THE CENTURY OF SPACE SCIENCE

The Century of Space Science

Volume II

Edited by

JOHAN A.M. BLEEKER
SRON – National Institute for Space Research, Utrecht, The Netherlands

JOHANNES GEISS
ISSI – International Space Science Institute, Bern, Switzerland

MARTIN C.E. HUBER
ESA – European Space Agency, Paris, France

History consultant

ARTURO RUSSO
Università degli Studi di Palermo, Italy

KLUWER ACADEMIC PUBLISHERS
DORDRECHT / BOSTON / LONDON

A C.I.P. Catalogue record for this book is available from the Library of Congress.

ISBN 0-7923-7196-8

Published by Kluwer Academic Publishers,
P.O. Box 17, 3300 AA Dordrecht, The Netherlands.

Sold and distributed in North, Central and South America
by Kluwer Academic Publishers,
101 Philip Drive, Norwell, MA 02061, U.S.A.

In all other countries, sold and distributed
by Kluwer Academic Publishers,
P.O. Box 322, 3300 AH Dordrecht, The Netherlands.

Printed on acid-free paper

Cover figure:
The nucleus of Comet Halley (rectangular frame;
copyright Max-Planck-Institut für Aeronomie) and the cosmic
microwave background (oval frame; image credited to NASA and
the Cosmic Background Explorer Team)

www.thecenturyofspacescience.com

All Rights Reserved
© 2001 Kluwer Academic Publishers
No part of this work may be reproduced, stored in a retrieval system, or transmitted in any form or by any
means, electronic, mechanical, photocopying, microfilming, recording, or otherwise,
without written permission from the Publisher, with the exception of any material supplied specifically for the purpose
of being entered and executed on a computer system, for exclusive use by the purchaser of the work.

Printed in The Netherlands.

Contents

Volume I

List of contributors	vii
Foreword	xiii
L Woltjer	

Introduction

1. The century of space science — 3
   JAM Bleeker, J Geiss and MCE Huber

The Beginnings

2. The space age and the origin of space research — 25
   A Russo
3. Enabling technology for space transportation — 59
   E Stuhlinger

The Early Epoch of Space Science

4. The cosmic radiation — 117
   JA Simpson
5. Magnetospheric physics — 153
   JA Van Allen
6. Barium cloud experiments in the upper atmosphere — 179
   R Lüst
7. Alkali metal cloud experiments in the upper atmosphere — 189
   J-E Blamont
8. Early solar space research — 203
   C de Jager
9. A history of the solar wind concept — 225
   EN Parker
10. The terrestrial planets at the dawn of the space age — 257
    WK Hartmann
11. The Moon before Apollo — 271
    JR Arnold
12. From the ionosphere to high energy astronomy – a personal experience — 277
    H Friedman
13. Early ultraviolet spectroscopy from space — 287
    BD Savage
14. The early days of infrared space astronomy — 301
    M Harwit

Fundamental Science in Space

Cosmology and Gravitational Physics

15. Verification of general relativity: tests in the Solar System — 335
    K Nordtvedt
16. Verification of general relativity: strong fields and gravitational waves — 353
    CM Will
17. The cosmological constants — 373
    GA Tammann
18. COBE, dark matter and large-scale structure in the Universe — 399
    KM Górski and AJ Banday
19. The origin of the light elements in the early Universe — 423
    H Reeves
20. Gravitational lensing — 441
    J Surdej and J-F Claeskens

Extragalactic Astronomy

21. Clusters of galaxies — 473
    RF Mushotzky
22. Gamma-ray bursts — 499
    RAMJ Wijers
23. Quasars — 529
    M Elvis
24. Blazars — 549
    G Ghisellini
25. X-ray and infrared properties of normal galaxies — 561
    G Fabbiano and MF Kessler

The Milky Way

26. The hot part of the interstellar medium — 581
    SL Snowden
27. Space-borne observations of the life cycle of interstellar gas and dust — 607
    EF van Dishoeck and AGGM Tielens
28. The interstellar medium of our Galaxy — 647
    PC Frisch
29. Galactic cosmic rays — 677
    FB McDonald and VS Ptuskin
30. Stellar populations and dynamics in the Milky Way galaxy — 699
    GF Gilmore

#	Title	Page
31.	Pulsars and isolated neutron stars *W Becker and G Pavlov*	721
32.	Evolutionary concepts of binaries with compact objects *EPJ van den Heuvel*	759
33.	White dwarf binaries *PA Charles*	791
34.	Low mass X-ray binaries *J van Paradijs and M van der Klis*	811
35.	High-mass X-ray binaries *NE White*	823
36.	Black-hole binaries *Y Tanaka*	839
37.	The formation of stars and protoplanetary disks *C Waelkens*	857
38.	High-energy radiation from outer stellar atmospheres *R Pallavicini*	875
39.	Mass loss from stars *JP Cassinelli*	895
40.	Planetary nebulae *SR Pottasch*	913
41.	Supernovae and supernova remnants *H Tsunemi*	937

Volume II

The Solar System

#	Title	Page
42.	Acceleration processes of heliospheric particle populations *G Gloeckler and K-P Wenzel*	963
43.	Reconnection *G Haerendel*	1007
44.	The solar interior *DO Gough and PH Scherrer*	1035
45.	The solar atmosphere *SK Solanki and R Hammer*	1065
46.	The active Sun *D Alexander and LW Acton*	1089
47.	The solar wind *M Neugebauer and R von Steiger*	1115
48.	The heliosphere *A Balogh and LA Fisk*	1141
49.	The dusty heliosphere *E Grün*	1163
50.	The interaction of the heliosphere with the interstellar medium *R Lallement*	1191
51.	Comets: coma and beyond *K Szegö*	1217
52.	The morphology of cometary nuclei *HU Keller and L Jorda*	1235
53.	The constituents of cometary nuclei *K Altwegg and WT Huntress*	1277
54.	The Moon and terrestrial planets: geology and geophysics *JW Head III*	1295
55.	Radiometric chronology of the Moon and Mars *LE Nyquist, DD Bogard and C-Y Shih*	1325
56.	Chemical evolution of the Moon and the terrestrial planets *H Wänke*	1377
57.	The atmospheres of the terrestrial planets *FW Taylor*	1405
58.	Jupiter *DM Hunten*	1425
59.	The planets beyond Jupiter *T Encrenaz*	1431
60.	The satellites of the outer planets *N Thomas*	1451
61.	Planetary and lunar magnetism *NF Ness*	1479

The Earth and its Plasma Environment

#	Title	Page
62.	The magnetosphere as a plasma laboratory *RA Treumann and M Scholer*	1495
63.	Earth's magnetosphere *B Hultqvist*	1529
64.	Earth's ionosphere *T Hagfors and K Schlegel*	1559
65.	Oceanography *JA Johannessen, S Sandven and D Durand*	1585
66.	Satellite geodesy and geosciences *F Barlier and M Lefebvre*	1623
67.	Chemistry and physics of the atmosphere *JE Harries*	1653

Appendices

A basic chronology of the space age *A Russo*	1671
Catalog of space science launches 1957–2000 *J McDowell*	1677

Indices

Abbreviations and acronyms	1711
Index of cited authors	1719
Name index	1791
Subject index	1801

The Solar System

42

GEORGE GLOECKLER* AND KLAUS-PETER WENZEL**

Acceleration processes of heliospheric particle populations

Through space exploration an amazing variety of populations of energetic particles have been discovered in our Sun's surroundings. Wherever we look in the heliosphere, we find processes that can accelerate ions and electrons of the local plasma to energies of ~0.01 to ~1000 MeV, or even more.

The heliosphere is that volume of space surrounding the Sun and filled by the solar wind, the ionized gas expanding from the Sun at supersonic speed and embedding all planets. This region was named the 'heliosphere' by Dessler (1967) and originally defined 'as the region of interplanetary space where the solar wind is flowing supersonically'. The current definition refers to that region of space where the solar plasma dominates: the heliosphere extends beyond the heliospheric (solar wind) termination shock, believed to lie about 100 AU from the Sun. Beyond this is the heliopause, the division between solar and interstellar plasma and magnetic fields.

That the space between Sun and Earth (and beyond) is not completely a vacuum has been recognized for nearly a century (or even slightly more), ever since discussions started about the observed relationships between solar activity and geomagnetic variations (see e.g. historical sources in Chapman and Bartels 1940). That the heliosphere is populated with 'energetic particles' – electrons, protons, and heavier ions with a few tens of keV energy to tens of MeV energy – was not known until the advent of the flight of appropriate detectors on spacecraft that travelled beyond the boundaries of the Earth's magnetosphere, beginning in the 1960s.

In the early years of space exploration, the heliosphere was viewed as a passive domain for energetic particles. Solar energetic particles were known to be injected periodically at its centre, galactic cosmic rays continually penetrated from outside, and occasional energetic particles leaked from the Earth's magnetosphere. The heliosphere was considered to be responsible for merely dispersing and decelerating (cooling) these particle populations. It was a degrading receptacle.

However, in the last thirty years our view of the heliosphere has changed dramatically. The discoveries of several new energetic particle populations in the 1960s and early 1970s produced convincing evidence that the interplanetary medium is the site of large-scale and nearly continuous acceleration of charged particles to energies as high as ~30 to ~100 MeV per atomic mass unit (amu). The heliosphere is not simply a mostly passive medium that transports solar particles, modulates galactic cosmic rays and occasionally accelerates particles to modest energies by interplanetary shocks.

The great variety of new energetic particle populations accelerated at different sites throughout the heliosphere is illustrated in Figure 1: solar energetic particles (SEPs); particles associated with travelling shocks (the discovery of 'energetic storm particles' goes back to the earliest era of space science); 'upstream particles' accelerated at the bowshocks of Earth and other planets; 'corotating' ion events associated with the forward and reverse shocks bounding 'corotating interaction regions' (CIRs) in the

---

* University of Maryland, College Park, MD, USA
** ESA – European Space Research and Technology Centre, Noordwijk, The Netherlands

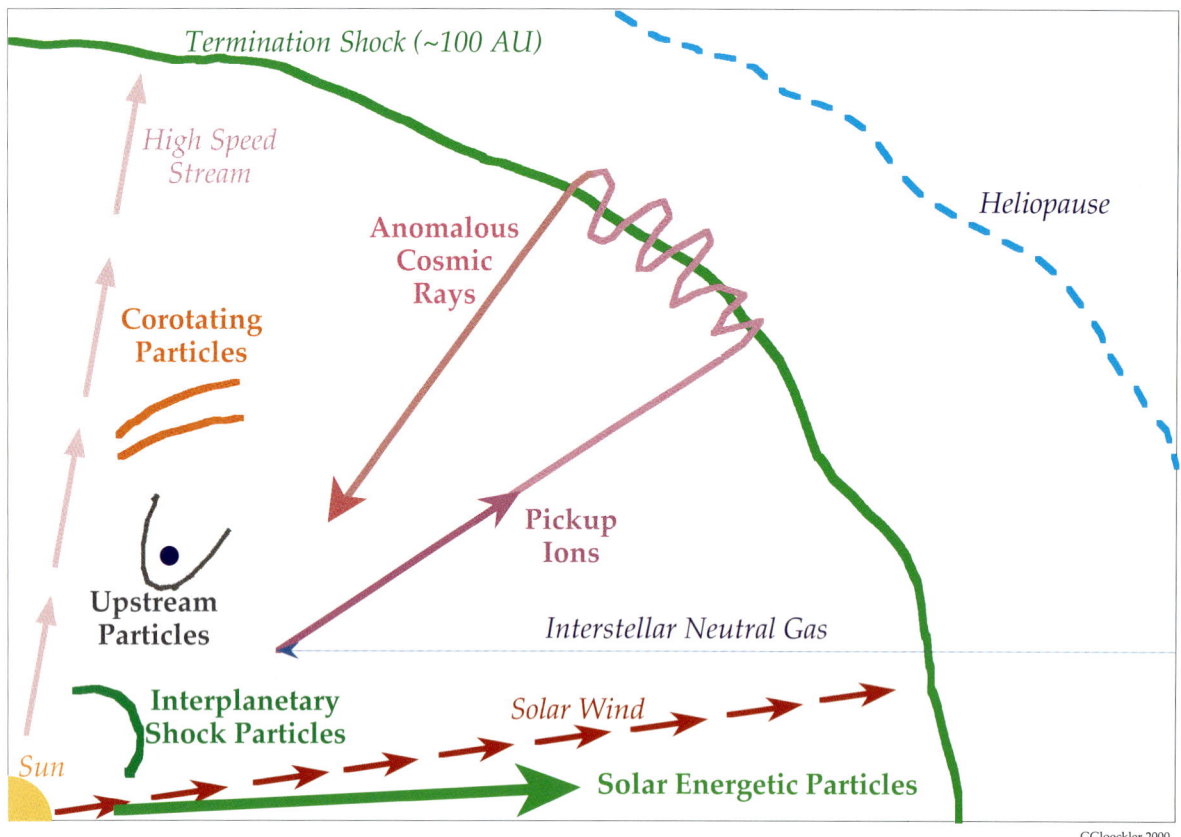

Figure 1 Schematic illustration of heliospheric energetic particle populations (bold labels). Solid curves indicate shocks.

solar wind; the anomalous cosmic ray (ACR) component; and pickup ions. These particles often propagate over great distances, carrying information on the nature, location and composition of their sources and about the physics of particle acceleration in their energy spectra, ionization states, and abundance of elements and isotopes.

Most of these heliospheric energetic particle populations are directly associated with collisionless shock waves. The prime acceleration site of the ACR component is thought to be the solar wind termination shock. With the exception of impulsive SEP events and interstellar pickup ions, there is nearly a one-to-one correspondence between shocks and energetic (upwards of several hundred keV) particle populations in the heliosphere. Since the solar wind is supersonic, it is not surprising that it is riddled with shock waves. But originally it came as a surprise that the shocks are such prolific particle accelerators. Thus, for more than three decades, the heliosphere has provided a gigantic and effective astrophysical laboratory for intense studies of particle acceleration processes by means of direct measurements. These *in situ* studies of particle acceleration have formed a basis for our understanding of acceleration processes throughout the Universe, for example, for the problem of the origin of galactic cosmic rays (see Chapter 29).

Typical intensity spectra as a function of energy per amu (energy/nucleon) for the different populations of energetic particles are illustrated in Figure 2. We will discuss these, with the exception of galactic cosmic rays, in later sections.

SPACECRAFT AND INSTRUMENTATION

The first energetic particle signatures that were found in the heliosphere were detected as a 'target of opportunity' by spacecraft that had another prime objective, including Mariner 2, Explorer 12 and Explorer 14. Later spacecraft that played a key role in the discovery and study of the various heliospheric ion populations included the Interplanetary Monitoring Platform (IMP) series, in particular IMPs 7 and 8; Helios 1 and 2 in the inner heliosphere; the International Sun–Earth Explorers (ISEE 1, 2, 3); Pioneers 10 and 11 and Voyagers 1 and 2 in the outer heliosphere; Ulysses, exploring the three-dimensional heliosphere; and Wind and the Advanced Composition Explorer (ACE). Most of

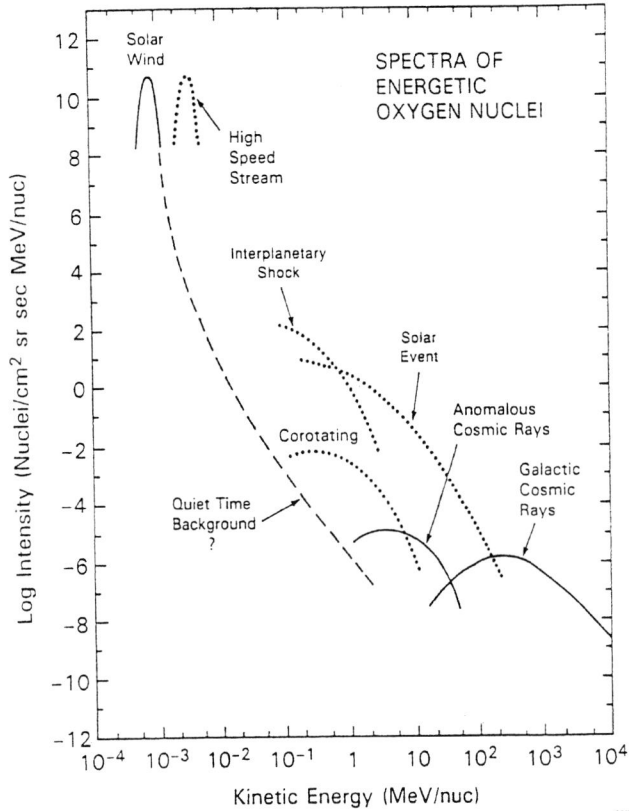

Figure 2 Typical oxygen differential spectra for different populations of heliospheric particles. (After Figure 2.1 of von Rosenvinge et al. 1995.)

these spacecraft were spinning, thus being able to provide information on the arrival direction of the particles measured.

By the end of the twentieth century we had explored the heliosphere in three dimensions, mapping the spatial distributions of the solar wind, of CIRs and of coronal mass ejections (CMEs) as Ulysses flew over the solar poles. Energetic particles accelerated at CIR shocks have been followed to latitudes far higher than the shocks themselves, thus serving as probes of the magnetic topology in the solar corona and high-latitude heliosphere. The Voyager spacecraft have tracked the modulation of the spectra of the ACRs out beyond 75 AU. Observations with the large-area detectors on Wind and ACE near 1 AU provide new insights into the physical processes of acceleration of SEPs. And we must not forget those spacecraft that continue to operate for long periods of time, such as IMP 8, which has been providing data for 30 years. These give us a complete perspective on solar-cycle variations that underlie the physical processes we study.

Table 1 lists the space missions that have made major contributions to the study of heliospheric particle populations.

Detectors for energetic particles used in the very early days of space exploration included Geiger counters, nuclear emulsions, proportional counters and scintillators with photomultiplier tubes. Semiconductor (solid-state) detector developments, reducing weight and enhancing reliability of instruments, began in the late 1950s and early 1960s. At lower energies space plasma detectors used Faraday cups. These

Table 1 Space missions that played a key role in studies of heliospheric particle populations

Spacecraft	Operation	Orbit	Achievements
Explorer 12	1961–1962	Elliptical Earth orbit	First detection of 'energetic storm particles'
Mariner 2	Aug–December 1962	Reached Venus in Dec. 1962	First observation of an interplanetary shock
IMP 7	1972–1978	Circular, ~30 R_E	Discovery of ACR oxygen; first measurement of ionization states of energetic particles
IMP 8	1973–now	Circular, ~30 R_E	Discovery of ACR He; 1 AU reference mission due to long orbital life
Helios 1/2	1974/76–1986/1980	Inner heliosphere to 0.3 AU	Spatial structures of SEP events; CIR radial gradients from 1 to 0.3 AU
Pioneer 10/11	1972–1999	Outer heliosphere	ACRs; CIR radial gradient beyond 1AU
Voyager 1/2	1977–now	Outer heliosphere	ACRs
ISEE 1/2	1977–1987	Elliptical Earth orbit	Earth's bow shock and upstream particles
ISEE 3	1978–1982	215 R_E Sunward of Earth	Interplanetary shock particles; two classes of SEP events
AMPTE/IRM	1984–1986	Elliptical Earth orbit	Discovery of He^+ pickup ions
Ulysses	1990–now	Solar polar orbit (1.3 to 5.4 AU)	Discovery of H^+ and heavier pickup ions; CIRs at high latitudes
SAMPEX	1992–now	Earth polar orbit	Charge states of ACRs
Wind	1995–now	Upstream of Earth	Large area detector systems
ACE	1997–now	215 R_E Sunward of Earth	High-resolution composition (elemental and isotopic and charge states) with large collection area

early instruments provided a measure of the total ion flux and could only resolve protons from helium and heavier ions.

In the course of time, instruments have improved enormously, in sensitivity (e.g. larger collecting areas and lower energy thresholds), in resolution (e.g. better separation of different species) and in high-speed on-board processing. Where we once measured event-averaged abundances and energy spectra in SEP events or across CIR shocks, we can now probe the time-dependent spectral evolution in such events for many particle species over 4–5 orders of magnitude in energy.

What led to the many discoveries and provided progress in the detailed understanding of the different particle populations were – besides new spacecraft on different trajectories probing new regions in the heliosphere – new space-borne instruments capable of measuring the composition of particles at increasingly lower energies. The most common composition-resolving particle detectors were so called 'energy loss v. energy' (dE/dx v. E) telescopes, using stacks of solid-state detectors (see Gloeckler (1970) for descriptions of these and other particle detection techniques). In these particle telescopes, the amplitudes of the coincident signals from the thin front dE/dx detector and the thick back E detector (or stack of detectors) were digitized (or pulse-height analysed) and transmitted to Earth. When the pulse heights of both detectors were plotted against each other for many recorded particles, distinct tracks would appear for each resolved nuclear species, with the separation between tracks depending on the combination of nuclear charge (atomic number) and mass. In this way the atomic number and mass of the nucleus were determined. Ever lower energy limits were reached by decreasing the thickness (but not the area) of the dE/dx detectors. This was done by, for example, using thin-window proportional counters of the Ultra-Low Energy Telescope (ULET) (Hovestadt and Vollmer 1971) provided by the Max-Planck-Institut (Garching) as part of the University of Maryland Experiments on IMPs 7 and 8 (Tums et al. 1974), or a mosaic of ultra-thin solid-state detectors in the Low Energy Charged Particle (LECP) experiment on Voyagers 1 and 2 (Krimigis et al. 1977). Large-area, position-sensitive solid-state detectors made it possible to resolve individual isotopes of many of the elements in the various particle populations (e.g. von Rosenvinge et al. 1995, Stone et al. 1998).

To reach even lower energies (tens of kilo-electronvolts), large-area plastic sheets were exposed in space and then returned to Earth for analysis (Price et al. 1967). This technique makes use of the fact that highly ionizing, heavy particles will damage the plastic material along their track. Once retrieved from space, these tracks become visible after chemical etching, and from the size and shape of the cones etched into the material, information on the nuclear charge and energy of the particle is obtained. The main disadvantage of plastic track detectors is their lack of good time resolution and the need to recover them.

A different approach to measure the composition of particles in the low-energy range between about 0.1 to 10 MeV/amu uses the 'time-of-flight v. energy' (TOF–E) technique (Gloeckler and Hsieh 1979). The start detector, a thin foil, is separated from the stop detector (usually a solid-state detector) by 10–50 cm. Secondary electrons, emitted from the respective surfaces of the start and stop detectors, are guided, typically by electric fields, to microchannel plates that detect these fast electrons and produce the start and stop signals for the TOF analysis. Measurements of the speed of the particle using TOF analysis combined with measurement of its energy (E) in the stop detector yields its mass. Composition instruments of this type were flown on the Wind (Gloeckler et al. 1995) and ACE spacecraft (Mason et al. 1998).

Even lower energies are reached by combining TOF–E with electrostatic analysis, as is done in the Solar Wind Ion Composition Spectrometer (SWICS) instrument on Ulysses, Wind and ACE (Gloeckler et al. 1992, 1995, 1998). Such instruments provide composition measurements from less than 1 to more than 100 keV/charge and measure, in addition to the mass, the charge or ionization state of the particle. This class of instruments is used to measure solar wind composition and the velocity distribution of pickup ions.

The most recent set of instruments to measure the composition of the solar wind, solar energetic particles and galactic cosmic rays with large collection areas and excellent elemental and isotopic resolution are assembled on ACE. It includes, besides SWICS (essentially the same instrument as its counterpart on *Ulysses*; see below), the Ultra Low Energy Isotope Spectrometer (ULEIS), the Solar Energetic Particle Ionic Charge Analyzer (SEPICA), the Solar Isotope Spectrometer (SIS) and the Cosmic Ray Isotope Spectrometer (CRIS). These state-of-the-art ACE instruments are described in Russell et al. (1998).

The acceptance of novel and technically advanced experiments for spaceflight has not always been easy, and the road to the pioneering measurements that were eventually made was often tortuous and lengthy. For example, there was considerable resistance to the use of ultra-thin solid-state detectors that were eventually flown in the LECP experiment (Krimigis et al. 1977) on Voyagers 1 and 2, yet these detectors make it now possible to measure the lowest energy ACRs in the outer heliosphere.

The TOF technique found it particularly difficult to gain acceptance in flight experiments. TOF instruments were proposed for an Explorer mission and again for Galileo, but were not selected by NASA because the technology was considered unproven and too risky for flight. However, in

1977 a joint ESA–NASA committee selected a TOF instrument (SWICS) for flight on the ESA spacecraft of the ESA–NASA International Solar Polar Mission (ISPM), then scheduled for launch in 1983. SWICS was a wide-field-of-view, collimator-type energy/charge analyser followed by a TOF spectrometer using 30 kV post-acceleration to allow measurement of the residual energy of the ions with solid-state detectors, and incorporated a number of new technologies (TOF, ultra-high voltages, thin carbon foils and an unconventional electrostatic analyser). The primary aim of SWICS (see Gloeckler et al. 1992 for details) was to measure the elemental and charge-state composition of the solar wind (see Chapter 47), but the energy range of SWICS was chosen to be sufficiently large to also detect and study the very rare pickup ions as secondary objectives of the investigation.

The ISPM mission, consisting of two spacecraft which were to be launched together towards Jupiter (~ 5 AU), and then sent over the north and south poles of the Sun respectively, was to be the first to explore the third dimension of the heliosphere. Had ISPM been launched as scheduled, all pickup ions would have been discovered a decade earlier. However, ISPM encountered several difficulties during its development. First was the cancellation of the NASA spacecraft in 1980 caused by budgetary problems at NASA. Then there were difficulties with the launch system, and in 1986 came the *Challenger* disaster, which placed space shuttle launches of all spacecraft on hold. Explorer class spacecraft such as the Active Magnetospheric Particle Tracer Explorer (AMPTE), and the Phobos mission to Mars, whose development started after ISPM, continued to be launched by expendable vehicles, and carried TOF instruments modelled after SWICS long before the ESA-built surviving ISPM spacecraft finally started its high-latitude journey in October 1990 (Wenzel et al. 1992), some 25 years after the initial concept for the exploration of the three-dimensional heliosphere was first proposed. Among the many achievements of the Ulysses mission, as ISPM was renamed, was the discovery and detailed study of all interstellar pickup ions (with the exception of He^+, discovered almost a decade earlier with AMPTE by Möbius et al. (1985), and Ar^+, which is outside the mass range of SWICS): H^+ (Gloeckler et al. 1993), $^3He^+$ (Gloeckler and Geiss 1996), $^4He^{2+}$ (Gloeckler et al. 1997), N^+, O^+ and Ne^+ (Geiss et al. 1994). Using data from their GAS instrument on *Ulysses*, Witte et al. (1993) discovered interstellar atomic He and provided the first direct measurements of its density, temperature and flow direction.

There were other instances, however, when instruments employing new and presumably risky technologies were selected for flight the first time they were proposed. An example of this was the Maryland Ion–electron Experiment (MIE) proposed for IMPs 7 and 8. This instrument, which uses ± 10 kV deflection supplies and an unconventional collimator-type deflection system, is still operating flawlessly on IMP 8 after 30 years in space, and in 1975 was the first to provide measurements of the ionization states of energetic particles. A proposed enhancement to the original MIE experiment was also accepted for flight. This was the ULET sensor (Hovestadt and Vollmer 1971) with the flow-through, thin window gas proportional counter never flown before. Data from this sensor led to the discovery of ACR oxygen in 1973.

SOLAR ENERGETIC PARTICLES

High-energy particle acceleration at the Sun is associated with the most energetic astrophysical phenomenon occurring in the Solar System: the explosive release of magnetic energy in the corona. By observing the energetic particles accelerated in SEP events, we can obtain information about the physical mechanisms causing the acceleration, the source site conditions and the mechanisms of transport of the particles from the acceleration site to the observer in interplanetary space.

Classification of SEP events

There is a rich history of the study of solar flares, dating back to when the first observations were reported by Carrington (1860). Although solar flares were discovered primarily by their visible light, H-alpha (656 nm) component, it became clear over the years that a flare has important high-energy aspects, as is apparent from the X-ray (0.1–1 nm) and microwave radiation (both discovered at the end of the 1950s) and gamma-ray emission (discovered in 1972).

The first evidence of high-energy particle emission from the Sun dates back to the pre-space era. The history began with the report by Forbush (1946) of a sudden transient increase in the counting rate of his ionization chamber on 25 July 1946 in association with a large solar flare, observed in H-alpha at Meudon. Lange and Forbush had previously observed two similar counting rate increases, occurring in association with solar activity in early 1942, but Forbush did not make the connection between counting rate increases and the simultaneous flaring until he observed the 1946 event.

The first SEP event to be analysed in detail was the famous ground-level event of 23 February 1956 that was observed by the neutron monitors developed by Simpson and located at Chicago and five other sites. Meyer et al. (1956) drew attention to the short duration (~ 45 minutes) of the flare (marked with a heavy line in Figure 3) and the long

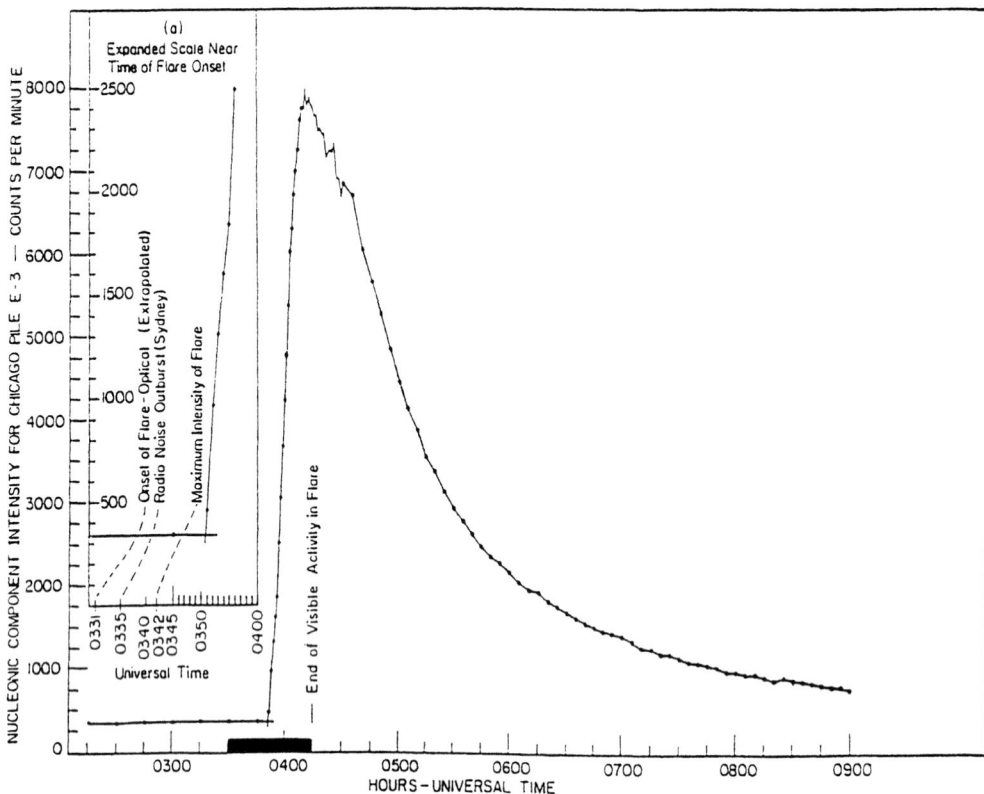

Figure 3 Chicago neutron monitor record of the ground level event of 23 February 1956. (Figure 2 of Cliver 2000; adapted from Meyer et al. 1956.)

duration (~15 hours) of the particle event. They concluded that a short-lived acceleration was followed by diffusive propagation. This observation led to the then prevailing paradigm for particle acceleration in association with solar flares which was that all particles were rapidly accelerated at the flare site.

Over the next 20 years, observations of these solar particles using ground-based neutron monitors and riometers (that measure radio opacity of the ionosphere), and later space-borne instruments on balloons and satellites, led to an extensive body of knowledge on the time profiles, spectra and particle abundances in large SEP events. The general wisdom was to assume that the acceleration of the particles somehow occurred in spatial and temporal conjunction with the solar flare itself. The prime reason why flares have been related to energetic particle events was that most particle events are preceded by a flare.

It was, however, not so easy to explain all SEP observations in terms of a flare source. Flare activity at the Sun lasts, at most, for hours while SEP events can persist for many days. Particles could be seen from events anywhere on, and sometimes behind, the visible disk; yet it was well known that particles could not cross the interplanetary magnetic field lines over such large distances.

The proposed explanation, which dominated the study of energetic particles for more than two decades, was the Reid–Axford model (Reid 1964, Axford 1965). The key to this model was a solar corona that stored the particles, somehow allowing them to diffuse in the corona easily in longitude, and then to leak outward from the corona along the interplanetary magnetic field. No mechanism for the 'coronal diffusion' (or coronal propagation) was ever identified. However, the model offered enough adjustable parameters to fit some of the well-behaved time profiles of the particle intensities in large SEP events.

At about the same time, Wild et al. (1963) provided a much different picture to explain their radio observations. The emission frequency in radio bursts is related to the local plasma frequency, which varies as the square root of the electron density. Thus the fast frequency drift of type III bursts was ascribed to 10–100 keV electrons streaming out of the corona from a flare through plasma of decreasing density. On the other hand type II bursts (meter wavelengths) had a much slower drift rate that corresponded to local acceleration at a ~1000 km s^{-1} shock wave moving out through the corona. This led to the idea of two-phase acceleration, with electron acceleration primarily early in the impulsive phase of an event, followed by proton acceleration later at the expanding coronal shock. While we now know that protons and electrons are accelerated in both 'phases', these Australian radio astronomers emphasized the presence of two acceleration mechanisms and, prophetically, suggested that particle abundances might distinguish them. Evidence for the two-phase picture was also provided by Lin (1970) who reported on the basis of space observations that pure electron events were preferentially associated with flares that only exhibited type III emission while events with protons and relativistic electrons tended to follow flares with type II/IV radio events. Early indications for proton acceleration by type II burst producing shock waves came for example from Svestka and Fritzova-Svestkova (1974).

The accelerating source for SEP events remained a topic of debate and studies for many years. Two main factors contributed in the 1980s and first half of the 1990s to a gradual change of our picture of SEP events from the flare as the only source. One was the realization of the significance of coronal shocks driven by coronal mass ejections (CMEs) from the Sun. CMEs were detected on the Skylab mission about a decade after the 'coronal diffusion' models and were the last major solar transient phenomena to be discovered (e.g. Gosling et al. 1974). They can account for at least as much or even more, of the energy released in an associated flare. These sudden, violent ejections are capable of producing the high-speed flows required to drive travelling interplanetary shocks. That CMEs are the drivers of interplanetary shocks was shown without a doubt by the Solwind coronagraph observations (e.g. Sheeley et al. 1985, Cane et al. 1987). Shocks ahead of a CME can accelerate particles over a large spatial region and are today generally accepted to be the source of the large interplanetary (proton) events that scientists had studied for many years.

The other factor was the improvement in instrumentation: more sensitive detector systems were flown with larger collection areas and capabilities to measure the SEP composition, i.e. heavier elements and not only protons and electrons (see above). Particle abundances (elemental and isotopic) and ionization state measurements have become important tools that can distinguish flare-accelerated particles from those accelerated at shock waves (Mason et al. 1984) driven into interplanetary space by a huge CME (e.g. Reames 1988).

Two solar cycles after the paper of Meyer et al. (1956), Kahler et al. (1978) presented the first evidence indicating that a CME rather than a flare was the essential ingredient for a large proton event. These authors also noted that the sizes of CMEs were compatible to the helio-longitude range over which particle events did not show delays to onset, the so-called 'fast propagation region' in which coronal diffusion had been assumed to occur rapidly. Domingo et al. (1979) first discussed an energetic proton event which was not associated with a flare, but only with an erupting prominence and an interplanetary shock. Cliver et al. (1983) reported a number of energetic proton events in which the impulsive phase was weak. Mason et al. (1984) showed that the particle abundances of events did not vary as a function of longitude of the event. This observation put severe constraints on the 'coronal diffusion' process, which implied that different ions diffuse at different rates, but was consistent with particles being accelerated over a range of heliolongitudes at a large shock.

The connection between soft X-ray (~ 0.1 keV) events and energetic particles in space was made when Cane et al. (1986), following the suggestion by Wild et al. (1963), showed that SEP events come in two basic classes that were distinguished by both the characteristics (duration and spectra) of their associated flare electromagnetic emissions and their particle composition (proton/electron ratio) in space. The large, long-duration events were associated with long-duration flares, with CMEs, with coronal and interplanetary shocks, and extended to much higher proton energies than the impulsive events. They could originate anywhere on the solar disk. In contrast, the other class of events, the impulsive particle events were associated with impulsive flares, without CMEs, were only observed following flares in regions that were magnetically well connected to the observer, suggesting a much smaller source region, and had high electron-to-proton ratios. The authors deduced that the two classes of particle events arise because different acceleration processes operate preferentially for the two classes of flares. The fact that soft X-ray events could be divided into two classes, namely 'impulsive' events and long-duration or 'gradual' events, had earlier been recognized by Pallavicini et al. (1977). The terms 'gradual' and 'impulsive' have stuck. They now distinguish the time scales of the SEP events themselves rather than the duration of soft X-ray emission (Figure 4).

A different line of evidence came from particle abundances. Small SEP events with greatly enhanced abundance of the rare isotope ^{3}He were discovered by Hsieh and Simpson (1970). The ^{3}He-rich events were subsequently found to have ~ 1000-fold enhancements in ^{3}He/^{4}He and ~ 10-fold enhancements in Fe/O relative to coronal abundances. The average abundances in large SEP events were known to reflect coronal abundances. Abundance variations were often explained in terms of rigidity-dependent transport in the corona from the flare site ('coronal diffusion'), but it was impossible to explain the huge enhancement in ^{3}He/^{4}He in this way. It became clear that two different mechanisms of acceleration were required. By tying low-energy (~ 10 keV) electron events and ^{3}He-rich SEP events together, Reames et al. (1985) provided compelling evidence that ions, accelerated in the flare's impulsive phase and made manifest by gamma-ray lines, can escape to space.

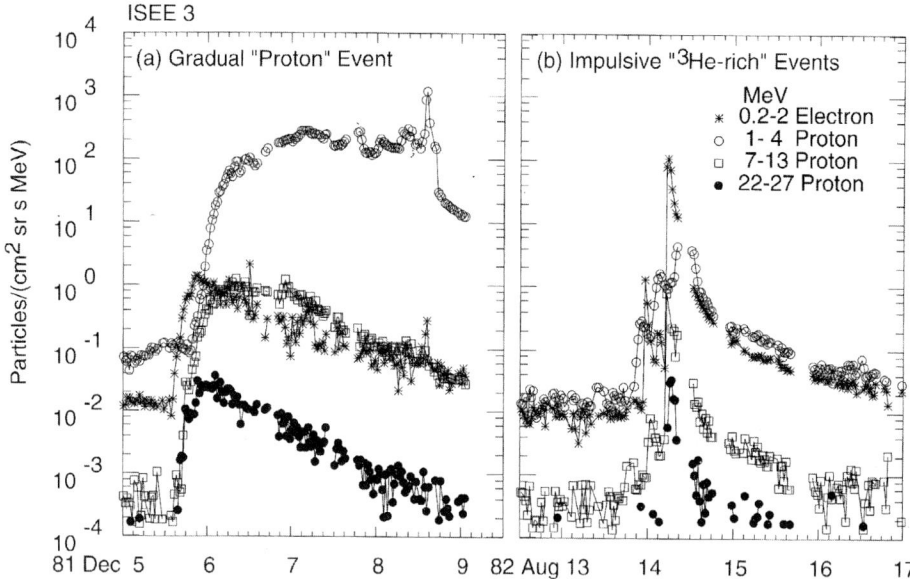

Figure 4 Intensity–time profiles of electrons and protons in (a) a gradual SEP event and (b) impulsive SEP events. The gradual event is a disappearing-filament event with a CME, but no impulsive flare. It is dominated by protons; a small peak at ~ 1 MeV appears near the time of shock passage. The impulsive events come from a series of flares with no CMEs. Electrons dominate the impulsive events, with peak intensities exceeding that in the gradual event. (Figure 2.2 of Reames 1999.)

Another compelling line of evidence for the different origin of impulsive and gradual SEP events came from measurements of the ionization states of the energetic ions (Klecker et al. 1984, Luhn et al. 1984, 1987). In gradual events, none of the elements above He were fully ionized; the ionization state of Fe (Q_{Fe}) was found to be 14 ± 1, on average. This indicated source material with an electron temperature of $\sim 2 \times 10^6$ K, a typical temperature of the ambient corona (Figure 5). Since Fe would be rapidly stripped of additional electrons at coronal densities where flares occur, Fe ions with low charge state must have been accelerated high in the corona where low-density material is traversed by a collisionless shock wave. In contrast, in ^{3}He-rich events, all elements up to Si were nearly fully ionized; Q_{Fe} was found to be 20.5 ± 1, on average, typical of flare-heated material of $\sim 10^7$ K.

Thus the initial paradigm of flares being the only source of SEPs has gradually changed. This change, illustrated in Figure 6, has provided an awakening in understanding the physical mechanisms of particle acceleration in SEP events. Today's widely accepted two-class paradigm of solar particle acceleration, advocated largely by D.V. Reames and colleagues (e.g. Cane et al. 1986, Reames 1988, 1995, 1999), is that in most of the large (proton) events a CME-driven shock wave accelerates particles from the ambient plasma of the corona and solar wind as it propagates over a large region of space and time ('gradual events') (e.g. Mason et al. 1984, Cane et al. 1988). The particles which are seen in small-intensity, but more numerous, events in a limited longitude range, are actually accelerated in impulsive flares and have unusual ^{3}He-rich and heavy-ion-rich abundances ('impulsive events') (e.g. Cane and Reames 1990). Recent composition measurements at low

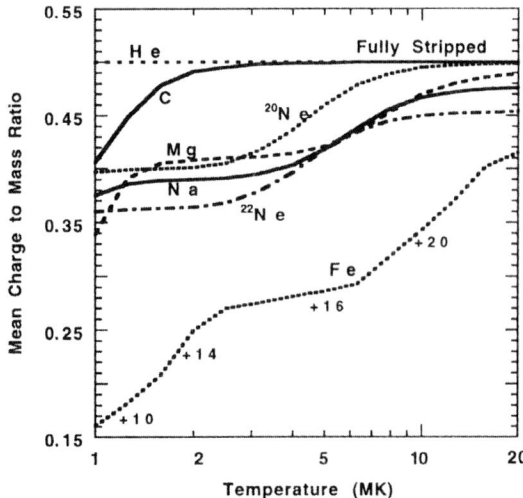

Figure 5 The mean charge-to-mass ratios of several ion species in SEP events as a function of electron temperatures ranging from 1 to 20 MK. These curves, which assume thermal equilibrium, are based on calculations. The line 'fully stripped' is correct only for nuclei with an equal number of protons and neutrons. Note that the ionization state of Fe (indicated below the Fe curve) is expected to be an especially good thermometer. (Figure 3 of Mewaldt 2000.)

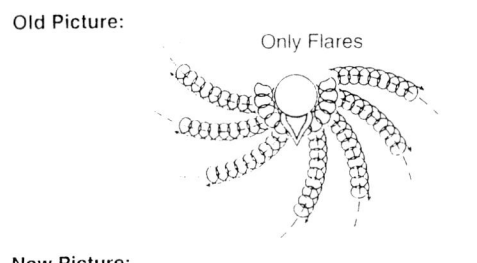

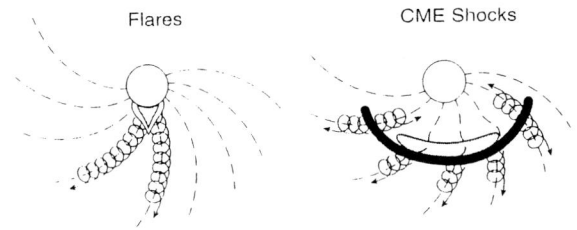

Figure 6 Schematic illustrating the paradigm shift for the source of solar energetic particles. (Figure 2.1 of Reames 1999.)

Table 2 Characteristics of SEP events (adapted from Reames 1995)

Characteristics	Impulsive	Gradual
General event size	Small	Larger
Abundance	Electron-rich (proton-poor)	Proton-rich
^{3}He/^{4}He	~1	0.001–0.1
Fe/O	~1	~0.1
H/He	~10	~100
Fe charge state Q_{Fe}	$+20 \pm 1.2$	14 ± 1
Duration of event	Hours	Days
Longitude cone	<30°	~180°
Radio type	III, V (II)	II, IV
Duration of soft X-rays	Impulsive (<1 h)	Gradual (>1 h)
Coronagraph	—	CME (96%)
Solar wind	—	IP shock
Events/year	~1000	~10

(<1 MeV/amu) energy on ACE have provided 'graphical' evidence that the particles are accelerated impulsively from the same solar active region and all travel the same distance to the spacecraft showing the effects of time dispersion, i.e. higher energy particles arrive first, followed by progressively lower energy ions. ^{3}He-rich events are explained in terms of resonant wave–particle interactions in the source plasma (e.g. Fisk 1978, Temerin and Roth 1992).

The main characteristics that led to this classification of the two basic SEP populations are given in Table 2.

The new paradigm constituted a declaration of independence of the energetic particle community from the flare community that was not accepted easily by solar scientists. CMEs came late to the domain of known solar eruptive phenomena. Although various aspects of this new paradigm had been proposed since the early 1980s, a heated debated arose following Gosling's (1993) paper 'The solar flare myth'. This paper produced hostile dismay that 'Jack Gosling and a few other revisionists' would 'wage an assault on over 30 years of solar-flare research' based on the 'low-grade optical data the CME people use' (Zirin 1994) and led to a 'flares v. CMEs' controversy in the solar–terrestrial community (e.g. Hudson et al. 1995, Miller 1995).

The motivation to distinguish flares and CMEs goes far beyond semantics. The term 'flare' evokes the idea of a limited spatial and temporal extent. Flares also imply chromospheric and coronal heating and are classically observed by photons from a hot gas. The 100-year history of flare observation had produced a significant photon bias that tended to discount interplanetary observations of plasma processes that are not visible using photons. Energetic ions tell us about the conditions in the source plasma, but they have also opened a new window on the complex plasma physics of particle acceleration. The existence of resonant wave–particle interaction that enhances ^{3}He by many orders of magnitude could not be inferred from photon observations of flares.

Apart from the magnetic energy source, the physical processes producing 'impulsive' and 'gradual' events are different. In the case of impulsive events and flares, detailed plasma physics involves pulsed electron beams with spatial scales as small as 10 km; multiple injections combine to form both type III radio and hard X-ray bursts, and beam-generated waves that are resonantly absorbed to enhance energetic ^{3}He. In the case of gradual events, shock waves are the prime acceleration source. Large solar flares and CMEs can produce coronal shocks. The detailed relationship between flares and CMEs is, however, not yet fully established. The shocks driven by the CME, crosses field lines to accelerate the particles at significant distances from the Sun over vast regions of space and for long periods of time in the largest SEP events.

A prime purpose in distinguishing gradual and impulsive SEP events is to permit separate studies of the two physical mechanisms of particle acceleration involved (see below). Of course there are also 'mixed' or 'hybrid' cases where both mechanisms appear to contribute. This was recognized since the first statistical studies (e.g. Cane et al. 1986). In these events, one mechanism operates in the flare while the other operates independently at the CME-driven shock. However, one population or the other seems to dominate in a surprisingly large fraction of the events. The best evidence of a gradual event is the presence of a fast CME, while the opposite is the case for an impulsive event (e.g. Cane et al. 1990). Only when the shock transit speeds exceed 500 km s^{-1} do SEP events become likely, while speeds >750 km s^{-1} always produce SEP events (Reames *et al.* 1997a).

Cliver *et al.* (1982) reviewed the ground-level events observed from 1942 to 1978. On occasions, solar energetic protons have been detected with energies beyond ~20 GeV (e.g. Simnett 1995). The famous 23 February 1956 event of the pre-space era and the 29 September 1989 event, detected both on spacecraft and by ground and underground monitors, are well-studied examples. A recent high-energy event occurred on 6 November 1997. In these events the protons arrived promptly, i.e. shortly after the electromagnetic signature of the solar parent flare. This strongly suggests that these high-energy protons were accelerated during the impulsive phase of the flare at or at least near the Sun (e.g. Kahler 1994). Whether these rare very high-energy particles are accelerated primarily in flares or coronal shocks is still not fully resolved (see the following section).

The most unambiguous signature of energetic (>30 MeV) protons in the solar atmosphere comes from the observation of gamma-ray lines produced through nuclear reactions. Following the first detection of gamma-ray lines from the Sun in 1972 (Chupp *et al.* 1973), manifesting ion acceleration in the flare impulsive phase, gamma-ray observations have provided powerful diagnostics of particles that interact at the Sun. Comparisons of the interacting (gamma-ray producing) and escaping interplanetary particles or SEPs (e.g. Cliver *et al.* 1989, Murphy *et al.* 1991) have largely supported the two-class picture: (1) they show a clear lack of correlation between the numbers of interacting and interplanetary particles, consistent with a poor connection between flares and large SEP events; and (2) they reveal a similarity between the composition of SEPs following impulsive flares and the composition of accelerated beam particles in gamma-ray flares of all durations.

SEP composition

Elemental composition

The abundance of the elements C through Si in an SEP event were first measured on sounding rockets in 1960 (Fichtel and Guss 1961) and extended to elements up to Fe during the 1960s. Measurements on spacecraft (e.g. Mogro-Campero and Simpson 1972, Teegarden *et al.* 1973, Cook *et al.* 1980) improved during the next decades. But in order to get abundances or abundance ratios, it was necessary to average the data over an event or even several events. The latest generation of large-geometry instruments launched on the Wind and ACE spacecraft now allows us to measure the time variations of abundances within an event with increasing detail. Element abundances have proved to be a most powerful tool in identifying particle populations and the nature and properties of the source of the plasma where they originate.

SEP abundances show characteristic signatures for the two classes of SEP events (Figure 7). Two well-resolved populations for gradual and impulsive events are found. Variations of different elements tend to be correlated in gradual events and uncorrelated in impulsive events. For gradual events the SEP abundance ratios cluster around the coronal abundance. For impulsive events the heavy ion abundances are enriched compared to the corona.

SEP abundances are known to vary with energy. Most of the abundance measurements made in the past were around 5 MeV/amu. At a different energy one can expect to find a different dependence of abundance on the charge to mass ratio Q/M. Mazur *et al.* (1992) studied the energy dependence of abundances and found that the abundances approach coronal values with minimal event-to-event variation at low energies near ~1 MeV/amu. At higher energies (>10 MeV/amu) they show an increasing divergence. At even higher energies (>100 MeV/amu) a new domain of variations may begin.

Once we average over the variations seen within an event and between different events, the gradual SEP events provide at present the most complete information we have on element abundances in the solar corona. There are two established differences between the elemental composition of gradual SEP events and that of the photosphere:

1. The abundances of elements with low (<10 eV) first ionization potential (FIP) are enhanced in SEPs by a factor of ~4 (e.g. Figure 8). This fractionation process occurs during transport of material from the cool photosphere, where elements with high FIP exist primarily as neutrals and those with low FIP are mostly singly ionized. The ions are more efficiently transported to the hot corona than the neutrals, resulting in an over-abundance of low FIP elements in the corona.

2. The measured abundances in any single event as compared to coronal values are enhanced or depleted by an amount that is well correlated with the element's ionic charge to mass ratio, Q/M (e.g. Figure 9). This second fractionation process has been empirically determined from SEP measurements during many large events. It is presumed that this fractionation, typically expressed as a power law in Q/M, results from acceleration processes that depend on the particles' rigidity (momentum per charge) (Breneman and Stone 1985).

Taking these two processes to be the dominant ones affecting relative abundances, one can relate the observed SEP abundances to the solar coronal and photospheric abundances. The element abundances in gradual events, reflecting coronal

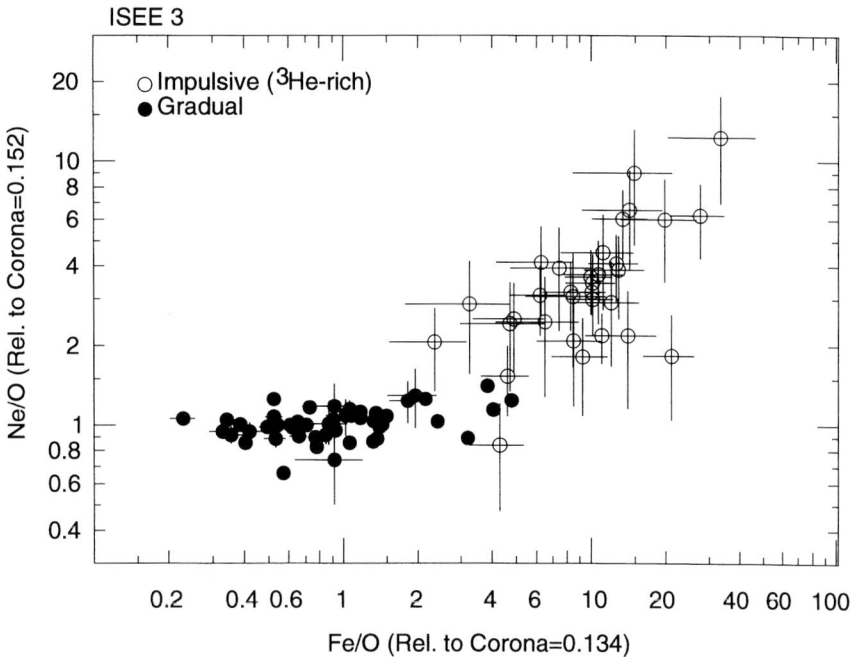

Figure 7 Plot of abundance ratios Ne/O v. Fe/O at ~5 MeV/amu for gradual and impulsive events. Each point represents abundances averaged over one SEP event. The abundance ratios are normalized to coronal values. (Figure 2.4 of Reames 1999.)

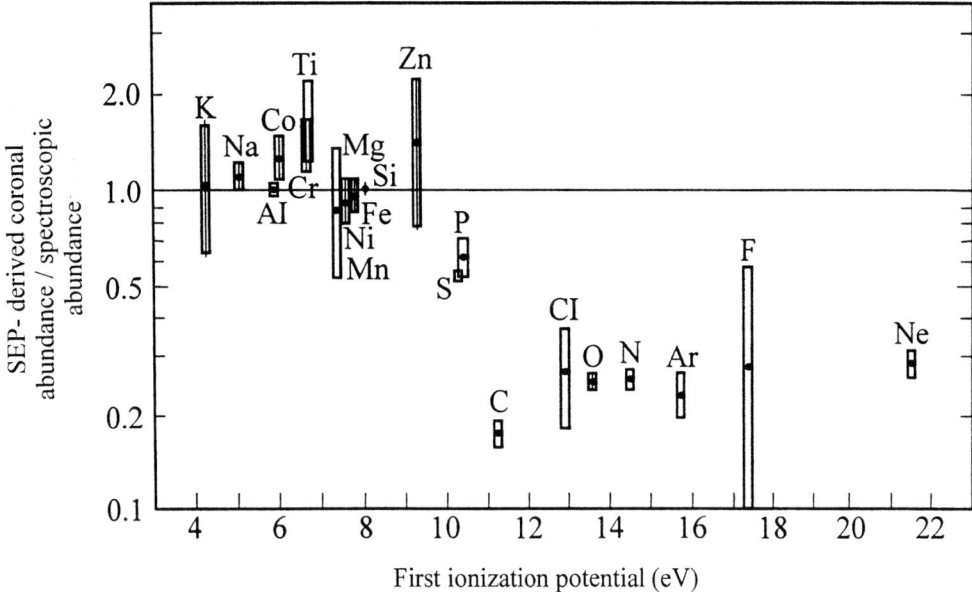

Figure 8 SEP-derived coronal abundances relative to spectroscopic photospheric abundances plotted v. first ionization potential. (After Figure 4 of Breneman and Stone 1985.)

values, are compared in Table 3 (later in this chapter) along with other energetic particle populations.

In the two-class picture, high ionization states and enhanced abundances of heavy elements are attributed to stochastic acceleration at the flare site. Recent ACE observations (e.g. Cohen et al. 1999) have revealed that these characteristics are not solely limited to small impulsive events. In response to the new observational results, a more sophisticated shock model has been developed (Ng et al. 1999) that indicates that in large SEP events both high ionization states and enhanced heavy element abundances might be explained in terms of shock acceleration with electron stripping taken into account. Because gradual events involve both flares and CME-driven shocks, the possibility remains that some SEP events with favourably located (i.e. western-hemisphere) eruptive flares may be hybrids (e.g. Cane et al. 1991). In order to address the evolution of composition from an 'impulsive-like' composition in the early stage to a 'normal' composition in later stages of such events and to include the characteristics of the particles that interact at the Sun to produce gamma-ray emission, Cliver (1996) expanded Reames' tabular summary of the two-class picture presented in Table 2.

Ionization states

The ionization states (often also referred to as 'ionic charge states' or 'charge states') of SEPs provide evidence for the site of origin and initial acceleration of the SEP population. They are expected to reflect the temperature of the material from which they were accelerated, as well as subsequent processes that may remove additional electrons. ACE is now providing direct measurements of SEP ionic charge states, with significantly improved resolution and collecting power, compared to the initial charge state measurements with IMP 7 in 1976 (Gloeckler et al. 1976) and ISEE 3 in the early 1980s (Luhn et al. 1984, 1987). Figure 10 shows an example from a large SEP event observed by ACE indicating that the ionization states of several elements increase with increasing energy (Möbius et al. 1999). At the higher energies measured on ACE there is general agreement with the average ionic charge states reported from ISEE 3. Instruments on the polar-orbiting Solar Anomalous and

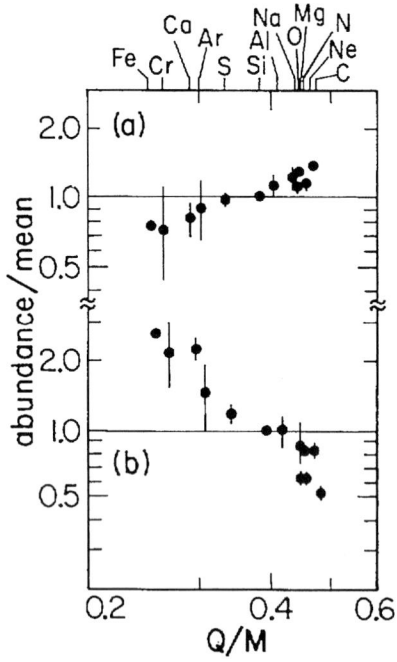

Figure 9 Abundances, relative to the mean SEP abundance, for two particle events v. charge-to-mass ratio. The Q/M-dependent abundance fractionation can be represented by a power law. (Figure 2 of Breneman and Stone 1985.)

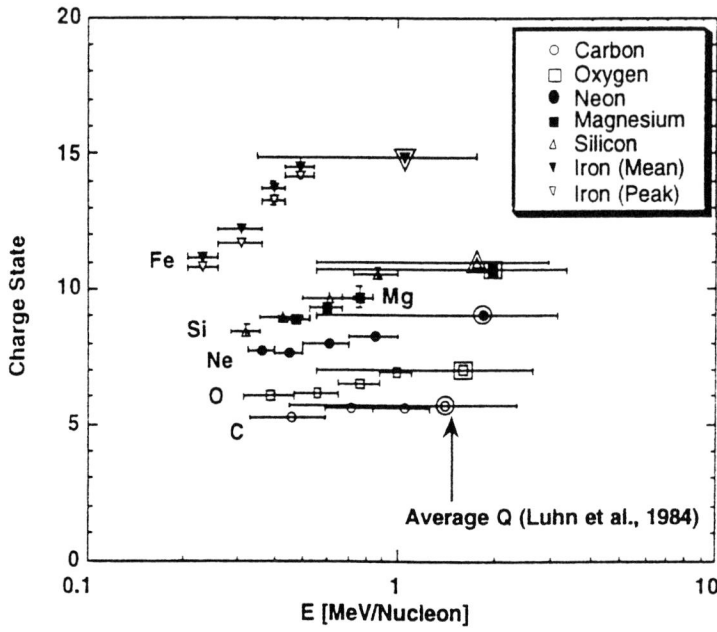

Figure 10 ACE measurements of ionic charge states in an SEP event (7–9 November 1997), compared with the average charge states measured ISEE 3 (Luhn et al. 1984). (Figure 3 of Möbius et al. 1999.)

Magnetospheric Particle Explorer (SAMPEX) spacecraft also find an energy-dependent ionization state in the same event using a geomagnetic technique to measure the ionic charge-to-mass ratio over a wide energy range, with an Fe charge state of ~20 at >30 MeV/amu (Mazur et al. 1999). The interpretation of these energy-dependent charge-state distributions is not yet clear. Suggestions centre on the timescales for acceleration and ionization equilibrium, on effects of energy-dependent electron stripping or on a mixture of different particle populations.

ACE has also measured the charge-state distributions in individual events (Popecki et al. 1999). Figure 11 shows the distributions for Fe measured only days apart, one with a mean of ~10 from a small gradual event, and one with a mean of ~17 from an impulsive event. It is evident that the Fe charge-state distribution is considerably broader than expected from thermal equilibrium, indicating possibly that material from regions of differing temperature was accelerated, or that additional ionization processes have occurred. The ACE data suggest that non-thermal processes may be important in some of the SEP events.

Isotopes

One of the intriguing characteristics of impulsive events is that the rare isotope ^{3}He is enhanced by factors of around 10^2 to 10^4 or more over its abundance in the photosphere and solar wind (^{3}He/^{4}He ~0.0004). Studies with the previous generation of instruments led to a convention of labelling ^{3}He-rich events to be those with ^{3}He/^{4}He >0.1, because small ^{3}He enhancements could not be reliably identified. ACE observations have revealed many SEP events (including large, presumably gradual events) with ^{3}He/^{4}He ratios from ~0.001 to ~0.1. It has been suggested that the ^{3}He-rich events can form a seed population for SEP particles accelerated by shocks that accompany large SEP events (Mason et al. 1999).

Prior to ACE there were only limited measurements of the isotopic composition of SEP ions with $Z \geqslant 6$. ACE has made it possible to measure the isotopes of a number of heavy elements (e.g. O, Ne, Mg, S, Fe). A primary objective of such measurements is to determine the isotopic composition of the solar corona, since it is difficult to do this with spectroscopic techniques. First results have shown that the isotopic composition varies considerably from event to event (Leske et al. 1999a). Although it was well known that the SEP elemental composition varies from event to event, the large variations in the isotopic composition of elements other than He (e.g. ^{22}Ne/^{20}Ne by a factor of 5) were not expected. It is an open question whether these isotope variations are due to the same acceleration and transport processes that fractionate the elemental composition of SEPs, or whether they result from some additional mass-dependent fractionation process.

Summary

SEP observations from ACE are beginning to affect some of the views on SEP events that had evolved in the decade before ACE. Initial results have blurred the distinctions between the two classes of SEP events. They have shown overlap in the ^{3}He/^{4}He ratio, broader distributions of the Fe charge states, charge-state increases with energy, provided evidence for non-equilibrium ionization processes and composition changes during events (e.g. Cliver 1996, Mewaldt 2000).

Acceleration of SEPs

Two generic acceleration processes have been identified for accelerating SEP particles:

1. *Shock acceleration* Here the particle's energy is derived from the relative motion between scattering centres (e.g. waves) on either side of a strong shock (diffusive acceleration), or between such centres and the shock itself (first-order Fermi acceleration). It occurs in high-β plasma, including CME-driven coronal shocks, i.e. it is the dominant process for gradual SEP events; β is the ratio of plasma pressure to magnetic pressure. Shock acceleration also occurs in other heliospheric settings: at interplanetary propagating shocks (see the following section), planetary bow shocks, at CIR shocks (see section on 'Corotating ion events') and at the heliospheric termination shock (see section on 'Anomalous Cosmic rays').

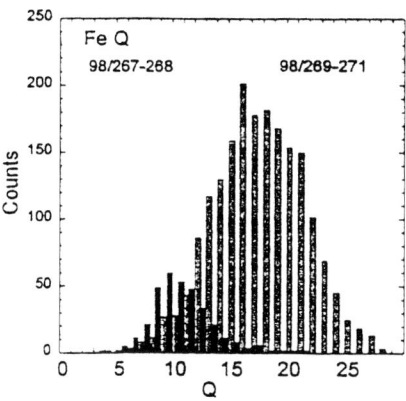

Figure 11 Charge-state distributions for Fe measured by ACE during a small gradual event (mean charge = 10) and an impulsive event several days later (mean charge = 17). (After Popecki et al. 1999.)

2. *Stochastic acceleration* Here energy is transferred from waves to particles, involving resonant wave–particle interactions. It occurs in regions of high magnetic field and low-β plasma, notably in solar flares, i.e. it is the dominant process for impulsive SEP events.

Gradual events

Numerous observations support the scenario that the acceleration in large, gradual SEP events takes place at a shock wave driven out from the Sun by a fast CME. Supra-thermal solar-wind ions (i.e. coronal material) serve as the seed population. Protons, the dominant particle species, streaming away from the strong shock generate Alfvén waves (they represent a mode of transmitting information and electromagnetic energy through magnetized plasmas) that trap particles in the acceleration region and act as scattering centres for the particles in a diffusive acceleration process.

Self-generated waves were first used as the basis of a shock-acceleration theory for galactic cosmic rays (e.g. Axford *et al.* 1977). This theory was then applied by Lee (1983) to locally accelerated particles seen in conjunction with interplanetary travelling shocks (see the following section). Over the years it has been recognized that interplanetary shocks seen at 1 AU accelerate particles all the way out from the Sun's corona. Near the Sun SEPs are strongly focused into beams which rapidly amplify Alfvén waves to great intensities, so that the waves in turn quickly scatter the SEPs. The enhanced scattering near a shock dramatically increases its acceleration efficiency, so that a quasi-equilibrium is soon established between ion densities and streaming on the one hand and wave intensities and growth rates on the other.

Alfvén wave generation at a distance is unobservable. However, SEPs arriving at a spacecraft carry information on their past interaction with the waves. Careful study of the SEPs has thus provided indirect evidence for wave excitation and diffusive shock acceleration near the Sun. The key observation here is the SEP abundance histories. Two ion species with unequal charge-to-mass ratios Q/A and the same velocity interact resonantly with Alfvén waves of different wavelengths. The resulting difference in their acceleration and transport is sensitively reflected in their abundance ratio history.

The recent model of the coupled evolution of energetic ions and Alfvén waves (Ng *et al.* 1999) has been tested against two of the gradual SEP events observed by Wind and ACE. Although it does not fit all observations precisely, it describes the qualitative behaviour of several newly observed abundance variations. For example, the flattening of energy spectra at low energies or the remarkable abundance evolution (e.g. Fe/O v. He/H) in gradual events can be understood.

Impulsive events

Impulsive ^{3}He-rich events must involve acceleration physics that differs greatly from that in gradual events. The unique charge-to-mass ratio of ^{3}He and its gyro-frequency uniquely situated between those of dominant species, H and ^{4}He, led to the suggestion of a selective enhancement by resonant wave absorption in the source plasma during an impulsive flare (Fisk 1978). The strong association between ^{3}He-rich events, streaming 10–100 keV electron and type III ratio bursts has been exploited in the theory of Temerin and Roth (1992). They noted an analogy with electron-beam generated waves that couples energy to ions in the Earth's aurora, and suggested that a similar process might occur in solar flares. Electromagnetic ion cyclotron waves produced below the H gyro-frequency (Ω_H) by downward streaming electrons, can resonate directly with ^{3}He ($\Omega_3 = \frac{2}{3}\Omega_H$). This process accelerates ions to MeV energies, without requiring a second process. Ions heavier than He can be accelerated, as they interact through the second harmonic of their gyrofrequencies. An extensive review of the status of various theories of particle acceleration in impulsive flares has been given by Miller *et al.* (1997).

Our views of both acceleration processes have evolved considerably. Element abundances have become one of the most powerful tools to study the physics of acceleration of solar energetic particles. In gradual SEP events, they probe the transparency of the proton-generated wave spectrum as a function of rigidity. In impulsive events, they tell us the average accelerating wave spectrum as function of gyro-frequency. The new measurements and the contemporary insight into the physical processes through theory have brought promising progress in our understanding of SEP acceleration. Models for the two underlying processes of acceleration have emerged.

INTERPLANETARY SHOCK PARTICLES AND PARTICLES UPSTREAM OF PLANETARY BOW SHOCKS

Wherever shocks are observed or inferred to exist in the tenuous plasma of the solar corona, in interplanetary space or around planets, energetic charged particles are observed or inferred to exist as well. Shocks can both efficiently accelerate particles and affect their subsequent transport.

Interplanetary shock particles

Large solar flares and coronal mass ejection (CME) events produce shock waves, which propagate out through the corona into interplanetary space. The existence of shocks in the solar corona is inferred from Type II radio bursts. As coronal shocks they are an important source for accelerating solar energetic particles close to the Sun (see the previous section). Interplanetary shocks are observed in association with energetic particle enhancements throughout the heliosphere, and they manifest themselves at Earth as magnetic storm 'sudden commencements'. They often play a major role in the time-evolution of the interplanetary transport of energetic particles between the Sun and Earth and beyond. They are a source for accelerating particles from the ambient interplanetary medium and are responsible with the associated CME for Forbush decreases in the galactic cosmic radiation.

Sonett et al. (1964) were the first to directly identify shocks in interplanetary space. They used data from Mariner 2 and Explorer 14, and magnetic field measurements in the magnetosphere of the Earth. The first observations of energetic particles accelerated by a travelling interplanetary shock also goes back to the earliest era of space discoveries. (For a review of observations and theory in the USSR, see Toptyghin (1980).) One example of these early observations comes from Bryant et al. (1962), who reported Explorer 12 measurement of an 'energetic storm particle' (ESP) event, so named because of the close temporal correspondence of the event with a sudden commencement at Earth. The ESP enhancement, superposed on the decay phase of a solar proton event, occurred at the same time as a Forbush decrease in the galactic cosmic ray count rate, then recorded by a ground-based neutron monitor. The origin of the ESP event, as was soon realized, was a travelling interplanetary shock. The energy spectrum of the ESP event was clearly steeper than that of the solar particles. The generic name 'ESP event' for the intensity peak near a shock has been maintained in the literature up to the present time.

Further evidence that interplanetary shocks accelerate particles was found, for example, by Rao et al. (1967). It was believed that the maximum energy achievable was about 10 MeV and that the whole effect was seen only locally as the shock passed. It was thought that the particles from the flare provided an energetic seed population on which the shock operated.

The 1970s and early 1980s witnessed an explosive growth in research concerned with understanding this close association between shocks and energetic particles within the heliosphere. The rapid growth has been spurred by the wealth of comprehensive and detailed observations of shocks and associated particle events, made in interplanetary space by a multitude of spacecraft ranging in heliocentric distance from 0.3 AU (Helios) to over 75 AU (Voyager).

Pulse-like increases in which the intensity of low-energy particles (0.3–5 MeV/nucleon) increased abruptly (for a few tens of minutes) at the front of propagating interplanetary shock waves were reported, for example, by Armstrong et al. (1970). This encouraged Fisk (1971) to suggest his model for producing these increases (see below).

Extensive measurements of ESP and other particles events associated with travelling interplanetary shocks were made during the 1980 solar maximum by the International Sun Earth Explorer (ISEE) spacecraft, with which the present authors were involved (e.g. Scholer et al. 1983, Van Nes et al. 1984, Sanderson et al. 1985a,b) (see Figure 12). These studies greatly contributed to the current understanding that diffusive acceleration of solar wind ions dominates ESP events at quasi-parallel shocks, whereas shock drift acceleration of energetic solar flare ions dominates ESP events at quasi-perpendicular shocks (see below).

Strong shocks can occasionally accelerate >100 MeV protons out to 1 AU. An example of these rare events is the 19 October 1989 event, the largest of solar cycle 22 (Figure 13). The 110–500 MeV protons behaved just as ~5 MeV protons do in the numerous small events observed: they showed a similar intensity–time profile with an intensity peak near the time of shock passage, as do lower-energy particles.

ESP events continue to be present in the outer heliosphere although their occurrence rate is reduced by decay and coalescence of shock waves. Pyle et al. (1984) showed, using energetic particle events measured by Pioneer 10 between heliocentric radial distances of 24–28 AU, that periods of solar activity produce a coalescence of flare and/or shock produced energetic particles in the outer heliosphere (Figure 14). Decker et al. (1981) showed that ions above ~10 keV are accelerated by interplanetary shocks out to at least 40 AU and over 30° in latitude in the outer heliosphere.

With the advances in instrumentation in the 1990s, the measurements of the composition of the shock accelerated particles have become an important tool to determine their source. In general, observations have shown that the high-energy tails of solar wind and of the ambient interplanetary energetic particle population are the source of particles accelerated by interplanetary shocks. Whether shocks can accelerate solar wind particles to energies of tens or even hundreds of mega-electronvolts locally in the interplanetary medium, or whether the higher-energy ions are initially accelerated close to the Sun where shock acceleration is more efficient, is an open question.

In summary, the classification of the enhancements of energetic charged particle intensities produced by the interplanetary shock acceleration processes established in the 1970s and early 1980s (e.g. Pesses et al. 1982) still holds. This classification accounts for three types:

1. Enhancement, produced by travelling interplanetary shock waves and observed during the decay phase of a SEP event are called energetic storm particle (ESP)

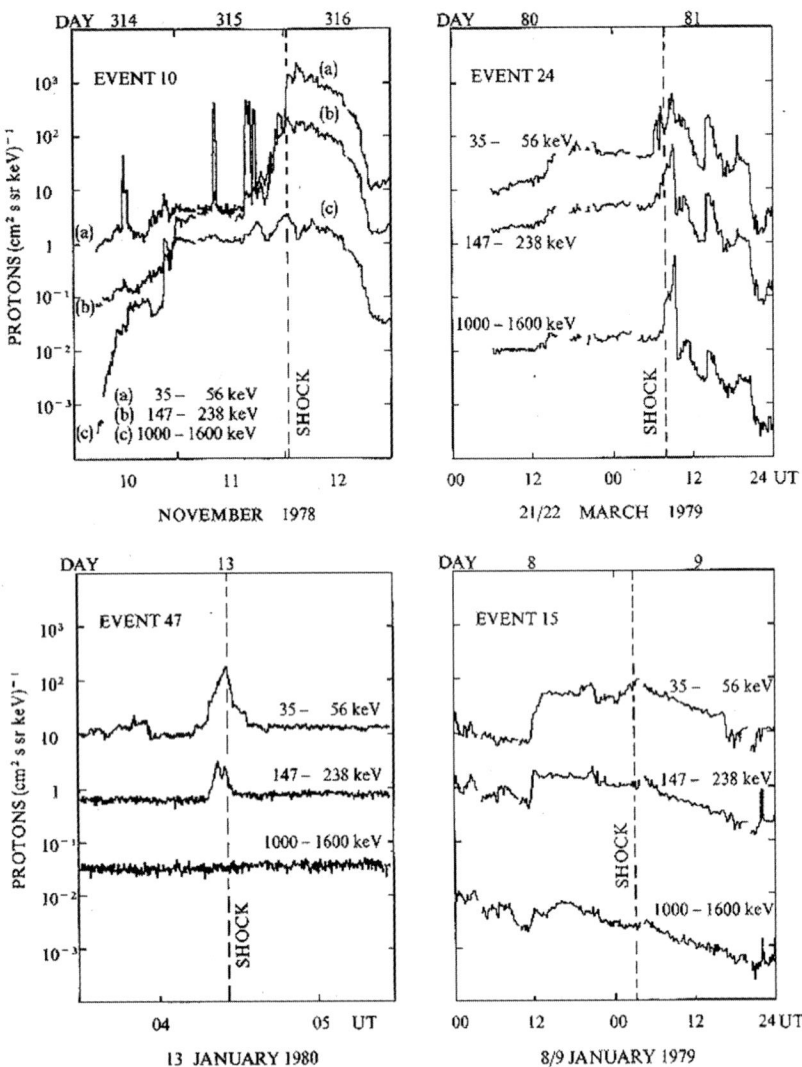

Figure 12 Four examples of ion intensity enhancements associated with the passage of an interplanetary shock at the ISEE 3 spacecraft. (Figure 1 of Van Nes et al. 1984.)

events. These events are generally ion enhancements of energies up to a few mega-electronvolts, occurring for a few hours prior to and after shock passage.

2. Enhancements that are characterized by relatively short-lived ion intensity spikes (0.1–1 h) at the time of shock passage, are called 'shock spike' events.
3. Enhancement produced by the shock waves that bound the solar wind stream–stream Corotating Interaction Regions (CIRs) (Smith and Wolfe 1976). These enhancements are usually observed at both the leading and trailing edges of the CIR (see the following section).

Theory

Fisk (1971) was the first to provide a quantitative theory for the shock origin of ESP events. He connected a solution of the energetic particle transport equation across the shock and derived an expression for the particle differential intensity. If he had not assumed a power-law energy spectrum and not allowed a discontinuity in particle intensity at the shock, he would have discovered the mechanism of 'diffusive shock' acceleration in which particles scatter across the shock many times and are accelerated by being coupled to the shock compression. Scholer and Morfill (1975) also came close with their work on ESP events using a Monte Carlo numerical method. However, the discovery of that mechanism and its ability to produce power-law energy spectra had to await the work of Axford et al. (1977), Blandford and Ostriker (1978), Bell (1978) and others on the origin of galactic cosmic rays (see also Chapter 29).

Independent of diffusive shock acceleration is the theoretical work of, for example, Decker (1981) and Pesses

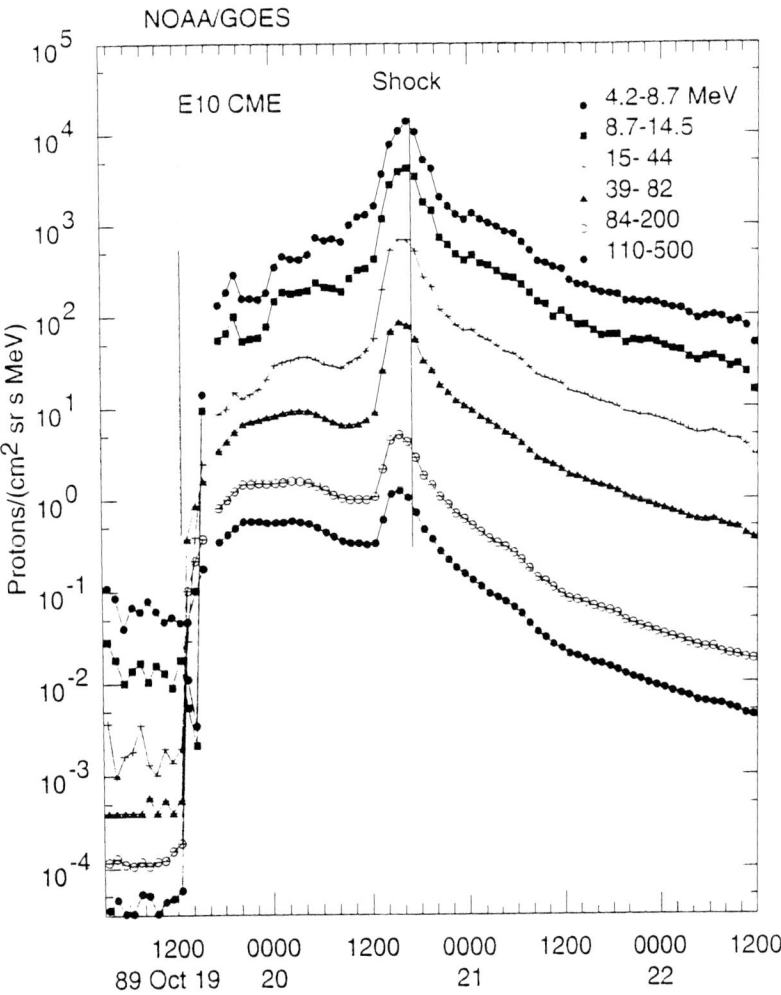

Figure 13 Intensity–time profiles at different energies for a large SEP event, showing flat time profiles with intensity peaks near the time of shock passage even at very high energies. (Figure 3.2 of Reames 1999.)

et al. (1982) which attempted to understand ESP and 'shock spike' events. They interpreted the short-duration shock spike events at the time of shock passage as transient reflections of solar ions from the shock front whenever the shock is locally quasi-perpendicular (i.e. the normal of the shock front is nearly perpendicular to the local magnetic field). Particles are mirrored and reflected from the increased field strength at a fast shock during a single encounter and gain energy by drifting parallel to the $V_{sh} \times B$ electric field in the shock frame. This acceleration mechanism became known as 'shock drift' acceleration.

In the early 1980s there was much controversy between the proponents of the two types of shock acceleration, diffusive and drift, concerning which type properly described ESP events. A major step forward was made by Jokipii (1982), who showed that both mechanisms are included in the energetic particle transport equation if the drift term is kept, and that the relative contribution of each is reference-frame dependent.

The particles seen in an ESP event are trapped near the shock by self-generated waves. Lee (1983) developed a quasi-linear theory in which the particle and wave intensities are in equilibrium, and which thus describes a plasma shock peak of infinite extent after an infinite time. Lee found the equilibrium distribution and the spectra of both particles and waves and their spectral distribution as a function of distance from the shock – in other words, he described the structure of an ESP event. At the shock, the equilibrium spectrum is a power law with a spectral index that depends upon the shock compression ratio. The predicted relationship between particle and wave spectra at shocks was confirmed by comparing the predictions with observational data (e.g. Kennel *et al.* 1986). Lee (1983) also estimated the maximum energy of accelerated particles.

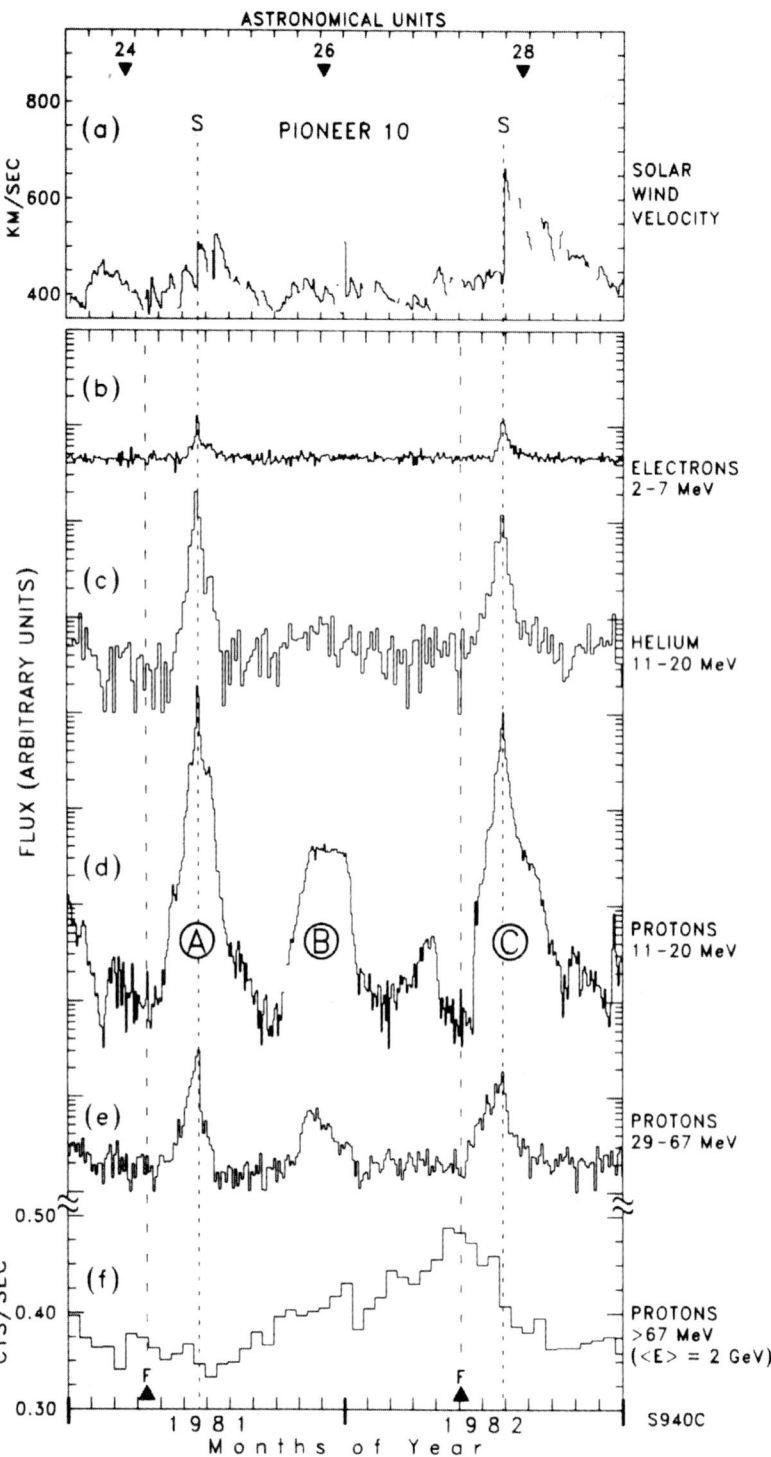

Figure 14 Energetic particle increases associated with shocks in the outer heliosphere as observed by Pioneer 10. The three ringed letters A, B, C refer to three different particle events. (Figure 3 of Pyle *et al.* 1984.)

Since he considered only background turbulence in his estimate, he obtained a low value (see, however, Lee 1997). Including low-frequency intensity fluctuations in B in the estimate will create waves generated by low-energy protons to scatter protons of higher energy more efficiently.

The strong shock created by the huge flare-induced blast wave of 4 August 1972 played a special role in the study of shock acceleration in the heliosphere. Eichler (1981) showed that the shock was a 'mediated' shock in which the energetic particle pressure is comparable to the thermal plasma pressure and modifies the structure of the shock. Mediated shocks are very rare in the heliosphere but are expected to be important in accelerating galactic cosmic rays in interstellar space.

Particles upstream of planetary bow shocks

Planets, and their interaction with the interplanetary medium, can be a rich source of energetic particles. In general there are three distinct regions around planets where energetic particles can be found: planetary bow shocks; inner magnetospheres, where particles can be stably trapped in well-defined radiation belts; and the dynamic regions of the outer magnetospheres, magnetosheaths and magnetotails. We will focus here on planetary bow shocks that are relevant to other heliospheric particle populations. Most important for the studies of particle acceleration has been the Earth's bow shock, which provides a stable structure where energetic particles and the spectrum of their self-generated waves have been studied together with the properties of the shock by many space missions.

Although energetic particles in the energy range 10–100 keV/charge and often associated ultra-low-frequency (ULF) waves had been routinely observed in the vicinity of Earth's bow shock since the 1960s (e.g. Asbridge et al. 1968, Greenstadt et al. 1968), the ISEE mission was the first one to be dedicated to understanding their behaviour. ISEE 1 and ISEE 2, launched in 1977, could measure particle and wave intensities along the sub-solar bow shock under a variety of solar wind conditions and unravel temporal and spatial structure. These observations, especially of the particle's angular distributions, led to the classification of the energetic particle distribution function into three classes (Hoppe et al. 1981).

Typically, as the solar wind flows into the shock, magnetic flux tubes at the nominal spiral direction first encounter a quasi-perpendicular shock on the dusk side of the Earth. They are then convected across to the quasi-parallel region on the dawn side. A reflected beam of ions is seen upstream of the shock, streaming back along the magnetic field ('reflected (R) ions' or 'field aligned beams'). As the flux tubes are convected across to the quasi-parallel region on the dawn-side, resonant waves generated by this beam are seen and the distribution of back-streaming ions begins to broaden from interaction with the waves ('intermediate (I) ions'). Finally, a diffusive region occurs with a nearly isotropic distribution of particles extending up to ~ 100 keV ('diffusive (D) ions') and a related complex pattern of wave packets (Figure 15).

The picture that has emerged from the ISEE observations is that these distributions represent temporal phases of shock acceleration as a given flux tube first contacts and then sweeps across the nose of the shock toward the flanks. Thus the Earth's bow shock has provided a laboratory to study shock acceleration and injection under conditions of limited geometry and finite acceleration time.

These spatial and temporal limitations play the role of adiabatic deceleration for the corotating ion events (see the following section) in limiting the acceleration of upstream ions to exponential spectra (in this case in energy/charge). However, ironically it is not yet understood how the exponential spectrum is formed. Finite time of connection to the bow shock is reasonable but need not yield energy/charge dependence and cannot account for the exponential spectra when the magnetic field is radial.

Acceleration at the Earth's bow shock was studied theoretically by Lee (1982). Wave generation by the particles was included in this model, as it was in the interplanetary shock model (Lee 1983) (see above). The principal difference for the bow shock is the shorter acceleration time as field lines are rapidly convected past the shock with a timescale of ~ 10 min.

Monte Carlo simulations of the particle acceleration have been performed more recently (e.g. Ellison et al. 1990, Scholer and Burgess 1992). Ions that undergo a few traversals of the shock contribute to increasingly higher energies in the spectrum. These simulations do not explicitly include wave generation. However, the use of a short scattering mean free path, λ, of only a few gyro-radii, presumes the presence of significant wave growth. These small values of λ are required for acceleration of particles directly from the solar wind. Unfortunately, no spatial dependence is assumed for λ in the Monte Carlo calculations; one would expect small values of λ to exist only near the shock, as found by Lee (1982).

Energetic ion events seen upstream of the Earth's bow shocks typically last between 10 minutes and 3 hours. The occurrence rate of these 'upstream events' increases when both the solar wind speed and the geomagnetic activity index are enhanced and when the interplanetary magnetic field at the spacecraft is directed so as to intercept the bow shock. While most observations have been made within about 25 Earth radii of the bow shock, upstream events have also been observed far upstream of the Earth, for example by ISEE 3, the first spacecraft in an orbit around the sunward Lagrangian point, approximately 215 Earth radii upstream of the Earth. Here most of the events exhibit large (100:1) field-aligned anisotropies, implying scatter-free transport, and little or no velocity dispersion during their onset or decay phases, indicating the convection of spatial structures populated with upstream particles over the spacecraft (e.g. Sanderson et al. 1981). Recent dual-spacecraft measurements show a large spatial extent of these events (Dwyer et al. 2000).

The origin of energetic ions (a few kilo-electronvolts to 1–2 MeV) observed upstream and within ~ 25 Earth radii of the Earth's bow shock has been a matter of controversy for more than two decades. It has been explained in terms

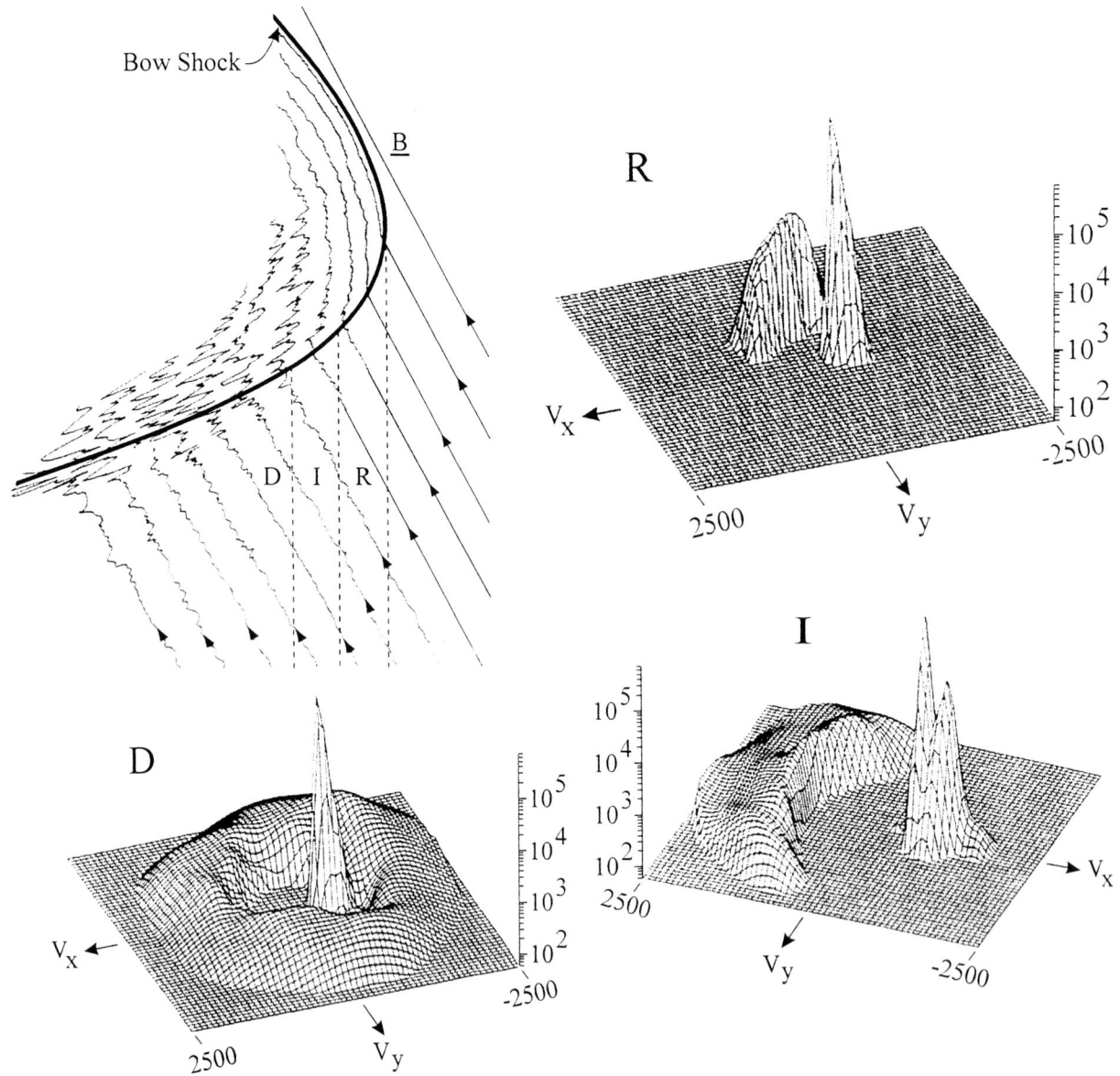

Figure 15 Reflected (R), intermediate (I) and diffuse (D) ion distributions and their locations upstream of the Earth's bow shock. (After Figures 4, 7 and 13 of Paschmann *et al.* 1981 and Figure 33 of Hoppe *et al.* 1981.)

of either (1) the leakage of magnetospheric particles accelerated inside the magnetosphere (e.g. during substorms) into the upstream region (e.g. Sarris *et al.* 1976) or (2) the acceleration of solar wind ions, either via reflection at the quasi-perpendicular portions of the bow shock or via scattering by the waves, i.e. first-order Fermi process, prevalent near the quasi-parallel portion of the shock (e.g. Lee 1982).

A key measurement that has not been available until the launch of high-sensitivity mass spectrometers on the Wind spacecraft is the energetic ($\geqslant 75$ keV/amu) heavy ion composition of upstream events. The elemental abundances should provide new insights into the origin of these particles, since the composition of the two potential sources of the parent populations (i.e. the Earth's magnetosphere and the solar wind) are considerably different. The analysis available so far, shows that the ion composition of the upstream events observed at Wind is similar to that measured in CIRs formed when high-speed (~ 600–750 km s^{-1}) solar wind streams interacted with slower (~ 300–450 km s^{-1}) solar wind streams (see the following section). This could be taken as evidence for further acceleration of an energetic seed population (provided by CIRs or SEPs) at the bow shock. A recent extensive study of the characteristics of more than 1200 upstream events observed in more than four years (Desai *et al.* 2000) concludes, however, that neither the main predictions of Fermi acceleration models nor those of magnetospheric leakage can satisfactorily account for the results of this study. The origin of the upstream ion population therefore remains a challenge.

Upstream particles have also been observed at other planets. Magnetospheric origin rather that than shock acceleration was suggested by Krimigis (1992) for the origin of upstream ions observed by Voyager at Jupiter, Saturn and Uranus. Upstream accelerated particles at Venus were reported by Williams et al. (1991). Hoppe and Russell (1982) concluded that ion acceleration is a characteristic of all planetary bow shocks.

COROTATING ION EVENTS

Discovery of 27-day recurrent particle increases

At the beginning of the 1965 solar minimum Bryant et al. (1965) and Fan et al. (1965) reported observations of modest increase in the intensity of ~ 1 to ~ 20 MeV/amu protons and alpha particles that lasted for three to six days and often reappeared several times at ~ 27-day intervals as shown in Figure 16. These 27-day recurrent events were not associated with solar flares and did not exhibit the velocity dispersion, characteristic of impulsive acceleration, often observed at the onset of solar energetic particle (SEP) events. Thus, they represented a new particle population, distinct from SEPs. It was initially believed that these were solar particles, continuously accelerated in active regions of the Sun, which after escaping into open magnetic field regions would then propagate along, and be confined within, stable interplanetary magnetic regions that corotated with the Sun. The recurrence of the increases at 27-day intervals, the rotation period of the Sun, was interpreted to be related to the long-term coherence of these magnetic field configurations. Based on this picture it was expected that the intensity of the recurrent particle increases would decrease with increasing distance from the Sun.

Our ideas about the origin of particles in the 27-day recurrent increases changed dramatically with new observations of these events during the next solar minimum in 1974–76. Advancements in instrumentation that allowed composition measurements down to sub-MeV energies, and most important the launches of Mariner 10, Helios 1 and 2, and the Pioneer 10 and 11 space probes exploring for the first time the inner (~ 0.3–1 AU) and the outer (>1 AU) heliosphere were crucial for the dramatic advances in our knowledge of this new heliospheric particle population.

Non-solar origin

Pioneer 10 observations (McDonald et al. 1976) revealed that the peak intensity of ~ 1 MeV protons in 27-day recurrent particle streams measured between 1 and 4 AU increased with increasing distance from the Sun. This was entirely inconsistent with a solar origin, and implied that these particles were accelerated in the interplanetary medium beyond several AU. There were earlier indications of positive gradients of few MeV particles. Vernov et al. (1970) using data from Zond 3 and Venus 2 between 0.7 and 1 AU found a 200% per AU increase with heliocentric distance in the 1.5 MeV protons associated with Forbush decreases during 1965–66, and raised the possibility of acceleration of mega-electronvolt particles in the inhomogeneities of the solar wind. Roelof and Krimigis (1973) observed events with positive gradients using Explorer 35 and Mariner 5. But it was only with the Pioneer 10 gradient observations when the non-solar origin of the recurrent events became to be generally accepted.

The location where this interplanetary acceleration took place and the solar wind structure responsible for this acceleration were first established by Barnes and Simpson (1976) and Smith and Wolfe (1976). In Figure 17 is shown the spatial–temporal correlation between the ~ 1 MeV corotating particle fluxes, the interplanetary magnetic field strength and the solar wind speed observed with Pioneers 10 and 11 at 4.7 and 4.4 AU, respectively. The magnetic field and solar wind variations (shown in the bottom two panels of Figure 17) led Smith and Wolfe (1976) to suggest that an interplanetary structure was formed by a forward–reverse shock pair with a compression region of strong and turbulent magnetic field in between. This interaction region bounded by the shock pair, would then corotate with the Sun and be observed by a nearly stationary spacecraft every 26 to 27 days. The formation of the shock pair mostly beyond 1 AU was a consequence of the interaction between fast and slow wind streams, which are a stable feature of the heliosphere around solar minimum.

The particle intensity profiles show a remarkable correlation with the solar wind and magnetic field structures of the CIR, with the peaks coinciding within about one day with the forward and reverse shocks. There are other repeatable features seen in these two CIR associated increases representative of the average characteristics of some 50 CIR events in that study. The time-integrated flux as well as the peak intensity is usually greater at the reverse shock than at the forward shock. The differential energy spectra are steeper (characterized by a higher spectral index) at the forward shock than at the reverse shock, and the proton to helium ratio is over 100 at the forward shock and only ~ 10 to 20 at the reverse shock. Taken together, these observations provided strong evidence for the existence of large-scale, continuous particle acceleration in these CIR structures, and that the corotating shocks, especially the reverse shock accelerated particles to at least several mega-electronvolts.

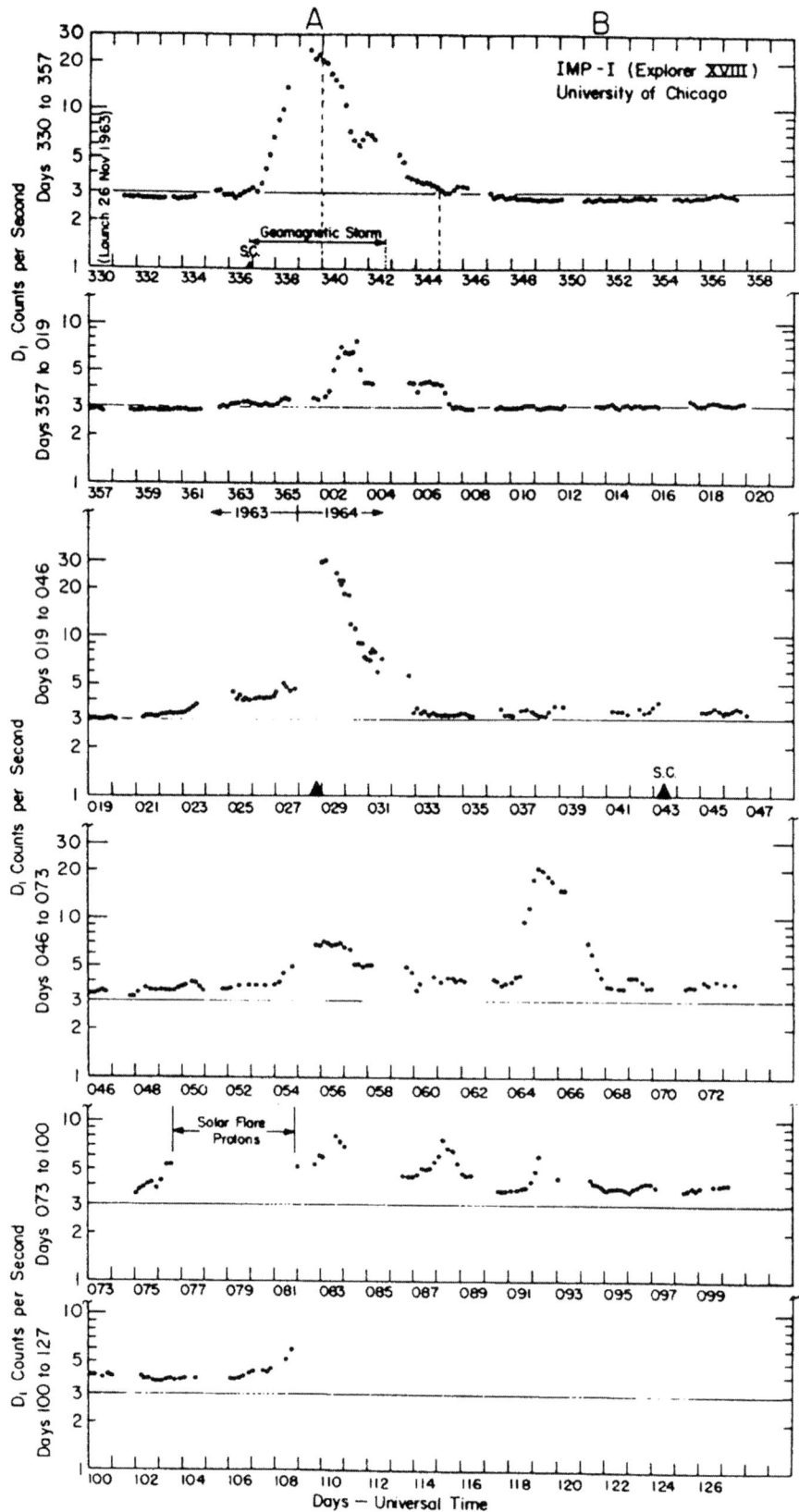

Figure 16 Intensities of 20 MeV protons during several solar rotations in 1963–64, showing a series of corotating regions under A, and intermittent series under B (Fan *et al.* 1965).

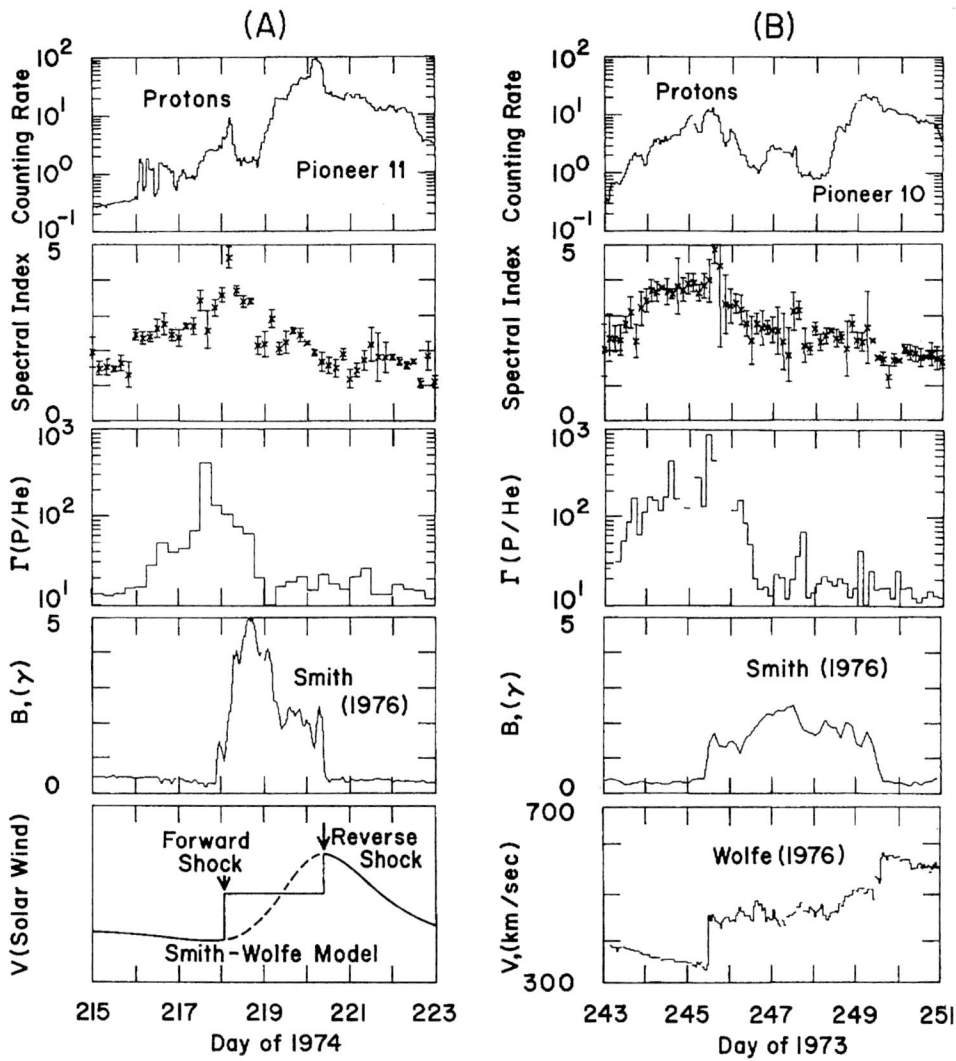

Figure 17 Two CIR proton events observed at 4.4 and 4.7 AU. Note the peaks of proton intensity at the boundaries of the CIR, coinciding with the observed jump in solar wind speed (bottom panels). Smith (1976) and Wolfe (1976) refer to private communications. (Figure 3 of Barnes and Simpson 1976.)

There were many studies that followed which cemented these ideas of interplanetary acceleration of CIR particles. Careful measurements of the radial gradients of corotating particle streams using Helios 1 and 2, IMP 7 and Pioneers 10 and 11 by Van Hollebeke et al. (1978) indicated a strong positive gradient from 0.3 to ~3 to 4 AU, followed by a decrease at 10 AU as shown in Figure 18. These observations established that the acceleration process was dominant between 3 and about 5 or 6 AU. Using Mariner 10, IMP 7 and Pioneer 11 data, John Simpson's group at Chicago found that the gradients of solar energetic particles (SEPs) and those of corotating particles were dramatically different. While SEPs had the expected negative gradient (-350% per AU), CIR particles had a positive gradient of $\sim 300\%$ per AU between ~ 0.5 and 2.2 AU. The positive gradient measurements for CIR particles were also supported by observations of anisotropies and flow directions of corotating particles at 1 AU. For example, Ed Stone's group at Caltech using IMP 7 data showed that on average, the few MeV/amu proton and alpha particles in corotating events streamed diffusively toward the Sun, indicating a source of particles beyond the orbit of Earth.

What ambient material was accelerated and the nature of the acceleration mechanism were far from being resolved. Acceleration mechanisms discussed most frequently included both statistical processes driven by magnetic and plasma turbulence which are observed in the CIRs (e.g. Fisk 1976a) and acceleration at the forward and reverse

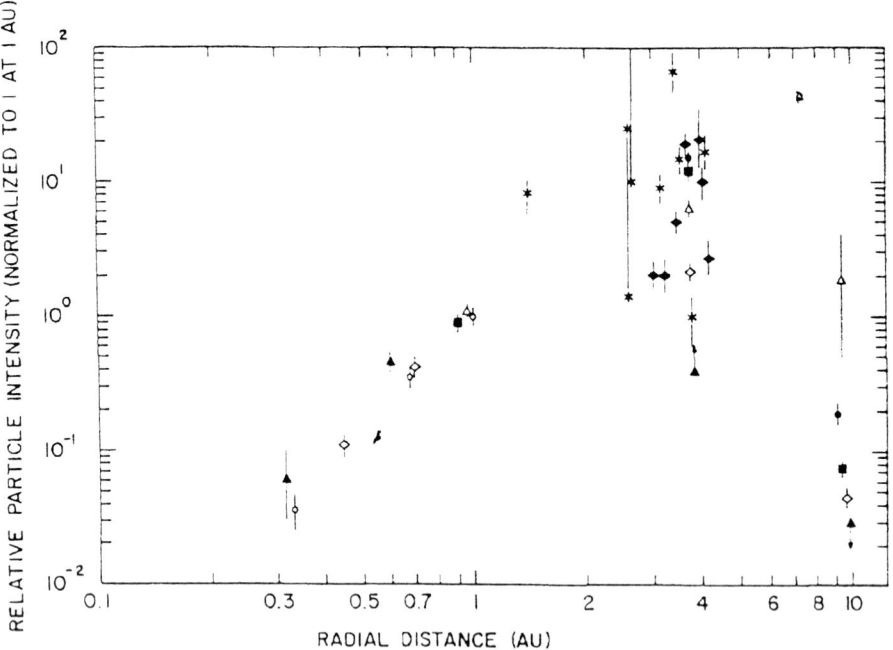

Figure 18 Relative intensity of ~1 MeV protons as a function of radial distance from the Sun; the observations have been normalized to the intensity at 1 AU. (Figure 5 of Van Hollebeke *et al.* 1978.)

shocks of the CIRs (e.g. Palmer and Gosling 1978). Based on the then available composition measurements the most plausible source seemed to be the solar wind. However, Gloeckler (1979) also suggested another source, an 'as yet undetected, low-energy (<200 keV/amu) residual particle population presumably of solar origin' in addition to 'the high-energy tail of the solar wind.' It is interesting to note that this other potential source of CIR particles, namely the 'inner source' pickup ion population (Geiss *et al.* 1995), was discovered some 20 years later.

Composition and energy spectra of particles in corotating events

Having established the location of the acceleration regions to be primarily between ~3 and ~6 AU, the next task was to pinpoint the source of the CIR accelerated particles. Detailed measurements of the velocity distributions and composition of corotating particles were made during the 1974–76 solar minimum using composition spectrometers on IMP 7 and 8, at 1 AU, Helios 1 and 2 in the inner heliosphere and Pioneers 10 and 11 in the outer heliosphere. It was found that the velocity distributions (phase space density v. speed) at 1 AU for all measured elements had the same exponential shape with the same slope of $\sim 3 \times 10^8$ cm s^{-1} for each species. These characteristic spectra did not change significantly with time following the onset of the CIR event (Gloeckler *et al.* 1979). Furthermore, these same spectral forms with the same slopes were observed by Frank McDonalds's group at the Goddard Space Flight Center both in the inner and outer heliosphere. The exponential velocity spectra were consistent with a predominantly statistical acceleration process, such as transit-time damping (Fisk 1976a), in or near the CIRs.

The average composition of corotating particles (Table 3) was also distinct in several respects. Unlike the case of solar energetic particles, the relative abundance of CIR particles appeared to be less variable from one event to the next. Compared to the SEP composition, corotating particles had a large amount of He and C relative to O. On the whole the composition in recurrent events reflected most closely the relative abundance of the solar corona, and, by inference, that of the solar wind, whose composition was not well measured at that time. Based on the then available composition measurements it was believed that the high-velocity tail of the solar wind was the most likely source material for corotating particles.

Additional observations of the composition of particles in corotating events during the 1993–95 solar minimum at both 1 AU (compiled by Reames 1999) and ~5 AU (Keppler *et al.* 1995), and the comprehensive SWICS Ulysses measurements of the solar wind composition which then became available for the first time (von Steiger *et al.* 2000) made it possible to re-examine the hypothesis that the solar wind was the only source of CIR particles. Comparing relative abundances of the solar wind with CIR particles (Table 3) one finds that in corotating events at

Table 3 Abundance relative to O of the solar wind, solar energetic particles, corotating ion events at 1 AU, anomalous cosmic rays, interstellar pickup ions and inner source pickup ions

Element or isotope ratio	Average of the fast and slow solar wind (von Steiger et al. 2000)	Gradual solar energetic particle events (Reames 1999)	CIR events (Gloeckler 1979, Reames 1999)	Anomalous cosmic rays (Cummings et al. 1999, Stone and Cummings 1997, Leske et al. 1999b)	Interstellar pickup ions (Gloeckler and Geiss 2001) Inner source pickup ions (Gloeckler et al. 2000a)
H	3069 ± 215	1570 ± 220	2300 ± 200	$(1.22 \pm 0.07) \times 10^1$	$(1.83 \pm 0.40) \times 10^3$
He	85 ± 18	57 ± 3	165 ± 15	6.27 ± 0.36	$(3.02 \pm 0.60) \times 10^2$
C	0.69 ± 0.04	0.465 ± 0.009	0.97 ± 0.10	$(6.25 \pm 0.74) \times 10^{-3}$	$(2 \pm 1) \times 10^{-3}$
N	0.093 ± 0.022	0.124 ± 0.003	0.140 ± 0.014	$(1.39 \pm 0.09) \times 10^{-1}$	$(1.47 \pm 0.36) \times 10^{-1}$
O	1.000	1.000 ± 0.01	1.000 ± 0.037	1.00	1.00
Ne	0.088 ± 0.014	0.152 ± 0.004	0.185 ± 0.017	$(7.25 \pm 0.44) \times 10^{-2}$	$(1.43 \pm 0.36) \times 10^{-1}$
Na		0.010 ± 0.001	—	$(1.57 \pm 0.63) \times 10^{-4}$	—
Mg	0.128 ± 0.025	0.196 ± 0.004	0.135 ± 0.013	$(4.71 \pm 2.03) \times 10^{-4}$	$(6.6 \pm 3.5) \times 10^{-4}$
Si	0.135 ± 0.026	0.152 ± 0.004	0.094 ± 0.010	$(9.99 \pm 2.05) \times 10^{-4}$	$(4.3 \pm 2.2) \times 10^{-4}$
S	0.050 ± 0.011	0.0318 ± 0.0007	0.055 ± 0.011	$(2.15 \pm 1.01) \times 10^{-4}$	—
Ar	—	0.0033 ± 0.0002	—	$(6.40 \pm 0.65) \times 10^{-3}$	—
Fe	0.101 ± 0.021	0.134 ± 0.04	0.096 ± 0.012	$(1.0 \pm 0.3) \times 10^{-3}$	—
^{3}He/He				—	$(2.48 \pm 0.66) \times 10^{-4}$
^{15}N/N				$< \sim 0.02$	—
^{18}O/^{16}O				~ 0.002	
^{22}Ne/^{20}Ne				~ 0.07	

1 AU both He and Ne are about a factor of two, and C and N are about 40–50% more abundant than in the solar wind, while the abundance of H and Si is lower by approximately 30–40% compared to that of the solar wind. The CIR particle composition at 5 AU is similar to that at 1 AU.

Observations of corotating particles with Ulysses

With the launch of Ulysses in 1990 came new advances as well as some surprises concerning the origin, acceleration and propagation of corotating particles. Ulysses is well equipped to measure spectra and composition of particles starting with the solar wind (see Chapter 47) and extending to high energies, well beyond tens of MeV/amu. Ulysses' orbit and time of launch were such that it would spend much of its time between ~ 3 and 5.4 AU during the 1993–96 solar minimum, in the region where, and at a time when, well-developed CIRs appeared regularly.

One of the surprising results was the persistence of recurrent events to high latitudes, long after the disappearance of the CIRs that accelerated these particles (see Keppler (1998) and Kunow et al. (1999) for a summary of high-latitude observations). These crucial observations led Fisk (1996) to propose a new global magnetic field model (see Chapter 47) which provided in a natural way for direct magnetic connection across a large latitude range, allowing particles accelerated in CIRs at low latitudes to reach the high latitudes where they were observed. This was, however, not the only way to transport energetic particles to high latitude. An alternative transport method suggested by Jokipii et al. (1995) was by diffusion perpendicular to the average magnetic field due to random walk of field lines.

Additional sources for corotating particles

Ulysses observations of the solar wind and pickup ions in CIRs at 4 to 5 AU showed that pickup He$^+$ ions were far more easily injected and accelerated in these turbulent regions than the solar wind He^{2+} ions (Gloeckler et al. 1994). Thus, after protons, which are very likely a mixture of solar wind and pickup hydrogen, pickup He$^+$ was the second most abundant species from ~ 10 keV to ~ 60 keV. Presumably most of the several MeV/amu CIR accelerated helium whose ionization state is not measured is also singly charged. This result demonstrated that pickup ions were injected far more efficiently than solar wind ions for further acceleration. Interstellar pickup ions are then clearly an important source for CIR accelerated particles, and when available are preferentially accelerated compared to solar wind ions. The efficient acceleration of interstellar pickup

He, and presumably of other pickup ion species, provided a natural explanation for the overabundance of He, Ne and N. But the high abundance of C could not be explained in this manner because interstellar pickup carbon was absent. It was found, however, that the C/O, as well as the Ne/O ratios in corotating events at 1 AU were each well correlated with solar wind speed (Mason et al. 1997), reaching the highest values at the highest speeds. The positive correlation with speed argued for the preferential injection of pickup ions at large solar wind speeds since, with their high density at speeds of up to twice the solar wind speed, most of them would have the highest speed and thus a more efficient injection. This explained the high Ne/O ratio at high speed because the interstellar pickup ion ratio at several AU is about 0.4. However, compared to O there is less than 1% of interstellar pickup C (Gloeckler and Geiss 1998), so that the large abundance of C in CIR events in the high-speed wind had to come from another source.

The suggestion that the C found in the inner source could account for the anomalous cosmic ray carbon was first made by Geiss et al. (1995) (see the following section). For this to occur, inner source pickup ions would have to be accelerated inside the heliosphere to overcome their strong adiabatic cooling by the time they reached the termination shock at about 80–110 AU. Gloeckler (1999) proposed that the inner source pickup ions were another plausible source of CIR particles. Compared to the other sources it had the largest C/O ratio ($\sim$1). Inner source pickup ions were thus the best candidate to supply the large amounts of C often observed in corotating particles. This source also contained relatively large amounts of Ne, Mg and Si (Gloeckler et al. 2000a) and could thus contribute these elements as well.

Acceleration of corotating particles

As soon as it was established that corotating particles were not solar, but accelerated in the heliosphere, various acceleration processes for these particles began to be discussed. Prominent among these were both statistical processes, such as transit-time damping (Fisk 1976a) and second-order Fermi acceleration by Alfvén waves (e.g. Jokipii 1971), and shock acceleration (e.g. Palmer and Gosling 1978). All these mechanisms are still being considered to either accelerate the ambient material first to some intermediate energies (several hundred kilo-electronvolts), from these intermediate energies to the full energies observed for CIR particles (tens of mega-electronvolts), or to the top energies by a single process.

In the left-hand panel of Figure 19 numerically calculated distribution functions (at various heliocentric distances) based on the diffusive shock acceleration model of Fisk and Lee (1980) are compared with observations at 1 AU of CIR spectra of different elements normalized to protons (Gloeckler et al. 1979). The calculated velocity distributions at the forward and reverse shock respectively are shown in the right-hand panel. In this model, which not only describes the acceleration but also the propagation of the CIR particles back to, for example, 1 AU, the acceleration takes place at the forward and reverse shocks of a CIR by multiple scattering between the shock and the upstream turbulence. As in many other shock models, some injection energy threshold is assumed. With reasonable model parameters it is possible to match the data well down to about 100 keV/amu.

The most comprehensive measurements of spectra for H^+, He^+ and He^{2+} in and outside CIRs are available from Ulysses. These observations extend from below solar wind energies to about 10 to 50 times solar wind energies. In the left-hand panel of Figure 20 (from Gloeckler 1999) the velocity spectra of the three ion species downstream of the forward shock are compared with those in the downstream region of the reverse shock, while in the right-hand panel the spectra of the three ion species, all normalized to H^+, downstream of the reverse shock are compared to the upstream H^+ distribution. The various spectra are normalized (as indicated by the numbers in parenthesis) to reveal the following remarkable features. First, more pickup He^+ than solar wind He^{2+} is accelerated even though solar wind He^{2+} is at least 1000 times more abundant than pickup He^{2+}. Second, the spectral shapes of each of the three species behind the forward shock is the same as that behind the reverse shock, even though the two shocks had quite different parameters. The stronger reverse shock heated the solar wind far more than the weaker forward shock, yet both produced identical accelerated spectra. Finally, as shown in the right-hand panel, all three species had identical spectra. This implies that the acceleration mechanism depends primarily on the ion speed. Furthermore, the spectral shapes are not power laws as predicted by simple shock acceleration mechanisms.

Observations such as these in the critical suprathermal velocity range where injection and initial acceleration occur, place important constraints on possible acceleration mechanisms. Observations of suprathermal particles, as those shown here, are most consistent with statistical acceleration, such as transit-time damping, at least to energies of 100 keV or more. Schwadron et al. (1996) found evidence for such acceleration by comparing suprathermal ion fluxes with magnetic turbulence levels, and were able to reproduce the observed spectral shapes using transit-time acceleration. A two stage acceleration process for CIR particles was proposed by Gloeckler et al. (1994). In the initial step, particles are accelerated by some statistical mechanism in the turbulence between the forward and reverse shocks of

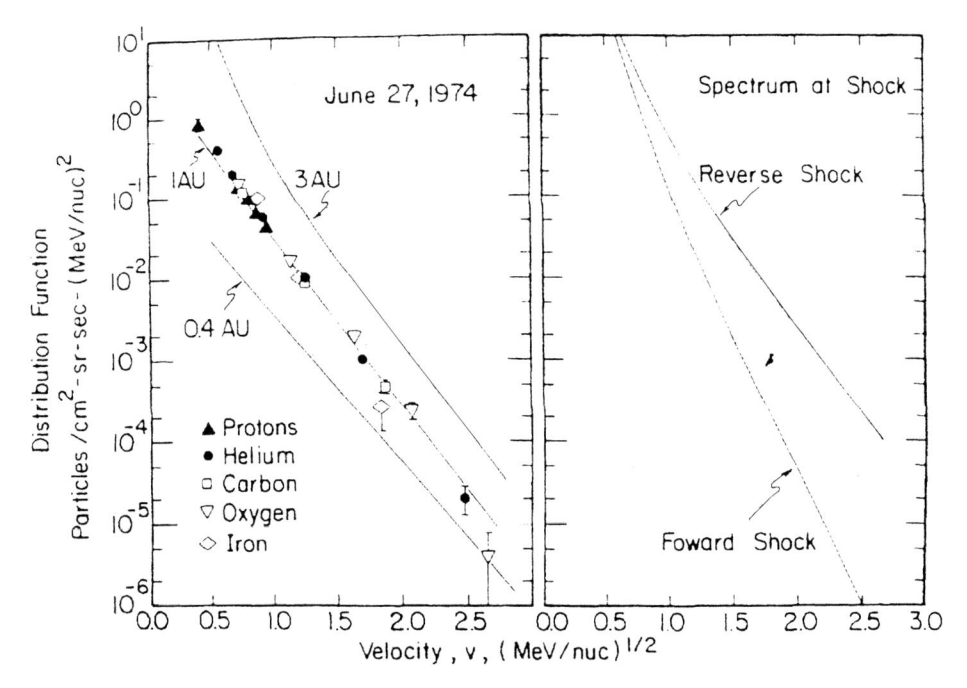

Figure 19 Numerical calculation of distribution functions, using the Fisk and Lee (1980) model. Left: Functional forms at three heliocentric radii, and comparison with 1 AU data (Gloeckler et al. 1979). Right: Spectra at the shocks, assumed to be at 4 AU. (Figure 2 of Fisk and Lee 1980.)

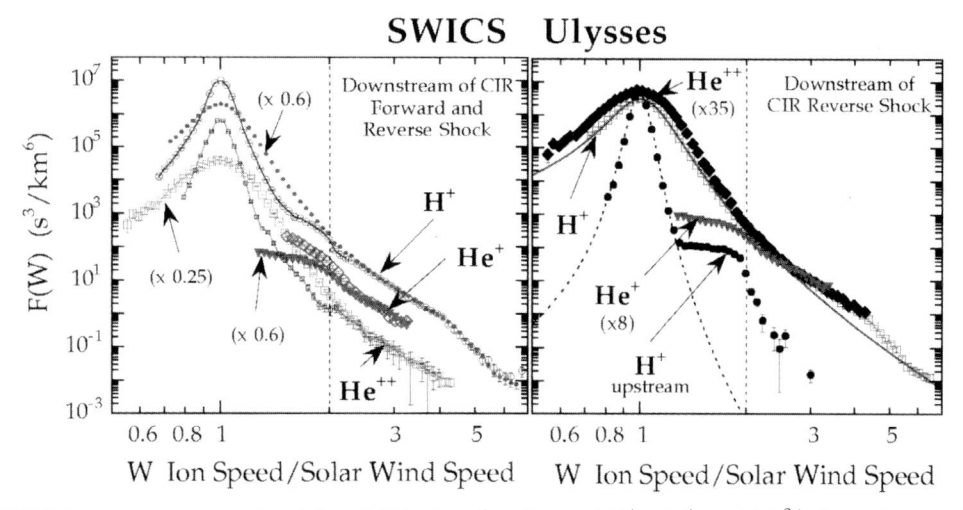

Figure 20 SWICS/Ulysses measurements of the distribution functions of H^+, He^+ and He^{2+} downstream of a CIR forward shock (open symbols) and reverse shock (filled symbols). Left: The respective spectra downstream of the reverse shock have been normalized to those downstream of the forward shock to reveal the remarkable similarity in the spectral shape of each of the three species in the high-velocity tail region. Right: The spectra of He^+ and He^{2+} downstream of the reverse shock have been normalized to that of H^+ to reveal the identical shapes of all three species in the tail region. The upstream H^+ spectrum is shown for comparison. (Figure 7 of Gloeckler 1999.)

the CIR. Once these ions were accelerated to these intermediate energies they would have sufficient energy to be injected and then accelerated to much higher energy by diffusive shock acceleration at the forward and reverse shocks of the CIR.

The CIRs become more complex in the outer heliosphere as the forward and reverse shocks spread and cross each other and parcels of solar wind plasma are multiply shocked. The coalescence of interaction regions beyond several astronomical units form merged interaction regions, which are regions of relatively intense magnetic fields with a radial extent of at least several astronomical units (Burlaga *et al.* 1985). These turbulent regions do not only effect the galactic cosmic ray intensity, but are likely to accelerate particles as well.

ANOMALOUS COSMIC RAYS AND THEIR PICKUP ION SOURCES

Shortly before the 1974–76 solar minimum several research groups reported observations of energetic particles with unusual composition. These tens-of-MeV particles were seen during solar quiet times, in the absence of solar flare particles. In rapid succession came reports by the University of Chicago group of helium fluxes exceeding proton fluxes in the energy range between about 20 and 30 MeV/amu (Garcia-Munoz *et al.* 1973), and by the Max-Planck-Institut (Garching) and the University of Maryland groups of oxygen fluxes substantially higher than carbon between about 1 and 10 MeV/amu (Hovestadt *et al.* 1973). The true significance of these independently obtained results was not immediately apparent. Then the Goddard Space Flight Center and the University of New Hampshire group reported N/He as well as O/He flux ratios about 10 times higher than normal around 10 MeV/amu (McDonald *et al.* 1974) (Figure 21). Taken together, these observations indicated a highly anomalous composition, unlike anything seen before for either galactic cosmic rays or solar energetic particles. The ability to observe these low-energy, heavy nuclei for the first time was the result of new particle detector technology that was coming into use, flown on spacecraft such as IMP 7 and IMP 8 and Pioneer 10 that spent most of their time in interplanetary space.

By the beginning of 1974, there was a considerable amount of information available concerning this new anomalous component and Stone's group at Caltech as well as Price's Berkeley group confirmed its existence. The oxygen hump at ~5 MeV/amu seen by Hovestadt *et al.* (1973) and McDonald *et al.* (1974) persisted. The information then available on the anomalous component or the anomalous cosmic rays (ACRs) as they were soon called (the term 'anomalous modulation' was used by John Simpson, while 'anomalous component' was first used by McDonald *et al.* 1974), was summarized in an invited talk by Gloeckler given in 1974 at the Spring Meeting of the American Physical Society in Washington, DC. In addition to the anomalies in the spectral shapes (Figure 21) and abundances (Figure 22) of O and N between ~2 and ~30 MeV/amu, and of He between ~10 and ~60 MeV/amu, Gloeckler listed the following experimental results available at that time:

1. Variations of intensities of the ACRs with both time (reported by both the Caltech and Goddard groups) and distance (reported by the Goddard group) were similar to those seen for galactic cosmic rays and thus ruled out a solar origin (see Figure 23 for more extensive observations of ACR oxygen).
2. Elements such as ~10 to 30 MeV/amu B, Ne, Mg, Si, and Fe had an abundance roughly similar to that of galactic cosmic rays at higher energies (reported by the Goddard and Berkeley groups).
3. The major isotopes of O and N were ^{16}O and ^{14}N (reported by Caltech).

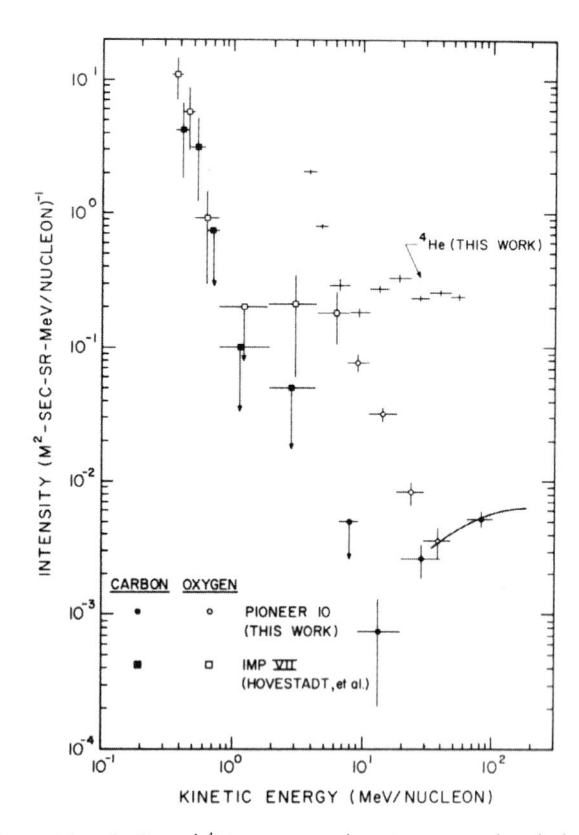

Figure 21 C, O and ^{4}He spectra, showing anomalously high O and ^{4}He abundance at ~2 to 20 and ~10 to 40 Mev/amu respectively. The C and O spectra at higher energies derived from the Goddard cosmic-ray telescope on IMP 7 are indicated by the solid line. (Figure 2 of McDonald *et al.* 1974.)

One of the burning questions at that time concerned the origin of the ACRs. The traditional approach, given the extra-solar origin, was to postulate nearby stellar sources with highly abnormal composition. Indeed, Hoyle and Clayton (1974) proposed such a scenario. They started with the usual nova phenomena in which material of approximately solar composition was transferred from a giant to a white dwarf companion of the binary system. Then they assumed a carbon- and oxygen-rich surface for the white dwarf with sufficient surface mixing to produce hydrogen

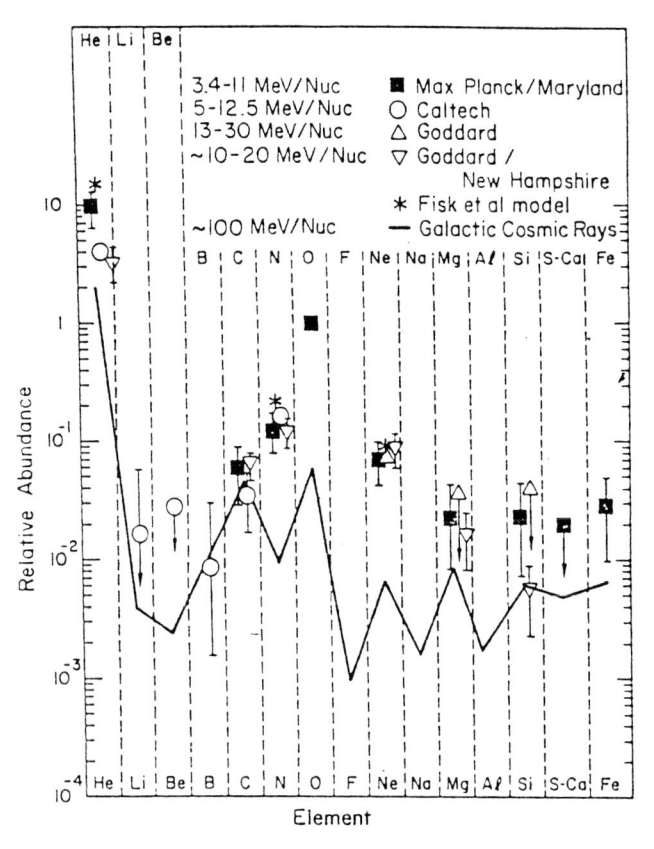

Figure 22 Comparison of the relative abundances in the anomalous component between ~3 and 30 MeV/amu with the composition of galactic cosmic rays at ~100 MeV/amu (solid curve) and abundances (*) deduced from the model of Fisk et al. (1974) for He, N, O and Ne. The elements N, O and Ne in the anomalous component are over-abundant by about a factor of 10 compared with the respective elements in galactic cosmic rays. (Figure 3 of Gloeckler 1975.)

to carbon ratios of order unity. Under these special conditions nuclear burning of carbon and the infalling hydrogen will produce large amounts of nitrogen. Finally, the acceleration of this enriched material to tens of MeV/amu was proposed to be by conventional nova shocks or turbulent magnetic fields generated in the post-explosion phase nova shocks. There were difficulties with this and other such nucleosynthesis models. All such models predicted enrichments of rare isotopes such as ^{13}C, ^{15}N and ^{17}O, that were not observed in the 1972–74 ACR data, but could not be entirely ruled out. More serious objections came from considerations of power requirements necessary to maintain the observed oxygen intensity. Only by changing drastically the fundamental assumptions of cosmic ray modulation theories was it barely possible to reconcile the power requirements of these nucleosynthesis models with the spectra observed for ACR oxygen (Fisk 1976b).

There were several other suggestions for the origin of the anomalous component. For example, Fowler et al. (1979) proposed a cometary source whereby cometary molecules, released at distances between 0.5 and 4 AU from the Sun, would eventually break up, become ionized and accelerated. Biswas et al. (1981) invoked stellar winds of O stars to supply the material for the ACRs. Particles released from such stars would presumably be enriched in O and N, would be partially ionized and accelerated by interstellar shocks.

Fisk et al. (1974) proposed an entirely different explanation for the origin of the anomalous component. They noted that elements that were highly overabundant in the ACRs (O and N) were precisely those that had a high first ionization potential (FIP). They then argued that a source for the ACRs that would fit the unusual abundance pattern was the neutral component of interstellar gas devoid of low-FIP atoms that were already all ionized in the interstellar medium. Several years earlier the presence of interstellar atomic hydrogen well inside the heliosphere was inferred from all-sky Lyman-α maps (Bertaux and Blamont 1971) and the first rudimentary theory for the entry and evolution of interstellar neutral gas in the solar system was proposed by Blum and Fahr (1970). Fisk et al. (1974) argued that ionized interstellar gas would be prevented from entering the Solar System by the magnetic field configuration at the interface between the heliosphere and the interstellar medium. However, the neutral, high-FIP elements of the interstellar gas would penetrate deep into the Solar System because of the relative motion of the Sun with respect to the local interstellar cloud. This simple process thus supplies the high-FIP (He, N, O, Ne), but not the low-FIP (C, Mg, Si, Fe) material, matching the anomalous composition of the ACRs. Of course acceleration is still required to take the few-eV neutrals and convert them to the tens-of-MeV particles of the anomalous component. As described in Fisk et al. (1974), the first step in this acceleration process is straightforward. Once the interstellar neutrals came sufficiently close to the Sun they would be ionized by solar UV and/or by charge exchange in the solar wind. The newly created ions, called pickup ions, would be immediately picked up by and carried outward with the solar wind. Pickup ions will have energies up to twice the solar wind speed, or roughly over 1 keV/amu. So now one had a source of ~1 keV/amu, singly charged ions, inside the heliosphere with a composition that is very similar to the tens-of-MeV ACRs. Fisk et al. (1974) then assumed that while being convected outward with the solar wind some fraction of these singly ionized pickup ions would be preferentially accelerated, and by the time they reached the outer heliosphere at a distance of about 100 AU they would have gained energies of tens of MeV. The power required for this acceleration was found to be a small fraction of the solar wind

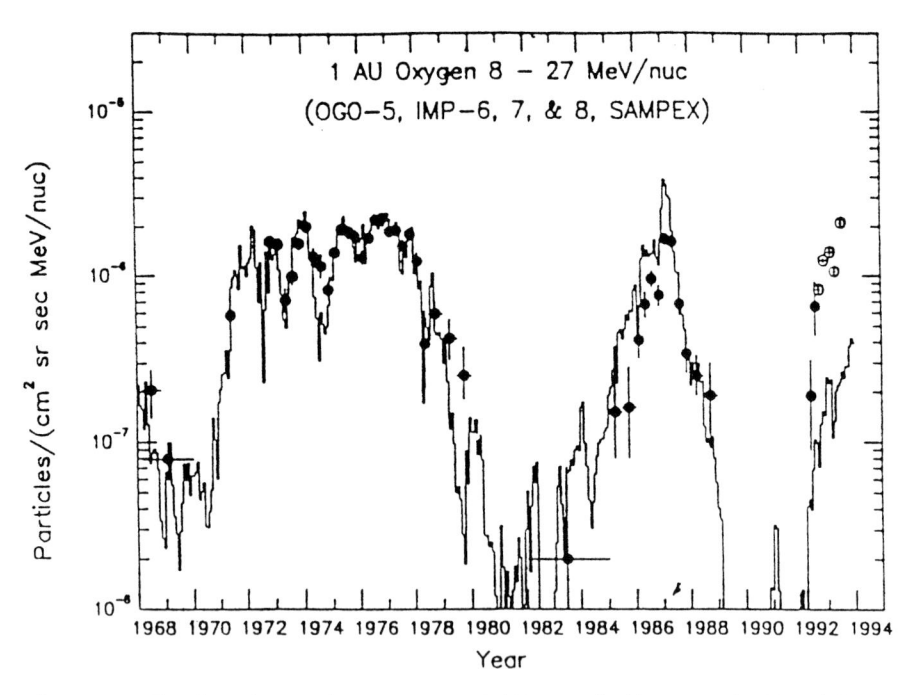

Figure 23 The flux of anomalous oxygen at 1 AU compiled between 1968 and 1994.

flow energy. The energetic singly charged ions would then be subject to solar modulation, just like galactic cosmic rays, as they diffused inward as the anomalous component. However, unlike the fully ionized low-energy galactic cosmic rays, the ~ 10 MeV/amu singly charged ACRs would suffer relatively little modulation because of their large rigidities.

Ten years after their discovery it was still not possible to settle entirely the question of the origin of the ACRs. Even though the Fisk *et al.* (1974) concept of interstellar origin became the most favoured scenario, the other ideas could not be entirely ruled out on the basis of experimental knowledge of ACRs then available.

Current state of knowledge

ACR Ne, Ar and H

One of the predictions of the interstellar gas origin theory (Fisk *et al.* 1974) for the ACRs was that, in addition to O, N and He, the elements Ne and H should also be overabundant in the anomalous component. The early composition measurements were not sensitive enough to detect such enhancements. However, with additional observations the increased Ne abundance predicted by Fisk *et al.* (1974) was indeed observed by von Rosenvinge and McDonald (1975) as shown in Figure 22. Cummings and Stone (1987) found Ar to be enhanced in the ACRs. This was perfectly consistent with the Fisk *et al.* (1974) idea because some Ar, like Ne, was expected to be neutral in the interstellar medium and thus be able to enter the heliosphere (e.g. Vasyliunas and Siscoe 1976).

To detect anomalous hydrogen proved to be far more difficult, in part because, having the same charge state as the galactic cosmic ray hydrogen, both anomalous and galactic protons were strongly modulated at low energies. The first report of the detection of ACR protons was based on the observation of a hump between ~ 30 and ~ 300 MeV in the modulated H spectrum (Christian *et al.* 1988). But other explanations were suggested, such as modulation effects of galactic cosmic rays that could reproduce the observed spectrum (Reinecke and Moraal 1992). Other methods for detecting anomalous protons proved inconclusive. These were based on measurements of the D/H ratio which for the ACR was expected to be $\sim 2 \times 10^{-5}$, and ~ 0.05 in the galactic cosmic rays. Although D/H at ~ 100 MeV was found to be more than a factor of 2 lower than the galactic cosmic ray value (Lopate and McKibben 1991), modulation models could be adjusted to accommodate this decrease assuming a purely galactic component. Later came the Ulysses measurements of interstellar pickup hydrogen (Gloeckler *et al.* 1993) providing clear evidence that the source material for ACR H did indeed exist. But the definitive measurements that showed the existence of ACR protons were not made until the 1994–96 solar minimum. McDonald *et al.* (1995), using Voyager 1 (58 AU) and Pioneer 10 (61 AU) measurements, showed conclusively the presence of anomalous hydrogen. In 1996 both Voyager and Pioneer observations showed clearly that the hump in the proton spectrum below ~100 MeV was quite pronounced and the proton intensity about five times above the expected galactic cosmic rays both at Pioneer 10 at 64 AU, Voyager 1 at 63 AU and Voyager 2 at 49 AU (see Klecker et al. (1998) for details). There was now no way to adjust modulation parameters to explain the bulge in the proton spectrum assuming only a galactic component. The Voyager 1 and 2 proton spectra in the distant heliosphere averaged from 1993 to 1999 (Cummings et al. 1999) had the unmistakable bulge characteristic of ACRs at ~30 to ~40 MeV. The elusive anomalous protons were finally positively identified.

Charge states of the anomalous component

Almost immediately after their discovery it was recognized that measurements of charge states of the ACRs would be decisive in establishing their origin. The interstellar gas origin of Fisk et al. (1974) made a firm prediction that ACRs would be predominantly singly charged. Direct measurements of charge states of tens-of-MeV particles are difficult and have not yet been made. However, indirect evidence indicated as early as 1977 that ACR He was most likely singly ionized (McKibben 1977) and that the ionization state of oxygen was less than 3 (Klecker et al. 1980). These studies examined either the temporal or spatial variations of ACRs resulting from modulation, assuming either low or high ionization states for these energetic particles and used modulation parameters derived from galactic cosmic rays at higher energies. Using a variant of this technique and assuming that modulation depends on a combination of velocity and rigidity (momentum per charge), Cummings et al. (1984) inferred that ACR He, N, O and Ne are singly ionized by examining spectral features such as energies at which the flux for each of these species peaked.

A method independent of assumptions regarding modulation made use of ACR measurements on low-altitude, high-latitude Earth-orbiting spacecraft. With such orbits it was possible to use the Earth's magnetic field to determine the rigidity of the particle. Early results based on ACR measurements on Skylab 3 (Biswas et al. 1977) and on Spacelab 1 and 3 (e.g. Oschlies et al. 1989) were limited by counting statistics but indicated the presence of singly charged O, N and Ne. Later, Adams et al. (1991) derived a mean charge of about 1 for ~10 MeV/amu ACR O. Using advanced instrumentation on the 82° inclination, Earth-orbiting SAMPEX satellite, Klecker et al. (1995) found a large number of singly charged ACR N and Ne and placed an upper limit of 15% on the O^{2+}/O^+ ratio. SAMPEX measurements not only showed that the ACRs were predominantly singly ionized at around 10 MeV/amu, but found in addition that the fraction of singly charged to the total for each of ACR N, O and Ne decreased with increasing energy above ~10 MeV/amu (Klecker et al. 1997, Selesnick et al. 1997) as is shown in Figure 24. The SAMPEX observations, in particular, provided the decisive confirmation for the interstellar origin of the anomalous component. Furthermore, charge state measurements with SAMPEX showed that at higher energies ($\geq$15–20 MeV/amu) multiply-charged ACRs become increasingly more abundant. It had been pointed out earlier by Jokipii (1992) and Klecker (1995) that the relatively low abundance of multiply-charged ions below ~10 MeV/amu places limits on acceleration time-scales in the outer heliosphere, since the acceleration must be sufficiently rapid to compete with charge exchange processes that would further strip the source ions. The dominance of multiply-charged ACRs at higher energies, on the other hand, explains ACR energies close to the observed 100 MeV/amu because multiply-charged ions gain proportionally more energy in the acceleration process at the termination shock (Mewaldt et al. 1996).

Elemental and isotopic abundance of the anomalous component

The interstellar origin theory of Fisk et al. (1974) for the anomalous component predicted that elements with high

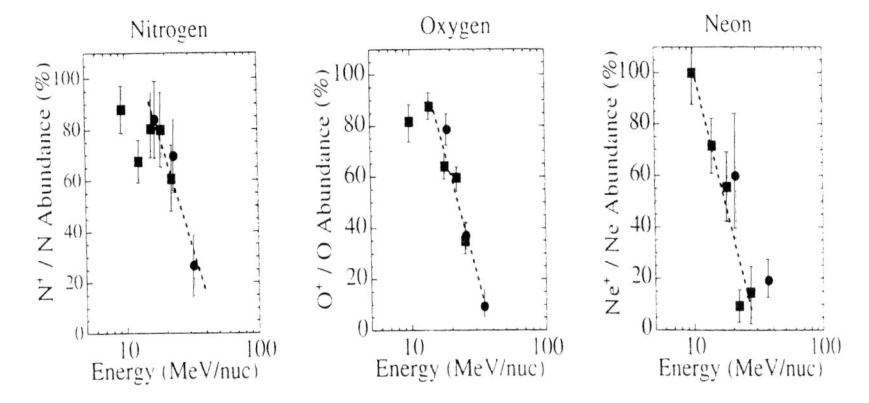

Figure 24 Relative abundance of singly ionized ACRs (N, O and Ne) as a function of energy. (From Klecker et al. 1997 (squares), and Selesnick et al. 1997 (circles).)

first ionization potential that had a relatively large neutral component in the interstellar medium would be the only elements present in the ACRs. All these elements, H, He, N, O, Ne and Ar were indeed found in the anomalous component. The upper limits placed by early measurements on the rare isotopes of Ne and O were consistent with an interstellar origin and excluded unusual galactic sources for their origin.

Isotopes

The most recent ACR isotopic measurements of N, O and Ne from ACE (Leske *et al.* 1999b) are relatively close to solar or solar wind ratios and thus provide further convincing evidence for the local interstellar origin of ACRs (Figure 25). In fact, measurements of the ACR isotopic composition of Ne and O have now advanced to the point that they may be used to provide information of the evolution of the local interstellar medium since the formation of the Solar System, 4.5 Gyr ago.

Minor ACRs

An interstellar gas origin for the ACRs implies that low FIP elements such as C, Mg and Si would be absent in the anomalous component. Thus the surprising discovery of ACR C by Cummings and Stone (1987) presented a real challenge for explaining its origin. Subsequently, measurements made with instruments on Wind (Reames *et al.* 1997b) and Geotail (Takashima *et al.* 1997) at 1 AU reported ACR S, another low-FIP element not present as neutrals in the interstellar medium. Voyager 1 and 2 observations (e.g. Cummings *et al.* 1999) and Wind measurements (Reames 1999) have provided clear evidence for ACR Na, Mg, Si and Fe in addition to C and S. Geiss *et al.* (1995) proposed that the source of ACR carbon could be accelerated pickup C from the inner source they had just discovered in the inner heliosphere. Mewaldt (1999) suggested re-accelerated remnants of solar energetic particles could be responsible for the ACR Si, S and Fe. While the existence of the minor ACR component is now well established its origin is less certain. Definitive measurements of the charge states of the minor ACRs would discriminate between the two alternatives that have been proposed – re-accelerated remnants of solar energetic particles or accelerated inner source pickup ions (Geiss *et al.* 1995, Gloeckler *et al.* 2000a). Current observations, however, seem to support the idea that inner source pickup ions are the primary source of the minor ACRs. For example,

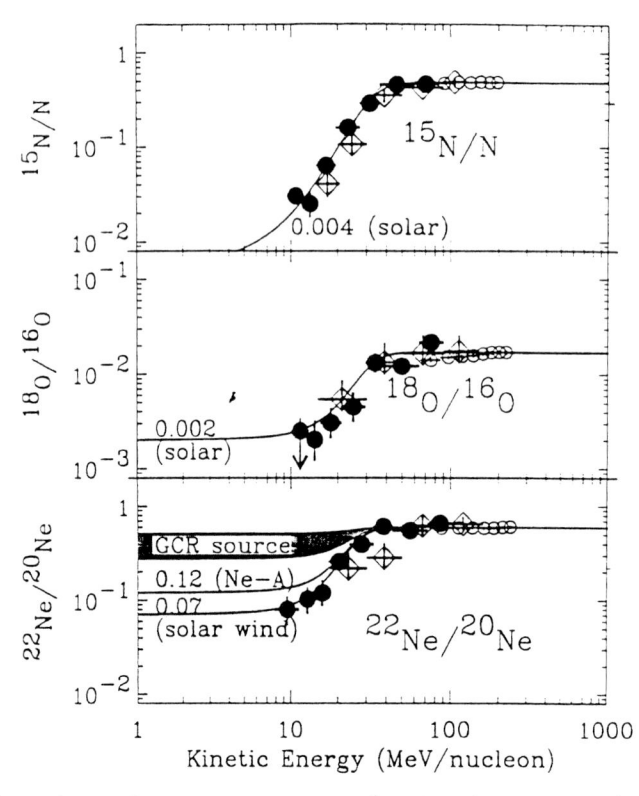

Figure 25 Observed energy dependence of isotopic composition of N, O and Ne compared with expected values (curves and shaded region). The observations are from the Solar Isotope Spectrometer (SIS; filled circles) and the Cosmic Ray Isotope Spectrometer (CRIS; open circles) on ACE as well as from SAMPEX (diamonds). (Figure 2 of Leske *et al.* 1999b.)

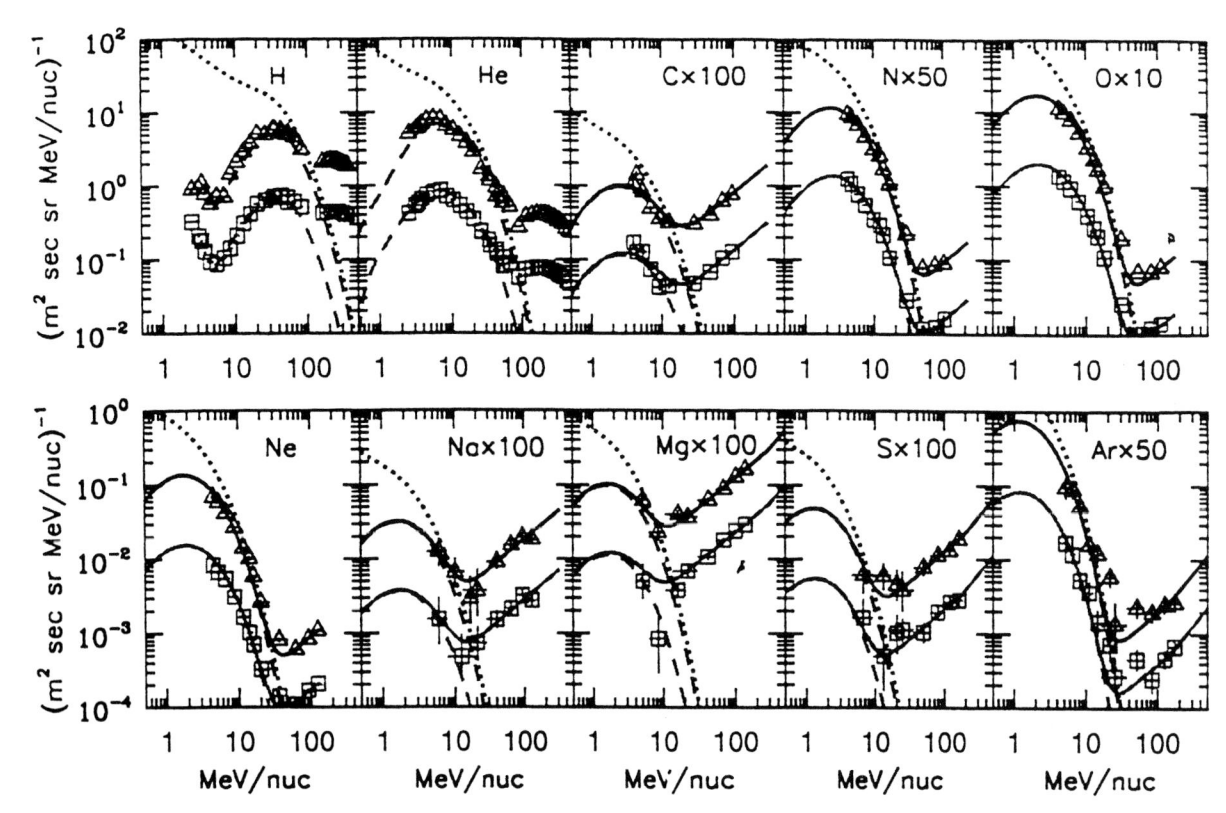

Figure 26 Energy spectra of 10 elements at Voyager 1 (V1, triangles) and Voyager 2 (V2, squares) for 1993/53–1998/365. In some panels factors are shown which have been applied to the V1 observations before plotting. The V2 observations use the same factors for V1 plus an additional factor of 0.2 for clarity. The dashed curves are model calculations using the model and parameters from Steenberg et al. (1999), except for the intensity normalization. The dotted curve is the spectrum at the shock at a polar angle of 60°. The solid curves are the sum of ACRs and GCRs. (Figure 1 from Cummings et al. 1999.)

Mewaldt et al. (1996) found singly and low-charged ACR C, Cummings et al. (1999) showed indirect evidence for low-charge Si, and Gloeckler et al. (2000b) estimated that the density of inner source pickup C is sufficient to account for the ACR C (see also Table 3).

Our current state of knowledge of the energy spectra and composition of the anomalous component is summarized in Figure 26 and Table 3.

Pickup ions: The source population of ACRs

The plausible suggestion by Fisk et al. (1974) that the most likely source of the ACRs were interstellar pickup ions preceded their discovery by more than a decade. The discovery of pickup He by Möbius et al. (1985) was almost accidental, made possible by a TOF instrument (SULEICA) on the board the German satellite (Ion Release Module, IRM) of the magnetospheric US–German AMPTE mission. The payload of the US satellite of AMPTE (Charge Composition Explorer, CCE) contained CHEM (Charge Energy and Mass), a TOF spectrometer based, as was SULEICA, on the SWICS design. However, because the CCE spacecraft, unlike IRM, never left the Earth's magnetosphere, CHEM could not detect pickup He. When singly charged He ions were detected by SULEICA (Suprathermal and Low Energy Ion Composition Analyzer) it was first thought that they were of magnetospheric origin (D. Hovestadt, personal communication). But careful analysis of their energy spectrum showing the predicted sharp drop-off of flux at twice the solar wind speed, and the expected seasonal variation of the He^+ intensity proved that these ions were interstellar pickup ions.

The discovery of all the other pickup ions was not possible with IRM because of all the neutrals in the interstellar gas, from which pickup ions are created only He survives at 1 AU. All other components of this neutral interstellar gas have been nearly completely converted to pickup ions and swept away by the solar wind before reaching 1 AU. To detect pickup ions other than He required spacecraft reaching heliocentric distances of at least several AU. The Ulysses mission had the right orbit, and the SWICS instrument had the capability to make these measurements.

Detailed studies of interstellar pickup ions with SWICS (Gloeckler *et al.* 1993, Geiss *et al.* 1994; see also Gloeckler and Geiss 1998, 2001 and Gloeckler *et al.* 2001 for details) revealed many of the expected characteristics of pickup ions. Except for the rare He^{2+}, they were all singly charged and had energy spectra that showed a pronounced drop in density at twice the solar wind speed. They were adiabatically cooled in the expanding solar wind. The variations of the pickup ion fluxes with radial distance (radial gradients) were consistent with the expected spatial distribution of the parent interstellar gas. There were, however, several surprises. It was found that, contrary to expectation, the scattering mean free path of pickup ions was large (~ 1 AU), there were large, short-term variations in the intensity of pickup H and He (and presumably for the less abundant heavier species as well, although this could not be established) that were correlated. But of most significance for the acceleration of pickup ions on their way to become the ACRs it was found (Gloeckler 1999, Gloeckler *et al.* 2000b) that even in the most quiet and least disturbed low speed solar wind, the energy spectra of pickup ions had significant density above the cut-off at twice the solar wind speed (so called high-velocity tails), which indicated continuous acceleration of pickup ions in the unperturbed slow solar wind as they were carried outward to the termination shock.

The densities of interstellar atomic H, He, N, O and Ne and the intensities of the corresponding pickup ions near the heliospheric termination shock have been deduced from SWICS/Ulysses measurements of the distribution functions of these ions between about 3 and 5 AU. The ratios at the termination shock of interstellar pickup ion densities relative to oxygen are given in Table 3. The most abundant elements in the ACRs are the same as those observed in the pickup ion population, with the relative abundance of the ACR and pickup ions being almost identical for the heavy elements N, O and Ne. The relative abundance of pickup H and He is, however, much larger than that of the ACR H and He. This is attributed to a much smaller injection and/or acceleration efficiency for these light ions at the termination shock.

Inner source as a possible primary source of the minor ACRs

The discovery of interstellar pickup ions was anticipated and not much of a surprise. In fact, there were expectations that the plasma instruments on Voyagers 1 and 2 and Pioneers 10 and 11 would find pickup hydrogen. Eventually, a careful examination of Pioneer data revealed signatures of pickup H (Intriligator *et al.* 1996), and recent analysis of the Voyagers' LECP data provides evidence for the high-velocity tails of pickup oxygen in the distant heliosphere (Krimigis *et al.* 2000).

However, the discovery with Ulysses of singly charged C in abundance comparable to O^+ (Geiss *et al.* 1995) came as a complete surprise. After all, C was believed to be fully ionized in the interstellar medium and as such it could not enter the heliosphere. It was quickly established that the differential energy spectrum and radial gradient of C^+ were respectively different from those of interstellar pickup O^+, and based on this evidence, Geiss *et al.* (1995, 1996) proposed that pickup C^+, unlike the other pickup ions, came from an extended 'inner source'. Further work (Gloeckler *et al.* 1998, 2000a,b, Schwadron *et al.* 2000) showed that inner source H^+, N^+, O^+, Ne^+, Mg^+ and Si^+ were also present. The velocity spectra of these inner source pickup ions as well that of C^+ peaked somewhat below the solar wind bulk speed and showed no clear evidence of a cut-off at twice the solar wind speed. The peaked spectra most likely resulted from extensive adiabatic cooling indicating that the source of neutrals from which these ions originated was well within 1 AU (Schwadron *et al.* 2000). Production of slow neutrals deep inside the heliosphere required the presence of grains or dust close (a few tens of solar radii) to the Sun. Because inner source Ne was clearly observed, the primary mechanism proposed for generating inner source ions involved reprocessing or recycling the solar wind (Gloeckler *et al.* 1998, Schwadron *et al.* 2000). In this process, dust grains near the Sun act as reservoirs in which solar wind ions become embedded. Solar wind material is then released, primarily as slow-moving neutrals, which then become ionized by photoionization and charge exchange with the solar wind.

The composition of inner source pickup ions (C, Mg and Si) relative to interstellar O^+ is given in the last column of Table 3. Considering possible systematic uncertainties of about a factor of two (only statistical errors are given in the table) the resemblance between the relative abundances of the minor ACRs and the inner source pickup ions is striking. While not proof by itself, this similarity in the compositions suggests that these singly charged inner source pickup ions could well be the main source of the minor ACRs.

Acceleration of ACRs

While the origin of the dominant component of the ACRs is now well established, the mechanism responsible for their acceleration and the location where the acceleration takes place is still under debate. The requirements for the acceleration process are substantial – the few-keV pickup ions have to have their energy increased to about 10 MeV or more. There were three different types of acceleration mechanisms proposed: (1) acceleration by transit-time damping (Fisk 1976a, Klecker 1977), (2) acceleration by the polar termination shock (Pesses et al. 1981), and (3) acceleration by the ecliptic termination shock (Fisk 1986).

Transit-time damping

This was the first theory for the acceleration of the several-keV pickup ions to the tens-of-MeV ACRs. Fisk (1976a) pointed out that turbulence, such as in the fast–slow wind stream interaction regions in the outer heliosphere, could lead to large-scale magnitude fluctuations in the magnetic field. These fluctuations or plasma waves would then transfer energy to pickup ions which would resonate with the waves, and which, in turn, would be damped in this process by a transit-time damping mechanisms. Using calculated acceleration rates and taking the magnitude of the spectrum as a free parameter, Klecker (1977) was able to reproduce the spectra and intensities of ACR O, N and Ne in a spherically symmetric, steady-state model for acceleration and modulation. Supporting this mechanism, observations in the outer heliosphere (e.g. Burlaga et al. 1996) revealed that large-scale variations in the field magnitude do indeed occur, and high-velocity tails in the distribution functions of pickup ions, presumably a consequence of transit-time damping, are ubiquitous in the low speed solar wind (Gloeckler et al. 2000b) being especially intense in the corotating interaction regions (Gloeckler 1999).

Polar termination shock

A first-order Fermi mechanism was proposed by Pesses et al. (1981) to accelerated ACRs at the heliospheric termination shock at ~ 100 AU where the supersonic solar wind is expected to transition to subsonic flow. This shock, surrounding the entire heliosphere, would then accelerate the ACR in a process similar to that at the Earth's bow shock or at interplanetary shocks. The particles are scattered back and forth across the shock front, gaining energy by being compressed in the converging solar wind flow. To be accelerated, the pickup ions must first be reflected at the shock with sufficient velocity to be able to propagate upstream in the solar wind. This injection of ions was thought to be most effective at the termination shock over the solar poles where the magnetic field is nearly radial. Indeed, Jokipii (1986) using quantitative two-dimensional numerical simulation to solve the full transport equation showed that the observed spectrum and gradients of ACRs could be obtained with reasonable values for the input parameters. Because of their low charge, ACRs have high rigidity, and the gradient and curvature drifts of these particles both along the shock and inside the heliosphere are important in producing the spectral shapes of the ACRs and the intensity gradients. An important result of this model, which included gradient and curvature drifts, is the prediction of a 22-year cycle based on the reversal of polarity of the solar magnetic field. There is observational evidence to support this. Cummings et al. (1995) found a difference in ACR oxygen spectral shapes and intensities for comparable levels of solar modulation but opposite magnetic field polarities.

Ecliptic termination shock

Acceleration of the ACRs at the heliospheric termination shock near the equatorial regions has been suggested by Fisk (1986). This would be an attractive alternative should gradient and curvature drifts prove to be less important. Because the average magnetic field in the distant equatorial heliosphere is nearly azimuthal and thus parallel to the shock front, particles would have to have substantially higher velocity to be able to propagate upstream, resulting in a much higher injection threshold. However, Pioneer 11 observations at ~ 32 AU showed large directional fluctuations ($\sim 35°$ FWHM) around the average azimuthal direction (Smith 1993). Assuming a similar spread in the direction of the magnetic field at the termination shock, the fraction of time when the injection efficiency is low turns out to be only about 20%. Furthermore, the persistence of high-velocity tails observed for pickup ions at low latitudes but not at high latitudes (Gloeckler et al. 2000b), would compensate for the higher injection threshold. Thus it appears that injection near the equatorial plane is not a problem, and in fact injection may take place at all heliolatitudes.

What then is the most likely scenario for the acceleration of the ACRs by more than three orders of magnitude from the few-keV pickup ions? It is very likely a combination of all three mechanisms proposed. Transit-time damping of the turbulence in the slow solar wind, especially in corotating interaction regions and at the interface between the fast and slow wind around solar minimum that probably extends right up to the termination shock will accelerate pickup ions to moderate (perhaps hundreds of keV/amu) energies as these populations are convected towards the termination shock. Other mechanisms, such as a second-order Fermi process or shock surfing acceleration (Lee et al. 1996) may contribute as well to pumping up the energy of pickup ions. But to attain the really high energies of the ACRs almost certainly requires efficient first-order Fermi acceleration by the termination shock. It is likely that this acceleration takes place both at the polar shock and the near equatorial shock.

CONCLUDING REMARKS AND OUTLOOK

The era of *in situ* spacecraft observations has revealed a rich variety in the populations of energetic particles in the heliosphere. The discoveries of the many heliospheric populations could not have been made without the technological advances that allowed us to reach distant regions of our heliosphere, and new observational tools that allowed us, at the same time, to delineate the individual characteristics of the various particle populations. Driven by observations using these advanced instruments and new spacecraft trajectories has come progress in theory, simulation and modelling. Our understanding of the acceleration and transport processes of energetic particles, and of the sites and composition of the source populations for these particle populations has dramatically increased. We may be nearing the end of this rapid discovery phase. More attention is now focussing on the detailed physics of particle acceleration in the heliosphere, stimulated by the refined observations that are coming from new instrumentation. What will be the future prospects in this field, what are the unresolved questions and the theoretical and observational challenges?

Unanswered questions

Among the unresolved questions is the origin of the minor ACRs. Inner source pickup ions are the most likely possibility, but it is not clear how these strongly cooled ions can be efficiently injected and accelerated. The origin of particles accelerated in CIRs is also not entirely clear, and very likely depends on the heliocentric distance where acceleration takes place. Interstellar pickup ions undoubtedly contribute, especially in the outer heliosphere, as does the inner source closer in. The suprathermal tails of the solar wind are clearly an important source as well. Measurements of charge states of the accelerated particles of these populations, especially at distances of a few AU where CIR acceleration dominates, would go a long way to settle these questions.

Another important question that remains concerns the location of the acceleration regions for the ACRs. Initially it was believed that the acceleration took place continuously inside the heliosphere; later, that all the acceleration would take place at the solar wind termination shock. Probably it is a combination of both, although there is no general agreement on this.

Finally, the mechanism producing the quiet-time suprathermal tails of distribution functions remains a mystery. Are these tails produced by a statistical mechanism, or by compressions in the solar wind, or by some other process? Understanding the origin of these tails is clearly important because these highly suprathermal particles will be most efficiently injected for further acceleration by other mechanisms.

Theoretical challenges

Our main theoretical challenge is then to describe the most plausible acceleration process that can best account for the observed characteristics of the various heliospheric particle populations. Can shock acceleration models alone accommodate current observations or are other processes, such as statistical mechanisms required? If so, what are the respective roles of each of these mechanisms in producing each of the observed populations?

Our views of the acceleration processes of SEPs have evolved considerably. We once saw stochastic acceleration as the random transfer of energy to particles from an intense, but simple power-law Alfvén-wave spectrum. We now believe that the spectrum must be complex with resonant peaks of electromagnetic ion cyclotron waves generated by electrons and damped by ^{3}He. We saw shock acceleration in terms of particle scattering against ambivalent turbulence. We now see it as a dynamic process where wave generation by low-energy particles traps them near the shock, increasing the efficiency of their acceleration to higher energy, a process that repeats to giga-electronvolt energies in the strongest shocks near the Sun. We have made remarkable progress in understanding the acceleration of solar energetic particles. But with new observations arriving, new questions will arise and the full test of the two-class paradigm for SEP acceleration is still outstanding.

We need to understand acceleration injection efficiencies. For example, how can we account for the observed inefficient acceleration of ACR H and He compared with the heavier elements? Intuitively, one expects that the injection efficiency increases with particle energy or speed. But what is the dependence of the injection efficiency on particle velocity? It is very difficult to answer these questions from observations alone because one cannot easily separate the source from the accelerated population. Much more work needs to be done to describe acceleration of electrons, for example in CIRs.

Related to the question of the acceleration of the ACRs is their modulation that determines the energy spectra observed deep in the heliosphere. Modulation models should be improved to be both three-dimensional (3D) and time-dependent. Similarly, 3D and time-dependent models of the heliosphere and its interaction with the interstellar gas and magnetic field surrounding it would lead to a better understanding of the remote regions, including the termination shock of the heliosphere.

Observational challenges

To make further substantial progress in our understanding of the acceleration and transport of heliospheric particles, and their interaction with the solar wind, it will be necessary to measure the 3D velocity distribution functions of various ion species of these populations on time-scales much shorter than presently available. The 3D distribution functions should be obtained over a wide energy range, extending from typical energies of the source populations (few tens of electronvolts) to that of the full energy of the accelerated particles (tens of mega-electronvolts), and will require high-sensitivity instruments that are now being developed.

With these same high-sensitivity instruments, capable of determining the composition and ionization states of the particles, it will also be possible to characterize far better than now the various source populations and the resulting accelerated particles. For instance, the composition of interstellar pickup ions, including isotopic ratios of key elements, will be obtained with high precision, along with better charge state and isotopic measurements of the ACRs, including the minor ACRs. Such observations will not only bring significant advances in our knowledge of the local interstellar medium, but will also shed new light on the locations as well as mechanisms for the injection and acceleration of ACRs, and the origin of the minor ACRs. In a similar fashion, the detailed characterization of the composition of inner source pickup ions will be made possible, along with more extensive measurements of CIR accelerated particles, including their charge states. Such observations will be decisive in settling the question of the origin of corotating particles.

There is as yet scarce information on molecular ions in the source populations, let alone of the particles accelerated from this material. *In situ* measurements of molecules over a wide energy range pose a real experimental challenge for the twenty-first century, but carry with it rich rewards. The study of molecular composition in the coming years will provide us with a powerful tool for tracing the sources of energetic particles, and will undoubtedly lead to unanticipated discoveries.

New missions, New discoveries

Gamma-ray observations have opened up an important window to study particle acceleration on the Sun. This will likely stimulate the next advances in this field with the advent of observations from the High Energy Solar and Spectroscopic Imager (RHESSI) spacecraft which is for the first time providing high-resolution images of solar gamma rays. These observations may also help to shed more light on the acceleration site of the very high-energy solar particles.

To meet the observational challenges discussed above will require missions to place advanced, high-sensitivity instruments in strategic locations in the heliosphere. For example, to study in detail the composition of the interstellar pickup ions and at the same time measure the characteristics of inner source pickup ions a spacecraft with a 1 by ~3 to 4 AU heliocentric, low inclination orbit would be sufficient. Technology for achieving this orbit exists today and this mission could be accomplished within the next 10 years.

More challenging is the concept for the *in situ* exploration of nearby interstellar space starting with a detailed survey of the outer heliosphere and the termination shock region. A 30-year mission that could reach a distance of 200 to 400 AU would require new propulsion technology, such as solar sails. With a comprehensive payload of advanced, low-mass instruments, this mission would resolve a number of outstanding questions of heliospheric particle acceleration that would otherwise go unanswered. But even more important would be the discoveries this pioneering mission would make.

Acknowledgments

The authors thank H. V. Cane, L. A. Fisk and G. M. Mason for critical reading of the draft manuscript. The authors acknowledge the support of Ms S. Ihaddadene and C. Nilsson (Space Science Department of ESA) for the preparation of the manuscript.

REFERENCES

Adams, J.H., Garcia-Munoz, M., Grigorov, N.L., Klecker, B., Kondratyeva, M.A., Mason, G.M., McGuire, R.E., Mewaldt, R.A., Panasyuk, M.I., Tretyakova, Ch.A., Tylka, A.J. and Zhuravlev, D.A. (1991). The charge state of the anomalous component of cosmic rays. *Astrophysical Journal*, **375**, L45–L48.

Armstrong, T.P., Krimigis, S.M. and Behannon, K.W. (1970). Proton fluxes at 300 keV associated with propagating interplanetary shock waves. *Journal of Geophysical Research*, **75**, 5980–5988.

Asbridge, J.R., Bame, S.J. and Strong I.B. (1968). Outward flow of protons from the Earth's bow shock. *Journal of Geophysical Research*, **73**, 5777–5782.

Axford, W.I. (1965). Anisotropic diffusion of solar cosmic rays. *Planetary and Space Science*, **13**, 1301–1309.

Axford, W.I., Leer, E. and Skadron, G. (1977). The acceleration of cosmic rays by shock waves. In Proceedings of the 15th Interational Cosmic Ray Conference, Plovdiv, Vol. 11, pp. 132–137.

Barnes, C.W. and Simpson, J.A. (1976). Evidence for interplanetary acceleration of nucleons in corotating interaction regions. *Astrophysical Journal*, **210**, L91–L96.

Bell, A.R. (1978). The acceleration of cosmic rays in shock fronts. *Monthly Notices of the Royal Astronomical Society*, **182**, 147–156.

Bertaux, J.L. and Blamont, J.E. (1971). Evidence for a source of an extraterrestrial hydrogen Lyman-alpha emission: The interstellar wind. *Astronomy and Astrophysics*, **11**, 200–217.

Biswas, S., Durgaprasad, N., Nevatia, J., Sarkar, S. and Venkatavaradan, V.S. (1977). Energy spectra and abundances of carbon to nickel ions in the low energy 8–100 MeV/amu cosmic rays observed in the Skylab experiment. In Proceedings of the 15th International Cosmic Ray Conference, Plovdiv, Vol. 2, pp. 327–332.

Biswas, S., Durgaprasad, N. and Trivedi, S.S. (1981). On the origin of the low energy rays from stellar sources. *Earth Planet Science (Indian Academy of Sciences)*, **90**, 337–344.

Blandford, R.D. and Ostriker, J.P. (1978). Particle acceleration by astrophysical shocks. *Astrophysical Journal*, **221**, L29–L32.

Blum, P.W. and Fahr, H.-J. (1970). Interaction between interstellar hydrogen and the solar wind. *Astronomy and Astrophysics*, **4**, 280–290.

Breneman, H.H. and Stone, E.C. (1985). Solar coronal and photospheric abundances from solar energetic particle measurements. *Astrophysical Journal*, **299**, L57–L61.

Bryant, D.A., Cline, T.L., Desai, U.D. and McDonald, F.B. (1962). Explorer 12 observations of solar cosmic rays and energetic storm particle events after the solar flare of September 28, 1961. *Journal of Geophysical Research*, **67**, 4983–5000.

Bryant D.A., Cline, T.L., Desai, U.D. and McDonald, F.B. (1965). Continual acceleration of solar protons in the MeV range. *Physical Review Letters*, **14**, 481–484.

Burlaga, L.F., McDonald, F.B., Goldstein, M.L. and Lazarus, A.J. (1985). Cosmic ray modulation and turbulent interaction regions near 11 AU. *Journal of Geophysical Research*, **90**, 12027–12039.

Burlaga, L.F., Ness, N.F., Belcher, J.W., Lazarus, A.J. and Richardson, J.D. (1996). Voyager observations of the magnetic field, interstellar pickup ions and solar wind in the distant heliosphere. *Space Science Reviews*, **78**, 33–42.

Cane, H.V., McGuire, R.E. and von Rosenvinge, T.T. (1986). Two classes of solar energetic particle events associated with impulsive and long-duration soft x-ray flares. *Astrophysical Journal*, **301**, 448–459.

Cane, H.V. and Reames, D.V. (1990). The relationship between energetic particles and flare properties for impulsive solar flares. *Astrophysical Journal*, **73**, 253–258.

Cane, H.V., Reames, D.V. and von Rosenvinge, T.T. (1988). The role of interplanetary shocks in the longitude distribution of solar energetic particles. *Journal of Geophysical Research*, **93**, 9555–9567.

Cane, H.V., Reames, D.V. and von Rosenvinge, T.T. (1991). Solar particle abundances at energies of greater than 1 MeV per nucleon and the role of interplanetary shocks. *Astrophysical Journal*, **373**, 675–682.

Cane, H.V., Sheeley, Jr., N.R. and Howard, R.A. (1987). Energetic interplanetary shocks, radio emission and coronal mass ejections. *Journal of Geophysical Research*, **92**, 9869–9874.

Cane, H.V., von Rosenvinge, T.T. and McGuire, R.E. (1990). Energetic particle observations at the Helios 1 spacecraft of shocks associated with coronal mass ejections. *Journal of Geophysical Research*, **95**, 6575–6579.

Carrington, R.C. (1860). Description of a singular appearance seen on the sun on September 1, 1859. *Monthly Notices of the Royal Astronomical Society*, **20**, 13–15.

Chapman, S. and Bartels, J. (1940). *Geomagnetism*. Oxford University Press, London.

Christian, E.R., Cummings, A.C. and Stone, E.C. (1988). Evidence for anomalous cosmic-ray hydrogen. *Astrophysical Journal*, **334**, L77–L80.

Chupp, E.L., Forrest, D.J., Higbie, P.R., Suri, A.N., Tsai, C. and Dunphy, P.P. (1973). Solar gamma ray lines observed during the solar activity of August 2 to August 11, 1972. *Nature*, **241**, 333–335.

Cliver, E.W. (1996). Solar flare gamma-ray emission and energetic particles in space. In R. Ramaty, N. Mandzhavidze and X.M. Hua, (eds), *High Energy Solar Physics, AIP Conference Proceedings*, **374**, pp. 45–60.

Cliver, E.W. (2000). Solar flares photons and energetic particles in space. In R.A. Mewaldt, J.R. Jokipii, M.A. Lee, E. Möbius and T.H. Zurbuchen (eds), *Acceleration and Transport of Energetic Particles observed in the Heliosphere*, ACE–2000 Symposium, *AIP Conference Proceedings*, **528**, pp. 21–31.

Cliver, E.W., Forrest, D.J., Cane, H.V., Reames, D.V., McGuire, R.E., von Rosenvinge, T.T., Kane, S.R. and MacDowall, R.J. (1989). Solar flare nuclear gamma-rays and interplanetary proton events. *Astrophysical Journal*, **343**, 953–970.

Cliver, E.W., Kahler, S.W. and McIntosh, P.S. (1983). Solar proton flares with weak impulsive phases. *Astrophysical Journal*, **264**, 699–707.

Cliver, Kahler, S.W., Shea, M.A. and Smart, D.F. (1982). Injection of $\sim 2\,\text{GeV}$ protons, $\sim 1\,\text{MeV}$ electrons, and $\sim 100\,\text{keV}$ electrons in solar cosmic ray flares. *Astrophysical Journal*, **260**, 362–370.

Cohen, C.M.S., Mewaldt, R.A., Leske, R.A., Cummings, A.C. and Stone, E.C. (1999). New observations of heavy ion-rich solar particle events from ACE. *Geophysical Research Letters*, **26**, 2697–2700.

Cook, W.R., Stone, E.C. and Vogt, R.E. (1980). Elemental composition of solar energetic nuclei. *Astrophysical Journal*, **238**, L97–L101.

Cummings, A.C., Mewaldt, R.A., Blake, J.B., Cummings J.R., Fränz, M., Hovestadt, D., Klecker, B., Mason, G.M., Mazur, J.E., Stone, E.C., von Rosenvinge, T.T. and Webber, W.R. (1995). Anomalous cosmic ray oxygen gradients throughout the heliosphere. *Geophysical Research Letters*, **22**, 341–344.

Cummings, A.C. and Stone E.C. (1987). Elemental composition of the anomalous cosmic ray component. In Proceedings of the 20th International Cosmic Ray Conference, Moscow, Vol. 3, pp. 413–416.

Cummings, A.C., Stone, E.C. and Steenberg, C.D. (1999). Composition of anomalous cosmic rays and other ions from Voyager observations. In Proceedings of the 26th International Cosmic Ray Conference, Utah, Vol. 7, pp. 531–534.

Cummings, A.C., Stone, E.C. and Webber, W.R. (1984). Evidence that the anomalous cosmic-ray component is singly ionized. *Astrophysical Journal*, **287**, L99–L103.

Decker, R.B. (1981). The modulation of low-energy proton distributions by propagating interplanetary shock waves. A numerical simulation. *Journal of Geophysical Research*, **86**, 4537–4554.

Decker, R.B., Pesses, M.E. and Krimigis, S.M. (1981). Shock-associated low-energy ion enhancements observed by Voyagers 1 and 2. *Journal of Geophysical Research*, **86**, 8819–8831.

Desai, M.I., Mason G.M., Dwyer, J.R., Mazur, J.E., Von Rosenvinge, T.T. and Lepping, R.P. (2000). Characteristics of energetic ($\geqslant 30\,\text{keV}\,\text{nuc}^{-1}$) ions observed by the Wind/STEP instrument upstream of the Earth's bow shock. *Journal of Geophysical Research*, **105**, 61–78.

Dessler, A. (1967). Solar wind and interplanetary field. *Review of Geophysics*, **5**, 1–41.

Domingo, V.R., Hynds, R.J. and Stevens G. (1979). A solar proton event of possible non-flare origin. In Proceedings of the 16th International Cosmic Ray Conference, Kyoto, Vol. 5, pp. 192–197.

Dwyer, J.R., Mason, G.M., Desai, M.I., Mazur, J.E. and von Rosenvinge, T.T. (2000). The spatial size of ion events measured far upstream of the earth's bow shock by ACE/ULEIS and WIND/STEP. *Geophysical Research Letters*, **27**, 65–68.

Eichler, D. (1981). A cosmic-ray-mediated shock in the solar system. *Astrophysical Journal*, **247**, 1089–1092.

Ellison, D.C., Möbius, E. and Paschmann, G. (1990). Particle injection and acceleration at Earth's bow shock: comparison of upstream and downstream events. *Astrophysical Journal*, **352**, 376–394.

Fan, C.Y., Gloeckler, G. and Simpson, J.A. (1965). Protons and helium nuclei within interplanetary magnetic regions which corotate with the Sun. In Proceedings of the 9th International Cosmic Ray Conference, London, Vol. 1, pp. 109–111.

Fichtel, C.E. and Guss, D.E. (1961). Heavy nuclei in solar cosmic rays. *Physical Review Letters*, **6**, 495–497.

Fisk, L.A. (1971). Increases in the low-energy cosmic ray intensity at the front of propagating interplanetary shock waves. *Journal of Geophysical Research*, **76**, 1662–1672.

Fisk, L.A. (1976a). The acceleration of energetic particles in the interplanetary medium by transit time damping, *Journal of Geophysical Research*, **81**, 4633–4645.

Fisk, L.A. (1976b). Solar modulation and a galactic origin for the anomalous component observed in low-energy cosmic rays. *Astrophysical Journal*, **206**, 333–341.

Fisk, L.A. (1978). ^{3}He-rich flares: A possible explanation. *Astrophysical Journal*, **224**, 1048–1055.

Fisk, L.A. (1986). The anomalous component, its variation with latitude and related aspects of modulation. In R.G. Marsden (ed.), *The Sun and the Heliosphere in Three Dimensions*, Reidel, Dordrecht, pp. 401–411.

Fisk, L.A. (1996). Motions of footpoints of heliospheric magnetic field lines at the Sun: Implications for recurrent energetic particle events at high heliographic latitudes. *Journal of Geophysical Research*, **101**, 15,547–15,554.

Fisk, L.A., Kozlovsky, B. and Ramaty, R. (1974). An interpretation of the observed oxygen and nitrogen enhancements in low-energy cosmic rays. *Astrophysical Journal*, **190**, L35–L37.

Fisk, L.A. and Lee, M.A. (1980). Shock acceleration of energetic particles in corotating interaction regions in the solar wind. *Astrophysical Journal*, **237**, 620–626.

Forbush, S.E. (1946). Three unusual cosmic-ray increases possibly due to charged particles form the Sun. *Physical Review*, **70**, 771.

Fowler, P.H., Redfern, R.M. and Swordy, S.P. (1979). Do comets provide material for the anomalous component of the cosmic rays? *Nature*, **279**, 622–624.

Garcia-Munoz, M., Mason, G.M. and Simpson, J.A. (1973). A new test for solar modulation theory: the 1972 May–July low-energy galactic cosmic-ray proton and helium spectra. *Astrophysical Journal*, **182**, L81–L84.

Geiss, J., Gloeckler, G., Fisk, L.A. and Von Steiger, R. (1995). C$^+$ pickup ions in the heliosphere and their origin. *Journal of Geophysical*, **100**, 23,373–23,377.

Geiss, J., Gloeckler, G., Mall, G., Von Steiger, R., Galvin, A.B. and Ogilvie, K.W. (1994). Interstellar oxygen, nitrogen and neon in the heliosphere. *Astronomy and Astrophysics*, **282**, 924–933.

Geiss, J., Gloeckler, G. and von Steiger, R. (1996). Origin of C$^+$ ions in the heliosphere. *Space Science Reviews*, **78**, 43–52.

Gloeckler, G. (1970). The measurements of non-relativistic charged particles of extra-terrestrial origin. In H. Ögelman and J.R. Wayland (eds), *Introduction to Experimental Techniques of High-Energy Astrophysics*, NASA Publ. SP-243, pp. 1–61.

Gloeckler, G. (1975). Low energy particle composition. In Proceedings of the 14th International Cosmic Ray Conference, Munich, Vol. 11, pp. 3784–3804.

Gloeckler, G. (1979). Composition of energetic particle populations in interplanetary space. *Reviews of Geophysics and Space Physics*, **17**, 569–582.

Gloeckler, G. (1999). Observation of injection and pre-acceleration processes in the slow solar wind. *Space Science Reviews*, **89**, 91–104.

Gloeckler, G., Balsiger, H., Bürgi, A., Bochsler, P., Fisk, L.A., Galvin, A.B., Geiss, J., Gliem, F., Hamilton, D.C., Holzer, T.E., Hovestadt, D., Ipavich, F.M., Kirsch, E., Lundgren, R.A., Ogilvie, K.W., Sheldon, R.B. and Wilken, B. (1995). The solar wind and supra-thermal ion composition investigation on the Wind spacecraft. *Space Science Reviews*, **71**, 79–124.

Gloeckler, G., Cain, J., Ipavich, F.M., Tums, E.O., Bedini, P., Fisk, L.A., Zurbuchen T.H., Bochsler, P., Fischer, J., Wimmer-Schweingruber, R.F., Geiss, J. and Kallenbach, R. (1998). Investigation of the composition of solar and interstellar matter using solar wind and pick up ion measurements with SWICS and SWIMS on the ACE spacecraft. *Space Science Reviews*, **86**, 497–538.

Gloeckler, G., Fisk, L.A. and Geiss, J. (1997). Anomalously small magnetic field in the local interstellar cloud. *Nature*, **386**, 374–377.

Gloeckler, G., Fisk, L.A., Geiss, J., Schwadron, N.A. and Zurbuchen, T.H. (2000a). Elemental composition of the inner source pickup ions. *Journal of Geophysical Research*, **105**, 7459–7463.

Gloeckler, G., Fisk, L.A., Zurbuchen, T.H. and Schwadron, N.A. (2000b). Sources, injection and acceleration of heliospheric ion populations. In R.A. Mewaldt, J.R. Jokipii, M.A. Lee, E. Möbius and T.H. Zurbuchen (eds.), *Acceleration and Transport of Energetic Particles Observed in the Heliosphere*, ACE-2000 Symposium, AIP Conference Proceedings, **528**, 221–228.

Gloeckler, G. and Geiss, J. (1996). Abundance of ^{3}He in the local interstellar cloud. *Nature*, **381**, 210–212.

Gloeckler, G. and Geiss, J. (1998). Interstellar and inner source pickup ions observed with SWICS on Ulysses. *Space Science Reviews*, **86**, 127–159.

Gloeckler, G. and Geiss J. (2001). Heliospheric phenomena deduced from pickup ion observations. In R.G. Marsden (ed.), *The 3-D Heliosphere at Solar Maximum*. Proceedings of the 34th ESLAB Symposium. *Space Science Review*, **97**, 169–181.

Gloeckler, G., Geiss, J., Balsiger, H., Bedini, P., Cain, J.C., Fischer, J., Fisk, L.A., Galvin, A.B., Geiss, J., Gliem, F., Hamilton, D.C., Hollweg, J.V., Ipavich, F.M., Joos, R., Livi, S., Lundgren, R., Mall, U., McKenzie, J.F., Ogilvie, K.W., Ottens, F., Rieck, W., Tums, E.O., von Steiger, R., Weiss, W. and Wilken, B. (1992). The solar wind ion composition spectrometer. *Astronomy and Astrophysics* Suppl., **92**, 267–289.

Gloeckler, G., Geiss, J., Balsiger, H., Fisk, L.A., Galvin, A.B., Ipavich, F.M., Ogilvie, K.W., von Steiger, R. and Wilken, B. (1993). Detection of interstellar pick-up hydrogen in the solar system. *Science*, **261**, 70–73.

Gloeckler, G., Geiss, J. and Fisk, L.A. (2001). Heliospheric and interstellar phenomena revealed from observations of pickup ions. In A. Balogh, E.J. Smith, and R.G. Marsden (eds), *The Heliosphere near Solar Minimum: The Ulysses Perspective*. Springer-Praxis, Berlin, 287–326.

Gloeckler, G., Geiss, J., Roelof, E.C., Fisk, L.A., Ipavich, F.M., Ogilvie, K.W., Lanzerotti, L.J., von Steiger, R. and Wilken, B. (1994). Acceleration of interstellar pickup ions in the disturbed solar wind observed on Ulysses. *Journal of Geophysical Research*, **99**, 17,637–17,643.

Gloeckler, G., Hovestadt, D. and Fisk, L.A. (1979). Observed distribution functions of H, He, C, O, and Fe in corotating energetic particles streams: Implications for interplanetary acceleration and propagation. *Astrophysical Journal*, **230**, L191–L195.

Gloeckler, G. and Hsieh, K.C. (1979). Time-of-flight technique identification at energies from 2 to 400 keV/nucleon. *Nuclear Instruments and Methods*, **165**, 537–544.

Gloecker, G., Sciambi, R.K., Fan, C.Y. and Hovestadt, D. (1976). A direct measurement of the charge states of energetic iron emitted by the sun. *Astrophysical Journal Letters*, **209**, L93–L96.

Gosling, J.T. (1993). The solar flare myth. *Journal of Geophysical Research*, **98**, 18937–18949.

Gosling, J.T., Hildner, E., MacQueen, R.M., Munro, R.H., Poland, A.I. and Ross, C.L. (1974). Mass ejections from the Sun: a view from Skylab. *Journal of Geophysical Research*, **79**, 4581–4587.

Greenstadt, E.W., Green, I.M., Inouye, G.T., Hundhausen, A.J., Bame, S.J. and Strong, I.B. (1968). Corrrelated magnetic field and plasma observations of the Earth's bow shock. *Journal of Geophysical Research*, **73**, 51–60.

Hoppe, M.M. and Russell, C.T. (1982). Particle acceleration at planetary bow shocks waves. *Nature*, **295**, 41–42.

Hoppe, M.M., Russell, C.T., Frank, L.A., Eastman, T.E. and Greenstadt, E.W. (1981). Upstream hydromagnetic waves and their association with backstreaming ion populations: ISEE 1 and 2 observations. *Journal of Geophysical Research*, **86**, 4471–4492.

Hovestadt, D. and Vollmer, O. (1971). A satellite experiment for detecting low energy heavy cosmic rays. In Proceedings of the 12th International Cosmic Ray Conference, Hobart, Vol. 4, 1608–1613.

Hovestadt, D., Vollmer, O., Gloeckler, G. and Fan C.Y. (1973). Differential energy spectra of low energy (<8.5 MeV per nucleon) heavy cosmic rays during solar quiet times. *Physical Review Letters*, **31**, 650–653.

Hoyle, F. and Clayton, D.D. (1974). Nucleosynthesis in white-dwarf atmospheres. *Astrophysical Journal*, **191**, 705–710.

Hsieh, K.C. and Simpson, J.A. (1970). The relative abundances an energy spectra of ³He and ⁴He from solar flares. *Astrophysical Journal*, **162**, L191–L196.

Hudson, H., Haisch, B. and Strong, K.T. (1995). Comment on 'The solar flare myth' by J.T. Gosling. *Journal of Geophysical Research*, **100**, 3473–3477.

Intriligator, D.S., Siscoe, G.L. and Miller, W.D. (1996). Interstellar pickup H^+ ions at 8.3 AU: Pioneer 10 plasma and magnetic field analyses. *Geophysical Research Letters*, **23**, 2181–2184.

Jokipii, J.R. (1971). Deceleration and acceleration of cosmic rays in the solar wind. *Physical Review Letters*, **26**, 666–669.

Jokipii, J.R. (1982). Particle drift, diffusion, and acceleration at shocks. *Astrophysical Journal*, **255**, 716–720.

Jokipii, J.R. (1986). Particle acceleration at the termination shock 1: Applications to the solar wind and the anomalous component. *Journal of Geophysical Research*, **91**, 2929–2932.

Jokipii, J.R. (1992). Constraints on the acceleration of anomalous cosmic rays. *Astrophysical Journal*, **393**, L41–L43.

Jokipii, J.R., Kota, J., Giacalone, J., Horbury, T.S. and Smith, E.J. (1995). Interpretation and consequences of large-scale magnetic variances at high heliographic latitude. *Geophysical Research Letters*, **22**, 3385–3388.

Kahler, S.W. (1994). Injection profiles of solar energetic particles as functions of coronal mass ejection heights. *Astrophysical Journal*, **428**, 837–842.

Kahler, S.W., Hildner, E. and Van Hollebeke, M.A.I. (1978). Prompt solar proton events and coronal mass ejections. *Solar Physics*, **57**, 429–443.

Kennel, C.F., Coroniti, F.V., Scarf, F., Livesey, W.A., Russell, C.T., Smith, E.J., Wenzel, K.P. and Scholer, M. (1986). A test of Lee's quasi-linear theory of ion acceleration by interplanetary travelling shocks. *Journal of Geophysical Research*, **91**, 11,917–11,928.

Keppler, E. (1998). The acceleration of charged particles in corotating interaction regions (CIR): A review with particular emphasis on the Ulysses mission. *Surveys in Geophysics*, **19**, 211–278.

Keppler, E., Fränz, M., Korth, A., Krupp, N., Reuss, M.K., Wilken, B., Balogh, A., Forsyth, R.J., Quenby, J.J. and Blake, B. (1995). Energetic particle observations at high heliographic latitudes. *Space Science Reviews*, **72**, 285–290.

Klecker, B. (1977). The anomalous component of low energy cosmic rays: A comparison of observed spectra with model calculations. *Journal of Geophysical Research*, **82**, 5287–5291.

Klecker, B. (1995). The anomalous component of cosmic rays in the 3-D heliosphere. *Space Science Reviews*, **72**, 419–430.

Klecker, B., Hovestadt, D., Gloeckler, G. and Fan, C.Y. (1980). On the charge state of the anomalous oxygen component. *Geophysical Research Letters*, **7**, 1033–1036.

Klecker, B., Hovestadt, D., Gloeckler, G., Ipavich, F.M., Scholer, M., Fan, C.Y. and Fisk, L.A. (1984). Direct determination of the ionic charge distribution of helium and iron in ³He-Rich solar energetic particle events. *Astrophysical Journal*, **281**, 458–462.

Klecker, B., McNab, M.C., Blake, J.B., Hamilton, D.C., Hovestadt, D., Kästle, H., Looper, M.D., Mason, G.M., Mazur, J.E. and Scholer, M. (1995). Charge states of anomalous cosmic-ray nitrogen, oxygen, and neon: Sampex observations. *Astrophysical Journal*, **442**, L69–L72.

Klecker, B., Mewaldt, R.A., Bieber, J.W., Cummings, A.C., Drury, L., Giacalone, J., Jokipii, J.R., Jones, F.C., Krainev, M.B., Lee, M.A., Le Roux, J.A., Marsden, R.G., McDonald, F.B., McKibben, R.B., Baring, M.G., Ellison, D.C., Lanzerotti, L.J., Leske, R.A., Mazur, J.E., Moraal, H., Oetliker, M., Ptuskin, V.S., Selesnick, R.S. and Trattner, K.J. (1998). Anomalous cosmic rays. *Space Science Reviews*, **83**, 259–308.

Klecker, B., Oetliker, M., Blake, J.B., Hovestadt, D., Mason, G.M., Mazur, J.E. and McNab, M.C. (1997). Multiply charged anomalous cosmic ray N, O, and Ne: observations with HILT/SAMPEX. In Proceedings of the 25th International Cosmic Ray Conference, Durban, Vol. 2, 273–276.

Krimigis, S.M. (1992). Voyager energetic particle observations at interplanetary shocks and upstream of planetary bow shocks: 1977–1990. *Space Science Reviews*, **59**, 167–201.

Krimigis, S.M., Armstrong, T.P., Axford, W.I., Bostrom C.O., Fan, C.Y., Gloeckler, G. and Lanzerotti, L.J. (1977). The low energy charged particle (LECP) experiment on the Voyager spacecraft. *Space Science Reviews*, **21**, 329–354.

Krimigis, S.M., Decker, R.B., Hamilton, D.C. and Gloeckler, G. (2000). Observations of pick-up ions in the outer heliosphere by Voyagers 1 and 2. In R.A. Mewaldt, J.R. Jokipii, M.A. Lee, E. Möbius, and T.H. Zurbuchen (eds), *Acceleration and Transport of Energetic Particles observed in the Heliosphere*, ACE-2000 Symposium, AIP Conference Proceedings, **528**, 333–340.

Kunow, H., Lee, M.A., Fisk, L.A., Forsyth, R.J., Heber, B., Horbury, T.S., Keppler, E., Kóta, J., Lou, Y.Q., McKibben, R.B., Paizis, C., Potgieter, M.S., Roelof, E.C., Sanderson, T.R., Simnett, G.M., von Steiger, R., Tsurutani, B.T., Wimmer-Schweingruber, R.F. and Jokipii, J.R. (1999). Corotating interaction regions at high latitudes. *Space Science Reviews*, **89**, 221–268.

Lee, M.A. (1982). Coupled hydromagnetic wave excitation and ion acceleration upstream of the Earth's bow shock. *Journal of Geophysical Research*, **87**, 5063–5080.

Lee, M.A. (1983). Coupled hydromagnetic wave excitation and ion acceleration at interplanetary traveling shocks. *Journal of Geophysical Research*, **88**, 6109–6119.

Lee, M.A. (1997). Particle acceleration and transport at CME-driven shocks. In N. Crooker, J.A. Jocelyn and J. Feynman (eds), *Coronal Mass Ejections*. AGU Press, pp. 227–234.

Lee, M.A., Shapiro, V.D. and Sagdeev, R.Z. (1996). Pickup ion energization by shock surfing. *Journal of Geophysical Research*, **101**, 4777–4789.

Leske, R.A., Mewaldt, R.A., Christian, E.R., Cohen, C.M.S., Cummings, A.C., Stone, E.C., von Rosenvinge, T.T. and Wiedenbeck, M.E. (1999b). Measurement of the isotopic composition of anomalous cosmic ray N, O, and Ne from ACE. In Proceedings of the 26th International Cosmic Ray Conference, Utah, Vol. 7, pp. 539–542.

Leske, R.A., Mewaldt, R.A., Cohen, C.M.S., Cummings, A.C., Stone, E.C., Wiedenbeck, M.E., Christian, E.R. and von Rosenvinge, T.T. (1999a). Event-to-event variations in the isotopic composition of neon in solar energetic particle events. *Geophysical Research Letters*, **26**, 2693–2696.

Lin, R.P. (1970). The emission and propagation of ~ 40 keV solar flare electrons. *Solar Physics*, **12**, 266–303.

Lopate, C. and McKibben, R.B. (1991). The ratio ²H/¹H as a test for the presence of anomalous protons at Pioneer-10 near solar minimum in 1987. In Proceedings of the 22nd International Cosmic Ray Conference, Dublin, Vol. 3, pp. 390–393.

Luhn, A., Klecker, B., Hovestadt, D., Gloeckler, G., Ipavich, F.M., Scholer, M., Fan, C.Y. and Fisk, L.A. (1984). Ionic charge states of N, Ne, Mg, Si and in solar energetic particle events. *Advances in Space Research*, **4**(2–3), 161–164.

Luhn, A., Klecker, B., Hovestadt, D., and Möbius, E. (1987). The mean ionic charge of silicon in ³He-rich solar flares. *Astrophysical Journal*, **317**, 951–955.

Mason, G., Gold, R.E., Krimigis, S.M., Mazur, J.E., Andrews, G.B., Daley, K.A., Dwyer, J.R., Heuerman, K.F., James, T.L., Kennedy, M.J., Lefevere, T., Malcom, H., Tossman, B. and Walpole, P.H. (1998). The ultra-low-energy isotope spectrometer (ULEIS) for the ACE spacecraft. *Space Science Reviews*, **86**, 409–448.

Mason, G.M., Gloeckler, G. and Hovestadt, D. (1984). Temporal variations of nucleonic abundances in solar flare energetic particle events: II.

Evidence for large-scale shock acceleration. *Astrophysical Journal*, **280**, 902–916.

Mason, G.M., Mazur, J.E. and Dwyer, J.R. (1999). ³He enhancements in large solar energetic particle events. *Astrophysical Journal*, **525**, L133–L136.

Mason, G.M., Mazur, J.E., Dwyer, J.R., Reames, D.V. and von Rosenvinge, T.T. (1997). New spectral and abundance features of interplanetary heavy ions in corotating interaction regions. *Astrophysical Journal*, **486**, L149–L152.

Mazur, J.E., Mason, G.M., Klecker, B. and McGuire, R.E. (1992). The energy spectra of solar flare hydrogen, helium, oxygen, and iron: evidence for stochastic acceleration. *Astrophysical Journal*, **401**, 398–410.

Mazur, J.E., Mason, G.M., Lopper, M.D., Leske, R.A. and Mewaldt, R.A. (1999). Charge state of solar energetic particles using the geomagnetic cutoff technique: SAMPEX measurements in the 6 November 1997 solar particle event. *Geophysical Research Letters*, **26**, 173–176.

McDonald, F.B., Lukasiak, A. and Webber, W.R. (1995). Pioneer 10 and Voyager 1 observations of anomalous cosmic-ray hydrogen in the outer heliosphere. *Astrophysical Journal*, **446**, L101–L104.

McDonald, F.B., Teegarden, B.J., Trainor, J.H., von Rosenvinge, T.T. and Webber, W.R. (1976). The interplanetary acceleration of energetic nucleons. *Astrophysical Journal*, **203**, L149–L154.

McDonald, F.B., Teegarden, B.J., Trainor, J.H. and Webber, W.R. (1974). The anomalous abundance of cosmic-ray nitrogen and oxygen nuclei at low energies. *Astrophysical Journal*, **187**, L105–L108.

McKibben, R.B. (1977). An experimental test for the charge state of the anomalous helium component. *Astrophysical Journal*, **217**, L113–L116.

Meyer, P., Parker, E.N. and Simpson, J.A. (1956). Solar cosmic rays of February, 1956 and their propagation through interplanetary space. *Physical Review*, **104**, 768–783.

Mewaldt, R.A. (1999). Solar and interplanetary particles re-accelerated at the solar wind termination shock. *Advances in Space Research*, **23**(3), 541–545.

Mewaldt, R.A. (2000). New views of solar energetic particles from the Advanced Composition Explorer. In B.L. Dingus, D.B. Kieda and M.H. Salamon (eds), 26th International Cosmic Ray Conference, Invited, Rapporteur and Highlight Papers, AIP Conference Proceedings, **516**, 265–273.

Mewaldt, R.A., Cummings, J.R., Leske, R.A., Selesnick, R.S., Stone, E.C. and von Rosenvinge, T.T. (1996). A study of the composition and energy spectra of anomalous cosmic rays using the geomagnetic field. *Geophysical Research Letters*, **23**, 617–620.

Miller, J.A. (1995). Much ado nothing. *EOS*, **76**, 401–407.

Miller, J.A., Cargill, P.J., Emslie, A.G., Holman, G.D., Dennis, B.R., LaRosa, T.H., Winglee, R.M., Benka, S.G. and Tsuneta, S. (1997). Critical issues for understanding particle acceleration in impulsive solar flares. *Journal of Geophysical Research*, **102**, 14,631–14,659.

Möbius, E., Hovestadt, D., Klecker, B., Scholer, M., Gloeckler, G. and Ipavich, F.M. (1985). Direct Observation of He$^+$ pick-up ions of interstellar origin in the solar wind. *Nature*, **318**, 426–429.

Möbius, E., Popecki, M., Klecker, B., Kistler, L.M., Bogdanov, A., Galvin, A.B., Heirtzler, D., Hovestadt, D., Lund, E.J., Morris, D. and Schmidt, W.K.H. (1999). Energy dependence of the ionic charge state distribution during the November 1997 solar energetic particle event. *Geophysical Research Letters*, **26**, 145–148.

Mogro-Campero, A. and Simpson, J.A. (1972). The abundances of solar accelerated nuclei from carbon to iron. *Astrophysical Journal*, **177**, L37–L41.

Murphy, R.J., Ramaty, R., Kozlovsky, B. and Reames, D.V. (1991). Solar abundances from gamma-ray spectroscopy: Comparisons with energetic particle, photospheric, and coronal abundances. *Astrophysical Journal*, **371**, 793–803.

Ng, C.K., Reames D.V. and Tylka A.J. (1999). Effect of proton-amplified waves on the evolution of solar energetic particle composition in gradual events. *Geophysical Research Letters*, **26**, 2145–2148.

Oschlies, K., Beaujean, R. and Enge, W. (1989). On the charge state of anomalous oxygen. *Astrophysical Journal*, **345**, 776–781.

Pallavicini, R., Serio, S. and Vaiana, G.S. (1977). A survey of soft x-ray limb flare images: The relation between their structure in the corona and other physical parameters. *Astrophysical Journal*, **216**, 108–122.

Palmer, I.D. and Gosling, J.T. (1978). Shock associated energetic proton events at large heliocentric distances. *Journal of Geophysical Research*, **83**, 2037–2046.

Paschmann, G., Sckopke, N., Papamastorakis, I., Asbridge, J.R., Bame, S.J. and Gosling, J.T. (1981). Characteristics of reflected and diffuse ions upstream from the Earth's bow shock. *Journal of Geophysical Research*, **86**, 4355–4364.

Pesses, M.E., Decker, R.B. and Armstrong, T.P. (1982). The acceleration of charged particles in interplanetary space. *Space Science Reviews*, **32**, 185–204.

Pesses, M.E., Jokipii, J.R. and Eichler, D. (1981). Cosmic ray drift, shock wave acceleration, and the anomalous component of cosmic rays. *Astrophysical Journal*, **246**, L85–L88.

Popecki, M.A., Möbius, E., Klecker, B., Kistler, L.M., Galvin, A.B., Heirtzler, D., Morris, D., Siren, C., Bogdanov, A. and Hovestadt, D. (1999). Time profiles of ionic charge states for rapidly rising solar active periods. In Proceedings of the 26th International Cosmic Ray Conference, Utah, Vol. 6, 187–190.

Price, P.B., Fleischer, R.L., Peterson, D.D., O'Ceallaigh, C., O'Sullivan, D. and Thompson, A. (1967). Identification of isotopes of energetic particles with dielectric track detectors. *Physical Review*, **164**, 1618–1620.

Pyle, K.R., Simpson, J.A., Barnes, A. and Mihalov, J.D. (1984). Shock acceleration of nuclei and electrons in the heliosphere beyond 24 AU. *Astrophysical Journal*, **282**, L107–L111.

Rao, U.R., McCracken, K.G. and Bukata, R.P. (1967). Cosmic ray propagation processes: 2. The energetic particle event. *Journal of Geophysical Research*, **72**, 4325–4341.

Reames, D.V. (1988). Bimodal abundances in the energetic particles of solar and interplanetary origin. *Astrophysical Journal*, **330**, L71–L75.

Reames, D.V. (1995). Solar energetic particles: a paradigm shift. *Review of Geophysics* Suppl., **33**, 585–589.

Reames, D.V. (1999). Particle acceleration at the sun and in the heliosphere. *Space Science Reviews*, **90**, 413–491.

Reames, D.V., Barbier, L.M. and von Rosenvinge, T.T. (1997b). Wind, EPACT observations of anomalous cosmic rays. *Advances in Space Research*, **19**(5), 809–812.

Reames, D.V., Kahler, S.W. and Ng, C.K. (1997a). Spatial and temporal invariance in the spectra of energetic particles in gradual solar events. *Astrophysical Journal*, **491**, 414–420.

Reames, D.V., Meyer, J.P. and von Rosenvinge, T.T. (1994). Energetic-particles abundances in impulsive solar flare events. *Astrophysical Journal Suppl.*, **90**, 649–667.

Reames, D.V., von Rosenvinge, T.T. and Lin, R.P. (1985). Solar ³He-rich event and nonrelativistic electron events: A new association. *Astrophysical Journal*, **292**, 716–724.

Reid, G.C. (1964). A diffusive model for the initial phase of a solar proton event. *Journal of Geophysical Research*, **69**, 2659–2667.

Reinecke, J.P.L. and Moraal, H. (1992). On the form of the 1987 hydrogen spectrum in the outer heliosphere. *Astrophysical Journal*, **392**, 272–276.

Roelof, E.C. and Krimigis, S.M. (1973). Analysis and synthesis of coronal and interplanetary energetic particle, plasma, and magnetic field observations over three solar rotations. *Journal of Geophysical Research*, **78**, 5375–5410.

Russell, C.T., Mewaldt, R.A. and von Rosenvinge, T.T. (eds) (1998). The Advanced Composition Explorer mission. *Space Science Reviews*, **86**, 1–4.

Sanderson, T.R., Reinhard, R. Van Nes, P. and Wenzel, K.P. (1985a). Observations of three-dimensional anisotropies of 35- to 1000-keV protons associated with interplanetary shocks. *Journal of Geophysical Research*, **90**, 19–27.

Sanderson, T.R., Reinhard, R. Van Nes, P. and Wenzel, K.P. (1985b). Observations of 35- to 1600-keV protons and low-frequency waves upstream of interplanetary shocks. *Journal of Geophysical Research*, **90**, 3973–3980.

Sanderson, T.R., Reinhard, R. and Wenzel, K.P. (1981). The propagation of upstream protons between the earth's bow shock and ISEE 3. *Journal of Geophysical Research*, **86**, 4425–4434.

Sarris, E.T., Krimigis, S.M. and Armstrong, T.P. (1976). Observations of high energy protons and electrons at $\sim 35\,R_E$ with IMP 7. *Journal of Geophysical Research*, **81**, 2341–2355.

Scholer, M. and Burgess, D. (1992). The role of upstream waves in supercritical quasi-parallel shock re-formation. *Journal of Geophysical Research*, **97**, 8319–8326.

Scholer, M. Ipavich, F.M. Gloeckler, G. and Hovestadt, D (1983). Acceleration of low-energy protons and alpha particles at interplanetary shock waves. *Journal of Geophysical Research*, **88**, 1977–1988.

Scholer, M. and Morfill, G. (1975). Simulation of solar flare particle interaction with interplanetary shock waves. *Solar Physics*, **45**, 227–240.

Schwadron, N.A., Fisk, L.A. and Gloeckler, G. (1996). Statistical acceleration of interstellar pick-up ions in co-rotating interaction regions. *Geophysical Research Letters*, **23**, 2871–2874.

Schwadron, N.A., Geiss, J., Fisk, L.A., Gloeckler, G., Zurbuchen, T.H. and von Steiger, R. (2000). Inner source distributions: Theoretical interpretation, implications, and evidence for inner source protons. *Journal of Geophysical Research*, **105**, 7465–7472.

Selesnick, R.S., Mewaldt, R.A. and Cummings, J.R. (1997). Multiply charged anomalous cosmic rays above 15 MeV/nucleon. In Proceedings of the 25th International Cosmic Ray Conference, Durban, Vol. 2, pp. 269–272.

Sheeley Jr, N.R., Howard, R.A., Koomen, M.J., Michels, D.J., Schwenn, R., Mühlhäuser, K.H. and Rosenbauer, H. (1985). Coronal mass ejections and interplanetary shocks. *Journal of Geophysical Research*, **90**, 163–175.

Simnett, G.M. (1995). Protons in flares. *Space Science Reviews*, **73**, 387–432.

Smith, E.J. (1993). Magnetic fields throughout the heliosphere. *Advances in Space Research*, **13**(6), 5–14.

Smith, E.J. and Wolfe, J.H. (1976). Observations of interaction regions and corotating shocks between one and five AU: Pioneer 10 and 11. *Geophysical Research Letters*, **3**, 137–140.

Sonett, C.P., Colburn, D.S., Davis, L., Smith, E.J. and Coleman, Jr, P.J. (1964). Evidence for a collisionfree magnetohydrodynamic shock wave in interplanetary space. *Physical Review Letters*, **13**, 153–156.

Steenberg, C.D., Cummings, A.C. and Stone, E.C. (1999). Drift calculations on the calculations of anomalous cosmic rays during the 1998 solar minimum period. In Proceedings of the 26th International Cosmic Ray Conference, Utah, Vol. 7, pp. 593–596.

Stone, E.C., Cohen, C.M.S., Cook, W.R., Cummings, A.C., Gauld, B., Kecman, B., Leske, R.A., Mewaldt, R.A., Thayer, M.R., Dougherty, B.L., Grumm, R.L., Milliken, B.D., Radocinski, R.G., Wiedenbeck, M.E., Christian, E.R., Shuman, S. and von Rosenvinge, T.T. (1998). The solar isotope spectrometer for the Advanced Composition Explorer. *Space Science Reviews*, **86**, 357–408.

Stone, E.C. and Cummings, A.C. (1997). Evidence for anomalous cosmic ray S, Si, and Fe in the outer heliosphere and for a non-ACR source of S at 1 AU. In Proceedings of the 25th International Cosmic Ray Conference, Durban, Vol. 2, pp. 289–292.

Svestka, Z. and Fritzová-Svestková, L. (1974). Type II bursts and particle acceleration. *Solar Physics*, **36**, 417–431.

Takashima, T., Doke, T., Hayashi, T., Kikuchi, J., Kobayashi, M., Shirai, H., Takehana, N., Ehara, M., Yamada, Y., Yanagita, S., Hasebe, N., Kashiwagi, T., Kato, C., Munakata, K., Kohno, T., Kondoh, K., Murakami, H., Nakamoto, A., Yanagimachi, T., Reames, D.V. and von Rosenvinge, T.T. (1997). The first observation of sulfur in anomalous cosmic rays by the Geotail and the Wind spacecraft. *Astrophysical Journal*, **477**, L111–L113.

Teegarden, B.J., von Rosenvinge, T.T. and McDonald, F.B. (1973). Satellite measurements of the charge composition of solar cosmic rays in the $6 \leqslant Z \leqslant 26$ interval. *Astrophysical Journal*, **180**, 571–581.

Temerin, M. and Roth, I. (1992). The production of ^{3}He and heavy ion enrichments in ^{3}He-rich flares by electromagnetic hydrogen cyclotron waves. *Astrophysical Journal*, **391**, L105–L108.

Toptyghin, I.N. (1980). Acceleration of particles by shocks in a cosmic plasma. *Space Science Reviews*, **26**, 157–213.

Tums, E., Gloeckler, G., Fan, C.Y., Cain, J. and Sciambi, R. (1974). Instrument to measure energy and charge of low energy interplanetary particles. *IEEE Transactions in Nuclear Science.*, **NS-21**(1), 210–217.

Van Hollebeke, M.A.I., McDonald, F.B., Trainor, J.H. and von Rosenvinge, T.T. (1978). The radial variation of corotating energetic streams in the inner and outer solar system. *Journal of Geophysical Research*, **83**, 4723–4730.

Van Nes, P., Reinhard, R., Sanderson, T.R., Wenzel, K.P. and Zwickl, J. (1984). The energy spectrum of 35- to 1600-keV protons associated with interplanetary shocks. *Journal of Geophysical Research*, **89**, 2122–2132.

Vasyliunas, V.M. and Siscoe, G.L. (1976). On the flux and the energy spectrum of interstellar ions in the solar system. *Journal of Geophysical Research*, **81**, 1247–1252.

Vernov, S.N., Chudakov, A.E., Vakulov, P.V., Gorchakov, E.V., Kontor, N.N., Logatchev, Y.I., Lyubimov, G.P., Pereslegina N.V. and Timofeev, G.A. (1970). Propagation of solar and galactic cosmic rays of low energies in interplanetary medium. In V. Manno and D.E. Page (eds), *Intercorrelated Satellite Observations Related to Solar Events*, Reidel, Hingham, MA, pp. 53–89.

Von Rosenvinge, T.T., Barbier, L.M., Karsh, J., Liberman, R., Madden, M.P., Nolan, T., Reames, D.V., Ryan, L., Singh, S., Trexel, H. and Winkert, G. (1995). The Energetic Particles: Acceleration, Composition and Transport (EPACT) investigation on the Wind spacecraft. *Space Science Reviews*, **71**, 155–206.

Von Rosenvinge T.T. and McDonald, F.B. (1975). IMP 6, 7, and 8 observations of the composition and time variations of low energy cosmic rays. In Proceedings of the 14th International Cosmic Ray Conference, Munich, Vol. 2, pp. 792–797.

Von Steiger, R., Schwadron, N.A., Fisk, L.A., Geiss, J., Gloeckler, G., Hefti, S., Wilken, B., Wimmer-Schweingruber, R.F. and Zurbuchen, T.H. (2000). Composition of quasi-stationary solar wind flows from SWICS/Ulysses. *Journal of Geophysical Research*, **105**, 27,217–27,238.

Wenzel, K.P., Marsden, R.G., Page, D.E. and Smith, E.J. (eds) (1992). Ulysses instruments. *Astronomy and Astrophysics*, Suppl., **92**, 207–440.

Wild, J.P., Smerd, S.F. and Weiss, A.A. (1963). Solar bursts. *Annual Review of Astronomy and Astrophysics*, **1**, 291–366.

Williams, D.J., McEntire, R.W., Krimigis, S.M., Roelof, E.C., Jaskulek, S., Tossman, B., Wilken, B., Stüdemann, W., Armstrong, T.P., Fritz, T.A., Lanzerotti, L.J. and Roederer, J.G. (1991). Energetic particles at Venus: Galileo results. *Science*, **253**, 1525–1528.

Witte, M., Rosenbauer, H., Banaszkiewicz, M. and Fahr, H. (1993). The Ulysses neutral gas experiment: Determination of the velocity and temperature of the interstellar neutral helium. *Advances in Space Research*, **13**(6), 121–130.

Zirin, H. (1994). Solar storminess. *Sky & Telescope*, November, p. 9.

FURTHER READING

Balogh, A., Gosling, J.T., Jokipii, J.R., Kallenbach, R. and Kunow, H. (eds) (1999). Corotating interaction regions. *Space Science Reviews*, **89** (1–2).

Mewaldt, R.A., Jokipii, J.R., Lee, M.A., Möbius, E. and Zurbuchen, T.H. (eds) (2000). *Acceleration and Transport of Energetic Particles observed in the Heliosphere*, ACE-2000 Symposium. *AIP Conference Proceedings*, **528**.

Simpson, J.A. (1995). The anomalous nuclear component in the three-dimensional heliosphere. *Advances in Space Research*, **16**(9), 135–149.

Zank, G.P. and Gaisser, T.R. (eds) (1992). *Particle Acceleration in Cosmic Plasmas*. *AIP Conference Proceedings*, **264**.

43

GERHARD HAERENDEL*

Reconnection

The reconnection of magnetic fields in a conducting fluid or gas is one of the most important processes in cosmical plasmas. It converts energy stored in magnetic fields into kinetic energy, transforms magnetic configurations, and enables interactions between two magnetized plasma regimes which otherwise would be much weaker. The recognition of the existence of this process grew half a century ago out of the attempt to understand the dramatic energy releases in the solar corona and chromosphere during flares. Processes on the Sun on all scales of energy release have the been targets of reconnection studies ever since. With the beginning of space research and the discovery of the magnetosphere it soon became clear that the fundamental process underlying solar wind–magnetosphere interactions and the ensuing internal dynamics of the magnetosphere must be reconnection. Furthermore, there was the chance to explore the process by extensive *in situ* and global response studies.

Reconnection is a process that breaks the magnetic connection of plasma elements in situations in which the magnetic field can be generally considered as frozen in. It requires high current densities which are concentrated in thin sheets or filaments and are thus exceptional regions in a much larger quasi-inert environment. The formation of reconnection regions – that is, of the necessary thin current sheets – may occur over long build-up times. The onset of reconnection can thus have a quasi-explosive character and lead to a rapid release of magnetic energy stored during the preparatory process. However, in some situations reconnection can also be essentially stationary. Since cosmic plasmas outside stellar interiors, dense atmospheres, and dense molecular clouds are essentially collisionless and highly conducting, reconnection is the most striking transformation process of magnetic configurations and the most powerful conversion process of magnetic into kinetic energy in the magnetic regimes around stars and in interstellar space. Magnetic reconnection is, however, also a necessary ingredient of the dynamos that act in the slow plasma flows in stellar interiors.

Reconnection is conceptually not a difficult process. The main controversy has been not so much its existence in highly conducting fluids, but its speed or efficiency. Postulated in the late 1940s, it took three decades to verify its existence in space and two decades more to derive an understanding of the microphysical origin of its efficiency. Although it was first proposed in an attempt to explain the sudden energy release in solar flares, it was the direct access to reconnection situations in the laboratory and in space that laid the foundations of its acceptance by the wider community of plasma physicists and astrophysicists, and it was thanks to the rapidly developing art of computer simulations during the last two decades of the twentieth century that led to a deeper understanding of the microprocesses involved. Reconnection is fundamental to the energetics and dynamics of the solar corona, which is the source of the plasma and magnetic fields that fill the heliosphere. Reconnection also plays a key role in transferring energy from the solar wind flow to planetary magnetospheres and atmospheres. It is thus the most important ingredient of solar–terrestrial physics. Although it is most familiar in the high-beta plasmas of corona, magnetopause, and magnetotail, a modified form of reconnection occurs in the very low-beta plasmas ($\leq 10^{-4}$) at the bottom of the corona or planetary magnetospheres. However, in this context the term "reconnection" has been mostly avoided: although magnetic connections are broken and magnetic energy is released, terms such as "auroral acceleration," "thawing" of field lines, and "magnetic fractures" tend to be used instead.

---

*International University Bremen, Germany

This chapter attempts to trace the evolution of the understanding of the concept of reconnection from the first conjectures, via the growing body of evidence, to the present-day analysis of the relevant microprocesses. After looking at the early history and the development of basic reconnection configurations on the Sun and in the magnetosphere, we turn to the growth of indirect and direct evidence of its existence and to the insights gained from specifically designed reconnection experiments in the laboratory and from numerical simulations. Finally, we return to our point of origin, the application to the Sun. There are still many unanswered questions, and controversies are still debated. However, the foundations for a successful interpretation of some of the most striking phenomena on the Sun and in near-Earth space have become increasingly solid, and applications to distant cosmic phenomena are gaining credibility. The author is grateful for having been an observer and participant, although not in the front row, of the exciting quest for a thorough understanding of this powerful and yet elusive process.

THE EARLY HISTORY: FROM GIOVANELLI TO PETSCHEK

In a *Nature* article of 1946, Donald G. Giovanelli (1946) proposed a theory of solar flares according to which electrons accelerated by induction electric fields near magnetic neutral points excite the optical emissions of chromospheric atoms. He assumed that a discharge takes place because, with increasing energy, the electrons undergo fewer collisions. Consequently, he proposed the existence of very high current densities. Cowling (1953) heavily criticized Giovanelli's and Hoyle's (1949) subsequently developed ideas on the grounds that this implied current sheets a few meters in width "which cannot by any stretch of

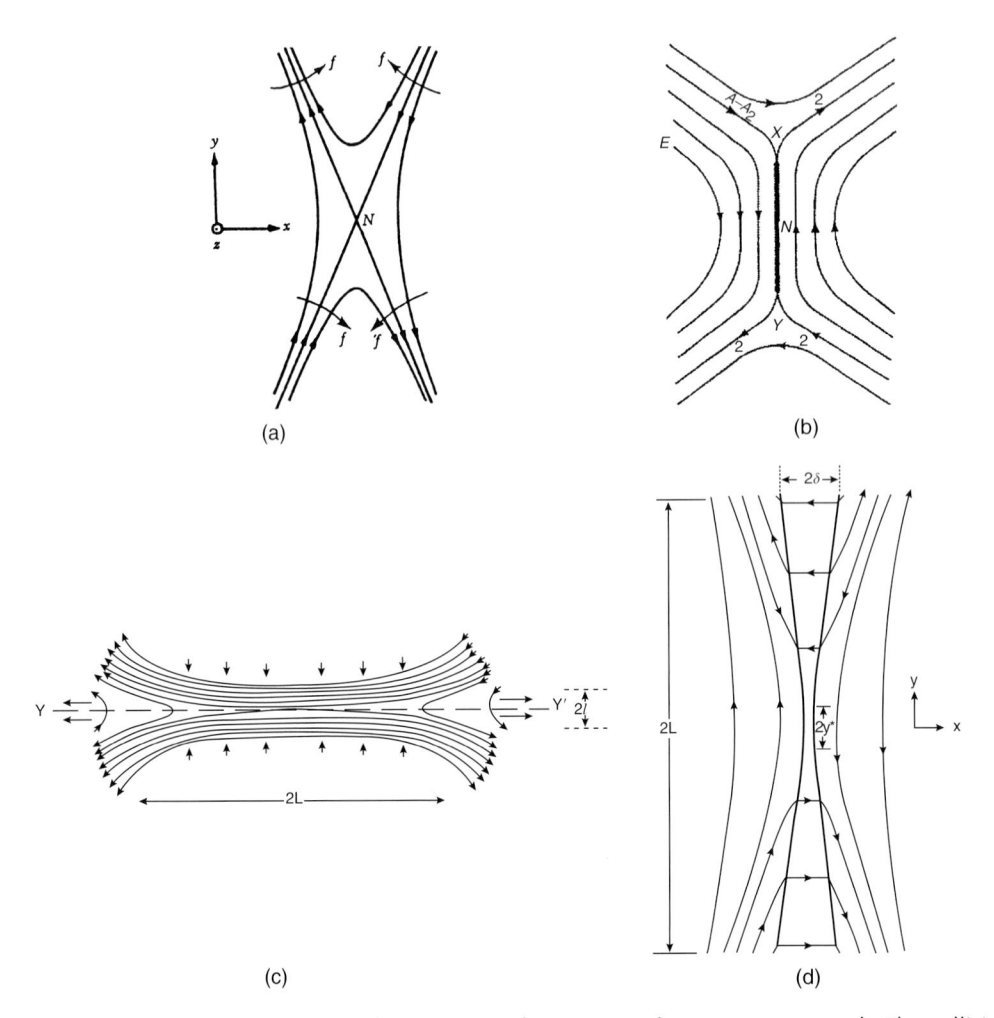

Figure 1 (a) The direction of the magnetic force, f, near a neutral point, N. (After Dungey 1958.) (b) The collision layer. (After Sweet 1958.) (c) Schematic of a reconnection configuration. (After Parker 1963.) (d) Petschek's solution: a small diffusion region with standing slow shocks attached. (After Petschek 1964.)

imagination be regarded as a possible thickness of a flare layer." He further concluded that "the discharge is not a regular one, but an irregular dissipation of energy in a violently twisted field." He then asked how the field can become so heavily twisted. The electromagnetic forces may well be there. "But induced currents, whose effect is always to oppose the changes to which they are due, smother the effect which was sought and leave only an unsatisfactory vestige of it remaining." This argument of Cowling, in which he invoked Lenz's law, was disputed by Hoyle's student Jim Dungey (1958), who showed that near neutral points Lenz's law is reversed: that the magnetic force density, $f = \mathbf{j} \times \mathbf{B}$ (Figure 1), "tends to compress the material and field in the x-direction and stretch them in the y-direction. Since this motion reduces the acute angle between the limiting lines of force at N, it seems probable that it increases the current density." He added, "While a rigorous investigation is lacking, then, the indications are that the pressure gradient cannot prevent the discharge."

Dungey's ideas were taken up by Sweet (1958), who concluded that the magnetic forces would flatten the field and form a thin "collision layer," as shown in Figure 1b. In analogy to the hydromagnetic situation, he considered two plates being forced together, calculated correctly the outflow velocity, and investigated the dissipation of magnetic energy by Joule heating in the current layer of decreasing thickness. It was Parker (1957, 1963) who, inspired by Sweet's work, posed the reconnection problem in the simplest possible terms, at the same time drawing attention to the fact that none of the then "known mechanisms are sufficiently rapid to account for the solar flare from the annihilation of magnetic fields." And he was concerned that "there appears to be some popular belief that the necessary annihilation of magnetic fields has been accounted for by quantitative theory." In his papers he elaborated quantitatively on the existing difficulties, the most striking of which was the *rate* of reconnection. This can be derived immediately from Parker's famous set of relations. With $2l$ as the width of the current sheet in Figure 1c and $2L$ its length, conservation of mass yields

$$vL = Vl \quad (1)$$

where v is the speed of magnetic field merging and V the efflux velocity, which from pressure equilibrium across the neutral sheet must be the Alfvén speed:

$$V = \frac{B_0}{\sqrt{\mu_0 \rho}} = V_A \quad (2)$$

where B_0 is the upstream magnetic field. According to Sweet and Parker, magnetic diffusion controls the merging speed v:

$$v = \frac{D_m}{l} \quad (3)$$

where D_m is the magnetic diffusivity,

$$D_m = \frac{\eta}{\mu_0} \quad (4)$$

and η is the electrical resistivity. Combining eqns (1)–(4) provides an expression for the merging or reconnection rate in terms of the Alfvénic Mach number:

$$M_A = \frac{v}{V_A} = R_m^{-1/2} \quad (5)$$

where $R_m = V_A L / D_m$ is the magnetic Reynolds or Lundquist number. The difficulty, which Parker pointed out and which exists still today (see, e.g., Biskamp 1993), is that for current sheets of macroscopic length L, R_m tends to be a very large number in space and solar plasmas, and the reconnection rate v or M_A is very small, much too small to account for the fast energy release in solar flares.

The solution came from Petschek (1964), in a paper he presented at the AAS–NASA symposium on the physics of solar flares in 1963. As shown in Figure 1d, he attached standing hydromagnetic waves to a diffusion region of length y^*, much shorter than the length L of the overall reconnection situation. He determined y^* to be

$$y^* = \frac{D_m}{v_A M_A^2} \quad (6)$$

whereby the merging rate, M_A, is considered as an externally controlled parameter. His theory then provides a maximum possible flow rate, which (with a correction by a factor of $\frac{1}{2}$, found by Vasyliunas (1975)) is

$$M_{A,max} = \frac{\pi}{8 \ln(2 M_{A,max}^2 R_m)} \quad (7)$$

So, Parker's merging rate, which depends inversely on the square root of R_m, has now been replaced by a logarithmic dependence. Even with $R_m = 10^{10}$, maximum merging rates of 0.05 appear to be possible. "The principal effect of including the wave propagation mechanism is to reduce the length over which the diffusion mechanism must operate," wrote Petschek. With this reduced length, Parker's relation (eqn (3)) is still valid.

The two waves attached to the diffusion region are slow mode shocks which nearly switch off the tangential field component. They propagate with the Alfvén speed based on the normal magnetic field component. Thus

$$\frac{B_n}{B_0} = M_A \quad (8)$$

These waves are dissipative, and the pressure increases across the shock so as to maintain continuity of the total (gas plus magnetic) pressure.

As pointed out by Vasyliunas (1975), Petschek's solution requires a convergent flow toward the diffusion region (Figure 2). The magnetic field strength is thus lowered in the region immediately upstream of the diffusion region. Here the Alfvén speed is lower than it is farther away. Since the Alfvén Mach number must not exceed unity just outside the diffusion region, it follows that the merging rate must be less than unity. This conclusion, however, depends on Petschek's assumption that the magnetic field far upstream is homogeneous. Other boundary conditions can produce even higher merging rates.

Petschek's (1964) paper constituted a major breakthrough in the theory of reconnection and in plasma astrophysics generally. Several alternate models producing even higher reconnection rates were subsequently presented, for instance by Sonnerup (1970) and Yeh and Axford (1970). Their solutions are basically similarity solutions in that, at scales between that of the diffusion region (considered to be nearly zero) and the external scale L, the magnetic field and flow velocity depend only on the ratio x/y of the two relevant coordinates. These and other models have been critically analyzed by Vasyliunas (1975). The outcome was:

> Neither Petschek's model nor the similarity model predicts a definite merging rate. In both models the speed of plasma inflow (v) is, within limits, a free parameter whose value is assumed to be determined by the boundary conditions; the merging rate fixes the dimensions of the diffusion region in terms of the appropriate microscopic length scale, but neither resistivity nor inertial effects have any influence upon the models outside the reconnection region.

The microscopic length scales Vasyliunas was referring to arise from the various terms of the generalized Ohm's law and are the resistive length, D_m/v_A, and the ion inertial length, c/ω_{pi}, or the electron inertial length, c/ω_{pe}. In a later section we return to the role of these scales in the so-called diffusion region according to present-day understanding.

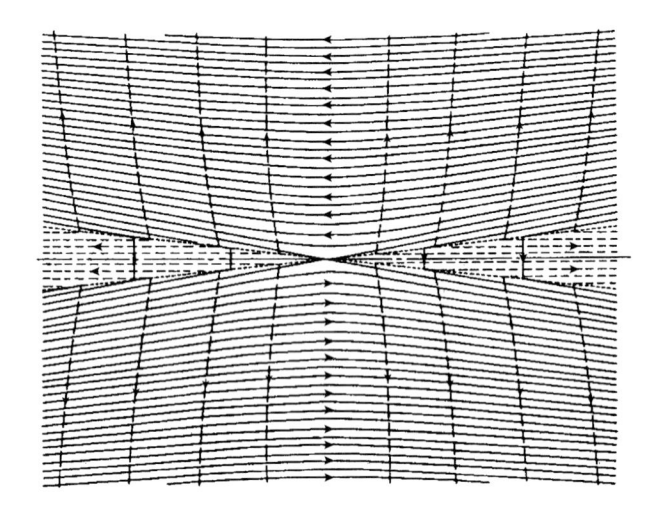

Figure 2 Convergent flows in the Petschek solution. (After Vasyliunas 1975.)

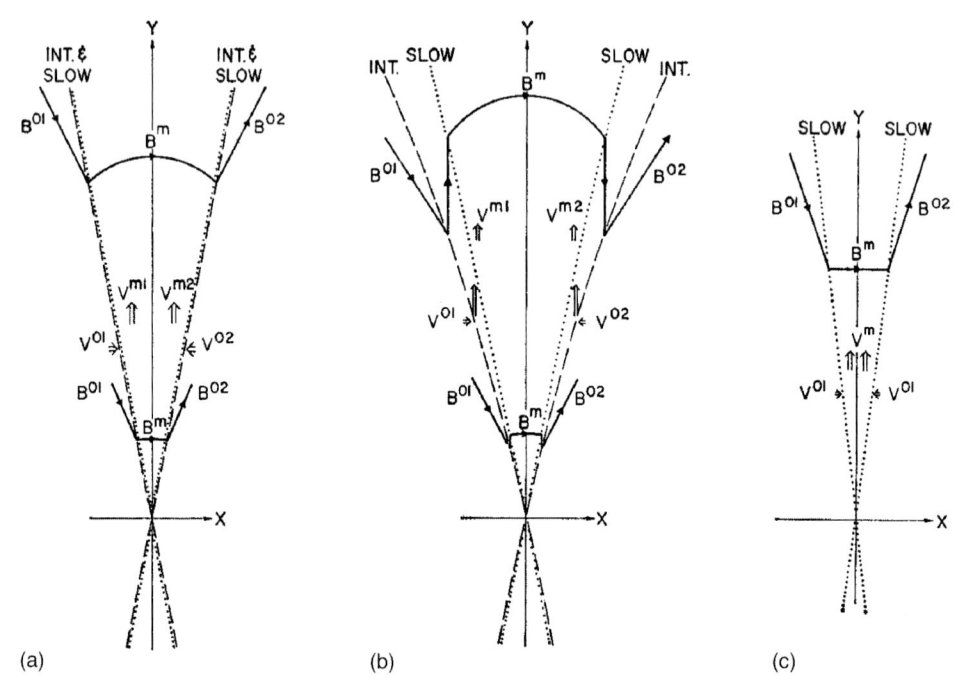

Figure 3 Standing wave solutions for an external field with spatial gradient: (a) for an incompressible fluid, (b) for a compressible one. In (c) the external field is homogeneous. (After Petschek and Thorne 1967.)

Even when processes other than magnetic diffusion are considered to act near the X-line, one conventionally refers to this region as the "diffusion region." We shall see later that, at least on scales of c/ω_{pe}, some dissipative diffusive process may accomplish the field-line breaking.

Priest and Forbes (1986) later derived a unified model for steady-state reconnection in which the solutions of Petschek (1964) and Sonnerup (1970) appeared as special cases. Like Petschek, they used an expansion method and classified the solutions by a parameter b which is related to the flow velocity at the inflow boundary. All of these solutions applied to an extremely symmetric and homogeneous plasma and field with strictly antiparallel fields on either side of and at some distance from the reconnection layer. In practise there may be variations of the magnetic field magnitude away from the neutral line and a rotation of the upstream field vectors relative to a strict antiparallel orientation. Slow shocks suffer from the restriction that the tangential components of **B** must have the same sign on either side of them. The more general situations can, however, be accommodated by splitting the standing waves into pairs of intermediate waves and slow shocks, as shown in Figure 3 (Petschek and Thorne 1967). The intermediate waves travel at exactly the Alfvén speed, and in the compressible case stand somewhat farther upstream and in the incompressible case coincide with the slow shocks. They rotate the tangential field vector, without change in magnitude, by 180°, while the slow shocks produce a change in magnitude.

Another variant, particularly important for reconnection at the front-side magnetopause, deals with situations in which the plasma density and temperatures change drastically across the reconnection layer. This was considered by Levy et al. (1964). Their solution involves an intermediate wave on the high-density side through which the thermodynamic properties remain constant, followed by a slow expansion wave through which the magnitude of **B** increases in order to balance the higher plasma pressure on the upstream side (Figure 11).

After the seminal work of Sweet, Parker, and Petschek, the research on reconnection took essentially three directions:

(1) the application to space and astrophysics, in particular solar magnetic configurations, assuming essentially Petschek's maximum merging rate;
(2) the search for evidence of reconnection, both by indirect and by direct *in situ* signatures;
(3) the pursuit of a deeper understanding of the controlling processes, both by specifically designed laboratory experiments and by numerical simulations.

All of these areas will be discussed subsequently.

Before leaving the subject of the early history of reconnection it is appropriate to refer briefly to a somewhat related process discovered during the same period – the

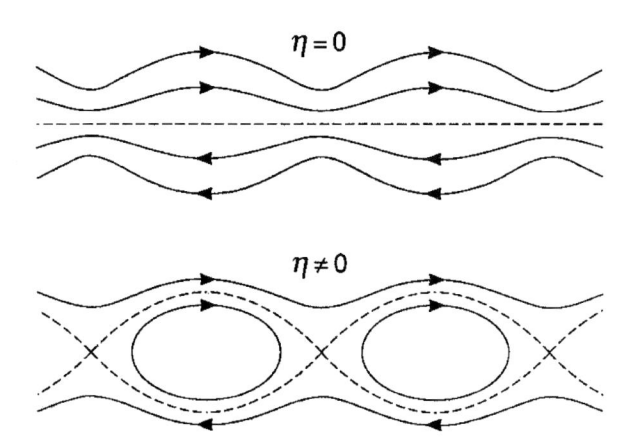

Figure 4 Periodic filamentation of sheet current and plasma leads to tearing mode instability for $\eta \neq 0$. (After Furth et al. 1963.)

tearing mode instability (Furth et al. 1963). It occurs in resistive current sheets, leads to current filamentation, and converts magnetic energy ohmically into heat (Figure 4). For wavelengths larger than the width of the current sheet the growth time is the geometric mean of the resistive diffusion time, $\tau_d = d^2/D_m$, and the Alfvénic transit time, $\tau_A = d/v_A$. Beyond structuring the current sheet into elongated magnetic islands, the tearing mode is important because of coalescence instability (Finn and Kaw 1977). Because parallel currents attract each other, an array of small islands may coalesce to form a larger plasmoid. Biskamp (1993) treated this instability extensively, and in numerical simulations studied, among other things, the repetitive plasmoid generation and ejection from a tearing, unstable Sweet–Parker current sheet. This process can be of relevance in magnetotails.

For several decades the relevance of the tearing mode to space and astrophysical situations, in which η essentially vanishes, has been a matter of debate. While the inertia of the ions appears to be the non-ideal process responsible for the instability, the adiabatic magnetized electrons turn out to have a stabilizing effect, through electrostatic coupling to the ions. However, there may be non-adiabatic processes, for instance pitch-angle scattering, which allow the electrons to cross field lines and thus can render the system unstable (Coroniti 1980). Another cause of non-adiabaticity and therefore destablization of the collision-free tearing mode can be the chaotization of the electron orbits when traversing thin current sheets with a finite normal field component, B_n (Büchner and Zelenyi 1987). Chaotization sets in when the parameter $\kappa = (B_n/B_0)(L/\rho_{eo})^{1/2}$ tends to unity (<1.6), where L is the thickness of the current sheet and ρ_{eo} the electron gyro-radius. The relevance in the context of reconnection arises from the possibility that destabilization of the tearing mode via island coalescence may lead to the spontaneous onset of reconnection.

RECONNECTION CONFIGURATIONS

On the Sun

As the reconnection process was conceived in the search for a powerful energy conversion process operating in the solar atmosphere, it is not surprising that the elaboration of the basic concepts by Sweet, Parker, Petschek, and others was followed by a vast amount of literature attributing the solar flare essentially to a reconnection of opposing magnetic fields in the corona or chromosphere. Also, less dramatic solar events down to the scale of micro- and nanoflares (Parker 1988) were analyzed in the framework of reconnection. No attempt is made in this chapter to summarize the wide spectrum of ideas about specific magnetic configurations and quantitative assessments of how reconnection may provide the energy released in a flare and account for the wide range of associated phenomena. We focus instead on the principal reconnection situations which are likely to operate in the solar atmosphere and which underlie radiative outbursts, from the largest flares (10^{33} erg) to nanoflares ($\leq 10^{24}$ erg).

There seem to be only three categories of magnetic field configurations and driving forces:

(1) those in which opposing fields are *pushed* together by plasma flows or magnetic buoyancy;
(2) those in which fields are *pulled* and stretched;
(3) those in which flux tubes are *twisted* and, owing to the onset of internal kink modes, are pressed against neighboring flux tubes.

We illustrate these situations by reference to some of the early papers dealing with each of these mechanisms.

Pushed fields

In his first paper on reconnection, Sweet (1958) proposed a way of producing narrow current sheets with neutral points by displacing two bipolar spot groups towards each other (Figure 5). "If the medium is a perfect electrical conductor the lines of force through the neutral point move with the medium and the two fields A and B do not interpenetrate further. The fields may be said to collide." In the discussion

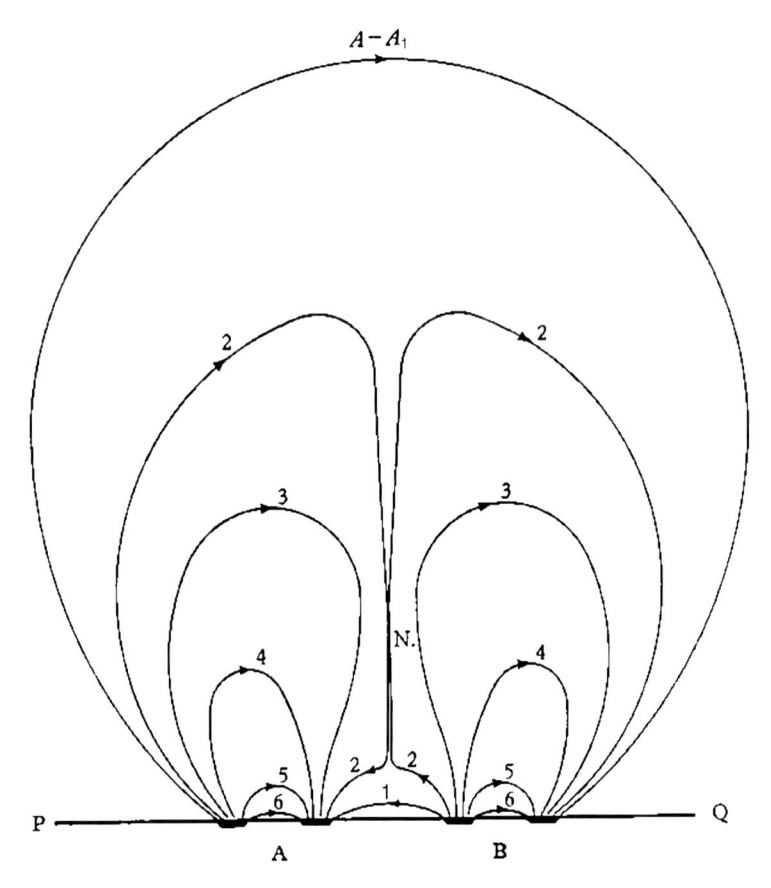

Figure 5 The field of displaced bipolar systems in a conducting medium. (After Sweet 1958.)

following Sweet's presentation of this paper at the IAU Symposium No. 6 at the Royal Institute of Technology in Stockholm in August 1956, Cowling asked how motions in the photosphere would be transmitted into the corona. If by hydromagnetic waves, would differences in neighboring propagation paths upset the formation of the current layer? Sweet replied that the motions were so slow that the transmission of forces could be worked out without reference to hydromagnetic waves. This discussion illustrates well the novelty of the whole subject and the need for pioneering thinking during the early conception phase of reconnection and flare theories.

A second way of pushing fields together results from the emergence of magnetic loops from underneath the photosphere in an established spot group (Figure 6). Heyvaerts et al. (1977) considered three phases:

(1) the preflare phase with continuous reconnection and some energy release leading, among others, to soft X-ray emission;

(2) an impulsive or flash phase, where the configuration gets out of equilibrium; high-current densities, microinstabilities, and anomalous resistivity are generated; and particles can be accelerated to very high energies;

(3) the main phase, in which a new equilibrium is established and reconnection is still proceeding.

No detailed conclusions will be given here on electric currents, fields, effective resistivity, and overall energy release, except to note that the width of the current sheet, which evolves through the various stages of the rising field, is calculated on the basis of eqn (3) with different assumptions for the resistivity, η, and externally impressed velocity, v. By invoking a very high turbulent resistivity, thicknesses of 10 km are quoted, which is much wider than Cowling's estimates and wider than is implied by nonturbulent, inertia-based reconnection. Turbulent resistivities have played a large role in the development of theoretical as well as computational reconnection models; we return to this topic in a later section.

Pulled fields

The first suggestion of how thin current sheets might be created by stretching a dipolar field upward was made by Carmichael (1964) at the famous AAS–NASA symposium on solar flares in 1963. He referred to the solar wind, just proposed by Parker, as the driving agent, but without making any quantitative assessment of the strength of the driving force. Sturrock (1966) suggested that the flare reconnection site is located in the current sheet or sheet pinch separating antiparallel fields in coronal streamers (Figure 7). He also considered the set-up of the streamer configuration by the distortion of closed field lines and their upward motion through the sheet pinch due to excess pressure in the cusp. He then described the onset of reconnection somewhere in the sheet pinch as a nonlinear evolution of the tearing mode instability. Once established, "the upper parts of the U-shaped field lines now exert a tension on the plasma associated with it, so that this plasma will be ejected from the Sun as from a slingshot." He concluded, "As the 'magnetic pocket' is ejected form the surface of the Sun it moves into a region of lower magnetic pressure and lower gas pressure, and for this reason is likely to expand transversely." With this sentence, Sturrock outlined a model of coronal mass ejections (CMEs) long before their discovery. But more important is the conclusion, not explicitly drawn, that because of the absence of any obstacles in the upward outflow, the reconnection process can last for an extensive period. As reconnected flux (and plasma) is removed quickly (at the Alfvén speed) and is not piling up, a small width of current sheet can be sustained until much of the free energy of the configuration is released. And the rate of reconnection is likely to be maintained at the maximum possible rate – that is, the Petschek rate.

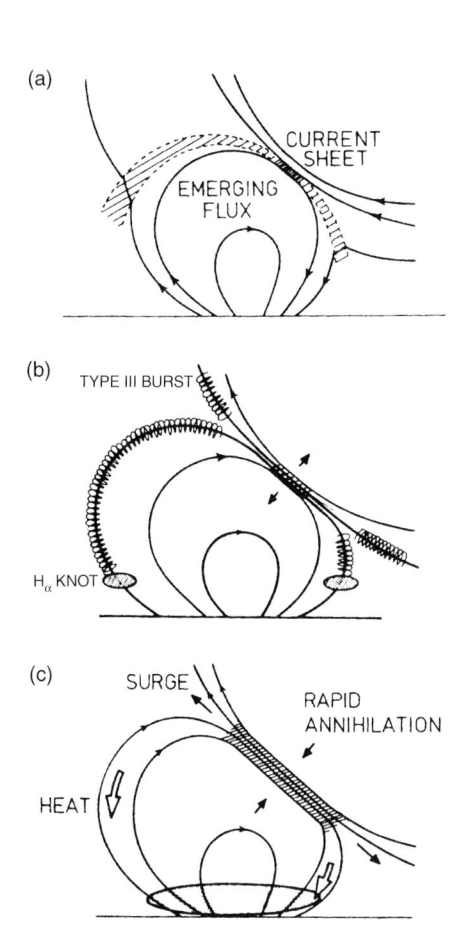

Figure 6 Three phases of the emerging flux mechanism for a subflare: (a) preflare heating, (b) the impulsive phase, and (c) the main phase. (After Heyvaerts et al. 1977.)

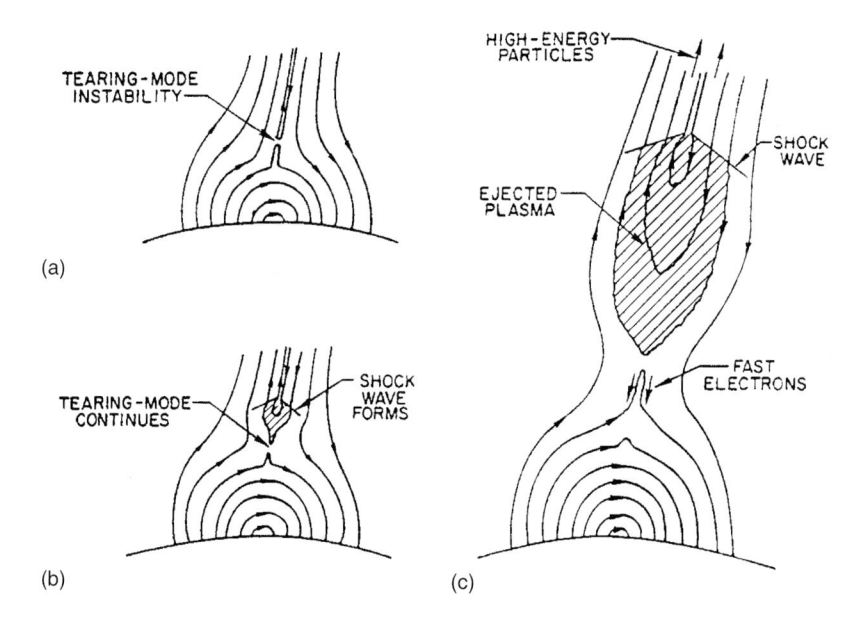

Figure 7 Three phases of a high-energy flare. (After Sturrock 1966.)

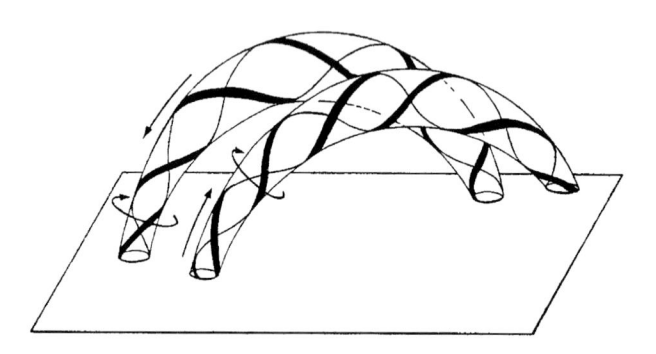

Figure 8 Two oppositely twisted flux tubes approaching each other. (After Gold and Hoyle 1960.)

Twisted fields

Interaction of twisted flux tubes was first considered by Gold and Hoyle (1960), as shown in Figure 8. Twisted bundles of magnetic field serve as stores of energy and, when pressed together, they provide regions with antiparallel field components prone to reconnect. With this model the authors tried to explain the long energy storage time due to continuous twisting at the photospheric feet of the field lines, with regard to the short, catastrophic energy release time when parallel currents in the boundary regions of two twisted filaments attract each other. In a series of papers, Parker (1972, 1983, 1988) investigated the equilibrium of magnetic fields in an infinitely conducting fluid subject to various kinds of motion at the boundaries (corresponding to the photospheric feet of the flux tubes). Rotations and random walks create twisting and braiding, such as shown in Figure 9. Parker concluded that static equilibrium of such configurations was not possible and that, by necessity, magnetic neutral sheets and reconnection occur. He considered the braiding and twisting as a mechanism of continuous energy transfer into coronal fields, and the ensuing reconnection as a continuous heating process. While the continuous energy input in an active region is of the order of $10^7 \, \mathrm{erg \, cm^{-2} \, s^{-1}}$, the output in individual reconnection events may range from less than 10^{24} to 10^{27} erg. Parker thus explained the X-ray corona as well as the observations of fast jets (Brueckner and Bartoe 1983) ejected from localized reconnection regions.

From these brief characterizations of solar reconnection configurations, we now turn to the magnetosphere, which offers the tremendous advantage of being accessible to *in situ* observations.

At the magnetosphere

After an early speculation by Hoyle (1949) that auroral particles may be accelerated at neutral points forming at interfaces of the interplanetary magnetic field (IMF) and the geomagnetic field, it was Hoyle's student Dungey (1961) who first applied the reconnection concept to the interaction of the solar wind and the magnetosphere. His famous diagram (Figure 10) showed that, with a southward IMF, magnetospheric field lines may merge with the IMF on the front side, become open, get pulled across the polar cap, and be transformed back into closed field lines at a reconnection site on the rear side. A magnetic circulation or plasma convection would thus be initiated inside the magnetosphere. Levy *et al.* (1964) evaluated the effect of magnetospheric reconnection more quantitatively and concluded that the

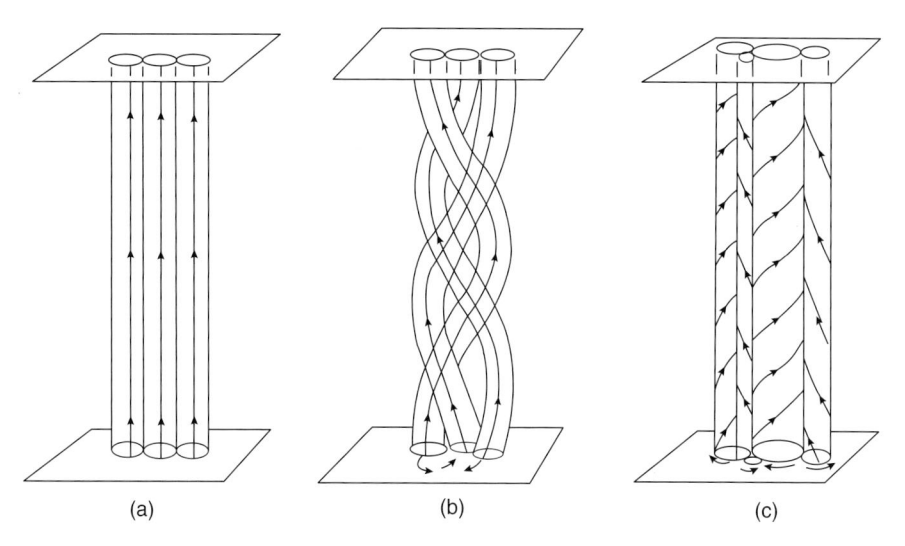

Figure 9 Schematic representation of (a) individual flux tubes, and (b) braiding and (c) twisting by fluid motions at one end. (After Parker 1983.)

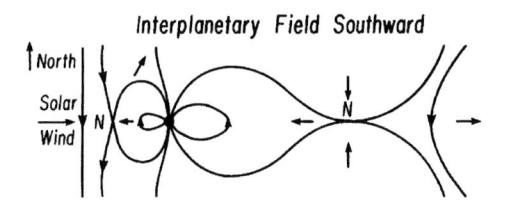

Figure 10 The magnetic topology of the reconnecting magnetosphere at the noon/midnight meridian, with southward interplanetary magnetic field. (Dungey 1961.)

openness of the magnetosphere was rather moderate. Only 10–20% of the incident southward magnetic flux would reconnect with the Earth's field, the remainder being carried past a well-defined magnetopause. They also considered the consequences of a strong density and plasma pressure drop across the magnetopause or, for that matter, across a reconnection site. Figure 11 shows their solution. The first discontinuity attached to a diffusion region is a rotational one or an intermediate wave. It rotates the magnetic field vector, ideally by 180°, and accelerates the inflowing plasma by an amount given by

$$[v_t] = \pm \frac{[B_t]}{\sqrt{\mu_0 \rho}} \quad (9)$$

if the plasma pressure is isotropic. The subscript t refers to the tangential component of **B**, and the square brackets denote the jump of the respective quantities across the discontinuity. Further inward, the plasma expands into the relative vacuum of the magnetosphere through an expansion wave or fan and is further accelerated. The real situation is more complex in that the internal plasma pressure is

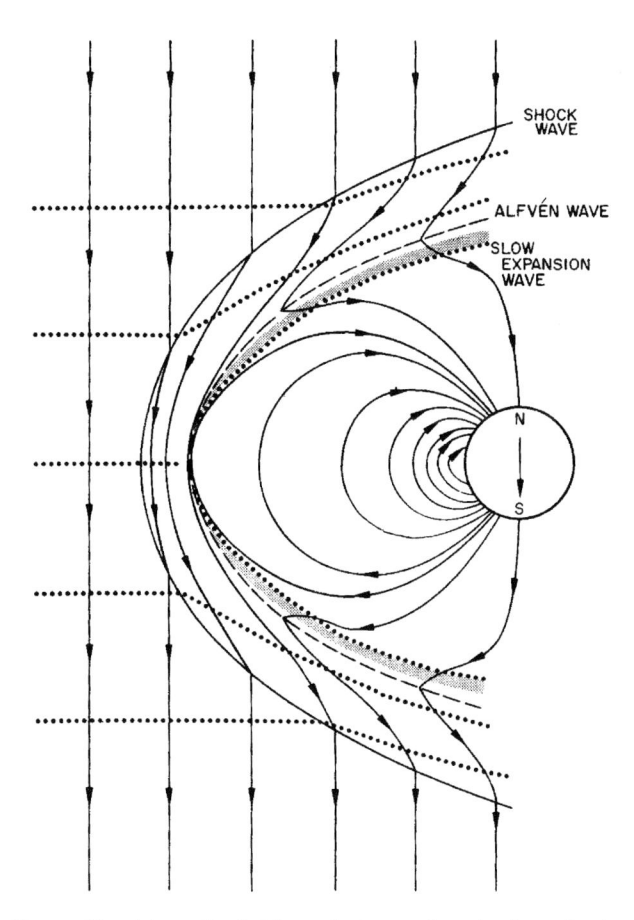

Figure 11 Magnetic fields and plasma flow in the subsolar region. The flow is decelerated by the bow shock. The magnetosphere boundary is resolved into an Alfvén wave and a slow expansion fan. (After Levy et al. 1964.)

not negligible and the plasma pressure may be anisotropic. In the latter case, eqn (9) has to be modified to

$$[v_t] = \left[\frac{(1-\alpha)^{1/2} B_t}{(\mu_0 \rho)^{1/2}}\right] \quad (10)$$

where

$$\alpha = \frac{p_\parallel - p_\perp}{B^2/\mu_0} \quad (11)$$

In either case there is a distinct flow enhancement just inside the rotational discontinuity, identified as the magnetopause.

Whereas the *front-side* reconnection may be classified as due to *pushed* fields, in the sense discussed in the previous section, *tail* reconnection may correspond more to the *pulled* field situation, although not to the same extent as in coronal streamers, since there is little external pressure drop along the tail current sheet. Hones (1979) proposed the existence of two neutral lines in the tail, one more permanent at large distances ($>100 R_E$), and the other forming intermittently after thinning of the near-Earth plasma sheet at typical distances of 20–30 R_E (Figure 12). It is the set-up of the latter reconnection site that, in the opinion of many researchers, is the origin of the magnetospheric substorm with all its consequences of inward plasma and field transport, intense magnetosphere/ionosphere currents and aurorae, and the formation of a plasmoid between the two neutral lines and its ejection in the downtail direction.

Almost all initial theories of reconnection have dealt with the merging of antiparallel fields. The situation at the front side of the magnetosphere is, however, characterized by steadily and irregularly changing orientations of the IMF. One orientation often lasts only for 10 or 20 minutes, and rarely for several hours. Therefore, it was necessary to explore the consequences of all possible clock angles, Θ, of the IMF at the nose of the magnetosphere, the ideal case being $\Theta = 180°$, that is, oriented due south. There were two approaches, the so-called "component merging" (Sonnerup 1974, Gonzalez and Mozer 1974) and "antiparallel merging" (Crooker 1979). The latter is easily described, the assumption being that reconnection occurs predominantly where the IMF draped over the front-side magnetopause is antiparallel to the magnetospheric field. This determines the site of the diffusion region – the "reconnection lines" – as shown in Figure 13. This has certain consequences for the predominant convection flow across the polar cap after reconnection and on the boundary layer flow, such as is shown in Figure 11, as a function of clock angle. In particular, the dividing line between the northward and southward flows may be far from the equator.

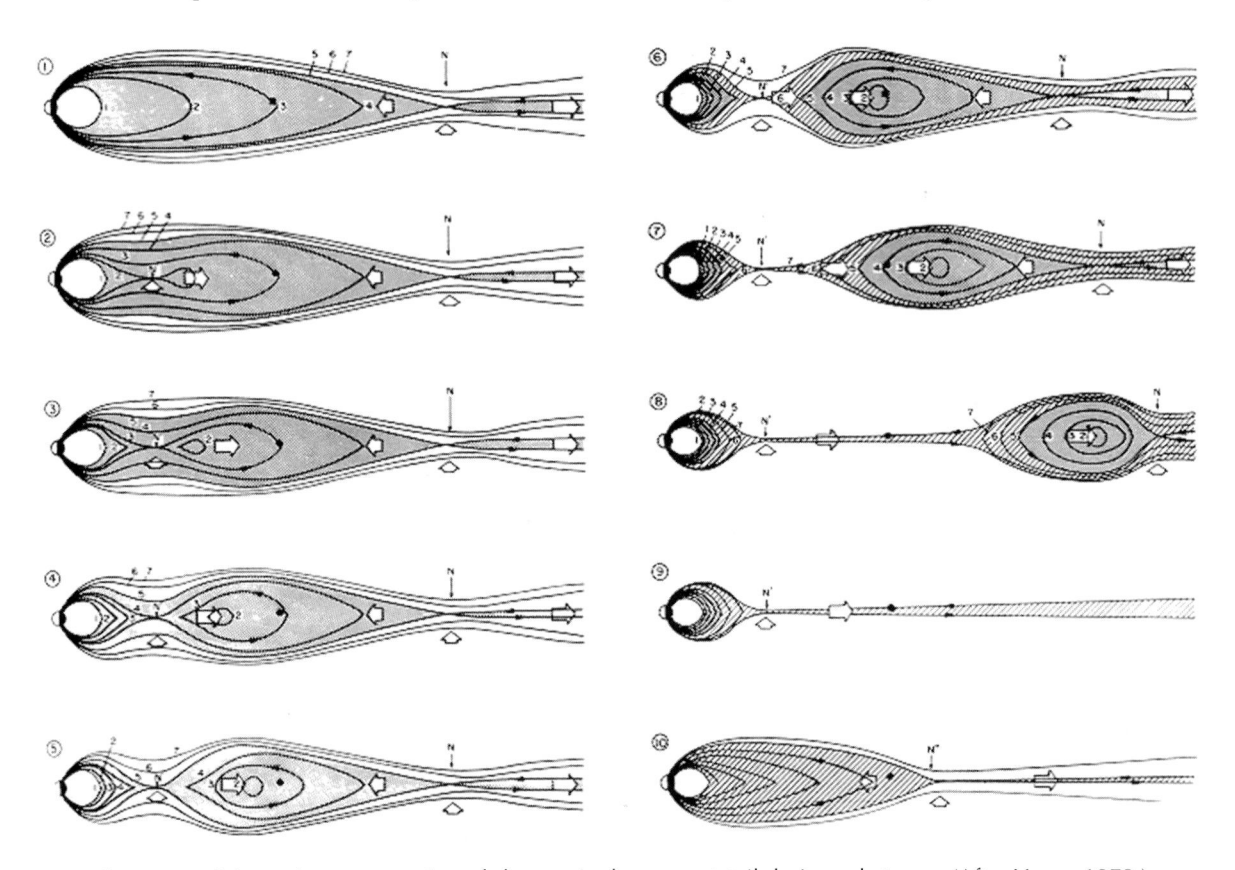

Figure 12 Schematic representation of changes in the magnetotail during substorms. (After Hones 1979.)

Figure 13 A model of antiparellel field merging. (After Crooker 1979.)

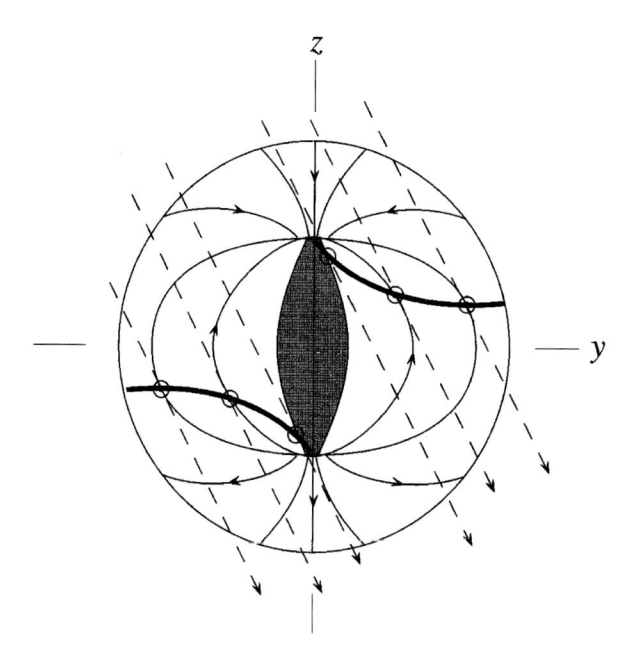

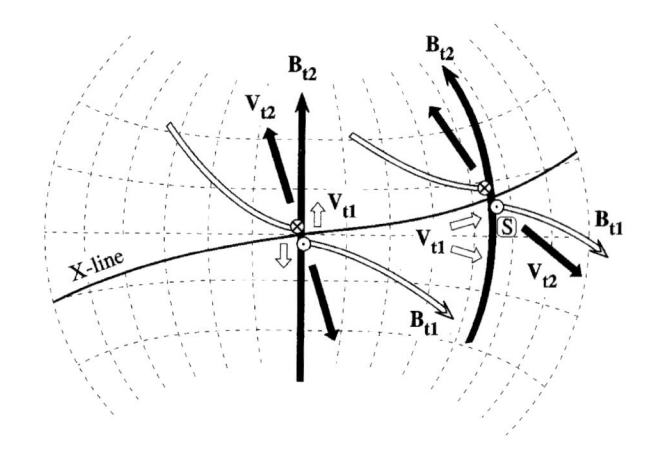

Figure 14 A front view of the magnetopause, showing the reconnection or X-line, the external (B_1) and internal (B_2) tangential magnetic fields, and the flow directions before and after reconnection. (Sonnerup et al. 1981.)

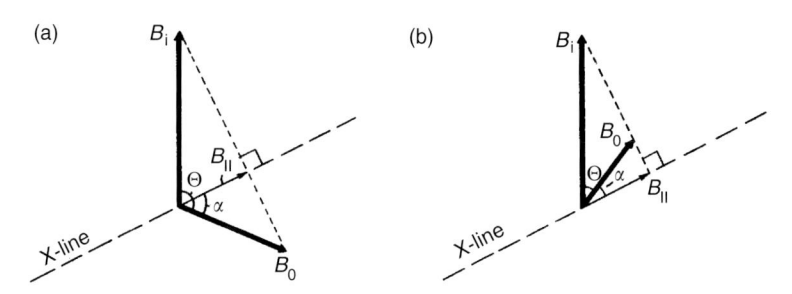

Figure 15 Orientation of the X-line (a) when $\cos \Theta < B_0/B_i$ and (b) when $\cos \Theta > B_0/B_i$ for $B_0/B_i < 1$. (After Sonnerup 1974.)

When considering component merging, one had to deal not only with various external field orientations, but also with various field strengths relative to that of the internal field. The basic assumption was that only the antiparallel field components mattered for the merging, and that the reconnection or X-line should follow essentially an external flow line; that is, by necessity it should intersect the subsolar point. This is shown in a diagram by Sonnerup et al. (1981), reproduced here as Figure 14. Locally, the speed of reconnection, v, was expressed as a tangential electric field component parallel to the X-line, $E_\parallel$. Since, according to Petschek (1964) and Levy et al. (1964), v was expected to be at best of the order of $0.1 v_A$, one calibrates in terms of $k v_A B_0$ with k of the order of 10^{-1}. Petschek (1964) had argued that in the incompressible situation the reconnection rate would not change if one added a constant magnetic field component parallel to the X-line. On the basis of this argument, component merging is evaluated such that the internal field, B_i, and the external field, B_0 have the same parallel component, as shown in Figure 15. If the magnitude of B_0 is smaller than that of B_i, one sees immediately that there are no antiparallel transverse components – that is, no reconnection – when $\cos \Theta > B_0/B_i$. The outcome of the analysis of Sonnerup (1974) is

$$E_\parallel \leq k v_{A0} B_0 \frac{((B_0/B_i) - \cos \Theta)^2}{1 + (B_e/B_i)^2 - 2(B_e/B_i) \cos \Theta} \quad (12)$$

In the case where $B_0 = B_i$, this reduces to

$$E_\parallel \leq k v_{A0} B_0 \sin^2 \tfrac{1}{2}\Theta \quad (13)$$

These relations lend themselves to experimental verifications, as we shall see in the next section. If the reconnection line has a total length L, then $E_\parallel L = V$ is the total potential along this line or the total transport rate of open magnetic flux across the polar cap. V can be measured in various ways in the polar cap ionosphere and compared with the predictions of eqns (12) and (13), in particular with respect to the Θ dependence and the value of k.

OBSERVATIONAL EVIDENCE IN THE MAGNETOSPHERE

Indirect evidence

Since this chapter focuses on reconnection as a process and not on the magnetosphere, we deal here only very briefly with its consequences for magnetospheric dynamics, which range from internal convection and ionospheric currents to aurorae, substorms, magnetic storms, and radiation belt replenishments. The temporary openness of magnetic field lines connected to the high-latitude ionosphere has been amply verified by the access of energetic solar electrons to the polar cap. The overall configuration of the geomagnetic tail shows that temporary connections occur with the interplanetary field.

More quantitative information can be derived from correlations of certain magnetospheric phenomena with properties of the solar wind and the IMF. Such phenomena include the position of the dayside boundary of the polar cusp, which can be located by certain particle precipitations. Reconnection leads to an erosion of the front-side closed field, and the equatorward border of the cusp at low altitudes moves closer to the equator by several degrees when the IMF turns south (Burch 1972).

A wide class of tests is based on the polar cap convection and the associated Hall currents, whose magnetic perturbation fields can be measured conveniently on the ground. Nishida (1971) observed a strong coherence of the high-latitude perturbation fields and southward components of the IMF. The concentration of convection and Hall currents on either the morning or afternoon side of the northern polar cap, depending on whether the B_y-component is positive or negative ($+y$ being eastward), was observed by Friis-Christensen et al. (1972) via the magnetic perturbation fields, while Heppner (1972) directly measured the convection electric field above the polar cap with a low-orbiting satellite. Figure 16 shows, on the basis of these measurements, how the electric equipotentials or convection streamlines are concentrated to one or the other side of the polar cap depending on the sign of the IMF B_y component. Diagrams like Figure 14 suggest immediately the reasons for this dependence.

We now jump a vast literature and quote a recent result of Burke et al. (1999). If the merging rate v/v_A is substantially smaller than unity, then, in accordance with eqn (7), this means that only a fraction of the magnetic flux transported toward the magnetopause connects with magnetospheric field lines, while most of the flux passes by without interaction. This is shown in Figure 17a. The width L_G, of the channel through which the reconnecting flux is transported, multiplied by the upstream solar wind electric field,

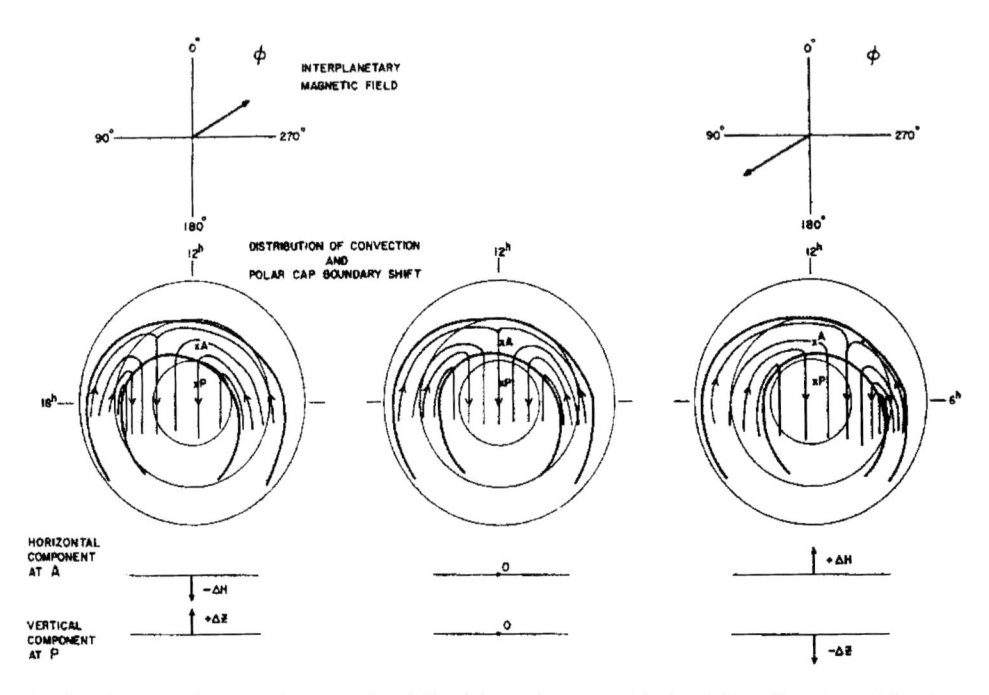

Figure 16 The distribution of convection and the shift of the polar cap with the IMF y-direction. (After Heppner 1972.)

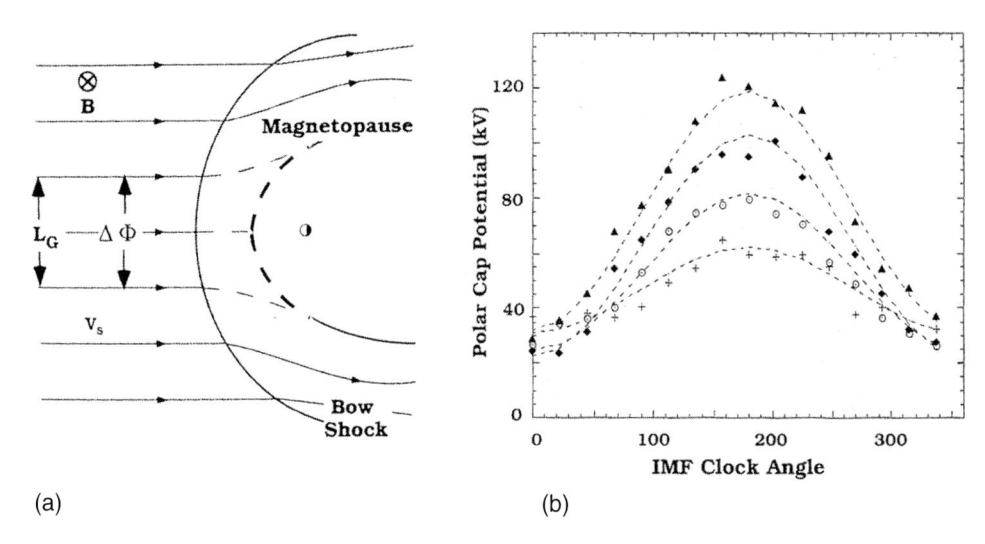

Figure 17 (a) Representation of the merging potential $\Delta\Phi$; that is, the magnetic flux transported towards the magnetopause which actually undergoes reconnection. (b) Scatter plot of polar cap potentials measured with the DE 2 satellite as a function of IMF clock angle for four ranges of the tangential magnetic field ($\langle B_T\rangle$ = 2.3, 4.3, 6.1, and 10.4 nT, respectively). (After Burke et al. 1999.)

yields the potential $\Delta\Phi$ which is to be found subsequently across the polar cap. Measurements from 720 passes of a low-orbiting satellite were sorted in groups of various IMF magnitudes and further sorted in 16 overlapping bins of Θ; Figure 17b shows the result. The dashed lines are fits to the equation

$$\Delta\Phi_{PC} = \Phi_0 + L\langle v_s\rangle\langle B_T\rangle \sin^2\langle \tfrac{1}{2}\Theta\rangle \qquad (14)$$

which is tailored on eqn (13). $\langle B_T\rangle$ is the group average of the measured tangential upstream magnetic field and $\langle v_s\rangle$ is the corresponding average of the solar wind speed. The main result of a regression analysis is that L turns out to be of the order of $4R_E$. As one can safely assume (since this is supported by *in situ* measurements) that the reconnection line extends across the whole front side toward the morning and afternoon flanks of the magnetopause – that is, it is not less than about $40R_E$ long – this statistical finding confirms the theoretical predictions from 1964 that $k \approx 0.1$, or $v \approx 0.1v_A$, or $B_n \approx 0.1B_0$ for antiparallel fields. Although such a reconnection rate may not hold locally at any time (see below), it may be safely used in astrophysical applications. The constant, Φ_0, in eqn (14), which turns out to be about 30 kV, may be due to other means of momentum transfer, like viscous interaction with the solar wind, and/or to delayed responses of the magnetospheric system.

Other means of testing indirectly the consequences of reconnection of the IMF with the Earth's field include measurements of the magnetic flux piling up in the tail when the IMF turns southward, and the correlation of the ensuing substorm activity with the reconnection rate predicted by the upstream conditions. Since the formation of a near-Earth X-line, as shown in Figure 12, is involved here, and this needs a build-up time of about 0.5 to 1 hour, this correlation is optimum if such a delay is incorporated. Summaries of such studies can be found in Nishida (1975) and Haerendel and Paschmann (1982).

In situ observations

By the 1970s the art of plasma and field measurements in space was sufficiently well developed that one could hope to determine the main indicators of an ongoing reconnection process with some degree of reliability. The most direct measurement would be that of the normal plasma velocity, v_n, or the normal component of the magnetic field, B_n. However, for realistic reconnection rates both are small quantities compared with the respective tangential components and, in addition, the magnetopause does not stand still, but more often than not has a much larger normal motion than the reconnection rate, v_n. In spite of that, the minimum variance technique developed by Sonnerup and Cahill (1967) allowed in quite a few instances the determination of B_n, which typically was a few nanoteslas. Figure 18 shows an example.

Much easier is the assessment of the tangential momentum gain as the shocked solar wind or, in magnetospheric terminology, the magnetosheath plasma traverses the leading standing wave, producing what is termed a rotational discontinuity (RD), as shown in the model of Levy *et al.*

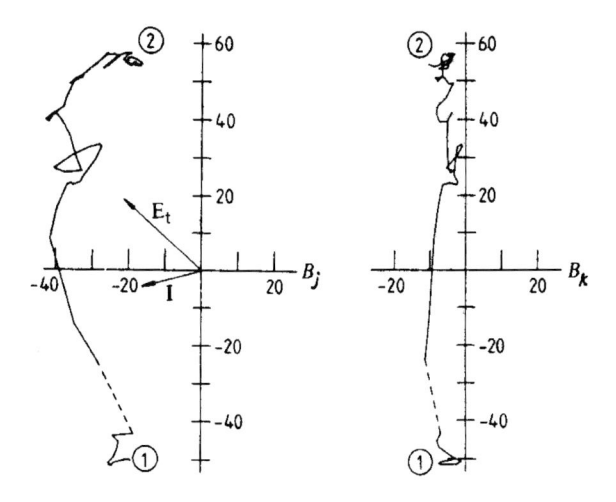

Figure 18 Tangential (B_j) and normal (B_k) components of **B** at the magnetopause from a minimum variance analysis during the crossing shown in Figure 19. (After Paschmann et al. 1979.)

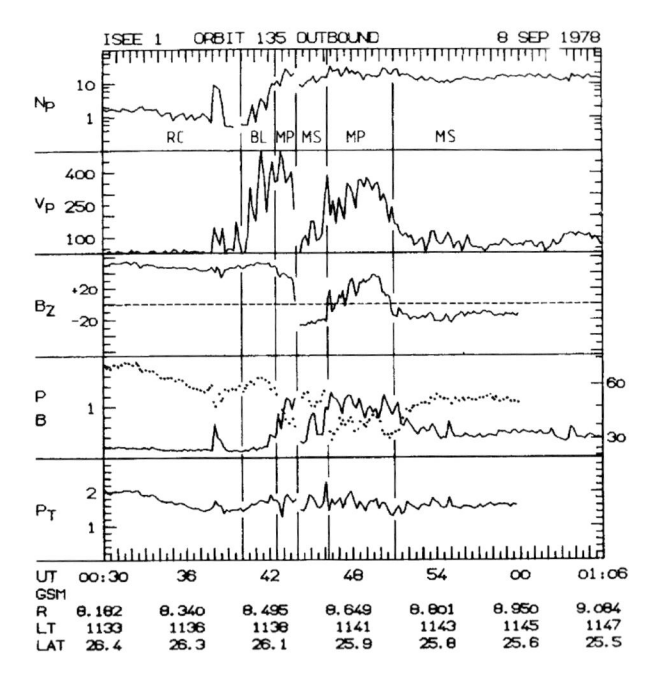

Figure 19 Plasma and field data from the first identification, by ISEE 1, of a rotational discontinuity and the related velocity increase (v_p) at the magnetopause. N_p = plasma density, P = plasma pressure, and P_T = total pressure. (After Paschmann et al. 1979.)

(1964) (Figure 11). The observable increase in the tangential velocity, v_t, should follow eqn (9) or (10), all quantities in which are well measurable, particularly if measurements are performed not too far from the stagnation point, where the external flow speeds are small. Figure 19 shows the original data of the first *in situ* confirmation of the existence of an RD at the magnetopause, obtained with the first International Sun–Earth Explorer probe (Paschmann *et al.* 1979). The plasma flow velocity, v_p, just inside the abrupt jump of B_z to negative values (at the magnetopause) is enhanced by a factor of 8 to 10. Application of the Walén condition (eqns (9) and (10)) revealed excellent agreement of data and theory within the experimental uncertainties. Also, the minimum variance technique yielded substantial values of B_n. Figure 20 shows a collection of subsequent determinations of Δv_t from the same mission. A horizontal arrow of length unity would signal perfect agreement with fluid theory, which actually is hardly to be expected in a non-Maxwellian anisotropic plasma composed of more than one species.

The art of testing the Walén relation has been well developed by plotting the flow velocity, **v**, in the so-called de Hoffmann–Teller frame, which moves along the magnetopause at the speed, $\mathbf{v}_{HT}$ (Sonnerup *et al.* 1990). In this frame, the plasma flows along the field at the local Alfvén speed:

$$\mathbf{v} - \mathbf{v}_{HT} = \pm \mathbf{v}_A \qquad (15)$$

and the electric field vanishes (on average); $\mathbf{v}_{HT}$ is determined by minimizing, for a certain time interval, the deviations of $[(\mathbf{v} - \mathbf{v}_{HT}) \times \mathbf{B}]^2$, where **v** and **B** are the

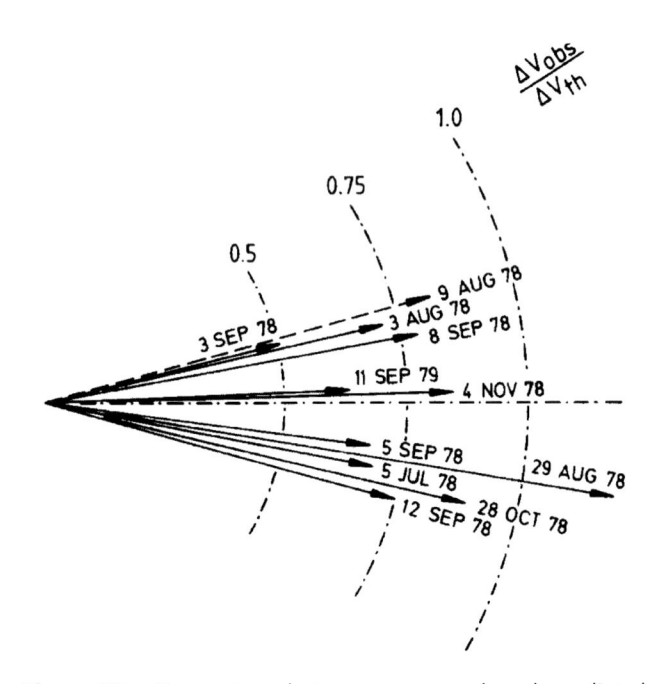

Figure 20 Comparison between measured and predicted tangential velocity enhancements across the magnetopause for 11 cases observed by ISEE 1 and 2. For perfect agreement, the vectors should be horizontal and of length 1.0. (After Sonnerup et al. 1981.)

instantaneous measured values. Figure 21 shows the result for one particular magnetopause crossing. Of course, it often happens that no such correlation, which ideally should have a slope of 1, is found at the magnetopause. In such a case, which can be described as a tangential discontinuity, the magnetopause is closed. In only 60% of the cases with a clock angle Θ larger than 45° could the existence of an RD be confirmed (Phan *et al.* 1996); that is,

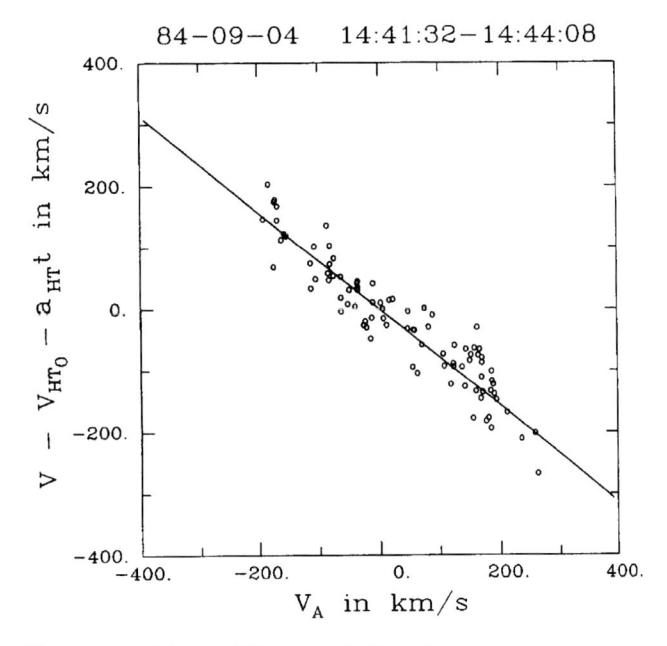

Figure 21 The Walén correlation for a magnetopause crossing. (After Sonnerup *et al.* 1990.)

reconnection often appears to be locally absent in spite of favorable field orientations.

Another clear indication of reconnection at the magnetopause is the distinctly different behavior of plasma and magnetic pressures for low and high magnetic shears (clock angle). Figure 22 shows the results of a superposed epoch analysis (Phan *et al.* 1994). For low shear, for which reconnection is absent or weak, the magnetic flux appears to pile up while the plasma is being depleted, probably by being squeezed out along **B** by the enhanced magnetic pressure. For high shear, reconnection and transit of the plasma into a fast boundary layer flow prevents the pile-up of **B**. Since in high shear β is greater than in low shear, the velocity gain according to eqns (9) and (10) is noticeably reduced.

In situ measurements offer very detailed diagnostics of the response of particles of different origin (magnetosheath or magnetosphere) and different mass and energy to the existence of a rotational discontinuity. The RD acts as a velocity filter creating peculiar velocity and spatial distributions depending on whether the particles are reflected or transmitted and, in the latter case, on their proper speed along the field lines. The predictions of simple kinematic considerations have been verified by observations (e.g., Gosling *et al.* 1990).

Finally, it must be stressed that *in situ* measurements have revealed a most important property of dayside (pushed) reconnection: its intermittency. While most theories consider stationary situations, most of the actual encounters of reconnection events have a transient character, the durations being of just a few minutes (Haerendel *et al.* 1978, Russell and Elphic 1978). The result of such short-duration reconnection

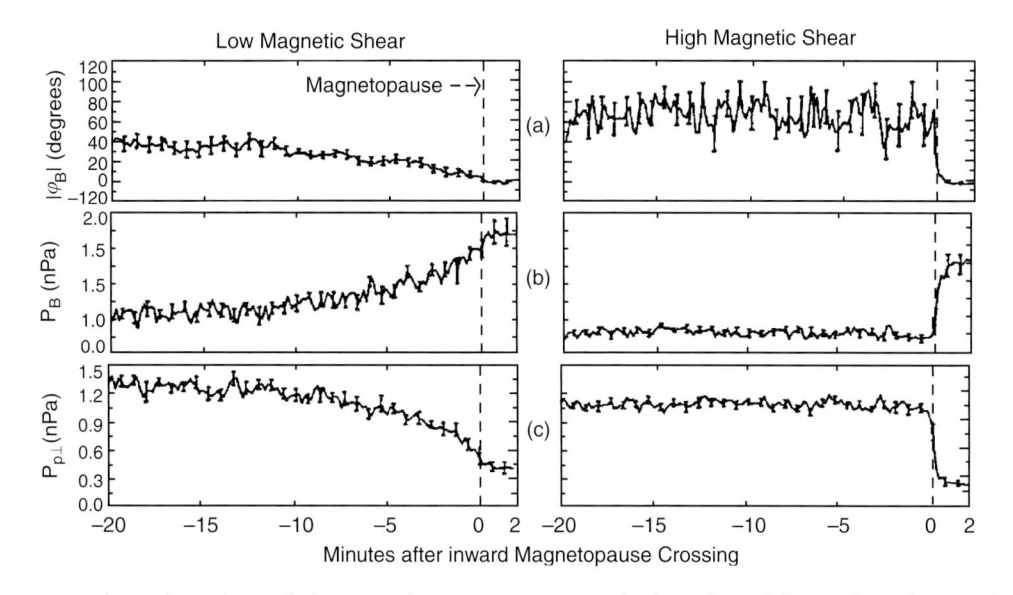

Figure 22 Superposed epoch analysis of plasma and magnetic pressures for low-shear (left) and high-shear (right) crossings of the magnetopause as recorded by the Ion Release Module on board the Active Magnetospheric Particle Tracer Explorer. (After Phan *et al.* 1994.)

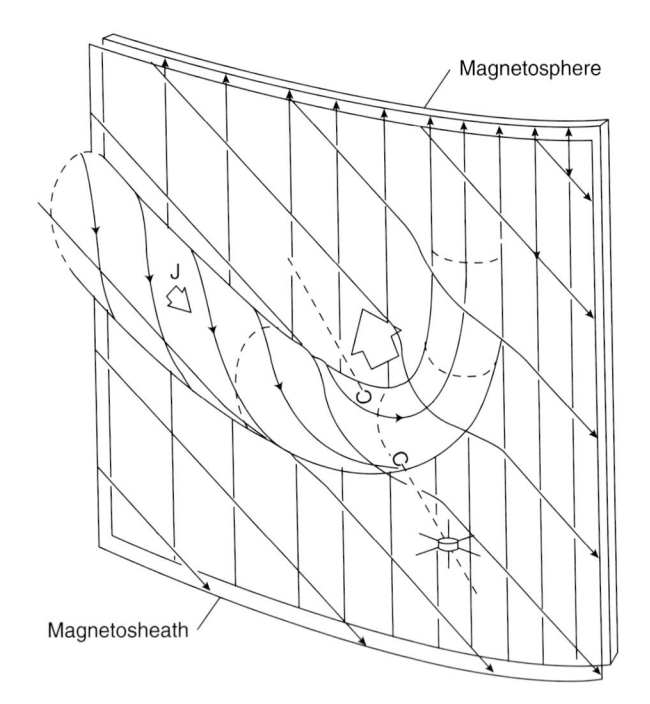

Figure 23 Schematic of the Russell – Elphic model of a flux transfer event with twisted magnetic field. (After Russell 1990.)

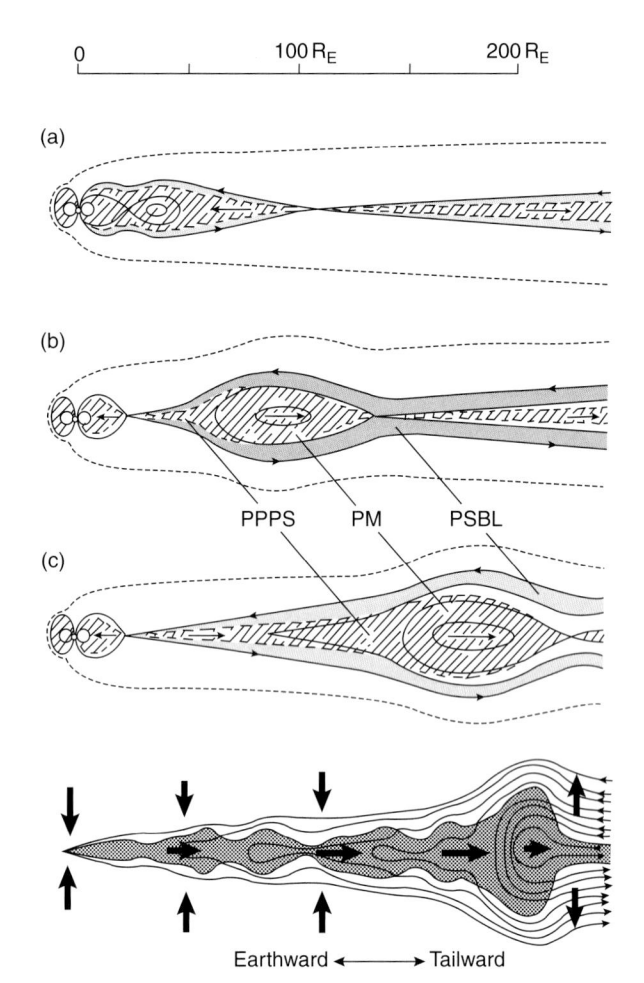

Figure 24 (a) The ISEE 3 picture of the magnetotail during plasmoid ejection. (After Richardson et al. 1987.) (b) The Geotail view. (After Mukai et al. 1998.)

events is flux tubes of finite extent, so-called flux-transfer events (FTEs). Figure 23 shows a schematic derived on the basis of their typical magnetic signatures, notably the bipolar variation of the normal B component. Their plasma signatures are also quite intriguing, as they exhibit a mix of magnetosheath and magnetospheric plasmas. These flux tubes have been found to be twisted, and the tangential stress of the twisted field balances an internal overpressure (Paschmann et al. 1982). Various interpretations have been given for the origin and characteristics of FTEs, as either spatial or temporal structures of the reconnection process (for a summary see Russell (1990)). The reason for such a spatial or temporal intermittency is, however, unclear. It seems as if the reconnection process has a self-regulating capability.

All of the above observations apply essentially to the wave region in Petschek's model. Only few claims for an encounter with the diffusion region have been made. Indeed, as is discussed below, it is not clear which particle and field structure should be expected for the region. However, as shown by Treumann et al. (1995; see also Chapter 62), the observed wave intensities at the magnetopause are incapable of producing diffusion coefficients consistent with the required merging rates and observed thicknesses of the magnetopause current layers. Therefore, if we continue to use the name "diffusion region" for the central part of a reconnection configuration, we must remember that diffusion is not the basic process that allows the cutting of field lines.

Not only the front-side magnetopause but also the Earth's magnetotail turned out to be an excellent laboratory for the study of reconnection. As already conjectured by Hones (1979) (Figure 12), magnetospheric substorms are due to the formation of a near-Earth neutral line, with a plasmoid (i.e., O-type magnetic field structures, disregarding any east–west core fields) building up between the inner and outer neutral lines and its subsequent downtail ejection. Two deep-tail missions (ISEE 3 in the early 1980s and Geotail in the 1990s) provided deep insights into the plasma and field structure of the outer reconnection regions. It was found (e.g., Scholer et al. 1984, Richardson et al. 1987, Mukai et al. 1998) that the first indication of near-Earth reconnection at large downtail distances was the arrival of energetic ions streaming tailward at high speeds. They constitute the population of a broad boundary layer originating from the near-Earth X-line, as shown in Figure 24. Subsequently, the magnetic field at the position of the spacecraft turns northward, indicating the arrival of closed-field structures (the plasmoid), to turn southward again after several minutes. The whole duration of the plasmoid phase is about 20 minutes. With flow speeds increasing outward from 400 to 700 km s^{-1}, this suggests a total length of about $80\,R_E$. After the plasmoid phase there is a long-duration but highly variable reconnection phase (Figure 24b) characterized by the mixing and counter-streaming of cold ions arriving from the tail lobes, (ultimately from the solar wind), and hot ions that have been heated upon entering the deeper layers of the plasmoid. This superposition of ion streams of various flow velocities and temperatures and the progressive mixing and isotropization toward the center of the reconnection layer are characteristic of collision-free reconnection and modify significantly the theoretical fluid models discussed so far. We return to this important topic in the discussion of the microscopic properties.

LABORATORY STUDIES OF RECONNECTION

Although this chapter deals with reconnection in the heliosphere and concentrates on solar and magnetospheric plasmas, a short excursion into the laboratory is appropriate and rewarding. There are two types of reconnection experiments: unwanted ones which occur in fusion machines, and ones specifically designed to study the reconnection process in support of space and astrophysical studies. Of the first type we mention only one, the internal disruption in tokamaks and its interpretation by Kadomtsev (1975). The reason for its inclusion in this chapter is that in tokamaks reconnection does not occur between antiparallel fields but with weakly sheared fields, similar to the configurations considered by Gold and Hoyle (1960) and Parker (1983) (Figures 8 and 9). However, there is a huge difference between the cosmic configurations and the tokamaks in that the former are subject to line-tying, while the latter have purely toroidal geometries. Therefore the core (toroidal) field can be neglected in the consideration of internal disruptions, and only the poloidal fields need to be taken into account (Figure 25). Kadomtsev (1975) explained the disruption as a resistive kink mode forming a magnetic island (Figure 25b and 25c) which grows at the expense of the central plasma column until it is completely consumed and the rotational symmetry is re-established. The basic process is, indeed, reconnection occurring at the radius where the safety factor $q = rB_z/RB_\Theta$ (R and r are the major and minor radii, and B_z and B_Θ the toroidal and poloidal fields) is unity; in other words, where field lines make one complete spiral turn around the plasma torus and are easily interchanged. The driver of the instability is the steepening of the current profile due to continued heating – that is, increase of the magnetic shear. Another difference with space situations is that the diffusion rate appears to be controlled by resistivity.

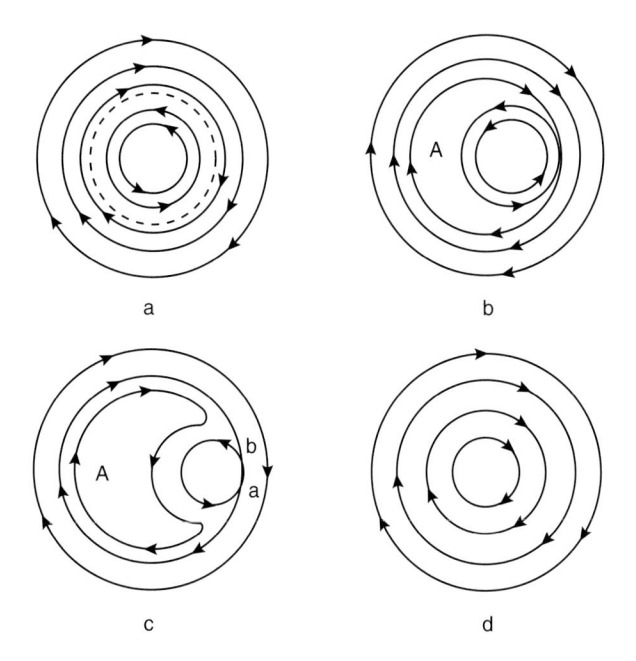

Figure 25 A model of internal disruption in a tokamak, showing the auxiliary field, B_ϕ, before, during, and after reconnection. (After Kadomtsev 1975.)

Of the second type of experiment we discuss just two: the famous reconnection experiments in the electron magnetohydrodynamic (EMHD) regime by Stenzel and Gekelman (1979) and Stenzel et al. (1982) in which only the electrons are magnetized, and the experiments of Yamada et al. (1997) in which even the ion gyro-radius is much smaller than the plasma device. While the detailed experimental set-ups cannot be described here, we mention a few striking results.

Gekelman et al. (1982) and Stenzel et al. (1982) found that in the long, flat reconnection region between driven antiparallel fields, pressure gradients, and thereby space charge fields in addition to the applied electric induction fields, were building up. This led to additional Hall currents. Extensive plasma diagnostics and careful evaluation of the terms in the generalized Ohm's law revealed a strong increase of the resistivity due to microturbulence. However, in contrast to most theoretical models that incorporate anomalous resistivity, also in the context of reconnection simulations (see below), regions of maximum resistivity and maximum current density did not coincide. Strong heating was observed, and much of the converted electromagnetic energy was fed not into fluid motion but into heat. However, after several Alfvénic transit times the plasma developed the classic flow pattern, with jetting from the edges of the neutral sheet at speeds close to v_A. Figure 26 shows the evolution of the ion flow vectors. Although no direct transfer of these results to the space plasmas is possible, there is an important hint that, on scales smaller than the ion gyro-radius, secondary charge separation fields and Hall currents are generated which have an impact

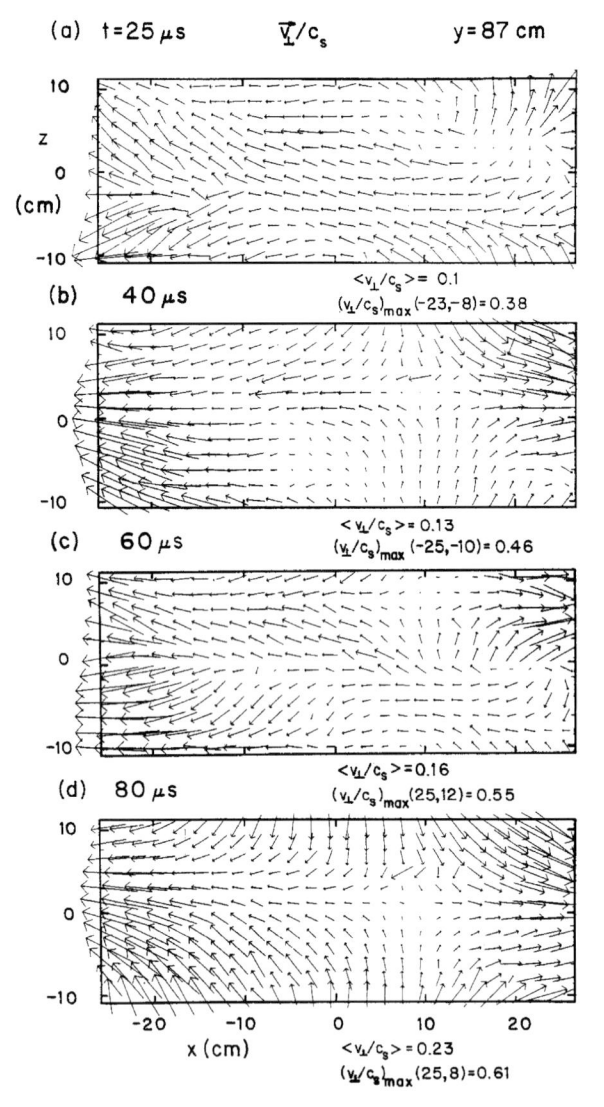

Figure 26 Vector fields of the measured transverse ion flow velocity, $v_\perp$, normalized to the sound speed, c_s, at different times during the external current rise. (After Gekelman et al. 1982.)

on the electron dynamics and magnetic field evolution. Furthermore, ion heating must not be neglected against acceleration.

In recent years, several new laboratory experiments have been designed for the purpose of investigating reconnection. Two prominent ones are the MRX (Magnetic Reconnection Experiment) at Princeton (Yamada et al. 1997) and the TS-3 (Tokyo University Spherical Torus) merging device (Ono et al. 1997). Both experiments use two toroidal plasma and field configurations pushed against each other, and subsequently, at decreasing external current, pulled away from each other (Figure 27). The magnetic fields in the reconnection region can be made antiparallel, or a component in the sheet current direction can be added (co-helicity). The plasma and field parameters are such that the Lundquist

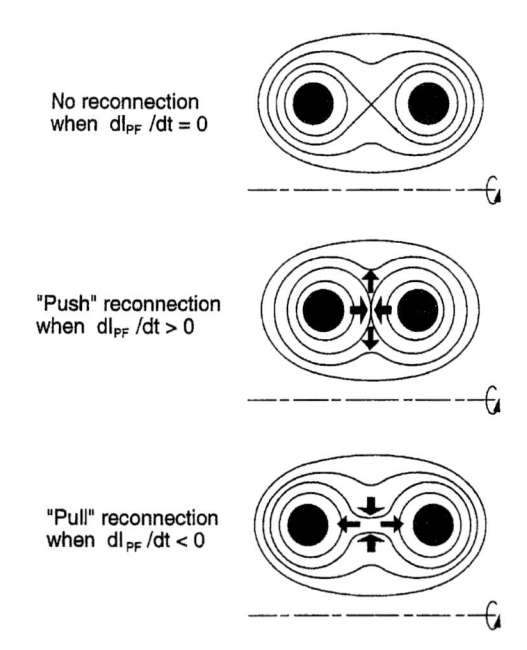

Figure 27 "Push" and "pull" reconnection in the double annular plasma configuration of the MRX experiment. (After Yamada et al. 1997.)

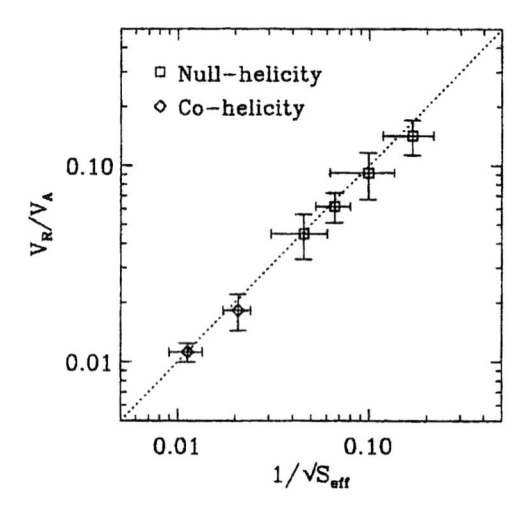

Figure 28 Observed reconnection rates, v_R, in the MRX experiments as functions of $S^{-1/2}$ as predicted by the Sweet–Parker model (dotted line). (After Ji et al. 1999.)

number, S, is large and the ion gyro-radius, ρ_i, is small. For antiparallel fields a thin Sweet–Parker current sheet develops with a width $\delta \approx \rho_i$. As in previous laboratory experiments, strong ion heating and enhanced resistivity are observed. Calculating the effective Lundquist number, S_{eff}, with the enhanced resistivity and correcting for compressibility and enhanced downstream pressure, Ji et al. (1999) have found a Sweet–Parker dependence of the reconnection rate on S_{eff} as shown in Figure 28.

In recent years, the dialogue between laboratory and space physicists has intensified, for instance at the International Conferences on Plasma Physics (ICPP) and the Symposia on Interrelationship Between Plasma Experiments in Laboratory and Space (IPELS). These are forums for comparing the results of experiments with observations made in vastly different environments, and attempting to separate the relevant effects from the incidental. Their implications with regard to Sweet–Parker or Petschek-type reconnection and the role of anomalous resistivity are still a matter of heated debate (see also Biskamp 1993, Kulsrud 1998).

MICROPHYSICS OF RECONNECTION

The concepts of reconnection and the interpretations of experimental data mostly follow frameworks developed in the magnetohydrodynamic (MHD) picture. This is quite natural since outside reconnection regions MHD appears to be a good context in which to describe a cosmic plasma. However, recently research on reconnection has focused increasingly on non-MHD effects, in particular with respect to the applicability of Ohm's law and the role of parallel electric fields (e.g., Hesse et al. 1990, Scudder 1997). During the 1970s and 1980s, numerical modeling employed (mostly incompressible) MHD approaches with various expressions for an enhanced resistivity. The insights obtained were essentially twofold. First, fast (Petschek) reconnection could not be achieved with uniform resistivity, but only when it was limited to small regions of current concentrations. Second, the dependences of the extent of the central current sheet and thus the reconnection rates on the boundary conditions were clearly recognized. Summaries of this work have been given by Scholer (1991) and Priest and Forbes (2000).

Further insights have come from non-MHD numerical simulations as well as from space observations. Both the diffusion region and the standing slow shocks begin to appear in a new light. Single-particle pictures with wave-induced scattering, Hall MHD, and non-isotropic pressure tensors have been used. Which description is the most appropriate is not yet clear; here we discuss only a few selected results.

A series of studies has concentrated on the question of which of the two terms on the right-hand side of the generalized Ohm's law in the form

$$\mathbf{E} + \mathbf{v}_e \times \mathbf{B} = -\frac{1}{ne}\nabla \cdot \mathbf{P}_e - \frac{m_e}{e}\frac{d\mathbf{v}}{dt_e} \quad (16)$$

contributes more to the reconnection electric field – the pressure tensor

$$E_p = -\frac{1}{ne}\left(\frac{\partial P_{exy}}{\partial x} + \frac{\partial P_{eyz}}{\partial z}\right) \quad (17)$$

or the electron inertia

$$E_{in} = -\frac{m_e}{e}\left(\frac{\partial v}{\partial t_e} + \mathbf{v}_e \cdot \nabla \mathbf{v}_e\right) \quad (18)$$

The answer, derived by employing a fully electromagnetic particle-in-cell code (Hesse and Winske 1998), is that for thicker current sheets of order c/ω_{pi} the off-diagonal terms of the electron pressure tensor clearly dominate. For thin current sheets, of order c/ω_{pe}, however, the time-dependent electron inertial term gives the highest contribution, but that of the non-gyrotropic pressure term is not negligible.

Studies of reconnection with a 2.5-dimensional hybrid code by Shay et al. (1998) revealed that a two-scale structure of the diffusion region develops: an outer ion region (c/ω_{pi}) and an inner electron region (c/ω_{pe}). The frozen-in magnetic flux constraint is broken in the inner region. In the outer region, the electrons are frozen to the magnetic field, Hall currents flow, and the electron dynamics determine the electromagnetic field evolution near the X-line. Here, whistler mode waves develop and the electron velocities are of the order of v_A^-, much higher than in normal MHD calculations. But within about $10c/\omega_{pi}$ in the outflow direction the ions become coupled to the electrons by charge separation fields, and the mass flow velocity approaches the Alfvén velocity. The most striking result is that with these dynamics the "diffusion region" appears to develop automatically a length-to-width ratio, L/l (eqn (1)), of the order of 10. Since $V = v_A$, the reconnection rate settles naturally at $v \cong 10^{-1} v_A$, consistent with the Petschek rate. Figure 29 (from Shay et al. 1998) shows the Hall and electron inertial regions and the magnetic field configuration, while Figure 30 shows the ion and electron flows for a mass ratio $m_i/m_e = 5$. The Hall currents driven by the electric polarization fields create out-of-plane magnetic field components whose tensions accelerate the electrons to energies comparable to those of the ions. Thus the zone near the X-line becomes whistler mediated. As convincing as the picture emerging from these studies is, caution is still necessary, because the resulting magnetic configuration, with its short current sheet, is very suggestive of Petschek reconnection. Also, the actual field-breaking processes, which mostly operate on the c/ω_{pe} scale (electron wave turbulence?) are still unclear.

Our second topic of the microphysics concerns the slow shocks. Here evidence from both observations in the deep tail and hybrid simulations presents a picture of slowly mixing plasma populations, which to call a "discontinuity" is a strong idealization. Observations of ion and electron velocity distributions following the passage of a plasmoid (see above), in particular in the trailing part, exhibit the dominance of two counter-streaming ion beams with $T_\perp > T_\parallel$. Obviously, ions entering from both sides of the tail lobe become heated, predominantly transverse to $\mathbf{B}$. The electrons are characterized

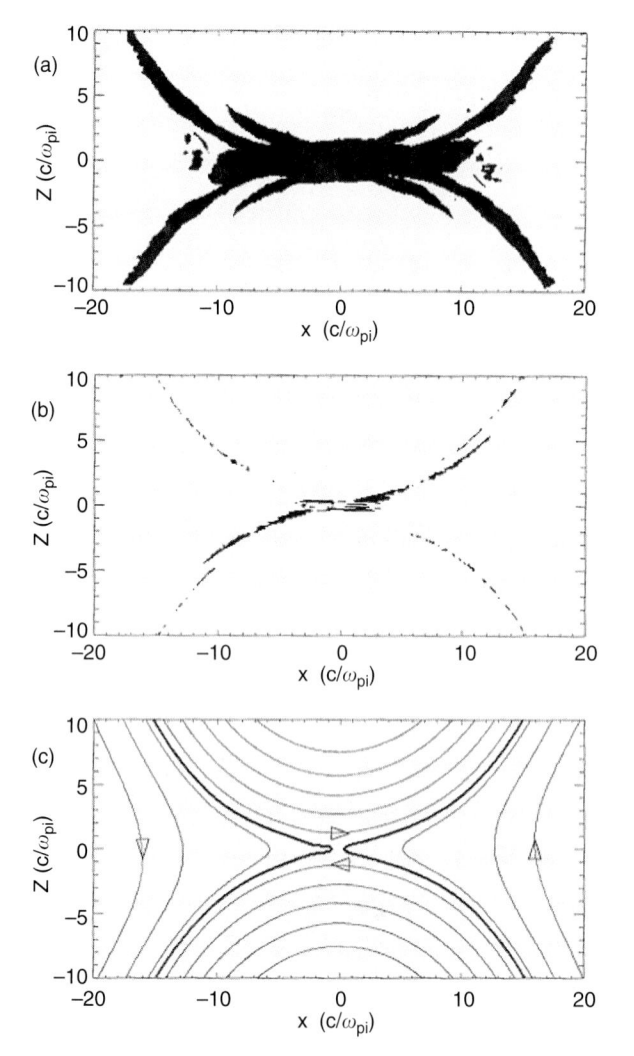

Figure 29 (a) The Hall region, (b) the electron inertial region, and (c) the magnetic field configuration in numerical computer experiments of reconnection using a 2.5 dimensional hybrid code. (After Shay *et al.* 1998.)

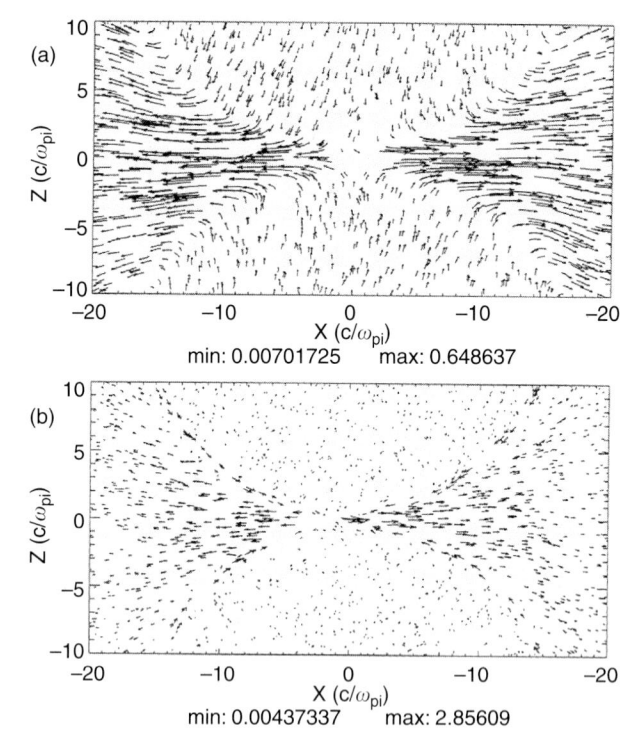

Figure 30 (a) Ion and (b) electron flow vectors for the same simulation and time as in Figure 29. The normalization in the two panels differs (compare the "max" values). (After Shay *et al.* 1998.)

by so-called flat-top distributions which indicate the shift of the whole distribution function by a fixed amount of energy, as for instance through the traversal of a potential drop, combined with energy redistribution by instabilities. Close to the neutral sheet, non-gyrotropic ion distributions prevail (Mukai *et al.* 1998).

Scholer and Lottermoser (1998) used a hybrid code to study reconnection on a large scale, as would occur in the magnetotail. They found that inward-drifting cold lobe ions perform meandering (moving in so-called Speiser orbits) after they arrive in the center of the plasma sheet, then become energized by the reconnection electric field, and are subsequently ejected on lobe field lines. Here they interpenetrate freshly arriving cold ions. The model predicts a concentration of the electric current in the center of the wedge-shaped plasma outflow region, and not at the edges, where slow shocks would be located. Figure 31, a set of ion distributions from this simulation, shows the increasing thermalization of the cold and hot ion distributions as the center of the reconnection wedge is approached. This is attributed partially to wave scattering, but mostly to phase mixing in the turbulent fields in the center of the wedge. Overall, the effect of heating by a slow-mode shock is achieved, not in a discontinuity but in a distributed fashion. However, as expected in a slow shock, a cross-shock electric potential drop seems to exist in the real situation as well, creating the flat-top electron distribution. This was already found at $-200 R_E$ by Feldman *et al.* (1985).

In summary, it appears that the "diffusion region" is of order c/ω_{pi} with a dissipative electron substructure of order c/ω_{pe}, requiring little noise or anomalous resistivity for the field-line cutting. The attached standing slow shocks seem to be realized not as discontinuities, but as broad regions of phase mixing of a cold incoming and hot beam population, energized in a thin central current sheet. The substantial width of plasmoids and of the post-plasmoid reconnection regions revealed by deep-tail observations would seem to favor fast, Petschek-type reconnection rates.

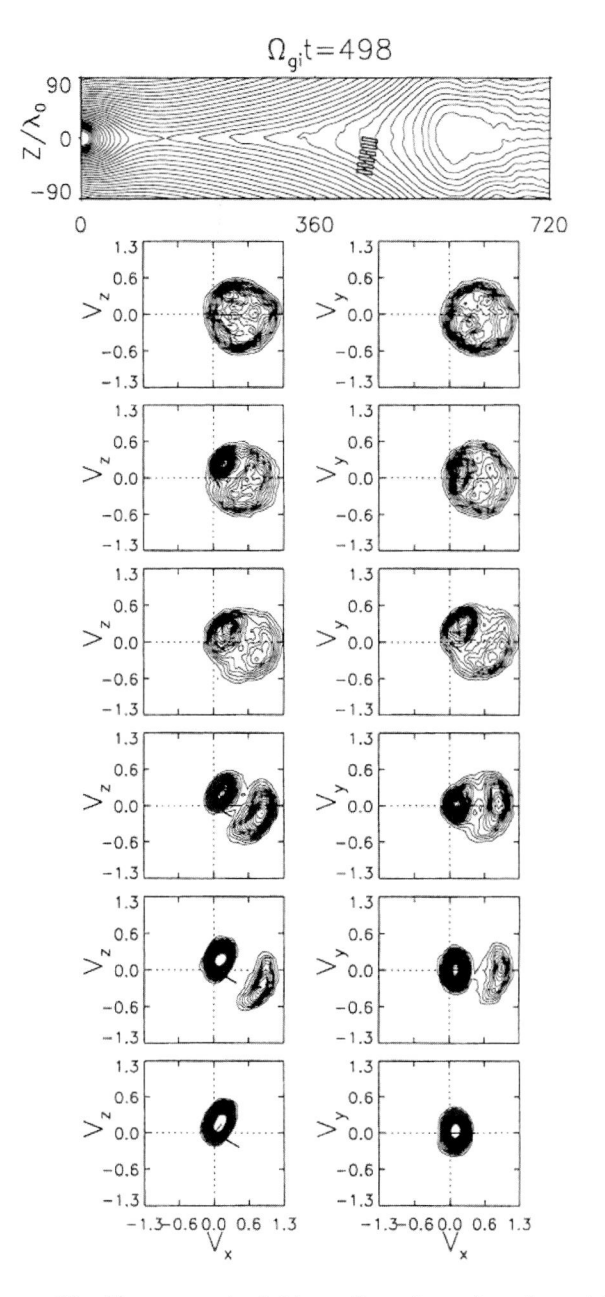

Figure 31 The magnetic field configuration of a plasmoid ejection (top) and ion velocity distributions between the center and edge (top to bottom) of the post-plasmoid reconnection flow. (After Scholer and Lottermoser 1998.)

SOLAR OBSERVATIONS

The concept of reconnection was developed in search of a mechanism of rapid release of magnetic energy on the Sun. Over the past half-century, confidence in the viability of this mechanism has been widely established. Applications are numerous, but many of them lack rigor with regard to the speed and availability of free energy – some skepticism still exists. On the other hand, the art of fitting complex three-dimensional coronal structures with theoretical magnetic fields has developed very well, making it possible to predict or identify potential sites of reconnection in complex spot groups (e.g., Priest and Démoulin 1995, Priest and Forbes 2000). However, unambiguous identifications of reconnection in the solar corona and chromosphere are difficult, mostly impossible, but there are various consistency checks. We conclude this chapter with four examples of phenomena which are naturally explained by the action of reconnection, and also discuss the quantities involved.

The first phenomenon is X-ray bright points (XBPs), discovered by Vaiana et al. (1970) by X-ray imaging from a sounding rocket and subsequently extensively explored (Golub et al. 1977). They are diffuse clouds 2000 to 20 000 km in size, with lifetimes ranging from several minutes to many hours. Typically a few hundred XBPs are visible at any one time. Flare phases of a few minutes with rapid intensity variations are quite common. They are more numerous and more uniformly distributed over the Sun during solar minima than during solar maxima. It has been found that two-thirds of the XBPs are related to canceling magnetic features in the photosphere, and one-third to newly emerging fields (Harvey 1985). Priest et al. (1994) presented a model of XBPs based on canceling magnetic fields of two approaching and reconnecting magnetic bipoles. Once two opposite polarities have come sufficiently close to each other to generate narrow current sheets, reconnection can proceed, hot plasma is ejected, and magnetic field is annihilated (Figure 32). In the final state, the coronal and photospheric fields disconnect; the latter submerge and disappear from magnetograms, while the former form an overlying arch. The lifetime of an XBP is interpreted as determined by the speed of the approaching fields, $\sim 0.5\,\mathrm{km\,s^{-1}}$, and their spatial extent. However, for the flaring phases one must instead take the reconnection speed. For Petschek reconnection – that is, for $M_A = 0.1$, and $B = 10$ G and $n = 10^9\,\mathrm{cm^{-3}}$ – it would amount to $\sim 70\,\mathrm{km\,s^{-1}}$. This means that in 5 minutes the total magnetic flux would be annihilated. While 5 minutes may be a long time for the actual flaring phase, the longer-term brightening and flux disconnection from the photosphere may actually occur in the temperature minimum, where the electrical resistivity maximizes and may be of Sweet–Parker type. Litvinenko and Martin (1999) derive reconnection speeds for this case of order about $0.2\,\mathrm{km\,s^{-1}}$, even smaller than typical magnetic transport speeds. These few notes on XBPs demonstrate the typical degree of uncertainty in explaining solar plasma processes.

Closely related to the previous topic, but not necessarily identical, are the "explosive events" observed by Dere et al. (1991). These are plasma jets observed in spectral lines formed in the transition region to the corona ($T \sim 10^5\,\mathrm{K}$). Typical speeds are $100\,\mathrm{km\,s^{-1}}$ and durations are 30–60 s. They seem to occur at the edges of high-field regions, often

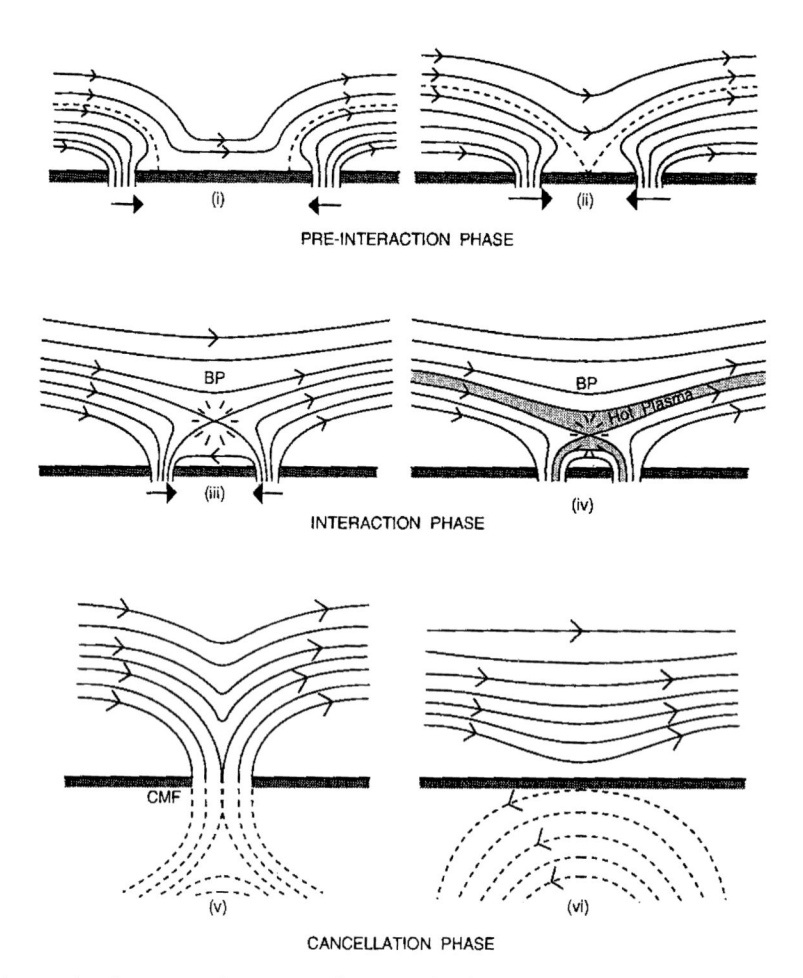

Figure 32 A model of X-ray bright points: the stages of approach of two equal magnetic fragments and their interaction by reconnection and creation of canceling magnetic features. (After Priest et al. 1994.)

associated with the nearby emergence of new magnetic flux. The interpretation, as for XBPs, is that reconnection occurs when pre-existing and newly emerging fields of opposing polarities meet (Figure 33). Dere et al. (1991) concluded that the reconnection occurs in bursts of typically 30 s duration, repeating over the time of flux cancellation, that is, over several hours. Since the observed jet speeds should be close to v_A, a magnetic field of 10 G would imply a not unreasonable density of 5×10^{10} cm^{-3}. With a Petschek reconnection rate of 10 km s^{-1} and $\tau = 30$ s, in each burst the flux of a slab of 300 km would be reconnected. Twenty such reconnection bursts would suffice to consume the entire approaching new flux of opposite polarity. Spatially resolved observations of bipolar jets of the same kind were reported by Innes et al. (1997), and strongly support the interpretation suggested in Figure 32. The burst-like and repetitive character of the explosive events seems to be a necessary feature of "pushed" reconnection events when the restoration of free energy (i.e., of sufficiently narrow current sheets) is significantly slower than the energy release.

Almost indisputable evidence of reconnection in the solar corona has been provided by the TRACE mission (Schrijver et al. 1999) when, about 10 hours after the emergence of new magnetic flux, new connections between that flux and previously existing fields where observed in the 171 Å passband, that is, with flux tubes filled with 10^6 K plasma. Of course, the actual reconnection process remains invisible, but chromospheric heating probably associated with the energy release during reconnection soon leads to an "illumination" of some of the reconnected loops.

Finally, the frequent observations of large loops with cusp-like tops obtained with the X-ray satellite Yohkoh have provided the most impressive evidence for the link between flares and reconnection (Shibata 1996, Tsuruda 1996). The very suggestive images have revealed temperatures of around 12–13 MK in what presumably are the outflow regions from between the slow shocks. Foot point separations with ~ 10 km s^{-1} are consistent with the classical model (Kopp and Pneuman 1976) of two-ribbon flares. The cusp-like shape of the loop top (Figure 34) is particularly

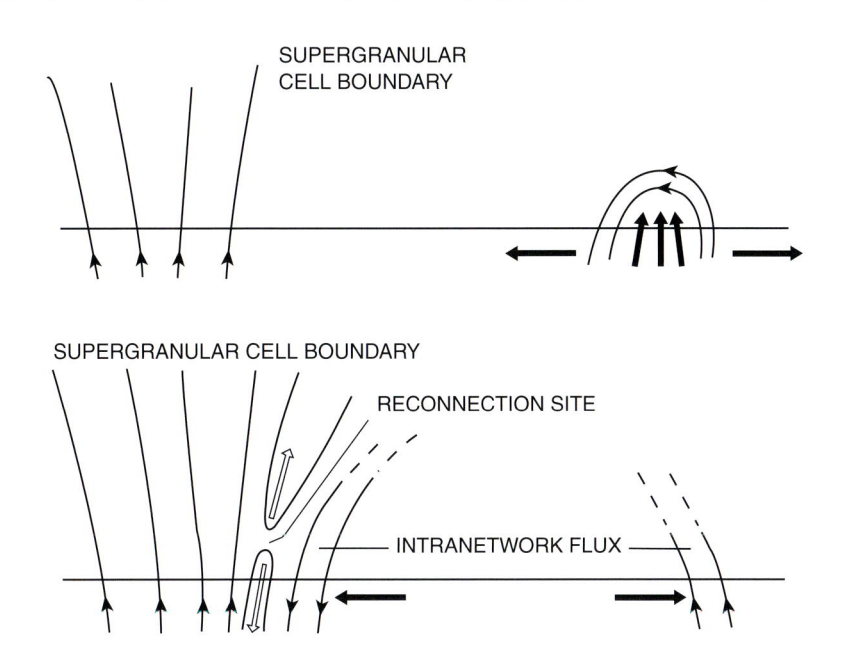

Figure 33 Intranetwork magnetic flux emergence and reconnection as the origin of explosive events with fast jets at transition region temperatures. (After Dere et al. 1991.)

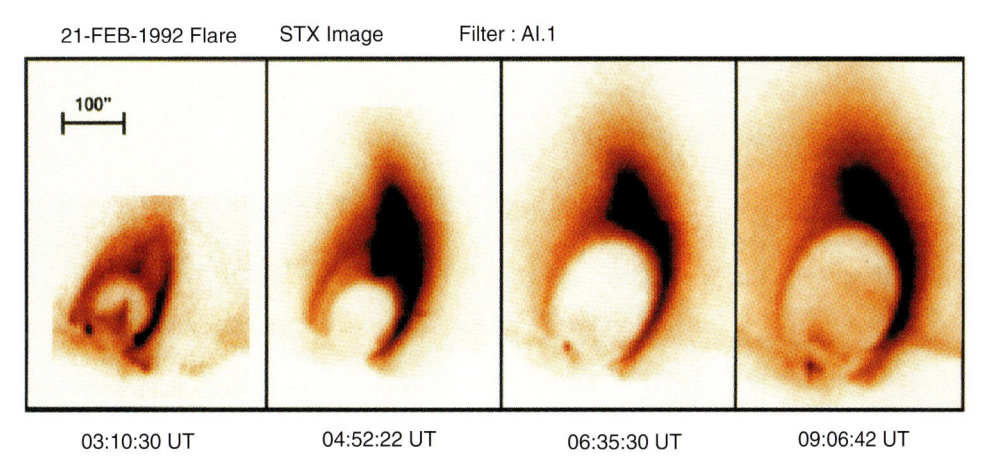

Figure 34 Images of a flare observed by the soft X-ray telescope (SXT) on Yohkoh with a clearly defined cusp structure. (Courtesy of The Institute of Space and Astronautical Science, 1996.)

suggestive of the braking of the downward reconnection jet, leading to strong heating and driving currents down to the chromospheric/photospheric feet. Somewhat different interpretations of the observed features have been given by different researchers. Figure 35 shows the interpretation by Shibata (1996) which contains, in addition, the unobserved element of an upward plasmoid ejection or filament eruption. Such a driver may be a necessary element for long-duration flares and long-duration reconnection, since it would allow quick removal of magnetic flux from the upper outflow region, and would also move the X-line upward, while the lower outflow leads to a pile-up of dipolarized flux, consistent with the observable height increase of the inner loop. A "pulled" reconnection explains not only the long duration but also the synchronization of the flash phase along lengths of 30 000 km or more, as witnessed by the fast onset of two-ribbon flares, which calls for an external driver. Remember that fast reconnection and breaking of field lines requires somewhere (on the X-line) a current width of order c/ω_{pi}, that is, ~10 m. How can such a narrow current sheet be established over a length of tens of thousand of kilometers within seconds, and be sustained for

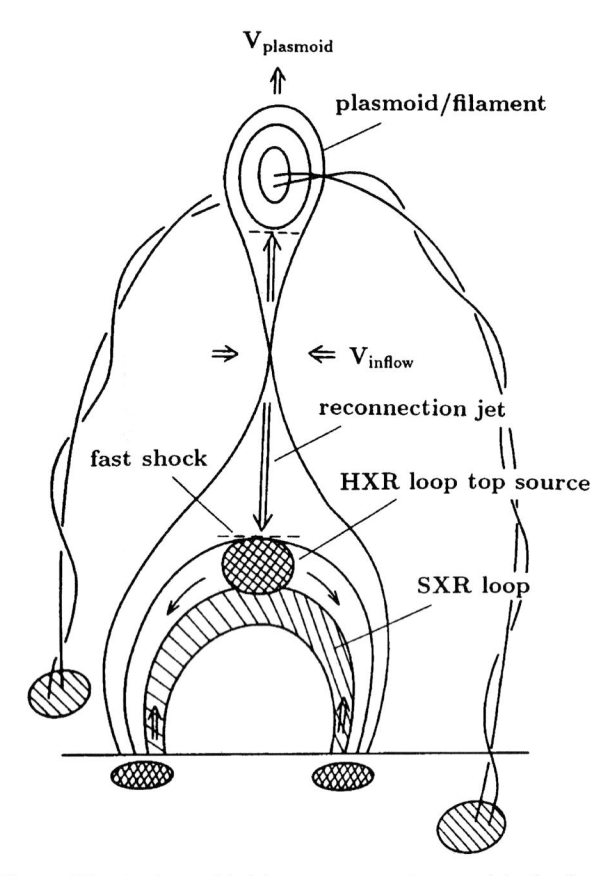

Figure 35 A plasmoid-driven reconnection model of a flare. (After Shibata 1996.)

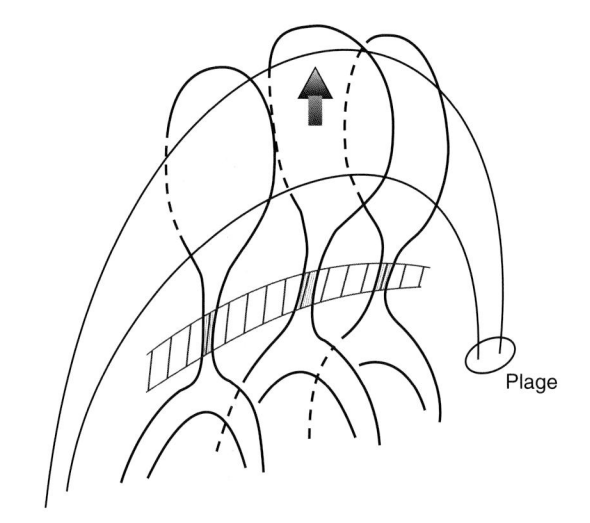

Figure 36 A flare as a "pulled" reconnection event above a magnetic arcade.

at least a few minutes, if the fields were just pushed toward each other? This requires a coordinating element. A rising core field intersecting the loop tops of an arcade-like field configuration, as shown in (Figure 36), could provide such a simultaneous onset, in addition to being an outflow driver.

Leaving aside many fascinating observational details we close the subject of flare reconnection with a simple assessment of the energy release. We assume antiparallel fields of $B = 30$ G, $n = 3 \times 10^9$ cm^{-3}, a horizontal length of $l = 30\,000$ km, and a vertical extent of the energy conversion region (slow shocks) of $h = 10\,000$ km. Then $v_A = 1200$ km s^{-1}, and the total energy conversion rate with a Petschek rate of reconnection becomes

$$\dot{W} = 2 \cdot \frac{B^2}{2\mu_0} \cdot 0.1v \cdot v_A \cdot l \cdot h = 2.6 \times 10^{28} \text{ erg s}^{-1} \quad (19)$$

During the impulsive phase of ~ 100 s, magnetic flux from 12 000 km on either side would become reconnected, and 3×10^{30} erg would be released with 7.5 keV per ion. This would not be a big flare. With higher fields and larger volumes, the energy can be greatly enhanced.

CONCLUDING REMARKS

This chapter has dealt with the principles of reconnection as they have been explored over the past five decades in the context of solar and magnetospheric physics, one a high- and the other a low-density plasma, but both, with regard to reconnection, collision free. Supporting or complementary evidence from laboratory and numerical experiments has been summarized. But expansion of the topic to reconnection situations elsewhere in the Solar System or the Universe, although there are many fascinating objects, has not been done. For instance, coronal mass ejections, which are now observed on a daily basis, are most likely to be accompanied by disconnections of magnetic flux from the Sun. The interaction of the solar wind and interplanetary magnetic field with the magnetic fields of other planets by necessity involves reconnection, with all its consequences. Particularly intriguing is the situation of Mercury, which lacks an insulating atmosphere, or of Neptune, which during the Voyager 2 encounter presented its polar cusp toward the oncoming solar wind. Little is known about reconnection with fields trapped in atmospheres of (essentially) non-magnetized planets (Mars, Venus) or comets, when the interplanetary field changes direction. Large-scale disconnection events in cometary plasma tails have been associated with reconnection. Reconnection must play a role in almost all interacting magnetized cosmic objects, as for instance in accretion disks and their interaction with stellar magnetospheres and rotating black holes. These few remarks not only give an idea of what is missing in this chapter, but also show how universal the process is, and how relevant near-Earth or solar studies are for successful applications elsewhere in the Universe.

So, what are the key insights? First, reconnection does exist and can proceed at a fast (Petschek) rate. However, this does not seem to happen in a continuous, stationary fashion, but rather as transient events. What regulates the duration of these events is not well understood. It may be a narrowing of the current sheet over a greater length and the transition from Petschek to Sweet–Parker reconnection. On other occasions it may be caused by changes of the upstream parameters. This applies to the situation at the Earth's magnetopause, where the inflow of magnetic flux is much faster than even the fastest reconnection rate. In other cases, for example on the Sun, temporary exhaustion of free energy may cause intermittency because the reconnection speed exceeds the convective speed with which opposite polarity is supplied. All of these situations can be described as "pushed" reconnection. "Pulled" reconnection, as happens for instance along coronal streamers or in the magnetotail, may last much longer and proceed at high rates. Very little explored, because of the intrinsic complexity, is the reconnection between neighboring twisted flux tubes, the balance of energy input and release, the effect of local untwisting, and the subsequent random walk of magnetic field lines.

The width of the magnetopause current sheet, as well as studies with laboratory experiments and hybrid simulations, demonstrate the necessity of thin current layers of order c/ω_{pi} or ρ_i somewhere between the reconnecting fields. As to the actual field-line breaking process, it is not clear whether anomalous resistivity is needed, or just some dissipation on the c/ω_{pe} scale. It may depend on the plasma parameters. Our understanding of the nature of the standing waves, idealized as slow shocks or combinations of intermediate and slow-mode waves, is also evolving away from the description as discontinuities. And the plasma behavior is found to deviate strongly from that of a fluid. Kinetic descriptions with interpenetrating ion beams, which originate from non-adiabatic acceleration in narrow current sheets, and cold bulk plasma subject to pitch-angle scattering and phase mixing, replace the original MHD concepts, at least for the dilute plasma of solar wind and magnetosphere. There are still many unknowns. However, we can say that the ideas developed between Giovanelli (1946) and Petschek (1964) have been placed on a solid foundation in subsequent decades, and can be safely applied to distant objects, as long as the sources of free energy are known and the requirements of thin current sheets and flux ejection are respected.

REFERENCES

Biskamp, D. (1993). *Nonlinear Magnetohydrodynamics*. Cambridge University Press.

Brueckner, G.E. and Bartoe, J.D.F. (1983). Observations of high-energy jets in the corona above the quiet Sun, the heating of the corona, and the acceleration of the solar wind. *Astrophysical Journal*, **272**, 329–348.

Büchner, J. and Zelenyi, L.M. (1987). Chaotization of the electron motion as the cause of an internal magnetotail instability and substorm onset. *Journal of Geophysical Research*, **92**, 13456–13466.

Burch, J.L. (1972). Precipitation of low-energy electrons at high latitudes: Effects of interplanetary magnetic field and dipole tilt angle. *Journal of Geophysical Research*, **77**, 6696–6707.

Burke, W.J., Weimer, D.R. and Maynard, N.C. (1999). Geoeffective interplanetary scale sizes derived from regression analysis of polar cap potentials. *Journal of Geophysical Research*, **104**, 9989–9994.

Carmichael, H. (1964). A process for flares. In W.N. Hess (ed.), *Physics of Solar Flares*, NASA SP-50, Washington, DC, pp. 451–456.

Coroniti, F.V. (1980). On the tearing mode in quasi-neutral sheets. *Journal of Geophysical Research*, **85**, 6719–6728.

Cowling, T.G. (1953). Solar electrodynamics. In G.P. Kuiper (ed), *The Sun*. University Chicago Press, pp. 532–591.

Crooker, N.U. (1979). Dayside merging and cusp geometry. *Journal of Geophysical Research*, **84**, 951–959.

Dere, K.P., Bartoe, J.-D.F and Brueckner, G.E. (1991). Explosive events and magnetic reconnection in the solar atmosphere. *Journal of Geophysical Research*, **96**, 9399–9407.

Dungey, J.W. (1958). The neutral point discharge theory of solar flares: A reply to Cowling's criticism. In *IAU Symposium No. 6*, Cambridge University Press, pp. 135–140.

Dungey, J.W. (1961). Interplanetary magnetic field and the auroral zones. *Physical Review Letter*, **6**, 47–48.

Feldman, W.C., Baker, D.N., Bame, S.J., Birn, J., Gosling, J.T., Hones, Jr., E.W. and Schwartz, S.J. (1985). Slow shocks: A semipermanent feature of the distant geomagnetic tail. *Journal of Geophysical Research*, **90**, 233–240.

Finn, J.M. and Kaw, P.K. (1977). Coalescence instability of magnetic islands. *Physics of Fluids*, **20**, 72–78.

Friis-Christensen, E., Lassen, K., Wilhjelm, J., Wilcox, J.M., Gonzalez, W. and Colburn, D.S. (1972). Critical component of the interplanetary magnetic field responsible for large geomagnetic effects in the polar cap. *Journal of Geophysical Research*, **77**, 3371–3376.

Furth, H.P., Killeen, J. and Rosenbluth, M.N. (1963). Finite resistivity instabilities of a sheet pinch. *Physics of Fluids*, **6**, 459–484.

Gekelman, W., Stenzel, R.L. and Wild, N. (1982). Magnetic field line reconnection experiments: 3. Ion acceleration, flows, and anomalous scattering. *Journal of Geophysical Research*, **87**, 101–110.

Giovanelli, R.G. (1946). A theory of chromospheric flares. *Nature*, **158**, 81–82.

Gold, T. and Hoyle, F. (1960). On the origin of solar flares. *Monthly Notices of the Royal Astronomical Society*, **120**, 89–105.

Golub, L., Krieger, A.S., Harvey, J.W. and Vaiana, G.S. (1977). Magnetic properties of X-ray bright points. *Solar Physics*, **53**, 111–131.

Gonzalez, W.D. and Mozer, F.S. (1974). A quantitative model for the potential resulting from reconnection with an arbitrary interplanetary field. *Journal of Geophysical Research*, **79**, 4186–4194.

Gosling, J.T., Thomsen, M.F., Bame, S.J., Onsager, T.G. and Russell, C.T. (1990). The electron edge of the low-latitude boundary layer during accelerated flow events. *Geophysical Research Letter*, **17**, 1833–1836.

Haerendel, G. and Paschmann, G. (1982). Interaction of the solar wind with the dayside magnetosphere. In A. Nishida (ed.), *Magnetospheric Plasma Physics*, Reidel, Dordrecht, pp. 49–142.

Haerendel, G., Paschmann, G., Sckopke, N., Rosenbauer, H. and Hedgecock, P.C. (1978). The frontside boundary layer of the magnetosphere and the problem of reconnection. *Journal of Geophysical Research*, **83**, 3195–3216.

Harvey, K.L. (1985). The hydromagnetics of the Sun. In *Proceedings of the 4th European Meeting on Solar Physics*, ESA SP-220, pp. 235–236.

Heppner, J.P. (1972). Polar-cap electric field distribution related to the interplanetary magnetic field direction. *Journal of Geophysical Research*, **77**, 4877–4887.

Hesse, M., Birn, J. and Schindler, K. (1990). A self-consistent two-dimensional resistive fluid theory of field-aligned potential structures including charge separation and magnetic and velocity shear. *Journal of Geophysical Research*, **95**, 18929–18938.

Hesse, M. and Winske, D. (1998). Electron dissipation in collisionless magnetic reconnection. *Journal of Geophysical Research*, **103**, 26479–26486.

Heyvaerts, J., Priest, E.R. and Rust, D.M. (1977). An emerging flux model for the solar flare phenomenon. *Astrophysical Journal*, **216**, 123–137.

Hones, E.W., Jr. (1979). Plasma flow in the magnetotail and its implications for substorm theories. In S.-I. Akasofu (ed.), *Dynamics of the Magnetosphere*, Reidel, Dordrecht, pp. 545–562.

Hoyle, F. (1949). Magnetic storms and aurorae. In *Some Present Researches in Solar Physics*, Cambridge University Press, pp. 102–104.

Innes, D.E., Inhester, B., Axford, W.I. and Wilhelm, K. (1997). Bi-directional plasma jets produced by reconnection on the Sun. *Nature*, **386**, 811–813.

Ji, H., Yamada, M., Hsu, S., Kulsrud, R., Carter, T. and Zaharia, S. (1999). Magnetic reconnection with Sweet–Parker characteristics in two-dimensional laboratory plasmas. *Physics of Plasmas*, **6**, 1743–1750.

Kadomtsev, B.B. (1975). Reconnection of field lines and disruptive instability in tokamaks. *Fizika Plazmy*, **1**, 710 [In Russian; English translation: *Soviet Journal of Plasma Physics*, **1**, 389 (1976)].

Kopp, R.A. and Pneuman, G.W. (1976). Magnetic reconnection in the solar corona and the loop prominence phenomenon. *Solar Physics*, **50**, 85–98.

Kulsrud, R.M. (1998). Magnetic reconnection in a magnetohydrodynamic plasma. *Physics of Plasmas*, **5**, 1599–1606.

Levy, R.H., Petschek, H.E. and Siscoe, G.L. (1964). Aerodynamic aspects of the magnetospheric flow. *AIAA Journal*, **2**, 2065–2076.

Litvinenko, Y.E. and Martin, S.F. (1999). Magnetic reconnection as the cause of a photospheric canceling feature and mass flows in a filament. *Solar Physics*, **190**, 45–58.

Mukai, T., Fujimoto, M., Hoshino, M., Kokubun, S., Machida, S., Maezawa, K., Nishida, A., Saito, Y., Teresawa, T. and Yamamoto, T. (1998). Structure and kinetic properties of plasmoids and their boundary regions. *Journal of Geomagnetism and Geoelectricity*, **48**, 541.

Nishida, A. (1971). Interplanetary origin of electric fields in the magnetosphere. *Cosmic Electrodynamics*, **2**, 350–374.

Nishida, A. (1975). Interplanetary field effect on the magnetosphere. *Space Science Reviews*, **17**, 353–389.

Ono, Y., Inomoto, M., Okazaki, T. and Ueda, Y. (1997). Experimental investigation of three-component magnetic reconnection by use of merging spheromaks and tokamaks. *Physics of Plasmas*, **4**, 1953–1963.

Parker, E.N. (1957). Sweet's mechanism for merging magnetic fields in conducting fluids. *Journal of Geophysical Research*, **62**, 509–520.

Parker, E.N. (1963). The solar flare phenomenon and theory of reconnection and annihilation of magnetic fields. *Astrophysical Journal*, Suppl., **8**, 177–211.

Parker, E.N. (1972). Topological dissipation and the small-scale fields in turbulent gases. *Astrophysical Journal*, **174**, 499–510.

Parker, E.N. (1983). Magnetic neutral sheets in evolving fields: I. General Theory; II. Formation of the solar corona. *Astrophysical Journal*, **264**, 642–647; 635–641.

Paschmann, G., Haerendel, G., Papamastorakis, I., Sckopke, N., Bame, S.J., Gosling, J.T. and Russell, C.T. (1982). Plasma and magnetic field characteristics of magnetic flux transfer events. *Journal of Geophysical Research*, **87**, 2159–2168.

Paschmann, G., Sonnerup, B.U.Ö., Papamastorakis, I., Sckopke, N., Haerendel, G., Bame, S.J., Asbridge, J.R., Gosling, J.T., Russell, C.T. and Elphic, R.C. (1979). Plasma acceleration at the earth's magnetopause: Evidence for reconnection. *Nature*, **282**, 243–246.

Petschek, H.E. (1964). Magnetic field annihilation. In W.N. Hess (ed.), *Physics of Solar Flares*, NASA SP-50, Washington, DC, pp. 425–439.

Petschek, H.E. and Thorne, R.M. (1967). The existence of intermediate waves in neutral sheets. *Astrophysical Journal*, **147**, 1157–1163.

Phan, T.-D., Paschmann, G., Baumjohann, W., Sckopke, N. and Lühr, H. (1994). The magnetosheath region adjacent to the dayside magnetopause: AMPTE/IRM observations. *Journal of Geophysical Research*, **99**, 121–141.

Phan, T.-D., Paschmann, G. and Sonnerup, B.U.Ö. (1996). Low-latitude dayside magnetopause and boundary layer for high magnetic shear: 2. Occurrence of magnetic reconnection. *Journal of Geophysical Research*, **101**, 7817–7828.

Priest, E.R. and Démoulin, P. (1995). Three-dimensional magnetic reconnection without null points: 1. Basic theory of magnetic flipping. *Journal of Geophysical Research*, **100**, 23443–23463.

Priest, E. and Forbes, T. (2000). *Magnetic Reconnection – MHD Theory and Applications*, Cambridge University Press.

Priest, E.R., Parnell, C.E. and Martin, S.F. (1994). A converging flux model of an X-ray bright point and an associated cancelling magnetic feature. *Astrophysical Journal*, **427**, 459–474.

Richardson, I.G., Cowley, S.W.H., Hones, E.W. and Bame, S.J. (1987). Plasmoid-associated energetic ion bursts in the deep geomagnetic tail: Properties of plasmoids and the post-plasmoid plasma sheet. *Journal of Geophysical Research*, **92**, 9997–10013.

Russell, C.T. (1990). The magnetopause. In C.T. Russell, E.R. Priest and L.C. Lee (eds), *Physics of Magnetic Flux Ropes*, Geophysical Monograph, **58**, American Geophysical Union, Washington, DC, pp. 439–453.

Russell, C.T. and Elphic, R.C. (1978). Initial ISEE magnetometer results: Magnetopause observations. *Space Science Reviews*, **22**, 681–715.

Scholer, M. (1991). Numerical models of magnetic reconnection. *Geophysical and Astrophysical Fluid Dynamics*, **62**, 51–68.

Scholer, M., Gloeckler, G., Klecker, B., Ipavich, F.M., Hovestadt, D. and Smith, E.J. (1984). Fast moving plasma structures in the distant magnetotail. *Journal of Geophysical Research*, **89**, 6717–6727.

Scholer, M. and Lottermoser, R.-F. (1998). Hybrid simulations of magnetotail reconnection: Plasmoids, the post-plasmoid plasma sheet, and slow mode shocks. In S. Kokubun and Y. Kamide (eds), *Substorms-4*, Terra Scientific, Tokyo, pp. 467–472.

Schrijver, C.J., Title, A.M., Berger, T.E., Fletcher, L., Hurlburt, N.E., Nightingale, R., Shine, R.A., Tarbell, T.D., Wolfson, J., Golub, L., Bookbinder, J.A., De Luca, E.E., McMullen, R.A., Warren, H.P., Kankelborg, C.C., Handy, B.N. and De Pontieu, B. (1999). A new view of the solar outer atmosphere by the Transition Region and Coronal Explorer. *Solar Physics*, **187**, 261–302.

Scudder, J.D. (1997). Theoretical approaches to the description of magnetic merging: The need for finite β_e, anisotropic, ambipolar Hall MHD. In B. Hultqvist and M. Øieroset (eds), *Transport Across the Boundaries of the Magnetosphere*, Kluwer, Dordrecht, pp. 235–267.

Shay, M.A., Drake, J.F., Denton, R.E. and Biskamp, D. (1998). Structure of the dissipation region during collisionless magnetic reconnection. *Journal of Geophysical Research*, **103**, 9165–9176.

Shibata, K. (1996). New observational facts about solar flares from Yohkoh studies: Evidence of magnetic reconnection and a unified model of flares. *Advances in Space Research*, **17**, 4/5, 9–28.

Sonnerup, B.U.Ö. (1970). Magnetic field reconnection in a highly conducting incompressible fluid. *Journal of Plasma Physics*, **4**, 161–174.

Sonnerup, B.U.Ö. (1974). Magnetopause reconnection rate. *Journal of Geophysical Research*, **79**, 1546–1549.

Sonnerup, B.U.Ö and Cahill, L.J., Jr. (1967). Magnetopause structure and attitude from Explorer 12 observations. *Journal of Geophysical Research*, **72**, 171–183.

Sonnerup, B.U.Ö., Papamastorakis, I., Paschmann, G. and Lühr, H. (1990). The magnetopause at large magnetic shear: Analysis of convection electric fields from AMPTE/IRM. *Journal of Geophysical Research*, **95**, 10541–10557.

Sonnerup, B.U.Ö., Paschmann, G., Papamastorakis, I., Sckopke, N., Haerendel, G., Bame, S.J., Asbridge, J.R., Gosling, J.T. and Russell, C.T. (1981). Evidence for magnetic field reconnection at the Earth's magnetopause. *Journal of Geophysical Research*, **86**, 10049–10067.

Stenzel, R.L. and Gekelman, W. (1979). Experiments on magnetic-field-line reconnection. *Physical Review Letter*, **42**, 1055–1057.

Stenzel, R.L., Gekelman, W. and Wild, N. (1982). Magnetic field line reconnection experiments: 4. Resistivity, heating, and energy flow. *Journal of Geophysical Research*, **87**, 111–117.

Sturrock, P.A. (1966). Model of the high-energy phase of solar flares. *Nature*, **5050**, 695–697.

Sweet, P.A. (1958). The neutral point theory of solar flares. In *IAU Symposium No. 6*, Cambridge University Press, pp. 123–134.

Treumann, R.A., LaBelle, J. and Bauer, T.M. (1995). Diffusion at the magnetopause: An observational perspective. In P. Song, B.U.Ö. Sonnerup and M.F. Thomsen (eds), *Physics of the Magnetopause, Geophysical Monograph 90*, American Geophysical Union, Washington, DC, pp. 331.

Tsuneta, S. (1996). Structure and dynamics of magnetic reconnection in a solar flare. *Astrophysical Journal*, **456**, 840–849.

Vaiana, G.S., Krieger, A.S., Van Speybroeck, L.P. and Zehnpfennig, T. (1970). *Bulletin of the American Physical Society*, **15**, 611.

Vasyliunas, V.M. (1975). Theoretical models of magnetic field line merging. *Reviews of Geophysics and Space Physics*, **13**, 303–336.

Yamada, M., Ji, H., Hsu, S., Carter, T., Kulsrud, R., Bretz, N., Jobes, F., Ono, Y. and Perkins, F. (1997). Study of driven magnetic reconnection in a laboratory plasma. *Physics of Plasmas*, **4**, 1936–1944.

Yeh, T. and Axford, W.I. (1970). On the reconnection of magnetic field lines in conducting fluids. *Journal of Plasma Physics*, **4**, 207 229.

44

DOUGLAS O. GOUGH* AND PHILIP H. SCHERRER**

The solar interior

The quest to learn the internal structure and dynamics of the Sun is motivated by several issues. For most people the most obvious is the desire to understand the source of energy for the Earth, which is essential for the maintenance of life. Astronomers are interested in the Sun because it is an example of a typical main-sequence star that can be studied in enormously greater detail than any other. And in addition, the Sun can be used to investigate fundamental physics: it is an important source of gravity, providing a testbed for the general theory of relativity, and it contains material at high pressure and temperature permitting us to study particle, nuclear and atomic physics, plasma physics, and fluid dynamics under conditions that cannot be achieved on Earth. Perhaps less fundamental but certainly more important to society, the solar interior is the source of both secular and cyclic variability in the electromagnetic and particle fluxes, and of all their effects on the Earth and human technological systems.

This chapter describes first our understanding of the solar interior prior to 1975. It then provides a narrative discussion of developments in each of several major topics since that time. These topics include neutrinos, shape, irradiance, calibration of stellar models, rotation, the dynamics of the convection zone and of large-scale features within it, magnetic field generation and the solar cycle, and the nature of active regions and active longitudes. Beginning in about 1975 a new tool for probing the interior of the Sun, and some day of other stars, was discovered. This tool is helioseismology. The impact of helioseismological inferences on our understanding of the solar interior is so dominant that a brief review of the techniques by which it is used will be provided prior to an examination of the development and present state of each of the above topics. For the most part the further development of the main scientific topics has followed the development of helioseismology, and therefore the discussion of those topics will be intertwined with the discussion of helioseismology.

STATUS PRIOR TO 1975

Before 1975 the only observable quantities that yield information about the solar interior were the global parameters – mass, radius, luminosity, and effective temperature – which were used to calibrate (spherically symmetrical) theoretical models. A measure of the neutrino flux, which reveals energy generation processes directly, was also available. The age of the Sun is inferred from the age of the Earth, and other interplanetary material. The observed surface differential rotation and the existence and organization of magnetic fields in the photosphere also provided information about processes assumed to be operating in the convection zone. There was also a measure of the oblateness of the photosphere, from which one should be able to infer a measure of the mean rotation of the interior and the oblateness of the exterior gravitational equipotentials. But the interpretation of some of those data was in question. Primarily, our understanding of the solar interior was based on the application of stellar models.

Prior to the observation that there were too few neutrinos coming from the Sun, theoretical models could be adjusted to match the global observables. Explanation of the low neutrino flux was a big problem. Although in principle the solar data were adequate to determine the uncertain parameters that specify the simplest theoretical models – the initial helium abundance Y, the total heavy-element abundance Z (the relative proportions of most of the heavy elements can be determined by spectroscopic analysis) and a scaling parameter α contained in the mixing-length theory used to model the convective heat flux – the outcome was

* University of Cambridge, United Kingdom
** Stanford University, CA, USA

generally not believed, for it yielded an implausibly low value of the helium abundance, namely $Y \simeq 0.16$. Even though it was in the Sun that helium was discovered, a direct abundance determination is not possible. The atmosphere is too cool for a spectroscopic analysis, and the theory of radiative transfer in the corona is too uncertain to yield a reliable value. Abundance measurements were available from *in situ* measurements of the solar wind, but these exhibited temporal and spatial variability, implying that chemical differentiation had taken place and that therefore none of the measured values necessarily represents the value in the Sun. However, there were astronomical estimates: values of Y determined in the atmospheres of hot stars believed to be of comparable age to the Sun and measurements in ionized interstellar gas clouds suggested a value of about 0.25. Moreover, calibrated theories of Big Bang nucleosynthesis yielded values of Y that exceeded 0.20 – comfortably less than the values observed in the stars which have condensed from gas clouds that have been enriched by nuclear processed material from supernova explosions, yet substantially greater than the value from the solar calibration. Therefore the value of 0.16 was not generally accepted; instead it was assumed that $Y \simeq 0.25$, and that it was the neutrino flux that could not be explained. Consequently the issue was dubbed "the solar neutrino problem." Nevertheless, there remained some doubt about Y.

MOST IMPORTANT ISSUES IN 1975

(i) Solar neutrino flux: Is it a problem with the structure of the Sun, or with nuclear physics or even particle physics? Spontaneous oscillations between neutrino flavors had been suggested by Pontecorvo (1968), raising the possibility that a substantial fraction of the electron neutrinos that are produced in the Sun have been transformed into μ or τ neutrinos by the time they reach the Earth, and thereby had evaded detection. However, for such a process to occur, neutrinos would need to have mass, and at the time it was almost universally believed that neutrinos were massless. There had been a great industry in adjusting parameters of standard solar models, and adding extra ingredients almost always in a spherically symmetrical fashion. That some neutrinos were detected at all was regarded as a triumph for nuclear physics. A crucial assumption in the construction of the standard models is that they are in thermal balance (the thermal relaxation time is very much less than the characteristic time of structural variations arising from chemical composition changes produced by nuclear transmutations). This implies that the integrated rate of generation of thermal energy by nuclear reactions is equal to the "observed" luminosity of the Sun (at least, the luminosity inferred from irradiance measurements, assuming spherical symmetry). A second assumption is that the core is motionless, and that therefore all but the slowest nuclear reactions have reached local equilibrium. Together, these assumptions impose tight theoretical constraints on the balance of reactions and the consequent neutrino production rates.

(ii) Oblateness of gravitational equipotentials and the history of solar spin-down: How much greater is the angular velocity Ω in the core than it is at the surface? There had been many studies of potential instabilities arising from shear presumed to be imposed by spin-down and to which the Sun was regarded as being neutrally stable. They all implied that Ω increases inwards, which would cause the oblateness of the Sun to be greater than one might suppose from the surface rotation alone. Dicke and Goldenberg (1967, 1974) had claimed such a greater oblateness which might have been compatible with some of these theoretical studies, but there were problems with interpreting the optical measurements of the shape of the Sun, owing to there being a greater emission of light from the solar atmosphere in equatorial regions making the solar disk appear to be more oblate than the matter distribution. Pole–equator variation in convective flow could also influence the measurement. The contribution to the oblateness from the centrifugal force due to the rotation of the photosphere needs to be subtracted from the raw measurement: the residual would be only about 4% of the total if Ω were uniform, rendering the corrected measurement uncertain. A measurement of the oblateness of the surface would have been fine if Ω were much larger in the interior, as Dicke had predicted to be required for the Brans–Dicke theory of gravity.

The interpretation of the measurements of Dicke and Goldenberg were doubted by the community, partly because they contradicted general relativity by implying too great a precession of the perihelion of the orbit of Mercury. A subsequent contradictory measurement by Hill and Stebbins (1975) was accepted immediately by most astronomers, however, perhaps not as critically as one should expect. Nevertheless, it still appears that any measurement of the shape of the photosphere is an extremely unreliable guide to the oblateness of the gravitational equipotentials, and that more direct methods are preferable. As we describe below, the inference from helioseismology is now by far the best, and is likely to remain so for a long time.

(iii) The source of magnetic fields and the activity cycle was also a mystery. While a number of competing models were put forward, the general belief was that a dynamo process operating in the convection zone and driven by rotational shear is the source of active-region fields and of the 22-year magnetic cycle. A phenomenological discussion invoking differential rotation, buoyancy, and supergranulation to distribute the fields into the observed patterns (Babcock 1961, Leighton 1969) has held considerable credence, and it is not unlikely that these processes form the

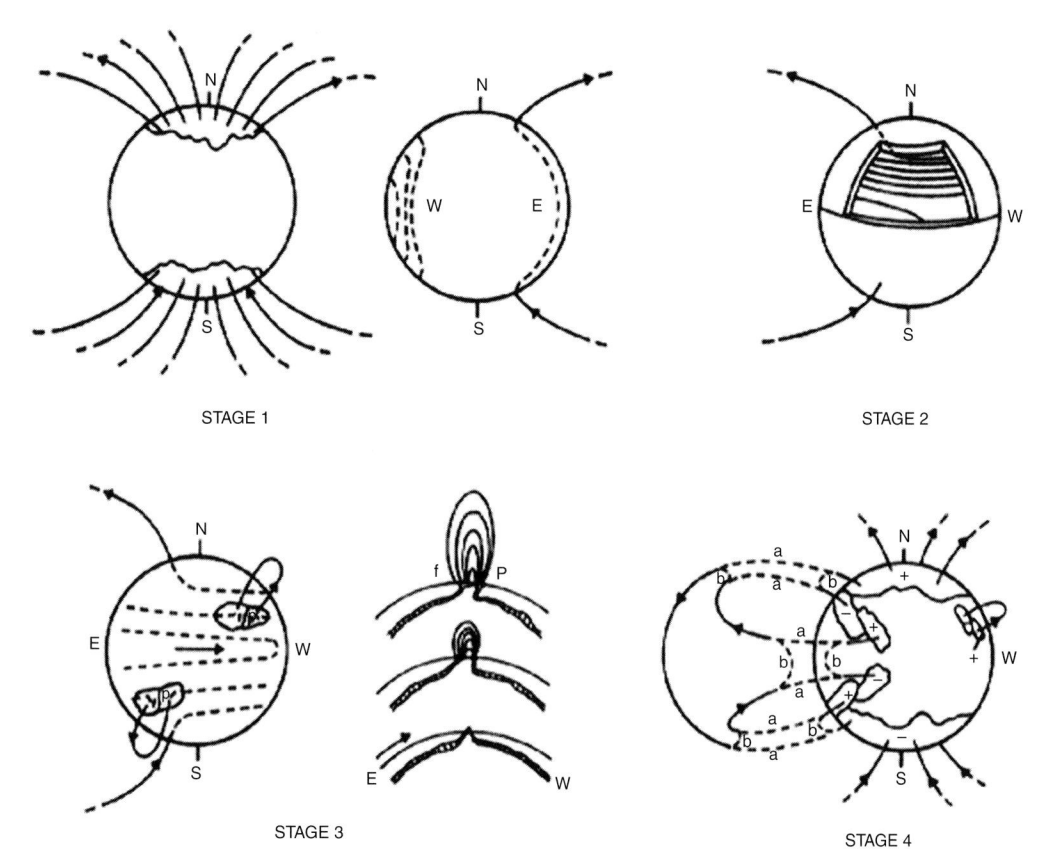

Figure 1 Schematic from Babcock (1961) via Rabin et al. (1991) showing the proposed evolution of the 22-year solar magnetic cycle, which still forms the basis of many modern views. Stage 1 is a state of basically poloidal field. In stage 2 the field lines are stretched into toroidal spirals by the differential rotation (principally in the tachocline, perhaps), and once the field has reached a strength sufficient to become buoyant, parcels of fluid rise through the convection zone drawing with them loops of field which erupt through the photosphere (stage 3) to form bipolar magnetic regions. In stage 4 advection by a combination of differential rotation and the (anisotropic) convection causes leading active regions to migrate equatorward and trailing regions to migrate poleward, creating new poloidal field of opposite polarity; the toroidal field dissipates to produce stage 5, which is like stage 1 but with the sense of the field reversed.

basis of much of what is happening. Figure 1 shows the essential characteristics of these models. But there was no precise model dynamo that could reproduce the observed magnetic cycle as shown in Figure 2. The observed fact of "active longitudes" or "zones" could not be accounted for in any model. Much progress had been made to understand the source of coronal and interplanetary fields in terms of the photospheric fields, but there was no understanding of the origin of the photospheric fields, nor of their organization.

(iv) The stability of the "solar constant," or total irradiance was unclear. There was historical evidence of secular changes, and there was uncertainty about the variability with activity. The correlation of solar-activity indices with northern European climate records suggested strongly that there is at least a long-term coupling of solar activity with climate (Eddy 1976). Attempts to measure total irradiance and its variations from the ground, under way since the beginning of the twentieth century, had not yielded convincing results. In fact, all observations were consistent with an unchanging Sun, and with measurement variations induced by the Earth's atmosphere through which the measurements had been made – thus the term "solar constant" was believed to be a reasonable connotation for the flux of radiation emitted by the Sun (e.g. Smith and Gottlieb 1973). We had to wait for measurements from space before variations in the solar constant could be measured reliably.

THE STUDY OF SOLAR OSCILLATIONS PRIOR TO 1975

In the summers of 1960 and 1961, Robert Leighton (1961) and his two students Robert Noyes and George Simon made observations that fundamentally altered the

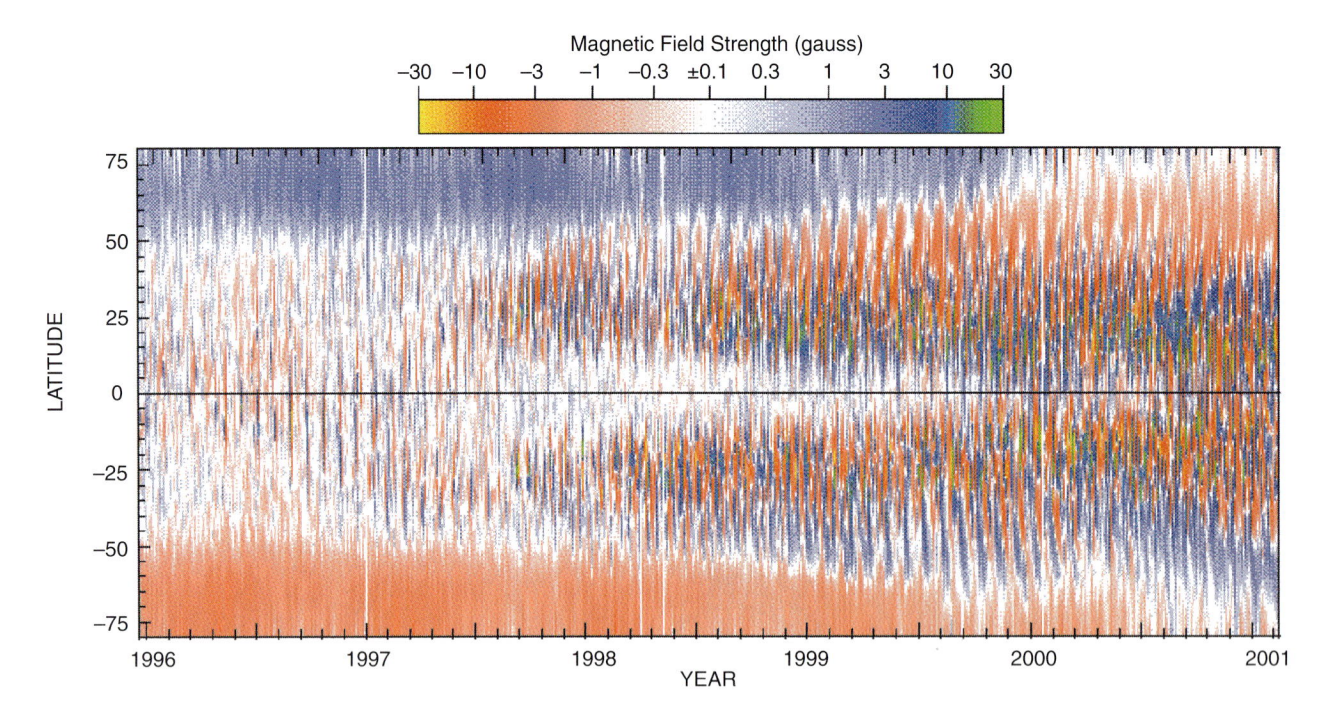

Figure 2 Solar magnetic flux during the beginning of the present activity cycle. The strong fields in darker colors can be seen to migrate toward the equator while the weaker fields share the structure of the large-scale polarity pattern which organizes the corona. Each Carrington rotation is evident in the vertical banded structure, which is a symptom of the variation in field strength (and activity) with longitude. (Courtesy of Dave Hathaway.)

study of solar physics (Leighton et al. 1962). By optical subtraction of spectroheliograms obtained in the wings of a spectrum line they discovered both supergranulation (Simon and Leighton 1964) and the 5-minute oscillations (Noyes and Leighton 1963). Although we still do not really understand supergranulation in detail, we do know that it structures and rearranges magnetic fields and plays a crucial role in the outer atmosphere. The discovery of the 5-minute oscillations spawned helioseismology and the modern study of the solar interior.

The spatial or temporal coverages of the early studies were limited, masking the essential nature of the oscillations. The true nature of the oscillations as the superposition of millions of trapped acoustic waves was at first neither understood nor exploited.

Frazier (1968) claimed that 5-minute solar oscillations were trapped acoustic modes. He suggested to Ulrich that he carry out a calculation which resulted in the understanding that the observations were of evanescent waves that are vestiges of waves propagating in the interior of the Sun and being reflected beneath the observing layer (Ulrich 1970). At about the same time Leibacher and Stein (1971) reached the same conclusion. This understanding was not fully accepted until Deubner (1975) observed the predicted ridges in power after making an observation with greater spatial and temporal resolution. These observations were soon followed by more detailed observations by Rhodes et al. (1977).

At about the same time H.A. Hill announced having seen oscillations in the solar diameter measurements which were interpreted as being global modes of oscillation (e.g. Hill and Caudell, 1979). The diagnostic potential was recognized immediately (Christensen-Dalsgaard and Gough 1976), and helioseismology was born. Hill et al. (1991) prepared a detailed review of these early observations.

HELIOSEISMOLOGY IN A NUTSHELL

The Sun is a ball of almost fully ionized gas. Its radius is about 700 Mm. The nuclear reactions that heat the Sun occur mainly in the inner 150 Mm. Up to about 500 Mm from the center, the energy is carried outward by radiation. An average photon takes about 30,000 years to traverse the radiative zone; the thermal cooling time is 1,000 times longer. The outer 200 Mm is unstable to convection, and there the energy is efficiently carried by convection up to the thin layer at the surface called the photosphere from where the energy is radiated into space. It takes several months for energy to get through the convection zone and about 8 more minutes to get to the Earth. Figure 3a is a schematic view of the main features of the Sun.

A number of types of waves can propagate in the Sun. Acoustic waves can propagate throughout the interior. The sound travel time through the Sun is about 2 hours. Acoustic

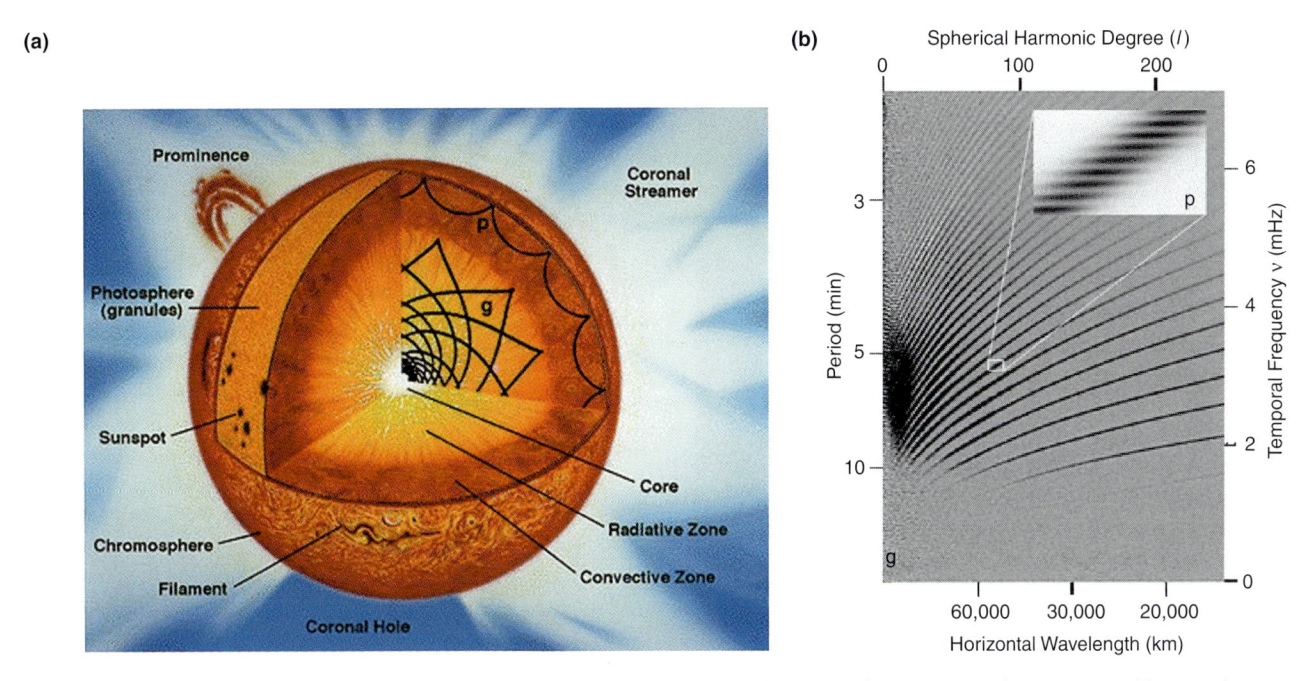

Figure 3 (a) A cutaway sketch of the Sun showing the principal regions. (b) A sample spectrum of acoustic oscillations showing p modes. (From the GONG project.)

waves with frequencies below about 5 mHz cannot propagate in the upper atmosphere, and are reflected back into the interior. The acoustic wave speed, c, is about $7\,\mathrm{km\,s^{-1}}$ near the surface and increases with depth to near $500\,\mathrm{km\,s^{-1}}$ at the center. The existence of the sound-speed gradient causes acoustic waves propagating in any direction other than exactly downward to be refracted away from the solar center before returning to the surface. The smallest observed scale of convection at the surface, called granulation, is believed to be the source of broadband acoustic noise with power peaking near (although somewhat below) 3 mHz (5-minute periods), which is about the observed lifetime of the granulation. Thus the interior is populated with sound waves traveling in all directions. For any particular wavetrain, the turning point $r = r_t$ at which a downward propagating wave is refracted back towards the surface is at a level where the characteristic angular frequency ω of the wavetrain divided by the horizontal component of the wavenumber k_h is the local sound speed, that is, where $\omega/k_h = c$.

It is reasonable to think of waves in the interior as the superposition of many individual wave pulses originating from near-surface disturbances. Many of the wave pulses live long enough to travel around the Sun many times, to interfere with themselves to form normal modes. When the waves reflect at the surface, the observable surface moves. We can detect the reflection by measuring either the motion of the surface or the brightness of the surface, which varies from the compression. By observing the surface on a regular grid we can selectively detect waves that follow particular paths through the interior and reflect at appropriate pairs of points, maintaining relative phase (see Figure 5). We use the terminology of spherical harmonics to describe the modes, and refer to the selected horizontal scale of the waves with the dimensionless spherical harmonic degree l, the east–west component of which is the azimuthal order m. Figure 4 shows a sample spherical harmonic function. For waves with more than a few hundred wavelengths around the Sun, we often speak instead of horizontal wavenumber k (in units of radians $\mathrm{m^{-1}}$).

For historical reasons we usually refer to the frequency of moderate- and low-degree modes ($l <$ few hundred) in cyclic frequency ν (Hz), and use the angular frequency ω (radian $\mathrm{s^{-1}}$) for high-k waves. Since there are many waves traveling in random directions and phases, they add up to some net signal at each location on the surface. We can describe these waves either as normal modes of oscillation or as traveling waves, depending on our purpose. A good analogy is a bell in a sandstorm. The bell will be heard to ring at some global frequencies with amplitudes changing randomly with time.

The Sun can be said to be ringing in several million modes with observable amplitude. A random place on the solar surface is found to be moving up and down with a period near 5 minutes and an amplitude of several hundred meters per second. This corresponds to velocities of about $10\,\mathrm{cm\,s^{-1}}$ per mode. We can measure the mode frequencies from the power spectrum (Figure 3b) and compare them to calculations of model stars, and learn which model

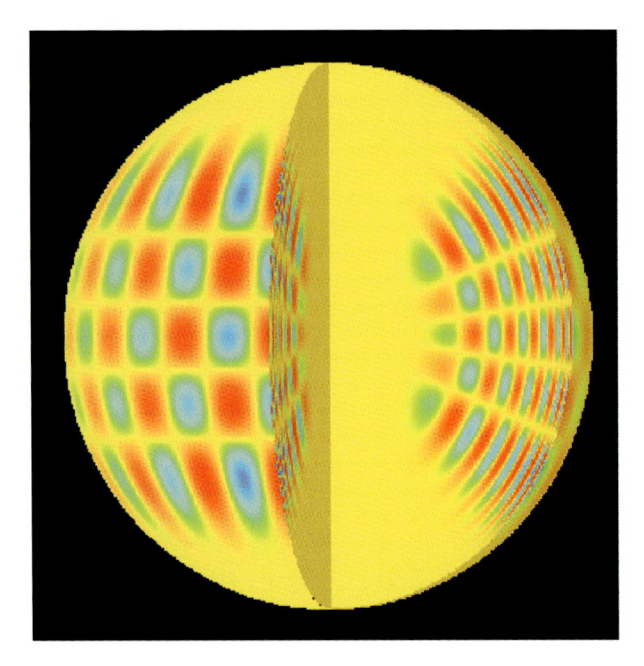

Figure 4 Example of spherical harmonic on the surface of the Sun, which represents the line-of-sight Doppler image of a single mode of acoustic oscillation: yellow represents zero velocity and the scale of the color variation has been magnified by a factor of about 10^9. The mode has $l = 20$, $m = 16$, $n = 14$. The color variation on the meridional cuts represents (right) the vertical and (left) the horizontal components of the velocity in the interior scaled with the square root of density. The degree l is a measure of the total horizontal wavelength: the azimuthal order m is the number of meridional nodal planes (half the number of nodes around the equator); there are $l-m$ nodal half-cones about the center, which intersect the surface in (yellow) lines of latitude; the (radial) order n is the number of spherical nodal surfaces in the vertical velocity, which intersect the right-hand meridional cut in yellow circles.

parameters best describe the Sun. This process has been remarkably successful. We now have models that describe a star which match the observed mass, radius, luminosity, and mode frequencies of the Sun very well. These models predict about two or three times as many neutrinos as are observed, which justifies the connotation "neutrino problem."

In addition to regarding the waves as normal modes, we can consider them to be waves propagating through the Sun along ray paths and use them to probe the wave travel time between multitudes of pairs of points on the surface by simply finding the time shift that allows the best correlation in wave forms observed at each point. Figure 5 shows some sample ray paths used for such time–distance studies. The difference in travel times for waves going in opposite directions between each pair of points provides information about the bulk motion of the interior along the ray path. The average travel time indicates the average wave speed, and thus the temperature (given the chemical composition), along the path.

For both of these approaches we can combine either many modes or many ray paths and determine what model properties are consistent with the observed data. This process of inverting the data is what finally yields maps of interior rotation and sound speed. Both space-based and ground-based observations have been crucial to the development of the subject.

SOLAR MODELS AND NEUTRINOS PRIOR TO HELIOSEISMOLOGY

Prior to helioseismology the desire of most theorists was to reduce theoretical values of the neutrino luminosity L_ν to a minimum, with the aim of simply finding a solar model that

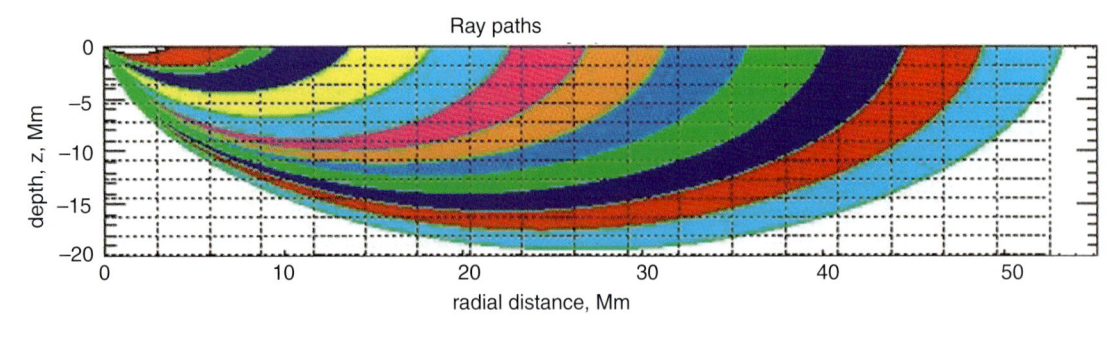

Figure 5 Ray paths in the upper convection zone that have been used for time–distance analysis on a computational grid indicated by the dotted lines (Kosovichev et al. 2001). Note that the horizontal and vertical scales are different; the depth of penetration of a ray is actually about $1/\pi$ of the distance between the two points at which it intersects the surface. The travel time for waves along these paths is measured by correlating signals at those two surface points.

was not inconsistent with the observations of the day. Since most of the neutrinos that were detected at the time – by capture by ^{35}Cl; Bahcall (1989) provides an extensive discussion of the measurements – were high-energy neutrinos emitted during the spontaneous decay of ^{8}B, which is produced mainly near the center of the solar core at the highest temperatures, the objective was to flatten the temperature profile in the core, reducing the central value yet increasing the values in the outer core in order to maintain the total rate of production of thermal energy at the observed luminosity. Devices for reducing gas pressure, and hence temperature, in the central core were introduced, such as imagining rapid rotation or intense magnetic fields to help support the star against gravity, but all to no avail. The only potentially viable idea was for the Sun to be composed of material with a low mean molecular mass, which could be achieved by imagining a low helium-to-hydrogen abundance ratio. Model computations indicated that along with that went a low heavy-element abundance, too low to explain photospheric spectroscopic analyses. However, that could perhaps have been overcome if subsequent to its formation the Sun had accreted heavy elements by infalling comets which have been mixed into the convection zone but have not settled beneath it. Indeed, such models have very shallow convection zones, containing a mass of perhaps no more than $10^{-5}M_\odot$, requiring the accretion of only $10^{-7}M_\odot$ or so of heavy material, which is certainly not out of the question. However, one would still have the cosmological problem with the helium abundance to contend with. Moreover, simulations by Gilman of the dynamics of the convection zone were unable to reproduce the equatorial acceleration of the rotation observed in the Sun's surface unless the convection zone were much deeper. Thus, the depth of the convection zone was an interesting quantity. Once its value had been ascertained by seismology, it became a commonly used gauge of solar models.

IRRADIANCE AND SHAPE SINCE 1975

Monitoring variations in the solar irradiance is an interesting issue. It is evident that in addition to providing a basic measure of the state of the convection zone, changes in the irradiance could have a profound effect on climate, and even perhaps on weather. Moreover, to make precise irradiance measurements requires going to space, because the irradiance variations are less than the variations in atmospheric transparency. A study in 1973 concluded that there is no variation at wavelengths greater than 1500 Å over the 11-year cycle at the limit of 1% (Smith and Gottlieb 1973). Thus in the pre-space era there was no measurement with sufficient precision or cadence to detect the variations we now know to be present.

It is very difficult to make an absolute measurement of total irradiance at all wavelengths, and two advances were necessary to solve the problem. These were the development of self-calibrating radiometers and the availability of platforms above the atmosphere. A number of high-altitude balloon and rocket measurements were made during these years, but without a firm conclusion about the reality of variations at the level of a few tenths of a per cent (e.g. see *Variations of the Solar Constant,* ed. S. Sofia, NASA Conference Publication 2191, 1981).

The first continuing irradiance observations began in 1978 with the Earth Radiation Budget experiment (ERB) on board the NIMBUS 7 spacecraft. This mission showed indications of variability at a few tenths of a per cent.

The first dedicated total solar irradiance experiment was the Active Cavity Radiometer Irradiance Monitor (ACRIM) aboard the Solar Maximum Mission (SMM) launched in 1980 (Willson *et al.* 1981). When the first data became available, irradiance variations over several days with amplitudes of several tenths of a per cent were found to be well correlated with the passage of sunspots. These results demonstrated that sense could be made of irradiance observations only if a continuous series of observations be made. The early ACRIM data also revealed variations with characteristic periods in the 5-minute band (Fröhlich 1981, 1984). A review of the development of instrumental techniques and the early results can be found in Livingston *et al.* (1991).

Early ACRIM and ERB observations suggested the existence of a decline in irradiance with time, which might be associated with the solar cycle, although there was the fear that it might instead have been the result of instrumental degradation. Not until the irradiance started to rise in the mid-1980s after the onset of a new sunspot cycle was the community convinced that the variation was solar. The increase in total irradiance with increasing activity seemed at first to be inconsistent with the observation of a decrease in irradiance when sunspots are present. However, although sunspots are darker than the average photosphere, the facular regions associated with activity are brighter. At least in the direction of the ecliptic the facular or other positive activity-related contributions outweigh the negative effect of sunspots. This result is at least of the right sign to be consistent with the long-term correlation of solar activity with terrestrial climate variations. However, the first models to compare the estimated irradiance increase from faculae combined with spots failed to explain the variations completely.

But already by about 1980 there were speculations about other sources of luminosity variation, and the extent to which they should be associated with changes in the Sun's radius. Jack Eddy's (1979) analysis of solar radius variations indicated an apparent decline, and there was considerable discussion of the suggestion that it might be a long-term secular variation. Now there was the possibility of substantial energy liberated by gravitational contraction. Could that have provided an explanation of the low neutrino flux? If a substantial fraction of the solar luminosity came from gravitational contraction, the nuclear reaction rates would be lower than those required of a star that is essentially in thermal balance. Evidently, the contraction would have had to be a transient, because the timescale was so short – very much less than the age of the Sun. But some of us wondered whether the apparent radius variation could be a solar-cycle effect, and we were interested in what might be the cause. The most common suggestion was a variation in the properties of convection on the granular scale, presumably resulting from the interaction with a changing magnetic field. What was evident to at least those who could carry out a numerical calculation was that perturbations to convection influenced only the structure of the very thin superadiabatic boundary layer of the convection zone, and produced an extremely small relative change in the total radius of the Sun. The publicity given by Eddy to the subject stimulated astrolabe measurements in Haut Provence and eventually the radius monitoring program set up by Tim Brown (the results of which were reported recently by Brown and Christensen-Dalsgaard 1998). All the measurements were fraught with problems produced by atmospheric refraction, as indeed had been Dicke and Goldenberg's attempt to measure the shape of the solar disk. The theoretical work in the late 1970s showed that the ratio $W = \Delta\ln R/\Delta\ln L$ of relative changes in radius R and luminosity L was on the whole an increasing function of the depth at which the principal perturbations to the structure of the Sun takes place. Therefore a measurement of W was a very interesting objective. One needed to go to space for measuring both $\Delta\ln L$ and $\Delta\ln R$. The latter was proposed by Sofia et al. (1991) with the solar disk sextant, but it has never been flown. Brown and Christensen-Dalsgaard obtained only an upper bound to $\Delta\ln R$. However, Emilio et al. (1998) report a possible detection at $\Delta\ln R = 2 \times 10^{-5}$, although they worry that that might be instrumental and therefore be only an upper bound too. Their result implies $W \lesssim 0.04$, which is possibly consistent with a modulation of the geometry of the convective motion being the immediate cause of both $\Delta\ln R$ and $\Delta\ln L$. The estimate of W presumes that relative changes in irradiance represent relative changes in L. But the uncertainty in the measurement of R leaves us still unsure of the value of W after more than 20 years.

It has been shown that most, if not all, of the variation in the observed total irradiance is associated with localized magnetic activity. It may therefore have little to do with global-scale changes in the dynamics of the convection. However, the match between irradiance and magnetic field observations is still not complete, and there have been suggestions that there are latitudinal variations in luminous flux that change with the cycle and are not the result of localized magnetic activity. Nevertheless, irradiance variations, which have been measured only in the plane of the ecliptic, may overestimate luminosity variations.

In 1979 a study of the proposed Solar Probe mission led one to believe that careful measurement of the gravitational potential could yield information about the internal rotation via the gravitational quadrupole moment J_2; that could be so provided any time dependence due to internal density-changing motion in the Sun could be removed from the data. The Solar Probe is still a mission for the future, but no longer is its goal J_2, for that has now been measured by seismology. More recent proposals for gravitational-wave detectors in solar orbit have raised the possibility of detecting internal gravity-mode oscillations (otherwise known as g modes, and for which buoyancy provides the restoring force) of the Sun, with perhaps more sensitivity than observations of velocity, brightness, or shape of the solar surface. These are also missions for the twenty-first century.

We have already reported above that early ground-based observations of the shape of the visible limb indicated significant oblateness – such that the inferred internal rotation would be several times the surface rate, and make enough of a bulge in the gravitational equipotentials to affect the precession of Mercury enough to make it inconsistent with the prediction of general relativity. Later results from the ground and from space with MDI (see below) have shown that there is no such bulge of a magnitude inconsistent with the surface rate being typical of the rotation of the entire body of the Sun. However, these observations did yield information on latitudinal variations in irradiance.

Ground-based observations of oblateness also suggested the detection of many instances of g modes and p modes. Solar p modes do have a detectable signal in the limb-position data (MDI), but at an amplitude less than the early reports. The claims of g mode detection have not been confirmed by later observers who have placed upper limits on g mode amplitudes that are significantly smaller than the early claims.

INTERIOR STUDIES – 1975 THROUGH 1983

Once it was realized that the trapping of the 5-minute waves in an acoustic cavity beneath the solar surface implied that the properties of the oscillations could be used to infer conditions in the Sun's interior, a number of new observations were made. The new understanding pointed to the need for long time series of images of brightness or motion of the photosphere, in order to obtain accurate oscillation frequencies. The first series of imaged observations (Deubner 1975) were made of trapped waves of high k, which produce constructive interference ridges corresponding to modes with one, two or maybe a few nodes in the oscillation wave function in the vertical direction, but with many in the horizontal direction (n low, l high). These observations were first used to obtain an estimate of the depth of convection zone (Gough 1977); it was obtained by scaling the frequencies to estimate the mean stratification of only the upper layers of the convection zone, and comparing the outcome with theoretical models to get the depth. The estimate was 200 km, in contrast to the depth of only 150 km in the favored models of the time. This suggested that Y is not low, and might even exceed the suspected 0.25 value. This was therefore further evidence that the neutrino problem was real. Subsequent observations (Rhodes et al. 1977, Deubner 1979) enabled the estimate to be refined.

Sun-as-a-star and annulus–center measurements

At the same time as the high-k observations were being made, other observers were developing methods to measure the average line-of-sight velocity of the entire Sun, that is, to observe the Sun as a star. These observations were made with a resonance scattering technique developed by Isaak (1961) and Fossat and Roddier (1971). In this case, in view of the roughly 2-hour sound travel time through the Sun, the resonant 5-minute modes are of high order, and except near the center of the Sun they behave nearly as radial modes with 15 to 25 wavelengths across the cavity. For these whole-Sun oscillations the theory of pulsating stars could be applied (Ledoux and Walraven 1958) whereas for the smaller patches the waveguide model of Ulrich and Leibacher and Stein is appropriate.

The power spectrum of the Birmingham whole-disk data (Claverie et al. 1979) suggested a sequence of approximately uniformly spaced peaks, which were interpreted as being produced alternately by groups of modes with $l = 0$ and 2 with m even and modes with $l = 1$ and 3 with m odd. The power in each group, although it suffered statistical fluctuations, was independent of the parity, in agreement with theory. Therefore it was impossible to distinguish between the even- and odd-degree groups. The spacing between the peaks (what is now called the large separation) is a measure of the acoustic travel time through the Sun; the precise location of the peaks depends principally on the conditions near the upper reflecting layers in the convection zone. Because turbulent convection is not well understood, those layers could not be reliably modeled, and it was therefore natural first to calibrate models solely according to large separation. That suggested that Y was anomalously low, in contradiction to the deduction from the high-degree modes, and some observers held the opinion for a while that the new result might alleviate or even solve the neutrino problem. The suggestion was important, for it resulted from a global measure of the solar interior, whereas the earlier contradictory calibration using Deubner's data, which effectively sampled merely the upper convective boundary layer, rested more heavily, it seemed, on theory.

The parity of the modes was determined soon afterwards from observations made by Grec, Fossat, and Pomerantz (1980), who achieved the required greater frequency resolution by observing the Sun continuously for five days. This they accomplished by observing from the South Pole. Using a superposed frequency analysis, the mean peaks of both odd and even parity in the power spectrum were resolved into two, but differently: modes of even parity produced two peaks of similar power, with a frequency difference of about 10 μHz which is now called the small separation, and modes of odd parity are split by about 5/3 as much, with most of the power in the $l = 1$ peak. The small separation contains information of a different kind, because it is determined predominantly by conditions in the energy-generating core. But it was the identification of the parity of the peaks that was used first: it showed that the previous calibration of the Birmingham data by large separation alone was incorrect, for that calibration had wrongly identified observed and theoretical peaks with opposite parity. A correct identification would require a relative frequency shift either upwards or downwards (with different assignments of the radial order n to the peaks), which would either reduce Y yet further, or raise it even above 0.25. Which was correct? If only n were known, the issue could be decided. But it was not possible to extrapolate the sequence to the fundamental mode and count down in order to determine n; the spacing was not precisely uniform and the nonuniformity was unknown, for it depends on the structure of the Sun that is being sought, and the mode of lowest frequency that it had been possible to observe was too far from the fundamental for the nonuniformity in the spacing to be of no consequence to the counting. Nevertheless, the exercise highlighted the importance of long continuous observations for resolving individual modes. Figures 6 and 7 show these modes as now observed from SOHO with GOLF (Lazrek et al. 1997, Gabriel et al. 1997).

Shortly afterwards, Scherrer et al. (1983) detected modes with $l = 4$, 5, and 6 by analyzing the difference in Doppler shifts in a central portion of the solar disk and the surrounding annulus. Comparison of the results with theory reinforced support for the high-Y solution. But because the theoretical models could not be calibrated to match the data precisely, there remained considerable uncertainty.

The next advances: modes of high and intermediate degree

The answer to the ambiguity in the calibration was provided by Duvall and Harvey (1983), who observed modes with degrees ranging from 1 to 139. The observations were made of zonal modes (i.e., modes with $m = 0$), and were made by

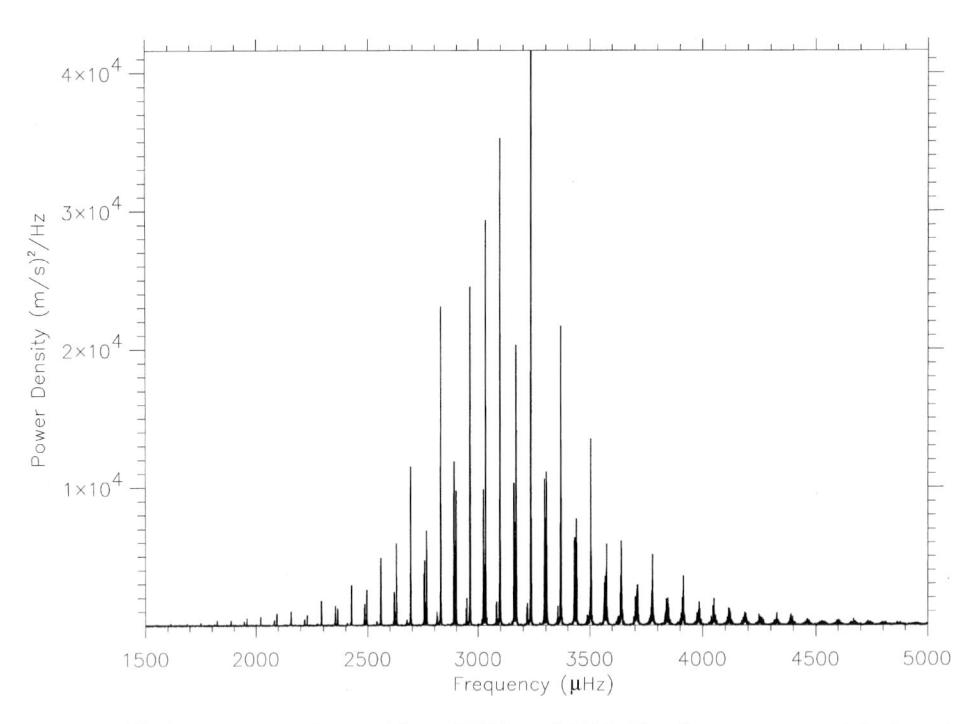

Figure 6 Sun-as-a-star oscillation spectrum observed from GOLF on SOHO. The discrete nature of the low-degree spectrum is quite obvious in modern observations. (From Lazrek *et al.* 1997.)

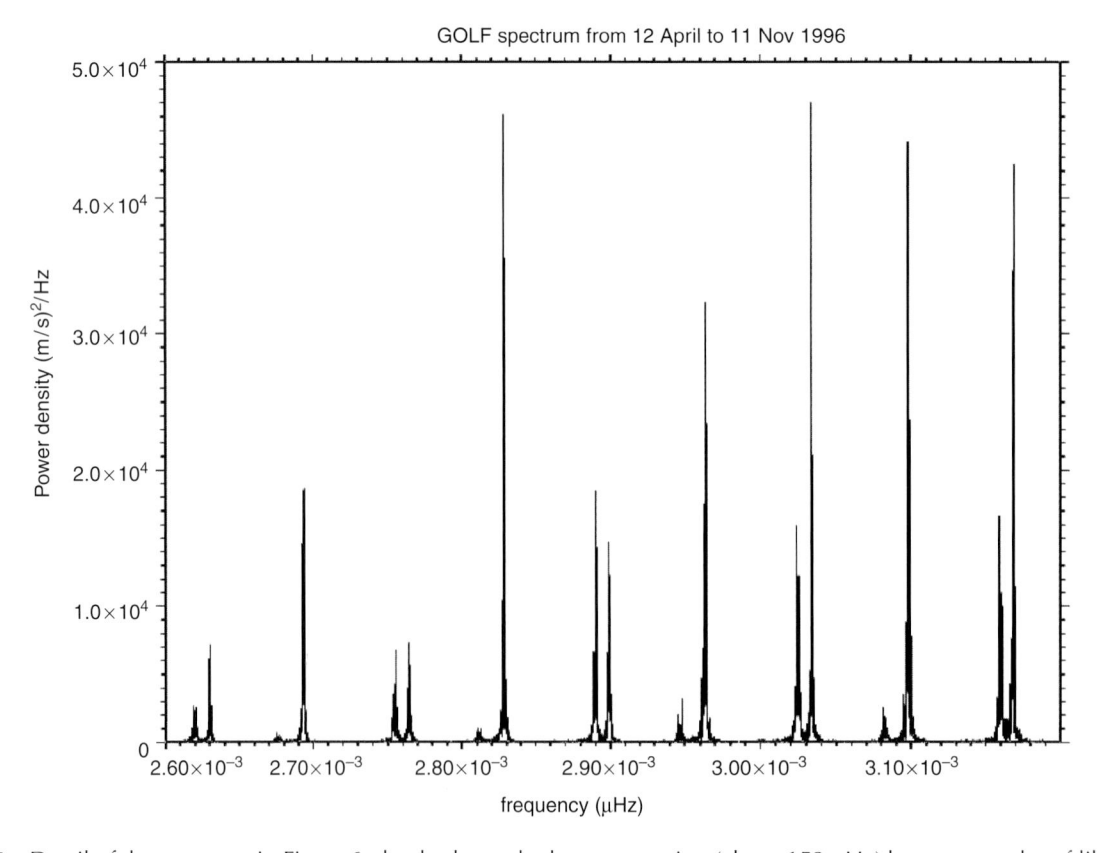

Figure 7 Detail of the spectrum in Figure 6 clearly shows the large separation (about 153 μHz) between modes of like degree *l* and small separations between peaks in a group (about 10 μHz between peaks associated with *l* = 0 and *l* = 2, and about 17 μHz between peaks with *l* = 1 and *l* = 3). The structure of the spectrum was clear in 1997 from these GOLF observations, but was difficult to discern in the late 1970s and early 1980s. (From Gabriel *et al.* 1997.)

contracting the solar image to a line along the rotation axis by passing the light through a cylindrical lens. At the higher degrees, the orders of the modes can be identified; in particular, the frequencies ω of observable modes are bounded from below by those of the f modes whose frequencies satisfy $\omega = \sqrt{(gk)}$ in the limit $k \to \infty$, irrespective of the structure of the Sun. Here g is the gravitational acceleration. By counting upwards one can identify n at high degree and then follow modes of like n down to low l. The outcome was to select the model calibration with the higher value of Y, which confirmed the original calibration against Deubner's high-degree data.

A second advance was made by Tom Duvall (1982), who discovered that the frequencies satisfied the law $(n + \alpha)/\omega = f(\omega/L)$ of a fixed waveguide, where $L^2 = l(l+1)$, L/r is the horizontal wavenumber k, and f is some function. Actually f is also a function of the penetration depth d, which is not fixed, but since $d = R - r_t$, where R is the radius of the Sun and r_t is the lower turning point, where the waves travel horizontally and therefore satisfy $\omega r_t/L = c(r_t)$, it too is a function of ω/L and the form of the law is preserved. Indeed, the functional form of f was invertible, and led immediately to the first determination of $c(r)$ to a depth of 80 Mm (Gough 1985a). Soon afterwards Duvall and Harvey obtained a more extensive data set by making measurements at the South Pole, and it was then possible to determine the sound speed almost to the center of the Sun (Christensen-Dalsgaard et al. 1985). Needless to say, the result favored the high-Y solar models, confirming convincingly that the resolution of the neutrino problem did not lie in adjusting the structure of simple spherically symmetrical solar models.

After measuring zonal modes, Duvall and Harvey (1984) had the idea of rotating their cylindrical lens through $90°$, in order to concentrate on sectoral modes. Although lines of latitude project onto the sky as straight lines, so that the cylindrical collapse to the rotation axis is fine, lines of longitude do not. But the greatest curvature is near the East and West limbs, where the signal is relatively weak. Therefore anyone like we, who considered the use of a cylindrical lens to measure zonal modes a clever idea to filter a class of modes according to their true structure, regarded the rotation of the lens, which does not provide a precise filter for sectoral modes, as absolutely brilliant. It permitted the measurement of the rotational splitting with respect to azimuthal order m, from which it was possible to infer the radial variation of the interior rotation of the Sun (Duvall et al. 1984).

The inferences from Duvall and Harvey's observations were stunning, for they provided detailed information about the structure and rotation of the solar interior, and resolved immediately some important issues. First, Ω does not increase dramatically with depth: in fact Ω (which was actually an average in the equatorial regions) was almost uniform, but with evidence that the region around the core actually rotates more slowly than does the photosphere at the equator, a feature which had been anticipated by no-one; there was also evidence for a slight rise in Ω with depth immediately beneath the photosphere in the outer layers of the convection zone (confirming early results from Deubner (1979) and Rhodes et al. 1983), and possibly a rapidly rotating core – but not sufficiently rapid to sustain the idea of a substantial Brans–Dicke effect (although by then the Brans–Dicke theory had already been rejected from planetary radar-ranging measurements).

An important inference that was made from the sound-speed determination was that in a localized region immediately beneath the convection zone the temperature of the Sun appeared to be greater than that of the models, and that that could be accounted for most plausibly if the opacity tables supplied by Los Alamos were too low by 10–20% in that region. The helioseismologists persuaded the Livermore researchers to do some spot checks, which confirmed the helioseismological inference; then a comparison of the Livermore calculations with those of Los Alamos revealed an error in the Los Alamos treatment of spin–orbit coupling in radiative transitions. Oblateness aside, this was the first, admittedly indirect, contribution of helioseismology to basic physics. In consequence, more extensive calculations were carried out to produce new tables of opacity (Iglesias and Rogers, 1987).

These dramatic conclusions, from data obtained using instruments that had originally been designed to make measurements of other kinds, transformed solar physics; yet they also made it abundantly clear that with a well-designed observing program improved data would provide an amazing degree of precision that would transform solar astrophysics yet further. What was required were long, continuous observations to obtain clean spectra with minimal sidelobes in order that precise eigenfrequencies could be obtained. From that requirement grew GONG and SOHO (GOLF, VIRGO, and SOI/MDI).

Even before the Duvall–Harvey observations, there were plans being laid in the USA for an imaged helioseismic space mission (Wellesley 1978). The observational requirements outlined for this proposed mission were, in rough terms, similar to the eventual requirements for SOHO (Newkirk 1980).

The DISCO mission (Bonnet et al. 1981) was proposed in Europe and would have contained instruments similar in capability to the SOHO GOLF and SOHO VIRGO instruments to observe solar irradiance and low-degree oscillations from the Sun–Earth L_1 vantage point. It would also have provided solar-wind observations. Owing to competition from other missions, DISCO was not selected for development, but the concepts formed part of the basis for the later SOHO mission.

In 1986 ESA and NASA agreed to co-sponsor the SOHO (SOlar and Heliospheric Observatory) mission. SOHO carries

12 experiments to study the Sun and heliosphere (Domingo et al. 1995). As finally selected, SOHO contained three instruments for studying the solar interior: GOLF – Global Oscillations at Low Frequency (Gabriel et al. 1995), VIRGO – Variability of solar IRradiance and Gravity Oscillations (Fröhlich et al. 1995), and SOI/MDI – Solar Oscillations Investigation with the Michelson Doppler Imager (Scherrer et al. 1995). The payload selection was complete in 1987, and launch was on 2 December 1995.

Why space for visible-light observations and instrumentation

Helioseismic observations are made in the part of the solar spectrum that is easily observable from the Earth. One might ask why spend the money and put up with the delays and inconveniences of observing from space if the observations could be made from the ground. It turns out that one is forced, by a combination of several factors contributed by the atmosphere, to observe from above the atmosphere in order to discover some important parts of the story.

In 1983 NASA sponsored a year-long study of the requirements for observations that could exploit the potential of helioseismology. The study culminated in a symposium (SNOWMASS 1984), and a report (Noyes and Rhodes, 1984) which outlined the conclusion that yes one can study part of the problem from the ground but yes one must go to space to exploit fully the opportunity to study the solar interior. This dual conclusion led to the parallel development of the ground-based international GONG project (see special issue of *Science*, **272**, 1281 ff, 1996) and the inclusion of helioseismic instrumentation on the SOHO spacecraft. At Snowmass we were still in the "interpreting causes of discrepancies between theory and observation" stage, and we were planning – or was it assessing the feasibility of? – a space mission. (It was both.)

The primary limitations to ground-based observations have different impacts on the attainment of different goals. The low-degree nearly radial oscillations that probe the energy-generating core of the Sun require continuous minute-by-minute observations for years at a time in order to measure the oscillation frequencies with enough accuracy to determine rotation and the thermal structure in the core. Observations from the ground suffer severely from differential transparency over the solar disk as the Sun moves across the sky and as winds move even imperceptible clouds across the Sun. These problems prevent total irradiance observations and limit oscillation observations to frequencies above about 1.5 mHz. Lower-frequency modes can be observed, but not with the required accuracy in a time less than the time over which the frequencies change with the solar cycle. In the presumed g-mode part of the spectrum, the noise from the ground is such that a 10-year ground observation still has more noise than a 2-year space-based observation. Nevertheless, if observations of many decades will be required to detect g modes, this may well be accomplished on the ground before continuous decade-long observations will be made from space. The VIRGO and GOLF instruments on SOHO are designed to observe irradiance and brightness oscillations and Sun-as-a-star oscillations, but only for as long as SOHO is operational. There are several continuing ground-based low-degree observing programs, including BiSON (e.g. Chaplin et al. 1996) and IRIS (e.g. Fossat 1995), which will last longer.

For slightly-higher-degree observations where a solar image, and from it a map of surface velocity, can be obtained, the transparency problems can be overcome. That is so up to the point where atmospheric seeing distorts the image by too much. This is the middle ground, containing what are referred to as "medium-l" modes, which can be effectively observed from the ground. These modes are the primary target of the GONG project and LOWL (Tomczyk et al. 1995). They require multi-year continuous observations with at least a one-minute cadence. This can be achieved from the ground only by combining observations made at a network of observatories judiciously sited around the world. The GONG project grew from the 1983 study and, with funding from the US National Science Foundation and support from Australia, India, Spain, and Chile, has been in continuous operation since October 1995.

At the high-wavenumber end of the spatial spectrum, solar acoustic waves are confined to the upper parts of the convection zone where the processes that generate solar activity are likely to be found. They require observations of high resolution over large portions of the visible disk for many days at a time. We do not know of any way to obtain the required image stability without atmospheric seeing distortions except by observing from above the atmosphere. The MDI instrument on SOHO was designed to observe velocity oscillations at all spatial frequencies (down to scales of a few Mm where small-scale granulation dominates) but with particular attention to the short-wavelength (high-l) range which cannot be observed accurately from the ground. In the range where MDI and GONG overlap, cross-calibration has proven useful to both programs and has led to improved confidence in the conclusions about rotation in the convection zone. MDI observes only in one spectral line, sampling only one height in the solar atmosphere, so there is continuing need for supplementary observations from the ground. Also, the SOHO spacecraft is expensive to operate and may not be available beyond its 6-year goal. There are plans for future missions to take over where SOHO leaves off. Improvements in detector technology have partly closed the gap between GONG and MDI with the deployment of GONG+ (new higher-resolution cameras in the GONG instruments) expected to become operational in the year 2001.

Details of the various types of instruments can be found in Hill et al. (1991) and in SOHO (1995).

WAITING FOR SOHO AND GONG – 1984 THROUGH 1995

The decade leading up to the commencement of GONG and the launch of SOHO was a period of consolidation and diversification. Libbrecht undertook an intense observing program at the Big Bear Solar Observatory. Together with his collaborators he produced new frequency information from $l = 4$ up to $l = 1300$ (Libbrecht and Kaufmann 1988; Duvall et al. 1988): only the modes with $l < 100$ were measured individually; the others were obtained from ridge fitting. Moreover, the whole-disk observations of the BiSON and IRIS networks provided frequencies of modes with $0 \le l \le 3$, and modes with l between 5 and 6 were measured by Scherrer and his collaborators at Stanford. Phobos data (Fröhlich et al. 1991) were used too. Coupling these data sets permitted improved inversions for the sound speed in the solar interior, which now extended into the energy-generating core. The results were more-or-less in accord with theoretical expectation, the model sound speed differing from that in the Sun by typically no more than about 0.5%. However, the magnitude of this discrepancy was many standard deviations of the estimated data errors; the time was ripe for taking into account physical processes in the Sun that had previously been thought to be minor.

Nuclear reaction cross-sections were continually being revised by those interested in neutrino production, and we have already mentioned the correction by Iglesias and Rogers to the opacity. However, a process that had not been incorporated into the models was gravitational settling of heavy elements. This had been discussed informally for many years, but no-one had previously judged the outcome to be great enough to warrant putting its investigation high on their priority list. The first estimates of helium settling were made by Bahcall and Loeb (1990), to assess the effects on neutrino production rates. Then Christensen-Dalsgaard et al. (1993) made a careful study of the influence on the sound speed. They found that the settling of helium moved the model sound speed to within a few tenths of a per cent of that of the Sun, although we had to wait for GONG and SOI/MDI to be sure that that was the case.

Prior to this work, an analysis of the second derivative of the sound speed inferred seismologically from data of Duvall et al. (1988) led to the first accurate determination of the depth of the convection zone: $d_c = 0.287R$ (Christensen-Dalsgaard et al. 1991). The analysis was repeated by Basu and Antia (1997), using more recent data from Libbrecht et al. (1990), yielding the same result. This global property of the Sun, which is just a single number, has provided a convenient first (and sometimes only) criterion against which to calibrate theoretical solar models.

Of great interest was the structure of the energy-generating core. This could be determined only by combining the low-degree whole-disk data with the rest, which possibly led to systematic errors arising from combining data from different instruments. Because the central sound speed is extremely sensitive to the relation between the low-l frequencies and the others, any systematic error could be disastrous. Even the sign of the difference between the central sound speed of the Sun and the reference models was disputed, and indeed it is still in doubt today (although the magnitude is now much smaller than it was).

Brown (1985), then Harvey (1988), Woodard and Libbrecht (1988), Morrow (1988), and Rhodes et al. (1988) measured degeneracy splitting in detail, from which it first became evident that the latitudinal variation of the differential rotation observed at the surface appears to extend throughout the convection zone, and that the interior appears to rotate uniformly, with a transition layer which is too thin to resolve (Brown et al. 1989).

This period also saw the beginning of new techniques for analyzing helioseismic data, designed particularly for diagnosing conditions locally. All of them regard the seismic disturbances as propagating waves rather than normal modes. Braun et al. (1987) compared a measure of wave flux into and out of a circle surrounding a sunspot, and discovered that within a given range of horizontal wavenumber the inward flux exceeds the outward flux; this indicated that waves incident on a sunspot are either absorbed or are converted into other waves, such as more deeply penetrating acoustic waves with smaller horizontal wavenumber or magnetoacoustic waves which might propagate either upwards into the atmosphere or downwards to regions of high density where they are probably dissipated. The result stimulated theoretical work on wave absorption and scattering. The hope is that in the fullness of time the processes responsible for the phenomenon will be understood well enough for diagnosing sunspot structure. That objective is still beyond our sight.

F. Hill and his collaborators have developed a procedure for measuring disturbances locally. The disturbances are considered as waves propagating in a waveguide bounded above and below by the boundaries of the acoustic cavity. By Fourier-analyzing the disturbances in a small region of the photosphere a local relation between (horizontal) wavenumber k and frequency ω can be obtained: a two-dimensional power spectrum with respect to k at fixed ω is greatest in a series of almost circular rings centered near the origin, each one corresponding to a set of waves of given (radial) order. The displacement of the center of each ring measures an average of the horizontal velocity that advects the waves in their appropriate acoustic cavity; in principle, the averages can be deconvolved (inverted) to yield horizontal velocity as a function of depth. The deviation of the shapes of the rings from pure circles measures local horizontal anisotropy of propagation, which may result from horizontal variations of the (scalar) wave speed or the vertical component of the velocity of convection motion, or from a magnetic field. Perhaps the most important result to arise from the early analysis was the discovery of a shear in the Sun's angular velocity immediately beneath the photosphere. The discovery was later confirmed by SOHO MDI, and is not understood.

More adventurous were the attempts by Duvall and by Lindsey and Braun to identify where and when an acoustic disturbance propagating downwards from the surface reappears later at a different location. These methods have become very powerful when applied to the data from GONG and MDI which came later; we shall therefore return to discuss them in a subsequent section.

We should not leave this summary of the decade without pointing out that an international team of scientists was established prior to GONG and SOHO to develop methods of analyzing the imminent superior seismic data. Since an essential ingredient of the art of inverting data is judgement, based on experience, several "hare-and-hounds" exercises were carried out to gain the expertise and experience, in which the hare constructed artificial data which the hounds then inverted. Carrying out the numerical work was relatively easy, once the computer codes had been written; what was difficult was deciding how to interpret the results. The interpretations were presented at meetings and then the "true" answers were revealed. The exercises were not only extremely instructive, but also the element of competitiveness amongst the hounds led to more work on developing inversion techniques and writing software than might otherwise have occurred, and consequently caused the subject to advance faster. Such hare-and-hounds exercises have now become commonplace within small research groups, although their utility on an organized international scale has probably come to an end.

It was in this decade that the first useful seismic space observations were made. The first was from the Active Cavity Radiometer built by R.C. Willson for the SMM spacecraft. Although designed to measure longer-term solar irradiance variations, Woodard and Hudson (1983) discovered that the short exposures were precise enough to measure oscillations in the Sun's radiative intensity. The observations provided frequencies of the modes of lowest degree: $l = 0, 1, 2$. Because the SMM spacecraft was eclipsed by the Earth for a substantial part of its orbit the power spectrum of the oscillations suffered sidelobes as do single-site ground-based observations. Nevertheless, the data appeared to be superior to the ground-based data of the time, and after some years of observation provided convincing evidence that the seismic frequencies change with time, presumably in association with the solar cycle, confirming an earlier report by BiSON (Elsworth et al. 1990). Continuous intensity data were subsequently obtained from a high-precision photometer called IPHIR (InterPlanetary Helioseismology by Irradiance measurements, Fröhlich et al. 1991), attached to the side of the ill-fated Phobos spacecraft and taking measurements during the flight to Mars. Inversions of those data, coupled with intermediate-degree frequencies from BBSO, indicated that the sound speed in the core of the Sun is somewhat smoother than that in the standard theoretical reference model, as had been indicated previously from inversions of ground-based data. Was this evidence that some degree of material redistribution in the core had taken place?

CONTRIBUTIONS OF GONG AND SOHO – 1996 ONWARDS

SOHO was launched in December 1995, and observations began early in 1996. Finally we were able to begin to exploit prospects for helioseismology without the atmosphere in the way. Also 1995 saw the deployment of the ground-based GONG project providing nearly continuous coverage of modes which sample most of the solar interior. There are to date more than 500 papers describing the results from analysis of these projects. The papers mentioned here are merely representative of the studies presently under way.

One of the first results from the new measurements was an improved measure of the global (spherically averaged) thermal and hydrostatic structure of the Sun. That structure can be determined by examining the multiplet average frequency of the normal modes, and is presented in Table 1. This means that waves traveling in all directions are averaged such that only the radial information is retained. There remained a discrepancy between the observations of sound speed and the best models, most notably just beneath the base of the convection zone and also near the core (Figure 8). The higher sound speed in the Sun just beneath the base of the convection zone is likely to be due to a deficit of helium at this depth, decreasing the mean molecular weight and increasing the sound speed (Kosovichev et al. 1996).

In addition to the globally averaged quantities are those properties obtainable by examining the small frequency splittings of the $2l + 1$ modes of different m for each degree l and order n. These are the frequencies of waves traveling in different directions in the Sun with the same magnitude of horizontal wavenumber and approximately the same frequency. Analysis of the differences in frequency for waves traveling in different directions offers the possibility of studying aspherical contributions to the structure by measuring the even splittings, and the angular velocity with the odd splittings. The techniques for analyzing the frequencies are described in some detail by, for example, Gough and Thompson (1991).

Table 1 Internal structure of the Sun. The variable τ is the propagation time at the adiabatic sound speed c from the center of the star to radius r, measured in units of 3523 s, the propagation time to the photosphere $r = R$; m is the mass enclosed within the sphere of radius r, M is the total mass of the Sun, and X is the abundance by mass of hydrogen; ρ is density, T is temperature and p is pressure, all in cgs units. The frequencies ν_L, ν_c and $N/2\pi$ are respectively the reduced Lamb frequency, the acoustic cutoff frequency and the cyclic buoyancy (Brunt-Väisälä) frequency, all in μHz. Acoustic modes of degree l with frequency ν penetrate to the depth where $\nu / (l + \frac{1}{2}) = \nu_L$, and are reflected near the surface where $\nu = \nu_c$; N is the characteristic g-mode frequency. The information in the first 8 columns was determined seismologically, using frequencies obtained from a 360-day data set from MDI. The quantities in the last three columns were obtained by perturbing the standard solar model of Christensen-Dalsgaard et al. (1996) to make it consistent with the seismic data; they are not purely seismic, and are therefore less reliable than the other quantities

τ	r/R	m/M	ρ	c	ν_L	ν_c	$N/(2\pi)$	X	T	p
0.000	0.000	0.000	1.53E+02	5.042E+02	∞	0.00E+00	0.00E+00	3.38E-01	1.57E+07	2.33E+17
0.010	0.026	0.002	1.45E+02	5.065E+02	4.53E+03	1.95E+02	2.24E+02	3.61E-01	1.54E+07	2.23E+17
0.020	0.051	0.013	1.27E+02	5.098E+02	2.27E+03	3.56E+02	3.76E+02	4.15E-01	1.48E+07	1.97E+17
0.030	0.077	0.040	1.05E+02	5.108E+02	1.51E+03	4.74E+02	4.47E+02	4.83E-01	1.40E+07	1.64E+17
0.040	0.103	0.082	8.49E+01	5.068E+02	1.13E+03	5.54E+02	4.60E+02	5.50E-01	1.30E+07	1.31E+17
0.050	0.128	0.137	6.81E+01	4.977E+02	8.87E+02	6.07E+02	4.50E+02	6.05E-01	1.19E+07	1.01E+17
0.060	0.153	0.201	5.44E+01	4.848E+02	7.24E+02	6.42E+02	4.35E+02	6.44E-01	1.10E+07	7.67E+16
0.070	0.177	0.269	4.34E+01	4.697E+02	6.06E+02	6.62E+02	4.25E+02	6.71E-01	1.01E+07	5.75E+16
0.080	0.201	0.338	3.46E+01	4.541E+02	5.17E+02	6.72E+02	4.25E+02	6.87E-01	9.34E+06	4.28E+16
0.090	0.223	0.405	2.75E+01	4.387E+02	4.49E+02	6.73E+02	4.27E+02	6.97E-01	8.66E+06	3.18E+16
0.100	0.245	0.468	2.19E+01	4.240E+02	3.95E+02	6.67E+02	4.28E+02	7.03E-01	8.06E+06	2.36E+16
0.125	0.297	0.601	1.25E+01	3.918E+02	3.02E+02	6.34E+02	4.26E+02	7.11E-01	6.86E+06	1.15E+16
0.150	0.345	0.702	7.30E+00	3.655E+02	2.43E+02	5.87E+02	4.06E+02	7.14E-01	5.96E+06	5.85E+15
0.175	0.389	0.775	4.43E+00	3.438E+02	2.02E+02	5.40E+02	3.79E+02	7.16E-01	5.27E+06	3.14E+15
0.200	0.432	0.828	2.79E+00	3.254E+02	1.72E+02	4.96E+02	3.51E+02	7.17E-01	4.72E+06	1.77E+15
0.225	0.472	0.867	1.82E+00	3.095E+02	1.50E+02	4.57E+02	3.25E+02	7.19E-01	4.27E+06	1.05E+15
0.250	0.510	0.897	1.23E+00	2.955E+02	1.32E+02	4.23E+02	3.00E+02	7.20E-01	3.89E+06	6.43E+14
0.275	0.547	0.919	8.52E-01	2.829E+02	1.18E+02	3.95E+02	2.77E+02	7.20E-01	3.56E+06	4.09E+14
0.300	0.582	0.936	6.07E-01	2.713E+02	1.07E+02	3.70E+02	2.55E+02	7.21E-01	3.27E+06	2.68E+14
0.325	0.615	0.949	4.43E-01	2.604E+02	9.67E+01	3.50E+02	2.34E+02	7.21E-01	3.01E+06	1.80E+14
0.350	0.648	0.959	3.30E-01	2.497E+02	8.82E+01	3.33E+02	2.11E+02	7.23E-01	2.77E+06	1.24E+14
0.375	0.679	0.968	2.52E-01	2.388E+02	8.05E+01	3.20E+02	1.76E+02	7.31E-01	2.52E+06	8.61E+13
0.400	0.708	0.974	1.98E-01	2.263E+02	7.31E+01	3.12E+02	1.01E+02	7.37E-01	2.25E+06	6.08E+13
0.425	0.736	0.980	1.61E-01	2.114E+02	6.57E+01	3.11E+02	-1.55E-01	7.37E-01	1.96E+06	4.30E+13
0.450	0.762	0.984	1.31E-01	1.973E+02	5.92E+01	3.13E+02	-2.08E-01	7.37E-01	1.71E+06	3.05E+13
0.475	0.786	0.988	1.06E-01	1.841E+02	5.36E+01	3.16E+02	-2.54E-01	7.37E-01	1.49E+06	2.15E+13
0.500	0.808	0.991	8.57E-02	1.716E+02	4.85E+01	3.21E+02	-3.03E-01	7.37E-01	1.30E+06	1.51E+13
0.525	0.829	0.993	6.91E-02	1.597E+02	4.40E+01	3.29E+02	-3.60E-01	7.37E-01	1.13E+06	1.06E+13
0.550	0.849	0.995	5.53E-02	1.484E+02	4.00E+01	3.39E+02	-4.29E-01	7.37E-01	9.73E+05	7.30E+12
0.575	0.867	0.996	4.40E-02	1.375E+02	3.63E+01	3.51E+02	-5.16E-01	7.37E-01	8.37E+05	4.99E+12
0.600	0.883	0.997	3.47E-02	1.270E+02	3.29E+01	3.66E+02	-6.26E-01	7.37E-01	7.15E+05	3.35E+12
0.625	0.899	0.998	2.70E-02	1.169E+02	2.97E+01	3.85E+02	-7.71E-01	7.37E-01	6.07E+05	2.21E+12
0.650	0.913	0.999	2.08E-02	1.071E+02	2.68E+01	4.07E+02	-9.63E-01	7.37E-01	5.11E+05	1.43E+12
0.675	0.926	0.999	1.57E-02	9.761E+01	2.41E+01	4.35E+02	-1.23E+00	7.37E-01	4.25E+05	8.97E+11
0.700	0.938	0.999	1.17E-02	8.833E+01	2.15E+01	4.68E+02	-1.60E+00	7.37E-01	3.50E+05	5.45E+11
0.725	0.948	1.000	8.42E-03	7.924E+01	1.91E+01	5.09E+02	-2.14E+00	7.37E-01	2.84E+05	3.17E+11
0.750	0.958	1.000	5.89E-03	7.033E+01	1.68E+01	5.60E+02	-2.96E+00	7.37E-01	2.26E+05	1.76E+11
0.775	0.966	1.000	3.96E-03	6.154E+01	1.46E+01	6.23E+02	-4.23E+00	7.37E-01	1.77E+05	9.14E+10
0.800	0.973	1.000	2.53E-03	5.291E+01	1.24E+01	6.99E+02	-6.27E+00	7.37E-01	1.36E+05	4.40E+10
0.825	0.980	1.000	1.50E-03	4.498E+01	1.05E+01	7.97E+02	-9.82E+00	7.37E-01	1.02E+05	1.93E+10
0.850	0.985	1.000	8.21E-04	3.793E+01	8.81E+00	9.48E+02	-1.70E+01	7.37E-01	7.34E+04	7.38E+09
0.875	0.989	1.000	3.96E-04	3.004E+01	6.95E+00	1.14E+03	-3.14E+01	7.37E-01	5.00E+04	2.33E+09
0.900	0.992	1.000	1.55E-04	2.309E+01	5.32E+00	1.34E+03	-5.96E+01	7.37E-01	3.48E+04	5.93E+08
0.925	0.995	1.000	4.69E-05	1.813E+01	4.17E+00	1.58E+03	-1.22E+02	7.37E-01	2.51E+04	1.19E+08
0.950	0.997	1.000	1.05E-05	1.444E+01	3.31E+00	1.87E+03	-2.74E+02	7.37E-01	1.86E+04	1.77E+07
0.975	0.999	1.000	1.65E-06	1.144E+01	2.62E+00	2.28E+03	-7.39E+02	7.37E-01	1.38E+04	1.80E+06
1.000	1.000	1.000	2.00E-07	7.893E+00	1.80E+00	4.52E+03	2.86E+03	7.37E-01	5.78E+03	7.61E+04

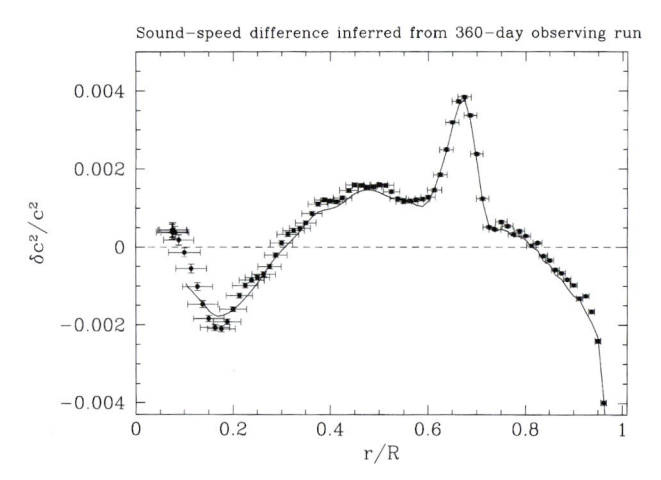

Figure 8 Relative differences between the square of the sound speed in the Sun, determined by seismology, and the square of the sound speed in a standard solar model of Christensen-Dalsgaard et al. (1996). The seismic measures are actually averages over localized regions whose characteristic widths (roughly the full width at half maximum of a smoothly varying localized averaging kernel) are indicated by the horizontal bars, with standard errors represented by the vertical bars. The continuous curve is a smoothed function that fits the seismic data by least squares. (From Kosovichev et al. 1997.)

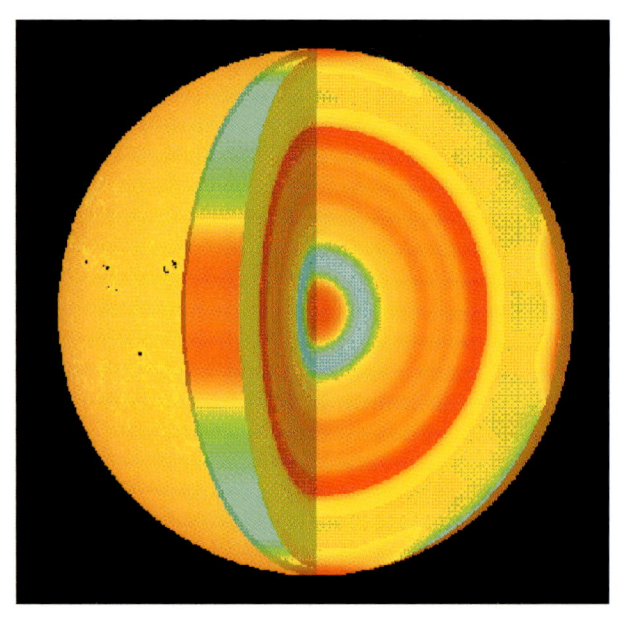

Figure 9 Color represenation of the relative deviation of sound speed throughtout the Sun from that of the (spherically symmetric) standard model. Significant latitudinally varying deviations can be seen only near the surface. The magnitude of the deviations can be gauged by comparing this with Figure 8. (Data from SOHO MDI and VIRGO, courtesy A.G. Kosovichev.)

The zonally averaged sound speed is found to vary with latitude, at least near the surface, being slightly lower at mid-latitudes, at least at solar minimum, as is shown in Figure 9 (Kosovichev et al. 1997). Detailed analysis of the contributions to this variation from each of the components of the even splittings and from the surface magnetic field confirms the earlier observation by Woodard and Libbrecht (1993) that the spatial organization of the thermal structure is the same as that of the magnetic field when viewed appropriately (Goode et al. 1998, Antia et al. 2001). It is now understood that these variations are the source of the correlation between the variation in global-mode frequencies and the solar activity levels which was discovered with the ACRIM observations on SMM in the 1980s. Basu and Antia (2000) have examined the bump in sound speed at the base of the convection zone, and found no evidence of variation with phase in the cycle. Thus the thermal cycle variations appear to be confined to a region near the top of the convection zone.

The ability to observe the Sun without an intervening atmosphere allowed the best direct determination to date of the oblateness of the Sun's surface (Kuhn et al. 1998). The conclusion is that there is no excess oblateness beyond what is consistent with the helioseismically determined rotation.

Other early results included a precise measurement of the seismic radius (Schou et al. 1997), the recognition of the effect of relativistic terms in the equation of state (Elliott and Kosovichev 1998), and an understanding of asymmetries of lines in the p-mode spectrum which implies that the source of oscillations is 50 to 100 km beneath the photosphere (e.g. Nigam et al. 1997, 1998; Kumar and Basu 1999). Kuhn et al. (1998) found evidence for variations in the shape of the Sun. It is not yet understood if these are related to the sound-speed-magnetic field correlation or with more fundamental aspects of convection.

As described above, analysis of the odd splittings from observations obtained prior to SOHO and GONG showed that the surface differential rotation extended down to about the base of the convection zone, with the suggestion of more uniform rotation beneath. The more detailed measurements from SOHO MDI confirm the earlier results and show that the hint of fast rotation in the upper layers is real and extends to high latitudes (Figure 10; Schou et al. 1998).

There is a local maximum near $r = 0.95R$ (at a depth of 35 Mm) at all latitudes, the rise with increasing depth beneath the surface being monotonic at low latitudes; at higher latitudes the rotation rate first declines with depth immediately beneath the photosphere before rising to its maximum. This can be seen more easily in Figure 11 (Basu and Antia 2001). Comparison with surface magnetic features which also rotate faster than the photosphere suggest that they are in fact rooted at a depth of 35 Mm.

Figure 11 shows a slow rotation near the poles. The slow polar rotation has been known for years, and early helio-seismological analysis of BBSO observations suggested

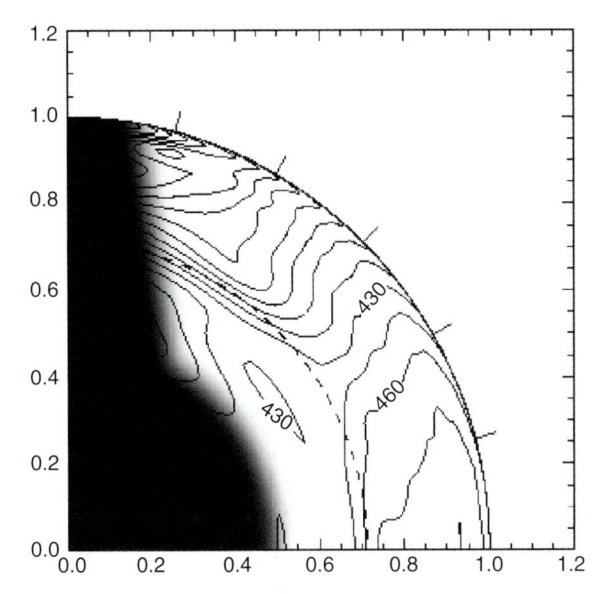

Figure 10 A depiction of the rotation rate (in nHz) on a meridional cut through the Sun, inferred by inverting the rotational splitting of normal modes of seismic oscillation. The normal-mode method tells us only about the north–south symmetrical component, so only a quarter section needs to be shown. With currently available data the inversion cannot yield well-defined results near the rotation axis or deeper than about half way to the center, where the figure has been blacked out. The nearly radial contours in the convection zone, above the dashed circle, indicate that the variation of the angular velocity that is seen directly at the surface does not change substantially through most of the convection zone. (From Schou et al. 1998.)

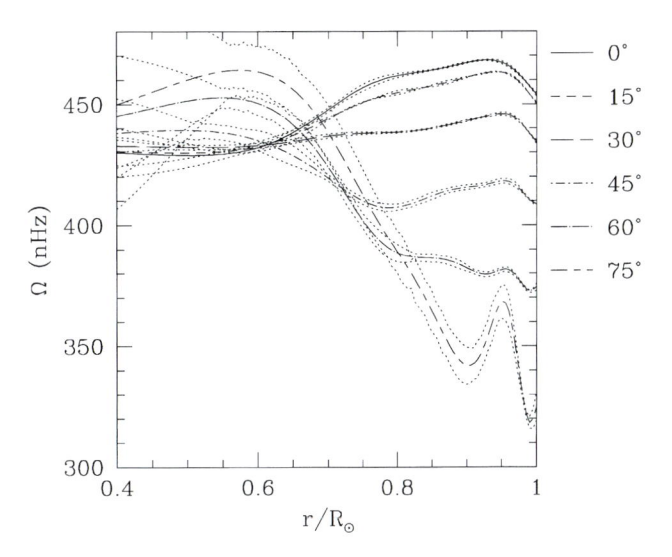

Figure 11 Rotation rate in nHz from MDI observations, plotted against radius for several latitudes; each curve is flanked by two others, indicating ± one standard error in the inference. The greater rotation speed at $0.95R$ is seen at all latitudes, and at high latitudes there is also a local minimum beneath. Below about $0.7R$ all latitudes appear to rotate at near 430 nHz, which corresponds to a period of 26.9 days sidereal (or 29.1 days as seen from the Earth). (From Basu and Antia 2001, Figure 1b.)

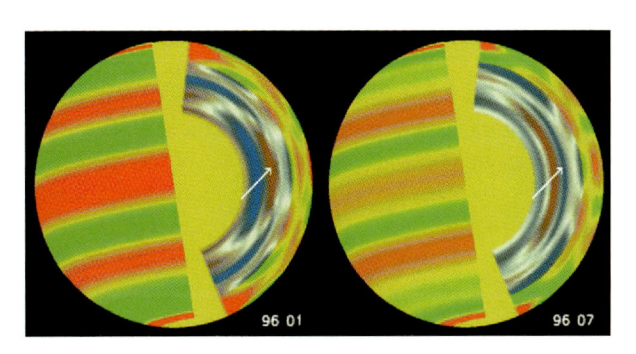

Figure 12 The cutaways illustrate the rotational shear in January 1996 and in the following July, somewhat less than half a period of the 1.3-year oscillation later. In the inner region, blue and red indicate faster and slower rotation, whereas in the upper half of the convection zone red represents faster and green slower. The amplitude of the oscillation in angular velocity near the base of the convection zone is about 10% of the mean jump in angular velocity across the tachocline. The equatorward migration of the pattern of zonal flow in the surface layers is evident from a comparison if the two panels. (After Howe et al. 2000a.)

that the polar rate may vary with the cycle. Birch and Kosovichev (1998) have found some evidence for change with MDI data, but a clear determination must wait a few more years until we pass the present maximum of the solar cycle. The rotation of the very polar regions is particularly interesting. It is common practice to expand the angular velocity in the form $\Omega = \Sigma_l \, \Omega_{2l} \cos^{2l}\Theta$, where Θ is colatitude. If a fit of the seismic inferences to the expansion is made only in the latitude range equatorward of 75°, the MDI data imply that at least for $r \geq 0.95R$ the coefficients ω_{2l} are extremely small for $l > 2$; poleward of latitude 75° the angular velocity appears to drop abruptly (by about 20 nHz) relative to the smooth expansion. Unfortunately, inversions have not been possible poleward of 80°; this inferred feature of the angular velocity profile is therefore only on the verge of detectability, and may not be real.

Figures 10 and 11 show the strong shear at the base of the convection zone near $0.7R$. This region where the rotation speed changes abruptly has been named the "tachocline." While one might expect the rotation of the Sun to be on the whole very steady, localized flows can be contained in the rotation signal. To examine local flow patterns and possible variations, an approach of subtracting the smoothed large-scale rotation has been developed. This has been used by Howe et al. (2000a) to examine the tachocline and the surrounding regions for stability. They found evidence that the strength of the shear varies with a period near 1.3 years, as illustrated in Figure 12. It is too soon to tell if this is periodic, a random variation, or the result of some elusive measurement error. If real, it will have important implications for theoretical models of the solar cycle that depend on the cycle fields being generated principally by this shear.

The technique of examining variations from a large-scale smoothed rotation profile has been used for some years to study the zonally averaged surface zonal flows. Howard and LaBonte (1980) first discovered alternating fast and slow bands that drift equatorward during the cycle. These were thought to be superficial effects of the presence of active-region fields or the result of random noise in the measurements. One of the early results from MDI demonstrated that these flow patterns are real (Kosovichev and Schou 1997; Schou et al. 1998), and later analysis revealed that they extend at least throughout the upper half of the convection zone (Schou 1999, Howe et al. 2000b). See also Figure 13.

The results above were determined by analysis of the normal-mode frequencies of the Sun. These frequencies provide information about only the north–south symmetric components of sound speed, density, and rotation, although the inversions are probably the most sensitive approaches to determining those components. Notwithstanding that deficiency, we can be reasonably sure on dynamical grounds that the north–south asymmetry is not large, and therefore we can calculate from the seismological inferences the small centrifugal distortion from spherical symmetry of the solar structure. Unlike the rotational splitting, which is only of first order in the small quantity Ω/ω, the distortion is of second order – actually $O(n^2\Omega^2/\omega^2)$ which is very much smaller, despite the factor n^2. This is part of the reason why the quadrupole moment J_2 of the gravitational equipotentials is determined more accurately by the seismologically determined angular velocity Ω/ω than it is by direct measurement. Eigenfrequency inversions do not provide information about north–south differences in the solar structure; nor do they provide detailed information about aspects of the structure that vary in longitude or that are actually local to the vicinity of, for example, a sunspot or active region. In the years before SOHO and GONG, other methods of analysis were developed that specifically examined local properties of the Sun by examining properties of wave propagation in localized regions. These studies can be roughly divided into two approaches.

The first to be developed is actually an extension of the very first analyses of the 1970s where the waves are considered to be nearly horizontally propagating waves bounded in a vertical cavity. In this approach the solar surface is mapped into rectangular patches each of which is examined separately for horizontal wave advection in what is known as "ring diagram analysis" (Hill 1989). As discussed above, this method first led to the firm confirmation of the rotational shear just beneath the surface before normal modes were observed with high enough degree l to enable near-surface information to be resolved. It has now been used to tile the solar surface systematically into 15-degree square patches and examine the flow across those patches as a function of depth in the upper 10–20 Mm. The method is sensitive to flows as small as a few meters per second.

The other "local helioseismological" approach achieves higher spatial resolution by considering, in several of the procedures, waves propagating to or between individual pairs or sets of observed locations (e.g. Duvall et al. 1993b). Variants of these methods have been called acoustic imaging, acoustic holography, heliotomography, time–distance seismology, and so on. The methods differ somewhat in the analysis procedures, in sensitivity to different phenomena, in computational complexity, and in nomenclature. Several of them are being studied in detail; the methods and the means to interpret the results are developing rapidly. Each approach has proven useful in making inferences about the solar interior.

One of the interesting results from local seismological approaches is the mapping of meridional circulation. A poleward flow of between 10 and 20 m s^{-1} was detected from surface velocity measurements in the 1970s and has

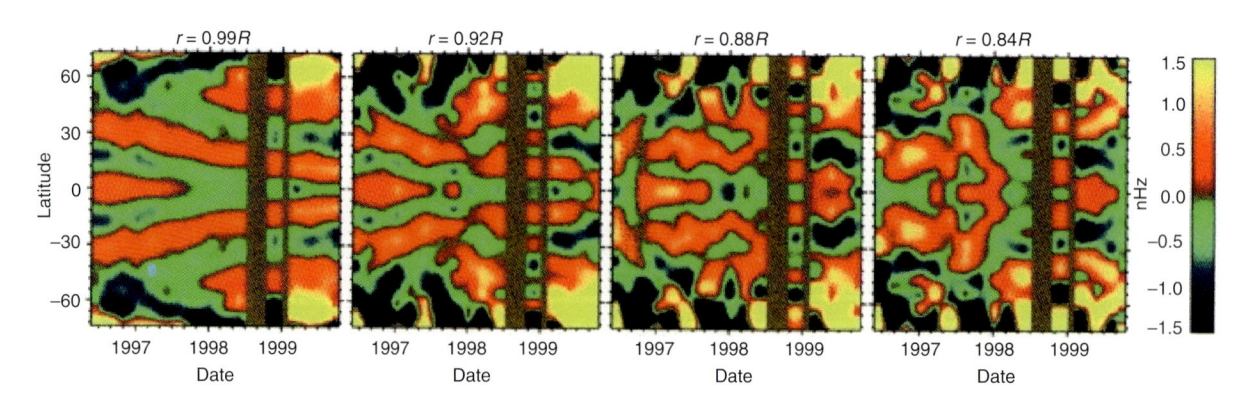

Figure 13 Zonal flow at four depths in the upper half of the convection zone. Red represents regions in which the flow is faster than the average rotation at that latitude, green and blue where it is slower. The alternating bands of faster and slower rotation are seen to migrate equatorward during the cycle. (From Howe et al. 2000b.)

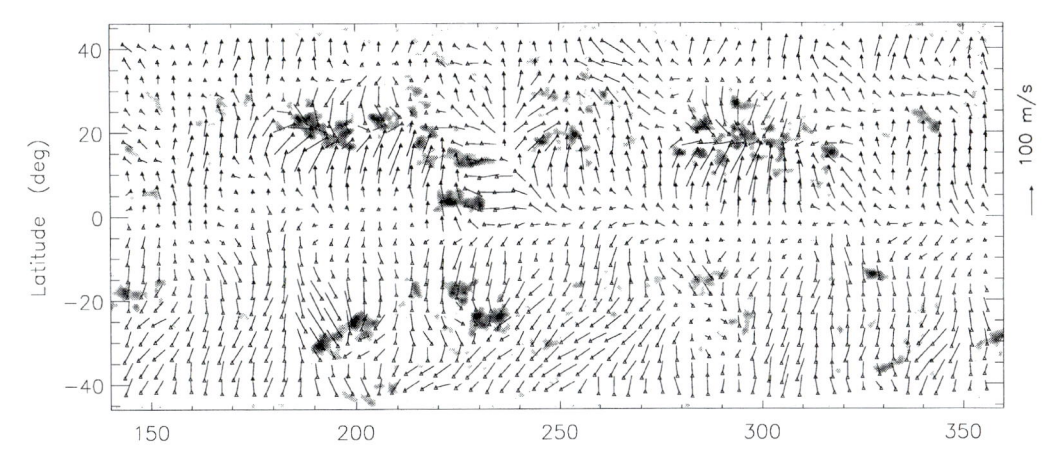

Figure 14 Residual horizontal flows from time–distance analysis of f mode wave advection obtained for part of Carrington rotation 1949 (in 1999) after subtraction of a smoothed rotation background. Arrows are plotted every 3.84°. The map includes about 40% of the surface of the Sun, and shows the poleward flow and the large-scale flow directed toward active regions. Mean travel times shorter than average (dark shade, up to 3% relative change) are associated with magnetic activity. (From Gizon et al. 2001.)

been included in phenomenological models of the solar magnetic cycle for many years. This flow is now observed with local seismological techniques, and is found to extend through most of the convection zone (e.g. Giles et al. 1998, González Hernández et al. 1998). It is presumed, but not yet observed, that there is an equatorward return flow lower in the convection zone or in the tachocline.

The poleward flow appears to vary during the cycle. It also appears to vary with longitude, as was suggested by direct ground-based velocity observations from Mt Wilson. Near the beginning of the MDI and GONG observations, at an epoch coincident with solar minimum, the flow extended to high latitudes – at least to 60°, which is about the present observational limit. There are indications that in 1999 the flow in the north terminated at the active-region band. Observations of this flow can presently be made by MDI only during the 2–3 months per year of continuous telemetry, so the persistence and evolution of these large-scale structures are uncertain. Figure 14 shows the flows in the top 2 Mm derived from the advection of f-mode waves.

As can be seen in Figure 14, not only very-large-scale ordered flows but also local flows toward active regions can be monitored by these methods. There is not yet sufficient coverage in time to see if these localized flows are always directly associated with the active regions, and are always directed toward the active-region centers.

Acoustic holography and time–distance methods can be used also to probe local variations in wave speed. It is found that there is an increase in sound-wave travel speed between about 3 and 15 Mm beneath sunspots, and more generally there is an increase beneath all surface regions in which there are concentrations of magnetic field. The large, well-isolated sunspot in the MDI high-resolution field of

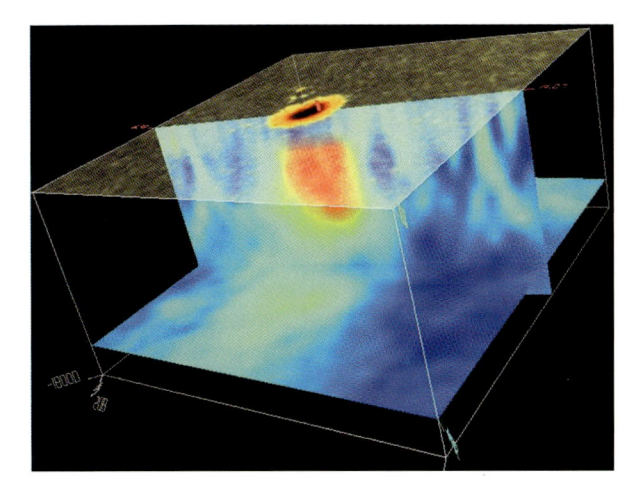

Figure 15 Acoustic wave speed beneath a sunspot. The analysis region is from the surface down to a depth of 20 Mm, with a horizontal extent of 38°, or about 460 Mm on a side. The three surfaces shown are white-light intensity at the Sun's surface, and wave speed on vertical and horizontal cuts, shown in color: red is fast, relative to the average in the magnetically quiet regions, and blue is below the average. There is a region of slight excess speed seen in the horizontal cut just southeast of the spot. A new spot emerged at that location a few days later, but it is too soon to know whether that is likely to have been fortuitous or whether it was the first detection of field emergence.

view in June 1998 has been particularly well studied. Figure 15 shows the wave-speed variations beneath this spot. The reddish regions are where the wave speed is higher than average, while the blue region just beneath the spot is a region of lower speed. The lower-speed region is almost certainly due to cooler material associated with the

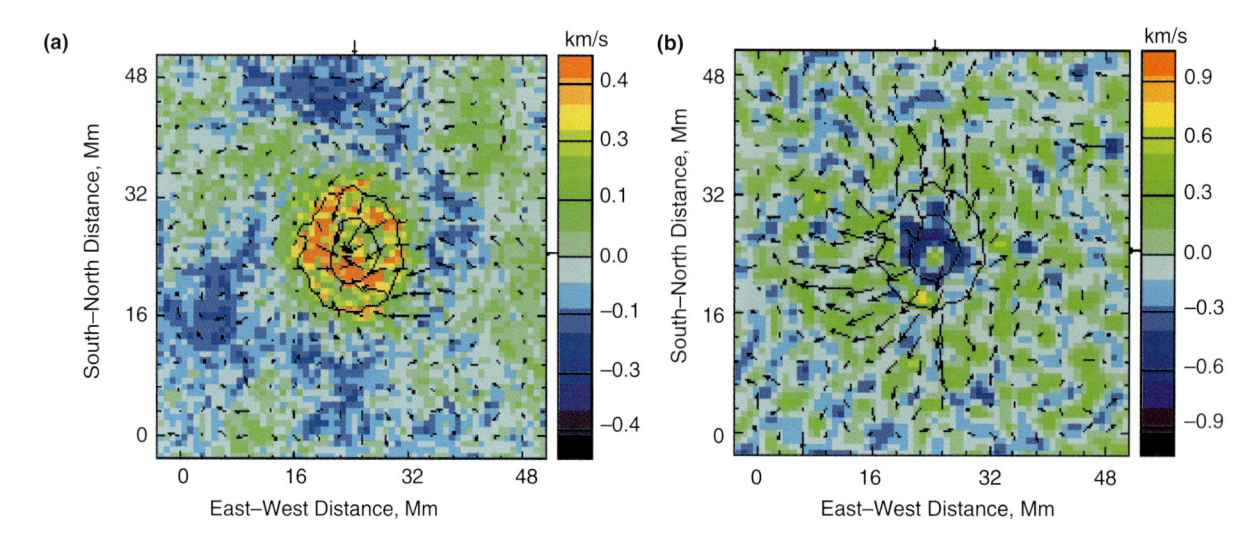

Figure 16 Motion beneath a spot. These figures illustrate how material flows in the depth range (a) 0–3 Mm and (b) at 6–9 Mm. The data are from the MDI high-resolution field of 19 June 1998. The outlines of the sunspot umbra and penumbra are shown. The colors represent vertical motion, positive values (red) corresponding to downflow and negative (blue) to upflow. The arrows denote the horizontal motion; the longest arrows correspond to 1.0 km s^{-1} in panel (a) and to 1.6 km s^{-1} in (b).

spot. The higher-speed regions appear to be due to some, yet to be determined, combination of magnetic field concentration and higher temperature (Kosovichev et al. 1997, Lindsey and Braun 1999).

Local flow beneath a sunspot can also be inferred by time–distance methods. Figure 16 shows vertical motion (colors) and horizontal motion (arrows) on two cuts beneath the sunspot shown in Figure 15. Analysis of this and other spots suggests a well-defined flow pattern, with converging flow toward the spot between the surface and about 6 Mm depth and with a downflow extending to about 6 Mm directly beneath the spot. At greater depths the flow diverges outside the spot, and there is upflow beneath the spot. There is also a region of upflow in the top 6 Mm in an extensive region around the spot (Zhao et al. 2001). The whole circulation pattern is similar to one hypothesized by Parker (1979), and has been reproduced in numerical models by Hurlburt and Rucklidge (2000).

Lindsey and Braun (2000) have demonstrated that phase-sensitive acoustic imaging techniques (essentially the same as the time–distance method) can be used to image magnetic regions on the far side of the Sun. Far-side flux estimates can be made on a daily basis using MDI medium-*l* continuous Dopplergram data. These data, available each day in near real time, can be used to detect changes in activity a week or more before the active region rotates onto the Earth-facing hemisphere of the Sun. The images here (Figure 17) show the whole Sun, 27 and 13.5 days before NOAA region number 9393 was at disk center. It is clear that the NOAA region, which produced a strong geoeffective CME and the largest flare in more than a decade (the aurora borealis was seen in Mexico), was evident in the seismic images for more than a week before it rotated into the visible disk. There was little hint of it when that longitude was Earth-facing in the prior rotation. Data like these are bound to become useful in predicting activity.

In addition to imaging the far side of the solar surface, these and related methods have been used to analyze MDI data to examine acoustic wave absorption by sunspots, excess acoustic emission at high frequencies (above 5 mHz) around active regions (e.g. Lindsey and Braun 1999, Donea et al. 2000), and acoustic "moats" around sunspots (Braun et al. 1998).

Acoustic waves emanating from the first strong flare after the launch of SOHO were studied by Kosovichev and Zharkova (1998) – see Figure 18. Their technique has yet to provide diagnostic information that can be used to obtain the detailed structure in the vicinity of active regions, but it may in the future if there are enough large flares with accompanying good data coverage.

A combination of direct Doppler velocity observations and seismic probing can also be used to study near-surface features such as supergranulation (Hathaway et al. 2000, Beck and Schou 2000, Beck and Duvall 2001). These studies show that supergranular cells rotate at different rates according to their size, the larger cells rotating faster than the smaller ones. It is interesting that supergranules rotate faster than any other solar phenomenon, including the peak in material rotation 5% into the Sun. These studies reveal that we really do not understand the true nature of supergranules.

Analysis of velocity residuals indicates the existence of so-called giant cells, which may simply be the large-sized,

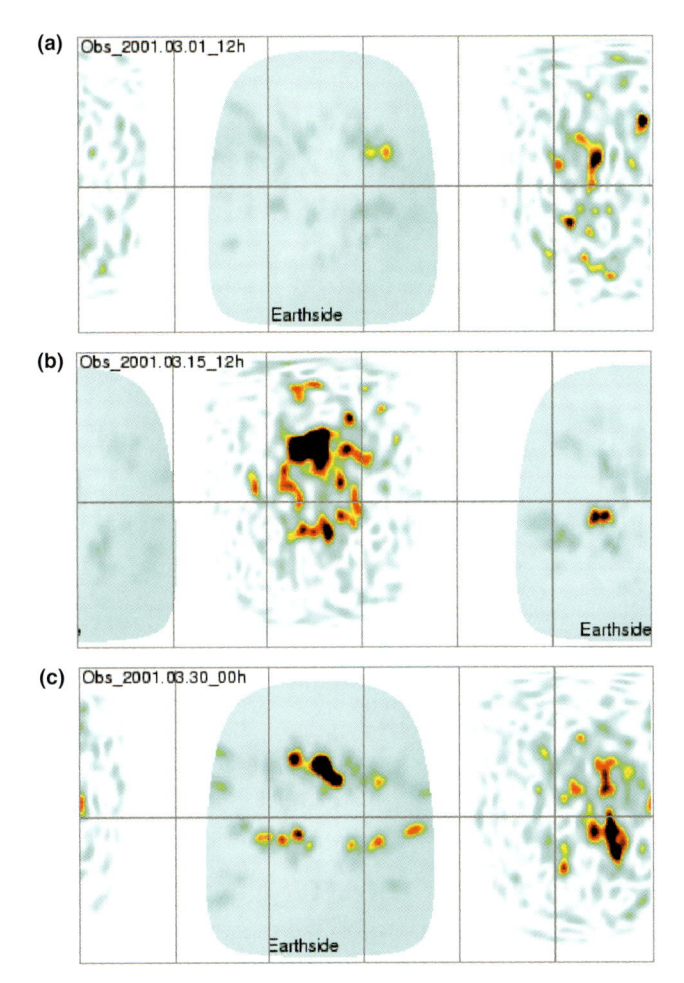

Figure 17 Carrington maps of magnetic flux (Earthside) and the flux inferred by acoustic imaging on the far side, on (a) 1 March, (b) 15 March, (c) 30 March 2001. Active Region 9393 is visible in the direct magnetic image in (c) near the center of the Earthside view on 30 March; it is centered at Carrington longitude 154, latitude 17N. The region was first detected on the far side by acoustic imaging on 12 March; it is prominent in the far-side image of 15 March above, in which it can be seen to be situated also near Carrington longitude 154, latitude 17N. There is no sign of the region in the direct flux measurements of 1 March.

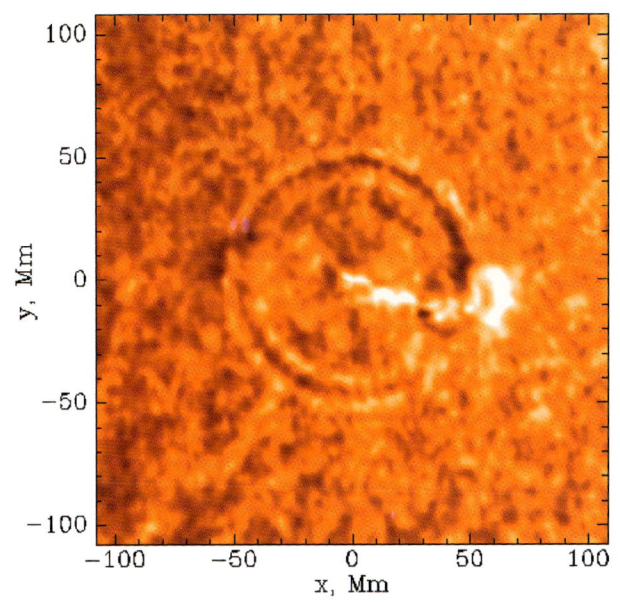

Figure 18 Seismic response of the solar atmosphere to the impulse of a solar flare in 1997. The ripple in line-of-sight velocity spreading from the flare site has been enhanced by a factor of 5. (From Kosovichev and Zharkova 1998.)

long-lived tail of the supergranule distribution (Beck et al. 1998). Kuhn et al. (2000) interpreted similarly sized variations implied from limb observations as Rossby waves.

One of the key goals for SOHO GOLF and SOHO VIRGO and a goal for SOHO MDI is the detection of the g modes. These are oscillations for which buoyancy provides the restoring force. They are essentially confined beneath the convection zone, hardly penetrating to the surface, which is why they are difficult to detect. They are computed to be the most sensitive possible measure of conditions in the energy-generating core. To date, while there have been published claims of g mode detection, there is no repeatable detection of these important signals from the interior. The upper limits to the amplitudes of g modes established with SOHO observations are well below the early ground-based claims, yet still well above theoretical estimates (Gough 1985b; Kumar 1997). The search for g modes must continue (Appourchaux et al. 2000).

The need to measure accurately the total irradiance has been met by the VIRGO experiment on SOHO. Figure 19 shows irradiance data combined from several missions. It shows a clear variation with the cycle. Models of irradiance based on spot and plage areas and empirical scaling laws fail to explain completely the observed variability, there remaining a few-per-cent unexplained residual. In particular, the "cycle" in irradiance started before the magnetic activity increased in 1996. To explain the residual few per cent is a challenge for the future.

FUTURE MISSION CONCEPTS

At the beginning of the new century a number of key questions about the solar interior remain. There are mission concepts in various stages of development designed to address some of these questions.

One of the outstanding questions is the existence and detectability of the g modes, which could inform us about the deepest part of the solar interior – the region where the

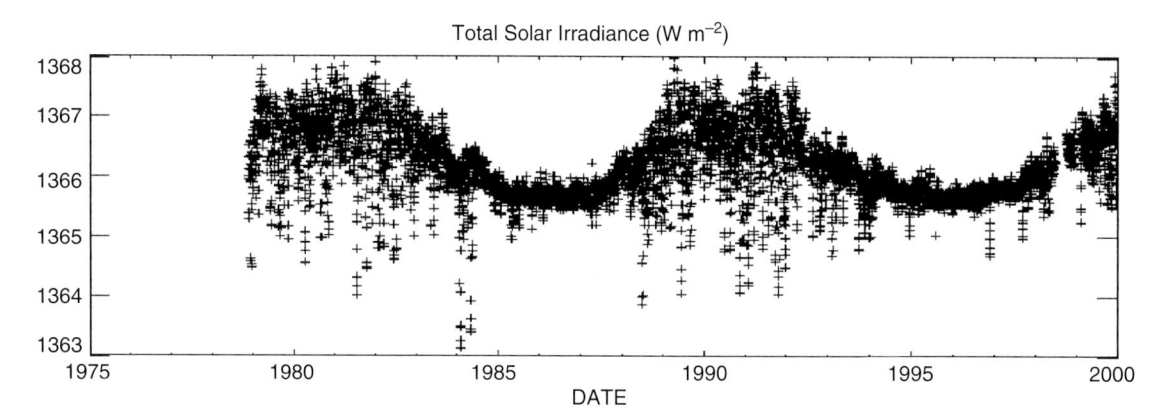

Figure 19 Data from measurements of total solar irradiance acquired from various spacecraft after making adjustments to normalize the absolute levels of the different data sets in such a way as to produce a coherent time sequence. (Courtesy C. Fröhlich and the VIRGO team, updated from Quinn and Fröhlich (1999).)

energy is generated. It may be that appropriate data will come from missions whose prime objective is not seismological. For example, there are currently several mission concepts under study, such as ASTROD and LISA, for the detection of gravitational radiation from cosmological sources. The detections would be made by determining extremely precisely the distance between several spacecraft separated by up to many million kilometers. Solar g modes would be a local source of noise for these experiments. Such noise, which must be removed for studying gravitational waves, would provide the data for the study of solar gravity waves.

Following the lead from SOHO, there are several mission concepts designed explicitly for studying the solar interior by means of helioseismology. One of these, the Solar Dynamics Observatory (SDO), is a key part of the new NASA "Living with a Star" program, which has the goal of learning about those aspects of the Sun and the solar–terrestrial system that can affect our home planet and human technological systems. For the first time a "systems approach" will be used to examine nearly all aspects of the Sun–Earth system, which can have economic effects on society. The SDO project will be tasked with learning enough about the solar cycle and the evolution and development of solar activity to pave a way to predicting solar activity, if at all possible.

In a more exploratory mission, ESA has selected the Solar Orbiter (SO) mission for launch between 2009 and 2013. It will probe the inner Solar System at high latitudes, and study the Sun and the inner heliosphere. Helioseismic instruments on SO could provide the first good look at the polar flows and magnetic field in the region that is the source of much of the solar wind.

NASA is studying missions that would examine the Sun continuously from far vantage points. If these missions include helioseismic instruments, then names such as Sentinel, Safari, and Farside will become part of the solar-interior vocabulary.

The continued operation of ground-based instruments such as those in the upgraded GONG, LOWL, and BiSON projects is also crucial for obtaining the continuous coverage which is so necessary for learning about the deep interior of the Sun and its variation over the solar cycle.

MAJOR ISSUES AT THE BEGINNING OF THE TWENTY-FIRST CENTURY

We have moved a very long way indeed in the last quarter of a century, from merely trying to calibrate simple standard models of the Sun in order to satisfy the few observational constraints that were available at the time to asking subtle questions about physics and the internal dynamics of a star. The original measure of the solar neutrino flux was the first observable quantity to be even half-way predicted by the early simple models, and with that the theory appeared to fail immediately once neutrino flux data had been obtained. Now it seems not unlikely that neutrino flavor is not conserved between the creation of the neutrinos in the solar core and their detection on Earth, so the apparent failure may not be real. Moreover, with the wealth of new seismic data that are now available, we can put very tight constraints on the spherically averaged structure of the Sun. It is now possible to reproduce with the standard models the observed sound speed to within a few tenths of a per cent, and the density to within about 1%. The simple standard model appears to fare very well indeed.

The recent observations at SuperKamiokande that at least μ neutrinos oscillate into other flavors (Kajita 1999), probably τ (Fukuda *et al.* 2000), renders it likely that electron neutrinos do too. Indeed, e-neutrino transitions are consistent with the recent neutrino counts announced by the Sudbury Neutrino Observatory (Ahmad et al. 2001). The SuperKamiokande measurements indicate that the e-neutrino oscillation wavelength exceeds the diameter of the Earth. Therefore the Sun will be the prime neutrino source for future e-neutrino physics, and it behoves helioseismologists to constrain the structure of its energy-generating core as tightly as possible.

It is widely believed that even knowledge of the frequencies of only a few g modes would substantially enhance the power of seismological diagnosis of the structure of the core. To find and measure g modes has been the principal goal of one of the investigations (GOLF) on SOHO, and one of the principal goals of another (VIRGO). Unfortunately, a robust identification of even a single g mode has not been achieved. The failure is not yet one of theory, because current observational upper bounds to the amplitudes exceed theoretical expectations by a substantial margin. Yet in principle the instruments should be sensitive enough to detect modes at least below the greater of the theoretical amplitudes – they were designed to do so. The impediment is what, for the purposes of g-mode detection, is the solar noise: direct contributions to the signal from convective motion. In straightforward Fourier power spectra of data of a few years' duration, the amplitude of the Doppler noise level in the frequency range of the gravest g modes is a few millimeters per second, which is to be compared with expected g-mode amplitudes of some tenths of millimeters per second or less. Evidently it is impractical to try simply to identify single peaks in the power spectrum – additional information must be brought to bear, such as expected theoretical constraints on the distribution of the g-mode frequencies in the seismic spectrum. It is also important to improve the techniques for determining the frequencies of the deeply penetrating low-degree p modes that have been detected, for we must recognize that g modes may not be detected with the current suite of instruments.

What are the issues that we wish to use g modes to address? The most obvious is to determine with greater accuracy the variation of the spherically averaged sound speed and density through the core, and, with the help of theoretical constraints on the thermal structure, determine the temperature, which is related to sound speed and density through the chemical composition. Then we need to ascertain whether the core is spherically symmetric and motionless, aside from rotation, as is presumed in the construction of standard models, or whether the products of the nuclear reactions are advected into regions of different temperature and density, thereby throwing the nuclear reactions out of balance: that would change the predictions of neutrino production rates. Addressing this issue will not be easy, for any such motion is likely not to produce a direct measurable seismic signature. It is likely that we shall need to bring indirect evidence into play, perhaps using dynamical arguments in association with information about the variation of angular velocity with both radius and latitude.

Further refinements in our knowledge of the hydrostatic stratification of the Sun will enable us to extend our use of the Sun as a laboratory for investigating physics. We have already mentioned the role played by the first sound-speed inversions in uncovering an error in the opacity calculations. But more powerful is diagnosis of the lower part of the convection zone, where the stratification is almost isentropic. The isentropic condition, coupled with the knowledge that the zone is very close to being in hydrostatic support and that thermodynamic fluctuations induced by the convective motion are small, enables us to set constraints on certain thermodynamic relations implied by the equation of state that are not dependent on the other assumptions of solar structure theory. The goal is to use those constraints to test theories of screening and compound–particle interactions in dense plasmas. A first step has been taken already, with the calibration of a simple theory to demonstrate that the mean apparent sizes of neutral hydrogen and neutral and singly ionized helium are accessible to helioseismology (Baturin et al. 2000). The future outcome will be not only useful to plasma physics, but also pertinent to calculating nuclear reaction rates, including those producing neutrinos.

An important issue to be resolved is why the sound speed in the radiative interior differs from that in standard theoretical models by so much. To be sure, theoretical models are much closer to reality than what a decade or so ago would hardly have been contemplated possible, but they deviate from the observational inferences by many standard errors (Figure 8). Is that real, or are the inferences biased by correlated observational (or data-reduction) error? Adjustments of the physical assumptions upon which the models have been built can bring those models more-or-less into line with current inference (e.g. Turck-Chièze et al. 2001), but the manner by which that can be accomplished is not unique. We require not merely to produce a model that is not in disaccord with available observations, but to determine which of the many, if indeed any, represents the truth. That will require additional, nonseismic, information, coupled, no doubt, with more sophisticated theory.

We have obtained an enormous amount of information about the Sun's rotation from seismology in the last years of the previous century, and also about other forms of motion in the convection zone. This has led to the rejection of previously held theories of solar spin-down and the role of convection in angular momentum transport. As a result, we now have a rather better understanding of the internal dynamics of the Sun. But there remain many unanswered questions; and yet more questions have been raised by the fine details of the new seismological inferences. Why is there a generally poleward meridional flow in the surface layers, at least at low latitudes? Why, in an epoch near the current solar maximum, does the flow seem to terminate at the active-region latitudes in the north? Is that primarily a consequence of the activity, or is the activity channelled by the flow? And does it really vary with the solar cycle? Why is there such a termination of the flow only in the northern hemisphere? Perhaps it has actually occurred in the southern hemisphere too, but started later and has not been observed because there has yet been no further long interval of continuous telemetry from SOHO. Is the temporal north–south asymmetry in the flow random, or does it alternate with the solar cycle? What causes the banded structure in the angular velocity of the upper layers of the convection zone? Is the dynamical nature of that structure similar to that of Jupiter? And why, unlike in Jupiter, does the structure migrate equatorward at low latitudes, against the meridional flow? Answering these, and many other questions that come immediately to mind, will no doubt form an essential step for increasing our understanding of the more prominent large-scale symptoms of the Sun's global dynamics: the pole–equator variation of angular velocity in the convection zone and the solar cycle itself.

There have been substantial advances in numerical simulations of the convection zone, but current numerical methods and computer resources are inadequate to cope with the full problem. A highlight of the simulations has been the successful modeling of the jump in entropy across the upper superadiabatic boundary layer of the convection zone without recourse to an adjustable parameter (Stein and Nordlund 2000); its correctness was established by matching the boundary layer to an adiabatically stratified interior, and thereby computing the depth of the convection zone, whose value we know from seismology. However, whereas it was possible to address this issue by simulating only the outer layers of the convection zone, to study the distribution of angular velocity requires taking the entire zone into account because large-scale currents redistribute angular momentum globally. This poses a much more difficult problem. It may be that a global model will first be achieved by studying local processes, as Brummell et al. (1998) and Tobias et al. (2001) have done, and then piecing them together into a whole. Out of such work might also emerge explanations to other intriguing properties of the observed angular velocity Ω, such as the fact that in the outer layers of the convection zone the latitudinal variation is well represented by only a three-term expansion in $\cos^2\Theta$ equatorward of latitude $75°$, and that near latitude $75°$ there appears to be an abrupt drop in Ω towards the poles. It is interesting that the anomalously slow polar rotation occupies the region threaded by open magnetic field lines, from which the slow component of the solar wind emanates. One is tempted to conjecture that this coincidence is not fortuitous, and that the same process controls the rotation, the field, and the wind.

Understanding the solar tachocline, the shear layer at the base of the convection zone, is of great current interest, particularly because it is not unlikely that the tachocline dynamics is intimately related to the solar cycle. Gough and McIntyre (1998) have argued that the transition from the differentially rotating convection zone to an almost uniformly rotating radiative zone beneath necessarily requires the presence of a large-scale magnetic field pervading the radiative interior to hold it rigid. The existence of the tachocline shear then implies the existence of a meridional circulation homogenizing the tachocline material with the convection zone, which is now deficient in helium as a result of gravitational settling. That provides a natural explanation of the positive sound-speed anomaly which is evident in Figure 8 immediately beneath the convection zone. However, the story may not be simple: the flow could vary with time, as Gilman and Dikpati (2000) have conjectured, which would have important repercussions on our ideas of magnetic field augmentation and transport in the tachocline, and hence on the mechanism of the solar cycle. Is the magnetic field sustained by a genuine dynamo, as most solar physicists believe? Or is the field in the outer layers of the Sun produced by magnetohydrodynamic stretching of a fossil field that is emerging from the radiative interior, while the whole system is running down on the global ohmic diffusion time?

The dynamical discussions of the tachocline raise several issues, some of which are to be goals of future helioseismology. What is its shape? Evidence is that the base of the convection zone and the surface of greatest rotational shear in the tachocline are both prolate (Gough and Kosovichev 1995; Charbonneau et al. 1999), with an eccentricity of about 2%, but this value has not been determined reliably because the structure of the tachocline has not been resolved. It is also important to determine whether the thickness of the tachocline varies with latitude, and whether it varies also with time. If the tachocline were much thicker near the poles, it could provide the flow to cycle convection-zone material to depths sufficient for lithium to be destroyed by nuclear reactions, and so explain the low abundance of lithium observed in the photosphere. It has been conjectured that such tachocline pits exist at the poles, which, unfortunately, would render them very difficult to detect from SOHO. Supplementary seismology from a future spacecraft flying over the Sun's poles, such as the Solar Orbiter, might answer the question. The detection of temporal variations of the tachocline would have a direct bearing on our ideas of the maintenance of the solar cycle.

The 1.3-year oscillation of the angular velocity immediately above and below the tachocline close to the equator (Howe et al. 2000a) was discovered in a search for a solar-cycle variation. Is that oscillation directly associated with the cycle or not? Its characteristic period might be considered to correspond to the fourth harmonic of the cycle, but the coincidence may be accidental. Evidently, much further observational investigation over a solar cycle or longer must be undertaken. This emphasizes the need for long continuous seismic observations if one is to entertain even the hope of understanding solar dynamics.

There remain many other issues related to the global dynamics about which we yet have little idea. For example, what is the reason for the active longitudes? What is the cause of other characteristic timescales of activity, such as the prominent one at 140 days? Can flares and coronal mass ejections be predicted? It is likely that, as with terrestrial weather, the detailed motion in the solar convection zone, which is no doubt responsible for these phenomena, cannot be predicted over a very long timescale, although with accurate observations useful medium-term predictions (over a very few rotation periods) may eventually become possible. But even now we are able to detect sunspot precursor signatures by acoustic imaging several days before the sunspots emerge at the surface. Refining the techniques will be important for giving reliable advance warning of inclement space weather and potential magnetic storms on Earth.

Although the study of the excitation of the seismic modes is still in its infancy (e.g. Goldreich *et al.* 1994), it is likely to provide a valuable diagnostic of solar convective flow. Attempts are being made to relate cycle variation of oscillation amplitudes and acoustic spectral line widths with the properties of granulation, so far with a modicum of success (Houdek *et al.* 2001). Asymmetry of spectral lines provides information about the level of excitation (Duvall *et al.* 1993a), although it appears to be contaminated by a correlated direct signal from the granulation which is presumed to excite the modes (Nigam *et al.* 1998). Acoustic generation in a stratified environment such as the solar convection zone, in which the waves return to the region of generation to interact with the exciting convective flow, is much harder to study than the almost homogeneous acoustically compact sources first considered by Lighthill (1952). Nevertheless, enormous advances have been made by aerodynamic engineers, and it is not unreasonable to anticipate analogous progress to be made by astrophysicists too.

Finally, what can be said of other stars? Recent observations by Martic *et al.* (1999) and Bedding *et al.* (2001) have provided strong evidence that stars roughly similar to the Sun oscillate in a similar fashion, with amplitudes not too dissimilar from those predicted by theory. They give us confidence that there are extremely useful data to be successfully gathered by the imminent asteroseismological space missions COROT (Catala *et al.* 1995), MOST (Matthews 1998), MONS (Kjeldsen and Bedding 1998), and, further in the future, the more ambitious ESA mission Eddington (Favata *et al.* 2000; Penny *et al.* 2001). All of these missions will be capable of measuring solar-like acoustic spectra, the outcome of which will be a profound transformation of stellar astrophysics.

FURTHER READING

Much of the work in the early years of the development of helioseismology has been reported in the proceedings of conferences and symposia. These proceedings and several key collections of papers published as books are listed here as an aid and to show the progression from studies of future missions to the results from those missions. The bold type will serve as citations of these volumes when needed in other references.

WELLESLEY 1979: *Proceedings of the Symposium on the Study of the Solar Cycle from Space*, Wellesley, MA, June 1979, NASA Conference Publication CP-2098, 1979.

TUCSON 1979: Hill, H. and Dziembowski, W. (eds) (1980), *Nonradial and Nonlinear Stellar Pulsation*, Proceedings of a workshop held in Tucson, Arizona, 12–16 March 1979, Lecture Notes in Physics 125, Springer-Verlag, Berlin.

GSFC 1980: S. Sofia (ed), *Variations of the Solar Constant*, Proceedings of a workshop held at Goddard Space Flight Center, 5–7 November 1980, NASA Conference Publication CP-2191, 1981.

SUN AS STAR 1981: Jordan, S. (ed) (1981). *The Sun as a Star*, Monograph Series on Nonthermal Phenomena in Stellar Atmospheres, NASA SP-450, Washington, DC.

CRIMEA 1981: Gough, D.O. (ed) (1983). *Problems of Solar and Stellar Oscillations*, Proceedings of the 66th IAU Colloquium held at the Crimean Astrophysical Observatory, USSR, 1–5 September 1981, Reidel, Dordrecht [also *Solar Physics*, **82**].

CATANIA 1983: Belvedere, G. and Paterno, L. (eds) (1984). *Oscillations as a Probe of the Sun's Interior*, EPS Study Conference held in Catania, Italy, 20–24 June 1983, Memorie della Societa' Astronomica Italiana, **55**, Nos. 1–2.

SNOWMASS 1983: Ulrich, R.K., Harvey, J., Rhodes, E.J., Jr. and Toomre, J. (eds) (1984). *Solar Seismology from Space*, NASA JPL 84–84, Pasadena, CA.

GARMISCH 1985: *Future Missions in Solar, Heliospheric and Space Plasma Physics*, Proceedings of an ESA Workshop, Garmisch-Partenkirchen, Germany, 30 April–3 May 1985 (ESA SP-235), 1985.

CAMBRIDGE 1986: Gough, D.O. (ed) (1986). *Seismology of the Sun and the Distant Stars*, NATO conference, Reidel, Dordrecht.

TENERIFE 1988: Rolfe, E.J. (ed) (1988). *Seismology of the Sun and Sun-Like Stars*, ESA SP-286, Paris.

AARHUS 1988: Christensen-Dalsgaard, J. and Frandsen, S. (eds) (1988). *Advances in Helio- and Asteroseismology*, IAU Symposium 123, Reidel, Dordrecht.

VERSAILLES 1989: Berthomieu, G. and Cribier, M. (eds) (1990). *Inside the Sun*, Proceedings of the 121st Colloquium of the IAU, held at Versailles, France, 22–26 May 1989, Kluwer, Dordrecht.

SANTA BARBARA 1990: Gough, D.O. and Toomre, J. (eds) (1991). *Challenges to Theories of Moderate Mass Stars*, Proceedings of a conference held in Santa Barbara, CA, 19–22 June 1990, Lecture Notes in Physics 388, Springer-Verlag, Berlin.

COX 1991: Cox, A.N., Livingston, W.C. and Matthews, M.S. (eds) (1991). *Solar Interior and Atmosphere*, Conference Proceedings, Tucson, AZ, 15–18 November 1988, University of Arizona Press, Tucson, AZ.

GONG 1992: Brown, T. (ed) (1993). *GONG 1992: Seismic Investigations of the Sun and Stars*, ASP Conference Series Vol. 42, Astronomical Society of the Pacific, San Francisco.

GONG 1994: Ulrich, R.K., Rhodes, E.J., Jr. and Däppen, W. (eds) (1995). *GONG '94: Helio- and Astero-Seismology from the Earth and Space*, ASP Conference Series Vol. 76, Astronomical Society of the Pacific, San Francisco.

ASILOMAR 1995: Hoeksema, J.T., Domingo, V., Fleck, B. and Battrick, B. (eds) (1995). *Fourth SOHO Workshop, Helioseismology*, held at Pacific Grove, CA, 2–6 April 1995, ESA SP-376, 2 vols.

SOHO 1995: Fleck, B., Domingo, V. and Poland, A. (eds) (1995). *The SOHO Mission*, Kluwer, Dordrecht [also *Solar Physics*, **162**(1–2)].

NICE 1996: Provost, J. and Schmider, F.-X. (eds) (1998). *Sounding Solar and Stellar Interiors*, Proceedings of IAU Symposium 181, Nice, France, 30 September–3 October 1996, Kluwer, Dordrecht.

SOHO 1997: Fleck, B. and Svestka, Z. (eds) (1997). *The First results from SOHO*, Kluwer, Dordrecht [also *Solar Physics*, **170**(1) and **175**(2)].

KYOTO 1997: Deubner, F.-L., Christensen-Dalsgaard, J. and Kurtz, D. (eds) (1997). *IAU Symposium 185: New Eyes to See Inside the Sun and Stars*, Proceedings from Symposium held in Kyoto, Japan, 18–22 August 1997, Kluwer, Dordrecht.

BOSTON 1998: *Structure and Dynamics of the Interior of the Sun and Sun-like Stars*, SOHO 6/GONG 98 Workshop, 1–4 June 1998, Boston, USA, ESA SP-418 (2 volumes).

STANFORD 1999: Duvall, T.L., Jr., Harvey, J.W., Kosovichev, A.G. and Svestka, Z. (eds) (2000). *Helioseismic Diagnostics of Solar Convection and Activity*, SOHO 9 Workshop held 12–15 July 1999 at Stanford, CA, Kluwer, Dordrecht [also *Solar Physics*, **192**(1–2) and **193**(1–2)].

MANCHESTER 2000: Brekke, P., Fleck B. and Gurman J.B. (eds) (2001). *Recent Insights into the Physics of the Sun and Heliosphere: Highlights from SOHO and other Space Missions*, Proceedings of IAU Symposium No. 203, Astronomical Society of the Pacific.

TENERIFE 2000: Wilson, A. and Palle, P. (eds) (2001). *Helio- and Astero-seismology at the Dawn of the Millennium*, SOHO 10/GONG 2000 Workshop, 2–6 October 2000, Santa Cruz de Tenerife, Tenerife, Spain, ESA SP-464, The Netherlands.

REFERENCES

Ahmad, Q.R. *et al.* (2001). Measurement of the rate of $v_e + d \rightarrow p + p + e^-$ interactions produced by ^8B solar neutrinos at the Sudbury Neutrino Observatory. *Physical Review Letters*, **87**, 071301.

Antia, H.M., Basu, S., Hill, F., Howe, R., Komm, R.W. and Schou, J. (2001). Studying asphericity in the solar sound speed from MDI and GONG data. In **TENERIFE 2000**, pp. 45–50.

Appourchaux, T., Fröhlich, C., Andersen, B., Berthomieu, G., Chaplin, W., Elsworth, Y., Finsterle, W., Gough, D.O., Hoeksema, J.T., Isaak, G.R., Kosovichev, A.G., Provost, J., Scherrer, P.H., Sekii, T. and Toutain, T. (2000). Observational upper limits to low-degree solar g modes. *Astrophysical Journal*, **538**, 401–414.

Babcock, H.W. (1961). The topology of the Sun's magnetic field and the 22-year cycle. *Astrophysical Journal*, **133**, 572–587.

Bahcall, J.N. (1989). *Neutrino Astrophysics*, Cambridge University Press.

Bahcall, J.N. and Loeb, A. (1990). Element diffusion in stellar interiors. *Astrophysical Journal*, **360**, 267–274.

Basu, S. and Antia, H.M (1997). Seismic measurement of the depth of the solar convection zone. *Monthly Notices of the Royal Astronomical Society*, **287**, 189–198.

Basu, S. and Antia, H.M. (2000). Possible solar cycle variations in the convection zone. In **STANFORD 2000**, pp. 449–458.

Basu, S. and Antia, H.M. (2001). A study of temporal variations of the tachocline. In **TENERIFE 2000**, pp. 297–300.

Baturin, V.A., Däppen, W., Gough, D.O. and Vorontsov, S.V. (2000). Seismology of the solar envelope: Sound-speed gradient in the convection zone and its diagnosis of the equation of state. *Monthly Notices of the Royal Astronomical Society*, **316**, 71–83.

Beck, J.G. and Duvall, T.L., Jr. (2001). Time–distance study of supergranulation. In **TENERIFE 2000**, pp. 577–581.

Beck, J.G., Duvall, T.L., Jr., Scherrer, P.H. and Hoeksema, J.T. (1998). The detection of giant velocity cells on the Sun. In **BOSTON 1998**, pp. 725–729.

Beck, J.G. and Schou, J. (2000). Supergranulation rotation. In **STANFORD 2000**, pp. 529–539.

Bedding, T.R., Butler, R.P., Kjeldsen, H., Baldry, I.K., O'Toole, S.J., Tinney, C.G., Marcy, G.W., Kienzle, F. and Carrier, F. (2001). Evidence for solar-like oscillaions in β Hydri. *Astrophysical Journal*, **549**, L105–L108.

Birch, A.C. and Kosovichev, A.G. (1998). Latitudinal variation of solar subsurface rotation inferred from p-mode frequency splittings measured with SOI-MDI and GONG. *Astrophysical Journal*, **503**, L187.

Bonnet, R., Crommelynck, D., Delaboudiniere, J., Fröhlich, C., Simon, P. and Thuillier, G. (1981). Disco Assessment Study, European Space Agency Report ESA SCI(81), 3 May 1981.

Bonnet, R., Crommelynck, D., Delaboudiniere, J., Fröhlich, C., Simon, P. and Thuillier, G. (1991). Disco Re-Assessment Study European Space Agency Report ESA SCI(81), 6, September 1991.

Braun, D.C., Duvall, T.L. and LaBonte, B.J. (1987). Acoustic absorption by sunspots. *Astrophysical Journal*, **319**, L27–L31.

Braun, D.C., Lindsey, C., Fan, Y., and Fagan, M. (1998). Seismic holography of solar activity. *Astrophysical Journal*, **502**, 968.

Brown, T.M. (1985). Solar rotation as a function of depth and latitude. *Nature*, **317**, 591–594.

Brown, T.M., Christensen-Dalsgaard, J., Dziembowski, W.A., Goode, P.R., Gough, D.O. and Morrow, C.A. (1989). Inferring the Sun's internal angular velocity from observed p-mode frequency splittings. *Astrophysical Journal*, **343**, 526–546.

Brown, T.M. (1998). Observational challenges in asteroseismology. In R. Donahue and J. Bookbinder (eds), *Cool Stars, Stellar Systems and the Sun*, Tenth Cambridge Workshop, ASP Conference Series Vol. 154, Astronomical Society of the Pacific, San Francisco, pp. 289–300.

Brown, T.M. and Christensen-Dalsgaard, J. (1998). Accurate determination of the solar photospheric radius. *Astrophysical Journal Letters*, **500**, L195–L198.

Brummell, N.H., Hurlburt, N.E., Toomre, J. (1998). Turbulent compressible convection with rotation. II. Mean flows and differential rotation. *Astrophysical Journal*, **493**, 955–969.

Catala, C., Mangeney, A., Gautier, D., Auvergne, M., Baglin, A., Goupil, M.J., Michel, E., Zahn, J.P., Magnan, A., Vuillemin, A., Boumier, P., Gabriel, A., Lemaire, P., Turck-Chieze, S., Dzitko, H., Mosser, B. and Bonneau, F. (1995). COROT: A proposal to study stellar convection and internal rotation. In **GONG 1994**, pp. 426–431.

Chaplin, W.J., Elsworth, Y., Howe, R., Isaak, G.R., McLeod, C.P., Miller, B.A., van der Raay, H.B., Wheeler, S.J. and New, R. (1996). BiSON performance. *Solar Physics*, **168**, 1–18.

Charbonneau, P., Christensen-Dalsgaard, J., Henning, R., Larsen, R.M., Schou, J., Thompson, M.J. and Tomczyk, S. (1999). Helioseismic constraints on the structure of the solar tachocline. *Astrophysical Journal*, **527**, 445–460.

Christensen-Dalsgaard, J. and Gough, D.O. (1976). Towards a heliological inverse problem. *Nature*, **259**, 89–92.

Christensen-Dalsgaard, J., Däppen, W., Ajukov, S.V., Anderson, E.R., Antia, H.M., Basu, S., Baturin, V.A., Berthomieu, G., Chaboyer, B., Chitre, S.M., Cox, A.N., Demarque, P., Donatowicz, J., Dziembowski, W.A., Gabriel, M., Gough, D.O., Guenther, D.B., Guzik, J.A., Harvey, J.W., Hill, F., Houdek, G., Iglesias, C.A., Kosovichev, A.G., Leibacher, J.W., Morel, P., Proffitt, C.R., Provost, J., Reiter, J., Rhodes, E.J., Jr., Rogers, F.J., Roxburgh, I.W., Thompson, M.J. and Ulrich, R.K. (1996). The current state of solar modeling. *Science*, **272**, 1286–1292.

Christensen-Dalsgaard, J., Duvall, T.L., Jr., Gough, D.O., Harvey, J.W. and Rhodes, E.J., Jr. (1985). Speed of sound in the solar interior. *Nature*, **315**, 378–382.

Christensen-Dalsgaard, J., Gough, D.O. and Thompson, M.J. (1991). The depth of the solar convection zone. *Astrophysical Journal*, **378**, 413–437.

Christensen-Dalsgaard, J., Proffitt, C.R., and Thompson, M.J. (1993). Effects of diffusion on solar models and their oscillation frequencies. *Astrophysical Journal*, **403**, L75.

Claverie, A., Isaak, G.R., McLeod, C.P., van der Raay, H.B. and Roca Cortés, T.R. (1979). Solar structure from global studies of the 5-minute oscillation. *Nature*, **282**, 591–594.

Deubner, F.L. (1975). Observations of low-wavenumber nonradial eigenmodes of the Sun. *Astronomy and Astrophysics*, **44**, 371–375.

Deubner, F.L. (1979). Five-minute oscillations on the Sun. *Reports on Astronomy, Transactions of the IAU*, **17A**, Part 2, 53–56.

Dicke, R.H. and Goldenberg, H.M. (1967). Solar oblateness and general relativity. *Physical Review Letters*, **18**, 313–316.

Dicke, R.H. and Goldenberg. H.M. (1974). The oblateness of the Sun. *Astrophysical Journal Supplement*, **27**, 131–182.

Domingo, V., Fleck, B. and Poland, A. (1995). The SOHO mission: An overview. In **SOHO 1995**, pp. 1–37.

Donea, A.-C., Lindsey, C. and Braun, D.C. (2000). Stochastic seismic emission from acoustic glories and the quiet Sun. In **STANFORD 2000**, pp. 321–333.

Duvall, T.L. (1982). A dispersion law for solar oscillations. *Nature*, **300**, 242–243.

Duvall, T.L., Jr., Dziembowski, W.A., Goode, P.R., Gough, D.O., Harvey, J.W. and Leibacher, J.W. (1984). Internal rotation of the sun. *Nature*, **310**, 22–25.

Duvall, T. L. and Harvey, J.W. (1983). Observations of solar oscillations of low and intermediate degree. *Nature*, **302**, 24–27.

Duvall, T.L. and Harvey, J.W. (1984). Rotational frequency splitting of solar oscillations. *Nature*, **310**, 19–22.

Duvall, T.L., Jr., Harvey, J.W., Libbrecht, K.G., Popp, B.D. and Pomerantz, M.A. (1988). Frequencies of solar p-mode oscillations. *Astrophysical Journal*, **324**, 1158–1171.

Duvall, T.L., Jr., Jefferies, S.M., Harvey, J.W., Osaki, Y. and Pomerantz, M.A. (1993a). Asymmetries of solar oscillation line profiles. *Astrophysical Journal*, **410**, 829–836.

Duvall, T.L., Jefferies, S.M., Harvey, J.W. and Pomerantz, M.A. (1993b). Time–distance helioseismology. *Nature*, **362**, 430–432.

Eddy, J.A. (1976). The Maunder minimum. *Science*, **192**, 1189–1202.

Eddy, J.A. and Boornazian, A.A. (1979). Secular decrease in the solar diameter 1863–1953. *Bulletin of the American Astronomical Society*, **11**, 437.

Elliott, J.R. and Kosovichev, A.G. (1998). Relativistic effects in the solar equation of state. In **BOSTON 1998**, pp. 453–456.

Elsworth, Y., Howe, R., Isaak, G.R., McLeod, C.P. and New, R. (1990). Variation of low-order acoustic solar oscillations over the solar cycle. *Nature*, **345**, 322–324.

Emilio, M., Leister, N.V. and Laclare, F. (1998). Solar diameter latitude dependence. In **BOSTON 1998**, pp. 457–460.

Favata, F., Roxburgh, I.W. and Christensen-Dalsgaard, J. (2000). Eddington: A mission to map stellar evolution through oscillations and to find habitable planets. ESA-SCI (2000) 8, ESTEC, Noordwijk.

Fossat, E. (1995) IRIS status report. In **GONG 1994**, pp. 387–391.

Fossat, E. and Roddier, F. (1971). A sodium experiment for photospheric velocity field observations. *Solar Physics*, **18**, 204–210.

Frazier, E.N. (1968). A spatio-temporal analysis of velocity fields in the solar photosphere. *Zeitschrift für Astrophysik*, **68**, 345–356.

Fröhlich, C. (1981). The variability of the solar output. In **GSFC 1980**. pp. 37–44.

Fröhlich, C. (1984). Wavelength dependence of solar luminosity fluctuations in the five-minute range. In **CATANIA 1983**, pp. 237–243.

Fröhlich, C., Romero, J., Roth, H., Wehrli, C., Andersen, B., Appourchaux, T., Domingo, V., Telljohann, U., Berthomieu, G., Delache, P., Provost, J., Toutain, T., Crommelynck, D., Chevalier, A., Fichot, A., Däppen, W., Gough, D., Hoeksema, T., Jimenez, A., Gomez, M., Herreros J., Roca Cortes, T., Jones, A., Pap, J. and Willson, R. (1995). VIRGO: Experiment for helioseismology and solar irradiance monitoring. In **SOHO 1995**, pp. 101–128.

Fröhlich, C., Toutain, T., and Schrijver, C.J. (1991). Helioseismology with the IPHIR instrument on the USSR Phobos mission. *Advances in Space Research*, **11**, 69–76.

Fukuda, S., Fukuda, Y., Ishitsuka, M., Itow, Y., Kajita, T., Kameda, J., Kaneyuki, K., Kobayashi, K., Koshio, Y., Miura, M., Moriyama, S., Nakahata, M., Nakayama, S., Obayashi, Y. and Okada, A. *et al.* (2000). Tau neutrinos favored over sterile neutrinos in atmospheric muon neutrino oscillations. *Physical Review Letters*, **85**, 3999–4003.

Gabriel, A., Grec, G., Charra, J., Robillot, J-M., Roca-Cortes, T., Turck-Chieze, S., Bocchia, R., Boumier, P., Cantin, M., Cespedes, E., Cougrand, B., Cretolle, J., Dame, L., Decaudin, M., Delache, P., Denis, N., Duc, R., Dzitko, H., Fossat, E., Fourmond, H.-J., Garcia, R., Gough, D., Crivel, C., Herroros, J., Lagardere, H., Moalic, J.-P., Palle, P., Petrou, N., Sanchez, M., Ulrich, R. and Van der Raay, H. (1995). Global oscillations at low frequency from the SOHO mission (GOLF). In **SOHO 1995**, pp. 61–99.

Gabriel, A.H., Charra, J., Grec, G., Robillot, J.-M., Cortes, T.R., Turck-Chieze, S., Ulrich, R., Basu, S., Baudin, F., Bertello, L., Boumier, P., Charra, M., Christensen-Dalsgaard, J., Decaudin, M., Dzitko, H., Foglizzo, T., Fossat, E., Garcia, R.A., Herreros, J.M., Lazrek, M., Palle, P.L., Petrou, N., Renaud, C. and Regulo, C. (1997). Performance and early results from the Golf Instrument flown on the Soho mission. In **SOHO 1997**, pp. 207–226.

Giles, P. M., Duvall, T.L., Jr. and Scherrer, P.H. (1998). Time–distance measurements of subsurface rotation and meridional flow. In **BOSTON 1998**, pp. 775–780.

Gilman, P.A. and Dikpati, M. (2000). Joint instability of latitudinal differential rotation and concentrated toroidal fields below the solar convection zone. II: Instability of narrow bands at all latitudes. *Astrophysical Journal*, **528**, 552–572.

Gizon, L., Duvall, T.L., Jr. and Larsen, R.M. (2001). Probing surface flows and magnetic activity with time–distance helioseismology. In **MANCHESTER 2000**, pp.189–191.

Goldreich, P., Murray, N. and Kumar, P. (1994). Excitation of solar p modes. *Astrophysical Journal*, **424**, 466–479.

González Hernández, I., Patrón, J., Bogart, R.S. and the SOI Ring Diagram Team (1998). Meridional flows from ring diagram analysis. In **BOSTON 1998**, pp. 781–786.

Goode, P.R., Dziembowski, W.A., DiMauro, M.P., Kosovichev, A.G. and Schou, J. (1998). Solar asymmetries from SOHO-MDI splitting data. In **BOSTON 1998**, pp. 887–891.

Gough, D.O. (1977). Random remarks on solar hydrodynamics. In R.M. Bonnet and P. Delache, (eds), *The Energy Balance and Hydrodynamics of the Solar Chromosphere and Corona*, Proceedings of IAU Colloquium No. 36, G. de Bussac, Clairmont-Ferrand, pp. 3–36.

Gough, D.O. (1985a). Recent advances in helioseismology. In B. Chen and C. de Jager, (eds), *Proc. Kunming Workshop on Solar Physics and Interplanetary Travelling Phenomena*, Science Press, Beijing. vol. **1**, pp. 137–164.

Gough, D.O. (1985b). Theory of solar oscillations. In E. Rolfe and B. Battrick (eds), *Future Missions in Solar, Heliospheric and Space Plasma Physics*, ESA SP-235, ESTEC, Noordwijk, pp. 183–197.

Gough, D.O. and McIntyre, M.E. (1998). Inevitability of a magnetic field in the Sun's interior. *Nature*, **394**, 755–757.

Gough, D.O. and Thompson, M.J. (1991). The inversion problem. In A.N. Cox, W.C. Livingston and M.S. Matthews (eds), *Solar Interior and Atmosphere*, University of Arizona Press, Tucson, AZ, pp. 519–561.

Gough, D.O. and Kosovichev, A.G. (1995). An attempt to measure the latitudinal variation of the depth of the convection zone. In **ASILOMAR 1995**, pp. 47–48.

Harvey, J.W. (1988). Solar internal rotation from helioseismology. In **TENERIFE 1988**, pp. 55–66.

Hathaway, D.H., Beck, J.G., Bogart, R.S., Bachmann, K.T., Khatri, G., Petitto, J.M., Han, S. and Raymond, J. (2000). The photospheric convection spectrum. In **STANFORD 2000**, pp. 495–508.

Hill, F. (1989). Solar oscillation ring diagrams and large-scale flows. *Astrophysical Journal*, **343**, L69–L71.

Hill, F., Deubner, F-L. and Isaak, G. (1991). Oscillation observations. In **COX 1991**, pp. 329–400.

Hill, H.A. and Caudell, T.P. (1979). Global oscillations of the Sun: Observed as oscillations in the apparent solar limb-darkening function. *Monthly Notices of the Royal Astronomical Society*, **186**, 327–342.

Hill, H.A. and Stebbins, R.T. (1975). The intrinsic visual oblateness of the Sun. *Astrophysical Journal*, **200**, 471–483.

Houdek, G., Chaplin, W.J., Christensen-Dalsgaard, J., Däppen, W., Elsworth, Y., Gough, D.O., Isaak, G.R., McLeod, C.P., New, R. and Rabello-Soares, M.C. (2001). On changes in convective properties over the solar cycle. *Monthly Notices of the Royal Astronomical Society*, **327**, 483–487.

Howard, R. and Labonte, B.J. (1980). The sun is observed to be a torsional oscillator with a period of 11 years. *Astrophysical Journal*, **239**, L33–L36.

Howe, R., Christensen-Dalsgaard, J., Hill, F., Komm, R.W., Larsen, R.M., Schou, J., Thompson, M.J. and Toomre, J. (2000a). Dynamic variations at the base of the solar convection zone. *Science*, **287**, 2456–2460.

Howe, R., Christensen-Dalsgaard, J., Hill, F., Komm, R.W., Larsen, R.M., Schou, J., Thompson, M.J. and Toomre, J. (2000b). Deeply penetrating banded zonal flows in the solar convection zone. *Astrophysical Journal*, **533**, L163–L166.

Hurlburt, N.E. and Rucklidge, A.M. (2000). Development of structure in pores and sunspots: flows around axisymmetric magnetic flux tubes. *Monthly Notices of the Royal Astronomical Society*, **314**, 793–806.

Iglesias, C.A., Rogers, F.J. and Wilson, B.G. (1987). Reexamination of the metal contribution to astrophysical opacities. *Astrophysical Journal*, **322**, L45–L48.

Isaak, G.R. (1961) An atomic beam spectrophotometer. *Nature*, **189**, 373–374.

Kajita, T. (1999). Atmospheric neutrino results from Super-Kamiokande and Kamiokande – Evidence for ν_μ oscillations. *Nuclear Physics B Proceedings Supplements*, **77**, 123–132.

Kjeldsen, H. and Bedding, T.R. (eds) (1998). *The First MONS Workshop: Science with a Small Space Telescope*, Aarhus Universitet, Aarhus.

Kosovichev, A.G., Duvall, T.L., Jr., Birch, A.C., Gizon, L., Scherrer, P.H. and Zhao, J. (2001). Heliotomography of the outer layers of the Sun. In **TENERIFE 2000**, p. 701.

Kosovichev, A.G., Nigam, R., Scherrer, P.H., Schou, J., Christensen-Dalsgaard, J., Dziembowski, W.A., Goode, P.H., Gough, D.O., Reiter, J. and Rhodes, E.J., Jr. (1997). Spherical and aspherical structure of the Sun: First year of SOHO-MDI observations. In **KYOTO 1997**, pp. 157–164.

Kosovichev, A.G. and Schou, J. (1997). Detection of zonal shear flows beneath the Sun's surface from f-mode frequency splitting. *Astrophysical Journal*, **482**, L207.

Kosovichev, A.G., Schou, J., Scherrer, P.H., Bogart, R.S., Bush, R.I., Hoeksema, J.T., Aloise, J., Bacon, L., Burnette, A., DeForest, C., Giles, P.M., Leibrand, K., Nigam, R., Rubin, M., Scott, K., Williams, S.D., Basu, S., Christensen-Dalsgaard, J., Däppen, W., Rhodes, E.J., Jr., Duvall, T.L., Jr., Howe, R., Thompson, M.J., Gough, D.O. Sekii, T., Toomre, J., Tarbell, T.D., Title, A.M., Mathur, D., Morrison, M., Saba, J.L.R., Wolfson, C.J., Zayer, I. and Milford, P.N. (1996). Internal structure and rotation of the Sun: First results from MDI data. In **NICE 1996**, pp. 203–210.

Kosovichev, A.G. and Zharkova, V.V. (1998). X-ray flare sparks quake inside Sun. *Nature*, **393**, 317–318.

Kuhn, J.R., Armstrong, J.D., Bush, R.I. and Scherrer, P.H. (2000). Rossby waves on the Sun as revealed by solar 'hills'. *Nature*, **405**, 544–546.

Kuhn, J.R., Bush, R.I., Scheick, X. and Scherrer, P.H. (1998). The Sun's shape and brightness. *Nature*, **392**, 155.

Kumar, P. (1997). Excitation of solar acoustic oscillations. In J. Provost and F.-X. Schmider (eds), *Sounding Solar and Stellar Interiors*, IAU Symposium 181, Kluwer, Dordrecht, pp. 287–305.

Kumar, P. and Basu, S. (1999). Line asymmetry of solar p modes: Properties of acoustic sources. *Astrophysical Journal*, **519**, 396–399.

Lazrek, M., Baudin, F., Bertello, L., Boumier, P., Charra, J., Fierry-Fraillon, D., Fossat, E., Gabriel, A.H., García, R.A., Gelly, B., Gouiffes, C., Grec, G., Pallé, P.L., Pérez Hernández, F., Régulo, C., Renaud, C., Robillot, J.M., Roca Cortés, T., Turck-Chièze, S. and Ulrich, R.K. (1997). First results on p modes from GOLF Experiment. In **SOHO 1997**, pp. 227–246.

Ledoux, P. and Walraven, T. (1958). Variable stars. *Handbuch der Physik*, **51**, 353–604.

Leibacher, J.W. and Stein, R.F. (1971). A new description of the solar five-minute oscillation. *Astrophysical Letters*, **7**, 191–192.

Leighton, R.B. (1961). Untitled comments on supergranulation and 5-min oscillations. In R.N. Thomas (ed), Aerodynamic Phenomena in Stellar Atmospheres, IAU Symposium 12, *Il Nuovo Cimento Supplement Series 10*, **22**, 321–327.

Leighton, R.B. (1969). A magneto-kinematic model of the solar cycle. *Astrophysical Journal*, **156**, 1–26.

Leighton, R.B., Noyes, R.W. and Simon, G.W. (1962). Velocity fields in the solar atmosphere. I. Preliminary report. *Astrophysical Journal*, **135**, 474.

Libbrecht, K.G. and Kaufman, J.M. (1988). Frequencies of high-degree solar oscillations. *Astrophysical Journal*, **324**, 1172–1183.

Libbrecht, K.G., Woodard, M.F. and Kaufman, J.M. (1990). Frequencies of solar oscillations. *Astrophysical Journal Supplement*, **74**, 1129–1149.

Lighthill, M.J. (1952). *Proceedings of the Royal Society of London, A*, **211**, 564–587.

Lindsey, C. and Braun, D.C. (1990). Helioseismic imaging of sunspots at their antipodes. *Solar Physics*, **126**, 101–115.

Lindsey, C. and Braun, D.C. (1999). Chromatic holography of the sunspot acoustic environment. *Astrophysical Journal*, **510**, 494–504.

Lindsey, C. and Braun, D.C. (2000). Seismic images of the far side of the Sun. *Science*, **287**, 1799–1801.

Livingston, W.C., Donnelly, R.F., Grigoryev, V., Demidov, M.L., Lean, J., Steffen, M., White, O.R. and Willson, R.L. (1991). Sun-as-a-star spectrum variability. In **COX 1991**, pp. 1109–1160.

Martic, M., Schmitt, J., Lebrun, J.-C., Barban, C., Connes, P., Bouchy, F., Michel, E., Baglin, A., Appourchaux, T. and Bertaux, J.-L. (1999). Evidence for global pressure oscillations on Procyon. *Astronomy and Astrophysics*, **351**, 993–1002.

Matthews, J.M. (1998). Asteroseismology from space: Getting the MOST science for the least money. In **BOSTON 1998**, pp. 395–398.

Morrow, C.A. (1988). Solar rotation models and the a_1, a_3 and a_5 splitting coefficients for solar acoustic oscillations. In **TENERIFE 1988**, pp. 91–98.

Newkirk, G., Jr. (1980). Chairman, Solar Cycle and Dynamics Mission, Final Report, July 1980, Goddard Space Flight Center, SCADM #3.

Nigam, R., Kosovichev, A.G., Scherrer, P.H. and Schou, J. (1997). Line asymmetry and excitation mechanism of solar oscillations. In **KYOTO 1997**, pp. 195–198.

Nigam, R., Kosovichev, A.G., Scherrer, P.H. and Schou, J. (1998). Asymmetry in velocity and intensity helioseismic spectra: A solution to a long-standing puzzle. *Astrophysical Journal*, **495**, L115–L118.

Noyes, R.W. and Leighton, R.B. (1963). Velocity fields in the solar atmosphere. II. The oscillatory field. *Astrophysical Journal*, **138**, 631–647.

Noyes, R.W., Rhodes, E.J., Jr. (1984) *Probing the Depths of a Star: The Study of Solar Oscillations from Space*. Report of the NASA Science Working Group on the Study of Solar Oscillations from Space, July 1984, NASA JPL, Pasadena, CA.

Parker, E.N. (1979). Sunspots and the physics of magnetic flux tubes. I. The general nature of the sunspot. *Astrophysical Journal*, **230**, 905–913.

Penny, A.J., Favata, F., Deeg, H.J., Eddington Science Team (2001). The Eddington planet-finding and asterseisomology mission. In *Planetary Systems in the Universe*, IAU Symposium 202, Manchester, England, August 2000. Astronomical Society of the Pacific, in press.

Pontecorvo, B. (1968). Neutrino experiments and the problem of conservation of leptonic charge. *Soviet Physics JETP*, **26**, 984–988.

Quinn, T.J. and Fröhlich, C. (1999) Accurate radiometers should measure the output of the Sun. *Nature*, **401**, 841–842.

Rabin, D.M., DeVore, C.R., Sheeley, N.R., Harvey, K.L. and Hoeksema, J.T. (1991). The solar activity cycle. In **Cox 1991**, pp. 781–843.

Rhodes, E.J., Cacciani, A. and Korzennik, S.G. (1988). Initial high-degree p-mode frequency splittings from the 1988 Mt. Wilson 60-foot Tower Solar Oscillation Program. In **TENERIFE 1988**, pp. 81–86.

Rhodes, E.J., Harvey, J.W. and Duvall, T.L. (1983). Recent observations of high-degree solar p-mode oscillations at the Kitt Peak National Observatory. In **CRIMEA 1981**, p. 111.

Rhodes, E.J., Ulrich, R.K. and Simon, G.W. (1977). Observations of non radial p-mode oscillations. *Astrophysical Journal*, **218**, 901–919.

Scherrer, P.H., Bogart, R.S., Bush, R.I., Hoeksema, J.T., Kosovichev, A.G., Schou, J., Rosenberg, W., Springer, L., Tarbell, T.D., Title, A., Wolfson, C.J., Zayer, I. and the MDI Engineering Team (1995). The Solar Oscillations Investigation – Michelson Doppler Imager. In **SOHO 1995**, pp. 129–188.

Scherrer, P.H., Wilcox, J.M., Christensen-Dalsgaard, J., Gough, D.O. (1983). Detection of solar five-minute oscillations of low degree. *Solar Physics*, **82**, 75–87.

Schou, J. (1999). Migration of zonal flows detected using Michelson Doppler Imager f-mode frequency splittings. *Astrophysical Journal*, **523**, L181–L184.

Schou, J., Antia, H.M., Basu, S., Bogart, R.S., Bush, R.I., Chitre, S.M., Christensen-Dalsgaard, J., DiMauro, M.P., Dziembowski, W.A., Eff-Darwich, A., Gough, D.O., Haber, D.A., Hoeksema, J.T., Howe, R., Korzennik, S.G., Kosovichev, A.G., Larsen, R.M., Pijpers, F.P., Scherrer, P.H., Sekii, T., Tarbell, T.D., Title, A.M., Thompson, M.J. and Toomre, J. (1998). Helioseismic studies of differential rotation in the solar envelope by the solar oscillations investigation using the Michelson Doppler Imager. *Astrophysical Journal*, **505**, 390–417.

Schou, J., Christensen-Dalsgaard, J., Howe, R., Larsen, R.M., Thompson, M.J. and Toomre, J. (1998). Slow poles and shearing flows from helioseismic observations with MDI and GONG spanning a year. In **BOSTON 1998**, pp. 845–849.

Schou, J., Kosovichev, A.G., Goode, P.R. and Dziembowski, W.A. (1997). Determination of the Sun's seismic radius from SOHO-MDI. *Astrophysical Journal*, **489**, L197.

Simon, G.W. and Leighton, R.B. (1964). Velocity fields in the solar atmosphere. III. Large scale motions, the chromospheric network, and magnetic fields. *Astrophysical Journal*, **140**, 1120–1147.

Smith, E.V.P. and Gottlieb, D.M. (1973). Solar flux and its variations. In W.R. Bandeen and S.P. Maran (eds), *Possible Relationships between Solar Activity and Meteorological Phenomena*, GSFC Symposium 74, Nov. 1973, NASA SP-366, Washington, DC, pp. 97–117.

Sofia, S., Maier, E. and Twigg, L. (1991). The Solar Disk Sextant – Monitoring the size and shape of the Sun. *Advances in Space Research*, **11**, 123–132.

Stein, R.F. and Nordlund, Å. (2000). Realistic solar convection simulations. *Solar Physics*, **192**, 91–108.

Tobias, S.M., Brummell, N.H., Clune, T.L. and Toomre, J. (2001). Transport and storage of magnetic field by overshooting turbulent compressible convection. *Astrophysical Journal*, **549**, 1183–1203.

Tomczyk, S., Streander, K., Card, G., Elmore, D., Hull, H. and Cacciani, A. (1995). A instrument to observe low-degree solar oscillations. *Solar Physics*, **159**, 1–21.

Turck-Chièze, S. Couvidat, S., Kosovichev, A.G., Gabriel, A.H., Berthomieu, G., Brun, A.S., Christensen-Dalsgaard, J., García, R.A., Gough, D.O., Provost, J., Roca Cortés, T., Roxburgh, I.W. and Ulrich, R.K. (2001). Solar neutrino emission deduced from a seismic model. *Astrophysical Journal Letters*, **555**, L69–L73.

Ulrich, R.K. (1970). The five-minute oscillations on the solar surface. *Astrophysical Journal*, **162**, 993–1002.

Willson, R.C., Gulkis, S., Janssen, M., Hudson, H.S. and Chapman, G.A. (1981). Observations of solar irradiance variability. *Science*, **211**, 700–702.

Woodard, M.F. and Hudson, H.S. (1983). Frequencies, amplitudes and linewidths of solar oscillations from total irradiance observations. *Nature*, **305**, 589–593.

Woodard, M.F. and Libbrecht, K.G. (1988). On the measurement of solar rotation using high-degree p-mode oscillations. In **TENERIFE 1988**, pp. 67–71.

Woodard, M.F. and Libbrecht, K.G. (1993). Solar activity and oscillation frequency splittings. *Astrophysical Journal Letters*, **402**, L77–L80.

Zhao, J., Kosovichev, A.G. and Duvall, T.L., Jr. (2001). Investigation of mass flows beneath a sunspot by time–distance helioseismology. *Astrophysical Journal*, **557**, 384–388.

45

SAMI K. SOLANKI* AND REINER HAMMER**

The solar atmosphere

As a typical star, and the only one that can be spatially resolved by direct means, the study of the Sun has provided an insight into many of the fundamental processes taking place in stellar atmospheres, often at small scales. A prime example is magneto-convection or the formation of coronae and the consequent emission of copious amounts of X-rays. In addition, the Sun's apparent brightness allows measurements with unprecedented accuracy. Thus the Sun is the standard against which cosmic abundances are compared. Its high apparent brightness also means that the Sun is a strong source at almost all wavelengths and thus detectable with simple, not particularly sensitive equipment such as the early instruments flown in space. Thus for many wavelengths the Sun was the first (or one of the first) cosmic source(s) detected.

However, only the lowest layers of the Sun's atmosphere, the photosphere and chromosphere, can be regularly observed from the ground over the solar disk. The transition region, corona and the solar wind are best studied from space, and even many properties of the photosphere (such as the variation of solar irradiance with time) had to await space-based observations for their determination or discovery.

1 OVERVIEW OF THE SOLAR ATMOSPHERE

Traditionally the atmosphere of the Sun is divided into four layers, starting with the photosphere at the bottom, moving up through the chromosphere and transition region to the corona. The photosphere is the layer in which the temperature drops outwards from around 5800 K at the solar surface to around 4000 K at the temperature minimum. Beyond that point it rises again, first relatively gently (forming the chromospheric plateau), but then very rapidly in the transition region (TR). The temperature profile becomes flatter again in the corona. The boundary between the corona and the TR is often drawn at approximately 10^6 K. This boundary, like that between chromosphere and TR, is not sharp or well defined. At still greater distances from the solar surface the temperature gradually decreases again, achieving values of approximately 10^5 K at 1 AU (whereby electrons and ions need not have the same temperature in the heliosphere). As we shall see in subsequent sections, the simple plane-parallel representation of the solar gas outlined above is not tenable in any layer of the atmosphere. At any given height more than one atmospheric component is present, each having its own temperature, density and velocity structure.

Features as diverse as granular convection cells in the photosphere (Figure 1) and magnetic loops in the corona (Figure 2) are now known to structure the respective layers of the atmosphere. In addition to being spatially inhomogeneous at almost all spatial scales, the solar atmosphere is also highly dynamic at almost all timescales. Much of the interesting physics to be learnt by studying the solar atmosphere is related to this structuring and dynamics and the associated heating of the chromosphere and corona.

In the following we discuss the various atmospheric layers, starting with the photosphere and moving outward. Particular emphasis is placed on the contributions made by space missions to our knowledge and understanding of the solar atmosphere. Since these contributions are largest for the transition region and corona our discussion of these layers will be more detailed than of the photosphere and chromosphere. Table 1 summarizes the space missions mentioned in this chapter.

2 THE PHOTOSPHERE

2.1 The plane-parallel photosphere

The solar photosphere is the layer that emits most of the solar radiative energy flux, with the emitted spectrum

---

* Max-Planck-Institut für Aeronomie, Katlenburg-Lindau, Germany
** Kiepenheuer-Institut für Sonnenphysik, Freiburg i/B, Germany

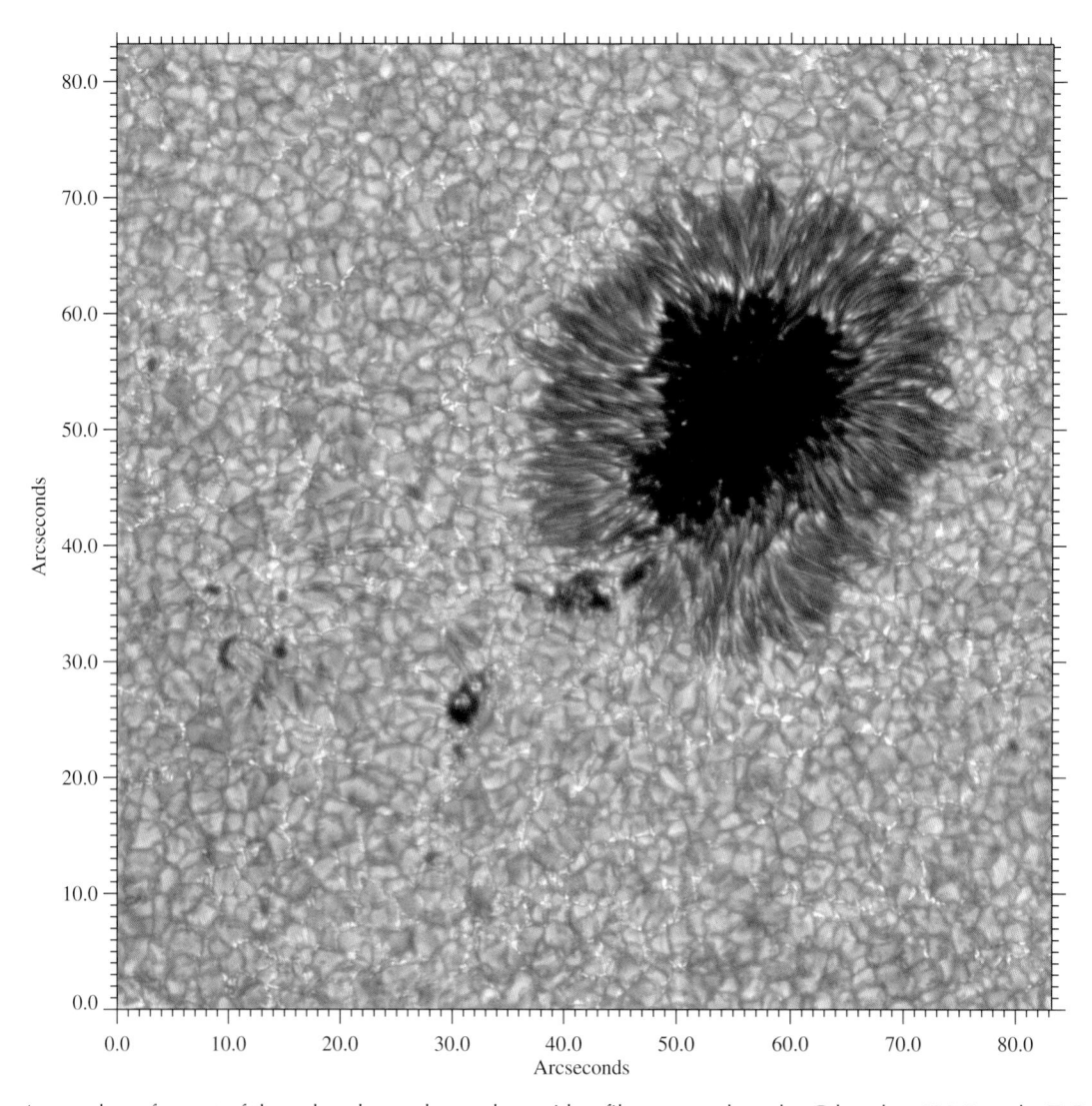

Figure 1 A snapshot of a part of the solar photosphere taken with a filter centred on the G band at 430.5 nm by T. Berger and G. Scharmer. The image covers 60 000 × 60 000 km on the Sun. The most prominent feature is a sunspot. The much smaller dark features are pores. Also visible are granules (bright cells surrounded by dark lanes) and bright points corresponding to magnetic elements. (Courtesy of T. Berger.)

having its peak in the visible (in the green part of the wavelength range). As such, the photosphere is the atmospheric layer most easily observed from the ground and consequently the one to whose investigation spacecraft have contributed the least. This, however, is changing at a rapid pace, with the ESA–NASA Solar and Heliospheric Observatory (SOHO; Fleck and Domingo 1995) providing the first glimpses of how space-based telescopes can revolutionize our understanding of the photosphere. The next major highlight is expected to be provided by the Japan–US–UK Solar B mission.

The brightness across the solar disk is not constant but rather decreases from the centre of the disk to its edge (the solar limb) at visible wavelengths. This is called limb darkening. Since at the limb the radiation is emitted at greater heights, limb darkening implies a decrease in the temperature with height. Furthermore, the spectral form of the limb darkening provides information on the continuum absorption coefficient. Such observations confirmed the proposal by Wildt (1939) that in the visible the absorption is dominated by the H^- ion in spite of its low abundance (Chalonge and Kourganoff 1946).

Traditionally the limb darkening and the shapes and strengths of absorption lines (Fraunhofer lines) have been employed to determine the temperature stratification in the solar photosphere. These diagnostics reveal that the temperature decreases outwards in the solar photosphere from over 6500 K at the deepest observable layers to around 4000 K at the temperature minimum (e.g. Holweger 1967). The advent of UV observations from space, in particular

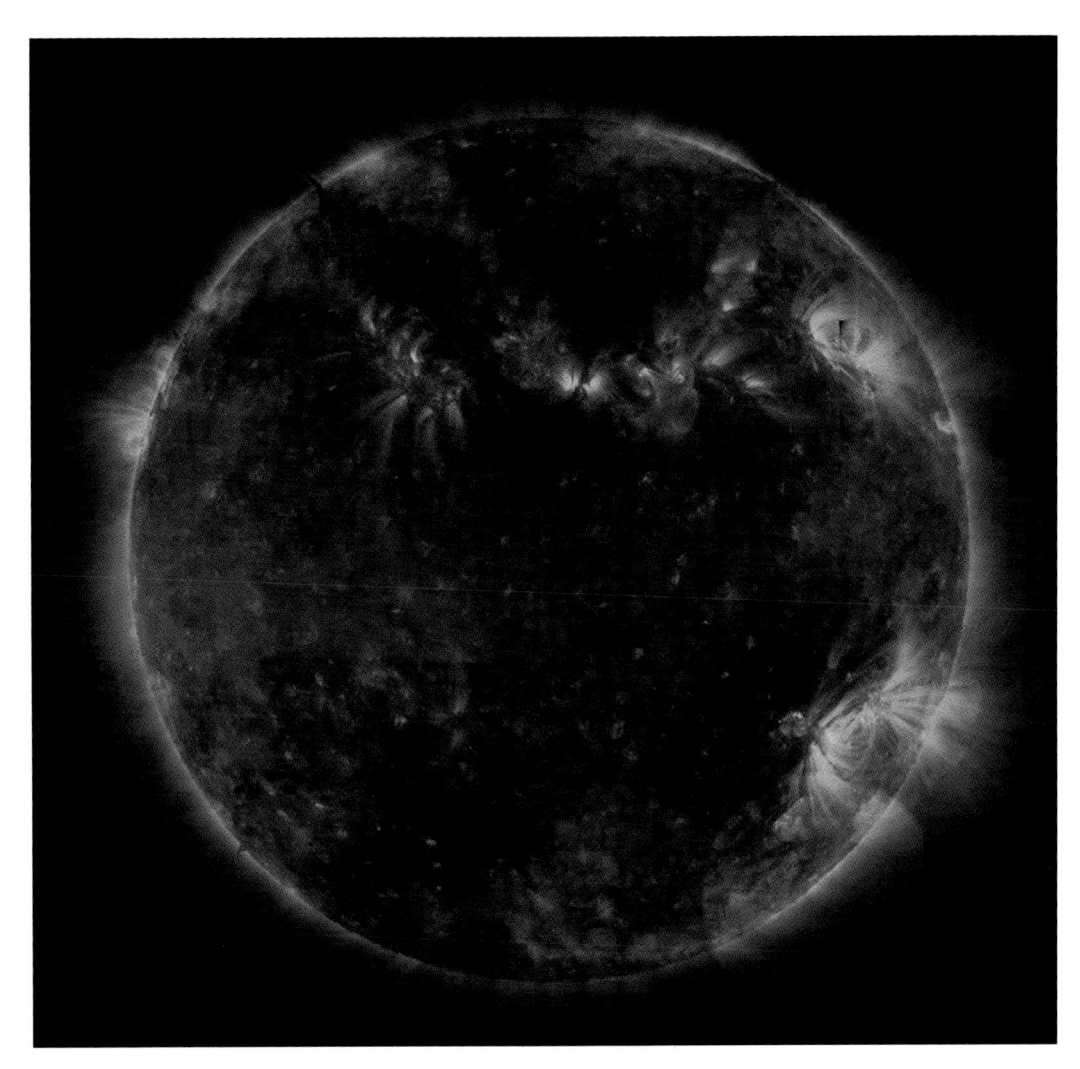

Figure 2 Composite of several high-resolution images taken with the Transition Region and Coronal Explorer (TRACE; Handy et al. 1999) in a spectral band near 171 Å, which is dominated by emission from eightfold ionized iron atoms (Fe IX) formed around 10^6 K. At these temperatures the network is no longer visible, and the disk emission is dominated by active regions, by coronal loops in the quiet corona (i.e. outside of coronal holes and active regions), and by numerous bright points. Plumes extend as ray-shaped density enhancements from the north and south polar coronal holes. (Courtesy of the TRACE team. TRACE is a mission of the Stanford–Lockheed Institute for Space Research, and part of the NASA Small Explorer program.)

Table 1 Space missions mentioned in this chapter

Mission	Operation period
Stratoscope	several balloon flights 1957 and 1959
OSO 4 (Orbiting Solar Observatory)	1967–69
OSO 6 (Orbiting Solar Observatory)	1969–72
Skylab	1973–74
Spektrostratoskop	balloon flight 1975
OSO 8 (Orbiting Solar Observatory)	1975–78
HRTS (High Resolution Telescope and Spectrograph)	rocket and shuttle flights since 1975
TRC (Transition Region Camera)	rocket flights 1979 and 1980
SMM (Solar Maximum Mission)	1980–89
SOUP (Solar Optical Universal Polarimeter)	experiment on Spacelab 2, 1985
NIXT (Normal-Incidence X-ray Telescope)	rocket flights, e.g. 1993
Yohkoh	since 1991
SOHO (Solar and Heliospheric Observatory)	since 1995
TRACE (Transition Region And Coronal Explorer)	since 1998
Solar B	launch scheduled for 2005

from Skylab (Tousey 1977), provided a new diagnostic, the wavelength dependence of the continuum intensity, since at shorter wavelengths the continuum radiation emanates from higher layers (e.g. Vernazza et al. 1973, 1981). The advantage of UV and EUV spectra is that they also contain emission lines belonging to different ions that carry information on the temperature in the solar chromosphere, transition region and corona.

A reliable knowledge of the thermal stratification is fundamental for the accurate determination of elemental abundances. The pioneering work by Russell (1929) and the seminal compilation by Goldberg et al. (1960) have been followed by increasingly detailed and accurate determinations of the abundances of ever more elements. The current status of our knowledge of solar abundances (from the solar core to its corona) is discussed in the volume edited by Fröhlich et al. (1998), with the photospheric abundances being reviewed therein by Grevesse and Sauval (1998). On the whole these abundances agree surprisingly well with the meteoritic values, although there are some minor deviations and some residual uncertainty. The latter is due partly to the inhomogeneity of the solar atmosphere (discussed in Sections 2.2 and 2.3), which has generally not been taken into account when determining abundances. However, at the level of accuracy currently being achieved such inhomogeneities begin to have a significant effect.

2.2 Convection

It was evident relatively early that a single atmospheric component cannot adequately describe the solar photosphere. The dark sunspots and the bright faculae (bright structures most prominent near the limb), already visible with a small telescope, highlight the need for multiple thermal components. Sunspots and faculae are associated with magnetic activity (Section 2.3), but even the quiet parts of the Sun are known to be inhomogeneous since the discovery by William Herschel of solar granulation, bright structures typically 1000 km in diameter separated by a dark network. Figure 1 shows a snapshot of solar granulation surrounding a sunspot. On a larger scale a bright network (most prominent in radiation coming from chromospheric and transition-region layers) is also known to exist. To account for such regions with different brightness, sets of plane-parallel models have been produced (e.g. Vernazza et al. 1981, Fontenla et al. 1993). Again, UV spectra taken outside the terrestrial atmosphere have played an important role in constructing such model families.

High-resolution observations and the modelling of spectral lines have shown that at least in the photospheric layers it is mainly inhomogeneities at scales smaller than approximately 1000 km on the Sun that are of physical relevance. For example, faculae, which have sizes of 10^4–10^5 km, are found to be composed of many small magnetic elements, each with a diameter of the order of 100 km.

The major inhomogeneity in photospheric layers is introduced by the granulation, which is the surface signature of overshooting convection. The bright granules identify hot upflowing gas overshooting from the convectively unstable layers below the solar surface into the stably stratified photosphere. These are surrounded by multiply connected cool and hence dark lanes of downflowing gas. Properties of the granulation have been deciphered using data obtained with balloon-borne telescopes (with the Stratoscope, Danielson 1961; and the Spektrostratoskop, Mehltretter 1978), in space (Solar Optical Universal Polarimeter (SOUP), Title et al. 1989) and from the ground (Muller 1999).

A particular success have been detailed two- and three-dimensional numerical simulations, that is computations of the radiation hydrodynamics under conditions corresponding as closely as possible to those present on the Sun, based on a minimum of simplifying assumptions. Such simulations have reproduced a wide variety of observations (e.g. Nordlund 1984, Lites et al. 1989), so that they are likely to include the main physical ingredients necessary to describe solar granulation. Mainly, however, they have led to a better physical understanding of solar convection and the influence of granulation on, for example, abundance determinations (e.g. Solanki 1998). Both observations and simulations suggest that the vertical velocity associated with granules decreases rapidly with height, while the horizontal velocity becomes increasingly strong, being supersonic over portions of the largest granules. This last fact is one of the rare predictions made by theory in solar physics that have been subsequently confirmed by observations.

An oscillatory velocity component is also present in the photosphere and chromosphere. In the photosphere its power peaks occur at a period of around 5 min, while in the chromosphere the power peak lies near 3 min. The 5 min oscillations are evanescent in the solar atmosphere, but propagate in the solar interior. They are used to probe the subsurface layers of the Sun (helioseismology). The amplitude of the vertical oscillatory velocity increases with increasing height and dominates over the vertical granular flow field at the top of the photosphere.

In addition to granulation three larger scales of convection are known to affect the solar atmosphere, mesogranulation (5–7 Mm in size) discovered by November et al. (1981), supergranulation (20–30 Mm) discovered by Simon and Leighton (1964) and giant cells (covering 40° in longitude and less than 10° in latitude) discovered by Beck et al. (1998) using Dopplergrams recorded by the Michelson Doppler Interferometer (MDI) on SOHO. Granulation has by far the most readily visible signature in the photosphere, followed by supergranulation, while the influence of the other scales of convection on the solar atmosphere is so subtle that it can only be detected with the help of special techniques. In addition to revealing giant cells, MDI has also provided the best images of supergranulation at the solar surface (from Doppler shifts). Yet another important contribution of MDI to the study of solar convection has come from the application of local helioseismic techniques to time series of MDI Dopplergrams. Such analyses of the solar oscillation spectrum have provided the first images of supergranular flows below the solar surface (Duvall et al. 1997). A comparison between the subsurface supergranulation (reconstructed from MDI local helioseismology) and MDI magnetograms provides direct confirmation of the traditional picture that the magnetic network is located at the convergence points of the (subsurface) supergranules, so that the magnetic features float in the downflow lanes of the supergranules (Duvall and Gizon 2000). This increases the confidence in the results of local helioseismology. The study of solar convection has thus been firmly catapulted into the space age by MDI, after the SOUP paved the way.

Simulations are now starting to move beyond granules to the larger convective cells (Ploner et al. 2000). They suggest that the large-scale convective phenomena observed at the surface are driven at or very close to the surface itself and are not due to the ionization of helium in deeper layers, as had earlier been suggested.

2.3 Magnetic fields

The strongest structuring agent of the photosphere besides granulation is the magnetic field. It is concentrated into flux tubes with a field strength of 1–1.5 kG at photospheric levels, but also has a weaker component, which contains the same order of magnitude of flux as the tubes, but only a small fraction of the magnetic energy. This has to do with the fact that whereas the flux is proportional to the field strength the magnetic energy scales with the square of the field strength.

In the photosphere the magnetic flux tubes are nearly vertical and can be considered to be vertical bundles of concentrated magnetic field lines surrounded by nearly field-free gas.

The largest flux tubes, or rather their intersections with the solar surface, are visible as dark sunspots, while the smallest ones are the magnetic elements, groups of which form faculae and the network. Sunspots have a diameter lying in the range 4000–60 000 km. They are dark and are distinguished from the generally smaller pores by the fact that sunspots have two main components, a darker umbra (with an effective temperature $T_{\rm eff} \approx 4500$ K) and a less dark penumbra ($T_{\rm eff} \approx 5500$ K), while pores have basically one, umbra-like component. Sunspots have peak field strengths of 2000–3500 G, increasing with size. The field strength averaged over the whole sunspot (i.e. over the cross-section of the flux tube) is 1000–1500 G and is very close to that found for small-scale magnetic elements, although the latter carry up to 10^6 times less magnetic flux. The field strength averaged over the flux-tube cross-section actually appears to be independent of flux-tube size at all heights in the photosphere (Solanki et al. 1999). An explanation for this result has so far not been given.

An additional basic property of sunspots is the presence of an outflow in the penumbra along thin horizontal flux tubes embedded in the generally inclined field. This outflow, termed the Evershed effect (Evershed 1909), decreases with height and finally turns into an inflow at chromospheric heights, where it is called the inverse Evershed effect (St. John 1913; see Solanki 1997 for a review). Observations with the spectrometers UVSP on the Solar Maximum Mission (SMM; Chipman 1981) and more recently with the Solar Ultraviolet Measurement of Emitted Radiation (SUMER) and the Coronal Diagnostics Spectrometer (CDS) on SOHO revealed that in the transition region the inverse Evershed effect continues into the umbra, where it manifests itself as a downflow that can reach supersonic velocities at transition region temperatures, in particular in bright structures called sunspot plumes (Kjeldseth-Moe et al. 1988, Brynildsen et al. 1999).

At the other end of the size scale of flux tubes are the magnetic elements, whose diameters are close to or less than the best currently achievable spatial resolution (which corresponds to approximately 150 km in visible light). These flux tubes are bright and harbour most of the magnetic energy in photospheric layers, although they probably carry less than half of the total magnetic flux (Meunier et al. 1998). They are constantly being moved around by granules and supergranules. Rapid jostling by granules may produce waves propagating upward along the flux tube (Roberts and Ulmschneider 1997, Grossmann-Doerth et al. 1998), which may contribute to the heating of the chromosphere and corona (e.g. Choudhuri et al. 1993). They also appear to periodically dissolve or fragment and form again later (Berger et al. 1998, Gadun et al. 2001). Schrijver et al. (1998) estimated from the high rate of magnetic flux emergence revealed by MDI magnetograms of the quiet Sun that the flux in the magnetic network is replaced every 40 hours. The flux emerges in the form of a small loop, whose footpoints move ever further apart with time and rarely come together again later. Thus, from the way that the emerged flux is seen to evolve it is clear that reconnection between field lines must be commonplace and must happen almost uninterruptedly (Section 4.2).

Magnetic elements and sunspots not only structure the photosphere, through their field lines they provide also links to the chromosphere and corona. Along these links energy can be transported from the solar interior (where it is present in abundance) to the outer atmosphere, where it needs to be deposited. Significant advances in the study of photospheric fine structure (granulation, sunspots, magnetic elements) and of its connection to chromospheric and coronal features are expected to follow from the Solar B mission, currently scheduled to be launched in 2005.

Magnetic elements and sunspots are also thought to be largely responsible for the observed fluctuations of the solar irradiance, that is the brightness of the whole solar disk as measured from above the Earth's atmosphere. All successful observations of solar irradiance variations have been carried out from space. In the meantime the irradiance could be monitored almost continuously for two full solar cycles, although with a variety of instruments whose records only partially overlap (Fröhlich 2000). Such measurements have led to the discovery of brightness dips lasting weeks, produced by the passage of sunspots across the solar disk in connection with solar rotation, as well as a brightening (by 0.1% in total irradiance) at solar activity maximum relative to activity minimum (Willson and Hudson 1991). Both these effects can be quantitatively reproduced on the basis of the evolution of the surface area and spatial distribution of sunspots and faculae on the solar surface (Solanki and Fligge 2000).

3 THE CHROMOSPHERE

3.1 The chromospheric spectrum

The solar chromosphere is visible without filters for a short time at the beginning and end of totality of a solar eclipse at which point the solar limb changes colour dramatically. Outside of eclipses it can be observed in the cores of strong absorption lines at visible or near-ultraviolet wavelengths. Alternatively many of the emission lines in the extreme ultraviolet (EUV) part of the spectrum arise in the chromosphere.

Prominent spectral lines of chromospheric origin are the Ca II H and K lines at around 390 nm, or the Mg II h and k lines at around 280 nm. The cores of these absorption lines show a central intensity peak, which indicates a reversal of the temperature gradient with height, that is that the temperature decrease with height in the photosphere is followed by a temperature increase in the chromosphere, although alternative explanations (assuming a time-dependent or spatially structured chromosphere) are also possible (e.g. Carlsson and Stein 1995).

The spectrum of the chromosphere and of the hotter transition region and corona is, however, much richer when observed from space for two reasons. Firstly, the shorter wavelengths at which most of the transitions from ionized species that sample higher temperatures take place can only be observed from above the Earth's atmosphere. Secondly, at increasingly shorter wavelengths the continuum is formed at ever greater heights, so that at wavelenghts below roughly 160 nm all the spectral lines must be formed in the chromosphere or above. Another advantage of the EUV is that in contrast to the cores of strong lines in the visible many of the lines in the EUV are optically thin and thus easier to analyse. It is therefore not surprising that a significant part of the effort in space-based solar physics had been invested into EUV spectroscopy, culminating in the two spectrometers CDS and SUMER on board the SOHO spacecraft. In particular the latter has provided extremely rich spectra of the chromosphere and the transition region at high spatial and spectral resolution. In Figure 3 a spectrum of the quiet Sun obtained by SUMER is plotted (Curdt et al. 1999). Note the large number of emission lines in the spectrum. The first- and second-order spectra of the SUMER grating overlap and refer to the lower and upper wavelength scales, respectively. Most prominent are the Lyα line of hydrogen at 1216 Å and the Lyman continuum starting near 912 Å. Figure 4 shows blow-ups of two spectral regions, with the identifications of the main spectral lines being indicated. Many transitions of neutral (C I, N I, Ne I, S I) and singly ionized species (Fe II, N II, C II, Ar II) are visible. Most of these are of chromospheric origin.

3.2 Standard chromospheric models

In standard, time-independent plane-parallel models of the solar atmosphere (such as those of Fontenla et al. 1993) the chromosphere covers the height range between the temperature minimum at the top of the photosphere and the bottom of the transition region, where the temperature increases rapidly with height. In the lower chromosphere of such a model the temperature increases outward appreciably but becomes reasonably height independent in the middle and upper chromosphere. The chromosphere is the lowest part of the atmosphere in which the temperature increases significantly outward from the solar surface and which therefore definitely cannot be in radiative equilibrium, but rather requires some mechanical or magnetic source of energy input. The current picture of the solar chromosphere is geometrically far more complex and more dynamic than portrayed by such 'standard' models (Section 3.4), but for many purposes the standard plane-parallel models are adequate and are still used.

The chromospheric layers of these models are in general based on EUV continua below 1600 Å and lines formed partly or completely in the chromosphere (Lyα, Mg II k, Ca II K). Of particular importance have been spectra obtained by the Orbiting Solar Observatories OSO 4 and OSO 6 (Vernazza et al. 1973) and by Skylab (Vernazza et al. 1981, Fontenla et al. 1993). A difficult problem within the context of these models has been posed by the hydrogen Lyα line, whose profile could not be reproduced by static models.

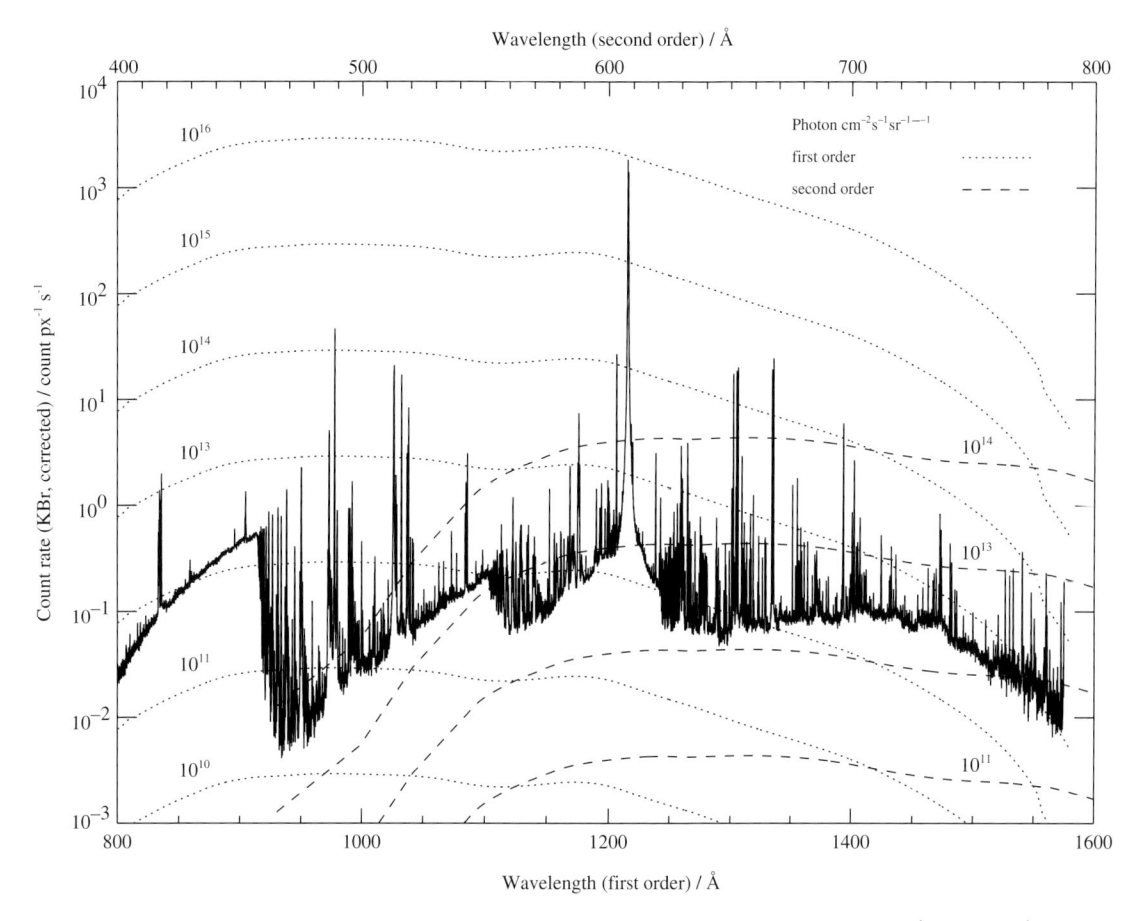

Figure 3 Quiet Sun spectrum of 12 August 1996 from 01:13 to 03:40 UT in first order from 800 Å to 1590 Å. The spectrum is corrected for detector dead time and local gain depression effects, and the attenuation at and near H I Ly α is compensated. Isoradiance contours render the radiometric calibration. (Courtesy of W. Curdt.)

Only the introduction of ambipolar diffusion by Fontenla et al. (1993, and reference therein) led to satisfactory results.

The presence of horizontal structuring in the chromosphere, clearly visible in ground-based images made in the core of the Ca II K line and in all chromospheric UV and EUV lines (Bonnet et al. 1980, Lemaire et al. 1997) is taken into account in a simple manner in these models by introducing a set of plane-parallel atmospheres, each describing different parts of the Sun (different atmospheric components) ranging from the interiors of network cells (darkest and coolest) to the network (brightest and hottest).

3.3 Chromospheric heating

Soon after it was realized that the solar corona is hot, Biermann (1946) suggested an explanation that evolved into the standard heating theory for both the solar chromosphere and corona over a period of three decades. This theory is based on acoustic waves that are generated abundantly in the turbulent flow field of the upper convection zone. As these waves propagate upward, their profile changes since in large-amplitude waves the wave peaks have a higher propagation speed than the valleys. Therefore the peaks attempt to overtake the valleys. This leads to the formation of shocks – thin zones in which the velocity and temperature change so rapidly that viscosity and thermal conduction convert the wave energy into heat, which is then available to sustain the elevated temperature of the upper solar atmosphere.

The formation of shocks out of sound waves is not an everyday experience. Imagine, for example, that we confine the sound of a tuning fork into a tube in order to enforce one-dimensional propagation. Such a small-amplitude wave must travel a distance of the order of the circumference of Earth before it shocks. The waves generated in the upper solar convection zone can form shocks a hundred times faster, despite their longer wavelengths. This is because of their large amplitudes – they are already large when the waves start in the convection zone, and they increase further as the waves propagate upward into a region of decreasing density.

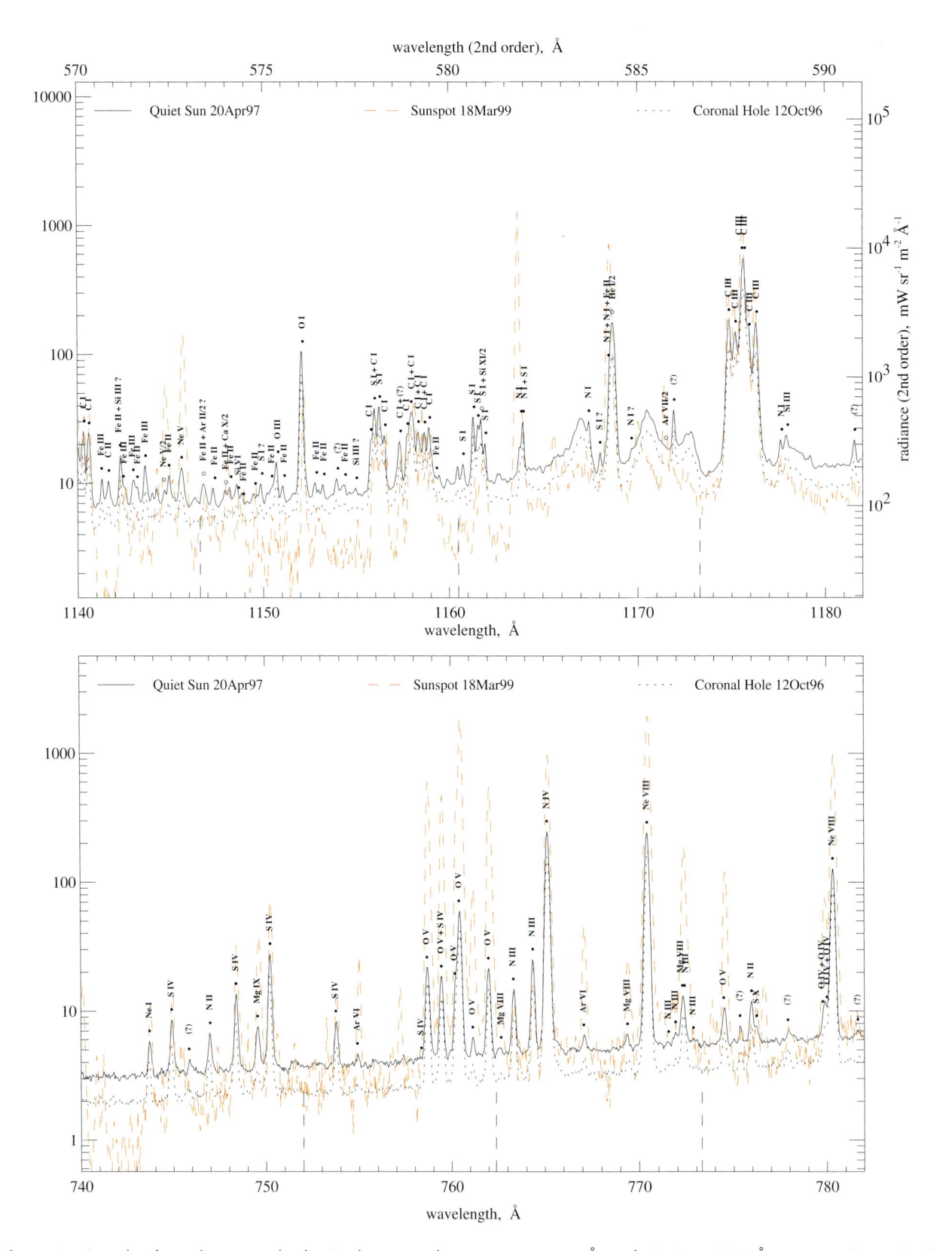

Figure 4 Samples from the spectral atlas in the spectral range 740 to 782 Å and 1140 to 1182 Å, representing quiet Sun (solid), sunspot (dashed), and coronal hole (dotted). (Courtesy of W. Curdt.)

When shocks have been formed, the wave amplitude is controlled by two competing effects: the outward density decrease in the solar atmosphere tends to increase it further, whereas the energy loss associated with shock dissipation tends to decrease it. Ultimately both effects balance each other; then the shock amplitude remains roughly constant. The heating associated with such a wave of constant amplitude is proportional to the density. This matches nicely the behaviour of the radiative output in standard (hot) chromosphere models (Ulmschneider 1970, Anderson and Athay 1989).

Therefore, chromospheric heating by acoustic shock waves has some attractive properties: it starts in the lower chromosphere, after shocks have been formed, and then it decays in a way consistent with the observed radiation loss. For these reasons, acoustic waves are still thought to be the main heating source of nonmagnetic parts of the solar chromosphere. They can also explain the so-called 'basal' emission from the chromospheres of stars with very low magnetic activity (Buchholz et al. 1998).

The same properties that make acoustic waves attractive for heating the chromosphere, however, make them unsuited for heating the corona. Their energy flux, decaying proportionally to the density, is virtually exhausted when they reach the corona. This was shown with data obtained by the OSO 8 satellite, which placed very severe constraints on the amount of upward propagating acoustic energy flux in the upper chromosphere and lower transition region (Athay and White 1978, Bruner 1981). Ground-based observations (e.g. Schmieder and Mein 1980) confirmed these constraints, and theoretical models showed that purely acoustic waves are unlikely to be able to produce a corona in either Sun-like or giant stars (Hammer and Ulmschneider 1991).

Sophisticated numerical simulations of shock waves in the nonmagnetic solar chromosphere (Carlsson and Stein 1995, 1997) reproduced very well the observed brightenings in the spectral lines H and K of singly ionized calcium and showed convincingly that the observed time-dependent variations in the line profiles are caused by the passage of large-amplitude shocks. The temperatures and densities behind these shocks were found to be so high that these postshock regions alone could produce the entire emission from the lower chromosphere, without an outward rise of the average chromospheric temperature. However, these simulations neglected short-period acoustic waves, which are also generated in the convection zone and could provide some background heating, as argued by Kalkofen et al. (1999). Therefore it is not yet clear if the entire lower chromosphere in nonmagnetic regions is cool and where the average temperature starts to rise. This problem will be discussed further in Section 3.4.

In magnetic parts of the chromosphere, in particular in the network, the magnetic field plays an important role in the heating process. In the simplest case, this role is rather passive, when magnetic flux tubes only act as quasi-static ducts that channel longitudinal wave motions. Such waves can be generated by squeezing flux tubes in convective flows. They can also be generated out of transverse flux tube waves, similar to water splashing out of a water hose that is shaken around. The presence of a magnetic field changes the propagation speed of longitudinal tube waves only slightly; however the onset of shock formation and heating can be delayed in flux tubes that expand rapidly because the wave amplitude grows slower when the wave energy is spread over an increasing area. Otherwise longitudinal tube waves have similar properties as regular acoustic waves; in particular they heat rapidly and are thus good candidates for heating magnetic parts of the chromosphere of the Sun and other stars (Cuntz et al. 1998).

The magnetic field can also play a much more active role in the heating process, for example when transverse or torsional wave motions of the magnetic field carry the main wave energy, or when stored magnetic energy is released when different magnetic structures collide. Such heating mechanisms will be discussed further in Section 5.1 in the context of coronal heating. Several of the mechanisms that might heat the corona could also contribute to the heating of magnetic parts of the chromosphere.

3.4 Thermal and dynamic structure of the chromosphere

An image taken in almost any spectral line formed at chromospheric temperatures reveals prominent spatial structure, the most dominant being the so-called chromospheric network. This consists of patches of enhanced brightness located along the boundaries of supergranulation cells (e.g. Bonnet et al. 1982). This network is quite inhomogeneous and patchy, with its brightness being related to the amount of magnetic flux concentrated at that particular location. In general, the chromospheric network looks qualitatively similar to that visible in lower transition-region lines, such as He II 304 Å, although the individual network features in the chromosphere are usually finer scale. In the chromosphere the enhanced brightness in the network is thought to be caused by the dissipation of waves travelling along the magnetic flux tubes (cf. Section 3.3). Magnetoacoustic waves, that is acoustic waves modified by the magnetic field and propagating along the field lines are the principal mode thought to be of importance for chromospheric heating. In contrast to other, incompressible, wave modes supported by magnetic flux tubes magnetoacoustic waves with reasonable amplitudes in the photosphere steepen to form shocks at chromospheric heights. Such shocks provide an efficient mechanism for the dissipation of the wave's energy, that is its conversion into thermal energy of the local gas.

Although in atomic lines the brightness contrast between network and cell interior is generally smaller in the chromosphere than in the transition region, there is evidence from molecular transitions that this is not the whole story.

The fundamental band of the rotational–vibrational transitions of the CO molecule is located around 4.8 μm in the IR. When observed near the solar limb or off the limb the cores of the strongest lines of this band are formed in the chromosphere and exhibit temperatures as low as 3500–3800 K, that is below the traditional temperature-minimum value (Ayres and Testermann 1981, Solanki et al. 1994). This led to the picture that much of the lower solar chromosphere is in a cool state, quite different from that described by standard chromospheric models (Section 3.2). Combining these observations with the picture revealed by atomic chromospheric lines such as the Ca II H and K lines (which show the bright chromospheric network) it was concluded that only a small fraction of the solar chromosphere is actually in a hot state. This consists mainly of the network, with the cell interiors being very cool according to this picture (Ayres et al. 1986). On the other hand, using data obtained by SMM, Athay and Dere (1990) deduced that at least in the layers at which the O I and C I lines at UV wavelengths are formed, 90% of the solar suface is covered by gas at chromospheric temperatures.

The theoretical ideas used to explain this complex and seemingly contradictory thermal structure have evolved considerably with time. Ayres (1981) first pointed out that the CO molecule itself may contribute to the high contrast in temperature between that deduced from atomic lines and the CO lines. If the heating rate is not sufficiently high and the chromospheric temperature drops below a given value (roughly 4000 K), CO begins to form rapidly. Since this molecule is an efficient radiative cooling agent it then lowers the temperature even further until a steady state is reached. Sufficiently above the critical temperature CO never forms and the atmosphere remains hot. This process leads to a thermal bifurcation.

More recently another, more dynamic scenario has emerged from the one-dimensional radiation-hydrodynamic simulations of Carlsson and Stein (1995). These indicate that the passage of shock waves through the chromosphere produces strong peaks in temperature that can be higher than the traditional chromospheric temperature. The chromospheric emission in atomic lines samples mainly these high temperatures, while the CO lines are a better diagnostic of the cool gas between the shocks, since CO gets dissociated at the shock temperatures. Interestingly, when averaged over time the chromospheric gas in these simulations is cool, with temperatures close to those deduced from CO (Section 3.3).

Finally, in three dimensions we expect interactions between acoustic waves propagating in different directions (Rutten and Uitenbroek 1991) as well as overshooting from granular convection in the lower chromosphere (e.g. Steffen and Muchmore 1988) to also play a role in shaping and structuring the chromosphere.

3.5 Magnetic canopy

Magnetograms obtained near the solar limb in chromospheric lines reveal the presence of large patches of almost horizontal magnetic field, which have been interpreted as the base of a magnetic canopy (Giovanelli and Jones 1982). This means that above a certain height, which is thought to be around 700–1000 km in the quiet Sun, the atmosphere is filled with magnetic field. Recently, independent Hanle effect measurements have confirmed the presence of such a magnetic canopy (Bianda et al. 1999). The Hanle effect is a quantum interference effect that allows the measurement of much weaker magnetic fields in the solar atmosphere than is possible with the generally used Zeeman effect.

The explanation of such a low-lying magnetic canopy makes use of the large temperature contrast between the hot magnetic elements and their cool surroundings since the pressure scale height is proportional to temperature. The magnetic field in a flux tube is confined by the excess of gas pressure in the surroundings. In the hotter flux tubes the gas pressure decreases less rapidly with height than in the cooler surroundings. Solanki and Steiner (1990) showed that for empirically derived temperature stratifications the internal gas pressure becomes larger than the external value at around 700–1000 km. At this height the magnetic field cannot be confined by the external gas any more and spreads out rapidly. Thus a canopy is formed, in agreement with the observations.

One interesting quantity describing the relative importance of gas and field for the dynamics is the plasma $\beta = 8\pi p/B^2$, which is simply the ratio of the thermal energy density of the gas to the magnetic energy density. In the photospheric layers of flux tubes the magnetic energy density dominates over that of the gas, that is, $\beta < 1$. In the canopy, however, β can locally become larger than unity, since due to the rapid expansion the field becomes rather weak. Higher in the atmosphere (e.g. the corona) $\beta \ll 1$ everywhere and the dynamics are to a large extent magnetically driven. This decrease in β is produced because the gas pressure decreases exponentially, whereas above the canopy the field strength decreases only slowly, roughly following a power law.

4 TRANSITION REGION

The transition region between the chromosphere and corona belongs to the most fascinating parts of the solar atmosphere. It separates two vastly different temperature regimes, in which the energy balance between heating and cooling processes operates in different ways. This thermal interface is thin, but highly structured and extremely dynamic. There

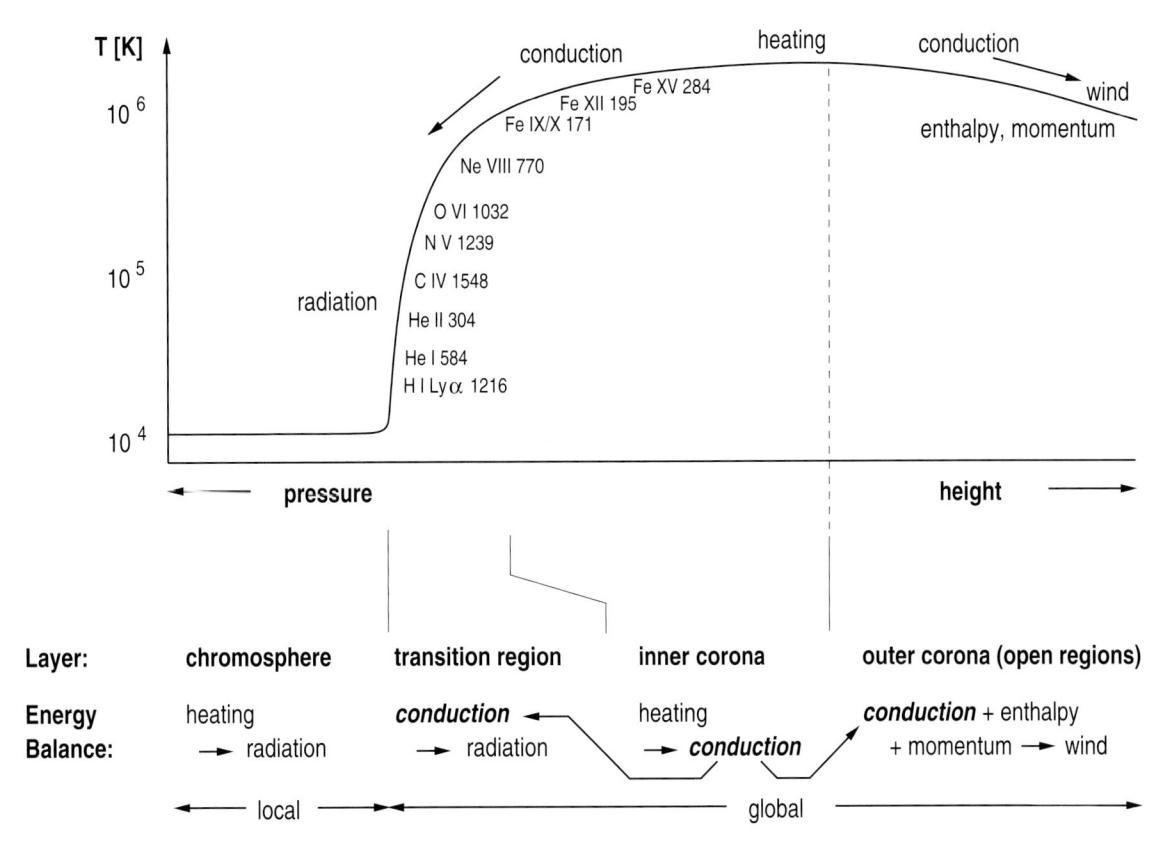

Figure 5 Schematic representation of the energy balance and average temperature along a magnetic flux tube in the outer solar atmosphere. The right-most part of the diagram (labelled outer corona) applies only to 'open' flux tubes that extend out to interstellar space. This part of the diagram does not apply to flux tubes belonging to a coronal magnetic loop; here the remaining part of the diagram describes one loop 'leg', from the solar surface up to the point near the loop top where the maximum temperature is reached. The chromosphere is characterized by a *local* balance between heat input and radiative output. Transition region and corona are characterized by a *global* energy balance, where the heat input is redistributed by thermal conduction to the places where it is needed. In both open and closed flux tubes, heat is conducted back into the inner corona and transition region, where the densities are high enough that the energy can be radiated away in strong emission lines of various ions. In magnetically open regions, heat is also conducted outward and helps to lift the solar wind out of the solar gravitational field and to accelerate it to its high speed. In addition to conduction, the wind is also powered by the energy set free by the cooling of the outflowing gas ('enthalpy') and by the direct energy and momentum input from waves.

are indications that significant variations occur on spatial and temporal scales that are smaller than could be resolved with the best current instruments. In fact, the transition region may be that part of the solar atmosphere that will ultimately impose the highest demands on the spatial and temporal resolution of future space observations if we ever want to understand the dominating physical processes in sufficient detail.

4.1 Energy balance

Before we discuss small-scale variations, a few simple theoretical considerations are in order, to explain the different character of the energy balance in the chromosphere and corona, why these two regimes are separated by a thin transition region and where this transition region is located. The essence of these theoretical considerations is illustrated in Figure 5.

Throughout the entire chromosphere, the Sun is able to adjust the temperature and ionization state of the gas in order to radiate away the energy that is locally deposited by the heating mechanisms. The further out we go, however, the lower the density becomes, and the longer it takes for the atmosphere to cool by radiation after a heating event. Ultimately radiation becomes so inefficient in the outer, tenuous parts of the solar atmosphere that any substantial amount of heating leads to the formation of a hot corona.

In the corona, the temperature is large enough that all atoms are highly ionized, and that the free electrons have such high thermal speeds that they can transport energy very efficiently from hotter to cooler places. This energy redistribution by thermal conduction changes the character of the energy balance fundamentally, since now the energy sources (where the heating occurs) and sinks (where the radiation is emitted or gas motions are powered) can be spatially separated. It is through this trick of globalizing its energy balance that the corona can once again achieve energy equilibrium. Thermal conduction (and to some extent also flows and waves) collects the heat input in the inner corona and transports it back towards higher densities where it can be radiated away more easily. In magnetically open regions, part of the energy is also transported outward, where it helps to lift the solar wind plasma out of the gravitational field of the Sun and to accelerate it to its final speed.

A major energy sink of the corona is the transition region to the underlying chromosphere. Within this thin layer the temperature jumps by two orders of magnitude, from around 10^4 K in the upper chromosphere to around 10^6 K in the corona (Figure 5). Many ions with strong spectral lines exist in this temperature range. Particularly large amounts of energy are emitted in the resonance lines of the most abundant elements, hydrogen (H I Ly α 1216 Å) and neutral (He I 584 Å) and singly ionized (He II 304 Å) helium, which are all formed at the foot of the transition region. Numerous other strong spectral lines from a variety of ions are also formed in the transition region. Since virtually all of these emission lines are located in the UV and EUV parts of the spectrum, which are absorbed by the Earth's atmosphere, any direct observational information on this layer must come from space instruments.

For the outer solar atmosphere, the total emitted energy per volume and time can be shown to vary as the square of the density times a function of temperature. This function peaks around 10^5 K; and moreover the density in the lower transition region is two orders of magnitude higher than in the corona. This explains why the transition region radiates much more efficiently than the corona. Plasma with a density and temperature typical of the lower transition region can radiate away its thermal energy in a so-called radiative cooling time of only a few seconds, while plasma with properties typical of the inner corona needs about an hour to cool by radiation.

For this reason a lot of energy is transported by thermal conduction from the corona back into the transition region, from where it is radiated away. At high temperatures, thermal conduction is mostly due to electrons, which are bound to follow the magnetic field lines. The conductive energy flux is then given by the temperature gradient along the magnetic field times the thermal conductivity, which is determined by the speed at which the electrons can move. As the conductive energy flows downward to lower temperatures, the electron speed decreases rapidly. The resulting decrease of the conductivity must be offset by a steepening of the temperature gradient in order that the energy can still be transported further down. As a result, the transition region is a thin layer with a particularly steep temperature gradient in its lower parts.

The location of the onset of the transition region within the chromosphere determines its pressure, or density, and thus the amount of energy that it can radiate away. This location is therefore adjusted to changes of the coronal heating rate. Suppose the energy input into the corona is temporally enhanced. Then the excess energy can no longer be radiated away either in the corona or the transition region at their current densities. It is therefore conducted right into the upper chromosphere, where it heats up plasma, which then expands into the corona. This process of 'chromospheric evaporation' effectively pushes the transition region downward until the density in the transition region and corona becomes high enough that the excess heat input can be radiated away. Conversely, when the heat input into the corona is temporally reduced, coronal plasma cools and flows back down into the chromosphere ('coronal condensation'). In this way fluctuations in the coronal heating rate generate flows through the transition region.

4.2 Structure

The transition region is not simply a thin, spherically symmetric shell around the Sun. It is highly structured by the magnetic field. The latter fills all available space already above the middle chromosphere (Section 3.5), but is still distributed inhomogeneously in the overlying transition region and corona. Even along a given magnetic field line the location of the transition region varies in time, in response to the coronal heat input, as discussed in Section 4.1.

On the disk, the most prominent feature seen in spectral lines from the lower transition region (Figure 6) is the network, an extension of the chromospheric network to higher temperatures. The brightest network elements are up to an order of magnitude brighter than the darkest points in the cell interior. However, not the entire network lane area is bright at any given point in time. The total network area accounts for about two-thirds of the total emission. The width of the network lanes is typically around 7 Mm (e.g. Patsourakos *et al.* 1999), while the cell diameters are of the order of 20–30 Mm. Beyond 2.5×10^5 K the network widens rapidly with increasing temperature, hence it becomes increasingly diffuse and is no longer recognizable above 10^6 K (Figure 2).

The ultimate reason for the enhanced emission in the network is that there the magnetic flux density is larger than in the cell interior. Horizontal convective flows transport newly emerging magnetic flux from the interior of supergranulation cells towards the borders, where it accumulates. Studies with the MDI instrument on the SOHO satellite showed that some of these flux elements disappear during this trip, for example by subduction under the solar

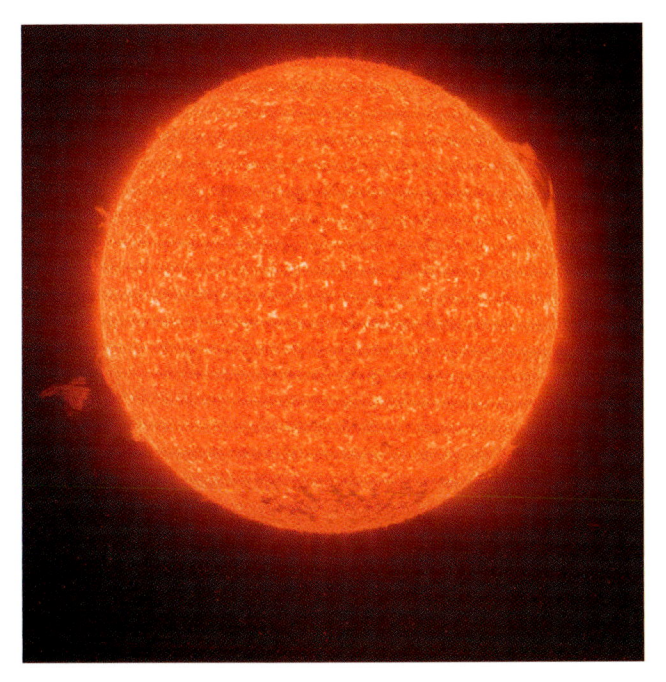

Figure 6 Solar image, taken with the Extreme Ultraviolet Imaging Telescope (EIT, Delaboudinière et al. 1995) on SOHO in a wave-length band dominated by the spectral line He II 304 Å of singly ionized helium, formed around 60 000 K in the lower transition region. On the disk, the most characteristic feature seen at these temperatures is the network structure. The limb shows several prominences, which consist of plasma much cooler than the surrounding corona, supported by a magnetic field. These structures can become unstable and erupt, as in the lower left. Other limb features are needle-like EUV macrospicules. The significantly reduced brightness in coronal holes (like the one at the bottom of the image) is not typical of spectral lines formed in the lower transition region, but a special property of helium lines. (Courtesy of the SOHO/EIT team. SOHO is a project of international cooperation between ESA and NASA.)

surface. Others, however, reach the borders of the supergranulation cells, at a rate sufficient to replenish the network magnetic field within 40 hours (Schrijver et al. 1998). Some of these magnetic flux elements collide with pre-existing network field, often causing reconnection events (Chapter 43). The latter might be responsible for most of the observed dynamics, which will be discussed later. Reconnection also leads to the reconfiguration of the magnetic field. As a result, the network field is a continuously changing mixture of magnetic flux tubes of various geometries and sizes, consisting mostly of a range of magnetic loops, from the smallest ones that cannot yet be resolved observationally to large ones that connect to the solar surface at large distances. In some areas there are also open flux tubes, which reach out all the way to the interstellar medium. The small loops do not extend to large heights.

Thus with increasing height the magnetic structure becomes simpler, and only the so-called coronal funnels (the legs of large loops and open regions) survive and expand into the available space above the cell interior. Models of this expansion (Gabriel 1976) describe the observed widening of the network in the upper transition region quite well. The emission from the lower transition region ($T < 2 \times 10^5$ K), however, is much harder to model. This will be discussed in more detail in Section 4.4.

While with increasing temperature the network fades, other features begin to dominate the appearance of the Sun (Figure 2). They outline the magnetic field structure on larger scales and show us the basic components of the corona: coronal holes are regions where the large-scale magnetic field is unipolar, while active regions consist of complex systems of magnetic loops with enhanced magnetic field strength, temperature and pressure. The corona outside of holes and active regions is called the quiet corona; it is composed of loops of lower temperature and pressure than in active regions, and perhaps some interspersed small open field regions.

Coronal holes are barely noticeable in the emission intensity from temperatures corresponding to the lower transition region. The small-scale magnetic network seems to be similar, although not identical (Huber et al. 1974), underneath coronal holes as under the quiet corona. With increasing temperature (beyond 5×10^5 K) the brightness contrast between quiet Sun and coronal holes increases, and the densities and temperature gradients are smaller in coronal holes. Near 10^6 K (Figure 2), finally, no emission is visible any more from coronal holes because their maximum electron temperature is somewhat smaller. Here we see only emission from the diffuse quiet corona (with maximum temperatures in the range $1-2 \times 10^6$ K), active regions (with temperatures of several million kelvin) and compact bright points.

Along the solar limb, several types of inhomogeneities can be seen at chromospheric and transition region temperatures, including prominences and various types of spicules. Prominences are relatively cool clouds of gas (at around chromospheric temperatures) that are embedded in the hot corona and usually supported against gravity by the magnetic field. They are often surrounded by, and interspersed with, plasma emitting at transition region temperatures. Spicules can be seen in strong chromospheric lines as columns of gas protruding out of the solar limb, 10 Mm high and less than 1 Mm thick. They consist of chromospheric matter that is ejected upward at speeds exceeding 20 km s^{-1} and then either falls back or disappears from the visible part of the spectrum after a total lifetime of 5–10 min. Spicules transport much more mass into the corona than is needed by the solar wind, so essentially all of it must flow back. Those spicules that are not seen to fall back at chromospheric temperatures are probably heated up to a few times 10^5 K before most of their gas flows down again and contributes to the transition region emission. On the other hand, as long as the spicule matter is cool, it absorbs part of the transition region emission at short wavelengths from the solar limb. These emission and absorption contributions by spicules are one reason why the transition region does not appear as an extremely thin spherical shell at the solar limb. Macrospicules are giant versions of spicules within coronal holes; they can be seen in EUV lines formed at temperatures up to 2×10^5 K. Fascinating movies of EUV macrospicule jets have been obtained with Skylab and with EIT and CDS on SOHO. The ultimate cause for the upward ejection of (macro)spicular plasma has not yet been identified, although several possible mechanisms have been proposed (including various types of waves as well as scenarios involving the buffeting of magnetic flux tubes by granular motions in the photosphere).

Unfortunately, the best spatial resolution (down to 1 arcsecond, or 700 km on the Sun) that has so far been achieved with space instruments (like the Transition Region Camera (TRC; Bonnet et al. 1980), the High Resolution Telescope and Spectrograph (HRTS; Bartoe and Brueckner 1975), the Normal-Incident X-ray Telescope (NIXT; Golub et al. 1990), SOHO/SUMER and TRACE) has turned out to be insufficient to resolve much of the fine structure that we now know to be important for the physics of the transition region. We are still far from understanding the complex magnetic topology of the lower transition region and the structure, acceleration mechanisms and ultimate fate of spicules. While recent space experiments have given us a glimpse of the amazing level of fine structure that governs this part of the solar atmosphere, they also uncovered a number of problems that can be solved only with future observations with even better resolution.

4.3 Dynamics

The transition region is not only complicated because of its fine structure, but also due to various types of motions and a generally high level of temporal variations. This is impressively illustrated by TRACE movies like those distributed on CD-ROM with the review article by Schrijver et al. (1999). They show variations and apparent motions on all spatial and temporal scales – including brightening, moving and oscillating magnetic loops; magnetic field reconfigurations with associated mass ejections; or up- and downflowing cool gas in spicules and active-region filaments.

From such movies, however, it is often not easy, or even feasible, to assess the speeds involved – some of the apparent motions might simply be moving wave fronts rather than moving gas. A more direct, but still not always unique, measurement of plasma velocities is based on the analysis of the positions and shapes of spectral lines recorded with instruments such as SO82 B (Bartoe et al. 1977) on Skylab, OSO 8, HRTS or the spectrometers SUMER and CDS on SOHO.

Such measurements identify motions both towards and away from the observer. The gas that contributes most to the emission from lower transition region temperatures is predominantly moving towards the Sun (Figure 7, left panel), at speeds reaching a maximum of $10 \,\mathrm{km\,s}^{-1}$ for temperatures near 2×10^5 K. The average downflow speed decreases both towards lower and higher temperatures and turns into upward speeds for $T > 5 \times 10^5$ K (Peter and Judge 1999; Figure 7). It has been suggested that the apparent down- and upflows represent the compression regions of downward and upward running acoustic waves that are generated by sudden heating events in magnetic loops. Alternatively the observed flow pattern could be caused by chromospheric gas (like spicules and similar phenomena) that was ejected upward at cooler temperatures and then heated to a few times 10^5 K. In this picture, the main part of the gas falls back towards the Sun and causes the downflow at lower temperatures, while the hottest parts of the gas might further expand into the corona, thus causing the outflow at higher temperatures. Flows initiated by highly asymmetric and temporally variable heating in magnetic loops have also been suggested to contribute to the observed velocities. The ultimate explanation of this phenomenon needs observations with even higher resolution as well as more sophisticated numerical simulations. The upflows at higher temperatures come mostly from network boundaries or intersections of boundaries, in particular in coronal holes, where the large-scale magnetic field is open. Here the measured outflows have been interpreted as a signature of the onset of the fast solar wind (Hassler et al. 1999, Peter and Judge 1999, Stucki et al. 1999, Wilhelm et al. 2000), which has long been known to emanate from coronal holes.

Wave motions have also been identified, either as periodic oscillations in individual spectral lines formed in the network and in active regions, or as a time delay between velocity fluctuations in lines formed in the chromosphere and the transition region. The latter provides evidence for upward travelling waves in the network cell interior (Wikstøl et al. 2000), presumably the transition region remnants of strong shocks in the chromosphere (Section 3.3).

The width of transition region spectral lines (with the exception of the strongest ones) is mainly determined by the thermal motion of the atoms and by macroscopic, but spatially unresolved gas motions. The latter, so-called nonthermal motions, are found to reach values up to $30 \,\mathrm{km\,s}^{-1}$, with a large scatter, and are even larger in coronal holes than in the quiet Sun. Spectral line shapes are often a superposition of multiple components (Kjeldseth-Moe and Nicolas 1977), most often consisting of a narrow main component and a weaker, but broader component that is slightly shifted in

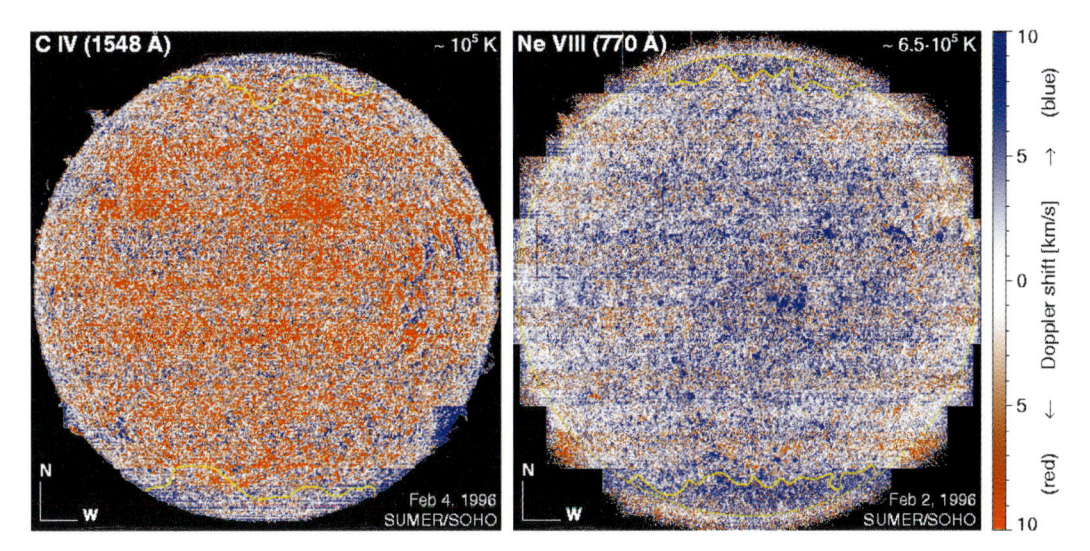

Figure 7 Doppler shifts of spectral lines in the transition region, measured by scanning with the SUMER instrument on SOHO over the Sun. On the disk, emission from lower transition region temperatures (left panel) is dominated by redshifts. They could be produced by flows and downward travelling waves generated in transient, asymmetric heating events in magnetic loops or by down-falling plasma at 10^5 K that has been lifted up at other (presumably lower) temperatures. In the upper transition region (right panel) blueshifts prevail. Coronal holes (bordered by the yellow lines near both poles) are the source regions of the fast solar wind, which is a possible cause of the dominance of blueshifts in these areas. (After Peter 1999.)

wavelength. According to a statistical analysis of SUMER data by Peter (2000), this phenomenon is restricted to bright network elements and can be attributed to the different magnetic structures that exist there: coronal funnels and small-scale magnetic loops (Section 4.2).

Temporal variations in the transition region occur on many different scales. Very common are sudden brightenings in the EUV intensity as detected with CDS and SUMER on SOHO. These events ('blinkers', e.g. Harrison 1997) were identified as density enhancements lasting about half an hour on average. Related brightenings have also been detected with EIT on SOHO. They might represent the reaction of the transition region to energy release events in the corona.

More violent explosive events have been observed with HRTS and SUMER. They occur mostly at network borders, like the brightenings (termed microflares) detected with SMM. Strong lineshifts indicate that gas is ejected from the explosion site as a bidirectional jet with speeds up to 150 km s^{-1} (Innes *et al.* 1997). Simultaneous magnetograph measurements show that explosive events are associated with magnetic field cancellation. Their most likely explanation is therefore that they are generated when small magnetic loops are transported by the convective flow to the network border, where they collide with a pre-existing field in such a way that magnetic field reconnection occurs, which reorganizes the field and converts magnetic energy into heat. At higher temperatures in the corona, Yohkoh (Ogawara *et al.* 1991) has also detected jets associated with reconnection.

These latter types of variability point towards a continuous spectrum of reconnection events, ranging from large flares (a flare is a strong, reconnection-driven explosion; cf. Chapter 46) to microflares and then down to nanoflares, the latter being small-scale heating events along individual field lines within magnetic loops that were suggested to be a basic heating mechanism of coronal loops (Parker 1988). There has been a lot of discussion if the combined effect of all these observed variations is already sufficient to heat the corona in magnetically closed regions. At the time of writing it appears that other heating mechanisms are operating as well (e.g. Aschwanden *et al.* 2000) – but this issue is not yet finally solved, mainly because it is very difficult to estimate the energy released in the smallest events, which are the most common ones.

4.4 Energetics of the lower transition region

As the previous discussion showed, the lower part of the transition region (below 2×10^5 K) plays a special role: it is structured on finer scales and is temporally more variable than the upper transition region, down to scales that lie beyond the capabilities of current instrumentation. Another problem is that the emission from the lower transition region is much more intense than can be explained by simple steady-state transition region models that are based solely on energy supply by classical electron thermal conduction along the magnetic field (Section 4.1). While such models can well reproduce the average emission of the upper transition region, they fail to explain the observed intensity from plasma below about 2×10^5 K by several orders of magnitude. One reason for this discrepancy is probably that this type of thermal conduction, which dominates in the transition region and inner corona, becomes inefficient at small temperatures, so that the temperature gradient becomes very steep, as explained in Section 4.1. In typical models of this kind, the thickness of the lower transition region is of the order of only 100 km. Such a thin layer does not contain much plasma, and its total emission is thus small – much smaller than observed.

In the real Sun other, more efficient types of conductive energy transport become important in the lower transition region. First, ions and neutral atoms contribute in addition to the electrons. Diffusion lets neutral atoms drift upward to higher temperatures, while ions and electrons drift downward. Second, it is likely that the plasma is strongly filamented by the magnetic field on small spatial scales. If neighbouring filaments have different temperatures, then thermal conduction *across* the magnetic field becomes important at low temperatures (Rabin and Moore 1984). Third, it is possible that the large observed line widths are produced by unresolved turbulent motions, as suggested by some heating theories. The associated energy transport by turbulent eddies enhances drastically the thermal conduction in the lower transition region (Cally 1990). Models incorporating these types of diffusion and enhanced thermal conduction could successfully account for the strong emission at low temperatures.

A number of other possible explanations have been suggested as well. Waves might deposit momentum and energy directly in the lower transition region, thus reducing the need for thermal conduction from above (e.g. Woods *et al.* 1990). Even without direct heating, waves increase the emission due to associated time-dependent effects (Wikstøl *et al.* 1998), although perhaps not to the required amount (Feldman 1998) – but this needs further clarification in terms of more detailed models. Similarly, time-dependent coronal heating also enhances the emission (Athay 1984, Sturrock *et al.* 1990, Roumeliotis 1991).

Moreover, it is important to realize that the fastest electrons have the lowest collision rates with other particles and could thus penetrate the thin transition region with few or no collisions. In particular, fast coronal electrons could reach the lower transition region (Shoub 1983). Even the possibility that the fastest of the chromospheric electrons could, without collisions, produce the entire transition region and corona without any further heating has been discussed (Scudder 1992) and found to produce the correct emission at low temperatures, but to be inconsistent, in its simplest form, with the emission and other properties at higher temperatures (Anderson *et al.* 1996). This latter suggestion needs to be studied further with more detailed theoretical models.

So there exist indeed a number of promising ideas for explaining the observed excess emission at temperatures corresponding to the low transition region (smaller than about 2×10^5 K). The most often discussed suggestion, however, is that a sizable fraction of this emission does not at all originate from a 'transition region' between chromosphere and corona, but rather from an ensemble of magnetic loops that do not reach coronal temperatures and are so small that they could not yet be resolved with current instruments (Feldman 1983, Antiochos and Noci 1986, Dowdy *et al.* 1986; Section 4.2).

The emission from the important spectral lines of helium, the second most abundant element of the Sun, poses additional problems. Both neutral and singly ionized helium requires particularly large amounts of energy to reach electronic states from where lines can be emitted. These large energies can be provided either by collisions with fast electrons in the transition region, or by coronal EUV and X-radiation that ionizes chromospheric helium atoms, which upon subsequent recombination reach the excited energy states from where the lines are emitted.

In relatively weak helium lines we can see through the transition region into the chromosphere, where the latter mechanism dominates. Therefore, such lines are sensitive to radiation from the overlying corona and have thus been used as proxies to infer the amount of coronal X-ray emission, both for the Sun and for other types of stars.

In the strongest helium lines (like the one shown in Figure 6), however, we can usually see only into the transition region, where electron collisions dominate (Andretta and Jones 1997). When compared to other lines excited by electron collisions in the transition region, however, these helium lines are much more intense, by typically an order of magnitude. In fact, the strongest helium line is usually even stronger than the available ionizing EUV and X-ray emission from the corona, which confirms that it is predominantly excited by electron collisions, not by photoionization. Jordan (1975) suggested that the enhancement could be caused by some process that mixes ions with hotter electrons. This would affect the emission from helium more than from other atoms because helium needs more energy for ionization and more time to adjust its ionization equilibrium. Several of the mechanisms discussed above could effectively provide such a mixing of helium ions with hot electrons: diffusion, turbulence, nonthermal electrons or time-dependent effects such as waves or 'bursty' heating. The enhancement factor is observed to increase from coronal holes to quiet regions. This suggests a dependence of the mixing process on the density, temperature gradient or the magnetic field structure of the overlying corona.

Many details of the formation of the helium spectrum are not yet understood, despite active ongoing research. Recent studies involved instruments like SUMER and CDS on the SOHO satellite, combined with simultaneous ground-based observations or rocket experiments. Several such studies attempted to compare the spatial and temporal behaviour of helium lines with other lines. But the data are complex, and their correct interpretation requires radiative transfer and hydrodynamic calculations, which need further refinements. And the small spatial and temporal scales that characterize the lower transition region call for observations with higher resolution than our current instruments have achieved.

Helium emission is very important for the energetics of the transition region and upper chromosphere, mainly because helium is the most abundant element after hydrogen. Quite surprisingly, its abundance is not constant. Precise values are known only for the solar interior (where the helium to hydrogen abundance ratio is 8.4% by number, as determined from the analysis of solar oscillations) and in the distant solar wind (where the abundances have been measured directly by spacecraft, as discussed in Chapter 47, and where the helium abundance is found to be only about half the surface value, and highly variable). In the entire solar atmosphere, however, the variation of the helium abundance as a function of height and magnetic structure is comparatively uncertain because it is very difficult to determine from spectroscopic observations, in particular in the upper chromosphere and lower transition region, where the important helium lines are formed. It is possible that small-scale abundance variations in space and time contribute to the observed anomalous behaviour of the helium lines.

5 CORONA

There is no unanimously accepted definition of a boundary temperature that separates the upper transition region and corona. An image like Figure 2, which probes plasma at temperatures near 1 million kelvin, illustrates only partial aspects of each of the basic components of the inner corona: coronal holes dominated by an open magnetic field configuration; active-region magnetic loop complexes and their small-scale counterparts, bright points; and finally the quiet corona consisting of more diffuse magnetic loops and small-scale open regions. Measurements from SOHO instruments showed that the electron temperature in coronal holes rises only up to around 8×10^5 K, while from Yohkoh we know that active region loops reach $2-5 \times 10^6$ K during quiet phases, and up to 10^7 K during heating bursts. An image like Figure 2 therefore outlines cooler coronal loops in the quiet corona and in the outer parts of active regions but does not contain significant emission from the coronal parts of coronal holes, while in the hottest parts of active regions it shows only the transition region footpoints of very hot loops. The coronae of magnetically closed and open regions in the solar atmosphere are fundamentally different, not only with respect to their magnetic topology, but also according to their energy balance, temperature, density and other physical properties. Therefore it is useful to discuss them separately.

5.1 Coronal loops

The presence of large magnetic loop structures in the outer solar atmosphere has long been known from ground-based observations such as those of loop prominences at the solar limb, or from images taken in the few coronal spectral lines that happen to fall in the visible part of the spectrum and are thus accessible from the ground. But only after the first rocket-borne X-ray observations, and in particular since the highly successful Skylab mission, did it become evident that essentially the entire X-ray emission from the Sun originates in magnetic loops, and that these loops are not restricted to a few activity centres, but dominate both the active and quiet corona outside of coronal holes. For a review of these early findings see, for example, Vaiana and Rosner (1978). The one-to-one correspondence between magnetic loops and X-ray emission makes X-ray observations an ideal tool for studying magnetic activity on stars other than the Sun (cf. Chapter 38).

A series of subsequent space missions led to a continuous refinement of our knowledge of solar coronal loops. Over a complete solar cycle, Yohkoh monitored the properties of loops at high temperatures, to which it is particularly sensitive. The rocket-borne NIXT experiment provided snapshots of the somewhat cooler corona with higher spatial resolution. EIT evolved into a major workhorse on SOHO, taking images of the entire Sun in four different temperature regimes of the transition region and corona, which proved indispensable as synoptic reference for observations with other SOHO instruments. The latter include the spectrographs SUMER and CDS, which measure flow speeds, temperatures and densities in the transition region and inner corona; and the coronagraphs UVCS (Ultraviolet Coronagraph Spectrometer) and LASCO (Large Angle and Spectrometric Coronagraph), which map the structure and flows of the large-scale corona and solar wind. Like EIT, TRACE obtains images at several different temperatures, but with much higher spatial resolution, at which it can observe only smaller sections of the solar disk at a time.

With respect to coronal loops, perhaps the single most important result of all these observations is the appreciation of the high level of structuring and dynamics in the corona.

For example, images with the so far highest resolution show thin magnetic loops that are only one pixel wide along their entire length. It is likely that these fine loops are still underresolved and in reality even thinner. Such images were taken by TRACE in a wavelength band sensitive to radiation from a limited temperature range, usually around 10^6 K. Does the fact that we recognize the complete loop then imply that it has a virtually constant temperature along its entire length? If so, this would impose severe constraints on the distribution of heating along the loop (Schrijver et al. 1999) and/or its variation with time. But alternatively it is also possible (Reale and Peres 2000) that even such thin loops consist of a bundle of even thinner magnetic threads, all having their own separate energy balance, temperature profile and time evolution. If this is the case, the emission from the cooler threads would contribute to the TRACE wavelength band predominantly near the centre part of the loop, while the footpoints of the hotter threads would dominate the image near the loop ends. Neither of these different types of threads needs to have constant temperature. Moreover, one should expect that such extremely thin filaments are also temporally variable, so that our observations average in both space and time. Thus the seemingly constant temperature within the observed loops could well be an illusion due to the fact that our spatial and temporal resolution is not yet high enough to resolve the important physical phenomena at their intrinsic scales. Only future improved measurements can help to resolve these issues.

A similar averaging effect could also explain why Yohkoh observes generally thicker loops than TRACE. Yohkoh samples a broader temperature range, to which more threads might contribute (Peres 1999). Another aspect to keep in mind is that Yohkoh senses higher temperatures, where the radiative cooling times are much longer (Section 4.1). Therefore, if the temperatures of individual loop threads are controlled by heating events followed by cooling, one observes at any given instant more threads at high temperatures than at low ones, so one should expect a general trend towards sharper contours at lower temperatures.

In active regions, TRACE observed in its 10^6 K band an irregular pattern of fine-scale structures that resembles, and was therefore termed, 'moss' (see the review by Schrijver et al. 1999). This phenomenon is restricted to regions in which simultaneous Yohkoh observations detect hot overlying loops of at least 3×10^6 K. Hence, moss has been interpreted as the upper transition region in hot loop threads. The moss pattern is highly variable, partly because of absorbing cool matter that moves up and down in front of the loop footpoints. This is possible because bright, hot loops have high pressures, which push their transition regions down to height levels below the top of the surrounding chromosphere.

The reason for the overall bright emission from loops is that they keep the gas trapped within them. Since the matter is highly ionized at coronal temperatures, all particles can only spiral around, and move along (but not across), the magnetic field lines. In the inner solar corona, the gas pressure is typically an order of magnitude smaller than the pressure associated with the magnetic field ($\beta \approx 0.1$; Section 3.5), so the gas in a coronal loop cannot normally open up its container (except for the outermost loops, which will be discussed in Section 5.2). As described in Section 4.1, all the energy that is deposited in a loop as heat is ultimately radiated away, after redistribution by thermal conduction and adjustment of the location of the transition region and the loop pressure. The more heat a loop receives, the larger its pressure becomes, and the brighter the emission. By contrast, magnetically open regions can also cool by outward thermal conduction and in particular by energy losses associated with the solar wind, which acts as a 'safety valve'. If a closed and an open flux tube are heated at the same rate, the open one has thus a lower temperature (because of additional cooling mechanisms) and density (because less energy needs to be radiated away) and therefore a much lower pressure.

But what *are* the mechanisms that heat coronal loops? It has become increasingly clear since the late 1970s that acoustic waves cannot transport enough energy beyond the chromosphere (Section 3.3). Therefore, the energy transport into, and very likely also the heating of, the corona must be mediated by the magnetic field. The footpoints of magnetic field lines are anchored in the deep photosphere, where convective and turbulent flows move them around. This gives rise to a large number of different means to transport energy into the upper layers. Unfortunately, we can address here only some of the most basic mechanisms. For comprehensive reviews, see, for example, Narain and Ulmschneider (1990, 1996).

Quite generally, rapid footpoint motions generate waves, which travel upward along the magnetic field; while slow motions let the atmosphere move through a sequence of quasi-equilibrium states, during which electric currents are important.

There exist many different types of waves that travel along the magnetic field. Slender magnetic flux tubes (where the physical quantities can be assumed constant over the cross-section) support three types of waves: *kink* waves when their footpoints are pushed back and forth by the convective motions; *torsional* waves when they are twisted; and longitudinal (or *sausage*) waves when they are squeezed. Additional types of waves are possible in thick magnetic flux tubes, for example modes that run predominantly along the surface of the tube and others that run in the interior. And outside of flux tubes, for example in the space-filling magnetic field above the canopy, there are again different magnetic wave modes possible. All these types of waves have been investigated in some detail, trying to answer questions like: how much wave energy is generated at which frequency in the convection zone; how does the wave propagate along the structured magnetic field; how much of the wave energy 'leaks' into the surrounding environment and how much is converted into other wave modes; what fraction of the energy is reflected off the transition region; and where and how is the mechanical energy ultimately converted into heat? Those wave types that have compression as their major restoring force, such as longitudinal tube waves, dissipate rapidly, like acoustic waves, and are therefore good candidates for heating the chromospheric network, but not the corona. Only waves where the magnetic field provides the main restoring force are viable candidates for heating the corona.

In addition to these wave heating mechanisms (sometimes called AC mechanisms) there are also several possibilities for non-wave heating caused by slow footpoint motions (DC mechanisms). Newly emerging magnetic flux appears all over the solar surface as small loops, which are subsequently swept by the supergranular flow towards the cell boundaries. When they collide with a pre-existing field in the network, discontinuous changes arise in the directions of neighbouring magnetic field lines. In such a situation the magnetic field can reconfigure itself; and part of the stored magnetic field energy is suddenly set free, causing heating, waves, plasma jets and brightenings. (For a more exhaustive description of such processes see Chapter 43.) Reconnection events can also occur when rising freshly emerged flux collides with overlying magnetic flux. Even without collisions between different loops, reconnection might also be important in individual coronal loops. This is because the continuous shuffling of magnetic field line footpoints causes a 'braiding' of the threads that compose the loop, and thus again differences in the direction of neighbouring magnetic field lines, associated with electric currents that may contribute to coronal heating (Parker 1979). If the braiding gets strong enough, the magnetic field again reconfigures itself in reconnection events called nanoflares by Parker (1988). TRACE has indeed observed indications of such a braiding of individual loop threads (Schrijver *et al.* 1999).

Braiding can lead to a state of small-scale turbulence in the corona, which can be maintained by photospheric motions (e.g. Heyvaerts and Priest 1992). In such a state, the conversion of mechanical energy into heat is much more efficient. New results from SOHO and TRACE indicate that at least parts of the corona are indeed in a state of turbulence. For example, TRACE observed coronal loops that oscillated back and forth a couple of times after being hit by a strong wave from a nearby flare. From the damping rate of this oscillation, Nakariakov *et al.* (1999) concluded that the corona is turbulent.

5.2 The open corona

Higher up in the atmosphere the magnetic structure becomes less complex, and magnetic loops disappear. The solar wind opens up and carries away the outermost loop field lines, leading to a configuration with a thin layer with oppositely directed magnetic field on both sides, which contains electric currents. The resulting helmet- and ray-shaped structures are called streamers (Figures 8 and 9). At solar activity minimum, they are mainly confined to the equatorial zone, while with increasing solar activity the streamer belt extends to higher latitudes and becomes more irregular. Except during high-activity phases, large coronal holes are usually located at the polar regions. Their field lines bend around the streamers and occupy at larger distances the entire heliosphere outside of a narrow zone near the equatorial plane (cf. Chapter 47).

The solar wind has been predicted theoretically and then measured by numerous satellite missions (Chapters 9 and 47). One of the main objectives of the SOHO mission was to identify the coronal origins of the solar wind and to study its initial acceleration. Coronal holes have long been identified as the main source of the fast, regular solar wind that dominates the heliosphere at higher latitudes. Coronal holes are structured by plumes, ray-shaped features that can be traced out to several solar radii. They have higher densities and lower temperatures than the interplume regions. Model calculations as well as observations with SOHO instruments (e.g. Wilhelm *et al.* 1998) have now firmly established that the outflow speeds are much lower within plumes, so that interplume regions are the genuine source regions of the fast solar wind. Therefore, the plasma that makes up the fast wind flows first through open funnels within the network (Section 4.2) and is then accelerated within the low-density interplume lanes.

The slower, more irregular wind near the equatorial plane could originate in several different types of solar regions (e.g. Wang 1994, Noci *et al.* 1997), the relative importance of which is still being debated and might depend on the phase in the solar cycle. The possible source regions of the slow wind include (i) the borders of coronal holes, with their larger areal expansion out to large distances (Figure 9); (ii) small open regions between streamers, with their early rapid expansion and subsequent constriction; and (iii) the outermost parts of closed regions, which occasionally open up. SOHO/UVCS measurements indicate that the legs of streamers have similar ion abundances as were measured by satellites in the slow-wind zone.

The SOHO coronagraphs discovered that the slow solar wind is accelerated slowly, while the fast wind is accelerated surprisingly rapidly (e.g. Antonucci 1999, and references therein). Time sequences from the LASCO coronagraph show elongated blobs of gas that appear above

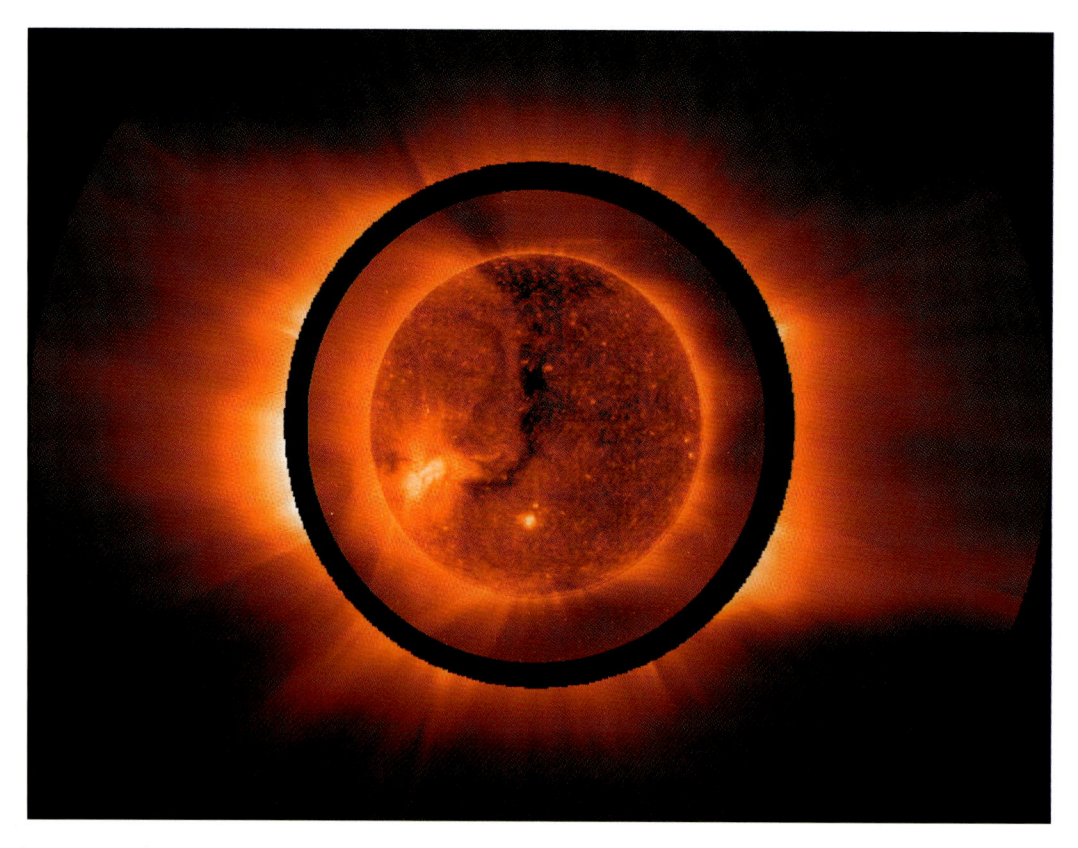

Figure 8 Composite of two SOHO images of the Sun and its corona. The inner part was taken with the EIT instrument and shows the solar disk and inner corona in a wavelength band dominated by iron lines formed near 2 million kelvin. Notice the long extension of the polar coronal hole, two active regions and the streamer regions of enhanced emission over parts of the solar limb. The streamers extend far away from the Sun, as shown in the outer image, taken with the UVCS instrument in the light of fivefold ionized oxygen (O VI). The overall structure of the corona is obviously controlled by the magnetic field. (Courtesy of the SOHO/EIT and SOHO/UVCS teams. SOHO is a project of international cooperation between ESA and NASA.)

the cusps of streamers and are then carried away by the slow wind. They can serve as tracers for measuring the wind speed. One finds that the slow wind is continuously accelerated throughout at least the first 20 solar radii ($R_\odot$), reaching speeds around $300\,\mathrm{km\,s^{-1}}$ there, about three-quarters of the average speed measured at the orbit of Earth. The observed acceleration is consistent with thermal wind expansion (Parker 1958) at a temperature of $1.1 \times 10^6\,\mathrm{K}$. By contrast, observations covering the radial range 1.6–$3\,R_\odot$ in coronal holes indicate that the fast solar wind is accelerated an order of magnitude more rapidly; presumably it reaches its final speed of up to $800\,\mathrm{km\,s^{-1}}$ already within the first $10\,R_\odot$.

It has long been known that the decrease of the density with increasing distance from the Sun must ultimately lead to a situation where collisions between different particles are so inefficient that electrons, protons and other species each have their own temperature and their own energy and momentum balance.

A major result of the SOHO mission was the discovery that the electron temperatures and densities in coronal holes are significantly lower than expected; hence the decoupling between species occurs much earlier than anticipated. The electrons were found to reach their maximum temperature of somewhat less than $10^6\,\mathrm{K}$ already low in the corona, some 10–20% of $R_\odot$ above the photosphere. Beyond this point, the electron temperature declines outward, more rapidly inside plumes than outside. As a result of this low maximum and early decline, and of the inefficient collisional coupling, outward electron heat conduction is lower than expected and may not be able to provide significant amounts of energy out to distances where the fast solar wind needs it: of the order of $R_\odot$ for its potential energy and (as just discussed) a few $R_\odot$ for its kinetic energy. Therefore, much of the wind energy must be supplied either by direct energy input from waves or by enthalpy redistribution (Section 4.1) – that is, by using up the energy from cooling particles that were heated to high temperatures at lower distances.

energy into the gyration motions of ions. The required waves have been suggested to be generated during reconnection events in the network (Axford and McKenzie 1997), but it is still debated if they can reach far enough into the corona or if they are damped so rapidly that they must be replenished continuously within the corona from the turbulent decay of other waves (Isenberg and Hollweg 1983, Cranmer et al. 1999).

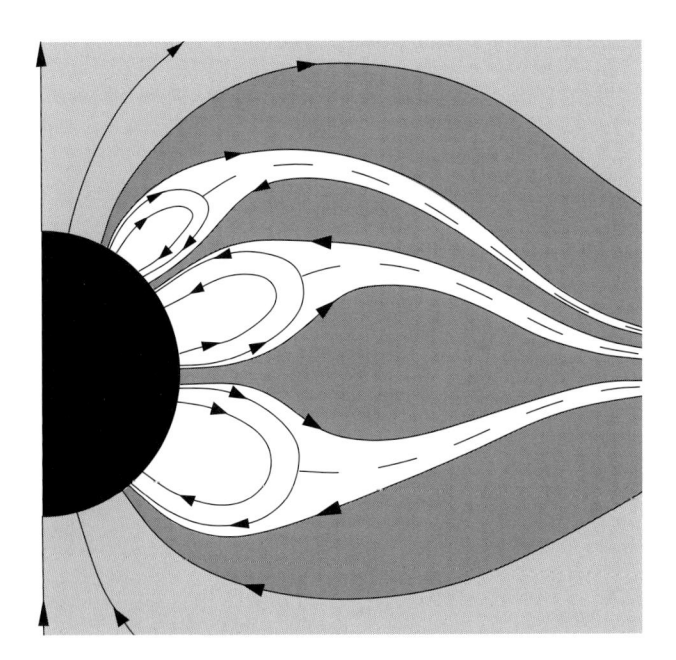

Figure 9 Schematic representation of the large-scale magnetic field structure of the solar corona. The direction of the magnetic field is indicated by arrows. Closed regions exist only up to a certain radial distance, beyond which their field lines are opened up by the solar wind. This leads to regions with oppositely directed magnetic field lines close to each other, and thus to current sheets (dashed lines) within streamers (white). The fast steady solar wind (light grey) comes mainly from large coronal holes, whereas the irregular slow solar wind has been suggested to come from small open regions with strong areal expansion and subsequent constriction (dark grey), with possible further contributions from the boundaries of coronal holes and from the opening of the outermost closed regions.

Both these possibilities appear to play a role. The temperature of protons and neutral hydrogen atoms increases to around $3-4 \times 10^6$ K near $3R_\odot$, where both species decouple and hydrogen starts to cool, while the proton temperature can no longer be determined (see the review by Antonucci 1999). Five-fold ionized oxygen ions were also measured with UVCS on SOHO. Their temperature decouples from that of the protons already at $1.5 R_\odot$ and then increases to extremely high values. Moreover, the velocity distribution of these ions along the magnetic field lines differs markedly from the velocity associated with their gyration motions around the field lines. When interpreted as kinetic temperature, these velocity distributions correspond to at least 10^7 K along the magnetic field direction, and 10^8 K perpendicular to it.

The inefficiency of collisions, and the fact that the temperatures of minor ions are much higher and anisotropic, has been attributed to ion-cyclotron heating by high-frequency Alfvén waves. Such waves can pump magnetic wave

REFERENCES

Anderson, L.S. and Athay, R.G. (1989). Chromospheric and coronal heating. *Astrophysical Journal*, **336**, 1089–1091.

Anderson, S.W., Raymond, J.C. and van Ballegooijen, A.A. (1996). Ultraviolet emission-line intensities and coronal heating by velocity filtration: Collisionless results. *Astrophysical Journal*, **457**, 939–948.

Andretta, V. and Jones, H.P. (1997). On the role of the solar corona and transition region in the excitation of the spectrum of neutral helium. *Astrophysical Journal*, **489**, 375–394.

Antiochos, S.K. and Noci, G. (1986). The structure of the static corona and transition region. *Astrophysical Journal*, **301**, 440–447.

Antonucci, E. (1999). Solar wind acceleration region. In J.-C. Vial and B. Kaldeich-Schürmann (eds), *Plasma Dynamics and Diagnostics in the Solar Transition Region and Corona* (Proceedings of the 8th SOHO Wksp.), ESA-SP 446, pp. 53–60.

Aschwanden, M.J., Tarbell, T.D., Nightingale, R.W., Schrijver, C.J., Title, A.M., Kankelborg, C.C., Martens, P. and Warren, H.P. (2000). Time variability of the "quiet" Sun observed with TRACE. II. Physical parameters, temperature evolution, and energetics of EUV nanoflares. *Astrophysical Journal*, **535**, 1047–1065.

Athay, R.G. (1984). The origin of spicules and heating of the lower transition region. *Astrophysical Journal*, **287**, 412–417.

Athay, R.G. and Dere, K.P. (1990). Bifurcation in the low chromosphere. *Astrophysical Journal*, **358**, 710–717.

Athay, R.G. and White, O.R. (1978). Chromospheric and coronal heating by sound waves. *Astrophysical Journal*, **226**, 1135–1139.

Axford, W.I. and McKenzie, J.F. (1997). The solar wind. In J.R. Jokipii, C.P. Sonett, M.S. Giampapa (eds), *Cosmic Winds and the Heliosphere*, University of Arizona Press, Tucson, pp. 31–66.

Ayres, T.R. (1981). Thermal bifurcation in the solar outer atmosphere. *Astrophysical Journal*, **244**, 1064–1071.

Ayres, T.R. and Testerman, L. (1981). Fourier transform spectrometer observations of solar carbon monoxide. I. The fundamental and first overtone bands in the quiet Sun. *Astrophysical Journal*, **245**, 1124–1140.

Ayres, T.R., Testerman, L. and Brault, J.W. (1986). Fourier transform spectrometer observations of solar carbon monoxide. II. Simultaneous cospatial measurements of the fundamental and first-overtone bands, and Ca II K, in quiet and active regions. *Astrophysical Journal*, **304**, 542–559.

Bartoe, J.-D.F. and Brueckner, G.E. (1975). New stigmatic, coma-free, concave-grating spectrograph. *Journal of the Optical Society of America*, **65**, 13.

Bartoe, J.-D.F., Brueckner, G.E., Purcell, J.D. and Tousey, R. (1977). Extreme ultraviolet spectrograph ATM experiment S082B. *Applied Optics*, **16**, 879–886.

Beck, J.G., Duvall, T.S. and Scherrer, P.H. (1998). Long-lived giant cells detected at the surface of the Sun. *Nature*, **394**, 653–655.

Berger, T.E., Löfdahl, M.G., Shine, R.A. and Title, A.M. (1998). Measurements of solar magnetic element dispersal. *Astrophysical Journal*, **506**, 439–449.

Bianda, M., Stenflo, J.O. and Solanki, S.K. (1999). Hanle effect observations with the Ca I 4227 Å line. *Astronomy and Astrophysics*, **350**, 1060–1070.

Biermann, L. (1946). Zur Deutung der chromosphärischen Turbulenz und des Exzesses der UV-Strahlung der Sonne. *Naturwissenschaften*, **33**, 118–119.

Bonnet, R.M., Bruner, E.C. Jr, Acton, L.W., Brown, W.A. and Decaudin, M. (1980). High-resolution Lyman-alpha filtergrams of the Sun. *Astrophysical Journal*, **237**, L47–L50.

Bonnet, R.M., Bruner E.C., Jr., Acton, L.W., Brown, W.A., Decaudin, M. and Foing, B. (1982). Rocket photographs of fine structure and wave patterns in the solar temperature minimum. *Astronomy and Astrophysics*, **111**, 125–129.

Bruner, E.C., Jr (1981). OSO 8 observational limits to the acoustic coronal heating mechanism. *Astrophysical Journal*, **247**, 317–324.

Brynildsen, N., Maltby, P., Brekke, P., Haugan, S.V.H. and Kjeldseth-Moe, O. (1999). SOHO observations of the structure and dynamics of sunspot region atmospheres. *Solar Physics*, **186**, 141–191.

Buchholz, B., Ulmschneider, P. and Cuntz, M. (1998). Basal heating in main-sequence stars and giants: Results from monochromatic acoustic wave models. *Astrophysical Journal*, **494**, 700–714.

Cally, P.S. (1990). Turbulent thermal conduction in the solar transition region. *Astrophysical Journal*, **355**, 693–699.

Carlsson, M. and Stein, R.F. (1995). Does a non-magnetic solar chromosphere exist? *Astrophysical Journal*, **440**, L29–L32.

Carlsson, M. and Stein, R.F. (1997). Formation of solar calcium H and K bright grains. *Astrophysical Journal*, **481**, 500–514.

Chalonge, D. and Kourganoff, V. (1946). Recherches sur les spectres continus du soleil. *Annals of Astrophysics*, **9**, 69.

Chipman, E.G. (1981). The Solar Maximum Mission. *Astrophysical Journal*, **244**, L113–L115.

Choudhuri, A.R., Dikpati, M. and Banerjee, D. (1993). Energy transport to the solar corona by magnetic kink waves. *Astrophysical Journal*, **413**, 811–825.

Cranmer, S.R., Field, G.B. and Kohl, J.L. (1999). Spectroscopic constraints on models of ion cyclotron resonance heating in the polar solar corona and high-speed solar wind. *Astrophysical Journal*, **518**, 937–947.

Cuntz, M., Ulmschneider, P. and Musielak, Z.E. (1998). Self-consistent and time-dependent magnetohydrodynamic chromosphere models for magnetically active stars. *Astrophysical Journal*, **493**, L117–L129.

Curdt, W., Brekke, P., Schühle, U., Wilhelm, K. and Dwivedi, B.N. (1999). The SUMER EUV atlas in the spectral range 670 Å–1498 Å. In: *Proceedings of the 8th SOHO Workshop Plasma Dynamics and Diagnostics in the Solar Transition Region and Corona*, Paris. ESA SP-446. Noordwijk: ESA Publ. Div., pp. 251–256.

Danielson, R.E. (1961). The structure of sunspot penumbras. I. Observations. *Astrophysical Journal*, **134**, 275–288.

Delaboudinière, J.-P., Artzner, G.E., Brunaud, J., Gabriel, A.H., Hochedez, J.F., Millier, F., Song, X.Y., Au, B., Dere, K.P., Howard, R.A., Kreplin, R., Michels, D.J., Moses, J.D., Defise, J.M., Jamar, C., Rochus, P., Chauvineau, J.P., Marioge, J.P., Catura, R.C., Lemen, J.R., Shing, L., Stern, R.A., Gurman, J.B., Neupert, W.M., Maucherat, A., Clette, F., Cugnon, P. and van Dessel, E.L. (1995). EIT: Extreme-Ultraviolet Imaging Telescope for the SOHO Mission. *Solar Physics*, **162**, 291–312.

Dowdy, J.F., Jr, Rabin, D. and Moore, R.L. (1986). On the magnetic structure of the quiet transition region. *Solar Physics*, **105**, 35–45.

Duvall, T.L. Jr and Gizon, L. (2000). Time–distance helioseismology with f modes as a method for measurement of near-surface flows. *Solar Physics*, **192**, 177–191.

Duvall, T.L. Jr, Kosovichev, A.G., Sherrer, P.H., Bogart, R.S., Bush, R.I., De Forest, C., Hoeksema, J.T., Schou, J., Saba, J.L.R., Tarbell, T.D., Title, A.M., Wolfson, C.J. and Milford, P.N. (1997). Time–distance helioseismology with the MDI instrument: Initial results. *Solar Physics*, **170**, 63–73.

Evershed, J. (1909). Radial movement in sun-spots. *Monthly Notices of the Royal Astronomical Society*, **69**, 454–457.

Feldman, U. (1983). On the unresolved fine structures of the solar atmosphere in the 30,000–200,000 K temperature region. *Astrophysical Journal*, **275**, 367–373.

Feldman, U. (1998). On the unresolved fine structures of the solar atmosphere. III. Elemental abundances consideration. *Astrophysical Journal*, **507**, 974–977.

Fleck, B. and Domingo, V. (eds) (1995). *The SOHO Mission*, Kluwer, Dordrecht.

Fontenla, J.M., Avrett, E.H. and Loeser, R. (1993). Energy balance in the solar transition region. III. Helium emission in hydrostatic, constant-abundance models with diffusion. *Astrophysical Journal*, **406**, 319–345.

Fröhlich, C. (2000). Observations of irradiance variations. In E. Friis-Christensen, C. Fröhlich, J.D. Haigh, M. Schüssler, R. von Steiger (eds), *Solar Variability and Climate*, Kluwer, Dordrecht. [*Space Science Reviews*, **94**, 15–24.]

Fröhlich, C., Huber, M.C.E., Solanki, S.K. and von Steiger, R. (eds) (1998). *Solar Composition and its Evolution – From Core to Corona*. Space Science Series of ISSI, Kluwer, Dordrecht. [*Space Science Reviews*, **85**.]

Gabriel, A.H. (1976). A magnetic model of the solar transition region. *Philosophical Transactions of the Royal Society A*, **281**, 339–352.

Gadun, A.S., Solanki, S.K., Sheminova, V.A. and Ploner, S.R.O. (2001). A formation mechanism of magnetic elements in regions of mixed polarity. *Solar Physics*, **203**, 1–7.

Giovanelli, R.G. and Jones, H.P. (1982). The three-dimensional structure of atmospheric magnetic fields in two active regions. *Solar Physics*, **79**, 267–278.

Goldberg, L., Müller, E.A. and Aller, L.H. (1960). The abundances of the elements in the solar atmosphere. *Astrophysical Journal Supplement Series*, **5**, 1–137.

Golub, L., Nystrom, G., Herant, M., Kalata, K. and Lovas, I. (1990). Sub-arcsecond observations of the solar X-ray corona. *Nature*, **344**, 842–844.

Grevesse, N. and Sauval, A.J. (1998). Standard solar composition. *Space Science Reviews*, **85**, 161–174.

Grossmann-Doerth, U., Schüssler, M. and Steiner, O. (1998). Convective intensification of solar surface magnetic fields: Results of numerical experiments. *Astronomy and Astrophysics*, **337**, 928–939.

Hammer, R. and Ulmschneider, P. (1991). On the intrinsic difficulty of producing stellar coronae with acoustic waves. In P. Ulmschneider, E.R. Priest and R. Rosner (eds), *Mechanisms of Chromospheric and Coronal Heating*, Springer-Verlag, Berlin, pp. 344–346.

Handy, B.N. and 47 coauthors (1999). The transition region and coronal explorer. *Solar Physics*, **187**, 229–260.

Harrison, R.A. (1997). EUV blinkers: The significance of variations in the extreme ultraviolet quiet sun. *Solar Physics*, **175**, 467–485.

Hassler, D.M., Dammasch, I.E., Lemaire, P., Brekke, P., Curdt, W., Mason, H.E., Vial, J.-C. and Wilhelm, K. (1999). Solar wind outflow and the chromospheric magnetic network. *Science*, **283**, 810.

Heyvaerts, J. and Priest, E.R. (1992). A self-consistent turbulent model for solar coronal heating. *Astrophysical Journal*, **390**, 297–308.

Holweger, H. (1967). Ein empirisches Modell der Sonnenatmosphäre mit lokalem thermodynamischen Gleichgewicht. *Zeitschrift für Astrophysik*, **65**, 365–417.

Huber, M.C.E., Foukal, P.V., Noyes, R.W., Reeves, E.M., Schmahl, E.J., Timothy, J.G., Vernazza, J.E. and Withbroe, G.L. (1974). Extreme-ultraviolet observations of coronal holes – Initial results from Skylab. *Astrophysical Journal*, **194**, L115–L118.

Innes, D.E., Inhester, B., Axford, W.I. and Wilhelm, K. (1997). Bi-directional plasma jets produced by magnetic reconnection on the Sun. *Nature*, **386**, 811–813.

Isenberg, P.A. and Hollweg, J.V. (1983). On the preferential acceleration and heating of solar wind heavy ions. *Journal of Geophysical Research*, **88**, 3923–3935.

Jordan, C. (1975). The intensities of helium lines in the solar EUV spectrum. *Monthly Notices of the Royal Astronomical Society*, **170**, 429–440.

Kalkofen, W., Ulmschneider, P. and Avrett, E.H. (1999). Does the Sun have a full-time chromosphere? *Astrophysical Journal*, **521**, L141–L144.

Kjeldseth-Moe, O., Brynildsen, N., Brekke, P., Engvold, O., Maltby, P., Bartoe, J.-D.F., Brueckner, G.E., Cook, J.W., Dere, K.P. and Socker, D.G. (1988). Gas flows in the transition region above sunspots. *Astrophysical Journal*, **334**, 1066–1075.

Kjeldseth-Moe, O. and Nicolas, K.R. (1977). Emission measures, electron densities, and nonthermal velocities from optically thin UV lines near a quiet solar limb. *Astrophysical Journal*, **211**, 579–586.

Lemaire, P., Wilhelm, K., Curdt, W., Schühle, U., Marsch, E., Poland, A.I., Jordan, S.D., Thomas, R.J., Hassler, D.M., Vial, J.C., Kühne, M., Huber, M.C.E., Siegmund, O.H.W., Gabriel, A., Timothy, J.G. and Grewing, M. (1997). First results of the SUMER telescope and spectrometer on SOHO. *Solar Physics*, **170**, 105–122.

Lites, B.W., Nordlund, Å. and Scharmer, G.B. (1989). Constraints imposed by very high resolution spectra and images on theoretical simulations of granular convection. In R.J. Rutten and G. Severino (eds), *Solar and Stellar Granulation*, Reidel, Dordrecht, pp. 349–357.

Mehltretter, J.P. (1978). Balloon-borne imagery of solar granulation. II. The lifetime of solar granulation. *Astronomy and Astrophysics*, **62**, 311–316.

Meunier, N., Solanki, S.K. and Livingston, W.C. (1998). Infrared lines as probes of solar magnetic features. XIII. The relative flux in weak and strong quiet-Sun magnetic fields. *Astronomy and Astrophysics*, **331**, 771–781.

Muller, R. (1999). The solar granulation. In A. Hanslmeier and M. Messerotti (eds), *Motions in the Solar Atmosphere*, Kluwer, Dordrecht, pp. 35–70.

Nakariakov, V.M., Ofman, L., DeLuca, E.E., Roberts, B. and Davila, J.M. (1999). TRACE observations of damped coronal loop oscillations: Implications for coronal heating. *Science*, **285**, 862–864.

Narain, U. and Ulmschneider, P. (1990). Chromospheric and coronal heating mechanisms. *Space Science Reviews*, **54**, 377–445.

Narain, U. and Ulmschneider, P. (1996). Chromospheric and coronal heating mechanisms. II. *Space Science Reviews*, **75**, 453–509.

Noci, G. and 24 coauthors (1997). The quiescent corona and slow solar wind. In A. Wilson (ed), *The Corona and Solar Wind Near Minimum Activity (Proceedings of the 5th SOHO Wksp.)*, ESA SP-404. Noordwijk: ESA Publ. Div. pp. 75–84.

Nordlund, Å. (1984). Modelling of small-scale dynamical processes: Convection and wave generation. In S.L. Keil (ed), *Small-Scale Dynamical Processes in Quiet Stellar Atmospheres*, National Solar Observatory, Sunspot, NM, pp. 181–221.

November, L.J., Toomre, J., Gebbie, K.B. and Simon, G.W. (1981). The detection of mesogranulation on the Sun. *Astrophysical Journal*, **245**, L123–L126.

Ogawara, Y., Takano, T., Kato, T., Kosugi, T., Tsuneta, S., Watanabe, T., Kondo, I. and Uchida, Y. (1991). The Solar-A mission – An overview. *Solar Physics*, **136**, 1.

Parker, E.N. (1958). Dynamics of the interplanetary gas and magnetic fields. *Astrophysical Journal*, **128**, 664–676.

Parker, E.N. (1979). *Cosmical Magnetic Fields*, Clarendon Press, Oxford, pp. 359–391.

Parker, E.N. (1988). Nanoflares and the solar X-ray corona. *Astrophysical Journal*, **330**, 474–479.

Patsourakos, S., Vial, J.-C., Gabriel, A.H. and Bellamine, N. (1999). Transition-region network boundaries in the quiet Sun: Width variation with temperature as observed with CDS on SOHO. *Astrophysical Journal*, **522**, 540–546.

Peres, G. (1999). Coronal heating and structuring. In J.-C. Vial and B. Kaldeich-Schürmann (eds), *Plasma Dynamics and Diagnostics in the Solar Transition Region and Corona (Proceedings of the 8th SOHO Wksp.)*, ESA SP-46. Noordwijk: ESA Publ. Div. pp. 43–52.

Peter, H. (1999). Analysis of transition-region emission-line profiles from full-disk scans of the Sun using the SUMER instrument on SOHO. *Astrophysical Journal*, **516**, 490–504.

Peter, H. (2000). Multi-component structure of solar and stellar transition regions. *Astronomy and Astrophysics*, **360**, 761–776, and erratum in *Astronomy and Astrophysics*, **364**, 933–934.

Peter, H. and Judge, P.G. (1999). On the Doppler shifts of solar ultraviolet emission lines. *Astrophysical Journal*, **522**, 1148–1166.

Ploner, S.R.O., Solanki, S.K. and Gadun, A.S. (2000). Is solar mesogranulation a surface phenomenon? *Astronomy and Astrophysics*, **356**, 1050–1054.

Rabin, D. and Moore, R. (1984). Heating the Sun's lower transition region with fine-scale electric currents. *Astrophysical Journal*, **285**, 359–367.

Reale, F. and Peres, G. (2000). TRACE-derived temperature and emission measure profiles along long-lived coronal loops: The role of filamentation. *Astrophysical Journal*, **528**, L45–L48.

Roberts, B. and Ulmschneider, P. (1997). Dynamics of flux tubes in the solar atmosphere: Theory. In C.E. Alissandrakis, G. Simnett and L. Vlahos (eds), *Solar and Heliospheric Plasma Physics, Proceedings of the 8th European Meeting on Solar Physics*, Springer, Berlin, pp. 75–101.

Roumeliotis, G. (1991). Joule heating as an explanation for the differential emission measure structure and systematic redshifts in the Sun's lower transition region. *Astrophysical Journal*, **379**, 392–400.

Russell, H.N. (1929). On the composition of the Sun's atmosphere. *Astrophysical Journal*, **70**, 11–82.

Rutten, R.J. and Uitenbroek, H. (1991). Ca II H_{2V} and K_{2V} cell grains. *Solar Physics*, **134**, 15–71.

Schmieder, B. and Mein, N. (1980). Mechanical flux in the solar chromosphere. II. Determination of the mechanical flux. *Astronomy and Astrophysics*, **84**, 99–105.

Schrijver, C.J., Title, A.M., Harvey, K.L., Sheeley, N.R. Jr, Wang, Y.-M., van den Oord, G.H.J., Shine, R.A., Tarbell, T.D. and Hurlburt, N.E. (1998). Large-scale coronal heating by the small-scale magnetic field of the Sun. *Nature*, **394**, 152–154.

Schrijver, C.J. and 16 coauthors (1999). A new view of the solar outer atmosphere by the Transition Region and Coronal Explorer. *Solar Physics*, **187**, 261–302.

Scudder, J.D. (1992). Why all stars should possess circumstellar temperature inversions. *Astrophysical Journal*, **398**, 319–349.

Shoub, E.C. (1983). Invalidity of local thermodynamic equilibrium for electrons in the solar transition region. I. Fokker-Planck results. *Astrophysical Journal*, **266**, 339–369.

Simon, G.W. and Leighton, R.B. (1964). Velocity fields in the solar atmosphere III. Large-scale motions, the chromospheric network, and magnetic fields. *Astrophysical Journal*, **140**, 1120–1147.

Solanki, S.K. (1997). Dynamics of flux tubes in the solar atmosphere: Observations. In G. Simnett, C.E. Alissandrakis and L. Vlahos (eds), *Solar and Heliospheric Plasma Physics*, Lecture Notes in Physics, Springer-Verlag, Heidelberg, pp. 49–73.

Solanki, S.K. (1998). Structure of the solar photosphere. In C. Fröhlich, M.C.E. Huber, S.K. Solanki and R. von Steiger (eds), *Solar Composition and its Evolution – From Core to Corona*, Kluwer, Dordrecht. [*Space Science Reviews*, **85**, 175–186.]

Solanki, S.K., Finsterle, W., Rüedi, I. and Livingston, W. (1999). Expansion of solar magnetic flux tubes large and small. *Astronomy and Astrophysics*, **347**, L27–L30.

Solanki, S.K. and Fligge, M. (2000). Reconstruction of past solar irradiance. In E. Friis-Christensen, C. Fröhlich, J.D. Haigh, M. Schüssler and

R. von Steiger (eds), *Solar Variability and Climate*, Kluwer, Dordrecht. [*Space Science Reviews*, **94**, 139–144.]

Solanki, S.K., Livingston, W. and Ayres, T.R. (1994). New light on the heart of darkness of the solar chromosphere. *Science*, **263**, 64–66.

Solanki, S.K. and Steiner, O. (1990). How magnetic is the solar chromosphere? *Astronomy and Astrophysics*, **234**, 519–529.

St. John, C.E. (1913). Radial motion in sun-spots. I. The distribution of velocities in the solar vortex. *Astrophysical Journal*, **37**, 322–353.

Steffen, M. and Muchmore, D. (1988). Can granular fluctuations in the solar photosphere produce temperature inhomogeneities at the height of the temperature minimum? *Astronomy and Astrophysics*, **193**, 281–290.

Stucki, K., Solanki, S.K., Rüedi, I., Stenflo, J.O., Brković, A., Schühle, U., Wilhelm, K. and Huber, M.C.E. (1999). Coronal holes versus normal quiet sun observed with SUMER. In *VII International Plasma Astrophysics and Space Physics Conference*, Kluwer, Dordrecht. [*Astrophys. Space Sci.*, **264**, 43–52.]

Sturrock, P.A., Dixon, W.W., Klimchuk, J.A. and Antiochos, S.K. (1990). Episodic coronal heating. *Astrophysical Journal*, **356**, L31–L34.

Title, A.M., Tarbell, T.D., Topka, K.P., Fergusen, S.H., Shine, R.A. and the SOUP Team (1989). Statistical properties of solar granulation derived from the SOUP instrument on Spacelab 2. *Astrophysical Journal*, **336**, 475–494.

Tousey, R. (1977). Apollo Telescope Mount of Skylab – An overview. *Applied Optics*, **16**, 825–836.

Ulmschneider, P. (1970). On frequency and strength of shock waves in the solar atmosphere. *Solar Physics*, **12**, 403–415.

Vaiana, G.S. and Rosner, R. (1978). Recent advances in coronal physics. *Annual Reviews of Astronomy and Astrophysics*, **16**, 393–428.

Vernazza, J.E., Avrett, E.H. and Loeser, R. (1973). Structure of the solar chromosphere. Basic computations and summary of the results. *Astrophysical Journal*, **184**, 605–632.

Vernazza, J.E., Avrett, E.H. and Loeser, R. (1981). Structure of the solar chromosphere III. Models of the EUV brightness components of the quiet Sun. *Astrophysical Journal Supplement Series*, **45**, 635–725.

Wang, Y.-M. (1994). Two types of slow solar wind. *Astrophysical Journal*, **437**, L67–L70.

Wildt, R. (1939). Negative ions of hydrogen and the opacity of stellar atmospheres. *Astrophysical Journal*, **90**, 611–620.

Wilhelm, K., Dammasch, I.E., Marsch, E. and Hassler, D.M. (2000). On the source regions of the fast solar wind in polar coronal holes. *Astronomy and Astrophysics*, **353**, 749–756.

Wilhelm, K., Marsch, E., Dwivedi, B.N., Hassler, D.M., Lemaire, P., Gabriel, A.H. and Huber, M.C.E. (1998). The solar corona above coronal holes as seen by SUMER on SOHO. *Astrophysical Journal*, **500**, 1023–1038.

Wikstøl, Ø., Hansteen, V.H., Carlsson, M. and Judge, P.G. (2000). Chromospheric and transition region internetwork oscillations: A signature of upward propagating waves. *Astrophysical Journal*, **351**, 1150–1160.

Wikstøl, Ø., Judge, P.G. and Hansteen, V. (1998). On inferring the properties of dynamic plasmas from their emitted spectra: The case of the solar transition region. *Astrophysical Journal*, **501**, 895–910.

Willson, R.C. and Hudson, H.S. (1991). The Sun's luminosity over a complete solar cycle. *Nature*, **351**, 42–44.

Woods, D.T., Holzer, T.E. and MacGregor, K.B. (1990). Lower solar chromosphere–corona transition region. II – Wave pressure effects for a specific form of the heating function. *Astrophysical Journal Supplement Series*, **73**, 489–512.

46

DAVID ALEXANDER* AND LOREN W. ACTON**

The active Sun

Following the sun we left the old world – Christopher Columbus

The title of this chapter presupposes that there exists a Sun that is Non-active or Quiet. Modern research, however, is discovering more and more that there is no such thing as a perfectly placid Sun. Activity occurs on all scales at all times in the solar atmosphere. As solar instrumentation improves, the subsequent advances in our knowledge of what physics controls this continuous activity will lead to a better understanding of how stars like the Sun work. Ever since Fabricius and Galileo first confirmed that sunspots were solar phenomena, the nature of our nearest star has been a subject of scientific investigation. The culmination of this line of inquiry is the present-day field of solar physics, the goal of which is to investigate the constantly changing Sun. Solar activity in all of its myriad forms, from cyclical variations of the whole Sun to small-scale transient phenomena, is ubiquitous in solar research. Improved instruments now demonstrate that even the quiet Sun is far more dynamic than once thought. The intrinsic variablity of the Sun provides a touchstone to the intricate workings of stars. In this chapter we will explore the active Sun and examine how our knowledge of this dynamic star has blossomed in this *Century of Space Science*.

An important early tool in the exploration of solar activity was the solar eclipse. The popular attraction of eclipses remains strong today but their scientific value has steadily declined since the invention of the coronagraph by French astronomer Bernard Lyot in 1930. The eclipse expeditions of the late nineteenth and early twentieth centuries are a testimony to the dedication of early solar astronomers. The word "expedition" itself conjures up the effort involved in observing an eclipse since, more often than not, the eclipse was best observed from a remote and inhospitable part of the planet. Despite all of the hardships, the astronomer was often rewarded with wonderful scientific data and occasionally an important discovery. Eclipse observations provided a strong foundation for the subsequent research into solar activity.

Eclipses always yielded spectacular displays in the darkened sky, but until 1836 the coronal phenomena observed were thought to originate on the Moon or even in the Earth's atmosphere and, therefore, merited little scientific interest. The last four decades of the nineteenth century, however, were witness to many discoveries which laid the foundation for the study of solar activity in the solar corona. Eclipse data, in particular, were responsible for a number of these discoveries and significantly advanced our knowledge of the Sun and its behavior: the solar origin of prominences (1860; Pietro Secchi, Warren De La Rue), the discovery of helium (1868; Jules Janssen), the existence of a chromosphere (1870; Charles Augustus Young), the unequivocal identification of the corona as solar (1871; Jules Janssen), and the relation between sunspot cycle and corona (1878; Jules Janssen). The eclipse of 1878 was particularly interesting as it occurred during solar minimum and exhibited a corona with a marked equatorial extension (in contrast with the nearly circular corona observed in 1871). It was noted that the coronal streamers, as the extensions off each limb were called, had a strong resemblance to magnetic lines of force and it was proposed that the Sun must, in fact, be a large magnet (Frank Bigelow in 1889 and Störmer in 1911). Subsequently in 1912,

---
* Lockheed-Martin Solar and Astrophysics Lab, Palo Alto, CA, USA
** Montana State University, Bozeman, MT, USA

Henri Deslandres suggested that the forms and motions of prominences seen during solar eclipse appeared to be influenced by a solar magnetic field. The link between magnetic field and the emitting plasma on the Sun was beginning to take shape.

The key to understanding solar activity is the Sun's ever-changing magnetic field. It is now virtually certain that all solar activity, and perhaps the solar atmosphere itself, is there because of solar magnetism. However, it is only in this last century that the fundamental importance of the magnetic field for solar phenomena has been realized. The epochal discovery of magnetic fields on the Sun by American astronomer George Ellery Hale in 1908 signalled the birth of modern solar physics. This realization led to fundamental progress in our understanding of many physical processes occurring on the Sun and set the foundation for most of the solar physics advances in the modern age. The discovery of solar magnetism arose out of Hale's supposition that the distinctive alignments of penumbral filaments in sunspots bore a remarkable resemblance to the iron filing patterns formed around the poles of a bar magnet. To test his hypothesis Hale looked for the line splittings expected from the newly-discovered Zeeman effect: the first application of this in astronomy. Using the spectrograph at the recently completed solar tower telescope at Mount Wilson, Hale was immediately rewarded with a clear line-triplet indicating the presence of a strong magnetic field.

The study of the Sun's magnetic field was limited to strong-field regions, such as those of sunspots, until the invention of the solar magnetograph in 1952 (see Figure 1). Harold Babcock and his son Horace used the first ever solar magnetograph to discover that the magnetic field existed outside of sunspots and was, in fact, distributed across the whole solar surface. The solar magnetograph, refinements of which are now used in virtually every major solar observatory in the world (there is even one currently in space), introduced an improvement of about two orders of magnitude in the sensitivity of magnetic field measurements over the contemporary visual or photographic techniques. Without these observational breakthroughs we would know very little of what turned out to be the active Sun.

Solar physics research was, literally, taken to new heights by the advent of observations from space. The age of space astronomy has its roots in the flight of a captured World War II German V2 rocket on 10 October 1946 from White Sands Missile Range in New Mexico. This rocket, carrying instruments to observe for the first time the far ultraviolet spectrum of the Sun, reached an altitude of 90 km above the Earth's surface. However, despite the increasing sophistication of rocket-borne experiments in the following decade and a half, discoveries were hampered by the very short flight durations attainable by these sounding rockets.

In March of 1953 the National Academy of Sciences appointed a National Committee to oversee US participation

Figure 1 Two of the original magnetograms of the Sun taken with the Babcock magnetograph in July 1953. (Reproduced with permission of H.W. Babcock and the American Astronomical Society.)

in the International Geophysical Year (IGY), a comprehensive series of global geophysical activities set up by the International Council of Scientific Unions the previous year. The IGY spanned 1957–58 and included investigations into solar activity and the upper atmosphere of the Earth. In connection with the upper atmosphere research, the USA undertook to develop an orbiting satellite program.

The advent of orbiting satellites opened the way for continuous solar observations at wavelengths unattainable from the Earth. A series of eight satellites known as the Orbiting Solar Observatories provided much of the space-based solar observations through the 1960s, although the OSO satellites were modest in size with relatively small instruments. Solar physics received an enormous boost in 1973 when NASA launched the Skylab space-station mission (Figure 2). Skylab allowed long-focal-length solar telescopes to be flown in space for the first time and the commensurate improvement in observations was so great that the data are still being used today.

In terms of versatility and reliability, the performance of the Skylab telescopes and instruments exceeded the highest aspirations of astronomers at the time. Skylab made a number of important discoveries which significantly enhanced our knowledge of solar activity and which changed the nature of solar physics research permanently. In the almost three decades since the launch of Skylab many solar observatories have been placed in space, each making a step towards a better understanding of the Sun. The physics and phenomena we will discuss here depend heavily on results from these space missions and the legacy left by sounding rockets, OSO, and Skylab.

Much of this chapter will center on what we have learned about the Sun through advances in our knowledge of the magnetic field, its variability, and its interaction with the solar plasma. While this is of intrinsic interest to the solar physics community, the problem posed by the observed magnetic activity of the Sun is of fundamental importance to all of astrophysics. Activity is observed in many distant stars and galaxies. However, the proximity of the Sun makes it the only star whose activity can be seen in enough detail to guide us toward an understanding. Naturally, to do justice to a century of research would tax even the space limitations of a single dedicated volume. To confine such a task to a single chapter necessitates a more circumspect approach. We attempt to highlight the major discoveries which have led us to where we are today in our

Figure 2 The Skylab space station seen from Skylab 4. (Reproduced with permission of Skylab, NASA.)

understanding of solar activity and to describe the phenomena which best exemplify solar variability.

The generation of the magnetic field in the interior of the Sun is the subject of a different chapter of this volume (Chapter 44) but its importance for the observed phenomena which have driven solar research over this last century provides us with a natural starting point for our discussion into the activity of the Sun. The solar dynamo is the machine that is responsible for this magnetism and whose most evident manifestation is the solar cycle.

THE CYCLICAL SUN

With orb and cycle girds the starry throng – Wordsworth

Some two hundred years after the death of Galileo, a pronounced quasi-regular period was discovered in the number of sunspots by a German apothecary and amateur astronomer, Heinrich Schwabe, in 1843. This period demonstrated that the Sun had an 11-year cycle of activity which had been repeated, with a couple of important interruptions, for about 250 years. It may seem surprising that it took approximately two centuries for this pattern to become apparent. However, it should be pointed out that the cyclical pattern in the sunspot behavior is most clearly recognizable after considerable averaging. Indeed, the oft-quoted cycle of about 11 years is far from regular, in either duration or form. Lengths of cycles vary from a minimum of about 8 to a maximum of about 15 years, the long-term average being about 11.1 years. While the solar cycle is a discovery of the nineteenth century, it has had a profound effect on the solar physics of the twentieth century, driving much of the advances in our current understanding of the Sun.

Essentially all solar phenomena exhibit 11-year cycles, including radiative outputs, particle and plasma emissions, interior oscillations, and perhaps even fundamental processes in the Sun's nuclear burning core. Diverse solar parameters such as the integrated X-ray emission, UV irradiance, and solar radio flux record this variability over wide ranging spectral and temporal scales. Like the sunspot number, the total radiative output and the entire solar spectrum exhibit pronounced quasi-11-year and 27-day cycles, linking all aspects of solar radiation variability to a common source: the magnetic activity of the Sun. The 27-day period corresponds to solar rotation. Detailed investigation of this activity led Hale to discover, in 1913, that a predominance of major spot groups exhibited a very distinctive pattern. He found that the preceding and following polarities were opposite in the northern and southern hemispheres and, more importantly, that this pattern was reversed from one 11-year cycle to the next. This ultimately led Hale to conclude (with Seth Barnes Nicholson some 25 years later) that the magnetic cycle of the Sun must have a duration of 22 years.

The variability best quantified observationally is that of the Sun's total (spectrally integrated) radiative output. The extremely small amplitude of this variability was undetectable until the development of radiometers of sufficient sensitivity in the 1970s. Overlapping cross-calibrated measurements made by active cavity radiometers since November 1978 compose a total irradiance database with sufficient long-term precision to identify an 11-year total irradiance cycle of about 0.1% amplitude during solar cycles 21 and 22 (1976–1986 and 1986–1996, respectively), in phase with solar activity. Although these variations in solar luminosity seem quite small they are of the same order of magnitude as the longer period insolation change that give rise to major ice ages. However, to date it has not been fully determined whether cyclical changes in the Sun's radiative output have a marked effect on the Earth's climate or not.

While the spectrally integrated solar output provides the best measure of the solar variability, specific wavelengths can provide some insight into the nature and cause of the variability. A prime example of this is the variation in the X-ray irradiance over the course of a solar cycle. The Soft X-ray Telescope (SXT) on board the Yohkoh spacecraft has demonstrated a variation by as much as a factor of 40–50 from solar maximum to solar minimum in the wavelength range 3–60 Å, which is to be compared with that seen at shorter wavelengths by the (Geostationary Operational Environmental Satellite (GOES) series of satellites (1–8 Å) of about a factor of 80–100. The advantage of modern grazing (and, more recently, normal) incidence optics at X-ray wavelengths is that the imaging allows us to determine the role played by different solar phenomena in the solar variability. The SXT has shown that the whole corona participates in the solar cycle not just the most active parts (see Figures 3 and 4).

The solar activity cycle is variable in several aspects. One obvious signature of this variability has shown up in the last century. The most recent series of solar maxima, spanning 30–40 years, exhibited the largest minimum to maximum amplitudes in the 400-year record, suggesting that the overall activity level is at a historical high. (At the time of writing, Cycle 23 appears to have entered its declining phase after peaking in March 2001 with a maximum which was significantly lower than early predictions.) There are some rules, however, that apply to the vast majority of sunspots, demonstrating large-scale patterns in the activity cycle, which reveal fundamental properties of the magnetic structure below the atmosphere. Studies of the active Sun can therefore complement the recent advances in helio-seismology (see Chapter 44) by providing important clues to the working of the solar dynamo.

Understanding the solar dynamo and its observational effects is now a full-time occupation for an increasing number of solar scientists. A network of ground stations (called GONG, the Global Oscillation Network Group) and a

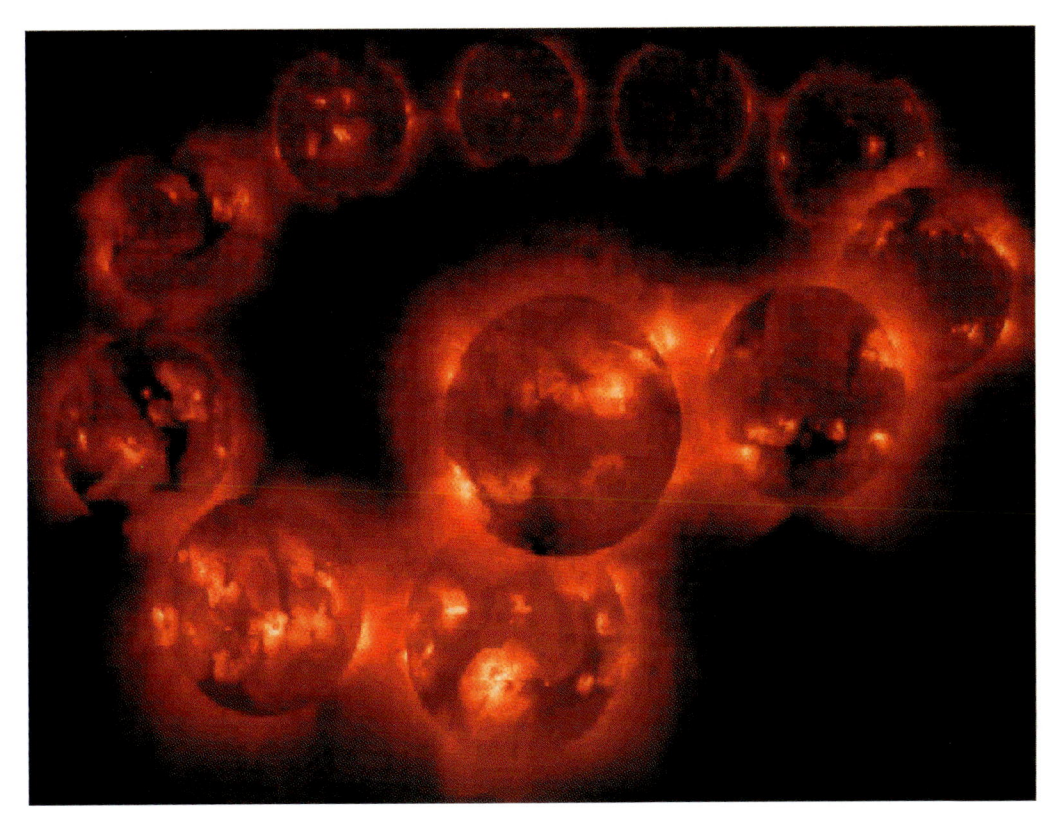

Figure 3 The Yohkoh spacecraft has completed almost an entire solar cycle in orbit. These images taken by the Soft X-ray Telescope illustrate the cyclical nature of the solar atmosphere. (Reproduced with permission of Greg Slater, Lockheed Martin Solar and Astrophysics Lab.)

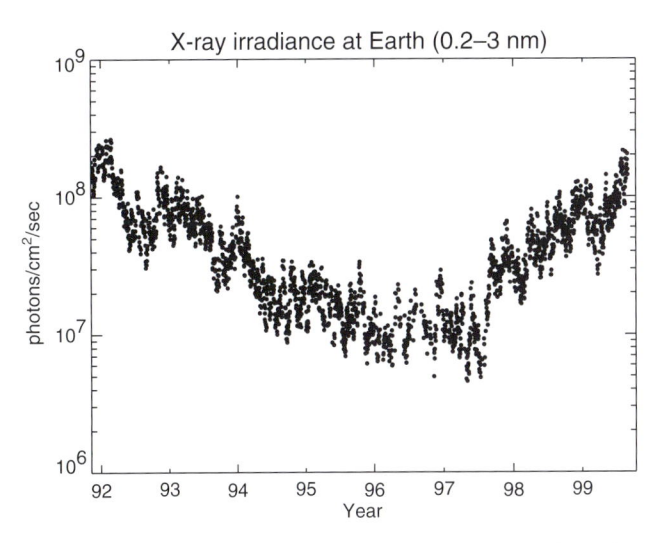

Figure 4 The X-ray irradiance of the Sun follows the solar cycle with a variation in the 3–60 Å wavelength range of almost two orders of magnitude between solar maximum and solar minimum.

spacecraft at the first Lagrangian point L_1 (SOHO, the Solar and Heliospheric Observatory) provide continuous monitoring of the solar surface with the single aim of collecting long-term time series data from which the oscillation spectrum of the Sun can be determined. The data collected from GONG and SOHO are revolutionizing the ways in which we view the solar dynamo (see Chapter 44). However, the seminal paper by Horace Babcock (1961) still gives a comprehensible glimpse of how magnetic field is generated at the Sun. This empirical dynamo model is able to explain many of the observational phenomena associated with the emergence, evolution, and decay of sunspots and active regions along with the larger scale cycle manifestations such as the Maunder butterfly diagram (Figure 5).

The majority of the bipolar active regions emerge with a preferred east–west orientation at a small inclination angle such that the preceding polarity is closer to the equator. In addition, active regions generally follow Hale's polarity law such that the preceding (defined relative to solar rotation) polarity spots in all sunspot groups that occur in the northern hemisphere have the same magnetic polarity, while the following spots have the opposite magnetic polarities. The polarities are exactly reversed for the southern hemisphere.

DAILY SUNSPOT AREA AVERAGED OVER INDIVIDUAL SOLAR ROTATIONS

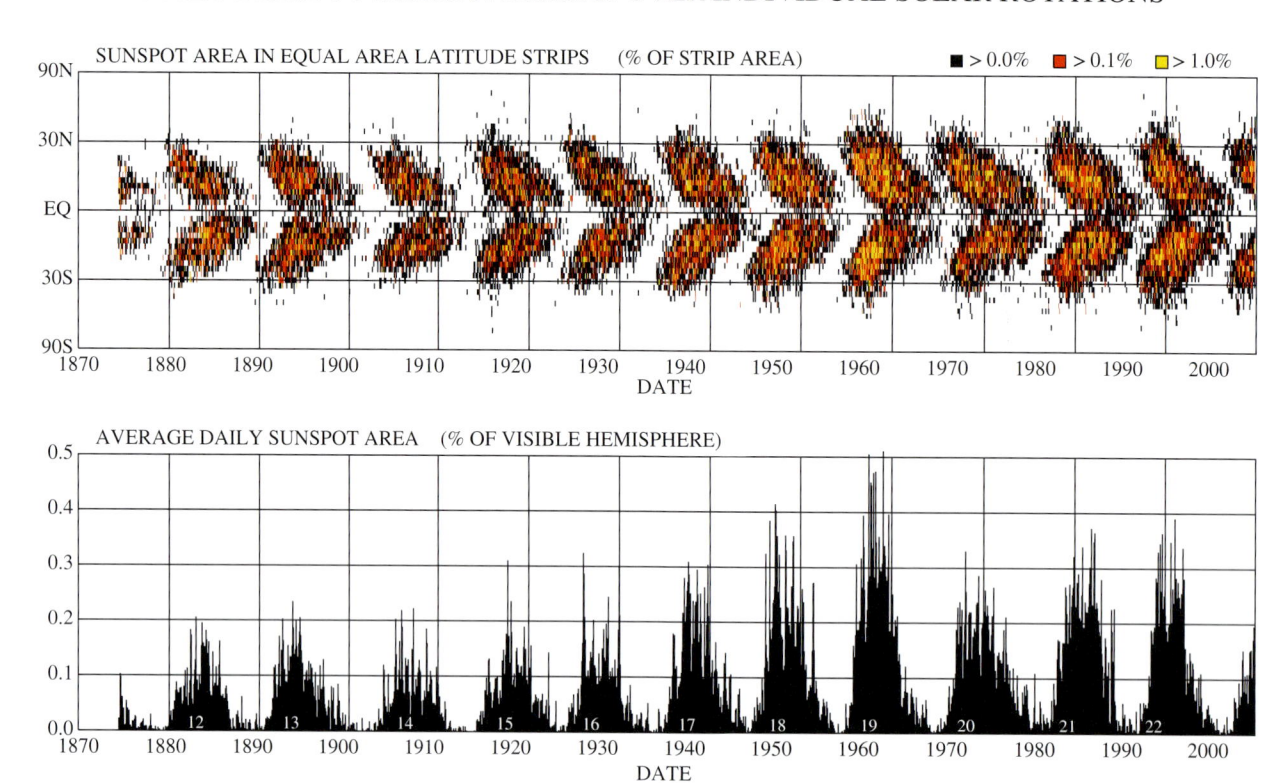

Figure 5 Butterfly diagram showing the latitudinal distribution of sunspots throughout the last 11.5 solar cycles. Also shown is the corresponding averaged daily sunspot areas. (Reproduced with permission of David Hathaway, NASA/MSFC.)

Deviations from this pattern usually indicate the emergence of new cycle flux within old cycle regions. The latitudinal dependence of sunspot emergence depending on when in the cycle they appear (known as Spörer's law after German astronomer Gustav Spörer although originally discovered by British astronomer Richard Carrington) provide further criteria required to be satisfied by models of the solar activity cycle. In particular, these rules suggest that active regions arise as Ω-shaped loops from a toriodal flux strand, formed at the base of the convection zone, and that this toroidal magnetic flux system is intrinsically strong. The rapid advance of computing power in the last few years has resulted in several numerical models capable of simulating many of these observed features.

In addition to large active regions with sunspots, smaller-scale activity features can also display a cyclical pattern and provide extra clues to the process behind solar activity. For example, it was argued based on results from Skylab and sounding rocket flights in the 1970s that X-ray bright points, small bipolar structures with enhanced X-ray emission, had an activity cycle 180° out of phase with that of active regions. This led to the interesting notion that the solar activity cycle reflected an oscillation in the spectrum of the emerging flux rather than a variation of the amount of flux. Subsequent observations of the dynamics of the network field, enhanced during sunspot minimum, were able to explain the predominance of X-ray bright points at solar minimum, thereby eliminating the basis for this contentious suggestion. (More recent results from the Japanese Yohkoh spacecraft have suggested that X-ray bright points do not show any pronounced variation with the solar cycle. This argues that the smaller-scale magnetic fields responsible for producing the bright points are independent of the global magnetic dynamo.)

An imbalance in the level and timing of activity between the northern and southern hemisphere has been recognized as a property of many recent solar cycles. This asymmetry was also detected in the activity of the total magnetic flux. Evident early in Cycle 21 (i.e., 1977) is a lag of several rotations in the onset of activity in the southern hemisphere relative to the northern hemisphere. The north–south asymmetry in the magnetic flux is much more pronounced in the strong field component (> 25 G), corresponding to active regions, but the asymmetry is still detectable in the weak-field observations. The asymmetry of the magnetic fields outside active regions is in the same sense as that for active regions but of a significantly smaller amplitude. The magnetic data also show pulses or episodes of activity in agreement with what is seen in the whole-Sun measures. Both hemispheres display this episodic behavior but are apparently uncorrelated for much of the cycle. Late in the cycle, however, the pulses of activity in the north and south synchronize suggesting that the sources of flux for the two hemispheres are interacting. The variability of this north–south asymmetry with the solar cycle provides a clue to the interaction of the dynamo field components in the convective zone.

A key problem with the Babcock picture of the solar cycle is the conversion of old-cycle flux to new-cycle flux. Babcock's basic idea is that the new-cycle flux is generated from the remnant field of the previous cycle. However, the long-term evolution of the solar corona requires removal of magnetic flux early in the solar cycle as the old-cycle flux is gradually replaced. Global flux observations demonstrate that the total magnetic flux on the Sun varies by only about a factor of 3 between solar minimum and solar maximum, some two orders of magnitude lower than the variation expected from the observed rate of flux emergence. This means that the flux removal rate must nearly match the emergence rate throughout the solar cycle. The removal of flux from the photosphere can occur either via the submergence of the magnetic field or by the reconnection and expulsion of magnetic flux into the interplanetary medium (e.g., via a coronal mass ejection). The latter process is constrained by the observation that the interplanetary magnetic field (IMF) is essentially constant over a solar cycle. This requires that the expulsion mechanism "severs" the magnetic connection of the ejected flux to the solar surface. Otherwise, the continual supply of magnetic flux to the interplanetary medium would result in the indefinite build-up of the IMF at 1 AU; it has been estimated that the rate of flux escape would be sufficient to double the magnetic field at 1 AU approximately every 100 days.

In addition, the reversal of the magnetic polarity of the Sun as it transits from one cycle to the next is manifested in the creation of large unipolar regions and mixed-polarity areas. All of these processes involve the emergence and cancellation of magnetic flux on a variety of temporal and spatial scales indicating that reconnection is a fundamental prerequisite in the global magnetic behavior of the Sun.

A discussion of the solar cycle would not be complete without, at least briefly, mentioning the apparent lack of solar activity in the seventeenth and the early eighteenth century. This period of severely depressed activity between 1645 and 1715, called the Maunder Minimum after Walter Maunder, although first discussed by Gustav Spörer in 1887, may indicate the presence of more complicated physics in the generation of the Sun's global magnetic field and be suggestive of additional cycles in the Sun's evolution. Indeed, comparisons of solar and stellar radiation suggest that the Sun is potentially capable of a wider range of variability than that which we are witnessing in the contemporary era – an idea supported by the presence of the Maunder Minimum (and similar phenomena). In twentieth-century investigations of solar activity over the millennia, the Maunder Minimum showed up in a diversity of signatures such as high cosmogenic isotope concentrations (determined from tree ring and ice core samples), decreased solar rotation (determined from contemporary sunspot observations), and increased solar diameter (determined from timings of Mercury transits and durations of solar eclipses) relative to contemporary levels (see the review by Lean 1997). Determining the long-term, large-amplitude variability of the solar cycle is important for the development of solar dynamo models. As our observational knowledge of the solar interior increases with the progress of helioseismology, the theory of magnetic field generation and transport through the convective zone will need to be refined not only to explain the helioseismology data but also the long-term manifestations of solar variability.

THE MAGNETIC SUN

If the Sun had no magnetic field, it would be as uninteresting as most astronomers think it is – R.B. Leighton

Solar activity takes many forms, from the large-scale, long-term patterns of the solar cycle to the relatively short-term transients of solar flares and coronal mass ejections. Even the quiet Sun displays a startling level of activity and dynamics. *All of this diverse activity is magnetic in origin.* As mentioned above, the Sun's magnetic nature has been known for almost a century and the dramatic improvement in observational techniques can now use information about the magnetic field and its changes to probe the transient phenomena displayed by the Sun. The most obvious manifestation of the magnetic field is that of the sunspot. It was suggested as early as 1941 by German astrophysicist Ludwig Biermann that the photosphere is dark in the sunspot locations because of the magnetic inhibition of convection there, leading to cooler temperatures. A sunspot is one of several magnetic features which make up what is called an active region. This term dates back to a review by German solar physicist Karl Otto Kiepenheuer (1953), who coined the phrase "centre of activity". A centre of activity was described as a "limited and coherent region on the Sun with an area usually smaller than about one-tenth of that of the visible hemisphere". Active regions have since proven to be critical to the investigation of the magnetic nature of the Sun. The key to understanding active regions, and therefore much about solar activity, lies in the relationship between the magnetic field and the coronal plasma.

Active regions are comprised of a number of distinct magnetic flux systems which exhibit a variety of characteristics. The majority of the magnetic flux passing through the photosphere is concentrated into elements of locally high field strength on a variety of scales, from the quiet network to sunspots. Active regions can display the whole hierarchy of solar magnetic flux elements, including compact plages of relatively high filling factor, sunspots, pores, and micropores. These are all formed during the emergence phase over a period of several days and are believed to be fabricated in or just below the convection zone. It is found that the number of active regions present on the Sun as a function of area increases rapidly with decreasing area. Small, short-lived ephemeral active regions are very numerous and can appear well outside the activity belts defined by the large active regions. These ephemeral regions have fluxes as low as 10^{18}–10^{20} maxwells and lifetimes as short as hours, each some two to three orders of magnitude lower than observed in large active regions. The dynamic range of solar magnetic activity, all of which must be related to the generic process of field generation and evolution, suggests a similarity of magnetic development on a range of scales.

Dividing observed photospheric magnetic fields into strong and weak components demonstrates a clear distinction between the temporal behavior of active-region and quiet-Sun fields. Over Cycle 21 the Sun's total magnetic flux increased by at least a factor of 5, peaking some 1.7 years after the sunspot maximum. The strong-field component, however, showed a variation of a factor of 15 from cycle minimum to cycle maximum while the quiet-Sun fields increased by no more than a factor of 2 during the same interval. The conclusion from this is that most of the flux emerging in active regions disappears before it is able to disperse, with the estimation that more than 70% of the magnetic field disappears *in situ* while only a small fraction disperses into the background. Large active regions occupy only a small fraction of the photosphere, and they are not observed outside the activity belts. Magnetic field is found everywhere on the Sun and while the active regions are key signatures in the solar variability, the network and intranetwork field is of no less importance in our understanding of solar activity. The general network field is found to be concentrated on the boundaries and, in particular, the vertices of supergranular velocity cells, while the intranetwork field consists of small-scale mixed polarity magnetic flux within the network cells. Intranetwork field is qualitatively different from network field and ephemeral regions. An important consequence of the intrinsic weakness and mixed polarity of the intranetwork field is that it does not penetrate into the outer atmosphere. The magnetic field in the quiet network, however, is found to display a dynamic behavior which allows it to couple to the larger-scale field. Recent analysis of the quiet solar network field by Karel Schrijver and colleagues, using data from SOHO/MDI (Michelson Doppler Imager), has shown that there is a recycling of small-scale network magnetic flux on a timescale of 1.5–3 days. In other words, the whole network field is constantly changing, emerging, submerging, and cancelling, with the whole network being replaced in a little under 2 days.

The dynamical behavior of the many small magnetic concentrations which make up the quiet network (dubbed "magnetic carpet," Figure 6) suggests that there is a continuous process of flux emergence and cancellation which may conspire to transfer substantial magnetic energy upward into the solar corona. The presence of the magnetic carpet also implies that the quiet network field must be generated locally and that very little of it is due to flux dispersal from active regions. This implies that the dynamics of this field is independent of the overall solar dynamo process. The foregoing results, among others, have led modern researchers to raise the possibility that a second "surface" dynamo might be at work on the Sun. This dynamo would operate just below the photosphere and be a result of turbulent flows in the magnetoconvection process. However, there is some evidence to indicate that the magnetic network varies in polarity characteristics and in mean flux density with the phase in the activity cycle, suggesting that any surface dynamo is not completely isolated from the global solar dynamics.

The discovery of the Sun's magnetic nature by Hale in 1908 and the invention of the magnetograph by the Babcocks in 1953 are pivotal points in the history of solar physics. The legacy of these events is thriving in the present day. Each advance in solar instrumentation yields further proof of the pervasiveness of the magnetic field and its variability. The Sun's transition region and corona not only attest to the dominance of the magnetic field but may actually exist because of it. The interaction between the magnetic field and the plasma which it confines is most evident in the array of energetic transient activity exhibited by the Sun. Phenomena such as solar flares and coronal mass ejections provide the means by which changes in the topology of the magnetic field at the Sun can be experienced at the Earth.

THE TRANSIENT SUN: THE EARLY YEARS

One swallow does not make a summer – Richard Carrington quoting Aristotle (*from Nicomachean Ethics*) when reporting his observations of the first flare to the Royal Astronomical Society

An important consequence of solar activity is the effect it has on the Earth. That long-term variations of the solar radiative output can have an effect on terrestrial climate is clear, as attested by the Little Ice Age associated with the

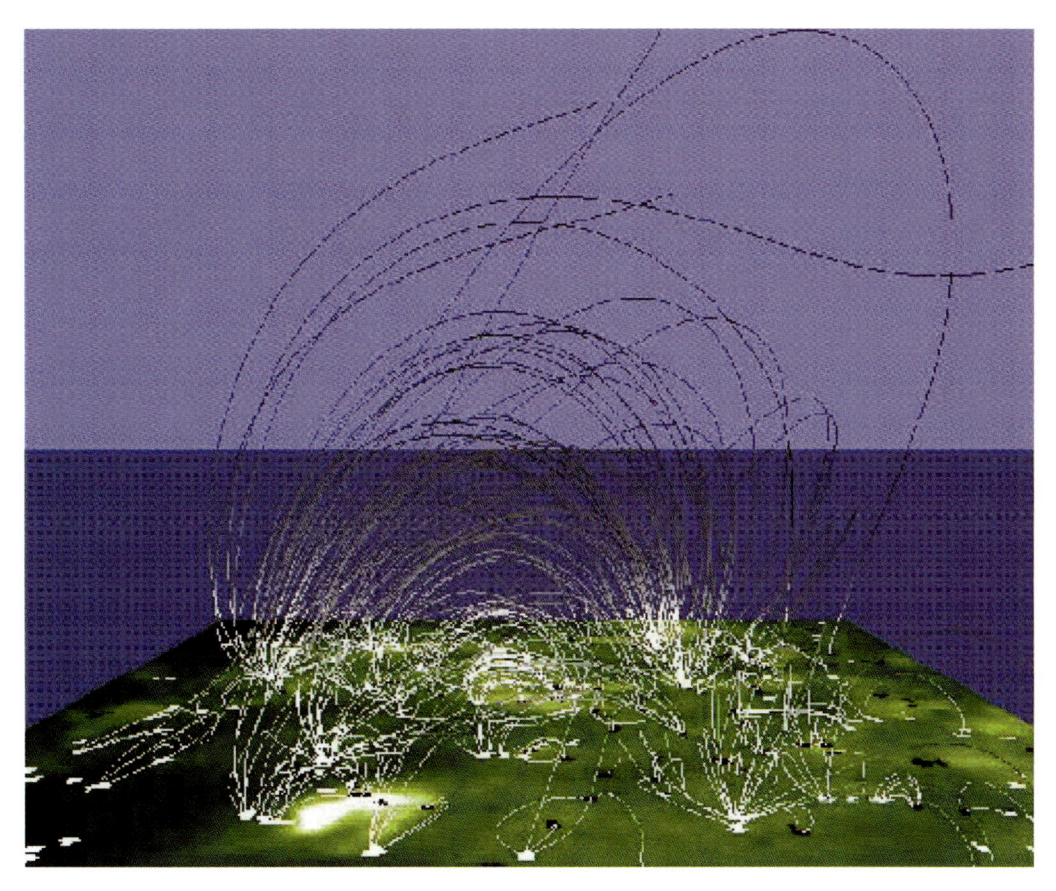

Figure 6 Theoretical predictions for the structure and heating of the magnetic field above the solar surface. The white magnetic fieldlines emanate from the magnetic carpet and form arches from one magnetic polarity (white) to the other (black). The image underlying the arches is the heating observed at the same time by EIT in the iron line at 195 Å, with bright green corresponding to relatively hot regions and dark green corresponding to cool ones. (Reproduced with permission of Neal Hurlburt, Lockheed Martin Solar and Astrophysics Lab.)

Maunder sunspot minimum. On a much shorter timescale, transient activity at the Sun creates dramatic responses at the Earth which were "mostly harmless" until relatively recently. The modern demands for electric power in our largest cities and the increasing reliance on satellite communications for our daily activities have made us far more susceptible to the capriciousness of the Sun.

The first recognizable solar–terrestrial phenomena were geomagnetic disturbances, discovered to be solar in origin in the middle of the nineteenth century by Edward Sabine, Director of the British Colonial Observatories, who noted that the frequencies of both geomagnetic storms and sunspots followed the 11-year cycle. The more marked association of a 27-day recurrence of large storms was first discovered at the beginning of the twentieth century by Maunder. This firmly placed sunspots as the sources of intense geomagnetic activity. In the late 1920s a number of researchers, most notably Harold Newton, deduced that under such circumstances the most probable spot position was one day west of the central meridian and that large storms lasted about a day. This direct relationship pointed to a Sun–Earth travel time of around 1.5 days, suggesting that the disturbances responsible for the largest geomagnetic storms travelled at speeds in excess of $1000\,\mathrm{km\,s^{-1}}$. In addition to large geomagnetic storms, a great number of smaller storms were observed. However, no clear association could be made between these small storms and the passage of a sunspot across the solar disk, strongly indicating the existence of more than one solar source of geo-effective disturbances.

The first step in associating geomagnetic storms with what later became known as solar flares rather than the associated spot regions was the memorable observations in 1859 by British amateur astronomers Richard Carrington and Richard Hodgson, who independently witnessed a rapid intense flash of two bright ribbons on the Sun in visible light. This short-lived transient was followed two minutes later by a marked disturbance of the Earth's magnetic field, detected by the geomagnetic instruments at Kew Observatory in London, and then some 16 hours later by one of the largest magnetic storms on record. Carrington immediately made the far-reaching assumption that the terrestrial events were caused by the strange solar spectacle he had observed. The direct impact on the Earth by an event over 150 million km away had a profound influence on the study of solar–terrestrial relations which reaches to the present day.

The flare observed by Carrington and Hodgson was an example of a relatively rare event, a large white-light flare, in which the optical continuum is enhanced sufficiently over the background solar surface to be visible in contrast. Most flares are not so conspicuous in visible light, reserving their strongest emission for spectral lines such as hydrogen alpha (Hα) and higher energy EUV and X-ray radiation. Before the advent of space observations flares were best observed in the Hα line and as such were regarded as purely chromospheric phenomena. As observations improved and space astronomy developed, flare phenomena grew to include EUV, X-ray, radio, γ-ray, and energetic particle emission, all emitted as transient enhancements, either impulsively or gradually, over the background levels. We will return to these other observations later.

The importance of the chromospheric eruptions, as the early flares were known, for the Earth's space environment came through the study of these events and their apparent association with geomagnetic storms: the first tentative steps towards what today is called "space weather." The early observations of Newton and Maunder were improved upon by Hale's application of the spectrohelioscope at the Mount Wilson Observatory. However, it was Newton again who laid a solid foundation for the statistical association of large flares and storms. Newton surveyed all the large flares observed since 1892 and found a significant correlation between those flares and subsequent geomagnetic storms. With the development of radio astronomy in the 1930s, large solar flares came to be associated with more geomagnetic phenomena such as sudden ionospheric disturbances and ground-level enhancements. These indicated the presence of higher energy radiations and accelerated ions, paving the way for the magnetic theories of flare energization which were to follow. An Hα flare is a short-lived, sudden increase of intensity in the neighborhood of sunspots (see Figure 7). Because of the short duration of these events, many early flare observations were hampered by incomplete coverage making measurements of the time development and brightness determination extremely difficult. The first systematic photometry of solar flares was carried out in the late 1940s by Helen Dodson of the McMath-Hulbert Observatory in Michigan. The flare light curves determined from Hα spectroheliograms demonstrated the complexity of the solar flare phenomena. The most significant feature observed was the immense broadening of the hydrogen emission lines which takes place catastrophically at the time of the flare flash phase. Mervyn Ellison of the Royal Observatory in Edinburgh, attributed

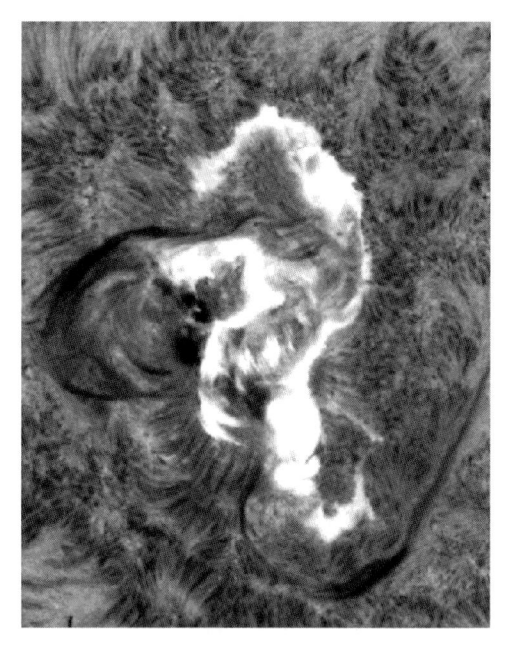

Figure 7 The development of a classic two-ribbon flare as observed in the blue wing of Hα (the "seahorse flare" of 7 August 1972). (Reproduced with permission of Big Bear Solar Observatory/New Jersey Institute of Technology.)

these great line widths to three potential causes: thermal Doppler broadening, Zeeman splitting, and the Stark effect. The first of these was ruled out since the flare spectrum was regarded as essentially one of low-temperature lines, while the lack of large line widths in the emission lines of metals (e.g., Ca, Fe) ruled out the Zeeman effect. This led Ellison and Fred Hoyle to conclude that the Stark effect was responsible for the observed line widths. In order to generate sufficient Stark broadening the presence of directed streams of electrons was invoked, with the electron beams being accelerated in electric fields near the sunspots. The presence of energetic particles in flares had been postulated earlier by William Swann in a model where particles were accelerated via the betatron process. These early suggestions pointed to an active role for the magnetic field in the flare process.

Statistically, the frequency of occurrence of Hα flares was found to vary markedly with the magnetic type of sunspot, increasing as we progress from the simple unipolar spot (α-type) through simple bi-polar types (β and βγ) to the magnetically complex type (γ). The αβγ sunspot magnetic classification, introduced by Hale, was extended to include a δ configuration by Horst Künzel in 1960. This new sunspot type defines a bipolar spot group with separate umbra but common penumbra, a configuration which was found to favor the production of large solar events such as flares. Modern theoretical studies invoke the magnetic complexity of the δ configuration to provide the necessary conditions for solar eruptive phenomena. This suggests that an important role is played by the magnetic field in the generation of flare activity. Indeed, the first magnetic interpretation for solar flares appeared as early as 1946 when Ron Giovanelli of the Commonwealth Solar Observatory in Canberra, Australia proposed the idea that flares were electrical discharges caused by changing magnetic fields in a high-conductivity medium. The reconfiguration of the magnetic field has been central in understanding flare phenomena ever since and it is now universally accepted that the whole flare phenomenon is critically governed by the properties of magnetized plasma.

In association with the Hα ribbons it was noted that the expulsion of hydrogen was observed near the peak intensity of the majority of bright flares. These emissions were found to occur in specific directions, usually along nearly vertical trajectories. Whether seen in elevation at the limb or in Doppler shifted Hα against the disk, this expulsion of material exibited all the characteristics of the well-known eruptive prominences (Figure 8). The initial velocity of a mass expulsion was around $500 \, \text{km s}^{-1}$ and while its Hα brightness was several times that of normal quiescent prominences it was still much fainter than the flare emission itself.

The first observations of prominences on the limb, outside of a solar eclipse, were those of Joseph Lockyer and

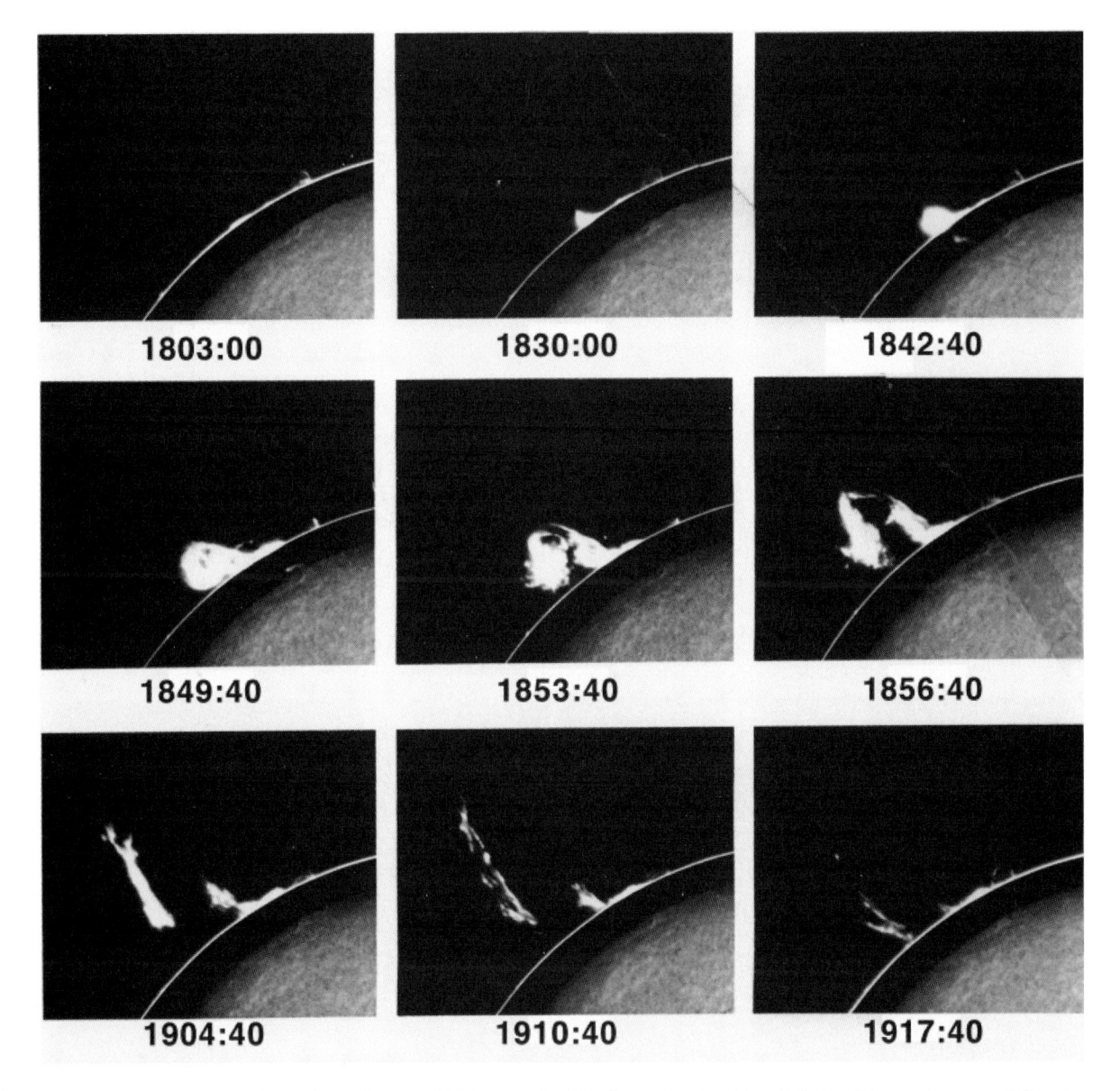

Figure 8 Prominence eruption of 11 August 1963 seen in Hα from the Lockheed Solar Observatory on Brier Summit, California. (Reproduced from Solar Filtergrams of the Lockheed Solar Observatory with permission of Lockheed-Martin Solar and Astrophysics Lab.)

Jules Janssen in 1868 using spectroscopes to measure prominence emission lines in full daylight. (It was from observations of prominences that the element Helium was discovered on the Sun). It was not until 1903, with the aid of photographic techniques that Hale and Ferdinand Ellerman discovered that the dark filaments observed in spectroheliograms were merely the projection of prominences on the disk (the term "filament" is due to French astronomer, Henri Deslandres). The invention of the coronagraph by French astronomer Bernard Lyot in 1930 made day-to-day observations of the prominences, and more generally the outer corona, possible. Prominences were classified into five types by American astronomer Edison Pettit based on their dynamical behavior. These included prominences described as active (type 1), containing streaming material, eruptive (type 2), containing ascending material, sunspot (type 3), found near sunspots, tornado (type 4), showing evidence for a spiral structure, and quiescent (type 5), relatively steady long-lasting structures. A second classification was derived by Harold Newton who used spectrohelioscope observations to categorize prominences into two main types: Type I prominences avoid the neighborhood of sunspots and consist of long well-defined filaments lasting several days (these correspond to Pettit's quiescent prominences); Type II prominences are associated with sunspots or with plage, are generally smaller than Type I and have lifetimes of the order of minutes or hours (this class corresponds to the other four types of prominence defined by Pettit). Nowadays, solar researchers refer to quiescent and active (or eruptive) prominences which relate more or less to Newton's Type I and II, respectively. Newton further subdivided his class II prominences into those which show large radial velocities and occur after the appearance of a localized emission and prominences which were originally of class I but became activated by the sudden appearance of an emission object. The latter category describes an interesting flare-prominence relationship. For the interested reader, an excellent history of prominence observations can be found in *Solar Activity* by Einar Tandberg-Hanssen (1967).

The relationship between solar flares and prominences goes back several decades to the phenomenon of *disparitions brusques*, catalogued in the late 1940s by researchers at Meudon Observatory in France. Nearly one-half of all low-latitude filaments were found to disappear temporarily at least once during their lifetime. When one makes allowance for the time the filaments are on the invisible hemisphere of the Sun this sudden disappearance phase appears to be an intrinsic property of all filaments. However, it was apparently caused by an external disturbance either from a flare or from a sunspot perturbation. This situation is complicated by the fact that in some cases the *disparitions brusques* are followed by a chromospheric brightening and therefore precede the flaring activity. Both quiescent and active prominences can be subject to *disparitions brusques*. The factors that cause this are important since filament eruptions appear to have a role in many of the coronal transients which make up the most energetic of solar activity (Figure 9). However, despite the fact that solar prominences have been known for several hundred years, and were easily observed, they were not thought to play a role in geomagnetic storms. A relationship was suggested by Harold Newton and W.M.H. Greaves in 1928 but Hale disagreed, pointing out three years later that erupting prominences generally fall back to the Sun. Newton, himself, later dismissed erupting prominences as the sources of high-speed streams because they rarely achieved escape velocity. The connection between prominence eruptions and geomagnetic storms was not fully appreciated until the work of JoAnn Joselyn and Patrick McIntosh in 1981. The flare was regarded as the primary agent for geo-effectiveness although it was clear that there was a strong correspondence with filament activity and flaring in solar active regions.

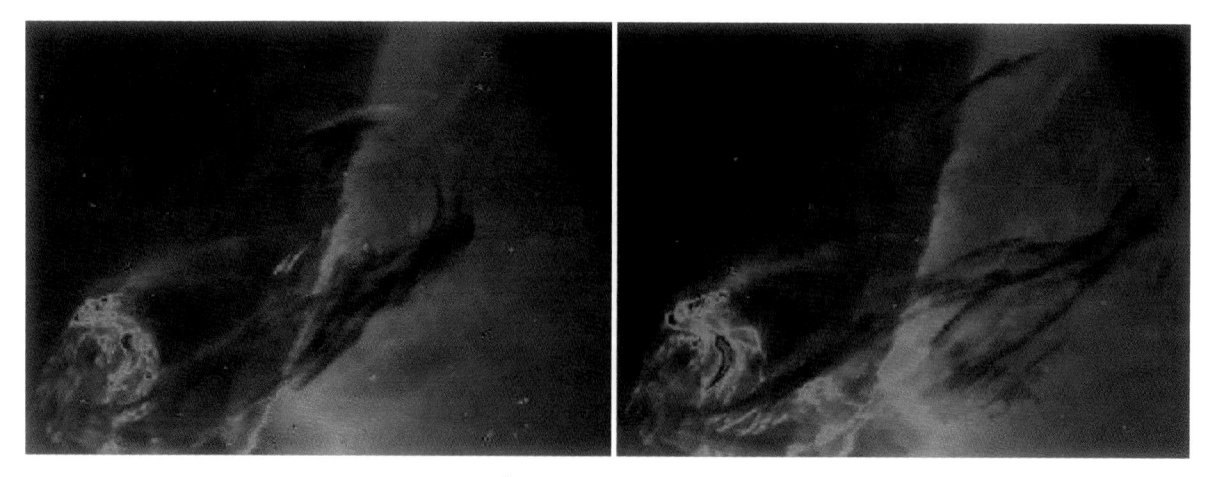

Figure 9 A solar filament eruption seen in the 195Å channel of the Transition Region and Coronal Explorer on 11 July 1998.

Karl Otto Kiepenheuer wrote in 1961:

Those who have seen in an accelerated movie the brightening of a flare out of a dark filament, and the almost chaotic interaction of bright and dark structures, will not doubt the existence of a causal relation between the activation of a dark filament and the formation of a flare.

Kiepenheuer further pointed out that the sudden disappearance of a prominence may occur in a number of ways: (1) the prominence flows down into the chromosphere, (2) the prominence shrinks, and (3) the prominence rises into the corona with increasing velocity that may eventually exceed the velocity of escape. The first of these was identified with the existence of magnetic fields which determine the streaming motion of the prominence material and that the triggering for this process was related to flaring activity in the sunspot. The third type of disappearance had been studied in detail with the conclusion that the ejected plasma is accelerated as it rises.

Despite the apparent association with flares it was also possible that the triggering agency was provided by sunspots. Anton Bruzek of the Fraunhofer Institute found that the development of magnetic field in and around sunspots generated a disturbance which propagated outward at $\sim 1\,\mathrm{km\,s^{-1}}$. Such a sunspot-induced disturbance should be distinguished from the similar flare-induced phenomenon where the triggering disturbance travels at about $100\,\mathrm{km\,s^{-1}}$. These studies were the precursors to present-day investigations into the relationship between filament eruptions and flares and preceded by three decades the discovery of coronal mass ejections.

Combined with the apparently clear association between geomagnetic disturbances and solar flares, the observed acceleration of material associated with prominence eruptions suggested a physical mediator for the transfer of energy from the solar atmosphere to the Earth. Statistical studies of the occurrence of geomagnetic storms following intense flares performed by Newton (1950) strongly suggested that solar flares resulted in the expulsion of a cone-shaped beam of particles having its origin at the time and place of the flare. Given the incontrovertible evidence for the existence of corpuscular radiation from the Sun, a major effort to detect the particles in transit was performed. The technique to do this relied on the observation of Doppler shifted absorption lines in the Fraunhofer spectrum. The results proved inconclusive since the absorption measured was of the same order as the errors inherent to the photographic photometry technique used. A more promising approach was the attempt to measure the absorptions at or shortly after the occurrence of the flare when the absorption would be at its most intense. Max Waldmeier in 1941 and Mervyn Ellison in 1943 independently detected a strong asymmetry in the wings of the Hα emission line of flares. Ellison interpreted this as being due to the absorption by hydrogen atoms which were being expelled in all directions from the flare site. This asymmetry was subsequently confirmed with spectrohelioscopes at observatories around the world. Ellison did caution, however, that:

While these asymmetric profiles provide the strongest possible evidence for the general expulsion of hydrogen during flares, we must await further work in order to prove that this constitutes the initial departure of the geomagnetic storm particles.

Compounding the mystery, sudden increases in cosmic ray intensity coinciding with large flares were detected, suggesting that in addition to plasma ejection the flare also contrived to accelerate charged particles to energies in excess of 5 GeV. The conclusion was that some unknown mechanism of the sunspot magnetic field was accelerating these particles to velocities close to the speed of light at the time of the flare. The flares for which the cosmic ray enhancements were detected were among the greatest flares recorded up to that time (1950). A problem remained, however, in that the delay between the flash phase of the flare and the peak of the cosmic ray intensity was about 30–60 minutes, yet the particle energies suggested a travel time of about 10–15 minutes. Two possibilities to account for the long delays were proposed: either the particles were accelerated at the time of the flare and then stored in the corona before escaping to the Earth or they were accelerated by some magnetic field mechanism which normally operates near the end of the flare and not at its beginning. The notion that the particles could be generated en route did not occur to the researchers at the time.

Additional evidence for the production of energetic particles in flares was provided by an entirely different part of the electromagnetic spectrum: solar radio emission. The possibility of the existence of solar radio waves was realized soon after Heinrich Hertz's pioneering experiments. In the early 1890s H. Ebert suggested that the solar corona was a visible electric discharge and concluded that it must therefore be the source of electromagnetic radiation. Attempts to detect this radiation at the turn of the century were unsuccessful, primarily due to the inadequate radio techniques available at the time. The development of radio advanced significantly in the 1920s but the notion of radio emission from the Sun had apparently been forgotten.

In 1942, however, two astronomers on two different continents independently observed and identified solar radio waves. In England James Stanley Hey, Superintendent of the Army Operational Research Group, noted that interference in his radar equipment operating at meter wavelengths was coming from the direction of the Sun. At the same time in the USA, George Clark Southworth of the Bell

Telephone Laboratories, using newly developed microwave radar receivers, found steady radiation at wavelengths of 3 and 10 cm coming from the Sun. Both astronomers noted that the intensity of the radiation far exceeded that expected from a black body of 6000 K, as the Sun was thought to be. Further analysis of Hey's data showed that the time of maximum effect coincided with the visual observation of a great flare over a large sunspot which was then on the Sun's central meridian.

In Australia, radio astronomers Ruby Payne-Scott and A.G. Little, were able to localize the source points of the radio outbursts accompanying flares and noted that the source location moved rapidly upwards through the corona. They suggested that the physical agency producing the radio emission might be identified with the departing corpuscular streams postulated to generate the terrestrial magnetic storms. This hypothesis was also supported by the fact that the time lag observed by Hey between flare and radio noise becomes longer with decreasing radio frequency. Thus, by about 1960 there was little reason to doubt that all three solar–terrestrial disturbances (geomagnetic storms, sudden ionospheric disturbances and ground-level enhancements) were directly caused by solar flares at the Sun. The expulsion of high-velocity particles necessary to cause the geomagnetic disturbances, therefore, required a physical explanation. Developments in theoretical solar physics were starting to attack this problem as we will see in a later section.

Before moving on to the modern era in which space-based astronomy became pre-eminent in solar activity research, we should pause to discuss a second type of geomagnetic storm, unconnected with solar flares but crucial, nonetheless, to the understanding of solar activity. These numerous small geomagnetic storms occur preferentially near the time of sunspot minimum. In contrast to flare-related storms these small storms displayed a marked 27-day recurrence. The 27-day periodicity detected in geomagnetism by Charles Chree and James Stagg in 1927 was convincing evidence for a repeatable source of magnetic storms, even though their data referred partly to minor disturbances below storm intensity. Recurrent sequences in storms were clearly defined and illustrated by "Bartels diagrams," so called from their first use in a paper by German scientist Julius Bartels in 1932 (see Figure 10). Bartels associated recurrent storms with hypothetical active regions which he denoted M regions. The problem was that he was unable to find any correlation with visible solar features. Statistical studies of geomagnetic storms indicated that the M regions tended to avoid the neighborhood of sunspots and the Sun–Earth travel time of the associated corpuscular streams was about 3 days, indicating an average velocity of approximately 600 km s^{-1}. In the late 1940s and early 1950s there was much discussion about the possible nature of these M regions with suggestions varying from possible connections to the long coronal streamers, seen on eclipse photographs, to regions which were first occupied by sunspots then free from photospheric disturbances for several months afterwards.

A pronounced decrease in the frequency of the M disturbances was found to occur three days after the central meridian passage of sunspots. Thus, it was concluded that M regions avoided spot groups within a distance of about 40° and were even destroyed by them. It was argued by Kiepenheuer in 1947 that this demonstrated a relationship between M disturbances and filaments. An important piece of evidence regarding the nature of the M regions came from a study of eight M storms which occurred in 1950–1951. Michael Smyth of the Royal Observatory, Edinburgh found that the onset of each of these storms followed the central meridian passages of unusually faint regions of the monochromatic corona. This was confirmed some 20 years later when Art Hundhausen of the High Altitude Observatory in Boulder determined that recurrent geomagnetic storms were associated with "inactive" solar regions, while Ludwig Oster and colleagues associated the storms with solar regions of open magnetic field. The picture was still not clear though with Constance Sawyer and Shirley Hansen arguing that the solar sources of recurrent storms were coronal equatorial arches. The true nature of these mysterious M regions was soon to be determined, however, by the combination of high-resolution X-ray imaging and observations from space.

The primary reason for the disparity in viewpoints about the source of recurring magnetic storms was the plain fact that the M regions could not be directly observed and that all studies were based on inferrences gleaned from those phenomena which could be observed. What was clear was that geomagnetic disturbances, large and transient or small and recurring, required the production of particles (the "corpuscles") at the Sun and their subsequent transport to the Earth. From this relatively simple fact the concept of a "corpuscular wind" was developed. The solar wind as it is now known is an important component of present-day solar–terrestrial research and is discussed in detail in Chapters 9 and 47.

THE TRANSIENT SUN: THE SPACE YEARS

There is a stability in the Universe because of the orderly and balanced process of change – Heraclitus

Our knowledge of the basic physics of solar activity has increased dramatically since we started taking observations from space. The original space instrumentation, however, was fairly primitive and was initially flown on sounding rockets. Following the success of the rocket program and the flight

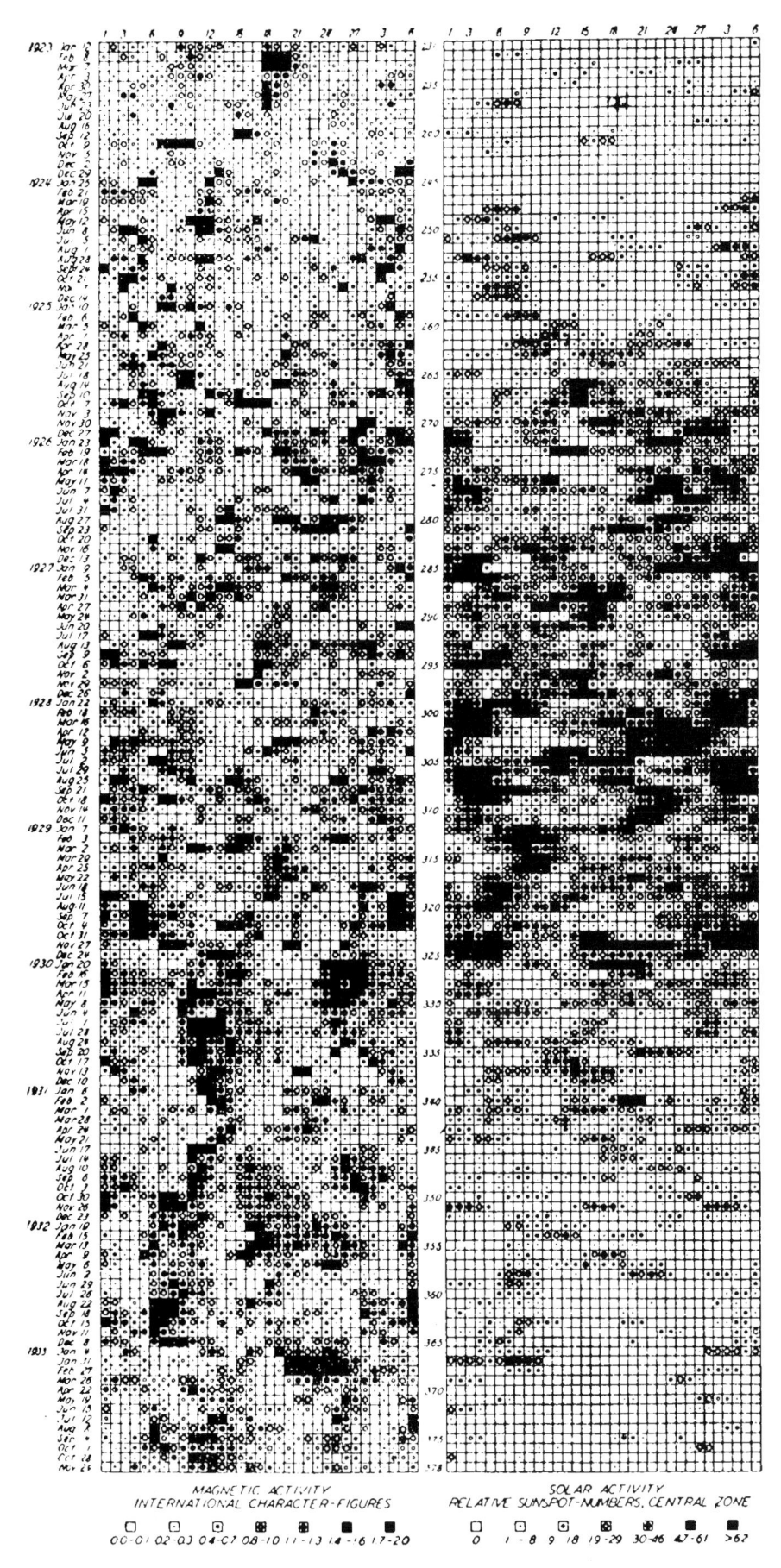

Figure 10 Bartels diagram for 1923–1933. (Reproduced with permission of Gesellschaft für wissenschaftliche Datenverarbeitung mbH, Göttingen.)

of the first satellites, space instrumentation developed at an increasing pace. Within the field of solar physics alone, several missions have been flown to study solar activity and, in particular, solar flares. These include NASA missions such as the OSO series of satellites in the 1960s and early 1970s, Skylab in 1973–4, and the Solar Maximum Mission from 1980 to 1989. More recently other nations have taken the lead in the study of solar activity (albeit with significant contributions from the USA via NASA). Japanese missions such as Hinotori in 1981 and Yohkoh, which was launched in 1991 and is still in operation today, were designed specifically to study solar flares. Predominantly European missions such as Ulysses and SOHO, operating currently, target the solar wind/interplanetary space and the global Sun from its core to far beyond the Earth's orbit, respectively. The Ulysses spacecraft was the first to leave the ecliptic plane since the Voyager spacecraft. Its orbit takes it over both solar poles to sample the slow and fast solar wind. The SOHO spacecraft, on the other hand, maintains a position on the Sun–Earth line by being in orbit about the L_1 Lagrangian point, about 1.5 million km from the Earth. NASA has returned to solar activity studies with the TRACE (Transition Region and Coronal Explorer) satellite launched in 1998 and the RHESSI (Ramaty High Energy Solar Spectroscopic Imager) mission launched on 5 February 2002. These solar observatories have been joined by many more solar–terrestrial missions over the years, e.g. Wind, ACE and Cluster II, and will be further complemented by the STEREO (Solar TErrestrial RElations Observatory) and Solar-B missions in 2004–5, providing a fleet of spacecraft all studying some aspect of solar activity.

A prime advantage of space observation is the ability to observe the Sun at wavelengths unobservable from the ground. This is particularly relevant for solar flares which were predominantly regarded prior to 1960 as chromospheric phenomena observed mainly in Hα. University of Chicago physicist Eugene Parker has been quoted as saying that "to construct a flare from its Hα image is a task which reminds me of the reconstruction of a dinosaur from its footprints." While this should not be used to denigrate the importance of observing the footprints, it does serve to emphasize the wealth of phenomena exhibited by even the smallest of flares and which can only be accessed via observations from space.

Developments in ground-based radio observations and the advent of space-based solar physics completely changed solar flare research to the extent where modern approaches treat flares as primarily a coronal manifestation with enhanced emission of high temperature radiation, accelerated particle production, mass ejection, and rapid morphological changes. However, space-based astronomy, in particular, resulted in a couple of discoveries which were to call into question the geo-effectiveness of solar flares; a spirited debate which continues today.

It had long been known that clouds of particles were expulsed from the Sun at the time of large solar flares. These particles resulted in the dramatic response of the Earth's magnetic environment in a geomagnetic storm. The problem was that this plasma cloud could not be detected directly on its journey from the Sun to the Earth, despite some ingenious ideas and observations. If a solar flare was to be accompanied by a stream of accelerated charged particles, then a mechanism was required to generate the expulsion. In 1961, Eugene Parker demonstrated that a large solar flare could drive a hydrodynamic blast wave to the Earth in 1–2 days. This idea was subsequently "confirmed" by a series of calculations on interplanetary shocks and was even supported by observation. However, Art Hundhausen in 1972 noted a number of apparent discrepancies between the shock wave models and observations, expressing some reservations about the association between large flares and the interplanetary shocks. First, there was an imperfect correlation between energetic flares and the shocks. Second, the masses and energies of the shocks were extremely large compared with the flare energies. Third, the occurrence of flares in closed magnetic field regions were unrelated to the solar wind flow along open magnetic field lines. Thus, one year prior to the launch of Skylab the physics of storm-causing interplanetary shocks was understood but the shocks themselves could not be directly related to any coronal events.

The source of the interplanetary shocks came into startling relief with the Skylab coronagraph observations. Indications of large, transient disturbances travelling through the Sun's outer corona had been noted in solar radio records and found in coronagraph observations from earlier unmanned spacecraft (most notably OSO-7). However, the as-then unprecedented sensitivity of the Skylab instrument was to create the dawn of a new era in solar physics research. The astronauts on the space station witnessed gargantuan loops rushing outwards from the Sun at fantastic speeds. Astronomers on the ground were elated by the first detailed pictures of the expulsion of erupting material which ultimately grew to be bigger than the disk of the Sun. Because of Skylab's extensive, concurrent coverage at ultraviolet and X-ray wavelengths, discovery was followed, in subsequent analysis, by a full description of the features and causes of coronal transient disturbances. It was evident that some of the coronal transients were initiated lower in the solar atmosphere and accompanied by flares while another, larger, class was associated with prominence eruptions. The disturbances of this latter class carried too much material to be simple expansions of the prominence and were, therefore, recognized as coronal phenomena. The first summary of the Skylab coronal disturbances by Jack Gosling and colleagues in 1974 left little doubt that these transients were the long-sought eruptions

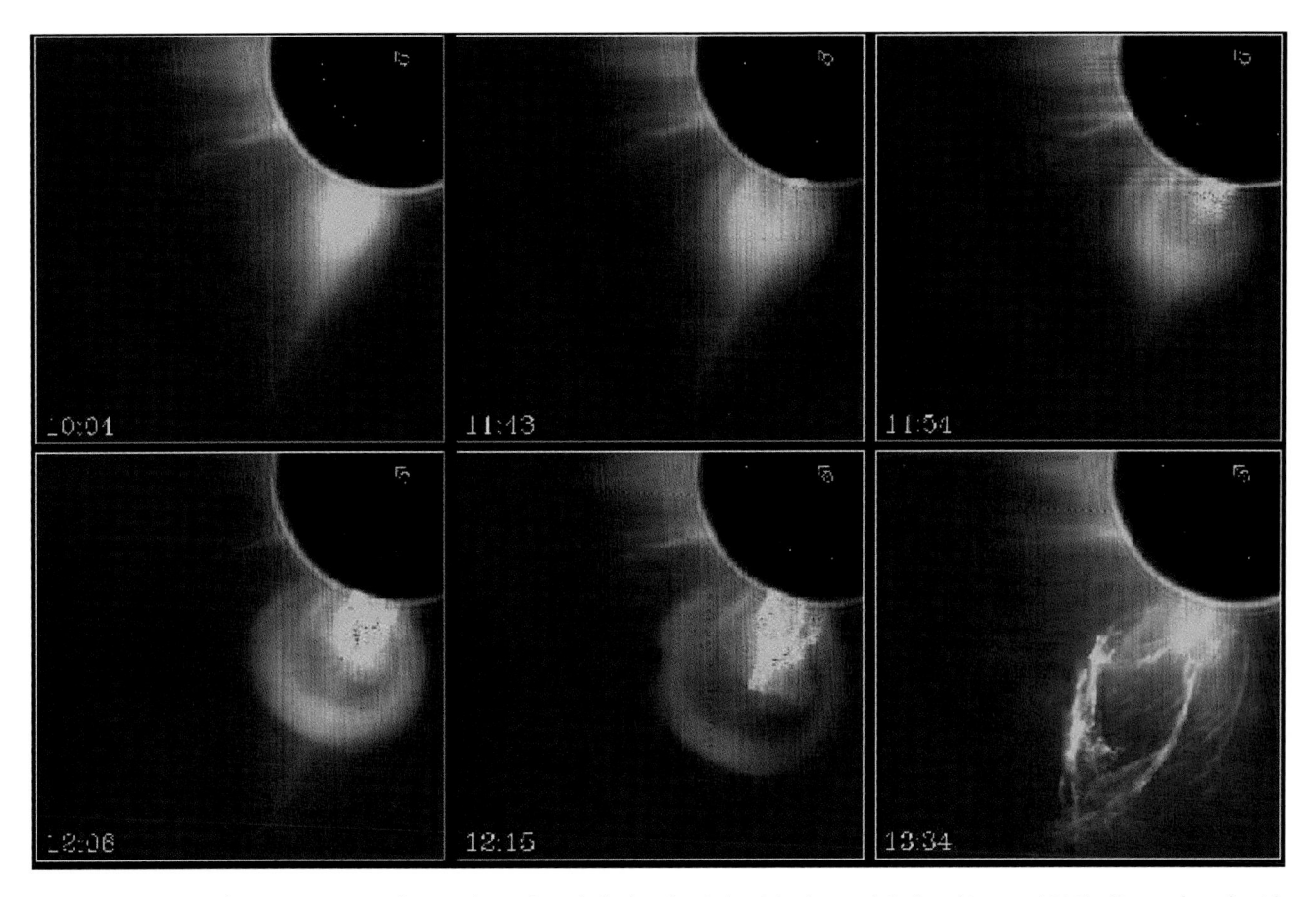

Figure 11 Coronal mass ejection observed in white light by the Solar Maximum Mission (August 1980). (Reproduced with permission of High Altitude Observatory.)

of coronal material required to produce the high speed transient flows of solar wind which, in turn, produce geomagnetic storms (Figure 11). Over the succeeding few years these events came to be known by a variety of noms-de-plume such as "plasma clouds," "solar mass ejections," "mass ejection coronal transients," and "coronal mass ejection events." Finally, in 1982, the abbreviation CME for Coronal Mass Ejection was coined in a paper by Len Burlaga and colleagues. This term is now widely and readily used to denote not only the coronagraph observed phenomena but also for X-ray dimming events at the Sun and particle events in interplanetary space. Art Hundhausen, who became the leading authority on CMEs, defined these events "to be an observable change in coronal structure that (1) occurs on a time-scale between a few minutes and several hours and (2) involves the appearance of a new, discrete, bright, white-light feature in the coronagraph field of view."

Coronal mass ejections have been a topic of extensive study since their discovery and have proved to be scientifically interesting in many different ways. A key feature is their transport through interplanetary space and their subsequent interaction with the Earth. It was estimated that a typical mass ejection contains approximately 4×10^{15} g of material and that the outward flow of this material drove pressure waves into the surrounding corona and solar wind. When the outward speed was sufficiently high, the ejections produced shock waves in the solar wind far from the Sun. The most intriguing aspect of these revolutionary observations was the number of surprises that they generated. First, far more CMEs were observed than expected from the frequency of occurrence of large geomagnetic storms or solar wind disturbances. Second, individual ejecta exhibited a wide range of outward speeds ranging from $<100\,\mathrm{km\,s^{-1}}$ to $>1200\,\mathrm{km\,s^{-1}}$, thereby explaining the first mystery since only the fast eruptions caused the interplanetary and geo-effective events. Third, it was nigh impossible to detect magnetic disconnection of the ejecta from the Sun even though this seemed to be required by the roughly constant magnetic flux in interplanetary space. Fourth, the ejections were found to be more commonly associated with prominence eruptions than with impulsive solar flares. Thus, the remote sensing capabilities of Skylab together with the *in situ* observations of spacecraft like Mariner 2 profoundly affected the way solar physicists viewed the

Sun–Earth connection. No longer could flares be regarded as the primary source of geo-effective disturbances. Even today the observational evidence supports the notion that CMEs are the primary cause of both energy release in the corona and the interplanetary disturbances. Flares, are not required to produce a CME and are arguably only secondary phenomena when they do occur with CMEs. Solar flares do, however, have terrestrial consequences, as evidenced by the detection of the sudden ionospheric disturbances and ground level enhancements discussed earlier. Such terrestrial effects were major disruptors of radio communications prior to the satellite era. However, it was observations from space which categorically proved that these disturbances were due to flare-generated hard radiation interacting with the upper layers of the Earth's atmosphere. Today these atmospheric effects are regarded as secondary to the major disruptions of the Earth's magnetosphere following a CME.

Before leaving Skylab we should return briefly to the mysterious M regions discussed in the last section. The 27-day recurrence of geomagnetic storms discovered in 1927 firmly suggested that the source was the Sun and led to much speculation on the nature of the solar phenomenon responsible. One idea put forward in the 1950s by Babcock and Babcock was that the hypothetical M regions could be identical with unipolar magnetic regions seen in their magnetograph observations. Regions of supressed emission were discovered in optical data by Max Waldmeier around the same time (and actually called *Koronalöcher*, i.e., coronal holes; see Chapter 8), but the link to the M disturbances was not apparently made. It was not until observations from outside the Earth's atmosphere were possible that the true nature of the M regions was discovered. Areas of strongly reduced emission were recorded on XUV spectroheliograms obtained from sounding rockets flown in 1967. These regions were subsequently observed with coarse spatial resolution by the Harvard instruments on the OSO 4 and OSO 6 spacecraft, leading to the definition of "coronal holes" – apparently independently from Waldmeier – as "being characterized by significant deficiency in the intensity of coronal emission lines" by Richard Munro and George Withbroe in 1972.

This discovery was the final piece in the puzzle of the mysterious recurrent geomagnetic storms. High-speed streams of particles detected in the solar wind by the Mariner 2 spacecraft were found to be associated with the recurrent storms and suggested that the Sun was unleashing bursts of particles from localized sources that rotated with the Sun (the M regions). The observed high speed streams were traced back to the Sun and were, without exception, found to originate from a coronal hole (once corrections had been made for the travel time of the stream and solar rotation). The mysterious M regions first postulated in 1932 had finally been discovered. The launch of Skylab in 1973 was soon to demonstrate that coronal holes were one of the most terrestrially important of all solar phenomena (Figure 12).

It should be noted that coronal holes are not completely devoid of emission. At temperatures associated with EUV and X-ray emission the radiative output of a coronal hole

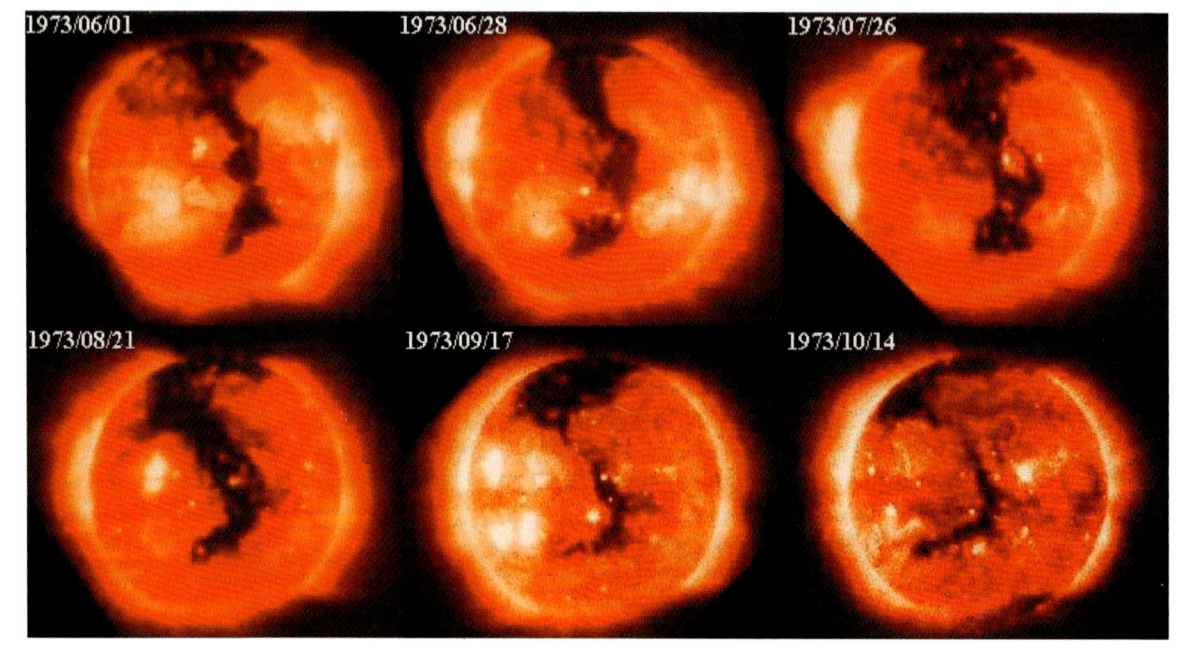

Figure 12 Development of the "Boot of Italy" coronal hole seen by Skylab in the summer of 1973. (Reproduced with permission of David Hathaway, MSFC/NASA.)

region is extremely small but detectable. The low level of radiative output is primarily a result of low density due to the fact that the magnetic field in the holes is open to interplanetary space allowing gases to escape the low corona. We refer the interested reader to Chapters 9 and 47 for a more complete discussion of the solar wind and the role played by coronal holes.

Following on the success of Skylab, several space missions were launched to study the transient Sun, many of them flying a coronagraph in combination with instruments designed to observe the high-energy radiation emitted from solar flares. The Solar Maximum Mission (SMM), launched on 14 February 1980, significantly advanced our knowledge of solar flares and coronal mass ejections. SMM was a multi-instrument orbiting observatory built to fly during the maximum of Cycle 21. Prior to SMM, the radiation from energetic ions and relativistic electrons had been measured for only a handful of flares and as a consequence, our understanding of flare particle acceleration was inferred largely from measurements of charged particles in interplanetary space and by extrapolating X-ray and radio observations.

Hard X-ray radiation ($E > 10$ keV) from solar flares was first observed as far back as 1959 by Lawrence Peterson and John Winckler using a balloon-borne X-ray detector. Subsequent, rocket, balloon and satellite observations improved our knowledge of the hard X-ray emission which was thought to be bremsstrahlung radiation from energetic electrons interacting with the ambient solar atmosphere. However, it wasn't until the flight of SMM that this energy regime was fully investigated. In the hard X-ray range, the electron bremsstrahlung component is significant in the overall flare energy budget reaching 10% or more of the total.

The most direct observational evidence for the acceleration of particles at the Sun came with the first signs of radio emission produced by fast electrons interacting with solar plasma and magnetic fields. This was subsequently followed by the discovery of radiations at higher energies (e.g., hard X-rays and γ-rays), implying particles with energies in the several tens of MeV range, and direct measurement of energetic electrons and ions in interplanetary space. These energetic radiations are all presumed to be generated by the interaction of energetic particles accelerated in the solar flare with the ambient solar medium. Many populations of energetic (i.e., non-thermal) particles created by solar flares have been identified covering several decades in energy. These populations include deka-keV electrons (10–100 keV) responsible for the bulk of the flare hard X-ray emission, relativistic electrons (>10 MeV) contributing to the γ-ray continuum, accelerated ions (10–100 MeV) which generate the rich nuclear de-excitation γ-ray line spectrum, >100 MeV ions which result in pion production in the solar chromosphere, and high-energy neutrons which result from the interaction of energetic protons and α particles in the flaring solar atmosphere. To varying extents, particle acceleration appears to occur in all flares, including relatively minor ones. Knowledge of the particle numbers, energy distributions and relative abundances provide the key to unlocking the secrets of the flare energy release.

The SMM yielded significant advances in all of these areas providing confirmation of several theoretical ideas developed from early observations. SMM also raised a number of questions which provided the impetus for the Yohkoh mission launched at the maximum of solar cycle 22. The many discoveries and scientific endeavors of the SMM mission is the subject of *The Many Faces of the Sun* edited by Keith Strong, Julia Saba, Bernie Haisch, and Joan Schmelz. The SMM satellite finally fell from orbit into the Indian Ocean in November 1989 having completed nine years of duty observing the most energetic phenomena of the Sun and redefining the Sun–Earth connection paradigm (Figure 13).

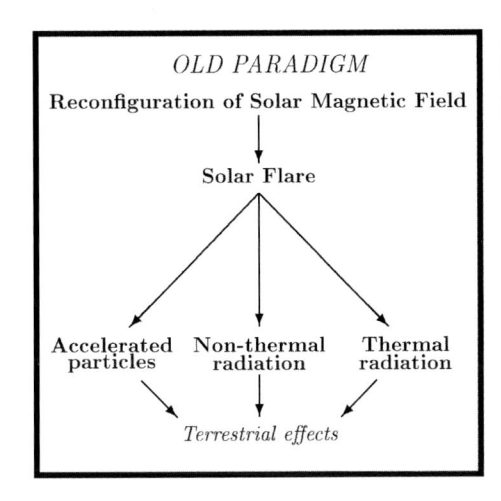

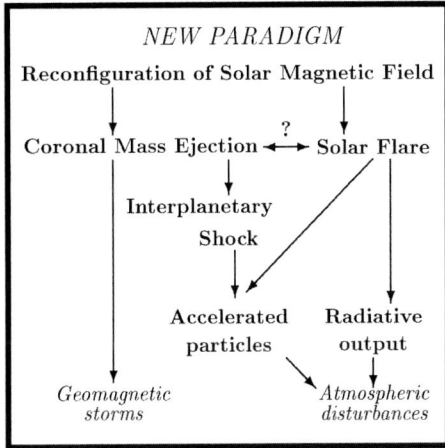

Figure 13 The discovery of coronal mass ejections, the solar wind and coronal holes led to a change in the solar–terrestrial paradigm. (After a figure by Jim Ryan in Strong *et al.* (1998).)

In addition to the scientific discoveries it made, SMM has the distinction of being the first satellite ever repaired on orbit. This occurred in April of 1984 when the crew of the Space Shuttle *Challenger* flew a mission (mission 41-C) to capture and replace the gyros which had failed after only nine months in space. The pilot and commander of the SMM repair mission was Francis R. Scobee, who was also the commander of *Challenger's* tragic last flight.

The relationship between solar flares and coronal mass ejections was a persistent theme during the 1990's with data from the Yohkoh and SOHO satellites being paramount. The Japanese satellite Yohkoh was launched as a solar maximum mission with a complement of instruments designed to study solar flares from soft X-ray to gamma-ray energies. The Hard X-ray Telescope (HXT) provided imaging capability at energies between 14 and 100 keV which confirmed the SMM discovery of localized hard X-ray emission at the footpoints of magnetic loops. The surprise discovery from the HXT was that localized hard X-ray emission was also found in the corona above some flaring loops. This has shed the first observational light on reconnection and particle acceleration in these events. More importantly for the flare/CME problem has been the Soft X-ray Telescope (SXT), the dynamic range of which has allowed Yohkoh to be a critical component in all solar flare investigations for almost a complete solar cycle. Observations by the SXT have blurred the distinction between the classical description of a solar flare and that of a classical CME. The SXT has demonstrated that CMEs typically have a response in the hot corona even when this response does not include typical flare emissions. The eruption of a filament or filament channel, a phenomenon which is found to be closely related to CMEs, often produces an X-ray arcade reminiscent of post-flare loop systems. The spatial scales of these CME-related arcades tend to be significantly larger than those in flares although they contain similar total energies. The large volumes involved tend to limit the production of hard X-rays and other non-thermal flare emissions and some would question whether this can rightly be called a flare. An intriguing observation which has recently been found to precede the arcade formation is the "dimming" of the X-ray corona, suggesting that a significant volume (and mass) of gas has been ejected from the flare site consistent with coronagraph observations in white light. The dimming events discovered by Yohkoh have been confirmed subsequently in EUV observations by the Extreme Ultraviolet Imaging Telescope (EIT) on SOHO and by TRACE. The quantitative relationship between this ejected mass and that seen in the CME has, however, yet to be established.

A joint European and American mission launched late in 1995 has once again brought CMEs to the fore in solar activity research. The ESA's SOHO spacecraft contains an impressive package of solar instrumentation which includes an EUV imager, UV/EUV spectrographs, UV and white-light coronagraphs, and *in situ* particle-and-field detectors. This suite of instruments originally designed to study the dynamics of the quiet Sun (local and global) and its effects on interplanetary space has been a great success and at the time of writing continues to yield spectacular science as we move towards solar maximum. The combination of a full disk EUV imager (EIT) and the three white light coronagraphs, collectively known as the Large Angle Spectrometric Coronagraph (LASCO), has demonstrated the coronal consequences of these large-scale magnetic reconfigurations, both in the hot lower corona and the extended outer corona. LASCO represents a significant advance over previous coronagraphs primarily in its expanded field of view (out to some 30 solar radii, i.e., about one-seventh of an astronomical unit), increased sensitivity and increased dynamic range. While the CMEs observed by LASCO are similar to those observed in previous coronagraphs, there are several new aspects: (i) many are accompanied by a global response of the solar corona, (ii) many show acceleration to the edge of the LASCO field of view, (iii) disconnection is a frequent occurrence, (iv) CMEs are occurring more frequently than had been expected at solar minimum, and (v) CMEs undergo extensive internal evolution as they move outward.

For all the information and detailed observations we have on flares and CMEs, we are still a long way from an understanding of the basic physical process leading up to an energetic solar eruption. How strongly these two diverse phenomena are inter-related and what physics governs this relationship is still unclear. However, it is becoming increasingly apparent that the magnetic reconfigurations associated with large flares and, more particularly, CMEs, have a role to play in mitigating the effects of the solar cycle. It has been suggested, recently, by B.C. Low of the High Altitude Observatory that the process by which the polarity of the Sun's large-scale field reverses over an 11-year cycle is mediated in the solar atmosphere by CMEs.

There is still much we do not know about solar eruptive events and it is clear that flares and CMEs will continue to inspire solar research for a long time to come. Of particular interest is the ability to predict these events and their potential terrestrial impact. Predicting the behavior of something as physically complex as the Sun, like terrestrial weather, is an extremely difficult endeavor. However, new and future missions (e.g., RHESSI, Solar-B, and STEREO) will significantly improve our chances of attaining this goal.

We have chosen to concentrate on the major phenomena which signify the activity of the Sun. However, it should always be borne in mind that activity is everywhere on the Sun. Although the term "quiet Sun" is commonly used, it is a misnomer. The observable Sun is dynamic at all scales, in all wavelengths and at all times. As instrumentation improves and higher spatial resolution becomes coupled to higher sensitivity detectors, the low-energy end of the transient-event spectrum is attracting more and more attention.

One satellite which is currently providing new observations of solar activity on the finest scales yet is NASA's Transition Region and Coronal Explorer. Launched in April 1998, TRACE observes the solar atmosphere at a spatial resolution unprecedented at EUV and UV wavelengths (1 arcsec, or 700 km, at the Sun). In addition, the extensive temperature coverage of TRACE enables us to observe transients throughout the solar atmosphere. The wealth of data and the remarkable results from TRACE are worthy of a chapter of their own. The reader is referred to two recent volumes of *Solar Physics*, Vols 192 and 196, which contain a number of papers highlighting a subset of the TRACE results. On the subject of transient events in the solar atmosphere, TRACE has observed, in remarkable detail, solar transients on scales some seven to eight orders of magnitude lower than the largest flares. This is a burgeoning area of solar research and promises to yield a more definitive picture of the active Sun than we have at present.

The opening up of space for scientific investigation of the Sun has led to a remarkable increase in our knowledge of astrophysics. The interplay of magnetic field and plasma occurs throughout the Universe and the proximity of the Sun makes it a fundamental proving ground for these astrophysical processes. Space missions have enabled us to investigate the Sun in more detail than we could have imagined at the end of the nineteenth century. Remote sensing observations in wavelengths unobservable from the Earth have shown us an object of inexpressible beauty and wonder. *In situ* observations by spacecraft travelling through the interplanetary medium have physically sampled the outer reaches of the solar atmosphere investigating the tendrils which connect the Sun to the Earth. What was once the final frontier is now a common destination, a location from which to learn the secrets of the cosmos. The next millenium promises to be a rich and exciting one.

THE PHYSICAL SUN

Quis, quid, ubi, quibus auxiliis, cur, quomodo, quando? Who, what, where, with what, why, how, when? – Unknown

As the nearest source of cosmic rays in the Universe, solar activity is an important phenomenon in astrophysics. Processes similar to solar flares and coronal mass ejections almost certainly occur in other stars and perhaps also in a wide range of astrophysical phenomena. In previous sections we have concentrated on the historical development of the observations of solar activity culminating in our present interest in solar flares, CMEs and related activity. However,

observational developments, as in any scientific field, progressed hand in hand with theoretical considerations. Sometimes observations led to new theories (e.g., the interpretation of Ludwig Biermann that comet tails were generated by a solar wind ultimately paved the way for what we now call space weather) while at other times theoretical developments led the quest for new and better observations (e.g., James Dungey's reconnection and instability ideas generated substantial observational interest in determining the changes in magnetic configuration during a flare). The development of theoretical models of solar activity has as rich a history as the observational side of solar physics. However, it was realized very early that most solar phenomena had something to do with the magnetic field. Consequently, the major improvements in our theoretical understanding of solar activity has come about through our ability to investigate the interplay between the plasma and the magnetic field, either as an ideal MHD process or as a dissipative process such as magnetic reconnection.

As in the case of observational studies, much of the theoretical interest in solar activity stemmed from its potential impact on the Earth space environment. As a consequence, solar flares dominated the theoretical landscape for the first 70 years of the twentieth century. Future developments, in particular the identification of the CME as the primary geoeffective solar event, provided a richer and more diverse arena in which to explore many of the same physical processes thought to be important for flares. Theoretical ideas, therefore, concentrated on the conversion of magnetic to thermal and kinetic energy.

For many years at the beginning of the 1900s the dictum of Lord Kelvin, that the Sun could not possibly be responsible for geomagnetic disturbances, influenced, and in fact hindered, the theory of solar activity. Kelvin's calculations were correct given the knowledge then available: he did not know about the solar wind which would have altered his reasoning considerably. Lord Kelvin was so revered that three decades later George Ellery Hale was still referring to the *apparent* terrestrial effects of solar eruptions. The author Arthur C. Clarke provides a cautioning motto: "If a famous scientist says something is impossible, he is probably wrong."

A major difficulty in the idea of a Sun–Earth connection was the lack of an apparent mechanism by which the Sun could influence the Earth's magnetic environment. The potential influence of the magnetic field on all forms of solar activity led William Swann to suggest in 1933 that energetic particles might be produced in solar flares by betatron acceleration thereby pointing to a possible means by which flares could cause geomagnetic disturbances. The first truly magnetic theory for a solar flare was proposed in 1946, some 13 years later, by Ron Giovanelli, who considered the high conductivity of the solar atmosphere and suggested that flares were electrical discharges caused by changing magnetic fields. Building on these ideas James Dungey introduced two concepts which still dominate flare research to this day: reconnection (see Chapter 43) and instability. In 1953 he showed that the electric field generated by changes in magnetic field at an X-type neutral point leads to a current which serves to further enhance the change in the magnetic field.

Several models, proposed in the late 1950s and early 1960s, concentrated on the specification of a magnetic field configuration that was to be metastable and therefore prone to instability. The sudden onset of an instability would serve to release free energy associated with coronal currents. It should be pointed out here that earlier models, in accord with observations at the time, assumed that flares were chromospheric phenomena and as such required extremely large fields to produce the observed energies. It was not until the rocket flights in the late 1950s demonstrated the existence of X-rays from solar flares that these phenomena were found to be primarily coronal in origin. The magnetic dominance over the plasma in the solar corona makes the magnetic reconfiguration required for the solar flare energy release more attainable theoretically.

The main challenge of solar flare theory in the 1950s and 1960s was to produce the large energies associated with solar flares and to do this fast enough. The requirements of flare theory at that time could be summarized briefly as:

1. the supply of $\sim 10^{32}$ erg by pre-flare energy storage, or by changes in the magnetic field during the flare
2. the triggering of the release of the stored energy
3. the explosive phase transformation of half the total energy supply into mass motion at $1500\,\text{km s}^{-1}$ in less than 300 s
4. the acceleration of 10^{36} electrons to energies of the order of 100 keV in 1 second, repeated several times during the explosive phase
5. the acceleration of 10^{36} protons in the range 0.5 MeV–30 GeV in less than 1000 seconds
6. the decay stage transformation of the other half of the energy supply into optical radiation on a decay time scale of 3000 seconds.

Some four decades later these basic requirements are essentially the same (additional requirements are necessary in large γ-ray flares where protons with energies as high as tens of mega-electronvolts are produced in a matter of seconds). Observations during this time have refined the details somewhat but modern theories still have to contend with all of the above. Developments in reconnection theory, in particular, have led to a more complete understanding of flare dynamic phenomena. However, even with the recent advances in computing power, a fully three-dimensional reconnection theory of solar flares still eludes us. In addition, the discovery of coronal mass ejections have added a new dimension to the "flare" problem. Whether we are studying a CME or a solar flare the basic physics of their cause is essentially the same; a sudden conversion of magnetic energy in the solar corona. Central to both is the reconfiguration of the magnetic field via reconnection and instability as originally proposed by Dungey in the 1950s. It is worthwhile, in this historical review, to consider a model proposed in 1960 by Tommy Gold and Fred Hoyle since it is one of the earliest models that satisfied many of the requirements outlined above. In some ways it is the "parent" of modern flare models. The basic idea of Gold and Hoyle is the interaction of two flux bundles (loops) each exhibiting a twisted magnetic field. The toroidal field component in each fluxtube is assumed to be in the same direction. It is argued that, if the two fluxtubes are in close proximity for an extended length of time then they will begin to penetrate into each other, thereby, enhancing the toroidal component of the combined field while destroying the longitudinal component. In this way a linkage of the toroidal field occurs creating a pinch effect and setting up a current sheet between the two fluxtubes. It was recognized by Gold and Hoyle that a process must exist whereby a catastrophic effect can occur. In other words, the axial parts of the fluxtubes must approach each other at an increasing rate. They further realized that this would require a rapid increase in magnetic diffusivity which is equivalent to a rapid decrease in the electrical conductivity.

The formula used by Gold and Hoyle for the conductivity of a magnetized plasma clearly suggests that they were treating the chromosphere as the source of the flare energy release in that it assumes an electric field purely orthogonal to the magnetic field. However, the linkage of magnetic field inherent to this model results in neutral points where the magnetic field is zero. The occurrence of reconnection at $\mathbf{B} \neq 0$ results in a current parallel to $\mathbf{B}$, driven by an electric field parallel to the magnetic field. This is where the Gold and Hoyle models differs from modern reconnection theories. The Gold and Hoyle model, for all its limitations, was partially successful in meeting many of the requirements for a solar flare. Thus, it was an important advance on previous ideas in suggesting a specific magnetic configuration and emphasizing the role of explosive reconnection of magnetic field as a mechanism for solar flare energy release; a suggestion which is very much with us today, some four decades later.

Following the success of the Gold and Hoyle model, a number of theoretical models were proposed which invoked field reconnection in a variety of magnetic configurations. The Gold and Hoyle configuration of an arched current-carrying flux tube was developed further by Hannes Alfvén and Per Carlqvist in 1967. This latter model invoked the notion of current interruption to explain the rapid release of energy in the flare. Another configuration considered was that of the large-scale current sheet put forward by Peter Sweet in 1958, Eugene Parker in 1963 and subsequently developed by Peter Sturrock in the late 1960s and early 1970s. Such a configuration was proposed to explain the observation of energetic particles and ejected plasma from a flare in the interplanetary medium, suggesting that the field had to be open in the vicinity of the flare. While there were both observational and theoretical reasons for either of the closed arch or current-sheet topologies both shared magnetic reconnection as the prime energy release mechanism.

Magnetic reconnection requires the diffusion of the magnetic field which in the solar corona is an incredibly slow process. It is therefore difficult to envisage how the reconnection process can occur fast enough to explain the flare phenomena. Harry Petschek in 1964 was the first to propose a fast reconnection model where the short reconnection timescales were accomplished by assuming that the region over which the magnetic field diffused was very small. While there are several problems with the Petschek mechanism as formulated in 1964, developments since then, by a number of theoreticians, have shown the basic premise to be sound and that some form of the Petschek scenario must be present for fast steady-state reconnection to be viable in the solar atmosphere. An important series of models appeared in the 1960s and 1970s which involved an open-field configuration. The first of these, by Hugh Carmichael in 1964, proposed that magnetic field lines high above the photosphere could be forced open by the solar wind. Developments of this line of thinking appeared from Peter Sturrock (1966), Tadashi Hirayama (1974) and Roger Kopp and Gerald Pneuman (1976) earning this class of the models the sobriquet of the CSHKP model. The later models departed from the notion that the field could be opened by the solar wind because it was argued that such a scenario could only happen where the plasma pressure was non-negligible relative to the magnetic pressure – that is, much higher in the solar atmosphere than where solar flares were observed to occur. The alternative approach was to consider the evolution of a closed, force-free magnetic field configuration as photospheric motions stress the field. The stressing of the force-free field builds up energy till it becomes unstable or metastable; energy would be released if the field were to make a transition to a different configuration (within the confines of MHD theory). Such an energy release, or eruption, would result in an open-field configuration. The flare model, accordingly consists of a pre-flare magnetic topology comprised mainly of closed magnetic flux which is subsequently stressed by photospheric motions. The energy of the stressed field steadily increases, remaining force-free as it does so, until an MHD instability occurs which leads to an eruption of some of the closed magnetic flux creating current sheets. Reconnection in the current sheets then provides much of the energy and dynamics required by the solar flare observations.

The prime strength of the above class of models is the ease with which filament eruptions are incorporated, thereby extending the relevance from flares to coronal mass ejections (which were discovered only after the basics of this theory had been developed). Filament eruptions are found to accompany the majority of flares and CMEs and fit nicely into the description above. The magnetic free energy associated with coronal currents is thought to be responsible for the ejection of the filament which is then free to open part of the field and form current sheets that reconnect to produce the flare or post-CME arcade.

It should be noted that the Kopp–Pneuman model, which has become ubiquitous in recent years, is a variant of the above scenario in that it considers the hydrodynamic effects of the closing of a system of open field lines by reconnection. A key difficulty of this model is that it requires a very large non-thermal energy flux into the corona, some one to two orders of magnitude larger than that of current estimates for coronal heating.

Other popular flare models include variants on the reconnection theme, some invoking emerging flux to form current sheets when interacting with ambient field, others involving current-carrying loops. Many of these models are now being applied to various aspects of coronal mass ejections and are, in some form, still being applied to solar flares. While observationally very different phenomena, solar flares and CMEs share much physics in common. This physics generally involves a reconfiguration of the magnetic field which results in the release of energy and the eruption of magnetic flux.

Reconnection has a variety of roles to play in the solar flare prior to its occurrence: enabling the storage of magnetic energy to be released in the flare, triggering of the energy release (e.g., tearing mode), and energy conversion from magnetic into thermal and kinetic. Modern approaches utilize the increasing power of computers to investigate the three-dimensional aspect of magnetic reconnection. These provide a rich variety of solutions which are now probing the intricacies of the flare/CME phenomena. A number of interesting effects arise in three-dimensional modeling of reconnection, which are not present in lower dimensions. The extremely dynamic nature of the reconnection process creates significant computational problems when carried into three dimensions. In addition, when the magnetic configuration of an active region is simulated in three dimensions, magnetic structures such as separator surfaces and quasi-separatrix layers appear, especially in those active regions with the magnetic complexity required for flare production. The role of these complex magnetic topologies in solar activity are only now being realized.

In the CME arena, the role of the open-field is under close scrutiny given the limitations set by the Aly–Sturrock conjecture after independent work by Jean-Jacques Aly and Peter Sturrock in 1991, which states that a completely open field has more energy than a closed force-free field with the same boundary conditions. Thus, the CSHKP class of models have to contend with the very simple fact that their proposed final state has more energy than any assumed initial state, challenging the ability of these models to explain adequately the energy released in the eruptive process.

The Aly–Sturrock conjecture can be circumvented in a number of ways, the most popular of which is to open only part of the magnetic field. The magnetic break-out model of Spiro Antiochos and collaborators at the Naval Research Laboratory in Washington, DC, provides the intriguing possibility that a CME can result from a multi-flux system via the eruption of sheared field and the partial opening of overlying closed field. Another approach is to avoid the force-free assumption of the initial magnetic configuration and assume a non-force-free corona. Such a system is not subject to the limitations of the Aly–Sturrock arguments and can provide sufficient energy to power an eruption.

Much of our discussion on the physics of solar activity has centered on the more energetic signatures. The spatial, temporal and energy scales of flares and CMEs make them more amenable to observation and we are more likely to discern the underlying processes behind these phenomena. Their potential for causing geomagnetic effects also adds to their intrinsic interest. However, as observations have improved in the spatial, temporal and spectral regimes and as the sensitivity of detectors increases, solar activity has been found to be ubiquitous on the Sun, consisting of phenomena on all scales. A burgeoning area of research as we move into the twenty-first century is the study of small-scale transient events and their role in heating the solar atmosphere, a long-standing problem discussed in the previous chapter. New instruments are allowing us to observe events on smaller and smaller scales both in the spatial dimension and in energy. The similarity between these phenomena and flares/CMEs suggest that similar physical processes are responsible although they may be occurring in different ambient environments.

Observations of solar activity over the past decade or so of the last century have improved dramatically and the amount of data has increased many-fold. Such detailed information creates significant challenges to the theoreticians and the need for new ideas is as great as ever. The marked improvement in computers has been of great service to solar modelers and has allowed the complex set of equations describing the interactions between magnetic field and plasma to be explored with ever increasing detail and accuracy. However, the computer developments and, indeed the developments in numerical techniques, have been no match for the developments in observations with enormous quantities of high spectral, spatial, and temporal resolution data being collected by instruments on spacecraft such as TRACE, SOHO, Yohkoh, ACE, Wind and RHESSI. The launch of solar observatories such as Solar-B and STEREO in the near future will only exacerbate the divisions between theory and observation.

As described above, most theories of solar activity involve the processes of reconnection and instability, processes which are thought to be governed by physics on scales of centimeters to meters. Observations on such physical scales are well beyond our current capabilities and so the challenge is to understand how the physical effects of these processes relate to the detailed observations on larger scales. The development of theoretical models is a small but vibrant area of solar research and the synergy with observation only helps to improve the subtlety and relevance of the theoretical ideas. The twenty-first century should show a steady progress in our understanding of the physics of the solar atmosphere and how it relates to the Sun as a star, the Sun as a driver of Space Weather and the Sun and heliosphere as a single physical system.

THE FUTURE

There is nothing permanent except change – Heraclitus

The twentieth century has been a remarkable one in our attempts to understand our neighborhood star. From the first magnetic observations of George Ellery Hale to the positioning of a magnetograph in space, 1.5 million km from the Earth, we have recognized the Sun as a magnetic object with almost every facet of its activity governed by its magnetic field. From the first recognition of a regular pattern in the Sun's activity by Heinrich Schwabe to the development of instruments which actually probe the inside of the Sun, we have learned that the turbulent solar interior generates magnetic field which migrates to the surface causing much of the solar activity we observe. From the discovery of the solar cause of geomagnetic storms by Edward Sabine and Walter Maunder to the discovery of coronal mass ejections we have found that the Sun and the Earth are inextricably linked and that an understanding of the solar–terrestrial environment is crucial to our society at the end of the twentieth century.

Our increasing reliance on space technology for communications, information, and entertainment has made our daily lives more vulnerable to the machinations of our nearest star. Understanding the nature of the Active Sun and its effects on our planet is no longer regarded merely as an academic pursuit. The direct impact of the Sun on the environment and its import for society makes it of relevance to everyone on Earth. The study of mass ejections and their relationship to other forms of solar activity provides an important link in the chain of evidence connecting all solar activity to its physical cause; the structure and evolution of the solar magnetic field. The *Century of Space Science* is marked with milestones on a path of discovery which embodies science, technology, education, human endeavor and a quest for knowledge unsurpassed at any time in the history of the human race. It is an often neglected fact that the Sun is an astrophysical body, a star, a fundamental building block of the Universe. In the last hundred years we have learned more about this star than any other object in the heavens. The complex physical phenomena exhibited by the Sun adds to its luster and the more we think we know, the more we find we have to learn. As we embark on a new century, the Sun remains an object of fascination and wonder. The discoveries of the last century serve to raise questions to be "answered" in the next. The first decade of the twenty first century alone will see great advances in our understanding of solar activity and the Sun as a whole. HESSI will explore the high-energy radiation emanating from solar flares and provide knowledge about some of the highest energy emissions of the Sun. The Japanese Solar-B satellite is planned for launch in 2004 as is the US STEREO mission. The former will investigate how the solar atmosphere responds to changes in the small-scale magnetic field, while the latter will provide stereoscopic observations of coronal mass ejections, investigating their morphology and their transport to the Earth.

Many more missions are proposed and new technologies are being developed to provide more efficient and cheaper access to deep space. Solar sails and electric ion propulsion will take us from the space age into the realm of what was once science fiction. Advances in ground-based instrumentation, with improvements in active optics already yielding results, will complement the observations from space, while continuing to make discoveries of their own. Ultimately, we will be exploring the activity of stars other than the Sun with as much detail as we currently observe our own. The next century promises to be an exciting one for solar physics and astronomy in general.

Acknowledgments

We would like to thank Tom Metcalf, Dick Shine, and Wendy Alexander for comments and suggestions.

REFERENCES

Aly, J.J. (1991). How much energy can be stored in a three-dimensional force-free magnetic field? *Astrophysical Journal Letters*, **375**, L61–L64.

Babcock, H.D. and Babcock, H.W. (1952). Mapping the Magnetic Fields of the Sun. *Publications of the Astronomical Society of the Pacific*, **64**, 282–287.

Babcock, H.W. (1961). The topology of the Sun's magnetic field and the 22-year cycle. *Astrophysical Journal*, **133**, 572–587.

Bartels, J. (1932). Terrestrial-magnetic activity and its relations to solar phenomena. *Terrestrial Magnetism and Atmosphric Electricity*, **37**, 1–52.

Burlaga, L.F., Klein, L., Sheeley, N.R., Michels, D.J., Howard, R.A., Koomen, M.J., Schwenn, R. and Rosenbauer, H. (1982). A magnetic cloud and a coronal mass ejection. *Geophysical Research Letters*, **9**, 1317–1320.

Carmichael, H. (1964). A process for flares. In *AAS/NASA Symposium. on Physics of Solar Flares*, NASA SP-50, Washington, DC, pp. 451–456.

Dungey, J.W. (1953). Conditions for the occurrence of electrical discharges in astrophysical systems. *Philosophical Magazine*, **44**, 725–738.

Giovanelli, R.G. (1946). A theory of chromospheric flares. *Nature*, **158**, 81–82.

Gold, T. and Hoyle, F. (1960). On the origin of solar flares. *MNRAS*, **120**, 89–105.

Gosling, J.T., Hildner, E., MacQueen, R.M., Munro, R.H., Poland, A.I. and Ros, C.L. (1974). Mass ejections from the Sun. *Journal of Geophysical Research*, **79**, 4581–4587.

Hale, G.E. (1908). Solar vortices and the Zeeman effect. *Publications of the Astronomical Society of the Pacific*, **20**, 220–223.

Hale, G.E. (1908). On the probable existence of a magnetic field in sunspots. *Astrophysical Journal*, **28**, 315–343.

Hale, G.E. (1913). Preliminary results of an attempt to detect the general magnetic field of the sun. *Astrophysical Journal*, **38**, 27–98.

Hundhausen, A.J. (1972). *Coronal expansion and solar Wind*, Springer-Verlag, New York.

Joselyn, J.A. and McIntosh, P.S. (1981). Disappearing solar filaments: A useful predictor of geomagnetic activity. *Journal of Geophysical Research*, **86**, 4555–4564.

Kiepenheuer, K.O. (1953). Solar activity. In G.P. Kuiper (ed.). *The Sun*, University of Chicago Press, pp.322–465.

Kopp, R.A. and Pneuman, G.W. (1976). Magnetic reconnection in the corona and loop prominence formation. *Solar Physics*, **50**, 85–98.

Low, B.C. (1996). Solar activity and the corona. *Solar physcis*, **167**, 217–265.

Lean, J. (1997). The Sun's variable radiation and its relevance for Earth, *Annual Review of Astronomy and Astrophysics*, **35**, 33–67.

Lyot, B. (1930). La couronne solaire étudiée en dehors des éclipses. *Comptes Rendues de l'Académie Science Paris*, **191**, 834.

Newton, H.W. (1950). A significant time-distribution of great solar flares and great geomagnetic storms. *The Observatory*, **70**, 233–234.

Parker, E.N. (1961). Sudden expansion of the corona following a large solar flare and the attendant magnetic field and cosmic-ray effects. *Astrophysical Journal*, **133**, 1014–1033.

Petschek, H.E. (1964). Magnetic field annihilation. In *AAS–NASA Symposium on Solar Flares*, NASA SP-50, 425–439.

Schrijver, C.J., Title, A.M., van Ballegooijen, A.A., Hagenaar, H.J., Shine, R.A. (1997). Sustaining the quiet photospheric network: The balance of flux emergence, fragmentation, merging, and cancellation. *Astrophysical Journal*, **487**, 424–436.

Sturrock, P.A. (1987). Solar flares and magnetic topology. *Solar Physics*, **113**, 13–30.

Sturrock, P.A. (1991). Maximum energy semi-infinte magnetic field configurations. *Astrophysical Journal*, **380**, 655–659.

Sweet, P.A. (1969). Mechanisms of solar flares, *Annual Review Astronomy and Astrophysics*, **7**, 149–176.

FURTHER READING

Kuiper, G.P. (1953). *The Sun*, University of Chicago Press, Chicago.

Tandberg-Hanssen, E. (1967). *Solar Activity*, Blasidell, Waltham, MA.

Noyes, R.W. (1982). *The Sun, Our Star*, Harvard University Press, Cambridge.

Strong, K.T., Saba, J.L.R., Haisch B.M. (eds.) (1998). *The Many Faces of the Sun*, Springer-Verlag, New York.

47

MARCIA NEUGEBAUER* AND RUDOLF VON STEIGER**

The solar wind

PARKER'S THEORY AND THE EARLY MEASUREMENTS OF THE SOLAR WIND

Shortly before the beginning of the space age, Eugene N. Parker of the University of Chicago predicted that interplanetary space would be filled with a plasma flowing rapidly outward from the Sun (Parker 1958). The likelihood that the Sun ejects charged particles that cause auroral and magnetic activity on Earth was generally accepted by that time. The observation that the plasma tails of active comets always point almost radially away from the Sun led Ludwig Biermann (1951) to postulate that the solar corpuscular radiation is continuous, rather than intermittent. It was also known that the outer atmosphere of the Sun, the solar corona, was extremely hot, with a temperature exceeding a million degrees. Sidney Chapman (1957) calculated that if the corona was in hydrostatic equilibrium, it must extend throughout the Solar System and cool off to only $\sim 2 \times 10^5$ K at the orbit of Earth. Parker (1958) put all these ideas together, explaining that the inward pressure of the interstellar medium was too weak to allow the solar atmosphere to be in hydrostatic equilibrium. He coined the phrase "solar wind" to describe the outward flowing solar corona which supplies the pressure required to stand off the local interstellar medium, to exert the necessary force on cometary plasma tails, and to transmit solar disturbances to the geomagnetic field. For a more complete theoretical explanation of Parker's prediction of the solar wind, see Chapter 9.

Parker's theoretical prediction was not uncontested, however. Most notably, Joseph Chamberlain (1960) proposed that rather than Parker's solar wind caused by the hydrodynamic outflow of the solar corona, there was merely a solar breeze,

consisting of plasma thermally escaping from the corona. Many of the early space investigations therefore included attempts to determine whether interplanetary space was filled with Parker's supersonic 500 km s^{-1} solar wind or with Chamberlain's subsonic 10 km s^{-1} solar breeze. A summary of those early missions and experiments is given in Table 1.

Not surprisingly, the Soviets were the first in space with instruments capable of measuring the interplanetary plasma. Their "ion traps" were simple Faraday cups with an inner grid held at -200 V to repel interplanetary electrons and to prevent the escape of photoelectrons from the cup and an outer grid at a positive potential to define the minimum energy of the ions entering the cup. Lunik 2 was the most successful of four missions, determining that there was indeed a flux of $\sim 2 \times 10^8$ cm^{-2} s^{-1} of positive ions with energy/charge >15 eV/charge (Gringauz et al. 1960). Because the speed of a proton with energy >15 eV is >53 km s^{-1}, the Lunik 2 measurements favored Parker's theory over Chamberlain's, but questions of the extent to which the speed exceeded that limit, the direction of the flow, and its persistence were left unanswered.

With the Explorer 10 mission in 1961, a group from the Massachusetts Institute of Technology (MIT) made the first US measurements of the solar wind (Bonetti et al. 1963). Their instrument was an advance over the Soviet ion traps in that it had an additional grid which carried a positive square-wave potential to allow measurement of the ion energy spectrum without confusion between the flux of ions entering the detector (an AC signal) and the flux of photoelectrons knocked out of the negative inner grid (approximately a DC signal). Before the spacecraft batteries died at a distance of ~ 34 Earth radii, the instrument measured an intermittent flux of ions from a direction within a 20° by 80° window which included the direction from the Sun. When the ions (assumed to be protons) were

---
* Jet Propulsion Laboratory, Pasadena, CA, USA
** International Space Science Institute, Bern, Switzerland

Table 1 The earliest attempts (1959–62) to observe the solar wind

Launch date	Spacecraft	Institution	Instrument*	Result
2 January 1959	Lunik 1	USSR	Four ion traps −10 to +15 V	No publishable data
12 September 1959	Lunik 2	USSR	Four ion traps −10 to +15 V	39–60 R_E Flux > 15 eV ≈ 2 × 10^8 cm^{-2} s^{-1}
4 October 1959	Lunik 3	USSR	Four ion traps −19 to +25 V	One observation of flux > 20 eV ≈ 4 × 10^8 cm^{-2} s^{-1} Other data < threshold (∼10^8 cm^{-2} s^{-1})
12 February 1961	Venus Probe	USSR	Ion traps 0 and 50 V	Very intermittent data One observation of flux ≈ 4 × 10^8 cm^{-2} s^{-1}
25 March 1961	Explorer 10	MIT	Modulated FC	Skimmed magnetopause flank Consistent with flow from Sun Measured n, v, T Supersonic and super-Alfvénic
16 August 1961	Explorer 12	NASA Ames	CPA	Dayside magnetosheath Did not detect any ions
22 August 1961	Ranger 1	JPL	6 CPAs	Failed to get out of parking orbit
18 November 1961	Ranger 2	JPL	6 CPAs	Failed to get out of parking orbit
22 July 1962	Mariner 1	JPL	CPA	Destroyed by range safety
27 August 1962	Mariner 2	JPL	CPA	113 days of data Continuous radial flow High-, low-speed streams n, v, T relations $v_\alpha \approx v_p$; n_α/n_p variable; $T_\alpha \approx 4\,T_p$
2 October 1962	Explorer 14	NASA Ames	CPA	Mostly magnetosheath UV interference

* FC is Faraday cup and CPA is curved-plate analyzer.

detected, their flux was in the range 1.0–2.5 × 10^8 cm^{-2} s^{-1}, their speed was ∼280 km s^{-1}, and their flow was supersonic, as predicted by Parker's theory. In retrospect, the ion fluxes detected by Explorer 10 were not in the solar wind proper, but downstream of the Earth's bow shock, in the flank of the magnetosheath.

A group at the NASA Ames Research Center attempted to measure the solar wind with instruments on Explorers 12 and 14 in 1961 and 1962 (Bader 1962, Wolfe and Silva 1965). These instruments were curved plate analyzers with a voltage applied perpendicular to the ions' direction of motion to bend their trajectories onto a detector. On Explorer 12 there was a problem that the field of view of the instrument did not include the solar direction, and on Explorer 14 there was a problem with contamination of the ion signal by solar ultraviolet radiation when the instrument did face the Sun. Furthermore, on both missions, the spacecraft trajectories were almost entirely downstream of the bow shock.

About the same time as the unsuccessful attempts by Bader and Wolfe at NASA Ames, one of the authors (MN) and her colleague, Conway W. Snyder, flew solar wind detectors on four different missions. The first two of those spacecraft, Rangers 1 and 2, failed to get out of low-Earth orbit, while the third spacecraft, Mariner 1, went astray and was destroyed by ground command. Finally, after some hair-raising misadventures (Neugebauer 1997), Mariner 2 was safely placed on a trajectory to Venus. The Mariner 2 instrument was a curved-plate analyzer which measured the ion current reaching a collector at each of 10 voltages on the deflection electrodes. Mariner 2 obtained a spectrum of the solar wind every 3.7 minutes almost continuously for 113 days. There was no longer any doubt that Parker had been correct; the solar wind exists.

Although the ion spectra obtained by Mariner 2 were very crude by today's standards, with measurable currents in no more than five energy/charge channels at any time, a lot of information about the properties of the solar wind could be gleaned from the data (Neugebauer and Snyder 1966). The solar wind blew continuously from within a 10° cone centered on the Sun. The wind was organized into low- and high-speed streams (velocities of ∼350 and 700 km s^{-1}, respectively), each of about 7 days' duration. The speed *versus* time profiles were steepened on the leading edges of the fast streams where the increased density indicated a snowplow effect. The proton temperature varied directly with the speed. These features are illustrated in Figure 1,

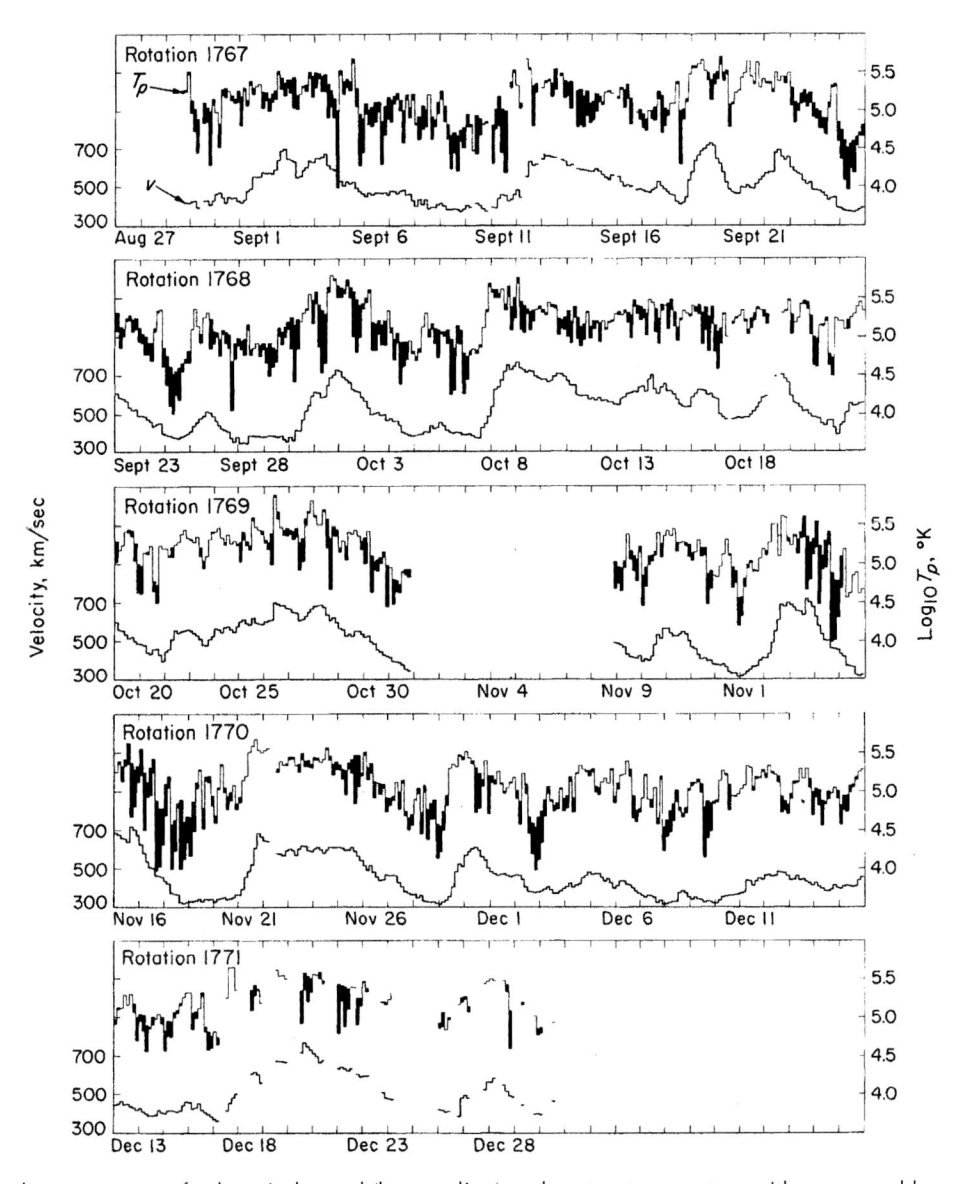

Figure 1 Three-hour averages of solar wind speed (bottom line) and proton temperature with upper and lower limit bars (top) observed by Mariner 2 in 1962. (From Neugebauer and Snyder 1966.)

which is a plot of three-hour averages of the solar wind speed and temperature over five 27-day rotations of the Sun. The pattern roughly repeated from one rotation to the next. On average, the ion flux and density varied as the inverse square of heliocentric distance between 1.0 and 0.7 AU.

It was often possible to detect a second spectral peak which was interpreted as being caused by alpha particles (helium nuclei) moving with approximately the same speed as the protons. This second peak could not, however, be fit to a model in which the alpha particles had the same temperature as the protons; instead, equal thermal speeds were indicated. The abundance of the alpha particles relative to the protons was sometimes highly variable from day to day.

Parker predicted not only the existence of the solar wind, but also the configuration of the interplanetary magnetic field (Parker 1958). Because of the very high electrical conductivity of the solar corona, the plasma and the magnetic field must move together. That is, the solar field is frozen into the solar wind. But at the same time that the field is being dragged nearly radially into space by the solar wind, it is still tied to the rotating Sun, with the result that the interplanetary field should have a spiral pattern with an angle to the radial direction of ~45° near 1 AU. The predicted spiral pattern of the field could be discerned in the data of the magnetometer on Mariner 2; this is illustrated in Figure 2, where each point represents a running average of five-hour averages. Although there is a great deal of scatter, the points are distributed in the quadrants predicted by the Parker spiral model. The properties of the fluctuations about the spiral direction continue to be studied intensively

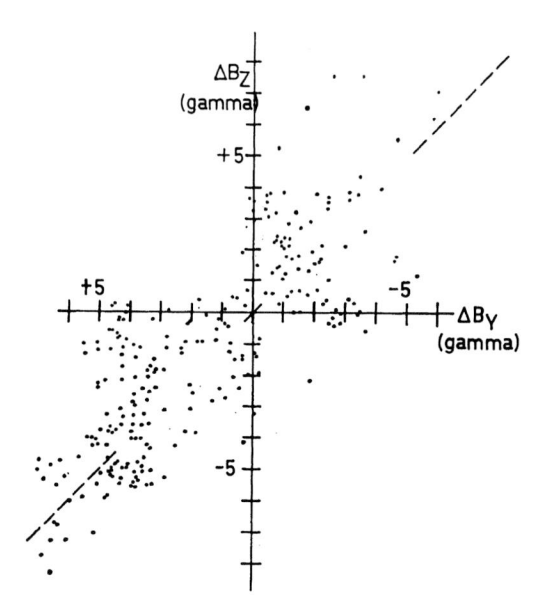

Figure 2. Five-hour sliding averages of hourly averages of the radial (ΔB_Z) and in-ecliptic transverse (ΔB_Y) components of the interplanetary magnetic field measured by Mariner 2. The dashed line shows the expected relation for the Parker spiral. (From Smith 1964.)

to reveal some of the fundamental processes occurring in the solar wind. A change in the direction of the interplanetary field from the first to the third quadrant in Figure 2, or the reverse, indicates a reversal of the polarity of the interplanetary field with the field sometimes pointing in toward and sometimes pointing out from the Sun. Week-long periods of persistent polarity were named "magnetic sectors" by Wilcox and Ness (1965).

MORPHOLOGY

The solar wind has probably been blowing for at least the past 3×10^9 years with essentially the same strength, as can be estimated by comparing the flux of xenon ions in today's solar wind with that deduced from the xenon content of the lunar regolith (Geiss 1973). Observations of comet tails reveal that the solar wind did not stop blowing even during the Maunder minimum, from about 1645 to 1715 when there were essentially no sunspots.

It thus seems that the solar wind is a ubiquitous and continuous phenomenon, but it is not a structureless one. Its density, speed, temperature, ion charge states, elemental composition, and other properties all vary with time and position on timescales from minutes (or less, but knowledge of fast fluctuations is limited by the typical time resolution of today's ion sensors) up to decades (or more, limited by the short duration of the space era). The large-scale structures of the solar wind are conveniently divided into recurrent or quasi-stationary streams and transient flows.

The discovery of a 27-day (the synodic period of solar rotation) modulation of cosmic rays by Forbush in 1938 was conclusively traced to dynamical phenomena in the interplanetary medium and related to recurring coronal "active regions" (in the terminology of those days) by Simpson in 1954 (Simpson 1998). As shown in Figure 1, such recurrent structure was indeed found in interplanetary space in the form of alternating high- and low-speed streams, each lasting several days. The polarity of the interplanetary magnetic field tended to remain constant throughout each of the high-speed streams, with consecutive streams having opposite polarities. It is important to note that there is not a one-to-one correspondence between fast streams and magnetic sectors. There need not be a fast stream within every magnetic sector, and the position of a fast stream relative to its magnetic sector boundary does not remain fixed in interplanetary space. The leading edge of each fast stream, where the solar wind speed increases, is now commonly called a corotating interaction region (CIR). Such interaction regions are an inevitable consequence if streams of sufficiently different speeds are emitted from the Sun at the same heliographic latitude. The effect of solar rotation is to eventually ram fast solar wind into slower wind emitted from more westerly heliographic longitudes. Figure 3 shows an early schematic and a newer version of this scenario. The newer version also shows how the CIR develops in interplanetary space to engulf the magnetic sector boundary.

As a general rule, two magnetized plasmas cannot intermix without the benefit of magnetic reconnection or other types of plasma instability. Therefore, the fast and the slow solar wind streams remain separated out to large heliographic distances. Discontinuities separating the two wind types were first studied by Belcher and Davis (1971) using Mariner-5 data.[1] Burlaga (1974) introduced the term "stream interface" for this boundary which is characterized by a decrease in density by a factor of ~2, accompanied by a similar increase in kinetic temperature. Sometimes, in order to enhance the signal, these two signatures are conveniently combined into the specific entropy argument, $T/n^{1/2}$, where T is temperature and n is density. As the solar wind expands to 1 AU and beyond, the stream interaction becomes progressively more pronounced. The leading (slow) plasma becomes accelerated and the trailing (fast) plasma becomes decelerated, building up hydromagnetic stresses which ultimately lead to the development of a pair of interplanetary shocks, a forward shock at the leading edge of the CIR and a reverse shock at the trailing edge.

---

[1] Mariner 5: a NASA Venus flyby mission launched on 14 June 1967 and operated to 21 November 1967.

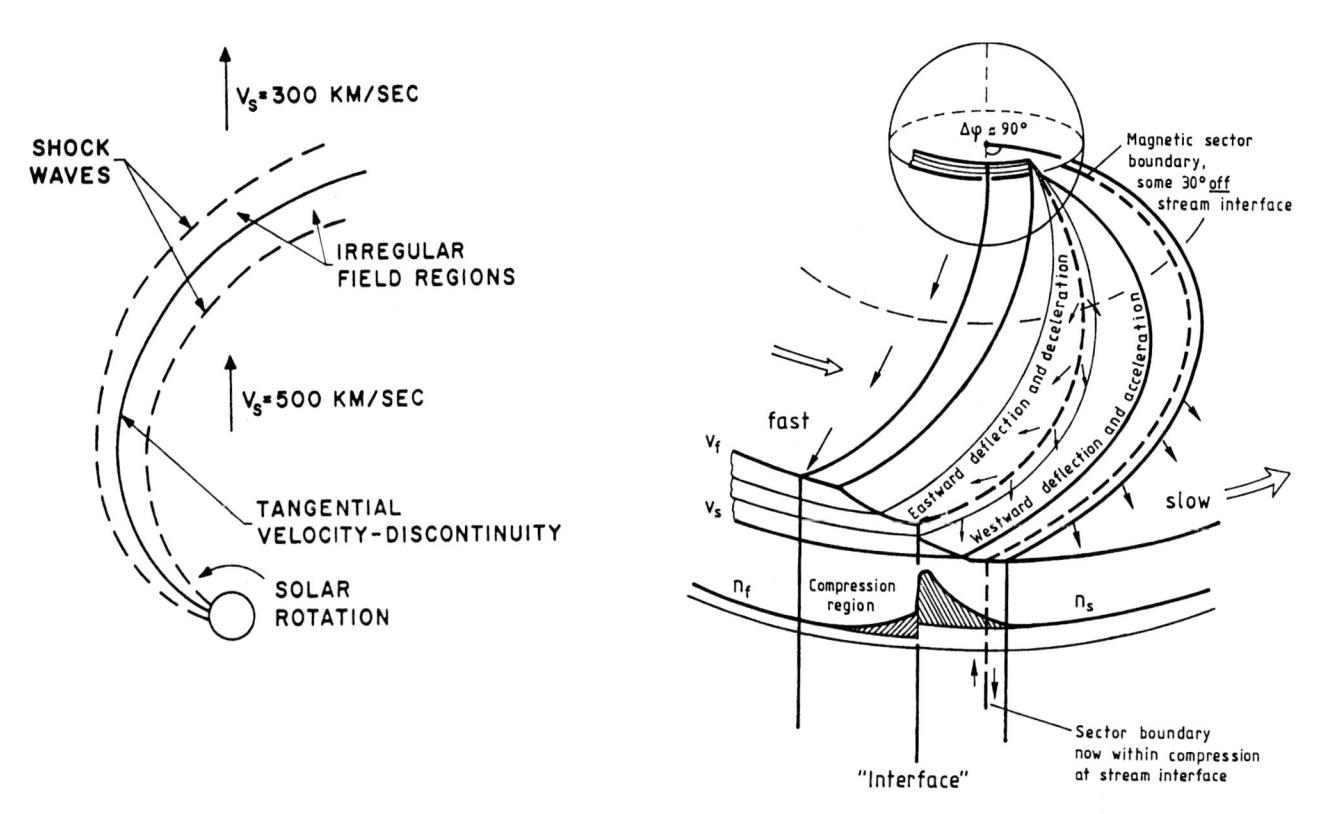

Figure 3 The first schematic of a CIR (left) by Dessler and Fejer (1963). The authors realized that "the collision of these plasmas will lead to the formation of two shock waves..." but nevertheless they chose to extend the shock all the way back to the Sun. A more modern version (right) shows the complex CIR morphology, including how the magnetic sector boundary, well outside the high-speed stream near the solar surface, will be engulfed into the CIR in interplanetary space. (From Schwenn 1990.)

These shocks normally do not develop within 1 AU of the Sun. They were first identified between 1 and 5 AU by Smith and Wolfe (1976) using data from Pioneers 10 and 11.[2] It is these corotating shocks that cause the 27-day modulation of cosmic rays mentioned above.

In 1973, the Skylab mission[3] obtained images of the solar corona in soft x-ray wavelengths. These images demonstrated that Simpson's "active regions" were in fact x-ray-dark regions, now called coronal holes, with lower-than-average density and temperature. (In modern usage, an active region is a group of sunspots with flare activity.) Krieger et al. (1973) showed that large coronal holes were the source of the quasi-stationary high-speed streams, thus confirming Simpson's conjecture. From the data from the Helios missions[4] obtained between 0.3 and 1 AU, Rosenbauer et al. (1977) recognized that fast streams are really distinct from the slow solar wind in their kinetic properties, thus defining the quasi-stationary solar wind as a two-state phenomenon. Using data from the IMP spacecraft,[5] Bame et al. (1977) quite accurately called the fast streams a "structure-free state" of the solar wind. Large-area coronal holes exist in the solar atmosphere mainly during the declining and minimum phases of the solar activity cycle, when they are more or less centered around the poles, but often develop large extensions to lower latitudes which are the main sources of fast streams near the ecliptic plane. Thus the fast streams and the CIR structure they generate are typical of solar minimum and cause an increase in the average solar

---

[2] Pioneers 10 and 11: NASA spacecraft launched on 3 March 1972 and 6 April 1973, respectively. Pioneer 10 flew by Jupiter and Pioneer 11 flew by Jupiter and Saturn before they both proceeded to the outer heliosphere and returning data until 31 December 1995 and 1 August 1992, respectively.

[3] Skylab: a crewed NASA mission, launched on 14 May 1973; it carried the Apollo Telescope Mount which was used for extensive observations of the Sun until 8 February 1974.

[4] Helios 1 and 2: a German/US mission consisting of a pair of spacecraft launched on 10 December 1974 and 15 January 1976, respectively, placed in heliocentric orbits with solar distances ranging between 0.3 and 1.0 AU; solar wind data are available through 4 March 1980.

[5] IMP: the Interplanetary Monitoring Platforms were NASA spacecraft in highly eccentric Earth orbits; the periods of operation for IMPs 6, 7, and 8 were 13 March 1971–2 October 1974, 23 September 1972–31 October 1974, and 26 October 1973– the present, respectively.

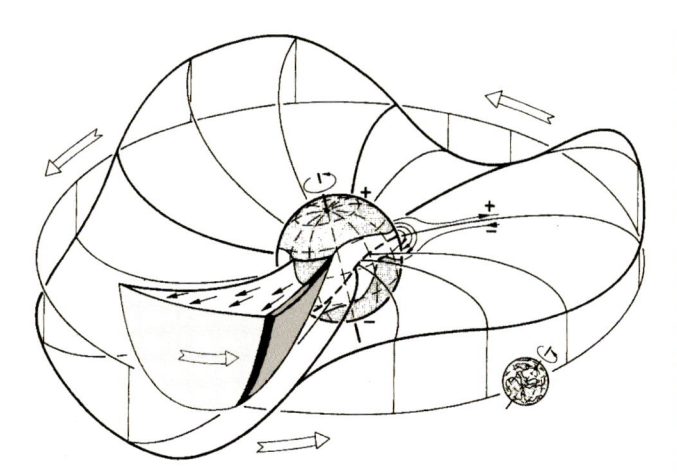

Figure 4 Schematic of the inner heliosphere during declining and minimum solar activity in terms of the "ballerina skirt" model proposed by Alfvén (1977). The coronal holes around each pole emit magnetically unipolar fast streams of opposite polarity, between which the slow wind from the belt around the heliomagnetic equator is confined. The warp of the current sheet separating fields of opposite polarity makes it loosely resemble a ballerina's skirt. (From Schwenn 1990.)

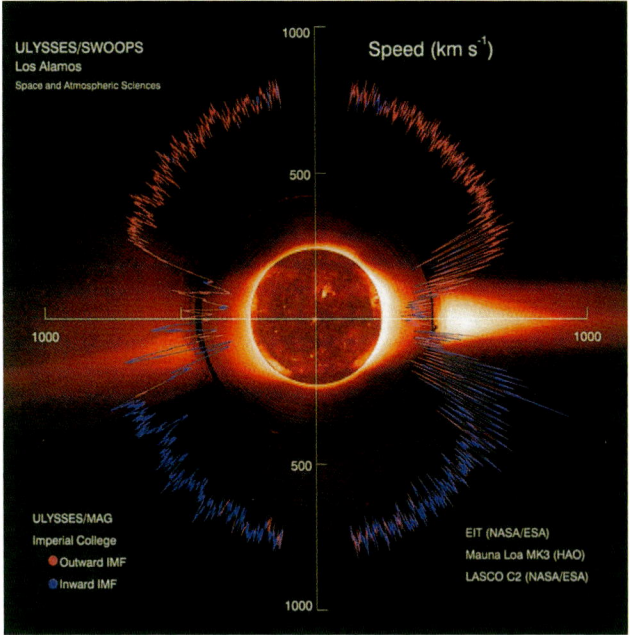

Figure 5 Bulk speed of the solar wind plotted *versus* heliographic latitude in a polar diagram. The color code indicates magnetic polarity; both fast streams are in fact magnetically unipolar, as brief reversals near the poles are spuriously caused by large-amplitude Alfvén waves. The fast streams occupy a much larger solid angle in the heliosphere than the coronal holes do on the Sun (visible on the inset image from SOHO/EIT), indicating a strong superradial expansion. The boundary between the stream types is very sharp, as seen in the left half of the figure obtained during Ulysses' fast latitude scan; it was crossed multiple times due to the warp of the "ballerina skirt" on the distant portion of the orbit (right half). (From McComas et al. 1998.)

wind speed during that period, as first demonstrated in a survey of a good part of solar activity cycle 20 by Feldman *et al.* (1978a).

When propagating out to larger heliocentric distances, the solar wind velocity structure behaves as if passed through a low-pass filter, as was observed and modeled by Gosling *et al.* (1976) using data from the first radial line-up of Pioneer 10 with the Earth. The formation of forward–reverse shock pairs is accompanied by smoothing of the smaller-speed structures present at 1 AU. Ultimately even the shocks disappear. A CIR's forward shock will eventually hit the reverse shock of the previous CIR – which might well be its own if there is only one CIR per solar rotation – and cancel into a tangential discontinuity. At large distances, CIRs are thus expected to coalesce into corotating merged interaction regions (CMIRs). These can produce successive increases and decreases of cosmic ray intensity over a period of several solar rotations, but generally little or no long-lasting or net modulation (Burlaga *et al.* 1985, McDonald and Burlaga 1997). The lack of net modulation is due to the C(M)IRs being quite limited in their latitudinal extent.

Smith *et al.* (1978) observed that the magnetic sector structure seen by Pioneer 11 gradually disappeared with increasing heliographic latitude and was nearly unipolar at merely 16° north in February 1976, that is around the minimum phase of solar cycle 20. From observations of the solar magnetic field, it is possible to infer a heliomagnetic coordinate system with its equator coinciding with the heliospheric current sheet. If the solar wind speed is plotted *versus* heliomagnetic latitude, an orderly picture emerges; slow wind is observed in a ±20° equatorial belt, while outside that belt there is only fast wind. This can be visualized in the "ballerina skirt" picture originally proposed by Alfvén (1977) and illustrated in Figure 4.

This general picture of the regions of fast and slow wind was confirmed impressively and extensively two solar cycles later by the Ulysses[6] spacecraft (Wenzel *et al.* 1992) on its first polar orbit around the Sun, in 1992–98. Figure 5 is a polar plot showing fast streams in the polar regions bounding a band of slow wind between them, separated by remarkably sharp boundaries. This once again emphasizes the two-state property of the quasi-stationary solar wind. The sources of the two stream types, polar coronal holes for the fast streams and coronal streamers for the slow wind, can easily be seen on the superposed pictures of the solar

---

[6]Ulysses: an ESA/NASA mission launched on 6 October 1990; after a gravitational assist at Jupiter, Ulysses has been in a near-polar orbit about the Sun, and as of 1 March 2002 it was still in operation.

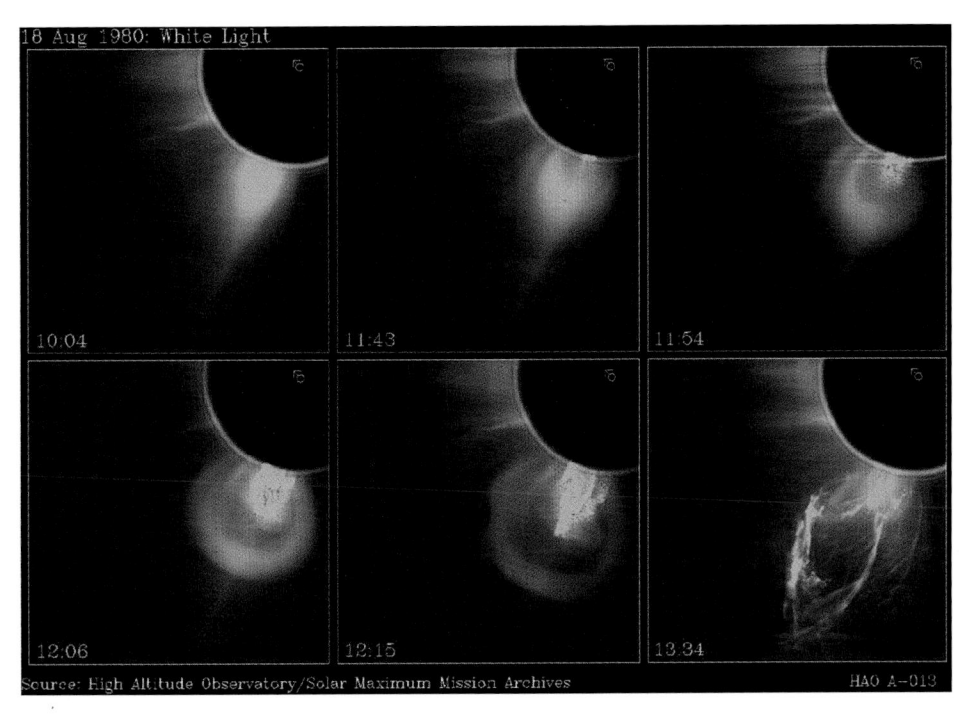

Figure 6 Development of a large CME observed by the coronagraph on the Solar Maximum Mission. (Courtesy of HAO/SMM C/P project team and NASA. HAO is a division of the National Center for Atmospheric Research, which is supported by the National Science Foundation.)

disk and corona. Note the strong superradial expansion of the fast streams which reach down to about ±20° latitude whereas the boundaries of the coronal holes lie near ±60°.

Long before the existence of the solar wind was established, it was known that times of high solar activity (e.g., as indicated by sunspot number), recurring every 11 years, coincided with high geomagnetic activity, as conjectured by Lord Carrington in the 1850s and established by Birkeland around the turn of the century. Since CIRs are recurrent and characteristic of the minimum heliosphere, a different interplanetary structure must account for those geomagnetic disturbances that are transient and characteristic of maximum activity. Obvious candidates are solar flares, which occur in solar active regions (i.e., regions of complex magnetic fields and sunspot groups) and were thought to eject streams of plasma into interplanetary space (Chapman and Ferraro 1931) that in turn would disturb the geomagnetic field several days later.

It was not until the Skylab era, when the outer corona became observable for longer periods than just the few minutes of a total solar eclipse, that such events were indeed observed by a coronagraph (MacQueen et al. 1974), and the term coronal mass ejection (CME) was coined to describe them. Since then, thousands of CMEs have been observed with coronagraphs on Solwind,[7] the Solar Maximum Mission,[8] and the Solar and Heliospheric Observatory.[9] An example from SMM is given in Figure 6, which shows the development of a particularly large CME in several steps. Initially, only an outer rope or bubble of enhanced density is visible. The loop then expands, and an apparent void – a region of low particle density and high magnetic field strength – appears within. Some CMEs also show an erupted prominence within the dark region. However, the relation between CMEs, erupting prominences, and solar flares is complex. Flares and erupting prominences are neither necessary nor sufficient conditions for a CME. Moreover, even when a flare appears in close connection with a CME, the start of the CME often precedes the flaring activity. This led Gosling (1993) to attribute the central role in the chain of transient events leading from the Sun to near-Earth space to CMEs rather than to flares (which is not to say – despite the mildly polemic title of Gosling's paper – that flares are irrelevant).

The discovery of CME plasma in interplanetary space predates the discovery of their optical counterparts at the

---

[7] Solwind: a US Department of Defense Earth-orbiting spacecraft launched on 24 February 1979, with an instrument complement including a solar coronagraph; operation ceased on 13 September 1985.

[8] SMM: a NASA spacecraft in low Earth orbit, launched from a Space Shuttle on 14 February 1980 and later refurbished by shuttle astronauts; the last data were acquired on 24 November 1989.

[9] SOHO: an ESA/NASA mission launched on 2 December 1995 into a halo orbit about the Sun–Earth Lagrangian libration point L1; SOHO carried instruments for both remote sensing of the Sun and in situ measurements, and was still in operation as of 1 March 2002.

Sun. Whereas at declining to minimum solar activity the time profile of the solar wind speed is dominated by recurrent fast streams, transient (non-recurrent) disturbances become more important or even dominant during solar maximum conditions. It is not, however, a trivial matter to identify CMEs in interplanetary space. Because of the wide range of velocities with which CMEs leave the Sun ($<50\,\mathrm{km\,s^{-1}}$ to $>1200\,\mathrm{km\,s^{-1}}$), their interplanetary signatures may differ widely. Moreover, the observing spacecraft may encounter different parts of a CME depending on its relative position to the ejected material, adding to the variability of the observations. It is therefore not possible to define the interplanetary signature of CMEs unambiguously, but only to give a list of characteristics of which some, but rarely all, apply to an individual event.

- The most energetic CMEs drive shocks into the preceding solar wind, as shown schematically in Figure 7, much like fast streams from coronal holes. Sheeley et al. (1985) found that only 2% of the shocks observed in 1979–82 (solar maximum 21) by Helios were clearly *not* associated with CMEs. The converse is not true, though, as numerous (slow) CMEs were not associated with shocks.
- It is usually quite simple to distinguish CME-associated shocks from CIR-driven shocks based on the kinetic temperature of the driving material. Both the proton temperature (Gosling et al. 1973) and the electron temperature (Montgomery et al. 1974) within a CME tend to be unusually low for a given solar wind speed. The cause is probably the expansion of the ejecta into a larger volume than they would otherwise occupy.
- CME ejecta often contain an anomalously high helium abundance (Hirshberg et al. 1972) which can easily reach twice the average solar wind value of $n_\alpha/n_p = 0.04$–0.05 and may go up to 0.30 in extreme cases. The source of these enrichments, which often are very patchy, may be pockets of dynamically accumulated helium that were left behind in the corona due to insufficient Coulomb drag with the solar wind protons (Geiss et al. 1970a).
- Bidirectional streaming of suprathermal electrons is an important indicator of CMEs (Montgomery et al. 1974). It is clear from Figure 7 that both ends of the solar magnetic field lines carried out in the ejecta may be rooted in the corona, thus allowing heat flux carried by the electrons to flow either way around the loop. Gosling et al. (1992) surveyed more than a full solar cycle (1978–90) of near-Earth bidirectional streaming events and found a variation of a factor of at least seven between solar maximum ($72\,\mathrm{yr}^{-1}$) and minimum ($<12\,\mathrm{yr}^{-1}$), again underlining the importance of CMEs structuring the heliosphere at solar maximum.
- In about one-third of the CMEs observed near Earth, the magnetic field inside the ejecta region undergoes a smooth rotation in at least one component, accompanied by an increase of the field strength and a temperature decrease. The magnetic field in such events, called magnetic clouds, have the configuration of a twisted flux rope (Burlaga et al. 1981, Lepping et al. 1990).
- Finally, the relative abundances and the charge states of heavy elements (carbon and heavier) are often significantly different within CME ejecta than in the surrounding solar wind. Charge states are generally higher, sometimes extremely so (Galvin 1997). Henke et al. (1998) noted a correlation between CMEs with a magnetic cloud topology and high charge states. This signature makes it simple to distinguish (fast) CMEs from recurrent fast streams, in which the heavy ions invariably have relatively low charge states. In very rare cases, towards the end of a CME event, the charge states become extremely low, which may be interpreted as the passage of cool prominence material (Gloeckler et al. 1999).

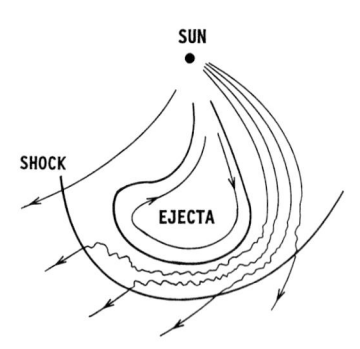

Figure 7 Schematic of a CME in interplanetary space, consisting of a bubble of ejecta driving a shock into the preceding solar wind (which is only the case if the ejection speed is sufficiently high), with a sheath of compressed plasma and magnetic field in between. (From Cane 1997.)

Not all of these features that distinguish CMEs from the ambient, quasi-stationary solar wind apply to each event, which makes CMEs a highly variable class of events. These tell-tale properties are almost entirely based on observations near the ecliptic plane. To date, only a small number of CMEs have been observed outside the range of the equatorial streamer belt, and all of them were fully immersed in coronal hole-associated fast wind near solar minimum (Gosling et al. 1994a). The plasma in all those CMEs had approximately the speed of the ambient (fast) solar wind of $\sim 750\,\mathrm{km\,s^{-1}}$. Nevertheless, most of them were preceded by a forward shock and, unlike at low latitudes, followed by a reverse shock. The presence of a shock pair indicates that they were not generated by the CME overtaking the wind, but by its over-expansion into the ambient plasma, that is, by the same process responsible for the low kinetic temperatures within. Near solar maximum, high-latitude CMEs may look much like their low-latitude cousins if slow solar wind dominates at all latitudes, but only the polar passes of the Ulysses mission during the solar maximum in 2000 or 2001 will tell.

The high rate of CME occurrence around solar maximum has a strong effect on the outer heliosphere. CME-generated transient interaction regions coalesce and form merged interaction regions (MIRs), which may be so numerous and extended at solar activity maximum that they cover the full solid angle around the Sun to become a global MIR, or GMIR (McDonald and Burlaga 1997). It is these GMIRs that are responsible for the attenuation of galactic cosmic ray fluxes during solar maximum. When a GMIR reaches the heliopause, it creates radio emission of about a year's duration, as discovered by Gurnett and Kurth (1996) from Voyager 1 data[10] both for cycles 21 and 22. From the travel times of these radio signals it is possible to estimate the overall size of the heliosphere, which these authors give as 110–160 AU.

COMPOSITION

Some of the motivation for studying the abundances of heavy ions in the solar wind is summarized in Figure 8. The protosolar nebula formed from the interstellar medium $\sim 4.6 \times 10^9$ years ago. Its composition is best preserved in the outer convective zone (OCZ) of the Sun, which can be regarded as well mixed and unaffected by fractionation processes (with a few exceptions). In the meantime, the local interstellar medium (LISM) has evolved chemically due to many generations of stars, and has been mixed continuously due to the 250×10^6 year galactic rotation. Differences in the compositions of the present OCZ and LISM can therefore be used to estimate the average evolution of the Galaxy over the lifetime of the Sun (Geiss and Gloeckler 1998). Similarly, differences in the compositions of the OCZ and various planetary materials provide clues to the evolution of the Solar System. Remote sensing can provide some elemental abundances of the OCZ (Grevesse and Sauval 1998), but most solar isotopic abundances are not yet known. The elemental and isotopic compositions of the solar wind are therefore highly relevant to understanding the evolution of the Galaxy and the Solar System. The solar wind, however, provides a biased sample of the OCZ because the mechanisms that generate the solar wind do not operate equally strongly on all elements. A fractionation is imposed that depends on elemental parameters such as ion mass, mass per charge ratio, first ionization potential, and so on. The resulting fractionation can only be assessed and characterized by comparing *in situ* measurements of the solar wind with remote observations of the photosphere. But once the fractionation mechanisms are understood, the

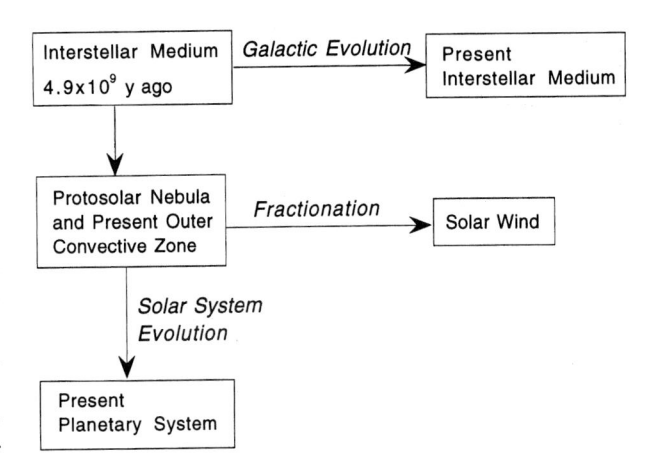

Figure 8 Logic diagram showing the relation between galactic and Solar System evolution and the solar wind.

process can be reversed and solar wind measurements can be used to infer the OCZ composition by theoretical modeling.

The second motivation for studying heavy ions is that the abundances, charge states, and kinetic properties of these ions provide information about the processes of coronal heating and solar wind acceleration. Heavy ions are ideal for this purpose because they act as tracers which have little effect on the large-scale dynamics and because, as shown later, the different types of solar wind flow have different heavy-ion properties.

Solar wind composition instruments of increasing complexity, resolution, and dynamic range have been flown on many spacecraft. Three generations of sensors can be distinguished.

The first solar wind instruments were either Faraday cups combined with retarding potential grids or curved-plate analyzers. Such electrostatic analyzers measure the energy per charge, E/q, of the incident ions, where $E = mv^2/2$ is the kinetic energy, m is ion mass, v is speed, and q is charge. By stepping the analyzer through a series of (usually logarithmically spaced) voltage steps, the distribution function (phase space density as a function of energy or velocity) of the ion beam is obtained. If all ions in the solar wind have approximately the same flow speed and if the velocity spreads due to thermal motions of the ions are small enough to avoid overlap of neighboring peaks, an E/q spectrum can be interpreted as a m/q spectrum. The very early solar wind measurements resolved helium from hydrogen in this way (Neugebauer and Snyder 1966). Later sensors with better E/q resolution and sensitivity were able to record the low charge states of O, Si, and Fe at times when the solar wind kinetic temperature was sufficiently low (Bame 1975a). At high kinetic temperatures, however, the individual peaks could not be resolved because they overlapped each other. Moreover, the high charge states of C and O always remained hidden behind the large He peak.

---

[10] Voyager 1: a NASA mission launched on 5 September 1977 which flew by Jupiter and Saturn before continuing into the outer heliosphere, where it is still in operation.

The second generation of sensors, first used by Ogilvie and Wilkerson (1969) on Explorer 34,[11] had a velocity selector (a Wien filter) added after the E/q analyzer. By stepping both the analyzer and the selector voltages, it was possible to obtain v and m/q of the incident ions independently. With the Ion Composition Instrument (ICI) on ISEE 3[12] (Coplan et al. 1978), it was possible to verify the equal velocity hypothesis and to find deviations therefrom, as well as to determine the abundances and the kinetic properties of ^{3}He, ^{4}He, O, Si, and Fe under most solar wind conditions. That experiment additionally measured Ne (Kunz et al. 1983), but other important elements such as C and Mg still remained hidden due to m/q overlap with more abundant species. A different type of second-generation instrument, such as the Ultra Low Energy Charge Analyzer (ULECA) on ISEE 3 (Hovestadt et al. 1978), consisted of an E/q analyzer with an array of solid-state energy detectors (SSD) added to obtain measurements of total energy E. It was thereby possible to determine the charge states of the CNO group and Fe (Ipavich et al. 1986), but measurements were limited to high-speed solar wind due to the inefficiency of the SSD at low energy.

The third generation of instruments uses the time-of-flight (ToF) technique (Gloeckler 1990). These sensors, of which the Solar Wind Ion Composition Spectrometer (SWICS) on Ulysses is the first one to be flown in interplanetary space (Gloeckler et al. 1992), combine a classical E/q analyzer with a ToF measurement giving v, and a total energy E measurement in a solid-state detector. Together, the three measurements provide energy (or speed), mass, and charge of each incoming ion separately. The main advantages of such an instrument are (1) a true mass measurement, thus resolving important m/q overlaps such as C^{6+}–He^{2+} or Mg^{10+}–C^{5+} and the high charge states of Si and Fe; and (2) low background, due to the triple coincidence technique used in registering the start and stop pulses of the ToF path and the energy measurement. Because solid-state detectors are inefficient at solar wind energies (~ 1 keV amu^{-1}), a post-acceleration of $\gtrsim 20$ keV is needed to boost the ion energy. This also guarantees that solar wind of very different speeds is measured under nearly the same conditions inside the sensor. Comparisons of high-speed and low-speed solar wind composition can thus be made reliably. With SWICS/Ulysses, the abundances of C, N, Mg, and S were measured for the first time, as were the high charge states of O, Si, and Fe.

A different type of ToF sensor, such as SOHO/CELIAS/MTOF (Hovestadt et al. 1995), makes use of a harmonic retarding potential in which the ToF is proportional to $\sqrt{m}$. These sensors reach a very high mass resolution of $m/\Delta m \gtrsim 100$ (at the expense of some detector efficiency and of the charge state information), allowing determination of elemental and isotopic abundances of several previously unmeasured elements such as Na, Al, and Ca (Bochsler 1998).

A completely different type of solar wind sensor was carried on the Apollo missions (Geiss et al. 1970b). Thin Al and Pt foils were exposed to the solar wind on the lunar surface. A large, well-known fraction of the ions became trapped in the foils, which were returned for analysis by laboratory mass spectrometers. Noble gas abundances and highly accurate isotopic abundances could be obtained by this technique. It is planned to extend this technique with the Genesis mission.[13] Finally, solar wind ions are also implanted in the lunar regolith, and solar wind abundances, particularly the heavy noble gases Kr and Xe, can be obtained by analysis of the lunar material (Wieler et al. 1996).

An overview of a few selected solar wind parameters obtained by Ulysses over an eight-year period is given in Figure 9. The data were collected during the post-maximum phase of solar cycle 22 and the onset of cycle 23 and thus predominantly represent conditions of low solar activity. Two features are readily apparent from the figure. First, the bimodal property of the solar wind already noted in the bulk speed in Figure 5 is also apparent in the composition parameters. The Si/O ratio is clearly enhanced in the slow solar wind, as are the other elements (Fe, Mg) with a low first ionization potential (FIP). This is believed to be caused by FIP fractionation in the chromosphere (Geiss 1982). The bimodality is even stronger in the charge-state ratios, which are expressed here as equivalent temperatures for collisional equilibrium between ions and electrons – the so-called freezing-in temperatures. The conclusion is that the fast and slow winds originate in regions with different coronal temperatures. Second, all parameters are much more variable in the slow solar wind, which validates the view of Bame et al. (1977) of the fast solar wind as a structure-free state.

In 1993–94 and once again in 1996–97, the slow wind observed at Ulysses alternated with a fast stream once every solar rotation (Figure 9), yielding an ideal period for direct comparison of the two quasi-stationary stream types. Geiss et al. (1995a,b) performed a superposed epoch analysis of that period, with the results shown in Figure 10. The figure

---

[11] Explorer 34, also known as IMP 4: a NASA mission launched on 24 May 1967 into a highly eccentric Earth orbit from which it returned data until 3 May 1969.

[12] International Sun-Earth Explorer 3: a NASA spacecraft launched on 12 August 1978 into a halo orbit about the Sun–Earth Lagrangian libration point L1; the spacecraft was later sent on a trajectory to explore the geomagnetic tail, and then, renamed the International Cometary Explorer, flew through the tail of Comet Giacobini-Zinner, and ended up in a heliocentric orbit, returning data until 5 May 1997.

[13] Genesis: a NASA spacecraft launched in August 2001 is collecting solar wind samples from a halo orbit about the Sun–Earth Lagrangian libration point L1 and will return the samples to Earth in September 2004.

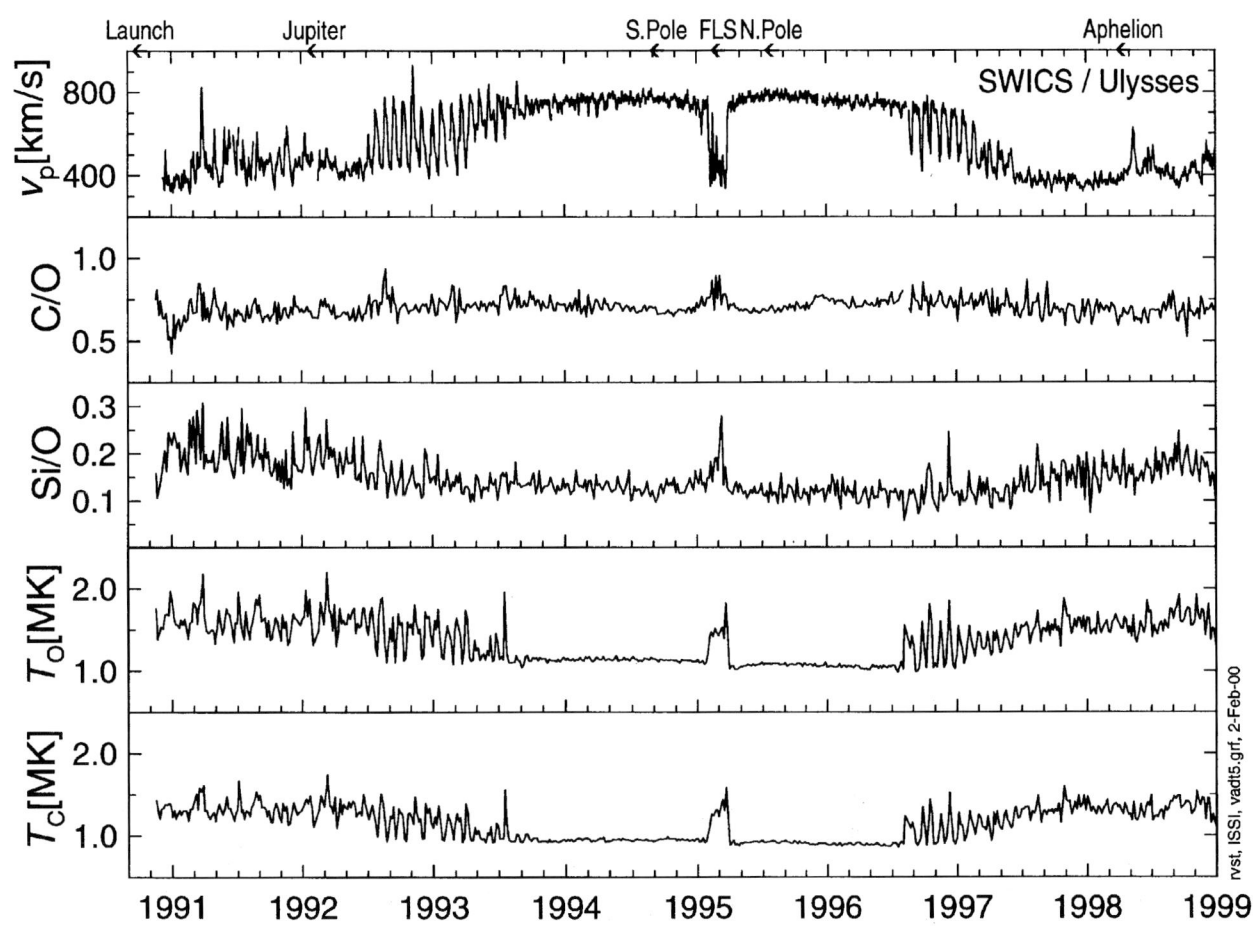

Figure 9 Overview of selected solar wind parameters obtained on Ulysses. From top to bottom: bulk speed of protons; C/O and Si/O abundance ratios; and freezing-in temperatures from the O^{7+}/O^{6+} and C^{6+}/C^{5+} charge-state ratios. All parameters are plotted at a time resolution of 5 days. The bimodal property of the solar wind speed in Figure 5 can also be seen in some of the other parameters: A strong anticorrelation of the freezing-in temperatures with speed is obvious, as is an enhanced abundance of the low-FIP element silicon in the slow wind. The Ulysses orbit remained near the ecliptic plane from launch until its gravitational assist maneuver at Jupiter and was tilted by some 80° thereafter, diving through the ecliptic during the fast latitude scan in early 1995. (Adapted from von Steiger et al. 2000.)

summarizes in a compact way one of the most essential results of SWICS regarding solar wind composition. There are two quasi-stationary types of solar wind that differ in speed (a heliospheric signature), in charge-state composition (a coronal signature), and in elemental composition (a chromospheric signature), and a sharp boundary separates the two states in the heliosphere, through the corona, and all the way down into the chromosphere. It is therefore quite likely that different physical processes must be invoked for the generation and acceleration of the two types of wind. From Figure 10 it is also apparent that the compositional jump is sharper at the leading side of the fast stream, that is, within the CIR, than in the trailing rarefaction region. Wimmer-Schweingruber et al. (1997, 1999) have investigated exactly how sharp the jumps are and where they occur. Not surprisingly they identified the location of the change in composition with the stream interface, and they showed that the jumps in the charge-state ratios usually occur faster than the time resolution of the sensor (~1 hour, depending on the particular ion species). The charge state ratio of O^{7+}/O^{6+}, for example, is thus established as a powerful diagnostic tool for telling the two solar wind types apart. Such a tool is particularly useful for identifying cases of multiple stream interface crossings, or for telling fast quasi-stationary streams from fast CMEs.

The composition measurements are summarized in Figure 11 (elements) and Figure 12 (charge states). The elemental data are plotted as the ratio of each element's abundance relative to oxygen observed in the solar wind to the same ratio observed in the photosphere *versus* the first ionization potential. The difference between the fast and slow solar wind again stands out. It can also be seen in Figure 11 that elements with low FIP ($\lesssim 10$ eV) lie on a plateau a

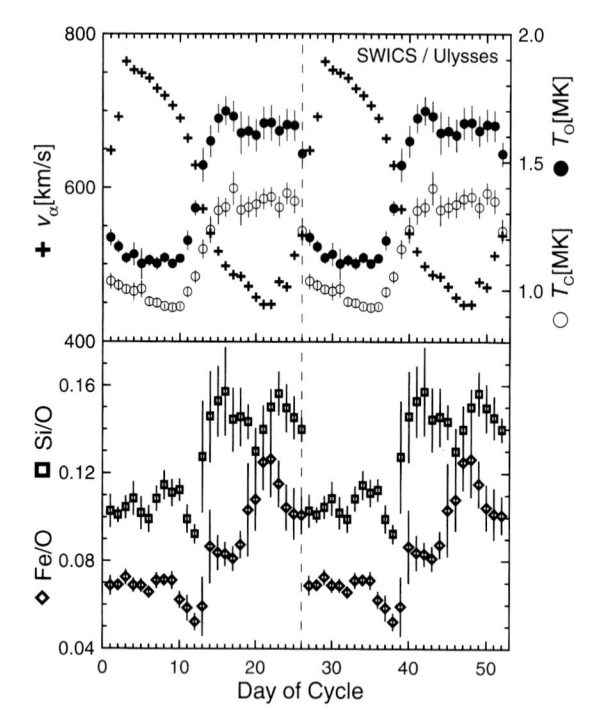

Figure 10 Superposed epoch plot of slow solar wind alternating with a fast stream once per solar rotation in 1992–93. The average profiles of the parameters in Figure 9 are superposed over nine recurrences of an average duration of 26 days, and the data are repeated twice to emphasize the periodicity. The anticorrelation of the speed with the charge-state temperatures and their positive correlation with the compositional signature is evident. Note also that the jump is sharp at the leading (CIR) side of the fast stream, but extended over 4–5 days at the trailing side. (After Geiss et al. 1995a.)

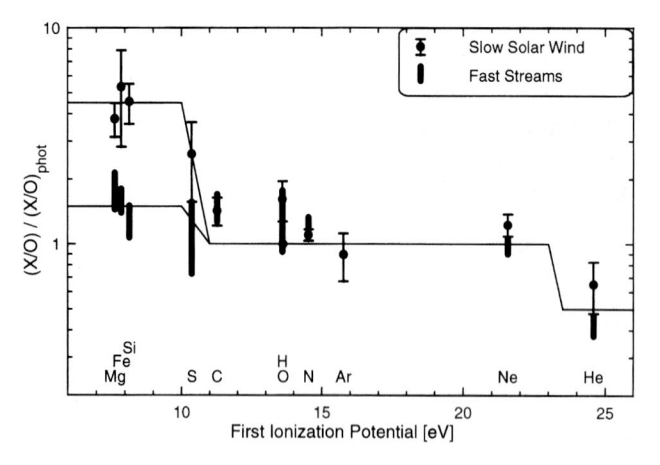

Figure 11 Solar wind abundances of heavy ions relative to oxygen, X/O, related to the photospheric ratios, $(X/O)_{phot}$, as a function of first ionization potential. (Adapted from von Steiger et al. 1997.)

factor of ~3–5 above the elements with high FIP. The intermediate elements C and S with FIPs near 10 eV are also enriched in the solar wind, but by smaller factors. Up to Ne, the high-FIP elements also lie on a plateau, but He is underabundant by a factor of ~2 relative to them. This basic FIP pattern was first described for the solar energetic particles by Hovestadt (1974), while Meyer (1981) first noted the same pattern in the slow solar wind near the ecliptic plane. In the fast streams, the fractionation factor between low- and high-FIP elements is reduced to a factor of ~2, but is still clearly present. Recent reanalysis of the Ulysses/SWICS data (von Steiger et al. 2000) indicates that the FIP bias of the slow wind is near the low end of the cited range, yet still significantly higher than in fast streams.

The observed compositional signature of fast streams must be imposed by a mechanism operating on the FIP of the elements. It is generally accepted that the FIP fractionation occurs by atom–ion separation at an altitude where the solar atmosphere is partially neutral, that is, in the upper chromosphere and the lower transition region of the Sun (Geiss 1982). A broad variety of models has been proposed to explain the FIP effect (von Steiger et al. 1997, Hénoux 1998). Each model can be characterized by the mode of ionization and by the mode of separation invoked. Most of the models take UV and EUV photons as the agent of ionization, but little consensus exists about the mode of separation. Atom–ion separation across magnetic field lines seems a good possibility, but there are several different ideas regarding the driving force (gravity, a density gradient, wave pressure, etc.). The different strengths of the FIP fractionation in the fast and in the slow wind imply, perhaps, that rather than a single mechanism, two (or more) mechanisms are at work.

Average charge-state spectra of the two solar wind types are shown in Figure 12. An obvious trend is a shift toward higher charge states in the slow wind compared to fast streams. Considering Figure 9, this comes as no surprise for C and O, since that shift translates directly into a higher freezing-in temperature. The situation for the elements that are spread over many charge states is less simple, however. The distributions of the low-charge states of Si and Fe are quite similar for both flow types, but there is an excess of high-charge states in the slow wind. Only the fast-stream charge-state distribution can be represented quite well by a single freezing-in temperature for each element (Geiss et al. 1995a).

The charge-state distribution of an element in the solar wind freezes in at that altitude in the corona where the ionization/recombination timescale exceeds the expansion timescale of the flow (Hundhausen 1972). In fast streams, all charge states of an element indicate approximately the same temperature; the freezing-in temperatures for C, O, Fe, and Si are 1.0, 1.2, 1.25, and 1.45 MK, respectively

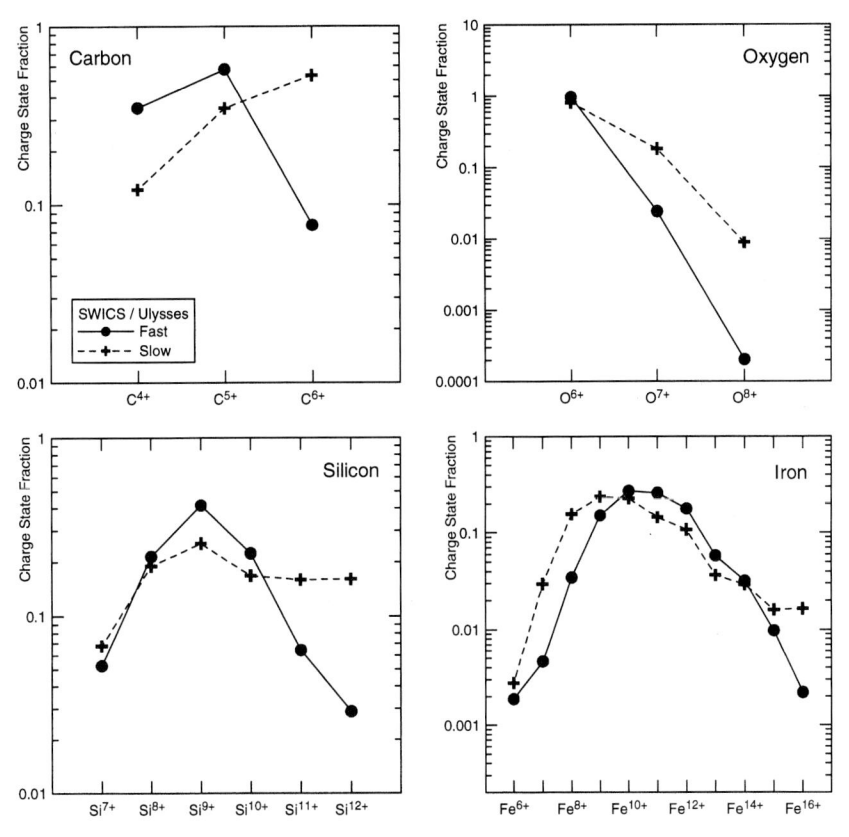

Figure 12 Average charge-state spectra of C, O, Si, and Fe from SWICS/Ulysses, of fast and slow solar wind. Each spectrum was obtained from a long-term accumulation over ~600 days. (Adapted from von Steiger et al. 2000.)

(Geiss et al. 1995a). Combination of these temperatures with ionization and recombination coefficients can yield a rough radial profile of electron temperatures in the corona. It is sufficient to assume a monotonically decreasing density as a function of heliocentric distance to show that the electron temperature must have a maximum of about 1.5 MK at a distance of a few solar radii (Ko et al. 1997).

PLASMA PROPERTIES

The previous sections deal with the macroscopic properties of the solar wind, including its organization into high- and low-speed streams and magnetic sectors and its chemical composition. A closer look, however, reveals that the solar wind is not an equilibrium plasma that can be described by locally unique values of velocity and temperature. Neither the ion nor the electron distributions are well described by Maxwell–Boltzmann functions, and several different streaming velocities and temperatures often exist simultaneously. These non-equilibrium distributions lead to some fascinating questions about plasma instabilities and wave–particle interactions, many of which are not yet well understood.

Consider first the solar wind electrons. The range of temperatures of solar wind electrons is less than that of the ions. This is shown in Figure 13, where one-year averages of the electron and proton temperatures observed by the Earth satellite Vela 4[14] are plotted *versus* solar wind speed. Early measurements by Vela 4 and by OGO 5[15] also showed that the electron distribution is anisotropic and provides an average net heat flux of $\sim 10^{-2} \mathrm{erg\, cm^{-2}\, s^{-1}}$ along the magnetic field and outward from the Sun (Montgomery et al. 1968, Montgomery 1972, Ogilvie et al. 1971). This is much lower than the heat conduction expected from classical collision-dominated thermal conductivity.

Later observations of solar wind electrons with the IMP 6–8 and Helios 1 and 2 spacecraft revealed an even more complex picture. The electron distribution can be usefully classified as consisting of a "core" population of thermal electrons, a faster, hotter "halo" population with a density

---

[14] Vela 3 and 4: pairs of US Department of Defense spacecraft operating from 20 July 1965–28 April 1967 and 28 April 1967–1 August 1972, respectively, in high-apogee Earth orbits.

[15] Orbiting Geophysical Observatory 3 and 5: NASA spacecraft operating 7 June 1966–29 February 1972 and 4 March 1968–6 August 1971, respectively, in highly eccentric Earth orbits.

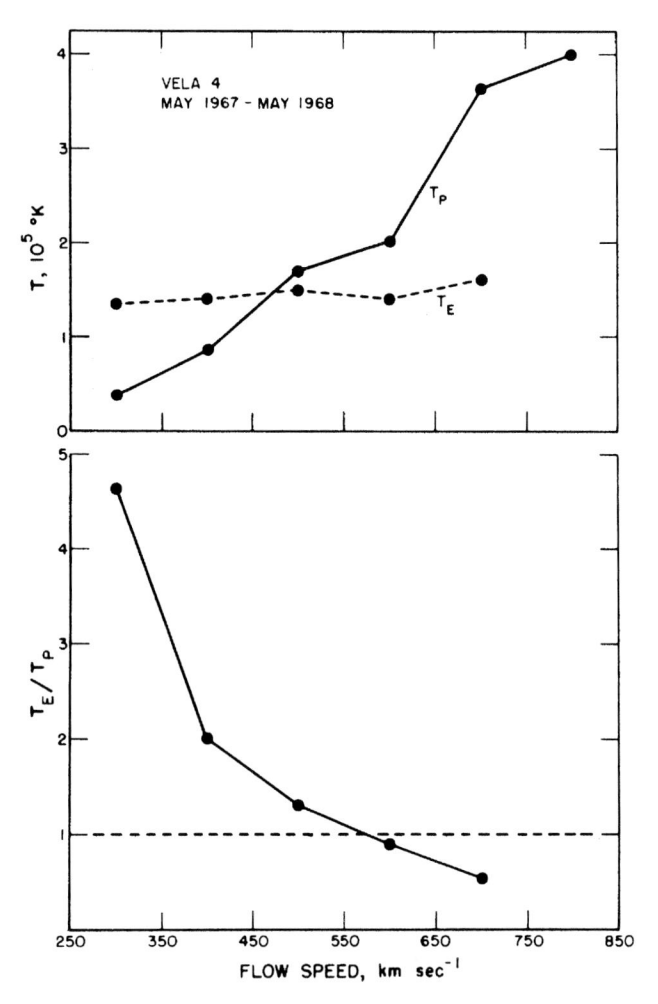

Figure 13. Top: dependence of electron and proton temperatures on the flow speed of the solar wind. Bottom: the ratio of the electron to proton temperatures. (From Montgomery 1972.)

~5% of the density of the core (Feldman et al. 1975, Pilipp et al. 1987a), and sometimes a narrower, higher-energy beam called a "strahl" (Rosenbauer et al. 1977). Each of these electron populations is anisotropic with higher temperatures parallel to the magnetic field than perpendicular to it. The heat conduction occurs through the drift of the strahl and halo populations relative to the bulk speed. The details of the distribution functions depend strongly on the solar wind stream structure, with the anisotropy being greatest and the strahl being most prominent in the fast wind from coronal holes and with collisions being most important in the slow, dense wind near the heliospheric current sheet (Feldman et al. 1978b, Pilipp et al. 1987b). Near the heliospheric current sheet, the heat flux sometimes disappears, which is taken as evidence that magnetic reconnection has occurred across the current sheet so that the plasma is no longer magnetically connected to the Sun. At other times, double strahls are observed, with one beam of suprathermal electrons moving out from the Sun along a magnetic field line and the other beam headed back toward the Sun (Gosling et al. 1987, Pilipp et al. 1987c). As discussed earlier, such events are interpreted as evidence for transient flow with a closed magnetic configuration in which both ends of the magnetic field lines are rooted in the Sun.

The variation of the electron distributions with distance from the Sun, R, could be studied by use of observations with Mariner 10[16] (0.45–1 AU), Helios 1 and 2 (0.29–1 AU), Voyager 2[17] (1.0–4.8 AU), and Ulysses (1.0–5.4 AU). The temperature of the electron core component is usually fit to a power law, with the exponent ranging widely, from $R^{-0.24}$ to $R^{-1.26}$; these results differ from both adiabatic expansion, for which the power-law exponent would be -1.33, and dominance by Coulomb collisions with an exponent of -0.33. The Helios and Ulysses measurements together indicate that the electron heat flux drops off approximately as $R^{-3.0}$ between 0.3 and 5 AU (Scime et al. 1994).

The velocity distribution of the protons in the solar wind is also anisotropic, usually with higher temperatures along the magnetic field than perpendicular to it (Hundhausen et al. 1967). At 1 AU, the proton heat flux is typically two or three orders of magnitude less than the electron heat flux. By the time of the IMP 6 mission, the energy resolution of electrostatic analyzers had improved to the point that two separate proton streams with different velocities could be discerned (Feldman et al. 1973). Helios 1 brought the first three-dimensional measurements of the solar wind plasma, which enabled mapping of the distribution functions parallel and perpendicular to the field. Figure 14 shows samples of some of the variety of proton distribution functions observed by Helios 2. The inner contours indicate that in the core of the distribution the temperature perpendicular to the field is usually greater than the parallel temperature, but the reverse is true for the outer contours. The top and bottom diagrams in the center column are examples of the presence of secondary proton beams streaming along the magnetic field with higher speeds than the core of the proton distribution. The average velocity difference between the primary and secondary proton beams increases with increasing wind speed and with decreasing distance from the Sun (Marsch et al. 1982a). Radial gradients of the temperature of solar wind protons have been measured with Helios 1 and 2, Pioneers 10 and 11, Voyager 2, and Ulysses. An approximately adiabatic expansion, with $T \propto R^{-4/3}$, was found in the slow solar wind (Marsch et al. 1982a, Liu et al. 1995), but in other flow regimes the proton temperature falls off more slowly than adiabatically,

---

[16]Mariner 10: a NASA spacecraft launched on 3 November 1973 for flybys of Venus and Mercury; the last data were obtained on 16 March 1975.

[17]Voyager 2: a NASA spacecraft launched on 20 August 1977; after flying by Jupiter, Saturn, Uranus, and Neptune, Voyager 2 continues through the outer heliosphere.

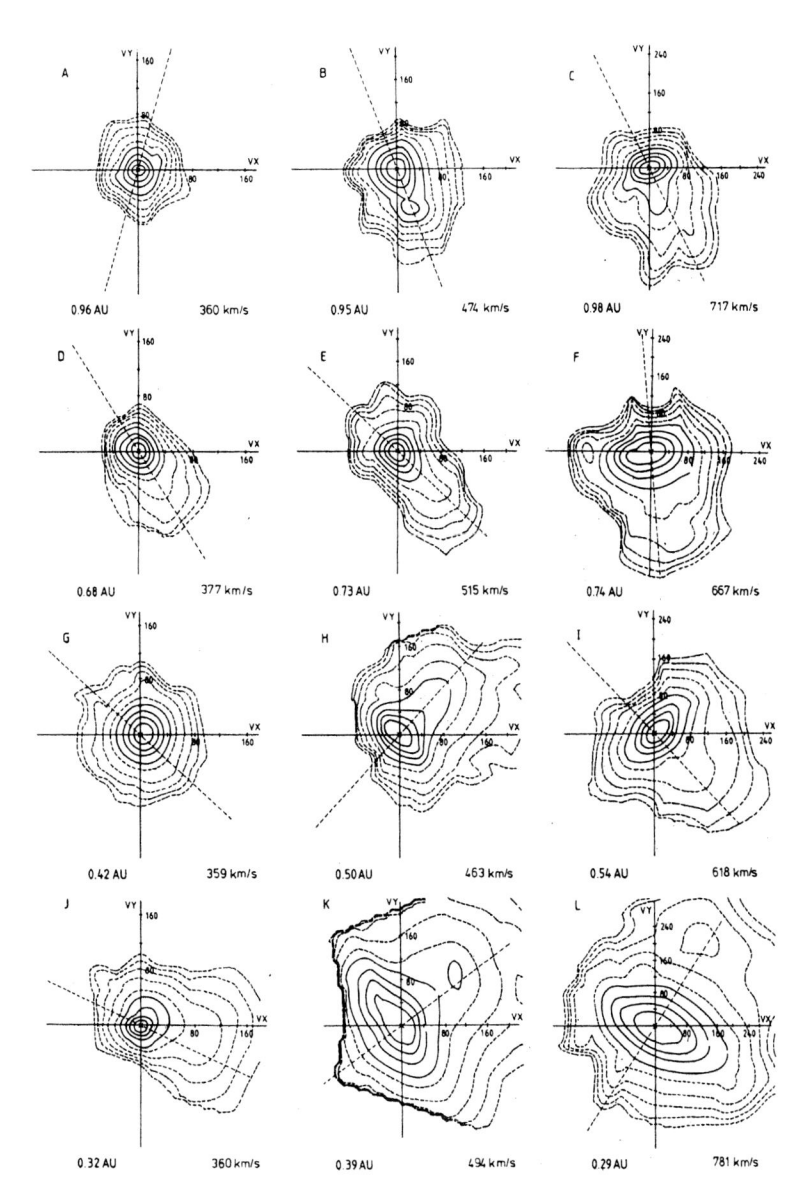

Figure 14 Proton velocity distributions measured over a range of solar wind speeds and solar distances by Helios 2. The two-dimensional plots are cuts through the three-dimensional distributions in a plane defined by the velocity vector (VX axis) and the magnetic field (direction indicated by a dashed line). Scales are in km s^{-1}. Continuous contour lines correspond to fractions 0.8, 0.6, 0.4, and 0.2 times the peak phase space density and dashed contours are logarithmically space to fractions of 0.1, 0.032, 0.01, 0.0032, and 0.001. (From Marsch et al. 1982a.)

with power-law exponents in the range −0.7 to −1.0 (Marsch et al. 1982a, Gazis et al. 1994, McComas et al. 2000). With the SWICS instrument on Ulysses which can separate the distributions of different ion species, it was discovered that both the protons and the heavy ions have high-energy tails, extending to at least 10 thermal widths above the bulk speed, and containing up to 1% of the ion density (Ogilvie et al. 1993). These approximately exponential-shaped tails are most prominent following the passage of interplanetary shocks.

It has already been mentioned that the Mariner 2 data indicated that the alpha particles in the solar wind were ∼4 times hotter than the protons, but later observations showed further peculiarities of the alpha distributions relative to that of the protons. Starting with the data from the Vela 3 spacecraft, it became apparent that, despite their greater mass, the alphas were, on average, moving away from the Sun faster than the protons (Robbins et al. 1970, Neugebauer 1981). The vector velocity difference between the alphas and the protons, $v_{\alpha p}$, is parallel to the magnetic field. In the dense, slow solar wind, occasional Coulomb collisions limit the velocity difference to values close to zero. In the high-speed wind at 0.3 AU, Helios found values of $v_{\alpha p} \geq 150$ km s^{-1}, approaching the Alfvén speed v_A

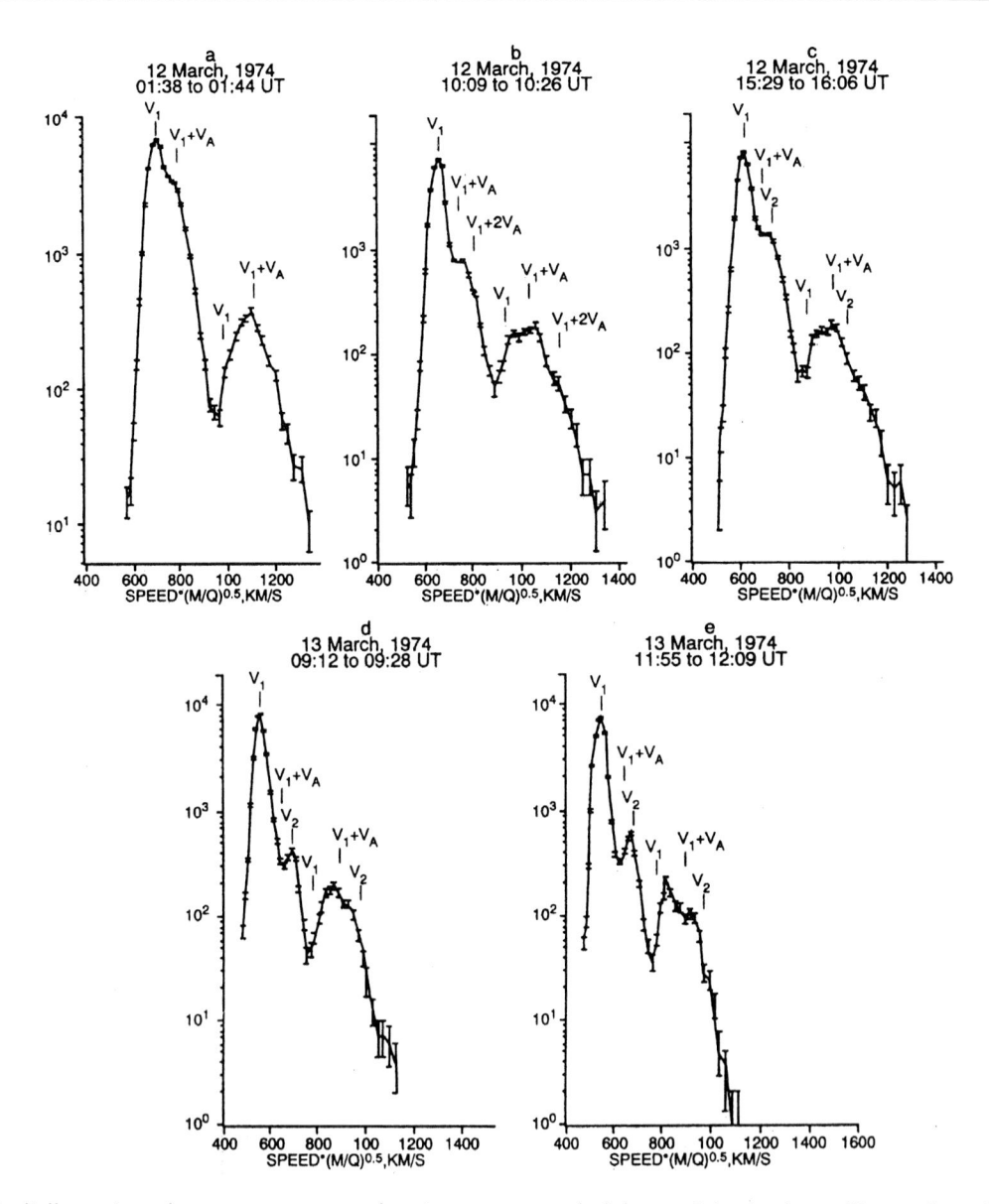

Figure 15 Radially projected count-rate spectra showing protons and alpha particles in the trailing region of a high-speed stream. The protons are at the lower values of $v(M/Q)^{0.5}$, while the alpha particles are at the higher values. V_1 is the speed of the primary proton peak, V_A is the Alfvén speed, and V_2 is the speed of a secondary proton peak. (From Feldman et al. 1993.)

(Marsch et al. 1982b). Helios, Voyager 2, and Ulysses data all indicate that both $v_{\alpha p}$ and $v_{\alpha p}/v_A$ decrease with decreasing speed and with increasing distance from the Sun (Neugebauer et al. 1996). To complicate matters further, the alpha particle distributions also sometimes exhibit two beams (Feldman et al. 1973; Asbridge et al. 1976). Some of the variety of proton and alpha beams are illustrated in the one-dimensional spectra plotted in Figure 15.

The rare ions heavier than the alpha particles, from carbon to iron, tend to obey rather simple rules with surprising accuracy: the heavy ions all flow with the same bulk speed as the alphas, $v_i = v_\alpha$, and they all have equal thermal speeds, that is, their kinetic temperatures are proportional to their masses, $T_i = m_i T_\alpha / 4$. Adherence to these rules is illustrated in Figure 16 which shows data obtained near 5 AU by Ulysses/SWICS. The bulk speeds are equal to a very high degree of accuracy ($\lesssim 1\%$) in both the slow and the fast solar wind. Closer to the Sun, near 1 AU, measurements of Si and Fe with the ISEE/ICI and the SOHO/CELIAS instruments showed those elements lagging behind He at the higher speeds (up to $600\,\mathrm{km\,s^{-1}}$) (Schmid et al. 1987; Bochsler 1989; Hefti et al. 1998). A possible interpretation is that the acceleration up to the He speed is completed somewhere between 1 and 5 AU. The equality of the

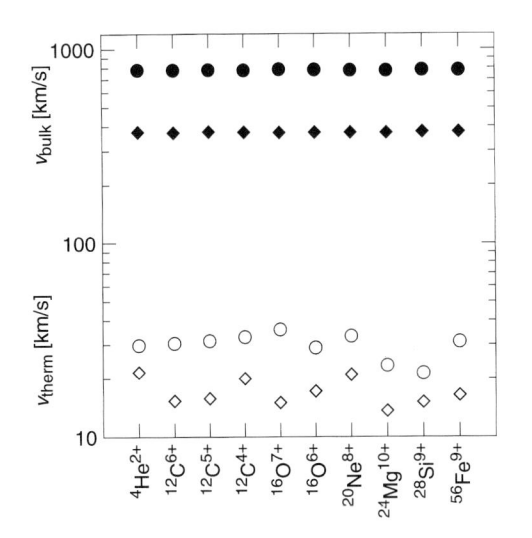

Figure 16 Bulk speeds (filled symbols) and thermal speeds (open symbols) of heavy solar wind ions obtained with SWICS/Ulysses during two 5-day time periods in fast (circles, days 1–5 of 1994) and in slow (diamonds, days 1–5 of 1998) solar wind.

thermal speeds shown in Figure 16 is obeyed much less accurately, but because there is no systematic trend with either mass or mass per charge they may be considered equal despite the scatter. Most measurements at 1 AU, from ISEE/ICI (Bochsler *et al.* 1985) and SOHO/CELIAS (Hefti *et al.* 1998), also show equal thermal speeds of the heavy ions. An important exception to the rule is that when the solar wind is very slow and rather dense, Coulomb collisions tend to equalize the temperatures rather than the thermal speeds (Bochsler *et al.* 1985).

Conformance with these rules under most solar wind conditions suggests a rather simple physical interpretation, although a fully satisfying model of the underlying mechanism has not yet been developed. There is some consensus that interactions of the ions with Alfvén waves traveling outward on the bulk solar wind accelerate all heavy ions, irrespective of their mass, to a maximum velocity $v_i \leq v_p + v_A \mathbf{b}$, where $\mathbf{b}$ is a unit vector parallel to the magnetic field. At the same time the heavy ions scatter in velocity space on a sphere around v_i with radius v_A, thus causing equal thermal speeds. A model by Isenberg and Hollweg (1983) reproduces the temperature rule for heavy ions quite well, but fails for their relationship to protons.

When examined on a finer temporal scale than that of the stream structure, the solar wind shows continuous fluctuations in nearly all parameters. There are many types of variations, arising from many different sources or processes, and exhibiting many different modes of propagation or evolution. Because these disturbances propagate at speeds well below the solar wind speed, instruments on spacecraft sense only one-dimensional cuts through what are usually three-dimensional structures or waves being carried outward from the Sun with the solar wind fluid.

Perhaps the simplest variations to understand are those caused by changes in the solar source of the wind. Neighboring streamlines or magnetic flux tubes may contain plasmas with slightly different speeds, densities, temperatures, elemental compositions, and magnetic field strengths and directions. The flux tubes with higher pressure than their neighbors expand until, as discovered with the instruments on Voyager, pressure-balance structures are a common feature of the solar wind far from the Sun (Vellante and Lazarus 1987). The boundaries between plasmas from different sources are often quite sharp; these structures, called tangential discontinuities, can pass a spacecraft in a matter of seconds.

More prevalent than pressure-balance structures are waves, of which many different kinds have been observed in the solar wind – electromagnetic, magnetohydrodynamic (MHD), magnetosonic, and electrostatic. The first waves to be studied in any detail were the low-frequency, transverse MHD waves called Alfvén waves. Alfvén waves can be thought of as propagating kinks in magnetic field lines. These waves were first detected in the Mariner 2 data (Unti and Neugebauer 1968), but a study based on Mariner 5 data by Belcher and Davis (1971) led to a more complete characterization of their properties. These large-amplitude, non-sinusoidal waves dominate the microscale structure of the solar wind at least half the time. They propagate outward from the Sun with wavelengths in the range 10^3 to 5×10^6 km. They are most prominent in the fast solar wind and in the trailing edges of high-speed streams and have the greatest amplitude in interaction regions. The MHD equations which describe a magnetized plasma such as the solar wind have solutions corresponding to longitudinal magnetosonic waves as well as to the transverse Alfvén waves. The magnetosonic waves are seldom observed (Belcher and Davis 1971), however, because they are rapidly damped (Barnes 1966).

There is another family of solutions of the MHD equations that corresponds to sharp jumps. The non-propagating tangential discontinuities (TDs) mentioned above are one solution, while propagating rotational discontinuities (RDs), which are sometimes thought of as steepened Alfvén waves, are another. Both types are seen in the solar wind at a rate of ~1 per hour (Siscoe *et al.* 1968, Burlaga 1969). There is a long-standing controversy, which is not yet resolved, about whether TDs or RDs occur more frequently. Other discontinuous solutions of the MHD equations correspond to shocks. The concept of collisionless shocks was first suggested by Gold (1959) as an explanation for the sudden commencement of geomagnetic storms, and, using data from Mariner 2, Sonett *et al.* (1964) were the first to demonstrate their existence.

Many high-frequency waves are present in the solar wind in addition to the Alfvén waves and MHD discontinuities.

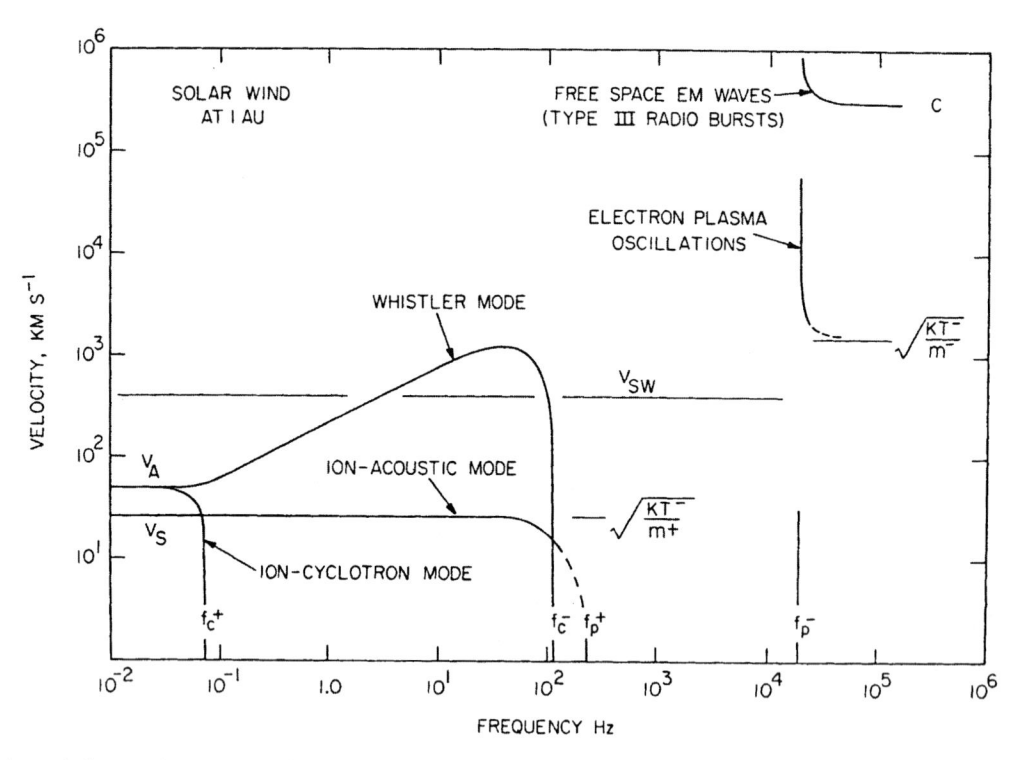

Figure 17 Plots of phase velocity *versus* frequency for selected plasma-wave modes that propagate parallel to the interplanetary magnetic field. (From Gurnett 1991.)

Figure 17 displays the phase velocity *versus* wave frequency of some of the plasma-wave modes seen in the solar wind near 1 AU; for simplicity the figure is limited to waves propagating along the magnetic field. Whereas the MHD waves with frequencies below the ion cyclotron frequency (f_c^+) are studied through their effects on the ions and the magnetic field, the higher-frequency plasma waves require different instrumentation. Rapidly varying magnetic fields are measured by the voltage induced in a coil wound around a high-permeability core, while electric fields are measured by the voltage difference between the two ends of long dipole antennas (often tens of meters long). Starting in the late 1960s, plasma-wave instruments have been flown on a large number of spacecraft, including OGO 3 and 5, Pioneers 8 and 9,[18] IMP 6, 7, and 8, Helios 1 and 2, ISEE 1, 2,[19] and 3, Voyagers 1 and 2, and Ulysses.

The interesting questions are: Where do all these different waves come from? What causes them? How do they evolve in the solar wind? How do they affect the physics of the solar wind?

Some of the waves and other disturbances are generated at the Sun, or in the solar corona, while others are generated *in situ* in the solar wind. The observation that the great majority of the Alfvénic waves propagate outward from the Sun is the basis for the argument that these waves must be generated closer to the Sun than the so-called critical point where the flow speed first exceeds the Alfvén speed at which the waves propagate. These waves are perhaps a remnant of the processes that heat the corona and accelerate the wind. Other waves are created by either dynamic or kinetic effects in the interplanetary medium. Regions where dynamic interactions generate waves include stream-interaction regions and other shear zones where the plasma may be subject to the Kelvin–Helmholtz instability.

Another burgeoning area of theoretical space plasma physics is MHD turbulence. As early as 1968, Coleman (1968) pointed out that the slope of power spectra of magnetic fluctuations in the solar wind was similar to the slope expected for isotropic, homogeneous fluid turbulence, and suggested that similar processes may occur in the solar wind. In the MHD turbulence view of the solar wind, the low-frequency Alfvén waves which propagate outward from the Sun or which are generated at shear regions in interplanetary space interact nonlinearly with each other or

---

[18] Pioneers 8 and 9: NASA spacecraft operating in heliocentric orbits from 13 December 1967 to 5 January 1975 and from 8 November 1968 to 18 August 1974, respectively; their heliocentric distances ranged from 0.99 to 1.09 AU for Pioneer 8 and from 0.75 to 0.99 AU for Pioneer 9.

[19] ISEE 1 and 2: a pair of closely spaced NASA spacecraft operating in highly eccentric Earth orbits from 22 October 1977 to 26 September 1987.

with a small population of inward-propagating waves to produce higher-frequency waves. The process continues, with the wave energy cascading downward to shorter wavelengths (higher frequencies). Eventually, when the wavelength reaches the ion cyclotron radius, the wave energy is absorbed by resonant ion heating. One such model by Tu (1988) successfully predicts both the change of slope and the variation of the proton magnetic moment as functions of distance from the Sun; the agreement between Tu's model and Helios measurements of the proton magnetic moment is shown in Figure 18. There continue to be many such studies using data from Pioneer, Voyager, Helios, and Ulysses which focus on distinguishing non-propagating structures from radially evolving turbulence, taking the three-dimensional expansion and dynamics of the solar wind into account.

Almost all the non-equilibrium distributions discussed above are either the cause of or are limited by kinetic effects and instabilities. In a nearly collisionless plasma like the solar wind, waves play a role similar to that played by collisions in ordinary fluids. The dynamics of the solar wind can result in unstable, non-equilibrium distributions, which in turn result in the generation of waves. The waves then interact with the ions or electrons to return them toward an equilibrium configuration. The net effect is that the free energy present as a result of multiple beams or other anomalies is transferred, via wave–particle interactions, to more stable, but hotter distributions. As the wind flows out through interplanetary space, this heating results in radial temperature profiles that are less steep than expected for adiabatic flow. Studies of these instabilities and wave–particle interactions comprise a very large subdiscipline within space physics; only a few examples can be mentioned here.

It was mentioned previously that the electron heat flux is much less than that expected from collisional processes alone. There have been theoretical investigations of several instabilities that might limit this heat flux. Ulysses measurements showed that the upper limit of the electron heat flux decreases with solar distance as R^{-3}, which supports the whistler (electromagnetic) heat-flux instability proposed by Gary et al. (1994). The Ulysses data also showed a rough anticorrelation of the electron heat flux with the simultaneously measured plasma-wave amplitude detected by the radio/plasma wave experiment (Scime et al. 1996). The whistler waves had more than enough power to scatter electrons out of the halo, which carries most of the heat flux.

Another broad category of non-equilibrium wave–particle interactions involves ion beams, such as interstellar pickup ions or the streaming of a secondary beam of protons or a beam of heavy ions relative to the main proton fluid. Observations close to the Sun by the Helios spacecraft indicated that both the differential streaming between the primary and secondary proton beams and between the protons and alpha particles increase or decrease with increases or decreases in the Alfvén speed (Marsch et al. 1982a,b); although Alfvén waves must play a role in either accelerating or limiting the speed of the secondary beam, the details of the interactions are still an area of active investigation. Farther from the Sun, beyond 1 AU, where the relative velocities of the several beams become substantially less than the Alfvén speed, other processes must be at work. Knowledge of the generation of waves by and the pitch-angle scattering of pickup ions in the solar wind was greatly advanced by measurements of waves and particle distributions in the outer comas of comet Halley and other comets where the solar wind picked up freshly ionized cometary ions (Tsurutani 1991).

The limitation of ion anisotropy is yet another example of important wave–particle interactions. In a nearly collisionless plasma such as the solar wind, one might expect an ion's magnetic moment (proportional to $T_\perp/B$) to be conserved, which would imply that as the solar wind expands and the magnetic field becomes weaker, the velocity component perpendicular to the field would rapidly decrease to yield a ratio of $T_\parallel/T_\perp$ greatly in excess of the observed anisotropy. Using data from IMP 6, Bame et al. (1975b) showed, however, that in high-speed streams, the core of the proton distribution has $T_\perp > T_\parallel$, which the authors interpreted as the result of local, interplanetary heating.

SOLAR WIND SOURCES AND ACCELERATION MECHANISMS

Parker's original theory, which successfully predicted the existence of the solar wind, assumed a thermally driven, spherically symmetric, time stationary, nonmagnetic, single fluid. We have already seen that the solar wind consists of several fluids (electrons, protons, and heavy ions), each

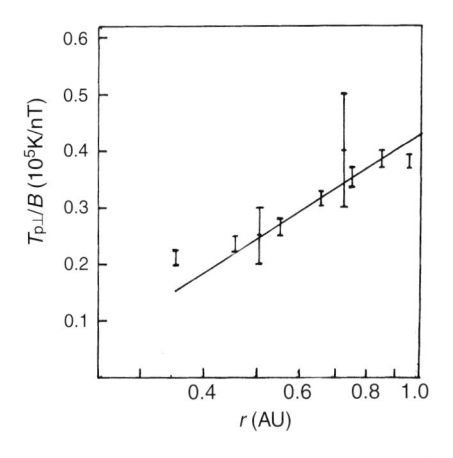

Figure 18 Plot of the proton magnetic moment $T_{p\perp}/B$ as a function of solar distance. (From Tu 1988.)

with their own temperatures and speeds, and often with multiple beams of a single particle type. The observed structure of fast, slow, and transient streams also tells us that the solar wind is neither spherically symmetric nor time stationary. It carries a magnetic field, which is too weak to affect the dynamics of the wind in interplanetary space, but which can control the motion of the plasma close to the Sun where the field is stronger. Finally, it turns out that the solar wind is not driven by thermal pressure gradients alone. This section provides a summary of the continuing effort to develop realistic models of the sources and the acceleration of the solar wind.

The early observations that the protons and electrons have different temperatures led to the development of two-fluid models of the solar wind (Sturrock and Hartle 1966). Use of two-fluid or multi-fluid models is justified by the low rate of Coulomb collisions between ions and electrons in interplanetary space. The two-fluid model predicted too high an electron temperature (3.4×10^5 K) and too low a proton temperature (4.4×10^3 K) at a solar distance of 1 AU. One must appeal to instabilities and wave–particle interactions to explain the redistribution of thermal energy from the electrons to the protons to provide the temperatures shown in Figure 13.

As mentioned earlier, it is well established that the fast wind emanates from coronal holes (Krieger et al. 1973). In coronal holes, the fields are obviously open, the expansion geometry appears to be simple, the acceleration occurs rapidly (Kohl et al. 1997), the charge states of heavy ions indicate a rapid freezing-in process, and the elemental abundances are close to solar values. However, bright features within coronal holes called polar plumes have been reported to have different elemental abundances than the fast solar wind. At least the prominently visible plume observed by Widing and Feldman (1992) appears to be enriched in low-FIP elements (such as in the Mg/Ne ratio) even more strongly than the slow solar wind. This is surprising because plumes are believed to be open-field structures within the coronal holes. It therefore seems that polar plumes are not a major source of the fast streams, and the absence of compositional variability in fast streams makes it questionable whether plumes contribute to the solar wind at all. Plumes might be essentially static structures which do not participate in the outward mass flow; recent SOHO observations of flow speeds in plumes suggest that this might be the case (Giordano et al. 2000). Moreover, a systematic search for compositional fine structures as possible interplanetary remnants of polar plumes within the largely uniform fast streams, for example in microstreams or in pressure-balanced structures, has revealed only a possible small signature in charge-state ratios, but no composition anomalies (von Steiger et al. 1997).

The single- and multi-fluid models all result in solar wind speeds well below the values observed in flows emanating from coronal holes. As early as Mariner 2 it was clear that the fastest wind did not come from the hottest parts of the Sun (Snyder and Neugebauer 1966), as would be implied by a thermally driven model. In fact, the temperatures in coronal holes are lower than the temperatures of the rest of the corona. These facts led to a search for nonthermal methods of accelerating the solar wind. Leer and Holzer (1980) showed theoretically that the final state of the solar wind depends strongly on where energy or momentum is added to the flow. The observed high speeds can be achieved only if energy or momentum is added to the flow after it has reached supersonic speed; energy input at lower altitudes increases the density and the mass flux, but not the speed. One possible source of this additional energy or momentum is waves created close to the Sun that propagate out into the wind. The high flux of outward-propagating Alfvén waves in the high-speed solar wind may be a remnant of such waves. The dependence of the differential flow between the protons and the alpha particles (or heavier ions) on the Alfvén speed also suggests a role for Alfvén waves in the dynamics of the solar wind.

There may also be geometric effects on the acceleration of the solar wind. The polar coronal holes occupy only a small fraction of the solar surface, but the fast wind diverging from those holes occupies about five times the solid angle of the holes themselves. Wang and Sheeley (1990) derived an empirical relation between solar wind speed and the divergence of the coronal magnetic field in the region where the wind originates; fast wind comes from slowly diverging field lines and the slow wind can be mapped back to regions of highly diverging fields at the boundaries of coronal holes.

Another possibly important geometric effect is the lengthening of the path the electrons take from the Sun to the point of detection in the solar wind. The electron heat flux is not conducted radially away from the Sun, as assumed in some of the early models, but along the magnetic field lines. The spiral winding of the field due to the Sun's rotation slightly lengthens the path between the Sun and 1 AU, but the smaller-scale twists, turns, and kinks in the field due to waves can lengthen the path considerably.

Probably all these effects – wave acceleration and heating, geometric effects, and a longer conduction path – play some role, but the entire process may be even more complicated than that. The heavy-ion data demonstrate that it is not realistic to calculate the properties of the solar wind starting with some temperature and density at a given level in the corona. There are strong correlations between heavy-ion abundances (set in the chromosphere or transition region), the ion charge states (set in the corona at distances of a few solar radii), and the speeds and temperatures of the heavy ions (probably set at altitudes where the wind has become supersonic). The three problems of how solar material is fed into the wind, how the corona is heated, and how the solar wind is accelerated are closely linked and must be solved together. This has not yet been done.

The energy required to heat the corona and to accelerate the wind must come from the turbulent convection in the outer layers of the Sun. The convection is thought to generate both waves which are damped in the chromosphere and corona and magnetic structures which relax by reconnection, but many questions remain about the nature of the waves and how they are damped and on the relative importance of magnetic reconnection. Some valuable information on coronal processes in the region of acceleration of the fast wind from coronal holes was obtained by SOHO. The Ultra-Violet Coronagraph Spectrograph (UVCS) instrument measures the intensities and profiles of the hydrogen Lyman α line and two lines of the O^{5+} ion in the corona from 1.5 to 5 solar radii (Kohl et al. 1995). The data provide information on the outflow velocities as well as on the random ("thermal") motions in both the solar radial and latitudinal directions. When combined with an empirical model of the coronal hole, the following results were obtained (Kohl et al. 1998): (1) the "thermal" velocities of the O^{5+} ions are much greater perpendicular to the magnetic field than parallel to it; (2) the component of the "thermal" velocity perpendicular to the magnetic field is much greater for O^{5+} ions than for protons; and (3) The outflow speed of O^{5+} ions is greater than the outflow speed of protons, which in turn is greater than that expected for electrons. These observations are inconsistent with any common motions of the O^{5+} ions and the protons; that is, the "thermal" speeds cannot be due to transverse waves such as Alfvén waves or to turbulent motions. Kohl et al. (1998) suggest that the ion motions may be a signature of ion-cyclotron waves. Lee and Wu (2000), however, believe that the proton and minor-ion dynamics are signatures of fast shocks moving through the corona.

New questions arise. If these ion motions are a signature of ion-cyclotron waves, how are the waves generated? If they are shocks, what is the source of the shocks? McKenzie et al. (1995) suggest that waves are generated by "microflares" caused by magnetic reconnection at the boundaries of convection cells on the solar surface and that the heavier ions are preferentially heated by resonant dissipation of the resulting high-frequency ion-cyclotron waves. But there is currently no detailed theory or numerical model of how such microflares generate the spectrum of waves required to match both the SOHO UVCS observations in coronal holes and the properties of the solar wind ions from those holes observed in interplanetary space. Both the microflare model of McKenzie et al. and a model by Parker (1987, 1990) in which twisting of magnetic fields by turbulent motions of their footpoints on the solar surface yield reconnection in "nanoflares" suppose the acceleration and heating to occur in discrete bursts or jets rather than smoothly and continuously throughout the corona as a whole. Short-lived jets of material have, in fact, been observed to move up into the corona from the solar surface (Brueckner and Bartoe 1983). Feldman et al. (1997) have proposed a composite model in which the bulk of the proton flux results from acceleration by thermal pressure gradients while discrete, transient jets contribute the higher speed secondary proton peak, the alpha particles, and the heavier ions.

Three sites have been suggested for the source(s) of the slow solar wind. Some of the slow wind can be traced back to small, low-latitude coronal holes (Neugebauer et al. 1998). Some of it probably originates in the bright, dense coronal streamers like those shown in Figure 5 (Feldman et al. 1981). It is not clear, however, how the material escapes from the obviously closed magnetic structures at the base of the streamer. Two proposed explanations are (a) quasi-stationary reconnection at the top of the closed field lines near the base of the current sheet, or (b) flow from the roots of the streamer along the open field lines at its periphery. The two scenarios can be tested by examination of the composition of the slow solar wind, in particular around the times of current sheet crossings which are thought to map back to the tip of the closed-field region. Oxygen is observed to have close to its normal abundance at times of sector boundary crossings (von Steiger et al. 1995). This clearly favors scenario (b) because SOHO/UVCS observations show that the abundances of oxygen and other high-FIP elements are depleted by an order of magnitude relative to photospheric values in the core of a quiescent equatorial streamer but resemble abundances measured in the slow wind along its periphery (Raymond et al. 1997).

In all its parameters, the slow solar wind is more highly variable than the fast solar wind. Because this variability includes elemental abundances, the fluctuations must originate at the Sun. Time-lapse sequences of coronagraph images acquired by the LASCO experiment on SOHO revealed discrete blobs of plasma which originate ~3–4 solar radii above the cusps of the closed helmet-shaped structures and which move radially outward as they accelerate up to speeds of 300 km s^{-1} near 25 solar radii (Sheeley et al. 1997).

How can this variability of the slow wind be explained? Fisk (1996) has proposed a new model of the interplanetary magnetic field to explain the latitudinal extent of the recurrent acceleration of energetic particles observed by Ulysses. In that model (Fisk et al. 1999), the open field lines at the edges of coronal holes continuously reconnect with neighboring closed-field loops. When reconnection occurs, the plasma originally in the loop joins the slow solar wind, whose properties then vary according to the properties (such as size, density, and temperature) of sequentially opened loops. Within this framework, Schwadron et al.

(1999) have suggested a way in which wave heating on coronal loops may explain both the stronger average FIP fractionation and the broader distribution of charge states (i.e., loop temperatures) in the slow solar wind compared to the fast wind.

Sequences of coronagraph images from Skylab (Gosling *et al.* 1974), Solwind (Howard *et al.* 1985), SMM (Hundhausen 1988), and SOHO (Howard *et al.* 1997) leave no doubt about the source of the transient solar wind; it is associated with the eruption of previously closed magnetic structures such as solar prominences. One type of such an event has been descriptively named "streamer blow-out." There is debate whether prominence eruptions cause CMEs or whether CMEs cause prominence eruptions (Hundhausen 1988). There are several suggestions concerning the cause(s) of the sudden release of energy in CMEs; these include the emergence of additional magnetic flux from beneath the solar surface (Feynman and Martin 1995) and the shear of previously emerged magnetic flux (Mikic and Linker 1997).

Observations with the Ulysses spacecraft provided new insight into the interplanetary acceleration of the transient solar wind. When seen moving through the corona, most CMEs have speeds well below typical speeds of CMEs detected in the solar wind at 1 AU or beyond. Near Earth, CME speeds typically range from 350 to $500\,\mathrm{km\,s^{-1}}$. However, every one of the six CMEs detected by Ulysses in the fast polar solar wind, at latitudes between 31°S and 61°S, had speeds greater than $650\,\mathrm{km\,s^{-1}}$ (Gosling *et al.* 1994b). The conclusion is that processes must be at work to accelerate the transient plasma up to approximately the speed of the ambient, quasi-stationary wind with which it interacts. Numerical models suggest a process in which a plug of slow CME plasma inserted into a fast wind is pushed on by the fast plasma behind it and will also be accelerated by the pressure gradient at its leading edge caused by the plasma ahead running away from it (Gosling and Riley 1996). Similar types of forces act to slow down fast CME material inserted into a slow ambient wind.

FUTURE DIRECTIONS

Since the early 1960s, solar wind studies have evolved from questioning the very existence and nature of the solar wind to seeking to understand some of the processes and mechanisms responsible for its properties.

- Perhaps the greatest unknown is the acceleration mechanism. For example, is the fast solar wind accelerated continuously by pressure gradients and waves or is it accelerated almost exclusively in discrete jets? For another example, what destabilizes large magnetic structures to yield CMEs? Addressing such questions probably requires sending a well-instrumented spacecraft extremely close to the Sun, within a few solar radii of its surface.
- There is another large set of questions associated with wave–particle interactions and solar wind turbulence. Progress in these areas would benefit from multi-spacecraft observations with very high time resolution of variations of the ion and electron distribution functions.
- Although great progress has been made in characterizing the fractionation of heavy elements between the solar surface and the solar wind, this work is not yet sufficiently definitive to relate the composition of the solar wind to the average composition of the Sun's outer convective zone. Determination of the isotopic abundances of the solar wind to an accuracy relevant to studies of Solar System evolution (see Figure 8) must await the return of solar wind samples by the Genesis mission.

There are other challenging problems involving the solar wind in areas within the scope of other chapters in this book. Of special interest to solar wind physicists are the acceleration of energetic particles in the solar wind and the interactions of the solar wind with planetary bodies and with the local interstellar medium.

FURTHER READING

A history of solar wind research, 1957–1970: Hufbauer, K. (1991). *Exploring the Sun*, Johns Hopkins University Press, Baltimore, pp. 213–258.

Broad overview of solar wind physics through the Helios missions: Schwenn, R. and Marsch, E. (eds) (1991). *Physics of the Inner Heliosphere*, Vols 1 and 2, Springer Verlag, Berlin.

Corotating interaction regions: Balogh, A., Gosling, J.T., Jokipii, J.R., Kallenbach, R. and Kunow, H. (eds) (1999). *Corotating Interaction Regions, Space Science Series of ISSI, Vol. 7*, Kluwer, Dordrecht.

Coronal mass ejections: Crooker, N., Joselyn, J.A. and Feynman, J. (eds) (1997). *Coronal Mass Ejections*, American Geophysical Union, Washington, DC.

Solar composition and heavy ions in the solar wind: Fröhlich, C., Huber, M.C.E., Solanki, S.K. and von Steiger, R. (eds) (1998). *Solar Composition and its Evolution – From Core to Corona, Space Sciences Series of ISSI, Vol. 5*, Kluwer, Dordrecht.

Most recent compendium of solar wind research: Habbal, S.R., Esser, R., Hollweg, J.V. and Isenberg, P.A. (eds) (1999). *Solar Wind Nine*, Conf. Proc. 471, American Institute of Physics, Woodbury, NY.

REFERENCES

Alfvén, H. (1977). Electric currents in cosmic plasmas. *Reviews of Geophysics*, **15**, 271–284.

Asbridge, J.R., Bame, S.J., Feldman, W.C. and Montgomery, M.D. (1976). Helium and hydrogen velocity differences in the solar wind. *Journal of Geophysical Research*, **81**, 2719–2727.

Bader, M. (1962). Preliminary Explorer 12 data on protons below 20 keV. *Journal of Geophysical Research*, **67**, 5007–5011.

Bame, S.J., Asbridge, J.R., Feldman, W.C., Gary, S.P. and Montgomery, M.D. (1975b). Evidence for local ion heating in solar wind high-speed streams. *Geophysical Research Letters*, **2**, 373–375.

Bame, S.J., Asbridge, J.R., Feldman, W.C. and Gosling, J.T. (1977). Evidence for a structure-free state at high solar wind speeds. *Journal of Geophysical Research*, **82**, 1487–1492.

Bame, S.J., Asbridge, J.R., Feldman, W.C., Montgomery, M.D. and Kearney, P.D. (1975a). Solar wind heavy ion abundances. *Solar Physics*, **43**, 463–473.

Barnes, A. (1966). Collisionless damping of hydromagnetic waves. *Physics of Fluids*, **9**, 1483–1495

Belcher, J.W. and Davis, L. Jr. (1971). Large-amplitude Alfvén waves in the interplanetary medium. *Journal of Geophysical Research*, **76**, 3534–3537.

Biermann, L. (1951). Kometenschweife und solare Korpuskularstrahlung. *Zeitschrift für Astrophysik*, **29**, 274–286.

Bochsler, P. (1989). Velocity and abundance of silicon ions in the solar wind. *Journal of Geophysical Research*, **94**, 2365–2373.

Bochsler, P. (1998). Structure of the solar wind and compositional variations. *Space Science Reviews*, **85**, 291–302.

Bochsler, P., Geiss, J. and Joos, R. (1985). Kinetic temperatures of heavy ions in the solar wind. *Journal of Geophysical Research*, **90**, 10,779–10,789.

Bonetti, A., Bridge, H.S., Lazarus, A.J., Rossi, B. and Scherb, F. (1963). Explorer 10 plasma measurements. *Journal of Geophysical Research*, **68**, 4017–4063.

Brueckner, G.E. and Bartoe, J.-D.F. (1983). Observations of high-energy jets in the corona above the quiet Sun, the heating of the corona, and the acceleration of the solar wind. *Astrophysical Journal*, **272**, 329–348.

Burlaga, L.F. (1969). Directional discontinuities in the interplanetary magnetic field. *Solar Physics*, **7**, 54–71.

Burlaga, L.F. (1974). Interplanetary stream interfaces. *Journal of Geophysical Research*, **79**, 3717–3725.

Burlaga, L.F., Pizzo, V., Lazarus, A. and Gazis, P. (1985). Stream dynamics between 1 AU and 2 AU: A comparison of observation and theory. *Journal of Geophysical Research*, **90**, 7377–7388.

Burlaga, L., Sittler, E., Mariani, F. and Schwenn, R. (1981). Magnetic loop behind an interplanetary shock: Voyager, Helios, and IMP 8 observations. *Journal of Geophysical Research*, **86**, 6673–6684.

Cane, H. (1997). The current status in our understanding of energetic particles, coronal mass ejections, and flares. In N. Crooker, J.A. Jocelyn and J. Feynman (eds), *Coronal Mass Ejections, Geophysical Monograph 99*, American Geophysical Union, Washington, DC, pp. 205–215.

Chamberlain, J.W. (1960). Interplanetary gas II. Expansion of a model solar corona. *Astrophysical Journal*, **131**, 47–56.

Chapman, S. (1957). Notes on the solar corona and the terrestrial ionosphere. *Smithsonian Contributions to Astrophysics*, **2**, 1–11.

Chapman, S. and Ferraro, V.C.A. (1931). A new theory of magnetic storms. *Terrestrial Magnetism and Atmospheric Electricity*, **36**, 77–97.

Coleman, P.J. Jr. (1968). Turbulence, viscosity, and dissipation in the solar wind plasma. *Astrophysical Journal*, **153**, 371–388.

Coplan, M.A., Ogilvie, K.W., Bochsler, P. and Geiss, J. (1978). Ion Composition Experiment. *IEEE Transactions on Geoscience Electronics*, **GE-16**, 185–191.

Dessler, A.J. and Fejer, J.A. (1963). Interpretation of Kp index and M-region geomagnetic storms. *Planetary and Space Science*, **11**, 505–511.

Feldman, W.C., Asbridge, J.R., Bame, S.J., Fenimore, E.E. and Gosling, J.T. (1981). The solar origins of solar wind interstream flows: Near-equatorial coronal streamers. *Journal of Geophysical Research*, **86**, 5408–5416.

Feldman, W.C., Asbridge, J.R., Bame, S.J. and Gosling, J.T. (1978a). Long-term variations of selected solar wind properties: Imp 6, 7, and 8 results. *Journal of Geophysical Research*, **83**, 2177–2189.

Feldman, W.C., Asbridge, J.R., Bame, S.J., Gosling, J.T. and Lemons, D.S. (1978b). Characteristic electron variations across simple high-speed solar wind streams. *Journal of Geophysical Research*, **83**, 5285–5295.

Feldman, W.C., Asbridge, J.R., Bame, S.J. and Montgomery, M.D. (1973). Double ion streams in the solar wind. *Journal of Geophysical Research*, **78**, 2017–2027.

Feldman, W.C., Asbridge, J.R., Bame, S.J., Montgomery, M.D. and Gary, S.P. (1975). Solar wind electrons. *Journal of Geophysical Research*, **80**, 4181–4196.

Feldman, W.C., Gosling, J.T., McComas, D.J. and Phillips, J.L. (1993). Evidence for ion jets in the high-speed solar wind. *Journal of Geophysical Research*, **98**, 5593–5605.

Feldman, W.C., Habbal, S.R., Hoogeveen, G. and Wang, Y.-M. (1997). Experimental constraints on pulsed and steady state models of the solar wind near the Sun. *Journal of Geophysical Research*, **102**, 26,905–26,918.

Feynman, J. and Martin, S.F. (1995). The initiation of coronal mass ejections by newly emerging magnetic flux. *Journal of Geophysical Research*, **100**, 3355–3367.

Fisk, L.A. (1996). Motion of the footpoints of heliospheric magnetic field lines at the Sun: Implications for recurrent energetic particle events at high heliographic latitudes. *Journal of Geophysical Research*, **101**, 15,547–15,553.

Fisk, L.A., Zurbuchen, T.H. and Schwadron, N.A. (1999). On the coronal magnetic field: Consequences of large-scale motions. *Astrophysical Journal*, **521**, 868–877.

Galvin, A.B. (1997). Minor ion composition in CME-related solar wind. In N. Crooker, J.A. Jocelyn and J. Feynman (eds), *Coronal Mass Ejections, Geophysical Monograph 99*, American Geophysical Union, Washington, DC, pp. 253–260.

Gary, S.P., Scime, E.E., Phillips, J.L. and Feldman, W.C. (1994). The whistler heat flux instability: Threshold conditions in the solar wind. *Journal of Geophysical Research*, **99**, 23,391–23,399.

Gazis, P.R., Barnes, A., Mihalov, J.D. and Lazarus, A.J. (1994). Solar wind velocity and temperature in the outer heliosphere. *Journal of Geophysical Research*, **99**, 6561–6573.

Geiss, J. (1973). Solar wind composition and implications about the history of the solar system. *Proceedings of the 13th International Cosmic Ray Conference*, **5**, 3375–3398.

Geiss, J. (1982). Processes affecting abundances in the solar wind. *Space Science Reviews*, **33**, 201–217.

Geiss, J., Eberhardt, P., Bühler, F., Meister, J. and Signer, P. (1970b). Apollo 11 and Apollo 12 solar wind composition experiments: Fluxes of He and Ne isotopes. *Journal of Geophysical Research*, **75**, 5972–5979.

Geiss, J. and Gloeckler, G. (1998). Abundances of deuterium and helium-3 in the protosolar cloud. *Space Science Reviews*, **84**, 239–250.

Geiss, J., Gloeckler, G. and von Steiger, R. (1995b). Origin of the solar wind from composition data. *Space Science Reviews*, **72**, 49–60.

Geiss, J., Gloeckler, G., von Steiger, R., Balsiger, H., Fisk, L.A., Galvin, A.B., Ipavich, F.M., Livi, S., McKenzie, J.F., Ogilvie, K.W. and Wilken, B. (1995a). The southern high-speed stream: Results from the SWICS instrument on Ulysses. *Science*, **268**, 1033–1036.

Geiss, J., Hirt, P. and Leutwyler, H. (1970a). On acceleration and motion of ions in corona and solar wind. *Solar Physics*, **12**, 458–483.

Giordano, S., Antonucci, E., Noci, G., Romoli, R. and Kohl, J.L. (2000). Identification of the coronal sources of the fast solar wind. *Astrophysical Journal*, **531**, L79–L82.

Gloeckler, G. (1990). Ion composition measurement techniques for space plasmas. *Review of Scientific Instruments*, **61**, 3613–3620.

Gloeckler, G., Fisk, L.A., Hefti, S., Schwadron, N.A., Zurbuchen, T.H., Ipavich, F.M., Geiss, J., Bochsler, P. and Wimmer-Schweingruber, R.F. (1999). Unusual composition of the solar wind in the 2–3 May 1998 CME observed with SWICS on ACE. *Geophysical Research Letters*, **26**, 157–160.

Gloeckler, G.L., Geiss, J., Balsiger, H., Bedini, P., Cain, J.C., Fischer, J., Fisk, L.A., Galvin, A.B., Gliem, F., Hamilton, D.C., Hollweg, J.V., Ipavich, F.M., Joos, R., Livi, S., Lundgren, R., Mall, U., McKenzie, J.F., Ogilvie, K.W., Ottens, F. and Rieck, W. (1992). The solar wind ion composition spectrometer. *Astronomy and Astrophysics Supplement*, **92**, 267–289.

Gold, T. (1959). Plasma and magnetic fields in the solar system. *Journal of Geophysical Research*, **64**, 1665–1674.

Gosling, J.T. (1993). The solar flare myth. *Journal of Geophysical Research.*, **98**, 18,937–18,949.

Gosling, J.T., Baker, D.N., Bame, S.J., Feldman, W.C., Zwickl, R.D. and Smith, E.J. (1987). Bidirectional solar wind electron heat flux events. *Journal of Geophysical Research*, **92**, 8519–8535.

Gosling, J.T., Bame, S.J., McComas, D.J., Phillips, J.L., Goldstein, B.E. and Neugebauer, M. (1994b). The speeds of coronal mass ejections in the solar wind at mid heliographic latitudes: Ulysses. *Geophysical Research Letters*, **21**, 1109–1112.

Gosling, J.T., Hildner, E., MacQueen, R.M., Poland, A.I. and Ross, C.L. (1974). Mass ejections from the sun: A view from Skylab. *Journal of Geophysical Research*, **79**, 4581–4587.

Gosling, J.T., Hundhausen, A.J. and Bame, S.J. (1976). Solar wind stream evolution at large heliocentric distances: Experimental demonstration and the test of a model. *Journal of Geophysical Research*, **81**, 2111–2122.

Gosling, J.T., McComas, D.J., Phillips, J.E., Weiss, L.A., Pizzo, V.J., Goldstein, B.E. and Forsyth, R.J. (1994a). A new class of forward–reverse shock pairs in the solar wind. *Geophysical Research Letters*, **21**, 2271–2274.

Gosling, J.T., McComas, D.J., Phillips, J.L. and Bame, S.J. (1992). Counterstreaming solar wind halo electron events: Solar cycle variations. *Journal of Geophysical Research*, **97**, 6531–6535.

Gosling, J.T., Pizzo, V. and Bame, S.J. (1973). Anomalously low proton temperatures in the solar wind following interplanetary shock waves – Evidence for magnetic bottles? *Journal of Geophysical Research*, **78**, 2001–2009.

Gosling, J.T. and Riley, P. (1996). The acceleration of slow coronal mass ejections in the high-speed solar wind. *Geophysical Reserach Letters*, **23**, 2867–2870.

Grevesse, N. and Sauval, A.J. (1998). Standard solar composition. *Space Science Reviews*, **85**, 161–174.

Gringauz, K.I., Bezrukikh, V.V., Ozerov, V.D. and Rybchinskii, R.E. (1960). A study of interplanetary ionized gas, energetic electrons, and solar corpuscular radiation using three-electrode charged particle traps on the second Soviet Cosmic Rocket. *Soviet Physics: Doklady*, **5**, 361–364.

Gurnett, D.A. (1991). Waves and Instabilities. In R. Schwenn and E. Marsch (eds), *Physics of the Inner Heliosphere. 2. Particles, Waves and Turbulence*, Springer-Verlag, Berlin, pp. 135–157.

Gurnett, D.A. and Kurth, W.S. (1996). Radio emissions from the outer heliosphere. *Space Science Reviews*, **78**, 53–66.

Hefti, S., Grünwaldt, H., Ipavich, F.M., Bochsler, P., Hovestadt, D., Aellig, M.R., Hilchenbach, M., Kallenbach, R., Galvin, A.B., Geiss, J., Gliem, F., Gloeckler, G., Klecker, B., Marsch, E., Möbius, E., Neugebauer, M. and Wurz, P. (1998). Kinetic properties of solar wind minor ions and protons measured with SOHO/CELIAS. *Journal of Geophysical Research*, **103**, 29,697–29,704.

Henke, T., Woch, J., Mall, U., Livi, S., Wilken, B., Schwenn, R., Gloeckler, G., von Steiger, R., Forsyth, R.J. and Balogh, A. (1998). Differences in the O^{7+}/O^{6+} ratio of magnetic cloud and non-cloud coronal mass ejections. *Geophysical Research Letters*, **25**, 3465–3468.

Hénoux, J.-C. (1998). FIP fractionation: Theory. *Space Science Reviews*, **85**, 215–226.

Hirshberg, J., Bame, S.J. and Robbins, D.E. (1972). Solar flares and solar wind helium enrichments: July 1965–July 1967. *Solar Physics*, **23**, 467–486.

Hovestadt, D. (1974). Nuclear composition of solar cosmic rays. In C. T. Russell (ed), *Solar Wind Three*. Edited by University of California, Los Angles, pp. 2–25.

Hovestadt, D., Gloeckler, G., Fan, C.Y., Fisk, L.A., Ipavich, F.M., Klecker, B., O'Gallagher, J.J., Scholer, M., Arbinger, H., Cain, J., Hofner, H., Kunneth, E., Laeverenz, P. and Tums, E. (1978). The nuclear and ionic charge distribution particle experiments on the ISEE-1 and ISEE-C spacecraft. *IEEE Transactions on Geoscience Electronics*, **GE-16**, 166–175.

Hovestadt, D., Hilchenbach, M., Bürgi, A., Klecker, B., Laeverenz, P., Scholer, M., Grünwaldt, H., Axford, W.I., Livi, S., Marsch, E., Wilken, B., Winterhoff, H.P., Ipavich, F.M., Bedini, P., Coplan, M.A., Galvin, A.B., Gloeckler, G., Bochsler, P., Balsiger, H., Fischer, J., Geiss, J., Kallenbach, R., Wurz, P., Reiche, K.-U., Gliem, F., Judge, D.L., Ogawa, H.S., Hsieh, K.C., Möbius, E., Lee, M.A., Managadze, G.G., Verigin, M.I. and Neugebauer, M. (1995). Celias – Charge, Element, and Isotope Analysis System for SOHO. *Solar Physics*, **162**, 441–481.

Howard, R.A., Brueckner, G.E., St. Cyr, O.C., Biesecker, D.A., Dere, K.P., Koomen, M.J., Korendyke, C.M., Lamy, P.L., Llebaria, A., Bout, M.V., Michels, D.J., Moses, J.D., Paswaters, S.E., Plunkett, S.P., Schwenn, R., Simnett, G.M., Socker, D.G., Tapin, S.J. and Wang, D. (1997). Observations of CMEs from SOHO/LASCO. In N. Crooker, J.A. Jocelyn and J. Feynman (eds), *Coronal Mass Ejections, Geophysical Monograph 99*, American Geophysical Union, Washington, DC, pp. 17–26.

Howard, R.A., Sheeley, N.R. Jr., Michels, D.J. and Koomen, M.J. (1985). Coronal mass ejections: 1979–1981. *Journal of Geophysical Research*, **90**, 8173–8191.

Hundhausen, A.J. (1972). *Coronal Expansion and Solar Wind*, Springer-Verlag, New York.

Hundhausen, A.J. (1988). The origin and propagation of coronal mass ejections. In V. J. Pizzo, T. E. Holzer and D. G. Sime (eds), *Proceedings of the Sixth International Solar Wind Conference*, NCAR/TN-306+Proc, National Center for Atmospheric Research, Boulder, CO, pp. 181–214.

Hundhausen, A.J., Bame, S.J. and Ness, N.F. (1967). Solar wind thermal anisotropies: Vela 3 and IMP 3. *Journal of Geophysical Research*, **72**, 5265–5274.

Ipavich, F.M., Galvin, A.B., Gloeckler, G., Hovestadt, D., Bame, S.J., Klecker, B., Scholer, M., Fisk, L.A. and Fan, C.Y. (1986). Solar wind Fe and CNO measurements in high-speed flows. *Journal of Geophysical Research*, **91**, 4133–4141.

Isenberg, P.A. and Hollweg, J.V. (1983). On the preferential acceleration and heating of the solar wind. *Journal of Geophysical Research*, **88**, 3923–3935.

Ko, Y.-K., Fisk, L.A., Geiss, J., Gloeckler, G. and Guhathakurta, M. (1997). An empirical study of the electron temperature and heavy ion velocities in the south polar coronal hole. *Solar Physics*, **171**, 345–361.

Kohl, J.L., Esser, R., Gardner, L.D., Habbal, S., Daigneau, P.S., Dennis, E.F., Nystrom, G.U., Panasyuk, A., Raymond, J.C., Smith, P.L., Strachan, L., van Ballegooijen, A.A., Noci, G., Fineschi, S., Romoli, M., Ciaravella, A., Modigliani, A., Huber, M.C.E., Antonucci, E., Benna, C., Giordano, S., Tondello, G., Nicolosi, P., Naletto, G., Pernechele, C., Spadaro, D., Poletto, G., Livi, S., von der Luhe, O., Geiss, J., Timothy, J.G., Gloeckler, G., Allegra, A., Basile, G., Brusa, R., Wood, B., Siegmund, O.H.W., Fowler, W., Fisher, R. and Jhabvala, M. (1995). The ultraviolet coronagraph spectrometer for the Solar and Heliospheric Observatory. *Solar Physics*, **162**, 313–356.

Kohl, J.L. Noci, G., Antonucci, E., Tondello, G., Huber, M.C.E., Cranmer, S.R., Strachan, L., Panasyuk, A.V., Gardner, L.D., Romoli, M., Fineschi, S., Dobrzycka, D., Raymond, J.C., Nicolosi, P., Siegmund, O.H.W., Spadaro, D., Benna, C., Ciaravella, A., Giordano, S., Habbal, S.R., Karovska, M., Li, X., Martin, R., Michels, J.G., Modigliani, A., Naletto, G., O'Neal, R.H., Pernechele, C., Poletto, G., Smith, P.L. and Suleiman, R.M. (1998). UVCS/SOHO empirical determination of anisotropic velocity distributions in the solar corona. *Astrophysical Journal*, **501**, L127–L131.

Kohl, J.L., Noci, G., Antonucci, E., Tondello, G., Huber, M.C.E., Gardner, L.D., Nicolosi, P., Strachan, L., Fineschi, S., Raymond, J.C., Romoli, M., Spadaro, S., Panasyuk, A., Siegmund, O.H.W., Benna, C., Ciaravella, A., Cranmer, S.R., Giordano, S., Karovska, M. and Martin, R. (1997). First results from the SOHO ultraviolet coronagraph spectrometer. *Solar Physics*, **175**, 613–644.

Krieger, A.S., Timothy, A.F. and Roelof, E.C. (1973). A coronal hole as the source of a high velocity solar wind stream. *Solar Physics*, **29**, 505–525.

Kunz, S., Bochsler, P., Geiss, J., Ogilvie, K.W. and Coplan, M.A. (1983). Determination of solar wind elemental abundances from M/Q observations during three periods in 1980. *Solar Physics*, **29**, 359–376.

Lee, L.C. and Wu, B.H. (2000). Heating and acceleration of protons and minor ions by fast shocks in the solar corona. *Astrophysical Journal*, **535**, 1014–1026.

Leer, E. and Holzer, T.E. (1980). Energy addition in the solar wind. *Journal of Geophysical Research*, **85**, 4681–4688.

Lepping, R.P., Jones, J.A. and Burlaga, L.F. (1990). Magnetic field structure of interplanetary magnetic clouds at 1 AU. *Journal of Geophysical Research*, **95**, 11,957–11,965.

Liu, S., Marsch, E., Livi, S., Woch, J., Wilken, B., von Steiger, R. and Gloeckler, G. (1995). Radial gradients of ion densities and temperatures derived from SWICS/Ulysses observations. *Geophysical Research Letters*, **22**, 2445–2448.

MacQueen, R.M., Eddy, J.A., Gosling, J.T., Hildner, E., Munro, R.H., Newkirk, G.A. Jr., Poland, A.I. and Ross, C.L. (1974). The outer solar corona as observed from Skylab: Preliminary results. *Astrophysical Journal*, **187**, L85–L88.

Marsch, E., Mühlhäuser, K.-H., Rosenbauer, H., Schwenn, R. and Neubauer, F.M. (1982b). Solar wind helium ions: Observations of the Helios solar probes between 0.3 and 1 AU. *Journal of Geophysical Research*, **87**, 35–51.

Marsch, E., Mühlhäuser, K.-H., Schwenn, R., Rosenbauer, H., Pilipp, W. and Neubauer, F.M. (1982a). Solar wind protons: Three-dimensional velocity distributions and derived plasma parameters measured between 0.3 and 1 AU. *Journal of Geophysical Research*, **87**, 52–72.

McComas, D.J., Bame, S.J., Barraclough, B.L., Feldman, W.C., Funsten, H.O., Gosling, J.T., Riley, P., Skoug, R., Balogh, A., Forsyth, R., Goldstein, B.E. and Neugebauer, M. (1998). Ulysses' return to the slow solar wind. *Geophysical Research Letters*, **25**, 1–4.

McComas, D.J., Barraclough, B.L., Funsten, H.O., Gosling, J.T., Santiago-Muñoz, E., Skoug, R.M., Goldstein, B.E., Neugebauer, M., Riley, P. and Balogh, A. (2000). Solar wind observations over Ulysses' first full polar orbit. *Journal of Geophysical Research*, **105**, 10,419–10,434.

McDonald, F.B. and Burlaga, L.F. (1997). Global merged interaction regions. In J. R. Jokipii, C. P. Sonett and M. S. Giampapa (eds), *Cosmic Winds and the Heliosphere*, University of Arizona Press, Tucson, AZ, pp. 581–616.

McKenzie, J.F., Banaszkiewicz, B. and Axford, W.I. (1995). Acceleration of the high speed solar wind. *Astronomy and Astrophysics*, **303**, L45–L48.

Meyer, J.-P. (1981). A tentative ordering of all available solar energetic particle abundance observations. *Proceedings of the International Cosmic Ray Conference*, **17**(3), 145–152.

Mikic, Z. and Linker, J.A. (1997). The initiation of coronal mass ejections by magnetic shear. In N. Crooker, J.A. Jocelyn and J. Feynman (eds), *Coronal Mass Ejections, Geophysical Monograph 99*, American Geophysical Union, Washington, DC, pp. 57–64.

Montgomery, M.D. (1972). Average thermal characteristics of solar wind electrons. In C. P. Sonett, P. J. Coleman Jr. and J. M. Wilcox (eds), *Solar Wind*, NASA SP-308, Washington, DC, pp. 208–218.

Montgomery, M.D., Asbridge, J.R., Bame, S.J. and Feldman, W.C. (1974). Solar wind electron temperature depressions following some interplanetary shock waves: Evidence for magnetic merging? *Journal of Geophysical Research*, **79**, 3103–3110.

Montgomery, M.D., Bame, S.J. and Hundhausen, A.J. (1968). Solar wind electrons: Vela 4 measurements. *Journal of Geophysical Research*, **73**, 4999–5003.

Neugebauer, M. (1981). Observations of solar wind helium. *Fundamentals of Cosmic Physics*, **7**, 131–199.

Neugebauer, M. (1997). Pioneers of space physics: A career in the solar wind. *Journal of Geophysical Research*, **102**, 26,887–26,894.

Neugebauer, M., Forsyth, R.J., Galvin, A.B., Harvey, K.L., Hoeksema, J.T., Lazarus, A.J., Lepping, R.P., Linker, J.A., Mikic, Z., Steinberg, J.T., von Steiger, R., Wang, Y.-M. and Wimmer-Schweingruber, R.F. (1998). The spatial structure of the solar wind and comparisons with solar data and models. *Journal of Geophysical Research*, **103**, 14,587–14,599.

Neugebauer, M., Goldstein, B.E., Smith, E.J. and Feldman, W.C. (1996). Ulysses observations of differential alpha-proton streaming in the solar wind. *Journal of Geophysical Research*, **101**, 17,047–17,055.

Neugebauer, M. and Snyder, C.W. (1966). Mariner 2 observations of the solar wind, 1: Average properties. *Journal of Geophysical Research*, **71**, 4469–4484.

Ogilvie, K.W., Geiss, J., Gloeckler, G., Berdichevsky, D. and Wilken, B. (1993). High-velocity tails on the velocity distribution of solar wind ions. *Journal of Geophysical Research*, **98**, 3605–3611.

Ogilvie, K.W., Scudder, J.D. and Sugiura, M. (1971). Electron energy flux in the solar wind. *Journal of Geophysical Research*, **76**, 8165–8173.

Ogilvie, K.W. and Wilkerson, T.D. (1969). Helium abundance in the solar wind. *Solar Physics*, **8**, 435–449.

Parker, E.N. (1958). Dynamics of the interplanetary gas and magnetic fields. *Astrophysical Journal*, **128**, 664–676.

Parker, E.N. (1987). Magnetic reorientation and the spontaneous formation of tangential discontinuities in deformed magnetic fields. *Astrophysical Journal*, **318**, 876–887.

Parker, E.N. (1990). Intrinsic magnetic discontinuities and solar x-ray emission. *Geophysical Research Letters*, **17**, 2055–2058.

Pilipp, W.G., Miggenrieder, H., Montgomery, M.D., Mühlhäuser, K.-H., Rosenbauer, H. and Schwenn, R. (1987a). Characteristics of electron velocity distribution functions in the solar wind derived from the Helios plasma experiment. *Journal of Geophysical Research*, **92**, 1075–1092.

Pilipp, W.G., Miggenrieder, H., Montgomery, M.D., Mühlhäuser, K.-H., Rosenbauer, H. and Schwenn, R. (1987b). Variations of electron distribution functions in the solar wind. *Journal of Geophysical Research*, **92**, 1103–1118.

Pilipp, W.G., Miggenrieder, H., Montgomery, M.D., Mühlhäuser, K.-H., Rosenbauer, H. and Schwenn, R. (1987c). Unusual electron distribution functions in the solar wind derived from the Helios plasma experiment: Double-strahl distributions and distributions with an extremely anisotropic core. *Journal of Geophysical Research*, **92**, 1093–1101.

Raymond, J.C., Kohl, J.L., Noci, G., Antonucci, E., Tondello, G., Huber, M.C.E., Gardner, L.D., Nicolosi, P., Fineschi, S., Romoli, M., Spadaro, D., Siegmund, O.H.W., Benna, C., Ciaravella, A., Cranmer, S., Giordano, S., Karovska, M., Martin, R., Michels, J., Modigliani, A., Naletto, G., Panasyuk, A., Pernechele, C., Poletto, G., Smith, P.L., Suleiman, R.M. and Strachan, L. (1997). Composition of coronal streamers from the SOHO ultraviolet coronagraph spectrometer. *Solar Phys.*, **175**, 645–665.

Robbins, D.E., Hundhausen, A.J. and Bame, S.J. (1970). Helium in the solar wind. *Journal of Geophysical Research*, **75**, 1178–1187.

Rosenbauer, H., Schwenn, R., Marsch, E., Meyer, B., Miggenrieder, H., Montgomery, M.D., Mühlhäuser, K.H., Pilipp, W., Voges, W. and Zink, S.M. (1977). A survey on initial results of the Helios plasma experiment. *J. Geophys.*, **42**, 561–580.

Schmid, J., Bochsler, P. and Geiss, J. (1987). Velocity of iron ions in the solar wind. *Journal of Geophysical Research*, **92**, 9901–9906.

Schwadron, N.A., Fisk, L.A. and Zurbuchen, T.H. (1999). Elemental fractionation in the slow solar wind. *Astrophysical Journal*, **521**, 859–867.

Schwenn, R. (1990). Large-scale structure of the interplanetary medium. In: R. Schwenn and E. Marsch (eds), *Physics of the Inner Heliosphere. 1. Large-Scale Phenomena*, Springer-Verlag, Berlin, pp. 99–181.

Scime, E.E., Bame, S.J., Feldman, W.C., Gary, S.P., Phillips, J.L. and Balogh, A. (1994). Regulation of the solar wind electron heat flux from 1 to 5 AU: Ulysses observations. *Journal of Geophysical Research*, **99**, 23,401–23,410.

Scime, E.E., Gary, S.P., Phillips, J.L., Balogh, A. and Lengyel-Frey, D. (1996). Electron energy transport in the solar wind: Ulysses observations. In D. Winterhalter, J. T. Gosling, S. R. Habbal, W. S. Kurth and M. Neugebauer (eds), *Solar Wind 8*, American Institute of Physics, Woodbury, NY, pp. 210–213.

Sheeley, N.R. Jr., Howard, R.A., Koomen, M.J., Michels, D.J., Schwenn, R., Mühlhäuser, K.H. and Rosenbauer, H. (1985). Coronal mass ejections and interplanetary shocks. *Journal of Geophysical Research*, **90**, 163–175.

Sheeley, N.R. Jr., Wang, Y.-M., Hawley, S.H., Brueckner, G.E., Dere, K.P., Howard, R.A., Koomen, M.J., Korendyke, C.M., Michels, D.J., Paswaters, S.E., Socker, D.G., Cyr, O.C.S., Wang, D., Lamy, P.L., Llebaria, A., Schwenn, R., Simnett, G.M., Plunkett, S. and Biesecker, D.A. (1997). Measurements of flow speeds in the corona between 2 and 30 Rs. *Astrophysical Journal*, **484**, 472–478.

Simpson, J.A. (1998). A brief history of recurrent solar modulation of the galactic cosmic rays (1937–1990). *Space Science Reviews*, **83**, 169–176.

Siscoe, G.L., Davis, L. Jr., Coleman, P.J. Jr., Smith, E.J. and Jones, D.E. (1968). Power spectra and discontinuities in the interplanetary magnetic field: Mariner 4. *Journal of Geophysical Research*, **73**, 61–82.

Smith, E.J. (1964). Interplanetary magnetic fields. In D. P. LeGalley and A. Rosen *Space Physics*, Wiley, New York, pp. 350–396.

Smith, E.J., Tsurutani, B.T. and Rosenberg, R.L. (1978). Observations of the interplanetary sector structure up to heliographic latitudes of 16 degrees, Pioneer 11. *Journal of Geophysical Research*, **83**, 717–724.

Smith, E.J. and Wolfe, J.H. (1976). Observations of interaction regions and corotating shocks between one and five AU; Pioneers 10 and 11. *Geophys. Res. Lett.*, **3**, 137–140.

Snyder, C.W. and Neugebauer, M. (1966). The relation of Mariner-2 plasma data to solar phenomena. In: R. J. Mackin and M. Neugebauer (eds), *The Solar Wind*, Pergamon Press, New York, pp. 25–32.

Sonett, C.P., Colburn, D.S., Davis, L. Jr., Smith, E.J. and Coleman, P.J. Jr. (1964). Evidence for a collisionfree magnetohydrodynamic shock in interplanetary space. *Phys. Rev. Let.*, **13**, 153–160.

Sturrock, P.A. and Hartle, R.E. (1966). Two-fluid model of the solar wind. *Phys. Rev. Let.*, **16**, 628–631.

Tsurutani, B.T. (1991). Cometary plasma waves and instabilities. In: R. L. Jr. Newburn, M. Neugebauer and J. Rahe (eds), *Comets in the Post-Halley Era*, Kluwer, Dordrecht, pp. 1171–1210.

Tu, C.-Y. (1988). The damping of interplanetary Alfvénic fluctuations and the heating of the solar wind. *Journal of Geophysical Research*, **93**, 7–20.

Unti, T.W.J. and Neugebauer, M. (1968). Alfvén waves in the solar wind. *Physics of Fluids*, **11**, 563–568.

Vellante, M. and Lazarus, A.J. (1987). An analysis of solar wind fluctuations between 1 and 10 AU. *Journal of Geophysical Research*, **92**, 9893–9900.

von Steiger, R., Geiss, J. and Gloeckler, G. (1997). Composition of the solar wind. In: J. R. Jokipii, C. P. Sonett and M. S. Giampapa (eds), *Cosmic Winds and the Heliosphere*, University of Arizona Press, Tucson, AZ, pp. 581–616.

von Steiger, R., Schwadron, N.A., Fisk, L.A., Geiss, J., Gloeckler, G., Hefti, S., Wilken, B., Wimmer-Schweingruber, R.F. and Zurbuchen, T.H. (2000). Composition of quasi-stationary solar wind flows from Ulysses/SWICS. *Journal of Geophysical Research*, **105**, 27217–27238.

von Steiger, R., Wimmer-Schweingruber, R., Geiss, J. and Gloeckler, G. (1995). Abundance variations in the solar wind. *Advances in Space Research*, **15**(7), 3–12.

Wang, Y.-M. and Sheeley, N.R. Jr. (1990). Solar wind speed and coronal flux-tube expansion. *Astrophysical Journal*, **355**, 726–732.

Wenzel, K.-P., Marsden, R.G., Page, D.E. and Smith, E.J. (1992). The Ulysses Mission. *Astronomy and Astrophysics Supplement Series*, **92**, 207–219.

Widing, K.G. and Feldman, U. (1992). Elemental abundances and their variations in the upper solar atmosphere. In: E. Marsch and R. Schwenn (eds), *Solar Wind Seven*, Pergamon Press, Oxford, pp. 405–410.

Wieler, R., Kehm, K., Hohenberg, C.M. and Meshik, A.P. (1996). Secular changes in the xenon and krypton abundances in the solar wind recorded in single lunar grains. *Nature*, **384**, 46–49.

Wilcox, J.M. and Ness, N.F. (1965). Quasi-stationary corotating structure in the interplanetary medium. *Journal of Geophysical Research*, **70**, 5793–5805.

Wimmer-Schweingruber, R.F., Bochsler, P. and Wurz, P. (1999). Isotopes in the solar wind: New results from ACE, SOHO, and Wind. In: S. R. Habbal, R. Esser, J. V. Hollweg and P. A. Isenberg (eds), *Solar Wind Nine*, AIP Conf. Proc. 471, American Institute of Physics, Woodbury, NY, pp. 147–152.

Wimmer-Schweingruber, R.F., von Steiger, R. and Paerli, R. (1997). Solar wind stream interfaces in corotating interaction regions. *Journal of Geophysical Research*, **102**, 17,407–17,417.

Wolfe, J.H. and Silva, R.W. (1965). Explorer 14 plasma probe observations during the October 7, 1962, geomagnetic disturbance. *Journal of Geophysical Research*, **70**, 3575–3579.

48

ANDRÉ BALOGH* AND LENNARD A. FISK**

The heliosphere

The heliosphere is the vast region of space surrounding the Sun and the Solar System that is filled with the solar wind. Its inner boundary is the outer atmosphere of the Sun, the solar corona. Its outer boundary is defined, to a first approximation, as a surface across which there is a balance of pressure between the solar wind and the local interstellar medium (LISM). This outer boundary is to be found at probably about two and a half to three times the orbital distance of the furthest planet, Pluto. The main regions of the heliosphere, described in this chapter, are illustrated schematically in Figure 1, although there are many uncertainties concerning this simple picture. The objective of this chapter is to provide an overview of the heliosphere as understood, at the beginning of the twenty-first century, complete with our very comprehensive understanding of the solar wind and its large-scale dynamic structures that shape the heliosphere. Together with the development of solar wind studies, the study of the propagation of cosmic rays in the heliosphere has also been of great importance in shaping our understanding. These topics are reviewed at some length. The space missions that particularly contributed to the exploration of the heliosphere are also summarized.

The many topics related to the heliosphere have been the subject of several books, as well as numerous review articles. Of particular interest and relevance are the books edited by Schwenn and Marsch (1990, 1991) on the results of the Helios mission in the inner heliosphere; four books related to the three-dimensional heliosphere and the Ulysses mission (Marsden 1986, 1995, 2001; Balogh *et al.* 2001); a volume on cosmic winds and the heliosphere (Jokipii *et al.* 1997); three volumes published under the imprint of the International Space Science Institute, Bern, on the interstellar medium and the heliosphere (von Steiger *et al.* 1996),

on cosmic rays in the heliosphere (Fisk *et al.* 1998), and on corotating interaction regions (Balogh *et al.* 1999). A volume that gathered his many contributions to heliospheric physics was published by Burlaga (1995). The outer heliosphere has been the subject of a volume edited by Grzedzielski and Page (1990). In addition, there have been, over the past 15 years, numerous review papers on aspects of the heliosphere; these, together with many original references are listed in this chapter.

THE EXISTENCE OF THE HELIOSPHERE AND ITS MAIN PROPERTIES

The existence of the heliosphere, as a volume of space controlled by the Sun, was first proposed by Davis (1955), who based this concept on the modulation of cosmic rays in anti-phase with the solar activity cycle. Cosmic rays are very high-energy particles which are accelerated (by vast shockwaves generated by supernova explosions) in the Galaxy, at considerable distances from the Solar System. The intensity of cosmic rays observed at the Earth is nearly constant, but is a few percent higher at sunspot minimum than at sunspot maximum. This 11-year modulation cycle can only be explained by assuming that the sunspot cycle has an effect on the space surrounding the Sun which, in turn, affects the propagation of cosmic rays into the inner Solar System, so that it is easier for cosmic rays to reach the Earth near solar minimum than around solar maximum. This explanation therefore implies that the medium surrounding the Sun undergoes changes between solar maximum and minimum. Given that these changes affect the propagation of high-energy cosmic rays (which are electrically charged particles), the changes in the medium around the Sun need to involve magnetic fields and, in effect, changes in the magnetic field itself, in response to the solar

* Imperial College, London, United Kingdom
** University of Michigan, Ann Arbor, MI, USA

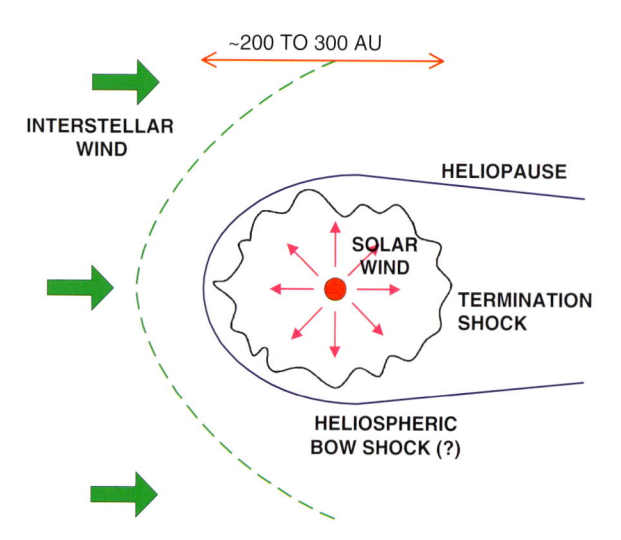

Figure 1 A schematic view of the heliosphere and its boundaries.

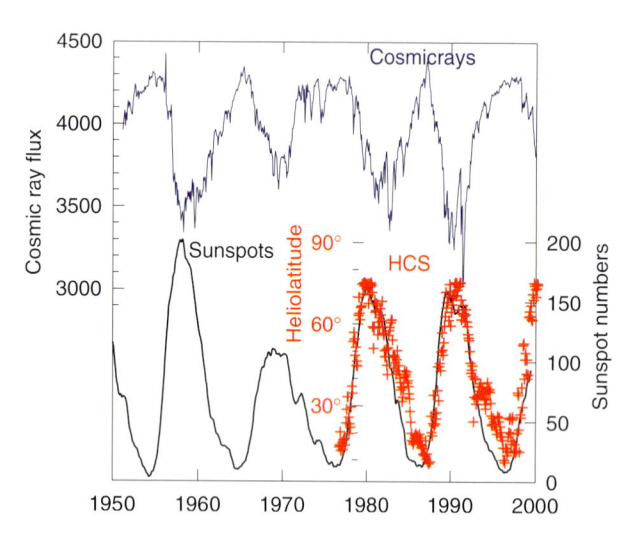

Figure 2 A composite plot showing the monthly number of sunspots (black trace), the daily average intensity of cosmic rays measured on the ground (blue trace) and the maximum heliolatitude of the coronal magnetic neutral line, representing the coronal imprint of the HCS (red symbols). There is clearly a very close correspondence between sunspots and the extension of the coronal magnetic neutral line; the anticorrelation between cosmic-ray intensity and sunspot numbers is indicative of the modulation of galactic cosmic rays by solar activity in the heliosphere.

shown. The historical background to the early cosmic-ray modulation studies has been reviewed recently by Simpson (1998) and McDonald (2000). It is remarkable that the recognition of the relationship came so early, in the mid-1950s, even before it had been observed for a complete sunspot cycle. There is also a third quantity plotted in Figure 2; this quantity, labelled HCS in the figure, is the calculated maximum heliolatitude of the heliospheric current sheet, in effect the physical extension of the Sun's magnetic equator into the heliosphere. The nature and importance of the HCS in structuring the heliosphere are described below. However, its is clear from the close match of this quantity to the sunspot numbers (once the two are superimposed, as in the figure) that the variation in the heliolatitude of the Sun's magnetic equator through the solar cycle is an important indicator of solar activity and, given the need for a physical description of the solar modulation of cosmic rays, the extension of the magnetic equator into the heliosphere is at least a part of the physical mechanism of the modulation process.

The necessary advance in understanding the magnetized medium surrounding the Sun assumed by Davis (1955) came about at the same time, from the recognition that the Sun continually emits a 'corpuscular' radiation. The existence of such a radiation, to explain the geomagnetic effects that could be related to activity on the Sun, had been assumed much earlier by Chapman (see Chapman and Bartels 1940, and references therein). The existence of the solar wind, as a constant outflow of plasma from the solar corona was first suggested by Biermann (see Biermann, 1957, and references therein) from the study of the ion tails of comets. The current understanding of the reason for the existence of the solar wind as a permanent outflow of supersonic plasma originated with Parker's theoretical model (Parker 1958, 1963). Within four years of Parker's first theoretical predictions, the existence of the solar wind was confirmed by the early interplanetary space missions (for a summary of the early results, see e.g. Lüst 1967). This confirmation (that included the charting of the interplanetary magnetic fields necessary for affecting the propagation of cosmic rays) finally placed on a firm foundation the concept of the heliosphere (e.g. Axford 1972) and opened the way for its exploration in the space age, from the early 1960s.

The phenomenology of the heliosphere covers numerous space plasma phenomena on all scales, from the kinetic plasma scale where the dynamics of particle distributions dominate the physical processes, to the large magnetohydrodynamic (MHD) scales where the large-scale dynamics of the solar wind and the magnetic field define the global properties of the heliosphere. Space plasma processes on many intermediate scales provide a constantly changing, dynamic environment in the different regions of the heliosphere. The relative accessibility of the heliosphere to direct observations by deep-space probes has made it the

activity cycle. Given also the large energies of cosmic-ray particles that are affected, the volume involved in the modulation also has to be large on the scale of interplanetary distances.

It is this relationship between cosmic-ray fluxes and sunspots that is illustrated in Figure 2, in which 50 years of observations, covering close to five solar cycles, are largest astrophysical plasma regime that can be studied in detail. Heliospheric physics is now a discipline in its own right, with obvious and important links to solar physics, space plasma physics and, by extrapolation, with many topics in astrophysics in general.

The size of the heliosphere is not yet known precisely. The outer boundary is currently estimated to be somewhere between 80 and 120 AU from the Sun. Both the size of the heliosphere and the nature of the outer boundary depend on the properties of the solar wind at large distances from the Sun, on the properties of the LISM and on the physical processes involved in their interaction. Even though these properties are not known in sufficient detail to draw definitive conclusions on either the size of the heliosphere or the nature of the boundary, there are several estimates based on indirect observations which, surprisingly, agree quite well with each other. Such indirect evidence has also made it possible to introduce useful models that can be tested as further observations become available.

A great deal is known about the heliosphere from the Sun out to the distance of the furthest direct, *in situ* observations, currently to 74 AU by the Voyager 2 spacecraft. The *in situ* observations, made since the early 1960s by deep-space missions as well as by the continuous monitoring of the solar wind in the vicinity of the Earth, have provided a vast, even if in some respects selective database for determining the structure and dynamics of the inner and middle heliosphere. At the same time, progress in understanding the physics of the solar corona, the inner boundary of the heliosphere, has led to a good understanding of the important relationships between solar and heliospheric phenomena. Space missions have also played the major role in increased understanding of the complex dynamics of solar and coronal phenomena relevant to heliospheric physics.

The role of space missions in exploring the heliosphere has led to phenomenological descriptions that emphasize the dependence of heliospheric phenomena on distance from the Sun. Although any division of the heliosphere into different regions is somewhat arbitrary, the inner, the middle and the outer heliospheric regions have been distinguished, based mostly on the dominant dynamic features observed.

The inner heliosphere is the region from the solar corona to about the orbit of the Earth. In this region the coronal sources of the different solar wind streams recognizably dominate the dynamics of the medium. The middle heliosphere, from about the Earth's orbit to Saturn's orbit at 10 AU, is the region where the dynamic evolution of solar wind structures forms large-scale structures that begin to mask the solar origin of the different solar wind streams. In the outer heliosphere, the structures formed in the middle heliosphere continue to evolve, but dissipative processes and the intrusion of material from the LISM make the link between the observed structures and their solar origin recognizable only on the largest temporal and spatial scales.

There is, however, another equally important way to divide the heliosphere. For as much as five or six years around solar minimum in each solar cycle, the polar regions of the heliosphere have significantly different properties from the equatorial region. Relatively uniform, high-speed solar wind fills both polar regions, extending down to between 20 or 30 degrees within the equator. In the equatorial region, fast and slow solar wind streams intermingle and interact, making this region more structured and dynamic: it is this region which can be conveniently subdivided into the three regions, inner, middle and outer heliospheric regions described above. Due to the absence of direct observations in the polar regions at different heliocentric distances, its structure and evolution with heliocentric distance can only be extrapolated from the unique set of observations made at high heliolatitudes by the Ulysses spacecraft at distances of about 2 AU.

For several years in each 11-year solar activity cycle, around maximum activity, the heliosphere becomes considerably more complex than around solar minimum. This is due in part to the fragmentation of solar wind streams and in part to the considerable increase in the occurrence rate and intensity of solar transients, in particular coronal mass ejections (CMEs). Instead of the relatively simple distribution of the solar wind sources in the corona at solar minimum when they are divided into large-scale coronal holes emitting fast solar wind and the equatorial belt of slow solar wind, non-uniform but mostly slow solar wind is emitted from the whole corona at solar maximum. Short-lived, small-scale coronal holes still emit fast solar wind, but at lower speeds than wind from the large polar coronal holes near solar minimum. Superimposed on the slow solar wind from the corona, CMEs occur not only more frequently than at solar minimum, but are distributed more or less evenly at all latitudes. Until Ulysses explores the polar regions of the heliosphere during the current solar maximum activity period in 2000–02, it is possible only to extrapolate the near-ecliptic conditions to those regions.

The terms used in this chapter, referring to the heliospheric medium and the heliospheric magnetic field, intend to underline the three-dimensional nature of the heliosphere and its phenomena. Historically, the terms 'interplanetary medium' and 'interplanetary magnetic field' have been extensively used to describe phenomena in the space between the orbits of the planets. This use of the terms was justified as most of the space missions that made observations in the solar wind remained in orbits restricted to close to the ecliptic plane, the Earth's orbital plane. The main impetus for change came from the Ulysses mission, described below, with its objective to explore the three-dimensional properties of the heliosphere in its unique, polar orbit around the Sun. However, other missions, Pioneer 11 first, when in transit between Jupiter and Saturn, and, later on, the two Voyager spacecraft left the ecliptic plane, following their final planetary encounters, reaching now far beyond the furthest planets in the Solar System. It has become therefore more appropriate to use the "heliospheric" terminology when describing phenomena in the three-dimensional volume around the Sun.

THE EXPLORATION OF THE HELIOSPHERE

Space probes have played a key role in the exploration of the heliosphere. Since the early 1960s, numerous missions have explored different regions in the vicinity of the Earth; in the distant, outer heliosphere; between the Sun and the Earth; and above the poles of the Sun. In the following, a few key missions that contributed in a significant way to our knowledge of the properties and structures of the heliospheric medium are described, highlighting their contributions to our knowledge of the heliosphere.

The Soviet space missions to the Moon (Lunik 1 and 2) carried plasma detectors that first observed the solar wind. However, the first extensive set of data confirming the existence of the solar wind and its main properties, such as its velocity, density, temperature, as well as its variability came from the Mariner 2 mission to Venus in 1962, launched on 27 August 1962 (Neugebauer and Snyder 1962). Some of the properties of the embedded magnetic field were also observed on this mission (Coleman *et al.* 1962), in particular the general agreement of the orientation of the magnetic field with Parker's predicted spiral geometry (Davis *et al.* 1966).

In the 1960s, early Earth-orbiting spacecraft, with apogees reaching beyond the magnetosphere, provided further evidence concerning many of the basic properties of the solar wind and the magnetic field embedded in it. All the most important phenomena that shape and characterize the heliospheric medium were first noted by Earth-orbiting spacecraft. In particular, good agreement was found in general between the Archimedean spiral structure of the magnetic field lines proposed by Parker (1958, 1963) and the observed orientation of the magnetic field (Davis *et al.* 1966; Ness and Wilcox 1964) on the Interplanetary Monitoring Platform (IMP 1) mission. On the same mission, the sector structure of the magnetic field, showing a recurring pattern of alternating polarities of the field as a function of solar longitude, was also identified (Wilcox and Ness 1964). A later spacecraft in the same series, the remarkable IMP 8, launched in 1974, has continued to provide observations in the solar wind for nearly 30 years. These observations now constitute a historic data set that is used for studying long-term trends in solar wind and cosmic ray parameters.

The first space mission that provided a constant monitoring of the solar wind, the heliospheric magnetic field, as well as energetic charged particles and cosmic rays in the vicinity of the Earth, yet removed from the influence of terrestrial effects, was the International Sun-Earth Explorer (ISEE 3) mission. This was the first spacecraft to be launched to orbit one of the Sun–Earth system's gravitational 'neutral' points, named after Lagrange. The Lagrange point L1 is situated at about 1% of the distance (or about 1,500,000 km) between the Sun and the Earth, ahead of the Earth. An orbit around this point in space allows the spacecraft to monitor the solar wind, and even to provide a warning of 30 to 60 minutes if a solar storm (in effect a CME) is about to hit the Earth's magnetosphere. The ISEE 3 spacecraft, launched on 12 August 1978, remained in this orbit for four years, through the solar maximum in 1979–80, before becoming the prototype of space wanderers by being retargeted first to the Earth's distant magnetospheric tail and then, following some spectacular manoeuvres, to the first spacecraft encounter with a comet in 1985.

The L1 orbit, first used by ISEE 3, has been used very successfully, more recently (since the mid-1990s) by the Solar and Heliospheric Observatory (SOHO) and the Advanced Charge Composition (ACE) spacecraft. SOHO was launched to provide a comprehensive and often spectacular monitoring of the Sun and its corona, and ACE is following up, two solar cycles after ISEE 3, and with a more up-to-date instrumentation, the monitoring of the solar wind and its magnetic field.

The Pioneer series of missions, from the early 1960s, have explored the region of space between the planets Mars and Venus, as well as the distant heliosphere past Jupiter and Saturn. The first spacecraft to the outer planets were the Pioneer 10 and 11 missions. Pioneer 10 was launched on 2 March 1972; it reached Jupiter on 3 December 1973 to make the first close-up observations of the giant planet. After its encounter with Jupiter, the spacecraft followed a trajectory to the outer reaches of the heliosphere. Pioneer 11 was launched on 5 April 1973 and it reached Jupiter on 2 December 1974. At Jupiter, the spacecraft was targeted in such a way that the flyby would allow it to reach Saturn. This involved a trajectory that, for the first time, reached a heliolatitude of 16° above the ecliptic plane. The two missions were terminated in March 1997 and November 1995, respectively. At the time when the missions ended, Pioneer 10 was at a heliocentric distance of 67 AU, while Pioneer 11 was at 43 AU.

These two spacecraft were the first to make detailed observations of the heliospheric medium out to Saturn and beyond. The first phase of the Pioneer mission took place around the minimum activity period in solar cycle 21, in the mid-1970s. They were the first to observe the recurring sequence of corotating interaction regions (CIRs) and their development with heliocentric distance (Smith and Wolfe 1977); these large-scale structures that consist of successively compressed and rarified solar wind and magnetic field are the most important features of the inner and middle heliosphere around solar minimum.

The inner heliosphere, between the Sun and the Earth's orbit, was the target of the joint German/NASA Helios mission. The two Helios spacecraft were launched on 10 December 1974 and 15 January 1976, respectively, into high-eccentricity heliocentric orbits with a perihelion of 0.3 AU and an aphelion of 1 AU. A few consecutive orbits of the two Helios spacecraft are shown in Figure 3, in a coordinate system fixed with respect to the line joining the Sun and the Earth. These two spacecraft collected data on heliospheric processes in the region between the Sun and the Earth, in particular on the early evolution of heliospheric structures as a function of distance from the Sun (Schwenn and Marsch 1990, 1991; Musmann *et al.* 1977).

The Voyager 1 and 2 spacecraft, launched in September 1977 and August 1977, respectively, still provide the only opportunity in the next decade or two to reach the outer boundary of the heliosphere. Since their launch, the two spacecraft visited Jupiter and Saturn, and Voyager 2 also flew by the two outer gas giants, Uranus and Neptune. Both spacecraft are now on trajectories that approach the outer boundaries of the heliosphere: Voyager 1 at a speed of about $3.5\,\text{AU}\,\text{yr}^{-1}$, Voyager 2 at about $3.1\,\text{AU}\,\text{yr}^{-1}$. The best chance to reach and detect at least the termination shock of the heliosphere rests with these two spacecraft. The orbits of the two Voyager spacecraft are illustrated in Figure 4 (together with the trajectories of the Pioneer spacecraft to the end of their mission). The race is clearly on, and there are good arguments (see section on 'The boundaries of the heliosphere') that show that the termination shock may well be reached before the fuel needed to stabilize the spacecraft runs out. This new phase in the long journey away from the Sun has been called the Voyager Interstellar Mission; this implies the possibility that the Voyager spacecraft will be

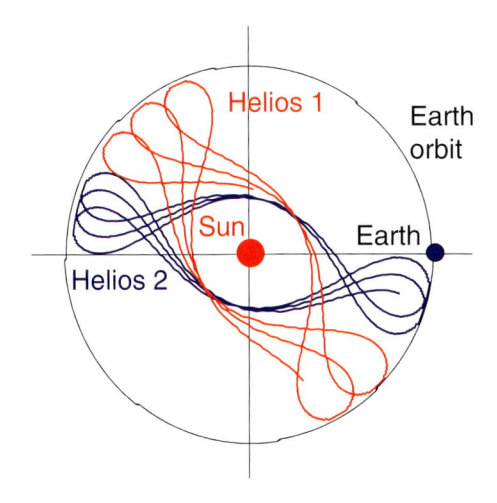

Figure 3 Orbits of the two Helios space probes in the inner heliosphere, in a coordinate system fixed with respect the Sun–Earth line.

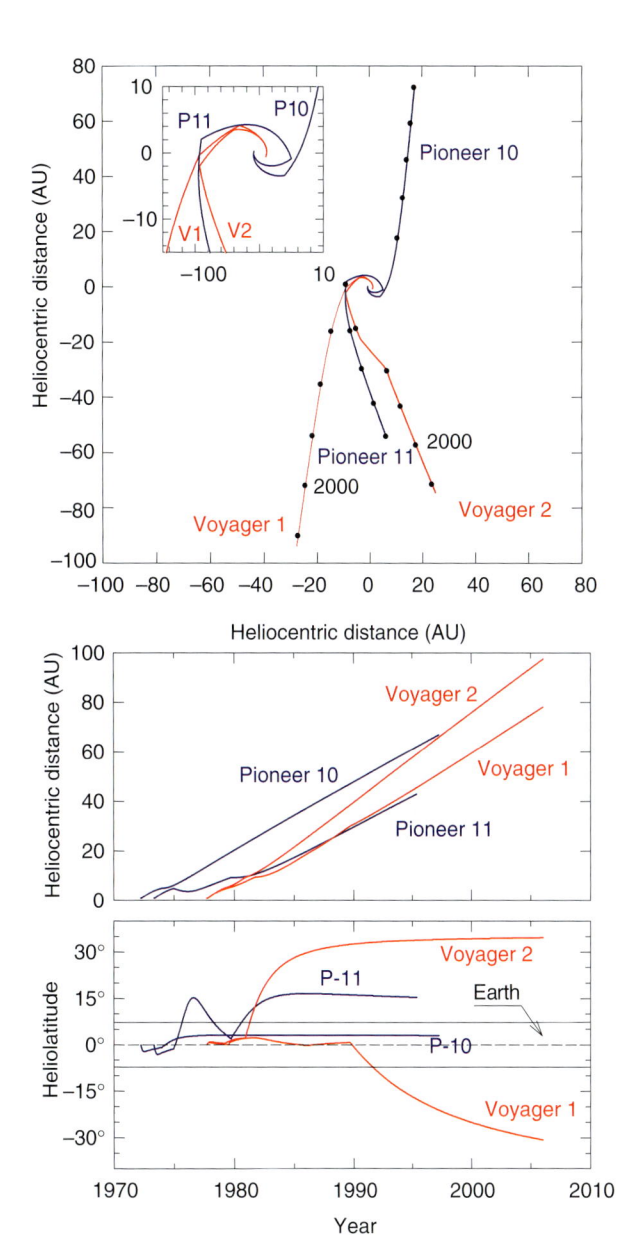

Figure 4 The race to the edge of the heliosphere. The upper panel shows the ecliptic projections of the orbits of the four spacecraft that have explored the outer regions of the heliosphere. The lower panels show the heliocentric distance and heliolatitude, respectively, of the four spacecraft. Note that the range of heliolatitudes covered by the Earth is also indicated in the lower panel. The two Voyager spacecraft are still functioning and are expected to survive long enough to observe the termination shock in the next few years.

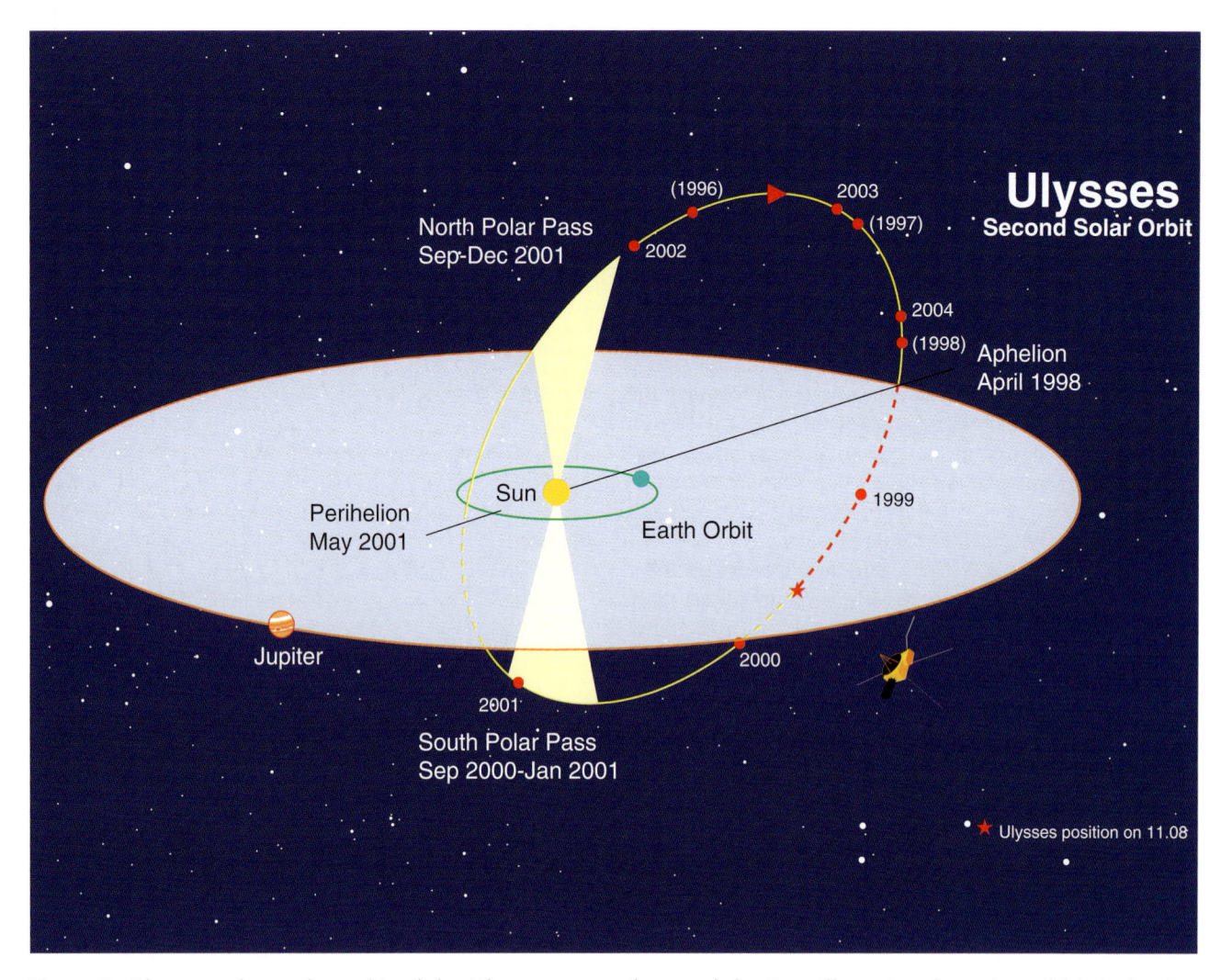

Figure 5 The second complete orbit of the Ulysses spacecraft around the Sun, illustrating the unique, high inclination reached by the mission to study the heliospheric medium over the polar regions of the Sun.

able to reach the heliopause, the boundary between the solar wind plasma and the interstellar medium. In the meantime, the Voyager spacecraft are returning valuable data, destined to remain unique for a long time, on conditions in the outer heliosphere.

The first mission targeted specifically to explore the third dimension of the heliosphere is the Ulysses spacecraft, launched on 6 October 1990 and foreseen to operate at least until September 2004. As celestial mechanics makes it difficult to place a spacecraft in an orbit with a significant inclination to the ecliptic plane, Ulysses used a gravitational swing-by at Jupiter to reach an orbit with a heliographic inclination of 80°. This mission has provided a significant step forward in our understanding of the heliosphere by charting its structure as a function of heliolatitude. Much of what we now understand of the three-dimensional structure of the heliosphere comes from the observations made by Ulysses at high heliolatitudes, first under solar minimum conditions around 1994 to 1996, and more recently, since 1998, under conditions of solar maximum activity. Although the many earlier observations by Earth-orbiting spacecraft, by the Helios, Pioneer and Voyager spacecraft identified many properties of the heliosphere, it was through the observations of Ulysses that these earlier data could be fully integrated into a three-dimensional picture. The second solar orbit of Ulysses is shown in Figure 5.

THE SOLAR WIND AND THE LARGE-SCALE STRUCTURE OF THE HELIOSPHERIC MAGNETIC FIELD

The fundamental reason for the existence of the heliosphere is the solar wind. It is therefore appropriate to review some of its basic characteristics that are important in shaping the structure of the heliosphere. (For more detailed accounts, see Chapters 9 and 47.) By correctly identifying the appropriate solution to the hydrodynamic equations of a steadily expanding, uniform and isothermal corona, the existence of the solar wind flowing radially from the corona at an asymptotically constant, supersonic velocity is easily deduced. Some of the simplifying assumptions used in this first derivation were already recognized by Parker (1963) as the sources of likely discrepancies between observations and theory. After nearly 40 years of measurements of the solar wind, its characteristics are much better known, and in fact are much better understood. Nevertheless, some basic questions concerning coronal heating mechanisms and the acceleration of the solar wind are still unanswered, although these questions, strictly without any answers 40 years ago, are now much better understood.

Since the identification of coronal holes, it has been clearly established that these regions of open magnetic fields in the corona are the sources of fast ($>600\,\text{km s}^{-1}$) solar wind streams. The origin of the slow ($<500\,\text{km s}^{-1}$) solar wind is less well understood, although it is clearly associated with regions of the corona in which magnetic fields are predominantly in the form of closed loops and streamers. The dynamics of magnetic fields in the boundary regions between coronal holes and closed field regions play a significant role in the formation of the solar wind (Fisk et al. 1999, Fisk and Schwadron 2001). The two kinds of solar wind, fast and slow, also have different characteristics in terms of temperature, density and elemental composition.

An accurate determination of the distribution of the solar wind as a function of time and location around the Sun is not possible. There is no direct way to measure the solar wind speed and density close to the Sun; it is possible, however, to infer the speed of the outflow at least occasionally. Two, mostly indirect means have been used. The first, the more direct of the two which relies on spectral imaging of coronal emission lines, has been implemented by instruments flown on the SOHO spacecraft as well as on the Spartan mission flown from the Space Shuttle. Doppler measurements have allowed estimates of the outflow velocities and associated densities to be determined, in particular in coronal holes where the fast solar wind originates. Other techniques, based on measurements of interplanetary scintillations (IPS), have been used to provide estimates of the solar wind speed close to the sun; these measurements are, in general, difficult to interpret as the speed profile has to be deduced from an integral of plasma in motion measured along the line of sight.

It is generally accepted that, whatever processes are responsible for the acceleration of the solar wind, the two kinds, fast and slow, originate in different coronal regions, with a sharp transition between source regions. The best evidence for this is provided by measurements of the composition of the solar wind. The main compositional differences are a drop in the abundance ratio of elements with low first ionization potential (FIP) to those with a higher FIP, and a drop in the abundance ratio of the high to low ionization state of given elements, such as oxygen, in fast solar wind streams. These indicators imply a lower coronal temperature for the regions of origin of the fast solar wind, in agreement with their identification with coronal holes. It important for the dynamics of solar wind streams that the compositional signatures show sharp boundaries between streams in the corona.

If it is assumed that, at some height in the corona, the solar wind accelerates away from the Sun uniformly, and if it is further assumed that the coronal magnetic fields are radially oriented at that height, the global equations that define the magnetic field in the solar wind can be easily deduced. The magnetic field lines are frozen into the plasma and convected in the solar wind. These equations that define idealized magnetic field lines in the heliosphere have been extensively used as the basic frame for organizing the actual observations. Parker's conceptual model is based on a uniform, radial solar wind velocity, and a uniform solar rotation.

The real solar wind differs in a number of significant respects from the ideal, uniform solar wind. Similarly, the coronal magnetic field which is convected in the solar wind, is also highly non-uniform and non-radial in the acceleration region, and therefore it is unsurprising that the Parker model of the heliospheric magnetic field is at best an approximation to what is actually observed in space. There are in fact many reasons for the differences between the ideal Parker geometry of the magnetic field and the observations. Some of these reasons are related to the propagation of non-uniform solar boundary conditions, such as differences in the rotation rates of the photosphere and coronal holes, and the motion of the footpoints of the magnetic field lines in the photosphere. Other departures from the ideal configuration are related to the dynamic evolution in the solar wind as it travels out into the heliosphere.

We can define, conceptually, source functions for both the solar wind, $\mathbf{V}_{SW}(r_0,\varphi,\vartheta)$, and the magnetic field, $\mathbf{B}_{SW}(r'_0,\varphi,\vartheta)$, on the surface of spheres (not rotating with the Sun) of heliocentric radius r_0 and r'_0, respectively, with φ and ϑ being the heliographic longitude and latitude. For the solar wind, we can take, again conceptually, the radius r_0 to be close to the sun, but at a distance where most of the acceleration of the solar wind has already taken place. For the magnetic field, r'_0 is normally taken to be between two and three solar radii, based on potential magnetic field models of the corona (Wilcox et al. 1980; Hoeksema 1991, 1995).

The general agreement of the magnetic field direction with the Parker model can be judged by the observations made by several space missions. The analysis was performed in each case using the actually measured solar wind speed, to calculate the difference of the magnetic field azimuth angle with respect to the expected Parker direction. Such analysis has been performed, for example, by Thomas and Smith (1980), Smith *et al.* (1986) and Smith (1997), using Pioneer data out to 35 AU; Behannon *et al.* (1989), using Voyager data; Forsyth *et al.* (1996a,b), using Ulysses data, in particular as a function of heliolatitude; and Lepping *et al.* (1996), near the Earth. The Ulysses results obtained during the fast latitude scan of the mission (when Ulysses moved from 80° south to 80° north in heliolatitude in 1994–95) are shown in Figure 6.

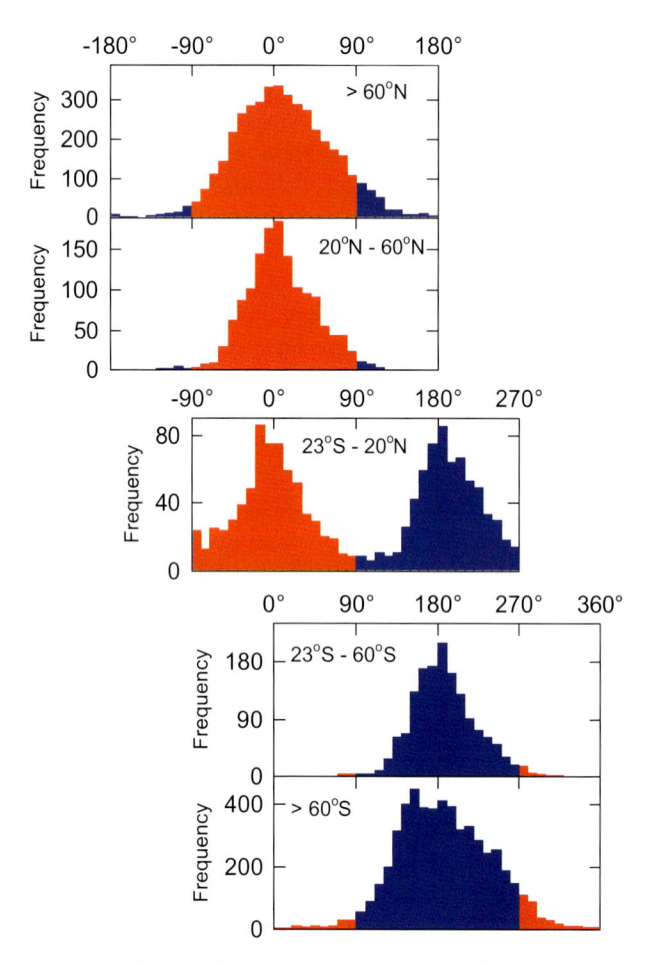

Figure 6 The distribution of the azimuthal direction of the heliospheric magnetic field as a function of heliolatitude, measured by the Ulysses spacecraft during its first fast latitude scan. These plots demonstrate that the sector structure of the magnetic field is restricted to the equatorial regions during solar minimum; away from the equatorial regions, the magnetic field is unpolar, with opposite polarities in the northern and southern hemispheres. The plots also show that the distributions are peaked around the expected direction, according to Parker's simple model. (Courtesy of ESA.)

The source functions of the solar wind and the heliospheric magnetic field are, in fact, strongly dependent on time on many scales. Along a radius vector from the Sun (i.e. for a given ϑ and φ), the time dependence is introduced not only by the solar rotation, but also, on a range of both longer and shorter timescales, by the evolution of the source regions of the solar wind, and on short timescales by dynamic phenomena such as CMEs.

In the solar wind, both the source magnetic field and the solar wind speed can undergo temporal changes. Neglecting temporal variations in the magnetic field, the time dependences in $\mathbf{V}_{SW}(r_0,\varphi,\vartheta)$ introduce a range of dynamic effects in the heliosphere. The evolving interactions of the non-uniform flows along a streamline of the solar wind introduce a corresponding structuring in the heliospheric magnetic field. In these interactions, the magnetic field plays only a peripheral role on the large scale (because of the dominance of the flow energy of the solar wind), but it is an essential ingredient of the small-scale processes that provide the physical basis for the MHD treatment of interacting solar wind streams.

Two timescales of major importance for both the solar wind and the HMF are the solar rotation period and the solar cycle. Around solar minimum, the corona is in a relatively stable state, with coronal holes, the regions of origin of the fast solar wind, covering both polar regions. The coronal streamer belt, a region of mostly closed magnetic field lines, is restricted to within about 15° to 30° of the solar equatorial plane. Slow solar wind streams are associated with the streamer belt. The relative stability of the coronal configuration, which persists over several solar rotations, introduces a periodicity in the function $\mathbf{V}_{SW}(r_0,\varphi,\vartheta)$ through its dependence on φ. For an interplanetary observer within about 15° to 25° of the solar equatorial plane, $\mathbf{V}_{SW}$ at the source surface switches from high speeds (above 700 km s^{-1}) to low speeds (below 500 km s^{-1}), corresponding to solar longitudes with coronal holes and the streamer belt, respectively. This periodicity generates a quasi-steady pattern of fast solar wind streams interacting with the preceding slow solar wind streams which result in the formation of the CIRs, as discussed below.

The source functions, simplified but appropriate to solar minimum conditions and in agreement, to the first approximation, with the Ulysses observations (Phillips *et al.* 1995a; Balogh *et al.* 1995), are illustrated in Figure 7. In particular, the existence of a sharp boundary between fast and slow solar wind streams has been shown through the analysis of freezing-in temperatures and abundance ratios of the solar wind ions (Geiss *et al.* 1995). For interplanetary observers within the ecliptic, the source functions in the lower panels of Figure 7 generate the alternating fast and slow solar wind streams and magnetic polarities with the period of the solar rotation.

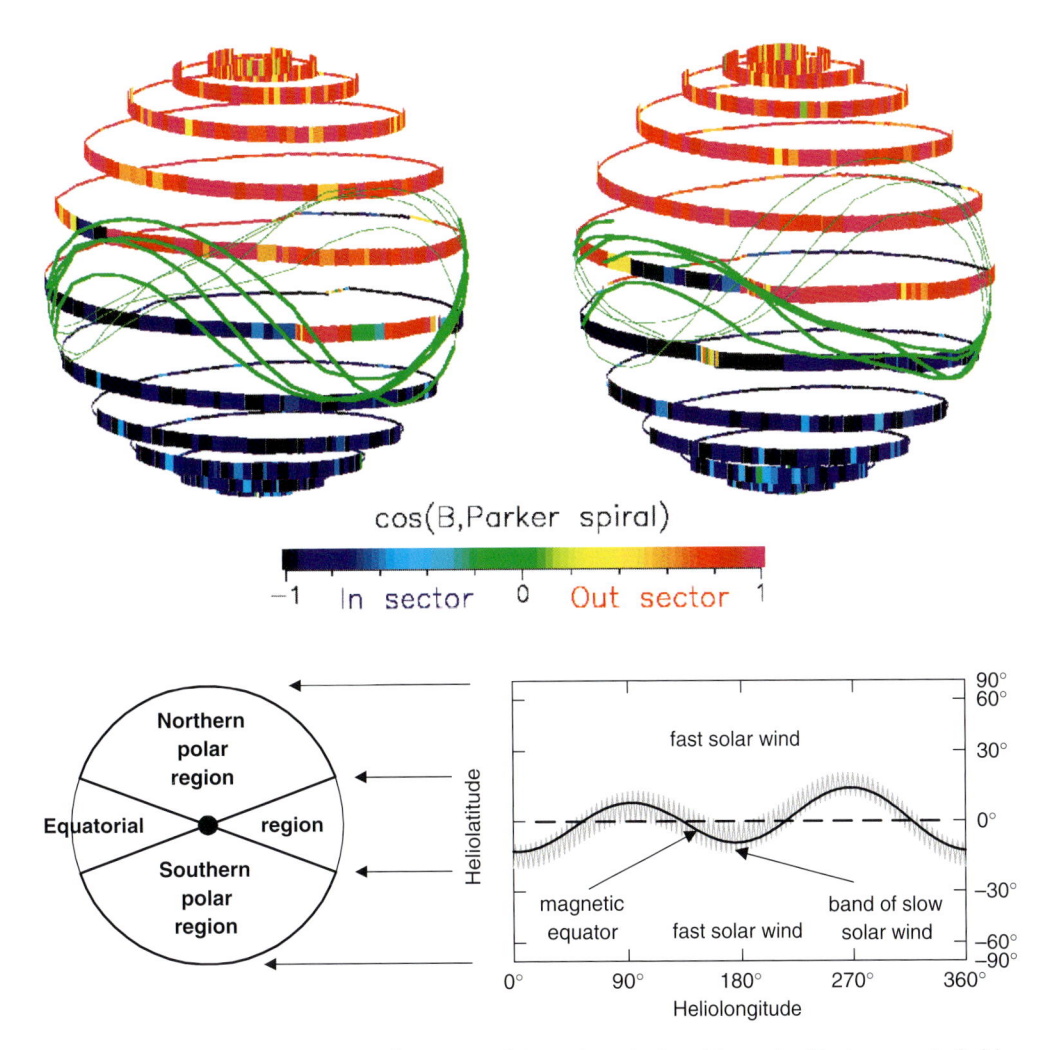

Figure 7 A schematic illustration of the source functions of the solar wind and its embedded magnetic field near solar minimum. In the upper panel, the magnetic polarities of the Sun's two hemispheres are illustrated, as observed by the Ulysses spacecraft in 1994–95. (Courtesy of G. Erdős, KFKI/RMKI, Budapest.) The green lines separating the two hemispheres represent the Sun's magnetic equator, as determined by coronal magnetic field modelling (Wilcox *et al.* 1980). The lower panel shows schematically the magnetic equator, the coronal streamer belt surrounding it and the coronal source regions of the fast solar wind. Although coronal holes only fill a small area in the corona, due to the non-radial expansion of the fast solar wind, most of the heliosphere is filled with fast solar wind at solar minimum.

The regions of origin of the solar wind undergo a significant change between solar minimum and solar maximum. At solar maximum, there are no dominant large-scale coronal holes and therefore the details of the source function are far more difficult to discern than at solar minimum; in terms of the function $\mathbf{V}_{SW}$ there are no clear periodicities at the solar rotation period. Although stream–stream interactions may still occur, persistent, corotating patterns no longer form at solar maximum. The somewhat surprising result, nevertheless, is that at least on average, extensive observations have shown that the magnetic field configuration in the heliosphere remains close to the simple Parker model. Despite its obvious simplicity and the many modifications that need to be made to make it conform to the actual observations, it has remained a useful frame of reference for organizing the observations.

Other than the large-scale dynamic processes that modify significantly the structure of the heliospheric medium (which are discussed below), there are two ways in which even a relatively steady solar wind and magnetic field can modify the structure of the Parker geometry. The first of these is due to the evolution of transverse components of the magnetic field as a function of heliocentric distance. Transverse components are due to non-radial contributions at the source surface that may arise from the random motion of the footpoints of the magnetic field. These are propagated out into the solar wind, at least partly in the form of long-wavelength, large-amplitude Alfvén waves. As the basic magnetic field equations show, these components depend on the inverse of the heliocentric distance, while the radial component of the magnetic field decays as the square of the distance from the Sun. Near the solar equatorial plane, the magnetic field is twisted into an increasingly tight Archimedean spiral (so that the magnetic field becomes close to perpendicular to the radial direction), and the effect of the slowly decaying random components in the transverse direction to the solar wind flow make a relatively small contribution to the structure of the field. However, sufficiently far from the Sun, at high heliolatitudes, where the Parker geometry implies relatively radial fields, the transverse component can, in principle, become the dominant one, as pointed out by Jokipii and Kóta (1989).

The second process that brings about a potentially significant modification of the Parker geometry of the magnetic field was identified by Fisk (1996a). This effect arises from the modification of the magnetic field line geometry by the difference in rotation rates between the solar photosphere and the overlying coronal holes. The footpoints of the magnetic field lines are anchored in the photosphere that rotates considerably slower at high heliolatitudes than near the equator. The fast solar wind that transports the magnetic field from the corona expands non-radially from the polar coronal holes. These coronal holes rotate faster than the underlying high-latitude photosphere, in fact at approximately the equatorial rotation rate of the photosphere. At the same time, the coronal holes rotate around an axis (notionally, the magnetic dipole axis) of the Sun that is somewhat offset from the rotation axis near solar minimum. These processes are schematically illustrated in Figure 8. This figure is drawn in the frame of reference that corotates at the equatorial rotation rate. The axis M is the axis of symmetry of the expansion of the fast solar wind from the Sun's polar coronal hole. This axis is fixed in this frame of reference and is offset from the rotation axis Ω of the Sun. The magnetic field line that is anchored in the photosphere at the heliographic pole undergoes non-radial expansion and moves to point p on a concentric outer surface. The footpoints of other field lines rotate around Ω at the slower rate appropriate to the high-latitude photosphere but are entrained through the faster rotating coronal hole. The circles on the outer surface in Figure 8 represent the footpoints of the magnetic field lines that are carried out into the solar wind, representing, at that latitude, the function $\mathbf{B}(r_0',\varphi,\vartheta)$.

The compound effect of the difference in the rotation of the coronal holes and the footpoints of the magnetic field is that field lines originating at high latitudes are emerging

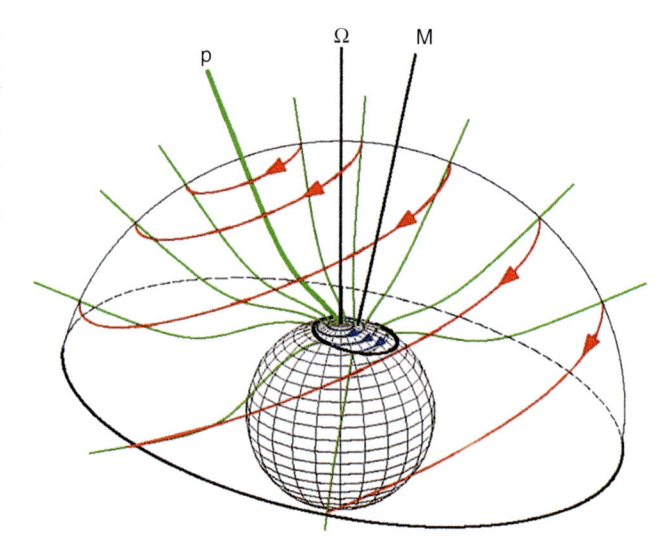

Figure 8 The model of the origin of the heliospheric magnetic field proposed by Fisk (1996a). This model takes into account the differential rotation of the photospheric footpoints of the magnetic field lines and their rigid rotation in the coronal holes; as a result, there is a transport of the magnetic field lines to latitudes considerably lower than their origin in the corona.

in the solar wind at lower latitudes. This effect thus modifies the Parker configuration that implies that the heliolatitude of the magnetic field lines is constant (in other words, that magnetic field lines are draped on the surface of cones axially aligned with the solar rotation axis). Simulations (Fisk 1996a) have shown that magnetic field lines originally at 70° heliolatitude in the inner heliosphere can provide a direct magnetic connection to much lower heliolatitudes at distances of 15 to 20 AU from the Sun.

This effect has been introduced to explain the propagation of energetic charged particles accelerated by shockwaves associated with CIRs from mid- to high heliolatitudes. This structuring of the magnetic field lines away from high latitudes is firmly based on the observation of the non-radial expansion of the solar wind and the faster rotation rate of the polar coronal holes. The extent to which it occurs, and can be discerned in the observations of the magnetic field direction, depends on the detailed geometry of the coronal holes and the stability of the source regions of the fast solar wind. It is likely that this effect is significant at times of low solar activity. Even in the absence of a steady-state modification of the overall structure of the heliospheric magnetic field, this effect probably introduces a significant level of magnetic flux transport away from the polar regions and contributes to the transport of energetic particles in heliolatitude an effect that is difficult to explain otherwise.

LARGE-SCALE DYNAMIC PHENOMENA IN THE HELIOSPHERE

The solar wind emitted from the corona is not uniform either spatially or temporally. The sources of the solar wind and the associated coronal magnetic structures vary considerably, in particular over the solar activity cycle, but also on the timescale of the solar rotation period. Along any given radial direction from the Sun, in a non-rotating frame, the solar wind streams can have different speeds, densities and other characteristics. The interaction between such solar wind streams of different velocities, densities, temperatures and composition along a streamline generates a wide range of dynamic structures in the heliosphere.

CIRs are a subset of large-scale structures in the heliosphere which arise from the dynamic interaction between fast and slow solar wind streams (Balogh *et al.* 1999). The conditions for the formation of CIRs are that both fast and slow solar wind streams be simultaneously present at a given (low-to-mid-) heliolatitude, and that the coronal sources remain approximately stable for several solar rotations. Slow streams are normally confined to the vicinity of the HCS. This holds demonstrably in the declining phase of the solar cycle and at solar minimum. Fast streams originate in the polar coronal holes and their equatorward extensions in the declining and minimum phases of the solar cycle; the coronal evolution of the source regions is slow on the timescale of the solar rotation.

It is not yet clear what are the characteristics of fast solar wind streams at solar maximum when all coronal holes are small and, because of the implicitly large spreading factor of magnetic field lines in the corona, cannot generate high speed streams (Wang and Sheeley 1994). The observations made by Ulysses in the rising phase of the current solar cycle show that the solar wind becomes very disturbed and variable, generally slow ($\sim 350\,\mathrm{km\,s^{-1}}$) with occasional, non-recurring periods of higher speeds (up to 550 to 600 $\mathrm{km\,s^{-1}}$).

The interaction between high- and low-speed streams operates from close to the Sun. The resulting structures and their evolution have been extensively observed by the two Helios spacecraft between 0.3 and 1 AU (for a review, see Schwenn 1990). There is also a wealth of data, covering more than three decades of observations, at 1 AU. CIRs evolve as a function of heliocentric distance; their evolution and eventual merging has been observed by the Pioneer and Voyager probes into the outer heliosphere (Smith 1989; Burlaga *et al.* 1984). In particular, the characteristic forward–reverse shock pair delineating the CIRs tends to develop only beyond 1 AU.

The three-dimensional characteristics of CIRs at midlatitudes were observed by Ulysses in 1992–93 (Bame *et al.* 1993). These observations were found to match well (Gosling *et al.* 1993) the three-dimensional model proposed by Pizzo (1991). This series of CIRs is illustrated in Figure 9. A more detailed description, together with a discussion of the shockwaves associated with them has been given by Gonzalez-Esparza *et al.* (1996). The Ulysses observations clearly established the three-dimensional nature of CIRs, their tilted propagation and the asymmetry in the propagation of the associated shockwaves (Riley *et al.* 1996).

CIRs therefore can be regarded as three-dimensional shell-like structures extending to about 30° to 40° away from the ecliptic plane during the declining phase of the solar cycle, but probably only to 15° to 30° at solar minimum. However, the actual latitudinal extent is likely to be longitude dependent, due to the longitude dependence of solar wind streams from the northern and southern hemispheres and to the (related) longitude dependence of the HCS.

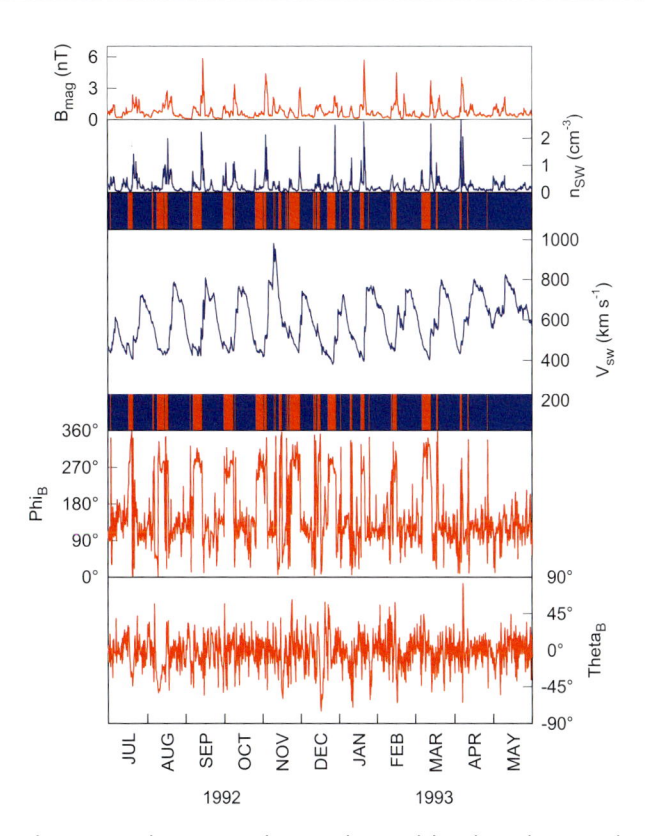

Figure 9 The series of CIRs observed by the Ulysses solar wind and magnetic field instruments, as the spacecraft moved from near the solar equatorial plane towards the south pole of the Sun. The large excursions in solar wind velocity in each solar rotation cause the compression of the slower solar wind stream ahead of the fast stream, resulting in a periodic increase in solar wind density and in the strength of the magnetic field. The magnetic sectors corresponding to the series of CIRs is also shown, as is the meridional angle of the magnetic field (in the lowermost panel) which shows the field deflections associated with the CIRs. (Balogh 1998.)

CMEs introduce a significant perturbation in the structure of the inner heliosphere in the form of interplanetary plasma clouds (IPCs). CMEs carry a large amount of mass and momentum capable of distorting the ambient solar wind and magnetic field, and can drive large-scale shockwaves which cause the characteristic Forbush decreases in the cosmic-ray flux in the inner heliosphere. Their plasma and magnetic structures are in general complex and differ significantly from the ambient HMF. While the signatures of many CMEs are readily identifiable in solar wind and magnetic field data, many critical questions remain concerning the contribution of CMEs to the structure and dynamics of the heliospheric medium. The observational database is far from complete for both CMEs near the sun and the IPCs. There is no comprehensive survey of CMEs available which could be in any way comparable to those which exist for instance for flares or sunspots. The existing and foreseeable heliospheric data set, even including the magnetosphere as an 'IPC detector', is even less comprehensive, based as it is on a very incomplete temporal and spatial sampling.

While case studies exist that link reliably some individual CMEs with their heliospheric IPC counterparts (see e.g. Weiss *et al.* 1996), in general the link between solar and heliospheric observations is difficult to establish. The lack of comprehensive heliospheric coverage, even if more comprehensive solar data become available, will remain a limitation.

Information on the temporal, heliographic and size distribution of CMEs as a function of the solar cycle, even if somewhat patchy, is available over the past 20 years (e.g. Hundhausen 1993). However, there is no single set of necessary and sufficient parametric signatures in heliospheric data to recognize the presence of CME ejecta or structures unambiguously. Solar wind data normally furnish the primary information, recognizing CMEs most commonly by the presence of bidirectional streaming of electrons, interpreted as tracers of magnetic structures that may be closed in the heliosphere or connected at both ends of the field lines to the corona. Given the identification of the CME signatures in the solar wind data, in a fraction of the cases (up to 50% or so) it is possible to identify an associated magnetic signature, based on smooth (low variance) and smoothly rotating magnetic field structures overlapping the solar wind signatures.

Magnetic clouds form an important and relatively easily recognizable subset of magnetic signatures of IPCs (for a review, see Burlaga 1991). These belong to a more general class in which large-scale magnetic flux ropes (Gosling 1990) with a more complex geometry, resulting from three-dimensional reconnection (Gosling *et al.* 1995), may be present. Considerable complexity, indicating the presence of both open and closed magnetic structures in CMEs, has been identified through the study of magnetic polarity reversals and associated bidirectional electron streams (Kahler *et al.* 1996).

There is a large uncertainty concerning the heliospheric extent of CMEs. At the Sun, their average latitude and longitude scales appear to be about 40°, but with a significantly long tail of the distribution to larger values (Hundhausen 1993). A key question for the structure of the HMF particularly near solar maximum (when CMEs may dominate the dynamics even in the inner heliosphere, see below) is the distribution of CME sizes and locations. Most CMEs appear to originate from the streamer belt (which therefore gives a good indication for their expected location). It has been suggested (Crooker *et al.* 1996) that the HCS acts as a conduit for CMEs of all sizes; in that case, both the detailed structure of the HCS and its large-scale properties may well be significantly modified by the frequency and size distribution of CMEs. However, CMEs may well be emitted from regions less clearly identified with the large-scale streamers. Ulysses in particular detected CMEs at relatively high latitudes, well polewards of the HCS (Gosling *et al.* 1994). These were also remarkable for being associated with forward–reverse shock pairs, apparently driven by the internal pressure of the CMEs.

Given the non-uniform nature of the heliospheric medium, disturbed both by the interaction of non-uniform solar wind flows and by transient phenomena, the interacting regions of plasma are continually transported away from the Sun, shaping in a complex way the heliosphere. This is recognizable both in the non-uniform properties of the solar wind and of the heliospheric magnetic field, as observed by the spacecraft in the different locations in the heliosphere. A particularly important question for the large-scale structure of the heliosphere is the dynamic evolution of the large-scale non-uniformities as a function of heliocentric distance.

The CIRs observed in the inner heliosphere evolve as they are convected out to 10 AU and beyond (Burlaga *et al.* 1984, 1996; Burlaga and Ness 1994, 1998, 2000; Ness and Burlaga 1996; Smith 1989). The evolution is accompanied by merging of the interaction regions as successive CIRs widen with heliocentric distance, and the leading forward shockwaves interact with the trailing reverse shockwaves of preceding CIRs. This leads to a coalescing of the compressed regions of magnetic field with heliocentric distance, as is shown in Figure 10, in which the magnetic field strengths measured simultaneously at 1 AU and at 16 AU are shown for a whole year, as the solar cycle was approaching its minimum activity phase. At the distance of the Earth, an evolving four-sector magnetic field structure was observed at the time, together with two CIRs in each solar rotation period. This structure is caused by fast solar wind streams generated in near-equatorial coronal holes,

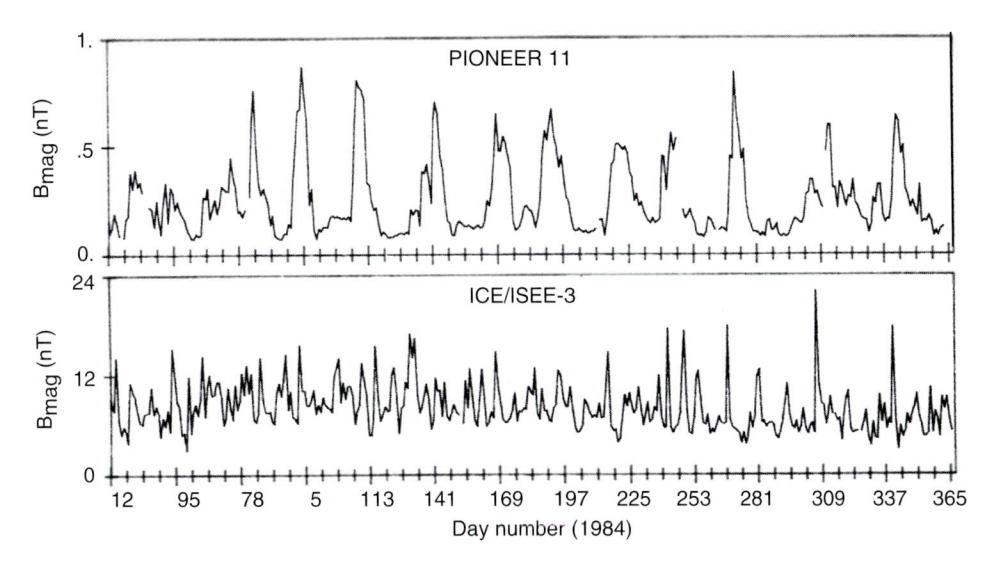

Figure 10 Comparison of the magnetic field magnitude observed in 1984, during the declining phase of the solar cycle, by ICE (ISEE 3) at 1 AU and Pioneer 11 at 16 AU, showing the coalescence of CIRs into merged interaction regions. (After Smith 1989.)

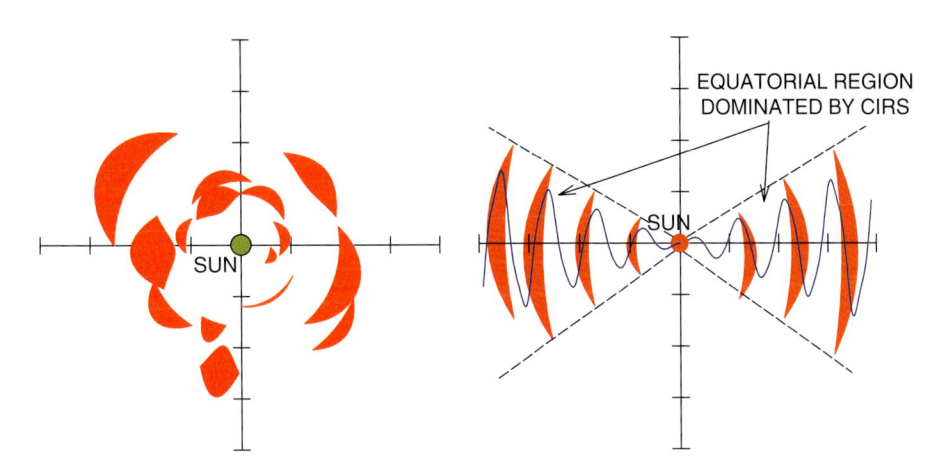

Figure 11 The two states of the inner and middle heliosphere, near solar maximum and solar minimum. The left panel shows schematically the situation near solar maximum when the Sun is surrounded by several CMEs of different sizes, emitted at all latitudes; the right panel shows the situation in the late declining and minimum phases of the solar cycle when CIRs and the wavy HCS dominate the low heliolatitude region.

located both north and south of the Sun's magnetic equator. These lead to the complex pattern of compressed magnetic fields through the formation of the CIRs at 1 AU, where the finer details of the non-uniform solar wind streams are still discernible. However, as the interaction regions are convected away, the compressed regions in each solar rotation period merge and lead to the simpler, larger scale structure observed by Pioneer 11 at 16 AU, where only one compression region was seen in each solar rotation period.

In general, these merged interaction regions, born of CIRs, dominate the regions around the solar equatorial plane out to beyond 15 or 20 AU around solar minimum as illustrated schematically in Figure 11. The periodicity at the solar rotation period in solar wind parameters, as well as in the magnetic field strength, remains a recognizable feature of the equatorial region around solar minimum at least to 50 AU, as observed by Voyager 2 up to 1995 (Lazarus *et al.* 1998).

The structures that dominate the heliosphere around solar maximum are the heliospheric counterparts of the CMEs. As already mentioned, the frequency, average size and heliolatitude range of CMEs all increase around solar maximum. At the same time, coronal holes, the sources of the high-speed solar wind, only exist as relatively small, transient features. Although the interaction between the generally slow solar wind and the rare, fast solar wind from the transient coronal holes remains a recognizable feature of the heliospheric medium, the resulting pattern is not stable on the scale of the solar rotation. The situation generally prevailing near solar maximum is also illustrated schematically in Figure 11.

COSMIC-RAY MODULATION AS A PROBE OF THE HELIOSPHERE

The phenomenon of cosmic-ray modulation, as outlined in the section on 'The existence of the heliosphere and its main properties', was at the origin of the naming of the heliosphere as a volume of space controlled by solar phenomena. Its study remains central to understanding the large-scale processes that shape the heliosphere, although the physical processes that control the access of cosmic rays into the inner heliosphere are now understood to be considerably more complex than originally assumed. The following comments relate the propagation and access of cosmic rays to the properties of the heliospheric medium. In the first instance, the basic transport processes are outlined; these processes are always active, but the properties of the medium change with the phase of the solar cycle. Around solar maximum, however, other, more extensive large-scale structural changes occur, as a result of an increased frequency and spatial extent of CMEs. These structures have a considerable effect on the propagation of cosmic rays and are thought to be a major factor in the decrease in cosmic-ray intensity in the inner heliosphere at solar maximum activity.

The intensity of cosmic rays in the heliosphere at a given location, in the relatively steady state of propagation conditions in the years around solar minimum, is the result of four different effects, as identified by Parker (1965). The first is diffusion in the irregular magnetic field, both parallel and perpendicular to the average magnetic field. An important question is the relative importance of parallel and perpendicular diffusion terms. Although early models of cosmic-ray propagation assumed that diffusion along the magnetic field was the dominant motion of the particles, the measurement of the latitude gradient of cosmic-ray intensity by Ulysses has highlighted the likelihood that diffusion perpendicular to the magnetic field lines is more important than expected (McKibben 1998; Potgieter 1998). The investigation of the diffusion coefficients applicable to cosmic rays in the heliosphere is a difficult task, but it is central to the understanding of the nature of the statistical and turbulent properties of the heliospheric magnetic field (e.g. Giacalone 1998).

The second effect is the drift of particles in the large-scale average magnetic field. As was noted by Jokipii et al. (1977) and Jokipii and Thomas (1981), electrically charged energetic particles undergo a significant drift motion in the large-scale heliospheric magnetic field that follows the Parker geometry described earlier. The drift of particles depends on the sign of their charge: electrons and positive ions drift in opposite directions. If A denotes the projection of the Sun's magnetic dipole axis, the sign of A reverses when the solar magnetic fields reverse near solar maximum. Consecutive solar cycles have therefore alternating signs of A. The direction of particle drift velocities depends on the sign of the product qA, where q is the charge (positive for ions, negative for electrons) of the particles. When qA is positive (as was the case during the last solar activity minimum in 1995–96), ions are expected to enter the inner heliosphere over the solar poles, whereas electrons enter along the Sun's equatorial plane. This preferred entry for oppositely charged particles reverses in consecutive solar cycles. When particles enter in the equatorial region, the wavy HCS, together with the CIRs present up to mid-latitudes, makes their access more difficult than over the solar poles. Evidence for drifts being an important large-scale effect on cosmic rays comes from the measurement of the recovery from the lower intensities at solar maximum. If particles enter preferentially over the solar poles, then the cosmic-ray intensities increase earlier in the inner heliosphere (e.g. near the Earth) than in the outer heliosphere (e.g. at Voyager). In the next solar cycle, the process is reversed: particles enter preferentially in the equatorial region, so that the recovery after solar maximum is observed first in the outer heliosphere, before it is observed near the Earth. This process is illustrated in Figure 12.

Convection away from the Sun and adiabatic deceleration in the expanding solar wind are the third and fourth effects of the heliospheric medium on cosmic rays. As the cosmic-ray particles gyrate around the magnetic field, the field lines are convected outwards in the solar wind, away from the Sun, carrying with them the particles. At the same time, the expansion of the solar wind leads to an adiabatic deceleration of the particles; this can also be interpreted as a rarefaction of the scattering centres as the solar wind expands.

The basic elements of the modulation process by the heliospheric medium and its structural changes as a function of the solar cycle are quite well understood (e.g. Potgieter 1998). However, one of the objectives of the Ulysses mission was to verify that the intensity of cosmic rays was, as expected on the basis of modulation theory, significantly higher over the polar regions of the Sun than near the ecliptic plane, at least at solar minimum. This expectation was effectively proved wrong: the heliolatitude gradient of cosmic-ray intensity was found to be very much smaller than predicted by theoretical models (e.g. McKibben 1998). The cause of this effect, implying a more isotropic propagation of cosmic rays than expected in the heliospheric medium, is not yet well understood. The most important property of the

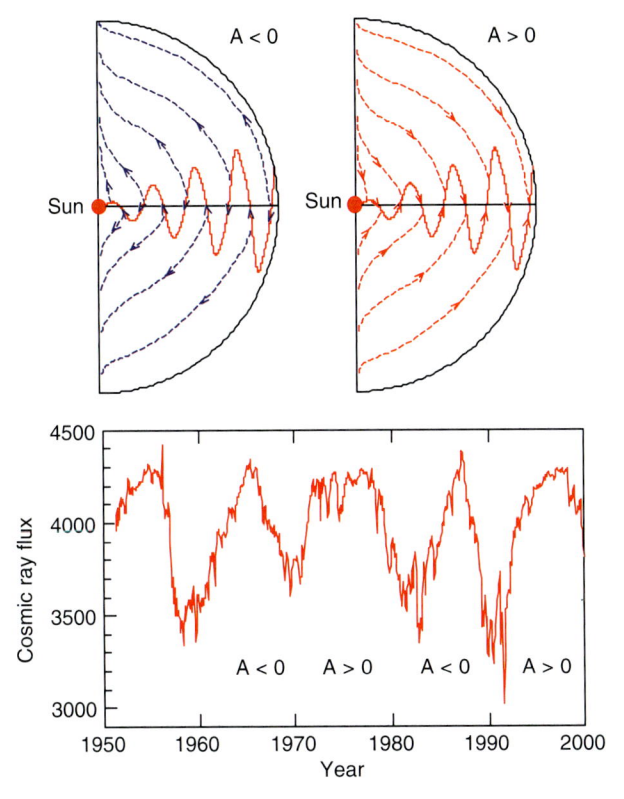

Figure 12 The preferential drift pattern of cosmic rays in the heliosphere during alternate solar cycles. The upper two panels show the drift paths for positively charged particles, the lower panel shows the different cosmic-ray intensity patterns corresponding to the alternating magnetic polarities of the polar regions of the Sun.

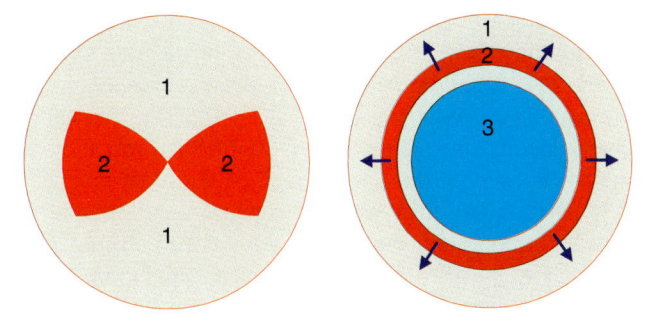

Figure 13 A schematic summary of the properties of the heliospheric medium that control cosmic-ray modulation. The diagram on the left illustrates the status around solar minimum, when cosmic-ray access into region 1 is through almost isotropic diffusion, whereas in region 2 access is restricted and controlled by CIRs. The diagram on the right shows the status around solar maximum, when the propagation conditions in the outer heliosphere (region 1) remain probably similar to those near solar minimum, region 2 represents the GMIRs, the major obstacles to cosmic-ray access into the inner heliosphere, and region 3 which is highly disturbed by the presence of frequent CMEs of all sizes.

medium that may influence the propagation of cosmic rays at high heliolatitudes is the much higher than expected level of fluctuations in the direction of the magnetic field in the uniformly high-speed solar wind (Smith *et al.* 1995; Horbury *et al.* 1996; Forsyth *et al.* 1996b). These fluctuations can contribute significantly to the diffusive propagation of cosmic rays perpendicular to the average magnetic field direction; the effect of an increased perpendicular diffusion is a reduction in the heliolatitude gradient in cosmic-ray intensity, in agreement with the observations of Ulysses.

In addition to the effects of the magnetic fluctuations on the cross-field diffusion of cosmic-ray particles, the transport of magnetic field lines from high to medium heliolatitudes proposed by Fisk (1996a), described in the section on 'The exploration of the heliosphere', may also contribute to an evening out of cosmic-ray fluxes away from the solar equatorial plane. At larger distances over the solar poles, the growth of large transverse fluctuations in the magnetic field may also be present, as proposed by Jokipii and Kota (1989), making the access of cosmic rays more difficult in the high-latitude heliosphere.

Given that the presence of CIRs and the wavy HCS in the equatorial region of the heliosphere influence strongly the access of cosmic rays in that region, and that the propagation of cosmic rays is nearly isotropic at high heliolatitudes, a simplified overall picture of the heliosphere, as seen by cosmic rays around solar minimum is as shown in the left panel of Figure 13.

As already mentioned, the principal cause of the decrease in cosmic-ray intensity in the inner heliosphere is the increase in frequency and spatial distribution of CMEs. CMEs introduce major structural disturbances into the solar wind; when these disturbances are spread in all directions around the Sun, the access of cosmic rays is significantly impeded from the outer into the inner heliosphere. There is, in addition, a further effect: in each solar cycle, there are a few very large CMEs, associated usually with significant flaring activity on the Sun. The solar disturbances that give rise to such large CMEs often produce in fact a series of outbursts, closely following each other. CMEs are usually associated with a decrease in cosmic-ray intensity (Forbush decreases) lasting several days, as the magnetic configuration of the CME limits their access into the volume occupied by the CME. It is natural to suggest that as the frequency of CMEs increases, the relatively small cosmic-ray decreases are added together to lead to the more general decrease in cosmic-ray intensity observed as solar activity increases towards maximum (for a review of this suggestion, see e.g. Cane 2000).

An alternative view, based on the cosmic ray and magnetic field observations made by the Voyager spacecraft, is that the decrease in cosmic-ray intensity occurs in steps, coincidently with the passage of strong magnetic fields through the heliosphere, resulting from the merging of large CMEs in the middle heliosphere (McDonald et al. 1981; see also, for a review, e.g. McDonald 1998). These so-called global merged interaction regions (GMIRs) form moving shells surrounding the Sun, formed of turbulent magnetic fields, with larger than average magnitude, and provide an outward propagating obstacle to the galactic cosmic rays (Burlaga et al. 1985). In the right panel of Figure 13, a schematic view of the heliosphere is shown, valid around solar maximum activity. In this schematic representation, the effects of both frequent, smaller CMEs are shown in the inner heliosphere, together with the outward propagating GMIRs in the outer heliosphere.

Anomalous cosmic rays (ACRs) represent a special component of the high-energy particles observed in the heliosphere. This component of the high-energy particle population (with energies from about 15 MeV to a few hundred MeV) consists of (mostly) singly ionized He, N, O and Ne ions, as well as protons (e.g. Cummings and Stone 1996, 1998). Their characteristic energy spectrum and elemental composition led to their identification as a population originating in the heliosphere, by acceleration at the termination shock (Pesses et al. 1981) from a seed population of interstellar atoms ionized in the heliosphere (Fisk et al. 1974; see also, for a recent review, Jokipii and Giacalone 1998). The study of the radial gradient of ACRs as the Voyager spacecraft has travelled out towards the outer boundary of the heliosphere has provided two important suggestions concerning the termination shock. The first is the likely distance of the termination shock: two different methods of extrapolating the Voyager observations have led to estimates between about 70 and 95 AU, at different epochs in the solar cycle (Cummings and Stone 1998). The second, based on the measured spectrum of the ACRs and its variation with heliocentric distance, implies a lower than expected Mach number for the solar wind in the outer heliosphere, and therefore a relatively weak termination shock (Fisk 1996b).

THE BOUNDARIES OF THE HELIOSPHERE

In simple terms, it can be stated that the outer boundary of the heliosphere is defined as the surface across which there is a pressure balance between the supersonic solar wind and the LISM. However, as the presence of the LISM cannot be sensed as the solar wind flows outwards at supersonic speeds, the solar wind suddenly has to slow down as its dynamic pressure becomes equal to the static pressure of the LISM. At this point, mass balance also has to be preserved. This can only happen through a shockwave, as the supersonic flow meets a static 'obstacle'. The shockwave is facing inward, and must surround the heliosphere globally, as the LISM envelops the Solar System and the region filled with the solar wind. The shockwave at which the solar wind thus slows down is the termination shock of the heliosphere, probably its most important boundary. In the frame of the solar wind, the termination shock is a reverse shock, somewhat similar to the reverse shocks encountered at the trailing edges of CIRs.

The properties of the LISM that are relevant for estimating the extent of the heliosphere have been extensively reviewed in articles in von Steiger et al. (1996); of particular relevance to establishing the properties of the LISM and their impact on the termination shock are the contributions by Fisk (1996b), Frisch (1996), Geiss and Witte (1996) and Lallement (1996).

The details of the termination shock, such as its distance from the Sun, its large-scale geometry, its thickness and its dynamics are very largely speculative at this stage (e.g. Axford 1972, 1996; Lee 1996; Suess and Nerney 1997). The dominant pressure term inside the termination shock is the dynamic pressure of the supersonic solar wind; other terms contributing to the total pressure in the solar wind, such as the pressure of the magnetic field and the thermal pressure of the plasma are much smaller. Estimating the external pressure is a considerably more difficult task, as the parameters of the LISM are only known indirectly, with large uncertainties. The basic equation used to estimate the distance to the termination shock is relatively simple:

$$\frac{1}{R_{TS}^2} (\rho_{SW} V_{SW}^2)_{1AU} = p_{LISM}$$

The term on the right-hand side of this equation is made up of a number of contributions which are difficult to estimate. In this equation, R_{TS} is the distance of the termination shock, measured in AU, ρ_{SW} and V_{SW} are the density and velocity of the solar wind, respectively, and p_{LISM} is the total pressure in the LISM. The values used for the solar wind density and pressure in this equation are taken from their observations at 1 AU, at the orbit of the Earth. This formula assumes that the solar wind dynamic pressure scales as the inverse square of heliospheric distance. The general validity of this approach, using Voyager observations in the outer heliosphere, was examined by Belcher et al. (1993).

The most important contributions to the external pressure are the thermal and magnetic pressure of the LISM; and its dynamic pressure, as the Sun and the Solar System move through the medium surrounding it at a speed about 20 to 25 km s^{-1}. In addition, the pressure of the neutral component of the LISM may well be important, and even the pressure exercised by the galactic cosmic rays has been considered as a term that cannot be neglected. An important question, but one that is difficult to answer, is whether the flow speed of the LISM is supersonic. The diagram used to introduce the heliosphere (Figure 1) was drawn assuming a supersonic flow past the heliosphere. In this case, the boundary between the shocked (subsonic) solar wind plasma outside the termination shock and the LISM, the heliopause, is a surface similar to the magnetopause around the Earth's magnetosphere.

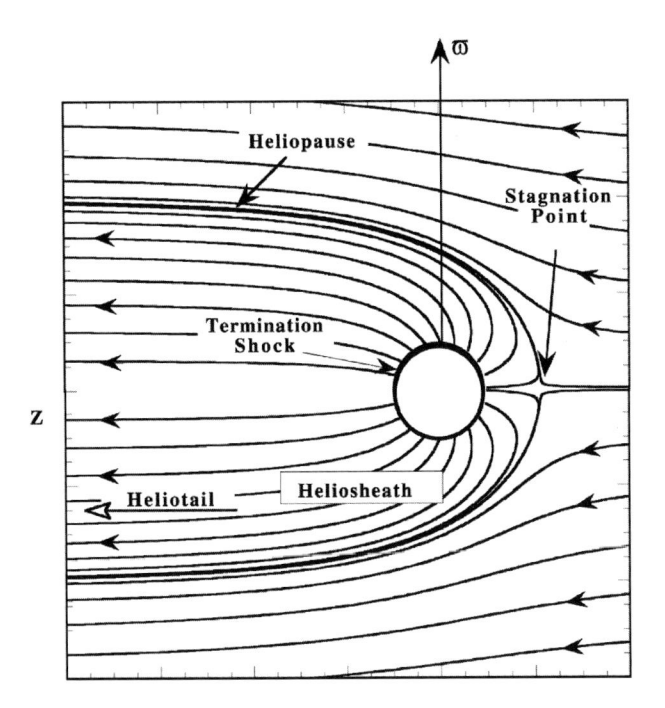

Figure 14 A model of the heliosphere assuming that the LISM is not supersonic. This picture represents an alternative view of the heliosphere to the schema shown in Figure 1. (By courtesy of S.T. Suess, NASA/Marshall Space Flight Center, Huntsville, Alabama, USA.)

There is also, in this case, a heliosheath, in which the subsonic LISM, slowed down at a heliospheric bow shock, flows past the heliosphere. The geometry shown in Figure 1 is, in very simplified terms, the picture that is generally assumed when describing the heliosphere.

An alternative view of the heliosphere is shown in Figure 14. The main difference in this view is the absence of a bow shock; the flow speed of the LISM is assumed to be subsonic. While this figure represents a possible particular case, it indicates the basic features that may be expected if a bow shock is not formed in front of the heliosphere, upstream, by a supersonic LISM. The heliopause remains a necessary feature in this model, as the surface separating the plasmas of interstellar and solar origins. In this case, the heliosheath is the region outside the termination shock, but inside the heliopause. The idealized streamlines of the subsonic solar wind plasma show that it is in fact turned towards the direction of the interstellar flow to form the heliotail (Suess and Nerney 1990).

It is important to revisit the idealized picture drawn of the termination shock above. Although the basic elements of the interaction are correctly represented by the pressure balance equation described above, a first necessary correction is the three-dimensional nature of the shock, as it envelops the heliosphere. The observations in the heliosphere show that the solar wind is present in all directions around the Sun; however, it is very non-uniform, both in density and speed, so that there is, at any time, a very large range of variability in the dynamic pressure of the solar wind as a function of heliographic latitude and, in fact, longitude. The processing of the solar wind in the outer heliosphere modifies, through the dynamic interaction of the different solar wind streams, the non-uniform distribution of the dynamic pressure observed in the inner heliosphere. This means that it is somewhat unclear to what extent the simplest assumption, that of spherical symmetry of the termination shock, is in error. It is nevertheless highly likely that there is at least a strong latitudinal dependence in the termination shock distance; this asymmetry may well be dependent on the phase of the solar cycle. The first complete latitude survey of the dynamic pressure of the solar wind by Ulysses showed (Phillips *et al.* 1995a) that the solar wind dynamic pressure was in fact higher at high heliolatitudes than in the equatorial regions, at least around solar minimum when the high-speed solar wind streams from the polar coronal holes dominated the heliosphere away from the equatorial plane. This observation led to the suggestion that the termination shock was at a greater distance from the Sun over the polar regions, when compared to its distance in the equatorial plane.

Another factor in the estimates of what happens near the boundary of the heliosphere is the role played in the dynamics by the neutral gas originating in the LISM, but streaming unimpeded into the heliosphere at the relative speed between the heliosphere and the LISM. Neutral particles are eventually ionized by either UV radiation from the Sun or by the charge exchange mechanism with the solar wind plasma (e.g. Isenberg 1999, and references therein). It is these particles, once picked up by the solar wind that are then convected outwards towards the termination shock and so constitute the seed population to the so-called anomalous component of the cosmic rays. The analysis and extrapolation of the ACR observations by Voyager have placed potentially important constraints on both the strength of the termination shock and on its heliocentric distance as already described in the section on 'Cosmic-ray modulation as a probe of the heliosphere'.

The pressure of pickup ions in the outer heliosphere may in fact contribute to a slowdown of the solar wind (Richardson *et al.* 1995; Burlaga *et al.* 1996). A study based on the comparison of solar wind speeds at Ulysses in the inner heliosphere and Voyager 2 in the outer heliosphere (Wang *et al.* 2000) has yielded an estimate of a slowdown of up to 10% out to about 60 AU. Such a change in the solar wind speed clearly needs to be taken into account when examining the pressure balance that is the basis of estimating the distance of the termination shock.

In addition, the potentially large variability of the termination shock distance, in response to varying conditions in the solar wind in phase with the solar activity cycle, remains a topic of great interest as the Voyager spacecraft approaches this first boundary of the heliosphere. The dynamic pressure of the solar wind may vary by as much as a factor of two;

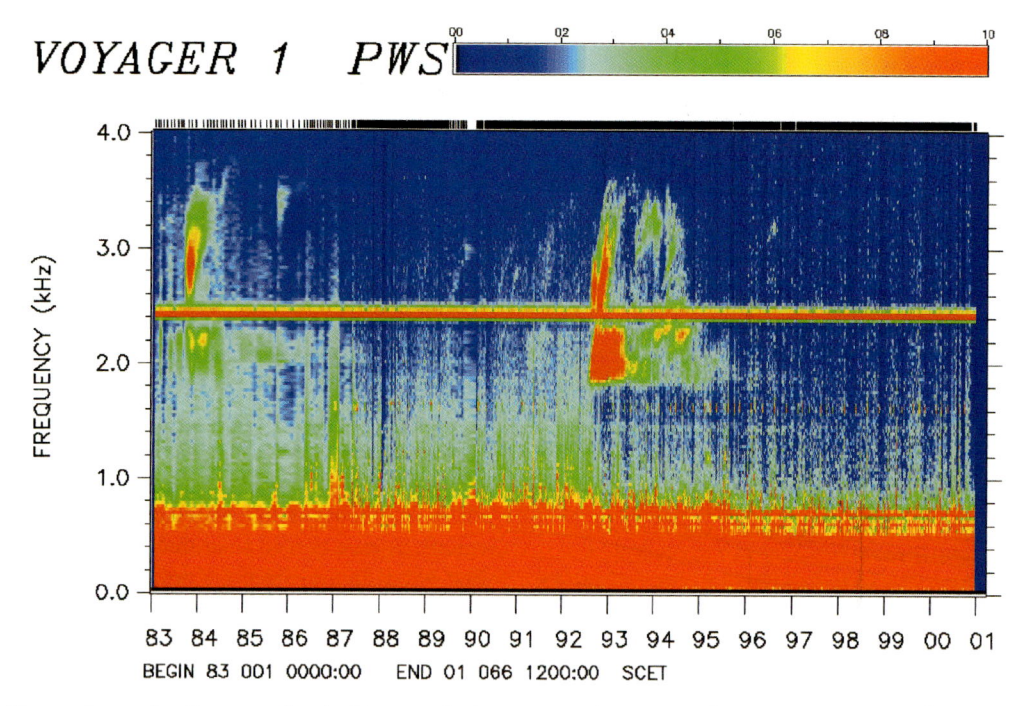

Figure 15 Observations of radio waves by the Voyager 1 plasma wave instrument from 1983 to 2000. The events in 1983 and 1992–93 have been interpreted as radio waves generated at the heliopause, in response to strong shockwaves that travelled out from the Sun at times of exceptionally high solar activity. (By courtesy of D.A. Gurnett and W.S. Kurth, University of Iowa.)

this figure leads to a variation in the termination shock distance of about 20 AU (Wang and Belcher 1999). Other efforts at modelling (e.g. Zank and Pauls 1996), taking into account the effects of the interstellar neutral atoms penetrating into the heliosphere and their conversion into a population of pickup ions, as well as the variability in solar wind dynamic pressure clearly show that both the location and structure of the termination shock are subject, at present, to uncertainties (see also review by Lee 1996). The only resolution will come from the direct observation of the termination shock itself. Even given the uncertainties, the Voyager spacecraft is expected to provide this vital evidence of the outer boundary of the heliosphere within the next decade.

Another estimate of the location of the boundary of the heliosphere comes from the remarkable observations made by the plasma wave instrument on the two Voyager spacecraft (Kurth et al. 1984; Gurnett and Kurth 1996), illustrated in Figure 15. First in 1983, then again in 1992–93, significantly enhanced wave activity was detected by the plasma wave instruments between 2 and 3 kHz. These long-lasting events have been interpreted (see review by Gurnett and Kurth 1996) as radio waves generated far in the outer heliosphere, in the vicinity of the heliopause, in response to shockwaves propagating from the Sun. The waves are likely to be emitted as a result of a shock-driven Langmuir-wave mode conversion mechanism at (or twice) the plasma frequency.

In both cases, there was very intense solar transient activity, about 400 days prior to the onset of the radio wave events. Several very large CMEs were observed in 1982 and in the first half of 1991 which, judging by their effects on cosmic-ray propagation in the heliosphere, created major, shock-led disturbances in the heliosphere as a whole. It is these two systems of shockwaves, reaching the heliopause, that are thought to have been at the origin of the events detected by both Voyager spacecraft. Given the delay between the solar events that gave rise to the disturbances, and given a range of assumptions concerning the propagation of the shock waves through the heliosphere, it is possible to estimate the distance to the heliopause to be between about 110 and 160 AU. Although the range of the estimate is very wide, the remarkable fact remains that the observations represent a remotely sensed, but nevertheless relatively direct detection of the heliospheric boundary.

REFERENCES

Axford, W.I. (1972). The interaction of the solar wind with the interstellar medium. In C.P. Sonnett, P.J. Coleman, Jr. and J.M. Wilcox (eds), *Solar Wind*, NASA-SP-308, Washington DC, p. 609.

Axford, W.I. (1996). The heliosphere. *Space Science Reviews*, **78**, 9–14. [Also in R. von Steiger, R. Lallement and M.A. Lee (eds) (1996). *The Heliosphere in the Local Interstellar Medium*, Kluwer, Dordrecht, pp. 9–14.]

Balogh, A. (1998). Magnetic fields in the inner heliosphere. *Space Science Reviews*, **83**, 93–104. [Also in L.A. Fisk, J.R. Jokipii, G.M. Simnett, R. von Steiger and K.P. Wenzel (eds) (1998). *Cosmic Rays in the*

Heliosphere, Space Science Series of ISSI, Kluwer Academic Publishers, Dordrecht, pp. 93–104.]

Balogh, A., Gosling, J.T., Jokipii, J.R., Kallenbach, R. and Kunow, H. (eds) (1999). *Corotating Interaction Regions*, Kluwer, Dordrecht.

Balogh, A., Marsden, R.J. and Smith, E.J. (eds) (2001). *The Heliosphere Near Solar Minimum: The Ulysses Perspective*, Springer-Praxis, London.

Balogh, A., Smith, E.J., Tsurutani, B.T., Southwood, D.J., Forsyth, R.J. and Horbury, T.S. (1995). The heliospheric magnetic field out of the ecliptic plane. *Science*, **268**, 1007.

Bame, S.J., Goldstein, B.E., Gosling, J.T., Harvey, J.W., McComas, D.J., Neugebauer, M. and Phillips, J.L. (1993). Ulysses observations of a recurrent high speed solar wind stream and the heliomagnetic streamer belt. *Geophysical Research Letters*, **20**, 2323–2326.

Behannon, K.W., Burlaga, L.F., Hoeksema, J.T. and Klein, L.W. (1989). Spatial variation and evolution of heliospheric sector structure. *Journal of Geophysical Research*, **94**, 1245–1260.

Belcher, J.W., Lazarus, A.J., McNutt, R.L., Jr. and Gordon, G.S., Jr. (1993). Solar wind conditions in the outer heliosphere and the distance to the termination shock. *Journal of Geophysical Research*, **98**, 15177–15183.

Biermann, L. (1957). Solar corpuscular radiation and the interplanetary gas. *The Observatory*, **107**, 109–110.

Burlaga, L.F. (1991). Magnetic clouds. In R. Schwenn and E. Marsch (eds), *Physics of the Inner Heliosphere*, Vol. 2, Springer-Verlag, Berlin, p. 1.

Burlaga, L.F. (1995). *Interplanetary Magnetohydrodynamics*, Oxford University Press, New York.

Burlaga, L.F., Klein, L.W., Lepping, R.P. and Behannon, K.W. (1984). Large scale magnetic fields: Voyager 1 and 2 observations between 1 AU and 9.5 AU. *Journal of Geophysical Research*, **89**, 10659–10668.

Burlaga, L.F., McDonald, F.B., Goldstein, M.L. and Lazarus, A.J. (1985). Cosmic ray modulation and turbulent interaction regions near 11 AU. *Journal of Geophysical Research*, **90**, 12027–12039.

Burlaga, L.F. and Ness, N.F. (1994). Merged interaction regions and large-scale magnetic fluctuations during 1991: Voyager 2 observations. *Journal of Geophysical Research*, **99**, 19341–19350.

Burlaga, L.F. and Ness, N.F. (1998). Voyager observations of the magnetic field in the distant heliosphere. *Space Science Reviews*, **83**, 105–121.

Burlaga, L.F. and Ness, N.F. (2000). Merged interaction regions observed by Voyagers 1 and 2 during 1998. *Journal of Geophysical Research*, **105**, 5141–5148.

Burlaga, L.F., Ness, N.F., Belcher, J.W., Lazarus, A.J. and Richardson, J.D. (1996). Voyager observations of the magnetic field, interstellar pickup ions and solar wind in the outer heliosphere. *Space Science Reviews*, **78**, 33–42. [Also in R. von Steiger, R. Lallement and M.A. Lee (eds), *The Heliosphere in the Local Interstellar Medium*, Space Science Series of ISSI, Kluwer, Dordrecht, pp. 33–42.]

Cane, H.V. (2000). Coronal mass ejections and Forbush decreases. *Space Science Reviews*, **93**, 55–77.

Chapman, S. and Bartels, J. (1940). *Geomagnetism*, Vols 1–2, Oxford University Press, Oxford.

Coleman, P.J., Jr., Davis, L., Jr., Smith, E.J. and Sonett, C.P. (1962). Interplanetary magnetic fields. *Science*, **138**, 1099–1100.

Crooker, N.U., Burton, M.E., Siscoe, G.L., Kahler, S.W., Gosling, J.T. and Smith, E.J. (1996). Solar wind streamer belt structure. *Journal of Geophysical Research*, **101**, 24331–24342.

Cummings, A.C. and Stone, E.C. (1996). Composition of anomalous cosmic rays and implications for the heliosphere. *Space Science Reviews*, **78**, 117–128. [Also in R. von Steiger, R. Lallement and M.A. Lee (eds) (1996). *The Heliosphere in the Local Interstellar Medium*, Kluwer, Dordrecht, pp. 117–128.]

Cummings, A.C. and Stone, E.C. (1998). Anomalous cosmic rays and solar modulation. *Space Science Reviews*, **83**, 51–62.

Davis, L.E., Jr. (1955). Interplanetary magnetic fields and cosmic rays. *Physical Review*, **100**, 1440–1444.

Davis, L., Jr., Smith, E.J., Coleman, P.J., Jr. and Sonett, C.P. (1966). Interplanetary magnetic measurements. In R.J. Mackin, Jr. and M. Neugebauer (eds), *The Solar Wind*, JPL Technical Report 32-360, Pergamon, London, pp. 35–52.

Fisk, L.A. (1996a). Motion of the footpoints of heliospheric magnetic field lines at the Sun: Implications for recurrent energetic particle events at high heliographic latitudes. *Journal of Geophysical Research*, **101**, 15547–15554.

Fisk, L.A. (1996b). Implications of a weak termination shock. *Space Science Reviews*, **78**, 129–136. [Also in R. von Steiger, R. Lallement and M.A. Lee (eds) (1996). *The Heliosphere in the Local Interstellar Medium*, Space Science Series of ISSI, Kluwer Academic Publishers, Dordrecht, pp. 129–136.]

Fisk, L.A. and Schwadron, N.A. (2001). The behavior of open magnetic field of the Sun. *Astrophysical Journal*, **560**, 425–438.

Fisk, L.A., Jokipii, J.R., Simnett, G.M., von Steiger, R. and Wenzel, K.-P. (eds) (1998). *Cosmic Rays in the Heliosphere*, Kluwer, Dordrecht.

Fisk, L.A., Kozlovsky, B. and Ramaty, R. (1974). An interpretation of the observed oxygen and nitrogen enhancements in low-energy cosmic rays. *Astrophysical Journal*, **190**, L35.

Fisk, L.A., Zurbruchen, T.H. and Schwadron, N.A. (1999). On the coronal magnetic field: consequences of large-scale motions. *Astrophysical Journal*, **521**, 868–877.

Forsyth, R.J., Balogh, A., Horbury, T.S., Erdos, G., Smith, E.J. and Burton, M.E. (1996b). The heliospheric magnetic field at solar minimum: Ulysses observations from pole to pole. *Astronomy and Astrophysics*, **316**, 287–295.

Forsyth, R.J., Balogh, A., Smith, E.J., Erdos, G. and McComas, D.J. (1996a). The underlying Parker spiral structure in the Ulysses magnetic field observations 1990–1994. *Journal of Geophysical Research*, **101**, 395–404.

Frisch, P.C. (1996). LISM structure – fragmented superbubble shell? *Space Science Reviews*, **78**, 213–222. [Also in R. von Steiger, R. Lallement and M.A. Lee (eds) (1996). *The Heliosphere in the Local Interstellar Medium*, Space Science Series of ISSI, Kluwer Academic Publishers, Dordrecht, pp. 213–222.]

Geiss, J., Gloeckler, G. and von Steiger, R. (1995). Origin of the solar wind from composition measurements. *Space Science Reviews*, **72**, 49.

Geiss, J. and Witte, M. (1996). Properties of the interstellar gas in the heliosphere. *Space Science Reviews*, **78**, 229–238. [Also in R. von Steiger, R. Lallement and M.A. Lee (eds) (1996). *The Heliosphere in the Local Interstellar Medium*, Kluwer, Dordrecht, pp. 229–238.]

Giacalone, J. (1998). Cosmic-ray transport coefficients. *Space Science Reviews*, **83**, 351–363.

Gosling, J.T. (1990). Coronal mass ejections and magnetic flux ropes in interplanetary space. In C.T. Russell, E.R. Priest and L.C. Lee (eds), *Physics of Magnetic Flux Ropes*, AGU Monograph 58, American Geophysical Union, Washington, DC, pp. 343–364.

Gosling, J.T., Birn, J. and Hesse, M. (1995). Three-dimensional magnetic reconnection and and the magnetic topology of coronal mass ejection events. *Geophysical Research Letters*, **22**, 869–872.

Gosling, J.T., McComas, D.J., Phillips, J.L., Weiss, L., Pizzo, V.J., Goldstein, B.E. and Forsyth, R.J. (1994). A new class of forward-reverse shock pairs in the solar wind. *Geophysical Research Letters*, **21**, 2271–2274.

Gonzalez-Esparza, J.A., Balogh, A., Forsyth, R.J., Neugebauer, M., Smith E.J. and Phillips, J.L. (1996). Interplanetary shock waves and large-scale structures: Ulysses' observations in and out of the ecliptic plane. *Journal of Geophysical Research*, **101**, 17057–17072.

Grzedzielski, S. and Page, D.E. (eds) (1990). *Physics of the Outer Heliosphere*, Pergamon Press, Oxford.

Gurnett, D.A. and Kurth, W.S. (1996). Radio emissions from the outer heliosphere. *Space Science Reviews*, **78**, 53–66. [Also in R. von Steiger, R. Lallement and M.A. Lee (eds) (1996). *The Heliosphere in the Local Interstellar Medium*, Space Science Series of ISSI, Kluwer Academic Publishers, Dordrecht, pp. 53–66.]

Hoeksema, J.T. (1991). Large scale solar and heliospheric magnetic fields. *Advances in Space Research*, **11**, 15–24.

Hoeksema, J.T. (1995). The large scale structure of the heliospheric current sheet during the Ulysses epoch. *Space Science Reviews*, **72**, 137. [Also in R.G. Marsden (ed) (1995). *The High Latitude Heliosphere*, Kluwer, Dordrecht, p. 137.]

Horbury, T.S., Balogh, A., Forsyth, R.J. and Smith, E.J. (1996). The rate of turbulent evolution over the Sun's poles. *Astronomy and Astrophysics*, **316**, 333–341.

Hundhausen, A.J. (1993). Sizes and locations of coronal mass ejections: SMM observations from 1980 and 1984–89. *Journal of Geophysical Research*, **98**, 13177–13200.

Isenberg, P.A. (1999). Interstellar pickup ions in the solar wind. In S.R. Habbal, R. Esser, J.V. Hollweg and P.A. Isenberg (eds), *Solar Wind Nine*, American Institute of Physics, Woodbury, NY, pp. 189–193.

Jokipii, J.R. and Giacalone, J. (1998). The theory of anomalous cosmic rays. *Space Science Reviews*, **83**, 123–136.

Jokipii, J.R. and Kota, J. (1989). The polar heliospheric magnetic field. *Geophysical Research Letters*, **16**, 1–4.

Jokipii, J.R., Levy, E.H. and Hubbard, W.B. (1977). Effects of particle drift on cosmic ray transport I. General properties, application to solar modulation. *Astrophysical Journal*, **213**, 861–868.

Jokipii, J.R., Sonett, C.P. and Giampapa, M.S. (eds) (1997). *Cosmic Winds and the Heliosphere*, University of Arizona Press, Tucson, AZ.

Jokipii, J.R. and Thomas, B.T. (1981). Effects of drifts on the transport of cosmic rays, IV. Modulation by a wavy interplanetary current sheet. *Astrophysical Journal*, **243**, 1115–1122.

Kahler, S.W., Crooker, N.U. and Gosling, J.T. (1996). The topology of intrasector reversals of the interplanetary magnetic field. *Journal of Geophysical Research*, **101**, 24373–24382.

Kurth, W.S., Gurnett, D.A., Scarf, F.L. and Poynter, R.L. (1984). Detection of a radio emission at 3 kHz in the outer heliosphere. *Nature*, **312**, 27–31.

Lallement, R. (1996). Relations between ISM inside and outside the heliosphere. *Space Science Reviews*, **78**, 361–374. [Also in R. von Steiger, R. Lallement and M.A. Lee (eds) (1996). *The Heliosphere in the Local Interstellar Medium*, Kluwer, Dordrecht, pp. 361–374.]

Lazarus, A.J., Belcher, J.W., Paularena, K.I. and Richardson, J.D. (1998). Voyager 2 solar wind observations in the outer heliosphere. *Space Science Reviews*, **83**, 87–92. [Also in L.A. Fisk, J.R. Jokipii, G.M. Simnett, R. von Steiger and K.-P. Wenzel (eds) (1998). *Cosmic Rays in the Heliosphere*, Kluwer, Dordrecht, pp. 87–92.]

Lee, M.A. (1996). The termination shock of the solar wind. *Space Science Reviews*, **78**, 109–116. [Also in R. von Steiger, R. Lallement and M.A. Lee (eds) (1996). *The Heliosphere in the Local Interstellar Medium*, Kluwer, Dordrecht, pp. 109–116.]

Lepping, R.P., Szabo, A., Peredo, M. and Hoeksema, J.T. (1996). Large scale properties and solar connection of the heliospheric current and plasma sheets: WIND observations. *Geophysical Research Letters*, **23**, 1199.

Lüst, R. (1967). The properties of interplanetary space. In J.W. King and W.S. Newman (eds), *Solar-Terrestrial Physics*, Academic Press, London, New York, p. 1.

Marsden, R.G. (ed) (1986). *The Sun and the Heliosphere in Three Dimensions*, D. Reidel, Dordrecht.

Marsden, R.G. (ed) (1995). *The High Latitude Heliosphere*, Kluwer, Dordrecht.

Marsden, R.G. (ed) (2001). *The 3-D Heliosphere at Solar Maximum*, Kluwer, Dordrecht.

McDonald, F.B. (1998). Cosmic ray modulation in the heliosphere: A phenomenological study. *Space Science Reviews*, **83**, 33–50.

McDonald, F.B. (2000). Integration of neutron monitor data with spacecraft observations: A historical perspective. *Space Science Reviews*, **93**, 263–284.

McDonald, F.B., Lal, N., Trainor, J.H., van Hollebeke, M.A.I. and Webber, W.R. (1981). The solar modulation of galactic cosmic rays in the outer heliosphere. *Astrophysical Journal*, **249**, L71–L75.

McKibben, R.B. (1998). Three-dimensional solar modulation of cosmic rays and anomalous components in the inner heliosphere. *Space Science Reviews*, **83**, 21–32.

Musmann, G., Neubauer, F.M. and Lammers, E. (1977). Radial variation of the interplanetary magnetic field between 0.3 and 1.0 AU. *Journal of Geophysics*, **42**, 591–598.

Neugebauer, M. and Snyder, C.W. (1962). The mission of Mariner II: Preliminary observations. *Science*, **138**, 1095–1097.

Ness, N.F. and Burlaga, L.F. (1996). Merged interaction regions and large scale fluctuations observed by Voyagers 1 and 2 in the distant heliosphere. In D. Winterhalter, J.T. Gosling, S.R. Habbal, W.S. Kurth and M. Neugebauer (eds), *Solar Wind Eight*, American Institute of Physics, Woodbury, NY, p. 591.

Ness, N.F. and Wilcox, J.M. (1964). Solar origin of the interplanetary magnetic field. *Physical Review Letters*, **13**, 461–464.

Parker, E.N. (1958). Dynamics of the interplanetary gas and magnetic field. *Astrophysical Journal*, **128**, 664.

Parker, E.N. (1963). *Interplanetary Dynamical Processes*, Wiley-Interscience, New York.

Parker, E.N. (1965). The passage of energetic charged particles through interplanetary space. *Planetary and Space Science*, **13**, 9–49.

Pesses, M.E., Jokipii, J.R. and Eichler, D. (1981). Cosmic ray drift, shock acceleration and the anomalous component of cosmic rays. *Astrophysical Journal Letters*, **246**, L85–L88.

Phillips, J.L., Bame, S.J., Barnes, A., Barraclough, B.L., Feldman, W.C., Goldstein, B.E., Gosling, J.T., Hoogeveen, G.W., McComas, D.J., Neugebauer, M. and Suess, S.T. (1995a). Ulysses solar wind plasma observations from pole to pole. *Geophysical Research Letters*, **22**, 3301.

Phillips, J.L., Bame, S.M., Feldman, W.C., Goldstein, B.E., Gosling, J.T., Hammond, C.M., McComas, D.J., Neugebauer, M., Scime, E.E. and Suess, S.T. (1995b). Ulysses solar wind plasma observations at high southerly latitudes. *Science*, **268**, 1030.

Pizzo, V.J. (1991). The evolution of corotating stream fronts near the ecliptic plane in the inner solar system: 2. Three-dimensional tilted dipole fronts. *Journal of Geophysical Research*, **96**, 5405–5420.

Potgieter, M.S. (1998). The modulation of galactic cosmic rays in the heliosphere: Theory and models. *Space Science Reviews*, **83**, 147–158.

Richardson, J.D., Paularena, K.I., Lazarus, A.J. and Belcher, J.W. (1995). Evidence for a solar wind slowdown in the heliosphere? *Geophysical Research Letters*, **22**, 1469.

Riley, P., Gosling, J.T., Weiss, L.A. and Pizzo, V.J. (1996). The tilts of corotating interaction regions at midheliographic latitudes. *Journal of Geophysical Research*, **101**, 24349–24358.

Schwenn, R. (1990). Large scale structure of the interplanetary medium. In R. Schwenn and E. Marsch (eds), *Physics of the Inner Heliosphere*, Vol. 1, Springer-Verlag, Berlin, p. 99.

Schwenn, R. and Marsch, E. (eds) (1990). *Physics of the Inner Heliosphere*, Vol. 1, Springer-Verlag, Berlin.

Schwenn, R. and Marsch, E. (eds) (1991). *Physics of the Inner Heliosphere*, Vol. 2, Springer-Verlag, Berlin.

Simpson, J.A. (1998). A brief history of recurrent solar modulation of the galactic cosmic rays (1937–1990). *Space Science Reviews*, **83**, 169–176.

Smith, E.J. (1989). Interplanetary magnetic field over two solar cycles and out to 20 AU. *Advances in Space Research*, **9**, 159.

Smith, E.J. (1997). Solar wind magnetic fields. In J.R. Jokipii, C.P. Sonett and M.S. Giampapa (eds), *Cosmic Winds and the Heliosphere*, University of Arizona Press, Tucson, AZ, p. 425.

Smith, E.J., Balogh, A., Neugebauer, M. and McComas, D.J. (1995). Ulysses observations of Alfven waves in the southern and northern solar hemispheres. *Geophysical Research Letters*, **22**, 3381.

Smith, E.J., Slavin, J.A. and Thomas, B.T. (1986). The heliospheric current sheet: 3-dimensional structure and solar cycle changes. In R.G. Marsden (ed), *The Sun and the Heliosphere in Three Dimensions*, D. Reidel, Dordrecht, p. 267.

Smith, E.J. and Wolfe, J.H. (1977). Pioneer 10, 11 observations of evolving solar wind streams and shocks beyond 1 AU. In M.A. Shea, D.F. Smart and S.T. Wu (eds), *Study of Travelling Interplanetary Phenomena*, D. Reidel, Dordrecht, p. 227.

Suess, S.T. and Nerney, S. (1990). Flow downstream of the heliospheric termination shock, 1: Irrotational flow. *Journal of Geophysical Research*, **95**, 6403–6412.

Suess, S.T. and Nerney, S. (1997). The termination shock and the heliosheath. In J.R. Jokipii, C.P. Sonett and M.S. Giampapa (eds), *Cosmic Winds and the Heliosphere*, University of Arizona Press, Tucson, AZ, p. 759.

Thomas, B.T. and Smith, E.J. (1980). The Parker spiral configuration of the interplanetary magnetic field between 1 and 8.5 AU. *Journal of Geophysical Research*, **85**, 6861–6867.

von Steiger, R., Lallement, R. and Lee, M.A. (eds) (1996). *The Heliosphere in the Local Interstellar Medium*, Kluwer, Dordrecht.

Wang, C. and Belcher, J.W. (1999). The heliospheric boundary response to large-scale solar wind fluctuations: A gasdynamic model with pickup ions. *Journal of Geophysical Research*, **104**, 549–556.

Wang, C., Richardson, J.D. and Gosling, J.T. (2000). A numerical study of the evolution of the solar wind from Ulysses to Voyager 2. *Journal of Geophysical Research*, **105**, 2337–2344.

Wang, Y.-M., Hawley, S.H. and Sheeley, N.R., Jr. (1996). The magnetic nature of coronal holes. *Science*, **271**, 464–469.

Wang, Y.-M. and Sheeley, N.R., Jr. (1994). Global evolution of interplanetary sector structure, coronal holes and solar wind streams during 1976–1993: Stackplot displays based on solar magnetic observations. *Journal of Geophysical Research*, **99**, 6597–6608.

Weiss, L.A., Gosling, J.T., McAllister, A.H., Hundhausen, A.J., Burkepile, J.T., Phillips, J.L., Strong, K.T. and Forsyth, R.J. (1996). A comparison of interplanetary coronal mass ejections at Ulysses with Yohkoh soft X-ray coronal events. *Astronomy and Astrophysics*, **316**, 384–395.

Wilcox, J.M. and Ness, N.F. (1964). Quasi-stationary corotating structure in the interplanetary medium. *Journal of Geophysical Research*, **70**, 5793–5805.

Wilcox, J.M., Scherrer, P.H. and Hoeksema, J.T. (1980). Origin of the warped heliospheric current sheet. *Science*, **209**, 603–605.

Zank, G.P. and Pauls, H.L. (1996). Modelling the heliosphere. *Space Science Reviews*, **78**, 95–106. [Also in R. von Steiger, R. Lallement and M.A. Lee (eds) (1996). *The Heliosphere in the Local Interstellar Medium*, Space Science Series of ISSI, Kluwer Academic Publishers, Dordrecht, pp. 95–106.]

49

EBERHARD GRÜN*

The dusty heliosphere

BEGINNINGS

Zodiacal light observations and meteor studies

Dust in space has been recognized for centuries. Three dusty phenomena in space can be observed by the naked eye: comets, meteors and zodiacal light. In the past the sporadic appearance of bright comets received much attention, and not only from astronomers, because it was linked to exceptional events in human history such as victory or defeat in war. The physical nature of comets was illuminated by the second dusty phenomenon: meteors. Although several meteors can be observed every clear night, there are special periods, so called meteor showers or meteor storms, when the rate is greatly enhanced. It was the coincidence between such meteor showers and the simultaneous apparitions of comets that suggested their relationship.

Triangulation already proved two hundred years ago that meteors are the luminous trail of pea-sized meteoroids that enter the Earth's atmosphere at about a height of 100 km. The speed of meteoroids was determined to range from $11\,\mathrm{km\,s^{-1}}$, the escape speed from the Earth, to $72\,\mathrm{km\,s^{-1}}$ which is the maximum collision speed with the Earth that a meteoroid can have on a bound orbit about the Sun. This speed range clearly identified meteoroids as members of the Solar System. The scattering of radio waves by the ionization trail in the atmosphere allows us to observe meteors even during the day. Such radar meteoroids can be as small as 0.1 mm in size. Football-sized and bigger meteoroids may survive the entry into the atmosphere, and some residual material may fall to the ground as meteorites, which can be picked up and examined. Up to now only a handful of meteorite falls have been observed with sufficient accuracy that an orbit could be established. All these cases were ordinary stony meteorites which had orbits with aphelia in the asteroid belt.

Because of their exotic nature, meteorites belong to the best-analysed samples on Earth. While some resemble ordinary stones, others consist of almost pure iron–nickel and a third type consists of a mixture of elements that is believed to be the mean composition (cosmic abundance) of the whole planetary system except for some volatile elements, like hydrogen and helium. The latter meteorites (carbonaceous chondrites) are considered to consist of primitive material that has not been modified by thermal processes like those that occur in the interior of planets. Meteoritic material has been used as a model for the material of which interplanetary dust particles are made.

From measurements of the entry trajectory and models of the upper atmosphere it became clear that meteoroids experience quite variable decelerations. While some slowed within a short time, others kept their speed for much longer times. The concept of loosely bound aggregate particles with large cross-section/mass ratios (i.e. low bulk density) was developed for those that decelerated rapidly; others behaved more like normal compact stones and even iron particles. Meteor physicists like Zdenek Ceplecha (1977) found a correlation between meteoroid densities and their orbits; namely, some meteoroids that have eccentric, comet-like orbits consist of low-density material, while higher-density meteoroids have orbits that just reached the asteroid belt.

The asteroid belt has always been suspected to be a source of meteoroids. At the beginning of the twentieth century, Max Wolf detected from the Sternwarte above Heidelberg such a large number of asteroids that he had a hard time to

---

*Max-Planck-Institut für Kernphysik, Heidelberg, Germany and University of Hawaii, Honolulu, HI, USA

find names for all of them. Since then the number of asteroids for which precise orbits are known has grown to over 20,000. Most asteroids have orbits between Mars and Jupiter. In 1918 Kiyotsugu Hirayama suggested that asteroids with very similar orbits form families that have been generated by the break-up of a larger asteroid in a recent (about 100 million years ago) catastrophic collision. Julius Dohnanyi (1970) developed a collision model of asteroids in which smaller asteroids can break up bigger ones. In each collision a flood of smaller fragments is generated. He showed that the size distribution of observed asteroids is compatible with a population of objects in collisional equilibrium. In these collisions a large amount of dust is generated.

The third dusty phenomenon is the zodiacal light, visible to the naked eye in the morning and evening sky in areas free of light pollution. As early as 1683, Cassini presented the correct explanation of this phenomenon: it is sunlight scattered by dust particles orbiting the Sun. A relation to other "dusty" interplanetary phenomena, like comets, was soon suspected. Comets shed large amounts of dust during their passage through the inner Solar System, which is most visible in their dust tails. The larger of these dust grains stay for extended periods close to the orbit of their parent comet, and are recognized as a meteor shower when they cross the orbit of the Earth. The finer dust gets injected into the zodiacal cloud. Astronomical observations showed that this dust is widely distributed in the planetary system, at least out to the asteroid belt. However, the situation of a zodiacal light observer is similar to that of a person in the midst of a cloud who is trying to determine the extent and the distribution of the cloud. Many particles along the line of sight which are at different distances from the Sun and have different scattering angles (i.e. the angle between Sun and observer) contribute to the zodiacal light brightness. The zodiacal light brightness, therefore, depends both on the dust density and the scattering function, both of which are unknown functions of the distance from the Sun. A way to resolve this problem, at least partially, was developed by René Dumont (1983). He used the fact that an observer (on the Earth or on a spacecraft) who is moving through the dust cloud will observe in the direction of motion different brightnesses at different times. The difference between two brightness values corresponds to the amount of dust along the line between the two observations. Thereby, the observer is able to derive the spatial density along the line of sight.

The dust community – astronomers and space physicists

In the early days of space research, the space agencies, mainly in the USA, organized conferences and workshops to present and discuss results from space measurements of dust, laboratory simulations and theoretical studies. International exchange of latest results in dust research took place at COSPAR (Committee on Space Research) conferences. It was Curt Hemenway, from the Dudley Observatory in Albany, USA, who founded the Cosmic Dust Panel within COSPAR. At the annual (and, later, biannual) meetings, researchers from around the world met and exchanged their results and ideas on dust in space. Initially, astronomers met in parallel at meetings of the International Astronomical Union (IAU). Soon it was recognized that it was mutually advantageous to bring together these partially overlapping communities of scientists. In 1967 Jerry Weinberg organized the first international dust meeting at the University of Hawaii in Honolulu, USA, that produced a significant proceedings volume (Weinberg 1967). After that, specialized international dust meetings took place every four to five years under the sponsorship of IAU and COSPAR. The first of these meetings, IAU Colloquium No. 13, was organized in 1971 by Curt Hemenway, Peter Millman and Alan Cook (Hemenway et al. 1973) in Albany, USA. It was followed in 1975 by IAU Colloquium No. 31 at the Max-Planck-Institut für Kernphysik in Heidelberg, Germany (Elsässer and Fechtig 1976; the attendees of this meeting are shown in Figure 1), in 1979 by IAU Symposium No. 90 at the Herzberg Institute of Astrophysics in Ottawa, Canada (Halliday and McIntosh 1980), by IAU Colloquium No. 85 in 1984 at the Laboratoire d'Astronomie Spatiale in Marseille, France (Giese and Lamy 1985), IAU Colloquium No. 126 in 1990 at the Kyoto University in Kyoto, Japan (Levasseur-Regourd and Hasegawa 1991), IAU Colloquium No. 150 in 1995 at the University of Florida in Gainesville, USA (Gustafson and Hanner 1996), and IAU Colloquium No. 181 organized in 2000 by Ian Williams and Tony McDonnell at the University of Kent in Canterbury, England. Proceedings of these meetings are a good record of the knowledge and an indicator of the advancements in the field.

Besides these international specialist dust meetings, there are many national meetings or meetings on broader topics in which dust research plays some role. Among the regular meetings are meetings of the Meteoritical Society at which analyses of meteoritic and other extraterrestrial material are discussed; the Lunar and Planetary Science Conferences in Houston, which focus on the study of the lunar, Martian and Venusian samples and related topics; AGU (American Geophysical Union) and EGS (European Geophysical Society) meetings at which a broad range of topics in geophysics and space plasma physics are discussed; DPS (Division of Planetary Science of the American Astronomical Society) meetings at which discussions on a wide range of planetary topics take place, from observations of planetary phenomena to theories of planetary formation; and the Asteroid, Comets, Meteors (ACM) meetings, which were founded in 1983 by Hans Rickman and Karl Lagerkvist in Upsala, Sweden, and later held at

1.	L. Kohoutek	16.	J.L. Weinberg	31.	J.W. Rhee	46.	J.R. Roach	61.	H.U. Keller
2.	P. Proisy	17.	Z. Ceplecha	32.	J. Trulsen	47.	D.A. Tomandl	62.	J. Rosinski
3.	Mrs. Proisy	18.	J.G. Sparrow	33.	Z. Sekanina	48.	S. Röser	63.	?
4.	K.W. Michel	19.	W. Kempe	34.	H.J. Staude	49.	V. Stähle	64.	D.S. Hallgren
5.	D.E. Brownlee	20.	G. Eichhorn	35.	J.M. Alvarez	50.	O.E. Berg	65.	L.W. Bandermann
6.	A. Mujica	21.	E. Grün	36.	E. Pitz	51.	J. Kissel	66.	D.W. Hughes
7.	A. Llebaria	22.	B. Donn	37.	T. Nishimura	52.	R. Soberman	67.	C.F. Lillie
8.	W. Kokott	23.	D.A. Morrison	38.	S. Drapatz	53.	J. Hartung	68.	H. Lee
9.	Mrs. Gehrels	24.	B. Marsden	39.	H. Elsässer	54.	H. Wolf	69.	R.A. Howard
10.	C. Leinert	25.	B.K. Dalmann	40.	P. Szkody	55.	J.C. Mandeville	70.	R.D. Wolstencroft
11.	T. Gehrens	26.	W. Gentner	41.	J.A.M. McDonnell	56.	R. Dumont	71.	G. Morfill
12.	J.G. Delcourt	27.	H.J. Völk	42.	E. Schneider	57.	B. König	72.	R.H. Munro
13.	S. Ternesvary	28.	M.S. Hanner	43.	V. Vanysek	58.	J.E. Blamont	73.	G. Schwehm
14.	R. Bloch	29.	F. Link	44.	J. Rahe	59.	A. Levasseur	74.	S. Hayakawa
15.	N. Pailer	30.	R. Robley	45.	F.E. Roach	60.	G.B. Burnett	75.	K.D. Schmidt
76.	H. Link								
77.	M. Kröger								
78.	R.H. Giese								
79.	G. Braun								
80.	P.W. Blum								
81.	H. Tanabe								
82.	M. Alexander								
83.	A.H. Delsemme								
84.	H. Fechtig								
85.	F.L. Whipple								
86.	J.S. Dohnanyi								
87.	C.L. Hemenway								
88.	P.M. Millman								
89.	R. Wlochowicz								

Figure 1 Participants of the 1975 Dust Meeting in Heidelberg (Elsässer and Fechtig 1976).

different places around the world. There are one or more meetings each year to which an active researcher in the dust field could go in order not to lose contact with the latest developments in the field.

In 1978 Tony McDonnell took up the task to assemble a comprehensive review of the "cosmic dust" field (Mc Donnell 1978). With contributions from Fred Whipple, Jerry Weinberg, David Hughes, Mayo Greenberg, Don Brownlee, D.G. Ashworth, Julius Dohnanyi and Hugo Fechtig as lead authors, a wide range of topics was covered: from zodiacal light over lunar craters to dust particles collected in the atmosphere; from dust dynamics to laboratory simulations.

It was more than 20 years before a new version of a comprehensive review of the field was published. The author, Bo Gustafson, Stan Dermott and Hugo Fechtig (Grün et al. 2001), edited reviews of topics ranging from near-Earth dust to circumplanetary to cometary and interstellar dust, from instrumentation over properties of interplanetary dust to orbital evolution of dust.

Table 1 lists major research centres and schools at which space dust was or is being studied. Key persons, relevant methods and space missions are given. Many early and recent dust researchers who made an impact on the field can be identified in the photograph (Figure 1), taken in 1975 at the IAU Colloquium No. 31. Since then some key figures have moved to other institutions, changed to different fields or terminated their dust research. Long-term involvement in the field and several researchers at one location have been used as criteria for inclusion in the table; however, the selection is subjective and is not complete in neighbouring fields, like comets, zodiacal light, theory, meteoritics, meteors, laboratory work, and others.

THE EARLY YEARS: FROM TOO MUCH DUST TO VERY LITTLE

Space dust: Dangerous, dirty and difficult

The danger from meteoroids to crewed and uncrewed activities was obvious. Centimetre-sized meteoroids entering the Earth's atmosphere at several tens of kilometres per second will certainly damage any spacecraft they strike. Smaller particles could also be dangerous to not-well-shielded satellites and to crewed extravehicular activity. Therefore, all space agencies considered it necessary to understand the hazard that comes from meteoroids. There were two aspects to consider: (1) the flux of meteoroids, which was a question of observations and measurements in space; and (2) the effect of hypervelocity impacts on space systems, which was best studied in the laboratory. Therefore, as one of the first space activities, vigorous research programmes were instigated to characterize the meteoroid hazard in space.

In everyday terms, "dust" is often a synonym "dirt" – something to be avoided. This is true as well for interplanetary dust. Astronomers who want to observe extra-Solar System objects have to contend with foreground obscuration by the zodiacal light. In the extremely empty space environment a few specks of dust, like paint flakes, cause significant problems to highly sensitive measurements; for example, infrared observers have to struggle with debris particles floating around their telescope. Physicist who want to study plasma effects in space have to modify their "clean" theories if dust is around. Dust is not easily controlled: it follows its own dynamics and disperses rapidly from its source, like smoke from a fire. This aspect, however, has a positive side: dust conveys messages from remote processes and objects by which it was generated.

Dust has many aspects and is thus difficult to quantify. An observer who determines one aspect of a dust particle in space, like its size, will find it difficult to determine other parameters. Dust particles come in many sizes, compositions and shapes. In all cases there is no single size, composition or shape that represents a certain space-dust environment. Therefore, many different parameters have to be measured in order to comprehensively characterize dust grains. Even if a dust particle is predominantly made of a single material, minute surface contamination changes its optical appearance so that it can no longer be easily recognized. If one wants to know the properties of an ensemble of dust particles, distributions in all relevant parameters have to be determined. Theoreticians modelling interplanetary dust have the difficulty of representing these particles by simplified models, for example a spherical particle of uniform composition and having the optical properties of a pure material. The description of the dynamics of dust involves many disciplines: Keplerian dynamics, interactions with the radiation field, and the plasma and magnetic environment. Some of the dust theories involve detailed assumptions about particle properties and the space environment that cannot be derived from first principles or that are not discernible from direct observations, and hence cause unease in theoreticians who are trained mostly in, for example, "pure" celestial mechanics. Interplanetary dust particles are difficult to characterize and to quantify, and only a small number of scientists, both experimenters and theoreticians, have dealt with them. However, it is the belief of the author that recent observational and theoretical methods are mature enough for us now to consider dust as an important and exciting subject of astrophysical research. Progress in the field is made by taking a multidisciplinary approach involving *in situ* space measurements, astronomical observations, theoretical studies and modelling, and laboratory investigations of basic processes. Close cooperation between astronomers, cosmochemists, dynamicists and experimental physicists has, in fact, been beneficial in solving dusty problems.

Table 1 Major space dust research centres (SC = Space Centre; Univ. = University) and schools, listed by location. Key figures (the numbers refer to Figure 1), space missions and relevant methods are included

Location and institution	Key figures	Space missions, topics, and relevant techniques
Albany, USA: Dudley Observatory and Univ.	Hanner (28), Hemenway (87), Weinberg (16)	Rockets, balloons, Skylab, Pioneers 10 and 11
Bochum, Germany: Univ.	Giese (78), Mann, Schwehm (73)	Microwave/light scattering, Ulysses
Boulder, USA: Univ.	Horanyi, Robertson	Experimental and theoretical studies of dust–plasma interactions
Bratislava, CS: Univ.	Kresak, Pittich, Porubcan	Meteors, theory
Cambridge, USA: Harvard-Smithsonian and Air Force Cambridge Research Laboratories (AFCRL)	Sekanina (33), Skrivanek, Southworth, Soberman (52), Whipple (85)	Comets, meteors, Pioneers 10 and 11
Canterbury, UK: Univ.	Green, Jennison, McBride, McDonnell (41)	Electrostatic dust accelerator, Ariel 2, Giotto, debris
Charlotteville, USA: Univ.	Singer	Theory
Chicago, USA: Univ.	Simpson, Tuzzolino	VeGa, Stardust
College Park, USA: Goddard SC	Alexander (82), Berg (50)	Electrostatic dust accelerator, Mariner 4, Lunar Explorer 35, OGO, Pioneers 8 and 9, LEAM
Gainesville, USA: Univ.	Dermott, Gustafson	Microwave/light scattering laboratory, theory
Hampton, USA: Langley SC	Alvarez (35), Humes, Kinard	Explorers 16 and 23, Lunar Orbiters, Pioneers 10 and 11, MTS, LDEF
Heidelberg, Germany: MPI-K and MPI-A	Auer, Elsässer (39), Fechtig (84), Grün (21), Jessberger, Kissel (51), Leinert (10), Lemke, Morfill (71)	Electrostatic dust accelerator, HEOS 2, Helios, Giotto, VeGa, Galileo, Ulysses, ISO, Cassini, Stardust, zodiacal light, interplanetary dust particles
Houston, USA: Johnson SC	Cour-Palais, Hörz, Kessler, Zook	Light gas gun, Gemini, Apollo, space shuttle
Huntsville, USA: Marshall SC	Naumann	Pegasus
Ithaca, USA: Univ.	Burns, Hamilton, Showalter	Galileo, rings, theory
Kobe, Japan: Univ.	Mukai	Muses, Hiten, zodiacal light
Marseille, France: Univ.	Lamy	D2A, zodiacal light
Moffet Field, USA: Ames SC	Farlow, Gault, Vedder	Light gas gun, electrodynamic dust accelerator, aircraft collections
Moscow, Russia: Acad. Science	Mazets, Nazarova, Vaisberg	Cosmos, Intercosmos, and Mars missions, VeGa
München, Germany: Univ.	Igenbergs	Plasma drag accelerator, Hiten, Planet B
Ondrjov, Czech Republic: Observatory	Ceplecha (17), Pecina	Meteors
Ottawa, Canada: Univ.	Millman (88)	Meteors
Prague, Czech Republic: Univ., Observatory	Solc, Svestka, Vanysek (43)	Comets
Redondo Beach, USA: TRW company	Früchtenicht	Dust accelerators, impacts
San Diego, USA: Univ.	Mendis	Theory of dust–plasma interactions
Seattle, USA: Univ.	Brownlee (5)	Stardust, interplanetary dust particles
St. Louis, USA: Univ.	Walker, Zinner	Interplanetary dust particles
Tokio, Japan: Univ.	Ohashi, Sasaki, Tanabe (81)	Nozomi, Electrostatic dust accelerator, zodiacal light
Verrières, France: Univ.	Blamont (58), Levasseur-Regourd (59)	D2A, zodiacal light lemage

A "Dust Belt" around the Earth?

Simple microphones were the first type of dust detector flown, in 1950 by Bohn and Nadig, on board a V2 rocket. It was easily verified by dropping glass beads on microphones that they responded to impacts of small particles. What the early investigators did not appreciate was that there were a variety of other effects to which microphones respond. Nilsson (1966) showed by using dual-method detectors that only a small fraction of the recordings were detected by the two methods simultaneously and hence were caused by dust impacts. Because of this effect, most modern dust detectors are "multi-coincidence", allowing dust impacts to be detected by several independent measurements.

Several events could mimic dust impacts: thermal cracks (such as in a heated kitchen oven) caused by temperature variations during a satellite's passage through the Earth's shadow; micro-vibrations caused by other systems on board the space vehicle; and electromagnetic interference caused by switching high currents or voltages. Thorough ground testing of the complete spacecraft helped to identify and prevent most of these interferences. Of the natural causes for fake impacts are cosmic rays hitting sensors or the sensitive first stages of the amplifier, and the time-variable ambient plasma parameter can interfere with sensitive dust detectors. One of the more intricate interference effects was observed on board the Ulysses spacecraft, where the sounder of the plasma wave instrument excited the ambient plasma by radio waves that in turn modulated the plasma currents to the spacecraft, which caused some noise in the dust detector.

Wire board detectors utilized the effect that an impact would have on an electronics board, namely cutting an electric connection. As with many early dust detectors, their sensitive area and exposure time did not match their sensitivity. Large particles, to which these detectors were sensitive, are rare and require large detector areas or long exposure times. Both were often unavailable, and hence no significant results were obtained. Sophisticated and highly sensitive early impact ionization detectors on the OGO (Orbiting Geophysical Observatory) satellites, that were developed by Alexander *et al.* (1971), produced no or only a few impacts, because – as we know today – their sensitive area was too small ($5\,cm^2$).

In the 1960s early microphone data from Explorer 8 and some Russian Cosmos and Salyut satellites were interpreted as near-Earth dust enhancements by factors of 1000 and higher than in interplanetary space. Measurements of the zodiacal light brightness set stringent constraints on the total dust density in interplanetary space. The near-Earth dust enhancement was supported by the first high-altitude dust collections using rockets, which had large numbers of particles on the collectors after their return to Earth. In a theoretical analysis, Shapiro *et al.* (1966) concluded that a near-Earth enhancement of at most a factor of 10 could reasonably be explained, but not enhancement factors of 100 and more. Indeed, the current best estimate of the gravitational enhancement is only a factor of 2. Simultaneously with the theoretical arguments, Nilsson (1966) expressed "some doubts about the Earth's dust cloud" on experimental grounds. He used "witness" microphone detectors that were shielded from dust impacts but otherwise identical to active ones. Both sensors recorded about the same event rates as the active ones, contradicting the dust hypothesis. Later, more sophisticated instruments showed that the initial high fluxes reported were erroneous and caused by the combined effect of immature dust detectors and the harsh near-Earth environment.

Very large dust detectors record very few meteoroids

From the beginnings of spaceflight, there were engineering-types of dust experiments that were designed not to address the meteoroid flux as a function of size, but to measure the resulting effect of the meteoroid flux on space systems. Micrometre-sized dust was of no interest since it could easily be shielded against. It was the milligram particles ($100\,\mu m$ and bigger) that the engineers were worried about. Such particles could penetrate millimetre-thick spacecraft walls and harm the sensitive systems inside. The typical dust detectors were of $1\,m^2$ area and employed robust detection methods like penetration detectors or capacitor sensors which were either immune to, or only slightly susceptible to, environmental interference.

Detectors of the "beer can" type were successfully flown on several early satellites and space probes (e.g. Explorers 16 and 23, Pioneers 10 and 11). These detectors consisted of a large number of pressurized cells that recorded the decrease in gas pressure that occurred when a wall was punctured by a meteoroid. The walls were metal sheets (stainless steel or copper–beryllium) of 25 or $50\,\mu m$ thickness that had penetration limits of 10^{-9} and $10^{-8}\,g$ at $20\,km\,s^{-1}$ impact speed. The Pegasus detectors were large-area (about $200\,m^2$) detectors that recorded penetrations by the discharge of a capacitor. Detectors like these determined the flux of meteoroids in near-Earth space in the $10\,\mu m$–$1\,mm$ size range. This range was important for the assessment of the meteoroid hazard of typical satellites and the crewed missions that lay ahead. Cour-Palais (1969) published the official NASA meteoroid model that described the Earth's natural meteoroid environment. It showed that no special or demanding precautions had to be taken to protect space systems against the meteoroid hazard. As a consequence, the interest in meteoroids shifted from technological to astrophysical questions and the amount of money to support dust studies declined. From the major groups at the NASA centres, only a few individual scientists continued to work in the meteoroid field; in Russia almost all meteoroid work ceased. It was universities and research institutions that maintained meteoroid research.

Revival of interest in the near-Earth dust environment

It was recognized by Kessler and Cour-Palais (1978) that steadily increasing space activity could potentially lead to hazardous conditions that would jeopardize any further space activities in low-Earth orbit. For example, explosions were observed in space of abandoned upper rocket stages

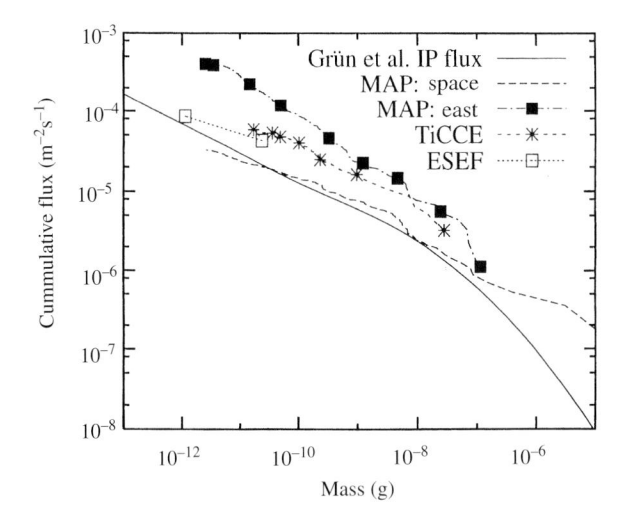

Figure 2 Natural dust v. space debris flux in low Earth orbit. Solid line is the interplanetary flux (Grün et al. 1985a) increased by a factor of two to account for gravitational focusing near Earth. This flux is compared with fluxes measured at two different faces of NASA's LDEF satellite (MAP space and MAP east), ESA's Eureca platform (TiCCE) and the Russian Mir space station (ESEF). The differences are due to different exposure geometries (McDonnell and Gardner 1998).

altitude in order to study the effects on materials during prolonged exposure to the space environment. Of primary interest was the effect of natural meteoroids and debris. Six years after launch, LDEF was retrieved by space shuttle and brought back to the ground. The study of the near-Earth dust environment was also the objective of the European Eureca satellite, which, like LDEF, was returned to Earth. The near-Earth meteoroid flux is about a factor of 2 higher than in deep space because of gravitational concentration by the Earth (Figure 2). Depending on the particle size, natural meteoroids are outnumbered by synthetic space debris. This is true for debris particles smaller than about 10 μm, which are about three times more abundant than natural meteoroids. Craters produced by space debris particles were identified by chemical analyses of residues in the craters. Residues have been found from space materials and signs of human activities in space like, for example, aluminium oxide spheres, aluminium, stainless steel, plastics and paint flakes. Monitoring the near-Earth dust environment is nowadays a routine activity of all major space agencies.

GAINING CONFIDENCE

Plenty of microcraters on lunar samples

In 1969 when NASA's astronauts returned samples from the Moon it was immediately recognized that they were peppered by micrometeoroid impacts (Figure 3). This demonstrated that the interplanetary dust flux had a significant effect on surfaces in space. Joseph Zähringer, a cosmochemist from the Max-Planck-Institut für Kernphysik, was a member of the team that had the first look at the lunar samples while they were still held in quarantine. Like the other team members, he immediately recognized the scientific potential of studying microcraters on lunar rocks. He suggested that his colleague Hugo Fechtig write a proposal to NASA to obtain lunar material for laboratory analysis. Like him, a large number of researchers started to study impacts on the Moon from the largest impact basins of thousands of kilometres in diameter to the smallest impact craters on surfaces of lunar rocks. On individual rocks, microcraters were found from millimetres down to submicrometre sizes. Several effects, however, made the comparison difficult. Since the exposure time of a given surface on a lunar rock cannot reliably be determined, no absolute flux values can be derived, only relative values for different crater sizes. Another difficulty is the calibration of microcrater dimensions with respect to meteoroid sizes. Hörz et al. (1975) reviewed all available calibration data and arrived at a size-dependent crater-to-projectile-diameter ratio ranging from a value of 2 for 10^{-18} g meteoroids to a value of 9 for 1 g meteoroids, respectively. It was recognized by Morrison and Zinner (1977) that only the steepest

which had not completely used up their liquid fuel. Explosions like this, as well as collisions of satellites over the Earth's poles, where many orbits of polar satellites cross one another, shed large amounts of debris in Earth orbit. Some of these are in high-altitude orbits with lifetimes of many years, and therefore pose significant hazards to future satellites. Kessler noted that the increased population of satellites will lead eventually to increased amounts of debris by accidental collisions. Shielding satellites by massive shields will eventually worsen the situation because even more mass needs to be transported into Earth orbit, which will eventually become debris. Kessler predicted that after 50 to 100 years, depending on the assumed growth rate of mass transportation to Earth orbit, a runaway process will commence that will shatter all matter in low-Earth orbit. An artificial collisional ring will be formed until atmospheric drag clears this region of space again. Because of this potentially disastrous outcome, most space agencies adopted measures to minimize the detrimental effects of spaceflight. These measures include emptying liquid fuel from used rocket stages; bringing inoperational satellites to orbits where the risk of collisions is low, or "de-orbiting" them to burn up in the Earth's atmosphere; and monitoring closely the amount of debris in space caused by human activities.

In 1984 NASA released the Long Duration Exposure Facility (LDEF) into near-Earth space at about 450 km

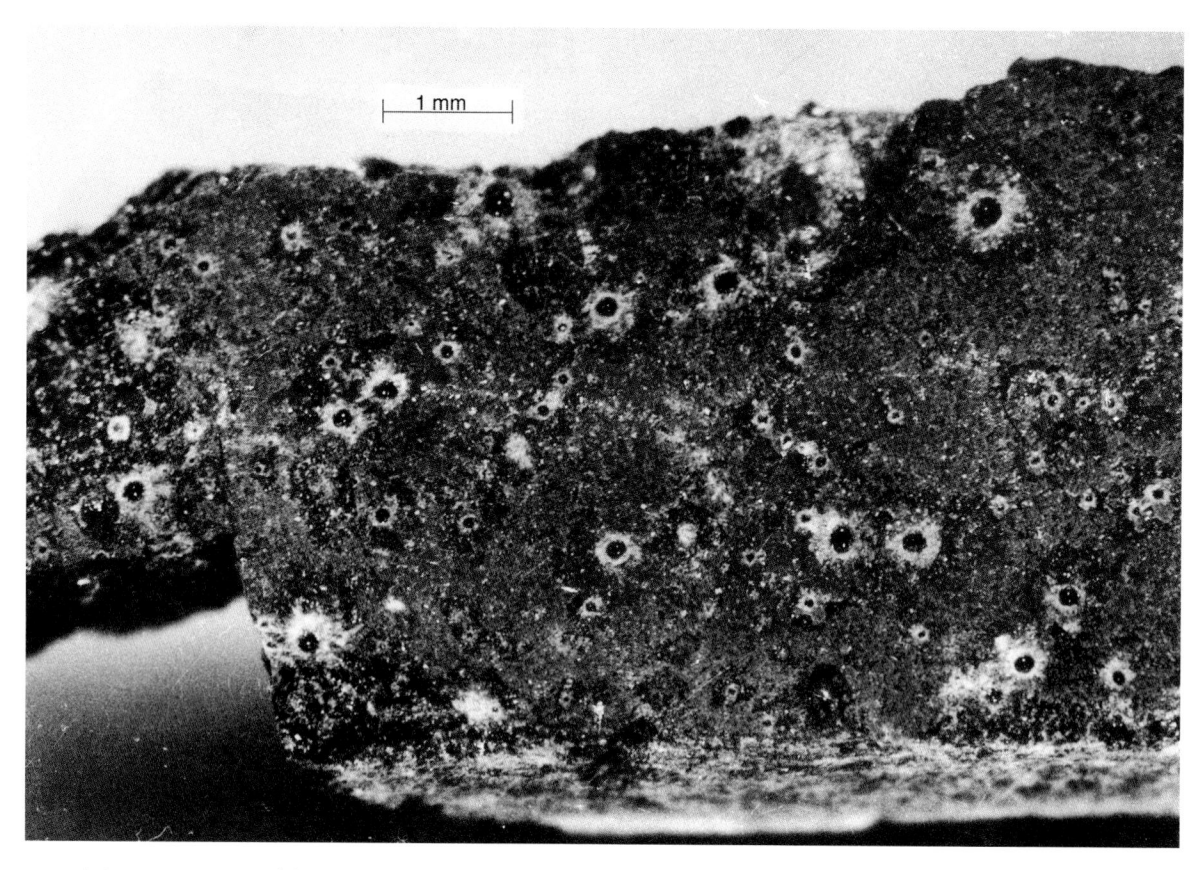

Figure 3 Microcraters caused by impacts of cosmic dust on a piece of lunar rock. Around the dark central pit there is the bright spallation zone. Microcraters have been found down to the resolution of optical microscopes and below (NASA).

slope of the crater size frequency measured on a single surface was least vulnerable to the effects of variations in exposure conditions and possible shielding by thin coatings of dust on the rock. Therefore, the author (Grün *et al.* 1985b) used their results and combined them with *in situ* measurements in order to derive the present microcrater production flux on the lunar surface. Studies of the effects of secondary microcraters produced by ejecta from primary impact craters on lunar samples showed that the number of microcraters is significantly increased for crater diameters below 10 μm. It was concluded that the interplanetary flux of meteoroids, $<10^{-9}$ g, is up to two orders of magnitude smaller than the ejecta flux on the lunar surface.

The flux of interplanetary meteoroids is generally given in the form of the cumulative meteoroid flux, which is the number of meteoroids with masses greater than or equal to mass m which impact on $1\,\mathrm{m}^2$ each second. Figure 2 shows the flux measured by a flat-plate sensor perpendicular to the ecliptic plane. From this size-dependent flux it can be calculated that most cross-sectional area is in particles of about 10^{-6} g which, therefore, contribute most to the zodiacal light brightness.

Reliable identification of collected extraterrestrial dust grains

Mineralogical, chemical and isotopic analyses of meteorites have provided deep insights into the physical and chemical conditions during their formation, and thus into the history of the Solar System as a whole. It is obvious that analyses of micrometeoroids could give similar information; in addition, they can be used as probes of their parent bodies. Early collections of dust in the atmosphere did not provide conclusive evidence of the extraterrestrial nature of the particles collected because of overwhelming contamination by terrestrial dust. A solution was to use rockets that exposed collection surfaces at heights of about 60 km above the Earth. Even in this method, severe contamination problems occurred: during the long path of the meteoroid collectors from their preparation in the clean room – through their installation on the rocket, the launch, exposure of the collection surfaces and closure at high altitudes, descent on a parachute, recovery and transport back to the laboratory – to the analysis of the collected particles, there were many leaks that contaminated the collectors with terrestrial dust, as was evidenced by "witness" samples that followed the same path but were not exposed to space. Ingenious methods like in-flight shadowing were developed in which particles were marked during their space exposure. Only a few probable micrometeoroids were collected this way. It was then recognized that cosmic dust collection by rocket flights (which at best lasted only a few minutes, and could carry small-area collectors only) was too expensive, difficult and inefficient.

Long exposures of large-area collectors on high-flying balloons seemed more promising, but still contamination by volcanic ashes was severe and analysis of a few dust grains scattered over collector surfaces of several square metres was tedious and labour-intensive. The breakthrough in micrometeoroid collection brought stratospheric collections by aircraft. High-flying aircraft like the U2 spy plane can cruise at 20 km altitude for many hours. On its wings it carries several 100 cm^2 flat-plate dust collectors which sweep huge amounts of air because of the high speed of the aircraft, thereby providing a strong concentration effect. Dust particles stick to the collector surfaces that are coated with silicone oil. After several hours of exposure the collector is retracted into a sealed storage container for return to the laboratory. After removal of the particles from the collector plate, the silicone oil is washed off and the particles can be examined (Figure 4).

Extraterrestrial grains (IDPs, interplanetary dust particles) of about 5–50 μm in diameter are collected this way. The lower size limit is determined by contamination by smaller terrestrial particles. Micrometre- and submicrometre-sized particles from volcanic eruptions can reach these altitudes in significant amounts. Another type of interference is caused by human contamination: about 90% of all collected particles in the 3–8 μm size range are aluminium oxide spheres that are products of solid fuel rocket exhausts. The upper limit is set by the low abundance of bigger particles, for example only about 10 IDPs more than 10 μm in size are collected during one hour of aircraft flight. The extraterrestrial nature of the collected particles was demonstrated in two steps. The first step was their chemical composition, which often resembles that of chondritic meteorites. The ratios of the elements magnesium, silicon, sulphur, iron and nickel are characteristic and distinguish the terrestrial or extraterrestrial nature of the collected particles. The final

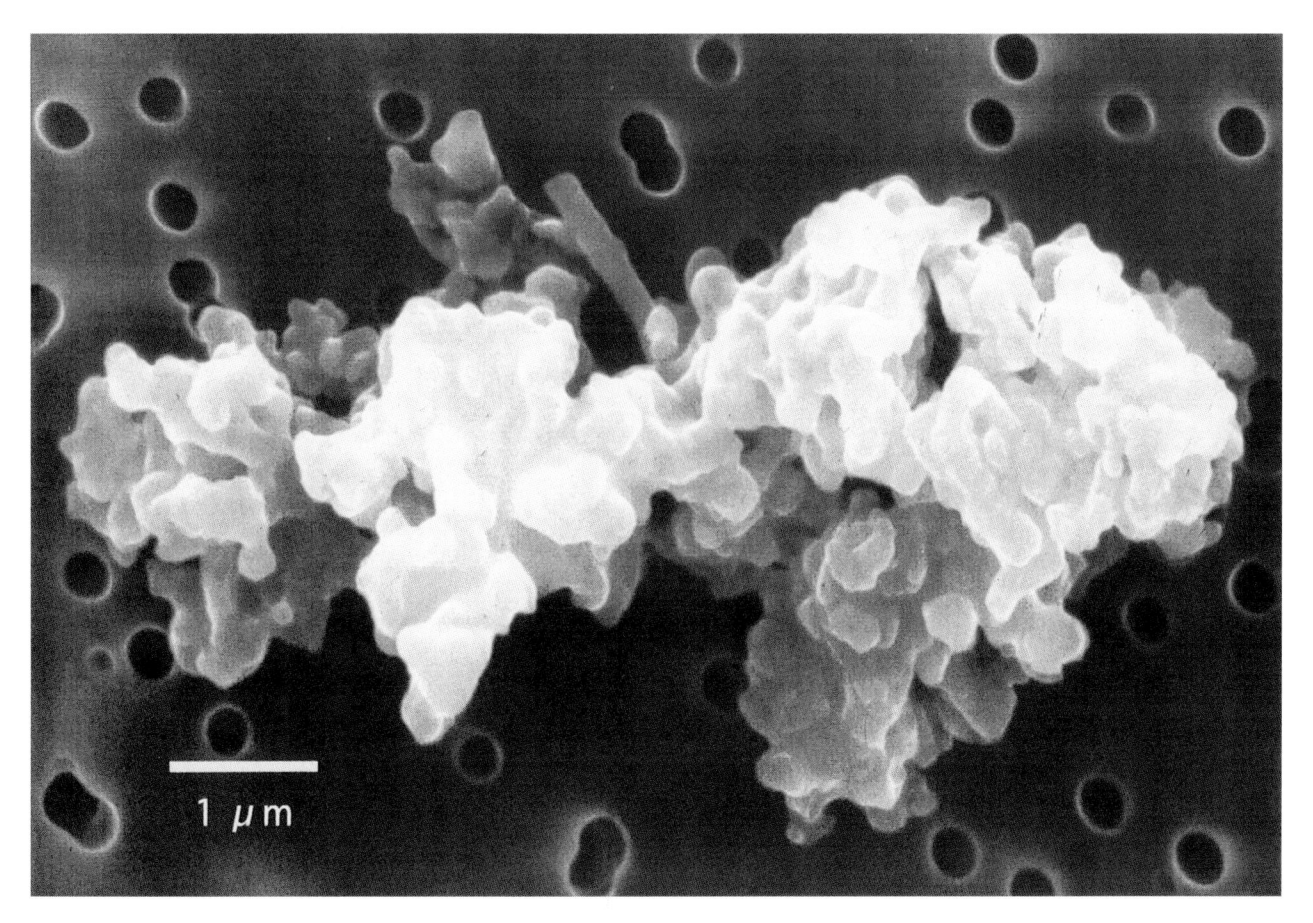

Figure 4 Interplanetary dust particle recovered by an aircraft from the stratosphere (NASA).

proof for the extraterrestrial origin of some classes of collected particles came after traces of solar wind helium and tracks from the exposure to highly energetic ions in space had been identified in the minerals by Brownlee et al. (1976).

Since 1981 NASA has performed routine cosmic dust collections by aircraft. The collected particles are isolated, chemically characterized, catalogued and distributed for further scientific investigation. Several types of IDPs have been identified: porous aggregates of chondritic composition, iron–sulphur–nickel grains and iron–magnesium-rich silicates, such as olivine and pyroxene. Chondritic aggregates may contain varying amounts of carbonaceous material and show a significant enrichment in volatile (low-condensation-temperature) elements compared with meteorites of C1 chondrites. This observation is used as an argument for the pristine nature of these particles that had never experienced temperatures above about 500°C. A remarkable feature of IDPs is their large variability in isotopic composition. Extreme isotopic anomalies have been found in some IDPs (e.g. factors of 1000 off the solar hydrogen isotope ratio). In typical Solar System material (from the Earth, the Moon and from meteorites), isotopic ratios vary by only fractions of 1%. The huge isotopic variations indicate that some grains are not homogenized with other Solar System material but have preserved much of their presolar character. This property, and their compositional similarity with comets, argues for a relationship between comets and IDPs. IDPs are an important source of primitive extraterrestrial material that is not available in any other way at the present time.

Like meteorites, IDPs do not represent an unbiased sample of meteoroids in space. They penetrate the atmosphere in a more or less disturbed state. Small IDPs of $<10\,\mu m$ diameter are decelerated in the tenuous atmosphere above 100 km, especially if their entry speed is low. At this height the deceleration is gentle, and the grains do not reach the temperature of substantial evaporation ($\sim 800°C$), because small particles have a high surface-area-to-mass ratio which enables them to radiate away excessive heat effectively. These decelerated dust particles subsequently float down through the atmosphere and become accessible to collection and scientific examination. At 20 km the concentration of $10\,\mu m$ diameter particles is about a million times higher than in space. Depending on the entry speed into the atmosphere, very volatile materials and loose aggregate particles are not accessible through this method: particles on low-eccentricity orbits which have low collision speeds with the Earth are favoured over those on high-eccentricity orbits.

Related methods which have even stronger biases are cosmic dust collections in deep-sea sediments, and in Arctic and Antarctic ices where extraterrestrial grains are recognized by their unusual chemical compositions. This material, which exists in abundance, originates mostly from ablation products of larger meteoroids. Analysis reveals some (but not very specific) cosmochemical information. Another approach is to study meteoroid residuals in impact craters generated during long-term exposure of collectors in space. A prominent example are surfaces on the LDEF, which characterized the near-Earth dust environment.

Laboratory impact experiments showed that impacts in low-density targets preserved much of the characteristics of the projectile's material. An application is the exposure of very low-density ($\ll 100\,kg\,m^{-3}$) aerogel in space that is able to collect little-modified projectiles with impact speeds of up to $10\,km\,s^{-1}$. This material is currently exposed by the Stardust mission, which is planned to collect cometary dust samples during its flyby of Comet 81P/Wild 2 in January 2004 (Brownlee et al. 1996).

Dust in the outer Solar System

Pioneers 10 and 11 explored the outer regions of the Solar System and made flybys of Jupiter and Saturn. The penetration "beer can" detectors they carried were designed to record any hazard from impacts of $10\,\mu m$ and bigger meteoroids in the asteroid belt. The implementation of this design goal caused difficulties in the more dusty environments of Jupiter and Saturn. It was realized only after Voyager's discovery of the Jovian ring that Pioneer 11 had passed through the outer fringes of this ring during its very close flyby of Jupiter several years earlier. However, only two impacts were recorded close to the ring position. As the instrument had a built-in dead time of 87 minutes after each detection, more penetrations may have happened during this passage. This dead time had been introduced in order to avoid multiple counts from marginal penetrations for which the pressure release from the cell may be very slow. The same effect may have caused also too few detections during the Pioneer 11 fly-through of Saturn's ring system (Humes 1980). A principal way to check this effect was to see whether the total number of recorded penetrations eventually reached the number of pressurized cells (108) of this detector. However, this check was prevented because beyond 20 AU the fill gas (argon and nitrogen) of the detectors froze in the very low ambient temperatures (Don Humes, personal communication). The pressure in the cells thus dropped, and the electrical discharges triggered multiple times. Nevertheless, reliable dust data out to 20 AU were obtained from the penetrations recorded by the Pioneer 10 and 11 detectors.

Another dust instrument on board Pioneers 10 and 11 was the Asteroid Meteoroid Detector (AMD). This instrument consisted of four telescopes that were designed to detect scattered sunlight from meteoroids passing through their field of view. From the timing and the amplitudes of the received light signals the trajectory of the meteoroid can be reconstructed. Because of a high noise level on individual channels (Auer 1974) no single trajectory could be uniquely identified. Some signals were later reinterpreted by Dubin and Soberman (1991) as light flashes from exploding meteoritic particles, which they called "cosmoids". However, since this cosmoid hypothesis is in direct conflict with zodiacal light observations and other *in situ* meteoroid measurements, it will not be considered here any further.

There was another aspect of the AMD story: Soberman's AMD instrument was preliminarily selected for inclusion in the Voyager mission. A high-sensitivity impact detector was considered as an addition to the AMD instrument. Both Joe Früchtenicht and the author designed an impact ionization mass analyser to be added to AMD. However, the failure of AMD on Pioneers 10 and 11 to generate reliable results caused the elimination of both dust instruments from the Voyagers. The main instruments that obtained information on dust rings was the camera on board (Figure 5). However, there was a surprise dust instrument on board that no one had expected. It was after the Saturn flyby that Gurnett *et al.* (1983) realized that the strange noise which the plasma wave instrument had recorded during the Voyagers' passage through Saturn's ring plane close to the newly discovered G-ring, could be interpreted as impacts of dust particles on the spacecraft skin. During the subsequent flybys of Uranus and Neptune, this instrument recorded dust impacts as well.

In 1974 the Pioneer 10 and 11 spacecraft had passed Jupiter en route to the outer Solar System. During the passage of the Jovian environment the "beer can" detectors on board had recorded only 11 and 2 impacts, respectively. These events nevertheless constituted a flux about 1000 times the interplanetary dust flux. The subsequent discussion was about whether these particles were gravitationally concentrated (focused) interplanetary meteoroids or whether there was a dust source within the Jovian system. Axford and Mendis (1974) described the theory of the interaction of dust particles with the Jovian magnetosphere. However, nothing of the suggested phenomena was seen in

Figure 5 Neptune with its ring system. The two photographs were taken by Voyager 2 shortly after its flyby of Neptune, looking back towards the planet (in the centre). Several diffuse rings circle Neptune (NASA/JPL).

the dust data from Pioneer 10 and 11, but a few years later the TV cameras on board the Voyager spacecraft had discovered two dusty phenomena in the Jovian system: a complex ring system inside about $3R_J$ from Jupiter, and the powerful volcanoes on Io that spew plumes of ash as high as 300 km above the surface.

The Voyager observations spawned a series of papers that tried to explain the observed phenomena. It was first suggested by Johnson *et al.* (1980) that dust from Io's volcanic plumes might leak into the magnetosphere. The suggested mechanism was that dust gets charged by the interaction with the ambient plasma and is subsequently picked up by Jupiter's magnetic field, which sweeps past at a speed of 57 km s^{-1}. Follow-on papers by Morfill *et al.* (1980) discussed the further fate of the ash specks and suggested that the ring may be connected to the volcanoes on Io. Nanometre-sized charged dust particles gyrate about magnetic field lines, just as electrons or ions would. Morfill suggested the "guiding-centre" diffusion process that would distribute such particles through the magnetosphere. Because of the smallness of the particles, statistical discrete charge fluctuations (i.e. the gain or loss of a single electron from the particle) will be enough to vary the gyro-radius of the particle so that it does not stay bound to a single magnetic field line, its guiding centre, but jumps from one field line to another, thereby diffusing throughout the magnetosphere. Eventually these particles reach the region of the dust ring where 10 km satellites were discovered by the Voyagers. It was suggested that impacts of nanometre-sized particles on the observed ring moons release micrometre-sized grains making up the observed ring. The major uncertainty of this theory lies in the largely unknown plasma environment in the inner magnetosphere, which is not resolved up to now.

An alternative proposal for the formation of the Jovian ring was made by Burns *et al.* (1980). They suggested that impacts of interplanetary meteoroids on the satellite cause the generation of the visible ring particles. Although uncertainties of the interplanetary meteoroid flux at Jupiter are considerable, this mechanism is generally accepted as an important source of micrometre-sized dust in the Jovian system.

More information on these dusty phenomena was expected from the follow-on mission to the Jovian system. In 1976, just after both Pioneers 10 and 11 had successfully passed Jupiter and about a year before both Voyagers flew through the same region of space, NASA issued an Announcement-of-Opportunity for the in-depth study of the Jovian system by the Galileo mission. It was Ian Axford who encouraged the author to propose for this mission one of the highly sensitive impact ionization dust detectors that were developed by the Heidelberg group. The objectives of the dust investigation were to search for dust rings and other dusty phenomena in the Jovian system. A year later the TV cameras on board the Voyagers discovered a wealth of dusty phenomena in the Jovian system. A twin instrument of the Galileo dust instrument was proposed and selected for the Ulysses mission. Both instruments were to bring new and exciting information on dust in the outer Solar System.

Development of highly sensitive dust detectors

A precondition for the development of advanced space dust detectors was the availability of appropriate simulation techniques. In order to simulate impact and collision phenomena in space, it was necessary to accelerate projectiles to speeds of over 10 km s^{-1}. Since projectiles from military guns reached speeds of only a few kilometres per second, new developments were necessary. The workhorse of accelerators for millimetre and bigger projectiles was the light gas gun, which consists of a conventional powder gun that pushes a piston into a barrel containing hydrogen gas. This barrel is sealed by a steel diaphragm which breaks when the pressure exceeds several kilobars. In front of the diaphragm sits the projectile, which is accelerated by the expanding gas. Because of the low molecular mass of the compressed and heated gas its expansion speed is high, much higher than that of the exhaust from the explosive. By this method projectile speeds up to 12 km s^{-1} were reached. Because of wear and tear the gun barrel had to be replaced frequently, making the method very expensive. Light gas guns are not too uneconomical to operate for speeds of up to about $5–7 \text{ km s}^{-1}$. A wide range of projectile materials can be accelerated by this method, especially when a sabot is used. A sabot is a small plastic (Teflon) bucket which holds the actual projectile. The sabot is accelerated by the gas and stopped at the muzzle of the barrel by a blind with a circular hole. The projectile keeps on flying through the hole and is used for impact experiments. Projectiles generally survive the acceleration process, and hence only a speed measurement is necessary to characterize the projectile. For millimetre-sized projectiles this is easily done by time-of-flight (TOF) measurement between two light curtains.

For smaller (0.1 mm) projectiles, the plasma drag gun was developed by Igenbergs and Kucera (1979). In this gun an electric discharge vaporizes a metal film, the ionized vapor (plasma) of which is compressed and accelerated by high current flowing through a metal coil. Plasma speeds of several 10 km s^{-1} are reached. In front of the coil is a thin plastic film at which several $100 \,\mu\text{m}$ projectiles are loosely attached. When the plasma beam hits the film it vaporizes and the projectiles are accelerated by plasma drag. Projectile speeds of up to 20 km s^{-1} can be reached. Because of the enormous thermal stress during the interaction with the plasma, only transparent glass beads (glass efficiently radiates excess heat away) can be successfully accelerated. However, even glass beads are often fragmented during acceleration, and an efficient diagnosis is necessary to obtain information on the projectile speed and mass. The speed is obtained by the TOF measurement between the discharge and the impact on a target. Because of the wide scatter, only a few of the glass beads accelerated hit the target. For the determination of the size and impact location a thin witness film is placed in front of the target. This film is optically inspected, and the diameter of the projectile is determined from the dimension of the penetration hole.

Micrometre-sized and smaller particles are accelerated by electrostatic accelerators that were developed by Shelton et al. (1960) and Friichtenicht (1962; see also Fechtig et al. 1978). These accelerators are similar to nuclear physics devices that accelerate charged particles by high electric voltages. Voltages of several megavolts are obtained by van de Graaff generators. In an electrostatic dust accelerator the ion source is replaced by a dust source in which dust is electrically charged. High charges on electrically conducting particles are obtained by bringing these particles into contact with the fine tip of a tungsten needle which is at a high electric potential. As the field strength for electron field emission is smaller than the field strength for ion emission by a factor of 10–100, positive potentials are used in a dust accelerator. Spherical iron particles can be charged so that the surface electric field reaches values close to the ion field emission limit of $10^{10}\,\mathrm{V\,m^{-1}}$. In a 2 MV accelerator, micrometre-sized iron particles reach speeds of $12\,\mathrm{km\,s^{-1}}$, and 0.1 µm particles reach speeds of $35\,\mathrm{km\,s^{-1}}$. Speeds as high as $70\,\mathrm{km\,s^{-1}}$ have been achieved for very small grains. Because of the charging process, only conducting particles can be accelerated. Up to now particles consisting of iron, carbon, aluminium, and metallic coated glass and plastic spheres have been used in these accelerators.

Another type of dust accelerator was developed by Vedder (1963). It used a quadrupole trap to suspend and charge dust particles by an ion beam. By this method almost any material can be charged. However, the charging process must be controlled by an operator who observes the particle through a microscope in order to keep it stable in the trap. Because of this control method the charging has to occur at ground potential, and a multiple-stage linear accelerator has to be used for the acceleration. However, the timing of such an accelerator is delicate, and speeds of up to only about $10\,\mathrm{km\,s^{-1}}$ have been reached. However, a range of non-conducting projectile materials (aluminium oxide, quartz, kaolin, soda-lime glass and polystyrene) were accelerated and used for the calibration of the Helios dust analysers (Dietzel et al. 1973).

The initial development of highly sensitive dust detectors went hand in hand with the development of dust accelerators. Within a short time after an electrostatic dust accelerator had been successfully developed at TRW, Friichtenicht and Slattery (1963) reported the experimental verification of a new impact phenomenon: impact ionization, which had been predicted theoretically by Raizer (1960). A dust particle that hits a solid target at speeds above a few kilometres per second produces an impact crater. During this process part of the projectile and target material are strongly compressed and heated, and some of the vapour is partially ionized. By an electric field applied to the impact area positive and negative charges (ions and electrons) can be separated and recorded. It was Alexander et al. (1971) from the Goddard group who applied this technique in a novel dust detector on the OGO satellites and on the Lunar Explorer 35 that were launched in 1966 and 1967, respectively. These sensors were TOF systems consisting of a thin-film front sensor and a rear sensor 10 cm apart. Both detectors employed impact ionization to detect the dust penetration of the front film and impact on the solid rear sensor, respectively. However, misled by the early false reports of a very high microparticle flux in the Earth's environment, Alexander selected only $5\,\mathrm{cm^2}$, as the sensitive area of a single detector. This had the consequence that no or only very few impacts were recorded. Berg and Richardson (1969) extended this idea by combining 16 TOF tubes into a single $100\,\mathrm{cm^2}$ dust detector, flown on Pioneers 8 and 9. This detector made the first important dust observation in interplanetary space and discovered new dusty phenomena (see below). Table 2 lists successful dust instruments on interplanetary missions.

The Heidelberg group took a slightly different route. They tried to avoid films as much as possible because they knew from laboratory experiments that a film, even if it is much thinner than the diameter of the projectile, disrupts a projectile that penetrates it at high speeds. Their first dust detector was flown on the Second Highly Eccentric Orbit Satellite (HEOS 2) (Dietzel et al. 1973) and employed an open impact ionization detector of about $100\,\mathrm{cm^2}$ sensitive area. It was launched in 1972 and took measurements in and outside the Earth's magnetosphere. From measurements of the signal amplitude and the rise-time, the mass and speed of the impacting micrometre and submicrometre-sized particle were derived. A major improvement of this simple dust impact detector was the tenfold increase of sensitive area ($1000\,\mathrm{cm^2}$) of the dust instruments (Figure 6) on the Galileo and Ulysses missions that took measurements in the outer Solar System.

The next step was to combine an impact ionization detector with a mass spectrometer and thus to analyse the ions released. In the laboratory the first compositional measurements using a TOF mass spectrometer were reported by Auer and Sitte (1968) and by Hansen (1968). The mass resolutions of the spectra obtained in the laboratory were low: only elements up to mass 50 amu could be resolved in the best cases. Compositional analysers for space application

Table 2 Characteristics of *in situ* dust instruments on interplanetary missions. Distance ranges are those in which dust measurements were obtained. The missions to Comet Halley carried several dust instruments. The mass threshold refers to the most sensitive instrument at 20 km s^{-1} impact speed

Mission	Launch year	Distance range (AU)	Dust instrument Sensitive area (cm^2)	Mass threshold (g)	Dust science
Pioneer 8	1967	1–1.1	94	2×10^{-13}	Interplanetary dust at 1 AU
Pioneer 9	1968	0.7–1	74	2×10^{-13}	Inner Solar System dust
Pioneer 10	1972	1–18	2600*	2×10^{-9}	Outer Solar System dust
HEOS 2	1972	1	100	2×10^{-15}	Earth system and interplanetary dust
Pioneer 11	1973	1–10	2600*	1×10^{-8}	Outer Solar System dust
Helios 1, 2	1974/76	0.3–1	120	9×10^{-15}	Inner Solar System dust
VeGa 1, 2	1984	0.8	5–500	4×10^{-15}	Comet Halley dust
Giotto	1985	0.9	5–2 $\times 10^4$	4×10^{-15}	Comet Halley dust
Galileo	1989	0.7–5.4	1000	4×10^{-15}	Jupiter system and interplanetary dust
Hiten	1990	1	100	2×10^{-15}	Earth system and interplanetary dust
Ulysses	1990	1–5.4	1000	4×10^{-15}	Three-dimensional dust distribution
Cassini	1997	0.7–10	1000	1×10^{-15}	Saturn system and interplanetary dust
Nozomi	1998	1–1.5	140	2×10^{-15}	Mars system and interplanetary dust
Stardust	1999	1–2.8	90	1×10^{-14}	Comet Wild 2 and interstellar dust

*Initial area; actual area decreased as cells were punctured.

were described by Dietzel *et al.* (1973) and Früchtenicht *et al.* (1973). Both instruments were of the 100 cm^2 sensitive area class. While Früchtenicht described an analyser with a 1 m static TOF tube, the instrument described by Dietzel employed time-lag focusing by which the spread in the ion energy was partially compensated and the mass resolution somewhat improved. The latter instrument was flown on Helios and was able to classify the observed meteoroids into chondritic and iron-rich categories (Leinert and Grün 1990). A major step forward in the direction of making impact ionization mass spectrometers an analytic tool was made by Kissel (1986) with his instruments for the Halley missions, Giotto and VeGa. He included an electrostatic reflector to improve the mass resolution beyond mass 100. With these instruments even isotopic analyses of cometary dust could be performed. An updated version of this mass analyser flies on the Stardust mission. This instrument has an enlarged sensitive area of ~ 100 cm^2 compared with to the 5 cm^2 of the Halley instruments. The Cassini instrument is a combination of the Galileo-type dust detector with a linear impact ionization mass analyser.

ASTROPHYSICAL STUDIES

Distribution of zodiacal dust in the Solar System

It is quite obvious that zodiacal light observations have many advantages when they are done from an Earth satellite or even from a deep-space probe (Figure 7): no detrimental weather effects, no dust and light pollution in the atmosphere and no airglow disturb the measurements. There is the possibility to observe in a wide range of directions and from different positions in space. Nevertheless, there are severe problems to solve in order to perform space-based zodiacal light observations: there is the significant variation of the background brightness from stars and galaxies, and there are instrumental stray light problems, especially for observations close to the Sun. Leinert *et al.* (1981) succeeded with baffled photometers on board Helios to measure the zodiacal light between 0.3 and 1 AU at brightness levels of 10^{-11} times of those of the Sun only 15° away from the Sun. The radial brightness profile was determined, and an inclination of 2° of the symmetry plane of the zodiacal cloud was found. Outside the Earth's orbit, the zodiacal light was observed by photometers on board Pioneers 10 and 11 (Weinberg *et al.* 1974). The zodiacal light brightness was found to exceed the background out to 3.3 AU from the Sun (Hanner and Weinberg 1973). Plans for the measurement of the zodiacal light outside the ecliptic plane were abandoned when NASA cancelled plans for the second Out-of-Ecliptic spacecraft. The only spacecraft that remained from the two-spacecraft mission was ESA's Ulysses, which carried an *in situ* dust detector.

Less than 10% of incident sunlight is scattered by interplanetary dust and contributes to the zodiacal light; the rest of the absorbed energy (>90%) is re-emitted as thermal infrared radiation mostly in the 10–50 μm wavelength range. Because of this, zodiacal infrared emission is a much

Figure 6 Galileo's dust detector with its electronics box. The diameter of the sensor is 40 cm, with a 0.1 m^2 sensitive impact area. The instrument can detect 0.1 μm dust grains at impact speeds above 30 km s^{-1}. An identical detector was also flown on the Ulysses mission.

more prominent astronomical phenomenon than the zodiacal light – much to the dismay of astronomers who are interested in more distant objects that are blinded by the foreground zodiacal emission. Once the technology for space-based infrared observations was developed in 1984, the IRAS satellite and later the COBE satellite obtained unprecedented information on the overall structure of the zodiacal cloud. Besides the large-scale structure of the zodiacal cloud, broad asteroidal bands as well as narrow comet trails were discovered. Kelsall *et al.* (1998) have

Figure 7 The zodiacal light behind the Moon, photographed by the Clementine spacecraft. Light from the Sun is blocked by the Moon. The ecliptic plane is delineated by the positions of the planets Mercury, Mars and Saturn (from left to right) (NASA).

presented a model that describes the infrared brightness of these phenomena as observed by COBE with an accuracy of better than 3%.

Zodiacal light observation has shifted from being an object of astronomical research to a method of characterizing the disturbing foreground. However, the detection of dust rings around other stars like β Pictoris by IRAS renewed the interest of astronomers in the study of the Solar System dust cloud, especially in the region of the long-suspected Edgeworth–Kuiper Belt. Outside Neptune's orbit a rapidly increasing number of comet- and asteroid-sized objects are being found, which should be accompanied by dust generated by mutual collisions of Kuiper Belt objects. The total amount of this dust and the signature of the outer planets on the structure of the dust cloud are of high astrophysical interest because structures observed in extrasolar dust rings could be indicative of planetary systems like ours.

Several properties of the zodiacal dust cloud can be understood by consideration of the dynamics of meteoroids in interplanetary space. For example, the radial structure of the zodiacal cloud is largely determined by the dynamics of dust in interplanetary space. All dust particles in space feel the gravitational pull of the Sun. For particles with masses of $>10^{-11}$ kg, solar gravity is by far the most dominating force. As a consequence they move on Keplerian orbits which are conic sections with the Sun at one focus – other forces are only small disturbances. Certainly, all observations of sizeable meteoroids are compatible with such orbits. However, micrometre-sized particles feel, in addition the repulsive force of solar radiation pressure and electromagnetic interactions with the interplanetary magnetic field.

The pressure exerted on dust in interplanetary space by solar radiation decreases with the inverse square of distance from the Sun – the same dependence as the gravitational force. Therefore, the ratio of gravitational and radiation-pressure forces is a constant for each particle everywhere in the heliosphere, and it depends only on the particle's size and material properties. This ratio is generally reffered to as β, which is inversely proportional to the size of the particle for particles bigger than the effective wavelength of sunlight. As a consequence, β increases for smaller sizes and reaches maximum values between 0.1 and 1 μm. The maximum value is about 0.5 for dielectric (transparent) materials, and reaches values of 3–10 for strongly light-absorbing particles.

There are important consequences for the dynamics of small particles because of the radiation pressure. Small particles which are generated from big particles (e.g. by emission from comets or by impact ejection from meteoroids or asteroids) carry the specific kinetic energy of their "parents". However, because of the radiation pressure they feel the reduced attraction of the Sun, and they therefore move on different orbits than their parents. For example, a dust particle with radiation pressure constant β>0.5 that is released from a large parent object on a circular orbit will leave the Solar System on a hyperbolic orbit.

Besides the direct effect of radiation pressure on the trajectories of small dust grains, there is also the more subtle Poynting–Robertson effect. This is caused by radiation pressure acting on a moving dust particle not perfectly radially outwards, but with a small component opposing the particle's motion. This drag force leads to a loss of angular momentum and orbital energy of the particle. The effect is strongest when the particle speed is highest, at its perihelion. The particle's orbit thus get slowly circularized, while it spirals towards the Sun. For a centimetre-sized particle on a circular orbit at the Earth's distance, the time to spiral into the Sun is 7×10^6 years.

In interplanetary space, solar effects determine the dynamics of dust grains. However, close to planets gravitational effects become important as well. Jackson and Zook (1989) found that for 10–100 μm-sized particles, planetary disturbances may stop the inward drift and the particle may be temporarily trapped in a resonance with the planet until its eccentricity is pumped up and the particle is released from the trap. Subsequent to this theoretical prediction, a heliocentric Earth-shepherded ring was found by Dermott *et al.* (1994) in infrared observations by IRAS.

This spiralling of micrometeoroids towards the Sun leads to a well-defined radial dependence of the spatial dust density with an increase towards the Sun proportional to the inverse radial distance (r^{-1}). Dust from a source in the asteroid belt at 3 AU is driven inwards by the Poynting–Robertson effect. As a consequence, at the Earth the dust density is three times the density in the asteroid belt. Leinert *et al.* (1981) found from zodiacal light measurements with Helios a radial density increase proportional to $r^{-1.3}$. He concluded that an asteroidal source is not sufficient to explain this slope but that a distributed source of dust is required, in addition, in the inner Solar System. This distributed source is mutual collisions between meteoroids, as we will discuss in the next section.

Solar System contribution to interstellar dust

Otto Berg had developed a multi-coincidence dust sensor for the Pioneer 8 and 9 missions that required three signals from each single dust impact: two from the front-film sensors and one from the rear sensor after the penetration of the thin front film. However, during the mission time of several years only 20 impacts showed the triple coincidence of signals as required. But several hundred events had only two simultaneous signals from the front-film sensors and none from the rear sensor. Worried by the noisy results from previous dust instruments, initially the investigator cautiously interpreted these signals as solar disturbances that were mostly recorded when the detectors faced the Sun. It took a careful analysis to realize that these events were signatures of reliable impacts – it was just that the small dust particles did not penetrate the front film. Consequently a prevailing flux of grains from the solar hemisphere was postulated by Berg and Grün (1973), who argued that these particles must be on trajectories that leave the Solar System. Dust grains on bound orbits about the Sun, as recorded in most triple coincidence events, should produce an equal number of impacts from the solar hemisphere as from the opposite hemisphere. Zook and Berg (1975) interpreted this outward dust flow as small grains being generated by collisions of meteoroids near the Sun and being expelled by the prevailing action of solar radiation pressure. Zook and Berg coined the name β-meteoroids for these particles because the β value of these particles had to be large ($\geqslant 0.5$). Whipple (1975) pointed out that the observed symmetry of impacts about the solar direction required that these particles had a similar angular momentum as the spacecraft by which they were detected. Since both Pioneers 8 and 9 orbited the Sun close to the Earth's orbit, most β-meteoroids observed by these spacecraft had to originate from distances just inside 1 AU. Since their discovery, β-meteoroids have been observed by dust instruments on Helios 1, Hiten and Ulysses.

The astrophysical significance of β-meteoroids is derived by theory. It is generally believed that β-meteoroids are generated by collisions between two meteoroids. In such collisions the impact speed is usually high enough ($>10\,\text{km}\,\text{s}^{-1}$) to result in fragmentation of one or both particles (the smaller of which is always shattered). A catastrophic collision (i.e. the shattering of both particles) occurs if the mass ratio of target and projectile does not exceed a certain value. This mass ratio was found experimentally in the laboratory to be 5×10^4 for 10 km s^{-1} impacts into basalt, and this value is assumed to be typical also for interplanetary meteoroids. If the target particle is bigger, it is only eroded by the impact cratering process. Dohnanyi (1970) showed that for meteoroids of masses <1 g – those of concern to us in the present context – erosive collisions are much less important than catastrophic collisions, so we can limit our discussions to the latter case.

From the meteoroid flux at 1 AU Grün et al. (1985a) calculated the rate of catastrophic collisions of meteoroids and their collisional lifetimes. At 1 AU, 10^{-5} kg (millimetre-sized) particles have the the shortest collisional lifetime, namely about 10,000 years. Bigger and smaller particles have longer lifetimes because the relative number of projectiles that can destroy them decreases. For smaller particles (with masses $>10^{-8}$ kg) the Poynting–Robertson effect becomes important and the Poynting–Robertson lifetime is shorter that the collisional lifetime: small particles spiral faster towards the Sun and evaporate there, rather than being destroyed by a collision.

The mass of meteoroids destroyed by collisions is converted to the form of smaller fragments, increasing particle numbers in the small size ranges. Comparing gains and losses, the net effect of collisions is to produce small dust particles ($m < 10^{-8}$ kg, $r < 100$ μm) at the expense of the larger meteoroids (Figure 8). Most of the interplanetary dust is produced by collisions of larger meteoroids, which form a reservoir that itself is continuously replenished by the disintegration of comets and asteroids. The majority of the material (90%) is transformed by collisions to β-meteoroids that are blown out of the Solar System. A relatively small portion (10%) is driven by the Poynting–Robertson effect towards to the Sun, where it is lost by evaporation. After being ionized the evaporated meteoroid material is swept by the solar wind out of the heliosphere as well. Estimates by the author indicate that inside 1 AU about 10 t s^{-1} of bigger meteoroids is ground up by collisions, of which about 9 t of small β-meteoroids leave the Solar System every second and mix in with interstellar dust. Other direct contributions to the heliospheric dust cloud come from comets, from interstellar dust that sweeps through the Solar System and from planets that shed dust from their environment into interplanetary space (see below).

Cometary dust: Size and compositional analyses

Comets have long been recognized to be carriers of pristine, unheated and unaltered material from the early stages of formation of the Solar System, and at the same time to be significant sources of dust in the Solar System. However, because of the high eccentricity of cometary orbits they are difficult to reach and to rendezvous with from Earth. Cheaper ways are flyby missions that just cross the orbit of a comet close to the nucleus. Even before 1970, comet flyby missions were studied by NASA. The scientists promoting missions to comets were cosmochemists whose background was laboratory studies of meteorites, where most information is gained from the analysis of trace elements and isotopic compositions. Hence, the conduction of similar analyses of cometary material was their expectation. Because cometary flyby occurs generally at high speeds (10 to 80 km s^{-1}) dust cannot be captured and analysed by methods like electron microscopes and ion probes that could be adopted from laboratory methods. Only impact ionization mass spectroscopy was capable of providing

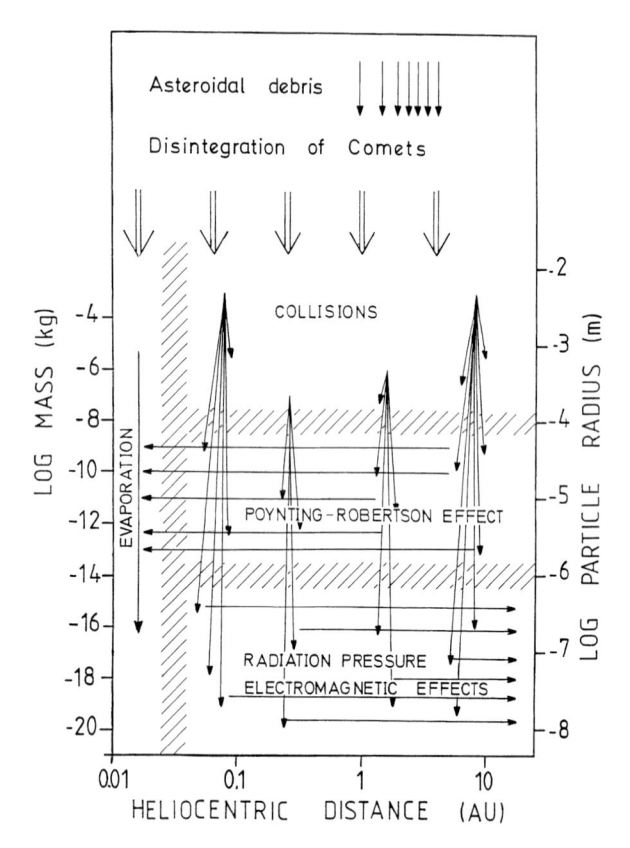

Figure 8 Dust evolution as function of particle size and distance from the Sun – the flow of meteoritic matter through the Solar System. Released from comets and asteroids, meteoroids are ground down by mutual collisions, until they are small enough that either the Poynting–Robertson effect transports them towards the Sun, where they evaporate, or radiation pressure drives them out of the Solar System. In this way the Solar System sheds more than 10 t s^{-1} of fine-grained dust into the interstellar medium (after Grün et al. 1985b).

chemical information of cometary dust during a flyby mission. This method had just been developed in some laboratories when the first comet missions were studied. Initially, the mass resolution of impact spectra was low, and the laboratory calibration data needed for the interpretation of the spectra were sparse. Mainly because of the limitations of the methods of dust analysis, early plans for cometary missions failed.

Nevertheless, the first space mission to get close to a comet had been planned by 1971. This was the ISEE 3 mission, the third space probe in the International Sun–Earth Explorer series that was launched in 1978 and was later renamed ICE (International Cometary Explorer). The original objectives of this mission were to study solar–terrestrial relations at the interface between the Earth's magnetosphere and the solar wind, and hence it carried mostly plasma and field instruments. It was the ingenuity of the mission designer Bob Farquar at Goddard Space Flight Center who recognized that through multiple fly-bys of the Moon the space probe could be directed to a close flyby of Comet Giacobini–Zinner in 1985. However, the comet flyby would occur only during a late extension of the mission, which would be executed only if all previous phases were successful. Therefore, preliminary ideas of putting a dust instrument on this mission because of the comet flyby opportunity were abandoned. Ironically, as with Voyager, it was discovered that the plasma wave instrument on board had some limited capabilities to detect dust impacts during the comet flyby.

Ready for the return to the inner Solar System of Comet Halley in 1986 were three space probes (two Russian VeGa craft and ESA's Giotto) that carried a suite of specifically designed dust instruments close to the comet. Most of these instruments were impact ionization detectors and mass spectrometers, but also piezoelectric microphones and thin-film detectors were flown. Jochen Kissel (1986) provided the impact mass spectrometers for all three missions. Tony McDonnell, Engeny Mazets, Oleg Vaisberg and John Simpson provided dust detectors for the determination of the dust flux and the size distribution. Besides the dedicated *in situ* dust instruments, many other instruments on board recorded the effects of the dusty cometary environment.

Significant results were the discovery of

(1) both smaller and bigger grains in the coma than the 1–10 μm grains that had been anticipated from astronomical observations
(2) the existence of a significant carbonaceous component in Halley dust
(3) a much wider scatter in some isotopic ratios than has been found in any other extraterrestrial material.

It was found that the dust-to-gas mass ratio is of the order of 1 and that most mass is in the biggest particles. Cometary particulates are an intimate mixture of two end-member components, refractory carbonaceous (rich in the elements H, C, N, and O) and stony (rich in rock-forming elements such as Si, Mg, Fe) material. The stony component comprises silicates, metals, oxides, sulphides and others. Both end-member components do not occur as pure components but are mixed to the finest scale. Figure 9 shows an impact spectrum that was recorded by the VeGa 2 dust spectrometer and contained much H, C and O and relatively little Mg, Si and Fe.

Comets are an important source of the interplanetary dust complex. But, contrary to the appearance, it is not the dust in the coma and tail of the comet that is the main contribution, but rather the millimetre-sized and bigger particles that initially stay close to the comet's orbit and that contribute significantly to the meteoroid complex. Particles in the cometary tail are strongly affected by radiation pressure and, therefore, become immediately β-meteoroids that leave the Solar System on fast hyperbolic trajectories. Bigger than millimetre-sized particles are only occasionally observable in edge-on views of the comet orbit (visible as the so-called antitail), when the path length through these particles is longest. Infrared observations by IRAS revealed the big particles released from comets as dust trails (Sykes *et al.* 1986). Comet trails form when big particles are released with low emission speeds (meters per second) and spread along the comet's orbit. After several revolutions of the Sun they fill a complete torus, the trail of Comet Encke being an example. When the Earth passes through such a trail a spectacular meteor storm is observed, as in 1966 and 1999 when the Earth passed through the trail of Comet Tempel–Tuttle, and Leonid meteor storms were observed. Gradually, gravitational scattering by the planets disperses such a dust torus in interplanetary space, and the meteoroids become part of the zodiacal cloud. This process constitutes the major mass input from comets to the meteoroid complex.

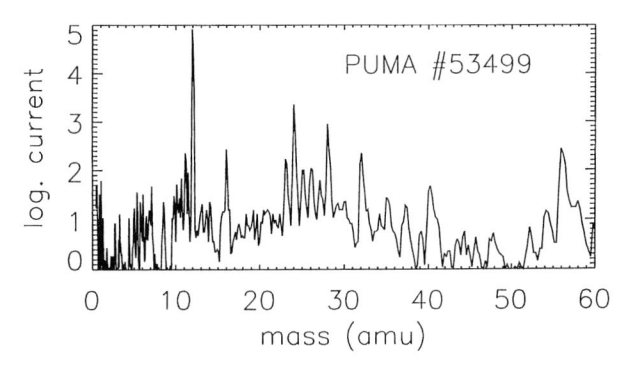

Figure 9 Mass spectrum of a dust particle from Comet Halley obtained by the impact ionization mass spectrometer on board the VeGa spacecraft. The particle is rich in carbon (12 atomic mass units, amu), oxygen (16 amu) and some magnesium (24, 25, and 26 amu) (Courtesy E. Jessberger).

Electromagnetically dominated dust: Dust streams from Io's volcanoes

For a long time, electromagnetic interactions of dust particles has been a topic of theoretical discussions only. Observational evidence for such effects in the Solar System had been absent – they were expected only for submicrometre-sized grains in strong magnetic fields, but most easily observed grains were much bigger. The situation changed in 1992 when Ulysses flew by Jupiter, the planet with the strongest magnetic field, and the dust detector on board recorded intense streams of tiny particles leaving the Jovian system (Figure 10). Before we present details of this observation we have to understand the basics of electromagnetic interactions.

All meteoroids in space are electrically charged. Several competing charging processes determine the actual charge of a meteoroid. Electrons and ions are collected from the ambient solar wind plasma. Irradiation by solar UV light liberates photoelectrons, which escape from the grain. Energetic ions and electrons cause the emission of secondary electrons. Whether electrons or ions can reach or leave the grain depends on their energy and on the polarity and electrical potential of the grain. Because of the predominance of the photoelectric effect in interplanetary space, meteoroids are mostly charged to a potential of a few volts positive. In some parts of planetary magnetospheres charging by plasma dominates, leading to negative charges because of the high electron flux. The potential depends on the energy of the plasma. If the plasma energy is higher than about 50 eV, secondary electron emission sets in and the particles lose more electrons than they collect, leading to a positive charge. A good understanding of the charging environment is required in order to determine the grain charge and its polarity. The timescale for charging is seconds to hours depending on the size of the particle and the plasma density; small particles charge slower.

Charged dust grains interact with the ambient magnetic field. Magnetic fields exist everywhere in the Solar System. In interplanetary space the outward streaming solar wind carries a magnetic field from the Sun. The polarity of the magnetic field is either positive or negative, depending on the polarity at the base of the field line in the solar corona, which varies spatially and temporary. The Lorentz force on a charged dust particle near the ecliptic plane is either upward or downward, depending on the polarity of the solar wind magnetic field. Near the ecliptic the polarity changes due to the sector structure, and the net effect of the Lorentz force on micrometre-sized interplanetary dust particles is small. Only secular effects on the bigger zodiacal dust particles are expected to occur which may affect the symmetry plane of the zodiacal cloud close to the Sun. Observations of the inner zodiacal light show such an effect of the symmetry plane, but there are other explanations, for example, planetary interactions.

When Ulysses flew by Jupiter in February 1992, streams of submicrometre-sized dust particles were discovered which showed impact rates up to a thousand times higher than the average rate in interplanetary space (Figure 10). The distribution of observed impact directions suggested monodirectional streams of particles that arrived from about the direction of Jupiter. Such dust streams were observed only within a distance of 2 AU from Jupiter. Remarkable was the observed periodicity of the streams, of 2 and 4 weeks. Measurements by the Galileo dust instrument confirmed the Ulysses findings and extended the analysis of dust streams during its orbital tour of Jupiter which began in 1996.

Electromagnetic interactions of the particles in the dust streams were evident in both the Ulysses and Galileo data: the deviations of the arrival direction from Jupiter's direction showed a clear correlation with the ambient interplanetary magnetic field which was simultaneously measured on board the spacecraft. Zook et al. (1996) demonstrated convincingly that only charged grains in the 10 nm size range would couple strongly enough to the interplanetary magnetic field to show the observed periodicity and directional variations. The corresponding impact speeds were in excess of $200\,\mathrm{km\,s^{-1}}$ – well beyond the speed range the instruments could be tested in the laboratory. Luckily, theory provided appropriate calibration for these particles.

Dust measurements by Galileo within the Jovian magnetosphere displayed fluctuations of the impact rates by up to two orders of magnitude with a main period of about 10 hours. Frequency analysis of the observed impact rate

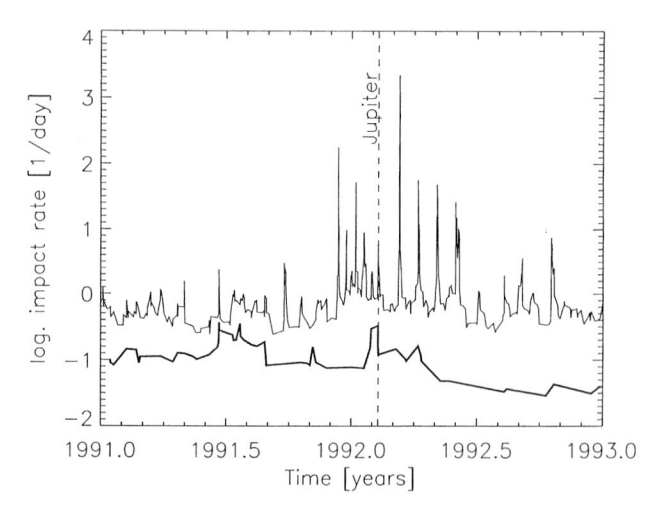

Figure 10 Dust impact rates recorded by Ulysses en route from Earth to Jupiter and beyond until the end of 1992. Jupiter flyby occurred on 10 February 1992. Thick line: rate of particles with mass exceeding 10^{-12} g; thin line: all particles.

showed also strong peaks at 5 and 42 hours. The impact direction of dust particles measured along Galileo's trajectory through the Jovian system was compatible with a dust source in the inner Jovian system inside about $6R_J$.

It was immediately recognized that Jupiter's magnetosphere would eject submicrometre-sized dust particles if they existed at all in the magnetosphere. At two places small dust in abundance had been recognized by Voyager's cameras before: in the Jovian ring at $1.8R_J$ (Jupiter radius, $R_J = 71,492$ km) and its weak extension out to $3R_J$, and in Io's volcanic plumes, which reach heights of about 300 km above the surface. Both phenomena have been suggested as the source of the dust streams: Horanyi et al. (1993) proposed an Io source, whereas Hamilton and Burns (1993) favoured a ring source.

Small dust particles are released from a larger body by impact processes or are emitted by the volcanic plumes on Io. Because of secondary electron emission and the photoelectric effect, in most of the Jovian magnetosphere dust grains get a charge corresponding to about $+3$ V surface potential, i.e. a 10 nm radius particle has lost about 20 elementary charges. Only in the dense plasma torus inside Io's orbit do dust particles rapidly get negatively charged.

Jupiter's magnetic field can be represented by a dipole magnetic field that is tilted by $9.6°$ with respect to the rotation axis and that rotates rigidly with Jupiter. The field strength decreases with the inverse third power of the distance from Jupiter. In the equatorial plane, where most dust sources are located, the magnetic field is roughly perpendicular to Jupiter's equatorial plane. Due to its rigid rotation with Jupiter, magnetic field lines sweep by a body with a speed that increases with distance to Jupiter. For example, at Io's distance ($5.9R_J$) the rotation speed is 74 km s^{-1} whereas Io's Keplerian orbit speed is only 17 km s^{-1} which results in a relative speed of 57 km s^{-1} between Io and Jupiter's magnetic field. Outside the "co-rotational distance" ($2.24R_J$) – the distance at which the Keplerian orbit period equals Jupiter's rotation period – positively charged particles feel an outward directed Lorentz force opposite to the gravitational pull of Jupiter. The effect of the moving magnetic field can be represented by a co-rotational electric potential that has a value of $+60$ million V at the position of Io. For particles smaller than about 200 nm the electromagnetic force is bigger than the gravitational force of Jupiter. The Jovian magnetosphere thus acts as a giant electromagnetic dust accelerator.

Charged dust grains in a certain size range are driven out of the magnetosphere by the co-rotational electric field. At Io's distance particles from about 5 to 200 nm radius are accelerated outward and leave the Jovian system. The escape speed from the Jovian system of the smallest particles is in the range of 300 km s^{-1}. Smaller particles remain tied to the magnetic field lines and gyrate around them, as do ions. Bigger particles move on gravitationally bound orbits that are more or less modified by the Lorentz force. The size range of expelled particles gets narrower when the source is located closer to Jupiter, with the upper size limit staying at about 200 nm. Due to the increased magnetic field strength at the distance of the tenuous Gossamer Ring ($3R_J$), the lower size limit is 28 nm (Hamilton and Burns 1993); smaller grains are magnetically bound.

Besides the outward acceleration there is a significant out-of-plane component of the electromagnetically induced force. Depending on the phase of the inclined magnetic field (with respect to Jupiter's rotation) at the position of the particle, its trajectory is deflected up or down from Jupiter's equatorial plane, where most particles originate. Dust particles that are continuously released from a source escape in a warped sheet of dust. Observers in Jupiter's equatorial plane would record dust particles when the warped dust sheet passes over their position. This occurs twice per Jupiter rotation, and a periodic variation of the flux results with a 5 hour periodicity.

There are several lines of evidence that the source of the dust streams is Io and not the Gossamer Ring: (1) 10 nm sized particles cannot escape the magnetic field at the ring distance, but they can escape from Io's distance, (2) the dense plasma torus at about $5R_J$ cannot easily be traversed by positively charged grains that originate in the ring, and (3) the observed periodicity of 42 hours of the impact rate, which is also the orbital period of Io, identifies this satellite as the source of the dust streams.

The source of dusty rings

All objects in the Solar System are constantly bombarded by meteoroids. While atmospheres around the planets and some moons mitigate and shield these effects, atmosphereless bodies are exposed to this bombardment. As an example, impact craters of all sizes have been found on lunar rocks; rocks from asteroids and most planetary satellites are excepted to show a similar cratering population. The material excavated during the impact process is ejected as particles in a wide range of fragment masses. Impact experiments in the laboratory have shown that the total mass ejected from a high-velocity impact can be more than 1000 times bigger than the projectile mass, depending on the impact speed. Some of the smallest ejecta particles have emission speeds in excess of the impact speed – the total kinetic energy in all ejecta particles, of course, is always less than the projectile kinetic energy. The gravity of the target body determines how much of the impact-generated fragments will escape. Targets of only a few kilometers in size will hardly retain any fragments. Some of the ejecta particles will even escape from moon-sized bodies; the rest will fall back to the surface, where it will accumulate with time. The lunar regolith, a fine-grained soil, is ejecta material retained from impacts of all magnitudes onto the lunar surface.

Although there are no dust storms on the Moon, there is a significant migration of dust due to electrostatic levitation and transport. There are a number of observations of dust above the surface of the Moon, either by instruments on automated lunar probes or by astronauts circling the Moon. Just before sunrise or after sunset the cameras on the Surveyor probes recorded a subtle horizon glow, and the photometers on board one of the Lunokhod rovers saw a post-sunset brightness overhead which was due to some dust still illuminated by the Sun. More direct evidence of dust migration on the lunar surface came from Otto Berg's Lunar Ejecta and Meteoroids (LEAM) instrument. This instrument was placed by the Apollo 17 astronauts on the lunar surface. It consisted of three sensors similar to the Pioneer 8 and 9 sensors that faced different directions: up, east and west. The three sensors recorded different flows of charged low-velocity dust, especially when the instrument site went in and out of shadow. This phenomenon is qualitatively explained by different electrical dust charging in sunlight and in shadow. During the day the lunar surface charges positively to about $+5$ V, due to the photoeffect from solar UV light. During the night, in the absence of solar UV photons, energetic electrons can charge the surface to several kilovolts negative. At the Sun–shadow interface high electric fields are set up which facilitate the transport of charged dust. This phenomenon requires further study because a future lunar station may become covered by dust if no precautions are taken.

The question of how much material leaves the Moon in comparison to the influx of meteoroids can be answered only by an experiment that monitors the flux towards and away from the Moon. Early attempts with dust detectors on Lunar Orbiter and Lunar Explorer 35 (Alexander *et al.* 1971) gave some indications of dust leaving the Moon, especially during meteor shower periods, but the data had severe statistical limitations. The HEOS 2 satellite recorded "groups" of impacts (two or more impacts in a few hours) which were suggested to be ejecta particles originating from impacts of large meteoroids on the Moon. Recently, the Japanese Hiten spacecraft orbited the Moon for a full year, but because of the small number of recorded particles and because of its wide field of view it could not accurately determine the ratio of incoming to outgoing particles.

The unique identification of ejecta particles leaving a satellite came from the Galileo spacecraft while orbiting Jupiter. During its flyby of the Galilean moons Callisto, Ganymede and Europa, the dust detector recorded a significantly enhanced flux of dust particles arriving from the direction of the satellite, whereas the background flux of particles arrived from the opposite direction. Krüger *et al.* (1999) showed that this flux was compatible with the production of ejecta particles from interplanetary meteoroids impacting the surface of the satellites. The measured speeds matched the flyby speed satisfactorily. The escaping particles could be followed during their dispersion throughout the Jovian system, forming a tenuous dust ring in the region of the Galilean satellites.

Dust production at small satellites is easily observable in the dust rings of Jupiter and Saturn. Images taken by Galileo's cameras showed that the sharp outer edges of the Gossamer Ring of Jupiter coincided with the position of small satellites. Ockert-Bell *et al.* (1999) interpret this behaviour as being due to the inward motion of ejecta particles once they are released from the satellite. Small satellites are more effective dust generators than are big satellites because their escape speeds are lower. For example, a rocky satellite of 10 km radius has an escape speed of about 10 m s^{-1} whereas the escape speed from the Moon, or any of the Galilean satellites, is more than 2 km s^{-1}. Since the ejecta speed distribution is strongly peaked at small values, many more particles can leave a small satellite than can leave a big one, despite the difference in cross-section. All the dusty rings of the giant planets must have satellites associated with them from which they originate. However, some of the ring source satellites have not yet been discovered, e.g. no satellite has been found that feeds Saturn's G ring.

Interstellar dust in the heliosphere

In 1977 the proposal for the Ulysses dust investigation was submitted to ESA and NASA. One of its objectives was the search for interstellar dust (ISD). It was hoped that a minute contribution of ISD could be identified above or below the Sun's poles where the interplanetary dust density was expected to be very low. It came as a big surprise that after Ulysses flew by Jupiter, the dust detector recorded impacts of interstellar grains that arrived predominantly from a direction that was opposite to the expected flow direction of interplanetary dust grains (Grün *et al.* 1993). It was found that on average the impact velocities exceeded the local Solar System escape velocity.

The motion of ISD through the Solar System was found to be parallel to the flow of neutral interstellar hydrogen and helium gas, with a speed of 26 km s^{-1} both for gas and dust. This proves that local interstellar dust and gas are nearly at rest with respect to each other. The ISD flow was continuously monitored by Ulysses, and persisted at a constant level at all latitudes above the ecliptic plane, even over the poles of the Sun, whereas interplanetary dust was strongly depleted away from the ecliptic plane. Starting in mid-1996 the flux of ISD began slowly to decrease and, in the year 2000, was about a factor of 3 lower (as will be explained below, this is related to the reversal of the magnetic field over in the course of the solar cycle).

Measurements in the ecliptic plane by Galileo confirmed that outside about 3 AU the ISD flux exceeds the flux of micrometre-sized interplanetary grains. Interstellar grains observed by Ulysses and Galileo range from 10^{-18} kg to above 10^{-13} kg. If compared with the ISD mass distribution derived by astronomers, the mass distribution observed *in situ* overlaps only with the biggest masses observed by remote sensing. More recently, even bigger (10^{-10} kg) interstellar meteoroids have been reliably identified by their hyperbolic speed (>100 km s^{-1}) at 1 AU (Taylor *et al.* 1996). The flow direction of these big particles varies over a much wider angular range than that of small grains observed by Ulysses and Galileo.

The deficiency of measured small grain masses is not solely caused by the detection threshold of the *in situ* instrumentation, but indicates a depletion of small interstellar grains in the heliosphere. Model calculations by Frisch *et al.* (1999) of the filtering of electrically charged grains in the heliospheric bow shock region and in the heliosphere itself show that 0.1 μm and smaller particles are strongly impeded from entering the Solar System by the interaction with the solar wind/magnetic field.

In 1976 Eugene Levy and Randy Jokipii suggested that small (<0.1 μm) interstellar grains will not reach the inner Solar System because they would interact strongly with the interplanetary magnetic field, and therefore would be carried out of the heliosphere by the solar wind. As a consequence no small grains (most astronomers believed that there are no big grains in the diffuse interstellar medium) could be found in the Solar System. However, Morfill and Grün (1979) and Gustafson and Misconi (1979) found, by looking into the details of this interaction, that there are exceptions to this rule. The overall polarity of the solar magnetic field changes with the solar cycle of 11 years. For one solar cycle, positive magnetic polarity prevails in the northern solar hemisphere, and negative polarity in the southern solar hemisphere. Interstellar particles that enter the Solar System are either deflected towards the ecliptic plane or away from it depending on the overall polarity of the magnetic field. Therefore, submicrometre-sized interstellar particles are either prevented (during one solar cycle) from reaching the inner Solar System or are concentrated (in the other solar cycle) near the ecliptic plane. In the period from 1989 to 2000 the overall magnetic field has the deflecting configuration. However, interstellar grains need about 20 years (almost two solar cycles) to traverse the distance from the heliospheric boundary (assumed to be at about 100 AU) to the Sun, and therefore the magnetic field change does not immediately affect the dust flow: there is a delay. Indeed, it took about six years until the filtering of small particles took effect, so that only big (micrometre-sized) interstellar particles reached the positions of Galileo and Ulysses.

Once it became evident that galactic ISD is accessible to *in situ* detection, NASA selected the Stardust mission to analyse and return to Earth samples of cometary and interstellar dust at the distance of the asteroid belt (Brownlee *et al.* 1996). Several times during its eccentric orbit about the Sun (out to about 3 AU) Stardust will capture interstellar dust, together with dust from Comet Wild 2, by impacts into aerogel, and return it to Earth in 2006. In addition, *in situ* detection and compositional measurements of cometary and interstellar grains will be performed.

FUTURE PROSPECTS

Despite the great advances made in the understanding of the heliospheric dust environment, there remain many important questions to be answered. Below is an incomplete collection of dusty problems (in the order of distance from the Sun) that may be addressed in the foreseeable future, or are already on their way at the beginning of the new century. We briefly discuss dust measurements in the F corona and in Earth orbit, the search for the Martain dust ring, analyses of dust from a variety of comets, exploration of the dust environments of Saturn, Uranus, Neptune and Pluto, the Kuiper Belt as a model for extrasolar dust clouds, and finally, the analysis of interstellar dust outside the heliosphere. This collection of problems demonstrates that dust analysis will remain an important topic of planetary and astrophysical research.

In situ measurements of dust in the F corona of the Sun (inside about $20 R_\odot$) get special attention because observations from the Earth during solar eclipses indicate transient peaks in the infrared radiation on top of a smooth increase towards the Sun. This is explained by material transported by the Poynting–Robertson effect towards the Sun until it sublimates. Dust of different chemical composition sublimates at different distances. At $4 R_\odot$ the black body temperature is about 2000 K, at which almost any meteoritic material sublimates. Inside this distance a dust-free zone is expected at the border of which a peak of the thermal emission from dust is located. Because meteoritic material is made of constituents of different volatility, these peaks are expected to depend on the material and the temperature at which it sublimates. This way nature separates meteoroid material near the Sun according to its volatility. Analysis of the spatial distribution of matter close to the Sun by *in situ* instruments that are much more sensitive than astronomical methods would immediately give us information on the volatility of their constituents.

Closer to home, the dust environment of the Earth is of interest, because humans are affecting this environment through their space activities. There is an increasing need to closely monitor the non-natural contribution in order to control its detrimental effects. However, the natural dust environment is also of technological and scientific interest. Hazards from meteor streams (like the Leonids) will require continuous attention. Meteoroids that pass the Earth are of scientific interest because they are messengers from distant worlds: asteroids, comets and even other stars. Once we know where dust grains originated, compositional analysis of grains can tell us many things about these worlds. For example, the identification of diamond, silicon carbide, graphite, corundum and silicon nitride in interstellar grains would provide the missing link between presolar dust grains that have been identified in meteorites and their stellar parents. Therefore the goal of such dust studies is to identify the sources of dust particles together with their in-depth analysis. Theoretical studies of dust dynamics together with accurate measurements of dust trajectories will enable us to identify their place of birth. Proposals for such missions carrying a dust telescope are currently under discussion. Although significant compositional information can be gained by *in situ* measurements, even more can be learned if this dust is collected and brought to the laboratory where the most advanced instrumentation can be used for its analysis. Sample return form Earth orbit is, of course, much easier than sample return from distant worlds. In summary, the future of dust measurements in Earth orbit is seen to lie in three areas: (1) environmental monitoring, (2) use of dust telescopes to separate and analyse dust populations of different origin, and (3) collection and sample return of dust from various sources for in-depth analysis in laboratories.

Enhanced dust densities in the Martian environment have long been suspected. Impacts of interplanetary meteoroids on the Martian satellites, Phobos and Deimos, generate ejecta particles that escape the weak gravity of the moons and reside for some time in the Martian environment. Attempts with the Viking cameras to observe a faint ring failed. The first indications of a Martian dust ring came from solar wind measurements with the magnetometer on board the Russian Phobos 2 mission, which observed magnetic field deflections similar to the ones observed in the dusty coma of Comet Halley. Since July 1998 the Japanese Nozomi spacecraft carrying Eduard Igenbergs' highly sensitive dust detector has been on its way to Mars to make the first *in situ* measurements of the Martian dust environment, in 2003. The comparison of the dust observations at Mars with those in the dusty rings of Saturn and the other giant planets will tell us what effects solar radiation pressure (strongest at Mars) and planetary magnetospheres (negligible at Mars) have. In addition, during a future human colonization of Mars dust from the Martian satellites may play a similar hazardous role as space debris in the Earth environment.

Analysis of particulates from Comet Halley brought us new and important information that has relevance to the understanding of the formation of the Solar System. Currently the Stardust mission is on its way to analyse, collect and return dust from Comet Tempel 2. This will give us a second example from the large variety of comets. Astronomical observations tell us that there are significant differences among comets: there are fresh comets that arrive for the first time from cold storage in the Oort Cloud, and there are others that have survived for a long time as short-period comets in the planetary system; there are very dusty comets and others that contain larger amounts of ice and gas; and there are comets that are more carbon-rich than others. It is the analysis of this variety that will tell us about the spatial and compositional variations in the protoplanetary nebula through the comets that are representative of different regions of this nebula.

The satellite and ring systems of the outer planets are models for the early Solar System, with satellites and ring particles in intimate interactions. There are rings that are fed from satellites, and others that may be the result of a break-up of a satellite; still others show signs of re-agglomeration and may eventually become satellites again. The diversity of ring structures – from the diffuse and voluminous E ring of Saturn to the extremely fine structure of Saturn's main rings, and the complex variability in many rings, exemplified by the F ring with its kinks, braids and multiple strands – is one of the challenging problems in planetary physics today. While Saturn's ring system is of high complexity, the rings of Uranus and Neptune show other features that have not yet been found elsewhere. The Pluto–Charon system may have rings with even other features. To understand the common characteristics of all these rings and the reasons why they are so different requires detailed measurements of the rings and their environments. Only the combination of *in situ* and remote observations of the physical, chemical and dynamical properties of ring particles will provide this crucial information. Currently, the Cassini mission is on its way to Saturn where it will address questions like mapping the size and spatial distributions of ring material throughout the Saturnian system, determining the particle dynamics that is manifested in the orbits in the diffuse rings, defining the role of meteoroid impacts as mechanism of ring erosion, analysing the chemical composition of ring particles, investigating interactions of the satellites with the ring system and determining their role as sources for ring particles, and studying interactions between the ring and the magnetosphere. Similar missions to the other outer planets are envisioned.

Detection of Kuiper Belt objects (KBOs) of up to a few hundred kilometres in diameter confirmed the existence of objects outside what had been traditionally thought of as the planetary system. Such objects had been predicted by Kenneth Edgeworth in 1949, and by Gerard Kuiper in 1951, in order to explain the frequency of occurrence of short-period comets. Also, models of the evolution of the disk of planetesimals by Julio Fernandez (1980) suggest the survival of planetesimals in the outer Solar System. About 540 objects have been found by April 2002 outside about 30 AU from the Sun. Theories predict an extension of kilometre-sized primordial objects out to 3,000 AU from the Sun. Mutual collisions among KBOs as well as impacts of interstellar grains generate dust locally. In the absence of big planets, the Poynting–Robertson effect is the most important dynamical effect on dust in the Kuiper Belt. However, Liou et al. (1996) have shown that during their inward orbital evolution micrometre-sized grains are trapped in resonances with the outer giant planets, which results in some structure of the radial and azimuthal distribution of dust at the edge of the planetary system.

The detection of an infrared excess at main sequence stars sparked renewed interest in the outer extensions of our own Solar System dust cloud; in particular the observation of dust disks around β Pictoris stimulated interest in the outer zodiacal cloud. In situ dust detectors on board the Pioneer 10 and 11 probes measured a constant dust flux of 10 μm particles out to 20 AU (Humes 1980). These may be the signature of the inner part of the Kuiper dust belt. Beyond that distance the plasma wave experiment on board the Voyager spacecraft suggested that there are dust grains out to about 50 AU. Observations of the Kuiper dust belt can, therefore, be used as a model for extrasolar dust clouds and can help to reveal information about other planetary systems.

The effects of the heliosphere reach into interstellar space out to about 300 AU from the Sun. This is the region where small (<0.1 μm) particles are affected and may be prevented from entering the heliosphere. A future mission to measure the undisturbed local interstellar medium, or one going even beyond there, to reach the next solar neighbour will deal – indeed, for technological reasons has to deal – with interstellar dust. We know that small (<0.1 μm) interstellar grains are filtered out before reaching the inner Solar System. Their origin, however, may be different from that of bigger grains accessible at Earth orbit. We know that evolved stars continuously lose mass. A large fraction of the stellar mass loss is provided by cool high-luminosity stars. As the ejected material cools in the expanding stellar envelope, solid particles condense out of the gas. This "stardust" provides the seeds for ISD grains that grow in cool interstellar clouds by accretion of atoms and molecules and by agglomeration. Interstellar shock waves provide an effective destruction mechanism for ISD grains. In diffuse interstellar clouds, the grains lose their volatile constituents owing to galactic UV irradiation. An unbiased look into this interstellar dust factory will provide us with information on processes that are difficult to quantify by astronomical observations alone. Thereby, in situ dust analysis will still be an important method when automated probes leave our Solar System.

REFERENCES

Alexander, W.M., Arthur, C.W. and Bohn, J.L. (1971). Lunar Explorer 35 and OGO 3: dust particle measurements in selenocentric and cislunar space from 1967 to 1969. In K. YA. Kondratyev, M.J. Rycroft and C. Sagan (eds), *Space Research XI*, Akademie-Verlag, Berlin pp. 279–285.

Auer S. (1974). The asteroid belt: doubts about the particle concentration measured with the Asteroid/Meteoroid detector on Pioneer 10. *Science*, **186**, 650–652.

Auer, S. and Sitte, K. (1968). Detection technique for micrometeoroids, using impact ionization. *Earth and Planetary Science Letters* **4**, 178–183

Axford, W.I. and Mendis, D.A. (1974). Satellites and magnetospheres of the outer planets. *Annual Review of Earth and Planetary Sciences*, **2**, 419.

Berg, O.E. and Grün, E. (1973). Evidence of hyperbolic cosmic dust particles. In M.J. Rycroft and S.K. Runcorn (eds), *Space Research XIII*, Akademie-Verlag, Berlin, 1047–1055.

Berg, O.E. and Richardson, F.F. (1969). The Pioneer 8 cosmic dust experiment. *Reviews of Scientific Instruments*, **40**, 1333–1337

Brownlee D.E., Burnett, D., Clark, B., Hanner, M.S., Horz, F. Kissel, J., Newburn, R., Sandford, S., Sekanina, Z., Tsou, P. and Zolensky, M. (1996). Stardust: Comet and interstellar dust sample return mission. In B.A.S. Gustafson and M.S. Hanner (eds), *Physics and Chemistry and Dynamics of Interplanetary Dust*, ASP Conf. Series 104, pp. 223–226.

Brownlee, D.E., Tomandl, D., Blanchard, M.B., Ferry, G.V. and Kyte, F. (1976). An atlas of extraterrestrial particles collected with NASA U-2 aircraft, 1974–1976, NAS TMX.

Burns, J.A., Showalter, M.R., Cuzzi, J.N., Pollack, J.B. (1980). Physical processes in Jupiter's ring: Clues to its origin by Jove. *Icarus*, **44**, 339–360.

Ceplecha, Z. (1977). Meteoroid populations and orbits. In A.H. Delsemme (ed), *Comets Asteroids Meteorites. Interrelations, Evolution and Origins* University of Toledo, pp. 143–152.

Cour-Palais, B.G. (ed) (1969). Meteoroid environment Model-1969. Near-Earth to Lunar Surface, NASA SP 8013.

Dermott, S.F. Jayaraman, S., Xu Y.L., Gustafson, B.A.S. and Liou, J.C. (1994). A circumpolar ring of asteroidal dust in resonant lock with the Earth. *Nature*, **369**, 719–723.

Dietzel, H., Eichhorn, G., Fechtig, H., Grün, E., Hoffmann, H.J. and Kissel, J. (1973). The HEOS A-2 and Helios micrometeoroid experiments. *Journal of Physics E*, **6**, S. 209–217.

Dohnanyi, J.S. (1970). On the origin and distribution of meteoroids. *Journal of Geophysical Research*, **75**, 3468–3493.

Dubin M. and Soberman R.K. (1991). Cosmoids: Solution to the Pioneer 10 and 11 meteoroid measurement enigma. *Planetary and Space Science*, **39**, 1573–1590.

Dumont, R. (1983). Zodiacal light gathered along the line of sight: The vicinity of the terrestrial orbit studied with photopolarimetry and with Doppler spectrometry. *Planetary and Space Science*, **31**, 1381–1387.

Elsässer, H. and Fechtig, H. (eds) (1976). *Interplanetary Dust and Zodiacal Light*, Lecture Notes in Physics 48, Springer-Verlag, Berlin.

Fechtig, H., Grün, E. and Kissel, J. (1978). Laboratory simulation. In J.A.M. McDonnell (ed), *Cosmic Dust*, Wiley, Chichester, pp. 607–669.

Fernandez, J.A. (1980). On the existence of a comet belt beyond Neptune. *Monthly Notices of the Royal Astronomical Society*, **192**, 481–491.

Friichtenicht, J.F. (1962). Two-million-volt electrostatic accelerator for hypervelocity research, *Reviews of Science Instruments*, **33**, 209–212.

Früchtenicht, J.F., Roy, N.L. and Becker, D.G. (1973). The cosmic dust analyzer: Experimental evaluation of an impact ionization model. In *Evolutionary and Physical Properties of Meteoroids*, NASA SP-319, 299.

Früchtenicht, J.F. and Slattery J.C. (1963). Ionization associated with hypervelocity impact. NASA Technical Note D-2091.

Frisch, P., Dorschner, J., Greenberg, M., Grün, E., Landgraf, M., Hoppe, P., Jones, A., Krätschmer, W., Linde, T., Morfill, G.E., Reach, W., Svestka, J., Witt, A. and Zank, G., (1999). Dust in the local interstellar wind. *Astrophysical Journal*, **525**, 492–516.

Giese, R.H. and Lamy, P. (eds). (1985). *Properties and Interactions of Interplanetary Dust*, Reidel, Dordrecht.

Grün, E., Zook, H.A. Baguhl, M. Balogh, A. Bame, S.J. Fechtig, H. Forsyth, R. Hanner, M.S. Horanyi, M. Kissel, J. Lindblad, B.-A. Linkert, D. Linkert, G. Mann, I. McDonnell, J.A.M. Morfill, G.E. Phillips, J.L. Polanskey, C. Schwehm, G. Siddique, N. Staubach, P. Svestka J. and Taylor A. (1993). Discovery of Jovian dust streams and interstellar grains by the Ulysses spacecraft, *Nature*, **362**, 428–430.

Grün, E., Zook, H.A., Fechtig, H. and Giese, R.H. (1985a). Collisional balance of the meteoritic complex *Icarus*, **62**, 244–272.

Grün, E., Zook, H.A., Fechtig, H. and Giese, R.H. (1985b). Mass input into and output from the meteoritic complex. In R.H. Giese and P. Lamy (eds), *Properties and Interactions of Interplanetary Dust*, Reidel, Dordrecht, pp. 411–415.

Grün, E., Gustafson, B.A.S., Dermott, S.F., Fechtig, H. (eds.) (2001). *Interplanetary Dust*, Springer, Berlin, Heidelberg, New York.

Gurnett, D.A., Grün, E., Gallagher, D., Kurth, W.S. and Scarf, F.L. (1983). Micron-sized particles detected near Saturn by the Voyager plasma wave instrument. *Icarus*, **53**, 236–254.

Gustafson, B.A.S. and Hanner, M.S. (eds) (1996). *Physics, Chemistry and Dynamics of Interplanetary Dust*, Astronomical Society of the Pacific Conference Series, Volume 104, San Francisco.

Gustafson, B.A.S. and Misconi, N.Y. (1979). Streaming of interstellar grains in the solar system. *Nature*, **282**, 276–278.

Halliday, I. and McIntosh, B.A. (eds) (1980) *Solid Particles in the Solar System*. Reidel, Dordrecht.

Hamilton, D.P. and Burns, J.A. (1993). Ejection of dust from Jupiter's gossamer ring. *Nature*, **364**, 695–699.

Hanner, M.S. and Weinberg, J.L. (1973) Gegenschein observations from Pioneer 10. *Sky and Telescope*, **45**, 217–218.

Hansen, D.O. (1968). Mass analysis of ions produced by hypervelocity impact. *Applied Physics Letters*, **13**, 89.

Hemenway, C.L., Millman, P.M. and Cook, A.F. (eds) (1973). *Evolutionary and Physical Properties of Meteoroids*, NASA SP-319, Washington.

Horanyi, M., Morfill, G.E. and Grün, E. (1993). Mechanism for the acceleration and ejection of dust grains from Jupiter's magnetosphere. *Nature*, **363**, 144–146.

Hörz, F., Brownlee, D.E., Fechtig, H., Hartung, J.B., Morrison, D.A., Neukum, G., Schneider, E., Vedder, J.F. and Gault, D.E. (1975). Lunar microcraters: Implications for the micrometeoroid complex. *Planetary and Space Science*, **23**, 151–172.

Humes, D.H. (1980). Results of Pioneer 10 and 11 meteoroid experiments: Interplanetary and near-Saturn. *Journal of Geophysical Research*, **85**, 5841–5852.

Igenbergs, E. and Kuczera, H. (1979). Micrometeoroid and dust simulation. In *Proceedings of the Comet Halley Micrometeoroid Hazard Workshop*, ESTEC, 18–19 April 1979, ESA SP-153, pp. 109–114.

Jackson, A.A. and Zook, H.A. (1989). A solar system dust ring with the earth as its shepherd. *Nature*, **337**, 629–631.

Johnson, T.V., Morfill, G. and Grün, E. (1980). Dust in Jupiter's magnetosphere: An Io source. *Geophysical Research Letters*, **7**, 305–308.

Kelsall, T., Weiland, J.L., Franz, B.A., Reach, W.T., Arendt, R.G., Dwek, E., Freudenreich, H.T., Hauser, M.G., Moseley, S.H., Odegard, N.P., Silverberg, R.F. and Wright, E.L. (1998). The COBE diffuse infrared background experiment search for the cosmic infrared background: II. Model of the interplanetary dust cloud. *Astrophysical Journal*, **508**, 44–73.

Kessler, D.J. and Cour-Palais, B.G. (1978). Collision frequency of artificial satellites: The creation of a debris belt. *Journal of Geophysical Research*, **83**, 2637–2646.

Kissel J. (1986). The Giotto particulate impact analyzer, ESA SP-1077, pp. 67–83.

Kissel, J., Brownlee D.E., Büchler, K., Clark, B.C. Fechtig, H., Grün, E., Hornung, K., Igenbergs, E.B., Jessberger, Krüger, F.R., Kuczera, H., E.K., Mc Donnell, J.A.M., Morfill, G.E., Rahe, J., Schwehm, G.H., Zekanina, Z., Utterback, N.G., Völk, H. and Zook, H.A. (1986). Composition of comet Halley dust particles from Giotto observations. *Nature*, **321**, 336–338.

Krüger, H., Krivov, A.V., Hamilton, D.P. and Grün, E. (1999). Detection of an impact-generated dust cloud around Ganymede. *Nature*, **399**, 558–560.

Leinert, Ch. and Grün, E. (1990). Interplanetary dust, In R. Schwenn and E. Marsch (eds). *Physics of the Inner Heliosphere I.*, Springer-Verlag, Berlin, pp. 207–275.

Leinert, Ch., Richter, I., Pitz, E. and Planck, B. (1981). The zodiacal light from 1.0 to 0.3 AU as observed by the Helios space probes. *Astronomy and Astrophysics*, **103**, 177–188.

Levasseur-Regourd, A.C. and Hasegawa, H. (eds) (1991). *Origin and Evolution of Interplanetary Dust*, Kluwer, Dordrecht.

Levy, E.H. and Jokipii, J.R. (1976). Penetration of interstellar dust into the solar system. *Nature*, **264**, 423–424.

Liou, J.C., Zook, H.A. and Dermott, S.F. (1996). Kuiper belt dust grains as a source of interplanetary dust particles. *Icarus*, **124**, 429–440.

McCracken, C.W., Alexander, W.M. and Dubin, M. (1961). Direct measurements of interplanetary dust particles in the vicinity of earth. *Nature*, **192**, 441–442.

McDonnell, J.A.M. (ed) (1978). *Cosmic Dust*, Wiley, Chichester.

McDonnell, J.A.M. and Gardner, D.J. (1998). Meteoroid morphology and densities: Decoding satellite impact data. *Icarus*, **133**, 25–35.

Morfill, G.E. and Grün, E. (1979). The motion of charged dust particles in interplanetary space – I. The zodiacal dust cloud. *Planetary and Space Science*, **27**, 1269–1282.

Morfill, G.E., Grün, E. and Johnson, T.V. (1980). Dust in Jupiter's magnetosphere. *Planetary and Space Science*, **28**, 1087–1123.

Morrison, D.A. and Zinner, E. (1977). 12054 and 76215: New measurements of interplanetary dust and solar flare fluxes. In *Proceeding of the 8th Lunar Science Conference*, pp. 841–863.

Nilsson, C. (1966). Some doubts about the Earth's dust cloud. *Science* **153**, 1242–1246.

Ockert-Bell, M.E., Burns, J.A., Daubar, I.J., Thomas, P.C., Veverka, J., Belton, M.J.S. and Klaasen, K.P. (1999). The structure of Jupiter's ring system as revealed by the Galileo imaging experiment. *Icarus*, **138**, 188–213.

Raizer, Yu.P. (1960). Residual ionization of a gas expanding in vacuum. *Soviet Physics JETP*, **10**, 411–416.

Shapiro, I.C., Lautman, D.A. and Colombo, G. (1966). The earth's dust belt: fact or fiction? 1. Forces perturbing dust particle motion. *Journal of Geophysical Research*, **71**, 5695–5704.

Shelton, H., Hendricks Jr., C.D. and Wuerker, R.F. (1960). Electrostatic acceleration of microparticles to hypervelocities. *Applied Physics*, **31**, 1243–1246.

Sykes, M.V., Lebovsky, L.A., Hunten, D.M. and Low, F. (1986). The discovery of dust trails in the orbits of periodic comets. *Science*, **232**, 1115–1117.

Taylor, D.A., Baggaley, W.J. and Steel, D.I. (1996). Discovery of interstellar dust entering the Earth's atmosphere. *Nature*, **380**, 323–325.

Vedder J.F. (1963). Charging and acceleration of microparticles. *Reviews, of Scientific Instruments*, **34**, 1175.

Weinberg, J.L. (ed) (1967). The zodiacal light and the interplanetary medium, NASA SP-150, Washington, DC.

Weinberg, J.L., Hanner, M.S., Beeson, D.E., DeShields, L.M. and Green, B.A. (1974). Background starlight observed from Pioneer 10. *Journal of Geophysical Research*, **76**, 3665–3670.

Whipple, F. (1975). Sources of interplanetary dust. In H. Elsässer and H. Fechtig (eds), *Interplanetary Dust and Zodiacal light*, Lecture Notes in Physics 48, Springer-Verlag, Berlin, pp. 403–415.

Zook H.A. and Berg, O.E. (1975). A source for hyperbolic cosmic dust particles. *Planetary and Space Science*, **23**, 183–203.

Zook H.A., Grün, E., Baguhl, M., Hamilton, D., Linkert, G., Liou, J.C., Forsyth, R. and Phillips, J.L. (1996). Solar wind magnetic field bending of Jovian dust trajectories. *Science*, **274**, 1501–1503.

50

ROSINE LALLEMENT*

The interaction of the heliosphere with the interstellar medium

Far out in the uncharted backwaters of the unfashionable end of the western spiral arm of the Galaxy lies a small unregarded yellow sun.

This first sentence of *The Hitch Hiker's Guide to the Galaxy*, Douglas Adams' fiction book, could serve as an introduction to this chapter, and paraphrasing his style (without his talent, I am afraid) I could add:

This small and unregarded yellow sun has encountered a tiny and insignificant interstellar cloud, which is so tenuous that the extremely weak wind of the yellow sun blows in the tiny cloud a cavity, which extends far beyond the planetary system attached to the yellow sun. Amazingly, trying to cross the edge of this cavity is one of the favorite games of a small group of obstinated individuals, who belong to the very primitive life forms on one of the orbiting planets.

Here I try to relate the story of the discovery of the small cavity carved by the solar wind in the local interstellar cloud, our heliosphere. Being in the middle of this structure, i.e. with a very limited and biased view of it, it has not been easy to infer its existence, and the first steps in the understanding of its structure in the 60's and 70's have required a strong imagination of the pioneers in this field. Then, during the last thirty years, the high number of new observations have made the subject more and more fascinating.

1 INTRODUCTION

1.1 A multidisciplinary field

The research area described in this chapter is by essence a multidisciplinary field, dealing with the properties of gases, radiation, and fields at distances from the Sun varying between a few solar radii and tens of parsecs, that is from the solar corona and the birth of the solar wind, to the edges of the heliosphere at about 100 AU, where the solar wind stops and comes into equilibrium with the galactic interstellar medium, and finally to the stars which are used as targets for measurements of the local interstellar medium properties, located at distances ranging between 1 and 100 pc from the Sun (1 pc = 200,000 AU, 1 AU = 215 solar radii, 1 solar radius = 700,000 km).

This multidisciplinary aspect is reflected in the large number of spacecraft whose instruments have recorded relevant data. I remember that during the first meeting held in the newly created International Space Science Institute in Bern in 1986, a meeting entitled "The Heliosphere in the Local Interstellar Cloud," I counted more than 20 spacecraft whose data were presented or discussed: Ulysses, Prognoz 5 and 6, Voyager 1 and 2, Pioneer 10 and 11, Pioneer-Venus, Galileo, IUE, HST, ISEE 3, IMP 8, AMPTE, Copernicus, EUVE, SOHO, ROSAT, WISCONSIN survey, DXS, and SAMPEX (and I must certainly have forgotten some others). Results obtained from ground using neutron monitors, telescopes and radio telescopes are also used in conjunction with space data. Such a variety (and such a "flotilla") has made this field so interesting during the last few decades. The expectation of the crossing of the boundary of our helio-sphere by one of the deep-space probes adds to the suspense.

An additional proof of the variety of the field is given by the (almost) contemporary historical overviews of S. Suess (1990), T. Holzer (1989), V. Baranov (1990), and I. Axford (1990). They give an idea of how different the descriptions by experts in solar wind, in plasma physics, and in cosmic

---

*CNRS – Service d'Aéronomie, Verrières-le-Buisson, France

rays can be. At variance with these reviews, the present chapter will have a neutral gas/interstellar medium "flavor."

Over the years, the number of erroneous interpretations in this field has been quite large, but this happens commonly, and we know from K. Popper that empirical disproof is a seminal event in the scientific process. All these errors have indeed stimulated the next advances. Also, while in some cases new instruments have been built specifically for a defined goal, for example the GAS experiment on board Ulysses for the "*in situ*" detection of interstellar neutral helium, other findings have occurred in a totally unexpected way. One of the best examples was the first detection of the "echoes" from the heliospheric boundary by the Voyager radio experiments in 1983. These echoes today remain, to a large degree, mysterious.

1.2 The heliosphere: a solar wind "bubble" in the ambient interstellar gas

Our Sun, an ordinary late-type star in the Local Arm (or Orion Arm), is presently traveling through a small interstellar cloud, the Local Cloud, whose size is of the order of a few parsecs (Figure 1). This cloud belongs to a small group of partially ionized clouds whose temperatures and densities are of the order of 5,000–10,000 K and 0.1 particles per cm^3. This group of clouds is embedded in a 100 pc wide volume, the Local Bubble (LB), believed to be filled with an extremely tenuous ($0.002\,cm^{-3}$) and hot ($10^6\,K$) gas. The origin of the LB is still a matter of much debate.

The relative motion between the Sun and the interstellar medium (ISM) creates a permanent flow of interstellar gas and dust around our star, the so-called interstellar wind (Figure 2). The flow velocity is of the order of $25\,km\,s^{-1}$ (4 AU per year). When the gas approaches the Sun, the ionized part of the interstellar gas and the interstellar neutrals are expected to behave differently: according to models, the plasma is decelerated, heated, and deviated, and flows around the heliopause, the contact discontinuity between the confined solar wind and the interstellar plasma. At variance with charged species that behave collectively, neutrals, to a first approximation, flow freely across the interface between the two plasmas. This behavior is due to their large mean free path with respect to neutral–ion (and –electron) interactions (of the order of tens of AU) and neutral–neutral collisions (hundreds of AU). Neutral atoms having successfully entered the heliosphere approach the Sun, and are increasingly affected by the solar wind and the solar radiation fluxes.

Around the Sun the so-called ionization cavity is formed, a volume devoid of neutrals, much smaller (about 10 AU for hydrogen and less than 1 AU for He) than the heliosphere. Nevertheless, a fraction of the atoms can approach close enough to scatter the solar UV lines by resonance, creating the so-called interstellar glow, which allows them to be detected from Earth orbit. The glow is like the patch of light surrounding a street lamp on a foggy night, except that it is not spherically symmetric, but has a conspicuous maximum on the so-called upwind side, where the interstellar wind originates, for the hydrogen Lyman-alpha glow (Figure 3), and on the contrary on the downwind side for the helium 58.4 nm glow (Figure 5). These differences arise because helium is ionized much closer to the Sun as compared to hydrogen, and thus can be focused by the Sun's gavitational field on the downwind side before becoming ionized. Actually, the helium density reaches its maximum in the focusing cone at about 1 AU. Historically, such a focusing of interstellar matter was first proposed by Lyttleton in 1950, not for atoms but for dust particles, to explain the formation of comets. Danby and Camm (1957) refuted this idea by showing that random thermal motion in the interstellar dust flow prevents the required "perfect" focusing. Contrary to the case of helium, the ionization time for hydrogen atoms as a function of the distance r to the Sun is such that hydrogen atoms traveling at about 20 $km\,s^{-1}$ have a 50% chance of surviving ionization at about 5 to 10 AU. Thus there is no focusing cone, and instead a cavity of this order of magnitude is formed. The maximum emissivity at Ly-alpha, the resonance wavelength for hydrogen, which varies roughly as the product of the density and r^{-2}, is upwind at 1.5–2.0 AU, that is, very close to the orbit of the Earth (Figure 3). After ionization, the newly formed particles (the so-called pickup ions) are convected outward by the solar wind electromagnetic field.

While interstellar helium is mainly photoionized, hydrogen photoionization is a minor effect and hydrogen atoms are preferentially ionized through charge exchange with the solar wind protons. It is important to note that charge exchange between a hydrogen atom and a proton (H^+) produces a "new" proton and a "new" atom, each one having essentially the same momentum as the former particle. In other words, it is almost exactly as if the electron had just left the hydrogen atom to become attached to the proton, while the two nuclei continue on their way. This means that all newly formed neutral atoms flow radially at the solar wind speed (300–800 $km\,s^{-1}$) and leave the inner Solar System very rapidly. This is why charge exchange is a loss process for the neutral flow. Moreover, these fast atoms no longer scatter the solar radiation, since, due to their high velocity, the resonance wavelength in their rest frame is shifted out from the solar Ly-alpha line which has a finite width of about 1 Å. They become invisible. From the dynamical point of view (trajectories of neutrals having escaped ionization), we know now that even in the absence of a strong ionization, hydrogen atoms would not be gravitationally focused, since radiation pressure due to the solar resonant photons balances the gravitation. All these effects

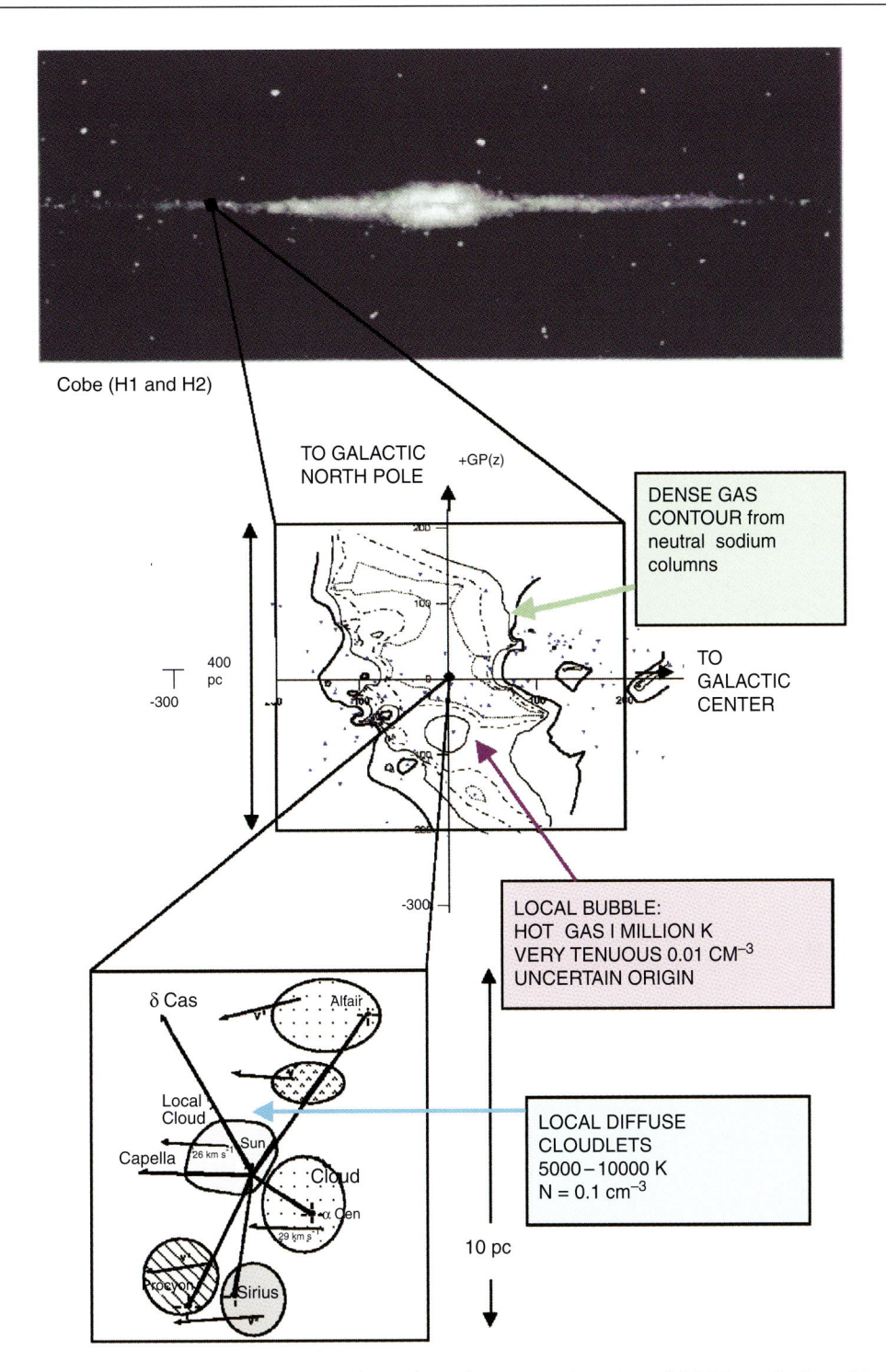

Figure 1 The solar environment at different scales: the Galaxy (here a combination of COBE results from Hauser *et al.* 1995), the local arm (local bubble contours from interstellar neutral sodium absorption, results of Sfeir *et al.* 1999), and the local cloudlets (derived from absorption lines towards very nearby stars (Lallement *et al.* 1995)).

were modeled for the first time in a realistic way by T.E. Holzer (1977). As we will see, charge exchange plays a fundamental role in the structure of the heliosphere, not only close to the Sun, but also at the heliospheric interface itself.

1.3 Four coincidences concerning the Sun's motion through the interstellar medium

Despite our ordinary Sun and ordinary Local Cloud, nothing is actually simple in our galactic environment and there

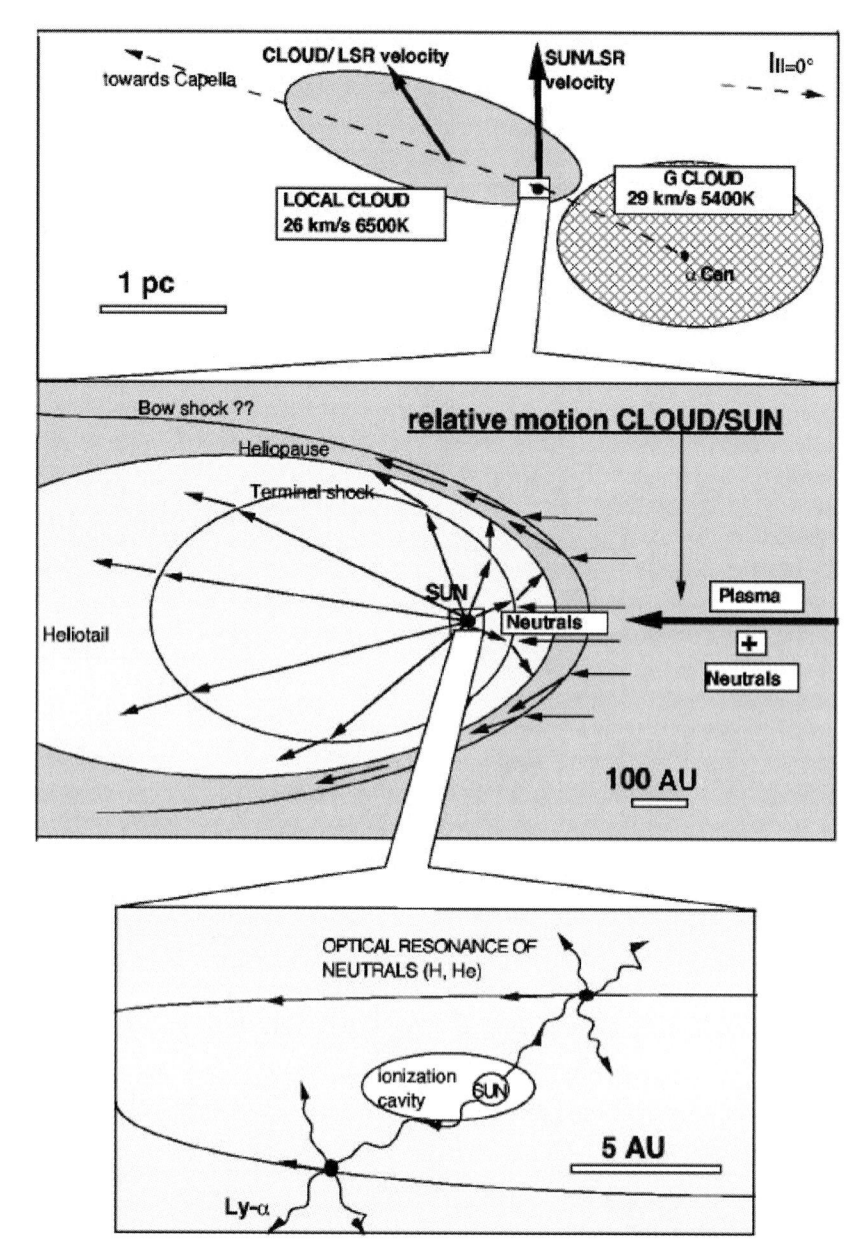

Figure 2 The solar environment at different scales: schematics of the two local clouds, the heliosphere, the interstellar hydrogen flow around the Sun and the "H glow." (From Lallement 1998.)

are (at least) four peculiarities about the Sun's motion in the Local Cloud. The first three have caused problems and have somewhat delayed the progress in the field during the last 30 years.

(1) In the rest frame of the Sun, the interstellar gas of the Local Cloud flows from a direction which points at 15° from the galactic center. In other words, for an observer at the Sun, the interstellar wind seems to flow from the central part of the Galaxy (the Sagittarius area). This is purely coincidental. The Sun is too old to be still embedded in its parent gas (the primordial cloud), it left it billions of years ago, and has already circled a few times around the galactic center. For the Sun, the Local Cloud is nothing more than one of the numerous interstellar galactic clouds it has already traveled through. The motion of the Local Cloud and the motion of the Sun are thus totally unrelated. It just happens that, by chance, the difference between the two velocity vectors points close to the galactic center (Figure 2). But this also implies that the interstellar gas emission glow has a maximum in the direction of the galactic center. As we will see, this has had (and still has) some consequences.

The emission is a maximum on the so-called upwind side, simply because it corresponds to the location of closest approach to the Sun of the inflowing interstellar atoms,

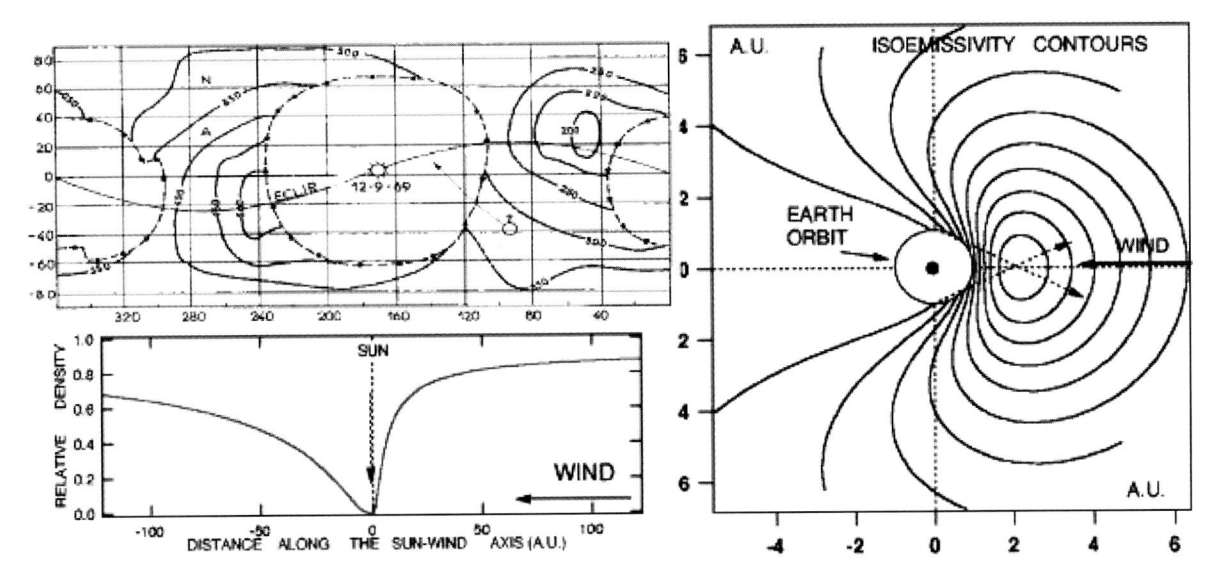

Figure 3 Top left: an example of one of the first maps of the interstellar hydrogen glow recorded on board OGO 5 in 1969. Isophotes of the Lyman-alpha emission are shown in equatorial coordinates. There is no enhancement in the direction of the solar Apex (direction A on the map). On the contrary there are a maximum and a minimum of intensity in two oppposite directions located close to the ecliptic plane. Bottom left: the modeled neutral hydrogen density along a Sun–wind axis in units of the density "at infinity." The ionization cavity created by solar EUV and solar wind is elongated on the downwind side. Right: the Lyman-alpha emissivity is a maximum on the upwind side at about 2 AU from the Sun. This explains why the direction of the emission maximum depends on the Earth location: the "parallax effect" illustrated in the figure and observed with OGO 5. (From Thomas 1972.)

before they are ionized. In terms of ecliptic configuration, the upwind direction is only very slightly inclined (by about 7°) on the ecliptic plane, and the ecliptic longitude of the incoming flow is about 254° (the Earth longitude at the beginning of June). As a consequence of the small inclination, the relative motion of the Earth and the interstellar flow varies strongly with time. In March the Earth is moving against the interstellar flow, and the relative motion is a maximum (about 50 km s^{-1}), while in September the Earth and the interstellar flow are in the same direction and their relative motion is only about 5 km s^{-1}.

(2) The Sun is by chance located at the very edge of the Local Cloud, which means that, using stellar spectroscopy as a tool for studying its properties, it can only be detected in one half of the celestial sphere, and not in all directions. In the opposite direction, another mass of interstellar gas with different properties (what we call another cloud) is detected. The exact geometry and the distance between the boundaries of the two clouds are still unclear. It has been suggested that the two clouds are a unique volume of gas, divided into two parts by a traveling shock. Except for the relative abundances, the properties of the two clouds point to such a situation (Grzedzielski and Lallement, 1996). This has delayed the determination of the physical properties of the medium surrounding the Sun.

(3) The modulus of the relative velocity of the Sun and the Local Cloud (25 km s^{-1}) is about twice the thermal velocity in the Local Cloud (about 10 km s^{-1} for a temperature of 7000 K). This corresponds, unfortunately, to the most ambiguous situation for the solar wind/interstellar plasma interface. Assuming neutrals and charged particles are decoupled on scales of the order of the heliospheric boundary size, which is reasonable, the speed of sound in the interstellar plasma is of the order of 14 km s^{-1} and, in the absence of any interstellar magnetic field, we could conclude that the flow is definitely supersonic, that is the interstellar plasma can not "feel" the solar atmosphere before being abruptly decelerated by a bow shock. If, as is likely, the local interstellar magnetic field (unmeasured yet) is within the range of observed values in the general interstellar medium, the fastest plasma waves (the fast magnetosonic mode) are traveling at a velocity which exceeds the speed of sound by a non-negligible amount. And here is the difficulty. The expected interval for the amplitude of the interstellar magnetic field (between 1 and 6 μG) is such that in the minimum field case, the relative velocity of the Sun and the ambient interstellar gas is still higher than the maximum velocity of traveling perturbations, and there must be a shock to decelerate the interstellar gas before it comes into equilibrium with the solar wind boundary. However, in the maximum field case, the flow is slower than the fast wave speed and there should be no shock transition and a very different structure for the interface. This remaining ambiguity has led to the existence of two competing types of models

and two groups of research teams. With the specificity that the Mach number for the local interstellar gas is higher by eastern researchers..., because the "Russian group" have developed the "supersonic flow" model, while theoreticians from Europe and the United States have essentially considered the Mach < 1 solution

(4) The fourth coincidence is more fortunate! By pure chance, the two Voyager probes, whose trajectories have been determined from planet positions of their consecutive visits, are now heading in the general direction of the incoming interstellar wind, that is, where the dynamic pressure of the interstellar medium adds to the other factors and the heliosphere is the most compressed. The distance the spacecraft have to cover before reaching the interface is thus the smallest in this direction. According to the results detailed below, there is a fair chance that one of the Voyagers will still be able to transmit data when it crosses the inner transition region, the solar wind termination shock.

2 THE CONCEPT OF THE HELIOSPHERE: WHERE AND HOW DOES THE SOLAR WIND STOP ITS EXPANSION?

There was certainly much progress in our understanding of the heliosphere in the twentieth century, simply because the concept of a heliosphere was born only 50 years ago! From then, one can distinguish two phases in the evolution of the field. The first phase was a series of major qualitative steps: pioneering theoretical works and observational discoveries that built the general picture that prevails today. As for quantitative aspects and detections of new species, they have progressed at a considerably increased rate during the last two decades, which constitutes the second phase. The way advances have been accomplished has been sometimes tortuous, being under the influence of very different types of diagnostics (but this is characteristic of the subject, as has already been noted), from energetic particles to stellar spectroscopy or radio waves.

2.1 The cosmic ray phase

The concept of the heliosphere was originally linked to the understanding of cosmic rays and solar energetic particle propagation. After the discovery of the solar wind (that is, the description by Biermann (1951) of the solar corpuscular radiation responsible for comet tails), Davis (1955, 1962) was the first to discuss the nature of the heliosphere as a volume filled by the solar wind and bounded by an external galactic medium, at this time thought to be a galactic magnetic field. The argument was that there is a need for a volume from which the galactic field is excluded, to account for the free propagation of solar particles between the Sun and the Earth. Otherwise they would be channeled along the field. Davis also suggested that galactic energetic particles (cosmic rays) could be trapped in this solar cavity and that changes in the size of the cavity could account for the modulation effect, that is, the observed temporal variation of the cosmic ray flux. Meyer et al. (1962) extended this work and were the first to consider the nature of the transition between the cavity blown by the solar wind and the ambient galactic magnetic field. At this period of time the estimated size of the heliosphere was highly variable (from five to hundreds of AU).

2.2 First UV glow measurements

At about the same time the first rocket measurements in the ultraviolet (UV) were obtained. In an attempt to detect the UV radiation from hot stars, Kupperian et al. (1959) detected, instead of point sources, a strong diffuse radiation in the spectral interval 1050–2025 Å. This signal was recorded from altitudes between 120 and 150 km, with almost no changes in intensity with rocket altitude. For this reason it was erroneously interpreted as an external emission (not linked to the Earth), arising from neutral gas filling the interplanetary space (essentially hydrogen emitting at the reso-nant transition, that is, 121.6 nm Lyman-alpha). We know now that this strong emission is due to the geocorona, the extended envelope of atomic gas around the earth, which resonantly scatters the solar UV radiation, and that only a negligible fraction of the recorded signal was of interplanetary origin. But this was the first time mention was made of the possible existence of an interplanetary gas emission. The first doubts about the actual origin of this diffuse UV emission were expressed by Schklovsky (1959) who put forward the idea that there could be a gaseous envelope around the Earth extending up to tens of thousand kilometers – the geocorona.

While the existence and the characteristics of the geocorona were progressively understood (thanks to the work of Chamberlain in particular), it became clear that spectroscopic measurements were needed to better understand the actual origin of the diffuse emission and to determine if all of it was due to the geocorona. Jacques Blamont suggested to American teams to apply to atomic hydrogen the technique he had used for atomic sodium, in the group of A. Kastler in France; that is, the use of an absorption cell. This idea was taken up by Morton and Purcell (1963) who found a way to produce atomic hydrogen in a cell using the dissociation of molecular hydrogen by contact with a heated tungsten filament (a method invented by Langmuir). Such a cell produces a deep and narrow absorption line and thus eliminates the signal within a wavelength range of the order of 0.01–0.1 Å depending on the cell regime, acting as a negative spectrometer. We will come back to the results of Morton and Purcell later.

2.3 New actors on stage: the solar wind termination shock and charge-exchange reactions with neutral interstellar gas

In the meantime theoretical work on the solar wind expansion was progressing well: Clauser (1960) and Weynman (1960) pointed out that the supersonic solar wind (flowing at velocities of a few hundred kilometers per second) must undergo a shock transition (we now call it the termination shock) to become a subsonic flow and be able to "feel" the pressure exerted by the external galactic medium. Also, at this time, in his description of the ultimate fate of the solar wind, Parker (1963) developed models of the heliosphere including such a solar wind termination shock, and distinguished different situations according to the different types of circumsolar interstellar media: (i) a pure galactic magnetic field, (ii) a fully ionized nonmagnetized galactic plasma, at rest with respect to the Sun, and (iii) a magnetized, ionized medium moving with respect to the Sun. In these cases he could establish analytical solutions.

For the first time the interaction between the solar wind and interstellar neutral atoms, that is, the expansion in a partially neutral galactic medium, was considered. Axford *et al.* (1963) estimated the consequences of charge-exchange reactions between neutral galactic hydrogen and solar wind protons beyond the termination shock front, where the solar wind becomes subsonic, with still high velocities but randomized directions. Recall here that the charge exchange reaction can be essentially represented as the transfer of an electron from a hydrogen atom to a proton, which means that the newly created proton has the motion of the former atom, and vice versa. Axford and co-workers inferred that these charge-exchange processes should lead to the entry of fast hydrogen atoms into the inner heliosphere (these atoms being formed after electron capture by hot solar wind protons in the subsonic region), and that most of the solar wind energy is transferred to the external medium through this charge-exchange mechanism. This is in fact now well established, as we will see below. Still, the distance to the termination shock was thought at this time to be essentially governed by the intensity of the galactic magnetic field. Axford and co-workers estimated that the shock, where the solar wind momentum flux would equal the galactic pressure, could be at a distance of 10–50 AU from the Sun.

2.4 What is the nature of the extra-geocoronal Lyman-alpha glow?

One consequence of solar wind charge exchange with galactic hydrogen is that some of the neutralized solar wind protons (that is, the newly formed hydrogen atoms first invoked by Axford and co-workers) can come back in the inner Solar System with a high velocity, be illuminated by solar radiation at Lyman-alpha, and can resonantly scatter this UV radiation. This mechanism was proposed by Patterson *et al.* (1963), and they modeled it to explain the Lyman-alpha rocket observations of Morton and Purcell (1963) made on 17 April 1961. Launched during the night, the experiment was a far-UV photometer placed behind a hydrogen absorption cell (the first one in space). When looking upward, the cell absorbed 85% of the signal, that is those 85% come from the geocoronal night glow which has a narrow enough emission line; the remaining 15% was some Lyman-alpha emission, but outside the 0.08 Å absorption linewidth. This fraction of the signal was therefore strongly Doppler shifted or characterized by a spectral linewidth broader than 0.08 Å, i.e. originating in a gas at high temperature, such as the population of fast neutrals created from neutralized solar wind protons, suggested by Axford *et al.* (1963).

We now know that this interpretation of the residual Lyman-alpha emission was (again!) not correct, since the emission is due to hydrogen atoms coming directly from interstellar space and approaching the Sun, and not to neutralized solar wind protons. The authors were mislead by the fact that the residual 15% seemed to be constant over the whole sky, while in the case of a directed flow, the hydrogen cell effect would have to depend on the angle with the Earth's orbital motion, modulating the Doppler shift and the absorption. In fact, this is due to the date of the rocket launch: it happened during the time when the Earth is moving upwind, adding to the interstellar wind velocity. In such a case, the region where there is some absorption of the interplanetary emission is concentrated in a very narrow band (a few degrees) along a great circle of zero Doppler shift, and if by chance the field of view (FOV) crossed this band, it affected only a few data points, going unnoticed by Morton and Purcell. If the launch had been repeated at another date, they would have noticed the variation with direction of the residual absorption, yielding other theories, in particular a preferential direction of the flow of emitting atoms.

While the interpretation of Patterson *et al.* (1963) was incorrect, it should be recognized that their paper was the first to mention a possible observation of the interaction of the solar wind with the surrounding interstellar medium, through Lyman-alpha measurements.

Later, several space missions again identified the presence of an extra-geocoronal UV radiation. However, the previous misinterpretation (fast hydrogen atoms from neutralized solar wind) was almost immediately replaced by a new one: the UV signal was assigned to a remote galactic source, rather than scattering by interstellar hydrogen atoms flowing in the inner Solar System as we know it is now. The reason was that it was thought at this time that the solar EUV radiation ($\lambda < 91.2$ nm) was ionizing totally hydrogen atoms in the interplanetary/interstellar medium, up to a distance of about 1500 AU: this is the calculated radius of the Strömgren sphere around the Sun, the distance at which the lifetime of one hydrogen atom v. ionization is equal to the probability of recombination (for a density of 0.1 atom/cm^3). While such Strömgren spheres can be huge (up to tens of parsecs) around hot early-type stars, creating the so-called hydrogen II regions filled by fully ionized interstellar gas and well observed through the H-alpha recombination line, a cool star like our Sun has, in principle, its own sphere, but smaller.

The "galactic" interpretation was strongly sustained by the fact that apparently the maximum of broad-band emission was reached close to the galactic plane and in particular in the galactic center region, according to the photometric results of Kurt (1965, 1967) from Zond 1 and later Venera 2, 3, and 4, and of Barth (1970) from Mariner 5 and 6. But this is simply the coincidental identity between the cloud/sun relative motion direction and the galactic center direction we have mentioned! Actually the galactic center has no role here (however the galactic story does not end quite yet and we will come back to the hypothesis of a galactic contribution to the Ly-alpha background emission in Section 4.3).

Blum and Fahr (1970) investigated how the Strömgren sphere of a cool star is modified when there is a relative motion between the star and the surrounding interstellar medium (or ISM), and understood that the concept of the ionization sphere completely collapses, even if this motion is as slow as a few km/s (random relative motions between stars are of the order of 20 km/s): as a matter of fact, the lifetime against ionization must be compared to the travel time, and if the ionization is weak, an atom can reach a much smaller distance than the Strömgren radius without being ionized. In such conditions, Blum and Fahr predicted that the neutral interstellar gas could easily produce a glow around the Sun. Indeed, the solar EUV output and the relative velocity are such that one interstellar hydrogen atom can penetrate down to 2–4 AU from the Sun before being ionized. The Blum and Fahr prediction was then extraordinarily rapidly followed by its observational confirmation, as we will see in the next section.

2.5 The parallax effect: the emission comes from the inner Solar System and not from the galactic disk

At the same time, the OGO 5 spacecraft was put into a spinning mode when being at its apogee (150,000 km), outside of the geocorona. Two Lyman-alpha photometers allowed mapping of the sky pattern of the extra-geocoronal emission for the first time (Figure 3): there was a broad maximum in the general direction of the galactic center (still the pure, but confusing, coincidence), and a minimum in the opposite direction. Fortunately the spin-up operation, though somewhat risky, was repeated three times. When compared, the maps recorded in September 1969 and April 1970, at two opposite positions of the Earth in its orbit, showed that the position of maximum was seen in a different direction, by about 30° (Figure 3, right). This parallax effect was the definite proof that the emission was coming from a distance of a few AU, and not from the Galaxy: the interstellar wind was discovered (Bertaux and Blamont, 1971; Thomas and Krassa, 1971), as a brilliant confirmation of the Blum and Fahr theoretical prediction.

In their modeling, Blum and Fahr predicted a maximum of Lyman-alpha emission in the direction of arrival of the flow of interstellar hydrogen atoms in the Solar System, because this is the place where hydrogen atoms can approach nearest to the Sun before being destroyed by ionization. They had predicted that this maximum would be in the direction of the Apex, or the direction of the motion of the Sun with respect to the surrounding stars forming the local standard of rest (LSR), at $\lambda = 271°$, $\beta = +53°$ in ecliptic coordinates. This was implicitly implying that the local ISM is at rest with respect to the LSR. But when measured, the direction of the maximum emission indicated a flow direction at 50° from the Apex direction ($\lambda = 252°$, $\beta = 7°$). This meant that the interstellar material in which the Sun is embedded was in reality moving through the local stars, a second reason to call this flow the interstellar wind.

While the description by Parker of the termination of the solar wind was a pure plasma description (solar wind and fully ionized interstellar plasma), the description of Blum and Fahr considered only the neutral part of the ISM. It had the feature of providing something to be observed from the inner Solar System, the Lyman-alpha emission. The direction of maximum emission indicated immediately the direction of maximum interaction between the solar wind and the ISM, where could be found the nearest locations of the solar wind termination shock and heliopause. As mentioned above, the two Voyager probes are heading in this general direction.

2.6 Neutrals or not neutrals: the dialog of the deaf

When it became clear that the ISM surrounding the Sun contained a substantial quantity of neutrals (or at least that a large quantity of neutrals was entering the heliosphere), it was a kind of bad news for plasma physicists, since hydrodynamic or magnetohydrodynamic (MHD) models of the heliosphere were no longer totally realistic. It was, however, good news for the observers, since neutrals can enter the heliosphere, scatter solar photons, and then become ionized, and thus there are glows to image and particles to measure. This is why, during the following years, a kind of dialog of the deaf was established. On one hand, theoreticians were building more and more sophisticated models of the heliospheric interface, still with no neutrals in the surrounding ISM (e.g. see Baranov et al. (1970) for the supersonic model,

Axford (1972) for the subsonic configuration). On the other hand, the new glow data on interstellar hydrogen and later on interstellar helium were analyzed with the help of much simpler models based on the assumption that the circumsolar ISM is totally neutral and that the neutral gas reaches the inner heliosphere and enters the supersonic solar wind with exactly the same characteristics as in the ISM. Based on this assumption, first estimates of the velocity V and the temperature T of the neutral hydrogen were obtained from the location of the maximum emission region (MER) on the upwind side through the parallax effect (Thomas 1972, Bertaux et al. 1972), and from glow and hydrogen cell data (Meier 1977, Bertaux et al. 1977, Wu and Judge 1979, Holzer 1977). As a matter of fact the distance between the Sun and the MER depends mainly on the velocity of the flow and on the total ionization rate: the faster the atom, the closer it can approach before being ionized by charge exchange with solar protons (more than 80% of the total ionization) and the solar EUV radiation. The linewidth of the glow emission line reflects the velocity dispersion of the atoms, and thus at first order the kinetic temperature. The favored values ($T = 7000$ K and $V = 20$ km s^{-1}) were in rough agreement with the first spectrum of the background emission directly recorded from space with a spectrograph (and not with a simple absorption cell), obtained with the Copernicus satellite (Adams and Frisch 1977).

As we said, the entry of neutrals deep inside the heliosphere is due to the relatively fast motion of the Sun in the Local Cloud. But the resulting inability of the Sun to ionize at distance the ISM through which it flows has another important consequence: it means that all the information on the interstellar matter which surrounds the heliosphere, based on observations made "inside," is applicable as such to the ISM, and does not need to be corrected for solar influence. This applies to ISM physical properties, isotopic and elemental abundances, and ionization state. It would not be true if our Sun were a hotter, UV-emitting star.

2.7 The discovery of anomalous cosmic rays (ACRs): how energy comes to neutrals

The neutral gas received some attention when the IMP 5 and Pioneer 10 observations of energetic particles (Garcia-Munoz et al. 1973, McDonald et al. 1974) led to the discovery of a new, low-energy component of cosmic rays (CRs), the so-called anomalous cosmic rays (ACRs). The energy of ACRs is too low to allow their entrance into the heliosphere from galactic space (they would be stopped by the interplanetary magnetic field), and thus they must originate from inside the heliosphere. Soon after their discovery, Fisk et al. (1974) searched for an explanation of their nonsolar relative abundances and rapidly identified them as the final stage of a fraction of the interstellar neutral atoms, after they have been ionized in the vicinity of the Sun, become the "pickup" ions, and then been convected in the solar wind and preferentially accelerated in the interplanetary medium. Although this "social ascension" from an ordinary neutral to a VIP (very important particle) is restricted to a small number of the atoms, while most atoms having entered the heliosphere leave it unchanged, interstellar neutrals are now definitely considered to be the seed population for ACRs. Since different types of ACRs have been recorded routinely by the Voyager probes, as He, C, N, O, Ne, Ar, and now Mg and Ca, ACRs have been providing a new probe of the local interstellar matter.

2.8 How strongly ionized is the surrounding interstellar medium? What confines the heliosphere?

Between one extreme aspect (a fully ionized ambient ISM) and the other extreme (a neutral surrounding ISM, with the corresponding neutral gas flow sweeping around the Sun, as revealed by the hydrogen glow and the ACRs), there is room for an intermediate situation. As a matter of fact, if the Sun produces EUV radiation too weak to create a Strömgren sphere and ionize the ambient gas, then it is not the only source of "hard" photons. Indeed, it was soon noticed that the diffuse EUV flux, essentially due to the hot stars in the vicinity of the Sun, is strong enough to ionize partially the local ISM (e.g. Cheng and Bruhweiler, 1990). Thus there must be an ionized component in the surrounding gas. But the problem is, this ionized part, as also the interstellar galactic field and the low-energy cosmic rays, are all excluded from the heliosphere. As a consequence, while producing pressure and being the confining agents, by definition they cannot be observed from the inside. So the questions appeared to be: what are the respective roles of plasma, magnetic field, and low-energy cosmic rays in the confinement of the solar wind? And how to measure them from inside the heliosphere?

The first theoretical description of neutral/plasma at the heliospheric interface in the context of a partially ionized surrounding ISM is due to Wallis (1975), who predicted that the neutral hydrogen of interstellar origin, when entering the heliosphere, charge exchanges with the interstellar ions, and, because those ions are slowed down and deviated due to their encounter with the solar wind (while neutrals are not), the net result of these charge-exchange reactions is not zero. On the contrary, the coupling between the neutral atoms and the plasma, due to these charge-exchange processes, produces perturbations of both the neutrals and the plasma flow (Figure 6).

The "eastern" Russian group then began a sophisticated modeling of the neutral–plasma interaction. Baranov and colleagues from Moscow (Baranov et al. 1979), calculated the feedback reaction, that is, the effect of the neutrals on the structure of the "supersonic" interface due to charge-exchange with the plasma. Following this work, Baranov *et al.* (1981) produced the first description of the heliospheric interface taking into account the mutual influence of plasma and neutrals, and indeed models of the neutral flow through the boundary region in the cases of the one-shock and two-shocks heliosphere confirmed that substantial modifications of the neutral hydrogen flow do occur.

At this time, again two diverging "schools" began to develop in parallel. The Russian group, arguing that the mean free path of the neutrals with respect to interactions with the charged particles (charge exchange with the protons and electron impact collisions) is of the order of the estimated size of the heliosphere (of the order of 100 AU), concluded that a fluid description of their behavior is precluded and that kinetic models are required. That is, the interface description requires a combination of fluid and kinetic codes. However, the computational time needed for a combined hydrodynamic (plasma)–kinetic (neutrals) model was at that time exceedingly large and the first descriptions produced were approximate not self-consistent models. The first real self-consistent solution was published much later (Section 4).

In the "West", fluid codes for both neutrals and plasma remained the new theoretical works, which in post-talk discussions during meetings had the effect to stimulate quick tempered reactions from Vladimir Baranov, for the pleasure of those who knew him (and the surprise of the others).

2.9 The early 1980s and the first evidence from ACRs for the presence of a solar wind termination shock around the heliosphere

The first observational evidence for the presence of a solar wind shock came from the analysis of the behavior of ACRs. Pesses *et al.* (1981) demonstrated that most likely these particles were accelerated at a relatively strong shock in the outer heliosphere, and that the solar wind termination shock could play this role. This was a fundamental finding, since for the first time there was evidence for what had been up to this point a theoretical construction only: the termination shock.

Of course this did not give information on the nature of the interface beyond the shock. As a matter of fact, the solar wind termination shock structure is not strongly dependent on the nature of the heliospheric interface on the interstellar side. It thus remained unclear at this time (and is still unclear) whether the interstellar flow is decelerated by a shock or not, and whether the interstellar magnetic field plays a dominant role or not. Nonetheless, from this date sophisticated models of the pickup ions' acceleration were built, and applied in particular to the Voyager ACR data, and ACR spectra and their gradients (Figure 4) began to give new clues as to the nature and location of the solar wind shock (e.g. Cummings *et al.* 1984, 1987 for recent data analysis and Chapter 4 for the full story).

3 THE 1980S AND THE FIRST CONTRADICTIONS BETWEEN NEUTRAL FLOW MODELS AND DATA; THE FIRST DETECTION OF PICKUP IONS

As long as interstellar hydrogen was the only observed neutral species, things were rather simple for data interpretation. The only slightly difficult point was the evaluation of the effect on the hydrogen atoms of the solar radiation pressure. This repulsive force is due to the cumulative effect of solar Ly-alpha photon absorptions by the atoms, which produces a radially oriented force, which is not compensated by the rediffusions of the photons because these rediffusions are isotropic and have a null resulting effect. It became likely that depending on solar activity (and thus solar UV output) radiation pressure varies and, depending on the activity, can overcome or not overcome the gravitational force which is in the opposite direction. In other words, hydrogen atoms are attracted to the Sun or repulsed according to the level of activity of the Sun. Indeed, spectroscopic observations have shown that, at solar minimum, atoms are slightly gravitationally focused, as derived by Bertaux *et al.* (1985). However, contrary to the case of helium, the strong ionization prevents the formation of a focusing cone.

Then, with the first measurements of the neutral helium glow, recorded by the US Army STP72 spacecraft (Weller and Meier 1981) and the Russian satellites Prognoz 5 and 6 (Dalaudier *et al.* 1984), the situation changed. Close to the Sun, neutral helium behaves very differently from hydrogen. Due to its longer lifetime before ionization (charge exchange is negligible, UV photons are the most significant source of ionization for helium atoms beyond 1 AU), and because radiation pressure is small and totally exceeded by gravitation, helium is concentrated on the downwind side of the Sun in the so-called "focusing cone" (Figure 5). The location of the narrow cone gives the direction of the interstellar flow (like a flag in the wind), while the shape of the cone is sensitive to the flow velocity and temperature. After analyses of the helium glow data, strong contradictions appeared. The helium velocity (27 km s^{-1}) and temperature (17,000 K) were found to be significantly above the values previously found for hydrogen, at this time determined from hydrogen cell data (Bertaux *et al.* 1985). That helium enters the heliosphere at a larger velocity than hydrogen was possibly consistent with some of the predictions of Wallis (1975) about the effect of the coupling with the plasma, leading to hydrogen deceleration, but in this case why was hydrogen much colder than helium, contrary to

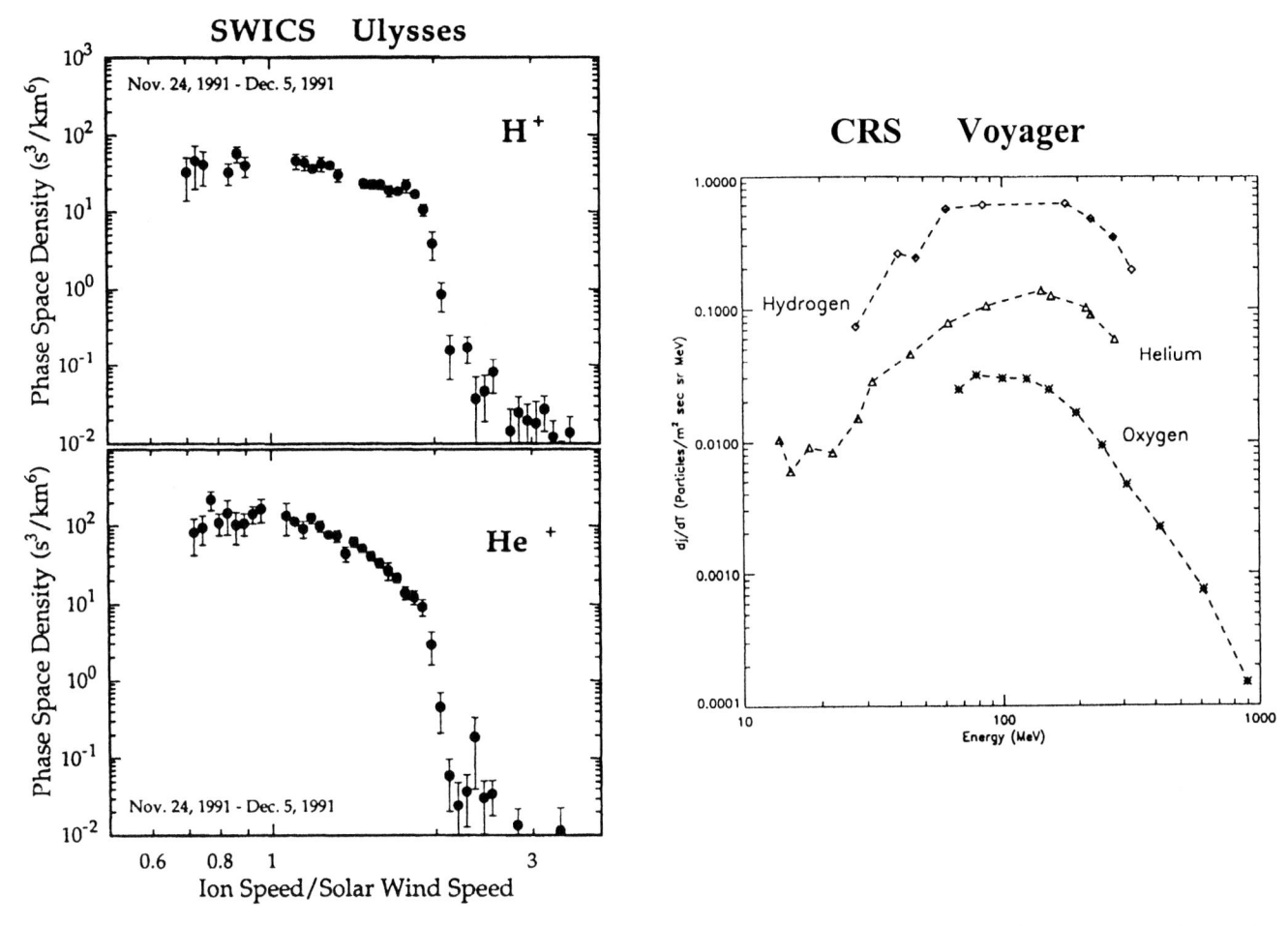

Figure 4 Interstellar neutral derivatives. Left: pickup ions are produced in the inner heliosphere after ionization of neutrals through solar EUV photoionization or charge exchange with solar wind ions. Their ring-shape velocity distribution and the sharp cutoff at twice the solar wind speed makes them distinguishable from ambient solar wind ions. They become a seed population for energetic particles accelerated in corotating shocks and at the termination shock, where they become ACRs. Right: ACR spectra detected by Voyager. Note that pickup ions have been detected much later than ACRs (Section 3). (From Gloeckler et al. 1993 and Cummings and Stone 1987, 1988.)

what was predicted? Explanations were actively searched for. The influence of the shape and the shift of the illuminating solar helium line at 58.4 nm, the role of elastic collisions with the protons, and so on were carefully considered (Chassefiere and Bertaux 1987, Chassefiere et al. 1988). Some effects were found, but nothing could satisfyingly reproduce temperature and velocity to be possibly consistent with hydrogen parameters, that is, simultaneously a larger velocity and a smaller temperature than for hydrogen. Who was right?

At the same time, the influence on the hydrogen flow of solar wind departures from sphericity and cyclic changes was investigated. Ly-alpha experiments, first on board Mariner 10 (Kumar and Broadfoot 1978, Ajello et al. 1979) and later on board Prognoz and Pioneer-Venus, had shown strong departures from axisymmetry of the hydrogen flow.

Anisotropies of the solar wind were shown to be responsible. That is, latitudinal variations of the solar wind, and thus of ionization of hydrogen by charge exchange, have an effect on the hydrogen distribution, an effect predicted by Joselyn and Holzer (1975). However, this had again been a surprise. While one was expecting more solar wind from the open field lines above the polar holes, and thus more ionization, the observed effect was exactly the opposite. A smaller solar wind flux was found at high solar latitude. It was indeed found to be smaller by as much as 40% (Lallement et al. 1985) above the solar poles. This has now been directly proved since Ulysses has flown over the solar poles, and the effect is clearly seen with the SOHO/SWAN Ly-alpha images. However, accounting for anisotropies, and also for the temporal evolution along the solar cycle, did not solve the temperature contradictions.

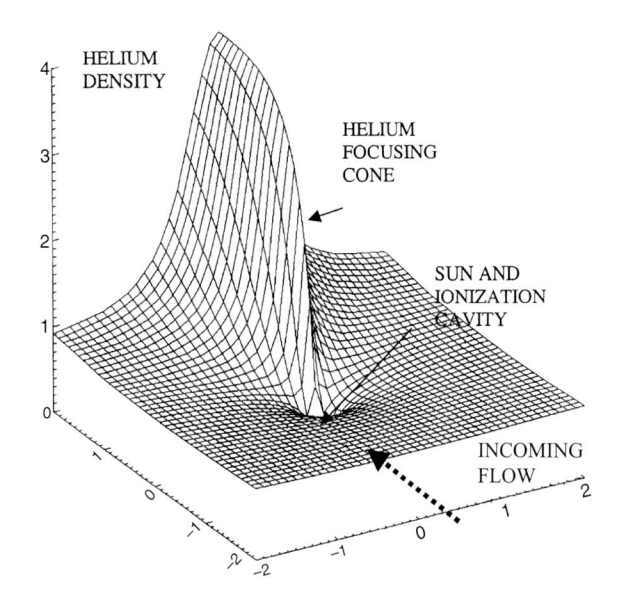

Figure 5 Density of interstellar neutral helium as predicted by the models. The density is in units of the interstellar density (scale along the Z axis). The X and Y axes are cartesian coordinates in AU in a plane containing the Sun and the interstellar wind vector. Helium is much less ionized than hydrogen and is concentrated in a prominent dense cone on the downwind side. The resulting glow of the helium cone at 58.4 nm gives very nicely the direction of the interstellar flow, like a flag in the wind.

Nor did the analysis of the first measurements of interstellar pickup ions, He^+ ions, by the ion spectrometer using on board the AMPTE (Mobius et al. 1985). These ions were detected mainly in the helium focusing cone on the downwind side, where helium is dense and more ions are created by photoionization. This first detection was fundamental in the sense that it definitely showed that pickup ions are not assimilated to the solar wind immediately after they have been picked up, but on the contrary they can be distinguished from the solar wind particles due to their ring-shaped velocity distributions (Isenberg 1986). New theoretical investigations followed, on the influence of the pickup ions on the solar wind by means of hydromagnetic wave excitation (Lee and Ip 1987), and the pickup acceleration to energetic particles and ACRs and the consequences at the termination shock (Lee and Axford 1988).

4 THE EARLY 1990S: A MULTI-SPACECRAFT STUDY OF THE HELIOSPHERE

The 1990s was a decade particularly rich in new and varying investigations. This was because of the number of sophisticated experiments simultaneously operating in space, the unprecedented conditions for the Voyager and the Pioneer probes now well beyond the planets, the first years of the very successful and rich Ulysses mission, the Hubble Space Telescope (HST) era, the Extreme Ultraviolet Explorer (EUVE), and the conjunction of efforts by a number of teams.

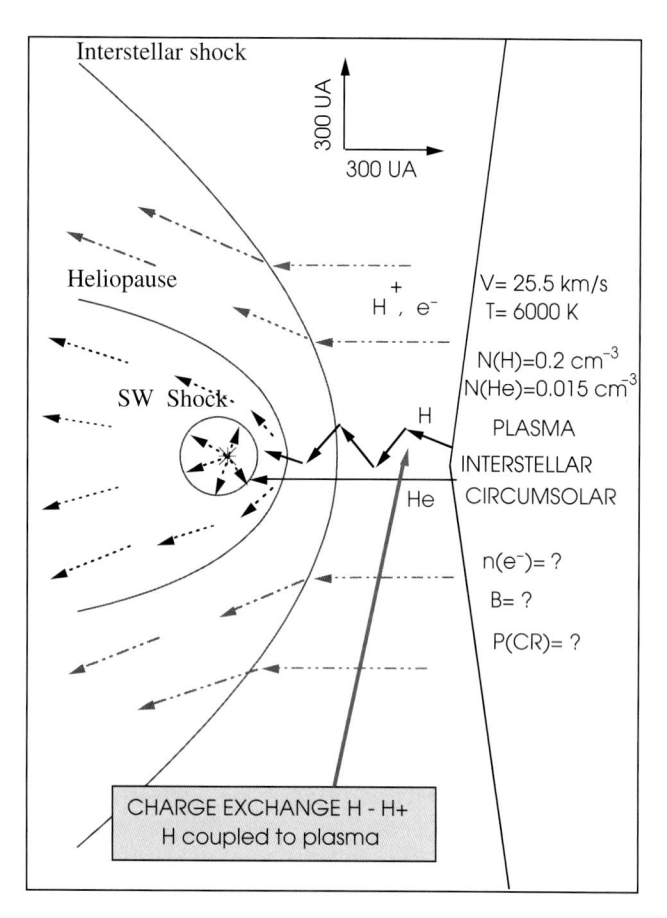

Figure 6 Theoretical discontinuity surfaces in the heliosphere, and hydrogen and helium differentiation. Contours are drawn according to the results of Baranov and Malama (1993) and Izmodenov et al. (1999a) in the two-shock case. The differentiation of interstellar helium and hydrogen atoms at the entry into the heliosphere is illustrated by the black arrows, and the "known" and unknown (as at 2000) parameters of the surrounding ISM are indicated on the right.

4.1 The interplanetary–interstellar connection

In order to define the boundary conditions for the heliosphere (Figure 6), one needs to know the physical characteristics and the abundances in the circumsolar ISM. A large part of ISM studies is based on stellar spectroscopy, that is, on the absorption lines of interstellar matter located along the line-of-sight between the Sun and the target star that are superposed on the star spectrum (Figure 7). Paradoxically, the local interstellar gas is not the easiest to observe, and is certainly not the easiest to interpret. This is due to the very

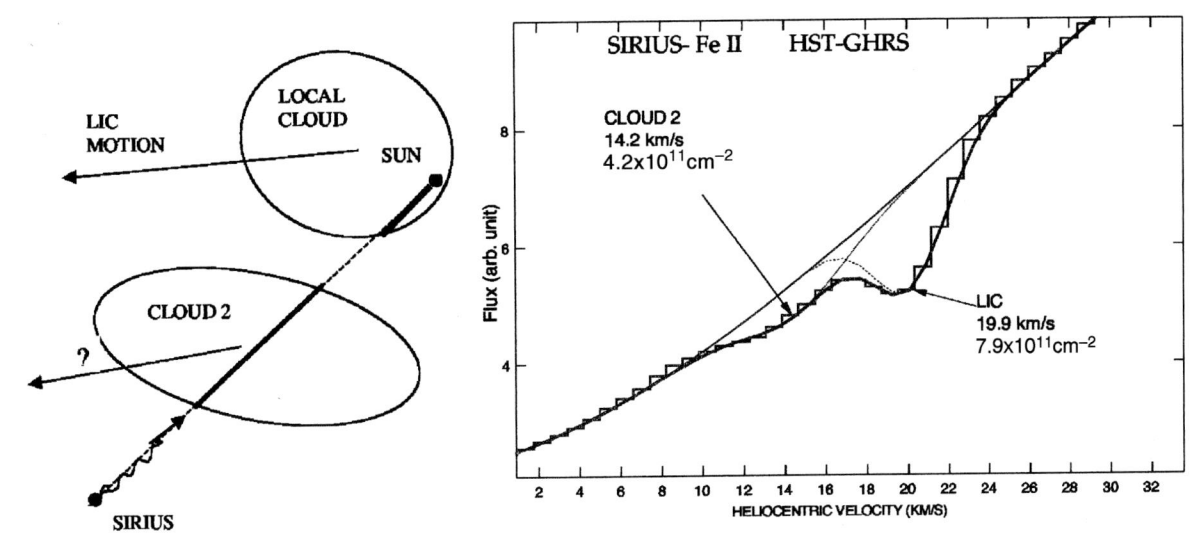

Figure 7 Spectroscopy of the nearby stars and Local Cloud absorption: the example of Sirius. There are two absorption lines detected towards Sirius (here ionized iron lines), one is due to the LIC material, the second to another mass of gas closer to Sirius. Doppler shifts and linewidths are used to derive gas motions and temperatures of the clouds. (From Lallement *et al.* 1994.)

low density of the surrounding medium, and thus the extreme weakness of the absorption, at least in the visible. Disentangling individual clouds along a line-of-sight can be achieved for short line-of-sights only; otherwise there are overlaps by the numerous absorbers.

The so-called "Doppler triangulation" method has been applied to the search for the local cloud properties. The principle of the method is extremely simple: if the mass of gas around us (the LIC) moves in first order as a solid body, the Doppler shifts of the absorbing lines produced by this cloud in the line-of-sight (LOS) towards different targets stars all around the Sun are simply the projections of the velocity vector onto the directions of the targets. One can then solve for such a velocity vector using a set of Doppler shifts for objects in different target directions (Figure 8) (Crutcher 1982). In reality, the problem is complicated by the fact that in general, more than one cloud is detected and one does not know which one is the local one (Figure 2, bottom). But using iterative methods and trying all combinations can solve this. Chronologically, the first determinations of very nearby gas motion were done from the ground, and very rapidly confirmed with HST. A set of absorption lines was used by Lallement and Bertin (1992) which revealed the presence of a mass of gas in the vicinity of the Sun moving at 25.7 km s^{-1} (with respect to the Sun) towards the direction (l,b) = (186.1°, −16.4°) in galactic coordinates, which corresponds to (λ, β)=)=(74.9°,−7.8°) in ecliptic coordinates. Unfortunately, the target stars producing this velocity vector were covering about half of the full celestial sphere sky only, and as a consequence it could not be deduced with certainty that the Sun is embedded in this mass of gas, and not simply

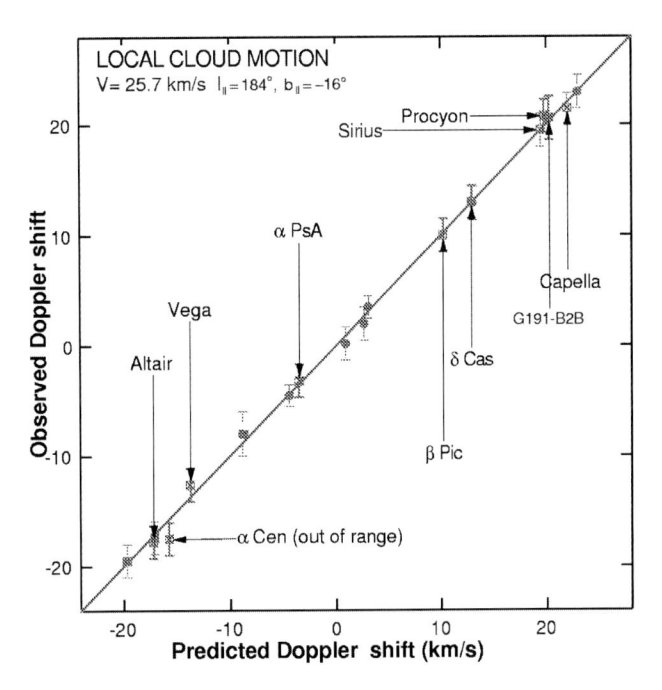

Figure 8 Local Cloud velocity vector determination: Doppler triangulation results. Agreement between predicted and measured Doppler shifts of the interstellar lines are found for the parameters shown. (From Lallement *et al.* 1995.)

very close to it. A few years before, another mass of gas had been found by Lallement *et al.* (1990), with a slightly different velocity vector (29.3 km s^{-1}, towards (l,b) = (184.5°, −20.5°) (or equivalently (λ, β)=)=(70.5°, −8.5°)) for target stars in the opposite hemisphere. Put together, these results

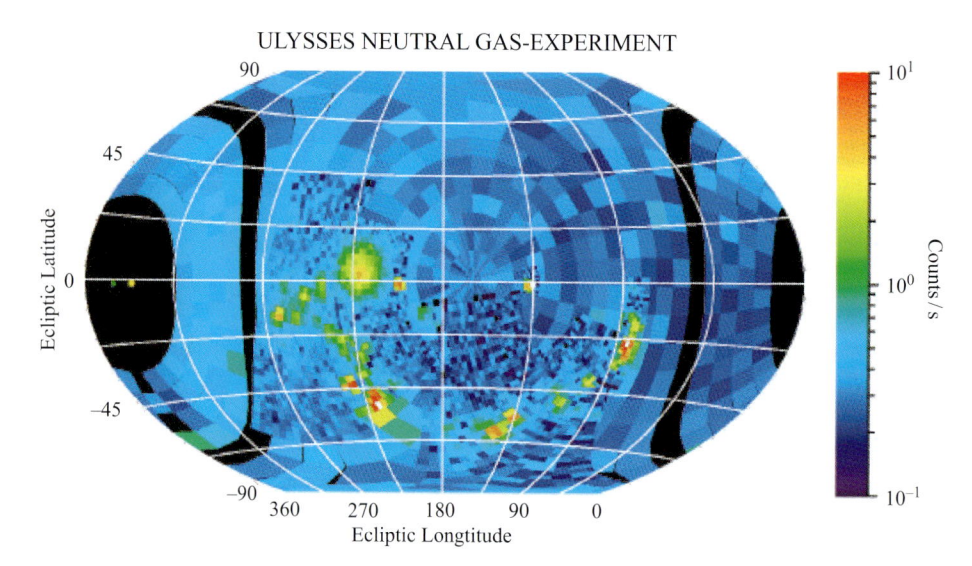

Figure 9 A typical map recorded by the GAS experiment on board Ulysses: helium Atoms from interstellar space correspond to the conspicuous bright spot at (220°, +3°). The bright stars of the Milky Way are seen as a half ring in the southern hemisphere, and neutrals around Jupiter produce a spot at (136°, −2°). (Courtesy M. Witte, Max-Planck, Lindau.)

suggested that either the Sun is by chance located between two slightly different masses of gas, called AG (for antigalactic) and G (for galactic), and nothing more precise could be said about which one is surrounding the Sun, or either something was wrong with the analysis. Thus, this was somewhat discouraging, and my colleagues from the Astrophysical Institute in Paris thought that we definitely had to wait for better data in the UV from the HST. However, convinced that the new (AG) vector, based on very nearby stars, and obtained after many hours spent recording, correcting, and adding spectra, had real significance, I presented a synthesis of the results for the two clouds at the Cospar meeting in 1992 in Washington. In the same session, and just before my talk was scheduled, was the presentation of the results of the GAS experiment on board the ESA/NASA Ulysses spacecraft, that is, the direct detection *in situ* of the interstellar helium atoms (Figure 9). The analysis was presented by Manfred Witte and resulted in the determination of the heliospheric helium flow velocity (both modulus and direction) and temperature (Witte *et al.* 1993). To my great pleasure, the numbers announced by my German colleague were similar to our results for the AG cloud, which I could in turn immediately comment about in my talk. Some of the participants to the session (as one of them confessed later) believed that we had prepared our presentations in conjunction, while already knowing the results of the other, but this was not true. While establishing for the first time a kind of bridge between the heliosphere and interstellar space, and showing that the Sun is embedded in the external part of the AG cloud (now the Local Cloud), these results also confirmed that, according to the theory, neutral helium is flowing freely through the heliospheric interface, and has the same physical state in our Solar System (before approaching too close to the Sun and suffering EUV ionization) as in interstellar space. This provided a kind of "zero perturbation" state for the interstellar gas and a new hope that, from this point, the rest of the puzzle would be solved gradually.

The results obtained with the HST, based on the strong UV lines of ionized iron and magnesium, essentially confirmed the two masses of gas and showed they have different temperatures and different relative abundances (Linsky *et al.* 1993, Lallement *et al.* 1995, Linsky and Wood 1996). It was very surprisingly found that the Sun must be extremely close to the boundary of the Local Cloud, that is, absorptions by the two clouds are never seen together.

The low temperature (7000 K) and high velocity (26 km s^{-1}) found for the ambient gas, from measurements both inside and outside the heliosphere, made sense, since this left room for a deceleration and even some small heating of hydrogen (Lallement *et al.* 1993), because the velocity and temperature inside the heliosphere had been found to be 20 km s^{-1} and about 8000 K (Bertaux *et al.* 1985). There was no longer a need for a cooling of hydrogen on entry to the heliosphere, which has no physical sense.

Still, there remained a series of unanswered questions. Why such a discrepancy between the high (17,000 K) "helium glow" temperature and the low (7000 K) helium particle kinetic temperature? This is still unclear, although it can be partially explained by a particular shape of the solar line, following the suggestion of Chassefiere *et al.* (1988). Also, why such a strong deceleration of hydrogen (from 26 to 20 km s^{-1}) and such a small associated heating (by only 1000 K)?

A third inconsistency also remained about the number densities of helium and hydrogen. The measurements were suggesting a H I/He I (that is, neutral hydrogen to neutral He) density ratio of about 10 in the inner heliosphere, that is, about the same value as the cosmological abundance ratio H/He in the ISM. At this time one was assuming that helium is fully neutral in the ISM, and hydrogen partly ionized, and thus H I/He I should be smaller than 10 outside the heliosphere. If a significant fraction of hydrogen atoms is prevented from entering the heliosphere due to the coupling with the excluded plasma, which models predict along with the deceleration, H I/He I should be significantly smaller than 10 inside the heliosphere. Thus from the measured numbers, hydrogen was apparently not (at least not significantly) filtered, which was inconsistent with the deceleration. Although we knew that (paradoxically) density measurements are the most difficult ones, since they are very sensitive to solar conditions (which vary) and the inaccuracy of UV instrument absolute calibrations (which do not vary), this was also a point of concern.

4.2 The echoes from the heliosphere boundary: Voyager radio data

The story of the Voyager radio emission, one of the most exciting stories in space science (and the suspense continues), began in 1983. Emissions in the range 2–3 kHz were detected at this time by the Plasma Wave Subsystem (PWS) of the two probes and analyzed by Kurth *et al.* (1984). A heliospheric origin was suspected; however, the role of Jupiter could not be discounted at this time. Then, after years of silence, new and stronger signals were detected again in 1992–93, shedding a new light on the phenomenon. As a matter of fact, from the comparison of the two events and the analysis of the solar wind conditions, Gurnett *et al.* (1993) suggested that the emissions are generated at the heliopause and could convincingly show that the two series of signals were triggered by two strong solar wind ejecta at the maxima of activity in 1981 and 1991. The idea of a triggering of the emission by coronal mass ejections (CMEs) had first been suggested by McNutt (1988). These new results definitely established that the 3 kHz radiation is generated at the boundary of the heliosphere, and that Voyager receives the echoes from the impact of strong solar wind on this boundary. The timed delay of a few hundred days between the solar ejection and the radio emission allowed Gurnett and co-workers to estimate the heliopause at 110–160 AU. The time interval length is linked to the uncertainty in the propagation speed of the corotating shock beyond the termination shock in the subsonic wind.

Figure 10 shows the spectrogram of the emission recorded by Voyager 1. Two different frequency bands characterize the emissions, the upper one between 2.5 and 3.5 kHz, and the lower one between 1.8 and 2.5 kHz.

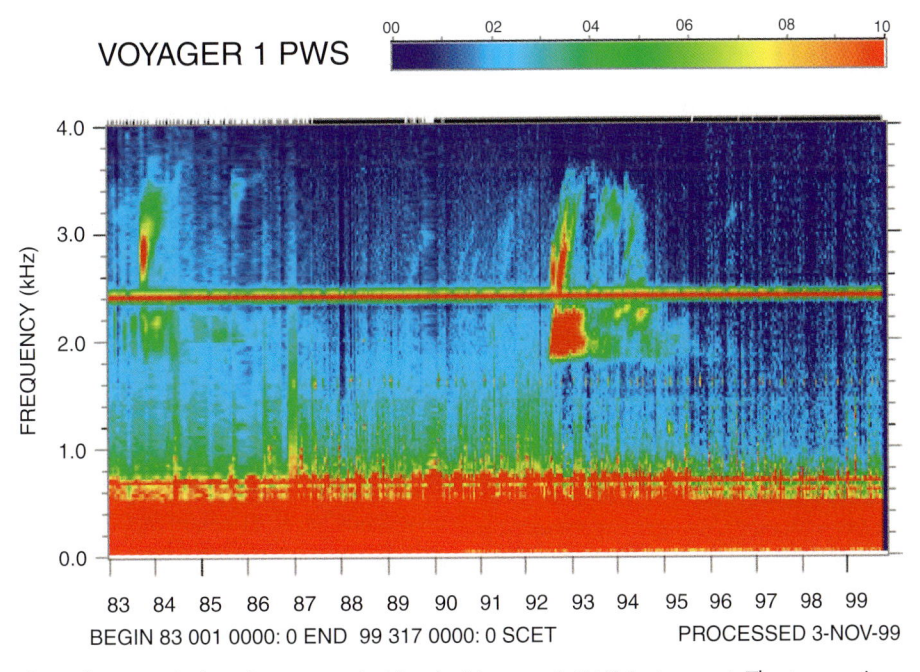

Figure 10 Echoes from the outer heliosphere recorded by the Voyager 1 PWS instrument. The two major events, in 1983 and 1993, correspond to the arrival times at about 150 AU of the strongest solar wind events which have marked the two consecutive solar activity maxima. Similar echoes were recorded by the PVS instrument on Voyager 2. (Note that the conspicuous sharp horizontal feature is of instrumental origin.)

Emissions appear almost simultaneously in the two bands, but there are some differences. The 2 kHz band seems to have a rather sharp and well-defined low cutoff at 1.8 kHz. Also, there is a striking difference in that, at variance with the upper band which originates from a preferential direction, that is, the upwind direction, the lower band seems to arise from all directions (Gurnett *et al.* 1998).

Sophisticated mechanisms for these two emissions have been proposed by different groups, and involve the interaction of shocks and density enhancements with the interface discontinuities (e.g. Zank *et al.* 1994, Czechowski and Grzedzielski 1994, Whang and Burlaga 1994). The periodic disappearance of the interstellar shock in response to global solar wind cyclic variations has also been considered (Richardson 1997). Gurnett and Kurth (1996) have suggested that the cutoff frequency could be the plasma frequency in the interstellar medium beyond the heliopause. The existence of a cutoff linked to the interstellar medium beyond the heliopause had been proposed by Axford (1985). However, there is no consensus yet on the origin of the 2 kHz emissions and of the cutoff.

4.3 The outer heliosphere Lyman-alpha emission: the Galaxy strikes back

Voyager UVS maps of the Lyman-alpha diffuse emission were analyzed by different groups, including our group in France, led by Jean-Loup Bertaux, who was a co-investigator on the UVS. We were doing comparisons with classical models for the hydrogen flow. Having noticed that the signal was slightly higher on the upwind side than it should have been according to our models, we were wondering whether we were beginning to see the emission from the region of high density associated with the plasma pile-up at the front of the heliopause, the "H wall," predicted to be formed by coupling with the plasma (Baranov and Malama 1993; Figure 11). We asked NASA to authorize and organize special maneuvers of the platform of the UVS, in order to make very long exposures towards a series of directions around the upwind region.

Thanks to NASA, maneuvers have been done regularly since then, on the basis of about two per year. Although the data gathered are not yet fully explained, as we will see below, they contain precious information, which we hope will be used in future (as with other Voyager experiments, it will be many years before new probes reach these distances and information of this type can be recorded again). The reasons for the delay in the analysis are mainly due to a new surprise, namely the detection of an additional Lyman-alpha emission, one which does not resemble what one would expect from the outer heliosphere, but has the characteristics of galactic radiation. Figure 12 (Quémerais *et al.* 1995) shows the recorded intensity at Lyman-alpha by Voyager 2 during one of the maneuvers, together with the expected emission from a simple model without heliopsheric interface perturbations of the hydrogen flow. Clearly the signal is higher than expected on the upwind side, revealing a higher density, but there is an additional feature in the form of a "bump," which can hardly be due to a local source. As a matter of fact, radiative transfer effects, at these distances

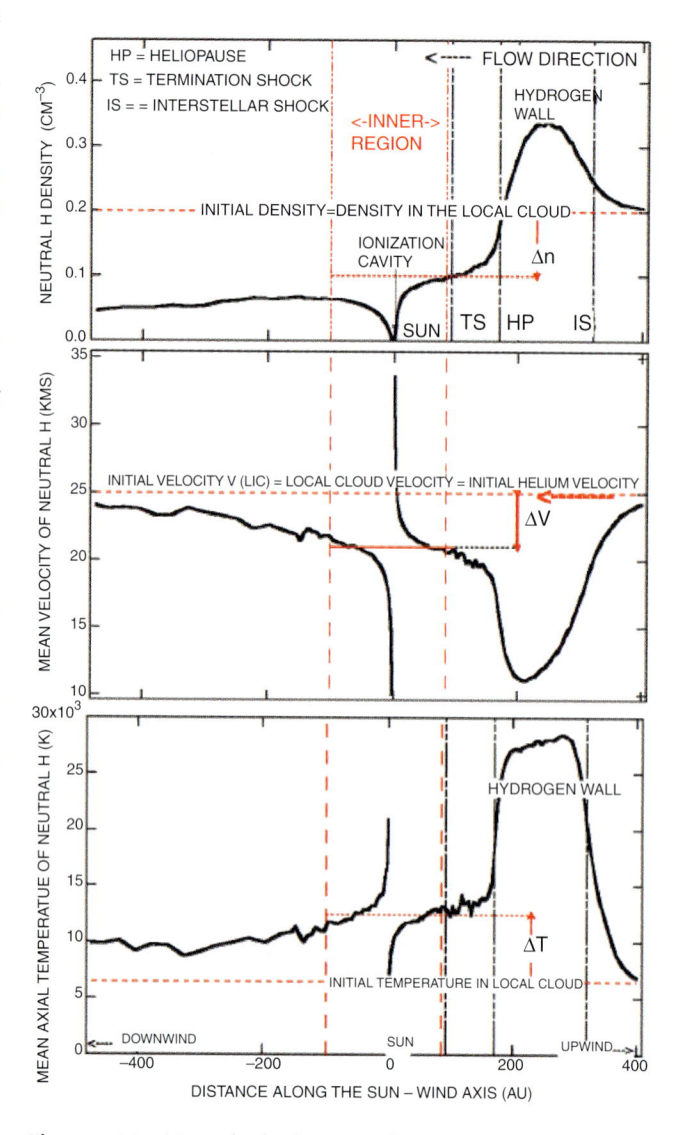

Figure 11 Neutral hydrogen characteristics along the Sun–wind axis according to Baranov and Malama (1993) and Izmodenov *et al.* (1999a). The density, average velocity, and apparent temperature are shown as a function of distance to the Sun for the "average" neutral flow. There are two components, primary atoms which have not suffered any charge exchange with the plasma and secondary atoms produced by charge exchange. When entering in the supersonic solar wind (left of the TS transition in red), the parameters N, V, and T have changed from their initial values in the unpertubed interstellar medium (right-hand side). The hydrogen wall, conspicuous in this two-shock model computed for a plasma density of $0.07\,cm^{-3}$, is seen between the bow shock and the heliopause.

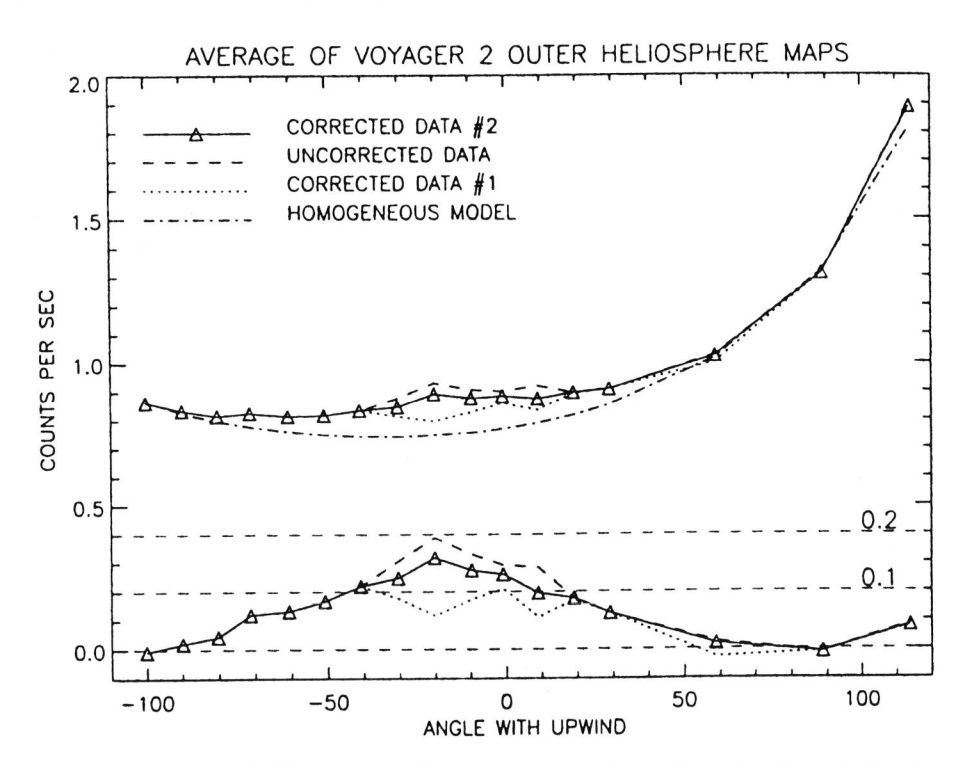

Figure 12 Lyman-alpha intensity recorded by Voyager 2 between 1993 and 1995 from the distant heliosphere as a function of line-of-sight angle with the upwind direction. Also shown is the expected intensity from a model without hydrogen compression in an "H wall." Data–model differences are shown at the bottom. In the galactic plane area, data have been corrected for a possible small contamination by diffuse emission from stars. Two extreme cases (dashed and dotted lines) are shown. (From Quémerais *et al.* 1995.)

from the illuminating source (the Sun), preclude angular variations at such small angular scales. Also, the locations of the maxima of this additional feature always coincide with the galactic plane, and not the upwind direction exactly. If one fits the data with the best-fitting heliospheric model, the additional intensity is of the order of about 10–20 Rayleigh, which is very small compared to the Lyman-alpha intensity at Earth orbit of a few hundred Rayleighs (Quémerais *et al.* 1996). The additional signal likely to be of galactic origin is probably originating from distant gaseous condensations. As a matter of fact, Lyman-alpha is produced not only by resonance of UV photons on hydrogen atoms, but also when protons recombine with electrons in the ionized regions around young stars (the H II regions), at the same time as H-alpha emission. Contrary to H-alpha, which can freely propagate through large quantities of gas, the Lyman-alpha radiation is strongly scattered by neutral hydrogen, and multiple scattering prevents photons from propagating unless they are gradually Doppler shifted from the resonance wavelength. Those photons which are not shifted have random motions from atom to atom and will finally disappear after encounters with dust particles.

For those who participated to the pioneering work on interplanetary emission, the detection of a galactic emission 25 years after the discovery of the Lyman-alpha glow would make them smile. As a matter of fact at the time of the discovery of the "interstellar glow," Jacques Blamont and co-workers had been disappointed when learning that the source of the emission was "local," that is, from gas inside the Solar System. They would have prefered a more "exotic" galactic source. Twenty-five years later, searching for a heliospheric emission from the "H wall," that is, a local source, a galactic emission is likely, but it complicates the diagnostic and it is not welcome. In both cases, scientists seem to be unhappy with what they find.

The existence of a new, probably galactic counterpart has indeed greatly complicated the determination of the intensity enhancement due to the hydrogen wall and the comparisons with model predictions. Up to now, although it is clear that the intensity pattern at Lyman-alpha as seen from increasing distances is not compatible with a uniform flow encountering the Sun and that there is a hydrogen wall (Hall *et al.* 1993, Quémerais *et al.* 1996), the actual level of the galactic counterpart and the actual additional intensity due to the compressed and heated hydrogen in the hydrogen wall are not yet resolved. More work is needed, and one expects some help due to the fact that the solar UV flux which illuminates the hydrogen wall varies with the solar cycle, as does the hydrogen wall emission, while the galactic counterpart is constant.

4.4 The boundary of the heliosphere seen in absorption towards nearby stars

The hydrogen wall of heated interstellar hydrogen was searched for in emission, with the problems of contamination discussed above, but again the unexpected occurred: it was detected very nicely in absorption! Interestingly, this discovery was done "live." At the IAGA conference in Boulder, I was responsible for a session entitled "The Heliosphere in the Local Interstellar Cloud," and, among other speakers from both heliospheric and interstellar communities, I had invited Vladimir Baranov from Moscow, Gary Zank and Lance Williams from Bartol, and Jeff Linsky from LASP in Boulder. The Baranov talk was a presentation of his latest theoretical results from the self-consistent model, and for the first time he had clearly presented separately the characteristics of the different species of neutral hydrogen in the outer heliosphere, and in particular the secondary atoms, that is, charge-exchanged interstellar protons, created mainly between the interstellar shock and the heliopause (Baranov and Malama 1993), and which are the main constituents of the hydrogen wall. These neutrals are slower and hotter than the unperturbed interstellar flow, as shown in Figure 11, since the initial protons are heated, compressed, and deviated. Other similar results were presented by Zank and Williams (Pauls *et al.* 1995, Zank *et al.* 1996b).

In the audience was Jeff Linsky, who had reduced and preliminarily analyzed the first high-resolution spectrum at Lyman-alpha of the star alpha Centauri, recorded with the Goddard High Resolution Spectrometer on board the HST. Linsky was planning to present the results on the interstellar gas intervening between the Sun and alpha Cen deduced from profile-fitting results of the spectrum (among other results). On the spot, hearing the results of Baranov and colleagues, he immediately did a "back of the envelope" calculation of the total column of heated and decelerated atoms, and he realized that the yet unexplained additional hydrogen absorption he had found when fitting the spectrum, which could not be due to the 5,000–10,000 K gas from the local interstellar clouds (otherwise the corresponding deuterium line would have been also detected), was very nicely explained by the hot gas of the heliospheric hydrogen wall. The target star alpha Cen is so close that the column of galactic gas along the line-of-sight between the Sun and the star is small and cold enough to allow the simultaneous detection, on the side of the main line, of the additional absorption due to the hydrogen wall gas, even if the hydrogen wall column is less than 0.1% of the total column.

Figure 13 (Linsky and Wood 1996) shows the alpha Cen spectrum with the best fit without hydrogen wall gas and the new fit after inclusion of the additional absorption by a column of gas having characteristics similar to those predicted by Baranov and Malama (1993). Clearly the

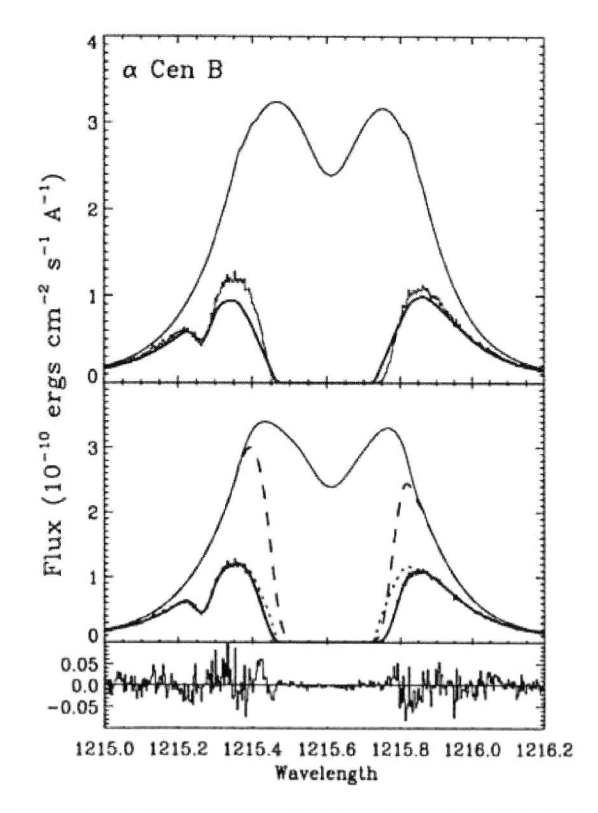

Figure 13 HST spectrum of alpha Centauri B (full line) at hydrogen Lyman-alpha and its modeling without (top) and with (bottom) absorption by the heliospheric "H wall." The fit is much better if the additional absorption by the hydrogen wall (dashed lines) is added to the cloud absorption (dotted line). The top curve is the predicted intrinsic stellar line before absorption by the two media. (From Linsky and Wood 1996.)

heliospheric absorption of H is not negligible and as a consequence the determination of the column density of H is strongly dependent on its inclusion (or not) in the fitting process, although, as we said, the heliospheric column is negligible compared to the interstellar column. This is not true for deuterium, because the transition is much weaker and heliospheric deuterium is not detectable. With a significant error on H (if one neglects the H wall), and no error on D, there are biases on the D/H ratio in the interstellar medium measured in this way using Ly-alpha absorption lines.

Since this discovery, absorptions due to hydrogen walls around a series of nearby stars have been detected from analyses of their Lyman-alpha (Wood *et al.* 1996, Piskunov *et al.* 1997). Also, both heliospheric and asterospheric hydrogen wall absorptions are simultaneously detected now around alpha Cen and Sirius (Lallement 1999, Izmodenov *et al.* 1999c). Because the presence of a hydrogen wall requires a stellar wind and a partially ionized interstellar gas around the star, this type of result has important implications: it shows for the first time that stars similar to the Sun have winds (although it seems natural, such winds had not been detected before). Also, it constrains the distribution and ionization of the gas in the local interstellar medium from *in situ* measurements at the locations of the stars. Besides these new aspects, the contribution of the hydrogen walls to the absorption has completely modified the interpretation of the spectra in terms of D to hydrogen ratios (Linsky and Wood 1996).

4.5 New messengers from interstellar space: the detection of pickup ions of C, O, N, and H, and the first isotopic ratios in the local ISM

In addition to the detection of neutral helium quoted above, the Ulysses mission produced new and fundamental data from the SWICS (Solar Wind and Interstellar Composition Spectrometer) experiment, a very sensitive plasma composition instrument implying time-of-flight technology. The unprecedented capabilities of this instrument in terms of charge per mass resolution and sensitivity have led to the detection of new pickup species, that is, those particles formed after ionization of interstellar neutral atoms. The ions H^+, $^4He^{2+}$, N^+, O^+, Ne^+ were successively discovered using SWICS (Gloeckler *et al.* 1993, Geiss *et al.* 1994).

From these pickup ions fluxes, it is possible to infer the fluxes of the "parent" neutral species and, for the first time, to connect relative abundances inside the heliosphere, measured *in situ*, and interstellar abundances deduced from astronomical measurements. As an example, the comparison between heliospheric and ISM neutral hydrogen, oxygen, and helium has shown that both neutral oxygen and neutral hydrogen are filtered at their entry into the heliosphere (Izmodenov *et al.* 1999a,b).

Only neutral species can enter the heliosphere, and one does not expect pickup ions from species which are fully ionized in the ISM. This is why the detection of the ion C^+ came as a surprise since there is no carbon in the neutral form in the ISM. This led to the discovery of the so-called "inner source," that is, the outgassing of neutral species by dust in the inner heliosphere, which produces its own pickup ions (Geiss *et al.* 1995).

Probably the most spectacular result was the first precise measurement of the $^3He/^4He$ isotopic ratio in the local ISM (Gloeckler and Geiss 1996), from the $^3He^+$ and $^4He^+$ fluxes. This ratio is of fundamental interest for cosmological models, and certainly nobody had expected 10 years ago to measure it directly inside the heliosphere.

On the heliospheric side, one of the consequences of the pickup measurements has been the improved accuracy in the determination of the densities of the interstellar neutrals (Section 4.6), compared to the use of optical data. But, mainly, the pickup measurements have stimulated numerous improvements of the propagation models, including wave–particle interactions (e.g. Isenberg and Lee 1998), and the study of the so-called "pressure-balance structures" (solar wind isobaric discontinuities), to infer the pickup pressure (Burlaga *et al.* 1994). Also, since the pickup ions account for a considerable fraction of the total energy at large heliocentric distances, they modify the solar wind termination shock (e.g. Zank *et al.* 1996a, Lee *et al.* 1996, Chalov *et al.* 1997).

4.6 Local interstellar helium is ionized: the contribution from the EUVE

An important and totally unexpected contribution to the field has also been brought by data from the Extreme Ultraviolet Explorer (EUVE), a satellite mainly devoted to the study of the EUV emission from hot or active stars, a topic previously without any relationship to the heliosphere.

As already said, the most accurate numbers for the density of heliospheric neutral helium have been obtained from both the direct detection of helium atoms and from pickup helium ions flux measurements by AMPTE and the SWICS instrument on board Ulysses. Modeling of He, He^+, and He^{2+} data provided $n(He)$ of the order of 0.014 cm^{-3} (Gloeckler 1996, Möbius 1996, Witte *et al.* 1996). Remember that helium is not filtered, and that this number density also applies to the local ISM.

Using the classical cosmological abundance ratio H/He = 10 and assuming that helium, the ionization potential of which is about twice that of hydrogen, is predominantly neutral in the local ISM, one infers from $n(He^0) = 0.14$ cm^{-3} that the total number of hydrogen nuclei $n(H^0) + n(H^+)$ (or equivalently $n(H\,I) + n(H\,II)$ with astronomical notation) is of the order of 0.14 cm^{-3}. The value 0.14 cm^{-3} is very close to the number density of 0.15 cm^{-3} derived for hydrogen far from the Sun from Voyager data (Quémerais *et al.* 1996), which is the largest value ever found for hydrogen. For comparison, pickup H^+, also detected by SWICS, leads to $n(H) = 0.12$ cm^{-3} (Gloeckler 1996). Under these conditions the hydrogen density decrease at the interface is smaller than 0.02 cm^{-3} (from 0.14 down to 0.12 cm^{-3}). In turn this implies that the plasma density in the ISM is very low, and that there is little confinement of the solar wind by the interstellar plasma. As a matter of fact, a plasma density as low as 0.05 cm^{-3} already implies a filtering as large as 50% in the frame of the two-shocks model. This was the situation in 1994.

However, surprisingly, the EUVE spectra of nearby white dwarf stars (Figure 14) revealed that interstellar helium is significantly ionized in the local ISM. Because EUV spectra of hot stars are strongly modified due to absorption by H^0, He^0, and He^+ along the line-of-sight, the analysis of the spectra allows one to infer the total column densities of these species. Dupuis *et al.* (1995) found that

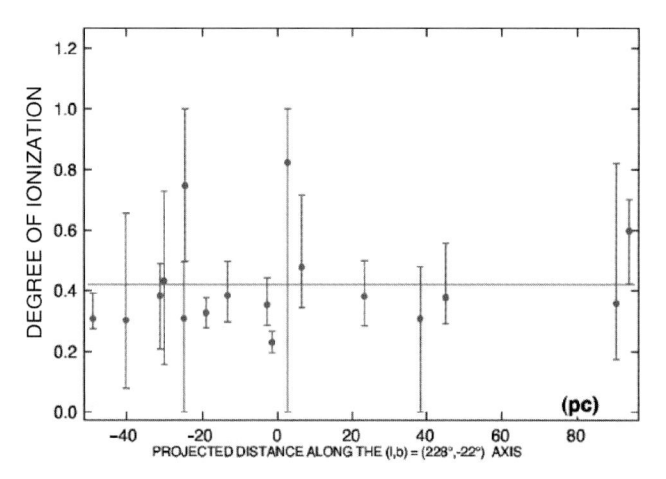

Figure 14 Interstellar helium ionization from the spectra of nearby white dwarfs recorded by the EUVE. Helium is found to be on average about 40% ionized. (From Wolff et al. 1999.)

along the line-of-sight towards some of the brightest white dwarfs helium is between 20 and 50% ionized. This result has been confirmed recently by a different team and an apparently constant value of 40% ionization has been derived for helium from 17 different targets (Wolff et al. 1999). Thus, the following new picture arises. Since in our local cloud helium is probably 40% ionized, the neutral helium density of $0.014\,\text{cm}^{-3}$ corresponds to a total helium density (neutral + ionized) of the order of $0.023\,\text{cm}^{-3}$ (remember that ionized helium does not penetrate), implying in turn a total hydrogen density of $0.23\,\text{cm}^{-3}$, and leaving room for a non-negligible filtering at the heliospheric interface. For example, if $n(\text{H}^+) = 0.05$ and $n(\text{H}^0) = 0.18\,\text{cm}^{-3}$ in the local ISM, a filtering of 30 to 40% is allowed (in other words, only 60 to 70% of hydrogen enters the heliosphere), which is roughly consistent with the filtering predicted by the models for a plasma density of the order of $0.05\,\text{cm}^{-3}$. As a conclusion, the EUVE results helped to solve one of the remaining inconsistencies about hydrogen and helium densities and the interface.

4.7 Neutral hydrogen flow heating and deceleration: a new determination of the plasma density

At the beginning of the twenty-first the SWAN experiment on board SOHO has now probably solved the last remaining inconsistency about the heliospheric hydrogen and He measurements. The instrument, whose primary objective is the determination of the solar wind mass flux as a function of cyclic activity and heliolatitude, is also addressing the question of the perturbations of neutral hydrogen at heliosphere entry. For this purpose, spectroscopic measurements of the Ly-alpha background are routinely obtained by using a hydrogen absorption cell (again). Using the motion along Earth orbit, that is, the $\pm 30\,\text{km s}^{-1}$ Doppler shift as a natural scanning mechanism, cell absorptions can be obtained at different locations of the Ly-alpha emission line, and the line profiles can be reconstructed. This was recently completed for the first year of data by Quémerais et al. (1999), providing the first map of the line-of-sight velocities of neutral hydrogen (Figure 15) while at the same time the absorption data were compared with a forward model for the hydrogen flow (Costa et al. 1999), leading to estimates of temperature and velocity of the flow before its interaction with the supersonic solar wind. The exceptional quantity of data recorded with SWAN has allowed the derivation of the velocity distribution of hydrogen without assumptions on the hydrogen cell parameters and has shown that its velocity dispersion corresponds to a temperature of about 10,000–11,000 K; that is, significantly larger than the temperature of the helium flow (and the LIC), and also larger than the temperature of 8000 K deduced earlier from the much smaller set of Prognoz data. This means that the hydrogen flow velocity distribution is broadened by interaction with the plasma through charge exchange, according to models.

The derived temperature and velocity can be compared with the same parameters outside the heliosphere. Figure 11 shows the velocity, temperature, and density of the neutral hydrogen along the Sun–wind axis as resulting from the self-consistent plasma/neutral model of Izmodenov et al. (1999a). What can be deduced from the SWAN measurements, using models for the flow in the inner heliosphere, are the parameters V, T and N at about 80–100 AU, that is, after the flow has crossed the termination shock. As shown in Figure 11, T and V are different from T(LIC) and V(LIC), the temperature and velocity of the Local Cloud, the asymptotic values on the right of Figure 11, which are identical to helium parameters. The temperature, velocity, and density jumps, ΔV, ΔT, ΔN, are a function of the coupling with the plasma. A preliminary comparison between the Baranov and Malama model and the data shows that the three gaps are best fitted for a plasma density of $0.04\,\text{cm}^{-3}$ (Costa et al. 1999, Lallement 1999). At variance with the 40% ionization of helium, the degree of ionization of hydrogen corresponding to such an electron density, that is, about 15% ionization, is in reasonable agreement with the expected ionizing field from the brightest hot stars (Vallerga 1996, Vallerga and Welsh 1995).

This result is to be compared with the radio data. In fact, $0.04\,\text{cm}^{-3}$ is the density corresponding to the plasma frequency of 1.8 kHz, that is, the exact cutoff observed for the low frequency of the Voyager data. This strongly suggests that the 2 kHz emission originates on the ISM side of the heliospheric interface, and that $0.04\,\text{cm}^{-3}$ is indeed the interstellar electron density outside the heliosphere in the Local Cloud, in agreement with one of the suggestions of Gurnett and Kurth (1996).

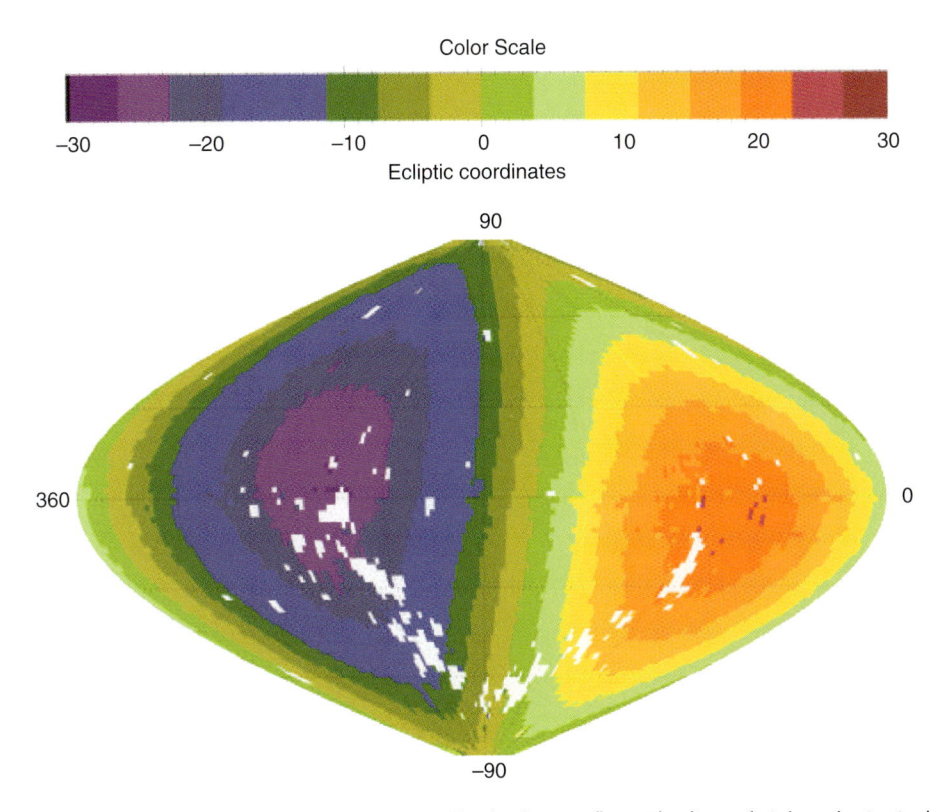

Figure 15 First map of line-of-sight velocities of the interstellar hydrogen flow. The line-of-sight velocity is the Doppler shift of the Ly-alpha emission, here shown as a function of the ecliptic coordinates of the line-of-sight. The hydrogen flow seems to move from the upwind direction (252°, 7°) at about 25 km s^{-1} and leave the Solar System in the opposite direction (72°, −7°) at about 20 km s^{-1}. These Doppler shifts contain information on the line-of-sight integrated properties of the flow close to the Sun and allow one to derive, with the help of flow models, the properties of the gas before the interaction with the Sun. (SOHO/SWAN results from Quémerais et al. 1999.)

However, for such a low plasma density, the two-shock models with no magnetic field or cosmic ray pressure predicts a termination shock at about 95 AU and the heliopause at 185 AU (Izmodenov et al. 1999a). This latter value is larger than the distance to the heliopause derived from the "echoes" of the solar wind events we have mentioned, suggesting something is missing in the model and probably an additional ISM pressure is required to further compress the heliosphere. But this is another matter of debate, since too strong an ISM magnetic field is incompatible with the supermagnetosonic character implicitly assumed in the two-shock model. Comparisons should be done with the "subsonic" models, such as the multi-fluid models of Williams et al. (1997). If the magnetic pressure is not negligible, MHD models (Ratkiewicz et al. 1998) are required. However, they need to include self-consistently the neutrals, which is one of the next achievements at which theoreticians are aiming (e.g. McNutt et al. 1998, Linde et al. 1998).

5 WHAT NEXT? UNANSWERED QUESTIONS AND PREDICTIONS

Can we really say now that our Sun's environment is no longer unknown territory? I would optimistically answer, "Yes." We have good ideas of temperature, velocity, densities, even turbulence (which is rather low, less than 1.8 km s^{-1} (Linsky et al. 1995)). But still there are two fundamental parameters which are not really constrained yet, the interstellar magnetic field and the low-energy cosmic ray flux (and pressure).

Can we reasonably expect to know them within the next five years? We have seen that the termination shock is very probably closer than about 95 AU and it is likely that the crossing of the shock by Voyager 1 will occur within the next 5–10 years (Figure 16). The predictions are made complicated by solar cycle effects. The solar wind pressure variation by a factor of two over the cycle is expected to produce a periodic oscillation of the shock with an amplitude of the

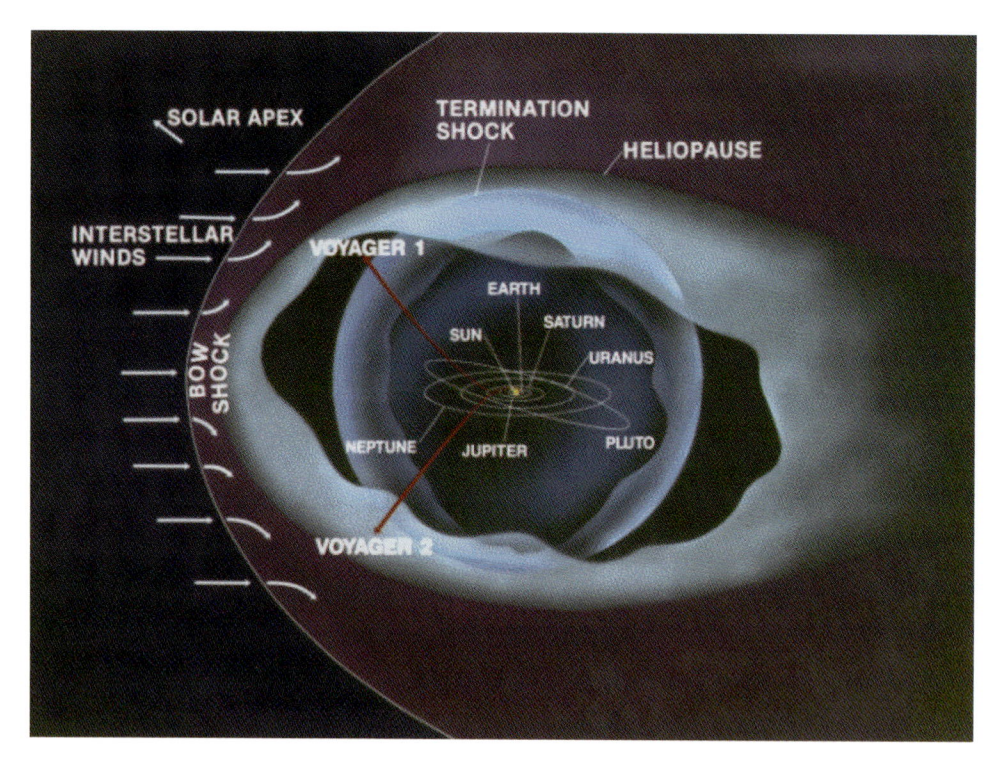

Figure 16 Voyager heading towards the edge of the heliosphere.

order of about 10 AU (e.g. Barnes 1994, Whang *et al.* 1999, Wang and Belcher 1999). Better than that, Voyager data will certainly warn us before the crossing: ACR gradients should tell us in advance at what distance is the shock (Stone *et al.* 1996, Stone and Cumming 1999) and hopefully this will allow the start of the countdown. (Marty Lee should then organize one of his famous pools which have marked the ends of scientific meetings on the heliosphere: " bet the distance of the termination shock." Up to now the "winning" distance has been increasing with time at roughly the speed of Voyager... maybe this will no longer be the case!)

Hopefully some of the instruments will detect the crossing and from the shock location we will get a better idea of the actual pressure in the surrounding ISM. This in turn should bring some constraints on the missing parameters. Still, it is very likely that the measurements will be totally different from those expected.

In addition to being patient, what can be done in the mean time?

- Narrow the constraints on the heliosphere structure by simply using the "inside" measurements, as neutrals and pickup-refined observations, to get velocity distribution departures from axisymmetry, distortions, and filtrations for such species. This is certainly achievable, if one takes into account carefully all the physical processes and solar cycle effects (e.g. Rucinski *et al.* 1996).

- Do high-quality and high-resolution direct spectroscopy of the Ly-alpha glow to disentangle galactic and heliospheric emissions. This is also achievable.
- Detect energetic neutral atoms created by charge exchange and measure their distribution. This should bring new clues on the heliospheric structure and is already in progress with the detection on board SOHO of their high-energy tail in the 40 keV range (Hilchenbach *et al.* 1998), that is, charge-exchanged accelerated pickup ions.
- Get new data and better understand the details about the origin of the Voyager radio emissions. This will probably be the "holy grail" since it is certainly one of the keys to the heliospheric structure.
- Wait for unexpected findings. They always happen, as shown again recently by Flynn *et al.* (1998) who used the neutral helium in the Earth exosphere as an absorption cell to derive the EUV helium glow background spectrum, and obtain new measurements of the helium flow velocity and temperature.

In future, hopefully all the objectives will be reached after the launch of an interstellar probe, traveling away from our Sun at 15–20 AU per year, and catching the old and brave Voyagers. Such a probe will tell us what are the intensity and the direction of the magnetic field in the LIC, what are the fluxes of the low-energy cosmic rays from which our heliosphere is protected, what is the total pressure in our small Local Cloud (is it really at a much lower pressure than the surrounding hot gas filling the "Local Bubble"?), what is the ionization balance for the different species (has a recent event modified the ionization of the ISM?), what is the dust size distribution in the LIC, how does the dust flow couple with the gas, and so on.

I would like to conclude this brief history with less timely questions, a kind of "interstellar forecast" paragraph. We are beginning to know our solar environment better, that is, the characteristics of the interstellar matter in the sun vicinity of the Sun, as illustrated at the beginning of this chapter. This means that we can begin to figure out which interstellar regions our Sun will travel through in the future. When will our Sun leave the Local Cloud? Probably in less than a few thousands years, according to the upper limit on the column of gas similar to the surrounding gas in the direction of its motion. When will it enter the next cloud (the so-called "G cloud"?). Within 50,000 years, certainly, since it is the distance towards alpha Cen, which is beyond the boundary of the G cloud, but maybe within 2000 years, if the clouds are contiguous, and a simple shock separates them. What kind of transition can we expect if the clouds are not contiguous, and tenuous hot gas of the Local Bubble separates them? The pressure could be higher, and the heliosphere could shrink substantially, when the Sun enters this hot gas. How will the voyage of the Sun through the different phases of the interstellar medium affect our heliosphere and our cosmic ray environment? The effects should be small, unless we are by chance going to encounter one of the small (a few hundred AU) and very dense ($10^5 cm^{-3}$) interstellar clouds claimed to fill our galaxy randomly. Zank and Frisch (1999) predict in this case significant effects on our heliosphere, if the temperature of these small condensations is about 10,000 K. But how important are these effects if the gas is extremely cold, which is likely if there is pressure balance between the different phases?

What will be the increase of both galactic and ACR fluxes reaching the Earth if the heliosphere shrinks in such a way our planet is no longer shielded? Did such an event already happen? Are the beryllium-10 enhancements recorded in the Antarctic ice, which suggest there have been periods of increased fluxes of energetic particles, linked to such hypothetical encounters? Or are they linked to more realistic astronomical events like supernovae explosions happening in our corner of the Galaxy, or to other phenomena? Will our interstellar environment suddenly suffer strong ionization and heating by a powerful beam of gamma or x-rays from a magnetar or a collapsing star which by chance will arise close to us? Life may be full of surprises, even for a small unregarded yellow Sun.

Acknowledgments

I wish to thank all the colleagues who participated in heliospheric studies and with whom I have always enjoyed lively discussions. It includes a very large number of the authors I quoted here and I had the chance to meet. I thank also particularly Sir Ian Axford, who read the manuscript and provided additional and precious information.

Finally, a special acknowledgement is due to Jean-Loup Bertaux, for his useful comments on the manuscript, and because he played a fundamental role both in the subject and in my personal involvement in this field.

REFERENCES

Adams, T.F. and Frisch, P.C. (1977). High-resolution observations of the Lyman alpha sky background. *Astrophysical Journal*, **212**, 300–308.

Ajello, J.M., Witt, N. and Blum, P. (1979). Four UV observations of the interstellar wind by Mariner 10 – Analysis with spherically symmetric solar radiation models. *Astronomy and Astrophysics*, **73**, 260–271.

Axford, W.I. (1972). The interaction of the solar wind with the interstellar medium. In C.P. Sonett, P.J. Coleman, Jr., and J.M. Wilcox (eds.), *Solar Wind*, NASA SP-308, Washington, DC, p. 609.

Axford, W. I. (1985), *Solar Physics*, Vol. 100, p. 575–586.

Axford, W.I. (1990). Introductory lecture – The heliosphere. In S. Grzedzielski and D.E. Page (eds), *Physics of the Outer Heliosphere*, COSPAR Colloquia Series, Vol. 1, Pergamon Press, London, pp. 7–15.

Axford, W.I., Dessler, A.J. and Gottlieb, B. (1963). Termination of solar wind and solar magnetic field. *Astrophysical Journal*, **137**, 1268.

Baranov, V.B. (1990). Gasdynamics of the solar wind interaction with the interstellar medium. *Space Science Reviews*, **52**, 89–120.

Baranov, V.B., Ermakov, M.K. and Lebedev, M.G. (1981). Three-component gasdynamic model of the interaction of the solar wind with the interstellar medium. *Soviet Astronomy Letters*, **7**, 206.

Baranov, V.B., Krasnobaev, K. and Kulikovsky, A. (1971). A model of the interaction of the solar wind with the interstellar medium. *Soviet Physics Doklady*, **15**, 791–793.

Baranov, V.B., Lebedev, M.G. and Ruderman, M.S. (1979). Structure of the region of solar wind – Interstellar medium interaction and its influence on hydrogen atoms penetrating the wind. *Astrophysics and Space Science*, **66**, 441–451.

Baranov, V.B. and Malama, Y.G. (1993). The model of the solar wind interaction with the local interstellar medium: Numerical solution of the self-consistent problem. *Journal of Geophysical Research*, **98**, 15157–15163.

Barnes, A. (1994). Motion of the heliospheric termination shock. 2: Energy loss due to energetic particle acceleration. *Journal of Geophysical Research*, **99**, 6553–6560.

Barth, C.A. (1970). Mariner 6 measurements of the Lyman-alpha sky background. *Astrophysical Journal*, **161**, L181.

Bertaux, J.L., Ammar, A. and Blamont, J.E. (1972). OGO-5 determination of the local interstellar wind parameters. *Space Research*, **12**, 1559.

Bertaux, J.L. and Blamont, J.E. (1971). Evidence for a source of an extraterrestrial hydrogen Lyman alpha emission: the interstellar wind. *Astronomy and Astrophysics*, **11**, 200.

Bertaux, J.L., Blamont, J.E., Mironova, E.N., Kurt, V.G. and Bourgin, M.C. (1977). Temperature measurement of interplanetary–interstellar hydrogen. *Nature*, **270**, 156–158.

Bertaux, J.L., Lallement, R., Kurt V.G. and Mironova, E.N. (1985). Characteristics of the local interstellar hydrogen determined from Prognoz 5 and 6 interplanetary Lyman-alpha line profile measurements with a hydrogen absorption cell. *Astronomy and Astrophysics*, **150**, 1–20.

Biermann, L. (1951). Kometenschweife und solare Korpuskularstrahlung. *Zeitschrift für Astrophysik*, **29**, 274.

Blum, P.W. and Fahr, H.J. (1970). Interaction between interstellar hydrogen and the solar wind. *Astronomy and Astrophysics*, **4**, 280.

Burlaga, L.F., Ness, N.F., Belcher, J.W., Szabo, A., Isenberg, P.A. and Lee, M.A. (1994). Pickup protons and pressure-balanced structures: Voyager 2 observations in merged interaction regions near 35 AU. *Journal of Geophysical Research*, **99**, 21511–21524.

Chalov, S., Fahr, H.J. and Izmodenov, V. (1997). Spectra of energized pickup ions of the two-dimensional heliospheric termination shock. II. Acceleration by Alfvenic turbulence and by large-scale solar wind turbulences. *Astronomy and Astrophysics*, **320**, 659–671.

Chassefiere, E. and Bertaux, J.L. (1987). Heating of helium of interstellar origin through elastic collisions with solar wind protons inside the heliosphere. *Astronomy and Astrophysics*, **176**, 121–130.

Chassefiere, E., Dalaudier, F. and Bertaux, J.L. (1988). Estimate of interstellar helium parameters from Prognoz 6 and Voyager 1/2 – EUV resonance glow measurements taking into account a possible redshift in the solar line profile. *Astronomy and Astrophysics*, **201**, 113–122.

Cheng, K.R. and Bruhweiler, F.C. (1990). Ionization processes in the local interstellar medium – Effects of the hot coronal substrate. *Astrophysical Journal*, **364**, 573–581.

Christian, E.R., Cummings, A.C. and Stone, E.C. (1988). Evidence for anomalous cosmic-ray hydrogen. *Astrophysical Journal*, **334**, L77–L80.

Clauser, T. (1960). 4th Symposium Cosmical Gas Dynamics, Varenna (1960).

Costa, J., Lallement, R., Quémerais, E., Bertaux, J.L., Kyrola E. and Schmidt W. (1999). Heliospheric interstellar hydrogen temperature from SOHO/SWAN hydrogen cell data. *Astronomy and Astrophysics*, **349**, 660–672.

Crutcher, R.M. (1982). The local interstellar medium. *Astrophysical Journal*, **254**, 82–87.

Cummings, A.C. and Stone, E.C. (1999). AGU meeting, December 1999, San Francisco.

Cummings, A.C., Stone, E.C. and Webber, W.R. (1984). Evidence that the anomalous cosmic-ray component is singly ionized. *Astrophysical Journal*, **287**, L99–L103.

Cummings, A.C., Stone, E.C. and Webber, W.R. (1987). Latitudinal and radial gradients of anomalous and galactic cosmic-rays in the outer heliosphere. *Geophysics Research Letters*, **14**, 174–177.

Czechowski, A. and Grzedzielski, S. (1994). Can a charge-exchange induced density rise at the heliopause explain the frequency drift of the 3 kHz Voyager signal? *Geophysical Research Letters*, **21**, 2777–2780.

Dalaudier, F., Bertaux, J.L., Kurt, V.G. and Mironova, E.N. (1984). Characteristics of interstellar helium observed with Prognoz 6 58.4 nm photometers. *Astronomy and Astrophysics*, **134**, 171–184.

Danby, J.M.A. and Camm, G.L. (1957). Statistical Dynamics and Accretion, *Monthly Notices Roy. Astron. Soc.* **117**, 50–71.

Davis, L.E., Jr. (1955). Interplanetary magnetic fields and cosmic-rays. *Physical Review*, **100**, 1440–1444.

Davis, L.E., Jr. (1962). The effect of solar disturbances and the galactic magnetic field on interplanetary gas. *Journal of the Physical Society of Japan*, **17**, Supplement A2, 543–545.

Dupuis, J., Vennes, S., Bowyer, S., Pradhan, A.K. and Thejll, P. (1995). Hot white dwarfs in the local interstellar medium: Hydrogen and helium interstellar column densities and stellar effective temperatures from extreme-ultraviolet explorer spectroscopy. *Astrophysical Journal*, **455**, 574–589.

Fisk, L.A., Kozlovsky, B. and Ramaty, R. (1974). An interpretation of the observed oxygen and nitrogen enhancements in low-energy cosmic-rays. *Astrophysical Journal*, **190**, L35.

Flynn, B., Vallerga, J., Dalaudier, F. and Gladstone G.R. (1998). EUVE measurement of the local interstellar wind and geocorona via resonance scattering of solar He I 584-A line emission. *Journal of Geophysical Research*, **103**, 6983.

Garcia-Munoz, M., Mason, G.M. and Simpson, J.A. (1973). A new test for solar modulation theory: The 1972 May–July low-energy galactic cosmic-ray proton and helium spectra. *Astrophysical Journal*, **182**, L81.

Geiss, J., Gloeckler, G., Fisk L.A. and Von Steiger R. (1995). C^+ pickup ions in the heliosphere and their origin. *Journal of Geophysical Research*, **100**, 23373–23378.

Geiss, J., Gloeckler, G., Mall, U., von Steiger R., Galvin, A.B. and Ogilvie, K.W. (1994). Interstellar oxygen, nitrogen and neon in the heliosphere. *Astronomy and Astrophysics*, **282**, 924–933.

Gloeckler, G. (1996). The abundance of atomic ^1H, ^4He and ^3He in the local interstellar cloud from pickup ions observations with SWICS on Ulysses. *Space Science Reviews*, **78**, 335–346.

Gloeckler, G. and Geiss, J. (1996). Abundance of ^3He in the local interstellar cloud. *Nature*, **381**, 210–212.

Gloeckler, G., Geiss, J., Balsiger, H., Fisk, L.A., Galvin, A.B., Ipavich, F.M., Ogilvie, K.W., von Steiger, R. and Wilken, B. (1993). Detection of interstellar pickup hydrogen in the solar system. *Science*, **261**, 70–73.

Grzedzielski, S., Lallement, R. (1996). Possible shock wave in the local interstellar plasma, very close to the heliosphere. *Space Science Reviews*, **78**, 247–258.

Grzedzielski, S. and Ratckiewicz, R. (1975). Asymmetric distant solar wind. *Acta Astronomica*, **25**, 177–204.

Gurnett, D.A., Allendorf, S.C. and Kurth, W.S. (1998). Direction-finding measurements of heliospheric 2–3 kHz radio emissions. *Geophysical Research Letters*, **25**, 4433.

Gurnett, D.A., and Kurth W.S. (1996). Radio emissions from the outer heliosphere. *Space Science Reviews*, **78**, 53–66.

Gurnett, D.A., Kurth, W.S., Allendorf, S.C. and Poynter, R.L. (1993). Radio emission from the heliopause triggered by an interplanetary shock. *Science*, **262**, 199.

Hall, D.T., Shemansky, D.E., Judge, D.L., Gangpadhyay, P. and Gruntman, M.A. (1993). Heliospheric hydrogen beyond 15 AU: Evidence for a termination shock. *Journal of Geophysical Research*, **98**, 15185–15192.

Hauser, M.G., Kelsall, T., Leisawitz, D. and Weiland, J. (1995). *COBE Diffuse Infrared Background Experiment Explanatory Supplement, Version 2.0*, COBE Ref. Pub. No. 95-A, NASA/GSFC, Greenbelt, MD.

Hilchenbach, M., Hsieh, K.C., Hovestadt, D., Klecker, B., Gruenwaldt, H., Bochsler, P., Ipavich, F.M., Buergi, A., Moebius, E., Gliem, F., Axford, W.I., Balsiger, H., Bornemann, W., Coplan, M.A., Galvin, A.B., Geiss, J., Gloeckler, G., Hefti, S., Judge, D.L., Kallenbach, R., Laeverenz, P., Lee, M.A., Livi, S., Managadze, G.G., Marsch, E., Neugebauer, M., Ogawa, H.S., Reiche, K.-U., Scholer, M., Verigin, M.I., Wilken, B. and Wurz, P. (1998). Detection of 55–80 keV hydrogen atoms of heliospheric origin by CELIAS/HSTOF on SOHO. *Astrophysical Journal*, **503**, 916.

Holzer, T.E. (1977). Neutral hydrogen in interplanetary space. *Reviews of Geophysics and Space Physics*, **15**, 467–490.

Holzer, T.E. (1989). Interaction between the solar wind and the interstellar medium. *Annual Review of Astronomy and Astrophysics*, **27**, 199–234.

Isenberg, P.A. (1986). Interaction of the solar wind with interstellar neutral hydrogen: Three-fluid model. *Journal of Geophysical Research*, **91**, 9965–9972.

Isenberg, P.A. and Lee, M. (1998). Transport of anisotropic interstellar pickup ions on bent flux tubes. *Journal of Geophysical Research*, **103**, 12037–12048.

Izmodenov, V.V., Geiss, J., Lallement, R., Gloeckler, G., Baranov, V.B. and Malama, Y. (1999a). Filtration of interstellar hydrogen in the two-shock heliospheric interface: Inferences on the local interstellar cloud electron density. *Journal of Geophysical Research*, **104**, 4731–4742.

Izmodenov, V.V., Lallement, R. and Geiss J. (1999b). Interstellar oxygen in the heliospheric interface: Influence of electron impact ionisation. *Astronomy and Astrophysics*, **344**, 317–321.

Izmodenov, V.V., Lallement, R. and Malama, Y. (1999c). Heliospheric and astrospheric neutral hydrogen absorption towards Sirius: No need for interstellar hot gas. *Astronomy and Astrophysics*, **342**, L13–L16.

Joselyn, J.A. and Holzer, T.E. (1975). The effect of asymmetric solar wind on the Lyman alpha sky background. *Journal of Geophysical Research*, **80**, 903–907.

Kumar, S. and Broadfoot, A.L. (1978). Evidence from Mariner 10 of solar wind flux depletion at high ecliptic latitudes. *Astronomy and Astrophysics*, **69**, L5.

Kupperian, J.E., Byram, E.T., Chubb, T.A. and Friedman, H. (1959). Far ultraviolet radiation in the night sky. *Planetary and Space Science*, **1**, 3.

Kurt (1965). Nasa techn. Transl. (translated from russian), F-389.

Kurt (1967). Kosmicheskiye Issledovaniga (in russian), Vol 5 (6).

Kurt V.G. (1965). In Skuridin et al. (eds), NASA technical translations, F-389, 769.

Kurt V.G. (1967). Kosmicheskiye Issledovaniga, 5, N°6.

Kurth, W.S., Gurnett, D.A., Scarf, F.L. and Poynter, R.L. (1984). Detection of a radio emission at 3 kHz in the outer heliosphere. *Nature*, **312**, 27–31.

Lallement, R. (1998). Observations of the local interstellar cloud. In D. Breitschwerdt, M.J. Freyberg and J. Truemper (eds), *The Local Bubble and Beyond*, Proceedings of IAU Colloquium 166, Lecture Notes in Physics Vol. 506, Springer-Verlag, Berlin, pp. 19–28.

Lallement, R. (1999). Global structure of the heliosphere: Optical observations. In S.R. Habbal, R. Esser, J.V. Hollweg and P.A. Isenberg (eds), *Solar Wind Nine*, AIP Conference Proceedings Vol. 471, American Institute of Physics, Washington, DC, pp. 205–210.

Lallement, R., Bertaux, J.L. and Clarke, J.T. (1993). Deceleration of interstellar hydrogen at the heliospheric interface. *Science*, **260**, 1095–1098.

Lallement, R., Bertaux, J.L. and Kurt, V.G. (1985). Solar wind decrease at high heliographic latitudes detected from Prognoz interplanetary Lyman-alpha mapping. *Journal of Geophysical Research*, **90**, 1413–1423.

Lallement, R. and Bertin, P. (1992). Northern-hemisphere observations of nearby interstellar gas: possible detection of the local cloud. *Astronomy and Astrophysics*, **266**, 479–485.

Lallement, R., Bertin, P., Ferlet, R., Vidal-Madjar, A. and Bertaux, J.L. (1994). GHRS observations of Sirius-A: Interstellar clouds towards Sirius and local cloud ionization. *Astronomy and Astrophysics*, **286**, 898–908.

Lallement, R., Ferlet, R., Lagrange, A.M., Lemoine, M. and Vidal-Madjar A. (1995). Local cloud structure from HST-GHRS. *Astronomy and Astrophysics*, **304**, 461.

Lallement, R., Ferlet R., Vidal-Madjar A. and Gry C. (1990). Velocity structure of the local interstellar medium. In S. Grzedzielski and E. Page (eds), *Physics of the Outer Heliosphere*, Proceedings of the 1st COSPAR Colloquium, Pergamon Press, London, pp. 37–42.

Lee, M.A. and Axford, W.I. (1988). Model structure of a cosmic-ray mediated stellar or solar wind. *Astronomy and Astrophysics*, **194**, 297–303.

Lee, M.A. and Ip, W.-H. (1987). Hydromagnetic wave excitation by ionized interstellar hydrogen and helium in the solar wind. *Journal of Geophysical Research*, **92**, 11041–11052.

Lee, M.A., Shapiro, V.D. and Sagdeev, R.Z. (1996). Pickup ion energization by shock surfing. *Journal of Geophysical Research*, **101**, 4777–4790.

Linde, T.J., Gombosi, T.I., Roe, P.L., Powell, K.G. and Dezeeuw, D.L. (1998). Heliosphere in the magnetized local interstellar medium – Results of a three-dimensional MHD simulation. *Journal of Geophysical Research*, **103**, 1889.

Linsky, J.L., Brown, A., Gayley, K., Diplas, A., Savage, B.D., Ayres, T.R., Landsman, W., Shore, S.N. and Heap, S.R. (1993). Goddard high-resolution spectrograph observations of the local interstellar medium and the deuterium/hydrogen ratio along the line of sight toward Capella. *Astrophysical Journal*, **402**, 694–709.

Linsky, J.L., Diplas, A., Wood, B.E., Brown, A., Ayres, T.R. and Savage, B.D. (1995). Deuterium and the local interstellar medium properties for the Procyon and Capella lines of sight. *Astrophysical Journal*, **451**, 335.

Linsky, J.L. and Wood, B.E. (1996). The alpha Centauri line of sight: D/H ratio, physical properties of local interstellar gas, and measurement of heated hydrogen (the hydrogen wall) near the heliopause. *Astrophysical Journal*, **463**, 254.

McDonald, F.B., Teegarden, B.J., Trainor, J.H. and Webber, W.R. (1974). The anomalous abundance of cosmic-ray nitrogen and oxygen nuclei at low-energies. *Astrophysical Journal*, **187**, L105.

McNutt, R.L., Jr (1988). A solar-wind 'trigger' for the outer heliosphere radio emissions and the distance to the terminal shock. *Geophysical Research Letters*, **15**, 1307–1310.

McNutt, R., Lyon, J., Goodrich, C.C. (1998). Simulation of the heliosphere – Model. *Journal of Geophysical Research*, **103**, 1905.

Meier, R.R. (1977). Some optical and kinetic properties of the nearby interstellar gas. *Astronomy and Astrophysics*, **55**, 211–219.

Meyer, P., Parker, E.N. and Simpson, A.J. (1962). Solar cosmic rays of February, 1956 and their propagation through interplanetary space. *Physical Review*, **104**, 768–783.

Moebius, E. (1996). The local interstellar medium viewed through pickup ions, recent results and future perspectives. *Space Science Reviews*, **78**, 375–386.

Moebius, E., Hovestadt, D., Klecker, B., Scholer, M. and Gloeckler, G. (1985). Direct observation of He(+) pickup ions of interstellar origin in the solar wind. *Nature*, **318**, 426–429.

Morton, D.C. and Purcell, J.D. (1963). Observations of the extreme ultraviolet radiation in the night sky using an atomic hydrogen filter. *Planetary and Space Science*, **9**, 455.

Parker, E.N. (1963). *Interplanetary Dynamical Processes*, Interscience Publishers, New York.

Patterson, T.N.L., Johnson, F.S. and Hanson, W.B. (1963). The distribution of interplanetary gas. *Planetary and Space Science*, **11**, 767.

Pauls, H.L., Zank, G.P. and Williams, L.L. (1995). Interaction of the solar wind with the local interstellar medium. *Journal of Geophysical Research*, **100**, 21595–21604.

Pesses, M.E., Jokipii, J.R. and Eichler, D. (1981). Cosmic ray drift, shock wave acceleration, and the anomalous component of cosmic-rays. *Astrophysical Journal*, **246**, L85–L88.

Piskunov, N., Wood, B.E., Linsky, J.L., Dempsey, R.C. and Ayres, T.R. (1997). Local interstellar medium properties and deuterium abundances for the lines of sight toward HR 1099, 31 Comae, beta Ceti, and beta Cassiopeiae. *Astrophysical Journal*, **474**, 315.

Quémerais, E., Bertaux, J.L., Lallement, R., Kyrola, E., Schmidt, W. and Berthé M. (1999). Interplanetary Lyman-alpha line profiles derived from SWAN/SOHO hydrogen cell measurements: Full-sky velocity field. *Journal of Geophysical Research*, **104**, 12585–12604.

Quémerais, E., Malama, Y.G., Sandel, W.R., Lallement, R., Bertaux, J.-L. and Baranov, V.B. (1996). Outer heliosphere Lyman-alpha background derived from two-shock model hydrogen distributions: Application to the Voyager UVS data. *Astronomy and Astrophysics*, **308**, 279–289.

Quémerais, E., Sandel, B.R., Lallement, R. and Bertaux, J.-L. (1995). A new source of Ly-alpha emission detected by Voyager UVS: Heliospheric or galactic origin? *Astronomy and Astrophysics*, **299**, 249.

Ratkiewicz, R., Barnes, A., Molvik, G.A., Spreiter, J.R., Stahara, S.S., Vinokur, M. and Venkateswaran, S. (1998). *Astronomy and Astrophysics*, **335**, 363–369.

Richardson, J. (1997). The heliosphere-interstellar medium interaction: One shock or two? *Geophysical Research Letters*, **24**, 2889.

Rucinski, D., Cummings, A.C., Gloeckler, G., Lazarus, A.J., Mobius, E. and Witte, M. (1996). Ionization processes in the heliosphere – Rates and methods of their determination. *Space Science Reviews*, **78**, 73–84.

Sfeir, D., Lallement, R., Crifo, F. and Welsh, B.Y. (1999). Mapping the contours of the Local Bubble: Preliminary results. *Astronomy and Astrophysics*, **346**, 785–797.

Shklovsky, J.S. (1959). On hydrogen emission in the night glow. *Planetary and Space Science*, **1**, 63.

Stone, E.C., Cummings, A.C. and Webber, W.R. (1996). The distance to the solar wind termination shock in 1993 and 1994 from observations of anomalous cosmic-rays. *Journal of Geophysical Research*, **101**, 11017–11026.

Stone, E.C. and Cummings, A.C. (1999). Solar Wind Nine. In Habbal *et al.*, (eds), AIP Conference Proceedings, Vol. 471, 201–204.

Suess, S.T. (1990). The heliopause. *Reviews of Geophysics*, **28**, 97–115.

Thomas, G.E. (1972). Properties of nearby interstellar hydrogen deduced from Lyman-alpha sky background. In C.P. Sonett, J.M. Wilcox and P.J. Coleman (eds), *Solar Wind*, NASA SP-308, p. 668.

Thomas, G.E., and Krassa, R.F. (1971). OGO-5 measurements of the Lyman-alpha sky background. *Astronomy and Astrophysics*, **11**, 218.

Vallerga, J. (1996). Observations of the local interstellar medium with the Extreme Ultraviolet Explorer. *Space Science Reviews*, **78**, 277–288.

Vallerga, J.V. and Welsh, B.Y. (1995). Epsilon Canis Majoris and the ionization of the local cloud. *Astrophysical Journal*, **444**, 702–707.

Wallis, M.K. (1975). Local interstellar medium. *Nature*, **254**, 202–203.

Wang, C. and Belcher, J.W. (1999). The heliospheric boundary response to large-scale solar wind fluctuations: A gasdynamic model with pickup ions. *Journal of Geophysical Research*, **104**, 549–556.

Weller, C.S. and Meier, R.R. (1981). Characteristics of the helium component of the local interstellar medium. *Astrophysical Journal*, **246**, 386–393.

Weymann, R. (1960). Coronal evaporation as a possible mechanism for mass loss in red giants. *Astrophysical Journal*, **132**, 380–403.

Whang, Y.C. and Burlaga, L.F. (1994). Interaction of GMIR shock with the heliopause and its relation to the 2- and 3-kHz radio emissions. *Journal of Geophysical Research*, **99**, 21457–21465.

Whang Y.C., Lu, J.Y. and Burlaga, L.F. (1999). The termination shock: 1979–1995. *Journal of Geophysical Research*, **104**, 28255.

Williams, L.L., Hall, D.T., Pauls, H.L. and Zank, G.P. (1997). The heliospheric hydrogen distribution: A multifluid model. *Astrophysical Journal*, **476**, 366–384.

Witte, M., Banaszkiewicz, M. and Rosenbauer, H. (1996). Recent results on the parameters of the interstellar helium from the Ulysses/GAS experiment. *Space Science Reviews*, **78**, 289–296.

Witte, M., Rosenbauer, H., Banaszkiewicz, M. and Fahr, H. (1993). The Ulysses neutral gas experiment: Determination of the velocity and temperature of the interstellar neutral helium. *Advances in Space Research*, **13**, 121–130.

Wolff, B., Koester, D. and Lallement, R. (1999). Evidence for an ionization gradient in the local interstellar medium: EUVE observations of white dwarf. *Astronomy and Astrophysics*, **346**, 969–978.

Wood, B.E., Alexander, W.R. and Linsky, J.L. (1996). The properties of the local interstellar medium and the interaction of the stellar winds of epsilon Indi and lambda Andromedae with the interstellar environment. *Astrophysical Journal*, **470**, 1157.

Wu, F.S. and Judge, D.L. (1979). Temperature and flow velocity of the interplanetary gases along solar radii. *Astrophysical Journal*, **231**, 594–605.

Zank, G.P., Cairns, I.H., Donohue, D.J. and Matthaeus, W.H. (1994). Radio emissions and the heliospheric termination shock. *Journal of Geophysical Research*, **99**, 14729–14735.

Zank, G.P. and Frisch, P.C. (1999). Consequences of a change in the galactic environment of the Sun. *Astrophysical Journal*, **518**, 965–973.

Zank, G.P., Pauls, H.L., Cairns, I.H. and Webb, G.M. (1996a). Interstellar pickup ions and quasi-perpendicular shocks: Implications for the termination shock and interplanetary shocks. *Journal of Geophysical Research*, **101**, 457–478.

Zank, G.P., Pauls, H.L., Williams, L.L. and Hall, D.T. (1996b). Interaction of the solar wind with the local interstellar medium: A multifluid approach. *Journal of Geophysical Research*, **101**, 21639–21656.

51

KÁROLY SZEGÖ*

Comets: coma and beyond

There is no clear definition of what the coma of a comet is. Most frequently it means the faint visible halo around comets, and in other context it denotes material that reflects or emits electromagnetic radiation (including light) back to observers. In this chapter we shall use the word 'coma' to describe all materials around the cometary nucleus, solids and volatiles, neutral and charged, irrespective how effectively a certain component is able to scatter or emit radiation. The origin of the coma is the nucleus itself, and we shall discuss the processes that eject or emit materials from the surface to the cometary atmosphere and beyond. We consider only physical processes; coma chemistry is not the topic of this chapter. It is unavoidable to take into account in this context the structure and the physics of the cometary surface layer as well, because this is the source of the coma. Comets are believed to contain pristine materials from the period when the Solar System was born; the cometary surface, however, has been exposed to many perturbations (cosmic rays, solar wind, heat cycles, etc.) and it definitely cannot be considered as pristine. How deep we have to dig to find unprocessed material is an open question. There is only limited experimental evidence on surface evolution obtained in simulations, and we have to rely mostly on guesses.

It is also important to note that *in situ* experimental investigation of comets is limited to a few flyby missions; most of our knowledge comes from astronomical observations. It is a matter of debate, however, as to what extent the observed coma features can be directly related to the properties of the nucleus. If a jet is seen in the coma, astronomers prefer to connect it with a source below it on the nucleus surface. However, as Crifo and Rodionov (1999a) have pointed out, two active sources on an aspherical nucleus can easily mimic the appearance of a jet between the sources, and it is much stronger than those visible above the actual active regions. This should make us cautious with inferences, but such caution is lacking in many publications. The complicated rotational motion of the generally irregularly shaped nucleus also renders inferences difficult. Even the free rotation of an irregular body strictly speaking is aperiodic in the ecliptic frame of reference; and in the frame of reference attached to the angular momentum vector the time variation of two of the Euler angles is characterized by two different periods, and the variation of the third Euler angle is aperiodic in general. The rotational motion under the effect of the torque due to nuclear activity can be even more complex. Filtering out all these details from ground-based observations is almost an impossible task, especially if the direct inference between coma and fixed sources on the rotating nuclear surface is also in question.

If we browse publications of the 1960s to find out what they contained about planets, we frequently find quite basic statements that were disproved later, after *in situ* measurements were done. When considering comets, we have to bear in mind that nobody has ever landed on a comet, and no nucleus has ever been explored from a close distance. At comet Halley, the Giotto probe took the highest resolution images from a distance of about 1200 km, with a nominal resolution of about 27 m; these images covered part of the nucleus. The two Vega craft imaged the nucleus from about 8500 km with a resolution of about 150 m. In these missions the high relative velocity between the nucleus and the spacecraft made dust analysis difficult, because no dust detector could be appropriately calibrated before launch. Formation of a plasma cloud due to the high impact velocity on the target surface, and its influence on data analysis in certain cases is still an open issue. Dust size measurements

*Research Institute for Particle and Nuclear Physics, Budapest, Hungary

made by different techniques differ in fine details. Modelling of the coma and the nucleus surface received a new impetus when the Rosetta mission of the European Space Agency was approved to explore and land on comet Wirtanen. The spectacular appearance of comet Hale-Bopp in 1997 also initiated new research.

In this chapter we first review models of cometary surfaces, to the level necessary to understand how the coma develops and what the boundary constraints are for the models for the acceleration of dust/gas mixtures. Next we discuss cometary atmospheres, both the regions where the properties of neutral gas are dominated by collisions, and where collisions cease to be present. We review the results of in-flight dust experiments, and we finish with a brief overview of the charged particle environment of comets.

THE SURFACE OF COMETS

It is the 'dirty snowball' vision of Whipple (1950) that lies at the heart of all comet models. Accordingly, to a zero-order approximation the surface is like the surface of a dirty snowball. The 'dirt' is not homogeneous, the chemical composition of the dirt varies depending on location, which is convincingly proved by the existence of chemically different jets in the coma (A'Hearn et al. 1986, Cosmovici et al. 1988, Clairemidi et al. 1990), and it is conceivable that there are inhomogeneities in other physical quantities as well. This entails that cometary activity might not be uniform, and though we cautioned against a direct inference from coma to surface, we are certain that this inhomogeneity is reflected in coma properties as well. Comets are irregularly shaped. The three-dimensional shape of comet Halley could be reconstructed based on the Vega flyby images, and Giotto contributed with a single view only (Szegö et al. 1995). The area of the sunlit nucleus surface varies as the object rotates, this is a natural cause of anisotropy.

Comets are always exposed to cosmic rays, which is the dominant radiation when they are in the Oort cloud. Their interaction with the local interstellar material also affects surface evolution, but as we do not have much knowledge of the details, we shall not discuss it here. The penetration depth of cosmic rays depends on their energy and on the local density; moderately relativistic charged particles lose energy in matter primarily by ionization. From ground experiments it is known (Caso et al. 1998) that the energy loss of protons more energetic than 1 GeV is a few $MeVg^{-1}cm^2$. For less energetic particles it grows considerably more; for example, 100 MeV protons penetrate a few centimetres into water ice. Cosmic radiation may trigger chemical reactions leading to surface differentiation, among others to the creation of impermeable chemical layers (Moore et al. 1983).

The solar wind can reach the surface of comets when cometary activity is low, that is at large distances from the Sun. It is known that dust can absorb solar wind ions and re-emit them as neutrals; this process is also likely to occur in comets. However, it is unlikely that the solar wind causes profound physical changes, and therefore we do not discuss this effect any further.

As a preparatory activity to a landing on comet 46P/Wirtanen, researchers have worked out a nucleus reference model (Möhlmann 1999), which is the best summary of our current knowledge of the thermal, mechanical and electromagnetic properties of the cometary surface. Ground simulations of the properties of ice–dust mixtures (e.g. Thiel et al. 1995) significantly contributed to this. Our focus here, however, is more limited; we are interested only in models that can describe how dust and gas can be released from the surface layer.

The first models concentrated on reproducing the inbound and outbound brightness curves of comets often showing an asymmetry at the same heliocentric distance. Whipple (1950) was the first to introduce the mantle as the likely surface layer of the nucleus, consisting of solid materials after the volatile component evaporated. Mendis and Brin (1977) suggested that erosion takes place, with particles smaller than a critical size being carried away by the outflowing gas. This was developed further by Horanyi et al. (1984) in the framework of the friable sponge model. The key assumptions of this model are that the dust loss rate is proportional to the momentum flux of the outflowing vapour, and that erosion takes place only in a thin surface layer (Figure 1). The gas is released from the surface of the icy core (a frozen dust–ice mixture), covered or uncovered by the dust mantle. This implies that the size distribution of the grains remains constant in the mantle with time,

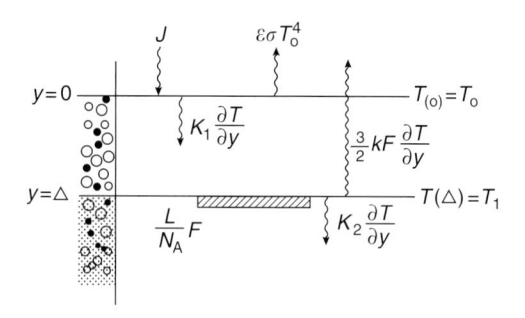

Figure 1 The 'friable sponge' surface model. The upper layer, the dust mantle, consists of degassed dust particles. Under the dust layer there is a dust–ice mixture. The heat flows are indicated, the sublimated vapour is in thermal equilibrium with the local mantle. L is the latent heat, N_A is the Avogadro number, F is the flux of sublimated vapour and K_1 and K_2 are is the thermal conductivities of the mantle and the core, respectively. (After Horanyi et al. 1984, Figure 1.)

independent of the thickness of the mantle. As this requires that all sizes are removed at the same rate, it is allowed that larger grains be broken into smaller pieces (friability).

The thermal balance on the dusty surface is given by the following equation:

$$(1-a)Je^{-\tau} - \varepsilon\sigma T_o^4 + D = -K\frac{\partial T \text{ (at } y = 0)}{\partial y} \quad (1)$$

where a is the surface albedo, J is the solar energy flux, τ is the optical depth of the coma, ε is the thermal emissivity of the surface, σ is the Stephan–Boltzmann constant, T_o is the temperature at the surface, D is the heat due to diffuse radiation from the coma, K is the thermal conductivity, and y is the distance measured from the surface downward. In this model the thermal conductivity can be expressed as $K = a + bT^3$; for the meaning of a and b see Horanyi et al. (1984).

There is a general agreement that the heat flux, J, due to solar irradiation is attenuated by the already existing coma. Its intensity depends on the heliospheric distance, the local incidence angle and that part reflected back in proportion to the surface albedo (first term of eqn (1)). The nucleus as a black body also irradiates heat proportional to its thermal emissivity (second term of eqn (1)), and the existing coma not only attenuates the incoming flux, but due to scattering and re-radiation, it contributes to local heating. The amount of heat reaching the surface due to the coma is a matter of debate. The 'rule of thumb' accepted nowadays is that this balances the attenuation (Salo 1988). The back-scattered heat from the coma is definitely not enough to maintain sizeable surface activity on the dark side; at least there is no evidence for such processes on the images taken during the Halley flybys.

In all surface models it is of paramount importance as to how heat is conducted inside, this accounting for the various gas production rates (Rickman 1991). The net heat flux on the surface is conducted inside through the degassed mantle, warming up the upward-flowing vapours. Models differ as to whether there is a local thermal equilibrium between the local mantle and vapour temperature. At the bottom of the degassed mantle the heat flow reaches the pristine core where the ice–dust mixture resides and is sublimated, the sublimation rate is governed by the Clausius–Clapeyron equation. In the simple friable sponge model all the heat is used on sublimation. The gas production rate remains constant until the mantle thickness reaches a value of about 10^{-2} cm, and then it decreases sharply with increasing mantle thickness. The surface temperature for 'pristine ice–dust' is the sublimation temperature, and as the mantle thickness increases it reaches the black-body temperature. The mantle thickness varies dynamically, and this can account for by the observed inbound/outbound asymmetry of the coma brightness.

In this model there is only one volatile component, and whereas it reproduces the basic dust-emission process, it evidently cannot elucidate the flux variation of the different gases as a function of heliocentric distance. A wide variety of models has been developed to remedy this and some other simplifications in the thermal properties. The most important modification is that gas is released not only from the ice surface, but from deeper layers as well. An excellent summary of these is given by Klinger et al. (1996). We follow this paper when summarizing the new features: (a) the gas phase contributes to the heat transfer to deeper layers; (b) volatiles are able to diffuse into deeper layers where they recondense; (c) sublimation can occur at various depths depending on the volatility of the different ices; (d) there is more than one ice phase, the water ice initially is in an amorphous phase that will become a crystalline phase; and (e) the pore size varies during the ougassing process. Such a surface model is shown in Figure 2. Most of these assumptions were verified by the KOSI comet simulation experiments (see references in Klinger et al. 1996).

In the following we discuss specifically the model developed by Podolak and Prialnik (1996). They assumed that the mantle layer is composed of dust, amorphous ice, crystalline ice, water vapour and gases such as CO and CO_2 trapped in the amorphous ice. The trapped gases are released when the transition from the amorphous to the crystalline phase takes place, but the gases released can recondense on pore walls and their sublimation may occur at a later stage. The gas flow is a free molecular flow, because the mean free path in the mantle is much larger than the pore sizes. The dust in this model is not only liberated from the surface, but may slowly move through the pores in the vertical direction,

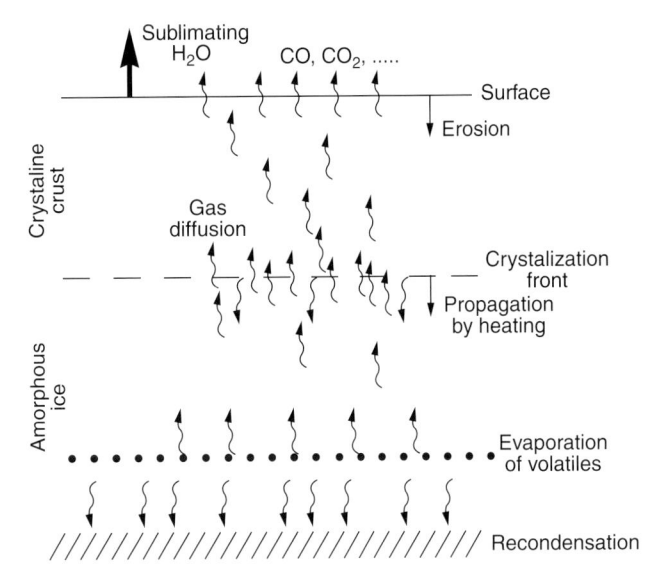

Figure 2 Schematic of a possible surface structure with different ice phases. (After Rickman 1991, Figure 10.)

though it is allowed that dust can move horizontally when the vertical path is blocked. All the pores are permeable to gas. Based on these and on some more specific assumptions, the probability of a dust particle leaving the mantle can be calculated. The one-dimensional mass and energy equations are solved only for the gas–ice mixture, the dust basically is treated separately, and only the dust and gas fluxes are coupled. These considerations lead to a stratified mantle structure: there is a highly porous dust layer on the top, followed deeper by a dense layer of crystalline ice and dust and, below that there is a layer of amorphous ice and dust, including other frozen-in volatile components at various depths (the frozen CO_2 is closer to the surface than the frozen CO). This model reproduces very well the observed gas emissions; namely that the CO production rate is higher at large heliospheric distances than that of water, but this ratio changes dramatically as the comet approaches the Sun. This proves that the coma composition does not reflect the composition of the nucleus.

These models are one-dimensional, with variations only in the radial directions being considered. This was remedied by Enzian et al. (1999) who developed a multidimensional rotating nucleus model, though still a spherical one, made up of a porous dust–ice matrix composed of water and CO. Heat and gas diffusion was allowed both in the radial and meridional directions. They found a near-uniform CO production, with water production and surface temperature showing local variations connected with the incoming heat flux.

Whereas these models are significant steps towards an understanding of cometary activity, there are still unresolved problems and controversies; for a review see, for example, Crifo and Rodionov (1999b). The model of Enzian et al. (1999) can barely account for the observed water production rate, even assuming that the surface is pure ice, and the radius is at the upper limit of the observations. As a pure ice surface is very unlikely, one tends to assume that bigger chunks of surface material can be released emitting gas in flight as well. This scenario is seemingly supported by radar observations (e.g. Harmon et al. 1997) indicating that the surfaces of several comets are rough on the centimetre scale; however, how such large pieces can be carried away is still very much a matter of debate.

CHARGING OF THE DUST IN FLIGHT AND THE COMETARY SURFACE

Dust particles and the surface of a cometary nucleus can be charged due to the ultraviolet radiation of the Sun and the escaping flux of photoelectrons. The current carried by the charged components of the solar wind is important when it can reach the cometary surface. In general, secondary ion currents are negligible. Horanyi (1996) gives an excellent review of charged dust dynamics in the Solar System. The motion of a dust particle is determined by the forces acting on it: the Lorenz force due to its own charge, gravitational forces and light pressure. In a coordinate system attached to the rotating nucleus, inertial forces should also be taken into account. Charging is relevant for those cases when the force due to charging is of the same order of magnitude as the other forces acting on the dust particle; typically for particles of a few micrometres in the vicinity of a few tens of metres of the nucleus.

The evolution of the charge created on the surface of an insulator is described by the following equation (Horanyi 1996):

$$\frac{dQ}{dt} = \sum_k J_k \qquad (2)$$

where J_k represents the charging currents. In a stationary case the left-hand side is zero.

In general, the flux of current density of particles, characterized by a distribution function $f(\mathbf{v})$, bombarding a surface with potential Φ, is given as

$$J = \int_{v^*}^{\infty} v^2 \, dv \int_0^{\pi/2} \sin\vartheta \, d\vartheta \int_0^{2\pi} d\psi \, v \cos\vartheta \, f(\mathbf{v}) \qquad (3)$$

where v^* is chosen for each plasma species so that $mv^{*2}/2 - e\Phi > 0$; that is, the integration volume in velocity space depends on the surface potential. It is easy to see that the limit of the integration is different if the Φ surface potential is positive or negative. If the size, a, of the dust particles is smaller than the characteristic Debye lengths, the surface we need to take into account for grain charging is the total surface, which is $4\pi a^2$ in the case of a spherical grain.

The actual formulae specifying the different charging currents are quite complicated, and are not given here. The potential distribution in the dusty plasma sheath above the nucleus can be obtained from the Poisson equation. At infinity the potential and the electric field should be zero.

The nightside surface potential differs: obviously there is no UV flux and the effect of the solar wind is also different. Electrons, due to their high thermal velocity, do reach the nightside surface, and the form of the charging current does not differ from that of the dayside. For protons, however, the situation is modified. Though the diameter of the nucleus is negligible compared to the proton gyroradius, since the flow velocity is higher than the proton thermal velocity, only a fraction of the proton distribution reaches the surface, those for which the gyration results in effective backward motion in the cometocentric system. As the nightside surface potential becomes negative, more protons can reach the surface. The estimated potential on the nightside can be as large as $-1\,\text{kV}$, deflecting the bulk kinetic energy of the solar wind protons moving past the comet.

However, if we assume that the proton distribution on the nightside is Maxwellian, the equilibrium nucleus surface potential would be about -20 V (Mendis *et al.* 1981). The gap between these two values is huge, and it cannot be resolved within the framework of current models.

The charging processes are fast relative to the rotation period of the nucleus, the characteristic timescale for charging being less than 1 s. Therefore, the plasma sheath of the nucleus stays static relative to the solar direction, it is not modified significantly by rotation.

Dust charging differs from surface charging, because any dust grains moving in the dayside plasma sheath are charged in addition by the photoelectron flux emitted from the surface. Calculations have been carried out only for spherical grains, approximating the relationship between the grain potential and its total charge as if it were a spherical conductor. Under these conditions a dust grain would levitate above the surface under the balancing force of electrostatic and gravitational forces. Below the region where levitation occurs, the grain would fall back, above it would escape. The existence of a dust-free region above the surface was also indicated by lunar data.

For all practical cases the electric field in the vicinity of the surface is of the order of V m^{-1}. Because, for example, the gravitational acceleration on the surface of comet P/Wirtanen is of the order of 10^{-2} cm s^{-2}, the electrostatic force acting on a dust particle is of the order of the gravitational force when the particle size is less than a few micrometres. This sets the limit on the importance of dust charging. For greater distances the potential gradient, that is, the electric field, is negligibly small; for larger size particles the electric force is negligible, and only gravity and solar radiation pressure should be considered.

As an example, Figure 3 shows the motion of 21 dust grains above the surface of comet P/Wirtanen, at 3 AU when the solar wind still reaches the surface (Juhasz and Szegö 1997). The ejecta are 2 μm dust grains with 0.2 m s^{-1} eject velocity, in a cone with a 60° opening angle. The forces acting on the dust particles are the gravitational force, the electrostatic Lorenz force and the light pressure. The centrifugal force should be taken into account if we describe the grain motion in a frame attached to the rotating nucleus. As can be seen, the grains form a levitating cloud above the surface, most of them settling quite far away

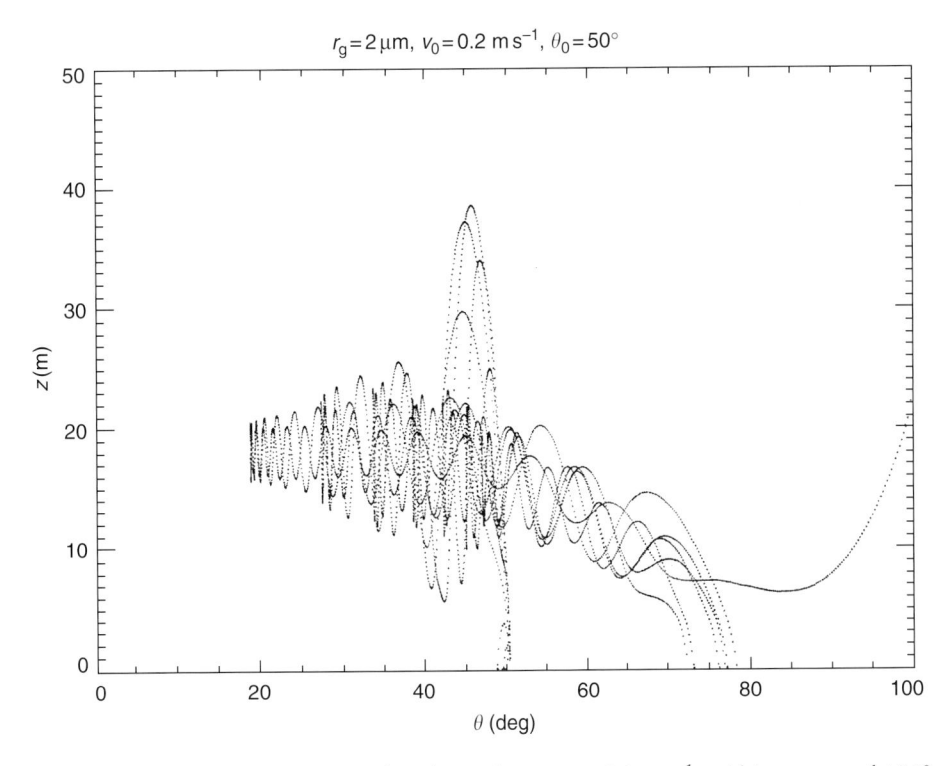

Figure 3 Motion of 2 μm size dust particles, emitted with a velocity $v_o = 0.2$ m s^{-1}, within a cone of $\pm 30°$, at 3 AU from the surface of comet Wirtanen, under nominal conditions (see text for details). The horizontal axis shows the angle between the Sun–comet line and the emission point in a plane perpendicular to the ecliptic; the vertical axis is distance above the surface. The grains form a cloud levitating about 20 m above the surface, and most of them settle between 70° and 80° solar zenith angle. Grains that enter into the shadow region escape tailward. (After Juhasz and Szegö 1997, Figure 7.)

from the ejection point. Grains that enter into the shadow region escape tailward.

The effect of charging is different for an active nucleus, because the solar wind no longer reaches the surface and only the solar UV radiation causes charging. In this case the nucleus surface potential is about $+2\,\text{V}$. We believe, however, that the qualitative picture does not change significantly.

THE COLLISION-DOMINATED COMA

When the dust and gas mixture leaves the surface it is dense enough to be dominated by collisions. For neutral gas, the mean free path λ is given approximately by $\lambda = (n\sigma)^{-1}$, where n is the gas number density and σ is the cross-section; its value is about $10^{-15}\,\text{cm}^2$. The gas production rate close to the surface is of the order of $10^{17}\,\text{mol}\,\text{cm}^{-2}\,\text{s}^{-1}$ at 1 AU, and if we take the sound velocity as an upper limit for the gas velocity, the mean free path is of the order of 10 cm. (The real value of the gas velocity differs from the sound velocity in some of the surface models described above, and molecular effusion theory also yields lower values (e.g. Gombosi 1994). For our qualitative description, however, the difference is not important.) The fluid regime prevails if the Knudsen number ($Kn = \lambda/L$, where L is the characteristic linear size of the problem, e.g. $L = n/|\text{grad}\, n|$) is less than 0.1. However, there is a region close to the surface where the gas distribution is not Maxwellian, simply because the outflowing gas distribution is not symmetric with respect to the upward and downward directions. The width of this layer for comet Wirtanen at 1 AU is about 1 m and at 2.5 AU is about 50 m (Crifo et al. 1999). At an altitude of about 100 km at 1 AU (and a few kilometres at 2.5 AU) the gas density becomes so low and the free mean altitude path so large that the fluid approximation becomes invalid.

Within the collision-dominated coma the fluid equations describing mass, momentum and energy conservation are assumed to be operational for the dust and gas interaction. Computer power has allowed only recently a full, time-dependent, three-dimensional code to be worked out (Crifo and Rodionov 1999a, Crifo et al. 1999b, and references therein). This approach, however, still requires certain simplifications. In the model only one gas component is taken into account, the dust particles are collected into several classes depending on their mass and no collision is allowed between dust particles. This is not a trivial limitation, because, as we shall see below, experimental data support the idea of dust fragmentation, that is a certain mixing between dust classes. The dust temperature is taken to be constant in flight, and equal to the black-body equilibrium temperature. No in-flight gas emission is allowed. During flight gas molecules do not pick up energy from the Sun. The boundary conditions at the lower end of the collision-dominated coma are obtained by solving together the fluid equations and the equations for the non-Maxwellian regime, they cannot be derived strictly from the surface models presented above.

The basic difficulty, however, lies in the gas–dust interaction. The only description available comes from Probstein (1969) who assumed that the dust grains are spherical, the grains reflect gas molecules diffusely with a Maxwellian velocity distribution close to the dust temperature and the molecular mean energy absorbed by the grain is completely converted to the thermal energy of the grain. The first problem is with the shape of the dust grain. Crifo and Rodionov (1999a) have pointed out that grain motion strongly depends on its shape, and in general the grain's trajectory is not linear. Similarly, the dust terminal velocity also exhibits shape dependence exceeding one order of magnitude variations in some cases. It is also very difficult to generalize the other assumptions of Probstein's model.

Previous one- and two-dimensional models confirmed that in the collision-dominated coma the water vapour cools down to such an extent that recondensation and icy grain formation is likely. The three-dimensional models cannot treat this question yet. It has not been clearly demonstrated whether the collision-dominated coma rotates with the nucleus or not, though the nucleus rotation might deform the dust structure.

Despite of all these limitations, the new models have yielded very important conclusions:

- Jet formation in the coma is not directly connected to source regions on the surface; it is sensitive to nucleus asphericity and inhomogeneity.
- Jet formation is possible without assuming active and inactive surfaces; actually even in the case of uniform surface activity.
- The gas distribution, due to collisions, shows broad features and does not reflect the chemical composition of the nucleus.
- The gas and dust in general do not move radially outwards, their velocity does not necessary increase monotonically and their density does not necessary decrease monotonically outwards. (In particular, under certain conditions, the nightside CO density can be higher than the dayside density!)

To illustrate this, Figure 4 shows the dust column density distribution around the nucleus of comet Halley (Rodionov et al. 2002). The image was produced by a three-dimensional gas dynamic model using the three-dimensional shape of comet Halley as derived from the flyby images (Merenyi et al. 1990). The nucleus attitude, illumination and view directions correspond to the Giotto encounter. It is assumed

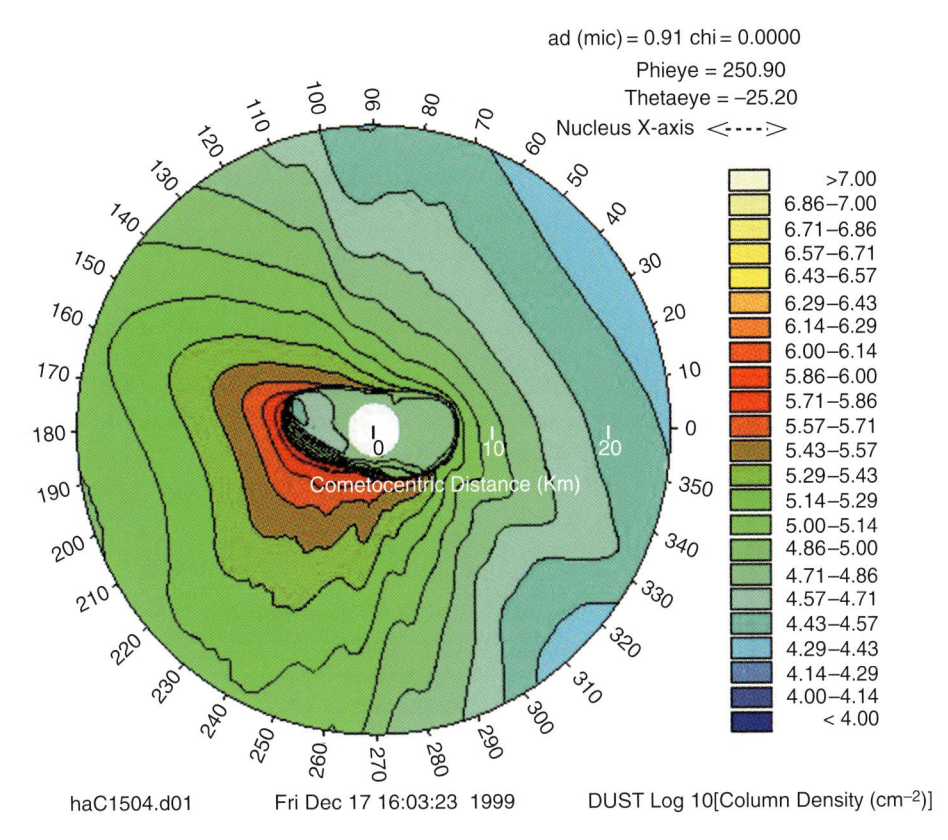

Figure 4 Colour-coded dust column densities, produced by a three-dimensional gas dynamic model using the three-dimensional shape of comet Halley as derived from the flyby images, and assuming *homogeneous* surface activity; that is, there are no active or inactive regions on the surface. The nucleus attitude, illumination and view directions correspond to the Giotto encounter. Due to the gas dynamics, the dust particles form clear jet features, corresponding very well to the actual measurements (c.f. Keller *et al*. 1994, Figure 77) (after Rodionov *et al*. 2002).

that the nucleus surface activity is homogeneous, that is there are no active or inactive regions and that the ice and dust ratio is the same everywhere. Despite this, due to the gas dynamics the dust particles form clear jet features, the pattern obtained corresponding very well to the jet features seen by the Giotto camera. This result clearly proves that the current theory on the existence of active and inactive surface regions is not necessary to account for coma observations.

The dust–gas coupling becomes unimportant further than about 10 nucleus radii from the surface. The dust particles start to move independently from the environment, under the effects of radiation pressure and possible internal processes (such as gas release and fragmentations). Though dust grains are very likely to be charged, no evidence has yet been found that dust charging influences their motion in this region or further outwards, and no interaction has yet been observed with ions. It has also been conjectured that the dust particles emit gas in flight, especially that the CN gas jets might be due to in-flight emission. The details of these models are not yet fully consolidated, though there is a general consensus that distributed sources do exist in the coma (Festou 1999).

The gas flow escaping from the surface is still in the collision-dominated region after having decoupled from the dust. The outflow velocity is of the order of $1\,\text{km}\,\text{s}^{-1}$ at 1 AU, the thermal velocity decreasing as the gas cools down. The gas composition, however, cannot be obtained from the composition at the surface; chemistry becomes more and more important, and gas released from dust also becomes a significant component. Collision between gas particles are important until the mean free path is commensurable with the volume in question, this being closer than a few thousand kilometres of the nucleus. Within this distance, due to scattering, the gas should fill the available volume more or less isotropically; this might be the reason why inhomogeneity of cometary activity is not seen in plasma processes. However, gas jets of different ions have been clearly observed (A'Hearn *et al*. 1986, Cosmovici *et al*. 1988, Clairemidi *et al*. 1990). Therefore, isotropization is either rather incomplete, or the sources of the gas jets are the not the nucleus, but the dust grains in flight, beyond the collision-dominated region. This controversy has not yet been really resolved.

Despite the fact that we do not discuss chemistry in this chapter, the formation of the hydrogen cloud cannot be left unmentioned. At about 1 AU the hydrogen corona radius is of several millions of kilometres around the nucleus. The neutral density depends on the type of ion, because different ions react differently to solar radiation pressure; this interaction is effective for hydrogen. There are several processes leading to hydrogen production. One is the photodissociation of water in the cometary atmosphere, which yields a molecular hydrogen velocity of about $20\,km\,s^{-1}$ relative to the nucleus. Photodissociation of OH gives a hydrogen component with a velocity of $8\,km\,s^{-1}$. There are also other non-molecular but dynamic processes (such as collisional thermalization) leading to slower components (Shimizu 1991). The hydrogen corona can only be observed from outside of the Earth's atmosphere, because the hydrogen atoms are detected by resonant scattering of the solar Lyman alpha radiation at 121.6 nm, and the atmosphere absorbs this radiation.

The neutral gas outflow was measured *in situ* aboard Giotto by the neutral mass spectrometer (Lämmerzahl *et al.* 1987). Between 900 and 2000 km the velocity is constant having values around $800\,m\,s^{-1}$. After this the velocity increases monotonically to about 40,000 km, reaching values between 1.2 and $1.6\,km\,s^{-1}$. Further details of the measurement cannot be explained without discussing coma chemistry. The neutral particle density was also measured aboard the Vegas (Gringauz *et al.* 1986a). The radial distribution could be fitted by an $r^{-2}\exp(-r/r_o)$ curve, but there were several local variations attributed to local effects. The inbound and outbound density values were somewhat different. The Giotto probe and the Vegas found different total neutral outflow, if the measurements taken along the orbit were extrapolated to the whole comet. It is difficult to assess whether the differences indicate local variations or a really rather different overall activity.

IN SITU DUST MEASUREMENTS DURING FLYBY MISSIONS

Several dust detectors operated on various comet flyby missions: the SP-1, SP-2 and DUCMA detectors on the Vegas (Vaisberg *et al.* 1986, 1987; Mazets *et al.* 1986, 1987; Simpson *et al.* 1986, 1987, 1989) and DIDSY on Giotto (McDonnell *et al.* 1987). The overall picture was that comet Halley produced about 2.5 more dust counts during the first flyby than during the second. However, on Vega 1 dust particles were only recorded closer than 260,000 km to the nucleus, whereas aboard Vega 2 dust particles were recorded from 320,000 km and closer. The Giotto encounter was the least dusty. It should be noted that as dust emission is very anisotropic, the amount of dust encountered on one specific orbit does not necessarily characterize the total dust output. McDonnel and Pankiewitz (1990) and Vaisberg (1990) have published excellent summaries. Below we focus on a few questions concerning dust production.

Table 1 Position of the boundaries for comet Halley

Spacecraft	Leg	P1 (km)	P2 (km)	P3 (km)	P4 (km)
Vega 1	Inbound	41,400	21,300	10,800	8,400
Vega 2	Inbound	42,700	24,500	11,200	7,900
Vega 2	Outbound	45,300	30,600	16,500	10,700

Existence of dust boundaries

The cumulative dust flux distribution as measured by SP-2 for both encounters and the dust counts as measured by SP-1 show steep changes in certain mass ranges at different distances. The SP-1 experimenters explicitly identified four boundaries, as shown in Table 1. The steep variations at the boundaries are not compatible with a simple fountain model, because in different mass ranges the boundary appears in the same place.

Mass flux distribution

The measured mass flux distribution, contrary to the pre-encounter models, increases for small dust masses throughout the measured range. A dip is observed for masses 10^{-11} to 10^{-14} g, is attributed to the increased radiation pressure for particles in Mie resonance with the radiation wavelength. The distributions cannot be fitted to a function of the form m^x with a single exponent.

The radial distribution profile is not uniform. In the plot of the SP-1 experiment between P2 and P3 the count rate varies as r^{-1}; beyond P3 it shows an $r^{-4.5}$ dependence. The SP-2 experimenters, however, found that the overall dust flux distribution is reconcilable with an r^{-2} variation.

Very fine dust

Measurements indicate that fine dust is plentiful in the cometary environment. Sagdeev *et al.* (1990), analysing unidentified peaks in the spectra of the PUMA dust impact mass analyser, came to the conclusion that the peaks are very likely associated with impacts of very fine dust, in the mass range 10^{-17} to 10^{-20} g. A further conclusion is that the very fine dust may contribute a few percent to the total dust mass around comet Halley. Actually, as molecules are of the order of 10^{-23} g, it is a question as to whether the distribution increases continuously to the molecular regime or not.

Large particles

The Vega spacecraft did not register hits caused by large dust particles, this differing from the Giotto experience. Therefore, the estimate of the total dust production rate is significantly different. It should be noted, however, that large dust particles are not distinguishable from cometary debris that stays in the vicinity of the nucleus for a long period of time; accordingly, the detection of large dust particles does not necessary mean detection of freshly emitted cometary material. Richter et al. (1991) argued that such particles could be emitted from the nucleus a long time previously relative to the orbital timescale.

Observation of jets

The dust distribution in the inner regions of the coma exhibits sharp local variations, which are attributed to dust jets. The intensity and the location of the variations change with dust mass, the least variable being the larger particles. Features of the spatial distribution can be enhanced if ratios of different mass distributions are generated (Vaisberg et al. 1987, Figure 3). During the encounters a few major jet events were registered.

During the Vega 2 flyby at about 50 s after closest approach the spacecraft crossed a prominent jet as derived from the imaging experiment aboard (Sagdeev et al. 1987). Both the three-dimensional reconstruction of the prominent jets and the general look of the image exposed during the jet crossing support this view. The dust counters also registered these jets, clearly showing dispersion in mass, the centre of the jet in each mass range was somewhat shifted, the larger particles arriving first (Vaisberg et al. 1987).

Dust fragmentation

Several features in the dust distribution data point to dust fragmentation, and we believe that the existence of this process is a very firm conclusion of the flyby missions.

One indicator is connected with dust boundaries. Oberc (1996) investigated this problem in depth, analysing the most pronounced enhanced gradients in the fluxes, and concluded in favour of dust disintegration. In many cases, especially in the DUCMA instrument, but in the other dust detectors as well, clusters of dust particles were frequently observed, with a non-Poisson mass distribution, totally different from the expectation on the basis of random arrival (Simpson et al. 1987). There is no other current explanation than fragmentation (Boehnhardt 1989).

It was mentioned already that depletion was observed in the mass distribution corresponding to an enhanced radiation pressure effect. However, a more detailed analysis has revealed that the dependence of the depletion on distance is weak, indicating that other particles, presumably coming from fragmentation, fill up the gap (Vaisberg et al. 1987).

In the mass spectrum of jets, low-mass dust particles were conspicuously missing, indicating that their source might not be directly the nucleus (Vaisberg et al. 1987).

Dust fragmentation can be caused by dust charging, an unavoidable phenomena. Fomenkova and Mendis (1992) pointed out that the electric forces generated could be higher than the tensile strength. The critical dust radius a_c (in micrometres) is $a_c = 7.75 |\Phi| / \sqrt{F_t}$, where F_t is the tensile strength in dyn cm^{-2} and Φ is the grain potential in volts. For a radius smaller than a_c, fragmentation is likely. Greenberg et al. (1995) derived a tensile strength of 2.7×10^3 dyn cm^{-2} for fluffy grains; as Φ is in general a few volts, a_c is in the regions of a few tens of micrometres.

Oberc (1996) elaborated a physical mechanism for slow fragmentation. In his model the fragmentation is the result of the slow sublimation of the gas holding the dust together. A fast fragmentation can be caused by a heat shock the dust particle undergoes after leaving the surface. On the surface the sublimating ice temperature is of the order of 200 K, whereas the dust temperature in flight very like reaches the black-body value.

Unresolved discrepancies between in-flight experiments

There are still a few unresolved discrepancies between the different dust instruments flown on board of the Vegas, especially between the DUCMA and the SP detectors. The position of most of the events correlates well, with a few exceptions. These are unaccounted for. Post-encounter calibration resolved the differences in the integral flux values. The major disagreement between the Vega and Giotto experiments is the total mass production, a number that heavily depends on the lower end of the dust distribution.

THE PLASMA ENVIRONMENT OF THE NUCLEUS

The neutral particles, mostly water molecules, are slowly ionized when flying away from the nucleus, and these newly formed ions interact with the solar wind flow creating a special plasma environment of the comet. We shall call this environment the 'cometospace' (coined in analogy to the term 'geospace' that denotes that volume around the Earth where the solar wind is disturbed due to the presence of the Earth). The cometospace differs from the plasma regions around planets, because at comets gravity does not confine the ions to the close vicinity of the centre. Accordingly, the cometospace is very much extended, and the cometary ions play a specific role in all plasma features formed. The velocity of the neutrals is about $v_g \sim 1\,\mathrm{km\,s^{-1}}$ in the radial direction, this being the velocity they acquired when they left the collision-dominated regime; their thermal velocity is negligible. (As was mentioned before, some of the neutrals have a higher velocity due to chemical reactions, and there is no reason to assume that the neutral environment is isotropic. These specifics will not be considered here.) We shall discuss only the effects of the water ions because they are the dominant species; their ionization time is $\tau = 10^6\,\mathrm{s}$. Accordingly, the ion density can be easily expressed through the total gas outflow Q, the cometocentric distance r and a mean free path for photoionization $v_g\tau$:

$$n_i = \frac{Q - Q\exp(-r/v_g\tau)}{4\pi r^2 v_g} \qquad (4)$$

The overall ion production measured on board was different for the different flybys of P/Halley. However, these numbers were extrapolated from measurements carried out along the spacecraft trajectories assuming isotropic ion distribution, an assumption that cannot be substantiated by observations. This might be a reason why the on board and ground-based ion production values differed.

The literature on cometospace is very broad. We give two general references in addition to a few specific ones cited below: Johnstone (1991) and Newburn et al. (1991). In the cometospace no specific plasma phenomena or feature could be connected with the dust–plasma interaction.

UPSTREAM OF COMETS

The neutrals fill a huge volume around comets due to their slow ionization time, and within this volume water ions are continuously generated by solar UV radiation. Because the neutrals are cold, the newly formed ions also form a cold beam in the velocity space relative to the solar wind. The ions are immediately picked up by the interplanetary magnetic field, retaining their velocity parallel to the magnetic field, whereas they start to gyrate around the magnetic field with their perpendicular velocity. Relative to the solar wind plasma they form a ring in the velocity space, in the configuration space the ions move on cycloid trajectories perpendicular to the magnetic field. This plasma distribution is unstable, and a collective plasma interaction sets in, leading to intensive wave generation. The most important waves far away from the nucleus are Alfven waves propagating, in general, along the magnetic field lines in opposite directions in the solar wind frame of reference. This wave excitation has been observed millions of kilometres upstream of the nucleus in the case of P/Halley and P/Giacobini-Zinner. The wave excitation is a nonlinear and turbulent process, and the characteristics of the waves observed at different cometary flybys significantly differ from each other (Glassmeier et al. 1997); the variation is connected to the cometary activity. However, this connection is not simple. P/Halley, which had the highest production rate at the time of encounter relative to the two other flybys (P/Giacobini-Zinner and P/Grigg-Skjellerup), showed the highest turbulence, though lowest nonlinearity. The waves observed at P/Giacobini-Zinner had a complicated structure composed of different types of waves (magnetosonic waves and whistler packets); at P/Grigg-Skjellerup, which had the lowest activity, the waves showed a fairly regular pattern.

The excited waves interact with the cold ions, transferring energy and momentum from the solar wind to the ions, and making them accommodate to the solar wind flow; consequently the solar wind slows down. The ion accommodation has several phases. The fastest process is that the waves scatter the ions at the pitch angle in the velocity space, generating two partial shells, centred at $V_{SW} \pm V_A$, where V_{SW} is the solar wind velocity, and V_A is the phase velocity of the Alfven waves relative to the solar wind flow. The two partial shells are connected at a circle, the position of which position is determined by the angle between the magnetic field vector and the velocity vector of the solar wind flow. The time required for pitch-angle scattering is a few ion gyration periods. The isotropization of the distribution function competes with the continuous generation of newly formed ions; therefore, during the flyby experiments the full (bispherical) shell was observed only near the cometary bow shock. A much slower process is the energization, that is, the pickup ions diffuse energy as well as shrinking and expanding the shell radii (the radius is proportional to the ion energy). Energy diffusion requires the presence of counter-propagating waves. The whole process of ion accommodation and solar wind deceleration is termed in the literature 'mass loading' (Szegö et al. 2000).

It is an intriguing question as to whether there is evidence for inhomogeneous cometary activity in the plasma processes of these distant regions. The TUNDE energetic ion spectrometer aboard the Vega space probes detected time intervals with recurring high count rates. One interpretation of the observations is that they are due to the inhomogeneous neutral outflow associated with the rotation of the nucleus. Whereas this interpretation has not been broadly accepted, no other has been put forward.

THE BOW SHOCK

Mass loading reduces the solar wind speed as its momentum is distributed to newly formed ions. Before the bow shock its velocity is still supersonic, and therefore a bow shock is formed around the comet. This shock, however, differs from the average planetary shocks in that it is much broader, and in some cases such a smooth transition is observed that it is questioned whether a shock is really formed. Such smooth transitions are termed bow waves, and observed, for example, when the Giotto probe encountered comet Grigg-Skjellerup. There are two reasons why the bow shock is so broad. The first one is associated with the constant production of ions even inside the shock (at the shock not only photoionization plays a role, but charge exchange also becomes more important), which changes the nature of the shock; these types of shocks are called mass-loaded shocks (Zank et al. 1993). The second reason is that whereas at 'ordinary' shocks their width is defined in terms of the actual proton gyroradius, at cometary shocks the ion gyroradius takes over this role, which by itself causes a factor of 16 increase in width. At Halley during the 1986 flybys the shock was about 100,000 to 150,000 km wide and it was located at about 1 million km from the nucleus, measured along trajectories almost perpendicular to the comet–Sun line. Along the Sun–comet line the bow shock was formed at about 350,000 km from the nucleus.

BEHIND THE BOW SHOCK

Behind the bow shock the shocked and loaded solar wind circumflows the nucleus, as indicated by the plasma flow vectors obtained during the SUISEI encounter with comet Halley, shown in Figure 5 (Mukai et al. 1986). The closest distance of the flow to the nucleus is about the location of the 'contact discontinuity', a boundary that plays a similar role to that of the ionopause of non-magnetic planets, and separating the flow from the cometary ionosphere. During the Halley flyby the distance of the contact discontinuity was about 5000 km from the centre. As the nucleus of comet Halley is non-magnetic, inside the contact discontinuity there was no magnetic field detected, this being the reason why the volume inside is also termed the magnetic cavity.

The loaded, shocked solar wind behind the bow shock has a complicated structure. This is well illustrated by the plasma measurements of the ICE spacecraft (Figure 6) that crossed the tail region of comet Giacobini-Zinner on 11 September 1985, at a distance of about 7800 km from the nucleus, on a path almost perpendicular to the ecliptic plane (Bame et al. 1986). At this comet the bow shock was particularly weak. Behind the bow shock the major regions

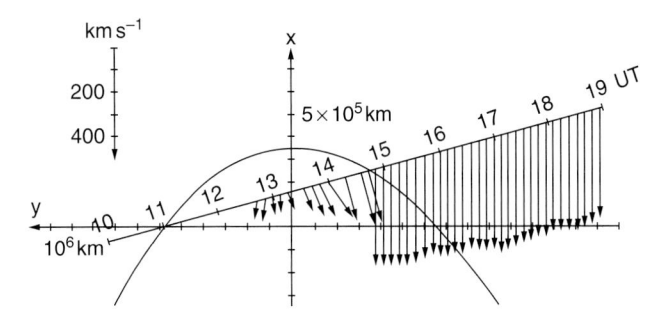

Figure 5 The solar wind velocity vectors obtained by the Japanese SUISEI spacecraft on 6 March 1986 at comet Halley before and after the bow shock crossing. The bow shock itself is denoted by a parabolic line; the x-axis points sunward, the y-axis is in the orbital plane of the spacecraft (almost identical to the ecliptic plane). Before the bow shock the velocity vectors are more or less parallel but decreasing in magnitude, because water ions originating from the comet load the solar wind, and the unperturbed solar wind momentum will be shared between the old and new ion components. The bow shock is a discontinuity: the sudden change in the magnitude of the velocity vectors clearly indicates this. Behind the bow shock the streaming plasma circumflows the obstacle, as indicated by the changing direction of the velocity vectors. The SUISEI probe made its closest approach of 151,000 km to the nucleus at 13:06 UT. (After Mukai et al. 1986, Figure 1.)

identified were the ionosheath, a boundary layer, a certain intermediate coma and the plasma tail.

In the ionosheath, as the plasma analysers carried aboard the flyby missions revealed, the flowing plasma has several components. One is the warm, Maxwellian solar wind, and the shell of the upstream pickup ions was also clearly discernible. A separate population is the freshly picked up ions behind the bow shock. These components have, in principle, different bulk velocities, and the velocity spread around the bulk also varies. The electrons, corresponding to these ions, also do not have a simple distribution. This is shown in Figure 7, in which electron plasma parameters are shown, as measured by the RPA-COPERNIS experiment aboard Giotto in the ram direction, during the last 6.2 hours before the encounter (Reme 1991). Several boundaries are seen mostly in the electron distribution, this region being called the 'mystery region' because no theoretical model has been devised to account for them (Reme et al. 1987). A similar region was also found when Giotto explored comet Grigg-Skjellerup. The difficulty in understanding all these features – in the opinion of the author – is based on the fact that our concept of the solar wind–planetary interaction is based on simple one-fluid magnetohydrodynamic (MHD) picture. At comets the shocked and loaded solar wind is far from being a one-fluid, MHD-type flow; therefore, the simple picture is not valid. However, no other theoretical tools

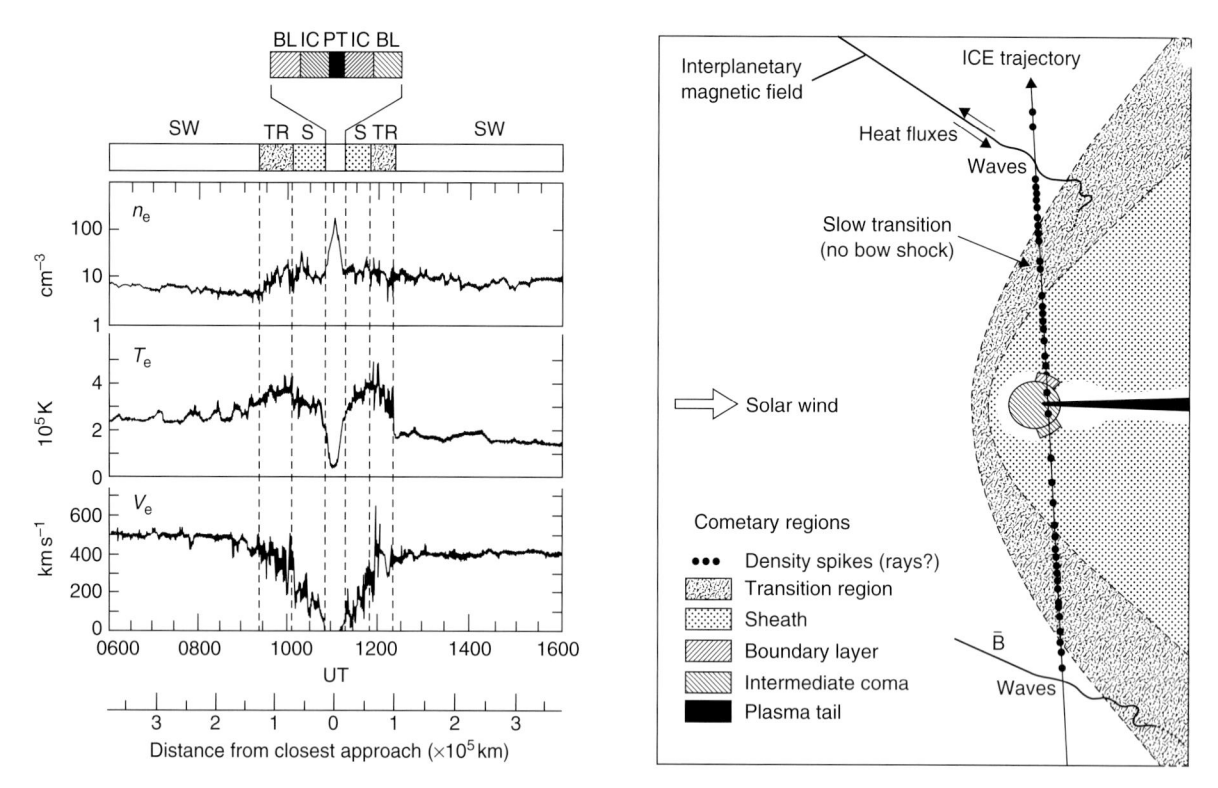

Figure 6 Left: overview plot of three running averages of electron density, temperature and velocity, measured during the encounter of ICE with comet Giacobini-Zinner. Plasma regions identified at the top are SW: solar wind; TR: transition region; S: sheath; BL: boundary layer; IC: cold intermediate coma; and PT: plasma tail. Right: the possible configuration of the plasma layers around the nucleus (After Bame et al. 1986, Figures 1 and 2.)

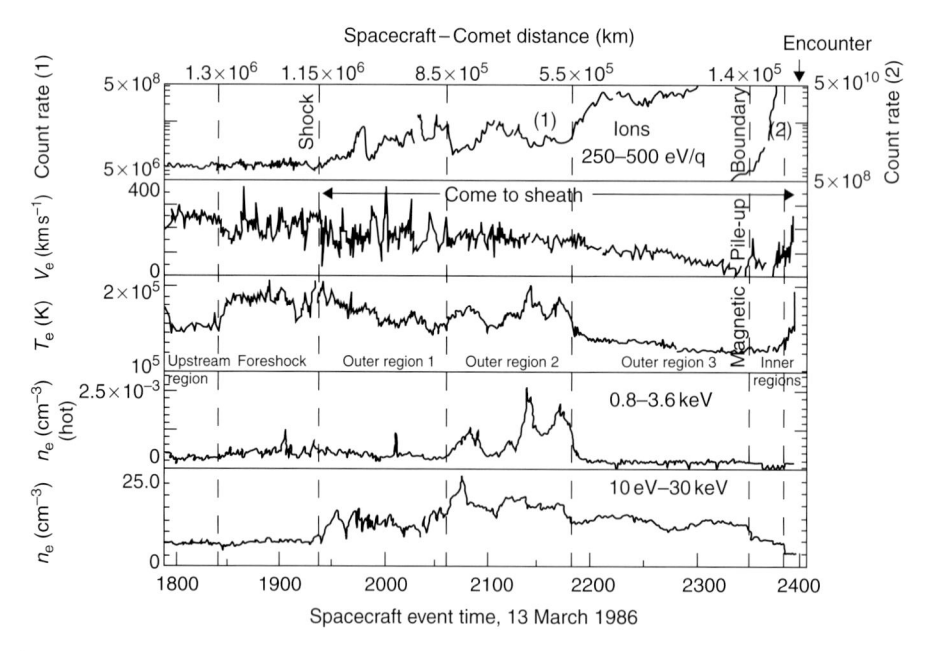

Figure 7 Plasma data collected a board the Giotto probe over a time interval that starts before bow shock crossing and ends at the closest approach to comet Halley. The complicated plasma regions surrounding a cometary nucleus are illustrated. The horizontal axis below the figure is the time (UT) on the day of the encounter. The upper plot shows the count rate of 250–500 eV/q ions. The plots below show electron velocity, temperature and densities for two electron components, respectively. The vertical dashed lines delimit the different plasma regions around the comet; these are identified by the changing properties of the electron and ion data. (After Reme 1991, Figure 1.)

are at hand to model this interaction, because of the physical and numerical complexity involved.

Therefore it is surprising that the one-fluid MHD picture, being a very crude approximation, is still working quite well. A recent comprehensive comparison of the Giotto data and the results of a three-dimensional adaptive MHD model has been published (Israelevich *et al.* 1999); this simulation reasonably reproduced the observed magnetic fields and ion velocities.

Closer to the nucleus a new type of boundary layer was reached. An analogy taken from non-magnetic planetary magnetospheres such as those of Venus and Mars indicates the presence of a special region above the dayside ionosphere where both the shocked solar wind and cometary ions play an equally important role. There are two candidates at comets for the upper boundary of this region: the first is the so-called 'cometopause' observed by the PLAZ-MAG charged particle analyser on board the Vega 2 probe (Vega 1 was not switched on at that location) at about 160,000 km from the nucleus (Gringauz *et al.* 1986b), and the second is the 'magnetic pile-up boundary' detected by the Giotto magnetometer at 135,000 km from the nucleus accompanied by a change in the electron spectrum (Neubauer *et al.* 1986). The cometopause is a boundary where the ion composition of the shocked solar wind changes. Above it protons are the dominant component, below it the cometary ion density takes over. The surprise is the sharpness of the transition, not seen by Giotto. The cometopause is associated with lower hybrid wave excitations, but only the Vega probes carried wave instruments, and no similar data are available from Giotto. The magnetic pile-up boundary is a sharp increase in the magnetic field value, but it was not detected by the Vega magnetometers during the two flybys. Therefore, the nature of these boundaries is still a matter of debate. We note, however, that the solar wind conditions were significantly different during the different flybys.

There might also be indirect confirmation of the existence of a cometopause at other comets. This is related to the cometary X-ray emission, first observed by Lisse *et al.* (1996), followed by several observations by ROSAT and EUVE. This confirmed that X-ray and UV emissions are general properties of comets; the origin of this radiation is a large volume on the dayside, inside the bow shock, as shown in Figure 8. It is generally believed that the emission is the result of the interaction of the solar wind with cometary neutrals, and the most likely scenario is that, due to charge exchange between high Z solar wind minority ions and cometary neutrals, highly excited atoms are produced that emit line radiation in the UV and soft X-ray energy range (Cravens 1997). There is also, however, another mechanism that can produce bremsstrahlung and K-shell radiation at the cometopause, which is due to lower

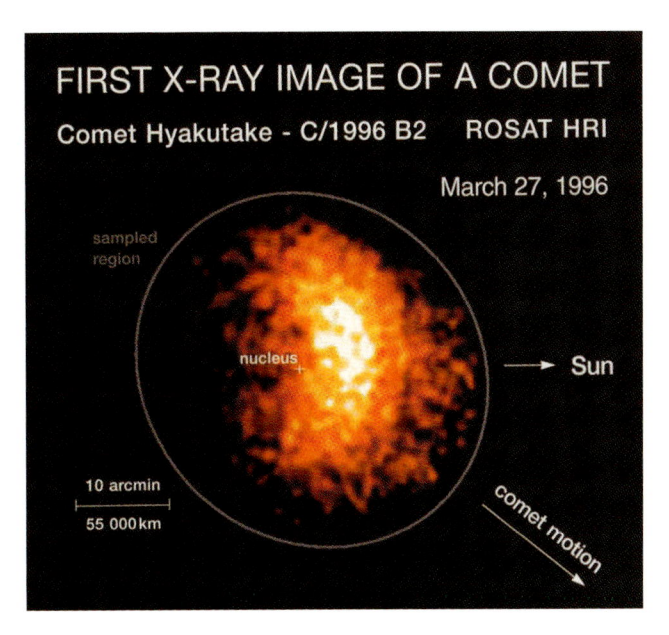

Figure 8 First observation of X-ray emission from comet Hyakutake. (From COSPAR Information Bulletin, No. 136, p. 2, August 1996, Elsevier Science; courtesy of NASA Goddard Space Flight Center.)

hybrid waves observed there. These waves are very efficient accelerators of electrons, and these accelerated electrons, in turn, may hit neutrals exciting K-shell X-ray and bremsstrahlung radiation (Shapiro *et al.* 1999). At the cometopause the accelerated electrons are the most effective source of X-ray emission. The observation of the soft X-ray and UV radiation of comets was one of the most surprising discoveries, resulting mainly from the observations of space-based telescopes.

CLOSER TO THE NUCLEUS

The region between the magnetic cavity and the cometopause is also very much structured; the draped magnetic field directions during the Giotto encounter are shown in Figure 9, in the 60,000 km vicinity of the nucleus of comet Halley (Neubauer 1991). The plasma here comprises of the shocked, warm solar wind loaded with cometary ions from the more distant regions, and a broad variety of energetic and cold cometary ions and electrons. Collisions become more important, and charge exchange plays a significant role in ionization. Ion-neutral friction, resistivity, thermal cooling and chemical reactions cannot be neglected. Sometimes it is customary to talk of the 'collisionopause', the upper boundary of the region where collisions cannot be neglected. The collisionopause boundary for the different

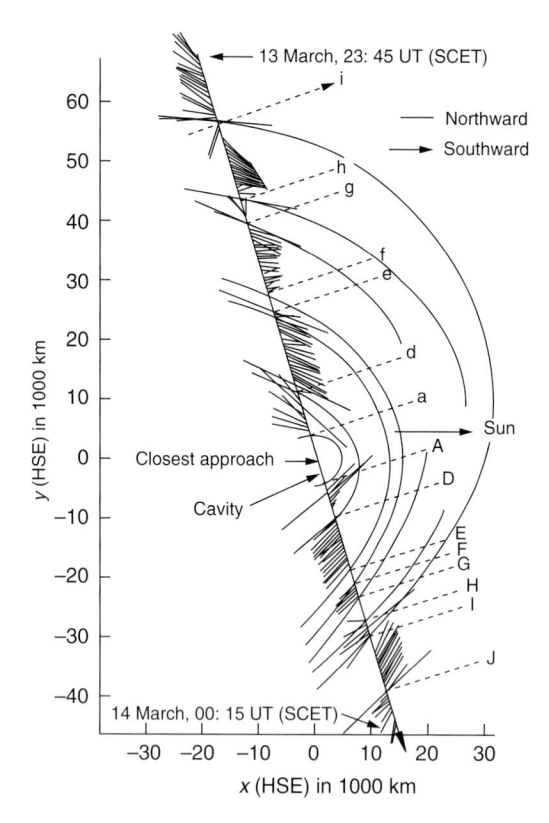

Figure 9 The structure of the magnetic field directions around the nucleus of comet Halley during the Giotto flyby. The discontinuities in the field directions before the encounter are denoted by lower-case letters and the corresponding ones after by upper-case letters. The possible shapes of the discontinuities are also indicated. (After Neubauer 1991, Figure 5.)

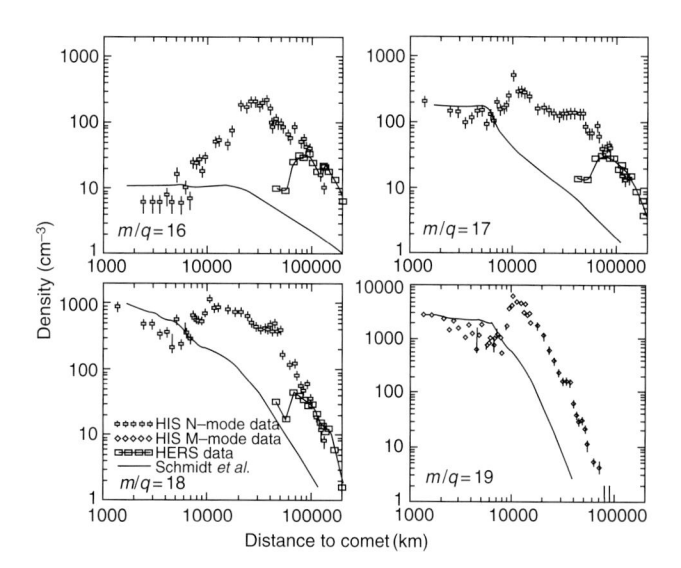

Figure 10 Ion density profiles for mass/charge of 16, 17, 18 and 19 amu/e as a function of distance from the nucleus. Stars and crosses indicate data from the IMS/HIS mass analyser, squares indicate data from the IMS/HER spectrometer and the solid lines represent theoretical data from an MHD model. (After Altwegg et al. 1993, Figure 1.)

processes has different locations; for charge exchange the collisionopause more or less coincides with the cometopause/magnetic pile-up layers.

We have learned from the investigation of the dayside mantle of non-magnetic planets that these regions are also structured. Specific features in certain plasma parameters (such as peaks, jumps, etc.) are not co-located, but rather ordered in altitude going away from the planets (Strangeway and Russell 1996, Szegö et al. 1998). By analogy, we cannot expect co-located structures in the different plasma parameters in this cometary region.

We describe the Halley flyby results next, following mostly Reme (1991). In this region the solar wind further decelerates, gets more and more loaded, and at about 20,000–40,000 km from the nucleus it disappears both from the field of view and the energy range of all the charged particle detectors. The ions, outflowing from the nucleus, leave the magnetic cavity with a radial velocity of about $1\,\mathrm{km\,s^{-1}}$. Outside the cavity the velocities of all the different ion components drop to close to zero, the densities are increased by a factor of about 10 and an ion pile-up region is formed.

This stagnation region extends to about 20,000 km from the nucleus. Its discovery was totally unexpected. Around other Solar System bodies no similar stagnation region was observed in analogous regions. The stagnation and ion pile-up regions are definitely permanent cometary features, because both the Vega (Gringauz et al. 1986) and the Giotto (Balsiger et al. 1986) instrumentation detected them. The PLAZMAG spectrometer aboard Vega 2 detected only cold cometary ions from the ram directions at about 15,000 km from the nucleus; the proton spectra clearly indicated that both the thermal and the bulk velocity of the H^+ component was low. This enabled PLAZMAG to be used as a mass spectrometer, and H^+, C^+, CO_2^+ and Fe^+ ions were identified with confidence. During the Giotto encounter the data obtained by the HIS sensor of the ion mass spectrometer allowed a much finer analysis of this region (Altwegg et al. 1993). The ion density profiles for mass/charge of 16, 17, 18 and 19 amu/e as a function of distance from nucleus are shown in Figure 10. Comparison with the MHD model of Schmidt et al. (1988), denoted by solid lines in the figure, clearly shows that MHD models cannot account for this. The steep increase of the ion temperatures outside the magnetic cavity in the pile-up region has led Altwegg et al. (1993) to conjecture that due to the elevated temperature the ion reaction rates could be modified, hence the apparent pile-up would be the result of 'chemical' effects. However, as the ion radial velocities are around zero, dynamic causes might well also be possible.

The pile-up region is structured in itself, different ions pile-up differently, and, for example, the $M/Q = 16$ ions had three peaks inside (Balsiger *et al.* 1987), as measured by IMS-HERS on Giotto. Before the pile-up region (that is, closer to the nucleus) the density drops as $1/r$. As follows from eqn (4) for small r values, after the pile-up region it drops as r^{-2}. In the ion stagnation region the ions are decoupled from the neutrals which are accelerated there. The total electron density in the $10\,eV–30\,keV$ energy range has three major discontinuities. The ion pile-up terminates at about 20,000 km from the nucleus, and the ion radial velocity starts to grow to a few kilometers per second on leaving this region going outward.

The planetary analogy of the magnetic cavity around the nucleus is the ionosphere. However, whereas at the ionopause the thermal pressure of the ionosphere balances the magnetic pressure, and the ionospheric ion distribution falls sharply at the boundary (that is, the ionospheric ions 'pause'), the physics at the cavity boundary is different (Cravens 1991). The ionospheric thermal pressure is replaced by ion-neutral drag force, since the cometary ionosphere is tenuous and cold. This force is the result of collisions between the outward accelerating neutrals and ions. It transfers momentum to the ions, hence creating kinetic pressure to balance the magnetic pressure. As the cavity is relatively small in size, the force due to the curvature of the magnetic field may also contribute to the magnetic pressure. Global MHD simulations indicate that the cavity is very likely tear-drop in shape and elongated in the direction of the tail. Accordingly, along the Giotto trajectory the curvature played a more important role outbound than inbound.

It is an important difference with respect to planetary ionospheres that the cometary ions do not pause at the cavity boundary, and the total ion density is observed to vary as $1/r$. Chemistry very likely plays a more important role here. The HIS sensor, carried aboard Giotto, revealed the details of the ion composition and the complicated chemical reactions taking place both inside and the near-outside vicinity of the cometary ionosphere; however, chemistry is beyond the scope of this chapter.

The plasma, flowing behind the nucleus, forms the plasma tail. The broad varieties of tail plasma structures, however, are beyond the scope of this chapter.

CONCLUSIONS

Comets are intriguing Solar System bodies, and our knowledge of them is very much incomplete. In the description given here we have concentrated on our lack of knowledge and on some of the controversies present in the current literature. We do not believe that these can be resolved without *in situ* observations. Astronomical observations might provide very useful hints and ideas that grasp the very essence of the questions (such as the dirty snowball model itself), but remote sensing cannot prove them conclusively.

We need to fly there and land.

REFERENCES

A'Hearn, M.F., Hoban, S., Birch, P.V., Bowers, C. and Klingesmith, D.A. (1986). Cyanogen jets in comet Halley. *Nature*, **324**, 649–651.

Altwegg, K., Balsiger, H., Geiss, J., Goldstein, R., Ip, W-H., Meier, A., Neugebauer, M., Rosenbauer, H. and Shelley, E. (1993). The ion population between 1300 km and 230 000 km in the coma of comet P/Halley. *Astronomy and Astrophysics*, **279**, 260–266.

Balsiger, H., Altwegg, K., Bühler, F., Geiss, J., Ghielmetti, A.G., Goldstein, B.E., Goldstein, R., Huntress, W.T., Ip, W.-H., Lazarus, A.J., Meier, A., Neugebauer, M., Rettenmund, U., Rosenbauer, H., Schwenn, R., Sharp, R.D., Shelly, E.G., Ungstrup, E. and Young, D. T. (1986). Ion composition and dynamics at comet Halley. *Nature*, **321**, 330–334.

Balsiger, H., Altwegg, K., Bühler, F., Fuselier, S.A., Geiss, J., Goldstein, B.E., Goldstein, R., Huntress, W.T., Ip, W.H., Lazarus, A.J., Meier, A., Neugebauer, M., Rettenmund, U., Rosenbauer, H., Schwenn, R., Shelley, E.G., Ungstrup, E. and Young, D.T. (1987). The composition and dynamics of cometary ions in the outer coma of comet P/Halley. *Astronomy and Astrophysics*, **187**, 163.

Bame, S.J., Anderson, R.C., Asbridge, J.R., Baker, D.N., Feldman, W.C., Fuselier, S.A., Gosling, J.T., McComas, D.J., Thomsen, M.F., Young, D.T. and Zwickl, R.D. (1986). Comet Giacobini-Zinner: plasma description. *Science*, **232**, 356–361.

Boehnhardt, H. (1989). Clusters and packets of grains in comet Halley and the fragmentation of dust. *Earth, Moon, and Planets*, **46**, 221.

Caso, C., Conforto, G., Gurtu, A., Aguilar-Benitez, M., Amsler, C., Barnett, R.M., Burchat, P.R., Carone, C.D., Dahl, O., Doser, M., Eidelman, S., Feng, J.L., Goodman, M., Grab, C., Groom, D.E., Hagiwara, K., Hayes, K.G., Hernandez, J.J., Hikasa, K., Honscheid, K., James, F., Mangano, M.L., Manohar, A.V., Mönig, K., Murayama, H., Nakamura, K., Olive, K.A., Piepke, A., Roos, M., Schindler, R.H., Shrock, R.E., Tanabashi, M., Törnqvist, N.A., Trippe, T.G., Vogel, P., Wohl, C.G., Workman, R.L., Yao, W.-M., Armstrong, B., Casas Serradilla, J.L., Filimonov, B.B., Gee, P.S., Lugovsky, S.B., Mankov, S., Nicholson, F., Babu, K.S., Besson, D., Biebel, O., Cahn, R.N., Crawford, R.L., Dalitz, R.H., Damour, T., Desler, K., Donahue, R.J., Edwards, D.A., Erler, J., Ezhela, V.V., Fass, A., Fetscher, W., Froidevaux, D., Gaisser, T.K., Garren, L., Geer, S., Gerber, H.-J., Gilman, F.J., Haber, H.E., Hagmann, C., Hinchliffe, I., Hogan, C.J., Höhler, G., Jackson, J.D., Johnson, K.F., Karlen, D., Kayser, B., Kleinknecht, K., Knowles, I.G., Kolda, C., Kreitz, P., Langacker, P., Landua, R., Littenberg, L., Manley, D.M., March-Russell, J., Nakada, T., Quinn, H., Raffelt, G., Renk, B., Ronan, M.T., Rosenberg, L.J., Schmitt, M., Schramm, D.N., Scott D., Sjöstrand, T., Smoot, G.F., Spanier, S., Srednicki, M., Stanev, T., Suzuki, M., Tkachenko, N.P., Valencia, G., van Bibber, K., Voss, R., Wolfenstein, L. and Youssef, S. (1998). Review of particle physics. *The European Physical Journal C*, **3**, 1–794, see also http://pdg.lbl.gov/

Clairemidi, J., Moreels, G. and Krasnopolsy, V.A. (1990). Spectro-imagery of P/Halley's inner coma in the OH and NH ultraviolet bands. *Astronomy and Astrophysics*, **231**, 235–240.

Cosmovici, C.B., Schwarz, G., Ip, W-H. and Mack, P. (1988). Gas and dust jets in the inner coma of comet Halley. *Nature*, **332**, 705–709.

Cravens, T.E. (1991). Plasma processes in the inner coma. In R.L. Newburn, M. Neugebauer and J. Rahe (eds), *Comets in the Post-Halley Era*, vol. 2, Kluwer, Dordrecht, pp. 1211–1258.

Cravens, T.E. (1997). Comet Hyakutake X ray source: Charge transfer of solar wind heavy ions. *Geophysical Research Letters*, **24**, 105–109.

Crifo, J.F., Lagerros, J., Rodionov, A.V. and Szegö, K. (2000). Comet Halley nucleus shape and activity I. First attempt at interpreting the *in situ* observed (1986) near-nucleus coma on the basis of a plausible nucleus shape model and of a three-dimensional coma gasdynamic model. *Submitted*.

Crifo, J.F. and Rodionov, A.V. (1999a). Modelling the circumnuclear coma of comets: Objectives, methods and recent results. *Planetary and Space Science*, **47**, 797–826.

Crifo, J.F. and Rodionov, A.V. (1999b). Modelling the surface activity of cometary nuclei. *Publications of the Astronomical Society of the Pacific*, in press.

Crifo, J.F., Rodionov, A.V. and Bockelee-Morvan, D. (1999). The dependence of the circumnuclear coma structure on the properties of the nucleus III. First modelling of a CO-dominated coma, with application to P/Wirtanen beyond 3 AU from the Sun. *Icarus*, **138**, 85–106.

Enzian, A., Klinger, J., Schwehm, G. and Weissman, P.R. (1999). Temperature and gas production distribution of the surface of a spherical model comet nucleus in the orbit of 46P/Wirtanen, *Icarus*. **138**, 74–78.

Festou, M.C. (1999). On the existence of distributed sources in comet comae. *Space Science Reviews*, **90**, 53–67.

Fomenkova, M.N. and Mendis, D.A. (1992). A note on the very small grains (VSGs) observed at Halley's comet. *Astrophysics and Space Science*, **189**, 327.

Glassmeier, K-H., Tsurutani, B.T. and Neubauer, F.M. (1997). Adventures in the parameter space, a comparison of low-frequency plasma waves at comets. In T. Hada and H. Matsumoto (eds), *Nonlinear Waves on Chaos in Space Plasmas*, Terra Science, Tokyo, pp. 77–113.

Gombosi, T.I. (1994). *Gaskinetic Theory*. Cambridge University Press.

Greenberg, J.M., Mizutani, H. and Yamamoto, T. (1995). A new derivation of the tensile strength of cometary nuclei: application to comet Shoemaker-Levy 9. *Astronomy and Astrophysics*, **295**, L35.

Gringauz, K.I., Gombosi, T.I., Remizov, A.P., Apathy, I., Szemerey, I., Verigin, M.I., Denchikova, L.I., Dyachkov, A.V., Keppler, E., Klimenko, I.N., Richter, A.K., Somogyi, A.J., Szegö, K., Szendro, S., Tatrallyay, M., Varga, A. and Vladimirova, G.A. (1986a). First *in situ* plasma and neutral gas measurements at comet Halley. *Nature*, **321**, 282–285.

Gringauz, K.I., Verigin, M.I., Remizov, A.P., Gombosi, T.I. and Tatrallyay, M. (1986b). Detection of a new 'chemical' boundary at Comet Halley. *Geophysical Research Letters*, **13**, 613–616.

Harmon, J.K., Ostro, S.J., Benner, L.A.M., Rosema, K.D., Jurgens, R.F., Winkler, R., Yeomans, D.K., Choate, D., Cormier, R., Giorgini, J.D., Mitchell, D.L., Chodas, P.W., Rose, R., Kelley, D., Slade, M.A. and Thomas, M.L. (1997). Radar detection of the nucleus and coma of comet Hyakutake. *Science*, **278**, 1921.

Horanyi, M. (1996). Charged dust dynamics in the solar system. *Annual Review of Astronomy and Astrophysics*, **34**, 383–418.

Horanyi, M. Gombosi, T.I., Cravens, T.E., Korosmezey, A., Kecskemety, K., Nagy, A.F. and Szegö, K. (1984). The friable sponge model of a cometary nucleus. *Astrophysical Journal*, **278**, 449–455.

Israelevich, P.L., Gombosi, T.I., Ershkovich, A.I., DeZeeuw, D.L., Neubauer, F.M. and Powell, K.G. (1999). The induced magnetosphere of comet Halley 4. Comparison of *in situ* observation and numerical simulation. *Journal of Geophysical Research*, **104**, 28, 309.

Johnstone, A. (ed) (1991). *Cometary Plasma Processes*, Geophysical Monograph Series 61, American Geophysical Union, Washington, DC.

Juhasz, A. and Szegö, K. (1997). Charged dust dynamics above the surface of a comet far from the Sun. *Journal of Geophysical Researches*, **103**, 12015.

Keller, H.U., Curdt, W., Kramm, J-R. and Thomas, N. (1994). *Images of the Nucleus of Comet Halley*, vol. 1, ESA SP-1127, ESTEC, Noordwijk.

Klinger, J., Levasseur-Regourd, A-C., Bouziani, N. and Enzian, A. (1996). Towards a model of cometary nuclei for engineering studies for future space missions to comets. *Planetary and Space Science*, **44**, 637–653.

Lämmezahl, P., Krankowsky, D., Hodges, R.R., Stubbemann, U., Woweries, J., Herrwerth, I., Berthelier, J.J., Illiano, J.M., Eberhardt, P., Dolder, D., Schulte, W. and Hoffman, J.H. (1987). Expansion velocity and temperatures of gas and ions measured in the coma of comet P/Halley. *Astronomy and Astrophysics*, **187**, 169–173.

Lisse, C.M., Dennerl, K., Englhauser, J., Harden, M., Marshall, F.E., Mumma, M.J., Petre, R., Pye, J.P., Ricketts, M.J., Schmitt, J., Trumper, J. and West, R.G. (1996). Discovery of X-ray and extreme UV emission from Comet C/Hyajutake. *Science*, **274**, 205–209.

Mazets, E.P. *et al.* (1986). Comet Halley dust environment from SP-2 detector measurements. *Nature*, **321**, 276.

Mazets, E.P. *et al.* (1987). Dust in comet P/Halley from Vega observations. *Astronomy and Astrophysics*, **187**, 699.

McDonnell, J.A.M., Evans, G.C., Evans, S.T., Alexander, W.M., Burton, W.M., Firth, J.G., Bussoletti, E., Grard, R.J.L., Hanner, M.S. and Sekanina, Z. (1987). The dust distribution within the inner coma of comet P/Halley: Encounter by Giotto's impact detectors. *Astronomy and Astrophysics*, **187**, 719–741.

McDonnell, J.A.M. and Pankiewitz, G.S. (1990). Comet Halley's dusty coma: *In situ* exploration with dust impact detector. In J. Mason (ed), *Comet Halley, Investigations, Results, Interpretations*, Ellis Horwood, New York, pp. 15–32.

Mendis, D.A. and Brin, G.D. (1977). Monochromatic brightness variations of comets. II – Core–mantle model. *Moon*, **17**, 359–372.

Mendis, D.A., Hill, J.R., Houpis, H.L.F. and Whipple, E.C. (1981). On the electrostatic charging of the cometary nucleus. *Astrophysical Journal*, **249**, 787–797.

Merenyi, E., Foldy, L., Szegö, K., Toth, I. and Kondor, A. (1990). The landscape of comet Halley. *Icarus*, **86**, 9–20.

Mohlmann, D. (1999). Activity and nucleus properties of 46 P/Wirtanen. *Planetary and Space Science*, **47**, 971.

Moore, M.H., Donn, B., Khanna, R. and A'Hearn, M.F. (1983). Studies of proton-irradiated cometary-type ice mixtures. *Icarus*, **54**, 388–405.

Mukai, T., Miyake, W., Terasawa, T., Kitayama, M. and Hirao, K. (1986). Plasma observation by Suisei of solar-wind interaction with comet Halley. *Nature*, **321**, 299–303.

Neubauer, F.M. (1991). The magnetic field structure of the cometary plasma environment. In R.L. Newburn, M. Neugebauer and J. Rahe (eds), *Comets in the Post-Halley Era*, Kluwer, Dordrecht. pp. 1107–1124.

Neubauer, F.M., Glassmeier, K.H., Pohl, M., Raeder, J., Acuna, M.H., Burlaga, L.F., Ness, N.F., Musmann, G., Mariani, F., Wallis, M.K., Ungstrup, E. and Schmidt, H.U. (1986). First results from the Giotto magnetometer experiment at Comet Halley. *Nature*, **321**, 352–355.

Newburn, R.L., Neugebauer, M. and Rahe, J. (eds) (1991). *Comets in the Post-Halley Era*, Kluwer, Dordrecht.

Oberc, P. (1996). Disintegration of dust aggregates as origin of the boundaries in Halley's coma: derivation of the sublimation parameters. *Icarus*, **124**, 195.

Podolak, M. and Prialnik, D. (1996). Models of the structure and evolution of comet P/Wirtanen. *Planetary and Space Science*, **44**, 655–664.

Probstein, R.F. (1969). The dusty gasdynamics of comet heads. In M.A. Lavrentiev (ed), *Problems of Hydrodynamics and Continuum Mechanics*, Society for Industrial and Applied Mathematics, Philadelphia, pp. 568–583.

Reme, H., Sauvaud, J.A., D'Uston, C., Cros, A., Anderson, K.A., Carlson, C.W., Curtis, D.W., Lin, R.P., Korth, A., Richter, A.K. and Mendis, D.A. (1987). General features of comet P/Halley: Solar wind interaction from plasma measurements. *Astronomy and Astrophysics*, **187**, 33.

Reme, H. (1991). Cometary plasma observations between the shock and contact surface. In A. Johnstone (ed), *Cometary Plasma Processes*.

Geophysical Monograph Series 61, American Geophysical Union, Washington, DC, pp. 87–105.

Richter, K., Curdt, W. and Keller, H.U. (1991). Velocity of individual large dust particles ejected from comet P/Halley. *Astronomy and Astrophysics*, **250**, 548.

Rickman, H. (1991). The thermal history and structure of cometary nuclei. In R.L. Newburn, M. Neugebauer and J. Rahe (eds), *Comets in the Post-Halley Era*, Vol. 2, Kluwer, Dordrecht, pp. 733–760.

Rodionov, A.V., Crifo, J.F., Szegö, K., Lagerros, J. and Fulle, M. (2002). An advanced model of cometary activity: Description and examples of application to comets Hyakutake and Halley. *Planetary and Space Science*, submitted.

Sagdeev, R.Z. *et al.* (1987). The spatial distribution of dust jets seen during the Vega 2 flyby. *Astronomy and Astrophysics*, **187**, 293.

Sagdeev, R.Z., Evlanov, E.N., Zubkov, B.V., Prilutskii, O.F. and Fomenkova, M.N. (1990). Detection of very fine particles near the nucleus of comet Halley. *Sov. Astron. Lett.*, **16**, 315.

Salo, H. (1988). Monte-Carlo modelling of the net effects of coma scattering and thermal reradiation of the energy inputs to cometary nucleus. *Icarus*, **76**, 253–269.

Shapiro, V.D., Bingham, R., Dawson, J.M., Dobe, Z., Kellet, B.J. and Mendis, D.A., (1999). Energetic electrons produced by lower hybrid waves in the cometary environment and soft X-ray emission: Bremstrahlung and K-shell radiation. *Journal of Geophysical Research*, **104**, 2537–2554.

Shimizu, M. (1991). The hydrogen clouds of comets. In R.L. Newburn, M. Neugebauer and J. Rahe (eds), *Comets in the Post-Halley Era*, vol. 2, Kluwer, Dordrecht, pp. 897–905.

Simpson, J.A. *et al.* (1986). Dust counter and mass analyser (DUCMA) measurements of comet Halley's coma from Vega spacecraft. *Nature*, **321**, 278.

Simpson, J.A., Rabinowitz, D., Tuzzolino, A.J., Ksanfomality, L.V. and Sagdeev, R.Z. (1987). The dust coma of comet P/Halley: measurements on the Vega-1 and Vega-2 spacecraft. *Astronomy and Astrophysics*, **187**, 742.

Simpson, J.A., Tuzzolino, A.J., Ksanfomality, L.V. and Sagdeev, R.Z. (1989). Ducma measurements of comet Halley dust mass spectra based on post-encounter dust calibrations. In *Asteroids, Comets, Meteors III.*, Proc. of the Uppsala conference, p. 345.

Schmidt, H.U., Wegmann, R., Huebner, W.F. and Boyce, D.C. (1988). Cometary gas and plasma flow with detailed chemistry. *Computer Physics Communications*, **49**, 17–59.

Strangeway, R.J. and Russell, C.T. (1996). Plasma waves and field aligned currents in the Venus plasma mantle. *Journal of Geophysical Research*, **101**, 17313–17324.

Szegö, K., Sagdeev, R.Z., Whipple, F.L., Abergel, A., Bertaux, J.-L., Merenyi, E., Szalai, S. and Varhalmi, L. (1995). *Images of the Nucleus of Comet Halley,* vol. 1, ESA SP-1127, ESTEC, Noordwijk.

Szegö, K., Klimov, S., Kotova, G.A., Livi, S., Rosenbauer, H., Skalsky, A. and Verigin, M.I. (1998). On the dayside region between the shocked solar wind and the ionosphere of Mars. *Journal of Geophysical Research*, **103**, 9101–9111.

Szegö, K., Glassmeier, K.-H., Bingham, R., Bogdanov, A., Fischer, C., Haerendel, G., Brinca, A., Cravens, T., Dubinin, E., Sauer, K., Fisk, L., Gombosi, T., Schwadron, N., Isenberg, P., Lee, M., Mazelle, C., Möbius, E., Motschmann, U., Shapiro, V.D., Tsurutani, B. and Zank, G. (2000). Physics of mass loaded plasmas. *Space Science Reviews*, **94**, 429–671.

Thiel, K., Kölzer, G., Kochan, H., Lämmerzahl, P. and Lorenz, E. (1995). Phenomenology and dynamic behavior of the dust component in the KOSI experiments. *Planetary and Space Science*, **43**, 375.

Vaisberg, O.L. *et al.* (1986). Dust coma structure of comet Halley from SP-1 detector measurements. *Nature*, **321**, 274.

Vaisberg, O.L., Smirnov, V., Omelchenko, A., Gorn, L. and Iovlev, M. (1987). Spatial and mass distribution of low-mass dust particles ($m < 10^{-10}$ g) in comet P/Halley's coma. *Astronomy and Astrophysics*, **187**, 753.

Vaisberg, O.L. (1990). The dust coma structure of comet Halley. In J. Mason (ed), *Comet Halley: Investigations, Results, Interpretations*, Ellis Horwood, New York, pp. 33–44.

Whipple, F.L. (1950). A comet model. I. The acceleration of comet Encke. *Astrophysical Journal*, **111**, 375–394.

Zank, G.P., Khabibrakhmanov, I.Kh. and Story, T.R. (1993). The structure of mass loading shocks. *Journal of Geophysical Research*, **98**, 5645–5649.

HORST UWE KELLER* AND LAURENT JORDA*

The morphology of cometary nuclei

The sudden appearance of a bright comet stretching over a large part of the night sky must have been one of the most awesome phenomena for early humans watching the sky. The nature of comets remained obscure well into the Middle Ages. Only with the introduction of astronomical techniques and analyses in Europe was the parallax of a comet determined by Tycho Brahe for the first time. He proved that comets are not phenomena of the Earth's atmosphere but are farther away than the Moon; in other words they are interplanetary objects. Later Kepler first predicted that comets follow straight lines, then Hevelius suggested parabolic orbits roughly a hundred years later. It was Halley who suggested that the comets of the years 1531, 1607 and 1682 were apparitions of one and the same comet that would return again in 1758. The success of this prediction made it clear that comets are members of our Solar System.

While it was now established that periodic comets are objects of the planetary system, their origin and nature continued to be debated. Were they formed together with the planets from the solar nebula (Kant) or were they of extra-solar origin as suggested by Laplace? This debate lasted for 200 years until well into the second half of the last century. Öpik (1932) suggested that a cloud of comets surrounded our Solar System. This hypothesis was quantified and compared to the observed distribution of orbital parameters (essentially the semi-major axes) of new comets by Oort (1950) (Section 2.1). Comets are scattered into the inner Solar System by perturbations caused by galactic tides, passing stars and large molecular clouds.

The Oort cloud would have a radius of 2×10^5 AU, a dimension comparable to the distances of stars in our neighbourhood. The lifetime (limited by decay due to activity and by perturbations caused by encounters with planets) even of the new comets on almost parabolic orbits and typical periods of the order of 10^6 years is short compared to the age of the planetary system (4.5 Gy). Therefore, observed comets could only recently have arrived on their orbits dipping inside the inner Solar System.

This reservoir of comets must have been established during the formation process of the planetary system itself. Cometesimals were agglomerated from interstellar/interplanetary gas and dust and scattered out of the inner Solar System by the giant outer planets (Section 2.3). This scheme implies that a central part of a comet, its nucleus, is stable enough to survive these perturbations. It must also be stable enough to pass the vicinity of the sun for many times in the case of a short-period comet.

Comets are bright and large when they are close to the sun and fade quickly when they recede beyond about 2 AU. Only with the advent of photography and large astronomical telescopes could a comet be followed until it becomes a star-like point source. What makes comets active near the Sun, blowing their appearances up to the order of 10^5 km? Bright comets often develop tails two orders of magnitude longer.

In an attempt to explain the cometary appearance, Bredichin (1903) introduced a mechanical model where repulsive forces drive the particles away from a central condensation. Spectroscopy revealed that dust grains reflect the solar irradiation. In addition, simple molecules, radicals and ions were found as constituents of the cometary coma and tail. The nature of the central condensation remained mysterious for a long time because of the observational dilemma. When the comet is close to the Earth and therefore to the Sun the dense coma obscures the view into its centre. When activity recedes the comet is too far away and too dim for detailed observations of its central condensation. During the middle of the nineteenth century the connection between

---

*Max-Planck-Institut für Aeronomie, Katlenburg-Lindau, Germany

comets and meteor streams was established. Schiaparelli (1866) calculated the dispersion of cometary dust within the orbital plane. From this time on the perception that the central condensations of comets were agglomerations of dust particles prevailed for about a century. The gas coma was explained by desorption of molecules from dust particles with large surfaces (Levin 1943). The storage of highly reactive radicals (most observed species (CN, CH, NH_2, etc.) were of this category) posed a major difficulty to be explained. The inference that these radicals should be dissociation products of stable parent molecules (such as $(CN)_2$, CH_4, NH_3, etc.) by Wurm (1934, 1935, 1943) led to our present understanding that these molecules are stored as ices within the central nucleus of a comet. Whipple (1950a,b) combined the astrometrical observations of changes of the orbital periods of comets with the existence of an icy cometary nucleus. The sublimation of ices cause reactive (rocket) non-gravitational forces that increase or decrease the orbital period of an active comet according to the sense of rotation of its nucleus.

Evidence in support of the icy conglomerate nucleus became more and more compelling by the derived high gas production rates that could not be stored by adsorption on dust grains (Biermann and Trefftz 1964, Huebner 1965, Keller 1976a,b) and by the same account by the large quantities of dust moving into the cometary tail (Finson and Probstein 1968b). The 'sand bank' model (Lyttleton 1953) was clearly dismissed in favour of a solid icy nucleus. Its formation and origin could now be explored.

While there was some knowledge about the chemical composition of the nucleus, its physical properties, even the basic ones like size, shape and mass, remained largely unknown because the nucleus could not be observed. Early attempts to derive the nucleus size from the 'nuclear' magnitudes of comets at large heliocentric distances while they are inactive (Roemer 1966a,b) led to a systematic overestimation of the size because their residual activity could not be eliminated.

The advent of modern detectors and large ground-based telescopes revealed that most comets display residual activity or clouds of dust grains around their nuclei. Taking the residual signal into account (mostly using simple models for the brightness distribution) the size estimates of the nuclei could be improved. The (nuclear) magnitude of a comet depends on the product of its albedo and cross-section. Only in a few cases could the albedo and size of a cometary nucleus be separated by additional observation of its thermal emission at infrared wavelengths. By comparison with outer Solar System asteroids Cruikshank *et al.* (1985) derived a surprisingly low albedo of about 0.04. A value in clear contradiction to the perception of an icy surface but fully confirmed by the first resolved images of a cometary nucleus during the flybys of the Vega and Giotto spacecraft of comet Halley (Sagdeev *et al.* 1986, Keller *et al.* 1986).

The improvements of radar techniques led to the detection of reflected signals and finally to the derivation of nuclear dimensions and rotation rates. The observations, however, are also model dependent (rotation and size are similarly interwoven as are albedo and size) and sensitive to large dust grains in the vicinity of a nucleus. As an example, Kamoun *et al.* (1982) determined the radius of comet Encke to 1.5 (+2.3, −1.0) km using the spin axis determination of Whipple and Sekanina (1979).

The superb spatial resolution of the Hubble Space Telescope (HST) is not quite sufficient to resolve a cometary nucleus. The intensity distribution of the inner coma, however, can be observed and extrapolated toward the nucleus based on models of the dust distribution. If this contribution is subtracted from the central brightness the signal of the nucleus can be derived and hence its product of albedo times cross-section (Lamy and Toth 1995, Rembor 1998, Keller and Rembor 1998; Section 4.3).

It has become clear that cometary nuclei are dark, small, often irregular bodies with dimensions ranging from about a kilometre (comet Wirtanen, the target of the Rosetta comet rendezvous mission) to about 50 km (comet Hale-Bopp, comet P/Schwassman-Wachmann 1). Their albedos are very low, about 0.04. Their shapes are irregular, axes ratios of 2:1 are often derived. Even though comets are characterized by their activity, in most cases only a small fraction of the nuclear surface (in some cases less than 1%) is active. An exception seems to be comet P/Wirtanen where all its surface is required to be active in order to explain its production rates (Rickman and Jorda 1998). The detection of trans-Neptunian objects (TNOs) in the Kuiper belt (Jewitt and Luu 1993) reveals a new population of cometary bodies with dimensions an order of magnitude bigger (100 km and larger) than the typical comet observed in the inner planetary system. Little is known about the extent, density, size distribution and physical characteristics of these objects. This region is supposedly the reservoir for short-period comets, manly those controlled by Jupiter (Jupiter family comets).

Our present concept of a cometary nucleus has been strongly influenced by the first pictures of the nucleus of comet Halley achieved during the Giotto flyby in 1986. While this revelation seems to be confirmed as typical by modern observations it carries the danger of prototyping new observational results and inferences. Missions and spacecraft are already on their way (Deep Space, Contour, Stardust, Deep Impact) or in preparation (Rosetta) to diversify our knowledge.

The morphology of cometary nuclei is determined by their formation process in the early solar nebula, their dynamics and evolution. The physics of the processes leading to their apparent activity while approaching the Sun are still obscure in many details but determine the small- and intermediate-scale morphology. The large-scale morphology, the shape, of a cometary nucleus is determined by its fragility and inner structure and by its generally complex rotational state. These topics will be reviewed in the following sections. Chemical and compositional aspects will be only discussed where they are important in the framework of the physical evolution of cometary nuclei. More details are given in Chapter 53. A brief survey of the current modelling efforts is given. The fate of cometary nuclei and their decay products follows. A summary and outlook ends this chapter on the morphology of cometary nuclei.

1 FROM DUST TO COMETS

1.1 Planet formation in the early Solar System

In Whipple's icy conglomerate model of a cometary nucleus, the ices were mixed with differentiated refractory matter deduced by analogy to meteoritic materials. The nucleus could be quite inhomogeneous, and large refractory boulders or even a refractory core were conceivable (Sekanina 1972). About 20 years after Whipple's model of a solid nucleus the concept that planet formation was triggered by gravitational instabilities of the dust component of the rotating solar nebula made comets to become building blocks of the planetary system. The growing dust grains settle towards the centre plane of the rotating disk until the density reaches a critical value where gravitational instability occurs (Safronov 1969, Goldreich and Ward 1973). The timescales for formation of these building blocks are short: 10^5 to 10^6 years. The size distribution of the resulting nuclei could be estimated based on the scale lengths for the gravitational instability to a few kilometres (Biermann and Michel 1978). Once a larger body is formed it grows very fast by gravitationally attraction (gravitational runaway) to form a planet. The ices of the cometary volatile components (predominantly water) require formation of the comets outside Jupiter's orbit. The sizes of the homogenous cometary nuclei remain small enough that gravitational compaction is unimportant and the grains from the molecular cloud are hardly altered. The degree of processing of these grains before they agglomerate depends on the physical parameters of the molecular cloud at the location of planetesimal formation such as the optical depth (shielding the dust from the central early Sun) and the resulting local temperature. A conceivable extreme is the formation directly from interstellar grains (Greenberg 1977, 1998).

1.2 Accreation of building blocks

In contrast to the formation of comets by gravitational instabilities, Weidenshilling (1995) shows that for plausible parameters of the solar nebula the presence of gas induces drag forces on the dust particles and prevents local gravitational instability. Submicrometre- to micronmetre-sized particles entrained in the gas of the contracting solar nebula grow by coagulation due to Brownian motion and settle toward the central plane. The larger grains decouple from the gas motion and sweep up the smaller grains to grow fast to centimetre sizes. This growth is based on interlocking molecular forces. Before the particles reach the critical density for gravitational instability to occur they would decouple from the gas and follow Keplerian orbits. The presence of the gas still influences the motion of the particles by inducing a drag force that is size dependent. The differential rotation relative to the gas causes shear forces that induce turbulence preventing the grain density increasing further (Weidenshilling 1980, Cuzzi et al. 1993). Thus cometesimals cannot form by collapse of a cloud of centimetre-sized particles but they have to grow by coagulation and agglomeration to metre size before they decouple from the shear-induced turbulence. But growth does not stop there because the gas drag-induced radial velocity dispersion decreases relatively fast for larger bodies (Figure 1). Once they reach dimensions of tens or hundreds of metres gas drag becomes insignificant. The lag of damping prevents local gravitational collapse to form solid planetesimals (Weidenshilling 1995).

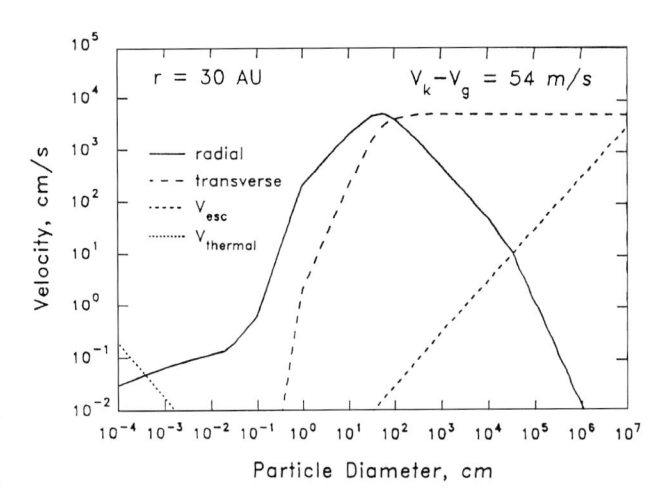

Figure 1 Particle velocities as a function of size in the model nebula at 30 AU. Particles are assumed to have a fractal structure at size 10^{-2} cm, and constant density of 0.7 cm^{-3} at $d > 1$ cm. Dotted line: thermal velocity at $T = 50$ K. Solid line: radial velocity, with peak value equal to $\Delta V = 54$ cm s^{-1} at $d \approx 10^2$ cm. Changes in slope are due to variation of particle density ($d \leq 1$ cm) and transition from Epstein to Stokes drag law ($d \approx 10^5$ cm). Dashed line: transverse velocity relative to pressure-supported gas. Short dashed: escape velocity from the particle's surface. (From Weidenshilling 1997.)

The radial velocity distribution as a function of particle size controls the evolution of the growing bodies. Figure 1 depicts a typical scenario in the solar nebula at a radial distance of 30 AU from the Sun (Weidenshilling 1997). Small particles rotate with the gas velocity but drift radially inward, while large particles follow Keplerian orbits and plough through the slower rotating gas. The peak velocity is reached for particle sizes for which the drag-induced response time $t_e = mv/F_D$ (m is particle mass, v is relative velocity and F_D is drag force) is comparable to the orbital period at the radial distance in question. This velocity distribution controls the agglomeration of the bodies. As long as gravitational attraction is unimportant the large bodies grow relatively slowly because of their small velocities. They typically grow from bodies at factors 3 to 5 times smaller that still have higher speeds. For example, after 8×10^4 s the largest bodies of 70 m accrete from 20 m 'particles'; at 1.5×10^5 s, 500 m from 200 m; at 2×10^5 s, 6 km from 1 km. Now the velocity becomes gravity influenced and at 2.5×10^5 s, 80 km bodies are growing by gravitational accretion. At any given time most of the mass is concentrated in a narrow size range.

1.3 Dust coagulation

These model calculations obviously require the dust particles to coagulate, to stick to each other at velocities up to several metres per second and agglomeration to prevail over destruction for metre-sized bodies at speeds of 50 m s^{-1}. Then the typical timescales for formation of cometesimals is a few thousand orbital periods so that comets could form within 10^6 years even in the Kuiper belt at 40 or 50 AU.

The formation of cometesimals by coagulation implies that the bodies are physically not homogenous but built of subnuclei of various sizes and typically about 3 to 10 times smaller than the body itself. Theses subnuclei themselves show a similar structure relative to their overall size. The speed of collisions are high enough that the building blocks may be partly shattered or have penetrated each other. The details will depend on physical parameters such as density, fluffiness (fractal dimension), stickiness, tensile strength, and so on. These parameters will vary as a function of body size. One expects voids between building blocks and volumes of increased density where penetration took place.

The timescale for accretion and the considerable radial velocities of the bodies imply radial migration over substantial heliocentric distances. Origination of particles from different heliocentric distances at different times during the formation of a cometesimal could lead to chemical differentiation, in particular in the dust to gas ratio.

2 RESERVOIRS OF COMETS

2.1 The Oort cloud

Comets are rather artificially divided into two classes, the short-period (SP) comets with orbitial periods of less than 200 years and the long-period (LP) comets. The activitity of comets near the Sun removes about 0.1 to 1% of the mass of the nucleus per orbit, so that the lifetime of a cometary nucleus is less than 1000 orbital periods. Obviously even LP comets with orbital periods of 10^6 years would not have survived on their present orbits from the beginning of the Solar System. Consequently comets are either formed or captured episodically. If of primordial origin they could not have formed on their present orbits. Hypotheses based on episodical events include gravitational focusing of passing interstellar cloud material (Lyttleton 1948), compression of interstellar clouds by shocks (McCrea 1975), or formation in giant molecular clouds and subsequent capture by the Solar System (Clube and Napier 1982) or even formation by eruption of the giant planets and their satellites (Vsekhsvyatskii 1967). These ideas have found little support.

Oort (1950) analysed the distribution of orbital energies of new (LP) comets, characterized by $1/a$ where a is the semi-major axis of the orbit. The more recent compilation by Marsden (1989b) shown in Figure 2 confirms the very strong peak with $1/a < 10^{-4}$ AU^{-1}. New (in the dynamical sense) comets come from distances almost comparable to the distances of nearby stars. Once penetrating the inner Solar System their orbits are strongly perturbed mainly by Jupiter with $\Delta(1/a) \approx \pm 6 \times 10^{-4}$ AU^{-1} (Everhart 1968) and rapidly diffuse to small semi-major axes if not expelled from the Solar System. Only a minute fraction ($<10^{-3}$) of these new comets will become SP comets, not enough by far to explain the presently known SP comets (about 600; Williams 2000). To explain the few (less than 10 per year) observed newly detected comets the spherical reservoir of randomly distributed comets must entail more than 10^{12} (Weissmann 1980) comets. Oort demonstrated that passing stars perturb the cloud of comets repeatedly and change the velocities of the comets (change of momentum) but change their orbital energies very little. Most of the comets, therefore, stay bound to the Solar System even though their orbital energies are very small. The probability of a comet in the Oort cloud being directed into the inner Solar System (<30 AU) is controlled by the very small solid angle the inner Solar System encompasses seen from the fringes of the Solar System at about 10^5 AU.

2.2 The Edgewood–Kuiper belt and TNOs

While Oort's hypothesis explains the currently observed numbers of new and LP comets it does not account for the

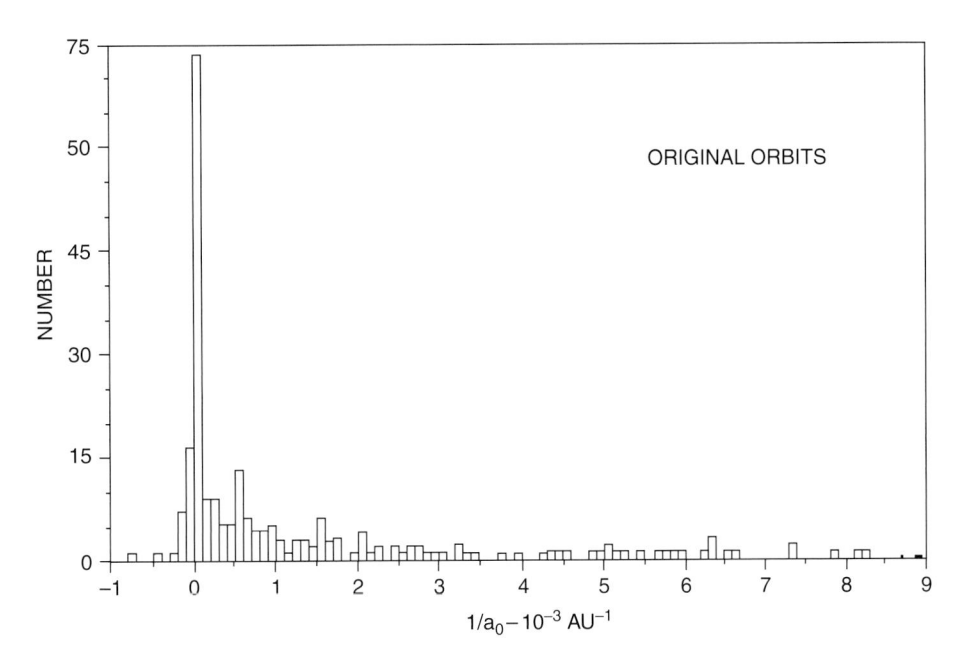

Figure 2 The distribution of original inverse semi-major axes for 264 long-period comets as found by Marsden (1989). The large spike at very small positive values of $1/a_0$ corresponds to comets on very long-period orbits, extending to interstellar distances. These are 'dynamically new' comets from the Oort cloud. The low continuous distribution with $1/a_0 > 1 \times 10^{-4}\,\mathrm{AU}^{-1}$ are returning comets which have been scattered in $1/a_0$ by planetary perturbations, primarily by Jupiter.

relatively large number of SP comets whose orbits are concentrated towards the ecliptical plane in contrast to the randomly distributed LP comets. Estimates and calculations of orbital evolutions (Everhart 1972, Joss 1973, Delsemme 1973) were not convincing. Fernandez (1980) showed that a transition to SP comets is more than two orders of magnitude more efficient from a belt of comets beyond the orbit of Neptune than from the Oort cloud. The existence of such a belt was suggested by Edgeworth (1949) and Kuiper (1951) and is now called the Edgeworth–Kuiper belt. This belt of comets with mostly low-inclination orbits was finally detected (Jewitt and Luu 1993). Since then, observations by several teams led to the discovery of more than 300 comets beyond the orbit of Neptune. The discovery rate is increasing rapidly: more than 100 new objects were discovered in 1999. About one-third of the TNOs are locked into a 3:2 resonance with Neptune, like Pluto. The other TNOs are separated into two families. The 'scattered TNOs' have high eccentricities ($e \gtrsim 0.5$) and a perihelion distance close to that of Neptune (q of about 30 AU). They have dynamical lifetimes comparable to the age of the Solar System (Duncan and Levison 1997). A prototype scattered TNO is 1996 TL66 ($q = 35$ AU, $e = 0.59$). Finally, the objects with semi-major axis ≥ 42 AU are called 'classical TNOs' (Jewitt 1999). They have orbits stable on a timescale larger than the age of the Solar System.

The size distribution of the TNOs is only determined from and for objects much larger than a typical comet. The equivalent radii are greater than or approximately equal to 100 km if a geometric albedo of 0.04 is assumed for all objects. The HST observation of smaller objects by Cochran et al (1995) has been questioned in several articles (Brown et al. 1997, Luu and Jewitt 1998, Jewitt 1999). A recent deep imaging survey with the Keck 10 m telescope (Luu and Jewitt 1998), combined with larger surveys (Jewitt et al. 1996, 1998; Gladman et al. 1998) yield a slope $q = 4.0 \pm 0.5$ of the size distribution of the TNOs (Jewitt et al. 1998, Jewitt 1999), a value compatible with both the collisionally relaxed equilibrium value $q = 3.5$ (Dohnanyi 1969) and the value $q \approx 4.5$ predicted by aggregations models (Yamamoto and Kozasa 1988). Further observations will allow one to reduce the uncertainty and reliably measure q for the three families of TNOs.

The size distribution of cometary nuclei in the Oort cloud should reflect the distribution during aggregation of the bodies because collisions do not play a role far away from the Sun. Hence it should be considerably steeper than that of the TNOs. Comets originating in the Oort cloud should therefore be larger on average than the SP comets from the EK belt.

2.3 Population of the Oort cloud

An even more intricate question is how the Oort cloud was populated by comets. Near the giant planets and particularly near Jupiter the density in the early solar nebula was high enough to produce large numbers of cometesimals. The scattering efficiency of Jupiter is so high that most cometesimals would be thrown out of the Solar System before their orbits could evolve to aphelia ($<5 \times 10^3$ AU) that would allow gravitational perturbations by the galactic tides to increase their perihelia so that they could escape the influence of the giant planets and evolve to members of the Oort cloud (Duncan *et al.* 1987). Early estimates and calculations (Fernandez and Ip 1981, 1983; Duncan *et al.* 1987) indicated that the placement efficiency of Neptune and Uranus is much higher and that in this case a strongly populated inner Oort cloud (extension about $<2 \times 10^4$ AU) is formed. The very same perturbations that erode the outer (classical) Oort cloud, namely galactic tides, passing stars and in particular giant molecular clouds, pump comets from the much more stable inner cloud into the outer Oort cloud.

2.4 The Sun born in a star cluster

The hypothetical inner Oort cloud is required for rejuvenation because the lifetime of the Oort cloud is smaller than the age of the Solar System (e.g. Hut and Tremaine 1985; Bailey 1983, 1986). Recently Fernandez (1997) showed that Neptune and Uranus are not efficient in scattering the cometesimals into highly elliptical orbits. They are rather passed to Saturn and Jupiter. This means that an inner Oort cloud cannot be built up as calculated by Duncan *et al.* (1987) because Jupiter throws most of the cometesimals out of the Solar System (see above). Explicit numerical calculations by Eggers (1999) confirm the approximations by Fernandez (1997). Eggers *et al.* (1999) suggest that the Sun was born in a star cluster (as most stars are) rather than as an isolated star. They investigate the evolution of the cometesimals in the early Solar System under the influence of slow encounters close to passing stars. The relative velocity of stars within the cluster is typically $1\,km\,s^{-1}$, an order of magnitude less than that of stars in the solar neighbourhood at present. The enhanced perturbations influence the orbits of cometesimals with relatively small aphelia. These are more often decoupled from the giant planets by increasing their perihelia appropriately before they can be scattered out of the Solar System. The placement efficiency of Jupiter and Saturn is thus sufficiently enhanced to form an inner Oort cloud (Figure 3). The important conclusion is that most of the comets in the Oort cloud are scattered by Jupiter and Saturn (and originate there) rather than by Neptune and Uranus. This has implications for the composition of the cometary nuclei.

The strong perturbations of the cluster stars also influence the comets outside Neptune's orbit. The model calculations show that the inclinations and eccentricities of TNOs far away from any influence of Neptune ('scattered TNOs', Section 2.2) are enhanced (Eggers *et al.* 1997). This is in good agreement with current observations for which no other sound explanation has been offered. Of course, this scenario requires that the TNOs formed before the star cluster disperses. Otherwise, the perturbations of the cometesimals will prevent the formation once the eccentricity becomes larger than 0.01 (Stern 1996). Just after the formation of the planetary system, however, there

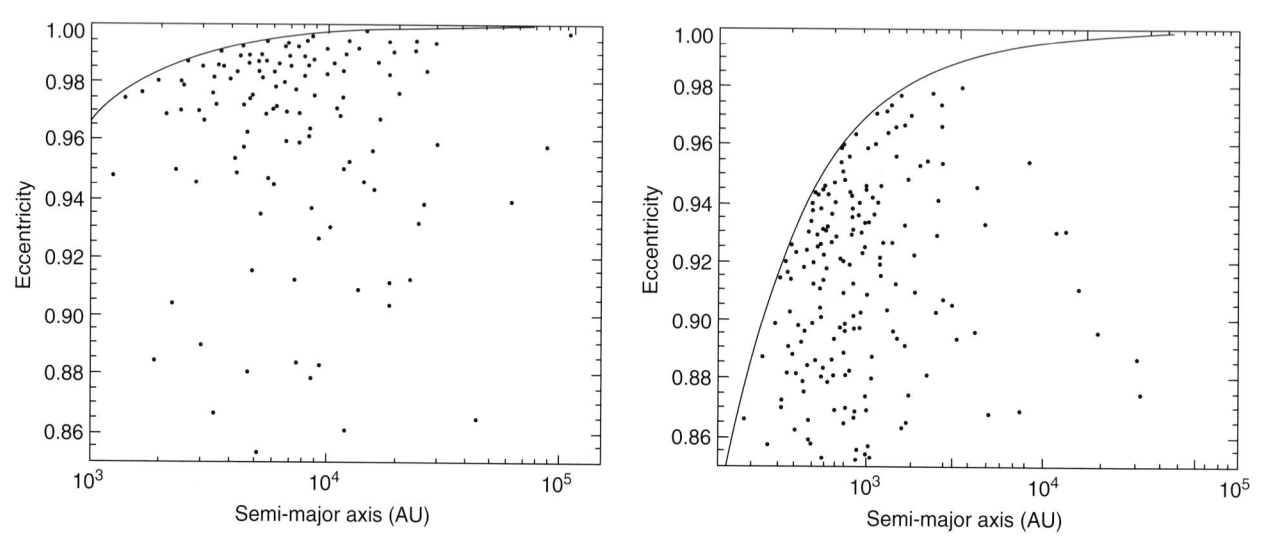

Figure 3 Development of cometary dynamics for the case that the early Sun is a member of a star cluster. Eccentricity v. final semi-major axis of Oort cloud comets for a low (left) and high (right) density star cluster after 20 million years. The line marks a perihelion distance of 33 AU. At smaller distances cometary orbits are not stable because of the perturbations of the planets. (After Eggers 1999 and private communication.)

may have been enough gas drag to compensate. There are no appropriate model calculations available.

Capture of comets from intra-cluster space that have been lost from other solar systems of the cluster is rather infrequent. The fraction of 'foreign comets' in the Oort cloud should be less than 10% even under favourable assumptions for the physical parameters.

3 CONSEQUENCES FOR THE NATURE AND MORPHOLOGY OF COMETARY NUCLEI

3.1 The icy conglomerate model

When Whipple explained the systematic changes of the orbital periods of SP comets by reactive forces caused by activity (sublimation of ice) from a rotating solid cometary nucleus the physical and chemical nature of this nucleus was described as a conglomerate of different ices and refractory particles. These particles refer to the observed dust grains of micrometer size in cometary tails (Finson and Probstein 1968a,b) but could also mean boulders by analogy to meteoroids (Sekanina 1972). These particles are embedded in the prevailing ices leading to the popular metaphor of the dirty snowball. This model could explain the high gas (water) production rates mainly derived from the first UV observations (see e.g. Keller 1976b for a review). The picture changed in the 1980s when the albedo of cometary nuclei was observed (from the ground (Cruikshank and Brown 1983) and from space (Sagdeev et al. 1986, Keller et al. 1986)) and found to be very dark, not resembling an icy surface.

3.2 Icy grains

The formation of cometary nuclei by agglomeration and their storage in the outer Solar System was discussed briefly following the most widely accepted solar nebula theory in Sections 1 and 2. What are the consequences for the physical nature and morphology of cometary nuclei? The agglomeration from interstellar dust at low velocities and low or moderate temperatures will lead to subnuclei of homogeneously mixed ice and dust particles. Whether the smallest units are unaltered icy interstellar grains (Greenberg 1984) or dust particles on which the ambient gas in the protosolar nebula condensed, obviously has a major influence on the composition and is discussed in Chapter 53 of this book but is of less importance for the question of nuclear structure and morphology. In the latter case it is conceivable that the ice to dust (here we mean all components that are less volatile than water ice) ratio varies from cometesimal to cometesimal depending on the place of formation and time of formation because the gas to particle ratio in the solar nebula changes with heliocentric distance and also perpendicular to the ecliptic plane. The amount of coating of the dust particles will depend on the surrounding vapour (gas) density and temperature but will also be proportional to the length of time of exposure.

3.3 Grain formation and growth

Laboratory experiments (Keller et al. 1992, Blum 1995) and modelling show that, in particular, cluster to cluster agglomeration leads to extremely fluffy structures with fractal dimensions below two. All simulations and laboratory experiments assume certain environmental conditions and parameters. The real parameters are not known and in addition it is highly probable that the particles pass through different regimes during their growth, such as accretion shocks, that in some model calculations are so strong that all material could be sublimated (Hayashi et al. 1985). Even with much weaker interactions that are appropriate to maintain the apparent pristinity of cometary material there will be processes such as restructuring (Dominik and Tielens 1997) increasing the density of the particles. Of course this extreme fluffiness cannot be maintained once the size of growing particles reaches a few thousand scale lengths of the original interstellar/interplanetary grains. The porosity of the small grains (up to several tens of micrometres) can be as high as 98%. Depending on the environmental conditions, if gas is still condensing, the empty cavities could be partially filled with ice. Even minor ambient gas pressures as well as moderate heating will lead to sintering. Gas predominantly condenses at areas with strong curvature and hence strengthens the joints between dust particles and hence the tensile strength of the cluster. Agglomeration by body-to-body collisions will prevail over accretion to large dimensions (Weidenshilling 1997). A 'moving size distribution' of the building blocks creates bodies with a fractal-like size distribution of the building blocks. The varying block size is reflected by a corresponding physical inhomogeneity of bodies and voids between them. With growing block size the relative velocities will increase and collisions will become more and more violent (Figure 1). One can expect physical destruction of the fragile and brittle building blocks at their surfaces of contact leading to zones of compaction. Metre and larger sized blocks could penetrate each other doubling the density in this volume. In the zones of destruction by impact the rubble will have less tensile strength and lead to weak junctures. The partial heating during impact could lead to diffusion of gas out of the dust matrix into the voids between the blocks forming pockets of volatile compounds.

3.4 Particle size distribution and subnuclei

In this scenario a cometary nucleus consists of subnuclei (building blocks) not much smaller than the nucleus itself.

The subnuclei themselves have a similar block distribution, and so on. The fine scale is determined by a matrix of fluffy dust particles coated with ice and sintered together. The tensile strength is rather high considering the high porosity of the matrix. The volatile to refractory (gas to dust) ratio can vary from building block to building block and volatiles could be concentrated in pockets. The average of the ratio will depend on the origin and the details of the formation process but a value around unity or a predominance of the non-volatile fraction are indicated (Greenberg 1998). The refractory components include not only silicates but complex organic molecules that only sublime when temperatures increase to 350 to 400 K.

3.5 Collisions

The relatively fluffy structure of low density will prevail (Donn 1991) until gravitational attraction becomes important for the largest blocks and the collisional speeds lead to more severe compaction and temperature increases that will change the composition of the ice and dust mixture. The size of the bodies has then reached dimensions considerably larger than the size of typical comets (of the order of ten kilometres). The TNOs with dimensions of hundreds or thousands (Pluto!) of kilometres (Stern and Colwell 1997) will suffer impact heating and compaction towards their centres. Finally their growth will be limited by surface erosion. Their number density in the EK belt is high enough so that their size distribution should be collision dominated (Farinella *et al.* 2000, Stern 1995). The timescale of collisions between large TNOs is comparable or longer than the age of the Solar System (Ipatov 1999). These, therefore, could escape catastrophic destruction.

3.6 Origins and evolution

The orbits of the SP comets of the Jupiter family (with orbital periods of about 5 years) strongly correlate to the ecliptic. They are prime candidates to have an origin in the EK belt from within 45 AU. Wether they are recent collisional fragments of larger TNOs or are only perturbed by them to reach the influence of Neptune and finally that of Jupiter is an open question as long as we know so little about the size distribution of objects in the EK belt. This effective handing down of comets from Neptune to Jupiter that allows for the transition of TNOs to SP comets prevents Neptune from being effective in the formation of the Oort cloud (Section 2.4). Over the lifetime of the Solar System, objects with diameters $\lesssim 100$ km will have anyhow suffered several collisions (Davis and Farinella 1997). Impact speeds will be determined by the gravitational attraction of the large TNOs, once they have a size of more than ~ 100 km, and reach 100 m s^{-1} and more.

A considerable fraction of the Halley-type SP comets with periods longer than 20 years orbit the Sun with large inclinations. Dynamical investigations show that only few comets change the group or family. Levison and Duncan (1994) used the Tisserand parameter, T, an approximate constant of motion in the restricted three-body (Sun–Jupiter–comet) problem, to characterize the Jupiter family ($T > 2$). Whether the Halley-type SP comets with $T < 2$ are end members of the Oort cloud comets (Quinn *et al.* 1990) or are connected to the TNOs on high-inclination orbits remains an open question.

Three different classes of comets can be identified according to their origin and evolution:

- Jupiter family
- Halley type
- Long-period and new comets

The Jupiter family comets were formed in the EK belt at low temperature but probably have suffered a rather violent collisional evolution and have therefore lost most of their original pristinity. The Halley types originating in the outer part of the EK belt could be the most pristine bodies considering that the new and LP comets from the Oort cloud were formed near Neptune or even as close in as Jupiter (Eggers *et al.* 2000). These comets have suffered very few collisions. One could expect that the chemical composition of the LP comets contains less of the more volatile species and have intact nuclei while the SP comets originally contain more volatile compounds but may have lost them during their further violent evolution by which their nuclei may have been shattered, ending up as rubble piles of loosely bound fragments.

4 NUCLEUS PROPERTIES FROM TELESCOPIC OBSERVATIONS

4.1 Derivation of the nucleus size and albedo

While comets exhibit spectacular tails and coma structures observable with the naked eye, their nuclei are among the most difficult objects to observe in our Solar System. Their apparent motion in the sky requires telescopes with good pointing capabilities (fast differential tracking rates and guiding). The coma brightness and morphology experience fast changes within a few hours or even less. Another difficulty arises from the bright dust coma which surrounds the nucleus when the comet becomes active. At the resolution achieved from the ground, the brightness contribution of the inner coma usually largely exceeds that of the nucleus. Last but not least, comet nuclei are too small to be spatially resolved by ground-based telescopes. As an example, the nucleus of comet 1P/Halley has a maximum dimension (length) of about 15 km which converts into an angular diameter of only 0.02 arcsec at a distance of 1 AU. It could be spatially resolved only during a close approach to the Earth at a geocentric distance of less than 0.01 AU. Several methods, described in Sections 4.2. and 4.3, allow one to overcome these difficulties and measure the signal from the nucleus.

When photometrically detected, the 'equivalent radius' of the nucleus is calculated from the following relationship (Russell 1916):

$$A_p r_n^2 = 2.235 \times 10^{22} r_h^2 \Delta^2 \times 10_\odot^{0.4(m_\odot - m)} \quad (1)$$

where A_p is the geometric albedo of the nucleus, m is the observed nuclear magnitude at zero phase angle, $m_\odot$ is the solar magnitude in the same bandpass as m, and r_h and Δ are the heliocentric and geocentric distances of the comet at the time of the observation (in AU), respectively. The correction of m to zero phase of m is performed by subtracting $\beta \cdot \alpha$ from the observed magnitude at phase angle α, where $\beta = 0.04$ mag deg^{-1} (Jewitt and Meech 1987). The equivalent radius in eqn (1) is generally not the physical radius of the nucleus because all bodies smaller than about 200 km radius have irregular shapes. In the case of a prolate ellipsoid, r_n^2 is the product of the projected semi-major axis and the semi-minor axis. If a temporal light curve can be obtained, its amplitude allows one to determine a lower limit for the axis ratio. When combined with optical measurements the mean geometric albedo of the nucleus can be derived from thermal infrared observations. A model is used to calculate a temperature map $T(\theta, \phi)$ of the surface as a function of the Bond albedo A_B and the thermal emissivity ε (e.g. Lebofsky and Spencer 1989). The mean geometric albedo A_p is related to A_B by an integral over the nucleus phase function. More sophisticated models involve additional parameters, such as the nucleus rotational parameters, thermal conductivity and thermal inertia (see e.g. Enzian et al. 1998). When $T(\theta, \phi)$ has been modelled, the energy balance is given by (Lebofsky and Spencer 1989):

$$\frac{\pi r_n^2 (1 - A_B) C_\odot}{r_h^2} = \eta \varepsilon \sigma r_n^2 \int_0^{2\pi} \int_{-\pi/2}^{\pi/2} T^4(\theta, \phi) \cos \phi \, d\phi \, d\theta \quad (2)$$

where $C_\odot$ is the solar constant, η is the 'beaming factor', an empirical normalization factor which takes into account the increase of thermal emission at small phase angles and σ is the Stefan–Boltzmann constant. Assuming that the geometric albedo and nucleus radius are the two unknown parameters, eqns (1) and (2) lead to the determination of both of them. When at least two measurements at different wavelengths are performed, the physical surface temperature of the nucleus can be calculated by comparing the data points with predictions of thermal models.

4.2 Observations of inactive comets

Modern optical detectors, such as CCDs, permit visible imaging of comets at heliocentric distances well beyond 10 AU (Hainaut et al. 1993). At these distances from the Sun, the cometary activity subsides and the bare nucleus is observed. When no coma is detected after a careful analysis with dedicated image-processing methods, an upper limit for the contamination by the coma in a small diaphragm is calculated. When this limit is small compared to the observed signal, the bare nucleus is being measured and the nucleus equivalent radius and elongation can be derived from the light curve. This method has been used to observe the nucleus of comets 1P/Halley at 18.8 AU from the Sun (Hainaut et al. 1995), 55P/Tempel-Tuttle at heliocentric distances between 3.5 and 6.7 AU (Hainaut et al. 1998) and 2P/Encke near aphelion at 4 AU (Jewitt and Meech 1987, Luu and Jewitt 1990). The derived equivalent radii range from 1.8 km (55P) to 5.5 km (1P) and the maximum axis ratio is greater than 2 (1P and 2P). The radius for 1P/Halley is in good agreement with the observations taken during the Giotto flyby (Keller et al. 1986). New detectors at wavelengths between 1.2 μm (near infrared) and 3 mm (millimetre and sub-millimetre ranges) were used for multi-wavelength observations of 95P/Chiron, a comet with a perihelion distance of 8.4 AU. Observations at 10–20 μm (Campins et al. 1994) and 1.2 mm (Altenhoff and Stumpff 1995) at 8.8–16 AU yield an estimate of its radius of 88 ± 10 km. A comparison of these observations with visible measurements implies a geometric albedo of 0.14 ± 0.01 (Campins et al. 1994, Altenhoff and Stumpff 1995), compatible with the presence of water ice at the surface (Luu et al. 2000).

This method seems promising since comets will be observable at larger heliocentric distances when the giant telescopes become fully operational. Future observation and reduction techniques for this type of observation were thoroughly described by Hainaut et al. (1994). However, some difficulties remain. First of all, the range of heliocentric distances is limited by the appearance of the activity on one hand, and by the limiting magnitude of the telescopes on the other. SP comets may well be active up to aphelion if very volatile ices such as CO or CO_2 sublimate. Indeed, comet 1P/Halley was active at 8.5 and 10 AU (West and Jorgensen 1989, West 1990), and a violent outburst occurred at more than 14 AU (West et al. 1991). Comets C/1980 E1 (Bowell) and C/1995 O1 (Hale-Bopp) were observed embedded in an extended coma at more than 13 AU (Meech and Jewitt 1987, McNaught 1995). 29P/Schwassmann-Wachmann 1 experiences continuous activity at more than 6 AU (Jewitt 1990). Dust around 95P/Chiron was first detected in the near infrared in 1988 (Tholen et al. 1988) and monitored later on in the visible

(Luu 1993a). Debris ranging from 10 cm to several metres lifted off by the gas drag at perihelion can reach stable orbits around the nucleus and form a 'dust debris cloud' of 100 to 1000 km in diameter (Richter and Keller 1995, Fulle 1997). This cloud could give a significant contribution to the signal attributed to the bare nucleus and lead to an overestimate of its equivalent radius.

Some comets – 10P/Tempel 2, 28P/Neujmin 1, 49P/Arend-Rigaux and 107P/Wilson-Harrington – show no or very little activity within 2 AU from the Sun. The bare nucleus of these comets can be observed close to the Earth without major coma contamination. These objects are probably 'transition objects' between active and extinct comets at the end of their active life: they are probably on the way to lose all their volatile materials and become inactive. Observations of comet 10P/Tempel 2 were carried out in the visible and thermal infrared at $r_h = 1.70$–1.92 AU. No coma was detected beyond 1.85 AU, and a nucleus equivalent radius of 4.7 km was derived from CCD images and photoelectric aperture measurements (Wisniewski 1990). A'Hearn et al. (1989) and Tokunaga et al. (1992) measured the nucleus thermal flux and determined a geometric albedo of 0.02–0.04 and an axes ratio greater than 1.9. Comet 28P/Neujmin 1 was observed at 1.7 AU from the Sun both in the visible and thermal infrared by Campins et al. (1987). Its equivalent radius amounts to 1.5 km and its geometric albedo is equal to 0.02–0.03. Comet 49P/Arend-Rigaux displayed a faint coma in the visible at 1.5–1.9 AU, but no coma was detected in the near and thermal infrared by Tokunaga and Hanner (1985) maybe because small micrometer-size grains invisible in the thermal infrared have a much larger cross-section than grains with a radius of 10 μm. A rough coma subtraction on the CCD images yielded a nucleus radius of 4.4–4.8 km and an albedo of 0.02–0.06 (Brooke and Knacke 1986, Birkett et al. 1987, Veeder et al. 1987, Millis et al. 1988). The axes ratio for this comet is greater than 1.6 (Veeder et al. 1987, Millis et al. 1988). Comet 107P/Wilson-Harrington was first named 'asteroid 1979 VA' before Bowell et al. (1992) reported a thin tail observed on several Schmidt plates. This object has been characterized by Campins et al. (1995), who measured a nucleus equivalent radius of 1.3–2.0 km and a geometric albedo of 0.05–0.10, depending on the thermal model used to interpret the data.

The physical properties of the largest TNOs in the Edgeworth–Kuiper belt have been studied for only a few years with the largest ground-based telescopes available and with space-based telescopes. Two TNOs were detected at 2.7-σ (1993 SC) and 3.4-σ (1996 TL66) levels by the Infrared Space Observatory (Thomas et al. 2000). Interpreted with the standard thermal model used for asteroids (Lebofsky and Spencer 1989), these measurements correspond to low geometric albedos of 0.02–0.03 (Thomas et al. 2000).

Photometry of five TNOs in several bands in the visible and near infrared (Jewitt and Luu 1998) shows different spectral reflectivities. The colour diversity among the largest TNOs still remains to be explained. It could be due to the development of an irradiation mantle combined with resurfacing during collisions with smaller objects (Jewitt 1999).

4.3 Observations of active comets

When a coma is present, several techniques can be used to separate the signal of the nucleus from that of the coma. Image-processing techniques were developed by West and Jorgensen (1989) and West (1990). They were able to identify three components on their CCD images of comet 1P/Halley: the outer and inner coma and the contribution from the nucleus. They subtracted a Gaussian profile to account for the nucleus contribution. The remaining image of the coma was smoothed and subtracted from the original frames in order to get 'nucleus frames', from which an equivalent radius of 5.5–7.0 km was measured. These values are compatible with the estimates of the nucleus size calculated from the Giotto/HMC images (Keller et al. 1987). However, this technique does not eliminate the risk of contamination by the coma, which peaks at the nucleus.

Meech et al. (1993) use a different approach in their analysis of photometric CCD measurements of comet 29P/Schwassmann-Wachmann 1. They measure the signal from the dust coma and the nucleus in 12 apertures with diameters ranging from 1.4 to 10 arcsec. In the diaphragms of smaller radii, they observe a modulation of the signal attributed to the rotation of the nucleus. They fit the observed rotational range of magnitudes as a function of the radius ρ of the diaphragms with a model in which the coma brightness varies as ρ^{-n} and the nucleus is a point-like source. They finally find a radius of 15 km assuming a geometric albedo of 0.04 and of only 8.6 km assuming a geometric albedo of 0.13 (Cruikshank and Brown 1983). The ratio of the major to the minor axis is found to be at least 2.6.

Finally, a new technique of nucleus–coma separation was introduced recently by Lamy and Toth (1995). It is applied to high-resolution images of comets observed at geocentric distances of 0.17–0.67 AU with the Wide-Field Planetary Cameras (WFPC and WFPC2) of the HST. The authors perform a fit of the images with a two-dimensional model of the coma and nucleus contributions (Lamy et al. 1999):

$$F(\rho, \theta) = \left(k_n + \frac{A(\theta)}{\rho^{n(\theta)}} \right) \otimes \text{PSF}(\rho) \quad (3)$$

where (ρ, θ) are polar coordinates centred on the nucleus, $k_n = \pi A_p r_n^2$ is the nucleus effective cross-sectional area and

PSF is the point-spread function of the WFPC and WFPC2. As a result of the fit, the nucleus is photometrically separated from the coma. Comets 4P/Faye (Lamy and Toth 1995), 19P/Borrelly (Lamy et al. 1998a), 46P/Wirtanen (Lamy et al. 1998b) and 45P/Honda-Mrkos-Pajdusakova (Lamy et al. 1999) have been observed so far. The nucleus properties derived for these comets are summarized in Table 1. This technique has also been used by Lisse et al. (1999) to interpret thermal infrared images of comet C/1996 B2 (Hyakutake) and determine a nucleus radius of 2.4 ± 0.5 km.

Adaptive optics is potentially a powerful tool to retrieve information from ground-based observations of small Solar System bodies. Resolved images of large asteroids with adaptive optics systems were successfully acquired and analysed (Saint-Pé et al. 1993, Drummond et al. 1998). The first observations of a comet with the ADONIS adaptive optics system were carried out at the European Southern Observatory by Marco et al. (1998). Comet C/1996 B2 (Hyakutake) was observed on 8 March 1996, at a geocentric distance of 0.59 AU. The achieved angular resolution is of the order of 0.2 arcsec, which converts into a spatial resolution of 85 km at the distance of the comet: although very useful information was retrieved about the inner dust coma the nucleus was not detected. Observations in late March 1996, when the comet approached the Earth to as close as 0.1 AU would have allowed a much better projected spatial resolution of less than 15 km with this system. Another attempt with adaptive optics was made for comet Hale-Bopp (C/1995 O1) in November 1997 and January 1998 (Marchis et al. 1999). The angular resolution was 0.1–0.2 arcsec, which converted into a projected spatial resolution of 280–470 km. Two bright knots separated by 0.2–0.4 arcsec were observed, but their nature (two nuclei, or a nuclei and a dust cloud) could not be determined unambiguously.

Table 1 Summary of nucleus equivalent radii, axes ratios and geometric albedos from visible, infrared, radio and radar observations

Comet	r_n^*(km)	Axes ratio[†]	A_p[‡]	Wavelength[¶]	Technique(s) used[§]
1P/Halley	5.5	2.0	0.04	VIS	SPC/DNM/SCM
2P/Encke	3.0–4.1	1.8		VIS/RAD	DNM/SRE
4P/Faye	2.7	1.2		VIS	SCM
10P/Tempel 2	4.5	1.5	0.02–0.04	VIS/TIR	DNM/MSD/SCC
19P/Borrelly	2.8	2.5		VIS	SCM
28P/Neujmin 1	9.7	1.2	0.02–0.04	VIS/NIR	MSD
29P/Schwassmann-Wachmann 1	8.6–15	2.6	0.13[**]	VIS/TIR	SCM
31P/Schwassmann-Wachmann 2	3.4	1.6		VIS	MSD
45P/Honda-Mrkos-Pajdusakova	0.34	1.3		VIS	SCM
46P/Wirtanen	0.6	1.2		VIS	SCM
49P/Arend-Rigaux	4.7	1.6	0.02–0.06	NIR/TIR	DNM/MSD/SCC/SCM
55P/Tempel-Tuttle	1.8	1.5		VIS	DNM
95P/Chiron	90	1.1	0.13–0.14	VIS/TIR/RAD	DNM/OCC
107P/Wilson-Harrington	1.3–2.0[††]		0.05–0.10	NIR/TIR	DNM
C/1983 H1 (IRAS-Aracki-Alcock)	5			TIR/RAD	MSD/SRE
C/1995 O1 (Hale-Bopp)	30–40[‡‡]			VIS[¶¶]	SCM
C/1996 B2 (Hyakutake)	2–3			VIS[¶¶]/TIR/RAD	SCM/SRE

* Nucleus equivalent radius, as defined in eqn (1).
† Lower limit for the ratio of the projected semi-major axis to the semi-minor axis of an elongated prolate spheroid.
‡ Mean geometric albedo of the nucleus.
¶ VIS = visible (0.3–1 μm), NIR = near infrared (1–3 μm), TIR = thermal infrared (3–100 μm), RAD = radio (>100 μm).
§ DNM = direct nucleus measurement – no coma was detected, MSD = measurement in a small diaphragm – the coma is faint enough to allow an estimate of the signal from the nucleus in a small diaphragm centred on the nucleus, SCC = subtraction of the coma contribution – a coma was detected and its contribution was estimated and subtracted, SCM = subtraction of a coma model – a coma model was subtracted to the image/spectrum, SRE = study of radar echo, SPC = in situ Giotto/HMC observations.
** Possible coma contamination.
†† Depending on the thermal model used to interpret the data.
‡‡ Lamy, personal communication.
¶¶ Adaptive optics observations available.

Occultations of stars by comet nuclei could provide high-accuracy measurements of their sizes and shapes. However, this technique requires a high-accuracy ephemeris of the comet in order to have a non-negligible chance of observing an occultation. Unfortunately, the position of the comet nucleus is difficult to measure when the nucleus is embedded in the coma. Furthermore, comets are perturbed by non-gravitational forces which are difficult to predict. The observation of a stellar occultation by a comet nucleus remains a highly challenging task. The only successful observation of a stellar occultation by a comet is that of a 14th magnitude star by 95P/Chiron (Bus et al. 1996), observed from five sites in California. The occultation has been detected from two sites, although only marginally (one data point) in the temporal series of photometric measurements obtained from 'site 2'. The derived nuclear radius was 90 ± 7 km, a value in fair agreement with those obtained with other techniques (Section 4.2).

Several attempts were made during the 1980s and 1990s to detect a radar echo from a cometary nucleus. The first echo was detected in the S-band at 12.6 cm from Arecibo when comet 2P/Encke passed at less than 0.3 AU from the Earth in November 1980 (Kamoun et al. 1982). The analysis of the echo led to a nuclear radius of 0.5–3.8 km. The measured value was affected by a large error bar because the angle between the nucleus spin axis and the line of sight affected the determination of the radius from the radar echo. Furthermore, assumptions need to be made on the spectral shape of the echo when the signal-to-noise ratio (S/N) is insufficient to measure it accurately.

In May 1983, comet C/1983 H1 (IRAS-Araki-Alcock) passed within 0.031 AU of the Earth, and radar echoes in the S-band (12.6–12.9 cm) and X-band (3.5 cm) have been easily detected by JPL's Goldstone antenna (Goldstein et al. 1984) and from Arecibo (Harmon et al. 1989). The high S/N achieved in the spectra allows one to identify a narrow-band component attributed to the nucleus (Harmon et al. 1989). The second component is attributed to a cloud of large dust particles around the nucleus. The derivation of the nucleus size from radar observations is difficult because the spin orientation of the nucleus is usually unknown. In the case of C/1993 H1, the echoes are compatible with a nuclear radius of about 5 km (Hanner et al. 1985) for a 'low-reflectivity, low-density surface, such as a deep layer of packed snow' (Harmon et al. 1989). A larger body of 8 km radius would be required to explain the echoes if the surface is constituted of loosely packed snow. The spectra in the S- and X-band are consistent with a very rough surface at centimetre to metre scales (Harmon et al. 1989, Goldstein et al. 1984). A more realistic dust-dominated surface was not considered.

Harmon et al. (1997) identified a radar echo from the nucleus of comet C/1996 B2 (Hyakutake) using the Goldstone antenna. Assuming a radar albedo of 0.039, they measured a nucleus radius of 0.4–2.5 km, a result in agreement with the infrared observations of Lisse et al. (1999). An echo was received during radar observations of comet 1P/Halley's dust coma from Arecibo in 1985 (Campbell et al. 1989), but the signal from the nucleus was too weak to be detected.

4.4 Summary of nucleus properties

Table 1 summarizes the most reliable nucleus parameters determined so far. The corresponding references are all given in Sections 4.2 and 4.3 and are not repeated here. The two last columns indicate the wavelength range used to derive the parameters and the applied technique(s). The nucleus equivalent radius has been determined for a total of 12 short-period comets, mainly for low-activity comets (28P, 31P, 49P, 107P), comets at large heliocentric distances (1P, 10P, 55P) and using coma subtraction techniques (1P, 4P, 19P, 45P, 46P). The nuclei of 29P and 95P have perihelia beyond 6 AU (8 AU for 95P) which results in a better contrast between nucleus and coma. Finally, the nuclei of only three long-period comets have been observed. Two of them were observed during a close approach to the Earth (C/1983 H1 and C/1996 B2) and the large nuclei of comet Hale-Bopp (C/1995 O1) could be separated from the coma on HST images. Most comet nuclei in Table 1 have been observed in the visible, and half of them *only* in the visible. The combination of observations at visible and infrared wavelengths provided the determination of the geometric albedo of six cometary nuclei.

When discussing the sizes of nuclei in this selection of comets one has to keep in mind that periodic comets are evolved bodies. Part (if not most) of their mass has been ejected during the years of activity spent in the inner Solar System. Therefore, for most of them, the measured sizes and shapes are probably very different from those they had when they were injected into the inner Solar System. All bodies listed in Table 1 are likely to have irregular shapes, as evidenced by the radar observations of comet C/1983 H1 (Harmon et al. 1989), which means that the equivalent radii can be slightly different from the actual physical radii. The measured nucleus equivalent radii range from 0.34 km (45P) to 90 km (95P). Four comets have a radius lower than 2 km, eleven have a radius between 2 and 10 km (if we choose to include 29P into this category), and only two (C/1995 O1 and 95P) have a radius larger than 10 km. The average radius of Table 1 is 11 km, but this value is only 6 km without 95P. In this sample, 1P/Halley is therefore a comet of typical size. Finally, it is important to notice that for all recent radius determinations of periodic comets obtained from HST images (4P, 19P, 45P and 46P) the measured value is <3 km. More observations of comet nuclei using different techniques will be required to get a size distribution and eliminate the biases associated with the various techniques. This is especially true for the long-period and new comets for which our sample amounts to only three comets.

The observed cometary geometric albedos are very low (0.02–0.14), placing comets among the darkest bodies in the Solar System. Comet nuclei are also red (except 95P), with reflectivity gradients of 10–20% per 100 nm (Campins et al. 1987, Thomas and Keller 1989, A'Hearn et al. 1989, Luu 1993b, Lamy and Toth 1995, Lamy et al. 1998a,), although some comets (45P, 21P/Giacobini-Zinner) show a decrease of this gradient in the near infrared (Luu 1993, Lamy et al. 1999). If one excepts the peculiar objects 29P and 95P, the measured geometric albedos are always < 0.10 (Table 1). This is indicative of a very porous surface containing carbonaceous material; speculatively due to the presence of aromatic kerogen-like compounds or other carbon-rich organic compounds (Gradie and Veverka 1980).

All the axes ratios in the third column of Table 1 are lower limits (Section 4.1). The comparison between the visible and infrared light curves (10P, 18P, 49P) shows that the photometric variations are produced by the rotation of the nuclei not by albedo spots (Jewitt 1992). Therefore, cometary nuclei are nearly prolate, very elongated bodies (Jewitt 1992) with axes ratios sometimes larger than 2 or even 2.5 (1P, 19P, 29P) and very often larger than 1.5 (2P, 10P, 31P, 49P, 55P). This seems difficult to reconcile with a strengthless model of comet nuclei in which at least several tens of fragments (small grains or larger cometesimals) are only bound by their self-gravity (Weissmann 1986, Asphaug and Benz 1996, Schenk et al. 1996). The fragments would tend to 'fall' into the self-gravitational well and form a nearly spherical body. Collisions would tend to spread them apart, but internal dissipation would dampen this effect in a short timescale. The effect of the rotation would be to create oblate rather than prolate bodies. Therefore, the tensile strength between the fragments (grains or cometesimals) must be large enough to prevent self-gravitation from making the elongated body collapse into a nearly spherical prolate body.

5 THE TIDAL DISRUPTION OF COMETARY NUCLEI

5.1 Splitting of cometary nuclei

According to recent observations, splitting of cometary nuclei occurs at a high rate of about 0.01 per year per comet (Chen and Jewitt 1994). Tidal breakups and splitting depend on several parameters characterizing the bulk nucleus properties, mainly the nucleus size, density and effective strength. The study of tidal breakups provides a unique opportunity to constrain these parameters. This is especially true for tidal breakups by Jupiter because the cometary activity is likely to be weak during the breakup at the heliocentric distance of 5 AU. Two comets are known to have experienced tidal breakups during a close approach to Jupiter: 16P/Brooks 2 and D/Shoemaker-Levy 9 (SL9). Several other comets have broken up after a close approach to the Sun (for a recent overview on split comets, see Sekanina 1997). Furthermore, catenae (chains of craters) created by the impact of fragments tidally disrupted by Jupiter have been identified on the moons Ganymede and Callisto (Figure 4; Melosh and Schenk 1993, Schenk et al. 1996). Similar chains have also been identified on the Moon (Melosh and Whitaker 1994).

5.2 Models of tidal disruption and application to SL9

On 24 March 1993, C.S. Shoemaker, E.M. Shoemaker and D.H. Levy (Shoemaker et al. 1993) discovered a diffuse object of magnitude 14 close to Jupiter on films taken at the 0.46 m Palomar telescope. This object was soon called comet Shoemaker-Levy 9 (hereafter SL9) and its past and future orbit around Jupiter calculated. SL9 had split into 21 pieces during a close approach at only 1.31 Jupiter radii (91 600 km) in July 1992. The length, position and spectacular appearance of the train of fragments, 'strung out like pearls on a string' (Luu and Jewitt 1993), has been carefully monitored by a multitude of observes (Weaver et al. 1995;

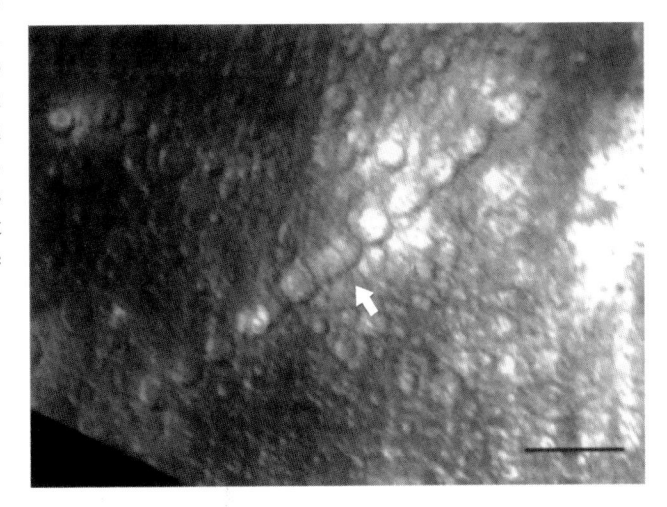

Figure 4 Gipul Catena on the surface of Jupiter's moon Callisto. This chain of 18 impact craters is about 625 km long. The arrow points to a lateral arcuate ridge bisecting a crater in Gipul Catena. The craters are elongated with aspect ratios up to 1.5. The average crater diameter is 29 km. (From Schenk et al. 1996.)

Figure 5). The fragments plunged into the Jovian atmosphere in July 1994, creating impacts of several tens of thousands kilometers observable with the naked eye.

Two main models of tidal breakup for SL9 were applied to interpret its splitting. The first model was introduced by Greenberg *et al.* (1995), and was discussed in more details by Asphaug and Benz (1996). When the comet penetrates within the distance of Jupiter where the cometary material fails to withstand the forces caused by the gravitational gradient across the nucleus, tidal stress creates a crack starting at the weakest flaw of the solid body. The crack grows on a timescale of about 1 s (the fracture timescale for a comet of ~1 km radius) separating the body into two pieces. When the smaller pieces come within the new distance of fragmentation, they split again, creating four fragments. The process repeats 4–5 times until 21 fragments are created. This phenomenon is called a 'fracture hierarchy'. The fragments then move apart assuming that their density is low enough to overcome self-gravitation. The time between two successive splittings is of the order of 1 h, implying that the overall process started well before perijove, possibly beyond 3 Jovian radii. At this distance, the tidal stress, and therefore the tensile strength of cometary material, would be lower than 10 Pa (0.1 mbar), an extremely low value, 10^5 times lower than the tensile strength of solid water ice. Such a low value is more than 10 times lower than the tensile strength of the porous aggregate interstellar dust model (Greenberg *et al.* 1995).

A second model was first applied to SL9 by Solem (1994). It was refined and further discussed by Asphaug and Benz (1996). The progenitor was assumed to be a strengthless self-gravitating aggregate of 20–2000 spherical 'grains' (Scotti and Melosh 1993). This means that the cohesive forces are small compared to the gravitational forces between the grains. The volume is equivalent to that of a sphere of 1.5 km diameter while the bulk density is a free parameter varied during the simulation. A radially symmetric repulsive potential is introduced to avoid the grains overlapping. An elastic regime is assumed: the energy dissipation during the collisions between grains is neglected. The trajectories of individual grains are integrated in the gravity field of the other bodies (N-body problem). Regardless of the nucleus density and the number of grains introduced, several clumps tend to form during the tidal disruption. The modelled chain has the same length and position angle as that of SL9 in 1993–1994 if the non-rotating progenitor is formed by more than 50 grains corresponding to a bulk density of $600 \pm 100 \mathrm{kg\,m^{-3}}$. The largest clumps contain 20% of the mass of the progenitor. For a body in prograde rotation, the observed chain of fragments could be explained only if the progenitor had a smaller radius (0.5 km) and a higher bulk density $10^3 \mathrm{kg\,m^{-3}}$. For retrograde rotation, the splitting is incomplete and a large central body survives.

The sublimation of very volatile species like CO or CO_2, which could be exposed to sunlight as a result of the breakup and accelerate the debris in the anti-solar direction, has not been taken into account in these models. The nucleus is considered inactive, although CO is probably produced by cometary fragments as fresh material becomes illuminated. Note that the outgassing – and the resulting nongravitational force – would tend to play against self-gravity, and therefore would imply a higher nucleus density for SL9. No activity could be detected by any observation.

Figure 5 CCD image of comet D/Shoemaker-Levy 9 obtained with the Hubble Space Telescope (WFPC2) in May 1994, about 14 months after the discovery of the comet by E. Shoemaker, C. Shoemaker and D. Levy (Shoemaker *et al.* 1993). A total of 22 fragments have been identified and labelled from A to W. The original comet nucleus split during a close approach at less than two Jovian radii (Section 5.1) in July 1992. The fragments G1, G2, P1, P2, Q1 and Q2 are the result of the splitting of three big fragments, originally labelled G, P and Q, well after the close encounter with Jupiter. The fragments produced spectacular dark spots observable with the naked eye when they plunged into the atmosphere of the giant planet two years later. (Courtesy of H. Weaver (Johns Hopkins University) and T.E. Smith (STScI) and NASA.)

Therefore it is unlikely that fresh (with substantial amounts of ices) interior was exposed during the breakups.

In the future, it would be interesting to study the size and mass distribution of the aggregates within the chain of fragments and compare them to what can be inferred from the observations of the visible images of SL9 and from the effects of the impacts in the atmosphere of Jupiter.

5.3 The formation of catenae by tidally disrupted comets

A total of 116 craters distributed in 11 crater chains on Ganymede and Callisto have a morphology consistent with impacts of fragmenting objects (Schenk *et al.* 1996). The lengths of the chains show that the impacts must have taken place just after the tidal breakup by Jupiter (Melosh and Schenk 1993). The ratios between the diameters of the largest and the smallest craters in a chain range from 1.4 to 4 with an average of 2.2. The craters are well aligned and do not have a preferred orientation on the surface. Gipul Catena, one of the longest chains on Jupiter's moon Callisto, is depicted in Figure 4. McKinnon and Schenk (1995) determined the mass of the bodies that created the crater chains on Ganymede and Callisto from the mass excavated during the formation of the craters. Assuming that the fragments were from Jupiter-family comets (and assuming a nucleus density of 10^3 kg m^{-3}), they found that the radii of the fragments were smaller than 2 km and the sizes of the progenitors ranged from ~0.5 to 4 km. Schenk *et al.* (1996) found a strong correlation (Figure 6) between the average fragment mass and the mass of the progenitor. In other words, a larger progenitor mass does not result in a chain with *more* craters but rather a chain with *larger* craters. This means that comets cannot be composed of cometesimals of uniform size, unless larger comets are composed of larger cometesimals. The building blocks of cometary nuclei must be smaller than the fragments responsible for the chains of craters. This is compatible with a nucleus model comprising a large number of loosely bound subnuclei. A concept of the nucleus structure is depicted in Figure 7 developed by Hughes (1988). The nucleus is built by a hierarchy of subnuclei similar to a fractal. The subnuclei are of low density and fluffy structure and only loosely bound (Section 1.3).

5.4 The implication of tidal breakups on the internal structure of SL9

It seems established that the progenitor of the 21 fragments of SL9 observed in 1992–1993, and more generally the progenitors of the fragments which created the chains of craters on Ganymede and Callisto, are weakly bound aggregates of at least several tens of 'particles'. Models that describe the nucleus as a collection of cometesimals with a characteristic size independent of the nucleus size are not compatible with the crater size distribution in the crater chains (Schenk *et al.* 1996). The bulk density of the progenitor ($500-1000 \text{ kg m}^{-3}$) is tightly constrained by the observed length and orientation of the chain of SL9 fragments. However, the particle size distribution is not constrained by the model calculations of Asphaug and Benz (1996). Finally, SL9 had not been observed before it broke up and finally disappeared. We have no indication of cometary activity of SL9 until it plunged into the atmosphere of the planet. The loss of volatile material before the breakup might have reduced the tensile strength of the nucleus if it was a 'dead comet'. The low tensile strength implied by the model calculations by Asphaug and Benz (1996) might therefore not be a value representative for active comets.

6 COMET HALLEY – A PROTOTYPE NUCLEUS?

6.1 Flyby missions of comet Halley

The first interplanetary mission of the European Space Agency (ESA) was a flyby of comet Halley in 1986. The reappearance of this famous comet after 75 years (its period) triggered the release of an armada of spacecraft: two Japanese passing the comet far outside its coma, two international Vega missions by the Soviet Union diving into the inner coma and the above-mentioned mission by ESA with the objective to come as close as possible to the still hypothetical nucleus (Reinhard 1986). The minimum approach distance of the Japanese missions of 150 000 km permitted the study of some interactions of the outer coma with the solar wind but not the comet itself. These

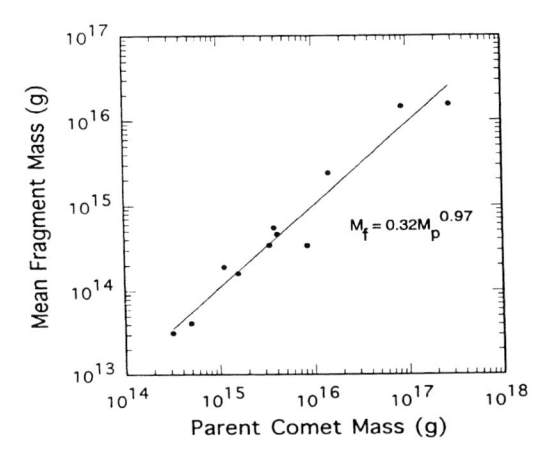

Figure 6 Correlation of mean fragment mass for individual crater chain comets and mass of the associated parent comet (Schenk *et al.* 1996; data from McKinnon and Schenk 1995). Bigger comets break up in bigger subnuclei.

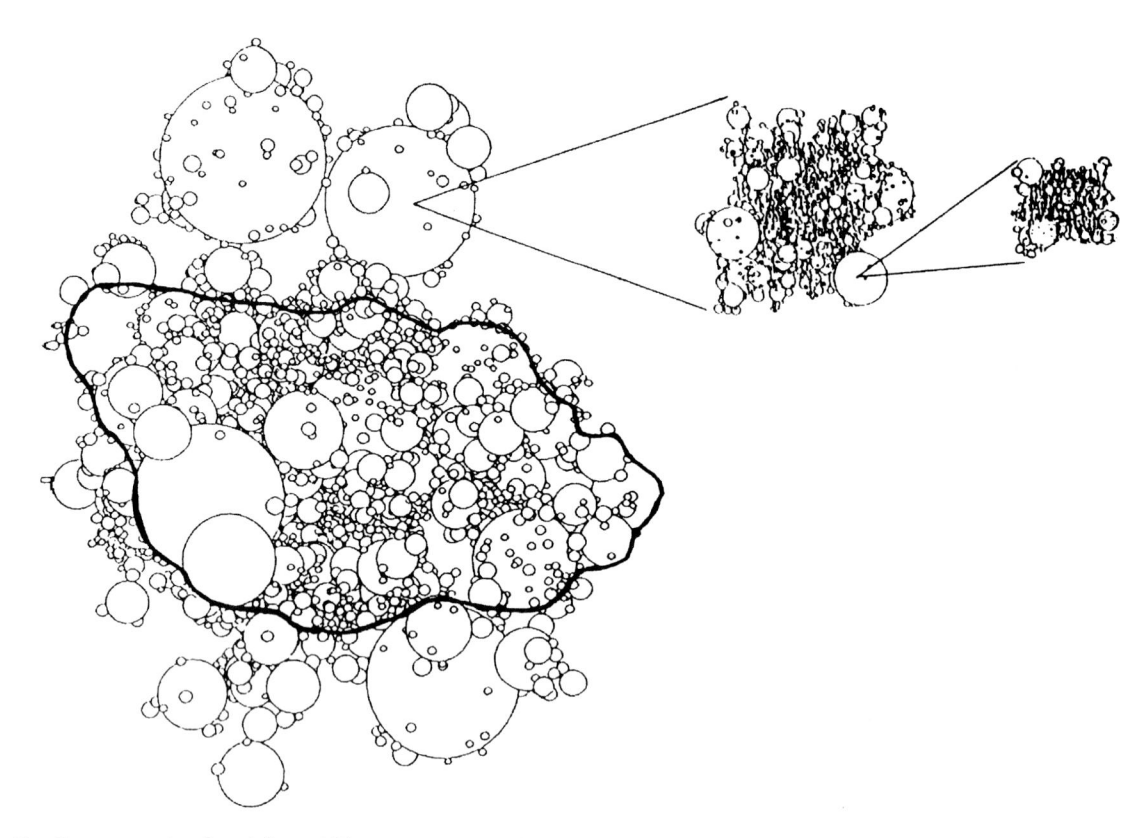

Figure 7 Computer-simulated fractal-like cometary nucleus of fluffy aggregates developed by Hughes (1988). The outline (thick line) depicts the approximate shape of comet 1P/Halley after the loss of loosely bound subnuclei during earlier orbits.

Table 2 Vega and Giotto encounter parameters

	Date (1986)	Time (UTC)	Distance of closest approach (km)	Phase angle* during approach (°)	Azimuth† of target point (°)	Relative velocity (km^{-1})	r (AU)	Δ (AU)
Vega 1	6 March	07:20:06	8889	112.2	19.2	79.222	0.7923	1.153
Vega 2	9 March	07:20:00	8030	123.4	15.0	76.785	0.8341	1.073
Giotto	14 March	00:03:01.84 ± 0.20‡	596 ± 2‡	107.2	−30.6 ± 0.1‡	68.373	0.9023	0.960

* Angle between Sun–comet line and relative velocity vector.
† Angle between the spacecraft–comet line at closest approach and the plane defined by the Sun–comet line and the relative velocity vector (positive: ecliptic north).
‡ Determined by Halley Multicolour Camera.

spacecraft are therefore not listed in Table 2 that provides details of the cometary encounters.

Vega 1 was the first to fly by on 6 March 1986. The images from a range of ~9000 km showed a diffuse intensity maximum with a secondary bright spot. Vega 2 images, taken three days later, looked similar. An immediate interpretation and determination of the nucleus size and shape was not possible. It took the Halley Multicolour Camera (HMC) on board ESA's space probe Giotto passing the nucleus at a distance of less than 600 km to reveal the size and shape of comet Halley's nucleus. The approach geometry was similar for all three missions. The visibility of the nucleus was slightly impaired by enhanced dust production during the Vega 1 flyby. The cameras on both Soviet spacecraft had major technical problems severely limiting the information in the transmitted images. The Giotto spacecraft and HMC were hit by large dust particles shortly before closest approach. HMC transmitted its last image of the nucleus from a distance of 1400 km. While the quality of the more than 2000 images were highly satisfactory a major drawback of the picture series was that the aspect angle started to change only just before the contact with the spacecraft was lost. Consequently the nucleus is only seen from one side, from one direction.

6.2 Images of comet Halley taken by HMC

In the following we will describe the imaging results and some of the images in more detail. They are the only views of a cometary nucleus and most of what we have learned about cometary nuclei is at least related to these images.

The cometary nucleus could be seen at first from a distance of 124 000 km, about half an hour before closest approach (Figure 8). The angular resolution of HMC was 22.4 μrad per picture element (pixel) so that the resolution (2 pixel) at the comet was 2.8 km. From the strong contrast gradient at the origin of the bright inner coma it was clear that this must be the nucleus of comet Halley. Unfortunately, the geometry during approach was unfavourable for all spacecraft. The Sun–comet–Giotto phase angle was 107°. The Sun was 17° below (behind) the image plane, that is HMC was looking toward the Sun and most of the nucleus surface was unilluminated. All HMC images shown are oriented in such a way that the plane defined by the relative velocity vector and the nucleus is parallel to a horizontal line. The direction towards the Sun varies from 25 to 30° above the horizontal. Celestial north is close to the up-direction of the images. All details of the images and their presentation are given in Keller *et al.* 1995).

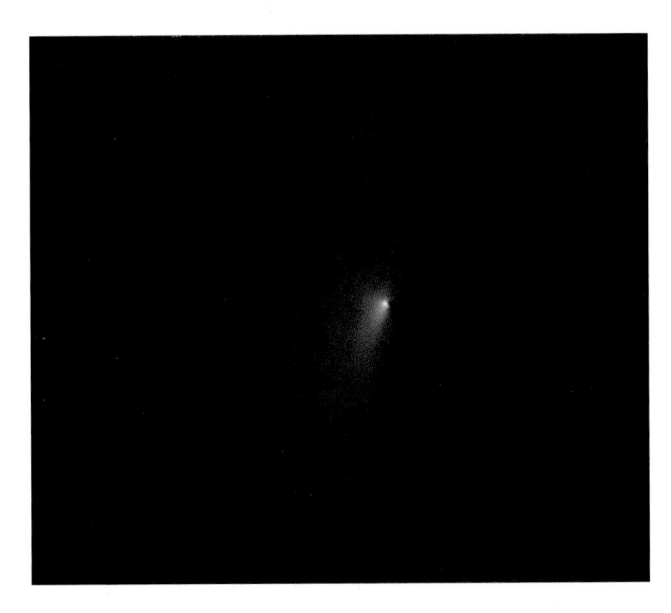

Figure 8 Halley Multicolour Camera (HMC) image 3056 of comet 1P/Halley taken at a distance of 124 000 km. The nucleus is resolved. North is approximately up and the Sun is approximately to the left. (All images taken from the HMC Atlas (Keller *et al.* 1995) are copyright MPAE.)

The series of images depicts the approach to the nucleus. The camera concentrates on the brightest parts of the scene. This is the jet activity near the sub-solar point at the northern tip. Image #3502 is one of last data transmitted and shows parts of the active region with a resolution of 50 m px^{-1} but somewhat reduced signal to noise because of the short exposure time.

Sixty-eight images are combined in a composite shown in Figure 9. Images with higher resolution replace the corresponding parts of images taken from a larger distance. The resolution of this composite image varies from 300 m px^{-1} at the southern end to 60 m px^{-1} at the active area in the north. The aspect angle changes also by 11°, particularly for the last images included. These changes cause some artificial effects that have to be considered when the image is interpreted.

6.3 Shape and orientation of the nucleus

Only about 25% of the visible surface is illuminated by the Sun. The terminator is running from the northern tip down to the southern end. Strong activity is visible on the sunward part of the surface concentrated near the sub-solar area. Surprisingly the outline of the bright limb is more difficult to define than the dark limb of the unilluminated surface. This is due to the fact that the dark nucleus is silhouetted against the illuminated dust of the inner coma in the background. Only because of the excellent performance of HMC and this fortuitous configuration are we in a position to determine the size and the shape of the nucleus of comet Halley.

The projected surface area seen by HMC corresponds to an ellipsoid of 14×7 km^2. The volume and three-dimensional shape of the nucleus cannot be determined from HMC images. Based on the knowledge of the full outline for one position of the nucleus (Figure 10), the images (Figures 11 and 12) of the Vega cameras (Szegö *et al.* 1995) can be used to derive the overall size and volume and some large figurative features. Extensive work in this area was done by Merényi *et al.* (1990) and Stooke and Abergel (1991). Even including in addition a discussion of the rotation period and position of the rotation axis (e.g. Belton *et al.* 1991) has not produced an unambiguous and generally agreed model. From the maximum length of the nucleus of 15.3 km observed by Vega one can estimate that the long axis in the HMC images points about 30° out of the image plane. Is it towards the observer or away? Which one is the big end on the Vega images? The comet Halley results demonstrate how difficult it is to determine the volume of an object with a flyby only. Merényi *et al.* (1990) give a size described by the best fit triaxial ellipsoid of 7.2, 7.22 and 15.3 km with an error of ±0.5 km on each axis. Stooke and Abergel (1991) found 6.7, 8.0 and 15.1 km for the three axes and estimate the uncertainty more realistically to 1 to 2 km. We rendered

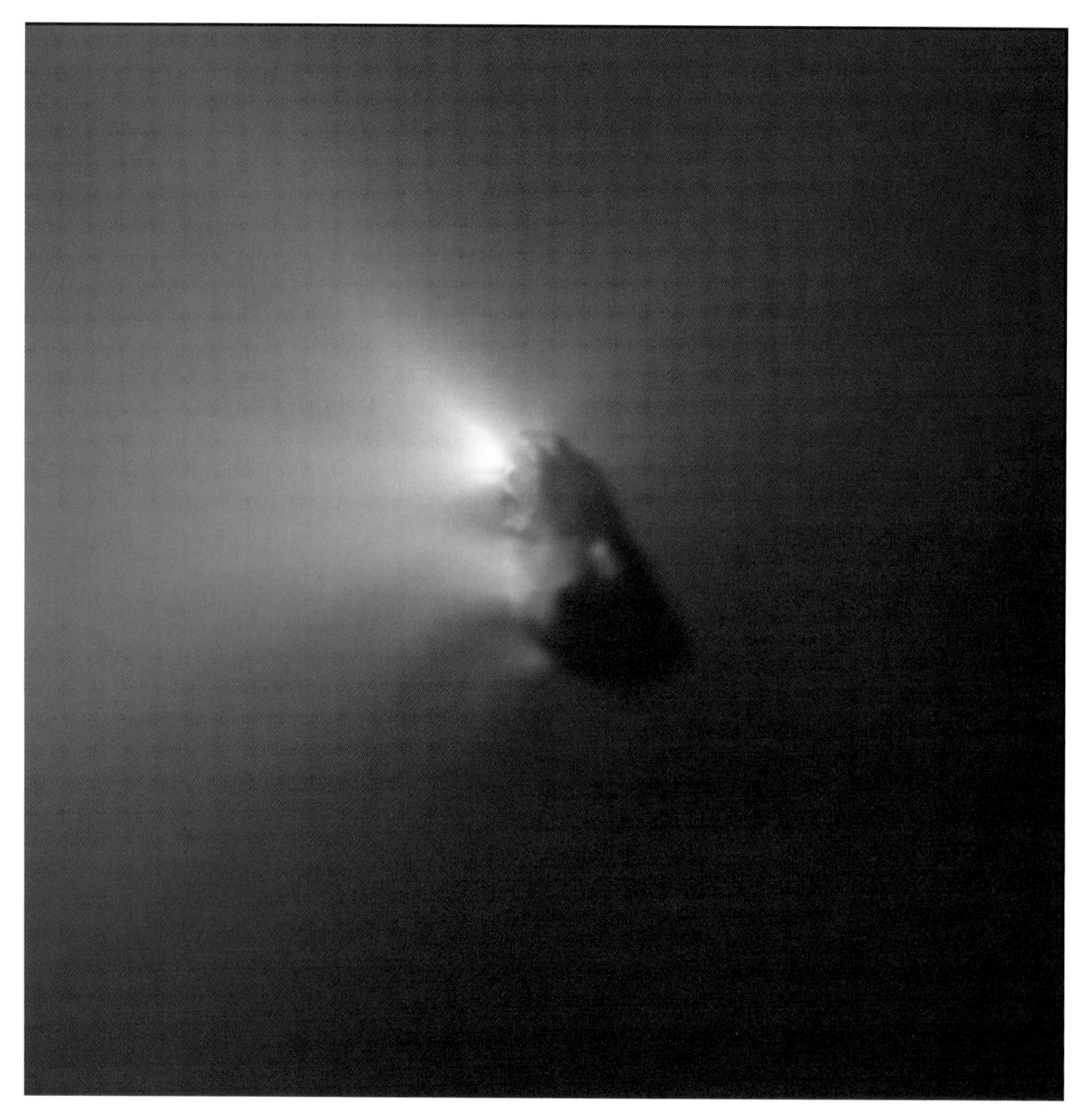

Figure 9 A composite image of the nucleus of comet 1P/ Halley composed of 68 images ('Best Image'). The resolution varies from 320 m px^{-1} at the south end to 50 m px^{-1} near the active region.

their model of the nucleus (Figures 13 and 14; see also figures in Stooke and Abergel 1991). Their shape cannot reproduce the slanted illumination of the Central Depression and the Mountain (see below) and hence the characteristic course of the terminator seen on the HMC images. Therefore the rendering of the aspect angle seen by HMC does not produce an image with high definition.

The overall shape could be described by two spheres, the one at the south end more massive, stuck together at the 'waist' as indicated in all models. The rotation axis (Section 8.2) is of course in the middle of the 'dumbbell'. Taking the deviations from the triaxial ellipsoid (420 km^3) into account leads to an estimated volume of 365 km^3 and an overall surface of 294 km^2 (Merényi *et al.* 1990).

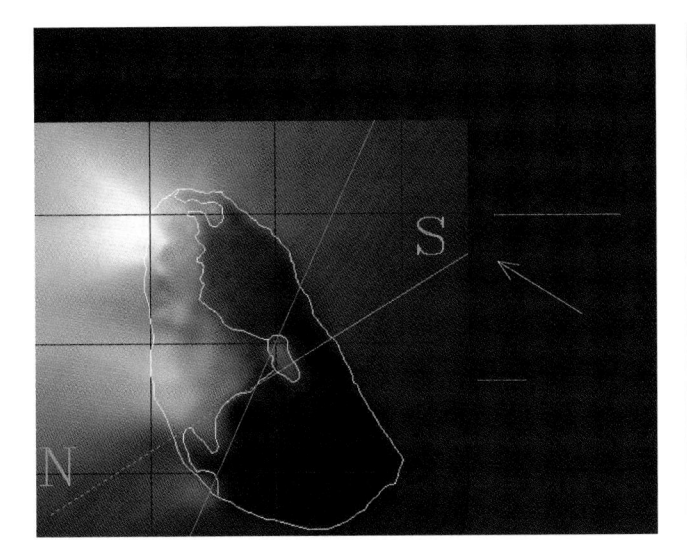

Figure 10 A composite image showing the rotation axis derived by Wilhelm et al. (1986) projected in the HMC image plane (yellow line). Note that the axis is almost perpendicular to the long axis of the comet. The projection of the angular momentum vector of Belton et al. (1991) is shown in green.

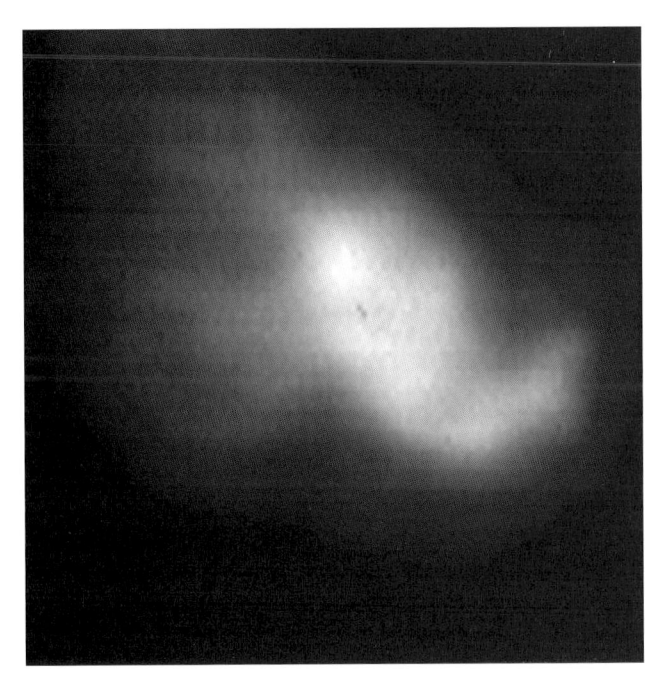

Figure 11 The Vega 1 image (6135) from the Vega Atlas of comet 1P/Halley (Szegö et al. 1995) was cleaned and geometrically corrected. It shows the nucleus and its vicinity from a distance of 8904 km at a phase angle of 29° on 6 March 1986. The view is onto the big end (following Merényi et al. 1990).

Rickman (1989) derived the mass of the nucleus from a study of the non-gravitational forces to 1 to 3×10^{14} kg and derived a density of 550 ± 250 kg m^{-3}.

Figure 12 The Vega 2 image (6190) from the Vega Atlas of comet 1P/Halley (Szegö et al. 1995) was cleaned and geometrically corrected. It shows the nucleus and its vicinity from a distance of 8030 km (near closest approach) at a phase angle of 28° on 9 March 1986. Strong jet activity (towards the lower left direction) is visible. Viewing direction is close to perpendicular to the long axis, that is, the full length of the nucleus is visible.

The geometric albedo of 0.04 (+0.02, −0.01) derived from Vega images (Sagdeev et al. 1986) assuming a Moon-like phase function is darker than that of the Martian moons Phobos and Deimos but similar to C-type asteroids and other nuclei determined from IR observations (Bowell and Lume 1979, Hartmann et al. 1982, Cruikshank et al. 1985). This makes comets one of the darkest objects in the Solar System. Figure 15 shows isophotes of the reflectivity. Note the brightest point at the foot of the most intense dust jet corresponds to only 0.8%. The illuminated surface at the phase angle of 107° is darker than 0.6% (Keller et al. 1986). The colour of the surface is slightly reddish with a gradient of 6 ($\pm$3) % (100 nm)$^{-1}$ over the wavelength range of 440 to 880 nm (Thomas and Keller 1989), similar to P-type asteroids (Figure 16). The bulk properties of the nucleus of comet Halley are summarized in Table 3. The surface of the nucleus is extraordinarily uniform in

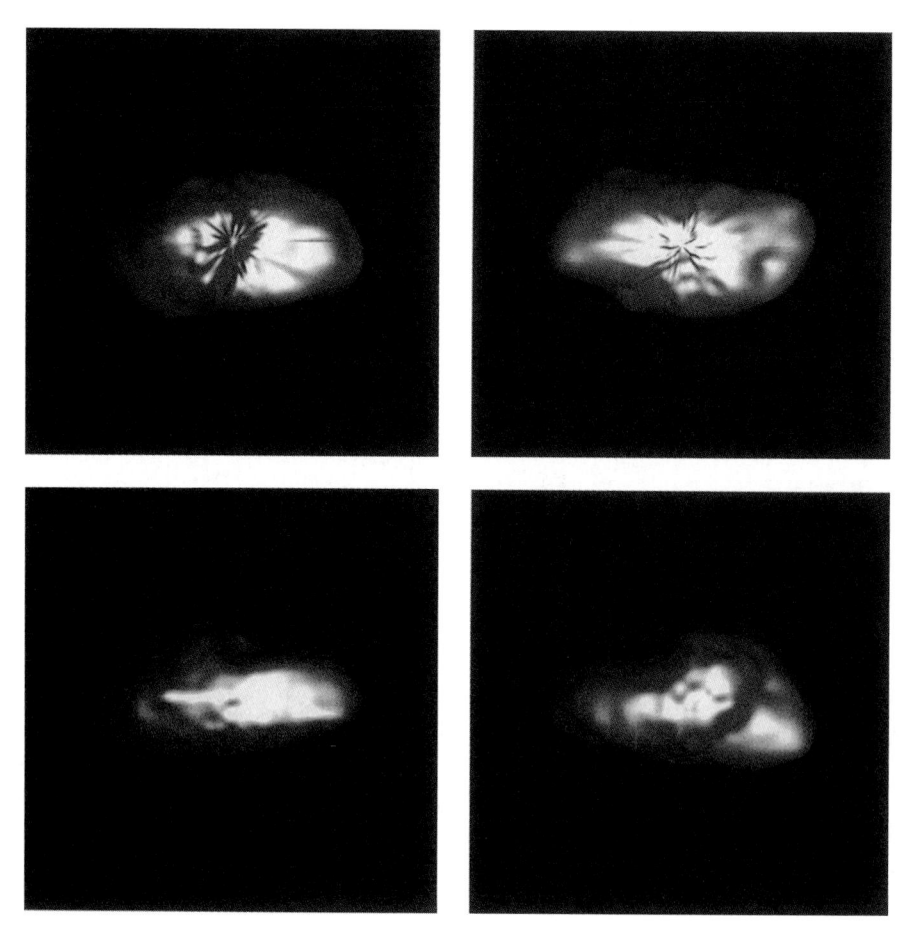

Figure 13 Stooke and Abergel (1991) proposed a shape model for 1P/Halley's nucleus based on images taken during the flybys of the nucleus by the Vega 1, Vega 2 and Giotto spacecraft. The top images are pictures of the nucleus as seen from the south (left) and north (right) poles. The bottom images are two opposite side views of the model, corresponding to longitudes of 0–180° (left) and 180–360° (right). The model has been rendered assuming a Lambertian surface. The 'rays' close to the poles are artifacts which appear during the resampling of the model data.

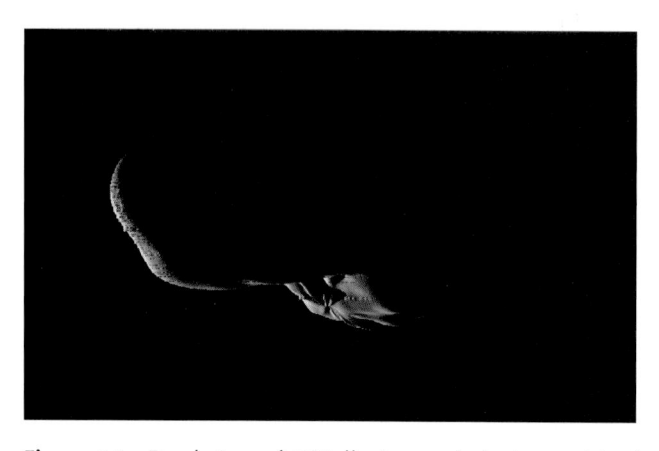

Figure 14 Rendering of P/Halley's morphologic model of Stooke and Abergel (1991) as seen from HMC assuming a Lambertian surface. The orientation is based on the rotational parameters of Sagdeev et al. (1989). It is similar to that of Wilhelm et al. (1986) (Figure 10). This image looks different from the HMC image (Figure 9) because the orientation relative to the Sun is incorrect, as discussed by Stooke and Abergel (1991) and in Section 8.2.

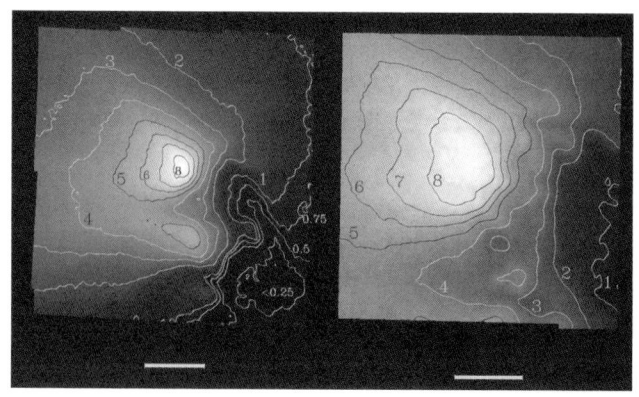

Figure 15 Two clear-filter images of the nucleus of 1P/Halley with intensity contours. The contour levels are marked and are in units of reflectivity ×1000. The highest marked level corresponds to a reflectivity of only 0.8%. (From Thomas and Keller 1989.)

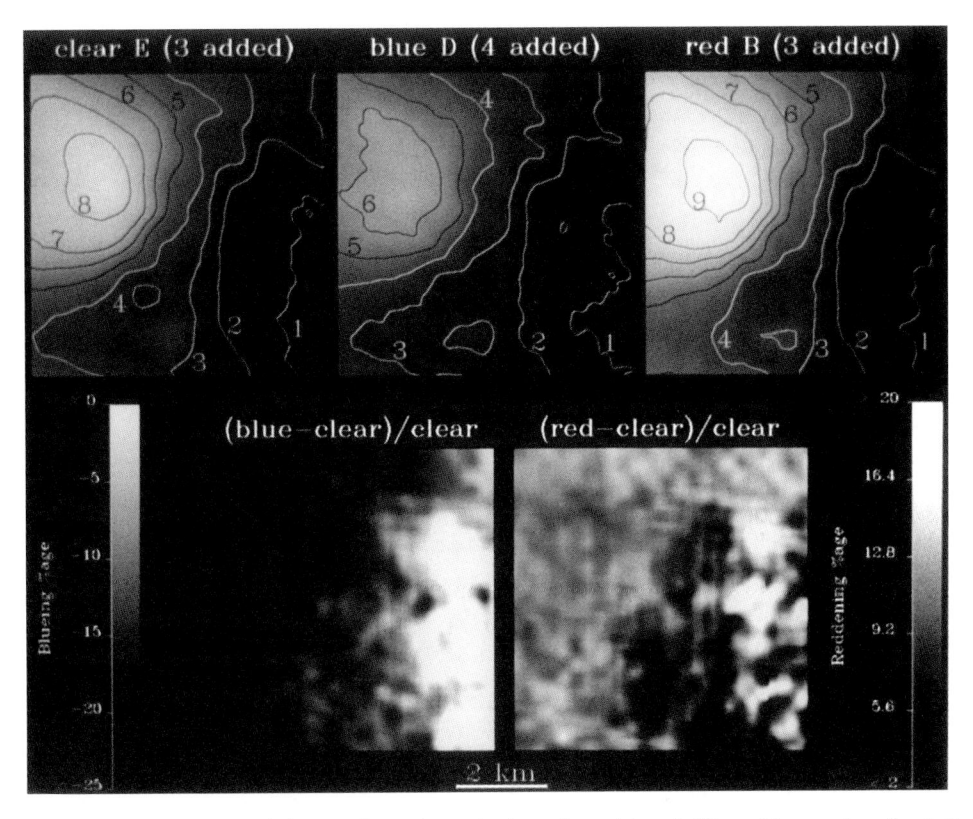

Figure 16 Reflectivity contour maps of the nucleus through three broad-band filters. The peak reflectivity in the blue is considerably below that in the red. Shown below are colour difference maps. The nucleus is less red than the surrounding dust. (From Thomas and Keller 1989.)

Table 3 Comet Halley's nucleus properties

Projected shape (full outline)	Max. length 14.2 ± 0.3 km Max. width 7.4 ± 0.2 km	HMC
Model body	15.3 × 7.2 × 7.22 km^3	Merényi et al. (1990)
Volume	420 ± 80 km^3 triaxial ellipsoid 365 km^3 model body	Merényi et al. (1990)
Surface	294 km^2	
Topography	Mountains, ridges, terraces	HMC
Activity	Concentrated in three major areas, ≤10% of surface	
Geometric albedo	0.04 (+0.02, −0.01)	Sagdeev et al. (1986)
Colour (reddish)	Reflectivity gradient: 6 ± 3% (100 nm)$^{-1}$ from 440 to 810 nm	Thomas and Keller (1989)
Mass	1–3 × 10^{14} kg	From non-gravitational forces Rickman et al. (1987)
Density	550 ± 250 kg m^{-3}	Rickman et al. (1987)
Rotation (complex)	Spin period 2.84 d 7.1 d around long axis 3.7 d nutation	Belton et al. (1991)

reflectance (Keller 1989), no bright spots (icy patches) were detected, neither in Vega nor in HMC images.

6.4 Topography and morphology

The fully visible outline of the HMC images shows a body the shape of which strongly deviates from a sphere ('snowball') and even from the best fit ellipse. The backside (unilluminated north-eastern) limb follows a straight line for about 10 km in length to terminate in an almost rectangular corner called the Duck Tail. This corner protrudes by $\Delta R/R = 0.3$ above the radius of the best fit ellipse. The relatively smooth course of the terminator north and south of the Central Depression being paralleled by a bright band (Ridge) at the morning southern illuminated side indicates large-scale features such as a terrace. The Central Depression tapers towards the Mountain. Its illuminated tip is estimated to protrude to $\Delta R/R = 0.25$ or 900 m above the best fit ellipse. Most features are identified in Figure 17.

These large-scale topographic features are counterbalanced by structures with typical dimensions of the order of 500 m to 1 km. The most obvious example is seen at the Chain of Hills. This scale is also found at places where the resolution of HMC images is smaller (down to about 50 m px^{-1}) near the bright activity and inside the Crater. This depression covers a projected area of 12 km^2 near the illuminated limb and is therefore distorted by the viewing geometry. All the observed topographic features (with the exception of the Mountain) could be shallow because the solar zenith angle is large. The depth of the crater was estimated at 200 m by Schwarz *et al.* (1986). Shallow, roundish crater-like features may be more common. At least one more 'crater' with a diameter of 600 m can be seen south of the Finger (the dark valley on the northern tip, the cutout is labelled Hill on Figure 17).

The global smoothness of the surface is contrasted by the highly irregular overall shape with its 2:1 axes ratio and the small radius (300 m) of curvature seen at the Duck Tail

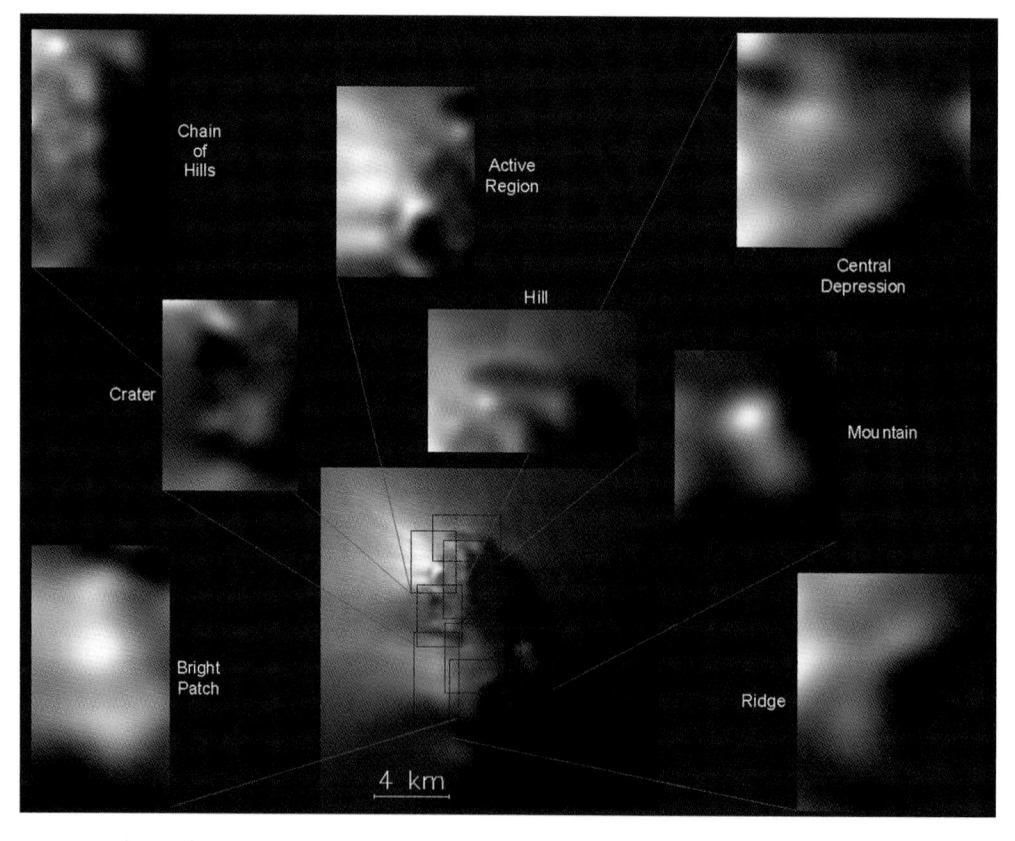

Figure 17 Features on the surface of the nucleus of comet 1P/Halley. Sections of the composite image (centre bottom) have been extracted and expanded by a factor of three to show, in detail, notable features on the nucleus mentioned in the text. The position of each expanded section is marked as a box on the composite and a corner of each section is linked to its counterpart by a line. Nonlinear enhancement has been applied to provide improved contrast within the cutouts. (From Keller *et al.* 1988.)

and high Mountain. Some sharp protrusions exist on the surface. Active areas may possibly be slightly brighter than the surroundings.

6.5 Activity and topography

The brightest spot on HMC images is near the northern tip. Dust cones are emitted from the areas at the illuminated limb. Several small filaments are visible, some of them originating from the visible surface. In two cases filaments in the inner coma could be traced to small active areas on the surface with diameters of about 500 m (Thomas and Keller 1987; for an explanation of the filaments see Knollenberg et al. 1996). The total extent of the active area is about 3 km along the limb. This active area is not the strongest source of dust. Most of the dust seen from far and intermediate distances comes from a source on the side turned away from HMC. This explains why the strongest dust feature changes its direction from south to north of the Sun–comet line when the close-up images start to resolve the optically thicker foot points of the dust jets at their origin. Taking also Vega and ground-based observations into account about five active areas are identified (Belton et al. 1991). The areas cover about 10% of the total or 20% of the illuminated surface at the time of the Giotto flyby (Huebner et al. 1986).

The amount of material lost from an active area during one appearance of comet Halley depends on the sublimation rate. Comparing the total active area to an estimate of the total gas production shows that roughly $30\,km^2$ of activity are sufficient if the water sublimation rate is not strongly quenched by a dust crust or mantle (Huebner et al. 1986). Depending on the assumed density and water to dust ratio this corresponds to a layer of roughly 6 m in depth. If the Crater is an area of activity (not yet activated on HMC images early in the morning) then it may have been just visible 2000 years ago.

There is no indication of any substantial nightside activity of the nucleus. The ratio of dust in the sunward to that in the anti-sunward hemisphere is 3.2:1. This very low value still awaits a realistic explanation. Assuming a triaxial ellipsoid for the nucleus and uniform dust production from all illuminated parts one would expect a ratio 9:1 for the HMC observational geometry. If the absorbed power is taken into account this ratio increases to 40:1. Lateral flows derived from hydrodynamic calculations of the gas–dust expansion above active surfaces do not provide a strong enough flow to explain the observations (Kitamura 1986, 1987). Keller et al. (1990) suggested that a major part of the sublimation takes part in a distributed source of ice particles lifted from the surface by gas drag. The subliming ice particles increase the lateral flow. Knollenberg (1994) showed that even under the most favourable circumstances the 3.2 ratio cannot be explained. At the same time these gas-dynamic calculations set an upper limit to the activity ratio on the nightside of less than 10^{-3}.

Two-thirds of all dust observed at a distance of a few hundred kilometres originate from the two major sources discussed above. The upper limits for the halfwidths of the jets are between 30 and 40°. These rather narrow jets can be modelled in axisymmetric gas dynamics of the acceleration of dust particles by subliming water (Knollenberg 1994) assuming a dust particle size distribution that is optically characterized by about $10\,\mu m$ particles. Even shallow indentations such as the Crater (200 m) will focus the jets by 25%. The model calculations based on three jets characterized in Table 4 represent the observations quite well (Figure 18).

Table 4 Parameters of major jets derived from the intensity distribution of dust near the nucleus of comet Halley

Direction (°)	Halfwidth (°)	Fraction (%)*
137	37	47
198	31	17
273	44	11

*A constant background comprising 25% supplements to total dust production.

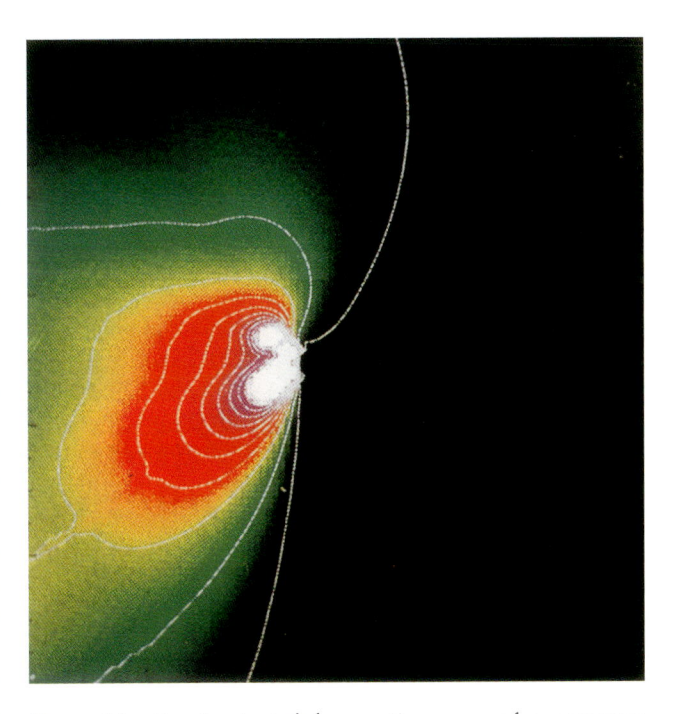

Figure 18 Gas is ejected from active areas of a cometary nucleus and drags dust particles along. This gas-dynamic interaction can be modelled. Three dust jets are superimposed to simulate the appearance of the dust coma at the time of the Giotto encounter with comet 1P/Halley. (From Knollenberg et al. 1996.)

The observed fine filaments (Thomas and Keller 1987) cannot be explained by collimation of dust, not even from deep, active craters (the bottom of which would not be reached by sunshine anyhow). Rather than by an enhancement of activity, the filaments can be formed by reduction of activity across areas of a few hundred metres across lying within a larger area of activity (Keller *et al.* 1994).

It is clear that the observed rather narrow jets are connected to large-scale (several kilometres) topographic features of the nucleus, mostly concave contributing to their collimation. Crifo and collaborators (Crifo and Rodionov 1997a,b, Crifo *et al.* 1999) have in a series of complex multidimensional gas-dynamic model calculations shown that most observed features can be reproduced by appropriately sculpturing the active areas. Crifo suggests that the jets are only formed by surface landscape (topography) and disputes the concept of active and inactive areas assuming uniform activity proportional to instantaneous insolation. The good fit based on three active sources (Figure 18) and in particular the observed low ratio of dust in the sunward and anti-sunward hemispheres of 3.2 are strong arguments against a uniformly active surface. In addition the filaments clearly argue for spots of reduced (or in-) activity, that is, for a heterogenous surface. A further argument is derived from the actual existence of such large-scale protrusions as the Duck Tail and the Mountain. A uniform surface activity should eliminate the 'sharp' edges of the nucleus first.

Vega images did not reveal any details of the nucleus surface even though it was claimed that linear features could be identified on two heavily stretched and processed Vega 2 images (Möhlmann *et al.* 1986, 1987).

6.6 Conclusions from nucleus imaging

The spacecraft observations and in particular the HMC images reveal an evolved, solid nucleus built of porous, dark, non-volatile matrix material. Its surface reaches temperatures of 400 K (Emerich *et al.* 1987a,b) much higher than the sublimation temperature of water (200 K). Active and inert surfaces can hardly be discerned by their reflectivity. The interior (material containing water ice) revealed within the active areas that lose many metres of material during one orbital revolution of comet Halley looks the same as the depleted inert surface. The frequent fragmentation of dust particles near the surface (Thomas and Keller 1990) underlines the low tensile strength and porosity and hence low density of the nuclear matrix.

Layers of 5 to 10 m (depending on the density of the material and the volatile to non-volatile ratio) are removed from the surface of active areas during each orbital revolution. Averaged over the whole surface the loss is of the order of 1 m. The strongly elongated body shape (2:1 ratio of main axes) could not be formed from an originally spherical body just by sublimation and erosion of surface material. This rough body must reflect the interior structure of the nucleus put together from a few large building blocks several kilometres in size. However, it is more probable that the present nucleus of comet Halley is the product of frequent splitting rather than of sublimation. The splitting would be facilitated by the inhomogeneity of its interior formed by subnuclei and their mutual penetration during the agglomeration of comet Halley. Estimates of the mass of the meteoric debris in the associated meteor streams (Section 10.3) lead to a 5 to 10 times bigger original mass some 2000 or 3000 orbits ago. This makes it clear that comet Halley was originally considerably larger and that its present nucleus has very little to do with its original shape.

Even if comet Halley's original surface was covered by a layer mutated by its long storage in the outer EK belt or Oort cloud, as discussed for the surface of new comets (Johnson *et al.* 1987) and for the surface of the EK belt objects (Jewitt and Luu 1998), nothing of that is left now.

The frequently observed scale of features around 0.5 to 1 km (Chain of Hills, filaments, structures within the Crater, etc.) indicates substructures of this size referring to subnuclei of the larger subnuclei determining the dumbbell shape of the nucleus.

Large parts of the present nucleus seem to be depleted of volatiles. The Mountain and the Duck Tail protrude so far above the reference ellipsoid that they are illuminated by the Sun from three sides. Volatiles are sufficiently enough depleted to suppress any activity and only the non-volatile bulk (matrix) material is left. Temperatures reach more than 400 K around perihelion and the heat wave can penetrate deeply once the volatiles in the matrix are lost.

The existence of these protrusions make a thin mantle of regolith on top of an essentially pristine icy interior rather improbable. The present loss mechanisms do not flatten protrusions as would be expected in the case of a uniform (icy) body.

Related to this topic are the most surprising observations of the restricted surface activity of the 'most active' short-period comet Halley. As a consequence, the surface and hence the nucleus are much larger than was originally expected. Many of the short-period comets display activity levels one or two orders of magnitude smaller than the level of comet Halley. In cases where information of the nuclear sizes could be achieved from ground-based remote sensing the volume of the nuclei (and in several cases the extreme axes ratio of 2:1) is comparable to that of comet Halley (A'Hearn 1988) resulting in an active surface fraction of less than 1% and smaller. How can comets maintain such low levels of activity over many orbital revolutions around the Sun? Comet Halley (or comet Encke) has hardly changed in activity since its first recordings some 30 revolutions or 2200 years ago (Yeomans and Kiang 1981). The physical explanation of this phenomenon is one of the keys to our understanding of the physics and nature of comets.

Compared to the pre-Halley era of cometary research when dust to gas ratios of 0.1 to 0.3 (Divine 1981) were generally 'observed' or assumed the role of the non-volatile component has strongly increased. The spacecraft observations revealed a large fraction of big dust particles. The images show that the non-volatile (matrix) component dominates the consistency of the nucleus. The nucleus of comet Halley does not look like a uniformly shrinking ice ball. It is not the ice that determines the appearance of the nucleus but rather the solid material. 'Icy dirt ball' may be more descriptive than 'dirty snowball' (Keller 1989).

7 MODELLING THE NUCLEUS

Assuming that typical comets (LP and SP) in the size range $0.5 < R < 50\,\mathrm{km}$ are not pieces produced by a breakup of larger cometary bodies (observed TNOs) the heating during the agglomeration process when larger cometesimals collide should be only a few tens of degrees. Locally of course somewhat higher temperatures could occur when bodies penetrate each other, in particular because the heat conductivity of the very porous dust matrix is very low. The matrix will be destroyed and highly volatile compounds could be sublimated and may recondense in the debris volume or diffuse into the voids between the individual subnuclei. Large bodies like the observed TNOs ($R > 100\,\mathrm{km}$) will be heated up by 50 K and more. But even in this case some signatures of the original material from which the cometesimals were agglomerated will be preserved.

One of the most important and debated properties of a cometary nucleus is the state of its water ice component. Is it crystalline or amorphous? Physical properties of the two states of water ice are considerable different. Of particular interest is the heat conductivity that is $K(T) = 567/T\,\mathrm{W\,m^{-1}\,K^{-1}}$ for crystalline ice and $K(T) = 2.34 \times 10^{-3}\,T + 2.8 \times 10^{-2}$ for amorphous ice (Klinger 1980, 1981, Kühmt 1984), both in compact form. The conductivity is orders of magitude smaller for amorphous ice. More recent laboratory experiments suggest even much lower values (Kouchi et al. 1992). Even more important for the activity of cometary nuclei is the fact that the transition from amorphous to crystalline ice is exogenic and takes place above about 135 K, on reasonable timescales at 153 K (Smoluchowski 1981). Under very favourable circumstances (compact, pure amorphous ice) a runaway transition into crystalline ice can occur. But for a porous matrix the transition wave triggered by heat from the surface dies after a few metres.

The interior of a cometary nucleus is only heated up after many revolutions around the Sun. Amorphous ice prevents the nucleus from reaching its equilibrium temperature, T_e:

$$T_e = \frac{1}{\tau}\int_0^\tau T_s\,dt$$
$$= \frac{1}{\varepsilon\sigma\tau}\int_0^\tau \left[\frac{C_s(1-A_s)}{4r_h(t)^2} - (1-f_d)LZ(T_s) - K\frac{dT}{dr}\bigg|_{r=R}\right]^{1/4}dt \quad (4)$$

where T_s is the surface temperature, σ the Stefan–Boltzmann constant, ε the emissivity, C_s the solar constant, A_s the surface albedo, r_h the heliocentric distance, f_d the fraction of inactive area, L the latent heat of water sublimation and Z the sublimation flux.

Figure 5 of Kühmt (1984) shows that a nucleus of crystalline ice with $R = 3\,\mathrm{km}$ on a Halley-type orbit would reach its equilibrium temperature of about 70 K after more than 100 revolutions around the Sun, a time much longer than Halley's comet is estimated to have been on its present orbit. The temperature in an amorphous ice nucleus does not change at all over this time interval.

The surface temperature T_s reaches more than 400 K once the comet is within 1 AU from the Sun (Emerich et al. 1987a,b). Within a few tens of metres the temperature drops down to the storage temperature of the nucleus (Oort cloud about 10 K, EK belt about 40 K). This temperature gradient in the radial direction but also along the surface (details depend on the orientation of the spin vector and the period) causes strong stresses within the near surface layers. Estimates and calculations (Kühmt 1984) show that these stresses can reach values orders of magnitude higher than the tensile strength of compact ice (about 2 MPa). Metre-sized pieces can break up. The strength of compact ice itself is orders of magnitude larger than what is estimated from splitting of cometary nuclei (Sekanina 1982). From the tidal break-up of comet Shoemaker-Levy values of about 10 Pa were derived (Section 5.2). This, however, may refer to the strength between building blocks rather of the subnuclei themselves. The frequency of observed cometary splitting (Section 5.1) clearly indicates that often subnuclei are only loosely bound. In this respect it is important to remember that the gravitational forces on the surface of a typical cometary nucleus are four orders of magnitude smaller than on Earth. Centrifugal forces at the tips of elongated nuclei in addition weaken the attractive forces. The 54 h spin period of comet Halley is, however, significantly longer than the period of 5–13 h required to balance gravity at the ends of the nucleus (Section 8.2).

The high-resolution observations of comet Hyakutake when it was as close as 0.1 AU to Earth revealed a series of small knots of activity (Desvoivres et al. 1999) that had

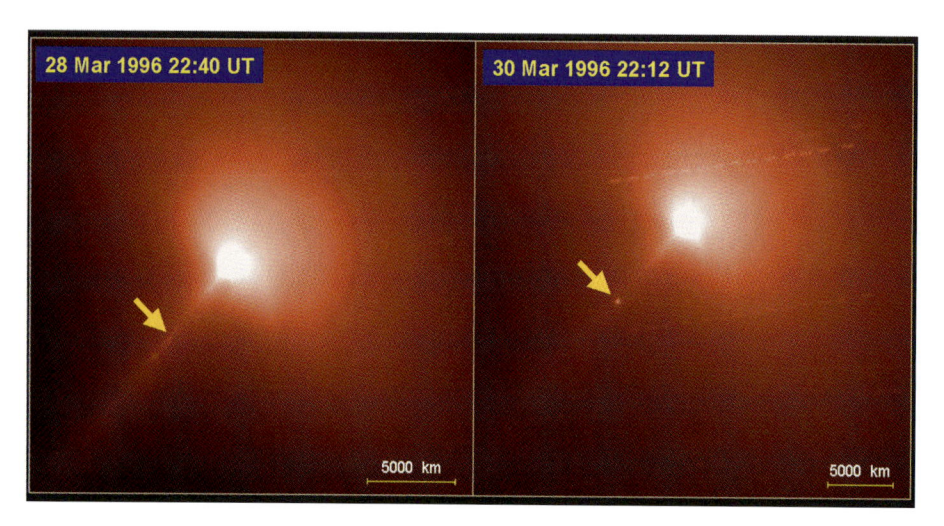

Figure 19 Images of comet C/1996 B2 (Hyakutake) obtained with the 1 m telescope at Pic du Midi Observatory on 1996 March 28.944 UT (left) and March 30.925 UT (right) with a broad-band filter. Both images show fragments (arrows) of several metres (Desvoivres *et al.* 1999) ejected by the nucleus several days before. (Courtesy of J. Lecacheux and F. Colas, Societe de Planetologie des Pyrenees.)

short lifetimes of a few days (Figure 19). The shedding of pieces of surface layers and smaller subnuclei during periods of activity is a common process and probably the rule. The lost pieces are normally just too small to be observed. The activity in the inner Solar System is a relatively violent process for the fragile cometary matrix material and inter-subnucleus bonds. Weak links (cracks) may be created by thermal stresses, and material strength and properties will vary with the content of volatiles. Recondensing ice strengthens the dust matrix by sintering and by increasing the Hertz factor enhances the heat conductivity.

7.1 Amorphous ice

Amorphous ice in cometary nuclei does not only provide an effective means of storing internal energy but can also store very volatile compounds such as CO that are trapped during the sublimation of the water ice and released during the crystallization process. The phase transformation takes place some 10 m below the nucleus surface and once initiated could affect several hundred metres of the amorphous ice (e.g. Prialnik and Bar-Nun 1987). How strong the runaway is depends on the temperature gradient inside the nucleus and hence of the amorphous ice. If the released transition energy suffices to heat the lower layers above the transition temperature the process continues into the nucleus.

The presence of amorphous ice has been used to explain the outbursts of cometary activity (West *et al.* 1991, Klinger 1993), the storage and release of highly volatile compounds (Bar-Nun *et al.* 1987) and the asymmetry of cometary activity relative to perihelion. The modelling of the (erratic) outbursts of comet 29P/Schwassmann-Wachmann 1 that orbits the Sun on an almost circular orbit at 6 AU shows that the CO production and its sudden variations can be understood based on a nucleus of amorphous water ice and dust (Enzian *et al.* 1997). The small variations of the insolation combined with the orientation of a rotation pole towards the Sun during perihelion suffice to trigger the amorphous ice transition and hence an outburst. Most of the physical processes explained by the presence of amorphous ice have also been simulated or explained without its presence. In fact the formation of comets from grains covered with amorphous ice is not probable.

Kouchi *et al.* (1994) discussed the conditions for condensation and preservation of amorphous ice and the cristallinity of astrophysical ices. During vapour deposition on the surface of a dust grain the condensation has to proceed fast enough that water molecules cannot rearrange themselves on the crystal lattice. The condition $F > F_c = D_s a^{-4}$, where D_s is the surface diffusion coefficient and a is the lattice constant of crystalline ice, requires that the flux of condensing water molecules, F, is larger than a critical flux, F_c, defined by physical properties of the ice. Figure 20 shows that this condition is neither fulfilled in the primordial solar nebula nor in circumstellar envelopes. Amorphous ice could only develop in molecular clouds. Even if comets were agglomerated from original material of a molecular cloud the question remains as to whether the amorphous ice could survive this agglomeration and formation of the Solar System. The timescale of crystallization is $t_c \propto e^{E/kT}$, an exponential function of temperature, T, and E, an experimentally determined

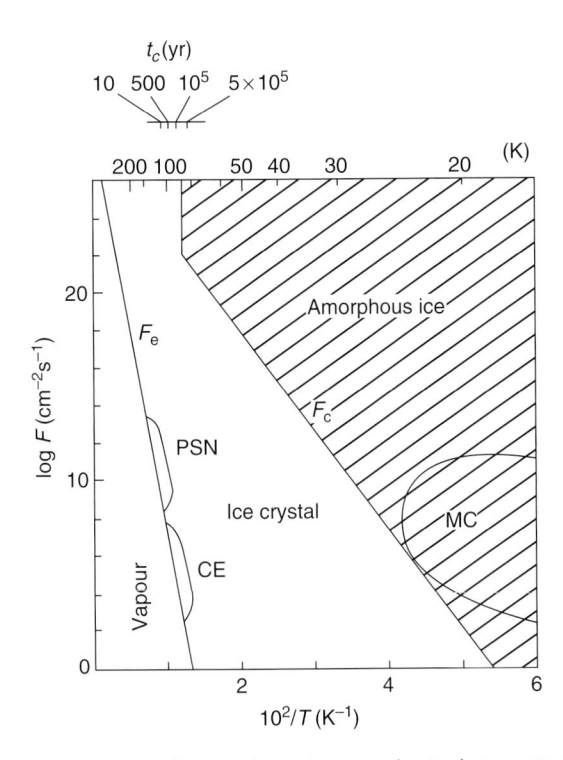

Figure 20 Crystallinity of ices in astrophysical sites. F_c^* is the critical flux and t_c is assumed to be 10^7 years. On the upper abscissa the timescale for crystallization is shown. PSN, CE and MC denote the primordial solar nebula, circumstellar envelope and molecular cloud, respectively. (After Kouchi et al. 1994.)

energy. Any, even short, heating (e.g. by shocks during collapse) will lead to conversion of the amorphous ice. In the later stages of the Solar System nebula temperatures at Uranus and beyond may have been low enough to sustain amorphous ice but not to recondense water in amorphous form (Kouchi et al. 1994).

IR observations of circumstellar envelopes (Omont et al. 1990) argue for condensation of crystalline ice (Kouchi et al. 1994). The striking similarity of IR spectra of cometary dust and crystalline silicate dust in disks around young stars (Crovisier et al. 1997, Ehrenfreud et al. 1997) indicate a high-temperature history at some stage. Klinger (1983) and Herman and Weissman (1987) come to the conclusion that Jupiter family comets should not contain amorphous ice and model calculations for a Jupiter family comet by Espinasse et al. (1991) lead to a runaway conversion of amorphous ice. In summary, it seems rather improbable that cometary nuclei contain amorphous ice. A further argument against amorphous ice in comets derives from internal heating by radioisotopes. It need not only be ^{26}Al that heats the central nucleus so that liquid water is produced (Wallis 1980) but even longer lived isotopes may produce enough energy to induce crystallization.

7.2 Physical properties of cometary nucleus material

Early models of the nucleus and its activity assumed a rather compact conglomerate of water ice and refractory components similar to what is known from meteorites. Sublimation from a pure water ice surface was calculated to determine the gas (e.g. Huebner 1965) and dust production (Finson and Probstein 1968a,b) rates of active comets. Based on these rates the sizes of cometary nuclei were estimated; in a few cases also their albedo (Delsemme 1985), leading to small nuclei with high albedo appropriate for pure ice. The dust to gas ratio was assumed and observed (due to dominating visibility of small grains (10 μm)) to be as small as 0.1 and in extreme cases as large as 0.3. A cometary nucleus is predominantly made from water ice with some impurities. Larger dust particles could gather on some parts of the surface and quench the sublimation of water.

The images of comet Halley showed that its nucleus was much larger than estimated by the best engineering model (Divine et al. 1986), that its surface had a very low reflectivity and that most of it was inactive. As a consequence the density of the nucleus should be smaller than $1000 \, \text{kg} \, \text{m}^{-3}$ (Rickman et al. 1987, Rickman 1989). The uncertainty of this determination based on nongravitational forces still includes higher values (Sagdeev et al. 1988, Peale 1989). This nucleus does not resemble a 'snowball' with impurities (icy conglomerate).There is now general agreement that

- the material of cometary nuclei is extremely porous and fluffy;
- gas diffusion is the main physical transport process of energy and mass;
- the refractory (non-volatile) to ice (volatile) ratio is at least unity; and
- most of the surface is inactive.

Several types of model calculations have been developed (for an overview see e.g. Klinger et al. 1996) and with the improvement of computer power multi-dimensional calculations have become available (Enzian et al. 1997, 1999). We will not review these papers further but discuss the consequences of comet Halley and later observations showing dust rather than ice is the dominant constituent of cometary nuclei. This fact has finally been recognized by the modellers and is reflected in the compositional mixture where now a typical dust to gas ratio of unity is assumed (Enzian et al. 1998, Orosei et al. 1999). Greenberg (1998) discusses formation of a cometary nucleus from interstellar material and finds a refractory to volatile (dust to gas) ratio of 1 and a dust to water ratio of about 2. He counts about 10% of PAH molecules to the volatile fraction. The dominance of refractory material leads also to physical consequences for the nucleus. The flow of gas and dust through porous matrix material has been modelled by Shoshany et al. (1999), Skorov and Rickman (1995) and Skorov et al. (1999).

Stimulated by the HMC images Keller (1989) suggested the concept of a cometary nucleus with a microstructure characterized by refractory material rather than by ice ('icy dirt ball' model). In contrast to cohesive bonding between ice grains, dust agglomerates are not eroded by thermal energy. This has major consequences particularly for the development of activity and the erosion processes on the nuclear surface.

7.3 The cohesive nucleus model (icy dirt ball)

The significance of cohesive forces becomes readily apparent if their strengths are compared to those of gravity and vapour pressure forces. Strength values due to van der Waals forces between grains in the primordial nebula are derived (Chokshi et al. 1993, Kührt and Keller 1996) for micrometre- to millimetre-sized grains ranging from 10^2 to 10^5 Pa in agreement with laboratory measurements (Saunders et al. 1986, Storrs et al. 1988). This range of values covers estimates for the tensile strength of comets (Whipple 1983) and for the strength of lunar regolith (Mitchell et al. 1973). Fireballs that probably originate from comets have a mechanical strength of 10^3 to 10^6 Pa (Wetherill and ReVelle 1982). The cohesive strength within the matrix structure of refractory material exceeds the vapour pressure of water ice subliming from comets. The vapour pressure even for a high sublimation temperature of 250 K does not exceed 100 Pa (e.g. Kührt and Keller 1994).

7.4 The icy conglomerate model and dust mantles

Early models (Whipple 1950a,b) and more quantitative ones (Shul'man 1972) used 'left over' grains, that were too large to be removed by the free ice sublimation, to cover parts of the nucleus surface and finally to quench the activity. The 'loose lattice' model by Mendis and Brin (1977) uses a fully permeable layer of loose non-cohesive grains to inhibit the insolation (Prialnik and Bar-Nun 1988). Brin and Mendis (1979) showed that such a layer would not be stable around perihelion for a comet on a Halley-type orbit. Rickman et al. (1990) included grain trapping within a 'heavy' dust grain layer (held on the surface by gravity) and in this way reduced the diffusion coefficient. As a result they find that for a Halley-type perihelion only a few percent of the surface is covered by a stable mantle of dust. Here the expression 'mantle' is used to describe a layer of refractory grains (cohesive or not) on top of an icy surface. In a review of the nucleus models Houpis (1990) emphasizes the discrepancy between model results and observations.

7.5 The cohesive model and crusts

Models based on the icy conglomerate concept using reasonable thermophysical parameters do not produce stable inert areas covering most of the nucleus surface. Models based on a cohesive refractory matrix of the nucleus (Kührt and Keller 1994) can produce a stable 'crust' on the surface that explains the observations of restricted activity. The word 'crust' describes the depleted uppermost part of the cometary refractory matrix. This crust as part of the matrix is cohesively connected to the inner nucleus. This concept was not recognized by Klinger et al. (1996) in their review of cometary nucleus models.

The physical processes involved in the formation of a crust (a layer of nucleus matrix material depleted from ice) are parameterized in the model calculation of Kührt and Keller (1994) with the purpose of investigating the stability of such a crust to breakup by vapour pressure by subliming water underneath the crust. The effective heat conductivity of the crust, K_{eff}, has a major influence on the thickness of the crust and the physical conditions. Details such as convective heat transport by gas (Horanyi et al. 1984, and many later papers) are not explicitly considered. The chosen range from 10^{-3} to $1\,\text{W}\,\text{m}^{-1}\,\text{K}^{-1}$ should cover all realistic situations. The lower value corresponds to that of a loose mantle (Brin 1980) and is widely used (Fanale and Salvail 1984, Rickman 1989). Higher values are similar to the conductivity of porous stony material. The extreme upper value is considered as an upper limit and is most favourable for the formation of a crust. The thermal parameters used are listed in Tables 1 and 5.

The development of the crust is followed for many orbits of typical comets such as P/Encke, P/Halley and P/Grigg-Skjellerup. The calculations show that for all reasonable conditions the vapour pressure underneath the crust exceeds the gravitational attraction of the depleted material generally by orders of magnitude at the early stages of crust formation. Figure 21 depicts the result for comet Halley for the two extremes of K_{eff}. For the extreme low value only a very thin crust a few centimetres thick develops (Figure 22). The crust reaches a thickness of metres if the high extreme for the crust heat conductivity is assumed (Figure 23). The pore size and porosity of the crust determine the pressure underneath

Table 5 Standard thermal parameters used for the calculations of crust formation on the surface of cometary nuclei

Parameter	Symbol	Value
Albedo	A	0.05
Specific heat	c	$600\,\text{J}\,\text{kg}^{-1}\,\text{K}^{-1}$
Emissivity	ε	1
Latent heat	H	$2.66 \times 10^6\,\text{J}\,\text{kg}^{-1}$
Crust porosity	p	0.5
Crust density	ρ	$500\,\text{kg}\,\text{m}^{-3}$
Effective pore size	l	1 mm
Crust heat conductivity	K	10^{-3} or $1\,\text{W}\,\text{m}^{-1}\,\text{K}^{-1}$
Core heat conductivity	K_{core}	$1\,\text{W}\,\text{m}^{-1}\,\text{K}^{-1}$
Dust-to-ice mass ratio	X	1

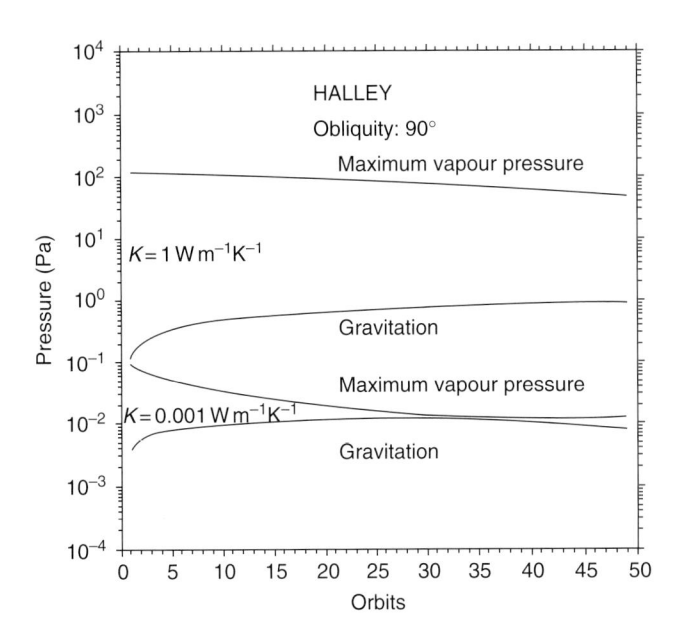

Figure 21 The development of the maximum vapour pressure (per orbit) under a growing crust of a a cometary nucleus. The resulting gravitational pressure of the crust is also shown. A Halley-like orbit and high and low thermal conductivities are used (obliquity 90°). (After Kührt and Keller 1994.)

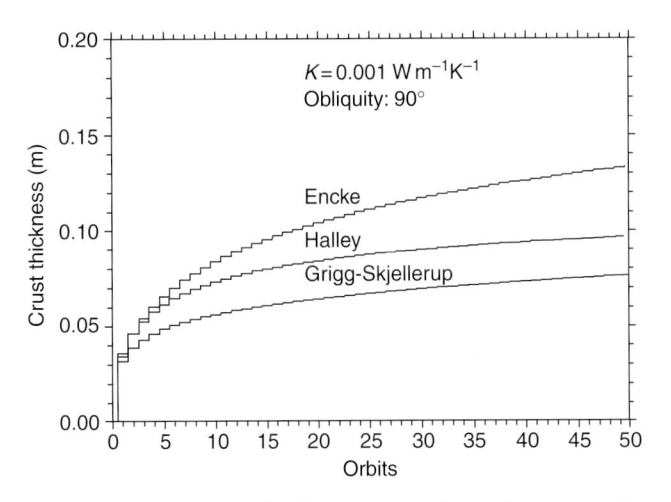

Figure 22 The crust development on the polar region for low heat conductivity of the crust and the case of obliquity 90°. Orbits corresponding to those of comets Encke, Grigg-Skjellerup and Halley have been chosen. (From Kührt and Keller 1994.)

the crust and, therefore, its stability. Both parameters are of minor importance for the sublimation flux through the crust. This flux is reduced by orders of magnitude compared with free sublimation of water ice and hence the crust is an inactive surface area. The calculations show that the crust (the depleted matrix of the nucleus) can withstand the vapour pressure under all physically plausible conditions.

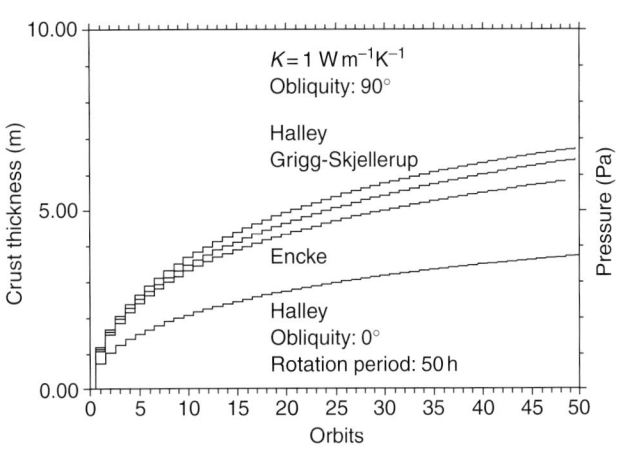

Figure 23 See Figure 22. A high thermal conductivity (upper three curves) is used in the model calculations. The lowest curve corresponds to the orbit of comet Halley (rotation period of 50 h) and depicts a point on the equator, assuming zero obliquity of the rotation axis. (From Kührt and Keller 1994.)

7.6 The quest for restricted (localized) but persistent cometary activity

While this model provides an explanation for the development of most of the inactive nucleus surface it does not describe the nucleus activity. Obviously, for a few percent of the surface, conditions must exist that allow the dust to be removed during the sublimation process. Possible explanations include thermal stresses (see above) but more generally some heterogeneity of the cometary nucleus:

(a) the content of ice within the lattice of a refractory material may be locally enhanced,

(b) vents of more volatile compounds such as CO or CO_2 help to remove the dust matrix, and

(c) the refractory matrix is physically disturbed and hence weakened.

(a) Inhomogeneities in the dust to gas ratio within limits can be expected from the formation process. It can be argued that colliding grains can be classified as icy or dusty depending on which component prevails. Then smaller ice clusters can be embedded in larger dust clusters and vice versa. Monte Carlo simulations can provide the size distribution as a function of the volume ratio of the components. Horanyi and Kecskemety (1983) modelled the random accretion for a dust concentration near the percolation limit. Clusters with enhanced ice content of several tens of metres are needed to maintain activity over several orbits. Following this idea, some regions of the nucleus contain enough ice to maintain activity (an icy conglomerate) whereas the majority of the nucleus volume is occupied by clusters corresponding to an icy dirt ball and are inactive. Active zones can become inactive by consuming the volatiles and inactive regions can become active by shedding depleted clusters (parts of the crust).

(b) The voids between clusters formed during the agglomeration process could be filled by ices recondensing after sublimation caused by the associated heating, particularly near the contact surfaces of the clusters. Here the water ice content could be enhanced so recondensing more volatile compounds could be concentrated. These regions of high porosity and hence low heat conductivity could be the preferred volume for recondensation of volatiles liberated by the internal heating of the nucleus due to impacting clusters but also due to radioactivity. This hypothesis would naturally explain the very restricted areas of activity.

(c) The refractory matrix could have lower than average tensile strength in regions that have been destroyed during the violent agglomeration process. This would help in combination with (b) to maintain activity over extended timescales.

At this point it has to be stressed again that the process of the highly restricted but persistent activity of cometary nuclei is not understood. Understanding the physics of this process will reveal a basic aspect of the nature of comets. This will probably have to wait for a rendezvous mission such as the Rosetta mission.

8 ROTATIONAL PROPERTIES OF COMETARY NUCLEI

8.1 Excited spin states

The topic of comet rotation has received fresh attention since the discovery of P/Halley's 'excited spin state' from the study of ground-based photometric results and Vega and Giotto images (Kerr 1986, Peale and Lissauer 1989). An excited spin state is a rotational state which involves several rotations and oscillations of the nucleus instead of a single rotation around one of the principal axes of inertia ('pure spin state'). The study of excited spin states and their relaxation by internal dissipation can allow us to assess some fundamental properties of the nuclei, which would otherwise remain unconstrained.

A fast-rotating body can lose material at the equator or even break up (Sekanina 1997) as a result of the centripetal acceleration. According to Chen and Jewitt (1994), at least one comet splits every year on average. Neglecting the tensile strength of the nucleus, one gets the following condition for the density ρ_n of a prolate nucleus to become unstable at the apex (Jewitt and Meech 1988):

$$\rho_n < \frac{\Omega_0^2}{2\pi G \, I(a_n/b_n)} \quad (5)$$

where G is the gravitational constant, Ω_0 is a typical angular velocity and $I(a_n/b_n)$ is a dimensionless shape-dependent integral (Jewitt and Meech 1988). The relationship shows that fast-rotating comets can become unstable and break up under the effect of the centripetal acceleration. Jewitt and Meech (1988) calculated that eqn (5) gives $\rho_n < 0.1\,\text{g cm}^{-3}$ for a comet with a rotation period of 12.7 h and $a_n/b_n \sim 2$, where a_n/b_n is the ratio of the semi-major to the semi-minor axis. The shortest rotation periods measured so far are $P_n = 5.2$–6.0 h for comets 31P/Schwassmann-Wachmann 2 (Luu and Jewitt 1992), 6P/d'Arrest (Fay and Wisniewski 1978) and 46P/Wirtanen (Lamy et al. 1999), suggesting a density $\rho_n > 0.5$–$0.7\,\text{g cm}^{-3}$ for these objects, assuming that their tensile strength is negligible compared to the centripetal fource. However, no direct observational link has been established between fast rotations and breakups. The statistical sample of rotation periods remains limited to 10–15 reliable values (Jewitt 1997). It has also been suggested that thermal stresses (Kührt 1984, Tambovtseva and Shestakova 1999) can contribute to, or even trigger, non-tidal splitting.

Most asteroids are believed to be very close to a pure spin state around their axis of maximum moment of inertia, close to their lowest possible rotational energy. For comets, the situation is radically different. Is this because of the comets' activity?

The rotational state can become excited when the torque of the tidal gravitational force becomes large enough to induce a nutation. The timescale for this effect is $\sim$10–100 years outside close approaches, but can become much shorter for Sun-grazing comets or during a close approach to Jupiter.

The sublimation and recondensation of water molecules (Crifo 1987) produces a non-gravitational force which has a measurable effect on the orbit of the comet (Whipple 1950a). The torque of this force can modify the rotational state of the nucleus on a timescale of 0.1–100 years, depending on the nucleus properties (size, shape, density and activity level). This timescale is usually shorter than the lifetime of comet nuclei in the inner Solar System. Many comets are expected to reach excited spin states when they leave the Kuiper belt or the Oort cloud.

When a body is in an excited rotational state, deformations occur as centripetal and transversal accelerations create stresses inside the body. The work associated with these deformations is dissipated into heat until the body comes back to its rotational state of lowest energy (pure spin). This effect has long been considered negligible, based on the work of Burns and Safronov (1973), because its timescale was believed to be much longer than the orbital period of any known comet (e.g. Peale and Lissauer 1989 for 1P/Halley). However, Efroimsky and Lazarian (2000) pointed out that the dissipated energy is an integral of the product of the strain tensor and the stress tensor.

Since both tensors contain terms proportional to $\cos(\omega t)$ and $\sin(\omega t)$, where ω is the frequency of nutation, their product involves a second harmonic. A detailed calculation for an oblate body yields an additional term in the equation of relaxation, which leads to a timescale for damping by internal dissipation several orders of magnitude lower than that of Burns and Safronov (1973):

$$\tau_{ID} \approx k \frac{\mu_n Q}{\rho_n a_n^2 \Omega_0^3} \quad (6)$$

where μ_n is the coefficient of rigidity or shear elastic modulus, Q the specific dissipation function or 'quality factor', a_n the semi-major axis of the oblate nucleus and the dimensionless factor k ($=1$–5) mainly depends on the initial rotational state. For a nearly prolate three-axial body, higher harmonics appear because the strain and stress tensors involve the Jacobian elliptic functions and the relaxation is even faster. The relationship shows that the timescale for damping is proportional to two unknown parameters, Q and μ_n. The value of the shear modulus is $\mu_n \sim 10^{10}$ Pa for carbonaceous and silicate rocks (Efroimsky and Lazarian 2000) and is estimated to be $\mu_n \sim 10^9$ Pa for dirty ice (Peale and Lissauer 1989). The 'quality factor' Q (or 'Q-factor') reflects the ability of the body to dissipate kinetic energy into heat. This parameter depends on the frequency of nutation ω and on the local temperature and pressure. Unfortunately, no measurement is available for comet- or asteroid-like material at low frequency ($<10^{-3}$ Hz), low pressure and low temperature. Efroimsky and Lazarian (2000) reviewed possible values for this parameter in the geological literature and concluded that $Q \sim 100$ or lower might be a good approximation for carbonaceous material at low temperature (20 K) and low frequency (10^{-5} Hz). However, this is an extrapolation of a measurement on lunar basalt at a frequency of 20 kHz. The value for cometary nuclei might be even smaller. As the authors pointed out, further laboratory measurements at frequencies of 10^{-3} to 10^{-6} Hz are needed to get measurements of Q appropriate to comets and asteroids.

8.2 Rotation states of comets 1P/Halley and C/1995 O1 (Hale-Bopp)

What do we know about the rotational state of the best observed comets: 1P/Halley and C/1995 O1 (hereafter Hale-Bopp)? Comet 1P/Halley was extensively observed during its last return in 1985–1986. A fleet of spacecraft was launched to study its nucleus and a worldwide campaign of ground-based observations was carried out. Unfortunately, the space-based observations by Vega and Giotto were essentially snapshots and yielded only *one* measurement of the nucleus orientation per flyby. Ground-based observations covered a wide temporal range, but did not have the spatial resolution required to trace the orientation of the nucleus. Colom and Gérard (1988), Schleicher *et al.* (1990) and several other authors (this topic has been thoroughly discussed by Belton *et al.* 1991) observed a clear periodicity with a period of 7 days. Dust and CN jets (e.g. Schulz and Schlosser 1989, Larson *et al.* 1987), as well as the study of the streamers observed in the dust tail (Grothues and Schmidt-Kaler 1996) showed a 2.2 day periodicity. It is now widely accepted that the nucleus is in an excited spin state (Sagdeev *et al.* 1989, Samarasinha and A'Hearn 1991, Belton *et al.* 1991, and references therein). Samarasinha and Belton (1995) demonstrated that a mechanism of excitation based on the outgassing from several point-like active areas could explain it. However, there is no common understanding of the actual spin state of 1P/Halley, and there is little hope that a rotational ephemeris will become available before its next return.

Comet Hale-Bopp is the brightest comet observed so far in modern history. It was followed from its discovery in 1995 (Hale *et al.* 1995) to its spectacular perihelion passage in 1997. A periodicity of about 11.5 h (Lecacheux *et al.* 1997), later refined to (11.35 ± 0.04) h (e.g. Jorda *et al.* 1997, Licandro *et al.* 1998, Samarasinha 2000), was determined from the analysis of dust and gas jets observed at visible wavelengths in January–April 1997. The dust features observed in 1995–1996 still need to be analysed in detail.

Comet 1P/Halley is in an excited spin state, which is most likely due to outgassing by an irregularly shaped nucleus with active areas (Samarasinha and Belton 1995). Internal dissipation can apparently not damp completely the excitation during a perihelion passage, although eqn (6) gives a timescale of 5 years for 1P/Halley. The angular velocities are such that the nucleus is far from the instability condition (eqn (5)). Despite the huge observational and theoretical efforts made to try to understand 1P/Halley's spin state, there is no consensus concerning the exact rotational state of the comet, which in fact illustrates the difficulty of determining cometary rotational states. Hale-Bopp is a faster rotator and a larger nucleus. Even if a feebly excited spin state would appear during a perihelion passage, it would be likely to be damped by internal dissipation according to eqn (6). Comets beyond or close to the condition of eqn (5) can split up or lose material near the instantaneous equator when the angular momentum increases under the effect of the torque of the non-gravitational force. This might be an important mechanism leading to reshape and finally destroy cometary nuclei (Chen and Jewitt 1994). Contrary to what has been believed until very recently, internal dissipation of rotational energy into heat might damp excited spin states between two perihelion passages, at least for large comets with long orbital periods (Jewitt 1997, Efroimsky and Lazarian 2000). Observations of the relaxation of an excited spin state could allow us to derive internal parameters, such as the coefficient of rigidity and the 'quality factor' of cometary material, which remain otherwise unconstrained. A long-term time series of high-quality measurements of the nucleus orientation is a good way to test the existing nucleus models by modelling the temporal evolution of the rotational state. The Rosetta mission to comet 46P/Wirtanen will give us a unique opportunity to do so.

9 SMALL-SCALE MORPHOLOGY OF THE SURFACE

9.1 Material loss rates

The large-scale topography of a cometary nucleus is determined by its largest subnuclei. These again by their largest subunits, and so on. While the loss averaged over the whole surface is only of the order of a few metres in radius the loss at active sites is an order of magnitude larger. In addition the overall loss of cometary material is probably strongly underestimated. Losses are usually estimated from the observed gas production (water) rates integrated over the active part of the orbit (typically for $r_h < 3$ AU). The dust to gas mass ratio is determined from observations of the tail mainly in the visible. Particles larger than $100\,\mu$m are difficult to assess. Most observations, and in particular the *in situ* observations during the comet Halley missions show that the distribution of the particle mass is such that the largest (centimetre size) particles contribute most. The extrapolations of the power law are usually truncated not later than 10 cm. This value corresponds to the maximum radius of particles that can be lifted off an active cometary nucleus by gas-dynamic drag forces (Keller 1990). IRAS observations (Sykes *et al.* 1986), however, show that dust trails of SP comets contain 'particles' with typical sizes of 60 cm. The sizes of fireballs are in the range of metres.

The observations of comet Hyakutake revealed several spots of activity lasting for short time intervals of the order of days only. The size estimates for these centres of activity were in the range from metres to few tens of metres (Desvoivres *et al.* 1999). These shed blocks of cometary material were only visible because the comet approached the Earth to within 0.1 AU making observations of exceptionally high resolution possible. One should assume that this process is common for most active comets. Finally, splitting of cometary nuclei is a frequently observed phenomenon (e.g. Sekanina 1982, Chen and Jewitt 1994; Section 5.1). Usually, lost pieces are an order of magnitude smaller (hundreds of metres in radius) than the dimension of the nucleus itself.

We can conclude that the mass loss of cometary nuclei is appreciably higher than estimated from their 'visible' activity alone. Depletion of volatiles near the nucleus surface seems less likely the fate of a cometary nucleus than its breaking up in pieces. This may be the reason why only few dormant (inactive) cometary nuclei are known or suspected (e.g. Hartmann *et al.* 1987). There are hardly any asteroid-like bodies observed on typical SP elliptical orbits.

9.2 Surface erosion

Erosion (and mass loss) of cometary nuclei does not take place by smooth and gentle sublimation of 'dirty' water ice. The very low gravity aided by centrifugal forces makes gas streams with densities orders of magnitude smaller than at the Earth's surface become violent fountains. Trapped gas can build up additional pressure and blast away pieces of the crust. Thermal stresses in the depleted surface layers produce cracks and loosen plates of metre thickness. Inter-block regions become weaker when they are depleted in ice that had stabilized the areas of contact by sintering.

While the cometary refractory matrix material of the nucleus is very porous and fluffy it is strong enough to withstand this violent environment during the periods of activity. In some areas (or volumes) the sublimation process may grind up the matrix and eject it as fine dust particles. Other parts of the nucleus are strong enough to withstand the treatment as large chunks. These chunks can survive in the inner Solar System for millions of years (much longer than the active comet itself) before they, for example, finally hit the Earth as fireballs. The tensile strength of these boulders is very low on the scale of objects on Earth but probably somewhat higher than the values derived from observed and analysed nucleus tidal breakups and splittings, some 10 to 10^4 Pa (Sekanina 1977, 1978, Asphaug and Benz 1996; Section 5.2). In the low-gravity environment this strength is enough to prevent even larger structures from collapse. The gravitational pressure $P_g = g_{comet}\, \rho\, \Delta$ of a layer Δ of 1 m thick, a density of 1000 kg m^{-3} and for $g_{comet} = 10^{-3}$ m s^{-2} exerts a pressure of only 1 Pa.

Some (large) dust particles will be lifted off the nucleus surface by the sublimation in active areas but will not reach a terminal speed to overcome the gravitational attraction of the nucleus. These will fall back like 'hail' onto the surface. Numerical calculations for a Jupiter family comet with a radius of 2 km and 25% of its surface covered by active areas showed that only about 10 to 20% of the total surface would be covered during one perihelion passage. This is less than the total extent of all active areas. And in addition most of the particles are found in and near the areas of activity trailing them due to the spin of the nucleus (Figure 24). These calculations by Kührt *et al.* (1997) show that the landscape of a cometary nucleus is not covered and certainly not overcast by regolith of pieces fallen back out of the jets of activity.

The nucleus surface should be covered by a bizarre landscape for which the Badlands (a heavily eroded terrain in Badland National Park) in North America can only be a

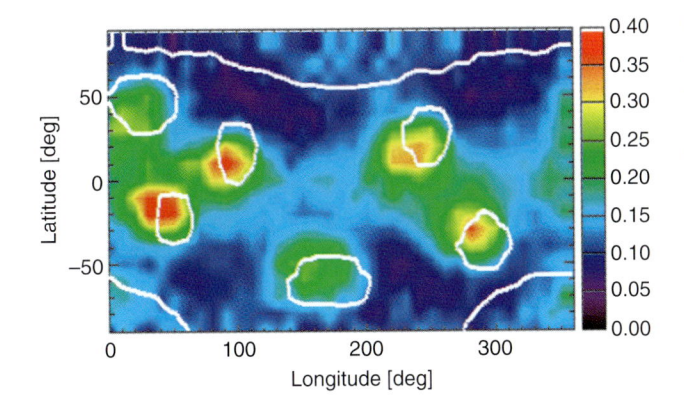

Figure 24 Area fraction of a P/Wirtanen-like nucleus covered by newly ejected dust integrated from 3 AU to perihelion. A size distribution index $\gamma = 3.8$, an intrinsic dust-to-ice ratio $\kappa = 1$ and a particle size range from 0.5 μm to 1 m are assumed. (From Kührt *et al.* 1997.)

rough approximation. On smaller scale the nucleus surface could look like certain limestone rocks at the seaside washed out by water and covered by acute pinnacles and little cavities on the scale of decimetres and centimetres. Areas of activity may be somewhat smoother because the tensile strength may here be low enough that the sublimation process could destroy the bonds of the matrix (refractory) material at least on the smallest sub-millimetre scale.

9.3 The cometary nucleus surface – a prediction

Let us describe how a typical (Halley-like) cometary nucleus will appear to a high-resolution camera on board a spacecraft approaching it at a phase angle of about 90°. As long as the resolution of the camera is worse than the dimension of the nucleus the nucleus will remain hidden by the dust in its vicinity (we assume a heliocentric distance of about 1 AU). The centre of brightness (the innermost coma) will become more and more acute and suddenly a dark spot will become visible in its centre giving rise to maximum contrast between the foot points of the jet activity and the unilluminated nucleus surface (Figure 8). An irregular shape will grow from the first signature of the nucleus, partly illuminated by the Sun, partly seen as outline against the illuminated background. The shape will deviate more and more from the ellipsoidal form. The limb becomes more and more rugged, and outcrops of several hundreds of metres will become visible at first. Parts of the nucleus body seem to be only loosely attached, large subnuclei gain shape. All the illuminated surface looks extremely rough. The interplay of light and shadows creates a bizarre maze of pinnacles of all sizes. Images of the surface with higher resolution look very much like those taken from 10 times the distance, the features seem homologous. Although the contrast between illuminated surface and shadows is extreme, photometry of the surface shows that it is very dark and uniform in reflectivity. Changes in surface brightness are due to variations of the phase function rather than caused by intrinsic variability of composition. No icy patches of enhanced albedo are visible.

Fountains of dust jets shoot from large patches of the surface predominantly near the sub-solar surface area. They obscure the view of the areas in the background, the scene looks foggy but the outlines of large surface features are still visible. The optical thickness is less than unity; however, the fine dust in the jets is bright in the sunshine, brighter than the average surface under the foot points of the jets where shadows tell of its roughness. It remains unclear why certain parts of the surface are active and spout the dust fountains. The active areas look very much like the inert majority of the nucleus surface. Possibly the mean roughness is somewhat smaller. The fringes of the areas of activity are irregular, islands of inactivity or reduced activity are visible on all scales, the biggest hundreds of metres in diameter. Looking closer at the dust jets, stretching the contrast, makes smaller structures visible. Fine filaments originating at spots of enhanced activity crisscross each other reflecting the inclinations of surface normals. Other filaments form by interactions of jet streams only a few tens or hundreds of metres above the surface. Some of them are curved.

Dust particles only slightly larger than the wavelength in the 1 to 10 μm range dominate in visibility. Careful processing of short exposures reveals large particles of centimetre and decimetre size. With some luck a trajectory of a larger piece could be followed while it falls back onto the surface at the fringes of a major jet area. While it impacts with a speed of a few metres per second it shatters into several pieces. Some of these larger particles reach highly elliptical orbits after barely avoiding an impact on the surface. Instruments on board the orbiting spacecraft ought not be hit. The probability of this happening is fortunately negligible. The number of 'orbiters' (large dust particles) will increase within the time interval of cometary activity. Some of these chunks were already elevated during previous perihelion passages (Richter and Keller 1995). Do they form a cloud of centimetre- or decimetre-sized particles as inferred from radar observations (e.g. Harmon *et al.* 1989, 1997)?

Some activity seems to follow boundaries between building blocks, coming to life when the Sun rises above them. It is not obvious which parts on the morning side will become active. Where should the lander touch down? There are areas where the large-scale roughness seems reduced, pinnacles with heights of tens of metres are less frequent. This seems to be more typical for the boundaries between obvious building blocks, but also for areas of activity (which often are one and the same). But close-ups reveal that the surface is everywhere rough on metre and sub-metre scales. Statistically distributed in between are patches with pinnacles and crevices of a few decimetres only, each covering a few square metres, too small to target the lander safely.

9.4 Landing on a cometary surface

The (Rosetta) lander approaches the surface and touches down softly next to a small area of activity. Its harpoon anchors it immediately. One of its three legs slips into a crevice, the lander leans to this side and the motion is stopped when the underside of the lander crushes the top of a pinnacle. The settling motion comes to a halt. Later when the experiments are switched on it turns out that the mechanical feeding mechanism for the analytic instruments is partly jammed. The high-resolution images transmitted show that the roughness of the surface continues to submillimetre scales. The nucleus matrix material is very porous. One of the feet penetrates into the surface indicating a very low tensile strength. The upper side of the foot is covered by dust from the cloud of 'ejecta' produced when the foot crashed into the surface. On the image it is not easy to discern the 'wounds' of the surface. There is no obvious surface taint or even crust. The material is very brittle. There is no firm contact, not even for the harpoon and the penetrators. Will the lander survive the cometary night? More quantitative descriptions of the hazards of landing on a comet can be found in Kührt *et al.* (1997).

10 FROM COMETS TO DUST

10.1 When comets die

The fate of an ageing comet depends on its physical nature and can follow three main routes:

(a) it sublimes if it is dominated by ice, shrinks, and becomes invisible and ceases to exist;
(b) it disintegrates by splitting and shedding of subnuclei and chunks, finally ending up as dust within the planetary system; or
(c) it becomes dormant when its surface becomes inert; it may remain as a body near its original size.

Only case (c) will lead to dormant bodies that will in some respects resemble asteroids. In cases (a) and (b) the cometary nucleus will finally cease to exist. We have given many arguments that show that cometary nuclei are not dominated by water ice and therefore will not follow case (a).

Comets with a nucleus of a cohesive dust matrix can form inert surfaces, crusts, if the depletion process is faster than the erosion. Cometary nuclei can become extinct or inactive while most of their interior is still pristine. They could be reactivated on a new orbit with a smaller perihelion or by a change of their rotation axis, or by splitting caused by thermal stresses. In the meantime they would not be classified as comets which are characterized by their activity in the form of a coma. However, TNOs from the EK belt do not show any activity (with one or two exceptions) but are generally considered to be comets. Obviously activity is a sufficient arguement but not a required condition to characterize a comet.

10.2 Relationship with asteroids

How many of the asteroids are dormant comets? How can we discern them from real asteroids? Means of selecting candidates are the orbital parameters (Weissman *et al.* 1989) or their surface finish such as low albedo and reddening similar to the asteroids of the outer Solar System (Hartman *et al.* 1987). About 20 to 30 candidates can be listed from the point of view of dynamics. Examples are comet P/Wilson-Harrington (1949 III) or asteroid 1979 VA.

As discussed in Section 2.2, SP comets may well be the remanents of bigger TNOs and have a collisional history similar to asteroids. Is there a generic relationship between asteroids and comets? Are they two extremes of one generic process? In other words are there intermediate bodies (Trojans, outer belt asteroids), half asteroid half comet, or are they all members of two distinct groups. Asteroids have formed in the inner Solar System and stayed there, while comets come from the outer Solar System (EK belt or Oort cloud) but originate from the planetary regime.

It is obvious that activity alone cannot be the only decisive argument in distinguishing comets from asteroids. A new classification scheme is required.

10.3 Final decay

Splitting seems to be the main route of decay (Section 5.1) for comets. Considering that the dynamical lifetime of a SP comet is about 5×10^4 years and that it may split once per 100 years or 20 orbital periods (Chen and Jewitt 1994) it will most often disintegrate into chunks rather than become dormant. Dormancy could only be considered as an intermediate state while it is 'bouncing' around in the inner Solar System perturbed by the planets. Its debris will at first form cometary trails then slowly be distributed within its orbital plane and possibly become visible as meteor showers. It is not only Jupiter family comets that decay in this way. The mass estimate of the meteoretic debris of the meteor streams η Aquarids and Orionids connected with comet Halley (Hughes 1987, Hajduk 1987) lead to an original mass of Halley's nucleus about 5 to 10 times what is presently observed. The dynamic age of the meteor streams (their dispersion in the orbital plane) implies that comet Halley attained its present orbit 2000 to 3000 revolutions ago.

The debris is not all small particles. Sykes et al. (1986) found a typical size of 60 cm for the debris in cometary trails, a dimension comparable to fireball meteoroids which have been related to comets because of their orbital dynamics and because of their very low density (200 kg m^{-3}; Ceplecha and McCrosky 1976) and friability. Their mass is in the region of 10^6 g. They consist of depleted matrix material. This is another hint that the tensile strength of the cometary matrix material on the level of small subunits can withstand the strong thermal stresses. If the fireball does not hit the Earth's atmosphere it will sooner or later collide in the inner planetary system and be shattered to interplanetary dust.

The cometary interplanetary dust particles (IDP) will resupply the zodiacal cloud (Whipple 1967) and some of them can be collected in the Earth's upper atmosphere after being strongly baked by solar radiation, sputtering, and the aero-braking during entry. Even these skeletons of silicate-rich material are of relatively low density and hence high porosity (Love et al. 1994).

Comets were made from dust (richly mixed with volatiles) and end up as dust (deprived of all compounds more volatile than silicates).

REFERENCES

A'Hearn, M.F. (1988). Observations of cometary nuclei. *Annual Review of Earth and Planetary Sciences*, **16**, 273–293.

A'Hearn, M.F., Campins, H., Schleicher, D.G. and Millis, R.L. (1989). The nucleus of comet P/Tempel 2. *Astrophysical Journal*, **347**, 1155–1166.

Altenhoff, W.J. and Stumpff, P. (1995). Size estimate of 'asteroid' 2060 Chiron from 250 GHz measurements. *Astronomy and Astrophysics*, **293**, L41–L42.

Asphaug, E. and Benz, W. (1996). Size, density, and structure of comet Shoemaker-Levy 9 inferred from the physics of tidal breakup. *Icarus*, **121**, 225–248.

Bailey, M.E. (1983). The structure and evolution of the Solar System comet cloud. *Monthly Notices of the Royal Astronomical Society*, **204**, 603–633.

Bailey, M.E. (1986). The mean energy transfer rate to comets in the Oort cloud and implications for cometary origins. *Monthly Notices of the Royal Astronomical Society*, **218**, 1–30.

Bar-Nun, A., Dror, J., Kochavi, E., and Laufer, D. (1987). Amorphous water ice and its ability to trap gases. *Physical Review B: Condensed Matter*, **35**, 2427–2435.

Belton, M.J.S., Julian, W.H., Anderson, A.J. and Mueller, B.E.A. (1991). The spin state and homogeneity of comet Halley's nucleus. *Icarus*, **93**, 183–193.

Biermann, L. and Michel, K.W. (1978). The origin of cometary nuclei in the presolar nebula. *Moon and Planets*, **18**, 447–464.

Biermann, L. and Trefftz, E. (1964). Über die Mechanismen der Ionisation und der Anregung in Kometenatmosphären. *Zeitschrift für Astrophysik*, **59**, 1–28.

Birkett, C.M., Green, S.F., Zarnecki, J.C. and Russell, K.S. (1987). Infrared and optical observations of low-activity comets, P/Arend-Rigaux (1984k) and P/Neujmin 1 (1984c). *Monthly Notices of the Royal Astronomical Society*, **225**, 285–296.

Blum, J. (1995). Laboratory and space experiments to study pre-planetary growth. *Advances in Space Research*, **15**, (10)39–(10)54.

Bowell, E. and Lume, K. (1979). Colorimetry and magnitudes of asteroids. In T. Gehrels (ed), *Asteroids*, University of Arizona Press, Tucson, AZ.

Bowell, E., West, R.M., Heyer, H.-H., Quebatte, J., Cunningham, L.E., Bus, S.J., Harris, A.W., Millis, R.L. and Marsden, B.G. (1992). (4015) 1979 VA = Comet Wilson-Harrington (1949 III). *IAU Circular* 5585.

Bredichin, T. (1903). *Mechanische Untersuchungen über Cometenformen in systematischer Darstellung*, G. Hassel, Leipzig.

Brin, G.D. (1980). Three models of dust layers on cometary nuclei. *Astrophysical Journal*, **237**, 265–279.

Brin, G.D. and Mendis, D.A. (1979). Dust release and mantle development in comets. *Astrophysical Journal*, **229**, 402–408.

Brooke, T.Y. and Knacke, R.F. (1986). The nucleus of comet P/Arend Rigaux. *Icarus*, **67**, 80–87.

Brown, M.E., Kulkarni, S.R. and Liggett, T.J. (1997). An analysis of the statistics of the Hubble Space Telescope Kuiper Belt object search. *Astrophysical Journal of Letters*, **490**, L119–L122.

Burns, J.A. and Safronov, V.S. (1973). Asteroid nutation angles. *Monthly Notices of the Royal Astronomical Society*, **165**, 403–411.

Bus, S.J., Buie, M.W., Schleicher, D.G., Hubbard, W.B., Larcialis, R.L., Hill, R., Wasserman, L.H., Spencer, J.R., Millis, R.L., Franz, O.G., Bosh, A.S., Dunham, E.W., James, D.H., Young, J.W., Elliot, J.L. Meserole, R., Olkin, C.B., McDonald, S.W., Foust, J.A., Sopata, L.M. and Bandyopadhyay, R.M. (1996). Stellar occultation by 2060 Chiron. *Icarus*, **123**, 478–490.

Campbell, D.B., Harmon, J.K. and Shapiro, I.I. (1989). Radar observations of comet halley. *Astrophysical Journal*, **338**, 1094–1105.

Campins, H., A'Hearn, M.F. and McFadden, L.-A. (1987). The bare nucleus of comet Neujman 1. *Astrophysical Journal*, **316**, 847–857.

Campins, H., Osip, D.J., Rieke, G.H. and Rieke, M.J. (1995). Estimates of the radius albedo of comet–asteroid transition object 4015 Wilson-Harrington based on infrared observations. *Planetary and Space Science*, **43**, 733–736.

Campins, H., Telesco, C.M., Osip, D.J., Rieke, G.H., Rieke, M.J. and Schulz, B. (1994). The color temperature of (2060) Chiron: A warm and small nucleus. *Astronomical Journal*, **108**, 2318–2322.

Ceplecha, Z. and McCrosky, R.E. (1976). Fireball end heights: A diagnostic for the structure of meteoric material. *Journal of Geophysical Research*, **81**, 6257–6275.

Chen, J. and Jewitt, D. (1994). On the rate at which comets split. *Icarus*, **108**, 265–271.

Chokshi, A., Tielens, A.G.G.M. and Hollenbach, D. (1993). Dust coagulation. *Astrophysical Journal*, **407**, 806–819.

Clube, S.V.M. and Napier, W.M. (1982). Spiral arms, comets and terrestrial catastrophism. *Quarterly Journal of the Royal Astronomical Society*, **23**, 45–66.

Cochran, A.L., Levison, H.F., Stern, S.A. and Duncan, M.J. (1995). The discovery of Halley-sized Kuiper belt objects using the Hubble Space Telescope. *Astrophysical Journal*, **455**, 342–346.

Colom, P. and Gérard, E. (1988). A search for periodicities in the OH radio emission of comet P/Halley (1986 III). *Astronomy and Astrophysics*, **204**, 327–336.

Crifo, J.F. (1987). Improved gas-kinetic treatment of cometary water sublimation and recondensation: Application to comet P/Halley. *Astronomy and Astrophysics*, **187**, 438–450.

Crifo, J.F. and Rodinov, A.V. (1997a). The dependence of the circumnuclear coma structure on the properties of the nucleus. I. Comparison between a homogeneous and an inhomogeneous spherical nucleus, with application to P/Wirtanen. *Icarus*, **127**, 319–353.

Crifo, J.F. and Rodinov, A.V. (1997b). The dependence of the circumnuclear coma structure on the properties of the nucleus. II. First investigation of the coma surrounding an homogeneous, asperical nucleus. *Icarus*, **129**, 72–93.

Crifo, J.F., Rodinov, A.V. and Bockelee-Morvan, D. (1999). The dependence of the circumnuclear coma structure on the properties of the nucleus. III.

First modeling of a CO-dominated coma with application to comets 46 P/Wirtanen and 29 P/Schwassmann-Wachmann 1. *Icarus*, **138**, 85–106.

Crovisier, J., Brooke, T.Y., Hanner, M.S., Keller, H.U., Lamy, P.L., Altieri, B., Bockelee-Morvan, D., Jorda, L., Leech, K. and Lellouch, E. (1997). The infrared spectrum of comet C/1995 O1 (Hale-Bopp) at 4.6 AU from the Sun. *Astronomy and Astrophysics*, **315**, L385–L388.

Cruikshank, D.P. and Brown, R.H. (1983). The nucleus of comet P/Schwassmann-Wachmann 1. *Icarus*, **56**, 377–380.

Cruikshank, D.P., Hartmann, W.K. and Tholen, D.J. (1985). Colour, albedo and nucleus size of Halley's comet. *Nature*, **315**, 122–124.

Cuzzi, J.N., Dobrovolskis, A.R. and Champney, J.M. (1993). Particle–gas dynamics in the midplane of a protoplanetary nebula. *Icarus*, **106**, 102–134.

Davis, D.R. and Farinella, P. (1997). Collisional evolution of Edgeworth–Kuiper belt objects. *Icarus*, **125**, 50–60.

Delsemme, A.H. (1973). Gas and dust in comets. *Space Science Reviews*, **15**, 89–101.

Delsemme, A.H. (1985). The nature of the cometary nucleus. *Publications of the Astronomical Society of the Pacific*, **97**, 861–870.

Desvoivres, E., Klinger, J., Levasseur-Regourd, A.C., Lecacheux, J., Jorda, L., Enzian, A., Colas, E., Frappa, E. and Laques, P. (1999). Comet C/1996 B2 Hyakutake: Observations, interpretation and modelling of the dynamimcs of fragments of cometary nuclei. *Monthly Notices of the Royal Astronomical Society*, **303**, 826–834.

Divine, N. (1981). Numerical models for Halley dust environments. In B. Battrick and E. Swallow (eds), *The Comet Halley Dust & Gas Environment*, ESA SP-174, ESA Scientific & Technical Branch, ESTEC, Noordwijk, pp. 25–30.

Divine, N., Fechtig, H., Gombosi, T.I., Hanner, M.S., Keller, H.U., Larson, S.M., Mendis, D.A., Newburn, R.L., Jr Reinhard, R., Sekanina, Z. and Yeomans, D.K. (1986). The comet Halley dust and gas environment. *Space Science Reviews*, **43**, 1–104.

Dohnanyi, J.W. (1969). Collisional models of asteroids and their debris. *Journal of Geophysical Research*, **74**, 2531–2554.

Dominik, C. and Tielens, A. (1997). Coagulation of dust grains and the structure of dust aggregates in space. *Astrophysical Journal*, **480**, 647–673.

Donn, B. (1991). The accumulation and structure of comets. In R. Newburn and J. Rahe (eds), *Comets in the Post-Halley Era*, Vol. 1, Kluwer, Dordrecht, pp. 335–359.

Drummond, J.D., Fugate, R.Q., Christou, J.C. and Hege, E.K. (1998). Full adaptive optics images of asteroids Ceres and Vesta: Rotational poles and triaxial ellipsoid dimensions. *Icarus*, **132**, 80–99.

Duncan, M.J. and Levison, H.F. (1997). A scattered comet disk and the origin of Jupiter family comets. *Science*, **276**, 1670–1672.

Duncan, M.J., Quinn, T. and Tremaine, S. (1987). The formation and extent of the Solar System comet cloud. *Astronomical Journal*, **94**, 1330–1349.

Edgeworth, K.E. (1949). The origin and evolution of the solar system. *Monthly Notices of the Royal Astronomical Society*, **109**, 600–609.

Efroimsky, M. and Lazarian, A. (2000). Inelastic dissipation in wobbling asteroids and comets. *Monthly Notices of the Royal Astronomical Society*, **311**, 269–278.

Eggers, S. (1999). Cometary dynamics during the formation of the Solar System. PhD thesis, Georg-August-Universität, Göttingen, Germany.

Eggers, S., Keller, H.U., Krupa, P. and Markiewicz, W.J. (1997). Origin and dynamics of comets and star formation. *Planetary and Space Science*, **45**, 1099–1104.

Eggers, S., Keller, H.U. and Markiewicz, W.J. (1999). Dynamics of the Kuiper belt during the formation of the Solar System. In L.M. Celnikier and T.T. Vân (eds), *Planetary Systems: the Long View (IX, Recontres de Blois)*. Edition Frontières, pp. 79–80.

Ehrenfreud, P., D'Hendecourt, L. Dartois, E. Jourdain de Muizon, M., Breitfellner, M. Puget, J.L. and Habing, H.J. (1997). ISO observations of interstellar ices and implications for comets. *Icarus*, **130**, 1–15.

Emerich, C., Lamarre, J.M., Gispert, R., Coron, N., Combes, M., Encrenaz, T., Crovisier, J., Rocard, F., Bibring, J.P., Moroz, V.I., Sanko, N.F. and Nikolsky, Yu.V. (1987a). Temperature of the nucleus of comet Halley. In E.J. Rolfe and B. Battrick (eds), *Symposium on the Diversity and Similarity of Comets*, ESA-SP 278, ESA Publications Division, ESTEC, Noordwijk, pp. 703–706.

Emerich, C., Lamarre, J.M. Moroz, V.I., Combes, M., Sanko, N.F., Nikolsky, Y.V., Rocard, F., Gispert, R., Coron, N., Bibring, J.P., Encrenaz, T. and Crovisier, J. (1987b). Temperature and size of the nucleus of comet P/Halley deduced from IKS infrared Vega-1 measurements. *Astronomy and Astrophysics*, **187**, 839–842.

Enzian, E., Cabot, H. and Klinger, J. (1996). A $2\frac{1}{2}$ thermodynamic model of cometary nuclei. I. Application to the activity of comet 29P/Schwassmann-Wachmann 1. *Astronomy and Astrophysics*, **319**, 995–1006.

Enzian, A., Cabot, H. and Klinger, J. (1998). Simulation of the water and carbon monoxide production rates of comet Hale-Bopp using a quasi 3-D nucleus model. *Planetary and Space Science*, **46**, 851–858.

Enzian, A., Klinger, J., Schwehm, G. and Weissman, P.R. (1999). Temperature and gas Production distributions on the surface of a spherical model comet nucleus in the orbit of 46P/Wirtanen. *Icarus*, **138**, 74–84.

Espinasse, S., Klinger, J., Ritz, C. and Schmitt, B. (1991). Modeling of the thermal behavior and of the chemical differentiation of cometary nuclei. *Icarus*, **92**, 350–365.

Everhart, E. (1968). Change in total energy of comets passing through the Solar System. *Astronomical Journal*, **73**, 1039–1052.

Everhart, E. (1972). The origin of short-period comets. *Astrophysical Journal Letter*, **10**, 131–135.

Fanale, F.P. and Salvail, J.R. (1984). An idealized short period comet model: Surface insolation, H_2O flux, dust flux and mantle development. *Icarus*, **60**, 476–511.

Farinella, P., Davis, D.R. and Stern S.A. (2000). Formation and collisional evolution of the Edgeworth–Kuiper belt. In V. Mannings, A.P. Boss and S.S. Russel (eds), *Protostars and Planets IV*, University of Arizona Press, Tucson, p. 1255.

Fay, T.D., Jr and Wisniewski, W. (1978). The light curve of the nucleus of Comet d'Arrest. *Icarus*, **34**, 1–9.

Fernández, J.A. (1980). On the existence of a comet belt beyond Neptune. *Monthly Notices of the Royal Astronomical Society*, **192**, 481–491.

Fernandez, J.A. (1997). The formation of the Oort cloud and the primitive galactic environment. *Icarus*, **129**, 106–119.

Fernández, J.A. and Ip, W.-H. (1981). Dynamical evolution of a cometary swarm in the outer planetary region. *Icarus*, **47**, 470–479.

Fernández, J.A. and Ip, W.-H. (1983). On the time evolution of the cometary influx in the region of the terrestrial planets. *Icarus*, **54**, 377–387.

Finson, M.L. and Probstein, R.F. (1968a). A theory of dust comets. I Model and equations. *Astrophysical Journal*, **154**, 327–380.

Finson, M.L. and Probstein, R.F. (1968b). A theory of dust comets. II. Results for comet Arend-Roland. *Astrophysical Journal*, **154**, 353–380.

Fulle, M. (1997). Injection of large grains into orbits around comet nuclei. *Astronomy and Astrophysics*, **325**, 1237–1248.

Gladman, B., Kavelaars, J.J., Nicholson, P.D., Loredo, T.J. and Burns, J.A. (1998). Pencil-beam surveys for faint trans-Neptunian objects. *Astronomical Journal*, **116**, 2042–2054.

Goldreich, P. and Ward, W.R. (1973). The formation of planetesimals. *Astrophysical Journal*, **183**, 1051–1061.

Goldstein, R.M., Jurgens, R.F. and Sekanina, Z. (1984). A radar study of comet Iras-Araki-Alcock 1983d. *Astronomical Journal*, **89**, 1745–1754.

Gradie, J. and Veverka, J. (1980). The composition of Trojan asteroids. *Nature*, **283**, 840–842.

Greenberg, J.M. (1977). From dust to comets. In A.H. Delsemme (ed), *Comets, Asteroids, Meteorites*, University of Toledo Press, OH.

Greenberg, J.M. (1984). The Structure and evolution of interstellar grains. *Scientific American*, **250**, 124–135.

Greenberg, J.M. (1998). Making a comet nucleus. *Astronomy and Astrophysics*, **330**, 375–380.

Greenberg, J.M., Mizutani, H. and Yamamoto, T. (1995). A new derivation of the tensile strength of cometary nuclei: Application of comet Shoemaker-Levy 9. *Astronomy and Astrophysics*, **295**, L35–L38.

Grothues, H.-G. and Schmidt-Kaler, T. (1996). The dust tail of comet 1P/Halley after its perihelion in 1986 and the rotation of the nucleus. *Monthly Notices of the Royal Astronomical Society*, **282**, 547–562.

Hainaut, O., Boehnhardt, K.J. and West, R.M. (1998). Early recovery of comet 55P/Tempel-Tuttle. *Astronomy and Astrophysical*, **333**, 746–752.

Hainaut, O., West, R.M., Marsden, B.G., Smette, A. and Meech, K. (1995). Post-perihelion observations of comet P/Halley IV, r=16,6 and 18,8 AU. *Astrophysical Journal*, **293**, 941–947.

Hainaut, O., West R.M., Smette, A. and Marsden, B.G. (1993). Imaging of very distant comets: present experience and future expectations. In W.F. Heubner, H.U. Keller, D. Jewitt, J. Klinger and R. West (eds), *Proceedings of the Workshop on the Activity of Distant Comets, Lenggries, Germany, October*. Southwest Research Institute, San Antonio, Texas, pp. 54–63.

Hainaut, O., West, R.M., Smette, A. and Marsden, B.G. (1994). Imaging of very distant comets: Current and future limits. *Astronomy and Astrophysical*, **289**, 311–324.

Hajduk, A. (1987). Meteoroids from comet P/Halley. The comet's mass production and age. *Astronomy and Astrophysical*, **187**, 925–927.

Hale, A., Stevens, J. and Bopp, T. (1995). Comet 1995 O1. *IAU Circular* 6178.

Hanner, M.S., Aitken, D.K., Knacke, R. Mccorkle, S. Roche, P.F. and Tokunaga, A.T. (1985). Infrared spectrophotometry of comet Iras Araki-Alcock (1983d): A bare nucleus revealed. *Icarus*, **62**, 97–109.

Harmon, J.K., Campbell, D.B., Hine, A.A., Shapiro, I.I. and Marsden, B.G. (1989). Radar observations of comet Iras-Araki-Alcock (1983d). *Astrophysical Journal*, **338**, 1071–1093.

Harmon, J.K., Ostro, S.J., Benner, L.A.M., Rosema, K.D., Jurgens, R.F., Winkler, R., Yeomans, D.K., Choate, D., Cormier, R., Giorgini, J.D., Mitchell, D.L., Chodas, P.W., Rose, R., Kelley, D., Slade, M.A. and Thomas, M.L. (1997). Radar detection of the nucleus and coma of comet Hyakutake (C/1996 B2). *Science*, **278**, 1921–1924.

Hartmann, W.K., Cruikshank, D.P. and Degewij, J. (1982). Remote comets and related bodies: VJHK colorimetry and surface materials. *Icarus*, **52**, 377–408.

Hartmann, W.K., Tholen, D.J. and Cruikshank, D.P. (1987). The relationship of active comets, 'extinct' comets, and dark asteroids. *Icarus*, **69**, 33–50.

Hayashi, C., Nakazawa, K. and Nakagawa, Y. (1985). Formation of the Solar System. In D.C. Black and M.S. Matthews (eds), *Protostars & Planets II*, University of Arizona Press, Tucson, AZ.

Herman, G. and Weissman, P.R. (1987). Numerical simulation of cometary nuclei. III. Internal temperatures of comertary nuclei. *Icarus*, **69**, 314–338.

Horanyi, M., Gombosi, T.I., Cravens, T.E., Körösmezey, A., Kecskeméty, K., Nagy, A.F. and Szegö, K. (1984). The friable sponge model of a cometary nucleus. *Astrophysical Journal*, **278**, 449–455.

Horanyi, M. and Kecskeméty, K. (1983). Percolation theory and the origin of comets. In T.I. Gombosi (ed), *Cometary Exploration*, part I, Central Research Institute for Physics, Budapest, pp. 21–25.

Houpis, H.L.F. (1990). Models of cometary nuclei. In J. Mason and P. Moore (eds), *Comet Halley, Investigations, Results, Interpretations Volume 2: Dust, Nucleus, Evolution*, Ellis Horwood, London, pp. 173–188.

Huebner, W.F. (1965). Über die Gasproduktion der Kometen. *Zeitschrift für Astrophysik*, **63**, 22–34.

Huebner, W.F., Delamere, W.A., Reitsema, H., Keller, H.U., Wilhelm, K., Whipple, F.L. and Schmidt, H.U. (1986). Dust–gas interaction deduced from Halley Multicolour Camera observations. In B. Battrick, E.J. Rolfe and R. Reinhard (eds), *20th Eslab Symposium on the Exploration of Halley's Comet*, ESA SP-250 II, ESA, Publications Division, ESTEC, Noordwijk, pp. 363–364.

Hughes, D.W. (1987). P/Halley dust characteristics: A comparison between Orionid and Eta Aquarid meteor observations and those from the flyby spacecraft. *Astronomy and Astrophysics*, **187**, 879–888.

Hughes, D.W. (1988). Cometary magnitude distribution and the ratio between the numbers of long- and short-period comets. *Icarus*, **73**, 149–162.

Hut, P. and Tremaine, S. (1985). Have interstellar clouds disrupted the Oort comet cloud? *Astronomical Journal*, **90**, 1548–1557.

Ipatov, S.I. (1999). Migration of trans-Neptutian objects to the Earth. *Celestial Mechanics and Dynamical Astronomy*, **73**, 107–116.

Jewitt, D. (1990). The persistent coma of comet P/Schwassmann-Wachmann 1. *Astrophysical Journal*, **351**, 277–286.

Jewitt, D. (1992). Physical properties of cometary nuclei. In A. Brahic, J.-C. Gerard and J. Surdej (eds), *Proceedings of the 30th Liege International Astrophysical Colloquium on Observations and Physical Properties of Small Solar System Bodies*, Universite de Liege.

Jewitt, D. (1997). Cometary rotation: An overview. *Earth, Moon, Planets*, **79**, 35–53.

Jewitt, D. (1999). Kuiper belts objects. *Annual Review of Earth and Planetary Sciences*, **27**, 287–312.

Jewitt, D. and Luu, J. (1993). Discovery of the candidate Kuiper belt object 1992 QB_1. *Nature*, **362**, 730–732.

Jewitt, D. and Luu, J. (1998). Optical–infrared spectral diversity in the Kuiper belt. *Astronomical Journal*, **115**, 1667–1670.

Jewitt, D., Luu, J. and Chen, J. (1996). The Mauna-Kea-Cerro Tololo (MKCT) Kuiper belt and Centaur survey. *Astronomical Journal*, **112**, 1225–1331.

Jewitt, D., Luu, J. and Trujillo, C. (1998). Large Kuiper Belt Objects: The Mauna Kea 8k Ccd Survey. *Astronomical Journal*, **115**, 2125–2135.

Jewitt, D.C. and Meech, K.J. (1987). Ccd photometry of comet P/Encke. *Astronomical Journal*, **93**, 1542–1548.

Jewitt, D. and Meech, K.J. (1988). Optical properties of cometary nuclei and a preliminary comparison with asteroids. *Astrophysical Journal*, **328**, 974–986.

Johnson, R.E., Cooper, J.F., Lanzerotti, L.J. and Strazzula, G. (1987). Radiation formation of a non-volatile comet crust. *Astronomy and Astrophysics*, **187**, 889–892.

Jorda, L., Rembor, K., Lecacheux, J., Colom, P., Colas, F., Frappa, E. and Lara, L.M. (1997). The rotational prameters of Hale-Bopp (C/1995 O1) from observatioons of the dust jets at Pic du Midi Observatory. *Earth, Moon, Planets*, **77**, 167–180.

Joss, P.C. (1973). On the origin of short-period comets. *Astronomy and Astrophysics*, **25**, 271–273.

Kamoun, P.D., Campbell, D.B., Ostro, S.J., Pettengill, G.H. and Shapiro, I.I. (1982). Comet Encke: Radar detection of nucleus. *Science*, **216**, 293–296.

Keller, H.U. (1976a). Thermalization of cometary hydrogen. *Astronomy and Astrophysics*, **38**, 150–152.

Keller, H.U. (1976b). The interpretations of ultraviolet observations of comets. *Space Science Reviews*, **18**, 641–684.

Keller, H.U. (1989). Comets – Dirty snowballs or icy dirtballs? In J. Hunt and T.D. Guyeme (eds), *Proceedings of an International Workshop on Physics and Mechanics of Cometary Materials*, ESA SP-302, ESA Publications Division, ESTEC, Noordwijk, pp. 39–45.

Keller, H.U. (1990). The nucleus. In W.F. Huebner (ed), *Physics and Chemistry of Comets*, Springer-Verlag, Berlin, New York, Tokyo, pp. 13–68.

Keller, H.U., Arpigny, C., Barbieri, C., Bonnet, R.M., Cazes, S., Coradini, M., Cosmovici, C.B., Delamere, W.A., Huebner, W.F., Hughes, D.W., Jamar, C., Malaise, D., Reitsema, H.J., Schmidt, H.U., Schmidt, W.K.H., Seige, P., Whipple, F.L. and Wilhelm, K. (1986). First Halley Multicolour Camera imaging results from Giotto. *Nature*, **321**, 320–326.

Keller, H.U., Blum, J., Donn, B., El Goresy, A., Fechtig, H., Feuerbacher, B.P., Grün, E., Ip, W.-H., Kochan, H., Mann, I., Markiewicz, W.J., Metzler, K., Morfill, G.E., Ratke, L. Rott, M., Schwehm, G. and Weidenschilling, S.J. (1992). Columbus-Proto-Planetesimal Dust Aggregation Experiment. In B. Kaldeich (ed), *Proceedings of the VIIIth European Symposium on Materials and Fluid Sciences in Microgravity*, ESA SP-333, ESA Publication Division, ESTEC, Noordwijk, pp. 839–844.

Keller, H.U., Curdt, W., Kramm, J.R. and Thomas, N. (1995). Images obtained by the Halley Multicolour Camera (Hmc) on board the Giotto spacecraft. In R. Reinhard, N. Longdon and B. Battrick (eds), *Images of the Nucleus of Comet Halley, Volume 1*, ESA Publication Division, Estec, Noordwijk, pp. 1–252.

Keller, H.U., Delamere, W.A., Huebner, W.F., Reitsema, H.J., Schmidt, H.U., Whipple, F.L., Wilhelm, K., Curdt, W., Kramm, J.R., Thomas, N., Arpigny, C., Barbieri, C., Bonnet, R.M., Cazes, S., Coradini, M., Cosmovici, C.B., Hughes, D.W., Jamar, C., Malaise, D., Schmidt, K., Schmidt, W.K.H. and Seige, P. (1987). Comet P/Halley's nucleus and its activity. *Astronomy and Astrophysics*, **187**, 807–823.

Keller, H.U., Knollenberg, J. and Markiewicz, W.J. (1994). Collimation of cometary dust jets and filaments. *Planetary and Space Science* **42**, 367–382.

Keller, H.U., Kramm, R. and Thomas, N. (1988). Surface features of the nucleus of comet Halley. *Nature*, **331**, 227–231.

Keller, H.U., Marconi, M.L. and Thomas, N. (1990). Hydrodynamic implications of particle fragmentation near cometary nuclei. *Astronomy and Astrophysics*, **227**, L1–L4.

Keller, H.U. and Rembor, K. (1998). Modeling cometary dust production rates post perihelion. *Bulletin of the American Astronomical Society*, **30**, 1090–1090.

Kerr, R.A. (1986). Halley's confounding fireworks. *Science*, **234**, 1196–1198.

Kitamura, Y. (1986). Axisymmetric dusty gas jet in the inner coma of a comet. *Icarus*, **66**, 241–257.

Kitamura, Y. (1987). Axisymmetric dusty gas jet in the inner coma of a comet. II. The case of isolated jets. *Icarus*, **72**, 555–567.

Klinger, J. (1980). Influence of a phase transition of ice on the heat and mass balance of comets. *Science*, **209**, 271–272.

Klinger, J. (1981). Some consequences of a phase transition of water ice on the heat balance of a comet. *Icarus*, **47**, 320–324.

Klinger, J. (1983). Classification of cometary orbits based on the concept of orbital mean temperature. *Icarus*, **55**, 169–176.

Klinger, J. (1993). Are outbursts of comets at large heliocentric distances due to internal or due to external causes? In W.F. Huebner, H.U. Keller, D. Jewitt, J. Klinger and R. West (eds), *Proceedings of the Workshop on the Activity of Distant Comets*, Southwest Research Institute, San Antonio, TX, pp. 110–113.

Klinger, J., Levasseur-Regourd, A.-C., Bouziani, N. and Enzian, A. (1996). Towards a model of cometary nuclei for engineering studies for future space missions to comets. *Planetary and Space Science* **44**, 637–654.

Knollenberg, J. (1994). Modellrechnungen zur Staubverteilung in der inneren Koma von Kometen unter spezieller Berücksichtigung der Hmc-Daten der GIOTTO-Mission. PhD thesis. Georg-August-Universität, Göttingen, Germany.

Knollenberg, J., Kührt, E. and Keller, H.U. (1996). Interpretation of HMC images by a combined thermal and gasdynamic model. *Earth, Moon, Planets*, **72**, 103–112.

Kouchi, A., Greenberg, J.M., Yamamoto, T. and Mukai, T. (1992). Extremely low thermal conductivity of amourphous ice: Relevance to comet evolution. *Astronomical Journal*, **388**, L73–L76.

Kouchi, A., Yamamoto, T., Kozasa, T., Kuroda, T. and Greenberg, J.M. (1994). Conditions for condensation and preservation of amorphous ice and crystallinity of astrophysical ice. *Astronomy and Astrophysics*, **290**, 1009–1018.

Kührt, E. (1984). Temperature profiles and thermal stresses in cometary nuclei. *Icarus*, **60**, 512–521.

Kührt, E. and Keller, H.U. (1994). The formation of cometary surface crusts. *Icarus*, **109**, 121–132.

Kührt, E. and Keller, H.U. (1996). On the importance of dust in cometary nuclei. *Earth, Moon, Planets*, **72**, 79–89.

Kührt, E., Knollenberg, J. and Keller, H.U. (1997). Physical risks of landing on a cometary nucleus. *Planetary and Space Science*, **45**, 665–680.

Kuiper, G.P. (1951). On the origin of the Solar System. In J.A. Hynek (ed), *Astrophysics*, McGraw-Hill, New York.

Lamy, P.L., A'Hearn, M.F. and Weaver, H.A. (1999). Hubble Space Telescope observations of the nucleus of comet 45P/Honda-Mrkos-Pajdusakova and its inner coma. *Icarus*, **140**, 424–438.

Lamy, P.L. and Toth, I. (1995). Direct detection of a cometary nucleus with the Hubble Space Telescope. *Astronomy and Astrophysics*, **293**, L43–L45.

Lamy, P.L., Toth, I., Jorda, L., Weaver, H.A. and A'Hearn, M.F. (1998a). The nucleus and inner coma of comet 46P/Wirtanen. *Astronomy and Astrophysics*, **335**, L25–L29.

Lamy, P.L., Toth, I. and Weaver, H.A. (1998b). Space telescope of the nucleus and inner coma of comet 19P/1904 Y2 (Borelly). *Astronomy and Astrophysics*, **337**, 945–954.

Larson, S., Sekanina, Z., Levy, D., Tapia, S. and Senay, M. (1987). Comet P/Halley near-nucleus phenomena in 1986. *Astronomy and Astrophysics*, **187**, 639–644.

Lebofsky, L.A. and Spencer, J.R. (1989). Radiometric and thermal modeling of asteroids. In R.P. Binzel, T. Gehrels and M.S. Matthews (eds), *Asteroids II*, University of Arizona Press, Tucson, AZ, pp. 128–147.

Lecacheux, A., Jorda, L. and Colas, F. (1997). Comet C/1995 O1 (Hale-Bopp). *IAU Circular* 6560.

Levin, B.J. (1943). Gas evolution from the nucleus of a comet as related to the variations in its absolute brightness. *Compt. Rend. Acad. Sci., USSR (NS)*, **38**, 72–74.

Levison, H.F. and Duncan, M.J. (1994). The long-term dynamical behavior of short-period comets. *Icarus*, **108**, 18–36.

Licandro, J., Bellot, R., Luis, R., Boehnhardt, H., Casas, R., Goetz, B., Gomez, A., Jorda, L., Kidger, M.R., Osip, D., Sabalisck, N., Santos, P., Serr-Ricart, M., Tozzi, G.P. and West, R. (1998). The rotation period of C/1995 O1 (Hale-Bopp). *Astrophysical Journal*, **501**, L221–L225.

Lisse, C.M., Fernández, Y.R., Kundu, A., A'Hearn, M.F., Dayal, A., Deutsch, L.K., Fazio, G.G., Hora, J.L. and Hoffmann, W.F. (1999). The nucleus of comet Hyakutake (C/1996 B2). *Icarus*, **140**, 189–204.

Love, S.G., Joswiak, D.J. and Rownlee, BD.E. (1994). Densities of stratospheric micrometeorites. *Icarus*, **111**, 227–236.

Luu, J.X. (1993a). Cometary activity in distant comets: Chiron. *Publications of the Astronomical Society of the Pacific*, **105**, 946–950.

Luu, J.X. (1993b). Spectral diversity among the nuclei of comets. *Icarus*, **104**, 138–148.

Luu, J. and Jewitt, D.C. (1990). The nucleus of comet P/Encke. *Icarus*, **86**, 69–81.

Luu, J.X. and Jewitt, D.C. (1992). Near-aphelion CCD photometry of comet P/Schwassmann-Wachmann 2. *Astronomical Journal*, **104**, 2243–2249.

Luu, J. and Jewitt, D.C. (1993). Comet Shoemaker-Levy (1993e). *IAU Circular* 5730.

Luu, J.X. and Jewitt, D.C. (1998). Deep imaging of the Kuiper belt with the Keck 10-meter telescope. *Astrophysical Journal*, **502**, L91–L94.

Luu, J.X., Jewitt, D.C. and Trujillo, C. (2000). Water ice in 2060 Chiron and its implications for Centaurs and Kuiper belt objects. *Astrophysical Journal*, **531**, L151–L154.

Lyttleton, R.A. (1948). On the origin of comets. *Monthly Notices of the Royal Astronomical Society*, **108**, 465–475.

Lyttleton, R.A. (1953). *The Comets and Their Origin*, Cambridge University Press.

Marchis, F., Boehnhardt, H., Hainaut, O.R. and LeMignant, D. (1999). Adaptive optics observations of the innermost coma of C/1995 O1. Are there a 'Hale' and a 'Bopp' in comet Hale-Bopp? *Astronomy and Astrophysics*, **349**, 985–995.

Marco, O., Encrenaz, T. and Gendron, E. (1998). First images of a comet with adaptive optics. *Planetary and Space Science*, **46**, 547–554.

Marsden, B.G. (1989a). The sungrazing comet group. II. *Astronomical Journal*, **98**, 2306–2321.

Marsden, B.G. (1989b). *Catalogue of Cometary Orbits*, 6th edn, IAU.

McCrea, W.H. (1975). Solar System as space-probe. *Observatory*, **95**, 239–255.

McKinnon, W.B. and Schenk, P.M. (1995). Estimates of comet fragment masses from impact crater chains on Callisto and Ganymede. *Geophysical Research Letters*, **22**, 1829–1832.

McNaught, R.H. (1995). Comet C/1995 O1 (Hale-Bopp). *IAU Circular* 6198.

Meech, K.J., Belton, M.J.S., Mueller, B.E.A., Dicksion, M.W. and Li, H.R. (1993). Nucleus properties of P/Schwassmann-Wachmann 1. *Astronomical Journal*, **106**, 1222–1236.

Meech, K.J. and Jewitt, D. (1987). Comet Bowell at record heliocentric distance. *Nature*, **328**, 506–509.

Melosh, H.J. and Schenk, P. (1993). Split comets and the origin of crater chains on Ganymede and Callisto. *Nature*, **365**, 731–735.

Melosh, H.J. and Whitaker, E.A. (1994). Split comets and crater chains on the Moon. *Nature*, **369**, 713–714.

Mendis, D.A. and Brin, G.D. (1977). The monochromatic brightness variations of comets – II. The core-mantle model. *Moon and Planets*, **17**, 359–372.

Merényi, E., Földy, L., Szegö, K., Tóth, I. and Kondor, A. (1990). The landscape of comet Halley. *Icarus*, **86**, 9–20.

Millis, R.L., A'Hearn, M.F. and Campins, H. (1988). An investigation of the nucleus and coma of comet P/Arend-Rigaux. *Astrophysical Journal*, **324**, 1194–1209.

Mitchell, J.K., Carrier, W.D., Costes, N.C., Houston, W.N., Scott, R.F. and Hovland, H.J. (1973). Soil mechanics. *NASA Conference Publication*, NASA SP-330.

Möhlmann, D., Börner, H., Danz, M., Elter, G., Mangoldt, T., Rubbert, B. and Weidlich, U. (1986). Physical properties of P/Halley – Derived from Vega Images. In B. Battrick, E.J. Rolfe and R. Reinhard (eds), *20th ESLAB Symposium on the Exploration of Halley's Comet*, ESA-SP 250 II, ESA, Publications Division, ESTEC, Noordwijk, pp. 339–340.

Möhlmann, D., Danz, M. and Börner, H. (1987). Properties of the nucleus of P/Halley. In E.J. Rolfe and B. Battrick (eds.), *Symposium on the Diversity and Similarity of Comets*, ESA-SP 278, ESA Publications Division, ESTEC, Noordwijk, pp. 481–492.

Omont, A., Forveille, T., Moseley, S.H., Glaccum, W.J., Harvey, P.M., Likkel, L., Loewenstein, R.F. and Lisse, C.M. (1990). Observations of 40–70 micron bands of ice in IRAS 09371+1212 and other stars. *Astrophysical Journal*, **355**, L27–L30.

Oort, J.H. (1950). The structure of the cloud of comets surrounding the Solar System and a hypothesis concerning its origin. *Bulletin of the Astronomical Institutes of the Netherlands*, **IX**, 91–110.

Öpik, E.J. (1932). Note on stellar perturbations of nearly parabolic orbits. *Proceedings of the American Academy of Arts and Sciences*, **67**, 169–183.

Orosei, R., Capaccioni, F., Capria, M.T., Coradini, A., De Sanctis, M.C., Federico, C., Salomone, M. and Huot, J.-P. (1999). Numerically improved thermochemical evolution models of comet nuclei. *Planetary and Space Science*, **47**, 839–853.

Peale, S.J. (1989). On the density of Halley's comet. *Icarus*, **82**, 36–50.

Peale, S.J. and Lissauer, J.J. (1989). Rotation of Halley's comet. *Icarus*, **79**, 396–430.

Prialnik, D. and Bar-Nun, A. (1987). On the evolution and activity of cometary nuclei. *Astrophysical Journal*, **313**, 893–905.

Prialnik, D. and Bar-Nun, A. (1988). The formation of a permanent dust mantle and its effect on cometary activity. *Icarus*, **74**, 272–283.

Quinn, T., Tremaine, S. and Duncan, M. (1990). Planetary perturbations and the origin of short-period comets. *Astrophysical Journal*, **355**, 667–679.

Reinhard, R. (1986). The Inter-Agency Consultative Group (IACG) and its associated working groups. In R. Reinhard and B. Battrick (eds), *Space Missions to Halley's Comet*, ESA Publications Division, ESTEC, Noordwijk, pp. 199–216.

Rembor, K.-M. (1998). Ableitung physikalischer Parameter von Kometenkernen aus Beobachtungen der Staubkoma. PhD thesis, Georg-August-Universität, Göttingen, Germany.

Richter, K. and Keller, H.U. (1995). On the stability of dust particle orbits around cometary nuclei. *Icarus*, **114**, 355–371.

Rickman, H. (1989). The nucleus of comet Halley: Surface structure, mean density, gas and dust production. *Advances in Space Research*, **9**, 59–71.

Rickman, H., Fernandez, J.A. and Gustafson, B.A.S. (1990). Formation of stable dust mantles on short-period comet nuclei. *Astronomy and Astrophysics*, **237**, 524–535.

Rickman, H. and Jorda, L. (1998). Comet 46P/Wirtanen, the target of the Rosetta mission. *Advances in Space Research*, **21**, 1491–1504.

Rickman, H., Kamél, L., Festou, M.C. and Froeschlé, Cl. (1987). Estimates of masses, volumes and densities of short-period comet nuclei. In E.J. Rolfe and B. Battrick (eds), *Symposium on the Diversity and Similarity of Comets*, ESA-SP 278, ESA Publications Division, ESTEC, Noordwijk, pp. 471–481.

Roemer, E. (1966a). The dimensions of cometary nuclei. *Nature et origine des comètes*, 23.

Roemer, E. (1966b). 1. Cometary nuclei. Introductory report. *Nature et origine des comètes*, 15–22.

Russell, H.N. (1916). On the albedo of the planets and their satellites. *Astrophysical Journal*, **43**, 173–196.

Safronov, V.S. (1969). Evolution of the protoplanetary cloud and formation of the Earth and planets. *NASA Technical Memorandum*, TT F-677.

Sagdeev, R.Z., Blamont, J., Galeev, A.A., Moroz, V.I., Shapiro, V.D., Shevchenko, V.I. and Szegö, K. (1986). Vega spacecraft encounters with comet Halley. *Nature*, **321**, 259–262.

Sagdeev, R.Z., Elyasberg, P.E. and Moroz, V.I. (1988). Is the nucleus of comet Halley a low density body? *Nature*, **331**, 240–242.

Sagdeev, R.Z., Szegö, K., Smith, B.A., Larson, S., Merenyi, E., Kondor, A., Toth, I. (1989). The Rotation of P/Halley. *The Astronomical Journal*, **97**, 546–551.

Saint-Pé, O., Combes, M. and Rigaut, F. (1993). Ceres surface properties by high-resolution imaging from Earth. *Icarus*, **105**, 271–281.

Samarasinha, N.H. (2000). Coma morphology due to an extended active region and implications for the spin state of comet Hale-Bopp. *The Astrophysical Journal*, **529**, L107–L110.

Samarasinha, N.H. and Belton, M.J.S. (1995). Long-term evolution of rotational states and non-gravitational effects for Halley-like cometary nuclei. *Icarus*, **116**, 340–358.

Saunders, R.S., Fanale, F.P., Parker, T.J., Stephens, I.B. and Sutton, S. (1986). Properties of filamentary sublimation residues from dispersions of clay in ice. *Icarus*, **66**, 94–104.

Schenk, P.M., Asphaug, E., Mckinnon, W.B., Melosh, H.J. and Weissman, P.R. (1996). Cometary nuclei and tital discruption: The geologic record of crater chains on Callisto and Ganymede. *Icarus*, **121**, 249–274.

Schiaparelli, G. (1866). Notice of paper, 'Sur la relation qui existe entre les Comètes et les Etoiles filantes'. *Monthly Notices of the Royal Astronomical Society*, **27**, 246

Schleicher, D.G., Millis, R.L., Thompson, D.T., Birch, P.V., Martin, R., Tholen, D.J., Piscitelli, J.R., Lark, N.L. and Hammel, H.B. (1990). Periodic variations in the activity of comet P/Halley during the 1985/1986 apparition. *Astronomical Journal*, **100**, 896–912.

Schulz, R. and Schlosser, W. (1989). CN-shell structures and dynamics of the nucleus of comet P/Halley. *Astronomy and Astrophysics*, **214**, 375–385.

Schwarz, G., Craubner, H., Delamere, W.A., Goebel, M., Gonano, M., Huebner, W.F., Keller, H.U., Kramm, R., Mikusch, E., Reitsema, H.J., Whipple, F.L. and Wilhelm, K. (1986). Detailed analysis of a surface feature on comet Halley. In B. Battrick, E.J. Rolfe and R. Reinhard (eds), *20th ESLAB Symposium on the Exploration of Halley's Comet*, ESA-SP 250 II, ESA, Publications Division, ESTEC, Noordwijk, pp. 371–374.

Scotti, J.V. and Melosh, H.J. (1993). Estimate of the size of comet Shoemaker-Levy 9 from a tidal breakup model. *Nature*, **365**, 733–735.

Sekanina, Z. (1972). A model for the nucleus of Encke's comet. In G.A. Chebatorev, E.I. Kazimirchak-Polonskaya and B.G. Marsden (eds), *The Motion, Evolution of Orbits and Origin of Comets*, Reidel, Dordrecht, p. 301.

Sekanina, Z. (1977). Relative motions of fragments of the split comets. I. A new approach. *Icarus*, **30**, 574–594.

Sekanina, Z. (1978). Relative motions of fragments of the split comets. II. Separation velocities and differential decelerations for extensively observed comets. *Icarus*, **33**, 173–185.

Sekanina, Z. (1982). The problem of split comets in review. In L.L. Wilkening (ed), *Comets*, University of Arizona Press, Tucson, AZ.

Sekanina, Z. (1997). The problem of split comets revisited. *Astronomy and Astrophysics*, **318**, L5–L8.

Shoemaker, C.S., Shoemaker, E.M. and Levy, D.H. (1993). Comet Shoemaker-Levy (1993e). *Iau Circular* 5724.

Shoshany, Y., Podolak, M., Prialnik, D. and Berkowitz, B. (1999). A Monte Carlo model for the flow of dust in a porous comet nucleus. *Icarus*, **137**, 348–354.

Shul'Man, L.M. (1972). The evolution of cometary nuclei. In G.A. Chebatorev, E.I. Kazimirchak-Polonskaya and B.G. Marsden (eds), *The Motion, Evolution of Orbits and Origin of Comets*, Reidel, Dordrecht, pp. 271–276.

Skorov, Yu.V. and Rickman, H. (1995). A kinetic model of gas flow in a porous cometary mantle. *Planetary and Space Science*, **43**, 1587–1594.

Skorov, Yu.V., Kömle, N.I., Mariewicz, W.J. and Keller, H.U. (1999). Mass and energy balance in the near-surface layers of a cometary nucleus. *Icarus*, **140**, 173–188.

Smoluchowski, R. (1981). Amorphous ice and the behaviour of cometary nuclei. *Astrophysical Journal*, **244**, L31.

Solem, J.C. (1994). Density and site of comet Shoemaker-Levy 9 deduced from a tidal breakup model. *Nature*, **370**, 349–351.

Stern, S.A. (1995). Collisional time scales in the Kuiper disk and their implications. *Astronomical Journal*, **110**, 856.

Stern, S.A. (1996). The historical development and status of Kuiper disk studies. In T.W. Rettig and J.M. Hahn (eds), *Completing the Inventory of the Solar System, Astronomical Society of the Pacific Conference Proceedings*, Vol. 107, pp. 209–232.

Stern, S.A. and Colwell, J.E. (1997). Accretion in the Edgeworth–Kuiper belt: Forming 100–1000 km radius bodies at 30 Au and beyond. *Astronomical Journal*, **114**, 841–849.

Stooke, P.J. and Abergel, A. (1991). Morphology of the nucleus of comet P/Halley. *Astronomy Astrophysics*, **248**, 656–668.

Storrs, A.D., Fanale, F.P., Saunders, R.S. and Stephens, J.B. (1988). The formation of filamentary sublimate residues (FSR) from mineral grains. *Icarus*, **76**, 493–512.

Sykes, M.V., Lebofsky, L.A., Hunten, D.M. and Low, F.J. (1986). The discovery of dust trails in the orbits of periodic comets. *Science*, **232**, 1115–1117.

Szegö, K., Sagdeev, R.Z., Whipple, F.L., Abergel, A., Bertaux, J.-L., Merényi, E., Szalai, S. and Varhalmi, L. (1995). Images obtained by the Television System (TVS) on board the Vega spacecraft. In R. Reinhard and B. Battrick (eds), *Images of the Nucleus of Comet Halley, Volume 2*, ESA Publications Division, ESTEC, Noordwijk, pp. 1–255.

Tambovtseva, L.V. and Shestakova, L.I. (1999). Cometary splitting due to the thermal stresses. *Planetary and Space Science*, **47**, 319–326.

Tholen, D.J., Hartmann, W.K. and Cruikshank, D.P. (1988). (2060) Chiron. *Iau Circular* 4454.

Thomas, N., Eggers, S., Ip, W.-H., Lichtenberg, G., Fitzsimmons, A., Jorda, L., Keller, H.U., Williams, Ip., Hahn, G. and Rauer, H. (2000). Observations of the trans-Neptunian objects, 1993 SC and 1996 TL_66, with the Infrared Space Observatory. *Astronomical Journal*, **534**, 446–455.

Thomas, N. and Keller, H.U. (1987). Fine dust structures in the emission of comet P/Halley observed by the Halley Multicolour Camera on board Giotto. *Astronomy and Astrophysics*, **187**, 843–846.

Thomas, N. and Keller, H.U. (1989). The colour of comet P/Halley's nucleus and dust. *Astronomy and Astrophysics*, **213**, 487–494.

Thomas, N. and Keller, H.U. (1990). Interpretation of the inner coma observations of comet P/Halley by the Halley Multicolour Camera. *Annales Geophysicae*, **8**, 147–166.

Tokunaga, A.T. and Hanner, M.S. (1985). Does comet P/Arend-Rigaux have a large dark nucleus? *Astrophysical Journal*, **296**, L13–L16.

Tokunaga, A.T., Hanner, M.S., Golisch, W.F., Griep, D.M., Kamisky, C.D. and Chen, H. (1992). Infrared monitoring of Comet P/Tempel 2. *Astronomical Journal*, **104**, 1611–1617.

Veeder, G.J., Hanner, M.S. and Tholen, D.J. (1987). The nucleus of comet P/Arend-Rigaux. *Astronomical Journal*, **94**, 169–173.

Vsekhsvyatskij, S.K. (1967). Physical characteristics of comets observed during 1961–1965. *Soviet Astronomy*, **10**, 1034–1041.

Wallis, M.K. (1980). Radiogenic heating of primordial cometary interiors. *Nature*, **284**, 431–433.

Weaver, H.A., A'Hearn, M.F., Arpigny, C., Boice, D.C., Feldmann, P.D., Larson, S.M., Lamy, P., Levy, D.H., Marsden, B.G., Meech, K.J., Noll, K.S., Scotti, J.V. and Sekanina, Z. (1995). The Hubble Space Telescope (HST) observing campaign on comet Shoemaker-Levy 9. *Science*, **267**, 1282–1288.

Weidenschilling, S.J. (1980). Dust to planetesimals: Settling and coagulation in the solar nebula. *Icarus*, **44**, 172–189.

Weidenschilling, S.J. (1995). Can graviational instability form planetesimals? *Icarus*, **116**, 433–435.

Weidenschilling, S.J. (1997). The origin of comets in the solar nebula: A unified model. *Icarus*, **127**, 290–306.

Weissman, P.R. (1980). Stellar perturbations of the cometary cloud. *Nature*, **288**, 242–243.

Weissman, P.R. (1986). Are cometary nuclei primordial rubble piles? *Nature*, **320**, 242–244.

Weissman, P.R., A'Hearn, M.F., McFadden, L.A. and Rickman, H. (1989). Evolution of comets into asteroids. In R.P. Binzel, T. Gehrels and M.S. Matthews (eds), *Asteroids II*, University of Arizona Press, Tucson, AZ, pp. 880–920.

West, R.M. (1990). Post-perihelion observations of comet P/Halley II. r=10.1 AU. *Astronomy and Astrophysics*, **228**, 531–538.

West, R.M., Hainaut, O. and Smette, A. (1991). Post-perihelion observations of P/Halley. III. An outburst at r = 14.3 AU. *Astronomy and Astrophysics*, **246**, L77–L80.

West, R.M. and Jorgensen, H.E. (1989). Post-perihelion observations of comet P/Halley at r=8.5 AU. *Astronomy and Astrophysics* **218**, 307–316.

Wetherill, G.W. and Revelle, D.O. (1982). Relationships between comets, large meteors, and meteorites. In L.L. Wilkening (ed), *Comets*, University of Arizona Press, Tucson, AZ, pp. 297–319.

Whipple, F.L. (1950a). A Comet model I. The acceleration of comet Encke. *Astrophysical Journal*, **111**, 375–394.

Whipple, F.L. (1950b). A comet model II. Physical relations for comets and meteors. *Astrophysical Journal*, **111**, 464–474.

Whipple, F.L. (1967). On maintaining the meteoritic complex. *The Zodiacal Light and the Interplanetary Medium*, NASA SP-150, pp. 409–426.

Whipple, F.L. (1983). Comets – Nature, evolution and decay. *Highlights of Astronomy*, **6**, 323–330.

Whipple, F.L. and Sekanina, Z. (1979). Comet Encke: Precession of the spin axis, non-gravitational motion and sublimation. *Astronomical Journal*, **84**, 1894–1909.

Williams, G. (2000). *Catalogue of Cometary Orbits*. 13th Edn, IAU.

Wisniewski, W.Z. (1990). Rotation of comet P/Tempel 2 from CCD and photoelectric photometry. *Icarus*, **86**, 52–57.

Wurm, K. (1934). Beitrag zur Deutung der Vorgänge in Kometen. I. *Zeitschrift für Astrophysik*, **8**, 281–291.

Wurm, K. (1935). Beitrag zur Deutung der Vorgänge in Kometen. II. *Zeitschrift für Astrophysik*, **9**, 62–78.

Wurm, K. (1943). Die Natur der Kometen. *Mitt. Hamburger Sternwarte*, **8**, 57–92.

Yamamoto, T. and Kozasa, T. (1988). The cometary nucleus as an aggregate of planetesimals. *Icarus*, **75**, 540–551.

Yeomans, D.K. and Kiang, T. (1981). The long-term motion of comet Halley. *Monthly Notices of the Royal Astronomical Society*, **197**, 633–646.

KATHRIN ALTWEGG* AND WESLEY T. HUNTRESS, Jr**

The constituents of cometary nuclei

Close to the edge of our Solar System is a cloud containing many tiny objects only a few kilometers in diameter. Even though their number has been estimated to be as large as $\sim 10^{13}$, their total mass is still negligible compared to that of the major planets. Their existence would go unnoticed were it not that, from time to time, some of them are gravitational by pertured and move inward, towards the Sun, and can then be observed as comets. Several space missions have been aimed uniquely at comets, among them the first European deep space mission, Giotto. Rosetta, a cornerstone mission of the Horizon 2000 program of the European Space Agency (ESA), is one of several comet missions being planned for the first decade of the twenty-first century. Why do comets deserve this special attention?

Comets streak brilliantly across the night sky, their tiny nucleus obscured behind an immense coma. Their long tails set them apart from all other celestial bodies, and their sometimes sudden, unexpected appearance and unusual paths across the night sky have stimulated the imaginations of skywatchers for thousands of years. In ancient China details of comets were meticulously recorded – their time of appearance, location and shape. For a long time the origin of comets was believed to be divine. Comets were usually associated with major happenings in the history of nations: lost battles, the death of a king, natural disasters. In the seventeenth century the English astronomer Edmund Halley proved that comets are not doomsayers dispatched by a supernatural being, or atmospheric phenomena, but bodies in our Solar System following their own trajectories.

The myth of the special significance of comet apparitions, however, has survived into modern times: when the large and brilliant comet Hale–Bopp appeared in 1998, there were some who saw the apparition of this comet as a divine sign, and hid in shelters or even committed mass-suicide.

But although comets were reduced from divine objects to members of the Solar System by the work of Edmund Halley, they have retained their special significance for scientists. It has been recognized that, while comets cannot tell us about our future, they can at least provide a wealth of information about the history of our Solar System. Comets provide insights into the origin of planets and the fate of the dark molecular cloud, long since disappeared, from which the Solar System – including our Earth – emerged.

COMETARY CONSTITUENTS AND THE ORIGIN OF THE SOLAR SYSTEM

The origin of solid matter can be divided into three phases: the synthesis of the atomic nuclei, the formation of molecules from atoms, and the condensation of gaseous atomic and molecular matter into solids. The history of cometary matter is complex, and is shown schematically in Figure 1. Atomic nuclei are synthesized in the interiors of stars and released into the interstellar medium at particular stages of stellar evolution (red giants, supernovae and novae). Condensation into solid grains occurs when densities are high and temperatures fall, as happens in stellar envelopes or remnants of stellar explosions (producing high-temperature condensates), molecular clouds (producing low-temperature condensates) and protostellar disks.

---
* Universität Bern, Switzerland
** Carnegie Institution of Washington, Washington, DC, USA

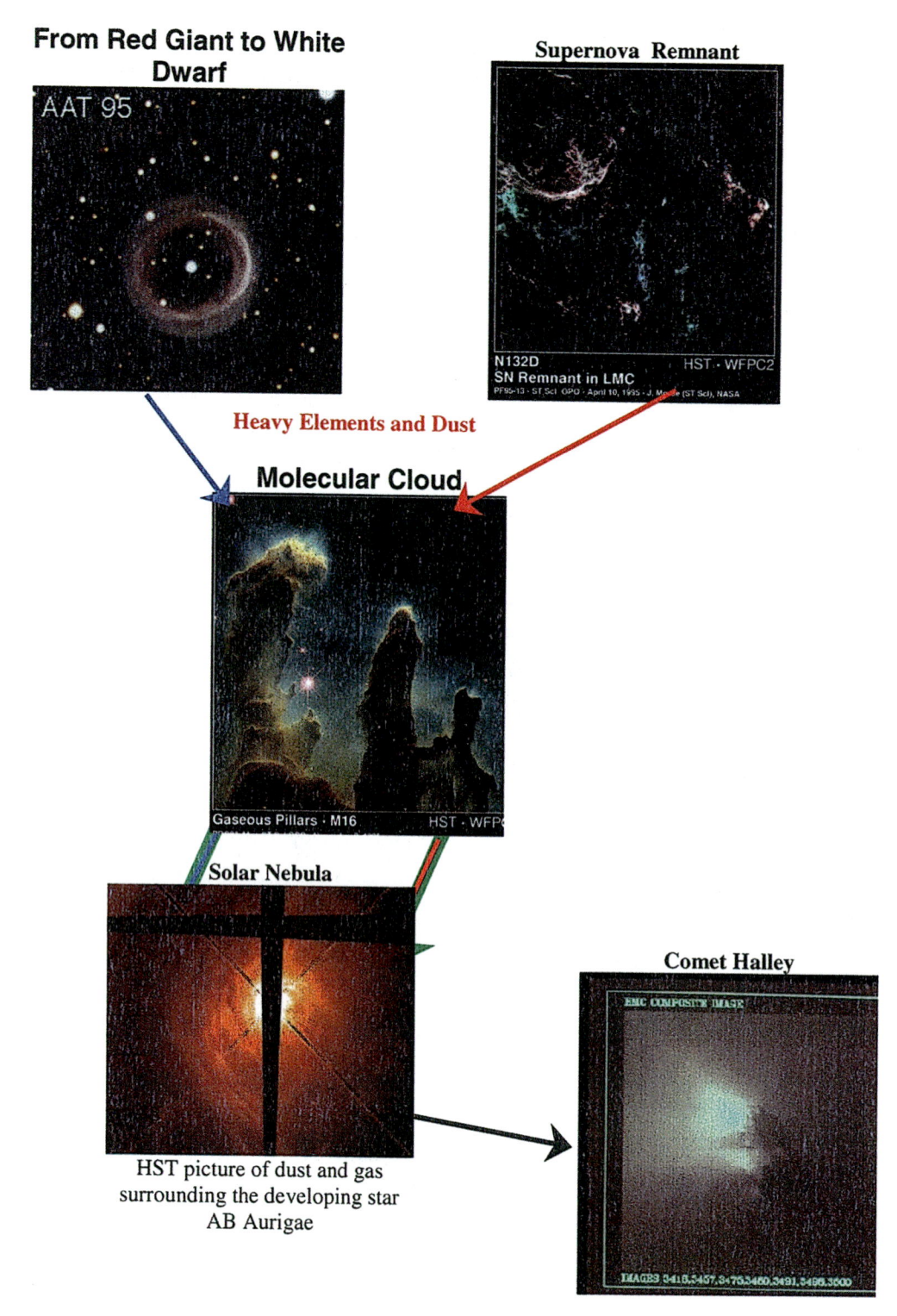

Figure 1 The origin of molecules, ice and dust in comets (after Geiss and Altwegg 1998).

Different stellar sources release matter with very different isotopic signatures, which are preserved in grains. The stellar origin of grains can therefore be identified from isotope abundance determinations, a method that has been applied with great success to certain refractory grains isolated from meteorites (e.g. Hoppe *et al.* 1996). Comets are expected to have preserved a larger variety of grains with the original stellar signature than meteorites. So far, however, such isotopic evidence is scarce. A few grains with high and variable $^{13}C/^{12}C$ ratios were found in the coma of

Comet 1P/Halley (Jessberger and Kissel 1989) indicating multiple stellar origins. More such results should be obtained by the Stardust and Rosetta missions.

Grains of stellar origin are mixed into the interstellar medium. When molecular clouds are formed, the grains may serve as nuclei for the condensation of more volatile molecules. Upon the dispersion of such clouds into the diffuse interstellar medium, the mantles of the grains can be processed by ultraviolet and particle radiation (Greenberg 1989). Material can be cycled many times between molecular clouds and the diffuse interstellar medium. About 4.6 Gy ago a galactic local protostellar molecular cloud partially collapsed to form several protostellar disks, among them our own protosolar cloud. During the collapse phase, and in the solar nebula, many grains and their volatile content were evaporated and recondensed prior to the accretion of the solid bodies of the Solar System.

Fred Whipple (1950) was the first to recognize the importance of comets for the history of the Solar System. In the last decades of the twentieth century it became accepted that comets contain the most pristine material to be found in any of the bodies in the Solar System. They were created far from the early Sun, and spent almost their entire lifetime at the edge of the Solar System, where the influence of the Sun was minimal, so they provide a reservoir of well-preserved and minimally altered material from the solar nebula. In studying this material we can deduce physical and chemical boundary conditions for the accretion of cometary material in the solar nebula. We can trace the origin of some cometary molecules to the dark molecular cloud from which the solar nebula emerged, and thus study the processes that led from the molecular cloud, through accretion into the solar nebula, to the present constituents of cometary nuclei.

COMETARY SPACE MISSIONS

The Halley fleet

In the 1970s scientists began to prepare for the 1985/6 apparition of Comet 1P/Halley. Halley is the most famous of all comets, and there were many who still remembered its last magnificent apparition in 1910. Since the 1960s, space missions had explored many of the planets and the technology for mounting a mission to a comet was available. It was therefore obvious that Halley would present a scientifically worthwhile and a technically challenging goal for a space mission, which at the same time would stimulate the interest of the public. ESA and NASA planned a joint mission in which a NASA-built spacecraft would head for a rendezvous with Comet Tempel 2, releasing en route an ESA-built probe that would be directed to encounter Halley. However, when NASA failed to obtain support for this mission from the US Government, ESA proceeded alone, and the mission was redefined from a rendezvous to a flyby mission to be launched on the European rocket Ariane 1. The time between the approval of this mission and the launch date was only five years. It was the first European planetary mission, and it took a great effort from both ESA and the different experiment teams to build the spacecraft and its payload on time. The spacecraft was named after the Italian painter Giotto di Bondone, whose frescoes of the adoration of the Magi painted in 1301 at the Scrovegni Chapel, Padua, used Comet Halley as the Star of Bethlehem (the comet was visible in the night sky at that time).

The Giotto spacecraft, with a sophisticated payload of ten instruments, was launched in June 1985 from Kourou in French Guiana. Giotto was not the only spacecraft to make the journey to Halley. A whole fleet, completed by two Japanese and two Russian spacecraft, was launched – other nations had recognized the scientific and public interest of comets. In addition, NASA reprogrammed its International Cometary Explorer (ICE) spacecraft (formerly ISEE 3, the third of the International Sun–Earth Explorer series) towards the tail of a comet. ICE went through the tail of Comet 21P/Giacobini–Zinner in 1985, becoming the first spacecraft ever to probe a comet *in situ*, and detect ions from the water group (Ogilvie 1985). It was then redirected towards Halley, passing upstream around 0.2 AU from Halley on 27 March 1986, and gathering data on the interaction of the comet with the solar wind.

The exploration of Halley's Comet led to an unprecedented intensive international cooperation. The International Halley Watch, co-ordinating the activities of astronomers and telescopes from all over the world, was initiated. Furthermore, a 'pathfinder' concept was established between Russia and ESA. The two Russian spacecraft, Vega 1 and Vega 2, flew past the comet at distances of 9000 and 8000 km, respectively, a few days before Giotto. They transmitted to ESA the exact position of Halley's nucleus, allowing Giotto to execute the very precise manoeuvre that would enable it to pass the comet's nucleus at the predefined distance of 600 km.

During the night of March 13/14 1986, people interested in cometary science either sat in the control centre in Darmstadt intently watching the computer screens as the Giotto data were received, or followed the events on television. But only the succeeding months and years revealed the wealth of data acquired during the few hours of the flyby.

Giotto survived this encounter, although its subsystems and the payload were damaged by dust impact. It was decided to redirect Giotto towards a second comet, 26P/Grigg–Skjellerup. Giotto became the first European spacecraft to enter a long hibernation phase and to make use of a planetary gravitational assist. In 1990, after four years of hibernation, it passed the Earth at a distance of ~23,000 km, and finally in July 1992 it passed comet Grigg–Skjellerup at a distance of 95 km. Another 'night of the comet' was celebrated in Darmstadt and on television. Grigg–Skjellerup is a short-period comet, but not very active. The flyby geometry was very different from that at Comet Halley, and the scientific results were somewhat less, but the data collected led to some interesting and novel findings about the interaction between comets and the solar wind. Analysis of the Giotto data from the Halley and Grigg–Skjellerup encounters continued throughout the 1990s.

Giotto's flyby of Comet Halley was certainly one of the highlights of ESA's space program in the twentieth century. But the Russian spacecraft also proved very successful. For Japan, the two spacecraft Suisei and Sakigake were the first successful planetary missions. The absence of NASA's participation in the Halley space armada was notable, and leadership in the space exploration of comets passed to ESA through the Giotto spacecraft and its particularly close flyby of Comet Halley.

Future space missions

In the 1990s the importance of cometary science was definitely established. The two great comets of that decade – Hyakutake (C/1996 B2) in 1996 and Hale–Bopp (C/1995 O1) in 1997 – triggered further interest in comets. This led to a big step forward in remote sensing of cometary constituents. But it also became evident that *in situ* measurements were far superior for a full picture of the nature of comets. New comet missions were planned in the 1990s which have not yet been launched or have not yet reached their goal. The US Comet Rendezvous and Asteroid Flyby (CRAF) mission, planned just after Giotto, was cancelled in 1991. The European Rosetta mission, originally planned as a sample-and-return mission jointly with NASA, was redefined to an ESA-built comet rendezvous mission with a German-built lander. It will be launched in January 2003 for a 2011 rendezvous with Comet 46P/Wirtanen, which will then be near aphelion. The spacecraft will accompany the comet in close vicinity through its perihelion passage in 2013. NASA launched the Stardust mission in 1998; a small mission aimed at collecting dust in the coma of Comet 81P/Wild 2. Stardust will return its sample to Earth in January 2006. In addition, NASA is redirecting its Deep Space 1 spacecraft to flyby the comets 19P/Borrelly and 107P/Wilson–Harrington. NASA is planning two additional missions, the CONTOUR mission to flyby three comets – 2P/Encke in 2003, 73P/Schwassmann–Wachmann 3 in 2006 and 6P/d'Arrest in 2008 – and Deep Impact, which will direct an impactor at comet 9P/Tempel 1 and examine the results during a flyby mission in 2005.

KNOWLEDGE OF COMETARY CONSTITUENTS BEFORE THE HALLEY ENCOUNTER

Before 1900

The first measurements of comet composition were made by astronomical spectroscopy in the first half of the twentieth century, but speculation on the composition of comets has a longer history. Ancient popular traditions held that comets are responsible for major effects, generally negative, on earthly life during their apparitions. Isaac Newton – whose *Principia* created a major scientific breakthrough in understanding the motion of heavenly bodies, including comets – was himself influenced by these beliefs, and was convinced that emanations from comets could lead to the spontaneous generation of plant life (Oparin 1938). Newton's more certain connection with comets was through his relationship with Edmund Halley, who prompted Newton to write *Principia* and also funded its publication.

Speculation related to the composition of comets also appeared during the nineteenth-century debate over Darwin's theory of the evolution of life. Unlike the rival theory of spontaneous generation, Darwin's concepts of natural selection and common ancestry required a singular and ancient origin for all life. The concept of panspermia was brought to bear on this issue by von Helmholtz (1871), who wrote, 'Who can say whether the comets and meteors which swarm everywhere through space, may not scatter germs wherever a new world has reached the stage in which it is a suitable place for organic beings.'

1900–1950

At the opening of the twentieth century the idea of panspermia, that the Earth was seeded with life from space, particularly by comets, was one of the leading explanations for the origin of life on Earth and had many proponents, including Lord Kelvin and Svante Arrhenius. Thomas Chamberlin (1911) attacked the idea that spores could be driven across the Universe to transport life from planet to planet, and proposed an alternate hypothesis that 'planetesimals', the small bodies out of which the planets formed, could have been a source of organic molecules on the early Earth.

According to Chamberlin and Chamberlin (1908), organic molecules imported to the early Earth by infalling planetesimals could have been the basis for abiotic chemical evolution on the young planet that would produce the chemicals necessary for life. Chamberlin was clearly influenced by the nineteenth-century discoveries of organic compounds in meteorites (Berzelius 1834, Wöhler 1858, Wöhler and Hoernes 1859). Chamberlin's work was aimed more towards an understanding of the origin of the Solar System, but included the original insight linking the origin of Earth with that of life on the planet. This work went essentially unnoticed until the end of this century (Oró and Lazcano 1996).

By the end of the first quarter of the twentieth century, the first detection had been made of molecular species in comets. The new technique of astronomical spectroscopy in the visible showed the presence of C_2 and CN radical species in the coma. While their origin was unknown, their presence led Oparin (1924) to the conclusion that hydrocarbons were present in comets, and hence to the suggestion that these species were the precursors of the first organic compounds on the Earth. The identification of CN in the coma of Comet Halley just prior to its apparition in 1910 also led to widespread public fear of cyanide poisoning when the Earth passed through Halley's tail.

1950–1986

Origin

Scientific investigation of the nature and composition of comets did not really begin in the twentieth century until it was half over. The first suggestions on origins followed rapidly after 1950. Oort (1950) proposed that 'new' comets came from an extensive and distant spherical cloud of comets that had been ejected to that distance from the outer Solar System in the early stages of planetary formation. This extension of the Solar System was subsequently named after him as the 'Oort Cloud'.

Oort's suggestion was supported by Öpik (1963), who demonstrated how the giant planets could have ejected comets into an extensive and distant spherical cloud about the Sun. This idea was picked up by Safronov (1972) and developed using numerical codes a few years later. The Oort Cloud is an inevitable result of the formation of the giant planets. During the growth of the cores of the giant planets, many of the icy planetesimals that approach them are ejected to large heliocentric distances rather than captured, and it is possible to calculate the relationship between the mass of the giant planet and the mass of icy planetesimals ejected. The ejections are in random directions, and inclination, outward ejections yielding an extended spherical cloud, and a significant fraction of inward ejections through the inner Solar System bombarding the young terrestrial planets. Duncan et al. (1987), Mumma et al. (1993) and Fernandez (1997) have advanced and secured this hypothesis with modern observational data on comet orbits and determined the relative contributions of Jupiter, Saturn, Uranus and Neptune to the formation of the Oort Cloud.

The Oort Cloud hypothesis was very successful in explaining the long-period and new comets, which exhibit a random distribution of inclinations. It did not, however, explain the short-period comets, which exhibit a strong preference for prograde orbits and a very flat distribution of inclinations from the ecliptic (with some notable exceptions). This disparity was solved by Fernandez (1980) and Duncan et al. (1988), who showed that the short-period comets could be objects detached from a flattened ring of small, icy objects beyond Pluto and interior to the spherical Oort Cloud.

This result revived the forgotten prediction by Gerard Kuiper (1951), based on the mass distribution in the Solar System, that there should be an extended belt of small bodies beyond Neptune and Pluto. The existence of the Kuiper Belt was confirmed in the 1990s (Jewitt and Luu 1995) by the detection of many of these objects in trans-Neptunian orbits. The few Halley-type short-period comets with high inclinations could be explained as former long-period comets that had been captured into the Solar System by close passage to Jupiter.

Structure and composition

The two 'giants' who laid the foundation for understanding the nature of the cometary nucleus in the twentieth century are Fred Whipple and Armand Delsemme; Whipple for structure of the nucleus and Delsemme for the chemical composition. Shortly after Oort's suggestion for the origin of comets, Whipple (1951) proposed his famous 'icy-conglomerate' model for the structure of the cometary nucleus. This remarkably prescient model remains in favour today, and Whipple (1991) presents a poignant review of the revelations from the 1986 Halley apparition which add the detail to his 1951 construction.

Delsemme (1998) has provided a history of his involvement in cometary science, and traced the development of our understanding of cometary composition in the second half of the twentieth century. At the time of Whipple's icy-conglomerate proposal, what was known of the composition of comets came from the identification of a few radical and ionic fragments in spectra of the coma. There was no understanding of the origin of the species seen in comet spectra, nor was there any identification of the nature of the parent material in the nucleus. Delsemme and Swings (1952) proposed, after Whipple's icy-conglomerate idea, that the nucleus was composed of a large amount of water-ice containing smaller amounts of carbon and nitrogen-containing gases, such as HCN, CH_4 and NH_3, as clathrates in water. Thus, water would control the sublimation rates of all species and explain the appearance of the photodissociation and photoionization products as a function of heliocentric distance.

Whipple and Delsemme's ideas were not immediately accepted. Lyttleton's earlier 'sandbank' model remained in the forefront, and the need was questioned for large amounts of water in the nucleus in order to explain C_2, CN, NH_2 or even OH in the coma. Delsemme reasoned that the consequences of solar action on large amounts of water in the nucleus would be to sublime water into a gaseous coma, and to form a halo of icy grains around the nucleus.

The released water would be photodissociated to form OH and high-velocity H atoms, and then the OH would in turn be photodissociated to form O(1D) and low-velocity H atoms. Models of the coma produced by these processes predicted a large Lyman-alpha coma, which could only be seen from space, and they predicted a -6 power for the dependence on heliocentric distance of the brightness of the OH and H coma fluorescence.

Two bright comets appeared in 1970, Tago–Sato–Kosaka and Bennett. The former was observed by the OAO-2 satellite (the second of the Orbiting Astronomical Observatories), which revealed an immense hydrogen halo. Observations also confirmed a -6 power for the dependence of both OH and H halos on heliocentric distance, and the brightness profiles in the coma of Bennett were shown to be consistent with a large halo of sublimating ice grains. These observations confirmed the Whipple icy-conglomerate model of cometary nuclei, and its chemical consequences as described by Delsemme (1977). The Whipple and Delsemme ideas are sometimes referred in combination as the 'dirty snowball' model. This model also predicted the observed steep drop in coma brightness for short-period comets at the distance from the Sun where water remains frozen on the nucleus.

RESULTS FROM THE HALLEY MISSIONS AND FROM REMOTE SENSING OF HYAKUTAKIE AND HALE–BOPP

Prior to the Halley apparition in 1986, almost all evidence for the nature of cometary nuclei was obtained through remote sensing from ground-based observations. The dirty-snowball model for the structure of the nucleus was well established, but our understanding of the chemical composition was still fragmentary. Most readily observable species in cometary comae were fragments, such as C_2, C_3, CN, OH, NH and NH_2, derived from the invisible parent species by solar photo-processing. By the time of the Halley apparition, only H_2O, CO, CO_2 (indirectly from CO_2^+), HCN and CH_3CN had been clearly identified as parent molecules in the coma. A broad feature due to silicates had been observed in the infrared spectrum of the dust.

The nature of the parent nuclear ice and dust was still mysterious. It was clear, however, that the nuclear volatiles were neither totally reduced nor totally oxidized, and that there was some organic material present. It had also been concluded by that time that because of the highly volatile nature of comets, they were the most primitive of Solar System objects. Comets clearly preserved a large fraction of the volatile material, other than hydrogen and helium, that was present in the solar nebula from which the planets formed. They were composed of three to ten times more volatile material containing, H, C, N and O than were the carbonaceous chondrite meteorites. A complete understanding of comet composition was frustrated by the fact that the coma is very bright at distances where ground-based spectroscopy is practical. This bright coma obscures the inner coma where the parent material might be observed before it is chemically modified by solar radiation.

Also, all comets seemed to be chemically very similar to one another, because with minor exceptions their emission line spectra all appeared very similar throughout the spectrum, from the ultraviolet through the near infrared. The same features seemed to be present for almost all comets. The largest spectral differences between comets were the line-to-continuum emission ratio, indicating a large spread in dust-to-gas ratios. This ratio could be interpreted in many ways, including age difference.

The Halley fleet was well equipped with instruments to study the coma composition. Giotto alone carried five mass spectrometers: a dust mass spectrometer, a neutral mass spectrometer and several ion mass spectrometers adapted to different energy ranges (ESA SP-1077). This payload was complemented by dust mass spectrometers (PUMA-1 and PUMA-2) on the Vega spacecraft (Grard *et al.* 1986). The main goal of the Halley missions was to detect the solid nucleus of the comet and to determine its constituents, with emphasis on the water group. An enormous wealth of data was collected in spite of the high relative speed (68 km s^{-1}) of the encounter with Halley.

The icy-conglomerate model survived the test at Halley, with the exception that the nucleus was found to have a much lower albedo than would be expected for ice alone. Features on the nucleus, including craters, clearly showed that it was structurally discrete, as would be expected for more strongly bound material. The nuclear material was light, however, with a low density on the order of 1.0 g cm^{-3}. The low albedo, ~4%, indicated a strong component of carbonaceous, non-volatile material.

It was immediately obvious that comets are much richer in organic components than had been expected, in the volatile component as well as in the dust. It took many years to fully decipher the data gathered by the mass spectrometers. This was due partly to the limited mass resolution and mass range of the instruments built at that time, which called for a significant amount of supplementary modelling, and partly to the problem of distinguishing between different species with the same mass. Also, chemical models are required in order to deduce the abundance of some parent molecules in the nucleus from measurements of daughter species in the coma. These models were not developed until after the Halley encounter, and are still being refined. Data from Giotto on changes in the mass spectrum with distance from the nucleus do provide clues that assist in unravelling these issues.

The number of molecules accessible to remote sensing was quite limited, even at the time of the Halley encounter. However, this changed drastically in the 1990s with the advancement of technology and the opening up of new spectroscopic wavelength ranges. The Hubble Space Telescope as well as the International Ultraviolet Explorer and the Infrared Space Observatory were responsible for a big step forward when in 1996 Comet Hyakutake, and in 1997 Comet Hale–Bopp, became easy targets for remote sensing. In the following section the most important results from the Halley missions and from remote sensing of Hyakutake and Hale–Bopp are summarized.

Composition of cometary dust

Giotto and the two Vega spacecraft each had a similar dust mass spectrometer on board. Due to the high relative velocity between spacecraft and comet, the impacting dust was ionized and the mass of its fragments determined by a time-of flight mass spectrometer (Kissel *et al.* 1986). For the first time it was possible to analyse a large number of particulates from a comet *in situ*. The results are very well summarized by Jessberger (1999) and Fomenkova (1999).

Excellent data were obtained on the composition of the fine dust (10^{-11}–10^{-16} g in mass, or 0.5–0.05 μm in size). The total dust/gas ratio was not well measured for Halley because of the lack of data on large dust grains, but is inferred to be about dust/gas = 1.7 by mass.

The fine dust was found to have two components: a silicate-rich portion of about $3\,\text{g cm}^{-3}$, and a refractory organic rich portion of about $1\,\text{g cm}^{-3}$. These components were found as nearly pure phases and as a third mixed phase, all highly variable in their relative abundance at any point in the coma, but overall present in about equal abundance. The refractory organic material was composed almost entirely of C, H, O and N. Labelled 'CHON', this material carries a considerable fraction of the nuclear volatiles and was identified as a distributed source of some of the coma gas species, particularly CO and formaldehyde. It also seems likely that CHON contains a large measure of unidentified non-volatile organic material, particularly unsaturated hydrocarbon species and CN-containing compounds. The average composition of the dust shows rock-forming elements in the same order of abundance as in the Sun and in C1 carbonaceous chondrites, with H, C and N enriched over their levels in C1 carbonaceous chondrites.

Figure 2 shows the elemental composition of Halley's dust, which shows a solar-like assemblage with the exception of the light elements, H, C, N and O. The missing fractions of C and O are contained in the volatile ice fraction of comets as measured from the major coma gases H_2O, CO and CO_2. Hydrogen is massively underabundant in comets because its principal form in the primitive solar nebula is molecular H_2, which condenses at a temperature well below the range of temperatures where comets are believed to have formed. The temperature in the Uranus–Neptune zone of the solar nebula, where the Oort Cloud comets are believed to have formed, can be as high as 60 K. The temperature in the solar nebula where the Kuiper Belt comets are believed to have formed ranges from 20 K at Neptune to 6 K at the outer edge of the belt, near 10,000 AU.

Particulate masses and densities were estimated to lie in the range 10^{-16}–10^{-11} g (approximate diameters 0.02–2 μm) with densities of 0.3–3 g cm^{-3} (Maas *et al.* 1989). The bulk abundances of the rock-forming elements in Halley's dust agree with the solar and chondritic abundances to within a factor of 2 (Jessberger *et al.* 1988). With the plausible assumption that the whole comet – dust and ice – has approximately solar composition with the exception of hydrogen and nitrogen, an overall dust/ice ratio of 2 was inferred (Grün and Jessberger 1990).

One of the most interesting findings is the fact that cometary dust contains two types of material: a rocky type which is rich in Mg, Fe, Si, etc., and a refractory organic type which contains mostly the elements C, H, O and N. The latter material is defined by having a ratio of carbon to any rock-forming element between 0.1 and 10 (Fomenkova *et al.* 1992). Most particulates measured at Comet Halley are of a mixed type. CHON- and rock-rich particulates each comprise about 25% of the particulates, while 50% are mixed. The number of particles in each group represented in the PUMA-1 and PUMA-2 data are rock, 430 and 161;

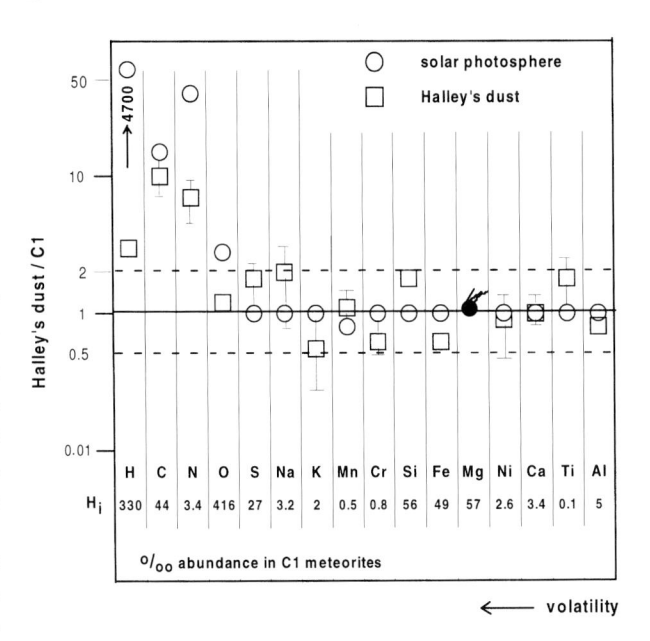

Figure 2 Elemental abundances in Comet Halley's dust (E.K. Jessberger, personal communication, 1999).

CHON, 464 and 51; mixed, 974 and 288. The overall mass ratio of silicates to organics in Halley's dust is between 2 and 1 (Fomenkova and Chang 1993). Grains composed almost entirely of carbon, or containing pure carbon inclusions, are typically among the smallest found, with masses $<10^{-15}$ g.

The refractory organic CHON component probably forms rims around rock-rich cores, an interpretation of PUMA-1 results (Kissel and Krueger 1987) that is in accordance with models of cometary grains by Greenberg (1982). The isotope ratios for the elements that could be analysed (e.g. Mg, S) show no deviations from the Solar System value except for carbon. An unequivocal detection of isotopically light carbon was reported with $^{12}C/^{13}C$ ratios as high as 5,000 in a particle composed of almost pure carbon. These high ratios link cometary matter to graphite grains extracted from carbonaceous chondrites (Zinner et al. 1990, Anders and Zinner 1993). There are two important implications. First, there is no characteristic *cometary* carbon isotopic composition (added to which the above result is based on data from only one comet, Halley). Second, the existence of such a wide range of isotopic compositions of carbon excludes the possibility that an equilibration processes affected the carbon carrier during the comet's formation or later in its history.

The extraction of mineralogical information from grain mass spectrometry is hampered by the inherent absolute uncertainty of the ion-to-atom yields, which can be as large as a factor of 2. Therefore, the following statements require the assumption that the rock-forming elements in Halley's dust are indeed present in their cosmic abundance. Clues to the mineralogical composition can be found by inspecting the variation of elemental composition from particle to particle. Mg/Fe ratios display a rather wide range, while that of the Si/Mg ratios is very narrow. Mg-rich, Fe-poor silicates constitute at least 40% and perhaps more than 60% of the particles. The next largest (~10%) group of particles is Fe(+Ni) sulphides, while Fe oxides are only a very minor constituent (<1%). The presence of unequilibrated high-temperature minerals like Mg-rich silicates and Fe sulphides (formed above 600 K) is evidence that equilibration at low temperatures is too slow a process to have affected the cometary dust particles (Schulze et al. 1997). Carbonates, sulphates or particles rich in refractory elements and resembling Ca,Al-rich inclusions known from chondrites have not been unambiguously identified (Jessberger et al. 1988, Schulze et al. 1997).

Whereas the investigations of the nature of the rocky dust component has been assisted by remote sensing studies (Hanner 1999), the nature of the CHON type dust has proved more difficult to clarify through remote sensing. Spectroscopic signatures of the organic component have been difficult to identify unambiguously by remote sensing.

To date, the Halley measurements remain the only means of assessing the nature of this component. However, with any interpretation of the results from the flyby of Halley, one has to be aware that grains measured by the PUMA-1 and PUMA-2 instruments had spent at least a few hours in the coma. It is possible, therefore, that during that time their original composition had been altered to some extent by heating and/or UV irradiation.

Cometary solids represent the most carbon-rich material in the Solar System. A few interplanetary dust particles (IDPs) exhibit a carbon abundance comparable to that in cometary dust (Thomas et al. 1993), and it is possible that these IDPs are of cometary origin. On average, CHON and mixed grains containing organic materials are heavier than rock particles (Fomenkova et al. 1992). It is possible that the structure of larger grains is similar to that of some IDPs – a carbonaceous matrix in which is embedded small mineral particulates. Particles containing organics (CHON and mixed) are more abundant closer to the nucleus in the central part of the coma, while rock particles dominate in the outer regions of the coma. These observations suggest that more volatile components of the grains gradually evaporate, releasing small silicate grains and providing an extended source for some of the gaseous species observed (A'Hearn et al. 1986, Eberhardt et al. 1987).

If one assumes that the overall dust/gas ratio in Comet Halley is 2, then the overall (dust+gas) abundance of C and O is solar, while H and N are depleted by factors of 600 and 2, respectively. Carbon and nitrogen occur predominantly in the dust, the partition between a solid phase and a gas phase being 2:1. On the other hand, H and O are twice as abundant in the gas than in the dust – as expected for a water-rich object. The bulk abundance of O in the cometary dust is about the same as its chondritic abundance, while the abundances of H, C and N in cometary dust are higher than in carbonaceous chondrites. The higher abundance of volatiles in cometary dust points to its more pristine nature.

Various unidentified organic hydrocarbon species of high molecular weight have been found in the pure carbon grain component of Halley dust, including polycyclic aromatic and highly branched aliphatic hydrocarbons ((H,C) grains), and polymers of carbon suboxide and of cyanopolyynes ((C,O) and (C,N) grains). The majority of grains ((H,C,N), (H,C,O) and (H,C,N,O) groups) contain heteropolymers and/or variable mixtures of carbon phases and complex organic compounds – compounds consistent with alcohols, aldehydes, ketones, acids and amino acids, and their salts. The exact make-up of these mixtures cannot be unambiguously identified from the available data. The simplest members of these homologous series occur in the interstellar medium (Irvine et al. 1998), and members of a higher molecular weight were found in the Murchison meteorite (Cronin and Chang 1993) and observed in laboratory experiments simulating the origin of comets ('cometary ice tholins'; McDonald et al. 1996).

The proportion of the pure carbon phases decreases when interstellar dust is being incorporated and processed into planetary materials. In the interstellar medium, up to 50% of the solid carbon is locked up in graphite or amorphous carbon. Halley's carbon grains contain ~10% of the total solid carbon, and in carbonaceous chondrites carbon phases account for only ~2% of the total carbon content. The proportion of carbon bound in complex organic material increases with the degree of processing: ~50% in interstellar grains and their mantles, 70–80% in Halley and 90–95% in carbonaceous chondrites. The observed diversity of types of cometary organic compounds is consistent with the interstellar dust model of comets (Greenberg 1982), and probably reflects differences in the history of the precursor dust.

Composition of cometary volatiles

The gas coma of Halley was measured *in situ* at distances as close 1100 km to the nucleus. These measurements included the neutral and ionic composition of the gas, photochemical products and dynamics. The apparitions of Hyakutake and Hale–Bopp in 1996 and 1997 provided an opportunity to train an advanced suite of ground-based spectroscopic technologies on two very bright comets. New spectroscopic windows were available in the radio region of the spectrum, and more sensitive detectors were available in both the infrared and the radio. These allowed the identification of many new parent and daughter species in Hyakutake and Hale–Bopp.

Gas production was observed in Hale–Bopp out to 7 AU, and the dominant coma gas beyond 4 AU was CO. Water became dominant in the coma only inside 3.5 AU (inbound). This clearly showed that cometary ices are not homogeneous and that other volatile species, present in lesser amounts but more volatile than water, can control sublimation at large heliocentric distances.

Table 1 is a summary of measurements of cometary volatile material measured in the gas coma of the three bright comets of the last two decades: Halley, Hale–Bopp and Hyakutake. The remote observations of the second and third of these comets added a number of organic and interstellar-like molecules to the list of detected parent volatiles, including formic acid, methyl formate, acetaldehyde, acetylene, ethane, HNC (probably synthesized in the coma), HNCO, HC$_3$N, NH$_2$CHO, and the sulphur-containing species H$_2$CS, OCS, CS$_2$ and SO$_2$.

Elemental abundance in cometary volatiles

Comets are likely to be the most pristine bodies of our Solar System. They were formed at large distance from the Sun, and have remained for most of their lifetimes outside Neptune's orbit in the Kuiper Belt, or at much greater distances in the Oort Cloud. Their size is quite small, so they have experienced almost no internal heating. They should therefore still consist of the original material from which they condensed.

Figure 3 (Geiss 1988) shows the relative abundance in Comet Halley of the volatile elements C, N and O, including both dust and ice fractions, for an assumed dust/gas ratio of 1.7. The observations that H$_2$O is the major repository for oxygen, and that O has a solar abundance, together argue for low-temperature accretion, in order to have preserved the original water abundance in the formation region, consistent with the high volatile content of comets. The water molecule is the major repository for oxygen in chemical models of interstellar clouds, of dense cloud cores and of protostellar nebulae. The fact that CO is present in large abundance relative to H$_2$O (about 7% in Halley) argues for a very low formation temperature, below the temperature at which CO is clathrated by H$_2$O.

The fact that C is also present in solar abundance, and that the major volatile repository for C is CO, argues for formation from material with the composition of an interstellar cloud, or in a kinetically controlled solar nebula, where CO is the major carbon-containing volatile. In the interstellar medium the CH$_4$/CO ratio is very small and controlled by chemical kinetics. Thermodynamic equilibrium would otherwise predict a very large ratio. Methane is therefore a minor species in interstellar clouds, and CO is the major repository of carbon. Similarly, the CH$_4$/CO ratio in solar nebula models is highly dependent on what assumptions are made about the relative influences of thermal and kinetic equilibria, and can be strongly controlled by various chemical kinetic factors including temperature, pressure and mineral abundance on grains.

The depletion of nitrogen (Figure 3) is significant and consistent with the formation of comets under chemically controlled conditions in either the interstellar medium or a solar nebula with abundant N$_2$ and CO relative to NH$_3$ and CH$_4$. The major reservoir of nitrogen is N$_2$, both in interstellar clouds and in modern chemical models of the early solar nebula in the region of the outer planets. Ammonia is present in dense, dark interstellar clouds, but N$_2$ remains the major repository of nitrogen. Likewise, solar nebula models for the low-temperature, outer regions of the solar nebula predict N$_2$ to be the major repository of nitrogen, with a low NH$_3$/N$_2$ ratio. The abundance of ammonia, as determined from *in situ* mass spectrometry in the coma of Comet Halley at a heliocentric distance of 0.9 AU, is 1.5% relative to water (Altwegg et al. 1993, Meier et al. 1994). This is about double the amount reported in the comae of Comets Hyakutake and Hale–Bopp. However, Bird et al. (1997b) reported an abundance of 1–1.8% of NH$_3$ in

Table 1 Abundances of cometary volatile material relative to water

Species	Comet	Source strength	References
H_2O	Halley	100	Krankowsky *et al.* (1986)
	Hyakutake	100	Mumma *et al.* (1996)
	Hale–Bopp	100	Weaver *et al.* (1997), Dello Russo *et al.* (1999)
CO	Halley	17	Eberhardt *et al.* (1987)
	Hyakutake	6–30	Mumma *et al.* (1996), Lis *et al.* (1997a), Weaver *et al.* (1996)
	Hale–Bopp	20	Weaver *et al.* (1997), Biver *et al.* (1997), DiSanti *et al.* (1999)
CO_2	Halley	3.5	Krankowsky *et al.* (1986)
	Hale–Bopp	20^*	Crovisier *et al.* (1997)
H_2CO	Halley	3.8	Geiss *et al.* (1991), Meier *et al.* (1993)
	Hyakutake	0.2–1	Lis *et al.* (1997a)
	Hale–Bopp	1	Biver *et al.* (1997)
HCOOH	Hale–Bopp	0.06	Bockelée-Morvan *et al.* (1998)
CH_3CHO	Halley	0.5	Altwegg *et al.* (1999)
$HCOOCH_3$	Hale–Bopp	0.06	Bockelée-Morvan *et al.* (1998)
NH_3	Halley	1.5	Allen *et al.* (1987), Meier *et al.* (1994)
	Hyakutake	0.5	Palmer *et al.* (1996)
	Hale–Bopp	0.7	Bird *et al.* (1997a), Bird *et al.* (1997b)
HCN	Halley	0.1	Geiss *et al.* (1991)
		0.1	Lis *et al.* (1997a)
		0.25	Weaver *et al.* (1997), Biver *et al.* (1997)
HNCO	Hyakutake	0.07	Lis *et al.* (1997a)
	Hale–Bopp	0.06	Lis *et al.* (1997b), Bockelée-Morvan *et al.* (1998)
HNC	Hyakutake	0.01	Eberhardt *et al.* (1987), Irvine *et al.* (1996)
	Hale–Bopp	0.04	Biver *et al.* (1997), Irvine *et al.* (1998)
CH_3CN	Halley	0.14	Geiss *et al.* (1999)
	Hyakutake	0.01	Bockelée-Morvan (1997)
	Hale–Bopp	0.02	Biver *et al.* (1997)
HC_3N	Hale–Bopp	0.02	Lis *et al.* (1997b), Bockelée-Morvan *et al.* (1998)
C_2H_5CN	Halley	0.028	Geiss *et al.* (1999)
NH_2CHO	Hale–Bopp	0.01	Lis *et al.* (1997b), Bockelée-Morvan *et al.* (1998)
CH_3OH	Halley	1.25; 1.7	Altwegg (1996), Eberhardt *et al.* (1995)
	Hyakutake	2	Lis *et al.* (1997a)
	Hale–Bopp	2	Biver *et al.* (1997), Lis *et al.* (1997b)
CH_2	Halley	0.27	Altwegg *et al.* (1994)
CH_4	Halley	<0.8	Altwegg *et al.* (1994)
	Hyakutake	0.7	Mumma *et al.* (1996)
	Hale–Bopp	0.6	Weaver *et al.* (1997)
H_2S	Halley	0.15; 0.4	Eberhardt *et al.* (1995), Altwegg (1996)
	Hyakutake	0.8	Bockelée-Morvan (1997)
	Hale–Bopp	1.5	Biver *et al.* (1997)
OCS	Halley	0.27	Altwegg (1996)
	Hyakutake	0.1	Woodney *et al.* (1997)
	Hale–Bopp	0.3	Lis *et al.* (1997b), Dello Russo *et al.* (1998)
SO	Hale–Bopp	0.2–0.8	Lis *et al.* (1997b), Bockelée-Morvan *et al.* (1998), Kim *et al.* (1997)
CS_2	Halley	0.1	Altwegg (1996)
	Hyakutake	0.1	Bockelée-Morvan (1997)
	Hale–Bopp	0.2	Biver *et al.* (1997)

Table 1 *Continued*

Species	Comet	Source strength	References
SO_2	Hale–Bopp	0.1	Bockelée-Morvan *et al.* (1998), Kim *et al.* (1997)
H_2CS	Hale–Bopp	0.02	Woodney *et al.* (1997)
S_2	Hyakutake	0.02	Weaver *et al.* (1996)
C_2H_2	Halley	0.3	Reber (1997)
	Hyakutake	0.5	Brooke *et al.* (1996)
	Hale–Bopp	0.1	Dello Russo *et al.* (2000)
C_2H_4	Halley	0.3	Reber (1997)
C_2H_6	Halley	0.4	Reber (1997)
	Hyakutake	0.2	Mumma *et al.* (1996)
	Hale–Bopp	0.4	Weaver *et al.* (1997)
C_3H_2	Halley	0.1	Altwegg *et al.* (1999)

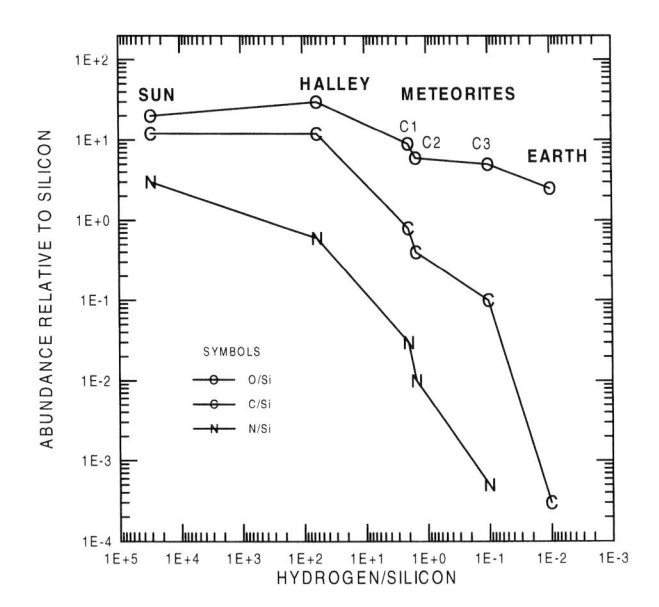

Figure 3 Relative abundances of elements released by Comet Halley (gas and grains) (Geiss 1988).

Hale–Bopp, more consistent with the value for Halley. This means that for Hyakutake the nitrogen depletion in the gas phase may be even more severe, or more of the nitrogen may still be in the form of N_2. However, the abundance of N_2 has not been directly measured. Its presence is deduced from observations of the bands of N_2^+, which are indirect indicators of N_2. For Halley the optically observed value for N_2 is 0.2% (Wyckoff *et al.* 1991). N_2 is hard to detect by remote sensing and by *in situ* mass spectrometry. The fact that nitrogen is cosmically underabundant and that $NH_3/N_2 > 1$ in comets implies that a significant amount of N_2 has escaped being frozen into the cometary nucleus during formation or has escaped subsequently.

Water-ice forms clathrates with gases such as CO, CH_4, N_2, Ne and Ar, trapping these more volatile gases in the ice phase at temperatures above their freezing points. Both CO and CH_4 are efficiently trapped by H_2O; the rare gases and N_2 are trapped less efficiently. The N_2/CO ratio in cometary volatiles is probably highly sensitive to formation temperature and subsequent thermal processing. There are as yet no good measurements of Ne and Ar in comets, but the abundance of these gases relative to water would also be excellent proxies for studying the thermal history of comets. The composition of Halley clearly showed that the region of the solar nebula in which comets formed was richer in the mildly reducing forms of C and N (CO and N_2) than the fully reduced forms (CH_4 and NH_3), although the latter are not completely absent.

Meteorites, which are also rather pristine bodies, are more highly depleted in light elements relative to Halley and the Earth. Meteorites are more clearly processed than Halley, so comets are indeed very pristine bodies.

Molecular composition of cometary volatiles

Whipple's icy-conglomerate model predicts that the main volatile molecule in cometary nuclei is water-ice. This was confirmed by the Halley measurements. The abundance of water in Halley's coma is approximately 70–80%, depending on the heliocentric distance. The list of additional molecules found in Halley's ice is long. Hale–Bopp and Hyakutake were both sufficiently active for several additional new minor species to be identified by remote sensing, such as OCS, CH_4, C_2H_2, and C_2H_6. Table 1 lists all the molecules that have been identified in the comae of these three comets; there is a wide variety of organic molecules, and at least 24 parent molecules have been identified (Crovisier and Bockelée-Morvan 1999). Results from *in situ* mass spectrometry of Halley and remote sensing are in most cases compatible (Altwegg *et al.* 1999, Crovisier and Bockelée-Morvan 1999, Eberhardt 1999). It is surprising that the oxidized molecules are more abundant than the reduced hydrocarbons. The most abundant molecules after water are carbon monoxide and carbon dioxide, followed by formaldehyde and methanol. The reduced species methane and ethane have been identified but are not very abundant. This observation is consistent with an origin from material with an interstellar-like composition. Organic molecules with masses as large as 100 amu or higher have been shown to exist in Halley's coma, which is also consistent with an interstellar origin.

It has been suggested that comets of dynamically different classes may be chemically different (A'Hearn et al. 1995, Feldman et al. 1997). The statistics are still poor, and an answer to this question has to await further remote sensing and space missions (e.g. Rosetta). Sulphur can be found as H_2S as well as in the form of OCS and CS_2, and all three of these molecules have sublimation temperatures between 25 and 80 K. As well as stable molecules, some radicals have been identified. Most of them are daughter products of the photodissociation of heavier parent molecules. However, at least two radicals have been identified which seem to be present as such in the cometary ice, namely CH_2 and C_4H (Altwegg et al. 1994, Geiss et al. 1999). C_4H is well known from radio astronomy to be present in molecular clouds, and CH_2 is a relatively abundant species in chemical models of dense interstellar clouds. These radicals are unlikely to have been present in sufficient quantities to be incorporated in ices frozen from a hydrogen-rich solar nebula – they are much more likely to have been incorporated into ices frozen on interstellar grain mantles at very low temperatures. This hypothesis requires these icy mantles to survive the accretion shock during their accumulation into the solar nebula.

Isotopic abundances in cometary volatiles

Isotopic ratios can provide additional clues to the physical conditions that prevailed during the formation of cometary volatiles. The most important is the D/H ratio in different molecules. A summary of all deuterium measurements in comets can be found in Meier and Owen (1999). Mass spectrometers aboard Giotto obtained the first accurate D/H ratios in the water of Comet Halley (Balsiger et al. 1995, Eberhardt 1996). Ground-based observations of HDO in Hyakutake (Bockelée-Morvan et al. 1998) and Hale–Bopp (Meier et al. 1998a), the detection of DCN in Hale–Bopp (Meier et al. 1998b), and upper limits for several other D-bearing molecules comprised the limited sample of D/H measurements. All three comets exhibit a similar D/H ratio in H_2O, enriched by about a factor of 2 relative to terrestrial water and by approximately one order of magnitude relative to the protosolar value. On this basis, comets cannot be the only source for the Earth's oceans.

The D/H ratio in cometary HCN is 7 times higher than the value in cometary H_2O. Species-dependent D-fractionations occur at low temperatures and low gas densities via ion–molecule or grain-surface reactions and cannot be explained by pure solar nebula chemistry. Cometary volatiles appear, therefore, to preserve the interstellar D fractionation. The observed D abundances set a lower limit to the formation temperature of 30 ± 10 K. Similar numbers can be derived from the *ortho*-to-*para* ratio in cometary water, from the absence of Ne in cometary ices and the presence of S_2. So far, all cometary D/H measurements refer to bulk compositions, and it is conceivable that significant departures from the mean value could occur at the grain-size level. Strong isotopic effects as a result of coma chemistry can be excluded for H_2O and HCN. A comparison of the cometary D/H ratio in water with values found in the atmospheres of the outer planets is consistent with the long-held idea that the gas planets formed around icy cores with a high cometary D/H ratio, and subsequently accumulated significant amounts of H_2 from the solar nebula with a low protosolar D/H.

The measurements of isotopes other than deuterium in the volatile part of cometary comae are rare. So far, all isotopic ratios ($^{18}O/^{16}O$, e.g. Balsiger et al. 1995, Eberhardt et al. 1995; $^{13}C/^{12}C$, e.g. Kleine et al. 1995; and $^{34}S/^{32}S$, Altwegg 1996) reported for comets, except for D/H, are compatible with the Solar System.

Dust as a source for cometary volatiles

Most of the volatile material detected in comets comes directly from the nucleus. These gases may be stored as a separate frozen phase or trapped in amorphous water-ice. Ice and refractory organic materials (CHON particles or hydrocarbon polycondensates) in the dust also contribute to the coma gas (Eberhardt 1999, Huebner and Benkhoff 1999). However, the unusual behaviour with heliocentric distance of most molecular species in the coma (in particular H_2O) may be related to seasonal effects because of the orientation of the spin axis of the nucleus (Kuehrt 1999). The organic material in dust particles is a distributed source for coma gas (radicals as well as other molecules). Among them is a distributed source for H_2CO and CO. Boice et al. (1990) and Eberhardt et al. (1995) conclude that in Comet Halley about two-thirds of the CO observed at a heliocentric distance $r = 0.9$ AU can be explained by the dissociation of H_2CO, which itself seems to come primarily from the dust (Eberhardt 1999). However, this is contested for other comets (Crovisier and Bockelée-Morvan 1999).

It seems that nuclear sources of coma gas dominate at large heliocentric distances r, while nuclear and distributed sources coexist at small r. DiSanti et al. (1999) indicate that the distributed source is effective for $r < \sim 2$ AU. Festou (1999) suggests that the observed density profile of CO can be explained by the outgassing of inhomogeneous regions of the comet, for example as 'jets' from CO-rich ice. The nature of the distributed sources of H_2CO and CO and of the release mechanisms from dust remain unexplained in the models by Greenberg and Li (1998). A similar situation exists for the observed CN. There is no doubt that some of the CN is related to the dissociation of HCN and CH_3CN. These two molecules may be the only parents of CN at large heliocentric distances; however, an additional source seems to be required at smaller distances, and it has been suggested that it is related to dust (A'Hearn et al. 1995).

The diversity of comets

Comets originated in the outer portion of the newly forming solar nebula, outside the ice condensation boundary and in the region of the developing giant planets and beyond. Comets might therefore be expected to exhibit some compositional differences reflecting their place of origin along the temperature gradient from the ice boundary near Jupiter to within the Kuiper Belt. The long-period comets ejected into the Oort Cloud from the Jupiter–Neptune region might preserve a higher-temperature, more processed composition, perhaps even retaining some vestige of their interaction with a giant planet sub-nebula, than do comets that formed at greater distances. The short-period comets of the Jupiter family probably originated at these greater distances in the Kuiper Belt. The Halley family of short-period comets is more likely composed of long-period comets that have been captured into short-period orbits by encounters with Jupiter. Thus, it might be expected that long-period and Halley family comets would have similar compositions, except as modified in Halley family comets by the longer time they have spent in orbits closer to the Sun. Another expectation would be that Jupiter family comets would be different from the both long-period and Halley family comets due to their origin at larger distances from the Sun in the solar nebula. Differences in short-period comets would be due to their differing histories of perihelion passages. Additional causes for differences among comets include large-scale radial mixing in the solar nebula, as well as collisional disruption, mixing and re-accretion of comet nuclei in the Kuiper Belt (Weissman 1999).

On this basis we might be expected Comet Halley to be compositionally more closely related to Comets Hale–Bopp and Hyakutake than to comets in the Jupiter family. These three comets do appear to be similar in composition, but composition measurements have not been sufficiently precise, and there is as yet no strong statistical basis for identifying large composition distinctions between comets. Compositional diversity has been studied in only one sample that has a statistically significant number of comets, all coming from the same database and acquired by the same observers (A'Hearn et al. 1995). They showed that about 30% of comets were deficient in C_2 and C_3 relative to CN. Most of the deficient comets were Jupiter family comets. A'Hearn et al. (1995) suggest that this deficiency is associated with the region of formation of these comets. Since we do not know the parent molecules of C_2, C_3 and CN (particularly since C_2 may be associated with more than one parent molecule), it is difficult to say whether this deficiency arose during comet formation or was caused by evolution.

COMETS AND THE INTERSTELLAR MEDIUM

The detection of CHON organic grains in comets, and the identification of interstellar-like organic and other exotic gases sublimated from cometary nuclei, give the definite impression that comets preserve a record of interstellar composition. Figure 4, based on Crovisier (1998), illustrates a quantitative similarity between interstellar ice and cometary ice. The interstellar data are from infrared spectra of icy grains in dense clouds, and the cometary data are from coma gases sublimed from the nucleus and dust.

The gas phase composition in the interstellar medium depends on many variables, but almost every parent species observed in comets has also been detected in the interstellar medium, either in molecular clouds or in dense cloud cores where star formation is taking place. N_2 and CO_2 have not been observed directly in the interstellar medium, but their presence is inferred from the detection of the protonated ions N_2H^+ and CO_2H^+. The detection of C_2H_6 in

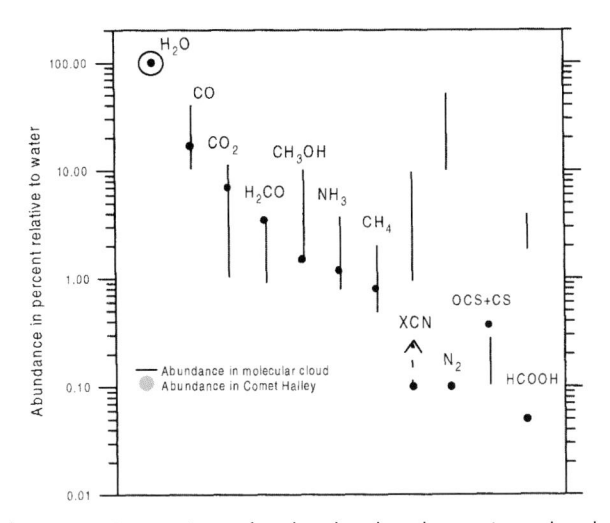

Figure 4 Comparison of molecular abundances in molecular clouds and comets (after Crovisier 1998).

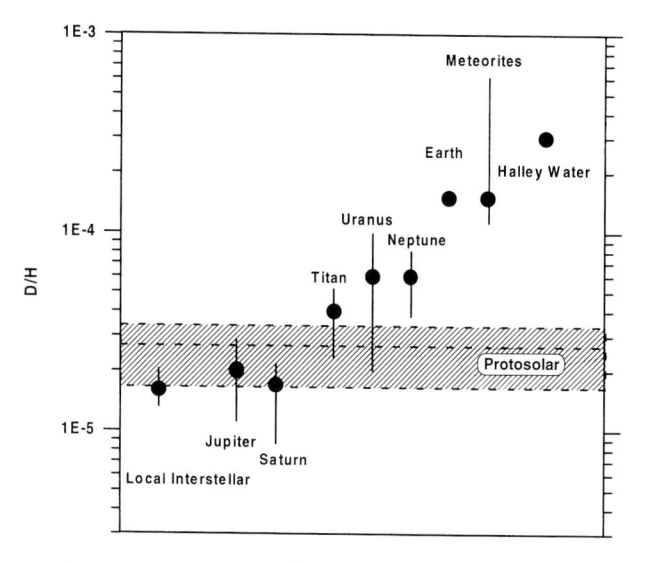

Figure 5 D/H ratio in the bodies of the Solar System.

Hale–Bopp and Hyakutake was unexpected, and holds a clue either to later processing of interstellar material in comets or to the composition of the interstellar medium itself. Alkanes have not yet been detected in the interstellar medium.

Chemical abundance and isotopic ratios of key elements in comets hold the clues to their origin and evolution. The D/H ratio is very large in the interstellar medium and can approach unity in some interstellar molecules. In interstellar water, D/H ranges from 3×10^{-4} to 4×10^{-3}. Figure 5 shows the D/H ratio as measured in the Sun, planets (including Earth), meteorites and Halley water. The local interstellar value is for hydrogen atoms, and is consistent with the protosolar value 4.5 Gyr ago as determined from the Sun.

A consistent picture emerges from Figure 5 if primordial solar nebula water contained elevated D/H ratios remaining from an origin in the interstellar medium. The giant planets Jupiter and Saturn are dominated by nebular H_2 and exhibit the solar value for D/H; Uranus and Neptune contain a significant addition of nebular ices added after the planets condensed, which accounts for their elevated D/H ratios relative to Jupiter and Saturn. The Earth's water retains memory of its D/H ratio in the same ice as formed the comets and meteorites. The range of D/H values exhibited by the meteorites, which encompasses the slightly different values in comets and the Earth, can be rationalized if their volatiles originated from interstellar material that was partially processed in the nebula and/or diluted with nebular material.

The value of D/H in comets is consistent with their formation at temperatures of 10–50 K, and with several other indicators of formation temperature. Measurements of the *ortho/para* hydrogen atom spin ratio (spin temperature) in cometary water indicate that it was last processed at about 30 K. The presence of CO in cometary ices indicates a formation temperature less than 60 K if CO was trapped in water as a clathrate, and 25 K for direct condensation of CO. The presence of N_2, even in small amounts, indicates a formation temperature near the freezing point of N_2, around 20 K. N_2 is not clathrated as strongly in water-ice as is CO.

Evidence for interstellar composition in the refractory component of comets comes from observations of cometary dust. In addition to the presence of CHON in Halley, in infrared spectra the dust appears similar to circumstellar grains. The strong 3.3–3.4 μm band indicative of C—H stretch transitions has features that can be explained by very small particles similar to the polyaromatic hydrocarbons (PAHs) observed in the interstellar medium. The infrared continuum of cometary dust also indicates the presence of small carbon grains. Some higher-temperature processing is indicated by the detection of crystalline as well as amorphous silicates, including both olivines and pyroxenes, in the 10 and 20 μm silicate bands of comet dust.

Greenberg (1982) has been a strong proponent of the idea that comets are formed by the aggregation of ice-covered interstellar grains. These grains are mantled by interstellar gas frozen as ice and photo-processed into more complex organic and non-volatile material. Support for this idea comes from the similarity of the 3.4 μm bands to what would be expected from such material, and the identification of CHON grains in the coma of Halley. Carbon isotope measurements of individual CHON grains reveal highly variable $^{12}C/^{13}C$ isotopic ratios, showing them to be highly chemically disequilibrated and consistent with multiple stellar origin.

The dust and gas measurements are therefore consistent with an interstellar origin, but indicate that some of the material has been processed at higher temperature in the early solar nebula. Some cometary material must therefore have survived the accretion shock in the solar nebula, and then been partially processed in the solar nebula and/or partially diluted with more highly processed solar nebula material. For extensive reviews of the relationship of cometary material to the interstellar medium, see Mumma *et al.* (1993) and Irvine *et al.* (2000). Table 2, taken in part from Mumma *et al.* (1993), lists some of the key indicators for an interstellar origin.

The results to date clearly indicate that comets are the product of chemical disequilibrium, and resemble in intricate detail the material observed in interstellar clouds. Abundances of the elements, their distribution in molecular species and their isotopic fractionation are all indicative of the retention of a large component of interstellar origin. This suggests that this material must somehow have survived accretion into the solar nebula, which is difficult to reconcile with the shock heating expected in current models of this process. Likewise, cometary composition appears to be remarkably consistent with what is observed in dense

Table 2 Indicators for interstellar origin

Indicator	Evidence
H_2O	Low hydrogen spin temperature in Halley indicates origin at about 30 K.
DHO	Enriched D/H in water is consistent with unmodified interstellar material.
C inventory	Significant proportion of carbon is present as complex organic material, consistent with interstellar origin. Inventory of organic compounds is consistent with interstellar origin. Distribution of unsaturated v. saturated species is consistent with interstellar origin. Distribution between dust and gas is consistent with interstellar material.
CO	Abundance is consistent with interstellar origin and condensation at 25–50 K. Readily clathrated with water at the higher range of temperatures. Variability from comet to comet may indicate processing history.
CH_4	Low abundance is consistent with interstellar origin.
CO/CH_4	High ratio is consistent with interstellar origin.
H_2CO	Abundance is consistent with interstellar gas and higher than for solar nebula. Polymer source in CHON grains is more consistent with interstellar grains.
CH_3OH	Abundance is more consistent with interstellar than with solar nebula origin.
CHON grains	Very high C ratio, volatility, $^{12}C/^{13}C$ ratios and indications of organic refractory component are indicative of an interstellar origin.
N inventory	Low total N inventory appears to be consistent with an interstellar origin where most N is in the form of N_2. N_2 is not easily incorporated into cometary ices due to condensation above the freezing point of N_2. Some N_2 is incorporated due to trapping (clathrate formation) in water-ice.
NH_3	Low abundance is consistent with interstellar origin.
NH_3/N_2	Ratio is higher than interstellar due to loss of N_2 in comet formation.
HCN	Abundance is consistent with interstellar gas and higher than for solar nebula.

'hot' cloud cores where stars are forming. This observation is also consistent with current theories of the origin of comets. These theories place the origin of long-period comets in the outer Solar System between Jupiter and Neptune, with subsequent ejection to the Oort Cloud, and place the origin of short-period comets in the Kuiper Belt, outside the orbit of Neptune, and extending to the inner edge of the Oort Cloud.

REFERENCES

A'Hearn, M.F., Hoban, S., Birch, P.V., Bowers, C., Martin, R. and Klinglesmith, D.A. III (1986). Cyanogen jets in Comet Halley. *Nature*, **324**, 649–651.

A'Hearn, M.F., Millis, R.L., Schleicher, D.G., Osip, D.J. and Birch, P.V. (1995). The ensemble properties of comets: Results from narrow band photometry of 85 comets, 1976–1992. *Icarus*, **118**, 223–270.

Allen, M., Delitsky, M., Huntress, W., Yung, Y., Ip, W.-H., Schwenn, R., Rosenbauer, H., Shelley, E. and Geiss, J. (1987). Evidence for methane and ammonia in the coma of comet P/Halley. *Astronomy and Astrophysics*, **187**, 502–512.

Altwegg, K. (1996). Sulfur in the coma of comet Halley from in situ measurements. Habilitationsschrift, University of Bern.

Altwegg, K., Balsiger, H., Geiss, J., Goldstein, R., Ip, W.-H., Meier, A., Neugebauer, M., Rosenbauer, H. and Shelley, E. (1993). The ion population between 1300 km and 230 000 km in the coma of comet P/Halley. *Astronomy and Astrophysics*, **279**, 260–266.

Altwegg, K., Balsiger, H. and Geiss, J. (1994). Abundance and origin of the CH_n^+ ions in the coma of Comet Halley. *Astronomy and Astrophysics*, **290**, 318–323.

Altwegg, K., Balsiger, H. and Geiss, J. (1999). Composition of the volatile material in Halley's coma from in situ measurements. *Space Science Reviews*, **90**, 3–18.

Anders, E. and Zinner, E. (1993). Interstellar grains in primitive meteorites: diamond, silicon carbide, and graphite. *Meteoritics*, **28**, 490–514.

Balsiger, H., Altwegg, K. and Geiss, J. (1995). D/H and $^{18}O/^{16}O$-ratio in the hydronium ion and in neutral water from in situ ion measurements in Comet Halley. *Journal of Geophysical Research*, **100**, 5827–5834.

Berzelius, J.J. (1834). Über Meteorsteine: 4. Meteorstein von Alais. *Annalen der Physik und Chemie*, **33**, 113–123.

Bird, M.K., Huchtmeier, W.K., Gensheimer, F., Wilson, T.J., Janardhan, P. and Lemme, C. (1997a). Radio detection of ammonia in Comet Hale–Bopp. *Astronomy and Astrophysics*, **325**, L5–L8.

Bird, M.K., Janardhan, P., Wilson, T.L., Huchtmeier, W.K., Gensheimer, P. and Lemme, C. (1997b). K-band radio observations of Comet Hale–Bopp: Detections of ammonia and (possibly) water. *Earth, Moon, and Planets*, **78**, 21–28.

Biver, N., Bockelée-Morvan, D., Colom, P., Crovisier, J., Germain, B., Lellouch, E., Davies, J.K., Dent, W.R.F., Moreno, R., Paubert, G., Wink, J., Despois, D., Lis, D.C., Mehringer, D., Benford, D., Gardner, M., Phillips, T.G., Gunnarsson, M., Rickman, H., Winnberg, A., Bergman, P., Johansson, L.E.B. and Rauer, H. (1997). Long-term evolution of the outgassing of Comet Hale–Bopp from radio observations. *Earth, Moon, and Planets*, **78**, 5–11.

Bockelée-Morvan, D. (1997). Cometary volatiles: The status after comet C/1996 B2 Hyakutake. In E.F. van Dishoeck (ed.), *Molecules in Astrophysics: Probes and Processes*, IAU Symposium 178, Kluwer, Dordrecht, pp. 219–235.

Bockelée-Morvan, D., Gautier, D., Lis, D.C., Young, K., Keene, J., Phillips, T., Owen, T., Crovisier, J., Goldsmith, P.F., Bergin, E.A., Despois, D. and Wootten, A. (1998). Deuterated water in comet C/1996 B2 (Hyakutake) and its implication for the origin of comets. *Icarus*, **133**, 147–162.

Boice, D.C., Huebner, W.F., Sablik, M.J. and Konno, I. (1990). Distributed coma sources and the CH_4/CO ratio in Comet Halley. *Geophysical Research Letters*, **17**, 1813–1816.

Brooke, T.Y., Tokunaga, A.T., Weaver, H.A., Crovisier, J., Bockelée-Morvan, D. and Crips, D. (1996). Detection of acetylene in the infrared spectrum of Hyakutake. *Nature*, **383**, 606–608.

Chamberlin, T.C. (1911). The seeding of worlds. *Journal of Geology*, **19**, 175–178.

Chamberlin, T.C. and Chamberlin, R.T. (1908). Early terrestrial conditions that may have favored organic synthesis. *Science*, **28**, 897–910.

Cronin, J. and Chang, S. (1993). Organic matter in meteorites: Molecular and isotopic analysis of the Murchison meteorite. In J.M. Greenberg, V. Pirronello and C. Mendoza-Gomez (eds), *The Chemistry of Life's Origin*, Kluwer, Dordrecht, pp. 209–258.

Crovisier, J. (1998). Physics and chemistry of comets: Recent results from Comets Hyakutake and Hale–Bopp. Answers to old questions and new enigmas. *Faraday Discussions*, **109**, 437–452.

Crovisier, J. and Bockelée-Morvan, D. (1999). Remote observations of cometary volatiles. *Space Science Reviews*, **90**, 19–32.

Crovisier, J., Leech, K., Bockelée-Morvan, D., Brooke, T.Y., Hanner, M.S., Altieri, B., Keller, H.U. and Leilouch, E. (1997). The spectrum of Comet Hale–Bopp (C/1995 O1) observed with the Infrared Space Observatory at 2.9 AU from the Sun. *Science*, **275**, 1904–1907.

Dello Russo, N., DiSanti, M.A., Mumma, M.J., Magee-Sauer, K. and Rettig, T.W. (1998). Carbonyl sulfide in comets C/1996 B2 (Hyakutake) and C/1995 O1 (Hale–Bopp): Evidence for an extended source in Hale–Bopp. *Icarus*, **135**, 377–388.

Dello Russo, N., Mumma, M.J., DiSanti, M.A., Magee-Sauer, K., Novak, R. and Rettig, T.W. (2000). Water production and release in Comet C/1995 O1 Hale–Bopp. *Icarus*, **143**, 324–337.

Delsemme, A.H. (1977). *Comets, Asteroids and Meteorites: Interrelations, Evolution, and Origins*. University of Toledo, OH.

Delsemme, A.H. (1998). Recollections of a cometary scientist. *Planetary and Space Science*, **46**, 111–124.

Delsemme A.H. and Swings P. (1952). Hydrates de gaz dans les noyeaux cometaires et les grains interstellaires. *Annales d'Astrophysique*, **15**, 1–6.

DiSanti, M.A., Mumma, M.J., Dello Russo, N., Magee-Sauer, K., Novak, R. and Rettig, T.W. (1999). Half the carbon monoxide of Comet Hale–Bopp originates from nuclear ices. *Nature*, **339**, 662–665.

Duncan, M., Quinn, T. and Tremaine, S. (1987). The formation and extent of the solar system comet cloud. *Astronomical Journal*, **94**, 1330–1338.

Duncan, M., Quinn, T. and Tremaine, S. (1988). The origin of short-period comets. *Astrophysical Journal Letters*, **328**, L69–L73.

Eberhardt P. (1998). In *Proceedings 1996 COSPAR Colloquia Series 10: Asteroids, Comets, Meteors*, Elsevier, Amsterdam.

Eberhardt, P. (1999). Composition of comets: The in situ view. In M.F. A'Hearn (ed.), *Cometary Nuclei in Space and Time*, IAU Colloquium 168, Astronomical Society of the Pacific, San Francisco.

Eberhardt, P., Krankowsky, D., Schulte, W., Dolder, U., Lämmerzahl, P., Berthelier, J.-J., Woweries, J., Stubbemann, U., Hodges, R.R., Hoffman, J.H., and Illiano, J.M. (1987). The CO and N_2 abundances in Comet P/Halley. *Astronomy and Astrophysics*, **187**, 481–484.

Eberhardt, P., Reber, M., Krankowsky, D. and Hodges, R.R. (1995). The D/H and $^{18}O/^{16}O$-ratios in water from comet P/Halley. *Astronomy and Astrophysics*, **302**, 301–306.

Feldman, P.D., Festou, M.C., Tozzi, G.P. and Weaver, H.A. (1997). The CO_2/CO abundance ratio in 1P/Halley and several other comets observed by IUE and HST. *Astrophysical Journal*, **475**, 829–834.

Fernandez, J.A. (1980). On the existence of a comet belt beyond Neptune. *Monthly Notices of the Royal Astronomical Society*, **192**, 481–491.

Fernandez, J.A. (1997). The formation of the Oort cloud and the primitive galactic environment. *Icarus*, **129**, 106–119.

Festou, M.C. (1999). On the existence of distributed sources in comet comae. *Space Science Reviews*, **90**, 53–67.

Fomenkova, M. (1999). On the organic refractory component of cometary dust. *Space Science Reviews*, **90**, 109–114.

Fomenkova, M. and Chang, S. (1993). Mass and spatial distribution of the carbonaceous components in Comet Halley. Lunar Plan. Sci. Conf. 24, pp. 501–502.

Fomenkova, M., Kenidge, J., Marti, K. and McFadden, L. (1992). Compositional trends in rock-forming elements of Comet Halley dust. *Science*, **258**, 266–269.

Geiss, J. (1988). Composition in Halley's Comet. *Reviews in Modern Astronomy*, **1**, 1–27.

Geiss, J. and Altwegg, K. (1998). *Giotto: A Unique ESA Science Mission*. ESA-SP 431.

Geiss, J., Altwegg, K., Anders, E., Balsiger, H., Ip, W.-H., Meier, A., Neugebauer, M., Rosenbauer, H. and Shelley, E.G. (1991). Interpretation of the ion mass spectra in the mass per charge range 25–35 amu/e obtained in the inner coma of Halley's comet by the HIS-sensor of the GIOTTO IMS experiment. *Astronomy and Astrophysics*, **247**, 226–234.

Geiss, J., Altwegg, K., Balsiger, H. and Graf, S. (1999). Rare atoms, molecules and radicals in the coma of P/Halley. *Space Science Reviews*, **90**, 253–268.

Grard, R., Gombogi, T. and Sagdeev, R.Z. (1986). The Vega Mission, ESA-SP 1066.

Greenberg, J.M. (1982). What are comets made of – a model based on interstellar dust. In L.L. Wilkening (ed.), *Comets*, University of Arizona Press, Tucson, AZ, pp. 131–163.

Greenberg, J. (1989). The core–mantle model of interstellar grains and the cosmic dust connection. In L.S. Allamandola and A.G.G.M. Tielens (eds), *Interstellar Dust*, Kluwer, Dordrecht, pp. 345–355.

Greenberg, J.M. and Li, A (1998). From interstellar dust to comets: The extended CO source in Comet Halley. *Astronomy and Astrophysics*, **332**, 374–384.

Grün, E. and Jessberger, E.K. (1990). Dust. In W.F. Huebner (ed.), *Physics and Chemistry of Comets*, Springer, Berlin, p. 113–176.

Hanner, M.S. (1999). The silicate material in comets. *Space Science Reviews*, **90**, 99–108.

Hoppe, P., Strebel, R., Eberhardt, P., Amari, S. and Lewis, R. (1996). Small SiC grains and a nitride grain of circumstellar origin from the Murchison meteorite: Implications for stellar evolution and nucleosynthesis. *Geochimica et Cosmochimica Acta*, **60**, 883–908.

Huebner, W.F. and Benkhoff, J. (1999). In M.F. A'Hearn (ed.), *Cometary Nuclei in Space and Time*, IAU Colloquium 168, Astronomical Society of the Pacific, San Francisco.

Irvine, W.M., Bergin, E.A., Dickens, J.E., Jewitt, D., Lovell, A.J., Matthews, H.E., Schloerb, F.P. and Senay, M. (1998). Chemical processing in the coma as the source of cometary HNC. *Nature*, **393**, 547–550.

Irvine, W.M., Bockelée-Morvan, D., Lis, D.C., Matthews, H.E., Biver, H.E., Crovisier, J., Davies, J.K., Dent, W.R.F., Gautier, D., Godfrey, P.D., Keene, J., Lovell, A.J., Owen, T.C., Phillips, T.G., Rauer, H., Schloerb, F.P., Senay, M. and Young, K. (1996). Spectroscopic evidence for interstellar ices in Comet Hyakutake. *Nature*, **383**, 418–420.

Irvine, W.M., Schloerb, F.P., Crovisier, J., Fegley, B., Jr. and Mumma, M.J. (2000). Comets: A link between interstellar and nebular chemistry. In V. Mannings, A. Boss and S. Russell (eds), *Protostars and Planets IV*, University of Arizona Press, Tucson, AZ, pp. 1159–1200. Invited review.

Jessberger, E.K. (1999). Rocky cometary particulates. *Space Science Reviews*, **90**, 91–97.

Jessberger, E.K., Christoforidis, A. and Kissel, J. (1988). Aspects of the major element composition of Halley's dust. *Nature*, **332**, 691–695.

Jessberger, E.K. and Kissel, J. (1989). Chemical properties of cometary dust and a note on carbon isotopes. In R.L. Newburn, M. Neugebauer and J.H. Rahe (eds), *Comets in the Post-Halley Era*, Kluwer, Dordrecht, pp. 1075–1092.

Jewitt, D. and Luu, J. (1995). The solar system beyond Neptune. *Astronomical Journal*, **109**, 1867–1876.

Kleine, M., Wyckoff, S., Wehinger, P.A. and Peterson, B. (1995). The carbon isotope abundance ratio in Comet Halley. *Astrophysical Journal*, **439**, 1021–1033.

Kim, S.J., Bockelée-Morvan, D., Crovisier, J. and Biver, N. (1997). Fluorescence and collisional processes of SO and SO2 in Comet Hale–Bopp (C/1995 O1). *Earth, Moon, and Planets*, **78**, 65–66.

Kissel, J., Brownlee, D.E., Buchler, K., Clark, B.C., Fechtig, H., Grün, E., Hornung, K., Igenbergs, E.B., Jessberger, E.K., Krueger, F.R., Kuczera, H., McDonnell, J.A.M., Morfill, G.M., Rahe, J., Schwehm, G.H., Sekanina, Z., Utterback, N.G., Völk, H.J. and Zook, H.A. (1986). Composition of Comet Halley dust particles from Giotto observations. *Nature*, **321**, 336–337.

Kissel, J. and Krueger, F.R. (1987). The organic component in dust from Comet Halley as measured by the PUMA mass spectrometer on board Vega 1. *Nature*, **326**, 755–760.

Krankowsky, D., Lammerzähl, P., Herrwerth, I., Woweries, J., Eberhardt, P., Dolder, U., Herrmann, U., Schulte, W., Berthelier, J.J., Illiano, J.M., Hodges, R.R. and Hoffman, J.H. (1986). In situ gas and ion composition measurements at Comet Halley. *Nature*, **321**, 326–329.

Kührt, E. (1999). H_2O activity of Comet Hale–Bopp. *Space Science Reviews*, **90**, 79–82.

Kuiper, G.P. (1951). On the origin of the solar system. In J.A. Hynek (ed.), *Astrophysics*, McGraw Hill, New York, pp. 357–424.

Levison, H. (1996). Comet taxonomy. In T.W. Rettig, J.M. Hahn (eds), *Completing the Inventory of the Solar System*, Astronomical Society of the Pacific, San Francisco, pp. 173–177.

Lis, D.C., Keene, J., Young, K., Phillips, F.G., Bockelée-Morvan, D., Crovisier, J., Schlike, P., Goldsmith, P.F. and Bergin, E.A. (1997a). CSO observations of comet C/1996 B2 (Hyakutake). *Icarus*, **130**, 355–372.

Lis, D.C., Mehringer, D., Benford, D., Gardner, M., Phillips, T.G., Bockelée-Morvan, D., Biver, N., Colom, P., Crovisier, J., Despois, D. and Rauer, H. (1997b). New molecular species in comet C/1995 O1 (Hale–Bopp) observed with the Caltech Submillimeter Observatory. *Earth, Moon, and Planets*, **78**, 13–20.

Maas, D., Krueger, F.R. and Kissel, J. (1989). Mass and density of SILICATE- and CHON-type dust particles released by comet P/Halley. *Asteroids Comets Meteors*, **III**, 389–392.

McDonald, G.D., Whited, L.J., Deruiter, C., Khare, B.N., Patnaik, A. and Sagan, C. (1996). Production and chemical analysis of cometary ice tholins. *Icarus*, **122**, 107–117.

Meier, R., Eberhardt, P., Krankowsky, D. and Hodges, R.R. (1993). The extended formaldehyde source in Comet P/Halley. *Astronomy and Astrophysics*, **277**, 677.

Meier, R., Eberhardt, P., Krankowsky, D. and Hodges, R.R. (1994). Ammonia in Comet P/Halley. *Astronomy and Astrophysics*, **287**, 268–278.

Meier, R. and Owen, T.C. (1999). Cometary deuterium. *Space Science Reviews*, **90**, 33–43.

Meier, R., Owen, T.C., Jewitt, D.C., Matthews, H.E., Senay, M., Biver, N., Bockelée-Morvan, D., Crovisier, J. and Gautier, D. (1998a). Deuterium in comet C/1995 O1 (Hale-Bopp): Detection of DCN. *Science*, **279**, 1707–1710.

Meier, R., Owen, T.C., Matthews, U.E., Jewitt, D.C., Bockelée-Morvan, D., Biver, N., Crovisier, J. and Gautier, D. (1998b). A determination of HDO/H2O in comet C/1995 O1 (Hale–Bopp). *Science*, **279**, 842–844.

Mumma, M.J., DiSanti, M.A., Dello Russo, N., Fomenkova, M., Magee-Sauer, K., Kaminski, C.D. and Xie, D.X. (1996). Detection of abundant ethane and methane, along with carbon monoxide and water, in comet C/1996 B2 (Hyakutake): Evidence for interstellar origin. *Science*, **272**, 1310–1314.

Mumma, M.J., Weissman, P.R. and Stern, S.A. (1993). Comets and the origin of the solar system: Reading the Rosetta Stone. In E. Levy and J.I. Lunine (eds), *Protostars and Protoplanets III*, University of Arizona Press, Tucson, AZ, pp. 1177–1252.

Ogilvie, K.W. (1985). The plasma regime at Comet Giacobini–Zinner. *Advances in Space Research*, **5**, 27–35.

Oort, J.H. (1950). The structure of the cloud of comets surrounding the Solar System and a hypothesis concerning its origin. *Bulletin of the Astronomical Institute of the Netherlands*, **11**, 91–110.

Oparin, A.I. (1924). The origin of life [Proiskhozhdenie zhizny]. In J.D. Bernal (1967). *The Origin of Life*, Weidenfeld & Nicolson, London, pp. 199–234.

Öpik, E.J. (1963). The stray bodies in the Solar System: Part 1. Survival of cometary nuclei and the asteroids. *Advances in Astronomy and Astrophysics*, **2**, 219–262.

Oro, J. and Lazcano, A. (1996). Comets and the origin and evolution of life. In P.J. Thomas, C.F. Chyba and C.P. McKay (eds), *Comets and the Origin and Evolution of Life*, Springer, New York, pp. 3–27.

Palmer, P., Wootten, A., Butler, B., Bockelée-Morvan, D., Crovisier, J., Despois, D. and Yeomans, D.K. (1996). Comet Hyakutake: First secure detection of ammonia in a comet. *Bulletin of the American Astronomical Society*, **28**, 927–928.

Reber, M. (1997). Die D/H- und $^{18}O/^{16}O$-Verhältnisse in Wasser sowie die Formaldchyd-Häufigkeit im Kometen P/Halley: Messungen mit dem Giotto Neutralmassenspektrometer. PhD Thesis, University of Bern.

R. Reinhard and B. Battrick (eds) (1986). *The Giotto Mission – Its Scientific Investigations*, ESA SP-1077.

Safranov, V.S. (1972). Ejection of bodies from the Solar System in the course of the accumulation of the giant planets and the formation of the cometary cloud. In G.A. Chebotarev, E.I. Kazimirchak-Polonskaya and B.G. Marsden (eds), *The Motion, Evolution of Orbits, and Origin of Comets*, IAU Symposium 45, Reidel, Dordrecht, pp. 329–334.

Schulze, H., Kissel, J. and Jessberger, E.K. (1997). Chemistry and mineralogy of Comet Halley's dust. In Y.J. Pendleton and A.G.G.M. Tielens (eds), *From Stardust to Planetesimals: Review Papers*, Astronomical Society of the Pacific, San Francisco, Vol. 122, 937–414.

Thomas, K., Blanford, G., Keller, L., Klock, W. and McKay, D. (1993). Carbon abundance and silicate mineralogy of anhydrous interplanetary dust particles. *Geochimica et Cosmochimica Acta*, **57**, 1551–1566.

von Helmholz, H. (1871), as quoted in Oró and Lazcano (1996) [they give it as von Helmholz, H. (1871), The Origin of the Planetary System. In *Selected writings of Hermann von Helmholtz* (Wesleyan University Press, 1971, p. 284). Quotation and reference are from J. Farley (1977), *The Spontaneous Generation Controversy: From Descartes to Oparin* (Johns Hopkins University Press, Baltimore), p. 142].

Weaver, H.A., Brooke, T.Y., Chin, G., Kim, S.J., Bockelée-Morvan, D. and Davies, J.K. (1997). Infrared spectroscopy of Comet Hale–Bopp. *Earth, Moon, and Planets*, **78**, 71–80.

Weaver, H.A., Feldman, P.D., McPhate, J.B., A'Hearn, M.F., Arpigny, C., Brandt, J.C. and Randall, C.E. (1996). Ultraviolet spectroscopy and optical imaging of Comet Hyakutake (1996 B2) with HST. In *Asteroids, Comets, Meteors* 1996, COSPAR Colloquium 10 (book of abstracts).

Weissman, P.R. (1999). Diversity of comets. *Space Science Reviews*, **90**, 301–311.

Whipple, F.L. (1950). A comet model: I. The acceleration of Comet Encke. *Astrophysical Journal*, **111**, 375–394.

Whipple, F.L. (1951). A comet model: II. Physical relations for comets and meteors. *Astrophysical Journal*, **113**, 464–474.

Whipple, F.L. (1991). The forest and the trees. In R.L. Newburn, M. Neugebauer and J.H. Rahe (eds), *Comets in the Post-Halley Era*, Vol. 2, Kluwer, Dordrecht, pp. 1259–1278.

Wöhler, M.F. (1858). Über die Bestandteile des Meteorsteines von Kaba in Ungarn. *Sitzungsberichte Akademie der Wissenschaften in Wien, Mathematische-Naturwissenschaftliche Klasse*, **33**, 205–209.

Wöhler, M.F. and Hörnes, M. (1859). Die organische Substanz im Meteorsteine von Kaba. *Sitzungsberichte Akademie der Wissenschaften in Wien, Mathematische-Naturwissenschaftliche Klasse*, **34**, 7–8.

Woodney, L.M., McMullin, J. and A'Hearn, M.F. (1997). Detection of OCS in Comet Hyakutake (C/1996 B2). *Planetary and Space Science*, **45**, 717–719.

Wyckoff, S., Tegier, S.C. and Engel, L. (1991). Ammonia abundances in four comets. *Astrophysical Journal*, **368**, 279–286.

Yamamoto, T. (1985). Formation environment of cometary nuclei in the primordial solar nebula. *Astronomy and Astrophysics*, **142**, 31–36.

Zinner, E., Wopenka, B., Amari, S. and Anders, E. (1990). Interstellar graphite and other carbonaceous grains from the Murchison meteorite: Structure, composition, and isotopes of C, N, and Ne. *Abstracts of the Lunar and Planetary Science Conference*, **21**, 1379–1380.

54

JAMES W. HEAD III*

The Moon and terrestrial planets: geology and geophysics

Before the Space Age, the study of the planets and their satellites was the domain of astronomers. Telescopic observations of planetary bodies focused on determining their size, position, orbit, density, average surface composition, and the physical state of their surface through photometric analysis. With few exceptions, geologists and geophysicists were occupied with the analysis of Earth, the very complex planet beneath their feet. But even in the latter part of the nineteenth century and the first half of the twentieth, a few geoscientists puzzled over the surface of the Moon, the nature of meteorites, and the geochemistry of the Solar System. G. K. Gilbert, Harold Urey, Ralph Baldwin, Gilbert Fielder, and Eugene Shoemaker each contributed to the foundation that was to become the basis for modern planetary geoscience (Stevenson 2000). Convergence between these disciplines began with the Space Age, and the ability to see the surfaces of other planetary bodies up close, to probe their interiors, and to analyze them from orbiters and flybys. More sophisticated and higher-resolution observations from spacecraft led to an understanding of the mineralogic makeup of their crusts through spectroscopic observations, and the distribution of their surface features through geological analysis of images taken at visible and radar wavelengths. Sophisticated tracking of spacecraft and direct deployment of instruments led to new insights into planetary geophysics. In the last half of the twentieth century, intense exploration of the Solar System changed planetary bodies from solely astronomical objects to geological and geophysical entities, and the picture that began to emerge was similar to that derived from the examination of any population of things: a tremendous diversity of characteristics, a handful of emerging themes, and a host of new questions.

How did planetary geologists and geophysicists approach these problems? Imagine observing a group of human beings. You might initially sort them by size and shape, and then distinguish them by various other physical attributes. But when the time came to understand the factors and processes that were responsible for these characteristics, you would need to look more closely, and to understand what was going on inside. Beneath the superficial variations in surface skin and in hair color and tone, what were the processes that were responsible for the activity of each of these organisms, its origin and its evolution? What internal structure and processes were responsible for its present state? How did the organism regulate its internal heat in the light of such extreme external variations? How did it give birth, how has it aged, and how will it die? Similar questions are key to understanding the birth and evolution of the array of terrestrial planetary bodies. As with humans, planets are complex systems in which most of the driving forces and regulating processes are hidden below the surface. Surface features can give clues as to how the interior works, but a detailed examination of the inside of these bodies is necessary before any real picture of the evolution of the planet as a whole can emerge. Of course, planets are not simple organisms that can be brought into the laboratory, studied, and dissected, allowing us to map out the anatomy of the interior and the role this plays in the evolution of its external features. Instead, indirect measurements are required, and indeed the surface features must be studied in detail if we are to infer the nature of the interior and how it may have changed with time.

*Brown University, Providence, RI, USA

Thus, among the most fundamental areas of analysis are the nature of planetary surfaces, the processes and sequence of events implied by their geologic records, and the structure and state of their interiors and how they have evolved with time. One of the supreme achievements of the Space Age has been the links that have been forged between geology and geophysics to address these issues and to sketch out the story of planetary evolution.

What was the intellectual basis for this phase of planetary exploration? We had many questions about planetary interiors: What is their basic structure; are they homogeneous, an even mixture of planetary ingredients throughout their interior? If not, are they like a plum pudding, or layered, perhaps in several large layers, or like a torte, with multiple thin layers? And what are these layers made of? Are their compositions sorted by density, and if so, how does the increase in pressure with depth inside planets influence the changes in the state of these materials? When and why did this structure develop, and how does it change with time? Volcanic activity shows that planets are hotter in their interior than at the surface. Where does this heat come from, how is it distributed in the interior, and what are the processes by which the planet redistributes and gets rid of heat? And how has this changed with time, in the course of the planet's history? Once this type of knowledge is to hand, we can ask even more sophisticated questions: How do planets differ in their basic internal structure, and how does this determine their evolution? What role does size, and position in the solar nebula during planetary formation, play in the further evolution of the planets?

The approach to addressing these questions was neither intellectual nor systematic. Missions and experiments were undertaken or denied for a variety of reasons: national goals, national security, proximity to Earth, international competition, international cooperation, technological sophistication, financial constraints, politics, professional advocacy, personal advocacy, and scientific rationale. We explored the Moon before we analyzed the surfaces of Venus and Mercury. The Soviet Union sent many missions to Venus, while the United States sent only a few. After the crowning achievements of Apollo, most of the Moon's surface remained unstudied by spacecraft for over twenty years. The sequential exploration of the Solar System is a complex story of the historical interplay of these many factors, and has been told elsewhere (Stevenson 2000, Chapman 1988, Cruikshank 1983, Colin 1983, Vaniman et al. 1991, Kieffer et al. 1992, Snyder and Moroz 1992, Morrison 1999).

In this chapter we integrate these important historical steps into an overview of the geology and geophysics of terrestrial planetary bodies, as revealed by exploration of the Solar System in the last half of the twentieth century. We first briefly outline the basic processes that may have been involved in the formation and evolution of planetary interiors, and then look at the types of measurements and observations that might lead us to an understanding of their present and past states. Armed with the basic physical properties of the terrestrial or Earth-like planets (size, density, position in the Solar System), we turn to a brief description of the nature and ages of the surfaces of each terrestrial planetary body, and what is now known about their interiors. We start with the smallest, the Moon, and proceed upward in size, via Mercury, Mars, and Venus, to Earth. What were the major external (e.g., impact cratering) and internal (e.g., volcanism) processes that were responsible for their evolution, and when did most of this activity occur? What do we know about each planetary body's interior, and how does this relate to what we see on the surface? Once we have this basic information, we can explore some themes and processes that help to explain the observed characteristics, and address such questions as: What are the major factors that cause the differences in the interiors and geological histories of the planets? How much of a planet's history is predetermined by its starting conditions (its "genetic" makeup), and how much is determined by its later history (its "environment")? We conclude with several outstanding questions that need to be investigated in the exploration of the Earth and terrestrial planetary bodies in the twenty-first century.

THE NATURE OF PLANETARY INTERIORS AND GEOPHYSICAL PROCESSES

Present planetary interior structure can be viewed from two perspectives: that of internal compositional variations, usually configured in layers such as crust, mantle, and core; and that of variations in internal temperature and state, producing layers on Earth such as the lithosphere, the asthenosphere, and the outer molten and inner solid core. Internal compositional variations are produced by differentiation – that is, the segregation of materials of different composition from more primitive and homogeneous parent materials. Differentiation can be rapid and catastrophic, as in the case of core formation, a process in which denser iron-rich material sinks to the deepest interior. Because planets are hotter in their early history, core formation is thought to occur in the first few percent of a planet's history, and the amount of gravitational potential energy it releases is so high that surface melting is implied, at least for the larger planets. Differentiation can take place over increasingly longer periods of time as heat in the interior causes partial melting of the mantle and the ascent of the hotter, less dense melt products (magma) toward the surface to form crustal materials (intrusions, and extrusions such as volcanoes). Depending on the amount of energy involved, and the way it is distributed in the interior, melting and differentiation can be global (e.g., if the energy is from a high influx of globally distributed random impacts), or local to regional (e.g., in the case of mantle plumes producing circular hot spots, or a global system of linear cracks at mid-oceanic ridges). Variations in the relative proportions and timing of these factors can lead to major differences in the nature and age of the outer differentiated layer, the crust. Energy from early intense bombardment can produce a primary crust, later internal heating and melting can produce materials for a secondary crust, and reworking of these earlier crusts can yield tertiary crusts (Taylor 1989).

One of the most fundamental aspects of understanding planetary interiors comes from knowledge of heat. What are the sources of heat, how much of it does a planet have at any given time, what is its internal distribution, and how does a planet transfer and get rid of heat over time? These simple questions are the keys to understanding planetary evolution. How much heat derives from initial accretional energy, position relative to the Sun, electromagnetic heating, core formation and other density instabilities, large impacts, short and long-term radioactive decay of minerals, and tidal interactions? How is this heat distributed over time in what is known as a planet's thermal history? What is the rate of change of temperature as a function of depth, and how does this relate to the physical state of material (e.g., liquid, partially molten, solid)? What is the nature and stability of thermal boundary layers, the transitional layers separating materials with different temperatures? How is heat transferred through the course of the planet's evolution? What role does conduction play? How important is convection? How significant is advection – the direct transfer of heat by movement of molten material from the interior to the surface as in the volcanic flooding of a planetary surface? How does material behave under the tremendous temperatures and pressures typical of planetary interiors? What materials change phases (rearrange their internal structure), and how do they then behave? And how does all this add up? How do different planets lose heat as a function of time in what is known as their thermal evolution? Are there many paths or only a few? And what are the factors that determine this? And where are we going – if the planets are indeed evolving, where is the Earth heading?

Our present knowledge of the interior of most planets is based on their bulk density and on information obtained from measurements of surface geochemistry, moment of inertia, gravity, and present and fossil magnetic fields. Also, very significant are inferences made from geologic structure, topography, and geologic history, analogies with Earth, and assumptions about starting conditions. Deep drilling, and tectonic uplift and exposure of rocks from depth, provide some information about the upper few hundredths of a percent of the Earth's radius, and impact cratering can expose material from greater depths, perhaps even below the crust, on some planets. Detailed assessment of the interior of a planet requires surface seismic networks and heat flow probes, which have so far been emplaced only on the Earth and Moon.

INTERNAL STRUCTURE AND GEOLOGICAL HISTORY OF THE TERRESTRIAL PLANETARY BODIES

The Moon

The Moon, as our closest neighbor in space, was the first body to attract the attention of geologists studying other planetary surfaces. Eugene Shoemaker and his co-workers (Don Wilhelms, Jack McCauley, Baerbel Lucchitta, Elliot Morris, Farouk El Baz, and others) applied the basic principles of terrestrial stratigraphy to lunar surface features and geologic structures, and this enabled them to define geologic units and produce geologic maps, and thus to delineate the major surface processes and the sequence of events in lunar history. Similar mapping techniques have been applied to each successive planet explored over the last 25 years, and the collective maps (Carr *et al.* 1984, Head 1999) provide the basis for understanding the history of each planet and comparative planetary evolution. The will and determination of a handful of geoscientists in the USA (Harold Urey, James Arnold, Gerald Wasserburg, Robert Walker, Paul Gast, and George Wetherill and colleagues) and the Soviet Union (M. V. Keldysh, A. P. Vinogradov, Roald Sagdeev, Valery Barsukov, Mikhail Marov and colleagues) convinced their respective governments that international competition could also produce significant scientific results.

Pluto and Charon excepted, the Moon is the largest satellite, relative to its parent body, in the Solar System, and its mode of formation has captivated scientists for years. Current thinking is that the Moon formed very early in Solar System history when a Mars-sized object, one-half the diameter of the Earth, impacted the proto-Earth, ejecting crust and upper mantle material which re-accreted in Earth orbit to form the Moon (e.g., Hartmann and Davis 1975). Soon afterwards a global crust formed, under a bombardment that lasted for several hundred million years in which a massive influx of projectiles impacted the newly formed surface at several kilometers a second, producing impact craters of many sizes. This bombardment fragmented and fractured the upper few kilometers of the Moon's crust to form a thick soil layer (megaregolith) and produced interfingering global geologic units of ejecta representing the first few hundred million years of lunar history (e.g., Wilhelms 1987). This so-called late heavy bombardment ended about 3.8 billion years ago (Wasserburg *et al.* 1977), but not before the largest projectiles had excavated huge depressions (as large as 2,000 km in diameter) and spread ejecta over immense areas (Spudis 1993), producing the extremely

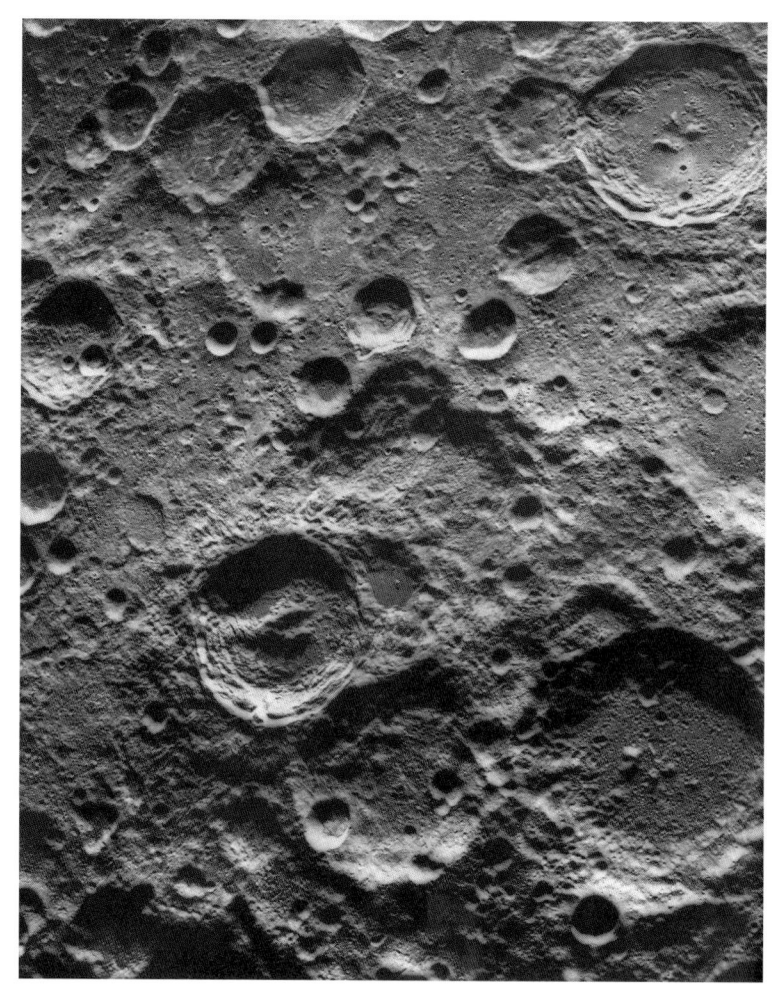

Figure 1 The heavily cratered lunar farside. The 75 km diameter King crater, with its lobster-claw-like central peaks, is the sharp-rimmed crater just to the lower left of the center. (NASA Apollo 16 image.)

rough surface topography typical of the lunar highlands that we see today (Figure 1). Data from the US missions Galileo, Clementine, and Lunar Prospector have provided a global view of the topography and mineralogy of the lunar crust (Spudis 1999), and have revealed details of a huge impact basin on the lunar farside that excavated to lower crustal and perhaps mantle depths (e.g., Pieters *et al.* 2000).

Volcanic flooding of the surface of the Moon became evident during the waning stages of the late heavy bombardment. By about 2.5–3.0 billion years ago, basaltic lavas had covered approximately 17% of the lunar surface, preferentially filling in the nearside, low-lying basin interiors to the lunar maria (Figure 2). Volcanic eruptions on the Moon were volumetrically significant, but far and few between, and occurred predominantly in the first half of its history, under a rapidly decreasing flux (e.g., Head and Wilson 1992). Tectonic activity on the Moon stands in stark contrast to that on our own planet. The limited array of lunar tectonic features occurs predominantly in and near the maria: linear rilles and graben, formed by crustal extension, were followed by sinuous (wrinkle) ridges formed by contraction (Solomon and Head 1980) (Figure 3). Virtually no major internally generated geologic activity for the last 2.5 billion years is manifested on the lunar surface. The Moon thus provides a picture of the first half of Solar System history (characterized by impact bombardment and early volcanism), and serves as a benchmark for the interpretation of the records preserved on other terrestrial planets. How do these features relate to the nature of the Lunar interior? Deployment of seismic instruments on the Moon and analysis of the results by Frank Press, Nafi Toksoz, Gary Latham, and others have helped to address this question.

Lunar samples collected by US Apollo and Soviet Luna missions (e.g., Papike *et al.* 1998), remote sensing, and surface seismic data show that the Moon has been internally differentiated into a crust, mantle, and possibly a small core (Figure 4). The feldspar-rich crust is thinner on the central nearside, about 55 km, but may reach thicknesses of 100 km

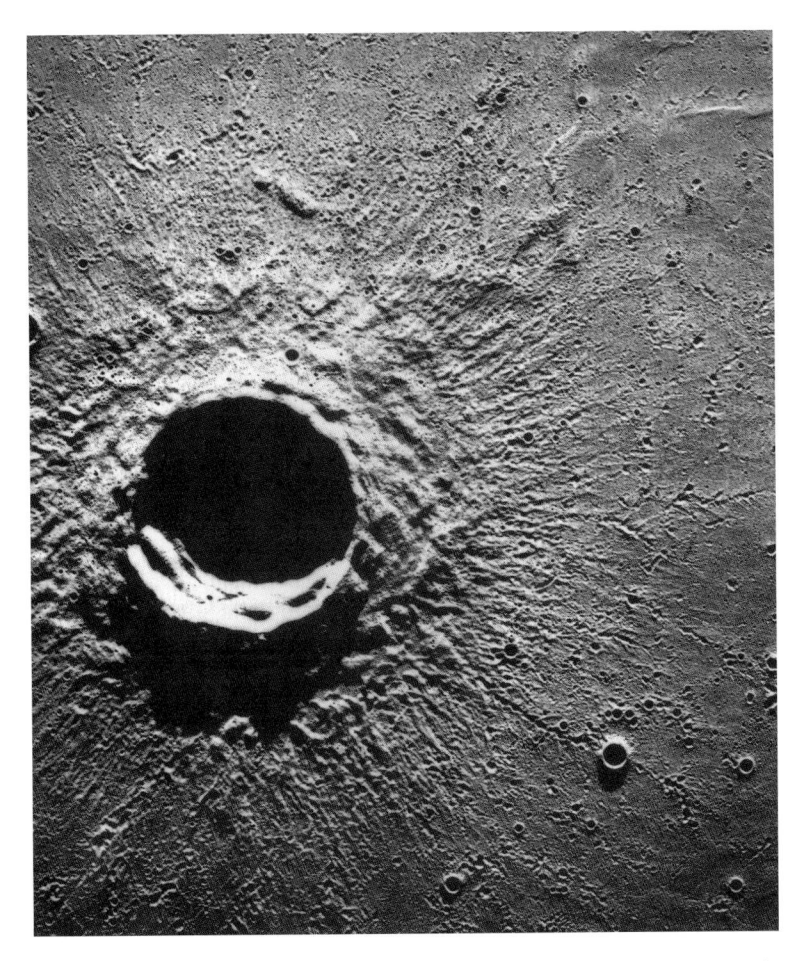

Figure 2 The relatively smooth lava-covered surface of Mare Imbrium. The impact crater Timocharis, 33 km in diameter, is surrounded by textured ejecta material. (NASA Apollo 15 image.)

on the farside. Seismic data and geologic mapping show that the lunar maria are relatively thin (a few kilometers thickness at most) and perched on a globally continuous feldspar-rich crust. The highland crust has been called a "primary" crust, derived from widespread melting associated with the energy of early impact cratering (Taylor 1989). Widespread melting was accompanied by fractional crystallization and separation of low-density plagioclase and its flotation to the surface, to create the crust and the residual upper mantle layers. Whether the melting was globally extensive (a magma "ocean") or regional (magma "lakes") is still a matter of controversy. In either case, the residual layers below the low-density crust were denser than the underlying mantle, probably leading to their gravitational collapse and sinking toward the interior, perhaps to form a core (e.g., Hess and Parmentier 1995). During this latter time period, partial melting of the mantle led to the formation of magmas which were emplaced as secondary crust – the lunar maria – collecting in low-lying craters and basins (e.g., Head and Wilson 1992).

Measurement of the present Lunar heat flow during the Apollo missions with instruments designed by Marcus Langseth and colleagues showed values much less than those of Earth, and consistent with a body losing heat by conduction. Seismic data indicate that at present the outer 800–1000 km of the Moon acts as a relatively rigid shell, or lithosphere. But the presence of the highland crust, mare basalts, and related sequential tectonic features show that the Moon was hotter in its interior earlier in its history. What was the thermal evolution of the Moon? How did the Moon lose heat as a function of time?

The formation of the globally continuous low-density buoyant crust apparently precluded the development of plate tectonics early in the history of the Moon. This led to a dominance of conductive cooling through this continuous layer, producing a globally continuous lithosphere, in contrast to the multiple, laterally moving and subducting plates on Earth. Thus the Moon quickly became a one-plate planet, losing heat primarily by conduction throughout the rest of its history (Solomon 1978). The large surface area to volume ratio meant that cooling by conduction was very efficient (the "radiator" effect) and that the Moon's lithosphere thickened rapidly with time. The small percentage of the

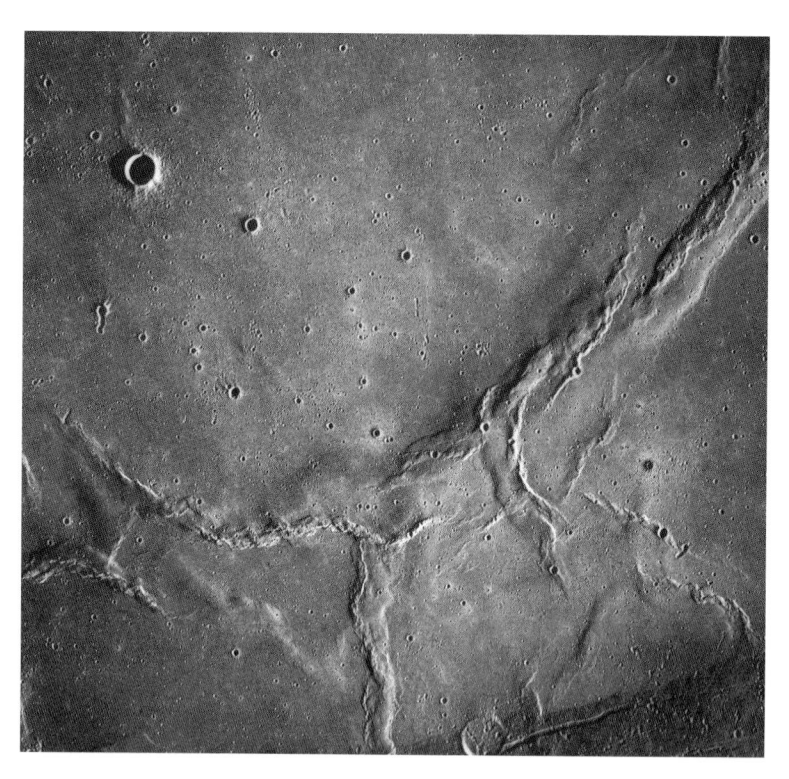

Figure 3 Lunar linear rilles and wrinkle ridges in southern Mare Serenitatis. Linear rilles or troughs (extensional structures) in the bottom of the image cut the darker surface, while wrinkle ridges (contractional structures) deform the lighter and younger mare of central Serenitatis. The crater is 15 km in width. (NASA Apollo 17 image.)

surface area formed by the volcanic maria indicates that advective cooling played a minor role. Evidence from tectonic features suggests that the Moon underwent a change from net global expansion before about 3.6 billion years ago to net global contraction from that point until today (e.g., Solomon and Head 1980). Gravity data obtained by William Sjogren and his colleagues showed that there are large positive anomalies (mascons) over the youngest mare-filled basins, and further details have been revealed by Clementine data (e.g., Neumann *et al.* 1996). The spacing and type of tectonic features around the mare margins provide evidence that the surface was flexing and subsiding under the load of the lavas on the lithosphere (Solomon and Head 1980). These data show that the lithosphere thickened rapidly with time early in the history of the Moon, and that the mascons plausibly represent the last mare load, emplaced on a lithosphere that was so thick that it supported this load. The lack of a significant dipole magnetic field at present, combined with evidence for a fossil magnetic field in some lunar samples, is consistent with this internal thermal evolution.

Summary

The Moon apparently formed from accretion of the ejecta of the impact of a Mars-sized object into early Earth. Continued heavy bombardment caused extensive melting and formation of an impact-energy related, global, low-density primary crust (now the highlands). Dense residual layers sank toward the interior, heated up, and later rose again. Conductive cooling created a globally continuous lithosphere that thickened rapidly with time. Secondary crust formation (the lunar maria) was minimal; the ascent of magma was inhibited by the thick, light crust causing the maria to form preferentially in basins on the thinner nearside highland crust. The high surface area to volume ratio meant that the Moon underwent rapid cooling, and little to no evidence of internal activity (e.g., magnetic field, surface volcanism) is seen in the last half of Solar System history.

Mercury

Mercury is one of the most poorly studied and enigmatic planets of the Solar System (Chapman 1988). Clues to its unusual nature come from the fact that it is about one-third the diameter, but about the same density, as Earth. Its proximity to the Sun seems a reasonable explanation for its lack of a significant atmosphere, and gives it the highest surface temperatures of any planet in the Solar System, and also the widest temperature range. In spite of this, it has radar-reflective materials located near its poles (Harmon

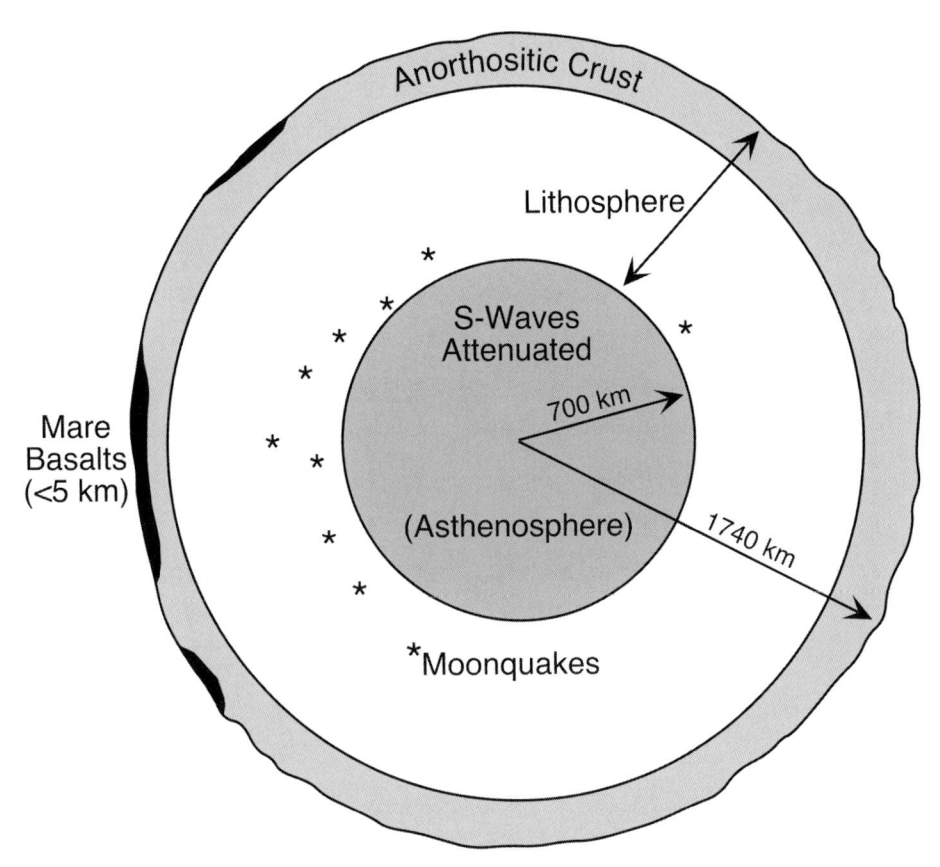

Figure 4 The interior of the Moon is known from Apollo seismic data and other geophysical measurements. The globe-encircling anorthositic highland crust has the later lunar basaltic maria superposed on it, largely filling impact basins and craters on the near-side, and is thicker on the farside (right) than the nearside (left). A small lunar core may be present (not shown). The presence of deep moonquakes (stars) is interpreted to mean that the outer rigid layer, or lithosphere, is almost 1000 km thick at present, and consists of the crust and much of the mantle. The attenuation of S (shear) waves provides evidence for an asthenosphere.

and Slade 1992) that have been interpreted as water-ice deposits in permanently shadowed areas known as "cold traps". With a size near that of the Moon, and a density that of the Earth, Mercury offered an opportunity to study the influence of size and internal structure on the geological history and thermal evolution of planetary bodies, and raised the question of initial starting conditions in determining planetary evolution.

In 1974 Mariner 10 provided the first close views of the surface of Mercury, returning images of about 35% of the planet's surface. Interpretation of these data by Bruce Murray, Robert Strom, and John Guest led to significant new insights. At first glance, these images revealed a remarkably Moon-like terrain (Figure 5). Detailed geologic mapping of the surface, however, has shown that Mercury differs from the Moon in several important respects (e.g., Spudis and Guest 1988). Large areas of relatively ancient, so-called intercrater plains may indicate that more extensive volcanism accompanied the period of heavy cratering on Mercury than on the Moon. Large extensive scarps (linear, asymmetrical drops in topography) on Mercury attest to episodes of regional shortening and, perhaps even the global contraction that would result from a modest decrease in the planet's circumference during solidification (Figure 6). Areas of smooth plains have nearly the same reflectivity as that of the heavily cratered regions, which has led to controversy over the origin (e.g., volcanic or ponded impact ejecta?) of the smooth regions (Vilas 1988). Mariner 10's low-resolution images make it difficult to resolve the uncertainty, but analysis of color data shows evidence for different units of probable volcanic origin in the plains (e.g., Robinson and Lucey 1997). With a few exceptions, Mercury generally resembles the Moon on its surface (e.g., Spudis and Guest 1988, Head et al. 2000), but its high density ($5.4 \, \text{g} \, \text{cm}^{-3}$) suggests that it bears more of a resemblance to the Earth in the interior.

Mercury's high density, together with its high-albedo surface, which in its reflectance properties is similar to the lunar regolith (e.g., Vilas 1988), shows that it must be internally layered, with iron likely comprising about 60–70% of the interior by mass. This, together with the presence of a dipolar magnetic field (Connerney and Ness 1988) thought by many to be internally generated, has been interpreted to

Figure 5 The surface of Mercury: heavily cratered terrain, with significant expanses of smooth intercrater plains. (NASA Mariner 10 image.)

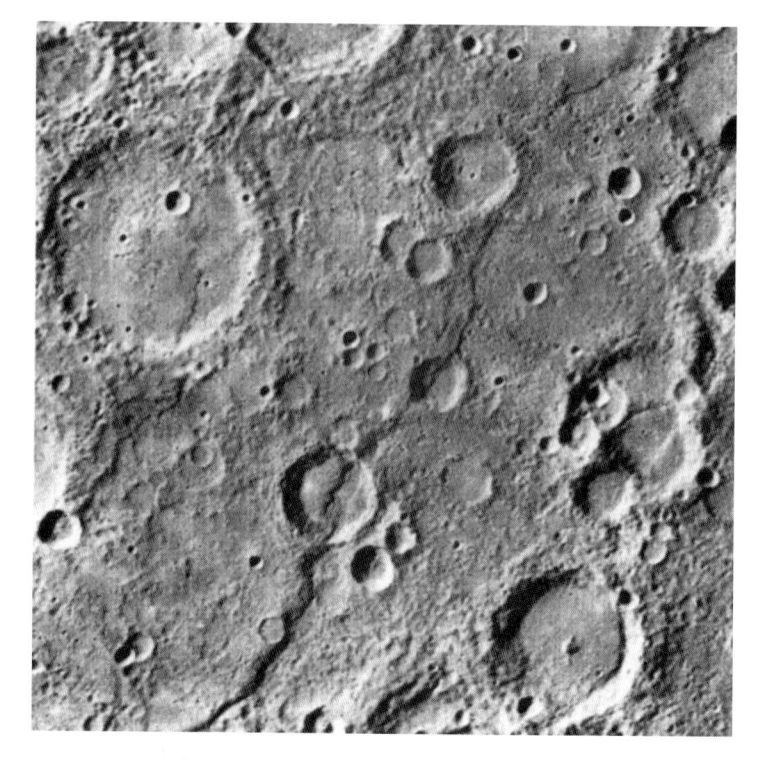

Figure 6 Large scarps (linear asymmetrical drops in the topography) on the surface of Mercury discovered by Mariner 10. The large crater at the upper left is about 130 km in diameter. A scarp is seen extending across its floor and into the adjacent terrain. The Discovery scarp, locally over 2 km in height, crosses the image in the middle in a parallel direction, cutting and shortening two impact craters. (NASA Mariner 10 image.)

mean that Mercury has a large, perhaps partially molten iron core approximately the size of the Earth's Moon (Figure 7). Why is there such a large core relative to those of other planetary bodies? Two competing ideas have emerged. Early studies of the formation of the Solar System emphasized the strong gradient in temperature and pressure as a function of increasing distance from the center of the collapsing solar nebula (Lewis 1988). To a first approximation, this trend explained the primary distinction between the inner and outer planets, with the solid silicate-rich terrestrial planets forming in a higher-temperature/pressure realm, and the large gas giants condensing and accreting at lower temperatures and pressures typical of the outer reaches of the nebula. It was viewed as logical that the planet closest to the Sun would have an unusually high abundance of more refractory materials such as iron. Increasing appreciation of the role of large impacts in the early history of the Solar System, brought about by the possibility that the Earth's Moon was formed as a result of the impact of a Mars-sized object, raised the question about similar events elsewhere in the Solar System. Could Mercury have been struck by a large object, stripping away large portions of its original outer crust (Vityazev et al. 1988, Benz et al. 1998) and leaving behind an abnormally iron-rich planet as a result? These two hypotheses can be tested by further exploration and analysis of Mercury, and – as with many scientific controversies – the answer may lie somewhere in between the two extremes.

How has Mercury evolved with time (Strom and Neukum 1988, Vilas 1999)? The presence of the unusual lobate scarps raises important issues about internally and externally driven processes (Melosh and McKinnon 1988, Thomas et al. 1988, Schubert et al. 1988). Some have hypothesized that these features are a consequence of Mercury's orbital evolution into its presently observed spin–orbit resonance. This phase of tidal despinning is predicted to cause equatorial contraction and high-latitude extension, but incomplete photographic coverage of the surface makes testing this hypothesis difficult. Alternatively, global contraction accompanying the thermal evolution of the planet could also cause shrinkage of the surface and deformation into large global-scale scarps (Figure 6).

Predictions about the thermal evolution of a body with the internal structure of Mercury (Solomon 1976, 1977, 1978; Schubert et al. 1988) show that core formation must have occurred before the end of the late heavy bombardment, because this event causes an intense release of gravitationally induced energy, producing large-scale melting and planetary expansion. Subsequent to this early phase of large-scale planetary differentiation, cooling and contraction of the interior and surface occurred, and part of the immense core solidified. The predicted amount of global contraction is equivalent to a decrease in planetary radius of several kilometers, somewhat more than the ~1–2 km determined from the geometry and distribution of the large scarps. The surface geology of Mercury suggests that, like the Moon, it has been characterized by a globally continuous crust and lithosphere throughout the history recorded in its surface units (Solomon 1978). Mercury has lost heat primarily by conduction (Solomon and Head 1982) (Figure 8), its lithosphere at least trebling in thickness with time. In

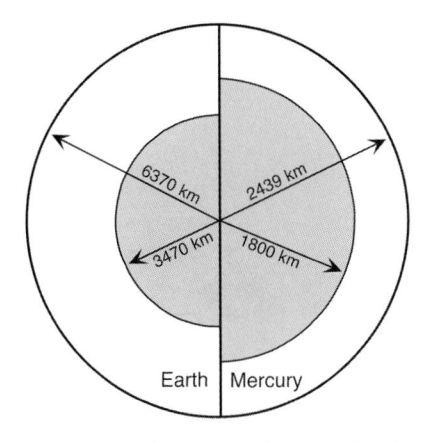

Figure 7 The interior of Mercury, showing the size of its core in relation to that of Earth. Mercury's dense iron core is slightly larger than the radius of the Earth's Moon (about 1740 km), and it makes up a significantly larger portion of the planetary interior than does the core of the Earth. The thickness of the crust of Mercury is unknown and is not shown here.

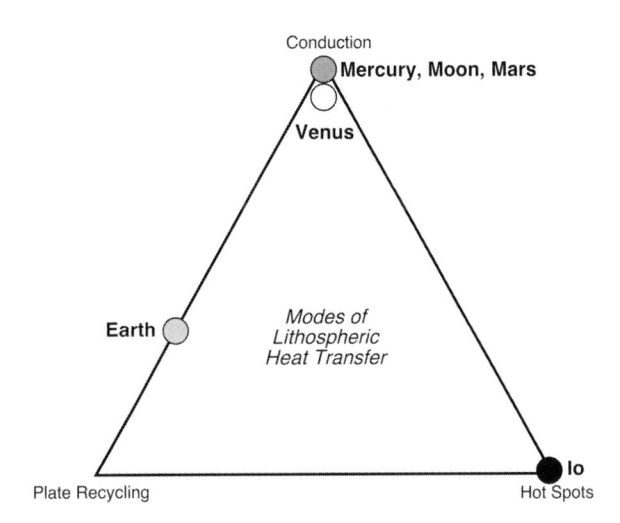

Figure 8 Schematic diagram showing various lithospheric heat transfer mechanisms. The Moon, Mars, and Mercury are dominated by conductive heat loss, while advective cooling through constant volcanic eruptions characterizes the innermost of the Galilean satellites, Io. Plate recycling, and radioactive decay of elements sequestered in the continental crust, dominate Earth. Venus is presently characterized by conductive heat loss, but its youthful surface and aspects of its geological record suggest that its heat loss may have been episodic. (After Solomon and Head 1982.)

addition, if cooling were sufficient to cause these large contractional structures, then the global state of stress in the lithosphere must have been contractional from this point on, which perhaps explains why secondary crustal formation (mare-like lava plains) is even more limited than on the Moon. Although the outer core of Mercury is probably still molten, and possibly the source of the observed magnetic field, this relatively near-surface internal dynamism is not reflected in the observed geology of the planet in the one-third of the surface observed in detail to date.

Summary

Mercury is a one-plate planet with a Moon-like surface which has been modified by global tectonic features linked to large-scale evolution of its Earth-like interior and probably its orbital evolution. Despite its high levels of initial heat, its Earth-like interior, and probable present-day molten outer core, there is no evidence of internal activity over the last several billion years. Are its enigmatic radar-reflective polar materials actually water-ice, and if so, where did they come from? Global geochemical and geologic mapping of Mercury and further determination of its internal structure will permit us to assess the relative roles of tidal despinning, and cooling-induced changes in radius, as factors in the formation of observed tectonic features, and to understand why (and if) its surface has been devoid of evidence of internal activity for the last half of Solar System history. Mercury provides an excellent example of the plausibility of modest changes in planetary radius resulting in large-scale changes in surface tectonics, an idea popular for the Earth in the first part of the twentieth century, but one which was eclipsed by the plate tectonics perspective. Among the exciting prospects in planetary exploration in the coming decade is the US Discovery MESSENGER mission, headed by Sean Solomon, which is designed to obtain comprehensive and global information about Mercury and its environment. Together with Bepi Colombo, a two-spacecraft mission of the European Space Agency, these results should ensure that Mercury takes its place among the better-known terrestrial planets.

Mars

Mars is twice the diameter of the Moon and about one-half that of Earth, but its density is closer to that of the Moon. Earth-based observations (Martin *et al.* 1992), extensive spacecraft exploration (Snyder and Moroz 1992), and meteorites ejected from Mars by impacts and transported to Earth (Longhi *et al.* 1992, McSween and Treiman 1998), provide the basis for understanding the surface (Tanaka *et al.* 1992, Carr 1999) and interior (Schubert *et al.* 1992). The US Mariner, Viking, and Mars Global Surveyor spacecraft have returned images revealing that Mars is more geologically diverse and complex than either the Moon or Mercury. Imaging teams led by Michael Carr, Thomas Mutch, and Michael Malin showed that Mars has a distinct hemispheric asymmetry in the distribution of its geologic units: the rough and often densely cratered southern hemisphere stands several kilometers above the sparsely cratered northern hemisphere (Tanaka *et al.* 1992, Smith *et al.* 1998). The southern hemisphere has two main components: a very ancient crust nearly saturated with large craters and cut by abundant small channels, apparently formed by groundwater sapping and runoff (Figure 9), and younger intercrater plains which appear to be ancient, but less modified. A variety of cratered plains cover the northern hemisphere. Volcanic flows characterize the plains surrounding large volcanoes (Figure 10) (Greeley *et al.* 2000), but elsewhere the plains are featureless except for craters, mare-like ridges, and an assortment of fluvial, eolian, and permafrost-like features. At intermediate to high northern latitudes the plains contain various kinds of patterned and striped ground, scarps, and irregularly shaped mesas. Their complexity may represent the influence of volatiles and changing temperature conditions (Squyres *et al.* 1992). At the highest latitudes the polar ice deposits and their enigmatic swirls are surrounded by seas of dunes (Thomas *et al.* 1992). In the equatorial regions the boundary between northern and southern hemispheres is extremely complex. Near the boundary the older, higher terrain of the southern hemisphere contains numerous channels tens of kilometers wide and hundreds of kilometers long (Figure 11) that are reminiscent of those formed on Earth by catastrophic flooding (Baker *et al.* 1992). Some scientists believe that these channels debouched enough water into the northern lowlands to produce temporary oceans (Parker *et al.* 1989, 1993).

The Tharsis region represents a major departure from the general surface geology of Mars. The Tharsis rise is a broad (8000 km wide) topographic bulge exhibiting ancient, heavily cratered units and young shield volcanoes (Figure 12). The region, centered at 14°S latitude and 101°W longitude, stands about 10 km above the surrounding terrain; some of its volcanic shields extend another 15 km in altitude. The vast majority of the linear rilles and fractures seen on Mars surround the Tharsis region (Banerdt *et al.* 1992). Valles Marineris, an enormous equatorial canyon system, extends radially away from Tharsis and is probably related to faulting that accompanied the evolution of Tharsis. Valles Marineris has been extensively modified by the collapse of its walls and also by channel formation (Lucchitta *et al.* 1992). The origin of Tharsis is uncertain; some believe that it is predominantly a massive uplift of the crust caused by some dynamic process deep within the planet's mantle, while others propose that the Tharsis region was the focus

Figure 9 Impact craters and plains in the cratered uplands of Mars. A high density of impact craters is testimony to the ancient age of this terrain, and destruction and partial erosion of parts of the craters show the importance of weathering. The superposed craters to the right of this image are filled with eolian material and volcanic plains. Small channels in the region are testimony to the work of water-related processes in the modification of the terrain. The image shows a region about 500 km across. (NASA Viking image mosaic.)

Figure 10 Perspective view of the volcano Arsia Mons, formed by "draping" a Viking image mosaic over topography derived from the Mars Orbiter Laser Altimeter (MOLA) instrument. The summit of the massive 500 km diameter volcano reaches elevations of over 17 km above the mean planetary radius of Mars. The large summit caldera structure is about 120 km in diameter. After the formation of the main edifice, lava flows began to erupt from the flanks, building the major protuberances seen in the foreground. Thousands of lava flows were erupted from these zones, spreading to distances of over a thousand kilometers from the vents. Together, these types of summit and flank eruptions contributed to the building of the massive Tharsis province. (NASA Viking image mosaic and MOLA data.)

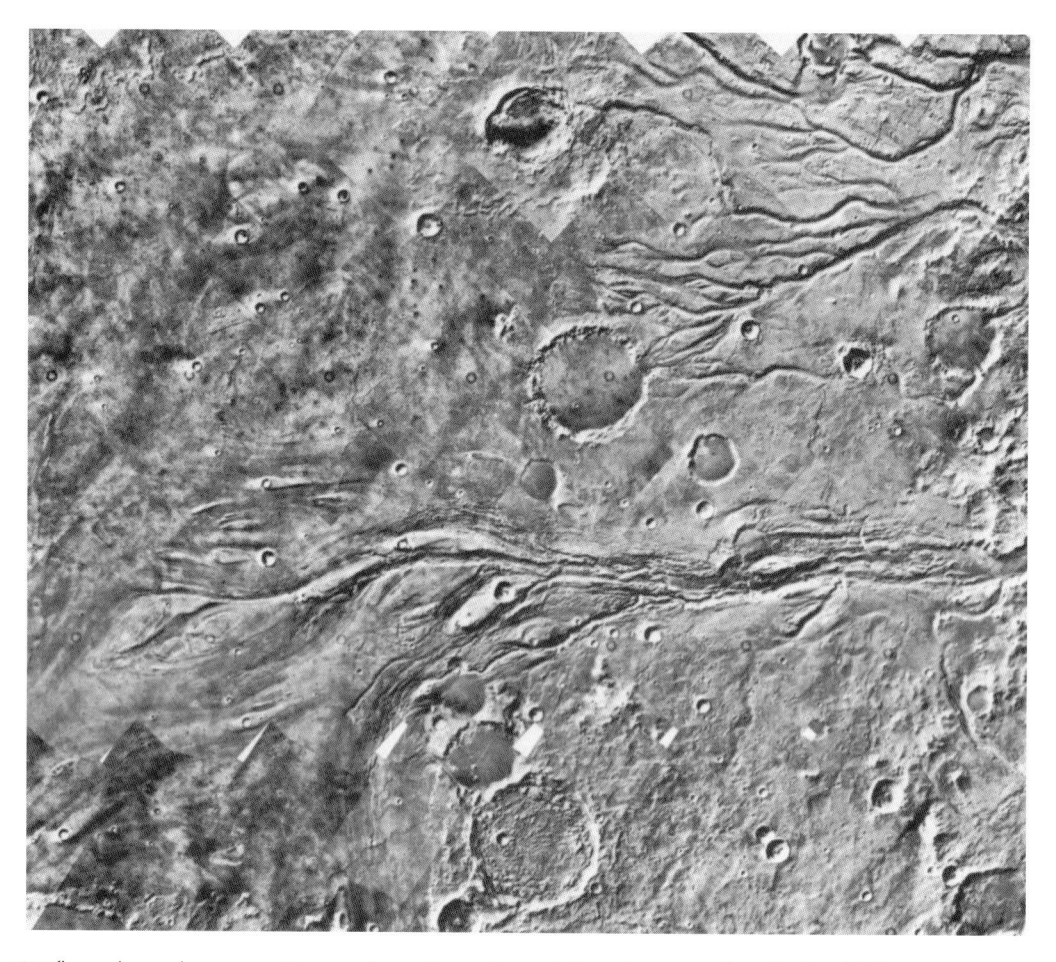

Figure 11 Outflow channels emerge from collapsed terrain and flow for many hundreds of kilometers across the Martian surface. In this Viking mosaic, two channel systems, Maja Valles to the south and Vedra Valles to the north, cut deeply into the old cratered terrain as they flow from Lunae Planum to the left down into Chryse Planitia to the right. The channels of Maja Valles converge (left) to form a distinctive gorge where it approaches Chryse Planitia (right). The region shown in this image is about 300 km wide. (NASA Viking image mosaic.)

of volcanic outpouring, topographic buildup, and related stresses, with uplift being minimal. Recent Mars Global Surveyor gravity and topography data lend support to the latter idea (Smith *et al.* 1999, Zuber *et al.* 2000).

Additional features unique to Mars include craters with lobate ejecta patterns that suggest a layer of ice or water in parts of the upper crust (Strom *et al.* 1992). Much of Mars is covered with wind-blown deposits (Greeley *et al.* 1992), which may mix with volatiles at the poles (Thomas *et al.* 1992). The surface units thus reveal an early chronology for Mars much like that of the Moon and Mercury (Tanaka *et al.* 1992), but with volcanism, particularly in the Tharsis region, extending well into the last half of Solar System history, perhaps even up to the present. However, the absolute chronology is not known because of the lack of documented samples from the surface units on Mars.

One of the most exciting aspects of Mars is the abundant evidence of changing climate (Figure 13) (Thomas *et al.* 1992, Owen 1992). What was the early atmosphere of Mars like? Under what conditions could runoff channels have formed, and how did the fluids in the upper crust get there? Did oceans once exist? Further exploration of Mars will provide important data for understanding the nature of the atmosphere, the interaction of the atmosphere and the surface, the nature of climate change, and the possible origin and evolution of life (McKay *et al.* 1992).

On the basis of remote sensing data and analysis of Martian meteorites, Mars is known to be differentiated into a crust, mantle, and core (Longhi *et al.* 1992, Schubert *et al.* 1992) (Figure 14). Gravity data suggest a relatively low-density crust characterized by variations in thickness and density, which by virtue of isostasy in the ancient cratered uplands must have formed in its earliest history (Esposito *et al.* 1992, Smith *et al.* 1999, Zuber *et al.* 2000). Crustal thicknesses are now known to be greater in the southern cratered terrain and to decrease toward the

Figure 12 The Tharsis region dominates the global topography of Mars. In these global views, compiled from Mars Orbiter Laser Altimeter (MOLA) data, the Tharsis region is seen in the lower image as a 10 km high bulge several thousand kilometers across, topped by three major edifices (the Tharsis Montes) rising about 14–18 km above mean planetary radius (MPR). Flanking Tharsis to the west is the huge Olympus Mons edifice, rising over 20 km above MPR, and to the north is Alba Patera, a gigantic volcanic edifice extending into the northern lowlands. To the east, the giant rift valley complex Valles Marineris extends for several thousand kilometers. In the other hemisphere, the heavily pitted cratered terrain is seen, as well as the huge ancient impact basin Hellas. (NASA MOLA data.)

boundary with the northern lowlands, which appear to be thinner and more uniform in thickness. This early crustal formation must have influenced subsequent thermal evolution. Mean density and moment of inertia data dictate the presence of a distinct dense central core ranging from 1300 to 2000 km in radius, and observations from recent Mars Pathfinder data constrain it even further. Spacecraft observations show little evidence for an intrinsic magnetic field, but analysis of Martian meteorites and recent Mars Surveyor data indicate that Mars had a magnetic field early in its history (Acuña et al. 1999). Broad, linear magnetic anomalies have even been interpreted as evidence of crustal spreading and plate tectonics in the early history of Mars (Connerney et al. 1999; see also Sleep 1994), but little supporting geologic evidence for plate tectonics has been found.

How do thermal evolution models compare with the observed geological record? In many ways Mars appears similar to the Moon: an ancient heavily cratered highland crust, a one-plate planet characterized by a globally continuous lithosphere, and volcanic resurfacing smoothing out lowlying areas and producing volcanic edifices. Thickening of the lithosphere with time is shown by the presence of some positive gravity anomalies similar to Lunar mascons, and the support of huge loads represented by the large volcanic shields formed in the intermediate to late period of Martian history. Although there is evidence for heterogeneities in the thickness of the lithosphere with time (Zuber et al. 2000), the general global cooling and thickening trend is clear.

But Mars is also fundamentally different from the Moon. First, the extremely large topographic rises of Tharsis and Elysium and their associated tectonic and volcanic features have no parallel on the Moon. If the Moon can be characterized as a one-plate planet with vertical tectonics linked to loading, flexure, and downward subsidence in the mare basins, then Mars has this, plus interior upwelling, vertical uplift, and extensive associated volcanism. Second, there is a significant amount of evidence pointing to the presence of a warmer, wetter climate earlier in the planet's history (e.g., Baker et al. 1991). Runoff and outflow channels, and evidence for permafrost, suggest that the Martian megaregolith initially collected water, occasionally catastrophically released water, and is now a reservoir of water (Clifford 1993). But where did the water come from? Did it emerge during volcanism as part of the heating and degassing of the interior, or was it a late addition from a veneer of cometary impacts? And where did it go?

Mars's greater distance from the Sun suggests that it would have accreted from more volatile-rich materials than the Earth did, a view supported by analysis of the SNC meteorites, which are almost certainly Martian in origin (Longhi et al. 1992). Late-stage accretion and initial outgassing of the interior would have concentrated water at or near the surface, but differentiation, core formation, and enhanced solar radiation would have radically altered its presence and abundance (Schubert et al. 1992). Most workers agree that the abundance of water would have been radically decreased by these processes, but opinion differs as to whether a very little or a lot remained, and where it was (Carr 1996). Was some water left in the mantle, to be released in the earliest history by the steep thermal gradient? Was a surface veneer of water delivered by comets late in accretion, and not admixed with the interior because of the stable lithosphere? Could Mars have been dry on the inside and wet on the outside in its early history? Could interior and exterior volatile reservoirs have remained relatively isolated from each other by the globally continuous lithosphere?

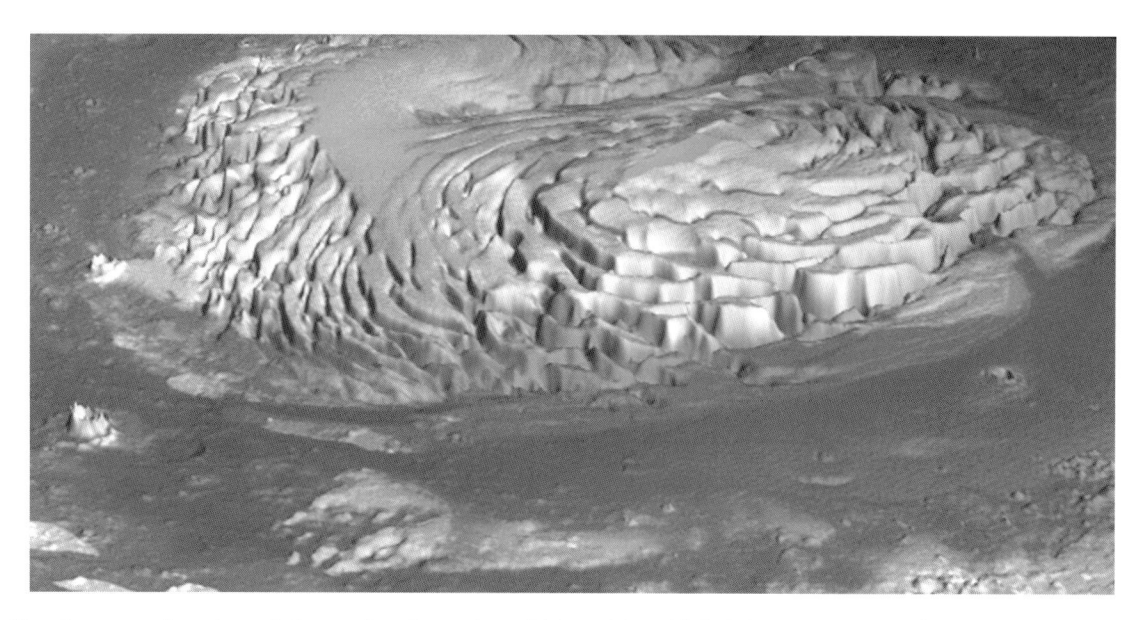

Figure 13 A perspective view of the north polar region of Mars. Mars Global Surveyor Mars Orbiter Camera (MOC) images have been "draped" over MOLA topography. Readily seen is the topography of the north polar cap, rising almost 3 km above the surrounding terrain. The permanent polar ices are seen in white, while the polar layered terrain is seen in gray on the slopes of the cap, cut by spiral troughs. The surrounding region is dominated by dunes and sedimentary deposits of the northern lowlands. (NASA MOC image and MOLA data.)

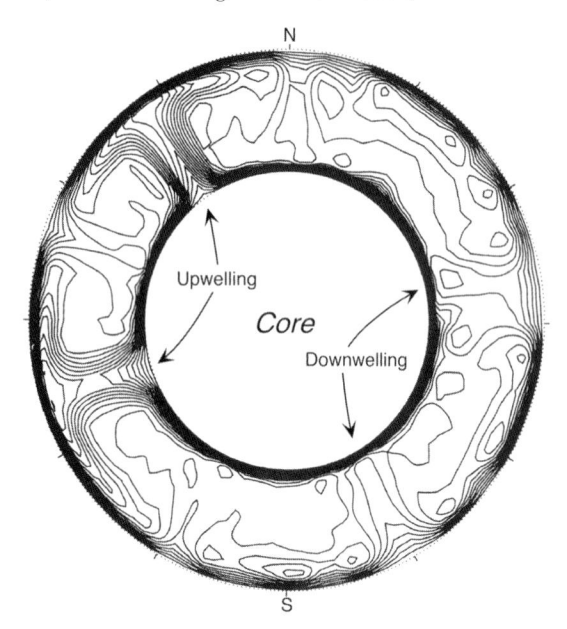

Figure 14 The interior of Mars. On the basis of Viking era geophysical and geochemical data, the core could have a radius ranging from 1300 to 1900 km. Recent Mars Global Surveyor data are refining this value and showing that the crust of Mars is thicker and more variable in the southern uplands (about 40–90 km) and thinner and more regular in the northern lowlands (about 35–40 km) (Zuber *et al.* 2000). This diagram shows a model of convective patterns in the Martian mantle with the assumption of over 90% heating from below the mantle. Several large upwelling regions are apparent. (After Schubert *et al.* 1992.)

The presence of Tharsis has puzzled planetary scientists for years (Banerdt *et al.* 1992). How could interior processes produce what appears to be about 10 km of uplift, and how could convective upwelling in the interior have consistently lasted over what appears to be several billion years? And indeed, how could such large amounts of melting continue for billions of years to produce the tremendous shield volcanoes? What type of chemical and thermal anomalies could be responsible, why do they occur only in certain places on Mars, and why not on other planets?

New insight into the formation of Tharsis has been provided by recent studies of the interior of Mars. In most planets convection in the mantle is thought to assume relatively regular patterns of upwelling and downwelling whose number and dimensions are linked to the thickness of the layer, and whether the layer is being heated from below or within. In some cases temperature and compositional differences in the mantle and underlying layers, and related density differences, can cause instabilities and anomalous departures from simple patterns of convective heat transfer. Subduction of lithospheric plates back into the Earth's mantle, for example, is thought to modify mantle convective patterns, and instabilities in the thermal boundary layer at the Earth's core/mantle boundary often produce mantle plumes and surface hot spots on a variety of scales. Even so, these types of perturbations are thought to be regional and relatively transient when considered with respect to the history of the planet.

How, then, could such a gargantuan anomaly occur and be sustained over such a long period of time? Recent studies of the interior of Mars suggest that it might be much different than previously thought in some very significant ways (e.g., Harder and Christensen 1996, Harder 1998, Schubert *et al.* 1990, 1995). This new insight comes from an understanding of the importance of changes in the phases of minerals as a function of changing temperature and pressure with depth, and whether these phase changes release heat (exothermic) or absorb it (endothermic). One such phase change of the mineral olivine within the Earth's interior has been shown to be endothermic, and the effect may be significant, shutting off the transfer of material and heat across the phase boundary. The lower gravity of Mars means that this transition would occur much deeper in the interior of Mars than it would in Earth. Analysis of the behavior of the thermal boundary layer close to the phase transition shows that the endothermic phase change suppresses instabilities in the hot thermal boundary layer and leads to only one or two major upwellings, or megaplumes, instead of a whole series distributed evenly over the planet. Further work predicted that at the onset there would have been a very small number of plumes, quickly changing to just one or two plumes, and eventually only one major plume would survive. Studies indicate that this convective state would stabilize in the first third of the history of Mars, and that after one-half of Mars's history, two plumes would dominate, and eventually only one would survive. This scenario provides a very close approximation to the observed history of volcanism on Mars, with dominance of two major plume-like loci of volcanism and tectonism (Elysium and Tharsis), and the continuation of activity in the larger area, Tharsis, well beyond that of Elysium. If this explanation is correct, it shows the importance of variations in the properties of planetary interiors under different conditions for the overall evolution of planetary surfaces and mechanisms of heat loss. Recent results from Mars Global Surveyor, however, raise new questions about the internal thermal structure of Mars and where such phase changes might occur, and show that the centers of volcanism in Tharsis are varied in space and time (e.g., Zuber *et al.* 2000). Models for Tharsis are currently being reevaluated with these new data.

Summary

Mars is an example of a one-plate planet bearing underlying similarities to the surfaces of the Moon and Mercury, but with an atmosphere, hydrosphere, and cryosphere. A further difference is the unusual concentration of volcanic and tectonic activity (in Tharsis and Elysium, respectively) which may be caused by the specific depth of phase changes, in turn related to the size of Mars. The mode of acquisition of water on Mars (internal or external), its subsequent history, and its relation to environments for the formation and evolution of life are some of the most exciting questions before the scientific community today, and a concentrated international exploration program is under way to address these questions.

Venus

The geology of Venus is of extreme interest because of the similarity of Venus to Earth in size, density, and position in the Solar System, and its dissimilarity from smaller terrestrial planets. Its dense carbon dioxide atmosphere (Donahue and Pollack 1983) has served, however, to obscure the surface of the planet from ground-based and spacecraft cameras. For many years our view of the surface consisted of radar images of small areas obtained from Earth through the pioneering efforts of Donald Campbell, Richard Goldstein, Ray Jurgens, and colleagues, and surface panoramas from Soviet Venera landers (Florensky *et al.* 1983, Moroz 1983). Near-global radar altimetry data returned in the late 1970s by the US Pioneer Venus spacecraft experiment headed by Gordon Pettengill allowed initial characterization of global topographic provinces: about 60% of the surface lies within 500 m of the most common radius, and only 5% lies more than 2 km above it (McGill *et al.* 1983). This is in contrast to the distinctly bimodal distribution of the Earth's topography, which represents the density differences between the lighter continents and the heavier ocean basins, and the thicker continental crust. Despite the distinctly unimodal distribution of altitudes on Venus, the range of elevations is comparable to that of Earth.

What do the detailed surface features of Venus look like, and what do they tell us about the tectonics, volcanism, atmospheric, and impact cratering processes operating to form and modify the planet's surface? What mechanism has Venus chosen to transfer its internal heat across its lithosphere and into space? Does it get rid of its heat by conduction, as do the smaller terrestrial planets? If this is so, we might expect to see an ancient Moon-like surface dominated by impact cratering and volcanism. Or does it get rid of its heat primarily by advection at a series of hot spots, similar to the innermost of the Galilean satellites, Io? If that were the case, we would expect to see extensive volcanic deposits and abundant centers of volcanism. Does Venus get rid of its heat in the same way as Earth does, predominantly through creation, lateral movement, subduction, and recycling of lithospheric plates? In this case we might expect to see evidence for the geologic features associated with divergent and convergent plate margins on Earth (Solomon and Head 1982). Or perhaps on Venus there operates a combination of these mechanisms, representing styles characteristic of each of these other planetary bodies – or indeed, perhaps something entirely different.

High-resolution imaging data obtained in 1983–84 by the Venera 15 and 16 spacecraft (designed under the guidance of V. Kotelnikov and O. Rhziga), and the global Magellan mission (designed under the guidance of Gordon Pettengill with a science team headed by R. Stephen Saunders) permitted the Venus-wide assessment of features and processes related to tectonism, volcanism, and impact cratering (Barsukov et al. 1986, Basilevsky et al. 1987, Head et al. 1992, Solomon et al. 1992, Phillips et al. 1992, Schaber et al. 1992, Saunders 1999). Surprisingly, it was found that Venus was quite different from Earth and the other terrestrial planets. Instead of major terrain types with two different ages (e.g., the equivalent of lunar highlands and maria, Mars heavily cratered terrain and northern lowlands, or Earth continents and seafloor), these images showed a global, relatively young surface, just a few hundred million years old, with only about a thousand impact craters over the entire planet. Even more surprisingly, the areal distribution of impact craters was initially found to be indistinguishable from a random one. This suggested that the entire surface may have been renewed by some process (tectonic, volcanic?) over a very short period of geologic time relatively recently (see discussion in Basilevsky et al. 1997). Furthermore, the resurfacing had to have been rapid, compared with the rate of accumulation of the impact crater population, in order to maintain the apparent randomness of areal crater distribution and to account for the lack of embayment of the vast majority of its craters. And of course, the questions of the fate of the earlier geologic record, and the apparent rapidity of the resurfacing, have very important geophysical implications (Phillips et al. 1997, Schubert et al. 1997, Hansen et al. 1997) and are still a subject of debate.

Because the morphological signatures of earlier times are not preserved, regional and global geological mapping and stratigraphic analyses could address the geologic history of only the last 10–20% of the history of Venus (e.g., Basilevsky and Head 1998, 2000; Basilevsky et al. 1997; Tanaka et al. 1997). The beginning of the observed history is characterized by intensive near-global tectonic deformation, forming the tessera terrain which today comprises less than 10% of the exposed surface area of Venus (Figure 15). Following tessera formation, several stages of extensive volcanism occurred, burying vast areas of tessera and forming what we see now as regional plains (Figure 16). Regional plains-forming materials can be subdivided and are separated from one another, from oldest to youngest, by tessera-forming deformation, dense fracturing, broad ridging, and finally, wrinkle ridging. These tectonic episodes are interpreted by some to be generally globally synchronous and to represent successive episodes characterized by the dominance of contraction, then extension, then contraction again, and finally extension once more. The last broadly distributed global tectonic episode, extensive wrinkle-ridging, occurred very close in time to the emplacement of the most areally abundant plains unit. This period marked the transition to the present stage of the history of Venus, which is characterized by a predominance of regional rifting and related volcanism (Figure 17). This latter stage appears to have gone on for longer than the other stratigraphic intervals (Basilevsky et al. 1997), although the resulting tectonic and volcanic features and deposits cover only 10–20% of the surface of Venus. This suggests that the general intensity of tectonics and the flux of volcanism in this latest period may have been much lower than in earlier times.

Summary

The observable part of the history of Venus is characterized by several key features that stand in contrast to the comparable period of Earth history (approximately the Phanerozoic), when global geodynamic processes were dominated by plate tectonics. First, Venus shows no sign of plate tectonics at present. Instead, global tectonism appears to have moved back and forth between periods of contraction and extension, the density of deformational structures and the strain rate declining with time. Also in contrast to the Earth's plate boundary volcanism, Venusian plains-forming volcanism occurred at a rate comparable to that of terrestrial volcanism at mid-oceanic ridges, but was emplaced in an entirely different style, as extensive flood plains. And then the style changed: for the last few hundred million years, Venus has been dominated primarily by rift-related volcanism emplaced at rates comparable to or even lower than present intra-plate volcanism on Earth (Head et al. 1992). At present, Venus appears to be a one-plate planet losing heat largely by conduction (Figure 8). But what about the first 80–90% of the history of Venus? How could two planets be so similar in so many ways, but so different in others? What aspects of the nature and evolution of the interior might have caused these differences? What are the implications for the early history of the Earth?

The similar size and mean density of Earth and Venus, and their close proximity in the Solar System, has been used to argue that their starting conditions were similar. One difference was revealed when spacecraft measurements showed that Venus has essentially no intrinsic magnetic field. An immediately appealing idea is that the lack of a magnetic field may be due to Venus's very slow rotation rate, but theory suggests that these factors are not related. Venus could, of course, have had a magnetic field in the past, but unfortunately the surface temperatures are currently above the point (the Curie point) at which the record of magnetic fields would be preserved in rocks.

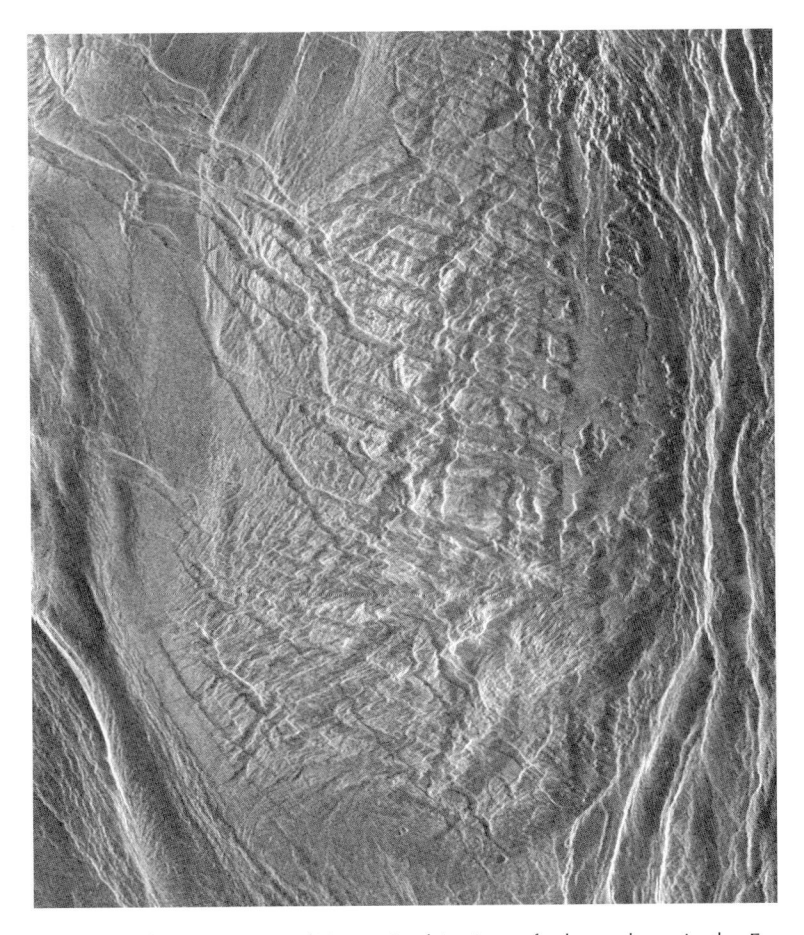

Figure 15 Tectonic deformation in the mountains of Venus. In this view of a huge dome in the Freya Montes region of Ishtar Terra, numerous tectonic features are testimony to the intense deformation that accompanied the creation of this and adjacent tessera terrain. Ringing the dome to the east and west are broad folds caused by shortening and contraction. On top of the dome is a set of intersecting extensional structures (graben), indicating that the dome underwent stretching and collapse. The width of the region covered by this image is about 75 km. (NASA Magellan image.)

Soviet Venera and Vega lander experiments (1972–85) showed that the composition of most plains units measured were consistent with basalts, similar to volcanic surface unit morphologies observed in the Magellan data (Surkov 1983, Surkov et al. 1984). Venera 8 measurements suggest that in some areas compositions may even be more silica-rich, toward the granitic end of the spectrum. No measurements were made in the highlands (tessera), and some workers have suggested that these areas, like the continents on Earth, may be more differentiated. These surface compositions show that the interior of Venus is differentiated and that a basaltic crust has been extracted from the mantle. The crustal thickness can be estimated by comparing the gravity and topography of different regions (Sjogren et al. 1997), and making assumptions about average composition. Estimates made in this way suggest that crustal thickness averages 25–40 km, with local variations, particularly in tessera regions, that probably exceed 50–60 km. How these thicknesses have changed with time, and indeed how they came about in the first place, are not well understood.

Planetary gravity fields are an indicator of lateral variations in density related to surface topography, crustal structure and composition, and to other mechanisms in the interior that form and maintain density differences (e.g., broad convection or hot spots). On Earth, large-scale gravity anomalies are generally not correlated with topography and crustal density anomalies, being instead related to broad convective flow processes in the mantle. Venus appears markedly different than Earth in this respect; its large-scale gravity anomalies are extremely highly correlated with topography. Among the suggested explanations for this difference are that:

- Venus does not have the equivalent of the Earth's asthenosphere (a partially molten layer at depth), which serves to partially decouple lithospheric plate movement from deeper mantle convection, or

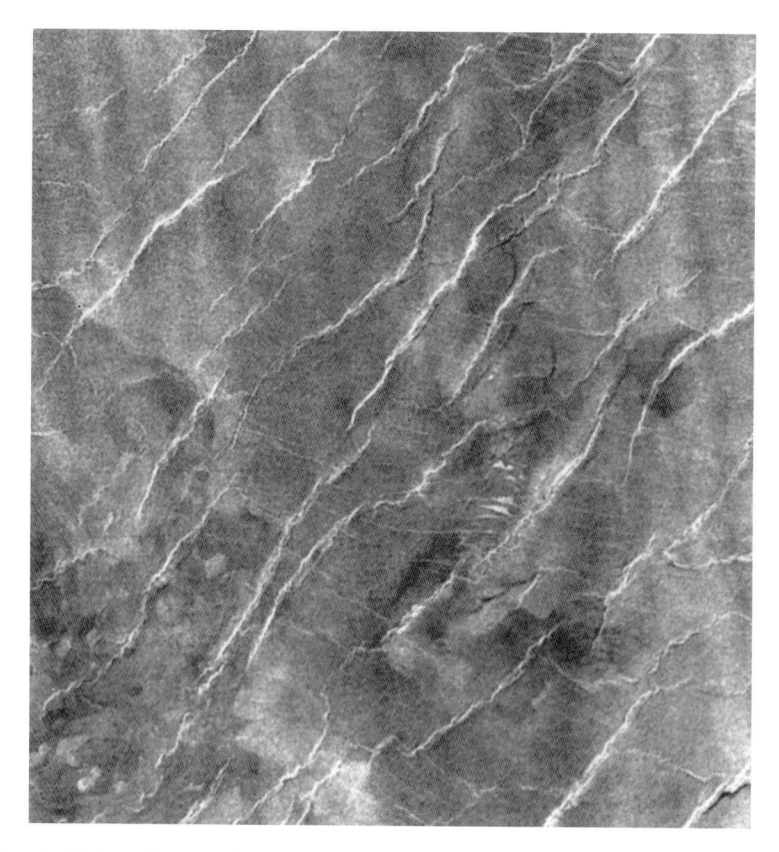

Figure 16 Regional plains in Helen Planitia. The majority of the surface of Venus is formed of dark, relatively smooth plains with superposed wrinkle ridges, known as regional or ridged plains. At the lower left, mottled material with some circular features thought to be small volcanic edifices are representative of an early phase of plains volcanism. The later regional plains (middle and upper part of the image) embay these features, and both units have subsequently been deformed by the ridges, which represent small folds and faults formed by regional contraction. The width of the region in this image is about 250 km. (NASA Magellan image.)

- the topography is supported by the finite strength of the interior; if this mechanism is not operating, the implication is that the topography is geologically very young and may be related to ongoing dynamic thermal convection and heat loss mechanisms.

No direct information exists about the deep interior of Venus, but because of the high, Earth-like mean density, virtually all workers assume, on the basis of comparisons with Earth, that Venus has differentiated a core (Schubert *et al.* 1997). Some investigators believe that the core may be frozen completely solid, while others suggest that core solidification has not yet commenced or is under way. In order to understand the present internal thermal structure and state without direct seismic and heat flow data, information must be gleaned from surface morphology and altimetry, gravity, topography, and Soviet Venera lander surface measurements of composition and heat-producing elements, coupled with an understanding gained from laboratory measurements of the strength of rocks under Venus conditions, and geophysical modelling. Estimates of the thickness of the lithosphere have been obtained by comparing actual topographic profiles to flexural estimates based on varying heat fluxes, and by comparing gravity/topography relationships to flexural models as a function of scale. The range of estimates of the effective elastic lithosphere thickness is 10–40 km, and from gravity/topography relationships, 20–40 km. New data on very dry rocks suggest that the crust of Venus may be almost as strong as the mantle. The implications of these data for the thermal flux of Venus are that the planet is presently losing less heat than the Earth is, perhaps because of the lack of the efficient terrestrial process of plate tectonics and subduction.

At present, mantle convection in the interior of Earth is dominated by the presence of a plate-tectonic-related, mobile lithosphere, a "lid" that moves laterally at high rates, flexes at plate margins, and is subducted back into the interior. However, the geologic record of Venus does not support the existence of plate tectonics in the recent past (Solomon *et al.* 1992). Thus, for a variety of reasons, Venus

Figure 17 A view of rifting and volcanism on Venus. Devana Chasma is a major north–south rift valley that crosses central Beta Regio. Over 1100 km long, it is comparable to the East African Rift Valley on Earth, and formed in a similar way, by east–west extension, stretching, and downfaulting. A 37 km diameter crater in the central part of the image has been cut by the faulting, and a portion of the crater has been destroyed. Lava flows are seen emerging from the central rift area and extending downslope in the lower left. The width of the area in this image is about 375 km. (NASA Magellan image.)

is thought of as being dominated by a stagnant lithospheric lid through which heat is lost primarily by conduction (Figure 8). In addition, layering of the mantle on Venus is predicted to be even more likely than on Earth. Mantle convection and heat loss models of Venus suggest that it is not now in thermal equilibrium with Earth-like abundances of radioactivity, and thus there appears to be a present lack of balance between internal heat production and heat loss. Geophysicists are presently investigating whether Venus may undergo episodic, and different, phases and styles of heat loss, and if so, what the mechanism of heat loss might be (e.g., Schubert *et al.* 1997, Hansen *et al.* 1997). Alternatively, there might be some more fundamental differences between Earth and Venus in abundance of heat-producing elements.

What models for the evolution of Venus have been proposed on the basis of these observations and data? One of the keys to the understanding of Venus is the origin of the distinct tectonic fabric associated with tessera, and the factors responsible for the transition to geologic processes dominated by volcanism and more focused deformation, such as rift zones. In some models, tessera terrain comes about through the normal evolution of crustal structure formed and modified by mantle convection patterns. In one model, tessera initially formed above large areas of mantle upwelling (hot spots or plumes) as regions of enhanced volcanism and crustal thickening; thermal decay and gravitational collapse then converted such volcanic plateaus into the highly deformed tessera terrain (e.g., Hansen et al. 1997). Alternatively, tessera terrain may have formed over zones of mantle downwelling when mantle flow patterns and crustal deformation coincided (e.g., Bindschadler et al. 1992). Another hypothesis cites the uniform decrease in heat output as a function of time and calls on the high surface temperature and the exponential relationship between temperature and rate of strain as a major factor in the evolution of the observed surface features. In this scenario, mantle convection is closely linked to the overlying lithosphere; over most of the history of Venus a weak lower crust deforms readily, resulting in very high levels of surface strain and production of the observed tessera deformation globally. At some point late in the history of Venus, the heat flux declines, the lower crust becomes more ductile, and rates of surface deformation decrease rapidly (e.g., Solomon 1993).

This last model illustrates how important surface and near-surface temperatures can be. Is there evidence for major changes in surface temperatures late in the history of Venus? Some recent calculations suggest that the release of gases associated with volcanism could significantly alter the surface temperatures and deformation style. In this scenario, gas loss associated with the extensive plains-forming volcanism (Bullock and Grinspoon, 1996) changes the atmosphere and surface temperatures sufficiently to propagate a thermal wave into the crust. This then influences the deformation style, favoring contraction, and possibly explaining the extensive, globally distributed wrinkle ridges formed on the regional plains (Solomon et al. 1999).

Some global geodynamic models assume a typical thermal evolution but predict evolutionary changes in mantle convection patterns. In these models, early very high rates of deformation are related to a highly convective interior and a highly deformable lithosphere capable of being incorporated into the convecting mantle. After a prolonged period of surface recycling into the mantle, the long-term planetary cooling trends result in diminished convective vigor, crustal and lithospheric stability, and a changeover to a one-plate, hotspot-dominated planet about 500 million years ago (e.g., Solomatov and Moresi 1996). Other models involve one or more periods of catastrophic resurfacing in the history of Venus. Analysis of the nature and distribution of impact craters has shown that the crater population represents an average crater retention age of about 300–750 million years, and that the distribution is difficult to distinguish from a completely spatially random population. These data, together with the very small number of craters that appear to have been modified by volcanic activity, have led some workers to propose that the surface of Venus underwent a global volcanic and/or tectonic resurfacing event about 300–750 million years ago which completely eradicated the crater population, and that the present crater population is a production population, with only a few craters modified by greatly reduced rates of volcanic activity.

What could account for such a global resurfacing? One possibility is episodic plate tectonics (Turcotte 1993). In this view, at several times in its history Venus has been a one-plate planet with a stable, conductively thickening lithosphere. This trend results in an increase in interior temperatures which leads to periods of enhanced mantle convection; these in turn lead to lithospheric foundering, periods of rapid lithospheric recycling and heat loss, and rapid resurfacing rates, prior to a return to stabilization, to be followed by another cycle. Another possible explanation is related to the style of crustal formation on Venus. Lunar secondary crust, the lunar mare, comprises only a small percentage of the surface and total crustal volume. But what happens when, unlike the Moon, a planet is large enough to produce vast quantities of secondary crust over geologic time, but does not recycle it, as Earth does? What happens when secondary crustal formation and destruction proceeds vertically, not laterally as in sea-floor spreading on Earth? Several investigators have modeled the sequence of events when vertical crustal accretion occurs on a one-plate planet, and what happens to the thickening basaltic crust and residual depleted mantle layer with time (e.g., Parmentier and Hess 1992). The results show that, over time, positive compositional buoyancy decreases in significance and negative thermal buoyancy increases, resulting in a net negative buoyancy for the depleted mantle layer. At this point the depleted mantle layer founders, deforming and delaminating both the depleted mantle and the overlying crustal material. As this material descends into the interior, hot fertile mantle material ascends from depth to replace it, undergoes pressure-release melting, and produces a phase of widespread surface volcanism. Following this overturn event, vertical crustal accretion continues at much reduced rates, and the process is predicted to repeat itself at intervals of 300–750 million years. Several aspects of the geological history of Venus are consistent with this vertical crustal accretion model, including:

- lack of preserved surface units from the first 75–85% of the history of Venus
- evidence for a vertically accreting crust over most of observed geologic time
- formation of tessera terrain as the first major unit of the present stratigraphic column
- major changes in the style and intensity of deformation following tessera formation
- emplacement of widespread regional plains over the vast majority of the surface of Venus closely following the period of tessera formation

- a substantial decrease in volcanic flux and a change from large-scale regional plains emplacement to focused local sources.

Presently, there is no firm consensus on this wide range of models, and each is being tested against the results of global geologic mapping, improved geophysical models of the interior, and the thermal evolution of Venus. Nonetheless, we now have a much clearer picture of the nature of the surface of Venus in the last few hundred million years of its geologic evolution, and its relation to Earth and the other terrestrial planetary bodies. Venus is a terrestrial planet very similar to Earth in size and mass, so comparisons between them offer a very promising way to understand some of the basic principles and trends in the evolution of terrestrial planetary bodies. During the part of Venus's history that is discernible from its present surface, Earth has been dominated by plate tectonic processes, with the most significant activity concentrated along the boundaries of lithospheric plates. The emerging picture of Venus reveals a very different situation. The majority of what we observe appears to have formed by vertical processes, not lateral movement as on Earth. Vertical crustal accretion dominates post-tessera time, and tectonic features reveal limited lateral movement. The general stratigraphic relations suggest an initial intense deformation, emplacement of extensive basaltic flood deposits over a geologically short period of time, and then relative quiesence, in contrast to the steady creation and destruction of lithospheric plates on Earth. No evidence of plate-tectonic-style recycling is observed to be presently operating on Venus. Yet the geological record provides important clues that Venus has changed its style of heat loss from the past to what we see today. Some models, in fact, predict that heat loss might have been not only catastrophic, but episodic (e.g., Parmentier and Hess 1992, Turcotte 1993). Future exploration of Venus should include surface stations to make long-term seismic and heat flow measurements, balloons and landers to assess the atmosphere and global surface composition, and eventual sample return.

Earth

The Space Age was a major boost to the study of terrestrial geology and geophysics. Advanced sensors that were products of the space and arms races allowed the perfection of techniques to study the Earth's interior in detail. Orbital remote sensing techniques provided global views of the Earth over wide spectral ranges and at very high spatial and spectral resolutions (e.g., the Landsat series, the first of which was launched in 1972). Satellite measurements provided evidence for crustal movements measured in centimeters (e.g., Lageos, launched in 1976, and its derivatives). Mapping of the sea floor for largely military reasons provided tremendous insight into the structure and morphology of the other two-thirds of the Earth's surface. The global view of this planet derived from the plate tectonics revolution and from seeing the Earth as a planet from space irrevocably changed our view of our own home planet. The Earth became known as a dynamic, changing planet, one that erased its early record as it created its new one. And in parallel, planetary exploration was revealing the nature of those missing chapters from the first half of Earth history.

The morphology and geology of the vast majority of the Earth's surface is concealed by liquid water and vegetation, and the atmosphere, hydrosphere, and biosphere are significant agents in the modification of the solid crust beneath us. The exploration of the ocean basins revealed their basic geologic and morphologic differences from the continents and paved the way for the development of the theory of plate tectonics. The Earth's surface is divided into a series of rigid lithospheric plates, and the formation, lateral movement and interaction, and destruction of these plates is responsible for most of the Earth's large-scale structural and topographic features. Plates collide to produce folded mountain belts and deep trenches (where plates flex, subduct into the mantle, and are destroyed). Continental rift valleys and vast plateaus of basalts accompany plate breakup. Strings of volcanoes appear where plates are consumed, and new crust forms at the gradually separating mid-oceanic ridges.

The distribution of terrestrial meteorite craters suggests that the ancient rocks of the Earth's surface (the 10% comprising the continental Precambrian shields) are the nearest terrestrial analog to cratered terrain (Figure 18), even though the density of craters is much less and the age much younger than for this surface unit on other planets. The basaltic plains of the ocean floor are the most pervasive terrain unit on the Earth, although their mode of origin is different than that of the basaltic plains on other terrestrial planets. These units are among the youngest of Earth's rocks, formed within the last 200 million years. Separate phases of volcanism have built broad basaltic flood plateaus and conical mountains over the last 65 million years. Polar caps like those on Mars occur on Earth, and they wax and wane with seasons and longer-term climatic fluctuations. Unlike the Moon, which has remained largely unchanged for the last half of Solar System history, the Earth has a surface constantly in motion, with the positions and relative abundances of terrain units constantly changing (Figure 18). Thus, the processes that are shaping the surface of the Earth can be studied as they occur. This high level of activity means, however, that a very significant fraction of the record of the history of the Earth has been destroyed. If we look at the percentage of the Earth's surface remaining from the first half of Solar System history, it is clear that most of that record has been destroyed (Figure 18).

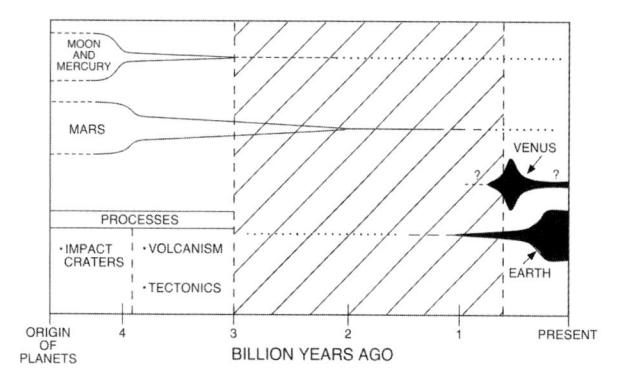

Figure 18 The geologic history of the terrestrial planets, showing the approximate distribution of ages of surface units presently exposed on planetary surfaces. On Earth, the very young ocean basins make up about two-thirds of the surface, while much of the rest of the exposed rocks are Phanerozoic platform sediments surrounding ancient continental cores. Few significant outcroppings of rock units from the first third of Earth's history are exposed. On the Moon, Mars, and Mercury, the record is dominated by heavily cratered terrain dating to the first third of planetary history, and volcanic resurfacing in early to mid history. On Venus, surface units are dominated by the regional plains, and the crater retention age is very young, although plate tectonics is not operating at present.

The present nature and structure of the Earth's interior is very well known relative to the other terrestrial planetary bodies (Anderson 1999). The Earth presently has a substantial intrinsic dipolar magnetic field due to dynamo activity in a convecting and conducting core, a field that has reversed its polarity abruptly over apparently random periods. Not only is there a global network of seismometers, but the extremely high level of activity on Earth means that there are ample seismic events that allow us to probe the interior. Samples from the crust and interior provide important information on differentiation and fractionation. The Earth's core consists of iron and admixed nickel and sulfur, and is about half the radius of the planet, with an inner solid portion and an outer liquid layer. The silicate-rich mantle is subdivided into an outer layer, extending to a depth of about 700 km, and an inner layer dominated by higher-density phases of mantle minerals. Two-thirds of the crustal surface area is made up of young, thin basaltic secondary oceanic crust, and the other third is made up of a much thicker and older tertiary continental crust. Although superficially similar to the lunar highlands and maria, the Earth's three-dimensional crustal structure is strikingly different.

The thermal structure is a key to the present dominant mechanism of heat loss (Figure 8). The outer thermal boundary layer, or lithosphere, averages about 100 km thick. Although it is relatively rigid, it is continuously laterally moving, being created at divergent boundaries, thickening progressively with age, and being flexed and subducted at convergent plate boundaries. Thus the plate tectonic process on the Earth provides a virtual conveyor belt of heat loss, in contrast to the smaller terrestrial planetary bodies whose lithospheres conductively thicken with time (Figure 8). Crustal spreading and lithospheric recycling are amazingly efficient; if present rates are typical of the last 4 billion years, the oceanic crust would have been renewed at least twenty times. Indeed, detailed studies of the eroded cores of ancient mountain belts show that they mark the location of earlier collision and subduction zones, indicating that the process has been repeated many times as rifting opened new oceans, leading to spreading, subduction, continental collision, and closure. It is tempting to think that plate tectonics has dominated throughout the entire history of Earth. The lithosphere overlies a partially molten layer (the asthenosphere) formed when the increase in temperature and pressure with depth temporarily cross into the region where upper mantle materials melt. This layer, typically at depths of between about 100 and 125 km, is partly responsible for the rapid lateral movement of lithospheric plates, and is a significant source of melt for surface volcanism.

Although manifestations of plate tectonic activity dominate the surface, evidence for mantle thermal anomalies are observed in the volcanoes produced over hot spots, or mantle plumes. As the resulting edifices grow, they load and flex the lithosphere in a manner similar to the maria loading on the Moon; on the Earth the rapidly laterally migrating lithosphere moves the volcano away from the source of the anomaly, resulting in the hot-spot traces seen in volcanic island chains. The sources of these plumes appear to be instabilities that develop in the thermal layer at the core/mantle boundary. Indeed, occasionally in Earth history, unusually large instabilities have produced megaplumes, vast outpourings of lava, and major modifications to the atmosphere, plate recycling patterns, oceanic chemistry, the biota, and the magnetic field (e.g., Coffin and Eldholm 1994).

Interior differentiation began early and continues today. Core formation and related mantle differentiation occurred in the earliest part of Earth history, possibly coincident with the late stages of accretion. No record of primary crust remains on the Earth and secondary crust produced before about 3.8 billion years ago has not been preserved. The earliest evidence of tertiary crust (the reworked primary and secondary crust seen in continents; Taylor 1989) at about this time suggests that no stable continents had emerged until over half a billion years after the Earth's formation. Continental crust has grown continuously in areal extent since that time, with a possible peak at about 3.0 billion years ago, and actual growth rates are a matter of controversy. Despite ample time and apparent mixing by convection and plate recycling, isotopically distinct volcanic source regions still persist in the mantle after billions of years. Still unknown is whether these represent chemical layering or smaller-scale heterogeneities.

Early heat sources for the Earth include accretion, with an unknown addition from the impact of the Mars-sized object thought to have produced the Moon, and core formation, which together suggest that early Earth was characterized by a vigorously convecting mantle and thin lithosphere. How did the Earth deal with this high heat flux in this early period? Was the planet completely covered by very rapidly moving oceanic crust, continuously recycling itself at very high rates? Did the high impact flux typical of this time play a role in crustal recycling, perhaps producing an early primary crust like the Moon? Or was there some sort of catastrophic event that caused the crust to founder during the first billion years, setting the stage for plate tectonics? Although plate tectonics is elegant in its global implications and simplicity, could the emerging picture of the surface and interior of Venus be telling us that early Earth history may have been characterized by catastrophism? None of these suggestions is completely satisfactory, and the history of the Earth in its infancy remains elusive. Indeed, observations of earlier Earth history (Nisbet, 1987) show that there are major phases and events that have not been repeated in later geologic history (e.g., a major phase of continent-building, a period of emplacement of iron-rich volcanic rocks known as komatiites, and a period of emplacement of feldspar-rich rocks known as anorthosites).

Whatever the heat loss mechanisms operating in early Earth history, the lithosphere certainly thickened with time as the planet cooled. After the first billion years, this was accompanied by increasingly larger portions of stable continental crust (Condie 1994, Goodwin 1991, Meissner 1986). Although plate tectonics is understood in principle, the process as we see it today requires a relatively rigid lithosphere several tens of kilometers thick, a condition likely to have characterized the Earth for the past several billion years. There is no consensus, however, on how or when plate tectonics might have been initiated. The first third of Earth history is very poorly known, but the other terrestrial planets offer evidence that core formation, high impact fluxes, early primary and secondary crustal formation, and thin lithospheres were characteristic of this time. When plate tectonics emerged, the Earth set out on a course of evolution that distinguishes it from any other planetary body observed to date. Lithospheric plate recycling, localization of geologic activity along plate boundaries, efficient mantle mixing, the formation of tertiary continental crust, and the production of vast quantities of secondary crust compared to the smaller terrestrial planetary bodies are unique to the Earth.

THEMES IN COMPARATIVE PLANETOLOGY

We return now to the questions we raised initially. What has the Space Age, the era of planetary exploration of the last half of the twentieth century, told us about these "organisms" called terrestrial planetary bodies? Beneath the surficial variations in geological features and crustal composition, what were the processes that were responsible for the internal activity of these bodies, their origin and evolution, and their present state? What are the basic themes and processes that help to explain the observed characteristics? What are the major factors that cause the differences in the interiors and geologic histories of the planets? How much of a planet's history is predetermined by its "genetic" makeup, or starting conditions, and how much is determined by its "environment," or later history? Of course, substantial answers to these questions will not be known until well into the present millennium, when sophisticated geophysical networks have determined the internal structure and are monitoring the conditions on all the terrestrial planetary bodies, and landers routinely make sophisticated *in situ* geochemical measurements and return samples to laboratories on Earth. Very significant progress towards these goals has been made, however (Head and Solomon 1981), and clear themes are beginning to emerge.

Initial chemical makeup

An important part of the "genetic" makeup of a planet is its initial compositional characteristics. The temperature/pressure distribution that must have existed during the final stages of collapse of the solar nebula and the formation of the planets must have insured significant sorting of volatile and less-volatile components of planetary starting materials as a function of distance from the proto-Sun. But further observations and more sophisticated theory have shown that there was very likely a significant amount of mixing of planetesimals from different parts of the Solar System during the rather chaotic period of the final accretion of the planets. And then there were the late-stage "singular" events, such as the impact of a Mars-sized object with the proto-Earth to produce the Moon, and the possible stripping of a low-density crust from Mercury by a very large impact event. So a challenge for the future lies in sorting out these several factors that clearly determine planetary starting conditions, and then tracking their influence on subsequent planetary evolution. For example, an unanswered – and almost unaddressed – question is, "How much of the Earth's seemingly 'unique' history (e.g., oceans, plate tectonics, life) was predetermined by the consequences of an impact of a Mars-sized object to form the Moon in the Earth's earliest history?" Could this fundamental event in the history of Earth explain the differences between Venus and Earth, so similar in so many other ways?

Differentiation and chemical layering

Once they had formed, temperature, compositional, and density variations caused planets to begin to undergo differentiation into internal layers of significant proportions (core, mantle, and crust), a process that has continued throughout their history. Although all the larger terrestrial planets underwent this process, the timing and duration of the formation of these layers differed considerably. Core formation appears to have occurred rapidly and earliest. The amount of gravitational potential energy released likely caused significant surface melting on the larger planets, but the geological record of this melting is obscured by impact crater deposits. Although the mantle remaining after core formation contemporaneously becomes a distinct layer, it continues to evolve throughout geologic time. Mantle differentiation continues as heat in the interior causes convective mixing and partial melting of the mantle, and the ascent of the hotter, less-dense melt products toward the surface to form crustal materials. And of course, changes in rates of heat production and patterns of heat loss with time help to modulate mantle differentiation.

Crustal formation and evolution

The outermost layer, the crust, has a very wide range of ages, rates, and styles of formation (Taylor 1989). Primary crust, as seen on the Moon in the form of the highlands, can result from melting associated with the impact energy of late heavy bombardment. The fate of primary crusts formed in this early period on Earth and Venus is currently unknown. Secondary crust, formed by partial melting of the mantle, can take several forms. The Lunar maria represent mantle partial melts that have been emplaced on the globally continuous highland primary crust, over a very small percentage of the surface, in the first half of Solar System history. Martian volcanic plains have similar origins, but their emplacement continued well into the second half of Solar System history and, in contrast to the Moon, secondary crust may comprise the vast majority of the thickness of the crust. In addition, phase changes in the mantle of Mars related to its size, compared to Earth, apparently resulted in the localization of upwellings into a very small number of focused centers, concentrating a significant part of secondary crustal formation in Tharsis and Elysium. The Earth's seafloor represents secondary crust that is very young, thin, actively forming today, and (in contrast to vertical crustal accretion typical of the Moon and Mars) moves laterally, much as a conveyor belt, and is thrust back into the mantle at subduction zones. Earth's secondary crust forms at such prodigious rates that the ocean floors are thought to have been renewed possibly as many as 20 times in the history of the Earth. Venus is covered with secondary crust whose age is also only a very small percentage of the total history of the planet. In contrast to the Earth, it appears to have formed by vertical crustal accretion late in Venus's history, by potentially catastrophic processes. Tertiary crust, formed by the reworking of primary and secondary crust, is best represented by the Earth's continental crust which has been forming for billions of years by erosion, intrusion, and tectonic reworking. The tessera of Venus and their associated deposits are thought by some scientists to be another example of tertiary crust.

Thermal and mechanical layering

Equally important to internal compositional layers are internal thermal structure and layering, sources of heat to determine and maintain them, their influence on the state of materials, and their changes with time. A major challenge for the future is to deconvolve the complex signal of the heat sources for a planet in its earliest history, which may include such diverse factors as accretional energy, position relative to the Sun, electromagnetic heating, core formation and other density instabilities, late-stage large impacts, short and long-term radioactive decay of elements, and tidal interactions. Once a planetary body has emerged from the first few hundred million years of its history, it will have acquired internal thermal layering, and convection patterns, to deal with these heat sources. An internal thermal gradient immediately evolves, depending on the total amount of heat, internal compositional layering, and the behavior of materials under different temperature/pressure conditions. Internal thermal boundary layers (e.g., at the core/mantle boundary) form and evolve, and convection patterns (localized, as in mantle plumes, or larger-scale) change.

The outer thermal boundary layer, the lithosphere, represents the interface between the interior and space, to which heat is lost, and its initial configuration is crucial to the subsequent evolution of the planet (Figure 8). The smaller terrestrial planetary bodies (the Moon, Mercury, and Mars) formed a globally continuous lithosphere in their earliest history and became "one-plate" planets, losing heat largely by conduction throughout the rest of their history. The large surface area to volume ratio of these bodies seems to be a reasonable explanation for this, with the well-known "radiator" effect causing a monotonic loss of heat as a function of time, a thickening of the lithosphere, and as an added benefit, the preservation of much of the record of the earlier history of the Solar System (Figure 18). This trend also had an influence on the surface volcanic record. As lithospheric thickness increased, and the mantle cooled efficiently, the formation of magmas became more difficult and their transport to the surface through this thick rigid layer was increasingly unlikely.

The Earth, however, adopted a different mode of heat transfer: seafloor spreading and plate tectonics (Figure 8). At divergent boundaries, new crust and lithosphere constantly forms. It then moves laterally away and cools conductively, and finally is swept back into the mantle at subduction zones. A small percentage of the heat loss is caused by segregation of heat-producing elements into the continental crust. This very efficient seafloor spreading, heat loss mechanism appears to have been operating on Earth for at least several billion years. How it originated, however, is currently unknown. Clearly the small thickness of the outer thermal boundary layer, relative to the total radius of the Earth, favors fracturing and potential destruction of the lithosphere, in contrast to the situation on the Moon, Mars, and Mercury. But how did plate recycling start?

Earth and its nearest planetary neighbor, Venus, are of approximately the same size and density, and have a surface geology that is comparable in average age (Figure 18). The unusual surface geology, the nature of the cratering record, and the presently globally continuous lithosphere (no active plate tectonics), however, all point to a geological and thermal history very different than that of the Earth. The geologic record has been interpreted by many workers to indicate that the surface has been catastrophically renewed by tectonism and volcanism in its recent geologic past. Venus, then, may indicate that there are exceptions to the examples of monotonic heat loss represented by the one-plate planets, and the "equilibrium" heat loss represented by Earth. Venus may undergo periods of episodic and globally catastrophic heat loss, separated by intervals of low heat loss and geological quiescence. Such an episodic heat loss history and thermal evolution is now thought more plausible, given several theoretical scenarios for the internal distribution of temperature and composition and their evolution with time. The implications of such events, however, are just now being considered. What is the influence of such an event on the atmosphere, its temperature, and the effects that increased surface temperatures would have on the deformation of rocks? Recent results suggest that the atmospheric and tectonic effects may be substantial (Bullock and Grinspoon 1996, Solomon *et al.* 1999). Have similar episodic and catastrophic changes occurred in Earth's past history? Planetary scientists are beginning to address this heretical view (e.g., Herrick and Parmentier 1994, Weinstein, 1993). And of course, the big question: Could such an event have been responsible for the initiation of subduction and plate tectonics early in Earth history?

INSIGHT INTO THE PAST, AND A VIEW OF THE FUTURE

The Space Age, with its remarkable ingenuity and advances in technology, has been responsible for the changing of our perception of planets from solely astronomical objects to dynamic geological and geophysical entities. In addition to the themes described above, the new perspectives provided by planetary exploration have given us major insight into our own home planet, Earth, and has raised a host of new questions and thoughts about the missing chapters of its history.

The record of the first half of Solar System history preserved on other planets, and the different evolutionary paths taken by these bodies, ultimately hold the keys to understanding the formative years of Earth. We now know that the following themes must be considered in the reconstruction of this hazy period:

- formation of the Earth from accretion of planetesimals (Melosh 1997)
- derivation of the Moon from the Earth as the result of a gigantic (Mars-sized) impact event
- late addition of a large amount of material from elsewhere in the Solar System
- stripping of the early atmosphere of the Earth, perhaps several times
- loss from the early Earth of a considerable amount of the solid upper layers
- massive changes in the internal constitution and thermal structure of the early Earth
- a continuing high impact flux subsequent to the Moon-forming event
- formation of large impact basins excavating deep into the planet and spreading ejecta widely, influencing the atmosphere and any biota
- delivery of rocks (and any available microbes) from the surfaces of other planetary bodies to Earth as meteorites
- launch of rocks (and any available microbes) from the Earth's surface to other planetary bodies as meteorites
- effusion of large volcanic outpourings over short periods of time, influencing the atmosphere, hydrosphere, and biosphere
- possible episodic (not just monotonic) heat loss from the interior, regional and perhaps global in scale (this would influence the atmosphere and surface temperatures, and perhaps the tectonic style)
- internal density instabilities causing global changes in volcanism and tectonism, and possible resurfacing of the entire planet
- planetary contraction due to thermal evolution (cooling or phase changes)
- a continuing role for impact events throughout the history of the Earth

- potential large-scale modification of the atmosphere and the biota at numerous but random times.

In the coming decades, these concepts derived from the last forty years of Solar System exploration will be folded into our ongoing reconstruction of the missing chapters of Earth history to produce a radically different picture of the formative years of our home planet. In the words of T. S. Elliot, "We will not cease from exploration, and the end of all of our exploring will be to arrive where we started and know the place for the first time."

And what about the future of the Earth? The lithosphere has thickened with time, and some scientists predict that plate tectonics will cease and the Earth will become a one-plate planet a few hundred million years from now (Takahashi 1990). And what will happen then? Will the Earth's surface become a stable recorder of impact events, like that of the Moon, Mars, and Mercury? Or will the buildup of heat in the interior lead to catastrophic overturn, as is thought to have happened on Venus? Solar System exploration in the Space Age of the last part of the twentieth century has created an emerging picture of the surfaces and interiors of the terrestrial planetary bodies. This perspective, whether related to the importance of impact events in Earth's geological and biological history, or to the origin and fate of plate tectonics, has provided us with a comparative basis on which to understand our own home planet. And ongoing and future exploration and perspectives during this time period will place our Solar System in the context of all of the others that are now known to exist around other stars, and the many more soon to be discovered (Goldsmith 1997). Similar scales of development in astrobiology will provide fundamental answers and the ultimate perspective.

In this first year of this third millennium, space has become permanently inhabited by humans in low Earth orbit. In the coming decades we will be returning samples from other planetary bodies, establishing geophysical networks and mobile geological laboratories, reaching new plateaus of understanding, and exploring new dimensions that we cannot even imagine today. Given the perspective of the events of the first fifty years of the Space Age, it is inevitable that humans will soon be back on the Moon, and will go on to Mars. And the women and men who will make these trips have already been born.

Acknowledgments

Thanks are extended to those many individuals who have enabled my participation in the exploration of the Solar System. Special thanks go to David R. Scott, John Young, and Jack Schmitt for taking me along for the ride, and to Noel Hinners, Farouk El Baz, Gordon Swann, and William Muehlberger for making it possible to explore the Moon. Tim Mutch taught me to keep asking why, and to have the resolve to get the answers. Thanks also to Anne Cote, Peter Neivert, and Nancy Carroll for their assistance in preparation of the manuscript.

REFERENCES

An italic reference of the form *Beatty et al. (1999)* indicates a multiauthor work which is listed in the Further Reading section that follows this list of references.

Acuña, M.H., Connerney, J.E.P., Ness, N.F., Lin, R.P., Mitchell, D., Carlson, C.W., McFadden, J., Anderson, K.A., Rème, H., Mazelle, C., Vignes, D., Wasilewski, P. and Cloutier, P. (1999). Global distribution of crustal magnetization discovered by the Mars Global Surveyor MAG/ER experiment. *Science*, **284**, 790–793.

Anderson, D.L. (1999). Planet Earth. In *Beatty et al. (1999)*, pp. 111–124.

Baker, V.R., Carr, M.H., Gulick, V.C., Williams, C.R. and Marley, M.S. (1992). Channels and valley networks. In *Kieffer et al. (1992)*, pp. 493–522.

Baker, V.R., Strom, R.G., Gulick, V.C., Kargel, J.S., Komatsu, G. and Kale, V.S. (1991). Ancient oceans, ice sheets, and the hydrological cycle on Mars. *Nature*, **352**, 589–594.

Banerdt, W.B., Golombek, M.P. and Tanaka, K.L. (1992). Stress and tectonics on Mars. In *Kieffer et al. (1992)*, pp. 249–297.

Basilevsky, A.T. and Head, J.W. (1995). Global stratigraphy of Venus: Analysis of a random sample of thirty-six test areas. *Earth, Moon and Planets*, **66**, 285–336.

Basilevsky, A.T. and Head, J.W. (1998). The geologic history of Venus: A stratigraphic view. *J. Geophys. Res.*, **103**, 8531–8544.

Basilevsky, A.T. and Head, J.W. (2000). Geologic units on Venus: Evidence for their global correlation. *Planetary and Space Science*, **48**, 75–111.

Basilevsky, A.T., Head, J.W., Schaber, G.G. and Strom, R.G. (1997). The resurfacing history of Venus. In *Bougher et al. (1997)*, pp. 1047–1084.

Basilevsky, A.T., Ivanov, B.A., Burba, G.A., Chernaya, I.M., Kryuchkov, V.P., Nikolaeva, O.V., Campbell D.B. and Ronca L.B. (1987). Impact craters of Venus: A continuation of the analysis of data from the Venera 15 and 16 spacecraft. *J. Geophys. Res.*, **92**, 12869–12901.

Barsukov, V.L., Basilevsky, A.T., Burba, G.A., Bobinna, N.N., Kryuchkov, V.P., Kuzmin, R.O., Nikolaeva, O.V., Pronin, A.A., Ronca, L.B., Chernaya, I.M., Shashkina, V.P., Garanin, A.V., Kushky, E.R., Markov, M.S., Sukhanov, A.L., Kotelnikov, V.A., Rzhiga, O.N., Petrov, G.M., Alexandrov, Yu.N., Sidorenko, A.I., Bogomolov, A.F., Skrypnik, G.I., Bergman, M.Yu., Kudrin, L.V., Bokshtein, I.M., Kronrod, M.A., Chochia, P.A., Tyuflin, Yu.S., Kadnichansky, S.A. and Akim, E.L. (1986). The geology and geomorphology of the Venus surface as revealed by the radar images obtained by Veneras 15 and 16. In Proceedings of the 16th Lunar and Planetary Science Conference, *J. Geophys. Res.*, **91**, Suppl., D378–D398.

Benz, W., Slattery, W.L. and Cameron, A.G.W. (1988). Collisional stripping of Mercury's mantle. *Icarus*, **74**, 516–528.

Bindschadler, D.L., Schubert, G. and Kaula, W.M. (1992). Coldspots and hotspots: Global tectonic and mantle dynamics of Venus. *J. Geophys. Res.*, **97**, 16135–16335.

Bullock, M.A. and Grinspoon, D.H. (1996). The stability of climate on Venus. *J. Geophys. Res.*, **101**, 7521–7530.

Carr, M.H. (1996). *Water on Mars*, Oxford University Press, New York.

Carr, M.H. (1999). Mars. In *Beatty et al. (1999)*, pp. 141–156.

Carr, M.H., Saunders, R.S., Strom, R.G. and Wilhelms, D.E. (1984). *The Geology of the Terrestrial Planets*, National Aeronautics and Space Administration, Washington, DC.

Chapman, C.R. (1988). Mercury: Introduction to an end-member planet. In *Vilas et al. (1988)*, pp. 1–23.

Clifford, S.M. (1993). A model for the hydrologic and climatic behavior of water on Mars. *J. Geophys. Res.*, **98**, 10973–11016.

Coffin, M.F. and Eldholm, O. (1994). Large igneous provinces: Crustal structure, dimensions, and external consequences. *Review of Geophysics*, **32**, 1–36.

Colin, L. (1983). Basic facts about Venus. In *Hunten et al. (1983)*, pp. 10–26.

Connerney, J.E.P., Acuña, M.H., Wasilewski, P.J., Ness, N.F., Rème, H., Mazelle, C., Vignes, D., Lin, R.P., Mitchell, D.L. and Cloutier, P.A. (1999). Magnetic lineations in the ancient crust of Mars. *Science*, **284**, 794–800.

Connerney, J.E.P. and Ness, N.F. (1988). Mercury's magnetic field and interior. In *Vilas et al. (1988)*, pp. 494–513.

Condie, K.C. (ed.) (1994). *Archean Crustal Evolution*, Elsevier, Amsterdam.

Cruikshank, D.P. (1983). The development of studies of Venus. In *Hunten et al. (1983)*, pp. 1–9.

Donahue, T.M. and Pollack, J.B. (1983). Origin and evolution of the atmosphere of Venus. In *Hunten et al. (1983)*, pp. 1003–1036.

Esposito, P.B., Banderst, W.B., Lindal, G.F., Sjogren, W.L., Slade, M.A., Bills, B.G., Smith, D.E. and Balmino, G. (1992). Gravity and Topography. In *Kieffer et al. (1992)*, pp. 209–249.

Florenskiy, K.P., Bazilevskiy, A.T., Burba, G.A., Nikolayeva, O.V., Pronin, A.A., Selivanov, A.S., Narayeva, M.K., Panfilov, A.S. and Chemodanov, V.P. (1983). Panorama of Venera 9 and 10 landing sites. In *Hunten et al. (1983)*, pp. 137–153.

Goldsmith, D. (1997). *Worlds Unnumbered: The Search for Extrasolar Planets*, University Science Books, Sausalito, CA.

Goodwin, A.M. (1991). *Precambrian Geology*, Academic Press, San Diego, CA.

Greeley, R., Bridges, N.T., Crown, D.A., Crumpler, L., Fagents, S.A., Mouginis-Mark, P.J. and Zimbelman, J.R. (2000). Volcanism on the red planet: Mars. In *Zimbelman and Gregg (2000)*, pp. 75–112.

Greeley, R., Lancaster, N., Lee, S. and Thomas, P. (1992). Martian aeolian processes, sediments, and features. In *Kieffer et al. (1992)*, pp. 730–766.

Hansen, V.L., Willis, J.J. and Banerdt, W.B. (1997). Tectonic overview and synthesis. In *Bougher et al. (1997)*, pp. 797–844.

Harder, H. (1998). Phase transitions and the three-dimensional planform of thermal convection on the Martian mantle. *J. Geophys. Res.*, **103**, 16775–16797.

Harder, H. and Christensen, U.R. (1996). A one-plume model of martian mantle convection. *Nature*, **380**, 507–509.

Harmon, J.K. and Slade, M.A. (1992). Radar mapping of Mercury: Full-disk images and polar anomalies. *Science*, **258**, 640–642.

Hartmann, W.K. and Davis, D.R. (1975). Satellite-sized planetesimals and lunar origin. *Icarus*, **24**, 504–515.

Head, J.W. III (1999). Surfaces and interiors of the terrestrial planets. In *Beatty et al. (1999)*, pp. 157–174.

Head, J.W., Crumpler, L.S., Aubele, J.C., Guest, J.E. and Saunders, R.S. (1992). Venus volcanism: Classification of volcanic features and structures, associations, and global distribution from Magellan data. *J. Geophys. Res.*, **97**, 13153–13197.

Head, J.W. and Solomon, S.C. (1981). Tectonic evolution of the terrestrial, planets. *Science*, **213**, 62–76.

Head, J.W. and Wilson, L. (1992). Lunar mare volcanism: Stratigraphy, eruption conditions, and the evolution of secondary crusts. *Geochimica et Cosmochimica Acta*, **56**, 2155–2175.

Head, J.W. III, Wilson, L., Robinson, M., Hiesinger, H., Weitz C. and Yingst, A. (2000). Moon and Mercury: Volcanism in early planetary history. In *Zimbelman and Gregg (2000)*, pp. 143–178.

Herrick, D.L. and Parmentier, E.M. (1994). Episodic large-scale overturn of two-layer mantles in terrestrial planets. *J. Geophys. Res.*, **99**, 2053–2062.

Hess, P.C. and Parmentier, E.M. (1995). A model for the thermal and chemical evolution of the Moon's interior: Implications for the onset of mare volcanism. *Earth and Planetary Science Letters*, **134**, 501–514.

Kieffer, H.H., Jakosky, B.M. and Snyder, C.W. (1992). The planet Mars: From antiquity to the present. In *Kieffer et al. (1992)*, pp. 1–33.

Lewis, J.S. (1988). Origin and composition of Mercury. In *Vilas et al. (1988)*, pp. 651–666.

Longhi, J., Knittle, E., Holloway, J.R. and Wänke, H. (1992). The bulk composition, mineralogy and internal structure of Mars. In *Kieffer et al. (1992)*, pp. 184–208.

Lucchitta, B.K., McEwen, A.S., Clow, G.D., Geissler, P.E., Singer, R.B., Schultz, R.A. and Squyres, S.W. (1992). The canyon system on Mars. In *Kieffer et al. (1992)*, pp. 453–492.

Martin, L.J., James, P.B., Dolfus, A., Iwasaki, K. and Beish, J.D. (1992). Telescopic observations: Visual, photographic, polarimetric. In *Kieffer et al. (1992)*, pp. 34–70.

McGill, G.E., Warner, J.L., Malin, M.C., Arvidson, R.E., Eliason, E., Nozette, S. and Reasenberg, R.D. (1983). Topography, surface properties, and tectonic evolution. In *Hunten et al. (1983)*, pp. 69–130.

McKay, C.P., Mancinelli, R.L., Stoker, C.R. and Wharton, R.A. Jr. (1992). The possibility of life on Mars during a water-rich past. In *Kieffer et al. (1992)*, pp. 1234–1248.

McSween, H.Y. Jr. and Treiman, A.H. (1998). Martian meteorites. In *Papike (1998)*, pp. 6-1–6-53.

Meissner, R. (1986). *The Continental Crust: A Geophysical Approach*, Academic Press, San Diego, CA.

Melosh, H.J. (ed.) (1997). *Origins of Planets and Life*, Annual Reviews Inc., Palo Alto, CA.

Melosh, H.J. and McKinnon, W.B. (1988). The tectonics of Mercury. In *Vilas et al. (1988)*, pp. 374–400.

Moroz, V.I. (1983). Summary of preliminary results of the Venera 13 and Venera 14 missions. In *Hunten et al. (1983)*, pp. 45–68.

Morrison, D. (1999). Exploring the solar system. In *Beatty et al. (1999)*, pp. 1–12.

Neumann, G.A., Zuber, M.T., Smith, D.E. and Lemoine, F.G. (1996). The lunar crust: Global structure and signature of major basins. *J. Geophys. Res.*, **101**, 16841–16863.

Nisbet, E.G. (1987). *The Young Earth: An Introduction to Archaean Geology*, Allen & Unwin, Boston.

Owen, T. (1992). The composition and early history of the atmosphere of Mars. In *Kieffer et al. (1992)*, pp. 818–834.

Papike, J.J., Ryder, G. and Shearer, C.K. (1998). Lunar samples. In *Papike (1998)*, pp. 5-1–5-234.

Parker, T.J., Gorsline, D.S., Saunders, R.S., Pieri, D.C. and Schneeberger, D.M. (1993). Coastal geomorphology of the Martian northern plains. *J. Geophys. Res.*, **98**, 11061–11078.

Parker, T.J., Saunders, R.S. and Schneeberger, D.M. (1989). Transitional morphology in West Deuteronilus Mensae, Mars: Implications for modification of the lowland/upland boundary. *Icarus*, **82**, 111–145.

Parmentier, E.M. and Hess, P.C. (1992). Chemical differentiation of a convecting planetary interior: Consequences for a one plate planet such as Venus. *Geophys. Res. Lett.*, **19**, 2015–2018.

Phillips, R.J., Johnson, C.L., Mackwell, S.L., Morgan, P., Sandwell, D.T. and Zuber, M.T. (1997). Lithospheric mechanics and dynamics of Venus. In *Bougher et al. (1997)*, pp. 1163–1204.

Phillips, R.J., Raubertas, R.F., Arvidson, R.E., Sarkar, I.C., Herrick, R.R., Izenberg, N. and Grimm, R.E. (1992). Impact craters and Venus resurfacing history. *J. Geophys. Res.*, **92**, 15923–15948.

Pieters, C.M., Head, J.W., Gaddis, L., Duke M. and Jolliff, B. (2001). Rock types of South Pole–Aitken Basin and distribution of ancient basalts. *J. Geophys. Res.*, **106**, 28001–28022.

Robinson, M.S. and Lucey, P.G. (1997). Recalibrated Mariner 10 color mosaics: Implications for Mercurian volcanism. *Science*, **275**, 197–200.

Saunders, R.S. (1999). Venus. In *Beatty et al. (1999)*, pp. 97–110.

Schaber, G.G., Strom, R.G., Moore, H.J., Soderblom, L.A., Kirk, R.L., Chadwick, D.J., Dawson, D.D., Gaddis, L.R., Boyce, J.M. and Russell, J. (1992). Geology and distribution of impact craters on Venus: What are they telling us? *J. Geophys. Res.*, **92**, 13257–13301.

Schubert, G., Anderson, C. and Goldman, P. (1995). Mantle plume interaction with an endothermic phase change. *J. Geophys. Res.*, **100**, 8245–8256.

Schubert, G., Bercovici, D. and Glatzmaier, G.A. (1990). Mantle dynamics in Mars and Venus: Influence of an immobile lithosphere on three-dimensional mantle convection. *J. Geophys. Res.*, **95**, 14105–14129.

Schubert, G., Ross, M.N., Stevenson, D.J. and Spohn, T. (1988). Mercury's thermal history and the generation of its magnetic field. In *Vilas et al. (1988)*, pp. 429–460.

Schubert, G. Solomatov, V.S., Tackley, P.J. and Turcotte, D.L. (1997). Mantle convection and the thermal evolution of Venus. In *Bougher et al. (1997)*, pp. 1245–1287.

Schubert, G. Solomon, S.C., Turcotte, D.L., Drake, M.J. and Sleep, N.H. (1992). Origin and thermal evolution of Mars. In *Kieffer et al. (1992)*, pp. 147–183.

Sjogren, W.L, Banerdt, W.B., Chodas, P.W., Konopliv, A.S., Balmino, G., Barriot, J.P., Arkani-Hamed, J., Colvin, T.R. and Davies, M.E. (1997). The Venus gravity field and other geodetic paramenters. In *Bougher et al. (1997)*, pp. 1125–1161.

Sleep, N.H. (1994). Martian plate tectonics. *J. Geophys. Res.*, **99**, 5639–5655.

Smith, D.E., Zuber, M.T., Frey, H.V., Garvin, J.B., Head, J.W., Muhleman, D.O., Pettengill, G.H., Phillips, R.J., Solomon, S.C., Zwally, H.J., Banerdt, W.B. and Duxbury, T.C. (1998). Topography of the northern hemisphere of Mars from the Mars Orbiter Laser Altimeter. *Science*, **279**, 1686–1691.

Smith, D.E., Zuber, M.T., Solomon, S.C., Phillips, R.J., Head, J.W., Garvin, J.B., Banerdt, W.B., Muhleman, D.O., Pettengill, G.H., Neumann, G.A., Lemoine, F.G., Abshire, J.B., Aharonson, O., Brown, C.D., Hauck, S.A., Ivanov, A.B., McGovern, P.J., Zwally, H.J. and Duxbury, T.C. (1999). The global topography of Mars and implications for surface evolution. *Science*, **284**, 1495–1503.

Snyder, C.W. and Moroz, V.I. (1992). Spacecraft exploration of Mars. In *Kieffer et al. (1992)*, pp. 71–119.

Solomatov, V.V. and Moresi, L-N. (1996). Stagnant lid convection on Venus. *J. Geophys. Res.*, **101**, 4737–4753.

Solomon, S.C. (1976). Some aspects of core formation in Mercury. *Icarus*, **28**, 509–521.

Solomon, S.C. (1977). The relationship between crustal tectonics and internal evolution in the Moon and Mercury. *Physics of the Earth and Planetary Interiors*, **15**, 135–145.

Solomon, S.C. (1978). On volcanism and thermal histories on one-plate planets. *Geophys. Res. Lett.*, **5**, 461–464.

Solomon, S.C. (1993). The geophysics of Venus. *Physics Today*, **46**, 48–55.

Solomon, S.C., Bullock, M.A. and Grinspoon, D.H. (1999). Climate change as a regulator of tectonics on Venus. *Science*, **286**, 87–90.

Solomon, S.C. and Head, J.W. (1980). Lunar mascon basins: Lava filling, tectonics, and evolution of the lithosphere. *Reviews of Geophysics and Space Physics*, **18**, 107–141.

Solomon, S.C. and Head, J.W. (1982). Mechanisms for lithospheric heat transport on Venus: Implications for tectonic style and volcanism. *J. Geophys Res.*, **87**, 9236–9246.

Solomon, S.C., Smrekar, S.E., Bindschadler, D.L., Grimm, R.E., Kaula, W.M., McGill, G.E., Phillips, R.J., Saunders, R.S., Schubert, G. Squyres, S.W. and Stofan, E.R. (1992). Venus tectonics: An overview of Magellan observations, *J. Geophys. Res.*, **97**, 13199–13255.

Spudis, P.D. (1993). *The Geology of Multi-Ring Impact Basins: The Moon and Other Planets*. Cambridge University Press.

Spudis, P.D. (1999). The Moon. In *Beatty et al. (1999)*, pp. 125–140.

Spudis, P.D. and Guest, J.E. (1988). Stratigraphy and geologic history of Mercury. In *Vilas et al. (1998)*, pp. 118–164.

Squyres, S.W. Clifford, S.M. Kuzmin, R.O. Zimbelman, J.R. and Costard, F.M. (1992). Ice in the Martian regolith. In *Kieffer et al. (1992)*, pp. 523–556.

Stevenson, D.J. (2000). Planetary science: A space odyssey. *Science*, **287**, 997–1005.

Strom, R.G., Croft, S.K. and Barlow, N.G. (1992). The Martian impact cratering record. In *Kieffer et al. (1992)*, pp. 383–423.

Strom, R.G. and Neukum, G. (1988). The cratering record on Mercury and the origin of impacting objects. In *Vilas et al. (1988)*, pp. 336–373.

Surkov, Y.A. (1983). Studies of Venus rocks by Veneras 8, 9, and 10. In *Hunten et al. (1983)*, pp. 154–158.

Surkov, Y.A., Barsukov, V.L., Moskalyeva, L.P., Kharyukova, V.P. and Kemurdzhian, A.L. (1984). New data on the composition, structure, and properties of Venus rock obtained by Venera 13 and Venera 14. Suppl., Proceedings of the 14th Lunar and Planetary Science Conference, *J. Geophys. Res.*, **89**, 393–402.

Takahashi, E. (1990). Speculations on the Archean mantle: Missing link between komatiite and depleted garnet peridotite. *J. Geophys. Res.*, **96**, 15941–15954.

Tanaka, K.L., Scott, D.H. and Greeley, R. (1992). Global stratigraphy. In *Kieffer et al. (1992)*, pp. 345–382.

Tanaka, K.L., Senske, D.A., Price, M. and Kirk, R.L. (1997). Physiography, geomorphic/geologic mapping and stratigraphy of Venus. In *Bougher et al. (1997)*, pp. 667–694.

Taylor, S.R (1989). Growth of planetary crust. *Tectonophysics*, **161**, 147–156.

Thomas, P.G., Masson, P. and Fleitout, L. (1988). Tectonic history of Mercury. In *Vilas et al. (1988)*, pp. 401–428.

Thomas, P., Squyres, S.W., Herkenhoff, K., Howard, A. and Murray, B. (1992). Polar deposits of Mars. In *Kieffer et al. (1992)*, pp. 767–798.

Turcotte, D.L. (1993). An episodic hypothesis for Venusian tectonics. *J. Geophys. Res.*, **98**, 17061–17068.

Vaniman, D., Dietrich, J., Taylor, G.J. and Heiken, G. (1991). Exploration, samples and recent concepts of the Moon. In *Heiken et al. (1991)*, pp. 5–26.

Vilas, F. (1998). Surface composition of Mercury from reflectance spectrophotometry. In *Vilas et al. (1998)*, pp. 59–76.

Vilas, F. (1999). Mercury. In *Beatty et al. (1999)*, pp. 87–96.

Vityazev, A.V., Pechernikova, G.V. and Safronov, V.S. (1988). Formation of Mercury and removal of its silicate shell. In *Vilas et al. (1988)*, pp. 667–669.

Wasserburg, G.J., Papanastassiou, D.A., Tera, F. and Huneke, J.C. (1977). Outline of a lunar chronology. *Philosophical Transactions of the Royal Society A*, **285**, 7–22.

Weinstein, S.A. (1993). Catastrophic overturns of the Earth's mantle driven by multiple phase changes and internal heat generation. *Geophys. Res. Lett.*, **20**, 102–104.

Wilhelms, D.E. (1987). The geologic history of the Moon. US Geological Survey Professional Paper 1348.

Zuber, M.T., Solomon, S.C., Phillips, R.J., Smith, D.E., Tyler, G.L., Aharonson, O., Balmino, G., Banerdt, W.B., Head, J.W., Johnson, C.L, Lemoine, F.G., McGovern, P.J., Neumann, G.A., Rowlands, D.D. and Shijie Zhong (2000). Internal structure and early thermal evolution of Mars from Mars Global Surveyor topography and gravity. *Science*, **287**, 1788–1793.

FURTHER READING

Beatty, J.K., Petersen, C.C. and Chaikin, A. (eds) (1999). *The New Solar System*, 4th edn, Sky Publishing Corp., Cambridge, MA.

Bougher, S.W., Hunten, D.M. and Phillips, R.J. (eds) (1997). *Venus II – Geology, Geophysics, Atmosphere, and Solar Wind Environment*, University of Arizona Press, Tucson, AZ.

Carr, M.H., Saunders, R.S., Strom, R.G. and Wilhelms, D.E. (1984). *The Geology of the Terrestrial Planets*. National Aeronautics and Space Administration, Washington, DC.

Chaikin, A. (1998). *A Man on the Moon: The Voyages of the Apollo Astronauts*. Penguin, New York.

Harland, D.M. (1999). *Exploring the Moon: The Apollo Expeditions*, Springer Praxis, London.

Hartmann, W.K. (1999). *Moons and Planets*, Wadsworth, New York.

Heiken, G.H., Vaniman, D.T. and French, B.M. (eds) (1991). *Lunar Sourcebook: A User's Guide to the Moon*, Cambridge University Press, New York.

Hunten, D.M., Colin, L., Donahue, T.M. and Moroz, V.I. (eds) (1983). *Venus*, University of Arizona Press, Tucson, AZ.

Kieffer, H.H., Jakosky, B.M., Snyder, C.W. and Matthews, M.S. (eds) (1992). *Mars*, University of Arizona Press, Tucson, AZ.

Murray, B., Malin, M.C. and Greeley, R. (1981). *Earthlike Planets: Surfaces of Mercury, Venus, Earth, Moon, Mars*, W.H. Freeman, San Franciso.

Papike, J.J. (ed.) (1998). *Planetary Materials*, Mineralogical Society of America, Washington, DC.

Shirley, J.H. and Fairbridge, R.W. (eds) (1997). *Encyclopedia of Planetary Sciences*, Chapman & Hall, London.

Spudis, P.D. (1993). *The Geology of Multi-Ring Impact Basins: The Moon and Other Planets*, Cambridge University Press.

Spudis, P.D. (1996). *The Once and Future Moon*, Smithsonian Institution Press, Washington, DC.

Taylor, S.R. (1992). *Solar System Evolution: A New Perspective*, Cambridge University Press, New York.

Vilas, F., Chapman, C.R. and Matthews, M.S. (eds) (1988). *Mercury*, University of Arizona Press, Tucson, AZ.

Weissman, P.R., McFadden, L. and Johnson, T.V. (eds) (1999). *Encyclopedia of the Solar System*, Academic Press, San Diego, CA.

Zimbelman, J.R. and Gregg, T.K.P. (eds) (2000). *Environmental Effects on Volcanic Eruptions: From Deep Oceans to Deep Space*, Kluwer Academic/Plenum, New York.

55

LAURENCE E. NYQUIST,* DONALD D. BOGARD* AND CHI-YU SHIH**

Radiometric chronology of the Moon and Mars

"How old is the Earth?" was a topic of scientific inquiry in the late nineteenth century. In the twentieth century the question became "How old are the Earth and other objects in the Solar System?" Related questions are: "How old is the Solar System?"; "How has the Earth changed over geologic time?"; and "How have the planetary bodies in the Solar System changed over time?" Also, "What has made Earth unique in our Solar System?" With the aid of spacecraft-acquired data and samples, and through the study of lunar and Martian meteorites, we are beginning to answer some of these questions as they relate to the Moon and Mars.

A century ago the age of the Earth was addressed by diverse approaches. These included calculations of how long it would take an initially hot Earth to lose its heat by thermal conduction and radiation to space; how long it would take the oceans to attain their current load of salt, assuming an estimated modern rate of weathering of surface rocks; and estimates of the rate of regression of the Moon away from the Earth. The ages obtained by such methods were often of the order of 10–100 Ma (million years). Unfortunately, these methods were based on faulty assumptions. For example, estimates of the age of the Earth based on observable geological processes failed to account for significant changes in the rate and nature of those processes acting over very long periods of time. Also, calculations of the thermal cooling of the Earth failed to consider the heat generated within the Earth by the natural radioactivity of potassium, uranium, and thorium.

The physical basis for determining the absolute ages of objects within our Solar System in the twentieth century is natural radioactivity, a phenomenon initially discovered in the closing years of the nineteenth century. A collection of radioactive atoms can be thought of as being like a clock, which is running at a constant and measurable speed. But unlike an ordinary clock which tells only the present time, the accumulated daughter products of radioactive decay give a measure of how long the radioactive clock has been running since it was initially "set."

USING NATURAL RADIOACTIVITY TO TELL GEOLOGIC TIME

Historical background

The discovery of natural radioactivity by H. Becquerel (1896) set scientists on the path to quantitative definition of geologic time. In an instance of scientific serendipity, Becquerel found that photographic plates kept near uranium-bearing minerals became darkened. Marie Sklodowska Curie pursued the observation for her PhD dissertation, noting that both uranium and thorium were active, and that two minerals of uranium, pitchblende and chalcolite, had greater radioactivity than uranium itself (Curie 1898). Pursuing the great activity of pitchblende, she and her husband, Pierre Curie, isolated the "new" element polonium from this mineral (Curie and Curie 1898). Next, the Curies, with G. Bèmont, also isolated radium from pitchblende (Curie et al. 1898). Although these shorter-lived daughter elements are far more radioactive than uranium, our interest here remains with uranium, their

* NASA – Johnson Space Center, Houston, TX, USA
** Lockheed–Martin, Houston, TX, USA

radioactive parent. Its decay period, or half-life, was soon found to extend over geologic time, and thus its radioactive decay was recognized as a possible basis for a geologic timescale (Rutherford 1906). For historical completeness, we note that Schmidt (1898) also discovered the radioactivity of thorium contemporaneously with M. Curie. Within a decade, two more long-lived naturally radioactive elements had been discovered which also could be used to tell geologic time: potassium (Campbell and Wood 1906, Campbell 1906) and rubidium (Campbell and Wood 1906, Campbell 1908). The radioactivity of a fourth element currently used for geologic age determinations, samarium, was discovered considerably later (Hevesey *et al.* 1933). Thus, the fundamental scientific discoveries leading to a quantitative geologic timescale were made near the beginning of the twentieth century.

The actual development of rigorous radiometric methods of telling geologic time occupied much of the first half of the century. Geochronology is the subject of a voluminous literature, and of a number of standard texts (e.g. Faure 1986). Here, we hope only to give enough background for the reader to gain an appreciation of the developments that permitted its application to the absolute chronology of the Moon and Mars. In preparing this overview, we rely heavily, but not exclusively, on "benchmark" papers assembled by Harper (1973), and his associated commentary.

The development of geochronology during the first three decades of the twentieth century was intimately intertwined with emerging views of the structure of the atom. The study of radioactive elements showed that atoms of the same chemical element could have different atomic weights. For example, we now know that "natural" uranium of atomic number 92 decays via α-particle emission with a very long half-life to the element of atomic number 90 (thorium), which, after two β-decays, becomes element 92 (uranium) again, but with a much shorter half-life than it had originally. However, its atomic mass is lighter by 4 mass units, the mass of the α-particle. This reasoning was expressed by Soddy (1913–14), who suggested the name isotopes for these two types of uranium. The name stems from the Greek *isos* (equal) and *topos* (place), meaning that two different kinds of atoms occupy the same place in the periodic table of the elements. J.J. Thomson also discovered that atoms of a stable element, neon, could have differing atomic weights (Thomson 1913). Later, F.W. Aston, a student of Thompson, discovered that many other elements also were composed of differing isotopes. Romer (1970) provides a historical account of the multiple contributions by many investigators to the "discovery" of isotopes.

A complete understanding of isotopes was not achieved, however, until the discovery of the neutron by Chadwick (1932). This discovery supported the view of the nucleus as composed of protons and neutrons. Atoms of a given element have a fixed number of protons, called the atomic number. However, atoms of a given element may have a variable number of neutrons. The atomic mass of an atom is approximately equal to the total number of protons and neutrons it contains. These concepts differ from those of Soddy (1913–14) principally in that he envisioned the nucleus to consist of electrons plus protons, and that the algebraic sum of charges in the nucleus gave the atomic number. Nuclei having differing arithmetical sums of particles would have differing atomic masses, but could nevertheless share atomic numbers, constituting his definition of isotopes. Rutherford later suggested that protons and electrons in the nucleus could be so closely bound that they would in effect constitute a new particle, the neutron. This suggestion was confirmed by Chadwick's work.

The modern term nuclide refers to an atom of specific atomic number and specific atomic mass. Radioactive decay results in conversion of a number, P, of radioactive parent nuclides into a number, D, of daughter nuclides. The daughter nuclide most often is an isotope of a different element than the parent, but sometimes branched-decay into more than one type of daughter nuclide also occurs. The activity, or decay rate, of P parent nuclides is

$$\frac{dP}{dt} = -\lambda P \qquad (1)$$

where λ is the decay constant, the probability per unit time that a parent-nuclide will decay. The decay constant is related to the half-life, τ, the time interval required for half the parent nuclides to decay by

$$\lambda = \frac{(\ln 2)}{\tau} \qquad (2)$$

The number of radiogenic daughter nuclides, D^*, produced in the time interval $(t-t_0)$ is given by the equation:

$$D^*(t) = P(t)\,[e^{\lambda(t-t_0)} - 1] \qquad (3)$$

where t_0 is the time when the volume containing the parent nuclides became closed, and the daughter nuclides began to be retained, and t is the time when the numbers of parent and daughter nuclides are measured. In geochronology, the volume of interest is a rock or mineral, and $(t - t_0) = T$, the age of that rock or mineral.

It often is convenient to normalize $D^*(t)$ to the number of nuclides, N, of a non-radiogenic isotope of the same element as the daugther to obtain:

$$\frac{D^*(t)}{N} = \frac{P(t)[e^{\lambda(t-t_0)} - 1]}{N} \qquad (4)$$

In natural systems, a number, D_0, of daughter nuclides may be present at time t_0. In this case, analytical measurements yield not $D^*(t)$, but $D(t) = D^*(t) + D_0$. Thus, in practice, eqn (4) is modified to

$$\frac{D(t)}{N} = \frac{D_0}{N} + \frac{P(t)[e^{\lambda(t-t_0)} - 1]}{N} \qquad (5)$$

which contains the measured quantities directly. Although eqns (4) and (5) are simple modifications of eqn (3), determining which equation best described various real systems played a significant role in the development of geochronology.

Development of U–Pb dating

Not surprisingly, the initial application of the above concepts involved the α-decay of uranium and thorium (e.g. Rutherford 1906). Uranium and thorium decay define separate α-decay series ending with three stable isotopes of lead: ^{206}Pb, ^{207}Pb, and ^{208}Pb. A fourth, minor, stable isotope, ^{204}Pb, is non-radiogenic. The two stable elements resulting from uranium and thorium decay are thus lead and helium. Historically, this gave rise to the "chemical lead method" and the "helium method" of telling geologic time, because only chemical methods were initially available to measure the products of radioactive decay. Thus, both methods initially attempted to apply eqn (3) with the simplifying assumptions that (a) the daughter element was initially absent in the rock or mineral analyzed, and (b) the contribution of thorium decay could be ignored, since with purely chemical means there was no way of telling which portion of the helium or lead came from thorium. These considerations initially led to application of the U–Pb method only to uranium-rich ores and minerals.

Arthur Holmes, in writing the first of several treatises dealing with the age of the Earth (Holmes 1913), placed the helium and lead ages of the day into the geological timescale, placing the oldest geological era, the Archean, at 1400–1600 Ma ago according to the lead method. A younger age for the Archean of about 700 Ma was indicated by the helium method, but was attributed to long-term leakage of some helium from such ancient minerals, a potential problem anticipated by Rutherford (1906). Holmes (1913) wrote: "Radioactive minerals, for the geologist, are clocks wound up at the time of their origin. After a few years' preliminary work, we now are confident that the means of reading these time-keepers is in our possession." Not all geologists of the time agreed with him, and indeed chemists and physicists were still struggling to better understand the phenomenon of radioactivity itself. Much more effort would go into refinement of the methods of telling geologic time.

The possibility of mass spectrographic determination of the isotopic composition of lead was realized within the next decade, and led to significant refinements of the U–Pb method. Aston (1927) reported the first determination of the isotopic composition of lead in collaboration with C.S. Piggot of the Geophysical Laboratory, Washington. Aston's instrument was called a mass spectrograph because it separated positively charged ions according to mass, in a manner similar to that in which a light prism separates sunlight into a spectrum of differing wavelengths, and because a photographic plate was exposed to the impact of electrically accelerated ions. The latter caused ionization of the photographic emulsion on the plate proportional to the number of incident ions. In later instruments, the magnitude of the positive ion current was measured directly. Such instruments are called mass spectrometers and in modern geochronology are used exclusively in preference to mass spectrographs.

Piggot (1928) records how he wrote to Aston, in October 1926, with the suggestion that several problems encountered in the chemical lead method, including the poorly known rate of thorium decay, could be overcome by means of mass spectrograph analysis, which would allow the identification of those lead isotopes produced by uranium decay only. He also wrote that it was he who supplied Aston with the sample of lead tetramethyl used for the analysis reported in the 1927 *Nature* article, and that he was "working up" a sample of radiogenic lead from some very pure Norwegian bröggerite, a mineral containing considerable proportions of uranium and lead, but a very small proportion of thorium. From this sample he hoped to determine directly the uranium-to-lead ratio, "and thereby secure a reliable estimate of its age." The next year, Aston (1929) reported the results of the isotopic analysis of the bröggerite with the comment: "These figures have been communicated to Mr Piggot, and when combined with the analyses of the mineral should enable its age to be fixed with considerable certainty."

Fenner and Piggot (1929) later reported the age of the bröggerite as between 920 and 908 Ma as determined from the ^{206}Pb and (^{206}Pb + ^{207}Pb) abundances, respectively, and 1313 Ma as determined from the ^{208}Pb abundance. Rutherford (1929), in a companion paper to the earlier paper by Aston (1929), had assumed the age to be 1000 Ma. He focused instead on interpreting the spectrographic line corresponding to mass number 207, rather than the mineral age. Rutherford, like Aston (1929), interpreted mass 207 as the end product of actinium decay, and further inferred its ultimate source as an isotope of uranium, of mass 235, which he called actino-uranium. Rutherford estimated "the period of transformation of the new isotope" to be 3.4 Ga. He interpreted this period as an upper limit to the age of the Earth, noting that it was "about twice the age of the oldest known radioactive minerals." In doing so, Rutherford (a) inferred the existence of undiscovered ^{235}U from the existence of ^{207}Pb, (b) estimated the half-life of ^{235}U, (c) utilized the knowledge that odd-mass isotopes are less abundant than the main even-mass ones to infer ^{235}U/^{238}U $\leq$ 1 at production, and (d) estimated an upper limit on the time of uranium decay, and by assumption on the age of the Earth, from the estimated present-day ^{235}U/^{238}U ratio. Rutherford's estimate was within 35% of the age presently accepted, but contained several errors, including the assumption that uranium was initially produced in the Sun. Furthermore, his estimate (a) used a present-day ^{235}U abundance that was more than twofold too low; and (b) used an estimated ratio of the decay constants for ^{235}U and ^{238}U, $\lambda_{235}/\lambda_{238} = 10.6$, which was too high by about 70%. This estimate nevertheless illustrates Rutherford's scientific insight.

The next advance in U–Pb dating was made without aid of a mass spectrograph. Rose and Stranathan (1936) determined the ratio of ^{207}Pb ("AcD") to ^{206}Pb ("RaG") for a number of radiogenic mineral leads from the hyperfine structure of their emission spectra. They assumed the actual progenitor of AcD was not known with certainty, calling it AcU to distinguish it from UI (^{238}U) and UII (^{234}U). They identified the probable mass of AcU as 235 or 239, most probably the former, following earlier suggestions (Russell 1924, Hahn 1925). Rose and Stranathan (1936) derived an expression for the age of the minerals as a function of ^{207}Pb/^{206}Pb and ^{235}U/^{238}U, but not explicitly dependent on the Pb/U ratio; that is, they determined the "Pb–Pb" ages of the minerals, and showed them to be in reasonable agreement with the U–Pb ages. However, they incorrectly assumed $\lambda_{235}/\lambda_{238}$ to lie between 10 and 11, from the earlier suggestions of Rutherford and others. Moreover, the ages of their oldest minerals were about 1 Ga, and did not improve estimates of the age of the Earth.

In a pair of classic papers, Nier (1939a,b) simultaneously reclaimed U–Pb dating for mass spectrometry, and put the method on a sound footing by positively identifying AcU as ^{235}U and by refining the values of the half-lives of both ^{238}U and ^{235}U. He also measured the isotopic composition of Pb in some U-rich minerals. Figure 1 is reproduced from Nier (1939b), and shows the relative abundance of ^{207}Pb and ^{206}Pb in pitchblende from Katanga, Africa. This "mass spectrum" shows that the Pb in this mineral, which Nier (1939b) reported as containing 72.2% U, was entirely radiogenic, and produced from U decay only, being free of both "common" ^{204}Pb and of ^{208}Pb from ^{232}Th decay. From the known α-disintegration rate of bulk uranium, Nier derived decay constants for ^{238}U and ^{235}U, respectively, that were within 2% of currently accepted values of $1.551 \times 10^{-10}\,\mathrm{a}^{-1}$ and $9.85 \times 10^{-10}\,\mathrm{a}^{-1}$ (Jaffey et al. 1971). Furthermore, the corresponding ratio of decay constants that he used was within 1% of the currently accepted value of 6.35, and significantly different from previous estimates. Nier (1939b) calculated ages from the ^{207}Pb/^{206}Pb ratios in the minerals he analyzed, and found that one uraninite from Manitoba, Canada, was 2200 Ma old, older than the then accepted age of the Earth.

Nier, a physicist, had entered the field of geochronology through work as a postdoctoral fellow at Harvard University in the early 1930s, and, after returning to the University of Minnesota, by building a mass spectrometer

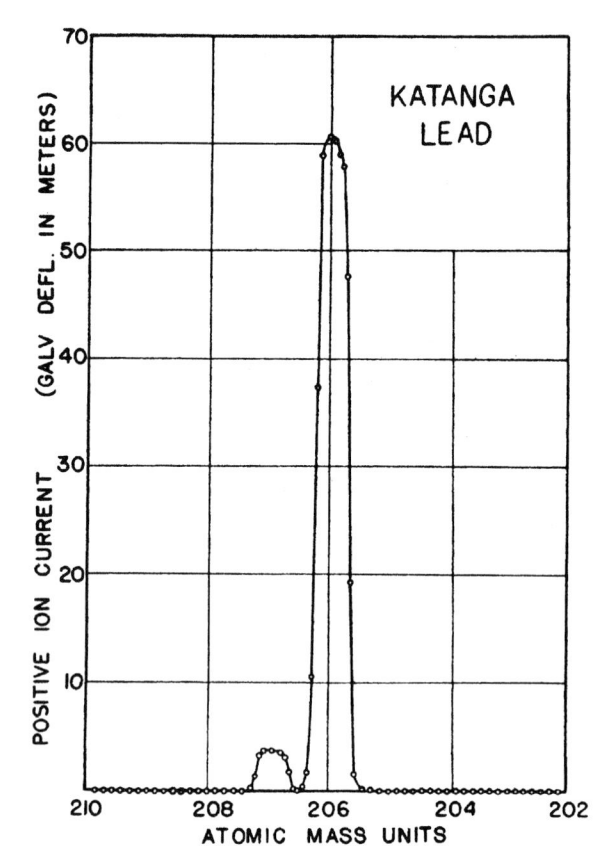

Figure 1 Mass spectrum showing the relative abundance of the isotopes ^{207}Pb and ^{206}Pb, daughter products of decay of ^{235}U and ^{238}U, in pitchblende from Katanga, Africa (Nier 1939b). The heights of the spectral peaks, and thus the abundances of the isotopes, are proportional to the positive ion current striking the ion-collecting electrode of a mass spectrometer. In this pioneering work, the positive ion current was measured with a ballistic galvanometer, as found in many elementary physics laboratories early in the twentieth century. The magnitude of the current was measured by the torsion of the galvanometer coil. A light beam was directed to a mirror attached to the coil, and the coil's torsion measured as meters of deflection of the reflected beam on an opposing laboratory wall (A.O.C. Nier, personal communication).

superior to other instruments of its day. Those interested in the historical development of mass spectrometry, the study of the isotopic abundances of the elements, and geochronology will find his short autobiographical sketch (Nier 1981) fascinating reading. His discovery of ^{235}U not only allowed it to be identified as the radioactive parent of ^{207}Pb, but also led to its identification as a fissionable uranium isotope (Nier et al. 1940). Nier often proudly referred to a handwritten note he had received from E. Fermi in October 1939, to the effect that "deciding whether the slow neutron fission (of uranium) is or is not due to the 235 isotope... is of considerable theoretical and possibly practical interest."

Later, during World War II, he became involved in the US Manhattan Project to achieve a uranium-fission chain reaction (Nier 1989).

U–^{4}He dating and cosmogenic nuclides

Whereas, the pre-World War II development of the "lead method" showed the age of the Earth to be ≥2.2 Ga, development of the "helium method" led to more equivocal results. Radioactive minerals often did not quantitatively retain helium, because helium is a light, non-reactive gas, which can readily diffuse out of mineral lattices. However, F. Paneth argued that helium would be reliably retained within metals and driven out of them when they were molten, making this method applicable to iron meteorites. The U–He ages of iron meteorites thus should be reliable against both inherited initial ^{4}He, and subsequent diffusive loss of radiogenic ^{4}Hc. By arguing that the Earth formed contemporaneously in the Solar System with "primitive" objects like meteorites, Paneth and his students introduced an important concept to the question of the Earth's age. Paneth (1931) summarized U–He ages for 26 iron meteorites. All were ≤2.8 Ga, an observation which he took as additional proof that meteorites originated in the Solar System, and not, as sometimes assumed previously, from fixed stars other than the Sun. However, Arrol *et al.* (1942) redetermined the uranium contents of several of the meteorites and found them to be lower than originally measured, causing the ages to increase to 6–7 Ga. Arrol *et al.* (1942) concluded that the age of the Solar System could not be less than 7 Ga, a conclusion that later drew criticism from F. Houtermans (1946). Indeed, Paneth and co-workers had ignored the possibility that the meteorites might contain helium of non-radiogenic origin, causing them to appear too old. The discovery of energetic cosmic rays led to suggestions that such cosmic rays could produce nuclear reactions when interacting with meteoritic matter. Among the products would be both ^{4}He and ^{3}He, which would be included in the volumetric helium measurements that Paneth and his co-workers had made. The validity of such suggestions was confirmed when isotopic analysis of helium liberated from five iron meteorites (Paneth *et al.* 1952) showed the helium to be composed of 18–32% ^{3}He.

It is worth noting at this juncture that the discovery of such cosmogenic helium in meteorites was one of the first instances in which laboratory analyses showed the effects of natural phenomena occurring outside Earth's atmosphere; that is, in space. "Space science" was in its infancy. The prior investigations of cosmic rays were done primarily by studying their effects in photographic emulsions attached to high-altitude balloons. Sounding rockets had probed the upper atmosphere, but in 1952 the first artificial satellite would not orbit the Earth for five more years. Now, while space itself was still foreign territory to humans and their machines, scientists turned to those things that come to us from space, cosmic rays and meteorites, to learn about Earth's environment in space. Also, at about this time, came a resurgence in interest in understanding how Earth and its siblings, the terrestrial planets, came to be. H.C. Urey, one of those responsible for rekindling this interest, sought to understand the accumulation and subsequent evolution of the Earth and other terrestrial planets. Urey (1951) wrote: "A two-stage process of accumulation into planetesimals is postulated, the first at low temperatures and the second at high temperatures, both followed by a period during which planetesimals accumulated to form the terrestrial planets." As evidence, he cited the Moon's surface, suggesting it was "formed by the collision of planetesimals up to approximately 100 km in radius." Although the lunar samples returned by the Apollo program disproved this particular suggestion, versions of the planetesimal accumulation hypothesis remain the most popular way of explaining the formation of the terrestrial planets.

Planetesimals are hypothesized to be more or less direct accumulations of primitive, chemically undifferentiated, solid matter as it condensed from a cooling solar nebula. Iron meteorites are composed predominantly of iron and nickel, and thus are strongly differentiated. How did they come to us? The concentrations of cosmogenic nuclides, especially the isotopes of neon and argon, as well as those of helium, were shown to vary as a function of depth within large iron meteorites (Signer and Nier 1960). This confirmed their production by cosmic-ray irradiation, and showed that the meteorites most recently were small, meter-sized, objects in space. Stone meteorites, too, were shown to contain cosmogenic noble gases. The period of cosmic-ray irradiation can be determined if the production rate of one of the cosmogenic nuclides can be determined. This was accomplished, for example, for the large stone meteorite, Norton County, by Begemann *et al.* (1957) who measured both its ^{3}He and ^{3}H (tritium) contents. Because the production and decay rates of tritium were in equilibrium, that is, equal to one another, the ^{3}He production rate could be determined from the tritium activity and the known ^{3}He/^{3}H production ratio. Begemann *et al.* (1957) calculated an irradiation age (also called a "cosmic ray exposure age") of 240 Ma for Norton County, much less than its K–Ar age of about 4400 Ma. The younger irradiation age was interpreted as giving the time when the meteorite was broken off from a much larger body. Subsequent work showed that many iron meteorites have somewhat longer irradiation ages, but many stone meteorites have shorter irradiation ages, typically ≤50 Ma.

The large irradiation ages of iron meteorites explain why Paneth's U–He ages for them were incorrect. Almost the entire amount of ^{4}He in iron meteorites is cosmogenic rather than radiogenic. Furthermore, the extremely low uranium contents were difficult to measure, and early-determined values had to be revised. However, the low uranium contents worked to advantage in determining the parameters that finally led to a precise U–Pb value for the age of the Earth.

U–Pb age of the Earth

Holmes (1946) and Houtermans (1946) independently developed an approach to determining the age of the Earth called the "common lead" method, which they applied to Nier's 1939 data. Houtermans' mathematical approach remains in current use. He noted that the $^{206}Pb/^{204}Pb$ and $^{207}Pb/^{204}Pb$ ratios of lead ores (galenas) of very low U/Pb ratios and common ages, and which were produced in a single stage of evolution from a common primordial isotopic composition, should lie along straight lines, which he called "isochrons," when plotted against one another. (This was the first use of the term isochron.) All such isochrons should pass through a common point corresponding to the initial lead composition, denoted by $(^{206}Pb/^{204}Pb)_0$ and $(^{207}Pb/^{204}Pb)_0$, and have a slope depending on the age of the Earth and the age of the ore. The latter must be known independently. Using Nier's data, Houtermans derived the age of the Earth to be 2.9 ± 0.3 Ga. He noted, however, that additional data were required, especially since the values of $(^{206}Pb/^{204}Pb)_0$ and $(^{207}Pb/^{204}Pb)_0$ were only approximated by the then available data.

A few years later, Patterson et al. (1953), using improved analytical techniques, solved the initial lead problem by measuring the isotopic composition of lead from iron meteorites. Measurements of lead and uranium extracted from troilite (iron sulphide) inclusions in the iron showed extremely low U/Pb ratios, and the lowest ratio of radiogenic to common, non-radiogenic, lead yet observed. This was "primordial" lead, separated from U near the time of the initial formation of the Solar System, as H. Brown (1947) had earlier suggested. Apparently, some early-formed planetesimals melted following accumulation, leading to chemical separation of metal and sulphides from silicates; that is, core formation in some planetesimals destined to be fragmented into iron asteroids.

In subsequent work, Patterson also measured the composition of lead from stone meteorites. Combining those data with data from iron meteorites and from terrestrial oceanic sediments (Patterson 1956), he showed that they all lay on the same isochron. Commenting that "One sample of oceanic sediment probably represents more material than a dozen galenas," Patterson interpreted the isochron age of 4.55 ± 0.07 Ga as simultaneously giving the age of the Earth and of meteorites. Arthur Holmes re-examined the issue following Patterson's publication, and concluded: "My own attempt to solve the problem from terrestrial evidence alone leads to essentially the same result (as Patterson's), which may be expressed as 4500 ± 100 million years" (Holmes 1956). That the age of the Earth, which occupied Holmes for more than 40 years, is a question of great durability is evident from the fact that it continues to be a topic for investigation. Allègre et al. (1995), for example, concluded that if the "age of the Earth" is interpreted to mean the end of accretion and core formation, accompanied by atmospheric extraction, that age is about 100 Ma younger than that of primitive meteorites, based on the terrestrial lead isotopic data.

The Rb–Sr dating method

The U–Pb method was found to be inappropriate for many problems of geochronological interest because the available samples had low uranium and lead abundances, making the lead isotopic analyses susceptible to chemical processing blanks. Thus, there was considerable interest in developing geochronological methods based on other parent nuclides, particularly isotopes of rubidium and potassium, which were known to undergo β-decay. Initially attention focused on rubidium, because the product of β-decay of potassium would be an isotope of calcium, one of the most abundant elements in the Earth's crust, and the probability of detecting the potassium decay product was judged to be nil.

Hahn and Rothenbach (1919) determined that the activity of natural rubidium corresponded to a half-life of about 100 Ga. Aston (1921) showed the existence of two isotopes of rubidium, ^{85}Rb and ^{87}Rb. Nier (1936, 1950) redetermined the isotopic abundances, and Hemmendinger and Smythe (1937) showed that the activity of rubidium was due solely to ^{87}Rb. Considerable development of the method occurred in Germany prior to World War II. Mattauch (1937) determined the isotopic composition of strontium separated from a rubidium-rich Canadian mica (lepidolite), and showed it to be nearly 100% ^{87}Sr. Based on this analysis, Hahn and Walling (1938) discussed the possibility of determining the geological ages of rocks and minerals by the "strontium method." They noted that the age of the mica as determined by the chemical lead method implied a corresponding ^{87}Rb half-life of 63 Ga. Although, in their words, this put the ^{87}Rb half-life on a sound basis, this value is about 29% greater than the value presently used. Furthermore, the method they proposed envisioned determination of rubidium and strontium abundances by chemical means, and thus would be limited to those minerals like mica and potassium feldspar that contained rubidium in high abundance and formed without inclusion of significant "common" strontium. Near the end of 1938, Otto Hahn artificially induced fissioning of uranium atoms, an event of immense political and historical impact. Nevertheless, Hahn continued to publish additional work on the strontium method well into the war years.

Ahrens (1949) resumed consideration of the strontium method after the war, utilizing the earlier work of Hahn and co-workers and comparing it to the lead and helium methods. Ahrens, like Hahn, considered it to be a chemical method, limited to rubidium-rich minerals. Although the "chemical strontium method" was practiced for a few years, the desirability of isotopic analysis to detect the possible presence of common strontium was clear. Mass spectrometric techniques for isotopic analysis had progressed significantly during World War II. Furthermore, nuclear research had resulted in the availability of isotopically enriched materials, which in turn led to the development of the isotope dilution method for measuring very low elemental abundances (Inghram 1954). Consequently, the "chemical strontium method" was replaced by the isotopic ^{87}Rb–^{87}Sr method. Schumacher (1956) at the University of Chicago, and Herzog and Pinson (1956) at the Massachusetts Institute of Technology applied the new method to stony meteorites. Schumacher (1956) calculated a meteorite age with the assumption that meteorites of two different types, one a chondrite and the second an achrondrite, were cogenetic, that is, formed contemporaneously. An unusually old age of between 5.46 and 5.80 Ga was obtained if a half-life for ^{87}Rb decay of 58.3 Ga was assumed, based on older direct half-life measurements. Ahrens (1949), for example, had recommended a value of 58 Ga based on the work of Eklund (1946). However, a significantly lower value of $(43 + 3/-2)$ Ga from direct counting had been reported by Geese-Bähnisch and Huster (1954). These workers noted the serious problem posed by self-absorption losses in counting the very low-energy β-particles from ^{87}Rb decay, which would cause the calculated half-life to be too high. Schumacher (1956) appealed to then ongoing work by Aldrich and co-workers (Aldrich et al. 1956) to suggest a "geological value" of the half-life of (49 ± 3) Ga leading to a Rb–Sr age of (4.7 ± 0.4) Ga for the meteorites. This value was in reasonable agreement with Patterson's (1956) age for meteorites.

Gast (1962) redetermined the ^{87}Rb–^{87}Sr ages of meteorites, analyzing four achondrites and five chondrites. He obtained an age of 4.67 Ga, in agreement with Schumacher's value, when using the geological determination of the ^{87}Rb half-life of (50 ± 2) Ga as reported by Aldrich et al. (1956). Flynn and Glendenin (1959) had reported an apparently precise determination of the half-life of (47 ± 1) Ga. In light of this direct half-life determination, Gast (1962) also calculated an age of 4.36 Ga for the meteorites relative to Flynn and Glendenin's value of the ^{87}Rb half-life. This meteorite age is considerably lower than that of Patterson (1956), and has not been substantiated by later work. Nevertheless, the half-life reported by Flynn and Glendenin (1959) had considerable impact on Rb–Sr geochronology for some years. Pinson et al. (1965), for example, noted that its widespread use in geochronology publications began with the 3 April 1961 meeting of the New York Academy of Sciences on the geochronology of rock systems, but that "It may well be in error." The small quoted error limit of 2% appears to have been overly optimistic, being based on the precision of determinations for five repeat experiments for which the standard deviation was 1%. Furthermore, the authors gave the "reliability" of the technique as only 2%, suggesting that the accuracy of their result might be expected to be in the range of 2–3%. Pinson et al. (1965) calculated an age of 4.52 ± 0.12 Ga for their suite of meteorites using 50 Ga as the ^{87}Rb half-life. More recently, Minster et al. (1982) determined Rb–Sr data for a large number of chondrites for which the Pb–Pb ages were known, and derived a "geological" value of (49.4 ± 0.3) Ga for the half-life. This value is used for Rb–Sr ages reported in this chapter. It differs by about 1.2% from the value of 48.8 Ga recommended by Steiger and Jäger (1977), but is nearly identical to 49.5 Ga attributed to Huster and Rausch (personal communication) by Aldrich and Wetherill (1958). A survey of literature values for direct determinations of the ^{87}Rb half-life shows that those determined by scintillator crystals are higher than those determined by other methods. Omitting such data, an average of values determined by Libby (1957), Flynn and Glendenin (1959), and McNair and Wilson (1961) is again 49.5 Ga, with a standard deviation of 2.9%. A more accurate determination of the "best" value of the half-life of ^{87}Rb remains an area for additional investigation (Begemann et al. 2000).

Two developments in the mid- to late 1960s affecting the Rb–Sr method significantly contributed to successful dating of the Apollo lunar samples in the 1970s. One was the application of the mineral or "internal" isochron technique to meteorites (Compston et al. 1965, Wasserburg et al. 1965). The second was the development of a new generation of automated mass spectrometers with digital data acquisition. The paper by Compston et al. (1965) is an early illustration of both trends; in it a precision of measurement of better than 0.05% is reported for normalized ^{87}Sr/^{86}Sr ratios. Wasserburg and Burnett (1969) reported the first data for a newly designed and constructed mass spectrometer (Wasserburg et al. 1969), later dubbed the Lunatic by the Caltech group. This instrument achieved a precision of measurement of ~0.01% for normalized ^{87}Sr/^{86}Sr ratios.

Figure 2 shows an example of an internal Rb–Sr isochron for lunar sample 10044. It is a representation of eqn (5) for $P(t) = \,^{87}$Rb, $N = \,^{86}$Sr, $D(t) = \,^{87}$Sr, and $D_0 = (^{87}\text{Sr})_I$, all expressed as atomic abundances in the various mineral phases in the rock. The ratio ^{87}Rb/^{86}Sr is plotted on the x-axis and the isotopic ratio ^{87}Sr/^{86}Sr is plotted on the y-axis. If all minerals in a rock started with identical initial $(^{87}\text{Sr}/^{86}\text{Sr})_I$ ratios, then with the passage of time the ^{87}Sr/^{86}Sr ratio in each mineral phase will increase in

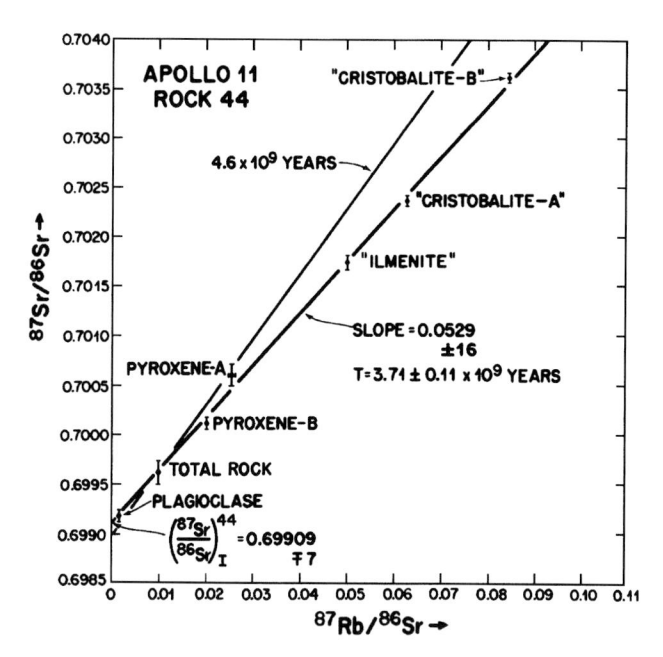

Figure 2 Rb–Sr isochron diagram (Papanastassiou et al. 1970) for lunar rock 10044 returned by the Apollo 11 astronauts. Values of ^{87}Sr/^{86}Sr increase regularly in proportion to the ^{87}Rb/^{86}Sr ratio in minerals separated from the rock. The mass spectrometer used for this work was designed in anticipation of the return of the lunar samples, and utilized what were then state-of-the-art electronics to measure the positive ion currents of ^{87}Sr and ^{86}Sr, for which the enrichment in ^{87}Sr/^{86}Sr was only 0.6%. Extrapolation of the isochron to ^{87}Rb/^{86}Sr = 0 yields the initial (^{87}Sr/^{86}Sr)$_I$ present in the rock when it solidified from a melt. Several other radiometric chronometers, for example Sm–Nd, also rely on similar isochron techniques.

proportion to the ^{87}Rb/^{86}Sr ratio in that phase in such a manner that data for all minerals will plot along a straight line of constant slope, that is, the isochron. The slope of the isochron is determined by the coefficient $[e^{\lambda(t-t_0)}-1]$ in eqn (5), and thus gives the age, $T = (t-t_0)$. The intercept of the isochron with the y-axis gives (^{87}Sr/^{86}Sr)$_I$, the value of which gives information about the Rb/Sr ratio in a possible precursor reservoir, and thus information about the composition of that reservoir. If the phases studied formed directly from the solar nebula, the value of (^{87}Sr/^{86}Sr)$_I$ thus obtained is the initial value for the nebula (Papanastassiou and Wasserburg 1969). The small enrichment in radiogenic (^{87}Sr)* of only ~0.6% required a measurement precision of ~0.01%, as achieved by the instrument described by Wasserburg et al. (1969), to determine the age precisely.

The K–Ar dating method

The K–Ar method of age determination was initially delayed longer than the Rb–Sr method, but rapidly caught up. The major isotopes of potassium are ^{39}K and ^{41}K. Nier (1935) also discovered the minor ^{40}K isotope, the existence of which had been previously suggested to account for the observed β-activity of potassium. Two years later, Smythe and Hemmendinger (1937) demonstrated that all the radioactivity of potassium was indeed associated with the ^{40}K isotope. Normal β-decay would lead to ^{40}Ca, the most abundant Ca isotope. Detection of radiogenic ^{40}Ca against a background of common Ca was considered impractical, because Ca is one of the most abundant elements in the Earth. However, von Weizsäcker (1937) reasoned that decay of ^{40}K to ^{40}Ar also would be energetically possible, and that this might occur via capture of one of the inner orbital electrons into the nucleus. Seeking empirical evidence of the process, he noted that it would account for both the γ-activity of potassium and also the high abundance of ^{40}Ar in the atmosphere. From the latter, he estimated a branching ratio of ^{40}Ar to ^{40}Ca of 1/3, with an uncertainty of a factor of 2 to 10. Thompson and Rowlands (1943) later detected the γ-radiation associated with electron capture, but were unable to quantify the rate of electron capture due to the low energy and extremely low intensity of the radiation. Aldrich and Nier (1948) found the isotopic ratio ^{40}Ar/^{36}Ar to be several times greater in argon extracted from potassium-bearing minerals than in the terrestrial atmosphere, which they took as evidence that much of the ^{40}Ar was radiogenic.

In the case of this branched decay, the age, T, of a potassium-bearing mineral is given by

$$T = (\lambda_e + \lambda_\beta)^{-1} \ln\left\{1 + \left(\frac{^{40}\text{Ar}}{^{40}\text{K}}\right)\left[\frac{1+R}{R}\right]\right\} \qquad (6)$$

where λ_e and λ_β are the decay constants for electron-capture and β-decay, respectively, $R = \lambda_e/\lambda_\beta$ is the branching ratio, and (^{40}Ar/^{40}K) is the ratio of the number of radiogenic ^{40}Ar atoms to ^{40}K atoms present in the mineral. In 1949 and 1950, the value of λ_β was determined to be about 0.55 Ga^{-1} by a number of workers, but the value of λ_e, or equivalently, R, was more uncertain. Direct counting measurements gave $R = 0.13$, but a number of early K–Ar age determinations on minerals of known age suggested a lower value. Wasserburg and Hayden (1954) made a careful determination of both the argon and potassium concentrations of a potassium feldspar mineral from the Bessner mine, Ontario, Canada, using the newly developed isotope dilution method. They compared the age calculated with the parameters determined by the counting experiments to that found by Nier (1939b) for Bessner uraninite, and found them to be in adequate agreement. However, in a more extensive work (Wasserburg and Hayden 1955a), they concluded that a smaller branching ratio, $R = 0.085 \pm 0.005$, was required to bring the K–Ar and U–Pb ages of coexisting feldspars and uraninites into agreement. Wasserburg and Hayden (1955b) reported the ages of two chondrites, Beardsley and Forest City, to be 4.82 ± 0.20 Ga and 4.58 ± 0.20 Ga, respectively, using $R = 0.085$. In an earlier application of the ^{40}K–^{40}Ar method to meteorites, Gerling and Pavlova (1951) had found the oldest age to be 3.03 Ga for $R = 0.124$. Wasserburg and Hayden (1955b) recalculated the age to 3.55 Ga for $R = 0.085$. Wetherill et al. (1955), however, noted that the ^{40}Ar/^{40}K ratios for feldspars from a given rock were often lower than for micas from the same rock, suggesting that some ^{40}Ar had been lost from the feldspars. They suggested, therefore, that R, as determined in the study of Wasserburg and Hayden (1955a) was likely to be too low, a suggestion confirmed by later work. Currently accepted values (Steiger and Jäger 1977) give $(\lambda_e + \lambda_\beta) = 0.5543$ Ga^{-1} and $R = 0.117$.

Geiss and Hess (1958) subsequently determined the K–Ar ages of 12 meteorites using the isotope dilution method and values of the decay parameters that are nearly the same as those used presently (Steiger and Jäger 1977). They recalculated the meteorite ages reported by Wasserburg and Hayden (1955a), reporting them to lie in the interval 4.15 to 4.35 Ga. Additionally, seven chondrites had ages in the interval 4.00 ± 0.12 to 4.40 ± 0.10 Ga and four achondrites had ages in the interval 3.23 ± 0.22 to 4.40 ± 0.70 Ga. One achondrite, Shergotty, had a calculated age of only 0.56 ± 0.11 Ga. As we will see later, Shergotty is one of the Martian meteorites, and this was the first indication of their unusually young ages.

The K–Ar chronometer has proven to be extremely useful. Because of the generally high abundance of K relative to other parent elements, it is often applicable where other chronometers are not. The conventional K–Ar method, however, has one serious drawback: It is prone to error due to diffusive loss of argon. The extent of this problem was delineated by an extensive study of 59 stony meteorites by Kirsten et al. (1963). A histogram of K–Ar ages showed a pronounced maximum at 4.5 Ga, but a wide distribution in ages with a secondary maximum at 0.5 Ga. The latter suggests a strong outgassing event at that time causing nearly complete loss of ^{40}Ar. The ^{39}Ar–^{40}Ar method established by Merrihue and Turner (1966) gives ways to recognize and compensate for ^{40}Ar loss. In this method, the sample is irradiated with fast neutrons and a portion of the ^{39}K is converted to ^{39}Ar. The argon is then extracted by stepped heating, and the ^{39}Ar is released along with the radiogenic ^{40}Ar. Provided that the proportion of ^{39}K converted to ^{39}Ar is known, apparent ages can be calculated for the argon released in each step. In practice, this conversion ratio is determined by irradiating a standard of known age with the samples of unknown age. Since the argon is first released from the less retentive mineral sites, the effect of argon loss can be detected, and the corresponding data discarded. Fortunately, this method also came "on line" in time for use with the Apollo lunar samples (Turner 1970a). All "K–Ar" data reported in this paper were obtained by this ^{39}Ar–^{40}Ar method.

Other chronometers

Although several other long-lived radioactivities have been applied in geochronology, only the α-decay of ^{147}Sm to ^{143}Nd will be mentioned here. The radioactivity of Sm was known early in the second quarter of the twentieth century (Hevesy et al. 1933, Hosemann 1936), but it was not utilized for geochronology until the last quarter of the century (Lugmair 1974). The Sm–Nd method is truly a product of the "space age." World War II, with all its attendant human suffering, had, nevertheless, given a tremendous boost to geochronology through advances in nuclear physics and the associated technology. But, even the technology of the late 1950s and early 1960s, by which time the preceding radiometric methods could be considered to have been established, was inadequate for the Sm–Nd system. The development of very stable electronics, digital data acquisition systems, and on-line computer control was required to obtain the very high analytical precision required for the Nd isotopic analyses. Such precision had been achieved for the instrument described, for example, by Wasserburg et al. (1969). Government funding made available during the US Apollo program allowed additional mass spectrometers to be upgraded to similar standards.

Development of the Sm–Nd method was also enabled by the development of chemical techniques for quantitative separation of rare earth elements from one another, as described, for example, by Eugster et al. (1970a). Quantitative separation is necessary to prevent isobaric interference in the mass spectrometer analysis. Isobaric interference can arise when two or more elements have isotopes of the same mass number, and thus interfere in the mass spectrum. Although the actual masses of the interfering nuclides will differ by a small amount, separation according to mass often requires a higher mass resolution than is commonly available with so-called isotope-ratio instruments. The requisite chemical separation methods were originally developed to enable measurement of neutron capture effects in Gd to study the irradiation history of meteorites and lunar rocks and soils (Eugster et al. 1970a,b, Lugmair and Marti 1971). Thus, the Sm–Nd method, now widely applied in terrestrial as well as in planetary geochronology, can be considered as a "spin-off" of the space program.

Some of the early concepts of geochronology are no longer applicable to the Sm–Nd method. One example is the concept of a "mineral age," which assumed that minerals exist which concentrate the element of the parent nuclide and exclude the element of the daughter nuclide. The two rare earth elements Sm and Nd resemble one another so closely in their chemical characteristics that such mineralogical segregation does not occur in nature. One must use the isochron approach, in which Sm and Nd isotopic data are obtained for at least two different mineral phases. Thus, there are no mineral ages, only rock ages. The same is true of the Rb–Sr method as applied to most extraterrestrial rocks. Most of these are basalts, which are not greatly different in chemical composition from chondritic meteorites, and for which the mineral phases do not efficiently separate parent and daughter elements. The concept of "radiogenic" and "common" components of the daughter element also loses significance, because even for radiogenic growth over the entire history of the Solar System, the "radiogenic" component of ^{143}Nd is tiny. For mineral isochron methods, the concept of common Sr and Nd is thus replaced by that of initial Sr and Nd, whereby the initial (^{87}Sr/^{86}Sr)$_I$ and (^{143}Nd/^{144}Nd)$_I$ ratios are determined by isochron methods (Figure 2). The initial Sr and Nd of two rocks may be distinguished, perhaps, by very small differences in isotopic composition, but such differences nevertheless record different paths of isotopic evolution in the sources from which the rocks were derived, as illustrated later in the chapter.

Because of the difficulty of chemically separating Sm from Nd, Sm–Nd isochron ages are very resistant to secondary events. They have proven invaluable for determining the ages of lunar highland rocks, and of other extraterrestrial rocks, which may have undergone impact metamorphism.

Our review of the historical development of geochronology in the twentieth century has now identified the primary methods that have been applied to lunar and Martian radiometric chronology: U–Pb, Pb–Pb, Rb–Sr, and Sm–Nd isochrons, complemented by ^{39}Ar–^{40}Ar step-heating analyses. These have been the "workhorse" methods for telling geological time for samples of our Solar System neighbors. To complement these, a number of "short-lived chronometers" based on now extinct radioactivities have also been developed to resolve events occurring close to the beginning of the Solar System. Some of these will be introduced, as needed, in the discussion to follow. However, the development of the first such "short-lived chronometer," the I–Xe chronometer (Reynolds 1960), significantly preceded that of the others, and belongs more properly to this historical record.

H. Brown (1947) suggested that meteorites could be used to determine the time of nucleosynthesis of the elements in stars if the daughter of an extinct natural radioactivity could be found in them. The nuclide ^{129}I decays to ^{129}Xe with a half-life of 17 Ma. This decay was recognized as potentially favorable for discovery of the effect, because xenon, a noble gas, should be present only in very low abundance in meteorites. However, initial searches for "fossil" ^{129}Xe in the Beardsley chondrite and the Nuevo Laredo achondrite failed (Wasserburg and Hayden 1955c, Reynolds and Lipson 1957). Nevertheless, Reynolds (1960) succeeded in finding clear evidence of such fossil ^{129}Xe in the Richardton chondrite. Assuming that ^{129}I and ^{127}I were produced in equal amounts in an hypothesized supernova explosion, he calculated that event occurred $\Delta t = 350$ Ma before the formation of the Solar System. Assuming the correctness of his model, an error of a factor of eight in either the I/^{129}Xe ratio in the meteorite, or in the initial ratio of ^{129}I/^{127}I, would cause an error of only three half-lives of ^{129}I, or 51 Ma (15%) in Δt. Wasserburg et al. (1960), however, argued that ^{129}I production more likely occurred continuously over the age of the Galaxy. They calculated a shorter interval, $\Delta t \sim 200$ Ma, since cessation of ^{129}I production for a galactic lifetime of ~ 10 Ga. More importantly in our context of planetary evolution, however, was Reynolds' comment that the I–Xe method would become important in dating meteorites because it dates from the beginning of the Solar System rather than the present. The same statement can now be made for a number of short-lived nuclides (Carlson and Lugmair 2000).

Application of radiometric methods

What does it mean to determine an isotopic age using one of the radiometric clocks? The common assumption is that some event reset the system and equilibrated the isotopic composition among all phases present. For example, if rock is totally melted to produce magma, then any differences in isotopic compositions resulting from prior radioactive decay are erased, and when the magma solidifies and cools all isotopic clocks are reset. However, if the rock cools slowly, various isotopic systems may start recording again at slightly different times and yield slightly different event ages. If the rock later experiences moderate heating, such as might occur through impact cratering or parent body metamorphism, various isotopic clocks may be partially or totally reset by solid-state diffusion (or gas loss if the daughter is a gas or volatile). Thus, if different isotopic clocks give different ages for a given sample, that could indicate slow cooling or a complex history involving several events. However, if different isotopic clocks all give the same age, the sample probably has had a simple history since all clocks were last reset. In those cases where parent and daughter elements have quite different geochemical properties, the isotopic clock is very sensitive to events that produce large-scale chemical segregation, such as separation of a planetary body into metallic core and silicate mantle.

The initial isotopic composition of the daughter within a sample at the time of a complete resetting event can give important information about the geochemical history prior to that event. For example, if we have reason to assume some specific initial parent/daughter ratio for a planetary body, but we find that a rock from that body gives an initial daughter isotopic composition which is inconsistent with that assumed parent/daughter ratio and the age of the parent body, then we may infer that an event occurred prior to the formation of the rock and produced geochemical separation of parent and daughter elements. Thus, isotopic clocks can have unusual characteristics in that we may be able to date events with clocks that have stopped (extinct radionuclides) and to characterize events that occurred before the clock began running (initial isotopic composition of daughter). For example, the isotopic data for lunar samples and Martian meteorites are used later in the chapter to characterize initial events of planetary differentiation that occurred before the formation of the rock samples themselves.

Other types of isotopic clocks are used to date geological events based on the production of nuclides by nuclear reactions, rather than by radioactive decay of naturally occurring nuclides. As we have seen, in the early 1950s analyses of meteorites showed the presence of nuclides that were produced from nuclear reactions caused by high-energy cosmic ray protons while the meteorite was exposed as a small body in space. An appreciable number of such "cosmogenic" nuclides were eventually identified in meteorites and later in lunar samples. Identification of a cosmogenic nuclide requires either that it be radioactive (e.g., ^{3}H, ^{22}Na, ^{26}Al, etc.) or, if stable, that it occur naturally in such a low abundance in the meteorite that the cosmogenic component can be identified (e.g., ^{3}He, ^{21}Ne, ^{38}Ar). Many analyses of cosmogenic nuclides in meteorites having different chemical compositions and different sizes while exposed in space, augmented by accelerator studies of the probabilities that energetic protons will create nuclear reactions with specific elemental targets (i.e., nuclear cross-section studies), have led to a rather sophisticated understanding of the production rates of many cosmogenic nuclides as a function of sample composition and size. Thus, for example, the production rate of cosmogenic ^{3}He in a silicate meteorite having a diameter of a fraction of a meter is approximately 4.5×10^5 atoms of ^{3}He per gram of meteorite per year. By dividing the measured amount of a cosmogenic product in a meteorite by its production rate, one determines the time of the impact event on the meteorite parent body that created the meteorite as a relatively small body and launched it into space. Through analogous studies of cosmogenic products in Apollo-returned lunar rocks, the formation times of several lunar craters have been determined. Furthermore, the lunar soils were found to typically contain higher concentrations of stable cosmogenic nuclides than did larger rocks, showing the effect of progressive comminution and "gardening" by meteorite bombardment of the lunar surface. We do not attempt to cover that aspect of lunar chronology in the present work, however.

FORMATION OF THE SOLAR SYSTEM: THE EARLIEST CHRONOLOGIES

Radiometric age dating of meteorites

As we have already seen, an upper limit to the formation time of the Earth, Moon, and Mars is obtained by determining the time of formation of primitive, small objects that were the first to form in the early Solar System. The larger planets and satellites are believed to have accreted from these smaller objects. Two kinds of early accretionary bodies are asteroids and comets, and at least samples of some asteroids fall to Earth as meteorites. Thus, determining the formation history of asteroids by studying meteorites gives us information about the very early period of the Solar System just prior to planet formation. These and related studies indicate that the first accretionary processes in the early Solar System occurred on a timescale of ~ 1 Ma, and incorporation of this early material into small bodies ~ 10–100 km in diameter perhaps occurred on a timescale of ~ 10 Ma (Podosek and Cassen 1994, Carlson and Lugmair 2000). Although most meteorite parent bodies formed very early, Solar System processes have affected them very differently. Some meteorites derive from primitive bodies that were little altered by internal heating and have retained the primitive nature of the material that formed them. Other meteorites derive from bodies that have been significantly heated, even melted, and have formed metallic cores and/or basaltic surfaces. Both types of objects, however, give important constraints on the timescales of early events and processes. This chronological information derives from three kinds of data: direct radiometric dating, use of extinct radionuclides to determine relative times, and measurements of the evolution of isotopic compositions of elements changed over time by radioactive decay.

In applying radiometric dating techniques to a meteorite, one has to consider what type of event is likely being dated. Primitive meteorites, for example carbonaceous and ordinary chondrites, contain phases that probably formed at different times. Some phases likely condensed from nebular gas in the early stages of Solar System formation. Other rare refractory phases contain isotopic signatures of individual supernova explosions, and these materials apparently formed in expanding supernova ejecta before formation of the Solar System (Anders and Zinner 1993). The existence of individual isotopic signatures is evidence that these grains did not equilibrate with other meteorite phases. (Unfortunately, determining the isotopic age of individual supernovae grains is a goal not yet attainable.) Studies of refractory inclusions (i.e., those that formed at high temperatures) in the Allende carbonaceous meteorite give the oldest precise age for dated meteoritic material. Using the Pb–Pb radiometric method, these inclusions have been dated at 4.566 Ga, with an uncertainty of only a few million years (Chen and Wasserburg 1981, Allègre *et al.* 1995). This age probably represents the earliest times of formation of solid materials within the solar nebula that became our Solar System.

The parent bodies of many meteorites were heated internally to such an extent that silicate was strongly altered. Even some primitive meteorites suggest alteration by low-temperature fluids. Some meteorite parent bodies experienced partial melting, which in some cases led to parent body differentiation. Many metallic meteorites are believed to derive from iron-rich cores that formed from early differentiation of some of these bodies. A few types of meteorites are basaltic, which implies that they were formed by partial melting and extrusion onto their parent body surfaces. The specific causes of this heating, which led some meteorite parent bodies to melt whereas others seem to have been little heated, is not fully understood, but accretional heating and decay of short-lived nuclides probably both contributed. In some respects these meteorite parent bodies show the same kinds of processes that occurred on a larger scale on bodies such as the Earth, Moon, and Mars. Because the planets are larger, however, it required longer times for them to form.

Most of the heating and differentiation within meteorite parent bodies appears to have occurred very early in Solar System history. One type of differentiated meteorite, the angrites, gives a Pb–Pb age of 4.558 ± 0.003 Ga (Lugmair and Galer 1992) and a less precise Sm–Nd age of 4.55 ± 0.03 Ga. Because angrites formed from a melt, this age would appear to date a specific formation event. A similar precise Pb–Pb age of 4.556 ± 0.006 Ga (Chen and Wasserburg 1985) was determined for a basaltic meteorite called a eucrite, which derives from a different parent body. In another study, a Pb–Pb age of 4.557 ± 0.002 Ga (Göpel *et al.* 1992) was reported for a phosphate mineral separated from the Acapulco meteorite, which was heated in still another parent body to temperatures approaching those required for initial melting. Because the phosphate mineral could have been substantially altered by this internal heating, the determined age may measure this heating episode and not the earlier accretion of parent body material. Pb–Pb dating of phosphate minerals from a suite of chondrites, meteorites from still other parent bodies that were heated but not melted, suggest a significant range in ages. Two chondrites gave precise Pb–Pb ages of 4.558 ± 0.006 Ga and 4.563 ± 0.001 Ga, but other chondrites gave younger ages to values as low as ~ 4.51 Ga (Göpel *et al.* 1994). The spread in ages presumably occurred because of an extended period of internal metamorphism caused by heating and slow cooling of the chondrite parent bodies. These and other studies of meteorites indicate that various meteorite parent bodies had substantially accreted by ~ 4.555 Ga ago, and in some cases these bodies had experienced significant internal heating (see review by Carlson and Lugmair 2000). Thus, in the relatively short time interval of ~ 10 Ma after formation of the earliest solid materials in the Solar System, bodies of tens to hundreds of kilometers in diameter had formed and differentiated. These bodies were the building blocks for the planets of the inner Solar System.

Extinct radionuclides and evolution of isotopic composition

Additional evidence for early and relatively rapid accretion of most meteorite parent bodies derives from certain short-lived isotopes, which were produced by the process of element nucleosynthesis in stars and were still present in small amounts when the Solar System formed. These extinct nuclides have long since decayed, but their decay products can give information about early planetary processes which occurred on timescales comparable to the half-lives of the extinct nuclides. We have already discussed the discovery by Reynolds (1960) of radiogenic ^{129}Xe*, the decay product of now extinct ^{129}I. Another extinct nuclide is ^{26}Al, which decays to stable ^{26}Mg with a half-life of 0.73 Ma. In primitive meteorites such as Allende there is abundant evidence for the existence of excess ^{26}Mg from ^{26}Al decay, but this evidence is much less common in meteorites such as eucrites and chondrites, which exhibit evidence of thermal processing in a parent body (MacPherson *et al.* 1995). Another extinct nuclide present in the early Solar System is ^{53}Mn (half-life of 3.7 Ma), which decays to ^{53}Cr. The decay products of ^{26}Al and ^{53}Mn in phases of some chondrites indicate that these meteorites formed a few million years after the Allende refractory inclusions. Several eucrites show evidence for ^{53}Mn, suggesting that these basalts also formed less than ~ 10 Ma after Allende, but to date only one eucrite suggests the presence of shorter-lived ^{26}Al (Carlson and Lugmair 2000). Time periods of ~ 1–10 Ma indicated by extinct nuclides are consistent with evidence noted above from radiometric dating for the time between formation of refractory phases in primitive meteorites and accretion and heating of meteorite parent bodies.

Yet a third extinct radionuclide is ^{107}Pd which decays to ^{107}Ag with a half-life of 6.5 Ma. This decay system is a sensitive indicator of early parent differentiation processes, which segregate metal from silicate, for example, formation of planetary cores. Enrichments of ^{107}Ag have been observed in several groups of iron meteorites, as well as stony-iron meteorites called mesosiderites and pallasites (Chen and Wasserburg 1996). Iron meteorites appear to have formed within a narrow time interval early in Solar System history, implying both rapid accretion and differentiation of the parent body and its rapid cooling. Some stony-iron meteorites suggest ages younger than iron meteorites by several million years, which may be due to their residence in larger parent bodies that cooled more slowly.

Chronology can also be determined by changes in the isotopic composition of elements containing the daughter products of radioactive decay. Examples are the elements Sr and Nd, where the isotopes ^{87}Sr and ^{143}Nd increase over time due to the decay of long-lived ^{87}Rb and ^{147}Sm. If the initial isotopic composition of Sr and Nd in the early Solar System can be determined, then the isotopic composition of Sr and Nd at the time the meteorite (or lunar or Martian rock) formed, that is, the initial ^{87}Sr/^{86}Sr and ^{143}Nd/^{144}Nd ratios contain time information. Two approaches to determining the initial isotopic compositions are to measure the element in a mineral that greatly concentrates the daughter element over the parent element (e.g., Sr is concentrated over Rb in plagioclase), or to measure the parent and daughter compositions in a suite of samples believed to have formed at the same time from a common parent. As an example of the latter approach, a suite of eucrite meteorites plotted on a ^{87}Rb–^{87}Sr age diagram gives an ^{87}Sr/^{86}Sr initial intercept at Rb/Sr = 0 of 0.69898 ± 0.00003 (BABI, basaltic achondrite best initial; Papanastassiou and Wasserburg 1969). An application of the evolution of isotopic compositions over time to the chronology of the Moon and Mars will be presented in a later section.

EARTH'S MOON

Space race – and the winner is: space science

On 25 May 1961, a young, newly elected US president declared "I believe that this nation should commit itself to achieving the goal, before this decade is out, of landing a man on the Moon and returning him safely to Earth." From the vantage point of century's end, John F. Kennedy's bold assertion is easily recognized as a defining moment for human endeavor. It was also a defining moment in space science. The ensuing US Apollo program directed public interest and resources to scientific investigations of the space environment of Earth, and of Earth's neighbors in space. Millions of television viewers worldwide tuned in to watch the activities of the astronauts on the lunar surface. For many, deployment of the US flag on the lunar surface symbolized victory in the space race and the ultimate achievement. However, the astronauts also deployed scientific packages and selected rocks to bring back to Earth. Each mission had its own drama, as the astronauts walked on the Moon, or rode their space buggy to lunar environs where human beings never had been before. As the astronauts got an "up close" look at those other-worldly environs, so did millions of the viewing public. Each mission ended in a suspenseful "spashdown" of the space capsule and astronauts in the Pacific Ocean. Television cameras searched the sky for the first appearance of the space capsule dangling from its colorful parachutes. They accompanied the personnel and helicopters that plucked the astronauts and capsule from the sea. In the democracies, public interest, combined with a sense of competition with the Soviet Union, were translated into legislative and financial support for space activities, including space science. For its part, the Soviet Union, proud of notable firsts in space, also pressed the race to the Moon, and, later, of automated exploration of the Moon. This competition, and the technological achievements accompanying it, enabled progress in many aspects of space science.

In all, 40 US and Soviet spacecraft visited the Moon between September 1959 and August 1976. The Soviet Luna and Zond series compiled a number of "firsts": The first impact on the Moon, the first uncrewed flyby photography, the first uncrewed landing, the first uncrewed orbital gamma-ray mapping, the first returned photographic film from an uncrewed flyby, the first uncrewed sample return, and the first uncrewed rover (Wilhelms 1987). The US Ranger, Surveyor, Lunar Orbiter, and Apollo series compiled its own firsts, many of them necessitated by the goal of a crewed landing and return. These included the first preimpact photography, the first uncrewed orbital photography, the first uncrewed landing in the rugged lunar highlands, the first crewed orbital mission, and the first crewed landing and return to Earth. John Kennedy's goal was achieved on schedule, but six years after his own death by an assassin's bullet.

In keeping with the relative openness of the Apollo program, the USA made the returned samples available to all qualified investigators, including scientists from around the world. Since Neil Armstrong, Buzz Aldrin, and Mike Collins of the Apollo 11 crew, and other Apollo crews to follow, came to the Moon "in peace, for all mankind," it seemed only natural that the best scientists in the world should participate in analyzing the samples, regardless of their workplace. Those participating showed a significant heritage from the meteorite studies of the 1950s and 1960s. These scientists, and many others, participated actively in landing site selection and in definition of the sampling program. Because of the intimate and dedicated involvement of the scientific community, as well as of the scientific knowledge gained, the investigation of the lunar samples is aptly considered a branch of "space science."

The Moon rocks: cornerstones for Solar System chronology

It is hard to overemphasize how meager was our understanding of the Moon prior to the return of lunar samples to

Earth in July 1969. Two very different scientific views of the Moon existed in the 1960s. One view was championed by Prof. Harold Urey, Nobel prize winner and a student of the Moon since the early 1950s. This view held that the Moon was a primitive body, possibly similar to a volatile-rich class of meteorites known as carbonaceous chondrites, and that lunar rocks would unlock the secrets of the earliest processes that formed the Earth and other planets (Urey 1951, 1952). The second view was that the Moon was a geologically active planet with abundant volcanic features, and that lunar rocks would reveal the parallel evolution of a body near the Earth. It was recognized that the age of the returned lunar samples would be critical in establishing which view was correct.

This basic question of the antiquity of the Moon was answered within six months of the first lunar landing (Albee et al. 1970, Turner 1970a). The maria were old, about 3.7 Ga, but not as old as primitive meteorites. Moreover, they contained basalts, the product of partial melting of the lunar interior. Thus, the Moon had once been a geologically active body. Such mixed characteristics of an ancient v. an evolved Moon emphasize its important place between the chronologies of meteorites and the Earth in our understanding of planetary evolution.

Some very early characteristics of the Moon have been preserved and reveal processes active during its formation and just afterward. These include evidence for major melting and differentiation to produce an early crust, and numerous impact craters dating from late stages of its formation. Similar features have been erased from the dynamic Earth. Lunar rocks give ages ranging from ~4.4 to ~3.1 Ga and reveal a complex evolution of the lunar mantle over significant time. Compared to studies of meteorites, whose parent bodies evolved quickly, if at all, before becoming thermally dead, the Moon tells us more about long-term thermal evolution of a planet.

Basic aspects of lunar thermal evolution were quickly established by the samples returned from Apollo 11. Rb–Sr dating of the basalt samples from Mare Tranquilitatis showed the mare had been filled by a sequence of lava flows (Papanastassiou et al. 1970), a finding supported by the ^{39}Ar–^{40}Ar data of Turner (1970b). The exact duration of magmatic activity at the Apollo 11 site was initially somewhat uncertain, however. The apparent range in ages of the analyzed basalts was comparatively narrow, and could have been increased by random errors and/or systematic biases, since the ages had been determined by several investigators using various radiometric techniques. To remove such uncertainites, Geiss et al. (1977) and Guggisberg et al. (1979) made a comprehensive study of the ^{39}Ar–^{40}Ar ages of most of the Apollo 11 basalts. They included analyses of whole rocks and separated mineral phases, especially of plagioclase. Plagioclase contains potassium in high abundance, and also is retentive of radiogenic argon. These analyses established the interval of volcanism at the Apollo 11 site as extending at least from 3.85 ± 0.03 Ga ago to 3.57 ± 0.05 Ga ago as calculated with the potassium decay parameters of Steiger and Jäger (1977). Guggisberg et al. (1979) summarized data for all the then recognized petrological subgroups at the landing site. They also reported data for several different mineral phases of some of the samples, allowing Ar diffusive losses to be detected. They concluded that at least five different lava flows, occurring over a period of about 300 Ma, were represented among the samples. This careful work confirmed the earlier conclusion (Geiss et al. 1977) that the maria formed in two stages: First, excavation of the mare basins by impact, and, second, their filling by successive lava flows. This work on the Apollo 11 samples also established that the Tranquilitatis Basin was formed prior to 3.85 ± 0.03 Ga ago, the age of the oldest basalt.

The oldest basalt included by Guggisberg et al. (1979) was sample 10003, for which the ^{39}Ar–^{40}Ar age had been determined by Turner et al. (1970b) and Stettler et al. (1974). An old Rb–Sr age of 3.81 ± 0.08 Ga, recalculated for $\lambda_{87} = 0.01402\,\mathrm{Ga}^{-1}$, was confirmed by Papanastassiou and Wasserburg (1975a). Sample 10003 had a particularly well-defined ^{39}Ar–^{40}Ar age spectrum as a function of argon released during step-heating, and thus significant prior argon loss could be excluded. The work of Guggisberg et al. (1979) also established ages unaffected by argon loss for samples 10044 and 10072, for which Rb–Sr ages had been measured by Papanastassiou et al. (1970, 1977). The ratio of ^{39}Ar–^{40}Ar ages to Rb–Sr ages for these three samples, for the decay parameters used here, is 0.998 ± 0.012 (1 standard deviation), showing that these parameters are self-consistent for these data. Later work has shown that samples 10062 (Papanastassiou et al. 1977), 10002 (Snyder et al. 1996), and 10085 (Shih et al. 1999) are as old as sample 10003.

As we have seen, several major questions about the Moon were answered with samples returned by the Apollo 11 mission from the first landing site. The samples from Apollo 12 extended the known period of mare volcanism to about 3.2 Ga ago. Later missions (Apollos 14, 15, 16, and 17) returned samples of the lunar highlands as well. The Apollo 17 mission occurred in December 1972. By the time of the Lunar Science Conference of March 1974, the outline of lunar chronology was evident: Early crustal formation in the time interval 4.56 to about 3.9 Ga ago, a short period of about 100 Ma years during which several of the lunar basins were formed (Tera et al. 1974, Jessberger et al. 1974), and mare volcanism in the time period about 3.9 to 3.2 Ga ago. Still, many details of the chronology remained to be explored: The range of rock types composing the lunar crust and their ages, whether there was pre-mare volcanism in the crust, the exact timing of formation of individual basins, the range of mare basalt types and whether there was a compositional dependence of mare basalt ages with basalt type, and so on. Addressing these questions requires the most comprehensive suite of samples possible, complemented by data of other types, especially global data obtained from spacecraft orbiting the Moon.

In all, some 382 kg of lunar rocks were returned to Earth by the Apollo program, the legacy of a tremendous national effort, and the foundation of lunar science. This resource was augmented by about 300 g of lunar soil and rock fragments returned by the uncrewed Soviet missions to three additional landing sites. The Apollo and Luna landing sites are shown in Figure 3. The last lunar sample return was by the Soviet uncrewed mission, Luna 24, in August 1976. More recently, the lunar scientific endeavor has been enriched by the discovery of lunar meteorites on Earth, and by data from the Clementine and Lunar Prospector orbital missions. By the end of the twentieth century, 13 lunar meteorites (excluding paired fragments of the same fall) had been recovered in the Antarctic by American and Japanese field teams (Lindstrom 1999). Three additional lunar meteorites have also been recovered elsewhere.

The returned Apollo and Luna samples represent six lunar mare (basaltic) sites and five highland (early crustal) sites. Apollo missions 15 and 17 are counted twice, each being to border areas between mare and highlands, and each returning both mare and highlands samples. Additionally, six of the lunar meteorites have mare compositions and seven have highland compositions. It has been argued that because some meteorites appear to have the same space exposure ages ("launch-pairing") the total number of lunar

Figure 3 Photograph of the lunar nearside with the locations of Apollo and Luna landing sites superimposed. Landing site locations are: Apollo 11, Mare Tranquilitatis; Apollo 12, Oceanus Procellarum; Apollo 14, Fra Mauro Formation; Apollo 15, Mare Imbrium near the Apennine Front Mountains; Apollo 16, Cayley Plains in the Descartes Mountains; Apollo 17, Mare Serenitatis, Taurus Littrow Valley between the North and South Massifs of the Taurus Littrow Mountains; Luna 16, Mare Fecunditatis; Luna 20, highlands between Mare Fecunditatis and Mare Crisium; Luna 24, Mare Crisium. The large crater Copernicus, referred to in the text, is visible to the north-northwest of the Apollo 14 landing site. (Photograph courtesy of NASA.)

sites represented among the lunar meteorites may be only five or six (Warren 1994, Lindstrom 1999). The lunar meteorites comprise a total mass of ~4 kg, with most of the mass in the four largest specimens. Although most of the data presented here are for Apollo samples, the Luna samples and lunar meteorites, respectively, help make the sampling more representative of the lunar surface.

The absolute radiometric ages of the lunar samples have been used to calibrate the cratering rate in locales that have been sampled, producing a sample-based chronology that can be extended globally. Furthermore, theoretical considerations can be used to extend cratering rates on the Moon to those elsewhere in the Solar System. Thus, the lunar chronology has in effect become the standard for chronologies throughout the Solar System, or at least throughout the inner Solar System (Basaltic Volcanism Study Project 1981).

In the following, we summarize the lunar radiometric age data, and discuss their implications for lunar evolution. We first categorize the ages according to compositional groups that are related to specific lunar processes, beginning with rocks from the lunar highlands. Highland rocks are related most fundamentally to processes of lunar crust formation, but also bear the imprint of a greatly enhanced lunar cratering rate early in lunar history. The isotopic data for a random sampling of lunar highland rocks would be dominated by impact-reset ages, or isotopic disturbance, as will be discussed later in the chapter. We seek to establish first the age of the protolith into which the impacts occurred. Our discussion relies heavily on earlier, more specialized reviews by Nyquist and Shih (1992) and Snyder et al. (2000a). An excellent summary of lunar highland rock types, accompanied by much valuable age information, also has been given by Papike et al. (1998).

Ferroan anorthosites (FANs): remnants of the earliest lunar crust

The lunar highland crust is composed dominantly of five rock groups and impact-generated mixtures thereof. The five rock groups are: (a) FANs; (b) Mg-suite rocks; (c) alkali-suite rocks; (d) evolved rocks; and (e) KREEP basalts. We use the descriptions of these rock groups given by Papike et al. (1998).

FANs contain >90% calcic plagioclase with anorthite contents >An_{96} and minor orthopyroxene, and are thought to have formed by flotation atop an initial global lunar magma ocean (LMO). Papike et al. (1998) summarize important LMO concepts and reproduce a schematic of a recent version of the model due to Hess and Parmentier (1995). Major features of the model are: (a) an initial magma ocean 800 km deep; (b) crystallization of mafic (i.e., Mg- and Fe-rich) cumulates that sink to the bottom of the ocean; (c) after the onset of plagioclase crystallization, flotation of an anorthositic lunar crust that reaches a final thickness of ~60 km; (d) the onset of gravitational instability as progressively denser cumulates are added to the top of the pile; (e) convective overturn, with late crystallizing ilmenite and a KREEP-rich residuum (i.e., a residuum rich in potassium (K), rare earth elements (REE), and phosphorus (P)) that sinks to deep mantle depths; (f) intrusion of Mg- and Al-rich magmas into the already formed anorthositic crust to form the Mg-suite rocks; (g) later generation of mare basalt magmas by remelting mafic cumulates "fertilized" by the addition of ilmenite and trapped KREEP-rich magmas.

According to this and similar LMO models, FANs should be the oldest lunar rocks, having formed after ~75–80% crystallization of the LMO (Snyder et al. 1992). Furthermore, if they formed from a Moonwide magma ocean, one might expect them to be abundant in the returned sample collection. In fact, they are relatively abundant only at the Apollo 16 site and among the lunar meteorites; they are rare at the Apollo 15 site, and nearly absent at the Apollo 14 and Apollo 17 sites. The largest Apollo 15 FAN, sample 15415, weighing 269 g, gained a degree of notoriety, being dubbed the "Genesis Rock" during the mission itself. Scientists present in the Mission Control Center of the Manned Spacecraft Center (now the Johnson Space Center) during the mission suspected that this rock might hold the key to the earliest stages of lunar formation. However, as early as the preliminary examination in the Lunar Receiving Laboratory (Lunar Sample Information Catalog, Apollo 15, 1971) it was noted that sample 15415 was actually a metamorphic rock, although probably originally "igneous." In a prescient observation, it was noted that "its complicated history suggests it should be interpreted with great care." Indeed, subsequent efforts to date this rock failed to prove its suspected antiquity, yielding an age of 3.89 ± 0.07 Ga (Stettler et al. 1973) rather than ~4.4–4.5 Ga as hoped. This failure is likely to be due to a characteristic of the applied ^{39}Ar–^{40}Ar dating method, which dates the time of last major Ar outgassing, which may differ from the time of igneous crystallization. Other age dating methods could not be applied because the rock is nearly monomineralic, being ~99% calcic plagioclase and only ~1% orthopyroxene. Nevertheless, the very primitive $^{87}Sr/^{86}Sr$ ratio of 15415 and other lunar anorthosites suggested a very old age, close to that of the Solar System itself. In later searches of the lunar sample collection, pristine FANs with higher pyroxene contents were sought, allowing application of Rb–Sr and Sm–Nd isochron techniques as well as ^{39}Ar–^{40}Ar. Many such samples are small (mm- to cm-sized) clasts in matrices of much larger masses of impact melt or otherwise brecciated material. The scarcity of datable clasts and the difficulty of the analyses has restricted the number of reliably determined ages of lunar anorthosites to only a few.

Figure 4a summarizes radiometric ages that we judge as reliably giving crystallization ages for FAN clasts extracted from highland breccias. The histogram includes four Sm–Nd ages and three ^{39}Ar–^{40}Ar ages. Because there is no evidence of a bias between ages determined by the two techniques, and for clarity of the figure, we do not distinguish the ages by technique. The distribution peak at ~4.4 Ga contains two ages determined by each technique. One age by each method lies at higher and lower values, respectively. This distribution probably reflects real variations among the FANs, since the oldest and youngest ages differ by considerably more than their respective error limits. Six of the seven ages are for samples returned by the Apollo 16 mission to the central lunar highlands. Four of the ages are for clasts from breccias collected at Station 11, Apollo 16, on the southeast quadrant of North Ray crater. North Ray is a moderately young (50 Ma) crater located just southwest of Smoky Mountain, which belongs to the Descartes formation. The Descartes formation is thought to be a large deposit of ejecta from the Nectaris Basin. The sampling site is only ~1 km from the base of Smoky Mountain and is the closest sampling site to any part of the Descartes formation. It is very likely that old "feldspathic fragmental breccias" (FFBs) collected at this site are representative of the Descartes formation (Norman *et al.* 2000). The oldest samples, 67016 (Alibert *et al.* 1994), 67215 (Norman *et al.* 2000), and 67435 (Dominik and Jessberger 1978), come from parent breccias that were collected within 20 m of one another on the rim of North Ray crater (Apollo 16 Preliminary Report 1972, Figure 6-65). The parent breccias probably were excavated from the megaregolith beneath the crater. The old ^{39}Ar–^{40}Ar ages of Luna 20 sample 22013 (Huneke and Wasserburg 1979) and lunar meteorite Yamato 86032 (Bogard *et al.* 2000) extend the geologic context of dated FANs beyond the Descartes formation and the central lunar highlands. Yamato 86032 also is an FFB, and has trace element and isotopic similarities to the FFBs at the North Ray site (Lindstrom *et al.* 1991, Premo and Tatsumoto 1991, Nyquist *et al.* 1996).

Two other ancient Apollo 16 anorthositic clasts were extracted from 60025 and 62236, collected near the Lunar Module (LM), approximately midway between Smoky Mountain to the north and Stone Mountain to the south. Stone Mountain is also an expression of the Descartes formation. The collection sites for these samples are closer to Stone Mountain, being within ~3–4 km of it. The collection sites likely contain ejecta from the young (2 Ma) South Ray crater, lying just 3 km west of Stone Mountain; however, the surface exposure age of 62236 is too old for it to have been derived from South Ray. These samples are true FANs, being mineralogically distinct from the North Ray FFB clasts, which Norman *et al.* (2000) call ferroan noritic anorthosites. It is likely that they represent the protolith of

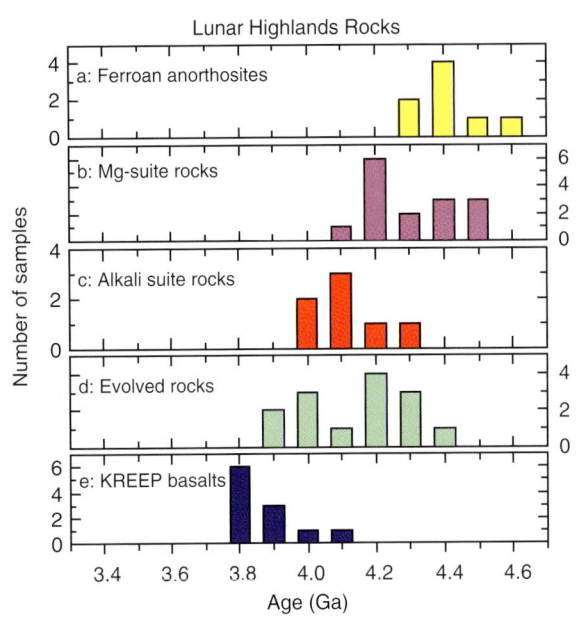

Figure 4 Radiometric ages for lunar highlands rocks. (a) Ferroan anorthosites. Ages include: (i) Sm–Nd internal mineral isochron ages of clasts that are "pristine" according to the definition of Warren and Wasson (1977); (ii) ^{39}Ar–^{40}Ar ages of anorthositic clasts for which a significant fraction of the total Ar released during stepwise heating shows an "ancient" age in excess of ~4.1 Ga. Literature references for five of the ages are given in Nyquist and Shih (1992). Sm–Nd ages for clasts from Apollo 16 samples 67016 (Alibert *et al.* 1994), 62236 (Borg *et al.* 1999a), and 67215 (Norman *et al.* 2000), as well as an ^{39}Ar–^{40}Ar age determination for lunar meteorite Yamato 86032 (Bogard *et al.* 2000) post-date that summary. (b) Mg-suite rocks. Ages are plotted for fifteen rocks or breccia clasts: ten norites, two gabbronorites, two troctolites, and a dunite. Six of the chosen ages were determined by Sm–Nd internal isochrons, five by Rb–Sr internal isochrons, four by U–Pb internal isochrons or ion probe zircon analyses, and two by ^{39}Ar–^{40}Ar stepwise analysis. There is substantial overlap of the ages determined by the different methods, each method yielding ages in the range 4.2–4.5 Ga. Several rocks have ages that are concordant by two or more methods. (c) Alkali-suite rocks. Three Apollo 14 samples have been dated either by Sm–Nd, or combined Sm–Nd and Rb–Sr isochrons. Three additional fragments from Apollo 14 breccias have been dated by U–Pb–zircon. A clast from Apollo 16 breccia 67975 has been dated by the ion probe U–Pb–zircon method as 4.339 ± 0.005 Ga, oldest among the alkali suite. (d) Evolved rocks. Eleven of the fourteen ages shown in the panel are by ion probe U–Pb–zircon analysis (Meyer *et al.* 1996). They show wide variation and an apparent bimodal distribution. The other ages are by Rb–Sr, Sm–Nd, and K–Ca isochron techniques. (e) KREEP basalts. The histogram includes samples from the Apollo 14, 15, and 17 sites. The youngest values are for Apollo 14 basalts, some of which have been questioned as possible impact melts. Also, the youngest ages are 3.84–3.85 Ga, which fall into the 3.8 Ga "bin."

Stone Mountain, and thus a geologically different portion of the Descartes formation than the North Ray crater FFBs.

One of the old ages plotted in Figure 4a is for an anorthositic rock fragment returned by the Luna 20 mission from a landing site on the northern border of Mare Fecunditatis with the adjacent highlands near the eastern limb of the Moon (Figure 3). Another of the old ages is for a lunar meteorite. Because of the preponderance of highlands area on the lunar farside, this meteorite has a high probability of being from the farside highlands. Thus the Luna 20 and lunar meteorite ages extend the area of known ~4.4 Ga FANs beyond the Descartes formation and probably to the lunar farside.

The comparatively young age, 4.29 ± 0.06 Ga and especially the positive initial $\varepsilon_{Nd} = +3.1 \pm 0.9$ of 62236 (Borg et al. 1999a) present a difficulty for "classical" LMO models. The positive ε_{Nd} implies a period of radiogenic growth of $^{143}Nd/^{144}Nd$ in a source region that was depleted in the light rare earth elements (LREE), La–Sm, compared to the heavy rare earth elements (HREE), Gd–Yb. Classical LMO models imply monotonically increasing LREE in the magma ocean as mafic cumulates settle out, and plagioclase flotation cumulates formed from them should have slightly *negative* values of ε_{Nd}. Instead, all of the FANs analyzed so far have slightly to significantly *positive* values of ε_{Nd} (Snyder et al. 2000a, Norman et al. 2000). The isotopic characteristics of 62236, and to a lesser extent the other FANs, imply radiogenic growth in an LREE-depleted environment like that of the hypothetical LMO mafic cumulates.

Models like that of Hess and Parmentier (1995) may be able to accommodate positive ε_{Nd} values, if they allow remelting of early-crystallized mafic phases and convective mixing with late-crystallizing plagioclase. Nevertheless, it is well to bear in mind that the extent of the verifiable presence of FAN on the lunar surface is limited. There is little direct evidence of an early, Moonwide, plagioclase-rich flotation crust, as hypothesized in the LMO model, among the lunar samples. In fact, the samples from two of the five "highland" sites (Apollo 14 and Apollo 17) contain no evidence of it at all. Furthermore, there is a paucity of anorthosite in soils from the Apollo 11 site, situated close to the Central Highlands. This may simply imply that lateral displacement of lunar material from its point of origin as ejecta from meteorite impacts is quite limited. Nevertheless, one might conclude that sources of FAN in the highlands neighboring the Apollo 11 site are limited.

Data from ground-based spectral reflectance studies and geochemical data obtained from the Apollo, Galileo, and Clementine spacecraft suggest that vast areas of the lunar highland surface consist of material with less than 4% FeO (Papike et al. 1998), however. Although this can be considered consistent with FAN composition, the "Genesis Rock" by comparison, contains >99% plagioclase, and essentially no FeO. Material with <1% FeO is exposed in the inner uplifted rings of some multi-ring basins. It might be supposed that the surface expression of FAN has been obscured by impact activity, but the orbital data suggest that the crust actually becomes more mafic with depth. Also, it has been argued that anorthosite may have been absent from the crustal segment that was the target of the Serenitatis basin-forming impact. Papike et al. (1998) concluded that: "Considerably more information is needed before the vertical and horizontal distribution of ferroan anorthosite in the lunar crust can be satisfactorily described and used in understanding lunar crustal evolution." Thus, the positive ε_{Nd} values of the FANs, although difficult to understand in the context of the LMO model, are not necessarily in conflict with other observations.

The crustal Mg-suite rocks: plutonic cumulates in a FAN Crust?

The Mg-suite rocks form the second major group of pristine lunar highland rocks. Unlike the FANs and ferroan noritic anorthosites, which are variations on a mineralogical theme, the Mg-suite rocks are mineralogically diverse. They include dunites (olivine cumulates), troctolites (olivine–plagioclase cumulates), norites (plagioclase–orthopyroxene cumulates), and gabbronorites (plagioclase–two pyroxene cumulates; Papike et al. 1998). The Mg-suite rocks are separated from FANs on the basis of plagioclase and mafic mineral compositions: Mg-suite rocks have higher values of Na/(Na + Ca) in plagioclase at a given ratio of Mg/(Mg + Fe) in the mafic minerals than do the FANs. These differences in major mineralogy correspond to generally higher trace element abundances in the Mg-suite rocks. This is important for radiometric age determinations because it means generally higher abundances of K, Rb, Sm, and U. Accordingly, age determinations for the Mg-suite rocks are easier than for the FANs, and a substantial number of ages were measured for these rocks in the decade following the end of the Apollo program. The compilation of Nyquist and Shih (1992) contains most of the available ages of Mg-suite rocks, and is the database for the age histogram shown in Figure 4b.

A widely held view, due primarily to James (1980), is that the Mg-suite rocks crystallized as plutons within an already-formed anorthositic crust. Mg-rich magmas may have intruded the lunar crust, forming layered mafic complexes similar to the Stillwater Complex of Montana, USA (Papike et al. 1998). The radiometric ages provide some support for this view in that the peak in the age distribution occurs at a lower value of ~4.2 Ga compared to the ~4.4 Ga age peak for the FANs. The statistical significance of this observation is limited, however, given the small number of measured ages, and the dominance of ages from the Apollo 16 site in the database for FANs. In our view, an equally important observation is that there is significant overlap between the ages of the two groups. If FANs did indeed crystallize directly from the magma ocean, then (a) this was an extended process, lasting ~200–300 Ma, and (b) intrusion of some Mg-rich magmas into newly formed crust must have begun almost immediately as the crust began to form.

Nine of the fifteen ages plotted in Figure 4b were obtained for rocks returned by the Apollo 17 mission to the Taurus Littrow valley located between the maria Serenitatis and Tranquilitatis (Figure 3). Two of the corresponding samples were collected at Station 2 on the flank of the South Massif, part of the Taurus Littrow Mountains. One of these was dunite 72417, dated at 4.51 ± 0.10 Ga (Papanastassiou and Wasserburg 1975b), and another was norite 72255, dated at 4.13 ± 0.05 Ga (Compston et al. 1975), both by Rb–Sr. These two samples were collected within 20–30 m of one another. Sample 72255 was a clast-rich impact melt sampled from Boulder 1, which probably rolled down from an upper portion of South Massif (Ryder 1993). The 4.13 Ga age was obtained for one of the clasts.

Astronaut H. Schmitt recognized sample 72417 as a clast in Boulder 3 at Station 2, which probably rolled down from near the top of the massif. Schmitt described it in the field as "olivine and something." Later examination showed it was originally a coarse-grained igneous rock consisting mainly of magnesian olivine (Ryder 1993). A sample of the adjoining matrix (72435) has been dated at ~3.9 Ga by both the Rb–Sr and ^{39}Ar–^{40}Ar methods. This clast appears to have been excavated from a deep crustal pluton by the Serentitatis basin-forming impact ~3.9 Ga ago. It seems to be direct evidence that some Mg-suite rocks were part of the earliest-formed lunar crust. The younger age of the norite clast in 72255 suggests that some crustal plutons formed much later, as well. A second norite clast, from polymict impact melt breccia 73215, was dated at 4.19 ± 0.01 Ga by the ^{39}Ar–^{40}Ar method (Jessberger et al. 1977). As for 72255 and other similar melt rocks at the site, the matrix was dated at ~3.9 Ga, establishing this as the age of the Serenitatis Basin.

The other old ages for Mg-suite rocks from the Apollo 17 site were found for a troctolite (76535) from Station 6 at the foot of the North Massif, a norite (77215) from Station 7, also at the foot of the North Massif, and two norites (78155 and 78235–78236) from the foot of the Sculptured Hills. The age of the troctolite has been a point of controversy. We have plotted the internal Sm–Nd age of 4.26 ± 0.06 Ga (Lugmair et al. 1976) in preference to the internal Rb–Sr age of 4.61 ± 0.07 Ga (Papanastassiou and Wasserburg 1976). The controversy was initially intensified by two questions: (a) Do any lunar samples have measurable ages within the "gap" between ~4.5 and ~4.0 Ga ago?

(b) Are any Mg-suite rocks old enough to date to the formation of the lunar crust itself? Later data seem to have answered both of these questions in the affirmative, lessening somewhat the significance of the exact crystallization age of 76535.

Nearly overlooked in the controversy over the 76535 age was the age of norite 77215, dated at 4.38 ± 0.03 Ga by the Rb–Sr internal isochron method (Nakamura et al. 1976). In part because of the discordance between the Rb–Sr and Sm–Nd ages of 76535, there is a tendency among lunar scientists to trust only the Sm–Nd ages of highland rocks. This viewpoint is in part justified by the possibility that Rb–Sr ages can be biased to erroneously high values by volatile loss of Rb accompanying post-impact thermal annealing. However, more generally "disturbed" isotopic systematics marked by failure of the Rb–Sr data to form an isochron is the most common expression of this phenomenon. We adopt the viewpoint that a linear alignment of data along an isochron within the error limits of the data should be considered a valid age in the absence of data to the contrary. The samples from Stations 6 and 7 discussed above appear to be remnants of plutonic rocks excavated into the Serenitatis rim by the Serenitatis Basin event, and that rolled downslope to their eventual collection site.

Norite 78155 was described in the Apollo 17 Preliminary Science Report as a friable feldspathic breccia that was "a particularly good candidate for Sculptured Hills material." Its ^{39}Ar–^{40}Ar and U–Pb ages of 4.17 ± 0.02 Ga (Oberli et al. 1979) place it among the "young," ~4.2 Ga plutonic Mg-suite rocks. The other Station 8 rock(s), 78235-8 were sampled from a half-meter-sized boulder described as having "originated from outside the sampling area" (Apollo 17 Preliminary Science Report 1973). This boulder, however, is important to overall lunar chronology. This is in part because of its size relative to the other samples we have discussed, many of which are severely crushed mm- to cm-sized clasts in polymict breccias. The 78235-8 boulder, although intensely shocked, has a well preserved, coarse-grained mineralogy, and in principle is well suited to age-dating methods. Indeed, it has ages determined by several methods which are concordant, or nearly so. It appears to belong to the "old" ~4.4 Ga group of highland rocks, although that assignment is somewhat uncertain. Carlson and Lugmair (1981) reported that the Sm–Nd isotopic systematics showed post-crystallization disturbance, presumably due to shock, but indicated crystallization in the lunar crust "about 4.34 AE (Ga) ago." Nyquist et al. (1981) also encountered disturbed isotopic systematics, but argued for an older age from the Rb–Sr and Sm–Nd data of the "most retentive" minerals, giving ages of 4.38 ± 0.02 Ga and 4.43 ± 0.05 Ga as preferred values from Rb–Sr and Sm–Nd, respectively. ^{39}Ar–^{40}Ar and U–Pb data are generally concordant with these values, but do not resolve the age within the ~4.34–4.43 Ga uncertainly. Samples 78235–8, 76535, and 72417 are our best and largest examples of lunar Mg-suite plutonic norites, troctolites, and dunites, respectively.

Three of the ages included in our compilation are from norite clasts from two Apollo 15 breccias, 15445 and 15455. The parent breccias were sampled at Station 7, Apollo 15, on the rim of Spur crater, at the foot of the Apennine Front Mountains, within about 20 m of the sampling site for the Genesis Rock, FAN 15415. Samples 15415 and 15455 were collected within a few meters of one another. Like the Taurus Littrow Mountains, the Apennine Front Mountains were uplifted by a basin-forming impact, in this case the impact that formed the Imbrium basin. Thus, it seems clear that these rocks were excavated into the basin rim from some depth in the lunar crust. They were either excavated from the megaregolith at the base of the Apennine Front by Spur crater, or rolled down the mountain and came to rest at the sampling site. In either case, one cannot distinguish a hypothetical magma ocean origin for the FAN from a hypothetical "crustal pluton" origin for the norite. An ancient Rb–Sr age of 4.54 ± 0.12 Ga was determined for the 15455 norite (Shih et al. 1993). Unfortunately, the Sm–Nd data were of inadequate quality for a precise age determination. An imprecise Sm–Nd age of 4.53 ± 0.29 Ga was inferred from the whole rock and pyroxene data only (Shih et al. 1993). Two discordant Sm–Nd ages of 4.46 ± 0.07 Ga, and 4.28 ± 0.03 Ga were obtained for different subsamples of the same norite clast in 15445. Because of the complex "breccia-in-breccia" textures of many highland clasts, Shih et al. (1993) suggested that the two ages probably represented two different, but similar, lithologies. If this is not so, the older age probably should be favored as the "true" age of the norite, which probably experienced a major thermal event at ~3.9 Ga, as recorded in the ^{39}Ar–^{40}Ar release pattern of the 15455 norite.

Shih et al. (1993) noted that the ε_{Nd} values of the norites appeared to trend from $+0.71 \pm 0.26$ at 4.46 ± 0.07 Ga ago to -0.35 ± 0.01 at 4.28 ± 0.03 Ga, coincident with the trend for KREEP basalts. This is consistent with the KREEP-like relative REE abundance patterns calculated for the parent magmas of norites by Papike et al. (1996). The absolute abundances calculated for the REE in the parent magmas range from ~200 to 500 times chondritic abundances, similar to those of so-called high-K KREEP basalts. Snyder et al. (1995a) modeled the fractional crystallization of a KREEP basalt magma, and found that magnesian plutonic norites like the Apollo 15 norites could be derived from the first ~55% of the crystallization sequence. Subsequent cumulates would be like the alkali norites to be discussed later. Thus, a strong case can be made that the norites were derived as cumulates of KREEP basalt magmas.

Papike et al. (1998) note that the largest bodies of igneous rocks actually observed by astronauts in the lunar highlands are cataclasized (impact-crushed) norites. Two meter-sized norites were sampled during the Apollo 17 mission. Nearly all the norites in the returned samples are from the Apollo 15 and 17 landing sites, none have so far been observed in the Apollo 16 samples. Papike et al. (1998) suggest that the pre-impact lunar crust in the Imbrium and Serentitatis impact areas may have been noritic. Whether these observations are consistent with the model of Mg-suite rocks forming as Stillwater-type intrusions in a plagioclase-rich flotation crust is unclear to the present authors. Like Shih et al. (1993) we believe that many observations concerning pristine highland rocks are difficult to explain by the LMO hypothesis alone. Some involvement of multiple petrogenetic processes of the types suggested by several authors seems more likely. Examples of processes which have been suggested are "serial magmatism" (Walker 1983) for Mg-suite rocks, and "multiple diapiric intrusions" (Longhi and Ashwal 1985) for FANs.

In any case, Mg-suite rock formation appears to have begun near the end of the magma-ocean era when REE abundances in the ocean had reached high levels because of prior fractionation of incompatible element-poor mafic minerals. We characterize this lunar era as the urKREEP era, following Warren and Wasson (1979). Although a rather broad operational definition of "KREEP" is often used by lunar scientists, we believe it advisable to distinguish the urKREEP, directly linked to heightened incompatible element abundances in a crystallizing LMO, from a later KREEP basalt, probably linked to remelting of urKREEP-enriched basalt source regions.

Evolved rocks: granites/felsites and quartz monzodiorites

During the interval ~4.3–4.0 Ga ago, the volume of lunar magmas crystallizing from the initial lunar magma ocean decreased, and the volume of new magmas generated within the Moon by radioactive decay of long-lived heat-producing elements, K, U, and Th, increased. The two eras probably overlapped, leading to magma production via a series of complex petrogenetic processes involving magma mixing, assimilation, and fractional crystallization. Some authors have suggested that metasomatic processes also played a role. Important in the thermal evolution of the Moon is the point at which it "turned on" to producing magmas from internal radioactive heat sources. Given the confusion of petrogenetic processes possibly occurring simultaneously, it appears difficult to identify when the "turn on" occurred. However, we have two clues: (a) An apparent change in the style of magmatism for incompatible element-rich lunar "granites"; and (b) the onset of mare basalt volcanism. Here we consider the lunar granite data, leaving the mare basalt data for later discussion.

The important role of incompatible element-rich lithologies in lunar evolution became increasingly evident as the Apollo missions progressed. The presence of such lithologies in the lunar soil returned by Apollo 11 was inferred from its chemical composition and its Rb–Sr model age of ~4.6 Ga. Chief among the lithologies suggested for the "cryptic," or "magic" component was "lunar granite"; a suggestion reinforced by return of granitic breccia 12013 by the Apollo 12 mission. The age of two granitic clasts in 12013, as determined by Rb–Sr internal isochrons, turned out to be ~4.0 Ga, however, not ~4.6 Ga as expected (Lunatic Asylum 1970). Nevertheless, a "whole rock" age of ~4.5 Ga was obtained from Rb–Sr data for separate clasts in the sample, giving it the Rb–Sr characteristics of the "magic component." However, impact melt glasses of KREEP composition (Meyer and Hubbard 1970) were discovered in the Apollo 12 soil at about the same time, and seemed to provide a more satisfactory fit in mixing models which sought to unravel the soil composition into end-member components (Meyer et al. 1971). The Apollo 14 mission to an unusual highlands area called the Fra Mauro Formation (Figure 3) confirmed an important role for KREEP in the lunar story. Although nearly every Apollo 14 sample seemed to be a KREEP-rich breccia, granites like 12013 were rare. Thus, it appeared that KREEP, not lunar granite, was the dominant incompatible element-rich end point of lunar differentiation. The Rb–Sr model ages of numerous samples of KREEP were ~4.3–4.4 Ga, leading to the conclusion that lunar differentiation may have required ~200 Ma to reach the stage where trace element abundances in the magma ocean were comparable to those for KREEP, or more precisely, urKREEP. The name urKREEP was coined by Warren and Wasson for "original KREEP," produced via differentiation of the magma ocean. But, how can one be sure that urKREEP actually was present ~4.3–4.4 Ga ago?

A few samples that are very enriched in trace elements hold the key. One sample in particular, the quartz-monzodiorite clast (QMD) in Apollo 15 breccia 15405, is one of the most incompatible element-enriched samples from the Moon, and also is large enough (2.5 g) that a number of geochemical studies have been made of it. Early attempts to date it by the Rb–Sr technique failed because of "recent" Rb loss, presumably at the time of major Ar outgassing and Pb volatilization ~1.3 Ga ago. Meyer et al. (1996), however, successfully dated it by applying the ion probe technique to the U–Pb system in zircon. The age they obtained, 4.294 ± 0.026 Ga, directly confirms the presence of urKREEP on the Moon ~4.3 Ga ago. Interestingly, this age is also the same as the Rb–Sr model ages of Apollo 15 KREEP basalts, implying little or no fractionation of the Rb/Sr ratio during petrogenesis of the basalts.

The ages found by U–Pb–zircon analyses of other evolved rocks (11 of the 14 ages shown in the "evolved rocks" panel in Figure 4d show wide variation and an apparent bimodal distribution. "Granophyre" 14303,29, with a KREEP-like REE pattern similar to the 15405 clast, has a similar age as well: 4308 ± 3 Ma (Meyer et al. 1996). Two other granites/felsites have bow-shaped REE patterns similar to those first reported for the 12013 granites, and their ages also are similar to those of the 12013 granites. One of these, the "large" (1.8 g) granitic clast in breccia 14321, was found to have concordant ages of 4.06 ± 0.10 Ga by three isochron techniques: Rb–Sr, Sm–Nd, and K–Ca. The U–Pb zircon age was found to be 4.010 ± 0.002 Ga, in agreement with the isochron methods. A second sample with a bow-shaped REE pattern, 12033,507, also a "large" (1.2 g) granite, has an age of 3.883 ± 0.003 Ga, apparently slightly younger than the 3.96 ± 0.05 Ga age of the 12013,29 granite. Although REE data are not available for all the "granites" included in the study, there is correspondence of young ages with bow-shaped REE patterns and old ages with KREEP-like patterns for those samples for which both types of data exist. Meyer et al. (1996) suggest an origin related to partial melting in the lunar interior for the young granites, and an origin related to the magma ocean for the old ones.

Meyer et al. (1996) also argue from their data for continuous granitic magmatism between ~4.4 and ~3.9 Ga ago. This conclusion is apparent from the data in Figure 4d, but the change in REE abundance patterns between the older QMD and younger granites/felsites also should be considered as indicating a change in the style of magmatism. The bow-shaped REE pattern of the granites/felsites is likely to be the result of phosphate fractionation at a higher degree of differentiation than reached by the QMD. This may be indicative of fractionation within a more confined volume of magma, consistent with more localized magmatism for the granites/felsites. If the bimodal distribution in Figure 4d is due to "old" QMD and young granites/felsites, the QMD may include ages as young as ~4.2 Ga. Those ages then would represent crystallization of the "last dregs" of the magma ocean. The oldest QMD ages are ~4.3 Ga, and provide an estimate of the time when the magma ocean first reached the stage of incompatible element enrichment consistent with their formation. The oldest age in the second peak in the age distribution, ~4.1 Ga, is that for the 14321 granite. This age probably represents the time of transition from magmatism associated with the lunar magma ocean to that augmented by heating from internal sources. Such heating might have begun in lower-crustal "pockets" highly enriched in the heat-generating elements during solidification of the last dregs from the magma ocean. Possibly, the young ages do not represent remelting, but rather failure of the last dregs of the magma ocean to solidify until intrusion into the crust cut them off from the heat of the main magma. In that case, heat generated in them by radioactive decay may simply have prolonged their solidification to younger times. This explanation, however, would not lead to a bimodal distribution of ages, but rather to a monotonic tailing off towards younger values. Thus, the bimodal distribution of ages for evolved rocks suggests "turning on" of internally generated lunar magmatism ~4.1 Ga ago.

KREEP basalts: descendants of urKREEP

As already noted, lunar "KREEP" was first discovered in Apollo 12 soils by Meyer and Hubbard (1970) and Hubbard *et al.* (1971). These co-workers of P.W. Gast, and Gast himself, were convinced that the material initially found only as impact-melted glass with a basaltic composition did indeed represent a lunar basalt type, produced by partial melting of the lunar interior. Subsequent geochemical analyses have shown that most of the incompatible elements in lunar rocks and soils seem to be associated with KREEP in one of its various forms. However, several of the many roles KREEP has been called on to fulfill require an early, Moonwide presence. "KREEP basalts" such as sample 14310, were discovered on the next successful mission, but were found to be only ~3.85 Ga old. Textural evidence suggested that these basalts might be impact melt rocks, however, and it was not until the Apollo 15 mission that basaltic-textured samples that everyone agreed were true basalts were returned. These, too, had ages near ~3.85–3.90 Ga. Thus, KREEP basalts seemed to be too young to fulfill the many roles assigned to "KREEP." This problem was solved by Warren and Wasson (1979) who introduced "urKREEP," a residuum of the magma ocean, to do early, Moonwide duty.

How are KREEP basalts related to urKREEP? Warren (1988) noted that the *mg*-number ($mg = 100 \times$ MgO/(MgO + FeO) in molar units) expected for urKREEP would be very low, whereas those of KREEP basalts are typical of other lunar basalts as well. To account for this and other chemical characteristics of KREEP basalts, he proposed that Mg-rich melts from lower in the crystallizing LMO passed through the developing crust/mantle boundary and assimilated or mixed with the urKREEP residuum. However, as shown in Figure 4e, there is no direct evidence that Mg-rich magmas were being intruded into the crust at times as late as required for the KREEP basalt magmas.

Figure 4e shows a peak in the KREEP basalt age spectrum at 3.8–3.9 Ga. No ages as low as 3.8 Ga have actually been measured for KREEP basalts; those 3.8 Ga ages shown in the figure are the result of 3.84 and 3.85 Ga ages falling into the 3.8 Ga "bin." The histogram also possibly is biased to "low" values by the inclusion of values for Apollo 14 basalts, some of which have been questioned as possible impact melts. We agree with Papike *et al.* (1998) that later remelting is required to generate the actual KREEP basalts, if their source regions formed in the manner suggested by Warren (1988).

There is widespread consensus that the Apennine Bench Formation is the origin of the Apollo 15 KREEP basalts (Spudis *et al.* 1988). This geological formation is interpreted as arising from a basalt flow into the Imbrium basin very shortly after its formation. Examples of slightly older KREEP basalts were found as two clasts in breccia sample 72275, from Boulder 1, Station 2, at the Apollo 17 landing site. These basalts probably were part of the pre-Serenitatis protolith. Thus, their ages of 3.98 ± 0.04 Ga (Rb–Sr) and 4.08 ± 0.05 Ga (Rb–Sr, Sm–Nd) combined with Apollo 15 KREEP basalt ages of 3.84 ± 0.05 Ga (^{39}Ar–^{40}Ar) to 3.90 ± 0.04 Ga (Rb–Sr, Sm–Nd) bracket the formation of the Serenitatis and Imbrium basins.

The onset of lower crustal remelting indicated at ~4.0–4.1 Ga required to produce the KREEP basalts appears to have occurred approximately simultaneously with remelting in the source regions of both rock types. It also is plausible that remelting would have "turned on" in these source regions before it did in the sources of most mare basalts.

Alkali-suite rocks: cumulates of uncertain origin

We have left the alkali-suite rocks (sometimes called highlands alkali suite or HAS) to the end of this discussion of the ages of lunar highland rock types because their context is poorly known. As noted by Papike *et al.* (1998), there is no formal definition of an alkali rock. Hubbard *et al.* (1971) are credited with identification of the first "rock" of this type, actually a fragment from the Apollo 12 soil, which they termed "KREEP anorthosite." The anorthite content of this fragment was low for a lunar anorthosite (An$_{88}$), while its REE contents were high compared to the FAN 15415. Hubbard *et al.* (1971) hypothesized that this fragment came from an anorthositic cumulate from a KREEP basalt magma. Other authors have also interpreted the alkali rock fragments as cumulates from KREEP basalt or QMD magmas (Snyder *et al.* 1995b). Warren (1993) described the alkali-suite rocks as probable intrusives in the lunar crust.

Most alkali-suite rocks have been found among the Apollo 12 and 14 samples, leading to their characterization as a "western" rock suite. Only one eastern rock clast, from breccia 67975 at Apollo 16, is considered to belong to the suite. The 67975 clast has been dated by the ion probe U–Pb–zircon method as 4.339 ± 0.005 Ga, oldest among the alkali suite (Figure 4c). Three other fragments also have been dated by U–Pb–zircon, and three additional samples have been dated either by Sm–Nd, or combined Sm–Nd and Rb–Sr isochrons. All of these are Apollo 14 samples. Shih *et al.* (2000) report Rb–Sr and Sm–Nd isochrons for two of the largest rock fragments, alkali anorthosite clast 14047c (~1.65 g) and olivine norite clast 14318c (~1.5 g). Although the weighted average age from the Rb–Sr and Sm–Nd isochrons is 4.04 ± 0.08 Ga, and is comparable to that of the KREEP basalts, the low initial $^{87}Sr/^{86}Sr = 0.699204 \pm 0.000037$ is too low for 14047c to be derived from a KREEP basalt magma, for which $^{87}Sr/^{86}Sr \sim 0.7003$ at ~4.0 Ga ago. A negative value of $\varepsilon_{Nd} = -1.06 \pm 0.52$ for 14047c is consistent with a KREEP-like LREE enrichment, however. This value is similar to that obtained for an age of 4.11 ± 0.05 Ga for alkali anorthosite 14304,267 by Snyder *et al.* (1995b), and also with their KREEP basalt accumulation model.

The weighted average age of the other alkali-suite rock dated by Shih *et al.* (2000), alkali norite 14318c, is 4.11 ± 0.05 Ga. This rock has a comparatively high initial $^{87}Sr/^{86}Sr = 0.700168 \pm 0.000066$, which could be consistent with crystallization from a KREEP basalt magma. However, $\varepsilon_{Nd} = -0.07 \pm 0.63$ is higher than expected for a source material with urKREEP or QMD characteristics, and Shih *et al.* (2000) conclude that this rock also could not have been derived from a magma with the chemical characteristics of Apollo 15 KREEP basalts. Rather, these isotopic results are consistent with the REE data and models of Shervais and McGee (1998). They modeled the equilibrium melt for 14318 as enriched in REE to ~500 times chondrites, but with chondritic, or possibly even an LREE-depleted abundance pattern instead of the characteristic LREE-enriched pattern of KREEP. These authors suggest that the REE data of western Mg-suite cumulates are best explained by mixing of residual magma ocean urKREEP with ultramagnesian komatiitic partial melts from the deep lunar interior.

Although there is apparent progress in deducing the origins of the alkali-suite rocks, we agree with Papike *et al.* (1998) that "their origins remain obscure." It is apparent, however, that a unique series of magmatic events occurred in the western lunar hemisphere in the pre-Imbrium era. Thus, it probably is not possible to think of early lunar evolution as a series of globally symmetric events. It is apparent that the LMO model forms a useful context in which to discuss early lunar evolution, but it is also apparent that the real petrogenetic environment was more complex than inferred from the simplest form of this model.

A lunar cataclysm: impact resetting of ages

Even a casual look at the Moon through a small telescope will reveal that it is covered with craters caused by the impact of objects of various sizes over the history of the Solar System. The violent, early meteorite bombardment that the Moon experienced was a major feature of lunar history, and it all but erased the evidence of early lunar evolution. The bombardment also made immeasurably more complicated the task of identifying early lunar rock types and deciphering early events. Painstaking efforts have been required to identify pristine rock samples, often tiny, that survived the bombardment, and to extract the information they hold. In addition, the nature of this bombardment and the time when it occurred is a question of considerable scientific interest.

Impacts produce heat, and in the case of large impacts that energy can be sufficient to reset radiometric ages, especially K–Ar ages. From the above discussion of formation times of lunar highland rocks, we might expect most highland samples to give radiometric ages of >4.1 Ga. However, most determined lunar highland ages are considerably less than 4.4 Ga, and many are in the range of ~3.8–4.1 Ga. A histogram of Rb–Sr and $^{39}Ar-^{40}Ar$ ages for highland samples returned by Apollo missions 14, 16, and 17 (Turner 1977, Nyquist 1977, Nyquist and Shih 1992, Bogard 1995) shows a broad distribution over ~3.8–4.1 Ga (Figure 5). In addition, analyses of the U–Th–Pb isotopic systems in lunar highland rocks indicate significant mobilization of Pb and resetting of this chronometer in a similar time period (Tera *et al.* 1974). It was recognized early that these younger ages were the result of impact resetting.

From photogeology it is recognized that the several large, nearside impact basins would have contributed varying amounts of impact ejecta to individual highland sites. Thus, to the extent that ejecta at a given locale can be attributed to a specific impact basin, the $^{39}Ar-^{40}Ar$ age of highland

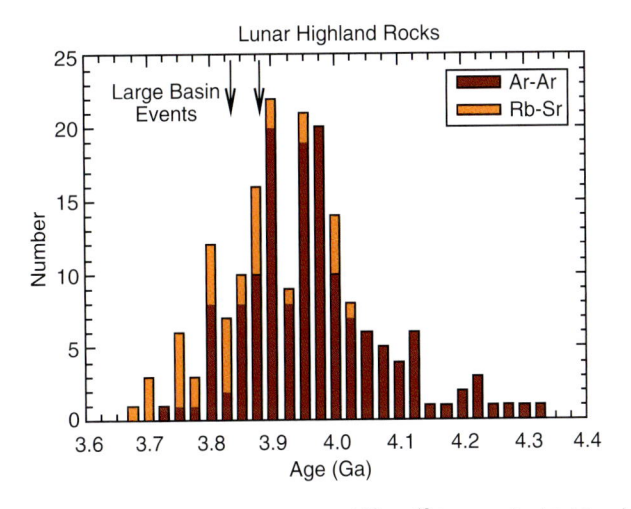

Figure 5 A histogram of Rb–Sr and $^{39}Ar-^{40}Ar$ ages for highland rocks returned by Apollo missions 14, 16, and 17 (Nyquist and Shih 1992, Bogard 1995).

rocks might date the formation time of large nearby basins. The many ^{39}Ar–^{40}Ar ages that have been reported for highland samples give a significant spread but with significant overlap in ages. For example, the Fra Mauro formation at Apollo 14 is believed to have derived dominantly from the Imbrium basin. But, the ^{39}Ar–^{40}Ar ages of Apollo 14 rocks commonly fall into two age groups: 3.86–3.89 Ga for rocks associated with Cone Crater and ~3.81–3.84 Ga, for rocks not associated with Cone Crater (Turner et al. 1971, 1972, Huneke et al. 1972). Is this difference the result of more than one resetting event, or incomplete resetting among samples with slightly different histories or composition, or some other cause? (All ^{39}Ar–^{40}Ar ages reported in the original references have been corrected here to the ^{40}K decay constant and isotopic abundance recommended by Steiger and Jäger (1977). Rb–Sr ages have been recalculated to $\lambda_{87} = 0.01402\,\text{Ga}^{-1}$ (Minster et al. 1982).)

Two small fragments returned in the soil from the Luna 20 highland site gave ^{39}Ar–^{40}Ar ages of 3.84 Ga, and were interpreted as possibly dating the formation of the Imbrium or Crisium impact basins (Podosek et al. 1973). In a following study, Jessberger et al. (1974) gave the time of major outgassing of KREEP-rich breccia 65015 as 3.92 Ga. Because this sample was neutron irradiated with the Luna 20 sample, a time resolution of 80 Ma between the two samples was established, showing that if each sample were outgassed in a basin-forming event, then two such events must have occurred within 80 Ma. Furthermore, the ^{39}Ar–^{40}Ar age was concordant with the Rb–Sr age of 3.89 Ga (recalculated to $\lambda_{87} = 0.01402\,\text{Ga}^{-1}$) previously determined by Papanastassiou and Wasserburg (1972) for 65015. The Rb–Sr age was interpreted as a secondary time of re-equilibration of the Sr isotopes in most of the mineral phases of the rock. Plagioclase in 65015, a remnant of the protolith from which the breccia was formed, was incompletely re-equilibrated, however. The interpretation that the breccia was formed by a basin-forming impact into an anorthositic crust follows readily.

Jessberger et al. (1974) appealed to calculations of basin ejecta thicknesses at the Apollo 16 site by McGetchin et al. (1973) to assign the Nectaris basin to this event. Maurer et al. (1978) determined the ^{39}Ar–^{40}Ar ages of many breccia fragments from the soil of the Apollo 16 highland site and found them to fall primarily into two groups. Samples in one group were anorthositic breccias and gave ages in the range 4.06–4.15 Ga; a second group of noritic anorthositic breccias gave younger ages of 3.83–3.96 Ga. The average age of this latter group was 3.92 Ga, which they attributed to Nectaris, in agreement with the age assigned by Jessberger et al. (1974). Maurer et al. (1978) interpreted their older age group to represent resetting in intermediate-sized impacts that produced craters ~10^2 km in diameter, and the younger age group to represent resetting by the Nectaris basin. However, in another study of Apollo 16 soil fragments, Schaeffer and Hussain (1974) found similar ^{39}Ar–^{40}Ar age groupings, but assigned the older group to Nectaris, placing its age at 4.18 Ga. Schaeffer and Husain (1974) appealed to the same ejecta thickness study by McGetchin et al. (1973) as did Jessberger et al. (1974), but made a different basin-age assignment. Chronologies of basin formation have also been proposed in several other age studies (e.g. Turner 1977, BVSP 1981, Deutsch and Stöffler 1987). Swindle et al. (1991) suggested an age of ~3.89 Ga for the Crisium basin. Some uncertainty still exists in the identification of the source basin for many dated samples, and in the exact age of the large basins in spite of the relatively large database. For example, ejecta identified with Imbrium often have been dated, yet, two relatively recent derivations of the age of Imbrium differ by ~80 Ma (Deutsch and Stöffler 1987, Dalrymple and Ryder 1993).

Dalrymple and Ryder (1993, 1996) determined ^{39}Ar–^{40}Ar ages of Apollo 15 impact melts and Apollo 17 breccia clasts, whose ages they argue have a high probability of dating the Imbrium and Serenitatis basin events, respectively. Their preferred ages are 3.85 ± 0.02 Ga for Imbrium and 3.893 ± 0.009 Ga for Serenitatis. In a detailed evaluation of rock type, lunar stratigraphy, and radiometric ages, Ryder and co-workers recently made an analysis of the likely ages of several large lunar basins (Ryder et al. 2000, Hartmann et al. 2000). The key to this analysis is establishing an age for Imbrium of 3.85 Ga and an age for Nectaris of 3.90–3.92 Ga. The Serenitatis and Crisum basins are argued from stratigraphy and crater density to be intermediate in age to Imbrium and Nectaris, and an age of 3.89 Ga is suggested for Crisium. Humorum is probably also intermediate in age between Imbrium and Nectaris. Based on stratigrapy and crater densities, the Orientale and Schrödinger basins are younger than Imbrium, and probably have ages between 3.80 and 3.85 Ga. Ryder suggests that several large lunar basins likely formed in a time period as short as 50–100 Ma.

How can the distribution of young reset ages for lunar highland samples be explained? One explanation offered is that the accretion of the Moon occurred over a significant period of time, with a steep decline in the rate of large impacts. This concept explains the concentration of rock ages in the ~3.8–4.1 Ga period as the result of continual resetting of radiometric ages due to the high impact rate, until ~4.1 Ga ago when the impact rate had fallen low enough for some rocks to escape resetting (Hartman 1980, Baldwin 1981, Deutsch and Stöffler 1987, Ryder 1990). The occurrence of some older rock ages is explained as a small, but nonzero probability that some rocks would escape being strongly broken and heated during this intense lunar bombardment.

A somewhat different interpretation for lunar highland age resetting is that it was a period of enhanced impacts long after formation of the Moon. This period has been called the lunar cataclysm (Tera et al. 1974), and is viewed as a significant and short-lived increase in the frequency of impactors over a relatively narrow period of time. It was recognized early in lunar studies that the reset ages of lunar highland rocks were not all the same, and thus that the period of formation of large basins possibly occurred over a period of at least ~0.1 Ga (Jessberger et al. 1974, Maurer et al. 1978, Turner 1977). Wetherill (1986) suggested that these impactors were the result of late (~4.0 Ga ago) breakup of a large object of ~10^{23} g, and that much of the mass of lunar impactors during this cataclysm was contained in a relatively few large objects that formed several of the large nearside basins. Thus, the cataclysm explanation would account for the apparent resetting of many highland samples by relatively few large basins in a relatively narrow interval of time. The lunar cataclysm concept, to the extent that it characterizes a short period of bombardment by large objects, may explain less well those reset ages of >4.1 Ga observed in some Apollo 16 samples. It is also conceivable that both concepts of the early lunar bombardment are valid to some degree.

Both interpretations of the nature of the early cratering history of the Moon are presented in the paper by Hartmann et al. (2000). Whichever the explanation of the early lunar bombardment, the consequences for the Earth seem profound. The Earth could not have escaped an even more intense bombardment, and this probably is an important factor in the observation that rocks older than ~3.8 Ga are very rare on Earth. An intense early bombardment of the Earth also could have affected the early evolution of the atmosphere (large impactors can eject atmospheric gases) and even the conditions for the early establishment of life.

Apollo 14 "mare" basalts: pre-Imbrium and mostly aluminous

The Imbrium basin is the largest, and one of the latest, well-preserved lunar basins. Samples from two Apollo landing sites, Apollos 14 and 15, transect in time the formation of the Imbrium basin. Pre-Imbrium samples from Apollo 15 are all highland crustal rocks. The samples at the Apollo 14 site have also allowed the identification of pre-Imbrium mare basalts, or at least mare-basalt-like samples. This conclusion has evolved mostly from investigations of the Apollo 14 samples done well after the missions as part of so-called "breccia pull-apart" studies. In these studies, several large football-sized samples returned by the Apollo 14 mission were dissected and "mined" for identifiable lithic clasts. These clasts were then studied petrographically, mineralogically, chemically, and chronologically by consortia of investigators. Ages determined for the mare basalt clasts identified in these studies, combined with those reported earlier from studies of larger samples, are shown in Figure 6a.

The three youngest Apollo 14 ages shown in the figure fall into the 3.8 Ga bin due to round-off from values of ~3.85 Ga. Two of these are so-called very high potassium (VHK)

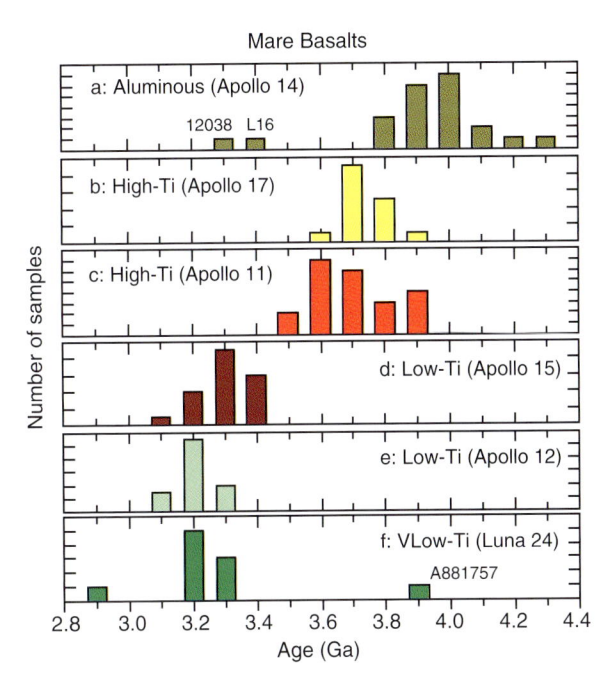

Figure 6 Histograms of the ages of lunar mare basalts categorized by chemical compositions and sampling site, where known. Ages are by the ^{39}Ar–^{40}Ar, Rb–Sr, and Sm–Nd methods. (a) Aluminous mare basalts. Ages are for two Apollo 14 basalts (14053 and 14072), and basaltic clasts in Apollo 14 breccias, Apollo 12 sample 12038 (3.31 ± 0.08 Ga), and a basaltic fragment from the Luna 16 landing site (3.41 ± 0.04 Ga). The Apollo 14 ages are all ≥3.85 Ga; three of the ~3.85 Ga ages fall in the 3.8 Ga bin. (b) High Ti basalts from Apollo 17. These basalts have been further divided into subgroups A, B, C, D, and U with average ages (Ga) as follows: A(6), 3.78 ± 0.01; B(5), 3.70 ± 0.02; C(2), 3.77 ± 0.07; D(1), 3.92 ± 0.04; U(2), 3.69 ± 0.01. Numbers in parentheses following subgroup type give the number of individual rocks analyzed. (c) High Ti basalts from Apollo 11. Subgroups and ages are: (i) High K: A(10), 3.58 ± 0.01; (ii) low K: B1(3), 3.67 ± 0.02; B2(3), 3.85 ± 0.02; B3(3), 3.71 ± 0.02; D(4), 3.87 ± 0.01. (d) Low Ti basalts from Apollo 15. Subgroups and ages are: quartz normative (8), 3.37 ± 0.01; olivine normative (10), 3.29 ± 0.01; picritic (1), 3.19 ± 0.04; ilmenite (1), 3.37 ± 0.03. (e) Low Ti basalts from Apollo 12. Subgroups and ages are: olivine (5), 3.22 ± 0.02; pigeonite (9), 3.16 ± 0.01; ilmenite (4), 3.17 ± 0.02. (f) VLT basalts: (i) Luna 24(8), 3.24 ± 0.01 Ga; (ii) lunar meteorite Asuka 881757, 3.87 ± 0.06 Ga. See Figure 7 for histograms of ages within the subgroups at each site.

basalts for which concordant K–Ar, Rb–Sr, and Sm–Nd ages were determined by Shih et al. (1986); the preferred age used for the diagram is 3.85 ± 0.02 Ga as determined by the ^{39}Ar–^{40}Ar ages. Thus, these precise ages are directly comparable to the majority of ages used to define the time of the Imbrium event (Figure 5), and imply these basalts solidified nearly simultaneously with the Imbrium impact. However, the petrographic studies of these samples showed them to be fragments of igneous rocks, and not impact melts. Their very high potassium contents were explained by Shih et al. (1986) as probably due to assimilation of granitic wall rock during magma genesis. These must have been some of the last pre-Imbrium basalt flows prior to formation of the Imbrium basin.

The same ^{39}Ar–^{40}Ar age of 3.85 ± 0.04 Ga was obtained for one of two clasts of lower, but nevertheless high, potassium content (HK basalts) from breccia 14304; no ^{39}Ar–^{40}Ar age was determined for the second clast. These two clasts have precisely determined Rb–Sr ages of 3.92 ± 0.04 and 3.96 ± 0.02 Ga, respectively (Shih et al. 1987); the older clast also has a concordant Sm–Nd age of 4.04 ± 0.11 Ga. The isotopic data for these two basalts can be explained by the assimilation model invoked for the VHK basalts. The younger ^{39}Ar–^{40}Ar age probably reflects Ar outgassing accompanying the Imbrium event. Shih et al. (1987) argued that aluminous mare basalt volcanism probably was derived from mantle sources underlying the highland crust, and that magmas from these sources were more likely to be contaminated by incompatible element-rich crustal materials than were other types of mare magmas. As discussed earlier in the chapter, very K-rich granitic clasts also are found in the Apollo 14 breccias, and have the same or older ages than these VHK and HK basalts. A probable lower crustal source for aluminous mare basalts was first suggested by Taylor (1975).

Two hand-specimen-sized samples, 14053 and 14072, are included in the compilation of Figure 6a. These samples have ages of 3.90 ± 0.02 Ga (^{39}Ar–^{40}Ar, Rb–Sr) and 3.97 ± 0.04 Ga (^{39}Ar–^{40}Ar, Rb–Sr, Sm–Nd), respectively. They are among those identified as pre-Imbrium by Wilhelms (1987). Another sample so identified is from the large "Big Bertha" breccia 14321, the source of most of the remaining aluminous mare basalt clasts included in the data compilation. Wilhelms (1987) presents the traditional interpretation of the Fra Mauro formation as fragmental ejecta from the Imbrium basin. Although others have challenged this view, it is consistent with the relative age relations among the samples. Pre-Imbrian ages are indicated especially strongly by ages of ~ 4.1–4.3 Ga for some of the aluminous mare basalt clasts in 14321 (Dasch et al. 1987). Taylor et al. (1983) had previously determined a Rb–Sr age of 4.20 ± 0.06 Ga for an olivine basalt clast in breccia 14305. In addition, many of the evolved and alkali-suite rocks included in the compilation of Figure 4 are from clasts in the Apollo 14 breccias, and have ages comparable to those of the Apollo 14 aluminous mare basalts. It seems clear that the Apollo 14 breccias contain a suite of rocks from the pre-Imbrium protolith that includes granites/felsites, alkali norites, anorthosites, aluminous mare basalts, and at least some more mafic rocks. Conspicuously lacking from this suite are the traditional lunar crustal rocks: FANs and magnesian-suite rocks. KREEP "basalts" from Apollo 14 are either examples of post-impact volcanism, as appears to be the case at Apollo 15, or they are impact melt rocks. In the latter case, their ages of $\sim 3.85 \pm 0.01$ Ga probably are a precise determination of the Imbrium basin age.

The above observations suggest a unique composition for the pre-Imbrium protolith. This in turn seems consistent with the high Th abundances observed for the Fra Mauro and adjacent areas by gamma-ray remote sensing on the Apollo 15, 16, and Lunar Prospector spacecraft. Thus, the need for caution when extrapolating conclusions based on Apollo 14 samples to Moonwide processes is evident. For example, we have argued from the ages of the evolved rocks (Figure 4) that internally generated magmatism appeared to "turn on" between ~ 4.1 and 4.2 Ga ago. Whereas that conclusion is supported by the existence of Apollo 14 mare basalts with ages in that time frame, perhaps it should not be extrapolated to other parts to the Moon. It is tempting to ascribe an "anomalous" lunar evolution in the pre-Imbrium region of the Moon to removal of much of the pre-Imbrium lunar crust by the hypothesized giant "Procellarum Basin," thought by some to have preceded the Imbrium basin, and to have included Oceanus Procellarum within its bounds. The existence of such a basin has been disputed by many, but geological arguments for its existence have been given by Wilhelms (1987) among others. In any event, the western part of the Moon, in the area now called Oceanus Procellarum, clearly was a topographic depression prior to the formation of the Imbrium basin. It seems reasonable to assume that it was filled by some kind of basalt flows prior to the Imbrium impact. It often has been assumed that these flows were of KREEP basalt. However, the age data provide no support for that assumption, and show instead that the sampled *pre-Imbrium* lithologies were aluminous mare basalts and the highlands alkali suite. The younger aluminous mare basalts for which ages are shown in Figure 6, 12038 and L16, are members of the low Ti group and will be discussed later with other members of that group.

High, low, and very low Ti basalts: remelting cumulates from the magma ocean?

The western, pre-Imbrium aluminous mare basalts may be atypical of the Moon as a whole. The same may be true of the western highlands alkali suite. Additionally, the magnesian-suite rocks and even the FANs appear to be asymmetrically distributed. There is an asymmetry in lunar crustal thickness, and even in the most obvious aspect of the Moon, its visible surface features. Nevertheless, lunar scientists continue to consider lunar evolution in terms of a spherically symmetric LMO model, which has among its features a radially stratified core, mantle, and crust. Why has this simple picture survived?

The reason, of course, is that the asymmetries cited above can all be viewed as surface features, whereas the lunar magma ocean, its crystallization products, and the ensuing lunar mantle processes may have extended to a depth of ~ 500 km in the Moon. Additionally, the LMO model yields important insights into lunar structure and evolution. Thus, for example, when lunar basalts are grouped according to chemical composition, and their ages are compared, important regularities appear. Chief among these is the apparent correlation between the ages and chemical compositions of lunar basalts as distinguished primarily by their TiO_2 content. The majority of lunar basalts can be grouped into one of three groups: high Ti basalts (9 to 14 wt% TiO_2), low Ti basalts (1 to 5 wt% TiO_2), and very low Ti (VLT) basalts (<1 wt% TiO_2). Basalt samples returned by Apollos 11 and 17 were primarily high Ti basalts, those returned by Apollos 12 and 15 and Luna 16 were primarily low Ti basalts, and those returned by Luna 24 were VLT basalts. Furthermore, the VLT type is more abundant on the lunar surface than its representation among the lunar samples indicates, as shown by Earth-based and spacecraft remote sensing and by the fact that the basaltic lunar meteorites known so far have been of the VLT variety. These groupings, although traditional, are nevertheless somewhat arbitrary. Apollo 12 ilmenite basalts are "low Ti basalts" of intermediate TiO_2 content (~ 5 wt%) that share some trace element and isotopic features with high Ti basalts, for example. The ages of lunar basalts are summarized according to landing sites and the traditional Ti groups in Figure 6.

One of the major triumphs of the LMO model has been its apparent ability to explain the primary geochemical and isotopic features of lunar basalts in terms of the "cumulate remelting model." In this model, cumulates from the magma ocean are thought to solidify in the lunar mantle, and to be remelted by the heat from decay of natural radioactivities contained within themselves. However, as noted by Ringwood and Kesson (1976), early-formed cumulates would be "barren" and never remelt because they would exclude the heat-producing elements during their crystallization. Only late-stage crystallization products from the magma ocean would be "fertile" and remelt. Detailed models for the thermal evolution of the Moon are in agreement with these predictions. Suggested solutions to this problem usually involve a variant on that suggested by Ringwood and Kesson (1976): By the time the magma ocean had evolved to ilmenite crystallization, the heat-producing elements, excluded from earlier phases, would have reached abundances high enough to cause remelting. Because ilmenite is heavy, ilmenite pods would sink to the bottom of the cumulate pile, accompanied by some trapped residual magma enriched in heat-producing elements, thereby fertilizing those cumulates where the ilmenite pod came to rest. Later versions of this idea call on mantle convection for ilmenite transport. This model seems to explain simultaneously the high TiO_2 contents of some mare basalts, the earlier remelting of high Ti basalts than of low Ti basalts, and the inferred depth of origin of the magmas. Snyder et al. (2000a) argue that, since ilmenite crystallization does not begin until the LMO is >95% crystallized, the most fertile site for partial melting is also the most Ti enriched. In this scenario, it follows that high Ti basalts should be the oldest lunar basalts.

High Ti basalts at the Apollo 11 and 17 sites do have relatively old ages up to ~ 3.9 Ga (Figure 6b,c). However, so does one VLT basalt, the lunar meteorite A881757. Further, although there is an apparent correlation between age and Ti group for the high and low Ti groups, this correlation ceases for the low Ti and VLT groups. Low Ti Apollo 12 ilmenite basalts (5 wt% TiO_2) have the same ages (~ 3.2–3.3 Ga) as VLT Luna 24 basalts (~ 1 wt% TiO_2). Finally, there are young aluminous mare basalts as well as old VLT basalts. Thus, the *apparent* correlation between the compositions and ages of the mare basalts is difficult to substantiate on the basis of the available lunar sample data. Furthermore, remote sensing studies suggest that among the many unsampled lunar maria, there are some with high Ti contents and young ages, as well as those with very low Ti contents and old ages. This point was made already by Geiss et al. (1977) and discussed more recently by Nyquist (1996).

As noted above, the environment most conducive to internal melt generation arose ~ 4.3 Ga ago at the base of the lunar crust where the last dregs of the magma ocean were concentrated. Under those circumstances, it is comparatively easy to envision generation of KREEP and aluminous mare basalt magma, for example, via the addition of an increment of heat due to decay of natural radioactivities. Later and deeper generation of mare basalt magmas apparently call for special circumstances. Perhaps, as favored by Snyder et al. (2000a, and preceding papers) basalt generation arises because "infertile" magma regions have been made fertile by the addition of ilmenite and trapped interstitial residual liquid from a nearly solidified magma ocean. Or, perhaps, as some authors have suggested, the Imbrium and other basin-forming impacts played a role in triggering mare basalt volcanism. This might happen via pressure-release melting, frictional heating as mantle rocks are forced past one another along shear planes, or simply via opening of magma channels to the surface. This might explain the apparent onset of mare volcanism at ~3.9 Ga, but would not explain the preponderance of low Ti and VLT volcanism in the ~3.2–3.3 Ga era. Clearly many things about the conditions leading to mare volcanism remain poorly understood.

The geologic context: mare basalt ages by landing site and basalt type

We have attempted to give the reader an appreciation of the global context of lunar chronology. Now we examine the local geologic context by taking a brief trip around the Moon, visiting the landing sites, identifying the rocks, saying a few things about their ages, and speculating on what they may be telling us. We will refer to Figure 7 as we go.

Like Neil Armstrong and Buzz Aldrin, the first men on the Moon, we begin at the Apollo 11 landing site. Early reports spoke of two basalt types: High K and low K. Additionally, all the basalts belong to the high Ti group. Later studies have subdivided the low K rocks into four subgroups: B1, B2, B3, and D. Groups B1 and B3 are most abundant and are nearly contemporaneous at 3.67 ± 0.02 and 3.71 ± 0.02 Ga, respectively. Groups B2 and D are slightly older, 3.85 ± 0.02 and 3.87 ± 0.01 Ga, respectively. Samples 10003 and 10062, discussed previously, as well as sample 10029 are members of Group B2, whereas samples 10002 and 10085 are members of Group D. Group D, at least, appears to be slightly older than the Imbrium basin. This is certainly possible. The Tranquilitatis Basin, where the landing site is, is pre-Nectarian in age, and is one of the oldest multi-ring basins (Taylor 1982). It is perhaps surprising that the basalts at the site are not even older. Group D basalt volcanism could have been triggered by the formation of

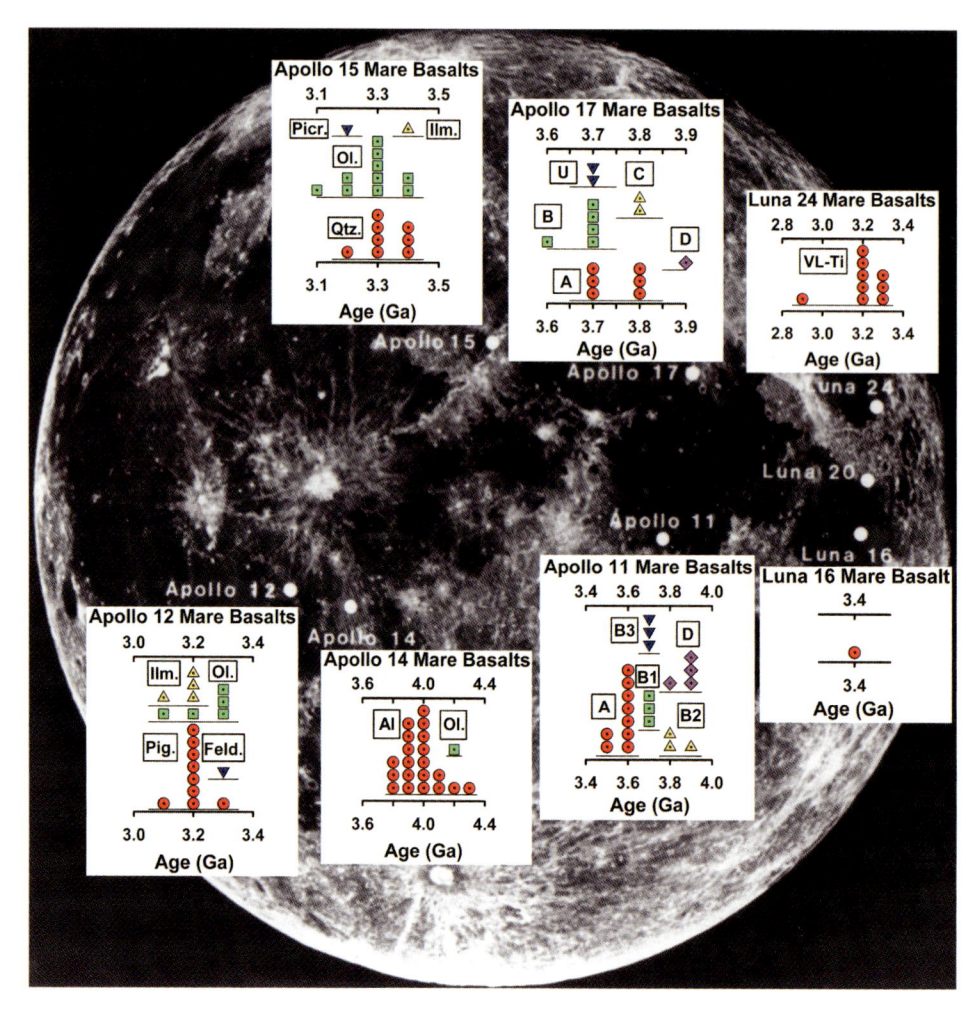

Figure 7 Lunar sampling sites with histograms of individual ages within each basalt subgroup at each site. See caption of Figure 6 for the average age of each subgroup.

the Nectaris basin to the south, or the Serentatis basin to the north. The high K rocks, now called Group A, are the youngest basalts at the site at 3.58 ± 0.01 Ga. Thus, the recorded duration of volcanism at this site is ~ 300 Ma, as concluded by Guggisberg et al. (1979).

Moving north, to the Apollo 17 site, we find a similar suite of basalts. Basalts here have even higher TiO_2 concentrations than at the Apollo 11 site. The majority of the dated basalts belong to Groups A, B, or U (ungrouped) with average ages of 3.78 ± 0.01, 3.70 ± 0.02, and 3.69 ± 0.01 Ga, respectively. Group C basalts have average ages of 3.77 ± 0.07 Ga. Only one Group D basalt has been identified and has an Rb–Sr age of 3.92 ± 0.04 Ga. This age is coincident within error limits with the preferred age of the Serenitatis Basin. Like the Apollo 15 KREEP basalts, this group D basalt apparently represents volcanic activity beginning immediately after basin formation. However, its $^{39}Ar-^{40}Ar$ age is younger at ~ 3.5 Ga, and the Sm–Nd data are disturbed (Nyquist et al. 2000). It is thus important to verify the ~ 3.9 Ga age for Group D for additional samples, if possible. Identical basin and basalt ages at Apollo 17 as well as at Apollo 15 would strengthen the evidence that volcanism can be triggered by basin-forming impacts.

Moving next to the Luna 24 landing site in Mare Crisium, we find that nine small VLT basalt fragments have been dated. All but one are $^{39}Ar-^{40}Ar$ ages only (Burgess and Turner 1998); the ninth was determined by Sm–Nd as well (Lunatic Asylum 1978). Eight samples have ages in the interval 3.2–3.3 Ga. The remaining sample has the youngest "reliable" age tabulated here: 2.93 ± 0.05 Ga. The Asuka 881757 VLT lunar meteorite has an age of 3.87 ± 0.06 Ga (Misawa et al. 1993). Thus, VLT volcanism apparently covered a time span of nearly 1.0 Ga. Because VLT basalts have low trace element abundances as well as low TiO_2 contents, this observation is hard to reconcile with the idea that the timing and duration of mare basalt volcanism is closely linked to the TiO_2 content of the source region. The Crisium Basin is Nectarian in age, next in sequence after the Serenitatis Basin (Taylor 1982). An $^{39}Ar-^{40}Ar$ age of a highlands sample from Luna 20 is consistent with near-simultaneity of these two basins. The sampled basalts at the landing site represent young flows occurring ~ 600 Ma after basin formation.

Aluminous soil and basalt fragments were found at the Luna 16 landing site in Mare Fecunditatis. One fragment was dated at 3.41 ± 0.04 Ga by $^{39}Ar-^{40}Ar$ (Huneke et al. 1972); a similar age of 3.39 ± 0.18 Ga was obtained for this fragment by Rb–Sr (Papanastassiou and Wasserburg 1972). This age is about 800 Ma younger than the oldest aluminous basalts from the Apollo 14 site. This single age is of great importance, since it comes from a known geological provenance of aluminous mare basalts, as inferred from the chemical composition of Luna 16 soil. The Fecunditatis Basin is pre-Nectarian in age, and is next youngest in the sequence of the major multi-ring basins to Mare Tranquilitatis (Taylor 1982). Thus, volcanism was also occurring in this basin at least ~ 500 Ma after its formation.

Mare volcanism at the Apollo 15 site was occurring as late as 700–800 Ma after basin formation. The geological context at this site is the most comprehensive of any of the sites. A stratigraphic sequence of mare basalt flows is exposed in the west wall of Hadley Rille, photographed by the astronauts. All of the mare basalts are low Ti basalts, but include as subgroups quartz normative basalts, olivine normative basalts, picritic basalts, and ilmenite basalts. Only one of each of the latter two groups has been dated. The quartz normative and olivine normative basalts show a nearly complete overlap in ages. The average age for quartz normative basalts is 3.37 ± 0.01 Ga and for olivine normative basalts is 3.29 Ga. The ilmenite basalt is slightly older. Spudis et al. (1988) interpret the stratigraphic sequence at Hadley Rille as olivine normative basalts overlying quartz normative basalts overlying Apollo 15 KREEP basalts in the Apennine Bench formation. The Apollo 15 site contains many of the major lunar rock types, including crustal, mare, and KREEP-rich basalts and quartz monzodiorites, but high Ti basalts are conspicuous by their absence.

The Apollo 12 site in Oceanus Procellarum is the westernmost site. Here, too, the mare basalts are low Ti, but contain four well-documented subgroups: Olivine basalts (3.22 ± 0.02 Ga), pigeonite basalt (3.16 ± 0.01 Ga), ilmenite basalts (3.17 ± 0.02 Ga), and one aluminous basalt, 12038 (3.31 ± 0.08 Ga). Sample 12038 may be unique to the site. Although its TiO_2 is $\sim 3.2\%$, and is typical of the other basalts at the site, its Al_2O_3 content of $\sim 12.5\%$ is similar to that of Luna 16 basalts ($\sim 13.6\%$) and the Apollo 14 aluminous basalts (also $\sim 13.6\%$). The REE pattern and abundances of 12038 match those of the Luna 16 regolith. The topographic low into which the Apollo 12 basalts flowed could have been produced ~ 1 Ga earlier. The ilmenite basalts here, as well as the ilmenite basalt at the Apollo 15 site, appear to have been produced by partial melting of a severely LREE-depleted source region, similar in this respect to the source regions of high Ti basalts.

The Apollo 14 landing site lies a relatively short distance east of the Apollo 12 site. As we have already seen, it is in a highland area, and the mare rocks and rock fragments are not indigenous to the site. The breccia clasts analyzed from this site may give us our best view of rocks in the pre-Imbrium terrain of Oceanus Procellarum. The clasts in question were mostly extracted from breccias collected near the rim of Cone crater, which is thought to have excavated through the local regolith to the Fra Mauro formation. Wilhelms (1987) presents the case that the Fra Mauro formation is primary ejecta from the Imbrium impact. Others have argued for a local origin for the Fra Mauro formation.

As we have seen, the ages of Apollo 14 aluminous mare basalts are consistent with a pre-Imbrium origin. They are readily consistent with the primary ejection model and the first such ages determined were so interpreted by Wilhelms (1987). The larger geological context for these ages is unclear to us if the Fra Mauro formation is of local origin. The aluminous mare basalts in the Apollo 14 breccia clasts are ~0.8–1.0 Ga older than the aluminous mare basalts at the Apollo 12 and Luna 16 sites. The single olivine basalt clast from breccia 14305 is also ~1.0 Ga older than the olivine basalts at the Apollo 12 site, which it closely resembles in major element composition. REE in the Apollo 14 olivine basalt are about a factor of two higher than in Apollo 12 olivine basalts, but are similar in abundance pattern. High Ti basalts are again absent, suggesting that they were absent from the Procellarum area before the Imbrium impact.

What have we learned from this tour around the landing sites? First, the sampled mare surface units vary in age from ~2.9 to ~3.9 Ga (Figure 6). These absolute ages have led to normalization of crater density ages to an "average lunar frontside maria" age of 3.45 Ga and an average crater density of 1.88×10^{-4} km^{-2} for craters >4 km in diameter (BVSP 1981). Second, it is difficult to say from the basalt ages when lunar magmatism "turned on." Most surface basalts flowed into basins already formed during the lunar cataclysm. Mare basalt volcanism may have been well underway by that time, but evidence of it may have been mostly obliterated. Only if the Fra Mauro formation is indeed primary ejecta from the Imbrium basin might we have samples from a "pre-basin" period of basaltic volcanism. Third, magmatism may at times (often?) have been triggered by basin-forming impacts, but once begun, it persisted for hundreds of millions of years. Fourth, volcanism involving several different magma types simultaneously was ongoing around the Moon ~3.2–3.3 Ga ago. Fifth, we have seen that the pre-Imbrium basalts at the Apollo 14 site, combined with those at the Apollo 12 and Luna 16 sites, suggest that the production of low Ti aluminous and olivine basalts began early (~4.2 Ga ago) and lasted for ~1.0 Ga. We have also seen from the Luna 24 data combined with the Asuka 881757 meteorite data that this is also the case for VLT basalts. We have no direct evidence in the lunar samples for a similarly long duration of high Ti volcanism, but the samples represent only about one-third to one-half the surface basalt types (Pieters 1978). Spectral reflectance studies (Pieters 1978, Sunshine et al. 1994) combined with crater densities (Boyce and Johnson 1978) suggest young (2.5 ± 0.5 Ga), high Ti volcanism in extreme western Oceanus Procellarum, however. In that case, high Ti volcanism was of comparably long duration to that of the other types of volcanism (Sunshine et al. 1994). The shorter span of volcanism indicated by the samples from a given site probably reflects the average depth of surface excavation due to impacts near the site. Sixth, at each landing site, there are variations in basalt composition. There is also a global scale heterogeneity in major basalt types. From west to east on the lunar nearside the dominant basalt types vary approximately from high Ti (from remote sensing), to low Ti mafic (Apollo 12), to high Ti (Apollo 11), to low Ti aluminous (Luna 16), to VLT (Luna 24). (See Pieters (1978) or BVSP (1981) for the distribution of lunar basalt types.)

Calibrating cratering chronologies

As apparent from the above discussion, mare surface ages determined from the density of impact craters per unit area provided an important augmentation to lunar chronology as determined from lunar sample ages. But, how comparable are the cratering ages to the absolute ages of lunar samples? We have already mentioned one calibration point: The average crater density on mare surfaces to the average age of mare surfaces as determined by sample analyses. It is also desirable to use radiometric ages and crater densities determined for surfaces of much younger age than the lunar mare in order to define the impactor size frequency curve for much younger times. Such young surfaces are smaller in area, and because no spacecraft has returned samples from such young surfaces, their radiometric age can only be inferred by indirect means. The best example of such a young feature is the large crater Copernicus (Figure 3), on whose extensive ejecta deposits crater densities have been measured. Radiometric ages of ~800 Ma have been determined on two rare types of samples recovered by Apollo 12 and believed ejected from Copernicus (Eberhardt et al. 1973, Bogard et al. 1994). These two samples are a small granitic clast (12033,507) and KREEP-rich material, both recovered from the returned soil. Although the ^{39}Ar–^{40}Ar age of the granitic clast is 0.8 Ga, its Rb–Sr age shows a disturbed age of ~2.2 Ga and the ^{40}K–^{40}Ca isochron age is 3.6 ± 0.1 Ga (Shih et al. 1994). Thus, the age of Copernicus is inferred from resetting of ^{39}Ar–^{40}Ar ages, presumably due to the thermal heat deposited in the ejecta. In a similar manner, an age of ~2.1 Ga was estimated for the large crater Autolycus based on ^{39}Ar–^{40}Ar age determination of glassy KREEP samples returned in the Apollo 15 soil (Ryder et al. 1991). Obviously the reliability of these dated surface features on the Moon critically rests with the interpretation that the Ar–Ar ages of the dated samples were reset by these large craters.

A heterogeneous lunar mantle: cumulates plus trapped magma from the LMO?

We have seen that lunar basalts vary in composition on both local and global scales. Because they are partial melts of the lunar mantle, the variation in their compositions has implications for the mantle composition as well. Is the mantle heterogeneous, and, if so, on what scale?

Basalt compositions are only indirect indicators of mantle composition. Compositional variations can arise during partial melting of the mantle to produce basaltic magmas, ascent of the magmas to the surface, and solidification at the surface. Because isotopic compositions are invariant to two of these three processes (partial melting and surface solidification), knowledge of the isotopic compositions of key elements during basalt genesis complements knowledge of chemical compositions of the basalts in an important way. The same isochron techniques are used to determine the initial isotopic composition of Sr and Nd, for example, as are used to determine ages by those methods (Figure 2). A review of lunar chronology would be incomplete without mentioning the initial radiogenic isotopic compositions that have been determined by this method, and their significance. The discussion to follow is of necessity abbreviated; a more complete review is given by Nyquist and Shih (1992).

In Figure 8, initial $^{143}Nd/^{144}Nd$ values are expressed in units of ε_{Nd}, following Lugmair et al. (1975). In this notation, values of $^{143}Nd/^{144}Nd$ at a particular time are expressed as deviations, in parts in 10^4, from the value in a "chondritic reservoir" at that time; that is, a reservoir that started with the same $^{143}Nd/^{144}Nd$ and Sm/Nd ratios as chondritic meteorites 4.56 Ga ago. Here we assume these values were also characteristic of the bulk Moon. Had the lunar mantle remained undifferentiated, values of $\varepsilon_{Nd} = 0$ would be obtained for all lunar basalts, regardless of age. This obviously has not been the case. Those source regions having Sm/Nd values differing most from those of chondritic meteorites are those for the Apollo 17 high Ti basalts, some Apollo 11 high Ti basalts, and some Apollo 12 and 15 low Ti basalts. High ε_{Nd} values of some of the low Ti basalts are especially surprising, since the low Ti basalts are more chondritic in bulk composition then the high Ti basalts. While it is true that comparatively large, positive ε_{Nd} values have only been observed for one Apollo 15 basalt and a few Apollo 12 basalts, the fact that such values are observed at all is significant. Also, there are large variations in ε_{Nd} values for basalts from a single landing site that correspond to the variations in bulk composition among the basalts.

Figure 9 shows a similar diagram for values of initial $^{87}Sr/^{86}Sr$ expressed in units of ε_{Sr} appropriate to the bulk Moon. The ε_{Sr} values are calculated as the deviations in $^{87}Sr/^{86}Sr$ from "bulk lunar" (BL) values, assuming the Moon began with $^{87}Sr/^{86}Sr = 0.69898$ (BABI; Papanastassiou and Wasserburg 1969) and $(^{87}Rb/^{86}Sr)_{BL} = 0.038$. The latter

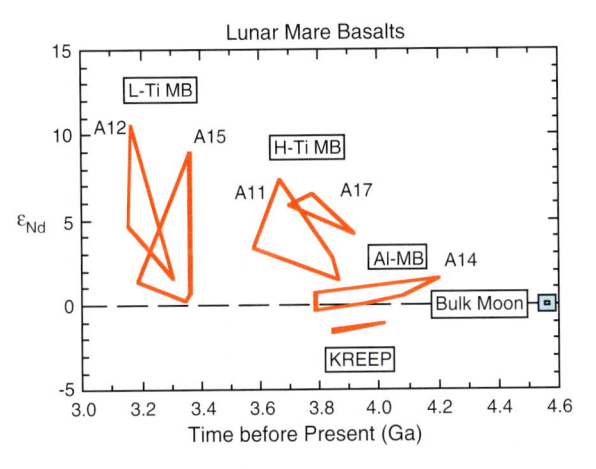

Figure 8 Initial $(^{143}Nd/^{144}Nd)_I$, defined analogously to initial $(^{87}Sr/^{86}Sr)_I$ (see Figure 2) expressed as ε_{Nd} values and plotted as a function of crystallization age for lunar basalts. Here $\varepsilon_{Nd} = 10^4 \times [(^{143}Nd/^{144}Nd)_I/(^{143}Nd/^{144}Nd)_{CHUR} - 1]$, where $(^{143}Nd/^{144}Nd)_{CHUR}$ is the value of $^{143}Nd/^{144}Nd$ in a chondritic uniform reservoir at the time of basalt crystallization. Values of $\varepsilon_{Nd} \sim 0$ imply a relatively undifferentiated mantle source region in which the relative REE abundances are similar to those in chondritic meteorites. Positive values of ε_{Nd} imply LREE depletion in the source regions, and negative values of ε_{Nd} imply LREE enrichment. Only KREEP basalts have negative ε_{Nd} values, indicative of involvement of urKREEP in their genesis, either via early incorporation into their source regions, or by assimilation of urKREEP during KREEP volcanism.

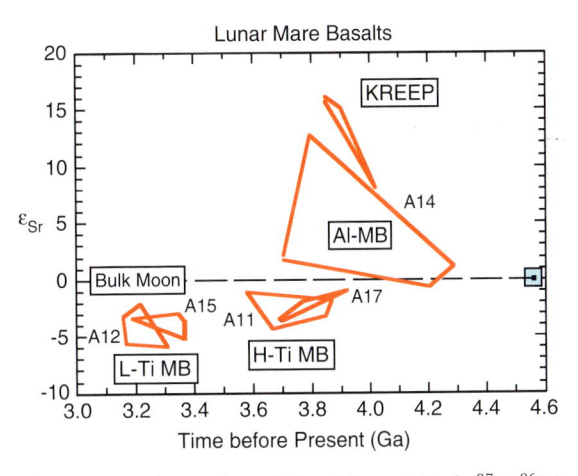

Figure 9 Values of ε_{Sr} defined from initial $(^{87}Sr/^{86}Sr)_I$ values relative to estimated values for the bulk Moon at the time of magma generation (see text). That is, $\varepsilon_{Sr} = 10^4 \times [(^{87}Sr/^{86}Sr)_I/(^{87}Sr/^{86}Sr)_{BL} - 1]$ where $(^{87}Sr/^{86}Sr)_{BL}$ is the estimated bulk lunar value. Only KREEP and some aluminous mare basalts have positive ε_{Sr} values, suggestive of evolution in a high Rb/Sr environment. High Rb/Sr would be characteristic of the urKREEP residuum from the lunar magma ocean, thus urKREEP is implicated to have been involved in the genesis of KREEP and aluminous mare basalts (Al-MB). The more common types of mare basalt are derivative from a lunar mantle in which Rb/Sr was depleted from its bulk lunar value.

value is chosen here to allow at least some members of each major basalt compositional group to have $\varepsilon_{Sr} \leq 0$. This assumption is made because Rb is more strongly excluded from typical mantle minerals than Sr. Thus, compared to Sr, Rb is preferentially partitioned into the crusts of rocky planets like the Earth. This and similar behavior for other incompatible trace elements leads to the concept of "enriched" crusts and "depleted" mantles, concepts and terminology borrowed from terrestrial geochemistry. The Apollo 14 mare basalts play a determining role in the choice of $(^{87}Rb/^{86}Sr)_{BL}$. Were they excluded from consideration, a value lower by perhaps a factor of two might be favored. Nyquist *et al.* (1977, 1979a) used an even higher value of $^{87}Rb/^{86}Sr = 0.05$ to model the isotopic evolution of the Apollo 12 basalts, however. A different choice of $(^{87}Rb/^{86}Sr)_{BL}$ will only cause the data field to rotate relative to the line $\varepsilon_{Sr} = 0$ in Figure 9. Although the value of $(^{87}Rb/^{86}Sr)_{BL}$ should be considered uncertain to about a factor of two, Figure 9 nevertheless illustrates some important conclusions.

First, KREEP basalts have positive ε_{Sr} and negative ε_{Nd}, thereby showing a "crustal" signature. (Here, by "crustal" we mean "non-mantle," since the isotopic signature of the anorthositic crust does not follow the terrestrial model. The presence of "non-mantle" Sr and Nd in KREEP basalts is consistent with terminology that refers to them as "non-mare" basalts.) This isotopic fingerprint results from the inclusion of an urKREEP component in the KREEP source, as discussed earlier. Next, the Apollo 14 aluminous mare basalts show both "mantle" and "crustal" isotopic signatures. This can also be interpreted as due to variable amounts of urKREEP, which for some samples totally dominates the isotopic signatures, but for others it does not, depending on the mass balance between Sr and Nd of "crustal" and "mantle" origin. We propose that the same pattern also holds for later, mare basalt volcanism, with some modifications. Thus, we propose that "mantle" Sr and Nd dominates the total inventory of these elements for most mare basalts, but there is an additional component of Sr and Nd that has a "crustal" isotopic signature. Like urKREEP, the latter component comes from the residual magma of the crystallizing magma ocean, but is variably enriched in incompatible elements according to the stage at which crystallization of mantle cumulates occurs. The mantle cumulates are, of course, the source of the "mantle" Sr and Nd. This scenario is similar to that presented by Snyder and Taylor (1993) who referred to the magma ocean component as "trapped instantaneous residual liquid" (TIRL) which characterizes its origin. In some later papers, these authors and their co-workers approximate the characteristics of the TIRL component by those of Apollo 15 KREEP basalt. This cannot be literally true, and we believe it introduces unnecessary confusion. These workers also introduce the concept of the "perfect adcumulate," also an oversimplification, since the compositions of the cumulate phases change with time. Nevertheless, we consider their basic model leading to mantle regions composed of (a) mafic cumulates, (b) trapped liquid from the magma ocean, and (c) possibly plagioclase and/or ilmenite and other phases co-crystallizing from the magma ocean at a particular stage of crystallization, to be consistent with the isotopic data.

That the isotopic data are qualitatively consistent with this model can be seen by inspection of Figures 8 and 9. Some basalts were derived from source regions in which the Rb/Sr and Sm/Nd ratios were highly fractionated relative to those in the bulk Moon. This can be accomplished via crystallization of mineral phases, but only if the parent magma is efficiently removed from the crystallizing phase. If some residual magma is left entrained among the crystals, for example, the resultant "mix" will have Rb/Sr and Sm/Nd ratios intermediate to those of the crystallizing phases and the residual magma. Since the concentrations of the incompatible elements are always very much higher in the residual magma than in the crystallizing phases, the residual magma stays close to bulk Moon values of Rb/Sr and Sm/Nd for much of the crystallization sequence. Thus, higher proportions of TIRL tend to produce mare source regions with isotopic parameters $\varepsilon_{Nd} \sim \varepsilon_{Sr} \sim 0$. As seen from the figure, each major group of mare basalts has some members which come from such sources. These sources would be the "fertile" ones containing relatively high concentrations of the heat-producing incompatible elements. Heat conduction, combined with heterogeneous distribution of TIRL within the mantle, prevents portions of the mantle which otherwise would be "barren" from being so. Thus, (a) the time of volcanism from a particular region of the lunar mantle depends on the *average* concentration of heat-producing elements in that portion of the mantle, (b) variations in major element compositions of the basalts depend on variations in the proportions of mineral phases crystallizing from the magma ocean when a particular source region formed, and (c) variations in Sr and Nd isotopic systematics of basalts from a particular region of the mantle depend on local variations in the proportion of TIRL in that region.

Today's Moon: frozen in time

We have seen that there are no lunar samples with formation ages <2.9 Ga. Furthermore, lunar surface crater densities suggest that most lunar volcanism ceased ~2.5 Ga ago, although areas on the Moon have been identified where volcanism may have persisted until only ~1.0 Ga ago. The cessation of volcanism probably occurred when the heat lost to space could not be compensated by radioactive heating in the lunar mantle. This is the subject of planetary thermal modeling, and the interested reader is referred to BVSP (1981, Chapter 9). There, it is concluded that "Virtually all lunar thermal history models that have high initial temperatures in the outer few hundred kilometers of the Moon maintain temperatures in the upper mantle sufficiently close to the solidus to sustain mare volcanism until times at least as recent as 3 b.y. ago." (1 b.y. = 1 Ga) Another conclusion is that "the retained heat and the average fraction of partial melt decrease more or less continuously with time" and "there should have been extensive lunar volcanism prior to 3.9 b.y. ago." As we have seen, the best candidates for this type of volcanism are KREEP basalts and the Apollo 14 aluminous mare basalts.

Figure 10 is a schematic representation of the genesis of igneous rocks on the Moon, after the manner of Nyquist and Shih (1992) and Snyder et al. (2000a). A stratified lunar crust, lower crust, and upper mantle is shown following Taylor (1982). Starting at ~4.5 Ga ago, FANs became a significant portion of the lunar crust. Nearly simultaneously,

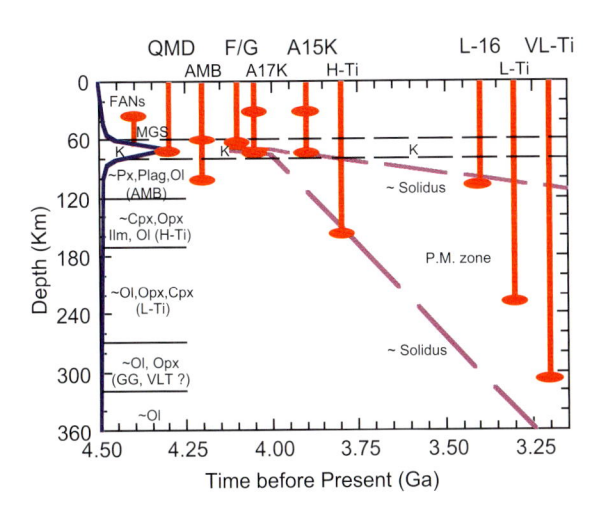

Figure 10 Schematic representation of the genesis of igneous rocks on the Moon, after the manner of Nyquist and Shih (1992) and Snyder et al. (2000a). A stratified lunar crust, lower crust, and upper mantle is shown following Taylor (1982). Ovals represent magma bodies. Starting at ~4.5 Ga ago, ferroan anorthosites (FANs) became a significant portion of the lunar crust. Mg-suite (MGS) magmas formed nearly simultaneously and ponded within the crust to form the Mg-suite plutons. Quartz monzodiorites (QMD) formed from the last dregs of the magma ocean as late as ~4.3 Ga ago. Aluminous mare basalt (AMB) magmas were generated shortly thereafter. Felsite/granites (F/G) formed nearly simultaneously. Apollo 17 and Apollo 15 KREEP basalts (A17K and A15K) formed in the same time frame, with volcanism lasting to ~3.9 Ga ago. High Ti basalts were produced next. Most low Ti and VLT basalts are among the youngest sampled basalts. The presence of aluminous mare basalts at the Luna 16 sampling site appears to require melting throughout a substantial portion of the lunar mantle ~3.0 Ga ago.

Mg-suite magmas were initially generated in the urKREEP zone near the base of the crust. These magmas assimilated less-differentiated materials and ponded within the crust to form the Mg-suite plutons. Quartz monzodiorites, probably our best example of urKREEP, formed from the last dregs of the magma ocean as late as ~4.3 Ga ago. Aluminous mare basalt magmas were generated shortly thereafter from the urKREEP layer near the base of the crust, but assimilated crustal anorthosites and some mafic magmas on their way to the surface. Felsite/granites formed nearly simultaneously via extensive differentiation of melts also derived from urKREEP. Apollo 17 and Apollo 15 KREEP basalts probably also formed from remelting within the urKREEP layer, but the magmas thus formed assimilated mafic and feldspathic components to yield overall basaltic compositions. High Ti basalts were produced next by the remelting of cumulates including ilmenite and clinopyroxene in addition to olivine and orthopyroxene formed from the LMO just prior to concentration of urKREEP. Later generation of low Ti and VLT basalts probably accompanied remelting of deeper, earlier cumulates. Simultaneous generation of aluminous mare basalts at the Luna 16 sampling site appears to require melting throughout a substantial portion of the lunar mantle at this time. Young aluminous mare basalts and old VLT basalts (not shown in Figure 10) are difficult to accommodate within this simple picture, and likely require large-scale heterogeneity in the lunar mantle to accommodate simultaneous generation of compositionally differing mare basalt magmas. Smaller-scale heterogeneity in the mantle source regions seems to be required by variations in chemical and isotopic compositions of the basalts at each landing site. The illustration of extensive melting throughout the lunar mantle at ~3.0 Ga ago in Figure 10 is the consequence of assuming a radially stratified mantle composition. A much more heterogeneous mantle that would allow melting to "pinch-out" within a more limited depth range ~3.0 Ga years ago as the Moon cooled is probably more realistic.

Thermal processes are not the only ones that may have played a role in the cessation of lunar volcanism, however. Again quoting BVSP (1981): "Global expansion early in lunar history should have aided magma ascent both during early mare volcanism and during a likely earlier period of volcanism as well. Global contraction starting at least 3.6 b.y. ago should have made magma ascent increasingly difficult... With time... local extensional stresses were likely exceeded by the increasingly compressive global surface stress... shutting off mare volcanism entirely." Thus, geophysical phenomena, unrelated to the composition of particular mare basalt source regions, may have been responsible for shutting off volcanism of all types in the ~2.5–3.0 Ga timeframe. Thereafter, endothermic magmatic activity on the Moon effectively ceased.

We note that the current state of knowledge of lunar history is far from complete. The magma ocean model, although extremely useful, requires special pleading in many instances. Lunar thermal history models do not, in fact, provide for cumulate remelting. The model of Solomon and Longhi (1977) is most closely tied to petrologic considerations. It rapidly solidifies the magma ocean, except for an urKREEP residuum, within ~100 Ma. Partial melting is initiated in the primitive lunar interior within ~100 Ma of lunar formation, but mafic cumulates from the magma ocean do not remelt. We suggest that this failure results from assuming an unwarranted "purity" of the mafic cumulates, thereby requiring too much of the heat-generating K, U, and Th to be removed from the cumulate phases. As presented above, the isotopic data are good evidence that some residual magma from the magma ocean was trapped in the mafic cumulates as they crystallized. The lowered heat-generating capacity in those phases would delay their melting relative to a "primitive" lower mantle, but melt they must have. Nyquist and Shih (1992) and Snyder *et al.* (2000a) have combined this line of reasoning with assumed versions of a solidified, stratified LMO to construct schematic representations of lunar evolution. These representations combine much observational data, more than can be given here, and we refer the interested reader to them.

We suggest that there is enough new data from a variety of fields to warrant another round of lunar thermal history models, which have not been pursued since the late 1970s. A study of the short-lived ^{146}Sm–^{142}Nd system, for example, showed that the cumulate source regions of mare basalts closed to Nd isotopic equilibration 240 ± 50 Ma years after Solar System formation (Nyquist *et al.* 1995a). Such constraints ought to be directly applicable to thermal models.

Finally, continued acquisition of new data is required to complete our knowledge of lunar evolution. Additional spacecraft-derived data are very helpful, but most helpful to solidifying the concepts discussed here would be isotopic analyses of additional lunar samples from known areas of the Moon. Automated sample returns that are cheap, frequent, and to areas of the Moon selected to answer specific questions would be the most useful. Such missions must have sufficient capacity to return samples large enough for complete isotopic study. Whether missions satisfying this requirement will be undertaken in the twenty-first century is unknown. Perhaps more likely are crewed missions of much greater complexity pursuing the vision that John Kennedy set before the generations to follow him. As exciting as establishing a lunar base, for example, would be, acquiring the answers to the remaining questions of lunar chronology would be best served by the ability to sample a number of areas distributed around the lunar globe. Until such time as lunar exploration is resumed, lunar scientists will continue extracting as much information as possible from samples already in the Apollo and Luna collections, and from lunar meteorites.

MARS

The century of space science has been marked by a number of serendipitous interactions between spacecraft missions and laboratory science. Probably nowhere has this been more evident than in the study of the chronology of planetary objects. The determination of the absolute ages of rocks, and therefore of planetary events, is a laboratory-based technique, which as yet has not been adapted to spacecraft instrumentation. Advances in laboratory techniques in the late 1960s enabled the determination of the absolute ages of the lunar samples returned by the Apollo missions. Also, serendipitously, additional "lunar samples" were recognized among meteorites found in the Antarctic. Nevertheless, as we have seen, the contribution to our overall knowledge of lunar chronology from the spacecraft-returned samples far outweighs that contributed by the lunar meteorites.

For Mars, the sample situation is reversed compared to the Moon. As we have not returned samples from Mars, all of our absolute ages relating to Martian chronology are derived from the Martian meteorites. Here the serendipity is that we have enough other spacecraft-derived data that we are reasonably sure that the meteorites are indeed from Mars. Furthermore, although the Martian surface is less quiescent than the lunar surface, meteor craters are nevertheless relatively reliably preserved, so that a cratering chronology has been derived for Mars. This cratering chronology is derived from orbital imaging data, mostly from the Viking 1 and 2 Orbiters (1976), and more recently from the Mars Global Surveyor (1997). The Viking 1 and 2 missions also included landers, which performed a number of geochemical analyses both during atmospheric entry and on the Martian surface. The Viking and Mars Global Surveyor missions, plus Pathfinder (1997) have provided the bulk of our current scientific information about Mars. Sample return is not scheduled until early in the twenty-first century. Relative crater densities per unit area give the relative ages of Martian surfaces, but determining the absolute ages of those surfaces awaits the return of Martian samples. Thus, here we concentrate on the meteorite data, since this chapter deals with absolute radiometric ages.

How do we know that "Martian meteorites" actually come from Mars? This question has been examined by a variety of authors during the past two decades, and the arguments most recently were summarized by Treiman *et al.* (2000). A prerequisite is to establish that they all belong to the same "oxygen isotope group," that is, that the oxygen isotopic composition in the meteorites in question can be related by isotopic mass fractionation, and thus that they probably were formed on the same planetary body. Direct evidence linking the group to Mars is: (a) the chemical compositions of the Martian meteorites are similar to the composition of Martian surface materials, as determined by Viking and Pathfinder, and show similarities in some ratios of elements that are markers of geochemical origin; (b) the generally young radiometric ages of the Martian meteorites (discussed below), in comparison to the generally old ages of other meteorites, indicate an origin from a large parent body, which cannot be the Earth or Moon; and (c) the observation that elemental and isotopic compositions of gases trapped within some of the meteorites match those of the Martian atmosphere, as measured by instruments carried on the Viking spacecraft. Because some of these trapped gases exhibit strong evidence of having been fractionated during residence in a planetary atmosphere, and because the Martian atmospheric composition is unique among known volatile reservoirs in the Solar System, Mars is the only reasonable source of these gases. In addition, the recognition that meteorites also come to us from the Moon has demonstrated that impact mechanics can distribute surface materials among planetary-sized bodies.

Historically, the first recognized Martian, or SNC, meteorites were classified into groups based on their mineralogy and named after the type example of each group: Shergotty, Nakhla, and Chassigny. Later, Antarctic meteorite Allan Hills 84001 was recognized as part of the Martian clan, but having a distinct mineral composition from the others. As it turns out, this mineralogical classification is also a good way to discuss the radiometric ages of Martian meteorites, because meteorites of a common classification tend to exhibit considerable similarity in their chronology. This similarity exists for both their radiometric, or formation, ages and for their space (cosmic-ray) exposure ages, and we shall use both types of ages in the following discussion.

Figure 11 shows the crystallization ages of Martian meteorites as determined from the radiometric data. Obviously our sampling is very incomplete. The data are much more limited than the lunar data, but cover an age span about three times larger, extending nearly to the present day. We have only a single sample older than 1.3 Ga, for example. The relative ages of various Martian surface units are estimated from crater densities, and extrapolation of the lunar cratering rate to Mars. The actual cratering rate at Mars may be uncertain to a factor of three. The surface ages of Mars so obtained are divided into three major epochs, Noachian (~4.5–3.5 Ga ago), Hesperian (~3.5–1.8 Ga ago), and Amazonian (~1.8 Ga–present), and these major units are further subdivided into early, middle, and late periods (Tanaka 1986, Tanaka et al. 1992). These age ranges are for the HT (Hartmann–Tanaka) scale. The age ranges for the NW (Neukum–Wise) scale are shifted to higher values.

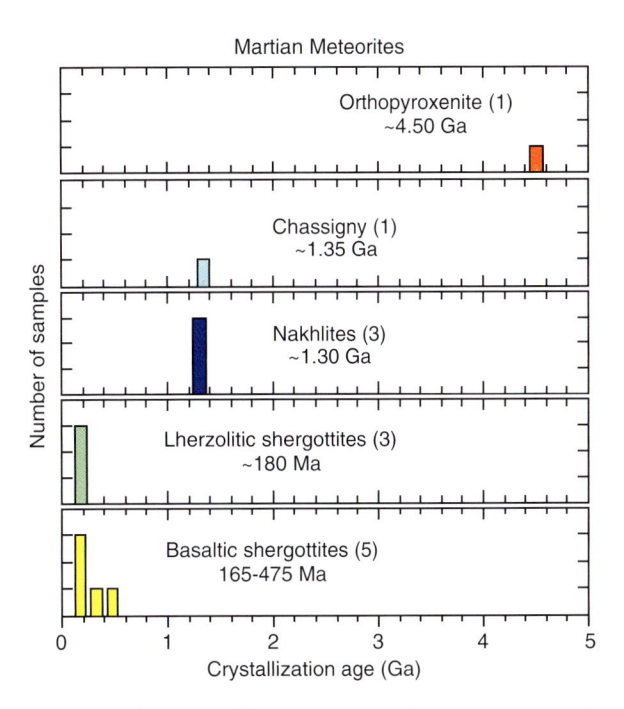

Figure 11 The crystallization ages of Martian meteorites separated by compositional groups as for the lunar samples. Martian volcanism appears to have extended over most of Solar System history, a conclusion also reached from crater retention ages (Hartmann and Berman 2000). Not unexpectedly, there are large gaps in the chronological record of the Martian meteorites.

In the lunar section of this chapter, we began the discussion with the oldest, crustal rocks and worked towards younger basaltic rocks. We hope thereby to have given the reader a "feel" for lunar evolution; that is, the changing nature of the Moon, especially its surface, with time. This approach is less fruitful for Mars because of the paucity of age and composition data, and because of lack of knowledge of the geological units represented. Nevertheless, we will adopt it here, beginning our discussion with the oldest, Noachian-age Martian meteorite. The gaps in our knowledge will be obvious.

Orthopyroxenite ALH84001: what have we learned from a Noachian-age Martian rock?

Antarctic meteorite ALH84001 is our only sample of the ancient Martian crust. This single sample has produced very interesting insights into Martian history. Perhaps its greatest fame derives from the report by McKay et al. (1996) of possible evidence of relic biogenic activity in secondary carbonates found within the meteorite. Its ancient age means it was present on the Martian surface when liquid water is thought by many to have also been present there. Thus, circumstances may have been conducive for life in the environment of ALH84001. The interpretation of evidence for fossil life has been challenged, but it has stimulated much research aimed at proving or disproving the life hypothesis, research that continues.

The old crystallization age of ALH84001 is direct evidence that portions of the Martian crust formed quickly after the planet accreted. There have been some variations in the reported ages. Jagoutz et al. (1994) reported an age of ~4.56 Ga by the Sm–Nd technique. Nyquist et al. (1995b) reported a conventional Sm–Nd age of 4.50 ± 0.13 Ga, and initial $^{146}Sm/^{144}Sm = 0.0022 \pm 0.0010$ relative to a chondritic uniform reservoir (CHUR). The latter result suggests formation more than one half-life of ^{146}Sm ($T_{1/2} = 103$ Ma) after the angrite meteorite LEW86010, for which $^{146}Sm/^{144}Sm$ lies in the range 0.0070–0.0076 (Lugmair and Galer 1992, Nyquist et al. 1994). A very precise Pb/Pb age of 4.558 Ga was determined for LEW86010 by Lugmair and Galer (1992). The $^{146}Sm/^{144}Sm$ ratio implies closure of the Sm–Nd system in ALH84001 no earlier than ~115 Ma after formation of the angrite. The long- and short-lived chronometers are thus concordant for an age of ~4.4 Ga. ALH84001 is contemporaneous with the oldest lunar crustal rocks, the FANs and anorthositic norites.

Unlike these lunar rocks, ALH84001 probably should not be considered a "typical" crustal rock. As we have seen, the hypothesis of a "magma ocean," as in the lunar case, would imply that the earliest crust should be feldspathic in composition, owing to the low density of feldspar minerals, which would allow them to float atop more mafic components as the magma crystallized. ALH84001 is itself mafic, and presumably would sink in a magma ocean. ALH84001 has much lower Al_2O_3, SiO_2, and K_2O and higher MgO and FeO than the Pathfinder "sulfur-free rock" (Rieder et al. 1997, Bell et al. 2000). Rocks at the Pathfinder site are thought to be derivative from the Southern Martian Highlands, but the Al_2O_3 content of the "sulfur-free rock" is much lower than that of lunar highlands soils, and lower even than that of lunar mare soils. We do not know how the Martian crust may have formed, but we can hypothesize that ALH84001 formed in a manner analogous to that of the lunar "Mg-suite;" that is, as a mafic cumulate in a layered igneous province within an earlier-formed crust.

Viewed in the context of the lunar samples, it seems fortuitous that one out of only 13 recovered Martian rocks would have preserved such an old age. The dated lunar FANs were small clasts extracted from lunar highland breccias, in which they often were surrounded by impact melt glass. As we have seen, only a few lunar crustal rocks have been reliably dated to have ages in the range of ~4.3–4.5 Ga, in spite of the fact that the Apollo 16 mission was specifically targeted to the central lunar highlands. The rarity of unadulterated "original" crustal material among the lunar highland rocks poses some questions: Did Mars experience the same high rate of meteor bombardment early in its history that the Moon did? Was the "late, heavy bombardment" as heavy or as late at Mars as at the Moon? Did Mars experience a "terminal cataclysm" of bombardment? ALH84001 has given us a few clues to the answers to these questions, but their resolution awaits combination of spacecraft orbital imaging and returned sample data, so that some heavily cratered areas on Mars can be absolutely dated.

One characteristic of ALH84001 that is consistent with a "terminal cataclysm," or, perhaps less likely, the end of the late, heavy bombardment, is an Ar outgassing age of ~4.0 Ga. This age is younger than the crystallization age, and corresponds to the hypothesized period of the "terminal lunar cataclysm." A plausible conclusion is that Mars experienced the same late-stage major bombardment that seems to be manifest throughout the inner Solar System. For Mars, this must be a tentative conclusion, based as it is on a single analysis. However, the case for a cataclysm can be made much more strongly for the Moon and the asteroid parent body of the HED meteorites (see discussion below). The conclusion concerning the "Martian cataclysm" is further qualified because both the value of the Ar outgassing age and the crystallization age are subject to alternative interpretations. Turner et al. (1997) prefer an Ar outgassing age of ~3.9–4.0 Ga, consistent with the time of the lunar cataclysm, but Bogard and Garrison (1999) suggest a wider range from 3.9 to 4.3 Ga. Furthermore, Wadhwa and Lugmair (1996) reported a Rb–Sr age of 3.84 ± 0.05 Ga for ALH84001. We concur with their interpretation that the Rb–Sr age most likely represents the time of intense shock recorded in the $^{39}Ar–^{40}Ar$ age. However, if it were not for the older Sm–Nd ages, the Rb–Sr age might well be interpreted as a crystallization age. Furthermore, the rock has undergone secondary mineralization, and now contains ~1% by volume of carbonates in cracks and shock-crushed areas. Thus, the possibility exists for disturbed isotopic systems due to the presence of secondary Sr or Nd.

Borg et al. (1999b) exploited the compositional zoning of the carbonate minerals in ALH84001 to dissolve selectively phases having different parent/daughter ratios for the Rb–Sr, U–Pb, and Sm–Nd systems. REE concentrations in the resultant solutions were too low for isotopic analysis, but the Rb–Sr and U–Pb isotopic analyses yielded concordant ages of ~3.9–4.0 Ga. Thus, carbonate formation appears to have been directly or indirectly linked to the impact event causing Ar outgassing. The mode of carbonate formation in ALH84001 has been a matter of debate, with both "impact metasomatism" and a "playa lake" model being suggested. The latter has gained favor, seeming to be more consistent with all the data. Although carbonates have not been detected on Mars from orbit, whitish deposits within possible ancient lake beds have been detected. The playa lake model is implied if outgassing were early, >4.1 Ga ago, the upper limit on the carbonate age. It is most reasonable to assume that Ar outgassing accompanied a crater-forming event. The carbonate age data would then imply carbonate formation a finite time interval later, for which the only identifiable mechanism is precipitation from a playa lake. (A second shock event, identified by Treiman (1995) and other authors, fractured carbonates already formed, and is thought to be connected with ejection of ALH84001 from the Martian surface.) The same scenario is possible for ^{39}Ar–^{40}Ar ages of ~3.9–4.0 Ga, because the hypothesized lake may have been a transient feature of duration much less than the precision of the radiometric dating methods.

Reset eucrite ages: evidence for an impact cataclysm on asteroid 4 Vesta and throughout the inner Solar System?

If the early bombardment that reset the isotopic ages of lunar highland rocks was the result of the final accretion of the Moon, other bodies in the Solar System (other than the Earth) might not have been similarly affected. In that case, the ages of early impact craters on Mars might be entirely different from those on the Moon, and the ~3.9–4.0 Ga ages for ALH84001 only a coincidence. However, radiometric ages of basalts believed to be from one of the minor planets, the large asteroid 4 Vesta, suggest that the impact cataclysm was real and affected the whole inner Solar System.

The specific parent bodies of most meteorites cannot be identified. However, Earth-based observations have led to the conclusion that several large asteroid fragments have been spalled from the ~520 km diameter Vesta and are still located in related orbits. Some of these fragments approach orbital resonances with Jupiter from which it is believed material can be readily perturbed by gravitational interactions into Earth-crossing orbits (Binzel and Xu, 1993). Further, spectra taken in the laboratory of a class of basaltic meteorites called eucrites closely match Earth-based spectra taken of Vesta and its fragments. These observations have been used to suggest that Vesta, the third largest asteroid, is the parent body of eucrites and the related differentiated meteorites called diogenites and howardites (Drake 1979, Binzel and Xu 1993). If Vesta is the parent body of eucrites, this permits us to relate the chronological history determined in the laboratory for eucrites to a specific object located in a known place in the Solar System, an important association for understanding Solar System evolution.

Radiometric age determinations of eucrites show that these basalts were produced very early, probably within 10 Ma after the formation of their parent body. If the basalts are from Vesta, these data imply that even a body as large as Vesta accreted relatively quickly and was heated sufficiently to differentiate internally and produce early basalts on its surface. Most eucrites are breccias, rocks broken apart and reassembled by impacts on the parent body surface. Because Ar diffuses relatively easily in silicate, the K–Ar age is more readily reset compared to other isotopic chronometers when a rock is subjected to moderate heating (Bogard 1995). The Sm–Nd chronometer is generally the most difficult to reset. The ^{39}Ar–^{40}Ar ages of most eucrites are either strongly disturbed or totally reset. The Rb–Sr and Pb–Pb ages of some eucrites also are disturbed, and a few apparently are reset.

The distribution of these reset Ar–Ar ages and a few reset Rb–Sr ages for eucrites is shown in Figure 12, and is similar to that for lunar highland rocks (Figure 5). It has been suggested that this resetting of eucrite ages is evidence for an impact cataclysm having occurred on the eucrite parent body (Bogard 1995). If this is indeed the case, then this impact cataclysm occurred throughout the inner Solar System and would have influenced the early chronology and evolution of other bodies such as Mars. The ~3.9–4.0 Ga ^{39}Ar–^{40}Ar and Rb–Sr ages of ALH84001 are consistent with the hypothesized impact cataclysm.

Because Vesta is a smaller body than the Moon, the gravitational energy released by large impacts on Vesta would be lower and the degree of material heating less.

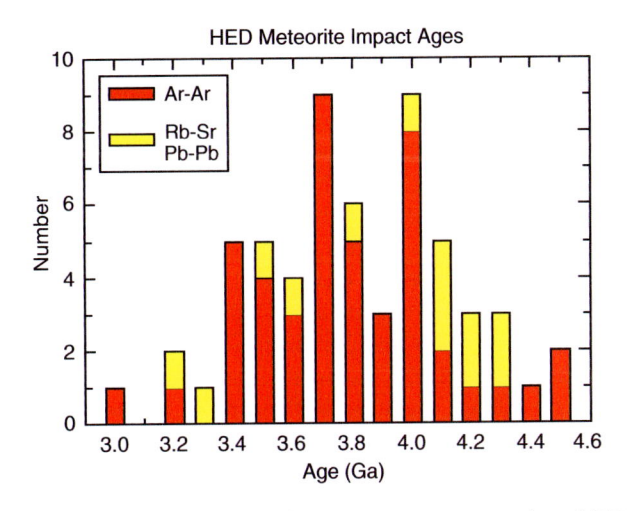

Figure 12 Histogram of impact-reset ages for HED (Howardite–Eucrite–Diogenite) meteorites, thought by many to be derived from the large asteroid 4 Vesta. The predominance of ages in the interval ~3.4–4.2 Ga is similar to that found for the Moon (Figure 5). This similarity suggests that the "lunar cataclysm" likely affected the entire inner Solar System. This also would have been true of Mars, explaining the Ar outgassing age of ~4.0 Ga of the ancient orthopyroxenite ALH84001.

This may be the explanation why, in contrast to the Moon, only the K–Ar chronometer, the one most easily reset, shows extensive impact resetting for Vesta. A few eucrites escaped brecciation, presumably because of deep burial, and these tend to show much older Ar–Ar ages (Bogard and Garrison 1995). Thus, the intermediate size of Vesta between the smaller meteorite parent bodies and planet-sized bodies such as the Moon, Earth, and Mars can be important in our understanding of those aspects of early planetary evolution involving both internal metamorphism and impact cratering.

The nakhlites: cumulate rocks from Early Amazonian volcanism?

So far, we have no Martian meteorites that are of Hesperian age; that is, ~ 3.5–1.8 Ga. Radiometric ages of the three nakhlite meteorites (Nakhla, Lafayette, and Governador Valadares), and Chassigny, as determined by the Rb–Sr, Sm–Nd, ^{39}Ar–^{40}Ar, and U–Pb methods are ~ 1.3 Ga (Figure 11). Close agreement of ages of these four meteorites by four dating techniques apparently unambiguously define their crystallization ages at ~ 1.3 Ga. The ~ 1.3 Ga ages of the nakhlites and Chassigny belong to either the Early (EA) or Middle Amazonian (MA) chronostratigraphic units as defined by Tanaka (1986) from crater densities, depending on whether one uses the HT or NW crater frequency curve. Type localities for rocks of this age are Amazonis Planitia (EA) and Acidalia Planitia (MA), respectively. Mineralogically, the nakhlites are clinopyroxenites, whereas Chassigny is a dunite. These mafic cumulates share similar trace element and isotopic characteristics, although trace element abundance patterns differ enough that the nakhlites and Chassigny probably are not comagmatic; that is, they did not form from a common magma. The crystallization history of Chassigny has been described as similar to that of the nakhlites, except that olivine and chromite are much more abundant (McSween 1994). The parental magma of the nakhlites has been modeled as similar to basaltic komatiites (Treiman 1986, Longhi and Pan 1989). Treiman (1986) concluded that the geologic settings that most fit these rocks are flood basalt provinces and shield volcanoes. To quote: "one may imagine flows of Nakhla parent magma coursing over the rough terrain left by previous flows, forming magma lakes and pools along their paths. Cumulates rich in augite and/or olivine settle out. The flows and pools cool according to their sizes ... Chassigny forms in a shallow intrusion."

Several authors have sought to identify Martian source terrains and candidate source craters for the SNC meteorites. Such attempts have relied primarily on the young crystallization ages of the meteorites (≤ 1.3 Ga) and the perceived characteristics of the ejecting impact and the resultant crater. Recent refinements in numerical simulations of near-surface spallation during impact seem to show that rock fragments can be ejected from even modest-size craters, so crater characteristics may not be a useful discriminator. Remotely sensed geochemical data from orbiting spacecraft are becoming increasingly available, and ultimately may help to delimit possible sources of the SNCs.

Shergottites: basaltic/lherzolitic rocks from Late Amazonian volcanism?

The preferred radiometric ages of basaltic and lherzolitic shergottites lie in the range ~ 175–475 Ma (Figure 11). The ~ 180 Ma ages of several shergottites belong to the EA chronostratigraphic unit according to the HT scale, or to the MA according to the NW scale. In seeking candidate source craters for the SNC meteorites, Mouginis-Mark et al. (1992) imposed the selection criterion that such craters must be ≥ 10 km in diameter. Prior to the orbital imagery from Mars Global Surveyor, this criterion and the age of candidate terrains appeared to restrict possible source craters for the ~ 180 Ma shergottites to the central Tharsis plains. Three later developments serve to loosen this restriction: (a) Mars Global Surveyor has shown that young, thin lava flows covering older flows exist at other localities; (b) older crystallization ages of ~ 330 Ma and ≥ 475 Ma have been found for some shergottites; and (c) the limiting crater size for meteorite ejection seems to have been pushed to lower values. Thus, it is unclear whether this restriction to the Tharsis region any longer applies. Nevertheless, the source region(s) for shergottites must be limited to very young basalt flows, probably in the Tharsis region, Amazonis Planitia, or Elysium Planitia. As for the nakhlites, improved orbital geochemical data may serve to further restrict the possible source areas.

Interpretation of some of the radiometric data for shergottites has been controversial because they often show complexities both within a single radiometric system and between systems. For example, the Rb–Sr ages for at least five of the eight shergottites, including Shergotty, cluster near 175 ± 10 Ma. These shergottites have similar Rb–Sr ages as the type example, Shergotty. Historically, it was the determination of the Rb–Sr age of Shergotty that gave rise to some of the early suggestions of a Martian origin for the shergottites (Wasson and Wetherill 1979, Nyquist et al. 1979b,c). However, the apparent ^{39}Ar–^{40}Ar ages of the shergottites are systematically older than the Rb–Sr ages. Many of the determined Sm–Nd ages are older than the Rb–Sr ages as well. Normally, Sm–Nd ages are the most robust against secondary disturbance, and thus these older Sm–Nd ages implied older crystallization ages. Although ^{39}Ar–^{40}Ar ages can be "too old" because of inherited ^{40}Ar, such situations are rare among meteorites. Rb–Sr ages can: (a) date the time of crystallization, in which case they should be concordant with the Sm–Nd age; (b) be partially reset to younger values by secondary major thermal events, in which case they should tend towards concordancy with the ^{39}Ar–^{40}Ar ages; and (c) be "disturbed" by volatile loss of Rb during secondary thermal events, in which case they either are undefined, or are apparently older than either the Sm–Nd or the ^{39}Ar–^{40}Ar ages. Yet the Rb–Sr ages of the shergottites are most often quoted as their crystallization ages. Can this be correct for meteorites that have been subjected to secondary shock and post-shock thermal annealing, as the shergottites apparently have?

The answer appears to be an equivocal "yes." Jones (1986) argued that preservation of elemental zoning patterns in major mineral phases of the shergottites, in spite of shock-induced mineral transformations, precluded identification of the Rb–Sr ages with the time of shock metamorphism: the earlier interpretation (Nyquist et al. 1979b, Shih et al. 1982). That earlier interpretation postulates that in a post-shock thermal regime Sr isotopic equilibration can be achieved more rapidly than chemical homogenization of major mineral components such as Ca. The issue remains unresolved, and is not addressable via in situ isotopic analyses by ion microprobe, for example, because the analytical precision required exceeds that provided by present instruments. Although, the majority of igneous petrologists side with Jones (1986), the issue has become a moot point due to additional isotopic measurements with improved instrumentation and techniques. More and more of the meteorites are being found to have Sm–Nd ages that are concordant with their Rb–Sr ages. The high ^{39}Ar–^{40}Ar ages are now explained as due to the ubiquitous presence of excess atmospheric ^{40}Ar shock-implanted into the shergottites and mantle ^{40}Ar present in the rocks when they crystallized (Bogard and Garrison 1999). Yet subtle isotopic inconsistencies are present in the other isotopic systems as well, including the U–Pb and Lu–Hf systems. These inconsistencies appear to require unusual petrogenetic processes accompanied by relict, unequilibrated isotopic effects in major mineral phases and secondary alteration/contamination processes that likely occurred on the Martian surface. Investigation of the ages of the shergottites remains an active area of research.

An initial controversy arose over the age of the Shergotty meteorite. Originally dated at 165 ± 11 Ma and 254 ± 10 Ma by the Rb–Sr and ^{39}Ar–^{40}Ar methods, respectively, (Nyquist et al. 1979b, Bogard et al. 1979), its Sm–Nd age was given as 620 ± 171 Ma by Shih et al. (1982) and as 360 ± 16 Ma by Jagoutz and Wänke (1986). The latter authors, however, actually reported two Sm–Nd ages. The ~360 Ma age was for a subset of the data that did not include the whole rock datum in the isochron regression. A second subset of the data that included the whole

rock datum gave an apparent age of 147 ± 20 Ma. We favor the younger age for several reasons: (a) for closed-system isotopic evolution, the whole rock datum must lie on the isochron defined by minerals from the rock; (b) the ubiquity of ages in the range of ~165–185 Ma among the shergottites; (c) the concordant Rb–Sr and Sm–Nd ages of ~175 Ma for Zagami, a near-twin of Shergotty; and (d) the possibility that the Shergotty sample was contaminated on Mars, or during handling on Earth. However, complex petrogenetic processes involving magma mixing might yield the two-age result in a non-trivial way, as suggested originally by Jagoutz and Wanke (1986).

Fewer ambiguities accompany the remaining data shown in Figure 11. The age data are incomplete for the lherzolitic shergottites, but the data which exist are consistent with the ~175 Ma age. For LEW88516, there is evidence of isotopic disturbance on Mars, apparently via surface alteration processes. Such processes may account for much of the aberrant data observed for other shergottites as well.

Because discordant ages were initially obtained for Shergotty and other shergottites by the various methods, Shih et al. (1982) sought an approach that would "see through" secondary events. They noted that the whole-rock Sm–Nd data for the basaltic shergottites Shergotty and Zagami, combined with that of the lherzolitic shergottite ALH77005, defined an apparent "isochron" of slope corresponding to an age of ~1.34 Ga, in remarkable agreement with the age of the nakhlites. This coincidence was reinforced by later data for basaltic shergottite EET79001. It seemed to suggest that the shergottites and the nakhlites might be related, if only indirectly via mantle processes. It also suggested that the ~1.3 Ga age might have significance for the shergottites. However, later interpretations have favored the view that this "isochron" is a mixing line, representing mixing among "crustal" and "mantle" end-members. As such, its only significance may be that all of the meteorites come from the same planet.

Ejection ages: how many ejection events?

In addition to their radiometric ages, so-called ejection ages (Eugster et al. 1997) provide useful information about the Martian meteorites. The ejection ages are conceptually the time since the meteorites' ejection from the Martian surface, and numerically equal to the sum of their cosmic-ray exposure ages and terrestrial ages. This definition assumes that the meteorites do not undergo secondary breakup in space. In practice, the cosmic-ray exposure ages are much longer than terrestrial ages, so "cosmic-ray exposure age" often is used synonymously with ejection age. Meteorites with similar ejection ages are interpreted as being ejected in a common event. However, some uncertainty exists in deriving these space exposure ages, especially when

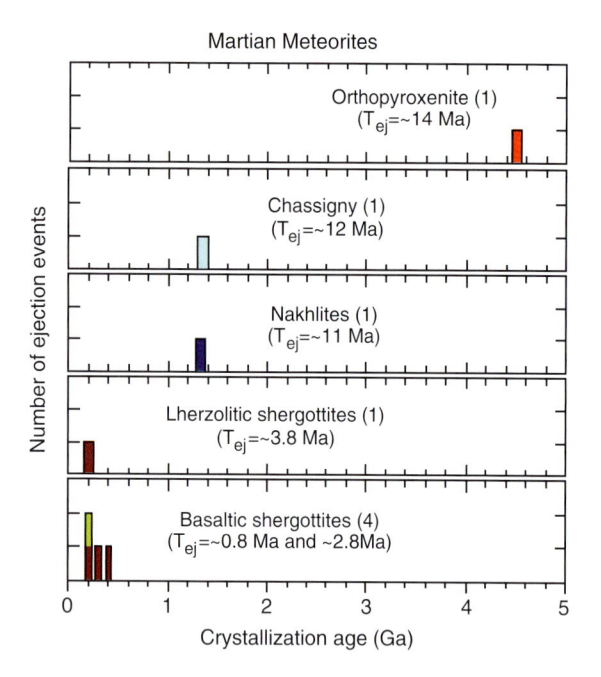

Figure 13 Histogram of "probable" ejection events from Mars. This histogram assumes that different events are indicated not only by different cosmic-ray exposure ages, but also by differences in other characteristics, such as chemical composition and different crystallization ages. The numbering of events follows that of Eugster et al. (1997), except that it counts a separate ejection event for basaltic shergottite QUE94201 because of its older crystallization age compared to Shergotty and Zagami, which have similar cosmic-ray exposure ages. Thus, this figure shows the "probable maximum" number of ejection events as eight. If QUE94201 and Chassigny are not given their own events, then a "probable minimum" number of events is six.

comparing ages of meteorites with significantly different compositions such as basaltic v. lherzolitic shergottites. Figure 13 compiles ejection events as inferred from the ejection ages, separated by meteorite type and crystallization age. To the extent that different Martian surface units are composed of rocks of distinct crystallization ages, crystallization age can be viewed as another "event discriminator." This is useful in identifying possible source areas for the meteorites, because much of the Martian surface has been classified according to relative age as determined from the density of meteorite impact craters per unit area. It should be noted, however, that these "crater retention ages" reflect the upper kilometer or so of near-surface layers in the areas where they are determined, and that lava flows of a variety of absolute ages may be present in a given area (Hartmann 1999, Hartmann and Berman 2000). The eight events illustrated in Figure 13 are the "probable maximum" needed to eject the thirteen meteorites. The "probable minimum" number of events is six (see figure caption).

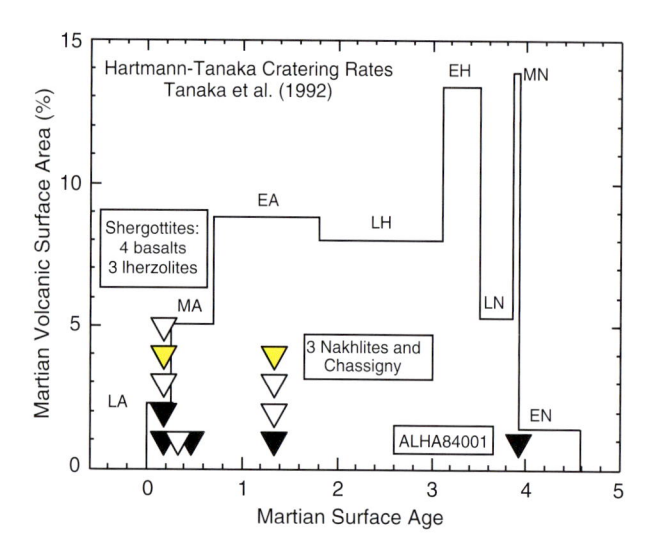

Figure 14 Comparison of the radiometric ages of Martian meteorites to Martian surface ages as inferred from crater counts. Absolute age boundaries for Martian cratering epochs and the percent of the total Martian volcanic surface area belonging to each epoch are taken from Tanaka et al. (1992). The Martian epochs are: EN, Early Noachian; MN, Middle Noachian; LN, Late Noachian; EH, Early Hesperian; LH, Late Hesperian; EA, Early Amazonian; MA, Middle Amazonian; LA, Late Amazonian. There are clearly a disproportionate number of young meteorites compared to the percentage of the Martian volcanic surface area that is equally young. The discrepancy is reduced when only the separate ejection events are considered. The number of definite ejection events for meteorites of a given crystallization age is represented by a black symbol. Possible additional ejection events are represented by a yellow symbol. Thus, for example, for the three nakhlites and Chassigny, there is one "definite" event for the nakhlites, and another "possible" event for Chassigny. The situation is complicated for the shergottites because some meteorites with the same ejection ages have different crystallization ages. The meteorite Dar al Gani 476 (crystallization age = 475 Ma, ejection age = 1.3 Ma) appears to represent an unique event. Meteorite QUE94201 (crystallization age = 330 Ma, ejection age = 2.8 Ma) has the same ejection age as other basaltic shergottites having crystallization ages of ~175 Ma. This event is registered as a black symbol at the younger age. A second event at that age is registered for EET79001, for which the crystallization age is ~175 Ma, but the ejection age is only ~0.8 Ma. A third event at ~175 for the llerzolitic shergottites is shown as "possible". Thus, by this accounting the total number of ejection events is 5–7.

Figure 14 seeks to combine information from Figures 11 and 13 with data on Martian crater retention ages. It is a histogram of the crystallization ages of Martian meteorites, superimposed on a histogram of the percentages of the Martian surface belonging to stratigraphic units of different ages, as determined by surface crater density. The figure presents a two-fold paradox: (a) the meteorites cluster very tightly at a few specific values of the radiometric ages; and (b) rocks of those ages are predicted to be relatively rare on the Martian surface. For example, there are many more meteorites with young ages, ≤1.3 Ga, than there are meteorites with old ages, ~4 Ga. Of the eight postulated events, only one yielded a meteorite of Noachian age, in spite of the fact that ~21% of the total volcanic surface area of Mars is Noachian in age. Noachian meteorites are thus under-represented by about a factor of two. This disparity can perhaps be rationalized by appealing to the "statistics of small numbers," or to other factors such as friability, which might discriminate against old, brecciated meteorites. However, as we have already noted, there are also no Hesperian-age (~3.5–1.8 Ga) meteorites, in spite of the fact that ~23% of the Martian surface area is that age. Since the Hesperian post-dates the era of heavy meteorite bombardment, rocks from that time period should resemble lunar mare basalts in terms of their bombardment history, state of brecciation, and friability. Among a comparable number of lunar meteorites there are several mare basalt breccia clasts, and one "hand specimen" mare basalt. Although only one has been dated (~3.9 Ga), all must be approximately "Hesperian" in age. Yet Hesperian age Martian meteorites are missing. In contrast, more than 90% of the meteorites are Amazonian in age, although only ~16% of the volcanic surface is that age. This imbalance does not significantly improve if ejection events are considered instead of simply the number of meteorites: Four to five out of six to eight ejection events must have occurred within the last ~4 Ma from ~16% of the volcanic surface area of Mars. These observations imply a Mars surface age dilemma. Either our assessment of the number of ejection events is wrong, or a much larger fraction of the Martian surface must consist of young volcanics, signifying that Mars has been more active geologically in the recent past than previously thought. How can these observations be reconciled?

One possible explanation is that the individual ejection events gave rise to a greater multiplicity of meteorites than illustrated in Figure 13. Some additional possibilities are illustrated in Figure 14. Although the three nakhlites were recovered at three widely separated localities, Egypt, Brazil, and Indiana, all three are clinopyroxenites, have the same radiometric crystallization ages, and the same ejection ages. We represent them by a single black triangle and two open ones in Figure 14, indicating three meteorites, but only one ejection event. This event is represented in the histogram of Figure 13 by a single block for nakhlites. The dunite, Chassigny, is a fourth meteorite with an ~1.3 Ga crystallization age, represented by a yellow triangle in Figure 14. It is mineralogically distinct from the nakhlites, but its ejection age is the same as that of the nakhlites within mutual error limits. The abundances of several key incompatible elements in Chassigny, as well as its radiogenic isotope composition, suggest a close relationship to the nakhlites. We suggest that it was ejected simultaneously with them, and thus that the EA has been sampled only once. This is commensurate with Martian meteorites having sampled Noachian terrain only once and Hesperian terrain not at all.

The shergottites present the greatest puzzle for the surface ages of Mars. Taken at face value, the ejection age data indicate four ejection events within 4 Ma on terrain of three different, young ages. Thus, more than half of the total number of ejection events apparently occurred on only ~7% of the Martian volcanic surface area, using the cratering statistics presented by Tanaka et al. (1992). Those statistics were based on Viking imagery, and, of course, may be revised based on newer data from Mars Global Surveyor. In this regard, observations of young, thin lava flows covering older terrain in broad regions of central Amazonis Planitia and southeast Elysium Planitia (Hartmann 1999, Hartmann and Berman 2000) are relevant. The ages of those lava flows have been estimated as no more than a few hundred Ma, and some areas are estimated to contain flows that are only ~10 Ma old. Thus, even acknowledging an uncertainty of about a factor of three in the absolute cratering rate, it is clear that lava flows as young as the ~180 Ma ages of many shergottites do exist on Mars. Nevertheless, it seems to us unlikely that four random events, at ~3.8, ~2.8, ~1.3, and ~0.8 Ma ago, would eject meteorites with crystallization ages within the narrow range of ~155–195 Ma allowed by the analytical uncertainties of the individual crystallization ages of the shergottites.

Because the exposure age calculations are subject to several sources of systematic bias, it may be that uncertainties in these ages are somewhat higher than currently estimated. Examples of such potential biases are diffusive loss of spallogenic noble gases related to particular orbital parameters, errors in correction of noble gas production rates for variations in chemical composition and variations in shielding from cosmic-ray protons, and contributions of solar protons to ^{21}Ne production. Such factors need to be carefully evaluated to see if the ~2.8 Ma and ~3.8 Ma ejection age groups, and the ~0.8 and ~1.3 Ma groups, respectively, are truly resolved from each other.

Nyquist et al. (1998) sought to evaluate whether the then known shergottites might have been ejected in a single event. Since then, new data have been added for DaG 476, which increase from two to three the number of distinct crystallization ages present among the shergottites. However, the probability that rocks of three different, but similar, crystallization ages might be present in the vicinity of a single impact is not significantly less than that rocks of two different crystallization ages be present. Basaltic shergottite QUE94201, with a crystallization age of ~330 Ma (Borg et al. 1997), appears to have been ejected in the same event as several ~175 Ma shergottites, for example (Eugster et al. 1997). Multiple crystallization ages at a given site might even be considered the rule, based on the recent observations of young lava flows overlying older ones in Amazonis Planitia and Elysium Planitia. The most difficult part of the single-impact scenario is explaining young, but different, ejection ages of ~0.8 Ma and ~1.3 Ma for two of the meteorites. Unless these two ages can be reconciled (one of the meteorites was heavily weathered and contaminated by terrestrial atmosphere), ejection from Mars in a single impact would require two secondary breakups in space, even if it were shown that the majority of basaltic and lherzolitic shergottites were ejected simultaneously. Nevertheless, the overall geochemical similarity of the meteorites involved, as well as their similar crystallization ages favor a single ejection event. In that case, the number of shergottite ejection events would be the same as the number of nakhlite and orthopyroxenite ejection events (one each).

Hartmann (1999) estimates the uncertainty in scaling the Martian crater density v. age curve to the lunar one to be of the order of a factor of three. If the Martian cratering rate were higher than assumed in Tanaka et al. (1992), and used in Figure 14, the ages of Martian surface units would be shifted to lower values. For example, surface terrain of age ≤ 0.5 Ga, now assumed to be present only in the LA and MA, respectively, would be found in the EA as well. Together these units represent ~16% of the Martian volcanic surface area, and the probability of an ejection event from within that area would be comparable to the probability of an event on older surfaces. Final resolution of the apparent paradox of having too many young meteorites from too little young surface will require absolute calibration of the Martian crater frequency curve via returned Martian samples.

Accretion and global geochemical differentiation of Mars

Theoretical estimates for the duration of planetary accretion have been made from statistical simulations resulting in "planetary systems" resembling our own in the number, location, and mass of the "planets." Wetherill (1986) estimated an accretion time of ~60 Ma for Mars, for example. Although the physical processes of accretion left no readable time markers, the abundance of certain radiogenic nuclides does provide a timescale for the chemical processes accompanying, or perhaps immediately following, accretion. What is evidently needed to measure the time interval of accretion are radionuclides which decayed to stable daughter nuclides during the time when accretion was occurring, and for which chemical fractionations of the parent/daughter ratio caused measurable variations in the abundance of the daughter product. Fortunately, a number of such radionuclides were present in the early Solar System as the result of element synthesis taking place in nearby stars, novae, and supernovae before the collapse of the interstellar cloud of matter which became the Solar System. We have mentioned a number of these "short-lived chronometers" in the introduction to this chapter. The two short-lived chronometers that have proven most useful for timing accretion of Mars and the accompanying global-scale chemical fractionations, or "geochemical differentiation," are ^{146}Sm and ^{182}Hf. This is not only because their half-lives for decay, 103 Ma, and 9 Ma, respectively, are of suitable length, but also because they partake geochemically in important aspects of global differentiation.

The decay of ^{146}Sm to ^{142}Nd is coupled to that of its long-lived sibling ^{147}Sm, which decays to ^{143}Nd. The ^{147}Sm–^{143}Nd decay is routinely used to infer the time-averaged Sm/Nd ratio in the mantle sources of rocks from their initial ^{143}Nd/^{144}Nd ratios. Several studies have shown that the time-averaged Sm/Nd ratio in nakhlite sources must have been ~17% greater than that in chondritic meteorites. Harper et al. (1995) measured the ^{142}Nd/^{144}Nd ratio in nakhlites, and found it to be ~70 parts per million higher than in chondrites. This enrichment in ^{142}Nd can only be explained if the enrichment in Sm/Nd occurred early in Solar System history, because ^{146}Sm would have completely decayed within ~300 Ma. Harper et al. (1995) modeled the growth of radiogenic ^{142}Nd in two stages: (a) in undifferentiated Martian material with the same ratio of Sm/Nd as chondritic meteorites, and (b) in a mantle reservoir with Sm/Nd ~17% greater than in chondrites. They concluded that Sm/Nd fractionation in the Nakhlite source must have occurred no later than 27 Ma after the origin of the Solar System. They further concluded that such fractionation, and therefore Martian differentiation, probably occurred within 15 Ma of Solar System formation. A three-stage model considered as a limiting case would have allowed Sm/Nd fractionation as late as ~100 Ma after Solar System formation.

Lee and Halliday (1997) pursued the question using the ^{182}Hf–^{182}W decay, which has a half-life of only 9 Ma. Whereas the ^{146}Sm–^{142}Nd system traces Sm/Nd fractionation among mineral phases in the planet's mantle and crust, the ^{182}Hf–^{182}W system traces Hf/W fractionation between a planet's metallic core and mantle, since W would be partitioned into the metal while Hf remained in the rocky mantle. The loss of W from the mantle would cause a large increase in the Hf/W ratio of the mantle to values much greater than in chondritic meteorites. If this happened before complete decay of ^{182}Hf, the measured ratio of radiogenic ^{182}W to a stable tungsten isotope, say ^{184}W, will exceed that in chondritic meteorites by an amount related to the time and degree of Hf/W fractionation. By analyzing the same samples as had previously been analyzed for ^{142}Nd, Lee and Halliday (1997) found that those samples with ^{142}Nd excesses, primarily the nakhlites and Chassigny, also had ^{182}W excesses. From the nakhlite data they concluded that core formation on Mars occurred within the first 15 Ma of Solar System history, in agreement with the preferred interpretation of the ^{146}Sm–^{142}Nd data by Harper et al. (1995). A further important conclusion they drew was that in order for a positive correlation to exist between the W and Nd data, the two parent/daughter ratios must have been fractionated together at an early stage. The simplest explanation is that silicate melting and metal segregation (core formation) occurred simultaneously. Their interpretation was that accretional energy combined with the decay of short-lived nuclides such as ^{26}Al would have quickly heated the planet and formed a shallow magma ocean from which metal could segregate.

The hypothesized Martian magma ocean is reminiscent of the popular lunar magma ocean model, and suggests that magma oceans may be a common feature of terrestrial planet formation. In a theoretical study, Tonks and Melosh (1992) concluded that by the time a planet grows to the size of Mars, it has a nearly 100% probability that a giant impact will have triggered core formation. Whether or not this is the case, it raises the question of how quickly such a magma ocean would cool to isotopic closure of the Sm–Nd system, and to what extent "core formation ages" from ^{182}Hf–^{182}W data and "mantle solidification" ages from ^{146}Sm–^{142}Nd data really ought to agree. Some geophysical models of planetary cooling suggest mantle solidification intervals of ~100 Ga, in better agreement with the three-stage model of Harper et al. (1995) for the Nakhla data than with their two-stage model. Shih et al. (1999) studied another nakhlite, Governador Valadares, and concluded that although a two-stage model with rapid formation of the nakhlite source provided an adequate fit to the data, a model allowing the source region to evolve to its pre-1.3 Ga value in two stages (three stages altogether) provided a still better fit. Research in this area will continue as new samples become available, and as new constraints are added from other isotopic systems.

We stated at the beginning of this section that the discussion assumed the Martian meteorites really are from Mars. If this is indeed true, we already have learned a great deal about Mars from them. We will be uncertain whether these conclusions are justified until samples of Mars are actually returned by spacecraft missions, a task for the new century of space exploration.

RADIOGENIC GROWTH OF ^{87}SR AND ^{143}ND IN PLANETARY MANTLES: EARTH, MOON, AND MARS

We have shown values of initial ^{87}Sr/^{86}Sr and ^{143}Nd/^{144}Nd in lunar basalts, expressed as $\varepsilon_{Sr}(T)$ and $\varepsilon_{Nd}(T)$, and plotted v. basalt age, T, in Figures 8 and 9. The Sr and Nd isotopic ratios are not changed by melting the host material, so, for simple petrogenetic processes, these also are the values of $\varepsilon_{Sr}(T)$ and $\varepsilon_{Nd}(T)$ in the lunar mantle just before and after basaltic magmas are formed by partial melting in the mantle. Because of the long half-lives of the radiogenic parents, ^{87}Rb and ^{147}Sm, thereafter the values of $\varepsilon_{Sr}(t)$ and $\varepsilon_{Nd}(t)$ in the mantle increase approximately linearly with time, t, to the present day, which by convention is taken to be $t = 0$. (Assuming no additional melts are extracted.) The rate of increase, or "radiogenic growth" of $\varepsilon_{Sr}(t)$ and $\varepsilon_{Nd}(t)$ is determined by the ^{87}Rb/^{86}Sr and ^{147}Sm/^{144}Nd ratios in the mantle source regions of the basalts. Furthermore, by assuming ^{87}Rb/^{86}Sr and ^{147}Sm/^{144}Nd was constant in the source regions during the entire time interval (4.56 Ga − T) *prior to* magma extraction, we can estimate the *time averaged* values of ^{87}Rb/^{86}Sr and ^{147}Sm/^{144}Nd in the source regions. Knowing the values calculated from such a "single-stage" model gives us useful insight into the chemical composition of the basalt sources. As we have seen, planetary mantles are typically depleted in certain incompatible elements, including K, Rb, LREE, and U, relative to the average concentrations of these elements in the bulk planet. Planetary crusts are therefore enriched in these same elements. Thus, it is common to speak of "the depleted mantle" of Earth, for example. In this section, we briefly compare what the isotopic data tell us about the lunar and Martian mantles to what we have learned about the Earth's mantle from similar data.

Figure 15 characterizes isotopic signatures, as expressed by present-day $\varepsilon_{Nd}(t=0)$ and $\varepsilon_{Sr}(t=0)$ values, in basalt source regions of Earth, Moon, and Mars; that is, in the planetary mantles. The values of these parameters were calculated by assuming that all three planetary bodies initially had chondritic ^{147}Sm/^{144}Nd and ^{143}Nd/^{144}Nd values and the bulk Solar System initial (BSSI) ^{87}Sr/^{86}Sr value (= BABI) at 4.56 Ga ago. However, the bulk ^{87}Rb/^{86}Sr values for these bodies were probably very different, as shown by characteristic volatile/refractory elemental ratios such as the K/La ratio; for example, K/La is ~400, ~70, and ~600 for Earth, Moon, and Mars, respectively. The (^{87}Rb/^{86}Sr)$_{BE}$ value for the bulk Earth is ~0.084 based on the correlation of ^{143}Nd/^{144}Nd and ^{87}Sr/^{86}Sr ratios for young terrestrial basalts. This correlation shows that ^{87}Sr/^{86}Sr = 0.7045 for the basalts at the chondritic value of ^{143}Nd/^{144}Nd; that is, for $\varepsilon_{Nd}(t=0) = 0$. Assuming the Rb/Sr ratio behaves similarly as the K/La ratio for these three planetary bodies, Rb/Sr for the Moon would be estimated to be ~1/6 the Rb/Sr value of the Earth and the Rb/Sr ratio for Mars would be estimated to be ~1.5 times the Earth value. As discussed earlier, we have used a higher value of (^{87}Rb/^{86}Sr)$_{BL}$ = 0.038 for the bulk Moon, as derived from the lunar isotopic data themselves. Because of the paucity of data for Mars, we use the K/La derived value (^{87}Rb/^{86}Sr)$_{BM}$ = 0.126 for bulk Mars. Other approaches lead to slightly different estimates. Borg et al. (1997), for example, estimated (^{87}Rb/^{86}Sr)$_{BM}$ = 0.16. The conclusions presented here

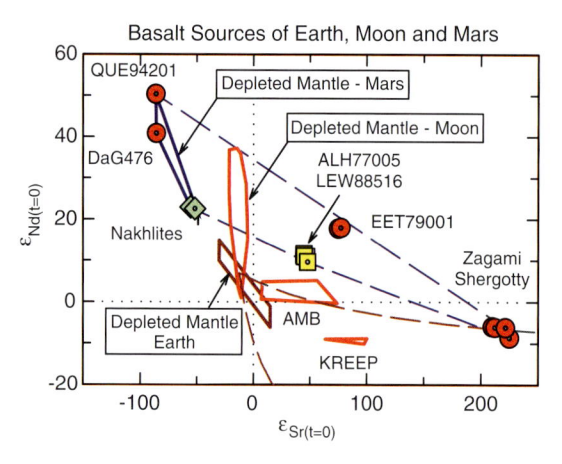

Figure 15 Variation of ε_{Nd} and ε_{Sr} for basalt sources on the Earth, Moon, and Mars. In each case, the undifferentiated body would have $\varepsilon_{Nd} = 0$, as in chondritic meteorites. "Bulk planet" values of $^{87}Rb/^{86}Sr = 0.084$, 0.038, and 0.126 are estimated to give $\varepsilon_{Sr} = 0$ for the Earth, Moon, and Mars, respectively. Variations in this parameter reflect different proportions of volatile (Rb) and refractory (Sr) elements in the three bodies. Interestingly, although the Earth is geologically most active, its mantle is most homogeneous, presumably due to re-homogenization due to mantle convection. The Martian mantle, in contrast, appears to be most heterogeneous, with the Moon's mantle a close second in this regard. The dashed curves show the approximate limits of ε_{Nd} and ε_{Sr} values for terrestrial mantle-derived magmas that have assimilated continental crustal components. Analogous rocks on the Moon are the KREEP basalts and aluminous mare basalts (AMB). For Mars, the ε_{Nd} and ε_{Sr} values of the "basaltic" meteorites Shergotty and Zagami clearly show evidence of a crustal component. (Multiple analyses are plotted for these two meteorites.) Basaltic shergottite EET79001 and lherzolitic shergottites ALH77005 and LEW88516 contain smaller crustal contributions. The isotopic systematics of the nakhlites appear to be those of "good" mantle-derived magmas, but these meteorites are curiously LREE-enriched, a feature difficult to reconcile with a LREE-depleted source. Thus, basaltic shergottites QUE94201 and DaG476 probably most reliably represent the isotopic compositions within the Martian mantle.

are insensitive to uncertainties in $^{87}Rb/^{86}Sr$ for bulk Moon and Mars.

The $\varepsilon_{Nd}(T)$ and $\varepsilon_{Sr}(T)$ values for the basalts were calculated from their respective $^{147}Sm/^{144}Nd$, $^{143}Nd/^{144}Nd$, $^{87}Rb/^{86}Sr$, $^{87}Sr/^{86}Sr$, and age (T) data. Then, their respective $^{147}Sm/^{144}Nd$ and $^{87}Rb/^{86}Sr$ values for the basalt sources were calculated from a single-stage model, that is, assuming the sources were formed by a single magmatic event at 4.56 Ga. The present-day $\varepsilon_{Nd}(t = 0)$ and $\varepsilon_{Sr}(t = 0)$ values plotted in the figure are ε_{Nd} and ε_{Sr} calculated for sources having the calculated single-stage values of $^{147}Sm/^{144}Nd$ and $^{87}Rb/^{86}Sr$ at $t = 0$.

The small parallelogram represents the so-called "mantle array" for basalt source regions of the Earth. The Earth's crust is mainly granitic in composition and crustal rocks plot towards the lower right corner of the diagram. The dashed curves represent approximate boundaries for mixtures of pristine mantle-derived magma and different granitic crustal materials. High Ti and low Ti mare basalts have values of $\varepsilon_{Nd}(t = 0)$ and $\varepsilon_{Sr}(t = 0)$ that plot in the upper left, "depleted mantle" quadrant of the figure, as do the nakhlites, Chassigny, and the two older shergottites, QUE94201 and DaG476. The younger shergottites, all with crystallization ages of ~180 Ma, with the possible exception of Shergotty, plot towards the lower right of the diagram. It has long been suspected that Shergotty and Zagami contain a "crustal" component. Whether this ought also to be concluded for lherzolitic shergottites ALH77005 and LEW88516, as well as basaltic shergottite EET79001 is at the moment unclear, and depends on where the $\varepsilon_{Sr}(t = 0) = 0$ line should be drawn; that is, on the value of $(^{87}Rb/^{86}Sr)_{BM}$. Lunar KREEP has a clear-cut crustal signature, as expected from the probable role of magma ocean residuum (urKREEP) in its genesis. Some of the Apollo 14 AMB (aluminous mare basalts), especially the VHK basalts, also display a generally "crustal" signature, but less associated with urKREEP. Crustal assimilation in this case probably involved granitic magmas, enriched in K and Rb, but depleted in REE relative to urKREEP.

Clearly, the magnitude of the isotopic compositions for basalt sources indicates the extent of planetary differentiation. If a planetary body did not differentiate at all, for example, the $\varepsilon_{Nd}(t = 0)$ and $\varepsilon_{Sr}(t = 0)$ values would remain at (0,0). Of the three planetary bodies considered here, the Martian mantle has by far undergone the most differentiation, followed by the mantles of Moon and Earth. This is apparently the result of mantle convection on Earth, and its lack on Moon and Mars. As a consequence, Earth has plate tectonics, the others do not.

CONCLUSIONS: WHAT HAVE WE LEARNED ABOUT THE CHRONOLOGY OF TERRESTRIAL PLANETS?

We have discussed some aspects of the understanding of Solar System chronology as the twenty-first century begins. Our emphasis has been the chronological evolution of Moon and Mars as planetary bodies. Meteorite chronology, which has been extensively discussed elsewhere, is presented here only to the extent that meteorite ages define upper limits for the ages of Moon, Mars, and other planets. Because Earth is a geologically active planet, the ages of terrestrial rocks define the complex ongoing evolution of Earth, but tell us little about its earliest history. Meteorites, deriving from small parent bodies that cooled relatively quickly, reveal processes of early accretion and heating of solid objects in the Solar System, but give little insight into long-term, planetary-scale processes. The Moon and Mars occupy intermediate positions in the size scale for rocky objects in the Solar System. The Moon is large enough to have extensively differentiated and produced rocks over a significant period of time. Yet it is small enough to have become thermally inactive part way through its evolution, and to preserve products of some of its earliest processes. Mars is closer to Earth in size and thermal history, and permits a comparison of the evolution of the two planetary bodies. In contrast to Earth, Mars apparently preserves elements of both its earliest history and of relatively recent thermal evolution. The concept of studying one planetary object in detail in order to understand other planetary bodies is called comparative planetology and is recognized as an important component of our investigation of the evolution of the Solar System.

We have presented considerable detail concerning the chronological evolution of Moon and Mars. These studies, along with studies of meteorites, have led both to a deeper understanding of major processes that have acted throughout Solar System history, and to formulation of other significant questions for future work. Some of the themes emerging from this study are summarized below.

The earliest solid materials formed in our Solar System about 4.566 Ga ago. Within a time period of ~10 Ma numerous small parent bodies had formed, and many had become strongly heated and differentiated. Several short-lived isotopes useful for assessing chronology were present at that time. Within ~50 Ma after formation of the earliest solid materials, isotopic systematics indicate that the Earth, Moon, and Mars had substantially formed and were experiencing major differentiation of components, including core formation. These findings derive from short-lived chronometers (e.g., ^{146}Sm and ^{182}Hf) and initial isotopic ratios (e.g., ^{87}Sr/^{86}Sr and ^{143}Nd/^{144}Nd). The oldest dated rocks from these bodies are ~4.5 Ga for Moon and Mars, but <4.0 Ga for the Earth. Most meteorite ages fall in the range 4.50–4.56 Ga.

Lunar studies have suggested that formation of an early magma ocean, representing melting to depths of many hundreds of kilometers, may have been a significant event on the inner planets and the Moon. Ca-rich feldspar solidified from the lunar magma ocean to form much of the crust of the Moon to depths of many tens of kilometers. Ages of a few individual rocks representing this early crust spread over a significant period, ~4.5–4.3 Ga, for reasons not fully understood. During the first few hundreds of Ma, this early feldspathic lunar crust was intruded by more mafic material from below as the magma ocean solidified throughout. A transition occurred from the magma ocean era to magmas generated by radioactive heating and melting. Some of these magmas interacted with and "contaminated" the lunar crust. This produced various lunar highland rocks spanning ages of ~4.5–4.0 Ga.

The depth profile through the lunar crust and upper mantle was compositionally very heterogeneous due to preferential flotation of feldspar and settling of Mg-rich components from the LMO. The early crust of the Moon also appears to have been compositionally heterogeneous in the lateral dimension. This heterogeneity apparently contributed to broad surface areas of rocks enriched in incompatible trace elements (i.e., urKREEP) and rare rock types such as granites that are also enriched in incompatible trace elements, but in different proportions than in urKREEP. These rocks have ages of ~4.3–3.8 Ga.

For the first ~700 Ma the Moon was strongly bombarded by impacting objects, which mixed the upper crust, brecciated many of the rocks, and reset most of the radiometric ages. The parent body of eucritic meteorites, probably the asteroid Vesta, apparently experienced a similar bombardment. Expectations are that an analogous bombardment occurred on Mars and the Earth, which would have greatly affected the early evolution of both the crust and atmosphere of these planets. The only old Martian meteorite known also experienced major impact heating at ~4.0 Ga.

This early lunar bombardment appears to have consisted of two components: an early residue from planet formation and a later (~3.8–4.0 Ga) period, called a cataclysm, in which infalling mass was concentrated into several large objects that created the nearside basins. Many of the ages of Apollo-returned highland rocks may have been reset by the Imbrium, Serenitatis, Crisium, and Nectaris basins. The ages of highland rocks from the lunar farside, where few young basins exist, could advance understanding of both the nature of the bombardment over time and the age distribution of highland rocks.

Radiometric ages for basalts from the lunar maria range from ~3.9 to ~2.9 Ga. An average age of ~3.45 Ga has been used to provide an "absolute" calibration for meteoroid cratering rates both elsewhere on the Moon and on other Solar System bodies. The upper limit to the ages of basalts currently filling the maria is provided by the ages of the mare basins. However, photogeologic studies have shown the presence of "cryptomaria" in the lunar highlands, which likely contain basalts older than those filling the current maria. Also, small fragments of pre-Imbrium aluminous mare basalts have been found within breccias excavated by the Imbrium impact. The ages of these fragments extend to ~4.2–4.3 Ga, dating the probable onset of internally driven volcanism. Photogeologic studies appear to extend the period of volcanism beyond the ~2.9 Ga minimum radiometric age of mare basalts, to as late as perhaps ~1.0 Ga ago.

Among the Apollo lunar samples, there is an apparent correlation between the composition and age of mare basalts. Aluminous mare basalts are generally oldest, followed by high Ti basalts, low Ti basalts, and VLT basalts, in that order. However, this apparent correlation is apt to be an artifact of limited sampling, since "young" aluminous mare basalts are present at the Luna 16 site, and a VLT basaltic lunar meteorite is as old as the oldest high Ti basalts. These observations are pertinent to inferences concerning the composition of the lunar mantle, and the way in which a melting zone may have migrated through it.

Of the 13 Martian meteorites for which chronologies have been determined so far, only one represents the early Martian crust, four give formation ages of ~ 1.3 Ga, and the rest indicate very young ages of ~ 175–475 Ma. Mars ejection times for Martian meteorites, as estimated from space exposure ages, suggest a maximum of six impact events at ~ 15, ~ 12, ~ 3.8, ~ 2.8, ~ 1.3, and ~ 0.8 Ma. The youngest four ejection ages involve shergottites, all of which also have young formation ages. The apparent situation that five out of six ejection events involve meteorites with ≤ 1.3 Ga formation ages contrasts with the observation from crater counts that $\sim 70\%$ of the basaltic surface of Mars is greater than ~ 1.3 Ga in age. The shergottite ages present a dilemma which implies either that the number of ejection events for shergottites is smaller than four (possibly only one), or that the cratering rate assumed for Mars is in error and the planet has been geologically more active in recent times than previously thought.

The usefulness of the initial $^{87}Sr/^{86}Sr$ and $^{143}Nd/^{144}Nd$ ratios as indicators of the degree of planetary differentiation has been demonstrated as a consequence of using high-precision isotopic ratio measurements for radiometric chronology. Compared to undifferentiated Solar System materials like chondritic meteorites, planetary crusts are enriched in refractory Nd over refractory Sm, whereas planetary mantles are depleted in Nd compared to Sm. Consequently crustal rocks have lower $^{143}Nd/^{144}Nd$ ratios than chondritic meteorites and mantle-derived rocks have higher $^{143}Nd/^{144}Nd$ ratios than chondrites. Because Rb and Sr differ significantly in volatility, the Rb/Sr ratios of bulk planets differ considerably, and the average $^{87}Sr/^{86}Sr$ ratios of rocks from their surfaces vary considerably. Relative to estimated averages for the bulk planets, Rb is enriched in crustal rocks and depleted in mantle-derived rocks compared to Sr. Among the three bodies Earth, Moon, and Mars the terrestrial mantle is least differentiated as a consequence of being constantly re-homogenized due to mantle convection. This re-homogenization is shown in only modest variability of initial $^{143}Nd/^{144}Nd$ and $^{87}Sr/^{86}Sr$ ratios among terrestrial basalts, compared to those from the Moon and Mars.

Finally, investigations of the chronological history of samples from the Moon, Mars, and meteorite parent objects emphasize the value of analyses of samples in terrestrial laboratories for increasing our understanding of the evolution of these Solar System objects.

Acknowledgments

We thank Prof. J. Geiss for offering us the opportunity to contribute to this volume, and Kluwer Academic Publishers for patiently seeing us through the writing of this paper. Comments by an anonymous reviewer were very helpful in redefining the content and tone of the paper. Discussions with F. McDowell and L. Long, and the loan of historical material by them, contributed significantly to the introductory sections on the development of geochronology. Discussions with F. Begemann and G. Wetherill helped to identify relevant references. An informal, but thorough, review of the "revised" manuscript by D. Papanastassiou improved it in several areas. Finally, J. Hultberg of the JSC Technical Library cheerfully ordered many ancient reference works, without which portions of the paper would not have been possible. Financial support for this work was provided via NASA RTOP 344-31-30.

REFERENCES

Ahrens, L. (1949). Measuring geologic time by the strontium method. *Geological Society America Bulletin*, **60**, 217–266.

Albee, A.L., Burnett, D.S., Chodos, A.A., Eugster, O.A., Huneke, J.C., Papanastassiou, D.A., Podosek, F.A., Russ, G.P. III., Sanz, H.G., Tera, T. and Wasserburg, G.J. (1970). Ages, irradiation history, and chemical composition of lunar rocks from the Sea of Tranquility. *Science*, **167**, 463–466.

Aldrich, L.T. and Nier, A.O. (1948). Argon 40 in potassium minerals. *Physical Review*, **74**, 876–877.

Aldrich, L.T. and Wetherill, G.W. (1958). Geochronology by radioactive decay. *Annual Review of Nuclear Science*, **8**, 257–298.

Aldrich, L.T., Wetherill, G.W., Tilton, G.R. and Davis, G.L. (1956). Half-life of Rb^{87}. *Physical Review*, **103**, 1045–1047.

Alibert, C., Norman, M.D. and McCulloch, M.T. (1994). An ancient age for a ferroan anorthosite clast from lunar breccia 67016. *Geochimica et Cosmochimica Acta*, **58**, 2921–2926.

Allègre, C.J., Manhès, G. and Göpel, C. (1995). The age of the Earth. *Geochimica et Cosmochimica Acta*, **59**, 1445–1456.

Anders, E. and Zinner, E. (1993). Interstellar grains in primitive meteorites: Diamond, silicon, carbide, and graphite. *Meteoritics*, **28**, 490–514.

Apollo 16 Preliminary Science Report (1972). NASA SP-315, Scientific and Technical Information Office, NASA, Washington DC.

Apollo 17 Preliminary Science Report (1973). NASA SP-330, Scientific and Technical Information Office, NASA, Washington DC.

Arrol, W.J., Jacobi, R.B. and Paneth, F.A. (1942). Meteorites and the age of the Solar System. *Nature*, **28**, 235–238.

Aston, F.W. (1921). The mass spectra of the alkali elements. *Philosophical Magazine Series 6*, **42**, 430–441.

Aston, F.W. (1927). The constitution of ordinary lead. *Nature*, **120**, 224.

Aston, F.W. (1929). The mass spectrum of uranium lead and the atomic weight of proactinium. *Nature*, **123**, 313.

Baldwin, R.B. (1981). On the origin of the planetesimals that produced the multi-ring basins. *Proceedings of the Lunar and Planetary Science conference*, **12A**, 19–28.

Becquerel, H. (1896). Sur les radiations èmises par phosphorescence. *Comptes Rendues de l'Académie des Sciences*, **122**, 420–421.

Begemann, F., Geiss, J. and Hess, D.C. (1957). Radiation age of a meteorite from cosmic-ray-produced He^3 and H^3. *Physical Review*, **107**, 540–542.

Begemann, F., Ludwig, K.R., Lugmair, G.W., Min, K., Nyquist, L.E., Patchett, P.J., Renne, P.R., Shih, C.-Y., Villa, I.M. and Walker, R.J. (2000). Call for an improved set of decay constants for geochronological use. *Geochimica et Cosmochimica Acta*, **65**, 111–121.

Bell, J.F., III, McSween, H.Y., Jr, Crisp, J.A., Morris, R.V., Murchie, S.L., Bridges, N.T., Johnson, J.R., Britt, D.T., Golombek, M.P., Moore, H.J., Ghosh, A., Bishop, J.L., Anderson, R.C., Brückner, J., Economou, T., Greenwood, J.P., Gunnlaugsson, H.P., Hargraves, R.M., Hviid, S., Knudsen, J.M., Madsen, M.B., Reid, R., Rieder, R. and Soderblom, L. (2000). Mineralogic and compositional properties of Martian soil and dust: Results from Mars Pathfinder. *Journal of Geophysical Research*, **105**, 1721–1755.

Binzel, R.P. and Xu, S. (1993). Chips off asteroid 4 Vesta: Evidence for the parent body of basaltic achondritic meteorites. *Science*, **260**, 186–191.

Bogard, D.D. (1995). Impact ages of meteorites: A synthesis. *Meteoritics*, **30**, 244–268.

Bogard, D.D. and Garrison, D.H. (1995). ^{39}Ar–^{40}Ar age of the Ibitira eucrite and constraints on the time of pyroxene equilibration. *Geochimica et Cosmochimica Acta*, **59**, 4317–4322.

Bogard, D.D. and Garrison, D.H. (1999). Argon-39–argon-40 "ages" and trapped argon in Martian shergottites, Chassigny, and Allan Hills 84001. *Meteoritics and Planetary Science*, **34**, 451–473.

Bogard, D.D., Garrison, D.H. and Nyquist L.E. (2000). ^{39}Ar–^{40}Ar ages of lunar highland rocks and meteorites (abstract). *Lunar and Planetary Science XXXI*, CD-ROM 1138.

Bogard, D.D., Garrison, D.H., Shih, C.-Y. and Nyquist L.E. (1994). ^{39}Ar–^{40}Ar dating of two lunar granites: The age of Copernicus. *Geochimica et Cosmochimica Acta*, **58**, 3093–3100.

Bogard, D.D., Husain, L. and Nyquist, L.E. (1979). ^{40}Ar–^{39}Ar age of the Shergotty achondrite and implications for its post-shock thermal history. *Geochimica et Cosmochimica Acta*, **43**, 1047–1056.

Bogard, D.D. and Johnson, P. (1983). Martian gases in an Antarctic meteorite. *Science*, **221**, 651–654.

Borg, L.E., Connelly, J.N., Nyquist, L.E., Shih, C.-Y., Wiesmann, H. and Reese, Y. (1999b). The age of the carbonates in Martian meteorite ALH84001. *Science*, **286**, 90–94.

Borg, L.E., Norman, M., Nyquist, L., Bogard, D., Snyder, G., Taylor, L. and Lindstrom, M. (1999a). Isotopic studies of ferroan anorthosite 62236: A young lunar crustal rock from a light rare-earth-element-depleted source. *Geochimica et Cosmochimica Acta*, **63**, 2679–2691.

Borg, L.E., Nyquist, L.E., Taylor, L.A., Wiesmann, H. and Shih, C.-Y. (1997). Constraints on Martian differentiation processes from Rb–Sr and Sm–Nd isotopic analyses of the basaltic shergottite QUE 94201. *Geochimica et Cosmochimica Acta*, **61**, 4915–4931.

Boyce, J.M. and Johnson, D.A. (1978). Ages of flow units in the far eastern maria and implications for basin-filling history. *Proceedings of the Lunar Science Conference*, **9**, 3275–3283.

Brown, H. (1947). An experimental method for the estimation of the age of the elements. *Physical Review*, **72**, 348.

Burgess, R. and Turner, G. (1998). Laser argon-40–argon-39 age determinations of Luna 24 mare basalts. *Meteoritics and Planetary Science*, **33**, 921–935.

BVSP (Basaltic Volcanism Study Project) (1981). *Basaltic Volcanism on the Terrestrial Planets*, Pergamon Press, New York.

Campbell, N. (1906). The radioactivity of potassium. *Proceedings of the Cambridge Philosophical Society*, **14**, 557–567.

Campbell, N. (1908). The radioactivity of rubidium. *Proceedings of the Cambridge Philosophical Society*, **15**, 11–12.

Campbell, N.R. and Wood, A. (1906). The radioactivity of the alkali metals. *Proceedings of the Cambridge Philosophical Society*, **14**, 15–21.

Carlson, R.W. and Lugmair, G.W. (1981). Time and duration of lunar highlands crust formation. *Earth and Planetary Science Letters*, **52**, 227–238.

Carlson, R.W. and Lugmair, G.W. (2000). Timescales of planetesimal formation and differentiation based on extinct and extant radioisotopes. In R. Canup and K. Righter (eds), *The Origin of the Earth and Moon*, University of Arizona Press, Tucson, AZ, pp. 25–44.

Chadwick, J. (1932). The existence of a neutron. *Proceedings of the Royal Society*, **A136**, 692.

Chen, J.H. and Wasserburg, G.J. (1981). The isotopic composition of uranium and lead in Allende inclusions and meteorite phosphates. *Earth and Planetary Science Letters*, **52**, 1–15.

Chen, J.H. and Wasserburg, G.J. (1985). U–Th–Pb isotopic studies on meteorite ALH81005 and Ibitira (abstract). *Lunar and Planetary Science XVI*, 119–120.

Chen, J.H. and Wasserburg, G.J. (1996). Live ^{107}Pd in the early Solar System and implications for planetary evolution. In A. Basu and S.R. Hart (eds), *Earth Processes: Reading the Isotope Code*, American Geophysical Union, pp. 1–20.

Compston, W., Foster, J.J. and Gray, C.M. (1975). Rb–Sr ages of clasts from within Boulder 1, Station 2, Apollo 17. *The Moon*, **14**, 445–462.

Compston, W., Lovering, J.F. and Vernon, M.J. (1965). The rubidium–strontium age of the Bishopville aubrite and its component enstatite and feldspar. *Geochimica et Cosmochimica Acta*, **29**, 1085–1099.

Curie, M.S. (1898). Rayons émis par les composes de l'uranium et du thorium. *Comptes Rendues de l'Académie des Sciences*, **126**, 1101–1103.

Curie, P. and Curie, M.S. (1898). Sur une substance nouvelle radio-active, contenue dans la pechblende. *Comptes Rendues de l'Académie des Sciences*, **127**, 175–178.

Curie, P., Curie, Mme.P. and Bémont, G. (1898). Sur une nouvelle substance fortement radio-active, contenue dans la pechblende. *Comptes Rendues de l'Académie des Sciences*, **127**, 1215–1217.

Dalrymple, G.B. and Ryder, G. (1993). $^{40}Ar/^{39}Ar$ age spectra of Apollo 15 impact melt rocks by laser step-heating and their bearing on the history of lunar basin formation. *Journal of Geophysical Research*, **98**, 13085–13095.

Dalrymple, G.B. and Ryder, G. (1996). $^{40}Ar/^{39}Ar$ age spectra of Apollo 17 highlands breccia samples by laser step-heating and the age of the Serenitatis basin. *Journal of Geophysical Research*, **101**, 26069–26084.

Dasch, E.J., Shih, C.-Y., Bansal, B.M., Wiesmann, H. and Nyquist, L.E. (1987). Isotopic analysis of basaltic fragments from lunar breccia 14321: Chronology and petrogenesis of pre-Imbrium mare volcanism. *Geochimica et Cosmochimica Acta*, **51**, 3241–3254.

Deutsch, A. and Stöffler, D. (1987). Rb–Sr analyses of Apollo 16 melt rocks and a new age estimate for the Imbrium basin: Lunar basin chronology and the early heavy bombardment of the Moon. *Geochimica et Cosmochimica Acta*, **51**, 1951–1964.

Dominik, B. and Jessberger, E.K. (1978). Early lunar differentiation: 4.42-AE old plagioclase clasts in Apollo 16 breccia 67435. *Earth and Planetary Science Letters*, **38**, 407–415.

Drake, M.J. (1979). Geochemical evolution of the eucrite parent body: Possible nature and evolution of asteroid 4-Vesta? In T. Gehrels (ed.), *Asteroids*, University of Arizona Press, Tucson, AZ, pp. 765–782.

Eberhardt, P., Geiss, J., Groegler, N. and Stettler, A. (1973). How old is the crater Copernicus? *The Moon*, **8**, 104–114.

Eklund, S. (1946) Studies in nuclear physics. Excitation by means of X-rays. Activity of Rb^{87}. *Arkiv för Matematik, Astronomi och Fysik*, **A33**(14).

Eugster, O., Tera, F., Burnett, D.S. and Wasserburg, G.J. (1970a). Isotopic composition of gadolinium and neutron-capture effects in some meteorites. *Journal of Geophysical Research*, **75**, 2753–2768.

Eugster, O., Tera, F., Burnett, D.S. and Wasserburg, G.J. (1970b). The isotopic composition of Gd and the neutron-capture effects in samples from Apollo 11. *Earth and Planetary Science Letters*, **8**, 20–30.

Eugster, O., Weigel, A. and Polnau, E. (1997). Ejection times of Martian meteorites. *Geochimica et Cosmochimica Acta*, **61**, 2749–2757.

Faure, G. (1986). *Principles of Isotope Geology*, 2nd edn, Wiley, New York.

Fenner, C.N. and Piggot, C.S. (1929). The mass spectrum of lead from bröggerite. *Nature*, **123**, 793–794.

Flynn, K.F. and Glendenin, L.E. (1959). Half-life and beta spectrum of Rb^{87}. *Physical Review*, **116**, 744–748.

Gast, P.W. (1962). The isotopic composition of strontium and the age of stone meteorites. *Geochimica et Cosmochimica Acta*, **26**, 927–943.

Geese-Bähnisch, I. and Huster, E. (1954). Neubestimmung der Halbwertszeit des ^{87}Rb. *Naturwissenschaften*, **41**, 495–496.

Geiss, J., Eberhardt, P., Grögler, N., Guggisberg, S., Maurer, P. and Stettler, A. (1977). *Philosophical Transactions of the Royal Society*, **A285**, 151–158.

Geiss, J. and Hess, D.C. (1958). Argon–potassium ages and the isotopic composition of argon from meteorites. *Astrophysical Journal*, **127**, 224–236.

Gerling, E.K. and Pavlova, T.G. (1951). On the age of two stone meteorites. *Doklady Akademii Nauk SSSR*, **77**, 85–86 (in Russian).

Göpel, C., Manhès, G. and Allègre, C.J. (1992). U–Pb study of the Acapulco meteorite (abstract). *Meteoritics*, **27**, 226.

Göpel, C., Manhès, G. and Allègre C.J. (1994). U–Pb systematics of phosphates from equilibrated ordinary chondrites. *Earth and Planetary Science Letters*, **121**, 153–171.

Guggisberg, S., Eberhardt, P., Geiss, J., Grögler, N., Stettler, A., Brown, G.M. and Peckett A. (1979). Classification of the Apollo-11 mare basalts according to Ar^{39}–Ar^{40} ages and petrological properties. *Proceedings of the 10th Lunar Planetary Science Conference*, pp. 1–39. *Geochimica et Cosmochimica Acta*, **Suppl. 11**, Pergamon.

Hahn, O. (1925). Die Isotopen des Urans. *Zeitschrift für Anorganisch und Allgemeine Chemie*, **147**, 16–23.

Hahn, O. and Rothenbach, M. (1919). Über die radioactivität des rubidiums. *Physikalische Zeitschrift*, **20**, 194–202.

Hahn, O. and Walling, E. (1938). Über die Möglichkeit Geologischer Altersbestimmungen Rubidiumhaltiger Mineralien und Gesteine. *Zeitschrift für Anorganisch und Allgemeine Chemie*, **236**, 78–82.

Harper, C.T. (ed.) (1973). *Geochronology: Radiometric dating of rocks and minerals*. *Benchmark Papers in Geology*, Dowden, Hutchinson & Ross, Stroudsburg, Pa.

Harper, C.L., Jr, Nyquist, L.E., Bansal, B.M., Wiesmann, H. and Shih, C.-Y. (1995). Rapid accretion and early differentiation of Mars indicated by $^{142}Nd/^{144}Nd$ in SNC meteorites. *Science*, **267**, 213–216.

Hartmann, W.K. (1980). Dropping stones in magma oceans: Effects of early cratering. *Proceedings of the Lunar Highlands Crust Conference*, 155–171. *Geochimica et Cosmochimica Acta*, **Suppl. 12**, Pergamon.

Hartmann, W.K. (1999). Martian cratering VI: Crater count isochrons and evidence for recent volcanism from Mars Global Surveyor. *Meteoritics and Planetary Science*, **34**, 167–177.

Hartmann, W.K. and Berman, D.D. (2000). Elysium Planitia lava flows: Crater count chronology and geological implications. *Journal of Geophysical Research*, **105**, 15011–15025.

Hartmann, W.K., Ryder, G. Dones, L. and Grinspoon, D. (2000). The time-dependent intense bombardment of the primordial Earth/Moon system. In R. Canup and K. Righter (eds), *Origin of the Earth and Moon*, University of Arizona Press, Tucson, AZ, pp. 493–512.

Hemmendinger, A. and Smythe, W.R. (1937). The radioactive isotope of rubidium. *Physical Review*, **51**, 1052–1053.

Herzog, L.F. and Pinson, W.H. (1956). Rb/Sr age, elemental and isotopic abundance studies of stony meteorites. *American Journal of Science*, **254**, 555–566.

Hess, P.C. and Parmentier, E.M. (1995). A model for the thermal and chemical evolution of the Moon's interior: Implications for the onset of mare volcanism. *Earth and Planetary Science Letters*, **134**, 501–514.

Hevesey, G., Pahl, M. and Hosemann, R. (1933). Die Radioaktivität des Samariums. *Zeitschrift für Physik*, **83**, 43–54.

Holmes, A. (1913). *The age of the Earth*, Harper & Brothers, London.

Holmes, A. (1946). An estimate of the age of the Earth. *Nature*, **157**, 680–684.

Holmes, A. (1956). How old is the Earth? *Transactions of the Edinburgh Geological Society*, **XVI**, Part III, 313–333.

Hosemann, R. (1936). Die Radioaktivität des Samariums. *Zeitschrift für Physik*, **99**, 405–427.

Houtermans, F.G. (1946). Die isotopenhaüfigkeiten im natürlichen Blei und das Alter des Urans. *Naturwissenschaften*, **33**, 185–186.

Hubbard, N.J., Meyer, C., Jr, Gast, P.W. and Wiesmann, H. (1971). The composition and derivation of Apollo 12 soils. *Earth and Planetary Science Letters*, **10**, 341–350.

Huneke, J.C., Podosek, F. and Wasserburg, G.J. (1972). Gas retention and cosmic ray exposure ages of a basalt fragment from Mare Fecunditatis. *Earth and Planetary Science Letters*, **13**, 375–383.

Huneke, J.C. and Wasserburg, G.J. (1979). Sliva iz piroga (plum out of the pie): K/Ar evidence from Luna 20 rocks for lunar differentiation prior to 4.51 AE ago (abstract). *Lunar and Planetary Science X*, 598–600.

Inghram, M.G. (1954). Stable isotope dilution as an analytical tool. *Annual Review of Nuclear Science*, **4**, 81–92.

Jaffey, A.H., Flynn, K.F., Glendenin, L.E., Bentley, W.C. and Essling, A.M. (1971). Precision measurement of half-lives and specific activities of ^{235}U and ^{238}U. *Physical Review C*, **4**, 1889–1906.

Jagoutz, E. and Wänke, H. (1986). Sr and Nd isotopic systematics of Shergotty meteorite. *Geochimica et Cosmochimica Acta*, **50**, 939–953.

Jagoutz, E., Sorowka, A., Vogel, J.D. and Wänke, H. (1994). ALH84001: Alien or progenitor of the SNC family? (abstract). *Meteoritics*, **29**, 478–479.

James, O.B. (1980). Rocks of the early lunar crust. *Proceedings of the Lunar and Planetary Science Conference*, **11**, 365–393.

Jessberger, E.K., Huneke, J.C., Podosek, F.A. and Wasserburg, G.J. (1974). High resolution argon analysis of neutron-irradiated Apollo 16 rocks and separated minerals. *Proceedings of the 5th Lunar Science Conference*, 1419–1449. *Geochimica et Cosmochimica Acta*, **Suppl. 5**, Pergamon.

Jessberger, E.K., Kirsten, T. and Staudacher, T. (1977). One rock and many ages – Further K–Ar data on consortium breccia 73215. *Proceedings of the 8th Lunar Science Conference*, 2567–2580. *Geochimica et Cosmochimica Acta*, **Suppl. 8**, Pergamon.

Jones, J.H. (1986). A discussion of isotopic systematics and mineral zoning in the shergottites: Evidence for a 180 Ma igneous crystallization age. *Geochimica et Cosmochimica Acta*, **50**, 969–977.

Kirsten, T., Krankowsky, D. and Zähringer, J. (1963). Edelgas- und Kalium-Bestimmungen an einer grösseren Zahl von Steinmeteoriten. *Geochimica et Cosmochimica Acta*, **27**, 13.

Lee, D.-C. and Halliday, A.N. (1997). Core formation on Mars and differentiated asteroids. *Nature*, **388**, 854–857.

Libby, W.F. (1957). Simple absolute measurement technique for beta radioactivity – Application to naturally radioactive rubidium. *Analytical Chemistry*, **29**, 1566–1570.

Lindstrom, M. (1999). Lunar and Martian meteorites: Suites, pairing, and implications. In *Papers Presented to the Twenty-fourth Symposium on Antarctic Meteorites*, 1–3 June 1999, National Institute of Polar Research, Tokyo.

Lindstrom, M.M., Mittlefehldt, D.W., Martinez, R.R., Lipschutz, M.E. and Wang, M.-S. (1991). Geochemistry of Yamato-82192, -86032 and -793274 lunar meteorites. *Proceedings of the NIPR Symposium on Antarctic Meteorites*, **4**, 12–32.

Longhi, J. and Ashwal, L. (1985). Two-stage models for lunar and terrestrial anorthosites: Petrogenesis without a magma ocean. *Proceedings of the Lunar and Planetary Science Conference*, **15**, C571–C584.

Longhi, J. and Pan, V. (1989). The parent magmas of the SNC meteorites. *Proceedings of the Lunar and Planetary Science Conference*, **19**, 451–464.

Lugmair, G.W. (1974). Sm–Nd ages: A new dating method. *Meteoritics*, **9**, 369.

Lugmair, G.W. and Galer, S.J. (1992). Age and isotopic relationships among the angrites Lewis Cliff 86010 and Angra dos Reis. *Geochimica et Cosmochimica Acta*, **56**, 1673–1694.

Lugmair, G.W. and Marti, K. (1971). Neutron-capture effects in lunar gadolinium and the irradiation histories of some lunar rocks. *Earth and Planetary Science Letters*, **13**, 32–42.

Lugmair, G.W., Marti, K., Kurtz, J.P. and Scheinen, N.B. (1976). History and genesis of lunar troctolite 76535 or: How old is old? *Proceedings of the 7th Lunar Science Conference*, 2035–2054. *Geochimica et Cosmochimica Acta*, **Suppl. 7**, Pergamon.

Lugmair, G.W., Scheinen, N.B. and Marti, K. (1975). Sm–Nd age and history of Apollo 17 basalt 75075: Evidence for early differentiation of the lunar exterior. *Proceedings of the Lunar Science Conference*, **6**, 1419–1429.

Lunar Sample Information Catalog, Apollo 15 (1971). MSC 03209, Manned Spacecraft Center, NASA, Houston, TX.

Lunatic Asylum (1970). Mineralogic and isotopic investigations on lunar rock 12013. *Earth and Planetary Science Letters*, **9**, 137–163.

Lunatic Asylum (1978). Petrology, chemistry, age and irradiation history of Luna 24 samples. In R.B. Merrill and J.J. Papike (eds), *Mare Crisium: The View from Luna 24*, Pergamon Press, New York, pp. 657–678.

MacPherson, G.J., Davis, A.M. and Zinner, E.K. (1995). The distribution of aluminum-26 in the early Solar System. *Meteoritics*, **30**, 365–386.

Mattauch, J. (1937). Das Paar ^{87}Rb–^{87}Sr und die Isobarenregel. *Naturwissenschaft*, **25**, 189–190.

Maurer, P., Geiss, J., Grögler, N., Stettler, A., Brown, G.M., Peckett, A. and Krähenbühl, U. (1978). Pre-Imbrian craters and basins: Ages, compositions, and excavation depths of Apollo 16 breccias. *Geochimica et Cosmochimica Acta*, **42**, 1687–1720.

McGetchin, T.R., Settle, M. and Head, W. (1973). Radial thickness variation in impact crater ejecta: Implications for lunar basin deposits. *Earth and Planetary Science Letters*, **20**, 226–236.

McKay, D.S., Gibson, E.K., Thomas-Keprta, K.L., Vali, H., Romanek, C.S., Clemett, S.J., Chillier, X.D.F., Maechling, C.R. and Zare, R.N. (1996). Search for past life on Mars: Possible relic biogenic activity in Martian meterorite ALH84001. *Science*, **273**, 924–930.

McNair, A. and Wilson, H.W. (1961). The half-life of rubidium-87. *Philosophical Magazine*, **6**, 563–572.

McSween, H.Y., Jr (1994). What we have learned about Mars from SNC meteorites. *Meteoritics*, **29**, 757–779.

Merrihue, C. and Turner, G. (1966). Potassium–argon dating by activation with fast neutrons. *Journal of Geophysical Research*, **71**, 2852–2857.

Meyer, C., Jr, Brett, R., Hubbard, N.J., Morrison, D.A., McKay, D.S., Aitken, F.K., Takeda, H. and Schonfeld, E. (1971). Mineralogy, chemistry, and origin of the KREEP component in soil samples from the Ocean of Storms. *Proceedings of the 2nd Lunar Science Conference*, 393–411. *Geochimica et Cosmochimica Acta*, **Suppl. 2**, Pergamon.

Meyer, C., Jr and Hubbard, N.J. (1970). High potassium, high phosphorous glass as an important rock type in the Apollo 12 samples (abstract). *Meteoritics*, **5**, 210–211.

Meyer, C., Jr, Williams, I.S. and Compston, W. (1996). Uranium–lead ages for lunar zircons: Evidence for prolonged period of granophyre formation from 4.32 to 3.88 Ga. *Meteoritics and Planetary Science*, **31**, 370–387.

Minster, J.-F., Birck, J.-L. and Allègre, C.-J. (1982). Absolute age of formation of chondrites studied by the ^{87}Rb–^{87}Sr method. *Nature*, **300**, 414–419.

Misawa, K., Tatsumoto, M., Dalrymple, G.B. and Yanai, K. (1993). An extremely low U/Pb sources in the Moon: U–Th–Pb, Sm–Nd, Rb–Sr, and ^{40}Ar/^{39}Ar isotopic systematics and age of lunar meteorite Asuka 881757. *Geochimica et Cosmochimica Acta*, **57**, 4687–4702.

Mouginis-Mark, P.J., McCoy, T.J., Taylor, G.J. and Keil, K. (1992). Martian parent craters for the SNC meteorites. *Journal of Geophysical Research*, **97**, 10,213–10,225.

Nakamura, N., Tatsumoto, M., Nunes, P., Unruh, D.M., Schwab, A.P. and Wildeman, T.R. (1976). 4.4 b.y.-old clast in Boulder 7, Apollo 17: A comprehensive chronological study by U–Pb, Rb–Sr, and Sm–Nd methods. *Proceedings of the 7th Lunar Science Conference*, 2309–2333. *Geochimica et Cosmochimica Acta*, **Suppl. 7**, Pergamon.

Nier, A.O. (1935). Evidence for the existence of an isotope of potassium of mass 40. *Physical Review*, **48**, 283–284.

Nier, A.O. (1936). The isotopic constitution of rubidium, zinc, and argon. *Physical Review*, **49**, 272.

Nier, A.O. (1939a). The isotopic constitution of uranium and the half-lives of the uranium isotopes. I. *Physical Review*, **55**, 150–153.

Nier, A.O. (1939b). The isotopic constitution of radiogenic leads and the measurement of geological time. II. *Physical Review*, **55**, 153–163.

Nier, A.O. (1950). A redetermination of the relative abundances of the isotopes of neon, krypton, rubidium, and mercury. *Physical Review*, **79**, 450–454.

Nier, A.O. (1981). Some reminiscences of isotopes, geochronology, and mass spectromeztry. *Annual Review of Earth and Planetary Science*, **9**, 1–17.

Nier, A.O. (1989). Some reminiscences of mass spectrometry and the Manhattan Project. *Journal of Chemical Education*, **66**, 385–388.

Nier, A.O., Dunning, J.R., Booth, E.T. and Grosse A.V. (1940). Nuclear fission of separated uranium isotopes. *Physical Review*, **57**, 546.

Norman, M., Nyquist, L., Bogard, D., Borg, L., Wiesmann, H., Garrison, D., Reese, Y., Shih, C.-Y. and Schwandt, C. (2000). Age and origin of the highlands crust of the Moon: Isotopic and petrologic studies of a ferroan noritic anorthosite clast from Descartes Breccia 67215 (abstract). *Lunar and Planetary Science XXXI*, CD-ROM 1552.

Nyquist, L.E. (1977). Lunar Rb–Sr chronology. *Physics and Chemistry of the Earth*, **10**, 103–142.

Nyquist, L.E. (1996). High-Ti volcanism and the lunar mantle. *Meteoritics and Planetary Science*, **31**, 319–320.

Nyquist, L.E., Bansal, B., Wiesmann, H. and Shih, C.-Y. (1994). Neodymium, strontium and chromium isotopic studies of the LEW86010 and Angra dos Reis meteorites and the chronology of the angrite parent body. *Meteoritics*, **29**, 872–885.

Nyquist, L.E., Bansal, B.M., Wiesmann, H. and Shih, C.-Y. (1995b). "Martians" young and old: Zagami and ALH84001 (abstract). *Lunar and Planetary Science XXVI*, 1065–1066.

Nyquist, L.E., Bansal, B.M., Wooden, J.L. and Wiesmann, H. (1977). Sr isotopic constraints on the petrogenesis of Apollo 12 mare basalts. *Proceedings of the Lunar Science Conference*, **8**, 1383–1415.

Nyquist, L.E., Bogard, D.D., Wooden, J., Wiesmann, H., Shih, C.-Y., Bansal, B.M. and McKay, G.A. (1979c). Early differentiation, late magmatism, and recent bombardment on the shergottite parent planet (abstract). *Meteoritics*, **14**, 502.

Nyquist, L.E., Borg, L.E. and Shih, C.-Y. (1998). The Shergottite age paradox and the relative probabilities for Martian meteorites of differing ages. *Journal of Geophysical Research*, **103**, 31,445–31,455.

Nyquist, L.E., Reimold, W.U., Bogard, D.D., Wooden, J.L., Bansal, B.M., Wiesmann, H. and Shih, C.-Y. (1981). A comparative Rb–Sr, Sm–Nd, and K–Ar study of shocked norite 78236: Evidence of slow cooling in the lunar crust? *Proceedings of the Lunar and Planetary Science conference*, **12B**, 67–97.

Nyquist, L.E. and Shih, C.-Y. (1992). The isotopic record of lunar volcanism. *Geochimica et Cosmochimica Acta*, **56**, 2213–2234.

Nyquist, L.E., Shih, C.-Y., Reese, Y., Wiesmann, H., Bogard, D., Ryder, G. and Garrison, D. (2000). Age and petrogenesis of Apollo 17 Group D basalt (abstract). *Lunar and Planetary Science XXXI*, CD-ROM 1667.

Nyquist, L.E., Shih, C.-Y., Wooden, J.L., Bansal, B.M. and Wiesmann, H. (1979a). The Sr and Nd isotopic record of Apollo 12 basalts: Implications for lunar geochemical evolution. *Proceedings of the Lunar Science Conference*, **10**, 77–114.

Nyquist, L.E., Wiesmann, H., Bansal, B., Shih, C.-Y., Keith, J.E. and Harper, C.L., Jr (1995a). ^{146}Sm–^{142}Nd formation interval for the lunar mantle. *Geochimica et Cosmochimica Acta*, **59**, 2817–2837.

Nyquist, L.E., Wiesmann, H., Shih, C.-Y. and Dasch, J. (1996). Lunar meteorites and the lunar crustal Sr and Nd isotopic compositions (abstract). *Lunar and Planetary Science XXVII*, 971–972.

Nyquist, L.E., Wooden, J., Bansal, B., Wiesmann, H., McKay, G. and Bogard, D.D. (1979b). Rb–Sr age of the Shergotty achondrite and implications for metamorphic resetting of isochron ages. *Geochimica et Cosmochimica Acta*, **43**, 1057–1074.

Oberli, F., Huneke, J.C. and Wasserburg, G.J. (1979). U–Pb and K–Ar systematics of cataclysm and precataclysm lunar impactites (abstract). *Lunar and Planetary Science X*, 940–942.

Paneth, F. (1931). Uber die Zuverlässigkeit der "Heliummethode" und über das Alter von Eisenmeteoriten. *Naturwissenschaften*, **19**, 164–165.

Paneth, F.A., Reasbeck, P. and Mayne, K.I. (1952). Helium 3 content and age of meteorites. *Geochimica et Cosmochimica Acta*, **2**, 300–303.

Papanastassiou, D.A., DePaolo, D.J. and Wasserburg, G.J. (1977). Rb–Sr and Sm–Nd chronology and genealogy of mare basalts from the Sea of Tranquility. *Proceedings of the 8th Lunar Science Conference*, 1639–1672. *Geochimica et Cosmochimica Acta*, **Suppl. 8**, Pergamon.

Papanastassiou, D.A. and Wasserburg, G.J. (1969). Initial strontium isotopic abundances and the resolution of small time differences in the formation of planetary objects. *Earth and Planetary Science Letters*, **5**, 361–376.

Papanastassiou, D.A. and Wasserburg G.J., (1972). Rb–Sr age of a lunar 16 basalt. *Earth Planetary Science Letters*, **13**, 368–374.

Papanastassiou, D.A. and Wasserburg, G.J., (1975a). A Rb–Sr study of Apollo 17 boulder 3: Dunite clast, microclasts, and matrix. *Lunar Science*, **VI**, 631–633.

Papanastassiou, D.A. and Wasserburg, G.J. (1975b). Rb–Sr study of a lunar dunite and evidence for early lunar differentiates. *Proceedings of the 6th Lunar Science Conference*, 1467–1489. *Geochimica et Cosmochimica Acta*, **Suppl. 6**, Pergamon.

Papanastassiou, D.A. and Wasserburg, G.J. (1976). Rb–Sr age of troctolite 76535. *Proceedings of the 7th Lunar Science Conference*, 2035–2054. *Geochimica et Cosmochimica Acta*, **Suppl. 7**, Pergamon.

Papanastassiou, D.A., Wasserburg, G.J. and Burnett, D.S. (1970). Rb–Sr ages of lunar rocks from the Sea of Tranquility. *Earth Planetary Science Letters*, **8**, 1–19.

Papike, J.J., Fowler, G.W., Layne, G.D. and Shearer, C.K., (1996). Ion microprobe investigation of plagioclase and orthopyroxene from lunar Mg-suite norites: Implications for calculating parental melt REE concentrations and for assessing postcrystallization REE redistribution. *Geochimica et Cosmochimica Acta*, **60**, 3967–3978.

Papike, J.J., Ryder, G. and Shearer, C.K. (1998). Lunar samples. In P.H. Ribbe (ed.), *Reviews in Mineralogy, Vol. 36: Planetary Materials*, Mineralogical Society of America, Washington, DC, Chapter 5.

Patterson, C.C. (1956). Age of meteorites and the Earth. *Geochimica et Cosmochimica Acta*, **10**, 230–237.

Patterson, C., Brown, H., Tilton, G. and Inghram, M. (1953). Concentration of uranium and lead and the isotopic composition of lead in meteoritic material. *Physical Review*, **92**, 1234.

Pieters, C.M. (1978). Mare basalt types on the front side of the Moon: A summary of spectral reflectance data. *Proceedings of the Lunar and Planetary Science Conference*, **9**, 2825–2849.

Piggot, C.H. (1928). Lead isotopes and the problem of geologic time. *Journal of the Washington Academy of Sciences*, **18**, 269–273.

Pinson, W.H., Jr, Schnetzler, C.C., Beiser, E., Fairbairn, H.W. and Hurley, P.M. (1965). Rb–Sr age of stony meteorites. *Geochimica et Cosmochimica Acta*, **29**, 455–466.

Podosek, F.A. and Cassen, P. (1994). Theoretical, observational, and isotopic estimates of the lifetime of the solar nebula. *Meteoritics*, **29**, 6–25.

Podosek, F.A., Huneke, J.C., Gancarz, A.J. and Wasserburg, G.J. (1973). The age and petrography of two Luna 20 fragments and inferences for widespread lunar metamorphism. *Geochimica et Cosmochimica Acta*, **37**, 887–904.

Premo, W.R. and Tatsumoto, M. (1991). U–Th–Pb isotopic systematics of lunar norite 78235. *Proceedings of the Lunar and Planetary Science Conference*, **21**, 89–100.

Reynolds J.H. (1960). Determination of the age of the elements. *Physical Review Letters*, **4**, 8–10.

Reynolds, J.H. and Lipson, J. (1957). Rare gases from the Nuevo Laredo stone meteorite. *Geochimica et Cosmochimica Acta*, **12**, 330–336.

Rieder, R., Economou, T., Wänke, H., Turkevich, A., Crisp, J.A., Brückner, J., Dreibus, G. and McSween, H.Y., Jr (1997). The chemical composition of Martian soil and rocks returned by the mobile alpha proton X-ray spectrometer: Preliminary results from the X-ray mode. *Science*, **278**, 1771–1774.

Ringwood, A.E. and Kesson, S.E. (1976). A dynamic model for mare basalt petrogenesis. *Proceedings of the Lunar Science Conference*, **7**, 1697–1722.

Romer, A. (1970). *Radiochemistry and the Discovery of Isotopes*, Dover, New York.

Rose, J.L. and Stranathan, R.K. (1936). Geologic time and isotopic constitution of radiogenic lead. *Physical Review*, **50**, 792–796.

Russell, A.S. (1924). Radio-active disintegration series and the relation of actinium to uranium. *Philosophical Magazine*, **46**, 642–656.

Rutherford, E. (1906). The production of helium from radium and the transformation of matter. In *Radioactive Transformations*, Yale University Press, pp. 187–193. Silliman memorial lectures, Yale University Press.

Rutherford, E. (1929). Origin of actinium and age of the Earth. *Nature*, **123**, 313–314.

Ryder, G. (1990). Lunar samples, lunar accretion and early bombardment of the Moon. *EOS, Trans. AGU*, **71**, 313–323.

Ryder, G. (1993). Catalog of Apollo 17 Rocks. Vol. 1 – Stations 2 and 3 (South Massif). *Office of the Curator #87*, NASA Lyndon B. Johnson Space Center, Houston, TX.

Ryder, G., Bogard, D. and Garrison, D. (1991). Probable age of Autolycus and calibration of lunar stratigraphy. *Geology*, **19**, 143–146.

Ryder, G., Koeberl, C. and Mojzsis, S.J. (2000). Heavy bombardment of the Earth at ~3.85 Ga: The search for petrographic and geochemical evidence. In R. Canup and K. Righter (eds), *Origin of the Earth and Moon*, University of Arizona Press, Tucson, AZ, pp. 475–492.

Schaeffer, O.A. and Husain, L. (1974). Chronology of lunar basin formation. *Proceedings of the 5th Lunar Science Conference*, 1541–1555. *Geochimica et Cosmochimica Acta*, **Suppl. 5**, Pergamon.

Schmidt, G.C. (1898). Über die vom Thorium und den Thorverbindungen ausgehende Strahlung. *Verhandlungen der Physik Gesellschaft zu Berlin*, **17**(13), 14–16.

Schumacher, E. (1956). Alterbestimmungen von Steinmeteoriten mit der Rb–Sr Methode. *Zeitschrift für Naturforschung*, **11a**, 206–212.

Shervais, J.W. and McGee, J.J. (1998). Ion and electron microprobe study of troctolites, norite, and anorthosites from Apollo 14: Evidence for urKREEP assimilation during petrogenesis of Apollo 14 Mg-suite rocks. *Geochimica et Cosmochimica Acta*, **62**, 3009–3023.

Shih, C.-Y., Nyquist, L.E., Bogard, D.D., Bansal, B.M., Wiesmann, H., Johnson, P., Shervais, J.W. and Taylor, L.A. (1986). Geochronology and petrogenesis of Apollo 14 very high potassium mare basalts. *Journal of Geophysical Research*, **91**(B4), D214–D228.

Shih, C.-Y., Nyquist, L.E., Bogard, D.D., Dasch, E.J., Bansal, B.M. and Wiesmann, H. (1987). Geochronology of high-K aluminous mare basalt clasts from Apollo 14 breccia 14304. *Geochimica et Cosmochimica Acta*, **51**, 3255–3271.

Shih, C.-Y., Nyquist, L.E., Bogard, D.D., McKay, G.A., Wooden, J.L., Bansal, B.M. and Wiesmann, H. (1982). Chronology and petrogenesis of young achondrites, Shergotty, Zagami, and ALHA77005: Late magmatism on a geologically active planet. *Geochimica et Cosmochimica Acta*, **46**, 2323–2344.

Shih, C.-Y., Nyquist, L.E., Bogard, D.D. and Wiesmann, H. (1994). K–Ca and Rb–Sr dating of two lunar granites: Relative chronometer resetting. *Geochimica et Cosmochimica Acta*, **58**, 3101–3116.

Shih, C.-Y., Nyquist, L.E., Dasch, E.J., Bogard, D.D., Bansal, B.M. and Wiesmann, H. (1993). Ages of pristine noritic clasts from lunar breccias 15445 and 15455. *Geochimica et Cosmochimica Acta*, **57**, 915–931.

Shih, C.-Y., Nyquist, L.E., Reese, Y. and Wiesmann, H., 2000. Rb–Sr and Sm–Nd isotopic studies of two pristine alkali suite rocks (abstract). *Lunar and Planetary Science XXXI*, CD-ROM 1698.

Shih, C.-Y., Nyquist, L.E. and Wiesmann, H. (1999). Samarium–neodymium and rubidium–strontium systematics of nakhlite Governador Valadares. *Meteoritics and Planetary Science*, **34**, 647–655.

Signer, P. and Nier, A.O. (1960). The distribution of cosmic-ray-produced rare gases in iron meteorites. *Journal of Geophysical Research*, **65**, 2947–2964.

Smythe, W.R. and Hemmendinger, A. (1937). The radioactive isotope of potassium. *Physical Review*, **51**, 178–182.

Snyder, G.A., Borg, L.E., Nyquist, L.E. and Taylor, L.A. (2000a). Chronology and isotopic constraints on lunar evolution. In R.M. Canup and K. Righter (eds), *The Origin of the Earth and Moon*, University of Arizona Press, Tucson, AZ, pp. 493–512.

Snyder, G.A., Hall, C.M., Lee, D.-C., Taylor, L.A. and Halliday, A.N. (1996). Earliest high-Ti volcanism on the Moon: ^{40}Ar–^{39}Ar, Sm–Nd, and Rb–Sr isotopic studies of Group D basalts from the Apollo 11 landing site. *Meteoritics and Planetary Science*, **31**, 328–334.

Snyder, G.A., Neal, C.R., Taylor, L.A. and Halliday, A.N. (1995a). Processes involved in the formation of magnesian-suite plutonic rocks from the highlands of Earth's Moon. *Journal of Geophysical Research*, **100**, 9365–9388.

Snyder, G.A. and Taylor, L.A. (1993). Constraints on the genesis and evolution of the Moon's magma ocean and derivative cumulate sources as supported by lunar meteorites. *Proceedings of the NIPR Symposium on Antarctic Meteorites*, **6**, 246–267.

Snyder, G.A., Taylor, L.A. and Halliday, A.N. (1995b). Chronology and petrogenesis of the lunar highlands alkali suite. *Geochimica et Cosmochimica Acta*, **59**, 1185–1203.

Snyder, G.A., Taylor, L.A. and Neal, C.R. (1992). A chemical model for generating the sources of mare basalts: Combined equilibrium and fractional crystallization of the lunar magmasphere. *Geochimica et Cosmochimica Acta*, **56**, 3809–3823.

Soddy, F. (1913–14). Intra-atomic charge. *Nature*, **92**, 399–400.

Solomon, S.C. and Longhi, J. (1977). Magma oceanography: 1. Thermal evolution. *Proceedings of the Lunar Science Conference*, **8**, 583–599.

Spudis, P.D., Swann, G.A. and Greeley, R. (1988). The formation of Hadley Rille and implications for the geology of the Apollo 15 region. *Proceedings of the Lunar and Planetary Science Conference*, **18**, 243–254.

Stettler, A., Eberhardt, P., Geiss, J., Grögler, N. and Maurer, P. (1973). Ar^{39}–Ar^{40} ages and Ar^{37}–Ar^{38} exposure ages of lunar rocks. *Proceedings of the 4th Lunar Science Conference*, 1865–1888. *Geochimica et Cosmochimica Acta*, **Suppl. 4**, Pergamon.

Stettler, A., Eberhardt, P., Geiss, J., Grögler, N. and Maurer, P. (1974). On the duration of lava flow activity in Mare Tranquilitatis. *Proceedings of the 5th Lunar Science Conference*, 1557–1570. *Geochimica et Cosmochimica Acta*, **Suppl. 5**, Pergamon.

Steiger, R.H. and Jäger, E. (1977). Subcommission on geochronology: Convention on the use of decay constants in geo- and cosmochronology. *Earth and Planetary Science Letters*, **36**, 359–362.

Sunshine, J.M., Pieters, C.M. and Head, J.W. (1994). New evidence for compositional diversity on the Marius Hills Plateau from Galileo multispectral imaging (abstract). *Lunar and Planetary Science XXV*, 1359–1360.

Swindle, T., Spudis, P.D., Taylor, G.J., Korotev, R., Nichols, R.H. and Olinger, C.T. (1991). Searching for Crisium basin ejecta: Chemistry and ages of Luna 20 impact melts. *Proceedings of the Lunar and Planetary Science Conference*, **21** 167–184.

Tanaka, K.L. (1986). The stratigraphy of Mars. *Journal of Geophysical Research Supplement*, **91**, E139–158.

Tanaka, K.L., Scott, D.H. and Greeley, R. (1992). Global stratigraphy. In H. Kieffer, C. Snyder and M. Matthews (eds), *Mars*, University of Arizona Press, Tucson, AZ, pp. 345–382.

Taylor, L.A., Shervais, J.W., Hunter, R.H., Shih, C.-Y., Bansal, B.M., Wooden, J., Nyquist, L.E. and Laul, J.C. (1983). Pre-4.2 AE mare-basalt volcanism in the lunar highlands. *Earth and Planetary Science Letters*, **66**, 33–47.

Taylor, S.R. (1975). *Lunar Science: A Post Apollo View*, Pergamon.

Taylor, S.R. (1982). *Planetary Science: A Lunar Perspective*, Lunar and Planetary Institute, Houston, TX.

Tera, F., Papanastassiou, D.A. and Wasserburg, G.J. (1974). Isotopic evidence for a terminal lunar cataclysm. *Earth and Planetary Science Letters*, **22**, 1–21.

Thompson, F.C. and Rowlands, S. (1943). Dual decay of potassium. *Nature*, **152**, 103.

Thomson, J.J. (1913). Some further applications of the method of positive rays. *Nature*, **91**, 333–337.

Tonks, W.B. and Melosh, H.J. (1992). Core formation by giant impacts. *Icarus*, **100**, 326–346.

Treiman, A.H. (1986). The parental magma of the Nakhla achondrite: Ultrabasic volcanism on the shergottite parent body. *Geochimica et Cosmochimica Acta*, **50**, 1061–1070.

Treiman, A.H. (1995). A petrographic history of Martian meteorite ALH84001: Two shocks and an ancient age. *Meteoritics*, **30**, 294–302.

Treiman, A.H., Gleason, J.D. and Bogard, D.D. (2000). The SNC meteorites are from Mars. *Planetary and Space Sciences*, **48**, 1213–1230.

Turner, G. (1970a). Argon-40/argon-39 dating of lunar rock samples. *Science*, **167**, 466–468.

Turner, G. (1970b). Argon-40/argon-39 dating of lunar rock samples. *Proceedings of the Apollo 11 Lunar Science Conference*, **2**, 1665–1684.

Turner, G. (1977). Potassium–argon chronology of the Moon. *Physics and Chemistry of the Earth*, **10**, 145–195.

Turner, G., Huneke, J.C., Podosek, F.A. and Wasserburg, G.J. (1971). ^{40}Ar–^{39}Ar ages and cosmic ray exposure ages of Apollo 14 samples. *Earth and Planetary Science Letters*, **12**, 19–35.

Turner, G., Huneke, J.C., Podosek, F.A. and Wasserburg, G.J. (1972). Ar^{40}–Ar^{39} systematics in rocks and separated minerals from Apollo 14. *Proceedings of the 3rd Lunar Science Conference*, 1589–1612. *Geochimica et Cosmochimica Acta*, **Suppl. 3**, Pergamon.

Turner, G., Knott, S.F., Ash, R.D. and Gilmour, J.D. (1997). Ar–Ar chronology of the Martian meteorite ALH84001: Evidence for the timing of the early bombardment of Mars. *Geochimica et Cosmochimica Acta*, **61**, 3835–3850.

Urey, H.C. (1951). The origin and development of the Earth and other terrestrial planets. *Geochimica et Cosmochimica Acta*, **1**, 209–277.

Urey, H.C. (1952). *The Planets*, Yale University Press.

Von Weizsäcker, C.F. (1937). Über die Möglichkeit eines dualen β-Zerfalls von Kalium. *Physikalische Zeitschrift*, **38**, 623–624.

Wadhwa, M. and Lugmair, G.W. (1996). The formation age of carbonates in ALH84001 (abstract). *Meteoritics*, **31**, A145.

Walker, D. (1983). Lunar and terrestrial crust formation. *Proceedings of the Lunar and Planetary Science Conference*, **14**, B17–B25.

Warren, P.H. (1988). The origin of pristine KREEP: Effects of mixing between urKREEP and the magmas parental to the Mg-rich cumulates. *Proceedings of the Lunar and Planetary Science Conference*, **18**, 233–241.

Warren, P.H. (1993). A concise compilation of petrologic information on possibly pristine nonmare Moon rocks. *American Mineralogist*, **78**, 360–376.

Warren, P.H. (1994). Lunar and Martian delivery services. *Icarus*, **111**, 338–363.

Warren, P.H. and Wasson, J.T. (1977). Pristine nonmare rocks and the nature of the lunar crust. *Proceedings of the Lunar Science Conference*, **8**, 2214–2235.

Warren, P.H. and Wasson, J.T. (1979). The origin of KREEP. *Reviews of Geophysics and Space Physics*, **17**, 73–88.

Wasserburg, G.J. and Burnett, D.S. (1969). The status of isotopic age determinations on iron and stone meteorites. In P.M. Millman (ed.), *Meteorite Research*, Reidel, Dordrecht, pp. 467–479.

Wasserburg, G.J., Burnett, D.S. and Frondel, C. (1965). Strontium–rubidium age of an iron meteorite, *Science*, **150**, 1814.

Wasserburg, G.J., Fowler, W.A. and Hoyle, F. (1960). Duration of nucleosynthesis. *Physical Review Letters*, **4**, 112–114.

Wasserburg, G.J. and Hayden, R.J. (1954). The branching ratio of K^{40}. *Physical Review*, **93**, 645.

Wasserburg, G.J. and Hayden, R.J. (1955a). A^{40}–K^{40} dating. *Geochimica et Cosmochimica Acta*, **7**, 51–60.

Wasserburg, G.J. and Hayden, R.J. (1955b). Age of meteorites by the A^{40}–K^{40} method. *Physical Review Letters*, **97**, 86–87.

Wasserburg, G.J. and Hayden, R.J. (1955c). Time interval between nucleogenesis and the formation of meteorites. *Nature*, **176**, 130.

Wasserburg, G.J., Papanastassiou, D.A., Nenow, E.V. and Bauman, C.A. (1969). A programmable magnetic field mass spectrometer with on-line data processing. *Review of Scientific Instruments*, **40**, 288–295.

Wasson, J.T. and Wetherill, G.W. (1979). Dynamical, chemical and isotopic evidence regarding the formation locations of asteroids and meteorites. In T. Gehrels (ed.), *Asteroids*, University of Arizona Press, Tucson, AZ, pp. 926–974.

Wetherill, G.W. (1986). Accumulation of the terrestrial planets and implications concerning lunar origin. In W.K. Hartmann, R. Phillips and G.J. Taylor (eds), *Origin of the Moon*, Lunar Planetary Institute, Houston, TX, pp. 519–550.

Wetherill, G.W., Aldrich, L.T. and Davis, G.L. (1955). ^{40}Ar and ^{40}K ratios of feldspars and micas from the same rock. *Geochimica et Cosmochimica Acta*, **8**, 171–172.

Wilhelms, D.E. (1987). The geologic history of the Moon. *USGS Prof. Paper 1348*, US Government Printing Office, Washington, DC.

56

HEINRICH WÄNKE*

Chemical evolution of the Moon and the terrestrial planets

Knowledge of the chemical evolution of the Moon, Mercury, Venus, and Mars and the contributions to this knowledge from space missions vary greatly from body to body. Without doubt the Moon is a very special case. The six Apollo missions that landed there returned a total of 381.7 kg of lunar material. The first of these missions was launched on 16 July 1969 and returned to Earth on 24 July 1969. It was followed by Apollo 12, 14, 15, 16, and 17, the last of which was launched on 7 December 1972. Apollo 15, 16, and 17 carried a rover vehicle that extended the range of the astronauts considerably. During the Apollo 17 mission the rover covered a total of 30.5 km. The Russian uncrewed missions Luna 16, Luna 20, and Luna 24 returned about 370 g of lunar material. The robot rover of the Luna 17 mission traveled 10.5 km in 322 days and that of Luna 21 traveled 37 km in 139 days. In addition, 13 lunar meteorites totaling 4.1 kg have been recognized and studied thoroughly.

It should be stressed that with respect to the chemical evolution of a planet the availability of samples increases possible insights both in quantity and quality. Hence, it is no surprise that, except for the Earth, the Moon is the best studied object in the inner Solar System. For the foreseeable future there is no chance that even the most sophisticated instruments flown to anywhere in the Solar System might return as precise and detailed information as that obtainable from investigations of samples in laboratories on Earth.

With respect to our understanding of the subject of this chapter, next to the Moon comes Mars. Viking 1 and 2 returned the first information on the chemical composition of the Martian soil which seems to be well mixed on a global scale by strong storms. Viking 2 was launched ahead of Viking 1 on 20 August 1975, landed on Mars on 3 September 1976 and transmitted data for 3.5 years. Viking 1 was launched on 9 September 1975, landed on 20 July 1976 and transmitted data for 6.5 years. Between 1962 and 1988, the Russians sent many missions to Mars which mostly failed or achieved only limited success. The exception was Phobos 2, which arrived at Mars orbit on 29 January 1989; it identified water vapor in the Martian atmosphere, returned images of Mars and Phobos as well as visual and infrared spectra, but stopped transmitting data before it was able to deploy surface stations on Phobos. A large increase in our knowledge of the geochemistry of Mars was made by the very successful NASA Pathfinder mission. Its rover Sojourner, traveling a total of 104 m, returned high-quality data on the chemistry of five rocks and six soil samples analyzed by the APX-spectrometer on board the rover in addition to a number of close-up images. The second source of information on the chemical evolution of Mars stems from Martian meteorites, of which 14 with a total weight of 81 kg have been recognized and studied.

Venus is a most difficult planet to study because of the very harsh conditions on its surface. Nevertheless valuable data on the chemical composition of the Venusian surface were obtained by the Soviet Venera and Vega missions, launched between 1972 and 1984, which transmitted data, Venera 13 relaying the first color pictures of the surface. However, the returned data do not allow reliable conclusions to be drawn on the chemical evolution of this planet.

Almost nothing is known of the chemistry of Mercury. This is very unfortunate as Mercury has the highest density of all planets and, hence, this must be reflected in its chemical composition. It is hoped that the missions planned both by ESA and NASA will fill this gap.

---

* Max-Planck-Institut für Chemie, Mainz, Germany

Considering the chemical composition of the Moon and the planets, we have to bear in mind that the Solar System formed from the solar nebula. To show the fractionation process involved it is useful to compare individual bulk compositions of Solar System objects, or parts of them such as individual rocks, to the abundances of elements to solar abundance or even better to C1 (carbonaceous chondrites type 1) abundances which reflect the primordial Solar System abundances of all condensable elements. Data for C1 or solar abundances are given in the classical paper by Suess and Urey (1956) or in later compilations by Cameron (1973), Palme et al. (1981), Anders and Ebihara (1982), and Anders and Grevesse (1989).

THE MOON

The pioneer of modern cosmochemistry, Harold Urey (who won the Nobel Prize for Chemistry in 1934 for the discovery of deuterium), suggested the Moon as a possible place of the origin of stone meteorites (Urey 1959). He was, at that time, convinced that the Moon was a "primary" object. To account for the apparent about 10% lower density of the Moon compared with that of ordinary chondrites, he assumed the Moon to contain several percent water and/or graphite. It was his belief in a primitive nature of the Moon formed at low temperature, which made the Moon for him central to the understanding of the formation of the Solar System.

I, being 35 years younger than Urey, have been from early on fascinated by the great old man and his reasoning. In the early 1960s, I had found conclusive evidence for the solar wind origin of the solar-type rare gases observed in several bronzite chondrites (Wänke 1963, 1965; with respect to solar wind implantation in meteorites a number of other papers have to be mentioned: Signer and Suess 1963, Suess et al. 1964, Eberhardt et al. 1965, Zähringer 1966). To account for this observation, it was a necessity that individual meteorite grains prior to compaction were exposed to solar wind irradiation at the surface of their parent body, which should not have an atmosphere or magnetic field to hinder the solar wind implantation. I thought the Moon would be an ideal body for this implantation and together with some weaker arguments on the distribution of cosmic exposure ages, I published a paper (Wänke 1966). Urey was quite convinced by my arguments and in his letter to me of 10 February 1967 he wrote: "In fact, I think I am probably your most important public relations man in the United States at the present time.' Later on in this letter: 'It looks as though I must now begin to caution people not to think that the whole case for meteorites from the moon is completely settled. I keep telling them that only samples from the moon will definitely settle the problem."

The first chemical data obtained *in situ* on the Moon from Surveyors 5, 6, and 7 by alpha-scattering analyses (Turkevich et al. 1970, and references therein), indicated a basaltic composition at the two maria locations and a low-iron, basalt-like composition at the third one for the area near Tycho. They proved that Urey and I were wrong. Urey had just written a paper in which he summarized all arguments for a lunar origin of some meteorites (Urey 1968). On 26 February 1968, he wrote to me with respect to the Surveyor results: "I have the feeling that the data would seem to indicate that no meteorites come from the Moon at all. Moreover, I am led to doubt the Mars origin also." In my paper "On the lunar origin of the bronzite chondrites" (H-chondrites), I had mentioned that the other major group of ordinary chondrites – the hypersthene chondrites (L-chondrites) – might come from Mars.

The possibility of ejection of rocks from the Moon or Mars was considered to be absolutely impossible by the experts on cratering mechanisms at that time. When the first samples from the Apollo 11 mission were analyzed, I had the satisfaction that at least my proposition that the lunar dust should be loaded with solar wind particles turned out to be correct. Today, meteorites from the Moon – although not of chondritic composition – is a not disputed fact. A Mars origin of a small group of meteorites (the SNC meteorites) is widely accepted.

When NASA asked for proposals for lunar sample investigations my colleagues and I at the Max-Planck-Institute for Chemistry in Mainz, Germany, wrote altogether 12 proposals addressing different areas of research. As result of these proposals, the Mainz laboratory became a major player in the Lunar Sample Analyses Program of NASA. Our laboratory has received the largest amount of lunar material, both in terms of mass and number of sample specimens, of any laboratory outside the USA.

Apart from rare gas measurements, the multi-element analyses of lunar samples at the Mainz laboratory became highly appreciated. All major and minor elements and many geochemically important trace elements (in total up to 54 elements) have been determined in this multi-element analysis program. This turned out to be especially valuable for the investigations of lunar breccias with their considerable heterogeneity. Using this large data set quite a number of element correlations have been observed first or confirmed.

To work with lunar samples was very exciting for all of us involved. Let me describe an incident when we worked on our first lunar sample, a soil sample no. 10084. Most of our analytical work was carried out using neutron activation techniques starting non-destructive by gamma-counting the samples irradiated by thermal neutrons in a nuclear reactor. The reactor, a TRIGA research reactor of the Gutenberg-University of Mainz, was located next to our building. When we applied the first 6 hour irradiation from 9 a.m. to

3 p.m., the sample was in our gamma-ray spectrometer for less than about 30 minutes, and we waited for the first spectra. Our technician, Mr. Bernhard Spettel, came to me highly excited and said: "Very, very strange. I have never seen such a spectrum. There are only three lines, I have to look up the isotope table because I do not know these lines." This was indeed very unusual as Mr. Spettel was a real expert in this field and had most of the gamma-lines in his head. A few minutes later, he said: "You would not believe, the lines indicate only one element, namely indium." I knew that indium, a trace element in rock or soil samples, has a very high cross-section and due to the short half-life of the isotope produced by neutron capture, ^{116}In of only 54 minutes, it would be produced in high quantities even if present only in trace amounts. The short half-life meant that the isotope decayed rapidly, and in the evening of the same day we could see the lines of those elements we actually expected. What had happened? In the overall extraordinarily successful Lunar Sample Analyses Program, NASA had decided to return the samples under vacuum. A sample box was designed into which the astronauts had to put the samples and close the lid, which was supposed to be sealed by an indium gadget between the edges of the lid and box. The whole sealing mechanism did not work properly, and most of the samples returned from the Moon were heavily contaminated with indium. In our sample there must have been a tiny indium grain of about 0.1 mg. In this respect it was an exception, but practically all soil samples showed indium excesses.

The Apollo missions

Apollo 11 – Mare Tranquillitatis

Soil samples and soil breccias
As expected (Wänke 1965, 1966), all soil samples contained large amounts of solar wind-implanted rare gases. ^{4}He concentrations up to 0.5 cm^3 STP g^{-1} have been observed. From the ratios ^{4}He/^{20}Ne = 91 and ^{4}He/^{36}Ar = 505 compared with the ratios of more than three times higher found in magnetically separated metal grains, which hold rare gases more strongly, a considerable loss of helium due to diffusion was evident for the bulk samples (Hintenberger et al. 1970). The amount of hydrogen in the lunar soil, in soil breccias, was found to exceed that of ^{4}He by about a factor of six with a D/H ratio of about three times less than the terrestrial ratio. Laboratories worldwide were engaged in the Lunar Sample Analysis Program. With respect to the solar wind-implanted rare gas isotopes, the studies in Bern (Eberhardt et al. 1970) were very comprehensive, while Epstein and Taylor (1970) analyzed hydrogen. The findings of these groups were identical to those obtained in Mainz as well as by other laboratories.

The chemical composition of the soil and soil breccias was found to mimic that of the igneous rocks from which they were obviously derived with only small additional components (Wänke et al. 1970). From the amount of hydrogen evolved during the treatment of lunar soil with diluted acid under vacuum, the presence of 0.6% metal in sample 10084–18 was calculated. Analyses of metal particles separated from the bulk soil by a hand magnet showed that practically all of the nickel, cobalt, gold, and iridium in the bulk soil resides in the metal particles, indicating a mainly meteoritic origin. An exception was observed for tungsten, for which high concentrations of 24 ppm in these metal particles proved their equilibration with matter from lunar basalts at elevated temperatures. (In Figures 1 and 2, instead of individual mare rocks the soil sample 10084 is used for comparison as in this way small differences from rock to rock are avoided.)

The meteoritic component was extensively studied by Ganapathy et al. (1970) who estimated a 1.9% admixture of carbonaceous chondrite-like material, corresponding to an average influx rate of meteoritic and cometary matter of 3.8×10^{-8} g cm^{-2} yr^{-1}.

Rock samples
The igneous rocks, the mare basalts from Mare Tranquillitatis (Wänke et al. 1970) compositionally reflected highly differentiated material of clearly basaltic nature (Figure 1). In contrast to terrestrial rocks, the high titanium concentration

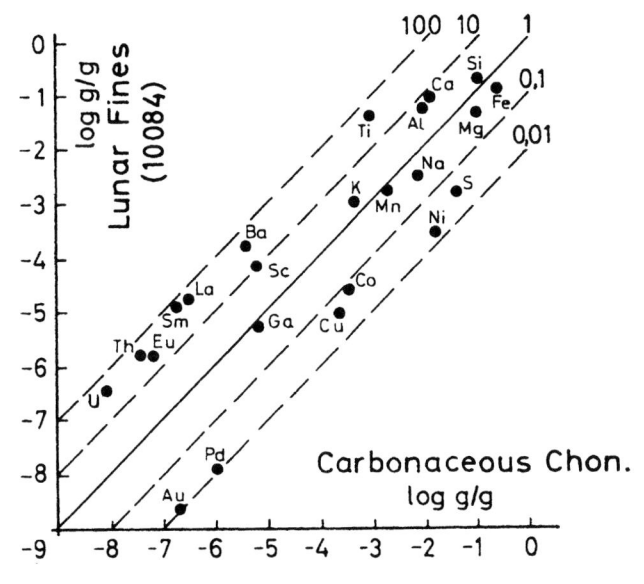

Figure 1 Comparison of the concentrations of various elements in lunar fines and carbonaceous chondrites type 1. The huge compositional differences clearly show that the Moon is not a primitive but a highly differentiated object. Data for lunar fines from Wänke et al. (1970). (After Wänke et al. 1970, reproduced with permission.)

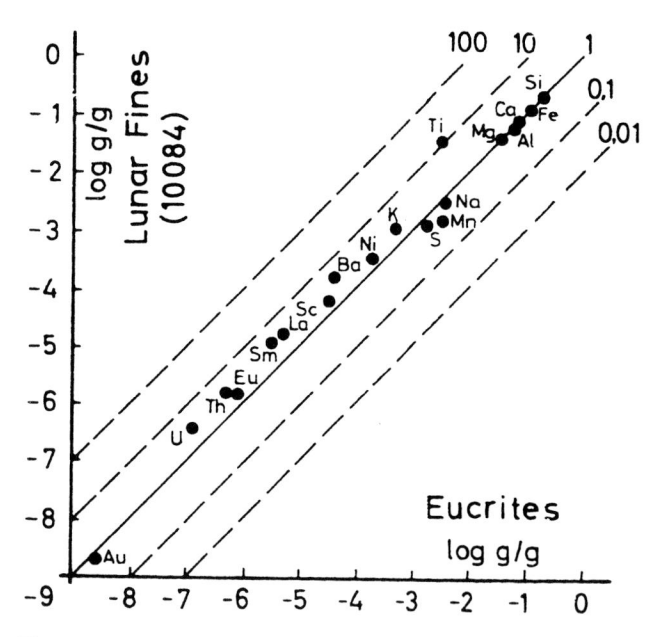

Figure 2 Comparison of the concentrations of various elements in lunar fines and eucrites. The higher differentiation of the Moon compared to the eucrite parent body is evident. (After Wänke et al. 1970, reproduced with permission.)

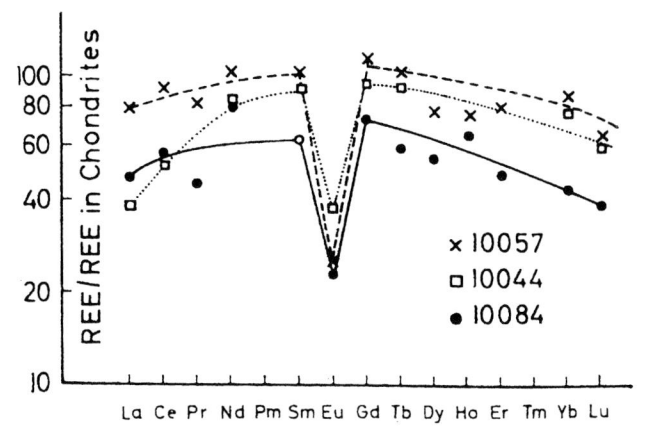

Figure 3 Rare earth concentrations in lunar rocks 10057 and 10044 and lunar fines 10084, normalized to chondritic values. Data for chondrites from Schmitt et al. (1963, 1964); data for lunar samples from Wänke et al. (1970). The negative Eu anomaly reflects the low oxygen fugacity of the Moon where Eu becomes divalent and does not fractionate with the other REEs that remain trivalent, but enters feldspar. The Eu anomaly has been reported by various authors (Gast and Hubbard 1970, Haskin et al. 1970, Philpotts and Schnetzler, 1970, Schmitt et al. 1970). (After Wänke et al. 1970, reproduced with permission).

in the lunar basalts was striking. Comparing the major and trace element abundances, it was shown that the lunar mare basalts are compositionally close to basaltic achondrites, the eucrite meteorites (Figure 2), although certain element ratios differ considerably excluding a lunar origin of eucrites. Striking were also the fractionated rare earth element (REE) patterns with a large negative europium anomaly (Figure 3), reflecting a lower oxygen fugacity on the Moon as compared to the Earth. As in the case of rare gases, the major and trace element compositions were also studied by many groups. The REEs, for example, were studied by Wakita et al. (1970). The paper by Gast et al. (1970) should also be mentioned in this respect. It contains, apart from data on REEs and some other trace elements, suggestions on the petrogenesis of the Apollo 11 basalts. In all cases the Mainz data agreed very well with those from other investigators.

One of the proposals for our investigations on lunar samples to be carried out at Mainz was the determination of radioactive isotopes produced by the interaction of cosmic-ray particles. For this work sample masses of about 100 g were required. It was necessary to decompose chemically the sample in order to extract the various radioisotopes to be studied. The first step in this procedure was reduction and pulverization of the sample. Because of the high value of the sample, my colleague, Prof. Friedrich Begemann, and I had decided to do this step ourselves. For the pulverization we used a steel mortar that we had put inside a small plastic tent so that particles thrown out of the mortar or the sieves could be caught. We started our work around 8 p.m. When we finished around 11 p.m., I walked back to my office on this rather cold but very clear late evening with the Moon high up in the sky. Suddenly I noticed an irritation in my nose and got out my handkerchief. When I put down the handkerchief I noticed two black dots on it, which obviously were due to some fine dust liberated during the sample preparation in spite of the plastic tent. I looked up to the Moon on which Mare Tranquillitatis was easy visible with the naked eye, thinking that a rock is missing from it now and a tiny, tiny fraction of it I have just removed from my nose. This was the most touching moment of my life.

Apollo 12 – Oceanus Procellarum

Soil and rock samples

Both the soil and the rock samples returned by the Apollo 12 mission showed larger variations in chemical composition than those from Apollo 11, which may just reflect the fact that all the Apollo 11 samples were collected much closer to the landing module.

Apollo 12 landed within 200 m of the Surveyor 3 spacecraft. Some distant ejecta rays from the large crater Copernicus, located 400 km to the north, crossed the site. However, none of the returned samples could be proven to contain material originating from Copernicus.

With respect to their major-element chemistry, the mare basalts from Oceanus Procellarum were divided into two

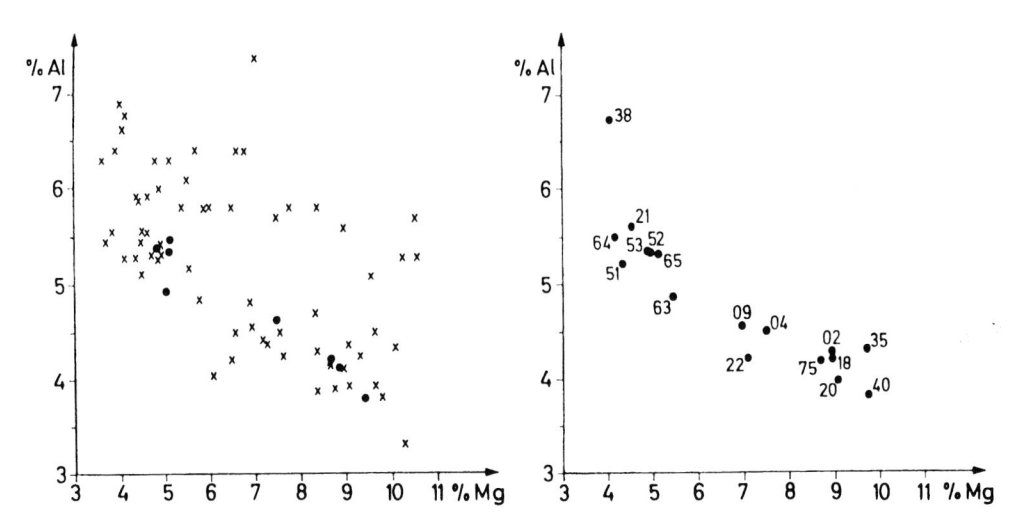

Figure 4 Al v. Mg in lunar igneous rocks from Apollo 12. Left: measurements of all investigators as individual entries (×); Mainz data (•). Right: the numbers refer to the last two digits of the five-digit sample numbers. Average values of all investigators. Notice the anticorrelation of Mg and Al as is expected in magmatic processes. The high quality of the Mainz data is evident comparing the two plots. Data from Willis *et al.* (1971), Compston *et al.* (1971), Ehmann and Morgan (1971), Morrison *et al.* (1971), Bouchet and Kaplan (1971), Cuttitta *et al.* (1971), and Wänke *et al.* (1971). (After Wänke *et al.* 1971, reproduced with permission.)

groups: low-Mg basalts with about 5 wt% Mg and high-Mg basalts with about 7–9 wt% Mg. As illustrated in Figure 4, magnesium is anticorrelated with aluminum. Figure 4 also shows the superiority of the Mainz data in respect of accuracy. Among the trace elements, nickel, chromium, and cobalt concentrations were similar to magnesium, while the REEs showed an opposite trend (Wänke *et al.* 1971).

The Apollo 12 soil samples showed a large range in their abundances of trace elements, in particular of the incompatible ones. It was shown that the observed range could be explained by the mixing of two major components. One of which could be assigned to the local high-Mg rocks, while the second component was a trace element-rich silicious or "granitic" component. The latter, named KREEP (rich in potassium (K), REEs, and phosphorus) by Hubbard and Gast (1971), was noticed for the first time at the Apollo 12 landing site; however, it was never observed in pure form but only in the form of admixtures with highland breccias and soils.

Detailed studies had shown that the chemical composition of lunar soils reflect their mixed origins. In fact, all soil samples returned from the Moon contain some components exotic to the site of collection. Even at the Apollo 11 site, located in the middle of a mare basalt plain, the presence of anorthositic material was the first indication of the anorthositic nature of the lunar highlands (Wood *et al.* 1970), while the Apollo 12 soils show an admixture of KREEP as explained above. In addition a small meteoritic component is present in all soil samples.

Apollo 14 – Fra Mauro

The samples returned from Apollo 14 differed from those of the previous missions as igneous rocks were almost absent, being only present in the form of lithic fragments. Most of the samples were breccias compositionally similar to the soil samples and very rich in KREEP elements. With these samples it became evident that many additional incompatible trace elements belonged to the KREEP component. Apart from U and Th, these were W, Ba, Zr, Ta, and Hf as well as the heavy alkaline earth elements Rb and Cs.

Apollo 15 – Hadley Rille, Mare Imbrium;
Apollo 16 – Descartes region (typical highland);
Apollo 17 – Taurus-Littrow region, eastern Mare Serenitatis

With respect to the chemical composition of the Moon, the last three Apollo missions as well as the Soviet Luna 16, 20, and 24 missions confirmed several of the earlier findings but also contradicted earlier conceptions. Among the latter, only mentioned here is the connection of the Moon to the eucrites, the basaltic meteorites as well as the idea of a highly refractory Moon. The Moon can, if at all, only be slightly enriched in refractory elements relative to the Earth.

As mentioned in the introduction, many element correlations were discovered which later on turned out to hold for the Earth and other planetary objects as well (see the section on "Martian meteorites and the bulk composition of

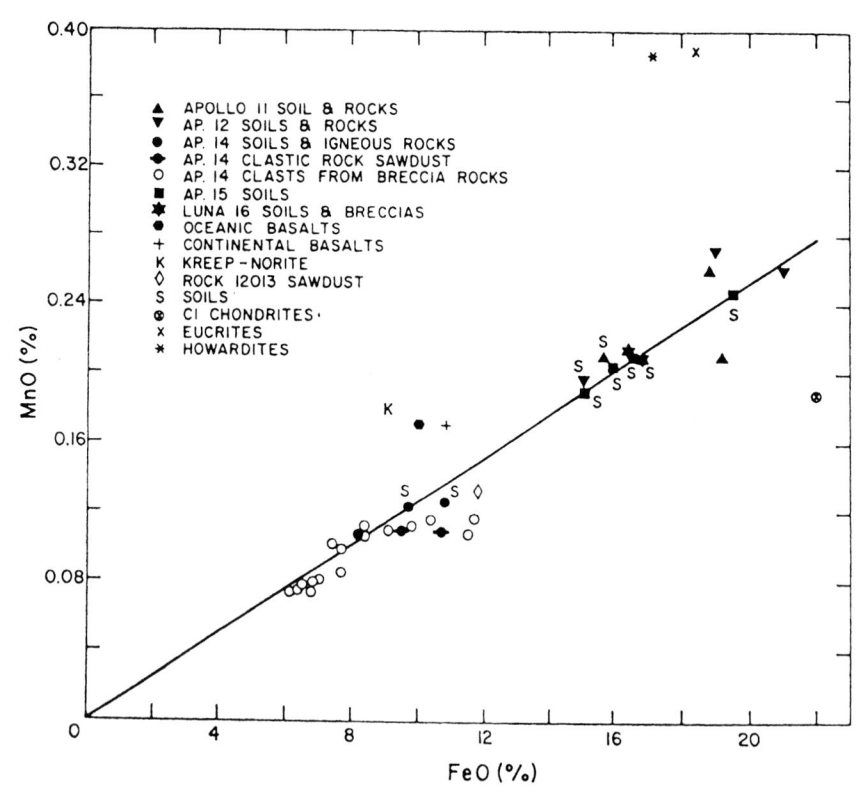

Figure 5 Abundances of FeO and MnO in lunar, meteoritic, and terrestrial samples. This was the first element correlation noticed. The correlation coefficient is 0.95 (>99% confidence level for all lunar samples). (After Laul et al. 1972, reproduced with permission.)

Mars"). The first of these correlations, namely MnO v. FeO (Figure 5), was discovered by Laul et al. (1972).

Although most of the Apollo 16 samples, like those from Apollo 14, were breccias, the anorthositic nature of the lunar highland could be confirmed. A number of real anorthosites were identified, although in some cases they were only found as clasts in breccias; these are samples 15415, 60015, 60025, 61016, and 65315. It was noticed that there obviously exists a lower limit of large-ion lithophile (LIL) elements in these anorthosites. With the help of the relevant partition coefficients the composition of the melt from which they crystallized could be derived. From these data it became evident that the refractory trace elements had about a 10-fold enrichment relative to type I carbonaceous chondrites in this magma (Palme et al. 1984), which was once part of a global magma ocean (see next section).

The command and service modules of the Apollo 15 and 16 missions carried an integrated geochemical package that included x-ray and gamma-ray spectrometers. These instruments were the first built for a global geochemical mapping applying these techniques. For the x-ray spectrometer, solar x-rays were used for exitation and, hence, it was restricted to the dayside area overflown. These restrictions did not apply for the gamma-ray spectrometer, using the emission of the natural radioactive elements K, Th, and U as well as cosmic-ray-induced nuclides. Large differences of the gamma-ray signals from K, Th, and U were observed between the mare and the highland areas. The highest concentrations of the natural radioelements were observed east, west, and south of Mare Imbrium, while the highland areas, especially those of the farside showed almost an order of magnitude lower abundances of natural radioelements (Metzger et al. 1973).

The x-ray spectrometer yielded good data for Mg, Al, and Si. As expected for the major elements, the variations were smaller but clearly showed higher Al/Si and lower Mg/Si ratios over highland areas while for the mare regions lower Al/Si and higher Mg/Si ratios were observed (Adler et al. 1973). These observations can be interpreted by the dominance of anorthosite in the highland crust and the mafic nature of the mare rocks. The highly incompatible trace elements K, Th, and U refer to the trace element-rich KREEP component.

The origin of KREEP and the Mg-rich matter in lunar highlands

Most samples originating from lunar highlands show identical patterns of incompatible elements (Palme and Wänke

1975). This is most easily explained by the addition of various amounts of a trace element-rich component (KREEP) to all these samples as originally proposed by Hubbard and Gast (1971) for the Apollo 12 samples.

The trace element patterns of mare basalts are in a certain sense complementary to the trace element patterns of the highland crust (Figures 6 and 7). Regarding the origin of the KREEP component, it is still not clear if it was formed as a residual liquid from the magma ocean confined to the outer 300 km of the Moon or formed by a small degree of partial melting of most of the Moon. Palme and Wänke (1975) favored a two-step model that links the formation of KREEP to that of the mare basalts. According to this model the KREEP component originates as an early partial melt from the deep interior of the Moon highly enriched in incompatible elements, including the heat-producing elements K, Th, and U. This liquid would concentrate in the crust and in some cases even penetrate the crust. The trace elements may be enriched further by fractional crystallization. After the separation of the first liquid, the still hot interior of the Moon would be heated further by the U and Th left in the residual interior. This could in a very short time lead to a second partial melting, which would be more localized depending on the degree of U and Th depletion in the first stage. In this way, magma chambers would form, out of which the mare basalts were later derived. Any fractional crystallization of KREEP or any dilution of the mare basalt magmas by melting the wall rocks of the magma chambers would, of course, influence the apparent degree of partial melting.

The main argument against any derivation of KREEP and mare basalts from the same source region is the higher Mg/Fe ratio of KREEP-rich highland rocks compared to the low

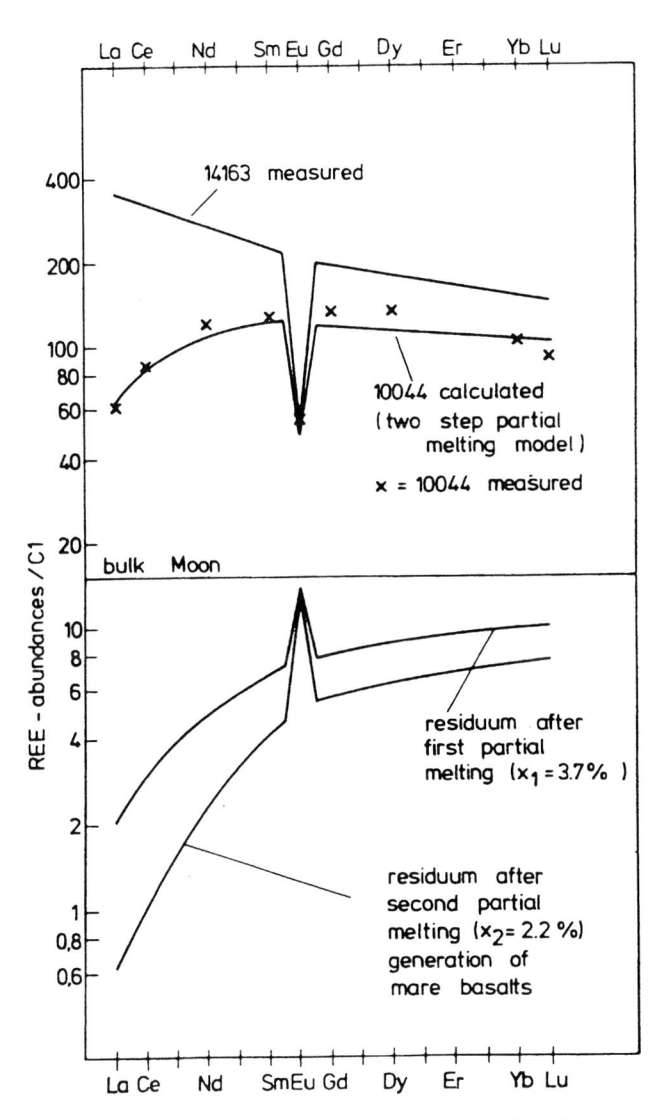

Figure 6 REE abundance patterns in the mare basalt 10044 calculated from a two-step partial melting model; x_1 and x_2 denote fractions of partial melting. The close match of the calculated REE abundance curve with the measured values is obvious. (After Wänke et al. 1974, reproduced with permission.)

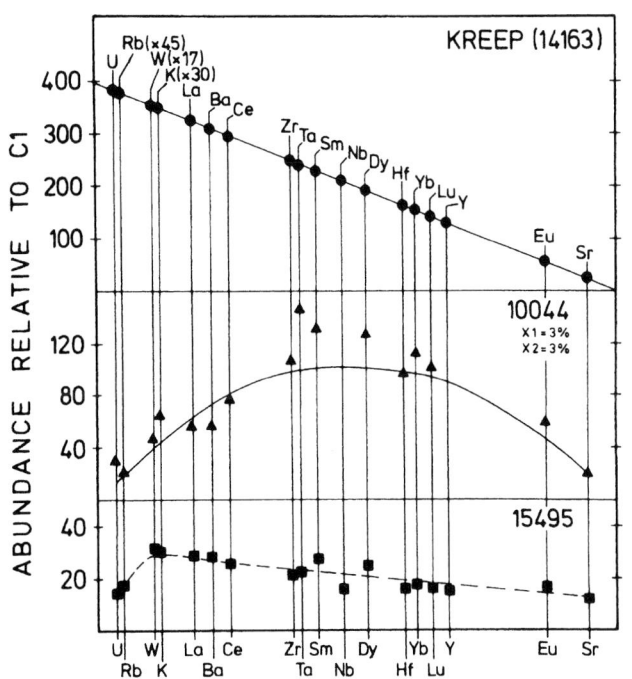

Figure 7 Enrichment factor relative to C1 for incompatible elements in KREEP-rich soil 14163, the Apollo 11 mare basalt 10044, and the Apollo 15 mare basalt 15495. For the model calculation for 10044 the same x_1 was used as for similar calculations for Apollo 17 basalts. A slightly lower x_1 would give a better fit, indicating that the primary depletion of incompatible elements was different for the source region of Apollo 11 and Apollo 17 mare basalts. The Apollo 15 basalts are certainly from a much less depleted source region; no attempt was made to calculate the pattern for 15495. (After Palme and Wänke 1975, reproduced with permission.)

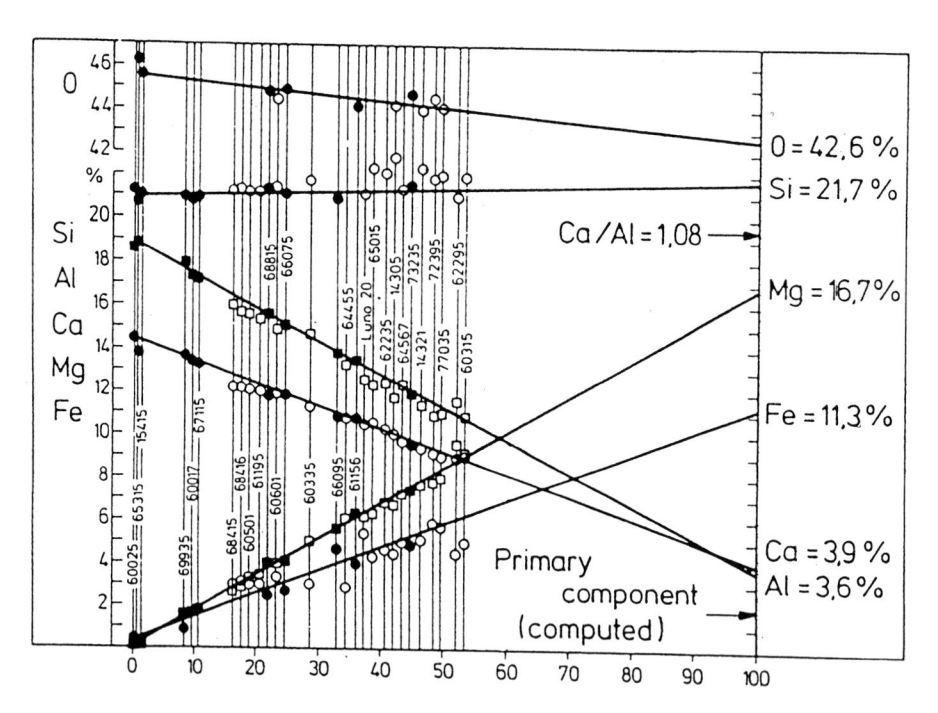

Figure 8 Mixing diagram. The sample positions along the x-axis were calculated from the data of Al, Ca, and Mg alone, without any assumption of possible endmembers. Samples with high portions of KREEP and samples of crystalline character (open symbols) were not included in the computation of the element lines. Most of the highland rocks fit well into this binary mixing diagram with pure anorthosite as the Ca,Al-rich endmember and a primitive Mg-rich endmember on the other side, which contains all major and trace elements in chondritic ratios, except the Mg/Si ratio. The Ca/Al ratio of 1.08 is the chondritic value. (After Wänke et al. 1976, reproduced with permission.)

Mg/Fe ratio of the mare basalts (Gast 1972, Hubbard et al. 1974). In the model advocated by Wänke et al. (1975) for the formation of the lunar crust, the highland rocks are viewed as mechanical mixtures of a primitive lunar plagioclase component derived from differentiation of an early magma ocean, a Mg-rich component (see below) and KREEP, the latter being only a minor constituent compared to the major element chemistry. The same holds for the small meteoritic component. If this model is correct, the mare basalts would have been generated after differentiation of KREEP. Allowing for the two minor components, KREEP and the meteoritic component, most of the highland rocks fit into a binary mixing diagram (Figure 8) with pure anorthosite as the Ca,Al-rich endmember and a Mg-rich endmember on the other side, called a "primary" component by Wänke et al. (1975). It was found that the Mg-rich primary component contained all major and trace elements in chondritic ratios except the Mg/Si ratio. The composition of the primary component was located along the mixing lines at a point where Ca and Al displayed the chondritic ratio. However, as seen from Figure 8, this location is not defined very well as the element lines for Ca and Al have similar slopes.

Subsequently, Wänke et al. (1977) slightly modified this mixing diagram by including Sc and extrapolating the compositional data to the point where the Al/Sc ratio became chondritic. (The elements Al, Ca, and Sc are all refractory lithophile elements and they can be expected to be present in chondritic abundance ratios in all objects of the Solar System.) The revised composition of the primary component was very similar to that obtained previously. According to their interpretation, the Mg-rich primary component represents material of the last accretion stage of the Moon when a solid anorthositic crust had already formed. However, to make the Mg/Si ratio of the primary component chondritic, it was necessary to add 31% olivine (Table 1).

As pointed out by Ringwood et al. (1986/87), there may be a simpler and perhaps more plausible interpretation of the

Table 1 Primary component in the lunar highlands

	Mg-rich component	Olivine 72417	Primary component = Mg-rich component + 31% olivine
Mg (%)	15.9	27.4	19.5
Al (%)	3.24	0.64	2.43
Si (%)	22.5	18.7	21.3
Ca (%)	3.57	0.82	2.71
Fe (FeO) (%)	10.5	9.1	10.1
Sc (ppm)	23	4.3	17.5

Table 2 Comparison of the composition of the bulk Moon with that of the unfractionated primary component observed in lunar highland samples (for references see Wänke and Dreibus 1986)

	Moon (silicate portion) (%)		Unfractionated primary component (%)	
			Mg/Si = 1	Mg/Si = Earth's mantle
	Ringwood and Kesson (1977)	Delano (1985)	Wänke et al. (1977)	Wänke and Dreibus (1982)
SiO_2	44.6	41.9	45.6	44.2
MgO	33.4	34.3	32.4	35.5
FeO	13.9	16.7	13.0	12.7
Al_2O_3	3.7	3.58	4.6	3.76
CaO	3.4	2.86	3.8	3.15
MnO	—	0.23	0.18	0.16
Cr_2O_3	0.4	0.38	0.4	0.37

nature of the primitive component than that originally proposed by Wänke et al. (1975). The Mg-rich component could also be recognized as a komatiitic magma, as it resembles terrestrial komatiites rather closely except for the higher FeO and lower Na_2O in the lunar composition. On Earth, komatiites are thought to be formed from magmas derived by large degrees of partial melting. Ringwood et al. (1986/87) took the similarity in composition between lunar and terrestrial komatiites as a strong indication of a similarity of the composition of the bulk Moon and the Earth's upper mantle.

Wänke et al. (1977) considered the composition of the primary component as representative of the lunar bulk composition. Later, instead of the chondritic Mg/Si ratio that of the Earth's mantle was used. Data for the bulk composition of the Moon obtained in this way are listed in Table 2 with data from other authors.

Differentiation of the Moon: the magma ocean concept

Large compositional differences of the Fe- and Mg-rich mare basalts and the Ca- and Al-rich highland rocks indicate large-scale differentiation processes of either the whole Moon or at least of the outer 300 km, corresponding roughly to one-third of the mass of the Moon.

The Ca-rich plagioclase ($CaAl_2Si_2O_8$), the dominant mineral observed in the highlands, has a relatively low density which led to the suggestion that the early lunar crust was formed by flotation of plagioclase on top of a 300–400 km deep magma ocean, covering the whole Moon. Apart from anorthosites (Ca-rich plagioclase), KREEP basalts and Mg-rich rocks have also been observed among the returned highland rocks. The latter, especially are all very old: 4.2 Gyr or older (Carlson and Lugmair (1979, 1981). This means that they are much older than the mare basalts that range between 2.9 and 3.9 Gyr. Thus the lunar crust was formed very early, just after the formation of the Moon. An early formation of the proposed magma ocean solves the problem of the heat source, as accretional heat might have played an important role, especially if the Moon was formed by an impact-induced splash-off of material from the Earth followed by its rapid accretion in orbit around the Earth.

The mare basalts with their negative Eu anomalies (as opposed to the positive Eu anomalies of most highland rocks) probably formed from sunken cumulates after plagioclase had crystallized. The partial melts of these cumulates filled the basins formed by large impacts.

The KREEP magmas with their extremely high concentrations of incompatible elements formed in this scenario from residual liquids after 99% of the magma ocean had crystallized. Nevertheless, Rb–Sr model ages cluster around 4.25–4.40 Gyr. (Palme 1977, Warren and Wasson 1979).

Estimates of the depth of the lunar magma ocean vary from about 100 km to a totally molten Moon. As summarized by Warren (1985), calculations of the depth of melting are based mostly on mass balance arguments. According to Warren, the best argument for a lunar magma ocean is the fact that the average plagioclase content of the lunar surface is about 75%, which for a 60 km thick layer corresponds to an amount of Al of 1.9% of the mass of the whole Moon. This is close to the bulk Al content of the Moon, for which estimates range from about 2 to 3%. Even if one assumes a crust of only 30 km thick and a lower plagioclase content, one still needs about one-third of the Moon to be molten, even making the assumption of an Al extraction of 100%.

Oxygen isotopes and other indicators for a genetic relationship between earth and moon

From the detailed work of Clayton and Mayeda (1975, and references therein), it became known that the lunar samples lie within the analytical uncertainties on the terrestrial oxygen fractionation line in the three-isotope plot of $^{17}O/^{16}O$ v. $^{18}O/^{16}O$ shown in Figure 9. This fact became even more

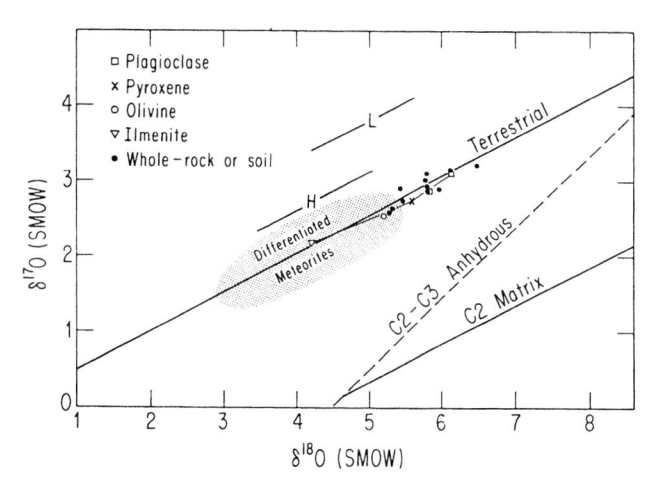

Figure 9 Three-isotope plot of $^{17}O/^{16}O$ v. $^{18}O/^{17}O$. The plotted data points are lunar rocks and soils from Apollo 11, 14, 15, 16, and 17; separated phases from rock 79155 are joined by a line. (After Clayton and Mayeda 1975, reproduced with permission.)

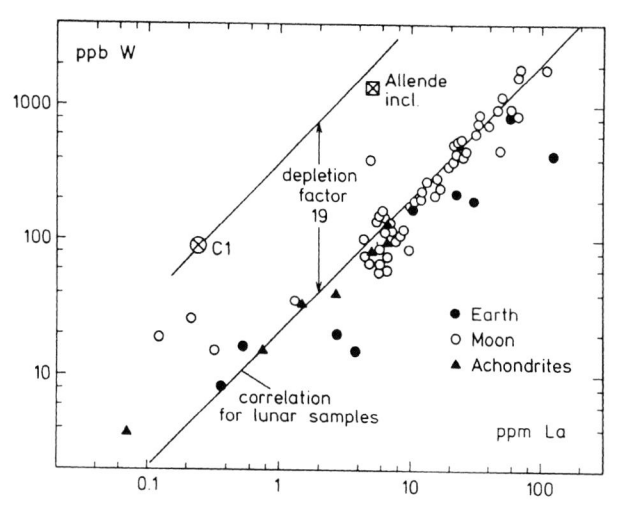

Figure 10 W v. La in lunar, terrestrial, and meteoritic samples. The strong correlation of these two elements indicate their presence in all types of lunar samples in a constant abundance ratio independent of their absolute concentration. The similarity of the abundance ratio of W and La for the Moon and the Earth points to a close genetic relationship of Earth and Moon. All data from the Mainz Laboratory.

significant when it became known that samples from the eucrite parent body, possibly the asteroid Vesta, as well as the Martian meteorites fall distinctly away from the terrestrial oxygen isotope fractionation line. Thus, among the rocks of these four objects that underwent magmatic differentiation, and of which we have samples, from two, Earth and Moon, the rocks have identical isotope patterns while the other two differ. The work of Clayton and his group was the first indication of a genetic relation between the Moon and the Earth.

A strong correlation of W with La was observed for all types of lunar samples (Figure 10). All other refractory lithophile elements with similar chemical fractionation behavior (LIL elements) are present in lunar samples in their chondritic abundance ratios (Wänke et al. 1973). This also holds to a lesser extent for rocks of the Earth because of more complex magmatic fractionation processes. Only W, a highly refractory and incompatible element, is depleted relative to other refractory elements to an almost identical extent in rocks from the Earth's mantle and in samples from the Moon. As was suggested by Wänke et al. (1973), the reason being that W is a moderately siderophile element of which only a small fraction stayed in the silicate phase where it fractionated together with the other incompatible elements while most of it went into the metal phase. Measuring metal–silicate partition coefficients for W at different temperatures and oxygen fugacities, Rammensee and Wänke (1977) concluded that about 26% metal had to be in chemical equilibrium with the silicate phase to account for the observed W depletion. For the Earth with a core mass of over 30% as well as the achondrites there is no problem; however, for the Moon with a core mass of 5% or less, the observed W depletion requires a scenario with a strong genetic relationship between Earth and Moon. Hence, the metal, in amounts required to explain its depletion in the Moon, obviously resides in the Earth's core. A similar case can be made for phosphorus. Later it was found that W is in fact more incompatible than La and correlates even better with U. The W and P depletions in the lunar samples were the basis of the proposition by Rammensee and Wänke (1977) of a close genetic relationship between Earth and Moon.

At the time when this proposition was put forward, there was no physical explanation for it. Hartmann and Davis (1975), Cameron and Ward (1976), and Ringwood (1979) suggested collision-induced fission of the Earth for the formation of the Moon from the Earth's mantle. In the model calculation of Wetherill (1976), the largest object impacting on the Earth during its accretion was of the order of several lunar masses. Today, the collision of the Earth with an object of the mass of Mars is the widely accepted model for the origin of the Moon. According to Melosh and Sonett (1986), jets of dense, hot vapor will form after such a collision which will expand and cool adiabatically around the Earth thus forming the Moon on a fast timescale. The incorporation of large amounts of Earth mantle material in the jets accounts for the geochemical similarity between Earth and Moon. Since jetting is a near-surface phenomenon, the vapor jet will not include core material. This can explain the lower density of the Moon caused by a lower abundance of metal. Some differences between Earth and Moon may be attributed to the admixture of projectile material in the jet. The depletion of the Moon in volatile and moderately volatile elements can also be explained straightforwardly.

For a detailed discussion of the geochemical evidence for such a scenario see Wänke and Dreibus (1986).

The elements Mn, Cr, and V are substantially depleted in the Earth's mantle. The most likely explanation for this depletion is the removal of these elements into the Earth's core, presumably in reduced form (Wänke 1981). All three elements show increasing siderophile tendencies with decreasing oxygen fugacities. In the case of Mn, the most volatile of these three elements, depletion by volatilization may also have played some role. In spite of the strong depletion of the moderately volatile elements Na and K on the Moon, the Earth, the eucrite parent body (EPB), and Mars, no depletion of Mn, Cr, and V is observed on Mars and the EPB (Figure 11). Depletion by volatilization can be excluded for Cr and V, the latter being a refractory element.

There is no general consensus about the reason for the depletion of Mn, Cr, and V in the Earth's mantle; however, whatever mechanism caused it, their depletion is a very characteristic feature of the Earth's mantle and is strongly coupled to the accretion mode of the Earth. Very similar depletion of these elements has been observed for the Moon (Wänke and Dreibus 1986). Removal of Mn, Cr, and V into the Earth's core in reduced form requires inhomogeneous accretion and core formation during accretion because these elements would not be stable in metallic form in equilibrium with larger amounts of FeO. Inhomogeneous accretion and core formation during accretion is favored in the model of Wänke (1981).

Another important element correlation is that of K v. La. K, like La, is a highly incompatible element. However, being moderately volatile it is subject to depletion by volatilization. Thus, in Figure 12 the observed depletion of K for the Earth, the achondrites (eucrites, howardites, and diogenites), and the lunar samples (most depletion) reflects its loss by volatilization. The abundance of the highly refractory element La can be assumed to be chondritic relative to silicon, and thus the lower K/La ratios are not, as originally thought by many workers, due to an overabundance of the refractory La, but merely due to the depletion of K. Similar arguments can be made from the K/U ratio.

Siderophile elements

Relative to the Earth's upper mantle the lunar mantle shows a slight overabundance of Fe, W, and Mn, a slight depletion of Co and P, and an increasing depletion going from Ni to Re. This depletion parallels the effective metal–silicate partition coefficients (Figure 13).

A very specific feature of the Earth's mantle is its high concentrations of moderately siderophile elements: 2110 ppm Ni and 105 ppm Co. Thus, if the Moon was formed from the Earth's mantle, one should expect that the

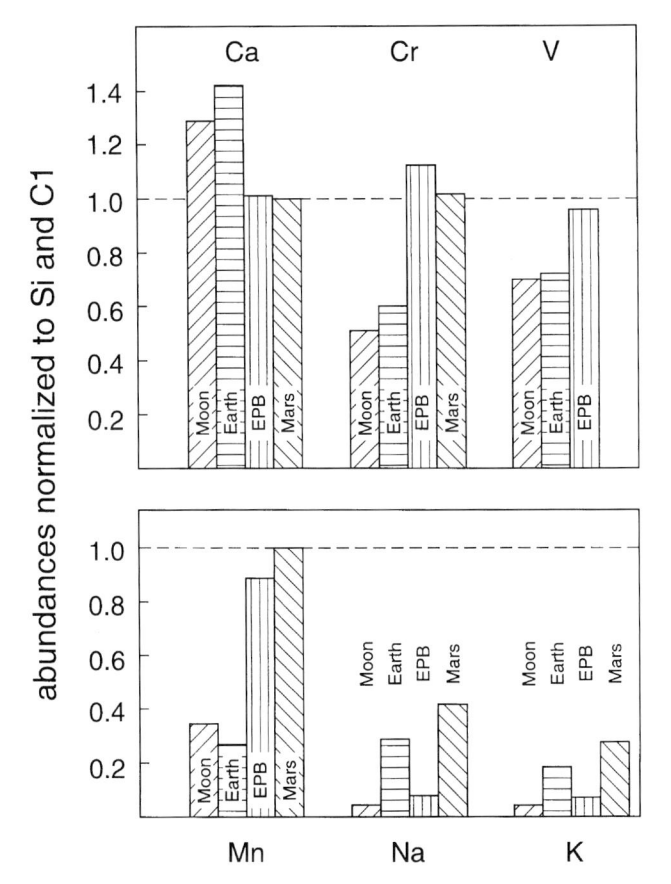

Figure 11 Abundances of Mn, Cr, and V on the Moon, Earth, EPB, and Mars, compared to the abundances of the refractory element Ca and the moderately volatile elements Na and K. The similar depletion of Mn, Cr, and V on the Moon and Earth is striking, while both Mars and EPB show no depletion of these elements. For explanation see text. Na and K show higher depletions for the Moon as compared to the Earth.

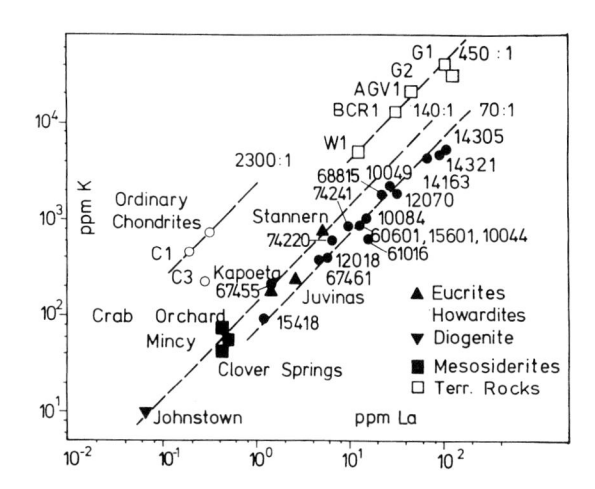

Figure 12 K v. La in lunar, terrestrial, and meteoritic samples. The ratio of K and La can be used as fingerprint of the object from which the samples are derived. The depletion of K is highest for the Moon. All data from the Mainz laboratory. (After Wänke et al. 1973, reproduced with permission.)

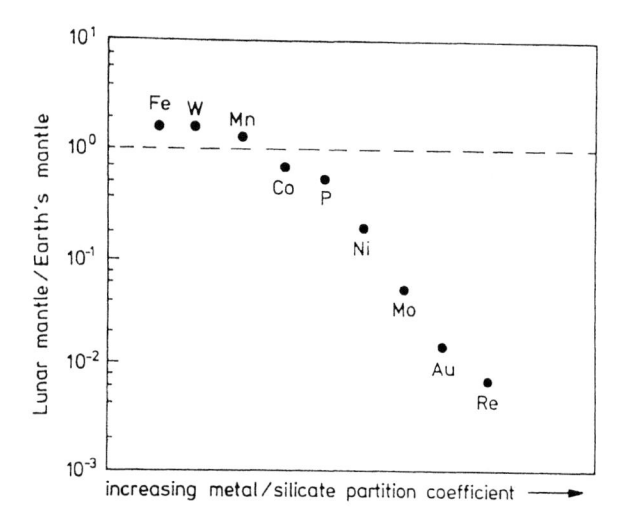

Figure 13 Abundances of siderophile elements in the lunar mantle relative to those in the Earth's mantle. Note the increasing depletion of elements in the lunar mantle with increasing siderophilicity. (After Wänke and Dreibus 1986, reproduced with permission.)

silicate matter out of which the Moon formed to have contained similar concentrations of these elements. However, we have to keep in mind that even small amounts of metal, indigenous or of meteoritic origin or formed by *in situ* reduction, would, after segregation, considerably reduce especially the amount of Ni in the silicates. Crystallization of olivine would further reduce the Ni content of the magma due to the large olivine–liquid partition coefficient of Ni. Indeed the amount of Ni is much lower in mare basalts, which generally contain less than 100 ppm Ni. Co will be significantly less affected by both processes. Both lunar mare basalts as well as pristine lunar non-mare basalts indeed contain up to 50 ppm Co. Wänke *et al.* (1978) put together evidence that led to their conclusion that large fractions of Ni and Co in lunar highland breccias are non-meteoritic in origin. Today, Ni and the other siderophile elements in the highland breccias reside almost exclusively in small metal grains. However, this does not exclude the possibility that a considerable portion of them were originally present in oxidized form in the silicates. It is easy to show that W derived originally from the KREEP component is now primarily concentrated in the metal grains of the highland breccias. As was also emphasized by Taylor *et al.* (1976), the present composition of metal grains in these samples only reflect the conditions of the last stage of equilibrium. Any record of previous events is erased.

Using different arguments, Delano and Ringwood (1978) also favored considerable amounts of Ni and Co and other siderophile elements in lunar highland rocks to be of lunar origin. This point of view was strongly opposed by Anders (1978), who advocated an almost exclusively meteoritic origin of these elements. The issue of siderophile elements in the lunar highlands concentrated on the question as to whether the meteoritic fraction of these elements should be expected in chondritic abundance ratios or not. Mare soil samples, for example, have Au/Ir ratios close to those of C1 chondrites, while highland soils and breccias have much higher Au/Ir ratios, sometimes reaching five times the C1 ratio. However, Au/Ir ratios lower than chondritic abundance are also observed. Wänke *et al.* (1978) suggested that the meteoritic component of siderophile elements would always be added in C1 abundance ratios. In the environment to which the lunar highland samples were exposed, mobilization of the less refractory elements occurred. Similarly to W moving from silicate host phases to metal grains, the more volatile siderophile elements like Ge or Au would move from metal to metal ending up where the temperatures were lowest. This idea was heavily criticized by Anders (1978).

In order to clarify this issue a series of laboratory experiments on the mobilization of Au and other siderophile elements at elevated temperatures were carried out at Mainz (Dreibus *et al.* 1981). Lunar highland soils and breccias, after neutron activation, were put into evacuated quartz tubes together with thin Ni respectively Fe foils, which were kept about 10 mm apart from the irradiated samples. After evacuation the samples were heated to temperatures between 400 and 1200°C for 1–50 hours. The oxygen fugacity was kept fixed by the FeO–Fe respectively FeO–Ni buffer. At the end of each experiment the metal foil was removed from the quartz tube, carefully cleaned of adhering dust, and analyzed by gamma-ray spectroscopy. After 50 hours at a temperature of 1000°C about 25% of Au and W activity was found in the metal foils accompanied by 92% Cu and 78% Zn and even 21% Fe and 12% Co; only Ir did not move but stayed in the lunar soil samples. Na and K were never observed in the metal foils; however, large amounts of these two elements entered the walls of the quartz tubes at temperatures above 800°C.

Dreibus *et al.* (1981) concluded from these experiments: "Even at temperatures below 800°C non-refractory siderophile elements can be redistributed in the layers of the lunar highlands." W is probably mobilized as an oxide and does not behave as a refractory element under these conditions. Under isothermal conditions mobilized siderophiles will only be trapped by metal phases. As most of the metal is of meteoritic origin, the amount of metal originally present in any sample is related to its Ir content. Hence, Au will always remain loosely correlated with Ir. Thus, it is the siderophile character of Au and Ir that keeps the two elements together and not a Maxwellian demon as inferred by Anders (1978). Since all but refractory metals can be redistributed at fairly low temperatures, it may be sometimes difficult to define a meteoritic or intrinsic component even for Co and Ni.

"Pristine" non-mare rocks – that is rocks with no or only a small meteoritic component – as recognized and thoroughly analyzed by Warren and Wasson (1978) represent the key for understanding siderophile elements in lunar highland rocks. According to Delano and Ringwood (1978) most pristine rocks come from impact melts, although there is no general consensus about their genesis. However, whatever their origin may be the pristine highland rocks crystallized from a melt, and, hence, must have been in chemical equilibrium with other phases present in the upper layers of the Moon. By definition, concentrations of strongly siderophile elements like Ir are very low in pristine highland rocks. High Co and low Ni concentrations in non-pristine highland rocks indicate equilibration with small amounts of metal.

An excellent correlation of Co with (MgO + FeO) was observed (Wänke and Dreibus 1986) for terrestrial rocks and pristine lunar highland rocks. The latter are depleted in Co by only a factor of 1.7 relative to terrestrial samples (Figure 14). With the assumption of chondritic Ni/Ir ratios in the meteoritic component, Ni concentrations in all samples of Figure 15 were corrected for the meteoritic contribution. As shown in Figure 15, the Ni/Mg ratio varies in terrestrial samples. However the data points follow quite closely a trend line that is mainly due to fractionation of olivine. As in the case of Co, it was found that many Apollo 16 breccias plot along the line representing the Ni/Mg ratio of the Earth's mantle. The Ni/Mg ratios of lunar mare basalts are about a factor of four lower than those of terrestrial samples with identical Mg concentrations. In accordance with these results of Wänke and Dreibus (1986), depletion factors of 1.2 and 4.0 have been obtained, respectively, for Co and Ni from investigations of lunar volcanic glasses (Delano 1985). The relatively high Ni/Mg ratios of lunar mare basalts are noteworthy in this respect.

Estimates of the average composition of the surface layers in the lunar highlands by Taylor (1982) are given in Table 3. The most recent space mission to the Moon, that of the Lunar Prospector, showed that the KREEP-rich material is concentrated around Mare Imbrium and the south pole – Aitken Basin (Binder 1998). In contrast, the highlands have relatively low and uniform KREEP abundances. Hence, it is likely that Taylor's estimate based on the KREEP-rich samples around the Imbrium Basin is not really representative of the whole Moon. Palme *et al.* (1991) have pointed out that lunar meteorites of highland origin are chemically different from the bulk of the Apollo 16 highland samples in having higher Fe/Mg ratios and lower contents and less fractionated patterns of incompatible and siderophile elements. The seven lunar highland meteorites represent probably four different events that delivered material from the lunar highlands to Earth. The chemical composition of these meteorites is to a first approximation very similar. Their mean average composition is also listed in Table 3; it may well give a more reliable data set on the average lunar highland composition.

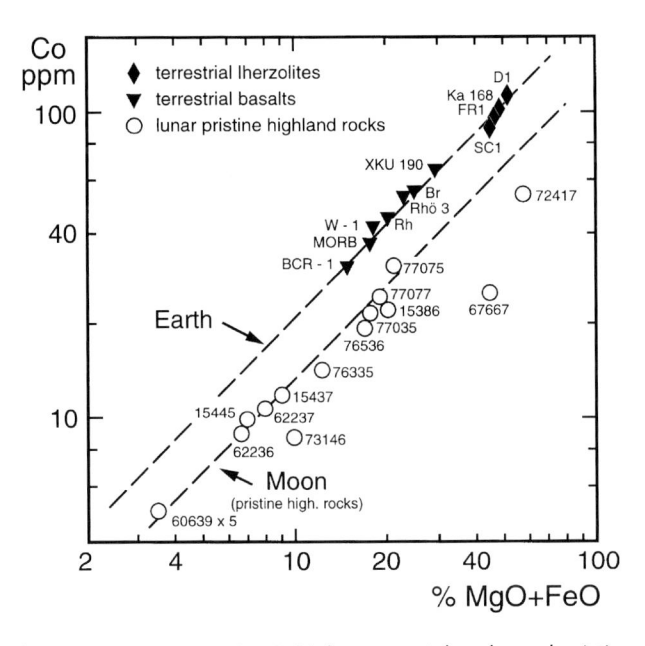

Figure 14 Co v. (MgO+FeO) for terrestrial rocks and pristine lunar highland rocks. As outlined in the text, Co is overabundant in the Earth in respect to its metal–silicate partition coefficient. Pristine highland rocks are depleted in Co only by a factor of 1.7.

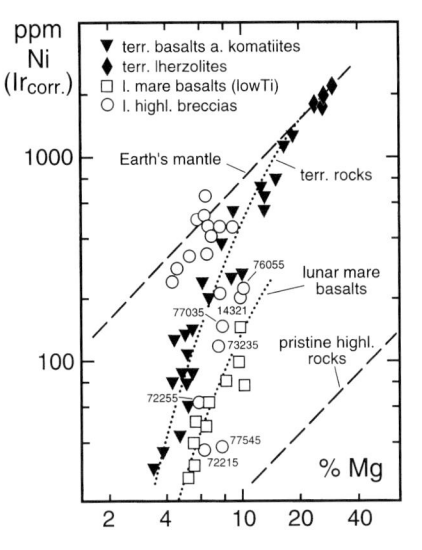

Figure 15 Ni v. Mg. The Ni concentrations have been corrected for the meteoritic contribution using the highly siderophile element Ir as a monitor. The relatively high Ni concentrations in lunar mare basalts are especially noteworthy. The dotted lines represent least squares fits of data points of terrestrial basalts and lunar mare basalts. The lower dashed line is a least squares fitted linear correlation line obtained for pristine lunar highland rocks. The upper dashed line indicates the Ni/Mg ratio of the Earth as obtained from most primitive mantle nodules.

Table 3 Average lunar highland composition (Palme et al. 1991)

	Average highland* (%)	Lunar meteorites† (%)
Mg	4.1	3.7
Al	13.0	13.8
Ca	11.3	11.0
Ti	0.34	0.16
Fe	5.13	4.07
	(ppm)	(ppm)
Na	3339	2484
P		103
K	600	203
Sc	10	11
V	24	30
Cr	680	843
Mn		574
Co	15	18
Ni	100	152
Zn		20
Ga		3.0
Br		0.07
Rb	1.7	0.69
Sr	120	138
Y	13.4	8.5
Zr	63	29
Nb	4.5	2.1
Sn	4.5	0.40
Cs	0.07	0.02
Ba	66	28
La	5.3	1.98
Ce	12	5.48
Nd	7.4	3.08
Sm	2.0	0.99
Eu	1.0	0.75
Gd	2.3	0.72
Tb	0.41	0.22
Dy	2.6	1.46
Ho	0.53	0.33
Yb	1.4	0.92
Lu	0.21	0.13
Hf	1.4	0.73
Ta		0.11
Ir		0.006
Au		0.002
Pb		0.62
Th	0.9	0.24
U	0.24	0.077

* Average highland composition (Taylor 1982).
† Average of five lunar meteorites.

The question of polar ice deposits

Deposits of frozen water ice inside permanently shaded craters at both poles of the Moon have been postulated (e.g. Arnold 1979). The presence of such deposits has been widely discussed in recent years in connection with their use for a lunar base planned to be built in the future. The search for water ice was the prime objective of the Lunar Prospector mission, which with its neutron spectrometer indeed observed significant reductions of the epithermal and thermal neutron fluxes above the lunar poles (Feldman et al. 1998). Feldman et al. (2000) inferred from these reductions that both poles contain about the same amount of total hydrogen. Unfortunately the data do not allow one to distinguish between atomic hydrogen and water ice. The concentration of atomic hydrogen amounts to about 50 ppm for the equatorial regions as known from returned lunar sample analyses and assigned almost exclusively to the implantations of the solar wind hydrogen. The observed reductions of the neutron fluxes above the polar areas point to an excess of about 100 ppm or to hydrogen concentrations of about a factor of three higher in the polar areas relative to the equatorial areas. In form of water ice, this would correspond to 880 ppm of H_2O in the polar regolith. With the assumption that the water ice at the south pole is restricted to the areas in permanent shade, Feldman et al. (2000) estimate a concentration of $1.5 \pm 0.8\%$ water. The maximum amount of water ice estimated from the neutron data of Lunar Prospector corresponds to 2.05×10^9 metric tons of H_2O at both poles of the Moon. This inventory is small compared to that estimated as possibly delivered and subsequently retained on the Moon from cometary and meteoritic impacts, and by accumulation of H_2O from interplanetary dust particles. The total inventory from these sources is estimated to be 10 to 100×10^9 metric tons (Arnold 1979). In any case, if the predictions of the amount of H_2O delivered to the Moon are correct most of the water has been lost to space.

Let me add a personal note. Threefold higher concentrations of hydrogen derived from solar wind implantation at the polar regions could easily be expected because of the reduced diffusion loss to space due to the lower temperatures near the poles compared to that at equatorial regions. At the beginning of the section on "The Apollo missions", I referred to observations of Hintenberger et al. (1970) of a substantial loss of solar wind implanted helium from lunar fines. Interestingly, this loss amounts to at least a factor of three. Hence, I have severe doubts on the reality of ice deposits at the lunar south pole. A number of other scientists have also expressed their doubts.

MARS

Viking 1 and 2

The first information on the chemistry of the Martian surface was obtained by the Viking mission from the two landers: Viking 1, which landed on 20 July 1976 at Chryse

Planetia and Viking 2, which landed on 2 September 1976 at Utopia Planetia. The landers carried, apart from a number of rather sophisticated instruments to search for biological activity, a special miniaturized energy-dispersive x-ray fluorescence spectrometer (XRFS), using two radioisotope sources, ^{55}Fe and ^{109}Cd, for exitation and proportional counters for analysis. Soil samples were collected with a scoop mounted on a steerable arm of about 3 m in length and transferred to the analysis chamber which was filled with about 25 cm^3 of soil. The construction of the sampling device allowed the passage of dust and pebbles of up to about 1 cm in diameter to the analysis chamber. It was possible to dump the sample from the chamber after analysis and fill it with new material. At Chryse Planetia, 13 samples were acquired and analyzed and 8 at Utopia. For the results of all 21 samples analyzed see Clark et al. (1982). Mean values of the analytical results are given in Table 4. In some cases soil samples were taken from beneath rocks after they had been pushed aside. In other cases samples were taken from holes dug by the sampling arm. These samples were called "protected samples". As can be seen from Table 4, there are no noticeable differences between protected samples and surface samples and no real differences between the two landing sites. The latter is especially remarkable as the two locations were about 6500 km apart. The pebbles acquired turned out to be duricrust not rock, that is plate-shaped fragments (clods) about 0.5 cm thick, lying in and on the regolith surface. At Chryse the clods had the same composition as fines except for somewhat higher content of sulfur. Clark et al. (1982) suggested that this might reflect the presence of cementing salts like MgSO$_4$. The low bulk density of the clods of only 1.2 g cm^{-3} also indicates that they consist of cemented fines. At the Utopia site attempts to sample pebbles or clods failed, apparently because of weaker cementation that caused break-up during sampling.

The results of the chemical analyses by Viking landers 1 and 2 did not lead to a unique interpretation for the origin of the fines. However, as stated by Clark et al. (1982), a source material of predominantly mafic to ultramafic nature is almost certainly required.

The high concentrations of sulfur and chlorine are especially noteworthy as terrestrial rocks contain only about 0.2% sulfur at maximum. The absence of detectable sulfur-containing volatiles in the gas chromatograph–mass spectrometer (GCMS) experiments on board the Viking landers, during which the samples were heated up to 500°C points towards sulfates as the most probable sulfur-containing species in the Martian fines. Sulfur and chlorine most likely originate as volcanic exhalation products in the form of SO$_2$, Cl$_2$, or HCl, as was proposed by Clark and Baird (1979). Alternatively they could have been supplied as efflorescence salts from leachates percolating through the upper layers of the Martian crust.

Martian meteorites and the bulk composition of Mars

Wasson and Wetherill (1979) and independently Nyquist et al. (1979) proposed that a small number of meteorites linked to each other by distinct trace element and oxygen isotope ratios might in fact represent rocks from the surface of Mars. It was thought that large impacts on Mars accelerated part of the ejecta material to velocities exceeding the escape velocity of this planet (around 5 km s^{-1}), delivering it into ecliptic orbits around the Sun. After a relatively short transfer time these rocks might cross the Earth's orbit and fall eventually as meteorites. These meteorites were called SNC meteorites relating to the known three subgroups

Table 4 Average compositions (wt%) of Mars soil classified by sample types. (After Banin et al. 1992; for the original data of all samples analyzed see Clark et al. 1982.)

Sample type	Protected fines		Surface fines		Clods
Samples	Chryse	Utopia	Chryse	Utopia	Chryse
Averaged	C-6, -11	U-2, -4, -6, -7	C-1, -7, -8	U-1, -3, -5	C-2, -5, -13
SiO$_2$	44	43	43	43	42
Al$_2$O$_3$	7.3	(7)*	7.3	(7)*	7
Fe$_2$O$_3$	17.5	17.3	18.5	17.8	17.6
MgO	6	(6)*	6	(6)*	7
CaO	5.7	5.7	5.9	5.7	5.5
K$_2$O	<0.15	<0.15	<0.15	<0.15	<0.15
TiO$_2$	0.62	0.54	0.66	0.56	0.59
SO$_3$	6.7	7.9	6.6	8.1	9.2
Cl	0.8	0.4	0.7	0.5	0.8

* Not measured because of instrument difficulties; assumed to be the same as in Chryse samples

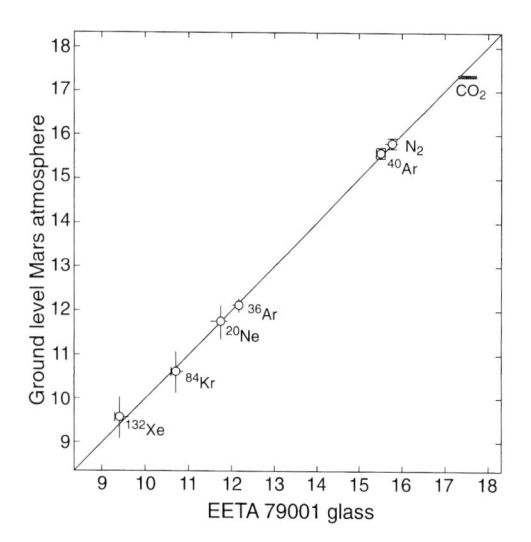

Figure 16 Comparison of the relative abundances of various gases trapped in EETA 79001 glass with samples of the Martian atmosphere taken by the Viking spacecraft. The number of particles per cubic centimeter is shown on a log–log plot. (After Pepin 1985.)

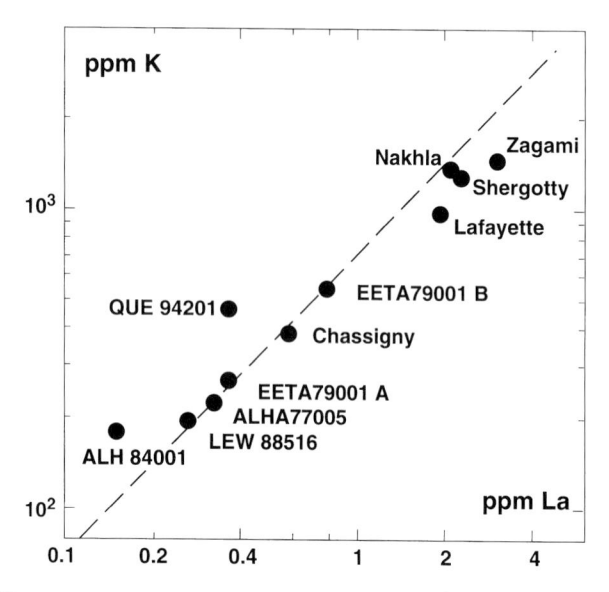

Figure 17 K v. La in Martian meteorites. The correlation of the moderately volatile element K relative to the refractory element La allows one to estimate the abundance of K on Mars. All data from the Mainz laboratory.

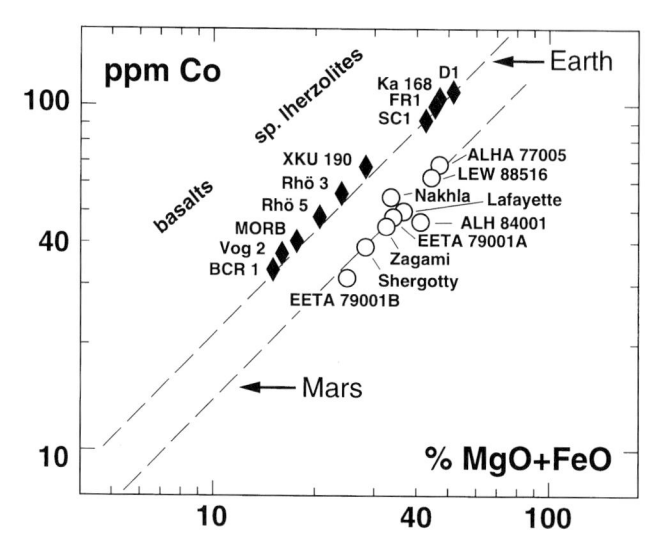

Figure 18 Co v. (MgO + FeO) in terrestrial rocks and in Martian meteorites.

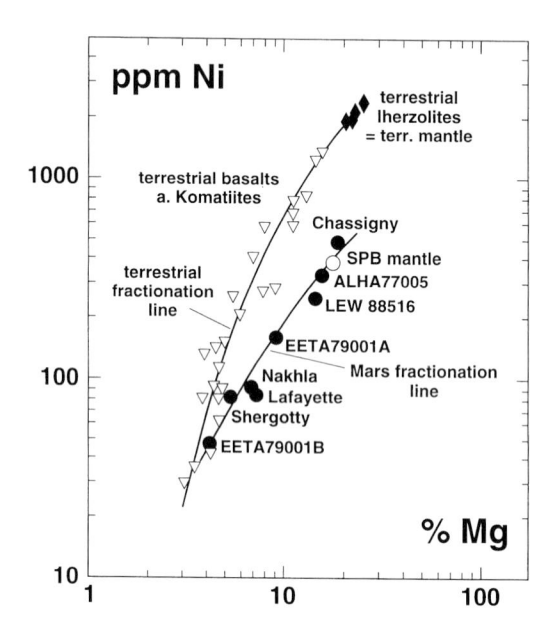

Figure 19 Ni v. Mg in terrestrial rocks and in Martian meteorites.

shergottites, nakhlites and chassigny. Their young crystallization ages of less than 1.3 Gyr and their fractionated REE patterns seemed to exclude an asteroidal origin. Strong additional evidence came from the discovery of a trapped rare gas and nitrogen component in shock glasses of shergottite EETA 79001, with element and isotope ratios very different from those observed in any other meteorites, but closely matching the highly characteristic ratios in the Martian atmosphere; for example, $^{40}Ar/^{36}Ar$, $^{14}N/^{15}N$, $^{14}N/^{40}Ar$, and $^{129}Xe/^{132}Xe$ (Bogard and Johnson 1983, Becker and Pepin 1984, Pepin 1985; Figure 16). Whereas all other differentiated meteorites have crystallization ages close to 4.5 Gyr, Jagoutz and Wänke (1986), for example, determined a Sm–Nd age for Shergotty of only 360 Myr.

Using various element correlations (Figures 17–19) as well as cosmochemical constraints, it was possible to make quite reliable estimates of the bulk composition of Mars, assuming Mars is indeed the parent body of the SNC meteorites (Dreibus and Wänke 1987). The results are given

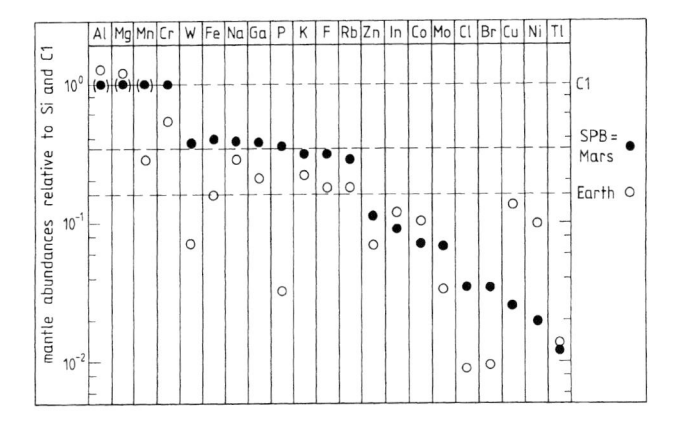

Figure 20 Element abundances in the mantles of the Earth and Mars normalized to Si and C1. Notice the depletion of the chalcophile elements Cu, Co, and Ni in the Martian mantle, which led Dreibus and Wänke (1987) to favour a homogeneous accretion of Mars. (After Wänke and Dreibus 1994, reproduced with permission.)

Table 5 Bulk composition (%) of the Earth and Mars according to Dreibus and Wänke (1987)

	Earth	Mars
Mantle + crust		
Na_2O	0.39	0.50
MgO	36.8	30.2
Al_2O_3	4.20	3.02
SiO_2	46.0	44.4
P_2O_5	0.015	0.16
K_2O	0.028	0.037
CaO	3.54	2.45
TiO_2	0.23	0.14
Cr_2O_3	0.44	0.76
MnO	0.13	0.46
FeO	7.58	17.9
Core		
Fe	78.6	78.1
Ni	4.9	7.8
Si	13.2	
S	1.6	14.0
Co, Mn, Cr	1.7	
Core mass	32.4	21.0

in Table 5 and Figure 20. As the SNC meteorites were thought to be magmatic fractionation products of the Martian mantle, the connection to the mantle is rather straightforward. To estimate the composition and mass of the Martian core, Dreibus and Wänke (1987) assumed that the whole planet has a Fe/Si ratio equal to that in C1 (carbonaceous chondrites type 1) or, in other words, abundances of Fe and Ni equal to 1.00 relative to Si and a S abundance of 0.35 (mean abundance of elements with moderately volatile character like S). It was further assumed that the portion of Fe not contained as FeO in the mantle as well as all Ni and S to be present in the elemental state in the core. In this way a core mass of about 21% was obtained.

The composition of the Martian mantle as obtained from SNC meteorite data listed in Table 5, seems rather reliable and is generally accepted. Compared to the Earth's mantle the Martian mantle contains more than twice the amount of FeO as also indicated by the high FeO abundance in the Martian soil as measured by the Viking landers. The higher abundance of moderately volatile elements like K is noteworthy. Furthermore the much higher concentration of P as reflected by similar high values in SNC meteorites compared to terrestrial basalts is of importance. On Earth, P is considerably depleted in the mantle, but the reason for this depletion is not yet fully understood. The depletion of Mn, Cr, and V in the Earth's mantle is discussed in the section on "The Moon". Contrary to the terrestrial situation, Mn and Cr are not depleted in the Martian mantle (Figure 11).

From Figure 20, one notices that the siderophile elements Co, Ni, and Cu, which exist in high abundances in the Earth's mantle, are much less abundant in the Martian mantle. To explain the observed elemental abundance patterns of the Earth's mantle the following two-component model has been proposed (Ringwood 1979, Wänke 1981):

- Component A: highly reduced and free of all elements with equal or higher volatility than Na, but containing all other elements in C1 abundance ratios. Fe and all siderophile elements are in metallic form, and even part of Si as metal.
- Component B: oxidized and containing all elements, including the volatiles, in C1 abundances. Fe and all siderophile and lithophile elements are present mainly as oxides.

For the Earth a mixing ratio of component A to component B of 85:15 was deduced as well as an inhomogeneous accretion scenario (Wänke 1981). According to this model the accretion of the Earth started with the highly reduced component A. It was thought that component A formed inside the present orbit of Mars where temperatures in the primordial solar nebula were high enough to reduce the matter and simultaneously to prevent condensation of volatile and moderately volatile elements. Segregation of the metallic FeNi phase, that is core formation, was supposed to be almost contemporaneous with accretion, leading to an FeO-free mantle during this stage. The high present-day mantle abundances of siderophile elements like Ni and Co, which in the case of Ni exceeds by far the amount to be expected from the metal–silicate partition coefficients, exclude equilibration with a pure FeNi phase. This is the reason for the conclusion in this model that the oxidized component B, which supplied FeO, NiO, CoO, and so on to the mantle, was added in substantial amounts only after the Earth had reached about two-thirds of its present mass. In this inhomogeneous accretion model the major portion of the matter from component B was never in equilibrium with larger amounts of metallic FeNi. Component B was thought to have formed outside the present orbit of Mars where water condensed as ice or formed hydrates. Due to perturbations by the growing Jupiter, the matter between about 1.5 and 4 AU was pushed inwards as well as outwards leading to the present mass distribution minimum in the asteroidal belt and adding part of this matter to the Earth during its final stage of accretion. Mars, located near the formation boundary of component A and component B, was fed from both regions almost simultaneously. However, this boundary is by no means meant to represent a sudden compositional change but a gradual transition with components A and B as endmembers.

The obvious depletion of all chalcophile elements in the Martian mantle as seen in Figure 20 was the reason that Dreibus and Wänke (1987) favored a homogeneous accretion for Mars with a mixing ratio of component A to component B of 60:40. It was assumed that sulfur supplied mainly by component B was responsible for the formation of sulfur-rich FeNi alloys, leading to sulfide–silicate equilibrium. During core formation extraction of elements from the mantle took place according to their sulfide–silicate partition coefficients.

The homogeneous accretion as inferred for Mars had another important consequence. Water being added in considerable amounts to the growing planet from component B reacted with the metallic iron from component A. The reaction $Fe + H_2O \rightarrow FeO + H_2$ generated huge amounts of H_2 which in a hydrodynamic escape mechanism carried large quantities of heavier species with it. The result was a planet with a FeO-rich mantle and, compared to the Earth, a smaller core.

Dreibus and Wänke (1987) estimated that a total of about 3.4% water was added to Mars during the main stage of accretion, but only about 36 ppm remained in the mantle. Making the highly unrealistic assumption of 100% release, this would yield a water layer of 130 m deep covering the whole planet. Thus, the Martian interior is very dry, as for the Earth's mantle about 1000 ppm water is estimated.

However, considerable amounts of water are required to explain the erosion features on the Martian surface. The water was obviously added to the crust in the last accretion stage of the planet. Hence, Mars was once wet on the outside but stayed dry in its interior. Today the Earth is both wet on the outside and in its interior However, it might well be that the Earth was also dry inside after accretion and the water was only brought to its interior after the onset of plate tectonics due to the subduction of oceanic plates. Plate tectonics and subduction of surface material seem to be absent on Mars.

Mars Pathfinder

New and important information about the chemistry of Mars was obtained by the very successful Mars Pathfinder mission. The lander, which represented the first return to the surface of Mars 21 years after Viking, carried, apart from a camera and a number of meteorological instruments, a small rover named Sojourner. This vehicle carried, in turn, an instrument for the chemical analysis of soil and rocks. This was an **a**lpha-**p**roton **x**-ray **s**pectrometer (APXS) designed and built at the Max-Planck-Institut for Chemistry, Mainz, Germany, especially for that purpose. The technique used allowed the chemical analysis of soils and rocks without any sample preparation. All what was required was to put the sensor head of the spectrometer against the sample to be analyzed, which was then exposed to irradiation of alpha particles from curium sources. In fact, the APXS was added to the Sojourner payload rather late so that it could not be mounted at the front of the vehicle where the camera was located, but was mounted to the rear.

The APXS sensor head was known as "Sojourner's nose". But contrary to normal living creatures with eyes and nose mounted in the same direction, this was not so in the case of Sojourner. Hence, when the project lead, using the stereo camera of the lander, had selected a certain rock of interest, Sojourner was directed toward this rock for a close-up image. In order to make a chemical analysis of the rock in question, the rover then had to reverse for a certain distance, turn around by 180° and move forward again. As one can imagine this was not an easy task for the rover team. As mentioned above, the Pathfinder mission was a big success, and the rover stayed active for 83 Mars days, much longer than planned. All the instruments worked to perfection. The APXS worked without a noticeable gain shift and returned high-quality spectra of five rocks and of five soil locations. In addition several more analyses were performed which for various reasons were of poorer quality.

Pathfinder was launched on 4 December 1996 from Cape Canaveral, Florida. After two delays of 24 hours each, the Delta rocket roared into the night sky. From the standpoint I had there watching the launch it actually looked as if the rocket was going directly to Mars which was clearly visible in the night sky. I kept my fingers crossed and wished success for the mission. The APXS was originally designed and built for the 1996 Russian Mars mission which just three weeks earlier had failed during launch. The counterparts of the APXS just on its way to Mars had come down in the eastern Pacific.

Originally, we had proposed our APXS for a planned ESA–NASA mission called "Intermarsnet". In that way, the technical performance of the APXS became known to American scientists. I remember that Steve Squyres got very exited when I presented the APXS at a joint

ESA–NASA Science Team meeting in Noordwijk, The Netherlands. To my regret, Intermarsnet never materialized, but I was asked by representatives of JPL (Jet Propulsion Laboratory, Pasadena, USA) if I would be willing to provide our APXS for NASA's Pathfinder mission, specifically to be mounted on the rover. I was more than happy as, according to the planning with the fixed small stations of the Russian mission, we would only have been able to analyze the soil at the two landing sites of these landers. With the Pathfinder rover, it would be possible to bring the APXS sensor head in contact with rocks on the Martian surface. A smooth and very satisfactory collaboration began between the APXS team with the Principle Investigator Rudolf Rieder from my department and the JPL engineers. The JPL engineers designed a simple but ingenious mechanism that allowed the APXS sensor head to be put against rock surfaces and also to turn it downwards to the soil.

After a seven month flight, Pathfinder entered the Martian atmosphere on 4 July 1997 directly after cruise-stage separation. The approach velocity of $7.5\,km\,s^{-1}$ was much reduced by the entry shield, before parachute deployment. At an altitude of 1.6 km the lander obtained radar ground contact. After inflation of the air bags, small landing rockets were fired and the lander surrounded by air bags hit the ground at a velocity of $14\,m\,s^{-1}$. The lander bounced up and down at least 15 times without air bag rupture demonstrating the robustness of this landing system (Golombek et al. 1997).

A full panorama of the surface was obtained at the end of the first day on Mars. On the next day the rover was driven down the rear ramp. After switch-on of the APXS its sensor head was directed into the atmosphere. The alpha backscatter spectrum of the Martian atmosphere (Figure 21) did not bring any new information, but the fact that it exactly matched laboratory CO_2 spectra was the first sign that also our APXS had survived and was in good health.

In the 83 Mars days until radio contact was lost, Sojourner drove altogether 104 m around the lander, staying within 12 m of it, and put the sensor head of the APXS to nine rocks and down to the soil at seven locations (Figure 22).

The chemical composition of the soil turned out to be again very similar to that at the landing sites of Viking 1 and 2 (Figure 23). The only surprise was the amount of potassium. While Clark et al. (1982) reported an upper limit of 0.15% K_2O for both Viking sites, the APXS found about 0.5% K_2O at Ares Vallis (Rieder et al. 1997), which is only 1500 km from the Viking 1 site. If Clark et al. (1982) have not overestimated their sensitivity considerably one would have to invoke a special process which adds potassium-rich material to Ares Vallis from the southern highlands.

In this respect, it is important to note that Surkov et al. (1989), based on data from the Phobos 2 mission, reported a mean concentration of the Martian surface of 0.3% K. I have to confess that up to the Pathfinder analyses, I did not believe

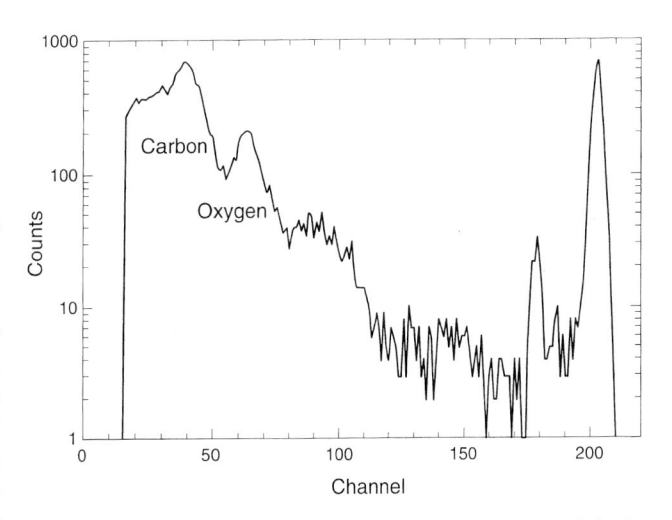

Figure 21 The first spectrum returned from the Pathfinder lander: an alpha backscatter spectrum of the Martian atmosphere.

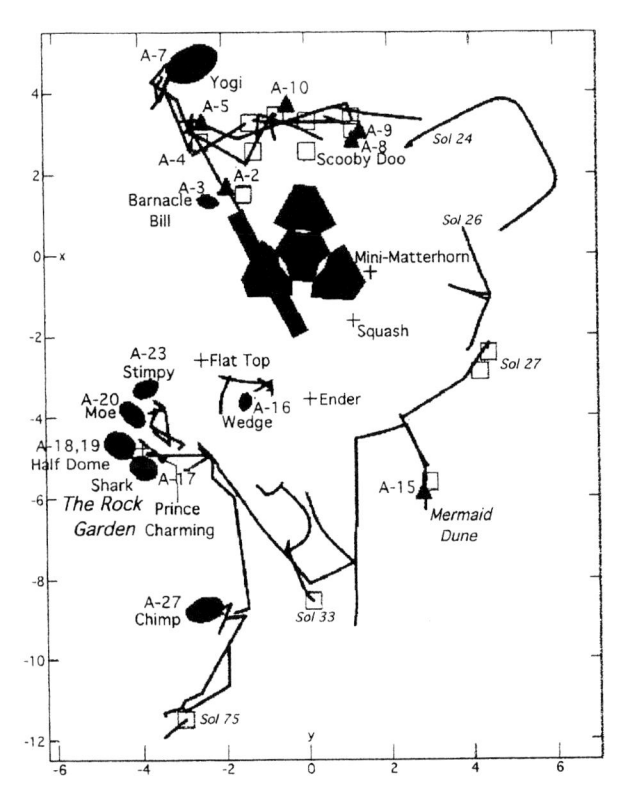

Figure 22 Map showing rover traverse (black lines). The gaps are due to dead reckoning errors that built up during each drive. (After Golombek et al. 1999, reproduced with permission.)

these data, which were obtained by gamma-ray spectrometry from orbit, because the Martian meteorites contain only up to 0.15% K: but obviously I was wrong. Surkov et al. (1989) have not only reported high K concentrations but also high

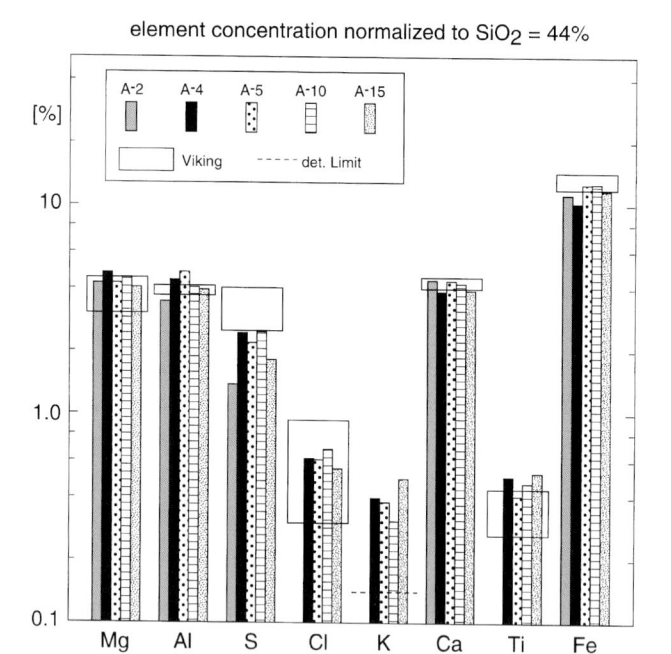

Figure 23 Chemical composition of five soil samples analyzed by the APXS mounted on the Sojourner rover of the Pathfinder mission. The range of the respective concentrations obtained by both Viking landers is also shown.

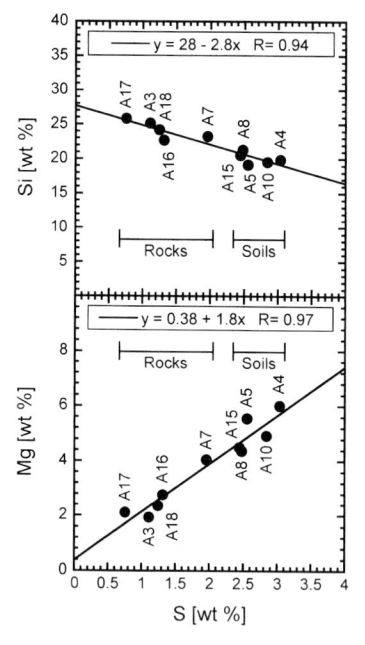

Figure 24 Concentrations of Si and Mg v. S in soils and rocks as measured on the Martian surface by the APXS.

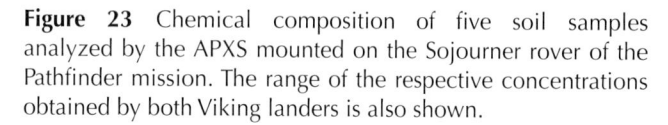

values for U and Th; relative to the concentrations in Martian meteorites the differences are even higher. This is clearly evident looking at the K/U ratio of 6000 (Surkov et al. 1989) and of the respective ratio of 18000 in Martian meteorites.

U, Th, and practically all LIL elements in Martian meteorites reside to a large extent in apatite, which dissolves easily in slightly acidic aqueous solutions. Hence, if the low values for the K/U or K/Th ratios of Surkov et al. (1989) are correct, one might speculate that water percolating to the upper crust extracted U and Th, which after evaporation of water were incorporated into the soil.

The biggest surprise of the Pathfinder APXS data was the observation that the composition of the rocks embedded in the soil at the Pathfinder site differed greatly from the soil composition. As all the rocks analyzed were covered to a variable extent by soil, this difference was not so clear at first. The APXS, due to the short range of the alpha particles, could not distinguish between bare rock and adherent fines. The fines had SO_3 contents of about 6%, while the SO_3 concentration in the rocks varied between less than 1% and about 4%. Such high values exceed by far what can be expected in igneous rocks and would even be unusual in sedimentary rocks. The fact that good linear correlations both positive and negative, were observed for many elements when plotted relative to the concentration of sulfur (Figure 24), led Rieder et al. (1997) to propose that all rocks analyzed have nearly identical compositions but the amount of soil adhering to the surface of the rock is variable. This adhered reddish soil on the surface of the gray rocks is clearly visible on the Pathfinder images. By extrapolation to zero sulfur the composition of a soil-free rock could be calculated (Table 6). The normative mineral composition of the soil-free rocks calculated from the chemical composition showed a dominance of feldspar, orthopyroxenes, and quartz with minor amounts of Fe–Ti oxides.

In the course of extensive and careful calibration work it was also possible to obtain reliable data on the abundance of P in the samples at the Pathfinder site. This was a difficult task due to the limited energy resolution of the x-ray detector of the APXS. Almost identical concentrations of P for both soil and rock samples (Table 6) were found which at a first glance might indicate a sedimentary origin of the rocks which contained more than twice as much K compared to the soil and also considerably higher SiO_2. Hence, one should also expect the rocks to have a higher abundance of P, which turned out not to be the case. The observed P concentrations of about 0.5% (Dreibus et al. 2000) are very high compared to concentrations in terrestrial rocks but they are consistent with the high P abundance derived from analyses of SNC meteorites and predicted for the Martian mantle by Dreibus and Wänke (1987).

According to the chemical classification of terrestrial volcanic rocks, the soil-free Pathfinder rocks fall in the range of terrestrial andesites or icelandites, while the shergottite meteorites are in the basaltic range (Table 6).

Table 6 Corrected APXS data (%) of Mars Pathfinder rocks and soil (Brückner et al. 2001) based on the analyses of Rieder et al. (1997) compared with other rocks

Element	Martian crust= soil average	Martian rock Shark A-17	"Soil-free" rock	Basaltic Martian meteorites Shergotty	Earth continental crust	Earth andesite SSK 1
Na_2O	1.1	2.0	2.5	1.29	3.0	3.57
MgO	8.7	3.5	1.5	9.28	3.1	1.96
Al_2O_3	8.0	10.0	11.0	7.06	15.2	14.1
SiO_2	42.3	55.2	57.0	51.4	60.2	60.5
P_2O_5	0.98	0.98	0.95	0.80	0.2	0.22
SO_3	6.8	1.88	0.75	0.33	0.2	0.008
Cl	0.55	0.38	0.32	0.01	0.2	0.15
K_2O	0.61	1.14	1.36	0.16	2.9	1.13
CaO	6.5	8.8	8.1	10.0	5.5	6.12
TiO_2	1.0	0.65	0.69	0.87	0.7	1.16
Cr_2O_3	0.3	0.05	0.05	0.20	—	—
MnO	0.5	0.49	0.55	0.53	0.1	0.19
FeO	22.3	14.8	15.7	19.4	6.1	10.7

The Pathfinder rocks represent highly fractionated crustal material rich in SiO_2 and K, but low in MgO. Hence, contrary to the mafic nature of the Martian soil and the even more mafic Martian meteorites they are felsic in composition. This holds independent of the not yet solved nature of these rocks, that is, igneous or sedimentary. Figure 25 illustrates the difference between the composition of the Pathfinder rocks and the composition of the soil in which they are embedded. The large differences clearly indicate that the soil cannot be made by reduction of these rocks even if weathering and the addition of SO_2 and HCl from volcanic gases are taken into account. Addition of material richer in Mg and Fe but poorer in K as observed in Martian meteorites seems to be an unavoidable necessity.

Taking the almost identical soil composition at the three landing sites as representative for the whole surface soil of Mars, Brückner et al. (1999) were able to show that all elements fit well into a two-component mixing diagram with the Pathfinder rocks on the high K–low Mg side and the Martian meteorites on the opposite side (Figure 26). Hence, it was concluded that large geologic units of andesitic (Pathfinder rocks) as well as of basaltic (Martian meteorites) composition must exist on Mars and cover about equal areas. The recently obtained thermal emission spectrometer (TES) data of the Mars Global Surveyor (MGS) confirmed this model.

Although the sensitivity for detecting carbon by the APXS was reduced due to the high contribution from the Martian atmosphere, the upper limit obtained of 0.8% C equal to less than 5% $MgCO_3$ clearly showed that

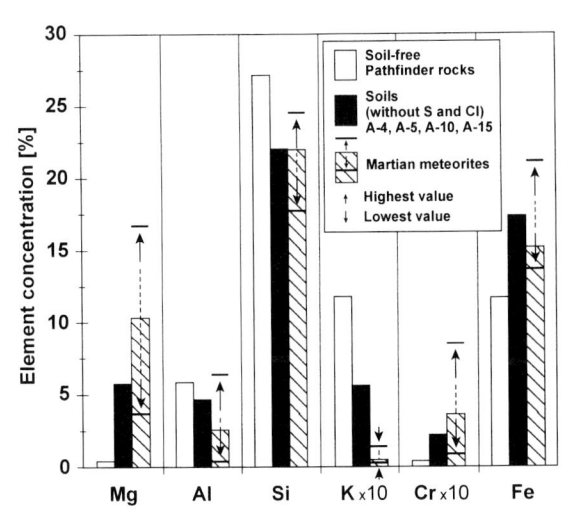

Figure 25 Chemical composition of soils and rocks at the Pathfinder landing site and the composition of Martian meteorites.

carbonates are low or absent in the Martian soil. Hence, the proposition of the presence of large amounts of carbonates at the Martian surface as the remnants of an early thick CO_2 atmosphere seems not to be justified.

Although Pathfinder was originally planned as a test for the airbag landing system and a test for a roving vehicle, this mission brought a wealth of information not only about its specific landing site but also about the global surface

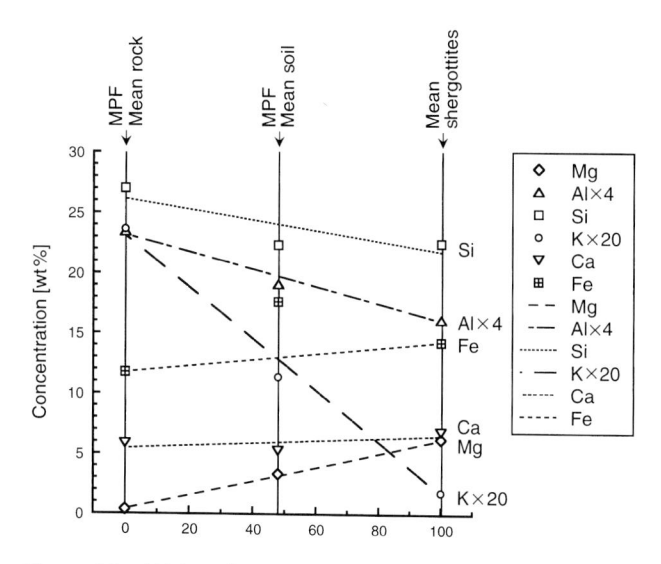

Figure 26 Mixing diagram containing the Pathfinder rocks as felsic endmember, the Martian meteorites, especially the shergottites, as basaltic endmember, and the Viking and Pathfinder soils as mixtures of these two major components. In addition, the soil contains $MgSO_4$ and Cl from evaporates. The contribution of haematite (Fe_2O_3) is clearly visible, because the soil data points for Fe fall above the line connecting the two endmembers.

Table 7 Chemical composition (%) of Venusian rocks (Surkov *et al.* 1986)

	Venera 13	Venera 14	Vega 2
MgO	11.4 ± 6.2	8.1 ± 3.3	11.5 ± 3.7
Al_2O_3	15.8 ± 3.0	17.9 ± 2.6	16.0 ± 1.8
SiO_2	45.1 ± 3.0	48.7 ± 3.7	45.6 ± 3.2
K_2O	4.0 ± 0.63	0.2 ± 0.07	0.1 ± 0.08
CaO	7.1 ± 0.96	10.3 ± 1.2	7.5 ± 0.7
TiO_2	1.59 ± 0.45	1.25 ± 0.41	0.2 ± 0.1
FeO	9.3 ± 2.2	8.8 ± 1.8	7.7 ± 1.1
SO_3	1.62 ± 1.0	0.88 ± 0.77	4.7 ± 1.5
MnO	(0.2 ± 0.1)	(0.16 ± 0.08)	(0.14 ± 0.12)

chemistry of Mars. In short these findings can be summarized in 10 points:

1. Martian soil can be considered to represent material from reduced rocks of very different composition and their reaction products with volcanic gases, especially SO_2 and HCl, mechanically mixed and distributed on a global scale by dust storms.
2. The felsic endmember rocks have high SiO_2, high K, but low Mg contents; chemically they correspond to terrestrial andesites or icelandites.
3. The mafic endmember rocks have a basaltic composition similar to shergottites, the most abundant subgroup of Martian meteorites.
4. The geologic units of andesitic and basaltic compositions cover similar portions of the Martian surface, recently confirmed by the TES data of the MGS orbiter.
5. The Martian soil contains about 10% $MgSO_4$ and about 1% chloride either in the form of NaCl or $MgCl_2$.
6. Carbonates are low or absent in the Martian soil as predicted by Wänke and Dreibus (1994), because of the dominance of sulfates.
7. Identical trace element patterns of Martian meteorites and Pathfinder rocks as well as similar high abundances of P confirm Mars to be the parent body of these meteorites.
8. The high concentrations of P and Fe in the Pathfinder rocks confirm the high abundances of these elements in the Martian mantle as predicted by Dreibus and Wänke (1987).
9. The K concentrations in the Martian soil are at 0.5% considerably higher than determined by the Viking missions, but are in line with gamma-ray data as reported by Surkov *et al.* (1989).
10. The K/U ratio reported by Surkov *et al.* (1989) of 6000 is low compared to a value of about 18000 in Martian meteorites. However, as pointed out by Dreibus *et al.* (1996), this discrepancy might be due to the fact that in the Martian environment U and Th reside mainly in apatite that dissolves easily in slightly acidified brines and thus could be redistributed within the surface layers. It is hoped that the **g**amma-**r**ay **s**pectrometer (GRS) of the 2001 orbiter will shed light on this issue.

VENUS

The chemical composition of Venusian surface material has been studied by several Russian landers at various locations on the surface of Venus.

The major element analyses listed in Table 7 were carried out by x-ray fluorescence (XRF), for which the samples from the Venusian surface obtained with a drill were brought through an air lock into the interior of the landers of the Venera 13, 14, and Vega 2 missions. The data reported by Surkov *et al.* (1986) include values for MnO. These data are in parentheses, because I consider them unreliable as they were obtained by XRF, using proportional counters that make it almost impossible to separate the weak Mn line from the much stronger line of Fe. The Venera 8, 9, 10, and the Vega 1 and 2 missions also returned data on the natural radioelements K, Th, and U obtained by gamma-ray spectroscopy (Table 8). According to the data the Venusian surface is of basaltic composition. The soils from the ancient

Table 8 Uranium, thorium, and potassium (%) in Venusian rocks (Surkov et al. 1987)

	K	U × 10^{-4}	Th × 10^{-4}
Venera 8	4.0 ± 1.2	2.2 ± 0.7	6.5 ± 0.2
Venera 9	0.47 ± 0.08	0.60 ± 0.16	3.65 ± 0.42
Venera 10	0.30 ± 0.16	0.46 ± 0.26	0.70 ± 0.34
Vega 1	0.45 ± 0.22	0.64 ± 0.47	1.5 ± 1.2
Vega 2	0.40 ± 0.20	0.68 ± 0.38	2.0 ± 1.0

upland rolling plains (Venera 8 and 13) are high in K and resemble terrestrial alkaline gabbros, while the soils of all other landing sites have low K contents and generally resemble terrestrial oceanic basalts in their composition. So far no areas have been found that would compositionally resemble terrestrial continental crust high in SiO_2.

The data from the Venusian surface rocks indicate that the mantle of Venus is probably similar in composition to that of the Earth. With the exception of the K-rich rocks from the Venera 8 site the K/U ratio is generally 30–40% less than that of terrestrial rocks. This might indicate a somewhat larger depletion of the moderately volatile K on Venus relative to the refractory elements. Such an increase of the depletion of moderately volatile elements with decreasing distance from the Sun is to be expected on theoretical grounds. For Mars a lower depletion of K relative to the Earth has been observed. The high SO_4 concentration in the Venusian soil probably reflects the presence of $CaSO_4$ formed by the reaction $CaCO_3 + SO_2 = CaSO_4 + CO$. With an abundance of about 150 ppm, SO_2 is the most abundant sulfur-containing molecule in the lower atmosphere of Venus; it exceeds by far H_2S and OCS.

The observed densities of all terrestrial planets (Mercury, Venus, Earth, and Mars) exceed a value that could be accounted for by an exclusively rocky composition. The existence of metallic cores in all planets is an unavoidable consequence. The variation of the core fractions explains the varying densities of the terrestrial planets. Thinking as a cosmochemist, I prefer to change this statement by saying: "The varying densities reflect the differences of oxidation. Oxygen is a light element and, hence, adding it makes the planet lighter."

If Venus were identical to the Earth in composition and structure, possessing the same core mantle crust ratio and the interior temperature were the same at corresponding pressures, then Venus would possess a mean density of $5.34\,\mathrm{g\,cm^{-3}}$ compared to the observed value of $5.24\,\mathrm{g\,cm^{-3}}$. In other words, the observed value of the density of Venus is 1.9% smaller compared to an Earth-like model. Although this difference is small it is nevertheless well determined since it is based on the application of the same empirically derived equation of state for both planets. Taking the higher near-surface temperature of Venus into account the density differential reduces to 1.7%.

To explain this density difference it has been suggested that the Venusian mantle contains twice as much FeO than the Earth's mantle, leading to a much smaller core fraction of only 0.23 compared to 0.325 for the Earth. However, in this case the FeO content of the Venusian surface rocks should also exceed the terrestrial values by a factor of two. As is evident from the XRF analyses of Venusian surface rocks by Venera 13 and 14 as well as by Vega 2 this is not the case and, hence, the lower bulk density of Venus remains surprising.

One might speculate that the higher density of the Earth is related to the formation of the Moon. In the now widely preferred giant impact model for the formation of the Moon, the Moon was formed from the mantle phases of the Earth and the impactor, while most of the core phases merged into the Earth's core. The mass of the Moon is 1.23% of the Earth's mass. As in all likelihood the total mass lost from the system was substantially higher, preferential loss of low-density mantle material might explain the density difference between Venus and Earth. Generally a somewhat lower core mass fraction of Venus than that of the Earth seems an unavoidable consequence of the density difference.

MERCURY

Despite many models developed in the last 30 years to infer the chemical composition of Mercury, one has to state that we know almost nothing about the planet's chemistry. Solid evidence is limited to two observations.

First, Mercury's high density. Although Johannes Keppler proposed that Mercury had the highest bulk density of all planets, it was only relatively recently that a precise figure of $5.43 \pm 0.01\,\mathrm{g\,cm^{-3}}$ was obtained (Anderson et al. 1987), from the encounters of Mariner 10 in 1974 and 1975. Hence, its density is considerably above the uncompressed density ($4.07\,\mathrm{g\,cm^{-3}}$) of the Earth leading to a core fraction about twice that of the Earth.

Second, Mercury's magnetic field. From the first and third close encounters of the Mariner 10 spacecraft it became evident that Mercury possesses a small but Earth-like magnetosphere, that is, a dipole field of internal origin, with an axis of symmetry aligned nearly perpendicular to the ecliptic. This dipole field in all likelihood is the result of an active dynamo in the core of the planet.

A core fraction twice of that of the Earth requires an increase in the Fe/Si ratio of about a factor of five. This number reduces somewhat if one assumes that the core contains, apart from the traditionally siderophile elements such as iron, nickel, and cobalt, metallic silicon. In other words, a highly reduced planet. Evidence of metallic silicon in the early solar nebula stems from the observation of a few percent silicon in the metal grains of enstatite chondrites.

The magnetic field of Mercury and its surface features seem to require a large, at least partially still molten core. In order to avoid solidification of the core and also in order to bring the thermal models of the planet into accordance with the lack of large-scale volcanism, an admixture of sulfur to the core material has been suggested. In order to lower the melting point of the core material by 200°C, an admixture of about 20% sulfur is required. Sulfur is more volatile than sodium or potassium, the two elements that are depleted on the Earth by a factor of about six. Hence, it is very difficult to explain such a high sulfur abundance on Mercury. However, as pointed out by Schubert et al. (1988), a total sulfur abundance of around 2–3% relative to iron allows for both a rapid growth of an inner core prior to cessation of early bombardment (to satisfy the geological constraint of little planetary contraction over geologic time) and the possibility of an ongoing convection in a thin outer core.

Today there is considerable evidence that the degree of oxidation increases going from the Earth to Mars; in the same manner the abundance of volatile elements increases. Going closer to the Sun, the opposite seems to be the case. Hence, even 2% sulfur in the original core of Mercury seems to be too high. Here one should remember that in fact any light alloying element in the core will lower its temperature of solidification. A few percent metallic silicon might be a better choice as the alloying element in Mercury's core.

As pointed out by Goettel (1988), the FeO abundances increase monotonically with decreasing condensation temperature in the solar nebula. Thus the bulk FeO content of a planet may be correlated with the abundance of species with lower condensation temperatures, such as sodium, potassium, sulfur (as FeS), and water. The FeO content of Mercury's crust is in principle observable from reflectance spectroscopy. Applying this technique, McCord and Clark (1979) estimated an FeO content of about 5.5% for the surface material of Mercury, while in a perhaps more precise study of the reflectance spectra by Vilas (1985), FeO bands were reported to be absent. Hence, it looks as if the abundance of FeO is indeed low in the surface layers of Mercury and in all likelihood also in the mantle of the planet. This fits well with a picture of decreasing FeO content going from Mars (mantle abundance 18% FeO) inward to the Earth (7.6% FeO) and to Mercury (<2% FeO).

There are two types of model to explain the high density of Mercury. The first involves the mechanical separation of high-density (metal) from low-density silicate material: (a) metal silicate fractionation in the solar nebula prior to accretion, and (b) blow-off of silicate mantle material from the planet due to impacts. The latter scenario attracted interest when the widely accepted consensus was reached that the most likely model for the origin of the Moon involved splash-off of Earth mantle material due to a collision with a Mars-sized object. The second model concerns the high-temperature loss of silicates: (a) Mercury accreted from material that was devoid of SiO_2 and all more volatile species, and (b) evaporation of a large fraction of the planet's silicate mantle. In addition a high degree of reduction – silicon partially in metallic form – should also be considered.

The high temperature loss of silicates as favored by Cameron et al. (1988) should lead to a planet in which the silicate phase is depleted in SiO_2 and FeO relative to MgO, Al_2O_3, and CaO. The amounts of Na and K should be very low. Of the three radioelements K, Th, and U observable by gamma-ray spectrometry from orbit as well as from a lander, the K/Th and K/U ratios should be very low. Even U should be considerably depleted. Only the highly refractory element Th should not be affected.

The indication of pure water ice in craters at both poles of Mercury by radar reflection measurements was quite a surprise (Harmon et al. 1994) at first sight: The axis of rotation of Mercury is perpendicular to the ecliptic. The absence of any noticeable atmosphere leads on the illuminated side to a temperature distribution that is a strong function of the latitude, with values up to 430°C at the equator and far below freezing point in the permanently shaded polar craters. The abundance of water in the interior of Mercury is probably very low. However, in addition to water vapor due to outgassing from the interior, water delivered by the impact of comets and also of water-rich asteroids has to be taken into account. Although the time that water molecules can spend in the gravity field before being carried away by the solar wind is limited, a small fraction of those which remain will be cold-trapped in the polar regions.

The discovery of water ice at the poles of Mercury triggered the search for water ice in lunar polar craters. The search for water ice was the prime goal of the recent Lunar Prospector mission. Radar measurements did not find conclusive evidence for water ice at the poles of the Moon nor did the neutron spectrometer of Lunar Prospector. In fact it has been speculated that the water ice observed within several craters near both of Mercury's poles might be the result of a recent impact of a moderate-sized comet on Mercury and thus be derived from a singular event.

The importance of meteorites from the Moon and Mars for the study of the chemical composition of these objects has been stressed in preceding sections. Hence, one might ask: Could we also receive rocks from Venus and Mercury? In the case of Venus, the thick CO_2 atmosphere would require a very large impact which could temporarily remove the atmosphere from the hemisphere of the impact. This and the high escape velocity of $10.4\,km\,s^{-1}$ makes ejection of surface rocks from Venus highly unlikely. Mercury has no atmosphere and its escape velocity is only

4.2 km s^{-1}. However, on average an additional velocity of 10 km s^{-1} is necessary to transfer a meteorite from the orbit of Mercury to that of the Earth. Thus we have to conclude that rocks from Mercury and Venus delivered in the form of meteorites to Earth are not impossible, but highly unlikely.

REFERENCES

Adler, I., Trombka, J.I., Schmadebeck, R., Lowman, P., Blodget, H., Yin, L., Eller, E., Podwysocki, M., Weidner, J.R., Bickel, A.L., Lum, R.K.L., Gerard, J., Gorenstein, P., Bjorkholm, P. and Harris, B. (1973). Results of the Apollo 15 and 16 x-ray experiment. *Geochimica et Cosmochimica Acta, Supplement 4*, 2783–2791.

Arnold, J.R. (1979). Ice in the lunar polar regions. *Journal of Geophysical Research*, **84**, 5659–5668.

Anders, E. (1978). Procrustean science: Indigenous siderophiles in the lunar highlands, according to Delano and Ringwood. *Geochimica et Cosmochimica Acta, Supplement 10*, 161–184.

Anders, E. and Ebihara, M. (1982). Solar-system abundances of the elements. *Geochimica et Cosmochimica Acta*, **46**, 2363–2380.

Anders, E. and Grevesse, N. (1989). Abundances of the elements: Meteoritic and solar. *Geochimica et Cosmochimica Acta*, **53**, 197–214.

Anderson, J.D., Colombo, G., Esposito, P.B., Lau, E.L. and Trager, G.B. (1987). The mass, gravity field and ephemeris of Mercury. *Icarus*, **71**, 337–349.

Banin, A., Clark, B.C. and Wänke, H. (1992). Surface chemistry and mineralogy. In H.H. Kieffer, B.M. Jakosky, C.W. Snyder and M.S. Matthews (eds), University of Arizona Press, *Mars*, Tucson AZ, pp. 594–625.

Becker, R.H. and Pepin, R.O. (1984). The case for a Martian origin of the shergottites. Nitrogen and noble gases in EETA 79001. *Earth and Planetary Science Letters*, **69**, 225–242.

Binder, A.B. (1998). Lunar Prospector: Overview. *Science*, **281**, 1475–1476.

Bogard, D.D. and Johnson, P. (1983). Martian gases in an Antarctic meteorite? *Science*, **221**, 651–654.

Bouchet, M. and Kaplan, G. (1971). Spark mass spectrometric analysis of major and minor elements in six lunar samples. *Proceedings of the 2nd Lunar Science Conference*, MIT Press, Vol. 2, pp. 1247–1252.

Brückner, J., Dreibus, G., Lugmair, G.W., Rieder, R., Wänke, H. and Economou, T. (1999). Chemical composition of the Martian surface as derived from Pathfinder, Viking, and Martian meteorite data (abstract no. 1250). *Lunar and Planetary Science XXX*, Lunar and Planetary Institute, Houston, TX (CD-ROM).

Brückner, J., Dreibus, G., Rieder, R. and Wänke, H. (2001). Revised data of the Mars Pathfinder alpha proton X-ray spectrometer: Geochemical behavior of major and minor elements. (abstract no.1293). *Lunar and Planetary Science XXXII*, Lunar and Planetary Institute, Houston, TX (CD-ROM).

Carlson, R.W. and Lugmair, G.W. (1979). Sm–Nd constraints on early lunar differentiation and the evolution of KREEP. *Earth and Planetary Science Letters*, **45**, 12–132.

Carlson, R.W. and Lugmair, G.W. (1981). Time and duration of lunar highland crust formation. *Earth and Planetary Science Letters*, **52**, 22–238.

Cameron, A.G.W. (1973). Abundances of elements in solar system. *Space Science Reviews*, **15**, 121–146.

Cameron, A.G.W., Fegley, Jr., B., Benz, W. and Slattery W.L. (1988). The strange density of Mercury: Theoretical considerations. In F. Vilas, C.R. Chapman and M.S. Matthews (eds), *Mercury*, University of Arizona Press, Tucson, AZ, pp. 692–708.

Cameron, A.G.W. and Ward, W.R. (1976). The origin of the Moon (abstract). *Lunar Science VII*, Lunar Science Institute, Houston TX, pp. 120–122.

Clark, B.C. and Baird, A.K. (1979). Is the Martian lithosphere sulfur rich? *Journal of Geophysical Research*, **84**, 8395–8403.

Clark, B.C., Baird, A.K., Weldon, R.J., Tsusaki, D.M., Schnabel, L. and Candelaria, M.P. (1982). Chemical composition of Martian fines. *Journal of Geophysical Research*, **87**, 10059–10067.

Clayton, R.N. and Mayeda, T.K. (1975). Genetic relations between the moon and meteorites. Proceedings of the 6th Lunar Science Conference, *Geochimica et Cosmochimica Acta, Supplement 6*, 1761–1768.

Compston, W., Berry, H., Vernon, M.J., Chappell, B.W. and Kaye, M.J. (1971). Rubidium–strontium chronology and chemistry of lunar material from the ocean of storms. *Proceedings of the 2nd Lunar Science Conference*, MIT Press, Vol. 2, pp. 1471–1485.

Cuttitta, F., Rose, H.J., Jr., Annell, C.S., Carron, M.K., Christian, R.P., Dwornik, E.J., Helz, A.W. and Ligon, D.T., Jr. (1971). Elemental composition of some Apollo 12 rocks and soils. *Proceedings of the 2nd Lunar Science Conference*, MIT Press, Vol. 2, pp. 1217–1229.

Delano, J.W. (1985). Mare volcanic glasses, II: Abundances of trace Ni and the composition of the Moon (abstract). *Lunar and Planetary Science XVI*, Lunar and Planetary Institute, Houston, TX, pp. 179–180.

Delano, J.W. and Ringwood, A.E. (1978). Siderophile elements in lunar highlands: Nature of the indigenous component and implications of the origin of the Moon. *Geochimica et Cosmochimica Acta, Supplement 10*, 111–159.

Dreibus, G., Brückner, J. and Wänke, H. (2000). Phosphorous in Martian rocks and soils and the global surface chemistry of Mars as derived from APXS on Pathfinder (abstract no. 1127). *Lunar and Planetary Science XXXI*, Lunar and Planetary Institute, Houston, TX, (CD-ROM).

Dreibus, G., Jagoutz, E., Spettel, B. and Wänke, H. (1996). Phosphate-mobilization on Mars? Implication from leach experiments on SNCs. *Lunar and Planetary Science XXVII*, Lunar and Planetary Institute, Houston, TX, pp. 323–324.

Dreibus, G., Palme, H., Rammensee, W., Spettel, B. and Wänke, H. (1981). On the mobilization and redistribution of Au and other siderophiles in lunar highland materials. *Lunar and Planetary Science XII*, Lunar and Planetary Institute, Houston, TX, pp. 240–242.

Dreibus, G. and Wänke, H. (1987). Volatiles on Earth and Mars: A comparison. *Icarus*, **71**, 225–240.

Eberhardt, P., Geiss, J., Graf, H., Grögler, N., Krähenbühl, U., Schwaller, H., Schwarzmüller, J. and Stettler, A. (1970). Trapped solar wind noble gases, exposure age and K/Ar age in Apollo 11 lunar fine material. *Geochimica et Cosmochimica Acta, Supplement 1*, 1037–1070.

Eberhardt, P., Geiss, J. and Grögler, N. (1965). Further evidence on the origin of trapped gases in the meteorite Khor Temiki. *Journal of Geophysical Research*, **70**, 4375–4378.

Ehmann, W.D. and Morgan, J.W. (1971). Major element abundance in Apollo 12 rocks and fines by 14 MeV neutron activation. *Proceedings of the 2nd Lunar Science Conference*, MIT Press, Vol. 2, pp. 1237–1245.

Epstein, S. and Taylor, H.P. (1970). The concentration and isotopic composition of hydrogen, carbon and silicon in Apollo 11 lunar rocks and minerals. *Geochimica et Cosmochimica Acta, Supplement 1*, 1085–1096.

Feldman, W.C., Lawrence, D.J., Elphic, R.C., Barraclough, B.L., Maurice, S., Genetay, I. and Binder, A.B. (2000). Polar hydrogen deposits on the Moon. *Journal of Geophysical Research*, **105**, 4175–4195.

Feldman, W.C., Maurice, S., Binder A B., Barraclough, B.L., Elphic, R.C. and Lawrence, D.J. (1998). Fluxes of fast and epithermal neutrons from Lunar Prospector: Evidence for water ice at the lunar poles. *Science*, **281**, 1496–1500.

Ganapathy, R., Keays, R.R., Laul, J.C. and Anders, E. (1970). Trace elements in Apollo 11 lunar rocks: Implications for meteorite influx and origin of the moon. *Geochimica et Cosmochimica Acta, Supplement 1*, 1117–1142.

Gast, P.W. (1972). The chemical composition and structure of the Moon. *The Moon*, **5**, 121–148.

Gast, P.W. and Hubbard, N.J. (1970). Abundance of alkali metals, alkaline and rare earths and strontium-87/strontium-86 ratios in lunar samples. *Science*, **167**, 485–487.

Gast, P.W., Hubbard, N.J. and Wiesmann, H. (1970). Chemical composition and petrogenesis of basalts from Tranquillity Base. *Geochimica et Cosmochimica Acta, Supplement 1*, 1143–1163.

Goettel, K.A. (1988). Present bounds on the bulk composition of Mercury: Implications for planetary formation processes. In F. Vilas, C.R. Chapman and M.S. Matthews (eds), *Mercury*, University of Arizona Press, Tucson, AZ, pp. 613–621.

Golombek, M.P., Anderson, R.C., Barnes, J.R., Bell, III, J.F., Bridges, N.T., Britt, D.T., Brückner, J., Cook, R.A., Crisp, D., Crisp, J.A., Economou, T., Folkner, W.M., Greeley, R., Haberle, R.M., Hargraves, R.B., Harris, J.A., Haldemann, A.F.C., Herkenhoff, K.E., Hviid, S.F., Jaumann, R., Johnson, J.R., Kallemeyn, P.H., Keller, H.U., Kirk, R.L., Knudsen, J.M., Larsen, S., Lemmon, M.T., Madsen, M.B., Magalhaes, J.A., Maki, J.N., Malin, M.C., Manning, R.M., Matijevic, J., McSween, H.Y., Jr., Moore, H.J., Murchie, S.L., Murphy, J.R., Parker, T.J., Rieder, R., Rivellini, T.P., Schofield, J.T., Seiff, A., Singer, R.B., Smith, P.H., Soderblom, L.A., Spencer, D.A., Stoker, C.R., Sullivan, R., Thomas, N., Thurman, S.W., Tomasko, M.G., Vaughan, R.M., Wänke, H., Ward, A.W. and Wilson, G.R. (1999). Overview of the Mars Pathfinder mission: Launch through landing, surface operations, data sets and science results. *Journal of Geophysical Research*, **104**, E4, 8523–8553.

Golombek, M.P., Cook, R.A., Economou, T., Folkner, W.M., Haldemann, A.F.C., Kallemeyn, P.H., Knudsen, J.M., Manning, R.M., Moore, H.J., Parker, T.J., Rieder, R., Schofield, J.T., Smith, P.H. and Vaughan, R.M. (1997). Overview of the Mars Pathfinder mission and assessment of landing site predictions. *Science*, **278**, 1743–1748.

Harmon, J.K., Slade, M.A., Vèlez, R.A., Crespo, A., Dryer, M.J. and Johnson, J.M. (1994). Radar mapping of Mercury's polar anomalies. *Nature*, **369**, 213–215.

Hartmann, W.K. and Davis, D.R. (1975). Satellite-sized planetesimals and lunar origin. *Icarus*, **24**, 504–515.

Haskin, L.A., Helmke, P.A. and Allen, R.O. (1970). Rare earth elements in returned lunar samples. *Science*, **167**, 487–490.

Hintenberger, H., Weber, H.W., Voshage, H., Wänke, H., Begemann, F. and Wlotzka, F. (1970). Concentrations and isotopic abundances of the rare gases, hydrogen and nitrogen in Apollo 11 lunar matter. *Geochimica et Cosmochimica Acta, Supplement 1*, 1269–1282.

Hubbard, N.J. and Gast, P.W. (1971). Chemical composition and origin of nonmare lunar basalts. *Proceedings of the 2nd Lunar Science Conference*, MIT Press, Vol. 2, pp. 999–1020.

Hubbard, N.J., Rhodes, J.M., Wiesmann, H., Shih, C.-Y. and Bansal, B.M. (1974). The chemical definition and interpretation of rock types returned from non-mare regions of the Moon. *Geochimica et Cosmochimica Acta, Supplement 5*, 1227–1246.

Jagoutz, E. and Wänke, H. (1986). Sr and Nd isotopic systematics of Shergotty meteorite. *Geochimica et Cosmochimica Acta*, **50**, 939–953.

Laul, J.C., Wakita, H., Showalter, D.L., Boynton, W.V. and Schmitt, R.A. (1972). *Geochimica et Cosmochimica Acta, Supplement 3*, 1181–1200.

McCord, T.B. and Clark, R.N. (1979). The Mercury soil: Presence of Fe^{2+}. *Journal of Geophysical Research*, **84**, 7664–7668.

Melosh, H. and Sonett, C.P. (1986). When worlds collide: Jetted vapor plumes and the Moon's origin. In: W.K. Hartmann, R.J. Phillips and G.J. Taylor (eds), *Origin of the Moon*, Lunar Planetary Institute, Houston, TX, pp. 621–642.

Metzger, A.E., Trombka, J.I., Peterson, L.E., Reedy, R.C. and Arnold, J.R. (1973). Lunar surface radioactivity: Preliminary results of the Apollo 15 and 16 gamma-ray spectrometer experiments. *Science*, **179**, 800–803.

Morrison, G.H., Gerard, J.T., Potter, N.M., Gangadharam, E.V., Rothenberg, A.M. and Burdo, R.A. (1971). Elemental abundances of lunar soil and rocks from Apollo 12. *Proceedings of the 2nd Lunar Science Conference*, MIT Press, Vol. 2, pp. 1169–1185.

Nyquist, L.E., Bogard, D.D., Wooden, J.L., Wiesmann, H., Shih, C.Y., Bansal, B.M. and McKay, G.A. (1979). Early differentiation, late magmatism and recent bombardment of the shergottite parent body (abstract). *Meteoritics*, **14**, 502.

Palme, H. (1977). On the age of KREEP. *Geochimica et Cosmochimica Acta*, **41**, 1791–1801.

Palme, H., Spettel, B., Jochum, K.P., Dreibus, G., Weber, H., Weckwerth, G., Wänke, H., Bischoff, A. and Stöffler, D. (1991). Lunar highland meteorites and the composition of the lunar crust. *Geochimica et Cosmochimica Acta*, **55**, 3105–3122.

Palme, H., Spettel, B., Wänke, H., Bischoff, A. and Stöffler, D. (1984). Early differentiation of the Moon: Evidence from trace elements in plagioclase. *Journal of Geophysical Research*, **89**, C3–C15.

Palme, H., Suess, H.E. and Zeh, H.D. (1981). Abundances of the elements in the solar system. In: K. Schaifers and H.H. Vogt (eds) *Landolt-Börnstein, Numerical Data and Functional Relationships in Science and Technology, Vol. 2 Astronomy and Astrophysics*, Springer Verlag, Heidelberg, pp. 257–265.

Palme, H. and Wänke, H. (1975). A unified trace element model for the evolution of the lunar crust and mantle. *Geochimica et Cosmochimica Acta, Supplement 6*, 1179–1202.

Pepin, R.O. (1985). Evidence of Martian origins. *Nature*, **317**, 473–475.

Philpotts, J.A. and Schnetzler, C.C. (1970). Potassium, rubidium, strontium, barium and rare earth concentrations in lunar rocks and separated phases. *Science*, **167**, 493–405.

Rammensee, W. and Wänke, H. (1977). On the partition coefficient of tungsten between metal and silicate and its bearing on the origin of the Moon. *Geochimica et Cosmochimica Acta, Supplement 8*, 399–409.

Rieder, R., Economou, T., Wänke, H., Turkevich, A., Crisp, J., Brückner, J., Dreibus, G. and McSween, H.Y., Jr. (1997). The chemical composition of Martian soil and rocks returned by the mobile alpha proton x-ray spectrometer: Preliminary results from the x-ray mode. *Science*, **278**, 1771–1774.

Ringwood, A.E. (1979). *On the Origin of Earth and Moon*, Springer, New York.

Ringwood, A.E. and Kesson, S.E. (1977). Composition and origin of the Moon. *Geochimica et Cosmochimica Acta, Supplement 8*, 371–398.

Ringwood, A.E., Seifert, S. and Wänke, H. (1986/1987). A komatiite component in Apollo 16 highland breccias: Implications for the nickel–cobalt systematics and bulk composition of the Moon. *Earth Planetary Science Letters*, **81**, 105–117.

Schmitt, R.A., Smith, R.H., Lasch, J.E., Mosen, A.W., Olehy, D.A. and Vasilevskis, J. (1963). Abundance of fourteen rare-earth elements, scandium and yttrium in meteorites and terrestrial matter. *Geochimica et Cosmochimica Acta*, **27**, 577–622.

Schmitt, R.A., Smith, R.H. and Olehy, D.A. (1964). Rare-earth, yttrium and scandium abundances in meteoritic and terrestrial matter–II. *Geochimica et Cosmochimica Acta*, **28**, 67–86.

Schmitt, R.A., Wakita, H. and Rey, P. (1970). Abundandes of 30 elements in lunar rocks, soil, and core samples. *Science*, **167**, 512–515.

Schubert, G., Ross, M.N., Stevenson, D.J. and Spohn, T. (1988). Mercury's thermal history and the generation of its magnetic field. In: F. Vilas, C.R. Chapman and M.S. Matthews (eds), *Mercury*, University of Arizona Press, Tucson, AZ, pp. 429–460.

Signer, P. and Suess, H.E. (1963). Rare gases in the sun, in the atmosphere, and in meteorites. In: J. Geiss and E.D. Goldberg (eds), *Earth Science and Meteoritics*, North-Holland Amsterdam, pp. 241–272.

Suess, H.E. and Urey, H.C. (1956). Abundances of the elements. *Reviews of Modern Physics*, **28**, 53–74.

Suess, H.E., Wänke, H. and Wlotzka, F. (1964). On the origin of gas-rich meteorites. *Geochimica et Cosmochimica Acta*, **28**, 595–605.

Surkov, Yu.A., Barsukov, V.L., Moskaleva, L.P., Kharyukova, V.P., Zaitseva, S.Ye, Smirnov, G.G. and Manvelyan, O.S. (1989). Determination of the elemental composition of martian rocks from Phobos 2. *Nature*, **341**, 595–598.

Surkov, Yu. A., Kirnozov, F.F., Glazov, V.N., Dunchenko, A.G., Tatsy, L.P. and Sobornov, O.P. (1987). Uranium, thorium, and potassium in the Venusian rocks at the landing sites of Vega 1 and 2. *Journal of Geophysical Research*, **92**, B4, E537–E540.

Surkov, Yu.A., Moskalyova, L.P., Kharyukova, V.P., Dudin, A.D., Smirnov, G.G. and Zaitseva, Ye. (1986). Venus rock composition at the Vega 2 landing site. *Journal of Geophysical Research*, **91**, B13, E215–E218.

Taylor, L.A., Misra, K.C. and Walker, B.M. (1976). Subsolidus reequilibration, grain growth, compositional change of native FeNi metal in lunar rocks. *Geochimica et Cosmochimica Acta, Supplement 7*, 3461–3477.

Taylor, S.R. (1982). *Planetary Science: A Lunar Perspective*, Lunar and Planetary Institute, Houston, TX.

Turkevich, A.L., Franzgrote, E.J. and Patterson, J.H. (1970). Chemical composition of the lunar surface in a region near the Crater Tycho. *Science*, **168**, 825–828.

Urey, H.C. (1959). Primary and secondary objects. *Journal of Geophysical Research*, **64**, 1721–1737.

Urey, H.C. (1968). The origin of some meteorites from the Moon. *Naturwissenschaften*, **55**, 49–57.

Vilas, F. (1985). Mercury: Absence of crystalline Fe^{2+} in the regolith. *Icarus*, **64**, 133–138.

Wakita, H., Schmitt, R.A. and Rey, P. (1970). Elemental abundances of major, minor and trace elements in Apollo 11 lunar rocks, soil and core samples. *Geochimica et Cosmochimica Acta, Supplement 1*, 1695–1717.

Wänke, H. (1963). Cosmic ray data derived from isotope studies in meteorites. *Proceedings of the International Conference on Cosmic Rays*, Jaipur, India, 2–14 Dec., Vol. 3, pp. 473–479.

Wänke, H. (1965). Der Sonnenwind als Quelle der Uredelgase in Steinmeteoriten. *Zeitschrift für Naturforschung*, **20a**, 946–949.

Wänke, H. (1966). Der Mond als Mutterkörper der Bronzit-Chondrite. *Zeitschrift für Naturforschung*, **21a**, 93–110.

Wänke, H. (1981). Constitution of terrestrial planets. *Philosophical Transactions of the Royal Society*, **A303**, 287–302.

Wänke, H., Baddenhausen, H., Blum, K., Cendales, M., Dreibus, G., Hofmeister, H., Kruse, H., Jagoutz, E., Palme, C., Spettel, B., Thacker, R. and Vilcsek, E. (1977). On the chemistry of lunar samples and achondrites. Primary matter in the lunar highlands: A re-evaluation. *Geochimica et Cosmochimica Acta, Supplement 8*, 2191–2213.

Wänke, H., Baddenhausen, H., Dreibus, G., Jagoutz, E., Kruse, H., Palme, H., Spettel, B. and Teschke, F. (1973). Multielement analyses of Apollo 15, 16 and 17 samples and the bulk composition of the Moon. *Geochimica et Cosmochimica Acta, Supplement 4*, 1461–1481.

Wänke, H. and Dreibus, G. (1982). Chemical and isotopic evidence for the early history of the Earth–Moon system. In: P. Brosche and J. Sündermann (eds), *Tidal Friction and the Earth's Rotation II*, Springer Verlag, Berlin, pp. 322–344.

Wänke, H. and Dreibus, G. (1986). Geochemical evidence for the formation of the Moon by impact-induced fission of the proto-Earth. In: W.K. Hartmann, R.J. Phillips and G.J. Taylor (eds), *Origin of the Moon*, Lunar Planetary Institute, Houston, TX, pp. 649–672.

Wänke, H. and Dreibus, G. (1994). Chemistry and accretion history of Mars. *Philosophical Transactions of the Royal Society*, **A349**, 295–293.

Wänke, H., Dreibus, G. and Palme, H. (1978). Primary matter in the lunar highlands: The case of the siderophile elements. *Geochimica et Cosmochimica Acta, Supplement 10*, 83–110.

Wänke, H., Palme, H., Baddenhausen, H., Dreibus, G., Jagoutz, E., Kruse, H., Palme, C., Spettel, B., Teschke, F. and Thacker, R. (1975). New data on the chemistry of lunar samples: Primary matter in the lunar highlands and the bulk composition of the Moon. *Geochimica et Cosmochimica Acta, Supplement 6*, 1313–1340.

Wänke, H., Palme, H., Baddenhausen, H., Dreibus, G., Jagoutz, E., Kruse, H., Spettel, B., Teschke, F. and Thacker, R. (1974). Chemistry of Apollo 16 and 17 samples: Bulk composition, late stage accumulation and early differentiation of the Moon. *Geochimica et Cosmochimica Acta, Supplement 5*, 1307–1335.

Wänke, H., Palme, H., Kruse, H., Baddenhausen, H., Cendales, M., Dreibus, G., Hofmeister, H., Jagoutz, E., Palme, C., Spettel, B. and Thacker, R. (1976). Chemistry of lunar highland rocks: A refined evaluation of the composition of the primary matter. *Geochimica et Cosmochimica Acta, Supplement 7*, 3479–3499.

Wänke, H., Rieder, R., Baddenhausen, H., Spettel, B., Teschke, F., Quijano-Rico, M. and Balacescu, A. (1970). Major and trace elements in lunar material. *Geochimica et Cosmochimica Acta, Supplement 1*, 1719–1727.

Wänke, H., Wlotzka, F., Baddenhausen, H., Balacescu, A., Spettel, B., Teschke, F., Jagoutz, E., Kruse, H., Quijano-Rico, M. and Rieder, R. (1971). Apollo 12 samples: Chemical composition and its relation to sample locations and exposure ages, the two component origin of the various soil samples and studies on lunar metallic particles. *Geochimica et Cosmochimica Acta, Supplement 2*, 1187–1208.

Warren, P.H. (1985). The magma ocean. Concept and lunar evolution. *Annual Review of Earth and Planetary Sciences*, **13**, 201–240.

Warren, P.H. and Wasson, J.T. (1978). Compositional–petrographic investigations of pristine nonmare rocks. *Geochimica et Cosmochimica Acta, Supplement 10*, 185–217.

Warren, P.H. and Wasson, J.T. (1979). The origin of KREEP. *Reviews of Geophysics and Space Physics*, **17**, 73–88.

Wasson, J.T. and Wetherill, G.W. (1979). Dynamical, chemical and isotopic evidence regarding the formation location of asteroids and meteorites. In: T. Gehrels (ed), *Asteroids*, University of Arizona Press, Tucson, AZ, pp. 926–974.

Wetherill, G.W. (1976). The role of large bodies in the formation of the Earth and Moon. *Geochimica et Cosmochimica Acta, Supplement 7*, 3245–3257.

Willis, J.P., Ahrens, L.H., Danchin, R.V., Erlank, A.J., Gurney, J.J., Hofmeyr, P.K., McCarthy, T.S. and Orren, M.J. (1971). Some interelement relationships between lunar rocks and fines, and stony meteorites. *Proceedings of the 2nd Lunar Science Conference*, MIT Press, Vol. 2, pp. 1123–1138.

Wood, J.A., Dickey, J.S., Jr., Marvin, U.B. and Powell, B.N. (1970). Lunar anorthosites and a geophysical model for the Moon. *Geochimica et Cosmochimica. Acta, Supplement 1*, 965–988.

Zähringer, J. (1966). Primordial helium detection by microprobe technique. *Earth and Planetary Science Letters*, **1**, 20–22.

57

FRED W. TAYLOR*

The atmospheres of the terrestrial planets

It has been known since the end of the nineteenth century that the Earth's nearest planetary neighbours, Venus and Mars, have extensive atmospheres. However, the nature of these atmospheres was not at all well understood until the advent of planetary missions by robot probes, beginning with Mariner 2 which encountered Venus in December 1962. Since then, the spurts of new information from missions like Pioneer Venus, and Mariner and Viking to Mars, plus radically improved Earth-based techniques including the use of infrared, microwave and radar sensing, have revealed most of the important basic atmospheric properties such as composition, temperature, surface pressure, cloud properties, dynamics and general circulation. Conditions on these planets are now known to be, for the most part, very different from even the most informed speculation that took place in the last few decades of the nineteenth century.

The bright and featureless aspect of Venus showed early observers that the planet was cloudy, and it was perhaps natural to assume that these were water clouds like those on the Earth, made more copious by the increased evaporation of surface water due to the greater proximity of Venus to the Sun. From this it seemed logical to expect rain and a swampy, tropical surface environment with carbonated-water oceans and at least a good chance of life. The searing temperatures and crushing pressures that we now know lie under clouds of concentrated sulphuric acid were recognized only after the dawn of the space age, and only fully accepted by some after the first landing on the planet by Venera 7 in 1970.

Mars has long enjoyed a reputation as a habitable sort of place. As early as 1774, Herschel concluded from his observations that it had a 'considerable' atmosphere containing clouds; Flammarion in 1896 noted the existence and variability of the polar caps. However, as late as the 1950s theoreticians analysing the carbon dioxide lines in the Martian spectrum were writing of the surface pressure in terms that were more than a factor of 10 too high. It was only with the development of high-resolution ground-based spectrometers and the arrival of Mariner 4 in the early 1960s that the surprisingly low value of around 6 mbar was derived and the composition of nearly pure carbon dioxide confirmed.

Mercury has always been elusive because its position close to the Sun means that telescopic observations from the Earth and spaceflight in its vicinity are both relatively difficult. Some putative spectroscopic evidence for a dense atmosphere was obtained in the nineteenth century, but this was later called into question on theoretical grounds when it was recognized that gases on such a small, hot body would readily escape to space. A thin atmosphere like that of Mars was still considered a possibility, for example by those who thought they could discern the occasional obscuration of surface features by airborne dust, well into the second half of the twentieth century. The arrival at Mercury of Mariner 10 in 1973 finally demonstrated that the surface pressure on the planet is less than the most perfect vacuum that can be created in laboratories on the Earth.

The fact that the terrestrial planet atmospheres are now well characterized in general terms does not of course mean that we fully understand them. The curious dynamics of Venus's cloud layers, the vivid evidence for massive climatic change on Mars and the possible existence of ice deposits near the poles on Mercury are just the tip of an enormous iceberg of ignorance and curiosity concerning our planetary neighbours. The flotilla of scientific probes being aimed towards Mars in the first decade of the twenty-first century is one testament to the fact that the magnificent journey of discovery that essentially began in the twentieth century is set fair to reach its zenith in the next.

---

*University of Oxford, United Kingdom

Table 1 Physical data for the terrestrial planets

	Mercury	Venus	Earth	Mars
Orbital and rotational data				
Mean distance from Sun (km)	5.79×10^7	1.082×10^8	1.496×10^8	2.279×10^8
Eccentricity	0.2056	0.0068	0.0167	0.0934
Obliquity (deg)	0	177	23.45	23.98
Siderial period (d)	87.97	224.701	365.256	686.980
Rotational period (h)	1407.5	5832.24	23.9345	24.6229
Solar day (d)	115.88	117	1	1.0287
Solar constant (kW m^{-2})	See text	2.62	1.38	0.594
Net heat input (kW m^{-2})	See text	0.367	0.842	0.499
Solid body data				
Mass (kg)	3.302×10^{23}	4.870×10^{24}	5.976×10^{24}	6.421×10^{23}
Radius (km)	2439	6051.5	6378	3398
Surface gravity (m s^{-2})	3.70	8.60	9.78	3.72
Atmospheric data				
Composition	He, Na	See Table 2		
Mean molecular weight	—	43.44	28.98 (dry)	43.49
Mean surface temperature (K)	~400	730	288	220
Mean surface pressure (N m^{-2})	$<10^{-12}$	92	1	0.007
Mass (kg)	$\sim 8 \times 10^3$	4.77×10^{20}	5.30×10^{18}	$\sim 10^{16}$

Table 2 Composition of the terrestrial planet atmospheres (for Mercury, see Table 4). Values are given as fractional abundances or as parts per million (ppm)

	Venus	Earth	Mars
Carbon dioxide	0.96	0.0003	0.95
Nitrogen	0.035	0.770	0.027
Argon	0.00007	0.0093	0.016
Water vapour	~0.0001(?)	~0.01	~0.0003
Oxygen	0.0013	0.21	~0
Sulphur dioxide	0.00015	0.2 ppb	~0
Carbon monoxide	0.00004	0.12 ppm	0.0007
Neon	5 ppm	18 ppm	2.5 ppm

Physical data for Mercury, Venus, Earth and Mars are given in Table 1, and data for the atmospheres of Venus, Earth and Mars are given in Tables 2 and 3.

MERCURY

Having already described the atmosphere of the innermost planet as resembling a perfect vacuum, it is perhaps capricious to go on at length about its interesting properties. Nevertheless, there are a few, and the story of how we got to our present understanding of this wispy envelope, and its associated mysteries, is engaging.

Most of the early observations of Mercury concentrated on trying to resolve details on the surface and on determining thereby the period of rotation. Maps were drawn, and features thought to be permanent were assigned romantic names. Maps by Schiaparelli in 1889 and Lowell in 1896 showed narrow, linear features reminiscent of those on their charts of Mars and probably of the same chimerical origin. Antoniadi, who published a book about Mercury in 1934 and made many observations of his own, was convinced that Mercury exhibited synchronous rotation, with the same face always directed at the Sun. He deduced that Mercury possessed an atmosphere thick enough to support dust storms, which seemed sometimes to obscure the dark markings he saw on the planet. In fact, he was viewing different parts of the disk, as Mercury rotated once every 58.6 days, rather than the 88 days that would have matched its annual journey around the Sun.

Table 3 Properties of clouds and dust in the terrestrial planet atmospheres*

	Venus	Earth	Mars
Fractional coverage	1.00	0.40	0.05 (cloud)
			0–1.0 (dust)
Typical optical depth	25–40	5–7	0.2–6 (dust)
Composition	$H_2SO_4 \bullet H_2O$	H_2O	Magnetite etc. (dust); H_2O, CO_2
Number density (cm^{-3})			
liquid	50–300	100–1000	0
solid	10–50	0.1–50	30–1000
Mass loading ($g\ m^{-3}$)	0.01–0.1	0.1–10	0.0002–0.1
Main production process	Chemistry	Condensation	Windblown (dust)
Equivalent depth (mm)	0.1–0.2	0.03–0.05	1–100
Effective radius (μm)	2–4	10	0.4–2.5 (dust)
Main forms	Stratiform, cumulus	stratiform, cumulus	stratiform, mixed (dust)
Temporal variability	Slight (haze), high (deep)	High	High

*The equivalent depth is the estimated thickness of the cloud material if it were deposited on the surface. The effective radius is the radius of the spherical particles having most nearly the same scattering properties as the cloud at visible wavelengths. After Esposito et al. (1983), with changes and additions

Table 4 Properties of the atmosphere of Mercury, from Mariner 10 results

Surface pressure	$<10^{-12}$ bar
Average surface temperature	440 K (590–725 K, sunward side)
Atmospheric composition	Detected compounds
	42% oxygen, 29% sodium, 22% hydrogen, 6% helium, 0.5% potassium
	Possible trace constituents
	Argon, carbon dioxide, water, nitrogen, xenon, krypton, neon

Another way to detect an atmosphere is by analysing the polarization of light reflected from the planet. Summarizing the evidence from a long sequence of such observations at several wavelengths, Dollfus in 1961 concluded that there was evidence for a pressure at the surface of up to about 1 mbar, one thousandth of that on Earth. Ingersoll in 1971 revised this to an upper limit of 0.28 mbar (for carbon dioxide; the value depends on the composition assumed). These researchers recognized that their analysis depended on the assumption that the surface itself does not contribute to the polarization of the reflected light; if it does, their values get smaller. In fact, the surface pressure is less than one billionth of a millibar.

Reviewing in 1988 the question of surface markings on Mercury, and the various maps that had been drawn over the years, Patrick Moore wrote 'My own conclusion is that although it is probable that a few albedo features were glimpsed occasionally, the errors in observation were so unavoidably large that, without the spacecraft, we would never have learned anything definite about Mercury's features.' The spacecraft he refers to is, of course, Mariner 10, the only one to visit Mercury to date.

It is to Mariner 10 that we also owe the resolution of the question of the surface pressure on Mercury. During three encounters, on 29 March 1974, 21 September 1974 and 16 March 1975, two ultraviolet spectrometers recorded atmospheric data. The first observed at a selection of wavelengths chosen to correspond to the emission lines of several candidate atmospheric species, specifically helium, neon, argon, xenon, hydrogen, carbon and oxygen. The second observed the Sun, and detected absorption by the atmosphere in four wavelength bands as the Sun went behind the limb of the planet. This is the so-called occultation technique, which is very sensitive to the presence of small traces of gas. The results are summarized in Table 4.

The total mass of Mercury's atmosphere has been estimated at only about 8 tonnes. The low surface pressure means that the atoms and molecules can escape from the planet, or collide with the surface, before they collide with each other: by analogy with the outermost layers of the thicker atmospheres of the other planets, this is called an exosphere. It follows that the atmosphere is transient; to exist at all, it must be constantly replenished. Since mixing is ineffective, the composition, and the density, may vary considerably over the globe, depending on the local balance between sources and sinks for each molecule.

Oxygen, sodium, and potassium in the atmosphere are probably baked out or otherwise derived from the minerals in the surface or the crust of Mercury. The hydrogen and helium come in as the solar wind and are temporarily trapped. It seems likely that trace amounts of argon, carbon dioxide, water, nitrogen, xenon, krypton and neon are also present.

The metallic ions sodium and potassium are probably produced by sputtering at the surface caused by micrometeorites, solar wind particles or ions from Mercury's magnetosphere. There is observational evidence for localized concentrations which may imply venting from subsurface sources or some other mechanism. Accumulations of radar-bright material in craters in the polar regions are thought by many to be deposits of water ice. They occur in locations that are permanently sheltered from sunlight that will, in the absence of a substantial atmosphere, be very cold. If substantial amounts of ice are indeed present (and it remains to be shown that the features are not due to some other radar-bright volatile, such as sulphur or even sodium) their origin is probably in meteoritic bombardment, although again outgassing from a source within the planet cannot be ruled out.

Mercury is a prime target for an orbiter mission in the first decade of the new century. NASA has announced plans for the Mercury Surface, Space Environment, Geochemistry and Ranging mission (Messenger), while the European Space Agency (ESA) has definite plans for a 'cornerstone' mission in 2008 or there abouts, albeit with an unfortunately puerile name 'Bepi Colombo'. ESA recently completed a technical and scientific study of a mission to return dust, rock and a core sample from the surface of Mercury, although no date has been set for this very ambitious mission. While primarily directed towards understanding the origin and evolution of Mercury as a planet, key atmospheric science questions can also be addressed. Firstly, better spectrometers can be used to confirm the composition given in Table 4, and to search for trace species like water vapour and carbon dioxide. Secondly, the variation in space and time of these species can be mapped, and any low-pressure 'volcanoes', fissures, or other regions where gases are issuing from the interior, located. Finally, the question of the nature and origin of the reflective deposits near the poles can be investigated with observations that include infrared temperature measurements. For a planet with a vacuum for an atmosphere, Mercury will remain an interesting place for many decades to come.

VENUS

The twentieth century was more than 60 years old before the technology was in place to probe beneath the all-enveloping cloud veil on Venus. Measurements of emission from the planet at microwave (centimetre) wavelengths, which penetrate all but the most massive clouds, showed a source temperature of more than 600 K, a value so high that most planetary scientists assumed at first that the source must be something other than the surface of the planet. It could, among other possibilities, be emission by some unidentified process in the planet's ionosphere, or from sustained lightning activity in the clouds. However, it was pointed out by Carl Sagan and others that the visible diameter of Venus is several tens of kilometers larger than its radar diameter, so the atmosphere below the cloud tops must be correspondingly deep. In an optically thick atmosphere, the vertical temperature gradient cannot deviate much from a value known as the adiabatic lapse rate, which on Venus is about $10\,\mathrm{K\,km^{-1}}$. The temperature at the cloud tops is relatively easy to measure, with infrared sensors on ground-based telescopes, and was known to be around 240 K. Thus, if this temperature was, say, 50 km above the ground, the latter had to be around 740 K, in agreement with the microwave emission temperatures. Sagan and colleagues also went on to show that it might, under certain conditions that remained to be proven, be possible to sustain such a high temperature by the greenhouse effect, a well-known atmospheric phenomenon without which our own planet would be an icy wasteland.

Still, the idea of as much as 500 K of greenhouse-induced heating was hard to believe at first. On the Earth, the effect is only about 35 K; the larger amounts of carbon dioxide on Venus and it being closer to the Sun imply a larger value but not, in most people's minds, so much. For one thing, the clouds of Venus are so reflective that about 80% of the Sun's heat is reflected back into space: the amount absorbed is less than that of the Earth and actually about the same as that of the relatively distant, dark and icy planet Mars.

The ensuing controversy was resolved only by the flight of Mariner 2, the world's first successful interplanetary probe. It had on board a microwave radiometer similar to those that had detected the hot emission from Venus, the difference being that, carried close to the planet, it could resolve the difference between heat emission from the surface and from the upper atmosphere or ionosphere. The results showed the intensity of the emission falling off from the centre of the planet to the limb, clear support for a source at the surface. Direct temperature sensors on the first successful Venera entry probe provided the final confirmation in 1967.

There followed a flurry of Venus investigation, led by the spacecraft of the Mariner, Venera and Pioneer programmes. Altogether there were 9 instrumented flybys, 15 landers and 3 orbiters at Venus between Mariner 2 in 1962 and Venera 16 in 1984. This 22-year period of Venus exploration concentrated on the massive and mysterious atmosphere of Venus, and constituted a golden age in this one key aspect of space science. The rest of the century was rather quiet by comparison, as attention and scarce resources swung away from Venus and towards the outer planets and Mars.

A large proportion of what we now know about Venus's atmosphere came from the highly successful Pioneer Venus (PV) project of the late 1970s. Consisting of four entry probes and a separately launched orbiter, with a co-ordinated theory programme and a data distribution system that was advanced for its time, the objective was to understand the detailed processes involved in the massive greenhouse effect, the circulation of the atmosphere, the production and maintenance of the cloud layers and the physics of the upper atmosphere. The investigators, around 100 of them all told, under the intellectual leadership of professors Richard Goody and Donald Hunten, whose influence had led NASA to mount the mission, gathered for project meetings at the NASA Ames Research Center near San Francisco. They drew up a long list of questions for the mission to target, most of them related to three impassioned areas of half knowledge, half ignorance where new measurements were thought to be the key.

The first of these, obviously, was to explain in full detail the mechanism behind the high surface temperature and pressure. Although this clearly was a version of the greenhouse effect so familiar on the Earth, scientists were still far from being able to show how the effect was able to become so large on Venus. Carbon dioxide alone cannot produce enough warming: the spectral regions where it absorbs cover only part of the infrared spectrum, and even the large amounts of the gas now known to be present on Venus leave 'windows' that have to be filled by other absorbers. Data were needed on the composition and energy balance of the atmosphere, obtained by radiometry and optical and mass spectroscopy on the orbiter and the descending probes. The modellers were particularly lacking basic parameters on the vertical extent, particle size, density and optical properties of the clouds. Carl Sagan by this time had moved on to other fields of endeavour, but his erstwhile student, the late Jim Pollack, had become the leading expert on the Venusian greenhouse effect and was prominent among the theory team.

The next question concerned the circulation of the atmosphere. Despite the Earth-like nature of the solid planet, and of some aspects of the vertical structure of the atmosphere, especially if the dense layers below the clouds are taken to be the Venusian analogue of the terrestrial oceans, the atmospheric motions of the two planets have very little in common. It had been known since the 1920s that markings in the cloud tops appear when Venus is viewed through an ultraviolet filter; these markings rotate rapidly around the planet in the zonal direction (parallel to the equator) suggesting very high winds. The PV scientists hoped to discover what drives this 'super-rotation' on such a slowly turning planet. As with the greenhouse effect, conceptual models existed that explained how the phenomenon might have originated, but quantitative numerical simulations could not come close to achieving the observed amplitude of the effect. Dynamical models of Venus's atmosphere, derived from terrestrial general circulation models, could produce slow zonal winds driven by the transfer of momentum from below, or from the overhead motion of the Sun, but had serious difficulty in driving the speed up to the observed $100\,\mathrm{m\,s^{-1}}$. Detailed wind profiles obtained by tracking the descent of the probes were expected to be the key to identifying the missing physics in the models.

Finally, there was the process of mapping. The PV orbiter was the first artificial satellite of Venus and therefore presented the first opportunity to characterize the whole planet in a systematic way. Radar mapping of the surface topography was a high priority; so was an exploration of Venus's particles and fields environment. For the atmospheric scientists, the infrared and ultraviolet mapping of the middle and upper atmosphere, including spectroscopic and polarization data, were the focus. It was hoped that the variable features in the temperature, cloud and composition fields would reveal previously unknown properties of the planet, and so it proved.

PV was in fact two missions, launched separately within a few months of each other. The first launch placed in orbit the first artificial satellite of Venus (the 'orbiter' mission), while the second delivered four probes, one large and three small, directly into the atmosphere (the 'multiprobe' mission). The orbiter mapped the atmosphere above the clouds at infrared wavelengths, the clouds themselves in the visible and ultraviolet and the surface using radar. Operating from a highly elliptical orbit, it also investigated the thermosphere of Venus, utilizing measurements of the drag on the spacecraft as it dipped down into the atmosphere, to altitudes as low as 150 km, and occasionally lower. Aided by the angular coverage that it obtained as a result of its spin-stabilized configuration, the orbiter also obtained the first comprehensive maps of Venus's magnetosphere.

The multiprobe mission was to address questions of the nature of the clouds on Venus, their layering, microstructure and composition; the solar heating of the atmosphere as a function of depth; the atmospheric circulation and its driving forces; the bulk composition of the atmosphere, the loss of water and stability of carbon dioxide; the deep atmosphere vertical temperature structure; and ionospheric turbulence, ion chemistry and exospheric temperature. The orbiter, to be in position before, and operating during the probe descent, was to address cloud-top morphology; the cause and extent of the four-day 'super-rotation' at the cloud tops; the vertical temperature structure, gravitational moments, exospheric temperature, composition and loss of water; and ionospheric temperatures and motions.

For investigators on a planetary mission, especially a complex, multidisciplinary and multi-spacecraft venture like PV, participation is a lengthy toil of instrument building and planning meetings, launch and its associated excitements, followed by a seemingly interminable period of waiting, and then a rush of new information. Sometimes, because a new technique is being applied to a relatively unexplored world, completely unexpected findings emerge almost immediately. The author's infrared radiometer experiment on the PV orbiter furnished several such examples.

The first came on the night that the first data arrived at the Earth. The PV project team was a compact, inexpensive organization, as space missions go, a model for today's 'faster, cheaper' approach which it anticipated by a quarter of a century. The investigators each had a room for their team that was close to the main operations centre, where the spacecraft was being controlled and monitored. When the circumstances required it, we could walk in and request that commands be sent to our instrument, and wait and watch for the response. Following the orbiter's arrival at Venus, and successful injection into orbit, we had just sent the once-only command to release the cover over the aperture of the instrument, which we called VORTEX (Venus Orbiter Radiometric Temperature-sounding Experiment). It was designed to measure the temperature of the cloud tops, and of successively higher layers of the overlying atmosphere, by sensing the infrared emission from them in carefully selected wavelength bands.

The instrument was already switched on and checked out, but could not see the planet until the cover, which protected the optics during launch and flight, was flipped out of the way. The paper streaming out of the line printer was full of zeros, and stubbornly remained that way until several minutes after the time we had calculated for the delay due to the communications link. Then, at last, real numbers of about the right magnitude appeared, everyone too tense to cheer. We took half an hour's worth of data off to our designated room to check it out.

One of the most important parameters for any instrument measuring weak signals is the background noise level: it governs the final accuracy of everything measured. We had measured this carefully before departure, of course, a year earlier, but we needed to know if it had got worse during the rigours of installation and launch at Cape Canaveral, or possibly even better, as a result of the vibrationless environment on the orbiter as it skimmed silently around the huge globe of Venus. In fact, it was about the same.

While the rest of us were looking at this, Dan McCleese of the VORTEX team took it on himself to plot the atmospheric temperatures as a function of time, marking off on the plot approximately where on the planet the instrument was looking at that time. He noticed straight away that Venus seemed to get hotter as the scan moved from the equator to the north pole – the opposite of what we expected. Venus has no axial tilt: the solar heating is always greatest at the equator. How could the atmosphere be warmer, the further north one went? We stared at this, tired in the early hours of the morning from lack of sleep and all the excitement, and knowing that we had a press conference the next day.

After looking at the in-flight calibration data from VORTEX we convinced ourselves that the effect was real, and duly reported it to a somewhat bemused press conference. A few days later, the instrument scanned the north pole itself, and the cloud-top temperature channel went off the scale! This meant that the instrument had recorded a temperature that was higher than the highest end of the expected range. Again, this was real. David Diner took the polar data and succeeded in synthesizing it into an image, so we could see how the high temperatures were distributed. The results showed three successive shocks. First, it confirmed the polar warming we had seen on that sleep-deprived first night. Second, the hot feature at the pole was curiously chevron-shaped, and double, with one chevron at each side of the north pole, surrounded by a 1000 km wide band of very cold air, like a cold river running around the pole with two bright eyes staring out from inside. (Figure 1) Finally, we observed the nature of the temperature variations around the equator of Venus. Such variations were not unexpected, of course; the atmosphere warms and cools during the day, depending on where the Sun is in the sky. On Venus, however, the simple pattern we are used to on the Earth (warmest in the early afternoon, coolest just after dawn) did not seem to apply. Instead, the atmosphere showed high temperatures twice a day, separated by two cold minima.

In the fullness of time, these surprises became part of the body of knowledge about Venus's atmosphere, and spawned a new set of questions. The high polar temperatures are forced dynamically, and are intimately linked to the rapidly rotating cloud tops. The polar dipole and cold polar collar are symptoms of a gigantic polar vortex, part of a planet-wide circulation system in which air rises near the equator and descends at the poles, spun rapidly by the global super-rotation (although why the eye of the vortex has this curious double structure is still unknown). The double maximum in the daily temperature cycle is due to the dominance of the diurnal tide, a familiar but less pronounced feature of Earth's atmosphere which is amplified on Venus.

Other features emerged from the data. The main cloud tops have an overlying veil of haze that extends high above the visible clouds, and they rise and fall (once per day,

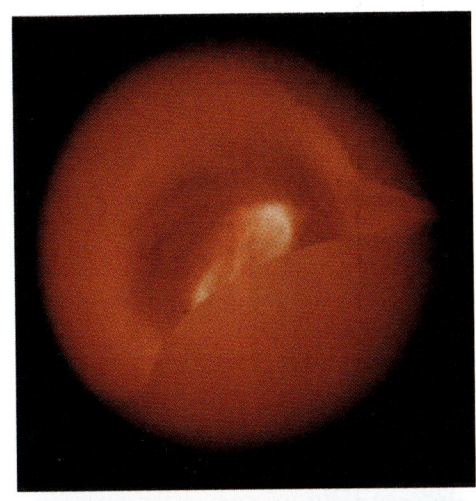

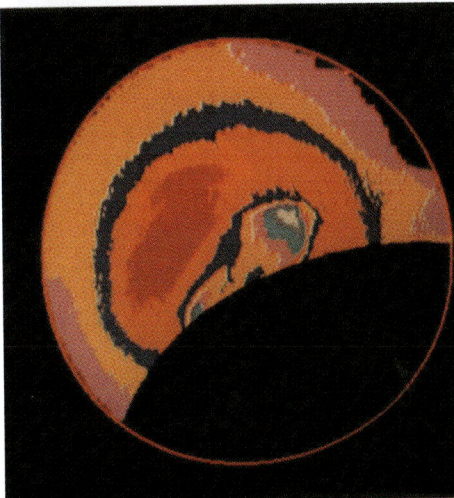

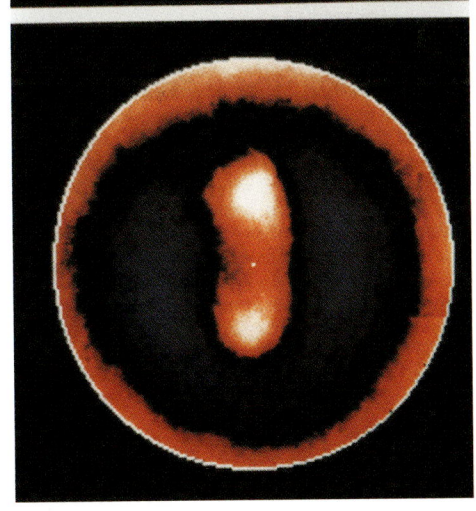

Figure 1 The Venusian polar dipole. Top: as observed by Pioneer Venus Orbiter OIR at 12.5 μm wavelength; middle: a contrast-stretched view; bottom: a time-averaged view, in a co-ordinate system that rotates with the dipole at a rate of once per 2.7 days.

unlike the temperature of the air) with the diurnal cycle. A patch of relatively wet air 1000 km across follows the Sun at the equatorial cloud-top; the heat of the Sun is either drawing the moisture up from below by convection, or baking it out of the cloud droplets. Either way, the feature has a very sharp boundary and needs to be studied more before it can be fully understood.

Remarkable features also emerged from the other instruments. The temperature of the thermosphere, the low-density region of the atmosphere above about 150 km altitude, is moderately warm (350 K) on the dayside, and extremely cold (100 K) on the nightside, with an almost unbelievably sharp transition between the two. The polarization measurements confirm that the clouds are concentrated sulphuric acid, and that the drop size is surprisingly constant at around 1 μm radius. Ultraviolet spectroscopy indicates that sulphur dioxide in the atmosphere at cloud-top level is very variable; declining by as much as a factor of 10 over a 10-year period. Perhaps this reflects the sporadic nature of the release of sulphurous gases from volcanoes on the surface.

The accumulated data from all the missions flown show that the lower atmosphere of Venus is composed mainly of CO_2 (96.5%) and N_2 (3.5%), with small amounts of the noble gases, principally argon, and of chemically active species including H_2O, CO, OCS, SO_2, HCl and HF. Variability in space and time is to be expected of reactive species, but the scale of the variability is virtually unknown in every case, and even the mean values are uncertain. For the best studied example, water vapour, values ranging from 20 to 25 000 ppmv have been published over the last 30 years. The highest of these came from direct sampling instruments on entry probes, and were in conflict with ground-based microwave observations. Measurements at frequencies near 1 cm wavelength placed upper limits on the water vapour abundance of 2–300 ppmv, assuming uniform mixing below the clouds.

The Venera 11, 12, 13 and 14 probes carried spectrophotometers that scanned the range from 0.45 to 1.2 μm with moderate spectral resolution (0.02 μm). Again using water vapour as the most important example of the results obtained from these early experiments, the probes found mixing ratios close to the modern value in the lower troposphere, with larger and more ambiguous values at higher levels near the cloud decks.

The Venusian clouds totally cover the planet at heights between about 45 and 70 km, with thinner hazes above and below. Instruments on the Venera 9 and 10 landers and the four PV entry probes provided simultaneous *in situ* measurements of the vertical structure and particle size distributions in the clouds. They found that the main cloud consists of three distinct layers, separated by relatively clear regions. These layers, known as the upper (~57–70 km altitude), middle (~49–57 km) and lower (~47–49 km) clouds, are bounded above and below by more diffuse haze layers. Sulphuric acid (H_2SO_4) aerosols are the principal constituent of all three layers, but the particle size distributions differ from layer to layer. The PV orbiter Cloud Photopolarimeter measurements showed that the upper haze layer is composed primarily of very small 'mode 1' particles, which have modal radii between 0.25 and 0.4 μm. The Large Probe Cloud Particle Spectrometer on the same spacecraft showed that the lower haze is also composed primarily of these particles. The upper cloud consists mainly of a second particle type, called mode 2, which has an equivalent radius of, ~1 μm but also includes significant numbers of mode 1 particles. A third particle type, mode 3, is the principal component of the middle and lower clouds. This consists of relatively large particles, in a distribution with a mean radius of 3.85 μm and some particles as large as 35 μm equivalent radius. Mode 3 may have a crystalline component of uncertain composition. The smaller modes are found mixed in with mode 3 at all levels where thelatter occurs.

The Vega 2 lander also carried a particle size spectrometer, the second of only two to be deployed on Venus to date. The Vega measurements revealed very few large particles at altitudes below 55 km, in marked contrast to the PV large probe, which of course entered at a different place and time. The measurements of cloud thickness by the nephelometers on the four Pioneer probes differed by less than ~20% in the middle cloud (50–57 km), but revealed much larger differences in the particle densities within the lower cloud (47–50 km). Similar results were obtained by the nephelometers aboard Venera 9, 10 and 11. It is now clear that the cloud opacity, the relative abundances of size modes and possibly the modes themselves vary considerably across the planet, and evolve in time.

Winds measurements below the cloud tops were obtained by tracking the PV, Venera and Vega entry probes as they descended through the atmosphere. These measurements show that the winds blow from east to west at all levels, with amplitudes that decrease almost monotonically with altitude. The descent probe measurements are again too limited in spatial and temporal sampling to discriminate between atmospheric waves and the time- and space-averaged motions. They also lack the accuracy needed (better than $\pm 1\,\text{m s}^{-1}$) to describe the much smaller zonal winds in the lowest scale height (~16 km), or the meridional winds at any level, well enough to understand the role that momentum transport by the meridional circulation and vertically propagating waves may play in the maintenance of the super-rotation.

Many other results from PV were realized, and are well documented in the literature, for example in the two books *Venus* and *Venus 2* published by the University of Arizona Press (see Further Reading). The Russian Venera series also had many successes which are reviewed in those publications.

The downside of this intense period of exploration of the Venusian atmosphere was that it was considered by many to have been 'solved' and its priority for future new missions fell. Since Venera 14 the only mission to Venus has been Magellan, in 1990. While this produced a spectacular study of the surface by radar, it had no atmospheric science component.

In the absence of any new focused mission, those still puzzling over the many questions raised by PV, like the nature of the polar vortex, the chemistry and variability of the cloud layers and, still, the forces driving the super-rotation, had a surprise bonus when astronomers studying Venus in the mid-1980s detected bright emission from the nightside at certain wavelengths in the near infrared. It soon became clear that this must be thermal emission from the lower atmosphere, diffusing through the clouds. In addition to providing a possible explanation for the long-standing mystery of the 'ashen light' on Venus, such a source offered a new and different approach to remote sensing of the composition and cloud structure of the deep atmosphere, previously accessible only at microwave and radio frequencies. The author had been fortunate enough to be asked to review the original manuscript by David Allen, in which he reported the discovery, for the journal *Nature*. In addition to providing an opportunity to suggest what turned out to be the correct origin for the emissions, this also led to calculations (by Lucas Kamp) of the emission from a model Venus atmosphere. This showed that it was in fact quite possible for infrared radiation to escape through the clouds at wavelengths between the strong absorption bands of carbon dioxide and water vapour, in particular at 1.7 and 2.3 μm, where Allen and colleagues had observed the highest fluxes. In not anticipating this sooner two things (at least) had been overlooked: firstly, that sulphuric acid droplets have a high single scattering albedo in the near infrared. This means that photons are more likely to be scattered than absorbed, and is one of the reasons that Venus has such a powerful greenhouse effect. That fraction of the incoming solar radiation that is not reflected can diffuse through the clouds towards the surface, just as the thermally emitted radiation can escape to be detected by Allen and his colleagues. The thermal flux peaks at slightly longer wavelengths, since the surface of Venus, although hot enough to glow, is of course much cooler than the Sun. The other key factor is the width of the atmospheric absorption bands: the simple Lorentz theory of molecular line shapes predicts that they should blend into each other and seal off the spectral windows, whereas in fact they clearly do not. A sub-Lorentzian line shape can explain the behaviour of carbon dioxide, in particular, at high temperatures and pressures.

Allen's measurements and Kamp's model also permitted an exciting prediction, namely that the 1.7 and 2.3 μm windows, plus a number of other windows not observable from the Earth because of absorption in the terrestrial atmosphere, could probe not only the cloud layers on Venus but all the way to the surface. To do this effectively would require a near infrared imaging spectrometer on a spacecraft making a close approach to, or better still in orbit around, Venus. Earth-based images of the dark side of Venus at 2.3 μm, which were soon being made by a number of observers, showed contrasts that appeared to be structure in the clouds. Nobody had seen the deep cloud structure on Venus before; it held important clues to the meteorological behaviour of the pressure-cooker environment on the Earth's twin.

At around this time (1989) the Galileo Jupiter orbiter spacecraft was at last being prepared for launch, having been seriously delayed by the accident to the space shuttle *Challenger* in 1986. As fate would have it, it was now programmed to reach Jupiter by means of close flybys of Venus and Earth, a consequence of the abandonment of the high-energy upper stage which had been declared unsafe for use in the Shuttle. Galileo carried the perfect instrument for observing Venus, the Near Infrared Mapping Spectrometer (NIMS). As a member of the NIMS team, the author made a half-serious attempt to persuade NASA to fire the orbit insertion motor at Venus, and forget about Jupiter! Although this proposal was treated with the expected disregard, permission was granted to turn NIMS on during the Venus encounter for scientific observations. Calculations showed that NIMS could achieve a spatial resolution on Venus of as good as 25 km, far better than Earth-based images. Also, it had a spectral range from 0.7 to 5.2 μm, covering all of the predicted windows.

The results, obtained in February 1991, were spectacular (Figure 2). Bright regions (white in the image) correspond to regions where the total cloud thickness is at least 20 times less than the thickest (dark blue) regions. The cloud veil on Venus is not the thick, uniform mass that it had for so long appeared to be. The morphology of the clouds speaks of intense weather activity, featuring large, cumulus-like clumps. Exactly what is going on is impossible to determine from the stationary snapshot that results from a flyby, but it can be said that the deep atmosphere of Venus must be a very active place.

In the brightest regions the spectral features of a number of atmospheric constituents show clearly. These include water vapour and carbon monoxide and, even at the low spectral resolving power of NIMS, it can be seen that the latter varies across the planet while water does not seem to. Andrew Collard showed in 1994 that the dominant trend is for carbon monoxide to increase from equator to pole, a surprising result since it implies a source for the gas at high latitudes. The author showed that this is unlikely to be from volcanism, since the amounts involved are much too large; instead, it seems most likely that the polar vortices are efficiently transferring CO-rich air from the upper atmosphere,

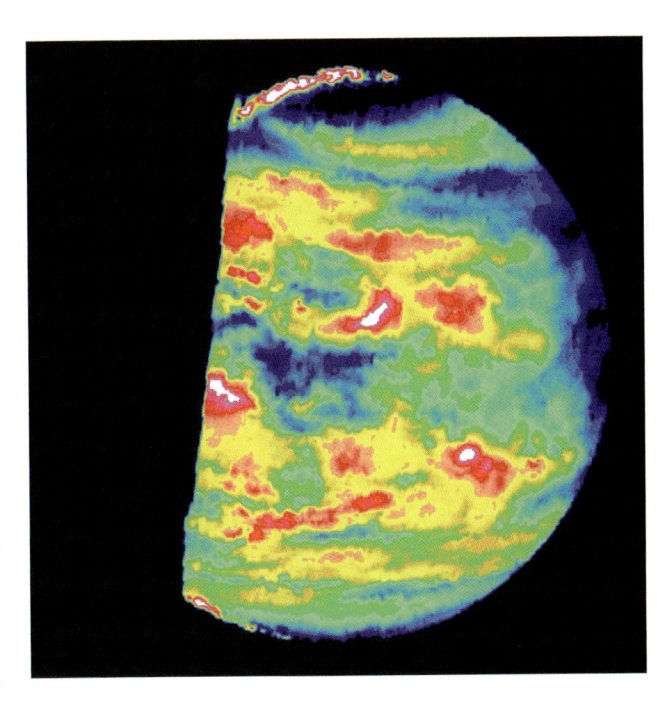

Figure 2 Venus at 2.3 μm wavelength, as observed by the Galileo Near Infrared Mapping Spectrometer in February 1991. (From Carlson *et al.* 1991.)

Table 5 Mixing ratios of minor gases measured from spectroscopic investigations of the nightside emission on Venus (After Taylor *et al.* 1997.)

Gas	Mixing ratio (ppmv)	Altitude (km)
H_2O	30 ± 10	26–45
	30 ± 10	15–30
	30 ± 15	0–15
D/H	120 ± 40 times terrestrial	26–45
CO	23 ± 7	30
	29 ± 7	40
HF	0.005 ± 0.002	30–40
HCl	0.5 ± 0.15	15–30
OCS	14 ± 6	30
	0.35 ± 0.1	38
SO_2	130 ± 40	35–45

where it is generated by the action of solar UV radiation on CO_2, to the layers near the surface. Systematic measurements of carbon monoxide from a future Venus orbiter should show the details of the general circulation at work for the first time.

Tables 1 and 5 show the accumulated knowledge of the composition of Venus's atmosphere from all sources, mostly PV direct measurements and Earth-based near infrared spectroscopy. Following the serendipitous opportunity afforded by Galileo, and the spectacular surface studies by Magellan, the drive to persevere with new studies of Venus declined sharply. Although understandable, this is not logical, and Venus enthusiasts left the twentieth century on a definite low. In part this is a consequence of the breathtaking success of the preceding decades, a breathing space to assimilate all that has been learned and to develop a fresh appreciation for the complexities represented by a planet and its atmosphere. In part it is a consequence of a renewed focus on Mars, which brings us to the next section.

MARS

Being almost cloud free, Mars is easier than Venus to study from the Earth, and its surface and atmospheric properties are considerably better known than those of Venus. The atmosphere has circulation and weather patterns similar to the Earth's in many ways, but also distinctively different in others. The planet-wide dust storms that occur every few years on Mars have no terrestrial counterpart. Mars shows dramatic evidence for major climate changes in the fluvial features on the surface, and isotopic evidence and solar wind interactions indicate that the composition of the atmosphere has changed with time. The implications of a warmer, wetter climate in the past lead us to consider the possibility that life could have started there, with the attendant hopes of the discovery of past or present indigenous lifeforms. The history of Mars exploration in the twentieth century was characterized by the repeated raising, and then dashing, of hopes on this front. Most notable in this respect was the canal hypothesis advanced by Lowell and others in the early years of the century, through the Viking landings with their life-detection instruments in 1976, right up to the very recent controversy over the nature of 'microfossils' in a meteorite believed to be of Martian origin.

Planetary missions have tended to alternate between Mars and Venus, as if paying attention to both simultaneously was too much of a strain. Although Venus was first with Mariner 2 in 1962, as the twenty-first century begins the emphasis is now very much on Mars. At the time of writing, Global Surveyor is in its main mapping phase, Pathfinder is dead but not forgotten, Mars Climate Orbiter and Mars Polar Lander are in flight and two more launches to Mars are planned for each of the next three windows of opportunity, in 2001, 2003 and 2005. The last of these may return samples to Earth. Even the ESA is at last making its first tentative move towards its planetary neighbours, lamentably late, with a mission called Mars Express, which, is now expected to launch in 2003.

The first spacecraft to explore Mars was Mariner 4, arriving on 14 July 1965. Its main objective was to photograph the surface, but for atmospheric scientists the key result came from the pioneering radio occultation experiment conducted under the leadership of Arvydas Kliore. This study of the effect of the atmosphere on the radio signal from Mariner as it passed behind the planet, and emerged, confirmed that the surface pressure was not more than about 0.7% of that of the Earth. This was tenuous indeed compared not only to the hopes and assumptions of those who dreamed about life, but also much less than some spectroscopists had inferred with some confidence from their telescopic observations only about a decade earlier.

Next came Mariners 6 and 7 in 1969. Apart from an erroneous detection of ammonia and methane, the most memorable atmospheric results from these flybys were the detection of carbon monoxide in the atmosphere, frozen carbon dioxide in the northern polar cap, and a dayside ionosphere. Mariner 9 became the first Mars orbiter on 14 November 1971, followed on 29 November and 2 December by the Soviet spacecraft Mars 2 and 3. One of the global dust storms occurred shortly before Mariner 9 acquired Mars orbit, obscuring the face of the planet, and did not clear until March 1972, when fortunately the spacecraft and its instruments were still fully operational. The ultraviolet spectrometer measured ozone for the first time, while the infrared spectrometer, a relatively high-resolution Michelson interferometer, detected no new species but performed the first global mapping of temperature and water vapour in the Martian atmosphere.

The definitive study to date of the atmosphere of Mars was accomplished by the Viking orbiter–lander missions launched in 1975. These were not the first spacecraft to land on Mars – Mars 2 crash-landed on 27 November 1971 – but they were the first to operate successfully and deliver detailed information about the atmospheric composition (Table 2). They also produced the only two detailed vertical temperature profiles for Mars that existed until the arrival of Pathfinder in 1998, and a global map of the water vapour distribution by latitude and season.

The great scientific success of Viking, with its long and fairly complete survey of basic conditions on Mars, combined with the disappointment of failing to detect any evidence for life, led Martian space science into the doldrums, much as Pioneer was to do for Venus five years later. Apart from the partial success of Phobos 2 in 1989, and some spectacular long-distance imaging by the Hubble Space Telescope, the investigation of Mars from space was essentially suspended for nearly 20 years.

This was not for want of effort in the scientific community. The author was a member of no less than eight study teams for Mars missions in the 1980s and early 1990s, all of which seemed to have good prospects at the time. On the American side, these included Mars Rover – Sample Return, which did not materialize because it was even more exciting financially than it was scientifically. Perhaps its main legacy was that it helped to usher in the era of low-cost missions advocated by Daniel Goldin at NASA. On the European side, the story was one of trying to initiate a planetary programme, which initially seemed a fairly easy task in the face of increased European unity and the desire for more, which had led to a successful and confident ESA. This reckoned, however, without the immensely powerful European astronomy lobby, who soon developed competing plans for a series of increasingly sophisticated optical, infrared, ultraviolet and gamma-ray observatories in space.

The first serious European attempt to mount a mission to Mars began in 1981, at the height of the success of the PV mission. Several of the investigators on the Venus mission were European, and there was a strong feeling among them that a Martian version had the potential not only to get Mars exploration going again after Viking, but also to remain within the less ambitious budget envelope of an ESA mission. A consortium led by Ulf von Zahn, who had developed the mass spectrometer experiment for PV, duly proposed a low-cost Mars orbiter called Kepler in one of ESA's regular competitions. Kepler won an industrial assessment study but lost out in the final round to SOHO/Cluster and then on the second attempt to the Huygens entry probe for Titan. At about the same time, NASA undertook a study of a similar mission called Mars Geoscience Climatology Orbiter, which later developed into Mars Observer. In 1985 a decision was made to link Mars Observer and Kepler to produce the first attempt at an international Mars mission, the so-called Mars Dual Orbiter.

Kepler was to have been a spin-stabilized orbiter in a highly eccentric (periapsis 150 km, apoapsis 7000 km) polar orbit about Mars with a period of 5.1 hours. Mars Observer was a three-axis stabilized polar orbiter in a close, circular orbit, 360 km above the surface. The Kepler orbit came so low that direct sampling of the upper Martian atmosphere would have been possible about five times per day. The latitude of periapsis changed by about 1° per day, to give good global coverage during the course of the mission, extending into both polar regions. Joint science aspects were defined, requiring two spacecraft operating simultaneously. Benefits would result for both missions because the two spacecraft could carry complementary instruments, they would be in different types of orbits and one spacecraft could observe the other. Launch of both spacecraft was planned for the September 1993 opportunity, which in the event only Mars Observer achieved.

A fresh European consortium was formed a few years later and began to consider a more ambitious Mars mission. In 1993 a new ESA phase-A study was completed jointly with NASA, on a mission, now called Marsnet, to place surface stations on Mars. The scientific objectives were to establish a network of small stations on the surface about 1000 km apart, from which the internal structure of the planet, the chemical and mineralogical character of Martian rocks and soils, the atmospheric circulation and weather patterns and the exobiological conditions existing on the surface could all be studied. An orbiter would act as a relay station and provide simultaneous data on the atmosphere, surface roughness and plasma environment.

Marsnet was a community proposal in Europe, which means it was submitted to ESA by a scientific consortium in response to a call for new mission proposals as part of the new selection cycle of ESA's long-term plan, known as Horizon 2000. After evaluating the proposals, ESA embarked on an industrial phase-A study of Marsnet. This was co-ordinated with NASA, in view of possible future co-operation on a joint ESA/NASA Mars network mission. NASA in parallel studied similar landers in the framework of the Mars Environmental Survey (MESUR) mission. The Marsnet and MESUR study teams worked together and exchanged representatives at their respective meetings. However, NASA aimed to deploy 18 probes on the surface of Mars and ESA only three. One mission priced itself out of the market, while at the same time making the affordable mission look inadequate, and neither MESUR nor Marsnet reached fruition.

Nevertheless, the concept of a network mission to Mars remained a high scientific priority; it still is. A revamped version of Marsnet was studied by ESA in 1992–93 under the name InterMarsnet, this time as a full-scale collaboration with NASA with the specific goal of a joint mission to be launched in June 2003. The NASA component was drawn from the Mars Surveyor programme, which already had Congressional approval. A joint ESA/NASA InterMarsnet Science Working Group was formed to support the phase-A activities from 1994 to 1996, together with engineering and industrial teams. An International Workshop on InterMarsnet was held in September 1995 on the island of Capri, a great success which was held by ESA to have demonstrated a fundamental and deep interest in Mars exploration among its wider international scientific communities.

The mission was again to consist of a network of stations on the Martian surface, but now with a more sophisticated orbiter capable of atmospheric sounding and imaging, roughness radar measurements and plasma environment monitoring. The main scientific goals of the mission were expanded somewhat, and given as: the study of the internal structure of the planet; the study of the surface morphology and geology at the landing sites; the geochemical and mineralogical analysis of Martian rocks, soils and volatiles; the study of the atmospheric circulation, structure and weather patterns; the study of the magnetic field and geodesy of the planet; and the study of exobiological conditions during the planet's history.

Before the conclusion of the phase-A study the ESA science programme was hit by a reduction of 10%, possibly to become 15% over five years. The financial pressure thus generated worked against InterMarsnet, which was deemed by ESA's analysts to be the most expensive of the competing missions. The less complex, cheaper and ESA-only COBRAS/SAMBA mission (now called Planck) was selected as the third medium-sized mission of the Horizon 2000 long-term programme, and once again ESA involvement in the exploration of Mars was postponed.

Over a period of about 15 years, three fully co-operative (in the sense of broadly equal proposed contributions) European–American missions to Mars were studied to a high level. In addition, there were several studies that either were carried out only at a very low level, or else were not truly joint because most of the funding was intended to come from a single international partner. Among these were Mars Aeronomy Orbiter, MESUR and Mars Rover Sample Return, all primarily American studies, with considerable guest involvement from Europe. A huge amount of energy was invested by all concerned, to no avail.

The frustrated European planetary scientists turned to the Americans and the Russians for flight opportunities. A number of European experiments were on board the Russian Mars 96 mission that failed shortly after launch. A 'silver lining' to this catastrophe was that it finally galvanised ESA into planning a Mars mission of its own: it commissioned a smaller, faster mission called Mars Express to carry spare or reconstructed versions of the instruments that formed the European part of the Mars 96 science payload to Mars in 2003. In contrast to the vigorous planetary programmes of the USA and the USSR, no European mission would reach any planet in the twentieth century, but the future looks brighter.

After an abortive attempt to study Mars with Mars Observer, discussed further below, success finally came for NASA in 1996 with the approval of the Mars Pathfinder mission, a very low-cost lander which also included a rudimentary roving vehicle called Sojourner. The community was excited: a new Mars mission at last! The following passages were written by the author in a 'Letter from America' to the members of the Royal Astronomical Society, just after Pathfinder landed. They hopefully capture some of the excitement that one feels during a planetary encounter. After a digression into the latest findings from the Galileo Jupiter orbiter, the piece goes on as follows:

> Anyway, I am supposed to be writing about Mars Pathfinder, and indeed at the moment the buzz here at NASA's 'Gateway to the Solar System' is all about that little shoebox-sized object crawling on our red neighbour. The Mall at JPL is blocked by press trailers, and the Laboratory has had to expand its Web site to handle the incredible public interest in Pathfinder – nearly 100 million 'hits' per day at its peak. Down the hill in Pasadena the film 'Contact' is playing; JPL is proud of the fact the cost of Pathfinder (somewhere between 150 and 200 million dollars, depending on how you count it) was no more than that of an expensive movie. The toy stores are sold out of model Sojourner rovers.
>
> I take a break from Galileo once a day or so to walk down to the Space Instrument Laboratory, where our own contribution to Mars exploration – a version of a weather satellite instrument developed at Oxford and now adapted to study the climate and dynamics of the thinner Martian atmosphere – is in final test before being shipped for launch next year. Today the engineers look gloomy, behind their cleanroom smocks and masks, because the instrument is malfunctioning and they don't yet know why. One of our team, a former student at Oxford and now one of the project scientists on Pathfinder, is just leaving and asks me if I want to come up with him to Mission Control.
>
> The first thing I notice is how crowded it is (how do they pay all these people on a shoestring budget?). They are mostly scientists, because it is night-time at the landing site on Mars and most of Pathfinder's equipment is turned off to conserve battery power. When sunlight again strikes the solar panels, the lander will wake up, the rover can move and data will again flow down from Mars. Meanwhile, the operations people have gone off to get some rest. Many of them hardly slept for days during the crucial landing sequence and its aftermath.
>
> The record of the landing is recorded in tables and diagrams strewn around a central table. One of them shows the accelerometer data from the impact: a series of big spikes, calibrated in g's, a few seconds apart but getting smaller and closer together as time goes on. Each spike marks a bounce of the whole spacecraft as it hit the surface, moments after cutting the parachute free and inflating the airbags. The airbags are developed from those we now have in our cars. They completely surrounded Pathfinder when it hit the ground, imparting the appearance of a giant raspberry to the latest visitor from Earth. It was still bouncing when the record relayed back to Earth comes to its planned end after a few minutes; whoever did the sums did not expect it to keep bouncing for such a long time.
>
> The 'geology' room is packed with people wearing red and green glasses and staring at the wall, which is completely filled with a huge, blurry picture barely recognizable as a 360-degree mosaic of the landing site in the ancient flood plain known as Ares Vallis. I slip on a pair of the 3-D glasses (smart plastic and glass spectacles, not the flat cardboard kind from comic books – this is the Space Programme) and the image leaps into startlingly sharp detail and relief, so realistic that I have to sit down

to look at it or risk falling over. Hills which I know are over a mile away seem close, the rubble on their slopes clearly visible. A shallow volcanic caldera on the horizon, over twenty miles away, can be dimly perceived through the thin veil of reddish dust which hangs always in the Martian atmosphere. The rover sits in the foreground, seeming to move as we look at it towards a large rock a few metres ahead. The whole scene is breathtaking, viewing it an experience which has almost nothing in common with the drab reproductions of parts of the same picture which I have seen in the newspapers.

As I leave the room and go back into the main area I bump into Jens-Martin Knudsen from the University of Copenhagen. He and I worked together on an ESA committee, and I know that he has an experiment on Pathfinder to investigate the magnetic properties of the Martian soil. This consists of a set of six permanent magnets of different strength, which the camera photographs every day to see how much dust is sticking. A similar experiment on the Viking Mars landers of two decades ago failed because the magnets were too strong, and they all became completely saturated with dust in the first day or so. He shows me the latest picture: this time they've got it right. After a couple of weeks, the strongest magnet has a thin coating and the weakest no noticeable deposit yet. Even the 'strong' magnet is very weak – the soil is very magnetic, he says.

In the 'atmosphere' paddock atmospheric modellers from Ames Research Center near San Francisco are poring over meteorological data from the lander. Plots show the temperature swinging up and down by more than 100 degrees from day to night. Yesterday the noontime maximum was about $-20°C$; on a warm day it gets above freezing. The pressure record shows a similar diurnal swing, and an overall downward trend. This is the effect of the carbon dioxide air freezing onto the south pole of Mars, which, it being midwinter there, is in permanent darkness and very cold. The wind is light at about 10 metres per second and very variable in direction, with no immediately obvious pattern to the swings from easterly to westerly. The meteorologists are anxious to get as long a record as they can, but, to keep it cheap, Pathfinder was only designed to work for a month and already its batteries are only accepting about half the charge they did at the beginning of the daily discharge cycle. Soon, other things may start to fail.

The geochemists are in a race against time to get as much data as possible, too. The power for the APX (alpha-proton-xray) spectrometer, which analyses the composition of rocks and soil, comes from a non-rechargeable battery. Unlike most of the systems on lander and rover, the APX can *only* work at night, because it has to be cold. Some of the early data was lost because it was turned on too soon after nightfall, and wasn't cold enough. A measurement takes several hours and the noise in the warm part of the integration swamped the information in the rest. Once they realized what was happening it was quickly fixed, and a string of large and small rocks and soil patches have now been analysed. The rocks have been given cute names – Barnacle Bill, Yogi, Cradle, Ginger – and the preliminary analyses are puzzling. The soil samples are all the same, not surprising for material which has been mixed around the planet as airborne dust, but each rock is different and, by some criteria, a couple of them have more Earth-like compositions than expected. This sort of data takes a long time to analyse properly, however.

In some ways the level of interest in Pathfinder is embarrassing to NASA, because it was never meant to be a mission of scientific exploration at all. With its short lifetime and just four very small scientific instruments, such new results as it gets will not answer any of the big questions about Mars, and were not intended to. Pathfinder is an engineering test bed, checking out cheaper ways of doing things like direct atmospheric entry and hard-landing safely on another world. Its original name – MESUR Pathfinder – reflected the fact that another sixteen or twenty identical landers were intended to follow once the test had proved successful. These would have been deployed all across Mars in a network in a mission called Mars Environmental Survey, or MESUR. This plan has now been abandoned, but the Mars programme still has a lot in store. Mars Global Surveyor, a large and heavily instrumented orbiter, is in space and approaching Mars as I write. Outside in the Mall a giant display board shows its trajectory, seeming to overshoot Mars as the planet moves in its orbit to catch up. Next year, a powerful orbiting station (with our weather and climate instrument on it, among others) called Mars Surveyor 98 will be launched along with a much more sophisticated lander/rover combination on a separate rocket which will eventually be deployed on the surface. No less than eight other Mars missions are in the pipeline between now and 2005, one of them the European Mars Express. This wave of activity will culminate in the return of selected Martian samples to Earth, and a new assault on the question of ancient warm climates and possible life forms. Everyone here hopes the wave of interest and hype in modest little Pathfinder will whet, and not saturate, the appetite of the public and the media for results from its more exotic siblings which will follow.

Before Pathfinder, NASA had launched Mars Observer in 1992 to study the present-day global climate, the environment on the surface of the planet and the geological evidence, clearly discernible even from space, that Mars used to be warmer and wetter, with rivers and seas. Disastrously, the spacecraft ceased transmission shortly before insertion into Mars orbit, resulting in a failure to acquire any data from the scientific payload. This included an instrument using techniques developed at Oxford, originally for the Earth's atmosphere, but also very similar to those employed at Venus on the PV mission discussed above. The new instrument, descended from, but much more sophisticated than, VORTEX was called the Pressure Modulator Infrared Radiometer, or PMIRR (MORTEX was thought to have unpleasant connotations!).

A new version of PMIRR was built for the Mars Climate Orbiter (MCO), a mission developed by NASA as part of its strategy to recover the scientific objectives of Mars Observer, once it was clear that the latter could not be rescued. The other parts are the 1996 Mars Global Surveyor and an orbiter mission in 2001. MCO was a small spacecraft (Figure 3), and the payload consisted entirely of the PMIRR atmospheric remote sounding instrument, together with a small camera, the Mars Color Imager, both dedicated to observing the lower and middle atmosphere of Mars globally and continuously. Launched at about the same time, the Mars Polar Lander and two 'microprobe' penetrators were to deploy on the surface of Mars to relay additional climate-related information.

PMIRR included channels to measure the global atmospheric and surface temperatures, water vapour profiles and airborne dust abundances on Mars. It had a limb and nadir scanning capability, intended to map the three-dimensional and time-varying thermal structure of the atmosphere from the surface to 80 km altitude, as a function of atmospheric pressure. At the same time, other channels would measure

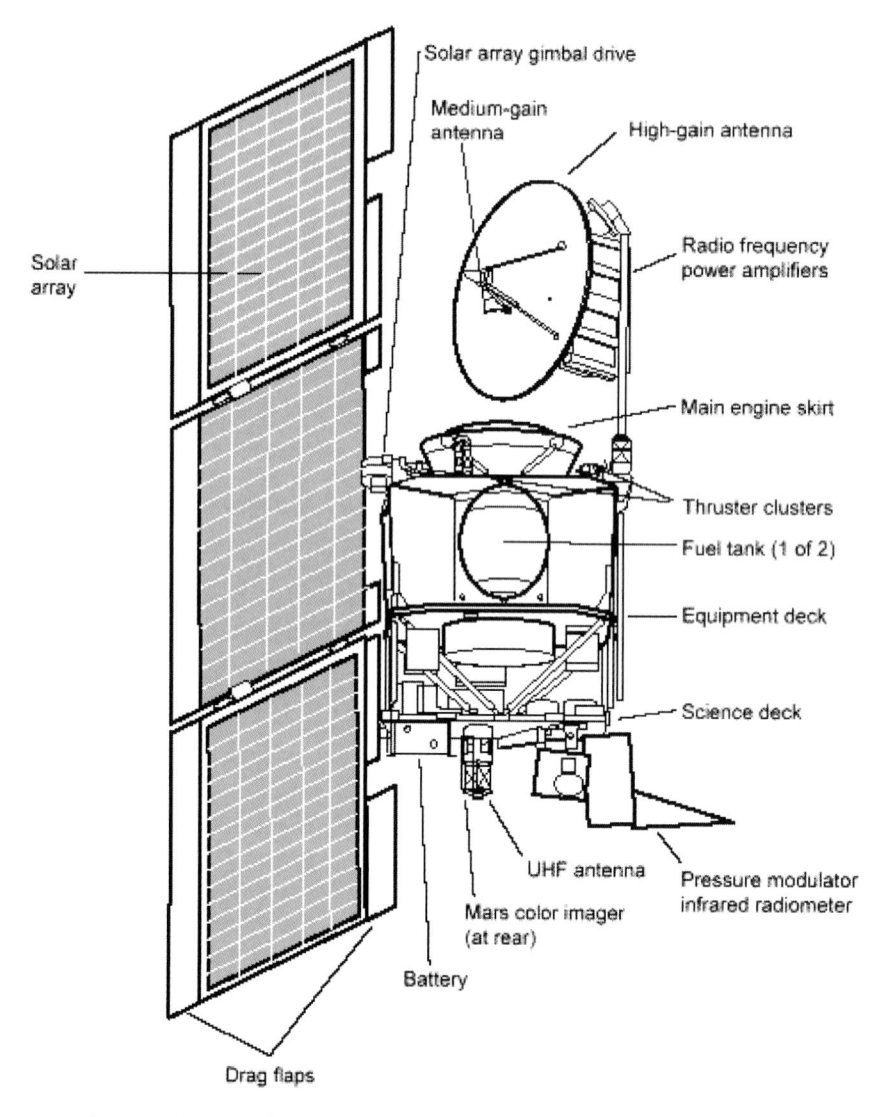

Figure 3 The 1999 Mars Climate Orbiter and its payload, the Pressure Modulator Infrared Radiometer (PMIRR) and the Mars Color Imager (MARCI). (NASA.)

the atmospheric dust loading and its global, vertical and temporal variation and the seasonal and spatial variation of the vertical distribution of atmospheric water vapour in the lower 30 km of the atmosphere. Multi-wavelength observations would be used to distinguish atmospheric condensates of H_2O and CO_2 from each other and from suspended dust, and to derive surface temperatures, albedo and thermal inertia to understand the atmosphere's interaction with the ground. The inclusion of some broad spectral bands permitted an investigation of the radiative balance of the polar regions, particularly in winter when the long period of total darkness leads to the condensation of about a third of the entire atmospheric inventory of carbon dioxide onto the polar caps.

The PMIRR instrument was large by the standards of most instruments on planetary probes, with a mass of 42 kg and using 41 W of power. It observes in a broadband visible channel and in eight spectral intervals in the 6–50 μm range in the thermal infrared. Spectral discrimination within this infrared range is achieved using multilayer interference filters and pressure-modulator units, the latter to provide high spectral resolution in the carbon dioxide (temperature-sounding) and water vapour bands.

At Mars, pressure-modulation radiometry offers the major advantage of being able to distinguish radiation originating from airborne dust from that emitted by atmospheric carbon dioxide and water vapour. This is especially important when limb sounding, a technique that is very sensitive to suspended dust, which is always present in significant amounts in the Martian atmosphere. Water vapour and carbon dioxide modulators are used to modulate the pressure of gas within a sealed optical cell, using dynamically balanced pistons driven at resonance. The signal-to-noise ratio in the pressure-modulator channels is improved by cooling the mercury cadmium telluride detectors and the focal plane assembly to the temperature of liquid nitrogen. The quantities of cryogen required for the long flight to Mars would be prohibitive, so a passive radiative cooler is used instead. This is basically a carefully insulated and shielded black plate, which faces cold space and cools by radiating heat away in the infrared.

The hardware was built by team members at the Jet Propulsion Laboratory in California under principal investigator Dan McCleese, in the author's department at Oxford University in England and in the Space Research Institute in Moscow, Russia, by Vasili Moroz and colleagues.

The general circulation of the Martian atmosphere, its meteorology, polar energy balance and conditions on the surface are all controlled by the annual cycles of carbon dioxide, water vapour and airborne dust, which the instrument observes as time-varying temperature–pressure, humidity, dust and derived wind fields. Most of what was previously known about the global dynamics of the atmosphere of Mars has been derived from general circulation models (GCMs) which use codes developed and tested on the Earth's atmosphere, suitably adapted to the Martian conditions of insolation, gravity, surface topology and so on. PMIRR aimed to obtain the first sufficiently detailed data to test these models, and to determine to what extent the atmospheric circulation on Mars is analogous to that on the Earth. Groups in Oxford and Paris developed a Mars GCM specifically for this purpose, including a facility for assimilating the experimental data from PMIRR and the surface station directly into the model. The assimilation procedure constrains the model towards agreement with the data, while producing diagnostics that indicate to what extent the measured properties of the atmosphere are consistent with the way the model has parameterized the physics. Terrestrial GCMs, for example those that are used in weather forecasting, are 'tuned' to give good results and it is unlikely that they will also do so under the different conditions found on Mars: the discrepancies are the result of interest.

The most dramatic and interesting weather phenomenon on Mars, apart perhaps from the massive cycling of much of the atmosphere between the polar caps, is the global dust storm. These occur in some years, yet for some reason not in others, and generally originate in the southern hemisphere during southern spring and summer, when Mars is closest to the sun and the rate of solar heating is greatest. Localized centres of dust-raising are common on Mars; sometimes they expand and coalesce until they form a band that fills a zone extending around the planet at heights up to about 50 km. This can then expand to other latitudes, including the northern hemisphere, over the course of a few days, resulting in a global storm. A few days later, clearing begins gradually and proceeds uniformly. The atmosphere takes from several weeks to a few months to return to normal conditions.

Airborne dust has a large effect on the atmospheric and surface temperatures, and on various components of the atmospheric circulation, even in the absence of global storms. The average amount of dust in the atmosphere is enough, through its effect on atmospheric solar and thermal radiation, to warm the air by 50 K or more. Such a large effect could be observed by PMIRR, along with the temperature fluctuations that accompany variations in the amount of dust and the large-scale thermal structure of the atmosphere prior to, during and after its onset. The vertical and meridional spreading and subsequent clearing of the dust itself, when mapped with good spatial and temporal resolution, should reveal its surface sources and sinks. With such data it will be possible to monitor the radiative forcing by dust, and to determine its effect on the residual-mean circulation and on stationary and travelling planetary waves, including atmospheric tides. These should in turn shed light on the dynamical processes at work during the onset, evolution and decline of dust storms.

There is also a seasonal water cycle in the Martian atmosphere. The evidence to date, mainly from Viking, suggests that this cannot be explained only in terms of the migration of water from the summer to the winter polar cap, via the atmosphere. Although this certainly occurs, on a grand scale, low-latitude sources apparently exchange water with the atmosphere as well. This is consistent with the belief that the water that once flowed on the surface of Mars is now frozen below the surface, with only about 10% or less in the polar caps. One interesting possibility is that the non-polar sources may be localized, especially if there is geothermal activity on Mars at the present time, and that these could be identifiable in the PMIRR water vapour maps. At any rate, PMIRR is capable of characterizing the loading and transportation of water by the different atmospheric regions as a function of location, altitude and time.

PMIRR can also detect water-ice hazes in the atmosphere, and enable estimates of water loss to the surface by condensation and snow. Co-located observations of temperature, water vapour, dust and ice hazes may also provide useful constraints on the horizontal transport of water ice, including ice condensed onto suspended dust particles.

The recession and growth of the Martian seasonal polar caps, due mainly to the annual cycle of condensation and release of atmospheric CO_2, is governed by their heat balance. The PMIRR wideband ultraviolet–visible–near infrared channel and radiometric filter channels in the thermal infrared measure the reflected solar and emitted thermal radiation emerging from the top of the atmosphere. These provide the radiative components of the energy balance of the north and south polar regions, while retrieved atmospheric temperatures and consistent model winds can be used to estimate the dynamical heat flux. Smaller terms include rates of subsurface heat conduction, determined through measurements of bare-ground and water-ice surface temperature variations, and the absorption or release of latent heat by ice and snow. The computation of these terms, constrained by the requirements of overall energy balance and mass conservation, will form a basis for a better understanding of the seasonal CO_2 cycle on Mars and the factors controlling it, including its interaction with the cycles of water and dust.

Thus, the Mars Climate Orbiter and the PMIRR sounder were built to carry out a major new study of Mars and its present-day climate. In addition to extending our knowledge and appreciation of terrestrial planet climate systems, it could also have led to a clearer idea of how Mars might have behaved in the past, when the climate may have been warmer and wetter. Several authors, most recently Haberle (1998), have commented on how difficult it is to model a more Earth-like early Mars without making unsatisfactory assumptions; better understanding of processes on present-day Mars is a prerequisite for further progress.

Another potential benefit is feedback into the modelling of Earth's present and future climate using physics-based computer codes, that is, GCMs. Important studies of global change on Earth are being carried out by many research groups but without a clear understanding of the domain in which the models are valid. Mars is in many ways a 'different' Earth which will provide a basic test of this kind of activity. But most of all we expect to learn a great deal about how Mars's atmosphere behaves currently, including its dramatic but poorly quantified dust, water and CO_2 cycles, and how these regulate the surface environment. This is not only of interest in its own right, it is of importance as a precursor for the success of future expeditions to our planetary neighbour.

The orbiter was launched on a nine-and-a-half-month journey to Mars on 10 December 1998 on a Delta II launch vehicle from Cape Canaveral in Florida. On arrival, the spacecraft fired its main engine intended to place it in a 14-hour elliptical orbit, so it could then use aerobraking to reduce the period of the orbit and make it more circular. This involves skimming through the upper atmosphere of Mars, using the drag on the solar panels to reduce the speed of the spacecraft. The final mapping orbit was to have been nearly circular at 421 km above the surface. Spacecraft and payload were designed to operate in this orbit for a full Martian year, 687 Earth days in length, which would have ended in January 2002 (Figure 4). In fact, the MCO crashed on the surface of the planet due to a miscalculation of the approach trajectory required to place it into Mars orbit. The result was the total loss, for the second time, of the scientific objectives and of PMIRR, which was the only current scientific space project investigating any terrestrial planet atmosphere, and certainly the most sophisticated of any of the twentieth century.

At the time of writing the full NASA report on the disaster is awaited and scientists are wondering when and how the instrument might be flown a third time. The cost of carrying out such a large experiment not once but three times is of course considerable, yet the rest of the Mars exploration programme to a greater or lesser extent relies on it. The global behaviour of the Martian atmosphere is still not well understood, because it has never been systematically observed in detail. There is no good data set with which to test and improve circulation and climate models, especially GCMs, and no plans to obtain such a fundamental data set except for PMIRR. The PMIRR instrument, although large and heavy by planetary instrument standards, is still state-of-the-art, especially if it could be combined with a nadir microwave sounder and a few surface stations. Without greatly improved knowledge of the atmosphere and its seasonal cycles, landing is more risky (because of uncertainty about

Figure 4 Mars Climate Orbiter as it should have been, orbiting Mars above the polar cap. (NASA.)

the winds), surface operations likewise (because of the threat from dust storms and so on) and attempts to model the past climate are less useful or plausible while we are unable to model the present one. Thus, the ambitious Mars programme that lies ahead needs MCO, or something rather like it, as an essential step on the way, quite apart from its intrinsic merits as stand-alone science, which are considerable.

Planetary research is a tough business for the experimentalist. PMIRR is one example of how not just years of hard work, but the hopes and aspirations of a number of committed individuals can be dashed forever by a trivial mistake or an unforeseen problem. *Say not the struggle naught availeth.*

THE FUTURE

In the first year of the twenty-first century, without MCO, there was no inner planet atmospheric investigation actually in space. What are the prospects for missions to these interesting and important, and still largely mysterious, neighbouring worlds in the years ahead? And what are the priorities following the accomplishments of the first 40 years of the space age?

For atmospheric science, the top priority must be to re-fly PMIRR or in some other way to address its atmospheric science objectives. The opportunity which has been selected is called Mars Reconnaissance Orbiter, now scheduled for launch in 2005. Plans for a network of very small orbiters, inspired by the Global Positioning System (GPS) network now in place around the Earth, may be brought forward to address the temperature sounding requirement by performing large numbers of satellite-to-satellite radio occultation soundings. These could be augmented by a number, possibly a large number, of small surface stations under a proposed successor to InterMarsNet now call Netlander. Both of these take advantage of the new opportunity to send multiple, small, low-cost missions to Mars as add-on or 'piggyback' payloads during Ariane 5 launches into Earth geostationary orbit. The water vapour and atmospheric dust studies might then be carried out by a smaller, simpler instrument than PMIRR using the latest detector technology to fly large infrared arrays.

In parallel with studies of these options, a lively programme is underway in the USA and France which is directed towards the return of selected samples of surface and sub-surface soil and rock from Mars for analysis in terrestrial laboratories. This is primarily directed towards the search for extraterrestrial life, as is the Beagle 2 lander attached to Mars Express, but new clues to the climate history of Mars, for example from isotopic analyses of the atmosphere and the surface, will also be obtained. Samples of the atmosphere and possibly of polar and sub-surface volatiles are likely to be returned at some stage. What happens after that will depend on what is gleaned from these samples about the climate history, and habitability and biology, if any, of Mars. The results from orbital missions like the successor to MCO and Mars Express on the general circulation and the present-day climate will also need to be assimilated. The biggest non-scientific question to be answered is whether crewed missions are essential and whether the political and popular will exists to face the costs and risks involved.

For Mercury, the initial reconnaissance is not yet complete. We can look forward to the first Mercury orbiter, which is likely to concentrate on mapping the surface and probing the interior but which also will look for volatiles in the polar regions, and hopefully will study the atmospheric variability and identify its sources and sinks. Both ESA and NASA have Mercury orbiters in their approved planning cycles and the launch of at least one of these in the first decade of the twenty-first century is reasonably assured. A sample return mission has already been the subject of a detailed study by ESA, published in 1999. Beyond that, we may anticipate that during the next 100 years, a crewed base (a solar observatory, for instance) will be set up on Mercury in the temperate region near the poles, perhaps drawing on the ice deposits.

The future for Venus is, like the planet itself, more cloudy. Most regrettably, no agency in the world has any mission planned to our intriguing neighbour. Although the Committee for Planetary and Lunar Exploration (COMPLEX) of the US National Academy of Science has recommended that 'the characterization of the basic structure, composition and

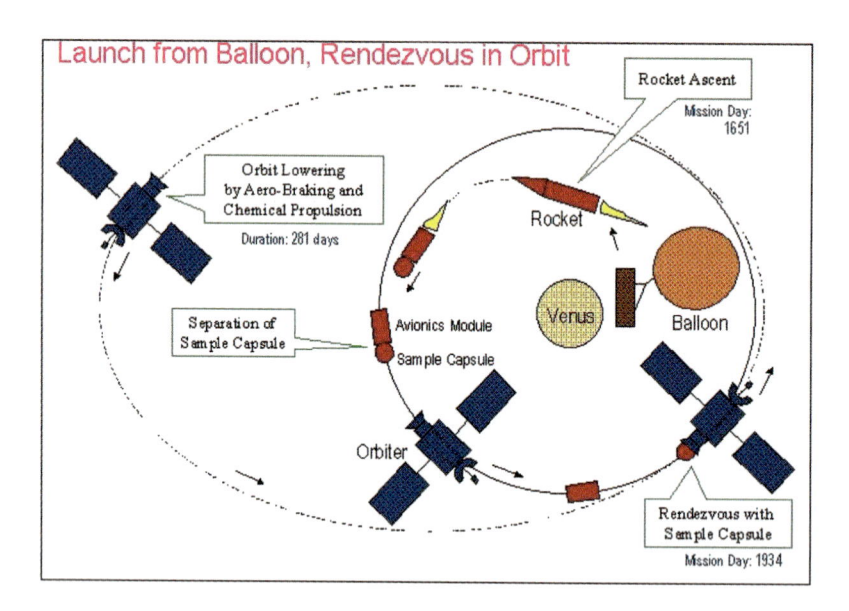

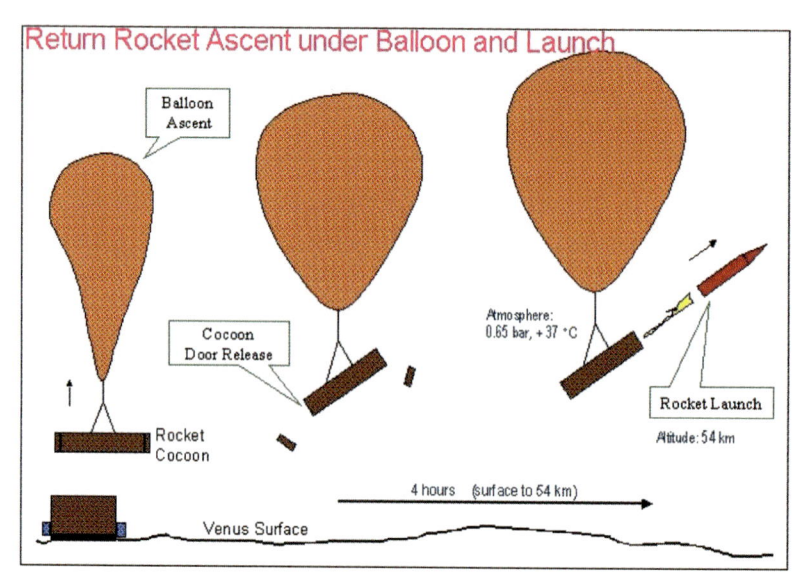

Figure 5 Concept for a Venus Sample Return Mission. (ESA.)

dynamics of Venus's atmosphere be a primary objective of planetary exploration', NASA has turned down a number of community proposals for low-cost Venus atmospheric missions. One of these will likely form part, sooner or later, of its Discovery programme.

A large number of key scientific questions about Venus remain unsolved. To list only the four most prominent, relating to the atmosphere, we have:

- Super-rotation. Why does the atmosphere of Venus rotate so much more rapidly than the planet? Observations of the ultraviolet markings alone cannot begin to answer this question, and theoretical models do not explain satisfactorily how the momentum required for a super-rotation at more than 50 times the surface period is transferred into the upper atmosphere. Global temperature and wind measurements as a function of height would seem to be a prerequisite for making progress on the problem.
- Deep meteorology. The complex cloud structures evident in the near-infrared window images, especially those at high spatial resolution from Galileo, are witness to a complexity of meteorological activity in the lower atmosphere of Venus that had previously not been suspected. The Galileo flyby provided only snapshots; systematic monitoring by a deep-atmosphere sounder is the next step.
- Polar dipole/collar. Nearly 20 years after its discovery, the polar dipole/collar complex at high latitudes on Venus remains a mystery. Again, systematic monitoring is essential; even a single wavelength 'movie' at high spatial resolution would begin to show its detailed structure and dynamical behaviour.
- Cloud chemistry. What maintains the dense sulphurous cloud layers on Venus? Is active vulcanism an essential part of the picture? What role do species such as CO and COS play in the chemical-dynamical cycle? A new orbiter could offer the first global, multilevel high-spatial-resolution views of clouds, cloud-tracked wind fields, surface and upper-level temperature fields and the distributions of several chemically active species, including the important greenhouse gases SO_2 and H_2O. Such maps and movies will permit an unprecedented understanding of the 'weather', the general circulation and the climate of Venus including the massive, greenhouse-produced troposphere below the clouds, the $200\,km\,h^{-1}$ winds and the mysterious polar vortices.

ESA has studied a Venus Sample Return project (Figure 5), although the main outcome of this was to reinforce the intuitive belief that such a mission would be extremely difficult, taking over six years and with many risky elements. The cost is well out of the reach of ESA's budget for the foreseeable future, and indeed probably of anyone else's. Like the crewed Mars mission, it may take major political reforms leading to a single space agency in a peaceful and united world before such an ambitious mission is readily taken on. It would take a brave soothsayer to put a date on when we will hold a piece of Venus's surface, or a bottle of its atmosphere and cloud material, in our hands. Looking on the bright side, it will certainly be before the turn of the *next* millennium.

FURTHER READING

A very comprehensive account of the scientific knowledge of each of the inner planets, with virtually all of the relevant references, up to the date of publication, is given in the publications produced by the University of Arizona Press. The relevant ones are listed here.

REFERENCES

Bougher, S.W., Hunten, D.M. and Phillips, R.J. (eds) (1997). *Venus 2*, University of Arizona Press, Tucson, AZ.

Carlson, R.W., Baines, K.H., Encrenaz, T., Taylor, F.W., Droissart, P., Kamp, L.W., Pollack, J.B., Lollouch, E., Collard, A.D., Calcutt, S.B., Grinspoon, D., Weisman, P.R., Smythe, W.D., Ocampo, A.C., Danielson, G.E., Fanale, F.P., Johnson, T.V., Keiffer, H.B., Matson, D.L., McCord, T.B. and Soderblom, L.A. (1991). Galileo infrared imaging spectroscopy measurements at Venus. *Science*, **253**, 1541–1548.

Carr, M.H. (1996). *Water on Mars*, Oxford University Press, New York.

Esposito, L.W., Knollenberg, R.G., Marov, M.Ya., Toon, O.B. and Turco, R.P. (1983). The clouds and hazes of Venus. In D.M. Hunten, T.M. Donahue and V.I. Moroz (eds), *Venus*, University of Arizona Press, Tucson, AZ., pp. 484–564.

Grinspoon, D.H. (1997). *Venus Revealed*, Helix Books, Addison-Wesley.

Haberle, R.M. (1998). Early Mars climatic models. *J. Geophys. Res.*, **103**, No. E12, 28467–28479.

Hunten, D.M., Donahue, T.M. and Moroz, V.I. (eds) (1983). *Venus*, University of Arizona Press, Tucson, AZ.

Kieffer, H.H., Jakosky, B.M., Snyder, C.W. and Matthews, M.S. (eds) (1992). *Mars*, University of Arizona Press, Tucson, AZ.

Strom, R.G. (1987). *Mercury, the Elusive Planet*, Cambridge University Press.

Taylor, F.W., Crisp, D. and Bèzard, B. (1997). Near-infrared sounding of the lower atmosphere of Venus. In S.W. Bougher, D.M. Hunten and R.J. Phillips (eds), *Venus 2*, University of Arizona Press, Tucson, AZ, pp. 325–351.

Vilas, F., Chapman, C.R. and Matthews, M.S. (eds) (1988). *Mercury*, University of Arizona Press, Tucson, AZ.

58

DONALD M. HUNTEN*

Jupiter

Jupiter is the largest planet and is the prototype of the giant planets, the others of which are Saturn, Uranus and Neptune. It possesses four large, icy moons, a huge and intense magnetosphere, and a unique plasma torus surrounding the orbit of the satellite Io. It has therefore been a prime target for exploration, and in addition, its huge mass has made it a valuable target for gravity assists. The Pioneers and Voyagers used Jupiter in this way to reach Saturn, Uranus and Neptune, and Ulysses used it to obtain a solar orbit of high inclination. The announcement of the Pioneer opportunities in 1968 led to the establishment of the first real Jupiter community, and the measurements by these spacecraft taught us a great deal about the environment of the asteroid belt and Jupiter's magnetosphere, as well as the planet itself. Once they had been launched, studies were initiated for an ambitious, highly reliable spacecraft called Grand Tour but they were never implemented; instead they were replaced by the less-expensive Voyagers. These two spacecraft made history as they visited all four giant planets and also made important measurements of Jupiter's Galilean satellites, as well as Titan and Triton. Galileo included an entry Probe to explore Jupiter's atmosphere and an Orbiter to define the magnetosphere and the satellites. In spite of a delay of many years before it could be launched, and a loss of the reflector of the high-gain antenna, it has been a spectacular success. Ulysses made some measurements of the environment, and Cassini will do the same, as well as providing remote sensing of the planet and satellites.

THE PIONEERS

In 1968, NASA issued an Announcement of Opportunity to solicit experiment proposals for the two spacecraft that would eventually be named Pioneers 10 and 11. At that time there was no real Jupiter 'community'; our body of knowledge was expressed in occasional papers but there was little or no discussion at scientific meetings. Jeffreys had shown that the temperature of the surface [or cloud tops] must be close to the solar-equilibrium value, and Menzel had confirmed this by analysis of thermal-infrared observations (see Wildt 1969). Jupiter's small mean density had led to the inference that hydrogen and helium were major constituents, and models of the interior structure had been worked out (DeMarcus 1958). Öpik (1962) had published a detailed analysis of the composition, leaning heavily on results of a pioneering observation of a stellar occultation by Baum and Code (1953). Unfortunately, the scale height obtained from these data was much too small, and Öpik obtained a helium abundance that was correspondingly much too large. Münch had pointed out the important role of pressure-induced absorption by H_2 in controlling the thermal opacity, and models of the thermal structure involving this opacity had been worked out (Trafton and Münch 1969).

Radio emission from the planet and its magnetosphere were early discoveries as radio astronomy underwent its rapid development after the Second World War. Convenient summaries appear in Berge and Gulkis (1969) and Carr and Desch (1969). Analysis of the synchrotron emission by electrons in the radiation belt allowed an estimate of the strength and orientation of Jupiter's magnetic field. Intense pulses at decametric wavelengths were eventually found to be linked to the position of the satellite Io, and Goldreich and Lynden-Bell (1969) proposed that the mechanism involves a unipolar generator as Jupiter's rapid rotation sweeps the magnetic field past the satellite. The thermal emission by Jupiter's atmosphere appears in the wavelength region from a few mm to tens of cm. The principal opacity source is ammonia in the pressure region of a few bars, and

*University of Arizona, Tucson, AZ, USA

allows estimates of the ammonia abundance and the temperature profile (Berge and Gulkis 1969, de Pater and Massie 1985). Measurement of the thermal radiation of the planet had shown that it radiates roughly twice as much energy as it absorbs from the Sun (Low 1976).

The principal objectives of the Pioneers would be to measure the zodiacal light, and therefore the density of interplanetary dust; to diagnose the potential hazard to spacecraft of the asteroid belt; and to make an initial reconnaissance of Jupiter and its radiation belts. One result of the announcement of opportunity was the formation of a Jupiter community, much of which was represented at the Third Arizona Conference on Planetary Atmospheres (1969); several of the papers presented there have been quoted in the preceding paragraph. By the time of the Pioneer encounters (December 1973, December 1974) a large body of Jupiter science had accumulated. This science and the Pioneer results are represented in the book 'Jupiter' (Gehrels 1976).

While Pioneers 1 and 2 were being built and on their way to Jupiter, a number of valuable results were obtained from Earth-based studies. First came the 1971 occultation of the star β Sco by Jupiter. It was successfully observed by three groups: Combes et al. (1971), Hubbard et al. (1972) and Veverka et al. (1974). A critical review by Hunten and Veverka (1976) concludes that the temperature is 170–200 K in the mesosphere (number density 5×10^{14} cm^{-3} or pressure near 3 μbar). Two years later (1973) Brown discovered strong emission of the yellow D lines of Na in the vicinity of Io (Brown and Yung 1976), and this was followed by the even more remarkable discovery of the Io Plasma Torus (Kupo et al. 1976). This object and its physics are described in several chapters of the book by Dessler (1993). One of the first discoveries of the Kuiper Airborne Observatory was that of Jovian water vapor (Larson et al. 1975). The absorptions are strong, but they can only be observed from high altitudes because those of the Earth's lower atmosphere are even stronger.

Returning to the Pioneers, the Jupiter-oriented instruments included an Imaging PhotoPolarimeter, a two-channel UltraViolet Photometer, a Magnetometer, and a 2-channel InfraRed Radiometer. Occultation by the ionosphere and atmosphere was observed by means of the telemetry carrier, and tracking of the spacecraft gave information on the mass and gravity field. Charged particles from plasma to cosmic rays were measured by four instruments, and dust particles by two. Although Pioneer 10 passed right through the Io Plasma Torus, its plasma instrument was oriented in the wrong direction to readily detect it, and a detection was not reported until several years after the encounter (Intriligator and Miller 1981). The principal discoveries concerning the planet were:

- The images showed that the well-known cloud structure of belts and zones is supplanted at latitudes above 45° by structures resembling large-scale convection.

- The IR Radiometer confirmed that the planet radiates approximately twice as much energy as it absorbs from the Sun, and showed that this total radiation is essentially independent of latitude (Ingersoll et al. 1976). Because the solar heat is deposited primarily at low latitudes, the internal heat must therefore come out preferentially at the higher latitudes, probably as a consequence of the special properties of convective heat transport (Ingersoll and Porco 1985).

- The radio occultation experiment measured the electron densities in the ionospheres of Io and Jupiter. Initial results for the density of Jupiter's neutral atmosphere were very strange; the anomaly was eventually traced to neglect of the planet's oblateness, and corrected temperature profiles are given by Kliore and Woiceshyn (1976). The exospheric temperature was found to be an astonishingly hot ~ 1000 K, instead of the <200 K expected for heating by solar UV. Primary candidates or the additional heat source are absorption of wave energy that has propagated up from the lower atmosphere (Yelle et al. 1996), and precipitation of ions and electrons from the magnetosphere (Hunten and Dessler 1977).

THE VOYAGERS

During 1976 studies were under way at the Jet Propulsion Laboratory for the 'Grand Tour'. These two spacecraft were designed for the long life and extreme reliability needed to take advantage of a forthcoming opportunity to visit all four of the Jovian planets by use of gravity assists. Accompanying the spacecraft studies were studies of potential instruments. Right at the end of 1976, the U.S. Congress withdrew support for these ambitious missions, and gave it instead to a pair of missions that would be based on the Mariner line that had successfully explored Venus, Mars and Mercury. Initially called 'MJS' (Mariner Jupiter-Saturn), these spacecraft were later re-named 'Voyagers 1 and 2'. Fortunately the planetary alignment did not go away and the Voyagers kept working, so both missions were eventually extended to carry out the exploration of the Uranus and Neptune systems.

The Voyagers arrived at Jupiter in March and July 1979. With 3-axis stabilization each spacecraft was able to carry a pair of cameras, much more powerful than the spin-scan instrument on the Pioneers. Figure 1 shows two false-color mosaics, one each for Voyagers 1 and 2, showing one complete rotation (actual colors are various shades of yellow). An UltraViolet Spectrometer (UVS) was used in a novel mode, in which it observed occultation of the Sun by the upper atmosphere. It was therefore possible to obtain height profiles of the major gases, especially H_2, H and CH_4. The Pioneer occultation experiments of the telemetry carrier

Figure 1 Comparison of two cylindrical projections of Jupiter, made from images taken by Voyagers 1 and 2. Longitudes are 400 to 0 degrees in the top image, and the bottom one is aligned with it. Colours have been enhanced in the processing. Relative motions of features can be seen; for example, the Great Red Spot has moved westward and the white ovals eastward. Regular plume patterns are equidistant around the northern edge of the equator, while a train of small spots at approximately 80 degrees south latitude has moved eastward. Significant changes are evident in the recirculating flow east of the Great Red Spot, in the disturbed region west of the Great Red Spot, and in the brightening of material spreading into the equatorial region from the more southerly latitudes. (P-21772 C.)

were repeated. Both measurements confirmed the very high exospheric temperatures, ~1000 K, found by the Pioneers, but the height of the ionosphere was much greater than expected (McConnell *et al.* 1982). Presumably the ionospheric plasma is raised by vertical drifts.

The UVS also observed the airglow from the same atmospheric region. Its most striking discovery was of intense far-ultraviolet emissions from the Io Plasma Torus, making it clear that this body is the seat of enormous energies, much greater than had been expected from the ground-based studies mentioned above. The principal constituents are ions of O and S, each in several charge states. The source is Io's atmosphere, of which SO_2 was found by IRIS, the InfraRed Interference Spectrometer to be a (or the) major constituent (Pearl *et al.* 1979). When this gas escapes from the satellite, it is dissociated and ionized by torus electrons, and the ions are then accelerated to co-rotation with Jupiter by the latter's magnetic field. This acceleration is a major source of energy, which probably powers the entire medium. The plasma temperature is typically several eV or several hundreds of thousands of degrees K, with a component of electrons at several hundred eV (Brown *et al.* 1983). After the ions are neutralized, their velocity is high enough to carry them to distances of ~100 Jovian radii, where they can again become ionized and join the magnetospheric population. As they work their way back inward, they are accelerated to energies in the millions of eV and contribute a major part of the magnetospheric population.

A detailed analysis of the infrared spectra has been presented by Carlson *et al.* (1993). Confirming earlier Earth-based work, they found the abundances of the ice-forming vapors CH_4, NH_3 and H_2O to be somewhat greater than 'solar', the values they would have if all the C, N and O were converted to these molecules. The NH_3 and H_2O were depleted at the higher altitudes and lower temperatures roughly as expected for condensation into the cloud particles that are inferred to be present. The excess is generally thought to be accounted for by late accretion of icy planetesimals which would be vaporized in the atmosphere. A careful analysis of the energy balance confirmed that Jupiter radiates 1.668 times as much as it receives from the Sun (Hanel *et al.* 1981). The excess heat from the interior is believed to arise from a combination of residual primordial heat and release of gravitational energy by a slight shrinkage of the planet.

Some of the images of the night side showed bright spots attributed to lightning flashes. Close analysis (Borucki and Williams 1986) showed that they had the size and elliptical shape expected if the flashes had occurred at the 3–5 bar level of the water clouds and the light was diffused by the ammonia clouds at ~0.6 bar.

GALILEO

The idea of a Jupiter Orbiter-Probe (JOP) mission received a great deal of study even before the launch of the Voyagers. The Jet Propulsion Laboratory had organized a Science Advisory Group as early as 1971 and JOP was recommended by the U.S. National Academy's Space Science Board (SSB) after consideration by its Committee on Planetary and Lunar Exploration (COMPLEX). It was proposed in the U.S. President's budget for consideration late in 1977, but was omitted from the version of the budget prepared by the Budget Committee of the House of Representatives. A major stated reason, which turned out to be all too valid, was that the Space Shuttle and the necessary Centaur upper stage would not be ready in time to launch the mission. In response, the U.S. planetary community staged a supporting campaign, which was successful in that the House rejected the recommendation of its own committee and appropriated the funds necessary to start the mission, which was soon re-named 'Galileo'. As the date of availability of the Shuttle slipped again and again, the same thing happened to Galileo. Development of the Shuttle version of the Centaur upper stage caused further delays, but in 1985 the launch system was declared ready and the spacecraft was shipped to Cape Canaveral for launch preparations. The JPL crew took time off from these preparations to watch the launch of Challenger and its disastrous breakup. Investigation of the reasons imposed a delay in the launch date of Galileo. Next, NASA cancelled the adaptation of the Centaur stage to the shuttle, on the grounds that it would be too dangerous to launch with a payload containing such large quantities of liquid hydrogen and oxygen. It appeared that there would be no way to get Galileo to Jupiter by means of the available solid-fueled upper stages, until it was realized that the necessary velocity could be gained by means of gravity assists, a flyby of Venus followed by two of the Earth. Such a trajectory required the addition of a system of sun shades to Galileo, which had not been designed to withstand the additional heating that it would encounter so close to the Sun. After a further delay of three years, the mission was finally launched in October 1989. Because of the three extra orbits, Galileo did not arrive at Jupiter until December 1995. During its transit, Ulysses obtained its Jupiter gravity assist in February 1992. It made some measurements of the plasma environment, but was not equipped with any instruments that could sense the planet itself.

Shortly after the first Earth encounter a command was sent to Galileo to unfurl the reflector of the High-Gain Antenna, which was a duplicate of one that had been used on several Earth satellites operating in geosynchronous orbit. Although the motor responded, something in the system stuck and the reflector only partially opened. In spite of many attempts to complete the deployment, none was successful and the mission had to be carried out with a low-gain antenna with a bit rate of only a few tens of bits per second. A massive re-programming of the spacecraft's central computer was successfully undertaken, providing the ability to compress the data to the greatest possible extent. There was no loss of the relayed Probe data, but considerable restriction in the amount of data that could be taken by and transmitted from the remote-sensing and magnetospheric instruments on the Orbiter.

The Jupiter objectives of Galileo, and the corresponding instruments, were reviewed by Hunten et al. (1986). Many of the instruments were carried into the atmosphere by the Probe; they comprised

- the Atmospheric Structure Instrument, measuring temperature, pressure and acceleration;
- the Mass Spectrometer, measuring composition throughout the descent;
- the Helium Abundance Interferometer, dedicated to an accurate measurement of this one gas;
- the Net Flux Radiometer, to measure this quantity in several IR and visible wavelength bands;
- the Lightning and Radio Detector, also including a detector for very energetic electrons and ions in the innermost magnetosphere;
- equipment to measure the Doppler shift of the Probe-Orbiter relay signal and therefore obtain information on the descent rate and winds.

The Orbiter carried a magnetometer and a suite of instruments to measure the fluxes of ions and electrons as a function of energy in the torus and magnetosphere. More closely related to the planet and the satellites were an imaging system, the Near Infrared Mapping Spectrometer (NIMS), and the Photo-Polarimeter Radiometer (PPR), with many channels from the visible to the far infrared.

These three instruments had an unexpected workout in 1994 when the fragments of Comet Shoemaker-Levy 9 entered Jupiter's atmosphere at locations visible from Galileo but not from the Earth (Carlson et al. 1997). Many other measurements of the impact sites were obtained from Earth after they had rotated into view. Here only three will be mentioned: detection of cometary water vapor from the KAO (Bjoraker et al. 1996, Sprague et al. 1996) and non-detection of seismic waves propagating away from the impacts, which would have given a probe of Jupiter's interior (Walter et al. 1996). Generally speaking, this event taught us far more about the comet and the nature of the impact phenomena than about Jupiter.

The entry site of the Probe was constrained to a very low latitude for various reasons, including the need to take advantage of Jupiter's rapid rotation to minimize the entry velocity relative to the atmosphere. This velocity, of order $50 \, \text{km s}^{-1}$, was much greater than the $11 \, \text{km s}^{-1}$ experienced by Apollo capsules returning from the Moon, and the energy per unit mass greater by a factor of 20. The challenge of designing a successful entry system was therefore a major one. Unfortunately, the equatorial region is the site of extensive areas with much less cloud than exists elsewhere. These areas are called 'hot spots' not because they are any warmer than other areas, but because they appear bright in 5 μm radiation, transmitted through holes or thin areas in the clouds from a deep level of the atmosphere that is hidden from view in normal cloudy areas. The Probe entered one of these regions and found no sign of the water-ice cloud whose study was a major objective of the mission. Fortunately, remote sensing by the NIMS confirmed earlier ideas that there is a gradient of water-vapor abundance at the edge of a hot spot, with much larger values outside.

Great emphasis was placed on an accurate measurement of the helium abundance, which at the time the payload was chosen was regarded as representing the primordial value. Later it was realized that some helium might sink into the interior through the formation of He-rich droplets at a pressure level of a few megabars, and this idea was supported by Voyager measurements suggesting that the Jovian abundance is actually significantly less than solar, and that it is greatly depleted in Saturn's atmosphere. The Galileo results obtained by the Mass Spectrometer and the Helium Abundance Interferometer agreed almost exactly, and the implications are discussed in detail by von Zahn *et al.* (1998). There is no reduction relative to the solar value, which is itself slightly reduced from the primordial one by diffusive separation in the radiative zone.

Much of the remote-sensing work from the Orbiter concerns the satellites; a collection of papers appears in the Special Issue of *Icarus* on Remote Sensing Results of the Galileo Orbiter Mission (1998). Here we shall allude only to the observations of Jovian auroras by the imaging subsystem (Ingersoll *et al.* 1998). This paper gives references to previous work in the UV by Voyager, the International Ultraviolet Explorer and the Hubble Space Telescope, in soft X-rays, and in the near IR by Earth-based telescopes. Galileo, with good views of the night side, was able to obtain sharp images in visible light. The auroral oval occurs on magnetic field lines that intersect the equatorial plane at a distance of 13 Jovian radii, well outside the Io Plasma Torus at 6 R_J.

CASSINI

In 1984 a joint committee of the U.S. Space Science Board and the European Science Foundation was formed to study potential joint missions of NASA and ESA, the European Space Agency. Their choice was a Saturn Orbiter-Titan Probe (SOTP, now called 'Cassini-Huygens'). The Orbiter would be (and now is) the responsibility of NASA and the Probe that of ESA. After Venus and Earth gravity assists, Cassini flew past Jupiter at the end of 2000 on its way to its orbital tour of the Saturn-Titan system. During the Jupiter approach and departure it transmitted far more images to Earth than the entire yield from Galileo and other experiments were just as successful. The reference item Cassini at Jupiter (2002) is a set of papers concerning Jupiter's magnetosphere and aurora.

This chapter has focused on the history and politics of the exploration of Jupiter and has mentioned the principal scientific results only briefly. More details may be found in the recent article by Encrenaz (1999).

CONCLUSION

The Century of Space Science has offered many opportunities for exploration of Jupiter and its system, and has stimulated many ideas and theoretical developments. Our knowledge of Jupiter is at a totally different level from what it was when Pioneers 10 and 11 were announced.

REFERENCES

Baum, W.A. and Code, A.D. (1953). A photometric observation of the occultation of σ Arietis by Jupiter. *Astronomical Journal*, **38**, 108–112.

Berge, G.L. and Gulkis, S. (1969). Earth-based radio observations of Jupiter: Millimeter to meter wavelengths. In: T. Gehrels (ed.), *Jupiter*, University of Arizona Press, Tucson, pp. 621–692.

Bjoraker, G.L., Stolovy, S.R., Herter, T.L., Gull, G.E. and Birger, B.E. (1996). Water after the collision of fragments G and K of Comet Shoemaker-Levy 9 with Jupiter. *Icarus*, **121**, 411–421.

Borucki, W.J. and Williams, M.A. (1986). Lightning in the Jovian water cloud. *Journal of Geophysical Research*, **91**, 9893–9903.

Brown, R.A. and Yung. Y.L. (1976). Io, its atmosphere and optical emissions. In: T. Gehrels (ed.), *Jupiter*, University of Arizona Press, Tucson, pp. 1102–1145.

Brown, R.A., Pilcher, C.D. and Strobel, D.F. (1993). Spectrophotometric studies of the Io torus. In: A.J. Dessler (ed.), *Physics of the Jovian Magnetosphere*, Cambridge University Press, Cambridge, pp. 197–225.

Carlson, B.E., Lacis, A.A. and Rossow, W.B. (1993). Tropospheric gas composition and cloud structure of the Jovian North Equatorial Belt. *Journal of Geophysical Research*, **98**, 5251–5290.

Carlson, R.W., Drossart, P., Encrenaz, Th., Weissman, P.R., Hui, J. and Segura, M. (1997). Temperature, size, and energy of the Shoemaker-Levy 9 impact fireball. *Icarus*, **128**, 251–274.

Carr, T.D. and Desch, M.D. (1969). Decametric and hectometric observations of Jupiter. In: T. Gehrels (ed.), *Jupiter*, University of Arizona Press, Tucson, pp. 693–737.

Combes, M., Vapillon, L. and Lecacheux, J. (1971). First results of the occultation of β Sco by Jupiter. *Astronomy and Astrophysics*, **49**, 399–403.

DeMarcus, W.C. (1958). The constitution of Jupiter and Saturn. *Astronomical Journal*, **63**, 2–56.

de Pater, I. and Massie, S.T. (1985). Models of the millimeter-centimeter spectra of the Jovian planets. *Icarus*, **62**, 143–171.

Encrenaz, T. (1999). The planet Jupiter. *The Astronomy and Astrophysics Review*, **9**, 171–219.

Goldreich, P. and Lynden-Bell, D. (1969). Io, a unipolar inductor. *Astrophysical Journal*, **156**, 59–78.

Hanel, R.A., Conrath, B.J., Hearth, L.W., Kunde, V.G. and Pirraglia, J.A. (1981). Albedo, internal heat, and energy balance of Jupiter: Results of the Voyager infrared investigation. *Journal of Geophysical Research*, **86**, 8705–8712.

Hubbard, W.B., Nather, R.E., Evans, D.S., Tull, R.G., Wells, D.C., van Citters, G.W., Warner, B. and Vanden Bout, P. (1972). The occultation of Beta Sccorpii by Jupiter and Io. I. Jupiter. *Astronomical Journal*, **77**, 41–59.

Hunten, D.M. and Veverka, J. (1976). Stellar and spacecraft occultations by Jupiter: A critical review of derived temperature profiles. In: T. Gehrels (ed.), *Jupiter*, University of Arizona Press, Tucson, pp. 247–83.

Hunten, D.M. and Dessler, A.J. (1977). Soft electrons as a possible heat source for Jupiter's thermosphere. *Planetary and Space Science*, **25**, 817–821.

Hunten, D.M., Colin, L. and Hansen, J.E. (1986). Atmospheric science on the Galileo mission. *Space Science Reviews*, **44**, 191–240.

Ingersoll, A., Münch, G., Neugebauer, G. and Orton, G.S. (1976). Results of the infrared radiometer experiment on Pioneers 10 and 11. In: T. Gehrels (ed.), *Jupiter*, University of Arizona Press, Tucson, pp. 197–205.

Ingersoll, A.P. and Porco, C.C. (1985). Solar heating and internal heat flow on Jupiter. *Icarus*, **35**, 27–43.

Ingersoll, A.P., Vasavada, A.R., Little, B., Anger, C.D., Bolton, S.J., Alexander, C., Klaasen, K.P., Tobiska, W.K. and the Galileo SSI Team (1998). Imaging Jupiter's aurora at visible wavelengths. *Icarus*, **135**, 251–264.

Intriligator, D.S. and Miller, W.D. (1981). Detection of the Io plasma torus by Pioneer 10. *Geophysical Research Letters*, **8**, 409–412.

Kliore, A.J. and Woiceshyn, P.M. (1976). Structure of the atmosphere of Jupiter from Pioneer 10 and 11 radio occultation measurements. In: T. Gehrels (ed.), *Jupiter*, University of Arizona Press, Tucson, pp. 216–246.

Kupo, I., Mekler, Y. and Eviatar, A. (1976). Detection of ionized sulfur in the Jovian magnetosphere. *Astrophysical Journal Letters*, **205**, L51–L53.

Larson, H.P., Fink, U., Treffers, R. and Gautier, T.N. (1975). Detection of water vapor on Jupiter. *Astrophysical Journal*, **197**, L137–L140.

Low, F.J. (1976). Discussion. In: T. Gehrels (ed.), *Jupiter*, University of Arizona Press, Tucson, pp. 203–204.

McConnell, J.C., Holberg, J.B., Smith, G.R., Sandel, B.R., Shemansky, D.E. and Broadfoot, A.L. (1982). A new look at the ionosphere of Jupiter in light of the UVS occultation results. *Planetary and Space Science*, **30**, 151–167.

Niemann, H.B., Atreya, S.K., Carignan, G.R., Donahue, T.M., Haberman, J.A., Harpold, D.N., Hartle, R.E., Hunten, D.M., Kasprzak, W.T., Mahaffy, P., Owen, T.C. and Way, S.H. (1998). Composition of the Jovian atmosphere as determined by the Galileo probe mass spectrometer. *Journal of Geophysical Research*, **103**, 22,831–22,845.

Öpik, E. (1962). Jupiter: Chemical composition, structure, and origin of a giant planet. *Icarus*, **1**, 200–257.

Pearl, J. Hanel, R., Kunde, V., Maguire, W., Fox, K., Gupta, C., Ponnamperuma, C. and Raulin, F. (1979). Identification of gaseous SO_2 and new upper limits for other gases on Io. *Nature*, **280**, 755–758.

Sprague, A.L., Bjoraker, G.L., Hunten, D.M., Witteborn, F.C., Kozlowski, R.W.H. and Wooden, D.H. (1996). Water brought into Jupiter's atmosphere by Fragments R and W of Comet SL-9. *Icarus* **121**, 30–37.

Trafton, L. and Münch, G. (1969). The structure of the atmospheres of the major planets. *Journal of Atmospheric Sciences*, **26**, 813–835.

Veverka, J., Wasserman, L.H., Elliot, J., Sagan, C. and Liller, W. (1974). The occultation of β Scorpii by Jupiter. I. The structure of the Jovian upper atmosphere. *Astronomical Journal*, **79**, 73–84.

von Zahn, U., Hunten, D.M. and Lehmacher (1998). Helium in Jupiter's atmosphere: Results from the Galleo probe helium interferometer experiment. *Journal of Geophysical Research*, **103**, 22,815–22,829.

Walter, C.M., Marley, M.S., Hunten, D.M., Sprague, A.L., Wells, W.K., Hoffmann, W.F., Sykes, M.V., Deutsch, L.K., Fazio, G.G. and Hora, J.L. (1996). Search for seismic waves from the impact of the SL-9 R fragment. *Icarus*, **121**, 341–350.

Wildt, R. (1969). The outer planets: Some early history. *Journal of Atmospheric Sciences*, **26**, 795–797.

Yelle, R.V., Young, L.A., Vervack, R.J. Jr, Pfister, L. and Sandel, B.R. (1996). Structure of Jupiter's upper atmosphere – Predictions for Galileo. *Journal of Geophysical Research*, **101**, 2149–2161.

FURTHER READING

Cassini at Jupiter (2002). *Nature*, **415**, 965–966; 985–1005.

Dessler, A. (ed.) (1993). *Physics of the Jovian Magnetosphere*, Cambridge University Press, Cambridge.

Galileo Probe (1996). *Science*, **272**, 837–860.

Galileo Orbiter (1996). *Science*, **274**, 377–403.

Gehrels, T. (ed.) (1976). *Jupiter*, University of Arizona Press, Tucson.

The Galileo Probe Mission to Jupiter (1998). *Journal of Geophysical Research*, **103**, 22, 775–23,069.

Remote sensing results from the Galileo mission (1998). *Icarus*, **135**, 1–376.

Science Advisory Group (1973). Outer solar system exploration – An overview. *Space Science Reviews*, **14**, 347–590.

Third Arizona Conference on Planetary Atmospheres (1969). *Journal of Atmospheric Sciences*, **26**, 795–1001.

The Voyager missions to the outer system (1977). *Space Science Reviews*, **21**, 75–276.

The Voyager mission through the Jupiter encounters (1981). *Journal of Geophysical Research*, **86**, 8123–8841.

59

THÉRÈSE ENCRENAZ*

The planets beyond Jupiter

INTRODUCTION

At large heliocentric distances, the giant planets of the Solar system – Jupiter, Saturn, Uranus and Neptune – have specific characteristics: with respect to the terrestrial planets, they have a large size, a low density, a fast rotation period around their axis; they also all have a ring system and a large number of satellites. The two brightest of them, Jupiter and Saturn, have been known since Antiquity. In contrast Uranus and Neptune, which are both smaller and farther away from the Sun, have been discovered relatively recently: Uranus was first observed by William Herschel in 1781, following the new developments of large telescopes, while the discovery of Neptune, by John Adams and Urbain Le Verrier in 1846, was the direct result of celestial mechanics calculations.

Both Jupiter and Saturn have been monitored from ground-based telescopes for over three centuries, after Galileo Galilei, in 1610, turned to the sky his newly built refractor and discovered, in particular, the four satellites of Jupiter later called galilean. Concerning the two major giant planets, we thus have a huge data base, first made of drawings, then completed with photographs and spectra. The first images revealed the latitudinal structure of belts and zones on both planets, the existence of the Great Red Spot on Jupiter, and the nature of Saturn's ring system. Photographs of improving quality provided a monitoring of the planets' meteorological features, showing, in particular, the stability of the Great Red Spot on Jupiter but the variability of the other smaller features. The development of ground-based spectroscopy, first in the visible, then in the infrared range, provided a quantitative information about the chemical composition of Jupiter and Saturn. Methane and ammonia were detected in 1932, and hydrogen, the main constituent of the giant planets, was observed for the first time in 1960. In the 1970s, the improvement of infrared spectroscopy techniques, coupled with the use of larger and larger ground-based telescopes, led to the discovery of a long list of minor atmospheric constituents, on both Jupiter and Saturn. We thus had already a good knowledge of the nature and composition of their atmospheres when the first spacecraft devoted to the exploration of the giant planets, Pioneer 10 and 11, were launched in 1972 and 1973 respectively.

Uranus and Neptune, in contrast, were very poorly known until the space era, which, in their case, started with the launch of the second Voyager spacecraft, in 1977. Prior to that date, the available information about these two planets was limited to poor-quality images, which, still, provided evidence for temporal variability and climatic changes in the case of Neptune. Hydrogen and methane were for long the only atmospheric species detected from ground-based spectroscopy.

Our knowledge of the giant planets has been entirely revised with the two sets of NASA spacecraft, Pioneer 10 and 11 and later Voyager 1 and 2. The four spacecraft encountered Jupiter in 1973 (Pioneer 10), 1974 (Pioneer 11) and 1979 (Voyager 1 and 2) respectively; Jupiter was later extensively studied between 1995 and 2000 by the Galileo mission, launched in 1989. Saturn was first encountered by Pioneer 11 in 1979, then by Voyayer 1 and 2 in 1980 and 1981. Voyager 2 continued its journey to encounter Uranus in 1986 and Neptune in 1989, completing the first *in situ* space exploration of the giant planets and their systems. This space program turned out to be an outstanding scientific success. In the future, scientists are looking for the long-term exploration of Saturn by the Cassini mission arriving there in 2004; for Uranus and Neptune, no further space exploration has yet been scheduled.

---

* Observatoire de Paris, Meudon, France

In addition to *in situ* space exploration, space observatories in Earth orbit have greatly contributed to our knowledge of the giant planets. The first of them was the International Ultraviolet Observer (IUE), a cooperative project of NASA, ESA and the UK, launched in 1977, which explored the ultraviolet sky for almost twenty years. The next step was the Hubble Space Telescope (HST), a NASA/ESA mission in operation since 1989, which has been providing unprecedented quality images, as well as UV spectra, of all sorts of astronomical sources. The third space observatory to be used for planetology was the Infrared Space Observatory (ISO), an ESA mission with NASA and Japanese participations, in operation between 1995 and 1998, which recorded high sensitivity images and spectra in the infrared range. Two major missions, currently under development, are expected to be the next observatories performing planetary observations: the Far Infra-Red and Submillimeter Telescope (FIRST), an ESA mission with a possible contribution from NASA, will explore the long-wavelength range of the electromagnetic spectrum with unprecedented sensitivity and spectral resolution, and the Next Generation Space Telescope (NGST), developed by NASA with a potential ESA participation, will be a follow-up of the HST with an extension of its spectral range toward the infrared.

THE GIANT PLANETS BEYOND JUPITER

It is possible to understand the general properties of the giant planets (such as their large size, low density and large number of satellites) in the light of their formation scenario. This so-called "nucleation scenario", now widely accepted by the scientific community, is based on a whole set of observational data concerning the dynamics of solar-system bodies and their chemical composition, as well as the analysis of star-forming regions and protoplanetary disks, which bring stringent constraints to theoretical models. Following the initial concepts proposed by Emmanuel Kant and Pierre-Simon de Laplace in the XVIIIth century, the current scenario assumes that the Solar system was formed after the gravitational collapse of a protosolar cloud into a disk, in which solid particles accreted together through collisions to form first planetesimals, and later larger bodies (Cassen and Woolum 1999). The chemical composition of this protosolar cloud, made of interstellar matter, was likely to reflect the cosmic abundances, i.e. include first hydrogen (about 75%), then helium (about 25%), then traces of other heavier elements (O, C, N, ...).

In the vicinity of the Sun, where the terrestrial planets formed, the temperature was probably such that only heavy elements (silicates and metals) were in solid form; these elements represent only a small fraction of the cosmic matter. In contrast, at large heliocentric distances, where the giant planets formed, most of the matter, apart from hydrogen and helium, was solid, in form of ices (H_2O, CH_4, NH_3, ...). These ices, incorporated into the planetesimals, were abundant enough to form massive cores (about ten to fifteen terrestrial masses, according to theoretical models), which could in turn accrete the surrounding protosolar nebula, mostly made of hydrogen and helium. This simple first-order scenario can account for the large sizes and the low densities of the giant planets, as well as their ring and satellite systems.

According to the nucleation scenario (Mizuno 1980), all giant planets should have formed from an initial core of comparable mass. Comparing this number (10–15 terrestrial masses) to the actual sizes of the giant planets immediately shows a distinction between the four giants. Jupiter and Saturn, whose initial core is 3 to 10% of their total mass, are mostly gaseous, while Uranus and Neptune, with an initial core of more than 50% of their total mass, can be called the icy giants. What can be the origin of this difference? It has been suggested that Uranus and Neptune, located farther from the Sun where the density of the disk was lower, took a longer time than Jupiter and Saturn to accrete their initial cores. In this case, the accretion of the surrounding gas around Uranus and Neptune may have happened after most of the gas was blown away by the strong solar wind expelled by the early Sun in its T-Tauri phase, and Uranus and Neptune would have found much less protosolar gas available for their accretion. This is no more than a possible explanation, however; there are still many open questions about the formation scenarios of the giant planets, and the recent discovery of many "giant" exoplanets in the close vicinity of their star raises new problems, currently unsolved, about their formation.

Apart from the question of their origin, giant planets, as observed today, are far from being fully understood. Among the major issues are their internal structure, the nature of their internal source of heat, the composition of their clouds and their meteorology. The four giant planets were considered for long as being similar to one another; however, recent observations, especially through space exploration, has revealed that the giant planets, like the terrestrial ones, are unique worlds by themselves, and the reasons for these differences remain to be solved. The main orbital and physical parameters of the giant planets are summarized in Table 1; their chemical composition is given in Table 2. In Table 3, abundance ratios in the giant planets are compared to solar and protosolar values.

PLANETS BEYOND NEPTUNE

Early in the XXth century, it was announced by some astronomers, including Percival Lowell, that the orbital motions of Uranus and Neptune could not been explained without the gravitational perturbations of another planetary body, located at further distances from the Sun. This new object called "Planet X" was unsuccessfully searched for during the following decades. As a result of this observing campaign, however, the small Pluto was discovered in 1930 by Clyde Tombaugh. With a diameter of about 2400 km and a density close to $2\,g\,cm^{-2}$, Pluto was not able to account for the orbit anomalies mentioned above. In 1978 however, a satellite, called Charon, was detected around Pluto by James Christy from a study of Pluto's orbital trajectory. Pluto has a distinctly eccentric orbit (e = 0.246), that is conspicuously inclined against the ecliptic plane (i = 17° 10′).

No bigger planet was found after Pluto, but a whole family of smaller objects, physically similar to Pluto and later called Trans-Neptunian Objects (TNOs) have been discovered during the past decade in the Edgeworth-Kuiper Belt, between 30 and 50 AU (Levison and Weissman 1999). The existence of this belt was predicted independently, on theoretical grounds, by Kenneth Edgeworth and Gerard Kuiper in the 1940–50s. They argued that the expected density of the protoplanetary disk, beyond Neptune, should have been too low to allow the formation of large bodies like the giant planets, but should have been sufficient for the formation of planetesimals, comets and possibly very small planets. It was later suggested by Julio Fernandez that this belt could be the reservoir for short-period comets which have a low inclination. These theoretical studies motivated new observing campaigns which benefited from the use of large ground-based telescopes and high-sensitivity detectors. The first TNO was detected in 1992 by David Jewitt and Jane Luu. As of 1 January, 2000, about 200 objects have been found with diameters larger than 100 km. As in the case of comets, their albedo is expected to be low (a few percent). It now appears that the TNOs are members of a new family of solar-system objects which could be as many as 100 000 (with a diameter above 100 km) in the Edgeworth-Kuiper Belt, with a total mass of a few tenths of a terrestrial mass. The existence of this new class has important implications on physico-chemical and dynamical models of solar-system formation. The search for TNOs and the spectroscopic study of the detected objects have become a new, rapidly exploding field of research.

Most of our knowledge regarding Pluto and the TNOs has come from ground-based astronomy (see Stern and Yelle 1999, for a review). In particular, in the case of Pluto, two methods have been very powerful. The first one is the observation of mutual occultations and transits of Pluto and Charon, in the beginning of the 1980s, which provided a determination of the masses and densities of both objects. The second method is the observation of a stellar occultation by Pluto in 1988, which showed evidence for a stable atmosphere around Pluto. This atmosphere, dominated by nitrogen N_2 with a minor contribution of CH_4, shows remarkable similarities with that of Neptune's satellite Triton. Finally, ground-based infrared spectroscopy has allowed the detection of several ices on Pluto's surface (N_2, CO, CH_4). Water ice, in contrast, was found on Charon, which confirms that Pluto and Charon are different in nature, as illustrated by their different densities. The orbital, physical and chemical properties of Pluto and Charon suggest an origin different from the planetary subnebular material from which the outer satellites formed. It has been suggested that Pluto and Charon accreted independently in the outer solar system and later collided, which led to the binary system observed today (Mueller and McKinnon 1997; Stern and Yelle 1999).

In addition to the ground-based exploration of Pluto, significant information has come from HST images of Pluto's surface which have shown evidence for a differentiation of its surface. The same conclusion came from observations of Pluto's far-infrared flux with ISO. Both HST and ISO have also been used for measuring the visible and far-IR fluxes of a few TNOs. There has been no *in situ* space exploration of Pluto so far, but a flyby mission (Pluto Fast Flyby) is currently under study at NASA.

THE SPACE MISSIONS

The flyby missions: Pioneer 10/11 and Voyager

Following a series of Pioneer spacecraft devoted to the exploration of the Moon and the solar wind, Pioneer 10 and Pioneer 11 were designed by NASA for a low-cost exploratory mission to Jupiter (Lasher 1997). In addition to the exploration of the jovian environment, the major objectives were to study the interplanetary medium beyond the Mars orbit and to investigate the possible hazards in crossing the asteroidal belt. After the success of the Pioneer 10 flyby of Jupiter (4 December 1973), Pioneer 11 was retargeted to allow a Saturn flyby after the Jupiter encounter (3 December 1974). The Saturn flyby took place on 1 September 1979. Among the 11 instruments of the scientific payload were an imaging photopolarimeter, a UV photometer, an IR radiometer, several plasma physics instruments and two impact detectors.

These flybys provided the first short-range images of Jupiter, Saturn and the Saturn rings, showing with unprecedented detail the structure of the jovian belts and zones, the Great Red Spot and Saturn's ring system. Another highlight of these missions was the first measurement of the jovian magnetic field (previously detected from ground-based radio-emission of Jupiter) and the first detection of Saturn's magnetosphere. Pioneer 10 and 11 also demonstrated that spacecraft could survive after crossing the asteroid belt, and thus opened the road to future more sophisticated missions.

Figure 1 The planet Saturn as seen from Voyager 1 on 18 October 1980, from a distance of 34 million km. Colors have been enhanced to better show the contrasts in the belt-zone structure (NASA).

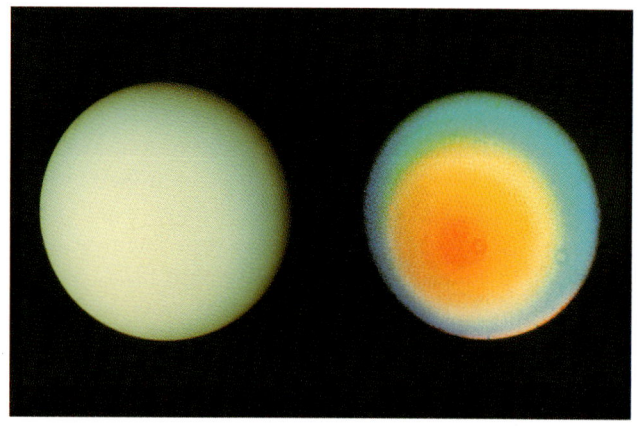

Figure 2 The planet Uranus as seen from Voyager 2, one week prior to the encounter (January 1986), from a distance of 9 million km. Left side: real colors; right side: false colors (NASA).

Following this success, Voyager 1 and 2, two Mariner-type vehicles, were launched by NASA in September 1977 for a flyby exploration of the four giant planets (Miner 1997). Voyager 1 encountered Jupiter on 5 March 1979 and Saturn on 12 November 1980. Voyager 2 encountered Jupiter on 9 July 1979, Saturn on 25 August 1981, Uranus on 24 January 1986 and Neptune on 25 August 1989. The two probes are expected to send data back to Earth until about 2015, where they will be at 130 AU from the Sun. Their objective is to detect the heliopause, defining the limit between the solar-system plasma and the interstellar medium, which is expected to occur, according to the Voyager data, between 110 and 160 AU from the Sun.

The Voyager spacecraft included in their scientific payload (11 instruments) a camera, IR and UV spectrometers, a magnetometer, and charged-particles, plasma and radio wave experiments. The camera observed the cloud structure and the dynamical phenomena of the atmospheres (Figures 1 to 3), whose temperature structure and chemical composition were inferred from the IR spectrometer IRIS; the UV spectrometer probed the stratospheres and the ionospheres.

A huge amount of new results and discoveries came out of the Voyager exploration of the giant planets. Among the highlights are the first detection of active volcanism on Io, the detection of Jupiter's rings and the fine structure of the other planetary rings, the composition and structure of Titan's and Triton's atmospheres, the first observation of Uranus and Neptune's magnetospheres, the first observation of the surfaces of Miranda and Triton, and the detection of many new small satellites around the four giant planets.

Figure 3 The planet Neptune as seen by Voyager 2 at the time of encounter (August 1989). The image is a false-color combination of 3 exposures in 3 different filters. The deep blue color (which corresponds to the real color) is due to large abundances of gaseous methane. White spots are believed to be high-altitude CH_4 cirrus. The red color, which is not real, corresponds to high-altitude haze in Neptune's stratosphere (NASA).

The space observatories: IUE, HST and ISO

The IUE satellite consisted of a 45-cm telescope equipped with two spectrographs, giving a resolving power as high as 10^4 over the 1150–3200 Å spectral range (Boggess and Wilson 1987). UV spectra of the giant planets recorded by IUE have allowed to probe their upper atmospheres, and in particular their aeronomy and photochemistry. Above about 1600 Å, the UV radiation comes from the

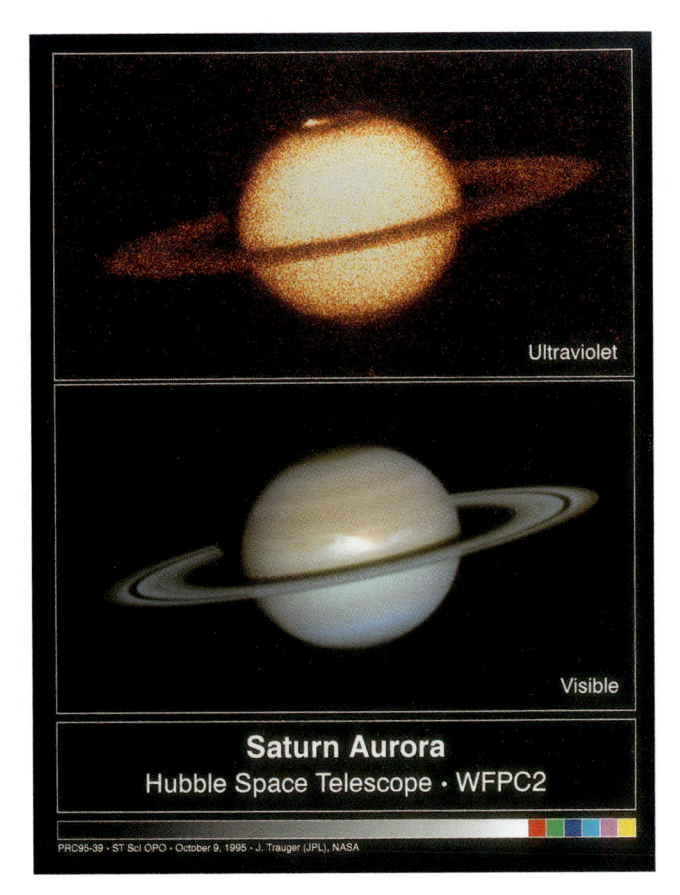

Figure 4 Two images of Saturn taken by the Hubble Space Telescope (HST). Top: UV image (9 October 1994) showing an aurora around Saturn's north pole. Bottom: Visible image (1 December 1994). The aurora is not detectable at visible wavelengths, but a bright meteorological feature, probably arising in the upper troposphere of Saturn, is present near the equatorial region (NASA-ESA).

upper levels of the homogenously mixed atmosphere, below the homopause level. Information has been retrieved on the presence of hydrocarbons in Saturn, Uranus and Neptune. At lower wavelengths, the upper stratosphere above the homopause is probed. The main result of IUE was the first detection of a bright and variable aurora on Uranus, implying the presence of a significant magnetic field; this result was later confirmed by Voyager UVS (Moos and Encrenaz 1987).

HST, a 2.4 m telescope equipped with cameras and UV spectrographs, operating in Earth orbit since 1989, has provided the scientific community with images of the giant planets of unprecedented quality in the near UV, visible and near-IR range. In particular, images of Saturn taken in 1995, when the rings were seen edge on from Earth (Figure 4), allowed the observation of small inner satellites within the rings (Nicholson *et al*. 1996). Images of Neptune (Figure 5) have revealed the temporal structures of the dark

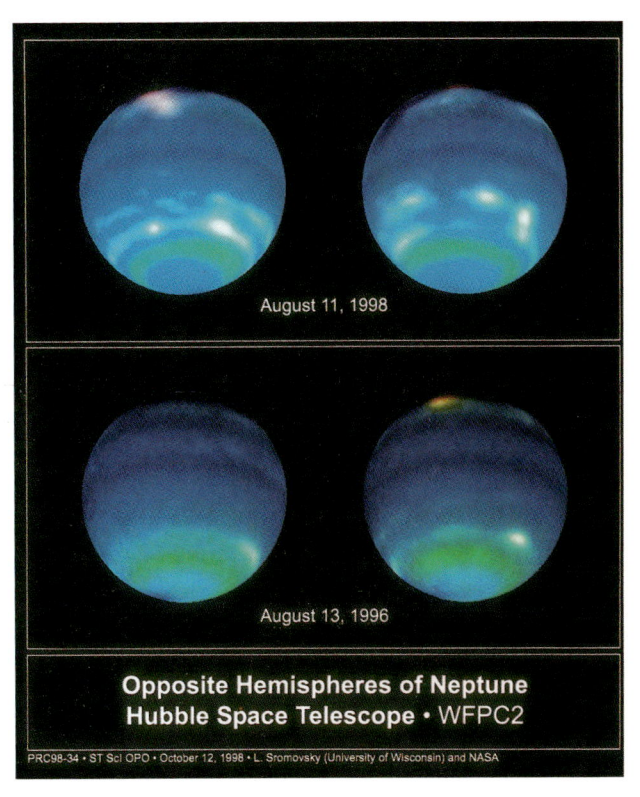

Figure 5 Images of Neptune taken by the HST on August 11, 1998 (top) and August 13, 1996 (bottom). They show the strong variability of Neptune's cloud structure (NASA-ESA).

spots, first identified by Voyager 2 (Nelson and Domingue 1999). HST also provided the best high-resolution images of the Pluto-Charon system (Figure 6).

The third space observatory used for planetary studies, ISO, consisted of a 60-cm telescope with a camera, a photometer and two spectrometers, all encased in a dewar containing superfluid helium. ISO covered the entire infrared range (2–200 µm) with a resolving power ranging from 50 to 30 000 (Kessler *et al*. 1996). Infrared spectroscopy of the giant planets with ISO has led to many new results and discoveries, including a new determination of the D/H ratio, the unexpected detection of an external oxygen source in their upper stratospheres, and the detection of several new hydrocarbon species (Lellouch 1999).

THE PLANET SATURN

Saturn shows several similarities with Jupiter. It is mostly a gaseous planet, as its initial core represents only about 10% of its total mass. As Jupiter, Saturn is in rapid rotation around its axis and accordingly, exhibits a parallel system of belts and zones (Figures 1 and 4). These features, however, are less contrasted than on Jupiter, possibly

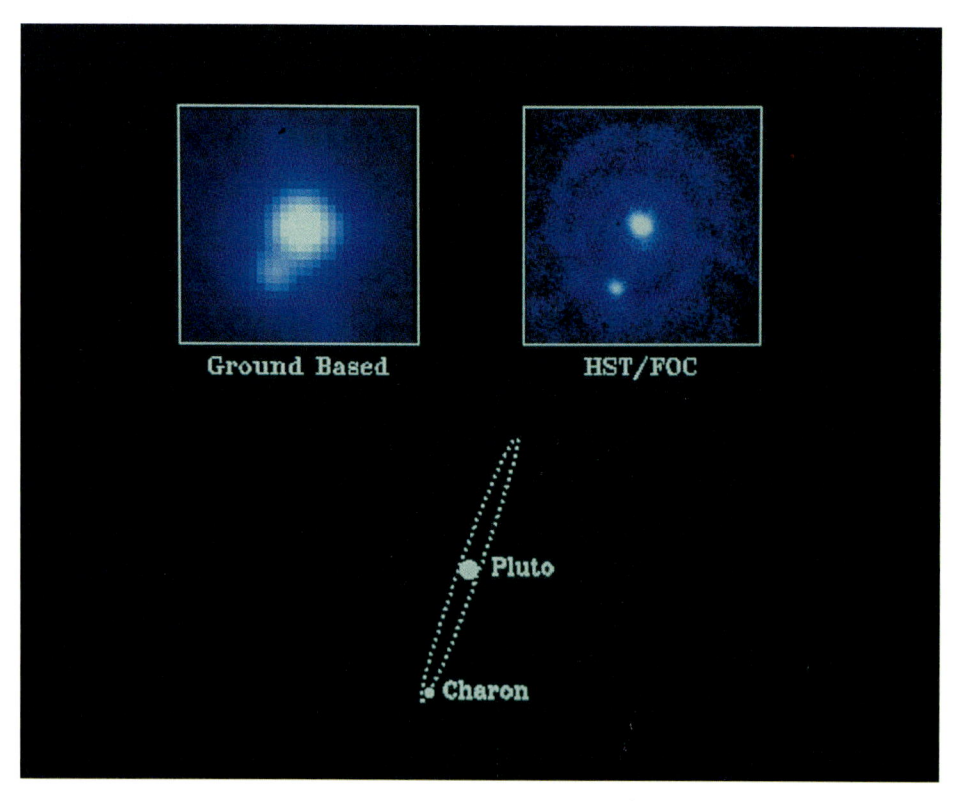

Figure 6 The Pluto-Charon system as observed by the Faint Object Camera of the HST. A ground-based image is shown for comparison (NASA-ESA).

because of the thicker NH_3 ice cloud which is seen in the visible light.

Internal structure

We only have indirect knowledge of the internal structures of the giant planets. Our information comes from the measurements of their oblateness and gravitational fields, coupled with data on their internal energy and their atmospheric chemical composition.

Saturn is more oblate than Jupiter and the other giant planets (Table 1), due to its low density and its fast rotation speed. Its gravitational potential coefficients were determined from the Pioneer 11 flyby (Null *et al.* 1981) which made a closer approach than the Voyager flybys. It has been known for several decades, from ground-based infrared observations, that Saturn posesses a significant internal source; its amount (1.78 times the solar absorbed energy) was later refined by Voyager (Hanel *et al.* 1983). With regard to Saturn's chemical composition, it has been believed for more than a decade, on the basis of Voyager data, that the helium content in Saturn was significantly lower than in the other giant planets (Conrath *et al.* 1984); however, a recent reanalysis of these data has led to a re-estimation of the He mixing ratio in Saturn, much closer to the Jupiter value (Conrath and Gautier 2000; Table 2).

Using these data, models have been developed to infer the distribution of hydrogen, helium and heavy elements, as well as the temperature distribution, down to the planet center. Recent models infer a central core of heavy elements of about one terrestrial mass, surrounded by a region made of a mixture of hydrogen and helium including about 25 terrestrial masses of heavy material, whose relative content decreases toward the outer layers (Hubbard 1997; Marley 1999). The temperature above the core is expected to be about 13 000 K and the pressure should be around 10^7 bars. If cosmic abundances were assumed, the total content of Saturn in heavy elements should only be 2 terrestrial masses; in the case of Jupiter, a content of about 20 terrestrial masses is found according to models, while it should be only 6 under the assumption of cosmic abundances. There is thus a significant excess of heavy elements on both planets, especially so on Saturn. This is consistent with the nucleation model which implies that Saturn's initial core (in which the heavy elements were incorporated) has a larger mass fraction than the jovian one.

Saturn's interior looks, at first approximation, similar to that of Jupiter, except that the temperatures and internal

Table 1 Orbital and physical parameters of the giant planets

	Jupiter	Saturn	Uranus	Neptune
ORBIT				
Distance from the Sun (AU)	5.2	9.6	19.2	30.1
Eccentricity	0.048	0.056	0.046	0.009
Inclination	1°19'	2°30'	0°46'	1°47'
Sidereal period (y)	11.86	29.45	84.02	164.8
PLANET				
Equ. radius (km)	71 492	60 268	25 559	24 764
Oblateness	0.062	0.091	0.06	0.02
Mass (terrestrial masses)	318.1	95.1	14.6	17.2
Mean density (g·cm^{-2})	1.24	0.63	1.21	1.67
Mean surface gravity (m·s^{-2})	24.8	10.5	8.5	10.8
Sidereal rotation (h)	9.9	10.7	17.4	16.2
Axial inclination (°)	3.1	26.8	97.9	28.8
Ratio of emitted energy to absorbed solar energy	1.7	1.8	<1.1	2.5
Magnetic field tilt (°)	9.4	0.0	58.6	46.9
Magn. field offset (planetary radius)	0.119	0.038	0.352	0.485

pressures are lower. In both Jupiter and Saturn, the pressure is high enough for hydrogen to be in metallic phase. The region of metallic hydrogen is expected to start under a depth, below the visible surface, of about 30 000 km in the case of Saturn, and 10 000 km in the case of Jupiter. This conductive region, very extended in the case of Jupiter, could be responsible for the magnetic fields of both planets, and in particular for the very strong jovian field.

In the metallic phase, helium is expected to condense; this differenciation mechanism might be responsible for the fact that the helium mixing ratio, on both Jupiter and Saturn, is lower than the protosolar value (Tables 2 and 3). It could also be at least partly responsible for the internal energy source which has comparable amounts on both planets (Gautier and Owen 1989; Table 1).

Atmospheric structure and composition

As in the case of the other giant planets, the thermal structure of Saturn (Figure 7) is characterized by a convective region. This region, the troposphere, where the temperature decreases with increasing altitudes, extends up to a level of about 0.1 bar (the tropopause). The stratosphere, above this level, is a radiative region where the temperature increases with altitude, due to the absorption of solar light by methane and aerosols. The temperature profile, together with the mean molecular weight (i.e. the helium abundance) has been simultaneously retrieved from the radio-occultation experiment and the infrared spectroscopy experiment (IRIS) of Voyager (Hanel et al. 1992; Conrath et al. 1984; Figure 8).

As mentioned above, the first results (Lindal et al. 1985) have been recently reanalyzed, leading to a small change of the temperature profile, but a drastic change of the helium abundance (Conrath and Gautier 2000).

From thermochemical models, one can predict, at least to first order, the cloud structure of the giant planets (Atreya 1986; West 1999; Figure 9). In the case of Saturn, an ammonia cloud is found around 150 K, near the 1-bar pressure level; this cloud, thicker than in the case of Jupiter, is mainly responsible for the visible aspect of the planet. At deeper levels, an NH_4SH cloud is expected around 210 K (pressure level of about 4 bars) and a thick water cloud at 220–270 K (P = 6–10 bars); these two clouds, however, are not directly accessible to observations. In the upper troposphere, Saturn exhibits various features, less numerous and less contrasted than in the case of Jupiter, which have been used to map the winds. These winds are predominantly east–west (zonal) and symmetrical with respect to the equator, as for the other giant planets, and show a very strong prograde jet around the equator (Gierasch and Conrath 1993; Figure 10); Jupiter, in contrast, shows a series of prograde and retrograde jets oscillating with latitude.

In the troposphere, in addition to methane, minor atmospheric species are found (Table 2). Most of them have been detected from ground-based infrared spectroscopy in the 4.5-μm region, which is a spectral window both in the Earth atmosphere (absence of H_2O and CO_2 absorptions) and in the giant planets (absence of methane absorption). These measurements allowed the detection of PH_3, CH_3D, CO, GeH_4 and AsH_3 (see Atreya et al. 1999 for a review).

Table 2 Abundances in the atmospheres of the giant planets

Species	Jupiter	Saturn	Uranus	Neptune
H_2	1	1	1	1
HD	1.8×10^{-5}	2.3×10^{-5}	5.5×10^{-5}	6.5×10^{-5}
He	0.136	0.11–0.16	0.18	0.23
CH_4 (trop)	2.1×10^{-3}	4.4×10^{-3}	2×10^{-2}	4×10^{-2}
CH_4 (strat)	2.1×10^{-3}	4.4×10^{-3}	3×10^{-5}–10^{-4}	7×10^{-4} (0.05–1 mb)
$^{13}CH_4$ (trop)	2×10^{-5}	4×10^{-5}		
CH_3D (trop)	2.5×10^{-7}	3.2×10^{-7}	10^{-5}	2×10^{-5}
CH_3D (strat)				2.2×10^{-7}
C_2H_2		3.5×10^{-6} (0.1 mb)	2–4×10^{-7} (0.1–0.3 mb)	1.1×10^{-7} (0.1 mb)
$^{12}C^{13}CH_2$	*	2.5×10^{-7} (mb)		
C_2H_6	4.0×10^{-6} (0.3–50 mb)	4.0×10^{-6} (<10 mb)		1.3×10^{-6} (0.03–1.5 mb)
CH_3C_2H	*	6.0×10^{-10} (<10 mb)		
C_4H_2		9.0×10^{-11} (<10 mb)		
C_2H_4	7×10^{-9}			
C_3H_8	6×10^{-7}			
C_6H_6	2×10^{-9}	*		
CH_3		0.2–1×10^{-7} (0.3 μb)		2–9×10^{-8} (0.2 μb)
NH_3 (trop)	2×10^{-4} (3–4 b)	2–4×10^{-4} (3–4 b)		
$^{15}NH_3$	8×10^{-7}			
PH_3 (trop)	6×10^{-7}	1.7×10^{-6}		
GeH_4	7×10^{-10}	2×10^{-9}		
AsH_3	3×10^{-10}	2×10^{-9}		
CO (trop)	1.5×10^{-9}	2×10^{-9}		
CO (strat)	1.5×10^{-9}	2×10^{-9}	10^{-6}	
CO_2 (strat)	4×10^{-10} (<10 mb)	3×10^{-10} (<10 mb)		5×10^{-10} (<5 mb)
H_2O (trop)	1.4×10^{-5} (3–5 b)	2×10^{-7} (>3 b)		
H_2O (strat)	1.5×10^{-9} (<10 mb)	2–20×10^{-9} (<0.3 mb)	5–12×10^{-9} (<0.03 mb)	1.5–3.5×10^{-9} (<0.6 mb)
HCN				3×10^{-10}
H_3^+	*	*	*	

*Detected.

In addition, tropospheric H_2O was detected using the ISO short-wavelength spectrometer (de Graauw et al. 1997). Many of these species, including CH_4, PH_3 and NH_3 (the latter derived from radio measurements) are found to be enriched with respect to cosmic values; the implications of this result will be discussed below. An important exception, however, is water vapor which was found to be strongly undersaturated in Saturn's troposphere (below the 3-bar pressure level). The same situation was observed in Jupiter, both from the Galileo Probe data and from 5-μm Voyager-IRIS spectroscopic measurements. The current explanation is that this anomaly is the result of convective motions: the oxygen depletion would only appear in specific cloud-free areas ("hot spots") which would be subsidence regions. The H_2O depletion would then be the result of dynamical activity, on both Jupiter and Saturn, and would not be representative of the oxygen abundance in the deep atmospheres of the two planets.

Another important result of ISO was the measurement of D/H from the identification of the R(1) HD line at 56 μm (Griffin et al. 1996). The derived result (Tables 2 and 3) is in good agreement with the D/H value inferred from

Table 3 Abundance ratios in the giant planets (adapted from Atreya et al. 1999)

	Jupiter	Saturn	Uranus	Neptune	Solar (S) or Protosolar (P)
He mass fraction (Y)	0.234 (0.8 × P)	0.18–0.25 (0.8 × P)	0.21–0.31 (0.8–1.0 × P)	0.27–0.37 (1.0–1.3 × P)	0.275 (P)
D/H	2.6×10^{-5} (1.2 × P)	2.3×10^{-5} (1.1 × P)	5.5×10^{-5} (2.6 × P)	6.5×10^{-5} (3.1 × P)	2.1×10^{-5} (P)
C/H	1.05×10^{-3} (2.9 × S)	2.25×10^{-3} (6 × S)	0.008 (22 × S)	0.011 (30 × S)	3.62×10^{-4} (S)
N/H	4×10^{-4} (3.6 × S)	$2–4 \times 10^{-4}$ (?) (2–4 × S?)			1.12×10^{-4} (S)
P/H	3.0×10^{-7} (0.82 × S)	$2.5–5 \times 10^{-6}$ (6–12 × S)			3.73×10^{-7} (S)
S/H	3.8×10^{-5} (16 bars) (2.4 × S)				1.62×10^{-5} (S)
O/H	3×10^{-4} (19 bars) (0.35 × S)				8.51×10^{-4} (S)

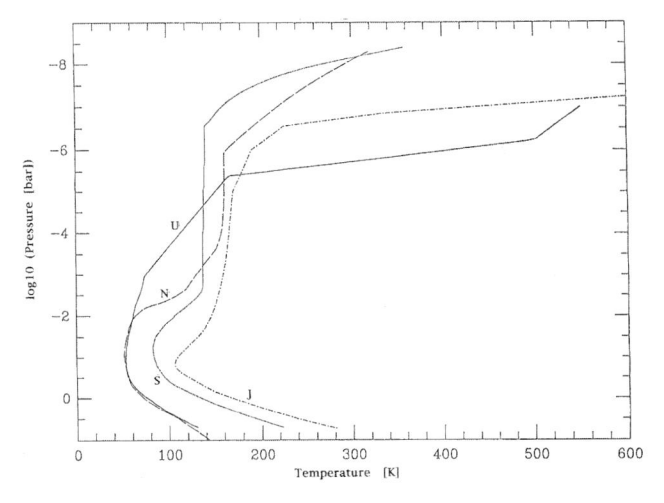

Figure 7 Thermal vertical profiles of the four giant planets. The profile of Jupiter is derived from the Galileo probe data. The other profiles are inferred from Voyager radio-occultation data. (From Encrenaz 1999a.)

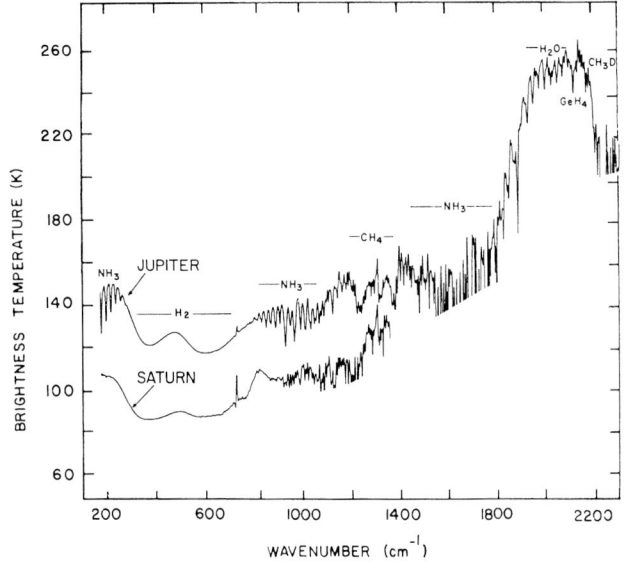

Figure 8 Spectra of Jupiter and Saturn recorded by the infrared interferometer (IRIS) of Voyager. In the 200–500 cm^{-1} range (20–50 μm), the continuum is dominated by the pressure-induced hydrogen absorption, due to H_2–H_2 and H_2–He collision. This dependence of the absorption on the helium abundance is used, together with the radio-occultation data, to infer simultaneously the thermal profile and the helium content. Emissions due to CH_4 and hydocarbons are also present on both Jupiter and Saturn, while absorption due to NH_3 and other tropospheric species are detected in the Jupiter spectrum. (From Hanel et al. 1992.)

CH_3D, and compatible with the estimate of the protosolar D/H value (Geiss and Gloeckler 1998).

In the stratosphere, many hydrocarbons are present, as a result of methane photodissociation. Acetylene C_2H_2 and ethane C_2H_6, the most abundant ones, were first detected at 12–14 μm from ground-based spectroscopy, and their vertical distributions were subsequently retrieved from the Voyager IRIS data (Atreya et al. 1999). More recently, observations of Saturn with the ISO short-wavelength spectrometer have led to the detection of several new hydrocarbons in the 14–17 μm range (Figure 11): methyl radical CH_3, methylacetylene CH_3C_2H, diacetylene C_4H_2 and benzene C_6H_6 (Bézard et al. 1999). In addition, the

CH_4 emission band at 7.7 μm was used for a simultaneous retrieval of CH_4 (constant with height in the probed region) and the temperature stratospheric profile; this study was done first on the IRIS data (Courtin et al. 1984) and later

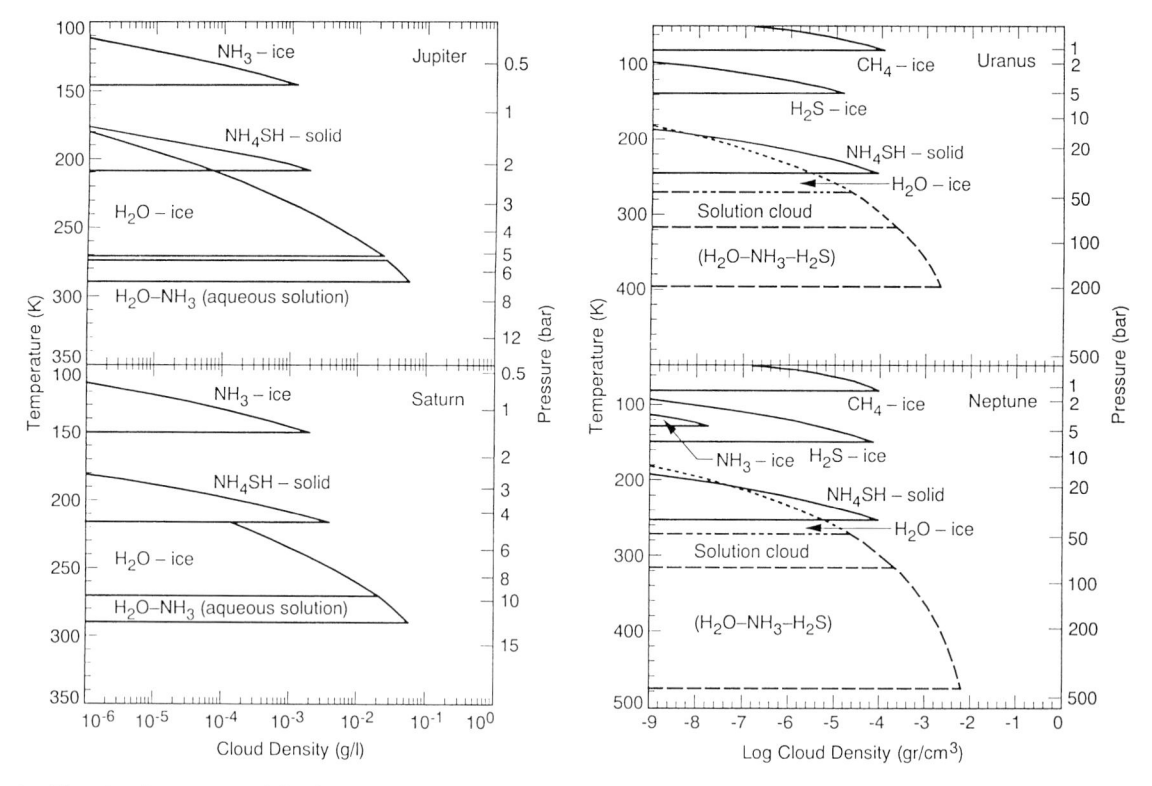

Figure 9 The cloud structure of the four giant planets as inferred from photochemical models and spectroscopic observations. This figure was constructed from models by Atreya and Romani (1985) for Jupiter and Saturn and de Pater et al. (1991) for Uranus and Neptune. Further information on the upper parts of Uranus and Neptune's atmospheres (not included in this figure) was provided by Baines et al. (1995). (From West 1999.)

applied to ISO data in order to study the temporal variations of the thermal profile (Bézard et al. 1999).

The analysis of the hydrocarbon distribution in Saturn's stratosphere, together with a study of a fluorescence band of methane at 3.3 μm (Drossart et al. 1999) has led to a determination of the eddy diffusion coefficient which characterizes the vertical mixing of the stratosphere. The value inferred for Saturn (about $10^8 \, cm^2 \, s^{-1}$ at the homopause level) confirms a previous determination from Voyager UVS data (Atreya et al. 1984) and is about hundred times higher than the Jupiter value (Atreya et al. 1999). The reason for such a high vertical mixing in Saturn is not yet understood.

Another major highlight of ISO was the detection of H_2O and CO_2 in the stratosphere of Saturn (Figure 12). The presence of water was suspected in Saturn's stratosphere because of the presence of the ring system which appeared as a potential oxygen source; also, water had been tentatively detected in the UV range by IUE and CO_2 had been detected on Titan from IRIS-Voyager. Water vapor was indeed observed in the four giant planets and Titan, from IR emissions lines between 28 and 45 μm; in addition, CO_2 was detected at 15 μm on Saturn and Neptune (Feuchtgruber et al. 1997, 1999). The origin of the external oxygen source will be discussed below.

The ionosphere and the interaction with the magnetosphere

At low pressure, above the μbar pressure level, the mean free path of molecules becomes large enough for each one to have its own scale height; this region is called the heterosphere. In this region, extreme UV radiation dissociates and/or ionizes molecules (first molecular hydrogen but also hydrocarbons) and atoms (mostly H and He). The most abundant ion in Saturn is H_3^+, detected from ground-based IR spectroscopy by Geballe et al. (1993); H^+ and H_2^+ were also detected by Voyager.

The ionospheric structure of Saturn was investigated by the radio-occultation experiments of Pioneer 11 (S band at 2.3 GHz) and Voyager (S band and X band at 8.6 GHz). Both the upper stratosphere and the ionosphere were found to be hotter than expected, which implies that, in addition to extreme UV radiation, other energy sources are implied (Atreya et al. 1984; Atreya 1986). Electron precipitation,

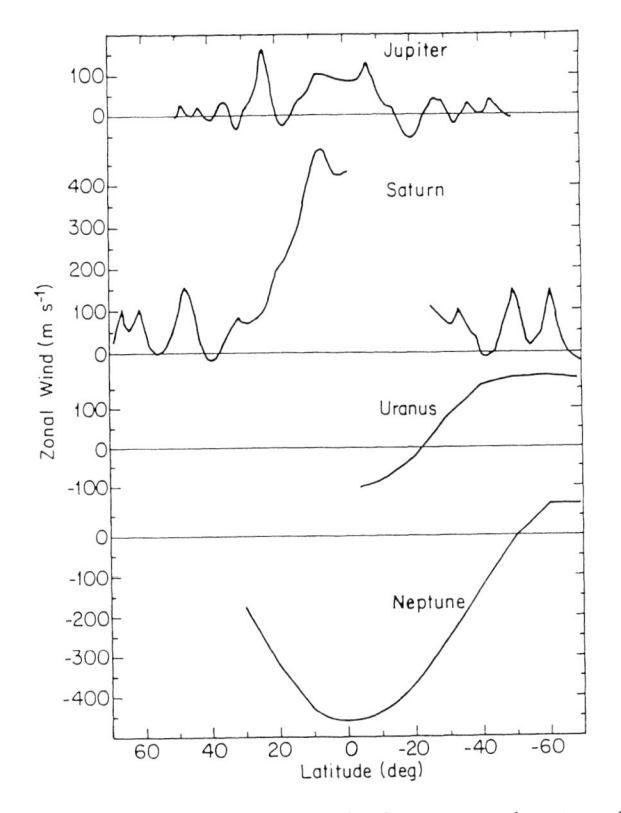

Figure 10 Zonal (east–west) wind velocities as a function of latitude for the four giant planets. All profiles look symmetrical with respect to the equator. Both Uranus and Neptune show retrograde winds at the equator while the opposite is observed on Jupiter and Saturn. (From Gierash and Conrath 1993.)

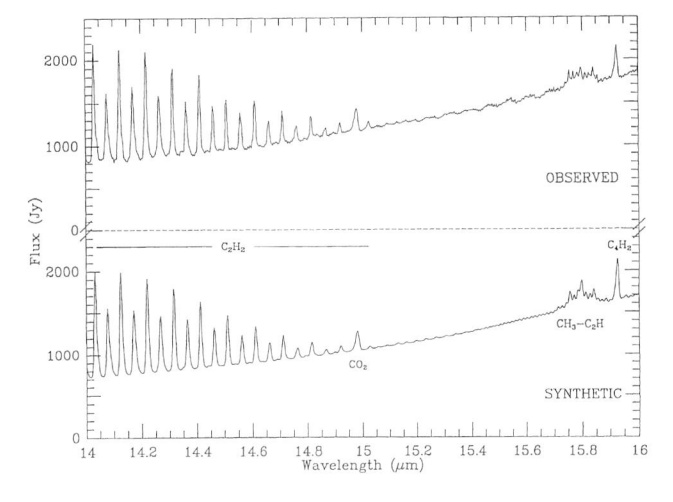

Figure 11 The 14–16 μm spectrum of Saturn, showing the emissions of CO_2, CH_3C_2H and C_4H_2 (de Graauw et al. 1997). Top: data from the ISO short-wavelength spectrometer; bottom: synthetic spectrum. (From Encrenaz et al. 1999.)

also invoked in the case of Jupiter on the basis of the Galileo probe data (Young 1998), appears as a plausible mechanism. Another possible contribution could come from breaking planetary gravity waves.

In the UV range, in addition to the global Ly α and H_2 emissions of Saturn, qualitatively similar to those of Jupiter, the Voyager UV spectrometer detected aurorae near 80° latitude, confirming previous tentative observations by IUE and Pioneer 11. A ring of hydrogen emission was also measured between 8 and 25 Saturnian radii (Broadfoot et al. 1981; Figure 13).

The magnetosphere of Saturn was first detected by Pioneer 11, and further investigated by the two Voyager probes. In contrast to the other giant planets, Saturn has almost no tilt nor radial offset (Table 1). As mentioned above, Saturn's magnetic field, as the jovian one, is likely to be generated through dynamo effect from convective motions in the internal metallic hydrogen layers. The Planetary Radio Astronomy experiment aboard Voyager recorded two types of radio bursts, the kilometric radiation and other sporadic bursts called Saturn electrostatic discharges, whose diurnal variations rather suggest an atmospheric origin (Warwick et al. 1981).

URANUS AND NEPTUNE

With an initial core representing more than half of their total mass, the icy giants Uranus and Neptune show significant differences with respect to Jupiter and Saturn. First, their lower masses and higher densities imply a different internal structure, more enriched in heavy elements; second, their larger heliocentric distances leads to lower temperatures and the condensation of many species, implying a different atmospheric composition and cloud structure.

As a unique particularity among the planets, Uranus's rotation axis is tilted by almost 90°, i.e. almost lays in the ecliptic plane. At the time of the Voyager 2 observations (1986), the sub-solar point was close to the south pole. This single characteristic of Uranus might be the result of a collision with a large body which occurred in the early history of the planet, before its satellite system (close to the planet's equatorial plane) was formed.

The visual aspect of the two planets is very different from the gaseous giants, and very different to each other. Uranus is pale green-bluish with very low contrast and very few features (Figure 2), while Neptune exhibits a deep blue color with some white and dark spots (Figure 3). On both planets, the color probably reflects the presence of gaseous methane in large, but different amounts; the main dark spot observed on Neptune by Voyager 2 might be an anticyclonic structure comparable to the Great Red Spot of Jupiter; it is not as stable as the GRS, however, since more

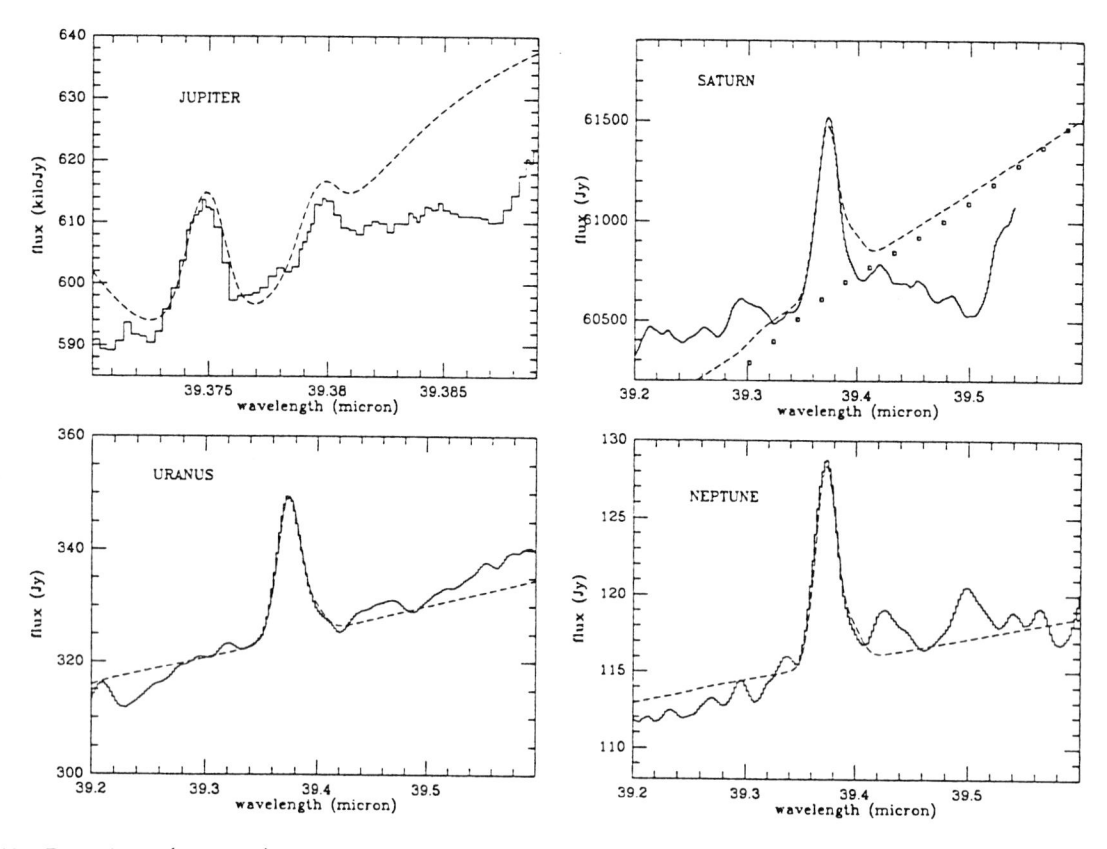

Figure 12 Detection of stratospheric water vapor in the four giant planets by the ISO short-wavelength spectrometer (Feuchtgruber et al. 1997, 1999). Several infrared H_2O lines were detected in each planet. (From Encrenaz et al. 1999.)

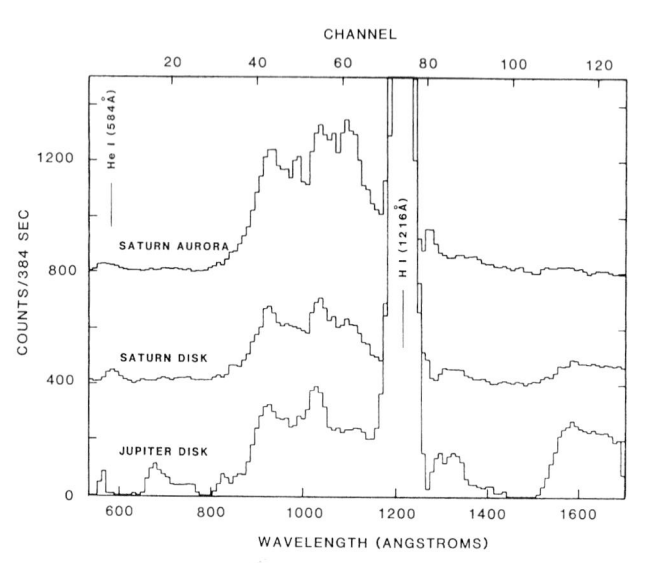

Figure 13 UV spectra of Saturn recorded by the UVS experiment of Voyager 1. A Jupiter spectrum is shown for comparison. The Saturn spectra have been multiplied by 3.3, the ratio of the solar flux at Jupiter to that at Saturn. Emissions on Saturn are due to H Ly α and to the Lyman and Werner bands of H_2 (900–1130 Å). (From Broadfoot et al. 1981.)

recent HST observations have shown that this feature has disappeared (Figures 3 and 5). The white spots are believed to be due to high cirrus of methane ice with high reflectivity. Zonal flows were measured from the motion of the features: they both exhibit, especially on Neptune, a strong retrograde jet around the equator (Gierasch and Conrath 1993; Figure 10). The situation is thus the opposite of the Saturn case, while Uranus and Neptune are very similar, in spite of the very different position of their polar axis with respect to the ecliptic plane; this similarity illustrates that solar radiation is not responsible for the wind field latitudinal distribution. There is currently no explanation of the differences between the wind fields of the four giant planets.

Internal structure

In spite of the similarity of their sizes, Uranus and Neptune have different internal structures, as already indicated by their different densities (Table 1) and also demonstrated by the Voyager measurements of their gravitational fields, which imply that Neptune is less centrally condensed than Uranus. Current models of the planets include the possible presence of a core, whose mass would be less than one terrestrial mass, and an icy mixture with a decreasing density (from about 5 to 1 g cm^{-3}) toward the external layers up to about 0.8 planetary radius; an hydrogen-rich envelope surrounds this level. As a result, the interior of Uranus and Neptune is mostly icy, and their is no metallic hydrogen. Following cosmic abundances, the ices should be mostly H_2O, with a contribution of CH_4 and possibly NH_3. The magnetic fields of Uranus and Neptune could be generated by conductive motions within the hot icy region, presumably ionized and dissociated. The temperature in the deep interior is expected to be several thousand K, and the pressure of several million bars (Hubbard 1997; Marley 1999).

Since their is no metallic hydrogen in the interiors of Uranus and Neptune, no helium condensation is expected. Indeed, for both planets, the helium abundance determined by Voyager IRIS and radio-occultation experiments (Figure 14) was compatible with the protosolar value (see Encrenaz 1999b for a review). However, there is one measurement which is still not understood: Neptune, like Jupiter and Saturn, has a significant internal energy source (2.5 times the absorbed solar energy) while Uranus doesn't (Gautier an Owen 1989; Table 1). The reason of this difference is not yet understood. In the case of Neptune, the origin of the internal energy could be gravitational contraction; this mechanism is also expected to take place in Jupiter. In the case of Uranus, it has been suggested that gradients in Uranus' internal composition might inhibit convection; this hypothesis, however, remains to be tested.

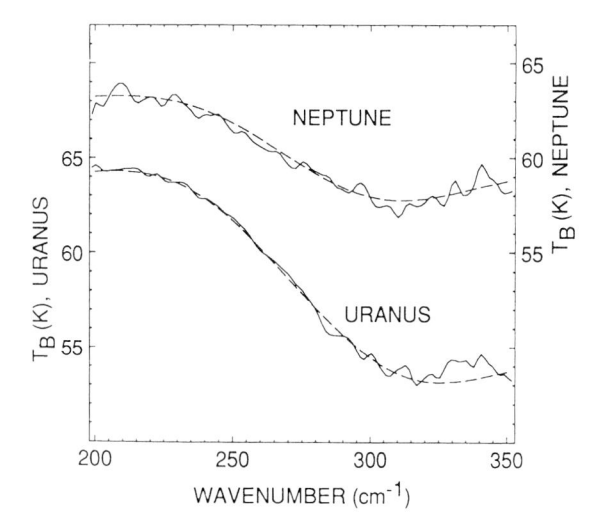

Figure 14 Infrared spectra of Uranus and Neptune, observed by IRIS-Voyager 2 between 200 and 350 cm^{-1} (28–50 μm). Spectra have been offset from one another by 5 K. Dashed curves show synthetic models corresponding to the retrieval (with the use of radio-occultation data) of the temperature profile and the helium content. (From Hanel et al. 1992.)

Atmospheric structure and composition

Because of their large heliocentric distances, Uranus and Neptune have thermal structures comparable to those of the gaseous giants (Figure 7), but a very low temperature at the tropopause (about 50 K). The temperature profiles were retrieved from the Voyager radio-occultation data. As a result of the low tropospheric temperature, many atmospheric species condense in the upper troposphere, including CH_4. Apart from hydrogen, methane and their isotopes, all the atmospheric species detected in Uranus and Neptune are stratospheric.

The troposphere of Uranus and Neptune has been probed by ground-based spectroscopy of H_2 and CH_4 bands in the visible and near-IR range (Baines et al. 1995) and also from radio measurements (de Pater et al. 1991). The inferred cloud structures (Figure 9) include a thick cloud, presumably due to H_2S ice, around T = 120 K (P = 3–4 bars), a CH_4 cloud between 50 and 80 K (0.1–1.2 bar), and several hydrocarbon clouds in the stratosphere. NH_3 condensation is expected to take place below or within the H_2S cloud, and HCN condensation, in the case of Neptune, occurs in the stratosphere (Baines et al. 1995).

The CH_4 mixing ratio inferred from the visible and near-IR spectra is 0.016 for Uranus and 0.022 for Neptune, which implies C/H enrichments of about 20 and 30 respectively as compared to the solar value (Baines et al. 1995; Table 3).

The D/H ratio in Uranus and Neptune was retrieved by two different means. The first method used ground-based spectroscopy of a CH_3D near-IR band, at 1.6 μm and, in the case of Neptune, Voyager and ground-based data at 8 μm; its accuracy, however, was limited by the uncertainty in the fractionation factor which is used to infer D/H from CH_3D/CH_4. The second method used the detection of the R(2) line of HD at 38 μm by ISO (Feuchtgruber et al. 1999; Figure 15). For both Uranus and Neptune, the inferred D/H ratio was about 3 times larger than the protosolar value (Table 3). The implications of the results concerning the abundance ratios will be discussed below.

The stratospheres of Uranus and Neptune show strong differences in their thermal structures (Figure 7), which illustrates that different heating mechanisms are involved. Uranus' lower stratosphere, at pressures of 0.1–1 mbar, is significantly colder than Neptune's, but its temperature increases at a faster rate as the altitude increases, so that the upper stratosphere of Uranus is warmer. As a result, there is no infrared detection of CH_4 nor hydrocarbons in the lower stratosphere of Uranus, but species are detected at higher altitudes, near the homopause at the μbar level. On Uranus, only C_2H_2 has been detected by ISO (Encrenaz et al. 1998; Figure 16), leading to a determination of the eddy diffusion coefficient which confirms earlier measurements by Voyager-UVS (Atreya et al. 1991); this coefficient is

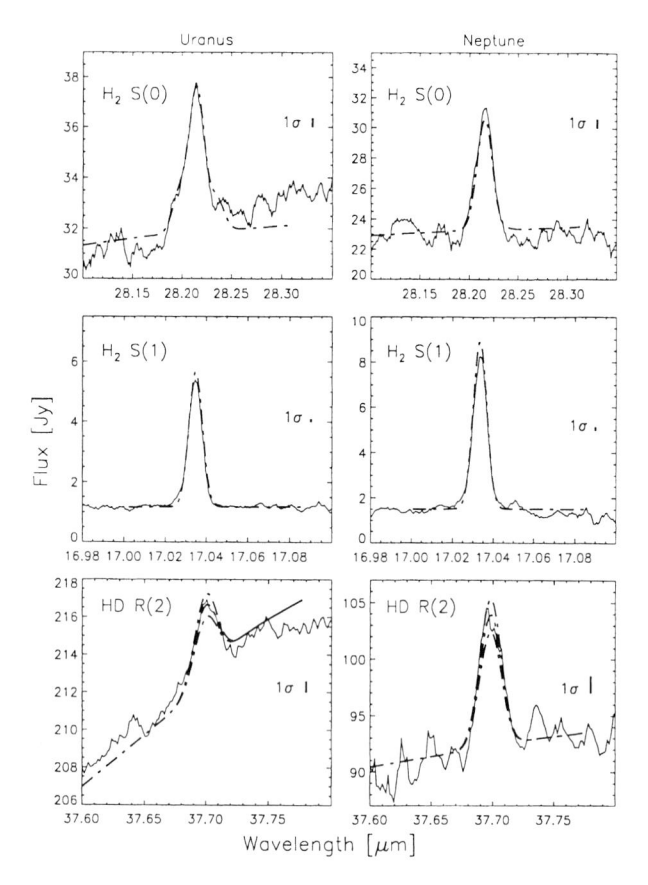

Figure 15 Detection of H_2 and HD infrared transitions on Uranus (left column) and Neptune (right column) with the short-wavelength spectrometer on ISO. The observations of the H_2 S(0) and S(1) quadrupole lines, at 28.2 μm (top) and 17.0 μm (middle) respectively, were used to retrieve the thermal profile in the lower stratospheres. The HD R(2) rotational transition was observed at 37.7 μm (bottom) and compared to several models, using the thermal profile inferred from the H_2 observations and different values of the HD mixing ratio. The HD mixing ratios used in the models are 9, 11, 13×10^{-5} for Uranus and 11, 13, 15×10^{-5} for Neptune (the HD mixing ratio is twice the D/H ratio). (From Feuchtgruber et al. 1999.)

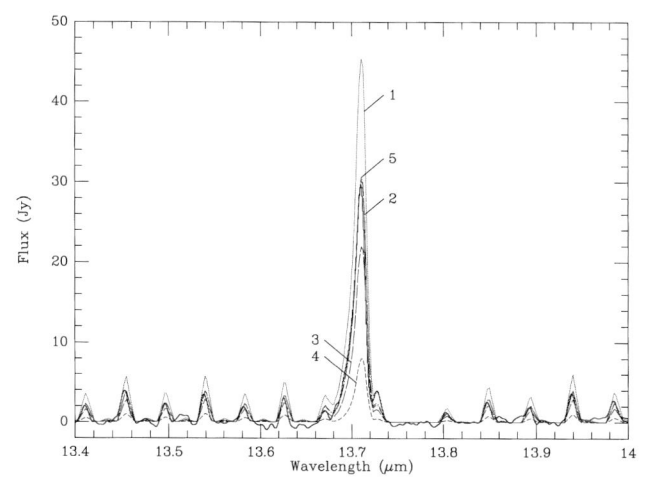

Figure 16 The ISO spectrum of Uranus in the C_2H_2 band (thick line) compared to synthetic models (thin lines). Best fits (Models 2 and 5) are provided for a methane stratospheric mixing ratio of $3-10 \times 10^{-5}$ and an eddy diffusion coefficient of $5-10 \times 10^{-3}$ cm$^2 \cdot$s^{-1}. (From Encrenaz et al. 1998.)

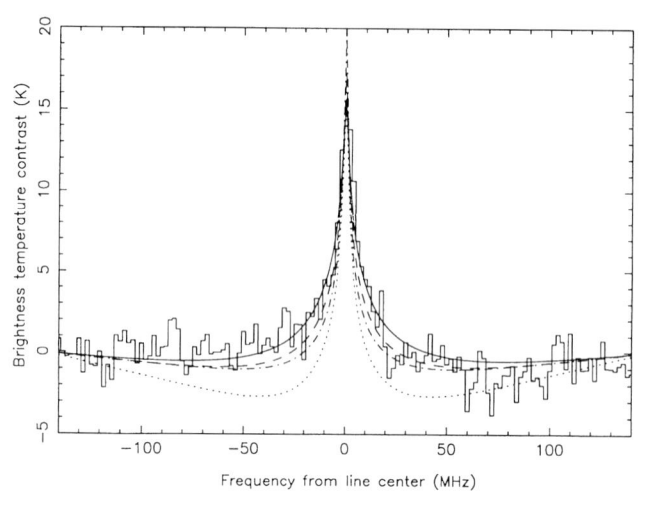

Figure 17 Observation of CO in Neptune's stratosphere with the 30-m IRAM antenna. (From Rosenqvist et al. 1992.)

very low (about 10^4 cm^2 s^{-1}) which implies that Uranus' stratosphere is very sluggish; its homopause is actually much lower than is the case with the other giants. On Neptune, the eddy diffusion coefficient derived from Voyager-UVS is at least 10^7 cm^2 s^{-1} (Romani et al. 1993) which implies a rapid vertical mixing. In addition to C_2H_2 and C_2H_6, previously detected from ground-based and IRIS-Voyager observations, two new hydrocarbons, the methyl radical CH_3 and ethylene C_2H_4, were detected by ISO (Bézard et al. 1999; Schulz et al. 1999).

An unexpected discovery was, in 1992, the ground-based detection of CO and HCN in Neptune's stratosphere from heterodyne spectrosocpy (Rosenqvist et al. 1992; Marten et al. 1993; Figure 17). The CO measured mixing ratio (about 10^{-6}) was about 1000 times larger than the value expected from thermochemical models which is in agreement with the values of Jupiter and Saturn (Table 2). Stringent upper limits (but still compatible with the Jupiter and Saturn values) were obtained for Uranus. Finally, as mentioned above, water vapor was detected by ISO in the stratospheres of Uranus and Neptune (Figure 12), showing evidence for an external source of oxygen; ISO also detected CO_2 on Neptune (Feuchtgruber et al. 1997, 1999). We will give below a comparative study of these results on the four giant planets and a discussion about the possible origin of this oxygen source.

Ionospheres and magnetospheres

Uranus and Neptune have extensive ionospheres which have been investigated by the Voyager 2 radio science and UVS experiments. They are dominated by the H^+ ion, but H_3^+ has also been detected on Uranus by use of IR ground-based spectroscopy, later confirmed by ISO. As is the case with Jupiter and Saturn, the temperatures of the ionospheres are higher than expected (Atreya et al. 1991), implying the existence of energetic processes in addition to EUV radiation.

The magnetic fields of Uranus and Neptune show strong tilts and offsets of their dipoles (Table 1); the reason for this peculiar geometry is still unclear. As a result, their magnetospheres are very different from that of Saturn and the aurorae appear far from the poles. These aurorae, first detected by IUE in the case of Uranus, were observed by Voyager 2-UVS on both planets; the proposed emission mechanism is excitation by charged particles (Connerney 1993).

CONSTRAINTS ON FORMATION AND EVOLUTION SCENARIOS

Abundance ratios

The measurement of elemental and isotopic ratios in Saturn, Uranus and Neptune can be used, by comparison with the Jupiter data, to derive information on formation and evolution scenarios. The C/H ratio, measured in the four giant planets, provides information on the abundance of CH_4 ice which was included in the initial cores of the planets; similar information comes from the determination of D/H in Uranus and Neptune, as deuterium is expected to be enriched, by ion-molecule reactions at low temperature, in the protoplanetary ices (Irvine and Knacke 1989). The D/H ratio in Saturn, in contrast, should be, as in the case of Jupiter, more representative of the protosolar D/H value. Finally, the He/H ratio appears to be a diagnostic of internal evolution for Jupiter and Saturn, while its value in Uranus and Neptune appears to reflect the protosolar helium abundance. Table 3 summarizes the abundance ratios in the four giant planets. In the case of Jupiter, we have a precise determination of these ratios from Galileo probe measurements; values quoted in Table 3 refer to the data recorded at the deepest atmospheric levels. They should be more representative of the interior values (see Atreya et al. 1999, for a review).

As shown in Table 3, there is a regular enrichment of the C/H ratio as a function of heliocentric distance: it ranges from 3 at 5 AU up to about 30 at 30 AU. This result is fully consistent with the nucleation model, since this model assumes an initial icy core of comparable mass for the four giant planets. When the primordial nebula accreted around these cores, the ices were heated and the carbon enriched the outer envelope; the enrichment was maximum for Uranus and Neptune, whose initial core represents more than half of their total mass. The measurements of C/H thus provide at least a qualitative agreement to the nucleation theory. The measurements of N/H and P/H on Jupiter and Saturn seem to show a similar trend; as mentioned above, due to dynamical effects, the O/H determination is probably not representative of its value in the interior of the planets.

The same trend is shown in the case of D/H (Table 3 and Figure 18). Jupiter and Saturn show comparable values,

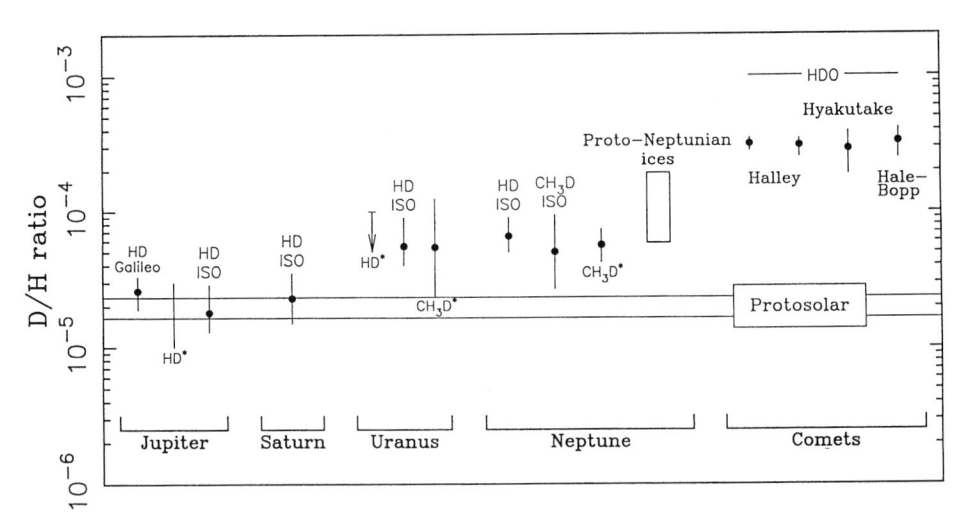

Figure 18 D/H ratios in the Solar system. Asterisks indicate ground-based measurements. (Adapted from Bockelée-Morvan et al. 1998, taken from Encrenaz et al. 1999.)

which are also consistent with the protosolar measurement derived from the solar wind. For Uranus and Neptune, the increasing enrichment of deuterium can be explained, as in the case of CH_4, by the larger fraction of ices in these planets. Models have been used to infer the D/H value of the proto-uranian and proto-neptunian ices. As shown in Figure 18, results seem to indicate that this value is lower than the D/H ratio inferred from comets (Bockelée-Morvan et al. 1998), which, if confirmed, might have implications on formation processes in the outer solar system.

Table 3 indicates that the helium mass fraction in Jupiter and Saturn is lower than the protosolar value, estimated from evolutionary models, while the values of Uranus and Neptune are compatible with it. It has to be noted that the new Saturn determination by Conrath and Gautier (2000) is not very different from that for Jupiter, while the previous value (Conrath et al. 1984), used as a reference in the literature until 2000, was about 4 times smaller. As mentioned above, a reasonable explanation for the helium depletion in Jupiter and Saturn is helium condensation which is expected to have taken place in the liquid hydrogen phase of their interiors during the cooling phase. As Saturn started from a cooler temperature than Jupiter, this condensation is expected to have started earlier than in Jupiter, which would imply a lower value of the helium mass fraction; this is not incompatible with the present results taking into account the uncertainty in Saturn's value.

CO and HCN in Neptune

As mentioned above, CO was detected in very large amounts in Neptune's stratosphere (about 10^{-6}), while its mixing ratio in Jupiter and Saturn (about 10^{-9}), compatible with Uranus' upper limit, is consistent with the values expected from thermochemical models (Prinn and Fegley 1989; Fegley et al. 1991). HCN, undetected so far in the other giant planets, was also detected in Neptune's stratosphere.

From a theoretical point of view, the low value of the CO/CH_4 ratio in the giant planets results from the thermochemical reaction

$$CO + 3H_2 \rightleftharpoons CH_4 + H_2O$$

which evolves toward CH_4 at low temperature and high pressure and toward CO at high temperature and low pressure. In the conditions of the giant planets formation, calculations show that the expected CO/H_2 in the sub-nebulae of the giant planets is about 10^{-9}. In the same way, the N_2/NH_3 ratio depends upon the reaction

$$N_2 + 3H_2 \rightleftharpoons 2NH_3$$

and increases for high temperature and low pressure. As a result, both CH_4 and NH_3 are expected to be the dominant species in the outer planets. This is confirmed by the low value observed for CO/CH_4 on Jupiter and Saturn (Table 2); however we have no indication of the N_2/NH_3 ratio, because the symmetrical N_2 molecule cannot be measured in the infrared rotation spectra.

What is the origin of CO and HCN in Neptune? Due to its long lifetime, CO is probably of internal origin. It has been suggested that HCN could be formed in Neptune's stratosphere from reactions between CH_3 and N atoms. Nitrogen could either come from Triton's atmosphere or from the dissociation by galactic cosmic rays of internal N_2, which, although undetected so far, could be present at the percent level. The question of the presence of nitrogen in Neptune's interior is to be raised after the detection of tropospheric CO in anomalously high abundance. Why are CO and HCN absent on Uranus? If the presence of tropospheric CO and N_2 is confirmed in Neptune, this would imply different formation and/or evolution scenarios for both planets. It has been suggested that the absence of CO and HCN in Uranus could be due to the lack of internal energy source in Uranus which would inhibit convection in the deep atmosphere. If this interpretation is correct, the difference in the interior structures of Uranus and Neptune remains puzzling, especially in view of the similarity of their magnetic fields (Table 1).

The external oxygen source

Recent observations by ISO have shown the presence of water vapor in the stratospheres of the giant planets and Titan; CO_2 has also been found in all objects but Uranus. Because of the low temperature of the tropopause which acts as a cold trap, water cannot be transported from the interior of the planets to the stratosphere, so it has to be of external origin; the same conclusion applies for carbon dioxide in Neptune and Titan.

Two major sources of external oxygen have been considered (Feuchtgruber et al. 1997, 1999): the local source, coming from the rings and/or the icy satellites, and the interplanetary source, which could consist of micrometeoritic ice particles or large objects like cometary nuclei. However, as mentioned above, the water flux inferred from the ISO data was in the same range (10^5–10^7 cm^2 s^{-1}) for all giant planets and Titan. This similarity seems to favor the second hypothesis, as the local sources are very different from one planet to another. A flux of micrometeoritic particles seems to be a plausible explanation, on the basis of data on interplanetary dust collected at 1 AU and at larger heliocentric distances by the Pioneer 10 and 11 probes. The comet source looks more unlikely, on the basis of the relatively low frequency of cometary collisions like that of Shoemaker-Levy-9 with Jupiter in 1994 – the so-called SL-9 impact – and considering the short lifetime of water after such events.

The question of the origin of CO_2 in Jupiter, Saturn, Neptune and Titan is also connected to the external oxygen source (Encrenaz et al. 1999). It is tempting to associate the presence of stratospheric CO_2 in Neptune (and its absence in Uranus) to the high content of stratospheric CO in Neptune and the lack of it in Uranus. In this case, CO_2 could be formed chemically in Neptune's stratosphere from CO and OH; the same scheme could take place in Titan (other possible photochemical reactions involve H_2O and CH_3 radicals). However, this explanation does not account for the presence of stratospheric CO_2 in Jupiter's and Saturn's stratospheres, where the CO content is very low. Another explanation might be that CO_2, as H_2O, comes from interplanetary particles; in this case, the absence of CO_2 in Uranus could possibly be explained by a lack of thermal emission in its colder lower stratosphere. Another puzzling effect is the fact that, in Jupiter's stratosphere, CO_2 has been detected in the south polar region, but not on the north pole; this latitudinal variability suggests that the observed CO_2 is a remnant of the SL9 impact; this interpretation seems compatible with photochemical models (Moses 1996). The question which remains open is then the origin of Saturn's stratospheric CO_2, observed in abundances comparable to the Jupiter (south) and Neptune values.

In conclusion, the exploration of the giant planets, using space-borne but also ground-based means, has led to a new understanding of their physico-chemical properties. To first order, the main similarities and differences between the four giant planets can be explained with the current scenarios of formation and evolution. However, as our knowledge has improved, new unsolved questions appear. Among the critical issues are the origin of the giant planet's internal heat sources, the nature of their dynamics, the heating mechanisms of their upper stratospheres and the large variety of their magnetospheres. Another major problem is the origin of the differences in atmospheric composition and internal energy between Uranus and Neptune, which probably provide a diagnostic of their formation scenarios.

THE FUTURE OF PLANETARY EXPLORATION BEYOND JUPITER

In situ space missions

After the success of the Voyager mission, the natural next step for *in situ* space exploration of the giant planets was long-term monitoring by orbiters and *in situ* analysis by descent probes. In the case of Jupiter, this step was successfully achieved by the NASA Galileo mission. Its atmospheric probe entered the jovian atmosphere in December 1995, and its orbiter monitored the jovian system from 1996 to 2000. In spite of the loss of the orbiter's high-gain antenna, both aspects of the mission were highly successful. In particular, the Galileo probe provided unique data about the elemental and isotopic composition of its atmosphere, the wind field, the thermal distribution and the cloud structure (Young 1998).

Following Jupiter's *in situ* exploration by Galileo, the NASA-ESA Cassini/Huygens mission, launched in 1997, will encounter Saturn in 2004. The Huygens probe will be sent into the atmosphere of its satellite Titan and will investigate the nature of its surface, while the orbiter will monitor the planet, its rings and its satellites during at least four years.

After Cassini, unfortunately, there is currently no follow-up project dealing with the exploration of the giant planets. An orbiter around Jupiter's galilean satellite Europa is under study by NASA, with the aim of exploring the water ocean which migh exist below its icy surface, but there is no planned mission toward Uranus and Neptune, mostly because of budget constraints. Still, the success of Galileo has illustrated the need for this type of mission, and in particular the need for atmospheric descent probes: these tools are essential to determine the atmospheric abundances (including the rare gases, not detectable by remote sensing spectroscopy) and thus to bring new information to the problem of planetary origin.

Ideally, two types of probe missions should be considered. On one hand, multi-probe missions should be sent to Jupiter and Saturn, in order to separate the dynamical local effects and to infer the abundance ratios in their deep atmosphere. Indeed, in the case of Jupiter, the Galileo probe entered a "hot spot", an atypical clear region very different from the rest of the planet, and some results acquired by the probe (in particular about the cloud structure and the oxygen abundance) were not representative of the entire planet. On the other hand, probes should be sent in the atmospheres of Uranus and Neptune, in order to search for Uranus' internal source and, in the case of Neptune, get a more accurate measurement of this source. In addition, these probes should measure the deep tropospheric thermal profile and, in the case of Uranus, search for convective inhibition, and measure abundance ratios, in particular nitrogen and rare gases. Such space measurements would provide a qualitative improvement in our knowledge of the giant planets, their origin and evolution.

As mentioned above, a space mission (Pluto Fast Flyby) presently under study at NASA, is expected to explore Pluto within the forthcoming decades. This mission should provide new answers (and probably also new questions) about the origin of Pluto itself, the origin of the Pluto-Charon system, and their link to the TNOs.

Space observatories

Following IUE, HST and ISO, a new generation of space observatories is under study, both within NASA and ESA.

In the near future, the Space Infra-Red Telescope Facility (SIRTF), to be launched by NASA in December 2001, will explore the infrared sky with an 85-cm telescope and several high-sensitivity instruments, including a medium-resolution spectrometer. This facility will be especially useful for measuring the infrared spectrum of Pluto and detecting the thermal emission of the TNOs, which was just beyond ISO's capabilities. The temperature measurements of the TNOs, coupled with their photometry in the visible range, will allow a simultaneous determination of their diameters and their albedos.

As a follow-up of HST, the New Generation Space Telescope (NGST), presently studied by NASA, should consist of a 8-m telescope with high-sensitivity cameras and spectrometers operating from the visible up to 30 μm. Its performances should be better than the existing facilities by several orders of magnitude. Its launch is planned around 2008. On the basis of the ISO results, one can expect new discoveries concerning the chemical composition of the giant planets from infrared spectroscopy. NGST should also obtain high-quality spectra of TNOs in the near-infrared range, and thus provide key information upon their chemical composition.

At even longer wavelengths, the Far Infra-Red and Submillimeter Telescope (FIRST), selected as the fourth cornerstone of ESA's "Horizon 2000" long-term program, will be a 3-m telescope with two imaging spectro-photometers and a heterodyne spectroscopy receiver, covering the submillimeter range with a resolving power as high as 10^6. FIRST, to be launched in 2007, will provide new information about the stratospheric composition of the giant planets. In particular, the vertical distribution of the stratospheric water vapor should be better constrained. This should bring more information on the nature of its external origin.

Ground-based observations

As an epilog, it is appropriate to stress that the exploration of the outer solar system has also largely benefited from ground-based astronomy, and will continue to take advantage of the instrumental developments, as the new generation of 8-m telescopes is beginning to operate. Several areas can be explicitly mentioned.

Over the past decade, the systematic monitoring of the Edgeworth–Kuiper Belt with large telescopes and sensitive cameras has led to the detection of the first samples of a new class of objects. This long-term program has to be performed from the ground and will continue to grow in the near future. In particular, the population of TNOs beyond 50 AU will be explored. In addition, spectrophotometry of the TNOs in the visible range, presently at the limit of sensitivity, should become accessible in the forthcoming decade, allowing a first classification as a function of their chemical composition.

The use of adaptive optics on large telescopes, in the near infrared range, provides a spatial resolution as high as that of the HST, if not higher, and has already given excellent results. In particular, edge-on images of Saturn's rings obtained in 1995, allowed a direct observation of inner satellites; in 1999, images of Neptune were used to analyse the meteorology of the planet.

In the millimeter range, we have seen that ground-based heterodyne spectroscopy had been successful in detecting CO and HCN in Neptune's stratosphere. This kind of study will develop in two directions in the forthcoming years. First, the extension of the spectral range in the submillimeter up to frequencies of about 1 THz will enlarge the number of possible species which could be detected. Second, the operation of large millimeter and submillimeter interferometers, like the ALMA project, will provide the advantage of a high spatial resolution and allow the mapping of planetary disks.

REFERENCES

Atreya, S.K. (1986). *Atmospheres and Ionospheres of the Outer Planets and their Satellites*. Springer-Verlag.

Atreya, S.K. and Romani, P.R. (1985). Photochemistry and clouds of Jupiter, Saturn and Uranus. In: G.H. Hunt (ed.), *Recent Advances in Planetary Meteorology*, Cambridge University Press, pp. 17–68.

Atreya, S.K., Waite, J.H., Donahue, T.M., Nagy, A.F. and McConnell, J.C. (1984). In: T. Gehrels and M.S. Matthews (eds.), *Saturn,* University of Arizona Press, pp. 239–277.

Atreya, S.K., Sandel, B.R. and Romani, P.N. (1991). Photochemistry and Vertical Mixing. In: J.T. Bergstralh, E.D. Miner and M.S. Matthews (eds.), *Uranus,* University of Arizona Press, pp. 110–146.

Atreya, S.K., Wong, M.H., Owen, T.C., Mahaffy, P.R., Niemann, H.B., de Pater, I., Drossart, P. and Encrenaz, Th. (1999). A comparison of the atmospheres of Jupiter and Saturn: deep atmospheric composition, cloud structure, vertical mixing and origin. *Planetary Space Science*, **47**, 1243–1262.

Baines, K.H., Mickelson, M.E., Larson, L.E. and Ferguson, D.W. (1995). The abundances of methane and ortho/para hydrogen on Uranus and Neptune: implications of new laboratory 4-0 H_2 quadrupole line parameters. *Icarus*, **114**, 328–340.

Bézard, B., Feuchtgruber, H. and Encrenaz, Th. (1999). Observations of hydrocarbons in the giant planets. In: Proceedings of the Conference "The Universe as seen by ISO", Paris, France, 20–23 October 1998, ESA SP-427, pp. 153–156.

Bockelée-Morvan, D., Gautier, D., Lis, D.C. *et al.* (1998). Deuterated water in comet C/1996 B2 (Hyakutake) and its implications for the origin of comets. *Icarus*, **133**, 147–162.

Boggess, A. and Wilson, R. (1987). The history of IUE. In: Y. Kondo (ed.), *Exploring the Universe with the IUE Satellite*, Reidel, pp. 3–20.

Broadfoot, A.L., Sandel, B.R., Shemansky, D.E. *et al.* (1981). Extreme Ultraviolet Observations from Voyager 1 Encounter with Saturn. *Science*, **212**, 206–211.

Cassen, P.M. and Woolum, D.S. (1999). The origin of the Solar System. In: P.R. Weissman, L. McFadden and T.V. Johnson (eds.), *Encyclopedia of the Solar System*, Academic Press, pp. 35–63.

Connerney, J.E.P. (1993). Magnetic fields of the outer planets. *Journal of Geophysical Research*, **98E**, 18659–18679.

Conrath, B.J., Gautier, D., Hanel, R.A. and Hornstein, J.S. (1984). The helium abundance of Saturn from Voyager measurements. *Astrophysical Journal*, **282**, 807–815.

Conrath, B.J. and Gautier, D. (2000). Saturn helium abundance: a reanalysis of Voyager measurements. *Icarus*, **144**, 124–134.

Courtin, R., Gautier, D., Marten, A., Bézard, B. and Hanel, R. (1984). The composition of Saturn's atmosphere at northern temperature latitudes from Voyager IRIS spectra: NH_3, PH_3, C_2H_2, C_2H_6, CH_3D, CH_4 and the saturnian D/H isotopic ratio. *Astrophysical Journal*, **287**, 899–916.

de Graauw, Th., Feuchtgruber, H., Bézard, B. *et al.* (1997). First results of ISO-SWS observations of Saturn: detection of CO_2, CH_3C_2H, C_4H_2 and tropospheric H_2O. *Astronomy and Astrophysics*, **321**, L13–L16.

de Pater, I., Romani, P.N. and Atreya, S.K. (1991). Possible microwave absorption by H_2S in Uranus and Neptune's atmospheres. *Icarus*, **91**, 220–233.

Drossart, P., Fouchet, Th., Crovisier, J., Lellouch, E., Encrenaz, Th., Feuchtgruber, H. and Champion, J.-P. (1999). In: Proceedings of the Conference "The Universe as seen by ISO", Paris, France, 20–23 October 1998, ESA SP-427, pp. 169–172.

Encrenaz, Th. (1999a). The planet Jupiter. *Astronomy and Astrophysics Review*, **9**, 171–219.

Encrenaz, Th. (1999b). Light elements in the Solar system. In: L. Maiani, F. Melchiorri and N. Vittorio (eds.), *CP476, 3 K Cosmology: EC-TMR Conference*, The American Institute of Physics.

Encrenaz, Th., Feuchtgruber, H., Atreya, S.K. *et al.* (1998). ISO observations of Uranus: the stratospheric distribution of C_2H_2 and the eddy diffusion coefficient. *Astronomy and Astrophysics*, **333**, L43–L46.

Encrenaz, Th., Drossart, P., Feuchtgruber, H. *et al.* (1999). The atmospheric composition and structure of Jupiter and Saturn from ISO observations: a preliminary review. *Planetary Space Science*, **47**, 1225–1242.

Fegley, B., Gautier, D., Owen, T. and Prinn, R.G. (1991). Spectroscopy and chemistry of the atmosphere of Uranus. In: J.T. Bergstralh, E.D. Miner and M.S. Matthews (eds.), *Uranus*, University of Arizona Press, pp. 147–203.

Feuchtgruber, H., Lellouch, E., de Graauw, T., Bézard, B., Encrenaz, T. and Griffin, M. (1997). External supply of oxygen to the atmospheres of the giant planets. *Nature*, **389**, 159–162.

Feuchtgruber, H., Lellouch, E., Encrenaz, Th., Bézard, B., Coustenis, A., Drossart, P., Salama, A., de Graauw, Th. and Davis, G.R. (1999). Oxygen in the stratospheres of the giant planets and Titan. In: Proceedings of the Conference "The Universe as seen by ISO", Paris, France, 20–23 October 1998, ESA SP-427, pp. 133–136.

Gautier, D. and Owen, T.C. (1989). The composition of outer planet atmospheres. In: S.K. Atreya, J.B. Pollack and M.S. Matthews (eds.), *Origin and Evolution of Planetary and Satellite Atmospheres*, University of Arizona Press, pp. 487–512.

Geballe, T.R., Jagod, M.-F. and Oka, T. (1993). Detection of H_3^+ infrared emission lines in Saturn. *Astrophysical Journal*, **408**, L109–L112.

Geiss, J. and Gloecker, G. (1998). Abundances of deuterium and helium-3 in the protosolar cloud. *Space Science Reviews*, **84**, 239–250.

Griffin, M.J., Naylor, D.A., Davis, G.R. *et al.* (1996). First detetion of the 56-μm rotational line of HD in Saturn's atmosphere. *Astronomy and Astrophysics*, **315**, L389–L392.

Hanel, R.A., Conrath, B.J., Kunde, V.G., Pearl, J.C. and Pirraglia, J.A. (1983). Albedo, internal heat flux, and energy balance of Saturn. *Icarus*, **53**, 262–285.

Hanel, R.A., Conrath, B.J., Jennings, D.E. and Samuelson, R.E. (1992). *Exploration of the Solar System by Infrared Remote Sensing*. Cambridge University Press.

Hubbard, W.B. (1997). Saturn: interior structure. In: J.H. Shirley and R.W. Fairbridge (eds.), *Encyclopedia of Planetary Sciences*, Chapman and Hall, pp. 717–718.

Irvine, W.M. and Knacke, R.F. (1989). The chemistry of interstellar gas and grains. In: S.K. Atreya, J.B. Pollack and M.S. Matthews (eds.), *Origin and Evolution of Planetary and Satellite Atmospheres*, University of Arizona Press, pp. 3–34.

Kessler, M.F., Steinz, J.A., Anderegg, M.E. *et al.* (1996). The Infrared Space Observatory (ISO) Mission. *Astronomy and Astrophysics*, **315**, L27–L31.

Lasher, L.E. (1997). Pioneer 10 and 11 missions. In: J.H. Shirley and R.W. Fairbridge (eds.), *Encyclopedia of Planetary Sciences*, Chapman and Hall, pp. 579–581.

Lellouch, E. (1999). Solar system observations. In: Proceedings of the Conference "The Universe as seen by ISO", Paris, France, 20–23 October 1998, ESA SP-427, pp. 125–132.

Levison, H.F. and Weissman, P.W. (1999). The Kuiper Belt. In: P.R. Weissman, L. McFadden and T.V. Johnson (eds.), *Encyclopedia of the Solar System*, Academic Press, pp. 557–583.

Lindal, G.F., Sweetnam, D.N. and Eshelman, V.R. (1985). The atmosphere of Saturn: an analysis of the Voyager radio occultation measurements. *Astronomical Journal*, **90**, 1136–1146.

Marley, M.S. (1999). Interiors of the Giant Planets. In: P.R. Weissman, L. McFadden and T.V. Johnson (eds.), *Encyclopedia of the Solar System*, Academic Press, pp. 339–355.

Marten, M., Gautier, D., Owen, T., Sanders, D.B., Matthews, H.E., Atreya, S.K., Tilanus, R.P.J. and Deane, J.R. (1993). First observations of CO and HCN on Neptune and Uranus at millimeter wavelengths and their implications for atmospheric chemistry. *Astrophysical Journal*, **406**, 285–297.

Miner, E.D. (1997). Voyager missions. In: J.H. Shirley and R.W. Fairbridge (eds.) *Encyclopedia of Planetary sciences*, Chapman and Hall, pp. 922–927.

Mizuno, H. (1980). Formation of the giant planets. *Progress of Theoretical Physics*, **64**, 544–557.

Moos, H.W. and Encrenaz, Th. (1987). Planetary atmospheres and aurorae. In: Y. Kondo (ed.) *Exploring the Universe with the IUE Satellite*, Reidel, pp. 45–65.

Moses, J.I. (1996). SL9 impact chemistry: long-term photochemical evolution. In: K.S. Noll, H.A. Weaver and P.D. Feldman (eds.), *The Collision of Comet Shoemaker-Levy 9 and Jupiter*, Cambridge University Press, pp. 243–268.

Mueller, S. and McKinnon, W.B. (1997). Pluto. In: J.H. Shirley and R.W. Fairbridge (eds.), *Encyclopedia of Planetary Sciences*, Chapman and Hall, pp. 645–650.

Nelson, R.N. and Domingue, D.L. (1999). The Solar System at Ultraviolet Wavelengths. In: P.R. Weissman, L. McFadden and T.V. Johnson (eds.), *Encyclopedia of the Solar System*, pp. 697–713.

Nicholson, P.D., Showalter, M.R., Dones, L. *et al.* (1996). Observations of Saturn's ring-plane crossing in August and November. *Science*, **272**, 509–516.

Null, G.W., Lau, E.L., Biller, E.D. and Anderson, J.D. (1981). Saturn gravity results obtained from Pioneer 11 tracking data and Earth-based satellite data. *Astronomical Journal*, **86**, 458–468.

Prinn, R.G. and Fegley, B. (1989). In: S.K. Atreya, J.B. Pollack and M.S. Matthews (eds.), *Origin and Evolution of Planetary and Satellite Atmospheres*, University of Arizona Press, pp. 78–136.

Romani, P.N., Bishop, J., Bézard, B. and Atreya, S. (1993). Methane photochemistry on Neptune: Ethane and ethylene mixing ratios and haze production. *Icarus*, **106**, 442–463.

Rosenqvist, J., Lellouch, E., Romani, P.N., Paubert, G. and Encrenaz, T. (1992). Millimeter-wave observations of Saturn, Uranus and Neptune: CO and HCN on Neptune. *Astrophysical Journal*, **392**, L99–L102.

Schulz, B., Encrenaz, Th., Bézard, B., Romani, P.N., Lellouch, E. and Atreya, S.K. (1999). Detection of C_2H_4 in Neptune from ISO/PHT-S observations. *Astronomy and Astrophysics*, **350**, L13–L17.

Stern, S.A. and Yelle, R.V. (1999). Pluto and Charon. In: P.R. Weissman, L. McFadden and T.V. Johnson (eds.), *Encyclopedia of the Solar System*, Academic Press, pp. 499–517.

Warwick, J.W., Pearce, J.B., Evans, D.R. *et al.* (1981). Planetary radio-astronomy observations from Voyager 1 near Saturn. *Science*, **212**, 239–243.

West, R.A. (1999). Atmospheres of the Giant Planets. In: P.R. Weissman, L. McFadden and T.V. Johnson (eds.), *Encyclopedia of the Solar System*, Academic Press, pp. 315–337.

Young, R.E. (1998). The Galileo probe mission to Jupiter: Science overview. *Journal of Geophysical Research*, **103E**, 22775–22790.

Gierasch, P. and Conrath, B.J. (1993). Dynamics of the outer planets – Post-Voyager measurement objectives. *Journal of Geophysical Research*, **98**, 5459–5469.

60

NICOLAS THOMAS*

The satellites of the outer planets

The discovery of the four large satellites of Jupiter by Galileo Galilei in January 1610 was one of the most significant events in the history of science. Galileo's observations of moons appearing to move from one side of the planet to the other provided support for the Copernican view of the Universe by showing that celestial objects did not all orbit the Earth. His claim as the first observer of what he called the 'Medicean planets' was hotly disputed by the German astronomer Simon Marius. Marius claimed to have observed the satellites in November 1609 and there is little doubt that he was using a telescope to study the heavens around the same time as Galileo. However, Galileo published his observations while there is no record of those of Marius. In Galileo's honour, the four large satellites of Jupiter are often now referred to as the Galilean satellites.

While the scientific and philosophical significance of Galileo's discovery cannot be underestimated, the observations also started an enormous expansion in the number of known objects in our Solar System. Prior to Galileo's discovery, eight Solar System objects were known: six planets (Mercury, Venus, Earth, Mars, Jupiter and Saturn), one moon (the Moon) and the Sun. His observations raised the number of known Solar System objects by 50% at a stroke. There are now 61 satellites of the 9 planets known (several other irregular satellites of Saturn await confirmation), 57 of them are satellites of the gas giant planets Jupiter, Saturn, Uranus and Neptune (Figure 1). Many of the satellites have remarkable phenomena associated with them. There are, for example, volcanoes on Io, geysers on Triton and a thick atmosphere on Titan. In some senses, therefore, Galileo's discovery was also the start of a new discipline in planetary sciences – the study of planetary satellites.

* Max-Planck-Institut für Aeronomie, Katlenburg-Lindau, Germany

The rest of this chapter is organized as follows. It begins with a historical overview of the discoveries of planetary satellites in the twentieth century. Included are brief anecdotes about some of the astronomers involved in these observations and a short description of the spacecraft that made many of the more recent discoveries. In the following sections the orbits of planetary satellites are discussed in more detail, including a look at the limited amount of information we have about surface composition. This is followed by a description of the atmospheres of planetary satellites which have become a major topic since the early 1980s. The availability of spacecraft imaging of outer Solar System objects has also led to an amazing growth in the field of planetary geology which in turn has given us clues as to the evolution of planetary satellites. The chapter concludes with a brief discussion of what the twenty-first century could hold.

HISTORICAL INTRODUCTION

Status in 1900

By the start of the twentieth century 22 planetary satellites had been discovered. After the discovery of the Galilean satellites, Christiaan Huygens (1629–1695) and Giovanni Domenico Cassini (1625–1712) found most of the larger moons of Saturn in the seventeenth century. The larger satellites of Uranus were found in the eighteenth and nineteenth centuries by William Herschel (1738–1822) and William Lassell (1799–1880) while the largest satellite of Neptune, Triton, was also discovered by Lassell in 1846.

In his introductory chapter to the book *Satellites*, Burns (1986) suggested a means of classifying planetary satellites. In his classification, regular satellites are large, spherical and

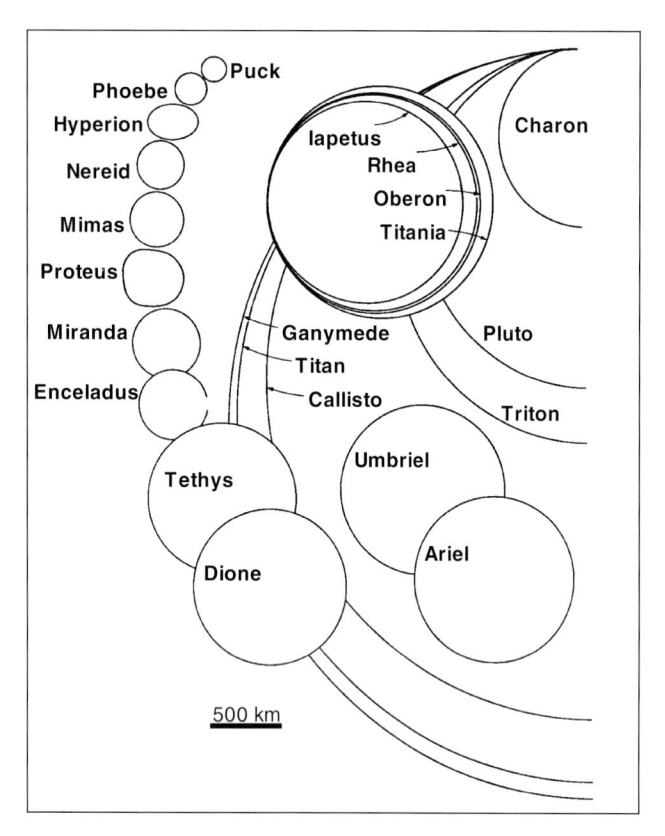

Figure 1 A montage of the relative sizes and shapes of some of the larger planetary satellites in the Solar System. (From Croft and Soderblom 1991, courtesy of University of Arizona Press).

relatively close to the parent planet. They are in prograde orbits with low inclination and eccentricity and are often in synchronous rotation (that is, the same hemisphere always faces the parent). Their masses are small compared to the parent. Collisional shards are small, irregularly shaped objects, extremely close to the planet or co-orbital with a regular satellite. They have prograde orbits with essentially zero inclination and eccentricity. They are thought to be the 'debris' left over after planet and regular satellite formation. Irregular satellites are far from the parent and have substantial inclination, eccentricity, or both. They are small. They are thought to have been captured into their present orbits (although it has to be said that a reasonable mechanism for this process remains to be demonstrated). A fourth group of unusual satellites, which do not fall into any of the above categories, was also suggested (Table 1).

By around 1850 all the regular satellites had been discovered. Two of the three satellites we now categorize as unusual (the Moon and Triton) had also been found (the third, Charon, would not be found until 1978). Shortly before the beginning of the twentieth century, telescopes became powerful enough to find some of the smaller collisional shards. Hyperion, which might be considered to be intermediate between a regular satellite and a collisional shard, was discovered in 1848. Phobos and Deimos were discovered by Asaph Hall (1829–1907) in 1877 and Edward Emerson Barnard (1857–1923) observed Amalthea, a satellite inside the orbit of Io, from the Lick Observatory in 1892 during a systematic search for new Jovian satellites.

Up to this point the discovered satellites were all relatively close to the parent (within 60 planetary radii of the centre of mass). However, the discovery of the irregular satellite Phoebe, in a retrograde orbit, 215 planetary radii from Saturn by William Henry Pickering (1858–1938) in April 1899 (on a plate obtained seven months previously) indicated that many objects at hundreds of planetary radii away from the parent planet could exist.

Initial discoveries

The first significant event of the twentieth century was the discovery in 1904 of the irregular satellite of Jupiter, Himalia, by the American astronomer Charles Dillon Perrine (1878–1951). Perrine worked at the Lick Observatory on Mount Hamilton in northern California and later became director of the Argentinian National Observatory in Cordoba. An 86 km diameter crater on the Moon (at 42.5°N, 127.8°W) is named after him.

Philibert J. Melotte (1880–1961) discovered the eighth moon of Jupiter, Pasiphaë, in 1908. While photographing Pasiphaë with the 36-inch (0.9 m) Crossley reflector at the Lick Observatory, Seth Barnes Nicholson (1891–1963) discovered a ninth moon which was named Sinope. Nicholson was born in Springfield, Illinois, and became interested in astronomy as an undergraduate at Drake University in Iowa. He graduated in 1911 and became a graduate student at the University of California from which he gained his doctorate in 1915. His dissertation was based on his discovery and orbit computation of Sinope.

Nicholson subsequently moved to the Mount Wilson Observatory near Pasadena, California, where he spent the rest of his career. His main job was to observe the Sun with the 150-foot (46 m) solar tower telescope. He produced annual reports on sunspot activity (with Hale and St John) and magnetism for decades. During this period, however, he also discovered three more Jovian satellites (Lysithea, Ananke and Carme) and a Trojan asteroid (Menelaus) as well as computing the orbits of several comets. He also worked with Edison Pettit to produce a vacuum thermocouple which was used to measure the temperatures of the Moon and planets in the early 1920s. He was twice president of the Astronomical Society of the Pacific and became the 56th winner of the Catherine Wolfe Bruce medal, awarded by the society, shortly before he died.

The Uranian satellite Miranda was discovered in 1948 by the Dutch-born astronomer Gerrit Pieter Kuiper (1905–1973).

Table 1 Classification of planetary satellites

	Earth	Mars	Jupiter	Saturn	Uranus	Neptune	Pluto
Regular			Io (Galileo 1610) Europa (Galileo 1610) Ganymede (Galileo 1610) Callisto (Galileo 1610)	Mimas (Herschel 1789) Enceladus (Herschel 1789) Tethys (Cassini 1684) Dione (Cassini 1684) Rhea (Cassini 1672) Titan (Huygens 1655) Iapetus (Cassini 1671)	Miranda (Kuiper 1948) Ariel (Lassell 1851) Umbriel (Lassell 1851) Titania (Herschel 1787) Oberon (Herschel 1787)		
Shards		Phobos (Hall 1877) Deimos (Hall 1877)	Metis (Synnott 1979) Andrastea (Jewitt) Amalthea (Barnard 1892) Thebe (Synnott 1979)	Atlas (Terrile 1980) Prometheus (Danielson 1979) Pandora (Collins et al. 1980) Epimetheus (Fountain, Larson/Walker 1966/1980) Janus (Dollfus 1966) Telesto (Yoyager 2, 1980) Calypso (Pascu et al.1980) Helene (Lecacheux, Laques 1980) Hyperion (W. and G. Bond/Lassell 1848)	Cordelia (Voyager 2, 1986) Ophelia (Voyager 2, 1986) Bianca (Voyager 2, 1986) Cressida (Voyager 2, 1986) Desdemona (Voyager 2, 1986) Juliet (Voyager 2, 1986) Portia (Voyager 2, 1986) Rosalind (Voyager 2, 1986) Belinda (Voyager 2, 1986) Puck (Voyager 2, 1986)	Naiad (Voyager 2, 1989) Thalassa (Voyager 2, 1989) Despina (Voyager 2, 1989) Galatea (Voyager 2, 1989) Larissa (Reitsema et al. 1981) Proteus (Voyager 2, 1989)	
Irregular			Leda (Kowal 1974) Himalia (Perrine 1904) Lysithea (Nicholson, 1938) Elara (Perrine 1905) Ananke (Nicholson, 1951) Carme (Nicholson, 1938) Pasiphaë (Melotte 1908) Sinope (Nicholson, 1914) S/1999 J1 (Scotti et al. 1999)	Phoebe (Pickering 1898) Several others await confirmation	Caliban (Gladman et al. 1997) Sycorax (Gladman et al. 1997)	Nereid (Kuiper 1949)	
Unusual	Moon					Triton (Lassel 1846)	Charon (Christy 1978)

Kuiper, who is better known by his Americanized name, Gerard Peter, turned to studies of the Solar System shortly before the end of World War II. He discovered the methane atmosphere of Titan in 1944 (see below), the presence of CO_2 in the atmosphere of Mars and the small satellite of Neptune, Nereid, in 1949 at the McDonald Observatory. He was also instrumental in setting up the first institute dedicated to the study of the Solar System. The Lunar and Planetary Laboratory (LPL) of the University of Arizona, at Tucson, remains one of the leading planetary research institutes in the world.

Kuiper not only made ground-based observations. He also pioneered the use of telescopes in high-altitude aircraft for infrared observations. The absorption by the Earth's atmosphere of infrared radiation often prevents the investigation of the compositions of gaseous planets and satellites, the interstellar medium and molecular clouds. Kuiper's idea led to a C-141 StarLifter jet aircraft being used as a flying observatory. The Kuiper Airborne Observatory (KAO) was put into service in 1975 and is operated by NASA from its Ames Research Center at Moffett Field near San Jose. It is expected to be replaced by a flying observatory called SOFIA, which is based upon a Boeing 747, being built in a collaboration between the USA and Germany. Kuiper was also involved in spaceflight and became chief experimenter for the Ranger series of missions to the Moon. He also assisted in the selection of Cerro Tololo in Chile and Mauna Kea in Hawaii as sites for new high-altitude observatories. His name is associated with craters on Mercury, the Moon and Mars. The most prestigious prize of the Division of Planetary Sciences of the American Astronomical Society is also named after him.

The first satellite of Saturn to be discovered in the twentieth century was first observed by Andouin Dollfus in 1966 from the Pic du Midi Observatory in France. Dollfus has had a long and distinguished career as a ground-based observer of planets and comets. Much of his more recent work has concerned the investigation of the polarization of light reflected by dust grains ejected from comets. The satellite of Saturn he observed is called Janus. However, there was considerable uncertainty about the orbit of the new satellite. In 1978 John Fountain and Stephen Larson clarified the uncertainty by re-analysing the results of Dollfus's observations and demonstrating that there were two objects in the same orbit about Saturn. The smaller of the two co-orbital objects was subsequently called Epimetheus. In March 1980, just prior to the arrival of Voyager 1 at Saturn, Lecacheux and Lacques who, like Dollfus, worked at the Pic du Midi Observatory detected a small satellite (Helene) in the same orbit as Dione. Hence, prior to the arrival of the Voyager spacecraft, Saturn was known to have at least 12 satellites.

In 1974 Charles Kowal discovered the 13th satellite of Jupiter (Leda). Kowal worked at the Lowell Observatory in Flagstaff, Arizona. This privately endowed observatory was founded by Percival Lowell (1855–1916) in 1894. In subsequent years, Lowell studied the orbit of Uranus and predicted that the perturbations in the orbit could be explained by a large planet outside the orbit of Neptune. According to Lowell, this planet, called Planet X, needed to be approximately seven Earth masses to fit his observations. The discovery of Pluto by Clyde W. Tombaugh (1906–1997) in 1930 merely ignited the controversy because Pluto appeared to be much fainter than would be expected of such a large object. In 1977 Kowal started a photographic search for other objects outside the orbit of Neptune. He studied a region 15° on either side of the ecliptic. He found nothing attributable to an object beyond Neptune but did discover the first Centaur, 2060 Chiron, which may indicate a link between asteroidal debris in the outer Solar System and short-period comets.

In astronomy, the United States Naval Observatory (USNO) is probably best known for being the organization that, together with the Royal Greenwich Observatory, produces the Astronomical Almanac. However, in 1978, while measuring the positions of Pluto on a series of images, the USNO's James W. Christy made an immensely important discovery. Christy saw that the image of Pluto was extended to one side. On examining other plates, he saw that the extension had moved to the opposite side of Pluto. Christy realized that he was observing a very close satellite of Pluto. By viewing a series of older plates he noticed that the extension seemed to move with a period of about a week. After being reminded by a colleague that Pluto had a light-curve with a period of 6.387 days, he concluded that the orbital period was identical to the period of the light-curve. His colleague Robert Harrington computed the orbit the next day from the detailed analysis of the observations and found excellent agreement with a 6.387-day period. This discovery was important because it allowed a determination of the mass of Pluto that demonstrated that Pluto's mass was very much less than that of the Earth and therefore could not be the Planet X predicted by Percival Lowell. Interestingly, Seth Nicholson was also involved these studies. He and others had investigated perturbations in Neptune's orbit and concluded that the mass of Pluto should be only 0.8–0.9 Earth masses. Although this was moving in the right direction, Christy's discovery led to a mass estimate of only 1/400th of the Earth's mass.

The space age

By the time of Christy's discovery the exploration of the outer Solar System with robotic spacecraft was already in full swing. On 3 March 1972 Pioneer 10 became the first probe to be launched to Jupiter. It made its closest approach to Jupiter (132 000 km) on 3 December 1973. Pioneer 11 was launched on 5 April 1973 and passed Jupiter (42 800 km) on

2 December 1974 and Saturn (20 800 km) on 1 September 1979. Although these spacecraft carried considerable instrumentation, they were not particularly sophisticated. Pictures of the motion in Jupiter's atmosphere were returned, for example. Information was also provided about the plasma environment of Jupiter and a possible ionosphere of the innermost Galilean satellite, Io (see below). However, the imaging systems did not identify any new satellites during the flybys and the images of the existing satellites were not particularly good either. The Voyager spacecraft were much better equipped to study the planets and their satellites and rings. Voyager 2 was launched on 20 August 1977. Voyager 1 followed just 16 days later. Over the next 12 years Voyager 2 would complete what was known as a Grand Tour of the outer Solar System by passing each of the giant planets in turn (Figure 2). This voyage of discovery found many new planetary satellites but, more importantly, revealed many of the previously known satellites to be remarkable objects which deserved in-depth study in their own right.

The team responsible for the imaging systems on the Voyager spacecraft was led by Bradford A. Smith from LPL in Tucson. In addition to his work with the Voyager mission, Smith was also involved in the Vega mission to comet Halley in 1986 and remains a first-class ground-based observer. Together with Richard Terrile, he acquired the first images of a proto-planetary disk (that of β Pictoris) in 1984 (Smith and Terrile 1984). The images from the Voyager spacecraft were carefully analysed for evidence of small satellites in orbits very close to the parent planets. The observations of the ring systems of the planets proved to be particularly fruitful. The data revealed that the dynamics of thin rings about the planets were controlled by tiny satellites often less than 60 km in diameter. These satellites were christened 'shepherds' and their discovery prompted considerable theoretical work about their orbits and their interaction with ring particles. Ten new satellites were discovered at Uranus alone by Voyager 2. The Voyager imaging data also resolved a long-standing controversy about a possible satellite of Neptune. In 1981 Harold Reitsema and colleagues from LPL observed a near-occultation of a star by Neptune. The observers were looking for evidence of rings. They noticed a dip in the signal from the star on one side of Neptune but this was not reproduced at the corresponding position on the other side of Neptune. There were two possible explanations. Either the ring was only partial (not completely surrounding Neptune) or there had been an occultation by a satellite. The chances of the latter were thought so improbable that most believed a ring-arc had been observed. Subsequent detections of rings using the occultation technique by Brahic and Hubbard seemed to support this hypothesis. However, Reitsema and his colleagues were not so convinced and continued to insist that they believed it to have been a satellite. The Voyager 2 results fully supported their assertions. Rings were clearly found but, at the distance from Neptune at which Reitsema and co-workers made their observation of a dip in the stellar signal, no ring was found. However, a small satellite, later named Larissa, was in a circular orbit about Neptune at precisely that distance. Reitsema and co-workers have been credited with its discovery.

Ground-based observations

At the end of the 1970s spacecraft were becoming much more sophisticated and plans for a major mission to Jupiter, to be called Galileo, were well underway. Ground-based observations were not standing still, however. There were two areas of development. Firstly, charge-coupled devices (CCDs) were beginning to become available. CCDs are silicon semiconductors that convert incoming photons into electrons in a linear way. The efficiency of this process can be very high. Modern devices operating in the optical can reach 90% efficiency. Even at the beginning of the 1980s, CCDs with peak efficiencies of the order of 30% were

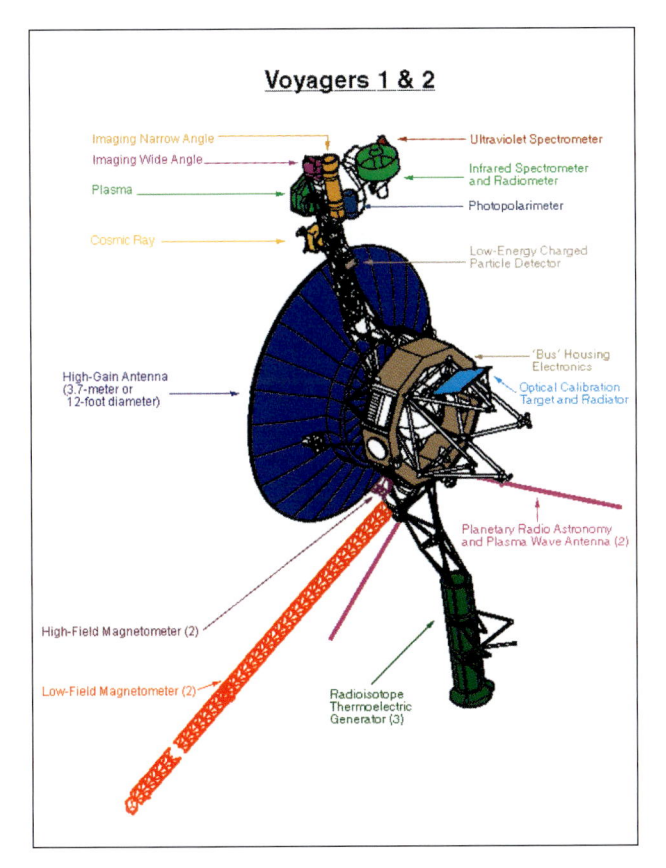

Figure 2 A drawing of the Voyager 1 and 2 spacecraft showing the positions on the instruments and some of the spacecraft sub-systems. Between 1979 and 1990, the Voyager 2 spacecraft visited all four giant planets and returned hundreds of images of their satellite systems. (Image: NASA/JPL.)

available. The first CCDs to be selected for flight on a spacecraft were ordered in 1981 and subsequently returned remarkable results during the Giotto flyby of comet Halley in March 1986. The high efficiency and linearity of these detectors make them ideal for use in ground-based astronomy where the faintest objects are often of the most significance.

The second development was the production of infrared imaging devices. Observations in the infrared offer many advantages over optical imaging. For the study of planetary satellites, one powerful advantage is the ability to image within one of the many methane absorption bands in the infrared. The gas giant planets all contain significant amounts of methane in their upper atmospheres. The methane absorbs solar illumination very strongly making Saturn, for example, appear almost black at a wavelength of $2.2\,\mu m$. For an observer of planetary satellites, this is extremely important. Searches for satellites near Saturn would normally confront a strong straylight background produced by scattering of the light from Saturn by both the atmosphere and the telescope itself. However, this scattered background is no longer present at $2.2\,\mu m$ because of the methane absorption, making satellites close to the planet more visible because of the increased contrast. Ground-based observers took advantage of these possibilities to detect new satellites and to confirm the existence and orbits of others detected by the Voyager spacecraft.

A good example of the power of CCDs came in 1997 when Brett Gladman and colleagues discovered two remote moons of Uranus. Caliban and Sycorax are vast distances from Uranus (Caliban and Sycorax have semi-major axes of 7.1×10^6 and $12.2 \times 10^6\,Km$–278 and 477 Uranian radii, respectively) and are closer to the edge of the sphere of influence of their parent planet than any other satellite (Gladman *et al.* 1998). The brighter of the two satellites is in a highly eccentric, retrograde orbit and is perturbed by the Sun and by the other planets to such an extent that simple Keplerian (elliptical) orbits completely fail to describe its motion. It is highly likely that Caliban and Sycorax have only recently (on geological timescales) been captured into orbit and they may be ejected from the Uranian system on similar timescales. This illustrates that planetary satellites can be transient. A more dramatic example of this phenomenon was comet Shoemaker-Levy 9 which was probably captured into a Jupiter-centric orbit and thus, temporarily, became a satellite of Jupiter before splitting into at least 21 pieces and crashing into Jupiter's atmosphere in July 1994.

SATELLITE DYNAMICS

Orbital and rotational characteristics

It is widely assumed that the planetary system formed in the outer regions of a disk-shaped nebula that comprised the proto-Sun at its centre and a mixture of gases, ices and dust that was slowly spiraling inwards towards it (see Dormand and Woolfson (1989) for an alternative view). In this scenario, the proto-planets began as small instabilities in the disk that then started to accumulate material themselves. In an analogous way, the proto-planets also comprised a central condensation and a surrounding disk. Out of this disk the regular satellites of the gas-giants formed (e.g. Pollack and Fanale 1982). This hypothesis naturally leads to the idea of the regular satellites being in nearly circular, low-inclination orbits having small eccentricity. However, satellites with orbits relatively close to the parent planet experience tidal forces that can have profound effects on their orbits.

It was established long before the start of the twentieth century that tidal forces between a planet and a satellite could modify the size of the satellite's orbit. George Howard Darwin (1845–1912), the son of Charles Darwin, took this further in 1880 by showing that the orbital elements of the satellite could also change. Darwin was particularly interested in the origin and evolution of the Sun–Earth–Moon system but his work was applicable to other planetary satellites. It was in the 1960s that more elaborate and general developments of the governing equations were made by W.M. Kaula and by G.J.F. MacDonald (Burns 1977).

The basic concept is straightforward. The planet and the satellite respond to the combined gravitational field. Hence, because the objects are not rigid, tidal bulges are raised on them. If the planet has a rotational angular velocity greater than the orbital angular velocity of the satellite about the planet (as is the case for Jupiter and Io, for example), then the tidal bulge on the planet will appear to rotate ahead of the satellite because of friction. Similarly, if the planet takes longer to complete one rotation than the satellite takes to complete one orbit (Mars and Phobos, for example), then the tidal bulge on the planet will follow the satellite around the planet.

The tidal bulge has mass. Therefore, if the satellite lags behind the bulge on the planet, the gravitational attraction towards the bulge will accelerate the satellite. Energy is in effect being transferred from the planet's rotation (reducing its rotation rate) to the satellite's orbit (which therefore increases). If the satellite is in front of the bulge, the satellite's orbit will contract and the planet's rotation rate will increase. The same forces tend to reduce the eccentricity of the orbit and lower the inclination. Thus, a single satellite orbiting about a planet will move towards a circular orbit of low inclination.

The planet also raises a tide on the satellite, of course. Here, the forces acting on the satellite bring it towards synchronous rotation so that the same hemisphere faces the planet throughout the satellite's orbit. The Moon is the classic example of this phenomenon but all of the Galilean satellites, all of the regular moons of Uranus and most of the regular moons of Saturn have the same property. For the Galilean satellites, it was William Herschel who first concluded in 1797 that they were in synchronous rotation. However, this was disputed for many years. Eventually Stebbins and Jacobsen (1928) proved it to be the case when they made the first photoelectric measurements of the satellites. All of the regular and unusual satellites, except Titan, are known to be in synchronous rotation. Because of its optically thick atmosphere, the rotation state of Titan has not yet been clearly established, although Hubble Space Telescope (HST) observations through atmospheric windows in the near-infrared appear to suggest that the rotation is indeed synchronous.

In a completely evolved system, the energy taken from the planet's rotation will be sufficient to reduce the rotational angular velocity of the planet so that it matches the orbital period of the satellite. The energy taken from the satellite's rotation will also bring it into synchronous rotation. Hence, the same hemisphere of the satellite always faces the same hemisphere of the planet. Ground-based and HST observations of Pluto and Charon appear to show that this system is completely evolved in this sense.

The tidal evolution of the Uranian satellite system may have been somewhat unusual. The orientation of the rotation axis of Uranus is very different from the other gas giants being tilted by 98° with respect to its orbit. This has led to speculation that the orientation of the spin axis was changed by a large impact. One must then ask how this change affected Uranus's regular satellite system, the orbital momenta of which are aligned with the planet's axis of rotation. It is possible that the impact occurred very early in the lifetime of the planet, before the satellites formed. In this case, the proto-satellite disk would have re-oriented itself about the new equatorial plane relatively quickly and the satellites could then have formed. However, if the impact occurred after original satellite formation, there may have been impacts between satellites as tidal forces acted to modify their orbits.

Voyager 2 images of the innermost of the five regular Uranian satellites, Miranda, in 1986 revealed that something remarkable may have happened to the satellite. Miranda exhibits an incredible diversity of surface structures (Greenberg *et al.* 1991). It has been suggested that Miranda was broken up by an impact into relatively large blocks and that the blocks then re-accreted by way of rather gentle collisions to produce the observed surface structures seen today. However, there are alternative explanations (see below).

All the irregular satellites are, by definition, remote from the parent planet and in orbits that are relatively eccentric and inclined to the equatorial plane of the planet. It is widely suspected that they are captured asteroidal objects but the exact mechanism by which capture occurs is unclear. The possible mechanisms include:

- capture from solar orbit resulting from gas drag;
- capture from solar orbit resulting from asteroidal (cometary) outgassing influencing the orbit;
- capture by tidal dissipation which, however, requires a close pass which might, in turn, lead to tidal disruption of the asteroid and also requires some means of raising the periapsis to the present value;
- capture following collision with another object;
- three-body capture using the fortuitous circumstance of a third large object in heliocentric orbit in the proximity of the planet; and
- mass change of the Sun and/or the parent planet at the time of capture, but this is only achievable during the accretionary epoch because the temporary capture time of an asteroid must be short (typically around 10^4 years).

An unusual characteristic of the outer satellites of Jupiter is that they fall into two distinct groups. The satellites that have names ending with an 'a' are in direct (prograde) orbits inclined by approximately 27° with respect to Jupiter's equator. The eccentricities of the orbits are around 0.16 and they are all about 165 Jovian radii (R_J) from the planet. The satellites with names ending with an 'e' are in retrograde orbits with inclinations of around 150°, eccentricities of around 0.25 and at a Jovicentric distance of about $320R_J$. A new member of the outer group has recently been discovered. It has been suggested that each group had a common precursor that broke up at some stage. Kuiper (1951), for example, proposed that two objects were captured by the proto-Jovian nebula and that gas drag with the denser equatorial parts of the nebula led to their break-up. There are, however, problems with this hypothesis because the outermost group is close to the limiting distance of gravitational stability. One way out of this dilemma would be to assume Jupiter to have been more massive in the past.

Sykes *et al.* (2000) have made a study of the spectrophotometric properties of the two groups and conclude that the inner (prograde) group are spectrally very similar, suggesting a common, homogeneous precursor. However, the outer (retrograde) group are spectrally dissimilar. Sykes *et al.* suggest that the precursor may have been extremely heterogeneous in this case.

Most of the collisional shards are between the parent planet and the innermost regular satellite of the system. They are usually associated with the ring systems of the planet and it is likely that they are the collisional remains after multiple impacts, fragmentation and possible re-accretion throughout the lifetime of the Solar System. Objects such as Amalthea and Janus may be just very small regular satellites. From observations of the planetary satellites and of asteroids, it is apparent that sphericity is a common property of all satellites larger than 200 km. Mimas is the smallest spherical satellite (radius $r = 196$ km) while Hyperion (semi-axes of 205 km × 130 km × 110 km) and Amalthea (135 km × 83 km × 75 km) are the two largest irregular satellites. Sphericity is a consequence of gravity overcoming the structural strength of the material that makes up the satellite. The pressure at the centre of a satellite, P, can be estimated by integrating the product of the mass and the gravational acceleration of material in a column above the centre of the satellite, leading to the equation

$$P = \frac{2}{3} \pi G \rho^2 r^2$$

where ρ is the bulk density. Table 2 gives four examples of the internal pressures computed from this formula.

A comparison of the estimated tensile strength of rocky material with the internal pressure suggests that typical tensile strengths of around 10 MPa are exceeded within bodies of $r \geq 150$ km. Bodies of this radius or greater therefore restructure and equilibrate themselves as the interior fractures and flows under pressure. This process may be influenced by the heat generated by radioactive decay or by the gravitational energy released during formation. There is a simple analytical solution for the kinetic energy released from the gravitational in-fall of material from infinity onto the surface taking into account the increasing gravitational acceleration as the body grows. For uniform density, the kinetic energy, E_T, gained is (Schubert et al. 1986)

$$E_T = \frac{16}{15} \pi^2 \rho^2 G r^5$$

from which one can determine the temperature rise expected using an estimate for the specific heat capacity, θ. A typical value for θ would be 1.7×10^3 J kg^{-1} K^{-1} (Lewis 1997) leading to very small temperature rises for bodies of size $r \approx 100$ km. Radioactive decay heating provides 4×10^{-11} W kg^{-1} which suggests that over periods of $>10^8$ years higher temperatures could be reached. However, the smaller the body, the larger the surface to volume ratio, and hence cooling via conduction and thermal emission should be rather efficient. Hence, it is rather straightforward to understand why small bodies are not spherical.

Tidal forces again play a role in modifying the rotational characteristics of the objects. The long axis of an irregularly shaped satellite is always aligned with the object–planet line (Hyperion is the only known exception). If a satellite were a perfect fluid, it would distort into an ellipsoid following the equipotential surface. However, at the Roche limit (a_R), no closed equipotential surface for the satellite exists. This would suggest that a perfectly fluid satellite should disintegrate inside the Roche limit. The limit is given by the equation

$$a_R = 2.456 \left(\frac{\rho_P}{\rho_S} \right)^{1/3} r_P$$

where ρ_P and ρ_S are the densities of the planet and satellite, respectively, and r_P is the radius of the planet.

The Roche limit is also important when an object (such as a comet) comes close to a planet. An example is comet Shoemaker-Levy 9 which is thought to have broken up during a very close pass to Jupiter two years before it actually impacted the planet. It needs to be stressed that the limit does not apply to small moons held together by, for example, van der Waals forces rather than self-gravity. Thus, Metis (J16), for example, either has a fairly high density (≥ 3.48 g cm^{-3}), which is somewhat unlikely, or it is held together by chemical forces. Several collisional shards inside the orbits of regular satellites are close to or inside the Roche limit. Hence, the collisional evolution and formation of these objects has been strongly affected by tidal forces and the balance between these forces and the chemical/structural forces that tend to hold the satellites together.

The rotational characteristics of Hyperion are a special case in the Solar System. Hyperion's orbit has an eccentricity of around 0.1 which is forced by its gravitational interaction with Titan and Iapetus. Furthermore, the satellite is highly asymmetric. Wisdom et al. (1984) showed that the combination of the forced eccentricity and the unusual shape produced a bizarre situation. As tidal dissipation drives Hyperion's spin toward a nearly synchronous value, Hyperion enters a chaotic state, becoming attitude-unstable and tumbling. It was therefore predicted that Hyperion

Table 2 Internal pressures of selected planetary satellites

Satellite	Radius (km)	Density (g cm^{-3})	Mass (10^{19} kg)	Internal pressure (Mpa)
Mimas	196	1.44	4.55	10.9
Enceladus	250	1.13	7.40	12.6
Iapetus	730	1.15	188	100
Europa	1569	3.01	4879	3180

would be found to be tumbling chaotically. Subsequent ground-based observations have supported this conjecture.

Resonances and shepherds

The orbital periods of Io, Europa and Ganymede were known to be close to the ratios 1 to 2 to 4 shortly after their discovery. It was Laplace (1805) who recognized that the mean orbital angular velocities, n, of the three inner Galilean satellites formed an exact resonant system described mathematically as

$$0 = n_I - 3n_{II} + 2n_{III}$$

As the orbits of other satellites in the Solar System became better defined, it became clear that resonances between satellites were extremely common. Roy and Overden (1954) showed that many more resonances existed than were possible by chance. In the Saturnian system, Enceladus and Dione (2:1), Mimas and Tethys (2:1) and Titan and Hyperion (4:3) are locked in orbital resonances. There are no simple resonances in the Uranian system but it is apparent from the inclination of Miranda and the non-negligible eccentricities of the other satellites that tidal effects have been present. Goldreich (1965) showed that resonances were stable against the tidal expansion of the orbits and, thus, satellites could be captured into resonances. Borderies and Goldreich (1984) subsequently produced a simple analytical determination of the capture probability.

Like the Galilean satellite system, the three inner large satellites of Uranus (Miranda, Ariel and Umbriel) are near the Laplace resonance. However, this resonance is not exact and there are certain orbital anomalies (Pollack et al. 1991). In particular, Miranda has a rather high inclination for a regular satellite ($>4°$). The lack of mean-motion commensurabilities among the Uranian satellites is somewhat of a puzzle but it appears to be caused by the fact that Uranus has a very small oblateness (compared to Jupiter, for example) which leads to resonant terms with comparable frequencies which, in turn, perturb each other. This can lead to a nonlinear system which has chaotic behaviour in certain configurations. It seems probable that the Uranian satellite system has this property and is consequently dynamically highly complex. This does exclude the possibility that Uranian satellites have passed through resonances during the lifetime of the Solar System.

Of further interest are the so-called co-orbital satellites. These small shards are in the same orbit as a regular satellite and are thus in a 1:1 resonance. Considerable progress on the dynamical behaviour of these objects has been made based on the similarity of the physics to that of the Trojan asteroids. Examples of Saturnian co-orbitals are Telesto and Calypso (which co-orbit with Tethys), Helene (which co-orbits with Dione) and Epimetheus (which co-orbits with Janus, although this system is unique in that Janus and Epimetheus are roughly the same mass). The Tethys and Dione co-orbitals librate (oscillate) about the stable stationary Lagrangian points, L_4 and L_5, 60° ahead and behind the regular satellites. In a reference frame with Saturn at its centre and that rotates with the regular satellite, the co-orbitals appear to follow 'tadpole'-shaped trajectories (Figure 3). For Janus and Epimetheus the situation is slightly different in that they are capable of perturbing each other. For example, as Janus approaches Epimetheus from behind, the mutual interaction drags it forward, which increases the angular momentum and hence increases the orbital radius. Conversely, Epimetheus loses angular momentum and its orbital radius decreases. With a lower orbital radius, Epimetheus's angular velocity increases and hence it moves away from Janus and starts to follow it around Saturn. In a rotating reference frame, Epimetheus appears to follow a trajectory shaped like a horseshoe.

It is also interesting to note that collisional shards may have resonances with the ring systems of the giant planets which may place constraints on the lifetimes of ring

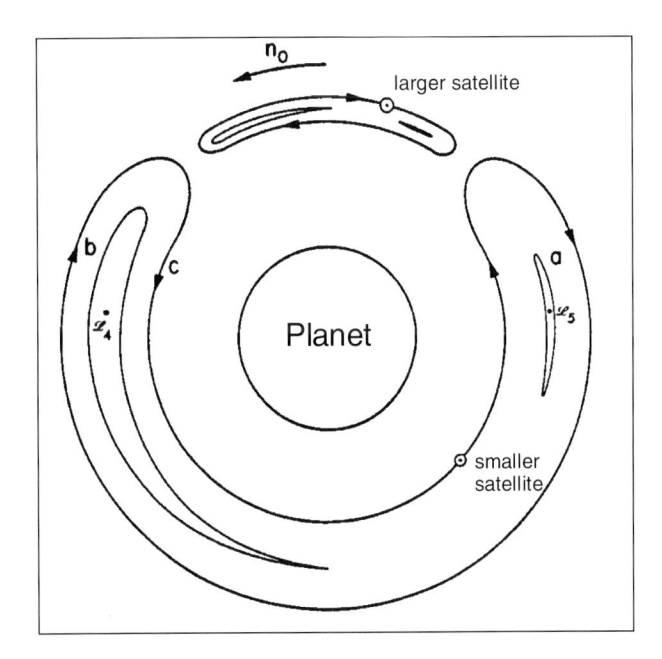

Figure 3 Examples of coorbital satellite orbits. The figure shows a reference frame which is rotating about the planet at a constant rate (referred to as the average mean motion). For one of the coorbitals, (a) shows an example of a small libration about the stable Lagrange L_5 point, (b) shows the limiting case for so-called "tadpole" orbits centred about the L_4 point, and (c) shows a horseshoe orbit. Examples of the motion of the second co-orbital are also shown. The magnitude of its motion depends upon the ratio its mass to that of the first co-orbital. (After Yoder et al. 1983.)

systems in their present states (Peale 1986). Cordelia and Ophelia in the Uranian system appear to have resonances associated with several rings. Galatea in the Neptune system may be another example. Satellites close to the rings can also control the properties of the rings. Cordelia and Ophelia act like 'shepherds' to confine particles in the ϵ ring, for example (Figure 4). The γ and δ rings appear to be similarly confined but no satellites were found near their orbits. (The satellites may have been below the detection threshold of the Voyager cameras; French *et al.* 1991.) Pandora and Prometheus in the Saturnian system perform a similar function and also appear to exhibit resonances with the regular satellites. From the orbital perturbations (Nicholson *et al.* 1996), the densities of these satellites were found to be rather low (0.65 g cm^{-3}). The satellites are expected to be icy and their growth from ring particles may resemble the comet growth process in the solar nebula.

Tidal flexing

It has been seen how important resonances have been in determining the orbits of the planetary satellites. Resonances also have effects on the surface properties of planetary satellites. The most spectacular example of this is Io. Seen from Earth, Io is about 1 arcsec in diameter. This is approximately the resolution limit one can obtain with conventional ground-based telescopes in reasonable weather.

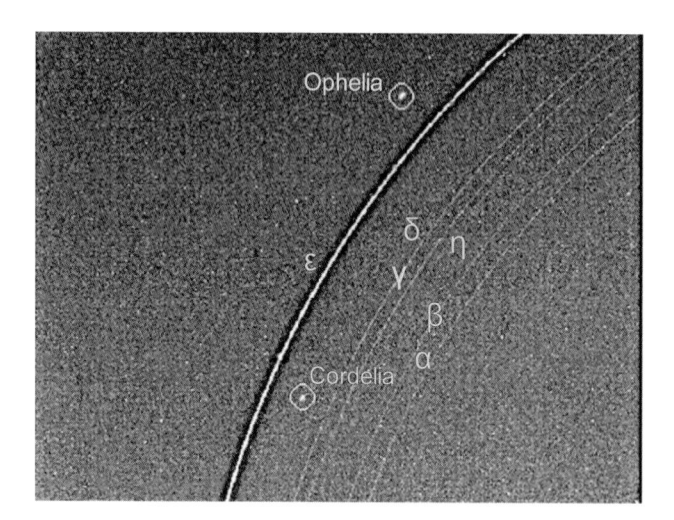

Figure 4 The satellites, Cordelia and Ophelia, orbit Uranus on either side of the ε ring. The satellites thereby confine the particles in the ring and keep them from spreading out and dissipating. This Voyager 2 image shows the two satellites and part of the Uranian ring system. Several other discrete rings can be seen. It is thought that shepherd satellites also shape the appearance of these rings. However, these satellites may have been so small and dark that Voyager's cameras were not able to detect them. (Adapted from a NASA/JPL image.)

Minton (1972) acquired a photograph of Io on a really exceptional night at the telescope and was actually able to resolve Io's disk and identify broad surface characteristics. He saw that the poles of Io were relatively dark, the midlatitudes were reddened and the equatorial region was bright in the infrared. Minton did not realize that he was observing, for the first time, the effects of volcanism on Io.

In 1979, just five days before the Voyager 1 encounter with Io, Peale *et al.* (1979) published a paper in Science in which they made predictions about the influence of tides on Io. If Io were the only satellite in the Jovian system, its orbit would be perfectly circular and the tidal bulge raised on the satellite would be in the direction towards Jupiter. However, Europa and, to a lesser extent, Ganymede introduce a forced eccentricity into Io's orbit. This has the effect that the tidal bulge on Io is not directed towards Jupiter all the time. Tidal forces act to correct this misalignment but these forces are resisted by friction which, in turn, produces heat. Peale *et al.* (1979) concluded that the total dissipation of energy within Io was so high that the Voyager spacecraft might see active volcanoes. When Voyager arrived, their calculations and predictions were fully borne out by the observations. The surface of Io was found to be almost devoid of impact craters with evidence of recent flows originating from dark calderas.

A little while after the first images were returned to Earth, Linda Morabito of the Jet Propulsion Laboratory was processing some of the images to improve their contrast. She was analysing an image of the whole disk taken at a phase angle of about 90° so that the terminator was roughly in the centre of the disk. She saw, however, that the terminator was broken by a round blob which was obviously material still in sunlight and therefore well above the surface. Morabito had identified a volcanic plume of dust that had been ejected by a volcano that was currently active (Figure 5).

Analysis of the Voyager data produced evidence of eight active volcanic plumes. From studies of changes in the surface appearance, it was clear that at least two other large volcanic events had occurred between the Voyager 1 and 2 encounters. An entire belt within 30° of the equator appeared to be active with at least eight long-lived plumes identified in Voyager 1 and 2 data. This was what Minton saw with his remarkable observation. The long-lived plumes ejected material to heights of around 70 km. However, there was also evidence of much larger, shorter-lived events, such as those originating from a vent called Pele, which could throw material up to 300 km above the surface. The ejection velocity at the mouth of the vent was over 1 km s^{-1}.

The shorter-lived, larger volcanoes did not appear to be equally distributed about Io nor were they confined to the equatorial belt like the longer-lived plumes. They seemed to

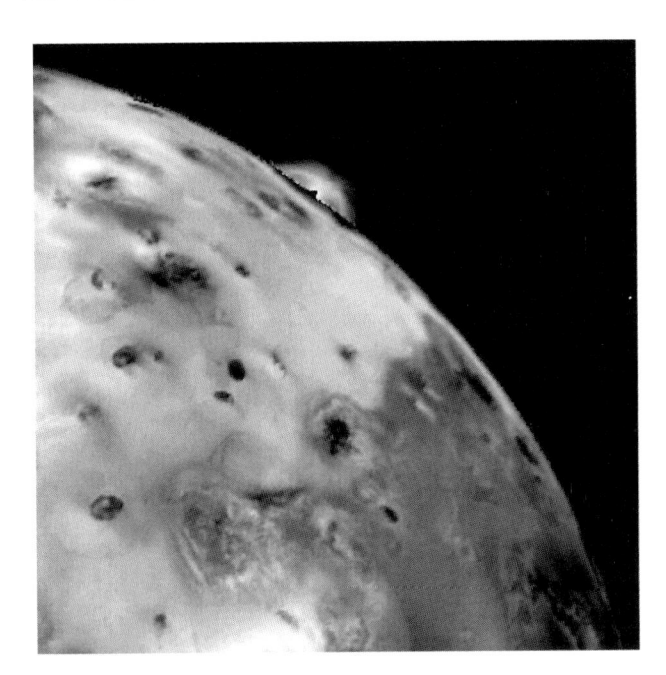

Figure 5 Enhanced Voyager image of Io showing the plume of the volcano, Prometheus (not to be confused with the Saturnian satellite of the same name). (Image: NASA/JPL.)

be only evident over a specific range of longitudes. Modelling of Io's internal structure and its response to tidal forces has attempted to explain the distribution of volcanic vents with some degree of success (e.g. Ross *et al.* 1990). It is now estimated that the tidal flexing mechanism produces at least 2×10^{14} W of heat inside Io. Some estimates are as high as 10^{15} W, or $24\,\text{W}\,\text{m}^{-2}$ averaged over the whole surface. For comparison, the measured heat flow for the Moon is around $0.017\,\text{W}\,\text{m}^{-2}$ (Kaula *et al.* 1986).

Io is the most spectacular example of a satellite with tidally-induced volcanism in the Solar System. However, there may be several other examples. Although Miranda is not presently active, there is evidence of geologically recent flow on its surface. It is difficult to estimate an absolute age for when this flow occurred because dating requires accurate knowledge of the cratering rate and evolution with time over the age of the Solar System. A possibility is that during the orbital evolution of the larger, outer satellites of Uranus, a resonance occurred that produced tidal heating in Miranda leading to internal melting, pressure build-up and outflow. Ariel is another candidate in the Uranian system (see below).

Voyager 2 imaging of Neptune's large, irregular satellite Triton revealed evidence of active plumes of material being ejected from the surface (Kirk *et al.* 1992). Triton's retrograde orbit around Neptune is slowly decaying because of tidal forces and, on a timescale of greater than 3×10^7 years, will crash onto Neptune. The eccentricity of Triton's orbit is essentially zero as a result of orbital decay. Tidal forces

may have provided an internal heat source in the past that explains Triton's current activity. However, there are other, more plausible explanations (for example, a solid-state greenhouse effect, see below).

Stevenson (1992) has shown that the eccentricity of Titan's orbit about Saturn provides a small heat source to the interior. However, this heat source appears to be insignificant compared to radioactive decay heating. It is unlikely that the small possibility of volcanic activity on Titan can be verified or rejected until the arrival of Cassini at Saturn in 2004.

SURFACE COMPOSITION

The absorption bands of many ices can be found in the near-infrared wavelength range between 1.0 and 5.0 μm. The Voyager spacecraft were not equipped with any instrumentation for detection in this range. The thermal infrared spectrometer (IRIS) provided excellent observations of the thermal emission from the satellites and the planets and, of course, identified atmospheric SO_2 on Io. However, most of what we know about the surface composition of outer planet satellites comes from ground-based observations. An exception is the Galilean satellite system because of the recent success of the near-infrared mapping spectrometer (NIMS) on board Galileo.

Most planetary satellites appear to have water ice on their surfaces in varying quantities. Darker material is evident on many surfaces which may be organic in origin. The group around Carl Sagan (1934–1996) at Cornell University made a number of experiments to investigate whether more complex organic molecules could be created by the irradiation of mixtures of varying amounts of methane, nitrogen, carbon dioxide and water. The experiments resulted in the formation of non-volatile organic residues that became known as 'tholins'. The spectral properties of these materials were found to match features around 3.4 μm in the spectra of comets, Titan and several other objects (see below).

In addition, rocky material must be present on the surfaces of most planetary satellites. However, identification of these materials is difficult without direct sampling. Hence, the details of planetary satellite surface composition remain somewhat obscure. In the following, some of the more interesting properties of several specific objects are addressed.

Io

Io's low ultraviolet reflectivity had been known since the 1920s. In the 1970s it was established that Io was very bright in the near-infrared. Wamsteker (1972) first pointed out that sulphur was one of the few materials that matched the main characteristics of Io's spectrum. It was clearly established prior to the Voyager flybys that Io was almost completely devoid of water but possessed a strong spectral feature near 4.08 μm. The Voyager spacecraft confirmed the importance of sulphur and demonstrated the presence of SO_2 (responsible for the 4.08 μm band) on Io. Sill and Clark (1982) gave a comprehensive review of the spectra of different allotropes of sulphur. Surface silicates were inferred to be present from the fact that steep slopes are seen on the surface in dormant vents. Slopes of pure sulphur collapse at lower slope angles than those seen on Io.

Subsequently, searches for other molecules were made. Although the detection of CO_2 was often reported, all of the bands were later attributed to weak SO_2 absorptions. It has been suggested that H_2S may be present at the poles to buffer the atmosphere rather in the way the polar caps buffer the CO_2 atmosphere on Mars. This remains uncertain despite detection of hydrogen emission by HST. The detection of the alkali metals sodium and potassium in the neutral clouds indicates that other materials are present and the recent detection of chlorine ions in the Io plasma torus suggests the possible presence of HCl or even NaCl either in the atmosphere or on the surface. However, the composition of non-sulphur-bearing molecules (for example, silicates) remains obscure.

Europa, Ganymede and Callisto

Kuiper first suggested the possible presence of water frost on Europa and Ganymede by comparison of the 1 and 2 μm surface reflectivities. By the early 1970s this analysis was supported by further observations with updated equipment and Callisto was also shown to have a significant covering of water ice. The Galileo NIMS experiment indicated the presence of CO_2 and SO_2 (McCord et al. 1998b), which are believed to exist as trapped molecules in the surface ice or in darker surface material. The trapping of O_2 in the surface ices on Ganymede has also been demonstrated (Spencer et al. 1995). The accumulation of detectable abundances of O_3, produced by the action of ultraviolet or charged-particle radiation on O_2, also appears to occur leading to an ultraviolet band at 0.26 μm. The Galileo NIMS experiment has also detected the presence of salts on all of the icy Galilean satellites (e.g. McCord et al. 1998a). Hydrogen peroxide has been detected on Europa following a prediction by Johnson and Quickenden (1997).

Iapetus

Cruikshank et al. (1999) suggested that 3 μm observations of the low-albedo hemisphere of Iapetus (see below) are well matched by a surface mixture that includes 'Triton' tholin, an involatile organic substance produced by the irradiation of a gaseous mixture of N_2 and CH_4 (McDonald et al. 1994). Inclusion of small quantities of this nitrogen-rich tholin in models of the other, higher-albedo satellite surfaces also produces good fits to the latest spectral data, particularly in the case of Rhea.

The icy Saturnian satellites

In the early 1980s it was established that the surfaces of Rhea and Dione are composed primarily of water ice. The spectrum of Dione's leading surface is characterized by being relatively blue at near-infrared wavelengths with deep absorption bands at 1.5 and 2.05 μm and shallow absorption bands at 1.05 and 1.3 μm. All properties of the spectrum are attributable to water ice. Following the detection of O_3 on Ganymede, Noll et al. (1997) made a search for O_3 on Rhea and Dione and made positive detections on both occasions. Hence, trapping of O_2 in surface ices and subsequent O_3 production may be a common phenomenon on icy satellites.

The Uranian satellites

All the Uranian satellites have relatively low albedoes and significant quantities of organic material on their surfaces is suspected. The surfaces of Titania and Oberon have been investigated using ground-based near-infrared spectroscopy (Roush et al. 1998). Combinations of water ice and varying amounts of spectrally neutral material provide reasonable fits to the spectra. Three broad features are evident (at 1.52, 1.65 and 2.05 μm). The 1.65 μm band may be attributable to hexagonal water ice at low temperatures which would allow an estimate of the surface temperature. No evidence of CO_2, CO, NH_3 or CH_4 has so far been found. The relatively neutral surface material may be related to tholins.

Triton and Nereid

Triton's spectrum has been known to contain evidence of methane for some years. In recent observations, vibrational bands of CH_4, N_2, CO and CO_2 have been identified (Quirico et al. 1999). There is likely to be significant amounts of rocky material in addition to the ices. The plumes or streaks seen in Voyager 2 data probably indicate hydrocarbons which may be more complex but they are likely to escape detection from ground-based observations for quite some time.

Spectra of Nereid are now possible through the use of the latest 10 m telescopes. The near-infrared spectra show absorptions near 1.54 and 2.03 μm which are well matched by a synthetic spectrum of an intimate mix of low-temperature, particulate water ice and dark, blue-coloured material (Brown et al. 1999).

SATELLITE INTERIORS

The densities of most of the regular satellites have been established by measurements of the perturbations to spacecraft trajectories. The knowledge of the density allows one to estimate the relative abundances of icy material and rocky (silicate) material within the satellite. This is only an estimate, however, because the density of the rocky material depends strongly upon its specific composition. Therefore, in general, the structure of satellite interiors is rather poorly constrained.

The best-studied satellites are in the Jupiter system because of the multiple flybys by the Galileo spacecraft. Here, the bulk density decreases with increasing Jovicentric distance from Io ($3.57\,\mathrm{g\,cm^{-3}}$) to Europa ($3.02\,\mathrm{g\,cm^{-3}}$) to Ganymede ($1.94\,\mathrm{g\,cm^{-3}}$) to Callisto ($1.85\,\mathrm{g\,cm^{-3}}$), which, implies an increasing fraction of lighter, icier material with Jovicentric distance. It is widely assumed that this density distribution was the result of Jupiter being much warmer during its formation. The higher temperature was sufficient to drive volatiles (principally water) from the inner regions of the proto-satellite disk. Hence, Io is almost completely devoid of water. Galileo measurements suggest that it has a two-layer structure. At the centre is a metallic core, probably made of iron and iron sulphide, about 900 km in radius. This is covered by a mantle of partially molten rock and crust.

There is still no clear conclusion on whether Io possesses an intrinsic magnetic field. The magnetometer aboard Galileo has recorded an intriguing set of phenomena at Io but the system is so complex that attribution of some of these phenomena to an intrinsic field has proved to be very difficult. However, the magnetometer clearly identified Ganymede as having a magnetic field. The tracking data from the spacecraft also suggest that Ganymede's interior is differentiated into a three-layer structure. At its centre is a metallic core surrounded by a rocky silicate mantle. An icy shell encapsulates this mantle. Ganymede's magnetic field may therefore be generated in an analogous way to that of the Earth with a fluid mantle rotating over a metallic core. Europa may well be similar in internal structure. A somewhat weaker magnetic field than at Ganymede was detected. Callisto appears to be undifferentiated. Given the similarities between Callisto and Ganymede (size, density, surface composition, and so on), this distinct difference in internal structure is extremely surprising.

SATELLITE ATMOSPHERES

Of the nine planets in our Solar System, only Mercury does not have a collisionally thick atmosphere (although the atmosphere of Pluto is a special case). Thus, atmospheres tend to be associated with large objects – objects with substantial gravitational fields – which is clear when one considers the Jeans formula for the rate of escape by thermal evaporation of a gas from a planetary atmosphere (e.g. Chamberlain and Hunten 1987). At the exobase of an atmosphere, the thermal velocity of the atmospheric gas must not exceed the escape velocity. This can be expressed approximately in the form

$$\frac{GM}{r_0 - H\log(1/n_0 HQ)} > \frac{kT}{m}$$

where m is the gas molecular weight, G is the gravitational constant, M is the mass of the planet or satellite, r_0 is the radius of the planet or satellite, n_0 is the surface gas density, Q is the molecular cross-section, T is the atmospheric temperature (assumed to be constant), k is Boltzmann's constant and H is the atmospheric scale height. This inequality shows that if one reduces the mass of a satellite the stability of an atmosphere can only be preserved if the atmospheric temperature is low and/or the molecular weight of the atmospheric gas is high. This favours large satellites, remote from the Sun, with heavy atmospheric molecules. Titan, the largest satellite of Saturn, matches this description.

Titan

In 1908 J. Cornas Solá noticed that Titan showed a pronounced limb darkening and suggested that this meant that it has an atmosphere (Leverington 1995). In 1944 Kuiper confirmed that a thick atmosphere was present when he detected methane lines in the spectrum. However, it was not clear at this stage whether the atmosphere was relatively thin (20 mbar) and composed mostly of methane or whether there were additional constituents that made the atmosphere much denser at the surface.

The Voyager missions to the outer planets were of tremendous importance as missions of discovery and once again they provided an answer. However, this time at a cost. It was planned that the Voyager 2 trajectory would take it via Jupiter and Saturn to Uranus and then Neptune. For Voyager 1 there were two options. It was possible to use Voyager 1's Jupiter encounter to steer the spacecraft towards Pluto for a flyby. However, there was considerable pressure to make a close approach to Titan during the Voyager 1 flyby specifically to look at what was expected to be an extremely interesting object and to carry out a radio occultation experiment to probe the atmosphere. A close flyby of Titan eliminated the chance of going to Pluto. Furthermore, the flight to Pluto required an early launch date. At the time, Pluto was thought to be rather dull, and hence the flyby of Titan was chosen. One can speculate on whether the same decision would be made in similar circumstances today.

The radio occultation experiment of Voyager 1 effectively resolved the issue of the atmospheric pressure. The mean molecular weight of the atmosphere was found to be 28 with a surface temperature of 94 K and a surface pressure of 1.5 bar. Thus, methane is a minor constituent in a molecular nitrogen atmosphere as predicted by D.M. Hunten in 1978. It should even be in a liquid state at the surface under these conditions, which is actually a significant problem for models of Titan's atmosphere because there is no evidence of departure from a dry adiabatic lapse rate ($-dT/dz = g/C_p$) in the radio occultation data.

The presence of several other organic molecules was suggested by ground-based observations of Titan's spectrum. The infrared spectrometer on Voyager 1 confirmed many of these and detected several more, including C_2H_2, C_2H_4, C_2H_6, HCN, C_3H_4 and C_4H_2. Between 1996 and 1998 the Infrared Space Observatory (ISO) undertook a dedicated programme to study Titan's organic inventory. At the time of writing, the results of these observations are just becoming available. An interesting one is the detection of H_2O in the stratosphere. Because the tropopause acts as a sort of cold trap, the H_2O observed above this level must be exogenic in origin, possibly the result of the in-fall of cometary material. This is further evidence for the importance of exogenic processes in the evolution of planetary atmospheres.

The surface of Titan was completely obscured from the Voyager cameras by a thick smog of brown hydrocarbons (Figure 6). The methane in the atmosphere undergoes photochemical reactions that produce more complex organics such as the brown smog and ethane. Methane should be lost in this way and must therefore be replenished from below. Ethane should condense out. If Titan's atmosphere is long lived compared to the age of the Solar System, a layer of liquid ethane several hundred metres thick should have formed leading to a global ethane ocean. However, there is now strong evidence that most of the surface is solid. There are windows in the 900 nm to 1 μm region of the spectrum in which Titan's aerosol atmosphere is not optically thick and hence the surface can be probed. HST images (Smith *et al.* 1997) suggest that there are variations in albedo across the surface that, although small, would rule out a global ocean (Figure 7).

The thermal structure of Titan's atmosphere is shown schematically in Figure 8. The optical haze is situated about 250 km above the surface. At lower altitudes several other cloud or haze layers should be present as the temperature at first decreases with decreasing altitude and condensation of specific organics occurs.

Titan invokes considerable interest because of the abundance of hydrocarbons and the atmospheric nitrogen. In some ways, this is reminiscent of the Earth. An intriguing possibility is that Titan's present state is somehow similar to the early Earth. Titan simply did not evolve because of the low temperatures and lack of liquid water at 10 AU from the Sun.

Three scenarios in which Titan could have acquired this dense atmosphere have been considered (Atreya 1986):

- direct capture of N_2 from the proto-Saturnian nebula;
- systematic loss of N_2 from a clathrate–hydrate molecule containing N_2; and
- photolysis of NH_3 during the early phases of accretion.

There remain, however, large uncertainties as to the exact nature of the processes involved. Hence, studies of Titan

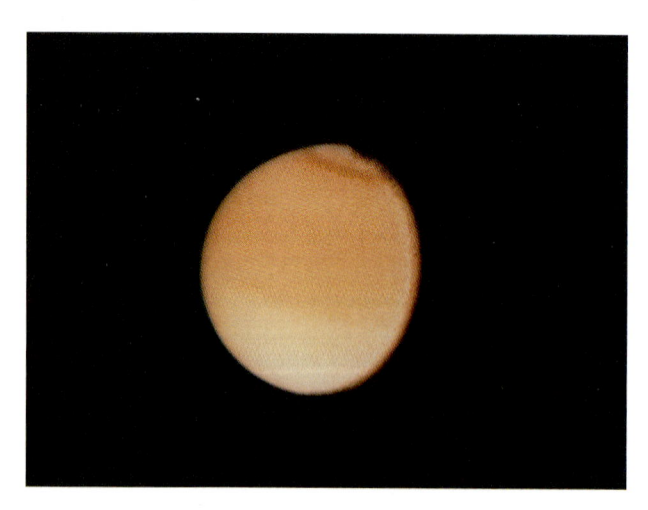

Figure 6 Enhanced Voyager image of Titan. The surface is obscured by an orange/brown haze in the thick atmosphere. (Image: NASA/JPL).

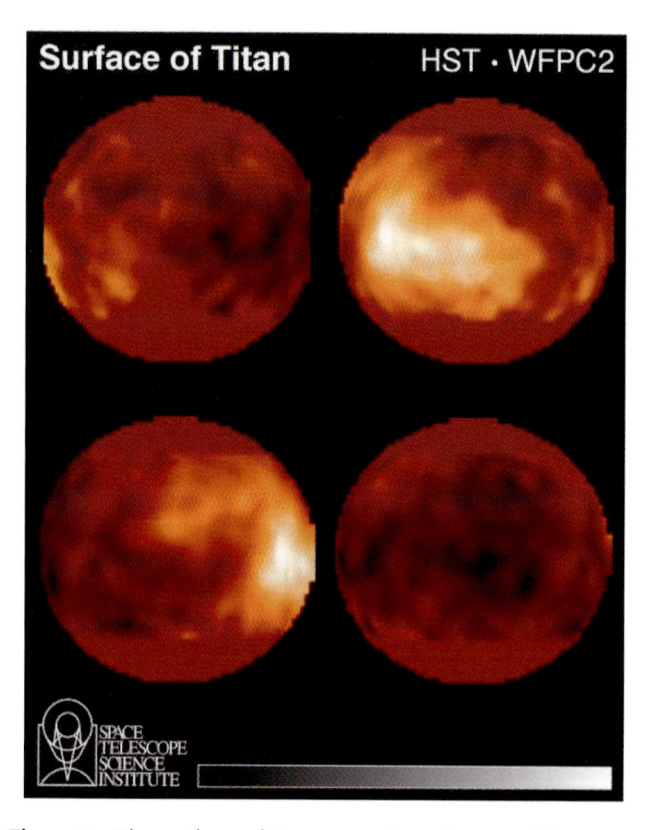

Figure 7 The surface of Titan seen through one of the near-infrared "windows" in the atmosphere. The images were obtained by the Hubble Space Telescope and show that Titan's surface does not have a uniform albedo and therefore is not completely covered by a global methane or ethane ocean. (Image: University of Arizona/NASA/STScI.)

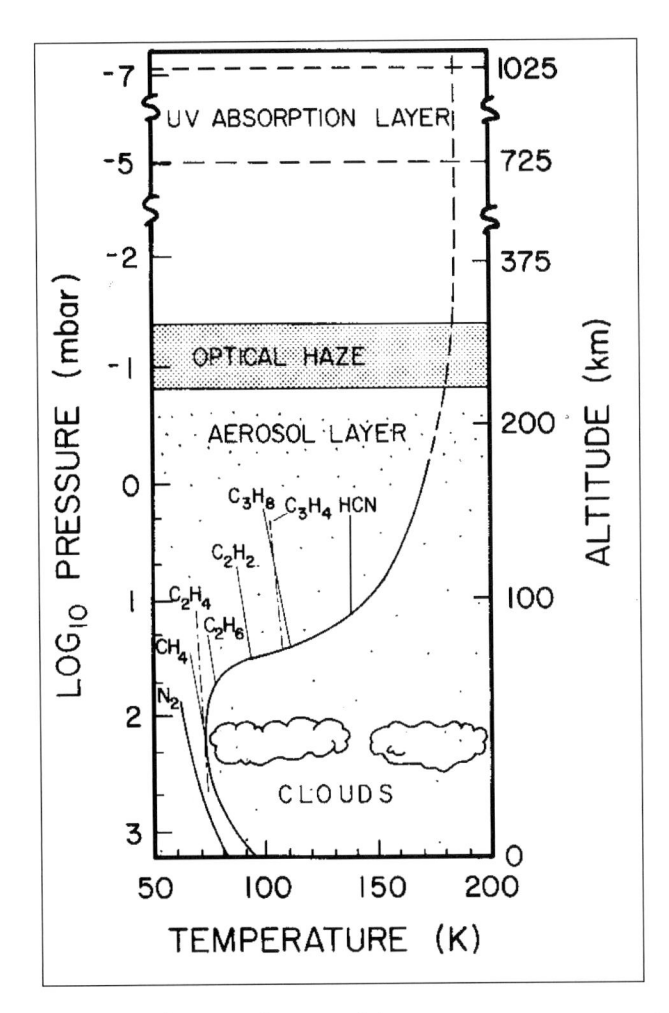

Figure 8 Schematic drawing of the temperature structure of Titan's atmosphere showing the altitudes of aerosol/cloud layers. (From Atreya 1986.)

will continue to be pursued. The first major event of the twenty-first century will be the entry of the European Space Agency's Huygens probe into its atmosphere in April 2005. Huygens, which will be ejected from NASA's Cassini spacecraft shortly after it enters orbit around Saturn, carries a payload that is dedicated to the investigation of the pressure, temperature, composition and dynamics of Titan's atmosphere. A remarkable camera system (the Descent Imager/Spectral Radiometer experiment) will provide the first images and visible and infrared spectra of the surface after Huygens penetrates the smog and cloud layers and will also determine aerosol particle sizes and their variation with height. A radar mapper aboard Cassini will provide the first global view of the structure of Titan's surface.

Tenuous atmospheres

Of the other regular satellites in the Solar System, only Io and Triton appear to possess atmospheres in which the exobase is above the surface. Even for Io this is somewhat uncertain as will be seen. The surface pressures are comparable to those found in good vacuum chambers on Earth. However, the complexity and importance of these atmospheres to the physics of the satellites far outweigh their tiny densities. Furthermore, although the atmospheres of Triton and Io are different from other atmospheres and, in most ways, from each other, they do possess one common theme. In both cases, the molecule making up the gaseous phase is also present in the solid phase on the surface. Hence, the properties of the atmospheres are linked to the equilibrium vapour pressure (EVP) of the respective gases (N_2 for Triton, SO_2 for Io) at the surface temperatures found on the satellites. However, over the years researchers have discovered that there are many twists in this particular tale.

Triton

The study of Triton's atmosphere began with the detection of methane in the infrared by Cruikshank and Silvaggio (1979). Nitrogen was detected five years later. It was quickly shown that the absorption bands seen in the spectra were produced by surface ices rather than by atmospheric gases. However, the expected surface temperatures were such that nitrogen was expected to be a major atmospheric constituent. In addition to CH_4 and N_2, the signatures of CO_2 ice and CO ice have also been detected.

The atmosphere was first unambiguously detected by the phase delay in the X-band radio signal from Voyager 2 when the spacecraft was occulted by the satellite during the Neptune flyby in August 1989. Refraction caused by N_2 (which was subsequently detected in the gas phase from the ground at 2.16 μm) gave a tangential column density at the surface of $8.8 \times 10^{26}\,m^{-2}$ and a scale height equivalent to a surface temperature of 38 K. The mean molecular weight of the atmospheric gases was 28. This is consistent with N_2 in equilibrium with the surface. Haze particles were also found to be present at low optical depths although they appear to have limited significance for the dynamics of the atmosphere.

The EVP of nitrogen is strongly temperature dependent and hence, if the gas has a density equivalent to the EVP at the local surface, one might expect large variations in the surface density from the equator to the poles because of the difference in the solar insolation. However, there is a subtle twist. The temperatures on Triton are so low that the latent heat released by condensation on the surface is important for the heat balance of the surface (Figure 9). The heat balance can be described by

$$m_s c \frac{dT}{dt} = \frac{S \cos\theta}{R_h^2}(1 - A_H) - \varepsilon \sigma T^4 + L\frac{dm}{dt} + K\frac{dT}{dz}$$

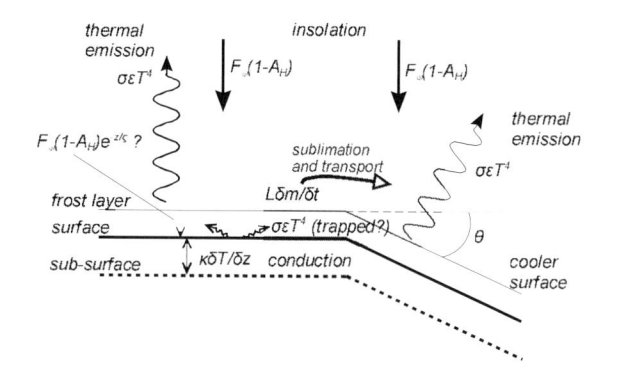

Figure 9 The physical processes affecting the thermal balance at the surface of a satellite include insolation, thermal emission, and conduction. In addition, surface frosts may sublime or condense (depending upon the temperature) and solid-state greenhouse effects may occur. (Figure modified from Hansen and Paige 1992.)

where m_s is the mass of the surface layer, T is the temperature, S is the solar constant at 1 AU, θ is the solar zenith angle, R_h^2 is the heliocentric distance in AU, A_H is the directional hemispheric albedo, ε is the infrared emissivity, σ is Stefan's constant, L is the latent heat of sublimation (typically $2 \times 10^6 \, \text{J kg}^{-1}$ for H_2O, for example), dm/dt is the condensation rate and K is the thermal conductivity. The magnitude of the condensation term is such that the N_2 frost on the surface is isothermal independent of position (Yelle et al. 1995) and hence the surface atmospheric density is constant over the entire globe.

In EVP systems there is a net transport of material towards the poles (assuming that the rotation axis is roughly perpendicular to the ecliptic plane). Triton is particularly interesting here because of the inclination and precession of its orbit which means that the sub-solar latitude wanders from 50 °S to 50 °N over timescales of a few hundred years (Harris 1984). The sub-solar latitude at the present time is very close to its southernmost extreme. Thus, the frost distribution on the surface should be rather mobile. However, the Voyager images of the frost distribution are not consistent with the simple EVP theory. Furthermore, recent evidence has been presented of a doubling of the surface pressure which may be caused by the motion of the sub-solar point or by changes in the frost distribution.

Io

The detection of decametric radio emission from Jupiter by Franklin and Burke (1956) was the first step towards finding out just how complex an object Io is. Much of the interesting physics associated with Io, the innermost Galilean satellite, is related in some way to its 'atmosphere', and hence several points are included under this one heading.

The first photoelectric measurements of Io and the other Galilean satellites were made by Stebbins (1927) and Stebbins and Jacobsen (1928) while the first evidence of a leading–trailing hemisphere asymmetry in the brightness of Io was presented by Harris (1961). In 1962 Binder and Cruikshank (1964) made the first observation of a phenomenon that was to remain controversial for many years. They observed a post-eclipse brightening of Io and suggested that this was evidence of a thin atmosphere which condensed out during the 2.25-hour eclipse behind Jupiter. This experiment was subsequently repeated by many observers with contradictory results. However, its scientific significance was overtaken by events.

In 1964 Bigg analysed an extensive set of decametric observations and saw that emissions were correlated with Io's orbital position. This suggested a strong electromagnetic interaction between Io and Jupiter's magnetic field. Piddington and Drake (1968) and Goldreich and Lynden-Bell (1969) considered the possibility that Io was a conductor moving through Jupiter's magnetic field and that a current system would be set up that could generate the power to supply the radio emissions. The problem was that if Io were a solid, rocky, cold body, the electrical conductivity of the satellite should be negligible. However, in 1973 the Pioneer 10 radio occultation experiment revealed a tantalizing possibility. Io was shown to have a small, asymmetric ionosphere (and hence an atmosphere) which could 'short circuit' the current system around Io.

Io was also shown to be unusual in other ways. Brown (1974) detected sodium D-line emission from the vicinity of Io. This observation implied that a gaseous cloud of sodium was accompanying Io in its orbit about Jupiter. Potassium emission was discovered just a few months later. Matson et al. (1974) suggested that high-energy ions in Jupiter's magnetosphere were sputtering material from Io's surface to produce the cloud. The nature of the sputtering ions was first revealed by Irene Kupo, a student of Yuri Mekler and Aharon Eviatar at Tel-Aviv University. She made the first detection of singly ionized sulphur (S^+) from the vicinity of Io. The ratio of the strengths of the forbidden doublet emissions at 6716 and 6731 Å is extremely sensitive to the electron density of the plasma (n_e). This revealed $n_e = 2000 \, \text{cm}^{-3}$. This was in direct contradiction to observations made by the Pioneer 10 spacecraft which were interpreted in terms of a dense plasma $n_e = 25\,000 \, \text{cm}^{-3}$ composed of protons. Further ground-based observations supported Kupo's work and Pilcher and Morgan (1979) added to the mystery by detecting O^+ emission from the same region. The Voyager 1 spacecraft provided the final confirmation by detecting ionized sulphur and oxygen species in a torus (known as the Io plasma torus) that completely surrounds Jupiter and is centred on Io's orbit.

The origin of the sulphur and oxygen species was also revealed by Voyager 1. Io's SO_2 atmosphere was first detected by the Infrared Imaging Spectrometer (IRIS) on Voyager 1 (Pearl *et al.* 1979) close to the volcanic vent Loki. The derived pressure was around 10^{-7} bar. This is remarkably close to the EVP of SO_2 at the temperature of the sub-solar point on Io (≈ 130 K). A logical conclusion was that the atmosphere was in equilibrium with surface ices over the whole satellite. However, IRIS measured the temperature of the nightside of Io to be around 87 K. The EVP of SO_2 decreases by seven orders of magnitude from 130 to 87 K and therefore huge pressure gradients and substantial transport of SO_2 from the sub-solar point towards the poles were predicted. It followed that ions in Jupiter's magnetosphere sputter material from the atmosphere of Io to produce the neutral clouds. Sodium and potassium are minor constituents (the oxygen neutral cloud was to be detected in 1981) but are easily detectable because these atoms are extremely efficient scatterers of sunlight. The neutrals undergo electron impact ionization to form the ions that are then swept up by Jupiter's co-rotating magnetic field. The neutral atmosphere would naturally form an ionosphere that would close the current loop and thereby produce the observed radio emissions.

However, this rather straightforward interpretation soon ran into problems as ground-based observers attempted to make direct detections of the SO_2 in the ultraviolet and microwave regions in the early 1980s. Observations suggested disk-integrated SO_2 densities far below that expected for an EVP model (e.g. Lellouch 1996). It was therefore suggested that the SO_2 detected by IRIS was actually in Loki's volcanic plume and therefore not representative of the ambient atmosphere. The ambient pressure might be much lower than predicted by the EVP model because of cold traps. Regional cold trapping of atmospheric gases can occur in two ways. Firstly, assume the presence of a naturally dark surface with a slightly brighter surface adjacent to it. The dark surface will be warmer because of its lower albedo. Atmospheric gases will therefore condense preferentially on the lighter surface so that the gas pressure will equilibrate at the temperature of the lighter surface. There is a potential positive feedback mechanism here because ice condensates tend to have very high albedos driving the surface temperature down further. The second type of cold trap is the sub-surface cold trap (e.g. Matson and Nash 1983). The sub-surface temperature is controlled by the thermal conductivity. If the surface layers are extremely porous and of low thermal conductivity, the sub-surface layers can be much colder than the illuminated surface. Because of the porosity, gas can flow into the sub-surface layers and condense there so that the equilibrium pressure is at the temperature of the sub-surface layer, not that of the surface.

HST observations (McGrath *et al.* 2000) have indeed detected higher densities of SO_2 and S_2 above known active regions while microwave observations continue to show low global densities. The present interpretation is that there is a global SO_2 atmosphere (together with its dissociation products) of considerably lower density than the EVP model would predict. Superimposed upon this global atmosphere are areas of higher, variable density above volcanic vents. Over most of the satellite the global atmosphere is thick enough to protect the surface from energetic charged particle sputtering. An exception may be the polar caps where the density may be significantly reduced because the surface is extremely cold. Here, energetic particles from Jupiter's magnetosphere could strike the surface directly leading to a phenomenon known as radiation darkening of the surface material and hence explaining why Io's polar caps are of lower albedo than its equatorial belt. However, this remains to be proved by experiment.

Since their discovery 30 years ago, the neutral clouds, which surround Io, and the Io plasma torus have been investigated in increasing detail. The improvement in observational quality, however, has not led to a clear understanding of their formation. Optical observations of the neutral sodium cloud (the brightest neutral emission and still widely assumed to be a tracer for other more numerous species such as sulphur and oxygen) in the last 10 years, for example, have frequently revealed structures in the neutral cloud that cannot be explained by a straightforward sputtering mechanism. At least two other mechanisms appear to be acting to remove material from Io's atmosphere in addition to the sputtering mechanism (Figure 10).

An ionospheric pick-up mechanism appears to be of particular importance. The field lines of Jupiter's magnetosphere pass through Io's ionosphere at a relative velocity of 57 km s^{-1} (Figure 11). Ions in the ionosphere 'see' these field lines. As a consequence of Faraday's law, the ions are accelerated to this relative velocity and begin to gyrate about the lines. This process is known as pick-up. Just before leaving Io at this high velocity, the ions may undergo a charge exchange reaction that produces a fast neutral. Sodium and potassium neutrals travelling at this velocity with respect to Io and in a directed jet consistent with this process have been observed. When they leave Io, the neutrals are travelling at around 74 km s^{-1} (the speed of the motion of the magnetic field lines at Io's orbit produced by Jupiter's rotation). This is far in excess of the escape velocity from the Jupiter system which is roughly 25 km s^{-1} from Io's orbital distance. Hence, the neutrals completely leave the Jupiter system and travel into interplanetary space if their lifetimes against electron impact ionization and photoionization are sufficiently long. An enormous cloud of sodium neutrals about Jupiter has indeed been detected.

The second potential mechanism is a variation on the first. In this mechanism, molecular ions are picked up in the ionosphere of Io and are trapped and held by Jupiter's magnetic field. The ions eventually undergo dissociation or

Figure 10 The innermost Galilean satellite of Jupiter, Io, loses around $4\,t\,s^{-1}$ of material. It is the major source of heavy ions in Jupiter's magnetosphere. Although sodium is only a minor component, it is an extremely efficient scatterer of sunlight at 5890 and 5896 Å, which makes it easily visible with current ground-based instrumentation. The lower image shows a great deal of structure in the emission of neutral sodium escaping from Io. The observed structures are characteristic of the escape mechanisms. Above is an image of S^+ emission from the Io plasma torus obtained from Catalina observatory in 1990. Ions in Jupiter's inner magnetosphere are confined to a torus shaped volume centred on an equilibrium position which is determined by the balance between centrifugal force and magnetic mirror force. As Jupiter rotates, the torus appears to a ground-based observer to open and close. This is a consequence of the 10° tilt of Jupiter's magnetic axis relative to its rotation axis. Notice the bright structure in the torus called the "ribbon". (Image: N.M. Schneider and J.T. Trauger.)

dissociative recombination to provide another source of fast neutrals. This process also appears to have been observed (through observation of the inferred dissociation product, sodium) but, remarkably, the signature of this process is not always observed and hence it appears to be highly time variable. A possibility is that the presence of sodium-bearing molecular ions in Io's ionosphere is related to volcanic activity.

Neutrals that leave Io at low velocities can undergo electron impact ionization and enter the Io plasma torus (IPT). The most populous species in the IPT have also been regularly observed over the past decade (Figure 10). The ions in the IPT oscillate about a plane that is defined by the equilibrium position between magnetic mirror force in Jupiter's offset tilted dipole magnetic field and the centrifugal force resulting from Jupiter's rapid rotation. The density, temperature and composition of the IPT vary strongly with the radial distance from Jupiter. Inside Io's orbit the plasma is cool and the composition is dominated by singly ionized species (mostly S^+ and O^+). The electron and ion temperatures, although not in equilibrium, rise rapidly between $5.4R_J$ and $6.0R_J$. The higher-energy electrons give rise to a more highly ionized plasma and S^{2+} becomes the most numerous sulphur ion around $5.8R_J$. The electron-density reaches a maximum of around $3000\,cm^{-3}$ at around $5.7R_J$ – just inside the orbit of Io. This region is seen to be very bright in optical emission and is referred to as the 'ribbon'. The origin of this spatial structure remains poorly understood. The estimated composition at $5.9R_J$ (at Io's orbit) is estimated to be 14% S^+, 48% O^+, 19% S^{2+}, 2% O^{2+}, 2% S^{3+}, 2% Na^+ and $\leq 2\%$ SO_2^+. A 10% mixing ratio of protons is usually assumed but this is uncertain. The estimated mass of the IPT is around $2\times 10^6\,t$ and the loss rate from Io is thought to be around $4\,t\,s^{-1}$. Significant variability over many different timescales has been observed. In particular, the 'ribbon' has exhibited remarkable variability over the years. In some observations it has been almost totally absent. It is widely assumed that there must be some link between the variation in volcanic activity on Io (with its inferred effect on the atmosphere) and the

Table 3 Atmospheric pressures on the Galilean satellites

	Species	Pressure (bar)	Peak electron density (cm^{-3})
Io	SO_2	10^{-7}*	$2-10 \times 10^4$
Europa	O_2	$0.4-2.5 \times 10^{-11}$	$5-14 \times 10^3$
Ganymede	O_2	$0.2-2 \times 10^{-10}$	5×10^3
Callisto	CO_2	Exosphere†	2×10^3

* Possible contribution from a volcanic plume
† Discovered by Carlson et al. (1999)

Figure 11 The Galileo spacecraft acquired this image of Io when the Galilean satellite was in eclipse behind Jupiter. It shows substantial emission from atmospheric constituents in a non-uniform distribution. Although the composition of the emitting species and the excitation mechanism remain controversial, electrons originating from Jupiter's magnetosphere or Io's own ionosphere are probably involved in exciting emissions of neutral species such as oxygen in Io's atmosphere. These electrons will also be involved in electron impact ionization which maintains an ionosphere on Io. The ions in the ionosphere could then be stripped from Io by Jupiter's co-rotating magnetic field. (Image: NASA.)

variations seen in the neutral clouds and the IPT. However, this has not yet been clearly demonstrated.

Titan, Triton, Europa and Enceladus are also known to influence the heavy ion populations of their parent planets and the processes described above could, to a greater or lesser extent, be operating there also (e.g. Shemansky et al. 1993, Ip et al. 1998). It is possible that Charon and Pluto have an even more complex interaction because of the potential for the flow of the atmosphere of Pluto to Charon. There has been a suggestion that Charon possesses a weak atmosphere following a stellar occultation experiment. However, the evidence is not strong at present. Charon itself has a lower albedo than Pluto and has a rather low-amplitude rotational light curve which again points to a lack of surface frosts, although it is known that water frost is present on the surface from infrared spectroscopy measurements.

Sputtered exospheres

Sputtering of planetary surfaces is a means of vaporizing material to form gases above the surface of a planetary satellite. Hence, if there are sputtering particles available that can make direct impacts on a planetary satellite, it is to be expected that gas will be present above the surface at some level. The column density of the gas cannot exceed a value that blocks the sputtering particles reaching the surface since this 'turns off' the source. For the Galilean satellites, this value corresponds to surface pressures of around 0.1 nbar. Although such densities are incredibly small, they are detectable from Earth. Hall et al. (1998), for example, have detected neutral oxygen emission at 1304 Å from Europa and Ganymede which arises from an aurora-type process (Table 3).

The radio occultation experiment aboard Galileo has recently measured the densities of the ionospheres of all of the Galilean satellites. The peak densities are shown in Table 3. The ionospheres are all affected by interaction with Jupiter's magnetosphere. In Ganymede's case, the intrinsic magnetic field, discovered by the magnetometer on Galileo, may limit this interaction (Kliore et al. 1998).

It is to be expected that the Saturnian satellites will also have sputtered exospheres of this type. Voyager observations of ions trapped in Saturn's magnetic field and HST observations of OH emission near the orbits of the icy Saturnian satellites (Shemansky et al. 1993) offer strongly supporting circumstantial evidence for this. While improved ground-based and Earth-orbiting observations and the forthcoming arrival of Cassini at Saturn in 2004 should lead to the detection of these exospheres in the near future, the detections of any sputtered exospheres at Uranus and Neptune may have to wait a little longer.

If a satellite's orbit lies within the magnetosphere of its parent planet, the sputtered material may be held within the magnetosphere. In this way, planetary satellites can be the principal sources of heavy ions in planetary magnetospheres. Satellites outside the magnetosphere of the parent planet interact with the solar wind and material sputtered from the surface will, in general, be lost to interplanetary space fairly quickly. Figure 12 shows the positions of many of the regular satellites with respect to the surface, Roche limit and magnetospheric bow shock of their parent planets.

SURFACE GEOLOGY

Cratering rates

Resurfacing rates calculated from the absence of craters critically depend upon knowledge of the cratering rate and its variation with time. It is generally assumed that after their initial formation, the planets and their satellites

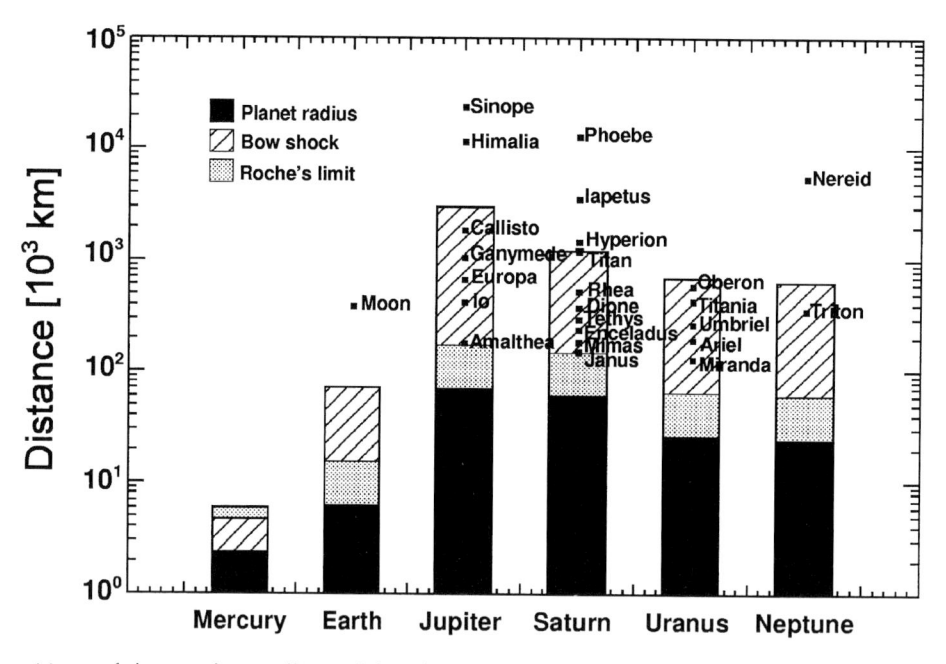

Figure 12 The positions of the regular satellites of the planets with respect to the planetary radius, the Roche limit of the planet, and the planetary magnetosphere.

experienced a period of heavy bombardment that ended around 3.5 billion years ago. During this time, the existing large bodies swept up small objects left over from planetary formation. The cratering rate subsequently slowed down to its present rate.

In addition to being time variable, the impact characteristics must be position dependent. In the outer Solar System, the relative velocities between the impactors and the impacted are slower. This reduces the impacting energy and affects the morphology of the resulting crater. However, the velocities are increased by the potential well of a planet. Shoemaker and Wolfe (1982) showed how impact velocities on the Galilean satellites rise steeply as one moves inward from Callisto to Io. The orbital velocities of the satellites also affect the crater statistics and produce asymmetries between the leading and trailing hemispheres.

The nature of the impacted surface also influences the structure of the craters produced by impacts. The surface properties must therefore be taken into account when interpreting the cratering record. Ganymede and Callisto, for example, show a sharp drop-off in the number of craters with diameter ≥ 60 km compared to the terrestrial planets. One possibility is that the large craters on these icy surfaces have relaxed. However, the overall crater distribution on these objects suggests that the entire impactor distribution on these satellites is incompatible with the modern terrestrial planet projectile population (Chapman and McKinnon 1986). Thus, entirely different populations of projectiles are required depending upon the position within the Solar System. Furthermore, collisional shards produced during satellite formation that ultimately impact satellites form an additional source of craters which may alter significantly the impactor distribution for the giant planet satellites. Therefore, estimates of the absolute ages of satellite surfaces from crater statistics remain extremely imprecise. However, cratering does allow a comparison of the relative ages of surface units.

Resurfacing processes

If surface units on the same satellite show markedly different crater statistics, one can infer that a resurfacing process has occurred at some time in the history of the body. If there are large differences between adjacent planetary satellites in a satellite system, one may also be able to draw such a conclusion and address the issue of global resurfacing events. In the following, several resurfacing mechanisms are discussed by referring to explicit examples.

Atmospheric effects

In the EVP atmospheric model, there is a net transport of ices from the equator to the poles. Thus, on Io, there may be continued deposition of SO_2 at the poles which obscures the surface. It is unlikely, however, that this dominates volcanic resurfacing (see below). A more interesting example is Triton.

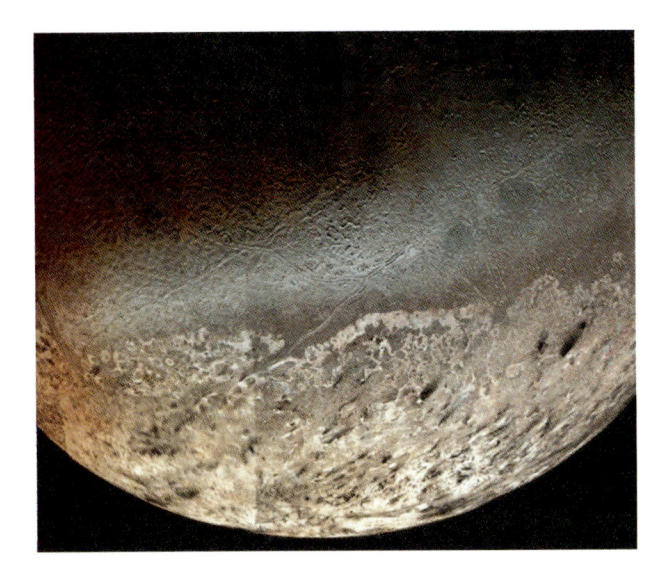

Figure 13 The surface of Triton observed by Voyager 2. (Image: NASA.)

The Voyager images of the surface indicate very clearly that extensive resurfacing has occurred. Water-rich materials with water-soluble admixtures have been suggested to explain the observed landforms. Stern and McKinnon (1999) have recalculated the cratering rates based upon the size distribution of objects in the trans-Neptunian region. They suggest that Voyager images show crater densities well below the values expected and that Triton must therefore be active now. Indeed, active geyser-like plumes were found by the Voyager 2 spacecraft.

The first images of Triton from Voyager 2 showed a series of dark streaks across the surface (Figure 13). Complex hydrocarbons were immediately suspected. However, the transport of nitrogen frosts across the surface should have buried such features on relatively short timescales. Hence, the streaks had to be geologically recent. Soderblom et al. (1990) searched the Voyager data for a source of the streaks using stereoscopic techniques and discovered active plumes that extended up to 8 km above the surface. The mechanism that drives the geyser activity is not known. One possibility is that the reservoir of cryomagma below the surface has an extremely large volume. A single body of cryomagma, which cools only very slowly, could then trigger several small local geysers. Some external factor, such as an impact, could then expose this volume leading to extensive resurfacing. However, there are many other possibilities only some of which have received detailed attention.

The orbital precession of Triton means that the position of the sub-solar latitude varies strongly over long timescales and hence the surface frost distribution should be very mobile. This may also lead to erosion processes which can affect the apparent crater distribution. Similar

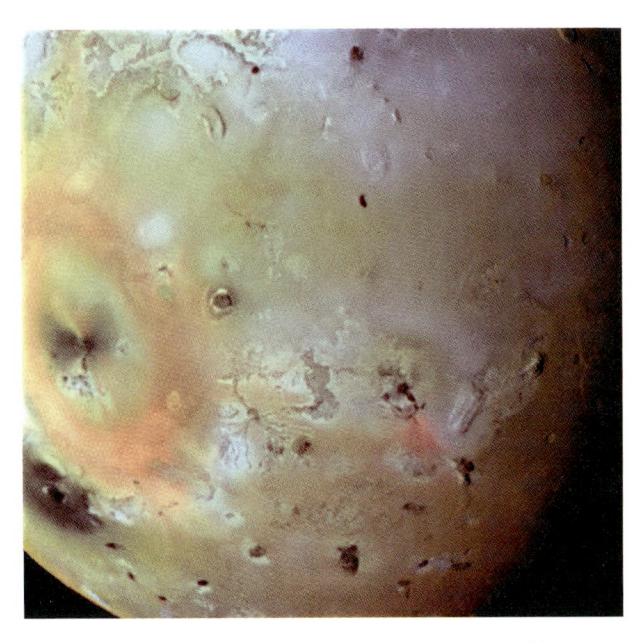

Figure 14 The surface of Io observed by Galileo. Low resolution color data have been combined with a high spatial resolution clear filter frame to emphasize the varied effects of volcanism on Io. To the left is the volcanic vent, Pele. It is surrounded by a bright orange ring which indicates extensive deposition of sulphurous material ejected from the vent. A second orange area is evident to the right near the vent, Marduk. The remarkable colors on the surface of Io were identified with sulphur by the late Carl Sagan. (Image: NASA.)

processes may be active on Pluto and, possibly, Charon. Although we know little about the surface appearance of Titan, it is almost certain that the atmosphere will play a strong role in modifying the surface through aeolian and 'rain'-type processes.

Volcanism

Volcanic activity can lead to lava flows which erase existing impact craters. The most striking example of this mechanism is, of course, Io (see above). It is estimated that the resurfacing rate is, on average over the whole surface, around 0.1–1 mm per year (Figure 14). The remarkable colours on the surface of Io (the Voyager imaging team leader, Bradford Smith, used to joke about Io looking like a giant pizza) were correctly interpreted by Carl Sagan as being indicative of a large sulphur content. However, steep slopes in the calderas indicate that silicates must also be present and temperatures in the active calderas may reach 2000 K. As sulphur cools from 450 to 420 K, its viscosity decreases by several orders of magnitude. Hence, pure sulphur flows on Io can expand away from the vent very quickly even when relatively cold. On Earth, volcanic plumes are driven by the explosive escape of gases (mostly

Figure 15 The surface of Ariel observed by Voyager 2 showing evidence of cryovolcanism. (Image: NASA.)

H_2O). Io is completely devoid of water but the presence of sulphur on the surface and SO_2 in the atmosphere suggests that these compounds (in gaseous phase) drive the eruptions.

Ariel has clearly had a complex geological history, especially when compared to its neighbour in the Uranian system, Umbriel. There appears to have been extrusive resurfacing which might be termed volcanic. In this case, however, the materials involved were probably composed of melted ices. Hence, this type of 'cold' volcanism is sometimes referred to as cryovolcanism and the resurfacing flows are called cryolavas. On Ariel the resulting deposits have a wide range of ages indicating that there have been multiple stages of resurfacing activity (Figure 15). By comparison with Umbriel, it appears that Ariel underwent a global volcanic resurfacing early in its history that erased the oldest craters. Subsequently, cryovolcanic flows occurred on at least two (and probably more) separate occasions. The flows were probably viscous as suggested by the heights of their margins. There is also evidence of flows on Miranda. Here, too, flows are characterized by distinct topographic margins, which indicate extrusion of a relatively viscous material. The topography of the flows can be used to estimate flow viscosity. The estimates range from 10 MP to 1 GP (Schenk 1991). The heat source for the occasional activity on Ariel is unknown but the complexity of the orbital resonances in the Uranian satellite system certainly seems to be a 'smoking gun'. Tittemore (1990) estimated that orbital evolution through the 4:1 resonance with Umbriel around 3.8 billion years ago could have raised the internal temperature of Ariel by as much as 20 K, thereby triggering resurfacing.

The cratering on Enceladus clearly indicates a range of different ages (Figure 16). Some plains terrains show no

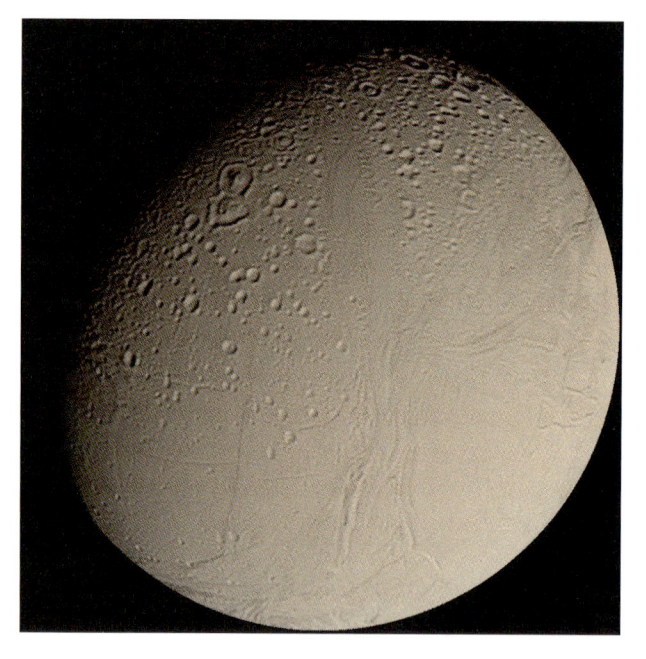

Figure 16 The surface of Enceladus appears to have a wide range of ages. Craters dominate a significant part of the surface and yet some plains terrains show no craters at all at the resolution limit of the Voyager cameras. Extensive re-surfacing must have taken place. (Image: NASA.)

craters at all at the limit of the Voyager image resolution. The abrupt change from cratered regions to plains suggests a widespread flooding by cryovolcanic flows possibly caused by geologically recent tectonic activity driven by tidal heating. It has even been suggested that Enceladus may be active now with this activity also leading (by an unspecified mechanism) to the production of Saturn's E-ring, the density of which is strongly peaked at the orbit of Enceladus (Squyres and Croft 1986). Dione and Tethys also show some evidence of liquid resurfacing.

In the Jovian system, Ganymede ought to have a higher crater density than its neighbour, Callisto, because of gravitational focusing. However, Ganymede is clearly less cratered and, in particular, there is a paucity of large craters. Ganymede exhibits two distinct types of terrain. A dark terrain is relatively old but still apparently younger than the surface of Callisto. The relatively youthful appearance of this terrain may be because viscous relaxation of large craters occurred more rapidly on Ganymede. A high-albedo terrain is obviously even younger and shows a range of ages consistent with emplacement over an extended period of time. There are also a series of bright belts that suggest flooding of fault-bounded valleys. Once again cryovolcanic flows are suspected. The fault system is extensive and may be related to phase changes in the ice-dominated lithosphere which has produced expansion as the lithosphere cooled after Ganymede's formation.

Mass movements

Mass movements include phenomena such as landslides which can alter the appearance of the surface and erase evidence of ancient craters. Galileo observations appear to show that this is an important process on Callisto (Figure 17). Craters can also be erased by viscous relaxation processes where crater rims slump. Subsidence of crater rim blocks can also occur particularly if the lithosphere material is icy rather than rocky.

Miranda may show evidence of another type of mass movement. Three geologic regions on the surface, referred to as 'coronae', appear to have similar squarish shapes bounded by a zone of concentric parallel lineaments that are rounded at the corners. The surfaces of the three coronae (called Arden, Inverness and Elsinore) are much less cratered than their surroundings. However, the albedos of the regions are distinctly different and the outer banded zones of the three also differ. One possible formation mechanism is that these young regions might have been produced by upwelling of lighter material as Miranda began to differentiate. These so-called 'diapirs' would have been large teardrop-shaped accumulations of ices that would have been surrounded by heavier rock–ice mixtures. There are, however, alternative hypotheses.

Collision – Miranda

It is widely thought that the change in the axial tilt of Uranus was produced by an impact. The satellite system that existed prior to the impact was then destabilized leading to collisions between the satellites and subsequent formation of the present system. The older surfaces on Miranda may be original surfaces while the grooved and faulted terrain may be connected in some way to the collision and re-equilibration process. The coronae would then be the youngest surfaces that would have formed the interior of the original satellites in the system. Such a process might explain the unusual geology of the satellite (Figure 18).

Marzari et al. (1998) simulated the breakup of Miranda by a high-velocity impact and computed the size and orbital distributions of the collisional fragments. They then simulated the reaccumulation of the satellite. They showed that reaccumulation occurs on a short timescale (10^3 years), in spite of the initial large dispersion of the ring debris and the presence of Ariel at the outer border of the ring. However, the process depends critically on the relative importance of fragmenting or accumulating collisions.

Gene Shoemaker suggested a combination of the diapir and the collisional theories. During the last phases of accretion, Miranda was impacted by several large silicate bodies.

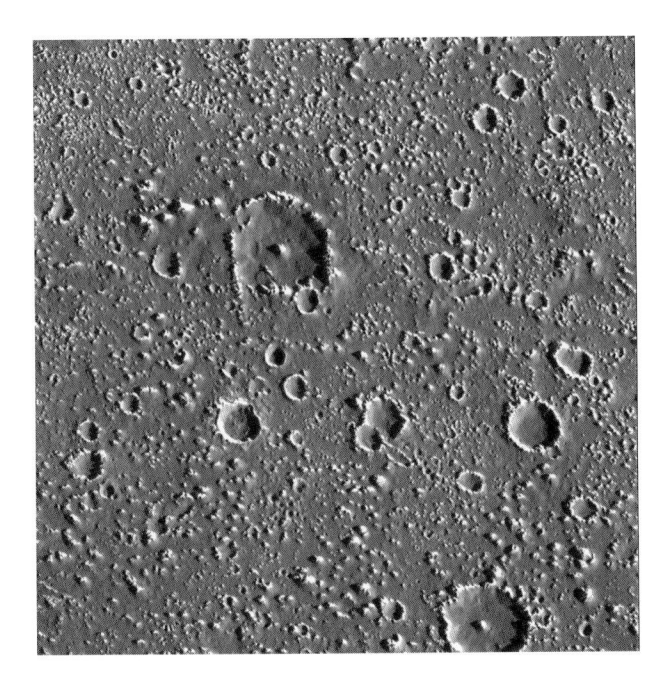

Figure 17 This moderately high resolution view of Jupiter's icy moon, Callisto, shows two, probably related, phenomena. Firstly, a dark, mobile blanket of material covers Callisto's surface. Movement of this material occurs on slopes, as seen here on some crater walls. Secondly, while Callisto has a significant number of large craters, it lacks the related number of small craters which are seen in the crater size distributions of other similar bodies in our solar system. Small craters near slopes would become filled in by the downward movement (landslides) of the dark material. (Image: NASA.)

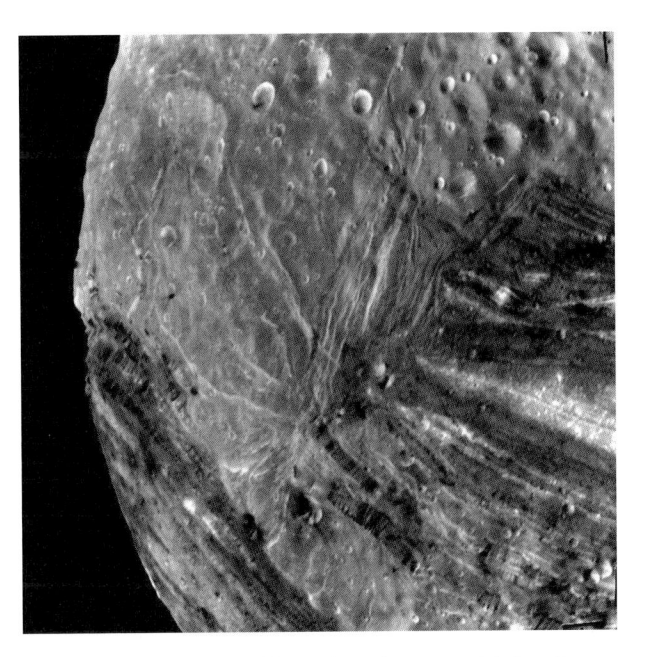

Figure 18 The remarkable surface features of Miranda seen by Voyager 2. To the right, one can see one of the coronae with its low crater density compared to the old surfaces (e.g. top). (Image: NASA.)

These bodies would have been heavy compared to their surroundings and would have sunk through an upper layer that had a lower density. The coronae are the remaining evidence of this sinking process. This process has the disadvantage of not giving a satisfactory explanation for the differences between the coronae. However, no existing theory offers an entirely satisfactory explanation for the surface features on Miranda.

Liquid–solid surface transitions

The entire surface of Europa is icy and perhaps, on average, only 100 million years old. There is increasing evidence that the surface of Europa comprises an icy shell that covers a salt-rich liquid water 'ocean'. For example, the magnetometer on Galileo recorded directional magnetic field changes with time consistent with a conductive liquid water layer less than 100 km below the surface. Furthermore, the surface of Europa also shows evidence of mobile 'iceraft'-type features (Figure 19). The detailed morphology of this terrain strongly suggests the presence of liquid water at shallow depths below the surface, either now or at some time in the past (Carr *et al.* 1998). Although this type of terrain is widespread, it is not global. Moore *et al.* (1998) have analysed impact craters on the surface and conclude that there are two basic types. 'Classical' impact craters are evident which look like lunar craters of similar size but are more topographically subdued. There are also very flat circular features that lack the basic topographic structures of impact craters (for example, raised rims, central depressions or central peaks). Moore *et al.* suggest that the type of crater depends upon the rheology of the surface. 'Classical' craters are found where the surface down to a depth of 10 km is essentially solid. Impact simulations suggest that the distinctive morphologies of the other type of crater would not be produced by impact into a solid ice target, but may be explained by impact into an ice layer 10 to 15 km thick overlying a low-viscosity material such as liquid water.

Europa is also in an orbital resonance that raises tides. Although the forces involved are not as strong as at Io, they may still have a profound effect on the surface. If there is an internal heat source, the ice shell above the liquid could melt. Greenberg *et al.* (1999) suggest that a region of chaotic terrain on Europa is evidence of this phenomenon. Furthermore, the tides in the hypothetical sub-surface ocean could rise and fall by as much as 30 m. This would put stress on the ice shell so that when the stress exceeds the tensile strength of the ice, a crack forms possibly exposing liquid underneath. Once exposed the liquid water would freeze and thereby close the crack in the ice shell. Hence, resurfacing occurs.

Solid-state greenhouse effect

One theory of the origin of Triton's plumes is the solid-state greenhouse effect. Ice is translucent at visible wavelengths. If an icy surface is also opaque in the thermal infrared then a 'greenhouse' effect is created (Matson and Brown 1989) and changes in the vertical temperature distribution of the surface layers can occur. This can result in sub-surface melting of ices and possible phase changes. If the surface ice forms a seal over the sub-surface layers then pressure can build up which may lead to an explosive fracture of the ice. Alternatively, if the surface ice is sufficiently porous, plume production can occur. There are many icy planetary satellite surfaces (for example, Europa and Ganymede) to which this concept can be applied.

External resurfacing – Iapetus

Cratering by a population of asteroidal objects is the most commonly seen effect of exogenic processes on planetary surfaces. Energetic ions impacting a satellite surface might be considered an exogenic process (although often the satellite is itself the source of the impacting ions). However, Iapetus shows evidence of a completely different exogenic process. The trailing hemisphere of Iapetus has an albedo of around 0.5 and looks like a typical icy satellite surface. The leading hemisphere, however, is extremely dark with an albedo of 0.05. Voyager images suggest that the dark material is overlying the icy surface (Figure 20). Iapetus is in synchronous rotation and the dark material lies exactly symmetrically about the apex of the leading hemisphere. The most plausible theory of the origin of this material is that the outer Saturnian satellite Phoebe was, at one point in its history, emitting dust (that is, some form of cometary emission). This material spiralled inwards towards Saturn under the effect of Poynting–Robertson drag and was swept up by the leading hemisphere of Iapetus. Numerical simulations have demonstrated the feasibility of this concept (D. Hamilton, personal communication).

This concept has not been completely accepted, however, because there are dark-floored craters in Iapetus's high-albedo hemisphere, and the north pole is brighter than parts of the trailing hemisphere. However, these problems

Figure 19 The highest resolution images of Europa from the Galileo spacecraft appear to show "icerafts" indicating fluid flow.

Figure 20 Iapetus is one of the most unusual objects in the Solar System. While its trailing hemisphere appears similar to many icy satellites, its leading hemisphere seems to be covered in a dark, possibly carbonaceous, material.

are greatly reduced when one considers that Iapetus took nearly a billion years to move into synchronous rotation with Saturn. Hamilton (1997) has suggested a scenario in which the inward spiralling of dust from Phoebe began before Iapetus was tidally locked and ended after synchronous rotation was reached. An ice covering on the trailing hemisphere then formed and the dark-floored craters seen there are simply relatively young craters exposing the early Phoebe deposits. Wilson and Sagan (1996), however, suggest that Iapetus's asymmetry is best explained by a thick, primordial low-albedo sub-surface layer of organics exhumed by impact erosion.

There remains much to be understood about these processes. Recent observations suggest a spectral similarity between Iapetus and Hyperion which is not understood. Several of the outer satellites of Jupiter and Uranus might also be captured comet-like objects and hence they may also have influenced the surfaces of regular satellites in a similar but less profound manner. It is hoped that the Cassini mission will provide a better understanding of the surface of Iapetus.

THE TWENTY-FIRST CENTURY

So what will happen in the next century of investigation of the satellites of the outer Solar System? The irregular satellites are destined to remain rather obscure objects for the foreseeable future. Although Cassini will take a quick look at Phoebe on its way into the Saturnian system, it is widely thought that these objects will offer little new insight into the major questions connected to Solar System formation and evolution. It is these questions that will drive the studies of outer planetary satellites and they can only be answered by studying the regular satellite systems.

The overall goal of spacecraft missions to outer planetary satellites will be to understand the planetary satellite formation process and to identify what processes have led to the diversity of the outer planetary satellites. Studies will also help us learn about processes that affected the early Earth. In many instances we are still rather ignorant of the physics behind some of the phenomena occurring today. Spacecraft missions can also clarify some of these issues. In the short term, there are several objects of immediate interest.

Europa has become a major priority for detailed investigation because of the increasing evidence that liquid water exists near its surface. It seems almost certain that a lander will be sent to the surface within the next two decades to investigate the surface composition and to probe the interior by seismometry. The need for liquid water is regarded as one of the precursors for life and, hence, the investigation of Europa will be a spur to the newly developing scientific field of astrobiology.

Titan is also a target for astrobiology. The vast quantities of organic material and nitrogen in Titan's atmosphere suggest similarities between the satellite and the early Earth. The investigation of the chemistry in this environment will receive a great boost from the results of the Huygens descent probe which will enter Titan's atmosphere in 2005. Subsequent missions will depend to a large extent on the results from this probe.

The Cassini spacecraft will do for the Saturnian satellite system what Galileo has done for the Jovian system. It is to be expected that the surface compositions of many satellite surfaces will be better understood and that some of the processes that have led to their current state will be more apparent from higher-resolution imaging. The Cassini spacecraft will also investigate the structure of the ring system and the influence of the shepherd satellites.

Triton is a fascinating world which demands a more detailed investigation. Interest in a Neptune orbiter mission is increasing in the USA. It is possible that such a mission could be launched early in the second decade of the twenty-first century. This mission would accomplish many of the goals outlined for Galileo and Cassini for the Neptune system. This would leave the Uranian satellite system as the only giant planet satellite system not visited by an orbiter. There are presently no plans for a mission of this type. This is rather unfortunate because, in 2008, the Sun passes through the plane containing the orbits of the satellites.

This means that all points on a satellite will be illuminated at some time during the orbit of the satellite about Uranus. This phenomenon occurs only every 42 years because of the orbit of Uranus about the Sun. This is particularly important because the tilt of the rotation axis of Uranus with respect to the ecliptic means that, for example, in 2030 only one hemisphere of a satellite can be observed. The other hemisphere is in permanent darkness.

Detailed surveys of most of the regular and unusual satellites could be completed by the middle of the next decade and in-depth studies of Titan and Europa will also have been performed. An investigation of the physical processes on Io is also high on most planetary scientists' wish-lists. However, the radiation from the Io plasma torus makes Io an extremely difficult target for *in situ* investigation. For many of the other regular satellites, however, orbiter or lander missions can already be conceived to study their properties in detail and one could easily imagine that by the end of the century objects such as Iapetus, Triton and Miranda will have been investigated by landers able to analyse surface composition.

The 20th century saw the study of planetary satellites grow from almost a 'hide and seek'-style game into a major scientific discipline with implications for our own origin. The twenty-first century will provide us with answers to many existing questions. Further exploration will undoubtedly raise new questions probing even deeper into the formation and evolution of our planetary system.

FURTHER READING

The University of Arizona Space Science Series is an excellent compendium of knowledge of planetary satellites. The books relating to the Jupiter system do not take into account the results from the Galileo mission, however. A specific overview of the physics of planetary satellites can be found in this series in Burns, J.A. and Matthews, M.S. (eds) (1986). *Satellites*, University of Arizona Press, Tucson, AZ. However, this book is becoming a little out of date.

A mathematical description of planetary atmospheres which refers to planetary satellite atmospheres can be found in Chamberlain, J.W. and Hunten D.M. (1987). *Theory of Planetary Atmospheres*, 2nd edition, Academic Press, San Diego, CA. A detailed description of the processes in Titan's atmosphere can be found in Atreya, S.K. (1986). *Atmospheres and Ionospheres of the Outer Planets and their Satellites*, Springer-Verlag, Heidelberg.

A non-mathematical description of planetary satellites can be found in Rothery, D.A. (1992). *Satellites of the Outer Planets*, Clarendon Press, Oxford. This is recommended for the non-specialist. It is rather slanted towards geology, however.

A good section on the volcanism of the outer satellites is included in Frankel, C. (1996). *Volcanoes of the Solar System*, Cambridge University Press.

A compendium of data on planetary satellites can be found in Bakich, M. (2000). *The Cambridge Planetary Handbook*, Cambridge University Press.

REFERENCES

Atreya, S.K. (1986). *Atmospheres and Ionospheres of the Outer Planets and their Satellites*, Springer-Verlag, Heidelberg.

Binder, A.B. and Cruikshank, D.P. (1964). Evidence for an atmosphere on Io. *Icarus*, **3**, 299–305.

Borderies, N. and Goldreich, P. (1984). A simple derivation of capture probabilities for the J+1:J and J+2:J orbit–orbit resonance problems. *Celestial Mechanics*, **32**, 127–136.

Brown, R.A. (1974). Optical line emission from Io. In A. Woszczyk and C. Iwaniszewska (eds), *Exploration of the Planetary System*, IAU Symp. 65, Reidel, Dordrecht.

Brown, R.H., Cruikshank, D.P., Pendleton, Y. and Veeder, G.J. (1999). Note: Water on ice on Nereid. *Icarus*, **139**, 374–378.

Burns, J.A. (1977). Orbital evolution. In J.A. Burns (ed), *Planetary Satellites*, University of Arizona Press, Tucson, AZ.

Burns, J.A. (1986). Some background about satellites. In J.A. Burns and M.S. Matthews (eds), *Satellites*, University of Arizona Press, Tucson, AZ.

Carr, M.H., Belton, M.J.S., Chapman, C.R., Davies, M.E., Geissler, P., Greenberg, R., McEwen, A.S., Tufts, B.R., Greeley, R. and Sullivan, R. (1998). Evidence for a subsurface ocean on Europa. *Nature*, **391**, 363.

Chamberlain, J.W. and Hunten, D.M. (1987). *Theory of Planetary Atmospheres*, 2nd edition, Academic Press, San Diego, CA.

Chapman, C.R. and McKinnon, W.B. (1986). Cratering of planetary satellites. In J.A. Burns and M.S. Matthews (eds), *Satellites*, University of Arizona Press, Tucson, AZ.

Croft, S.K. and Soderblom, L.A. (1991). Geology of the Uranian satellites. In J.T. Bergstralh, E.D. Miner and M.S. Matthews (eds), *Uranus*, University of Arizona Press, Tucson, AZ.

Cruikshank, D.P., Roush, T.L., Owen, T.C., Geballe, T.R., Dalle Ore, C.M., Khare, B.N. and de Bergh, C. (1999). Saturn's icy satellites: Infrared reflectance spectra of Rhea, Tethys, Dione, and Iapetus at $\lambda > 2.5\,\mu m$. American Astronomical Society, DPS meeting 31, 03.02 (abstract).

Cruikshank, D.P. and Silvaggio, P. (1979). Triton: A satellite with an atmosphere. *Astrophysical Journal*, **233**, 1016–1020.

Dormand, J.R. and Woolfson, M.M. (1989). *The Origin of the Solar System: The Capture Theory*, Ellis Horwood/Prentice Hall, Chichester.

Franklin, K.L. and Burke, B.F. (1956). Radio observations of Jupiter. *Astronomical Journal*, **61**, 177.

French, R.G., Nicholson, P.D., Porco, C.C. and Marouf, E.A. (1991). Dynamics and structure of the Uranian rings. In J.T. Bergstralh, E.D. Miner and M.S. Matthews (eds), *Uranus*, University of Arizona Press, Tucson, AZ.

Gladman, B.J., Nicholson, P.D., Burns, J.A., Kavelaars, J.J., Marsden, B.G., Williams, G.V. and Offutt, W.B. (1998). Discovery of two distant irregular moons of Uranus. *Nature*, **392**, 897–899.

Goldreich, P. (1965). An explanation of the frequent occurrence of commensurable motions in the solar system. *MNRAS*, **130**, 159–181.

Goldreich, P. and Lyndon-Bell, D. (1969). Io: A Jovian unipolar inductor. *Astrophysical Journal*, **156**, 59–78.

Greenberg, R., Croft, S.K., Janes, D.M., Kargel, J.S., Lebofsky, L.A., Lunine, J.I., Marcialis, R.L., Melosh, H.J., Ojakangas, G.W. and Strom, R.G.

(1991). Miranda. In J.T. Bergstralh, E.D. Miner and M.S. Matthews (eds), *Uranus*, University of Arizona Press, Tucson, AZ.

Hall, D.T., Feldmann, P.D. McGrath, M.A. and Strobel, D.F. (1998). The far-ultraviolet oxygen airglow of Europa and Ganymede. *Astrophysical Journal*, **499**, 475.

Hamilton, D.P. (1997). Iapetus: 4.5 billion years of contamination by Phoebe dust. American Astronomical Society, DPS meeting 29, 20.02 (abstract).

Hansen, C.J. and Paige, D.A. (1992). A thermal model for the seasonal nitrogen cycle on Triton. *Icarus*, **99**, 273–288.

Harris, A.W. (1984). Physical properties of Neptune and Triton inferred from the orbit of Triton. In J.T. Bergstralh (ed), *Uranus and Neptune*, NASA CP-2330.

Harris, D.L. (1961). Photometry and colorimetry of planets and satellites. In G.P. Kuiper and B.M. Middlehurst (eds), *Planets and Satellites*, University of Chicago Press.

Ip, W.-H., Williams D.J., McEntire, R.W. and Mauk, B.H. (1998). Ion sputtering and surface erosion at Europa. *Geophysical Research Letters*, **25**, 829–832.

Johnson, R.E. and Quickenden, T.I. (1997). Photolysis and radiolysis of water ice on outer solar system bodies. *Journal of Geophysical Research*, **102**, 10985–10996.

Kaula, W.M., Drake, M.J. and Head, J.W. (1986). The Moon. In J.A. Burns and M.S. Matthews (eds), *Satellites*, University of Arizona Press, Tucson, AZ.

Kirk, R.L., Soderblom, L.A., Brown, R.H., Kieffer, S.W. and Kargel, J.S. (1995). Triton's plumes: Discovery characteristics and models. In D. Cruikshank (ed), *Neptune and Triton*, University of Arizona Press, Tucson, AZ.

Kliore, A.J., Herrera, R.G., Hinson, D.P., Twicken, J.D., Flasar, F.M. and Schinder, P.D. (1998). The ionosphere of Io and the plasma environments of Europa, Ganymede, and Callisto. American Astronomical Society, DPS meeting 30, 55.P09 (abstract).

Kuiper, G.P. (1951). Satellites, comets, and interplanetary material. *Proceedings of the National Academy of Sciences*, **39**, 1153–1158.

Laplace, P.S. (1805). *Mechanique celeste 4*, Courcier, Paris. (Translation by N. Bowditch reprinted 1966, Chelsea, New York.)

Lellouch, E. (1996). Urey Prize Lecture. Io's Atmosphere. *Icarus*, **124**, 1–21.

Leverington, D. (1995). *A History of Astronomy*, Springer-Verlag, London.

Lewis, J.S. (1997). *Physics and Chemistry of the Solar System*, Academic Press, San Diego, CA.

Marzari, F., Dotto, E., Davis, D.R., Weidenschilling, S.J. and Vanzani, V. (1998). Modelling the disruption and reaccumulation of Miranda. *Astronomy and Astrophysics*, **333**, 1082–1091.

Matson, D.L. and Brown, R.H. (1989). Solid state greenhouses and their implications for icy satellites. *Icarus*, **77**, 67–81.

Matson, D.L., Johnson, T.V. and Fanale, F.P. (1974). Sodium D-line emission from Io: Sputtering and resonant scattering hypothesis. *Astrophysical Journal*, **192**, L43–L46.

Matson, D.L. and Nash, D. (1983). Io's atmosphere: Pressure control by regolith cold trapping and surface venting. *Journal of Geophysical Research*, **88**, 4771–4783.

McCord, T.B., Hansen, G.B., Clark, R.N., Martin, P.D., Hibbitts. C.A., Fanale, F.P., Granahan, J.C., Segura, M., Matson, D.L., Johnson, T.V., Carlson, R.W., Smythe, W.D. and Danielson, G.E. (1998b). Non-water-ice constituents in the surface material of the icy Galilean satellites from the Galileo near-infrared mapping spectrometer investigation. *Journal of Geophysical Research*, **103**, 8603–8626.

McCord, T.B., Hansen, G.B., Fanale, F.P., Carlson, R.W., Matson, D.L., Johnson, T.V., Smythe, W.D., Crowley, J.K., Martin, P.D., Ocampo, A., Hibbitts, C.A. and Granahan, J.C. (1998a). Salts on Europa's surface detected by the Galileo's Near Infrared Mapping Spectrometer. *Science*, **280**, 1242.

McDonald, G.D., Thompson, W.R., Heinrich, M., Khare, B.N. and Sagan, C. (1994). Chemical investigation of Titan and Triton tholins. *Icarus*, **108**, 137–145.

McGrath, M.A., Belton, M.J.S., Spencer, J.R. and Sartoretti, P. (2000). Spatially resolved spectroscopy of Io's Pele plume and SO_2 atmosphere. *Icarus*, **146**, 476–493.

Minton, R.B. (1973). The red polar caps of Io. *Communications of the Lunar and Planetary Laboratory*, **10**, 35–39.

Moore, J.M., Asphaug, E., Sullivan, R.J., Klemaszewski, J.E., Bender, K.C., Greeley, R., Geissler, P.E., McEwan, A.S., Turtle, E.P., Phillips, C.B., Tufts, B.R., Head, J.W., Pappalardo, R.T., Jones, K.B., Chapman, C.R., Belton, M.J.S., Kirk, R.L. and Morrison, D. (1998). Large impact features on Europa: Results of the Galileo nominal mission. *Icarus*, **135**, 127–145.

Nicholson, P.D., Showalter, M.R., Dones, L., French, R.G., Larson, S.M., Lissauer, J.J. McGhee, C.A., Sicardy, B., Seitzer, P. and Danielson, G.E. (1996). *Science*, **272**, 509.

Noll, K.S., Roush, T.L., Cruishank, D.P., Johnson, R.E. and Pendelton, Y.J. (1997). Detection of ozone on Saturn's satellites Rhea and Dione. *Nature*, **388**, 45–47.

Peale, S.J. (1986). Orbital resonances, unusual configurations, and exotic rotation states among the planetary satellites. In J.A. Burns and M.S. Matthews (eds), *Satellites*, University of Arizona Press, Tucson, AZ.

Peale, S.J., Cassen, P.M. and Reynolds, R.T. (1979). Melting of Io by tidal dissipation. *Science*, **203**, 892–894.

Pearl, J., Hanel, R., Kunde, V., Maquire, W., Fox, K., Gupta, S., Ponnamperuma, C. and Raulin, F. (1979). Identification of gaseous SO_2 and new upper limits for other gases on Io. *Nature*, **280**, 757–758.

Piddington, J.H. and Drake, J.F. (1968). Electrodynamic effects of Jupiter's satellite Io. *Nature*, **217**, 935–937.

Pilcher, C.B. and Morgan, J. (1979). Detection of singly ionized oxygen around Jupiter. *Science*, **205**, 297–298.

Pollack, J.B. and Fanale, F. (1982). Origin and evolution of the Jupiter satellite system. In D. Morrison (ed), *The Satellites of Jupiter*, University of Arizona Press, Tucson, AZ.

Pollack, J.B., Lunine, J.I. and Tittermore, W.C. (1991). Origin of the Uranian satellites. In J.T. Bergstralh, E.D. Miner and M.S. Matthews (eds), *Uranus*, University of Arizona Press, Tucson, AZ.

Quirico, E., Doute, S., Schmitt, B., de Bergh, C., Cruikshank, D.P., Owen, T.C., Geballe, T.R. and Roush, T.L. (1999). Composition, physical state, and distribution of ices at the surface of Triton. *Icarus*, **139**, 159–178.

Ross, M.N., Schubert, G., Spohn, T. and Gaskell, R.W. (1990). Internal structure of Io and the global distribution of its topography. *Icarus*, **85**, 309–325.

Roush, T.L., Cruikshank, D.P., Owen, T.C., Geballe, T.R., Benedix, G.K., de Bergh, C., Noll, K.S. and Khare, B.N. (1998). Titania and Oberon: Surface composition from new near-infrared observations and reflectance models. American Astronomical Society, DPS meeting 30, 44.04.

Roy, A.E. and Ovenden, M.W. (1954). On the occurrence of commensurable mean motions in the solar system. *MNRAS*, **114**, 232–241.

Schenk, P.M. (1991). Fluid volcanism on Miranda and Ariel – Flow morphology and composition. *Journal of Geophysical Research*, **96**, 1887–1906.

Schubert, G., Spohn, T. and R.T. Reynolds (1986). Thermal histories, compositions and internal structures of the moons of the Solar System. In J.A. Burns and M.S. Matthews (eds), *Satellites*, University of Arizona Press, Tucson, AZ.

Shemansky, D.E., Matheson, P., Hall, D.T., Hu, H.-Y. and Tripp, T.M. (1993). Detection of the hydroxyl radical in the Saturn magnetosphere. *Nature*, **363**, 329–331.

Shoemaker, E.M. and Wolfe, R.F. (1982). Cratering time scales for the Galilean satellites. In D. Morrison (ed), *Satellites of Jupiter*, University of Arizona Press, Tucson, AZ.

Sill, G.T. and Clark, R.N. (1982). Composition of the surfaces of the Galilean satellites. In D. Morrison (ed), *Satellites of Jupiter*, University of Arizona Press, Tucson, AZ.

Smith, B.A. and Terrile, R.J. (1984). A circumstellar disk around Beta Pictoris. *Science*, **226**, 1421–1424.

Smith, P.H., Lemmon, M.T., Lorenz, R.D., Sromovsky, L.A., Caldwell, J.J. and Allison, M.D. (1997). Titan's surface revealed by HST imaging. *Icarus*, **119**, 336–349.

Soderblom, L.A., Kieffer, S.W., Becker, T.L., Brown, R.H., Cook, A.F., II, Hansen, C.J., Johnson, T.V., Kirk, R.L. and Shoemaker, E.M. (1990). Triton's geyser-like plumes: Discovery and basic characterization. *Science*, **250**, 410–415.

Spencer, J.R., Calvin, W.M. and Person, M.J. (1995). CCD spectra of the Galilean satellites: Molecular oxygen on Ganymede. *Journal of Geophysical Research*, **100**, 19049–19056.

Stebbins, J. (1927). *Lick Observatory Bulletin*, **13**, 1–11.

Stebbins, J. and Jacobson, T.S. (1928). Further photometric observations of Jupiter's satellites and Uranus, with tests for the solar constant. *Lick Observatory Bulletin*, **13**, 180–195.

Stern, A.S. and McKinnon, W.B. (1999). Triton's surface age and impactor population revisited (evidence for an internal ocean). 30th Annual Lunar and Planetary Science Conference, Houston, TX.

Stevenson, D.J. (1992). Interior of Titan. In ESA, Symposium on Titan, ESA SP.

Sykes, M.V., Nelson, B., Cutri, R.M., Kirkpatrick, D.J., Hurt, R. and Skrutskie, M.F. (2000). Near infrared observations of the outer Jovian satellites. *Icarus*, **143**, 371–375.

Squyres, S.W. and Croft, S.K. (1986). The tectonics of icy satellites. In J.A. Burns and M.S. Matthews (eds), *Satellites*, University of Arizona Press, Tucson, AZ.

Tittemore, W.C. (1990). Tidal heating of Ariel. *Icarus*, **87**, 110–139.

Wamsteker, W. (1972). Narrow-band photometry of the Galilean satellites. *Communications of the Lunar and Planetary Laboratory*, **9**, 171–177.

Wilson, P.D. and Sagan, C. (1996). Spectrophotometry and organic matter on Iapetus 2. Models of interhemispheric asymmetry. *Icarus*, **122**, 92–106.

Wisdom, J., Peale, S.J. and Mignard, F. (1984). The chaotic rotation of Hyperion. *Icarus*, **58**, 137–152.

Yelle, R.V., Lunine, J.I., Pollack, J.B. and Brown, R.H. (1995). Lower atmospheric structure and surface–atmosphere interaction on Triton. In D. Cruikshank (ed), *Neptune and Triton*, University of Arizona Press, Tucson, AZ.

Yoder, C.F., Colombo, G., Synnott, S.P. and Yoder, K.A. (1983). Theory of motion of Saturn's coorbital satellites. *Icarus*, **53**, 431–443.

61

NORMAN F. NESS*

Planetary and lunar magnetism

In situ studies by spacecraft of the magnetic fields of Earth and all the planets except for Pluto began with the USSR's launch of Sputnik 3 in 1958. The study of the geomagnetic field by the USA followed with Vanguard 3 in 1959. Since then the US Explorer, Mariner, Pioneer, Voyager, Ulysses, and Galileo missions have surveyed all the planets from Mercury to Neptune as well as the Earth's Moon, the Galilean moons of Jupiter, and Saturn's largest moon, Titan. The USSR repeatedly studied Earth's Moon, Venus, and Mars with their Luna, Venera, Mars, and Phobos missions.

Mariner 10 discovered in 1974 and 1975 that Mercury was globally magnetized but with such a small intrinsic magnetic field that no durably trapped radiation belts were permanently associated with the magnetosphere, formed by the solar wind. Venus appears to be globally unmagnetized, according to results from Venera and Mariner spacecraft since 1961. The solar wind interaction with its atmosphere and ionosphere creates an induced magnetosphere, but devoid of any trapped radiation belts.

The magnetic field of the Moon has been studied by several spacecraft, especially Explorer 35 in 1968 (Ness 1969). Measurements on the lunar surface were conducted by the robotic rover Lunokhod in 1968 (Dolginov *et al.* 1975) and with the three Apollo Lunar Surface Experiments Packages (ALSEP) that were placed there during the Apollo program of the late 1960s and early 1970s (Dyal *et al.* 1972, 1974). The Apollo 15 and 16 missions each carried a sub-satellite which was launched into a 100 km altitude, low-latitude orbit. The data they gathered suggested the existence of magnetized crustal regions (Hood *et al.* 1981). More recent data provide much firmer evidence of localized regions of remanent crustal magnetization by the MAGER (MAGnetic field Electron Reflectometer) instrument on Lunar Prospector (Lin *et al.* 1998).

The latest detailed study of these lunar magnetized regions from Lunar Prospector data suggest a special geometrical relationship and correlation with impact craters which are antipodal in location to them (Halekas *et al.* 2001). The source of the magnetic field for these magnetized regions is unclear at present. They may indicate that there was an ancient lunar dynamo that had magnetized the lunar crust, or perhaps there is a magnetization process that occurs as a result of the high-velocity impacts which generate the observed craters. It is also possible that both processes are important.

In 1964 the US Mariner IV, and subsequently several Soviet Mars and Phobos probes, had placed upper limits on any global intrinsic Martian field, although the existence of a detached bow shock and magnetic tail were observed (Ness 1979a). Since achieving orbit in 1997, the US orbiter Mars Global Surveyor has finally clarified the long-standing enigma of the status of the Martian magnetic field with the discovery of localized intense remanent crustal magnetization but no significant global field (Acuña *et al.* 1999, Connerney *et al.* 1999).

Pioneers 10 (1974) and 11 (1975) and Voyagers 1 (1979) and 2 (1979) examined *in situ* the magnetic field and magnetosphere of Jupiter, which had been inferred from its non-thermal radio emissions, first detected in 1955. Jupiter's magnetic field is much stronger than Earth's and is distinctly non-dipolar close to the planet. More recently, the US Galileo and ESA Ulysses spacecraft have added to the database of *in situ* studies of the main field of Jupiter.

*Bartol Research Institute, Newark, DE, USA

Galileo has also studied the four Galilean moons during multiple close flybys, and results suggest that some of these moons possess an intrinsic or induced field.

In 1979 the US Pioneer 11 discovered a significant global magnetic field and trapped radiation belt at Saturn. This was studied in much more detail by Voyagers 1 and 2 in 1980 and 1981. Saturn was found to have a weaker field than Jupiter, and one that appears to be remarkably axisymmetric (to order and degree $n = 3$ in a spherical harmonic or multipole representation) about its rotation axis.

The Voyager 2 flybys of Uranus and Neptune in 1986 and 1989 discovered significant global magnetic fields and trapped energetic particle radiation belts. Both Uranus and Neptune display quite similar magnetic fields, but very different from those of Jupiter, Saturn, and Earth. As astrophysical objects, Uranus and Neptune are best described as oblique rotators because of the large angular offsets of their magnetic axes from their rotation axes (59° and 47°, respectively). Additionally, their *magnetic centers* are offset from their center of figure by substantial fractions of a planetary radius ($0.31R_U$ and $0.55R_N$, respectively).

This chapter summarizes our present knowledge of the quantitative characteristics of the magnetic fields of the planets and certain of their moons. An early review of space magnetometry by Ness (1970) summarized the first decade of *in situ* studies of the magnetic fields of Earth, Mars, Venus, and the Moon, discussed the technical problems of measuring weak interplanetary fields, and presented technical details of instrumentation. A more recent review of the author's role in space exploration has appeared (Ness 1996).

The dates of significant discoveries of planetary magnetic fields and the associated spacecraft are given in Table 1. An additional relevant mission parameter in these studies of solar wind interactions and the quantitative analysis of the intrinsic magnetic fields is the closest flyby or periapsis distance of the spacecraft trajectories relative to the center of the planet, in units of planetary radii.

GEOMAGNETISM AND PLANETARY DYNAMOS

The most plausible explanation for the origin of planetary magnetic fields, and the observed and well-documented terrestrial secular variation, is that there is a coherent regenerative dynamo motion of an electrically conducting fluid in a planet's interior. This motion is a result, most likely, of thermally driven convection. There are also suggestions that coupling of precessional torques, which arise in the differential motion between the cores and mantles of the planets, may drive this coherent motion. Finally, this coherent motion may also be driven by tidal forces associated with other celestial bodies in close proximity, for example Jovian tides in the Galilean moons, which display spin–orbit coupling as a result of tidal dissipation.

Study of the Earth's magnetic field has been under way ever since William Gilbert published his famous treatise, *De magnete*, in 1600. Geomagnetism is a well-established discipline with a long and distinguished history, but one that lies outside the scope of this chapter. Interested readers are referred to any of the many texts that address the many special problems of geomagnetism, such as Busse (1978), Campbell (1997), Backus *et al*. (1996), Jacobs (1994), Merrill *et al*. (1996), Stevenson (1983), and Soward (1992).

The internal motion in the Earth's electrically conducting core is regenerative in the sense that it leads to the maintenance of electrical currents and associated magnetic fields by the dynamo process. The observed secular variation of the geomagnetic field (Cain 1995) adds substantial support to this concept of dynamic fluid motion in the interior.

Table 1 Summary of US spacecraft, 1971–2004, which have discovered and investigated *in situ* the magnetic fields of the planets

	Year of encounter or orbit/periapis in R_p					
	Mercury	Mars	Jupiter	Saturn	Uranus	Neptune
Mariner 10	•1974/1.30					
	1975/1.13					
Pioneer 10			1973/2.84			
Pioneer 11			1974/1.31	•1979/1.35		
Voyager 1			1979/4.88	1980/3.07		
Voyager 2			1979/10.1	1981/2.69	•1986/4.18	•1989/1.18
Ulysses			1992/6.3			
Galileo (orbiter)			1995+/6–15			
Cassini (orbiter)				[2004/1+]		
Mars Global Surveyor		•1997/1.03				

•Discovery encounter.

The history of the Earth's magnetic field is recorded in magnetized rocks and sediments, which bear magnetizable materials. The study of these ancient samples, which is the discipline of paleomagnetism, indicates that the geomagnetic field has reversed itself many times in the Earth's geological past (McElhinney 1973, Piper 1987). These reversals are dated through the study of the relative abundances of certain isotopes of the radioactive elements.

The physical properties of the interiors of the other planets can only be estimated, whereas the study of earthquake generated seismic signals provides accurate estimates of the structure and physical properties of Earth's interior. There appears to be no obstacle to the assumption that all planetary global magnetic fields are generated within electrically conducting core or shell regions. However, a continuing challenge is the development of a comprehensive and quantitative theory of how rapidly rotating, self-gravitating, and highly condensed bodies throughout the Universe generate their magnetic fields.

Dynamo theory remains one of the continuing basic challenges in theoretical planetary physics and astrophysics (Roberts 1995, Proctor and Gilbert 1996). Recent supercomputer simulations of the Earth's dynamo process that reveal intrinsic reversals have been reported by Glatzmaier and Roberts (1995, 1996) and Glatzmaier et al. (1999). Interesting laboratory experiments using molten sodium have recently been conducted which demonstrate the dynamo effect (Gailitis et al. 2000).

THE GEOMAGNETIC FIELD

Studies of the geomagnetic field prior to the space era were conducted on land and sea, sometimes with specially constructed sailing vessels such as the one used in the seventeenth century by Edmund Halley in an investigation of the deviation of the compass direction from true north in the Atlantic Ocean near South America. The major advance in studies of the geomagnetic field resulting from land and sea measurements occurred just as the space age began. This was the development of the theory of plate tectonics in the early 1960s. This was based upon paleomagnetic studies of crustal rocks and a vast database of total field measurements by instruments towed behind oceanographic research vessels.

Paleomagnetic results had shown that the geomagnetic field had reversed its polarity many times in the past. These reversals are features which mark the time of these events and were accurately dated from radioisotopes present in common rock-forming minerals. The oceanographic data were combined to demonstrate conclusively that ocean floors were spreading apart from central regions, called ridges, as a result of convection in the mantle driving the relative motion of the continents.

Thus, the oceanographic magnetic data were analogous to a tape recording since the hot upwelling molten material from the mantle preserved a memory of the status of the geomagnetic field as it cooled below its Curie point. The concept and the theory of *continental drift* was proposed in 1912 by the German meteorologist Alfred Wegener. He became Professor of Geophysics and Meteorology at the University of Graz, Austria, from 1924 to 1930, the year of his death. He had based his theory originally on the geometrical similarities of the coastlines of the South American and African continents, and matching geological formations on opposite sides of the South Atlantic. Wegener's ideas and theory of such moving continents were initially met with skepticism, which continued for a long time, with some geophysicists "proving" that such motion was physically impossible.

The first orbital flight of a magnetometer was aboard the Soviet satellite Sputnik 3 in 1958; this was followed by the US Vanguard 3 in 1959. It was not until 1979 that the USA dedicated a special mission to study the geomagnetic field, MAGSAT. This spacecraft carried both an ultra-precise triaxial fluxgate vector magnetometer and a cesium vapor total field unit to measure accurately the magnitude of the field. Use of these spacecraft data and standard ground station magnetic observatories allowed the development of models of the Earth's magnetic field up to degree and order 13 and beyond (McLeod 1996).

GLOBAL MAGNETIC FIELDS AND SOLAR WIND RAMIFICATIONS

The existence of a global planetary magnetic field with sufficient strength to deflect the solar wind leads to several unique features of the planetary environment (see the reviews by Stern and Ness 1982, Schulz 1995). The main feature is the formation of a magnetic cavity, or *magnetosphere*, from which undisturbed solar wind plasma is excluded and within which a distorted planetary field controls the motion of energetic charged particles – mostly electrons and protons, but some heavier ions too. Examples of the latter are found in the Jovian magnetosphere, which is populated with heavier ions such as S^{2+} and SO^+ that originate in the SO_2 ejections from Io's active volcanoes. The neutral atoms become ionized by charge exchange with the co-rotating magnetosphere of Jupiter.

Outside this cavity a detached bow shock wave develops in the supersonic and super-Alfvénic solar wind flow. Finally, a *magnetotail* is formed which trails far behind the planet in the anti-solar-wind direction. The distance to the sunward stagnation point of solar wind flow, measured in units of planetary radii, varies from the low values of 1.2 at Mars (Vignes et al. 2000) and 1.4 at Mercury (Ness 1976) to as great as 70 or more at Jupiter. This parameter, and those describing the global field to lowest order and degree as a simple offset tilted dipole, are summarized in Table 2.

The traditional way of describing a planetary magnetic field quantitatively is based upon a method developed by Gauss. This uses spherical harmonic coefficients to represent magnetic multipoles. Modern methods of deriving planetary field representations from spacecraft data using generalized inversion techniques have been developed by Connerney (1981). The Gaussian coefficients for all of the planetary magnetic fields known to date are given in Table 3. The lowest-order magnetic field multipole moment ($n = 1$) is a dipole. Thus, a basic result of the solar wind interaction is that a bipolar magnetotail is developed. Within this magnetotail is a field reversal region, referred to as a plasma or neutral sheet, which contains an embedded plasma. The origin of this plasma is the atmosphere of the parent planet, any natural moons, and the shock-modified solar wind. The strength of the magnetic field in the magnetotails of planets depends primarily upon the properties of the solar wind. The magnetotail appears to be an important energy store, playing an important role in the overall dynamics of the magnetosphere. See Ness (1987) for a review of the discovery in 1964 and early space age studies of the Earth's magnetotail.

A unique feature of the magnetotails of Uranus and Neptune is the obliquity of their rotational axes combined with the large angular offsets of the magnetic axes from the rotational axes. This combination produces a unique magnetic pole-on interaction with the solar wind at certain periods during their orbits. As the planet rotates, the magnetic axis of the planetary dipole field becomes colinear with that of the solar wind velocity. This is theorized to lead to a magnetotail configuration in which the plasma sheet becomes cylindrical in shape, surrounding a unipolar flux tube (Voigt *et al.* 1987, Voigt and Ness 1990). This is the opposite of the case at Earth, where the plasma sheet develops as a transverse planar structure in the magnetotail, which separates the bipolar lobes or flux tubes of the magnetotail.

At Jupiter and Saturn a bipolar magnetotail and plasma sheet similarly develops. Important consequences of the solar wind interaction, trapped radiation belts, and plasma sheets are that there are a number of electrical current systems external to the planet. The magnetic fields of these currents have a significant influence on *in situ* measurements of the magnetic fields by flyby or orbiting spacecraft. These currents must be carefully considered and, if appropriate, explicitly modeled when attempts are made to extract accurate estimates of the intrinsic field of the planet.

MERCURY

Only one spacecraft, the US Mariner 10, has studied the planet Mercury. Launched in 1973, it used a gravity-assist maneuver at Venus to encounter Mercury. There were three encounters, due to a serendipitous exact-integer relationship between the heliocentric orbital periods of Mercury and Mariner 10. Successful encounters in March 1974 (Ness *et al.* 1974) and again in March 1975 provided direct observational evidence for a global magnetic field at Mercury (Ness 1978). The second one, in September 1974, was dedicated to imaging the surface and did not penetrate the magnetosphere or cross the magnetopause and detached bow shock wave.

Data from the magnetic field experiment on board Mariner 10 during the 1975 encounter are shown in Figure 1. The positions of the detached bow shock and magnetopause, scaled from Earth data, are shown superimposed on the plot of the magnetic field vectors projected on the plane of the spacecraft's orbit (Connerney and Ness 1988).

The discovery of an intrinsic magnetic field at Mercury was a great surprise, since there was neither any prior

Table 2 Summary of magnetic fields, solar wind sunward stagnation point, and rotation periods of the planets (Note: 1 G cm^3=10^{-3} A m^2; 1 nT=10^{-5} G=10^{-9} T)

Planet	Dipole moment (G cm^3)	Tilt and sense	Dipole equatorial field (nT)	Average stagnation point distance (R_p)	Rotation period
Mercury	5×10^{22}	+14°	330	1.4	58.7^d
Venus	$< 4 \times 10^{21}$	—	< 2	1.0+	−243^d
Earth	8.0×10^{25}	+11.7°	31 000	10.4	23.9^h
Moon	$< 1 \times 10^{19}$	—	< 0.2	none	
Mars	$< 2 \times 10^{20}$	—	< 0.5	1.29	24.6^h
Jupiter	1.6×10^{30}	−9.6°	428 000	65 ± 15	9.92^h
Saturn	4.7×10^{28}	−0.0°	21 200	20 ± 3	10.66^h
Uranus	3.8×10^{27}	−58.6°	23 000	20	17.24^h
Neptune	2.0×10^{27}	−46.9°	14 000	26	16.1^h

Table 3 Schmidt normalized spherical harmonic coefficients or multipole parameters for Earth and the four giant planets, Jupiter, Saturn, Uranus, and Neptune. UR denotes unresolved. (Parentheses give resolution matrix value, if less than unity)

Planet	Earth	Jupiter	Saturn	Uranus	Neptune
Radius (km)	6378	71 492	60 330	25 600	24 765
Model	IGRF 85	VIP 4	Z3	O3	O8
g_1^0	−0.29877	4.205	+0.21535	+0.11893	+0.09732
g_1^1	−0.01903	−0.659	0	+0.11579	+0.03220
h_1^1	+0.05497	0.250	0	−0.15685	−0.09889
g_2^0	−0.02073	−0.051	+0.01642	−0.06030	+0.07448
g_2^1	+0.03045	−0.619	0	−0.12587	+0.00664
h_2^1	−0.02191	0.497	0	+0.06116	+0.11230
g_2^2	+0.01691	−0.361	0	+0.00196	+0.04499
h_2^2	−0.00309	0.053	0	+0.04759	−0.00070
h_3^0	+0.01300	−0.016	+0.02743	+0.02705	−0.06592
g_3^1	−0.02208	−0.520	0	+0.01188	+0.04098
h_3^1	−0.00312	0.244	0	−0.07095	−0.03669
g_3^2	+0.01244	−0.176	0	−0.04808	−0.03581
h_3^2	+0.00284	−0.088	0	−0.01616	+0.01791
g_3^3	+0.00835	0.408	0	−0.02412	+0.00484
h_3^3	−0.00296	−0.316	0	−0.02608	−0.00770
g_0^4		−0.168(0.75)			
g_1^4		0.222(0.48)			
g_2^4		−0.061(0.89)			
g_3^4		−0.202(0.94)			
g_4^4		0.066(UR)			
h_1^4		0.076(UR)			
h_2^4		0.404(0.93)			
h_3^4		−0.166(0.89)			
h_4^4		0.039(UR)			
Dipole moment (G)	$0.304 R_E^3$	$4.28 R_J^3$	$0.215 R_S^3$	$0.228 R_U^3$	$0.142 R_N^3$
Dipole tilt	+11.7°	−9.6°	−0.0°	−58.6°	−46.9°
OTD offset	$0.08 R_E$	$0.07 R_J$	$0.04 R_S$	$0.31 R_U$	$0.55 R_N$

evidence nor speculation that there would be a planetary field of global extent. The surface of Mercury is similar to that of the Moon, with a history of impact bombardment clearly visible in its numerous craters. It was expected that Mercury's interior had cooled or frozen out any active dynamo motion. Since then, studies have revealed the presence of a tenuous atmosphere consisting of sodium and other elements sputtered from the surface by impacting energetic particles (Lammer and Bauer 1997).

The Mariner 10 magnetic field data set is a challenge for quantitative analysis, because of the significant contributions of electrical currents associated with the deflected solar wind flow (Ness 1979a,b, Connerney and Ness 1988). Figure 2 illustrates the general problem of non-uniqueness in these data inversions. Uncertainties in the derived harmonic coefficients are due to coupling among them, which is predetermined by the spatial coverage afforded by the spacecraft trajectory. This figure shows that a smaller dipole moment can be compensated for with a larger quadrupole moment and associated (and correlated) changes in the external fields.

The unavoidable conclusion is that Mercury possesses a global magnetic field with an equivalent dipole moment whose equatorial field is approximately 0.1% of Earth's. The dipole magnetic field axis is estimated to be tilted approximately 14° from the rotation axis, a value similar to Earth's 11.4° (Table 2). There is no permanently trapped radiation belt at Mercury (Simpson et al. 1974) because the magnetosphere is very small relative to the planet itself; for this reason, remote studies of the magnetic field at Mercury cannot be conducted by radio astronomers. There are currently two missions under careful but separate study by the USA and ESA to return to Mercury and investigate more fully its magnetic field, magnetosphere, and atmosphere, as well as study its surface and atmosphere. NASA has approved the Messenger mission to encounter and orbit Mercury in 2009 following launch in 2004. The ESA mission is still awaiting final approval.

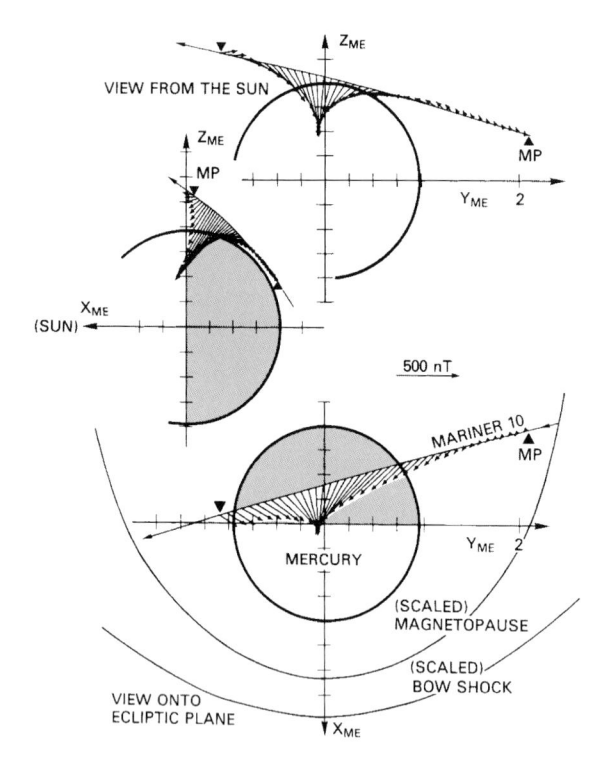

Figure 1 Summary projections of magnetic field data obtained by Mariner 10 during the third encounter with Mercury on 16 March 1975. The spacecraft passed close to and nearly over the north pole of the planet. The coordinate system employed has the positive x_{ME} axis pointing from Mercury toward the Sun, the positive y_{ME} axis parallel to the ecliptic plane and pointing in the opposite direction to the heliocentric motion of Mercury, and z_{ME} completing the system. The location of the magnetopause crossings are indicated by "MP". The relative positions of the magnetopause and bow shock, scaled from Earth, are projected onto the x_{ME}–y_{ME} plane. Forty-two measurements of orthogonal components of the magnetic field were made each second; samples for each component were averaged over 6 seconds, and the resultant vector average was used in the display and in the analyses (Connerney and Ness 1988).

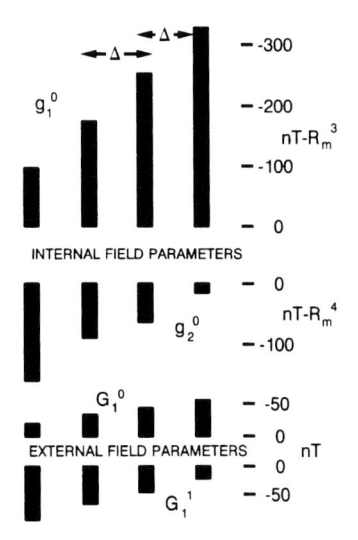

Figure 2 Correlated variations of the derived internal g_1^0 and g_2^0 Gaussian harmonic coefficients with the external terms G_1^0 and G_1^1 for Mercury. This illustrates how the paucity and location of data limits the determination of the low-order internal terms, as they and the external terms "masquerade" or "alias" each other in the least-squares analysis fitting model fields to observations (Connerney and Ness 1988).

Thermal models of Mercury's evolution predict too thin a sub-Curie-point shell of permanently magnetized material, which would be sufficient to carry the requisite terrestrial-like magnetization. Thus, the prevailing view has been that the planet's magnetic field is due to an active dynamo in the interior (Ness 1979a,b). However, with the discovery of intense crustal magnetic fields at Mars and the implied large magnetizations, it may be necessary to revise this view when new data are obtained and analyzed at Mercury.

Mariner 10 was the first flight of the dual magnetometer method which was developed to eliminate, while in-flight, time-varying magnetic fields generated by the spacecraft itself (Ness et al. 1971, Lepping and Ness 1978). The dual magnetometer method solved a long-standing problem with JPL's policy of not designing or building minimally magnetic spacecraft, unlike the Goddard Space Flight Center and the Ames Research Center. This was and continues to be especially important in the study of weak interplanetary magnetic fields.

MARS

Initial studies in the 1960s and 1970s by US and Soviet flyby spacecraft were able only to place upper limits on the global magnetic field (Ness 1979a). This is because they did not pass close enough to the planet to detect anything other than the detached bow shock wave. From the location of that feature it was possible to estimate the equivalent magnetic obstacle to solar wind flow by comparison with Earth. Mariner 4 set the first upper limit on the dipole field in 1964 (Smith et al. 1965). Unfortunately, there were no magnetometers selected for flight on board the twin US missions of Mariner 6 and 7, Mariner 8 and 9, and Viking 1 and 2, which flew past Mars or were placed into orbit around it in the 1960s and 1970s.

In the 1980s, the USSR began an ambitious Mars program named Phobos to study Mars and its moon Phobos. Only one of the two launched Phobos spacecraft achieved orbit. From that the lowest limit yet set for the global magnetic field was derived (2×10^{22} G cm^3) from analysis and interpretation of the structure, size, and characteristics of the magnetotail observed at the distance of the moon's orbit.

Following the failure of the US Mars Observer (MO) in 1993, the USA followed with the successful launch of Mars Global Surveyor (MGS) in 1996. It was placed into a highly elliptical Martian orbit in 1997. During the ensuing two years, aerobraking was used to achieve the final polar and Sun-synchronous circular mapping orbit at an altitude of ~400 km and at the local solar time of 0200–1400. During the aerobraking periapsis passes, MGS passed only 101–120 km from the surface and surveyed all latitudes from the north to the south pole. This was a very serendipitous aspect of the MGS mission which would not have occurred had Mars Observer been successful! This is so because the mapping orbit altitude of MO and MGS was planned to be 350–400 km, and the launch vehicle capability for MO was such that a direct injection into mapping orbit was planned. This eliminated the need for any aerobraking maneuvers for MO.

Data from the magnetic field experiment (Acuña et al. 1998, 1999; Connerney et al. 1999, Ness et al. 1999) showed that the crust of Mars is intensely magnetized, mainly in the southern hemisphere. This immediately implies that there was an active Martian dynamo in the planet's early history. There appears to be no significant global magnetic field discernable in the data at present, so the dynamo must have ceased to be active. Exactly when it died may eventually be determined. On the basis of our present-day understanding of the observed hemispherical contrast in the cratering record on the surface of Mars, it appears that the dynamo decayed shortly after the formation of the planet. The much older southern hemisphere is the location of the majority of the remanent crustal fields.

Interpretation of MGS data has lowered the equivalent dipole moment of Mars from the pre-MGS value by an order of magnitude to 2×10^{21} G cm^3. This corresponds to an equatorial field of less than 6 nT. There are readily evident patterns of lineations in the southern hemisphere which suggest the possibility of a process similar to ocean-floor spreading on Earth. But, as is always the case with such magnetic field data (and gravity data, too), there is no way to determine the distribution of magnetic sources below the surface.

Very large Martian fields were detected during periapsis passes during MGS's aerobraking phase, and this permitted detailed study of the magnetic lineations. The task of modeling the location and characteristics of the subsurface sources is work in progress. As at Earth, when the magnetic structure of the crustal fields is resolved and interpreted, it may well provide a history of the evolution of the planet. The low-altitude aerobraking periapis passes reveal fields as large as 1500 nT at the spacecraft's altitude. This is much larger than the magnitude of terrestrial magnetic anomalies observed by spacecraft at the same distance from the surface of Earth. The implication of these large values is that there is a massive amount of intensely magnetized material in the crust.

Figure 3 presents the results of an initial quantitative study of the magnetic lineations in which 20 infinitely long 30 km thick strips of 200 km width are assumed to exist in the subsurface. The individual magnetization values are represented by the tabular bar graphs in the middle of the figure (Connerney et al. 1999). The fit of observations to the model for the two field components perpendicular to the direction of lineation is quite good.

These exciting MGS data are being corrected for the contamination associated with spacecraft-generated magnetic fields. These range between 2 and 10 nT, and are generated primarily by the moving high-gain antenna and the associated TWT electronics as well as the solar arrays. On MGS there was no long boom to place the magnetometer sensors at remote distances from the spacecraft to reduce the contamination level.

The corrected MGS magnetic field data has been published and made available to the scientific community via the several NASA-sponsored archives (Acuña et al. 2001). The most puzzling aspect of these results is that the intensity of magnetization of the Martian crustal materials is inferred to be as high as 20–60 A^{-1}, many times larger than the most intensely magnetized Earth rocks. Since the exact chemistry of Martian crustal material is unknown at present, the answer to this puzzle may not be found until samples are returned (see Connerney et al. 2001).

JUPITER

Jupiter's non-thermal microwave radio emissions were discovered in 1955. They were interpreted as indicating the presence of a global magnetic field at Jupiter, and generated by trapped and precipitating electrons in Jupiter's radiation belts. Careful study of the polarization, frequency, and time-variable characteristics yielded good estimates of an equivalent dipolar magnetic field of Jupiter. *In situ* spacecraft observations since 1974 have considerably refined our knowledge of the Jovian main magnetic field and also the electrical currents flowing through the Jovian magnetosphere, which significantly distort the field in the magnetosphere close to the planet.

The most recent quantitative model of the main magnetic field of Jupiter is that developed by Connerney et al. (1998). Referred to as the VIP-4 model (Figure 4), it is based on *in situ* observations by Voyagers 1 and 2 and Pioneer 11 and the location of the Io associated fluxtube footprint. The latter parameter confines the geometry of the field lines threading the moon Io at $5.905 R_J$ to the trace of

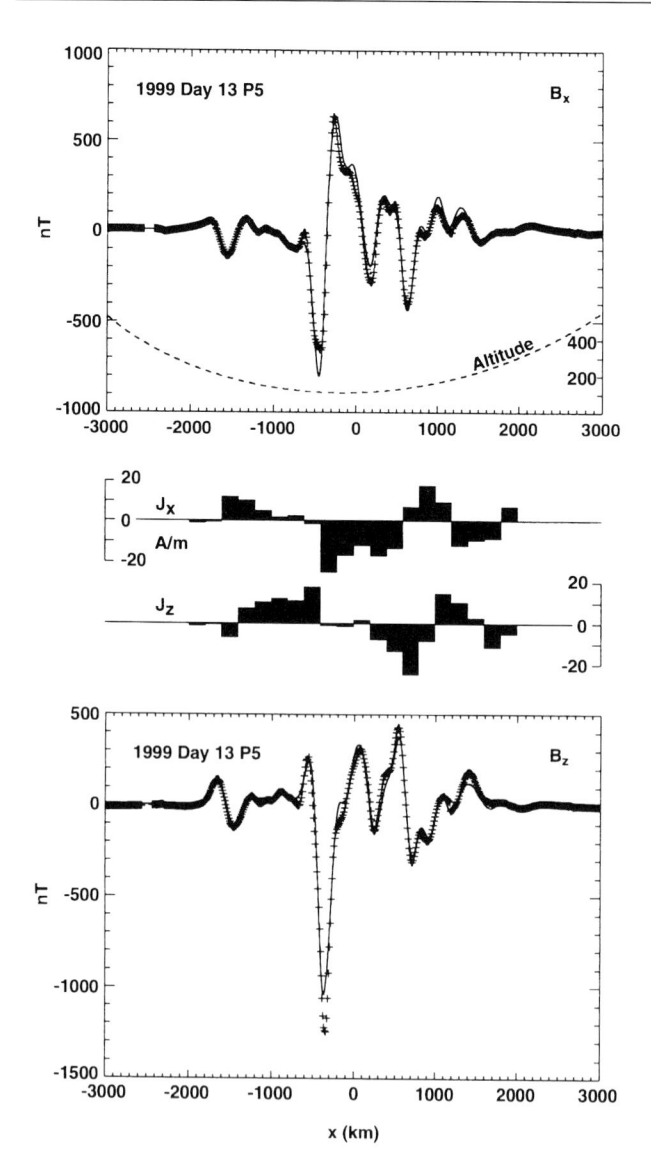

Figure 3 Mars Global Surveyor magnetic field data gathered during its passage over the magnetic lineations observed in the heavily cratered, older, southern hemisphere of Mars. The data are interpreted through a model in which a set of 20 thin two-dimensional plates represent the crustal magnetized regions, each of 30 km depth and 200 km width. The coordinate system is such that x is perpendicular to the assumed y-axis trend of the lineations, and the z axis is parallel to the local zenith. Derived parameters are the simultaneously determined J_x and J_z magnetization per unit volume (A m^{-1}) values required to provide the fit to the observations (+). The altitude of MGS during the overflight is shown in upper panel. The latitude and longitude of periapsis is 60°S, 180°W (Connerney et al. 1999).

the infrared aurora observed from the NASA Infrared Telescope Facility (IRTF) on Mauna Kea. The number on each isointensity line in Figure 4 indicates the upper limit

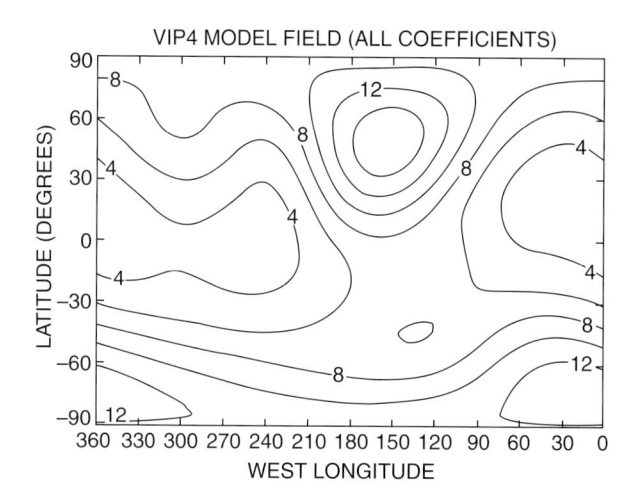

Figure 4 An isointensity Mercator plot of the Jovian magnetic field on surface of planet, based on the VIP-4 model of the field, which incorporates data obtained by Voyagers 1 and 2 and by Pioneer 11, as well Hubble Space Telescope observations of the Io fluxtube footprint locations in the polar regions (Connerney et al. 1998). The maximum field is slightly above 14 G near 55°N, 150°W.

of the order and degree of the internal field that has been determined with confidence.

For the giant planets we have introduced a shorthand nomenclature in which a letter specifies the degree of the highest multipole term determined with confidence, and a number indicates the highest degree of the multipole term employed in the analysis. Thus, for example, O6 refers to an octupolar representation based upon a sixth-degree (and order) analysis. An earlier O6 model was based upon the 1975 observations by Pioneer 11 and 1979 observations by the Voyager 1 spacecraft. No evidence for any secular change in the main magnetic field of Jupiter (the dipole term) was deduced during the interval between Pioneer 11 in 1975 and the Voyager observations in 1979. The 1992 encounter of Jupiter by the ESA-NASA spacecraft Ulysses was not at a close enough distance to contribute meaningfully to our knowledge of any secular change of the dipole moment of the planetary field.

The maximum field in the northern hemisphere is estimated from non-thermal radio emissions to be slightly above 14 G. This is in good agreement with the observed maximum frequency of microwave radio emissions, 42 MHz. The equatorial fields range from 4 to 8 G. This variation in field intensity and lack of polar symmetry (10.4 G. in the south) is due in large part to the presence of important higher-order multipole terms. The geometry of the Jovian and terrestrial fields appears similar when appropriately normalized, primarily because of the similar tilt of the axes of their magnetic fields (9.6° and 11.4°) from their rotation axes.

The US Galileo spacecraft has been contributing to studies of the inner magnetosphere since 1995. Data have been used primarily to model the time variations of the Jovian current sheet. Due to the hazards of radiation damage associated with the intense radiation belts of Jupiter, after its first periapsis at $5.905R_J$, the periapsis of Galileo was raised to $15R_J$. Following the end of the primary mission, the extended mission has included several close passes of the Galilean moons: Io (at $5.905R_J$), Europa (at $9.395R_J$), Ganymede (at $14.99R_J$), and Callisto (at $26.37R_J$). The ambient Jovian field at these moons was measured to be 1835, 420, 120, and 35 nT.

SATURN

Saturn possesses an unusual magnetic field. It was discovered by Pioneer 11 in 1979 and confirmed and further investigated by Voyagers 1 and 2 in 1980 and 1981. Saturn's magnetic field appears to be symmetric with respect to the rotation axis (Acuña et al. 1983, Connerney et al. 1982). The derived quantitative model of the magnetic field, Z_3, includes only the axially symmetric dipole, quadrupole, and octupole terms (g_1^0, g_2^0, g_3^0). Figure 5 shows the latitude variation in various models of the Saturnian main field.

There is an interesting technique for validating global field models which employs observations of the absorption of trapped radiation by natural moons. Moons behave like occulting disks and create *holes* in the radiation belts. Their observed locations can be used to test predicted locations of such holes using global field models. This has been done successfully at Saturn.

The enigma of Saturn is that its auroral radio emissions vary periodically. These time variations can be used to determine the rotation period of the interior of the planet, since there is no visible solid surface; the rotation periods of the other giant planets are determined by similar analyses. Detailed examination of the source location of these modulated radio emissions indicates that they are generated in a region close to the northern rotation pole. Thus, this Saturnian kilometric radiation is evidence for the departure of the magnetic field in the polar region, close to the planet, from axial symmetry. The ESA–NASA Cassini mission, successfully launched in 1997 (to arrive at Saturn in 2004) should provide the necessary observational data to improve our knowledge of Saturn's magnetic field and resolve this enigma.

URANUS

The only measurements of the magnetic field and magnetosphere of Uranus thus far were conducted by the discovery mission of Voyager 2 in January 1986 (Connerney et al. 1987, Ness et al. 1986, 1991). It was with considerable surprise that a global magnetic field was detected, as well as an associated and well-developed magnetosphere and radiation belt structure. This is because the planet Uranus is not exothermic – it does not radiate as much energy as it receives from the Sun. Thus, it was thought that there were no internal energy sources to drive and power a dynamo.

An even greater surprise at Uranus was that the magnetic dipole axis is offset from the rotation axis by 58.6°. It was initially speculated that this might be related to the large obliquity of the planet, 98.2°. The observation in 1989 of a similarly highly oblique magnetic field at Neptune meant that this hypothesis was not well founded. The combined obliquity and large tilt of the planetary magnetic field means that at times in its orbit the magnetosphere configuration of Uranus will periodically be pole-on (Table 2 and Figure 6). A pole-on magnetosphere would have a unique magnetic tail structure with a cylindrically shaped neutral sheet imbedded within the overall tail, not a planar feature crossing the entire tail. Another important feature of the magnetic field of Uranus is that its magnetic center is spatially displaced by a significant amount ($0.31R_U$) from the center of the planet. This means that there are large, higher-order multipole moments at Uranus (Table 3).

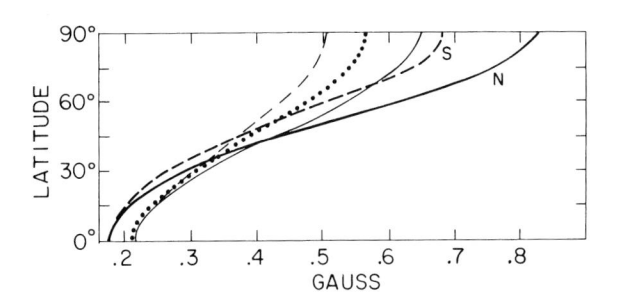

Figure 5 Variations with latitude of axisymmetric models of the Saturnian internal magnetic field (in Gauss) on surface of planet with an assumed flattening of 10.6%. The three models are: centered dipole (light dashed line) with field of 0.21 G at the equator; axially offset dipole (OD, dotted and light solid lines); and zonal harmonics only model (Z_3, heavy dashed and heavy solid lines), including axially symmetric dipole, quadrupole, and octupole terms. The Z_3 model is not equivalent to a single offset dipole. N and S represent the northern and southern hemispheres (Connerney et al. 1982, Acuña et al. 1983).

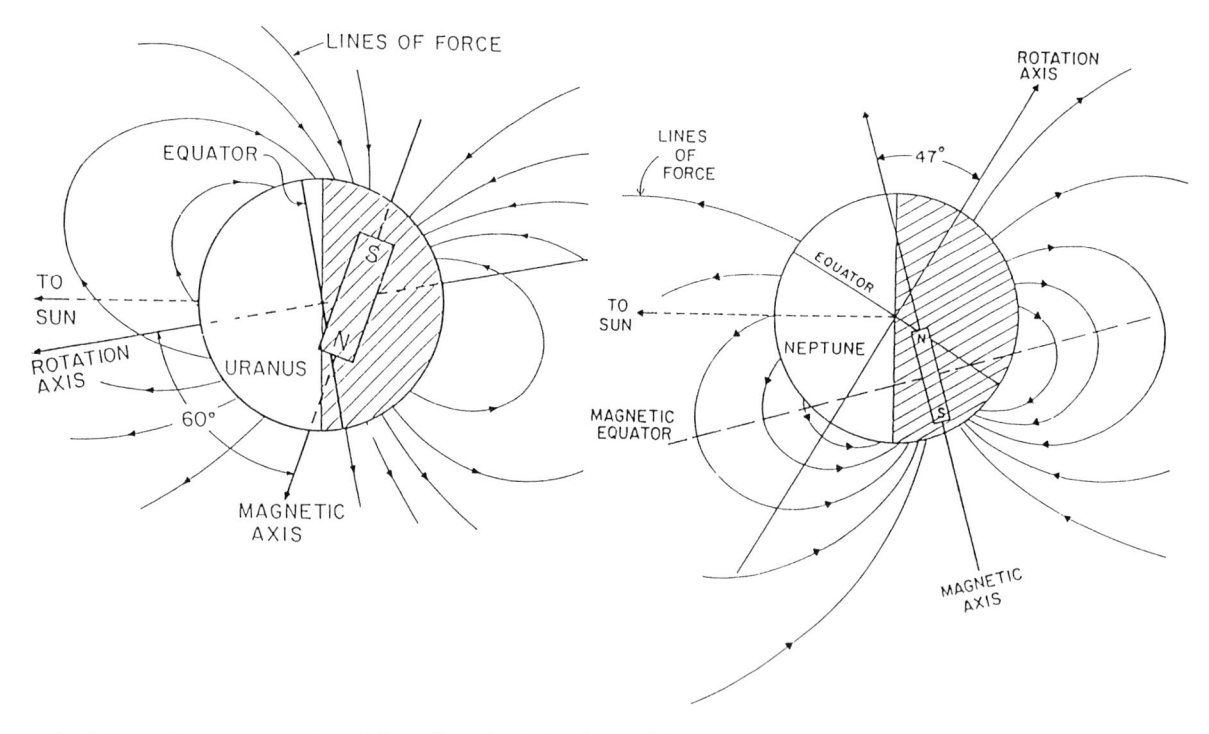

Figure 6 Schematic representation of the offset tilted dipole model magnetic fields of Uranus and Neptune. Note the large spatial offsets of the magnetic centers from the center of each planet as well as the large angular offsets of the magnetic axes from the rotational axes (Ness 1994).

At Jupiter and Saturn, the transit time for a flyby spacecraft while within their large magnetospheres is several to many times the rotation period of the parent planet. Thus, in planet-centered coordinates, the flyby trajectories appear to spiral radially inward and outward. This yields broad radial and longitudinal coverage and, for certain trajectories, a wide latitude range. At Uranus the flyby was more limited in longitude although it did have a wide latitude extent. These combine in the quantitative analysis to limit the ability to determine with confidence any multipole moments of order higher than 3. At present, the quantitative models for the Uranian field include dipole, quadrupole, and with less certainty, octupole terms. Figure 7 (upper panel) presents the Q3 representation of the Uranian field on its surface. Figure 7 (lower panel) presents the O3 model. It is seen that the magnetic field intensities, extrapolated to the surface of the planet, are rather similar, as are the locations of the auroral zones, for these two models.

The auroral zones at Uranus are displaced far from the polar regions of the planet. One of the "polar" regions is essentially equatorial. Unfortunately, due to the geocentric bias toward scheduling Voyager 2 UV instrument searches for aurora near the rotational polar regions in advance of the encounter, there are very sparse observations of actual aurora at Uranus.

NEPTUNE

Following the very surprising results at Uranus, it was with confidence that a global magnetic field was anticipated at Neptune, because this planet is exothermic (Curtis and Ness 1988). An even bigger surprise of the 1989 encounter by Voyager 2 was that Neptune also possessed a global magnetic field for which the dipole magnetic axis is offset far from the rotational axis (46.8°) (Ness et al. 1989). Moreover, its magnetic center is also offset, but even more so, from the center of the planet: 55% at Neptune, 31% at Uranus (see also Holme and Bloxham 1996).

Figure 8 presents the O8 model of the Neptunian magnetic field, as well as the location of the theoretically computed dip equator and auroral zones (Acuña et al. 1993, Ness et al. 1995). Both Uranus and Neptune are remarkable in that the aurora are expected to occur far from the polar regions. But as at Uranus, the Voyager 2 pre-encounter planning targeted its UV auroral observations for the polar regions. Thus, very limited data were obtained on auroral phenomena.

Associated with a large angular tilt between the magnetic and rotation axes at Neptune, its magnetosphere will, like Uranus, periodically present a pole-on configuration to the solar wind (Voigt and Ness 1990). During the 1990s,

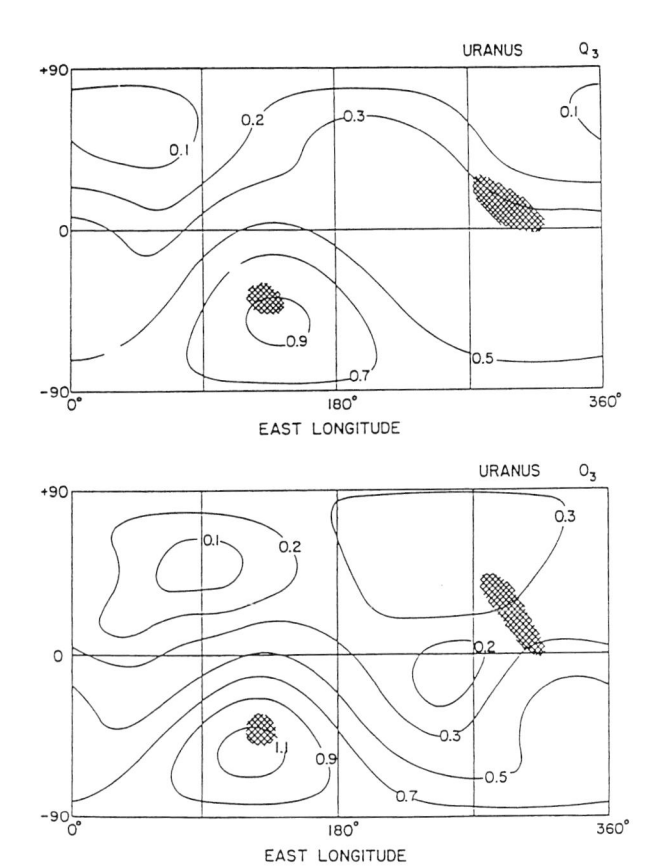

Figure 7 Isointensity Mercator plots of the surface magnetic fields at Uranus. The models employ terms up to and including quadrupole Q3 or octupole O3, although the degree and order of the terms used in the data fitting extend to 3 for both models. The hatched regions represent the location of the auroral regions where high magnetic latitude field lines intersect the surface of the planets. Note the low planetary latitude of the Uranian auroral oval (Connerney et al. 1987).

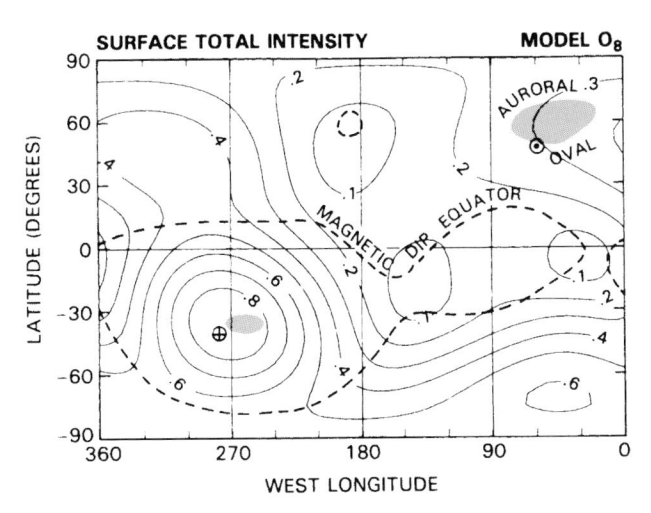

Figure 8 An isointensity Mercator plot of the surface magnetic field of Neptune, using the O8 representation with the location of the magnetic dip equators illustrated by dashed lines. The circled dot and cross represent the intersections of the model dipole magnetic axis of the planet with its surface (Acuña et al. 1993, Connerney et al. 1991).

during each planetary rotation, its magnetosphere alternated from being Earth-like to pole-on because of the planet's obliquity and the angular offset of its magnetic axis.

The time of the Neptune encounter was especially serendipitous. Although determined entirely by the desire to achieve a close post-Neptune Triton encounter, the spacecraft entered the magnetosphere when the polar cusp region of the solar wind interaction was pointed Sunward. This is the only pole-on magnetosphere configuration for which *in situ* observations have been made (Lepping et al. 1992).

MAGNETIC FIELDS OF MOONS

The Moon has been magnetically studied by a number of spacecraft since the initial measurements by the Soviet Luna 2. The results from Luna 10 suggested the presence of an intrinsic magnetic field, but this interpretation was shown to be in error by Explorer 35 in 1968 (Ness 1969). Luna 10 had actually made its observation near the time of full moon when the Moon was in the Earth's extended magnetotail. In the US Apollo missions of the late 1960s and early 1970s placed sub-satellites into close but low-inclination lunar orbits and detected several near-equatorial regions with localized surface magnetic fields. The US Lunar Prospector, in a close polar orbit, used a magnetometer and a low-energy electron detector to identify regions of crustal magnetization from reflected electrons. The term *mini-magnetosphere* was introduced to describe the geometry of these magnetic field regions (Lin et al. 1998).

Beginning in 1995 with its injection into Jovian orbit, the Galileo spacecraft made many close flybys of the four Galilean moons. These have led to interpretations indicating the possible existence of intrinsic global magnetic fields at some of these moons. Certain of these initial interpretations have not proved robust. Early interpretations ignored the process of electromagnetic induction in the interior of the moons; induction is now invoked to explain some of the observations without the need to include an intrinsic field. The frail nature of early interpretations is also due to the complex interaction of these moons with the co-rotating Jovian magnetosphere plasma. This leads to disturbances including Alfvén wings, which trail away from the moon along the distorted Jovian field line (Neubauer 1998b).

An internal magnetic field has been discovered at Ganymede (Gurnett et al. 1996, Kivelson et al. 1996, 1997a, Williams et al. 1997, 1998). The situation here is complex, with a moon's magnetosphere within a planetary magnetosphere controlling the plasma and energetic particle environment.

The study of the putative internal field at Io has continued, with data from many flybys (Kivelson et al. 1996, Khurana et al. 1997a, 1998a; Neubauer 1998a). It appears that by taking proper account of the interactions of its atmosphere and ionosphere with the co-rotating Jovian magnetosphere, observations can be explained without an internal field (Frank et al. 1996, Neubauer 1998a, Saur et al. 2002). The most convincing evidence of the important and dominant role played by interactions of the co-rotating Jovian magnetosphere with the atmosphere and ionosphere of Io has been the Hubble Space Telescope (HST) observations of UV aurora in the equatorial regions of Io. These data have been interpreted and modeled by Saur et al. (2000) as demonstrating that there is no need to invoke any appreciable intrinsic global field of Io to understand all the plasma, field, energetic particles, and UV observations by Galileo and HST to date.

The moon Europa is now thought to possess an induced magnetic field (Khurana et al. 1998b, Neubauer 1999) and not to be internally magnetized, as was originally considered (Kivelson et al. 1997b). The electromagnetic induction process is postulated to occur in an electrically conducting shell within the interior wherein the electolytic properties of dissolved salts in an aqueous solution provide the necessary electrically conducting material. The rotation of the non-axially symmetric Jovian field provides the time-varying inducing field. Callisto, like Europa, is now considered to be intrinsically non-magnetic but to possess a conducting interior so that the magnetic field signatures detected by Galileo are due to an induction process (Khurana et al. 1997b, 1998b; Kivelson et al. 1999).

All four of these moons are very complex objects (Stevenson et al. 1986, Showman and Malhotra 1999) with atmospheres and intrinsic or induced magnetic fields interacting with a dynamic plasma and field environment. It will take future missions to clarify and resolve the present-day ambiguities about the origins of the magnetic fields of Ganymede and Europa.

COMPARISONS OF PLANETARY MAGNETIC FIELDS

Initial *in situ* studies of the magnetic fields of all the planets, except for Pluto, are now complete after 40 years. Schulz and Paulikas (1990) have shown that the magnetic fields of the planets Earth, Jupiter, Uranus, and Neptune demonstrate a unique equi-partition of global magnetic energy, based upon the spectrum of the harmonic coefficients representing the internal magnetic fields. From this they have deduced that the region in which the magnetic field is generated has an effective normalized radius at Earth, Jupiter, Uranus, and Neptune of 0.432, 0.756, 0.464, and 0.662, respectively. These values are tantalizingly close to those for the radii of the conducting regions developed independently on the basis of plausible physical models of composition, rotation rate, and body figure. It will be most interesting to follow future developments of this concept and its refinement along with the investigations of the symmetry properties of the planets, as conducted by Rädler and Ness (1990).

SUMMARY AND QUESTIONS FOR THE FUTURE STUDY

During the last four decades of the twentieth century, exploration of the planets and their moons by spacecraft has led to discoveries and knowledge of the quantitative characteristics of the magnetic fields of the planets Mercury, Venus, Mars, Jupiter, Saturn, Uranus, and Neptune. Except at Jupiter, all of these observations represented significant discoveries. At Jupiter, discoveries of the magnetic fields of intrinsic or induced origins at the Galilean moons have also proved to be surprising and complex. Earth's Moon also presents an interesting puzzle, with the recently identified antipodal relationship of craters and crustal magnetized regions.

At certain positions in their orbits, the magnetotails of Uranus and Neptune have unique configurations. While the planets Earth, Jupiter, and Saturn possess "standard" magnetotail structures, with bipolar lobes separated by a transverse plasma sheet, Uranus and Neptune possess cylindrical plasma sheets in their pole-on configurations. This is an interesting problem for future study in planetary magnetospheres.

Additional measurements will be important in establishing the characteristics of secular variations of intrinsic planetary magnetic fields. At Mercury, a close-orbiting spacecraft will be required to investigate the magnetic field quantitatively, because the planet itself occupies a large volume fraction of its magnetosphere. Departures from axial symmetry of the Saturnian magnetic field are present but not yet quantitatively described. This is certainly a task for the Cassini mission, due to encounter Saturn in 2004. Finally, the recent discovery of the unique localized intense Martian magnetic fields of crustal origin will be studied for the insight they will provide into the history of evolution of this planet.

The existence of this new body of experimental observations on the magnetic fields of self-gravitating, rapidly rotating and condensed objects in the Solar System is having an important impact on our knowledge of planetary interiors and our comparisons with other objects in the Universe.

Acknowledgments

I appreciate the contributions to the research reported upon here by my colleagues at the NASA Goddard Space Flight Center: Mario H. Acuña, Kenneth W. Behannon, John E.P. Connerney, and Ronald P. Lepping; and at the University of Köln: Fritz M. Neubauer.

REFERENCES

Acuña, M.H., Connerney, J.E.P. and Ness, N.F. (1983). The Z_3 zonal harmonic model of Saturn's magnetic field: Analyses and implications. *Journal of Geophysics Research*, **88**, 8771–8778.

Acuña, M.H., Connerney, J.E.P. and Ness, N.F. (1993). Neptune's magnetic field: Calculation of field geometric invariants derived from the 18E1 GSFC/BRI model. *Journal of Geophysics Research*, **98**, 11275–11284.

Acuña, M.H., Connerney, J.E.P., Ness, N.F., Lin, R.P., Mitchell, D., Carlson, C.W., McFadden, J., Anderson, K.A., Rème, H., Mazelle, C., Vignes, D., Wasilewski, P. and Cloutier, P. (1999). Global distribution of crustal magnetization discovered by the Mars Global Surveyor MAG/ER experiment. *Science*, **284**, 794–798.

Acuña, M.H., Connerney, J.E.P., Wasilewski, P., Lin, R.P., Anderson, K.A., Carlson, C.W., McFadden, J., Curtis, D.W., Mitchell, D., Rème, H., Mazelle, C., Sauvaud, J.A., d'Uston, C., Cros, A., Médale, J.L., Bauer, S.J., Cloutier, P., Mayhew, M., Winterhalter, D. and Ness, N.F. (1998). Magnetic field and plasma observations at Mars: Preliminary results of the Mars Global Surveyor mission. *Science*, **279**, 1676–1680.

Acuña, M.H., Connerney, J.E.P., Wasilewski, P., Lin, R.P., Mitchell, D., Anderson, K.A., Carlson, C.W., McFadden, J., Rème, H., Mazelle, C., Vignes, D., Bauer, S.J., Cloutier, P. and Ness, N.F. (2001). Magnetic field of Mars: Summary of results from the aerobraking and mapping orbits. *Journal of Geophysics Research*, **106**, 23403–23417.

Backus, G., Parker, R. and Constable, C. (1996). *Foundations of Geomagnetism*. Cambridge University Press.

Busse, F. (1978). Theory of planetary dynamos. In C.F. Kennel, L.J. Lanzerotti and E.N. Parker (eds), *Solar System Plasma Physics – A 20th Anniversary Review*, North-Holland, Amsterdam, pp. 293–318.

Cain, J C. (1995). Main field and secular variation. *Reviews of Geophysics*, Supp., **33**, 145–152.

Campbell, W.H. (1997). *Introduction to Geomagnetic Fields*. Cambridge University Press.

Connerney, J.E.P. (1981). The magnetic field of Jupiter: A generalized inverse approach. *Journal of Geophysics Research*, **86**, 7679–7693.

Connerney, J.E.P., Acuña, M.H. and Ness, N.F. (1987). The magnetic field of Uranus. *Journal of Geophysics Research*, **92**, 15329–15336.

Connerney, J.E.P., Acuña, M.H. and Ness, N.F. (1991). The magnetic field of Neptune. *Journal of Geophysics Research*, **96**, 19023–19042.

Connerney, J.E.P., Acuña, M.H., Ness, N.F. and Satoh, T. (1998). New models of Jupiter's magnetic field constrained by the Io flux tube footprint. *Journal of Geophysics Research*, **103**, A6, 11929–11939.

Connerney, J.E.P., Acuña, M.H., Wasilewski, P., Ness, N.F., Rème, H., Mazelle, C., Vignes, D., Lin, R.P., Mitchell, D.L. and Cloutier, P. (1999). Magnetic lineations in the ancient crust of Mars. *Science*, **284**, 790–793.

Connerney, J.E.P., Acuña, M.H., Wasilewski, P.J., Kletetschka, G., Ness, N.F., Rème, H., Lin, R.P. and Mitchell, D.L. (2001). The global magnetic field of Mars and implications for crustal evolution. *Geophysical Research Letters*, **28**, 4015–4018.

Connerney, J.E.P. and Ness, N.F. (1988). Mercury's magnetic field and interior. In M. Matthews, C. Chapman and F. Vilas (eds), *Mercury*, University of Arizona Press, Tuscon, A.Z., pp. 494–513.

Connerney, J.E.P., Ness, N.F. and Acuña, M.H. (1982). Zonal harmonic model of Saturn's magnetic field from Voyager 1 and 2 observations. *Nature*, **298**, 44–46.

Curtis, S.A. and Ness, N.F. (1988). Remanent magnetism at Mars. *Geophysics Research Letters*, **15**, 737–739.

Dolginov, Sh.Sh., Yeroshenko, Ye.G., Zhuzgov, L.N., Sharova, V.A., Vnuchkov, C.A., Okulesski, B.A., Bazileuski, A.T., Vanyan, L.L., Egorov, I.V. and Fainberg, E.B. (1975). Magnetism and electrical conductivity of the Moon from Lunokhod 2 data. In *Cosmochemistry of Moon and Planets*, pp. 314–322.

Dyal, P., Parkin, C.W. and Cassen, P. (1972). Surface magnetometer experiments: Interval lunar properties and lunar surface interactions with the solar plasma. In Proceedings of the Third Lunar Science Conference, *Geochemica et Cosmochemica Acta*, Suppl. **3**, 2287–2307.

Dyal, P., Parkin, C.W. and Daily, W.D. (1974). Magnetism and the interior of the Moon. *Reviews of Geophysics and Space Physics*, **12**, 568–591.

Frank, L.A., Paterson, W.R., Ackerson, K.L., Vasyliunas, V.M., Coroniti, F.V. and Bolton, S.J. (1996). Plasma observations at Io with the Galileo Spacecraft. *Science*, **274**, 394–395.

Gailitis, A., Lielausis, O., Dement'ev, S., Platacis, E., Ceifersons, A., Gerbeth, G., Gundrum, T., Stefani, F., Christen, M., Hänel, H. and Will, G. (2000). Detection of a flow induced magnetic field eigenmode in the Riga Dynamo Facility. *Physics Review Letters*, **84**, 4365–4368.

Glatzmaier, G.A., Coe, R.S., Hongre, L. and Roberts, P.H. (1999). The role of the Earth's mantle in controlling the frequency of geomagnetic reversals. *Nature*, **401**, 885–890.

Glatzmaier, G.A. and Roberts, P.H. (1995). A three-dimensional self-consistent computer simulation of a geomagnetic field reversal. *Nature*, **377**, 203–209.

Glatzmaier, G.A. and Roberts, P.H. (1996). Rotation and magnetism of Earth's inner core. *Science*, **274**, 1887–1891.

Gurnett, D.A., Kurth, W.S., Roux, A., Bolton, S.J. and Kennel, C.F. (1996). Evidence for a magnetosphere at Ganymede from plasma-wave observations by the Galileo spacecraft. *Nature*, **384**, 535–537.

Halekas, J.S., Mitchell, D.L., Lin, R.P., Frey, S., Hood, L.L., Acuña, M.H. and Binder, A.B. (2001). Mapping of crustal magnetic anomalies on the lunar near side by the Lunar Prospector electron reflectometer. *Journal of Geophysics Research*, **106**(E11), 27841–27852.

Holme, R. and Bloxham, J. (1996). The magnetic field of Uranus and Neptune: Methods and models. *Journal of Geophysics Research*, **101**, 2177–2200.

Hood, L.L., Russell, C. and Coleman, P., Jr. (1981). Contour maps of lunar remanent magnetic fields. *Journal of Geophysics Research*, **86**, 1055–1069.

Jacobs, J.A. (1994). *Reversals of Earth's Magnetic Field*, 2nd edn, Cambridge University Press.

Khurana, K.K., Kivelson, M.G. and Russell, C.T. (1997a). Interaction of Io with its torus: Does Io have an internal magnetic field? *Geophysics Research Letters* **24**(19), 2391–2394.

Khurana, K.K., Kivelson, M.G., Russell, C.T., Walker R.J. and Southwood, D.J. (1997b). Absence of an internal magnetic field at Callisto. *Nature*, **387**, 262–264.

Khurana, K.K., Kivelson, M.G., Stevenson, D.J., Schubert, G., Russell, C.T., Walker, R.J. and Polanskey, C. (1998b). Induced magnetic fields as evidence for subsurface oceans in Europa and Callisto. *Nature*, **395**, 777–780.

Khurana, K.K., Linker, J.A., Kivelson, M.G. and Russell, C.T. (1998a). Reply. *Geophysics Research Letters* **25**(13), 2351–2352.

Kivelson, M.G., Khurana, K.K., Coroniti, F.V., Joy, S., Russell, C.T., Walker, R.J., Warnecke, J., Bennett, L. and Polanskey, C. (1997a). The magnetic field and magnetosphere of Ganymede. *Geophysics Research Letters*, **24**(17), 2155–2158.

Kivelson, M.G., Khurana, K.K., Joy, S., Russell, C.T., Southwood, D.J., Walker, R.J. and Polansk, C. (1997b). Europa's magnetic signature: Report from Galileo's pass on 19 December 1996. *Science*, **276**, 1239–1241.

Kivelson, M.G., Khurana, K.K., Russell, C.T., Walker, R.J., Warnecke, J., Coroniti, F.V., Polanskey, C., Southwood, D.J. and Schubert, G. (1996). Discovery of Ganymede's magnetic field by the Galileo spacecraft. *Nature*, **384**, 537–541.

Kivelson, M.G., Khurana, K.K., Stevenson, D.J., Bennett, L., Joy, S., Russell, C.T., Walker, R.J., Zimmer C. and Polanskey, C.L. (1999). Europa and Callisto: Induced or intrinsic fields in a periodically varying plasma environment. *Journal of Geophysics Research*, **104**, 4609–4625.

Lammer, H. and Bauer, S.J. (1997). Mercury's exosphere: Origin of surface sputtering and implications. *Planetary Space Science*, **45**(1), 73–79.

Lepping, R.P., Burlaga, L.F., Lazarus, A.J., Vasyliunas, V.M., Szabo, A., Steinberg, J., Ness, N.F. and Krimigis, S.M. (1992). Neptune's polar cusp region: observations and magnetic field analysis. *Journal of Geophysics Research*, **97**, 8135–8144.

Lepping, R.P. and Ness, N.F. (1978). An extension of the dual magnetometer method for use on a dual spinning spacecraft. *Journal of Geophysics Research*, **83**, 2211–2215.

Lin, R.P., Mitchell, D.L., Curtis, D.W., Anderson, K.A., Carlson, C.W., McFadden, J., Acuña, M.H., Hood, L.L. and Binder, A. (1998). Lunar surface magnetic fields and their interaction with the solar wind: Results from Lunar Prospector. *Science*, **281**, 1480–1484.

McElhinney, M.W. (1973). *Paleomagnetism and Plate Tectonics*. Cambridge University Press.

McLeod, M.G. (1996). Spatial and temporal power spectra of the geomagnetic field. *Journal of Geophysics Research*, **101**, 2745–2763.

Merrill, R.T., McElhinny, M.W. and McFadden, P.L. (1996). *The Magnetic Field of the Earth*, 2nd edn, Academic Press, San Diego, CA.

Ness, N.F. (1969). Lunar Explorer 35. *Advances in Space Research*, **9**, 678.

Ness, N.F. (1970). Magnetometers for space research. *Space Science Reviews*, **11**, 111–222.

Ness, N.F. (1976). The magnetosphere of Mercury. In D.J. Williams (ed.), *Physics of Solar Planetary Environments*, American Geophysical Union, Washington, DC, pp. 933–946.

Ness, N.F. (1978). Mercury: Magnetic field and interior. *Space Science Reviews*, **21**, 527–554.

Ness, N.F. (1979a). The magnetic fields of Mercury, Mars and Moon. *Annual Review Earth Planet Science* **7**, 249–288.

Ness, N.F. (1979b). The magnetosphere of Mercury. In C.F. Kennel, L.J. Lanzerotti and E.N. Parker, (eds), *Solar System Plasma Physics*, Vol. II, North-Holland, Amsterdam, pp. 285–286.

Ness, N.F. (1987). Magnetotail research: The early years. In T.Y. Lui (ed.), *Magnetotail Physics*, Johns Hopkins University Press, Baltimore, MD, pp. 11–22.

Ness, N.F. (1994). Intrinsic magnetic fields of the planets: Mercury to Neptune. *Philosophical Transactions of the Royal Society A*, **349**, 249–260.

Ness, N.F. (1996). Pioneering the swinging 1960s into the 1970s and 1980s. *Journal of Geophysics Research*, **101**, 10497–10509.

Ness, N.F., Acuna, M.H., Behannon, K.W., Buralga, L.F., Connerney, J.E.P., Lepping, R.P. and Neubauer, F.M. (1986). Magnetic fields at Uranus. *Science*, **233**, 85–89.

Ness, N.F., Acuña, M.H. and Connerney, J. (1995). Neptune's magnetic field and field geometric properties. In D.P. Cruikshank (ed.), *Neptune and Triton*, University of Arizona Press, Tucson, AZ, pp. 141–168.

Ness, N.F., Acuña, M.H., Connerney, J., Wasilewski, P., Mazelle, C., Sauvaud, J., Vignes, D., d'Uston, C., Rème, H., Lin, R., Mitchell, D.L., McFadden, J., Curtis, D., Cloutier, P. and Bauer, S.J. (1999). MGS magnetic fields and electron reflectometer investigation: Discovery of paleomagnetic fields due to crustal remanance. *Advances in Space Research*, **23**(11), 1879–1886.

Ness, N.F., Behannon, K.W., Lepping, R.P. and Schatten, K.H. (1971). Use of two magnetometers for magnetic field measurements on a spacecraft. *Journal of Geophysics Research*, **76**, 3564–3573.

Ness, N.F., Behannon, K.W., Lepping, R.P., Schatten, K.H. and Whang, Y.C. (1974). Magnetic field observations near Mercury: Preliminary results from Mariner 10. *Science*, **185**, 151–160.

Ness, N.F., Connerney, J.E.P., Lepping, R.P., Schulz, M. and Voigt, H. (1991). The magnetic field and magnetospheric configuration of Uranus. In J. Bergstralh, E.O. Miner and M.S. Matthews (eds), *Uranus*, University of Arizona Press, Tucson, AZ, pp. 739–779.

Ness, N.F., Neubauer, F.M., Acuña, M.H., Burlaga, L.F., Connerney, J.E.P. and Lepping, R.P. (1989). Magnetic fields at Neptune. *Science*, **246**, 1473–1478.

Neubauer, F.M. (1998a). Comment on "Interaction of Io with its torus: Does Io have an internal magnetic field?" by K.K. Khurana, M.G. Kivelson and C.T. Russell. *Geophysics Research Letters*, **25**(13), 2349.

Neubauer, F.M. (1998b). The sub-Alfvénic interaction of the Galilean satellites with the Jovian magnetosphere. *Journal of Geophysics Research*, **103**, 19843–19866.

Neubauer, F.M. (1999). Alfvén wings and electromagnetic induction in the interiors: Europa and Calisto. *Journal of Geophysics Research*, **104**, 28671–28684.

Piper, J.O.A. (1987). *Paleomagnetism and the Continental Crust*, Open University Press, Milton Keynes.

Proctor, M.R.E. and Gilbert, A.D. (1996). *Lectures on Solar and Planetary Dynamos*, Cambridge University Press, New York.

Rädler, K-H and Ness, N.F. (1990). The symmetry properties of planetary magnetic fields. *Journal of Geophysics Research*, **95**, 2311–2318.

Roberts, P.H. (1995). Dynamics of the core, geodynamo. *Reviews of Geophysics Suppl.* **33**, 443–450.

Saur, J., Neubauer, F.M., Strobel, D.F., Summers, M.E. (2002). Interpretation of Galileo's Io plasma and field observations: the J0, I24, I27 flybys, and close polar passes. *Journal of Geophysics Research*, in press.

Saur, J., Neubauer, F.M., Strobel, D.F., Summers, M.E. (2000). Io's ultraviolet aurora: Remote Sensing of Io's interaction. *Geophysics Research Letters*, **27**, 2893–2896.

Schulz, M. (1995). Planetary magnetospheres. *Earth, Moon and Planets*, **67**, 161–173.

Schulz, M. and Paulikas, G.A. (1990). Planetary magnetic fields: A comparative view. *Advances in Space Research*, **10**(1), 55–64.

Showman, A.P. and Malhotra, R. (1999). The Galilean satellites. *Science*, **286**, 77–84.

Simpson, J.A., Eraker, J.H., Lamport, J.E. and Walpole, P.H. (1974). Electrons and protons accelerated in Mercury's magnetic field. *Science*, **185**, 160–166.

Smith, E.J., Davis, L. Jr, Coleman, P.J. Jr and Jones, D.E. (1965). Magnetic field measurements near Mars. *Science*, **149**, 1241–1245.

Soward, A.M. (1992). Dynamo theory. *Advances in Space Research*, **12**(8), 257–263.

Stern, D.P. and Ness, N.F. (1982). Planetary magnetospheres, *Annual Review of Astronomy and Astrophysics*, **20**, 134–161.

Stevenson, D.J. (1983). Planetary magnetic fields. *Physics*, **46**, 555–620.

Stevenson, D.J., Harris, A.W. and Lunine, J.I. (1986). Origins of satellites. In J.A. Burns, and M.S. Matthews (eds), *Satellites*, University of Arizona Press, Tucson, AZ, pp. 39–88.

Vignes, D., Mazelle, C., Rème, H., Acuña, M.H., Connerney, J.E.P., Lin, R.P., Mitchell, D.L., Cloutier, P. and Ness, N.F. (2000). The solar wind interaction with Mars: Locations and shapes of the bow shock and the magnetic pile-up boundary from the observations of the MAG/ER experiment onboard Mars Global Surveyor. *Geophysics Research Letters*, **27**(1), 49–52.

Voigt, G-H, Behannon, K.W. and Ness, N.F. (1987). Magnetic field and current structures in the magnetosphere of Uranus. *Journal of Geophysics Research*, **92**, 15337–15346.

Voigt, G-H and Ness, N.F. (1990). The magnetosphere of Neptune: Its response to daily rotation. *Geophysics Research Letters* **17**(10), 1705–1708.

Williams, D.J., Mauk, B. and McEntire, R.W. (1998). Properties of Ganymede's magnetosphere as revealed by energetic particle observations. *Journal of Geophysics Research*, **103**, 17523–17534.

Williams, D.J., Mauk, B.H., McEntire, R.W., Roeof, E.C., Armstrong T.P., Wilken, B., Roederer, J.G., Krimigis, S.M., Fritz, T.A., Lanzerotti, L.J. and Murphy, N. (1997). Energetic particle signatures at Ganymede: Implications for Ganymede's magnetic field. *Geophysics Research Letters*, **24**(17), 2163–2166.

The Earth and its plasma environment

62

RUDOLF A. TREUMANN* AND MANFRED SCHOLER*

The magnetosphere as a plasma laboratory

INTRODUCTION

Most of the baryonic matter in the Universe is in the plasma state. This comes about when the temperature of the matter becomes so hot that the atoms spontaneously dissociate into positively charged ions and electrons. Because of charge conservation in this dissociation process, plasmas are usually quasineutral with equal numbers of negative and positive charges. In other words, the number densities, $n_{e,i}$, of the oppositely charged particles are equal, $n_e = Z_i n_i$, where the indices e, i indicate electrons and ions, and Z_i is the nuclear charge number. Since the overwhelming majority of baryonic matter is hydrogen, one usually has equal numbers of electrons and protons in the volume, $n_e = n_i$. Even when the fraction of the ionized component is small the ionized component shows a totally different behaviour than neutral matter. It is governed by the laws of electrodynamics and many particle theory rather than by gravitation. It is these interactions which lead to the particular properties of a plasma and distinguish it from other matter states.

Information about remote plasmas is obtained solely from the electromagnetic radiation that they emit. At temperatures on the order of eV (1 eV = 1.6×10^{-19} Joule corresponds to a temperature of ~11 000 K) and much higher, this radiation is in the UV, X-ray, and radio wave ranges, the latter indicating the presence of magnetic fields. Of all this radiation, only long-wavelength radio waves can reach the surface of Earth, because all short wavelengths are absorbed by the atmosphere. The study of collisionless processes in plasmas in space thus became possible only after the advent of the space era. For astronomy, which had been previously restricted to the optical and long-wavelength radio wavebands, the space era opened up the areas of UV-, X-ray, Gamma-ray and Infrared Astronomy. However, from the point of view of physics, the greatest success in space exploration must be attributed to Space Plasma Physics. This domain profited directly from the possibility of *in situ* observation of the processes, the structure, and the particle and field distributions in the plasma environment of Earth. It discovered its richness and beauty and uncovered a wealth of new and unknown physical phenomena and processes which could not have been produced in the laboratory but are of utmost importance for the understanding of the properties of remote plasmas in astrophysics. Because, however, space plasma physics as a field of research does not, by its very nature, produce a multitude of nice pictures and photographs which please a broad public, the success of space plasma physics has been by far less spectacular than that of both groundbased and spacebound astronomy, where each technical progress literally opens up new vistas out into the Universe.

The plasma in near Earth space, which is the only one accessible to *in situ* measurements, is, with a few exceptions, invisible. The most spectacular of these exceptions is the Aurora. As it turns out, however, this phenomenon provides only a marginal indication of other much more violent processes taking place in the depths of near Earth space which would never have been discovered nor understood without direct access to these regions. One may easily conclude from this fact that many astrophysical observations are nothing but marginal indications of processes to which we have not the slightest access and which are hidden forever to our knowledge, forcing us into interpretations which may have little to do with what actually takes place out there.

*Max-Planck-Institut für Extraterrestrische Physik, Garching bei München, Germany

In this respect, space plasma physics is of undeniable value. It provides the only possibility of testing models and processes which are believed to take place in plasma but cannot yet be directly measured. In fact, astrophysics has, probably more than any other science, benefited from the insights reached in space plasma physics. Many of its more successful models and descriptions are based on extensions and rescaling of the processes inferred from measurements in the plasma in near Earth space and in particular the Earth's magnetosphere, the easiest accessible part of all the cosmos.

To mention only one example, we refer to the very existence of Earth's magnetosphere (cf. Chapter 63). The possibility of a magnetosphere surrounding Earth and confining its magnetic field into a drop-like region was first suggested by T. Gold in 1949 as a theoretical idea lacking any experimental evidence and without any mathematical explication. It was picked up by the Russian plasma physics school at the end of the fifties, when it was already possible to perform measurements in space, using rockets and satellites. By the time when astrophysicists (Pacini 1967, Gold 1968, Goldreich and Julian 1969) made the first use of the notion of a magnetosphere in 1967–1969 in order to understand the physics of pulsars in analogy to Earth, the Earth's magnetosphere was already a well established and also, in its basic properties, a well understood fact.

The plasma in the magnetosphere and its closer environment constitutes an invaluable huge laboratory for the investigation of the processes taking place in a collisionless (or at the best weakly collisional) dilute warm plasma. This has been immediately recognized after the discovery of the magnetosphere. Thus most of the efforts to explore the processes taking place in the magnetospheric plasma are driven by two complementary intentions: the purely explorational wish to map and understand the space-time evolution of the near-Earth environment with the highest available precision and, at the same time, to use this environment in order to infer about the physical processes which take place in an extended dilute natural plasma. The former intention is the more geophysically and environmentally oriented one, which is of enormous importance in understanding the more extended human living space. The latter intention instead is entirely physically oriented and can be considered as the contribution of space physics to basic plasma research. These processes arise in the first place in the interaction between the high speed collisionless solar wind stream and the dipolar almost stationary geomagnetic field confined to the magnetosphere. They also arise in the dynamical mixing of plasma of solar wind origin with plasma of atmospheric origin. Thus the three main actors in the physical spectacle of space plasma physics are the solar wind, the geomagnetic field confined to the magnetosphere, and the electrically conducting upper layer of Earth's atmosphere, the ionosphere. With the solar wind, geomagnetic field and ionosphere at hand, one is in the fortunate position to monitor many processes to which we otherwise would not have any access.

In the present essay we review a number of the magnetospheric processes in the context of their history and their physical importance. When identifying such processes we have to be very selective. We therefore focus on only a few most obvious ones: the investigation of particle confinement in a magnetic mirror geometry as naturally given in the radiation belt dipolar geomagnetic field region, the formation of the Earth's bow shock wave, the surface of violent deceleration of the solar wind by the presence of Earth's magnetic field, on transport processes at the magnetospheric boundary, the magnetopause, on reconnection, on the generation of electric fields and particle acceleration, and on the most interesting processes of wave generation and radiation. Wherever possible, we also in passing attempt to point on the key historical events. However, as our historical knowledge and abilities are rather restricted we apologize for that the historical aspect will naturally fall short compared with the scientific description. Unfortunately, historical recollections by its main actors on these subjects are rather sparse. We refer the more historically interested reader to the article by Stern (1996) and references therein. Of interest in this respect are also the books by Van Allen (1983) and Kennel (1995), and an essay by Van Allen (1990). Additional historical information can be found in Gillmor and Spreiter (1997).

PLASMA CONFINEMENT: THE RADIATION BELTS

Historically, the phenomenon of plasma confinement in a magnetic mirror geometry was the first physical problem where space physics became aware of the existence of the magnetosphere.

The first few Russian spacecraft launched at the end of the fifties and in the early sixties of the twentieth century the first one, including the famous Sputnik I, launched on 4 October 1957 in honour of the fiftieth anniversary of the Bol'shevyk October Revolution (for a recollection of the shock on the American public and US administration and the immediate science-political activity following it see McDougall 1985), carried on board a few simple particle detectors which were sensitive to energies higher than a few $10\,\text{eV}$ (or temperature of a few $100\,000\,\text{K}$). One of us (R.T.) vividly remembers his feelings when, as a young high-school student during the autumn holidays in early October 1957, he saw the faint glimmering light of Sputnik I passing slowly and quietly over the dark late evening sky. No-one ever before had witnessed a man-made moon even as tiny as this one flying around Earth across the vast infinity of space. It was the sole privilege of our generation to be the first to visually experience and enjoy this demonstration and manifestation of the enormous achievements, dimensions and potentialities of human scientific and technological endeavour.

Very little was known at that time about the existence, density, particle flux and temperature of the plasma component in space. It was in fact widely believed that space was essentially empty, sometimes only crossed by high-energy cosmic rays, asteroids, quasi-regularly traversed by meteoric showers, occasionally by comets, and from time to time also by solar magnetic cloud ejections, which then caused strong geomagnetic storms the physics of which was largely unclear. Following the simple pioneering theory of Chapman and Ferraro, put forward nearly thirty years earlier in 1931, storms were boldly attributed to a brief compression of the geomagnetic field when one of these clouds, during its passage over Earth, transitionally compressed the geomagnetic field. The suddenness by which such storms begin did support this theory quite well, as did – in an entirely wrong interpretation – the long-lasting main phase of the storm, when the geomagnetic field on the Earth's surface falls below its undisturbed value, until it, finally, gradually recovers in the storm recovery phase. This undershoot of the field was, in its simplistic interpretation, attributed to an over-expanding geomagnetic field after the magnetic cloud had passed the Earth. One should note this disturbing and long prevailing lack of physical understanding of the storm mechanism and the even less well understood smaller sister of magnetic storms, the magnetic bay disturbance or, as it is called today, the *substorm*! The latter one was attributed solely to variations of currents flowing in the ionosphere. The only indication of space being filled continuously with plasma came from observation of comet tails, which are directed radially away from the Sun (Biermann 1951). This indication anticipated the solar wind.

Which instrument to put on a spacecraft depends on the expectation of what could be measured, and on the availability of instruments. The early Soviet spacecraft carried some low-energy particle traps. As it was known that space is continuously traversed by cosmic rays, the hope of the investigators was to measure *in situ* some of the lower energy cosmic rays which could not pass through the atmosphere, and possibly or occasionally some hypothetical medium energy solar particles, which, since the early theoretical endeavors of Störmer (1907), were thought to be responsible for the optical auroral emissions. They detected relatively strong charged particle fluxes at low altitudes and medium latitudes, the origin of which was difficult to explain (Gringauz *et al.* 1961, see also Lemaire and Gringauz 1998). The interpretation of these findings remained a mystery until the launch of the first tiny

Figure 1 William Pickering, James Van Allen, and Wernher von Braun (from left) proudly lifting a scale copy of the first U.S. satellite Explorer 1 and its launcher at a press conference after its successful launch into space by one of von Braun's Juno rockets a model of which is standing on the right of the table. (Courtesy, NASA, JPL photo, see also Murray 1990.)

American satellite Explorer 1, of which a scale model mounted on its carrier could be lifted by its three principle experimenters, Wernher von Braun, James Van Allen, and William Pickering (Figure 1). Explorer 1 was carried into space in 1958 by one of von Braun's Army rockets "Juno", which was developed from the Second War German V2/V3 family also designed by von Braun.

Explorer 1 had on board just one single Geiger counter, which was provided in the last minute by James Van Allen and his collaborators, Carl McIlwain, Georg Ludwig and Ernest Ray, as the sole scientific instrument available at that time expected to work under space conditions (for a personal recollection of the history of Van Allen's contribution, see Van Allen 1996). Each time when the spacecraft passed a certain range of low magnetic latitudes at an altitude of a few hundred kilometers the counter fell into saturation. High particle flux had been detected permanently in those latitudes. It was Van Allen who immediately realized that these latitudes corresponded to a particular restricted range of dipolar geomagnetic field lines. The high particle flux detected here corresponded to particles that had been about permanently trapped in the dipolar geomagnetic field, a property of charged energetic particles that had been theoretically predicted long ago by H. Alfvén, E. Teller, T. Northrop and others, based on the theory of adiabatic motion of charged particles (cf. Northrop 1963). These regions were later called the Earth's Radiation Belts or *Van Allen Belts*, and are since publicly known as some regions which are biologically dangerous for astronauts to pass through. This is partially true, but the radiation belts are of danger in the first place for modern spacecraft like Hubble, Chandra, Rosat, SOHO and others because the high intensity of particle radiation can affect the highly sensitive instrumentation on board.

The origin of the radiation belts can be understood resulting from an interplay between the Earth's dense atmosphere and the flux of energetic galactic cosmic rays coming from the depths of the Universe and hitting the atmosphere. The collisions between the cosmic radiation and the atmosphere cause the cosmic ray albedo neutron decay. Neutrons, produced in the nuclear interaction between the cosmic rays and the nuclei of the atmosphere are reflected upward from the atmosphere. These free neutrons are unstable and decay into energetic protons and electrons which become trapped in the closed geomagnetic field configuration near the Earth, forming an inner and an outer radiation belt. The spectrum of the trapped protons typically decreases approximately inversely with energy from ~ 1 MeV to ~ 1 GeV. A third innermost belt is produced by the energetic interstellar ions of the anomalous cosmic ray component in our solar system. These ions have a high rigidity, ignore the geomagnetic field and penetrate deep into it. In collisions with the upper atmosphere they lead to ionization and become themselves stably trapped. Radiation belts have also been artificially generated during the enormous charged particle releases from atmospheric nuclear tests like the "Starfish" nuclear explosion in 1962. Sometimes similar injections of particles have been observed during very strong magnetic storms. Thus the origin of the radiation belts is well understood. However their persistent confinement, spatial structure and temporal variation requires explanation and provides an important and interesting exercise in plasma physics.

Plasma confinement is particularly important in fusion research, where a hot plasma with a temperature of $\sim 10^7$–10^8 K (or energy ~ 1–10 keV) must be magnetically confined for a sufficiently long time in order to make possible the ignition and self-sustainment of the nuclear fusion reaction. At the high densities of a fusion plasma, these times are rather short and limited by collisions. At the low densities of the trapped particles in the Earth's (Van Allen) radiation belts, as these regions have been called, the trapping times can become very long, of the order of years, allowing for a detailed investigation of the injection and loss processes of trapped particles. Ideally, the particles can be trapped for infinitely long times because the trapped component does not experience any collisions. Thus the discovery of the radiation belts provided the immediate opportunity to study the properties of an ideally trapped plasma. This study led to a large number of hitherto unexpected results and to a deep understanding of the trapping process, one of the greatest successes in the early period of space science.

The early spacecraft carried instrumentation which was unable to distinguish between electrons and ions. Though it was clear that both kinds of particles could be trapped in the radiation belts, determination of the real distribution of electrons and ions in the belts had to wait until around 1975 for investigation and confirmation. At about that time, it was confirmed that protons above 100 keV energy populated one single large radiation belt, while the radiation belt electrons above 40 keV distributed themselves into two spatially separated belts, the inner and outer electron belts. The region between those two is called the *slot* of the radiation belts.

Pitch-angle diffusion

Stably trapped particles possess three adiabatic invariants: the magnetic moment, the bounce and the azimuthal drift invariant. Typical gyration times for electrons and protons

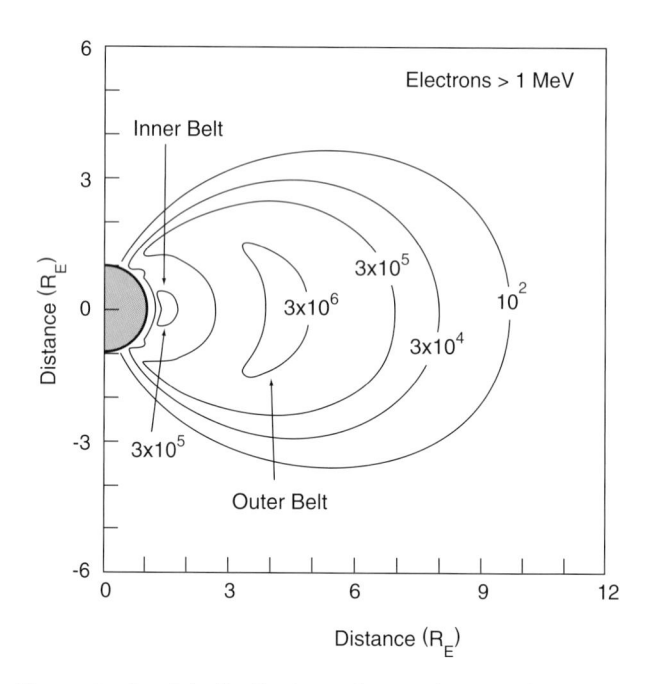

Figure 2 Spatial distribution of omnidirectional radiation belt electron flux (in units of electrons cm^{-2} s^{-1}) following the 1962 "Starfish" atmospheric nuclear test explosion. The flux distribution indicates the presence of an inner and an outer electron radiation belt. Constant flux contours follow the geomagnetic dipole field flux tubes. (After Walt 1994.)

in the radiation belts at about 4 R_E equatorial distance from Earth are on the order of $\sim 10^{-5}$ s and $\sim 10^{-2}$ s, respectively. Bounce periods depend on energy but are on the order of several seconds, while azimuthal drift times are much longer, in the tens-of-minutes to hours range, depending on particle energy and mass.

One can easily imagine that any time variations in the magnetic field which proceed in times shorter than one of these three time scales, the gyration time ω_{cq}^{-1}, bounce τ_b or azimuthal drift τ_ϕ times, will violate the constancy of one or the other adiabatic invariant, destroy the corresponding periodic motion of the particles and cause loss of radiation belt particles. This had already been recognized by H. Alfvén in the forties and has been mathematically evaluated by T. Northrop (1963).

Azimuthal drift violation is the most important natural injection or loss process of relatively low energy particles in the outer radiation belts. It is observed during magnetic storms and generates the magnetospheric *ring current*, an azimuthal current of trapped or quasi-trapped near-equatorial particles flowing around the Earth and being carried essentially by protons. The diamagnetic effect of this ring current provides the correct interpretation of the decrease of the magnetic field at low latitudes observed during the main phase of the magnetic storms, mentioned above.

The most important violation of adiabatic invariants relates to the magnetic moment. When the particle is exposed to waves with frequency ω close to the cyclotron frequency $\omega_{cq} = qB/m$, the magnetic moment is not conserved anymore. This leads to fast losses of particles from the radiation belts and is the paradigm for the so-called *pitch-angle diffusion*. It was one of the great successes of space plasma physics, in which space is conceived as a plasma laboratory (see the recollection of the history as described by Petschek 1996), when the very mechanism of pitch-angle diffusion and particle losses from the radiation belts had been clarified by Kennel and Petschek (1966). The equation describing pitch-angle diffusion is

$$\frac{\partial F(v,\alpha,t)}{\partial t} = \frac{1}{\sin\alpha}\frac{\partial}{\partial \alpha}\left[D_{\alpha\alpha}\sin\alpha\frac{\partial F(v,\alpha,t)}{\partial \alpha}\right] \quad (1)$$

which is the Fokker-Planck equation in velocity space and is of the same type as the heat conduction or any other diffusion equation in ordinary configuration space. The right-hand side describes the change in pitch-angle α under the action of the diffusion coefficient $D_{\alpha\alpha} = \langle (\Delta\alpha)^2/2\tau \rangle$, the left-hand side the corresponding local change of the phase-space distribution F in time. The average $\langle \rangle$ is over the gyration period, and τ is the diffusion time.

The problem consists in the determination of the pitch-angle diffusion coefficient. In interaction with a self-excited wave spectrum, it should be a function of the power spectral density $(\delta B)^2/\Delta\omega$, where δB is the magnetic fluctuation amplitude, and $\Delta\omega$ is the bandwidth of the wave in interaction with the particles. Kennel and Petschek (1966) derived for the diffusion coefficient the expression

$$D_{\alpha\alpha} \approx \frac{\pi}{4}\frac{\omega_{cq}^2}{\Delta\omega}\frac{|\delta B|^2}{B^2}\frac{1}{v_\parallel}\frac{\partial \omega}{\partial k_\parallel} \quad (2)$$

which explicitly shows the dependence on the power spectral density of the wave spectrum. The particles experiencing diffusion have resonant energies

$$W_{\parallel,res} \geq \frac{m}{2}v_{Aq}^2 G\left(\frac{p_\perp}{p_\parallel}\right) \quad (3)$$

larger than the magnetic energy per particle which is given by the Alfvén energy $(m/2)v_{Aq}^2 = B^2/2\mu_0 n$ and is inversely proportional to the density n. The function G is of the order of one and depends only weakly on the pressure anisotropy $p_\perp/p_\parallel$. Increasing the background plasma density will drastically reduce the resonant energy. As there are more particles at low than at high speeds, one recognizes that, in a denser background plasma, the resonant effect will be stronger than in dilute plasma, and consequently pitch-angle diffusion will be much stronger and particles will be confined for shorter times.

The clarification of the pitch-angle diffusion process and its experimental confirmation could be performed only with the help of observations of radiation-belt life times. In particular, the monitoring of the decay of the electron and proton fluxes injected into the radiation belts during the Starfish explosion enabled to determine the lifetimes of radiation belt particles under relatively quiet conditions.

Structure of the radiation belts

The pitch-angle diffusion equation has two families of solution, when the diffusion time τ is short and when it is long with respect to the loss time τ_{loss} of particles inside the loss-cone. For long diffusion times, the pitch-angle diffusion process cannot fill the loss cone. This limit corresponds to $|D_{\alpha\alpha}|\tau_{loss} \gg 1$. Hence the loss cone is void of particles. The distribution function drops steeply at its edge. Only few particles, making it into the loss cone are then precipitated into the ionosphere. Under quiet magnetospheric conditions the diffusion is weak. On the other hand, for $|D_{\alpha\alpha}|\tau_{loss} \ll 1$ pitch-angle diffusion is very strong, and the loss-cone is filled such that the distribution at $\alpha < \alpha_{loss}$ is practically flat. This happens under substorm injection conditions. The diffusion is strong then, leading to high and bursty precipitation fluxes of particles which are lost from the radiation belts and ring current to the ionosphere. This precipitation is very well correlated with local X-ray emissions originating from the thick-target bremsstrahlung emitted from the precipitating fast electrons. It also correlates

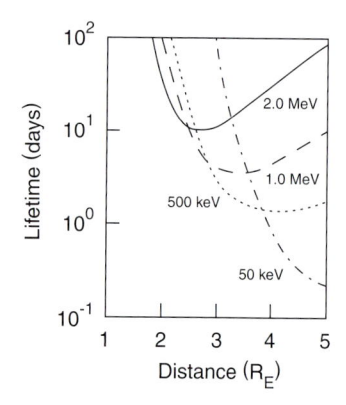

Figure 3 Field-line averaged lifetimes of radiation belt electrons calculated from whistler pitch-angle diffusion theory (after Lyons *et al.* 1972). Lifetimes depend on particle energy. The higher the energy the longer the particles remain trapped. Medium energy electrons have minimum lifetime between 3–4R_E distance from Earth corresponding to the radiation belt slot region.

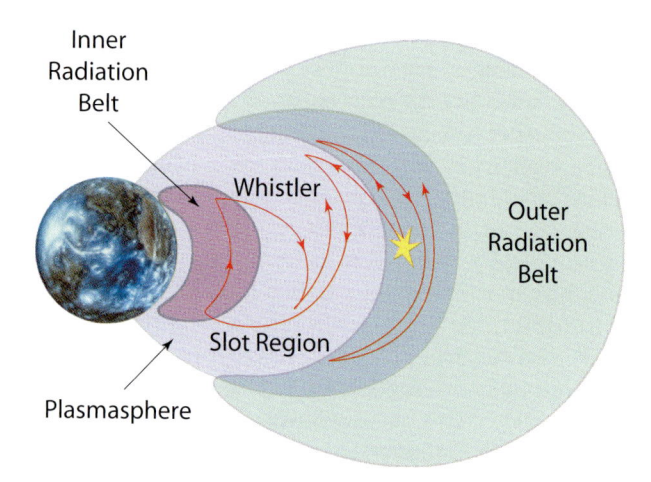

Figure 4 A schematic of the structure of the inner and outer radiation belts showing the position of the slot region separating the two belts. The irregular line shows the path of radiation belt excited oblique whistlers. These whistlers propagate away from the field line where they have been excited and become amplified in the slot region to a level where they strongly interact with the electrons until these undergo pitch-angle diffusion and precipitate. Much of this stronger amplification and interaction is due to the enhanced density of background plasma in the slot region, which causes stronger resonance with waves. (From Baumjohann and Treumann 1997.)

with high whistler mode turbulence in the region of the radiation belts. Moreover, proton precipitation from the ring current caused by ion whistlers (electromagnetic ion-cyclotron waves, which are the ion equivalent of electron whistler waves) leads to the long lasting mid-latitude stable auroral red arcs (SAR arcs) observed during magnetic storms.

A spectacular confirmation of this theory in space was provided when it was discovered that the radiation belts are radially structured. The outer radiation belt consists of two regions of high flux separated by a slot of very low particle intensities centered somewhere at about ~3–4R_E radial distance (see Figures 3 and 4). The slot region could be nicely understood when taking into account a broad and obliquely propagating whistler wave spectrum excited by the trapped radiation belt electrons themselves. The growth rate of the oblique waves maximizes in the slot region due to overlap of many harmonic resonances and locally increases the pitch-angle diffusion. Particles entering the slot have the shortest lifetimes against pitch-angle scattering, as clearly shown in Figure 3, and are readily lost into the loss-cone and to the ionosphere. The reason for this increase in pitch-angle diffusion can be found not only in the obliqueness of the propagation of the whistler noise but also in the radially inward increase of the background plasma density in the slot region, which approximately coincides with the position of the plasmapause. This increase lowers the resonance energy of the electrons according to eqn (3) and causes the strong growth of the whistler noise amplitude, which ultimately causes strong pitch-angle diffusion.

Let us come back to our initial remark that families of waves other than whistler may also cause pitch-angle diffusion. Indeed, it has been found that, for instance, electrostatic electron cyclotron waves cause very strong pitch-angle diffusion, this time on the low-energy electrons. Similarly, electrostatic ion-cyclotron waves lead to pitch-angle diffusion of low-energy ions.

The radiation belt and ring current physics provide an elegant example of how theoretical predictions of particle and plasma behaviour under ideal trapping conditions act are confirmed. The theory of confinement and particle loss by pitch-angle diffusion could be proved by experiments in space. Other applications have made use of injecting particle beams along the magnetic field from ground or spacecraft into the radiation belts and artificially exciting whistler mode waves, which then could be detected near the conjugate field line foot point or on spacecraft when taking advantage of the antiparallel propagation of electrons and whistlers in resonance. Moreover, injecting cold plasma into the radiation belt locally causes enhanced whistler mode growth and consequently increased precipitation levels of radiation belt electrons.

Historically, experiments of this kind have been performed in order to check Kennel and Petschek's (1966) mechanism of radiation belt electron depletion by self-generated whistler mode turbulence, something that could not have been performed in the laboratory simply because of the difficulty of preparing a laboratory plasma in an appropriate way. The huge dimension of the plasma in the magnetosphere was, however, ideally suited for these experiments. By now, it has been well established that the long-term stability of the radiation belts is a result of the interplay between injection of energetic particles by neutron albedo and/or geomagnetic storms and the excitation of whistler and electromagnetic ion-cyclotron turbulence by the freshly injected particles, which determines their longevity. At the same time, this mechanism explains the precipitation of energetic electrons and ions into the ionosphere.

Pitch-angle diffusion, as confirmed by radiation belt physics, has a wide range of applications outside the radiation belts. In cosmic ray physics, pitch-angle scattering by magnetic field fluctuations or in warped magnetic field flux tubes is very important in the theory of cosmic ray transport across the heliosphere and the galaxy. It scatters the particles out of their parallel motion. Since the energy is not changed in this process, the particles attain a larger perpendicular velocity which increases their gyroradii. This must necessarily cause diffusion across the magnetic field in addition to inhibiting parallel propagation by causing reduced parallel diffusion. The effect of pitch-angle diffusion on cosmic ray particles is thus to confine them locally in the heliosphere and to diffuse them perpendicular to the heliospheric magnetic field, which may explain the penetration of cosmic rays into regions of the heliosphere which otherwise should be inaccessible to them. The generalization of this mechanism to astrophysical conditions where cosmic rays diffuse across the galaxy is obvious.

Cosmic ray research is not the only astrophysical application of trapping, confinement, and pitch-angle diffusion. Other important applications include the trapping of charged particles in solar coronal loops, the atmospheres and magnetospheres of magnetised stars, white dwarfs and, in particular, in neutron star magnetospheres. All these objects are not accessible to direct investigation, and any conclusions about their behaviour and structure involve analogies to known systems of which the magnetosphere and its radiation belts are particular examples. In solar coronal loops, for example, enormous amounts of energy can be stored in the form of charged particles. Their presence is indicated by solar gyro-synchrotron radio-emission in the meter-to-decimeter wave-bands familiar as type-IV radio emission. During solar flares and solar coronal mass ejection (CME) events these particles are violently released into the Sun's environment, causing dense particle clouds to be ejected into interplanetary space which drive interplanetary shock waves and cause a complexity of disturbances in near-Earth space that has been summarised under the somewhat spectacular name of 'Space Wheather'. Pitch-angle diffusion and injection of these particles into the dense solar atmosphere is responsible for hard-X ray and even Gamma-ray emission during solar flares, indicating the enormous energy that can be stored in loops. Similar effects causing stellar winds are believed to occur and have been observed in magnetised stars. In particular, in pulsar magnetospheres particle confinement, trapping and pitch-angle diffusion are involved into processes controlling the accretion of matter onto neutron stars, quasi-periodic oscillations in pulsar emission, the stability of pulsar magnetospheres, its loading with charges particles and the appearance of charge separation electric fields, plasma convection motions resulting from these fields, and particle acceleration and injections into interstellar space in the form of cosmic rays.

PLASMA BOUNDARIES AND TRANSPORT

The discovery of the radiation belts was quickly followed by a hastened exploration of the remaining Earth's environment leading to the discovery of the *solar wind*, confirmation of the Parker spiral structure of the interplanetary magnetic field, and the discovery of the *magnetosphere* with its frontside boundary and its nightside long extended *magnetospheric tail* draped around Earth like as was originally believed, a huge tear drop. In particular, the discovery of the tail came as a great surprise. It was Ness (1965) who showed, from magnetic measurements obtained with the Imp 1 spacecraft, that the magnetosphere possessed a tail much longer than expected with nearly antiparallel field lines forming a long magnetically nearly neutral sheet in the center of the tail close to the ecliptic plane. This observation implied that the neutral sheet carried a narrow electric sheet current separating the antiparallel fields north and south of it, a conclusion that led Williams and Ness (1966) to propose their tail sheet magnetic model, or theta-model, of the magnetosphere and which ultimately led to the proposal (Speiser and Ness 1967) of so-called Speiser particle orbits in the neutral sheet (for a personal historical recollection of all these discoveries, see Ness 1996). These are strange meandering orbits that are performed by the particles in a magnetic null sheet.

The discovery of the solar wind cannot be attributed to one single person but is shared by different groups in the US and the group of K. Gringauz in the Soviet Union. In the US, the leading group was the group headed by Ness, who launched the Imp spacecraft series and, with Imp 1, discovered the interplanetary magnetic field and the changes in the magnetic field related to the bow shock and the magnetopause, thereby confirming the interaction between the solar wind and the geomagnetic field. Gringauz's group discovered plasma in space but were less successful in their correct interpretation.

Discovery of the solar wind quickly stimulated Parker's theory of the origin of the solar wind as a supersonic outgassing of the solar corona. It was immediately recognized that the shape of the magnetosphere was the result of the interaction of the geomagnetic field **B** with the fast supersonic and super-alfvénic (Alfvénic Mach number $\mathcal{M}_A \sim 4$–10) solar wind stream. The explanation was in terms of a simple pressure equilibrium $m_i n_{SW} v_{SW}^2 = B^2/2\mu_0$ of the specularly reflected solar wind of proton mass m_i, density n_{SW} and velocity v_{SW}. The boundary between the stream and the magnetic field, the so-called *magnetopause*, turned out to be an extremely narrow interface, a *discontinuity*.

Magnetohydrodynamics, assuming that the solar wind stream can be described in fluid terms, allows for five such discontinuities: *contact*, *tangential*, and *rotational discontinuities*, and two types of *shock waves*, the *fast* and the *slow* shock wave. Observation of the magnetopause suggests that, to the lowest approximation, it forms a tangential discontinuity with the magnetospheric magnetic field, being tangential to the surface and the solar wind plasma streaming tangentially over its surface around the magnetosphere. One of the main tasks of experimental and theoretical investigation of the processes in the magnetosphere has since been the clarification of the structure of this thin boundary. How do such boundaries form? What is the reason for their stability? In which way may plasma pass across such a boundary? These have been key questions which have only been partially answered.

Magnetopause transport

Most of magnetopause observations made over the last quarter of a century by a large fleet of spacecraft have supported the view that the gross shape of the magnetopause is indeed the result of a large-scale pressure equilibrium such as the one required by a tangential discontinuity. Among those spacecraft, the most successful in investigating the shape and structure of the magnetopause have been ISEE 1 and 2, the three AMPTE satellites IRM, UKS and CCE and, recently, Geotail, Equator-S, and Cluster. The ISEE spacecraft, launched in the seventies, and the AMPTE IRM and UKS spacecraft, launched in 1984, had comparable instrumentation to measure the plasma and field parameters in the solar wind, magnetosheath and the external magnetosphere at the magnetopause and in the plasma sheet. ISEE 1 and 2 as well as IRM and UKS were kept on relatively short mutual distance, allowing for the investigation of some of the spatial variations in the field and plasma parameters. Equator-S was a short lived European satellite launched in 1998 into a near magnetopause orbit intended to map the magnetopause variations. Geotail, launched in 1992, is on the other hand, a deep tail spacecraft with highly sophisticated instrumentation to measure the field and plasma parameters in the plasma sheet and plasma sheet boundary layer. Occasionally it, however, passed across the magnetopause and also the bow shock providing valuable information about these regions. Cluster is a 4-spacecraft ESA-mission intended to investigate the spatio-temporal structure of the outer boundaries near the Polar Cusp.

The magnetopause is not entirely opaque to the solar wind plasma. On the contrary, its inner side is populated by a broad layer of solar wind plasma, the *low-latitude boundary layer* on the front side and the *plasma mantle* at high latitudes. Hence, plasma may indeed penetrate across the magnetopause and supply the magnetosphere with plasma to add to the plasma supplied along the magnetic field by the Earth's ionosphere.

What is the mechanism of transport of plasma across the magnetopause, as a tangential discontinuity should theoretically be impenetrable for any particles except cosmic rays with their huge gyroradii on the order of the size of the entire magnetosphere? It is believed today that the main transport mechanism is *reconnection*, the discussion of which we delay until a later section. In the absence of reconnection, there is only one way of plasma transport across a tangential discontinuity: diffusion. Thus the observation of a discontinuous free boundary of the magnetosphere challenged plasma physics to understand the process of diffusion.

In the presence of collisions $\nu \neq 0$, it is easily understood that particles can be pushed across the magnetic field. Collisions, if they are frequent enough, violate the frozen-in condition by destroying conservation of the magnetic moment of the particles. In a dilute plasma like the solar wind collisions are however so infrequent that it is illusionary to think of any diffusive effect caused by them. More violent processes are required. These processes can be found in the plasma turbulence which we will discuss later. But let us consider at this point a very simple model of a diffusive process in a plasma. Assume that during each gyration around the magnetic field a particle is kicked out of its orbit just once. This implies that, with each kick, the particle starts meandering through the plasma thereby jumping from field line to field line. This is the fastest diffusive motion one can imagine; any faster motion corresponds to free streaming in which the magnetic field does not play any role. The collision frequency is $\nu \sim \omega_{cq}$ in this case, and the diffusion coefficient perpendicular to the magnetic field, and hence across the boundary, is given by the diffusivity $D_\perp = D_B \sim k_B T_q/m\omega_{cq}$. This formula, giving the dependence of the diffusivity on the thermal energy $k_B T_q$ and magnetic field (through $\omega_{cq} = B/mq$), had been proposed in 1949 by David Bohm in dealing with fission bomb plasmas and is thus called Bohm diffusion. Ordinary classical diffusion gives the stronger B-dependence $D_\perp^{cl} \propto k_B T/B^2$. It turns out that at the magnetopause the diffusivity required for populating the low-latitude boundary layer has Bohm-dependence, and is on the order of Bohm diffusion, numerically about $D_B \sim 10^9 \, m^2 s^{-1}$. Why is this so and how could one produce a collision frequency of a value comparable to the electron cyclotron frequency? The first part of this question remains unanswered even today.

Anomalous collisions

Processes of generating collision frequencies in a plasma that is otherwise collisionless on all relevant spatial and time scales must be of extremely violent nature. It can therefore be expected that any attempt of producing collision frequencies of this magnitude all over the boundary of the magnetopause will ultimately fail simply because it should be very difficult to sustain such processes over a whole surface of the size of the magnetopause. Experimental and theoretical investigations (e.g. Treumann et al. 1995) as well as numerical simulations (e.g. Winske et al. 1995) of the various proposals detailing how these collision frequencies can be produced have indeed confirmed this negative result. Nevertheless let us briefly review the various theories because they show how space plasma has served to clarify one of the most important and tantalizing questions in theoretical plasma physics dealing with collisionless processes.

Plasma boundaries, like the magnetopause, constitute current sheets because the magnetic field tangential to the boundary may rotate and experience a jump in its magnitude. When this current, which is usually carried by electrons, exceeds a certain threshold, it excites a plasma instability, i.e. the free energy in the current feeds a particular plasma wave. As a consequence of this instability, the current loses energy while at the same time the electrons are scattered by the self-generated waves. This is the basic mechanism for generating anomalous collisions and hence anomalous resistivity. The nature of this anomalous resistivity depends on the direction of the current, parallel or perpendicular to the magnetic field. It also depends on the kind of wave that is amplified. Field-aligned currents amplify other waves than transverse currents. The three most important wave types amplified by the former are the Buneman-wave, the ion-sound wave, and the electrostatic ion-cyclotron wave (among other less important ones). Transverse currents amplify a large class of waves known under the common name of drift waves.

In order to understand what the mechanism is, let us briefly consider the general definition of the resistivity η in fully-ionized plasma. This is given by the Spitzer-Braginskii formula $\eta = \nu/\epsilon_0 \omega_{pe}^2$, where ω_{pe} is the electron plasma frequency. The collision frequency can be written as

$$\nu = \omega_{pe} W_{th} \ll \omega_{pe}, \quad W_{th} \simeq (n\lambda_D^3)^{-1} \quad (4)$$

showing that it is directly proportional to the thermal fluctuation level W_{th} in the plasma oscillations. Hence, increasing the collisionality of the plasma as required by many orders of magnitude requires an increase in the fluctuation level by orders of magnitude above thermal. This is just what the instability does.

Buneman-waves run unstable when the electron current velocity $v_D > v_e$ in the ion frame exceeds the electron thermal speed. This is a very strong case because usually the thermal electron speed is large, a considerable fraction of the velocity of light. It may, however, happen. For instance, near the magnetopause and also in the magnetospheric tail the electron temperature is roughly seven times less than the ion temperature which implies that the electron thermal speed is only about 16 times the ion thermal speed. The Buneman-wave will rise very fast until its large amplitude traps electrons and heats them quickly to such a temperature that the electron thermal speed exceeds the electron drift speed. This kills the instability. However, at drifts $v_D \sim v_e$, when this stage is reached, the ion-sound wave will grow quickly. It is a resonant and thus much weaker instability. It produces an anomalous collision rate

$$\nu_a^s \simeq 0.01 \, \omega_{pi}(T_e/T_i)(v_D/c_s) \quad \text{for} \quad v_D/c_s < 6 \quad (5)$$

where $c_s = \sqrt{k_B(T_e + T_i)/m_e}$ is the ion-sound speed. Electrostatic ion-cyclotron waves have a lower threshold for going unstable but produce much lower collision frequencies, which are of little interest.

The strongest collision rates are generated by a particular drift mode, the lower-hybrid drift wave, that experiences instability when there are density gradients at the boundary. As this is the case at the magnetopause, it is probable that the lower-hybrid drift wave is excited there. The collision frequency produced by this wave is

$$\nu_a^{lh} \simeq \omega_{lh} \sqrt{\frac{\pi}{8} \frac{m_i}{m_e}} \langle W^{lh} \rangle \quad (6)$$

where $\omega_{lh} \sim \sqrt{\omega_{ce}\omega_{ci}}$ is the lower-hybrid frequency, which is the geometric mean of the electron and ion-cyclotron frequencies. This collision frequency just reproduces the desired Bohm dependence of the diffusivity.

Diffusion

Measurement of the wave spectral densities at the magnetopause allows to test the relative importance of the waves for the diffusion of plasma across the magnetopause. Figure 5 shows the application of the lower-hybrid collision theory to the reconnection region at the magnetopause in dependence on the measured lower-hybrid wave electric field intensity. Here we plotted the perpendicular diffusion coefficient

$$D_\perp^{lh} = \frac{k_B T_e}{m_e} \frac{\nu_a^{lh}}{\omega_{ce}^2} \quad (7)$$

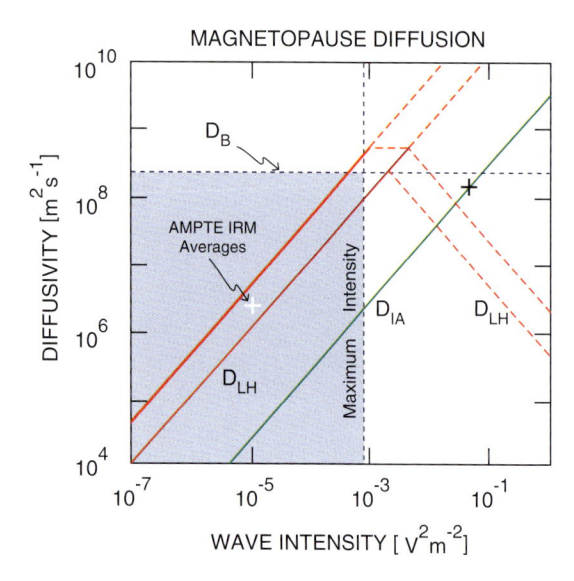

Figure 5 Lower-hybrid drift diffusion coefficient in dependence on the electric wave intensity for reconnection at the magnetopause. The horizontal dashed line indicates the Bohm diffusion D_B limit corresponding to about $\nu \sim \omega_{ce}$, as the upper limit of any reasonable anomalous collisional diffusivity for the magnetopause/low latitude boundary layer. The light cross marks the integrated electric wave field measurements at the magnetopause. The diffusivity is orders of magnitude below Bohm diffusion. For comparison, the linear ion-acoustic diffusivity D_{IA} has also been drawn. The cross on this curve marks the theoretical quasilinear estimate. It obviously overestimates the diffusivity. D_{IA} for all measured intensities turns out to be weaker than D_{LH}, which dominates in the magnetized case. The turnover of the diffusion coefficient at high wavefields indicated by the dashed lines is due to the diffusion becoming isotropic at large collision frequencies. Diffusivity then goes like $D \propto \nu_a^{-1}$, which causes its decrease. (After Treumann et al. 1995.)

as function of $\langle |\mathbf{E}^{lh}|^2 \rangle$. The cross marks the real measurement of wave intensity, while the horizontal dashed line indicates the theoretical Bohm diffusion limit, which corresponds to the upper limit on any reasonable diffusion process or collision frequency $\nu \simeq \omega_{ce}$ equal to the electron cyclotron frequency. The width of the band around the cross corresponds to the typical average fluctuation of the intensity. The vertical dashed line terminating the shadowed region indicates the extreme upper limit of wave intensities, which is occasionally measured. For comparison, the figure also shows the ion-sound diffusivity, which is much less than that generated by the lower-hybrid drift wave. The cross on the ion-sound curve marks the theoretical Sagdeev estimate, which obviously overdoes those limits given by the real measurements.

The lesson learned from this figure is that the measurements never ever, except for the very rare cases of occasionally very high intensities corresponding to the vertical dashed line, support collision frequencies high enough for *global* Bohm diffusion. The actual collision frequency is roughly two orders of magnitude lower than ω_{ce}, which is not enough for *global diffusive* transport across the magnetopause in order to populate the low-latitude boundary layer. Hence, the space laboratory was able to decide not only among the importance of the different wave modes for generating anomalous resistivity but also about the relevance of such anomalous resistivities for transport across the magnetospheric boundaries and thus very generally across plasma boundaries. It seems highly improbable that anomalous collisions will by themselves produce such a transport over an entire boundary, although they may be locally very important. One case when this importance becomes viable is *reconnection*, which we will discuss separately.

Diffusivity is just one of a whole set of problems related to what we call *transport* in plasma. A plasma is a multi-particle system with huge numbers of charged particles per Debye-sphere $N_D \equiv 4\pi n \lambda_D^3 / 3 \gg 1$, each particle acting on and reacting to the variations of the total electromagnetic field. A plasma must thus be described by kinetic theory. Since this is a very difficult task, one would like to treat it rather on the basis of one- or multi-fluid theories. Such theories assume that the kinetic equations can be replaced by transport equations which describe the macroscopic behaviour of the plasma while the microscopic interaction is buried in some phenomenological transport coefficients like the average collision frequency ν, anomalous resistivity η, viscosity ζ, heat conductivity κ, etc. The basic problem of anomalous transport has been reviewed by Dum (1995). In collisionless plasma, the existence of such transport equations is non-trivial. In particular, the jump conditions at discontinuities result from the requirement on the transport equations that these equations are valid on both sides of the discontinuity. The jump conditions are the boundary conditions which glue the two solutions to both sides together in a way which violates these equations only in a narrow region restricted to the interior of the discontinuity.

Transport equations make sense only on scales larger than the microscopic scale. In a magnetized plasma this scale is the gyro-radius. Because of the enormous scale lengths in space plasma physics – the ion gyro-radius in the solar wind is typically the order of several 1000 km – spacecraft are ideally suited for investigation of the internal structure of the natural plasma boundaries.

There has been enormous progress in recent years in the investigation of the inner structure of discontinuities. The best investigated discontinuities are the Earth's magnetopause, the Earth's bow shock, and the tail plasma sheet. The most valuable information about the structure of these discontinuities has been provided since the mid-seventies among others by the spacecraft ISEE 1 and 2 and ICE (formerly ISEE 3), AMPTE, Geotail, and Equator-S. Various other interplanetary discontinuities are traveling shocks in the solar wind, which have been detected by the three ISEE spacecraft, by the Wind spacecraft and in particular by the Ulysses and Galileo satellites, the solar wind current sheet, CME shock waves, corotating interaction region shocks, tangential discontinuities in the solar wind, all explored by the same spacecraft and in the first place by Ulysses. Around Jupiter and Saturn magnetospheres and their boundaries were discovered by the Pioneer 10 and 11 spacecraft launched in 1972 and 1973, respectively. The first detections of Jupiter's magnetosphere date back to the Pioneer 10 Jupiter encounter in November 1973. Saturn was passed by Pioneer 11 about six years later in August/September 1979. Jupiter's bow shock, about which the Galileo satellite provided information, has also been investigated *in situ*. In addition, remote information has been obtained from the boundary of the heliosphere from the two Voyager spacecraft which are now far out at distances beyond 60 AU and are expected to pass the terminal shock of our heliosphere believed to be located between 80–100 AU within the next few years and within another decade the heliopause, the outermost heliospheric boundary.

Structure of the magnetopause

From a naive point of view, the magnetopause is a discontinuity whose width is of the order of the ion gyroradius, roughly 1000 km thick in perpendicular direction. Since electrons penetrate much less deeply into the magnetic field, the magnetopause should contain a charge separation field in the normal direction. The near absence of plasma in the magnetosphere close to the magnetopause should in addition cause a steep outward radial gradient in the density and an inward radial gradient in both the magnetic field and the plasma temperature. Rarely, however, do measurements confirm this simple theoretical picture, not speaking about the difficulty or impossibility of measuring the space charge electric field inside the magnetopause. Measurements of this field have been reported or inferred but are still not reliable. One is desperately waiting for the *in situ* data from the electron drift experiment (EDI) on Cluster II, which is expected to provide the first reliable data on the electric field strength perpendicular to **B** across the magnetopause and in other region of the magnetosphere. This instrument (cf. Paschmann *et al.* 1993) is based on the simple principle that, under very wide conditions, a weak electron beam injected into the magnetic field in a direction nearly perpendicular to **B** will, in addition to its gyrational motion, perform a perpendicular drift $\mathbf{v}_{E\perp} = \mathbf{E}_\perp \times \mathbf{B}/B^2$ proportional to a possible transverse electric field $\mathbf{E}_\perp$ and inversely proportional to the magnetic field magnitude B. Knowledge of the latter and measuring the slight transverse displacement of the beam provides a direct measurement of the electric field which is free of all antenna and sheath effects. At the time when this book is published, we hopefully will have available the measurements of EDI's giving us the first unambiguous *in situ* measurements of convection electric fields in space.

The measurements of the particle and field quantities across the magnetopause exhibit a steep change in field direction when passing from the magnetosheath into the magnetosphere, a weak field gradient, a relatively flat density gradient over the whole width of the low-latitude boundary layer followed by a steep decay at the inner edge of this layer, and a susceptible temperature gradient indicating the presence of hot magnetospheric plasma and possibly heating of the entering sheath plasma.

This behaviour shows that the structure of the magnetopause is provided by microphysical processes and is by far not understood even in the simplest case when the magnetopause seems to satisfy most of the jump conditions imposed by the tangential discontinuity model. Using the magnetic field in three dimensions and applying a minimum variance technique, one can find the direction of the normal in the magnetopause in an average sense from a satellite path across the boundary. Such a procedure shows that the magnetopause usually possesses a consistent normal direction. Quasiperiodic variations of the direction of the normal are caused by either periodic surface waves which propagate along the magnetopause surface or by the internal structure of the magnetopause. Again, this cannot be decided by measurements of one single spacecraft. It awaits clarification by multi-spacecraft missions like Cluster II. Cluster II is the revived Cluster experiment which was unfortunately terminated in June 1996, when its launcher, the Ariane 5 rocket, exploded shortly after launch from Kourou. As had been planned for Cluster, Cluster II consists of four spacecraft which can be maneuvered and located at different mutual distances. Cluster II has, at about the time of writing of this article, been successfully launched in two separate groups of two spacecraft each on July 15 and August 9, 2000. It will start its mission near the dayside polar cusp of the magnetopause in autumn 2000 and, half a year later, will begin investigating the nightside magnetotail plasma sheet. Its four satellites will allow for a space-time resolution of many processes taking place at the magnetopause and in the plasma sheet.

In a bold two-dimensional fluid-dynamic model, assuming that the field structure changes smoothly across the tangential discontinuity magnetopause, the single-spacecraft magnetic field measurements along a slow crossing of the magnetopause can be used to reconstruct the field configuration. Assuming that the field configuration is stationary, the pressure equilibrium condition with transverse to the magnetopause pressure $P(A) = p(A) + B_z^2(A)/2\mu_0$, a sole function of the single remaining (out of plane) magnetic vector potential component $A(x, y)$ and normal magnetic field component $B_z(A)$ can be cast into an equation for

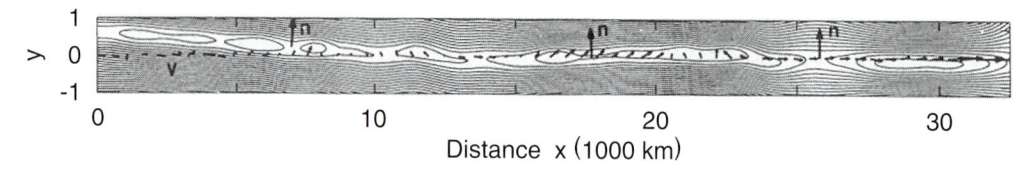

Figure 6 Reconstruction of the interior magnetopause structure by solving the Grad-Shafranov equation with input of a real magnetic field profile across the magnetopause measured by Ampte Irm (after Hau and Sonnerup 1999). Arrows labeled **n** show the local direction of the magnetopause normals used in the reconstruction model as determined from minimum variance analysis of the data. The small arrows show the measured transverse to the magnetic field velocity vectors in the deHoffmann-Teller frame along the path of the Irm spacecraft ($x = 0$-line) in relative units. The length of the velocity vector at the right boundary corresponds to a velocity or ~ 164 km s^{-1}. Scales at the x and y axes are measured in units of 1000 km.

$A(x, y)$. This so-called Grad-Shafranov equation (see Biskamp 1993, p. 30)

$$\nabla_\perp^2 A(x, y) = -\mu_0 \frac{\partial P(A)}{\partial A} \quad (8)$$

has been derived by several authors (Lüst and Schlüter 1957, Grad and Rubin 1958, Shafranov 1958). This equation is an inhomogeneous Laplace equation (or Poisson equation) with the right-hand side depending only on the unknown function itself, which is the vector potential $A(x, y)$. The Grad-Shafranov equation can be solved numerically with spacecraft measurements as input. An example of such a calculation is shown in Figure 6.

The surprising result of this calculation is that the interior of a tangential magnetopause apparently consists of a collection of magnetic islands forming long chains with changing thickness of the magnetopause transition region along the magnetopause. Occasionally, the magnetic field leaves some spaces open in this chain, small "wormholes", where field lines may leak out of the magnetopause to its environment and plasma may cross the magnetopause. Of course, the reality of this result may be questioned because of the many assumptions in this calculation, the validity of the two-dimensional model and the validity of the fluid and pressure balance approximation. However, it demonstrates that even in the most simple models of discontinuities, the use of real data not only probes the model assumptions but leads to entirely new views on the physics of the discontinuity. So far, such calculations have been performed for tangential discontinuities. Similar calculations will have to be performed also for rotational discontinuities which are expected in reconnection models.

RECONNECTION

Reconnection is widely believed to be the process which competes with diffusion in transport across boundaries. Reconnection as a fundamental plasma process was proposed in 1956/1958 by Sweet and Parker, following original

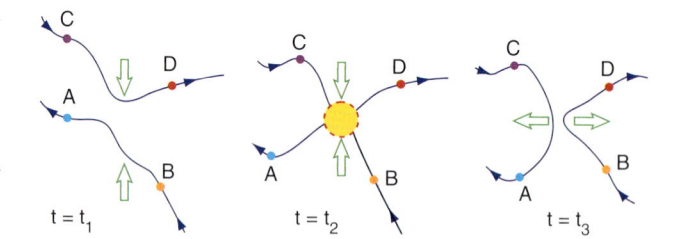

Figure 7 The mechanism of reconnection as seen from the point of view of plasma and field-line topology. At time t_1, two anti-parallel magnetic field lines (flux tubes) approach each other. A,B and C,D are four particles tied to each of the field lines. Open arrows indicate the flow direction. At t_2 the field lines meet, reconnect and reorder. At t_3 the reconnected field lines, having a new topology, have separated. Now the order of the particles on the field lines is A,C and B,D without any need for diffusion. The only region where diffusion is probably necessary is in the tiny circular region where the two flux tubes meet at time t_2. (From Treumann and Scholer 2001.)

suggestions by Giovanelli (1946) and Hoyle (1949), as a mechanism to explain the fast release of energy in solar flares by converting stressed magnetic energy into plasma kinetic energy. Though very simple, the basic idea of reconnection is ingenious. It is sketched in Figure 7. The problem of reconnection is that it supposes that even in plasma magnetic field lines can be rearranged to assume a different topology as shown in Figure 7. However, magnetic field lines cannot be cut apart in no way. A better physical picture is to think of magnetic flux density instead of field lines. The magnetic flux density **B** is continuous. This is prescribed by the condition of zero divergence $\nabla \cdot \mathbf{B} = 0$, which simply expresses the fact that magnetic monopoles have never been observed. Quantum theory suggests that the magnetic flux can be exchanged only in microscopically tiny quanta $\Delta \Phi_{mag} = 2\pi\hbar/e \approx 10^{-15}$ Vs. As the field lines have no end, they all and forever must close on all time scales. The sole possibility is that in some spatial point the magnetic flux is annihilated by some process. One can visualize this process by topologically letting two opposite field lines come into mutual contact. After the contact, they may have connected in different ways. This implies that at the contact point the flux has become annihilated, and magnetic energy has been converted into kinetic plasma energy. The effect is that the new field topology is highly bent and tends to release the magnetic stresses into plasma jetting as indicated by the arrows in Figure 7. The force is provided by the Maxwell stresses at the slow mode shocks. As the magnetic flux is negligible at the reconnection site, the new magnetic topology is transported away from the reconnection site by the bi-directional jets. Jetting in the direction of inflow is inhibited because of the strong magnetic field and plasma momentum there.

It has been reported that Parker, on the return flight from a meeting on solar eruptions where Sweet suggested the possibility that magnetic field lines may rearrange from one into another topology, did a short calculation based on the incompressible continuity of flow across an extended current sheet giving Sweet's idea a sound physical formulation. The paper was submitted shortly after Parker arrived home and was published immediately in 1958. Sweet and Parker's mechanism was applied by Dungey (1961) to the magnetosphere in a very simply topological model. Dungey (1995) remembered that he was sitting in a *Café Parisien* on Montparnasse when almost suddenly and unprepared the enlightening idea came to his mind that, by referring to reconnection, one could understand how the ionospheric current systems were generated by magnetospheric convection. More elaborated versions of his model have subsequently been developed (cf. Chapter 63). When the field lines of the solar wind and the magnetosphere interconnect at the dayside magnetopause, which can happen in the first place when the interplanetary magnetic field possesses a southward component, the bi-polar outflow from the magnetopause reconnection site transports the field lines over the poles into the magnetospheric tail. This causes a tailward motion of the magnetic field in the polar regions of the magnetosphere. In the far tail, the newly added magnetic flux reconnects in the center of the plasma sheet in the tail with the bi-polar outflows now being directed earthward and tailward. The tailward jet transports field and plasma downtail forming a *plasmoid*, while the earthward jet generates earthward return flow of field and plasma. This is shown in Figure 8. The model nicely explains the magnetospheric convection pattern and has by now become the standard magnetospheric plasma flow model. A collection of the wealth of the related magnetospheric effects is given in Chapter 43.

There is a large number of indirect confirmations of reconnection at the magnetopause and its pendant in the magnetotail. Plasma jetting in the low latitude boundary layer parallel to the magnetopause has been reported since

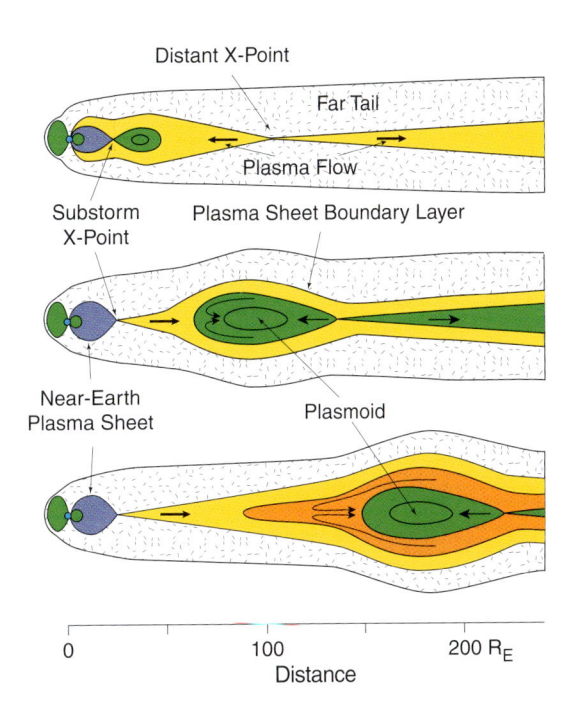

Figure 8 Three stages of plasmoid formation in the tail magnetosphere by reconnection. Initially, there is a distant X-point in the plasmasheet at about 100 R_E. During substorm, the added magnetic flux in the tail causes another X-point to form close to Earth. The reconnection at this point jets the tailward plasma in form of a plasmoid downtail.

1979 (e.g. Paschmann *et al.* 1979). Recently, using data from two nearly equatorial spacecraft (Equator-S and Geotail), both northward and southward jetting could be observed, suggesting that the reconnection site or X-line was somewhere between the two spacecraft which were separated by less than $3R_E$. Other indirect evidence includes the existence of a de Hoffman-Teller reference frame moving tangential to the magnetopause and identifying it as a rotational discontinuity. Transforming to this frame eliminates the perpendicular electric field. The residual plasma motion is parallel to the magnetic field which in the magnetopause is normal to the discontinuity in a rotational discontinuity and has the value of the Alfvén velocity based on the normal magnetic component. Hence, measuring the normal component of the magnetic field and comparing the normal flow with the Alfvén speed provides another test, the so-called Walén test. Figure 9 shows how well this test works in the case of reconnection at the magnetopause, when the magnetopause is a rotational discontinuity. Finally, some particular signatures in the magnetic field at the magnetopause have been interpreted as single reconnected magnetic flux tubes or "Flux Transfer Events". This must have been produced in "bursty" reconnection.

The magnetopause and dayside magnetospheric conditions are summarized in Figure 10 in a sketchy way. The

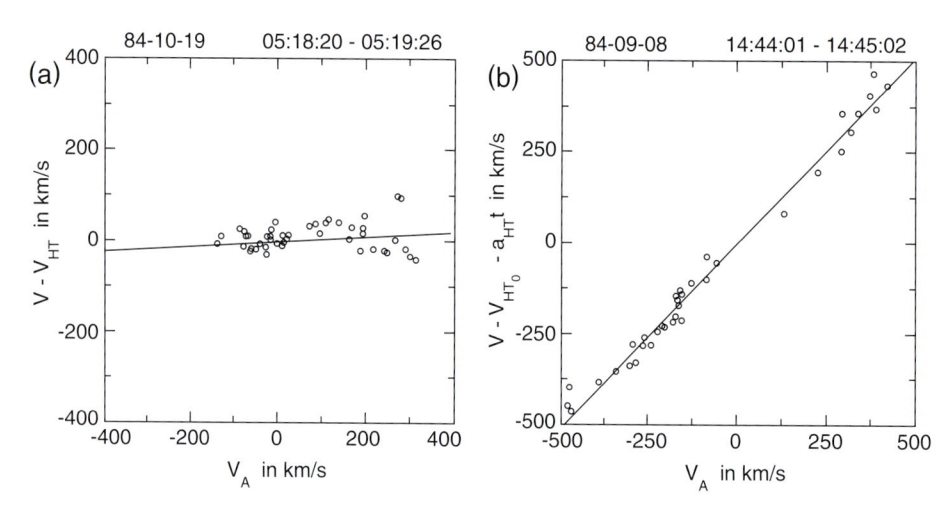

Figure 9 The Walén condition for two kinds of magnetopause crossing. (a) Tangential discontinuity. The Walén relation is not satisfied. (b) Rotational discontinuity. The Walén relation indicates the presence of reconnection at the magnetopause. (From Hultqvist et al. 1999.)

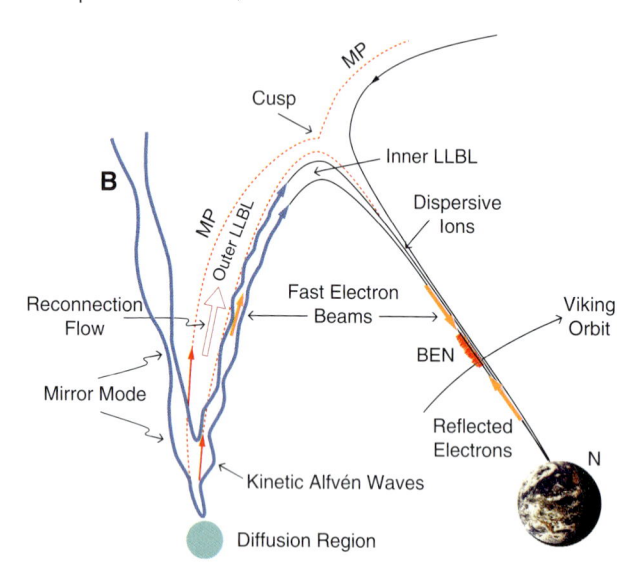

Figure 10 Schematic of the connection between the cusp dayside ionosphere and the reconnection region at the magnetopause (MP). The figure shows only the northern daylight quadrant. The real geometry has approximate north-south symmetry. The interconnected magnetosheath and magnetosphere field lines move slowly at Alfvén speed together with the bulk plasma-jet flow in the outer low-latitude boundary layer (LLBL) northward (in the southern quadrant, which is not shown, they move southward) while fast electron and ion beams accelerated in or close to the reconnection diffusion region escape along newly interconnected field lines in the inner low-latitude boundary layer down to the ionosphere. These beams form the velocity dispersed particle flux which is observed at low-altitude spacecraft altitude (e.g. by Viking whose orbit is indicated), where they excite broadband electric noise (BEN) on cusp field lines. BEN in itself is an indication of the presence of electron beams. (From Treumann and Pottelette 1999.)

reconnection in the diffusion region jets the plasma along the magnetopause to north and south forming the outer low latitude boundary layer while fast electrons and ions accelerated in the reconnection region escape along the newly reconnected field lines into the polar region forming the inner low latitude boundary layer. These beams enter the inner magnetosphere as dispersed particle fluxes in the polar cusp region, where they generate broadband electric noise (BEN) in medium altitudes, which is an indication of their presence. At the same time, kinetic Alfvén waves carry pulses of field-aligned currents which connect the ionosphere to the magnetopause.

The evidence for tail reconnection is not as numerous. Its most spectacular manifestation is the magnetospheric substorm, a description of which is beyond the scope of this essay. For our purposes, it suffices to note that reconnection in the tail plasma sheet has been indirectly confirmed by an overwhelming amount of observations, the interpretation of which is very difficult to reconcile in models that ignore the process of reconnection. The only disturbing fact is that nobody ever reported convincingly that he had crossed the reconnection site. What is so special about it? The Sweet-Parker model predicts that the very reconnection process, i.e. the annihilation of magnetic flux, takes place in a narrow but extended diffusion layer in the otherwise long and homogeneous current sheet separating the two oppositely directed magnetic fields. Reconnection in this case proceeds at a slow rate which can be expressed as the Mach-number constructed from the incoming flow speed and the Alfvén speed, and by the particle flux conservation. This limiting rate turns out to be $\mathcal{M}_{SP} \approx 1/\sqrt{Lu}$, where $Lu = \mu_0 L v_A / \eta$ is the Lundqvist number (or Reynolds number based on the Alfvén speed). Since η is very small, the Lundqvist number is very large, and thus the reconnection rate is very low causing very slow reconnection.

In 1964 Harry Petschek (see also Petschek's reminiscences 1995 and 1996) gave a talk in which he proposed an interesting variant of this process. His paper was published in the nearly inaccessible conference proceedings but nevertheless became one of the most cited papers in all of plasma physics and in particular in space plasma physics. Petschek realized that the restriction of the diffusion region to a small localized domain has the advantage that the magnetic fields need to diffuse only at a narrow reconnection site. This makes possible the configuration shown in Figure 7. The main transport of plasma as well as plasma heating then does not take place at the reconnection site but far outside in four long extended slow shock waves, which, Petschek proposed, emanated from the tiny diffusion region. This model has the advantage that anomalous diffusivity need not be generated all over the entire magnetopause but rather in a tiny spot within the current sheet, in the diffusion region. As we have explicated in the section on anomalous collision frequencies, the anomalous resistivity produced at the magnetopause is by far large enough to provide the diffusivity required for Petschek reconnection (or one of its more elaborate modern versions).

The reconnection rate in Petschek's model becomes essentially independent of resistivity $\mathcal{M}_{Pet} \sim 1/\log Lu$. As a very general rule in space and astrophysics, all logarithms of large numbers are less than 100. Hence, Petschek's reconnection rate is much larger than the above Sweet-Parker rate. By our argument, it should in fact be larger than 1%. The reason is the fact that rather a tiny fraction of all the plasma passes through the current layer. As the diffusion region is not extended spatially, it is much less surprising that the reconnection diffusion region has barely been detected by any spacecraft. Petschek theory predicts that its extension is the order of the ion inertial length $\lambda_i = c/\omega_{pi}$, which, at the magnetopause, is on the order of several ~ 100 km. Preliminary full particle simulations show that it can be comparable to the electron inertial length. Hence, assuming a homogeneous coverage of the dayside magnetopause by spacecraft orbits, the probability of detection of the diffusion site for stationary reconnection is a mere 10^{-6} or so.

On the other hand, it is barely known how the diffusion reconnection site would manifest itself in the data. Theories do not agree about its structure. A resistive diffusion region is expected to contain hot electrons and energetic ions in chaotic motion together with possibly strong lower-hybrid and ion-acoustic and sometimes even sporadic Buneman turbulence. Other theories, based on hybrid simulations or including finite electron mass, propose that Hall-current driven whistler mode waves should be an indication of the reconnection site, while anomalous resistivity should be negligible. The entire reconnection region then has the structure of one large localized nonlinear whistler wave similar to a whistler solitary wave. Purely inertial effects may play a role as well, giving rise to "inertial resistivity". Moreover, finite Larmor radius effects have been proposed. Finally, non-resistive non-turbulent reconnection has been suggested as well, based on observation of mirror mode generation in plasma exhibiting pressure anisotropy. This effect resembles the Meissner effect in superconductivity, where the magnetic field is annihilated by diamagnetism instead of by heating. The only fact that is certain is the conclusion that electrons play the important role in determining the length scale of the diffusion region, while the motion of the ions determines the reconnection rate. Here they must be liberated by some fast process from being enslaved by the magnetic field in the frozen-in condition. This liberation is the basic requirement for making reconnection possible.

We should not forget to mention a recent development in reconnection theory which is open to testing in space plasma. It concerns the proposal that the irregular magnetosheath magnetic field provides conditions at the magnetopause which support a collisionless tearing mode structure similar to that reconstructed in Figure 6. If the tearing islands are irregularly distributed in the magnetopause some magnetic field lines may wander through the magnetopause and connect with the magnetospheric field. There is no well-formulated theory for this proposal. However, it has been proposed that such a theory could be based on statistical mechanical grounds, if it became possible to write down the partition function for such a field line wandering process. The prediction in such a case would be that collisionless reconnection is not a simple instability but a phase transition which could be treated by renormalization theory. This is similar to the above notion that a plasma boundary, which goes pregnant with reconnection, must contain bubbles of low magnetic field and must thus have undergone a quasi-superconducting phase transition.

None of these mechanisms is convincing as yet. The field is widely open for speculation and will be clarified only by combination of measurements in the magnetopause and the tail plasma sheet and particle simulation models in as many dimensions as feasible (for a timely review, see Scholer 1991). Three-dimensional simulation calculations are most desirable. The available hybrid and quasi-full particle codes in two dimensions have already produced interesting and testable results about the structure and dynamics of the remote reconnection layer as well as the diffusion region. These results indeed indicate that the very reconnection region is dominated by electron dynamics. Moreover, the current sheet is highly unstable and seems to decay into filaments. Finally, for a small diffusion region slow shocks will develop near the separatrix but far (the order of 100 ion-inertial lengths) away from the reconnection site. These shocks act to thermalize the upstream plasma when it passes across them into the region of the jetting bulk plasma. All the mechanisms related to this kind of slow shock formation and heating are of utmost interest in the fundamental process of reconnection. One expects that in near future the field will become sufficiently experimentally mature to ultimately test the different theoretical proposals and to understand the physics of the reconnection diffusion region.

COLLISIONLESS SHOCKS: THE EARTH'S BOW SHOCK WAVE

Collisionless shock waves in plasma are a particularly important family of discontinuities. They have first been proposed by Gold (1953) realizing that information from the corona through interplanetary space to Earth can only be transmitted via a shock wave which is necessarily collisionless. Astrophysical application, when speaking about emission of radiation and particle acceleration, makes wide and frequently unreflected and unjustified use of the concept of a shock wave. It is another great advantage of space plasma physics that Nature provides us with an easily accessible sample of a collisionless shock, the Earth's bow shock wave which arises as the result of the impact of the super-alfvénic solar wind stream on the magnetosphere, here playing the role of a blunt obstacle. This fact allows to investigate the structure of collisionless shocks in great detail.

The bow shock is a supercritical shock of medium Mach number $M_A \lesssim 10$, where $M_A = v_{SW}/v_A$ is taken relative to the solar wind Alfvén velocity, which is the relevant unit of reference because the thermal velocity in the solar wind is unimportant. The bow shock is a fast shock. Its effect is to retard the solar wind and to compress the solar wind magnetic field. The bow shock is strongly curved. In addition, its overall size is small of the same order as the size of the magnetosphere. Considering the direction of the Parker spiral solar wind magnetic field, the bow shock consists of a region where it has the character of a quasi-perpendicular shock, while the more extended remaining shock surface is quasi-parallel, as is shown in Figure 11. Here parallel and perpendicular refer to the magnetic field-shock normal angle. Again, this multitude of different regions and properties makes the Earth's bow shock an ideal object of investigation of the various types of supercritical collisionless shock waves.

Many of the properties of supercritical and in particular high-Mach number shocks have not yet been clarified. One of the basic questions still remains the mechanism of plasma heating and entropy production which causes the dissipation of the energy of flow and the steepening of the

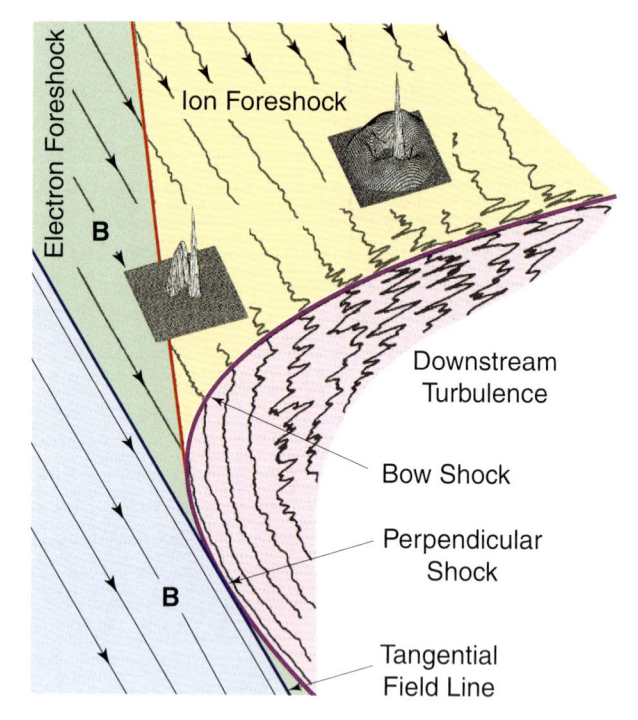

Figure 11 The bow shock geometry. At the tangential field line the bow shock is a perpendicular shock. Indicated are the electron and ion foreshock regions. The small inserts show the ion distribution functions close to the boundary of the ion foreshock and deep in the ion foreshock in front of the quasi-parallel shock region. At the boundary, the distribution exhibits the solar wind beam in the center and the reflected ion beam. In the foreshock, the distribution still shows the solar wind beam, but the reflected ion beam has dispersed into a ring shell distribution of hot ions. The magnetic field structure shows the distortion of the magnetic field due to the excitation of large amplitude waves by the ion beam-beam and ion beam-shell interaction in the foreshock region.

shock. In subcritical shocks, it is believed that the mixing of the upstream supersonic flow and the plasma reflected from the obstacle causes a beam-beam instability. This can be understood in terms of a temperature anisotropy with higher temperature in the direction of the flow, which in a quasi-perpendicular shock is perpendicular to the magnetic field. The anisotropy excites electromagnetic ion-cyclotron waves and electron whistlers very similar to the mechanism discussed above in the context of the stability of the radiation belts. As the anisotropy is constantly restored, these waves are driven very hard until reaching nonlinear amplitudes. The whistlers subsequently steepen, and a shock forms when, in addition, some mechanism generates an anomalous resistivity. Such a mechanism is provided by the transverse current instability in the steepened wave, as has been discussed above. This mechanism has been discovered in 1967 by Kennel and Sagdeev (1967), shortly after the confirmation of the existence of the bow shock in front of the magnetopause. For historical consistency, we should note that Parker (1961) had already proposed that the steepening and quasilinear saturation of an Alfvén wave excited by the fire-hose instability in a streaming plasma like the solar wind would also lead to the formation of a shock wave, while the general philosophy of a microscopic nonlinear theory of collisionless shocks had been developed by Sagdeev (1966).

This mechanism works only for shocks of Mach numbers below a critical Mach number of $M_c \sim 2.7$. As the collisionless heating by anomalous resistivity is a thermodynamic process and hence is a relatively slow process, faster flows leave not enough time for entropy production and braking of the flow. In supercritical quasi-perpendicular shocks with $M_A > M_c$, the lacking dissipation is provided by specular reflection of some of the upstream ions and electrons at the steepened waves back into the upstream flow, long before they can get into contact with the obstacle. Since not the whole flow is reflected, these particles do not carry magnetic flux back upstream. Thus, in the direction perpendicular to the upstream solar wind magnetic field they can propagate only a distance comparable to their gyroradius based on the streaming velocity. The gyrating reflected ions just in front of the quasi-perpendicular shock ramp constitute a local current that possesses a magnetic field, which adds to the upstream solar wind field increasing its strength. It was one of the important experimental verifications of shock theory when this "magnetic foot" was discovered in the seventies.

Faster high-Mach number quasi-perpendicular shocks reflect more upstream particles. In order to be as efficient as possible, the shock ramp must steepen ever more with increasing M_A. The more particles will be reflected as the foot becomes stronger and larger. At a certain not well defined Mach number, the momentum of the reflected ions will become large enough to carry some magnetic flux further upstream, until the foot region widens and another shock front is formed by beam-beam instability further upstream, such that a series of sub-shocks cascades the upstream flow stepwise down.

The foot region exerts other functions in collisionless shock formation as well. It is the electric current in the foot which provides the required dissipation for electron heating. When this current is strong enough, it excites electrostatic Buneman and ion-acoustic plasma waves as well as lower-hybrid drift waves, all of them contributing to anomalous resistivity and thus to dissipation of flow energy and electron heating in the foot region even before the shock ramp. This process broadens the transition region from up- to downstream and causes structuring of the quasi-perpendicular shock. In addition, nonlinear effects on the particles and waves appear in the foot region, which modify the wave spectrum and the electron distribution and may cause the generation of radiation. Such radiation is well known from solar type II radio bursts, which are associated with strong shocks in the solar corona.

Figure 12 shows a dynamic spectrum of plasma waves during a pass of the Wind spacecraft, launched in November 1994, across the quasi-perpendicular bow shock. These spectra were measured during the initial bow shock passes of Wind on its way toward the sun and into the solar wind, where it monitors the variations of the solar wind in order to have a pre-warning facility far out there to predict the space weather conditions near Earth. Wind is one of the cooperative scientific satellites which constitute the International Solar Terrestrial Physics (ISTP) programme, which has been set up by the space physics community to improve the understanding of the physics of solar-terrestrial relations. The other spacecraft belonging to this programme are Geotail, Polar, SOHO, and Cluster II.

The magnetic shock profile seen in Figure 12 is typical for the highly oscillatory nature of the field in and behind the shock ramp and the flat foot region. The ramp is the place of strong low frequency broadband wave emission. The narrow line at about 30 kHz in front of the shock comes from Langmuir waves excited near the solar wind plasma frequency by the reflected electron beams moving upstream along the electron foreshock field line. Two weak

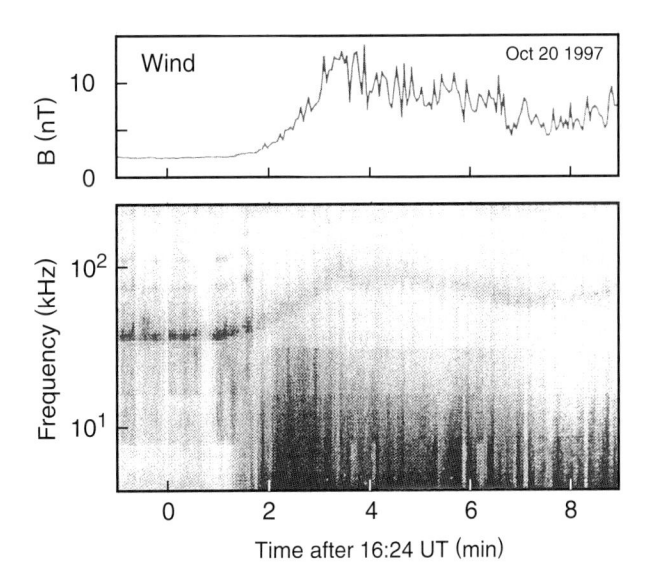

Figure 12 Plasma wave dynamical spectrum during a bow shock crossing of the Wind satellite (after Bale et al. 1999). *Top*: the profile of the magnetic field across the shock exhibiting the strong magnetic fluctuations. *Bottom*: Wave spectrum showing the plasma Langmuir wave fluctuations and the high intensity broadband emissions in the shock transition region.

emission lines at roughly the second and third harmonic of the plasma frequency appear in the foreshock. These are foreshock generated electromagnetic emissions characteristic of strong shock waves (see the section on radiation). No radiation is produced neither in the ramp nor in the downstream region. The fable continuation of the plasma frequency line across the shock ramp into the magnetosheath is caused by weak thermal fluctuations and maps the local plasma density profile across the shock. It thus provides an excellent measurement of the density compression. As predicted by theory, this compression nicely follows the magnetic field compression.

The second function of the foot region is to act as a particle source for the quasi-parallel part of the bow shock. Indeed, the reflected particles in the foot region have in addition to their perpendicular velocity component a component $v_\parallel$ parallel or antiparallel to the upstream magnetic field. When this component exceeds the upstream parallel flow speed, the reflected particles move away upstream along the solar wind magnetic field from the quasi-perpendicular shock into the undisturbed solar wind upstream of the quasi-parallel bow shock. At the same time the magnetic field line along which they move is convected towards the shock with the flow. Usually the reflected electrons are much faster than the reflected ions and therefore can overcome the upstream flow easier. Beams of such reflected electrons populate the solar wind in front of the quasi-parallel bow shock between the field line tangential to the quasi-perpendicular shock and the shock. The slower ions need more time. Hence the field line on which they are found first has moved some distance closer to the shock. The region between the tangential and the first field line populated by ion beams is called the *electron foreshock* while the whole region populated by ions in front of the quasi-parallel shock is the *ion foreshock*.

Figure 11 shows the different regions around the Earth's bow shock with the electron and ion foreshocks, the quasi-perpendicular region around the tangential field line, the large quasi-parallel region of the bow shock and the plasma turbulence indicated by the wave irregular field lines. This plasma turbulence is caused by the ion distribution functions which are shown close to the ion foreshock boundary (double beam) and deep in the foreshock as a ring distribution.

The role that foreshock particles play in the formation of quasi-parallel shocks is very important. Foreshock ion beams and diffuse ion ring distributions interact with the upstream flow causing a beam-beam and beam-ring interaction, which excites large amplitude magnetic fluctuations far upstream of the quasi-parallel shock. These fluctuations, which are polarized more or less perpendicular to the upstream magnetic field, isotropize the upstream particles heating them to temperatures higher than the solar wind. This causes a broad region of dissipation in front of the shock ramp of the quasi-parallel shock and thus an effective broadening of the shock dissipation region into the upstream region. The net effect is that the solar wind is already retarded to some degree before even reaching the very shock ramp. In addition, the downward convection of the large amplitude upstream waves modifies the shock in a number of ways, changing it, reforming it, becoming transformed at the shock and passing it into the region downstream of the shock, the magnetosheath (see e.g. Scholer and Burgess 1992). The most important modification of the shock is that locally along the quasi-parallel shock front the magnetic field direction is turned by the impacting wave packets to become about quasi-perpendicular. Hence, any curved quasi-parallel shock will locally and temporarily behave quasi-perpendicular and, to a lesser degree than a genuinely perpendicular shock, will itself reflect particles upstream until they form a foot in front of the quasi-parallel shock. These particles, due to the inclination of the magnetic field, will have substantial parallel speed to themselves, escape into the foreshock and add to the already present diffuse ion distribution there. This implies that practically any supercritical quasi-parallel shock behaves locally like a quasi-perpendicular shock, while possessing a broad turbulent foreshock. It can thus rightly be said that this foreshock constitutes the extended shock transition from upstream to downstream, and there is no really sharp boundary between the undisturbed upstream flow and the shock transition except for the dynamically unimportant field line that forms the foreshock boundary.

What concerns the further fate of reflected and isotropized ions when they are convected toward the shock, a number of investigations have verified that some of these ions traverse the shock into the magnetosheath. Their distribution is a shell or ring distribution which is unstable against electromagnetic ion-cyclotron waves. Such waves are excited just behind the shock in the magnetosheath, where they contribute to magnetosheath plasma heating. Their existence and further dissipation has been proven by direct observation.

Electrons behave differently from ions. Figure 13 shows three parallel electron distribution functions measured when they were crossing the perpendicular shock. The solar wind distribution, which is close to a Maxwellian, is transformed during shock crossing into a broad heated flat-top distribution. This effect is mostly due to the electric potential drop in the shock. However, plasma turbulence in the shock ramp itself contributes substantially to electron heating right in the shock ramp. Within the ramp, the electron distribution consists of a residual solar wind beam distribution and a beginning flat-top distribution indicating that the solar wind beam runs into the hot shock ramp electron plasma causing a beam-hot plasma interaction, which retards the beam and heats the plasma. The nature of this interaction is most interesting for shock physics. As,

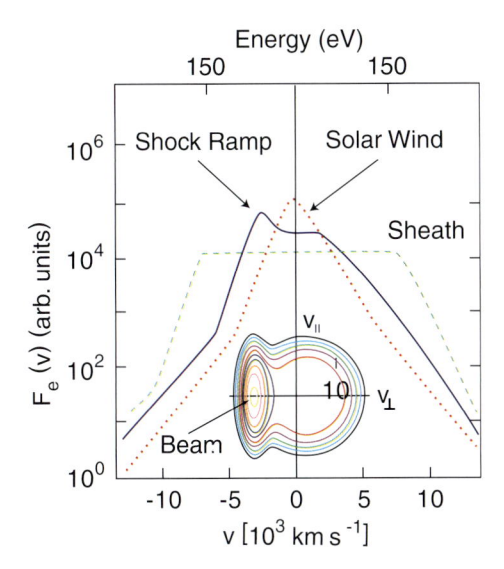

Figure 13 Electron distribution at different positions in the shock transition. The solar wind electron distribution becomes a mixed distribution in the shock ramp showing a residual of the solar wind electron stream moving into the hot shock ramp distribution. Behind the ramp in the adjacent sheath the electron distribution has been heated into a top-flat distribution. The insert shows the phase-space isodensity contours of the distribution in the shock ramp mixing region with cold electron beam and hot ramp plasma. (Adapted from Feldman 1985.)

discussed above, the electron dissipation generated in the course of this interaction in high-Mach number shocks does, however, not suffice for the entire deceleration of the upstream plasma beam, whose energy resides basically in the super-alfvénic upstream ions. We have already described how Nature is able to circumvent this hurdle and to generate the required increase in entropy in high-Mach number shocks.

Many of the processes mentioned are still the subject of experimental and simulation studies. Numerical simulations combined with observations of bow shock crossing have contributed to today's degree of understanding of collisionless shock wave physics.

PLASMA WAVES AND PLASMA TURBULENCE

Holding an electric antenna of a length longer than the Debye length into near-Earth space and measuring the potential difference along the antenna as a function of time will provide you with a wave form measurement. Depending on your time resolution, you cover a certain range of frequencies.

The collisionless plasma of near-Earth space is full of waves of all frequencies and wavelengths. Most of these waves are dynamically unimportant. They have very small amplitudes and, consequently, their energy density is much less than the thermal or kinetic energy density of the plasma. However, knowing the generation mechanism of the waves provides very useful information about the parameters and dynamics of the plasma as the generation of the waves depends on the plasma free energy, and the waves have characteristic frequencies which in many cases are directly related to the plasma parameters. A wave synoptic of the magnetosphere maps the spatial variation of the plasma parameters of the magnetosphere along the satellite path similar to Figure 12, where the Langmuir wave line maps the variation of the plasma density from the solar wind foreshock region across the shock ramp into the magnetosheath. In the magnetosphere, the function of the Langmuir wave line is taken by the upper hybrid wave. In addition, the strong magnetic field and the presence of temperature anisotropies in the electrons causes excitation of electron-cyclotron waves which map the magnetic field variation in several harmonics.

Other waves are lower-hybrid waves, whistlers and all kinds of low-frequency electromagnetic waves. Some of the latter have been indicated in Figure 11 as magnetic oscillations in the ion foreshock, where they are excited by ion-ion beam instability. In the magnetosheath, one has in addition one very important low frequency magnetic wave, the mirror-mode, the physics of which is barely understood even today. It is excited by an excess in perpendicular pressure over parallel pressure in the magnetosheath and accumulates as large amplitude compressive oscillations in the magnetic field just in front of the magnetopause. We already noted that this wave may be important in igniting reconnection. On the other hand, being a very low-frequency mode and often having been observed at extraordinarily large amplitude makes it very interesting in any low-frequency turbulence theory of collisionless plasmas.

Plasma turbulence can be divided into low-frequency turbulence and high-frequency turbulence. The former is treated mostly on the basis of magnetohydrodynamic theory by simulation technique. It seems that such simulations verify Kolmogoroff's spectral turbulent power law $W(k) \propto k^{-5/3}$. This dependence has been verified also in solar wind turbulence. It has not yet been checked in magnetosheath turbulence which may be closer to two-dimensional and thus should lead to another dependence suggested by Kraichnan.

High frequency plasma turbulence is basically kinetic plasma turbulence and therefore implicates the full kinetic theory. It is of utmost interest for all of plasma research that the confirmation of the pitch-angle diffusion theory by observation of the lifetime of radiation belt particles also validates the famous *quasi-linear plasma theory* at least for this particular type of wave-particle interaction.

Quasi-linear plasma theory, the lowest-order approximation to well-developed plasma turbulence, had originally not been invented in view of space application. It emerged in the fifties independently and at about the same time in both the Russian and US plasma physics schools at a time when the two-sided embargo on exchange of scientific information between the western capitalist and eastern communist societies was still active. When, in the late fifties and early sixties, this embargo fell and scientists from both parts of the world first met, they were very surprised that their ideas, theories and independently obtained results agreed so well. They took this fact not as a coincidence but as the confirmation of the independence of scientific thought on ideological influence and guidance.

We cannot however be as certain about the independent discovery and the shared priority of these research programs. In the former Soviet Union, it arose as a transformation of Landau mean-field theory in statistical mechanics to the classical field of plasma theory. It was Sagdeev and his coworkers, Vedenov and Velikhov, who became interested in the fifties in the reaction of a strong turbulent plasma-wave field on the behaviour of the plasma. In order to treat this problem analytically, they developed the quasi-linear formalism. During the first International Conference on Fusion Research in 1961, they presented their theory to the international plasma community. It then took only a few weeks until Drummond and Pines announced a similar and essentially identical theory, both getting published in the conference series. An application to shock formation (Parker, 1961) followed almost immediately. The above theory of plasma confinement by Kennel and Petschek (1966) is but the consequent application of quasi-linear theory to the evolution of whistler and ion-cyclotron turbulence. It is one of the most beautiful examples of this theory, which in many other occasions provides only a crude approximation.

The original target of quasi-linear theory concerned the evolution of the so-called gentle-beam instability. It arises when an electron beam of a certain speed v_b traverses an (unmagnetized) plasma. This happens in Nature when an electron beam flows parallel to the magnetic field, the nicest example being the electron-foreshock reflected electron beams. The free energy in the beam is used to generate Langmuir waves by means of the instability in the beam itself. The resonance condition is simply that $kv_b \approx \omega$, and the waves are excited with phase velocities slightly below beam speed, $\omega/k \leq v_b$, where the distribution function has positive slope in v-space. This was found as early as 1946 by Leo Landau. The interaction of the plasma and the beam with the excited Langmuir wave spectrum has two consequences: (1) the waves slow down the beam, and (2) the waves heat and accelerate the background electrons which have comparable velocities. The result is that the gap between the background electron distribution and the beam is replenished and filled with electrons until the instability is quenched, a process described by quasi-linear theory.

In 1972 Zakharov, working with Sagdeev, realized that an intense Langmuir wave field long before quasi-linear saturation should exert some radiation pressure on plasma which modulates the plasma density in the regions of large amplitude. The plasma wave field is trapped in the density modulations and decouples from the beam. In one dimension, the plasma wave radiation pressure can be balanced by the thermal pressure of the surrounding plasma. This process generates Langmuir solitons which are stationary solutions of a nonlinear Schrödinger equation. Zakharov found that in higher dimensions no steady state would be reached. The plasma collapses locally creating ever narrower holes filled with plasma waves and generates a spiky spectrum of ion-acoustic and Langmuir fluctuations of localized but enormous intensity. A quiescent plasma in this way evolves into a state of intense turbulence.

Zakharov's discovery ignited a vivid activity in investigation of this strongly turbulent state leading to a well developed technique of collapse theory. In space, this theory could be confirmed only in very rare cases. The most probable place where it may be valid is the electron foreshock. Here spikey Langmuir wave emissions have been observed frequently allowing the theory to be checked. In the magnetosphere, plasma collapse is rather improbable. The many other wave modes which can be excited in a strongly magnetized plasma carry away the free energy, and high-frequency turbulence stays mostly weak. The other case when it becomes important is in drift wave turbulence, which may contribute to plasma transport, as has been discussed above.

Strongly turbulent stationary wave solutions other than radiation pressure generated collapsing cavitons result from the interplay of nonlinear wave steepening and wave dispersion. Such processes have been theoretically predicted for instance for ion-acoustic waves. The dispersion relation of these waves allows for short wavelength waves to flow at ever decreasing phase velocities. Hence, when the ion-acoustic wave steepens in its nonlinear evolution, then newly generated short wavelength components stay behind the traveling wave until the steepening stops. This process is described by the Korteweg-de Vries equation and causes a large amplitude solitary wave or density ripple propagating at ion-acoustic speed in the plasma. Strongly turbulent plasmas are full of such ripples, as has been found by various space observations in the solar wind and in the auroral zone. Similar effects are also caused by other wave modes, like electron acoustic waves and drift waves of the lower-hybrid type. They all contribute to strong plasma turbulence and provide a broad research field in space plasma physics, the goal of which is the ultimate understanding of well developed plasma turbulence.

The most interesting recent observation of strongly nonlinear effects in high-frequency plasma modes has been restricted to the downward and upward current regions in the dilute plasma of the auroral magnetosphere and their connection to the plasma sheet boundary layer. Here the presence of strong electron beams leads to an evolution of highly nonlinear wave modes, which are of different nature in the two regions. In the plasma sheet boundary layer and the downward current region, it has been experimentally found that they form electron phase space holes which cause strong electron heating and localized wave fields typical for strong high-frequency plasma turbulence. Side effects of the presence of these structures are the acceleration and heating of electrons and possibly the generation of small-scale field-aligned electric potential drops. In the upward current region, the waves form very intense broadband wave packets, which probably relate to generation of radiation. This field of research is of undeniable importance, though it is still in its infancy.

PARALLEL ELECTRIC FIELDS

Evidence for the existence of magnetic-field aligned electric potential drops in the magnetosphere has come from observation of particle acceleration in the auroral region. Similar evidence has been provided by observation of very impulsive solar hard-X ray and Gamma-ray flares. In both cases the generation of high energy particles in very short time is hard to understand if not assuming that strong electric potential drops can exist at least for limited times in the plasma along the magnetic field.

Perpendicular electric fields generate plasma drift via the $\mathbf{E} \times \mathbf{B}$ motion and do not pose any problem in collisionless processes in plasma. *In situ* measurements have been provided by plasma injections (e.g. Haerendel *et al.* 1967, Föppl *et al.* 1968). On the other hand, field-aligned (parallel) electric fields $\mathbf{E}_\parallel$ can hardly be supported in collisionless plasma. They necessarily induce electron motion along the field which cancels out the potential drop. The observation of parallel electric potential drops thus indicates the presence of violent processes which violate charge neutrality. On a fast time-scale, the order of the electron plasma scale $\tau_{pe} \sim \omega_{pe}^{-1}$ non-neutrality poses no problem; local violation of charge neutrality is the reason for the existence of plasma waves on this time scale. On longer time scales, however, charge neutralization happens on electron flow times $\tau_e \sim L_\parallel/v_{e\parallel}$ along the magnetic field and is suppressed only in cold plasma. Parallel fields thus require some mechanism which, in hot plasma, inhibits free motion of electrons along the magnetic field. However, their direct observation is difficult. First attempts to measure them date back to Mende (1968).

In collisional plasma, resistivity provides a mechanism for generating magnetic field-aligned potential drops. Under certain conditions, anomalous resistivity η_a along the magnetic field replaces ordinary collisions in the way we have exemplified above. If this is the case, the parallel electric field is simply given by $E_\parallel = \eta_a j_\parallel$, where $j_\parallel$ is the field-aligned current density. In marginally stable conditions when the current just quenches itself to the onset of the most relevant instability threshold, the field can be estimated from $E_\parallel \approx \eta_a e n_e v_{c\parallel}$ where $v_{c\parallel}$ is the critical parallel electron current speed which, for ion-sound turbulence, is the ion-acoustic velocity c_s, and η_a is obtained from the Sagdeev collision frequency ν_a^s (eqn (5)). This marginally stable parallel electric field scales as

$$E_\parallel^{\mathrm{marg}} \simeq 0.01 \frac{m_e}{e} \left(\frac{m_e}{m_i}\right)^{\frac{1}{2}} \frac{T_e}{T_i} \frac{v_D}{c_s} \omega_{pe} v_D \quad \text{with} \quad v_D \sim v_{c\parallel} \sim c_s \tag{9}$$

In the auroral zone, these fields could reach values of $E_\parallel \sim$ up to 1 mV/m and thus could theoretically cause potential drops $\Delta\Phi \sim 10\,\mathrm{kV}$ along the magnetic field, if the parallel electric field is maintained over a distance of $>10\,\mathrm{km}$. Such fields are sufficient to accelerate $\sim 10\,\mathrm{keV}$ electrons during disturbed times. However, though typical auroral electrons have energies in this range, measurements have never been able to verify the presence of anomalous resistivities together with the required ion sound turbulence in the auroral magnetosphere. The situation is similar to that at the magnetopause, where searches for anomalous resistivity and diffusivity have been unsuccessful and the measured turbulence levels in all kinds of plasma waves are far below the quasi-linear estimates of the anomalous collision frequencies indicating again the general problems one encounters in collisionless plasma turbulence when trying to determine transport coefficients. We already mentioned in the previous section that instead of the expected broad-band ion-acoustic turbulence the observations rather reveal that the relevant plasma turbulence is highly localized. We will return to this surprising and important observation below.

The current continuity $\nabla \cdot \mathbf{j} = 0$ requires that any field-aligned current closes somewhere. In the magnetosphere, such a closure is possible via currents perpendicular to the magnetic field lines flowing either in the ionosphere or in the ring current or boundary layer regions. In the latter case, these currents resemble diamagnetic drift currents. Knowledge of the perpendicular currents $\mathbf{j}_\perp$ yields for the parallel current the expression

$$j_\parallel = -\int_s ds\, \nabla_\perp \cdot \mathbf{j}_\perp (s) \tag{10}$$

Thus, when the perpendicular current diverges in the direction perpendicular to the magnetic field, the above integral taken along the magnetic field (with field-aligned coordinate s) determines the parallel current density. Since, in the region of parallel potential drop $\Delta\Phi \neq 0$, the parallel current $j_\parallel$ is continuous, there exists a relation $j_\parallel = K\Delta\Phi$. Knowing K as a functional of the particle distribution one can determine the potential drop from the divergence of the perpendicular current density. Otherwise, measuring the latter and $\Delta\Phi$, one may find the coefficient K which contains all the dynamics of generation of field-aligned potential drops.

It is, however, not easy to determine whether and where the perpendicular currents diverge. Direct measurements of field aligned current are therefore sparse, though one may try to determine them from magnetic signatures when crossing field lines in perpendicular direction or from electron flux along the field. The latter method requires knowledge of the entire parallel distribution function of electrons which, again, is difficult to achieve. In general the low energy electrons are badly determined. Detectors are usually polluted by low-energy photo-electrons emitted by the body of the spacecraft. Moreover, the presence of field-aligned electric potentials modifies the particle distribution function in several ways. They accelerate ions and electrons in opposite directions forming opposing electron and ion beams. Usually the original parallel electron distribution is a Maxwellian. In a field aligned potential drop Φ, it becomes a shifted Maxwellian

$$f_{e\parallel}(v_\parallel) \propto \exp\left\{\frac{1}{k_B T_{e\parallel}}\left(\frac{m_e v_\parallel^2}{2} - e\Phi\right)\right\} \quad (11)$$

This suggests that the potential is important only for particles of parallel energy less than $e\Phi$. The more energetic particles ignore the field. This causes the distribution function to become asymmetric in energy, as is shown in Figure 14. The electric field shifts part of the parallel distribution function in energy space causing a beam-like distribution. Based on this expression a very simple functional form for the coefficient K in the relation between the parallel current and the potential drop was given first by Knight (1973). In collisionless plasmas, shear flows also generate field-aligned potential drops (see Figure 16 below). In constant magnetic fields and with no space charge, shear flows imply non-zero divergence

$$\nabla_\perp \cdot \mathbf{E}_\perp = -\nabla_\parallel E_\parallel = \nabla_\parallel^2(\Delta\Phi) \quad (12)$$

of the perpendicular electric field thus causing parallel electric fields and the corresponding field-aligned potential drops $\Delta\Phi$.

In a mirror magnetic field geometry with loss-cone, parallel fields deform the distribution function in a characteristic way as shown in Figure 15, where a measured electron distribution is given. This distribution has been observed in the auroral downward current acceleration region and has

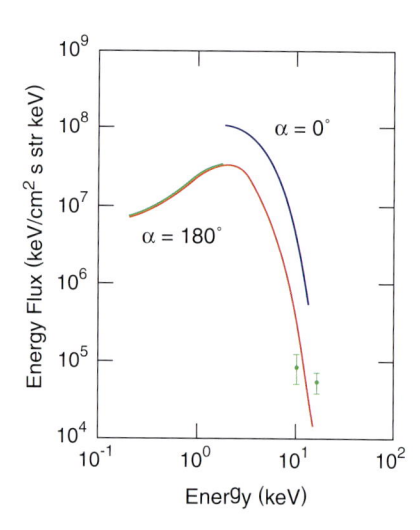

Figure 14 Deformation of a Maxwellian electron distribution with loss-cone in the presence of a field-aligned electric potential drop. Model calculation. (From Chiu and Schulz 1984.)

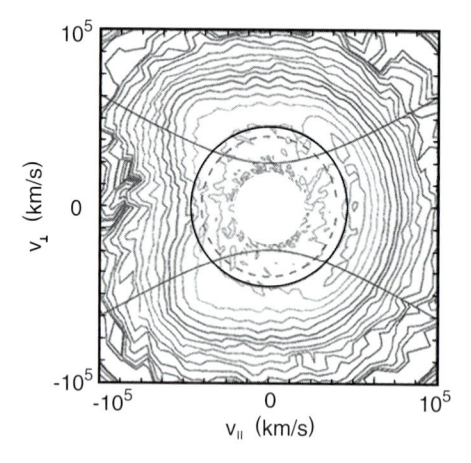

Figure 15 Measured electron phase-space distribution (isodensity contours) near the mirror point in the auroral zone (from Delory et al. 1998). The photoelectron contribution caused by electron emission from spacecraft at small velocities has been eliminated (empty inner region). The distribution has the shape of a "horseshoe" and must have been caused by a parallel electric potential drop. The overlaid curves (circle and hyperbolas, see text) show the regions of trapping and acceleration in the Chiu and Schulz model. The dashed circle indicates the relativistic-resonance line used below in the discussion of electron-cyclotron maser radiation.

the typical shape of a "horseshoe". The location of the parallel potential drop along the field line close to the mirror point of the particles does not simply shift the particles in parallel energy; is also widens the distribution to become a ring. The horseshoe is created by the presence of the loss-cone. Similar distributions had also been observed earlier

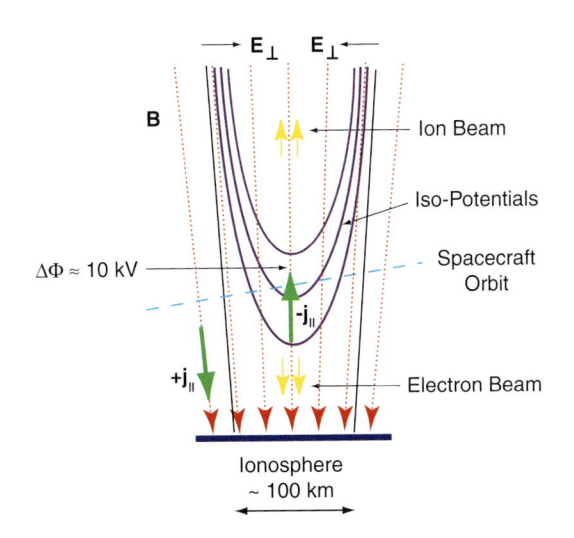

Figure 16 Electric field and current structure in the auroral magnetosphere in the upward current region $-j_\parallel$. The electric potential lines are bent due to the existence of a field aligned electric potential drop $\Delta\Phi$ generated by shear flows resulting from oppositely directed $E_\perp$ Ions are accelerated upward, electrons downward by the potential drop. Observations (see Figure 17) indicate that it is on the order of $\Delta\Phi \simeq 10$ keV.

and had been called electron conics, but the presence of the electric field is important in transforming them into a horseshoe. The circle and hyperbolae in the figure separate different particle species. In particular, the particles between the cirle and the hyperbola are trapped between mirror points along the field line. Particles below the hyperbola are backscattered from the ionosphere. Those above are of magnetospheric plasma sheet origin.

It is thus not decided yet what is the origin of the field aligned electric fields and electric potential drops. A schematic representation of the global field-aligned potential structure in the upward current region is shown in Figure 16. Their presence is well established now, at least in the Earth's auroral zone, but poses a challenge for explanation. Simple models, like double layers, have been developed in the past. These models have been proven inconsistent with observation and collisionless plasma behaviour. Also, the above mentioned mechanism based on anomalous resistivity has not been verified by observations. Observations instead indicate that any field aligned potential drops should be restricted to narrow localized regions along the magnetic field and of small transverse dimension. These structures are of meso-scale and seem to arise typically in collisionless plasma turbulence. More recent theoretical calculations suggest that most of the electric potential drop occurs in a large number of these small-scale structures which have been identified as ion phase-space holes (e.g. Gray *et al*. 1993) or electron phase-space holes (e.g. Omura *et al*. 1998, Muschietti *et al*. 1999), resulting from the graining of the particle phase-space. Small-scale potential drops like these are well understood by now and can be maintained by microscopic plasma equilibria for susceptible times. The many small-scale potential drops may add up to a large field-aligned large-scale potential drop. Their existence in the auroral zone has been convincingly demonstrated by recent observations of FAST, a small spacecraft launched in August 1996 by the Space Science Laboratory of the University of California Berkeley into a near-polar medium-altitude orbit with apogee of about 4000 km. The FAST instrumentation is particularly suited for unprecedented high-resolution measurements of plasma and field parameters, including three components of the electric wave fields. An example of FAST plasma and wave field data is shown in Figure 17. Electron holes have also been detected in the magnetotail plasma sheet boundary layer by the Japanese Geotail spacecraft and close to the bow shock by the Wind spacecraft, both members of the ISTP mission. Geotail (cf. Nishida *et al*. 1998), launched in 1992, can be considered as a follow-up and considerably more advanced mission of the deep tail probe ICE (which formerly was the ISEE 3 satellite). But the only hint about the effectiveness of electron holes in contributing to large-scale potential drops comes at present from observation of the fine structure of auroral radiation in the kilometer wave band. We will discuss this radiation below.

ACCELERATION

Space plasmas are particularly well suited for the investigation of particle acceleration. There are at least four mechanisms of particle acceleration in and near the magnetosphere. The most common one is shock acceleration at the Earth's bow shock. Investigation of the flux of energetic particles at the bow shock provides a good measure of this kind of acceleration. Other important mechanisms are acceleration in reconnection, acceleration in the tail and magnetopause current sheets, acceleration in wave-particle interaction, and direct acceleration in field aligned potential drops.

There is no need to review the various mechanisms of shock acceleration at this point (see e.g. Terasawa and Scholer 1989). Some of them, like shock drift acceleration of ions, Fermi acceleration, diffusive acceleration and the still debatable mechanism of particle injection have been verified and/or tested at the bow shock. Also, acceleration in current sheets is known to exist. It is famous under the name of Speiser-orbit acceleration, and is responsible for the presence of energetic ions escaping from the tail current sheet in the plasma sheet and forming ion beams along the magnetic field. They have been detected in connection with observation of plasmoids in the tail and are regularly observed in the plasma sheet boundary layer.

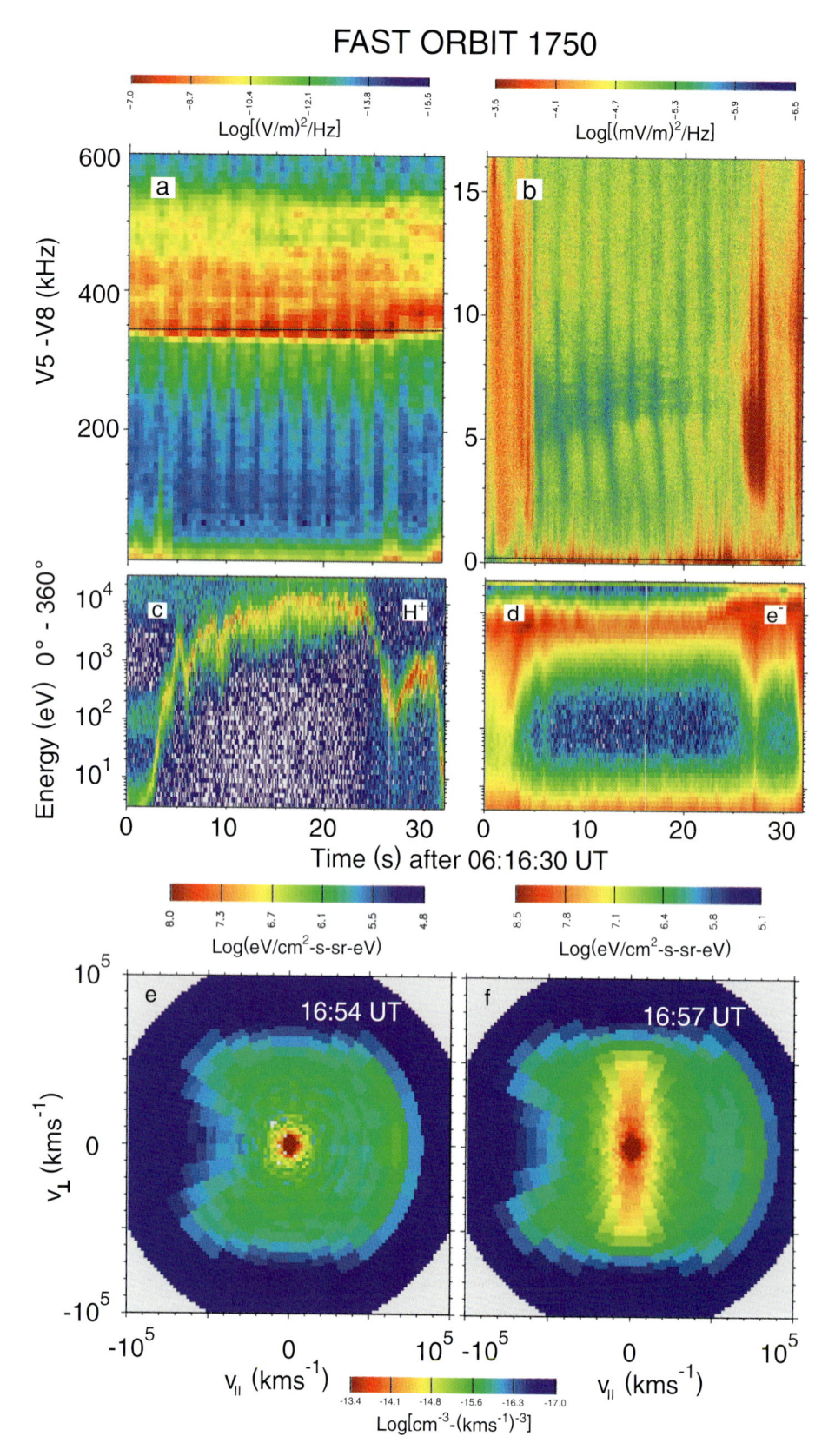

Figure 17 Plasma and field observations in the auroral upward field-aligned current region by FAST. Ion and electron energy spectra are shown in the two middle panels. The violent acceleration of both particle families is exhibited in the spectra when the spacecraft passes the boundary of the acceleration region. Particle energies jump up to ~10 keV. The electron pitch-angle distributions in the two lower panels show the formation of horseshoe distributions, an indication of the auroral electron beam and (unexplained) perpendicular heating of the electrons. (From Pottelette et al. 1999.)

Acceleration in wave-particle interaction should occur rather frequently. Lower-hybrid waves are believed to be the most efficient in accelerating ions perpendicular to the magnetic field, and electrons along the magnetic field. Detection of lower-hybrid waves in many places in the magnetosphere, starting from the foot of the perpendicular bow shock, the magnetopause, the auroral zone, plasma sheet, and the plasmapause region, has been frequently reported. It is however, difficult to measure both the waves and the particles simultaneously. Acceleration of ions by electrostatic ion-cyclotron waves is a much exploited mechanism. For a long time, it has been believed to be the dominant mechanism for generation of field-aligned potential drops. This belief has recently been damaged by the observation of ion and electron holes in the auroral acceleration region with weak ion-cyclotron wave activity. The latter may be a cause of the former. Nevertheless, it seems that ion-cyclotron waves are responsible for the existence of ion-conics.

The most important mechanism for accelerating particles in the auroral zone is direct acceleration by field-aligned electric potential drops. The first indirect hint of the existence of such magnetic field-parallel potential drops in the auroral zone at altitudes higher than a few thousand kilometers was obtained from observation of a barium cloud release in the auroral zone (Haerendel et al. 1976). During this release, it was observed that the ionized barium ions reaching a certain altitude suddenly started speeding up at high velocity along the magnetic field into the magnetosphere. The only reasonable explanation for such a velocity change is the presence of electric potential drops along the magnetic field line occupied by the particles. Later experiments have provided more indirect evidence for these electric potentials. The production and origin of such potential drops, especially when they are very strong, is still under debate and is very badly understood, as we have explained in the previous section. However, when they exist and can be sustained for sufficiently long time over some distance along the magnetic field lines, then particles can pick up the potential difference when passing across and become accelerated even to the full value $\Delta\Phi$. Figure 16 shows a sketch of the expected potential structure in the acceleration region.

Another very interesting and unexpected observation goes back to the early exploratory days of space plasma physics in 1968, when the tail plasma sheet was discovered. V. Vasyliunas, then working on his PhD thesis at MIT and analyzing the particle and field data provided by the OGO 1 and OGO 3 spacecraft monitoring the particle component in the near-Earth magnetospheric tail, was struck by the frequently observed strange particle distribution that he could not avoid to describe. These distributions deviated considerably from the ordinary equilibrium distribution functions which are usually expected to exist in systems under equilibrium. The observed distributions under quiet magnetospheric conditions exhibited the properties of Maxwellians at low energies, but turned into some kind of power law distributions at high energies. Distributions of this kind had been suspected as early as 1853 by the world-famous French mathematician Augustine Cauchy, and were observed already in 1926 by Lewis F. Richardson, who investigated scale-invariant turbulence, and by Paul Lévy (1886–1971) who was interested in the unusual dynamics of particles and fluid elements in turbulence. Vasyliunas fitted a power law index κ to the observed distributions, which later were called Kappa-distributions (or also Lévy distributions). Many such Kappa-distributions have since been measured (see e.g. Christon et al. 1991) in the tail of the magnetosphere, the solar wind and in other places in space. For instance, the undisturbed solar wind electron distribution function possesses, over many orders of magnitude in energy, a Kappa-distribution.

What process is responsible for generating of such distribution functions? Obviously, the plasma during times when Kappa-distributions are measured is in a quiescent state, apparently near some kind of equilibrium. In such a state, one expects that the distribution be Maxwellian (or in terms of statistical physics, Boltzmannian). The fact that it is not poses a very interesting problem for all of basic physics. Are there stationary states in equilibrium when the plasma (or another system) behaves collisionless and therefore is far from thermal equilibrium? A solution to this problem has recently been proposed in terms of *stationary states far from thermal equilibrium*. Such states may occur in collisionless but otherwise turbulent plasma which evolve into wave excitation on all scales and therefore behave scale-invariant over a large range of scales. Such systems have long been known to behave strangely, similar to that which was predicted in the presence of Lévy flights (Shlesinger et al. 1993). When this happens, it is demonstrated that the basic solution of the Liouville equation differs from the Boltzmann-Gibbs ansatz, suggesting the presence of a generalized collision integral. In equilibrium, this collision integral allows for a distribution function resembling a Kappa-distribution (Treumann 1999). The existence of a collision integral, including long-range interactions and differing from the well-known Boltzmann collision integral, has consequences for the derivation of transport equations and transport coefficients, a point that will have to be explored in more detail in the future. This finding may have consequences for quasi-thermal equilibria in plasma physics and in particular in potential applications to the Kappa-distributions measured in the solar wind (cf. Chapter 42) and in the magnetosphere. It may also be of interest in other fields of statistical physics.

RADIATION

Plasmas are hot, having kinetic temperatures well above $T > 10^4$ K or, in energy units $\gg 1$ eV. There temperatures can easily reach values of keV or even more. Such hot systems should emit radiation, which classically according to Wien's law, should have a total (black body) intensity proportional to the fourth power of the temperature $I \propto T^4$. This holds, when the plasma is collisional, which is the case for plasmas like the sun, stars etc. The radiation emitted is free-free emission, coming from the retardation of electrons when passing close to nuclei (protons). Most of the plasmas in near-Earth space are, however, non-collisional, which implies that their bremsstrahlung free-free emission is negligible. Immersing such plasmas into sufficiently strong magnetic fields causes the plasma to radiate synchrotron radiation, which comes from the cyclotron motion of the electrons around the magnetic field lines. The latter radiation is non-thermal in nature because it is not generated by collisions; however, here we consider it as thermal. The reason is that, in a plasma, the stationary external magnetic field provides a similar order to the plasma particles as collisions. These two types of thermal radiation have been known for a long time (cf. Jackson 1975). The first can be used to measure the temperature and density of the remote radiation source; the second allows for a determination of the plasma temperature and the magnetic field magnitude.

Space plasmas sometimes radiate in a way that is fundamentally different from either of these two types. Their emissions may be called nonthermal, sporadic, and irregular. When measuring the intensity of this sporadic radiation in terms of Wien's law, one very frequently obtains equivalent temperatures on the order of 10^{15}–10^{18} K, corresponding to energies of 100–10 000 GeV. Since plasmas in near-Earth space cannot have thermal energies higher than several MeV, the radiation is clearly nonthermal. Typical examples of this kind of radiation are the solar and interplanetary radio bursts of the type I to type V families (for a closer description see e.g. McLean and Labrum 1985). Planetary emissions belonging to this category are found in the intense auroral kilometric radiation (AKR) emitted from Earth's auroral zone, and some of Jupiter's radio radiation in the kilometric and hectometric wave band. The best investigated radiation of all these is the terrestrial AKR (Figure 18). It profits from the advantage of proximity to Earth and thus from relatively easy accessibility.

The electron cyclotron maser

Discovery of AKR about 40 years ago by Benediktov *et al.* (1968) was almost ignored until Gurnett (1974) rediscovered it and estimated its power. Benediktov *et al.* (1968) used data from a radio wave receiver on board the Soviet spacecraft

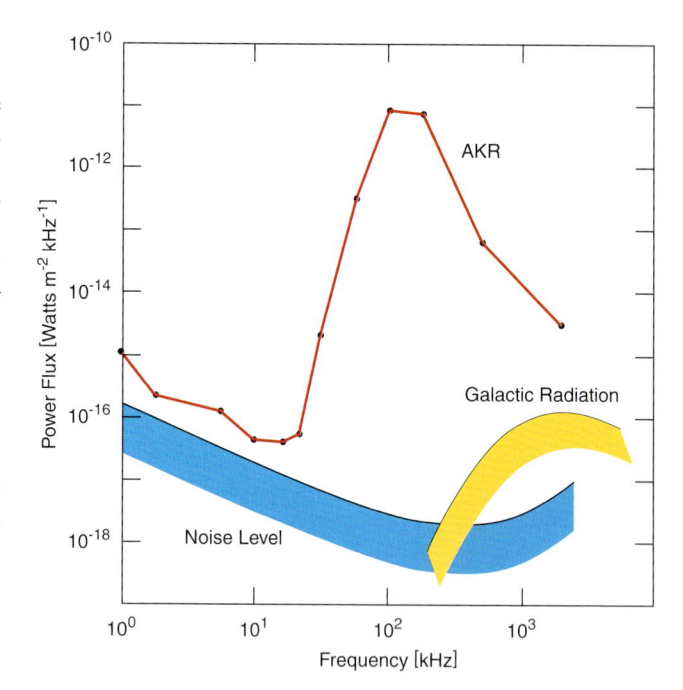

Figure 18 The first preliminary integrated spectrum of the auroral kilometric radiation (AKR) many orders above the galactic background radiation. (Gurnett 1974.)

Elektron-2, which was launched in 1964 into a relatively low orbit. It detected a broad band of electromagnetic noise which remained unexplained but was suspected to be ionospheric in origin. The issue was cleared only when, ten years later, Gurnett (1974) used the more sophisticated wave instruments on board the American Imp 6 and Imp 8 spacecraft launched in March 1971 and October 1973, respectively, into highly eccentric orbits. Gurnett was able to measure the spectrum of the radiation (Figure 18). Taking advantage of the large distance of Imp 8, which was located at 32.1 R_E geocentric distance, and of the directivity of the radiation, Gurnett determined the location of its source to be in the auroral ionosphere and having size $<6°$. He could also show that the radiation was strongly correlated with auroral substorm activity. This identified it as what it is known today: an auroral substorm phenomenon of high variability contributing to loss of energy into free space.

The estimated radiative power came as a big surprise and a challenge to our understanding about the kind of violent mechanism that could generate radio waves capable of carrying away between 1–10% of the estimated total energy set free in an auroral substorm. Gurnett's (1974) observation immediately ignited a vivid theoretical effort on the presumably nonthermal radiation mechanism. The first attempt was directed at exploiting the shortly before discovered plasma collapse (Zakharov, 1972). When Langmuir or upper-hybrid waves destabilized by the energetic auroral electron beam reach nonlinear amplitudes, their radiation pressure modulates the plasma, expelling it from the region of largest wave power. As this seemed to provide a violent process, it was natural to assume that the shaking of electrons, which interact strongly with the nonlinear wave packets formed in the collapse, could lead to intense radio wave emission. The radiation should then escape from the many little antennae created by the auroral beam along its path down the magnetic field lines. However, though calculation revealed that this mechanism yielded radiation wave power scaling as a low power with ratio of plasma wave energy density to kinetic plasma energy density, which itself is naturally only a small number, the order of 10^{-3}–10^{-5}, the source of radiation was readily pinned down to lie in a region where the plasma frequency is much below the electron cyclotron frequency. Collapse would thus be in a frequency range far away from the observed radiation frequency, which is in the range of the electron cyclotron frequency. Thus collapse radiation, though not improbable, turned out to be unable to explain the observed nonthermal AKR emission. Other attempts including upper-hybrid collapse lead to the correct frequency but yielded low emission rates.

Up to the present, AKR remained an unresolved mystery, even though when, in 1979, Wu and Lee (1979) realized that the weak relativistic effect of the few keV auroral electrons, when combined with their nearly empty loss-cones at auroral altitudes, provided a mechanism to directly destabilize the free-space electromagnetic modes which can propagate in a plasma, the X-, O-, and Z-modes. Excitation of waves by linear instability is a most attractive way because it is a first-order effect and therefore may have high efficiency. A mechanism of this kind has already been tested by Melrose (1973), who assumed temperature anisotropy of nonrelativistic auroral electrons, as the source of the radiation with perpendicular temperature larger than parallel temperature. However, entirely unreasonable anisotropies $T_\perp/T_\parallel \sim 100$ were needed for instability.

Wu and Lee (1979) understood that the weakly relativistic electrons that resonate with the unstable wave are located along a relativistically displaced resonance curve $k_{v\parallel} = \omega - \omega_{ce}/\gamma_r$. The inclusion of the relativistic factor γ_r, which is a function of the perpendicular $v_\perp$ and parallel $v_\parallel$ resonant electron speeds, transforms the resonance curve from a circle (at constant $v_\perp$ in the nonrelativistic case) into an ellipse. Since the instability is driven by the part on the resonance curve that falls into the steepest perpendicular velocity gradient $\delta_{v_\perp} f_e(v_\perp, v_\parallel)$ of the electron distribution $f_e(\mathbf{v})$, this ellipse could be placed into the loss-cone in such a way that it covered the steep transverse loss-cone gradient in the distribution function. This led to a susceptible increase in the growth rate of the instability. Because the free-space modes are usually absorbed in plasma, the inversion of the absorption rate into a growth rate led to call this mechanism the *electron cyclotron maser*.

The strongest emission was found just above the electron cyclotron frequency ω_{ce}, and to weaker extent at the next higher harmonic. It turned, however, out that to obtain reasonable agreement with observation and to avoid total reabsorption, the plasma had to be so dilute that $\omega_{pe}^2/\omega_{ce}^2 \ll 1$ which could not be found a good reason for though radiation was always strongest when this condition was satisfied. Nevertheless, this theory was a crucial step ahead towards an understanding of the nonthermal radio emission from collisionless plasma. It was immediately adopted by the scientific community and was applied to all possible other occasions where plasma radiation of similar signature as AKR was observed. The emission from Jupiter and Saturn, and in the first place solar type I, solar type IV, and even solar radio-spike emissions were proposed to be explained by the cyclotron maser mechanism. Moreover, slow temporal variations in the radiation of radio stars were attributed to observation of electron cyclotron maser radiation from such objects.

The loss-cone cyclotron maser had a number of disturbing disadvantages. It required a very steep loss cone for susceptible growth rates. It produced broadband emissions above the cyclotron frequency. Its escape from the radiation source remained unexplained. The observations instead indicated weak loss cones, very narrow-band emissions frequently below the cyclotron frequency and high temporal variability of these emissions. It was attempted to turn the disadvantage of trapping the radiation within the radiation source into an advantage by calculating the quasilinear reaction of the emitted radiation on the electron distribution. This attempt showed that the loss-cone could be readily depleted by the radiation. There was even an attempt to explain the relatively high radiation effects by further coherent amplification of the trapped radiation in a similar way to the one known from laboratory lasers. But none of these attempts could reproduce the most intense emissions below the cyclotron frequency.

This theoretical problem was resolved only when it was realized that the relativistic effect also reduced the lower cut-off frequency of X-mode radiation. Free-space electromagnetic waves propagate in two modes, the right-hand polarized or X-mode and the left-hand polarized of O-mode, where O stands for ordinary and X for extraordinary. The O-mode is reflected from the position where its frequency matches the local plasma frequency. The X-mode is reflected from a frequency which is above the electron cyclotron frequency and thus, in a dilute plasma, is much higher than the O-mode cut-off. However, when taking into account the relativistic decrease in the electron cyclotron frequency, this cut-off may fall below the nonrelativistic cyclotron frequency. In this case, findings even reveal that, for X-mode waves of such low frequency, the propagation angle is practically 90°, which implies $k_\parallel = 0$. Considering the resonance condition one then finds that the resonance ellipse becomes a resonance circle centered at the origin of velocity space.

For a loss-cone distribution, this would practically rule out any instability because at no place along the resonance curve is the condition $\delta_{v_\perp} f_e(v_\perp, v_\parallel) > 0$ even satisfied. However, Nature has found a way to exploit this effect. When the distribution function becomes a ring or "horseshoe" distribution, as shown in Figure 15, then the resonance circle passes a long way along positive perpendicular velocity gradients and may pick up all the contributions of the particles having excess perpendicular energy in order to transform this excess into X-mode radiation. Thus, the prerequisite of effective nonthermal plasma radiation in the radio frequency range is the formation of a horseshoe distribution function. We have shown that this requires the existence of a strong parallel potential drop and correspondingly a field-aligned electric field in otherwise collisionless plasma. The observation of this kind of non-thermal plasma radiation can therefore be taken as a strong indication of the presence of parallel electric fields. However, a high-time and high-frequency resolution of the AKR uncovered another mystery, as shown in Figure 19. It seems that the entire spectrum of AKR consists of very narrow-band elementary radiation sources which move with high velocity across the spectrum. The maximum of emission occurs when these elementary sources stop or even turn around. When one transforms their spectral displacement into spatial motion along the magnetic field, one finds that most of these sources move down the field lines obviously riding on the back of the auroral electron beam. This lets one suspect that these sources are electron phase space holes in the electron horseshoe distribution function and that another important modification of the nonthermal plasma maser radiation mechanism is required until the problem of radiation from AKR and the electron-cyclotron maser in general can be considered as settled.

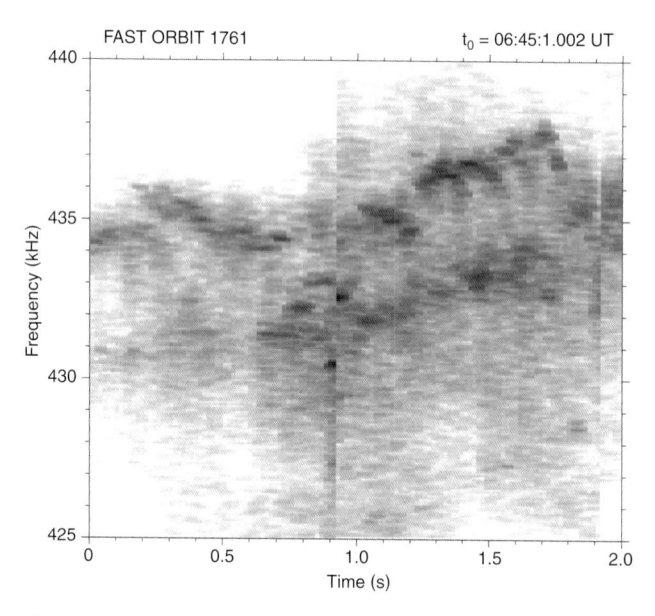

Figure 19 A high-time and frequency resolved observation of AKR fine structure in the auroral source region by the FAST spacecraft. The striking observation is that AKR consists of a multitude of very narrowband fast drifting emission sources which when retarded at a certain altitude contribute the most to radiation. Note that most of the structures move downward along the magnetic field as can be read from the increasing frequency of the emission. (From Pottelette et al. 2000.)

Fundamental-harmonic emission in collapse

Maser radiation works only in *underdense* plasma. In dense plasma, like the solar wind, the foreshock regions, the solar atmosphere, and Earth's plasma sheet and plasma sheet boundary layer $\omega_{ce} < \omega_{pe}$, and the maser is set out of work. Radiation in these circumstances is probably generated by electron beams and the subsequent collapse of nonlinear waves according to Zakharov's mechanism. The high concentration of Langmuir or upper-hybrid wave energy in the solitons and cavitons of plasma implies that electrons interacting with this high-frequency electrostatic wave field experience sufficient 'shaking' that they may efficiently emit radio waves due to continuous acceleration and deceleration by the localized wave fields. Radiation in these cases is close to the electron plasma frequency ω_{pe} and its harmonics. An example of an observation of these harmonic emissions in front of Earth's bow shock wave is shown in Figure 12. Other examples are known from solar type III and V radio bursts, which are related to propagating and scattered electron beams emanating from the solar corona and passing through interplanetary space. Similar emissions have also been observed in the plasmasheet boundary layer. All of them can be taken as proof that electron beams have been accelerated and propagate along the magnetic field with little scatter. Their existence challenges the dominance of quasi-linear theory, which predicts that Langmuir wave excitation would quickly retard the beam and smear it out.

On the other hand, observations *in situ* could not directly prove the collapse mechanism. The observations showed that spikey Langmuir waves occur in the source region in front of the bow shock in correlation with radio wave emission, but the correlation is not good enough. One tends to believe today that the collapse proceeds too fast in order to be able to produce enough radiation. It rather stabilizes the beam while the energy of the trapped waves is readily burnt out by transit time damping as a result of a few passing electrons. This is too weak for generation of susceptible radiation power. However, subsequent trapping of Langmuir waves in the slowly decaying density troughs, which are basically ion-acoustic waves, generates the observed radiation in a statistical incoherent way. Whether this mechanism is correct or is merely another way out of the dilemma of explaining the observation of fundamental, second and third harmonic radiation from electron beams cannot be decided at present. Further observation of wave fields and particle spectra in the coming century will certainly lead to the resolution of this question. It can be safely argued that we are already close to its final settlement.

ACTIVE EXPERIMENTS

The above recollection of the role of space plasma in understanding basic plasma processes important to astrophysics as well as laboratory plasma physics has concentrated mainly on the natural effects. Last but not least, we note that a limited number of active experiments have been performed in the past century in space plasma. Such experiments were designed to test the response of the different types of natural plasma in different regions of space to artificial distortions of conditions.

The first of such "experiments" were the atmospheric nuclear tests which injected large amounts of energetic particles into the Earth's radiation belts. Subsequent experiments were designed for answering more scientific questions. As already mentioned, injection of intense electromagnetic waves close to but below the electron cyclotron frequency in the whistler mode band along the magnetic field from ground stimulates electron-whistler resonance and causes enhanced mono-energetic electron precipitation, as has been shown by such frequency-tuned experiments. Similarly, injection of cold plasma from spacecraft into the trapping regions outside the plasmasphere lowers the electron resonance energy and stimulates electron precipitation as well as whistler noise, as experiments have also shown. Such injection of cold plasma has become a well-known technique using Barium clouds (see Chapter 6, where both the history and scientific results of the first Barium cloud experiments are described in detail; for a personal recollection of some of the history, see also Haerendel 1996) and injecting them either from rockets into both the auroral and equatorial ionospheres or from spacecraft into more remote regions of near-Earth space. In the equatorial ionosphere, the intention was to stimulate instability of the equatorial electrojet and F-layer by adding cold and heavy plasma. This causes the Rayleigh-Taylor and some other related instabilities to evolve there, the evolution of which could be observed from ground by incoherent radar backscatter. Moreover, optical emission observations provided information about the ionization process, while instrumentation on the rocket can measure the distortion of the ionosphere *in situ* in both the density and wave fields.

One particularly successful experiment of this kind was the neutral Xenon beam injection experiment (Häusler *et al.* 1986) during the Porcupine rocket campaign (Haerendel and Sagdeev 1981), which took place in March 1979 in Kiruna (Sweden). It was followed in the eighties by a number of similar experiments launched from Alaska into the aurora and from Brazil into the equatorial electrojet.

Historically, the Xenon beam experiment on Porcupine is of singular interest and importance as one of the very first common and coordinated efforts in Space Research between the Soviet Union and Western countries during the period of the Cold War. This is in contrast to all other active experiments flown later which were performed merely by the US or in bilateral Western cooperation. The Porcupine Xenon beam experiment fell right into the Cold War Brezhnev era, when there were still strong tensions between the US and the Soviet Union. When, in the later seventies, Gerhard Haerendel from the Max-Planck-Institute of Extraterrestrial Physics in Garching, Germany and Roald Sagdeev from the Institute of Space Research (IKI) in Moscow met at a scientific meeting, they ad hoc decided to give such an effort a try, and despite all contrary forecasts, this effort turned out to be successful. Though the political tensions remained, it was nevertheless possible to place a Soviet plasma gun on one of the Porcupine rockets. This gun was provided as a black box by the Institute of Space Research of the Soviet Academy of Sciences. It was accompanied merely by a brief description of its working parameters. Nobody of the Western scientists ever saw its interior or technical construction. Obviously it was the prototype of an instrument regularly built for and put on the fleet of Soviet military satellites. Professor Sagdeev, at that time Director General of IKI, managed to obtain permission to bring it into operation on the Porcupine rocket which, considered in retrospect, was indeed a heroic action. The Western countries involved in this experiment were France, Germany, Sweden, and the US with the Max-Planck-Institute for extraterrestrial Physics as the lead institution and Gerhard Haerendel as the Principal Investigator.

The Porcupine Xenon beam experiment allowed for the study of the ionization and neutralization processes of the injected dense and fast heavy ion beam, the diamagnetic disturbance caused by the beam, the excitation of field aligned electron flux and currents, and the generation of a multitude of instabilities of plasma waves like ion-acoustic, lower-hybrid and ion cyclotron waves. It provided insight into the formation of anomalous resistivity on the Xenon cyclotron-radius scale and on the penetration of an electrically neutral ion beam into the plasma (Häusler *et al.* 1986). Such penetration still remains to be a badly understood process. The main problem is that the space charge of the ions when penetrating into the plasma must necessarily be compensated either by electron fluxes into the beam from the environmental plasma or by the original beam electrons themselves. Under normal conditions electrons are frozen-in into the magnetic field and can move only along the magnetic field, however. Therefore the above problem is traced back to the general problem of how electrons can move across a magnetic field.

The neutralization of an overdense high-momentum ion beam is made possible by the diamagnetic effect of the beam. In other words, the enormous ram pressure of the beam compresses the ambient magnetic field and excludes it from the beam core. The original beam electrons are then able to follow the ions into the magnetic field-free region being coupled to them by the ambipolar space charge electric field. Hence, charge neutralization along the magnetic field by ambient electrons becomes necessary only in a narrow beam-surface layer the width of which is proportional to the electron skin depth. Here neutralization is provided solely by electrons from the environment plasma population which carry the strong field-aligned currents which are responsible for the observed ion-acoustic and high-harmonic ion-cyclotron waves. In addition the beam-density gradients generate the lower-hybrid wave spectrum responsible for anomalous transport, electron and background-ion heating and acceleration, and generation of anomalous electric fields.

The Xenon-beam experiment also contributed to the understanding of Alfvén's critical-velocity ionization effect (see Alfvén 1981). This effect was originally proposed by Alfvén in 1942 in order to explain the fast ionization of cosmic gas clouds which are not exposed to UV radiation such that radiative ionization does not work. Ionization of the neutral gas cloud was assumed to proceed when getting in contact with star wind plasma. When the difference in the velocities of the cloud and plasma exceeded the velocity needed for ionization, Alfvén claimed the gas should become spontaneously ionized in the contact layer. Under collisionless conditions the effect is rather complicated, as could also be demonstrated in the artificial Barium cloud experiments which will be mentioned below. It is believed today that seed ionisation of the neutrals is produced by the few superthermal electrons from the tail of the electron distribution having high enough energy to ionize some neutrals. The newly created ions behave like pick-up ions which form a ring-beam distribution in phase space. Whence the ring-beam enters the plasma on a scale of the ion gyro-radius, lower-hybrid wave turbulence is generated which further heats and accelerates electrons. This causes an avalanche of superthermal electrons and additional excess ionization. The result is an ionization explosion that ultimately transforms the neutral gas into plasma. The local time-scale of this process is of the order of several lower-hybrid times and is thus very short. It must, however, be admitted that this process is still barely understood and requires much more theoretical and experimental effort.

A number of low-momentum neutral-ion cloud-release experiments in the auroral magnetosphere provided information about the electric field structure of the upper ionosphere by observing the perpendicular drift of the visible emission of the Barium ions (see e.g. Föppl et al. 1968). One of the most interesting findings in these experiments was the indication of field aligned electric potentials which, in one of the Barium releases, accelerated the newly created Barium ions along the magnetic field. In addition, observation of the behaviour of the Barium cloud has been used also to infer about Alfvén's critical velocity ionization effect mentioned above, but the results were not unambiguous.

Releases at remote places in space have been used to create "artificial comets" (Valenzuela et al. 1986, Gurnett et al. 1985) in the solar wind and magnetosheath and large ion clouds in Earth's geomagnetic tail. These experiments were performed by the AMPTE IRM mission in fall 1984 and spring 1985 and involved releasing Lithium and Barium clouds into the near-Earth plasma environment. The cold neutral gas clouds released symmetrically with respect to the main spacecraft body were found to be readily ionized by solar UV radiation. The experiments were intended to investigate the erosion of this quickly solar-UV ionized plasma clouds by the solar wind stream as well as the generation of turbulence inside and in the boundary layers of such clouds. In the solar wind releases, similarities to cometary bow shock build-up and tail disruption were found. Shock formation, and generation of many instabilities have been observed in these experiments, giving insight into the dynamical processes of cometary environments, cometary tail disruption and generation of radiative processes. Moreover, experiments of this kind were able to prove the existence and to investigate the mechanism of pick-up ion acceleration and dynamics, as well as energetic tail formation on the ion distribution function in a streaming plasma, as had been theoretically proposed and is of particular importance in view of understanding the generation of the anomalous cosmic ray component observed in the heliosphere.

In addition to these efforts space is also used for more technically oriented experiments, the propagation of strong ion and electron beams across and along the magnetic field, the investigation of the charge neutralization processes in these cases. Discharge experiments between injected beams and the spacecraft body have been performed. Another famous experiment is the tether-experiment for which there is not enough space to describe here.

Finally, we should mention that high-power radio waves are continuously injected into the ionosphere from the ground (for instance, from the European Ionospheric Scatter radar facility (EISCAT) near Tromsø in Norway) in order to infer the nonlinear response of the ionospheric plasma to the radio wave energy input. The observations show all kinds of instabilities and waves being excited in these experiments, side-band decay of the radio waves, resonant absorption, and strong plasma heating which causes heat flux instabilites. This field of research is of great practical and theoretical interest for understanding the role of ionospheric heating by natural and anthropogenic electromagnetic waves on atmospheric and climatic variations and even for the understanding of some astrophysical problems. Under many astrophysical conditions, extraordinarily intense radiation is injected into gases and plasma and causes a wealth of effects impossible to investigate in the laboratory. Experiments of the above-mentioned kind provide pioneering information about the effects which may occur when plasmas are heated by extraordinarily large amplitude radio waves.

CONCLUSIONS AND FUTURE PROSPECTS

The present recollection of basic processes, which are related to the use of Earth's plasma environment as a convenient laboratory provided by Nature, must necessarily be colored by our personal preferences. It is thus neither complete nor objective. We have selected those problems which appeared to us to have the quality of being major achievements in twentieth century space science and could not have occurred in groundbased laboratory experiments. We have also concentrated more on those effects which space offered us for investigation than on man-made active experiments. This selection reflects the real division of the amount of activity invested into both directions of research: using space passively and actively as a laboratory. The passive role of man in exploring space and exploiting the accumulated knowledge has been much larger and much more efficient than its active role. This may change in near future, when active experiments will be performed in larger number. The information extracted about the behaviour of plasma under largely collisionless conditions has, however, already been enormous. It mostly brought up new problems, discovered unexpected plasma behaviour and faced us with an ever increasing number of unresolved questions.

Today we understand the gross properties of the magnetosphere, the cause of convection and magnetic variations in a system as large as the magnetosphere, we understand the formation of its boundary. We believe we have solved almost all the problems posed by the presence of the radiation belts. This part of space physics has largely been settled in the past century. However, a substantial number of problems remain to be solved in the coming decades. These problems relate to the very physical mechanism of reconnection, the generation of anomalous transport coefficients, the understanding of small-scale and large-scale turbulence, the problem of acceleration of particles in reconnection and in several spatial locations of the magnetosphere including and above all the auroral zones, the generation of strong electric potential drops along field lines, the generation of large amplitude waves of all kinds, and the problem of radiation.

Either of these questions has importance reaching farther than the very physics of the magnetosphere from which they emerge. Reconnection is one of the most important and most widely used concepts in plasma-astrophysics, where it is known that the whole of dynamo theory of generation of cosmic magnetic fields relies on it. It has important applications in laboratory plasma physics as well. Its understanding is crucial for the behaviour of fusion plasmas and losses of plasma and energy from fusion devices. Transport is another concept of extended importance both in astrophysics and in laboratory. Astrophysical application is boldly using estimates which in many cases are not sufficiently justified by basic plasma physics. Space plasma physics should be instrumental in providing justification and the most useful estimates. The physics of collisionless shocks is one of the most important fields of application in astrophysics. It has important applications as well in inertial heating of plasma and the attempts of igniting fusion by shock wave generation. A considerable amount of progress has been achieved in the past century observationally and theoretically, as well as by comparison with simulational work and by performing measurements on the Earth's bow shock and occasional interplanetary shocks. It seems as if this particular field will experience explosive progress in the next few decades, when larger and faster computer power will become available, real particle mass ratios can be used and simulations can be extended to three dimensions and run for longer times. The numerical technique is basically available already today. Simulations of this kind will also solve the problem of shock particle acceleration for both kinds of particles, ions and electrons. We believe that this will settle the debate about shock particle injection, the degree of how much electrons can be accelerated in shocks. It will also answer most of the questions arising in the physics of super-high Mach number shocks in astrophysics. Generation of large field-aligned electric potential drops is still barely understood. It is of enormous importance for the understanding of prompt acceleration of particles in solar flares, magnetized stars, and in very massive astrophysical objects, where one may expect that it generates extremely high-energy cosmic rays. The magnetosphere and, in particular, its auroral zone is the ideal laboratory to investigate just this exotic process and to infer about the conditions which must prevail for parallel electric fields to be maintained in plasma. These conditions are still barely known. They belong to the field of nonlinear plasma processes which can be studied solely by *in situ* spacecraft observations.

Acknowledgement

It has been a pleasure to accept the invitation to contribute to this volume. We thank the editors for giving us the opportunity to express our very personal feelings about the evolution of space plasma physics in the past century. We apologize to all those, whom we have not mentioned, and whose work and contributions to the development of our understanding of the physical processes in space plasmas have not been taken into account, or were unintentionally ignored. Much of the work by R.T. has been performed at ISSI. He thanks its directors J. Geiss, G. Paschmann, and R. von Steiger for their support as well as the ISSI staff for its hospitality.

BIBLIOGRAPHY: GENERAL LITERATURE

Alfvén, H. (1981). *Cosmic Plasma*, D. Reidel Publ. Comp., Dordrecht, Holland, 164 pp.

Biskamp, D. (1993). *Nonlinear Magnetohydrodynamics*, Cambridge University Press, Cambridge (UK). 348 pp.

Gillmor, C.S. and Spreiter, J.R. (eds) (1997). *Discovery of the Magnetosphere*, History of Geophysics, Vol. 7, AGU, Washington DC, 286 pp.

Hones, E.W., Jr (ed.) (1984). *Magnetic Reconnection in Space and Laboratory Plasmas*, Geophys. Monogr. 30, AGU, Washington DC, 386 pp.

Hultqvist, B., Øieroset, M., Paschmann, G. and Treumann, R. (eds) (1999). *Magnetospheric Plasma Sources and Losses*, Space Science Series of ISSI, Vol. 6, Kluwer Academic Publishers, Dordrecht, 482 pp.

Kennel, C.F. (1995). *Convection and Substorms: Paradigms of Magnetospheric Phenomenology*, Oxford University Press, New York, 408 pp.

Kivelson, M.G. and Russell, C.T. (eds) (1995). *Introduction to Space Physics*, Cambridge University Press, Cambridge (UK), 568 pp.

Lemaire, J.F. and Gringauz, K.I. (1998). *The Earth's Plasmasphere*, Cambridge University Press, Cambridge (UK).

Martin, J.C. and Fries, S.D. (eds) (1991). *A Spacefaring Nation*, Smithonian Inst. Press, Washington DC.

McDougall, W.A. (1985). *...the Heavens and the Earth. A Political History of the Space Age*, Basic Books, New York, 555 pp.

Michel, F.C. (1991). *Theory of Neutron Star Magnetospheres*, Chicago University Press, Chicago, 517 pp.

Murray, B. (1990). *Journey into Space, The First Thirty Years of Space Exploration*, W.W. Norton & Co, New York, 381 pp.

Nishida, A., Baker, D.A. and Cowley, S.W.H. (eds) (1998). *New Perspectives on the Earth's Magnetotail*, Geophys. Monogr. 105, AGU, Washington DC, 339 pp.

Northrop, T.G. (1963). *The Adiabatic Motion of Charged Particles*, Interscience, New York.

Parks, G.K. (1992). *Physics of Space Plasmas*, J. Wiley & Sons, New York, 538 pp.

Sagdeev, R.Z. (1966). Cooperative Phenomena and Shock Waves in Collisionless Plasmas. In: M.A. Leontovich (ed.), *Reviews of Plasma Physics*, Vol. 4, p. 23.

Sagdeev, R.Z. and Galeev, A.A. (1969). *Nonlinear Plasma Theory*, Benjamin, New York.

Song, P., Sonnerup, B.U.Ö. and Thomsen, M.F. (eds) (1995). *Physics of the Magnetopause*, pp. 447, Geophys. Monogr. 90, AGU, Washington DC.

Stone, R.G. and Tsurutani, B.T. (eds) (1985). *Collisionless Shocks in the Heliosphere. A Tutorial Review*, Geophys. Monogr. 34, AGU, Washington DC, 115 pp.

Tidman, D.A. and Krall, N.A. (1971). *Shock Waves in Collisionless Plasmas*, Wiley, New York, 175 pp.

Treumann, R.A. and Baumjohann, W. (1997). *Advanced Space Plasma Physics*, Imperial College Press, London, 387 pp.

Tsurutani, B.T. and Stone, R.G. (eds) (1985). *Collisionless Shocks in the Heliosphere. Reviews of Current Research*, Geophys. Monogr. 35, AGU, Washington DC, 303 pp.

Van Allen, J.A. (1983). *Origins of Magnetospheric Physics*, Smithonian Inst. Press, Washington DC, 144 pp.

Walt, M. (1994). *Introduction to Geomagnetically Trapped Radiation*, Cambridge University Press, Cambridge (UK), 168 pp.

REFERENCES

Bale, J., Kellogg, P.J., Larson, D.E., Lin, R.P., Goetz ,K. and Lepping, R.P. (1998). Bipolar electrostatic structures in the shock transition region: Evidence of electron phase space holes, *Geophysical Research Letters*, **25** 2929.

Baumjohann, W. and Treumann, R.A. (1997). *Basic Space Plasma Physics*, Imperial College Press, London.

Benediktov, E.A., Getmantsev, G.G., Sazonov, Y.A. and Tarasov, A.F. (1968). Preliminary results of measurements of the intensity of distributed extraterrestrial radio frequency emission at 725 and 1525 kHz frequency by the satellite Elektron-2. *Cosmic Research*, **3**, 492 [1965, *Kosm. Issled.* **3**, 614].

Biermann, L. (1951). Kometenschweife und solare Korpuskularstrahlung, *Zeitschrift für Astrophysik*, **29**, 274.

Chapman, S. and Ferraro, V.C.A. (1931). A new theory of magnetic storms. Part I. The initial phase. *Terrestrial Magnetism and Atmospheric Electricity*, **36**, 7.

Chiu, Y.T. and Schulz, M. (1978). Self-consistent particle and parallel electrostatic field distributions in the magnetospheric-ionospheric auroral region, *Journal of Geophysical Research*, **83**, 629.

Christon, S.P., Williams, D.J., Mitchell, D.G., Huang, C.Y. and Frank, L.A. (1991). Spectral characteristics of plasma sheet ion and electron populations during disturbed geomagnetic conditions. *Journal of Geophysical Research*, **96**, 1.

Delory, G.T., Ergun, R.E., Carlson, C.W., Muschietti, L., Chaston, C.C., Peria, W., McFadden, J.P. and Strangeway, R. (1998). FAST observations of electron distributions within AKR source regions. *Geophysical Research Letters*, **25**, 2969.

Dum, C.T. (1995). Can we find useful algorithms for anomalous transport?. In: *Cross-Scale Coupling in Space Plasmas*, Geophys. Monogr. 93, AGU, Washington DC, p. 1.

Dungey, J. (1961). Interplanetary magnetic field and the auroral zones, *Physical Review Letters*, **6**, 47.

Dungey, J.W. (1995) Origins of the concept of reconnection and its application to the magnetopause: A historical view. In: P. Song, B.U.Ö. Sonnerup, M.F. Thomsen (eds), *Physics of the Magnetopause*, Geophys. Monogr. 90, AGU, Washington DC, p. 17.

Ergun, R.E., Carlson, C.W., McFadden, J.P., Mozer, F.S., Delory, G.T., Peria, W., Chaston, C.C., Temerin, M., Elphic, R., Strangeway, R.J., Pfaff, R., Cattell, C.A., Klumpar, D., Shelley, E., Peterson, W., Moebius, E. and Kistler, L. (1998). FAST satellite wave observations in the AKR source region. *Geophysical Research Letters*, **25**, 2061.

Feldman, W.C. (1985). Electron velocity distributions near collisionless shocks. In: B.T. Tsurutani and R.G. Stone (eds.), *Collisionless Shocks in the Heliosphere: Reviews of Current Research*, Geophys. Monogr. 35, AGU, Washington DC, p. 195.

Föppl, H., Haerendel, G., Haser, L., Lüst, R., Melzner, F., Meyer, B., Neuss, H., Rabben, H.-H., Rieger, E., Stöcker, J. and Stoffregen, W. (1968). Preliminary results of electric field measurements in the auroral zone. *Journal of Geophysical Research*, **73**, 21.

Giovanelli, R.G. (1946). A theory of chromospheric flares. *Nature*, **158**, 81.

Gold, T. (1953). In: *Gas Dynamics of Cosmic Clouds*, IAU Symp. No. 2, p. 103.

Gold, T. (1968). Rotating neutron stars as the origin of the pulsating radio source. *Nature*, **218**, 731.

Goldreich, P. and Julian, W.H. (1969). Pulsar electrodynamics. *Astrophysical Journal*, **157**, 869.

Gringauz, K.I., Kurt, V.G., Moroz, V.I. and Shklovsky, I. (1961). Results of observations of charged particles observed out to 100,000 km with the aid of charged particle traps on Soviet space probes. *Soviet Astron. J.*, **4**, 680.

Gurnett, D.A. (1974). The Earth as a radio source: Terrestrial kilometric radiation. *Journal of Geophysical Research*, **79**, 4227.

Gurnett, D.A., Anderson, R.R., Häusler, B., Haerendel, G., Bauer, O.H., Treumann, R.A., Koons, H.C., Holzworth, R.H. and Lühr, H. (1985). Plasma waves associated with the AMPTE artificial comet. *Geophysical Research Letters*, **12**, 851.

Grad, H. and Rubin, H. (1958). Hydromagnetic equilibria and force-free fields. In: *Proceedings of the Second UN International Conference about Peaceful Uses of Atomic Energy*, Vol. 31, United Nations, Geneva, p. 190.

Gray, P.C., Hudson, M.K., Lotko, W. and Bergmann, R. (1991) Decay of ion beam driven acoustic waves into ion holes. *Geophysical Research Letters*, **18**, 1675.

Haerendel, G. et al. (1976). *First Observation of Electrostatic Acceleration of Barium Ions into the Magnetosphere*. ESA SP-155, p. 203.

Haerendel, G., Lüst, R. and Rieger, E. (1967). Motion of artificial ion clouds in the upper atmosphere. *Planetary and Space Science*, **15**, 1.

Haerendel, G. (1996). Curiosity and chance. *Geophysical Research Letters*, **101**, 10541.

Haerendel, G. and Sagdeev, R.Z. (1981). Artificial plasma jet in the ionosphere. *Advances in Space Research*, **1**, 29.

Hau, L.-N. and Sonnerup, B.U.Ö. (1999). Two-dimensional coherent structures in the magnetopause: Recovery of static equilibria from single-spacecraft data. *Journal of Geophysical Research*, **104**, 6899.

Häusler, B., Treumann, R.A., Bauer, O.H., Haerendel, G., Bush, R., Carlson, C.W., Theile, B., Kelley, M.C., Dokukin, V.S. and Ruzhin, Y.Y. (1985). Observations of the artificially injected Porcupine Xenon ion beam in the ionosphere. *Journal of Geophysical Research*, **91**, 287.

Hoyle, F. (1949). Magnetic storms and aurorae. In: *Some Recent Researches in Solar Physics*, Cambridge University Press, Cambridge, p. 92 and p. 102.

Jackson, J.D. (1985). *Classical Electrodynamics*, Wiley & Sons, New York, 848 pp.

Kennel, C.F. and Petschek, H.E. (1966). Limits on stably trapped radiation. *Journal of Geophysical Research*, **71**, 1.

Kennel, C.F. and Sagdeev, R.Z. (1967). Collisionless shock waves in high beta plasmas. *Journal of Geophysical Research*, **72**, 3303.

Knight. S. (1973). Parallel electric fields. *Planetary Space Science*, **21**, 741.

Lüst, R. and Schlüter, A. (1957). Axialsymmetrische magnetohydrodynamische Gleichgewichtskonfigurationen. *Zeitschrift für Naturforschung*, **12a**, 850.

Lyons, L.R., Thorne, R.M. and Kennel, C.F. (1973). Pitch-angle diffusion of radiation belt electrons within the plasmasphere. *Journal of Geophysical Research*, **77**, 3455.

McLean, D.J. and Labrum, N.R. (eds.) (1985). *Solar Radiophysics: Studies of Emission from the Sun at Metre Wavelengths*, pp. 516, Cambridge University Press, Cambridge.

Melrose, D.B. (1973). Coherent gyromagnetic emission as a radiation mechanism. *Australian Journal of Physics*, **26**, 229.

Mende, S.B. (1968). Experimental investigation of electric fields parallel to the magnetic field in the auroral ionosphere. *Journal of Geophysical Research*, **73**, 991.

Muschietti, L., Ergun, R.E., Roth, I. and Carlson, C.W. (1999). Phase-space electron holes along magnetic field lines. *Geophysical Research Letters*, **26**, 1093.

Ness, N.F. (1965). The Earth's magnetic tail. *Journal of Geophysical Research*, **70**, 2989.

Ness, N.F. (1996). Pioneering the swinging 1960s into the 1970s and 1980s. *Journal of Geophysical Research*, **101**, 10497.

Omura, Y., Matsumoto, H. and Kojima, H. (1996). Electron beam instability as generation mechanism of electrostatic electrostatic solitary waves in the magnetotail. *Journal of Geophysical Research*, **101**, 2685.

Pacini, F. (1967). Energy emission from a neutron star. *Nature*, **216**, 567.

Parker, E.N. (1957). Sweet's mechanism for merging magnetic fields in conducting fluids. *Journal of Geophysical Research*, **62**, 509.

Parker, E.N. (1961). A quasi-liner model of shock structure on a longitudinal magnetic field. *J. Nucl. Energy, Part C* **2**, 146.

Paschmann, G., Sonnerup, B.U.Ö., Papamastorakis, I., Sckopke, N., Haerendel, G., Bame, S.J., Asbridge, J.R., Gosling, J.T., Russell, C.T., and Elphic, R.C. (1979). Plasma acceleration at the Earth's magnetopause: Evidence for reconnection. *Nature*, **282**, 243.

Paschmann, G., Melzner, F., Haerendel, G., Bauer, O.H., Baumjohann, W., Sckopke, N., Treumann, R., McIlwain, C.E., Fillius, W., Whipple, E., Torbert, R.B. and Quinn, J.M. (1993). The electron drift instrument for Cluster. In: *Cluster: Mission, Payload and Supporting Activities*, ESA SP-1159, European Space Agency, Paris, p. 115.

Petschek, H.E. (1964). Magnetic field annihilation. In: *AAS-NASA Symposium on Physics of Solar Flares*, NASA Spec. Publ. 50, p. 425.

Petschek, H.E. (1995). The concept of rapid magnetic field reconnection: A retrospective view. In: P. Song, B.U.Ö. Sonnerup and M.F. Thomsen (eds.), *Physics of the Magnetopause*, Geophys. Monogr. 90, AGU, Washington DC, p. 21.

Petschek, H.E. (1996). Glimpses of space physics in the 1960s and 1990s. *Journal of Geophysical Research*, **101**, 10511.

Pottelette. R., Ergun, R.E., Treumann, R.A., Berthomier, M., Carlson, C. W., McFadden, J.P. and Roth, I. (1999). Modulated electron acoustic waves in auroral density cavities: FAST Observations. *Geophysical Research Letters*, **26**, 2629.

Pottelette, R., Treumann, R.A. and Berthomier, M. (2000). Auroral plasma turbulence and the cause of AKR fine structure. *Journal of Geophysical Research*, **105**, in press.

Pritchett, P.L., Strangeway, R.J., Carlson, C.W., Ergun, R.E., McFadden, J.P. and Delory, G.T. (1999). Free energy source and frequency bandwidth for the auroral kilometric radiation. *Geophysical Research Letters*, **104**, 10317.

Sagdeev, R.Z. (1979). The 1976 Oppenheimer lectures: Critical problems in plasma astrophysics, I. Turbulence and nonlinear waves, II. Singular layers and reconnection. *Reviews of Modern Physics*, **51**, 1.

Sckopke, N., Paschmann, G., Brinca, A.L., Carlson, C.W. and Lühr, H. (1990). Ion thermalization in quasi-perpendicular shocks involving reflected ions. *Journal of Geophysical Research*, **95**, 6337.

Scholer, M. (1991). Numerical models of magnetic reconnection. *Geophys. Astrophys. Fluid Dynamics* **62**, 51.

Scholer, M. and Burgess, D. (1992). The role of upstream waves in supercritical quasi-parallel shock re-formation. *Journal of Geophysical Research*, **97**, 8319.

Shafranov, V.D. (1958). Magnetohydrodynamical equilibrium configurations. *Sov. Phys. JETP*, **6**, 545.

Shlesinger, M.F., Zaslavsky, G.M. and Klafter, J. (1993). Strange kinetics. *Nature*, **363**, 31.

Speiser, T.W. and Ness, N.F. (1967). The neutral sheet in the geomagnetic tail: Its motion, equivalent currents and field line reconnection through it. *Journal of Geophysical Research*, **72**, 131.

Stern, D.P. (1996). A brief history of magnetospheric physics during the space age. *Reviews of Geophysics*, **34**, 1.

Störmer, C. (1907). Sur les trajectoires des corpuscules électrisés dans l'espace sous l'action du magnétisme terrestre, avec application aux aurores boréales. *Arch. Sci. Phys. Nat., Genéve*, **4**, period 5.

Sweet, P.A. (1958). The neutral point theory of solar flares. in: *Electromagnetic Phenomena in Cosmical Physics*, IAU Symposium (1956) No 6, Cambridge University Press, Cambridge, p. 123.

Terasawa, T. and Scholer, M. (1989). The heliosphere as an astrophysical laboratory for particle acceleration. *Science,* **244**, 1050.

Treumann, R.A. (1999). Generalized-Lorentzian thermodynamics. *Phys. Scripta,* **59**, 204.

Treumann, R.A., LaBelle, J. and Bauer, T.M. (1995). Diffusion at the magnetopause: An observational perspective. In: P. Song, B.U.Ö. Sonnerup, M.F. Thomsen (eds), *Physics of the Magnetopause*, Geophys. Monogr. 90, AGU, Washington DC, p. 331.

Treumann, R.A. and Scholer, M. (2001). Collisionless plasma processes. In: *Encyclopedia of Astronomy and Astrophysics*, Vol. 1, IoP, Bristol, pp. 398–408.

Treumann, R.A. and Pottelette, R. (1999). Auroral microprocesses, acceleration and radiation. *Advances in Space Research,* **23**, 1705–1720.

Valenzuela, A., Haerendel, G., Föppl, H., Melzner, F., Neuss, H., Rieger, E., Stöcker, J., Bauer, O., Höfner, H. and Loidl, J. (1986). The AMPTE artificial comet experiments. *Nature,* **320**, 700.

Van Allen, J.A. (1990). What is a space scientist? An autobiographical example. *Annual. Review of Earth and Planetary. Sciences,* **18**, 1.

Van Allen, J.A. (1996). Twenty-five milliamperes: A tale of two spacecraft. *Journal of Geophysical Research,* **101**, 10479.

Vasyliunas, V.M. (1968). A survey of low-energy electrons in the evening sector of the magnetosphere with OGO 1 and OGO 3. *Journal of Geophysical Research,* **73**, 2839.

Williams, D.J. and Ness, N.F. (1966). Simultaneous trapped electron and magnetic tail field observations. *Journal of Geophysical Research,* **71**, 5117.

Winske, D., Thomas, V.A. and Omidi, N. (1995). Diffusion at the magnetopause: The theoretical viewpoint. In: P. Song, B.U.Ö., Sonnerup and M.F., Thomsen (eds), *Physics of the Magnetopause*, Geophys. Monogr. 90, AGU, Washington DC, p. 321.

Wu, C.S. and Lee, L.C. (1979). A theory of terrestrial kilometric radiation. *Astrophysical Journal,* **230**, 621.

Zakharov, V.E. (1972). Collapse of Langmuir Waves. *Zh. Exp. Teor. Fiz.,* **62**, 1745 [*Sov. Phys. JETP,* **35**, 908].

63

BENGT HULTQVIST*

Earth's magnetosphere

INTRODUCTION

The magnetosphere was discovered in the early years of the space era. Magnetospheric physics became the first new scientific discipline conceived in the space age, although it was in the beginning called space plasma physics, auroral physics, or ionospheric physics, as the concept magnetosphere did not exist yet. Auroral and ionospheric physics were important items in the research program of the 'International Geophysical Year' (IGY), which, after long preparations, was implemented in the 18 months starting on 1 July 1957 and thereafter was followed up in a number of international research programs. Being the largest international co-operative research program ever, it resulted in important progress of magnetosphere-related research fields – such as the aurora – based on worldwide ground-based observations. IGY had also in its program the launching and operation of the first man-made satellites, both Soviet Sputniks and American Explorers and Vanguards, for *in situ* measurements in space. Still, Sputnik 1 caused an enormous sensation when it appeared in the sky on October 4, 1957.

The physics of near space came to dominate space research in the first decade or more after Sputnik. That had to do with the fact that the early satellites, which were stabilized by spinning, were well suited for accommodating the instruments of the space physicists, who were also very eager to send their instruments into space to measure directly on that hot plasma, the effects of which in the form of aurora, ionspheric variations and energetic particles at Earth's surface, they had studied for long from the ground. The astronomers could generally not be well served by the early satellites and in addition they were not used to build own instruments suitable for launch into space. Besides from auroral and ionospheric researchers, the magnetosphere recruited its workers from nuclear physics – generally via the cosmic rays and solar energetic particles fields – and laboratory plasma physics (ionized gases) groups. The magnetospheric physics community is still one of the largest in the solar system sciences.

In the four decades of the space era magnetospheric research has reached a certain degree of maturity. It is, however, still a young field in the sense that unexpected, 'surprising' results continue to constitute the most important new results from practically all new space missions. Earth's magnetosphere has become the main 'laboratory' for studies of space plasma physics. We rely on nature to make most of the 'experiments' in that laboratory, but some 'active' experiments have been made by researchers.

That the concept magnetosphere did not exist when Sputnik 1 was launched does not mean that no physical concepts of major importance in magnetospheric physics were known before the space age. A first attempt to describe what happens when a plasma cloud from the Sun reaches Earth with its strong internal magnetic field was made by Chapman and Ferraro already in the early thirties and in the late thirties and early forties Alfvén introduced large-scale electric fields around Earth and magnetic fields into the solar plasma interacting with Earth, as well as magnetohydrodynamics and the guiding center approximation.

Still, we must conclude that before the satellite era our knowledge of magnetospheric physics was very limited and many early ideas have turned out to be very far from reality. This is not astonishing considering what we now know of the complexity of magnetospheres. The complexity is so great that it is very difficult, if not impossible, to

---
* Swedish Institute of Space Physics, Kiruna, Sweden

derive more than a very limited amount of information from basic principles. Only *in situ* observations can generally tell what happens and theory needs a strong guidance from experiments in order to find its right directions. But the unavoidable limitations of observational possibilities in the huge magnetospheric system makes theoretical and numerical models of the system essential for the interpretation of the few scattered observations available when nature makes its 'experiments' around and within the magnetosphere. Thus, theory and experiments depend strongly on each other for progress. That the most important results of new magnetospheric missions are unexpected, as mentioned above, means that many theoretical and numerical models are in an early, incomplete, or even wrong state.

To learn to understand what are the prevailing processes among the many possible ones in the complex magnetosphere is, of course, an important task from a basic scientific point of view and it is important also because many of the basic magnetospheric processes, which can be studied in detail only in our own magnetosphere, certainly play important roles in the physics of the plasma universe in general, thus not only around the other bodies in our solar system and in the heliosphere as a whole, but around other stars and in the interstellar space. To the basic scientific interest adds that understanding of magnetospheric physics is a requirement for the exploration of space for various applications. 'Space weather' is affecting an increasing amount of human activities in the environment of humankind.

The present review deals with some of the most important results of magnetospheric research from the point of view not only of understanding the magnetosphere but for basic plasma physics and astrophysics in addition. The early part of the paper is organized mainly regionally. The first section describes the discovery of the magnetosphere, i.e. of the magnetopause and the bow shock and the physical processes there. Thereafter come in turn the inner magnetosphere, the high-latitude magnetosphere with the aurora-related phenomena, the magnetotail, the boundary layers. These regional sections are followed by some on processes: magnetospheric convection, substorms and storms, plasma sources and losses, and wave-particle interactions. Finally, future directions of research and the roles of magnetospheric research in the investigation of the universe are discussed briefly.

THE CELLULAR STRUCTURE

Chapman and Ferraro (1931, 1932), in a study of the cause of magnetic storms on Earth, had shown that a plasma cloud emitted by the disturbed Sun when reaching Earth would enclose the geomagnetic field and compress it in the direction of the Sun. This effect was only expected during magnetic storms. The existence of a permanently occurring solar wind was discovered by Biermann (1951) from his observations of comet tails and Parker (1958) gave a theory for it. From the presence of an always-present solar wind follows an always-present boundary to the geomagnetic field. The name 'magnetosphere' for the elongated volume within this boundary was used for the first time by Gold (1959). Several of the early US satellites with eccentric orbits showed a fluctuating and irregular magnetic field beyond certain distances (Pioneers 1 and 5 and Explorer 10), then considered as a broad, unstable boundary region. Explorer 12, launched in August 1961, was the first to provide a clear determination of the boundary of the magnetosphere in the subsolar region (Cahill and Amazeen, 1963). The sharp boundary, called the magnetopause, was found to be located in the range from 8 to 12 R_E near noon. The expected compression of the outer subsolar geomagnetic field was observed and the magnetic fluctuations were found to be larger outside the magnetopause than inside.

Less than two years after Cahill and Amazeen had made the first observations of the sharp boundary of the magnetosphere, Ness *et al.* (1964), in data from IMP-1, succeeded in interpreting the irregular variations in the magnetic field outside of the magnetopause and identified for the first time the bowshock of the magnetosphere in the solar wind. Their results are reproduced in Figure 1. That there would exist a bowshock in the collisionless plasma of the solar wind had been suggested by Axford (1962) and Kellogg (1962) shortly before, however without specifying the processes which replace the collisions in an ordinary gas. A collisionless shock wave had never been seen before. Collective plasma processes have in them taken the place of collisions in ordinary gases for the scattering and thermalization of the ions and electrons.

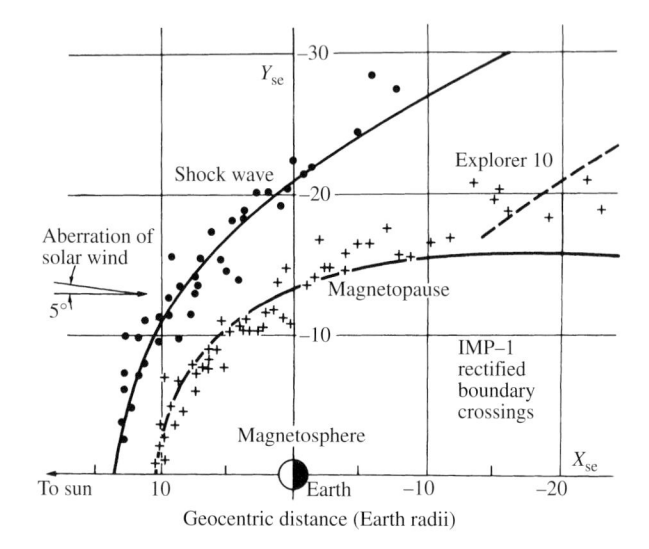

Figure 1 Observations of spacecraft crossings of the magnetopause and the standing shock wave in front of Earth's magnetosphere (after Ness *et al.* 1964).

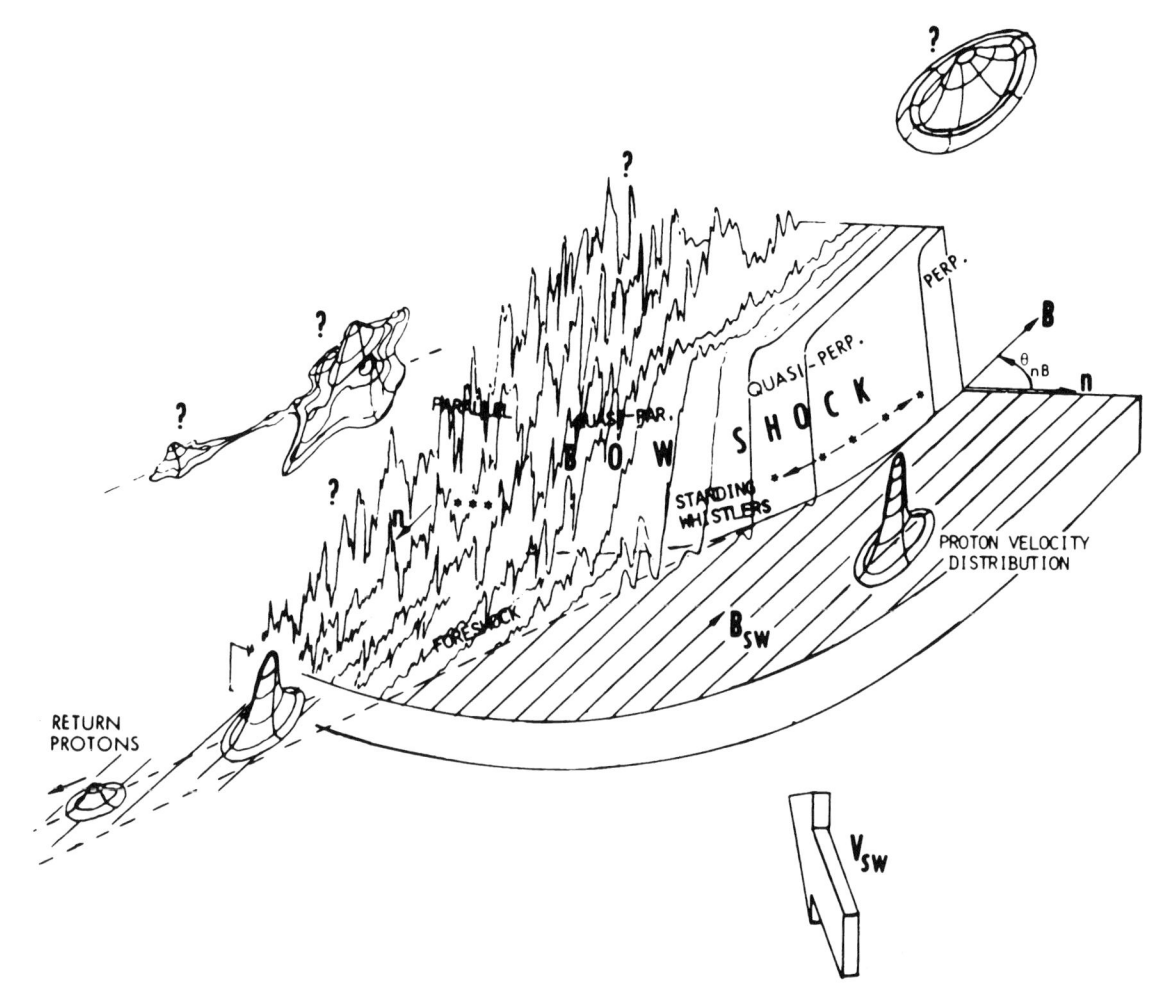

Figure 2 Schematic representation of magnetic field profiles at different locations of Earth's bowshock (after Greenstadt and Fredricks 1979). Field magnitude is plotted vertically. The superposed three-dimensional sketches represent solar wind proton thermal properties as number distributions in velocity space.

Collisionless shockwaves in hot magnetized plasmas are difficult to produce and investigate in the laboratory and most of what is known about them has been learnt from space measurements. Major efforts, both experimental and theoretical, have been devoted to the investigation of such shock waves since the middle of the sixties. A representation of observational results from Earth's bowshock is shown in Figure 2 (after Greenstadt and Fredricks 1979). It illustrates the complexity of the shock.

Collisionless shock waves are quite important for the acceleration of particles in the solar system and in the whole universe. Besides the bow shocks in front of planets, many travelling shocks generated near the Sun have been observed. There are still many aspects of the shock waves that are poorly understood and the experimental investigation of them in space will certainly go on for a long time. The reader is referred to e.g. Tsurutani and Stone (1985) for detailed information.

The magnetopause separates the comparatively dense and not too hot plasma of the solar wind from the thinner and hotter plasma in the magnetosphere. The transition takes place over a thin boundary layer. Also the magnetic fields in the two regions are different. An important result of space plasma physics research is the demonstration that this kind of thin boundary between plasmas of quite different properties characterizes the solar system as a whole and not only the vicinity of Earth. All planets have similar boundary layers and the interplanetary magnetic field sector boundaries are other examples. The solar system thus has a kind of cellular structure and it is most likely that this is true for the universe as a whole. The 'cell walls' are not possible to observe by remote sensing techniques but only by *in situ* measurements.

Such a cellular structure of the plasma universe may be of major astrophysical importance. Alfvén (1981) has even suggested that a consequence of the ability of plasma to concentrate differences between different populations into very thin boundary layers between them may be that a matter – antimatter symmetric universe may possibly exist in spite of lacking observational evidence, with matter and antimatter separated from each other by thin 'cell walls' and therefore the interaction between matter and antimatter in these thin boundaries being too weak to be observable from Earth.

The interaction of solar wind and magnetospheric plasmas and fields at the magnetopause is an interaction between two collisionless, high-temperature plasma populations containing different magnetic fields and moving relative to each other with supersonic and super-Alfvénic velocities. Such interaction is of basic interest and importance from both a general plasma physics and an astrophysics point of view. It is a complicated and difficult matter, still far from being understood in all its aspects. Practically all existing experimental knowledge about it has been obtained from space plasma physics investigations at the boundary between the solar wind and the magnetosphere of Earth. The investigations have shown that the magnetosphere is not closed but open in a way proposed by Dungey (1961) in the early space era and illustrated in Figure 3. How the interconnection of the magnetic fields of the solar wind and the magnetosphere really occurs, a process generally called reconnection and discussed in a special chapter of this book, is still a front line item of magnetospheric research. The dependence of the interaction on the direction of the interplanetary magnetic field being opposite to the direction of the magnetospheric field, or at least having a strong such component, was demonstrated observationally already a few years after the discovery of the magnetospause (Fairfield and Cahill 1966). The reconnection process has become an important tool in magnetospheric research as well as in many fields of astrophysics, although what really happens in the reconnection volume – the diffusion region – has not yet been demonstrated experimentally. Many conclusions about effects of reconnection can, however, be drawn, and are being drawn, without a detailed understanding of what happens in the diffusion region. That reconnection occurs under certain conditions must be considered as experimentally demonstrated beyond doubt, but a lot of work remains before the process is well understood. The magnetopause region of Earth will certainly be the main region for experimental investigations also in the future.

Another kind of interaction between the solar wind and the magnetosphere was proposed in the same year as Dungey published his reconnection-produced open magnetosphere, namely some kind of viscous processes at the magnetopause (Axford and Hines 1961). Although most evidence suggests that reconnection is quantitatively more important than viscous processes for the transfer of plasma and energy across the magnetopause, the latter processes are likely to occur and may even dominate the interaction when the interplanetary magnetic field is pointing northward. What micro-processes are most important for the viscous effects is still unknown.

That the differences between the properties of two plasma populations which are streaming relative to each other are concentrated to a thin boundary between them, does not mean that they do not affect each other beyond the boundary layer. All the phenomena to be reviewed below are in fact driven by the streaming of the solar wind around the magnetosphere. The characteristic solar wind speed of 400 km/s corresponds to a voltage difference over the width of the magnetosphere of several hundred kilovolts. Part of this voltage is transferred into the magnetosphere and causes convection and acceleration of the plasma. The solar plasma dynamic pressure on the magnetopause also has direct effects in the form of variable compression of the day-side magnetosphere caused by the variable solar wind and on the penetration of plasma in the polar cusp regions (see Figure 23) and possibly elsewhere.

The transient re-organizations of the inner magnetosphere that give rise to magnetospheric substorms and storms (see later section) involve power levels of the order of 10^{12} Watt or even more for the intense cases. Those processes use, at least partly, energy accumulated within the magnetosphere during longer periods than it is dissipated in, but the supply of energy by the solar wind is more than enough for running such processes continuously. That it does not happen is not because of too little energy in the streaming solar wind. In Earth's magnetosphere the rotation of the planet contributes significant energy to the magnetosphere but much less than the solar wind. For Jupiter, the opposite is true: the planetary rotation is a more important energy source for the magnetosphere than the solar wind.

A comparison: For an average rate of dissipation of energy within the near-Earth magnetosphere and the ionosphere of one tenth of the figure mentioned above for

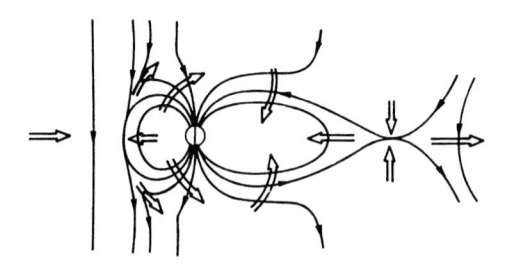

Figure 3 The 'open' magnetosphere proposed by Dungey (1961), in which the magnetic field lines of the solar wind plasma are connected with the geomagnetic field lines on the sunward side of Earth and they are disconnected again in the geomagnetic tail.

substorms, that average dissipation rate is of the order of one percent of the average total energy consumption rate for human activities in the world (assumed to be on the order 10^{20} Joule per year). The total average energy input rate to Earth's entire magnetosphere from the solar wind associated with the transfer of solar wind plasma to the magnetosphere is then at least of a similar order of magnitude as the global human energy consumption rate (see also later section on source and loss processes of magnetospheric plasma).

THE INNER MAGNETOSPHERE

The first big scientific surprise of the space era was that of the radiation belts, or Van Allen belts, by Van Allen's group at the University of Iowa, using a Geiger-Müller tube for energetic particle measurements on board the first US satellite, Explorer 1 (Van Allen et al. 1958). An early diagram showing the radiation belts can be seen in Figure 4. The discovery of the Van Allen radiation belts first of all demonstrated that large fluxes of energetic particles, contrary to what was expected, are naturally trapped in magnetospheres in forbidden orbits, i.e. orbits which cannot be reached by particles in a static magnetosphere either from inside or from outside. Such trapped particle populations are of general astrophysical importance as sources of electromagnetic radiation and of quasi-static magnetic fields. Strong radio emissions from Jupiter were among the first to be interpreted in terms of radiation from electrons trapped in a magnetic field (Warwick 1961).

The discovery of the Van Allen belts had an immediate effect on general studies of charged particle motion in magnetic and electric fields. Among the main contributors were members of the Iowa group, in particular Carl McIlwain. For these studies the guiding center approximation method, developed by Alfvén and published in 1940, was a basic tool. The radiation belt discovery also gave rise to many theoretical investigations and consequent improvements of the understanding of scattering and diffusion mechanisms for energetic particles in magnetospheres (e.g. Fälthammar 1966, 1968), of basic importance from plasma physics and astrophysics points of view.

In the space faring era good knowledge of the radiation belts quickly became a requisite for satellite engineering and orbit design. The radiation belts offer the most severe environment for spacecraft and space activities that can be found around Earth. It is generally unavoidable for satellites in orbits far from Earth to pass through them at least once, but that passage is made as short as possible, and low-orbiting application satellites are placed below them.

In spite of the fact that the radiation belts have been investigated since the early sixties, most intensely in the

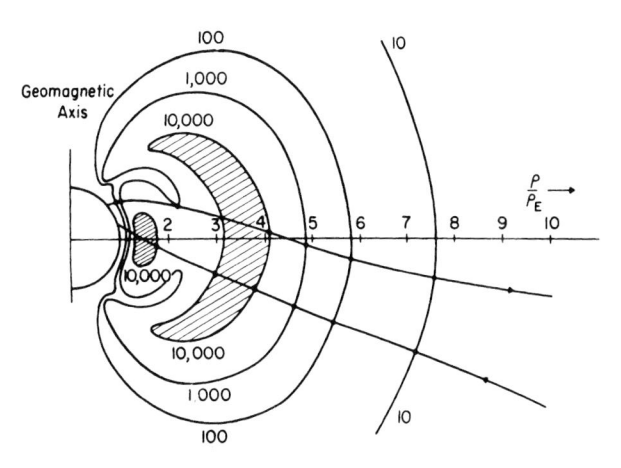

Figure 4 A meridian cross section of contours of equal intensity of geomagnetically trapped radiation based on data from Explorers I, III, and IV and Pioneer III. The semicircle at the left represents Earth and the two undulating curves that travers the diagram represent the outbound (upper curve) and inbound (lower curve) trajectories of Pioneer III. The labels on the contours are counts per second of the lightly shielded Geiger-Müller tube. The linear scale is in Earth radius (6371 km). The two cross-hatched regions of high intensity are the inner and outer Van Allen belts separated by the 'slot' (after Van Allen and Frank 1959).

sixties and seventies, surprising discoveries are still made. In the nineties a new radiation belt was found. It had quite an unusual composition, consisting of elements with high first ionization potential (He and O are present, C is not). The existence of a cosmic ray component, produced by acceleration of interstellar ions in the heliospheric boundary shock, was predicted by Fisk et al. (1974). That component, named Anomalous Cosmic Rays (ACR), would have the composition mentioned. It took almost 20 years until the trapped component in the magnetosphere was observed by means of the US spacecraft SAMPEX (Cummings et al. 1993). In the nineties it was also discovered that strong interplanetary shock waves when passing the magnetosphere penetrate all the way into the radiation belts and accelerates particles there (Li et al. 1993). We may still not know all important processes affecting the radiation belts and there are some characteristics of them which are not yet understood. An important example is a population of multi-MeV electrons in the heart of the outer belt, the origin of which remains to be understood.

The particle populations constituting the radiation belts and those of the ring current overlap to a large extent in energy. The radiation belts contain the high-energy ions and electrons that can penetrate into spacecraft shielding materials and eventually cause radiation damage to spacecraft instrumentation and to humans. The particles which contribute substantially to the ring current density are those in the medium-energy range of a few tens of keV to a few hundreds of keV, located mainly in the same regions of the inner magnetosphere as the radiation belts.

The ring current particle population is much more variable than the higher energy radiation belt particles. It gives rise to the global depression of the geomagnetic field, characteristic of the main phase of magnetic storms, when its density is rapidly increased and then decreases with a characteristic time of the order of day(s). This was known from ground-based magnetic observations long before the space age. The satellite measurements have provided *in situ* observations of the ions and electrons which constitute the ring current. Of the early satellite missions Explorer 45 (S^3) (e.g. Williams and Lyons 1974a,b) in the early seventies provided particularly important contributions to the understanding of the ring current. Its composition, which affects the magnetic effects, was not determined until the late eighties, when the AMPTE spacecraft for the first time brought ion mass spectrometers covering the entire energy range of interest through the whole volume of the ring current (between 3 and 9 R_E from Earth in the equatorial plane). Protons were found to carry the main part of the current except in strongly disturbed conditions when O^+ ions from the ionosphere may dominate (e.g. Hamilton *et al.* 1988). The contributions of various sources to the storm-time ring current is still a matter of some controversy. See also the section on substorms and storms below.

The inner magnetosphere does not contain only very energetic ions and electrons in the radiation belts and the ring current but also large amounts of cold and superthermal plasma. The ionosphere is the principal source of cold plasma (≤ 1 eV) throughout the magnetosphere. At low latitudes the ionosphere extends out along the magnetic field lines to several R_E with densities decreasing with altitude but still much higher than at similar altitudes along higher latitude field lines. Often the cold plasma density decreases sharply beyond a certain altitude. This was another astonishing result, first found by Gringauz *et al.* (1960) from data obtained with the Soviet Lunik rockets (see Figure 5). Not long thereafter the whistler method for observing plasma density near the equatorial plane, developed by Carpenter (1963), showed similar results. Carpenter could also determine a diurnal variation in the extension of the high-density region (called the plasmasphere, with the plasmapause as its boundary region). Nishida (1966) and Brice (1967) interpreted the plasma pause as the limiting surface within which the plasma co-rotates with Earth. Plasma moving along field lines outside of this surface would be convected away, but on the field lines in the co-rotating region, it would stay and develop some sort of diffusive equilibrium. The location of the limit to co-rotation depends on the large scale convection electric field within the magnetosphere. Since that field is strongly correlated with substorms and storms and other dynamic features of the magnetosphere, the size of the co-rotating region shrinks with increasing activity level. Equilibrium situations often have no time to develop. An illustration of the effect of a sudden increase of the convection field is shown in Figure 6. The model with the plasmapause being the limit of co-rotation (or 'last closed field

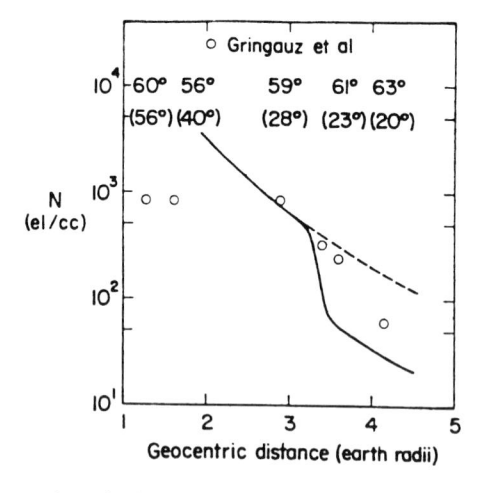

Figure 5 Plot of plasma density versus geocentric distance comparing an idealized equatorial density profile obtained from whistlers with measurements of ion density reported by Gringauz *et al.* (1960) from the Lunik 1 rocket (open circles), launched on January 19, 1959. The upper row of numbers shows the invariant latitude of the rocket (after Carpenter 1965).

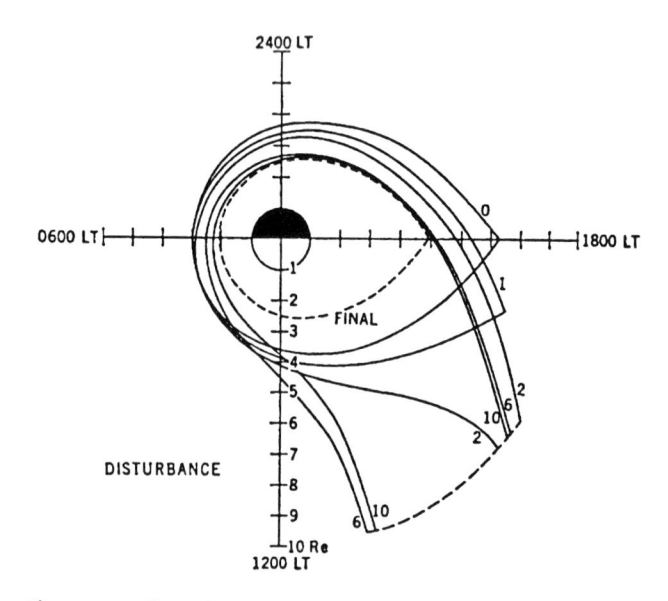

Figure 6 The plasmapause positions plotted for selected times (0, 1, 2, 6, and 10 hours) after the dawn-dusk electric field component has increased suddenly from an initial steady state to a value twice as large (after Grebowsky 1970).

line') is certainly an oversimplification, but it has been, and is, useful for the understanding of plasmasphere dynamics.

The inner magnetosphere also contains a superthermal population of electrons produced mainly as photoelectrons in the upper atmosphere and as secondary electrons from the interaction of precipitating energetic particles with the atmosphere. They have energies from a few to hundreds of eV and contribute in important ways to the heating of the cold plasma in the ionosphere and plasmasphere.

Despite nearly 40 years of study, our understanding of the plasma populations and the fields in the inner magnetosphere is still far from complete. The reader is referred to e.g. Carpenter and Lemaire (1997) or Chapter 4 of Hultqvist *et al.* (1999) for more information.

THE HIGH-LATITUDE MAGNETOSPHERE

The most spectacular sign of processes occurring in the high-latitude magnetosphere is the aurora (see Figure 7). It has always been a part of the human environment at high latitudes and has occasionally been seen also at middle to low latitudes. The mentioning of it in Aristotele's Meteorologica from about 340 before Christ is an early, but not the earliest documentation. Much older ones have been reported from China. More or less scientific studies of the aurora are known from the early 1700s. The French scientist de Mairan in the early years of that century carried out extensive statistical investigations of aurora in northern Scandinavia and claimed that there is a relation between aurora and disturbances on the Sun as well as an annual variation in auroral occurrence frequency. In the 1740s, Celsius and Hiorter in Upsala discovered that the geomagnetic field varies when there is aurora in the sky and, by correspondence with an English colleague, Graham, Celsius could conclude that the magnetic variations are not only local but at least regional. Sabine in 1852 demonstrated a close correlation between geomagnetic disturbances and solar activity as manifested by sunspots.

There were many more or less exotic generation mechanisms proposed during the 1800s, including various electric discharge phenomena around Earth. Alfvén carried the ideas involving electric fields forward in his 'Theory of Magnetic Storms', published in 1939–40. There he for the first time in a serious way introduced an electric field component along the magnetic field lines as a source of acceleration of aurora-generating particles. An important achievement of magnetospheric research has been the demonstration that a collisionless, magnetized plasma can sustain large potential differences in the direction of the magnetic field.

Several leading scientists were critical of Alfvén's parallel electric field. A common argument was that the conductivity along the magnetic field lines in and above the ionosphere is so high that any space charges and associated parallel electric fields would be eliminated very efficiently. The parallel electric field concept was more or less dismissed as being contrary to generally accepted theories and it was only the direct measurements in space in the 1970s, with release experiments on sounding rockets (Haerendel *et al.* 1976; Wescott *et al.* 1976) and also with different kinds of particle experiments on rockets and satellites that finally changed the situation completely. In the USA Review and Quadrennial Report to IUGG in 1979 Stern states that 'in the period 1975–78 ... it became increasingly evident that the condition $E_\parallel = 0$ is often grossly violated in the magnetosphere. It would be only fair to say that the period marked a transition from general scepticism concerning the role of $E_\parallel$ in the magnetosphere to the acceptance of $E_\parallel$ as an essential ingredient in such phenomena as discrete auroras ...'.

Among the many experimental results that paved the way for the general acceptance of parallel electric fields, some were of special importance in breaking the resistance. To these belong the earlier mentioned release experiments, which demonstrated the parallel acceleration in a very direct way, and also the discovery of the inverted V events by Frank and Ackerson (1971) which are illustrated in Figure 8.

Nearly monoenergetic electrons precipitating along the magnetic field lines into an aurora were observed already by McIlwain (1960) in the very first years after Sputnik by means of one of the early sounding rockets launched

Figure 7 Aurora over Kiruna, photographed by Torbjörn Lövgren.

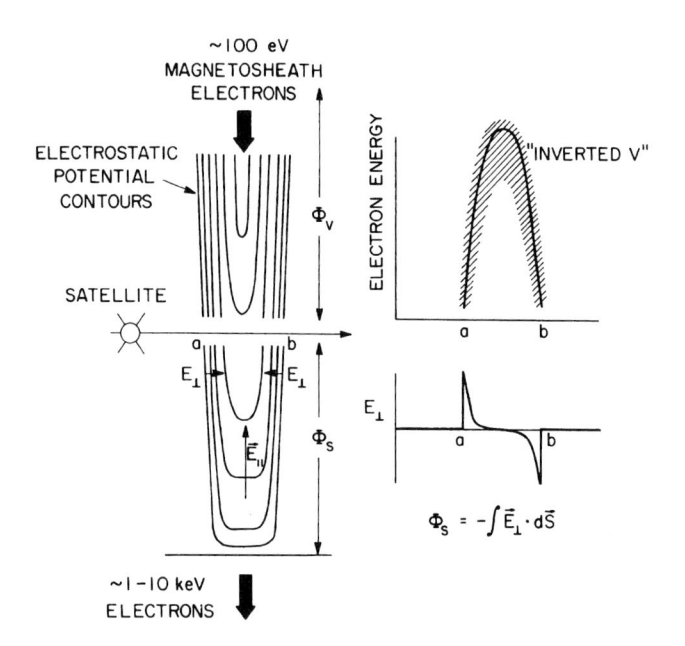

Figure 8 Qualitative form of the electrostatic equipotential contours associated with an inverted-V electron precipitation event (after Gurnett 1972).

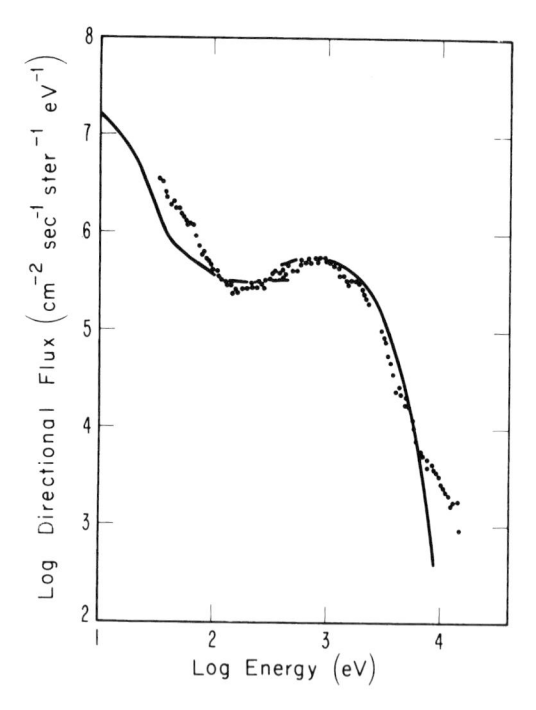

Figure 9 Comparison of Evans' (1974) model with observations by Frank and Ackerson (1971). The electrons were assumed to have originated from an 800 eV plasma of density 5 cm^{-3} and the field-aligned potential difference was taken to be 400 V (after Evans 1974).

into an aurora. One big stumbling block disappeared when Evans (1974) eliminated the problem that the existence of an intense low-energy tail of the observed peaked auroral electron spectra constituted (see Figure 9). He could explain that tail as composed of secondary electrons produced by the primary electrons in the ionosphere. The S3-3 satellite, launched in 1976, gave the final push in favor of the parallel electric field component, one of which was the new-discovered ion beams that move out of the upper ionosphere, first reported by the Lockheed group (Shelley et al. 1976). Another important result from S3-3 was from direct electric field measurements with a double probe experiment by Mozer and his group, which strongly suggested the existence of electric fields with a significant parallel component (Mozer et al. 1977). The current through the parallel potential difference has been found to be proportional to the voltage (Lyons et al. 1979). Knight (1973) gave a theoretical basis for the parallel fields.

The Swedish Viking satellite added new evidence for the existence of large potential differences along magnetic field lines. An important discovery by means of Viking was that the average field-aligned electric field component points upward for altitudes greater than 6000 km but downward below 4000 km (Marklund 1993; see Figure 10). While the observations of upward electric fields corroborated the earlier mentioned observations at medium altitudes, downward electric fields at lower altitudes came as a surprise, to the extent that it even raised the question of the reliability of electric field measurements. The Viking results have in the nineties been confirmed by the Swedish/German Freja and the US FAST satellites. They have shown that electrons in nearby regions to those with downward accelerated primary auroral electrons are accelerated upward to keV energies by parallel fields (Carlson et al. 1998). The upward electron acceleration and downward ion acceleration (observed with Freja) due to a downward directed electric field have been found at lower altitudes than the downward electron acceleration, as the average situation in Figure 10 indicates.

The argument against Alfvén's claim that magnetic field-aligned electric fields accelerate primary auroral electrons into the atmosphere – that the plasma conductivity (carried by electrons) along the magnetic field lines is too good to allow potential differences to build up – is obviously wrong at altitudes above several thousand kilometers and – an even greater surprise – it appears even to be wrong under certain conditions closer to the ionosphere, where the Freja and FAST results demonstrate that the electrons have been found to be accelerated in the upward direction and ions downward.

The high-latitude magnetosphere is not only characterized by quasi-static electric fields but also by fluctuations, irregularities, waves, and turbulence of many different kinds. The smallest spatial structures seen (such as auroral rays) are still not understood. The fluctuations and waves in

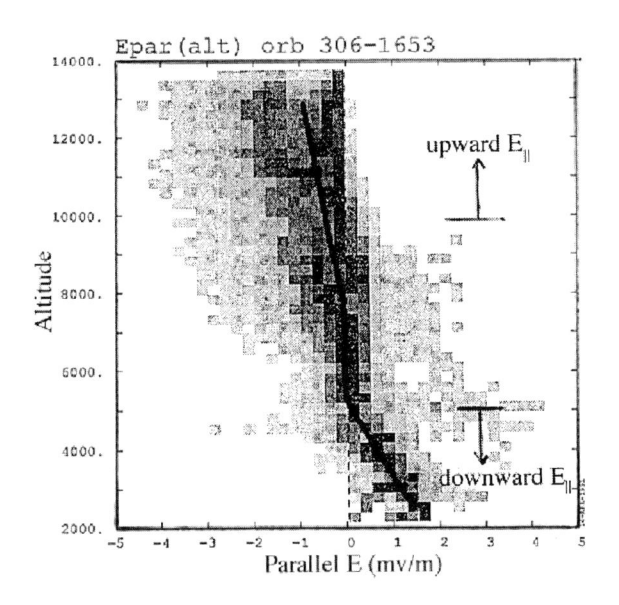

Figure 10 Statistical properties of the field-aligned electric field as determined from the Viking double probe experiment (after Marklund 1993).

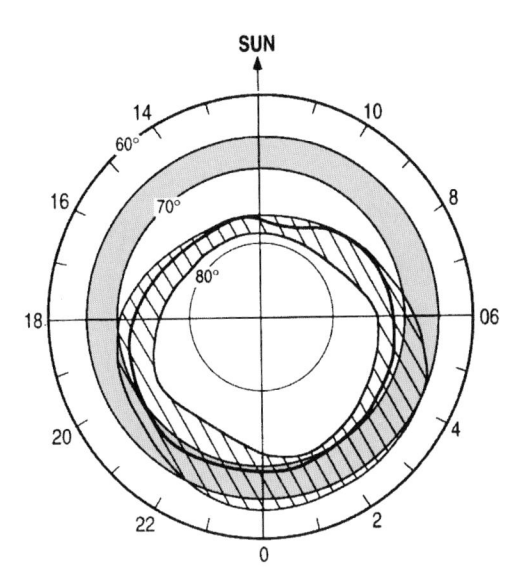

Figure 11 The auroral zone (shaded), the auroral oval (covered by slant lines) and the intersection line between the outer boundary of the outer radiation belt and Earth's surface (after Akasofu 1997).

a wide frequency band, starting at a fraction of a Herz, heat and increase the magnetic moment of the ions at medium altitudes above the high-latitude ionosphere so that they are driven out into the magnetosphere. Those processes have been intensely studies in the last two decades, but it is not yet quite clear what processes prevail under different conditions. For more information see e.g. André and Yau (1997) or Chapter 2 in Hultqvist et al. (1999) and references therein.

Not only plasma processes in the uppermost high-latitude ionosphere and at medium altitudes (of order 1 R_E) on auroral magnetic field lines are important for the generation of aurora. To reach such low altitudes charged particles in the equatorial regions of the magnetosphere have to be scattered into a narrow cone around the magnetic field lines within which the particles reach those altitudes and finally the atmosphere. The supply of the energetic electrons generating intense nighttime auroras is located in the tail of the magnetosphere and the electrons are scattered within a thin layer of very low magnetic field near the equatorial plane, where they undergo non-adiabatic acceleration in the large scale electric field within the magnetosphere so that some of them are scattered into the loss cone (so-called Speiser orbits; Speiser 1965). What process fills the loss cone on the day-side is still unknown in spite of many investigations. The cause of the most common kind of aurora, the diffuse one, which dominates on the day-side (but occurs also on the night-side), thus still remains to be identified.

For details the reader is referred to e.g. Lyons (1997) or Chapter 3 in Hultqvist et al. (1999) and references therein.

Some large-scale physical mechanisms associated with aurora have distributions in space and time, which were investigated from Earth' surface before the space era, in particular during the IGY, 1957–58. From all-sky camera recordings in more than 100 places, Feldstein (1963) was able to demonstrate that the aurora is not distributed along two roughly circular rings around Earth's two magnetic poles, of which only the night-side part was well observed, as had been generally believed since Fritz published his isochasms in 1881, but along ovals which are at much higher latitudes at noon than at midnight (see Figure 11). This could be verified by the first satellite that scanned the auroral emissions along the whole auroral oval in one passage over the polar regions, the Canadian ISIS-2 spacecraft in 1971 (Lui et al. 1975). The location of the oval on the day-side could soon be associated with the polar cusp regions of the magnetosphere, which separate magnetic field lines closing on the day-side from those closing on the night-side in the noon-midnight meridian plane of the magnetosphere (see Figure 23), and thereby later with the general gross features of the magnetosphere.

Also the major macroscopic relaxation process of the magnetosphere, the magnetospheric substorm (see later substorm section), as manifested in auroral dynamics, was very well described by Akasofu already in 1964 on the basis of IGY all-sky camera recordings (see Figure 12). Whereas the ISIS satellites produced one image of the

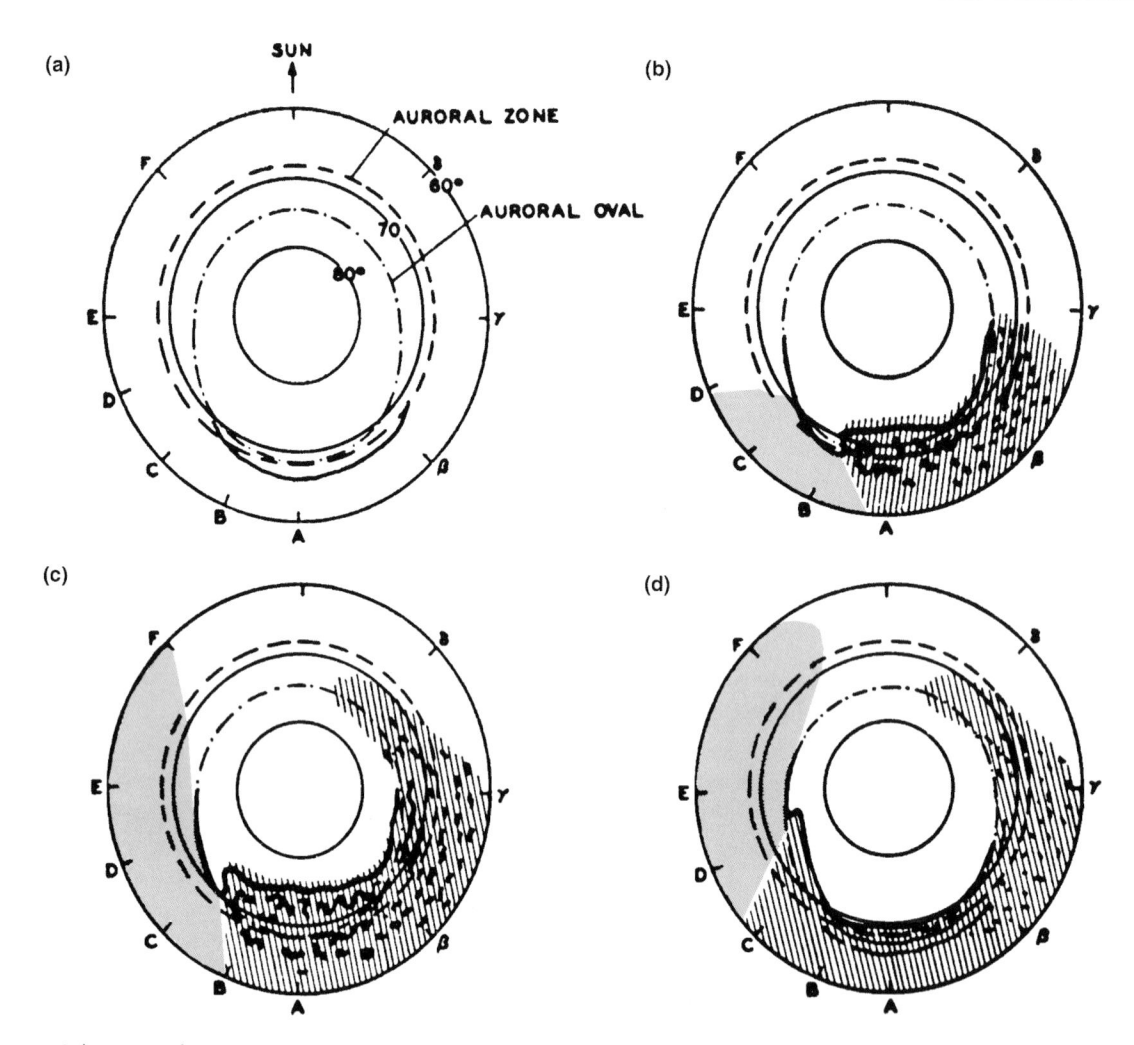

Figure 12 Schematic diagram to show the development of both the auroral and polar magnetic substorms, from a quiet situation (a), an early epoch of the expansive phase (b), the maximum epoch of the substorm (c), to an early epoch of the recovery phase (d). The region where a negative magnetic bay is observed is indicated by the line shade and the region of a positive bay by a dotted shade (after Akasofu 1968).

auroral oval per satellite orbit by using the spin to scan over all directions perpendicular to the spin axis, Dynamics Explorer 1 and Viking in the eighties used optical imaging systems and could follow the auroral dynamics in the entire oval in some detail. Viking produced one auroral image every 20 seconds and made the day-side aurora as available for studies as the night-side one. Although some new auroral distributions (like 'thaeta aurora' and 'pearl strings') were discovered by the space-born UV instruments and important complementary information was added on the day-side, the spacecraft mainly confirmed Akasofu's auroral substorm dynamics scheme. However, satellites provided data from various points in the magnetosphere without which the connection between substorm phenomena in the upper atmosphere and all over the magnetosphere would not have been possible to find.

Whereas IGY magnetic field measurements on Earth's surface were interpreted in terms of horizontal electric currents in the ionospheric E-layer, primarily because currents along the magnetic field lines cannot be measured from ground, early satellites demonstrated that magnetic field-aligned currents flow in and out of the auroral oval (Zmuda et al. 1966). The regions of upward field aligned currents were found to coincide partly with those of the auroral arcs and the aurora-generating electrons were shown to be important current carriers. Iijima and Potemra (1976) derived the current distributions shown in Figure 13 primarily from data obtained with the TRIAD satellites. Potemra also introduced the name Birkeland currents for the currents flowing along the magnetic field lines. The Birkeland currents are primarily carried by electrons and, as mentioned above, the current continuity condition frequently transfers

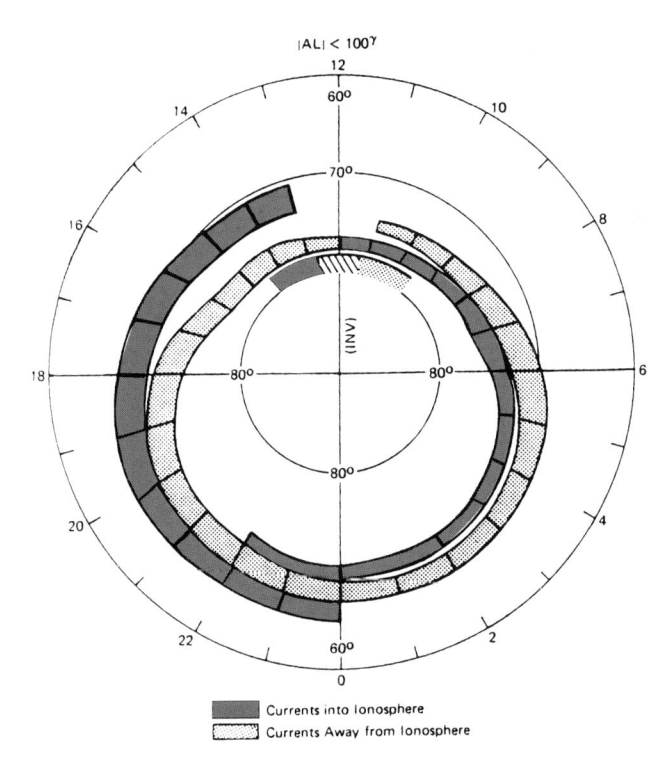

Figure 13 Field-aligned current regions in the ionosphere (after Iijima and Potemra 1976).

part of the voltage difference available within the magnetosphere to those parts of the auroral magnetic field tubes where the supply of current carriers in the circuit is lowest.

THE TAIL OF THE MAGNETOSPHERE

The tail of the magnetosphere, often called the magnetotail or simply tail, dominates the magnetosphere strongly in terms of volume and in terms of mass it is second only to the plasmasphere.

That the magnetosphere has a much larger extension in the anti-sunward than in the sunward direction was a conclusion of the Chapman and Ferraro (1931,1932) study already. In the first years after Sputnik theoretical gas-kinetic calculations of the balance between the dynamic pressure of the solar wind proposed by Parker (1958) and the geomagnetic field, representing a balancing pressure, were made by Zhigulev and Romishevskii (1959), Beard (1960), and Spreiter and Briggs (1961,1962) (see Spreiter and Stahara 1977, for details). The results showed what has later been found also from observations: that the magnetosphere in Earth's vicinity is shaped like a paraboloid of revolution, extending in the anti-sunward direction as a long cylinder. In the near-Earth region the tail diameter increases with downstream distance, being about 50 R_E at distances of 30–50 R_E from Earth. The flaring of the magnetospheric tail ceases beyond 100 R_E downstream, the tail diameter staying in the range 50–60 R_E.

A schematic tail cross section in the distance range of ten to several tens of R_E downtail is shown in Figure 14. Moving from the solar wind inward, the bowshock (BS) is the boundary in interplanetary space separating the shocked solar wind plasma and magnetic field in the magnetosheath (MS) and the un-shocked solar wind (as mentioned earlier). The magnetopause (MP) separates the magnetosheath from the magnetosphere (as also mentioned before). In the magnetospheric boundary layer (BL) magnetosheath-like plasma is found inside the magnetopause. The high-latitude boundary layer is called the plasma mantle (PM), or only mantle. It is on magnetic field lines which go out into the interplanetary space ('open' field lines). At lower latitudes, the low-latitude boundary layer (LLBL) is generally more magnetically turbulent than the mantle. It is still unclear to what extent it is on open and closed magnetic field lines. The plasma sheet (PS) is composed of hot, comparatively dense plasma, whereas the lobes (LB) are more or less empty of plasma and contain high, steady magnetic fields. The magnetic field within the plasma sheet is directed earthward north of the symmetry plane and anti-earthward south of it. The two regions of opposite magnetic field directions are separated by a thin layer with very low magnetic field, called the 'neutral sheet', which carries a strong current directed from the dawn to the dusk side and closing along the magnetopause. Most of the tail data within 23 R_E were obtained by means of the International Sun-Earth-Explorer spacecraft, ISEE 1 (NASA) and ISEE 2 (ESA), in the seventies.

In the deep tail the distributions of plasma and magnetic field are generally much more complex than shown in Figure 14. Cross sectional views of the topology of the magnetic field lines passing a tail cross section at 81 R_E downtail at two different times, derived by Frank et al. (1995), are reproduced in Figure 15. The position of the spacecraft providing the data, the Japanese-US Geotail, is indicated in the panels. As shown in the figure most field lines are not at all connected with Earth, and those which are connected at one end (open field lines) are not found within a cylindrical shell as in Figure 14 but in a sheath-like volume, strongly extended and inclined with regard to the equatorial plane. The closed field lines, corresponding to those in the plasma sheet in Figure 14, are found in a very irregular volume, very different from the regular shape of the plasma sheet closer to Earth.

Figure 15 could be obtained from the available measurements only by means of advanced numerical models of the magnetosphere. Only a few spacecraft have ever been in the very deep tail (ISEE 3 of NASA and the above-mentioned Geotail and, with one passage through the tail at 1000 R_E

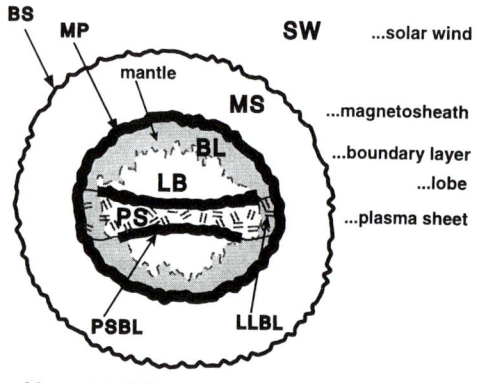

Figure 14 Schematic drawing of the magnetotail cross section at downtail distances from about ten to a few tens of Earth radii, showing basic plasma regimes (after Christon et al. 1998).

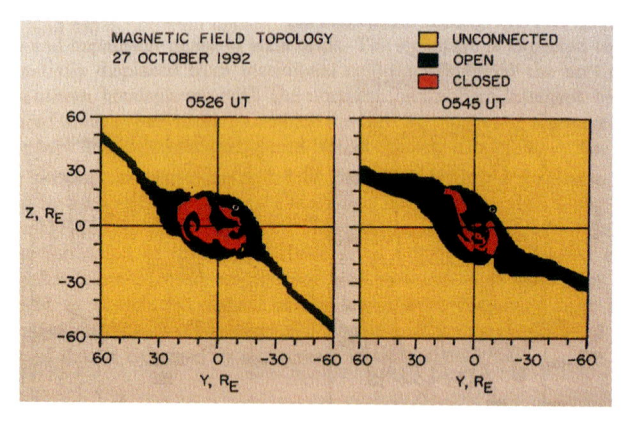

Figure 15 Cross-sectional views of the topology of the magnetic field lines at a downtail geocentric distance of 81 Earth radii. There are field lines not connected to Earth (unconnected), those with one end connected to Earth (open) and those with both ends connected to Earth (closed) (after Frank et al. 1995).

downtail, the American Spacecraft Pioneer 7). Both ISEE 3 and Geotail reached beyond 200 R_E. With the enormous volume of the tail, the direct measurements available are certainly much too few to derive anything like a self-consistent model on the basis of them alone. The tail is therefore a part of the magnetosphere where numerical modeling is absolutely necessary for interpreting the observations in space.

The global models of the magnetosphere are fluid magnetohydrodynamic models (MHD), with reconnection introduced at the magnetopause whenever certain conditions for the directions of the field lines on the two sides of the boundary are met. Those models are not fully self-consistent, but they are able to reproduce many of the known characteristics of the magnetosphere. The MHD approximation of the plasmas and fields is not valid everywhere and in particular not for small scale phenomena. The models are being continuously improved by modifications of the technique and by adding physical mechanisms as boundary conditions or by combining the large scale models with single particle theory calculations of particle trajectories in the model magnetic and electric fields and of the acceleration of the particles along the orbits. These latter combination models are often called 'Larger Scale Kinetic' models (LSK).

Although the MHD and LSK models have achieved remarkable successes in predicting e.g. the total flow of plasma through the tail, the formation of the plasma sheet, and variations of the magnetosphere with variations in the interplanetary magnetic field etc., there are a number of observational results that seem difficult to fit into the models. Perhaps the most important disagreement is found in the plasma density of the plasma sheet. Geotail has shown that the density has a maximum when the interplanetary field is northward and the input of plasma from the solar wind minimal according to models. The ions are then also colder than normally (Terasawa et al. 1997). Another observation that is difficult to interpret in terms of present models is the discovery by means of Geotail of cold ionospheric O^+ ions flowing along the tail with protons of solar wind origin as far away as 210 R_E from Earth (Mukai et al. 1994). Even if some ionospheric ions are seen in the distant tail, the composition of the tail is completely dominated by plasma of solar wind origin. For more information about the distant magnetotail, see e.g. Chapter 6 in Hultqvist et al. (1999) and references therein.

A special region of the tail of great importance for the dynamics of the magnetosphere near Earth is the plasma sheet. It is a main storage region for plasma and energy in the near-Earth part of the magnetosphere, providing particles for the ring current and the aurora. This near-Earth part is on closed magnetic field lines. Electrons and ions in the plasma sheet have average energies on the order 1 and 10 keV, respectively. The plasma density is generally of the order of 1 cm^{-3}. It contains ionospheric and solar wind ions in numbers of similar orders of magnitude, the ionospheric component increasing strongly with the disturbance level of the magnetosphere. The plasma sheet has a sharp cut-off near Earth at a distance of about 10 R_E, termed 'the inner edge of the plasma sheet' or the 'Alfvén layer' and caused by the combined effects of electric field and magnetic gradient drift. The solar wind component of the plasma is believed to be introduced and accelerated primarily on the field lines reconnecting in the neutral sheet of the tail. The flow of the plasma in the plasma sheet is earthward. The sections on convection, substorms and storms, and plasma sources and losses below contain some more information about the magnetotail.

The magnetotail, except for its part closest to Earth, is still quite poorly known experimentally and it is, as mentioned, the part where the dependence on models for interpreting data is most pronounced. The spatial/temporal complexity of it seems to increase with distance from Earth, which indicates that many surprises remain there for coming generations to experience.

THE BOUNDARY LAYERS OF THE MAGNETOSPHERE

Whereas classical fluid theory predicts that the magnetopause is impenetrable for plasma and magnetic fields, observations show that there exists a boundary layer containing magnetosheath-like plasmas just inside the magnetopause. This was first shown for the equatorial night-side magnetopause by Hones *et al.* (1972), for the low-latitude day-side magnetopause by Eastman *et al.* (1976) and for the high-latitude magnetopause by Rosenbauer *et al.* (1975). The boundary layer is generally divided up into the low-latitude boundary layer (LLBL), the entry layer near the polar cusps, and the plasma mantle (PM) along the high-latitude magnetotail.

Observations of ionospheric ions outside the magnetopause demonstrate that magnetospheric plasma is also lost across the magnetopause (e.g. Peterson *et al.* 1982). More energetic magnetospheric particles are also commonly seen outside the magnetopause (Meng and Anderson, 1970) and one talks about a magnetosheath boundary layer (MSBL) of such energetic particles. So the magnetopause obviously is not impenetrable. Solar wind particles, energy and momentum are transferred to the magnetosphere. The total amount of plasma entering the day-side magnetopause has been estimated to be on the order 10^{26} ions s^{-1} (Eastman *et al.* 1976, Hill 1979) and measurements within the magnetotail lobes indicate that plasma continues to enter all along the tail (e.g. Gosling *et al.* 1985).

The characteristics of the plasma in the LLBL and in the PM are quite different. These differences may be due to differences in the large-scale motions of the plasma in the regions and to differences in neighboring plasma populations. The plasma mantle particles are streaming tailward on field lines convecting inward into the lobes which are more or less empty, whereas the LLBL plasma is neighbor to the ring current and plasma sheet populations and rather moves towards the magnetopause. But it is also possible that the two layers form in different ways.

Many different physical mechanisms may be of importance for the transport of magnetosheath plasma, and of magnetospheric particles, across the magnetopause. Among these are magnetic reconnection, finite Larmor radius effects, diffusion, the Kelvin-Helmholtz instability, impulsive penetration and direct cusp entry. Of these the reconnection model has been developed furthest in terms of number of testable predictions, many of which have been confirmed by both case studies and statistical investigations. But the relative contribution of reconnection to the total amount of plasma transported across the magnetopause still remains to be determined. For reconnection theory makes no quantitative prediction about how much of the solar wind plasma incident on the cross section of the magnetosphere actually enters. This is because neither the rate at which magnetic flux is being reconnected, nor the extent of the reconnection region and its dependence on the interplanetary magnetic field are known a priori. Reconnection, contrary to all other mechanisms of possible importance for plasma entry, needs to take place only in a small (diffusion) region on the magnetopause but still has global effects. Once the interplanetary and terrestrial field become connected they remain interconnected while convected with the solar wind along the magnetopause, and plasma can continue to enter. The other processes mentioned operate locally.

For the other processes listed above the available observational evidence for their roles in the transport across the magnetopause is less than for reconnection both on microscales and macroscales. This does not mean that their unimportance is clear (see e.g. Lundin 1997). It is difficult to believe that diffusion does not occur across the magnetopause. It can supply plasma without any strong dependence on the IMF and can transport solar wind plasma onto possibly closed magnetic field lines in the LLBL. Present estimates of the diffusion coefficient fall substantially short of the required value, however.

For plasma entry through finite Larmor radius effects or impulsive penetration there are no measurement results yet that establish their occurrence, let alone determine the resulting flux across the magnetopause and the theories have not yet been developed to the state of quantitative predictions. There is evidence that plasma enters in the polar cusp regions, not only because subsolar reconnection populates cusp field lines, but also as a result of some unique process genuine to the cusps. But the nature of the process and its contribution to plasma transfer are not clear.

The boundary layers vary in thickness both with distance from the subsolar point of the magnetopause and with the IMF B_z. The thickness of the plasma mantle has been shown by Sckopke *et al.* (1976) to depend strongly on B_z. It has also been shown that there is no gap between the magnetopause and the outer boundary of the plasma mantle (Sckopke and Paschmann 1978). If the plasma entry were restricted to the region in the vicinity of the cusp, such a gap would develop as a result of the inward convection of the magnetic field lines. A model of the open magnetotail boundary has been developed by Siscoe and Sanchez (1987). Figure 16 shows a cross section of the magnetotail implied by this model. In this picture the magnetopause is a rotational discontinuity that maintains its distance from the center of the tail because it is standing in the flow, i.e. propagates upstream as fast as it moves downstream.

The continuous entry of solar wind plasma along the tail makes the plasma mantle thicker with increasing distance downtail. It fills the whole lobe and reaches the plasma sheet boundary in the distant tail, where the tail reconnection occurs. The lobe has been identified occasionally as far downtail as 200 R_E, but the observations are generally consistent with reconnection occurring at distances of order 100 R_E (when it occurs).

The experimental investigation of the boundary layers is in an early state. Their physics appears to be more complex than in most other magnetospheric regions and more complex than mostly used fluid models can simulate. For instance, plasma components of different mass and origin have been found to move differently (Lundin 1984). It is probably only in Earth's magnetosphere future detailed experimental studies of the boundary layers and of the interactions at the magnetopause that give rise to them can be made. This basic plasma physical and astrophysical matter will most likely stay on the agenda of space physics research for a long time. The ESA Cluster project will hopefully provide the next major step. For more information about present knowledge about the boundary layers and the lobes the reader is referred to e.g. Chapters 5 and 6 of Hultqvist et al. (1999) and references therein.

PLASMA CONVECTION IN THE MAGNETOSPHERE

Observations of aurora and ionospheric phenomena before and during the IGY had shown that structures generally moved westward before magnetic midnight and eastward after midnight and in the IGY it was found that there was in addition an equatorward motion at somewhat higher latitudes in the midnight region (e.g. Davis 1960). These and other low altitude observations led Axford and Hines (1961) to write their well known paper about the plasma convection in a closed magnetosphere. They assumed that there existed some sort of interaction at the magnetopause that could drive the magnetospheric plasma close to the magnetopause in the direction of the solar wind flow. The interaction was named 'viscous-like' but Dungey's reconnection process could do as well. Simple continuity considerations combined with the above-mentioned low-altitude observations necessitated a sunward flow in the interior of the magnetosphere. By adding the rotation of the plasma with Earth they arrived at the convection pattern shown in Figure 17. Although the magnetosphere has been demonstrated to be open, the pattern in Figure 17 is applicable to the region of the open magnetosphere earthward of the reconnection region in the tail. The earthward flow thus takes place within the plasma sheet (see Figure 23) and the downtail flow in the boundary layers. The low-energy plasma flows along the electrostatic equipotential surfaces within the magnetosphere.

A lot of observational efforts have been devoted to the convection and Axford's and Hine's pattern has been found to be a good basis for the understanding of important dynamical features of the magnetosphere. But, as commonly with complex systems, things have been found to be more complex than originally thought. A strong dependence on the interplanetary magnetic field (IMF) has been observed, as expected from the advanced reconnection model. Not only the north-south component of IMF (B_z), the sign of which controls whether reconnection occurs or not, but also the component of IMF perpendicular to the Sun-Earth direction and to Earth's dipole (B_y) affects the convection pattern strongly. The cross section of the tail shown in Figure 16, with the interplanetary magnetic field lines coming in on the dawn-side and going out on the

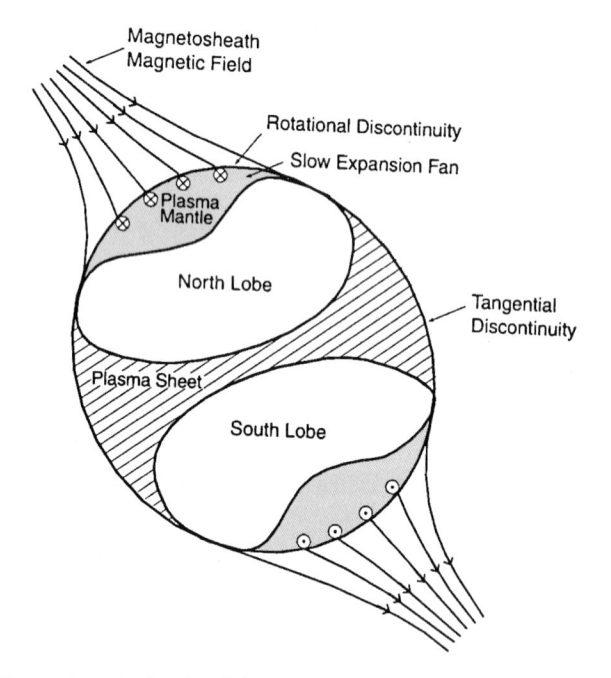

Figure 16 A sketch of the magnetotail cross section as seen from Earth, showing the magnetopause as a rotational discontinuity and the plasma mantle as a slow-mode expansion fan. The north-south and dawn-dusk asymmetries of the plasma mantle, as well as the tilted plasma sheet orientation are induced by a positive B_y component of the interplanetary magnetic field (after Siscoe and Sanchez 1987).

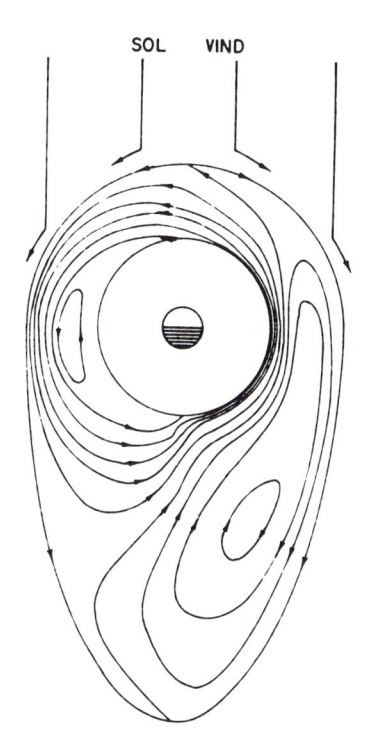

Figure 17 The convection pattern of the magnetospheric plasma expected from viscosity-like forces acting at the magnetopause and the rotation of Earth, according to Axford and Hines (1961).

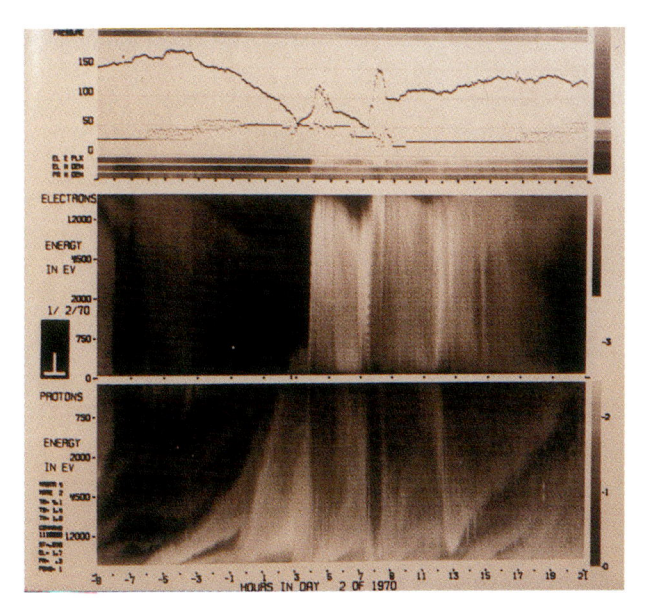

Figure 18 A 24-hours charged particle spectrogram sampled by the geosynchronous satellite ATS-6. Electron (above) and ion (below) energies go from 1 eV to 82 keV. Top panel shows magnetic field data. Whiter regions correspond to more intense fluxes of particles (after DeForest and McIlwain 1971).

dusk-side is an effect of IMF B_y pointing from dawn to dusk. The curvature of the reconnected field lines, carried downtail by the solar wind, brings the region with downtail convection into the dawn-side in the northern hemisphere and into the dusk-side in the southern hemisphere, as Figure 16 indicates. The regions with sunward motion are modified by this displacement. For northward IMF more complex and less clear convection patterns occur. The explanation of most convection observations is one of the observational supports for the reconnection model.

The supply of energetic particles to the ring current during magnetic storms has also been found to be largely interpretable in terms of inward convection of plasma sheet particles in an intensified convection electric field (Lyons and Williams 1984). The rapid change of composition of the ring current in the early phase of substorms and storms seems, however, to require also a significant input of accelerated ions from the ionosphere (e.g. Daglis *et al.* 1991).

That large scale convection cannot explain all plasma transport in the inner magnetosphere was demonstrated already by McIlwain (DeForest and McIlwain 1971; McIlwain 1972) from observations with the US geosynchronous ATS-6 satellite. He and his coworkers demonstrated that at the beginning of a substorm plasma very quickly is brought in to a sharp injection boundary (Figure 18). From there convection in the largescale electric and magnetic fields takes over. What produces the injection boundary and what brings the tail plasma to it very quickly is not yet well understood. The most promising attempt to explain these observations were made by Moore *et al.* (1981). They proposed that the plasma injection is caused by a compressional wave launched from the base of the magnetotail during substorm expansion. That wave would propagate earthward, steepening on its way and eventually break or dissipate. The inner limit of the wave's propagation would constitute the injection boundary. The nature of the processes that break and dissipate the wave is unknown.

Another problem with convective transport of plasma towards Earth in the tail was found by Erickson and Wolf (1980) and later by Schindler and Birn (1982). The pressure within the central plasma sheet was shown to rise so dramatically, adiabatic motion assumed, that the magnetic pressure of the lobes would be unable to contain the plasma sheet pressure. Continuous time-stationary earthward convection can therefore not exist. The many additional investigations of this problem that have been made have not solved the problem. More recent observations of transient or bursty flow within the magnetotail may indicate that a solution may be found in that direction. On the other hand, plasma bubbles (plasmoids) have been observed to move tailward from the inner part of the tail during substorms. The two phenomena may be connected. The observations of plasmoids are discussed in the substorm section below.

The convection in the distant tail, beyond the distant reconnection region, is everywhere downtail (see section below on sources and losses of magnetospheric plasma).

In conclusion, although convection has been studied since the early space era, there are quite a number of observational results which remain to be understood.

SUBSTORMS AND STORMS

Akasofu (1964) identified a sequence of dynamic events in auroral distributions on a global scale from the IGY all-sky camera recordings which he named auroral substorm, as mentioned in an earlier section (see Figure 12). The name substorm was coined by Chapman (see Akasofu 1968). That geomagnetic changes accompany aurora has been known, as mentioned earlier, since the eighteenth century. The magnetic equivalence to the auroral substorm was recognized by Birkeland at the end of the nineteenth century. He called it 'polar elementary storm'. Substorm effects have been found during the space era in practically all plasma and field variables in the magnetosphere. The magnetospheric substorm is a macroscopic instability by means of which the magnetosphere maintains its energy and mass content within certain limits and is probably of a similar nature as solar flares.

The ionospheric current system corresponding to a magnetic substorm's main phase is shown in Figure 19. It was derived by Fukushima (1953) from the large number of magnetometers distributed over the high-latitude and polar regions. The strongest current flows in the night-side of the auroral oval and is directed westward. It produces a global decrease of the horizontal component of the geomagnetic field at the surface of Earth. The region on Earth with such a decrease in the horizontal component, called negative bay, coincides with the region with active aurora from a westward limit in the evening hours defined by the so called westward travelling surge of the aurora, through midnight and the morning hours.

As mentioned in the previous section, spacecraft measurements within the magnetosphere have demonstrated that plasma is quickly injected into the inner magnetosphere from the tail to an injection boundary (Figure 18) when a substorm starts. From the injection boundary the injected plasma moves in the large-scale magnetic and electric fields, as expected. Not far from where the plasma injection occurs on the night-side, the AMPTE spacecraft have identified disruptions of the dawn-dusk directed neutral sheet current layer at a distance of about 9 R_E in the tail (Lui et al. 1988), which disruption appears to propagate tailward from there (Lopez et al. 1989). It remains to determine how these two kinds of observations are related.

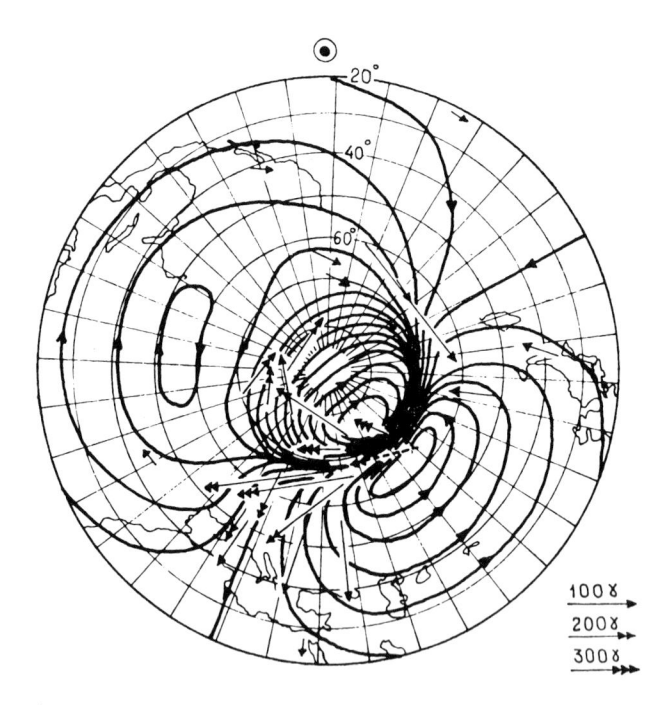

Figure 19 The current system for the polar magnetic substorm during the active main phase viewed from above the north pole; the current between adjacent streamed lines is 5×10^4 A (Fukushima 1953).

In the inner part of the tail the plasma sheet has been found to become very thin in the early phase of a substorm and in the main phase a large plasma element, 'plasmoid', is released from the inner tail in the direction of the distant tail and out of the magnetosphere (see Figure 20). These observations were first made by Hones and coworkers (e.g. Hones 1979) and the plasmoids have later been observed when passing out through the distant tail. A plasmoid is believed to be formed when the magnetic field lines in the very thin plasma sheet reconnect in the inner tail (at about $15R_E$). Its outer boundary is a distant reconnection region in the tail. Global MHD simulations of the magnetosphere have shown that the tail is driven to an unstable state by enhanced energy input with a quasi-stable magnetic neutral line forming in the tail and later a plasmoid moving away from Earth (Walker et al. 1993).

Many features of substorms are not understood and there is disagreement about how to interpret many experimental results. There is, however, a rather general agreement that at the beginning of the active phase of a substorm reconfiguration of currents and magnetic field occurs in the tail, with the cross-tail current partly being rerouted through the ionosphere in the auroral regions. What starts the substorm and what causes the disruption of the cross-tail current are matters of debate. Various plasma instabilities have

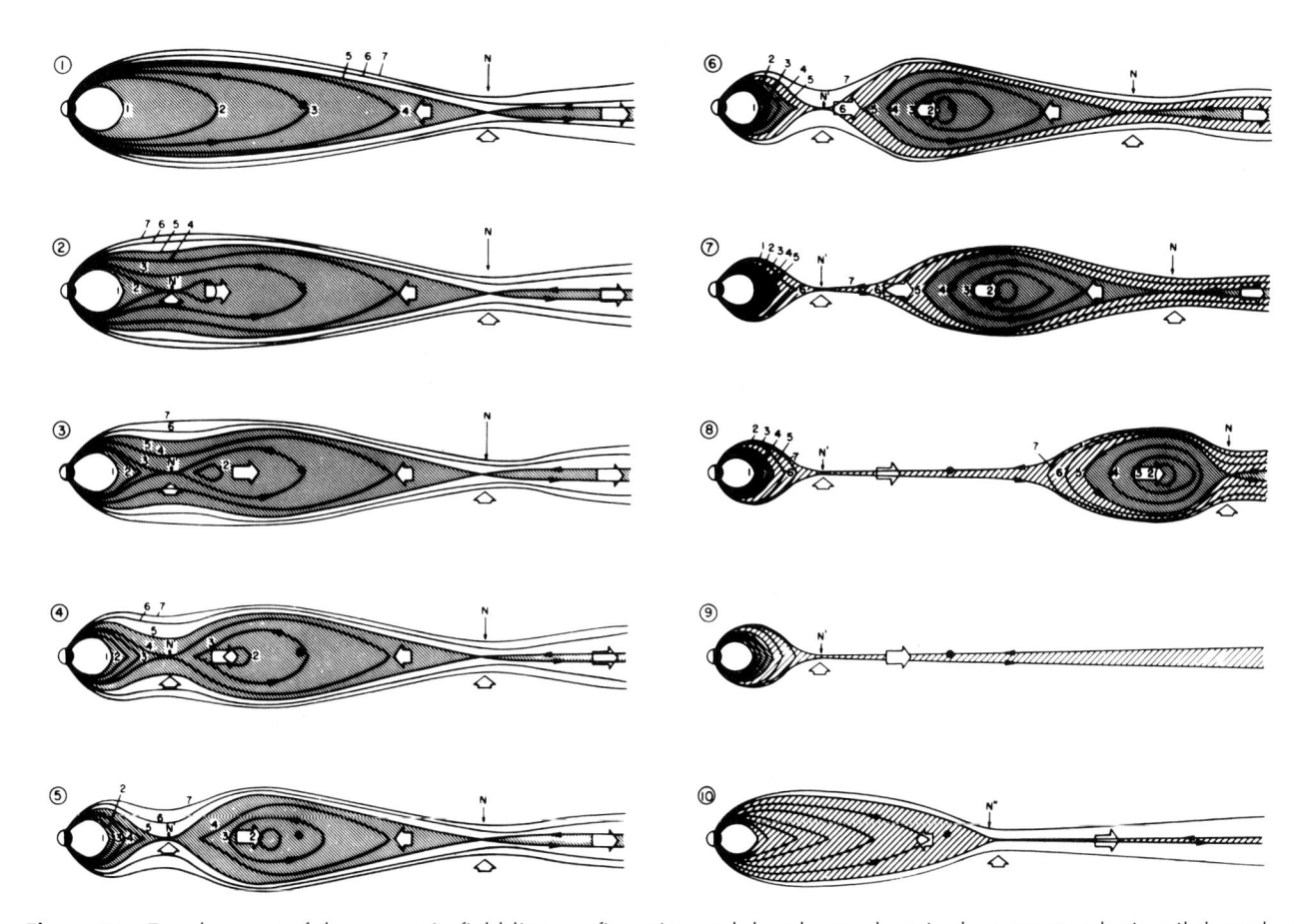

Figure 20 Development of the magnetic field line configuration and the plasma sheet in the magnetospheric tail through a substorm (after Hones 1979).

been suggested for the current disruption. Experimental data needed to solve the problem is, however, not available.

It was only in the space era that an initial 'loading' phase of substorms was detected, during which the magnetic flux in the tail is increasing by 30% or even more, and the amount of energy stored in the magnetosphere is growing (McPherron 1970). In terms of the reconnection model this is due to reconnection of southward directed interplanetary magnetic field lines with the northward directed geomagnetic field lines in the region of the subsolar point of the magnetosphere and subsequent transport of the field lines to the magnetotail. To what extent the substorm is a relaxation phenomenon, an unloading of flux and plasma through the tail, when the 'loading' has passed some limit, and to what extent it is directly driven by the solar wind and the interplanetary magnetic field, is another question central to present day research and discussion.

When the decrease of the horizontal component of the geomagnetic field at Earth's surface is very large and worldwide a disturbance is called a magnetic storm. A storm generally starts with an increase of the horizontal magnetic field at Earth's surface. This is an effect of the compression of the magnetosphere, sometimes to within the geostationary orbit, by the intense, fast moving plasma cloud from the Sun, as described by Chapman and Ferraro (1931, 1932). After a fraction of an hour to several hours, the magnetic field is depressed, sometimes by many hundred nanotesla (several percent) at the equator. This depression is caused by injection of hot plasma into the inner magnetospshere, in which a strong westward ring current is generated by the drift motions of the ions and electrons around Earth. The ring current populations are approaching Earth during strong storms and may reach 1–2 R_E altitude in the equatorial plane. The total current amounts to many million Amperes in a strong storm. The current decreases slowly over several days, due to the energetic ions in the ring current disappearing, probably mainly by charge exchange with neutral atoms in the upper atmosphere.

The contributions of spacecraft measurements to the understanding of macroscale characteristics of storms have been on the matter of composition and sources of the ring current ions as well as the spatial characteristics of various magnetospheric populations and processes during magnetic storms of various sizes. The contribution of ions by the high-latitude ionosphere to the ring current particle populations has been found to increase strongly with the magnitude of storms. Spacecraft measurements have also been imperative in timing processes all over the magnetosphere during various phases of storms and in determining relations between the various processes.

The relation between substorms and storms is still unclear. It appears that a storm is likely not to be only an accumulation of many substorms, although that has been believed by many researchers for many years.

Big magnetic storms is a space plasma phenomenon that has direct effects on human activities on Earth's surface. The strong magnetic fluctuations and strong electric fields induced by them have disrupted electric power and control systems and disturbed the operation of pipe lines.

SOURCE AND LOSS PROCESSES OF MAGNETOSPHERIC PLASMA

There are two and only two significant sources of magnetospheric plasma, the solar wind and the ionosphere. Until around 1970 there was a very general belief that there is only one source, the solar wind. Direct measurements in space during the seventies demonstrated that this belief was wrong. Even if the plasma sources are few, there are a large number of processes that may bring plasma across the inner and outer boundaries of the magnetosphere. Experimental data needed for identification of the prevailing processes is limited but have increased continuously in amount and relevance since the early seventies. Only spacecraft measurements have provided these data.

The energy transferred from the magnetosphere into the ionosphere and upper neutral atmosphere through energetic particle precipitation and ionospheric currents ends up practically completely as heat in the upper atmosphere. Only spacecraft measurements have demonstrated that large amounts of magnetospheric energy is also transferred more or less exclusively to the ionized part of the upper atmosphere, the ionosphere, giving rise to heating and expulsion of ionospheric plasma into the magnetosphere. A consequence of that it was taken more or less for granted by the space physics community that practically all plasma in the magnetosphere is of solar wind origin was that measurement of the composition was hardly considered worthwhile. The first measurements with ion mass spectrometers in the magnetosphere, which were carried out by the Lockheed group in the last years of the sixties, gave so astonishing results of precipitating energetic O^+ ions, i.e. ions which can only be of ionospheric origin, that the experimenters awaited the confirmation of the results on a second spacecraft before

publishing them (Shelley et al. 1972). The later discovery by means of the S3-3 satellite of outward flowing O^+ ions as beams along the magnetic field lines above the auroral zone ionosphere (Shelley et al. 1976) was not quite so unexpected, given the earlier observations of O^+ ions mentioned, but it was in some ways even more important. The first ion mass spectrometer sent into a highly eccentric orbit (apogee about 8 R_E) on GEOS-1 in 1977 by the Bern group demonstrated that the ionosphere is a source of plasma of comparable importance to solar wind in the regions of the magnetosphere reached by GEOS-1 (Geiss et al. 1978). Sometimes the ionospheric source dominates in the near-Earth magnetosphere, as is shown in Figure 21 for a magnetic storm in a period near solar maximum. The figure shows a strong majority of O^+ ions in the entire day-side of the magnetosphere. Chapell et al. (1987) have presented experimental evidence from Dynamics Explorer 1 indicating that the ionosphere may be the dominant plasma source not only during storms in solar maximum periods but always in the near-Earth magnetosphere.

There has thus been in the last decades a complete revolution of our knowledge and understanding of the interaction between the ionosphere and magnetosphere in regard of ion exchange. Earlier, the effectiveness with which the magnetosphere extracts plasma from the ionosphere was not understood at all. The processes giving rise to the pumping of the ionospheric plasma out into the magnetosphere have since the 1970s been, and still are, intensely studied.

The processes controlling the number and mass fluxes of ionospheric ions into the magnetosphere are those that work at the lowest altitudes, i.e. in the topside ionosphere. They define the source strength which cannot be surpassed at higher altitudes in a stationary situation, and in transient processes only for limited periods. Higher altitude processes are responsible for ions achieving energies high enough to overcome the gravitational field and escape out into the magnetosphere. Without such additional high-altitude acceleration many heavy ions would fall back into the ionosphere.

Important among the low-altitude processes is the polar wind, which occurs essentially at all times poleward of the plasmasphere and at all altitudes above a lower limit located in the upper atmosphere. It was predicted in the late sixties and was observed first by Brinton et al. (1971). The velocity of this bulk outflow increases with altitude and, on average, reaches 1 km s^{-1} at altitudes of about 2000, 3000, and 6000 km for H^+, He^+, and O^+, respectively (Abe et al. 1993). Typical thermal energies are around 0.3 eV. The polar wind outflow does not depend much on magnetospheric activity. The main forces causing the outflow are the vertical pressure gradients, ambipolar electric fields, and possibly centrifugal forces along rapidly convecting magnetic field tubes. There is still no quantitative theory for the

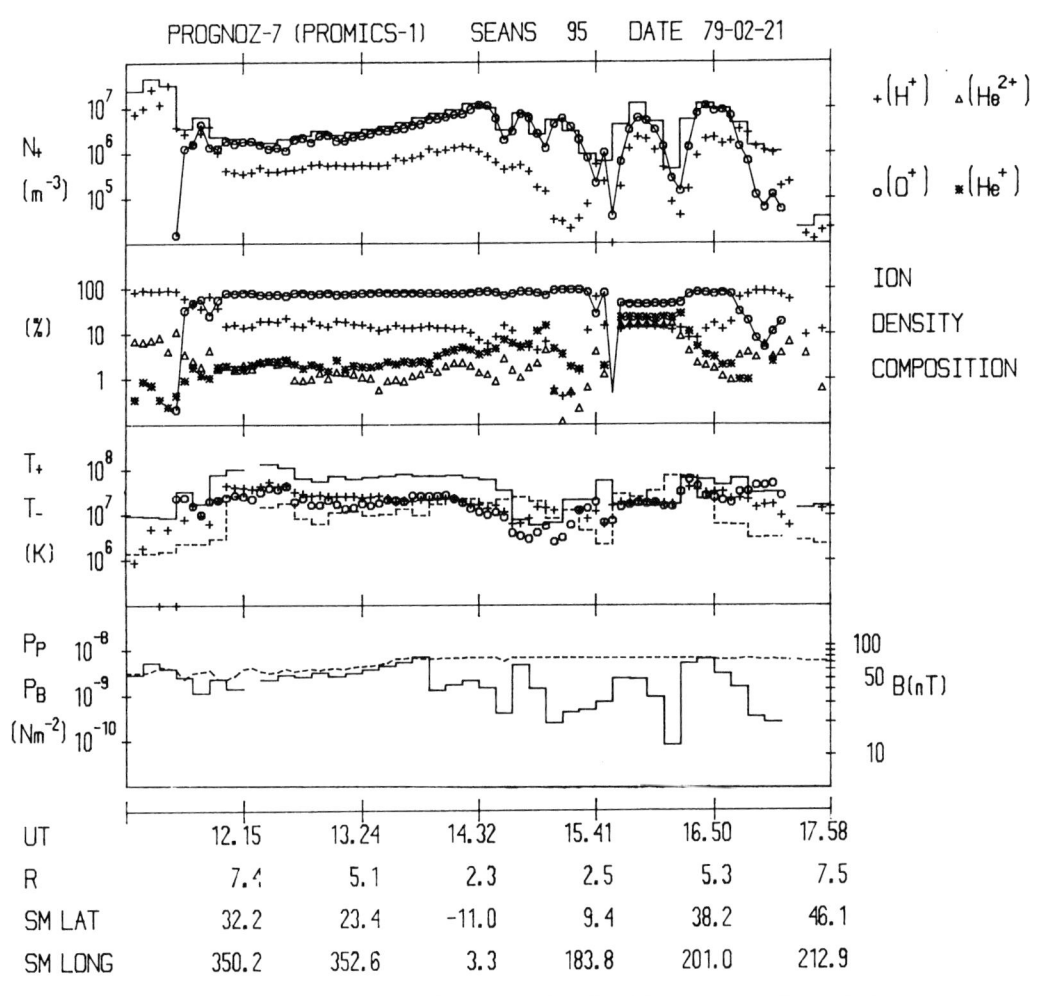

Figure 21 An example of strong dominance of O^+ ions in the dayside magnetosphere during a magnetic storm (after Hultqvist, 1982). Solid lines show the variables based on total ion measurements.

O^+ flow. The total polar wind outflow is on the order of 10^{25} ions s^{-1} both for H^+ and O^+ near solar maximum. The outflow of He^+ is generally smaller.

A bulk ion outflow in the O^+ dominated part of the upper ionosphere well below 1000 km altitude has been observed at auroral latitudes with flow velocities as high as 1 km s^{-1} and number fluxes important for the provision of the magnetosphere with plasma (Jones et al. 1988; Winser et al. 1988; Wahlund and Opgenoorth 1989). What causes this auroral bulk outflow is still not quite clear, but frictional heating, downward electron heat flux, and plasma processes producing an upward-directed electric field have been suggested.

A third kind of bulk upflow of importance for the supply is the so-called upwelling ion distributions in the day-side auroral oval (Lockwood et al. 1985). They have a conical pitch-angle distribution and a bulk upward velocity parallel to the magnetic field lines of about the same magnitude as the perpendicular velocity. The energy is in the range a few eV to a few tens of eV. The physical mechanisms involved in the acceleration certainly include perpendicular acceleration by some waves. The two last-mentioned kinds of ion outflow in the auroral oval are believed to be at least as important bottom-side suppliers of ions to the magnetosphere as the polar wind, with estimated total outflow of (1–2) 10^{25} ions s^{-1}.

Most of the heavy ions in the auroral bulk upflow would fall back into the ionosphere if they were not further accelerated at greater altitudes. The outflowing ions observed at altitudes of 1–2 R_E in auroral regions generally have much higher energies than the values associated with the bulk outflow lower down. Characteristic energies are of the order of a keV and the tails of energy distributions may reach many tens of keV. The additional acceleration is either perpendicular in wave fields – giving rise to conical pitch-angle distributions when the ions are forced upward in the diverging geomagnetic field by their increased magnetic moments that result from the acceleration – or as beams along the magnetic field lines caused by quasi-static parallel electric fields. For more information about outflow mechanisms the reader is referred to e.g. André and Yau (1997) or Chapter 2 in Hultqvist *et al.* (1999) and references therein.

On the basis of measurements by primarily the Japanese Akebono satellite (Yau *et al.* 1988) the combined total outflow of ion conics and beams has been found to be $2 \times 10^{25}\,\text{s}^{-1}$ to $10^{26}\,\text{s}^{-1}$ for H^+ and between $0.5 \times 10^{25}\,\text{s}^{-1}$ and $3 \times 10^{26}\,\text{s}^{-1}$ for O^+ for low and high levels of solar and magnetic activity, respectively. Its dependence on solar and magnetospheric activity levels, and in particular the increased outflow of O^+ with incresed magnetospheric disturbance level was first observed by means of the European GEOS satellites in the late seventies (Young *et al.* 1982). It is illustrated in Figure 22, which is from Yau *et al.* (1988). As can be seen there, the outflow of O^+ ions is larger than that of H^+ only for the highest disturbance levels near solar maximum. An outflow of $10^{26}\,\text{ions}\,\text{s}^{-1}$ of both O^+ and H^+ ions corresponds to a loss of 2.8 kg of the upper atmosphere per second. It is not known how much of other ion species is lost, but it is definitely much less than of O^+ and H^+ ions. The total values observed at low and high altitudes are consistent with the assumption that the low-energy upflow in the upper ionosphere is the source of the plasma accelerated further at altitudes of the order of Earth radii on auroral region field lines.

The medium and low-latitude ionosphere and the plasmasphere are generally negligible suppliers of ions to the magnetosphere (see e.g. Chapter 7 in Hultqvist *et al.* 1999).

Large statistical investigations of the loss of magnetospheric ions and electrons into the upper atmosphere in auroral and polar cap regions by Hardy *et al.* (1985, 1989) have given average values which are one order of magnitude less than the total outflow of ionospheric ions: 2×10^{24} to $1 \times 10^{25}\,\text{s}^{-1}$ for low and high magnetic activity, respectively. The processes scattering and accelerating particles into the atmospheric loss cone were discussed in the earlier section on the high-latitude magnetosphere. That the total precipitation rate is considerably smaller than the outflow rate is expected already because precipitating ions and electrons must have velocities nearly parallel to the geomagnetic field in order to reach the upper atmosphere, as they will otherwise be reflected by the magnetic mirror force above the atmosphere, whereas the outflowing ions will generally go out into the magnetosphere whatever their upward direction is.

Direct measurements of solar wind plasma transport across the magnetopause are exceedingly difficult to make because the plasma flow and the magnetic field are essentially directed tangential to the magnetopause. Any transport of plasma accross the magnetopause is only a small perturbation of the tangential transport. Even if it would be possible to measure the component normal to the magnetopause, each measurement would only apply locally and it seems fairly clear that the direct determination of the total transport of plasma into the magnetosphere over the entire magnetopause from *in situ* measurements is not a realistic goal. Instead indirect methods have to be used.

As mentioned in an earlier section on the boundary layers, estimates indicate that some 10^{26} ions cross the dayside magnetopause per second. About the same flux tailward is found in the magnetospheric boundary layers in the

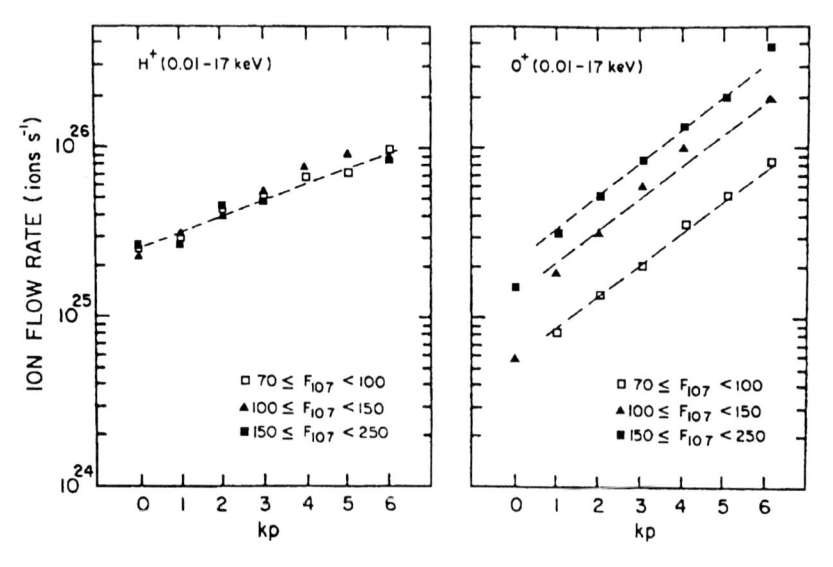

Figure 22 Outflow rates of H^+ and O^+ ions with energies in the range 0.01–17 keV observed on DE-1 at 16000–24000 km, integrated over all MLT and invariant latitudes above 56° in both hemispheres, as a function of K_p index, for different ranges of $F_{10.7}$ (after Yau *et al.* 1988).

near tail. Measurement results of plasma flow in the distant tail have only been published from the ISEE 3 and the Geotail spacecraft and from one passage of Pioneer 7 1000 R_E downtail. The amount of available plasma flow data in the distant tail is thus limited but the emerging picture is quite consistent. Using estimates of the magnetotail cross section and values reported from ISEE 3 measurements for the typical plasma density and velocity as a function of downstream distance, a flux of anti-sunward-moving ions in the magnetotail of $2 \times 10^{26}\,s^{-1}$ in the near-Earth tail (0–60 R_E), $7 \times 10^{26}\,s^{-1}$ in the distance range 60–120 R_E, $2 \times 10^{27}\,s^{-1}$ at 120–180 R_E, and somewhere between $3 \times 10^{27}\,s^{-1}$ and $3 \times 10^{28}\,s^{-1}$ beyond 180 R_E is found (Zwickl et al. 1984; Sibeck et al. 1985). Recently reported Geotail observations of magnetotail plasma parameters are entirely consistent with these estimates. About 1000 R_E downstream magnetotail densities and velocities observed with Pioneer 7 correspond to an anti-sunward flow of about $6 \times 10^{28}\,s^{-1}$ (Walker et al. 1975). As the total solar wind flow into the cross section of the magnetosphere is of the order of magnitude 10^{29} ions and electrons per second, the internal magnetospheric anti-sunward flow is more or less that of the solar wind at 1000 R_E.

The loss of magnetospheric plasma into the solar wind through the magnetopause is equally, or more, difficult to measure directly as the inflow, so no direct quantitative measurements have been made. However, there are arguments for the outflow being at least a factor of ten less than the inflow (see e.g. Paschmann 1997).

One clear conclusion can be drawn from the above figures, namely that the supply rate through the day-side magnetopause and from the ionosphere (both of order $10^{26}\,s^{-1}$) are orders of magnitude too small to account for the anti-sunward plasma flow through the deep tail. Therefore, plasma has to enter the magnetosphere elsewhere at high rate. Along the magnetotail appears to be the only remaining alternative and that is also supported by the large-scale simulations mentioned in the earlier section on the magnetotail. And plasma mantle observations indeed show that there is continuous entry along the tail magnetopause, consistent with models of the open tail boundary. In fact, the reconnection model qualitatively predicts many observations of the plasma mantle location and thickness as a function of the IMF B_y and B_z (but not all). It does, however, not predict the downtail evolution of the low-latitude boundary layer (LLBL), another possible region for solar wind plasma, and its dependence on IMF. The quantitative IMF dependence of the downtail flow remains unknown. That other mechanisms mentioned earlier, which possibly may contribute to the in- and outflow of plasma through the magnetopause can solve the above source problem at the tail magnetopause seems unlikely for simple supply rate reasons.

In conclusion, there is a situation where the strength of a plasma source, which has not been directly observed, is defined by the requirements of an observed plasma sink in the deep magnetotail. A scenario like in Figure 16 with regions along the entire magnetotail through which the interplanetary magnetic field lines and plasma flow are connected with field lines within the magnetosphere may match the loss of the plasma downtail (Sibeck and Siscoe 1984; Sibeck et al. 1985). It is also interesting to note that MHD/LSK simulations now yield numbers for the total inflow of plasma from the magnetosheath that seem consistent with the measured flow downtail (see e.g. Chapter 6 of Hultqvist et al. 1999).

Of the total flow of solar wind ions through the magnetosphere, amounting to the order of 10^{28}–$10^{29}\,s^{-1}$, only a small fraction affects the closed field line region in the magnetosphere near Earth, i.e. the plasma sheet earthward of the reconnection region in the tail and the day-side magnetosphere. The required source strength for this region is orders of magnitude less. Ionospheric and solar wind ions contribute on average similar orders of magnitude of ion fluxes to this near-Earth part of the magnetosphere, but the proportions vary with magnetospheric disturbance level and solar cycle phase. The ionosphere is likely to be the dominant plasma source in a region of the magnetosphere near Earth, sometimes referred to as the 'geosphere' (Moore and Delcourt 1995). The dominant sink for this region is probably the same as that for the solar wind outside the closed field line region, i.e. the downtail flow out of the magnetosphere.

A diagram summarizing some of the characteristics of the magnetosphere discussed above is shown in Figure 23. For more details about source and loss processes the reader is referred to e.g. Hultqvist et al. (1999).

WAVE-PARTICLE INTERACTIONS

Plasma waves play a major role in the redistribution of the magnetospheric plasma in phase space, for heating and acceleration of the plasma and for scattering of the plasma particles into the loss cone and thereby transferring magnetospheric energy into the upper atmosphere, as shown by Kennel and coworkers already in the 1960s (see e.g. Kennel 1969). The magnetosphere of Earth has become an important laboratory for investigations of wave-particle interactions in hot collisionless plasma.

Waves play a part in many phenomena discussed in the earlier sections. All the processes giving rise to randomization and heating of the solar wind plasma in the bowshock involve plasma waves. A wide variety of wave modes are believed to be involved (see e.g. Tsurutani and Stone 1985, for details). Scattering of the energetic particles in the inner magnetosphere into the loss cones is an important loss

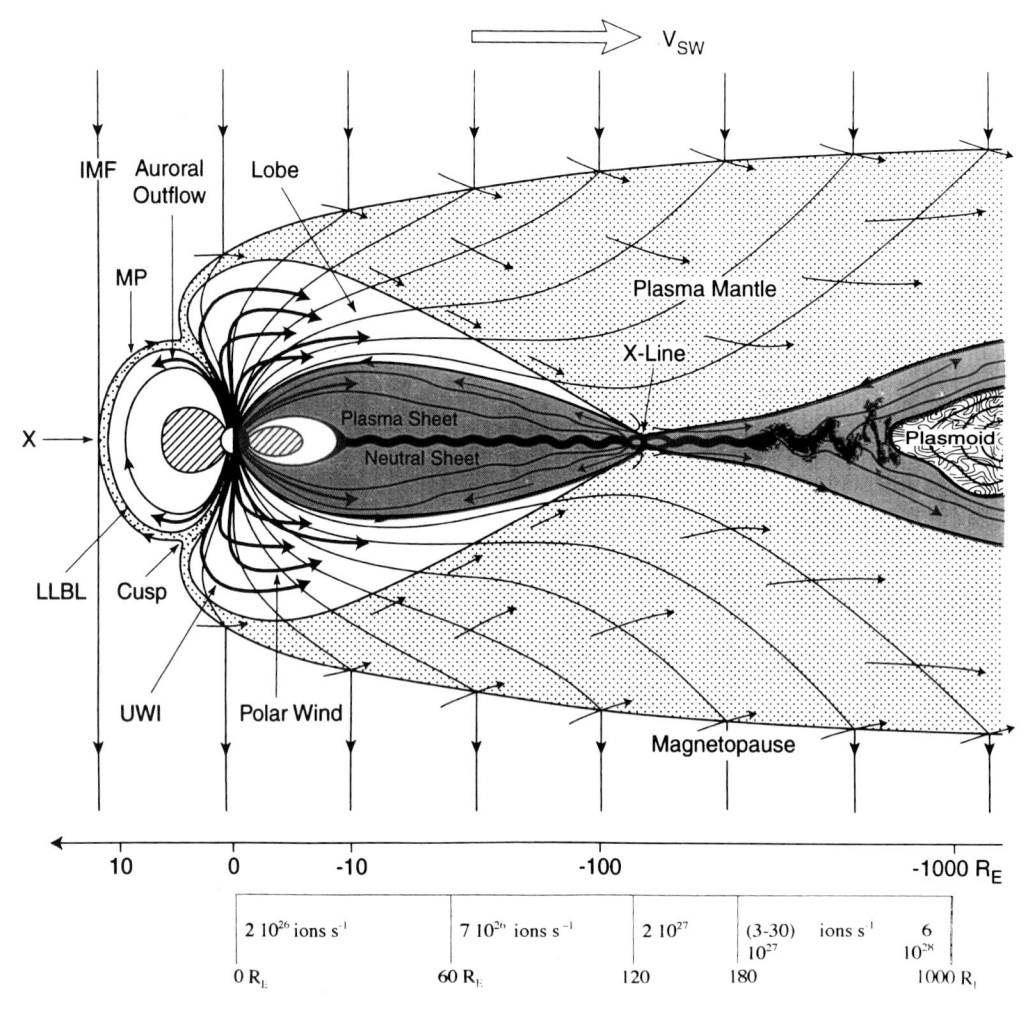

Figure 23 Summary figure for the flow of plasma into and within the magnetosphere (after Hultqvist et al. 1999). The figures at the bottom give the total observed flow in the downtail direction in different distance ranges.

process (see e.g. Lyons and Williams 1984, for theoretical background and discussions of examples). In the high-latitude magnetosphere wave-particle interactions are responsible for pitch-angle diffusion of auroral particles in some regions. However, what kind of waves are responsible for the scattering of the energetic electrons which cause the diffuse aurora is still not known (see e.g. Lyons 1997). Pitch-angle scattering generally does not change the energy of the particles, but waves may also accelerate charged particles along the magnetic field (e.g. kinetic Alfvén waves and lower-hybrid waves). Low-frequency variations in the magnetic and electric fields within the magnetosphere (magnetic pulsations) may modulate the precipitation of auroral electrons and produce pulsating aurora. Standing waves on inner magnetospsheric field lines modulate the plasma populations. The magnetotail is variable on many different time and space scales. Accumulating evidence indicates that the magnetotail, or at least some part of it, is in a permanently turbulent state. This challenges the long held view that magnetotail convection is basically ordered, driven by penetration of the interplanetary electric field into the magnetotail (see e.g. Chapter 6 of Hultqvist et al. 1999, and references therein).

Some sort of plasma instability is thought to be required in the diffusion region (to produce resistivity there) for reconnection to occur. The magnetospheric boundary layers are also unstable to certain surface waves (Kelvin-Helmholtz instability) which may affect some features of the solar wind-magnetosphere interaction.

Wave-particle interactions play several different roles in substorms: some kind of plasma instability is likely to play a decisive role in the initiation of substorms in the near tail (e.g. Lui et al. 1991), and in the reconnection process in the inner magnetospshere believed to give rise to the plasmoids and to the dipolarization of the magnetosphere on the Earth side of it. Wave-particle interactions cause most of the heating and ejection of ionospheric plasma into the magnetosphere. A very wide frequency range of waves is involved but the most important waves have frequencies from a fraction of one Herz to several hundred Herz, thus including the gyro frequencies of the major ion species at least for altitudes from 1000 km up to a few R_E. The spectral power density maximizes at the low end of the band mentioned. The average observed amplitudes can give rise to the generally observed ion energies. Sometimes the amplitudes are very large and mechanisms are known for production of even the highest observed energies, amounting to many tens of keV. There is no generally accepted theory for the generation of these waves but, without knowing the free energy, the original energy source can safely be predicted to be the interaction with the solar wind. For more information about the roles of wave-particle interactions in causing outflow of ionospheric plasma the reader is referred to e.g. André and Yau (1997).

Wave particle interactions are causing the strong radio emissions that several planets, including Earth, radiate. In the case of Earth this radiation cannot be observed from the surface because of the existence of the ionosphere. It was therefore discovered only when space probes looked back on Earth from large distances. The first evidence of radio emissions from Earth at frequencies less than one MHz was obtained from the Soviet Elektron 2 and 4 satellites by Benediktov *et al.* (1965) and there were also other indications later, but it was not until a more extensive study by Gurnett (1974) that it was clearly established that this radio emission is generated over the auroral regions in connection with the occurrence of auroral arcs. The wavelength is in the km range and the waves are called Auroral Kilometric Radiation (AKR). A power spectral density diagram observed in the generation region for AKR at an altitude of about $2R_E$ is shown in Figure 24. Gurnett also demonstrated that the total power emitted amounts to up to 10^9 W. This is by far the most powerful radio emission in the magnetosphere and it is comparable in many respects to the strong emissions from Jupiter and Saturn. Very high intensities are produced not only of propagating radio waves but also of non-propagating electrostatic waves at much lower frequencies.

The magnetosphere has become an increasingly important laboratory for basic plasma physics studies of waves in hot, collisionless, magnetized plasma. Here only one example will be mentioned, namely that the first direct quantitative comparison of experimental data with kinetic theory for generation of electrostatic waves in a collisionless hot plasma having a loss cone in its distribution function, was made by means of measurement results obtained with the GEOS 1 satellite (Rönnmark *et al.* 1978).

Practically all wave measurements in the magnetosphere give wave intensity as a function of frequency along the

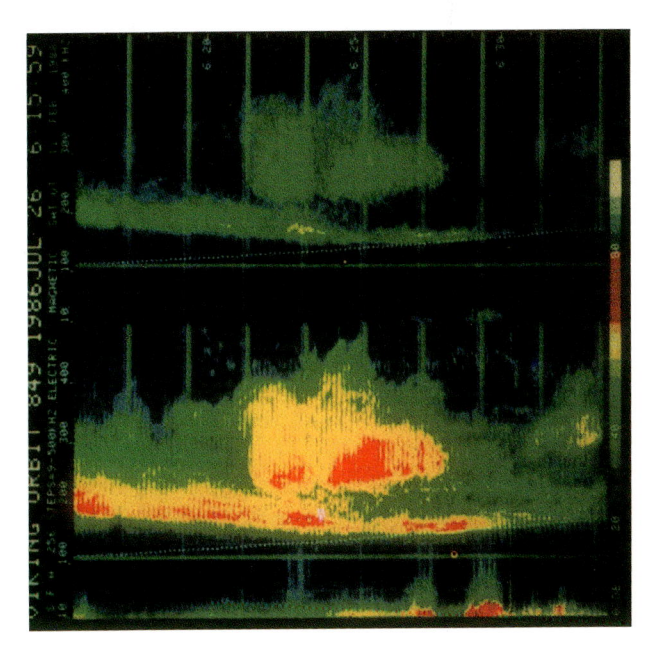

Figure 24 Color coded spectrograms of Auroral Kilometric Radiation (AKR) observed with the Swedish Viking satellite. Top panel shows magnetic field amplitude and the lower panel the electric field amplitude. The dotted lines below the frequency ranges with intense emissions indicate the gyrofrequency (courtesy A. Bahnsen). The high intensities are found at frequencies between 150 and 400 kHz.

spacecraft orbit (as in Figure 24). In some cases this is enough for identification of the wave mode with reasonable certainty but in many other cases it is not. The free energy available for the waves to grow is mostly unknown. The kind of comparison of theory with experiments mentioned above is only rarely possible because of lack of experimental data defining the plasmas and fields with sufficient detail and accuracy. The field of wave-particle interactions and their roles in magnetospheric physics is therefore in an early state of development.

FUTURE DIRECTIONS OF MAGNETOSPHERIC RESEARCH

In Earth's magnetosphere plasma conditions representing quite large parts of parameter space can be found, from cold collision-dominated molecular ion plasma in the lower ionosphere to hot collisionless light ion plasma with temperatures – or rather characteristic energies – of up towards hundred million degrees in the inner magnetosphere. All sorts of electromagnetic and electrostatic waves are generated and strong interactions occur between various parts of the magnetosphere. It is sometimes said that practically everything that can happen in a plasma happens in a magnetosphere, somewhere, sometime, which means that any proposal for a reasonable physical process to occur has a fair chance to be right, somewhere, sometime. The main task of magnetospheric research is to identify the most important processes, i.e. the prevailing processes among the many possible ones. It should not be astonishing that magnetospheric research, in spite of 40 years of efforts, is still a young research field, characterized by unforeseen surprises being the most important experimental results. The field is in an early state of development.

One of the most important future experimental lines of research is multi-point measurements, where the ESA Cluster project is the first major step, which will for the first time offer the possibility to separate time and space variations within certain scale ranges and to investigate spatial gradients and fine structure in various parameters, of basic importance for understanding the boundary regions of magnetospheres, for instance.

Another trend that we will see much more of in the future is to measure with strongly increased time resolution, thereby making possible investigations of new frequency ranges of kinetic plasma processes. The Freja and FAST satellites are the most important examples in this direction hitherto, but also Viking meant a major step in time resolution. Freja and FAST have opened up whole new fields of magnetospheric processes for experimental investigations and have produced very interesting new results, which are included in the above review.

A problem with practically all *in situ* measurements by spacecraft-born instruments, besides the problem of separating time and space variations, is that the measurements are very local. This is to some extent helped in high-latitude magnetospheric research by observations of the aurora, which give information about the distribution in space and time around the point of local measurement of the precipitation of energetic particles on a regional and even global scale. A new way of obtaining direct measurements of the distribution of energetic particles in whole magnetospheres has been developed in recent years and will be exploited in the future. The large scale distributions will be recorded by means of imaging instruments which measure energetic neutral particles instead of photons.

As emphasized in several places in this review, theoretical modeling and numerical simulations are playing very important roles in magnetospheric research and they will play even more important roles in the future. This is because of the complexity of magnetospheres. Models are necessary requisites for interpreting local *in situ* measurements, the more so the more we penetrate into the fine structure in time and space of the dynamics of the system.

Space physics research, including magnetospheric research, requires for progress to have well equipped spacecraft at the right place at the right time when nature (generally the Sun) makes its 'experiments'. The choice of complementary orbits for several complementary spacecraft is most important and the set of complementary instruments on the satellites has to be well chosen, but the satellite platforms in general need not be very large or very complex. This is the reason why several of the most important magnetospheric satellite projects in the last decade or two have been carried out within national space programs of even small European countries. The funding trends in Europe in the 90s have unfortunately reduced many national programs and the magnetospheric community in Europe is becoming more dependent on flight opportunities from the major agencies. Small satellite projects aimed e.g. for continued detailed investigations of particular physical processes in the magnetosphere using the only cosmic laboratory practically available, in a way somewhat similar to the way most physics research is carried out in ordinary laboratories on Earth's surface, often do not fit as well with the big space agencies, being oriented to major exploratory projects, as they do in small national programs. This is a problem that the magnetospheric community has to deal with.

Magnetospheric physics has many unsolved problems waiting for future research, as mentioned in the above review. Some important examples are the following.

– Search for and identification of reconnection site.
Reconnection has come to play a very important role not only in models of the magnetosphere but in astrophysics in general. Still it has not yet been possible to observe in detail what happens in the diffusion region, where the geomagnetic and interplanetary magnetic fields merge. Once the fields have merged, the interconnected field lines are convected along the magnetopause into the magnetotail. The tailward convection is a dissipation-free transport of the frozen-in magnetic field. Its global consequence is that the interconnected field lines allow for free access of solar wind plasma flowing along the magnetic field across the magnetopause. In this sense the microscale reconnection process has an enormous macroscale effect. All evidence for reconnection has when this is written been derived from measurements outside the diffusion region. No diffusion region has even been convincingly identified. As such identification will be the ultimate proof of reconnection, it is one of the most important tasks for the future. And it can only be done in the magnetosphere.

– Entry and transport processes.
The future multi-point measurement missions will hopefully throw light not only on reconnection but also on other processes that may lead to transfer of solar wind plasma across the magnetopause and its subsequent transport within the boundary layers, several of which have been mentioned in an earlier section. For none of these processes clear observational evidence exists yet and for most of them the theory has not been developed to the state where quantitative

predictions can be made. So future efforts have to go into both theory and observations.

– Mixing of the solar wind and ionospheric plasmas.

The question of the balance between the ionospheric and solar wind sources in different parts of the magnetosphere during different conditions of magnetospheric disturbance and solar cycle phase is still open. The solar wind obviously dominates as an ion source at large distances from Earth, while the ionosphere dominates close to Earth. In the inner magnetosphere two unsolved problems related to this question are the solar cycle dependence of the ring current composition and the dynamics of the plasmasphere, plasmapause, and detached plasma regions. Closely connected with the first problem is the question what energies ionospheric ions may achieve on their way to the equatorial region and what role these ions play in the dynamics of the ring current. Very interesting questions are also raised by the newly discovered downward pointing potential fields accelerating ionospheric electrons outward in auroral regions, which will certainly attract strong attention in planning future missions. Also the very large parallel electric fields, of the order of 1 V/m, concentrated in small-scale structures of the order of km extension along the field, first observed by the Freja and FAST satellites, will be important to study further and may possibly change some important features of the magnetospheric models in fundamental ways.

– The diffuse aurora.

The electrons causing the diffuse aurora have energies of the order of 1 keV and their latitudinal range of precipitation covers the foot points of the presumably closed field lines in the entire tail plasma sheet. The cause of this precipitation remains one of the major outstanding questions. The pitch-angle distribution of the precipitating electrons is close to isotropic at lower energies (<1 keV) but becomes increasingly anisotropic at energies of a few keV. Such behavior is typical for pitch-angle scattering in electrostatic high-frequency plasma waves. However, the intensity of such waves found in the magnetosphere is too low to account for the persistence of the precipitation. This indicates that either the conventional pitch-angle scattering theory does not apply or that other processes than wave- particle interactions dominate the plasma sheet electrons. Both alternatives will mean radical changes in the understanding of the magnetosphere and their may be others.

– Macroscopic dynamics.

Substorms and storms are frontlines of magnetospheric research since decades. Still there are many open questions left as indicated in an earlier section. As substorms affect the whole of the magnetosphere the observational problem is evident: it will probably never be possible to have suitably equipped spacecraft in all parts of the magnetosphere simultaneously and one is limited to using all new relevant data also for investigations of substorms and storms. That is what has been done and what is expected to be done in the future and a large part of the magnetospheric community is involved in an ongoing review and discussion of the development of the understanding of these phenomena which are very important, not only for understanding magnetospheric physics but also from the points of view of basic plasma physics and astrophysics.

– Global modeling.

Advanced magnetospheric models based on three-dimensional global MHD simulations have been developed in the last two decades. For the foreseeable future they will be the basic modeling tools. They are important because they provide the only consistent picture of the global topology of the magnetosphere as a function of IMF and solar wind parameters. The global MHD models give the electric and magnetic fields that can be used to determine the transport within the magnetosphere all the way to the inner magnetosphere. In a small number of large-scale kinetic simulations (LSK) transport of plasma has been derived from calculating the trajectories of ions from a measured distribution function backward in time to a source location at the magnetopause or in the ionosphere. Systematic LSK studies are required to determine how observed particle populations and fields depend on various sources for various solar wind and IMF conditions. The process of calibrating the MHD and LSK models by direct comparison with measured data will be an important part of the future modeling effort.

The existing MHD models are for fundamental reasons not applicable to regions with strong gradients, as in the near-Earth tail during substorms, and there are other limitations which should be eliminated, or at least reduced in future versions of the models. The goal is to achieve a full self-consistency that includes the transport in the global model. There is still some way to go before that goal is reached.

The above items are only a few examples of important future research directions. More open questions are indicated in earlier sections.

ON THE CONNECTIONS BETWEEN ASTROPHYSICS AND MAGNETOSPHERIC PHYSICS

Plasma systems organized by magnetic fields – intrinsic or induced – similar to Earth's magnetosphere in terms of physical processes, exist around all of the planets in the

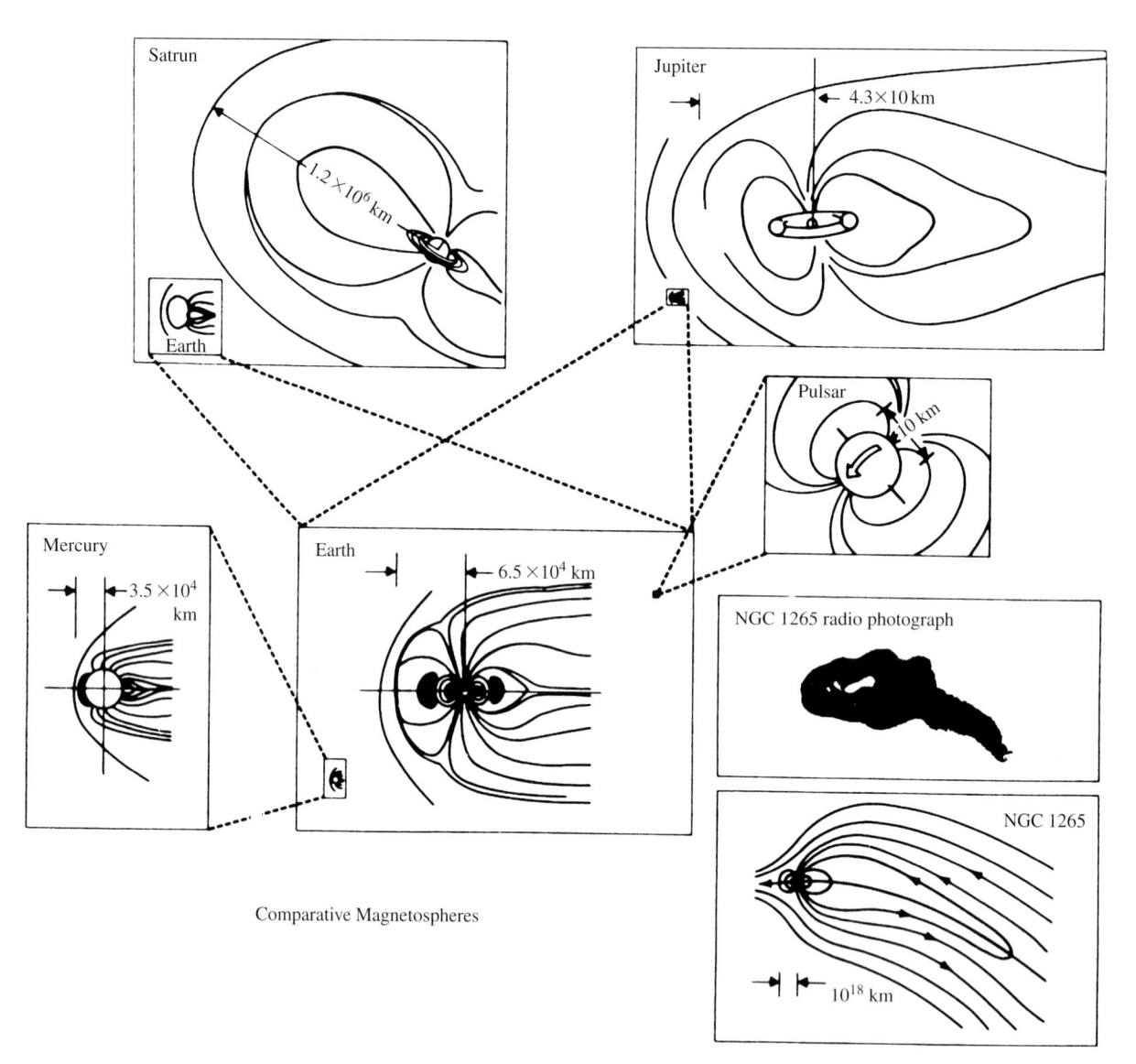

Figure 25 A number of different magnetospheres in the universe, from the smallest, that of Mercury, to the largest, that of galaxy no. 1265 (after Friedman 1986).

solar system and the entire heliosphere is a plasma system where magnetospheric physics rules. Although different magnetospheres have different sizes and partly different characteristics, most of the important physical processes appear to be the same. Many of the more exotic astronomical objects studied in recent decades contain hot plasma and strong magnetic fields and even some galaxies appear to be organized magnetically as are magnetospheres, thus indicating that another plasma system – an intergalactic wind – moves relative to the galaxy (see Figure 25). Magnetospheric physics in a wide meaning is thus expected to play a major role in the universe. The plasma state is by far the dominating aggregation state of the known matter in the universe, both in the stars and between them. Only where matter is concentrated and heated by gravitation does it radiate sufficiently intensely to be visible from Earth. All other plasma in the universe is invisible in practice. If, for instance, Jupiter's magnetosphere could be made visible it would be larger than the Sun as seen from Earth, although its distance from Earth is some five times larger.

Whereas astrophysicists have available as information source only electromagnetic radiation and sometimes energetic particles from the objects under study, space physicists have the possibility to investigate physical processes also by means of *in situ* measurements. But unfortunately we are then limited to the solar system and, for practical and economic reasons, mainly to our own magnetosphere. Earth's magnetosphere is used as a laboratory for detailed investigations of physical processes which are also likely to occur in other parts of the universe. Magnetospheric research can thus be said to be a sort of basic research also for the astrophysics field. However, there are of course many physical conditions in the universe, where gravitation dominates, about which nothing can be said from studies of magnetospheres.

Most of the physical concepts of space plasma physics, such as particle acceleration, convection, collisionless shocks, substorms, interactions at 'cell walls', non-linear plasma effects, turbulence, etc. are of astrophysical interest and the development of the understanding of them in magnetospheric research should therefore be of interest to astrophysicists. Whether the new knowledge from the magnetosphere will be used by astrophysicists or not is likely to depend primarily on how the interest develops in astrophysical research. It seems fairly safe to conclude by stating that the results of magnetospheric research will be used by astrophysicists when the time is ripe.

REFERENCES

Abe, T., Whalen, B.A.,Yau, A., Horita, R.E., Watanabe, S. and Sagawa, E. (1993). EXOS-D (Akebono) suprathermal mass spectrometer observations of the polar wind. *Journal of Geophysical Research*, **98**, 11191.

Akasofu, S-I. (1964). The development of the auroral substorm. *Planet. Space Sci*, **12**, 273.

Akasofu, S-I. (1968). *Polar Magnetospheric Substorms*, Dordrecht: D. Reidel Publ. Co.

Akasofu, S-I. (1997). Aurora research during the early space age: Personal account. In: C.S. Gilmore and J.R. Spreiter (eds.), *Discovery of the Magnetosphere History of Geophysics* 7, p. 13. Washington DC: American Geophysical Union.

Alfvén, H. (1939, 1940). Theory of magnetic storms, I, *Kungl. Svenska Vetenskapsakademiens Handlingar*, **3**, 18, No.3; II, III, ibid 18, No.9.

Alfvén, H. (1940). On the motion of a charged particle in a magnetic field. *Ark. mat. astron. fysik*, **27A**, No.22.

Alfvén, H. (1981). *Cosmic Plasma*. Dordrecht: D. Reidel Publ. Co.

André, M. and Yau, A.W. (1997). Theories and observations of ion energization and outflow in the high latitude magnetosphere. In: B. Hultqvist and M. Øieroset (eds.), *Transport Across the Boundaries of the Magnetosphere*, Space Science Series of ISSI, Vol 2, p. 27. Dordrecht: Kluwer Acad. Publ. (Also *Space Science Reviews*, **80**, issues 1–2, 1997.)

Axford, W.I. and Hines, C.O. (1961). A unifying theory of high-latitude geophysical phenomena and geomagnetic storms. *Canad. J. Phys.*, **39**, 1433.

Axford, W.I. (1962). The interaction between the solar wind and the earth's magnetosphere. *Journal of Geophysical Research*, **67**, 3791.

Beard, D.B. (1960). The interaction of the terrestrial magnetic field with the solar corpuscular radiation. *Journal of Geophysical Research*, **65**, 3559.

Benediktov, E.A., Getmansev, G.G., Sazonov, Yu.A., Tarasov, A.F. (1965). Preliminary results of measurements of the intensity of disturbed extraterrestrial radio-frequency emission at 725 and 1525 kHz frequencies by the satellite Elektron 2. *Kosm. Issled.*, **3**, 614.

Biermann, L. (1951). Kometenschweife und solare korpuskular Strahlung. *Z. Astrophys.*, **29**, 274.

Brice, N.M. (1967). Bulk motion of the magnetosphere. *Journal of Geophysical Research*, **72**, 5193.

Brinton, H.C., Grebowski, J.M. and Mayr, H.G. (1971). Altitude variation of ion composition in the mid-latitude trough region: Evidence for upward plasma flow. *Journal of Geophysical Research*, **76**, 3738.

Cahill, L.J. and Amazeen, P.G. (1963). The boundary of the geomagnetic field. *Journal of Geophysical Research*, **68**, 1835.

Carlson, C.W., Mc Fadden, J.P., Ergun, R.E., Temerin, M., Peria, W., Mozer, F.S., Cattell, D.M. and Pfaff, R. (1998). FAST observations in the downward auroral current region: Energetic upgoing electron beams, parallel potential drops, and ion heating. *Geophysical Research Letters*, **25**, 2017.

Carpenter, D.L. (1963). Whistler evidence of a 'knee' in the magnetospheric ionization density profile. *Journal of Geophysical Research*, **68**, 1675.

Carpenter, D.L. (1965). Whistler measurements of the equatorial profile of magnetospheric electron density. In: G.M. Brown (ed.), *Progress in Radio Science 1960–1963, Vol. III: The Ionosphere'* p. 76. Amsterdam: Elsevier.

Carpenter, D.L. and Lemaire, J. (1997). Erosion and recovery of the plasmapause region. In: B. Hultqvist and M. Øieroset (eds.), *Transport Across the Boundaries of the Magnetosphere*, Space Science Series of ISSI, Vol. 2, p. 153. Dordrecht: Kluwer Acad. Publ. (Also *Space Science Review*, **80**, issues 1–2, 1997.)

Chapman, S. and Ferraro, V.C.A. (1931, 1932). A new theory of magnetic storms, Part 1, The initial phase. *Terrestrial Magnetism and Atmospheric Electricity*, **36**, 77, 171, 1931 and **37**, 147, 1932.

Chappell, C.R., Moore, T.E. and Waite Jr., J.H. (1987). The ionosphere as a fully adequate source of plasma for the Earth's magnetosphere. *Journal of Geophysical Research*, **92**, 5896.

Christon, S.P., Eastman, T.E., Doke, T., Frank, L.A., Gloeckler, G., Kojima, H., Kokubun, S., Lui, A.T.Y., Matsumoto, H., McEntire, R.W., Mukai, T., Nylund, S.R., Paterson, W.R., Roelof, E.C., Saito, Y., Tsuruda, T.K., Sotirelis, T., Williams, D.J. and Yamamoto, T. (1998). Magnetospheric plasma regimes identified using Geotail measurements, 2: Statistics, spatial distribution, and geomagnetic dependence. *Journal of Geophysical Research*, **103**, 23521.

Cummings, J.R., Cummings, A.C., Mewaldt, R.A., Selesnick, R.S., Stone, E.C. and von Rosenvinge, T.T. (1993). New evidence for geomagnetically trapped anomalous cosmic rays. *Geophys. Res. Lett.*, **20**, 2003.

Daglis, I.A., Sarris, E.I. and Kremser, G. (1991). Ionospheric contribution to the crosstail current during the substorm growth phase. *Journal of Atmospheric Terrestrial Physics*, **53**, 1091.

Davis, T.N. (1960). The morphology of the polar aurora. *Journal of Geophysical Research*, **65**, 3497.

DeForest, S.E. and McIlwain, C.E. (1971). Plasma clouds in the magnetosphere. *Journal of Geophysical Research*, **76**, 3587.

Dungey, J.W. (1961). Interplanetary magnetic field and the auroral zones. *Physical Review Letters*, **6**, 47.

Eastman, T.E., Jones, Jr., E.W., Bame, S.J. and Astridge, J.R. (1976). The magnetospheric boundary layer: Site of plasma, momentum and energy transfer from the magnetosheath into the magnetosphere. *Geophysical Research Letters*, **3**, 685.

Erickson, G.M. and Wolf, R.A. (1980). Is steady convection possible in the Earth's magnetotail? *Geophysical Research Letters*, **7**, 897.

Evans, D.S. (1974). Precipitating electron fluxes formed by a magnetic field aligned potential difference. *Journal of Geophysical Research*, **79**, 2853.

Fairfield, D.H. and Cahill, Jr. L.J. (1966). Transition region magnetic field and polar magnetic disturbances. *Journal of Geophysical Research*, **71**, 155.

Fälthammar, C-G. (1966). On the transport of trapped particles in the outer magnetosphere. *Journal of Geophysical Research*, **71**, 1487.

Fälthammar, C-G. (1968). Radial diffusion by violation of the third adiabatic invariant. In: B.M. McCormac (ed.), *Earth's Particles and Fields*, p. 157. New York: Reinhold.

Feldstein, Y.I. (1963). Some problems concerning the morphology of auroras and magnetic disturbances in high latitudes. *Geomagnetizm in Aeronomiya*, **3**, 183.

Fisk, L.A., Kozlovsky, B., and Ramaty, R. (1974). An interpretation of the observed oxygen and nitrogen enhancements in low-energy cosmic rays. *Astrophysical Journal*, **190**, L 35.

Frank, L.A. and Ackerson, K.L. (1971). Observations of charged particle precipitation into the auroral zone. *Journal of Geophysical Research*, **76**, 3612.

Frank, L.A., Ashour-Abdalla, M., Berchen, J., Raeder, J., Paterson, W.R., Kokobun, S., Yamamoto, T., Lepping, R.P., Coroniti, F.V., Fairfield, D.H. and Ackerson, K.L. (1995). Observations of plasmas and magnetic fields in the Earth's distant magnetotail: Comparison with a global MHD model. *Journal of Geophysical Research*, **100**, 19177.

Friedman, H. (1986). *Sun and Earth*. New York: ScientificAmerican Library.

Fritz, H. (1981). *Das Polarlicht*. Leipzig.

Fukushima, N. (1953). *J. Fac. Sci. Tokyo Univ.*, Ser II, **8**, 293.

Geiss, J., Balsiger, H., Eberhardt, P., Walker, H-P., Weber, L., Young, D.T., and Rosenbauer, H. (1978). Dynamics of magnetospheric ion composition as observed by the GEOS mass spectrometer. *Space Science Reviews*, **22**, 537.

Gold, T. (1959). Motions in the magnetosphere of the Earth. *Journal of Geophysical Research*, **64**, 1219.

Gosling, J.T., Baker, D.N., Bame, S.J., Feldman, W.C., Zwickl, R.D., and Smith, E.J. (1985). North-south and dawn-dusk plasma asymmetries in the distant tail lobes: ISEE 3. *Journal of Geophysical Research*, **90**, 635.

Grebowsky, J.M. (1970). Model study of plasmapause motion. *Journal of Geophysical Research*, **75**, 4329.

Greenstadt, E.W. and Fredrics, R.W. (1979). Shock system in collisionless space plasma. In: L.J. Lanzerotti, C.F. Kennel and E.N. Parker (eds.), *Solar System Plasma Physics*, Vol. III, p. 3. Amsterdam: North-Holland Publ. Co.

Gringauz, K.I., Kurt, V.G., Moroz, V.I. and Shklovsky, I.S. (1960). Results of observations of charged particles up to R = 100000 km with the aid of charged particle traps on Soviet cosmic rockets. *Astron. Zh.*, p. 716; translated as *Soviet Astronomy A.J.*, **4**, 680.

Gurnett, D.A. (1972). Electric field and plasma observations in the magnetosphere. In: E.R. Dyer (ed.), *Critical Problems of Magnetospheric Physics*, p. 128. Washington DC: Nat. Acad. of Sci.

Gurnett, D.A. (1974). The Earth as a radio source: Terrestrial kilometric radiation. *Journal of Geophysical Research*, **790**, 4227.

Hamilton, D.C., Gloeckler, G., Ipovitch, F.M., Stüdemann, W., Wilken, B., and Kremser, G. (1988). Ring current development during the great geomagnetic storm of February 1986. *Journal of Geophysical Research*, **93**, 14343.

Hardy, D.A., Gussenhoven, M.S. and Holeman, E. (1985). A statistical model of auroral electron precipitation. *Journal of Geophysical Research*, **90**, 4229.

Hardy, D.A., Gussenhoven, M.S. and Brautigam, D. (1989). A statistical model of auroral ion precipitation. *Journal of Geophysical Research*, **94**, 370.

Hill, T.W. (1979). Rates of mass, momentum, and energy transfer at the magnetopause. In: B. Battrick (ed.), *Magnetospheric Boundary Layers*, p. 325. ESA SP-148.

Hones Jr, E.W., Asbridge, J.R., Bame, S.J., Montgomery, M.D., Singer, S., and Akasofu, S-I. (1972). Measurements of magnetotail plasma flow made with Vela 4B. *Journal of Geophysical Research*, **77**, 5503.

Hones Jr, E.W. (1979). Plasma flow in the magnetotail and its implications for substorm theories. In: S.I. Akasofu (ed.), *Dynamics of the Magnetosphere*, p. 545. Dordrecht: D. Reidel Publ. Co.

Hultqvist, B. (1982). Recent progress in the understanding of the ion composition in the magnetosphere and some major question marks. *Journal of Geophysical Research*, **93**, 9765.

Hultqvist, B., Øieroset, M., Paschmann, G. and Treumann, R. (1999.) *Magnetospheric Plasma, Sources and Losses*, Space Science Series of Issi, Vol. 6. Dordrecht: Kluwer Acad. Publ. (Also *Space Science Reviews*, **88**, issues 1–2, 1999.)

Iijima, T. and Potemra, T.A. (1976). The amplitude distribution of field-aligned currents at northern high latitudes observed by Triad. *Journal of Geophysical Research*, **81**, 2165.

Jones, G.O.L., Williams, P.J.S., Winser, K.J., Lockwood, M. and Suvanto, K. (1988). Large plasma velocities along the magnetic field lines in the auroral zones. *Nature*, **336**, 231.

Kellogg, P.J. (1962). Flow of plasma around the earth. *Journal of Geophysical Research*, **67**, 3805.

Kennel, C.F. (1969). Consequences of a magnetospheric plasma. *Rev. of Geophys.*, **7**, 379.

Knight, S. (1973). Parallel electric fields. *Planet. Space Sci.*, **21**, 741.

Li, X., Roth, I., Temerin, M., Wygant, J.R., Hudson, M.K. and Blake, J.B., (1993). Simulation of the prompt energization and transport of radiation belt particles during the March 24, 1991 SSC. *Geophysical Research Letters*, **20**, 2423.

Lockwood, M., Chandler, M.O., Horwitz, J.L., Waite Jr, J.H., Moore, T.E., and Chappell, C.R. (1985). The cleft ion fountain. *Journal of Geophysical Research*, **90**, 9736.

Lopez, R.E., Lui, A.T.Y., Sibeck, D.G., Takahashi, K., McEntire, R.W., Zanetti, L.J. and Krimigis, S.M. (1989). On the relationship between the energetic particle flux morphology and the changes in the magnetic field magnitude during substorms. *Journal of Geophysical Research*, **94**, 17105.

Lui, A.T.Y., Anger, C.D., Venkatesan, D., Sawchuk, W., abd Akasofu, S-I., (1975). A uniform belt of diffuse auroral emission seen by the ISIS-2 scanning auroral photometer. *Journal of Geophysical Research*, **80**, 1795.

Lui, A, T.Y. Lopez, R.E., Krimigis, S.M., McEentire, R.W., Zanetti, L.J., and Potemra, T.A. (1988). A case study of a magnetotail current disruption and diversion. *Geophysical Research Letters*, **15**, 721.

Lui, A., T.Y., Chang, C-L., Mankovsky, A., Wong, H-K. and Winske, D. (1991). A cross-field current instability for substorm expansions. *Journal of Geophysical Research*, **96**, 11389.

Lundin, R. (1984). Solar wind energy transfer regions inside the dayside magnetopause – II, Evidence for an MHD generator process. *Planet. Space Sci.*, **32**, 757.

Lundin, R. (1997). Observational and theoretical aspects of processes other than merging and diffusion governing plasma transport across the magnetopause. In: B. Hultqvist and M. Øieroset (eds.), *Transport Across the Boundaries of the Magneto-sphere*, Space Science Series of ISSI, Vol. 2, p. 269. Dordrecht: Kluwer Acad. Publ. (Also *Space Science Rev.*, **80**, issues 1–2, 1997.)

Lyons, L.R., Evans, D.S. and Lundin, R. (1979). An observed relation between magnetic field aligned electric fields and downward electron energy fluxes in the vicinity of auroral forms. *Journal of Geophysical Research*, **84**, 451.

Lyons, L.R and Williams, D.J. (1984). *Quantitative Aspects of Magnetospheric Physics*. Dordrecht: D. Reidel Publ. Co.

Lyons, L.R. (1997). Magnetospheric processes leading to precipitation. In: B. Hultqvist and M. Øieroset (eds.), *Transport Across the Boundaries of the Magnetosphere*, Space Science Series of ISSI, Vol. 2. p. 109. Dordrecht: Kluwer Acad. Publ. (Also *Space Science Reviews*, **80**, issues 1–2, 1997.)

Marklund, G.T. (1993). Viking investigations of auroral electrodynamical parameters. *Journal of Geophysical Research*, **98**, 1691.

McIlwain, C.E. (1960). Direct measurement of particles producing visible aurora. *Journal of Geophysical Research*, **65**, 2727.

McIlwain, C.E. (1972). Plasma convection in the vicinity of the geosynchronous orbit. In: B. McCormac (ed.), *Earth's Magnetospheric Processes,* p. 143. Dordrecht: D. Reidel Publ. Co.

McPherron, R.L. (1970). Growth phase of magnetospheric substorms. *Journal of Geophysical Research,* **75,** 5592.

Meng, C-I. and Anderson, K.A. (1970). A layer of energetic electrons (E > 40 keV) near the magnetopause. *Journal of Geophysical Research,* **75,** 1827.

Moore, T.E., Arnoldy, R.L., Feynman, J. and Hardy, D.A. (1981). Propagating substorm injection fronts. *Journal of Geophysical Research,* **86,** 6713.

Moore, T.E. and Delcourt, D.C. (1995). The geopause. *Rev. Geophys.,* **33** (2), 175.

Mozer, F.S., Carlson, C.W., Hudson, M.K., Torbert, R.B., Parady, B., Yatteau, J. and Kelley, M.C. (1977). Observations of paired electrostatic shocks in the polar magnetosphere. *Physical Review Letters,* **38,** 292.

Mukai, T., Hirahara, M., Machida, S., Saito, Y., Terasawa, T. and Nishida, A. (1994). GEOTAIL observation of cold ion streams in the medium distance magnetotail lobe in the course of a substorm. *Geophysical Research Letters,* **21,** 1023.

Ness, N.F., Scearce, C.S. and Seek, J.B. (1964). Initial results of the Imp 1 magnetic field experiment. *Journal of Geophysical Research,* **69,** 3531.

Nishida, A. (1966). Formation of plasmapause or magnetospheric plasma knee, by the combined action of magnetospheric convection and plasma escape from the tail. *Journal of Geophysical Research,* **71,** 5669.

Paschmann, G. (1997). Observational evidence for transfer of plasma across the magnetopause. In: B. Hultqvist and M. Øieroset (eds.), *Transport Across the Boundaries of the Magneto-sphere,* Space Science Series of ISSI, Vol. 2, p. 217. Dordrecht: Kluwer Acad. Publ., (Also *Space Science Reviews,* **80,** issues 1–2, 1997.)

Parker, E.N. (1958). Dynamical instability in an anisotropic ionized gas of low density. *Phys. Rev.,* **109,** 1874.

Peterson, W.K., Shelley, E.G., Haerendel, G. and Paschmann, G. (1982). Energetic ion composition in the subsolar magnetopause and boundary layer. *Journal of Geophysical Research,* **87,** 2139.

Rönnmark, K., Borg, H., Christiansen, P.J., Gough, M.P. and Jones, D. (1978). Banded electron cyclotron harmonic instability – A first comparison of theory and experiment. *Space Science Reviews,* **22,** 401.

Rosenbauer, G., Grünwaldt, H., Montgomery, M.D., Paschmann, G. and Sckopke, N. (1975). Heos 2 plasma observations in the distant polar magnetosphere: The plasma mantle. *Journal of Geophysical Research,* **80,** 2723.

Sabine, E. (1851, 1852). On periodical laws discoverable in the mean effects of the larger magnetic disturbances. *Philosophical Transactions of London,* 1851, pp. 123–39; 1852, pp. 103–24.

Schindler, K. and Birn, J. (1982). Self-consistent theory of time-dependent convection in the Earth's magnetotail. *Journal of Geophysical Research,* **87,** 2263.

Sckopke, N., Paschmann, G., Rosenbauer, G. and Fairfield, D.H. (1976). Influence of the interplanetary magnetic field on the occurrence and thickness of the plasma mantle. *Journal of Geophysical Research,* **81,** 2687.

Sckopke, N. and Paschmann, G. (1978). The plasma mantle: A survey of magnetotail boundary layer observations. *Journal of Atmospheric Terrestrial Physics* **40,** 261.

Shelley, E.G., Johnson, R.G. and Sharp, R.D. (1972). Satellite observations of energetic heavy ions during a geomagnetic storm. *Journal of Geophysical Research,* **77,** 6104.

Shelley, E.G., Sharp, R.D. and Johnson, R.G. (1976). Satellite observations on an ionospheric acceleration mechanism. *Geophysical Research Letters,* **3,** 654.

Sibeck, D.G. and Siscoe, G.L. (1984). properties of magnetic flux transfer events. *Journal of Geophysical Research,* **89,** 10709.

Sibeck, D.G., Siscoe, G.L., Slavin, J.A., Smith, E.J., Tsurutani, B.T. and Bame, S.J. (1985). Magnetic field properties of the distant magnetotail magnetopause and boundary layer. *Journal of Geophysical Research,* **90,** 9561.

Siscoe, G.L. and Sanchez, E. (1987). An MHD model for the complete open magnetotail boundary. *Journal of Geophysical Research,* **92,** 7405.

Speiser, T.W. (1965). Particle trajectories in model current sheets, 1, Analytical solutions. *Journal of Geophysical Research,* **70,** 4219.

Spreiter, J.R. and Briggs, B.R. (1961). Theoretical determination of the form of the hollow produced in the solar corpuscular stream by interaction with the magnetic dipole field of the earth. *NASA, TR-R-1120.*

Spreiter, J.R. and Briggs B.R. (1962). Theoretical determination of the form of the boundary of the solar corpuscular stream produced by interaction with the magnetic dipole field of the earth. *Journal of Geophysical Research,* **67,** 37.

Spreiter, J.R. and Stahara, S.S. (1997). Modeling solar wind flow past the magnetosphere. In C.S. Gillmore (ed.), *Discovery of the Magnetosphere, History of Geophysics,* **7,** p. 193. Washington DC: American Geophysical Union.

Stern, D.P. (1979). Electric fields in the Earth's magnetosphere. *Rev. Geophys. Space Phys.,* **17,** 626.

Terasawa, T., Fujimoto, M., Mukai, T., Shinohara, I., Saito, Y., Yamamoto, T., Machida, S., Kokubun, S., Lazarus, A.J., Steinberg, J.T., and Lepping, R.P. (1997). Solar wind control of density and temperature in the near-earth plasma sheet: Wind-Geotail collaboration. *Geophysical Research Letters,* **24,** 935.

Tsurutani, B.T. and Stone, R.G. (Eds.) (1985). *Collisionless Shocks in the Heliosphere: Current Research,* Geophys. Monogr. 35. Washington DC: Amer. Geophys. Union.

Van Allen, J.A. and Frank, A. (1959). Radiation around the Earth to a radial distance of 107400 kilometers. *Nature,* **203,** 1006.

Wahlund, J-E. and Opgenoorth, H.J. (1989). EISCAT observations of strong ion out-flows from the F-region ionosphere during auroral activity – preliminary results. *Geophysical Research Letters,* **16,** 727.

Walker, R.J., Ogino, T., Raeder, J. and Ashour-Abdalla, M. (1993). A global magneto-hydrodynamic simulation of the magnetosphere when the interplanetary magnetic field is southward: The onset of magnetotail reconnection. *Journal of Geophysical Research,* **98,** 17235.

Warwick, J.W. (1961). Dynamic spectra of Jupiter's decametric emissions. *Astrophysical Journal,* **137,** 41.

Williams, D.J. and Lyons, L.R. (1974). The proton ring current and its interaction with the plasmapause: Storm recovery phase. *Journal of Geophysical Research,* **79,** 4195.

Williams, D.J. and Lyons, L.R. (1974). Further aspects of the proton ring current and its interaction with the plasmasphere: Main and recovery phase. *Journal of Geophysical Research,* **79,** 4791.

Winser, K.J., Jones, G.O.L. and Williams, P.J.S. (1988). Large field-aligned velocities observed by EISCAT. *Journal of Atmospheric Terrestrial Physics* **50,** 379.

Yau, A.W., Peterson, W.K. and Shelley, E.G. (1988). Quantitative parametrization of energetic ionospheric ion outflow. In: T.E. Moore and J.H. Waite Jr, (eds.), *Modeling Magneto-spheric Plasma,* Geophys. Monograph **44,** p. 211. Washington DC: Amer. Geophys. Union.

Zhigulev, V.N. and Romishevskii, E.A. (1959). Concerning the interaction of currents flowing in a conducting medium with the earth's magnetic field. *Doklady Acad. Nauk SSSR,* **127,** 1001. (English translation: *Soviet Physic, Doklady,* 4, 859, 1960).

Zmuda, A.J., Martin, J.H. and Heuring, T.T. (1966). Transverse magnetic disturbances at 1100 km in the auroral region. *Journal of Geophysical Research,* **71,** 5033.

Zwickl, R.D., Baker, D.N., Bame, S.J., Feldman, W.C., Gosling, J.T., Hones Jr, E.W., McComas, D.J., Tsurutani, B.T. and Slavin, J.A. (1984). Evolution of the earth's distant magnetotail: ISEE 3 electron plasma results. *Journal of Geophysical Research,* **89,** 11007.

64

TOR HAGFORS* AND KRISTIAN SCHLEGEL*

Earth's ionosphere

1 INTRODUCTION

In this chapter we will deal with the ionosphere, a part of the terrestrial atmosphere ranging from about 80 km to more than 1000 km height. Its lower boundary lies in the mesosphere, at large heights a transition into the magnetosphere occurs. In altitude the ionosphere approximately coincides with the thermosphere a part of the neutral atmosphere (Figure 1). The term thermosphere is used when the properties of the neutral gas are under consideration, the term ionosphere when the electric properties and processes are regarded. Although the charged particle concentration in the ionosphere is at most around 1/1000 of the total particle density (around the F-layer peak, see Figure 1), the ions and electrons enable fundamentally different processes to take place in the ionosphere as compared to those occurring in the neutral atmosphere. Electric currents, electric fields, particle drifts in these fields and the reflection of radio waves are among these phenomena.

In Section 2 we shall first follow the exploration of ionospheric structure in its historical context. We will deal with the different layers (see Figure 1), will discuss peculiarities thereof and geographic as well as temporal anomalies. A separate subsection is devoted to ionospheric irregularities and some consequences of their existence.

Section 3 describes the development of our understanding of the fundamental physical processes leading to the formation and spatial and temporal behavior of the ionosphere. We shall deal with production and loss processes, chemistry, dynamics and transport, winds and waves in the ionosphere and thermosphere. Influences from above (of the magnetosphere) as well as from below (of the lower atmosphere) will be discussed as well. A brief summary of the current status of our knowledge of ionospheric processes completes this section.

The techniques for probing the ionosphere from the ground and *in situ* are the subjects of Section 4. Experiments by Appleton and Barnett (1925) and by Breit and Tuve (1926) designed to determine the height of reflection of radio waves from the ionospheric layers signaled the beginning of ionospheric research as a science. They in many ways can be said to represent the start of modern space science. They also, as it turned out, had an enormous influence on other fields. The development of the magneto-ionic theory to describe the propagation of radio waves in magnetized plasmas (Lassen 1927; Appleton 1932) marked a large step forward in plasma physics which before primarily had been confined to studies of gas discharges. The later studies of reflection and scattering from and transmission through irregular ionospheric layers and their interpretation had a profound influence on radio and radar astronomy (Ratcliffe 1948; Booker and Clemmow 1950; Booker *et al.* 1950). They helped to establish relations between the behavior of radio signals transmitted through the interplanetary and the interstellar medium, and led to increased understanding of the properties of plasmas in our solar system, in our galaxy and in other galaxies. Studies designed to detect small radio sources by observing scintillation caused by interplanetary plasma irregularities led to the discovery of pulsars (Hewish *et al.* 1968). They contributed to the interpretation of radar echoes from planets, asteroids and comets in terms of their degree of surface irregularity. The studies of the ionosphere even brought with them important technological advances in that all modern radar systems trace their roots to Breit

---

* Max-Planck-Institut für Aeronomie, Katlenburg-Lindau, Germany

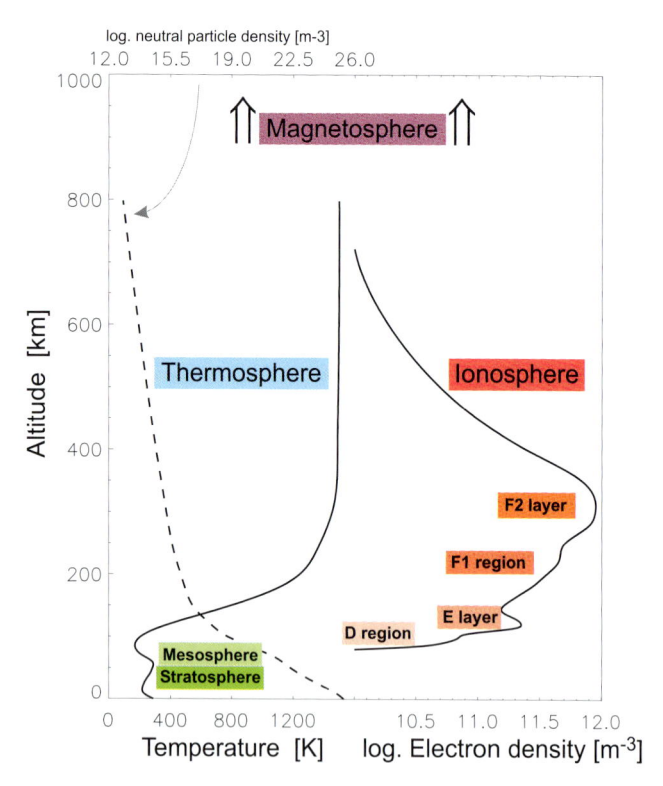

Figure 1 Schematic diagram of the ionosphere, one of the "spheres" of terrestrial atmosphere, between the mesosphere and the magnetosphere. It is characterized by its electron density height profile, subdivided into different layers and regions as described in the following sections (right hand part of the figure). The ionosphere covers approximately the same altitude range as the neutral thermosphere (left hand part of the figure), characterized by the neutral gas temperature. The neutral particle number density is plotted as a dashed line, its scale is on the top of the figure.

and Tuve's experiments with pulsed radar to determine the height of the ionosphere from the time delay of the echoes.

Finally, in Section 5 we briefly describe the role of the ionosphere and of ionospheric research in our modern society. Impacts on communication, geodesy and space technology are discussed in the context of solar-terrestrial relations and space weather.

The literature on the ionosphere is very extensive comprising several tens of thousands of papers, and it will be impossible to refer to them all in a book on the ionosphere, and let alone in such a brief chapter on the subject. The choice of references in this chapter is highly subjective and to some extent reflects our own interests and areas of research. We apologize for the omission of many important contributions to the field. These omissions must be ascribed to our ignorance of important contributions rather than to ill intent. We sincerely apologize for this limitation in acknowledging all those who have contributed to the field.

Speaking about important contributions, the most outstanding scientist in our field was certainly Professor Edward V. Appleton (1892–1965) as will become evident in the following sections. He received the Nobel Price for his achievements in 1947, the only scientist thus honored for ionospheric research.

2 EXPLORATION OF IONOSPHERIC STRUCTURE

2.1 Early ideas

The first ideas of an electrically conducting layer in the atmosphere were advanced by Gauss who in 1839 argued that small daily variations of the geomagnetic field could be explained by electric currents flowing in such a layer (see Kaiser 1962). Later Stewart (1882) suggested that these currents may be produced by the dynamo actions of winds blowing across the geomagnetic field. The basis of the electric conductivity however, remained unclear before the discovery of the electron in 1897 and ions around 1900.

After the successful reception of a radio signal transmitted from Pudhu/Cornwall at St. Johns/Newfoundland by Marconi on 12 December 1901, it became obvious that these waves did not propagate on a rectilinear path. Electromagnetic waves discovered by Hertz in 1887 were expected to travel along straight rays like visible light in vacuum, but could also undergo reflection and diffraction. Either one of these processes had to be involved to explain the propagation over such long distance. The first clear hypotheses of an electrical conducting layer and of a reflection of the signal by this layer were published independently by Kennelly (1902) and Heaviside (1902). The diffraction of radio waves around a conducting Earth was treated by MacDonald (1903), Rayleigh (1903), Poincare (1903), Zenneck (1907) and Sommerfeld (1909) and was discarded as possible explanation of Marconi's results.

Although radio transmissions were more and more extensively used for broadcasting and information transfer in the following years, it took more than two decades before the height of the layer could be established. Appleton's famous wavelength change experiment was performed in December 1924 (Appleton and Barnett 1925). It proved the existence of a ground wave (direct path) and a sky-wave (reflected at the layer) which are subject to interference at the reception point, causing a fading of the signal. From the fading characteristics the authors were able to derive a first height estimate of the conducting layer of about 90 km. From our present knowledge (see Figure 1) we know that this very low height had two causes. The sounding frequency of 0.75 MHz used was very low and therefore the reflection height was well below the E-region peak and, the measurements were made during night (because the BBC transmitter used was available for this research purpose only after the end of the normal broadcasting at midnight). About a year later the so-called Peterborough experiment, gave the first height variation: during the time one hour before sunrise to sunrise the layer descended from about 115 to 95 km. The interference pattern formed on the ground between transmitter and receiver was also utilized in another experiment to estimate the layer height. Hollingworth (1926) measured the variation of signal strength of a transmitter near Paris at various receiving locations between London and Edinburgh from which he deduced the height of reflection. At the same time, as mentioned in the introduction, Tuve and Breit (1925) used a transmitting method still applied today, the pulse technique. It gave similar results, namely around 100 km for the reflection height.

2.2 The ionospheric layers

The name of the layer was at that time still associated with researchers in the field, i.e. 'Heaviside layer' or 'Appleton layer'. After the British National Committee for Radio-Telegraphy in 1926 had suggested that a non-personal name should be found, the term 'ionosphere' was introduced by R.A. Watson-Watt and apparently also independently by Appleton. Appleton used the letter 'E' first to denote the electric field of signals reflected from the layer. Only later the letter was used to indicate the layer itself.

A thorough interpretation of these early radio experiments was only possible after a suitable theory of radio wave propagation was developed. Larmour (1924) was the first to realize that a plasma has a refractive index less than unity causing incoming waves to be bent away from the normal to the layer and consequently reflect radio waves. He derived the formula for the index of refraction

$$n^2 = 1 - \left(\frac{\omega_p}{\omega}\right)^2 = 1 - X \qquad (2.1)$$

with the plasma frequency

$$\omega_p = \sqrt{\frac{Ne^2}{m_e \varepsilon_0}}$$

where N and m_e are the electron density and electron mass, respectively, e the elementary charge, ε_0 the dielectric constant of free space, and ω the angular frequency of the transmitted signal. Appleton realized that the influence of the geomagnetic field on the motion of electrons and ions had to be taken into account. This finally led to the derivation of the so-called Appleton-Hartree formula (eqn (4.1)), (Appleton 1932; Hartree 1931) for the index of refraction. Lassen had derived the same formula even a few years earlier (Lassen 1926), but the name of the formula remained.

On the basis of eqn (2.1) it was obvious that the waves do not propagate with the velocity of light through the plasma and that measurements using the signal travel time only give a 'virtual' height h'. In the general case the virtual height is related to the true height h by

$$h' = \int_0^h \frac{c}{V_g} dz = \int_0^h \left(n + \omega \frac{\partial n}{\partial \omega}\right) dz \qquad (2.2)$$

where V_g is the group velocity of the wave and n the index of refraction (the simplified form is given in eqn (2.1)) depending on the height profile of the electron density N(h) and the sounding frequency ω.

Experiments with vertically reflected waves on a higher frequency led Appleton (1927) to the conclusion that another layer existed above the E layer. He called it F layer and found later that during day time it splits into an F1 and an F2 layer. Both layers (cf. Figure 1) were systematically studied in the following years making use of ionosondes developed at the beginning of the thirties (see Section 4). It was found that the ionization followed the solar zenith angle (Appleton and Naismith 1932) suggesting that the ionizing radiation came from the sun. The prevalence of the F layer ionization during night remained a puzzle for many years (see Section 3). Information about the layer thickness was first obtained in 1937, when Appleton showed that the E and F layer had a thickness ratio of 1:4.

A layer below the E layer escaped detection for quite a long time, because of measuring difficulties. The early ionosondes were not able to measure it, because of the low electron densities in this layer and the consequently necessary long wavelength to probe it (cf. eqn (1)). First indication of this 'D layer' came from fadeouts accompanying solar flares (Mögel 1931), later also from the study of the interference pattern between ground wave and sky wave near long-wavelength transmitters. Best and coworkers (1936) were able to derive reflection heights of 74 km during day and 92 km at night. More precise data about the electron density distribution were not obtained before the 1950s when the partial reflection (Gardner and Pawsey 1953) and the cross modulation techniques (Fejer 1955) were introduced (cf. Section 4).

The F1 and the D layers are not really layers with a well-defined peak as is the case for the E and the F2 layers (Figure 1). They form a ledge of enhanced ionization just below the E and the F2 layer, respectively. During the last twenty years, however, it became clear that they are really different from the layers above in terms of the physics involved (ionization, recombination, dynamics, see Section 3). Today we speak of D and F1 regions, rather than using the term layer.

It should be remembered that ionosondes do not give electron density data above the peak of a layer. This led to a gap in the measurements between the E and the F layer only closed after incoherent scatter facilities and rocket and satellite measurements became available. The electron density above the F2-layer peak was first observed with topside sounders from satellites in 1962 and also later with the incoherent scatter technique. The latter provides the most accurate measurements from the ground between about 80 km and several hundreds of km altitude (Figure 2 and Section 4).

At very large heights (above 1000 km) the ionospheric ion composition changes from O^+ to H^+ and He^+ ions, therefore this outer part of the ionosphere is sometimes called protonosphere. The whole near-Earth space filled with thermal plasma was named plasmasphere. This 'sphere' is corotating with the Earth and the thermal plasma within is transported upward and downward along magnetic field lines, even from one hemisphere to the other (Lemaire and Gringauz 1998). The plasmasphere exhibits a relatively sharp outer boundary, a steep drop in plasma density from about 10^9 to about 10^7m^{-3} at 3–5 Earth radii. This plasmapause was detected almost simultaneously from early satellite measurements (Gringauz et al. 1961) and from ground-based whistler observations (Carpenter and Park 1963). The plasmapause can also be regarded as the ultimate border of the ionosphere.

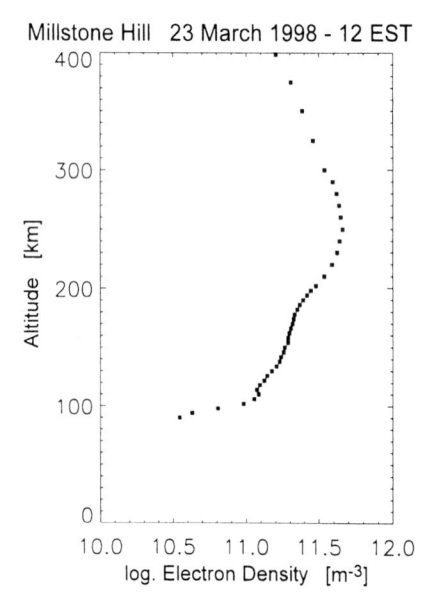

Figure 2 Example of an incoherent scatter measurement of the ionospheric electron density at mid latitudes with the Millstone Hill radar. It covers most of the ionosphere from the D layer to the topside above the F2 layer peak. The F1 layer is visible as well. With the help of advanced transmitter pulse schemes the height resolution can be kept variable, i.e. high in regions with small scale details (D, E, F1 region) and low in the F2 region and above. (J. Holt, private communication.)

2.3 Special effects and geographic peculiarities

The study of diurnal and seasonal variations of the F2 layer revealed several anomalies. The diurnal anomaly addresses the fact that the highest electron density does occur not at local noon (smallest solar zenith angle) but in the afternoon. The winter anomaly refers to the higher ionization in winter than in summer at mid latitudes (Berkner and Wells 1938). The former can be explained when dynamic effects are included in the physical description of the layers rather than only ionization, production and recombination (see Section 3), the latter by seasonal changes of the O/N_2 concentration ratio.

Already in the thirties a special form of the E layer was detected and later named 'sporadic E', or E_s (Appleton and Naismith 1940). It showed high electron densities with a peak often below the normal E-region peak, it was very narrow, and it occurred only sporadically. At mid latitudes observations showed a very pronounced maximum in the probability of occurrence during summer, whereas no clear seasonal variation was found at low and high latitudes (Smith 1957). Later, through measurements with sounding rockets, it was shown that E_s layers contain a considerable content of metallic ions, i.e. Fe^+ and Mg^+ besides O_2^+ and NO^+, the usual dominant ions at E-region height (Narcisi et al. 1967). In the sixties a wind-shear theory of sporadic E was developed by several authors (e.g. Whitehead 1961; Axford 1963). An altitude variation of the wind speed acts in the presence of the geomagnetic field to compress the ionization. This wind-shear theory is generally accepted today as one cause of mid latitude sporadic E.

A special form of E_s with peak electron densities well exceeding those of the F2 layer was observed at high latitudes (Appleton et al. 1933) and extensively studied particularly by Scandinavian scientists in the following years. After rocket and satellite measurements became available in the sixties, it was shown that these very pronounced layers (Figure 3) are created by electrons with energies of several keV precipitating from the magnetosphere along geomagnetic field lines. The particle energy determines the height of these layers, the particle flux its peak density (see Section 3). In recent years it became clear that the precipitation of these electrons (and sometimes protons) is controlled by the solar wind. The morphology of these layers is determined by magnetospheric processes also associated with geomagnetic disturbances. During these events significant changes in the F layer were also observed: so-called positive and negative storm effects, denoting an increase or decrease of the F-layer ionization, respectively (see Section 3).

Enhanced electron densities in the high latitude D layer were inferred from radio wave propagation disturbances (Mögel 1931; Dellinger 1937). These 'sudden ionospheric

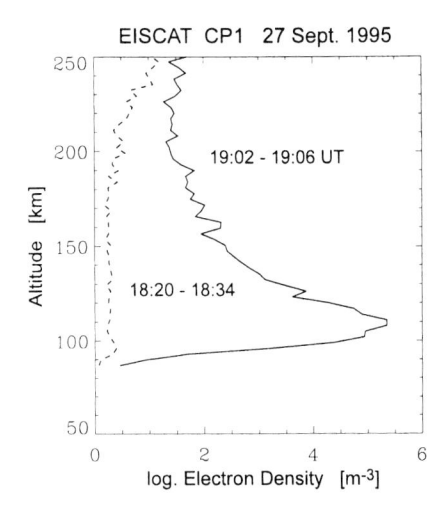

Figure 3 Example of an auroral E layer enhanced by particle precipitation recorded with the incoherent scatter facility EISCAT at Tromsö, Norway (solid line). Since particle precipitation events have time scales of a few minutes the electron density profile only about half an hour earlier looks completely different (dashed line). This latter profile characterizes the undisturbed auroral E region. In both cases there is no pronounced F region.

disturbances' (SID), or 'Mögel-Dellinger events' are caused by X-rays emitted during solar flares. Whereas a SID lasts only for a few tens of minutes, so-called 'polar cap absorption events' PCA, can go on for several days. This transpolar absorption of radio waves is caused by high-energy protons enhancing the D-region ionization (Hultqvist 1969). These MeV protons are also emitted during and after solar flares.

A peculiar phenomenon of the lower ionosphere/mesosphere was detected only recently. Strong continuous echoes in 50-MHz radar observations in the 80–90 km height range in summer were for the first time reported by Ecklund and Balsley (1981) from Alaska. These 'polar mesospheric summer echoes' (PMSE) were studied in the following years at many stations at high latitudes. Current explanations (e.g. Cho and Röttger 1997) indicate that they are a manifestation of a 'dusty plasma', since charged aerosols in the mesopause region play a strong role in their formation. They are closely related to noctilucent clouds.

Another geographic peculiarity is the 'equatorial anomaly' or 'fountain effect', experimentally described by Appleton (1947), but discovered by Japanese scientists already in 1939. This anomaly addresses the fact that the F-region electron density at the magnetic equator is smaller than that at about $10°$–$20°$ north and south of it. The zonal electric field at the magnetic equator creates a steady $\mathbf{E} \times \mathbf{B}$ plasma drift which is upward directed during the day. Therefore the dense equatorial plasma rises until the pressure forces are high enough that it starts to slide down along magnetic field lines, assisted by gravity, to the north and the south of the equator. This electromagnetic lifting therefore acts like a fountain.

2.4 Ionospheric irregularities

In the early years of ionospheric exploration by radio waves considerable effort was made to explain the fading of these waves. It was realized that they are often caused by irregularities of the medium. After Booker et al. (1950) provided a suitable theory of the diffraction of radio waves by a 'random screen' many techniques were developed to study the irregularities in detail, not only by remote sensing but also *in situ* by means of rocket and satellite experiments (see Section 4). During the 1960's it became clear that the irregularities are caused by a variety of plasma instabilities occurring in the ionosphere (Farley 1963; Buneman 1963; Maeda et al. 1963), for a review, see e.g., Fejer and Kelley (1980).

In the E region these instabilities are the so-called modified two-stream or Farley-Buneman instability and the gradient drift instability. The former is excited whenever the relative drift velocity between electrons and ions exceeds a certain threshold value, the latter if a small electron drift occurs together with a gradient in electron density (Simon 1963). Historically these instabilities were first studied at the equator where the differential drift of electrons and ions is caused by their different mobility under the influence of the geomagnetic field and neutral winds. Later it was realized that these instabilities are more strongly excited at high latitudes where the free energy of the instability is provided by large electric fields imposed on the ionosphere through magnetospheric convection. The corresponding irregularities were first described in the thirties (Harang and Stoffregen 1938), and their effect on radar signals later referred to as 'auroral backscatter', because of the occurrence together with visible aurora. Fairly recently it was found that the Farley-Buneman instability does not only create electron density irregularities, but that these interact themselves with the ambient electron gas and heat it well above normal levels. The correspondingly enhanced electron temperatures occurring in a relatively narrow height range around 110 km were first observed by Schlegel and St. Maurice (1981). They constitute an important heat source in the disturbed auroral zone (Figure 4).

Radio back-scatter experiments from mid latitude sporadic E performed since the fifties were later explained as resulting from the gradient drift instability (Ecklund et al. 1981) which was proven to act at mid latitudes, too. It was a great surprise, however, when clear evidence of Farley-Buneman instability excitation was reported (Schlegel and Haldoupis 1994) from the evaluation of spectra of 50-MHz back-scatter, because the relatively strong electric fields required to excite this instability were not expected at mid latitudes.

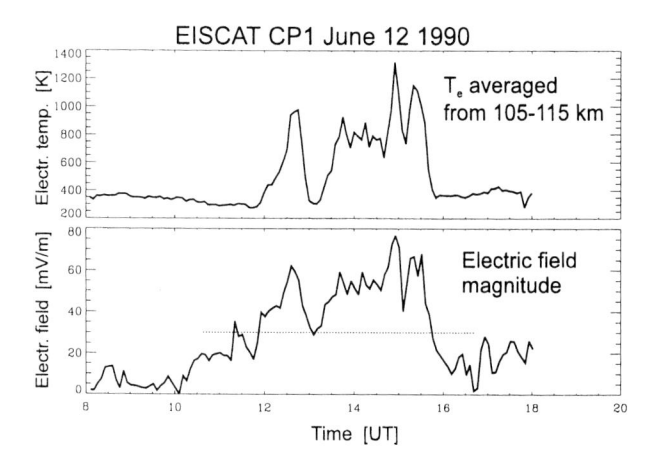

Figure 4 Enhanced electron temperatures in the auroral E region caused by the Farley-Buneman plasma instability (upper panel). The corresponding electric field magnitudes are plotted in the lower panel. When this electric fields exceeds a threshold of about 30 mV/m (dashed line) the instability is excited and T_e rises above the undisturbed levels of about 350 K. The T_e enhancement follows the temporal variation of the electric field, since the latter provides the free energy for the instability.

Similarly the so-called spread-F phenomenon, observed in the night-time ionosphere as a smearing and spreading of ionogram traces already in the 30's (Berkner and Wells 1934) which baffled the ionospheric physicists for decades is now associated with a broad range of plasma instabilities. They have scale sizes from tens of centimeters to hundreds of kilometers and result from multiple-step non-local processes involving collision and collision-less Rayleigh-Taylor and $E \times B$ instability and drift waves (Haerendel 1973). Satellite studies revealed that they are associated with large-scale density depletions aligned with magnetic meridians (Tsunoda et al. 1982) which are most probably seeded by gravity waves (Kelley and Hysell 1991).

Ionospheric irregularities seeded by neutral gas turbulence and affecting the spatial electron density structure have also been proposed for the lower ionosphere (Gurevich et al. 1996).

The investigations of instabilities led to a completely new field of research where the plasma physical rather than the geophysical aspects of the ionosphere are of prime interest (e.g. Fejer 1996). Practical consequences of the investigations were on the one hand better understanding of radio wave propagation through regions of instabilities affecting communication and navigation, on the other hand, and more important, the development of a whole new class of observational methods making use of the coherent back-scatter of radio waves from certain irregularities (Greenwald 1996).

Further steps in using the ionosphere as a plasma laboratory stem from the artificial modification of the ionosphere by powerful radio waves. It was observed that the modulation of the powerful Luxemburg radio transmitter caused this modulation to be imposed on radio transmissions from other stations. The theoretical explanation of this intermodulation was given by Bailey and Martin (1934). The strong Luxemburg transmissions modulate the plasma temperature causing the electron collision frequency and hence the absorption of other transmissions to be modulated in step. Proposals were also made to heat the ionosphere through the application of radio waves at the electron gyro-frequency to induce optical emissions, so-called air-glow (Bailey 1938). It was at one point even thought that such effects could be used for the night-time illumination of large areas. Air-glow phenomena have certainly been observed, but at a level many orders of magnitude weaker than required for illumination. The properties of this air-glow or artificial aurora have recently been studied by Kosch et al. (2000).

Systematic studies of the modification effects of strong electromagnetic radiation was begun in Colorado in 1971 (Utlaut 1970) and were followed by similar efforts in Puerto Rico, Russia, Norway and Alaska. Plasma turbulence associated with unstable ion-acoustic, lower-hybrid, Langmuir, upper hybrid and Bernstein waves have been the subject of these studies. The interpretation of the results of these observations have often led to animated discussions about their interpretation, whether to invoke weak nonlinear interactions associated with parametric decay (Stubbe et al. 1992) or strong nonlinear interactions described by the Zakharov-equation (DuBois et al. 1990). Field aligned irregularities are often created in heating experiments, and these irregularities can be used for ionospheric observation of plasma motion when naturally occurring field aligned irregularities are absent.

3 UNDERSTANDING IONOSPHERIC PROCESSES

3.1 Production and loss

By the second half of the 1920's it was firmly established that solar EUV radiation creates the ionosphere at altitudes above about 100 km. From our present knowledge and understanding the early ionospheric theorists faced the following problems in their attempts to describe the ionosphere:

1. The solar spectrum in the EUV range was not known. From the ground only radiation intensities above 289 nm were accessible due to the absorption by ozone.
2. Absorption coefficients and photo-ionization cross sections of the major atmospheric constituents were only poorly known.

3. Density, temperature, pressure and composition data of the neutral atmosphere above 100 km were not available.
4. Rates of chemical processes (e.g. charge exchange, recombination, and attachment) between ions, electrons and neutrals were not established.
5. The role of transport and dynamic processes (diffusion, electromagnetic drift, winds, waves) in shaping the ionosphere and controlling their temporal behavior was postulated, but the details were not known, either from their principle or from their mathematical treatment.

A milestone at that time was Chapman's (1931a) mathematical description of an ionospheric layer formed by ionizing radiation. The idea was that the production of charged particles would exhibit a maximum at a certain altitude. Descending in altitude this production will first grow because of the downward increase of atmospheric density, i.e. an increasing number of neutrals will be ionized. Lower down the production will eventually decrease because the ionizing radiation is more and more absorbed. Thus a maximum will be formed at some altitude. It should be noted that this treatment involved only relatively general principles (already formulated in part by Lenard, 1911) and did not require any of the above information. The most important restriction is the assumption of monochromatic EUV radiation and the existence of only one single atmospheric species. The production can then be expressed as

$$q(h) = q_0 \exp(1 - z - e^{-z} \sec\chi) \quad (3.1)$$

with

$$z = (h - h_0)/H \quad q_0 = S_\infty/(e^1 H) \quad h_0 = H \ln(Q_\alpha n_0 H)$$

Here H is the scale height of a neutral atmosphere with exponentially decreasing density, $n = n_0 \exp(-h/H)$, S_∞ the flux of incident EUV quanta at the top of the atmosphere, Q_α the absorption cross section of the single atmospheric constituent and χ the solar zenith angle. Figure 5 shows graphs of this function for different solar zenith angles. With a given scale height the height of the maximum depends only on the absorption cross section, and the maximal production q_0 only on the incident flux, whereas the shape of the layer is only defined by H.

In order to estimate the number of produced electrons one needs additionally a simplified continuity equation

$$\frac{dN}{dt} = q - \alpha N^2 \quad (3.2)$$

where α denotes the recombination coefficient. Assuming that α is height independent, the equilibrium electron density profile will be

$$N(h) = \sqrt{q(h)/\alpha} \quad (3.3)$$

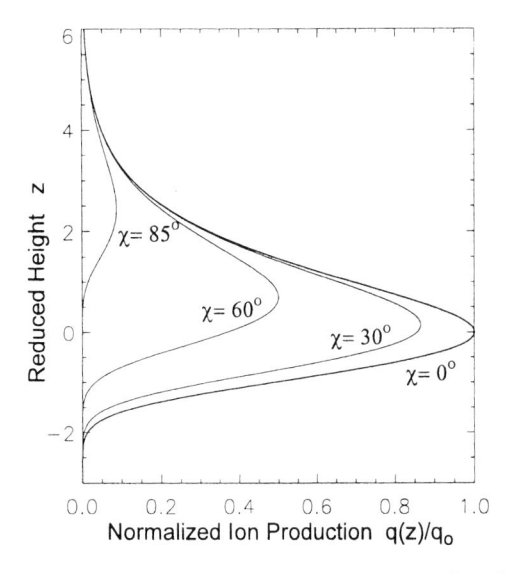

Figure 5 Curves describing a Chapman layer. The abscissa denotes the normalized ion production (eqn (3.1)), the ordinate the reduced height. The four curves are valid for different solar zenith angles.

Such a layer with a shape of $\sqrt{q}$ entered the scientific terminology as a Chapman layer. It was assumed in the early 1930's that the observed different ionospheric layers can just be attributed to different ionising wavelength bands (different S_∞) and different atmospheric constituents (different H, Q_α, α). The problem of calculating the actual electron density was, that these constants were only poorly known at that time (see items (1)–(3) of the above list).

The early estimates circumvented the lack of solar spectral data (1.) by assuming black body radiation of 6000 K and extrapolating ground based flux measurements at 289–3000 nm to values around 100 nm where the ionisation of oxygen and nitrogen was supposed to take place. In the early 1930's the development of quantum mechanics was sufficiently advanced to calculate values for the recombination rate α, and the absorption cross section Q_α for oxygen and nitrogen. The estimation of the scale height was more difficult. Since

$$H = \frac{kT}{mg}, \quad (3.4)$$

the neutral temperature T (k is Bolzmann's constant and g the gravitational acceleration) and the mass of the neutral gas molecules, m, is required. For the latter a mean value equivalent to 30 AMU was used and T was assumed to be around 300 K (a value far too low, from our present knowledge). Despite these imperfect values Chapman arrived at maximal electron densities close to $10^{12} m^{-3}$ at around 200 km, a value which was in reasonable agreement with observations from the F region (see Figure 1).

It is interesting to note that Chapman used a neutral temperature of 300 K in his estimates, although there were already hints of much higher values. Maris and Hulbert (1928) were the first who suggested that T could reach 1000 K above 150 km. They used extrapolations of a solar black body radiation and several different plausible assumptions of cooling. Later, more evidence for these high temperatures came from experimental estimates of the scale height (Martyn and Pulley 1936). They concluded that the atmosphere at about 300 km altitude either consists of He with a temperature of 300 K or of atomic oxygen with a temperature of 1200 K (cf. eqn (3.4)). Since there was no indication of Helium from spectroscopic measurements, it was concluded that the upper atmosphere was really hotter than so far assumed. Atomic oxygen was assumed to be the dominant atmospheric constituent at these heights, since Chapman (1931b) had shown that dissociation of molecular oxygen by solar UV radiation takes place above 100 km.

Chapman first assumed that his layer theory explains the F region. He had difficulties, however, to describe the E region. For the formation of the latter he therefore invoked particles from the sun. Later this was proven wrong: Measurements during a solar eclipse (Henderson 1933) provided clear evidence that both the E and the F1 layers were ionized by photons and therefore can be regarded as Chapman layers.

Shortly after Chapman's publication it was realized that recombination was not the only process to remove electrons from the ionosphere, electron attachment to neutrals was another possible mechanism. For attachment the continuity equation can be written

$$\frac{dN}{dt} = q - \beta N \qquad (3.5)$$

In equilibrium condition the electron density is therefore proportional to q, rather than proportional to $\sqrt{q}$ as in the case of recombination. By measuring the electron density at different values of the solar zenith angle and using eqn (3.1) it could therefore be experimentally tested which process was dominating. It turned out that the results favored the recombination mechanism in the E and F1 layer, but the results for the F2 layer were not conclusive. The 'anomalous' behavior of the F2 layer remained a puzzle for many years. Almost 20 years later Ratcliffe et al. (1956) arrived at the conclusion that the electron loss in the F2 layer can be described by a linear law as attachment suggests. A final consistent theory of the F2 layer was not achieved before transport processes were taken into account (see below).

Even the details of recombination processes invoked only formally in Chapman's theory proved elusive, and it took several years before they were clarified. Radiative recombination, i.e.

$$X^+ + e \rightarrow X + h\nu$$

could not account for the observed recombination rates (Bates et al., 1939). Therefore Bates and Massey (1947) proposed a two step process of charge exchange and dissociative recombination

$$O^+ + XY \rightarrow XY^+ + O \quad \text{and} \quad XY^+ + e \rightarrow X + Y$$

Together with diffusion (see 3.2) these reactions could explain an F2 layer above the F1 layer, even though there was only a peak of electron production at the F1-layer height. In addition the loss rate in the F2 layer becomes proportional to the electron density, thus simulating an attachment process.

An explosion of the insight in ionospheric processes occurred after World War II when *in situ* measurements by means of sounding rockets became possible. Suddenly several of the missing parameters listed above became available:

- A series of launches started with captured German V-2 rockets carrying EUV and X-ray spectrometers provided the first data of corresponding solar fluxes. The results led to the first reliable model of the production of the ionosphere in 1955. In the middle of the 60's the solar flux from the ozone cut-off at 289 nm down to a few nm was firmly established (Hinteregger 1970).
- Similarly, in early rocket experiments pressure, density and temperature were measured up to about 200 km altitude. In 1952 the first model atmosphere based on these data was prepared (Rocket Panel 1952). After the first density estimates in 1958 based on orbital changes of the Russian SPUTNIK-II (Jacchia 1959), this technique together with further rocket and ground-based data led to the first time dependent (season and solar activity) atmospheric model (CIRA 1961). This model eventually evolved into the MSIS model which is still in use (Hedin 1987).

These data enabled more accurate calculations of the mechanisms already known and led to the discovery of new processes important for the quantitative description of the ionosphere.

This was particularly true for the ion chemistry. The first reliable ion composition measurements by means of rocket-born mass spectrometers were made during the International Geophysical Year (IGY) (Johnson et al. 1958). One of the surprising results was that NO^+ was the most abundant ion in the E and F1 regions, although the primary produced ions were N_2^+, O_2^+ and O^+. A series of reactions like

$$N_2^+ + O \rightarrow NO^+ + N \quad O_2^+ + N \rightarrow NO^+ + O$$
$$O_2^+ + NO \rightarrow NO^+ + O_2 \quad O^+ + N_2 \rightarrow NO^+ + N$$

were later identified as sources (Bates 1974). With subsequent rocket flights metallic ions were detected and identified as meteoritic in origin (Narcisi et al. 1967). In addition it was demonstrated that Helium and Hydrogen ions were present in the topside ionosphere (Taylor et al. 1963).

In the following years the chemical models of the ionosphere became more and more complex. In present F-region models more than 20 reactions are taken into account (Mikhailov and Schlegel 1997). Current D-region models contain as much as 174 reactions among 24 positive and 11 negative ions (Burns et al. 1991).

3.2 Dynamics and transport

Parallel to the increase of knowledge of the ionospheric chemistry through experiments as outlined above, considerable progress was made in understanding the dynamics of the ionosphere. Already at the beginning of the 40's it became clear that purely static theories of production and recombination (see above) could explain several properties of the E and F1 layer, but failed to explain the behavior of the F2 layer. One of the first regarding diffusion in the ionosphere was Ferraro (1945), but he concluded that it should not be very important in the F2 layer. A few years later transport processes were treated by Martyn (1947, 1948). He proposed to add a transport term to the continuity equation used until then

$$\frac{dN}{dt} = q - L - \frac{\partial(NV)}{\partial z} \quad (3.6)$$

were L describes the different loss processes of (3.2) and (3.5). V is a vertical velocity which can generally be written as

$$V = V_{diff} + V_E + V_W \quad (3.7)$$

where V_{diff}, V_E, and V_W denote velocities caused by diffusion, electric fields and winds, respectively. Martyn (1947) treated the electric field case, whereas Yonezawa (1956) quantified the diffusion term on the basis of gas kinetics as

$$\frac{\partial}{\partial z}\left\{D_a\left[\frac{\partial N}{\partial z} + (1 + \alpha_T)\frac{N}{T_p}\frac{\partial T_p}{\partial z} + \frac{N}{H_p}\right]\right\} \quad (3.8)$$

where D_a is the so-called ambipolar diffusion coefficient, $T_p = (T_e + T_i)/2$ the plasma temperature, H_p the plasma scale height (which is different from the neutral scale height), and α_T the thermal diffusion coefficient.

Transport processes finally helped to solve the long-standing problem why the electron density reaches a peak at 250–350 km, i.e. the F2-region peak. Equation (3.6) without a transport term does not lead to a peak, since the loss term decreases faster with increasing height than the production term. Based on previous related work (Yonezawa 1956; Ratcliffe et al. 1956), Rishbeth and Barron (1960) made a detailed assessment how production, loss, drift, and diffusion interact to produce the daytime F2 peak.

During the 50's great progress was also made in the field which is now called ionospheric electrodynamics. Cowling (1945), and Baker and Martyn (1953) described the motion of charged particles under the influence of electric fields and collisions with neutral particles in the presence of a magnetic field which leads to the formulation of Ohm's law for the plasma:

$$\vec{j} = \sigma\vec{E} \quad (3.9)$$

The conductivity σ is a tensor in the ionosphere because of the different mobility of ions and electrons in different directions in relation to the magnetic field vector. The global current system in the E layer was experimentally explored through world-wide magnetometer records from the 30's on. The theory explaining this current system in terms of a wind-driven dynamo was developed by Martyn (1947, 1948, 1953) and Maeda (1955). The associated electric fields also affect the plasma in the F layer, leading for instance to the equatorial anomaly (see Section 2).

First data of atmospheric density and temperatures derived from orbital changes of satellites in the early 60's (see above) showed a region of high air density and temperature in the thermosphere around the subsolar point, and of low density and temperature in the early morning close to the equator. King and Kohl (1965) realized that such behavior corresponds to appreciable pressure gradients at F-region altitudes which, in turn, would drive a global thermospheric wind system. These winds will move F-region ionization up or down magnetic field lines, thereby changing the loss rates, since loss coefficients are always altitude dependent. In subsequent years the thermospheric wind system and/or its effect on the F region were investigated by many authors. Winds are also mainly responsible for the maintenance of the night-time F2-layer (Rishbeth 1968), a problem which took a long time to be solved.

3.3 Waves

Another manifestation of ionospheric dynamics is a quasi-periodic, wave-like change of plasma parameters called traveling ionospheric disturbance (TID). After their discovery by Munro (1948), Hines (1960) convincingly proved that TIDs are caused by atmospheric gravity waves (AGW).

With the derivation of a dispersion relation Hines showed two important properties of these waves:

- The wave frequency is limited by the so-called Brunt-Väisälä frequency, $\omega_g = \sqrt{(\gamma - 1)g/C}$ which describes the free oscillation of an air parcel under the influence of buoyancy and gravity (γ is the adiabatic exponent, g the acceleration of gravity, C the velocity of sound). The corresponding time period is between 5 and 15 min., depending on the temperature and on the mean mass of the neutral particles which are both height dependent.

- The directions of the phase and the group velocity (= direction of energy transport), are different in the atmosphere. While the phase shows a downward progression (see Figure 6) the energy transport is generally directed upward.

AGWs occur in the neutral atmosphere, but through the coupling between neutrals and ions, the ionization acts as a tracer for the neutral gas and therefore TIDs can be conveniently studied by ionospheric measurements techniques (Figure 6). In the 60's the propagation of TIDs was recorded by chains of ionosondes. By correlating the disturbances measured at different stations, the propagation velocity and direction could be estimated (e.g. Klostermeyer 1969). The observation showed that large-scale ($\lambda_h \approx 1000$ km) waves travel mostly equatorward from the auroral zone with velocities of 500–1000 m/s. Parallel to the observations many attempts have been made to develop analytical models of AGWs (Klostermeyer 1972; Richmond 1978). The interaction between the plasma and the neutral atmosphere proved to be very complex and difficult to solve, because continuity, momentum and energy equations have to be solved simultaneously for the ions and the neutrals to adequately describe the AGW – TID relationship. This problem has been properly solved (Kirchengast et al. 1996) only recently. This was only possible because of the revolution brought about in the study of TIDs by the incoherent scatter technique (see Section 4). With this method it became possible to study the influence of AGWs on several other ionospheric parameters besides electron density, namely on the electron and ion temperatures (Testud and Vasseur 1969) and on the ion drift (Schlegel 1986). Very valuable and precise data were gathered through four Worldwide Atmospheric Gravity Wave Studies (WAGS) between 1985 and 1990 when incoherent scatter stations, ionosondes and HF Doppler radar were used simultaneously. These measurements greatly advanced our understanding of the relevant processes (Williams et al. 1988, 1993). It also turned out that the waves proved to be very important for the energy transport in the thermosphere, considerable effort has therefore been made to assess their impact in a quantitative way. In recent complex models (e.g. Thermosphere-Ionosphere General Circulation Models – TIGCMs) the generation, propagation and dissipation of the waves can be modeled on a global scale with increasing confidence (Milward et al. 1993).

Another type of neutral atmospheric waves was named travelling atmospheric disturbances (TADs). They are soliton-like surges travelling from high latitudes towards the equator mainly as a consequence of local heating in the auroral zone during geomagnetic storms (see Section 3.3), but may also result from a superposition of AGWs (Richmond and Matsushita 1975). TADs in turn strongly affect ionospheric properties (Prölls et al. 1991).

Waves with a time period greater than that of AGWs (up to a few hours) also occur in the atmosphere. Humboldt recognized them at the beginning of the 19th century as a regular 12-hour variation in the barometric pressure. Following the solar day, they were identified as atmospheric tides (wave period of 24, 12, 8, 6 hours) and first described similarly to ocean tides (Laplace 1825). Their influence on the ionosphere was already in principle observed in the last century – before the discovery of the ionosphere – when corresponding regular variations of the geomagnetic field were observed (Schuster 1889). It was a long time, however, before these variations were attributed to ionospheric currents (a summary of contemporary knowledge at the end of the 30's is given in Chapman and Bartels 1940) and before they were quantitatively explained by Maeda (1955). Following an improved theory of tides (Chapman and Lindzen 1970) many studies showed their influence on other ionospheric parameters, in particular on the ion drift, strongly coupled to the neutral wind. Again it was the incoherent scatter technique which provided the most reliable data (Carru et al. 1967; Wand and Harper 1978). Tidal effects were also found and studied in electron and ion temperatures (Salah and Wand 1974). These studies also revealed tidal components affecting the height and the electron density of the F2-region. Advanced incoherent scatter facilities, like EISCAT (see Section 4), enabled continuous drift measurements in the lower E region where the measured ion drift is practically equal to the neutral wind because of the high ion-neutral collision frequency. With this technique tidal motions

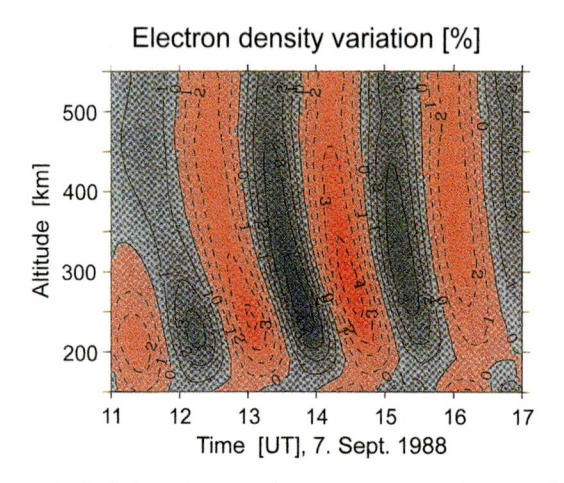

Figure 6 Relative electron density variations of a travelling ionospheric disturbance with a period of about 90 minutes. The forward slanted structures are caused by an upward propagating gravity wave. The plot was derived from incoherent scatter data (EISCAT), after background subtraction and filtering.

(e.g. Kunitake and Schlegel 1991; Virdi and Williams 1993) and their height dependence were studied. In recent years these and other measurements proved that continuous ionospheric data together with global satellite observations (e.g. UARS, McLandress *et al.* 1996) can make an important contributions to the study of tides in the upper atmosphere. Global models coupling the ionosphere and thermosphere (e.g. Fuller-Rowell *et al.* 1991) make extensive use of this data material.

A problem keeping the ionospheric researcher busy for many years was the day-to-day variability of the ionosphere. In recent years only, evidence mounted that planetary waves and their interaction with other atmospheric waves probably play a crucial role, particularly in the lower ionosphere. Planetary waves are free oscillations of the atmosphere with quasi periods of, for instance, 2 days, 5–6 days and 16 days. There is increasing evidence that non-linear interactions of the different atmospheric waves are very important for the atmospheric and in turn the ionospheric dynamics on time scales larger than one day. Effects of tidal – planetary wave interactions have been reported by Huuskonen *et al.* (1991) and Rüster (1994), the modulation of tides by AGWs by Fritts and Vincent (1987).

3.4 Forcing from above and below

It was already mentioned that Chapman invoked charged particles as a source for ionisation (see 3.1). Although some details of his ideas proved wrong, the effect of particle precipitation on the ionosphere, particularly at high latitudes became obvious after the first satellites discovered the particle environment of the earth (van Allen belts, van Allen and Frank 1959) and subsequently the magnetosphere.

After Birkeland's 'terrella' experiments (1901) and Størmers calculation of particle trajectories in the terrestrial magnetic field (1907) strong evidence existed that the aurora was caused by particles precipitating into the atmosphere, but the nature and source of these particles remained unclear. The first direct proof was delivered by means of a rocket fired directly into an aurora over Canada (Meredith *et al.* 1958). The results showed fluxes of electrons in the keV range. Subsequent rocket flights established the pattern and energy spectrum of theses auroral electrons (Hultqvist 1973) which are accelerated in the magnetosphere. It was soon realized that these particles not only created the splendid auroral forms, but provided in addition a new source of ionization, particularly in the E region. Rees (1963) developed an expression for the ion production by particle impact, which has to be included in the continuity eqn (3.6). It is a complicated expression depending on the flux of the particles, their energy, the neutral air properties, and the energy loss per ion pair formation. His calculations showed that precipitating electrons with energies of 1, 10,

100 keV created a maximal production at 160, 105, 83 km altitude. High fluxes of these particles create an auroral sporadic-E layer with electron densities much higher than those formed during the day by solar EUV radiation (Figure 3). In the eighties and nineties these auroral sporadic-E layers were extensively studied with the help of the incoherent scatter technique. In addition to ionization by auroral electrons, low energy electrons (supra thermal) also form a strong heat source for the upper ionosphere (Rees *et al.* 1983).

It was also realized that streams of particles spiraling down along a magnetic field line form the high-energy tail of the so-called Birkeland currents. Proposed by Birkeland as early as 1908, they were in fact not firmly established before the early seventies. At that time magnetic measurements on satellites showed a double sheet of field-aligned electric currents with opposite directions and magnitudes between 3×10^{-7} A/m^2 and 5×10^{-5} A/m^2 (Zmuda and Armstrong 1974). Birkeland currents are the field-aligned component of the ionospheric current system (see Section 3.2) and couple the ionosphere to the magnetosphere.

Since the conductivity along the magnetic field lines is high (because the charged particles can freely move along B) they form in fact equal potential lines thereby communicating electric fields induced in the magnetosphere by the solar wind magnetospheric dynamo to the ionosphere. These fields drive a complicated pattern of ionospheric ion drift in the high latitude ionosphere. Besides with satellites this drift pattern was again first studied with incoherent scatter radar (Banks *et al.* 1973; Foster 1983). It was already realized at that time that enhanced ion drifts have a profound influence on the ionospheric plasma. Large differences in the velocity of ions and neutrals cause frictional heating, which leads to enhanced ion and electron temperatures and ultimately also to increasing neutral temperature (St. Maurice and Hanson 1982). The enhanced plasma temperatures in turn affect many reaction rates thereby significantly changing the ion and neutral chemistry. Moreover, some reaction rates depend directly on the relative velocity of ions and neutrals. For instance the reaction

$$O^+ + N_2 \rightarrow NO^+ + N$$

is strongly enhanced during high electric fields thereby depleting the ionosphere of atomic oxygen ions in favor of nitric oxide ions which in-turn rapidly recombine with electrons (Banks *et al.* 1974). Through these two reactions the ionization of the high latitude ionosphere is effectively reduced in the course of geomagnetic storms, generally accompanied by large electric fields. The heating of the neutral upper atmosphere results in an upwelling and in turn in departures from diffusive equilibrium. This leads to a reduction in the thermospheric atomic oxygen density and to an increase of molecular nitrogen density (Prölls 1980). The expansion of the atmosphere also causes pressure gradients which modify the whole thermospheric circulation. Through ion drag the latter is also directly changed by large ion drifts. Both effects have a profound influence on the transport processes discussed above (Section 3.2). Together all these effects cause what is called an ionospheric storm. Many details of these storms have been unveiled in the last decades (see recent review by Buonsanto 1999), and are still under investigation since they are an important aspect of space weather (see Section 5). Many of these effects have been included in present-day ionospheric models describing ionospheric storm effects (Fuller-Rowell *et al.* 1994).

The forcing of the ionosphere and thermosphere from above through solar wind – magnetospheric processes is eminently important for the global energy budget of the whole atmosphere. Special efforts have therefore been undertaken to capture the global pattern of the ionospheric drifts, rather than measuring the drift very accurately at single stations with the incoherent scatter technique. This is, for instance, achieved by a chain of coherent backscatter radars surrounding the high latitudes: The SUPERDARN system presently provides the majority of data to study the ionospheric convection (Greenwald 1996) and has greatly advanced our understanding (see also Section 4).

A forcing of the ionosphere from below, i.e. from the mesosphere, stratosphere or troposphere is not so dramatic as the forcing from above, but may be similarly important. Already in the seventies it was realized that many of the gravity wave sources (severe weather, fronts, wind flowing over mountains, lee waves) are located in the troposphere (e.g. Richmond 1978). Tides and planetary waves are mainly excited in regions of high air density, too. In addition, effects of the global electric circuit on the ionosphere are discussed in recent years. The global electric circuit is driven by the global thunderstorm activity and is mainly confined to the region between the highly conducting earth and the sufficiently conducting ionosphere, called the earth – ionosphere wave guide (Volland 1984). That some energy of this circuit leaks into the ionosphere was already realized during the study of whistlers, low frequency electromagnetic waves emitted during lightning events (Storey 1953). They can interact with ionospheric structures and even cause particle precipitation (Inan *et al.* 1978). After the discovery of a new class of lightning events occurring between the top of thunderclouds and the ionosphere (sprites, blue jets, elves, Sentman and Wescott 1995) there is mounting evidence that these effects can cause transient disturbances of the ionospheric electric field and even in the electron density (Kelley *et al.* 1990). These new effects on the ionosphere are hot topics of present-day ionospheric physics. The controversial topic of earthquake-influences on the ionosphere also belongs to this field (e.g. Parrot 1995).

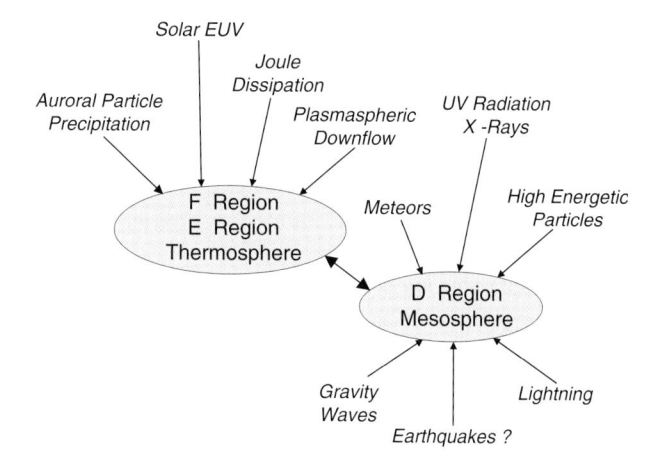

Figure 7 Schematic representation of processes, forces and radiation acting on the ionosphere-thermosphere system.

3.5 Current status

All the chemical and dynamic effects described above are taken into account in modern ionospheric models developed during the last two decades (Schunk and Sojka 1996, for a review). They solve a system of differential equations consisting of

- the continuity equation (conservation of mass, scalars):

  [Density change] = [production processes]
  $\quad$ − [loss processes] − [transport effects]

- the momentum conservation equation (vectors):

  [Acceleration per unit volume]
  $\quad$ = [force] − [drag] − [transport effects]

- the heat equation (a special form of the energy conservation equation, scalars):

  [Temperature change] = [heating processes] − [cooling processes] − [heat conduction]

The latter two equations of this system have to be solved simultaneously for the neutrals and for the ions, taking their interaction into account. Included are also many processes known to act on the ionosphere from outside (Figure 7). The mathematical treatment of these complex models was only possible after powerful computers became available.

4 IONOSPHERIC OBSERVATIONAL TECHNIQUES

4.1 Ground based radio wave propagation methods

The advances in determining the properties and understanding of the ionosphere have been driven by the development of methods to observe it. For this reason we give a brief description of some observational methods which have been of the greatest importance in the progress of ionospheric science and which have led to the status of our understanding as described in the previous two sections.

The most widely used technique for observations of the electron density in the ionosphere was, and still is, the swept frequency pulsed radar based on Breit and Tuve's (1926) original experiments. Short pulses centered on an angular frequency ω are transmitted vertically from the ground. The center frequency is then stepped through a certain frequency range, and the time delay of the reflected pulses is measured as a function of the pulse center frequency. The reflection occurs at a plasma cut-off frequency (Stix 1992) where the phase velocity of the wave approaches infinity and where the group velocity approaches zero. For a plasma without a magnetic field the cut-off frequency is given by eqn (2.1).

The presence of the magnetic field makes the plasma doubly refracting, and there will be two cut-off frequencies. The refractive index is given by the Lassen-Appleton-Hartree formula referred to in Section 2:

$$n^2 = 1 - \frac{X(U-X)}{U(U-X) - (1/2)Y^2\sin^2\theta \pm \sqrt{(1/4)Y^4\sin^4\theta + Y^2\cos^2\theta(U-X)^2}} \quad (4.1)$$

where the quantities occurring are defined by:

$$X = \frac{\omega_p^2}{\omega^2} \quad Z = \frac{\nu}{\omega}$$
$$Y = \frac{\omega_H}{\omega} \quad U = 1 - iZ \quad (4.2)$$

Here the undefined symbols are ν for the electron-neutral collision frequency and $\omega_H = eB_0/m_e$ for the angular gyro-frequency. B_0 is the geomagnetic field. For the ordinary mode the cut-off is given by $X = 1$, see eqn (2.1). Note that a true cut-off with $n^2 = 0$ only occurs when $Z = 0$. For the extraordinary wave mode it is:

$$X = 1 - \frac{\omega_H}{\omega} = 1 - Y \quad (4.3)$$

An example of an ionogram is shown in Figure 8. The ionograms which provide "virtual height", see eqn (2.2), can be inverted to give the electron density versus height relation, the so-called "true height" profiles, see also Figure 8 (Becker 1956; King and Cummack 1957; Titheridge 1959). The traces may become blurred due to irregular electron density structure, and the study of this blurring may reveal properties of the irregularities. The irregular diffraction pattern formed by interference of the many returned waves drifts over the ground at twice the velocity of the ionospheric plasma. The motions of this diffraction pattern can be used for studies of the dynamics and the structure of the electron density in the ionosphere (Briggs and Spencer 1954; Wright and Pitteway 1994; Reinisch 1996). In one model of the reflection from an

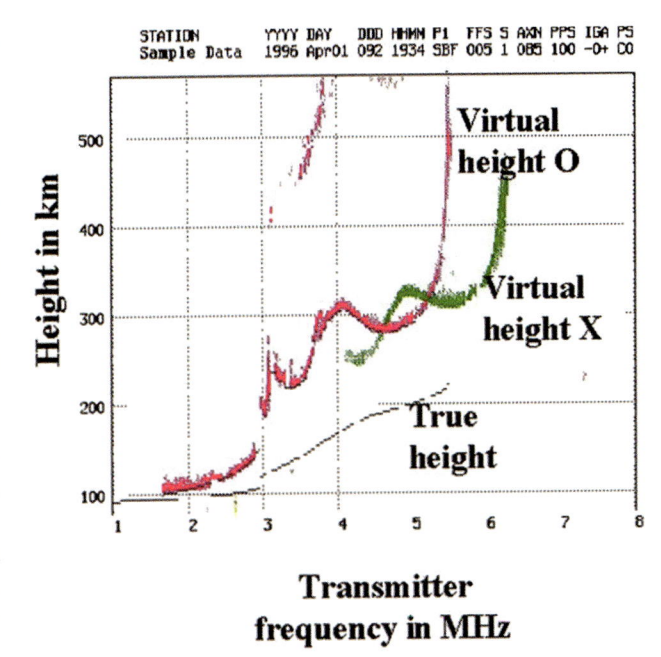

Figure 8 Example of virtual height versus frequency recorded with a modern ionosonde. The trace marked as "true height" indicates the actual height to that ionospheric level where the reflection, here of the ordinary or O-mode, takes place.

irregular ionosphere the level of constant electron density is considered to be a random function of position. The reflection properties depend on the depth of the phase modulation introduced by this diffraction screen and the angular spread of the return is determined jointly by the scale of the irregularities and the depth of the phase modulation (Booker and Clemmow 1950). A large number of ionosondes have been distributed over the globe and serve to provide a picture of the geographic distribution of electron density.

It was soon realized that the lack of a sharp cut-off makes ionosonde studies problematic in the D-region. In the partial reflection technique, introduced by Gardner and Pawsey (1953) and later improved by Belrose and Burke (1964) the scattering of radio waves by density gradients in the lower ionosphere rather then their reflection is utilized. Since the partial reflection is weak compared to total reflection strong transmitters (about 100 kW) have to be used. Another difficulty is the large antenna arrays necessary for efficiently transmitting and receiving at the required low frequencies (2–3 MHz). The electron density is computed from the ratio of the absorption coefficients for the ordinary and the extraordinary wave modes. These coefficients are given by the so-called Sen-Wyller (1960) formula, an extension of eqn (4.1) for high collision frequencies, as present at altitudes below 100 km. This technique provides reliable electron density data between about 70 and 95 km altitude and yields additionally an estimate of the electron-neutral collision frequency in this height range. Today the partial reflection technique with spaced antennas is widely used for drift measurements in the lower ionosphere/thermosphere (e.g. Manson and Meek 1984).

It was early discovered that scattering from elongated plasma density irregularities is much stronger when viewed in the direction perpendicular to the axis of elongation. Plasma density irregularities have a strong tendency to be elongated along the magnetic field, and for this reason strong back scattering from the irregular plasma shows preference for the direction normal to the magnetic field. With a VHF system, where the waves travel largely without bending, radar systems have been built in locations where the perpendicularity condition can be met at altitudes in the ionosphere where it is known that field aligned irregularities are formed, in the auroral and in the equatorial electrojets. The best known of these installations is the Scandinavian Twin Auroral Radar Experiment (STARE) operating from Malvik, Norway and Hankasalmi, Finland (Greenwald *et al*. 1978). With the two stations the measurements from overlapping areas can give two-dimensional vector drift velocities. Figure 9 shows the overlapping beam patterns of STARE used to produce two-dimensional maps of the plasma drifts over large areas. These maps can be used in studies of electric field distribution in the auroral E-region (Mazaudier *et al*. 1987) and to studies of the conductivity there (Nielsen *et al*. 1988). Similar studies have been conducted in the equatorial regions near Lima, Peru where the irregularities are associated with the equatorial electrojet and where the perpendicularity condition is met over a wide angular range in the magnetic meridian (Bowles *et al*. 1963).

At longer wavelengths, in the HF frequency range, similar enhancements of scattering from field aligned irregularities occur. In this frequency region with much smaller difference between the wave frequency and the plasma frequency there is considerable refraction and the waves are bent so as to become normal to the magnetic field in much larger spatial volumes, and in this case mostly in the F-region of the ionosphere. Chains of stations surrounding the two poles have been placed in operation under the name of Super Dual Auroral Network (SUPERDARN) and now regularly provide vector plasma velocities in polar regions (Greenwald 1996). Figure 10 shows an example of the coverage area of the northern stations in the SUPERDARN chain.

A great step forward in the observation of the ionosphere by radio waves was made with the suggestion of Gordon (1958) that modern radar systems could be made sensitive enough to detect Thomson scattering from free electrons in the ionosphere. As this technique does not depend on reflection at a cut-off frequency, it was initially thought that the density could be routinely monitored throughout the ionosphere, and not only below the maximum of the electron density. Measurement of the Doppler

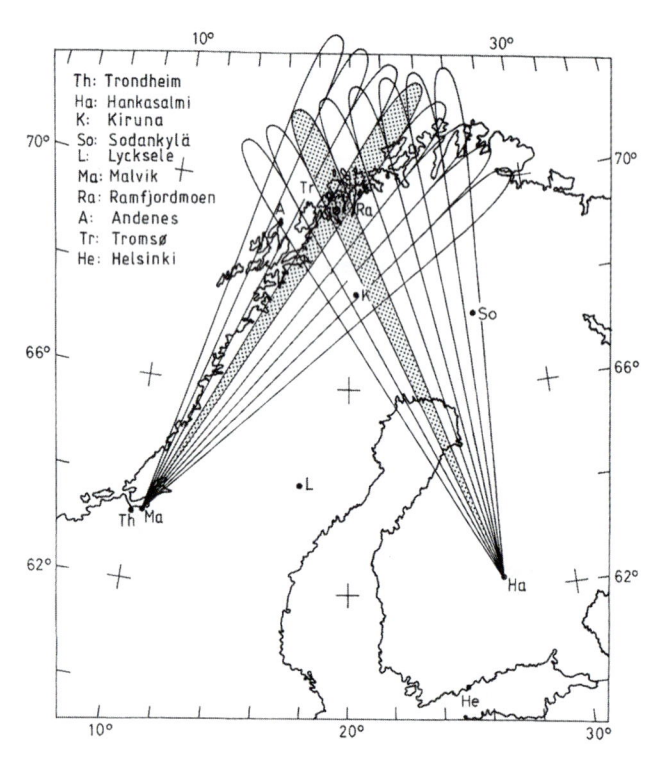

Figure 9 Plot of the overlapping multiple beam patters in the STARE VHF system which permits the measurement of two dimensional plasma drift over a large geographic area. The two beams crossing over the EISCAT area are shaded.

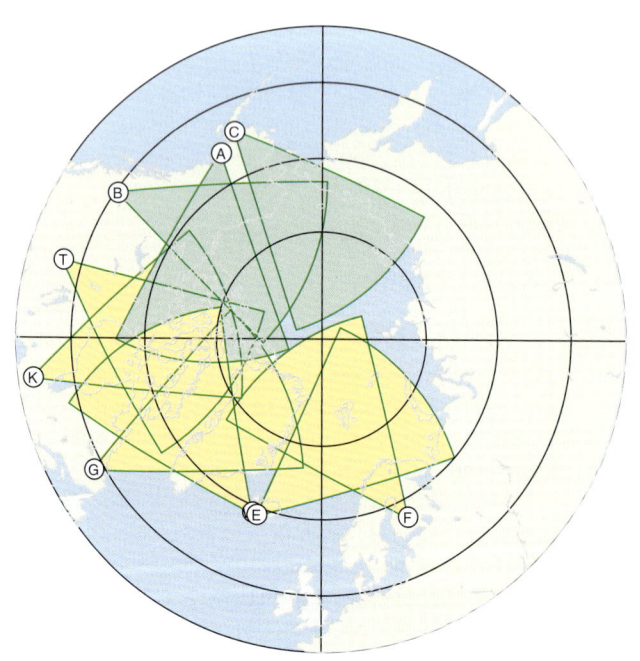

Figure 10 Diagram showing the pattern of the northern part of the SUPERDARN HF radar system which permits global coverage of the motions in the ionosphere.

spread of the returned signal, it was thought, would give information about the electron temperature. In a combination of unexpected experimental results obtained by Bowles (1958) in his first experiments, and a race by theoreticians to explain these results (Fejer 1961; Dougherty and Farley 1960; Salpeter 1960; and Hagfors 1961) the so-called incoherent scatter radar (ISR) technique was developed. What was most unexpected in the initial observations was the Doppler spread of the return, which appeared to show scattering by the ions rather than by the electrons, a physical impossibility. It turns out that the electrons are not really free, but tend to bunch together round the positive ions in "clouds" which move with the ions, and the motion of these clouds are with the ion thermal velocity, hence the appearance of the narrow spectra.

By the time this was understood two observatories had been funded, and were in the process of construction, the Jicamarca Observatory in Peru, and the Arecibo Observatory in Puerto Rico. This was the time when the United States was bent on catching up with the perceived scientific superiority of the Soviet Union as displayed by the launching of the first Sputniks. The time between the conception of an idea, and its realization was very short, and the review process now often causing long delays in scientific progress, was then minimal. The Arecibo Observatory was designed on the assumption of scattering from free electrons independent of the ions. The electrostatic forces between the electrons and the ions force the electron density to nearly completely follow the ion distribution at sufficiently large scales (scales larger than the Debye length, see below). The scattering from electrons thus bound to follow the motion of the ions will therefore possess the characteristics of the ion acoustic waves. Had the plasma physics been understood at the time, and had the review committees been allowed to ponder over the design for long, the observatory would probably have ended up with a 30 m rather than with a 300 m diameter antenna, and the world would have been without one of the world's primary tools for radio and radar astronomy in the frequency region from 50 MHz to 10 GHz.

Figure 11A shows a log-log plot of the theoretical spectra. The parameter is a measure of the ratio of the wavelength of the density fluctuations and the Debye length defined by

$$\lambda_D = \frac{v_{th}}{\omega_p} = \frac{\sqrt{k T_e/m}}{\omega_p} \qquad (4.4)$$

We see from the diagram how the spectra widen and become dominated by the electron thermal motion when the density fluctuation scale becomes smaller than the Debye length. With the radar systems used for ionospheric research and for the parameters which prevail in the ionosphere the spectra are nearly always dominated by the ions.

Inequality of electron and ion temperatures makes itself felt as a relative increase in the height of the shoulders of the spectra, and from parameter fitting routines it is possible to determine the ratio of temperatures, see also Figure 11B. When the ion population consists of two ions with distinctly different masses their relative abundance can again be determined from their spectra, see Figure 11C. In the lower part of the ionosphere collisions between ions and neutrals become important, and have a profound and predictable effect on the spectral shape, so much so that the collision frequency and the neutral density can be derived. Examples of these collision-dominated spectra are shown in Figure 11D.

A change in the ion mass has an effect in that the spectral width then changes approximately as $1/\sqrt{M_i} \propto \Delta\omega$. A bulk motion of the plasma with a velocity component in the direction of the line of sight is observed as a Doppler shift of the spectrum. The horizontal motion of the plasma is often determined by beam-swinging i.e. rotating the radar beam into different directions with a fixed elevation angle. In this case the assumption must be made that the velocity is constant, or systematically varying in some predictable fashion, over the space and the time of the beam-swinging experiment. A determination of the full three-dimensional velocity is only possible with a tri-static antenna configuration as employed in the EISCAT observatory (Rishbeth and van Eyken 1993). In regions of transitions between one type of major ion to another of different mass, the transition can be traced by studying the transition in the shape of the power spectra see Figure 11C.

The basic ISR measurements can be used to derive electron density, temperatures of electrons and ions, plasma velocities, collision frequencies and composition of the major ions. In addition a number of derived quantities can be obtained, such as electric field, neutral wind velocity, ohmic dissipation, spectrum of the energetic particle precipitation and several others. As we have seen in Sections 2 and 3 on the physical properties of the ionosphere the ISR technique has contributed in a major way to our present understanding of the physical state of the ionosphere (Evans 1969, 1974; Bauer 1975; Beynon and Williams 1978).

We have sketched some by now classical radio methods for the investigation both the dynamics and the physical state of the ionosphere. There are other new and promising methods under development to investigate ionospheric scattering structure with increasing resolution. A particularly interesting one makes use of TV transmitters in normal operation. The TV signals are received locally via a direct ray, and a scattered signal is extracted by computing a complex cross correlation with the direct signal used as a replica of the transmitted signal. The replica is then shifted in delay and offset in frequency in order to determine the scattered power as a function of time delay and Doppler

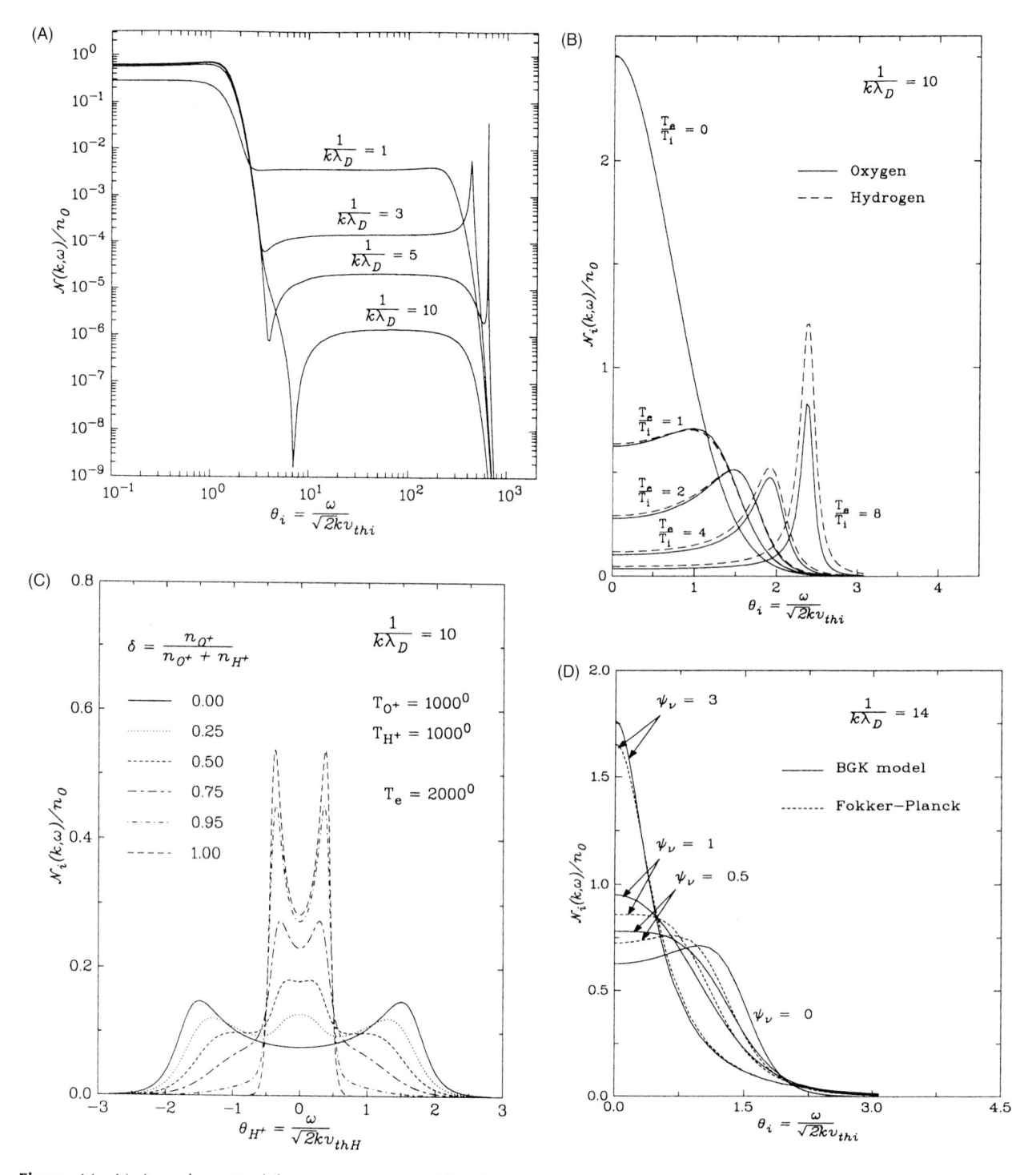

Figure 11 Various theoretical frequency spectra of incoherently scattered radio waves. The properties of these spectra find application for the determination of ionospheric parameters. The abscissa is normalized to the ion thermal speed in all panels. (A) Equilibrium spectra with $T_i = T_e$ for a range of Debye lengths showing the transition from ion-dominated to electron dominated spectra. Note also the feature due to the Langmuir oscillations when the Debye length is small. (B) Spectra showing the effect of variation of the T_e/T_i ratio for both hydrogen and oxygen ions for a fixed Debye length. (C) Spectra for a mixture of ions of different mass showing the changes in the spectral shape as the relative abundance of the two varies. (D) Variation of spectral shape associated with the changes of the ion-neutral collision frequency, allowing the neutral density to be determined.

shift (Sahr and Lind 1997). The scheme is particularly suitable for studies of irregularities. The method has not yet found wide application in ionospheric studies, but may be expected to do so in the future in the face of mounting difficulties in the allocation of transmitter frequencies and the interference from other services for the more classical methods of observations.

The use of and the importance of applying rockets as platforms for *in situ* measurements were realized immediately after World War II, and German V-2 rockets were used to carry EUV and X-ray spectrometers and pressure and temperature measuring instruments to heights of some 200 km. The consequences for our understanding of the ionosphere resulting from such measurements were described in Section 3 on ionospheric processes. We shall return to a description of some of the *in situ* measuring instruments presently.

4.2 Radio wave propagation experiments involving spacecraft

At about the time of the initial development of ISR and the construction of the first observatories to make use of this technique another event came to mark a milestone in the development of ionospheric observations, the launching of the first USSR Sputnik. At first the instrumentation was primitive, and the most important information was derived from the monitoring of continuous wave radio signals emitted by the satellites. The recording of the Doppler frequency versus time variations of these signals during satellite passes lead to accurate determination of orbital parameters, and from the change in these orbital parameters the satellite drag and the atmospheric densities could be derived (King-Hele 1963), and neutral predominant winds could even be derived (see Section 3.2).

Ionosondes, so successfully applied from the ground, were also installed in satellites and provided much new information about the structure of the topside of the ionosphere. New and interesting plasma effects were also studied in these experiments. Many new modes of plasma waves not observable from the ground have been identified. As the satellites are flying in a plasma environment both local and long distance plasma phenomena are excited (a collection of references may be found in Schmerling and Langille 1969).

Sophisticated methods for the measurement of ionospheric electron densities from radio propagation experiments were developed on the basis of the Navy Navigation Satellite (NSS) systems. These satellites transmit radio signals at two harmonically related frequencies for the purpose of removing the effects of the ionosphere in position determinations. Ionospheric investigators make use of the ability to remove ionospheric effects to obtain information about the ionosphere. For a particular angular frequency ω the phase of the signal received from a satellite beacon can be expressed as:

$$\phi = \frac{\omega}{c}\int_T^R \sqrt{1-X}\,ds + \omega t + \psi \approx \frac{\omega}{c}\int_T^R (1 - \frac{1}{2}X)\,ds + \omega t + \psi$$

$$= \frac{1}{c}\int_T^R (\omega - \frac{\omega_p^2}{2\omega})\,ds + \omega t + \psi \quad (4.5)$$

where X was defined in eqn (2.1), and where the integration is carried out between the satellite transmitter (T) and the receiver (R), where ψ is a phase constant and where it is assumed that the plasma frequency is much lower than the operating frequency. Obtaining the phase at the two operating frequencies and dividing by these frequencies we obtain a quantity which depends on the integrated electron density along the path:

$$\Phi = \frac{\phi_1}{\omega_1} - \frac{\phi_2}{\omega_2} = \text{const}\left(\frac{1}{\omega_2^2} - \frac{1}{\omega_1^2}\right)\int_T^R N_e\,ds + \psi_1 - \psi_2 \quad (4.6)$$

This integral provides what is called the total electron content (TEC) of the ionosphere, a quantity routinely derived at many stations world wide during the last two decenniums.

Observing these phase-terms at several different ground stations it is possible to obtain the crossing paths required to obtain two-dimensional cuts through the ionosphere necessary to derive the electron density distribution over large areas. Kunitsyn and his coworkers have particularly been active in perfecting such tomographic mapping of the ionosphere (see Kunitsyn *et al.* 1995, and the review by Leitinger 1996). Figure 12 shows an example of two consecutive reconstructions of the ionospheric profiles as a function of latitude and height along a path between Murmansk and Moscow. The method is well suited for mapping the behavior of the topside ionosphere in regions such as the polar cusp or the equatorial anomaly. Use of these satellite signals to study plasma density irregularities have also produced interesting results relating to both their amplitude and shape (Basu *et al.* 1988).

A remotely related method of observing the ionospheric electron density profile is to use occultation between satellites in navigation systems such as GPS or GLONAS (Jakowksi 1996). Such data might arise as a by-product of occultation observations of the neutral atmosphere, particularly with a view to monitor the water vapor content. Occultation methods have found wide application in studies of planetary atmospheres (Fjeldbo and Eshleman 1969), they have some promise for future work and bear a great potential for a world-wide mapping of the Earth's ionosphere particularly in combination with the tomographic techniques mentioned above.

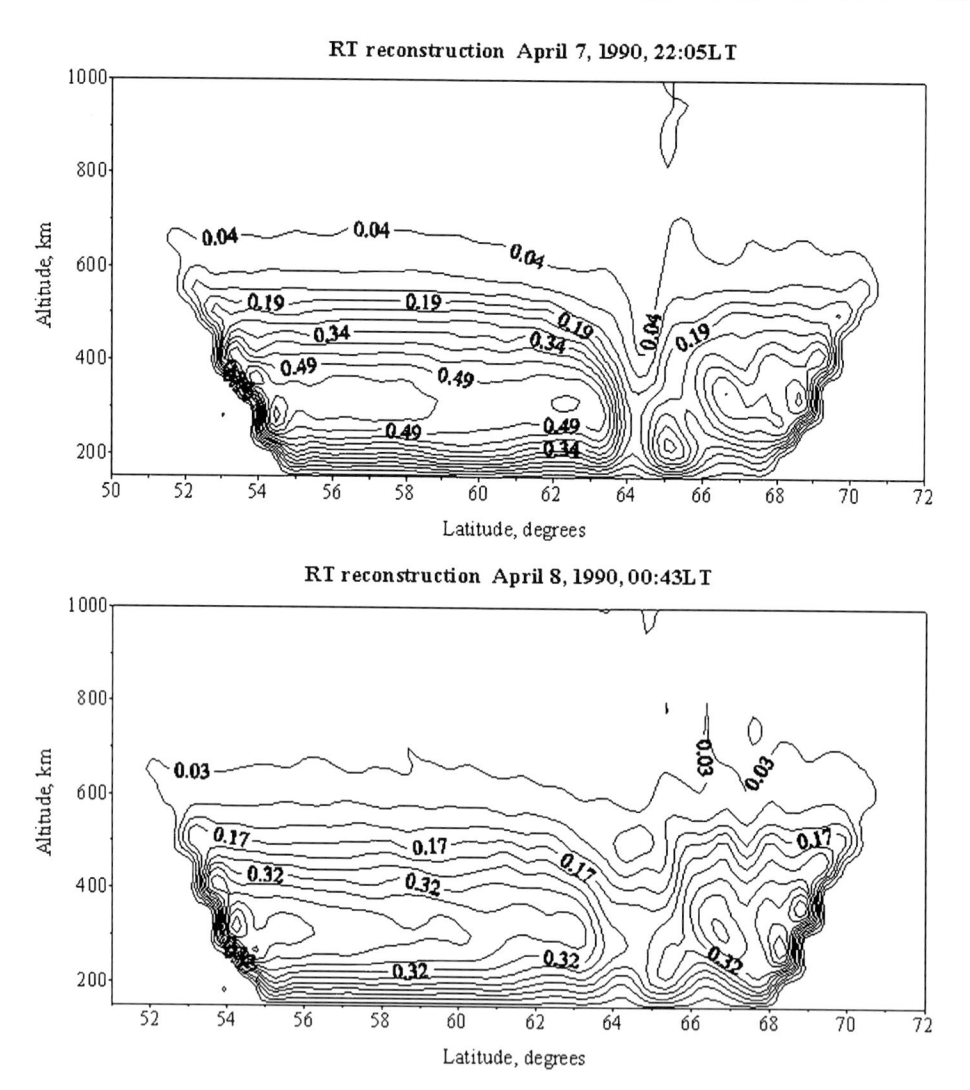

Figure 12 Electron density contours obtained with the radio tomography technique, 2 h 38 min apart. The contours are labeled in units of 10^{12} m^{-3}. The corresponding satellite signals have been received with a chain of three receivers between Murmansk and Moscow. This new method combines excellent height resolution, a large latitudinal and altitude coverage and a reasonable temporal resolution. The high density gradients at the northern and southern boundary of the field of view are artifacts of the reconstruction and have to be discarded (from V. Kunitsyn, private communication).

4.3 *In situ* measurements

The observational methods relying on remote observations are still developing and remain at the cutting edge of the scientific exploration of the ionosphere. However, when small scale phenomena need to be investigated, be they in electron density, temperature, composition, particularly related to trace gasses, or in electric and magnetic fields, charged particle velocity distribution, *in situ* measurements are indispensable.

Measurements of DC electric fields have been carried out in double probes both in satellites and in rockets. The double probe has a dipole-like structure, and the basic observation is the voltage between the two probes. As a rule the observing platform is rotating and the actual field must be determined by transformation first to a non-rotating frame of reference traveling with the platform, then by taking into account the electric field $\mathbf{V}_{s/c} \times \mathbf{B}$ due to the spacecraft motion in the geomagnetic field (e.g. Maynard 1996). The double probe technique does not only give the DC electric field, but is used also to measure the AC electric fields of electrostatic plasma waves up to a few kHz (Rinnert 1992). Another *in situ* measurement of a DC electric field is carried out by launching beams of charged particles perpendicular to the geomagnetic field and observe the amount of beam displacement during a gyro period. The drift velocity thus obtained is related to the electric field through the relation $|V_D| = E/B$. Such an experiment was flown successfully on the Freja satellite (Paschmann et al. 1994) and is also on board the Cluster mission.

Measurements of magnetic fields, of importance in studies of currents flowing in the ionosphere, are measured either by flux-gate magnetometers or by proton precession or alkali-vapor instruments. The former class of instruments can measure both magnitude and direction of the magnetic field, the latter provide an accurate value of the scalar field. The measured magnetic fields are related to the currents through Ampère's Law: $\nabla \times \mathbf{B} = \mu_0 \mathbf{J}$. The currents are not uniquely determined by the measured magnetic field. Additional assumptions must be made about the spatial and the temporal properties of the magnetic field. Magnetic field observations have been made on NASA's MAGSAT, and will be among the primary measurement parameter on the Danish ØERSTED and the German CHAMP satellites (Primdahl 1996).

Plasma waves most often cannot propagate away from the ionosphere and must be detected *in situ*. Electrostatic plasma waves and plasma structures are important because they transfer and dissipate free energy, modulate current systems, modify plasma gradients, scatter charged particles. They can only be observed indirectly on the ground after conversion to electromagnetic waves. Plasma waves in the ionosphere range in frequency from below 1 Hz to over 10 MHz. The most commonly used probe for the measurement of both density and temperature is the Langmuir probe. The probe assumes a voltage determined by the condition that the ion and the electron currents are equal. This voltage is given by:

$$V_f = -\frac{kT_e}{e}\ln\left(\frac{M_i}{m}\right)^{1/2} \quad (4.7)$$

The information about the physical state of the plasma is derived from the ion saturation region, from the electron retardation region and from the electron saturation region. These regions are characterized by:

$$\text{Ion saturation region}: I = \frac{NAeV_i}{\pi}\left(1 + \frac{kT_i}{M_iV_i^2} + \frac{2eV}{M_iV_i^2}\right)^{1/2}$$

$$\text{Electron retardation region}: I = NAe\left(\frac{kT_e}{2\pi m}\right)^{1/2}\exp\left(\frac{eV}{kT_e}\right) \quad (4.8)$$

$$\text{Electron saturation region}: I = \frac{NAe}{\pi}\left(\frac{kT_e}{m}\right)^{1/2}\left(1 + \frac{eV}{kT_e}\right)^{1/2}$$

Here A is the effective area of the probe and V_i the velocity of the plasma with respect to the spacecraft. Analysis of the data from such probes has provided useful data on electron temperature and density, and even on the ion mass. Great care must, however, be exercised in order to obtain reliable data. The geometry with respect to the velocity vector of the space craft, the probe surface material, the probe measurements in relation to the Debye length all must be taken into account for correct results to be obtained. For this reason incoherent scatter results and Langmuir probe measurements have often disagreed by significant amounts. A description of the intricacies of Langmuir probe measurements has recently been given by Brace (1996).

Other probes designed to explore the properties of the ionosphere depend on the measurement of the impedance of the probe when it is surrounded by plasma. The impedance is measured as a function of frequency and a number of resonances observed as a function of the frequency of the applied signal. Both the plasma temperature and the local plasma density can be determined this way. Advantages with impedance probes are that they provide good time resolution and that the plasma density can be measured independently of the satellite potential and the properties of the probe surface (Jensen and Baker 1992).

Ion drift, ion density and ion temperature have been derived using retarding potential analyzers. These analyzers collect ions through an aperture followed by a set of grids biased to different voltages thus passing ions through according to energy. The flow direction can also be determined by splitting the collector into different sectors and comparing the currents in the opposite portions of the detector. These devices have been described by Heelis and Hanson (1996) and in references in this paper.

The ion composition can be measured by ISR, but only the major ions and only in a few locations near one of the ISR observatories. *In situ* observations from satellites and rockets are therefore indispensable in ionospheric investigations. An early type of mass spectrometer making use of an electric filter is the quadrupole mass spectrometer described by Arnold and Viggiano (1986). The spectrometer consists of four parallel bars arranged in a quadrupole geometry. DC and AC voltages are applied to opposite bars. A charged particle traveling along the direction of the bar will only be passed for certain combinations of voltages and frequencies. As an example we show in Figure 13 the ion composition of the D and lower E region, measured by such an instrument (Steinweg et al. 1992). The threshold ion density was $10^{-6}\,\mathrm{m^{-3}}$ in this case. The height range covered here is particularly interesting because it contains a layer of enhanced metallic ion density between about 85 and 100 km. The density of Fe^+ can reach the same order of magnitude as the density of the major ion NO^+ at night. The metallic ions stem from meteors vaporized in this height region. Since their recombination rate ($\approx 10^{-18}\,\mathrm{m^3\,s^{-1}}$) is much lower than that of the gas ions, metallic ions have a long life time in the D region. Their abundance is therefore controlled mainly by dynamic processes, like wind and turbulence which are important at these heights. The influence of

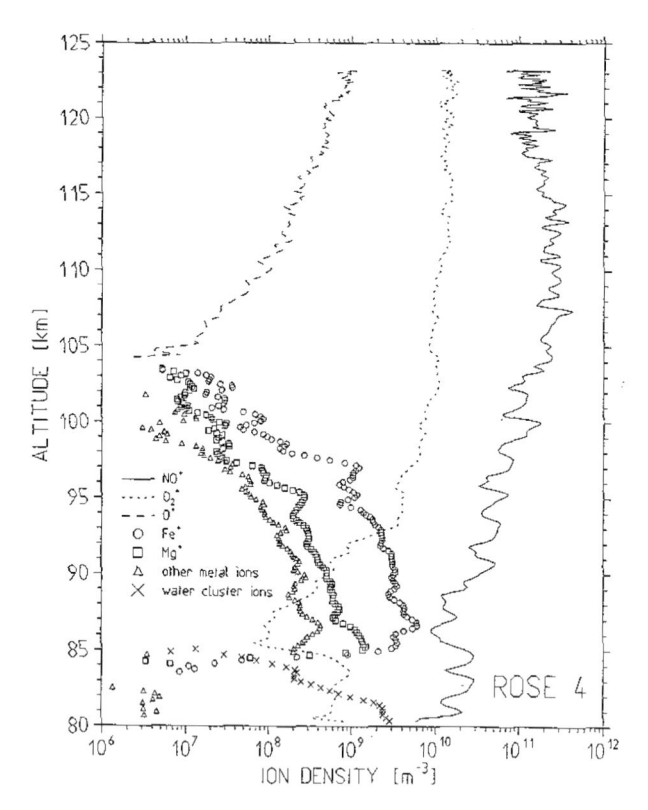

Figure 13 Height profiles of major and minor ions of the D and lower E region measured during the ROSE rocket project (Steinweg et al. 1992) above Kiruna/Sweden on Feb. 9, 1989 at 23:42 UT. Whereas the E region above 105 km is dominated by NO^+, O_2^+, and O^+, metallic ions of meteoric origin form a layer between 85 and 100 km ('other ions' contain mainly Na^+ and Ca^+). Below 85 km water cluster ions, $H^+(H_2O)_n$ are important.

metallic ions on sporadic E-layers was already mentioned (Section 2.3). Figure 13 also shows the density of water cluster ions of the form $H^+(H_2O)_n$ which are crucial for the ion chemistry of the lower D region (e.g. Burns et al. 1991).

For high time resolution measurement of the mass of the ions spectrometers with magnetic deflection have also found application. These instruments may be heavy because of the magnets required for the deflection, and there are also often problems with stray magnetic fields affecting other instruments onboard the spacecraft (Hoffman et al. 1973).

One very challenging problem in the study of charged particles in the ionosphere is to measure the tails in the distribution functions, tails representing deviations from the Maxwellian velocity distributions of the bulk of the particles. In the case of the electrons supra thermal tails in the distribution functions can under certain circumstances be studied by observing the "plasma lines" in the incoherent scatter spectrum corresponding to Langmuir waves. Energetic velocity tails can also be derived by studies of the D-region electron density profiles when these are controlled by energetic precipitating particles. For *in situ* observations the standard method of measurement is by means of retarding potential analyzers as mentioned above. The grid is then biased to accept only particles with energies above the bias level. Maier et al. (1980), Whalen et al. (1994) and Moore et al. (1995) have described such analyzers and their application to supra thermal particle measurements. For particles of relatively low energy in the range of 0.05 eV to a few eV such measurements are extremely difficult because of the satellite potential, modification of the distribution function by the geomagnetic field and because of photo emission from the spacecraft itself.

At higher particle energies, say 5 eV to 40 keV, some form of electrostatic plate analyzer is most often applied to determine both the energy spectrum and the angular distribution of velocities. These analyzers can either be of a cylindrical shape or a hemispherical shape, so called "top hat" analyzers. In these analyzers a potential difference is applied between the two plates, and by proper geometrical design particles entering within a narrow cone and having a particular energy can be brought to a focus at the output and detected by a channel electron multiplier. Such multipliers may have amplification factors as high as 10^7. Because of the large energy range, the high time resolution and the ease of programming them, the electrostatic analyzers have become one of the most versatile instruments for many *in situ* experiments (Carlson and McFadden 1996).

The electrostatic plate analyzer as such does not discriminate between different particle masses. Such discrimination can be achieved by adding a "time of flight" instrument. These instruments discriminate mass/charge ratios on the basis of the velocity acquired by a particle after being accelerated in an electrostatic potential (Wilken and Stüdemann 1984).

Other instruments apply magnetic deflectors after the electrostatic plate analyzer for this mass discrimination (Eliasson et al. 1994). For energies above 40 keV a plethora of instruments have found applications ranging from Geiger-Müller counters to solid-state detectors. As these higher energies are of greater interest in magneto-spheric studies than in ionospheric studies we shall not go into further detail here.

Measurements of the composition of the neutral gas usually relies on ionizing the gas and applying some form of mass spectroscopy of the ionized gas (for discussions of performance of some of these, see von Zahn 1974 and Offerman et al. 1981). Direct measurements of neutral wind and temperature have also been developed and are described by Hanson and Heelis (1975) and by Spencer et al. (1981). Measurement of the neutral wind can also be carried out by "coloring" the gas with a chemical such as strontium, lithium or tri-methyl-aluminum, depending on the height under investigation. The luminous trails formed after the release of these gases are traced by optical triangulation, and the wind vector derived (Larsen et al. 1995).

5 IMPACTS ON ASPECTS OF MODERN SOCIETY

Beginning with Marconi's telegraphy experiments (see Section 2) long distance communication depended nearly exclusively on ionospheric radio wave propagation. Communication circuits were set up between fixed points for telephone communication and for data transfer. Ionospheric radio wave propagation dominated broadcasting and communication with ships and airplanes.

The dominance of ionospheric radio wave propagation in long distance communication came to an end with the introduction of satellite communication. During the early 1960's satellites could be placed in geostationary orbits to serve as relay points for point to point communication at much higher frequencies and with much higher bandwidths than could be used on ionospheric paths. Satellite communication has in turn been placed in competition with optical fiber communication with even greater capability, but that belongs to a different story. Suffice it to say that the relative importance of the ionosphere as a communication medium is greatly diminished.

Rather than serving as a medium for communication, the ionosphere has become an impediment for satellite communication and for satellite navigation. In the former case ionospheric irregularities can cause enough stray scattering of the point to point signals that serious interference and multipath fading effects can arise. These irregularities most often occur in the auroral zones and near the equator. It is clearly important, therefore, to understand the ionospheric behavior in order to take evasive measures when the level of irregular structure is high. In the latter case the ionosphere can introduce extra phase delays, see Section 4, in particular eqn (4.5), and these phase delays can lead to inaccurate position determinations unless the ionospheric effects can be properly accounted for. Over the horizon radars used for early detection of missiles, airplanes and ships can also be seriously affected by changes in the ionospheric structure, and an understanding and monitoring of the ionosphere becomes important.

In case of very accurate position requirements in the cm range for instance for geodetic purposes not only irregularities but also the actual ionospheric influence has to be taken into account. If the model electron density often applied in the phase path calculations (eqn (4.5)) deviates too much from the actual density, systematic errors are introduced which limit the accuracy. The developments to use the signal of navigation satellites itself to determine the total electron content along the ray path (see Section 4.2) provides an iterative method to overcome this problem.

A number of satellite systems are in low orbit round the Earth. The various satellites monitoring environmental conditions are in low orbit because the observations are easier when the distance to the surface of the Earth is not so great. Observatory satellites such as that carrying the Hubble telescope are in low orbit because the instruments may from time to time have to be maintained by astronauts. The space stations MIR and the International Space Station (ISS) can only be assembled and serviced in low orbits. Satellites in low orbits are exposed to atmospheric drag requiring active correction to keep them on their intended track. The atmospheric density changes with temperature, largely under control by EUV radiation. Less predictable are the temperature changes caused by the ohmic heating due to the currents flowing in the ionosphere particularly near the equator and in the auroral zones. The monitoring of ionospheric conditions provides information on the atmospheric temperature and leads to an assessment of ionospheric drag. This is of great importance when orbit corrections must be made, and when planning new satellite missions.

Coronal mass ejection on the sun is well known to lead to enhanced particle precipitation in the Earth's vicinity, and to hazard for instrumentation onboard satellites, and the astronauts during extravehicular activity. Mass ejection can be monitored by spacecraft such as SOHO in near earth orbit, or from ground based observatories. The association of mass ejection and the association with particle precipitation are not a straightforward matter. The arrival of the particles in the vicinity of the Earth depends on the configuration of the interplanetary magnetic field, and cannot be predicted with certainty. The high-energy particles arriving near the Earth cause ionospheric effects, particularly enhanced ionization in the lower D-region. Studies of these ionization enhancements and their association with solar flares are, therefore, of interest for the understanding of the association of flares and particle precipitation.

The development of continent-wide distribution of electric power both in Europe and the Americas has brought with it very long transmission lines. The electric currents set up in the ionosphere can induce large currents in the power grid. These currents have from time to time caused damage to the distribution net, for instance by destroying transformer stations and causing blackouts. The best example is the Hydro-Quebec catastrophic failure in October 1989 due to geomagnetically induced currents (GIC). This failure caused damage in the $3 billion to $6 billion range, comparable to the damage caused by a major natural disaster such as Hurricane Hugo. Preventive measures can be taken both in design of circuits and in rerouting the currents in case of problem. Monitoring of ionospheric currents can help bring such measures about.

All these examples show that ionospheric investigations are still important, particular on a global scale. Therefore world-wide efforts and scientific programs have been launched to study the ionosphere as a link in the chain of solar wind – magnetosphere – ionosphere – thermosphere coupling. Nearly all industrialized nations participate in these efforts or have set up their own national programs in this field. Ionospheric studies in this framework of space weather will therefore continue in the future.

REFERENCES

Appleton, E.V. and Barnett, M.A.F. (1925). Local reflection of wireless waves from the upper atmosphere. *Nature*, **115**, 333.

Appleton, E.V. (1927). The existence of more than one ionized layer in the upper atmosphere. *Nature*, **120**, 331.

Appleton, E.V. (1932). Wireless studies of the ionosphere. *J. Inst. Elec. Engrs.* **71**, 642.

Appleton, E.V. and Naismith, R. (1932). Some measurements of the upper atmosphere ionization, *Proceedings of the Royal Society of London*, **A137**, 36.

Appleton, E.V., Naismith, R. and Builder, G. (1933). Ionospheric investigations in high latitudes, *Nature*, **132**, 340.

Appleton, E.V. and Naismith, R. (1940). Normal and abnormal region-E ionization. *Proceedings of the Physical Society of London*, **52**, 402.

Appleton, E.V. (1947). Geomagnetic control of F2-layer ionization. *Science*, **106**, 17.

Arnold, F. and Viggiano, A.A. (1986). Review of rocket-borne ion mass spectrometry in the middle atmosphere. In: R. Goldberg (ed.), *Handbook of MAP*, 19, ICSU Scientific Committee on Solar-Terrestr. Relations.

Axford, W.I. (1963). The formation and vertical movement of dense ionized layers in the ionosphere, due to vertical wind shears. *Journal of Geophysical Research*, **68**, 769.

Bailey, V.A. and Martyn, D.F. (1934). The influence of electric waves on the ionosphere. *Phil. Mag.* **18**, 369.

Bailey, V.A. (1938). Generation of auroras by means of radio waves. *Nature*, **142**, 613.

Baker, W.G. and Martyn, D.F. (1953). Electric currents in the ionosphere. I. Conductivity. *Philosophical Transactions of the Royal Society*, **A246**, 281.

Banks, P.M., Doupnik, J.R. and Akasofu, S.I. (1973). Electric field observations by incoherent scatter radar in the auroral zone. *Journal of Geophysical Research*, **78**, 6607.

Banks, P.M., Schunk, R.W. and Raitt, W.J. (1974). NO^+ and O^+ in the high latitude F region. *Geophysical Research Letters*, **1**, 239.

Basu, Su, Sa Basu, MacKenzie, E., Fougere, P.F., Coley, W.R.N., Maynard, C., Winningham, J.D., Sugiura, M., Hanson, W.B. and Hoegy, W.R. (1988). Simultaneous density and electric field fluctuation spectra associated with velocity shears in the auroral oval. *Journal of Geophysical Research*, **93**, 115.

Bates, D.R., Buckingham, R.A., Massey, H.S.W. and Unwin, J.J. (1939). Dissociation, recombination and attachment processes in the upper atmosphere. II. The rate of recombination. *Proceedings of the Royal Society of London*, **A170**, 322.

Bates, D.R. and Massey, H.S.W. (1947). The basic reactions in the upper atmosphere. *Proceedings of the Royal Society of London*, **A192**, 1.

Bates, D.R. (1974). Radiative and collision processes in the ionosphere. *Journal of Atmospheric Terrestrial Physics*, **36**, 2287.

Bauer, P. (1975). Theory of waves incoherently scattered. *Philosophical Transactions of the Royal Society*, **A280**, 167.

Becker, W. (1956). Über ein Verfahren zur routinemässigen Bestimmung der wahren Verteilung der Elektronendichte in einer Ionosphärenschicht aus den Durchdrehaufnahmen dieser Schicht. *Arch. Elektr. Übertr.* **10**, 207.

Belrose, J.S. and Burke, M.J. (1964). Study of the lower ionosphere using partial reflection. *Journal of Geophysical Research*, **69**, 2799.

Berkner, L.V. and Wells, H.W. (1934). Report of ionosphere investigations at the Huancayo magnetic observatory during 1933. *Proc. IRE* **22**, 1102.

Berkner, L.V. and Wells, H.W. (1938). Non-seasonal change of F2-region ion density. *Journal of Geophysical Research*, **43**, 15.

Best, J.E., Ratcliffe, J.A. and Wilkes, M.V. (1936). Experimental investigations of very long waves reflected from the ionosphere. *Proceedings of the Royal Society of London*, **A156**, 614.

Beynon, W.J.G. and Williams, P.J.S. (1978). Incoherent scatter of radio waves from the ionosphere. *Rep. Prog. Phys.* **41**, 909.

Birkeland, K. (1901). Expédition Norvégienne de 1899–1900 pour l'étude des aurores boréales. *Vid. Selsk. Skr. Math. Nat. Kl.* No 1.

Booker, H.G. and Clemmow, P.C. (1950). The concept of an angular spectrum of plane waves and its relation to that of polar diagram and aperture distribution. *Proceedings of the Institute of Electrical and Electronic Engineers*, **97**, 401.

Booker, H.G., Ratcliffe, J.A. and Shinn, D.H. (1950). Diffraction from an irregular screen with applications to ionospheric problems. *Philosophical Transactions of the Royal Society*, **242**, 579.

Bowles, K.L. (1958). Observation of vertical incidence scatter from the ionosphere at 41 Mc/s. *Physical Review Letters*, **1**, 454.

Bowles, K.L., Balsley, B.B. and Cohen, R. (1963). Field aligned E region irregularities identified with acoustic plasma waves. *Journal of Geophysical Research*, **68**, 2485.

Brace, L.H. (1996). Langmuir probe measurements in the ionosphere. In: R. Pfaff, J.E. Borovsky and D.T. Young (eds.), *Measurement Techniques in Space Plasmas*, American, Geophys. Union, Washington DC, p. 23.

Breit, G. and Tuve, M.A. (1926). A test of the existence of the conducting layer. *Physical Review*, **28**, 554.

Briggs, B.H. and Spencer, M. (1954). Horizontal movements in the ionosphere. *Reports on Progress in Physics*, **17**, 245.

Buneman, O. (1963). Excitation of field-aligned sound waves by electron streams. *Physical Review Letters*, **10**, 285.

Buonsanto, M.J. (1999). Ionospheric storms – a review. *Space Science Reviews*, **88**, 563.

Burns, C.J., Turunen, E., Matveinen, H., Ranta, H. and Hargreaves, J.K. (1991). Chemical modelling of the quiet summer D and E regions using EISCAT electron density profiles. *Journal of Atmospheric Terrestrial Physics*, **53**, 115.

Carlson, C.W. and McFadden, J.P. (1996). Design and applications of imaging plasma instruments. In: R. Pfaff, J.E. Borovsky and D.T. Young (eds.), *Measurement Techniques in Space Plasmas*, American Geophys. Union, Washington DC, p. 125.

Carpenter, D.L. and Park, C.G. (1963). Whistler evidence of a 'knee' in the magnetospheric ionization density profile. *Journal of Geophysical Research*, **68**, 1675.

Carru, H., Petit, M. and et Waldteufel, P. (1967). Mise en évidence de mouvements dans ionosphère au moyen de la diffusion incohérente, *Comptes Rendues de l'Académie des Sciences*, **264**, 560.

Chapman, S. (1931a). The absorption and dissociative or ionizing effect of monochromatic radiation in an atmosphere on a rotating earth. *Proceedings of the Physical Society of London*, **43**, 26.

Chapman, S. (1931b). Some phenomena of the upper atmosphere, *Proceedings of the Royal Society of London*, **A132**, 353.

Chapman, S. and Bartels, J. (1940). *Geomagnetism*, Vol. II, Chapter XXIII, Oxford.

Chapman, S.C. and Lindzen, R.S. (1970). *Atmospheric Tides*, D Reidel, Dordrecht.

Cho, J.Y.N. and Röttger, J. (1997). An updated review of polar mesospheric summer echoes: Observations, theory, and their relationship to noctilucent clouds and subvisible aerosols. *Journal of Geophysical Research*, **192**, 2001.

CIRA (1961). *Cospar International Reference Atmosphere*, North-Holland, Amsterdam.

Cowling, T.G. (1945). The electrical conductivity of an ionized gas in a magnetic field, with applications to the solar atmosphere and the ionosphere. *Proceedings of the Royal Society of London*, **A183**, 453.

Dellinger, J.H. (1937). Sudden ionospheric disturbances, *Journal of Geophysical Research*, **42**, 49.

Dougherty, J.P. and Farley, D.T. (1960). A theory of incoherent scattering of radio waves by a plasma. *Proceedings of the Royal Society of London*, **A259**, 79.

DuBois, D.F., Rose, H.A. and Russell, D. (1990). Excitation of strong Langmuir turbulence in plasmas near critical density: application to heating of the ionosphere. *Journal of Geophysical Research*, **95**, 21221.

Ecklund, W.L. and Balsley, D.B. (1981). Long-term observations of the Arctic mesosphere with the MST radar at Poker Flat, Alaska. *Journal of Geophysical Research*, **86**, 8775.

Ecklund, W.L., Carter, D.A. and Balsley, B.B. (1981). Gradient drift irregularities in midlatitude sporadic E. *Journal of Geophysical Research*, **86**, 858.

Eliasson, L., Norberg, O., Lundin, R., Lundin, K., Olsen, S., Borg, H., André, A., Koskinen, H., Rihelä, P., Boehm, M. and Whalen, B. (1994). The Freja hot plasma experiment – instrument and first results. *Space Science Reviews*, **70**, 563.

Evans, J.V. (1969). Theory and practice of ionosphere study by Thomson scatter radar, *Proc. IEEE* **57**, 496.

Evans, J.V. (1974). Some post-war developments in ground-based radiowave sounding of the ionosphere. *Journal of Atmospheric Terrestrial Physics*, **36**, 2183.

Farley, D.T. (1963). A plasma instability resulting in field-aligned irregularities in the ionosphere. *Journal of Geophysical Research*, **68**, 6083.

Ferraro, V.C.A. (1945). Diffusion of ions in the ionosphere. *Journal of Geophysical Research*, **50**, 215.

Fejer, J.A. (1955). The interaction of pulsed radio waves in the ionosphere. *Journal of Atmospheric Terrestrial Physics*, **7**, 322.

Fejer, J.A. (1961). Scattering of radio waves by an ionized gas in thermal equilibrium in the presence of a uniform magnetic field. *Canadian Journal of Physics*, **39**, 716.

Fejer, B.G. and Kelley, M.C. (1980). Ionospheric irregularities. *Reviews of Geophysics*, **18**, 401.

Fejer, B.G. (1996). Natural ionospheric plasma waves. In: H. Kohl, R. Rüster and K. Schlegel (eds.), *Modern Ionospheric Science*, European Geophysical Society, Katlenburg-Lindau, Germany, p. 216.

Fjeldbo, G. and Eshleman, V.R. (1969). The atmosphere of Venus as studied with the Mariner 5 dual radio-frequency occultation experiment. *Radio Science*, **4**, 879.

Foster, J.C. (1983). An empirical electric field model derived from Chatanika radar data. *Journal of Geophysical Research*, **88**, 981.

Fritts, D.C. and Vincent, R.A. (1987). Mesospheric momentum flux studies at Adelaide, Australia: Observations and a gravity-wave-tidal interaction model. *Journal of Atmospheric Science*, **44**, 605.

Fuller-Rowell, T.J., Rees, D., Parish, H.F., Virdi, T.S., Williams, P.J.S. and Johnson, R.G. (1991). Lower Thermosphere Coupling Study: Comparison of observations with predictions of the UCL-Sheffield Thermosphere-Ionosphere model. *Journal of Geophysical Research*, **96**, 1181.

Fuller-Rowell, T.J., Codrescu, M.V., Moffet, R.J. and Quegan, S. (1994). Response of the thermosphere and ionosphere to geomagnetic storms. *Journal of Geophysical Research*, **99**, 3893.

Gardner, F.F. and Pawsey, J.L. (1953). Study of the ionospheric D-region using partial reflections. *Journal of Atmospheric Terrestrial Physics*, **3**, 321.

Gordon, W.E. (1958). Incoherent scatter of radio waves by free electrons with applications to space exploration by radar. *Proc. IRE* **46**, 1824.

Greenwald, R.A., Weiss, W., Nielsen, E. and Thomson, N.R. (1978). STARE: A new radar auroral backscatter experiment in northern Scandinavia. *Radio Science*, **13**, 1021.

Greenwald, R.A. (1996). The role of coherent radars in ionospheric and magnetospheric research. In: H. Kohl, R. Rüster and K. Schlegel (eds.), *Modern Ionospheric Science*, European Geophysical Society, Katlenburg-Lindau, Germany, p. 391.

Gringauz, K.I., Kurt, V.G., Moroz, V.I. and Shklovsky, I. (1961). Results of observations of charged particles observed out to 100,000 km with the aid of charged particle traps on Soviet space probes. *Soviet Astronomy Journal*, **4**, 680.

Gurevich, A.V., Borisov, D.N. and Zybin, K.P. (1996). Ionospheric E-region turbulence induced by the turbulence of the neutral atmosphere. In: K. Schlegel (ed.), *Plasma Instabilities in the Ionospheric E-Region*, Cuvillier Verlag, Göttingen, Germany, p. 1.

Hanson, W.B. and Heelis, R.A. (1975). Techniques for measuring bulk gas-motions from satellites. *Space Sci. Instr.*, **1**, 493.

Haerendel, G. (1973). Theory of equatorial spread-F, Report, Max-Planck-Inst. für Extraterr. Physik, München, Germany.

Hagfors, T. (1961). Density fluctuations in a plasma in a magnetic field, with applications to the ionosphere. *Journal of Geophysical Research*, **66**, 1699.

Harang, L. and Stoffregen, W. (1938). Scattered reflections of radio waves from a height of more than 100 km. *Nature*, **142**, 832.

Hartree, D.R. (1931). The propagation of electromagnetic waves in a refracting medium in a magnetic field. *Proc. Camb. Phil. Soc.* **27**, 143.

Heaviside, O. (1902). Telegraphy, *Encycl. Brit.* 10th Ed, Vol **33**, 213.

Hedin, A.E. (1987). MSIS-86 thermospheric model. *Journal of Geophysical Research*, **92**, 4649.

Heelis, R.A. and Hanson, W.B. (1996). Measurement of thermal ion drift velocity and temperature using planar sensors. In: R. Pfaff, J.E. Borovsky and D.T. Young (eds.), *Measurement Techniques in Space Plasmas*, American Geophys. Union, Washington DC, p. 61.

Henderson, J.T. (1933). Measurements of ionization in the Kenelly-Heaviside layer during the solar eclipse of 1932. *Canadian Journal of Research*, **8**, 1.

Hewish, A., Bell, S.J., Pilkington, J.D.H., Scott, P.F. and Collins R.A. (1968). Observation of a rapidly pulsating radio source. *Nature*, **217**, 709.

Hines, C.O. (1960). Internal atmospheric gravity waves at ionospheric heights. *Can. J. Phys.*, **38**, 1441.

Hinteregger, H.E. (1970). The extreme ultraviolet solar spectrum and its variation during a solar cycle, *Ann. Géophys.*, **26**, 547.

Hoffman, J.H., Hanson, W.B., Lippincott, C.R. and Ferguson, E.E. (1973). The magnetic ion mass spectrometer on Atmsopheric Explorer. *Radio Science*, **8**, 315.

Hollingworth, J. (1926). The propagation of radio waves. *J. Inst. Electr. Engrs.* **64**, 579.

Hultqvist, B. (1969). Polar cap absorption and ground level effects. In: C. de Jager and Z. Svestka (ed.), *Solar Flares and Space Res.*, Amsterdam, p. 215.

Hultqvist, B. (1973). Auroral particles. In: A. Egeland, O. Holter and A. Omholt (eds.), *Cosmical Geophysics*, Oslo, Norway, p. 161.

Huuskonen, A., Virdi, T.S., Jones, G.O.L. and Williams, P.J.S. (1991). Observations of day-to-day variability in the meridional semi-diurnal tide at 70° north. *Annales Geophysicae*, **9**, 407.

Inan, U.S., Bell, T.F. and Helliwell, R.A. (1978). Nonlinear pitch angle scattering of energetic electrons by coherent VLF waves in the magnetosphere. *Journal of Geophysical Research*, **83**, 3235.

Jacchia, L.G. (1959). Two atmospheric effects in the orbital acceleration of artificial satellites. *Nature*, **183**, 526.

Jakowski, N. (1996). TEC monitoring by using satellite positioning systems. In: H. Kohl, R. Rüster and K. Schlegel (eds.), *Modern Ionospheric Science*, European Geophysical Society, Katlenburg-Lindau, Germany, p. 371.

Jensen, M.D. and Baker, K.D. (1992). Measuring ionospheric electron density using the plasma frequency probe. *Journal of Spacecraft and Rockets*, **29**, 91.

Johnson, C.Y., Heppner, J.P., Holmes, J.C. and Meadows, E.B. (1958). Results obtained with rocket-borne ion spectrometers. *Annales Geophysicae*, **14**, 475.

Kaiser, T.R. (1962). The first suggestion of an ionosphere. *Journal of Atmospheric Terrestrial Physics*, **24**, 865.

Kelley, M.C., Ding, J.G. and Holzworth, R.H. (1990). Intense ionospheric electric and magnetic field pulses generated by lightning. *Geophysical Research Letters*, **17**, 2221.

Kelley, M.C. and Hysell, D.L. (1991). Equatorial spread-F and neutral atmospheric turbulence: a review and a comparative anatomy. *Journal of Atmospheric Terrestrial Physics*, **53**, 695.

Kennelly, A.E. (1902). On the elevation of the electrically-conducting strata of the Earth's atmosphere. *Electr. World Engin.* **39**, 473.

King, G.A.M. and Cummack, C.H. (1957). Electron distribution in the ionosphere. *Journal of Atmospheric Terrestrial Physics*, **8**, 270.

King, J.W. and Kohl, H. (1965). Upper atmosphere winds and ionospheric drifts caused by neutral air pressure gradients. *Nature*, **206**, 699.

King-Hele, D.G. (1963). Improved formula for determining upper-atmosphere density from the change in a satellite's orbital period. *Planetary and Space Science*, **11**, 261.

Kirchengast, G., Hocke, K. and Schlegel, K. (1996). The gravity-wave-TID relationship: Insight via theoretical model-EISCAT data comparison. *Journal of Atmospheric Terrestrial Physics*, **58**, 233.

Klostermeyer, J. (1969). Gravity waves in the F region. *Journal of Atmospheric Terrestrial Physics*, **31**, 25.

Klostermeyer, J. (1972). Numerical calculations of gravity wave propagation in a realistic thermosphere. *Journal of Atmospheric Terrestrial Physics*, **34**, 765.

Kosch, M.J., Rietveld, M.T. and Hagfors, T. (2000). High latitude HF-induced air-glow displaced equator wards of the pump beam. *Geophys. Res. Lett.*, **27**, 2817.

Kunitake, M. and Schlegel, K. (1991). Neutral winds in the lower thermosphere at high latitudes from five years of EISCAT data. *Annales Geophysicae*, **9**, 143.

Kunitsyn, V.E., Tereshchenko, E.D., Andreeva, O.G., Khudukon, B.Z. and Melnichenko, Y.A. (1995). Radiotomographic investigations of ionospheric structures at auroral and middle latitudes. *Annales Geophysicae*, **13**, 1254.

Laplace, P.S. (1825). *Mécanique Céleste*, Paris.

Larmour, J. (1924). Why wireless electric rays can bend round the Earth. *Nature*, **114**, 650.

Larsen, M.F., Marshall, T.R., Mikkelsen, I.S., Emry, B.A., Christensen, A., Kayser, D., Hecht, J., Lions, L. and Waterscheid, R. (1995). Atmospheric response in aurora experiment: observations of E and F neutral winds in a region of postmidnight diffuse aurora. *Journal of Geophysical Research*, **100**, 17299.

Lassen, H. (1926). Über die Ionisation der Atmosphäre, und ihren Einfluß auf die Ausbreitung der kurzen elektrischen Wellen in der drahtlosen Telegraphie, II. *Z. Hochfrequenztechn*, **28**, 139.

Lassen, H. (1927). Über den Einfluss des Erdmagnetfeldes auf die Fortpflanzung der electrischen Wellen der drahtlosen Telegraphie in der Atmosphäre. *Elektrische Nachrichten-Technik*, **4**, 324–334.

Lemaire, J.F. and Gringauz, K.I. (1998). *The Earth's Plasmasphere*, Cambridge University Press, Cambridge.

Lenard, P. (1911). Über die Absorption der Nordlichtstrahlen in der Erdatmosphäre, Sitz. Ber. Heidelberger Akad. Wiss. 2A, 12.

Leitinger, R. (1996). Tomographie. In: H. Kohl, R. Rüster and K. Schlegel (eds.), *Modern Ionospheric Science*, European Geophysical Society, Katlenburg-Lindau, Germany, p. 346.

MacDonald, H.M. (1903). The bending of electric waves round a conducting obstacle: Amended results. *Proceedings of the Royal Society of London*, **A72**, 59.

Maeda, K. (1955). Researches on the geomagnetic distortion in the ionosphere. II Theoretical study on the geomagnetic distortion in the F2 layer. *Rep. Ionosph. Res. Japan* 9, 71.

Maeda, H. (1955). Horizontal wind systems in the ionospheric E region deduced from the dynamo theory of the geomagnetic Sq variation. *J. Geomag. Geoelectr.*, **7**, 121.

Maeda, K., Tsuda, T. and Maeda, H. (1963). Theoretical interpretation of the equatorial sporadic E layers. *Physical Review Letters*, **11**, 406.

Maier, E.J., Kayser, S.E., Burrows, J.R. and Klumpar, D.M. (1980). The suprathermal electron contributions to high-latitude Birkeland currents. *Journal of Geophysical Research*, **85**, 2003.

Manson, A.H. and Meek, C.E. (1984). Winds and tidal oscillations in the upper middle atmosphere at Saskatoon (52°N, 107°W, L=4.3) during the year June 1982–May 1983, *Planetary and Space Science*, **32**, 1087.

Maris, H.B. and Hulbert, E.O. (1928). Note on the ultraviolet light of the Sun as the origin of aurorae and magnetic storms. *Terrestinal Magnetism and Atmospheric Electricity*, **33**, 229.

Martyn, D.F. and Pulley, O.O. (1936). The temperature and constituents of the upper atmosphere. *Proceedings of the Royal Society of London*, **A154**, 455.

Martyn, D.F. (1947). Atmospheric tides in the ionosphere. I. Solar tides in the F2 region. *Proceedings of the Royal Society of London*, **A189**, 241.

Martyn, D.F. (1948). Atmospheric tides in the ionosphere. IV. Studies of the solar tide, and the location of the regions producing the diurnal magnetic variations. *Proceedings of the Royal Society of London*, **A194**, 445.

Martyn, D.F. (1953). Electric currents in the ionosphere III Ionization drift due to winds and electric fields. *Philosophical Transactions of the Royal Society*, **A246**, 306.

Maynard, N.C. (1996). Electric field measurements in moderate to high density space plasmas with passive double probes. In: R. Pfaff, J.E. Borovksy and D.T. Young (Eds.), *Measurement Techniques in Space Plasmas-Fields*, American Geophys. Union, Washington DC, p. 13.

Mazaudier, C., Senior, C. and Nielsen, E. (1987). Global convection electric field and current: Comparison between model's prediction and data from STARE, Saint-Santin, and magnetometers. *Journal of Geophysical Research*, **92**, 5991.

McLandress, C., Shepherd, G.G. and Solheim, B.H. (1996). Satellite-observations of thermospheric tides – Results from the WINDII Imaging Interferometer on UARS. *Journal of Geophysical Research*, **101**, 4093.

Meredith, L.H., Davis, L.R., Heppner, J.P. and Berg, O.E. (1958). IGY Rocket Rept, Ser. No. **1**, 169.

Mikhailov, A.V. and Schlegel, K. (1997). Self-consistent modelling of the daytime electron density profile in the ionospheric F region. *Annales Geophysicae*, **15**, 314.

Milward, G.H., Quegan, S., Moffet, R.J., Fuller-Rowell, T.J. and Rees, D. (1993). A modelling study of the coupled ionosphere and thermospheric response to an enhanced high-latitude electric field event, *Planetary Space Science*, **41**, 45.

Mögel, H. (1931). Über die Beziehungen zwischen Störungen des Kurzwellenempfangs und der erdmagnetischen Störungen. *Z. Geophys.* 7, 207.

Moore, T.E., Chappell, C.R., Chandler, M.O., Fields, S.A., Pollock, C.J., Reasoner, D.L., Young, D.T., Burch, J.L., Eaker, N., Waiter J.H., Jr, McComas, D.J., Nordholdt, J.E., Thomsen, M.F., Berthelier, J.J. and Robson, R. (1995). The thermal ion dynamics experiment and plasma source instrument. *Space Science Reviews*, **71**, 409.

Munroe, G.H. (1948). Short-period changes in the F region of the ionosphere. *Nature*, **162**, 886.

Narcisi, R.S., Baily, D.A. and Della L. Lucca (1967). Composition measurements of sporadic-E in the nighttime lower ionosphere. *Space Res VII*, Amsterdam, p. 123.

Nielsen, E., Senior, C. and Lühr, H. (1988). Ionospheric Hall conductivity deduced from ground-based measurements. *Journal of Geophysical Research*, **93**, 4125.

Offermann, D., Friedrich, V., Ross, P. and von Zahn, U. (1981). Neutral gas composition measurements between 80 and 120 km. *Planetary and Space Science*, **24**, 747.

Parrot, M. (1995). Electromagnetic noise due to earthquakes. In: H. Volland (ed.), *Atmospheric Electrodynamics*, CRC Press, Boca Raton, Vol. 2, p. 95.

Paschmann, G., Böhm, M., Höfner, H., Frenzel, R., Parigger, P., Melzner, F. Haerendel, G., Klitzing, A., Torbert, R.B. and Sartori, G. (1994). The electron beam instrument (F6) on Feja. *Space Science Reviews*, **70**, 447.

Poincare, H. (1903). Sur la diffraction des ondes electriques: À propos d'un article de M. MacDonald. *Proceedings of the Royal Society of London*, **A72**, 42.

Primdahl, F. (1996). Scalar magnetometers for space applications. In: R. Pfaff, J.E. Borovsky and D.T. Young (eds.), *Measurement Techniques in Space Plasmas-Fields*, American Geophys. Union, Washington DC, p. 85.

Prölls, G. (1980). Magnetic storm associated perturbations of the upper atmosphere: Recent results obtained by satellite-borne gas analyzers. *Rev. Geophys. Space Sci.* **18**, 183.

Prölls, G., Brace, L.H., Mayer, H.G., Carignan, G.R., Killen, T.L. and Klobuchar, J.A. (1991). Ionospheric storm effects at subauroral latitudes: A case study. *Journal of Geophysical Research*, **96**, 1275.

Ratcliffe, J.A. (1948). Diffraction from the ionosphere and the fading of radio waves. *Nature*, **162**, 9.

Ratcliffe, J.A., Schmerling, E.R., Setty, C.S.G.K. and Thomas, J.O. (1956). The rates of production and loss of electrons in the F region of the ionosphere. *Philosophical Transactions of the Royal Society*, **248**, 621.

Rayleigh and Lord (1903). On the bending of waves round a spherical obstacle. *Proceedings of the Royal Society of London*, **A72**, 40.

Rees, M.H. (1963). Auroral ionization and excitation by incident energetic electrons. *Planetary and Space Science*, **11**, 1209.

Rees, M.H., Emery, B.A., Roble, R.G. and Stamnes, K. (1983). Neutral and ion gas heating by auroral electron precipitation. *Journal of Geophysical Research*, **88**, 6289.

Reinisch, B.W. (1996). Modern Ionosondes. In: H. Kohl, R. Rüster and K. Schlegel (Eds.), *Modern Ionospheric Science*, European Geophysical Society, Katlenburg-Lindau, Germany, p. 440.

Richmond, A.D. (1978). Gravity wave generation, propagation and dissipation in the thermosphere. *Journal of Geophysical Research*, **83**, 4131.

Richmond, A.D. and Matsushita, S. (1975). Thermospheric response to a magnetic substorm. *Journal of Geophysical Research*, **80**, 2839.

Rinnert, K. (1992). Plasma waves observed in the auroral E region – ROSE campaign. *Journal of Atmospheric Terrestrial Physics*, **54**, 683.

Rishbeth, H. and Barron, D.W. (1960). Equilibrium electron distributions in the ionospheric F2-layer. *Journal of Atmospheric Terrestrial Physics*, **18**, 234.

Rishbeth, H. (1968). On explaining the behaviour of the ionospheric F region. *Reviews of Geophysics*, **6**, 33.

Rishbeth, H. and van Eyken, A.P. (1993). EISCAT: Early history and the first ten years of operation. *Journal of Atmospheric Terrestrial Physics*, **55**, 525.

Rocket Panel (1952). Pressures, densities, and temperatures in the upper atmosphere. *Physical Review*, **88**, 1027.

Rüster, R. (1994). VHF radar observations of non-linear interactions in the summer polar mesosphere. *Journal of Atmospheric Terrestrial Physics*, **56**, 1298.

Sahr, J.D. and Lind, F.D. (1997). The Manastash Ridge radar: a passive bistatic radar for upper atmospheric radio science. *Radio Science*, **32**, 2345.

Salah, J.E. and Wand, R.H. (1974). Tides in the temperature of the lower thermosphere at mid latitudes. *Journal of Geophysical Research*, **79**, 4295.

Salpeter, E.E. (1960). Electron density fluctuations in a plasma. *Physical Review*, **120**, 1528.

Schlegel, K. and St-Maurice, J.-P. (1981). Anomalous heating of the polar E region by unstable plasma waves, I Observations. *Journal of Geophysical Research*, **86**, 1447.

Schlegel, K. (1986). The study of tides and gravity waves with the help of field-aligned velocities measured by EISCAT. *Journal of Atmospheric Terrestrial Physics*, **48**, 879.

Schlegel, K. and Haldoupis, C. (1994). Observation of the modified two-stream plasma instability in the midlatitude E region ionosphere. *Journal of Geophysical Research*, **99**, 6219.

Schmerling, E.R. and Langille, R.C. (1969). (Guest Editors) Special Issue on Topside Sounding and the Ionosphere. *Proceedings of the IEEE* 57, Number 6.

Schunk, R.W. and Soika, J.J. (1996). Ionospheric models, In: H. Kohl, R. Rüster and K. Schlegel (eds.), *Modern Ionospheric Science*, European Geophysical Society, Katlenburg-Lindau, Germany, p. 181.

Schuster, A. (1889). The diurnal variation of terrestrial magnetism. *Philosophical Transactions of the Royal Society*, **A180**, 467.

Sen, H.K. and Wyller, A.A. (1960). On the generalisation of the Appleton-Hartree magnetoionic formulas. *Journal of Geophysical Research*, **65**, 3931.

Sentman, D. and Wescott, E. (1995). Red sprites and blue jets: Thunderstorm-excited optical emissions in the stratosphere, mesosphere and ionosphere. *Physics of Plasmas*, **2**, 2514.

Simon, A. (1963). Instability of a partly ionized plasma in crossed electric and magnetic fields. *Physics of Fluids*, **6**, 382.

Smith, E.K. (1957). World wide occurrence of sporadic E. *National Bureau of Standards Circ.*, 582.

Sommerfeld, A. (1909). Über die Ausbreitung der Wellen in der drahtlosen Telegraphie. *Ann. Phys.*, **28**, 665.

Spencer, N.W., Wharton, L.E., Niemann, H.B., Hedin, A.E., Garignan, G.R. and Maurer, J.C. (1981). The Dynamics Explorer wind and temperature spectrometer. *Space Sci. Instr.*, **5**, 417.

Steinweg, A.D., Krankowsky, D., Lämmerzahl, P. and Anweiler, B. (1992). Metal ion layers in the auroral lower E region measured by mass spectrometers. *Journal of Atmospheric Terrestrial Physics*, **54**, 703.

Stewart, B. (1882). Hypothetical views regarding the connexion between the state of the sun and terrestrial magnetism. *Encycl. Brit.* 9th Ed, Vol **16**, 181.

Stix, T.H. (1992). *Waves in Plasmas*, American Institute of Physics, New York.

St Maurice, J.-P. and Hanson, W.B. (1982). Ion frictional heating at high latitudes and its possible use for an *in situ* determination of neutral thermospheric winds and temperatures. *Journal of Geophysical Research*, **87**, 7580.

Störmer, C. (1907). Sur les trajectoires des corpuscules électrisés dans l'espace sous l'action du magnétisme terrestre avec application aux aurores boréales. *Arch. Sci. Phys. Nat.*, Genéve 4, period 5.

Storey, L.R.O. (1953). An investigation of whistling atmospherics. *Philosophical Transactions of the Royal Society*, **A246**, 113.

Stubbe, P., Kohl, H. and Rietveld, M.T. (1992). Langmuir turbulence and ionospheric modification. *Journal of Geophysical Research*, **97**, 6285.

Taylor, H.A., Brace, L.H., Brinton, H.C. and Smith, C.R. (1963). Direct measurements of helium and hydrogen ion concentration and total ion density to an altitude of 940 kilometer. *Journal of Geophysical Research*, **68**, 5339.

Testud, J. and et Vasseur, G. (1969). Ondes de gravité dans le thermosphère. *Annales Geophysicae*, **25**, 525.

Titheridge, J.E. (1959). The calculation of real and virtual heights of reflection in the ionosphere. *Journal of Atmospheric Terrestrial Physics*, **17**, 96.

Tsunoda, R.T., Livingstone, R.C., McClure, J.P. and Hanson, W.B. (1982). Equatorial plasma bubbles: vertically elongated wedges from the bottomside F layer. *Journal of Geophysical Research*, **87**, 9171.

Tuve, M.A. and Breit, G. (1925). Note on a radio method of estimating the height of the conducting layer. *Journal of Geophysical Research*, **30**, 15.

Utlaut, W.F. (1970). An ionospheric modification experiment using very high power, high frequency transmission. *Journal of Geophysical Research*, **75**, 6402.

Van Allen, J.A. and Frank, L.A. (1959). Radiation around the Earth to a radial distance of 107,400 km. *Nature*, **183**, 430.

Virdi, T.S. and Williams, P.J.S. (1993). Altitude variation in the amplitude and phase of tidal oscillations at high latitude. *Journal of Atmospheric Terrestrial Physics*, **55**, 697.

Volland, H. (1984). Atmospheric electrodynamics. In: L.J. Lanzerotti (ed.), *Physics and Chemistry in Space 11*, Springer, Berlin.

von Zahn, U. (1974). Composition studies in the thermosphere by means of mass spectrometers. In: F. Verniani (ed.), *Structure and Dynamics of the Upper Atmosphere*, Amsterdam, p. 389.

Wand, R.H. and Harper, R.H. (1978). Co-ordinated tidal observations at Arecibo. *Journal of Atmospheric Terrestrial Physics*, **40**, 887.

Whalen, B.A., Knudsen, D.J., Jau, A.W., Pilon, A.M., Cameron, T.A., Sebesta, J.F., McEwen, D.J., Koehler, J.A., Lloyd, N.D., Pocobelli, G., Laframboise, J.G., Li, W., Lundin, R., Eliasson, L., Watanabe, S. and Campbell, G.S. (1994). The Freja F3C cold plasma analyzer. *Space Science Reviews*, **70**, 541.

Whitehead, J.D. (1961). The formation of the sporadic E layer in the temperate zones. *Journal of Atmospheric Terrestrial Physics*, **20**, 49.

Wilken, B. and Stüdemann, W. (1984). A compact time-of-flight mass spectrometer with electrostatic mirrors. *Nuclear Instruments and Methods in Physics Research*, **222**, 587.

Williams, P.J.S., Crowley, G., Schlegel, K., Virdi, T.S., Mc Crea, I.W., Watkins, G., Wade, N., Hargreaves, J.K., Lachlan-Cope, T., Muller, H., Baldwin, J.E., Warner, P., van Eyken, A.P., Hapgood, M.A. and Roger, A.S. (1988). The generation and propagation of atmospheric gravity waves observed during the Worldwide Acoustic-Gravity wave Study (WAGS). *Journal of Atmospheric Terrestrial Physics*, **50**, 323.

Williams, P.J.S., Virdi, T.S., Lewis, R.V., Lester, M., McCrea, I.W. and Freeman, K.S.C. (1993). Worldwide Atmospheric Gravity-wave Study in the European sector 1985–90. *Journal of Atmospheric Terrestrial Physics*, **55**, 683.

Wright, J.W. and Pitteway, M.L.V. (1994). High-resolution velocity determinations from the dynasonde. *Journal of Atmospheric Terrestrial Physics*, **56**, 961.

Yonezawa, T. (1956). A new theory of formation of the F2 layer. *Journal of the Radio Research Laboratory*, **3**, 1.

Zenneck, J. (1907). Über die Fortpflanzung ebener elektromagnetischer Wellen längs einer ebenen Leiterfläche und ihre Beziehung zur drahtlosen Telegraphie. *Annals of Physics*, **23**, 846.

Zmuda, A.J. and Armstrong, J.C. (1974). The diurnal flow pattern of field-aligned currents. *Journal of Geophysical Research*, **79**, 4611.

JOHNNY A. JOHANNESSEN*, STEIN SANDVEN* AND
DOMINIQUE DURAND*

Oceanography

1 INTRODUCTION

At the onset of this new millennium increased awareness of the stresses being placed on the Earth system, often induced by human activities, has intensified the need for information on the present state of the Earth system and for enhanced capability to assess its evolution such as associated with environmental pollution, natural resource management, sustainable development, and global climate change.

This realisation has resulted in increased political and legal obligations on governments and on national and regional agencies to address Earth system topics of global concern. These obligations are often encapsulated within international treaties, whose signatories have explicit requirements placed upon them.

Many of these treaties call for systematic observations of the Earth to increase our understanding of its processes and our ability to monitor them:

- The UN Framework Convention on Climate Change (FCCC).
- The UN Convention to Combat Desertification in those Countries experiencing Serious Drought and/or Desertification.
- The Montreal Protocol of the Vienna Convention on the Protection of the Ozone Layer.

- Agenda 21 and the UN Commission on Sustainable Development.
- The Intergovernmental Panel on Climate Change.

These commitments require substantial economic, technical and scientific resources for their execution, and action at many levels, including significant programmes of global observations. In this context it is recognised that Earth observation satellites provide an important and unique source of information. The most well established international forum for coordinating the operational provision of data is without doubt the World Weather Watch (WWW) of the World Meteorological Organisation (WMO). Another prominent, though less established forum is the Global Ocean Observing System (GOOS). Common for these international forum and observing system is the role of satellite Earth Observation. Here it is worth mentioning the World Climate Research Program (WCRP), International Geopshere and Biosphere Program (IGBP) and Intergovermental Oceanographic Committee (IOC) in which research projects and observing systems highlight the importance of continuous and regular access to Earth Observation data.

Earth observations from satellite are highly complementary to those collected by *in-situ* systems. Whereas *in-situ* measurements are necessary for underwater observations, for high accuracy local observations, for the calibration of observations made by satellite and as input to numerical models, satellite observations provide an inherent wide area unique capability to obtain regular quantitative information of surface variables and upper layer phenomena at global, regional and local scales.

---

*Nansen Environmental and Remote Sensing Center, Bergen, Norway

Present-day applications of satellite data are widespread and cover research, operational and commercial activities. These activities are of interest in the global context and in the regional, national, and local context where Earth observation data are successfully applied in support of a range of different sectors, including (not exclusive):

- climate change research,
- stratospheric chemistry, particularly related to the ozone hole,
- weather forecasts based on Numerical Weather Prediction (NWP),
- agriculture and forestry services,
- resource mapping,
- hazard monitoring and disaster assessment,
- sea ice monitoring,
- coastal zone management,
- oceanographic applications.

Details of the outstanding scientific advances made possible by satellite observations of the ocean and the associated societal benefits are provided by Halpern (2000).

The number of Earth Observing satellites are growing rapidly for both scientific research and operational application within fields of land, atmosphere, marine meteorology and oceanography including sea ice covered regions. International investment in satellite platforms, instruments and associated ground segments is already substantial, and more investment is planned over the coming decade. There are currently over 45 missions operating, and around 70 more missions, carrying over 230 instruments, planned for operation during the next 15 years by the world's civil space agencies (CEOS 97). In addition, Space Agencies are currently implementing their new strategy for defining future Earth Observing satellites, both dedicated research missions and continuous (operational) monitoring missions.

In this article we will review the status of satellite oceanography at the onset of the new millennium. In Section 2 the principal methods, instruments and basic measured surface parameters are addressed. Examples of contribution to climate monitoring and operational oceanography are then given in Section 3 and Section 4, respectively. In Section 5 an outlook towards the near future satellite observing system is provided followed by a summary in Section 6.

2 PRINCIPAL METHODS, INSTRUMENTS AND SURFACE CHARACTERISTICS

Satellite oceanography is primarily using three domains within the electromagnetic (EM) spectrum, notably radiation in: – the visible/near-infrared (VNIR); – the thermal infrared (TIR); – and the microwave bands of the EM spectrum (Figure 1).

The visible and infrared channels utilize intervals of the EM spectrum with high atmospheric transmission, such as in the bands from 0.4–2.5 μm, 3.5–4.0 μm and 10–13 μm (Figure 1, top). The EM waves in these bands do not generally penetrate clouds, so remote-sensing observations of the Earth's surface in these bands can only be done satisfactorily under cloud free conditions. As such this is posing severe limitations in regions where clouds are frequently present. In the microwave area, on the other hand, at

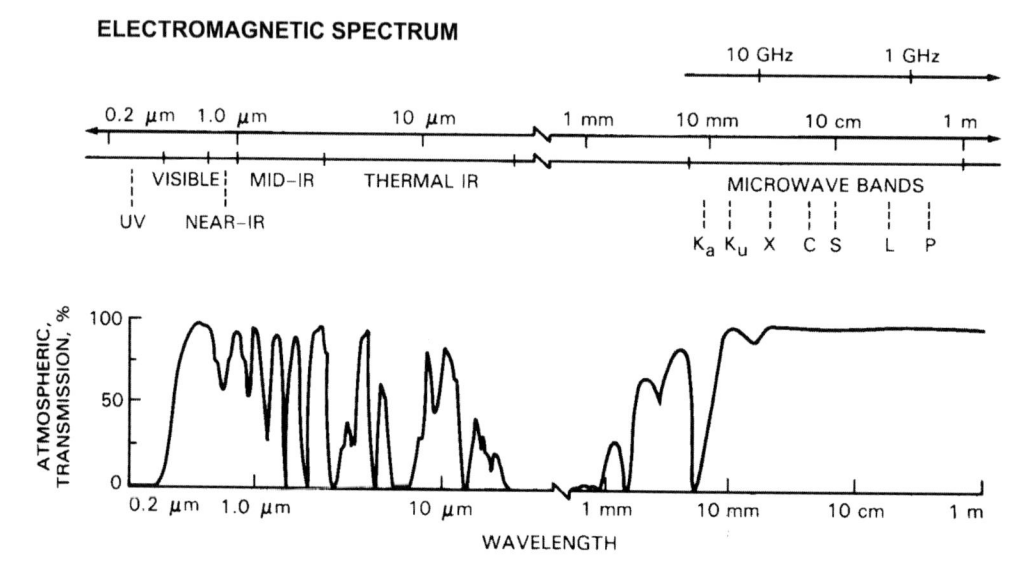

Figure 1 The electromagnetic (EM) spectrum showing the bands used in remote sensing together with the operating area for some sensors (upper graph). The atmospheric transmission of the EM spectrum is shown in the lower graph. Note that the operating areas for satellite oceanography are located in parts of the spectrum where atmospheric transmission is high.

wavelengths above 0.3 cm, EM waves generally penetrate clouds which makes it feasible to obtain regular, daily observations of ocean and sea ice surfaces (Figure 1, bottom). These characteristic spectral domains are further addressed in the following.

2.1 Visible/near infrared

The basic quantity observed in the VNIR domain is the albedo or alternatively the fraction of the incident sunlight that has been scattered and/or reflected in the atmosphere/ocean system. The incident solar radiation undergoes a number of interactions (absorption and scattering) with molecules and particles in the atmosphere and in the water, in addition to the reflection that occurs at the air – sea interface.

Only a fraction of the incident radiation penetrates the water body. Absorption by water molecules becomes critical at wavelength greater than 700 nm. Therefore water appears black at such wavelength, except when a high load of suspended particles is present near the surface. The visible light (400–700 nm) may propagate in the water medium and interact with water molecules, organic and inorganic particles in suspension, dissolved optically active substances, and possibly the sea-floor in optically shallow-waters. The penetration depth depends upon the wavelength and the water column absorption properties. Only a small fraction of the visible-light spectrum is scattered upwards to the surface giving rise to the so-called water-leaving radiance, L_w. This is expressed as:

$$L_w = (L_s - L_a - \alpha L_r)/\alpha \qquad (2.1)$$

where L_s is the radiance reaching the sensor, L_a is the atmospheric radiance, L_r is reflected radiation from the sea surface and α is the atmospheric diffuse transmittance. In most oceanic waters, L_w represents less than 10% of the total signal measured by a spaceborne sensor. Typically, 90% of radiation has been scattered in the atmosphere, without interaction with subsurface waters. Therefore any quantitative estimation of water-column optically active constituents depends upon an adequate retrieval of the water-leaving radiance L_w emanating from the water column. This requires reliable correction of the remotely sensed signal scattered in the atmosphere and reflected at the air–sea interface (Gordon, 1997).

Furthermore the capability of deriving the accurate quantity of a particular water quality parameter depends upon the complexity of the water column in terms of number and properties of components that interact with the electromagnetic signal. According to the optical complexity of the water body, two types of water are defined, namely case I and case II waters (Morel and Prieur, 1977).

Case I waters are natural open-ocean water bodies, for which water-leaving radiance measured by remote sensors are only dependent on chlorophyll pigment concentration. In such waters, which represent about 90% of the World ocean, the variation in the color of the upper water column could be related to the variation of the concentration of chlorophyll pigments contained in phytoplankton cells. The color of oceanic case I waters shifts from deep blue in oligotrophic waters (very low chlorophyll concentration) to dark green in eutrophic waters (high concentration). This shift results from the strong absorption by algae pigments (chlorophyll pigments and carotenoids) in the blue part of the visible spectrum, with a maximum around 445 nm, compared with the weak absorption around 550–580 nm (Morel, 1998). In case I type water, quite robust empirical relationships can be derived, linking the chlorophyll concentration to the ratio of water leaving radiance (and/or the reflectance of the sea) at these wavelengths.

Case II waters are more optically complex. In such waters the satellite derived water-leaving radiance, in addition to being dependent upon chlorophyll pigments and derived products, is also sensitive and modified by at least one other optically-active component, e.g., suspended sediment and/or colored dissolved organic matter. Most coastal waters are classified as case II waters. Furthermore the various optically active components do not display typical linear relationship. Therefore simple empirical models may no longer be used, and more sophisticated approach, such as inverse modeling must be considered.

2.2 Thermal infrared

For TIR (and PMW) the measured quantity is the emitted energy as function of surface temperature and emissivity. The emissivity is a dimensionless coefficient, e, and can be computed from the complex dielectric constant (or the relative permittivity) $e = e' - ie''$, which characterizes the electrical properties of the media. e' is referred to as the dielectric constant and e'' as the dielectric loss factor. Alternatively, e can be estimated from the complex index of refraction such as $n^2 = e$.

In the thermal infrared part of the spectrum, the surface signal expressed as the radiance observed by remote sensing can be used as input to Planck's law of radiation to find the sea surface temperature (SST) if the emissivity e of the surface is known. For water, the value of e in the most used thermal spectral band of 10 μm–12 μm is very high and stable, about 0.99. At a given wavelength λ the blackbody radiance, L, can be expressed as:

$$L = (2\pi hc^2/\lambda^5)/(hc/e^{\lambda kSST} - 1) \qquad (2.2)$$

The Planck's law is usually expanded in a Taylor series from which the linear term is maintained:

$$Ln(L) = aT_b^{-1} + b \qquad (2.3)$$

where T_b is the brightness temperature in the TIR part of the spectrum, and the coefficients a, b are constant values for each spectral band.

2.3 Passive microwaves

In the microwave domain the brightness temperature T_b provides the measurement of the microwave emission from the surface. The brightness temperature is defined by the real surface temperature T_s and the emissivity e by the relation

$$T_b = T_s * e \qquad (2.4)$$

Spatial variations in T_b observed over the surface of the Earth are due primarily to variations in the emissivity of the surface material and secondly to variations in surface temperature. For the most frequently used frequencies between about 6 GHz and 90 GHz, the emissivity of ice, snow and water show large variations allowing observations of a wide range of multidisciplinary parameters spanning land, ocean, cryosphere and atmosphere.

While e for calm water can be calculated quite accurately from the electric properties (Stogryn, 1971), the value and variation of e for the various forms of ice and snow is less accurately known, and therefore often has to be empirically measured. For sea ice, the dielectric constant e' is relatively constant with frequency above 1 MHz, but e'' is not. There is a minimum in e'' at 3–8 GHz with higher value for lower and higher frequencies.

For firstyear ice at 283 K and 8‰ salinity, the minimum e'' is approximately 0.3. As temperature decreases e'' will increase because precipitated salt will go back into solution. Furthermore, e'' will decrease with decreasing salt content. Multiyear ice has a lower e'' than firstyear ice and its temperature dependence is weaker. Thus, microwave radiation penetrates deeper into multiyear ice than firstyear ice.

2.4 Active microwaves

Radar instruments provide their own source of illumination in the microwave portion of the EM spectrum, at wavelengths on the order of 10^4 longer than those in the visible part of the spectrum (Figure 1). Because of this, radars can operate independent of solar illumination, cloud cover and precipitation conditions.

All radar measurements can be described by a basic equation, which relates transmitted power, distance, reflectivity and antenna characteristics. The equation can be formulated as

$$P_R = \frac{P_t}{4\pi R^2} G \frac{\sigma}{4\pi R^2} A \qquad (2.5)$$

where P_R is power received, P_t is power transmitted, G is the gain of the antenna, σ is the radar cross section, and A is antenna area. The energy of the outward propagating wave, which is spherically expanding, is given by the first ratio. This spherically expanding wave is focused down to an angular beamwidth by the antenna so that the fluxes becomes higher by a factor of G over that of a spherically expanding wave. The focused energy impinges on an object which has a radar cross section σ, which is defined as the equivalent of a perfectly reflecting object of a given area which reflects isotropically (spherically) as shown by the second ratio. Finally, the antenna area, A, term intercepts a portion of the reflected wave so that this portion of the flux defines the power received by the antenna.

The basic radar equation is general; it can be applied to any object of any shape or composition. For imaging over areas of terrain or ocean, a reflection coefficient is defined, σ_0, which is the radar cross section, σ, per unit area. The radar equation (2.5) can then be expressed as

$$P_R = \frac{\lambda^2}{(4\pi)^3} \int \frac{P_t G^2 \sigma_0}{R^4} dA \qquad (2.6)$$

The averaged received power for a radar can then be determined by examining the integral radar equation for distributed targets. The radar scattering coefficient, σ_0, also called backscatter coefficient, expresses a measure (usually in dB units) of the energy scattered back towards the antenna. It is a function of frequency, incidence angle, polarization and the scattering characteristics of the illuminated area.

Radar frequencies are identified by letter designations, and the most commonly used are K-band (30 GHz, 1 cm), X-band (9.4 GHz, 3.2 cm), C-band (5.3 GHz, 5.7 cm), L-band (1.25 GHz, 23.5 cm) and P-band (450 MHz, 62 cm). At these wavelengths the EM-waves are not appreciably attenuated by clouds, precipitation or the Earth's atmosphere (see Figure 1). Therefore, good quality radar data can be obtained during all kind of weather and light conditions. The three main classes of satellite radars (altimeter, synthetic aperture radar, and scatterometer) are further described in the following sections.

2.4.1 Radar altimeter

The radar altimeter measures the transit time and backscatter power of individual transmitted pulses. The transit time is proportional to the satellite's altitude above the ocean, land or ice surfaces. The pulse propagates toward the surface at time t_1 with a speed of light c, is backscattered by the surface, and an echo is received by the sensor at a time t_2. The time difference $t_d = t_2 - t_1$ is equal to the round trip distance to the reflecting surface divided by the propagation speed

$$t_d = \frac{2h}{c} \qquad (2.7)$$

The accuracy with which the distance h is measured is given by

$$\Delta h = \frac{\Delta c t_d}{2} + c \frac{\Delta t_d}{2} \qquad (2.8)$$

The time difference accuracy $\Delta t_d/2$ depends mainly on the sharpness of the pulse which is equal to $1/B$ where B is the signal bandwidth (Elachi, 1987). Over ocean surfaces the measured range is accurate to better than 2.5 cm at an along-track resolution of about 5 km. Higher range accuracy is achieved by detailed analysis of the received signal resulting from averaging a large number of echoes. Range errors and corrections may be associated with the instrument including the ultra-stable oscillator, and the environment such as the total electron content in the ionosphere, path delays within the atmosphere (dry and wet troposphere), and sea state bias. The latter results from the interaction of the altimeter's radar pulse with the sea surface. Simple parametric models are usually applied that express the sea state bias as a function of the wind speed and the significant wave height. The correction ranges between 0 to 20 cm and increases with wind speed and wave. As one typically lack precise knowledge of the wind speed and wave height along the altimeter ground track the sea state bias error is the major error source, and currently intensive investigations are conducted to develop more realistic models (Gaspar and Florens, 1998).

Over the ocean the magnitude and shape of the returned echoes also contain information about the characteristics of the reflecting surface, form which it is possible to retrieve geophysical parameters such as significant wave height, wind speed and sea ice edge location.

2.4.2 Synthetic aperture radar

The synthetic aperture radar (SAR) is a side-looking radar that forms an image by a series of electromagnetic pulses transmitted towards the Earth in a direction perpendicular to the satellite track and reflected back towards the antenna. As the incidence angle of a SAR is oblique to the local mean angle of the ocean surface, there is almost no direct specular reflection except at very high sea states in near range. It is therefore assumed that at first approximation Bragg resonance between the radar and ocean waves is the primary mechanism for backscattering radar pulses (Wright, 1978; Elachi, 1987). The Bragg equation is defined as:

$$\lambda_r = 2 \lambda_s \sin \theta \qquad (2.9)$$

where λ_r is the radar wavelength, λ_s is the sea surface wavelength and θ is the local angle of incidence. These short capillary-gravity waves at the Bragg-scale form in response to the wind stress and are further modulated by the longer waves.

Range is used to define the location of an object in the cross-track direction, which is perpendicular to the satellite flight direction, while azimuth defines the location in the along track direction. In imaging theory, *resolution* of a radar system is defined as the radar's ability to discriminate between two point targets, observable as bright spots in a particular image. Two different principles determine the resolution of a radar system: Azimuth resolution is defined by the beamwidth β of the radar, which is the angle by which the radar beam expands and is a function of the antenna size and range. The along track dimension, or azimuth resolution r_a is defined as

$$r_a = R\beta \qquad (2.10)$$

where R is the range distance to the illuminated object. The beamwidth of the radar is a function of the antenna length (D), with larger antenna producing a narrower beam. For a real aperture radar, such as SLAR, the only way to obtain fine azimuth resolution is to have a very short range or a very large antenna. Synthetic Aperture Radars (SAR) overcomes this problem by utilizing the Doppler shift induced by the along-track motion of the satellite (or aircraft) to simulate a very large antenna. The finest possible azimuth resolution is therefore approximated by:

$$r_a = D/2 \qquad (2.11)$$

The unique advantage of the SAR sensor is therefore that the azimuth resolution (in the order of 10 m) is independent of the range and the platform altitude.

The range resolution r_r of a radar system is

$$r_r = \frac{c t_p}{2 \sin \theta} = \frac{c}{2B \sin \theta} \qquad (2.12)$$

where t_p is the equivalent length of the transmitted radar pulse, c the speed of light, θ the incidence angle and B the pulse bandwidth. The range resolution is therefore corresponding to the minimum distance two objects must be separated to be detected by the SAR at these two unique ranges.

Imaging radars operate in the microwave frequencies between 400 MHz and 35.2 GHz, corresponding to wavelengths between 62 and 0.85 cm. As the bandwidth, $B = 1/t_p$ of the SAR transmitted signal pulses is large it is therefore possible to obtain a fine resolution (in the order of 10 m) in range. Note that this is achieved by modulating the transmitted pulse to obtain a short pulse length.

Another characteristic feature of imaging radar systems is polarization, which is defined as the direction of the electric vector of an EM wave. The electric vector is the plane in which the EM energy is transmitted. Visible light is unpolarized because the direction of the electric vector is randomly distributed. For radar systems it is common to emit pulses, which are either horizontally (H) or vertically (V), polarized. The radar antenna can receive either the horizontally- or vertically-oriented return signal, or both. Extensive description of radar remote sensing principles are found in Ulaby *et al.* (1982a, b) and Elachi (1987).

2.4.3 Scatterometer

Side looking scatterometers acquire wide swath coverage along the side of the satellite sub-track. Every strip at a distance y from the sub-track is observed at an incidence angle θ given by (neglecting the curvature of the Earth)

$$\theta = \tan^{-1}(y/h) \quad (2.13)$$

where h is the height of the satellite above the sub-track. The measurement principal is in general identical to the case of a real aperture imaging radar. With a fan-shaped beam the covered swath corresponds to the range footprint of the antenna. The azimuth resolution ($X_a = h\lambda/L \cdot \cos\theta$) is defined by the azimuth footprint width, and the range resolution ($X_r = c\tau/2\sin\theta$) by the pulse size in range direction (L is the antenna size, λ the scatterometer pulse length, and τ the transmitted pulse duration).

In contrast to the imaging SAR, scatterometers typically provide coarse resolution observations in the order of 50 km. Neither Doppler shift to obtain fine resolution in azimuth nor pulse modulation to obtain large bandwidth and hence gain fine range resolution are utilized as resolution is considered of secondary importance to the radiometric accuracy and extent of coverage.

In addition to the side looking configuration, scatterometers can be forward (or backward) looking, squint looking, or they can use pencil beam circular scanning principles. Hitherto the most common spaceborne scatterometers have used a combination of side looking and forward (backward) looking configurations (e.g. ERS scatterometer, and NSCAT). In so doing the wide swath capability (~500 km) is combined with the ability to observe the surface from different look directions. In turn, the angular signature (azimuthal behavior) of the surface scattering elements is acquired.

2.5 Parameter retrieval summary

From the electromagnetic measurements within these frequency domains and associated bands one can in turn derive a set of key physical and bio-chemical variables and fields based on dedicated retrieval algorithms. They include SST, surface roughness, sea surface height (or slope), sea ice field and so-called ocean color. As shown in Table 1 they can be further broken down into a large number of quantities in which the sea ice field can be related to ice extent, concentration, type and motion, and sea ice edge

Table 1 Overview of surface parameters derived from satellite based Earth Observation. The classes are: C – climate monitoring, OP – operational, and R – research. Not included in the table are atmospheric quantities such as column-integrated water vapor and liquid water derived by passive microwave observations

Parameter	Measurement principals				
	VNIR	TIR	PMW	AMW	
				Side-looking	Nadir
Sea ice extent	C,R		C,OP,R	C,OP,R	
Sea ice concentration & type			C,OP,R	C,OP,R	
Sea ice edge features	R		C,OP,R	C,OP,R	
Sea ice motion			C,OP,R	C,OP,R	
Sea surface temperature	C,OP,R		R		
Wave spectra				C,OP,R	
Significant waveheight					C,OP,R
Vector wind				C,OP,R	
Wind speed			R	R	C,OP,R
Current fronts & eddies			R	R	C,OP,R
Internal waves				R	
Sea level					C,OP,R
Surface geostrophic current					C,OP,R
Shallow water bathymetry	OP,R			OP,R	
Surface slicks (oil spill, film)	R			OP,R	
Chlorophyll concentration	OP,R			R	
Primary production	R,C				

features; surface roughness in different ways can be related to near surface vector wind, wind speed, wave spectra, significant wave-height, and surface current features, surface height can be related to surface geostrophic current and mean sea level, and ocean color or water leaving radiance can be related to chlorophyll concentration.

The most common satellite instrument used for measuring in the VNIR and TIR domains is the Advanced Very High Resolution Radiometer (AVHRR) which provide images with spatial resolution of about 1 km in 3–5 different frequency channels. Since the late 90's, a growing number of imaging spectrometers are also in operation (see Figure A1). Passive microwave (PMW) radiometers, such as SSM/I, observe in a similar manner the emitted radiation in the microwave domain at different wavelengths and polarizations. The ground resolution of PMW instruments varies in the range from 10 to 50 km, depending on the choice of frequency (or wavelength). Active microwave (AMW) observations can be classified in several types of instruments including synthetic aperture radars (SAR), side-looking radars (SLAR), scatterometers and radar altimeters. Whereas the three first AMWs are illuminating the surface at incidence angles typically varying from 20° to 50° the latter is nadir looking. The SAR instruments provide high resolution images, with pixel size down to about 10 m, while spaceborne SLARs provide images with medium resolution of 1–2 km. Scatterometer data have coarser resolution, varying from 10 km to about 50 km. Radar altimeters, on the other hand, are profiling the surface along the satellite ground track with a resolution of the order of 3–5 km.

A list of the most common satellites (past, present and future) and instruments of the classes presented above is provided in Annex B together with some relevant sensor and satellite characteristics such as swath width, spatial resolution, and launch date. In addition, Figures A1 and A2 provide a timeline of the most common VNIR/TIR and PMW/AMW sensors for oceanography operated in space since 1972 together with those approved and planned towards 2010.

Although there is some degree of overlap between the observed quantities and sensor types as indicated in Table 1 they are of complementary nature rather than redundant. In many cases this synergy of different sensors are allowing more reliable retrievals than otherwise would be the situation.

The separation into the three different classes, notably: C – climate monitoring, OP – operational and R – research is furthermore illustrating the multi-purpose use of the data in which, for instance, a derived quantity can be applied in an operational service and in parallel be used in research.

The scope of the paper is not to give a complete and comprehensive review of satellite oceanography within all the classes identified in the Table 1. In the remaining of the chapter we will, on the other hand, address the satellite retrieval capabilities and application in the context of Climate Research and Monitoring in Section 3 and Operational Oceanography in Section 4.

3 CLIMATE RESEARCH AND MONITORING

The past decade has seen increasing public concern about the Earth, its environment and mankind's impact upon it. Global threats such as climate warming, stratospheric ozone depletion, tropospheric pollution and more recent regional events such as the very intense El Niño, the fires in the S.E. Asia and the floods in many parts of the world have left public more concerned than ever about the need both to monitor and understand the Earth's environment.

At international level, agreements and treaties are signed by many countries encompassing climate change issues and at the highest level the Intergovernmental Panel on Climate Change (IPCC) was established under the auspices of the United Nations to advise governments on climate change and its implications. Priority issues, as identified by IPCC, are:

– sources, sinks and concentration of greenhouse gases,
– the Earth radiation balance,
– effect of ocean circulation on the timing and pattern of climate change,
– hydrological cycle,
– polar ice sheets, freshwater and changing sea level,
– ecosystem dynamics,
– large scale insertion of aerosols into the atmosphere.

Ultimately, our understanding of the Earth will improve by the development and elaboration of Earth system models into which data from various sources will be integrated. In view of satellite observations in oceanography this is further discussed in the context of sea ice, sea surface temperature, sea level and ocean circulation and carbon cycle. Note that a comprehensive overview of the new era for oceanography where the core needs for an integrated, coordinated and sustained global ocean observing system for climate and related physical oceanographic issues of research and of operational applications are provided by Smith and Koblinsky (2001).

3.1 Sea ice

Sea ice is defined as any ice which is formed as a result of freezing of sea water. Sea ice occurs at the surface of the ocean in areas where the surface temperature is cooled to the freezing point which is about $-1.8\,°C$ for sea water with salinity of about 35 parts per thousand (ppt).

Sea ice is a part of the cryosphere which interacts continuously with the underlying oceans and the atmosphere above. The growth and decay of sea ice occur on a seasonal cycle at the surface of the ocean at high latitudes. As much as 30 *mill.* km^2 of the Earth's surface can be covered by sea ice. In the Northern Hemisphere, sea ice extent (area enclosed by the ice boundary) fluctuates each year from a minimum in September, when most of the ice is confined to the central Arctic Ocean, Greenland Sea and Canadian Archipelago, to a maximum in March, when the sea ice covers almost the entire Arctic Ocean and many adjacent seas. In the Southern Hemisphere, the annual fluctuation is even greater, from a minimum in February to a maximum in September when the ice surrounds the Antarctic continent and extends equatorward to 55°–65° S (Gloersen *et al.*,

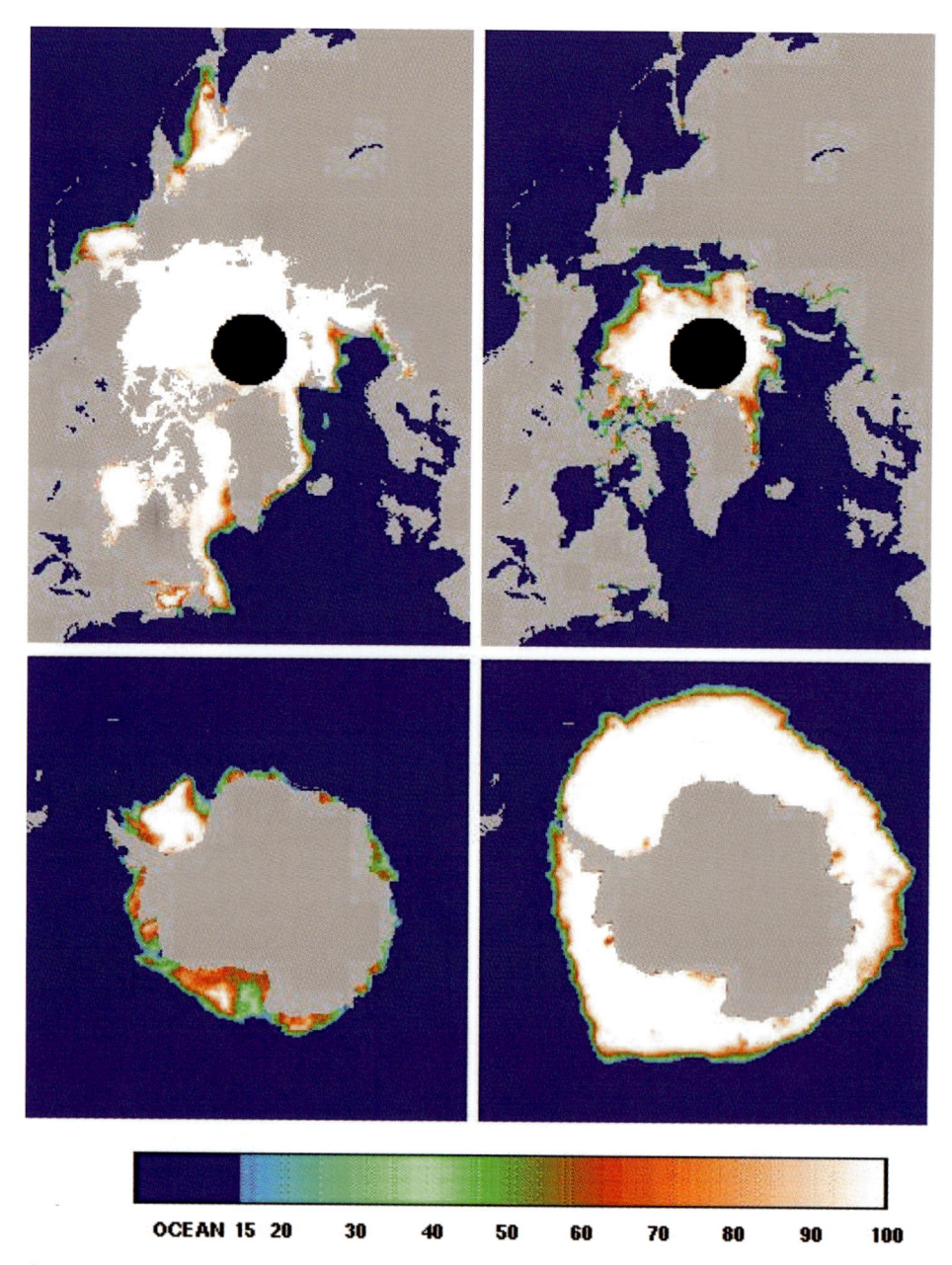

Figure 2 Maps of maximum and minimum sea ice extent and concentration in the Arctic and Antarctic obtained from passive microwave satellite data from the Special Sensor Microwave Imager (SSM/I) operated on the Defense Meteorological Satellite Program (DMSP) (after Bjørgo et al., 1997). The maps are from March 1993 (upper left and lower left figure) and September 1993 (upper right and lower right figure). Copyright American Geophysical Union. Ice concentrations, indicated by the colour bar with 100% in white, 15% in light blue and no ice in blue, were computed from SSM/I data using the Norwegian NORSEX algorithm (after Svendsen et al., 1983). Note that the two filled circles over the central Arctic marks the data-gap.

1992). Figure 2 shows example of maximum and minimum ice extent observed by passive microwave satellite data.

The largest volume of sea ice is found in the Northern Hemisphere in March, 0.05 *mill.* km^3, which is nearly twice the maximum sea ice volume in the Southern Hemisphere. The reason for this is the mean thickness of the Arctic sea ice which is about 3 m, whereas the mean thickness of the Antarctic sea ice is 1–1.5 m.

Sea ice has many roles in the global climate system. For one, it serves as an effective insulator between the ocean and the atmosphere, restricting exchange of heat, mass, momentum and chemical constituents. During winter when there is large temperature difference between the cold atmosphere and the relatively warm ocean surface, ocean-to-atmosphere heat transfer is essentially limited to areas of open water and thin ice within the pack. The winter flux of oceanic heat to the atmosphere from open water can be two orders of magnitude larger than the heat flux through an adjacent thick ice cover. As a result, the distribution of open water and thin ice is particularly important to the regional heat balance.

Another important role of sea ice in the global climate system is that it affects surface albedo. Ice-free ocean generally has albedo below 10–15%, whereas snow-covered sea ice albedo average to about 80%. A fresh snow cover on the ice can increase the surface albedo to values as high as 98%, whereas melt ponds can decrease the ice albedo to as low as 20%. Because the albedo of snow-covered sea ice is high, relative to that of open water, the presence of sea ice considerably reduces the amount of solar radiation absorbed at the Earth's surface. This is most significant in summer, when the insolation, or solar heating is high.

Sea ice processes also affect oceanic circulation directly by the rejection of salt to the underlying ocean during ice growth. This increases the density of the water directly under the ice, thereby inducing convection that tends to deepen the mixed layer. This convection contributes to driving the thermohaline circulation of the ocean. In regions with density structures that were initially weak or unstable, this can lead to overturning and deep water formation. Much of the world oceans' deep and bottom water is formed in polar regions by these mechanisms. Conversely, the input of relatively fresh water to the ocean during ice melt periods tends to increase the stability of the upper layer of the ocean, inhibiting convection. Furthermore, the net equator-ward transport of ice in each hemisphere produces a positive freshwater transport and a negative heat transport.

The first satellite sensors providing views of the large-scale structure and motion of sea ice utilized visible and infrared channels such as those onboard the early Nimbus, Tiros, and Earth Resources and Technology Satellite (ERTS, later renamed Landsat). By the late 1960s, it was apparent that the sequential synoptic observations needed for sea ice and climate studies could not be acquired by satellite-borne visible sensors, which are limited to cloud-free and well-illuminated conditions. Sea ice exists in regions which are dark for several months and are frequently cloudy in the remaining months (Gloersen et al., 1992).

Therefore, it has been necessary to develop observation methods using microwaves which are able to penetrate clouds and are not dependent on light conditions. The first passive microwave remote sensing systems for satellites were launched on the Russian Cosmos 243 and Cosmos 384 in 1968 and 1970, respectively. In the US, passive microwave technology was first used in remote sensing of sea ice during the late 1960s and early 1970s, when a prototype of the Electrically Scanning Microwave Radiometer (ESMR) was flown on Nimbus-5 over the Arctic (Campbell, 1973).

The period since 1970 has been one of great advancement in remote sensing of sea ice. After the ESMR period 1973–1976, a more advanced satellite instrument, the Scanning Multichannel Microwave Radiometer (SMMR) was operated on Nimbus-7 for nine years, from 1978 to 1987. Together with the DMSP SSM/I, which has been the "working horse" since 1987, the longest and most regular time series of global sea ice data has been provided at a resolution of typically 30 km.

In 1978, Seasat was the first satellite which provided high resolution SAR images of sea ice, but the satellite only operated for about three months. With spaceborne SAR data, which combines high spatial resolution with independence of cloud cover and light conditions, it is possible to observe sea ice with much better accuracy than visible and passive microwave methods. ERS-1 represented a major milestone in satellite SAR remote sensing of sea ice, because the satellite delivered tens of thousands of SAR images of sea ice from most ice covered regions in the world. The satellite SAR technology is being improved and SAR systems offered by the Canadian RADARSAT and the European ENVISAT satellites have wide swath, multimode, dual polarization. Also other microwave systems such as scatterometer and side-looking radar (SLR) data have shown promising results in sea ice observations (Gohin and Cavanie, 1994; Grandel et al., 1999).

Today, data from polar orbiting satellites are used extensively in research as well as monitoring of sea ice extent and other relevant ice parameters (Johannessen et al., 1992, 1995; Carsey et al., 1992). As general coupled atmosphere-ocean circulation models predict enhanced climatic warming in polar areas, it is expected that this should be reflected in reduced ice extent and concentration, a decreased mean sea ice thickness as well as possible changes in the intensity and number of pressure ridges. The regular use of satellite data is therefore also essential for model validation.

The microwave brightness temperature T_b of the Earth's surface depends on the electrical properties of the surface, embodied in its emissivity e and the physical temperature of the radiating portion of the surface T_s. From eq. 2.4, this may be expressed by the following relation in terms of the wavelength λ and polarization p:

$$T_b[\lambda, p] = e[\lambda, p] T_s \qquad (3.1)$$

This relationship is true only for e and T_s independent of depth, a typical assumption for sea ice (Steffen et al., 1992). The radiative transfer equation is the basis for the development of algorithms that convert the satellite radiance data into geophysical parameters. The microwave radiances received by the satellite are composed of various contributions from the Earth, atmosphere and space. The radiation received by the satellite, which is a function of wavelength and polarization, can be expressed by the equation:

$$T_b = e\,T_s\,e^{-\tau} + T_{up} + (1-e)T_{down}\,e^{-\tau} + (1-e)\,T_{sp}\,e^{-2\tau} \qquad (3.2)$$

where T_b, T_s and e are as before, and $e^{-\tau}$ represents atmospheric absorption, T_{up} is the atmospheric upwelling radiation, T_{down} is the atmospheric downwelling component, and T_{sp} is the cosmic background component.

There are several well documented algorithms for estimation of sea ice concentration based on brightness temperature observed in several frequencies and polarizations as suggested in Figure 3 (Steffen et al., 1992). Among the most common retrieval algorithms we can note the NASA team algorithm (Cavalieri et al., 1984), the Bootstrap algorithm (Comiso, 1986), and the NORSEX algorithm (Svendsen et al., 1983).

Analyses shows that the actual differences between these algorithms are in general small. In the following discussion we have chosen to use the NORSEX algorithm, which was developed after the NORSEX marginal ice zone experiment near Svalbard in 1979 conducted by the NORSEX Group (1983). The algorithm computes area concentration of total ice and two ice types: multiyear ice (MY) and firstyear ice (FY) using the 18 GHz and 37 GHz vertical polarization at a spatial resolution of 60 km.

An extension of the NORSEX algorithm (Svendsen et al., 1987) takes advantage of the improved spatial resolution of the 85 GHz channels provided by the SSMI system which is currently in operation. The 85 GHz channel, H polarization, (85H) is very sensitive to the difference between water and FY-ice as well as to the atmosphere. This channel is therefore used to sharpen the ice-water boundary.

Analysis of SMMR and SSM/I records taken separately revealed a greater reduction in Arctic sea ice area and extent during the SSM/I period as shown in Figure 4. The decreases from 1987–94 were ~4% per decade compared to ~2.5% per decade from 1978–87 (Johannessen et al., 1995), with no significant trends found in the Antarctic. Since then, merged SMMR-SSM/I time series have been produced and analyzed, establishing the trends more firmly. Bjørgo et al. (1997) showed that the trend in Arctic ice area and extent (1978–95) is about $-0.3 \times 10^6\,\text{km}^2$ per decade (Figure 4 a and b), corresponding to ~3% per decade. The 3% per decade decrease in the Arctic ice extent (1978–97) was subsequently corroborated in a separate analysis (Cavalieri et al., 1997) that also confirmed the hemispheric asymmetry seen earlier (Johannessen et al., 1995; Bjørgo et al., 1997). Cavalieri et al. (1997) found a slight (~1.5%)

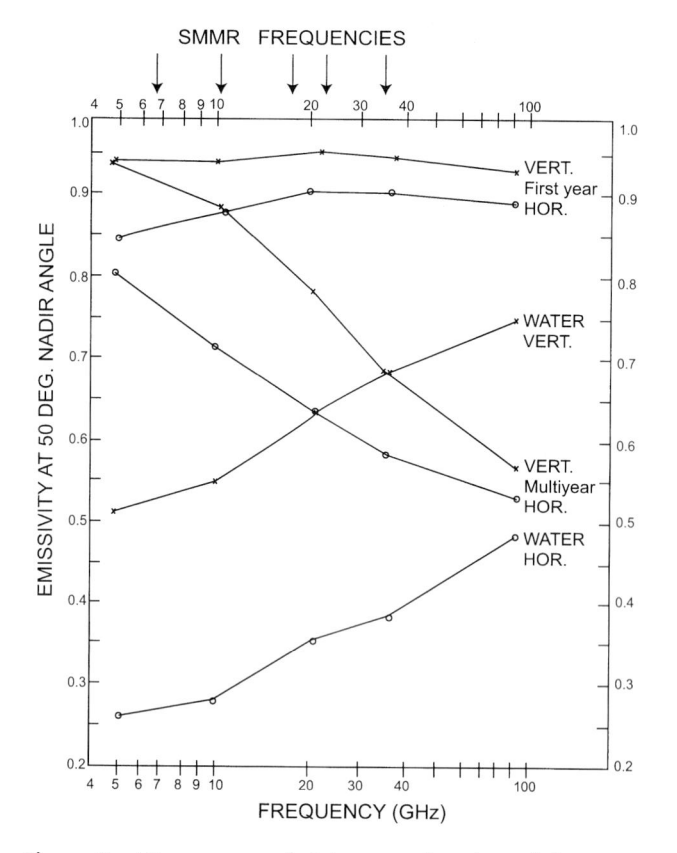

Figure 3 Microwave emissivity as a function of frequency and polarisation for open water, firstyear and multiyear ice, measured during the NORSEX experiment in 1979 (after Svendsen et al., 1983). Two curves are provided for each of the three classes as defined by the vertical (V-pol) and horizontal (H-pol) polarisation. Note that the spatial orientation of the electric (and magnetic) fields of an electromagnetic wave versus the surface of the medium that the wave is incident upon is normal for V-pol and parallel for H-pol. Copyright American Geophysical Union.

increase in the Antarctic, which may be considered significant. The hemispheric ice covers fluctuate quasi-periodically, with predominant periods between 3–5 years, though their variability is apparently not correlated (Cavalieri et al., 1997).

The capability to monitor interannual variations in multiyear ice area from SMMR and SSM/I data has recently been exploited using winter data, when firstyear and multiyear ice signatures permit their distinction (Johannessen et al., 1999). The analysis revealed a relatively large (~7% per decade) reduction in the multiyear ice area 1978–98 (Figure 4c), compared with an ~2% per decade decrease in the total ice area in winter. This finding is supported by a SMMR-SSM/I data analysis that found an 8% increase (5.3 days) in the length of the sea ice melt season in the Arctic from 1978–96 (Smith, 1998). It is also corroborated

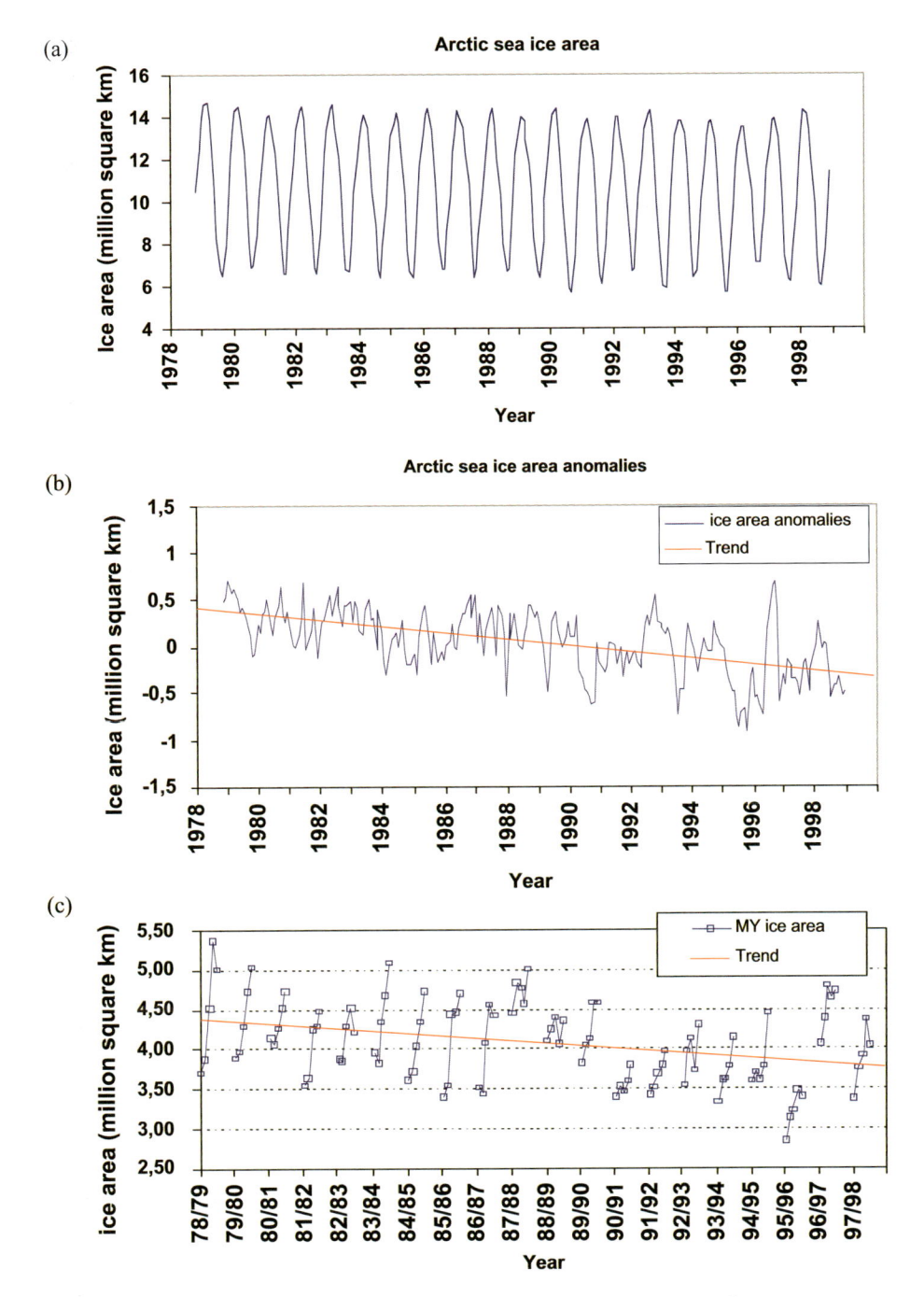

Figure 4 Time series of Arctic ice area derived from Nimbus-7 SMMR and DMSP SSM/I satellite passive microwave data: (a) is monthly mean and (b) area anomalies from 1978–98 where the linear regression indicates a ~31,000 km² yr⁻¹ decrease, corresponding to ~3% per decade (Bjørgo et al., 1997). (c) is the fraction of multi-year (i.e., having survived the summer melt) sea ice area in winter (November–March), 1978–98, where the linear regression indicates a ~30,000 km² yr⁻¹ decrease, corresponding to ~7% per decade (after Johannessen et al., 1999). (Copyright 2000 American Association for the Advancement of Science.)

by spatially- and temporally-fragmentary observations (from submarine sonar transects) of ice thickness decreases, as well as oceanographic data that have revealed changes in Arctic water masses since the 1970s that are reasoned to stem from a substantial (~2 m) melting of perennial MY ice. If this trend were to continue, it could eventually lead to a markedly different sea ice regime in the Arctic, altering heat and mass exchanges as well as ocean stratification.

3.2 Sea surface temperature

Knowledge of the ocean's central role in modifying climate, through its large heat capacity and transport of properties coupled with its complex interactions with the atmosphere and cryosphere, has long been insufficient for the accurate prediction of climate change resulting from fluctuations in natural or anthropogenic forcing. For example, it is known qualitatively that a large part of the excess energy input (the incoming solar radiation minus the infrared radiation to space) in tropical areas is carried by the oceans towards the poles, the other half being transported by the atmosphere (Peixoto 92). However, quantitative estimates are coarse, and predictions of how such fluxes would be modified by 'enhanced greenhouse forcing' are even more uncertain. Such uncertainties resulted in the formation of the World Climate Research Programme (WCRP) by the World Meteorological Organisation and the International Council of Scientific Unions and have been and are being addressed through very large oceanographic research programmes like WOCE and CLIVAR.

The monitoring of sea-surface temperature (SST) from Earth-orbiting infrared radiometers is the technique of marine remote sensing which has had the widest impact on oceanographic science (Robinson, 1985). The Advanced very High Resolution Radiometer (AVHRR) was first flown on TIROS-N in 1978, replacing the limited VHRR, which flew on the five first NOAA orbiting platforms. AVHRR can be seen as the first operational sensor for oceanography. Since 1978, there is always at least one functioning satellite in orbit, but the nominal configuration is two satellites working, in sun-synchronous orbits, separated by about 90° of longitude, so that one gives a morning and the other an afternoon daytime overpass at each location with a spatial resolution at nadir of 1 * 1 km. AVHRR is also part of the METOP/EPS payload as approved by Eumetsat Council in June 1996.

A list of TIROS–NOAA satellites carrying AVHRR are given in Table 2. The Geostationary Meteorological Satellite (GMS) programme also provides infrared imagery of the oceans, primarily covering low- and mid-latitudes, with a spatial resolution at nadir of 5 * 5 km and a repeat sampling capability of 30-minutes. The GMS programme is a worldwide co-operation which includes the American satellite series GOES and the European series Meteosat and MSG.

Table 2 TIROS -NOAA satellites carrying AVHRR sensors

Satellite	Operation dates	Split-channel
TIROS-N	Oct. 78–Jan. 80	No
NOAA-6	Jun. 79–Mar. 83	No
NOAA-7	Aug. 81–Feb. 85	Yes
NOAA-8	May 83–Oct. 85	No
NOAA-9	Feb. 85–Nov. 88	Yes
NOAA-10	Nov. 86–Sep. 91	No
NOAA-11	Nov. 88–Apr. 95	Yes
NOAA-12	Sep. 91–present	Yes
NOAA-14	Dec. 94–present	Yes
NOAA-15	May 98–present	Yes
NOAA-16	Sep. 00–present	Yes

The most popular algorithm for surface temperature retrieval from TIR sensor is based on the so-called Split-window method, which use two bands in the thermal infrared for correcting for varying atmospheric emission and transmission. This type of algorithm is fully operational and used with both NOAA-AVHRR and ERS-ATSR data.

The basic equation as given by McClain *et al.* (1985) is:

$$\text{SST} = a*T_4 + b_1*(T_4 - T_5) + c + b_2*(T_4 - T_5)*(\sec\theta - 1) + c \quad (3.3)$$

where $a = 1.02015$, $b_1 = 2.32$, $b_2 = 0.489$, $c = -278.52$, and θ is the scan angle: The sea surface temperature is given in degree Celsius.

By linear combination of brightness temperature T_4 of channel 4 (10–11 μm) and T_5 of channel 5 (11–12 μm) on AVHRR, the effect of varying atmospheric emission and transmission is largely removed.

Since moored buoys have been used to calibrate the algorithm effects of water emission coefficient and skin-to-bulk water temperature gradient can also been reduced or eliminated. The given coefficients are used globally for routine processing at NESDIS for daytime passes of NOAA-11. Slightly different values are used for night-time passes, and for instruments on other NOAA satellites. A distinction between ascending and descending pass data is often made, due to systematic differences in the algorithm and quality of day and night data, the night data being generally considered as the most accurate (less effect of the skin-temperature). Note that some instruments do not have split channels and as for Landsat-TM band 6, one are restricted to use single band algorithms.

Satellite infrared imagery play a major role in the study of horizontal structures in sea surface temperature thanks to its large coverage at fairly good resolution (~1 km) during cloud free conditions. In particular, these SST measurements are applied for many different purposes related to climate studies.

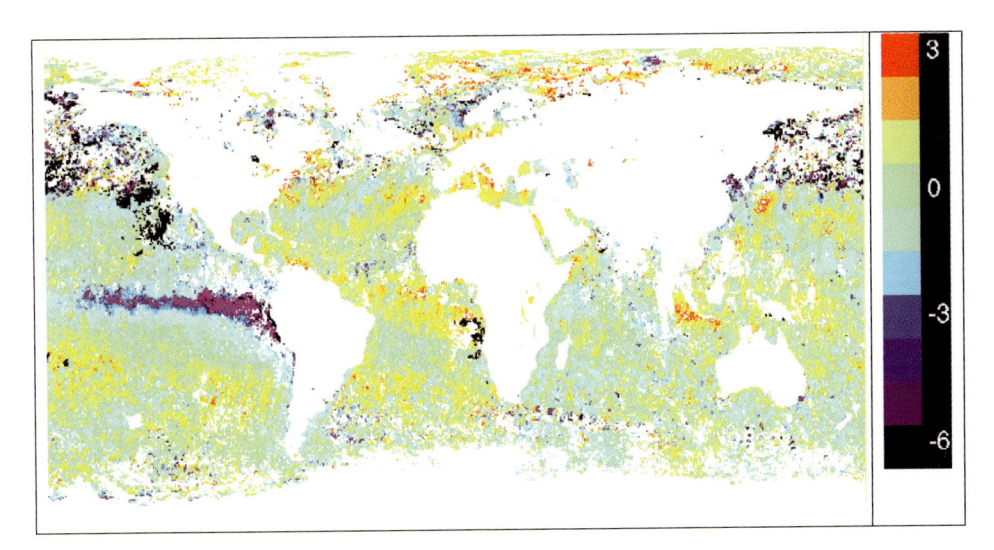

Figure 5 Monthly mean sea surface temperature (SST) differences between July 1995 and July 1997 derived from ERS-2 Along Track Scanning Radiometer (ATSR). Clearly noticed is the SST Expression of the 1997 El Niño with a sea surface temperature anomaly (the colour code is given in degrees Celsius with blue equal −3 °C). (Acknowledgement ESA-ESRIN.)

Applications concern for instance the tracking of long-term global changes in SST related to natural short term variation of the climate system and/or to man-induced changes in the composition of the atmosphere (greenhouse effect).

In Figure 5 the monthly mean sea surface temperature (SST) differences between July 1995 and July 1997 derived from ERS-2 Along Track Scanning Radiometer (ATSR) is shown. Clearly noticed is the SST expression of the 1997 El Niño with a positive sea surface temperature anomaly (with regards to the July 1995 situation) exceeding 5 °C in the eastern region. El Niño is a disruption of the ocean-atmosphere system in the tropical Pacific, which affects weather around the globe. The 1997–1998 El Niño is one of the strongest encountered in the last century causing destructive flooding in the US and Peru, and severe drought in the western Pacific, also associated with devastating fires. The phenomenon is characterised by a rise in the sea level (up to ∼40 cm) and in the sea surface temperature (up to 8 °C) in the eastern tropical Pacific. Such data are also important in research on equatorial dynamics and instability waves such as equatorial trapped large scale Kelvin waves and westward propagating Rossby waves. The interplay of these wave systems is considered to be important elements of the El Niño Southern Oscillation (ENSO). Further discussion of the sea level anomaly associated with the 1997–1998 El Niño is found in the next section.

As part of the ongoing 'AVHRR Oceans Pathfinder' project, NASA-JPL is tasked with reprocessing AVHRR to produce an accurate SST database especially suited for global climate studies (*http://podaac.jpl.nasa.gov/order/order_avhrr.html*). New processing procedures are used to improve the calibration accuracy and the number of valid retrievals (based on a quality flag). Weekly averaged data sets between November 1981 and August 1999 for both ascending (daytime) and descending (nighttime) orbits, are available on equal-angle latitude-longitude grids of 18 km squares at equator. An example of such a SST map is shown in Figure 6.

As well as playing a crucial role for the upper ocean thermal conditions the SST is also of fundamental importance in coupling the atmosphere and ocean. The skin of the ocean provides the lower boundary condition for the upwelling infrared radiation in the marine atmosphere. The air-sea temperature difference is a controlling factor in the exchange of heat and moisture (evaporation), and also determines the marine boundary layer stability, which has consequences on the air – sea fluxes of heat and momentum. The surface skin of the ocean is usually colder, by a few tenths of a degree, than the underlying bulk temperature of the water. This is due to the flow of heat from the ocean to the atmosphere by molecular processes through the upper millimeter or so of the water where turbulent transfer is suppressed by the density difference across the interface (Emery *et al.*, 1995).

Satellite infrared (TIR) SST retrievals have two significant limitations: proper retrievals cannot be obtained in cloudy conditions; and – atmospheric aerosols from volcanic eruptions and large fires can lead to significant cooling in the SST retrievals (Reynolds, 1993). As indicated in Table 1, on the other hand, passive microwave radiometry can offer a solution to the cloud and aerosol problem, provided the operating frequency is below 12 GHz (Wentz *et al.*, 2000). In that frequency domain the surface radiance is proportional to SST and the microwaves penetrate clouds with

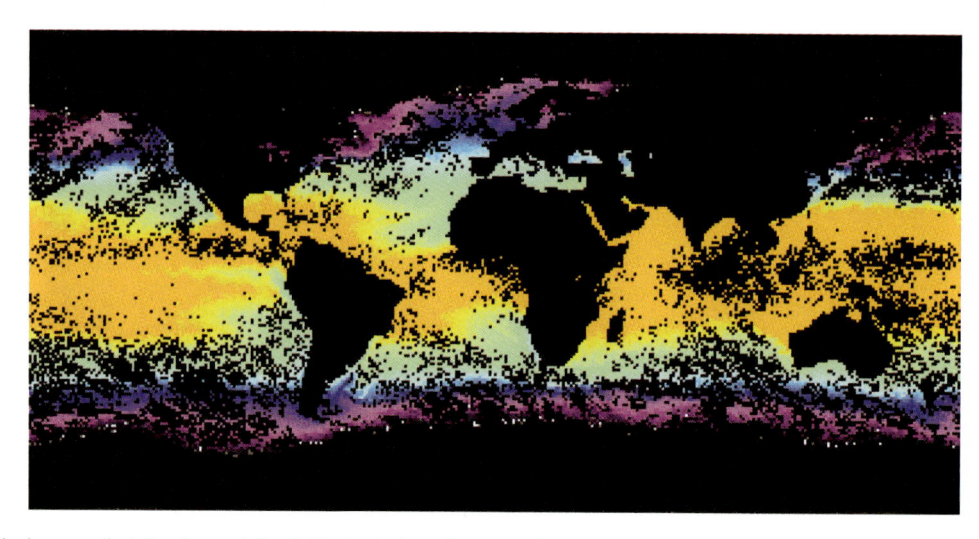

Figure 6 Global map of eight days of the PODAAC data showing the sea surface temperature distribution of warm equatorial and tropical waters (brown–yellow), extra-tropical and mid-latitude waters (yellow–green–blue) and the colder high latitude and polar waters (violet–pink). (Acknowledgement NASA/JPL.)

little attenuation as was demonstrated with the PMW operated on Seasat and Nimbus-7. Moreover, the impact on the brightness temperature from atmospheric aerosols can be ignored. The usefulness of the early radiometers was, however, limited by poor calibration and large spatial resolution (~150 km).

The Tropical Rainfall Measuring Mission (TRMM), launched in November 1997, operates the TRMM microwave imager (TMI) with frequencies ranging from 10.7 to 85 GHz. As the orbit inclination is only 30° it offers no global coverage, but nevertheless, its data collected in the tropical oceans have shown very promising results (Wentz *et al.*, 2000). Comparing their daily SST retrievals with buoy data they obtained a root mean square (rms) difference ranging from 0.5° to 0.7 °C. This difference is expected to be further reduced after more careful removal of retrieval errors and skin-to-buoy correction.

The TMI is the first in a series of satellite microwave radiometers. In the near future the Advanced Microwave Scanning Radiometer (AMSR) will also be operated on both a US and Japanese spacecraft. (See Figure A2 and subsequent Table.) The AMSR will even have a additional channel at 6.9 GHz that will further enhance the SST retrieval.

3.3 Sea level and ocean circulation

Sea-level change is one of the most important components of global climate research. Global-averaged sea-levels are considered to have risen by between 10 and 25 cm during the past century, and are predicted to rise by the order of half a meter in the next century (Warrick *et al.*, 1996). This will have important consequences for the 50–70% of the World's population that lives within coastal zones, which depend on the agricultural and industrial productivity of coastal regions, or which have to be protected from coastal flooding (Bijlsma *et al.*, 1996). Possible impacts on low-elevation coral islands (e.g. Maldives) and river deltas (e.g. Bangladesh) are obvious and have been well documented. Less well known is the fact that a rise of 50 cm along the coast of England and other parts of the European Atlantic coast will increase the probability of flooding by an order of magnitude (Coles and Tawb, 1990).

It is a challenge to come up with a full explanation for the sea level changes that has taken place in the 20th century since a number of competing geophysical processes, each of which is a complex process in itself, are contributing. Among these are interior Earth tectonics; the redistribution of water from ice sheet and glacier retreat; the rebound of the lithosphere and mantle and the affect of these on the Earth's gravity field; the thermal expansion of the oceans; the extraction of ground water; and changes in coastal sedimentation and erosion. The largest potential source is, nonetheless, the cryosphere. The rise in the 20th century corresponds to approximately 0.2% of the Antarctic Ice Sheet mass. However, little is known about the magnitudes of fluctuations in the ice sheets on this time-scale. In comparison, it appears that glacier retreat in Europe and North America may explain 4 cm of the present rise (Meier 84), whereas the thermal expansion of the ocean associated with global warming is also estimated to have contributed perhaps 4 cm during the last century.

Measurements of sea level are made from space via satellite radar altimetry and from *in-situ* devices such as coastal tide gauges, bottom pressure recorders and GPS-buoy systems. During the last decade, the technique of radar altimetry has become fully developed, enabling routine and precise, quasi-global measurements of mean sea level to be obtained. Analyses of TOPEX/POSEIDON altimetric data have provided observations of the ocean dynamic topography to an absolute accuracy of 2–3 cm. In comparison, the ERS-1 orbits are typically accurate to within 12–13 cm. However, since TOPEX/ POSEIDON and ERS-1 were flying simultaneously, the more precise TOPEX/POSEIDON data can be used to correct the ERS-1 orbit error as shown in Figure 7. The same is possible for ERS-2, but with the improved orbit determination from the PRARE (Precise Range and Range-Rate Equipment) the differences are less (Le Traon and Ogor, 1998).

Thanks to these simultaneously operating satellites, measurements of very precise, regular and quasi-global sea surface heights are obtained. As most changes in ocean surface currents (on timescales of a few days or longer) result in geostrophic balance, gradients of the sea surface pressure (or 'dynamic topography', the sea level above the geoid) as derived from radar altimetry can be employed almost directly as proxies for surface current information. Unlike *in-situ* measurements, they are global, synoptic and can be repeated for many years.

To advance our knowledge and prediction capabilities of the world climate at seasonal, interannual, and longer time scales, it is essential that ocean circulation processes are well observed, understood and simulated. Ocean thermodynamics has a stabilizing role on climate. The ocean and atmosphere together are responsible for the meridional heat transfers. Mechanical energy, mass and heat are exchanged at their interface and couple the two systems together. Therefore global, repeated, observations of the ocean topography are a critical element of the research on climate dynamics and on the perturbations to the coupled atmosphere/ocean system.

On basin to global scales, the primary application is climate monitoring. This includes phenomena such as the El Niño Southern Oscillation which is manifested in sea surface temperature and sea surface topography anomalies (Figures 5 and 8) which occurs on seasonal to inter-annual time scales. These observations are very complementary and as pointed out in the discussion of Table 1 they advance the interpretation and reliability of the data. For monitoring such large-scale features, the timeliness of the altimetric analyses can be relaxed to a few weeks. On the other hand, it is critical to have the most accurate orbit available in that time-frame in order to secure a height estimate with an accuracy of no less than a few centimeters.

It is not possible to optimize the sampling of any single satellite mission to observe all oceanic processes and regions. The sampling problem must therefore be thought of in terms of coincident access to a set of complementary altimetric missions such as the overlapping of ERS-1/2 on a 35-day, sun-synchronous, polar orbit and the 10-day orbit of TOPEX/POSEIDON. There is a large improvement in sampling characteristics when going from one satellite to two satellites. Compared to TOPEX/POSEIDON, the

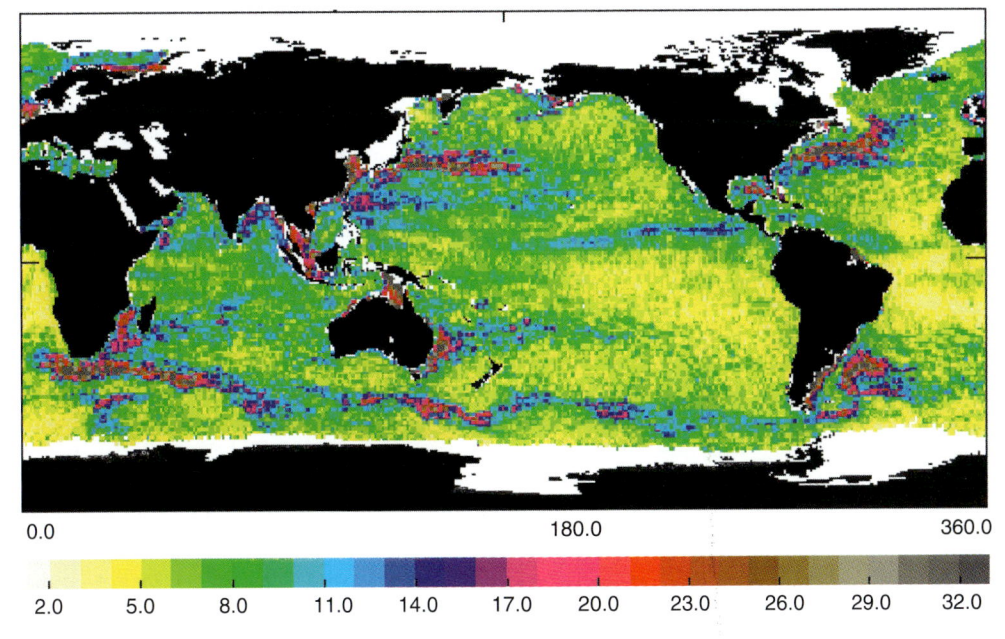

Figure 7 Global map of rms sea level variability (in centimeter height relative to the time-invariant mean surface) from ERS-1 Radar Altimeter obtained from cycles 6 to 18 (October 1992 to December 1993) after correction using TOPEX/POSEIDON precise orbit. The colour code represents height intervals from 2 cm (light yellow) to 25 cm (red). With the TOPEX/POSEIDON orbit as reference Traon and Oger (1998) calculated the ERS-1 orbit error to within about 2 cm rms. (Courtesy of P.-Y. Le Traon.)

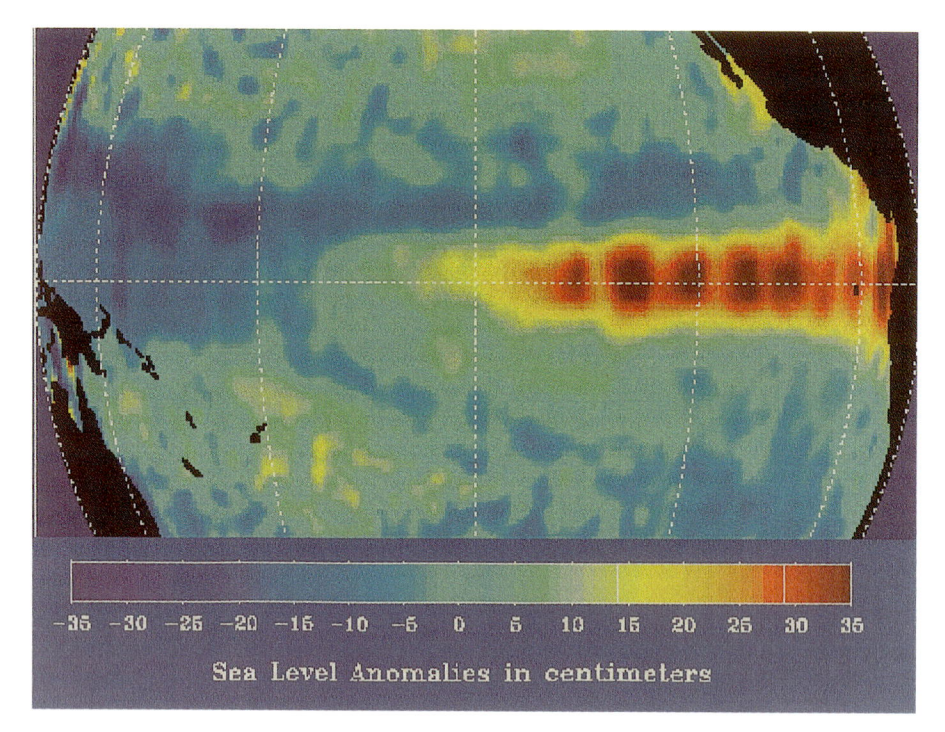

Figure 8 Sea surface topography anomaly ranging from about 30 to 10 cm (red-to-yellow colours) of the 1997 El Niño derived from the ERS-2 Radar Altimeter. (Acknowledgement Delft University of Technology/ESA-ESRIN).

combination of TOPEX/POSEIDON and ERS has, for example, a sea level mean mesoscale mapping error reduced by a factor of 4 and a standard deviation reduced by a factor of 5. The improvement in mesoscale sea level mapping is not as large when going from two to three or three to four satellites (Le Traon and Dibarboure, 1999).

On the other hand, they furthermore demonstrate that the mesoscale velocity field mapping is more demanding, in terms of sampling. The meridional and zonal mean velocity mapping errors are two to four times larger than the mesoscale sea level mapping error. The contribution from a third satellite altimeter is also more significant as only a combination of three satellites can actually provide a velocity field mapping error below 10% of the signal variance.

In summary, these results confirms what is generally agreed as being the main requirement for future altimeter missions (Guymer et al., 2000), i.e. that at least two (and preferably three) missions are needed with one very precise long-term altimeter system (such as T/P and later on Jason-1) to provide a reference for the other altimeter missions such as ERS, GEOSAT, and ENVISAT.

3.4 Carbon cycle and the biological pump

It is generally acknowledged that marine biochemical processes, and particularly the carbon cycle in the ocean, contribute to global climate variability and changes. The natural buffering capacity of the world-ocean with respect to carbon has a direct implication of the atmospheric concentration of carbon dioxide – CO_2, which is considered as the most important greenhouse gas in the atmosphere. It is commonly agreed that the world-ocean takes up as much as 2 Gt of carbon per year (Denman et al., 1996), hence contributing in the global decrease of atmospheric CO_2 of about 50%. The so-called biological pump is nowadays recognized as an essential component of the mechanisms of CO_2 exchange between the atmosphere and the upper ocean. Denman et al. (1996) report that the atmospheric concentration of CO_2 would have been 450 ppmv in absence of marine biota, instead of the current 280 ppmv.

Regarding this, satellite ocean color may provide valuable information on the state of the biological pump through the regular monitoring of chlorophyll pigment concentration as a proxy of phytoplankton biomass and primary production at global scale.

NIMBUS-7 (launched on 24 October 1978) operated, in addition to the SMMR, the Coastal Zone Color Scanner (CZCS); the first sensor specifically designed for ocean color surveillance. The CZCS was a scanning radiometer with five spectral bands at 443, 520, 550, 670, and 750 nm in the visible and NIR range, and a sixth thermal infrared band at 10.5–12.5 µm. The NIMBUS-7 orbit was placed at a height of 995 km, giving the CZCS a ground resolution at nadir of 825 * 825 m. It ceased operating after eight years in 1986. During this period, CZCS recorded approximately 250,000 minutes of data (about 125,000 two-minutes scenes) demonstrating that satellite ocean color sensing provide an invaluable data set for synoptic monitoring of open-ocean biochemical processes, in particular associated with chlorophyll concentration and distribution in the upper ocean.

After a data gap of about ten years a new generation of increasingly-sophisticated VNIR and imaging spectrometers has been developed including MOS, OCTS, POLDER, and SeaWiFS launched in 1996–1997, MODIS, MISR, OCM, OCI, OSMI launched in 1999, and, GLI, POLDER-2 and MERIS scheduled to be launched in 2000–2002 (See Annex B.) Presently more than six ocean color sensors are in operation. The instrument development and improvement have mainly consisted in better radiometric performances and increased number of spectral bands (from 5 for CZCS up to 36 for MODIS and GLI) (Morel, 1998). Other technological advancement have consisted in the introduction of multi-angle viewing capability (POLDER, MISR) and determination of the state of polarization of the reflected radiation (POLDER 1 and 2). In the past 5 years the status of ocean color sensing mission have shifted from experimental or scientific research missions to pre-operational demonstration missions.

As mentioned already in section 2 two main challenges are related to the processing of ocean-color data, namely: (1) atmospheric correction, and (2) the need to apply a bio-optical algorithm to derive water quality parameters from the satellite-measured radiation.

The atmospheric correction includes three major processes that effects the ocean color sensing, notably: (a) Rayleigh (molecule) scattering, (b) absorption by water vapor, ozone and other gases, and (c) absorption and scattering by suspended particles (aerosols). Whereas modeling of Rayleigh scattering and gas absorption is well established (Gordon et al., 1988) the main problem involves the correction from suspended particles in the atmosphere because their size and distribution are highly variable in space and time (Gordon and Wang, 1994; Fraser et al., 1997; Tanre et al., 1997).

Historically, empirical models have been developed relating a radiative quantity (water-leaving radiance, upwelling radiance, volume reflectance or remote sensing reflectance) derived from color sensing instruments and chlorophyll concentration. For case I water, the basic chlorophyll algorithms are based on relationships between chlorophyll concentration and the ratio of reflectance (R) or water leaving radiance (Lw) at two wavelengths (443–490 and 550–580 nm). As mentioned before, these two bands describe the impact of absorption and scattering of light by chlorophyll pigment. The choice of a ratio approach, rather than another kind of combination lies in the variability and the uncertainty affecting the absolute value of R or Lw.

The retrieval of the chlorophyll content is thus performed by a four step processing chain:

1. correction for the atmospheric contributions (Rayleigh scattering and gas absorption),
2. estimation of the contribution of atmospheric aerosols using NIR data,
3. derivation of a correction factor for aerosol scattering,
4. retrieval of the water leaving radiance and application of the empirical band-ratio model.

Variations in the concentration of chlorophyll in the upper ocean and consequent changes in the penetration of visible light have a fundamental impact on prediction of biological, physical, and geo-chemical oceanographic processes. Phytoplankton absorbs solar energy and converts it to organic matter, thus providing the basis for the world fisheries production. Solar irradiance absorbed by the phytoplankton also contributes to changes in the underwater visibility as well as to variations in the local heating rate and thus influences the development of the thermal structure and dynamics of the upper ocean.

An example of global seasonal distribution of chlorophyll, is shown in Figure 9. The deep blue and purple colors seen in most of the central ocean basins are oligotrophic waters resulting from selective absorption and scattering of pure seawater, unaltered from the presence of phytoplankton or other optically active substances. As one move closer to shore and shallow water regions nutrient inputs generally increase. The consequent development of higher concentrations of phytoplankton changes the color from blue to green.

Most of the empirical algorithms have been set up based on fitting ocean-color derived quantities to extensive worldwide in situ data set. However some algorithms have been developed for specific areas. Numerous empirical formulations have been proposed including Morel and Prieur (1977); Gordon et al. (1980); Gordon and Morel (1983); Morel (1988) for CZCS algorithms, and O'Reilly et al. (1998) for SeaWiFS and OCTS algorithms.

In the open ocean the complexity of the ocean optics is not as severe as in the coastal areas (case II waters), where the sediments and colored dissolved organic matter (CDOM) also significantly contribute to the signal measured by the satellite sensors. Deriving water quality parameters, with a sufficient accuracy, in case II waters is one of the main challenge of the coming years in marine optics. The current development includes new sensors with improved capability, i.e., more and narrower spectral bands, and new algorithms with improved atmospheric correction scheme and/or global approach (inverse methods). The most advanced work is undertaken in connection with the development of the next generation of optical ocean color sensor, i.e., the US MODIS and the European MERIS

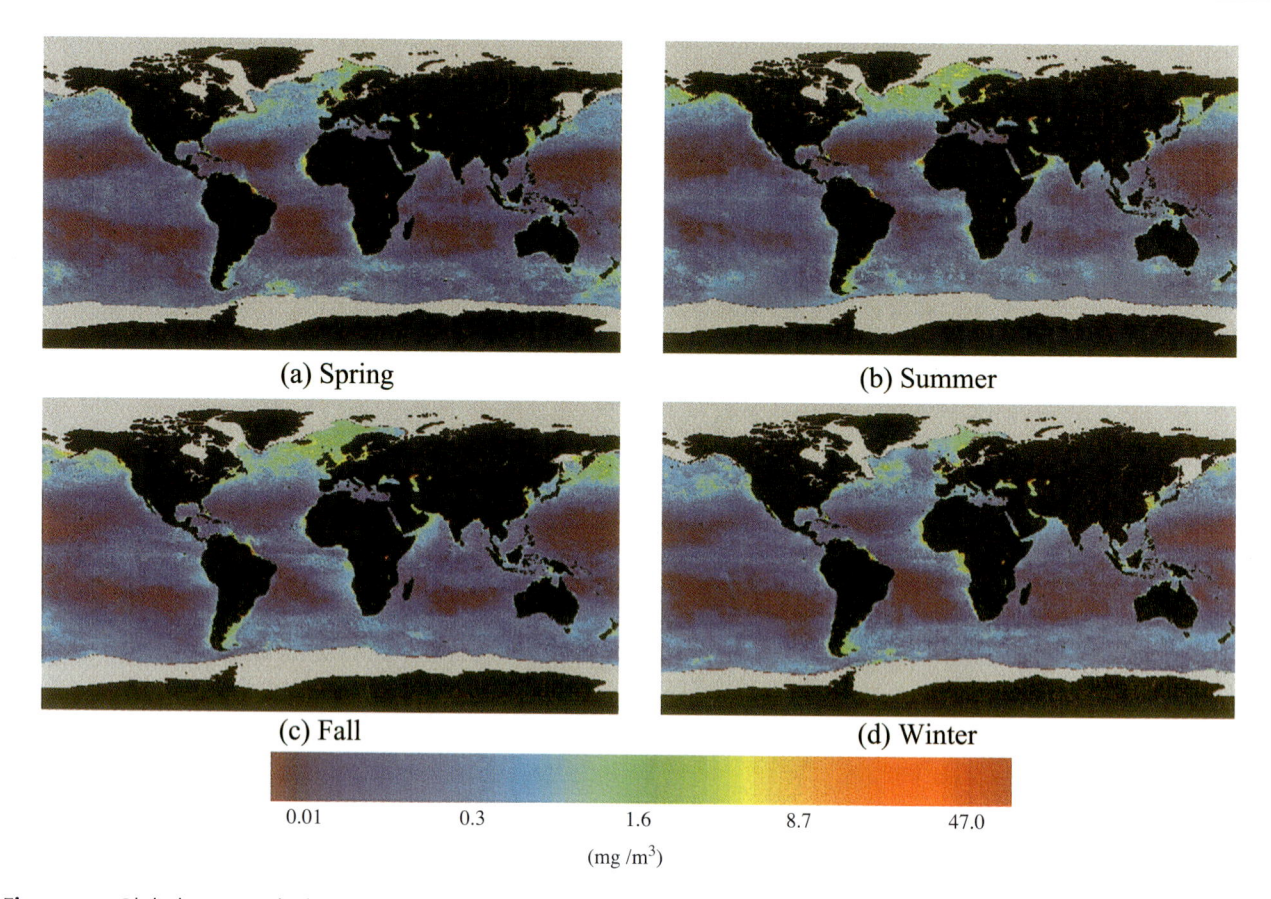

Figure 9 Global seasonal distribution of photosynthetic pigments (chlorophyll and phaeophytin) as derived from the processing of the four first years of the CZCS data acquisition from November 1978 to October 1981. Panel (a) shows the averaged pigment distribution from March to May. Panel (b) displays the same information for June to August. Panel (c) is for September to November, and Panel (d) for the period covering December to February. Among other, the highly productive coastal upwelling off West-Africa is clearly observed, as well as its seasonal variability, which is characterised by an increase of pigment concentration from spring to summer and a decrease in winter time. We clearly observe the variability of pigment concentration along the coast of Yemen and Oman, which can be related to the general moonson regime in the northern part of the Indian Ocean. (Acknowledgement NASA/GSFC.)

instruments. The state-of-the-art in chlorophyll concentration retrieval in case II waters can be found in e.g., Doerffer and Fischer, 1994; Schiller and Doerffer, 1999; Moore et al., 1999, which discuss novel approaches such as inverse method, including artificial neural network techniques, and improved atmospheric correction algorithms.

Algorithm for estimating the primary production rate from satellite-derived chlorophyll and surface temperature was first proposed by Platt and Sathyendranath (1988). The state-of-the-art primary production models can be found in e.g., Antoine et al. (1996) and Behrenfeld and Falkowski (1997). As phytoplankton absorb solar energy it removes inorganic carbon from the water, and reduce the partial pressure of carbon dioxide in the surface layers. In some cases, the partial pressure may be reduced below that in the overlying atmosphere resulting in a net flux of CO_2 from the lower atmosphere to the upper ocean. Global estimation of primary production has therefore become one of the main applications of ocean color data because of its use as a proxy of the carbon cycle in the ocean (Figure 10).

The current challenge regarding ocean color remote sensing is to enable the combined and simultaneous use of all the sources of ocean color data in order to make a significant increase in coverage of the global ocean, which is of paramount importance in the context of the limitations of the sensors in the presence of clouds. This requires that the data from multiple sensors can be merged into a single global product, providing regular fields in space and time. This has not yet been accomplished with ocean-color data and is potentially a difficult task (Yoder et al., 1998).

NASA's Sensor Inter-comparison for Marine Biological and Interdisciplinary Ocean Studies (SIMBIOS) project aims at determining how data merging can be accomplished by defining standards for data level, quality definition,

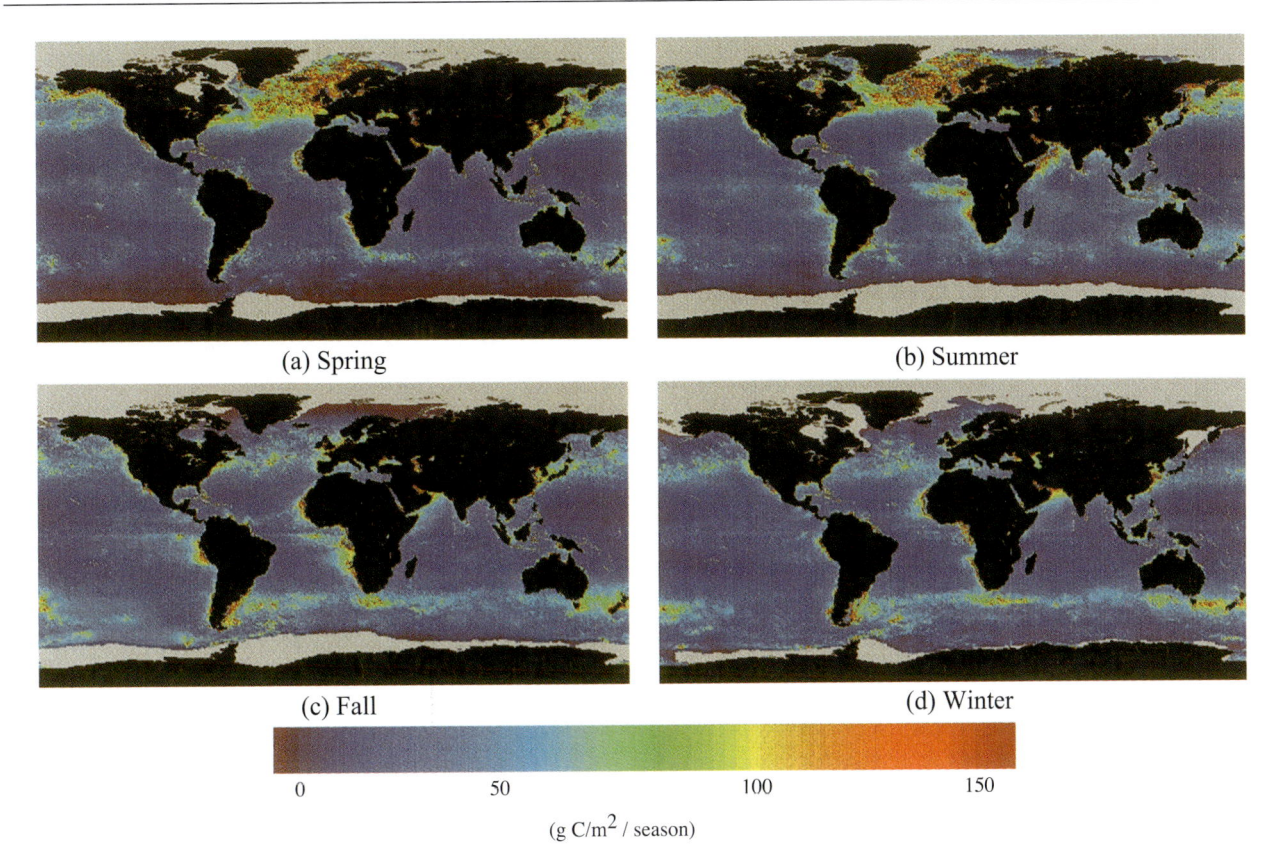

Figure 10 Seasonal averaged estimates of primary production using pigment concentration derived from the 1978–1981 CZCS data set and the so-called "Vertically Generalized Production Model" obtained by Behrenfeld and Falkowski (1997). In particular, seaonal variability between the northern and the southern hemisphere are well reproduced. (Courtesy of M.J. Behrenfeld and P.G. Falkowski; Copyright American Society of Limnology & Oceanography.)

radiometric constraints, and algorithm implementation. Regarding the latter, the main expected improvements in light of the many new sensors are based on a systematic use of non-linear models, including cubic polynomial, power and hyperbolic-power relationships, as well as multiple regression approach.

4 OPERATIONAL OCEANOGRAPHY

Operational ocean forecasting systems relies on an integrated use of observations of physical, biological, and chemical variables and coupled physical and marine ecosystem models. These observations and the model dynamics are integrated using so called data assimilation techniques. These are mathematical techniques, which are usually based on some prior statistical assumptions about the accuracy of the observations and dynamical models. Essentially, these techniques provide a mean for introducing the information about the true ocean-state into the models, hence limiting the model drift away from the real state of the ocean.

The new operational data assimilation systems, which are being developed at several centers also demand observations which are available in near real time. Thus, an extensive effort must be invested in the development of real time data analysis and processing, as well as in data distribution.

The Global Ocean Data Assimilation Experiment (GODAE), for instance, aims to implement and operate (between 2003 and 2007) a global system of observations, communications, modeling and assimilation, that will deliver regular, comprehensive information on the 3D state of the global oceans. Its objectives are two-fold, notably: (1) Apply state-of-the art ocean models and assimilation methods for short-range open-ocean nowcasts and forecasts, and for boundary conditions to extend predictability of coastal and regional subsystems, and for initial conditions of climate forecast models; and (2) provide global ocean analyses for developing improved understanding of the oceans, improved assessments of the predictability of ocean systems, and as a basis for improving the design and effectiveness of the global ocean observing system.

At the onset of the new millennium we are therefore gradually transferring from satellite oceanography to operational oceanography (Guymer et al., 2000). As indicated in Table 1 the suite of active microwave instruments currently in operation is providing for the first time consistent and regular global vector wind and wave fields, sea level and surface geostrophic height data in near real time (within three hours of observations), filling the gaps existing with conventional observation systems. This has opened up exiting new opportunities for wind, wave and current modeling, as well as assimilation into operational models for the production of marine weather, sea state and mesoscale current forecasts. Similarly, the large increase in the availability of optical sensors has lead to new opportunities for regular algae bloom monitoring which in turn has lead to pre-operational pilot demonstrations.

In this chapter we have selected to further address sea state, mesoscale current, sea ice and algae bloom in the context of operational oceanography.

4.1 Sea state

Satellite scatterometers provide a measure of both wind speed and direction over a swath of 500–1000 km width at a spatial resolution of 25–50 km. A complete global coverage is obtained in about three days. The quality of these wind field observations is widely recognized, and they enable for example small-scale low pressure systems and frontal lines to be identified properly compared with model background information. Provided the dual directional ambiguity of the vector wind field solutions are removed (Stoffelen and Andersen, 1997) the improvements in the initial wind field data provided by the scatterometer data have a beneficial impact on analyses and short-range forecasts, probably mainly from the improvements on the subsynoptic scales.

For example, the European Center for Medium-Range Weather Forecasting (ECMWF) provides operational meteorological forecast services, as well as sea-state forecasting services, to the national meteorological services of its seventeen participating states. The ERS scatterometer data are incorporated in such a way as to correct forecast surface wind fields continually over the oceans. These corrections are then propagated through the numerical model to provide corrections to other parameters such as atmospheric pressure, temperature and humidity. Trials have demonstrated that the incorporation of scatterometer data improve the accuracy of the short-range forecasts by approximately five percent over forecasts where scatterometer data were not included. In particular, the use of scatterometer data to improve the accuracy of the wind data for tropical cyclones and hurricanes has proven very useful, so that initial values of atmospheric parameters at the model grid points better match the actual values. Since January 1996 the assimilation of ERS-2 scatterometer data has been fully operational at ECMWF.

In addition the availability of wave height data from the radar altimeter has provided a strong influence on wave modeling and has also stimulated the development of wave height assimilation techniques. One of the principal motivations for developing the third generation wave model WAM (Komen et al., 1994) was to provide a state-of-the art model for the assimilation of global wind and wave data from satellites for improved wind and wave field analysis and forecasting. The WAM model is in use at a number of forecasting centers (NCEP, FNMOC, BMRC, and ECMWF to mention but a few) and altimeter wave height data are assimilated at MeteoFrance, KNMI, DNMI, UKMO and at ECMWF.

Moreover, as indicated in Table 1 the synthetic aperture radar (SAR) provides retrievals of the directional ocean wave spectrum either in wave mode operation (conducted in synergy with the scatterometer) or in full image mode. In the wave mode small SAR imagettes of 10 km × 5 km are acquired every 200 km along the scatterometer's near range coverage. These data are distributed to weather services for operational wave monitoring, analyses and forecasting (within three hours of observation) via the Global Telecommunication System (GTS). For the first time these data allow wave modelers to obtain global information on two-dimensional wave spectra.

In combination with the scatterometer wind field retrievals these data furthermore provide capabilities to separately study the wind sea, swell propagation and dissipation (Heimbach et al., 1998). This particularly has contributed to a provision of better open ocean wave forecasts since the ERS SAR wave mode data are allowing improved initialization of the swell field in the model. Moreover, the improved estimates of the wave spectrum (i.e., partitioning of the wave field into wind wave and swell components) are in turn used to refine wind field retrievals. Example of the direct application of such improved sea state prediction is associated with design of optimal ship routes.

In addition to the improvement in marine weather and sea state forecasting worldwide, the use of off-line products from ERS-1 is also providing benefits to the offshore industry as well as many other coastal activities. Time series of sea state information are being developed as a basis for predicting the sea state at a given time and location (Lasnier, 1996). This assists in a whole range of activities such as planning the timing and logistics of offshore activities to minimize risks to personnel, assessing marine risks, coastal defense planning, wave energy resource evaluations and setting engineering design parameters and is also helping insurance companies in risk analysis and the settlement of claims.

4.2 Mesoscale ocean current

Until 1996, operational ocean circulation models such as those run by the National Centres for Environmental Prediction (NCEP) and ECMWF (or several other numerical weather predictions centre) only assimilated sea surface temperature data from satellites. However, in recent years such models have began assimilating ERS-1/2 and T/P altimeter data on operational basis. The sea level heights from altimetry have been shown to increase the forecasting skill of the model, particular when salinity changes (in contrast to temperature) effect a large-scale change in sea level.

The need for a better monitoring and prediction of the marine environment has increased dramatically in recent years, in particular along the coastal boundaries and shelf regions where human activities are extensive and pollution has a significant impact. This has been documented in a number of unpredicted events along coasts worldwide, such as storm surges, harmful algae blooms and oil spills. A common parameter of significant importance in this context is the surface current; known to satisfactory level of accuracy at high spatial and temporal resolution.

The oceanographic mesoscale is comprised of highly energetic features, including western boundary currents and their eddy fields as well as narrow coastal current systems. Routine monitoring of strong current features may provide extremely valuable information to for instance fishermen, pleasure sailors, merchant ship crews and oil platform operators.

Mesoscale monitoring has been successfully demonstrated with ERS and TOPEX/POSEIDON data in the Gulf of Mexico and Gulf Stream current system, and in the North Atlantic Azores front region. Since 1991, the Service Hydrographique et Océanographique de la Marine (SHOM) (French Navy) has been conducting a long-term program – called SOAP – aiming at developing operational oceanography for mesoscale applications taking advantage of the availability of TOPEX/POSEIDON and ERS-1 altimeter data. The system has documented to be extremely useful at providing monitoring and prediction of the Azores front and its associated mesoscale variability. Comparison with *in-situ* data gathered during the Semaphore experiment demonstrated the quality of the system. The follow on system, called SOPRANE, is implemented on a larger area (North-East Atlantic) and is currently assimilating TOPEX/POSEIDON and ERS-2 data.

Coastal models could also utilize altimetry to provide oceanic boundary conditions on the open ocean side of the model domain. Altimetry does not give highly accurate data near land, but the combination of suitable models incorporating the altimetry at their open boundaries would enhance coastal current forecasting.

The need for high quality predictions of marine parameters and processes has been well identified in EuroGOOS and there is currently an extensive activity in developing systems for operational monitoring and prediction of the marine environment. Examples are a number of EC funded projects which includes among others the Mediterranean Forecasting System Pilot Project (MFSPP), and the DIADEM and TOPAZ projects. The latter two include development and implementation of an operational assimilation system for physical and ecosystem variables for the North Atlantic, Nordic Seas and the Arctic Ocean. The DIADEM system, for instance, couples physical and biochemical model modules and assimilates satellite derived sea surface temperature, sea surface height and ocean color in a model covering the North Atlantic, Nordic Seas and Arctic Ocean.

4.3 Sea ice

The possibilities to measure ice drift from satellites has been radically improved over the last ten years. Over that period it has been used in several pilot demonstration projects, notably in Canada, the Baltic Sea and on the Siberian Coast, and is currently used operationally.

Largescale sea ice drift, which determines the advective part of the ice balance, can be derived from several types of low- and medium-resolution satellite data: SSM/I, scatterometer and AVHRR data. Smaller scale ice motion including the formation of leads and polynyas, convergence with ridge formation and specific processes in the marginal ice zone such as eddies and ice tongues can only be measured by high-resolution Synthetic Aperture Radar (SAR) images, which have been available since the launch of ERS-1 in 1991 and further extended with RADARSAT in 1996.

The ice motion algorithm, also denoted ice motion tracker, is part of the RADARSAT Geophysical Processor System (RGPS) which is used routinely by Alaska SAR Facility to produce sea ice products from SAR images from the whole Arctic Ocean (Kwok, 1998).

The ice motion tracker is also used on SSM/I data to estimate large scale ice drift (Kwok *et al.*, 1998) as shown in Figure 11. With the 85 GHZ channel it is possible to estimate ice motion every three days for the whole Arctic Ocean. Liu *et al.* (1999) has demonstrated that also scatterometer data from NSCAT with 25 km resolution could be used successfully to calculate ice motion. Liu *et al.* (1999) has also shown that feature tracking using wavelet analysis can be a useful ice motion algorithm.

The presence of sea ice and its motion furthermore represents a major limitation for ships and offshore operations in at high latitudes in both hemispheres. The sea ice which is on average 2–3 m thick can only be penetrated by ice-strengthened vessels or icebreakers with sufficient ice class. Most ships and fishing vessels are not strengthened to sustain such conditions and must therefore avoid all ice areas. In many cases, when the ice concentration is 100% and the ice pressure is high, even the most powerful icebreakers can have problems to move forward through the icepack. Offshore platforms operating in ice-covered areas must have much stronger construction than what is required in ice-free waters. Also harbours and loading terminals on the coast require stronger construction in areas of sea ice. In countries and regions were sea ice occur it is therefore of primary importance to monitor the sea ice regularly and produce ice forecasts to assist ship traffic, fisheries and other marine operations (Sandven et al., 1998). In the Northern Sea Route, the longest ice navigation route in the world, Russia has build up an extensive ice service to support the ship traffic (Johannessen et al., 1996). The service includes ice monitoring and forecasting as well as icebreaker escorting of cargo ships (Figure 12).

The Canadian RADARSAT, operational from 1996, has sea ice monitoring as the prime objective and provides SAR data of ice covered areas on operational basis. With RADARSAT data, a new era in operational ice monitoring by satellites has begun, although there is still room for further research and development to advance ice scattering models and ice classification algorithms.

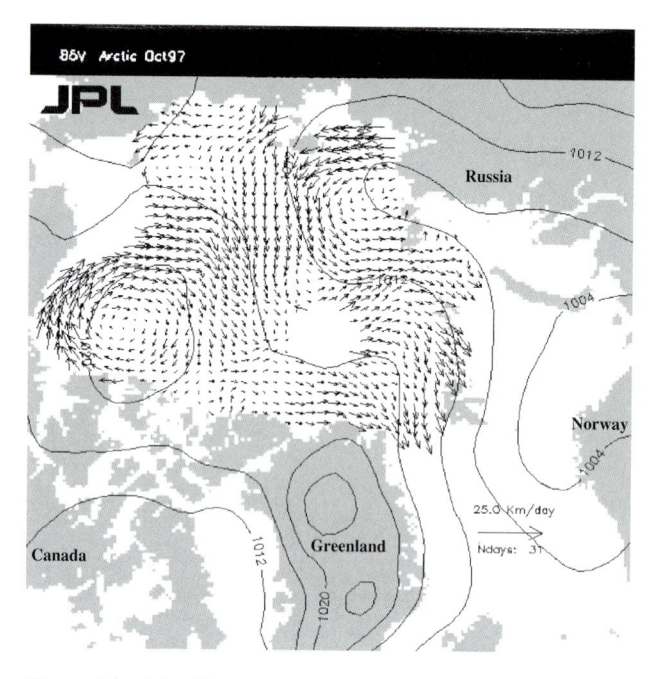

Figure 11 Monthly mean sea ice motion vectors delivered from RGPS using SSM/I data. The solid lines indicate the mean surface pressure (units of millibar) in October 1997, while the arrows give the monthly averaged sea ice drift speed and direction scaled to the 25 km/day (25 cm/s) marker. (Acknowledgement NASA/JPL.)

4.4 Algae bloom

One of the main issues of surveying phytoplankton distribution and concentration is the operational monitoring of harmful algae blooms (HAB) and potential fishery areas. The development of extreme algae bloom situations (harmful or not) generally depends on the following environmental conditions: (1) hydrodynamical conditions, (2) supply of macro-nutrients to the euphotic layer, (3) surface solar radiation, and (4) the optical properties of the water column. An

Figure 12 Picture of Russian nuclear powered icebreaker "Sibiria" leading a convoy of cargo vessels sailing through 2 m thick firstyear ice in the Northern Sea Route. (Courtesy Murmansk Shipping Company.)

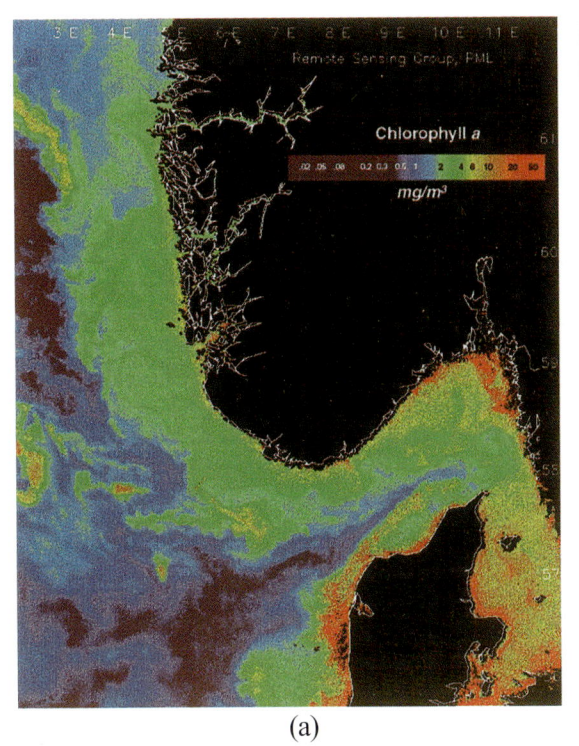

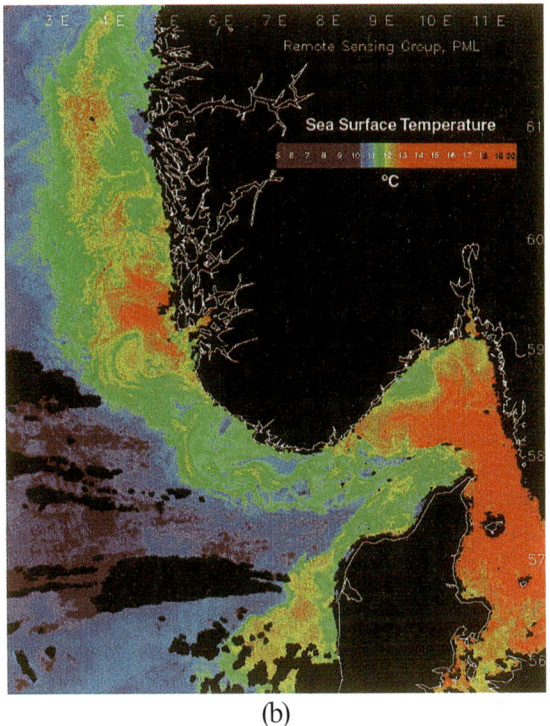

Figure 13 (a) Chlorophyll pigment concentration product from SeaWiFS over the North Sea and the Skagerrak on May 19th 1999 (© Orbital Imaging Corp. and NASA SeaWiFS project). (b) Sea surface temperature distribution as derived from NOAA-AVHRR for May 19th 1999 over the same region. These image data allow characterisation of the main water masses, circulation patterns and current fronts. One can observe the cool oligotrophic water coming from the Atlantic ocean into the North Sea and the Skagerrak Sea (in purple and blue). Warm waters from the Baltic (in red in the SST image) meet the cold waters in the Skagerak giving raise to turbulent processes, such as eddies. Along the western coast of Norway, the turbulent Norwegian current shows complex eddy circulation patterns, and is characterised by warm waters at this period of the year (13–15°), as well as somewhat high chlorophyll concentration (up to $5\,mg\cdot m^{-3}$). (Data courtesy: Steve Groom, Centre for Coastal and Marine Sciences – Plymouth Marine Laboratory.)

algae bloom may have its peak activity below the surface, and hence may not be detected by remote sensors. The alga themselves have limited mobility and hence a measure of the advection of an identified bloom may be indirectly done through monitoring of the currents and ocean circulation pattern.

The goal of most national HAB-monitoring plans is the effective management of fisheries, public health, and ecosystem problems related to marine biotoxins and harmful algae.

VNIR remote sensing data have already proved valuable for early warning of some algae blooms, and through use of information on front and gradients, a more efficient field sampling strategy can be implemented. In addition, routine monitoring data required for a forecast system provide a database containing information concerning the type, location, frequency, and duration of HABs. These data furthermore support studies on the impacts of HAB on the fisheries industry, on public health, and for basic algae and oceanographic research.

Figure 13 shows a SeaWiFS (left) and a NOAA-AVHRR (right) image from May 1999, covering the North Sea and Skagerrak area. The images resolve the pigment concentration in the water and the sea surface temperature, respectively. Major circulation and current patterns can be observed in both images. In particular the relatively cold and meso-trophic inflow of Atlantic water (in blue) in the Skagerrak is well observed, as well as the outflow of warm and eutrophic water from the Baltic.

In the same region, the outflow from the Oslofjord, characterized by high concentration of pigments, can be observed in the SeaWiFS image, as well as its spreading along the southeastern coast of Norway. High values are also observed, along the coast of Denmark. These originate from the high concentration of suspended matter, as well as the sea-floor contribution to the measured signal in this shallow water area. The mesoscale eddy variability along the western coast of Norway are also well captured in both images.

Despite the limitation in operational usage imposed by the presence of clouds, quality ocean color observation

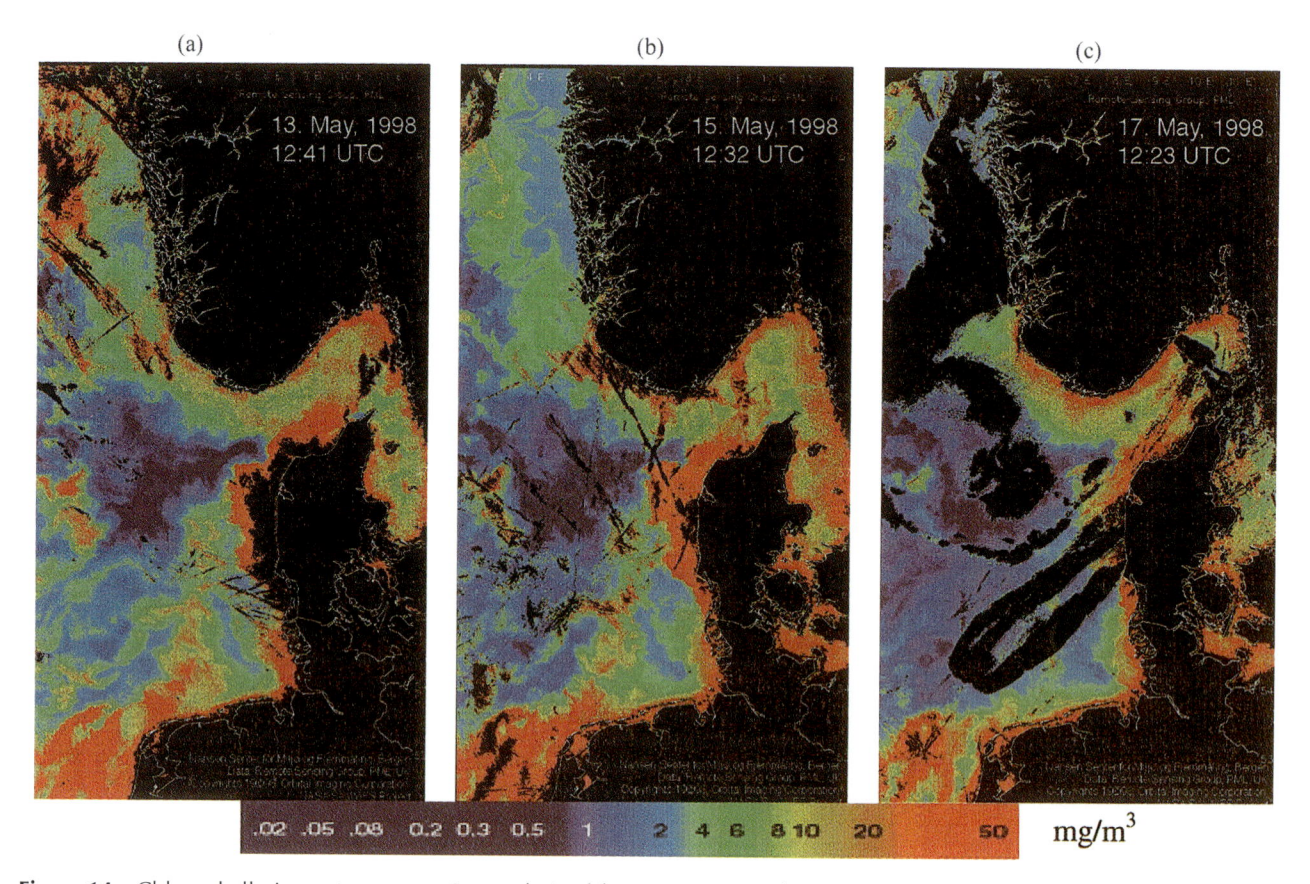

Figure 14 Chlorophyll pigment concentration as derived from processing of SeaWiFS ocean colour data for the North Sea and the Skagerrak, on (a) May 13, (b) May 15 and (c) May 17, 1998 (© Orbital Imaging Corp. and NASA SeaWiFS project). The time series shows the increases and demise of a *Chattonella* spp. bloom along the Danish coast, for the first time reported in this region of the European waters. The bloom of this toxic algae extended up to the Norwegian coast where intensive aquaculture of salmon is carried out, leading to the death of 350 tonnes of farmed salmon. Blue indicates low pigment; red indicates high concentration of pigment and thus of *Chattonella*. Black area over the sea indicate clouds or unprocessed pixels. (Data courtesy: Remote Sensing Group, Centre for Coastal and Marine Sciences – Plymouth Marine Laboratory.)

from space is valuable for early warning of algae bloom initiation and development in offshore areas. The derived products include chlorophyll a concentration, as an index of phytoplankton biomass (Figure 14), and the diffuse attenuation coefficient at 490 nm, as indicator of water clarity.

4.5 Emerging new applications and operational services using imaging SAR

Whereas the quantitative retrievals of wind waves and swell from SAR are operational, regular services for oil spill detection and shallow water bathymetry monitoring are now also relying on SAR image data, although the methods are only possible in a certain range of wind field and sea state conditions. Recently, moreover, quantitative wind field retrievals from SAR data have improved significantly (EOQ, 1998; Beal and Pichel, 2000). On the other hand better quantitative interpretations must be developed for retrievals of mesoscale ocean features in order to improve the use of SAR observations in mesoscale ocean current monitoring and prediction system.

A brief overview of the current state-of-the-art SAR image retrievals of wind, oil spill and natural films, shallow-water bathymetry, and current features and internal waves, is provided below.

Wind: Quantitative information of the near surface wind field is derived from synthetic aperture radar (SAR) backscatter intensity measurements and their spectral properties. This allows higher spatial resolution regional and local wind fields at about 2 to 10 km to be derived (see Figure 15a, b). Such fields are attractive for applications in coastal regions, smaller enclosed seas and in the vicinity of the marginal ice zones. This is not only for scientific interest (such as wind-wave coupling and air–sea interaction studies) but also for practical applications including towing

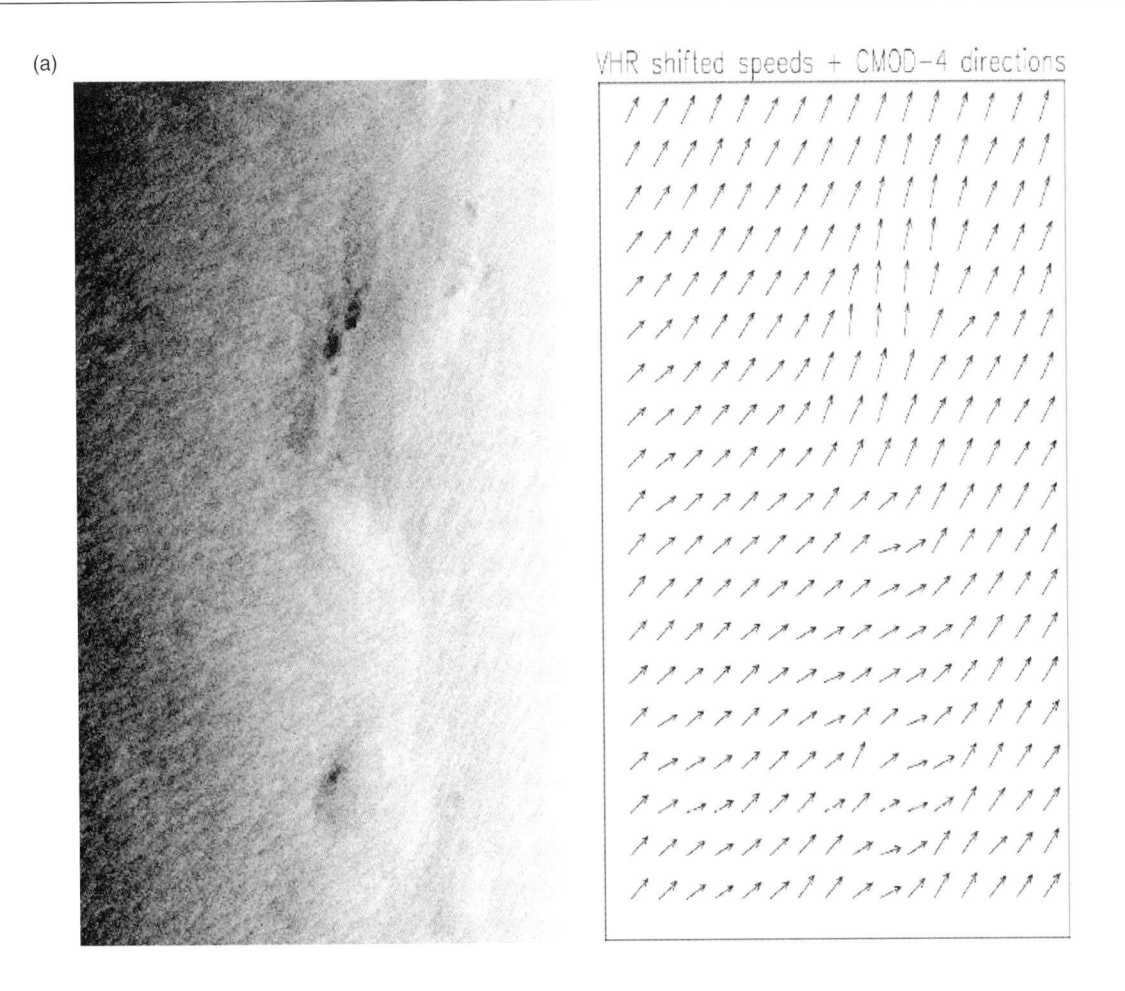

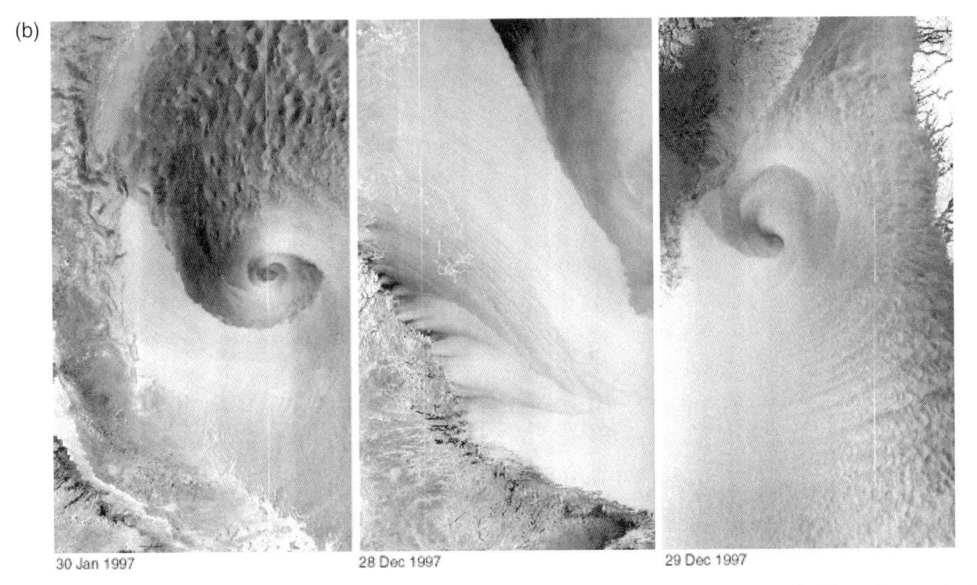

Figure 15 (a) SAR image (left) and example of optimized SAR derived mesoscale wind field (right) with the use of high-resolution wind field model (HIRLAM). Courtesy of M. Portabella; (b) RADARSAT ScanSAR wide images of the Labrador Sea, each showing mesoscale flow patterns, including synoptic and polar lows (after Vachon et al., 2000). Each image covers roughly 500 km from left to right.

operations, harbor and pilot boat operations, leisure boating and selection of sites for wind mills (Furevik and Espedal, 2002).

Following the launch of ENVISAT in March 2002 data from a new and advanced synthetic aperture radar (ASAR) will become available. In addition to ensuring the continuity of the ERS SAR and RADARSAT observations (Figure 15b) with a more versatile wave measurement mode, the ASAR instrument will offer new operating modes which look very promising for wind retrievals. In particular, the wide swath mode (150 m resolution) and global monitoring mode (1 km resolution) have the potential to deliver fine resolution wind field maps over a swath of more than 400 km.

This new and advanced capability to retrieve the fine resolution wind field from SAR clearly demonstrates the potential for new operational applications and research into coupled ocean-atmosphere interactive processes including atmospheric boundary and ocean mixed layer dynamics and exchanges of heat, momentum and gas across the ocean surface (Beal and Pichel, 2000).

Oil spills and natural films: Disaster management including flood (and sea level rise), toxic algae blooms and oil spill has been identified as a key area by the Committee on Earth Observation Satellites (CEOS) and Integrated Global Observing Strategy (IGOS). Extensive near real time use of SAR images for pollution monitoring along coastal waters are being conducted in several pre-operational projects (Pedersen et al., 1996).

It is clear that the imaging radars are very suitable for monitoring of oil spills notably in regions of intense ship traffic and in offshore areas of extensive oil and gas productions. Oil and natural slicks (film) dampens the capillary-short gravity waves on the ocean surface thus reducing or eliminating the presence of Bragg scattering mechanism. In turn, the radar cross section is low and the imaged area appears dark.

Dependent on the characteristics of the observed pollutant features (and its source) there are various ways of responding to the near real time information provided by the imaging satellite radar such as redirection of patrol aircraft, request for information from nearby ships, and formal requests to refineries, industrial plants, rig operators or ships.

The basic question for most end users is distinguishing man-made oil spills from natural films and other "lookalike" features. For practical determination of slick type (whether natural or man-made) and origin using a single frequency and polarization SAR, empirical and conceptual and interactive algorithms, employing wind history and drift models, appear to be most useful (Espedal et al., 1998).

Shallow water bathymetry: Knowledge of sea bottom topographic features and corresponding depths is of vital importance to ship routing in shallow water, for fisheries and for offshore operations. While traditional bathymetric surveys are time and cost consuming, the imaging capabilities with SAR have demonstrated a promising capability to retrieve bathymetry maps over large areas at relatively low costs, provided the average water depth does not exceed 20–30 m.

Off the coast of the Netherlands it has been shown that large scale topographic structures, such as the Bruine Banken, are clearly imaged under favorable moderate (3–7 m/s) wind conditions. At low and high winds, on the other hand, the images can be dominated by the presence of natural films and speckle. The magnitude of the surface current further modulates the backscatter expressions. For example, in areas of strong current (order of 1 m/s), such as off the coast of Zeeland and in the Waddenzee, sand waves (typically of 2–6 m height and crest to crest separation of about 500 m) and other permanent topographic features are regularly observed in the SAR images.

SAR images are now used to retrieve bathymetry information and subsequently to validate and initiate morphodynamic models which are being developed to forecast changes in bottom shapes linked with sediment transport, river deposition and coastal erosion. The results of studies of bathymetry estimation in the Plaatgat area show that if the number of ship survey tracks is reduced by a factor of 3–5, the rms error of the assessed depth map is still less than 30 cm (Hesselman et al., 1997), which is the accuracy required by the Dutch Ministry of Transport and Public Works (Rijkswaterstaat). This conclusion is corroborated by the results of depth assessments in other areas. It demonstrates that the use of SAR remote sensing mapping techniques can lead to a substantial reduction of the costs of traditional sounding campaigns.

Current shear, convergence/divergence: An excellent documentation that the frontal features associated with the mesoscale circulation patterns are expressed in SAR images is shown in Figure 16 (Johannessen et al., 1996), where the surface roughness anomalies detected by the radar (right) are compared to the SST distribution imaged by the AVHRR (left). In the latter the structure of the SST field with the curvilinear temperature fronts represents mesoscale variability of 10 to 50 km scale, characteristic of the unstable Norwegian Coastal Current (Johannessen et al., 1989). The frontal features contained in the radar image have a configuration and orientation that are in good qualitative agreement with those in the AVHRR image.

In spite of such fascinating images, quantitative interpretation of these image expressions are still limited. In particular, simulations of such SAR backscatter signatures associated with current fronts and eddies appear less intense than the observed signatures in satellite SAR images. The magnitude of the predicted perturbation depends strongly on the angular dependence of both the short wave equilibrium spectrum and the relaxation rate on the wind direction. Since the parameterization of these are uncertain it is difficult to draw definitive

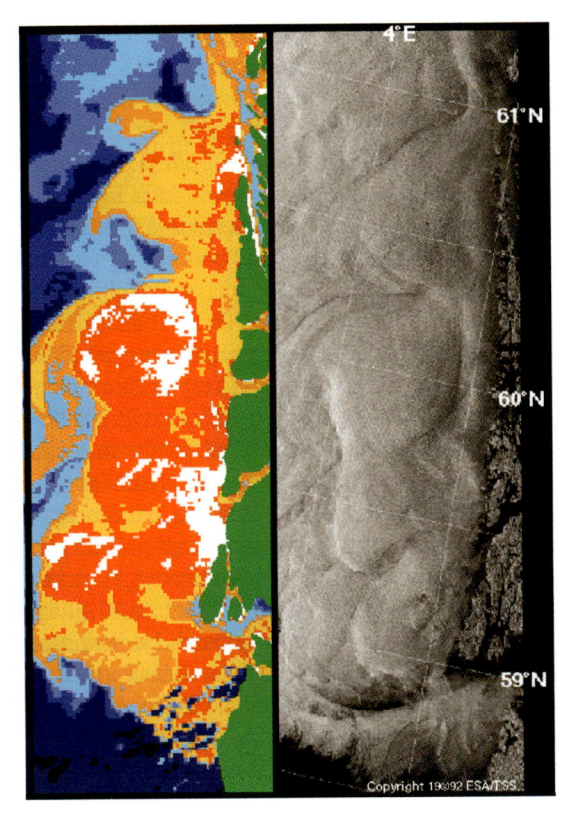

Figure 16 Manifestation of the Norwegian coastal current front off the west coast of Norway on October 3, 1992 in a NOAA-AVHRR 1-km resolution image of the sea surface temperature field (white is 14 °C and dark blue is 12 °C) acquired at 14:20 UTC (left) and a ERS-SAR 100-m resolution image of the surface roughness field acquired at 21:35 UTC (right) (after Johannessen et al., 1989). Both images cover approximately the same 100 km by 300 km region. In the left image the land is masked with green color. (© ESA/NERSC.)

conclusions regarding the quantitative accuracy of the models. To make further progress at a fundamental level, it may be necessary to use more sophisticated wave generation and growth, nonlinear wave-wave interaction, wave-current interaction (including the effect of wave breaking), and radar backscattering models (Komen et al., 1994).

Internal waves: There are several types of internal wave (IW) signature expressions in the SAR images such as alternatively bright to dark, single dark and single bright bands of radar backscatter. The generation mechanisms of these are mostly explained by tidal current interaction with bathymetry in a stratified ocean. The anomalous expressions of the IWs are dependent on the existence of film material in the upper ocean as well as the strength of the near surface wind.

Simulation models give estimates of SAR backscatter signatures of IW patterns which are of the correct order of magnitude. For practical purposes, if internal waves are used to determine the hydrographic structure, it may be good enough just to determine the wave locations, wavelengths and propagation speeds from available SAR imagery. However, if we actually wish to estimate the strengths of the currents associated with the waves, it may be possible to apply an inversion scheme based upon one of these forward models (Brandt et al., 1999).

5 NEAR FUTURE SATELLITE OBSERVATION

The need to explore new Earth Observation capabilities that aimed at bringing new data and scientific understanding, has been elaborated in ESA's Earth Explorer Programme (ESA 98) and the US Earth Observing System (EOS) Science Plan (EOS 99). These programs have been developed following extensive consultation with the Earth Observation communities, notably their concern about climate change and man's impact on it. Additional and complementary views and plans for Earth Observations are implemented by among other space agencies including CNES, NASDA, ISRO and Eumetsat (Lindstrøm et al., 2000; Guymer et al., 2000). Views and plans are also naturally discussed by the Committee on Earth Observation Satellites (i.e., CEOS 97).

Furthermore as indicated in the listing of existing (and planned) operating satellite sensors according to the four classes identified in Table 1, notably, VNIR, TIR, PMW and AMW (Annex B) ensure that the number of continuity missions appears relatively secure for the next 5 to 10 years. In some instances, i.e., METOP, it is even guaranteed that scatterometry and VNIR/TIR operations will continue for 15 years.

Three geophysical quantities, that in different ways are relevant for important processes within the fields of cryosphere and oceanography, and which hitherto has not been observed from satellites are: (a) the ice mass fluctuations; (b) the sea surface salinity; and (c) the marine geoid (and steady-state ocean circulation). It is generally agreed that the lack of these observations inhibits the development of scientific interpretation and understanding of basic processes that contribute to the ocean circulation and to the effects of the ocean on seasonal to multi-decadal climate change as indicated in Table 3.

It is therefore comforting to know that the space agencies have approved missions concepts specifically focussed to observe these quantities (Johannessen et al., 2001). These mission plans include: ESA's Cryosat mission complemented by NASA's plans for the Ice, Clouds and Land Elevation Satellite (ICESat); ESA's GOCE mission and NASA's Gravity Recovery and Climate Experiment (GRACE) mission; and ESA's SMOS mission which to some extent is complemented by the ongoing joint NASA/NASDA Tropical

Table 3 Connection between the three geophysical quantities, the oceanic processes that in different ways are contributing to climate and the candidate satellite missions

Processes	Parameters		
	Ice mass	Sea surface salinity	Marine geoid
Thermohaline circulation	ICESat, Cryosat	SMOS	GOCE (GRACE)
Sea level change	ICESat, Cryosat, GRACE, GOCE		GOCE
Air–Sea–Ice interaction (+albedo effect)	ICESat, Cryosat		
Evaporation minus precipitation		SMOS, +TRMM follow-on	
Mass and heat transport		(SMOS)	GOCE (GRACE)
Large scale frontal dynamics		SMOS	GOCE
Evolution of large scale salinity event	(ICESat, Cryosat)	SMOS, +TRMM follow-on	

Rainfall Measuring Mission (TRMM) and its planned follow-on mission. These missions are briefly discussed below.

Sea ice mass fluctuations: The cryosphere has a central role in the Earth's radiation budget imposed by the large albedo due to the presence of ice and snow. Loss of sea ice is predicted to cause a larger greenhouse-gas warming in the Arctic than elsewhere on the Earth, whereas uncertainty in ice sheet and glacier mass balances are the largest error sources on present sea level change (Wingham, 98). Moreover, thermohaline circulation and deepwater formation are affected by changes in sea ice and ice sheet masses.

The primary goal of the Cryosat mission, planned for launch in 2004, is to estimate trends in the ice masses of the Earth. This will be achieved by measuring the change in sea ice and ice sheet thickness with a radar altimeter using interferometric and synthetic aperture techniques for resolution enhancement. Cryosat will provide nearly complete and continuous coverage of the cryosphere, but is limited in its resolution. The US ICESat mission planned for launch in 2002 is of particular complementary importance to Cryosat. ICESat aims to provide details of the sea ice and land ice roughness spectrum by employing a fine resolution (100 m scale) laser altimeter technique. The limits imposed by dense cloud cover will reduce sea ice thickness measurements and change measurements at fixed ice sheet locations. This will make optimal combination with the Cryosat radar altimeter very attractive.

Sea surface salinity: The distribution of salt in the global ocean and its annual as well as interannual variability are crucial in understanding the role of the ocean in the climate system. *In-situ* salinity measurements are only scarcely distributed over the oceans. In fact, $1° * 1°$ boxes distributed over the global oceans show that for only about 70% of them a salinity measurement exists at all (Levitus and Boyer, 1994). A far smaller fraction of such areas has been monitored only once. This means that the average structures of the sea surface salinity (SSS) field are known to some extent, but details about its variability even on seasonal and interannual scales remain hidden.

The SSS varies as a result of exchange of water between the ocean and the atmosphere, via sea ice freezing and melting and from continental runoff. Salt affects the thermohaline circulation and therefore the distribution of mass and heat. Salinity may control the formation of water masses, which allows its use for tracer studies. Salinity is also thermodynamically important as salinity stratification can influence the penetration depth of convection at high latitudes and may determine the mixed layer depth in equatorial regions through the evaporation-precipitation coupling.

The sensitivity of L-band (1.4 GHz) passive microwave radiometer measurements of oceanic brightness temperature to SSS is well established (Lagerloef *et al.*, 1995; Kerr *et al.*, 1999). However, the sensitivity is a function of the sea surface temperature (SST) decreasing from 0.5 K/psu in 20 °C water to 0.25 K/psu for an SST of 0 °C. Hence, strong demands are put on the SSS retrievals from space in polar and sub-polar regions where the water masses are very sensitive to small changes in SSS (below 0.1 psu). Other oceanic factors which will influence the brightness temperature retrievals at L-band are surface roughness (wind speed and direction) and foam. Precise estimates for the uncertainties associated with these features are required in order to obtain sufficiently accurate SSS retrievals from space.

The ESA SMOS mission is planned for three years in order to cover two complete seasonal cycles, with a candidate launch date in 2005. Complementarity and synergy with other operating passive and active microwave systems are foreseen and in particular it will provide extremely valuable data for constraining the evaporation minus precipitation budget over the tropical oceans provided that a TRMM follow-on mission is flown simultaneously.

Marine geoid and steady-state ocean circulation: While variations of the sea level and thus of the ocean currents can be derived directly from satellite altimeter data, the absolute value of the ocean dynamic topography, and hence the absolute surface circulation, requires the independent determination of what would be the elevation of an ocean at rest, that is the geoid. The latter is not known at present with sufficient precision. The typical elevation scale of the dynamic topography is of the order of 0.1 to 1 m, while the precision of present geoid models is also multi-decimetric on the scale of many ocean circulation features.

The scientific objectives of the GOCE Mission are, based on the unique measurements by the gravity gradiometer and satellite-to-satellite tracking in high-low mode, to provide an accurate and detailed global model of the Earth's gravity field and geoid (ESA 99) with an accuracy of 1mG (1 milligal (Galileo) = 10^{-5} m/s^2) and 1 cm at a spatial resolution of 100 km. The mission duration is planned for about 20 months with candidate launch date in 2005.

GOCE is complementary with GRACE. GRACE will be the first gravity field mission using the principal of satellite-to-satellite low-low tracking mode. It will improve the accuracy of the spherical harmonic coefficients at the long and medium spatial scales ($<\sim 500$ km) by up to three orders of magnitude. This will allow measurement of the temporal variations in the gravity field to be recovered at 30–90 days interval over a period of about 5 years (planned to begin in 2002). Over the ocean this means that bottom pressure variations can be derived at a typical horizontal scale of 1000 km whereas changes in the ice masses can be studied over Antarctica and Greenland Ice Sheets. The aim (and challenge) is then to convert these sea floor pressure variations to changes in global ocean circulation.

6 SUMMARY

No *in-situ* observing system is presently conceivable on global scale, except perhaps underwater acoustic tomography. On the other hand no satellite remote sensing system provide sufficient subsurface observations. It is therefore only via the systematic combination of both observing systems that a 3D picture of the oceanic state can be drawn. Moreover, only with models able to assimilate data and simulate the 3D oceanic state can we conclude that we have a fully integrated system that can provide realistic forecasts.

Over the past two decades, the tools to observe, understand and model the Earth system have improved substantially through developments in observational technology, including remote sensing from space, through improvements in numerical simulation and high performance computing, and through new methods for assimilation of the time dependent atmospheric, oceanic and chemical data in an hierarchical set of dynamically evolving models. This has deepened our understanding of the complex interactions between the various components of the Earth system which together govern the evolution of the system as schematically illustrated in Figure 17.

In this perspective, satellites offer unique capabilities as they provide global, quasi-synoptic and repetitive data sets of homogeneous quality. The number of geophysical and bio-chemical variables that can be measured from space is large as indicated in Table 1, and addresses most components of the Earth System. All these characteristics of remote sensing from space are essential as the system itself is continuously changing over a wide range of space and time scales. Space observations have a key role to play in research and monitoring of the Earth system as they often provide the only means to acquire the relevant data.

However, it must be emphasized that the overall long term design criteria of the ocean observing system for climate, expressed by the Oceans Observation System Development Panel (OOSDP), is:

> to monitor, describe and understand the physical and bio-geochemical processes that determine ocean circulation and the effects of the ocean on seasonal to decadal climate change, and to provide the observations necessary for climate predictions

During the first decade of the new millennium relatively long time series (in some cases almost 30 years) of satellite derived quantities (including sea surface temperature, ocean wave field, near surface wind, ocean color, sea surface topography and sea ice extent, types and concentration) will become available. As an essential element of integrated ocean observing systems (in combination with *in situ* data collection system and model tools) continuous access to calibrated and corrected satellite data is indisputable. No space agency by itself can ensure to meet such a continuity requirement. International cooperation in implementation and operation of key missions is therefore highly necessary. This will also avoid duplication and ensure complementarity, and should lead to significant reductions in the agencies costs of addressing the objectives of the ocean observing system for climate. The challenge is furthermore,

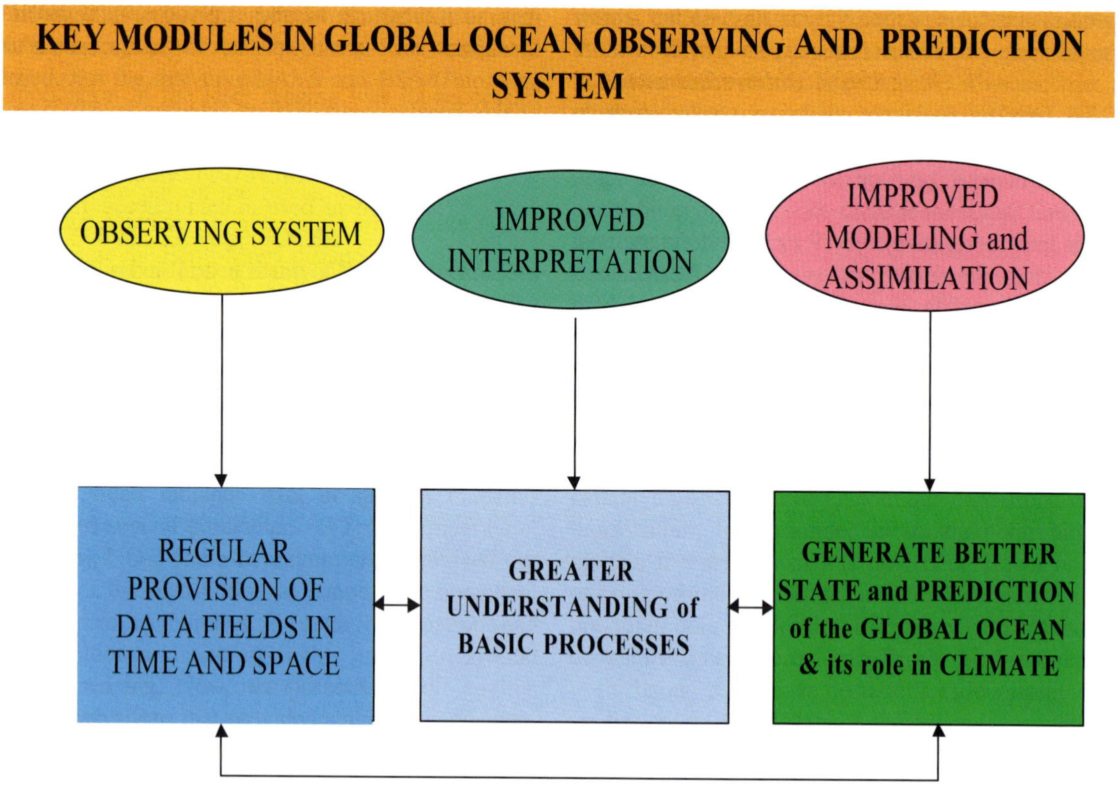

KEY MODULES IN GLOBAL OCEAN OBSERVING AND PREDICTION SYSTEM

Figure 17 Key modules of an integrated Earth system model applicable to the global ocean observing system consisting of an observation module, an interpretation module and a modeling and assimilation module. (After ESA SP-1227.)

to ensure the continuity of existing Earth observation data while at the same time developing new observation techniques. In the first decade of the new millennium Earth Observation from satellites, as an integrated part of the Global Ocean Observing System for Climate (*http://www.ocean.tamu.edu/OOSDP/FinalRept*) will be faced with exactly this challenging requirement.

Future plans and implementation of new Earth Observation satellite missions must also maintain a degree of flexibility to ensure optimum adjustment and complementarity with development and improvement of models and their subsequent need for data. The same is valid vis-à-vis technology development for *in situ* instruments. Regarding the latter, rapid development is undergoing for smart, autonomous expendable system and unmanned observing vehicles that return data via telemetry. The deployment, by 2004, of up to 3000 profiling Argo floats for salinity and temperature observations is one such example of a new and powerful element of a comprehensive international system for observing the global ocean. The expectation regarding future implementation of acoustic tomography and thermometry are other examples.

Acknowledgement

We are grateful to colleagues at the Nansen Environmental and Remote Sensing Center and at universities, research institutes and centers elsewhere for their valuable comments and support in making this article. Especially our thanks go to Ola M. Johannessen, Martin Miles, Elena Shalina, Lasse Pettersson, Einar Bjørgo, Pierre-Yves Le Traon, Paris W. Vachon, Steve Groom, and Marcos Portabello. We are also grateful to have received permissions to reproduce figure materials in this book-article. This is further specified in the figure captions. Financial support for this work has been obtained from Norwegian Space Centre, European Space Agency and Norwegian Research Council.

REFERENCES

Antoine, D., André, J.-M. and Morel, A. (1996). Oceanic primary production 1. Adaptation of a spectral light-photosynthesis model in view of application to satellite chlorophyll observations. *Global Biogeochem. Cycles*, **10**, 43–55.

Beal, R.C. and Pichel, W.G. (2000). Coastal and Marine Applications of Wide Swath SAR, Johns Hopkins APL, *Technical Digest*, vol. **21**, No. 1.

Behrenfeld, M.J. and Falkowski, P.G. (1997). Photosynthetic rates derived from satellite-based chlorophyll concentrations. *Limnol. Oceanogr.*, **42**, 1–20.

Bijlsma, L. (and 18 co-authors) (1996). Impacts, adaptions and mitigation of climate change: scientific–technical analysis. In: J.T. Houghton et al. (eds.), *Contribution to 2nd Assessment Report of the Intergovernmental Panel on Climate Change*, University Press, 572 pp.

Bjørgo, E., Johannessen, O.M. and Miles, M.W. (1997). Analysis of merged SMMR/SSMI time series of Arctic and Antarctic sea ice parameters. *Geophysical Research Letters*, **24**, 413–416.

Brandt, P., Romeiser, R. and Rubino, A. (1999). On the determination of characteristics of the interior ocean dynamics from radar signatures of internal solitary waves. *Journal of Geophysical Research*, **104**, 30039–30047.

Campbell, W.J. (1973). NASA remote sensing of sea ice AIDJEX. *Proceedings of the World Meteorological Organization Technical Conference*, Tokyo, Japan, WMO No. 350, pp. 55–56.

Carsey, F. (1992). *Microwave Remote Sensing of Sea Ice*, Geophysical Monograph 68, American Geophysical Union, p. 462.

Cavalieri, D., Gloersen, P. and Campbell, W.J. (1984). Determination of sea ice parameters with the Nimbus 7 SMMR. *Journal of Geophysical Research*, **89**(D4), 5355–5369.

Cavalieri, D.J., Gloersen, P., Parkinson, C.E., Zwally, H.J. and Comiso, J.C. (1997). Observed hemispheric asymmetry in global sea ice changes. *Science*, **278**, 1104–1106.

CEOS 97: Committee on Earth Observation Satellites: Toward an Integrated Global Observing Strategy, 1997 CEOS Yearbook, Published by Smith System Engineering Limited, UK.

Coles, S.G. and Tawb, J.A. (1990). Statistics of coastal flood prevention, *Philosophical Transactions of the Royal Society of London*, **A332**, 457–476.

Comiso, J.C. (1986). Characteristics of Arctic winter sea ice from satellite multispectral microwave observations. *Journal of Geophysical Research*, **91**(C1), 975–994.

Denman, K., Hofmann, E. and Marchant, H. (1996). Marine biotic responses to environmental change and feedbacks to climate. In: J.T. Houghton, L.G. Meira, B.A. Callander, N. Harris, A. Kattenberg and K. Maskell (eds.), *Climate Change 1995*, IPCC, pp. 487–516.

Doerffer, R. and Fischer, J. (1994). Concentration of chlorophyll, suspended matter, and glebstoff in case II waters derived from satellite coastal zone color scanner data with inverse modeling methods. *Journal of Geophysical Research*, **99**(C4): 7457–7466.

Elachi, C. (1987). *Spaceborne Remote Sensing: Applications and Techniques*, IEEE Press.

Emery, W.J., Wick, G.A. and Schluessel, P. (1995). Skin and bulk sea surface temperatures: satellite measurement and corrections. In: M. Ikeda and F.W. Dobson (eds.), *Oceanographic Applications of Remote Sensing*, CRC Press, pp.145–164.

EOS (January 1999). The US EOS Science Plan: The State of Science in the EOS Program, NASA NP-1998-12-069-GSFC.

EOQ (July 1998). *Earth Observation Quarterly*, No. 59, ESA Publication Division, Noordwijk, The Netherlands.

ESA SP-1227 (1998). The Science and Research Elements of ESA's Living Planet Program, ESA Publication Division, Noordwijk, The Netherlands.

ESA SP-1233 (1999). Report for Mission Selection, Gravity Field and Steady-State Ocean Circulation Mission, ESA Publication Division, Noordwijk, The Netherlands.

Espedal, H.A., Johannessen, O.M., Johannessen, J.A., Dano, E., Lyzenga, D. and Knulst, J.C. (1998). COASTWATCH'95, A tandem ERS-1/2 SAR detection experiment of natural film on the ocean surface. *Journal of Geophysical Research-Ocean*, **103**, 24969–24982.

Fraser, R.S., Matto, S., Yeh Eueng-Nan and McClain, C.R. (1997). Algorithm for atmospheric corrections of satellite measurements of ocean pigment. *Journal of Geophysical Research*, **102**(D14), 17,107–17,118.

Furevik, B. and Espedal, H. (2002). Wind energy mapping using Synthetic Aperture Radar. *Canadian Journal of Remote Sensing*. In press.

Gaspar, P. and Florens, J.P. (1998). Estimation of the sea state bias in radar altimeter measurements of sea level: Results from a new non-parametric method. *Journal of Geophysical Research*, **103**(C8), 15803–15814.

Gloersen, P., Campbell, W.J., Cavalieri, D.J., Comiso, J.C., Parkinson C.L. and Zwally, H.J. (1992). *Arctic and Antarctic Sea Ice, 1978–1987. Satellite Passive-Microwave Observations and Analysis*, NASA SP-511, 1992, 290 pp.

Gohin, F. and Cavanie, A. (1994). A first try at identification of sea ice using the three beam scatterometer of ERS-1. *International Journal of Remote Sensing*, **15**(6), 1221–1228.

Gordon, H.R. (1997). Atmospheric correction of ocean color imagery in the Earth Observing System era. *Journal of Geophysical Research*, **102**(D14), 17,081–17,106.

Gordon, H.R., Clark, D.K., Mueller, J.L. and Hovis, W.A. (1980). Phytoplankton pigments derived from the Nimbus-7 CZCS: initial comparisons with surface measurements. *Science*, **210**, 63–66.

Gordon, H.R. and Morel, A. (1983). *Remote Assessment of Ocean Color for Interpretation of Satellite Visible Imagery, a Review*. Lecture notes on coastal and estuarine studies. Springer-Verlag, 114 p.

Gordon, H.R., Brown, J.W. and Evans, R. (1988). Exact Rayleigh scattering calculations for use with the Nimbus–7 Coastal Zone Color Scanner. *Applied Optics*, **27**, 862–871.

Gordon, H.R. and Wang, M. (1994). Retrieval of water-leaving radiance and aerosol optical thickness over oceans with SeaWiFS: a preliminary algorithm. *Applied Optics*, **33**(3): 443–452.

Grandell, J., Johannessen, J.A. and Hallikainen, M. (1999). Development of a synergetic sea ice retrieval method for the ERS-1 AMI wind scatterometer and SSM/I radiometer, *IEEE Trans. Geoscience Remote Sensing*, **37**(2), 668–679.

Guymer, T., Font, J., Gaspar, P., Johannessen, J.A., van der Kolff, G. and Le Provost, C. (2000). Developing a European oceanographic satellite system. In: Report of the EuroGOOS Conference on Operational Ocean Observation form Space, Germany, 5–6 October, pp. 7–14.

Halpern, D. (ed.) (2000). *Satellites, Oceanography and Society*, Elsevier Oceanographic Series, 63, Amsterdam, The Netherlands.

Heimbach, P., Hasselmann, S. and Hasselmann, K. Statistical analysis and interpretation of WAM model data with global ERS-1 SAR wave mode spectral retrievals over 3 years, vol **103**, no. C4, pp. 7931–7979, April 15, 1998, Advances in Oceanography and Sea Ice Research using ERS Observations, *Journal of Geophysical Research-Ocean*.

Hesselmans, G.C. Calkoen and Wensink, H. Mapping of Seabed topography to and from SAR, ESA SP-414, May 1997, In Proceedings Third ERS Symposium, Space at the Service of our Environment, ESA Publication Division.

Johannessen, O.M., Campbell, W.J.R., Shuchman, S., Sandven, P. Gloersen, Johannessen, J.A., Josberger, E.G. and Haugan, P.M. (1992). Microwave study programs of air-ice-ocean interactive processes in the seasonal ice zone of the Greenland and Barents seas. In: F. Carsey (ed.), *Microwave Remote Sensing of Sea Ice*, AGU Geophysical Monograph 68, pp. 261–289.

Johannessen, O.M., Volkov, A.M., Grischenko, V.D., Bobylev, L.P., Sandven, S., Kloster, K., Hamre, T., Asmus, V., Smirnov, V.G., Melentyev, V.V. and Zaitsev, L. (1997). ICEWATCH – Ice SAR monitoring of the Northern Sea Route. In: J.H. Stel (ed. in chief), *Operational Oceanography. The Challenge for European Co-operation*. Proceedings of the First International Conference on EuroGOOS, 7–11 October 1996, The Hague, The Netherlands. Elsevier Oceanography Series, No. 62, pp. 224–233.

Johannessen, O.M., Miles, M. and Bjørgo, E. (1995). The Arctic's shrinking sea ice. *Nature*, **376**, 126–127.

Johannessen, O.M., Shalina, E.V. and Miles, M.W. (1999). Satellite evidence for an Arctic Sea ice cover in transformation. *Science*, **286**, 1937–1939.

Johannessen, J.A., Svendsen, E., Sandven, S., Johannessen, O.M. and Lygre, K. (1989). Three Dimensional Structure of Mesoscale Eddies in the Norwegian Coastal Current. *Journal of Phys. Oceanography*, **19**(1), 3–19.

Johannessen, J.A., Shuchman, R.A., Digranes, G., Lyzenga, D., Wkerman, Johannessen, O.M. and Vachon, P.W. (1996). Coastal ocean fronts and eddies imaged with ERS-1 SAR. *Journal of Geophysical Research*, **101**(C3), 6651–6668.

Johannessen, J.A., le Provost, C., Drange, H., Srokosz, M., Woodworth, P., Schlussel, P., Le Grand, P., Kerr, Y., Wingham, D. and Rebhan, H. (2001). Observing the Ocean from Space – Emerging Capabilities in Europe. In: C.J. Koblinsky and N.R. Smith (eds), observing the ocean in the 21st century, Bereau of Meteorology, Melbourne, Australia, pp. 198–208.

Kerr, Y. et al. (April, 1999). SMOS: Soil Moisture and Ocean Salinity, Proposal selected under the ESA Living Planet Program: Earth Explorer Opportunity Mission.

Komen, G.J., Cavaleri, L., Donelan, M., Hasselmann, K., Hasselmann, S. and Janssen, P.A.E.M. (1994). *Dynamics and Modelling of Ocean Waves*, Cambridge University Press, 532 pp.

Kwok, R. (1998). The RADARSAT Geophysical Processor System. In C. Tsatoulis and R. Kwok (eds.), *Analysis of SAR data of the Polar Oceans: Recent Advances*, Springer Verlag, pp. 235–257.

Lagerloef, G.S.E., Swift, C.T. and Le Vine, D.M. (1995) Sea surface salinity: The next remote sensing challenge, *Oceanography*, **8**, 44–50.

Lasnier, P. (1996). In: *ESA SP-1176/II, New Views of the Earth: Application Achievements of ERS-1*, ESA Publication Division, Noordwijk, The Netherlands.

Le Traon, P.-Y. and Oger, F. (1998). ERS-1/2 orbit improvement using the TOPEX/POSEIDON: The 2 cm challenge, Advances in Oceanography and Sea Ice Research using ERS Observations. *Journal of Geophysical Research*, **103**(C4), 8045–8058.

Le Traon, P.Y. and Dibarboure, G. (1999). Mesoscale mapping capabilities from multiple altimeter missions. *Journal of Atmospheric Oceanic Technology*, **16**, 1208–1223

Lindstrøm, E., Johannessen, J.A., Navalgund, R., Tanaka, T., Summerhayes, C., Fellous, J.-L. and Marra J. (2000). An Ocean Theme for the IGOS Partnership: Final report from the Ocean Theme Team, (http://www.igospartners.org).

Liu, A.K., Zhao, Y. and Wu, S.Y. (1999). Arctic sea ice drift from wavelet analysis of NSCAT and SSM/I data. *Journal of Geophysical Research*, **104**(C5), 11,529–11,538.

Levitus, S. and Boyer, T.P. (1994). *World Ocean Atlas 1994*, Volume 4: Temperature, NOAA Atlas NESDIS 4, NOAA, 117 pp.

McClain, E.P., Pichel, W.G. and Walton, C.C. (1985). Comparative performance of AVHRR-based multichannel sea surface temperatures. *Journal of Geophysical Research*, **90**(C6), 11587–11601.

Meier, M.F. (1984). Contribution of small glaciers to global sea level. *Nature*, **343**, 115–116.

Moore, G.F., Aiken, J. and Lavender, S.J. (1999). The atmospheric correction of water colour and the quantitative retrieval of suspended particulate matter in Case II waters: application to MERIS. *International Journal of Remote Sensing*, **20**, 1713–1733.

Morel, A. and Prieur, L. (1977). Analysis and variations in ocean color. *Limnol. Oceanogr*, 2(4):709–722.

Morel, A. (1988). Optical modeling of the upper ocean in relation to its biogenous matter content (case I waters). *Journal of Geophysical Research*, **93**(C9), 10,749–10,768.

Morel, A. (1998). Minimum Requirements for an Operational Ocean-color Sensor for the Open Ocean. Report of the IOCCG No. 1, Ed. IOCCG. Darthmouth, Canada, 46 p.

O'Reilly, J.E. et al. (1998). Ocean color chlorophyll algorithms for SeaWiFS. *Journal of Geophysical Research*, **103**(C11), 24,937–24,953.

Pedersen, J.P. et al. (February, 1996). Oil spill detection by use of ERS SAR data – From R&D towards pre-operational early warning detection service, ESA SP-383, *Proceedings Second International Workshop on ERS Applications*, ESA Publication Division, Noordwijk, The Netherlands.

Piexoto, J.P. and Oort, A.H. (1992). *The Physics of Climate*, American Institute of Physics, New York.

Platt, T. and Sathyendranath, S. (1988). Oceanic primary production: estimation by remote sensing at local and regional scales. *Science*, **241**, 1613–1620.

Robinson, I.S. (1985). *Satellite Oceanography: An Introduction for Oceanographers and Remote-sensing Scientists*. John Wiley & Sons, 455 p.

Sandven, S., Johannessen, O.M., Miles, M.W., Pettersson, L.H. and Kloster, K. (1999). Barents Sea seasonal ice zone features and processes from ERS-1 SAR: SIZEX92. *Journal of Geophysical Research*, **104**(C7), 15,843–15,858.

Sandven, S., Gronvall, H., Seina, A., Valeur, H., Nizovsky, H.,Steen, M., Andersen, H. and Haugen, V.E.J. (1998). Operational Sea Ice Monitoring by Satellites in Europe. OSIMS Final report. NERSC Technical Report no. 148, 91 pp.

Schiller, H. and Doerffer, R. (1999). Neural network for emulation of an inverse derivation of case II water properties from MERIS data. *International Journal of Remote Sensing*, 20(9), 1735–1746.

Smith, D.M. (1998). Recent increase in the length of the melt season of perennial Arctic sea ice. *Geophysical Research Letters*, **25**, 655–658.

Smith, N. and Koblinsky, C. (eds) (2001). Observing the Ocean for Climate in the 21st Century, Bureau of Meteorology, Melbourne, Australia.

Steffen, K., Key, J. Cavalieri, D. J., Comiso, J., Gloersen, P., St. Germain, K. and Rubinstein, I. (1992). The Estimation of Geophysical Parameters using Passive Microwave Algorithms. In: F. Carsey (ed.), *Microwave Remote Sensing of Sea Ice*, AGU Geophysical Monograph 68, pp. 201–231.

Stoffelen, A. and Andersen, D. (1997). Ambiguity removal and assimilation of scatterometer data. *Q.J.R. Meteorol. Soc.*, **123**, 491–518.

Stogryn, A. (1971). Equations for calculating dielectric constant of saline water. *IEEE Trans. Microwave Theory Tech.*, MTT-19, 733–736.

Svendsen, E.A., Kloster, K., Farrelly, B.A., Johannesen, O.M., Johannessen, J.A., Campbell, W.J., Gloersen, P., Cavalieri, D. and Matzler, C. (1983). Norwegian Remote Sensing Experiment: Evaluation of the Nimbus 7 Scanning Multichannel Microwave Radiometer for Sea Ice Research. *Journal of Geophysical Research*, **88**(C5), 2781–2791.

Svendsen, E., Mätzler, C. and, Grenfell, T.C. (1987). A model for retrieving total sea ice concentration from spaceborne dual-polarized passive microwave instrument operating near 90 GHz. *International Journal of Remote Sensing*, 8(10), 1479–1487.

Tanré D., Kaufman, Y.J. Herman, M. Mattoo, S. (1997). Remote sensing of aerosol properties over oceans using the MODIS/EOS spectral radiances. *Journal of Geophysical Research*, **102**(D14), 16,971–16,988.

Yoder, J.A. et al. (1998). Status for Satellite Ocean-color Missions: Considerations for Complementary Missions. Report of the IOCCG No. 2, Ed. IOCCG. Darthmouth, Canada, 46 p, ISSN 1098–6030

Ulaby, F.T., Moore, R.K. and Fung, A.K. (1982). Real Aperture Side-Looking Airborne Radars. In: *Microwave Remote Sensing – Active and Passive: Volume II – Radar Remote Sensing and Surface Scattering and Emission Theory*, Addison-Wesley Publishing Co., Reading, MA, pp. 562–629.

Ulaby, F. T., Moore R.K. and Fung, A.K. (1982). Synthetic Aperture Side-Looking Airborne Radar Systems. In: *Microwave Remote Sensing – Active and Passive: Volume II – Radar Remote Sensing and Surface Scattering and Emission Theory*, Addison-Wesley Publishing Co., Reading, MA, pp. 630–745.

Vachon, P.W., Adlakha, P. Edel, H. Henschel, M. Ramsey, B. Flett, D. Rey, M., Staples G. and Thomas, S. (2000). Canadian Progress Towards Marine and Coastal Applications of Synthetic Aperture Radar, Johns Hopkins APL, *Technical Digest*, **21**(1), 33–40.

Warrick, R.A., Le Provost, C., Meier, M.F., Oerlemans, J. and Woodworth, P.L. (1996). The science and climate change. In: J.T. Houghton *et al.*, (eds), Contribution to 2nd Assessment Report of the Intergovernmental Panel on Climate Change, University Press, 572 pp.

Wingham, D. *et al.* (1999). CRYOSAT, A mission to determine fluctuations in the mass of the Earth's land and marine ice fields, Proposal selected under the ESA Living Planet Program: Earth Explorer Opportunity Mission, April.

Wright, J.W. (1978). Detection of ocean waves by microwave radar: The modulation of short gravity-capillary waves, *Boundary Layer Meteorology*, **13**, 87–105.

Wentz, F.J., Gentemann, C., Smith, D. and Chelton, D. (2000). Satellite Measurements of Sea Surface Temperature through clouds. *Science*, **288**, 847–850.

ANNEX A ABBREVIATIONS AND ACRONYMS

AMW	Active Microwave
ASAR	Advanced Synthetic Aperture Radar
AVHRR	Advanced Very High Resolution Radiometer
CEOS	Committee on Earth Observation Satellites
CZCS	Coastal Zone Color Scanner
DMSP	Defense Meteorological Satellite Program
EC	European Commission
ECMWF	European Center for Medium-Range Weather Forecasting
EM	Electromagnetic
ENVISAT	Environmental Satellite
EO	Earth Observation (normally meaning the use of space-borne data)
ERS	European Remote Sensing Satellite
ERTS	Earth Resource Technology Satellite
ESA	European Space Agency
EuroGOOS	European Association for the Global Ocean Observing System
FCCC	UN Framework Convention on Climate Change
FY	Firstyear
GLI	Global Imager
GOCE	Gravity Field and Steady-state Ocean Circulation Explorer
GODAE	Global Ocean Data Assimilation Experiment
GOOS	Global Ocean Observing System
GPS	Global Positioning System
GRACE	Gravity Recovery and Climate Experiment
IGBP	International Geosphere and Biosphere Program
IOC	Intergovernmental Oceanographic Committee
IPCC	Intergovernmental Panel on Climate Change
IR	Infra-Red
LRI	Low Resolution Image
MERIS	Medium Resolution Imaging Spectroradiometer
METOP	Meteorological Operational
MISR	Multi-angle Imaging Spectroradiometer
MODIS	Moderate Resolution Imaging Spectroradiometer
MOS	Modular Opto-electronic Scanner
MY	Multiyear
NESDIS	National Environmental Satellite, Data & Information Service
NOAA	National Oceanic and Atmospheric Administration, USA
NORSEX	Norwegian Remote Sensing Experiment
NSCAT	NASA Scatterometer
OCI	Ocean Color Imager
OCM	Ocean Colour Monitor
OCTS	Ocean Color and Temperature Sensor
OOSDP	Ocean Observation System Development Panel
OSMI	Ocean Scanning Multispectral Imager
PMW	Passive Microwave
POLDER	Polarisation and directional
SAR	Synthetic Aperture Radar
SeaWiFS	Sea Wide Field of view Sensor
SIMBIOS	Sensor Inter-comparison for Marine Biological and Interdisciplinary Ocean Studies
SIZEX	Seasonal Ice Zone Experiment
SLR	Side-Looking Radar
SMMR	Scanning Multi-channel Microwave Radiometer
SMOS	
SSM/I	Special Sensor Microwave Imager
SSS	Sea Surface Salinity
SST	Sea Surface Temperature
TIR	Thermal Infrared
TMI	TRMM Microwave Imager
TRMM	Tropical Rainfall Microwave Measurements
VNIR	Visual and Near-Infrared
WAM	Wave Model
WCRP	World Climate Research Program
WMO	Word Meteorological Organization
WWW	World Weather Watch

ANNEX B   LIST OF SELECTED EO SATELLITES AND SENSORS FOR OCEANOGRAPHY: PAST, PRESENT, FUTURE

A list of the most common Earth observation satellites (past, present and future) and sensors is provided in this annex, together with some relevant sensor and satellite characteristics:

- Under the **satellite** column the name of the spaceborne platform is provided.
- In the **Instrument** column we make reference to the type of technology used and/or measurements performed by a given sensor.
- Under **acronym**, we give the common name used for a given sensor.
- The **swath width** is the size of the across track area on the ground which is covered by the instruments.
- The **spatial resolution** is approximately the size of the smallest feature on the ground that can be resolved with the instrument.
- The availability of data is given in terms of the **operating period** for past sensors, the start of the operating period for present sensors, and the approved/planned launch-date for future satellites.

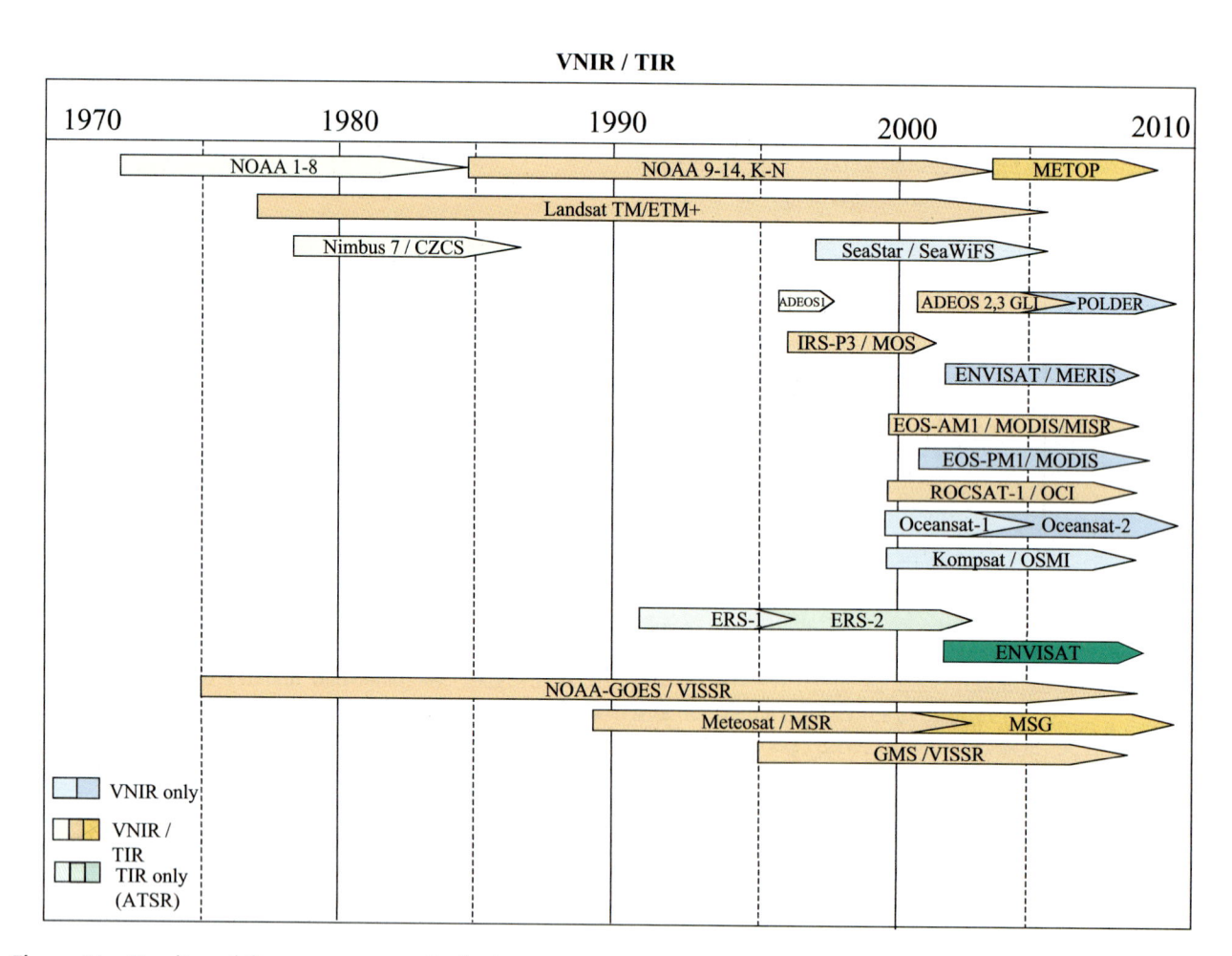

VNIR / TIR

Figure A1 Timeline of the most common Earth-observation VNIR/TIR passive sensors operated in space from 1972, together with those approved and planned towards 2010. The term "VNIR/TIR passive sensor" makes reference to instruments measuring either the radiation in the visible, near infrared part of the solar electromagnetic spectrum, which is reflected at the surface of the Earth (VNIR) or the radiation emitted by the Earth in the thermal infrared (TIR). VNIR sensors are used to derive biogeochemical parameters in the ocean upper layer, such as the concentration of chlorophyll pigments and suspended sediment. TIR instruments are mainly used to derive sea surface temperature.

PMW / AMW

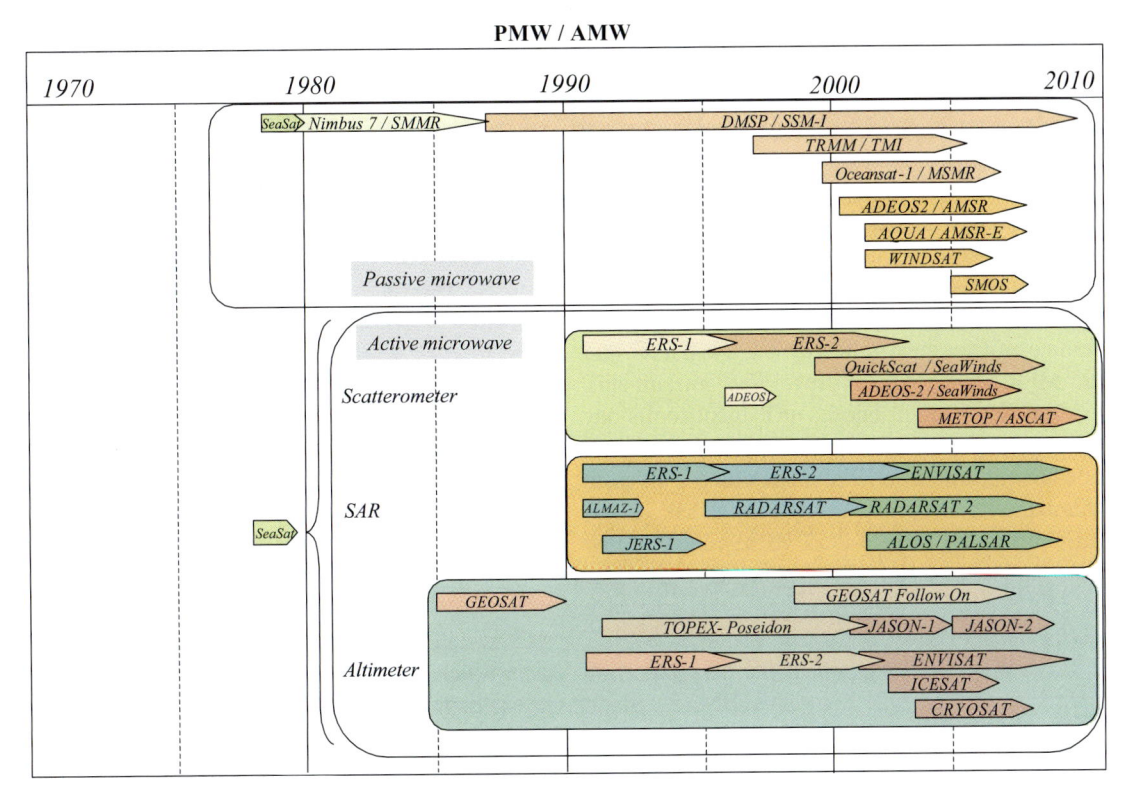

Figure A2 Timeline of the most common Earth-observation microwave sensors operated in space from 1978, together with those approved and planned towards 2010. The term "passive microwave sensors" makes reference to instruments that measure the microwave radiation emitted by the Earth (PMW). The term "active microwave sensors" refers to instruments transmitting a source of electromagnetic energy towards the Earth and measuring the portion of the signal that is scattered back to antenna (AMW). Active microwave sensors usually include three main types of instruments: scatterometers from which the wind field over the ocean can be derived, imaging synthetic aperture radars – SAR that give access to a number of physical processes and phenomena at the surface of the sea (current front, waves, wind, etc.), and altimeters from which the sea-surface height can be estimated in addition to significant wave-height and wind speed.

Past

Satellite	Instrument	Acronym	Swath width	Spatial resolution	Operating period
VNIR only					
ADEOS-1	Polarization and Directionality of the Earth's Reflectance	POLDER	1400 km	7×6 km	1996–1997
VNIR and TIR					
Nimbus-7	Coastal Zone Ocean Scanner	CZCS	1600 km	0.825 km	1978–1986
ADEOS-1	Ocean Color and Thermal Scanner	OCTS	1400 km	0.700 km	1996–1997
NOAA-1–8	Advanced Very High Resolution Radiometer	AVHRR/1	3000 km	1.1 km	1972–1984
ERS 1	Along-Track Scanning Radiometer	ATSR (TIR only)	500 km	1 km	1991–1996

Past Continued

Satellite	Instrument	Acronym	Swath width	Spatial resolution	Operating period
PWM					
SeaSat	Scanning Multichannel Microwave Radiometer	SMMR	3000 km	Function of the frequency (30–150 km)	1978
Nimbus-7	Scanning Multichannel Microwave Radiometer	SMMR	3000 km	Function of the frequency (30–150 km)	1978–1988
AMW					
SeaSat	Synthetic Aperture Radar	SAR	100 km	25 m	1978
	Wind-Scatterometer	SASS	750 km	50 km	
	Altimeter	RA		5–30 km	
ERS-1	Synthetic Aperture Radar	SAR	100 km	30 m	1991–1996
	Wind Scatterometer	AMI-WIND	500 km	50 km	
	Altimeter	RA		7 km	
JERS-1	Synthetic Aperture Radar	SAR	75 km	18 m	1992–1995
ALMAZ-1	Synthetic Aperture Radar	SAR	40–350 km	10–15 m	1991–1992
ADEOS-1	Scatterometer	NSCAT	600 km	50 km	1996–1997
GEOSAT	Altimeter	GEOSAT	NA	6.7 km	1985–1990

NA: not applicable.

Present

Satellite	Instrument	Acronym	Swath width	Spatial resolution	Operating period
VNIR only					
Orbview-2	Sea-viewing, Wide Field Sensor	SeaWiFS	1500–2800 km	1.1 km / 4.4 km	From 1997
Oceansat-I	Ocean Colour Monitor	OCM	1500 km	350 m	From 1999
TERRA (EOS-AM1)	Multi-angle Imaging Spectroradiometer	MISR	370 km	275 m	From 1999
Kompsat	Ocean Scanning Multispectral Imager	OSMI	800 km	850 m	From 1999
VNIR and TIR					
LANDSAT 5	Thematic Mapper	TM	185 km	120 m (TIR)	From 1976
LANDSAT-7	Enhanced TM	ETM+	705 km	60 m (TIR)	From 1999
NOAA-POES NOAA-9-14; K-N	Advanced very high resolution radiometer	AVHRR/2 AVHRR/3	approx. 3000 km	1.1 km	From 1984
ERS 2	Along Track Scanning Radiometer	ATSR (TIR only)	500 km	1 km	From 1995
PRIRODA (MIR)	Modular opto-electronic scanner	MOS	82 km	650 m	1995–1999
IRS-P3	Modular opto-electronic scanner	MOS	200 km	520	From 1996
ROCSAT-1	Ocean color imager	OCI	690 km	0.825 km	From 1999
TERRA (EOS-AM1)	Moderate Resolution Imaging Spectro Radiometer	MODIS	2300 km	250 m/1000 m	From 1999

Present Continued

Satellite	Instrument	Acronym	Swath width	Spatial resolution	Operating period
TERRA (EOS-AM1)	Advanced Spaceborne Thermal Emission and Reflection Radiometer	ASTER	60 km	90 m (TIR)	From 1999
NOAA-GOES 8–10	Visible and IR Spin Scanning Radiometer	VISSR	Horizon to horizon	1 km (VNIR.) 7 & 14 km (TIR)	From 1975
GMS Program GMS-5	Visible and IR Spin-Scan Radiometer	VISSR	Full Earth disk*	1.25 km (VNIR) 5 km (TIR)	From 1984
Meteosat program Meteosat-7	Visible and IR Spin-Scan Radiometer	MSR	Full Earth disk*	2.5 km (VNIR) 5 km (TIR)	From 1989
PWM					
DMSP	Microwave imager	SSM/I	1400 km	25–50 km	From 1987
TRMM	Microwave imager	TMI	760 km	4.4–45 km	From 1997
Oceansat-1	Microwave imager	MSMR	1360 km	40 to 120 km	From 1999
AMW					
ERS-2	Synthetic Aperture Radar	SAR	100 km	25 m	From 1995
	Wind Scatterometer	AMI-WIND	500 km	50 km	
	Altimeter	RA	3–5 km	7 km	
RADARSAT	Synthetic Aperture Radar	SAR	max. 500 km	28 m/50 m/ 100 m	From 1995
QuickScat	Wind scatterometer	SeaWinds	1800 km	25 km	From 1999
TOPEX-Poseidon	Altimeters	T/P	NA	6 km	From 1992
GFO	GEOSAT follow-on altimeter	RA	NA	5–10 km	From 1998

*: Swath width for geostationary satellite are indicated as full Earth disc.
NA: not applicable.

Future

Satellite	Instrument	Acronym	Swath width	Spatial resolution	Launch
VNIR only					
ENVISAT**	Medium Resolution Imaging Spectrometer	MERIS	1150 km	300 m/1200 m	2002
ADEOS-2	Polarization and Directionality of the Earth's Reflectance	POLDER-2	2400 km	6 km	2002
ADEOS-3	Super Global Imager	S-GLI (OCI)	1600 km	1 km	2004
VNIR and TIR					
ADEOS-2	Global Imager	GLI	1600 km	250 m–1 km	2002
NOAA-POES series	Advanced Very High Resolution Radiometer	AVHRR/3	3000 km	1 km	2001
METOP	Advanced Very High Resolution Radiometer	AVHRR/3	3000 km	1 km	2005
ENVISAT**	Advanced Along Track Scanning Radiometer	AATSR (TIR)	500 km	1 km	2002

Future Continued

Satellite	Instrument	Acronym	Swath width	Spatial resolution	Launch
AQUA (EOS-PM1)	MODIS	MODIS	2330	1 km	2002
NOAA-GOES GOES-L – M	Visible and IR Spin-Scan Radiometer	VISSR	Full Earth disc*	7, 14 km	2000/2001
MSG	Meteosat second generation	SEVIRI	Full Earth disk*	5 km	2003
PWM					
ADEOS-2	Advanced Microwave Scanning Radiometer	AMSR	1450 km	5–50 km	2002
AQUA (EOS-PM1)	Advanced Microwave Scanning Radiometer	AMSR-E	1450 km	5–50 km	2002
NPOESS	Passive microwave polarimetry	WindSat	1025 km	25 km	2002
SMOS	L-band microwave radiometer	SMOS	1200 km	50 km	2005
AMW					
ENVISAT**	Advanced SAR	ASAR	Max. 420 km	30 m/150 m/1 km	2002
	RA-2	RA-2		7 km	
RADARSAT2	Synthetic Aperture Radar	SAR	Max. 500 km	28 m/50 m/100 m	2003
ALOS	Phased array type L-band SAR	PALSAR	250–350 km	100 m	2003
ADEOS-2	Scatterometer	SeaWinds	1800 km	50 km	2002
METOP	Advanced Wind Scatterometer	ASCAT	1000 km	25 km	2005
JASON**	Altimeter	Jason	NA	6 km	2001
ICESAT	Laser altimeter	GLAS	NA	250 m	2002
CRYOSAT	Radar altimeter	–	NA	400 m	2004

*: Swath width for geostationary satellite is indicated as full Earth disc.
**: JASON and ENVISAT were successfully launched on 7th December 2001 and 1st March 2002 respectively. Note that GRACE 2002 and GOCE (2005) will acquire gravity measurements by combining accelerometers and GPS tracking.
NA: not applicable.

FRANÇOIS BARLIER* AND MICHEL LEFEBVRE**

A new look at planet Earth: Satellite geodesy and geosciences

INTRODUCTION

Yes! The Earth is also a planet! In the recent past, looking back at the long list of scientific space missions launched by space agencies, you have the impression that neither any celestial body, nor astronomical or geophysical theme has not been covered. But looking more carefully, you soon realize that something is missing... the Earth!

It is not the case that space is ignoring the Earth; indeed, huge programs are devoted to Earth observations, but they are designed as if there was no need for a scientific study of our planet, itself.

This equivocation is not just a semantic one. It seems that the Moon or other planets in the solar system deserve more attention than the Earth. As an example, the Magellan mission made a complete mapping of the surface of Venus which took five years with a SAR instrument (Synthetic Aperture Radar); several years before such an instrument could have very usefully flown around the Earth.

Now, there is a new appeal for understanding the Earth, and the situation has changed drastically. The Chernobyl accident caused a strong public reaction and created a widespread feeling that the atmosphere knows no borders. The recent large-scale ocean-atmosphere movements, like the ENSO (El Niño), widely discussed on by TV channels, contributes to public awareness.

The scientific community was perfectly aware of the lack of knowledge of the Earth as a planet and recognised that the main problem was the poor quality of observations. The need for a global perspective was such that a dedicated planetary program was undertaken and implemented in 1957–1958; the objective of this IGY (International Geophysical Year) was to collect as many geophysical measurements as possible from world-wide well distributed sites. Although successful in some fields it also showed the limits of this approach over the long term. As a coincidence, the first SPUTNIK satellite was launched in October 1957 and the government of USSR claimed officially that it had to be considered as a contribution to IGY, not a bad vision indeed.

The space agencies recently reconsidered their programs and made the knowledge of the Earth a top priority. ESA (European Space Agency) started a new program called "Earth Living Planet" while NASA (National Aeronautics and Space Administration) started a huge program devoted explicitly to Earth sciences. To be even more explicit, they named it "Destination Earth"; this movement was worldwide. Today the space agencies try to optimise their participation in a common endeavour through such ad hoc committees as CEOS (Committee on Earth Observation Satellites).

After 35 years of satellite geodesy and oceanography, our objective is to show that in spite of the absence of dedicated important programs in Earth physics from space, there was a de facto strategy. It began at the time when the necessity to undertake the study of the Earth became apparent and has produced significant results, as well as developed both tools and a living and multi-disciplinary community ready to go. We will show this development that through a historical overview of activities over the past 35 years, presenting the different phases and major turns in Earth-space science.

* Observatoire de la Côte d'Azur, Grasse, France
** Centre National d'Etudes Spatiales, Toulouse, France

1. FROM GEODESY TO SATELLITE GEODESY BEFORE 1957

Geodesy is one of the oldest disciplines in the IUGG (International Union of Geodesy and Geophysics) and has its roots in the early stages of civilization. When you live somewhere, the first thing to do is to know where you are, what is your local environment like, then to share your views with your neighbours, extend your perspective from local to regional and finally to planetary scales.

The Chamber's dictionary gives the following definition of geodesy: "it measures the Earth and its parts on a larger scale". The etymology from Greek tells us that geodesy comes from GE, the Earth and DAIEIN to share. Indeed what we have now to share is the EARTH.

When Sputnik was launched in October 4, 1957, geodesy was confined to improving the accuracy of measurements as well as increasing the resolution of the grid of networks. A high level of expertise existed and was exercised in specialized geodetic institutes, most of them sponsored by governments. The links with equivalent bodies working on military objectives have been varied but on average were well established, such that the release of data and results was often prevented. Concerning the shape of the Earth and its gravity field, numbers will give some ideas of the status. At the level of the continental formations, there were some efforts to get homogeneous sets of parameters representing at best the results obtained by triangulation and by astrogeodesy on the shape of the Earth and to get coherent geodetic systems like the NAD datum (North American Datum) or the EUR 50 datum (European datum). The relative accuracy, at least for the horizontal components, was acceptable inside a given system and when close to the main fundamental networks was of the order of 10^{-5} or 10^{-6} (1 meter over 1000 kilometers). However, it was uncertain in the limits of networks when no closure was available (as an example the south of Spain relative to EUR 50). The precision of vertical angle measurements was also limited by the atmospheric refraction.

The knowledge of the relationships between the origin of these systems and the center of mass of the Earth was poor, such that the absolute positions of the stations were maybe in error up to several hundred meters. No data were available over the oceans that are over 70% of the Earth surface. Over the continents, the data were sparse in many areas. Moreover, the networks were measured by discrete campaigns at different times with different instruments by different teams. Despite the high quality of the actors, there were some intrinsic limitations. The users had no other way than asking to their national geodetic institutes to make the geodetic links; the advantage was to have them made by professionals and well controlled. But the lack of a unified system was still a problem. Even in the 1960s the position of the same radar antenna given by 2 institutes from neighbouring countries differed in several cases by an amount larger than the internal error; it was simply coming from the use of different reference systems. The above comments do not aim at making any assessment of classical geodesy but rather at putting emphasis on the limitations of classical geodetic systems at the scale of the Earth. Now, satellite is there but how to use it at the best?

2. TRANSITION EXPLORATORY PHASE 1957–1970

2.1 The geometrical optical phase: A too obvious approach

First of all, how to observe? At first, the favoured method of observation was optical. Observatories had or developed big cameras and made photographs of artificial satellites lightened by the Sun, the spatial reference being provided by the surrounding stars. The first obvious idea was to use a satellite as a target, high enough to make intercontinental links that seem the most obviously missing element in geodesy and to have access to the 3D dimension. Everybody was enthusiastic; nobody realised that a satellite indeed ignores national boundaries: we remember the shock of some authorities, military or otherwise, when scientists published the co-ordinates of Malvern (UK) and Nice (FR), breaking for ever the long tradition to keep such data and positions secret.

The geometrical approach was limited by the magnitude of the satellite. A first solution was to launch some dedicated satellites shaped like balloons with a large diameter, about 25 meters, such as the ECHO 1 and 2 satellites (Note: it was the common approach used in a telecommunication experiment, the expectation being to use the satellite as a reflector for electromagnetic waves) The visual magnitude of ECHO satellites was around, 1, so that they were accessible to many small existing cameras, allowing the increase in the number of stations within the nets to link.

The two ECHO satellites were so bright that it was possible for anybody to watch them visually with the naked eye, and to follow their motion across the stars. It was attractive enough to drive the newspapers to publish the times that these "New Stars", passed over head.

Several million people watched and acquired a personal physical feeling for the existence of satellites and became aware that we were entering the space age. The visual observations were not just curiosities; networks of amateurs observed with some optical instruments and provided the directions, elevation and azimuth to some centers using these observations, especially from very low altitude satellites. The result was the first models of the Earth's gravity field and of the upper atmosphere density.

Several geodetic institutes were thus able, through intensive campaigns, to make geodetic intercontinental links. One on the most successful campaign was the geodetic connection between Europe and Africa.

Going beyond, a more optimized dedicated program was undertaken; PAGEOS, a better designed and more stable balloon satellite was put in orbit at a higher altitude. The existing BC – 4 cameras from the US Coast and Geodetic Survey were deployed in networks occupying 40 sites well distributed around the Earth; the geocentric positions of the 40 stations were published and were considered by this time as making one of the first homogeneous global Earth reference systems. It was in fact a dead-end and it is interesting to understand why. First the accuracy was not good enough. The best result obtained for the positioning was at the 10–15 meters level, but there is also a major disadvantage; this new reference system was not accessible to the common user.

2.2 The space geodetic scheme

The general principles of satellite geodesy may be depicted with an elementary diagram, the **GSM** scheme (Figure 1).

G is the center of mass of the Earth,
S is the position of a tracking station,
M is the position of the moving satellite.

At any time **GM = GS + SM**:

SM corresponds to the observation. It may be the measurement of range, range rate, directions whatever.
GS corresponds to the position of the tracking station. It is an unknown to be determined and it must be referred to a unified homogeneous Earth reference system.
GS moves (due to tides, tectonic motions, and local effects). The terrestrial reference frame, the position of the axis of rotation as well as the speed of rotation (the parameters of the Earth rotation) vary over time.
GM characterises the motion of the satellite.

The unknowns are:

– the initial conditions (position and velocity of the satellite) at a reference time.

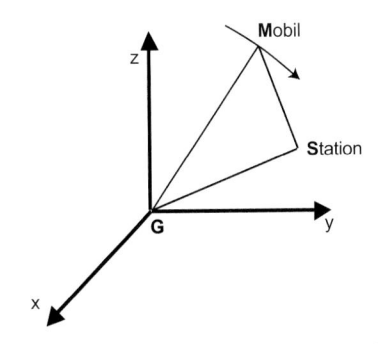

Figure 1 A simplified scheme for geodesy.

– the acting forces on the satellite (gravity field of the Earth, solid and ocean tides, air drag, direct and reflected solar radiation pressure …).

So the **GSM** game is easy to explain. Knowing the functions relating quantities (the measurements) to some parameters considered as unknowns, you have to compare your observation with a value computed by marking a first guess about the unknowns. In a linearized approach, you have to compute the partial derivatives of the observation relative to the unknowns. Then you have to minimize the differences between all computed and observed values using statistical assumptions and an algorithm of adjustment. The problem is in fact not linear, so you need to process by using iterations. Among the unknowns, there are possible systematic biases in your system or errors in the physics of your model. You expect to converge on accurate and reliable values.

In the dynamical approach, the motion of the satellite is an important component in many respects. Firstly, it is the natural way to scan any part of the planet. This sampling can be optimized using the orbital parameters that can be adjusted accordingly. Secondly, the perturbations of the satellite motion provide the determination of the acting forces and in priority the gravity field of the Earth. Conversely, a perfect knowledge of the forces provide a powerful constraint on the orbit determination. Then the accurate knowledge of the position and velocity of the satellite in a well-controlled reference system allows us to have the benefit of other measurements performed by onboard instruments; such as, for instance, those provided by radar altimeters in oceanography.

2.3 First look at observing technique candidates

2.3.1 Photographic observations

Photographic observations were the only precise material available for a while; nowadays, these types of observations are no longer used (except for specific applications, such the observation of space debris); but it is still interesting to examine all the efforts made.

Photographic Observations

The photography of satellites with respect to stars was the first type of precise measurements of the angular positions of satellites.

The procedure requires a capability of observing even satellites with faint magnitudes. The only way was to integrate enough light by a longer time of exposure, that means to use cameras able to track the satellite during its pass. The fully optimized system was the American network of BAKER-NUNN cameras that were modified by adding an extra degree of freedom. It was successful and a world-wide network was

(continued)

implemented. There was a large effort in automation not only to observe but also to measure the films and provide in a few weeks better right ascensions and declinations of the satellite, the reference being provided by the surrounding stars.

These photographic observations were the core of the first global determination of the gravity field (see Standard Earth below). The accuracy was limited to about 2 arc seconds, that is to say around 10 to 20 meters.

In parallel, other countries were developing their own systems, like the AFU 75 camera in USSR (aperture of 20 cm and focal length of 75 cm), the tracking camera ANTARES at the Nice Observatory, the HEWITT camera at the Malvern Observatory and the ZEISS camera in Germany.

For the purpose of developing the geometrical space geodesy with satellites like ECHO or PAGEOS, many smaller cameras were used: as examples the Wild BC4 camera of the Coast and Geodetic Survey in USA (aperture 11.7 cm, focal length 30.5 cm), the IGN camera in France (aperture 10 cm, focal length 30 cm). Schmidt Telescopes were also successfully used for observing flashing satellites (ANNA – 1B, GEOS-A and -B satellites launched by the USA) or for observing laser returns on the satellites and obtaining the 3 components of the station-to-satellite vector. But nowadays, all these techniques are generally abandoned, taking into account the progress realized with the laser techniques and the radio techniques. However, they played a major role in the beginning and old photographic data continued to be used in gravity field modeling to ensure a good decorrelation between the different harmonics.

2.3.2 Satellite laser ranging

Laser technology is able to emit highly concentrated phased optical energy in very narrow beams. This capability was used in transmitting energy from the ground towards the satellite equipped with corner cubes that reflect the light back in the same direction. The returned energy is detected and the time elapsed between emission and reception, after some corrections, provides the range. The first satellite equipped by the USA with laser corner cubes was BEB (1964). There was a competition to get first returns. The Goddard Space Flight Center (GSFC) team in the USA got the first ones in December 1964. The French CNRS team (Centre National de la Recherche Scientifique) at Verrières-le-Buisson obtained the first pass at the Haute Provence Observatory in January 1965. They only presented the first orbit computed with laser observations at the COSPAR meeting in Buenos-Aires in spring 1965. The claimed precision from the rms (root mean square) of the measurements was of about 1.0–1.5 meters.

Today, laser ranging is the most accurate technique, and it is still open to many improvements. One of the advantages is that the onboard equipment is light, cheap, has an infinite lifetime and does not consume any energy.

First Laser Returns, A Hunting Party

During the winter of 1964, we had the good luck to be involved in the first attempts to get some returns from BEB just to help R. and M. Bivas in charge of this experiment at the Haute Provence Observatory.

It was like a hunting party. At this time, the game was to view the satellite with binoculars, in a position as low as possible on the horizon, the best situation to get your prey in case of non-accurate predictions. So the game was to watch and, as soon as the satellite was in view, to transmit orally useful information to the Bivas. They were seated in an old turret that they manoeuvred around two axes to maintain the instrument in the direction of the satellite. In the meantime, they shot with the laser transmitter. The overall system was heating, requesting some cooling. The subsequent leakage of oil was evaporated with Mrs Bivas's hairdryer!

When you participate in such a venture, you become more respectful of the data, though without falling into exotic comments such as made by a newsman: shooting at a laser target is equivalent to firing at the eye of a bee flying around with a speed of 10 kilometers per second.

Thanks to celestial mechanics it was not as hard!

2.3.3 Radio frequency tracking data: The TRANSIT system

During the 60s, some radio-electric systems were developed and research undertaken to better understand the different components in order to identify the key points for the design of permanent and weather independent accurate system of tracking.

The TRANSIT system was developed very early by the US Navy to provide an improved navigation system for their fleet. The core was a one way Doppler downlink mode. In such a system, the transmitter is onboard the satellite, the receiver on the ground in tracking stations where the Doppler effect is measured and dated in the station time scale. The system is global.

The three main components are:

- a network of tracking stations well distributed around the Earth (TRANET network),
- a fleet of orbiting satellites, the TRANSIT satellites, with enough redundancy to provide a global coverage,
- a main operating center that collects the Doppler measurements from the ground stations, computes orbits and enters these coded orbits onboard the satellites.

In 1966, CNES (The French Space Agency), launched a small satellite named Diapason with a USO (ultra stable quartz oscillator) onboard to test accurate one-way Doppler downlink measurements. In 1967, two other satellites, named DIADEME 1 and 2, equipped with USO and laser reflectors were launched. The data from DIADEME were used in 1968 to make a pilot experiment with three ground stations equipped with laser tracking systems and Doppler receivers.

Notice that the Doppler measurements differ from the laser ones in that the former can only measure the relative speed ("range rate") between satellite and station, not the distance ("range") itself.

2.4 Dynamical approach and first zonal harmonics

At the surface of the Earth as well as in its environment, where the satellites revolve the value of the gravity, considerably varies. For representing such variations in space, the geodesists chose mathematical functions. They expanded the potential function of the gravity field with mathematical functions called spherical harmonics. They distinguished the functions only depending on latitude called zonal harmonics (characterised by a degree) and the others depending also on longitude called tesseral harmonics (characterised by a degree and an order). Determining the Earth gravity field is equivalent to determine the numerical values of the coefficients placed before each harmonic. Low degree coefficients describe long wavelengths, higher degrees increasingly shorter scales. Since gravity is the strongest force controlling the orbit of a satellite, one uses the best existent gravity model to predict the orbit, then the differences of the observed orbit are mapped back to changes in the gravity coefficients. The altitude effect is easily taken into account.

In space geodesy, the almost immediate result was the determination of the first coefficients of the zonal harmonics. The value of the first coefficient can be easily interpreted in terms of the position of the center of mass with respect to the crust. This position has been corrected by more than 100 meters. The value of the second coefficient (C_{20} or J_2) characterising the dynamic flattening of the Earth was computed with an accuracy 10 times better that the previous accepted value from classical geodesy. The value of the third coefficient (C_{30} or J_3), representing the gravity field asymmetry between the 2 hemispheres, was not known at all but was later discovered (the "pear" shape of the Earth!).

Several determinations were made, inter-compared and some initial attempts at geophysical interpretation published... In fact, It was rather easy; the amplitudes of the perturbations in the trajectory were comparatively quite large, up to several degrees/day on the ascending node (it defines the angular position of the orbital plane) and on the argument of perigee (the point of the orbit where the distance to the Earth center is minimal). But beyond these results, this dynamical approach provided the 2 objectives that one had in mind: tracking station and satellite positions using the gravity field.

2.5 Two major turns: The Standard Earth and the Williamstown Meeting

2.5.1. The Standard Earth

Following these preliminary studies, the first big step was when a first Earth model, the so-called Standard Earth (SE), was initiated in the mid-60s by the Smithsonian Astrophysical Observatory (SAO) at Cambridge, Massachusetts. The SE was a set of spherical harmonics coefficients depicting the gravity field and consistent station coordinates. There were a lot of new things. It was managed as a project with a clear global objective.

All the requirements were identified and actions were taken:

– for getting photographic observations from modified Baker Nunn cameras,
– for having a plan to select the satellites to track in order to have a maximized diversity of orbits,
– for deploying the network of these cameras to optimize the sampling,
– for developing analytical theories to describe the satellite motion and the statistical scheme for data assimilation with the recovery of unknowns,
– for creating and maintaining a unique atomic time scale, AS,
– for performing in parallel a model of atmospheric densities to filter the air drag.

This project was a full success. The Standard Earth I (SE I) was published and widely distributed during the COSPAR meeting in 1967. It was followed later by the SE II published by M. Gaposchkin and K. Lambeck in the Journal of Geophysical Research in 1971. This was the first civilian very significant project. The first solution provided co-ordinates of about 25 tracking stations and a gravity field to the degree and order 15. It was made in close co-operation with people coming from different countries; many of them were lucky enough to share the spirit of co-operation, the brain storming and hard work during summertime. SE was not an end but a basis for the future and had a strong impact on the future program.

2.5.2 The meeting in Williamstown – A vision

The expertise being acquired, it was time to express some views. A workshop was convened by NASA in July 1969, and was devoted to "Earth and Ocean Application Physics". W. Kaula chaired it. The main idea was, starting from existing technology, to present some initial results and emerging new objectives in order to set up a strategy with clear priorities expressed in terms of projects. The executive summary is one of the best visions in our discipline. It had a decisive

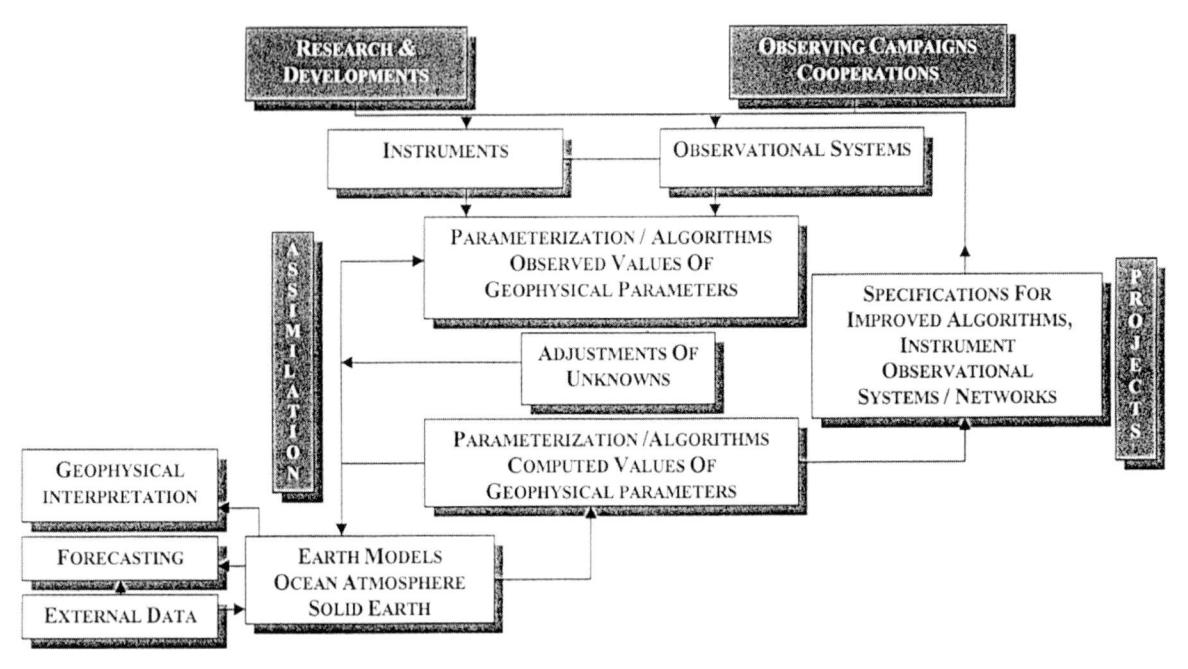

Figure 2 Lundquist's chart. (NASA document, 1969.)

impact at least on the 80 participants and is still considered as a foundation.

C. Lundquist, who was the manager of the Standard Earth and a driver in Williamstown, constructed a flow chart that seems very simple, but clearly put forward the relationships between the various tasks. This end-to-end strategy with an iterative process appeared as the key to success. We will use extensively an imitation of this chart to analyse the evolution (Figure 2).

Based upon the conclusions of Williamstown we wrote in 1970, a report called "Propositions à long terme en géodynamique". Our proposal was to-study "all that can be determined from accurate measurements of range, range rate, directions between ground stations and satellites, or between satellites". It concerns the study of the Earth and the Moon as elastic-viscous bodies; large motions of fluid envelopes like oceans, atmosphere or ice caps have to be considered. This proposal was criticized as being technical and not scientific. But 30 years later, it is recognized that it was a good approach. If we would like to have a more noble definition we could rename it the "Metrology of the Earth".

In conclusion, Williamstown was an opening and it is interesting that the program was aiming beyond scientific and technical objectives. It reveals a confidence in facing the challenge confronting scientists and humanity. This meeting was definitely a change of vision. The rationale was simple; there was enough scientific expertise to new possible applications of Earth and Ocean Physics. One of the major new outputs was the emergence of the radar altimetry as a new type of measurement and of a clear objective. It made a major contribution to understanding the Earth and its environment.

3. THE PREPARATORY PHASE 1970–1980

The flow chart described above provides a basis for the implementation of projects to be developed. The main difficulty is to maintain not only a high level of expertise in developing each component but also to maintain a high level of communication between the different teams. This requires a new way of involving scientists, and it is a major concern when the project is realized in an international cooperation in taking the existing cultural differences into account.

3.1 Observing systems and pilot experiments

3.1.1 Laser upgrading

First laser returns obtained from BEB then from D1C and D1D gave the impetus to put laser corner cubes on other satellites BEC, GEOS A, GEOS B, GEOS C. In parallel, several countries started building their own laser stations and, after testing, were ready to participate. The success of "Standard Earth" led to further improvements using more accurate laser data. That implied co-ordination of laser stations and the use of a larger diversity of satellites. In 1970, there was no satellite with inclination lower than 40°.

Therefore the CNES proposed to use a test launch of a new version of the DIAMANT launcher to put in orbit a flight model of the EOLE Spacecraft equipped with corner cubes; this so-called PEOLE satellite had an orbital inclination of 15°. CNES proposed to take this opportunity to co-ordinate an international program dedicated to the laser ranging use (adding optical direction measurements from cameras), that is a large observation campaign from the whole laser network to track 8 satellites. This program called ISAGEX, which was managed by G. Brachet, was quite successful. After 18 months, ISAGEX provided a unique new data set to undertake a new generation of gravity field models. (SE II and SE III at SAO, GEM series (Goddard Earth Model) at GSFC (Goddard Space, Flight Center) and GRIM series (GRgs/IMunich at Toulouse and Munich).

After ISAGEX, there were major improvements concerning the laser stations. They acquired the capabilities to make daylight measurements and to fire automatically at the satellites using an ephemeris, which could be improved locally. These developments were so promising that, in the 1970s, the idea came to have dedicated satellites fully optimized for laser tracking.

Two such satellites were launched:

- STARLETTE (CNES, 1975) in a low altitude orbit so as to be sensitive to gravity field and its temporal variations and even to the solid Earth and ocean tides,
- LAGEOS (NASA, 1976) in a very high altitude orbit to be as insensitive as possible to the gravity field and to the air drag effects, thus being a stable target to determine the Earth rotation motions and geocentric reference frames (i.e. in using the orbit of LAGEOS as a reference).

These two first dedicated satellites (followed later by similar satellites: STELLA (CNES), LAGEOS 2 (NASA and Italian Space Agency, ASI), AJISAI (Japan), ETALON (USSR), GFZ-1 (Germany)) were very useful, each having different but complementary characteristics. They all had a heavy weight in the gravity field solutions. As expected, LAGEOS was extremely useful in determining the polar motion and the rotation of the Earth. As expected, STARLETTE was also sensitive to the tidal potential and allowed researchers to determine the tidal potential parameters. Despite the limitations in coverage due to weather conditions and anisotropy of the laser network, the laser technique continued to be used and was upgraded again and again. One of the main interests of this kind of satellite is their lifetime: 1 million years for LAGEOS? 10000 years for STARLETTE? The Lifetime is an important factor for detecting and for determining long periodic perturbations. As an example, remember that the prime period of the nutation and associated tide is 18.6 years and the post-glacial rebound is of secular type. The latter is the main origin of the secular variation of the geodynamic flattening of the Earth.

Today, by using a spherical target, more homogeneous laser beams, and, most important, a new rapid and precise system of detection, it is possible to have laser echoes with only a few photons, thus providing a unique and proven accuracy. Moreover the laser observations are less sensitive to propagation delay errors due to the atmosphere than radioelectric measurements are therefore, the laser is used as a reference to calibrate the other systems and as a back up to continue to observe a spacecraft when it is no more transmitting, as happened to ERS 1 and GEOSAT FOLLOW ON.

> Story of STARLETTE
>
> *STARLETTE is a success story, which illustrates how fast a decision was sometimes taken in the 70s and then how drastically the situation changed later.*
>
> *There was a queue in the cafeteria of the CNES at Bretigny (1972). The queue was long enough to allow JC Husson, in charge of Earth and planetary programs, to inform us that a test of a new version of the Diamant Rocket was to be made but there was no payload and no money to do it.*
>
> *The project to build a small satellite optimized for laser tracking came back in to our minds. At the end of the queue, we had the orbital parameters and first draft of the scientific mission. After lunch a telex was mailed to SAO (Smithsonian Astrophysical Observatory) in Cambridge (USA). Their reply was enthusiastically positive and wise, and came back the same day. The complete dossier was ready in a few weeks and approved in a few months. Indeed the feasibility was confirmed, especially the core built with no radioactive Uranium by a mechanical department of CEA (the French Nuclear Agency). STARLETTE, only equipped with laser corner cubes, was successfully launched in 1975. To the big surprise of engineers asking for formats of telemetry, no format and even no telemetry were foreseen. Their comments were that a satellite with no telemetry is not really a satellite but a piece of "debris". We may recognise that, after the successful launch, we had some concerns because we got no returns at all until we realized that we were tracking the third stage of the Diamant launcher.*
>
> *STARLETTE was supposed to be both a perfect target for laser ranging and a perfect proof mass (physical realization of point M with mass m) and the laser observations were extensively performed and used as a core of gravity field improvements. STARLETTE has been tracked for more than 25 years now and will remain in the program for decades; it allowed us to determine perturbations like tidal potential over long periodic terms like those connected to the nutation.*
>
> *The last bit of trouble concerned administration which has to take account of any objects put in orbit. The case was applied to STARLETTE and it was difficult to explain that it is a "passive" satellite but we expect to track it for decades if not centuries to study the "dynamics" of the Earth. STARLETTE is not alone, the sister satellite STELLA was launched as a passenger of SPOT3.*

3.1.2 Other laser ranging applications: the Moon and time synchronisation

A new application of the laser ranging was its successful extension to the Moon. NASA put three panels of laser reflectors, on the landing sites of Apollo 11, 14 and 15. Two French panels were put on the Soviet vehicles Lunakhod. Today, only a few stations are able to get returns, but that is enough to measure the Earth-Moon distance with a one centimeter accuracy level. The lunar laser range root mean square deduced from a very precise trajectory is of about few centimeters for the last years (a result obtained at the JPL (Jet Propulsion Laboratory), the Côte d'Azur and the Paris Observatories). From these data, it is possible to determine the exact rate at which the Moon recedes from the Earth. This measurement is one of the ways to estimate the dissipation of energy within the Earth-Moon system. Several important results have been obtained concerning, for example, the properties of the Moon's rotation around itself, the Equivalence Principle, which was checked with a greater precision and the internal constitution of the Moon (Dickey et al. 1994, Samain et al. 1998).

Another application is the time synchronisation. The satellite laser ranging is also used to perform very accurate time synchronisation at intercontinental scales in combining ranges obtained by several stations from round trip measurements. It is necessary to determine the time differences of arrival of laser pulses emitted by the ground stations and detected by an optical sensor onboard a dedicated satellite. That was indeed realized by ESA on a geostationary telecommunication satellite (the Italian satellite SIRIO 2 with the LASSO package proposed by J. Gaignebet and M. Lefebvre in 1972 (Veillet et al. 1992) In the future, it could be possible to get this synchronisation with an accuracy of about a few tens of picoseconds (T2L2/PHARAO/ ACES experiment to be put onboard the International Space station in 2004).

Finally, note that new other applications from laser tracking emerge: the idea is to equip any launched bodies, including the stages of launchers put in orbit, with corner cubes allowing them to keep the orbital control of such objects in critical situations (ALOS project).

3.1.3 Radio-frequency tracking modes: TRANSIT, ARGOS, PRARE

Three different modes have been realized with TRANSIT, EOLE, and ARGOS satellites:

– One way Doppler downlink mode: TRANSIT system.

The TRANSIT system was operational from 1970 until 2000. Although designed for navigation and military purposes, it was used for such civilian objectives as positioning. It was continuously upgraded so it was able to give, for the first time, independent values of polar motion and new international reference frames. This includes a lot of international or regional campaigns: EDOC, EROSDOC, WEDOC, RETDOC, MEDOC, ADOS, a dedicated campaign in the African continent... (proceedings on Doppler positioning at Austin 1979). The same system was used in support of the precise orbit determination of satellites equipped with radar altimeters, like GEOS 3 (1975), SEASAT (1978), then GEOSAT (1985).

– Double way range and range rate: EOLE and PRARE systems.

The two-way range and range-rate systems were used on the EOLE satellite. One of the objectives was to locate a fleet of balloons moving in the stratosphere. The signal emitted onboard went to a transponder put on the balloon and, after a change of frequency, went back to the satellite. The received signal was directly compared to the transmitted one and allowed the Doppler measurement. Several hundred balloons were successfully located from EOLE. It was so promising that, as early as 1968, an extension of this concept was proposed to be implemented on a satellite in high orbit. This so-called GEOLE project was studied in detail but finally not approved. A similar project named POPSAT was later proposed by Germany to ESA but also not approved. The studies confirmed the value of the concept that was used for the implementation of the German PRARE (Precise Range And Range Rate) tracking system, which was flown on ERS 1 and 2 satellites. The accuracy was as expected but the system requires a good signal-to-noise ratio, which means that antennas must be pointed at the ground stations; thus, the system is difficult to operate and can present some difficulties when installed in remote areas is required.

– One way Doppler and one way uplink mode: the ARGOS system.

The third operational system was ARGOS. It was designed at the urging of oceanographers wanting to measure the eddies on the surface of the ocean by using buoy motions. The EOLE system was too complicated and not adequate for completely unmanned and remote ground stations. Therefore, the design of ARGOS was to minimize the complexity of the ground stations. The tracking mode was a one way Doppler uplink transmitter from the buoy receiver to the satellite, which measures the Doppler effect. The time scale is unique and provided by the onboard clock. Data are collected onboard and retransmitted by telemetry to a main center. The system was operational as soon as 1978 and is still in use.

> **ARGOS: Biodiversity and New Users**
>
> *A new class of users of ARGOS exists now: about 1700 animals of various species from whales to birds are permanently localized and their way of life recorded. Many lessons can be drawn from this new use of ARGOS: first the interest grows in a new scientific community to use new space systems. New ideas are proposed concerning biodiversity. Second, the biodiversity objective was considered in the beginning as being out of the specifications of ARGOS. It was wrong. Progress in small electronics allows us now to record permanently many biological parameters of animals and to transmit them later, thanks to ARGOS, alleviating the operational constraints. There is no more need to educate the animals to stay on the surface when the satellite is passing over!*

3.2 Altimetry emerging

Following the Williamstown's recommendations, a radar altimeter was placed in the space laboratory SKYLAB onboard a manned station (1973). It was switched-on by the astronauts and the first "around the world trip" was provided in real time. It demonstrated that there are bumps and hollows on the sea surface. The long wavelength features obtained in computing the geoid from the gravity field models were consistent with the amplitude of profiles. But at this time, it was amazing to see a lot of short wavelength details in whole coherence with tectonic features. This provided an extra impetus to get the decision for the launching of GEOS-3; the first satellite dedicated to satellite altimetry (the satellite orbits are not very sensitive to gravity features much shorter than the satellite height, but altimetry can see such scales on the sea surface).

However, the American Department of Defence (DOD) considered the altimetry as a very sensitive technique, and the overall experiment was classified. Indeed the altimetry provides deviations of the vertical as by-products, which can be useful for precisely launching rockets from the sea; at this time, these deviations were essential for cruise missiles as a part of the error in the initial conditions. Note that GEOS-3 had no onboard recorder, so the acquisition of telemetry was only possible in direct view of stations, what created large gaps. The orbit was computed from laser ranges and one-way downlink Doppler measurements, using a tracking network (the same network as used for TRANSIT). The accuracy of the orbit for the radial component was about 2 meters. Nevertheless, GEOS-3 was very useful. Indeed finally, there was some large release of data in delayed mode. These data were extensively used and a special issue of the Journal of Geophysical Research was published (Vol. 84, B8, July, 30, 1979). In the Kerguelen islands area, it was even possible to compute a marine geoid and to evidence strong correlations between significant oceans features and the sea-floor topography (Balmino et al. 1979). The second issue was the reaction of the scientific community frustrated to have been pulled out without any advertising. That triggers a renewal of interest and watching among the scientists.

GEOS-3 also triggered interest from the space agencies and centers. The JPL decided to prove definitively the potential of space for oceanography. What JPL did for other planets could not be a problem for the Earth, and it became SEASAT (a dedicated satellite for oceanography) with altimeter, scatterometer (to measure wind), SAR and passive radiometer. When the launch of SEASAT was certain, scientists undertook some preparatory actions without awaiting final official policy. Let us give an example: at the European level some high level scientists met at an informal meeting in London in July 1977 and wrote together an unsolicited proposal from their group, they named SURGE "SEASAT User Research Group in Europe". Their chairman P. Gudmandsen put the proposal on the desks of ESA and NASA. It was the right time to do it and it was very successful. At this time, ESA also took the decision to convene a workshop with European scientists, the main purpose being to make recommendations for solid Earth missions. The main themes were supposed to be applications in geodynamics in the broad sense and navigation.

Based on the SURGE activity, oceanography was added to the objectives. The workshop was called SONG (Space Oceanography Navigation and Geodynamics), and it took place at Schloss Elmau (Germany) in January 1978; 100 attendees during a full week made clear recommendations.

The proposed first priorities were as follows (Figure 3):

- (Figure 3A) A solid Earth program to deal with gravity field and precise positioning (crustal dynamics features) with satellites such as POPSAT or GEOLE. These satellites were not approved, but other solutions replaced these projects (DORIS, PRARE, GPS for the positioning and STELLA, LAGEOS2, GFZ-1 for the study of the gravity field).
- (Figure 3B) A so-called surface studies program to deal with ocean and ice dynamics, with two main satellites; (ice, ocean satellite: ERS-1 and ERS-2 satellites) and geoid satellites (today the GOCE satellite).

The Figures 3A and B give in a synthetic view the priorities and the schedule written by the executive committee. The reader may recognize in the first line what will be the ERS satellite panel. Moreover, other applications were envisioned; for example, the direct local determination of the Earth radiation budget was proposed. Indeed the CACTUS accelerometer built by ONERA, France (Office National d'Etudes et de Recherches Aéronautiques et

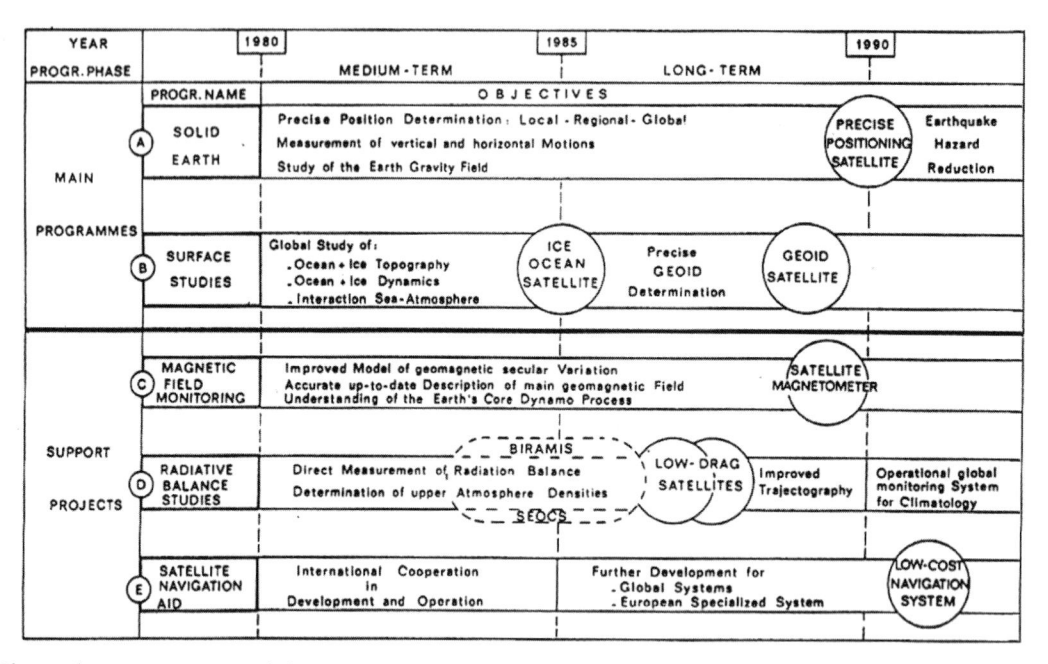

Figure 3 Chart of program proposal from the SONG workshop, Schloss Elmau, 16–31 January 1978. (From ESA-SP137 document, p33, 1978.)

spatiales) and launched by CNES in 1975 onboard the CASTOR satellite proved that such a system has the capability to measure many non-gravitational forces acting on a satellite. As a result, it can directly measure the local Earth radiation budget (the BIRAMIS project), as well as it can improve the upper atmosphere density models; it can also exhibit directly non-gravitational forces due to the Earth albedo (over ocean, land, and snow) and those due to the anisotropy of the satellite surface temperatures (it was the so-called thermal thrust effect or the photonic radiation thrust). However, the BIRAMIS project was not approved, but it was at the origin of many developments very crucial to geodesy; moreover, the Czech space project MIMOSA with the MACEK accelerometer onboard should be a very good opportunity to validate these interesting possibilities in the years to come (Sehnal 1994). Let us recall that ONERA micro-accelerometers will be used in the three gravity missions: CHAMP in 2000, GRACE in 2001, and GOCE in 2005. It will also used in the CNES/MICROSCOPE mission in 2004 to test the Equivalent Principle at the level of 10^{-15}. This illustrates very well the importance of developing new technologies a very long time in advance. It is of importance to recall that the non-gravitational forces are very probably today the limiting factor in the precise orbit determination of oceanographic satellites like TOPEX/Poseidon or JASON-1. New applications of accelerometry could be very usefully considered in the future.

SEASAT was launched in July 1978; the payload had all the major instruments for measuring the oceans from space. SEASAT had a short lifetime due to a satellite technical failure. In spite of this short lifetime and maybe thanks to that, SEASAT was really a starter. With only 3 months of data, a lot of investigations were successfully performed in many areas. The meeting "Ocean from Space" organized in Venice (1980), provided an opportunity to appreciate the results, and even the more sceptical attendees were convinced. So, for altimetry, the demonstration was convincing. The only missing component was the very precise orbit. SEASAT was put on a 3 days repeat orbit during 1 month: 9 collinear tracks were available and appeared to be extremely useful for learning a lot of things about the crossover technique and about mean sea surface features.

4. THIRD PHASE MATURITY AND DECISIONS 1980–1990

4.1 Time and frequencies

Extremely accurate time measurements are essential components of any space system involved in practical operational missions as well as in scientific research. Consequently a continuous effort has been made to improve the basic time technology and to transfer the improvements from ground to satellite systems.

This temporal accuracy is required not only for individual measurements but also throughout the duration of the mission at any time when observations are made. Moreover the time scale used has to be linked with the same accuracy to the international standard time scale, the TAI (International Atomic Time). The TAI is presently maintained at the "Bureau International des Poids et Mesures" (BIPM at Sèvres, France). There is also a need to know the position of the rotating Earth as a function of time (UT1 or the Universal Time1). These quantities are provided by international services, as the difference between UT1 and TAI. There was always a continuity and a partnership between the time and frequency laboratories and the various users and contributors. Global time synchronisation, which has been a nightmare in the past, is today easily achieved at a level of accuracy which satisfies most users, by employing space technologies. However, the constant progress in the atomic frequency standard requires further efforts to improve time comparisons. The frequency stability of oscillators is also a major requirement. All progress in space metrology (laser, radio tracking, and altimetry) is due to the progress in the time-frequency field (see Figure 4). The permanent dialogue between people developing new technologies and space project teams has also been very fruitful. As a last example, the time and frequency advances provided the basis for the successful GPS design (Global Positioning System).

4.2 Earth rotation

The way in which the Earth rotates depends on the way in which the Earth is constituted; core, mantle, crust, ocean and atmosphere are the basic constituents, the behavior of which allow us to interpret variations in Earth rotation. The progress concerning the determination of the Earth orientation parameter is illustrated in Figure 5, in which the results obtained in the beginning of the seventies by astrometric techniques can be compared 20 years later with new results obtained by using space based techniques. No erratic behavior is exhibited on the curve at the bottom of the figure. At the present time, the errors are of the order of a few

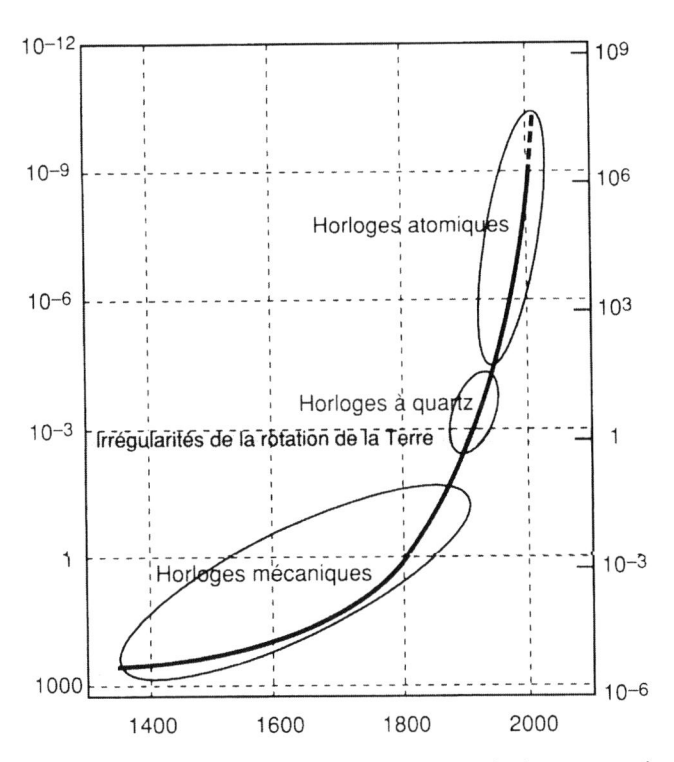

Figure 4 The y-axis gives the daily drift of clocks measured in seconds of time. The ordinate on the right gives the required number of years to obtain a drift of 1 second. On the x-axis the date is expressed in years.

Note: Improvements in the mechanical clock, quartz clock and atomic clock over half a millenium. A horizontal line gives the comparison with the rotation of the Earth considered as a clock. (From "Les fondements de la mesure du temps" by Claude Audouin and Bernard Guinot, Masson, Paris, 1998, p. 47.)

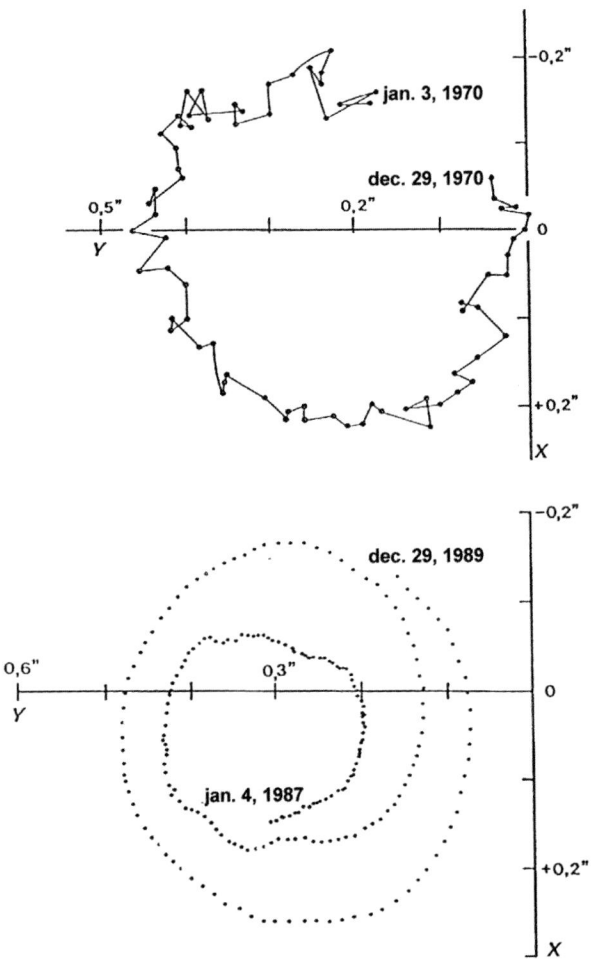

Figure 5 Comparison of the polar motion determined by the astrometric technique in 1970 and by space geodetic techniques in 1990. Note the improvement in precision. From the International Earth Rotation Service. (M. Feissel, Observatoire de Paris, 1990.)

tenths of milliarc second (less than 1 cm on the Earth crust). Anderle, in the U.S.A., formulated the first determination of polar motion in 1969, and then others in different countries undertook similar studies. The "Bureau International de l'Heure" (BIH) at the Paris Observatory started incorporating space geodesy results in 1974 for the determination of polar from motion, then UT1. The contributions of classical optical astrometry progressively decreased and was completely abandoned by the BIH in 1984.

Nowadays, the study of the Earth's rotation is based on GPS, SLR, VLBI and DORIS data. But the VLBI technique (Very Long Base Interferometry) has a specific role. It has no competitor for determining the direction of the rotation axis of the Earth in a celestial reference frame (quasi-inertial system defined by about 600 celestial radio-sources) and for linking terrestrial and celestial reference frames. This capability allows it to determine the length of day over long periods of time (a function equivalent to the determination by UT-1). VLBI is a fundamental and irreplaceable technique in the metrology of space.

4.3 The Earth gravity field status

It was important to better determine the zonal and tesseral harmonic coefficients in the Earth gravity field models, and regularly improved solutions were presented at the COSPAR meetings starting in the 60s. We have to keep in mind (cf. 2.5.1) that the first specific and historical civilian, effort was performed at the Smithsonian Astrophysical Observatory to arrive at a global solution, including determinations of the parameters in the gravity field model and the position of the observing stations, the so-called Standard Earth.

The basic principles of a dynamic solution were established during these first years of space geodesy and are still used today. In parallel to the pure dynamical solution, mixed methods using geometric constraints were developed. Then pure gravimetric methods were improved and finally also the mixed gravimetric, geometric and dynamic methods considered.

It is impossible to be exhaustive in this field but this period corresponds to a huge and fruitful international effort in America, in USSR, in Europe, in Japan, etc. ... In conclusion, it is clear that between October 4, 1957 and at the end of the sixties a big advance took place in this field, which has continued to progress for 30 years and yet is not finished. For the first period, references can be found in the Space Science Review (Kovalevsky and Barlier 1967).

Today, the principles used in the early times to mix data of various origins (orbit analysis, gravimetric data and geometric measurements) together with statistical constraints (Kaula's rule of thumb) continue to be used; new data types are now added and the models have become more and more sophisticated, coming to include relativistic corrections and non-gravitational forces modeling. The determination of the

Figure 6 The geoid.

Note: The geoid of the Earth look like a "patatoid". The variations of the geoid are given by colours and are in the range of about a hundred meters.

Avocado? "patatoid"
Who built that? When NASA released the GEOID corresponding to one of the GEM (Gravity Earth Models) of GSFC the sketch had exaggerated bumps and troughs. During the press conference it was that this Earth said appears as a non-fresh avocado or like a kind of potato, a "patatoid". This new nickname was used and popularized. Obviously, there was a reply in a newspaper an unhappy citizen "I am paying enough taxes to have a decent Earth. What NASA will plan to rectify it?"

gravity field by means of an orbital perturbations analysis is still in progress, due to the increased accuracy and coverage of tracking systems. The availability of a larger diversity of satellites being tracked by these systems is also an important factor; laser range measurements on STARLETTE, STELLA, LAGEOS, LAGEOS 2, Doppler measurements on GEOSAT, SEASAT and large sets of Doris data coming from the tracking of SPOT 2 and 3 are basic and crucial data. The geoid, such as it is determined from these gravity field models, is shown in Figure 6. Nevertheless, the upper limits of this approach is well known from the early beginning. A very high space resolution in the gravity field determination cannot be obtained only by orbital analyses. Therefore many projects have been studied at length both from a theoretical point of view and to take advantage of the new technologies.

The two main approaches and techniques are:

(i) Satellite to Satellite Tracking (SST Technique) in two possible modes:
   – the High–Low mode, where the low altitude satellite is sensitive to the small scale gravity anomalies and is tracked by a high altitude satellite,
   – The Low–Low mode, where two low altitude co-orbiting satellites measure their relative velocity or/and distance.

(ii) Gradiometry

Here the measurements are determined directly from the second derivatives of the gravitational potential in spacecraft axes; the technique is not only a challenge for the main instrument (a gradiometer) but imposes constraints on the satellite: it must be drag free, have attitude control contain orbit maintenance subsystems etc. But it is agreed that the scientific community accept the risk and promote advanced technologies. The proposing teams were never discouraged and thereforce improved their projects again and again for almost 20 years in Europe. The first GRADIO project was born in 1981 in France and at the same time, the GRM/SST gravity project was initiated in the U.S.A. They were in competition with TOPEX/Poseidon, which only measures the variations of ocean circulation, whereas one also needs the mean absolute geoid. As a matter of fact, the scientific output for geophysics was convincing enough per se, but it was not perceived as such by the decision-makers, and these first gravity field projects were not approved.

4.4 Altimetry

GEOSAT was launched by the US Navy in 1985 and its useful lifetime was 4 years. In a 2-year "classified" part of the mission, GEOSAT was put on a geodetic orbit with a dense grid. The corresponding data were classified until ERS 1 was put in a geodetic orbit as well, after which its data were released thank to the efforts of a few determined US Navy and NOAA oceanographers. GEOSAT was then placed in an exactly repeating "declassified" orbit. GEOSAT, although not fully optimized, had a major impact. In fact thanks to the active and efficient help of NOAA (Laboratory for Space Altimetry (R. Cheney)), a growing number of people and laboratories learned how to play with the data and in return to provid new insights.

One of the major was the initiation of operational activities including some consequences quasi real time pilot experiments. The repeat track of 17 days appeared to be a good compromise. More and more geophysicists realized that this was indeed a complete new system to study the ocean tectonics, and they analyzed altimetric profiles.

GEOSAT was transmitting a couple of coherent frequencies compatible with the TRANET network, allowing it to perform Doppler measurements. But tracking data were released with some reluctance. When they became available, they were used firstly to compute a tuned gravity field and then a precise orbit. The precision of the radial component was claimed to be of about 30–40 centimeters, one of the limitations coming from the Doppler measurements of the TRANET network, which were not precise enough by comparison with new modern systems.

4.4.1 Scientific requirements: DORIS tracking system, radar altimeter

In the early 80s, it the time had come to make key decisions about the undertaking of ambitious projects to provide data that would have a decisive impact on the models of ocean circulation, and also a unique set of data for the geophysicists studying ocean tectonics. The requirements for altimetry as a system were based on scientific objectives in the geophysical and oceanographic fields. Thus, we need to look at the two sets of data we have access to. Indeed, the sea surface shape is the result of the superposition of the two sets:

(i) The geophysical data set reports on the distribution of densities inside the Earth. The long wavelengths of the sea surface are due to large-scale convection inside the mantle and are comparatively well understood nowadays. The medium and short wavelengths are related to bathymetry density contrasts as parts of the ocean tectonic on these scales.
(ii) The oceanographic data set.

Let us assume we know the marine geoid perfectly; the circulation of the ocean modifies the sea surface topography in several ways: sea level variations due directly to the warming, and more important, due to the tides, the atmosphere and the variations of the ocean circulation (upper part and deep part). This is a major point to be better studied. Like the other instruments used in remote sensing altimetry to measure parameters at the surface. But this surface measurement is an integral of the field of densities from bottom to the surface. Thus, it is a quantitative measurement to be used to adjust models. The spectrum of this ocean data varies from meso-scale features like eddies to large western boundary currents in relationships with the climate.

A few bits of information, such as the seasonal effect between the two hemispheres, have amplitudes of 10–15 centimeters. The tropical variations linked to the ENSO-EL NINO phenomena have amplitudes of 20–25 centimeters on the scale of 10000 kilometers. To make some advances, it is necessary to determine these values with an error of 10% or less. If we want to give a specification to a system, we can use the following slogan:

> One Centimeter for a Monthly Value on an Ocean Basin Scale

To achieve these objectives, we have to look at the corresponding requirements for the system components. A precise orbit is not too critical for observing the eddies; their high frequency signature can be recovered in adjusting polynomial coefficients. It is a critical issue for large-scale variations, the most important for climate studies. At the time that the decisions for the DORIS system were being taken there was no adequate tracking system. We recall in a summary table (Table 1) the characteristics of existing systems and try to investigate where the source of error occurred and how to try to overcome them.

The main uncertainty was due to unknown parts of the gravity field and especially to those generating high

Table 1 Mean errors on different measurements over the years, which played a role in the determination of the sea surface (CNES document, Paris 1999). For the clocks, the best accuracy, which it was possible to get at a given time, is also indicated in brackets

	Accuracy of measurement			
	1975	1985	1995	2000
Laser	1.50 m	30 cm	3 cm	1 cm
Doppler	5 cm/s	1 cm/s	0.03 cm/s	0.01 cm/s
Altimeter	20 cm	5 cm	2 cm	0.5 cm
Accelerometer	10^{-9} m/sec^2		10^{-10} m/sec^2	10^{-13} m/sec^2
Clocks	10^{-11} (10^{-12})	10^{-12} (10^{-13})	$3 \cdot 10^{-13}$ (10^{-14})	10^{-14} ($5 \cdot 10^{-16}$)

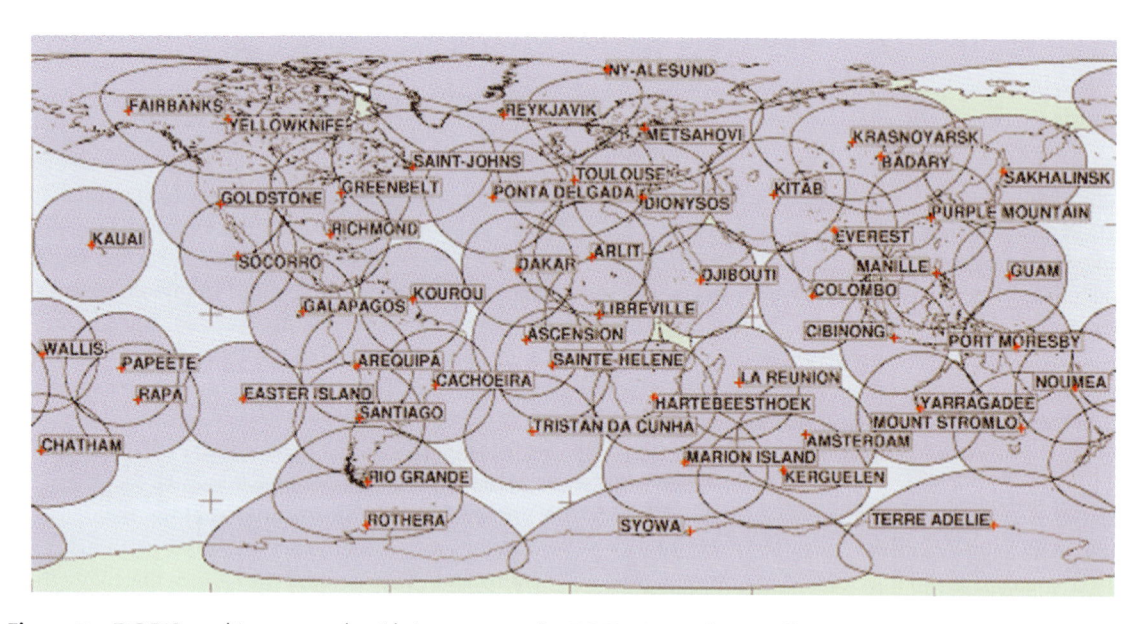

Figure 7 DORIS tracking network with its coverage for TOPEX/Poseidon satellite. (AVISO/CLS/CNES document.)

frequency perturbations, and to the unknowns coming from surface forces like air drag or solar radiation pressure, either direct or reflected. Both were had to be dealt with to increase sampling and accuracy.

Thus, it was decided to develop a new system taking the best points of previous systems and avoiding the worst ones: the DORIS system (Figure 7). It was based on a network of 50 tracking stations transmitting upward 2 frequencies driven by an USO (ultra stable oscillator) through a simple non-pointing antenna. The 2 frequencies are high enough and different enough to filter out the ionosphere effects. The transmitted frequencies are compared to the onboard receiver, also driven by an USO. The difference between then makes the Doppler measurements possible. The system operates in a one way up-link mode. The critical point is the stability of the USO frequencies. Research and technology were able to provide this new generation equipment even for remote sites. The short-term stability of the USO was 10^{-12} over 1000 seconds. The expected accuracy of the radial velocity measurement was about 0.5 mm/s. All the observations were dated in the time scale of the onboard clock, which was then compared with the TAI.

The design allowed ground stations to be almost fully automated and easy to install. The 50 stations were placed in well-distributed sites around the world and even in remote sites. More specifically, a well-balanced distribution between the 2 hemispheres has been achieved as well as extra coverage over stations oceans using island. This yields an orbital coverage of around 80% depending on the orbit height.

In the DORIS proposal, the ultimate accuracy was quoted as 5 cm for the radial orbital component. This value came from expertise acquired during several years of orbital analysis in Doppler tracking programs (MEDOC, GEOSAT

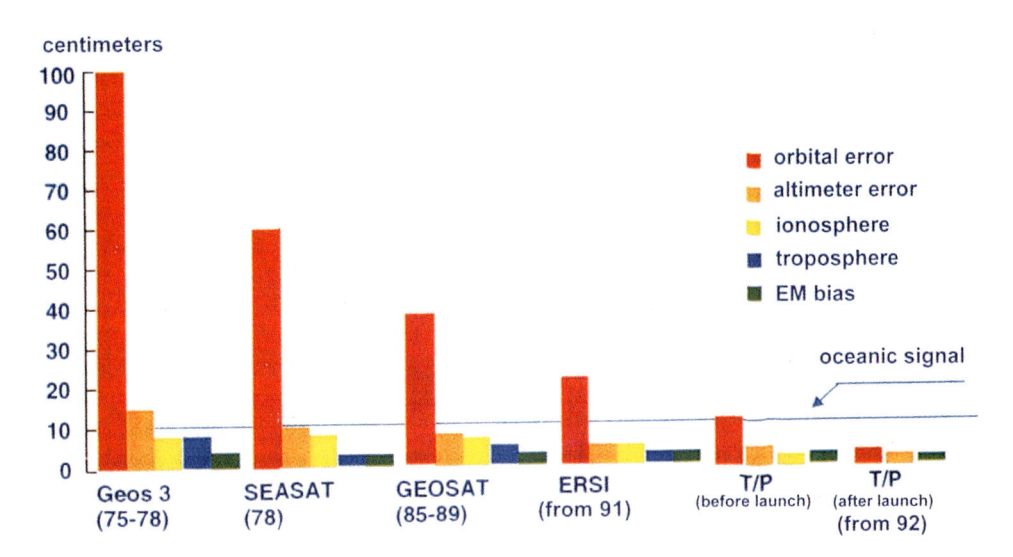

Figure 8 Shows progress in the orbit determination of oceanographic altimetry. Orbital error, altimeter error, ionospheric error, trosopheric error and electromagnetic bias of the altimeter are given by the colours ranging from red to blue, respectively. The blue line in the figure gives the amplitude of the oceanic signal to be detected. For TOPEX/Poseidon the status is given for the beginning of the mission and for the present. (GRGS/CNES document.)

and SEASAT). In these earlier projects Twenty stations were involved with a measurement precision of 7 mm/s. DORIS was based on fifty stations with a measurement precision of 0.5 mm/s. Today the accuracy of the radial orbital component is about 2 centimeters. The DORIS system was specified and designed for rapid, precise and automated orbit determination and navigation. It will be improved for the next missions. DORIS will play a major and irreplaceable role in space oceanography missions in the years to come (with JASON-1 and ENVISAT), which will also benefit greatly from laser tracking. Progress in orbit determination for altimetric satellites is shown in Figure 8.

The radar altimeter instrument measures the height range at nadir between the center of mass of the satellite and the center of the reflecting spot on the sea surface. Corrections are of three kinds, instrumental (e.g., drifts, center of mass), path delays (e.g., ionosphere, drag and wet troposphere), and surface effects (sea-state bias). The path delay corrections are made as follows:

- for the ionosphere, by combining the Doppler measurements of two coherent frequencies 13 GHz and 400 MHz,
- for the atmosphere "dry part", by using the fields of atmospheric pressure and temperature given by the meteorological centers,
- for the atmosphere "water vapour content part", by using data from the radiometer.

The problems of sampling and of determining a precise geoid to detect the mean ocean circulation have now to be considered. A single altimetric satellite provides insufficient sampling in space and time to cover the whole variability of the spectrum. Two satellites provide a much more adequate coverage, even if rapid phenomena are still inadequately observed.

Ideally speaking, it would be better to determine the geoid separately. Such a practice requires dedicated missions (see the planned gravity missions, discussed in this paper). In ensuring that the satellite has a repeat track, we restrict our access to variations alone but they provide nevertheless crucial and basic information.

4.4.2 ERS 1 and 2

Two majors programs were recommended to ESA during the SONG workshop in 1978:

- a surface study program with two complementary projects: an "Ice and Ocean satellite" (1985) followed by a "geoid satellite" (1989),
- a precise positioning satellite (1990).

Unfortunately, they were unexpectedly rejected due to the scientific structure of the ESA. At that time, the study of the Earth was strictly under the responsibility of the applications directorate, but fortunately though, ESA was able to take decisions for other satellites devoted to related applications: ERS 1 and his twin brother ERS 2.

As part of the payload, there was a radar altimeter designed just for sea state-wave height measurements. But in the first design, the satellite was at low attitude, 650 kilometers, and had only a laser system as a precise tracking device. However, step by step, actions were undertaken: the height was raised to 777 kilometers, the repeat orbit was planed to move to 35 days then to 168 days for a year (a so-called geodetic orbit with separated tracks of 8 km at the equator) and

Germany proposed a precise radio tracking system PRARE, doing double way range and range-rate measurements between satellite and ground pointing antennas. Our emphasis on this point is just to show that relatively minor modifications can greatly increase the scientific return of a mission (ESA proceedings, 1997).

4.4.3 TOPEX/Poseidon

Another decision was taken in 1987 by NASA and CNES to launch a fully dedicated satellite for satellite altimetry: TOPEX/Poseidon. NASA would build the satellite. CNES would launch it using an Ariane 42 P from Kourou (French Guyana) directly to the selected optimized orbit inclination of 65 degrees. The height of 1300 kilometers was chosen in order to avoid aliasing with solar tides, to decrease the air drag effect and the gravity field sensitivity, and to have a repeat period of 10 days. The payload was also shared:

- NASA took care of the 2 frequency radar altimeters, the microwave radiometer, the laser cubes and as an experiment, a GPS receiver,
- CNES took care of the fully dedicated tracking system DORIS, including the ground network of 50 stations, also providing, as an experiment, a new generation of radar and an altimeter using a solid state technology. This experimental altimeter was to use the new antenna for 10% of the time.

The expected lifetime was 3 years with a possible extension for two more years. The expected total error budget for the radial component was 13 centimeters, at least 3 times larger than the requirements. But there were some expectations that the main source of error was coming from the precision of the orbit and that this error could be later removed thanks to a better gravity field model and to better station positions. It proved to be true. TOPEX/Poseidon was decided mainly for scientific objectives and was supposed to be the starter for a long-range program. A Science Working Team (SWT), with all the selected Pi's (Principal Investigators) and project teams, was associated with the project to the end of the mission (in mid 2001 it is still operating). TOPEX/Poseidon was designed as the core for the program conceived as the WOCE (World Ocean Circulation Experiment) in 1988 at the International Oceanographic Conference at the UNESCO (Fu and Cazenave, 2001).

5. TIME OF RESULTS – TIME OF NEW DECISIONS 1990–2000

5.1 Some results – Scientific progress along the years

It is important to appreciate now how the space programs have totally changed and revolutionized our understanding of our planet (remember we are just looking at measurements of the metrologic type). There are several ways of gaining such an appreciation. We can first list some main results, along with a crude mechanism of comparison of what we knew before the space era and what we know today in terms of a series of progress that we can call gradients.

5.1.1 Gravity field

Before the space geodesy era, we had only an inaccurate idea of geodynamic flattening. Now, a complete determination of the gravity field with a space resolution up to 500 kilometers or even better has been made (Figure 9).

Recent models (from 1995) have been formulated at the Goddard Space Flight Center (GEM-T, JGM, EGM models), at Texas University (TEG models) and at GRGS (Toulouse, France) and GFZ (Potsdam, Germany) (GRIM Models).

5.1.2 Temporal variations of gravity field

This information was not accessible in the past. Now, accurate determinations of the C_{20} temporal variations (flattening of the Earth) with preliminary detection of variations of the first following terms (C_{30}, C_{40}, C_{50},) have been made. Seasonal and inter-annual variations of C_{20} are very important and are due to atmospheric and oceanic mass transports. They are not negligible in efforts to determine precise orbits. Secular variations of C_{20} are due to the postglacial rebound and can be decorrelated from the 18.6 year tide thanks to very long terms analyses of LAGEOS and STARLETTE orbits over 18 and 15 years, respectively. Averaging techniques have been employed very usefully in this field for the orbital analyses.

5.1.3 Plate tectonic motions

No direct geodetic measurements were possible before the space geodesy era. Now, the horizontal and vertical motion of geodetic stations can be detected by using space based techniques along with tracking stations on all the different tectonic plates (Figure 10). The accuracy is around 1–2 millimeters/year. The big effort carried out by NASA to co-ordinate international capabilities in the Crustal Dynamics Program between 1980 and 2000 has to be emphasized (D. Smith) as well as the Wegener program in Europe (P. Wilson).

5.1.4 Center of mass of the Earth

The location of the Earth's center of mass was only known within several hundred meters before the space geodesy era. The position of the center of mass can be located now with an accuracy of a few millimeters; some temporal variations are significant (Figure 11).

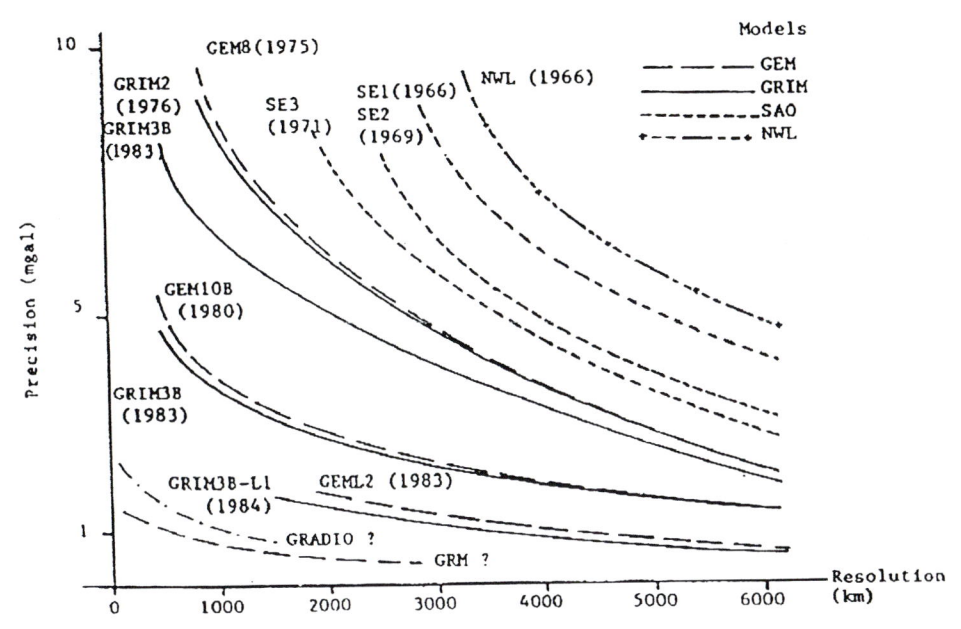

Figure 9 Precision of different gravity field models obtained from the space geodesy between 1965 and 1985, expressed in milligals as a function of the space resolution expressed in kilometers. This historical chart was constructed to gain approval for new missions.

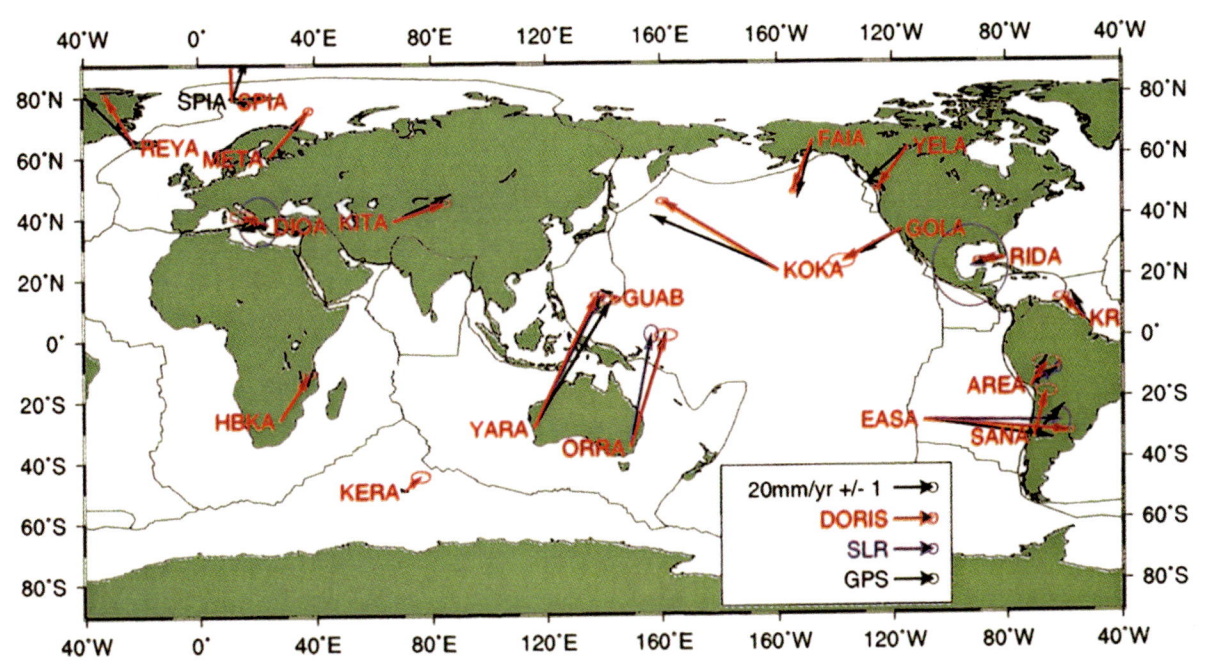

Figure 10 Velocities of plate tectonic motions deduced from three space geodetic techniques DORIS, SLR, GPS. (CNES/GRGS document, Toulouse, 1998.)

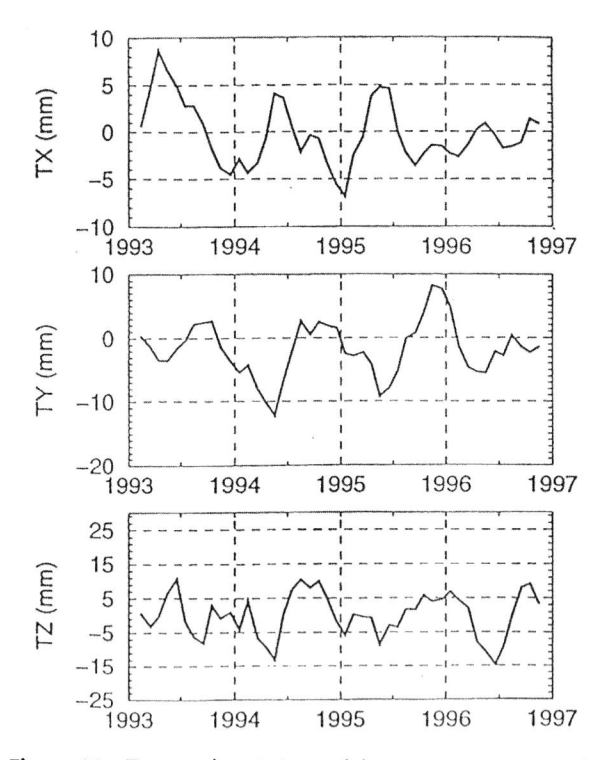

Figure 11 Temporal variations of the geocenter in a geodetic system defined by satellite laser ranging stations linked to the crust of the Earth. These variations are deduced from the LAGEOS Satellites. (From "l'étude des systèmes actuels et futurs de positionnement spatial DORIS" by Florence Bouillé, PhD-Université Paul Sabatier, Toulouse, 2000.)

5.1.5 Precise positioning

Within some continents, the relative precision of the positioning of geodetic stations was a little better than about 10^{-5} (10 meters over 1000 kilometers) before the space geodesy era.

Now, the terrestrial reference frames are coherent and controlled at the international level. Within these frames, the positions of individual stations are known with an accuracy of about one centimeter (Figure 12). The International Terrestrial Reference System (ITRS) has been developed at the "Bureau International de l' Heure" (Observatoire de Paris, B. Guinot, M. Feissel) in co-operation with the "Institut Géographique National" in France (C. Boucher and Z. Altamimi) in the 80s. The geodetic WGS 84 system was very closely linked to the ITRS from its beginning in 1984.

5.1.6 Earth rotation parameters

Before the space geodesy era, the determination of the Earth rotation parameters was based upon astronomical observations. They only permitted the detection of large-scale variations. Now, accurate determinations of the

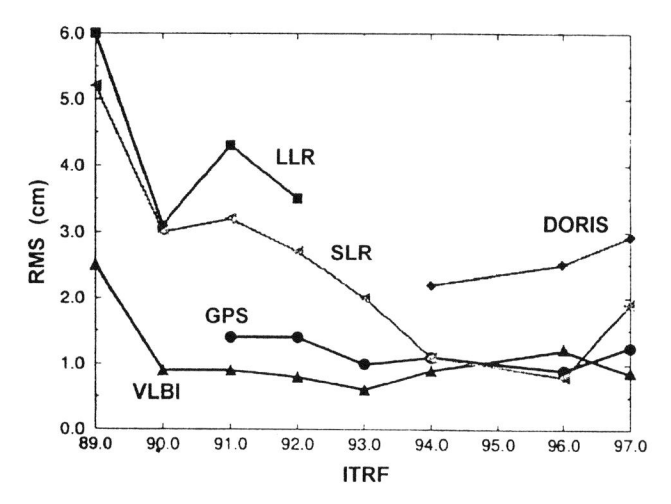

Figure 12 Precision of positioning by different space techniques over the years. (From the "modélisation des systèmes de références terrestres" by P. Sillard, PhD, Observatoire de Paris, 1999.)

Earth's rotation parameters can be made with a precision corresponding to errors less than 1 cm on the Earth's crust. Quantitative relationships between variations of speed of rotation and zonal variations of ocean and atmospheric mass displacements can also be determined.

It is the opportunity to emphasize the successful international effort performed in this field by the IERS under the co-ordination of the International Association of Geodesy (IAG) and the International Astronomical Union (IAU). Without such co-ordination, progress in space geodesy would have been impossible, nor would the progress accomplished in the various techniques have occurred (SLR, GPS, GLONASS, DORIS, VLBI), as they also require permanent co-ordination and dialogues. The progress much is shown in Figure 5.

5.1.7 Tides

Before the space geodesy era, a few long running series of tidal gauges and records existed but only involving a limited number of stations, as for example in Brest, France for more than 150 years. These records are performed along the coasts and yield the best estimate of the mean sea level that we can have for the last century. On the other hand, in the middle of oceans, no direct measurements could be performed except along the coasts of islands, and the modeling of the tides in the deep oceans was difficult. Now, the situation has drastically been improved with the availability of precise altimeter data. The improvement of hydrodynamic models is another factor for assimilating the altimeter data and for having now a very good model of the main tidal waves, with an accuracy of about 2 cm in many parts of the oceans. The tidal dissipation process and its location can also be precisely studied. The intercalibration between the altimetric systems and the tide gauge networks has to be undertaken on a continuous basis.

5.1.8 Ocean circulation

Before, there was a large set of descriptive knowledge about ocean circulation, but it was based upon discrete measurements. (Cruises of research vessels measuring accurate density profiles; these profiles were averaged in space and time, providing "climatic maps"). There was not a homogeneous accuracy on a planetary scale. TOGA and WOCE experiments were approved and set up to increase such worldwide coverage. Nowadays, the determination of the whole spectrum of sea surface topography has been completed thanks to altimetry. Models of the ocean circulation exist and first assimilation experiments of new space based data have been performed using these separate models. Nevertheless, the limitations of this approach have been well known from the early beginning of altimetry (for example the importance of knowing the geoid). Therefore many projects were studied at length both from a theoretical point of view and from a perspective established to take advantage of new technology (better and very precise

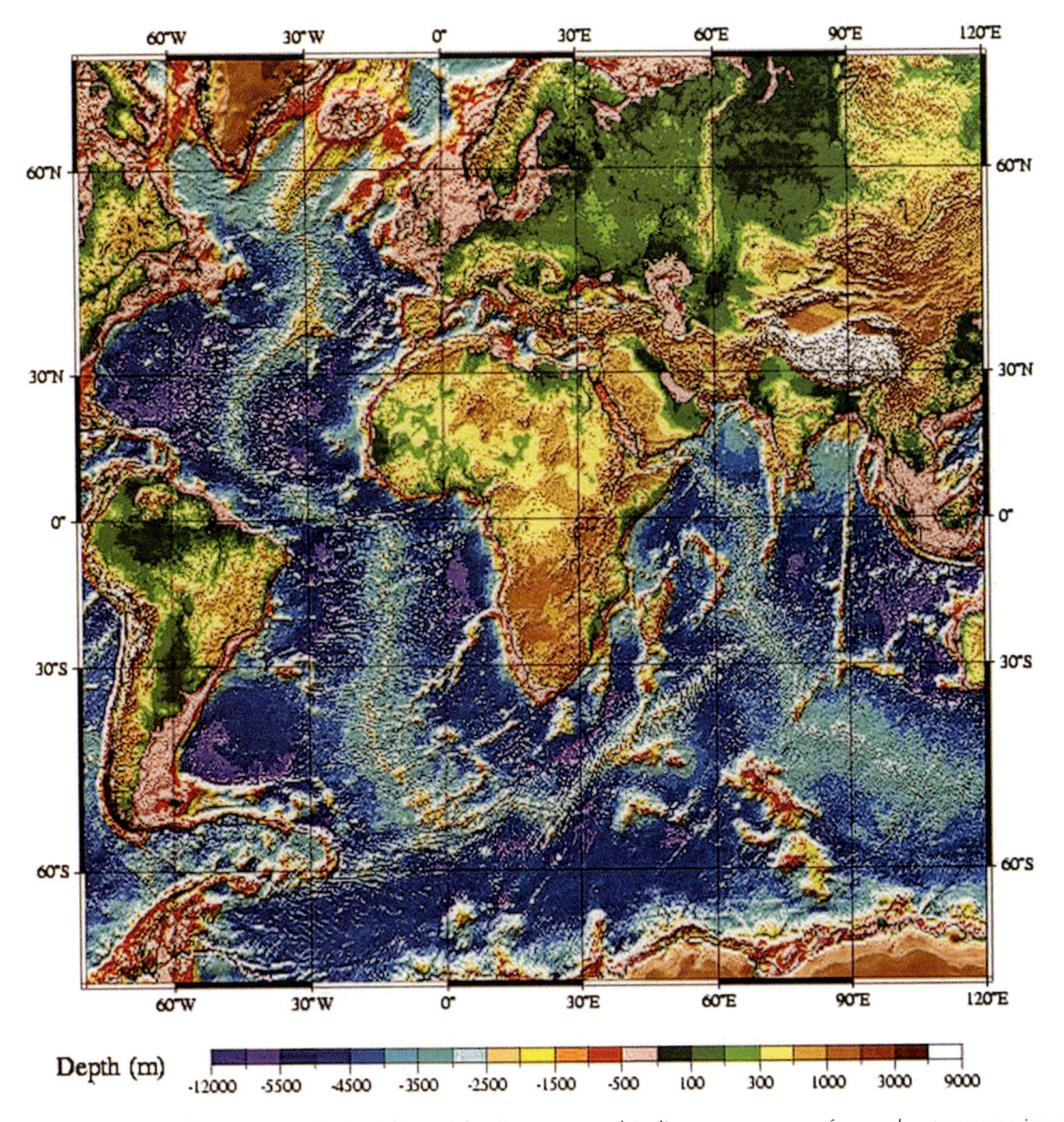

Figure 13.1 Global sea-floor topography in Africa, Atlantic ocean and Indian ocean zones from a least-squares inversion of altimetry-based high-resolution mean sea surface and shipboard soundings. (From S. Calmant, IRD, Noumea M. Bergé-Nguyen and A. Cazenave, LEGOS-GRGS/CNES, Toulouse, 2000.)

geoid, *in situ* measurements of various data in the deep oceans)

5.1.9 Sea-floor topography from satellite altimetry

New and original results have been obtained in research into the sea-floor topography. In the past, the determination of geoid was only possible on regional or local scales. Now, a global determination of the geoid can be computed from gravity field models with a space resolution of about 500 kilometers or a little better. These models, as explained above, combine much data of various origins including altimeter data.

Nevertheless, the long wavelengths of these models are well determined by space orbital data only and can be removed from the mean sea surface topography obtained from altimetric data. In the mean sea surface topography, the variable part of the oceanic signal (tide, oceanic circulation on different scales, atmospheric effects) can be removed directly when known, then the remaining variable part can be eliminated by averaging the sea surface topography over a certain period of time. Consequently, it is possible to determine a marine geoid. From this, it is then possible to inverse the problem of the geoid determination and to deduce the sea-floor topography generating the marine geoid. Much better results are obtained in combining bathymetric data obtained from shipboard soundings. Examples of accurate and high frequency sea floor topography for two large areas are given in the Figures 13.1 and 13.2. Dense altimetry data over the oceans from ERS1 data (geodetic

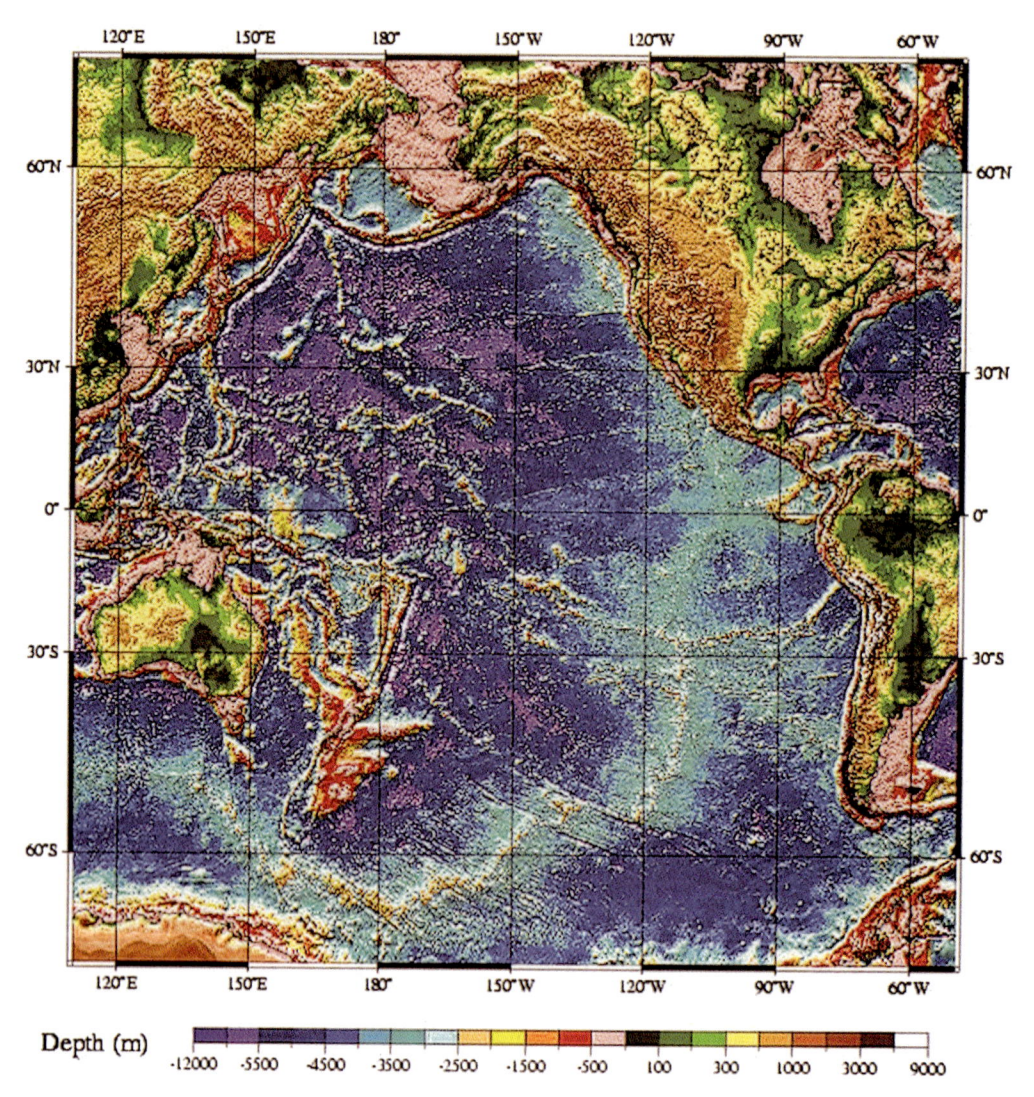

Figure 13.2 As in Figure 13.1 but for the Pacific area.

orbit of 168 days) were used to construct such images (CLS/ARGOS, Toulouse, Hernandez and Schaeffer 2000).

5.2 A Major turn: GPS, GLONASS, and GNSS/GALILEO

The DOD (Department Of Defense) initiated GPS (Figure 14) during the 70s in the USA. The US Navy had planned to replace their operating Navy Navigation Satellite System (NNSS) but had to join with the Army and the Air force to develop a common navigation system. This collaboration caused a few delays but augmented the capability of the system. The objective was clear: detection of any object, whatever its speed, in position in real time any where on the Earth.

To achieve this ambitious objective, the DOD decided to begin by setting up and maintaining a constellation of 24 satellites in high orbit (20,000 kilometers). Each satellite is equipped with atomic clocks and transmits downward 2 frequencies in the L-band (2 frequencies to filter the ionosphere effect). These two frequencies are coded: the first one by a Coarse Acquisition code (C/A) and the second one by a cryptic classified precise code (code-P). The idea was to make only the C/A code available for civilian uses. Some users (scientists being among them) were clever enough to perform precise positioning using differential methods and thus to eliminate the unknown part of the codes.

The DOD was worried about this fact and used other techniques to maintain the confidentiality of their system. But there was such pressure from non-military users that the US government took the decision to fully open the system to the public. The system, when used with highest quality receivers but not in real time, can provide positioning to an accuracy of one centimeter on even better. Progress in technology with miniaturized equipment made it easier. At this level, many applications are made and invented every week for the benefit of everybody. One of the most recent and spectacular uses is the decisive contribution to the discovery close to the coast of Egypt of two very old large cities (early 2000 B.C.). In this particular case, the essential role of GPS was to make possible the link between physical soundings and to establish precise geographical maps under the sea.

In science, the GPS outputs are obviously also very important. They concern the polar motion and the Earth rotation determined by the International Earth Rotation Service (IERS). They also include contributions to gravity missions (such as CHAMP, GRACE, and GOCE) and contributions to the International Terrestrial Reference Frame (ITRF) determination and to crust deformation studies.

It is important to draw attention to the fact that some other similar projects complement GPS, such as the existing Russian Global Navigation Satellite System (GLONASS), and that there are also new projects such as the

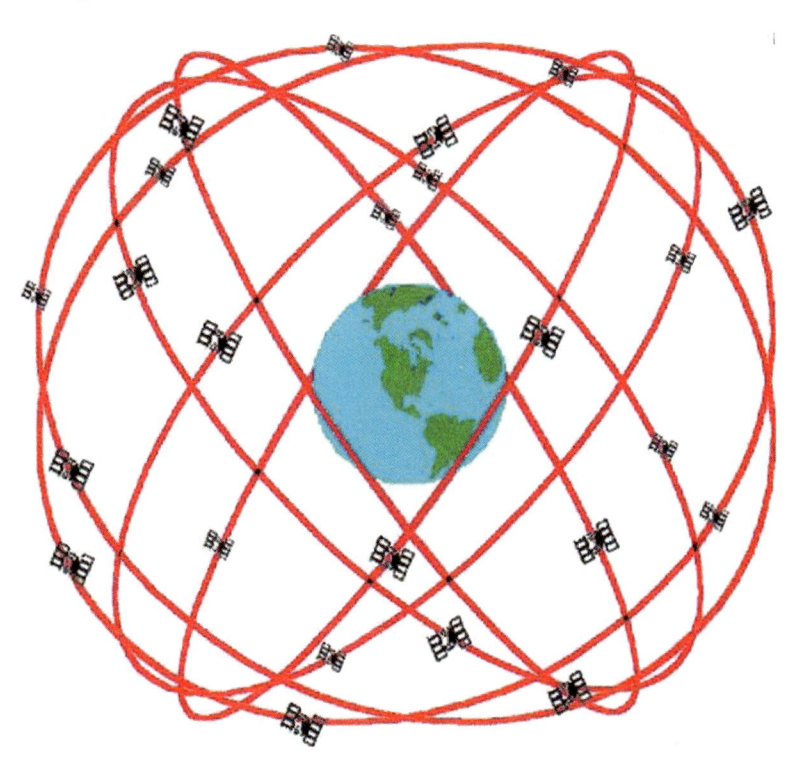

Figure 14 American GPS satellite constellation. There are 25 satellites orbiting at 20 200 km in 6 different orbital planes with an inclination of 55° with respect to the equator. (JPL document.)

GNSS/GALILEO system in Europe (Global Navigation Satellite System) for completing previous systems.

One of the potential customers will be international air-traffic control and also maritime and land navigation systems. Now, the scientists have the obligation to look carefully at the new coming systems and study whether they will have the same level of performance or not, a factor that can play a big role in the cost. As a result, scientists must have access to the data and be involved in the decisions. In other terms and based upon past experiments, one has to take care that, at this high level of metrology, one has the capability of controlling the systems by comparing them to other systems based on other techniques and by maintaining several teams working on them, so as not to have an isolated team of experts. This is especially important when faced with very large and superabundant sets of data in which too many parameters might be adjusted, such modifications hamper the detection of possible systematic errors. The international GPS service (IGS) plays an important role in these issues.

5.3 Gravity decisions

In the previous section, we were uncertain about decisions concerning the gravity missions. The ESA ARISTOTELES project was for a while the common candidate for this application nominated by both the European and the American geodesists, as discussed at the NASA workshop in Coolfont, USA (1989), but finally it was not approved. Fortunately, at the end of the decade 1990–2000, after many discussions and fights, three positive and very complementary decisions were taken:

– The first mission is CHAMP (CHAllenging Mini-satellite Payload), a German satellite successfully launched on July 16, 2000 in a polar orbit. The gravity mission is based on satellite-to-satellite tracking in the high-low mode, the high satellite being the satellites of the GPS constellation and the low satellite being the CHAMP satellite. This satellite was implemented by the GFZ (GeoForschungs Zentrum) Potsdam in Germany with some cooperation for providing science instruments by NASA (USA), CNES/ONERA (France), the Air Force Research Laboratories (USA).
– The second program is a joint project between NASA and the Deutsches Zentrum für Luft und Raumfahrt (DLR) led by the University of Texas and the GFZ Potsdam. It is based on the low – low satellite to satellite tracking mode. The two low satellites will measure their relative velocity with an accuracy of 1 micrometer/ second. In addition, the 2 low satellites will be tracked by GPS. GRACE will be launched in 2002; the lifetime is expected to be of 3–5 years. It will provide the static part and the monthly temporal variations of the Earth gravity field (Figure 15.1).
– The third project GOCE (Gravity Field and Steady-State Ocean Circulation Explorer) is an European ESA project, the first mission of the new Earth Explorer program. It is the more ambitious of all and is based upon an accurate gradiometer put inside a satellite equipped with an air drag free system. It will be put in a very low orbit of 250 kilometers. Launched in mid 2004, this mission should provide unprecedented accuracy of 2.5 millimeters for the geoid and of 0.08 milligal for the gravity field, with expected resolutions of 100 to 50 km (Figure 15.2)

Marketing Advice from Users!

In many cases including Earth space projects, you are required to have strong interest from so-called users. Although it seems to make sense it is not always so simple, as shown in a few past examples. CNES in the 60s proposed to implement the tracking of a fleet of stratospheric balloons in a two-way range and range rate measurement system on a satellite. This system, named EOLE, was finally successfully launched in co-operation with NASA.

Oceanographers were asked to express their possible interest and none was displayed. Two years later a group of oceanographers came back and made a strong pleu "we want a second EOLE". In the mean time, it happened that a major discovery, had occurred: the eddy activity was not limited to the meandering of large currents. Eddies were everywhere and their total energy was as large as the energy of involved in the main currents. There was a happy ending, and the ARGOS system was designed to respond to the request of oceanographers. Operational in 1978, it has been operating without any break from this time on:

Another example comes from the late 60s, when a rather negative attitude for measuring the plate tectonic motions was generally held, as the Earth rotation could only be determined with poor precision.

Consequently, we proposed altimetry to argue for ocean circulation. The reply was even more negative. The typical value of the amplitude of large scale data was in the range 10 to 20 centimeters and any significant progress would request a precision of 10%, that means an accuracy of one to two centimeters. What does it mean? Just that people with vision have a main part to play in satellite geodesy, as it is in many scientific domains. All scientists must be open to discussion but ready to defend their opinion with a lot of energy, persuasiveness and patience! Above all, they have to welcome the expertise of users; rather than asking them for their advices or requesting their support, they have to offer them opportunities.

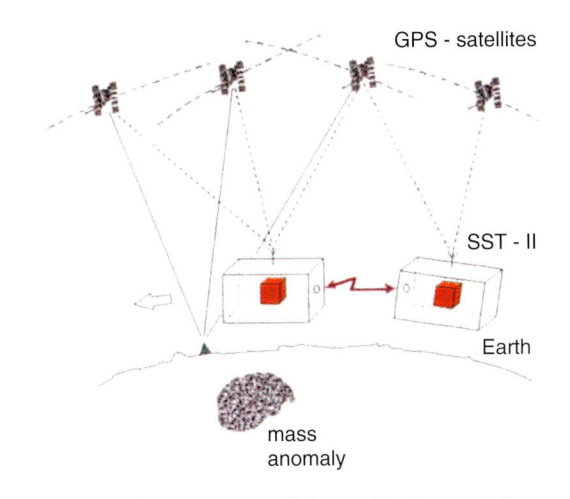

Figure 15.1 This mission will be realized as a joint project between NASA and DLR. (From ESA-SP1233 (1) document, p. 20, 1999.)

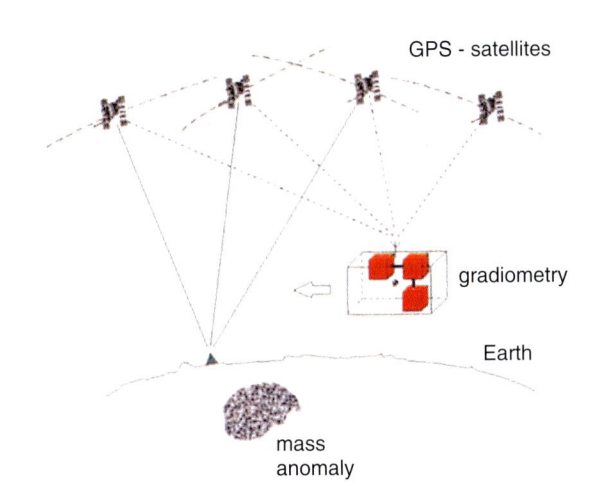

Figure 15.2 This mission will be realized in Europe by the ESA/GOCE satellite. (From ESA-SP1233 (1) document, p. 20, 1999.)

consequently, the decade 2000–2010 should be the decade of gravity.

One of the reasons for the positive decisions is the success of altimetric missions that obviously inspired further gravity missions. The GOCE mission will enhance altimeter results. The dialogue between oceanographers, geophysicists and geodesists has finally become fruitful. This dialogue originally involved altimetry programs, where the three fields of study are mixed together.

6. SUMMARY: SPACE SCIENCE AND EARTH

6.1 Metrology of the Earth

After almost 40 years, it is now possible to enhance the main features of this new geodesy with key words: global scale and sampling, observing system and system accuracy, redundancy, advanced technology, data management.

The first one is the access to the global scale. For the first time we can speak about the physics of the globe, and we are able to study the Earth on planetary scales. Regional studies are possible as well, but as part of global studies. However, the largest benefit from satellite missions is the unprecedented sampling capabilities both in space and time. The scientists must realize that they have some power, at least in the first phases of a project, to propose some sampling schemes. In some cases, it will be necessary to use a multi-satellite strategy to optimize the sampling.

Another specific feature is the observing system with its two parts: the onboard instrument and the ground network. The onboard instrumentation is the core component, which has the big advantage of using the same instruments throughout the mission, but such a practice requires calibration and a validation plan. The second component is a ground network, which is a prerequisite and has to be deployed all over the Earth providing data with an accuracy compatible with the onboard instruments. We have received from the analysis of existing data sets a better understanding of the type of observing systems we need; We need to infer the requirements for the observing system not only in terms of sampling but also in terms of accuracy. In fact, although very important, it is not easy to make assessment of the system accuracy. One of the methods used is to compare observations from satellites with *in situ* observations from calibration and validation sites to gain some insight into the observational errors but the real assessment has to be done on the global level. In fact, the next step is to use in the same scheme both ground based and satellite observations by assimilating them into the same model. The global sampling will give some hope to delineate systematic errors in the observing systems of inadequate modes resulting from the physics or in the assimilation methods. The final accuracy of a system will be the consequence of convergent independently gathered results.

Some redundancy on the metrological level is necessary. We need external checks using other systems of observations, other teams working independently with the same observations, or results coming from other disciplines. Ultimately, assessment can come later, when appropriate, by verifying the forcasting of specific events.

Speaking about the metrology of the Earth means there is a need to look carefully at new advances in technology and to investigate the possible benefit of medium or long range observing systems. Just to give a few significant examples:

– The triaxial microaccelerometer CACTUS onboard the CASTOR satellite (1975) drives currently the new generation of accelerometers used in the gravity space missions (CHAMP, GRACE, GOCE).
– The improved short-term stability of quartz oscillators was the basis of the DORIS system, baseline of a precise orbit for TOPEX/Poseidon, Jason1 and ENVISAT jointly with the laser technique.
– The progress in stability of the atomic clocks and the capability to put such devices onboard satellites is the basis of new navigation and positioning satellite systems such as GPS, GLONASS and GALILEO.
– The new device to detect very low levels of optical energy provides an impetus to the satellite laser ranging either in decreasing transmitted energy or in using very small array of reflectors onboard.

There are also major steps in computers, and other hardware and software to collect, distribute and archive the data. Data handling, management, formatting and access, along with time consumed in data processing were a big subject of discussion; it is a not only a domain where the techniques go fast. The real issue is or will be technical but also, it will be to find resources, mainly human means, and to watch the quality of the data including the feedback for reprocessing the data by taking observations from users and scientists into account.

Data exchange may be still difficult for political reasons but, sometimes, there is probably no way to prevent any scientists or users from acquiring a copy of a data set from a friend! An interesting point has to be noted. In the past, geodesy was in many countries more or less controlled by defense authorities, but any classification was circumvented very quickly; satellites ignore frontiers. The real problem is to archive the data, especially for future studies over very long periods of time. What will be left as data sets for our successors in 10, 20 or 50 years?

It is also necessary to take care to maintain a set of very long-term or even permanent well-equipped sites. These permanent sites, also called geodetic fundamental observatories, will be used as long-term references, or to construct images that act as anchors.

The outreach to the public is also very important and often occurs very quickly. We remember a time, when looking every hour at the occurrence and development of the oceanic phenomenon El Nino 97, we were very proud to see the images coming from TOPEX/Poseidon. We were happy to realize that in a few weeks the cumulative audience of people watching the event on TV channels would increase drastically, maybe to one billion around the world. But it remains an ethical problem especially in the sensitive areas like the rôle of oceans in possible climate changes.

6.2 Comments about scientific results

Before concluding, we want to clarify an important issue. Scientists often acknowledge the potential of satellite data but are reluctant to be fully committed to the space system and think that they have no way to change anything neither in design nor in data processing. They frequently think that future programs will be driven by the so-called applications without any focus on scientific objectives. We just want to give two examples here to highlight persistent scientific capabilities. Indeed, there are many examples where new results come out, although they were not a part of the original main objectives.

When we look at late 70s maps of the sea surface computed with GEOS 3 altimetric data, the input from geophysicists was rather minor. The first papers were considered as a curiosity. Ten years later we could personally remember an AGU session entirely devoted to ocean tectonics and altimetry. The chairman was obliged twice to change the session room to welcome an overcrowded audience of geophysicists. The marine geoid was thus considered as a prime objective interesting enough to convince ESA to move the ERS 1 satellite to a dedicated orbit allowing a grid of 8 kilometers at the equator and triggering the release of altimetric observations made by GEOSAT in the first part of its mission. This new set of data was the basis for computing the sea-surface topography used now as a reference. It was not a primary objective of the mission, but we can label that as one of its associated objectives (Figures 13.1 and 13.2). A message to the scientific community could be given: look at the design of a mission with enough vision to be able to discern such associated objectives so that minor modifications of design can provide a large scientific return.

As a second example, a recent major result was obtained from the study of the area of dissipation from the lunar tidal component on the Earth. The M2 tide (the role of other tides is smaller) provides an energy of about 3 Terawatts. The total amount of dissipation can be detected by a laser aimed at the Moon and the determination of the increase of the Earth–Moon distance by 3.8 centimeters per year and by the spin slow-down of the Earth or the increase of the length of day (about 2 ms per century). We have, with new space based techniques, more accurate determinations of the polar motion and the rotation of the Earth, making possible a tentative realistic scheme for the Earth-Moon system evolution. In the past, it was thought that most of this energy was dissipated in shallow seas or coastal zones.

However, in the 60s, best estimates from different authors were converging, claiming that the so dissipated energy is only 40 or 60% of the total energy and questions of how and where the remaining part was dissipated remained broadly open, and involved the possible role of the Earth's tides (Melchior 1973).

At the NASA Williamstown meeting (1969), this problem was also on the priority list of scientific questions, but no future program was expected to provide answers. Within the Science Working Team (SWT) of TOPEX/Poseidon, there was a subgroup in charge of the tidal model; the tidal model was considered as important and indeed was one of the drivers in the decision to put TOPEX/Poseidon in a non-heliosynchronous orbit; a prograde orbit with an inclination of 65 degrees was adopted.

Scientists started to develop new global tide models (see, e.g. Le Provost 2001). For most of the members of SWT, the tides were important enough to isolate tidal data, to have access to the filtered dynamical topography and to use it in models of ocean circulation. For example, Le Provost et al. (1994) set up a new global tide hydrodynamic model taking account of the new global map of bathymetry. Before any assimilation of altimetric measurements, it was clear from this model that, at least, a part of the dissipation occurs in the North Atlantic. However, Egbert and Ray (2000) went further in analyzing long term sequences (7 years) of the most accurate sets of TOPEX/Poseidon altimetric data. In addition to the well known areas, they found that dissipation also occurs in the areas of deep ocean. Consequently, they were able to construct a global map of the zones of dissipation. The proposed explanation is that the M2 tidal wave is reflected by tectonic ocean features, generating coherent energetic internal waves. The corresponding data is obtained from in the altimeter. As suggested in the past by Munk and Wunsch (1998), this tidal mixing provides the extra energy required to maintain circulation in the deep oceans (Wunsch 2000 and Kerr 2000). This result highlights the basic link between these questions and the exchanges of energy and momentum within the Earth-Moon system.

What we learn from this example is as follows: to get decisions, we need to have prime objectives that will be achieved if the baseline requirements are fulfilled. In addition, we should set associated (not secondary) objectives that may be more interesting but assume that we may be able to modify the specifications. In the example above (that we can call "tidal mixing"), it was possible to get a new result only because the system accuracy was 1–2 centimeters and the lifetime larger than specified. To provide a good image, we can use a sentence of Kerr (2000):

> Oceanographers needed a global tide gauge which they found in the TOPEX/Poseidon satellite.

To provide such a "tide gauge" it is essential to have scientists involved in all the phases of the project.

6.3 As a conclusion from "end-to-end" to "happy end"

Both of us started to work together in 1960 at the observatories of Paris and Meudon. At this time, we were sharing the first computer, the famous old IBM 650 called "the washing machine". One of us was involved in the determination of the Earth's rotation using the observation of stars with the Danjon astrolabe. The other one was already watching satellites and starting to study the density of the low atmosphere from visual observations.

40 years later, we recognize that we were lucky to have been involved in all the stages in this exciting period, when everything was new. The Earth as a living planet was the object of our research, an idea which was the dream of generations of geophysicists, geodesists, astronomers, and navigators.

We would like to share with the reader some major findings if not feelings. The major thing is the scale. Changing scale involves not just enlargement of each field. It produces global multi-disciplinary approach. We worked closely with teams from other fields, teams with new people and with expertise that they were anxious to share with others; this was really a new behavior. Until recently, each discipline of geosciences had enough interesting questions and observations for itself. But studying the Earth as a global planet makes progress only if scientists start a dialogue in depth.

The second finding is based upon the fact that space activity depends on projects having their own organization, schedule, rules, specifications, budget, and resources. In the beginning, it seems you have two different worlds and indeed, on the basis of efficiency, the relationships are organized through what we call interfaces.

We have experienced that real efficiency requires much more interactive relationships between the actors. The scientists have to know the technical constraints; project teams have to understand the impacts of decisions on scientific results. Such teamwork is even more crucial for data management. It is easier to ask the users about their requirements to build up a data processing system that will be stabilized after some period of validation. But the real success only comes if you have feedback, as for example, in reprocessing sets of data by taking account the findings of scientists which may require changes to the processing system. In the Earth sciences, it is necessary to understand the systems proposed for applications. In most cases, we have seen that the requirements in terms of sampling, continuity, mission parameters, data management are so close that differences are artificial. Again, the best is to have the active participation of scientists in the system.

Specific to Earth science is that a part of your system comes from the networks of *in situ* measurements and these ground networks are as important as the space instrumentation. The role of *in situ* is sometimes considered as one involving calibration of remote sensing. Such an idea is misleading; the different data have to be complementary and assimilated jointly in adequate models. Such a method is now favoured under the label of integrated programs.

In fact working together is the rule and requires scientific involvement in an "end-to-end" process. In our jargon, end-to-end means that one has to take care of all the tasks but that the links between the different tasks will be very important.

As a general conclusion, this contribution does not cover, by far, all the works made by the international community; but our goal was to testify through our limited experience that the improvement of scientific knowledge of the planet Earth is a reality, that a multi-disciplinary community exists and that a new way to associate the scientists is possible, necessary and efficient. New techniques, like the SAR interferometry, new themes like the ice cap dynamics, the extension of previous studies to planetary studies, are extremely promising and are in full development having produced with rewarding results. But we cannot present here all these new fields.

Let us borrow the words of our Australian colleague Neville Smith speaking for partnership in international programs: "You are a good player but nobody knows even you. Come and play with us and you will realize how good you are and how much we need you in the team. Play together and we will win and will have a lot of fun". We would like to finish with this remark: on our stage we play, we have a good casting, a good scenario called "The New Earth" but we have just to play at the same time, and so we will have a lot of fun and contribute to a "happy end for mankind". "We" means the geonauts.

We made some successful tests in proposing a new myth to express this feeling.

The GEONAUTS:

EARTH is a space station put successfully in orbit around the Sun 4.5 billions year ago. This station is manned and has welcomed 6 billion people; to emphasize our common venture we rename inhabitants of the Earth the GEONAUTS.

GE is EARTH, our ship; we are Nautes on it and navigate in the solar system (as Internautes are surfing on the web).

GEONAUTS are anxious to control the behavior of their Earth ship and to send around it automated and manned stations. This extra vehicular activity provides the requested observations to forecast their future.

6.4 As another conclusion: Plato

Research and applications from "The Republic" – Plato
EDUCATION OF THE PHILOSOPHER
RESEARCH AND APPLICATION, FOR.... MY OWN SATISFACTION?
...
And the third should be astronomy. Or don't you agree?
– Yes, I certainly agree. A degree of perception in telling the seasons, months and years, is useful not only to the farmer and sailor but equally to the soldier.

"You amuse me", I said, with your obvious fear that the public will disapprove if the subjects you prescribe don't seem useful. But it is in fact no easy matter, but very difficult for people to believe that there is a faculty in the mind of each of us, which these studies purify and rekindle after it has been ruined and blinded by other pursuits, though it is more worth preserving than any eye since it is the only organ by which we perceive the truth. Those who agree with us about this, we give your proposals unqualified approval, but those who are quite unaware of it will probably think you are talking nonsense, as they won't see what other benefit is to be expected from such studies. Make up your mind which party you are going to reason with-or will you ignore both and pursue the argument largely for your own satisfaction, though without grudging anyone else any profit he may get from it?

That's what I'll do, he replied; I'll go on with the discussion chiefly for my own satisfaction.

Acknowledgements

First of all, we are indebted to Prof. R. Rummel and Dr. G. Balmino; they were supposed to write this chapter and generously proposed that we should be substitutes in order to acknowledge our pioneering role. We were very sensitive to this and took this opportunity to give a personal account. We thank them very much for their advice and encouragement. We both were happy to provide our testimony on these 40 years, which was an exciting time. A testimony may contain some biases or misses. We take responsibility for them.

When 30 years ago we wrote a proposal after the Williamstown meeting, it was just a vision. We thought that after 30 years everything was going on except a dedicated mission to determine the gravity field of the planet Earth. But, fortunately, the decision to go ahead with GOCE was taken by ESA in 1999. We can testify that without the constant, competent and obstinate work over 20 years of our two friends and colleagues over 20 years, GOCE would be not here.

All that was undertaken in these fields was achieved in close co-operation with many colleagues during an exciting time. We can testify that the situation was stimulating, but also developed strong and friendly relationships. We would like to dedicate this article to our friend Dr. J.G. Marsh, who displayed a new spirit of cooperation. We want to thank again the numerous scientists and engineers with whom we worked, but also tell them that we deliberately chose to give a personal account rather than a compendium.

When accepting this task, we did not realize that the time was so short. Therefore we asked the DAG company in Toulouse (25, rue Saint Guilhem – http://www.dag.fr) at the last minute to help us. Without them and especially Lise Nobileau, Isabelle Labadiole and Emmanuel Ventura nothing would have been done. We shortened the possible delays thanks to the fact that DAG had organized several workshops for us and thus had acquired some basic knowledge of the subject.

Christiane Berger, Olivier Laurain, and Joëlle Nicolas from CERGA (Observatoire de la Côte d'Azur), Bernard Guinot, Philippe Gaspar, Raymond Zaharia, Victor Zlotnicki gave some extra help in reading and improving the text. We thank them very much. We are also indebted to the reviewers and to the editors for their help.

GENERAL REFERENCES

In space geodesy, looking at the number of symposia, books, reference papers over the first 40 years of this new field is amazing; several tens of thousand of references, hundreds of symposia, colloquia, many books... In the beginning, the leading rôle and enthusiasm were found in the USSR, USA, European countries, and Japan, but progressively most of the countries in the world became involved. Space geodesy has more and more applications in many fields including cartography, navigation, sea level research, etc. ... So space geodesy has become an important field of research and application. All continents and many islands now have, permanent geodetic stations equipped with either a GPS, GLONASS receiver, or and a DORIS beacon, or several of these.

How to describe all this enthusiastic effort in the beginning and this acceleration in the use of all these techniques? To us, it quickly appeared that it was impossible to give all the references. Moreover, very often many works have been duplicated or parallelled by several authors and teams in different countries. No doubt historians will have great difficulty describing precisely all this evolution of such a young discipline. Therefore, we decided to illustrate only a part of it and to select arbitrarily a certain number of books, proceedings and symposia in which many names of the first scientists can be read as well as the names of the new generation. We do that with a great feeling of modesty and humility, knowing that so many names and so many things will be forgotten. In any event, space geodesy appears as a very enthusiastic worldwide success story.

Proceedings

Baldi, P. and Zerbini, S. (1988). *Proceedings of the international conferences on the WEGENER/MEDLAS Project, Third international conference*. Bologna: University of Bologna.

Cospar (1960–1964). *Proceedings of the plenary meeting of the Committee on Space Research (COSPAR). Nice-Florence*. Amsterdam: North Holland.

ESA (1997). *Proceedings of the third ERS symposium: space at the service of our environment, volume I (Soil moisture, hydrology, land use, forestry, DEM, geology), volume II (GOME/ozone, atmosphere, ice properties, ice sheet, sea ice, ice dynamics, coastal zones, global change), volume III (Winds and waves, sea surface temperature and oceanography, ocean circulation, altimetry, marine geoid, meteorology, ATSR, instrument performances, orbits, SAR°Interferometry)*. Florence: ESA (SP-414).

Fedorov, E. and Bender, P. (1977). *Proceedings of the 78th IAU symposium, Nutation and the Earth's rotation*. Kiev: D. Reidel.

Gaposchkin, M. and Kolaczek, B. (1980). *Proceedings of the 56th colloquium of the IAU, Reference co-ordinate systems for Earth dynamics*. Warsaw: D. Reidel.

Giacagla, G.E.O. (1974). *The fourth COSPAR-IAU-IUTAM symposium, Satellite dynamics*. Sao-Paulo: Springer-Verlag.

Groten, E. and Strauss, R. (1988). *Lecture notes in Earth sciences, vol. 19, Proceedings of the international GPS workshop, Techniques applied to geodesy and surveying*. Darmstadt: Springer-Verlag.

Intercosmos co-operation (1965–1991). *Proceedings and publications of scientific results of the Intercosmos co-operation (1 or 2 meetings per year)*. Academies of the European Eastern Countries (in Russian and English).

Kovalevsky, J. (1965). *Trajectories of artificial celestial bodies, COSPAR-IAU-IUTAM symposium*. Paris: Springer-Verlag.

Levallois, J.-J. (1964). *Réseau géodésique européen par observation de satellites, symposium*. Paris: CNES and IGN.

Mc Carthy, D. and Pikington, J.D.H. (1978). *Proceedings of the 82nd IAU symposium, Time and the Earth's rotation*. San Fernando: D. Reidel.

McKinney, C.M., Kennedy, T.C. and Leroy, C. (1979). *Proceedings of the second International Geodetic Symposium on satellite Doppler positioning*. Austin (Applied Research Laboratories, Texas): Defense Mapping Agency and National Ocean Survey.

Melchior, P. and Yumi, S. (1972). *Proceedings of rotation of the Earth, IAU symposium 48*. Morioka (Japan): D. Reidel.

Roy, M. (1962). *Dynamics of satellites, IUTAM symposium*. Paris: Springer-Verlag.

Schwarz, K.P. (1999). *Geodesy beyond 2000. The challenges of the first decades (symposium 121), at the 35th general assembly of the International Association of geodesy*. Birmingham: Springer-Verlag.

Veis, G. (1963). *Proceedings of the international symposium on the use of artificial satellites for geodesy*. Amsterdam: North Holland.

Veis, G. (1965). *Proceedings of the second international symposium on the use of artificial satellites for geodesy*. Athens: The National Technical University of Athens.

Veis, G. (1973). *Proceedings of the international symposium on the use of artificial satellites for geodesy and geodynamics*. Athens: The National Technical University of Athens.

Veis, G. (1984). *Proceedings of the international symposium on the use of artificial satellites for geodesy and geodynamics*. Athens: The National Technical University of Athens.

Wytrzyszczak, I.M., Lieske, J.M. and Feldman, R.A. (1996). *Proceedings of IAU colloquium 165, Dynamics and astrometry of natural and artificial celestial bodies*. Poznan: Astronomical Observatory of A. Mieckiewicz University.

Contributions – Workshops

Schutz, B.E., Anderson, A., Froidevaux, C. and Parke, M. (1994). *Gravimetry and Space Techniques Applied to Geodynamics and Ocean Dynamics*. American Geophysical Union and International Union of Geodesy and Geophysics.

Smith, D. and Turcotte, D.L. (1993). *Contributions of Space Geodesy to Geodynamics, Crustal Dynamics, Geodynamics series, volume 23*. American Geophysical Union.

Smith, D. and Turcotte, D.L. (1993). *Contributions of Space Geodesy to Geodynamics, Earth Dynamics, Geodynamics series, volume 24*. American Geophysical Union.

Smith, D. and Turcotte, D.L. (1993). *Contributions of Space Geodesy to Geodynamic, Technology, Geodynamics series, volume 25*. American Geophysical Union.

Proceedings of 12 International Laser Workshops regularly held since 1973.

Earth and Ocean Application Physics (NASA Workshop), Williamstown, USA, July, 1969.

Space Oceanography, Navigation and Geodynamics (SONG Workshop), Schloss Elmau, Germany, January, 16–21, 1978, ESA-SP-137.

Solid Earth Science and Application Mission for Europe (SESAME Workshop), Chiemse, Germany, March, 4–6, 1986, ESA-SP-1080.

The Interdisciplinary rôle of Space Geodesy, International School of Geodesy "A. Marussi", Ettore Majorana center for scientific culture, E. Boshi and S. Zerbini, S.O.C., Erice, Italy, July, 22–29, 1988. Lecture notes in Earth Sciences, Eds; Mueller and Zerbini, Springer-Verlag.

Symposia, Cahiers

International Association of Geodesy, 1989–2000. *Symposia series from symposium 101, (Global and Regional Geodynamics) up to symposium 120 (Towards an Integrated Global Geodetic observation system)*. Springer-Verlag.

Paquet, P. and Elipe, A. (1997). *Conseil de l'Europe, Cahiers du Centre Européen de Géodynamique et de Séismologie, Proceedings of the Workshop: Accurate orbit determination and observations of high earth satellites for Geodynamics. COGEOS project chaired by A. Nobili*. Walferdange (Luxembourg): Centre Européen de Géodynamique et de Séismologie.

Books

Audoin, C. and Guinot, B. (1998). *Les Fondements de la Mesure du Temps. Comment les Fréquences Atomiques Règlent le Monde*. Paris: Masson.

Bertotti, B. and Farinella, P. (1990). *Physics of the Earth and the Solar System, Dynamics and Evolution, Space Navigation, Space-Time Structure, Geophysics and Astrophysics Monographs*. Kluwer Academic Publishers.

Bomford, B. (1962). *Geodesy*. Second ed. London: Oxford University Press.

Caputo, M. (1967). *The Gravity field of the Earth International Geophysics Series from Classical and Modern Methods*. Academic Press.

Cazenave, A. and Feigl, K. (1994). *Formes et Mouvements de la Terre, Satellites et Géodésie*. Paris: Croisée des Sciences, CNRS editions, Belin.

Dinescu, A. (1980). *Introducere in Geodezia Geometrica Spatiale*. Bucuresti: Editura Technica.

Fu, L.L. and Cazenave, A. (2001). *Satellite Altimetry and Earth Sciences. International geophysics series, volume 69*. San Diego: Academic Press.

Heiskanen, W.A. and Moritz, H. (1967). *Physical Geodesy*. San Francisco: W.H. Freeman and Co.

Kaula, W.M. (1966). *Theory of Satellite Geodesy*. Blarsdell Waltham, Mass.

King-Hele, D.G. (1964). *Theory of Satellite Orbits in Atmosphere*. London: Butterworths.

King-Hele, D.G. (1992). *Tapestry of Orbits*. Cambridge: Cambridge University Press.

Kovalevsky, J. (1963). *Introduction à la Mécanique Céleste*. Paris: Armand Colin.

Lambeck, K. (1980). *The Earth's Variable Rotation, Geophysical Causes and Consequences*. Cambridge: Cambridge University Press.

Lambeck, K. (1988). *Geophysical Geodesy*. Oxford: Clarendon Press.

Levallois, J.J. (1989). *Mesurer la Terre, 300 Ans de Géodésie Française, De la Toise du Châtelet au Satellite*. Paris: Ecole des Ponts et Chaussées, AFT.

Melchior, P. (1983). *Tides of the Planet Earth* (2nd edition). Oxford: Pergamon Press.

Milani, A., Nobili, A. and Farinella, P. (1987). *Non-gravitational Perturbations and Satellite Geodesy*. Bristol: Adam Hilger.

Montenbruck, O. and Gill, E. (2000). *Satellite Orbits, Models Methods Applications*, Springer-Verlag.

Moritz, H. (1989). *Advanced Physical Geodesy*. Wichman.

Moritz, H. and Hofman-Wellenhof, B. (1993). *Geometry, Relativity, Geodesy*. Karlsruhe: Wichman.

Mueller, I.I. (1964). *Introduction to Satellite Geodesy*. New-York: Frederick Ungar Publishing CO.

Munk, W.H. and Mac Donald, G.J.F. (1960). *The Rotation of the Earth*. Cambridge: University Press Cambridge.

Pick, M., Picha, J. and Wyskocil, V. (1973). *Theory of the Earth's Gravity Field*. Elsevier Scientific Publishing Company.

Rummel, R. (1992). *Geodesy, Encyclopaedia on Earth system, Sciences vol. 2*. Academic press, pp. 253–262.

Schneider, M. (1990). *Satellitengeodäsie*. Deutsche Forschungs Gemeinschaft, Technischen Universität München, Weinheim (Germany): VCH Verlagsgesellschaft.

Seeber, G. (1993). *Satellite Geodesy, Foundations, Methods and Applications*. Berlin. New York: Walter de Gruyter.

Zund, J. (1994). *Foundations of Differential Geodesy*. Springer-Verlag.

OTHER REFERENCES IN SPACE GEODESY

Balmino, G., Brossier, C., Cazenave, A., Nouël, F., Dominh, K. and Vales, N. (1979). Geoid of the Kerguelen islands area determined from GEOS 3 altimeter data. *Journal of Geophysical Research*, **84**, B8, 3827–3831.

Brachet, G. (1970). *International Satellite Geodesy Experiment Plan. Prepared by the Departement de Géodésie Spatiale*. Division Mathématique et Traitement, Centre spatial de Brétigny, CNES, 92 pp.

Brouver, D. (1959). Solution of the problem of artificial theory without drag. *Astronomical Journal*, **64**, 378–397.

Cazenave, A. and Forestier, F. (1971). Determination of the equations of condition for the zonal harmonics using the DIAL satellite. In: Kondratyev, Ya, M.J. Rycroft and C. Sagan (eds.), *Space Research XI*. Berlin: Akademie-Verlag, pp. 521–524.

Dickey, J.O., Bender, P.L., Faller, J.E., Newhall, X.X., Ricklefs, R.L., Ries, J.G., Shelus, P.J., Veillet, C., Whipple, A.L., Wiant, J.R., Williams, J.G. and Yoder, C.F. (1994). Lunar Laser Ranging: a continuing legacy of the Apollo program. *Science*, **265**, 482–490.

Egbert, G. and Ray, R.D. (2000). Significant dissipation of tidal energy in the deep ocean inferred from satellite altimeter data. *Nature*, **405**, 775–778.

Gaposchkin, E.M. (1966). Orbit determination. In: C. Lundquist and G. Veis (eds.), *Geodetic Parameters for a 1966 Smithsonian Institution Standard Earth*. Smithsonian Astrophys. Obs. Spec. Rep, **200**, 1, pp. 77–183.

Guier, W.H. and Newton, R.R. (1965). The earth's gravity field as deduced from the Doppler tracking of five satellites. *Journal of Geophysical Research*, **70**, 4613–4626.

Izsak, I.G. (1964). Tesseral harmonics of the geopotential and corrections to station co-ordinates. *Journal of Geophysical Research*, **69**, 2621–2630.

Jacchia, L.G. (1960). A variable atmospheric density model from satellite accelerations. *Journal of Geophysical Research*, **65**, 2775–2782.

Kaula, W.M. (1966). Tesseral harmonics of the earth's gravitational field from Camera tracking. *Journal of Geophysical Research*, **71**, 4377–4388.

Kaula, W. (1969). *The Terrestrial Environment, Solid-Earth and Ocean Physics, Application of Space and Astronomic Techniques, Report of a study at Williamstown, Mass to the NASA*. NASA, MIT, Cambridge, Mass, 155 pp.

Kerr, R.E. (2000). News of the week. *Science*, **288**, 1948–1949.

King-Hele, D.G., Cook, G.E. and Scott, D.W. (1969). Evaluation of odd zonal harmonics in the geopotential of degree less than 33 from the analysis of 22 satellite orbits. *Planetary and Space Science*, **17**, 629–664.

Kovalevsky, J. and Barlier, F. (1967). Géodésie terrestre et géodésie par satellites. *Space Science Review*, D. Reidel, 69–134.

Kozai, Y. (1961). The gravitational field of the earth derived from motions of three satellites. *Astronomical Journal*, **66**, 8–10.

Kozai, Y. (1962). Second order solution of artificial satellite theory without air drag. *Astronomical Journal*, **67**, 446–461.

Kozai, Y. (1963). *Effects of Solar Radiation Pressure on the Motion of an Artificial Satellite*. Smithsonian Contribution Astrophys, 6, pp. 109–112.

Lala, P. and Sehnal, L. (1969). The Earth's shadowing effects in the short-periodic perturbations of satellite orbits. *Bull. Astron. Czeck*, **20**, 327–329.

Le Provost, C., Genco, M.L., Lyard, F., Vincent, P. and Canceil, P. (1994). Tidal spectroscopy of the world ocean tides from a finite element hydrodynamic model. *Journal of Geophysical Research*, **99**, C12, 24777–24798.

Le Provost, C. (2001). Ocean tides. Fu and Cazenave (Eds.), *Satellite Altimetry and Earth Science*. International Geophysics series, volume **69**, San Diego Academic Press, pp. 267–303.

Lundquist, C. and Giacalia, G.E.O. (1969). *Possible Geopotential Improvement from Satellite Altimetry*. Smithsonian Astrophys. Obs. Spec Rep, 294, pp. 1–44.

Lundquist, C. and Veis, G. (1966). *Geodetic Parameters for a 1966 Smithsonian Institution Standard Earth*. Smithsonian Astrophys. Obs. Spec, 200 (3 vols.), 686 pp.

Marsh, J.G., Douglas, B.C. and Klosko, S.M. (1973). *A Global Station Co-ordinate Solution Based upon Camera and Laser Data*. Proceedings, the First International Symposium for the Use of Artificial Satellite for Geodesy and Geodynamics. Athens: National Technical University of Athens, pp. 749–799.

Melchior, P. (1973). *Physique et Dynamique Planétaire, Géodynamique*. Bruxelles, Vander, pp. 1–51.

Munk, W.H. and Wunsch, C. (1998). Abyssal recipes II: energetic of tidal and wind mixing. *Deep Sea Research*, **45**, 1977–2010.

Rapp, R.H. (1968). Gravitational potential of the earth determined from a combination of satellite observed and model anomalies. *Journal of Geophysical Research*, **73**, 6555–6562.

Rummel, R. (1986). Satellite gradiometry. In: H. Sünkel (ed.), *Mathematical and Numerical Techniques in Physical Geodesy*. Lecture notes in Earth sciences, 7, Springer, Berlin, pp. 318–363.

Samain, E., Mangin, J.F., Veillet, C., Torre, C., Fridelance, P., Chabaudie, J.E., Feraudy, D., Glentzlin, M., Pham-Van, J., Furia, M., Journet, A. and Vigouroux, G. (1998). Millimetric lunar laser ranging at OCA. *Astronomy and Astrophysics Suppl. Series*, **130**, 235–244.

Sehnal, L. and Pospisilova, L. (1994). *Lifetime of the CESAR satellite*. Publ. Astr. Inst., Academy of Sciences CR, **82**, 6–15.

Smith, D.E., Lerch, F.J. and Wagner, C.A. (1973). In M.J. Rycraft and S.K. Runcorn (eds.), *A Gravitational Field Model for the Earth. Space Research XIII*. Berlin: Akademie-Verlag, pp. 11–20.

Stanley, H.R. (1979). The GEOS-3 project. *Journal of Geophysical Research*, **84**, 3779–3783.

Veillet, C., Fridelance, P., Feraudy, D., Boudon, Y., Shelus, P.J., Ricklefs, R.L. and Wiant, J.R. (1992). *LASSO Observations at McDonald and OCA/CERGA: A Preliminary Analysis*. Meeting PTTI 113.

Wunsch, C. (2000). Moon, tides and climates. *Nature*, **405**, 743–744.

Yionoulis, S.A., Heuring, F.T. and Guier, W.H. (1972). A geopotential model determined from satellite Doppler data at seven inclinations. *Journal of Geophysical Research*, **77**, 3671–3677.

JOHN E. HARRIES*

Chemistry and physics of the atmosphere

1 INTRODUCTION

The advent of space techniques has had a profound effect on how mankind views the place of the Earth and of the human race in the universe. To see the planet Earth, in images taken from orbiting spacecraft such as the Space Shuttle, or from greater distances (as in the case of those famous images of the Earth from the surface of the Moon) tells us something about the fragility of the planet, and the importance of caring for our environment. Scientists were quick to recognise that this new perspective could be turned to great advantage in the drive to understand how the physical, chemical and indeed biological, systems of the Earth worked. From about 1960, a series of space missions began, aimed not outwards towards the stars, but inwards towards our atmosphere, oceans and land. In this chapter, we will follow the story of how space techniques have been applied in the 20th century to studies of:

- The meteorology and physics of the atmosphere and climate;
- The chemistry of the atmosphere.

We will also describe some of the scientific ideas that have been tested from space, as well as the satellites and instruments that have been developed: we will also mention some of the people involved in this exciting enterprise. It will be seen from this review that space has had a profound effect not only on mankind's perspective of the planet Earth, but also on the scientific understanding of the processes at work in the atmosphere and climate on the Earth. NASA captured just the right phrase when it created its "Mission to Planet Earth"[1] and ESA brought in the biological and anthropogenic aspects when it called the Earth science envelope programme "Living Earth."

2 HISTORICAL PERSPECTIVE

April Fool's day, 1960, was perhaps not the best choice of date for the launch of the World's great adventure in monitoring how the atmosphere of our planet Earth works, from a space platform. Nevertheless, on that date, the United States of America, acting through the new National Aeronautics and Space Administration (NASA), launched the TIROS 1 (the Television and InfraRed Observation Satellite).

The satellite was basically a cylinder with 18 flattened sides to mount solar power cells. The satellite was approximately 1.07 m in diameter, 0.56 m high (including the projecting television camera lens), and had a launch weight of approximately 128.4 kg, including fuel for small solid rockets to control the satellite's spin over time. (For comparison, the latest generation of Earth observing satellites often reach 4 m in length, and weigh several metric tonnes!). TIROS I ceased operating in mid-June 1960 due to an electrical failure. During the 77 days it operated, the satellite sent back 19,389 usable pictures that were used in weather operations. TIROS II was launched on November 23, 1960.

The main sensors that provided the cloud pictures were television cameras. The TIROS cameras were slow-scan devices that took snapshots of the scene below; one "snapshot" was taken every ten seconds. These were rugged, lightweight devices weighing only about 2 kg, including the camera lens. TIROS I was equipped with two cameras, wide angle and narrow angle. Progress since these simple early experiments has been enormous, as we shall see later.

NASA had been formed from the National Advisory Committee for Aeronautics (NACA) on October 1, 1958,

---

* Imperial College, London, United Kingdom
[1] Despite subsequently changing this to the more prosaic "Earth Science Enterprise".

and achieved many startling successes across the whole range of space activities, not only in studies of the Earth. Nevertheless, Earth observation was clearly one of the most important applications of the new space technology and systems that were being developed. In recognition of this, the National Oceanographic and Atmospheric Administration (NOAA) was formed, to become a major partner to NASA (primarily a technology agency) in the exploitation of the new observations. These two agencies between them would go on to dominate the observation, study and monitoring of planet Earth from space for the remainder of the century.

Elsewhere, other countries were quick to recognise the importance of Earth observation from space for studies of the environment. In Europe, the European Space Agency (ESA) was formed in 1975, and quickly developed the Meteosat series of spinning satellites, the first of which was launched in 1977 into an equatorial geostationary orbit at the Greenwich meridian position. This was just two years after the first NASA GOES (Geostationary Operational Environmental Satellite: a 3-axis stabilised spacecraft) had been placed into geostationary orbit over the Indian Ocean. ESA went on to develop the ERS-1 and -2 polar orbiting satellites, which carried a variety of visible, infrared and microwave instrumentation. Within Europe, 1986 saw the creation of what is in some ways a European equivalent of NOAA, the EUMETSAT (European Meteorological Satellite organisation), to exploit the new observations for meteorology and later for climate studies.

Across the world, other countries were becoming involved, including Japan, which through NASDA (National Aeronautics and Space Development Agency), and with the involvement of other national agencies, has developed a series of polar and geostationary satellites. Also, India, Brazil, Australia and other nations have set up national agencies and become active.

Now, at the beginning of the 21st century, a wide variety of observations of the Earth's atmosphere from space are available, including operational observations, made regularly, with good sampling over long periods; and also dedicated, more specific missions, designed to address specific issues. Progress has been enormous in just 40 years. Our understanding of the Earth and its atmosphere has advanced amazingly, often helped greatly by the new, global observations from space. In one way, this advent of global observations has come just in time as far as present worries over climate change are concerned: it is already the case that researchers are able to go back over much of those 40 years and study how planet Earth has actually developed, based on satellite observations. In another way, perhaps we have become more sensitive to global variations, partly by having more data available. Either way, space has made a major impact.

In the rest of this chapter we have been faced with a choice. We might either attempt to provide an absolutely comprehensive listing of all the missions that have occurred, along with relevant technical details; or to be more selective, and try only to illustrate the range of observations, and to pick out a few highlights of scientific advance that has come about, perhaps mentioning one or two individuals along the way, in an attempt to provide a little more colour to our story. Despite the fact that this latter approach will not be comprehensive, and will leave gaps (for which the author apologises right away), we have decided on the latter course. What is about to be served up is a subjective view of the highlights of efforts to study the atmosphere from space in the latter part of the 20th century.

3 THE METEOROLOGY AND PHYSICS OF THE ATMOSPHERE AND CLIMATE

3.1 Background

The measurement of basic atmospheric parameters, such as temperature and humidity, by instruments on the meteorological satellites flown by several agencies has been of great value in two ways. First, by providing images of cloud patterns and their developments, or by providing vertical profiles of temperature and humidity in the atmosphere, these data have had a direct and immediate impact on the accuracy of weather forecasts and extreme weather events. A second, extremely powerful use of the data has only more recently been appreciated, however. This is that, by providing a long term series of accurate, well inter-calibrated measurements, extending over several decades, these missions have provided data on the state of the atmosphere which is absolutely vital in studies of how the climate might be varying. Data from one-off dedicated "scientific" missions are of limited value in this sense, since they have very limited duration (with a few exceptions, such as UARS – see below). There is huge scientific value in this longevity and continuity of observation. The exceptional value of operational meteorological satellites in providing the basic, long term description of the state of the atmosphere necessary for a wide range of atmospheric problems has become apparent in the closing decades of the 20th century. In addition, the activities of other agencies such as NASA that have funded long-term observations by instruments measuring simple, but important physical parameters like the Earth Radiation Budget, have also been most important.

Geostationary satellites provide a continuous view of the earth disc from an apparently stationary position in space. The instruments on polar orbiting satellites, flying at a much lower altitude, provide higher resolution details about atmospheric temperature and moisture profiles, ozone amounts and radiation budget, although with a less frequent global coverage. The combination of the two types of measurement has proven to be very powerful.

We have already heard how TIROS-1 was put into orbit on April 1, 1960. There followed a rush of satellites and new developments of instruments, too numerous to account for fully here. Between 1960 and 1973, the ESSA series (first operational TIROS) was developed and flown; the Improved TIROS Operational System (ITOS) began in January 1970, carrying a second generation of visible and IR sensors; later in 1970, NOAA-1/ITOS-A heralded the first in the long series of NOAA satellites that operates up to the present day; and in 1973 the NOAA-3 spacecraft was the first satellite which provided direct broadcast VTPR (Vertical Temperature Profile Radiometer) data. The NOAA series has operated up to the end of the century, carrying instruments which have been gradually improved and developed. Then, in 1975 GOES-1 heralded the beginning of geostationary observations, joined by Meteosat-1 in 1977. Let us look a little more closely at some of these developments.

3.2 NOAA's geostationary weather satellites[2]

The idea of using the geostationary orbit generally has been ascribed to the science fiction writer, Arthur C Clarke, but the power of the geostationary orbit for meteorology, equipped with a variety of visible and infrared imagers and sounders was advocated strongly by Professor Verner Suomi[3] of the University of Wisconsin. His vision led to the enormous application that we see every day of geostationary observations to weather and climate research (Suomi and Vonder Haar 1969).

GOES satellites provide the kind of continuous monitoring necessary for intensive data analysis. They circle the Earth in a geosynchronous orbit, which means they orbit in the equatorial plane of the Earth at a speed matching the Earth's rotation. This allows them to remain over one position on the surface. The geosynchronous orbit is about 35,800 km above the Earth, high enough to allow the satellites a full-disc view of the Earth. Because they stay above a fixed spot on the surface, they provide a constant vigil for the atmospheric "triggers" for severe weather conditions such as tornadoes, flash floods, hail storms, and hurricanes. When these conditions develop, the GOES satellites are able to monitor storm development and track their movements.

GOES satellite imagery is also used to estimate rainfall during the thunderstorms and hurricanes for flash flood warnings, as well as to estimate snowfall accumulations

[2] For more information about NOAA geostationary and polar orbiting satellites, see: http:\\www.noaa.gov.
[3] For more about the life and work of Verner Suomi, see: http://profhorn.meteor.wisc.edu/wxwise/museum/a1main.html

Figure 1 GOES visible image of Hurricanes Madeline and Lester, off the coast of Mexico, October 1998. (Courtesy of NASA–Goddard Space Flight Center, data from NOAA GOES.)

and overall extent of snow cover. Such data help meteorologists issue severe weather warnings. Figure 1 shows a GOES image of hurricanes Lester and Madeline, off the coast of Mexico on October 17, 1998, illustrating the use of geostationary observations for one of these purposes, tracking storms.

The United States normally operates two meteorological satellites in geostationary orbit over the equator. Each satellite views almost a third of the Earth's surface: one monitors North and South America and most of the Atlantic Ocean, the other North America and the Pacific Ocean basin. GOES-8 (or GOES-East) is positioned at 75°W longitude and the equator, while GOES-10 (or GOES-West) is positioned at 135°W longitude and the equator. Coverage extends approximately from 20°W longitude to 165°E longitude.

The main mission is carried out by the primary instruments, the Imager and the Sounder. The imager is a multichannel instrument that senses radiant energy and reflected solar energy from the Earth's surface and atmosphere. The Sounder provides data to determine the vertical temperature and moisture profile of the atmosphere, surface and cloud top temperatures, and ozone distribution.

Other instruments on board the spacecraft include a Search and Rescue transponder, a data collection and relay system for ground-based data platforms, and a space environment monitor. The latter consists of a magnetometer, an X-ray sensor, a high energy proton and alpha detector, and an energetic particles sensor. All are used for monitoring the near-Earth space environment or solar "weather."

Table 1 GOES-10 characteristics

Main body	2.0 m by 2.1 m by 2.3 m
Solar array	4.8 m by 2.7 m
Weight at liftoff	2105 kg
Launch vehicle	Atlas I
Launch date	April 25, 1997, Cape Canaveral Air Station, FL
Orbital information	Type: Geosynchronous
	Altitude: 35,786 km
	Period: 1,436 minutes
	Inclination: 0.41 degrees
Sensors	Imager
	Sounder
	Space Environment Monitor (SEM)
	Data Collection System (DCS)
	Search and Rescue (S&R) transponder

Some typical characteristics of the GOES spacecraft are shown in Table 1.

Data from GOES satellites aid forecasters in providing better advanced warnings of thunderstorms, flash floods, hurricanes, and other severe weather. The GOES-I series provide meteorologists and hydrologists with detailed weather measurements, more frequent imagery, and new types of atmospheric soundings. The data gathered by the GOES satellites, combined with that from new Doppler radars and sophisticated communications systems make for improved forecasts and weather warnings.

3.3 NOAA's polar orbiting weather satellites

Complementing the geostationary satellites are two polar-orbiting satellites, constantly circling the Earth in an almost north–south orbit, passing close to both poles. The orbits are circular and sun synchronous, with an altitude between 830 (morning orbit) and 870 (afternoon orbit) km. One satellite crosses the equator at 7:30 a.m. local time, the other at 1:40 p.m. local time. The circular orbit permits uniform data acquisition by the satellite and efficient control of the satellite by the NOAA Command and Data Acquisition (CDA) stations located near Fairbanks, Alaska and Wallops Island, Virginia. Operating as pair, these satellites ensure that data for any region of the Earth are no more than six hours old.

A suite of instruments is able to measure many parameters of the Earth's atmosphere, its surface, cloud cover, incoming solar protons, positive ions, electron-flux density, and the energy spectrum at the satellite altitude. As a part of the mission, the satellites can receive, process and retransmit data from Search and Rescue beacon transmitters, and automatic data collection platforms on land, ocean buoys,

Table 2 NOAA-15 characteristics

Main body	4.2 m long, 1.88 m diameter
Solar array	2.73 m by 6.14 m
Weight at liftoff	2231.7 kg including 756.7 kg of expendable fuel
Launch vehicle	Lockheed Martin Titan II
Launch date	May 13, 1998, Vandenburg Air Force Base, CA
Orbital information	Type: sun synchronous
	Altitude: 833 km
	Period: 101.2 minutes
	Inclination: 98.70 degrees
Sensors	Advanced Very High Resolution Radiometer (AVHRR/3)
	Advanced Microwave Sounding Unit-A (AMSU-A)
	Advanced Microwave Sounding Unit-B (AMSU-B)
	High Resolution Infrared Radiation Sounder (HIRS/3)
	Space Environment Monitor (SEM/2)
	Search and Rescue (S&R) Repeater and Processor
	Data Collection System (DCS/2)

or aboard free-floating balloons. The primary instrument aboard the satellite is the Advanced Very High Resolution Radiometer or AVHRR.

Data from all the satellite sensors are transmitted to the ground via a broadcast called the High Resolution Picture Transmission (HRPT). A second data transmission consists of only image data from two of the AVHRR channels, called Automatic Picture Transmission (APT). For users who want to establish their own direct readout receiving station, low resolution imagery data in the APT service can be received with inexpensive equipment, while the highest resolution data transmitted in the HRPT service utilises a more complex receiver.

Table 2 contains some characteristics of the NOAA-15 satellite.

The polar orbiters are able to monitor the entire Earth, tracking atmospheric variables and providing atmospheric data and cloud images. The satellites provide visible and infrared radiometer data that are used for imaging purposes, radiation measurements, and temperature profiles. The polar orbiters' ultraviolet sensors also provide ozone levels in the atmosphere and are able to detect the "ozone hole" over Antarctica during mid-September to mid-November. These satellites send more than 16,000 global measurements daily via NOAA's CDA station to NOAA computers, adding valuable information for forecasting models, especially for remote ocean areas, where conventional data are lacking.

One development of major significance and importance in utilising the data from the polar orbiters was the idea of "inversion" of satellite-measured radiances as a function of wavelength across a spectral band (CO_2 for temperature measurement, H_2O for humidity) to yield vertical profiles of the required atmospheric parameter. Many people contributed to these ideas, notable amongst them Rodgers (1977) and Eyre (1989) in the UK, and Smith and others in NOAA in the USA.

Currently, NOAA is operating two polar orbiters: NOAA-14 launched in December 1994 and a new series of polar orbiters, with improved sensors, which began with the launch of NOAA-15 in May 1998. NOAA-12 continues transmitting HRPT data as a stand-by satellite.

The data from these satellites have been used in many and diverse scientific and meteorological applications. To mention only one, Geer *et al.* (1998) have used measurements of the radiance emerging from the atmosphere in the HIRS12 channel (HIRS, the High Resolution Infrared Sounder, is one of the instruments flown on the NOAA series: see Table 2). HIRS-12 radiances correspond to thermal radiation emitted from water vapour, in the mid-troposphere. The HIRS data set is reliably available over a roughly 20 year period (1979–present), and this was used in this work in order to develop a new method for detecting evidence for climate change using satellite radiances directly, without transforming into derived atmospheric variables, such as temperature or humidity. As Geer *et al.* show, this new method has met with some success, and indicates a new method for studying climate change using satellite radiance observations. Figure 2 shows a time series of HIRS-12 radiances (at a wavelength of about 6 μm) used in this study. The importance of a long, continuous data set, with careful attention being paid to inter-calibration at overlap points, is clear.

3.4 EUMETSAT's geostationary satellites

EUMETSAT's geostationary satellite programmes include the continuation of the current Meteosat system until at least the year 2002 with a Meteosat Second Generation (MSG) under development for the years until 2012. EUMETSAT has now launched several satellites of the operational Meteosat series, beginning in 1977. The last, Meteosat-7, was launched on 2 September 1997. These satellites enable operations to be assured until well after MSG becomes operational.

The Meteosat system provides continuous and reliable meteorological observations from space to a large user community. In addition to the provision of images of the Earth and its atmosphere every half an hour in three spectral channels (Visible, Infrared and Water Vapour), a range of processed meteorological parameters is produced. Meteosat also supports the retransmission of data from data collection platforms in remote locations, as well as the dissemination of meteorological information in graphical and text formats. Figure 3 provides one example of the (colour enhanced) data obtained in the visible channel, centred at 0.5 μm.

With the progression of science, and developments in the accuracy of numerical weather prediction, the need for more frequent and comprehensive data from space has evolved. This has led to the current work on the Meteosat Second

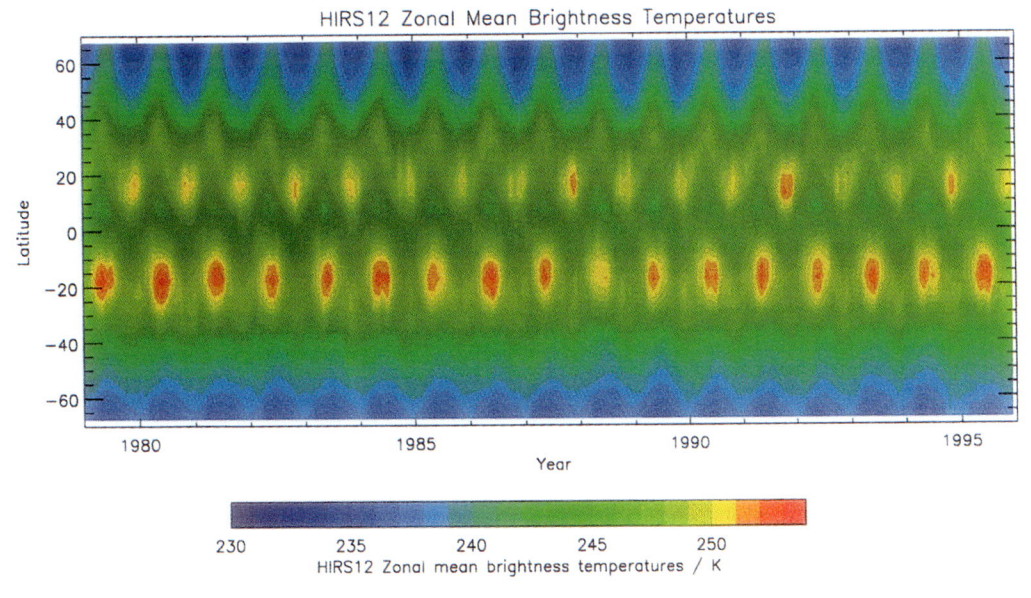

Figure 2 HIRS-12 (6 μm) top of the atmosphere radiances (in units of brightness temperature) plotted as a function of latitude and time, from 1979 to 1986. (Geer *et al.* 1998.)

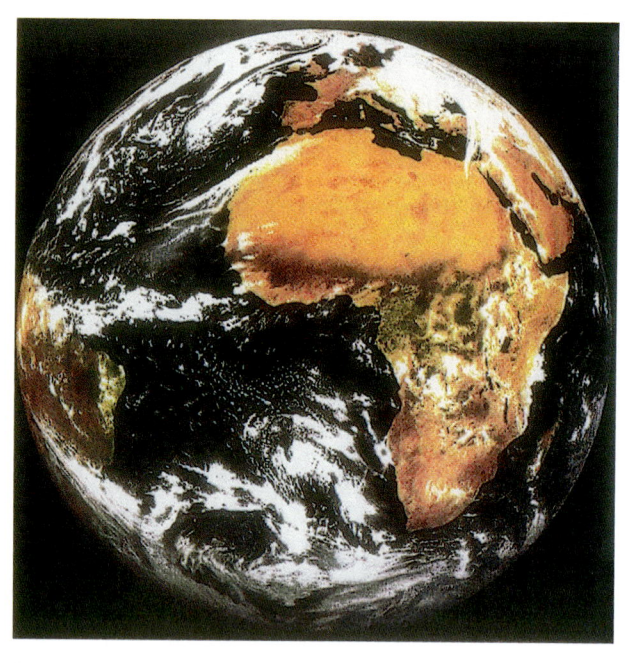

Figure 3 The Earth as seen from Meteosat (visible channel). (Courtesy of Eumetsat, Darmstadt, Germany.)

Generation system (MSG programme). The new satellites will be spin-stabilised like the current generation, but with many design improvements including a new radiometer, the Spinning Enhanced Visible Infrared Instrument, SEVIRI, which will produce images every fifteen minutes, in twelve spectral channels. MSG will also carry the first geostationary Earth Radiation Budget experiment, GERB (Harries and Cromelynck 1999).

The more frequent and comprehensive data collected by MSG will also aid the weather forecaster in the swift recognition and prediction of dangerous weather phenomena such as thunderstorms, fog and explosive development of small but intense depressions which can lead to devastating wind storms. In cooperation with EUMETSAT, ESA is responsible for the development of the first satellite planned for launch in the year 2002. Construction of follow-on models, their launches, development of a new ground system and operations are being implemented by EUMETSAT.

3.5 EUMETSAT Polar System (EPS)

Though Europe has not developed a polar orbiting meteorological stallite during the 20th century, a new system is envisaged for launch by about 2003. The lack of observational coverage in certain parts of the globe, particularly the Pacific Ocean and continents of the southern hemisphere, has led to the increasingly important role for polar orbiting satellite data in numerical weather prediction and climate monitoring. EUMETSAT is currently preparing the European component of a joint European/US polar satellite system. EUMETSAT plans to assume responsibility for the "morning" (local time) orbit and the US will continue with the "afternoon" coverage. It is planned to carry EUMETSAT instruments on the Metop satellite, developed in cooperation with ESA, for a launch in the year 2003. Metop-1 will be the first of a series of operational satellites providing service well into the second decade of the 21st century.

3.6 Earth radiation budget observations

NASA has been involved in the continued deployment of certain sensors for measurement of physical parameters, the most notable perhaps being those designed to measure the Earth radiation budget, or ERB. The ERB comprises two components, the shortwave solar energy absorbed by the Earth, and the longwave energy emitted by the Earth back to space. The former is measured, in practice, by measuring the reflected shortwave component, and the incoming solar flux (the "solar constant"), and taking the difference. The balance between these two fluxes of energy is expected to be maintained in the global average, over an annual cycle, but within these limits, considerable variation occurs as a function of time and space. By designing sensors that capture the whole spectrum of energy, no assumptions are made about the spectral distribution of incident or emergent energy, and small variations in either energy stream may be detected, as long as the sensors have the required sensitivity.

The first ERBE (Earth Radiation Budget Experiment) sensors were deployed on the Nimbus 6 and 7 satellites (launched in 1974 and 1978 respectively) as test flights. The ERBS satellite was deployed from the Space Shuttle Challenger in October 1984 and later launched into a 57 degree inclination precessing orbit with a period of approximately 72 days. The ERBS was the first of three ERBE platforms which would eventually carry the ERBE Instruments. The second ERBE Instrument was on board the NOAA-9 satellite when it was launched in January 1985, and the third was on board the NOAA-10 satellite when it was launched in October 1986. Although the scanning instruments on board all three ERBE satellites have failed, the non-scanning instruments are all presently functioning.

The ERBE system has produced top-of-the-atmosphere (TOA) fluxes from the ERBE scanning radiometers onboard the ERBS (Earth Radiation Budget Satellite), NOAA-9 and NOAA-10 satellites over a 5-year period from November 1984 to February 1990 (Barkstrom and Smith 1986).

ERBE consisted of two sensors, a scanning and a non-scanning radiometer. The scanning radiometer includes three detectors (thermistor bolometers for total, longwave, and shortwave radiation), which normally operate in a cross-track mode, scanning from above the horizon, down across the Earth disk, through nadir, continuing across the other

Earth limb up through space to the internal blackbody. The total detector measures radiation in the 0.2–50.0 μm wavelength band, the longwave detector measures radiation in the 0.5–50.0 μm band, and the shortwave detector measures radiation in the 0.2–5.0 μm band. The ERBE non-scanning instrument includes five detectors within the instrument, four of which normally operate in a nadir (Earth) "staring" mode. The fifth detector (the solar monitor) is used only for solar calibration measurements.

In Europe and Russia (previously the Soviet Union), the SCARAB project has also produced observations of the long- and short-wave components of the ERB (see Kandel et al. 1998).

As an example of the data obtained from these satellites, Figure 4 shows the longwave flux from the Earth averaged for the month of April 1985. This shows high fluxes in regions of low cloudiness, where the warmer surface and lower atmosphere are observed, and lower fluxes in regions of tropical high cloud, and at higher latitudes where the colder surface, or lower cloud tops, are observed. Figure 5 shows the measured variability in the solar constant, determined from the solar monitor channel of the ERBS project. Clearly, both the shortwave and longwave fluxes of the ERB demonstrate considerable variability in time and space. One question is how much any trends in climate may show up in these ERB signals: this is still an open question.

In a new development, in which Europe is taking a clear lead, the first ERB sensor for measurements of the radiation budget from geostationary orbit will be flown on the first Meteosat Second Generation satellite. The idea is to overcome the sampling problem of polar orbiting satellites. Thus, in the time between re-visits of the same scene by a polar orbiter, the cloud field and thence the radiation budget has changed considerably. From geostationary orbit, it will be possible, using the Geostationary Earth Radiation Budget experiment (GERB: see Harries and Crommelynck 1999, and also http://www.ssd.rl.ac.uk/) to measure the two components with a time resolution of as little as 15 minutes, allowing new studies of rapidly varying processes, such as the diurnal convection over Africa, dust storms, tropical storm generation, land surface radiation budget, and others.

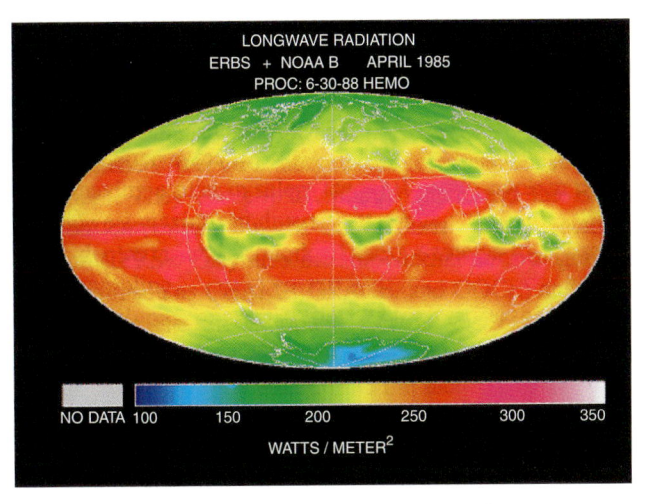

Figure 4 ERBS longwave radiative flux ($W m^{-2}$) averaged for April 1985. (Courtesy of Dr. Bruce Barkstrom, the ERBE Science Team, and NASA.)

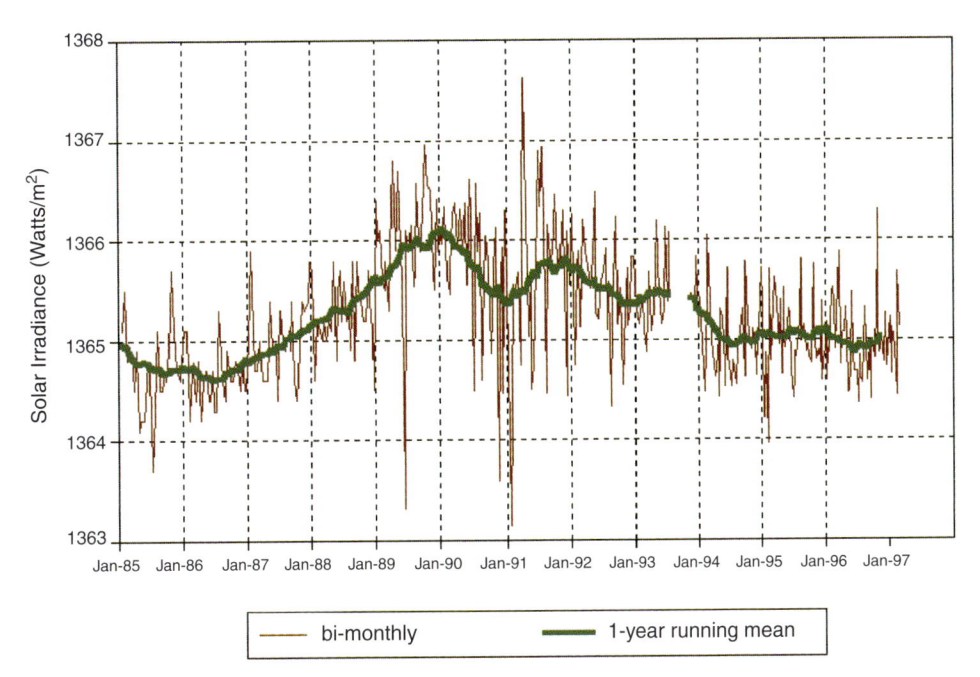

Figure 5 ERBS solar irradiance measurements ($W m^{-2}$), from 1985 to 1997. (Courtesy of Dr. Bruce Barkstrom, the ERBE Science Team, and NASA.)

In combination with continuing polar orbiting observations, the new data will allow a wide range of tests of our understanding of the coupling between radiation budget and other atmospheric and surface processes.

4 CHEMISTRY OF THE ATMOSPHERE

4.1 Background

A major area of scientific application of satellite Earth observation, since the inception of the NASA programme of Earth studies has been the study of the chemistry of the atmosphere, especially that of the stratosphere. Of course, this became an issue with concern during the 1970's about the potential for pollution of the stratosphere by the exhaust gases of high flying aircraft. This possibility was recognised following the Nobel Prize-winning work by Paul Crutzen on the catalytic destruction of ozone by nitrogen oxides in the stratosphere (Crutzen 1970).

During the 1980's this fear was increased through the work of Rowland and Molina (1975), who shared the Nobel Prize with Crutzen, showing that free chlorine in the stratosphere, released from sources such as fluorocarbons, could also severely deplete ozone. Concerns grew even greater when, in 1985, Farman and co-workers at the British Antarctic Survey showed evidence for the Antarctic ozone hole (Farman et al. 1985).

By this time, NASA had flown the last of the Nimbus series of satellites, Nimbus 7, launched in December 1978, which carried instruments capable of measuring a number of stratospheric constituents, including water vapour, ozone, nitrogen oxides, methane, as well as temperature. In planning was the successor to Nimbus, the Upper Atmosphere Research Satellite (UARS), a satellite that would carry a number of state-of-the-art instruments to monitor the state of the stratosphere. Planned many years earlier, and delayed by events such as the Challenger accident, UARS in the end was a timely project, which contributed considerably to understanding the stratosphere. While there have been other missions which have also contributed to stratospheric research, we will highlight these two NASA missions, Nimbus 7 and UARS, in what follows, as a reflection of their major importance.

One other unique feature of the era during which the Nimbus programme, the UARS programme, and others, were being developed was that it was a time of enormous creativity in terms of the development of original instrumental techniques by which to monitor the stratosphere. In view of the limitations here on space, we can only mention a few, in the following section. We will then describe Nimbus 7 and UARS.

4.2 An era of invention: New techniques and instruments

4.2.1 Nadir and limb sounding

The development of the meteorological sounding systems had led to the concepts of retrieving the profiles of temperature and humidity beneath an orbiting spacecraft through the inverse retrieval of these parameters from measurements at different wavelengths across a particular spectral feature (CO_2 for temperature sounding, H_2O for humidity sounding). The basic principles of the retrieval problem are described in a number of places: an accessible, not-too-mathematical treatment is given by Harries (1995: pp. 139–145), and will not be repeated here, interesting though it is.

Work by several researchers, notably Rodgers (1977) showed how, including the noise properties of the radiance signal detected by the satellite instrument, the accuracy of the retrieved atmospheric temperature or composition profile could be maintained, and a useful accuracy of temperature or composition obtained.

Such techniques worked well for the troposphere, where densities of gases were relatively high, and concentrations of water (and other trace gases) are high. In the stratosphere, however, where densities and temperatures are much lower, the weakness of the radiance signals makes the technique of nadir sounding less useful. With this in mind, Gille and House (1971) developed an alternative technique for the stratosphere, known as "limb sounding". This technique involved simply turning the instrument through 90°, so as to direct the field of view towards the horizon, or limb of the atmosphere. This involved observing through a much longer path length, thereby increasing the signal; it also meant that by keeping the field of view of the instrument narrow, the vertical resolution of the method could be reduced to just a few km at the limb. Finally, since the observations are made against the cold background of space, rather than the variable Earth's surface, the stability of the measurement is enhanced. This technique proved a great success, and has been used by generations of instruments since.

4.2.2 Selective and pressure modulation

Another characteristic difficulty associated with the stratosphere is that pressures are much lower than in the troposphere. Hence, a higher spectral resolution in the on-board instrument is required if the spectral lines emitted or absorbed by the atmosphere are to be resolved. This problem was overcome elegantly by the invention of two related techniques, by Houghton, Smith and colleagues (1966; see also Houghton et al. 1984). These involved carrying a sample of the gas to be observed (e.g., CO_2) within the instrument on the spacecraft. In the selective chopper technique, the beam of radiation from the atmosphere is chopped between a cell containing the gas, and a similar, but empty cell. In the pressure modulation technique, a single cell is used, but in this case the pressure of the gas is modulated. In both cases, demodulation at the chopping frequency yields a signal only at the frequencies of the absorption lines in the cell, and rejection of all other wavelengths: that is, automatic selectivity in wavelength, with the spectral resolution defined by the pressure of the gas in the cell. This pressure can be matched to pressures in the atmosphere, and very high effective spectral resolution obtained. Thus, high resolution and perfect spectral matching to only the desired radiation is possible. A number of not insignificant details need to be taken care of (e.g., Doppler shifts due to the high spacecraft velocity), but this technique has been well proven in instruments on Nimbus 6 and 7 and UARS.

4.2.3 Millimetre wavelength techniques

At the time of the Nimbus satellite series, heterodyne mixing techniques had been devleoped to the state where measurements with a reasonable sensitivity of the strong lines due to molecular oxygen, near 60 GHz frequency, were possible, making it possible to measure atmospheric temperature in the troposphere (e.g., Wilson and Schwartz 1981). In order to make the heterodyne systems sensitive enough to detect the much weaker emissions from the stratosphere, both from oxygen (to measure temperature) and from trace gases such as water vapour, ozone, chlorine monoxide, and others (to measure densities), a number of technical developments were required (Waters 1993), especially the development of low-noise, high frequency mixers (usually based on Schottky diodes). Suitable systems were developed in a number of centres, and Waters and co-workers at the Jet Propulsion Laboratory were able to successfully develop and fly a microwave radiometer on UARS which measured successfully up to about 300 GHz, and provided measurements of O_3, H_2O and ClO (e.g., Waters *et al.* 1993).

4.2.4 Back-scatter techniques

At much shorter wavelengths, in the visible and ultraviolet, progress in useful measurements of atmospheric state were hampered by a number of intrinsic problems. These included the low intensity of thermal emission at such short wavelengths; the problem of scattering by aerosols and dust; and the absence of strong absorption/emission features other than due to ozone. A novel way around at least the first of these problems was found by a number of workers, including McCormick and colleagues (1989). In this technique, which works only in sunlit conditions, the scattering of the incident light is used to provide a flux returning to the spacecraft from the atmosphere, with the spectral imprint of atmospheric absorption superposed on it. The technique has been successfully used in experiments such as the Solar Backscatter Ultra Violet (SBUV) spectrometer, and the Total Ozone Mapping Spectrometer (TOMS).

4.3 Nimbus

4.3.1 Background

The Nimbus series of spacecraft spanned the 1960's and 1970's and offered atmospheric scientists frequent access to space through the use of a well designed, modestly scaled spacecraft in a programme which, at the outset, recognised the need for a number of flights using a standardised "bus" spacecraft to achieve their goals. Unfortunately, neither NASA, nor any other agency, has had the wisdom since to recognise the common sense of the Nimbus approach: in contrast programme directors have swung from one extreme (the huge, expensive, "must carry everything" concept of the EOS programme) to the other (everything must be squeezed beyond the point of optimal returns, in order to make each mission "cheap and quick"). Moreover, the value of having a long term forward planned programme, yet one which retains some flexibility, has been ignored: planning swings between extremes instead of developing steadily. Programme planners should study the Nimbus programme as a model of "how to do it correctly"!

The Nimbus satellites, first launched in 1964, carried a number of instruments, including microwave radiometers, atmospheric sounders, ozone mappers, the Coastal Zone Color Scanner (CZCS), and infrared radiometers. Nimbus-7, the last in the series, provided significant global data on sea-ice coverage, atmospheric temperature, atmospheric chemistry, the Earth's radiation budget, and sea-surface temperature. The TOMS instrument aboard the NIMBUS satellite was able to map the areal extent of the Antarctic ozone hole. It also was used to monitor SO_2 in the atmosphere. After major volcanic eruptions, clouds of SO_2-enriched ash and gases ejected into the upper atmosphere can be tracked across much of the world until they dissipate.

NIMBUS-7, launched in October 1978, carried eight highly complex sensors which were all improved versions of sensors previously flown on NIMBUS satellites. They were a Limb Infrared Monitor of the Stratosphere, a High Resolution Infrared Radiation Sounder, an Earth Radiation Budget experiment, a Scanning Multichannel Microwave Radiometer (SMMR), a Pressure Modulated Radiometer, a Solar Backscatter UV/Total Ozone Mapping Spectrophotometer, a Temperature, Humidity Infrared Radiometer and a Tropical Wind, Energy Conversion and Reference Level experiment. The craft was powered by 10,500 solar cells and two SNAP-19 nuclear powered generators (NIMBUS was one of the first to use these).

The craft was placed in Sun-synchronous orbit and transmission of data from all of the experiments was completed as scheduled. For the first time NASA and ESA (European Space Agency which provided tracking support) were able to receive data concerning the global atmosphere in real time.

4.3.2 Nimbus science

Instruments on Nimbus 7 used many of the new techniques that were being developed at the time. As an example, we might quote the LIMS (Limb Infrared Monitor of the Stratosphere) experiment. This was a multi-channel filter radiometer, detecting thermal infrared atmospheric emission, employing solid cryogenically cooled detectors, and the limb-scanning technique (Gille and Russell 1984). This experiment provided radiance data in spectral channels corresponding to several parameters: CO_2 for temperature, H_2O, O_3, HNO_3, NO_2. These data quite revolutionised our understanding of the distribution of temperature and these constituents: prior to Nimbus 7, most information had come from balloon and aircraft flights and from ground based remote sensing. Now, we possessed high quality, near-global observations that provided a quite new perspective on the distributions of these species. As an example, Figure 6 shows a very exciting result, the latitude-height cross section of water vapour, for the month of May 1979, from about the tropopause (c. 100 mb, or hPa), up to the stratopause at about 1 mb (roughly 50 km altitude). This shows, for the first time in such detail, the source of dry air at the cold, tropical tropopause, with dry air rising vertically and towards the poles.

Higher latitudes show signs of moister air near the tropopause, and at very high altitudes. Subsequent work (e.g., Jones et al. 1986) showed how H_2O from LIMS could be combined with measurements of CH_4 by another instrument on Nimbus 7, the Stratospheric and Mesospheric Sounder: SAMS was the first instrument to use the pressure-modulation technique discussed earlier. Jones et al. were able to throw new light on the total hydrogen budget of the stratosphere.

No review would be complete without mention of the TOMS data. This instrument works on the solar backscatter principle, mentioned above. Thus, a disadvantage is that it cannot make measurements under conditions of no solar illumination, for example, as in the polar night. Nevertheless, the instrument has now been flown on a number of polar orbiting satellites, and the observations obtained of the total column density of ozone (the total amount of ozone in a vertical column, from the bottom to the top of the ozone layer) have been very widely used, in scientific studies and public debate of the ozone issue alike. Figure 7 shows a series of maps of total ozone in October each year, from 1979 to 1992, illustrating the appearance and growth of the Antarctic ozone hole (purple area over pole).

4.4 The Upper Atmosphere Research Satellite, UARS

4.4.1 Background

Following the termination of the Nimbus programme, planning was developed for a large "observatory" spacecraft that would, in principle, measure "everything" needed to understand the stratosphere. The idea that was prevalent at the time was that a complete set of measurements would lead to a complete understanding of the upper atmosphere (stratosphere and mesosphere) system. The difficulties with such a concept were essentially that of "all eggs in one basket", and also a lack of appreciation at the time of how much the properties of the atmosphere were variable in time, leading to the erroneous perception that one could mount a once-for-all experiment. Later, as we have noted above, the prevailing philosophy has switched (equally mistakenly, in this author's view) to the opposite view that everything must be done through small, cheap missions. Nimbus had already showed the value of something in between these extremes, blessed with some stability of planning.

Nevertheless, UARS was a milestone mission of great importance to atmospheric science. It was launched on September 15, 1991, and some of the instruments are still operational as this is written (early 2000). Due to its longevity, UARS has become a long term mission, considerably enhancing its value to science.

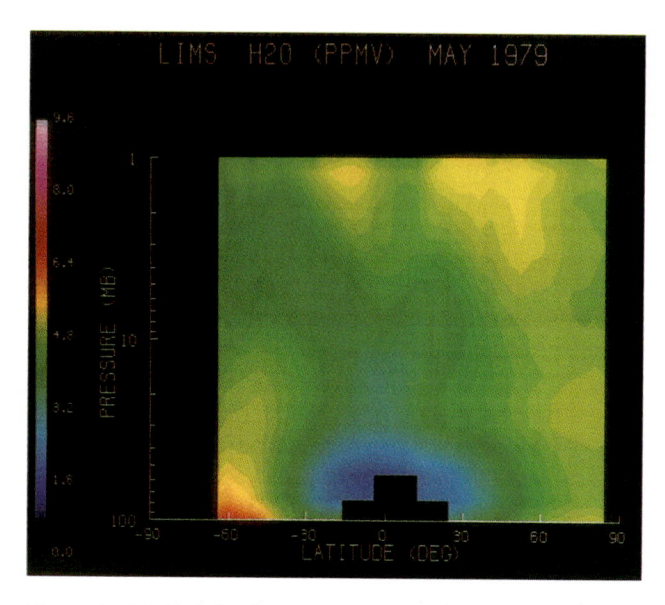

Figure 6 LIMS derived water vapour zonal mean cross section (parts per million by volume) for May 1979, plotted as function of latitude and height. The vertical scale runs from 100 hPa altitude to 1 hPa. The horizontal scale goes from 80°S on the left to 85°N on the right. The colour scale extends from a low mixing ratio value of about 1×10^{-6} by volume (dark blue) to a high of about 7×10^{-6} (red). (From Russell et al. 1984.)

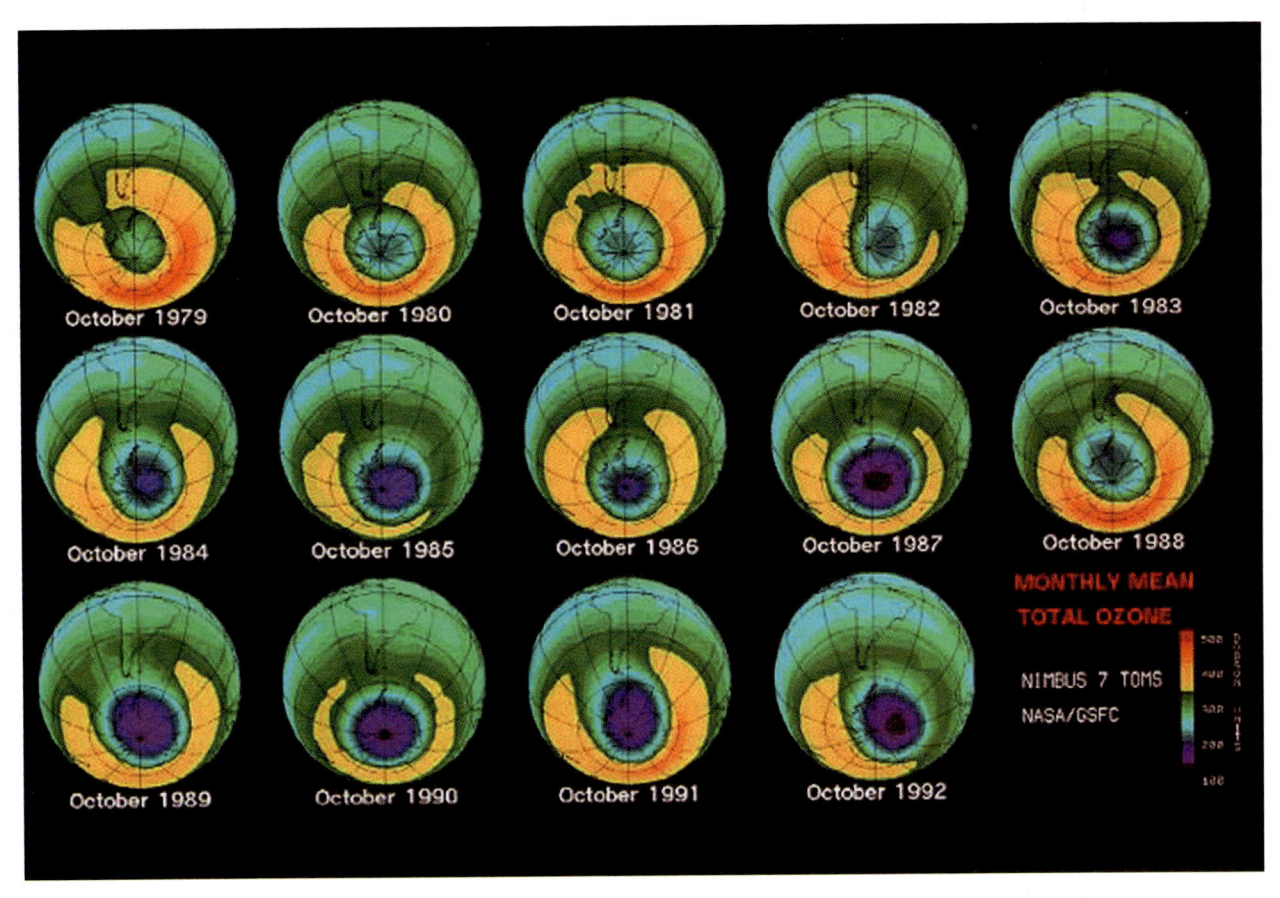

Figure 7 Nimbus 7 TOMS measurements of total ozone column for southern hemisphere for Octobers from 1979 to 1992. (Courtesy of NASA Goddard Space Flight Center.)

Figure 8 shows how the scale of spacecraft changed between Nimbus 7 and UARS (it also illustrates how the pre-occupation with large scale continued after UARS into the Earth Observing System epoch).

The UARS payload comprised nine complementary scientific instruments. A 10th instrument, the Active Cavity Radiometer II (ACRIM II), was not technically part of the UARS mission, but took advantage of a flight opportunity aboard UARS to study the Sun's energy output, an important variable in the study of the Earth's climate.

4.4.1.1 UARS chemistry payload
Four of UARS instruments measured the concentrations and distribution of gases important to ozone depletion, climate change and other atmospheric phenomena.

– Cryogenic Limb Array Etalon Spectrometer (CLAES). This was an infrared Fabry-Perot limb-sounding spectrometer, completely cooled for high sensitivity. Because of the use of expendable cryogen, the lifetime of CLAES was limited to about 3 years. CLAES determined concentrations and distributions by altitude of nitrogen and chlorine compounds, ozone, water vapour and methane, all of which take part in the chemistry of ozone depletion. The Principal Investigator for CLAES is Dr. Aidan E. Roche, Lockheed Palo Alto Research Laboratory.

– Improved Stratospheric and Mesospheric Sounder (ISAMS) studied atmospheric water vapour, carbon dioxide, nitrous oxide, nitric acid, ozone, methane and carbon monoxide. Like CLAES, ISAMS detected infrared radiation from the atmosphere and used it to derive information on atmospheric temperature and composition. Principal Investigator for ISAMS is Professor Fred W. Taylor, University of Oxford.

– Microwave Limb Sounder (MLS) measured microwave radiation from the atmospheric limb for the first time, and derived a global data set on chlorine monoxide, ozone and water vapour in the microwave spectral range. Principal Investigator for MLS is Dr. Joseph W. Waters, NASA's Jet Propulsion Laboratory (Waters *et al.* 1993).

– Halogen Occultation Experiment (HALOE) continues to observe absorption of solar radiation by the atmospheric limb during sunrise and sunset events (twice per orbit). From the measurements of atmospheric transmittance the

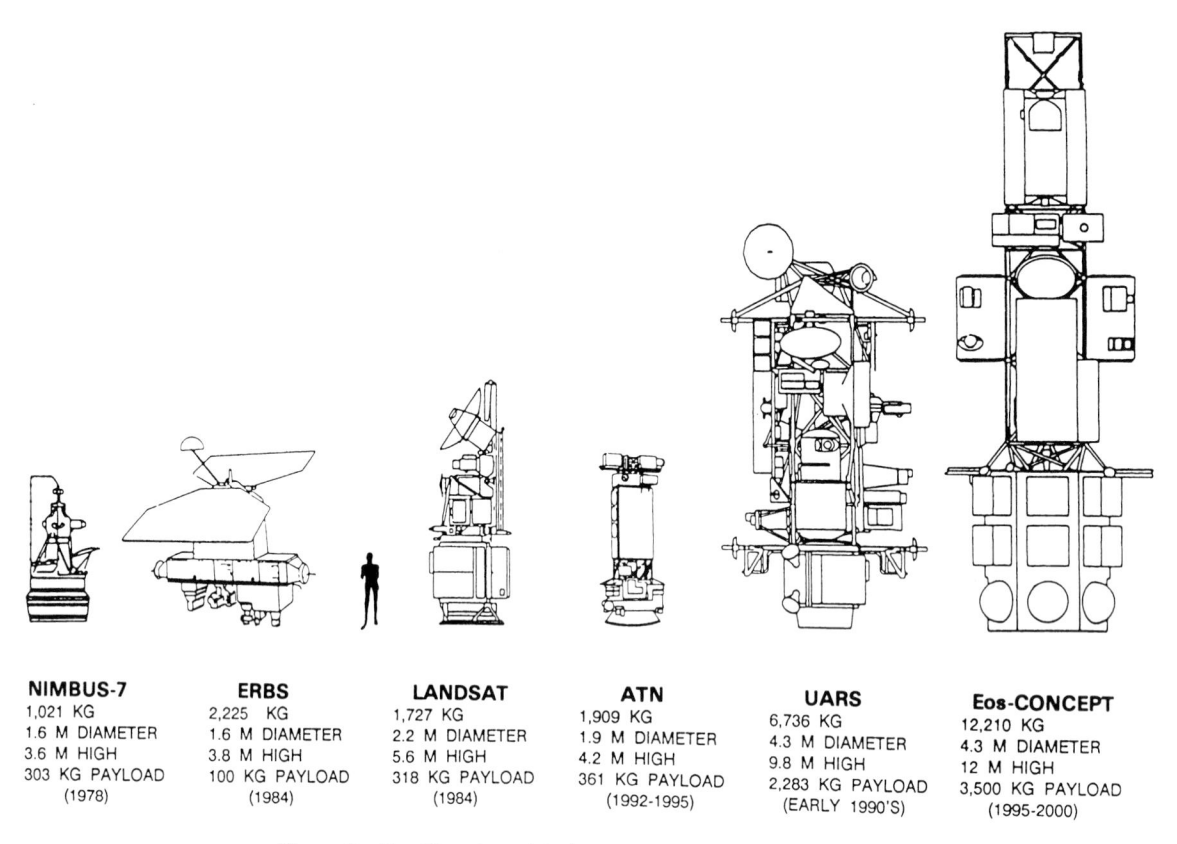

Figure 8 Satellites from Nimbus to EOS. (Courtesy of NASA.)

NIMBUS-7	ERBS	LANDSAT	ATN	UARS	Eos-CONCEPT
1,021 KG	2,225 KG	1,727 KG	1,909 KG	6,736 KG	12,210 KG
1.6 M DIAMETER	1.6 M DIAMETER	2.2 M DIAMETER	1.9 M DIAMETER	4.3 M DIAMETER	4.3 M DIAMETER
3.6 M HIGH	3.8 M HIGH	5.6 M HIGH	4.2 M HIGH	9.8 M HIGH	12 M HIGH
303 KG PAYLOAD	100 KG PAYLOAD	318 KG PAYLOAD	361 KG PAYLOAD	2,283 KG PAYLOAD	3,500 KG PAYLOAD
(1978)	(1984)	(1984)	(1992-1995)	(EARLY 1990'S)	(1995-2000)

vertical distribution of hydrofluoric acid, hydrochloric acid, methane, carbon dioxide, ozone, water vapour and members of the nitrogen family have been derived for almost 9 years. Each day, HALOE observes 28 solar occultations, measuring the energy absorption of the Sun's rays by these gases. Principal Investigator for HALOE is Dr. James M. Russell III, NASA's Langley Research Center (Russell *et al.* 1993).

4.4.1.2 UARS dynamics payload

Two instruments, the High Resolution Doppler Imager and the Wind Imaging Interferometer, will provide scientists with the first directly measured, global picture of the horizontal winds that disperse chemicals and aerosols through the upper atmosphere.

– High Resolution Doppler Imager (HRDI). By measuring the Doppler shifts of atmospheric chemicals, HRDI measured atmospheric winds between 10 and 45 km and above 54 km. These data are important to understanding the essential role of atmospheric motion on the distribution of chemicals in the upper atmosphere. Principal Investigator for HRDI is Dr. Paul B. Hays, University of Michigan.
– Wind Imaging Interferometer (WINDII) also used the Doppler shift measurement technique to develop altitude profiles of horizontal winds in the upper atmosphere. WINDII's measurements are sensitive to winds above about 80 km. Principal Investigator for WINDII is Dr. Gordon G. Shepherd, York University.

4.4.1.3 UARS energy payload

Three instruments, the Solar Ultraviolet Spectral Irradiance Monitor, the Solar Stellar Irradiance Comparison Experiment, and the Partial Environment Monitor, measured solar energy and energetic particles that reach the Earth.

– Solar Ultraviolet Spectral Irradiance Monitor (SUSIM). Measures ultraviolet light from the Sun, which is the driver of the ozone cycle, dissociating chlorine compounds into reactive chlorine atoms that in turn break up ozone molecules. Principal Investigator for SUSIM was Dr. Guenter E. Brueckner, Naval Research Laboratory.
– SOLar Stellar Irradiance Comparison Experiment (SOLSTICE) compared the Sun's ultraviolet energy to the UV radiation of bright blue stars, providing a standard against which the solar energy level can be measured in future long-term monitoring of the Sun. Principal Investigator for SOLSTICE is Dr. Gary J. Rottman, University of Colorado.
– Particle Environment Monitor (PEM). This instrument measured energetic particles from the Sun and allowed studies of the upper atmosphere, detecting and measuring the particles as they enter the atmosphere. Principal

Investigator for PEM is Dr. J. David Winningham, Southwest Research Institute.

- Solar Constant Active Cavity Radiometer Irradiance Monitor (ACRIM II). This was used to measure total solar irradiance, for long-term climate studies. ACRIM II was an instrument of opportunity, added to the UARS spacecraft after the engineering team determined that the spacecraft could fly a 10th instrument. Principal Investigator for ACRIM II is Dr. Richard D. Willson, Jet Propulsion Laboratory.

The UARS observatory consisted of a standard design Multi-mission Modular Spacecraft (MMS), coupled to a module that included the 10 instruments. As shown by Figure 8, the whole observatory measured over 4 m in length, and weighed over 6 tonnes.

4.4.2 UARS science

The UARS mission has yielded an enormous number of scientific papers and results. The official web site (*http://www.earth.nasa.gov/history/uars/uars.html*) lists a large number of key results, which include:

- Seasonal mapping of ClO radicals and reservoirs in the lower stratosphere: close association of ClO with polar stratospheric cloud formation temperatures.
- Containment of polar vortex chemistry within the vortex region.
- Descent in the center of the polar vortex: very low concentrations of the long lived trace gas methane (CH_4) in the center of the spring Antarctic polar vortex.
- Infrared mapping of aerosols from Mt Pinatubo (erupted on June 15, 1991) and Polar Stratospheric Clouds.
- First direct measurement of winds from space: measure the tropical quasibiennial oscillation (QBO) winds in the stratosphere.
- First global maps of chlorofluorocarbons and their products from space. Tropical transport in the stratosphere: annual cycle in the vertical transport of water vapour upward in the tropical stratosphere.
- Measurement of the UV and Visible component of solar variability.
- The role of energetic particles in stratospheric chemistry.
- Upper tropospheric water vapour in the presence of clouds, from microwave measurements.

Some of the most important discoveries achieved using UARS data have arisen from the long lifetime of the mission. As an example, Figure 9 shows a dramatic time series of HALOE measurements of water vapour at a pressure altitude of 130 hPa: these data are zonally averaged at each latitude band, and show details of the variability of water

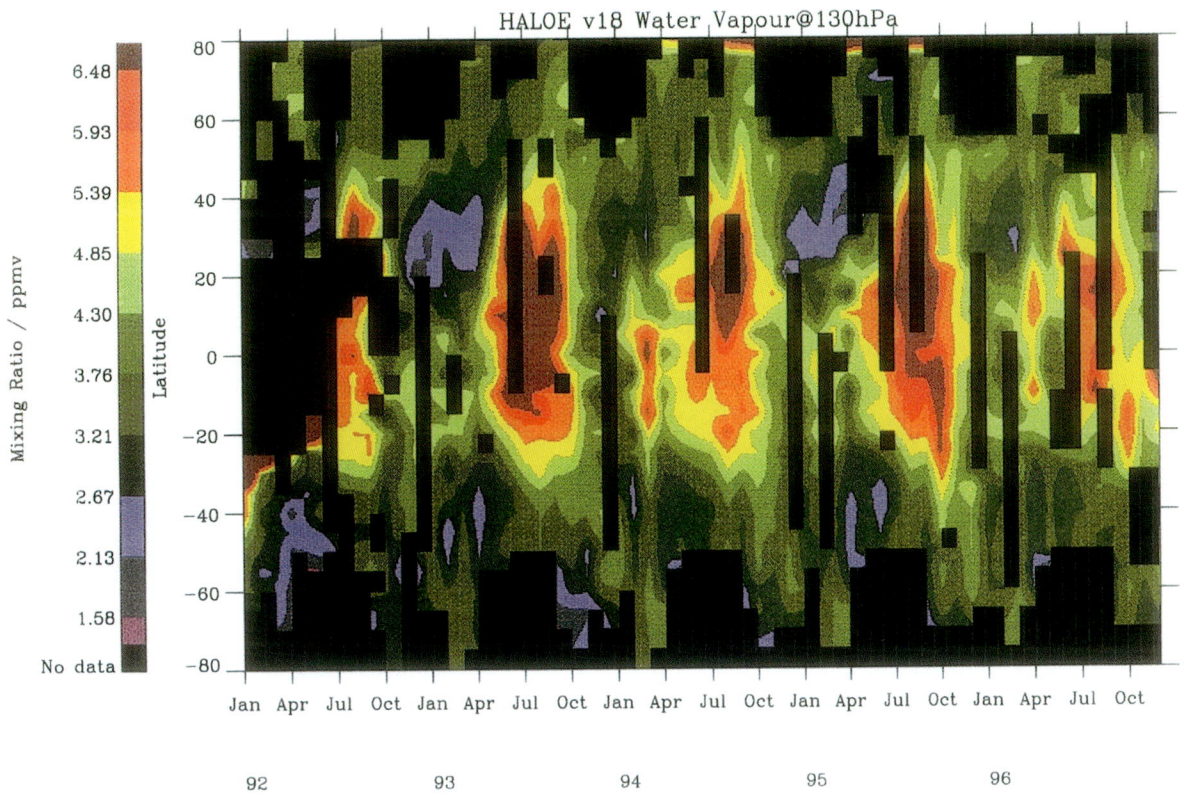

Figure 9 Time series (latitude-time) of HALOE measurements of H_2O at a pressure of 130 hPa, from January 1992 to October 1996. (Data courtesy of NASA Langley Research Center, Hampton, Va., USA.)

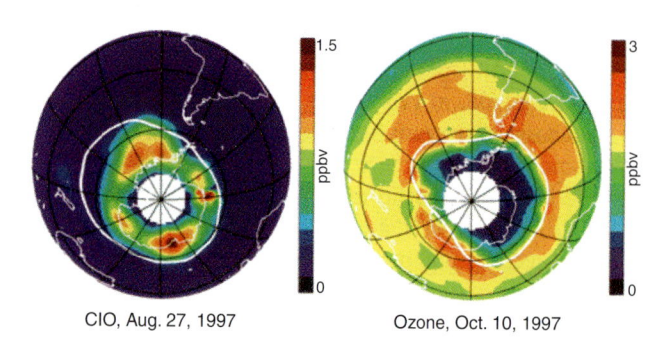

Figure 10 UARS MLS microwave measurements of ClO (August 1997) and O_3 (October 1997). (Courtesy of NASA Goddard Space Flight Center.)

vapour due to the annual cycle and inter-annual variations, on a scale never detected before. Observations such as this have provided us with new evidence for the processes which control the movement of moist/dry air horizontally and vertically, and have provided the underpinning for ideas such as the "downward control" exerted by the middle stratosphere on the lower stratosphere and the tropopause zone.

Figure 10 shows MLS microwave limb sounding measurements of the ClO and O_3 mixing ratios over the Antarctic, in August and October, 1997, respectively, clearly showing the complementarity between the two: high values of ClO in August are now known to be a pre-cursor to low values of ozone as the springtime sun returns to the Antarctic polar vortex. These observations, though at the limit of detectability of ClO for the MLS instrument, provided direct experimental evidence for the role of ClO (and by extension Cl) on ozone destruction in the Antarctic spring polar vortex.

Unfortunately, the UARS mission remains a "one-off" exercise, perhaps largely because of its very large cost. The Earth Observation System follows on, and will include a "chemistry" mission (EOS-CHEM, now renamed "Aura") which will fly in about 2003/5 or so. This, also, is a large satellite, containing generally large instruments. There is a clear dilemma here. On the one hand, the longevity and flexibility, coupled with the assurance of repeated access to orbit, offered for example, by the Nimbus mission provided atmospheric science with a valuable tool for atmospheric chemistry studies, which blossomed in the 1970s in particular. On the other hand, while providing only a single access to space, the UARS mission was of such a scale that it enabled large, complex, and highly sensitive instruments to be put into orbit, which have advanced our knowledge of the properties of the stratosphere and mesosphere enormously: it could be argued that such a major step up in knowledge could not have been obtained by a lower level but more continuous programme like Nimbus. So, which approach, Nimbus or UARS, should be adopted in the future? We will address this question later.

Finally, there is a more basic question. Now that we know so much more about the basic properties, physical and chemical, perhaps we do not need continuous monitoring of the chemical state of the stratosphere, but more occasional missions (say every decade), to check on that state periodically. This depends on whether we anticipate that sudden surprises (the ozone hole comes immediately to mind) are likely, and whether continuous monitoring of perhaps just a few basic variables (temperature, ozone and humidity), interspersed with occasional "chemical observatories" like UARS, is more appropriate.

One final comment concerning stratospheric chemistry. This concerns Europe and the rest of the world. To date, the monitoring of the chemical state of the atmosphere has been almost uniquely an American affair, through the Nimbus, UARS and in future the EOS programmes. However, Europe is implementing the ENVISAT mission, on which a number of European-developed instruments to measure the chemical state of the stratosphere will fly. Future European chemistry missions look to be way off in the future, so that ENVISAT would appear to be a rather isolated attempt to do so. In Japan, the ADEOS mission has provided an opportunity for Japanese scientists to gain experience in developing and flying instruments such as ILAS (Infrared Limb Atmospheric Sounder: operating on similar principles to the HALOE instrument).

4.5 The troposphere

So far, we have been concerned exclusively with the stratosphere, and to a lesser extent the mesosphere. This is, of course, not surprising, given the "ozone problem", which took the world by force in the 1970s, was accelerated by the ozone hole discovery of the 1980s, and probably peaked with the Nimbus 7-UARS period of global observation in the 1980s and 1990s. Now that the Montreal protocol limits the production of ozone-destroying chemicals released into the atmosphere, the politicians probably believe that this is one environmental issue "under control", and therefore not such a priority for expenditure.

However, the problem of chemical pollution of the troposphere has so far hardly been touched by space observations. For one thing, observations of trace chemicals in the shallow layers of the troposphere, in the presence of clouds, haze, and heavily saturated pressure-broadened lines, is experimentally extremely difficult. For another, surface based observations have been able to provide much of the needed information, at least over the land masses. Nevertheless, it is perhaps curious, given the very real problems of tropospheric chemistry, that more ingenuity has not been addressed to the exploitation of space measurements to study tropospheric chemistry. Why, for example, am I unable to report a similar "era of invention" in addressing this problem as was the case for the stratosphere (see Section 4.2)? The answer is probably multiple: certainly, the space measurement problem in the stratosphere is far easier than the troposphere; also, most "available" scientific expertise was, for the past 20 years, pre-occupied with the opportunities presented by the concerns over the stratosphere; maybe, also, the tropospheric question is really not appropriately addressed using space techniques. The question is, however, unresolved.

The first, serious attempts to study the chemistry of the troposphere from space will occur with the MOPITT (Measurements of pollution in the Troposphere: *http://eos-am.gsfc.nasa.gov/mopitt.html*) experiment on the NASA Terra mission, and the TES (Tropospheric Emission Spectrometer: *http://eos-chem.gsfc.nasa.gov/tes/*) experiment on EOS-CHEM/AURA.

5 SUMMARY

Space has had a profound effect on mankind's understanding and view of the Earth, its atmosphere and its climate systems. To begin with, the view of the planet Earth, hanging in space, covered with a fragile, thin blanket of atmosphere, has had a profound effect on the psyche of man. This one image may turn out to be one of the most significant images of the 20th century, and has surely contributed enormously to our sense of the unity but fragility of the planet which provides everything we need – other than the raw energy source – for survival.

More specifically, however, it is hoped that the review presented in the preceding pages has illustrated that space has given us a vast amount of new information about the atmosphere of the Earth, about its physical state, its variability, its clouds, humidity and composition. This information has helped the development of correspondingly large computer models of the atmosphere and climate which we now use to forecast our weather, and to predict the effects of anthropogenic and natural influences on the climate.

Seen in an historical perspective, these developments have been, perhaps somewhat erratic and spasmodic. Nevertheless, the direction has – nearly always – been forward, progressive and positive. Scientists and engineers have needed to develop elegant, sensitive new instruments, to detect the often feeble energy emerging from the planet. New techniques of data interpretation have been invented and developed: methods of inversion, by which parameters such as temperature or density may be retrieved from measurements of radiative intensities. On the one hand, repeated, systematic and regular measurements from space of a few basic properties of the atmosphere (temperature, humidity, ozone, winds) have proven to be of immense value in understanding the variability of our atmosphere, and brought us nearer to an understanding of how climate may develop. On the other hand, some space missions have shown the power and value of individual missions, aimed specifically at a sub-set of the problems facing us: an excellent example being the problem of stratospheric chemistry and the UARS mission.

What of the future? There will be a deluge of new data during the coming decade. In the USA, the Earth Observation System, part of the "Mission to Planet Earth" now re-named the "Earth Science Enterprise", has launched its first mission, called "Terra", right at the end of the century which we are reviewing here, on December 18, 1999. This mission will investigate properties of the land, oceans and atmosphere. The second mission will be "Aqua", to be launched about one year later, and will carry instruments to measure the atmospheric IR spectrum the radiation budget, clouds, sea surface temperature, and other parameters. This will be followed by EOS-CHEM (renamed "Aura"), as its original name suggests, a mission to study atmospheric chemistry, in a way a follow-on to the UARS mission. In Europe, ESA has launched its ENVISAT mission, which will study the chemistry and physical state of the atmosphere. EUMETSAT will fly its new operational polar orbiter, the METOP mission, as well as the Meteosat Second Generation series of satellites. In Japan, follow on missions to ADEOS are planned.

From this review we can, with some trepidation perhaps, venture to suggest some guidelines for the future. Thus, we need to move forward from the amazing developments that we have seen in space exploration of our own planet since the 1960's (a mere forty years!) with some principles in mind:

- The importance of long term planning and continuity of purpose is vital: extreme lurches of policy, from the huge mission to the tiny, from the vastly expensive to the cheap, (and back again?) do not help the cause one bit: in fact, they clearly hinder the development of a wise, measured response to mankind's needs as far as space is concerned. Leaders need to be chosen for this wisdom as much as any other quality. Certainly personal ego, the need to "make a mark" by changing things for its own sake should be "out".
- The importance of a balance between the long term monitoring and the short term focussed missions is very great. This requires close discussion and cooperation between operational, technology and science agencies, throughout the world.
- Space is extremely expensive. For this reason, the human race cannot afford duplication which arises because of international competition, lack of cooperation, or nationalistic (or, these days, regional) macho-ism. For this reason, too, international collaboration needs to be better than it is today.

– We have seen that, in the short history of our subject so far, there have been "spurts" of amazing creativity in developing new instruments and techniques, and other periods almost completely devoid of such inspiration. Policy, likewise, shifts from encouragement of new ideas for techniques to emphasis almost totally on data exploitation. Both are needed for a healthy development of space as a service to mankind and science.

Our understanding of the workings of the Earth's atmosphere and climate system is far from complete or adequate. We have, indeed, made enormous strides in this understanding over the past 30 years or so, and in this the advent of accurate, sensitive space measurements of atmospheric parameters has played a large role. The ingenuity shown by participating scientists and engineers in these developments has been superb, sometimes awe-inspiring. The work needs to continue: we need to monitor our Earth from space, and we need to address specific problems as they arise, and we need to do these things far more as an integrated global community than we do now. Of one thing there is no doubt: it will continue to be challenging, demanding and great fun to take part.

Acknowledgements

The author wishes to acknowledge with gratitude the information contained in the web sites of NASA, NOAA, ESA and EUMETSAT, which have been valuable in obtaining factual information used in this paper. The sites used are as follows:

http://www.eumetsat.de/en/
http://www.esa.int/presentation/
http://www.noaa.gov/
http://www.earth.nasa.gov/
http://www.earth.nasa.gov/history/uars/uars.html

REFERENCES

Barkstrom, B.R. and Smith, G.L. (1986). The Earth Radiation Budget Experiment: Science and Implementation. *Reviews of Geophysics*, **24**, 379–390.

Crutzen, P.J. (1970). The influence of nitrogen oxides on the atmospheric ozone content. *Quarterly Journal of Royal Meteorites Society*, **96**, 320–325.

Eyre, J.R. (1989). Inversion of cloudy satellite sounding radiance by non-linear optimal estimation. 1. Theory and simulation for TOVS. *Quarterly Journal of Royal Meteorites Society*, **115**, 1001–1026.

Farman, J.C., Gardiner, B.G. and Shanklin, J.D. (1985). Large losses of total ozone in Antarctica reveal seasonal clox/nox interaction. *Nature*, **315**, 207–210.

Geer, A., Harries, J.E. and Brindley, H. (1998). Spatial patterns of climate variability in upper-tropospheric water vapor radiances from satellite data and climate model simulations. *Journal of Climate*, **12**, 1940–1955.

Gille, J.C. and House, F.B. (1971). On the inversion of limb radiance measurements. I: Temperature and thickness. *Journal of Atmospheric Science*, **28**, 1427–1442.

Gille, J.C. and Russell, J.M. (1984). The Limb Infrared Monitor of the Stratosphere: Experiment description, performance and results. *Journal of Geophysical Research*, **89**, 5125–5140.

Harries, J.E. (1995). *Earthwatch: The Climate from Space*, Wiley-Praxis.

Harries, J.E. and Crommelynck, D. (1999). The Geostationary Earth Radiation Budget experiment on MSG-1 and its potential applications, *Advances in Space Research*, **24**, 915–919.

Houghton, J.T. and Smith, S.D. (1966). *Infrared Physics*, Oxford University Press.

Houghton, J.T., Taylor, F.W. and Rodgers, C.D. (1984). *Remote Sounding of Atmospheres*, Cambridge University Press.

Jones, R.L., Pyle, J.A., Harries, J.E., Zavody, A.M., Russell, J.M. and Gille, J.C. (1986). The water-vapour budget of the stratosphere studied using LIMS and SAMS satellite data. *Quarterly Journal of Royal Meteorites Society*, **112**, 1127–1143.

Kandel, R., Viollier, M., Raberanto, P., Duvel, J.P., Pakhomov, L.A., Golovko, V.A., Trishchenko, A.P., Mueller, J. Raschke, E. and Stuhlmann, R. (1998). The ScaRaB Earth radiation budget dataset. *Bulletin of the American Meteorological Society*, **79**, 765–783.

McCormick, P.M., Zawodny, J.M., Veiga, R.E., Larsen, J.C. and Wang, P.H. (1989). An overview of SAGE I and II ozone measurements. *Planetary and Space Science*, **37**, 1567–1586.

Rodgers, C.D. (1977). Statistical principles of inversion theory. In: A. Deepak (ed.), *Inversion Methods in Atmospheric Remote Sounding*, Academic Press.

Rowland, F.S. and Molina, M.J. (1975). Chlorofluoromethanes and the environment. *Reviews of Geophysics and Space Physics*, **13**, 1-end.

Russell, J.M., Gille, J.C., Remsberg, E.E., Gordley, L.L., Bailey, P.L., Fischer, H., Girard, A., Drayson, S.R., Evans, W.F.J. and Harries, J.E. (1984). Validation of water-vapor results measured by the limb infrared monitor of the stratosphere experiment on Nimbus-7. *Journal of Geophysical Research*, **89**, 5115–5124.

Russell, J.M., Gordley, L.L., Park, J.H., Drayson, S.R., Hesketh, W.D., Cicerone, R.J., Tuck, A.F., Frederick, J.E., Harries, J.E. and Crutzen, P.J. (1993). The Halogen Occultation Experiment. *Journal of Geophysical Research*, **98**, 10777–10797.

Smith, W.L. *et al.* (1990). GHIS – The GOEs High Resolution interferometer sounder. *Journal of Applied Meteorology*, **29**, 1189–1204.

Suomi, V.E. and Vonder Haar, T.H. (1969) Geosynchronous meteorological satellite. *Journal of Spacecraft and Rockets*, **6**, 342–34.

Waters, J.W. (1993). Microwave limb sounding. In: M.A. Janssen (ed.), *Atmospheric Remote Sensing by Microwave Radiometry*, Wiley and Sons.

Waters, J.W., Froidevaux, L., Read, W.G., Manney, G.L., Elson, L.S., Flower, D.A., Jarnot, R.F. and Harwood, R.S. (1993). Stratospheric ClO and ozone from the MLS on the UARS. *Nature*, **362**, 597–602.

Wilson, W.J. and Schwartz, P.R. (1981). Diurnal variations of mesospheric ozone using millimetre wave measurements. *Journal of Geophysical Research*, **86**, 7385–7388.

APPENDICES

Appendix A

ARTURO RUSSO

A basic chronology of the space age

1903		Publication of K. Tsiolkovsky's *Exploration of cosmic space by means of reactive devices*
1919		Publication of R. Goddard's *A method of reaching extreme altitudes*
1923		Publication of H. Oberth's *Die Rakete zu den Planetenräumen*
1926	*16 March*	First liquid-fuel rocket successfully launched in the United States by Goddard
1927		Creation of the *Verein für Raumschiffahrt* (VfR) in Germany
1928		Establishment of the *Gas Dynamics Laboratory* in Leningrad
1930		Publication of R. Esnault-Pelterie's *L'astronautique*
1931		Establishment of the *Groups for the Study of Reaction Motor* (GIRD) in Moscow and other towns in the Soviet Union
	March	First liquid-fuel rocket successfully launched in Germany
1932	*November*	W. von Braun recruited by the German Army to develop rockets for military use
1933	*17 August*	First liquid-fuel rocket successfully launched in USSR
1937		Establishment of the Peenemünde rocket research centre
1942	*3 October*	von Braun's A4 rocket (eventually V2) successfully launched for the first time from Peenemünde
1944	*September*	Launch of V2 offensive against England and Belgium
		Creation of the *Jet Propulsion Laboratory* for guided-missile research in USA
1945	*May–September*	End of World War II; V2 hardware shipped to the United States; von Braun and his team recruited by the US Army
1946	*16 January*	Establishment of the *V2 Rocket Panel* (eventually *Upper Atmosphere Rocket Research Panel*)
	16 April	First V2 launch in USA
		Scientific Research Institute 88 (NII-88) established in USSR, with S. Korolev in charge of developing long-range missiles.
1947	*18 October*	First V2 launch in USSR
1950	*June*	First launches from *Cape Canaveral*
	30 September	First International Astronautical Congress in Paris

1953	*15 March*	First launch of the Soviet *R-5* rocket
	20 August	First launch of the US Army's *Redstone* rocket
1956	*20 September*	First launch of the US Army's *Jupiter C* rocket
1957	*1 July*	Start of the *International Geophysical Year* (IGY), extending up to 31 December 1958
	21 August	First successful flight of the first ICBM, the Soviet *R-7* (*Semyorka*)
	20 September	First launch of the US Air Force's *Thor* rocket
	4 October	*Sputnik 1*, the first artificial satellite, successfully launched in USSR by a modified *R-7* rocket
	3 November	*Sputnik 2* successfully launched with dog *Laika* on board
	6 December	Launch failure of the American *Vanguard* satellite
1958	*31 January*	The first US satellite, *Explorer 1*, launched by a modified *Jupiter-C* rocket
	26 March	Launch of *Explorer 3*
	15 May	Launch of *Sputnik 3*, carrying about 1 ton of scientific instruments
	18 September	First successful flight of the first US ICBM, *Atlas*
	1 October	Creation of NASA
	18 December	Launch of the *Score* satellite (USA), the first telecommunication satellite
1959	*12 September*	The Soviet spacecraft *Luna 2* impacts on the Moon
	10 October	The Soviet spacecraft *Luna 3* sends the first images of the hidden face of the Moon
1960	*1 April*	Launch of *Tiros 1* (USA), the first meteorological satellite
	29 April	Space scientists from ten West European countries, meeting at the Royal Society in London, urge European cooperation in space research
	12 August	Launch of *Echo 1* (USA), a balloon satellite for passive space telecommunications
	19 August	The dogs *Strelka* and *Belka* successfully put into orbit aboard a Soviet spacecraft and returned safely to Earth
	28 November	Delegates from 11 West European countries, meeting at CERN in Geneva, agree to set up a European space research organisation
1961	*12 April*	Yuri Gagarin, a Soviet cosmonaut, is the first human being to orbit the Earth
	25 May	The US President John Kennedy commits his country to achieve a human Moon landing "before this decade is out" (*Apollo* mission)
	30 October	Delegates from six West European governments plus Australia, meeting in London, agree to undertake a joint program for a European heavy satellite launcher
1962	*20 February*	John Glenn is the first American astronaut to orbit the Earth
	7 March	Launch by NASA of the first *Orbiting Solar Observatory* (OSO). The last of the series, OSO-8, launched on 21 June 1975
	10 July	Launch of *Telstar* (USA), the first satellite for real-time space telecommunications
1963	*16 June*	Valentina Terechkova (USSR) is the first woman in space
	28 August	Launch of the *Nimbus 1* meteorological satellite (USA)
	27 November	Launch by NASA of the First Interplanetary Monitoring Platform (IMP). Ten IMPs were eventually launched, the last one on 25 October 1973

1964	*29 February*	Creation of the *European Launcher Development Organisation* (ELDO), with six European member states plus Australia
	20 March	Creation of the *European Space Research Organisation* (ESRO), with ten member states
	5 September	Launch by NASA of the first *Orbiting Geophysical Observatory* (OGO). The last of the series, OGO-6, launched on 5 June 1969
	15 December	Europe's first satellite, the Italian *San Marco 1*, successfully launched by an American *Scout* rocket
1965	*18 March*	Alekseï Leonov (USSR) is the first man to leave the spaceship and float freely in outer space
	6 April	Launch by NASA of *Early Bird* (*Intelsat 1*), the first telecommunication satellite in geostationary orbit. Commercial service inaugurated on 28 June
	23 April	Launch of the first *Molniya* communications satellite (USSR)
	14 July	The American spacecraft *Mariner 4* sends the first images of Mars
	26 November	The first French satellite, *Asterix 1*, successfully launched by a French *Diamant* rocket
	15 December	The American twin spacecraft *Gemini 6* and *7* achieve the first rendezvous in space
1966	*31 January*	The Soviet spacecraft *Luna 9* achieves the first successful soft landing on the Moon
	1 March	The Soviet spacecraft *Venera 3* impacts on Venus
	31 March	The Soviet spacecraft *Luna 10* is put into lunar orbit
	30 May	First soft landing on the Moon of an American spacecraft, *Surveyor 1*
	10 August	The American spacecraft *Lunar Orbiter 1* is put into lunar orbit
1967	*27 January*	The *Outer Space Treaty* signed simultaneously in London, Moscow, and Washington, DC, asserting the philosophy that the exploration of space should "contribute to broad international cooperation... [and] the development of mutual understanding... between States and peoples"
		American astronauts V. Grissom, E. White and R. Chaffee die in the fire of the *Apollo 1* flight simulator
	24 April	Soviet cosmonaut V. Komarov dies in crash while returning to Earth after first flight of *Soyuz 1* vehicle
	9 November	First launch of the *Saturn V* rocket
1968	*17 May*	First ESRO satellite, *Iris*, successfully launched
	15–21 September	The Soviet spacecraft *Zond 5* orbits the Moon and returns to Earth
	7 December	Launch by NASA of the first successful *Orbiting Astronomical Observatory* (OAO-2)
	21–27 December	The *Apollo 8* spacecraft, with three astronauts onboard, orbits the Moon
1969	*Spring*	Three *Intelsat 3* satellites provide for the first time global coverage in space telecommunications
	16–24 July	The *Apollo 11* mission achieves the first manned lunar landing: astronauts N. Amstrong and E. Aldrin walk on the Moon's surface
	31 July	First high-resolution pictures of Mars sent by *Mariner 6*
	8 November	First German satellite, *Azur 1*, launched by an American rocket
1970	*11 February*	First Japanese satellite launched by a national rocket (*Lambda 4S*)
	24 April	First Chinese satellite launched by a national rocket (*Long March 1*)

	12–24 September	Soviet *Luna 16* mission: automatic soft landing on the Moon and return to Earth with lunar material
	10 December	Launch by NASA of the *Uhuru* satellite, the first devoted to X-ray astronomy
	15 December	First soft landing on Venus by the Soviet spacecraft *Venera 7*
1971	*19 April*	Launch of the first space station, the Soviet *Salyut 1*. The last of the series (*Salyut 7*) launched in April 1982
	7–29 June	Soviet *Soyuz 11* mission: cosmonauts G. Dobrovolsky, V. Volkov, and V. Patasïev reach and live onboard the *Salyut 1* station but die during re-entry to Earth
	13 November	The American spacecraft *Mariner 9* put into Martian orbit
	27 November	The Soviet spacecraft *Mars 2* put into Martian orbit, a scientific capsule is released towards the surface but no results obtained
1972	*5 January*	The American President Richard Nixon launches the Space Shuttle program
	23 July	Launch by NASA of *Landsat 1*, the first satellite devoted to the study of Earth resources
	16 November	Launch by NASA of the SAS-B satellite, the first devoted to gamma-ray astronomy
	21 August	Launch by NASA of OAO-3 (*Copernicus*), carrying a special telescope for ultraviolet astronomical spectroscopy
	7–19 December	Last *Apollo* mission to the Moon (*Apollo 17*)
1973	*14 May*	Launch of the American manned space station *Skylab*
	3 December	The American spacecraft *Pioneer 10* flies close to Jupiter
1974	*29 March*	Mercury is visited by the American spacecraft *Mariner 10*
	17 May	Launch of *SMS-1* (USA), the first meteorological satellite in geostationary orbit
	3 December	*Pioneer 11* flies close to Jupiter
	18 December	Launch of the Franco-German *Symphonie* satellite, the first European communications satellite
1975	*30 May*	Creation of the European Space Agency (ESA)
	15–24 July	USA-USSR *Apollo-Soyuz* mission, with docking in orbit
	9 August	Launch of the first ESA satellite, COS-B, devoted to gamma-ray astronomy
	22 October	First images from Venus sent by *Venera 9*
1976	*20 July*	Soft landing on Mars achieved by the American mission *Viking 1*
1977	*12 August*	First manned free flight of the space shuttle Enterprise Launch by NASA of the first *High Energy Astronomical Observatory* (HEAO-1) satellite. Two other HEAO missions launched in 1978 and 1979
	23 November	Launch of the European meteorological satellite *Meteosat*
1978	*20 January*	First launch of the *Progress* automatic cargo spacecraft to the Salyut stations
	26 January	Launch of the joint NASA/ESA/UK *International Ultraviolet Explorer* (IUE)
	22 February	Launch by the US *Navstar 1*, the first GPS navigation satellite
	4–9 December	The American twin spacecraft *Pioneer Venus-1* and *2* reach Venus. The first is put into orbit, the second releases four probes to the planet

1979	*5 March, 9 July*	The American twin spacecraft *Voyager 1* and *2* fly close to Jupiter
	1 September	*Pioneer 11* flies close to Saturn
	24 December	First launch of the European launcher *Ariane*
1980	*18 July*	First Indian satellite launched by a national rocket (*SLV 3*)
	12 November	*Voyager 1* flies close to Saturn
1981	*12 April*	First launch of the space shuttle *Columbia*
	26 August	*Voyager 2* flies close to Saturn
1982	*13 May*	Two cosmonauts reach the newly launched *Salyut 7* station and will remain 211 days in orbit
1983	*25 January*	Launch of the *Infrared Astronomical Satellite* (IRAS), the first space mission devoted to infrared astronomy (NASA/Holland/UK collaboration)
	16 June	Launch by ESA of the first *Eutelsat* communications satellite
	28 November	First flight of the European space laboratory *Spacelab* onboard the shuttle *Columbia*
1984	*25 January*	The US President Ronald Reagan commits NASA to develop a space station
1985	*30–31 January*	Space ministers from ESA member states, meeting in Rome, approve plans for the long-term science programme 'Horizon 2000', the development of the *Ariane 5* launcher, the *Columbus* module for the space station, and the *Hermes* spaceplane
1986	*24 January*	*Voyager 2* flies close to Uranus
	28 January	Explosion of the shuttle *Challenger* soon after take-off, causing the death of its crew
	20 February	Launch of the Soviet space station *Mir*
	13–14 March	Close-up images of the Halley comet sent by the European spacecraft *Giotto*
1988	*15 June*	First launch of the *Ariane 4* rocket
	29 September	Signature of the international agreement between USA, Japan, Canada and ESA member states for the joint development of the Space Station
1989	*4 May*	Launch by NASA of the *Magellan* mission to planet Venus
	8 August	Launch by ESA of the astrometry satellite Hipparcos
	18 October	Launch by NASA of the *Galileo* mission to planet Jupiter and its moons
	18 November	Launch by NASA of the *COBE* mission, aimed at measuring the diffuse infrared and microwave radiation from the early universe
1990	*24 April*	Launch by NASA of the *Hubble Space Telescope*
	6 October	Launch of the ESA/NASA *Ulysses* mission, the first space mission outside the ecliptic plane
1991	*5 April*	Launch by NASA of the *Compton Gamma Ray Observatory*
	17 July	Launch by ESA of the first *European Remote Sensing* satellite (ERS-1). ERS-2 launched on 21 April 1995
	30 August	Launch of the Japanese *Yohkoh* satellite, devoted to X- and gamma-ray emissions from the Sun
1995	*March–July*	First American astronaut onboard the Russian *MIR* station
	27 June–7 July	First docking of the American *Space Shuttle* to the Russian *MIR* station

	17 November	Launch by ESA of the *Infrared Space Observatory* (ISO)
	2 December	Launch of the ESA/NASA *Solar and Heliospheric Observatory* (SOHO)
	7 December	The *Galileo* spacecraft arrives at Jupiter
1996	*7 November*	Launch by NASA of the *Mars Global Surveyor* mission. The spacecraft reaches Mars orbit on 11 September 1997
	4 December	Launch by NASA of the *Mars Pathfinder* mission to planet Mars
1997	*4 July*	Landing of *Mars Pathfinder* on Mars. The mini-rover *Sojourner* explores the landing region and sends images to Earth
	15 October	Launch of the NASA/ESA *Cassini-Huygens* mission to planet Saturn and its moon Titan. Arrival expected in July 2004
	30 October	First successful launch of the *Ariane 5* heavy satellite launcher
1998	*6 January*	Launch by NASA of the *Lunar Prospector* mission
	20 November	A Russian *Proton* rocket carries the International Space Station's first module into orbit
1999	*23 July*	Launch by NASA of the *Chandra X-ray Observatory*, named after the Nobel laureate astrophysicist Subrahmanyan Chandrasekhar
	10 December	Launch by ESA of the *X-ray Multi-Mirror* mission (XMM-Newton)
2000	*16 July, 9 August*	Launch of the ESA/NASA Cluster mission comprising 4-spacecraft to study space plasma physics in near-Earth space
	31 October	Launch by a Russian *Soyuz* vehicle of the first crew to the International Space Station. The three-man crew spent 138 days in the Station.

Appendix B

JONATHAN McDOWELL*

Catalog of space science launches 1957–2000

Most accounts of space science concentrate on a few famous spacecraft such as the Hubble Space Telescope or the Viking Mars missions. Here I present a comprehensive survey of space science satellites and probes. There are interesting stories behind even those satellites which carried quite minor or unsuccessful experiments, and I hope this will serve as a useful finding list for future historians, as well as providing an indication of the total extent of the space science effort as a function of nation and scientific discipline.

Unfortunately, the list does not include satellites which were lost in launch failures, and it does not include suborbital sounding rockets. It does include planetary probes, life science missions, microgravity missions, most geodesy flights, secondary experiments on piloted spaceships, secondary experiments on spy satellites, as well as the core space physics, astronomy, and atmospheric science automated missions in Earth orbit.

Constructing the list required significant subjective decisions. For example, the boundaries between

- atmospheric science and operational meteorology,
- infrared astronomy and missile early warning satellite background measurements,
- ionospheric research and communications satellite signal propagation technology,
- geophysical research and geodesy measurements for missile targeting,
- solar physics studies and solar flux monitoring for operational flare alerts

are all rather fuzzy, and I have exercised my best judgement.

*Harvard-Smithsonian Center for Astrophysics, Cambridge, MA, USA

Launch dates are Universal Time. Satellite and payload names are those assigned by the owner agency in their language, transliterated where the original is in a non-Latin alphabet. State is the nation-state in which the owner agency is located, and is usually the state which registered the satellite with the United Nations Office of Outer Space Affairs; "I/ESA" denotes International – European Space Agency. The orbital parameters are perigee and apogee in km with respect to a spherical Earth of radius 6378 km, and inclination with respect to the Earth's equator in degrees. "Operated until" represents the last day on which the satellite transmitted, as best as I have been able to determine it (this information is often hard to come by); the end of useful scientific operations may precede this date. "Primary Mission" indicates when the payload is a piggyback experiment on a vehicle with a different primary mission; in particular, the value "Spaceship" in this field indicates the payload was carried on a space vehicle designed to operate with a human crew (including, in a few cases such as Mercury MA-5, automated test flights of such vehicles). The mission of the scientific satellite is summarized in the "Discipline" column, which contains abbreviations (tabulated below) for the main subdisciplines. Where a brief (few words) summary of the mission is possible, this is included in the Comments. However, in many cases to expand on the list of disciplines in a useful way would require an extensive description of each satellite, which is beyond the scope of this table.

The table has been compiled using a large variety of sources, including especially the NASA Special Publications series, the TRW Space Log, the Russian astronautics magazine *Novosti Kosmonavtiki*, the *Journal of Geophysical*

Research, Spaceflight magazine, the *Journal of the British Interplanetary Society*, the journal *Kosmicheskie Issledovania*, as well as memos and primary unpublished materials in the libraries and archives of NASA HQ, NASA KSC, the Jet Propulsion Lab, and the US National Archives. The orbital data are ultimately from United States Space Command and are made public courtesy of the Orbital Information Group at NASA's Goddard Space Flight Center.

For up-to-date information on satellite launches, the reader is invited to visit http://hea-www.harvard.edu/~jcm/space

Key to disciplines:

A	Auroral studies
ALT	Altimeter missions
ATM	Atmospheric studies
BIO	Life sciences
ENV	Space environment
F	Fields
GEO	Geodesy
GRA	Gamma ray astronomy
ION	Ionospheric beacons
IRA	IR astronomy
M	Mixed cargo of experiments
MG	Microgravity
MM	Micrometeorites
OA	Optical astronomy
P	Particles
PL	Planetary and solar wind
RFA	Radio frequency astronomy
SMA	Submillimeter astronomy
SOL	Solar observations
SX	Solar X-ray observations
UVA	UV astronomy
XRA	X-ray astronomy
*	Indicates the more significant missions

CATALOG OF SPACE SCIENCE LAUNCHES 1679

Launch	Name	State	Peri ×	Apo (km) ×	Inc.	Operated until		Pri mission discipline	Comments
1957 Oct 4	1-y ISZ	USSR	214 ×	938 ×	65	1957 Oct 27		ENV*	First Earth satellite
1957 Nov 3	2-y ISZ	USSR	211 ×	1659 ×	65	1957 Nov 9		BIO	First animal payload
1958 Feb 1	Explorer 1	USA	359 ×	2542 ×	33	1958 Apr 1		P*	Discovered trapped radiation
1958 Mar 17	Vanguard I	USA	657 ×	3935 ×	34	1964 Mar		GEO	First geodetic satellite
1958 Mar 26	Explorer 3	USA	195 ×	2810 ×	33	1958 May 10		P*	—
1958 May 15	3-y Sovetskiy ISZ	USSR	207 ×	1247 ×	65	1958 Jun 5		P*	—
1958 Jul 26	Explorer 4	USA	258 ×	2233 ×	50	1958 Sep 19		P*	—
1958 Oct 11	Pioneer 1	USA	−100 ×	113 860 ×	29	1958 Oct 13		PL	—
1958 Dec 6	Pioneer 3	USA	−70 ×	102 200 ×	28	1958 Dec 7		PL	—
1959 Jan 2	AMS Luna-1	USSR	Solar orbit			1959 Jan 5		PL	First artificial planet
1959 Feb 17	Vanguard II	USA	564 ×	3304 ×	32	1959 Mar 7	Imaging	ATM	Cloud cover photos
1959 Mar 3	Pioneer 4	USA	Solar orbit			1959 Mar 6		PL	—
1959 Aug 7	Explorer 6	USA	250 ×	42 327 ×	46	1959 Oct 6		P/F*	—
1959 Sep 12	AMS Luna-2	USSR	Lunar impact					PL	First lunar impact
1959 Sep 18	Vanguard III	USA	521 ×	3758 ×	33	1959 Dec 12		MM/F*	—
1959 Oct 4	AMS Luna-3	USSR	500 ×	500 000 ×	55	1959 Nov 15		PL	Lunar farside photos
1959 Oct 13	Explorer 7	USA	560 ×	1087 ×	50	1961 Feb		P/F/ION*	—
1960 Mar 11	Pioneer V	USA	Solar orbit			1960 Jun 23		PL	—
1960 May 15	Korabl'-Sputnik	USSR	277 ×	674 ×	64	1960 May 19	Spaceship	BIO	—
1960 May 24	Midas 2	USA	482 ×	514 ×	33	1960 May 25	Warning	MM/P/ATM	—
1960 Jun 22	SR	USA	614 ×	1058 ×	66	1961 Apr 18	Sigint	SOL	Solar radiation
1960 Aug 10	Discoverer 13	USA	245 ×	614 ×	82	1960 Aug 11	Recon	BIO	—
1960 Aug 19	Korabl'-Sputnik-2	USSR	307 ×	307 ×	64	1960 Aug 20	Spaceship	BIO	—
1960 Nov 3	Explorer 8	USA	416 ×	2286 ×	49	1960 Dec 27		P/F*	—
1960 Nov 12	Discoverer 17	USA	182 ×	923 ×	81	1960 Nov 14	Recon	BIO	—
1960 Dec 1	Korabl'-Sputnik-3	USSR	171 ×	237 ×	65	1960 Dec 2	Spaceship	BIO	—
1960 Dec 7	Discoverer 18	USA	272 ×	536 ×	81	1960 Dec 10	Recon	BIO	—
1961 Feb 4	Tyazholiy Sputnik	USSR	179 ×	296 ×	64	1961 Feb 4?		PL	—
1961 Feb 12	AMS Venera	USSR	Solar orbit			1961 Feb 27?		PL	—
1961 Feb 16	Explorer 9	USA	635 ×	2581 ×	38	1961 Feb 16		ATM	Air density study
1961 Feb 22	Lofti	USA	167 ×	1002 ×	28	1961 Mar 30		F	VLF study
1961 Mar 9	Korabl'-Sputnik-4	USSR	173 ×	239 ×	64	1961 Mar 9	Spaceship	BIO	—
1961 Mar 25	Korabl'-Sputnik-5	USSR	163 ×	229 ×	64	1961 Mar 25	Spaceship	BIO	—
1961 Mar 25	Explorer 10	USA	221 ×	181 000 ×	33	1961 Mar 27		P/F*	Magnetic field probe
1961 Apr 12	Vostok	USSR	168 ×	314 ×	64	1961 Apr 12	Spaceship	BIO	First piloted spaceship
1961 Apr 27	Explorer 11	USA	497 ×	1777 ×	28	1961 Dec 6		GRA*	—
1961 Jun 16	Discoverer 25	USA	223 ×	362 ×	82	1961 Jun 19	Recon	MM/P	—
1961 Jun 29	Injun	USA	882 ×	996 ×	66	1963 Mar 6		P*	—
1961 Jun 29	Solrad 3	USA	879 ×	984 ×	66	1962 Aug	Sigint	SX	Solar X/signals intel.
1961 Jul 7	Discoverer 26	USA	229 ×	713 ×	82	1961 Jul 10	Recon	RFA/MM/P	—

Launch	Name	State	Peri ×	Apo (km) ×	Inc.	Operated until		Pri mission discipline	Comments
1961 Aug 6	Vostok-2	USSR	172 ×	218 ×	64	1961 Aug 7	Spaceship	BIO	—
1961 Aug 16	Explorer 12	USA	457 ×	77 170 ×	33	1961 Dec 6		P/F*	—
1961 Aug 23	Ranger 1	USA	179 ×	446 ×	32	1961 Aug 27		P	—
1961 Aug 25	Explorer 13	USA	364 ×	810 ×	36	1961 Aug 28		MM*	—
1961 Aug 30	Discoverer 29	USA	138 ×	511 ×	82	1961 Sep 1	Recon	A/P	—
1961 Sep 12	Discoverer 30	USA	231 ×	484 ×	82	1961 Sep 14	Recon	RFA/BIO	—
1961 Sep 17	Discoverer 31	USA	233 ×	380 ×	82	1961 Sep 19	Recon	A/P/RFA	—
1961 Oct 13	Discoverer 32	USA	233 ×	350 ×	81	1961 Nov 13	Recon	P/ION	—
1961 Nov 5	Discoverer 34	USA	225 ×	971 ×	82	1961 Nov 10?	Recon	P	—
1961 Nov 15	Traac	USA	934 ×	1129 ×	32	1962 Aug 14		P	—
1961 Nov 18	Ranger 2	USA	149 ×	169 ×	33	1961 Nov 19		P	—
1961 Nov 29	Mercury MA-5	USA	158 ×	237 ×	32	1961 Nov 29	Spaceship	BIO	—
1961 Dec 12	Discoverer 36	USA	223 ×	446 ×	81	1961 Dec 16	Recon	ION/BIO/P	—
1962 Jan 26	Ranger 3	USA	Solar orbit					PL	—
1962 Jan 26	Ranger Capsule 12	USA	Solar orbit					PL	Lunar lander capsule
1962 Feb 20	Friendship Seven (MA-6)	USA	149 ×	249 ×	32	1962 Feb 20	Spaceship	ATM	—
1962 Feb 27	Discoverer 38	USA	207 ×	357 ×	82	1962 Mar 3	Recon	A/P	—
1962 Mar 7	OSO 1	USA	546 ×	601 ×	32	1963 Aug 3		SOL/GRA*	—
1962 Mar 16	Kosmos-1	USSR	208 ×	731 ×	49	1962 Apr?		ION/P*	—
1962 Apr 6	Kosmos-2	USSR	218 ×	1449 ×	48	1962 Apr 17		ION/P*	—
1962 Apr 18	Discoverer 39	USA	201 ×	504 ×	73	1962 Apr 22	Recon	P	—
1962 Apr 23	Ranger 4	USA	Lunar impact			1962 Apr 26		PL	—
1962 Apr 23	Ranger Capsule 14	USA	Lunar impact			1962 Apr 26		PL	Lunar lander capsule
1962 Apr 24	Kosmos-3	USSR	226 ×	632 ×	49	1962 Jul?		P*	—
1962 Apr 26	Ariel 1	UK	397 ×	1202 ×	53	1964 Nov 9		P/SX*	—
1962 Apr 29	FTV 1125	USA	190 ×	414 ×	73	1962 May 5	Recon	P	—
1962 May 24	MA-7 Balloon Subsatellite	USA	154 ×	260 ×	32	1962 May 24	ATM	Atmospheric density, failed to deploy	—
1962 May 28	Kosmos-5	USSR	207 ×	1500 ×	49	1962?		ATM/MG*	—
1962 May 30	FTV 1128	USA	192 ×	315 ×	74	1962 Jun 2	Recon	P/RFA	—
1962 Jun 2	FTV 1127	USA	196 ×	373 ×	74	1962 Jun 5	Recon	P/RFA	—
1962 Jun 23	FTV 1129	USA	208 ×	267 ×	75	1962 Jun?	Recon	P/RFA	—
1962 Jun 28	FTV 1151	USA	193 ×	583 ×	76	1962 Jul 2	Recon	P/F/MM	—
1962 Jun 30	Kosmos-6	USSR	262 ×	304 ×	48	1962 Jul 6	Calibration	P	—
1962 Jul 21	FTV 1130	USA	200 ×	364 ×	70	1962 Jul 26	Recon	P/RFA	—
1962 Aug 2	FTV 1152	USA	193 ×	398 ×	82	1962 Aug 6	Recon	P/MM	—
1962 Aug 11	Vostok-3	USSR	158 ×	200 ×	64	1962 Aug 15	Spaceship	P/BIO	—
1962 Aug 12	Vostok-4	USSR	168 ×	221 ×	64	1962 Aug 15	Spaceship	P/BIO	—
1962 Aug 18	Kosmos-8	USSR	259 ×	569 ×	48	1962 Aug 23	Mil. tech.	MM	—
1962 Aug 25	[AMS Venera]	USSR	174 ×	248 ×	64	1962 Aug 25		PL	Venus probe, failed

CATALOG OF SPACE SCIENCE LAUNCHES

Date	Name	Country	Orbit				Notes
1962 Aug 27	Mariner 2	USA	Solar orbit			PL	Venus flyby
1962 Sep 1	[AMS Venera]	USSR	185 ×	246 ×	64	PL	Venus probe, failed
1962 Sep 12	[AMS Venera]	USSR	163 ×	195 ×	64	PL	Venus probe, failed
1962 Sep 17	FTV 1133	USA	192 ×	528 ×	81	P	Recon
1962 Sep 27	Kosmos-9	USSR	297 ×	339 ×	64	P	Recon
1962 Sep 29	Alouette 1	Canada	999 ×	1027 ×	80	ION/RFA*	
1962 Oct 2	Explorer 14	USA	915 ×	96 959 ×	33	P/F*	
1962 Oct 3	Sigma Seven (MA-8)	USA	156 ×	286 ×	32	P/ATM	Spaceship
1962 Oct 17	Kosmos-10	USSR	196 ×	368 ×	65	P	Recon
1962 Oct 18	Ranger 5	USA	Solar orbit			PL	Lunar probe, missed
1962 Oct 18	Ranger Capsule 18	USA	Solar orbit			PL	Lunar lander capsule
1962 Oct 20	Kosmos-11	USSR	249 ×	864 ×	48	ION/P*	
1962 Oct 24	[AMS Mars]	USSR	202 ×	260 ×	65	PL	Mars probe, failed
1962 Oct 26	Starad	USA	199 ×	5494 ×	71	P/F*	Studied artificial rad belt
1962 Oct 27	Explorer 15	USA	312 ×	17 619 ×	17	P/F*	Studied artificial rad belt
1962 Oct 31	ANNA 1B	USA	1081 ×	1179 ×	50	GEO	
1962 Nov 1	AMS Mars	USSR	Solar orbit			PL	Mars probe, missed
1962 Nov 4	[AMS Mars]	USSR	170 ×	170 ×	64	PL	Mars probe, failed
1962 Nov 5	FTV 1136	USA	194 ×	391 ×	75	P	Recon
1962 Nov 24	FTV 1135	USA	206 ×	317 ×	65	P	Recon
1962 Dec 13	Injun 3	USA	231 ×	2774 ×	70	P	
1962 Dec 16	Explorer 16	USA	759 ×	1170 ×	51	MM*	
1962 Dec 22	Kosmos-12	USSR	199 ×	387 ×	64	P	Recon
1963 Jan 4	[Luna-4]	USSR	178 ×	194 ×	64	PL	Lunar probe, missed
1963 Jan 4	ALS	USSR	180 ×	214 ×	64	PL	Lunar lander capsule
1963 Apr 2	Luna-4	USSR	200 ×	694 000 ×	65	PL	Lunar probe, missed
1963 Apr 2	ALS	USSR	200 ×	694 000 ×	65	PL	Lunar lander capsule
1963 Apr 3	Explorer 17	USA	246 ×	914 ×	57	ATM*	
1963 Apr 28	Kosmos-16	USSR	195 ×	384 ×	65	P	Recon
1963 May 9	DASH 1	USA	3611 ×	3676 ×	87	ATM	Calibration
1963 May 15	Faith Seven (MA-9)	USA	163 ×	265 ×	32	ATM	Spaceship
1963 May 15	MA-9 Balloon Subsatellite	USA	163 ×	265 ×	32	Atm. density study	ATM
1963 May 22	Kosmos-17	USSR	265 ×	751 ×	48	ION/P*	
1963 Jun 14	Vostok-5	USSR	158 ×	164 ×	64	P/BIO	Spaceship
1963 Jun 15	SR 6A	USA	165 ×	846 ×	69	SX*	Sigint
1963 Jun 16	Vostok-6	USSR	163 ×	191 ×	64	P/BIO	Spaceship
1963 Jun 27	Hitchhiker 1	USA	331 ×	4125 ×	82	P/F*	
1963 Jun 28	GRS	USA	421 ×	1298 ×	49	ATM/P*	
1963 Sep 28	Transit VE-1	USA	1071 ×	1129 ×	89	P*	
1963 Oct 17	ERS 12	USA	952 ×	102 375 ×	35	P	
1963 Oct 29	Hitchhiker 2	USA	290 ×	568 ×	89	P	

Launch	Name	State	Peri ×	Apo (km) ×	Inc.	Operated until		Pri mission discipline	Comments
1963 Nov 11	Kosmos-21	USSR	180 ×	219 ×	64	1963 Nov 11		PL	Venus probe, failed
1963 Nov 27	IMP 1 (Explorer 18)	USA	2072 ×	194 080 ×	35	1965 Mar 25		P/F*	—
1963 Dec 5	Transit VE-3	USA	1066 ×	1119 ×	89	1964 Mar 9		P*	—
1963 Dec 19	Explorer 19	USA	588 ×	2389 ×	78	1963 Dec 19		ATM*	—
1963 Dec 21	Hitchhiker 3	USA	312 ×	389 ×	64	1964?		P*	—
1964 Jan 11	EGRS 1	USA	905 ×	932 ×	69	1965?		GEO	—
1964 Jan 11	SR 7A	USA	909 ×	934 ×	69	1966 Jul		SX*	—
1964 Jan 30	Elektron-1	USSR	394 ×	7124 ×	60	1964 Mar		ION/MM/P*	—
1964 Jan 30	Ranger 6	USA	Lunar impact			1964 Feb 2		PL	Lunar probe, failed
1964 Jan 30	Elektron-2	USSR	408 ×	68 014 ×	60	1964 Jun		ION/MM/P/F*	—
1964 Mar 18	Kosmos-26	USSR	263 ×	367 ×	48	1964 Apr 1		F*	Magnetic field study
1964 Mar 27	Ariel 2	UK	289 ×	1343 ×	51	1964 Sep?		RFA/ATM/MM*	—
1964 Mar 27	Kosmos-27	USSR	197 ×	208 ×	64	1964 Mar 27		PL	Venus probe, failed
1964 Apr 2	Zond	USSR	Solar orbit			1964 Apr		PL	Venus probe, failed
1964 Apr 27	OPS 2921	USA	175 ×	426 ×	79	1964 May 26	Recon	ION	—
1964 Jun 6	Kosmos-31	USSR	218 ×	457 ×	48	1964 Oct 20		UVA*	UV background, failed
1964 Jun 13	Starflash 1A	USA	351 ×	357 ×	115	1964?		GEO	—
1964 Jul 10	Elektron-3	USSR	401 ×	7026 ×	60	1965?		P*	—
1964 Jul 10	Elektron-4	USSR	452 ×	66 264 ×	60	1965 Feb?		MM/RFA/SX*	—
1964 Jul 17	ERS 13	USA	319 ×	104 032 ×	38	1966 Jul 1		P	—
1964 Jul 28	Ranger 7	USA	Lunar impact			1964 Jul 31		PL	Lunar impact probe
1964 Aug 14	OPS 3316	USA	271 ×	3724 ×	95	1979 Mar 8		P/F*	—
1964 Aug 21	Starflash 1B	USA	329 ×	346 ×	114	1964?		GEO	—
1964 Aug 25	Explorer 20	USA	869 ×	1018 ×	79	1966 Jul		ION*	—
1964 Sep 5	OGO 1	USA	1289 ×	148 412 ×	32	1980 Aug 10		P/F/MM/RFA*	—
1964 Sep 13	Kosmos-45	USSR	200 ×	316 ×	64	1964 Sep 18	Recon	ATM	—
1964 Oct 4	IMP 2 (Explorer 21)	USA	666 ×	94 526 ×	33	1965 Oct 13		P/F*	—
1964 Oct 10	Explorer 22	USA	886 ×	1079 ×	79	1969 Apr 17		ION*	—
1964 Oct 24	Kosmos-49	USSR	259 ×	460 ×	48	1965 Aug 21		F*	Magnetic field study
1964 Nov 5	Mariner 3	USA	Solar orbit			1964 Nov 6		PL	Mars probe, failed
1964 Nov 6	Explorer 23	USA	464 ×	978 ×	51	1983 Jun 29		MM*	—
1964 Nov 18	OPS 3360	USA	174 ×	326 ×	70	1964 Dec 6	Recon	ION	Ionospheric beacon
1964 Nov 18	ORBIS	USA	174 ×	326 ×	70	1964 Dec 6?		ION	—
1964 Nov 21	Explorer 24	USA	542 ×	2480 ×	81	1968 Oct 18		ATM*	—
1964 Nov 21	Explorer 25	USA	528 ×	2492 ×	81	1966 Jul		P*	—
1964 Nov 28	Mariner 4	USA	Solar orbit			1967 Dec 20		PL	Mars flyby
1964 Nov 30	Zond-2	USSR	Solar orbit			1965 May 5		PL	Mars probe, failed
1964 Dec 9	Kosmos-51	USSR	259 ×	530 ×	48	1965 Nov 14		UVA*	UV background
1964 Dec 13	Transit VE-5	USA	1027 ×	1084 ×	89	1965 Jun 30		UVA/F*	UV background
1964 Dec 15	San Marco 1	Italy	198 ×	782 ×	37	1965 Sep 11		ATM/ION*	Air density/ionosphere

CATALOG OF SPACE SCIENCE LAUNCHES

Date	Name	Country	Orbit			Type	Category	Notes
1964 Dec 21	Explorer 26	USA	310 ×	26202 ×	20	1967 May 21		P/F*	—
1965 Jan 30	Kosmos-53	USSR	221 ×	1143 ×	48	1966 Aug 12		P*	—
1965 Feb 3	OSO 2	USA	549 ×	628 ×	32	1989 Aug 9		SOL/GRA*	—
1965 Feb 16	Pegasus 1	USA	500 ×	731 ×	31	1978 Sep 17		MM*	—
1965 Feb 17	Ranger 8	USA	Lunar impact			1965 Feb 20		PL	Lunar impact probe
1965 Feb 25	OPS 4782	USA	176 ×	350 ×	75	1965 Mar 18	Recon	A/P	—
1965 Mar 9	EGRS 3	USA	907 ×	939 ×	70	1960s?		GEO	—
1965 Mar 9	SR 7B	USA	909 ×	938 ×	70	1969 Jul		SX*	—
1965 Mar 11	EGRS 2	USA	282 ×	967 ×	89	1968 Feb 26		GEO	—
1965 Mar 12	Kosmos-60	USSR	194 ×	247 ×	64	1965 Mar 17	DRV	PL	Lunar lander, failed
1965 Mar 12	ALS	USSR	194 ×	247 ×	64	1965 Mar 17		PL	Lunar probe, failed
1965 Mar 18	Voskhod-2	USSR	167 ×	472 ×	64	1965 Mar 19	Spaceship	BIO	—
1965 Mar 21	Ranger 9	USA	Lunar impact			1965 Mar 24		PL	Lunar impact probe
1965 Mar 23	Gemini III	USA	161 ×	225 ×	32	1965 Mar 23	Spaceship	BIO	—
1965 Apr 3	EGRS 4	USA	1265 ×	1320 ×	90	1965 Apr 3		GEO	—
1965 Apr 17	Kosmos-65	USSR	205 ×	317 ×	65	1965 Apr 25	Recon	ATM	—
1965 Apr 29	Explorer 27	USA	936 ×	1316 ×	41	1973 Jul 20		ION*	—
1965 May 9	Luna-5	USSR	Lunar impact			1965 May 12	DRV	PL	Lunar lander, failed
1965 May 9	ALS	USSR	Lunar impact			1965 May 12		PL	Lunar landing capsule
1965 May 25	Pegasus 2	USA	508 ×	737 ×	31	1979 Nov 3		MM*	—
1965 May 29	IMP 3 (Explorer 28)	USA	195 ×	264252 ×	33	1968 Jul 5		P/F*	—
1965 Jun 3	Gemini IV	USA	165 ×	289 ×	32	1965 Jun 7	Spaceship	ATM/BIO/P	—
1965 Jun 8	Luna-6	USSR	Solar orbit			1965 Jun 11?		PL	Lunar lander, failed
1965 Jun 8	ALS	USSR	Solar orbit			1965 Jun 11?		PL	Lunar probe, failed
1965 Jul 2	Kosmos-70	USSR	225 ×	1113 ×	48	1966 Dec 18	DRV	PL	—
1965 Jul 16	Proton	USSR	177 ×	523 ×	63	1965 Oct 11		P*	Cosmic ray studies
1965 Jul 18	Zond-3	USSR	Solar orbit			1966 Dec		P*	Lunar flyby
1965 Jul 20	ERS 17	USA	566 ×	111790 ×	36	1968 Jul 1		PL	—
1965 Jul 30	Pegasus 3	USA	516 ×	536 ×	28	1969 Aug 4		GRA,P*	—
1965 Aug 10	Secor 5	USA	1136 ×	2424 ×	69	1960s?		MM*	—
1965 Aug 21	Gemini V	USA	170 ×	330 ×	32	1965 Aug 29	Spaceship	ATM, BIO	Long duration spaceflight
1965 Oct 4	Luna-7	USSR	Lunar impact			1965 Oct 7		PL	Lunar lander, failed
1965 Oct 4	ALS	USSR	Lunar impact			1965 Oct 7		PL	Lunar probe, failed
1965 Oct 5	OV1-2	USA	413 ×	3453 ×	144	1967 Apr		P/F*	—
1965 Oct 14	OGO 2	USA	418 ×	1509 ×	87	1981 Sep 17	DRV	P/F/SX/RFA*	Polar geophysics lab
1965 Oct 16	Kosmos-92	USSR	201 ×	334 ×	65	1965 Oct 24	Recon	BIO/ATM	—
1965 Oct 28	Kosmos-94	USSR	205 ×	271 ×	65	1965 Nov 5	Recon	BIO	—
1965 Oct 28	OPS 2155	USA	173 ×	432 ×	74	1965 Nov 17	Recon	A/P	Aurora-1965
1965 Nov 2	Proton-2	USSR	151 ×	277 ×	63	1966 Feb 6		P*	Cosmic ray studies
1965 Nov 6	Explorer 29	USA	1117 ×	2273 ×	59	1967 Dec 17		GEO	—
1965 Nov 12	Venera-2	USSR	Solar orbit			1965 Dec?		PL	Venus flyby
1965 Nov 16	Venera-3	USSR	Venus impact			1966 Jan?		PL	Venus impact
1965 Nov 16	Spuskaemiy apparat	USSR	Venus entry			1966 Jan?	DRV	PL	Venus lander capsule

Launch	Name	State	Peri ×	Apo (km) ×	Inc.	Operated until		Pri Mission Discipline	Comments
1965 Nov 19	SR 8 (Explorer 30)	USA	693 ×	898 ×	59	1967 Nov 20		SX*	–
1965 Nov 23	Kosmos-96	USSR	209 ×	262 ×	51	1965 Dec 9		PL	Venus probe, failed
1965 Nov 29	Alouette 2	Canada	502 ×	2986 ×	79	1975 Nov 29		ION/RFA*	–
1965 Nov 29	Explorer 31	USA	505 ×	2983 ×	79	1969 Jun 9		P	–
1965 Dec 3	Luna-8	USSR	Lunar impact			1965 Dec 6	DRV	PL	Lunar lander, failed
1965 Dec 4	Gemini VII	USA	299 ×	303 ×	28	1965 Dec 18	Spaceship	BIO/P/F	–
1965 Dec 6	FR-1	France	746 ×	759 ×	75	1966 Dec 31?		P/F*	Ionospheric VLF
1965 Dec 15	Gemini VI-A	USA	270 ×	274 ×	28	1965 Dec 16	Spaceship	ATM	–
1965 Dec 16	Pioneer 6	USA	Solar orbit			Active		PL*	Solar wind monitor
1965 Dec 21	OV2-3	USA	283 ×	34 649 ×	26	1975 Aug 17		P/F*	–
1966 Jan 31	Luna-9	USSR	Lunar impact			1966 Feb 3		PL	Lunar probe
1966 Jan 31	ALS Luna-9	USSR	Lunar landing			1966 Feb 3	DRV	PL	Lunar lander capsule
1966 Feb 11	Kosmos-108	USSR	190 ×	344 ×	48	1966 Nov 21		ATM/P*	–
1966 Feb 17	Diapason D-1A	France	502 ×	2735 ×	34	1967?		GEO	–
1966 Feb 19	Kosmos-109	USSR	204 ×	288 ×	64	1966 Feb 27	Recon	BIO	–
1966 Feb 22	Kosmos-110	USSR	192 ×	879 ×	51	1966 Mar 16	Spaceship	BIO	Long duration dog flight
1966 Mar 1	Kosmos-111	USSR	182 ×	194 ×	51	1966 Mar 3		PL	Lunar orbiter, failed
1966 Mar 16	Gemini VIII	USA	261 ×	270 ×	28	1966 Mar 17	Spaceship	BIO/ATM	–
1966 Mar 30	OV1-4	USA	888 ×	1014 ×	144	1966?		BIO	–
1966 Mar 31	Luna-10	USSR	Lunar orbit			1966 May 29		PL	Lunar orbiter
1966 Apr 8	OAO 1	USA	794 ×	804 ×	35	1966 Apr 10		UVA/GRA/XRA*	Multi-telescope package, failed
1966 Apr 22	OV3-1	USA	351 ×	5736 ×	82	1967 Feb		P*	–
1966 May 24	Kosmos-119	USSR	208 ×	1202 ×	48	1966 Nov 30		P/F*	VLF study, failed
1966 May 25	Explorer 32	USA	283 ×	2718 ×	64	1985 Feb 22		ATM*	–
1966 May 30	Surveyor 1	USA	Lunar landing			1966 Jun 2		PL	Lunar soft lander
1966 Jun 3	Gemini IX-A	USA	272 ×	274 ×	28	1966 Jun 6	Spaceship	MM/BIO	–
1966 Jun 7	OGO 3	USA	905 ×	121 519 ×	36	1981 Sep 15		P/F/RFA*	High altitude geophysics
1966 Jun 9	EGRS 6	USA	171 ×	3479 ×	90	1967 Jul?		GEO	–
1966 Jun 10	OV3-4	USA	644 ×	4730 ×	40	1960s?		P	Radiation dose study
1966 Jun 17	Kosmos-121	USSR	203 ×	325 ×	72	1966 Jun 25	Recon	P	Carried particle spectrometer
1966 Jun 24	Pageos	USA	3978 ×	4496 ×	87	1975 Jul 12		GEO	Balloon satellite for geodesy
1966 Jul 1	Explorer 33	USA	79 527 ×	477 380 ×	21	1971 May 31		P/F*	Intended for lunar orbit
1966 Jul 6	Proton-3	USSR	174 ×	490 ×	63	1966 Sep 16		P*	Cosmic rays
1966 Jul 18	Gemini X	USA	299 ×	757 ×	28	1966 Jul 21	Spaceship	P/BIO/MM	Docking and spacewalk test
1966 Jul 28	Kosmos-126	USSR	207 ×	338 ×	51	1966 Aug 6	Recon	P	Particle spectra

CATALOG OF SPACE SCIENCE LAUNCHES

Date	Name	Country	Orbit	Apogee	Incl.	End date	Type	Payload	Notes
1966 Aug 4	OV3-3	USA	358 ×	4479 ×	81	1968?		P/F*	Energetic particle study
1966 Aug 10	Lunar Orbiter 1	USA	Lunar orbit			1966 Oct 29		PL	Lunar mapping
1966 Aug 17	Pioneer 7	USA	Solar orbit			Active		PL*	Solar wind monitor
1966 Aug 19	EGRS 7	USA	3675 ×	3700 ×	90	?		GEO	—
1966 Aug 24	Luna-11	USSR	Lunar orbit			1966 Sep?		PL	Lunar orbiter
1966 Sep 12	Gemini XI	USA	298 ×	1368 ×	28	1966 Sep 15	Spaceship	UVA/BIO/ATM	
1966 Sep 20	Surveyor 2	USA	Lunar impact			1966 Sep 23		PL	Lunar lander, failed
1966 Oct 5	EGRS 8	USA	3682 ×	3704 ×	90	1960s?		GEO	—
1966 Oct 22	Luna-12	USSR	Lunar orbit			1967 Jan 19		PL	Lunar orbiter
1966 Oct 28	OV3-2	USA	316 ×	1589 ×	81	1967?		P*	Magnetosphere
1966 Nov 3	OV4-3	USA	277 ×	280 ×	32	1967 Jan 9	Spaceship	BIO/ION/MM	Space station mockup
1966 Nov 3	OV4-1R	USA	272 ×	272 ×	32	1967 Jan 5		P	—
1966 Nov 3	OV4-1T	USA	276 ×	300 ×	32	1967 Jan 11		P	—
1966 Nov 6	Lunar Orbiter 2	USA	Lunar orbit			1967 Oct 11		PL	Lunar mapper
1966 Nov 11	Gemini XII	USA	252 ×	291 ×	28	1966 Nov 15	Spaceship	BIO/MM/ F/UVA	Docking, spacewalk tests
1966 Dec 7	ATS 1	USA	35780 ×	35791 ×	0	1985 Apr	Comms	P/ATM	Geostationary environment
1966 Dec 11	OV1-9	USA	635 ×	769 ×	93	1967?		P	Radiation dose
1966 Dec 11	OV1-10	USA	634 ×	761 ×	93	1967 Jun		P/SX*	—
1966 Dec 12	Kosmos-135	USSR	251 ×	604 ×	48	1967 Apr 12		MM*	—
1966 Dec 14	Biosatellite 1	USA	281 ×	299 ×	33	1967 Jan 10		—	—
1966 Dec 14	Biosatellite 1 Capsule	USA	272 ×	286 ×	33	1967 Feb 15		BIO	—
1966 Dec 21	Luna-13 ALS	USSR	Lunar landing			1966 Dec 24	DRV	PL	Lunar lander capsule
1966 Dec 21	Kosmos-137	USSR	221 ×	1623 ×	48	1967 Nov 23		P*	—
1966 Dec 21	Luna-13	USSR	Lunar impact			1966 Dec 24		PL	Lunar lander probe
1967 Feb 5	Lunar Orbiter 3	USA	Lunar orbit			1967 Oct 9		PL	Lunar mapper
1967 Feb 8	Diademe D-1C	France	569 ×	1351 ×	39	1968?		GEO	—
1967 Feb 14	Kosmos-142	USSR	206 ×	1186 ×	48	1967 Jul 6		F/RFA*	—
1967 Feb 15	Diademe D-1D	France	590 ×	1882 ×	39	1967 Apr 5		GEO	—
1967 Mar 8	OSO 3	USA	537 ×	565 ×	32	1982 Apr 4		SOL/GRA*	Ionospheric VLF
1967 Mar 21	Kosmos-149	USSR	230 ×	264 ×	48	1967 Apr 7		ATM*	—
1967 Apr 6	ATS 2	USA	186 ×	10767 ×	28	1969 Sep 2	Comms	RFA	Carried radio experiment
1967 Apr 17	Surveyor 3	USA	Lunar landing			1967 Apr 20		PL	Lunar lander
1967 Apr 26	San Marco 2	Italy	210 ×	683 ×	2	1967 Oct 14		ATM/ION*	Air density/ ionosphere
1967 Apr 28	ERS 18	USA	8990 ×	110839 ×	33	1960s?		GRA*	Gamma ray background
1967 Apr 28	ERS 27	USA	9111 ×	110599 ×	34	1960s?		P/SX*	Solar flares
1967 May 4	Lunar Orbiter 4	USA	Lunar orbit			1967 Oct 6		PL	Lunar mapper

Launch	Name	State	Peri ×	Apo (km) ×	Inc.	Operated until		Pri mission discipline	Comments
1967 May 5	Ariel 3	UK	496 ×	600 ×	80	1970 Dec 14		RFA/ATM*	–
1967 May 16	Kosmos-159	USSR	671 ×	60294 ×	51	1977 Nov 11		PL	Lunar test launch
1967 May 24	IMP 4 (Explorer 34)	USA	242 ×	214382 ×	67	1969 May 4		P/F*	Magnetosphere
1967 Jun 5	Kosmos-163	USSR	251 ×	549 ×	48	1967 Oct 11		MM*	–
1967 Jun 12	Venera-4	USSR	Venus impact			1967 Oct 18	DRV	PL	Venus entry
1967 Jun 12	Spuskaemiy apparat Veneri-4	USSR	Venus entry			1967 Oct 18		PL	Venus lander capsule
1967 Jun 14	Mariner 5	USA	Solar orbit			1968 Nov 5		PL	Venus flyby
1967 Jun 16	Kosmos-166	USSR	277 ×	534 ×	48	1967 Oct 25		SOL*	Solar UV/X
1967 Jun 17	Spuskaemiy apparat	USSR	187 ×	262 ×	51	1967 Jun 25	DRV	PL	Venus entry capsule
1967 Jun 17	Kosmos-167	USSR	187 ×	262 ×	51	1967 Jun 25		PL	Venus probe, failed
1967 Jun 29	EGRS 9	USA	3797 ×	3946 ×	89	1970s?		GEO	–
1967 Jun 29	Aurora 1	USA	3797 ×	3946 ×	89	1960s?		A/P*	–
1967 Jul 14	Surveyor 4	USA	Lunar impact			1967 Jul 17		PL	Lunar lander
1967 Jul 19	Explorer 35	USA	Inclined LEO			1973 Jun 24		P/F*	Lunar orbit environment
1967 Jul 27	OV1-86	USA	475 ×	603 ×	101	1972 Feb 22		ATM	–
1967 Jul 27	OV1-12	USA	538 ×	555 ×	101	1980 Jul 22		P*	–
1967 Jul 28	OGO 4	USA	411 ×	896 ×	86	1972 Aug 16		P/F/RFA/SX*	–
1967 Aug 1	Lunar Orbiter 5	USA	Lunar orbit			1968 Jan 31		PL	Lunar mapper
1967 Aug 7	OPS 4827	USA	187 ×	343 ×	79	1967 Sep 1	Recon	SX/ATM	Background radiation
1967 Sep 7	Biosatellite 2	USA	289 ×	313 ×	33	1967 Oct 4		BIO	–
1967 Sep 8	Surveyor 5	USA	Lunar landing			1967 Sep 11		PL	Lunar lander
1967 Oct 18	OSO 4	USA	539 ×	572 ×	32	1982 Jun 15		SOL*	Solar observatory
1967 Nov 7	Surveyor 6	USA	Lunar landing			1967 Nov 10		PL	Lunar lander
1967 Nov 29	WRESAT	Australia	173 ×	1119 ×	83	1968 Jan 10		ATM/F*	Upper atmosphere study
1967 Dec 5	OV3-6	USA	403 ×	434 ×	90	1969 Mar 9		ATM*	Atmospheric composition
1967 Dec 13	Pioneer 8	USA	Solar orbit			Active		P/F*	Solar wind monitor
1967 Dec 19	Kosmos-196	USSR	220 ×	810 ×	48	1968 Jul 7		ATM/P*	–
1968 Jan 7	Surveyor 7	USA	Lunar landing			1968 Jan 10		PL	Lunar lander
1968 Jan 11	Geos 2 (Explorer 36)	USA	1084 ×	1573 ×	105	1973 Mar 27		GEO	–
1968 Feb 20	Kosmos-203	USSR	1185 ×	1201 ×	74	1968?		GEO	–
1968 Mar 4	OGO 5	USA	5075 ×	141939 ×	41	1972 Jul 14		P/F*	–
1968 Mar 5	SR 9 (Explorer 37)	USA	509 ×	882 ×	59	1990 Nov 16		SOL*	–
1968 Mar 21	Kapsula Kosmosa-208	USSR	195 ×	263 ×	64	1968 Mar 29		GRA	–
1968 Apr 6	OV1-13	USA	553 ×	9316 ×	99	1968 Jul 11		P*	–
1968 Apr 6	OV1-14	USA	554 ×	9938 ×	99	1968 Apr 13		P/F*	–
1968 Apr 7	Luna-14	USSR	Lunar orbit			1968 Jun 24		PL	Lunar orbiter
1968 Apr 15	Kosmos-213	USSR	193 ×	253 ×	51	1968 Apr 20	Spaceship	P/MM/ATM	–

CATALOG OF SPACE SCIENCE LAUNCHES

Date	Name	Country	Perigee ×	Apogee	Incl.	Decay	Code	Remarks
1968 Apr 18	Kosmos-215	USSR	248 ×	350	48	1968 Jun 30	UVA/XRA*	—
1968 Apr 26	Kosmos-219	USSR	214 ×	1647	48	1969 Mar 2	P/F*	—
1968 May 17	ESRO 2B	I/ESRO	329 ×	1075	97	1971 May 8	SX/P*	—
1968 Jun 4	Kosmos-224	USSR	200 ×	254	51	1968 Jun 12	ATM	—
1968 Jun 11	Kosmos-225	USSR	250 ×	492	48	1968 Nov 2	P*	Cosmic rays, SAA
1968 Jun 21	Kapsula Kosmosa-228	USSR	194 ×	213	51	1968 Jul 1	P	Recon
1968 Jul 4	RAE 1 (Explorer 38)	USA	5840 ×	5862	120	1972 Dec 25	RFA*	Low frequency radio observatory
1968 Jul 5	Kosmos-230	USSR	278 ×	518	48	1968 Nov 2	SOL*	Solar X/UV
1968 Jul 11	OV1-15	USA	147 ×	1544	89	1968 Nov 6	ATM*	—
1968 Jul 11	LOADS	USA	142 ×	530	89	1968 Aug 19	ATM	Dense sphere
1968 Aug 8	Explorer 39	USA	684 ×	2514	80	1981 Jun 22	ATM*	Air density balloon
1968 Aug 8	Injun 5 (Explorer 40)	USA	679 ×	2531	80	1970s?	P/F*	—
1968 Aug 27	Kosmos-237	USSR	197 ×	321	65	1968 Sep 4	P	Recon
1968 Sep 14	Zond-5	USSR	200 ×	385 000	51	1968 Sep 21	BIO/MM/PL	Spaceship, Circumlunar flight
1968 Sep 23	Kapsula Kosmosa-243	USSR	199 ×	274	71	1968 Oct 2	ATM	Microwave radiometer
1968 Sep 26	OV2-5	USA	35 064 ×	35 798	2	1968 Sep?	ION/P/F*	—
1968 Sep 26	ERS 28	USA	174 ×	34 315	25	1971 Feb 15	P/F*	—
1968 Oct 3	Aurorae	I/ESRO	258 ×	1490	93	1970 Jun 26	A/P*	—
1968 Oct 25	Soyuz-2	USSR	177 ×	196	51	1968 Oct 28	P	Spaceship
1968 Oct 26	Soyuz-3	USSR	180 ×	210	51	1968 Oct 30	ATM	Spaceship
1968 Nov 8	Pioneer 9	USA	Solar orbit			1983 May 18	P/F*	Solar wind monitor
1968 Nov 10	Zond-6	USSR	200 ×	400 000	51	1968 Nov 17	BIO/MM/PL	Spaceship, Circumlunar flight
1968 Nov 16	Proton-4	USSR	249 ×	447	51	1969 Jul 24	P*	Cosmic rays
1968 Nov 30	Kosmos-256	USSR	1173 ×	1225	74	1969?	GEO	—
1968 Dec 5	HEOS 1	I/ESRO	12 244 ×	210 550	52	1975 Oct 28	P/F*	Magnetosphere study
1968 Dec 7	OAO 2	USA	768 ×	778	35	1973 Feb 14	UVA*	Four 0.3 m telescopes
1968 Dec 14	Kosmos-259	USSR	212 ×	1210	48	1969 May 5	P/F*	—
1968 Dec 19	Kosmos-261	USSR	201 ×	611	71	1969 Feb 12	A/P*	Auroral particles at solar max
1968 Dec 26	Kosmos-262	USSR	255 ×	748	48	1969 Jul 18	SOL*	Solar UV/X
1969 Jan 5	Venera-5	USSR	Venus impact			1969 May 16	PL	Venus lander
1969 Jan 5	SA	USSR	Venus entry			1969 May 16	PL	Venus lander capsule
1969 Jan 10	Venera-6	USSR	Venus impact			1969 May 17	PL	Venus lander
1969 Jan 10	SA	USSR	Venus entry			1969 May 17	PL	Venus lander capsule
1969 Jan 14	Soyuz-4	USSR	206 ×	224	51	1969 Jan 17	P/ATM	Spaceship, DRV
1969 Jan 15	Soyuz-5	USSR	205 ×	227	51	1969 Jan 18	P/ATM	Spaceship, DRV
1969 Jan 22	OSO 5	USA	539 ×	563	32	1984 Apr 2	SOL*	Solar X/UV

1687

Launch	Name	State	Peri ×	Apo (km) ×	Inc.	Operated until		Pri mission discipline	Comments
1969 Jan 23	Kosmos-264	USSR	206 ×	297 ×	69	1969 Feb 5	Recon	GRA	Piggback gamma ray detector
1969 Jan 30	Isis 1	Canada	575 ×	3525 ×	88	1984 Mar		ION/RFA/P*	Ionospheric studies
1969 Feb 25	Mariner 6	USA	Solar orbit			1970 Dec		PL	Mars flyby
1969 Mar 17	Kosmos-272	USSR	1179 ×	1210 ×	73	1969?		GEO	–
1969 Mar 18	OV1-17	USA	386 ×	443 ×	99	1970 Mar 5		P/F/SX* study	Solar-terrestrial
1969 Mar 18	OV1-18	USA	461 ×	579 ×	98	1972 Aug 28		P/F*	Ionosphere
1969 Mar 18	OV1-19	USA	467 ×	5776 ×	104	1970?		P/F*	Trapped radiation
1969 Mar 18	OV1-17A	USA	176 ×	360 ×	99	1969 Mar 24		ION*	Ionosphere
1969 Mar 27	Mariner 7	USA	Solar orbit			1970 May 25		PL	Mars flyby
1969 Apr 14	Secor 13	USA	1072 ×	1131 ×	99	1980s?		GEO	–
1969 Apr 23	Kosmos-280	USSR	206 ×	246 ×	51	1969 May 6	Recon	P	–
1969 May 23	ERS 29	USA	16995 ×	111712 ×	33	1976 Aug		P/F*	VLF and particles
1969 May 23	ERS 26	USA	16977 ×	111583 ×	33	1976 Aug		SX/P*	Solar flare particles
1969 Jun 5	OGO 6	USA	396 ×	1085 ×	82	1979 Oct 12		P/F/SOL*	Geophysics
1969 Jun 21	IMP 5 (Explorer 41)	USA	640 ×	176219 ×	87	1972 Dec 23		P/F*	Solar wind
1969 Jun 29	Biosatellite 3	USA	352 ×	388 ×	33	1970 Jan 20		BIO	Monkey flight
1969 Jul 13	Luna-15 KT	USSR	Lunar impact			1969 Jul 21		PL	Lunar sample return, failed
1969 Jul 13	Luna-15 VA	USSR	Lunar impact			1969 Jul 21		PL	Lunar lander probe
1969 Jul 16	EASEP	USA	Lunar landing			1969 Aug 27		PL/P	Lunar science
1969 Aug 7	Zond-7	USSR	200 ×	400 000 ×	51	1969 Aug 14	Spaceship	BIO/P	Circumlunar flight
1969 Aug 9	OSO 6	USA	490 ×	556 ×	32	1981 Mar 7		SOL/XRA*	Solar observatory
1969 Aug 12	ATS 5	USA	35778 ×	35 791 ×	2	1984	Comms	P/F	Communications; particles
1969 Sep 23	Kosmos-300	USSR	175 ×	179 ×	51	1969 Sep 27		PL	Lunar sample return, failed
1969 Sep 23	VA	USSR	175 ×	179 ×	51	1969 Sep 27		PL	Lunar lander probe
1969 Oct 1	Boreas	I/ESRO	289 ×	378 ×	85	1969 Nov 23		A/P*	Auroral studies
1969 Oct 14	Interkosmos-1	USSR	250 ×	528 ×	48	1970 Jan 2		SOL*	Solar UV/X
1969 Oct 22	Kosmos-305	USSR	170 ×	210 ×	51	1969 Oct 24		PL	Lunar sample return, failed
1969 Oct 22	VA	USSR	170 ×	210 ×	51	1969 Oct 24		PL	Lunar return capsule
1969 Nov 8	Azur	Germany	388 ×	3142 ×	102	1970 Jun 29		A/P/F*	Aurora and radiation belts
1969 Nov 14	ALSEP	USA	Lunar landing			1977 Sep 30		PL	Lunar science
1969 Nov 24	Kosmos-312	USSR	1142 ×	1177 ×	74	1970?		GEO	–
1969 Dec 25	Interkosmos-2	USSR	200 ×	1078 ×	48	1970 Jun 7		ION/P*	Ionospheric research
1970 Jan 16	Kosmos-320	USSR	240 ×	298 ×	48	1970 Feb 10		ATM*	Atmospheric density
1970 Jan 20	Kosmos-321	USSR	259 ×	417 ×	70	1970 Mar 23		ION/F*	Ionosphere/fields

CATALOG OF SPACE SCIENCE LAUNCHES

Date	Name	Country				Date 2			Purpose
1970 Mar 4	OPS 0440	USA	180 ×	252 ×	88	1970 Mar 26	Recon	GEO	Doppler geodetic beacon
1970 Mar 10	Wika	France	313 ×	1607 ×	5	1978 Oct 5		P/F*	Ionospheric study
1970 Apr 8	TOPO 1	USA	1085 ×	1090 ×	99	1970s?		GEO	—
1970 Apr 11	ALSEP	USA	Cislunar orbit			1970 Apr 17		PL	Lunar science, failed
1970 Apr 24	Kosmos-335	USSR	247 ×	391 ×	48	1970 Jun 22		UVA*	UV background?
1970 May 20	OPS 4720	USA	171 ×	232 ×	83	1970 Jun 17	Recon	GEO	—
1970 Jun 1	Soyuz-9	USSR	236 ×	250 ×	51	1970 Jun 19	Spaceship	ATM/BIO/P	Long duration flight
1970 Jun 13	Kosmos-348	USSR	199 ×	589 ×	71	1970 Jul 25		A/P*	Aurora/ionosphere
1970 Aug 7	Interkosmos-3	USSR	201 ×	1154 ×	48	1970 Dec 6		P/F*	Trapped radiation
1970 Aug 10	Kosmos-356	USSR	226 ×	548 ×	81	1970 Oct 2		ION*	Ionospheric sounding
1970 Aug 17	Venera-7	USSR	Venus entry			1970 Dec 15		PL	Venus lander
1970 Aug 17	SA	USSR	Venus landing			1970 Dec 15		PL	Venus lander
1970 Aug 22	Kosmos-359	USSR	203 ×	685 ×	51	1970 Nov 6	DRV	PL	Venus lander capsule
1970 Sep 12	Luna-16 KT	USSR	Lunar landing			1970 Sep 20	DRV	PL	Lunar sample return
1970 Sep 12	Luna-16 VA	USSR	Cislunar orbit			1970 Sep 24	DRV	PL	Lunar probe
1970 Oct 8	Kosmos-368	USSR	205 ×	394 ×	64	1970 Oct 14		BIO	Radiation effects
1970 Oct 14	Interkosmos-4	USSR	254 ×	561 ×	48	1971 Jan 17		SOL*	Solar X/UV
1970 Oct 20	Zond-8	USSR	200 ×	400 000 ×	51	1970 Oct 27	Spaceship	PL/P	Lunar photos
1970 Nov 9	OFO	USA	287 ×	506 ×	37	1971 May 9		BIO	Frog experiment
1970 Nov 9	RM	USA	289 ×	475 ×	37	1971 Feb 7		P/MM	—
1970 Nov 10	Luna-17	USSR	Lunar landing			1970 Nov 17		PL	Lunar rover
1970 Nov 10	Lunokhod-1	USSR	Lunar landing			1970 Nov 17		PL	Lunar rover
1970 Nov 17	Kosmos-378	USSR	233 ×	1697 ×	74	1972 Aug 17		P*	Radiation study
1970 Nov 18	OPS 4992	USA	175 ×	216 ×	83	1970 Dec 11	Recon	GEO	Geodetic beacon
1970 Nov 24	Kosmos-379	USSR	171 ×	13 948 ×	51	1983 Sep 21	Spaceship	P	Lunar lander test
1970 Dec 2	Kosmos-381	USSR	966 ×	1012 ×	74	1971 Jan?		ION*	Ionosphere
1970 Dec 11	CEPE	USA	1423 ×	1474 ×	101	1970 Dec 11		P	—
1970 Dec 12	Uhuru (Explorer 42)	USA	534 ×	573 ×	3	1979 Apr 5		XRA*	X-ray sky survey
1970 Dec 12	Peole	France	509 ×	742 ×	15	1980 Jun 16		GEO	—
1971 Jan 31	ALSEP	USA	Lunar landing			1977 Sep 30		PL	Lunar science
1971 Feb 26	Kosmos-398	USSR	203 ×	10 870 ×	51	1995 Dec 10	Spaceship	P	Lunar lander test
1971 Mar 3	Shi Jian	China	264 ×	1799 ×	69	1979 Jun 17		P/F*	Test satellite
1971 Mar 13	IMP 6 (Explorer 43)	USA	1569 ×	203 825 ×	30	1974 Oct 2		P/F*	Magnetosphere/solar wind
1971 Mar 24	OPS 5300	USA	169 ×	226 ×	81	1971 Apr 12	Recon	GEO	—
1971 Apr 1	Isis 2	Canada	1356 ×	1428 ×	88	1984 Mar		ION/P/F/RFA*	Ionosphere
1971 Apr 15	Tournesol	France	457 ×	696 ×	46	1980 Jan 28		UVA*	Geocorona
1971 Apr 19	Salyut	USSR	223 ×	234 ×	51	1971 Oct 11	Spaceship	GRA/UVA/BIO	Space station
1971 Apr 22	Soyuz-10	USSR	204 ×	232 ×	51	1971 Apr 24	Spaceship	BIO	—
1971 Apr 24	San Marco 3	Italy	224 ×	678 ×	3	1971 Nov 29		ATM*	Air density
1971 Apr 28	Kosmos-409	USSR	1176 ×	1213 ×	74	1972?		GEO	—

Launch	Name	State	Peri ×	Apo (km) ×	Inc.	Operated until		Pri mission discipline	Comments
1971 May 6	Kapsula Kosmosa-410	USSR	195 ×	257 ×	64	1971 May 17		P	—
1971 May 10	Kosmos-419	USSR	126 ×	191 ×	51	1971 May 12		PL	Mars probe, failed
1971 May 19	Mars-2	USSR	Mars orbit			1972 Aug 22		PL	Mars orbiter
1971 May 19	SA	USSR	Mars impact			1971 Nov 27	DRV	PL	Mars lander capsule
1971 May 28	Mars-3	USSR	Mars orbit			1972 Aug?		PL	Mars orbiter
1971 May 28	SA	USSR	Mars landing			1971 Dec 2	DRV	PL	Mars lander capsule
1971 May 30	Mariner 9	USA	Mars orbit			1972 Oct 27		PL	Mars orbiter
1971 Jun 4	Kosmos-426	USSR	388 ×	1994 ×	74	1972 Jan 12		P*	Radiation studies
1971 Jun 6	Soyuz-11	USSR	250 ×	258 ×	51	1971 Jun 29	Spaceship	BIO	—
1971 Jun 8	P70-1	USA	542 ×	579 ×	90	1982 Jan 31		IRA	Military IR background survey
1971 Jun 24	Kapsula Kosmosa-428	USSR	199 ×	239 ×	51	1971 Jul 5		XRA	Hard X-ray bursts
1971 Jul 8	Solrad 10 (Explorer 44)	USA	435 ×	631 ×	51	1979 Dec 15		SOL*	Solar flux
1971 Jul 26	Endeavour	USA	Lunar orbit			1971 Aug 7	Spaceship	PL	Lunar remote sensing
1971 Jul 26	Apollo 15 Subsatellite	USA	Lunar orbit			1973 Aug 23		P/F	Lunar satellite
1971 Jul 26	ALSEP	USA	Lunar landing			1977 Sep 30		PL	Lunar science
1971 Aug 7	LOADS 2	USA	127 ×	1762 ×	92	1972 Jan 31		ATM	Air density
1971 Aug 7	RTDS	USA	131 ×	811 ×	87	1971 Sep 19		ATM	Air density
1971 Aug 7	OV1-21	USA	787 ×	915 ×	87	1971?		ATM/F*	Atm. composition and density
1971 Aug 7	AVL-802 Grid Sphere 7-1	USA	788 ×	916 ×	87	1971 Aug 7		ATM	—
1971 Aug 7	AVL-802 Grid Sphere 7-2	USA	778 ×	913 ×	87	1979 Mar 18		ATM	—
1971 Aug 7	AVL-802 Rigid Sphere	USA	778 ×	915 ×	87	1979 Nov 2		ATM	—
1971 Aug 7	AVL-802 Mylar Sphere	USA	784 ×	916 ×	87	1981 Sep 1		ATM	—
1971 Aug 12	Kosmos-434	USSR	188 ×	11 777 ×	51	1981 Aug 23	Spaceship	P	—
1971 Sep 2	Luna-18 KT	USSR	Lunar impact			1971 Sep 11		PL	Lunar sample return, failed
1971 Sep 2	Luna-18 VA	USSR	Cislunar orbit			1971 Sep 11		PL	Lunar probe
1971 Sep 10	OPS 5454	USA	146 ×	225 ×	74	1971 Oct 5	Recon	GEO	—
1971 Sep 28	Shinsei	Japan	869 ×	1867 ×	32	1972 Jan?		P/RFA*	Particles and solar radio bursts

CATALOG OF SPACE SCIENCE LAUNCHES

Date	Name	Country	Orbit			Launch date	Type	Payload	Purpose
1971 Sep 28	Luna-19	USSR	Lunar orbit					PL	Lunar orbiter
1971 Sep 29	OSO 7	USA	324 ×	569 ×	33	1972 Oct 20		SOL/XRA*	Solar physics and X-ray survey
1971 Oct 7	Kapsula Kosmosa-443	USSR	198 ×	288 ×	65	1971 Oct 19		P	Electron flux
1971 Oct 17	P71-2	USA	772 ×	800 ×	92	1974?		IRA?	Military IR mapping?
1971 Oct 28	Prospero	UK	545 ×	1580 ×	82	1989		MM*	Only satellite launched by UK
1971 Nov 15	SSS 1 (Explorer 45)	USA	234 ×	26780 ×	3	1992 Jan 10		P/F*	Magnetosphere
1971 Nov 20	Kosmos-457	USSR	1183 ×	1220 ×	74	1972?		GEO	–
1971 Dec 2	Interkosmos-5	USSR	198 ×	1044 ×	48	1972 Apr 7		P/F*	Magnetosphere
1971 Dec 2	Kosmos-461	USSR	486 ×	508 ×	69	1979 Feb 21		GRA*	Gamma ray bursts?
1971 Dec 11	Ariel 4	UK	473 ×	590 ×	82	1978 Dec 12		P/F/ATM*	Ionosphere
1971 Dec 27	Aureole 1	USSR	399 ×	2474 ×	73	1972 Sep 1		A/P*	Auroral particles
1972 Jan 31	HEOS 2	I/ESRO	994 ×	239 335 ×	90	1974 Aug 2		P/F/MM*	Magnetosphere
1972 Feb 14	Luna-20 VA	USSR	Cislunar orbit			1972 Feb 25	DRV	PL	Lunar sample return
1972 Feb 14	Luna-20 KT	USSR	Lunar landing			1972 Feb 21		PL	Lunar lander
1972 Mar 3	Pioneer 10	USA	Solar escape orbit			Active		Jupiter flyby	
1972 Mar 4	Kapsula Kosmosa-477	USSR	197 ×	291 ×	72	1972 Mar 14		P	Electron flux
1972 Mar 12	TD-1A	I/ESRO	525 ×	544 ×	97	1980 Jan 9		UVA*	UV telescope
1972 Mar 25	Kosmos-480	USSR	1173 ×	1201 ×	82	1973?		GEO	–
1972 Mar 27	Venera-8	USSR	Venus entry			1972 Jul 22	DRV	PL	Venus lander
1972 Mar 27	SA	USSR	Venus landing			1972 Jul 22		PL	Venus probe
1972 Mar 31	Kosmos-482	USSR	205 ×	9732 ×	52	1981 May 5	DRV	PL	Venus lander
1972 Apr 6	Kapsula Kosmosa-484	USSR	168 ×	177 ×	81	1972 Apr 18		A/P	Auroral particles
1972 Apr 7	Interkosmos-6	USSR	202 ×	244 ×	51	1972 Apr 11		P*	Cosmic rays
1972 Apr 14	Prognoz	USSR	468 ×	200 002 ×	64	1981 Mar 31		P/F/SX*	Magnetosphere
1972 Apr 16	Casper	USA	Lunar orbit			1972 Apr 27	Spaceship	BIO/UVA/PL	Lunar remote sensing
1972 Apr 16	Apollo 16 Subsatellite	USA	Lunar impact			1972 May 29		P/F	Lunar particles and fields
1972 Apr 16	ALSEP	USA	Lunar landing			1977 Sep 30		PL	Lunar science
1972 Apr 19	OPS 5640	USA	156 ×	247 ×	81	1972 May 12	Recon	ATM	Atmospheric density
1972 May 17	Kapsula Kosmosa-490	USSR	209 ×	258 ×	65	1972 May 27		P	Trapped and cosmic ray electrons
1972 May 25	OPS 6371	USA	157 ×	315 ×	96	1972 Jun 4	Recon	ATM	Atmospheric density
1972 Jun 29	Prognoz-2	USSR	222 ×	447 ×	64	1982 Dec 15		P/F/SX*	Magnetosphere
1972 Jun 30	Interkosmos-7	USSR	255 ×	508 ×	48	1972 Oct 5		SOL*	Solar UV/X
1972 Jul 13	Kosmos-502	USSR	200 ×	263 ×	65	1972 Jul 25	Recon	MM	–

Launch	Name	State	Peri ×	Apo (km) ×	Inc.	Operated until		Pri mission discipline	Comments
1972 Aug 13	Explorer 46	USA	494 ×	810 ×	37	1979 Nov 2		MM*	–
1972 Aug 19	Denpa	Japan	236 ×	6273 ×	31	1980 May 19		P/F*	Ionosphere/ magnetosphere
1972 Aug 21	OAO-3 Copernicus	USA	735 ×	747 ×	35	1981 Feb 15		UVA*	0.8-meter UV telescope
1972 Sep 15	Kapsula Kosmosa-518	USSR	196 ×	271 ×	72	1972 Sep 23		P?	–
1972 Sep 23	IMP 7 (Explorer 47)	USA	201100 ×	235 600 ×	17	1978 Oct 31		P/F/RFA*	Magnetosphere
1972 Oct 2	P72-1	USA	728 ×	749 ×	98	1973 May		GRA*	Gamma ray background
1972 Oct 18	Kapsula Kosmosa-525	USSR	191 ×	258 ×	65	1972 Oct 26		P?	–
1972 Nov 15	SAS 2 (Explorer 48)	USA	444 ×	632 ×	1	1980 Aug 20		GRA*	Gamma ray telescope
1972 Nov 22	ESRO 4	I/ESRO	240 ×	1124 ×	91	1974 Apr 15		A/P*	Ionosphere/aurora
1972 Nov 30	Interkosmos-8	USSR	197 ×	564 ×	71	1973 Mar 2		ION/P*	Ionosphere beacon
1972 Dec 7	America	USA	Lunar orbit			1972 Dec 19	Spaceship	BIO/UVA/PL	Lunar remote sensing
1972 Dec 7	ALSEP	USA	Lunar landing			1977 Sep 30		PL	Lunar science
1972 Dec 16	Aeros 1	Germany	216 ×	830 ×	96	1973 Aug 22		ATM/P*	–
1972 Dec 21	Kosmos-539	USSR	1342 ×	1381 ×	74	1974?		GEO	–
1973 Jan 8	Luna-21	USSR	Lunar landing			1973 Jan 15		PL	Lunar lander
1973 Jan 8	Lunokhod-2	USSR	Lunar landing			1973 Jan 16		PL	Lunar rover
1973 Feb 15	Prognoz-3	USSR	895 ×	199 442 ×	65	1976 Dec 31		SX/P*	Solar-terrestrial study
1973 Apr 6	Pioneer 11	USA	Solar escape orbit			1995 Nov		PL	Jupiter/Saturn flybys
1973 Apr 19	Interkosmos-9 Kopernik-500	USSR	193 ×	1343 ×	48	1973 Oct 15		P/RFA*	Solar radio bursts
1973 Apr 25	Kapsula Kosmosa-555	USSR	192 ×	204 ×	81	1973 May 4		P	–
1973 May 11	Kosmos-557	USSR	208 ×	237 ×	51	1973 May 22	Spaceship	P	Space station
1973 May 14	Skylab Orbital Workshop	USA	421 ×	444 ×	50	1979 Jul 11	Spaceship	SOL/ATM/ BIO	Space station
1973 May 25	Skylab SL-2	USA	425 ×	440 ×	50	1973 Jun 22	Spaceship	BIO	Station ferry
1973 Jun 10	RAE 2 (Explorer 49)	USA	Lunar orbit			1977 Apr 30		RFA*	LF radio astronomy
1973 Jul 21	Mars-4	USSR	Solar orbit			1974 Feb?		PL	Mars orbiter, failed
1973 Jul 25	Mars-5	USSR	Mars orbit			1974 Mar 1		PL	Mars orbiter
1973 Jul 28	Skylab SL-3	USA	425 ×	439 ×	50	1973 Sep 25	Spaceship	BIO	Station ferry
1973 Aug 5	Mars-6	USSR	Solar orbit			1974 Mar 12		PL	Mars lander, crashed
1973 Aug 5	SA Marsa-6	USSR	Mars impact			1974 Mar 12	DRV	PL	Mars lander capsule

CATALOG OF SPACE SCIENCE LAUNCHES

Date	Name	Country	Orbit			Arrival/End	Type	Category	Purpose
1973 Aug 9	Mars-7	USSR	Solar orbit			1974 Mar 9		PL	Mars lander, missed
1973 Aug 9	SA Marsa-7	USSR	Solar orbit			1974 Mar	DRV	PL	Mars lander capsule
1973 Sep 8	Kosmos-585	USSR	1374 ×	1406 ×	74	1974?		GEO	Station ferry test
1973 Sep 27	Soyuz-12	USSR	329 ×	341 ×	51	1973 Sep 29	Spaceship	ATM	Magnetosphere/solar wind
1973 Oct 26	IMP 8 (Explorer 50)	USA	215 092 ×	215 092 ×	19	2001 Oct 26		P/F/RFA*	Ionosphere
1973 Oct 30	Interkosmos-10	USSR	257 ×	1439 ×	74	1977 Jul 1		P/F*	—
1973 Oct 31	Kosmos-605	USSR	212 ×	397 ×	62	1973 Nov 22		BIO	Venus/Mercury probe
1973 Nov 3	Mariner 10	USA	Solar orbit			1975 Mar 24		PL	Station ferry
1973 Nov 16	Skylab SL-4	USA	420 ×	444 ×	50	1974 Feb 8	Spaceship	UVA, BIO	Seeds exposure
1973 Nov 30	Kosmos-613	USSR	251 ×	383 ×	51	1974 Jan 29	Spaceship	BIO	Atmospheric studies
1973 Dec 16	AE-C (Explorer 51)	USA	152 ×	4074 ×	68	1978 Dec 12		ATM*	Kohoutek observations
1973 Dec 18	Soyuz-13	USSR	221 ×	253 ×	51	1973 Dec 26	Spaceship	UVA, BIO	Auroral
1973 Dec 26	Aureole 2	USSR	399 ×	1971 ×	74	1974 Apr 30		A/P*	Atmospheric composition/density
1974 Feb 18	San Marco 4	Italy	236 ×	898 ×	2	1976 May 4		ATM*	—
1974 Apr 29	Kosmos-650	USSR	1368 ×	1401 ×	74	1975?		GEO	Solar UV and polarization
1974 May 17	Interkosmos-11	USSR	483 ×	511 ×	50	1979 Sep 6		SOL*	Lunar orbiter
1974 May 29	Luna-22	USSR	Lunar orbit			1975 Nov		PL	Neutral field points
1974 Jun 3	Hawkeye 1 (Explorer 52)	USA	4431 ×	122 973 ×	88	1978 Apr 30		P/F*	Space station
1974 Jun 24	Salyut-3	USSR	264 ×	270 ×	51	1975 Jan 24	Spaceship	BIO	Atmospheric studies
1974 Jul 16	Aeros 2	Germany	218 ×	829 ×	97	1975 Sep 25		P/ATM*	Dutch astronomical observatory
1974 Aug 29	Kosmos-675	USSR	1365 ×	1423 ×	74	1975?		GEO	British X-ray sky survey
1974 Aug 30	ANS	Netherlands	256 ×	1153 ×	98	1977 Jun 14		UVA/XRA*	
1974 Oct 15	Ariel 5	UK	506 ×	552 ×	2	1980 Mar 14		XRA*	
1974 Oct 22	Kosmos-690	USSR	362 ×		62			BIO	Lunar lander
1974 Oct 28	Luna-23 KT	USSR	Lunar landing			1974 Nov 12		PL	Lunar sample return, failed
1974 Oct 28	Luna-23 VA	USSR	Cislunar orbit			1974 Nov 6	DRV	PL	Atmosphere/ionosphere
1974 Oct 29	S3-1	USA	149 ×	3483 ×	96	1975 May 26		ATM/P/F*	Ionosphere
1974 Oct 31	Interkosmos-12	USSR	240 ×	662 ×	74	1975 Jul 11		ION/MM/P*	Ionosphere beacon
1974 Nov 15	Intasat	Spain	1442 ×	1460 ×	101	1976 Oct 6		ION*	Solar probe
1974 Dec 10	Helios 1	Germany	Solar orbit			1980s?		P/F/MM*	Space station
1974 Dec 26	Salyut-4	USSR	336 ×	347 ×	51	1977 Feb 2	Spaceship	ATM/XRA/SOL	Station ferry
1975 Jan 10	Soyuz-17	USSR	336 ×	348 ×	51	1975 Feb 9	Spaceship	BIO	—
1975 Feb 6	Starlette	France	804 ×	1108 ×	49	1975 Feb 6		GEO	

Launch	Name	State	Peri ×	Apo (km) ×	Inc.	Operated until		Pri mission discipline	Comments
1975 Feb 12	Kosmos-708	USSR	1369 ×	1410 ×	69	1976?		GEO	–
1975 Feb 24	Taiyo	Japan	249 ×	3115 ×	31	1980 Jun 29		P/SOL*	Solar-terrestrial studies
1975 Mar 27	Interkosmos-13	USSR	285 ×	1679 ×	82	1980 Sep 2		P/F*	Polar ionosphere
1975 Apr 9	Geos 3	USA	821 ×	855 ×	114	1981 May		ALT/GEO*	
1975 Apr 19	Aryabhata	India	568 ×	611 ×	50	1992 Feb 11		XRA/P*	First Indian satellite
1975 May 7	SAS 3 (Explorer 53)	USA	502 ×	509 ×	3	1979 Apr 9		XRA*	X-ray source location
1975 May 24	Soyuz-18	USSR	334 ×	348 ×	51	1975 Jul 26	Spaceship	BIO	Station ferry
1975 Jun 8	Venera-9	USSR	Venus orbit			1976 Jun 21		PL	Venus orbiter
1975 Jun 8	SA Veneri-9	USSR	Venus landing			1975 Oct 22		PL	Venus lander
1975 Jun 14	Venera-10	USSR	Venus orbit			1977 Mar 16		PL	Venus orbiter
1975 Jun 14	SA Veneri-10	USSR	Venus landing			1975 Oct 25		PL	Venus lander
1975 Jun 21	OSO 8	USA	540 ×	556 ×	32	1986 Jul 9		SOL/XRA*	Solar observatory
1975 Jul 15	Soyuz-19	USSR	217 ×	229 ×	51	1975 Jul 21	Spaceship	BIO	Docking with Apollo
1975 Jul 15	Apollo-Soyuz Test Project	USA	162 ×	224 ×	51	1975 Jul 24	Spaceship	BIO/ATM/UVA	Docking with Soyuz
1975 Aug 9	COS-B	I/ESA	2536 ×	96895 ×	92	1986 Jan 18		GRA*	Gamma-ray survey
1975 Aug 20	Viking Orbiter 1	USA	Mars orbit			1980 Aug 7		PL	Mars orbiter
1975 Aug 20	Mutch Memorial Station	USA	Mars landing			1976 Jul 20		PL	Mars lander
1975 Sep 9	Viking Orbiter 2	USA	Mars orbit			1978 Jul 24		PL	Mars orbiter
1975 Sep 9	Viking Lander 2	USA	Mars landing			1976 Sep 3		PL	Mars lander
1975 Sep 24	Kosmos-770	USSR	1165 ×	1207 ×	82	1976?		GEO	
1975 Sep 27	Aura	France	501 ×	712 ×	37	1982 Sep 30		UVA*	UV astronomy
1975 Oct 6	AE-D (Explorer 54)	USA	135 ×	3569 ×	90	1976 Mar 12		ATM/P*	Atmospheric studies
1975 Nov 17	Soyuz-20	USSR	340 ×	346 ×	51	1976 Feb 16	Spaceship	BIO	–
1975 Nov 20	AE-E (Explorer 55)	USA	142 ×	2819 ×	19	1981 Jun 10		ATM*	–
1975 Nov 21	Kapsula Kosmosa-780	USSR	197 ×	268 ×	65	1975 Nov 30		P?	
1975 Nov 25	Kosmos-782	USSR	216 ×	380 ×	62	1975 Dec 15		BIO	–
1975 Dec 4	S3-2	USA	233 ×	1544 ×	96	1978 May 1		P/F*	–
1975 Dec 11	Interkosmos-14	USSR	333 ×	1680 ×	73	1983 Feb 27		ION/P*	–
1975 Dec 22	Prognoz-4	USSR	231 ×	464 ×	65	1977 Dec 31		P/F*	–
1976 Jan 15	Helios 2	Germany	Solar orbit			1980 May 15		PL	Solar probe
1976 Feb 29	Ume	Japan	991 ×	1006 ×	69	1976 Apr 2		ION*	–
1976 Mar 15	SR 11A	USA	118 383 ×	119 180 ×	25	1977 Jun 12		SOL*	–
1976 Mar 15	SR 11B	USA	115 720 ×	116 645 ×	25	1979 Oct 31		SOL*	–
1976 May 4	Lageos	USA	5834 ×	5951 ×	109	1976 May 4		GEOD*	–
1976 Jul 8	S3-3	USA	232 ×	6038 ×	97	1986 Apr 24		P/F*	–
1976 Jul 21	Kosmos-842	USSR	966 ×	1007 ×	82	1978?		GEO	–

CATALOG OF SPACE SCIENCE LAUNCHES

Date	Name	Country	Orbit	Incl.	Status	Type	Purpose
1976 Jul 27	Interkosmos-16	USSR	464 × 518 ×	50	1979 Jul 10	SOL*	Solar photometer
1976 Aug 9	Luna-24 KT	USSR	Lunar landing		1976 Aug 18	PL	Lunar lander
1976 Aug 9	Luna-24 VA	USSR	Cislunar orbit		1976 Aug 23	PL	Lunar sample return
1976 Sep 22	Kapsula Kosmosa-856	USSR	199 × 290 ×	65	1976 Oct 3	GRA	Gamma spectrometer
1976 Nov 25	Prognoz-5	USSR	3869 × 195485 ×	66	1979 Jul 12	P*	
1977 Mar 29	Kosmos-900	USSR	455 × 519 ×	82	1979 Oct 11	A/P/F*	
1977 Apr 20	Geos 1	I/ESA	2083 × 38294 ×	26	1980 Jan?	P/F*	
1977 May 25	Kosmos-911	USSR	964 × 1001 ×	82	1979?	GEO	
1977 May 31	Kapsula Kosmosa-914	USSR	200 × 291 ×	65	1977 Jun 12	GRA	Gamma spectrometer
1977 Jun 17	Signe 3	France	458 × 518 ×	50	1979 Jun 20	GRA*	
1977 Aug 3	Kosmos-936	USSR	215 × 392 ×	62	1977 Aug 22	BIO	
1977 Aug 12	HEAO 1	USA	428 × 449 ×	22	1979 Mar 15	XRA*	X-ray survey
1977 Aug 20	Voyager 2	USA	Solar escape orbit		Active	PL	Outer planet probe
1977 Sep 5	Voyager 1	USA	Solar escape orbit		Active	PL	Outer planet probe
1977 Sep 22	Prognoz-6	USSR	1850 × 196379 ×	74	1978 Jan 26	UVA/GRB/P*	
1977 Sep 24	Interkosmos-17	USSR	465 × 509 ×	82	1979 Nov 8	P*	
1977 Oct 22	ISEE 1	USA	764 × 137531 ×	30	1987 Sep 26	P/F*	
1977 Oct 22	ISEE 2	I/ESA	780 × 137511 ×	30	1987 Sep 26	P/F*	
1977 Nov 24	Kosmos-963	USSR	1178 × 1207 ×	82	1979?	GEO	
1977 Dec 10	Soyuz-26	USSR	333 × 352 ×	51	1978 Jan 16	MG/BIO	Spaceship
1978 Jan 20	Progress-1	USSR	324 × 344 ×	51	1978 Feb 8	MG	Spaceship
1978 Jan 26	IUE	USA	25687 × 45892 ×	28	1996 Sep 30	UVA*	UV spectra
1978 Feb 4	Kyokko	Japan	642 × 3953 ×	65	1979 Nov 9	A/P/F*	
1978 Feb 16	Ume-2	Japan	974 × 1220 ×	69	1980s?	ION*	
1978 Mar 2	Soyuz-28	USSR	334 × 352 ×	51	1978 Mar 10	BIO	Spaceship
1978 Apr 26	HCMM	USA	610 × 633 ×	97	1981 Dec 22	ATM*	
1978 May 20	Pioneer Venus Orbiter	USA	Venus orbit		1992 Oct 22	PL	Venus orbiter
1978 Jun 27	Seasat 1	USA	775 × 797 ×	108	1978 Oct	ALT*	
1978 Jun 27	Soyuz-30	USSR	255 × 305 ×	51	1978 Jul 5	MG/BIO	Spaceship
1978 Jul 2	Kosmos-1026	USSR	206 × 244 ×	51	1978 Jul 6	P*	Cosmic rays
1978 Jul 7	Progress-2	USSR	327 × 330 ×	51	1978 Aug 4	MG/BIO	Spaceship
1978 Jul 14	Geos 2	I/ESA	35 801 × 35 768 ×	0	1984 Jan 25	P/F*	Magnetosphere study
1978 Aug 7	Progress-3	USSR	335 × 352 ×	51	1978 Aug 23	BIO	Spaceship
1978 Aug 8	Pioneer Venus Multiprobe	USA	Venus entry		1978 Dec 9	PL	Venus probe
1978 Aug 8	Sounder Probe	USA	Venus impact		1978 Dec 9	PL	
1978 Aug 8	North Probe	USA	Venus impact		1978 Dec 9	PL	
1978 Aug 8	Day Probe	USA	Venus landing		1978 Dec 9	PL	
1978 Aug 8	Night Probe	USA	Venus impact		1978 Dec 9	PL	

Launch	Name	State	Peri ×	Apo (km) ×	Inc.	Operated until		Pri mission discipline	Comments
1978 Aug 12	ISEE 3 (ICE)	USA	Earth-Sun L1			1997 May 5		P/F*	L1 and comet probe
1978 Sep 9	Venera-11	USSR	Solar orbit			1980 Feb		PL	Venus probe
1978 Sep 9	SA Veneri-11	USSR	Venus landing			1978 Dec 25		PL	Venus lander
1978 Sep 14	Venera-12	USSR	Solar orbit			1980 Apr		PL	Venus probe
1978 Sep 14	SA Veneri-12	USSR	Venus landing			1978 Dec 21		PL	Venus lander
1978 Sep 16	Jikiken	Japan	249 ×	28 876 ×	31	1980s		P/F*	–
1978 Oct 24	Interkosmos-18	USSR	402 ×	758 ×	82	1981 Mar 17		P/F*	–
1978 Oct 24	Magion	Czech SSR	402 ×	759 ×	82	1981 Sep 11		P/F*	–
1978 Oct 24	CAMEO	USA	926 ×	969 ×	99	1978 Nov 6		F*	Chemical release
1978 Oct 30	Prognoz-7	USSR	1127 ×	204 898 ×	65	1980 Sep 22		P/F/UVA/ GRA*	–
1978 Nov 13	Einstein Observatory	USA	520 ×	541 ×	23	1982 Mar 25		XRA*	First X-ray imaging observatory
1978 Dec 26	Kosmos-1067	USSR	1156 ×	1213 ×	82	1981?		GEO	–
1979 Jan 30	SCATHA	USA	27 580 ×	43 212 ×	7	1991 May 28		P/F*	–
1979 Feb 18	SAGE	USA	547 ×	659 ×	54	1989 Apr 11		ATM*	–
1979 Feb 21	Hakucho	Japan	540 ×	569 ×	29	1985 Apr 15		XRA*	–
1979 Feb 24	P78-1	USA	558 ×	601 ×	97	1992 Jul 20		SOL*	Solar observatory
1979 Feb 25	Soyuz-32	USSR	279 ×	307 ×	51	1979 Jun 13	Spaceship	BIO	–
1979 Feb 27	Interkosmos-19	USSR	474 ×	809 ×	73	1980s?		ION*	–
1979 Mar 12	Progress-5	USSR	278 ×	338 ×	51	1979 Apr 4	Spaceship	MG/GRA	–
1979 May 13	Progress-6	USSR	319 ×	337 ×	51	1979 Jun 9	Spaceship	MG/BIO	–
1979 May 25	Kapsula Kosmosa-1102	USSR	162 ×	180 ×	81	1979 Jun 6		P?	–
1979 Jun 2	Ariel 6	UK	599 ×	653 ×	55	1990 Sep 23		P/XRA*	–
1979 Jun 6	Soyuz-34	USSR	349 ×	355 ×	51	1979 Aug 19	Spaceship	BIO	–
1979 Jun 12	Kapsula Kosmosa-1106	USSR	197 ×	204 ×	81	1979 Jun 24		XRA	–
1979 Jun 28	Progress-7	USSR	323 ×	345 ×	51	1979 Jul 20	Spaceship	BIO	–
1979 Jun 28	KRT-10	USSR	359 ×	385 ×	51	1979 Aug 15		RFA	10-m radio telescope
1979 Sep 20	HEAO 3	USA	481 ×	497 ×	43	1981 Dec 7		GRA*	–
1979 Oct 30	Magsat	USA	347 ×	541 ×	96	1980 Jun 11		F*	–
1979 Nov 1	Interkosmos-20	USSR	454 ×	510 ×	74	1981 Mar 3		ATM*	–
1980 Feb 14	Solar Maximum Mission	USA	566 ×	569 ×	28	1989 Dec 2		SOL*	Solar observatory
1980 May 26	Soyuz-36	USSR	327 ×	341 ×	51	1980 Jul 31	Spaceship	MG/ATM	–
1980 Jun 29	Progress-10	USSR	321 ×	337 ×	51	1980 Jul 19	Spaceship	BIO	–
1980 Jul 23	Soyuz-37	USSR	324 ×	339 ×	51	1980 Oct 11	Spaceship	BIO/MG	–
1980 Sep 18	Soyuz-38	USSR	262 ×	313 ×	51	1980 Sep 26	Spaceship	MG/BIO/ ATM	–
1980 Sep 28	Progress-11	USSR	303 ×	318 ×	51	1980 Dec 11	Spaceship	F	–
1980 Dec 25	Prognoz-8	USSR	978 ×	197 369 ×	65	1984 Dec 28		P/F*	VLF experiment

Date	Name	Country	Orbit		Incl.	Reentry	Type	Class	Notes
1981 Feb 6	Interkosmos-21	USSR	467 ×	508 ×	74	1982 Jul 7		ATM*	—
1981 Feb 21	Hinotori	Japan	571 ×	636 ×	31	1991 Jul 11		XRA*	—
1981 Mar 22	Soyuz-39	USSR	335 ×	349 ×	51	1981 Mar 30	Spaceship	BIO/ATM/P	—
1981 Aug 3	Dynamics Explorer 1	USA	558 ×	23 300 ×	89	1991 Feb 28		P/F*	—
1981 Aug 3	Dynamics Explorer 2	USA	299 ×	990 ×	89	1983 Feb 19		P/F*	—
1981 Aug 7	IK Bulgaria-1300	USSR	793 ×	885 ×	81	1980s?		ION*	—
1981 Sep 19	SJ-2A	China	240 ×	1388 ×	59	1981 Sep 26		P/F*	—
1981 Sep 19	SJ-2B	China	233 ×	1527 ×	59	1982 Oct 6		ATM*	—
1981 Sep 19	SJ-2	China	231 ×	1503 ×	59	1982 Aug 17		P/F*	—
1981 Sep 21	Aureole 3	USSR	398 ×	1762 ×	82	1980s?		A/P/F*	—
1981 Sep 30	Kosmos-1312	USSR	1492 ×	1502 ×	82	1983?		GEO	—
1981 Oct 6	Solar Mesosphere Explorer	USA	481 ×	486 ×	97	1986 Dec 15?		ATM*	Mesosphere study
1981 Oct 30	Venera-13	USSR	Solar orbit			1982?		PL	Venus probe
1981 Oct 30	SA Veneri-13	USSR	Venus landing			1982 Mar 1		PL	Venus lander
1981 Nov 4	Venera-14	USSR	Solar orbit			1982?		PL	Venus probe
1981 Nov 4	SA Veneri-14	USSR	Venus landing			1982 Mar 5		PL	Venus lander
1981 Nov 12	Columbia	USA	254 ×	264 ×	38	1981 Nov 14	Spaceship	ATM	
1981 Nov 12	OSTA-1	USA	254 ×	264 ×	38	1981 Nov 14		ALT/ATM	SIR-A radar on Shuttle
1982 Mar 22	Columbia	USA	237 ×	245 ×	38	1982 Mar 30	Spaceship	BIO/MG	—
1982 Mar 22	OSS-1	USA	236 ×	249 ×	38	1982 Mar 30		ENV/P/SOL	Solar UV/X, plasma
1982 May 23	Progress-13	USSR	290 ×	346 ×	51	1982 Jun 6	Spaceship	MG/ BIO/OA	—
1982 Jun 24	Soyuz T-6	USSR	282 ×	305 ×	51	1982 Jul 2	Spaceship	MG/BIO/ XRA	—
1982 Jun 27	Columbia	USA	295 ×	304 ×	28	1982 Jul 4	Spaceship	BIO/MG	—
1982 Jul 10	Progress-14	USSR	303 ×	320 ×	51	1982 Aug 13	Spaceship	F	—
1982 Sep 18	Progress-15	USSR	299 ×	323 ×	51	1982 Oct 16	Spaceship	F/ATM	—
1982 Sep 24	Kosmos-1410	USSR	1493 ×	1502 ×	82	1984?		GEO	—
1982 Oct 31	Progress-16	USSR	352 ×	358 ×	51	1982 Dec 14	Spaceship	MG	—
1982 Nov 11	Columbia	USA	296 ×	318 ×	28	1982 Nov 16	Spaceship	BIO/MG/ENV	—
1983 Jan 26	IRAS	USA	893 ×	911 ×	99	1983 Nov		IRA*	IR sky survey
1983 Feb 20	Tenma	Japan	488 ×	501 ×	31	1988 Dec 17		XRA*	—
1983 Mar 23	Astron	USSR	25 129 ×	178 818 ×	79	1991 May 21		UVA*	—
1983 Apr 4	Challenger	USA	279 ×	289 ×	28	1983 Apr 9	Spaceship	MG/BIO/ATM	—
1983 May 26	Exosat	I/ESA	584 ×	191 510 ×	72	1986 May 6		XRA*	—
1983 Jun 2	Venera-15	USSR	Venus orbit			1985 Mar		PL	Venus radar
1983 Jun 7	Venera-16	USSR	Venus orbit			1985 Mar		PL	Venus radar
1983 Jun 18	Challenger	USA	296 ×	321 ×	28	1983 Jun 24	Spaceship	MG/BIO/UVA	—
1983 Jun 18	SPAS 1	Germany	291 ×	296 ×	28	1983 Jun 22		MG	—
1983 Jun 18	OSTA-2	USA	296 ×	321 ×	28	1983 Jun 22		MG	Materials experiments

Launch	Name	State	Peri ×	Apo (km) ×	Inc.	Operated until		Pri mission discipline	Comments
1983 Jul 1	Prognoz-9	USSR	380 ×	720000 ×	65	1980s?		SMA/GRA*	Big Bang study
1983 Aug 30	Challenger	USA	298 ×	309 ×	28	1983 Sep 5	Spaceship	P/MG/BIO	
1983 Oct 20	Progress-18	USSR	328 ×	344 ×	51	1983 Nov 16	Spaceship	F	VLF experiment
1983 Nov 24	Kosmos-1510	USSR	1480 ×	1524 ×	73	1985?		GEO	—
1983 Nov 28	Columbia	USA	241 ×	251 ×	57	1983 Dec 8	Spaceship	A/P	—
1983 Nov 28	Spacelab 1	I/ESA	242 ×	253 ×	57	1983 Dec 8		BIO/MG/UVA/P	First Spacelab
1983 Nov 28	Spacelab 1 Pallet	I/ESA	242 ×	253 ×	57	1983 Dec 8		ATM/P/UVA	—
1983 Dec 14	Kosmos-1514	USSR	212 ×	258 ×	82	1983 Dec 19		BIO	—
1984 Feb 3	Challenger	USA	305 ×	322 ×	28	1984 Feb 11	Spaceship	MG/BIO/P	—
1984 Feb 3	SPAS 1A	Germany	306 ×	321 ×	28	1984 Feb 11		MG	—
1984 Feb 14	Ohzora	Japan	356 ×	858 ×	74	1989 Jul 19		P/ATM*	Stratosphere/ mesosphere
1984 Feb 21	Progress-19	USSR	303 ×	305 ×	51	1984 Apr 1	Spaceship	ATM	—
1984 Mar 10	Kosmos-1543	USSR	216 ×	391 ×	62	1984 Apr 5		P*	Cosmic rays
1984 Apr 6	Challenger	USA	477 ×	481 ×	28	1984 Apr 13	Spaceship	BIO	—
1984 Apr 6	LDEF	USA	476 ×	478 ×	28	1990 Jan 20		ENV*	Materials exposure
1984 Aug 8	Kosmos-1589	USSR	1492 ×	1503 ×	82	1980s?		GEO	—
1984 Aug 16	Charge Composition Explorer	USA	1108 ×	49675 ×	4	1989 Jan		P/F*	Magnetosphere
1984 Aug 16	Ion Release Module	Germany	896 ×	113396 ×	27	1987 Dec 8		P/F*	Chemical release
1984 Aug 16	UK Subsatellite	UK	1127 ×	113291 ×	27	1988 Dec 8		P/F*	—
1984 Aug 30	Discovery	USA	296 ×	329 ×	28	1984 Sep 5	Spaceship	MG	—
1984 Oct 5	Challenger	USA	250 ×	261 ×	57	1984 Oct 13	Spaceship	MG/P	—
1984 Oct 5	ERBS	USA	596 ×	609 ×	57	Active		ATM*	—
1984 Oct 5	OSTA-3	USA	215 ×	228 ×	57	1984 Oct 13	Imaging	ALT/ATM	Imaging radar
1984 Nov 8	Discovery	USA	306 ×	313 ×	28	1984 Nov 16	Spaceship	MG/P	—
1984 Dec 15	Vega-1	USSR	Solar orbit			1987		PL	Comet probe
1984 Dec 15	SA Vega-1	USSR	Venus landing			1985 Jun 10		PL	Venus lander
1984 Dec 15	AZ Vega-1	USSR	Venus entry			1985 Jun 10		PL	Venus balloon
1984 Dec 21	Vega-2	USSR	Solar orbit			1987		PL	Comet probe
1984 Dec 21	SA Vega-2	USSR	Venus landing			1985 Jun 14		PL	Venus lander
1984 Dec 21	AZ Vega-2	USSR	Venus entry			1985 Jun 14		PL	Venus balloon
1985 Jan 7	Sakigake	Japan	Solar orbit			1999 Jan 7		PL	Comet probe
1985 Jan 24	Discovery	USA	334 ×	375 ×	28	1985 Jan 27	Spaceship	BIO/P/ATM	—
1985 Mar 13	Geosat 1	USA	784 ×	788 ×	108	1990 Jan 5		ALT*	—
1985 Apr 12	Discovery	USA	314 ×	461 ×	28	1985 Apr 19	Spaceship	MG/BIO	—
1985 Apr 16	Kosmos-1645	USSR	214 ×	388 ×	62	1985 Apr 29		MG	—
1985 Apr 26	Prognoz-10-IK	USSR	5975 ×	194737 ×	76	1994 Jan 12		P/F*	Magnetospheric study
1985 Apr 29	Spacelab 3	USA	345 ×	358 ×	56	1985 May 6	Spaceship	BIO/MG/UVA	—

CATALOG OF SPACE SCIENCE LAUNCHES   1699

Date	Name	Country	Orbit		Incl.	Reentry	Type	Payload	Notes
1985 Apr 29	SL 3 MPESS	USA	345 ×	358 ×	56	1985 May 6		ATM/P	Cosmic rays
1985 Jun 14	Kosmos-1660	USSR	1483 ×	1524 ×	73	1987?		GEO	—
1985 Jun 17	Discovery	USA	354 ×	380 ×	28	1985 Jun 24	Spaceship	MG/BIO	—
1985 Jun 17	Spartan 1	USA	354 ×	390 ×	28	1985 Jun 24		XRA*	—
1985 Jul 2	Giotto	I/ESA	Solar orbit			1992 Jul 23		PL	Comet probe
1985 Jul 10	Kosmos-1667	USSR	210 ×	268 ×	82	1985 Jul 17		BIO	—
1985 Jul 19	Kosmos-1669	USSR	353 ×	355 ×	51	1985 Aug 30		BIO	—
1985 Jul 29	Challenger	USA	311 ×	320 ×	49	1985 Aug 6	Spaceship	MG/BIO	—
1985 Jul 29	PDP	USA	310 ×	320 ×	49	1985 Aug 6		P/F	—
1985 Jul 29	Spacelab 2 PLT	USA	311 ×	320 ×	49	1985 Aug 6		SOL	Solar telescopes
1985 Jul 29	Spacelab 2 PLT	USA	311 ×	320 ×	49	1985 Aug 6		XRA	Hard X-ray imager
1985 Jul 29	Spacelab 2 PLT	USA	311 ×	320 ×	49	1985 Aug 6		IRA/P	IR telescope
1985 Jul 29	CRNE	USA	311 ×	320 ×	49	1985 Aug 6		P	Cosmic rays
1985 Aug 18	Suisei	Japan	Solar orbit			1991 Feb 22		PL	Comet probe
1985 Aug 27	Discovery	USA	351 ×	395 ×	28	1985 Sep 3	Spaceship	MG	—
1985 Oct 3	Atlantis	USA	476 ×	515 ×	28	1985 Oct 7	Spaceship	P/BIO	—
1985 Oct 30	Spacelab D-1	Germany	321 ×	333 ×	56	1985 Nov 6	Spaceship	MG/BIO	German Spacelab
1985 Oct 30	USS	Germany	321 ×	333 ×	56	1985 Nov 6		MG	—
1985 Nov 27	Atlantis	USA	368 ×	381 ×	28	1985 Dec 3	Spaceship	MG/BIO	—
1985 Dec 27	Kosmos-1713	USSR	216 ×	396 ×	62	1986 Jan 22		P*	—
1986 Jan 12	Columbia	USA	323 ×	346 ×	28	1986 Jan 18	Spaceship	MG/UVA	Halley monitor camera
1986 Jan 12	MSL-2	USA	323 ×	346 ×	28	1986 Jan 18		MG	—
1986 Jan 12	GBA-1	USA	323 ×	346 ×	28	1986 Jan 18		UVA/ATM	—
1986 Feb 11	Kosmos-1732	USSR	1479 ×	1526 ×	73	1980s?		GEO	—
1986 Feb 19	MAK-1	USSR	143 ×	151 ×	51	1991 Oct 18		ATM	—
1986 Feb 22	Viking	Sweden	821 ×	13 525 ×	98	1987 May 12		A/P/F*	—
1986 May 21	Kosmos-1744	USSR	218 ×	371 ×	62	1986 Jun 4		MG	—
1986 Aug 12	Ajisai	Japan	1479 ×	1497 ×	50	1986 Aug 12		GEO	—
1986 Dec 2	Kosmos-1803	USSR	1497 ×	1503 ×	82	1980s?		GEO	—
1986 Dec 18	Kosmos-1809	USSR	943 ×	966 ×	82	?		P/F*	—
1987 Feb 5	Ginga	Japan	509 ×	673 ×	31	1991 Nov 1		XRA*	—
1987 Feb 20	Kosmos-1823	USSR	1479 ×	1525 ×	73	1990s?		GEO	—
1987 Mar 3	Progress-28	USSR	353 ×	369 ×	51	1987 Mar 28	Spaceship	F	VLF experiment
1987 Mar 31	Kvant	USSR	343 ×	364 ×	51	2001 Mar 23	Spaceship	XRA*	Coded mask telescope
1987 Apr 24	Kosmos-1841	USSR	216 ×	379 ×	62	1987 May 8		MG	—
1987 Sep 23	Progress-32	USSR	298 ×	350 ×	51	1987 Nov 19	Spaceship	MG	—
1987 Sep 29	Kosmos-1887	USSR	215 ×	383 ×	62	1987 Oct 12		BIO	—
1987 Nov 20	Progress-33	USSR	325 ×	341 ×	51	1987 Dec 19	Spaceship	P	—
1988 Mar 25	San Marco 5	Italy	261 ×	598 ×	3	1988 Dec 6		ATM*	—
1988 Apr 14	Foton	USSR	216 ×	373 ×	62	1988 Apr 28		MG	—
1988 May 30	Kosmos-1950	USSR	1485 ×	1521 ×	73	1990s?		GEO	—
1988 Jul 5	Okean	USSR	632 ×	668 ×	82	1990 Jun 14	Recon	ATM*	—
1988 Jul 7	Fobos-1	USSR	Solar orbit			1988 Aug 28		PL	Phobos probe, failed

Launch	Name	State	Peri ×	Apo (km) ×	Inc.	Operated until		Pri mission discipline	Comments
1988 Jul 7	DPS	USSR	Solar orbit			1988 Aug 28		PL	Phobos lander
1988 Jul 12	Fobos-2	USSR	Mars orbit			1989 Mar 27		PL	Phobos probe, failed
1988 Jul 12	DPS	USSR	Mars orbit			1989 Mar 27		PL	Phobos lander
1988 Jul 12	PPS	USSR	Mars orbit			1989 Mar 27		PL	Phobos hopper
1988 Sep 29	Discovery	USA	302 ×	331 ×	28	1988 Oct 3	Spaceship	MG/ATM	—
1988 Nov 26	Soyuz TM-7	USSR	326 ×	354 ×	51	1989 Apr 27	Spaceship	BIO/MM	—
1988 Dec 2	Atlantis	USA	443 ×	454 ×	56	1988 Dec 6	Spaceship	A/P	—
1989 Jan 10	Kosmos-1989	USSR	19 102 ×	19 149 ×	64	1989 Jan 10		GEO	—
1989 Feb 21	Akebono	Japan	279 ×	10 434 ×	75	Active		ATM/P/A*	—
1989 Mar 13	Discovery	USA	301 ×	333 ×	28	1989 Mar 18	Spaceship	BIO/MG	—
1989 Apr 26	Foton	USSR	214 ×	369 ×	62	1989 May 11		MG	—
1989 May 4	Atlantis	USA	296 ×	332 ×	28	1989 May 8	Spaceship	ATM/MG	—
1989 May 4	Magellan	USA	Venus orbit			1994 Oct 12		PL	Venus radar mapper
1989 May 25	Pion	USSR	256 ×	269 ×	62	1989 Jul 23		ATM	—
1989 May 25	Pion	USSR	257 ×	268 ×	82	1989 Jul 24		ATM	—
1989 May 31	Kosmos-2024	USSR	19 099 ×	19 144 ×	64	1989 May 31		GEO	—
1989 Jul 18	Pion	USSR	254 ×	271 ×	82	1989 Sep 19		ATM	—
1989 Jul 18	Pion	USSR	254 ×	271 ×	82	1989 Sep 19		ATM	—
1989 Aug 8	Columbia	USA	294 ×	308 ×	57	1989 Aug 13	Spaceship	P/ENV	—
1989 Aug 8	Hipparcos	I/ESA	514 ×	35 890 ×	7	1993 Aug 15		OA*	Astrometric catalog
1989 Aug 28	Kosmos-2037	USSR	1484 ×	1525 ×	73	1990s?		GEO	—
1989 Sep 5	Soyuz TM-8	USSR	373 ×	389 ×	51	1990 Feb 19	Spaceship	MG	—
1989 Sep 15	Kosmos-2044	USSR	204 ×	261 ×	82	1989 Sep 29		BIO	—
1989 Sep 28	Interkosmos-24	USSR	500 ×	2490 ×	82	1995 Oct 11		P/F*	—
1989 Sep 28	Magion-2	Czech SSR	499 ×	2490 ×	82	1990 Nov 20		P/F*	—
1989 Oct 18	Atlantis	USA	299 ×	333 ×	34	1989 Oct 23	Spaceship	BIO/MG/ATM	—
1989 Oct 18	Galileo	USA	Jupiter orbit			Active		PL	Jupiter orbiter/probe
1989 Oct 18	Galileo Probe	USA	Jupiter entry			1995 Dec 7		PL	Jupiter probe
1989 Nov 18	COBE	USA	886 ×	896 ×	99	1997 May 1		SMA*	Big Bang study
1989 Nov 23	Discovery	USA	237 ×	558 ×	28	1989 Nov 28	Spaceship	A/P	—
1989 Dec 1	Granat	USSR	1814 ×	201 796 ×	53	1999 May 25		XRA*	Coded mask telescope
1989 Dec 20	Progress M-2	USSR	390 ×	393 ×	51	1990 Feb 9	Spaceship	MG/P	—
1990 Jan 9	Columbia	USA	319 ×	337 ×	28	1990 Jan 20	Spaceship	BIO/MG/ATM	—
1990 Jan 24	Hiten	Japan	Lunar orbit			1993 Apr 10		PL	Lunar orbiter
1990 Jan 24	Hagoromo	Japan	Lunar orbit			1990 Mar 18		PL	Lunar test satellite
1990 Feb 28	Okean-O1	USSR	637 ×	666 ×	82	1991 Jul 18		ATM*	—
1990 Feb 28	Atlantis	USA	238 ×	241 ×	61	1990 Mar 4	Spaceship	P	—
1990 Apr 5	Pegsat	USA	460 ×	675 ×	94	1998 Nov 14	Recon	F*	—
1990 Apr 11	POGS	USA	627 ×	745 ×	89	1990s?	Spaceship	F*	—

CATALOG OF SPACE SCIENCE LAUNCHES

Date	Name	Country	Orbit	Incl.	Status	Type	Instruments	Notes
1990 Apr 11	Foton	USSR	213 × 364	62	1990 Apr 27	Spaceship	MG	–
1990 Apr 24	Discovery	USA	614 × 618	28	1990 Apr 29	Spaceship	BIO/P/MG	–
1990 Apr 24	Hubble Space Telescope	USA	611 × 620	28	Active	Spaceship	OA/UVA*	Large space telescope
1990 May 31	Kristall	USSR	373 × 388	51	2001 Mar 23	Spaceship	MG	–
1990 Jun 1	ROSAT	Germany	562 × 584	52	1999 Jan		XRA*	Soft X-ray survey
1990 Jul 11	Gamma	USSR	411 × 430	51	1992 Feb 28		GRA*	–
1990 Jul 25	CRRES	USA	343 × 33 590	18	1990s		P/F*	–
1990 Jul 30	Kosmos-2088	USSR	1483 × 1524	73	1990s?		GEO	–
1990 Aug 15	Progress M-4	USSR	368 × 403	51	1990 Sep 20	Spaceship	P/F	Active plasma experiment
1990 Sep 3	Da Qui Weixing 1	China	843 × 901	98	1991 Mar 11		ATM	–
1990 Sep 3	Da Qui Weixing 2	China	860 × 900	98	1991 Jul 24		ATM	–
1990 Oct 6	Discovery	USA	283 × 302	28	1990 Oct 10	Spaceship	MG/BIO/P	–
1990 Oct 6	Ulysses	USA	Solar orbit		Active		PL	Solar polar survey
1990 Nov 15	Atlantis	USA	260 × 269	28	1990 Nov 20	Spaceship	A	–
1990 Dec 2	Columbia	USA	346 × 358	28	1990 Dec 11	Spaceship	ATM	–
1990 Dec 2	Soyuz TM-11	USSR	366 × 400	51	1991 May 26	Spaceship	BIO	–
1990 Dec 2	Astro 1 Fwd	USA	347 × 358	28	1990 Dec 11		UVA	Far UV observatory
1990 Dec 2	BBXRT	USA	347 × 358	28	1990 Dec 11		XRA	XR observatory
1991 Apr 5	Atlantis	USA	445 × 458	28	1991 Apr 11	Spaceship	A/MG	–
1991 Apr 5	Compton Observatory	USA	441 × 455	28	2000 Jun 4		GRA*	Gamma bursts and sources
1991 Apr 24	Meteor-3	USSR	1186 × 1213	82	1990s	Weather	ATM	–
1991 Apr 28	Discovery	USA	247 × 260	57	1991 May 6	Spaceship	ATM/P	–
1991 Apr 28	AFP-675	USA	247 × 260	57	1991 May 6		XRA/IRA	IR background
1991 May 18	Soyuz TM-12	USSR	390 × 392	51	1991 Oct 10	Spaceship	BIO	–
1991 May 30	Progress M-8	USSR	391 × 394	51	1991 Aug 16	Spaceship	P	–
1991 May 30	Naduvaniy gazovoy ballon	USSR	187 × 198	51	1991 Aug 29		ATM	–
1991 Jun 4	Okean	USSR	632 × 665	82	1994 Jan 4	Recon	ATM*	–
1991 Jun 5	Columbia	USA	278 × 296	39	1991 Jun 14	Spaceship	ATM	–
1991 Jun 5	Spacelab SLS 1	USA	279 × 299	39	1991 Jun 14	Spaceship	BIO	–
1991 Jun 5	GBA-2	USA	279 × 299	39	1991 Jun 14		MG/BIO	–
1991 Jun 29	REX	USA	771 × 875	89	?		P/F*	–
1991 Jul 17	ERS-1	I/ESA	776 × 778	98	Active		ATM/ALT*	–
1991 Jul 17	SARA	France	770 × 777	98	1990s?		RFA*	–
1991 Aug 2	Atlantis	USA	299 × 331	28	1991 Aug 11	Spaceship	ATM/A/MG	–
1991 Aug 15	Meteor-3	USSR	1186 × 1207	82	Active	Weather	ATM	–
1991 Aug 20	Progress M-9	USSR	379 × 396	51	1991 Sep 30	Spaceship	MG	–
1991 Aug 30	Yohkoh	Japan	521 × 788	31	Active		SX*	–
1991 Sep 12	Discovery	USA	563 × 575	56	1991 Sep 18	Spaceship	MG/BIO/P	–
1991 Sep 12	UARS	USA	573 × 580	56	Active		ATM*	–
1991 Oct 2	Soyuz TM-13	USSR	390 × 399	51	1992 Mar 25	Spaceship	BIO	–
1991 Oct 4	Foton	USSR	211 × 377	62	1991 Oct 20		MG	–

Launch	Name	State	Peri ×	Apo (km) ×	Inc.	Operated until		Pri mission discipline	Comments
1991 Nov 24	Atlantis	USA	361 ×	367 ×	28	1991 Dec 1	Spaceship	P	—
1991 Dec 18	Interkosmos-25	Russia	437 ×	3070 ×	82	?		P/F*	—
1991 Dec 18	Magion-3	Czech Rep.	437 ×	3071 ×	82	1992 Sep 9		P/F*	—
1992 Jan 22	Discovery	USA	292 ×	304 ×	56	1992 Jan 30	Spaceship	MG	—
1992 Jan 22	Spacelab IML-1	USA	293 ×	305 ×	56	1992 Jan 30	Spaceship	BIO/MG	—
1992 Jan 22	GBA-3	USA	293 ×	305 ×	56	1992 Jan 30		BIO/MG/UVA	Australian UV telescope
1992 Jan 25	Progress M-11	USSR	375 ×	394 ×	51	1992 Mar 13	Spaceship	MG/BIO	—
1992 Feb 11	Fuyo 1	Japan	569 ×	572 ×	97	Active		ALT/ATM*	—
1992 Mar 17	Soyuz TM-14	USSR	375 ×	396 ×	51	1992 Aug 10	Spaceship	BIO/MG	—
1992 Mar 24	Atlantis	USA	291 ×	300 ×	57	1992 Apr 2	Spaceship	ATM/MG/P	—
1992 Mar 24	Atlas 1 Fwd	USA	291 ×	301 ×	57	1992 Apr 2		ATM/P	—
1992 Mar 24	Atlas 1 Aft	USA	291 ×	301 ×	57	1992 Apr 2		ATM/UVA	FAUST UV telescope
1992 May 7	Endeavour	USA	275 ×	350 ×	28	1992 May 16	Spaceship	MG	—
1992 May 20	SROSS-C	India	250 ×	413 ×	46	1992 Jul 14		GRA*	—
1992 Jun 7	EUVE	USA	514 ×	527 ×	28	2001 Jan 31		UVA*	EUV sky survey
1992 Jun 25	Columbia	USA	299 ×	304 ×	28	1992 Jul 9	Spaceship	MG	—
1992 Jun 25	USML-1	USA	299 ×	304 ×	28	1992 Jul 9	Spaceship	MG/BIO	—
1992 Jun 30	Progress M-13	USSR	387 ×	410 ×	51	1992 Jul 24	Spaceship	MG/BIO	—
1992 Jul 3	SAMPEX	USA	514 ×	690 ×	81	Active		P*	—
1992 Jul 24	Geotail	Japan	3906 ×	104 552 ×	22	Active		P/F*	Explored magnetotail
1992 Jul 24	DUVE	USA	217 ×	1482 ×	27	1993 Mar 16		UVA	Diffuse UV background
1992 Jul 31	Atlantis	USA	421 ×	433 ×	28	1992 Aug 8	Spaceship	MG/BIO	—
1992 Jul 31	Eureca 1	I/ESA	500 ×	504 ×	28	1993 Jul 1		MG/XRA*	—
1992 Jul 31	TSS-1	Italy	295 ×	301 ×	28	1992 Aug 8		P/F	Tether
1992 Jul 31	TSS-1 PLT	USA	295 ×	301 ×	28	1992 Aug 8		P/F	—
1992 Jul 31	TSS-1 MPESS	USA	295 ×	301 ×	28	1992 Aug 8		P/F	—
1992 Jul 31	EOIM-3/TEMP2A-3	USA	227 ×	234 ×	28	1992 Aug 8		ENV/ATM	—
1992 Aug 9	FSW-2 1	China	170 ×	335 ×	63	1992 Sep 1		MG	—
1992 Aug 10	Topex-Poseidon	USA	1331 ×	1341 ×	66	Active		ALT*	—
1992 Aug 19	Pion-Germes-1	USSR	223 ×	233 ×	82	1992 Sep 25		ATM	—
1992 Aug 19	Pion-Germes-2	USSR	221 ×	232 ×	82	1992 Sep 24		ATM	—
1992 Sep 12	Endeavour	USA	298 ×	309 ×	57	1992 Sep 20	Spaceship	MG/BIO	—
1992 Sep 12	Spacelab J LM	Japan	298 ×	311 ×	57	1992 Sep 20	Spaceship	MG/BIO	Japanese Spacelab
1992 Sep 12	GAS Bridge	USA	298 ×	311 ×	57	1992 Sep 20		MG/BIO	—
1992 Sep 25	Mars Observer	USA	Solar orbit			1993 Aug 21		PL	Mars orbiter, failed
1992 Oct 6	Freja	Sweden	596 ×	1761 ×	63	1996 Oct 14		A/P/F*	—
1992 Oct 8	Foton	USSR	217 ×	352 ×	62	1992 Oct 24		MG	—

CATALOG OF SPACE SCIENCE LAUNCHES

Date	Name	Country	Perigee ×	Apogee ×	Inclination	Status/Date	Type	Payload	Notes
1992 Oct 22	Columbia	USA	285 ×	293 ×	28	1992 Nov 1	Spaceship	MG/ENV/ATM	—
1992 Oct 22	Lageos 2	Italy	5616 ×	5950 ×	52	1992 Oct 23		GEOD*	—
1992 Oct 22	USMP-1 Fwd	USA	285 ×	293 ×	28	1992 Nov 1		MG	—
1992 Oct 22	USMP-1 Aft	USA	285 ×	293 ×	28	1992 Nov 1		MG	—
1992 Oct 27	Progress M-15	USSR	396 ×	399 ×	51	1993 Feb 7	Spaceship	P	—
1992 Oct 27	MAK-2	USSR	389 ×	393 ×	51	1993 Apr 1		ATM	—
1992 Nov 15	Resurs-500	USSR	179 ×	238 ×	82	1992 Nov 22		MG	—
1992 Dec 2	Discovery	USA	317 ×	331 ×	57	1992 Dec 9	Spaceship	P/MG	—
1992 Dec 22	Kosmos-2226	Russia	1478 ×	1525 ×	73	1990s?		GEO	—
1992 Dec 29	Kosmos-2229	USSR	217 ×	373 ×	62	1993 Jan 10		BIO	—
1993 Jan 13	Endeavour	USA	300 ×	308 ×	28	1993 Jan 19	Spaceship	XRA/MG/BIO	Diffuse X-ray spectrometer
1993 Feb 20	Asuka	Japan	537 ×	646 ×	31	2001 Mar 2		XRA*	—
1993 Apr 8	Discovery	USA	292 ×	298 ×	57	1993 Apr 17	Spaceship	ATM/BIO/MG	—
1993 Apr 8	Spartan-201	USA	292 ×	298 ×	57	1993 Apr 13		SOL*	—
1993 Apr 8	Atlas-2	USA	292 ×	298 ×	57	1993 Apr 17		ATM/SOL	—
1993 Apr 25	Alexis	USA	747 ×	835 ×	69	Active		XRA*	—
1993 Apr 26	Columbia	USA	295 ×	305 ×	28	1993 May 6	Spaceship	MG	—
1993 Apr 26	Spacelab D-2 LM	Germany	296 ×	303 ×	28	1993 May 6	Spaceship	MG/BIO	German Spacelab
1993 Apr 26	USS	Germany	296 ×	303 ×	28	1993 May 6		MG/UVA	—
1993 May 22	Progress M-18	USSR	388 ×	390 ×	51	1993 Jul 4	Spaceship	BIO	—
1993 Jun 21	Spacehab SH-01	USA	472 ×	481 ×	28	1993 Jul 1	Spaceship	MG/BIO	—
1993 Jun 21	GBA-5	USA	472 ×	481 ×	28	1993 Jul 1		MG	—
1993 Sep 12	Discovery	USA	271 ×	307 ×	28	1993 Sep 22	Spaceship	BIO	—
1993 Sep 12	ORFEUS-SPAS	Germany	266 ×	342 ×	28	1993 Sep 22		UVA*	—
1993 Sep 26	Stella	France	797 ×	805 ×	98	1993 Sep 26		GEO	—
1993 Oct 11	Progress M-20	USSR	387 ×	397 ×	51	1993 Nov 21	Spaceship	MG	—
1993 Oct 18	Spacelab SLS 2 LM	USA	282 ×	289 ×	39	1993 Nov 1	Spaceship	BIO	—
1994 Jan 25	Meteor-3	Russia	1185 ×	1208 ×	82	1990s	Weather	ATM	—
1994 Jan 25	Clementine 1	USA	Lunar orbit			1995 May 4		PL	Lunar mapper
1994 Jan 25	ISA	USA	397 ×	126958 ×	65	1994 May 12		P/F*	—
1994 Feb 3	Discovery	USA	346 ×	358 ×	56	1994 Feb 11	Spaceship	A/P	—
1994 Feb 3	BREMSAT	Germany	339 ×	356 ×	56	1995 Feb 12		MM/ATM*	—
1994 Feb 3	Spacehab SH-02	USA	347 ×	358 ×	56	1994 Feb 11		BIO/MG	—
1994 Feb 3	Wake Shield Facility	USA	347 ×	358 ×	56	1994 Feb 11		MG	—
1994 Feb 3	GBA-6	USA	347 ×	358 ×	56	1994 Feb 11		MG	—
1994 Feb 8	Shi Jian 4	China	179 ×	36133 ×	28	1990s?		P/F*	—
1994 Mar 2	Koronas-I	Russia	486 ×	528 ×	82	2001 Mar 4		SOL*	—
1994 Mar 4	Columbia	USA	296 ×	305 ×	39	1994 Mar 18	Spaceship	ATM/MG/A	—
1994 Mar 4	USMP-2 Fwd	USA	296 ×	305 ×	39	1994 Mar 18		MG	—
1994 Mar 4	USMP-2 Aft	USA	296 ×	305 ×	39	1994 Mar 18		MG	—
1994 Mar 4	OAST-2	USA	296 ×	305 ×	39	1994 Mar 18		MG	—
1994 Apr 9	Endeavour	USA	211 ×	226 ×	56	1994 Apr 20	Spaceship	MG	—

Launch	Name	State	Peri ×	Apo (km) ×	Inc.	Operated until		Pri mission discipline	Comments
1994 Apr 9	MAPS	USA	211 ×	225 ×	57	1994 Apr 20		ATM	–
1994 Apr 9	SRL-1	USA	211 ×	225 ×	57	1994 Apr 20		ALT	Imaging radar
1994 May 4	SROSS-C2	India	431 ×	920 ×	46	2001 Jul 12		GRA/P*	–
1994 Jun 14	Foton	USSR	220 ×	358 ×	62	1994 Jul 2		MG	–
1994 Jul 3	FSW-2 2	China	207 ×	350 ×	62	1994 Sep 13		MG	–
1994 Jul 8	Columbia	USA	298 ×	302 ×	28	1994 Jul 23	Spaceship	BIO/MG	–
1994 Jul 8	Spacelab IML 2	USA	299 ×	303 ×	28	1994 Jul 23	Spaceship	BIO/MG	–
1994 Sep 9	Discovery	USA	253 ×	265 ×	57	1994 Sep 20	Spaceship	BIO/MG	–
1994 Sep 9	Spartan-201	USA	252 ×	265 ×	57	1994 Sep 20		SOL*	–
1994 Sep 9	LITE	USA	253 ×	265 ×	57	1994 Sep 20		ATM	Lidar experiment
1994 Sep 9	GBA-7	USA	253 ×	265 ×	57	1994 Sep 20		ATM/MG	–
1994 Sep 30	Endeavour	USA	212 ×	224 ×	57	1994 Oct 11	Spaceship	MG/BIO	–
1994 Sep 30	MAPS	USA	211 ×	223 ×	57	1994 Oct 11		ATM	Imaging radar
1994 Sep 30	SRL-2	USA	211 ×	223 ×	57	1994 Oct 11		ALT	–
1994 Nov 1	WIND	USA	186 ×	470 310 ×	28	Active		P/F*	Deep space explorer, lunar flybys
1994 Nov 3	Atlantis	USA	294 ×	311 ×	56	1994 Nov 14	Spaceship	ATM	–
1994 Nov 3	CRISTA-SPAS	Germany	294 ×	309 ×	57	1994 Nov 14		ATM*	–
1994 Nov 3	Atlas-3	USA	296 ×	310 ×	56	1994 Nov 14		ATM/SOL	–
1994 Nov 29	Geo-IK	Russia	1479 ×	1527 ×	73	?		GEO	–
1995 Jan 15	EXPRESS	Germany	110 ×	250 ×	33	1995 Jan 15		MG	–
1995 Jan 24	Astrid	Sweden	965 ×	1025 ×	82	1995 Mar 1		P/F*	–
1995 Feb 3	Spartan-204	USA	388 ×	388 ×	51	1995 Feb 11		UVA*	–
1995 Feb 3	Spacehab SH-03	USA	388 ×	389 ×	51	1995 Feb 11		MG	–
1995 Feb 3	CGP/ODERACS	USA	388 ×	389 ×	51	1995 Feb 11		MG	–
1995 Feb 16	Foton	USSR	217 ×	363 ×	62	1995 Mar 3		MG	–
1995 Mar 2	ASTRO-2 Fwd	USA	345 ×	359 ×	28	1995 Mar 18		UVA	–
1995 Mar 18	SFU	Japan	470 ×	492 ×	28	1996 Jan 20		MG/IRA/BIO*	–
1995 Apr 3	OrbView-1	USA	733 ×	748 ×	69	Active	Imaging	ATM	–
1995 Apr 9	GFZ-1	Germany	383 ×	393 ×	51	1999 Jun 23		GEO	–
1995 May 20	Spektr	USSR	392 ×	400 ×	51	2001 Mar 23	Spaceship	ATM/MG	–
1995 Jun 27	Spacelab-Mir LM	USA	393 ×	398 ×	51	1995 Jul 7	Spaceship	MG	–
1995 Jul 7	UPM/SAT 1	Spain	664 ×	676 ×	98	?		MG	–
1995 Jul 13	Discovery	USA	286 ×	315 ×	28	1995 Jul 22	Spaceship	MG/BIO	–
1995 Aug 2	Interbol-1	USSR	1443 ×	191 214 ×	63	2000 Oct 16		P/F*	–
1995 Aug 2	Magion-4	Czech Rep.	14 777 ×	178 122 ×	71	?		P/F*	–
1995 Aug 31	Fasat-Alfa	Chile	631 ×	668 ×	82	1995 Aug 31		ATM	–
1995 Sep 7	Endeavour	USA	369 ×	377 ×	28	1995 Sep 18	Spaceship	BIO	–
1995 Sep 7	Spartan 201	USA	369 ×	382 ×	28	1995 Sep 18		SOL*	–
1995 Sep 7	Wake Shield Facility	USA	396 ×	404 ×	28	1995 Sep 18		MG	–

CATALOG OF SPACE SCIENCE LAUNCHES

Date	Name	Country	Orbit	Incl.	Status	Type	Category	Notes
1995 Sep 7	GBA-8/CAPL	USA	396 ×	28	1995 Sep 18		MG	—
1995 Sep 7	IEH-1	USA	396 ×	28	1995 Sep 18		UVA	—
1995 Oct 8	Progress M-29	USSR	391 ×	51	1995 Dec 19	Spaceship	MG	—
1995 Oct 20	Spacelab USML-2	USA	267 ×	39	1995 Nov 5	Spaceship	MG	—
1995 Nov 12	Atlantis	USA	345 ×	51	1995 Nov 20	Spaceship	MG/ATM	—
1995 Nov 17	ISO	I/ESA	1110 ×	5	1998 May 16		IRA*	IR observatory
1995 Dec 2	SOHO	I/ESA	Earth-Sun L1		Active		SOL*	—
1995 Dec 18	Progress M-30	USSR	390 ×	51	1996 Feb 22	Spaceship	MG	—
1995 Dec 30	Rossi X-ray Timing Explorer	USA	564 ×	22	Active		XRA*	Time variability
1996 Feb 17	NEAR	USA	Eros orbit		2001 Feb 28		PL	Asteroid orbiter
1996 Feb 22	TSS-1R Deployer	USA	297 ×	28	1996 Feb 26		P/F	Tether experiment
1996 Feb 22	TSS-1R MPESS	USA	297 ×	28	1996 Mar 9		P/F	Tether measurements
1996 Feb 22	USMP-3 Fwd	USA	280 ×	28	1996 Mar 9		MG	—
1996 Feb 22	USMP-3 Aft	USA	280 ×	28	1996 Mar 9		MG	—
1996 Feb 24	Polar	USA	5157 × 50591	85	Active		P/F*	—
1996 Mar 9	REX II	USA	803 × 832	89	Active		P/F*	—
1996 Mar 21	IRS-P3	India	816 × 849	98	Active	Imaging	XRA	—
1996 Apr 23	Priroda	Russia	389 × 396	51	2001 Mar 23	Spaceship	ATM	—
1996 Apr 24	MSX	USA	896 × 906	99	?	Mil. tech.	IRA*	IR survey
1996 Apr 30	BeppoSAX	Italy	581 × 604	3	Active		XRA*	Broad band spectra
1996 May 19	Endeavour	USA	278 × 289	39	1996 May 29	Spaceship	BIO	—
1996 May 19	Spacehab SH-05 (CMAM-4)	USA	280 × 290	39	1996 May 29	Spaceship	MG/BIO	—
1996 May 19	GBA-9	USA	280 × 290	39	1996 May 29	Spaceship	MG/GRA	—
1996 Jun 20	Columbia	USA	264 × 274	39	1996 Jul 7	Spaceship	BIO	—
1996 Jun 20	Spacelab LMS	USA	268 × 278	39	1996 Jul 7	Spaceship	BIO/MG	Spacelab
1996 Jul 2	TOMS-EP	USA	493 × 511	97	Active		ATM*	—
1996 Jul 31	Progress M-32	USSR	375 × 391	51	1996 Nov 21	Spaceship	MG/BIO	—
1996 Aug 17	Midori	Japan	799 × 800	98	1997 Jun 30	Imaging	ATM*	—
1996 Aug 21	FAST	USA	351 × 4164	83	Active		A/P/F*	—
1996 Aug 29	Interbol-2	Russia	771 × 19207	62	Active		P/F*	—
1996 Aug 29	Magion-5	Czech Rep.	777 × 19205	62	Active		P/F*	—
1996 Sep 5	UNAMSAT-B	Mexico	966 × 1010	82	1996 Sep 7		ATM	Meteor echo
1996 Oct 20	FSW-2 No. 3	China	170 × 340	63	1996 Dec 3		BIO?	—
1996 Nov 4	HETE	USA	488 × 554	37	1996 Nov 4		GRA/XRA*	—
1996 Nov 4	SAC-B	Argentina	488 × 555	38	1996 Nov 4		XRA*	X-ray background, failed
1996 Nov 7	Mars Global Surveyor	USA	Solar orbit		Active		PL	Mars orbiter
1996 Nov 16	Mars-8	Russia	80 × 1500	51	1996 Nov 16		PL	Mars probe, failed
1996 Nov 16	MAS 1	Russia	80 × 1500	51	1996 Nov 16		PL	Mars lander, lost
1996 Nov 16	MAS 2	Russia	80 × 1500	51	1996 Nov 16		PL	Mars lander, lost
1996 Nov 16	Penetrator 1	Russia	80 × 1500	51	1996 Nov 16		PL	Mars penetrator, lost

Launch	Name	State	Peri ×	Apo (km) ×	Inc.	Operated until		Pri mission discipline	Comments
1996 Nov 16	Penetrator 2	Russia	80 ×	1500 ×	51	1996 Nov 16		PL	Mars penetrator, lost
1996 Nov 19	ORFEUS-SPAS	Germany	346 ×	358 ×	28	1996 Dec 7		UVA*	–
1996 Nov 19	Wake Shield Facility	USA	347 ×	358 ×	28	1996 Dec 7		MG	–
1996 Nov 19	Progress M-33	Russia	371 ×	388 ×	51	1997 Mar 12	Spaceship	BIO	–
1996 Dec 4	Sagan Memorial Station	USA	Mars landing			1997 Oct 7		PL	Mars lander
1996 Dec 4	Sojourner	USA	Mars landing			1997 Nov		PL	Mars mini-rover
1996 Dec 24	Bion No. 11	Russia	217 ×	378 ×	62	1997 Jan 7		BIO	–
1997 Feb 11	ORUC	USA	598 ×	608 ×	28	1997 Feb 21		UVA	STIS, FGS-1R for HST
1997 Feb 12	Haruka	Japan	574 ×	21401 ×	31	Active		RFA*	VLBI observatory
1997 Mar 4	Zeya	Russia	466 ×	478 ×	97	1999 Oct <		GEO	–
1997 Apr 4	Columbia	USA	298 ×	303 ×	28	1997 Apr 8	Spaceship	MG	–
1997 Apr 4	Spacelab MSL-1	USA	298 ×	303 ×	28	1997 Apr 8	Spaceship	MG	Spacelab MSL
1997 Apr 21	Minisat-01	Spain	562 ×	581 ×	150	Active		UVA/GRA*	–
1997 May 15	Spacehab SH-08	USA	384 ×	394 ×	51	1997 May 24	Spaceship	MG	–
1997 Jul 1	Columbia	USA	298 ×	303 ×	28	1997 Jul 17	Spaceship	MG	–
1997 Jul 1	Spacelab Long Module 1	USA	297 ×	303 ×	28	1997 Jul 17	Spaceship	MG	Spacelab MSL
1997 Aug 7	Discovery	USA	289 ×	300 ×	56	1997 Aug 19	Spaceship	UVA/MG/BIO	Hale-Bopp imaging
1997 Aug 7	CRISTA-SPAS	Germany	290 ×	299 ×	57	1997 Aug 19		ATM*	–
1997 Aug 7	TAS-1	USA	284 ×	295 ×	57	1997 Aug 19		ALT/SOL/MG	Laser altimeter
1997 Aug 7	IEH-2	USA	284 ×	295 ×	57	1997 Aug 19		UVA/SOL	Io torus imager
1997 Aug 23	Lewis	USA	294 ×	317 ×	97	1997 Sep 28	Imaging	UVA	–
1997 Aug 25	ACE	USA	Earth-Sun L1			Active		P*	Solar wind at L1
1997 Aug 29	FORTE	USA	800 ×	833 ×	69	Active		F*	Lightning studies
1997 Sep 26	Atlantis	USA	384 ×	389 ×	51	1997 Oct 6	Spaceship	MG	–
1997 Oct 9	Foton	Russia	218 ×	373 ×	62	1997 Oct 23		MG/BIO	Microgravity
1997 Oct 9	Mirka	Germany	2300 ×	360 ×	62	1997 Oct 23		MG	RV test
1997 Oct 15	Cassini	USA	Solar orbit			Active		PL	Saturn probe
1997 Oct 15	Huygens	I/ESA	Solar orbit			Active		PL	Titan probe
1997 Oct 22	STEP M4	USA	432 ×	501 ×	44	1997 Oct 22		ATM/P*	–
1997 Oct 30	TEAMSAT	I/ESA	530 ×	26633 ×	7	1997 Nov		ATM	–
1997 Nov 19	Columbia	USA	276 ×	282 ×	28	1997 Dec 5	Spaceship	MG/BIO	–
1997 Nov 19	Spartan 201	USA	277 ×	283 ×	28	1997 Dec 5		SOL*	–
1997 Nov 19	USMP-4 Forward	USA	279 ×	283 ×	28	1997 Dec 5		MG	–
1997 Nov 19	USMP-4 Aft	USA	279 ×	283 ×	28	1997 Dec 5		MG	–
1997 Nov 27	TRMM	USA	345 ×	347 ×	34	Active		ATM*	–

CATALOG OF SPACE SCIENCE LAUNCHES

Date	Name	Country	Orbit		Status		Type	Purpose
1997 Dec 2	Equator-S	Germany	564 ×	67070 ×	1998 May 1		P/F*	—
1998 Jan 7	Lunar Prospector	USA	Lunar orbit		1999 Jul 31		PL	Lunar mapping
1998 Jan 23	Endeavour	USA	377 ×	387 ×	1998 Jan 31	Spaceship	BIO/P	—
1998 Jan 23	Spacehab SH-10	USA	378 ×	384 ×	1998 Jan 31	Spaceship	MG	Mir cargo
1998 Jan 29	Soyuz TM-27	Russia	377 ×	384 ×	1998 Aug 25	Spaceship	BIO	—
1998 Feb 10	GFO	USA	781 ×	876 ×	Active		GEO	Altimeter satellite
1998 Feb 26	SNOE	USA	535 ×	579 ×	Active		ATM*	Nitric oxide survey
1998 Apr 2	TRACE	USA	597 ×	642 ×	Active		SOL*	Hi Res solar imaging
1998 Apr 17	Columbia	USA	252 ×	281 ×	1998 May 3	Spaceship	BIO	Spacelab mission
1998 Apr 17	Neurolab	USA	252 ×	281 ×	1998 May 3	Spaceship	BIO	—
1998 Jun 2	Discovery	USA	342 ×	373 ×	1998 Jun 12	Spaceship	MG/P	Mir cargo
1998 Jun 2	Spacehab SH-11	USA	342 ×	373 ×	1998 Jun 12	Spaceship	MG/BIO	Antimatter detector
1998 Jun 2	AMS	France	342 ×	373 ×	1998 Jun 12		P	Mars probe
1998 Jul 3	Nozomi	Japan	Solar orbit		Active		PL	—
1998 Jul 10	Resurs-O1	Russia	816 ×	818 ×	Active	Imaging	P/ATM	—
1998 Jul 10	WESTPAC	Australia	817 ×	818 ×	1998 Jul 10		GEO	—
1998 Oct 29	Spartan 201	USA	549 ×	560 ×	1998 Nov 7		SOL*	Research mission
1998 Oct 29	Spacehab SH-12	USA	550 ×	559 ×	1998 Nov 7	Spaceship	MG	Star formation studies
1998 Dec 6	SWAS	USA	638 ×	650 ×	Active		SMA*	Auroral studies
1998 Dec 10	Astrid-2	Sweden	978 ×	1013 ×	Active		P/F*	Mars lander crashed
1999 Jan 3	Mars Polar Lander	USA	Mars impact		1999 Dec 3		PL	Mars surface penetrator
1999 Jan 3	DS2 Scott Probe	USA	Mars impact		1999 Dec 3		PL	Mars surface penetrator
1999 Jan 3	DS2 Amundsen Probe	USA	Mars impact		1999 Dec 3		PL	Comet dust retrieval
1999 Jan 27	ROCSAT-1	Taiwan	586 ×	604 ×	Active		P/F/ATM*	Stardust return capsule
1999 Feb 7	Stardust	USA	Solar orbit		Active		PL	—
1999 Feb 7	Sample Return Capsule	USA	Solar orbit		Active		PL	—
1999 Feb 20	Soyuz TM-29	Russia	341 ×	357 ×	1999 Aug 28	Spaceship	BIO/MG	—
1999 Feb 23	ARGOS	USA	828 ×	845 ×	Active		A/P/F/XRA*	Magnetic field
1999 Feb 23	Orsted	Denmark	646 ×	864 ×	Active		F*	—
1999 Mar 5	WIRE	USA	539 ×	592 ×	Active		IRA*	IR survey, failed
1999 Apr 2	Progress M-41	Russia	336 ×	361 ×	1999 Jul 17	Spaceship	BIO	—
1999 Apr 28	ABRIXAS	Germany	554 ×	602 ×	1999 May 1		XRA*	Hard XR survey, failed
1999 May 10	Shi Jian 5	China	844 ×	868 ×	Active		P*	—
1999 May 18	TERRIERS	USA	540 ×	554 ×	1999 May 18		P/F*	—
1999 May 27	Starshine	USA	375 ×	393 ×	1999 Jun 5		ATM*	—
1999 Jun 20	QuikScat	USA	804 ×	807 ×	Active		ATM*	—
1999 Jun 24	FUSE	USA	753 ×	770 ×	Active		UVA*	Far UV spectra
1999 Jul 16	Progress M-42	Russia	340 ×	348 ×	2000 Feb 4	Spaceship	MG/BIO	—

Launch	Name	State	Peri ×	Apo (km) ×	Inc.	Operated until		Pri mission discipline	Comments
1999 Jul 23	Columbia	USA	272 ×	297 ×	28	1999 Jul 28	Spaceship	UVA/MG/BIO	–
1999 Jul 23	Chandra X-ray Observatory	USA	9999 ×	138826 ×	28	Active		XRA*	High res X-ray imaging
1999 Sep 9	Foton	Russia	215 ×	376 ×	62	1999 Sep 24		MG	–
1999 Oct 14	SACI-1	Brazil	733 ×	744 ×	98	1999 Oct		P/F/ATM*	–
1999 Dec 10	XMM	ESA	7079 ×	114028 ×	38	Active		XRA*	European X-ray Observatory
1999 Dec 18	Terra	USA	655 ×	686 ×	98	Active	Imaging	ATM*	Earth Observing System
1999 Dec 21	Arirang-1	KOREA	687 ×	708 ×	98	Active	Imaging	P*	–
1999 Dec 21	ACRIMSAT	USA	682 ×	727 ×	98	Active		SOL*	–
2000 Jan 27	ASUSAT	USA	750 ×	805 ×	100	2000 Feb		P/F	–
2000 Feb 3	Thelma	USA	752 ×	803 ×	100	2000 Feb 12		F	–
2000 Feb 3	Louise	USA	748 ×	804 ×	100	2000 Feb 12		F	–
2000 Feb 11	SRL-3 (SRTM)	USA	226 ×	241 ×	57	2000 Feb 22		ALT	Radar mapping
2000 Mar 12	MTI	USA	573 ×	613 ×	97	Active		SX	–
2000 Mar 25	IMAGE	USA	1055 ×	45925 ×	89	Active		P/F*	Magnetosphere studies
2000 Jul 15	MITA	Italy	415 ×	473 ×	87	Active		P*	Cosmic rays
2000 Jul 15	CHAMP	Germany	421 ×	475 ×	87	Active		P/F*	–
2000 Jul 16	Samba	I/ESA	17008 ×	120923 ×	90	Active		P/F*	–
2000 Jul 16	Salsa	I/ESA	16809 ×	120768 ×	90	Active		P/F*	–
2000 Aug 9	Rumba	I/ESA	17240 ×	120715 ×	90	Active		P/F*	–
2000 Aug 9	Tango	I/ESA	17313 ×	120961 ×	90	Active		P/F*	–
2000 Sep 8	Atlantis (STS-106)	USA	158 ×	324 ×	51	2000 Sep 20	Spaceship	MG	–
2000 Sep 26	Megsat-1	Italy	640 ×	646 ×	64	Active		A/MG	–
2000 Sep 26	Unisat	Italy	639 ×	644 ×	64	Active		ENV	–
2000 Oct 9	HETE 2	USA	595 ×	636 ×	1	Active		XRA/GRA*	Gamma ray bursts
2000 Nov 16	STRV 1c	UK	609 ×	39255 ×	6	Active		MM/P	–
2000 Nov 16	STRV 1d	UK	613 ×	39277 ×	6	Active		ATM/F	–
2000 Nov 21	SAC-C	USA	687 ×	707 ×	98	Active	Imaging	F*	–
2000 Nov 21	Munin	Sweden	697 ×	1800 ×	95	Active		A/P*	–

INDICES

Abbreviations and acronyms

There are numerous acronyms that are part of standard indentifications of astronomical objects. A complete catalogue can be found on a web site entitled "Astronomical Catalogue Designations" which is found under the address www.iee.org/publish/support/inspec/document/astron_ob.pdf, a summary of Astronomical Catalogues and Databases is also found on the web site www.eso.org/gen-fac/pubs/nra/section3_3.html

2dF	Two Degree Field
AAS	American Astronomical Society
AAVSO	American Association of Variable Star Observers
ABMDA	Advanced Ballistic Missile Defense Agency
ABRIXAS	A BRoad band Imaging X-ray All-sky Survey
AC118	designation for a cluster of galaxies
ACES	Atomic Clock Ensemble in Space
ACL	Advanced Charge Composition
ACCESS	Advanced Cosmic Ray Composition Experiment
ACE	Advanced Composition Explorer
ACIS	Advanced CCD Imaging Spectrometer
ACM	Asteroid, Comets, Meteors
ACO	Abell Corwin and Olowin, catalogue of clusters of galaxies
ACR	Anomalous Cosmic Ray
ACRIM	Active Cavity Radiometer Irradiance Monitor
ADAF	Advection-Dominated Accretion Flow
ADEOS	Advanced Earth Observing Satellite
ADONIS	ADaptive Optics Near Infrared System
ADS	Astronomical Data System
AESTUS	upper stage engine of the Ariane-5 rocket
AFCRL	Air Force Cambridge Research Laboratories
AFGL	Air Force Geophysical Laboratory
AGARD	Advisory Group for Aerospace Research and Development
AGASA	Akeno Giant Air Shower Array
AGB	Asymptotic Giant Branch
AGN	Active Galactic Nuclei
AGU	American Geophysical Union
AGW	Atmospheric Gravity Wave
AIS	American Interplanetary Society
AJISAI	Geodetic Satellite 'Ajisai'
AKG	Akademie für Kunst und Geschichte, Berlin
AKR	Auroral Kilometric Radiation
ALH	Allan Hills; finding place of meteorite
ALMA	Atacama Large Millimeter Array
ALOSI	a CNES project for satellite tracking
ALSEP	Apollo Lunar Surface Experiments Packages
AMB	Aluminous Mare Basalt
AMD	Asteroid Meteoroid Detector
AMPTE	Active Magnetospheric Particle Tracer Explorer
AMS	Alpha Magnetic Spectrometer
AMSR	Advanced Microwave Scanning Radiometer
AMW	Active Microwave
ANS	Astronomische Nederlandse Satelliet
AO	Announcement of Opportunity
APL	Applied Physics Laboratory
APT	Automatic Picture Transmission
APXS	Alpha-Proton X-ray Spectrometer
ARC	Atlantic Research Corp
ARGOS	Advanced Research and Global Observation Satellite
ARISTOTELES	an ESA study of a mission for measuring the Earth's gravity field
ARPA	Advanced Research Projects Agency
ARS	American Rocket Society
AS&E	American Science and Engineering
ASAR	Advanced Synthetic Aperture Radar
ASCA	Advanced Satellite for Cosmology and Astrophysics
ASI	Agenzia Spaziale Italiana
ASTRO	Shuttle missions devoted to astronomy (ASTRO 1 and 2)
ATM	Appollo Telescope Mount
ATN	Advanced TIROS-N
ATS	Application Technology Satellites
ATSR	Along Track Scanning Radiometer
AURA	Association of Universities for Research in Astronomy
AVHRR	Advanced Very High Resolution Radiometer
AXAF	Advanced X-ray Astrophysics Facility
AXP	Anomalous X-ray Pulsars
AXS	ASCA Satellite X-ray Source
AZUR	a German/US satellite for studying the magnetosphere and the Earth–Sun relationship
BABI	Basaltic Achondrite Best Initial
BAL	Broad Absorption Line Quasar
BATSE	Burst and Transient Source Experiment
BB	Black Body
BBN	Big Bang Nucleosynthesis
BBRC	Ball Brothers Research Corporation
BBSO	Big Bear Solar Observatory
BBXRT	The Broad Band X-ray Telescope
BD	Bonner Durchmusterung
BEN	Broadband Electric Noise
BESS	Balloon Borne Experiment with Superconducting Solenoidal Spectrometer
BH	Black hole
BHC	black hole candidate
BIB	Blocked Impurity Band
BIH	Bureau International de I'Heure
BIPM	Bureau International des Poids et Mesures
BIRAMIS	a (eventually not approved) project to measure directly the local Earth radiation budget
BIRAP	Balloon Infrared Platform
BiSON	Birmingham Solar Oscillations Network
BL	Boundary Layer
BMD	Ballistic Missile Defense
BMFT	Bundesministerium für Forschung und Technologie
BMRC	Bureau of Meteorology Research Centre, a research division of the National Meteorological Service in Australia
BNCSR	British National Committee for Space Research
BNKL	Becklin, Neugebauer, Kleinmann and Low
BOOMERanG	Balloon Observations Of Millimetric Extragalactic Radiation ANd Geophysics
BSSI	Bulk Solar System Initial
BUSS	Balloon Borne UV Stellar Spectrograph
BVSP	Basaltic Volcanism Study Project
CACTUS	The triaxial microaccelerometer flown aboard the CNES satellite CASTOR
CAK	Castor, Abbott, and Klein
CAL	Colombo Astrophysics Laboratory

CASDN	Comité d'Action Scientifique de la Défense Nationale
CASTLES	CfA-Arizona Space Telescope Lens Survey
CASTOR	a CNES satellite for geodesy flown in 1975
CBE	Collisionless Boltzmann Equation
CCD	Charge-Coupled Device
CCE	Charge Composition Explorer, pasrt of the AMPTE project
CDA	Command and Data Acquisition
CDM	Cold Dark Matter
CDOM	Colored Dissolved Organic Matter
CDS	Coronal Diagnostics Spectrometer
CE	Common Envelope
CEFSR	Committee for Evaluating the Feasibility of Space Rocketry
CELIAS	Charge, Element, and Analysis System
CEOS	Committee on Earth Observation Satellites
CERN	European Organization for Nuclear Research
CERGA	Centre d'Études et des Recherches en Géodynamique et Astronométrie
CFHT	Canada–France–Hawaii Telescope
CFSR	Committee for Evaluating the Feasibility of Space Rocketry
CGRO	Compton Gamma Ray Observatory
CHAMP	Challenging Mini-satellite Payload
CHEM	Charge Energy and Mass
CHON	Carbon, Hydrogen, Oxygen and Nitrogen containing material
CHUR	Chondritic Uniform Reservoir
CIAO	Chandra Interactive Analysis of Observations
CIE	Collisional Ionisation Equilibrium
CIR	Corotating Interaction Regions
CIRA	Cooperative Institute for Research in the Atmosphere
CL	Confidence Level
CLAES	Cryogenic Limb Array Etalon Spectrometer
CLASS	Cosmic Lens All Sky Survey
CLS	Collecte, Localisation, Satellites
CLIVAR	CLImate VARiability and Predictability
CMB	Cosmic Microwave Background
CMBR	Cosmic Microwave Background Radiation
CMD	Color–Magnitude Diagram
CME	Coronal Mass Ejections
CMIR	Corotating Merged Interaction Regions
CNES	Centre National d'Etudes Spatiales
CNM	Cold Neutral Material
CNO	Carbon-Nitrogen-Oxygen cycle
CNR	Consiglio Nazionale delle Ricerche
CNRS	Centre National de la Recherche Scientifique
CNS	Catalogue of Nearby Stars
COBE	Cosmic Background Explorer
COG	Curve Of Growth
COBRAS	COsmic Background Radiation Anisotropy Satellite; mission proposal to ESA, which – together with SAMBA – evolved into the Planck mission
COMPLEX	Committee for Planetary and Lunar Exploration
CONTOUR	COmet Nucleus TOUR, a NASA mission to fly by three comets
COPERS	Commission Préparatoire Européenne pour la Recherche Spatiale
COROT	COnvection et la ROTation des intérieurs stellaires
COS	Cosmic Origins Spectrograph
COSPAR	Committee on Space Research
CPA	Curved-Plate Analyser
CPC	Chopped Photometric Channel
CRAF	Comet Rendezvous and Asteroid Flyby
CRIS	Cosmic Ray Isotope Spectrometer
CSAGI	Special Committee for the International Geophysical Year
CSHKP	Carmichael-Sturrock-Hirayama-Kopp-Pneuman (model)
CTB	Cal Tech observation list B
CTT	Classical T Tauri stars
CV	Cataclysmic Variable
CV	Chaotic Variability
CVF	Circular Variable Filter and Continuously Variable Filter
CVF	Circular Variable Filter and Continuously Variable Filter
CVn	Canes [Standard Astronomical Designation for the 'Canes Venatici' (Hunting Dogs) Constellation]
CVZ	Continuous Viewing Zone
CXC	Chandra X-ray Center
CXO	Chandra X-ray Observatory
CZCS	Coastal Zone Color Scanner
DAC	Discrete Absorption Component
DAG	Agence DAG, Toulouse (http://www.dag.fr)
DAO	Dominion Astrophysical Observatory
DASA	Deutsche Aerospace AG, now part of Astrium
DEM	Differential Emission Measure
DENIS	Deep Near Infrared Sky Survey
DFVLR	Deutsche Versuchsanstalt für Luft- und Raumfahrt
DIADEM	an operational assimilation system for physical and ecosystem variables for the North Atlantic, Nordic Seas and the Arctic Ocean
DIADEME	two satellites, DIADEME 1 and 2, equipped with ultrastable oscillators and laser reflectors were launched by CNES in 1967
D/H	ratio of Deuterium to Hydrogen
DIDSY	Dust Impact Detection System
DIM	Disk Instability Model
DIRBE	Diffuse Infrared Background Experiment
DIRE	Direct Integration of Relaxed Einstein Equations
DISCO	Dual Irradiance and Solar Constant Observatory
DIVA	Deutsches Interferometer für Vielkanalphotometrie und Astrometrie
DL	Dividing Line
DMR	Differential Microwave Radiometers
DMR	Diffuse Microwave Radiometer
DMSP	Defense Meteorological Satellite Program
DNM	Direct Nucleus Measurement
DNMI	Norwegian Meteorological Institute
DOD	Department of Defense
DORIS	Doppler Orbitography and Radiopositioning Integrated by Satellite
DPS	Division of Planetary Science
DS 2	Deep Space 2
DTM	Department of Terrestrial Magnetism
DUCMA	Dust Counter and Mass Analyser
DXS	Wisconsin Bragg-crystal spectrometer
EA	Early Amazonian
EAS	European Astronomical Society
ECD	Electron Current Detector
ECHO	satellites with a diameter of about 25 meters and a reflective coating (ECHO 1 and 2)
EDI	Electron Drift Experiment
EETA	Elephant Moraine, Victoria Land, Antarctica; finding place of meteorite
EGB	Ellis, Grayson and Bond, designation for panetary nebula
EGRET	Energetic Gamma-Ray Experiment Telescope
EGS	European Geophysical Society
EISCAT	European Ionospheric Scatter radar facility
EIT	Extreme Ultraviolet Imaging Telescope
EK	Edgeworth–Kuiper (belt)
ELAIS	European Large-Area Infrared Space Observatory Survey
ELS	Eggen, Lynden-Bell and Sandage (model for the formation of the Galaxy)
ELDO	European Launcher Development Organization
ELOISE	European Large Orbiting Instrumentation for Solar Experiments
EM	Electromagnetic
EMHD	Electron Magneto-hydrodynamics
EMSS	Einstein Extended Medium Sensitivity Survey
ENSO	El Niño Southern Oscillation
ENVISAT	Environmental Satellite
EO	Einstein Observatory
EOQ	Earth Observation Quarterly

EOR	Earth Orbit Rendezvous Mode	GEO-600	a project aiming at the direct detection of gravitational waves by means of a laser
EOS	Earth Observing System		
EPB	Eucrite Parent Body	GEOS	Geodynamics Experimental Ocean Satellite (GEOS-1), magnetospheric research satellite (GEOS-2) interferometer of 600 m armlength (located near Hannover, Germany)
EPS	EUMETSAT Polar System		
ERB	Earth Radiation Experiment		
ERBE	Earth Radiation Budget Experiment		
ERBS	Earth Radiation Budget Satellite	GERB	Earth Radiation Budget experiment (on Meteosat Second Generation)
ERNO	Arbeitsgemeinschaft Entwicklungsring Nord, now Astrium G.m.b.H.		
		GfW	Gesellschaft für Weltraumforschung
EROS	Expérience de Recherche d'Objets Sombres	GFZ	GeoForschungs Zentrum Potsdam/Germany
ERTS	Earth Resources and Technology Satellite	GIC	Geomagnetically Induced Currents
ESA	European Space Agency	GINGA	ISAS X-ray satellite, formally known as ASTRO-C
ESLAB	European Space Research Laboratory, now Scientific Research and Support Department of ESA	GIRD	Group for the Study of Reaction Motion
		GIS	Gas Imaging Spectrometer
ESMR	Electrically Scanning Microwave Radiometer	GLA	Giant Luminous Arcs
ESO	European Southern Observatory	GLAST	Gamma-ray Large Area Space Telescope
ESOC	European Space Operation Center	GLONASS	Global Navigation Satellite System
ESP	Energetic Storm Particle	GMIR	Global Merged Interaction Regions
ESRIN	originally: European Space Research Institute, today it is referred to as 'European Space Agency establishment located in Frascati'	GMS	Geostationary Meteorological Satellite
		GNSS	Global Navigation Satellite System, also called GNSS/GALILEO
ESRO	European Space Research Organization	GOCE	Gravity Field and Steady-state Ocean Circulation Explorer
ESSA	Environmental Science Services Administration	GODAE	Global Ocean Data Assimilation Experiment
ESTEC	European Space Research and Technology Centre	GOES	Geostationary Operational Environmental Satellite
ETALON	geodetic, laser-tracked satellite of the USSR	GOLF	Global Oscillations at Low Frequency
EUMETSAT	European Organisation for the Exploitation of Meteorological Satellites	GOME	Global Ozone Monitor Experiment
		GONG	Global Oscillation Network Group
EUV	Extreme Ultraviolet	GOOS	Global Ocean Observing System
EUVE	Extreme Ultraviolet Explorer	GP-A	Gravity Probe-A
EVP	Equilibrium Vapour Pressure	GP-B	Gravity Probe-B
EXE	Eridanus Enhancement	GPS	Global Positioning System
EXOSAT	European X-ray Observatory Satellite	GR	General Relativity
FAIR	Fly-Along Infrared	GRACE	Gravity Recovery and Climate Experiment
FAST	Fast Auroral Snapshot Explorer	GRANAT	Venera-class spacecraft investigating gamma-ray bursts
FB	Flaring Branch	GRADIO	early studies for gradiometry satellites
FC	Fission Cell	GRASP	Gravitational Radiation Analysis and Stimulation Package
FCCC	Framework Convention on Climate Change	GRB	Gamma-Ray Bursts
F/G	Felsite/Granites	GRGS	Groupe de Recherché de Géodésie Spatiale
FGH	Field, Habing, Goldsmith	GRIM	GRgs/Imunich
FILM	Far Infrared Line Mapper	GRS	Great Red Spot
FIP	First Ionization Potential	GSFC	Goddard Space Flight Center
FIRAS	Far Infrared Absolute Spectrophotometer	GSPC	Gas Scintillation Proportional Counter
FIRSSE	Far Infrared Sky Survey Experiment	GTS	Global Telecommunication System
FIRST	Far Infrared and Submillimeter Telescope	GX5-1	Galactic X-ray Source5-1
FLRW	Friedmann–Lemaître–Robertson–Waker	HAB	Harmful Algae Bloom
FNMOC	Fleet Numerical Meteorology and Oceanography Forecasting Center	HAC	Hydrogenated Amorphous Carbon
		HALOE	Halogen Occultation Experiment
FOV	Field Of View	HAO	High Altitude Observatory
FP	Fabry-Perot	HAS	Highlands Alkali Suite
FPCS	Focal Plane Crystal Spectrometer	HB	Horizontal Branch
FR-I	Fanaroff–Riley-I (radio galaxies)	HBB	Hot Bottom Burning
FRI	Faranoff–Riley Class I (radio sources)	HBL	High-frequency peak BL Lac object
FSRQ	Flat Spectrum Radio Quasar	HBO	Horizontal Branch Oscillation
FTE	Flux-Transfer Events	HCS	the calculated maximum Heliolatitude of the heliospheric Current Sheet
FUSE	Far Ultraviolet Spectroscopic Explorer		
FUV	Far-Ultraviolet	HD	Henri Draper (catalogue)
FWHM	Full Width at Half Maximum	HDM	Hot Dark Matter
FY	First Year (ice)	HE	Hamburg-ESO Survey
GAIA	Global Astrometric Interferometer for Astrophysics	HE	1104–1805 quasar in HE survey
GALCIT	Guggenheim Aeronautical Laboratory of the California Institute of Technology	HEAO	High Energy Astronomical Observatory
		HED	Howardite–Eucrite–Diogenite
GALEX	Galaxy Evolution Explorer	HEOS	Highly Eccentric Orbit Satellite
GCM	General Circulation Model	HEOS 2	Second High Eccentricity Orbit Satellite
GCMS	Gas Chromatograph–Mass Spectrometer	HERS	High-Energy-Range Spectrometer
GCR	Galactic Cosmic-ray	HESSI	High Energy Solar Spectroscopic Imager
GCRS	Galactic Cosmic-Ray Sources	HETE	High Energy Transient Experiment
GDL	Gas Dynamics Laboratory	HH	Herbig-Haro object
GEM	Goddard Earth Model	HIFI	Heterodyne Instrument for FIRST
GEM	Gravity Earth Model	HIM	Hot Ionized Material
GEOSAT	altimetry satellite	HIMS	Hot Interstellar Medium Spectrometer

HIPPARCOS	HIgh Precision PARallax COllecting Satellite	IRAS	Infrared Astronomical Satellite
HIRLAM	High Resolution Wind Field Model	IRBM	Intermediate-range and Ballistic Missiles
HIRS	High resolution Infra Red Sounder	IRC	InfraRed Catalog
HIS	High-Intensity Spectrometer	IRIS	Infrared Imaging Spectrometer
HISTAR	an infrared-sky survey instrument launched on seven Aerobee 170 flights in 1971 and 1972, also flown, with closed-cycle cooler, on a satellite in 1971	IRM	Ion Release Module
		IRS	Infrared Spectrograph
		IRTF	Infrared Telescope Facility
HLQ	High-Luminosity Quasar	IRTS	Infrared Telescope in Space
HM	rocket engine burning hydrogen and oxygen, developed by MBB	ISAGEX	the first international laser-tracking campaign (involving 7 satellites)
HMC	Halley Multicolour Camera	ISAMS	Improved Stratospheric and Mesospheric Sounder
HMF	Heliospheric Magnetic Field	ISAS	Institute of Space and Astronautical Science
HMXB	High-Mass X-ray Binary	ISD	Interstellar Dust
HPQ	High-Polarization Quasar	ISEE	International Sun-Earth Explorers
HR	Hertzsprung-Russell	ISM	Interstellar Medium
HRC	High Resolution Camera	ISMIRAN	Institute of Earth Magnetism, Ionosphere and Propagation of Radiowaves of the Russian Academy of Sciences
HRDI	High Resolution Doppler Imager		
HRI	High Resolution Imager		
HRPT	High Resolution Picture Transmission	ISO	Infrared Space Observatory
HRTS	High Resolution Telescope and Spectrograph	ISOGAL	a 7 mm and 15 mm survey of the inner galactic disk performed with ISOCAM
HST	Hubble Space Telescope		
HUT	Hopkins Ultraviolet Telescope	ISPM	International Solar Polar Mission
HV	High velocity	ISR	Incoherent Scatter Radar
HVC	High-Velocity Clouds	ISRO	Indian Space Research Organisation
HXT	Hard X-ray Telescope	ISS	International Space Station
IAF	International Astronautical Federation	ISSI	International Space Science Institute
IAG	International Association of Geodesy	ISTI	International Space Technology, Inc.
IAGA	International Association of Geomagnetism and Aeronomy	ISTP	International Solar Terrestrial Physics
		ITOS	Improved TIROS Operational System
IAU	International Astronomical Union	ITRF	International Terrestrial Reference Frame
IAUC	International Astronomical Union Circular	ITRS	International Terrestrial Reference System
IBM	International Business Machines, Inc.	IUGG	International Union of Geodesy and Geophysics
ICBM	Intercontinental Ballistic Missiles	IVC	Intermediate-Velocity Clouds
ICM	IntraCluster Medium	JASON	altimetric satellite, follow-on of TOPEX/Poseidon
ICE	International Cometary Explorer	JATO	Jet-Assisted Take-off
ICESat	Ice Clouds and Land Elevation Satellite	JBIS	Journal of the British Interplanetary Society
ICI	Ion Composition Instrument	JFET	Junction Field Effect Transistor
ICPP	International Conferences on Plasma Physics	JHU	Johns Hopkins University
ICSU	International Council of Scientific Union	JILA	Joint Institute for Laboratory Astrophysics (University of Colorado/NIST)
IDP	Interplanetary Dust Particles		
IERS	International Earth Rotation Service	JOP	Jupiter Orbiter-Probe
IFCAI-Palermo	Istituto di Fisica Cosmica con Applicazioni all'Informatica, Palermo, Italy	JPL	Jet Propulsion Laboratory
		JVAS	Jodrell Bank VLA Astrometric Survey
IGBP	International Geosphere and Biosphere Program	KAMIOKA	location of the KAMIOKANDE detector
IGM	Intergalactic Medium	KAMIOKANDE	name of a neutrino detector derived from KAMIOKA, the location of the detector, and the name of the experiment, Nucleon Decay Experiment.
IGOS	Integrated Global Observing Strategy		
IGS	International GPS Service		
IGY	International Geophysical Year	KAO	Kuiper Airborne Observatory
IKI	Russian Space Research Institute in Moscow	KB-29	Design Bureau No. 29 in Moscow
ILMT	International Liquid Mirror Telescope	KBO	Kuiper Belt Object
IMF	Initial Mass Function	KOSI	KOmeten-SImulations experiment
IMF	Interplanetary Magnetic Field	KNMI	Royal Netherlands Meteorological Institute
IMP	Interplanetary Monitoring Platform	KPNO	Kitt Peak National Observatory
INTEGRAL	International Gamma-Ray Astronomy Laboratory	KREEP	K (potassium), Rare-Earth Elements and P (phosphorus)
IOC	Intergovernmental Oceanographic Committee		
IP	Intermediate Polar	LAS	Large Astronomical Satellite
IPAC	Infrared Processing and Analysis Center	LASCO	Large Angle and Spectrometric Coronagraph
IPC	Imaging Proportional Counter	LASER	Light Amplification through Stimulated Emitted Radiation
IPC	Interplanetary Plasma Clouds		
IPCC	Intergovernmental Panel on Climate Change	LASP	Laboratory for Atmospheric and Space Physics
IPELS	Interrelationship Between Plasma Experiments in Laboratory and Space	LBLS	Low-Frequency Peak BL Lacs
		LDEF	Long Duration Exposure Facility
IPHIR	Interplanetary Helioseismology by Irradiance Measurements	LEAM	Lunar Ejecta and Meteoroids
		LECP	Low Energy Charged Particle
IPLR	Interplanetary Laser Ranging	LECS	Low-Energy Concentrator Spectrometer
IPN	InterPlanetary Network	LEGOS-GRGS	Laboratoire d'Etudes en Géophysique et Océonographie Spatiales – Groupe de Recherche en Géodésie
IPS	Instrument Pointing System		
IPS	Interplanetary Scintillations		
IPT	Io Plasma Torus	LenGIRD	Leningrad GIRD
IR	InfraRed	LET	Low Energy Telescope
IRAC	Infrared Array Camera	LEW	Lewis Hills; finding place of meteorite
		LGM 1	Little Green Man 1

LHB	Local Hot Bubble	MIT	Massachusetts Institute of Technology
LIC	Local Interstellar Cloud	MJS	Mariner Jupiter–Saturn
LIGO	Laser Interferometer Gravitational Wave Observatory with 4 km armlength (two locations: Richland, Washington and Livingston, Louisiana, both USA)	MLS	Microwave Limb Sounder
		MMS	Multi-mission Modular Spacecraft
		MO	Mars Observer
		MOC	Mars Orbiter Camera
LIMS	Limb Infrared Monitor of the Stratosphere	MODIS	Moderate Resolution Imaging Spectroradiometer
LINER	Low-Ionization Nuclear Emission line Region	MODUST	a code treating the transfer of radiation through spherical distribution of dust grains
LISA	Laser Interferometer Space Antenna		
LISM	Local Interstellar Medium	MOLA	Mars Orbiter Laser Altimeter
LLBL	Low-Latitude Boundary Layer	MONS	Measuring Oscillations in Nearby Stars
LLR	Lunar Laser Ranging	MOPITT	Measurements of Pollution in the Troposphere
LMC	Large Magellanic Cloud	MosGIRD	the most important Group for the Study of Reaction Motion (GIRD), located in Moscow
LMO	Lunar Magma Ocean		
LMXB	Low-Mass X-ray Binary	MOUSE	Minimum Orbital Unmanned Satellite of the Earth
LMXRB	Low-Mass X-ray Binary	MP	Magnetopause
LOR	Lunar Orbit Rendezvous Mode	MPAe	Max-Planck-Institut für Aeronomie
LORAL	Loral Space Systems, Palo Alto, CA	MPE	Max Planck Institute for Extraterrestrial Research
LP	Long-Period comets	MPR	Mean Planetary Radius
LPL	Lunar and Planetary Laboratory	MRX	Magnetic Reconnection Experiment
LPQ	Low Polarization Quasar	MS	Magneto Sheath
LREE	Light Rare Earth Elements	MSBL	Magnetosheath Boundary Layer
LRI	Low Resolution Image	MSD	Measurement in a Small Diaphragm
LRS	Low Resolution Spectrometer	MSFC	Marshall Space Flight Center
LSD	Dock Landing Ship	MSG	Meteosat Second Generation
LSK	Large-Scale Kinetic simulations	MSIS	a thermospheric model
LSR	Local Standard of Rest	MSX	Midcourse Space Experiment
LSS	Large Scale Structures	MT	Main Telescope
LST	Lyman-Spitzer Telescope	MTOF	time of flight instrument designed to discriminate mass on CELIAS
LTE	Local Thermal Equilibrium		
LWP	Long-Wavelength Prime (spectrometer on IUE)	MTS	Meteorite Technology Satellite
		MX-774	a project for developing an intercontinental ballistic missile, studied by the Army Air Force in the immediate post-war period
LWR	Long-Wavelength Redundant (spectrometer on IUE)		
MACEK	an accelerometer in the Czech space project MIMOSA		
		MY	MultiYear (ice)
		NACA	National Advisory Committee for Aeronautics
LWS	Long-Wave Spectra	NAD	North American Datum
LWS	Long-Wavelength Spectrometer	NAL	Narrow Absorption Lines
MACHEK	an accelerometer in the Czech space project MIMOSA	NASA	National Aeronautics and Space Administration
		NASDA	National Aeronautics and Space Development Agency
MACHO	MAssive Compact Halo Object	NB	Normal Branch
MA	Middle Amazonian	NBO	Normal Branch Oscillation
MAD	Metal Abundance Deficiency	NBS	National Bureau of Standards
MAGER	MAGnetic field Electron Reflectometer	NCEP	National Centres for Environmental Prediction
MAGSAT	a NASA satellite measuring magnetic fields (1979–1980)	NEI	Non-Equilibrium Ionization
		NERVA	a nuclear-thermal rocket motor studied during the 1950s
MAP	Microwave Anisotropy Probe	NESDIS	National Environmental Satellite, Data and Information Service
MARCI	Mars Color Imager		
MARECS	Maritime European Telecommunications Satellite	NFI	Narrow Field Instrument
		NGC	New General Catalog
MAXIMA	Millimeter Anisotropy eXperiment IMaging Array	NGST	Next Generation Space Telescope
		NICMOS	Near Infrared Camera and Multi-Object Spectrograph
MBB	Messerschmitt-Boelkow-Blohm Corporation	NIMBUS	second-generation meteorological research and development spacecraft
MCO	Mars Climate Orbiter		
MDI	Michelson–Doppler Interferometer	NIMS	Near Infrared Mapping Spectrometer
MECS	Medium-Energy Concentrator Spectrometer	NIR	Near InfraRed
MER	Maximum Emission Region	NIVR	Nederlands Instituut voor Vliegtuigontwikkeling en Ruimtevaart
MERIS	Medium Resolution Imaging Spectroradiometer		
		NIXT	Normal-Incidence X-ray Telescope
MESUR	Mars Environmental Survey	NKVD	Russian initials for secret police
METOP	Meteorological Operational	NML	Neugebauer, Martz and Leighton
MFSPP	Mediterranean Forecasting System Pilot Project	NNSS	Navy Navigation Satellite System
		NOAA	National Oceanographic and Atmospheric Administration
MGS	Mg-Suite	NORSEX	Norwegian Remote Sensing Experiment
MHD	Magnetohydrodynamics	NPS	North Polar Spur
MICROSCOPE (CNES/...)	CNES Mission to test the Equivalent Principle, launch planned for 2004	NRAO	National Radio Astronomy Observatory
		NRL	Naval Research Laboratory
MIMOSA	MIcro Measurement Of Satellite Acceleration	NSA	Not Strongly absorbing
		NSC	Neutron Star Candidate
MIPS	Multiband Imaging Photometer for SIRTF	NSCAT	NASA Scatterometer
MIR	Merged Interaction Region	NSF	National Science Foundation

NTT	New Technology Telescope	POPSAT	study of two-way range and range-rate systems; the concept was implemented in the German PRARE tracking system flown on ERS 1 and 2
NWP	Numerical Weather Prediction		
OAO	Orbiting Astronomical Observatory		
OBC	Onboard Computer		
OCC	Occultation	POLDER	Polarisation and Drectional
OCZ	Outer Convective Zone	PPR	Photo-Polarimeter Radiometer
ODIN	an astronomy/aeronomy satellite for submm and mm wavelengths	PRARE	Precise Range and Range-rate Equipment
		PRODEX	PROgramme de Développement d'EXpériences Scientifiques
OGLE	Optical Gravitational Lensing Experiment	PS	Plasma Sheath
OGO	Orbiting Geophysical Observatory	PSAC	President's Science Advisory Committee
OGS	Objective Grating Spectrometer	PSC	Protostar cloud
OHB	OHB–System G.m.b.H., Bremen	PSF	Point Spread Function
OIMS	Russian initials for the Society for the Study of Interplanetary Communication	PSPC	Position Sensitive Proportional Counter
		PSR	Pulsar
OIR	(Pioneer Venus) Orbiter Infrared Radiometer	PUMA	dust impact mass analyser, flown on the Vega spacecraft
OIRO	Orbiting Infrared Observatory		
OKB-465	name of a rocket engine, whose chief designer was Glushko	PV	Pioneer Venus
		PWN	Pulsar-wind Nebulae
ONERA	Office National d'Etudes et de Recherches Aéronautiques et Spatialles	PWS	Plasma Wave Subsystem
		QBO	Quasi-biennial Oscillation
ONR	Office of Naval Research	QCC	Quenched Carbonaceous Carbon
OOSDP	Oceans Observation System Development Panel	QMD	Quartz-Monzodiorites
		QPO	Quasi-Periodic Oscillations
ORFEUS	Orbiting Retrievable Far and Extreme Ultraviolet Spectrometer	QSO	Quasi-Stellar Object
		QUE	Queen Alexandra Range, Victoria Land, Antarctica; finding place of meteorite
OSO	Orbiting Solar Observatory		
OSSE (in CGRO OSSE)	Oriented Scintillation Spectrometer Experiment	RADARSAT	Canadian synthetic-aperture Radar Satellite
		RAE	Royal Aircraft Establishment
OSTC	Federal Office for Scientific, Technical and Cultural Affairs, Belgium	RASS	Rosat All-Sky Survey
		RBCS	Rosat all-sky cluster survey purely based on X-ray data
OT	Optical Transient		
OTS	Orbiting Test Satellite	RBE	Region of Bizarre Emptiness
O VI	6th spectrum of Oxygen stemming from 5-fold ionised oxygen	RD	Rotational Discontinuities
		REE	Rare Earth Elements
OVV	Optically Violent Variable	REFLEX	ROSAT-ESO Flux-Limited X-ray galaxy cluster survey
OWL	Orbiting Array of Wide-Angle Light Collectors		
PACS	Photoconducting Array Camera and Spectrometer	RELIKT	Russian satellite mission to investigate the cosmic microwave background
PAGEOS	a balloon satellite used for ground-based geodetic measurements	RGPS	RADARSAT Geophysical Processor System
		RIT	an electric ion thruster
PAH	Polycyclic Aromatic Hydrocarbons	RL	a Pratt & Whitney rocket engine
PCA	Polar Cap Absorption Event	RLO	Roche Lobe Overflow
PDR	Photon-Dominated Regions	RMS	Root Mean Square
PDS	Photoswitch Detector System	RNII	Jet Scientific Research Institute
PEM	Particle Environment Monitor	ROSAT	Röntgen Satellite Trümper
PEOLE	Preliminary EOLE satellite, the latter named after named after Aeolus, the wind god	ROTSE	Robotic Optical Transient Search Experiment
PHARAO	Atomic clock based on laser cooling of atoms in orbit	RPA-COPERNIC	electron plasma experiment on Giotto
		RS CVn	RS Canum Venaticorum
PHEBUS	a Russian Payload for High Energy Burst Spectroscopy	RTG	Radioisotope Thermoelectric Generator
		RXJO5852. 0-4622	Rosat X-ray Satellite source position code, Julian 2002 epoch.
PHT	detector subsystem of the photo-polarimeter		
PHT-S	spectrometer detector subsystem of the photo-polarimeter	RXTE	Rossi X-ray Timing Explorer
		SAMBA	SAtellite to Measure Background Anisotropies; mission proposal to ESA, which – together with COBRAS – evolved into the Planck mission
PHOT (in ISO-PHOT)	Photo-polarimeter		
PL	Power Law		
PLANCK	ESA mission to map the cosmic microwave brackground (launch foreseen in 2007)	SAMPEX	Solar Anomalous and Magnetospheric Particle Explorer
PLANET	Probing Lensing Anomalies NETwork	SAMSO	Space and Missile Systems Organization
PLAZMAG	plasma analyser on board the Vega 1 and Vega 2 probes	SAO	Smithsonian Astrophysical Observatory
		SAR	Synthetic Aperture Radar
PM	Plasma Mantle	SAS	Small Astronomical Satellite
PM	Point-Mass	SAX	Satellite per Astronomia X (now called BeppoSAX)
PMIRR	Pressure Modulator Infrared Radiometer	SBBN	Standard Big Bang Nucleosynthesis
PMS	Pre-Main Sequence	SBUV	Solar Backscatter Ultra Violet
PMSE	Polar Mesospheric Summer Echoes	ScaRaB	observations of the long- and short-wave components of the Earth radiation budget
PMW	Passive Microwave		
PN	Planetary Nebula	SCORE	Signal Communication by Orbiting Relay Equipment
PODAAC	Physical Oceanography, Distributed Active Archive Center		
		SCR	Solar Corpuscular Radiation

SDI	Strategic Defense Initiative	SSC	Storm Sudden Commencement
SDO	Solar Dynamics Observatory	SSD	Solid-state Detectors
SDSS	Sloan Digital Sky Survey	SSMI	Special Sensor Microwave Imager
SE	Standard Earth	SSME	Space Shuttle Main Engine
SeaSat	Earth observing satellite with synthetic-aperture radar, scanning multichannel microwave radiometer, wind-scatterometer and altimeter (1978)	SSS	Sea Surface Salinity
		SSS	Solid State Spectrometer
		SSS	Super Soft Sources
SEC	Secondary-Electron Conducting	SST	Satellite to Satellite Tracking
SED	Spectral Energy Distribution	SST	Sea Surface Temperature
SEP	Société Européenne de Propulsion	STARE	Scandinavian Twin Auroral Radar Experiment
SEP	Solar Energetic Particles	STARLETTE	a passive satellite equipped with corner cubes for laser tracking
SEP	Strong Equivalence Principle		
SEPICA	Solar Energetic Particle Ionic Charge Analyzer	ST-ECF	Space Telescope European Coordinating Facility
SERC	Science and Engineering Research Council	STELLA	sister satellite to STARLETTE
SEREB	Société pour l'Etude et la Réalisation	STEREO	Solar TErrestrial RElations Observatory
SEVIRI	Spinning Enhanced Visible Infrared Instrument	STG	Space Task Group
SFR	Star-Forming Regions	STIS	Space Telescope Imaging Spectrograph
SG	Super Giant	STP	Solar-Terrestrial Physics (programme)
SGR	Soft Gamma-ray Repeater	STP72	a US Army spacecraft
SHOM	Service Hydrographique et Océanographique de la Marine	STScI	Space Telescope Science Institute
		STSP	Solar-Terrestrial Science Programme
SID	Sudden Ionospheric Disturbance	SUISEI	ISAS Halley probe, earlier called PLANET-A
SIM	Space Interferometry Mission	SULEICA	a time-of-flight instrument aboard the IRM of the AMPTE mission
SIMBIOS	Sensor Intercomparison for Marine Biological and Interdisciplinary Ocean Studies		
		SUMER	Solar Ultraviolet Measurement of Emitted Radiation
SIRTF	Space Infra-Red Telescope Facility		
SIS	Solar Isotope Spectrometer	SUPERDARN	Super Dual Auroral Network
SIS	Solid State Imaging Spectrometer	SURGE	SEASAT User Research Group in Europe
SISPRE	Società Italiana per lo Studio della Propulsione a Reazione	SUSIM	Solar Ultraviolet Spectral Irradiance Monitor
		SUVO	Space Ultraviolet-Visible Observatory
SIZEX	Seasonal Ice Zone Experiment	SWAN	Solar Wind ANisotropies
SKB	Russian initials of Special Design Bureau	SWAS	Submillimeter Wave Astronomy Satellite
SL (in SL9)	Shoemaker-Levy	SWICS	Solar Wind Ion Composition Spectrometer
SLAR	Side-Looking Radar	SWP	Short-Wavelength Prime (spectrometer on IUE)
SLEW	Slew Survey	SWS	Short-Wave Spectra
SLV	Strong Low-Velocity	SWS	Short-Wavelength Spectrometer
SMC	Small Magellanic Cloud	SWT	Science Working Team
SMEX	Small Explorer	SXRB	Soft X-ray Diffuse Background
SMM	Solar Maximum Mission	SXT	Soft X-ray Telescope
SMMR	Scanning Multichannel Microwave Radiometer	T2L2	Time Transfer by Laser Link
SMOS	Soil Moisture and Ocean Salinity (mission)	TAD	Travelling Atmospheric Disturbance
SN	Supernova	TAMA and TAMA 300	300 m ground-based interferometer for the detection of gravitational waves at the Mitaka campus of the National Astronomical Observatory in the Tama area of Tokyo, Japan
SNAP	Supernova Acceleration Probe		
SNC	Shergotty, Nakhla, Chassigny; the first recognised Martian meteorites		
SNOWMASS	see Ulrich, R.K., Harvey, J., Rhodes, E.J., Jr. And Toomre, J. (eds) (1984). *Solar Seismology from Space*, NASA JPL 84–84, Pasadena, CA.	TD	Tangential Discontinuities
		TEC	Total Electron Content
		TEGP	Theory and Experiment in Gravitational Physics
		TENMA	Japanese X-ray satellite, also known as Astro-B
SNR	Supernova Remnants	TES	Thermal Emission Spectrometer
SO	Solar Orbiter	TES	Tropospheric Emission Spectrometer
SOFIA	Stratospheric Observatory for Infrared Astronomy	TGS	Transmission Grating Spectrometer
SOHO	Solar and Heliospheric Observatory	TH	Thermal Component
SOI/MDI	Solar Oscillations Investigation with the Michelson Doppler Imager	TiCCE	Timeband Capture Cell Experiment on ESA's Eureca platform
SOLSTICE	Solar Stellar Irradiance Comparison Experiment	TID	Traveling Ionospheric Disturbance
SONG	Space Oceanography Navigation and Geodynamics	TIGCM	Thermosphere-Ionosphere General Circulation Model
SOPRANE	Système Océanique de PRévision en Atlantique Nord-Est		
		TIR	Thermal Infrared
SOTP	Saturn Orbiter-Titan Probe	TIRL	Trapped Instantaneous Residual Liquid
SOUP	Solar Optical Universal Polarimeter	TIROS1	Television and InfraRed Observation Satellite
SP	Short-Period comets	TIROS-N	Television InfraRed Operational Satellite – Next-generation
SPA	(in ORPHEUS-SPA) Shuttle Payload Pallet		
SPAS	Shuttle Payload Satellite		
Spartan	a small Shuttle-launched and -retrieved satellite	TNO	Trans-Neptunian Objects
SPIRE	Spectral and Photometric Imaging Receiver	TOA	Top-of the Atmosphere
SPT	Stationary Plasma Thruster	ToF	Time-of-Flight
SQUID	Superconducting Quantum Interference Device	ToF-E	Time-of-Flight v. Energy
SRC	Science Research Council	TOGA	Tropical Ocean Global Atmosphere
SRON	National Institute for Space Research	TOMS	Total Ozone Mapping Spectrometer
SRE	Study of Radar Echo	ToO	Target of Opportunity
SSB	Space Science Board		

TOPAZ	operational assimilation system for physical and ecosystem variables for the North Atlantic, Nordic Seas and the Arctic Ocean	VFW-Fokker	Zentralgesellschaft VFW-Fokker m.b.H
		VHF	Very-High Frequency
		VIP	Very Important Particle
TOPEX/Poseidon	Ocean TOPography EXperiment	VIRGO	Variability of solar Irradiance and Gravity Oscillations
TR	Transition Region		
TRACE	Transition Region and Coronal Explorer	VLA	Very Large Array
TRANET	a network of tracking stations well distributed around the Earth	VLBA	Very Long Baseline Array
		VLBI	Very Long Baseline Interferometry
TRANSIT	a fleet of orbiting US Navy satellites working in a one-way Doppler downlink mode and providing global coverage	VLF	Very-Low Frequency (radio waves)
		VLT	Very Low Ti
		VNIR	Visible/Near-Infrared
		VORTEX	Venus Orbiter Radiometric Temperature-sounding Experiment
TRC	Transition Region Camera		
TRIAD	an experimental U.S. Navy spacecraft	VTPR	Vertical Temperature Profile Radiometer
TRIGA	research reactor of the Gutenberg University in Mainz	WAC	Corporal rocket developed under Army Ordnance contract by the Jet Propulsion Laboratory (JPL)
TRMM	Tropical Rainfall Measuring Mission		
TRW	a global technology company, headquarters in Cleveland, Ohio	WAGS	Worldwide Atmospheric Gravity Wave Studies
		WAT	Wide-angle Tailed Source
TS	Thermal Soft	WCD	Wind Compressed Disk
TS-3	Tokyo University Spherical Torus	WCRP	World Climate Research Programme
TUNDE	Energetic Particle Analyser ion spectrometer aboard the Vega 1 space probe	WFC	Wide-Field Camera
		WFPC	Wide-Field Planetary Camera
UARRP	Upper Atmosphere Rocket Research Panel	WGS 84	geodetic system closely linked to the the International Terrestrial Reference System (ITRS)
UARS	Upper Atmosphere Research Satellite		
UBVRI	Ultraviolet-Blue-Visual-Red-Infrared		
UCLA	University of California at Los Angeles	WHIM	Warm Highly Ionized Medium
UDMH	Unsymmetrical Dimethyl Hydrazine	WHK	Welty, Hobbs, Kulkarni
UIB	Unidentified Infrared Bands	WiFS (in Sea WiFS)	Wide Field of view Sensor
UIR	Unidentified Infrared	WIM	Warm Ionized medium
UK	United Kingdom	WIMPs	Weakly Interacting Massive Particles
UKMO	Met Office UK	Wind	spacecraft to measure the solar wind, part of the Solar-Terrestrial Physics Programme (STP)
UKS	United Kingdom Satellite		
ULECA	Ultra Low Energy Charge Analyzer	WINDII	Wind Imaging Interferometer
ULEIS	Ultra-Low Energy Isotope Spectrometer	WIRE	Wide-field Infrared Explorer
ULET	Ultra-Low Energy Telescope	WISCONSIN	The Wisconsin survey (of the diffuse X-ray background)
ULF	Ultra-Low-Frequency		
ULIG	see ULIRG	WMH	Welty, Morton, Hobbs
ULIRGs	Ultraluminous Infrared Galaxies	WMO	World Meteorological Organisation
UPS	Unified Propulsion System	WNM	Warm Neutral Material
urKREEP	a lunar era preceding the KREEP (potassium, rare earth elements and phosphorus) era	WOCE	World Ocean Circulation Experiments
		WR	Wolf-Rayet
USAF	US Air Force	wTT	weak-lined T Tauri stars
USGS	U.S. Geological Survey	WUPPE	Wisconsin Ultraviolet Photo Polarimeter Experiment
USNO	United States Naval Observatory		
USO	Ultra Stable quartz Oscillator	WWW	World Weather Watch
UT1	Universal Time1	XBAC	X-ray-brightest Abell-type clusters of galaxies
UTC	Coordinated Universal Time		
UVAS	Ultraviolet Astronomical Satellite	XBP	X-ray Bright Points
UVCS	Ultraviolet Coronagraph Spectrometer	XEUS	X-ray Evolving Universe Spectroscopy Mission
UVS	Ultraviolet Spectrometer	XMM	X-ray Multi-Mirror Mission
UVSP	Ultraviolet Spectrometer Polarimeter	XRB	X-ray Binaries
UVSST	UV Sky Survey Telescope	XRF	X-ray Fluorescence
V2	Vergeltungswaffe 2	XRFS	X-ray Fluorescence Spectrometer
Vega	Russian Spacecraft that landed on the Moon, went Venus and flew by Comet Halley	X-ray UV	X-Ray and extreme UltraViolet spectral region
		YSO	Young Stellar Objects

Index of cited authors

Aarseth, S.J., 477, 495
Abbott, D.C., 834, 897, 902, 903, 908, 909, 910
Abe, K., 696
Abe, T., 1546, 1555
Abell, G., 475, 482, 495
Abergel, A., 612, 636, 637, 868, 869, 871, 872, 1233, 1251, 1252, 1254, 1274
Ables, E., 527
Abolins, J., 871
Abraham, P., 871
Abraham, R., 393
Abramovici, A., 357, 365, 370
Abshire, J.B., 1322
Abt, H.A., 58
Aceituno, J., 523
Achterberg, A., 696
Achtermann, J.M., 640, 642
Acker, A., 924, 935
Ackerson, K.L., 1491, 1535, 1536
Acosta-Pulido, J., 871
Acton, L.W., 1086
Acuña, M.H., 1307, 1320, 1321, 1232, 1479, 1485, 1487–1489, 1491, 1492
Adams, D.J., 498
Adams, F.C., 872
Adams, J.B., 260, 262, 268, 269, 270
Adams, J.H., 993, 999
Adams, T.F., 662, 671, 1199, 1213
Adams, W.S., 269, 649, 671, 895, 908
Adamson, A.J., 638
Ade, P.A.R., 394, 420, 466, 637, 638, 641, 869
Adelberger, K.L., 498, 526, 527
A Djie, H.R.E., 869
Adlakha, P., 1617
Adler, I., 528, 1382, 1401
Aellig, M.R., 1138
Agnese, P., 869
Agol, E., 465
Agrawal, P.C., 604
Aguilar-Benitez, M., 1231
Aharonson, O., 1322
A'Hearn, H.F., 1218, 1223, 1244, 1247, 1258, 1265, 1284, 1288
A'Hearn, M.F., 1231, 1232, 1269, 1272–1274, 1291–1294
Ahmad, Q.R., 1057, 1060
Ahrens, L.H., 1331, 1370, 1403
Aiken, J., 1616
Aitken, D.K., 1271
Aitken, F.K., 1373
Ajello, J.M., 1201, 1213
Ajhar, E.A., 393, 394, 467
Ajukov, S.V., 1060
Akasofu, S.-I., 156, 177, 1537, 1538, 1544–1556, 1580
Ake, T.B., 298, 299, 672
Akerlof, C., 518, 520, 523, 526
Akim, E.L., 1320
Albee, A.L., 1338, 1370
Albert, C.E., 604, 667, 671, 675

Albrow, M., 454, 465
Alcalà, J.M., 892
Alcock, C., 454, 465
Alcolea, J., 637
Al-Dargazelli, S.S., 693, 694
Aldcroft, T.L., 577
Aldering, G.S., 395, 420, 467, 560
Aldrich, L.T., 1331, 1332, 1370, 1376
Aleksandrov, S.G., 222
Aleshin, G.M., 275
Alexander, C., 1430
Alexander, W.M., 1168, 1175, 1184, 1187, 1188, 1232
Alexander, W.R., 1216
Alexandrov, Yu.N., 1320
Alfvén, H., 131, 149, 156, 159, 177, 231, 234, 237, 251, 1120, 1136, 1524, 1531, 1555
Alibert, C., 1341, 1370
Alibert, Y., 393
Alissandrakis, C.E., 1087
Alkemade, F.J.M., 910
Allamandola, L.J., 610, 617, 619, 620, 624, 626, 636, 639, 640, 642, 643, 644, 1274
Allamandola, L.S., 1274
Allegra, A., 1138
Allègre, C.J., 1330, 1336, 1370, 1372, 1373
Allen, C.W., 435
Allen, D.A., 301, 328, 612, 644
Allen, G.E., 689, 694
Allen, J.D., 270
Allen, M., 641, 645, 1286, 1291
Allen, R.O., 1402
Allen, S.W., 489, 490, 493, 495
Allendorf, S.C., 1214
Aller, H.D., 560
Aller, L.H., 673, 674, 1086
Aller, M.F., 560
Allison, M.D., 1478
Alloin, D., 547
Allsman, R.A., 465
Aloise, J., 1062
Alonso-Faus, A., 249, 251
Aloy, M.A., 553, 558
Alpar, M.A., 748, 752, 753, 756, 762, 784, 786–789, 812, 817, 820, 821
Alpher, R.A., 308, 328, 423, 436
Altenhoff, W.J., 1243, 1269
Althouse, W.E., 370
Altieri, B., 460, 466, 577, 638, 869, 1270, 1292
Altrock, R.C., 254
Altschuler, M.D., 220, 222
Altwegg, K., 1230, 1231, 1278, 1285–1288, 1291, 1292
Alvarez, L.W., 121, 122, 149, 678, 694
Alves, D.R., 465
Aly, J.J., 1113
Amari, S., 636, 1292, 1294
Amati, L., 525, 527, 754
Amazeen, P.G., 1530, 1555
Ammar, A., 1213

Amsler, C., 1231
Amstrong, E.B., 187, 201, 202
An, C.H., 910
An, P., 757
Anderegg, M.E., 328–330, 577, 870, 1449
Anders, E., 429, 434, 436, 479, 490, 495, 613, 636, 940, 959, 1284, 1291, 1292, 1335, 1370, 1378, 1388, 1401
Andersen, B., 1060, 1061
Andersen, D., 1604, 1616
Andersen, H., 1616
Andersen, J., 393
Andersen, M.I., 466, 520, 523, 527
Andersen, R.D., 757
Anderson, A., 1650
Anderson, A.J., 1269
Anderson, B.E., 298
Anderson, C., 1322
Anderson, C.M., 299
Anderson, D.L., 1316, 1320
Anderson, E.R., 1060
Anderson, F.W., 57, 57
Anderson, J.D., 349, 351, 1399, 1401, 1449
Anderson, J.T., 351
Anderson, K.A., 1232, 1320, 1491, 1492, 1541, 1557
Anderson, K.S.J., 547
Anderson, L.S., 1073, 1080, 1085
Anderson, N., 612, 638
Anderson, R.C., 1231, 1371, 1402
Anderson, R.R., 1527
Anderson, S., 397
Anderson, S.B., 748, 755, 784, 786, 788
Anderson, S.W., 1085
Andersson, B.G., 299
Ando, H., 438, 909
Ando, M., 754
André, A., 1581
André, J.-M., 1614
André, M., 858, 860, 862, 1537, 1548, 1551, 1555
André, P., 636, 637, 868, 869, 871, 872, 873, 891
Andreeva, O.G., 1582
Andretta, V., 1080, 1085
Andrews, G.B., 1002
Andrews, J.P., 299
Andriesse, C.D., 611, 636
Angelini, L., 755, 758
Anger, C.D., 1430, 1556
Angerhofer, P.E., 835
Anglin, J.D., 255
Angonin, M.C., 497
Angonin-Willaime, M.-C., 456, 466
Annell, C.S., 1401
Annis, J.A., 467
Anraku, K., 696
Antia, H.M, 1047, 1060, 1050, 1063
Antiochos, S.K., 893, 1080, 1085, 1088
Antoine, D., 1602, 1614
Antonelli, A., 527
Antonelli, L.A., 524, 527
Antoniadi, E.M., 268, 269, 1406
Antonucci, E., 222, 1083, 1085, 1137, 1138, 1139
Antonucci, R.R.J., 536, 537, 540, 546
Antunes, A., 886, 891
Anweiler, B., 1583
Aoki, T., 757, 954, 959
Apathy, I., 1232
Apostolatos, T.A., 370
Apparao, K.M.V., 604, 765, 786
Appenzeller, I., 299, 639, 861, 870
Applegate, J.L., 730, 756
Applegate, L., 440

Appleton, E.V., 1580, 1559–1563
Appourchaux, T., 1055, 1060, 1061, 1062
Aptekar, R.L., 525, 527, 755
Arbinger, H., 1138
Archinal, B.A., 351
Arendt, R.G., 329, 935, 1188
Arenou, F., 467
Arens, J.F., 640
Arevalo, M.J., 809
Argabright, V., 300
Arimoto, N., 569, 576
Arkani-Hamed, J., 1322
Armand, C., 636, 637, 638, 641, 869
Armandroff, T.E., 393
Armstrong T.P., 1492
Armstrong, B., 1231
Armstrong, E.B., 202
Armstrong, J.C., 1569, 1584
Armstrong, J.D., 1062
Armstrong, T.P., 177, 977, 999, 1002–1004
Armus, L., 523, 524, 526
Arnaboldi, M., 577
Arnaud, K.A., 478, 481, 484, 489, 495, 497, 498
Arnaud, M., 495, 498
Arndt, T.U., 904, 908
Arnett, W.D., 778, 786
Arnold, F., 1577, 1580
Arnold, J.R., 434, 436, 1390, 1401, 1402
Arnoldy, R.L., 1557
Arnould, M., 434, 436
Arons, J., 728, 752
Arpigny, C., 1271, 1272, 1274, 1293
Arrol, W.J., 1329, 1370
Arthur, C.W., 1187
Artzner, G.E., 1086
Aruga, Y., 577, 855
Arvidson, R.E., 1321
Arzoumanian, Z., 371, 753, 755
Asai, K., 577, 807, 808, 851, 852, 854, 855
Asanuma, T., 757, 960
Asaoka, I., 959
Asaoka, Y., 696, 959
Asbridge, J.R., 981, 999, 1003, 1032, 1033, 1130, 1136, 1137, 1139, 1231, 1527, 1556
Aschenbach, B., 597, 604, 605, 696, 719, 733, 737, 741, 752, 753, 756, 955, 959, 960
Aschwanden, M.J., 1079, 1085
Ash, M.E., 352
Ash, R.D., 1375
Ashby, M.L.N., 636, 639, 641, 642
Ashby, N., 351
Ashour-Abdalla, M., 1556, 1557
Ashwal, L., 1344, 1372
Asmus, V., 1615
Asphaug, E., 1247–1249, 1266, 1269, 1273, 1477
Aston, F.W., 1130, 1327, 1330, 1370
Astridge, J.R., 1555
Athay, R.G., 1073, 1074, 1080, 1085
Atlas, 57
Atreya, S.K., 437, 438, 1430, 1437, 1439, 1440, 1443, 1445, 1448, 1449, 1464, 1465, 1476
Attomici, R., 536
Attomici, S., 537
Au, B., 1086
Aubele, J.C., 1321
Aubourg, É., 394, 396, 397, 454, 466
Audard, M., 959
Audouin, C., 1633, 1650
Audouze, J., 438, 439
Auer, S., 1173, 1175, 1187

INDEX OF CITED AUTHORS

Auger, P., 123, 149
Augereau, J.C., 865, 868
Augueres, J.L., 869
Augusteijn, T., 524, 525, 817, 820
Augusto, P., 469
Aumann, H.H., 307, 312, 328, 329, 330, 577, 863, 868
Aussel, A., 466
Aussel, H., 576, 577, 869
Austin, S.M., 434, 436
Authier, B., 199, 201
Auvergne, M., 1060
Aviles, R., 394
Avni, Y., 543, 546, 548
Avrett, E.H., 1086–1088
Awaki, H., 498, 577
Axelrod, T.S., 465
Axford, W.I., 177, 246, 251, 690, 691, 694, 695, 968, 976, 978, 999, 1002, 1010, 1032, 1033, 1085, 1086, 1138, 1139, 1142, 1156, 1158, 1173, 1187, 1191, 1192, 1197, 1199, 1202, 1206, 1213–1215, 1530, 1532, 1542, 1543, 1555, 1562, 1580
Axon, D.J., 547
Ayani, K., 960
Aymon, J., 421
Ayon, J.A., 673
Ayres, T.R., 878, 891, 892, 904, 905, 908, 1074, 1085, 1088, 1215

Baade, D., 892
Baade, W., 694, 721, 752, 759, 786, 840
Baas, F., 640, 643
Babcock, H.D., 151, 229, 251, 1113
Babcock, H.W., 151, 229, 251, 1036, 1037, 1060, 1093, 1113
Babel, J., 909
Babler, B., 299
Babu, K.S., 1231
Bachiller, R., 637
Bachmann, K.T., 1062
Backer, D.C., 753, 755, 787
Backman, D.E., 560
Backus, G., 1480, 1491
Bacmann, E., 862, 868
Bacon, D., 460, 466
Bacon, L., 1062
Baddenhausen, H., 1403
Bader, M., 1116, 1136
Baggaley, W.J., 1188
Baggio, L., 371
Bagla, J.S., 528
Baglin, A., 1060, 1062
Baguette, J.M., 191, 193, 201
Baguhl, M., 1188, 1189
Bahcall, J.N., 478, 495, 497, 703, 711, 718, 787, 1041, 1047, 1060
Bahcall, N.A., 386, 387, 392, 393, 475, 476, 478, 479, 489, 495, 783
Bahnsen, A., 1551
Bailes, M., 753, 755, 757, 786–789
Bailey, M.E., 1240, 1269
Bailey, P.L., 1668
Bailey, V.A., 1564, 1580
Baily, D.A., 1582
Bailyn, C.D., 754, 806, 808, 841, 855
Baines, K.H., 1423, 1440, 1443, 1448
Baird, A.K., 1391, 1401
Baity, W.A., 528, 837
Baker, D.N., 167, 177, 1031, 1138, 1231, 1556, 1557
Baker, K.D., 1582
Baker, V.R., 1304, 1307, 1320
Baker, W.G., 1567, 1577, 1580
Bakes, E.L.D., 618
Bakes, E.L.O., 636, 675
Bakich, M., 1476
Bakker, E.J., 872

Balacescu, A., 1403
Balachandran, S.C., 427, 429, 436, 437
Balbi, A., 394, 420
Balbinot, R., 150, 547
Balbus, S.A., 794, 808
Baldi, P., 1649
Baldry, I.K., 1060
Baldwin, J.A., 536, 546
Baldwin, J.E., 651, 671, 1584
Baldwin, R., 243, 272, 273
Baldwin, R.B., 259, 261, 269, 1348, 1371
Baldwin, R.K., 275
Bale, J., 1511, 1526
Balick, B., 835, 917, 935
Baliunas, S.L., 881, 891, 892
Balkowski, C., 468
Ballester, G.E., 869
Ballet, J., 820, 959
Bally, J., 605
Balm, S.P., 870
Balmino, G., 1321, 1322, 1631, 1650
Balogh, A., 151, 255, 1001, 1002, 1005, 1138, 1139, 1140, 1141, 1148, 1151, 1158–1160, 1188
Balona, L.A., 909
Balonek, T.J., 438, 547, 560
Balsano, R., 523
Balser, D.S., 439
Balsiger, H., 1001, 1137, 1138, 1214, 1230, 1231, 1288, 1291, 1292, 1556
Balsley, B.B., 1580, 1581
Balsley, D.B., 1563, 1581
Baluteau, J.-P., 315, 328, 330, 627, 636–638, 641, 642, 869
Bame, S.J., 254, 255, 999, 1001, 1003, 1031–1033, 1113, 1119, 1123, 1124, 1133, 1136–1140, 1151, 1159, 1160, 1188, 1227, 1228, 1231, 1527, 1555–1557
Banaszkiewicz, B., 1139
Banaszkiewicz, M., 1004, 1216
Band, D.L., 505, 523, 524, 526, 527
Banday, A.J., 408, 413, 417, 419–421, 642
Bandeen, W.R., 202, 1063
Banderst, W.B., 1321
Bandyopadhyay, R.M., 1269
Banerdt, W.B., 1304, 1308, 1320–1322
Banerjee, D., 1086
Bania, T.M., 439, 547
Banin, A., 1376, 1391, 1401
Banks, P.M., 1569, 1580
Bansal, B.M., 1371–1375, 1402
Bappu, I., 907
Bappu, M.K.V., 910, 911
Baptista, R., 800, 808, 809
Baraffe, I., 393
Baranov, V.B., 1192, 1198, 1199, 1200, 1202, 1206, 1208, 1210, 1213–1215
Barat, C., 524, 526, 528
Barat, J., 192, 195–197, 201
Barban, C., 1062
Barbier, L.M., 1003, 1004
Barbieri, C., 1271, 1272
Barbuy, B., 393
Bardeen, J.M., 414, 419
Barger, A.J., 498
Baring, M.G., 689, 694, 1002
Barker, E.S., 547, 673
Barker, J.R., 636
Barker, P.J., 298
Barker, P.K., 909
Barkhouser, R.H., 299
Barkstrom, B.R., 1658, 1659, 1668
Barl, L., 869
Barlier, F., 1634, 1651

Barlow, M.J., 577, 633, 635–639, 641–643, 654, 671, 869, 903, 908, 935
Barlow, N.G., 1322
Barlow, T.A., 395
Barnard, E.E., 607, 636, 1452, 1453
Barnes, A., 139, 254, 1003, 1131, 1137, 1160, 1212, 1213, 1215
Barnes, C.W., 149, 983, 985, 999
Barnes, J.R., 1402
Barnett, M.A.F., 1559, 1560, 1580
Barnett, R.M., 1231
Barnothy, J.M., 444, 460, 465, 466
Bar-Nun, A., 1260, 1262, 1269, 1273
Baron, E, 808
Barr, P., 547
Barraclough, B.L., 254, 1139, 1160, 1401
Barrado y Navascués D., 865, 868, 872
Barreiro, R.B., 417, 419
Barret, D., 819, 820, 855
Barrett, A.H., 644
Barriot, J.P., 1322
Barron, D.W., 1567, 1583
Barry, D.C., 299
Barry, J.G., 122, 150
Barsony, M., 868
Barstow, M.A., 893
Barsukov, V.L., 1310, 1320, 1322, 1402
Bartelmann, M., 460, 465–467
Bartels, J., 142, 149, 156–158, 177, 251, 963, 1000, 1113, 1142, 1159, 1568, 1580
Barth, A.J., 526, 854
Barth, C.A., 1198, 1213
Barthelmy, S.D., 523, 525–527
Bartholdi, P., 639
Bartles, J., 229
Bartlett, D.F., 351
Bartman, F.L., 202
Bartoe, J.-D.F., 1014, 1031, 1078, 1085, 1087, 1135, 1137
Bartolini, C., 523
Barwick, S.W., 682, 694
Baryshev, Y.V., 528
Basile, G., 1138
Basilevsky, A.T., 270, 1310, 1320
Basri, G., 869
Bassani, L., 559
Bassner, B., 113
Bassner, H., 113
Basu, S., 1047, 1050, 1051, 1060–1062, 1063, 1575, 1580
Bates, B., 298, 672
Bates, D.R., 201, 1047, 1566, 1580
Bath, G.T., 834
Battrick, B., 187, 1270–1274, 1293
Baturin, V.A., 1057, 1060
Baud, B., 330, 577, 935
Baudin, F., 1061, 1062
Bauer, O.H., 638, 869, 1527, 1528
Bauer, P., 1573, 1580
Bauer, S.J., 1483, 1491, 1492
Bauer, T.M., 1033, 1528
Baugher, C.R., 352
Baum, S.A., 299
Baum, W.A., 290, 298, 381, 1425, 1429
Bauman, C.A., 1376
Baumjohann, W., 1032, 1500, 1526, 1527
Bauschlicher, C.W., 619, 636
Bauschlicher, Jr. C.W., 640
Bawn, C.E.H., 201
Baykal, A., 828, 834
Baym, G., 760, 787
Bazileuski, A.T., 1491, 1321
Beal, R.C., 1608, 1610, 1615
Beals, C.S., 261, 269, 895, 903, 908

Beard, D.B., 246, 251, 252, 1539, 1555
Beasley, A.J., 540, 546
Beatty, J.J., 694
Beatty, J.K., 113, 1322
Beaujean, R., 1003
Beaulieu, J-P., 465
Beaver, E.A., 298, 672
Beaver, M.L., 870
Bechtold, J., 546, 547
Beck, J.G., 1054, 1055, 1060, 1062, 1068, 1085
Beck, S.C., 617, 636
Becker, A.C., 465
Becker, C.M., 756, 959, 960
Becker, D.G., 1188
Becker, I., 300
Becker, R.H., 577, 754, 755, 819, 820, 822, 830, 834, 836, 893, 946, 951, 959, 1392, 1401
Becker, T.L., 1478
Becker, W., 724, 725, 727, 732–735, 737, 738, 740–742, 745–748, 750, 752, 753, 756, 785, 786, 1571, 1580
Becklake, J., 57
Becklin, E.E., 311, 312, 328, 526, 636, 643, 872, 873
Beckman, J.E., 301, 328, 439, 639, 673
Beckwith, S.V.M., 620, 629, 636, 657, 674
Beckwith, S.V.W., 636, 641
Becquerel, H., 149, 1325, 1371
Bedding, T.R., 1059, 1060, 1062
Bedford, D.K., 836
Bedinger, J.F., 190, 193, 201, 202
Bedini, P., 1001, 1138
Bedke, J., 379, 396
Beer, R., 641
Beer, W., 271, 275
Beers, T.C., 393, 396, 435, 437, 439
Beeson, D.E., 1189
Begam, M.C., 525
Begelman, M.C., 371, 558, 559
Begemann, B., 640
Begemann, F., 1329, 1331, 1371, 1402
Behannon, K.W., 999, 1148, 1159, 1492
Behr, A., 233, 235, 243, 251
Behrenfeld, M.J., 1602, 1603, 1615
Beichman, C.A., 329, 330, 577, 640, 868
Beier, E.W., 959
Beigman, I.I., 222
Beintema, D.A., 330, 437, 577, 638–640, 642, 644, 869, 870, 872, 873, 935
Beiser, E., 1374
Beish, J.D., 1321
Bekenkamp, H.E.G., 869
Belcher, J.W., 250, 251, 1000, 1118, 1131, 1137, 1156, 1158–1161, 1212, 1214, 1216
Belcourt, K., 641
Belczynski, K., 523
Belew, L.F., 90, 113
Bell, A.R., 690, 694, 978, 999, 1559
Bell, J.F., 722, 750, 753, 755, 950, 959, 1360, 1371, 1402
Bell, R.A., 427, 437
Bell, S., 787
Bell, S.J., 855, 1581
Bell, T.F., 755, 1581
Bellamine, N., 1087
Bell-Burnell, S.J., 835
Belloni, T., 757, 821, 884, 891
Bellot, R., 1272
Belmahdi, M., 870
Belrose, J.S., 1571, 1580
Belton, M.J.S., 1188, 1251, 1253, 1255, 1257, 1265, 1269, 1273, 1476, 1477
Bémont, G., 1371

Bender, K.C., 1477
Bender, P.L., 351, 1649, 1650
Bender, R., 394, 395, 546, 547
Benedettini, M., 639, 641, 642
Benedict, W., 269
Benediktov, E.A., 1520, 1526, 1551, 1555
Benedix, G.K., 1477
Benford, D., 1291, 1293
Benford, D.J., 637
Benit, J., 638
Benitez, E., 547
Benka, S.G., 1003
Benkhoff, J., 1288, 1292
Benn, C., 524
Benna, C., 222, 1138, 1139
Benner, L.A.M., 1232, 1271
Bennett, C.J., 120, 614, 636, 642
Bennett, C.L., 321, 328, 405, 419–421, 642, 645
Bennett, D.P., 465
Bennett, K., 286, 528, 753, 755, 758, 959
Bennett, L., 1491, 1492
Bennett, R.D., 149
Bentley, W.C., 1372
Benvenuiti, P., 298
Benz, W., 1247–1249, 1266, 1269, 1303, 1320, 1401
Béon, Y., 57, 57
Berchen, J., 1556
Bercovici, D., 1322
Berdichevsky, D., 1139
Berezhko, E.G., 690, 694
Berezinskii, V.S., 683, 685, 687, 688, 692, 693, 694
Berg, O.E., 1175, 1179, 1187, 1189, 1582
Bergaust, E., 57
Bergbusch, P.A., 391, 393
Berge, G.L., 1425, 1426, 1429
Berger, E., 526
Berger, T.E., 1032, 1085, 1066, 1069
Bergeron, J., 542, 546
Bergeron, L.E., 467, 524, 527
Berges, J.-C., 869
Berghöfer, T.W., 593, 603
Bergia, S., 150, 547
Bergin, E.A., 622, 636, 639–642, 1292, 1293
Bergman, M.Yu., 1320
Bergman, P., 1291
Bergmann, P.G., 499, 523
Bergmann, R., 1527
Bergstralh, J.T., 171, 177, 1476, 1477, 1492
Berkhuijsen, E.M., 663, 672, 694
Berkner, L.V., 149, 1562, 1564, 1580
Berkner, L.W., 57, 58, 136
Berkowitz, B., 1274
Berlind, P., 547
Berman, D.D., 1359, 1364, 1365, 1372
Bernadeau, F., 395
Bernal, J.D., 1293
Bernard, J.P., 189, 611, 636, 637, 868
Bernard, R., 189, 201
Bernardeau, F., 468
Bernas, R., 433, 437
Bernatowicz, J.J., 632
Bernatowicz, T.J., 636
Bernstein, G.M., 396, 397, 457, 466, 469
Bernstein, M.P., 638, 642
Berriman, R.W., 125, 149
Berry, H., 1401
Berry, W., 113
Bersolin, F., 394
Bertaux, J.L., 991, 999, 1062, 1198–1201, 1204, 1213–1215, 1233, 1274

Bertelli, G., 718
Bertello, L., 1061, 1062
Berthé M., 1215
Berthelier, J.-J., 1232, 1292, 1293, 1582
Berthias, J.-P., 371
Berthomier, M., 1527
Berthomieu, G., 1060, 1061, 1063
Bertin, E., 395, 468
Bertin, P., 438, 1203, 1215
Bertoldi, F., 617, 620, 630, 636, 638, 640, 642, 643, 863, 870
Bertotti, B., 150, 1650
Bertotti, R., 547
Bertout, C., 639, 862, 869
Bertram, D., 498
Bertram, P., 488
Bertram, R., 547
Bertsch, D.L., 696, 725, 753, 755, 757
Berzelius, J.J., 1280, 1291
Beskin, G.M., 757
Beskin, V.S., 725, 728, 753
Bessel, M.S., 439, 753, 755
Besson, D., 1231
Best, J.E., 1561, 1580
Bethe, H.A., 121, 125, 149, 230, 251, 436, 786
Betz, A.L., 329, 621, 636, 637
Beuermann, K., 687, 694, 802, 808, 809
Beust, H., 864, 869–871
Beynon, W.J.G., 222, 1573, 1580
Bézard, B., 437, 1423, 1439, 1440, 1444, 1448, 1449
Bezrukikh, V.V., 253, 1138
Bhat, P.N., 508, 523, 524, 526
Bhattacharjee, P., 692–695
Bhattacharya, D., 524, 748, 752, 762, 765, 769, 776, 785–787, 789, 810, 814, 819–822
Bianchi, L., 299, 855
Bianda, M., 1074, 1086
Bibring, J.P., 1270
Bickel, A.L., 1401
Biebel, O., 1231
Bieber, J.W., 1002
Bienayme, O., 718
Biermann, L., 123, 133, 149, 180, 187, 225, 232, 234, 243, 251, 896, 908, 1071, 1086, 1115, 1137, 1142, 1159, 1196, 1214, 1236, 1237, 1269, 1497, 1526, 1530, 1555
Biermann, P.L., 149, 488, 495, 690, 693, 696, 697
Biesecker, D.A., 1138, 1140
Bignall, H.E., 559
Bignami, G.F., 286, 744, 753–756, 758, 835
Bijlsma, J., 1598, 1615
Bildsten, L., 285, 370, 768, 786, 814–816, 819, 820, 828, 834, 855
Biller, E.D., 1449
Biller, S.D., 696
Billings, D.E., 221, 222, 234, 235, 251
Bills, B.G., 1321
Binder, A.B., 260, 269, 1389, 1401, 1466, 1476, 1491, 1492
Bindschadler, D.L., 1314, 1320, 1322
Binggeli, B., 379, 380, 393, 396
Bingham, R., 1233
Binney, J., 476, 495, 704, 718
Binns, W.R., 680, 689, 695–697
Binzel, R.P., 1272, 1361, 1371
Bionta, R.M., 527
Biraud, Y., 330
Birch, A.C., 1051, 1060, 1062
Birch, P., 465
Birch, P.V., 1231, 1273, 1291
Birck, J.-L., 1373
Bird, M.K., 1285, 1286, 1291
Birger, B.E., 1429
Birkeland, K., 229, 251, 1569, 1580

Birkett, C.M., 1244, 1269
Birkinshaw, M., 383, 393, 491, 498
Birn, J., 236, 251, 1031, 1032, 1159, 1543, 1557
Bischoff, A., 1402
Bishop, J., 1449
Bishop, J.L., 1371
Biskamp, D., 1009, 1011, 1025, 1031, 1032, 1506
Bisnovatyi-Kogan, G.S., 748, 753, 761, 786
Biswas, S., 991, 993, 999, 1000
Bittencourt, J.A., 238, 239, 251
Biver, H.E., 1292
Biver, N., 637, 1286, 1291, 1293
Biviano, A., 466, 577, 869
Bjoraker, G., 428, 437
Bjoraker, G.L., 1428, 1429, 1430
Bjørgo, E., 1592, 1594, 1595, 1615, 1616
Bjorkholm, P., 528, 1401
Bjorkman, J.E., 902, 908, 911
Bjorkman, K.S., 870, 909, 911
Björnsson, C.-I., 784, 786
Bjornsson, G., 523, 855
Blaauw, A., 702, 718, 760, 786
Black, D.C., 1271
Black, J.H., 615–617, 636, 639, 644, 645, 810, 835, 870, 872
Blackett, P.M.S., 123, 149
Blackwell, D.E., 233, 251
Blackwell, Jr. J.H., 547
Blades, J.C., 548, 670, 672, 673, 675
Blaes, O., 523
Blain, A.W., 462, 463, 466, 468
Blair, D.G., 365, 370
Blair, W.P., 299, 674, 809
Blake, B., 1002
Blake, G.A., 610, 621, 636, 638, 640, 643–645, 872
Blake, J.B., 1000, 1002, 1556
Blake, R.L., 285
Blakeslee, J.P., 394, 523
Blamont, J.E., 180, 187, 189, 191–202, 991, 999, 1196, 1198, 1207, 1213, 1273
Blanchard, A., 395
Blanchard, M.B., 1187
Blanchet, L., 357, 358, 367, 370
Blanco, P.R., 694
Blandford, R.D., 461, 465–467, 521, 523, 527, 544, 546, 552–559, 690, 695, 761, 788, 978, 1000
Bland-Hawthorn, J., 525
Blanford, G., 1293
Blasius, K., 270
Blau, M., 123, 125, 149
Bleeker, J.A.M., 497, 585, 589, 604, 605, 958, 959, 960
Bless, R.C., 287, 291–293, 298, 299, 609, 636, 753, 754
Bliokh, P.V., 465, 466
Blitz, L., 604, 860, 869
Bloch, J.J., 523, 582, 593, 603, 604
Blöcker, T., 935
Blodget, H., 1401
Bloemen, H., 959
Bloemen, J.B.G.M., 684, 688, 695
Blomberg, E.L., 352
Blommaert, J.A.D.L., 869, 871
Blommme, R., 894
Blondel, P.F.C., 862, 869, 870, 872
Blondin, J.M., 831, 834
Bloom, J.S., 511, 513, 516–518, 521, 523, 524–528
Bloom, S.D., 696
Blouke, M., 300
Bloxham, J., 1488, 1491
Blum, J., 1241, 1269, 1272
Blum, K., 1403
Blum, P.W., 991, 1000, 1198, 1213, 1214

Blumenthal, G.R., 505, 525, 552, 558
Blundell, K.M., 540, 546
Bober, A.S., 109, 113
Bobinna, N.N., 1320
Bobylev, L.P., 1615
Bocchia, R., 1061
Bocek, J., 525
Bochkarev, N.G., 547
Bochsler, P., 1001, 1124, 1130, 1131, 1137–1140, 1214
Bock, J.J., 394, 420, 466, 615, 637
Bockelée-Morvan, D., 626, 637, 638, 869, 1232, 1269, 1270, 1286–1288, 1291–1293, 1445, 1446, 1448
Bodenheimer, P., 777, 788
Bodo, G., 553, 558
Boehm, M., 1581
Boehnhardt, H., 526, 1225, 1231, 1272, 1273
Boehnhardt, K.J., 1271
Boehringer, H., 489
Boer, M., 525
Boesgaard, A.M., 429, 437, 872
Boffi, F.R., 396, 808
Bogaert, E., 871
Bogard, D.D., 1341, 1347, 1354, 1360–1363, 1371, 1373–1375, 1392, 1401, 1402
Bogart, R.S., 1061–1063, 1086
Bogdanov, A., 1003, 1233
Boggess, A., 287, 295, 296, 298, 300, 672, 870, 1434, 1448
Boggess, N.W., 321, 326, 328–330, 419–421, 577, 642, 645
Bogomolov, A.F., 1320
Bogun, S., 871
Bohlender, D., 395
Bohlin, J.D., 221, 222
Bohlin, R.C., 298, 299, 657–659, 664, 672, 674, 910
Bohm, A.B., 525
Böhm, K.H., 897, 908
Böhm, M., 1583
Bohn, H.G., 255
Bohn, J.L., 1187
Bohnhardt, H., 524
Böhringer, H., 489, 491, 493, 495–497
Boice, D.C., 1274, 1288, 1292
Boisson, C., 547
Bok, B.J., 704, 718, 860, 869
Boksenberg, A., 287, 294, 298, 300, 546, 547, 672
Bokshtein, I.M., 1320
Boland, W., 616, 637
Boldt, E.A., 497, 498, 543, 546, 577, 754, 820, 822, 834, 836, 959
Bolte, M.J., 547
Bolte, M.S., 396
Bolton, C.T., 760, 786, 811, 820, 823, 834, 840, 841, 854
Bolton, S.J., 1430, 1491
Boltwood, P., 560
Bomford, B., 1650
Bonaldi, M., 371
Bond, D., 547
Bond, G., 1453
Bond, H.E., 513, 523, 935
Bond, I.H., 696
Bond, J.R., 389, 393, 394, 410, 420, 466
Bondi, H., 829, 834
Bonestell, C., 262, 270
Bonetti, A., 247, 248, 251, 1115, 1137
Bonifacio, P., 393
Bonnal, J.F., 869
Bonneau, F., 1060
Bonnell, J.T., 467, 527, 560, 908
Bonnet, H., 467
Bonnet, R.M., 57, 58, 576, 1045, 1060 1071, 1073, 1078, 1086, 1271, 1272

Bonnet-Bidaud, J.M., 833, 837
Bontemps, S., 636, 637, 862, 868, 869, 871, 872
Boogert, A.C.A., 623–627, 637–640, 642, 644, 869, 872, 873
Bookbinder, J.A., 754, 891–894, 1032, 1060
Booker, H.G., 193, 202, 1559, 1563, 1571, 1580
Boonman, A.M.S., 626, 627, 637, 644, 872
Boonstra, A.-J., 869
Boornazian, A.A., 1061
Booth, E.T., 1373
Booth, L., 834
Booth, R.S., 672
Boothroyd, A.I., 436, 439
Bopp, T., 1271
Borderies, N., 1459, 1476
Boreiko, R.T., 621, 636, 637
Borg, H., 1557, 1581
Borg, L.E., 1341, 1342, 1360, 1366, 1367, 1371, 1373, 1375
Borgani, S., 498, 546
Borgeest, U., 456, 466, 468
Boriakoff, V., 782, 786
Borisov, D.N., 1581
Borken, R.J., 603, 605
Bornemann, W., 1214
Börner, G., 760, 786
Börner, H., 1273
Boroson, B., 832, 834
Boroson, T.A., 547
Borovicka, J., 525
Borovsky, J.E., 1580, 1581, 1583
Borrill, J., 394, 420, 466
Bortoletto, F., 869
Borucki, W.J., 1427, 1429
Boscaleri, A., 394, 420, 466
Bosch, J., 697
Bosh, A.S., 1269
Boshi, E., 1650
Boss, A.P., 640, 858, 869, 1270, 1292
Bostrom C.O., 177, 1002
Bottema, M., 300, 306, 328
Bottinelli, L., 382, 393, 395, 397
Bouchet, L., 820
Bouchet, M., 1381, 1401
Bouchy, F., 1062
Boudon, Y., 1651
Bougher, S.W., 1322, 1423
Boulade, O., 869
Boulanger, F., 611, 636–639, 869, 871, 872
Boumier, P., 1060–1062
Bourassa, R.R., 444, 449, 466
Bourgin, M.C., 1213
Bout, M.V., 1138
Bouvier, J., 869, 885, 891
Bouwman, J., 641, 644, 866, 869, 871–873
Bouziani, N., 1232, 1272
Bowell, E., 1244, 1253, 1269
Bowen, G.H., 904, 908, 911
Bowen, I.S., 119, 150
Bower, C.R., 694
Bower, G., 546, 547
Bowers, C., 1231, 1291
Bowers, C.W., 299, 300
Bowers, P.F., 330
Bowles, K.L., 1572, 1573, 1580
Bowyer, C.S., 582, 583, 593, 603
Bowyer, S., 285, 287, 298, 299, 603, 605, 669, 673, 722, 725, 753, 808, 811, 820, 823, 834, 1214
Boxhoorn, D.R., 638, 642, 644, 869, 872, 873
Boyarskiy, M., 960
Boyce, D.C., 1233
Boyce, J.M., 1322, 1354, 1371

Boyd, P.T., 190, 726, 733, 753, 754
Boyd, R.L.F., 51, 58
Boyd, R.N., 437
Boyd, T.J.M., 190, 201
Boyer, C., 265, 269
Boyer, T.P., 1612, 1616
Boyle, B.J., 395, 420, 452, 461, 466, 467, 525, 533, 540, 546
Boyle, P.J., 696
Boynton, P.E., 828, 834, 835
Boynton, W.V., 1402
Bózard, 1439, 1440, 1444
Brace, L.H., 1577, 1580, 1583
Brachet, G., 1650
Bradbury, S.M., 696
Bradt, H., 477, 496, 788, 789, 821, 822, 836
Bradt, H.L., 131, 150, 695
Bradt, H.V., 528, 753, 820 822
Bradt, H.V.D., 722, 725, 764, 786, 812, 817, 820, 821, 824, 834
Braes, L., 760, 786, 840, 854
Brainerd, J.J., 286
Brainerd, T.G., 466
Branch, D., 375, 393, 396, 516, 523, 808
Brand, P.W.J.L., 639
Brandt, J.C., 249, 251, 298, 669, 672, 1293
Brandt, N., 814, 820
Brandt, P., 1611, 1615
Brandt, S., 528
Brandt, W.N., 547
Branduardi, G., 497, 576, 754, 836, 909
Branduardi-Raymont, G., 755, 856
Brans, C., 345, 351, 461, 466
Brault, J.W., 437, 1085
Braun, D.C., 1047, 1048, 1054, 1060–1062
Brautigam, D., 1556
Brazier, K.T.S., 734, 737, 753, 755
Breakwell, J.V., 351
Brebenev, S., 960
Bredichin, T., 1235, 1269
Breen, F.H., 230, 252
Bregman, J.D., 637, 640, 643
Bregman, J.N., 58, 565, 576
Breidenthal, J.C., 352
Breit, G., 1559, 1561, 1571, 1580, 1584
Breitfellner, M., 638, 644, 1270
Breitschwerdt, D., 595, 599, 603, 605, 658, 672, 691, 695–697, 1215
Brekke, P., 1086, 1087
Bremer, M., 524, 525, 528
Brence, W.A., 185
Breneman, H.H., 972–974, 1000
Brenkle, J.P., 352
Breslin, A.C., 696
Bresolin, F., 393, 395–397
Bressan, A., 935
Brett, L., 602
Brett, R., 1373
Bretz, N., 1033
Breukers, R., 638
Brewer, J.P., 524, 526
Bricard, J., 189, 198, 202
Brice, N.M., 1534, 1555
Briceño, C., 885, 891, 893
Bridge, H.S., 251, 1137
Bridges, N.T., 1321, 1371, 1402
Bridle, S.L., 386, 393, 395
Briel, U.G., 496, 497, 576, 754
Briggs, B.H., 1571, 1580
Briggs, B.R., 246, 255, 1539, 1557
Briggs, M.S., 286, 504, 505, 518, 523–526, 758
Brighenti, F., 569, 576, 577
Brillet, A., 365, 370

Brin, G.D., 1218, 1232, 1262, 1269, 1273
Brinca, A.L., 1233, 1527
Brindley, H., 1668
Brinklow, A., 787
Brinkman, A.C., 526, 547, 809
Brinkmann, W., 724, 737, 752, 753, 837
Brinton, H.C., 1546, 1555, 1583
Bristol, R., 299
Britt, D.T., 1371, 1402
Brković, A., 1088
Broadfoot, A.L., 299, 1201, 1215, 1430, 1441, 1442, 1448
Broadhurst, T., 387, 393
Brock, M.N., 523, 527
Brocksopp, C., 788
Broeils, A., 527
Bromage, G.E., 546
Bromley, B.C., 387, 393
Brooke, T.Y., 638, 643, 869, 870, 1244, 1269, 1270, 1287, 1292, 1293
Brooks, C.G., 222
Brosch, N., 287, 298, 820
Brosche, P., 1403
Brossier, C., 1650
Broun, J.A., 228, 251
Brouver, D., 1650
Brown, A., 908, 1215
Brown, C.D., 1322
Brown, D.A., 836
Brown, E.F., 855
Brown, G.E., 526, 767, 778, 786
Brown, G.M., 222, 1372, 1373
Brown, H., 1330, 1334, 1371, 1374
Brown, J.W., 1615
Brown, L.M., 151
Brown, M.E., 1239, 1241, 1244, 1269
Brown, R.A., 1426, 1427, 1429, 1462, 1466, 1474, 1476
Brown, R.H., 299, 872, 1270, 1476–1478
Brown, T.M., 1042, 1047, 1060
Brown, W.A., 1086
Brown, W.R., 525
Browne, I.W.A., 469
Browne, W.A., 467
Brownlee, D.E., 1172, 1185, 1187, 1188, 1293
Brownsberger, K.R., 299
Brucato, R., 787
Bruce, D.E., 640
Brückner, J., 1371, 1374, 1397, 1401, 1402
Brueckner, G.E., 1014, 1031, 1078, 1085, 1087, 1135, 1137, 1138, 1140
Bruhweiler, F.C., 298, 661, 672, 673, 870, 1199, 1214
Brumberg, V.A., 371
Brumfield, M., 300
Brummell, N.H., 1058, 1060, 1063
Brun, A.S., 1063
Brunaud, J., 1086
Bruner, E.C., 1073
Bruner, E.C. Jr, 1086
Brunetti, M., 366, 370
Brusa, R., 1138
Bryam, E.T., 290, 298
Bryan, G.L., 477, 496
Bryant, D.A., 139, 149, 977, 983, 1000
Brynildsen, N., 1069, 1086, 1087
Buccheri, R., 286, 753, 755, 758, 786–789, 812, 820
Buchholz, B., 1073, 1086
Büchler, K., 1188, 1293
Büchner, J., 1011, 1031
Buckingham, R.A., 1580
Buckley, D.A.H., 754, 804, 808, 809, 822
Buckley, J.H., 696
Buckley, M.A., 869
Büdeler, W., 66, 68, 80, 88, 90, 113

Budich, W., 674
Buergi, A., 1214
Buff, J., 829, 830, 834, 836
Bühler, F., 437, 1137, 1231
Buie, M.W., 1269
Builder, G., 1580
Bukata, R.P., 1003
Bulanov, S.V., 694
Bulik, T., 521, 523, 730, 753
Bulkeley, R., 57, 58
Bullock, M.A., 1314, 1319, 1320, 1322
Buneman, O., 1563, 1580
Bunker, A.J., 460, 466, 523, 560
Bunn, E.F., 410, 421
Bunner, A.N., 582, 583, 603–605
Buonsanto, M.J., 1570, 1580
Buote, D.A., 488, 496, 567, 569
Buralga, L.F., 1492
Burba, G.A., 1320, 1321
Burbidge, E.M., 423, 433, 437
Burbidge, G.R., 437, 531, 533, 546, 559, 811, 820, 823, 836
Burbidge, M., 533, 546
Burch, J.L., 1018, 1031, 1582
Burchat, P.R., 1231
Burdett, A.M., 696
Burdo, R.A., 1402
Burenin, R.A., 508, 523
Burg, R., 547
Burgasov, M.P., 113
Burgdorf, M., 636–638, 641, 644, 869, 871, 872
Burger, M., 835
Burgers, E., 150
Burges, J.M., 608, 637
Burgess, D., 981, 1004, 1512, 1527
Burgess, R., 1353, 1371
Bürgi, A., 1001, 1138
Burke, B.F., 1466, 1476
Burke, M.J., 1571, 1580
Burke, W.J., 1018, 1019, 1031
Burke, W.L., 449, 466
Burkepile, J.T., 1161
Burkert, A., 810
Burlaga, L.F., 250, 251, 990, 997, 1000, 1113, 1118, 1120, 1122, 1123, 1131, 1137, 1139, 1141, 1151, 1152, 1156, 1157, 1159, 1160, 1206, 1209, 1214, 1216, 1232, 1492
Burles, S., 385, 393, 396, 428, 437, 440
Burnell, S., 497
Burnett, D.S., 1187, 1331, 1370–1372, 1374, 1376
Burnette, A., 1062
Burnight, T.R., 234, 251
Burns, C.J., 1567, 1578, 1580
Burns, J.A., 1174, 1183, 1187, 1188, 1264, 1265, 1269, 1270, 1403, 1423, 1451, 1456, 1476–1478, 1492
Burns, J.O., 488, 491, 496, 498
Burrows, C.J., 860, 861, 864, 869
Burrows, D.N., 590, 591, 597, 599, 603–605
Burrows, J.R., 1582
Burstein, D., 393, 497
Burstein, P., 589, 603
Burton, E.F., 677, 696
Burton F., 872
Burton, M.E., 1159
Burton, W.B., 587, 592, 604, 605, 607, 637, 639, 651, 663, 672, 707, 718
Burton, W.M., 294, 298, 1232
Burud, I., 456, 466, 523
Burwitz, V., 756
Bus, S.J., 1246, 1269
Buser, R., 396
Bush, R.I., 1062, 1063, 1086, 1527
Buss, R.H., 632, 637

Busse, F., 1480, 1491
Bussoletti, E., 328, 1232
Bussons, X.X., 696
Butcher, H., 475, 492, 496, 498
Butler, B., 1293
Butler, C.C., 125, 151
Butler, C.J., 892, 894
Butler, H.E., 298–300
Butler, R.C., 524, 527, 725, 753, 757
Butler, R.P., 864, 871, 1060
Butt, Y.M., 959
Butterworth, P.S., 523, 525–527
Buzzoni, A., 576
BVSP, 1354, 1357
Bybee, R., 300
Byers, N., 150
Bykov, A.M., 690, 695
Byram, E.T., 222, 285, 286, 475, 477, 496, 604, 753, 820, 834, 1215
Byun, Y.-I., 467

Ca, P.P.D, 298
Cabannes, J., 189, 202
Cabot, H., 1270
Cabrit, S., 189, 620, 637, 862, 869
Cacciani, A., 1063
Cahill, L.J., 1019, 1530, 1532, 1555
Cahill, L.J. Jr., 1033, 1555
Cahn, R.N., 1231
Caillault, J.-C., 893
Caillault, J.-P., 868, 877, 883, 891, 893
Cailloux, M., 468
Cain, D.L., 352
Cain, J.C., 1001, 1004, 1138, 1480, 1491
Cairns, I.H., 1216
Calcutt, S.B., 1423
Caldwell, J.A.R., 465
Caldwell, J.J., 1478
Callanan, P.J., 525, 821
Callander, B.A., 1615
Cally, P.S., 1080, 1086
Calvet, N., 630, 637, 861, 869
Calvin, W.M., 1478
Camerini, U., 754
Cameron, A., 827, 834
Cameron, A.G.W., 433, 437, 1320, 1378, 1386, 1400, 1401
Cameron, G.H., 120, 150
Cameron, T.A., 1584
Cami, J., 632, 634, 637, 640, 644, 873
Camilio, F., 371
Camilo, F., 748, 753, 755, 766, 767, 783, 784, 786
Camm, G.L., 1214
Camm, J.M.A., 1192
Campana, S., 524, 755
Campbell, A., 428, 437
Campbell, D.B., 352, 1246, 1269, 1271, 1320
Campbell, G.S., 1584
Campbell, N., 1326, 1371
Campbell, W.H., 252, 1480, 1491
Campbell, W.J., 1593, 1615, 1616
Campbell, W.W., 934
Campbell-Wilson, D., 855
Campins, H., 866, 869, 870, 1243, 1244, 1247, 1269, 1273
Campusano, L.E., 393
Canal, R., 810
Canceil, P., 1651
Candelaria, M.P., 1401
Cane, H.V., 969–971, 974, 1000, 1122, 1137, 1155, 1159
Canizares, C.R., 457, 458, 466, 567, 569, 576–578, 754, 755, 756, 960
Cannon, D.B., 497, 498
Cannon, R., 395, 525

Cantin, M., 1061
Canton, J.W., 226, 251
Canup, R., 1322, 1374, 1375
Caon, N., 523
Capaccioli, M., 380, 393
Capaccioni, F., 1273
Caplinger, J., 560
Cappa de Nicolau, C.E., 674
Cappellaro, E., 526
Cappi, M., 559
Capria, M.T., 1273
Caputo, F., 397
Caputo, M., 1650
Caraveo, P.A., 286, 524, 723, 737, 740, 753, 755, 756, 758
Card, G., 1063
Cardelli, J.A., 298, 615, 637, 660, 670, 672–674
Carey, S.J., 323, 328, 330
Cargill, P.J., 1003
Carignan, G.R., 1430, 1583
Carini, M.T., 547
Carlberg, R.G., 386, 393, 483, 496
Carlson, B., 428, 437
Carlson, B.E., 1427–1429, 1469
Carlson, C.W., 1232, 1320, 1491, 1492, 1526, 1527, 1536, 1555, 1557, 1578, 1580
Carlson, R.W., 1334–1336, 1343, 1371, 1385, 1401, 1413, 1423, 1429, 1477
Carlsson, M., 1070, 1073, 1074, 1086, 1088
Carlstrom, J., 645
Carmichael, H., 130, 149, 1013, 1031, 1111, 1113
Carole, 222
Carone, C.D., 1231
Carone, T.E., 547
Carpenter, D.L., 1534, 1535, 1555, 1562, 1580
Carpenter, G.F., 829, 837
Carpenter, K.G., 298, 672
Carpenter, R.L., 265, 269
Carpentier, G., 201
Carr, B., 386, 393
Carr, F.A., 298
Carr, J.S., 638, 640
Carr, M.H., 1297, 1304, 1307, 1320, 1323, 1423, 1474, 1476
Carr, R.E., 187
Carr, T.D., 1425, 1429
Carramiñana, A., 755
Carrier, F., 1060
Carrier, W.D., 1273
Carrillo, R., 547
Carrington, R.C., 229, 252, 967, 1000
Carroll, S.M., 390, 393
Carron, M.K., 1401
Carru, H., 1568, 1580
Carruthers, G., 291, 292, 298, 608, 637
Carsey, F., 1593, 1615
Carson, P.P.D., 672
Carswell, R.F., 469
Carter, D.A., 1581
Carter, T., 1032, 1033
Carter, W.E., 468
Carter-Lewis, D.A., 696
Casali, M.M., 872
Casanova, S., 891
Casares, J., 525, 841, 843, 854
Casas Serradilla, J.L., 1231
Casas, R., 1272
Caselli, P., 626, 637
Casey, S., 642
Cash, W.C., 299, 545, 546, 593, 604
Caskey, J.E., 201
Caso, C., 1218, 1231

Casperson, D., 523
Cassatella, A., 298
Cassé, M., 434, 437–440, 689, 695
Cassen, P.M., 1335, 1374, 1432, 1448, 1477, 1491
Cassinelli, J.P., 251, 253, 642, 872, 878, 879, 891–893, 896–899, 902, 903, 908–910
Cassisi, S., 391–393, 395, 1453
Castander, F.J., 461, 466
Castaneda, H., 871
Castelaz, M.W., 870
Castellani, V., 393
Castets. A., 637
Castor, J.I., 824, 829, 834–836, 897, 903, 908–910
Castro, P.G., 395, 420, 467
Castro-Tirado, A.J., 518, 523–525, 527, 528
Catala, C., 862, 869, 872, 1059, 1060
Catanese, M., 554, 558, 696
Catelli, J.R., 523
Cattaneo, F., 892, 910
Cattell, C.A., 1526
Cattell, D.M., 1555
Catura, R.C., 1086
Caudell, T.P., 1038, 1062
Caughlan, G.R., 430, 437
Caux, E., 314, 328, 636–638, 641–643, 869
Cavaleri, L., 1616
Cavaliere, A., 477, 496, 497, 558
Cavalieri, D.J., 1594, 1615, 1616
Cavallini, E., 371
Cavallo, G., 504, 509, 523
Cavanie, A., 1593, 1615
Cawley, M.F., 696
Cayrel, R., 391–393, 428, 437, 439, 908
Cazenave, A., 1638, 1650
Cazes, S., 869, 1271, 1272
Ceccarelli, C., 621, 626, 627, 637, 641, 869
Ceifersons, A., 1491
Celnikier, L.M., 1270
Celotti, A., 547, 556–560
Celsius, A., 226, 252
Cen, R., 392, 393, 395, 474, 477, 496
Cendales, M., 1403
Centurion, M., 524
Ceplecha, Z., 525, 1163, 1187, 1269, 1269
Cerdonio, M., 371
Cernicharo, J., 620, 621, 633, 634, 636–638, 644, 645
Cerulli, R., 641, 869
Cerutti, H., 437
Cesarsky, C.J., 570, 572, 573, 576, 577, 636, 637, 693, 695, 696, 862, 869, 871, 872
Cesarsky, D.A., 617, 620, 637, 869, 871
Cespedes, E., 1061
Cha, A.N., 299, 597, 604
Chabaudie, J.E., 352, 1651
Chaboyer, B., 391–393, 1060
Chadwick, D.J., 1322
Chadwick, J., 721, 1326, 1371
Chae, K.-Y., 457, 466
Chaffee, F.H., 468, 523, 526, 527, 548
Chaikin, A., 1322, 1323
Chakrabarty, D., 736, 753, 755, 786, 814, 820, 834
Challis, P., 395, 420, 468
Chalonge, D., 1066, 1086
Chalov, S., 1209, 1214
Chamberlain, J.W., 190, 202, 246, 247, 252, 1115, 1137, 1196, 1463, 1476
Chamberlain, V.D., 57, 58
Chamberlin, R.T., 1280, 1292
Chamberlin, T.C., 1280, 1292
Champion, J.-P., 1449
Champney, J.M., 1270
Chan, 642

Chan, H.F., 820
Chandler, M.O., 1556, 1582
Chandrasekhar, S., 132, 149, 230, 252, 653, 672, 721, 753, 895, 908
Chang, C.C., 255
Chang, C.-L., 1556
Chang, S., 247, 1284, 1292
Chanin, M.L., 198–202
Chapell, C.R., 1546
Chaplin, W., 1060
Chapline, G., 787
Chaplinne, W.J., 760, 1046, 1060, 1062
Chapman, C., 270, 1491
Chapman, C.R., 177, 1296, 1300, 1321, 1323, 1401, 1402, 1423, 1470, 1476, 1477
Chapman, G.A., 1063
Chapman, G.W., 190
Chapman, S., 133, 149, 156–158, 177, 190, 202, 228, 230, 234, 243, 252, 255, 963, 1000, 1115, 1121, 1137, 1142, 1159, 1497, 1526, 1530, 1539, 1545, 1555, 1565, 1566, 1568, 1580
Chappell, B.W., 1401
Chappell, C.R., 1555, 1556, 1582
Chapuis, C., 753
Chaput, C.J., 694
Char, S., 870
Charbonneau, P., 438, 1058, 1060
Charbonneau, R., 428
Charbonnel, C., 435, 437, 439, 440
Charles, P.A., 757, 801, 804, 808, 809, 821, 829, 834, 837, 854
Charles, P.H., 789
Charmandaris, V., 577
Charnley, S.B., 610, 621, 626, 637, 638
Charra, J., 1061, 1062
Charra, M., 1061
Chartas, G., 547
Chassefiere, E., 1201, 1204, 1214
Chaston, C.C., 1526
Chaussidon, M., 429, 437
Chauvineau, J.P., 1086
Chayer, P., 299
Che, A., 906, 909
Chebatorev, G.A., 1274, 1293
Chedin, A., 869
Chelebowski, T., 899, 910
Chelton, D., 1617
Chemodanov, V.P., 1321
Chen, H., 1274
Chen, J., 523, 1247, 1264–1266, 1268, 1269, 1271
Chen, J.H., 1336, 1371
Chen, K.Y., 784, 786
Chen, W., 597, 604, 641, 845, 849, 854, 856
Chenette, D.L., 147, 149, 250, 252
Cheng, A.F., 177, 752, 753, 786
Cheng, E.S., 419–421, 642, 645
Cheng, F.-H., 809, 810
Cheng, F.-Z., 547
Cheng, K.R., 1199, 1214
Cheng, K.S., 724, 728, 753, 820
Chereul, E., 718
Chernaev, V.I., 189, 202
Chernaya, I.M., 1320
Chernoff, D.F., 371, 367, 639
Chernych, N.S., 525
Chetin, T., 835
Cheung, A.C., 607, 637
Chevalier, A., 1061
Chevalier, R.A., 691, 695, 778, 786
Chèvreton, M., 468
Chew, G.F., 252
Chiang, E.I., 630, 637, 638
Chiang, J., 755
Chiappetti, L., 559

Chiappini, C., 435, 437
Chiar, J.E., 613, 614, 625, 638, 639, 642, 644, 869
Chieffi, A., 395
Chieppa, F., 351
Chihara, H., 640, 871
Chillier, X.D.F., 1373
Chin, G., 636, 639, 641, 642, 1293
Chincarini, G., 496, 497, 524
Chiosi, C., 935
Chipman, E.G., 1069, 1086
Chitre, S.M., 1060, 1063
Chiu, H.Y., 722, 753, 811, 820
Chiu, J., 786, 820, 834
Chiu, W.A., 387, 393
Chiu, Y.T., 1516, 1526
Chmielewski, Y., 429, 437
Cho, J.Y.N., 1563, 1581
Choate, D., 1232, 1271
Chochia, P.A., 1320
Chodas, P.W., 1232, 1271, 1322
Chodos, A.A., 1370
Chokshi, A., 1262, 1269
Choudhuri, A.R., 1069, 1086
Chree, C., 229, 252
Christen, M., 1491
Christensen, A., 1582
Christensen, J., 396
Christensen, J.A., 547
Christensen, L., 523
Christensen, U.R., 1309, 1321
Christensen-Dalsgaard, J., 1038, 1042, 1045, 1047, 1049, 1050, 1060–1063
Christensen-Dalsgaard, U.R., 1309
Christian, C.A., 394
Christian, E.R., 696, 697, 992, 1000, 1002, 1004, 1214
Christian, H.J., 524
Christian, R.P., 1401
Christiansen, P.J., 1557
Christoforidis, A., 1292
Christon, S.P., 1519, 1526, 1540, 1555
Christou, J., 871
Christou, J.C., 1270
Chu, Y., 673
Chu, Y.-H., 395, 917, 935
Chubb, T.A., 283, 285, 298, 753, 754, 820, 834, 1215
Chudakov, A.E., 1004
Chulkov, I., 820, 960
Chupp, E.L., 972, 1000
Churazov, E.M., 523, 820, 960
Church, S.E., 641, 869
Churchwell, E., 642, 860, 869, 904, 911
Chuvaev, K.K., 547
Chuvilgin, L.L., 696
Chwolson, O., 443, 466
Chyba, C.F., 1293
Ciaravella, A., 222, 1138, 1139
Ciardi, B., 520, 523
Ciardullo, R., 924, 935
Cicerone, R.J., 1668
Cimatti, A., 524
Cinti, M.N., 524, 525
Ciotti, L., 498, 567, 569, 576
Ciufolini, I., 347, 351
Cladis, J.B., 163, 177
Claeskens, J.-F., 454–458, 461, 465, 466, 468
Clairemidi, J., 1218, 1223
Clampin, M., 870
Claret, A., 523, 820, 869
Clark, B.C., 1187, 1188, 1293, 1376, 1391, 1395, 1400, 1401
Clark, B.G., 651
Clark, D.H., 834

Clark, D.K., 1615
Clark, F.O., 328
Clark, G., 577, 754
Clark, G.W., 285, 585, 586, 589, 604, 782, 786, 811, 812, 820, 821, 831, 833–836, 960
Clark, R.N., 1402, 1462, 1477, 1478
Clark, T.A., 560
Clarke, J.T., 299, 869, 1215
Clarke, T.E., 504, 523
Clasen, J.W., 466
Clauser, F., 242, 246, 252
Clauser, T., 1197, 1214
Clavel, J., 298, 329, 534, 546, 547, 577, 870
Claver, C.F., 523
Claverie, A., 1043, 1061
Clavier, J.-P., 870
Clay, J., 120, 149, 678, 695
Clayton, D.D., 902, 909, 990, 1002
Clayton, R.N., 1385, 1386, 1401
Clear, J., 723, 753
Cleaver, A.V., 114
Clegg, P.E., 330, 577, 636–639, 641, 642, 862, 869
Clements, S.D., 547
Clemett, S.J., 1373
Clemmow, P.C., 1559, 1571, 1580
Clette, F., 1086
Clifford, S.M., 1307, 1321, 1322
Cline, T.L., 149, 501, 502, 508, 523–528, 754, 1000
Cliver, E.W., 910, 969, 972, 975, 1000
Clocchiatti, A., 395, 420, 468
Cloutier, P.A., 1320, 1321 1491, 1492
Clow, G.D., 1321
Clube, S.V.M., 1238, 1269
Clune, T.L., 1063
Coble, K., 394, 420, 466
Coblentz, W.W., 304, 328
Coccia, E., 370, 371
Cochran, A.L., 547, 1239, 1269
Cocke, W.J., 722, 753
Code, A.D., 287, 289, 292, 293, 298, 299, 793, 809, 1425, 1429
Codrescu, M.V., 1581
Coe, R.S., 1491
Coffin, M.F., 1316, 1321
Cohen, C.M.S., 973, 1000, 1002, 1004
Cohen, D.H., 881, 891, 899, 901, 903, 908–910
Cohen, J.G., 527, 667, 672
Cohen, M., 636–638, 639, 641, 908
Cohen, M.H., 560
Cohen, R., 193, 202, 1580
Cohen, R.D., 547
Cohen, R.J., 577, 636
Cohen, R.S., 604
Colafrancesco, S., 497
Colas, E., 1270
Colas, F., 1271, 1272
Colbert, E.J.M., 540, 548, 854
Colburn, D.S., 1004, 1031, 1140
Cole, A., 525
Cole, S., 393
Coleman, C.I., 298
Coleman, P.C., 603
Coleman, P.J. Jr., 137, 139, 149, 151, 248, 252, 1004, 1132, 1137, 1139, 1140, 1144, 1158, 1159, 1216, 1491
Coleman, P.L., 604
Coleman, R.J., 252
Coles, P., 385, 393
Coles, S.G., 1598, 1615
Coletta, A., 524, 527
Coley, W.R.N., 1580
Colgate, S.A., 501, 516, 523, 524
Colin, L., 177, 1296, 1321, 1323, 1430

Collard, A.D., 1423
Colley, W.H., 459, 466, 467
Collins, C.A., 496, 497
Collins, R.A., 787, 855, 1581
Collin-Souffrin, S., 536, 546
Collmar, W., 560
Collura, A., 577, 893
Colom, P., 637, 1269, 1271, 1291, 1293
Colom, R., 1265
Colomb, F.R., 665, 672
Colombo, A., 371
Colombo, G., 268, 269, 1188, 1401, 1478
Colpi, M., 757
Colvin, R.S., 673
Colvin, T.R., 1322
Colwell, J.E., 1242, 1274
Comastri, A., 546–559
Combes, M., 869, 1270, 1273, 1426, 1429
Comeron, F., 860, 871, 882, 892
Cominsky, L., 285, 496, 754, 809, 835
Comiso, J.C., 1594, 1615, 1616
Compston, W., 1331, 1343, 1371, 1373, 1381, 1401
Compton, A.H., 120, 149, 678, 694
Conard, S.J., 299
Condie, K.C., 1317, 1321
Condon, J.J., 548
Conforto, G., 1231
Conlon, T.F., 149, 252
Connaughton, V., 286, 508, 524, 525
Connell, J.J., 689, 695
Connelly, J.N., 1371
Conner, J.P., 524
Connerney, J.E.P., 1301, 1307, 1320, 1321, 1445, 1449, 1479, 1482–1487, 1491, 1492
Connes, J., 269
Connes, P., 262, 269, 1062, 1445
Connors, A., 523
Conrath, B.J., 1429, 1436, 1437, 1441, 1442, 1446, 1449, 1450
Conselice, C., 525
Constable, C., 1491
Contaldi, C.R., 417, 420
Conti, L., 371
Conti, P.S., 829, 834
Cook, A., 1164
Cook, A.F., 1188
Cook, A.F. II, 1478
Cook, G.E., 1651
Cook, J.W., 1087
Cook, K.H., 465
Cook, R.A., 1402
Cook, W.R., 972, 1000, 1004
Cooke, B.A., 478, 496, 498, 836
Cooke, H.L., 677, 697
Cool, A.M., 719
Cooper, C.D., 202
Cooper, J.F., 1271
Cooper, R.G., 891, 900, 909
Cooray, A.R., 457, 466
Copet, E., 636, 869, 871, 872
Coplan, M.A., 1124, 1137–1139, 1214
Coppi, P.S., 854
Coradini, A., 1273
Coradini, M., 1271, 1272
Corbato, S.C., 695
Corbel, S., 736, 753
Corbet, R.H.D., 736, 753, 768, 769, 786, 828, 834, 835, 837
Corbin, M.R., 330
Cordes, J.M., 668, 672, 675, 755, 757, 761, 786
Cordier, B., 820
Córdova, F.A., 724, 753, 756, 793, 795–797, 808, 809
Corey, B.E., 467

Cormier, R., 1232, 1271
Cornell, E.A., 351
Cornett, R., 394
Cornwall, L., 871
Coron, N., 328, 330, 1270
Coroniti, F.V., 1002, 1011, 1031, 1491, 1556
Corradi, R.L.M., 935
Correia, J.C., 642
Cortes, T.R., 1061
Cortez, B.G., 959
Cosmovici, C.B., 1218, 1223, 1231, 1271, 1272
Cospar, 1649
Costa, E., 510, 523–528, 754
Costa, J., 1210, 1214
Costard, F.M., 1322
Costes, N.C., 1273
Coté, J., 869, 911
Cote, P., 523
Cotin, F., 869
Couch, W.J., 395, 420, 468, 498
Cougrand, B., 1061
Counselman, C.C. III, 777, 786
Coupiac, P., 870
Courbin, F., 466, 523
Cour-Palais, B.G., 1168, 1187, 1188
Courtès, G., 201
Courtillot, V., 261, 269
Courtin, R., 1439, 1449
Courts, G.R., 298
Coustenis, A., 1449
Coutu, S., 694
Couvidat, S., 1063
Covarrubias, R., 525
Covino, S., 518, 524, 755, 887, 891
Cowan, J.J., 391–393, 396, 397, 440, 547
Cowie, L.L., 299, 439, 481, 496, 661, 672, 673
Cowley, A.P., 807, 808, 814, 820, 841, 854, 855
Cowley, S.W.H., 1032
Cowling, T.G., 234, 236, 252, 1008, 1009, 1013, 1031, 1567, 1581
Cowsik, R., 636
Cox, A.N., 1060, 1061
Cox, D.E., 898, 909
Cox, D.P., 584, 589, 590, 595, 597, 601, 604, 605, 663, 665, 672
Cox, P., 559, 570, 576–578, 610, 633, 636–638, 640–645, 856, 869, 870, 872, 873
Craig, N., 675
Craig, W.W., 736, 754
Cram, L.E., 910
Crampton, D., 468, 547, 808, 820, 854, 855
Crane, P., 438
Cranmer, S.R., 222, 901, 909, 910, 1085, 1086, 1138, 1139
Craubner, H., 1274
Cravens, T.E., 1229, 1231–1233, 1271
Crawford, F., 755
Crawford, I.A., 667, 668, 671, 672, 792, 793, 804, 806, 811
Crawford, J.A., 808, 820
Crawford, R.L., 1231
Creech-Eakman, M.J., 638, 644, 872
Crenshaw, D.M., 547
Crespo, A., 1402
Cretolle, J., 869, 1061
Creze, M., 711, 718
Crifo, F., 605, 869, 1216
Crifo, J.F., 864, 1217, 1220, 1222, 1223, 1232, 1233, 1258, 1264, 1269
Crill, B.P., 394, 420, 466
Crips, D., 1292
Crisp, D., 869, 1402, 1423
Crisp, J.A., 1371, 1374, 1402
Cristiani, S., 466, 546
Crivel, C., 1061
Crivellari, L., 872

Crivelli Visconti, V., 371
Croft, S.K., 1322, 1452, 1472, 1476, 1478
Crommelynck, D., 1060, 1061, 1658, 1659, 1668
Cronin, J., 1284, 1292
Cronin, J.W., 123, 149
Crooker, N., 1002, 1088, 1137–1139
Crooker, N.U., 1016, 1017, 1031, 1152, 1159, 1160
Croom, S.M., 466
Cropper, M., 802, 809
Cros, A., 1232, 1491
Crosa, L., 810
Croswell, K., 702, 718
Crouch, T.D., 57, 58
Crovisier, J., 630–632, 637, 638, 640, 866, 869, 1261, 1270, 1286–1289, 1291–1293, 1449
Crowley, G., 1584
Crowley, J.K., 1477
Crown, D.A., 1321
Cruddace, R.G., 496, 497
Cruikshank, D.P., 171, 177, 260, 269, 300, 608, 642, 1236, 1241, 1244, 1253, 1270, 1271, 1274, 1296, 1321, 1462, 1465, 1466, 1476–1478
Cruise, A.M., 959
Crumpler, L.S., 1321
Crutcher, R.M., 666, 667, 672, 1203, 1214
Crutzen, P.J., 1660, 1668
Cruvellier, P., 869
Cugnon, P., 1086
Cui, W., 856
Cuillandre, J.C., 468
Culhane, J.L., 497, 755, 869, 893
Cullough, T., 270
Cullum, M., 757
Culter, C., 367
Cummack, C.H., 1571, 1582
Cummings, A.C., 142, 149, 696, 697, 785, 786, 987, 992–995, 997, 1000, 1002, 1004, 1156, 1159, 1200, 1201, 1212, 1214–1216, 1555
Cummings, J.R., 1000, 1003, 1004, 1533, 1555
Cunha, K., 425, 429, 434, 437
Cunningham, L.E., 1269
Cuntz, M., 909, 1073, 1086
Curdt, W., 1070–1072, 1086, 1087, 1232, 1233, 1272
Curie, M.S., 1325, 1371
Curie, P., 1325, 1371
Currie, B.W., 189, 202
Curtis, D.W., 1232, 1491, 1492
Curtis, H.B., 934
Curtis, H.D., 546
Curtis, S.A., 1488, 1491
Cushman, G.W., 210, 222
Cusumano, G., 524, 527, 755, 756
Cutler, C., 367, 370
Cutri, R.M., 546, 636, 1478
Cuttitta, F., 1381, 1401
Cuzzi, J.N., 1187, 1237, 1270
Cyr, O.C.S., 1140
Czarny, J., 869, 872
Czechowski, A., 1206, 1214
Czerny, B., 546

D'Alessio, P., 637
d'Odorico, V., 458
Da Costa, G.S., 394, 577
da Costa, L.N., 387, 393, 394
Daglis, I.A., 1543, 1555
Dahl, O., 1231
Dai, H.Y., 695
Daigneau, P.S., 1138
Daily, W.D., 1491
Dal Fiume, D., 524, 525, 527, 959
Dalal, N., 465

Dalaudier, F., 1200, 1214
Dalcanton, J.J., 457, 466, 458,
Dale, D.A., 378, 393
Daley, K.A., 1002
Dalgarno, A., 187, 201, 202, 617, 643, 656, 672
Dalitz, R.H., 1231
Dalle Ore, C.M., 1476
Dalrymple, G.B., 1348, 1371, 1373
Daltabuit, E., 898, 909
Dalton, J., 228, 252, 560
Dame, L., 1061
Dame, T.M., 597, 599, 601, 604, 605, 753
D'Amico, N., 755, 757, 787, 788
Daminelli, A., 909
Dammasch, I.E., 1086, 1088
Damour, T., 350, 351, 355–358, 361, 364, 365, 369–371, 1231
Danby, J.M.A., 1192, 1214
Danchin, R.V., 1403
Danielson, G.E., 1423, 1477
Danielson, R.E., 306, 328, 329, 1068, 1086, 1453
Danks, A.C., 300, 604
Danly, L., 498, 601, 604, 669, 670, 672
Danner, R., 736, 753
Dano, E., 1615
Dantel-Fort, M., 395, 467, 468
Danz, M., 1273
Danziger, I.J., 525, 755
Danziger, J., 527, 605
Danzmann, K., 365, 371
Däppen, W., 1060–1062
Dar, A., 525
Darrow, K.K., 117, 149
Dartois E., 626, 638, 869, 1270
Darwin, G.H., 272, 275, 777, 786, 1456
Das Gupta, M.K., 547
Dasch, E.J., 1350, 1371, 1374, 1375
da Silva, L., 397, 437–440
Datlowe, D.W., 528
Daubar, I.J., 1188
Daugherty, J.K., 728, 753
Davé, R., 392, 491, 496
David, L.P., 386, 393, 488, 496, 639
Davidsen, A., 589, 604, 823, 827, 833, 834
Davidsen, A.F., 298, 299
Davidson, C.R., 466
Davidson, G.T., 177
Davidson, J.A., 639, 643, 645
Davidson, K., 311, 329, 651, 672, 768, 772, 773, 786, 835, 903, 909
Davies, J., 636, 869, 871
Davies, J.K., 524, 637, 869, 1291–1293
Davies, M.B., 787
Davies, M.E., 1322, 1476
Davies, R.E., 828, 835
Davies, R.L., 393, 394
Davila, J.M., 1087
Davis, A.M., 1373
Davis, D.R., 1242, 1270, 1297, 1321, 1386, 1402, 1477
Davis, D.S., 488, 490, 496, 497
Davis, G.L., 1370, 1376
Davis, G.R., 437, 641, 869, 1449
Davis, J., 299
Davis, J.L., 467
Davis, L., 132–134, 149, 150, 231, 232, 246, 248–253, 255, 902, 905, 1004, 1118, 1131
Davis, L. Jr., 911, 1137, 1140, 1159, 1214, 1492
Davis, L.E., 1141, 1142, 1144, 1196
Davis, L.R., 1582
Davis, M., 393–396, 414, 420, 466, 526, 605
Davis, R.J., 292, 299, 810, 835
Davis, T.N., 1542, 1555
Davison, P.J.N., 497

Dawson, B.R., 695
Dawson, D.D., 1322
Dawson, J.M., 1233
Dawson, S., 560
Day, C.S.R., 789, 822, 824, 831, 835
Dayal, A., 1272
Dazeley, S.A., 697, 960
de Amici, G., 419, 421
de Bergh, C., 1476, 1477
de Bernardis, P., 394, 415, 420, 463, 466
de Boer, K.S., 294, 297–299, 668, 674
de Boula, O., 637
de Bruyn, A.G., 524, 757
De Campli, W.M., 884, 891
de Cuyper, J.P., 786
De Forest, C., 1086
De Gasperis, G., 420, 466
de Geus, E.J., 604
de Graauw, M.S., 637, 641, 871, 872
de Graauw, T., 437, 624, 637–642, 644, 645, 862, 863, 869–873, 1438, 1441, 1449
De Grandi, S., 485, 489, 496, 497
De Greve, J.P., 894
de Greve, J.R., 786
de Groene, P., 869
de Jager, C., 191, 193, 197, 201, 207, 210, 211, 214, 215, 218, 221, 222, 294, 295, 299, 554, 1581
De Jager, O.C., 554, 559, 757
De Jong, J.A., 813, 820, 909
de Jong, T., 330, 577, 616, 637, 640, 641, 643, 644, 868, 871, 873
de Jonge, G., 218, 222
De Keyser, J., 252
de Kool, M., 835, 836
de Korte, P.A.J., 584, 604
de Koter, A., 641, 644, 869, 871, 873
de La Reza, R., 870
de Loore, C., 761, 786, 789, 835
De Luca, E.E., 1032
de Mairan, J.J., 226, 235
De Marco, O., 935
De Maria, M., 120, 150
De Marian, J.J. d'Or, 252
de Martino, D., 908
de Medeiros, J.R., 397, 437, 438, 439, 440
de Muizon, J, 623, 632, 638, 640, 869, 870, 1276
de Muizon, M., 638
de Nolfo, G.A., 694
de Pater, I., 1426, 1429, 1440, 1443, 1448, 1449
De Pontieu, B., 1032
De Sabbata, V., 439
De Sanctis, M.C., 1273
de Sitter, W., 339
de Vaucouleurs, G., 269
de Vicente, P., 638
De Vorkin, D.H., 120
de Vries, C.P., 809, 959
de Wildt, P., 525
de Winter, D., 870, 872
Deane, J.R., 1449
DeBra, D.B., 351
Decaudin, M., 1061, 1086
Decher, R., 352
Decin, G., 864, 869
Decin, L., 637, 871
Decker, R.B., 977, 978, 1000, 1002, 1003
Decourchelle, A., 948, 949, 959
Deeg, H.J., 1063
Deepak, A., 1668
Deerenberg, A.J.M., 604

Deeter, J.E., 828, 834, 835
Defise, J.M., 1086
DeForest, C., 1062
DeForest, S.E., 1543, 1555
Degel, W., 525
Degewij, J., 1271
degl'Innocenti, S., 393
Deguchi, S., 643
Deharveng, J.M., 547
Dehnen, W., 718
Deinzer, W., 897, 908
Dekel, A., 387, 393, 396, 397, 440
DeKeyser, J., 249
deKoter, A., 910
del Sordo, S., 527
Delaboudinière, J.-P., 1060, 1077, 1086
Delache, P., 1061
Delamere, A., 300
Delamere, W.A., 1271, 1272, 1274
Delaney, M., 466, 577
Delano, J.W., 1385, 1388, 1389, 1401
Delcourt, D.C., 1549, 1557
Delgado-Martí, H., 825, 835
Delitsky, M., 1291
Deliyannis, C.P., 437, 439
Della Ceca, R., 498
Della L. Lucca, 1582
della Valle, M., 393, 396
Dell'Antonio, I., 397, 469, 498
Dellinger, J.H., 204, 222, 1562, 1581
Dello Russo, N., 1286, 1287, 1292, 1293
Delory, G.T., 1516, 1526, 1527
Delsemme, A.H., 1187, 1239, 1261, 1270, 1281, 1282, 1292
DeLuca, E.E., 1087
Delyannis, C.P., 436
DeMarcus, W.C., 1425, 1429
Demarque, P., 393, 394, 439, 1060
Dement'ev, S., 1491
Demidov, M.L., 1062
Démoulin, P., 1027, 1032
Dempsey, R.C., 886, 891, 908, 1215
Demyk, K., 612, 623, 638
den Boggende, A.J.F., 809
den Herder, J. W., 809
den Oord, G.H.J., 893
Dence, M., 270
Denchikova, L.I., 1232
Denis, N., 1061
Denissenko, D., 504, 524, 528
Denman, K., 1600, 1615
Dennerl, K., 1232
Dennis, B.R., 1003
Dennis, E.F., 1138
Dennison, B., 330
Densham, R.H., 834
Dent, W.R.F., 870, 1291, 1292
Denton, R.E., 1032
Depagne, E., 439
DePaolo, D.J., 1374
D'Ercole, A., 498, 576
Dere, K.P., 1027–1029, 1031, 1074, 1085–1087, 1138, 1140
Dermer, C.D., 518, 524, 552, 558, 689, 695, 697, 728, 757
Dermott, S.F., 1179, 1187, 1188
Deruelle, N., 358, 370
Deruiter, C., 1293
Desai, M.I., 982, 1000
Desai, U.D., 523, 524, 528, 1000
Desai, U.O., 149
Desch, M.D., 1425, 1429
Deschamps, A., 641

Désert, F.-X., 576, 612, 636–638, 869
DeShields, L.M., 1189
deSitter, W., 351
Deslanders, H., 229, 252
Desler, K., 1231
Despois, D., 637, 1291–1293
Dessler, A.J., 168, 177, 230, 246, 252, 269, 963, 1000, 1119, 1137, 1213, 1430
Desvoivres, E., 1259, 1260, 1266, 1270
Detal, A., 468
Deubner, F.L., 1038, 1042, 1043, 1045, 1061, 1062
Deustua, S., 395, 420, 467
Deutsch, A.J., 895, 896, 906, 909, 1348, 1371
Deutsch, E., 524
Deutsch, L.K., 1272, 1430
Deutschmann, W.A., 299
DeVore, C.R., 1063
DeVorkin, D.H., 57, 58, 120, 150
DeWarf, L.E., 394
Dewey, R.J., 761, 786
Dewi, J.D.M., 778, 786
Dey, A., 523
Dezalay, J.P., 508, 524, 528
Dezeeuw, D.L., 1215, 1232
d'Hendecourt, L.B., 610, 618, 623, 624, 638, 641, 863, 869, 1270
di Giorgio, A.M., 636–639, 641–643, 869
di Serego Alighieri, S., 546
Di Serego, S., 524
di Stefano, R., 788
Diachkov, A., 820
Diamond, P.J., 394, 665, 672
Diaz, J., 270
Dibarboure, G., 1600, 1616
Dicke, R.H., 345, 351, 400, 420, 461, 466, 1036, 1042, 1061
Dickens, J.E., 1292
Dickens, R.J., 497
Dickey, J.M., 590, 595, 601, 604, 663, 664, 672
Dickey, J.O., 346, 351, 352, 1630, 1650
Dickey, Jr. J.S., 1403
Dickinson, M., 498
Dicksion, M.W., 1273
Diego, F., 671
Diehl, R., 397, 959
Diercks, A., 395, 420, 468, 526
Dieter, N.H., 665, 672
Dieters, S., 526, 755, 758, 787, 821
Dietrich, J., 1322
Dietrich, M., 547
Dietz, R.S., 261, 269
Dietzel, H., 1175, 1176, 1187
Digiorgio, A., 636, 641
Digranes, G., 1616
Dikpati, M., 1058, 1061, 1086
Dillinger, W.H., 468
DiMauro, M.P., 1061, 1063
Dinescu, A., 1650
Ding, J.G., 1582
Ding, K.Y., 819, 820
Dingus, B.L., 523, 525, 696
Diplas, A., 664, 671, 672, 674, 1215
Diplock, B.R., 869
DiSanti, M.A., 1286, 1289, 1292, 1293
diSerego Alighieri, S., 540
Disney, M.J., 753, 755
Divine, N., 1261, 1270
Divine, R.A., 57, 58
Dixon, L., 386, 393
Dixon, W.V.D., 299
Dixon, W.W., 1088

Djorgovski, S.G., 511, 513, 515, 523, 524, 526, 527
Dmyk, K., 869
Dobe, Z., 1233
Dobrovolskis, A.R., 1270
Dobrzycka, D., 1138
Dobson, F.W., 1615
D'Odorico, V., 466
Dodson, R., 756
Doerffer, R., 1602, 1615, 1616
Dogiel, V.A., 684, 694, 695
Dohnanyi, J.W., 1164, 1180, 1239, 1270
Doi, Y., 641
Doke, T., 1004, 1555
Dokter, J.J., 547
Dokukin, V.S., 1527
Dolan, J.F., 753, 754, 811, 820
Dolan, T.E., 85, 113
Dolder, D., 1232
Dolder, U., 1292, 1293
Dolfus, A., 1321
Dolginov, S.S., 274, 275
Dolginov, Sh., 1479, 1491
Dollfus, A., 270, 1407, 1453, 1454
Domingo, C., 697
Domingo, V., 1000, 1046, 1061, 1066, 1086
Domingue, D.L., 1435, 1449
Dominh, K., 1650
Dominik, B., 1341, 1371
Dominik, C., 864, 869, 870, 1241, 1270, 1341
Donahue, M., 395, 473, 487, 496
Donahue, R., 1060
Donahue, R.A., 892–894
Donahue, R.J., 1231
Donahue, T.M., 177, 201, 438, 1309, 1321, 1323, 1423, 1430, 1448
Donald, E., 299
Donas, J., 299
Donati, J.F., 902, 909
Donatowicz, J., 1060
Dondi, L., 553, 558
Done, C., 852, 854
Donea, A.-C., 1054, 1061
Donelan, M., 1616
Dones, L., 1372, 1449, 1477
Donn, B., 291, 300, 1232, 1242, 1270, 1272
Donn, B.D., 611, 638
Donnelly, J., 680, 695
Donnelly, R.F., 1062
D'Onofrio, M., 393
Donohue, D.J., 1216
Dooling, D., 57, 58
Dorman, I.V., 688, 691, 695
Dorman, L.I., 691, 695
Dorman, V.L., 695
Dormand, J.R., 1456, 1476
Dorn, D., 300
Doroshenko, V., 547
Dörschner, J., 640, 673, 870, 1188
Doscheck, G.A., 893
Doser, M., 1231
Dotani, T., 577, 754, 808, 817, 820, 821, 836, 854, 855, 944, 959, 960
Dotson, J.L., 639
Dotto, E., 1477
Doty, J., 821
Doty, S.D., 637
Doublier, V., 524, 526
Dougherty, B.L., 1004
Dougherty, J.P., 1573, 1581
Dougherty, S.M., 644, 872
Douglas, A.E., 607, 638
Douglas, B.C., 1651

Doupnik, J.R., 1580
Doute, S., 1477
Dowdy, J.F., 1080, 1086
Dowell, C.D., 639
Downes, D., 548
Downes, R., 791, 809
Doxsey, R., 821
Doyle, J.G., 892, 894
Dragovan, M., 639
Draine, B.T., 515, 528, 612, 617, 620, 638, 643, 863, 866, 870
Drake, A.J., 465
Drake, J.F., 299, 300, 672, 674, 1032, 1466, 1477
Drake, J.J., 886, 891, 893
Drake, M.J., 1322, 1361, 1371, 1477
Drake, N., 809
Drake, S.A., 893
Drange, H., 1616
Drapatz, S., 636, 642–644, 869, 873
Drayson, S.R., 1668
Drechsel, G., 329, 577, 870
Drechsel, H., 809
Dreher, J.W., 547
Dreibus, G., 1374, 1385, 1387–1389, 1392–1394, 1396–1398, 1401–1403,
Dressler, A., 393, 394, 467, 475, 492, 496, 498, 546, 547
Dressler, K., 299
Drever, R.W.P., 370
Drew, J.E., 797, 809, 810, 908
Drinkwater, M.J., 548
Droissart, P., 1423
Dror, J., 1269
Drossart, P., 437, 1429, 1440, 1448, 1449
Drucker, A., 560
Drummond, D.L., 869
Drummond, J.D., 1245, 1270
Drury, L., 434, 438, 871, 1002
Drury, L.O'C., 695
Dryburgh, M., 671
Dryer, M.J., 1402
Du Vernois, M.A., 689
Dubin, M., 1173, 1187, 1188
Dubinin, E., 1233
Dubner, G.M., 597, 605
DuBois, D.F., 1564, 1581
Duc, P.-A., 577
Duc, R., 1061
Dudin, A.D., 1403
Dufay, J., 202
Dufour, R.J., 673
Dufton, P.L., 672
Duke M., 1321
Duley, W.W., 613, 617, 619, 638
Dultzin-Hacyan, D., 547
Dum, C.T., 1504, 1526
Dumont, A.-M., 546
Dumont, R., 1164, 1187
Duncan, D., 428, 434, 437, 439, 892
Duncan, M., 1273, 1281, 1292
Duncan, M.A., 644
Duncan, M.J., 1239, 1240, 1242, 1269, 1270, 1272
Duncan, R.C., 508, 524, 526, 742, 757, 758, 785, 789, 819, 822
Dunchenko, A.G., 1403
Dunford, E, 298, 298
Dungey, J.W., 249, 252, 1008, 1009, 1014, 1015, 1031, 1113, 1507, 1526, 1532, 1555
Dunham, E.W., 1269
Dunham, T., 259, 262, 269, 649, 672
Dunning, J.R., 1373
Dunphy, P.P., 1000
Dupke, R.A., 490, 496

Dupree, A.K., 299, 810, 832, 835, 891–893, 903, 906, 909
Dupuis, J., 299, 1209, 1214
Duquennoy, A., 859, 870
Dürbeck, H.W., 892
Durgaprasad, N., 999, 1000
Duric, 689
Durouchoux, P., 753
Durrance, S.T., 299
D'Uston, C., 1232, 1491, 1492
Dutrey, A., 630, 638
Duvall, T.L., 1043, 1045, 1047, 1048, 1052, 1054, 1059–1063, 1069, 1086
Duvall, T.S., 1085
Duvel, J.P., 1668
Duvernois, M.A., 694, 695
Duvert, G., 638
Duxbury, T.C., 1322
Dwek, E., 329, 420, 642, 1188
Dwivedi, B.N., 1086, 1088
Dwornik, E.J., 1401
Dwyer, J.R., 981, 1000, 1002, 1003
Dyachkov, A.V., 820, 1232
Dyal, P., 329, 1479, 1491
Dyce, R.B., 268, 270, 352
Dyck, H.M., 329
Dyer, C.C., 444, 466
Dyson, F.W., 442, 466, 538
Dziembowski, W.A., 1060–1063
Dzitko, H., 1060, 1061

Eaker, N., 1582
Eardley, D.M., 363, 372
Earl, T.A., 678, 695
Eastman, R.G., 528
Eastman, T.E., 1001, 1541, 1555
Ebbets, D.C., 298, 299, 637, 672
Ebel, D.S., 659, 661, 672
Ebeling, H., 485, 496, 497
Eberhard, N., 299
Eberhardt, P., 437, 1137, 1232, 1284, 1286–1288, 1292, 1293, 1354, 1371, 1372, 1375, 1378, 1379, 1401, 1556
Ebihara, M., 1378, 1401
Ebisawa, K., 577, 758, 808, 849, 854, 855
Eckart, A., 548
Eckart, P., 80, 113
Ecklund, W.L., 1563, 1581
Economou, T., 1371, 1374, 1401, 1402
Eddington, A.S., 340, 351, 442, 453, 466, 638, 672
Eddy, J.A., 608, 638, 1037, 1041, 1042, 1061, 1139
Edel, H., 1617
Edelstein, J., 299, 753
Edgar, R.J., 605, 664, 667, 672, 674, 756, 959
Edge, A.C., 481, 484, 495–497
Edge, D.O., 323, 329
Edlén, B., 217, 222, 252
Edlén, J.A., 234, 252
Edmonds, P.D., 786
Edmunds, M.G., 438
Edvardsson, B., 437
Edwards, B.C., 604, 605
Edwards, D.A., 1231
Edwards, H.D., 190, 202
Edwards, H.W., 189, 202
Edwards, P.G., 960
Edwards, R.T., 748, 754
Edwards, S., 862, 869, 870
Edy, P.B., 352
Eff-Darwich, A., 1063
Efremov, V., 960
Efremov, Y.N., 654, 672

Efroimsky, M., 1264, 1265, 1270
Efstathiou, G., 415, 420
Egami, E., 329, 576
Egan, M.P., 328, 330
Egbert, G., 1647, 1650
Eggelton, P.P., 771
Eggen, O.J., 701, 702, 718
Egger, R., 597, 603–605, 672, 756, 959
Eggers, S., 1240, 1242, 1270, 1274
Eggleton, P.P., 786
Eggleton, R., 270
Egorov, I.V., 1491
Egret, D., 395, 396, 891
Ehara, M., 1004
Ehlers, J., 465, 468
Ehmann, W.D., 1381, 1401
Ehrenfreund, P., 610, 623–625, 637–639, 642, 644, 869, 872, 1261, 1270
Eichelberger, A.C., 523
Eichhorn, G., 1187
Eichler, D., 254, 521, 524, 696, 980, 1000, 1003, 1160, 1215
Eidelman, S., 1231
Eikenberry, S., 527
Einasto, J., 495
Einstein, A., 227, 252, 336–338, 351, 352, 443, 466, 493
Eisenhardt, P., 560
Eisenhower, D.D., 328, 329
Eke, V.R., 393, 474, 477, 490, 496
Ekers, R.D., 526
Ekholm, T., 382, 393, 397
Eklund, S., 1331, 1371
Eklund, W.L., 1563
El Goresy, A., 1272
Elachi, C., 1589, 1590, 1615
Elbaz, D., 466, 574, 576, 577, 869
Elbert, D.D., 230, 252
Elbert, J.W., 695
Elcan, M.J., 528
Eldar, A., 396
Eldholm, O., 1316, 1321
Eliason, E., 1321
Eliasson, L., 1578, 1581, 1584
Elipe, A., 1650
Elitzur, M., 638
Eller, E., 1401
Eller, E.L., 528
Ellerman, F., 229, 253
Ellingsen, S.P., 855
Ellingson, E., 393, 496, 559
Elliot, H., 151
Elliot, J., 1430
Elliot, J.L., 753, 1269
Elliott, J.R., 1050, 1061
Ellis, G., 385, 393
Ellis, R.S., 393, 395, 420, 466, 467, 498
Ellison, D.C., 688, 690, 694–696, 981, 1000, 1002
Ellithorpe, D., 694
Elmegreen, B.G., 654, 672
Elmore, D., 1063
Elmsford, N.Y., 270
Elphic, R.C., 1021, 1032, 1401, 1526, 1527
Elsässer, H., 233, 243, 252, 1164, 1165, 1187
Elshin, V.K., 694
Elsner, R.F., 758, 827, 835, 836
Elson, L.S., 1668
Elster, J., 677, 695
Elston, E., 560
Elsworth, Y., 1048, 1060–1062
Elter, G., 1273
Elvey, C.T., 189, 202

Elvis, M., 532, 533, 539, 543, 546–548, 559, 577, 836, 893, 909
Elvius, A., 546
Elvius, T., 703, 718
Elwert, G., 210–212, 214, 222
Elyasberg, P.E., 1273
Emerich, C., 438, 1258, 1259, 1270
Emerson, B., 695
Emerson, J., 935
Emerson, J.P., 329, 330, 577
Emery, B.A., 1583
Emery, R.J., 636–638, 641, 869
Emery, W.J., 1597, 1615
Emilio, M., 1042, 1061
Emming, J.G., 299
Emry, B.A., 1582
Emslie, A.G., 1003
Encrenaz, P., 641
Encrenaz, T., 428, 437, 1270, 1273, 1423, 1429, 1435, 1439, 1441–1445, 1447–1449
Enge, W., 1003
Engel, L., 1294
Engelmann, J.J., 869
England, M.N., 547
Englhauser, J., 1232
Engvold, O., 1087
Enzian, A., 1220, 1232, 1243, 1260, 1261, 1270, 1272
Enzian, E., 1270
Eplee, R.E. Jr., 328, 420, 636, 642, 645
Epstein, A., 576, 578, 754
Epstein, G., 869
Epstein, R., 753
Epstein, R.I., 524, 526, 689, 695, 754
Epstein, S., 1379, 1401
Eracleous, M., 576, 793, 809
Eraker, J.H., 1492
Erben, T., 395, 468
Erdos, G., 1159
Ergma, E., 774, 778, 786
Ergun, R.E., 1526, 1527, 1555
Erice, 1650
Erickson, E.F., 639, 643, 645
Erickson, G.M., 1543, 1555
Erickson, N.R., 636, 639, 641, 642
Eriksen, K.A., 959
Eriksson, K., 642
Erlank, A.J., 1403
Erler, J., 1231
Ermakov, M.K., 1213
Ershkovich, A.I., 1232
Ertel, I.D., 222
Esa, 1649
Eshelman, V.R., 1449, 1575, 1581
Esin, A.A., 855
Eskridge, P.B., 568, 575, 576
Esnault-Pelterie, R., 64
Espedal, H.A., 1610, 1615
Espey, B.R., 547
Espinasse, S., 1261, 1270
Esposito, J., 689, 695, 696, 757
Esposito, L.W., 1407, 1423
Esposito, P.B., 1306, 1321, 1401
Esposito-Farése, G., 364, 365, 370
Esser, R., 253, 1138, 1140, 1160, 1215
Essling, A.M., 1372
Estaria, P., 329, 577, 870
Estulin, I.V., 528
et Vasseur, G., 1583
et Waldteufel, P., 1580
Etherington, I.M.H., 446, 466
Ettori, S., 489, 496

Eubanks, T.M., 343, 351
Eugster, O., 1333, 1363, 1364, 1366, 1370–1372
Evans, A., 809
Evans, A.G., 201
Evans, D.C., 298
Evans, D.R., 1450
Evans, D.S., 1430, 1536, 1555, 1556
Evans, G.C., 1232
Evans, I.N., 547, 577
Evans, J.U., 1573
Evans, J.V., 1581
Evans, N.J., 626, 638, 640, 643, 644
Evans, N.W., 386, 393, 397
Evans, R., 1615
Evans, R.G., 298
Evans, S.T., 1232
Evans, W.D., 508, 523, 524, 528
Evans, W.F.J., 1668
Everhart, E., 1238, 1239, 1270
Everitt, C.M.F., 347, 351
Evers, J., 869
Evershed, J., 1069, 1086
Eviatar, A., 249, 252, 1426, 1430
Evlanov, E.N., 1233
Evrard, A.E., 477, 489, 495, 496, 498
Ewart, D., 636–638
Ewart, J.D., 869
Ewen, H.I., 638
Ewen, H.J., 607
Eyre, J.R., 1657, 1668
Ezawa, H., 496, 497, 576
Ezhela, V.V., 1231

Fabbiano, G., 483, 498, 562–567, 569, 574–578, 836, 854, 872, 893, 910,
Fabbro, S., 395, 420, 467
Faber, S.M., 378, 393, 467, 546, 547
Fabian, A.C., 371, 474, 481, 489, 490, 495, 496, 498, 528, 533, 537, 539,
    542, 546, 548, 556, 558, 559, 569, 576, 577, 783, 786, 826, 831, 835,
    852, 854, 855, 891, 909
Fabian, D., 673
Fabricant, D., 489, 496, 497, 567, 576, 754
Fadda, D., 576
Faelker, J., 329, 577, 870
Fafone, V., 370
Fagan, M., 1060
Fagents, S.A., 1321
Fahlman, G.G., 395, 736, 754
Fahr, H.-J., 991, 1000, 1004, 1198, 1214, 1216
Fainberg, E.B., 1491
Fairbairn, H.W., 1374
Fairbridge, R.W., 1323, 1449
Fairchild, E.T., 298
Fairfield, D.H., 1532, 1555–1557
Fajardo-Acosta, S.B., 866, 870, 871
Falco, E.E., 456, 457, 465–468, 524
Falferi, P., 371
Falgarone, E., 636, 644, 668, 672, 869, 871, 872, 892
Falkowski, P.G., 1602, 1603, 1615
Faller, J.E., 351, 1650
Falomo, R., 528, 560
Fälthammar, C.-G., 156, 172, 177, 1533, 1555, 1556
Fan, C.Y., 137, 144, 150, 983, 984, 1000–1002, 1004, 1138
Fan, X., 393, 541, 546
Fan, Y., 1060
Fanale, F.P., 1262, 1270, 1273, 1274, 1423, 1456, 1477
Fang, L.Z., 546
Faraffiana, R., 299
Faraji, H., 640
Farese, P.C., 394, 420, 466
Farinella, P., 1242, 1270, 1650
Farley, D.T., 1563, 1573, 1581

Farman, J.C., 1660, 1668
Farrel, B.F., 352
Farrelly, B.A., 1616
Fass, A., 1231
Fassett, C.I., 394
Fassia, A., 524
Fauci, F., 786
Faulkner, J., 771, 781, 786
Faure, G., 1326, 1372
Favata, F., 888, 889, 891, 893, 947, 952, 959, 1059, 1061, 1063
Fawcett, B.C., 216, 222
Fay, T.D. Jr, 1264, 1270
Faye, G., 370
Fazio, G.G., 329, 870, 1272, 1430
Feast, M.W., 376, 393, 788
Febvre, P., 641
Fechtig, H., 1164, 1165, 1175, 1187, 1188, 1189, 1270,
    1272, 1293
Federico, C., 1273
Federman, S.R., 438
Federspiel, M., 378, 379, 382, 393, 396
Fedorov, E., 1649
Fedorov, R.E., 222
Fegan, D.J., 696
Fegley, B. Jr., 636, 640, 1292, 1401, 1446, 1449
Feigelson, E.D., 491, 496, 497, 576, 754, 884, 885, 891–893
Feigl, K., 1650
Feinberg, L., 300
Feissel, 1633
Fejer, B.G., 1581
Fejer, J.A., 1119, 1137, 1561, 1563, 1564, 1573, 1581
Feldman, H.A., 394
Feldman, P.D., 299, 1288, 1292, 1293
Feldman, R.A., 1649
Feldman, U., 1080, 1086, 1140
Feldman, W.C., 254, 1026, 1031, 1120, 1128, 1130, 1134–1140, 1160,
    1231, 1390, 1401, 1513, 1526, 1556, 1557
Feldmann, P.D., 1274, 1477
Feldmeier, A., 900, 909
Feldscher, L.R., 959
Feldstein, Y.I., 1537, 1556
Felenbok, P., 869, 872
Felgett, P.B., 304, 329
Felli, M., 869
Fellous, J.-L., 1616
Felten, J.E., 492, 496
Fender, R.P., 787, 788
Feng, J.L., 1231
Fenimore, E.E., 504, 518, 521, 524, 526–528, 736, 754, 1137
Fenner, C.N., 1327, 1372
Fenyves, E.J., 523
Feraudy, D., 352, 1651
Fergusen, S.H., 1088
Ferguson, A., 547
Ferguson, D.H., 526, 527
Ferguson, D.W., 1448
Ferguson, E.E., 1581
Ferguson, H., 524
Ferguson, H.C., 299, 396
Ferlet, R., 438, 440, 663, 668, 672–674, 864, 866, 869–871, 1215
Fermi, E., 131, 132, 149, 150, 653, 672, 678, 695
Fernandez, J.A., 867, 870, 1187, 1239, 1240, 1270, 1273,
    1281, 1292
Fernández, Y.R., 1272
Feroci, M., 523–528, 736, 754
Ferrando, P., 434, 437, 696
Ferrara, A., 577, 653, 665, 672
Ferrarese, L., 379, 393–397, 546
Ferrario, L., 800, 810
Ferraro, V.C.A., 133, 149, 230, 246, 252, 1121, 1137, 1497, 1526, 1530,
    1539, 1545, 1555, 1567, 1581

Ferreira, P.G., 394, 417, 420, 466
Ferriere, K., 695
Ferry, G.V., 1187
Festou, M.C., 1223, 1232, 1273, 1292
Fetscher, W., 1231
Feuchtgruber, H., 437, 638, 640, 641, 645, 869, 873, 1440, 1442–1444, 1446, 1448, 1449
Feuerbacher, B.P., 1272
Feynman, J., 1002, 1088, 1136–1139, 1557
Ficenec, D., 694
Fichett, M., 498
Fichot, A., 1061
Fichtel, C.E., 535, 546, 560, 691, 695, 696, 723, 753–757, 972, 1000
Fichtner, H., 691, 696
Field, G.B., 542, 546, 583, 601, 603–605, 608, 638, 647, 651, 653, 661, 665, 672, 674, 1086
Fielder, G., 261, 270
Fields, B., 439
Fields, S.A., 1582
Fienschi, S., 222
Fierro, J.M., 755
Fierry-Fraillon, D., 1062
Fieten, H., 935
Fikani, M.M., 526
Filimonov, B.B., 1231
Filippenko, A.V., 395, 420, 467, 468, 523, 526, 546, 547, 841, 843, 854
Filleux, C., 437
Fillius, R.W., 167, 169, 177
Fillius, W., 1527
Fimmel, R.O., 144, 150
Fineschi, S., 1138, 1139
Finger, M.H., 526, 758, 768–770, 786, 820, 834
Fink, R.L., 604
Fink, U., 1430
Finkbeiner, D.P., 396, 605
Finley, J.P., 605, 696, 754–757
Finn, J.M., 1011, 1031
Finn, L.S., 367, 370, 371
Finoguenov, A., 490, 496, 820
Finson, M.L., 1236, 1241, 1261, 1270
Finsterle, W., 1060, 1087
Finzi, A., 811, 820
Fiore, F., 524, 527, 535, 546, 559
Firth, J.G., 1232
Fischel, D., 298
Fischer, C., 1233
Fischer, H., 1668
Fischer, J., 636, 637, 641, 869, 1001, 1138, 1602, 1615
Fischer, P., 457, 466
Fish, A.C., 671
Fisher, A., 393
Fisher, K.B., 387, 393
Fisher, P., 396, 397
Fisher, R., 870, 1138
Fishman G.J., 286, 499, 500, 507, 523–528, 758, 787, 821
Fisk, L.A., 143, 150, 250, 252, 682, 695, 971, 976, 977, 985–988, 991–993, 995, 997, 1000–1002, 1004, 1135, 1137, 1138, 1140, 1141, 1150, 1155, 1156, 1159, 1160, 1199, 1214, 1233, 1533, 1556
Fitton, B., 298, 330
Fitzpatrick, E.L., 376, 394, 670–672, 675
Fitzsimmons, A., 1274
Fixsen, D.J., 328, 404, 420, 636, 642, 645
Fjeldbo, G., 1575, 1581
Flanagan, É.É., 367, 370
Flannery, B.P., 761, 772, 786
Flasar, F.M., 1477
Fleck, B., 1061, 1066, 1086
Fleischer, A.J., 908
Fleischer, R.L., 1003
Fleitout, L., 1322

Fleming, T.A., 882, 891, 893, 908
Fletcher, L., 1032
Fletcher, S., 523
Flett, D., 1617
Fligge, M., 1070, 1087
Floren, H.-G., 641
Florens, J.P., 1589, 1615
Florensky, C.P., 265, 270
Florensky, K.P., 1309, 1321
Florián, J., 525
Florkowski, D.R., 835
Flower, D.A., 1668
Flower, D.R., 638
Flynn, B., 1212, 1214
Flynn, K.F., 1331, 1372
Fockenbrock, R., 527
Foglizzo, T., 1061
Foing, B.H., 870, 1086
Földy, L., 1232, 1273
Folkner, W.M., 1402
Foltz, C.B., 468, 548
Fomalont, E.B., 453, 467, 496
Fomenkova, M., 1225, 1232, 1233, 1283, 1284, 1292, 1293
Fonger, W.H., 142, 150, 151, 255
Font, J., 1615
Fontana, A., 466
Fontenla, J.M., 1068, 1070, 1071, 1086
Föppl, H., 187, 1515, 1524, 1526, 1528
Forbes, T., 1011, 1025, 1027, 1032
Forbush, S.E., 120, 128, 130, 141, 150, 231, 233, 252, 967, 1001
Ford, E.C., 818, 820
Ford, H.C., 393–397, 546
Forestier, F., 1650
Forestini, F., 436
Forman, W., 285, 393, 478–480, 482, 483, 496–498, 567, 576–578, 723, 754, 800, 809, 823, 835, 910
Forrest, D.J., 1000
Forrest, W.J., 315, 329, 467, 612, 623, 639
Forrester, W.T., 299
Forsyth, R.J., 1002, 1138, 1139, 1148, 1155, 1159–1161, 1188, 1189
Fort, B., 395, 460, 465, 467, 468, 498
Fortini, P., 371
Fortner, B., 817, 820
Forveille, T., 642, 645, 870, 1273
Fosbury, R.A.E., 546
Fossat, E., 1043, 1046, 1061, 1062
Fossati, G., 547, 555, 558, 559
Foster, J.C., 1569, 1581
Foster, J.J., 1371
Foster, M.J., 524
Fouchet, Th., 1449
Fougere, P.F., 1580
Foukal, P.V., 1086
Fountain, G.H., 299
Fourmond, H.-J., 1061
Foust, J.A., 1269
Fowler, G.W., 1374
Fowler, P.H., 991, 1001
Fowler, R.G., 298
Fowler, W.A., 430, 433, 437, 439, 440, 1138, 1376
Fox, K., 1430, 1477
Frail, D.A., 505, 508, 513, 514, 523–527, 665, 672, 736, 755, 787, 821
Franceschini, A., 576, 869
Francis, P.J., 548
Francis, W.E., 252
François, P., 393
Frank, A., 918, 935, 1533, 1557
Frank, J., 532, 800, 809, 812, 820

Frank, L.A., 1001, 1490, 1491, 1526, 1535, 1536, 1539, 1540, 1555, 1556, 1569, 1584
Franka, S., 300
Frankel, C., 1476
Franklin, K.L., 1466, 1476
Fransson, C., 527, 695, 691, 788, 831, 835
Franx, M., 460, 467, 498
Franz, B.A., 329, 642, 1188
Fränz, M., 1000, 1002
Franz, O.G., 1269
Franzgrote, E.J., 275, 1403
Frappa, E., 1270, 1271
Fraser, R.S., 1601, 1615
Frazier, E.N., 1038, 1061
Frederick, C.L., 306, 309, 329
Frederick, J.E., 1668
Frederiks, D., 525
Frederiks, D.D., 754
Fredricks, R.W., 1531
Fredrics, R.W., 1556
Freedman, W.L., 376, 379, 393–397
Freeman, H.R., 298
Freeman, K.C., 465
Freeman, K.S.C., 1584
Freier, P., 125, 126, 131, 132, 150, 753
Freier, P.S., 678
Freire, R., 786
French, B.M., 1323
French, R.G., 1460, 1476, 1477
Frenk, C.S., 393, 420, 473, 496–498
Frenzel, R., 1583
Frericks, M., 869
Fresneau, A., 935
Freudenreich, H.T., 329, 1188
Freudling, W., 393, 394, 560
Freund, M.M., 330
Frey, H.V., 1322
Frey, S., 1491
Freyberg, M.J., 603, 605, 672, 719, 1215
Fricke, K.J., 525, 547
Fridelance, P., 352, 1651
Fridlund, C.V.M., 576
Fried, J.W., 560
Fried, P.M., 584, 589, 604, 605
Friedlander, M., 524, 525
Friedman, H., 217, 222, 277, 285–288, 298, 299, 475, 477, 496, 604, 753, 754, 820, 834, 1215, 1554, 1556
Friedman, P., 299
Friedman, S.D., 299, 674
Friedrich, V., 1583
Friend, D.B., 824, 835, 897, 902, 903, 909, 910
Frier, P.S., 695
Friichtenicht, J.F., 1175, 1176, 1187, 1188
Friis-Christensen, E., 1018, 1031, 1086
Frisch, P.C., 589, 590, 595, 597, 603, 604, 648, 653–655, 658, 662, 663, 667, 671–673, 675, 1159, 1183, 1188, 1213, 1216
Fritschel, P., 365, 371
Fritts, D.C., 1569, 1581
Fritz, G., 604, 722, 754
Fritz, H., 1537, 1556
Fritz, T.A., 1004, 1492
Fritzová-Svestková, L., 969, 1004
Froeschlé, Cl., 1273
Froeschlé, M., 453, 467
Fröhlich, C., 1041, 1046–1048, 1056, 1060, 1061, 1063, 1068, 1070, 1086, 1087
Froidevaux, C., 1650
Froidevaux, D., 1231
Froidevaux, L., 1668
Frondel, C., 1376

Frontera, F., 523–528, 559, 754
Fruchter, A.S., 395, 420, 467, 493, 511, 513, 515, 521, 524, 525, 527, 750, 754, 782, 786
Fruscione, A., 535, 546
Fry, W., 754
Frye, B., 393
Fryer, C.L., 507, 521, 524
Fryxell, B.A., 768, 786, 788, 828, 836
Fu, A., 836
Fu, L.L., 1638, 1650
Fucito, F., 370
Fuechsel, C.F., 298
Fuente, A., 617, 638
Fugate, R.Q., 1270
Fugmann, W., 460, 467
Fuhrmann, K., 329
Fujii, M., 524, 527
Fujikawa, M., 696
Fujimoto, M., 1032, 1557
Fujimoto, R., 809, 948, 959
Fukazawa, Y., 490, 496, 497, 566, 576–578
Fukuda, S., 1056, 1061
Fukuda, Y., 1061
Fukugita, M., 457, 467
Fukushima, N., 1544, 1556
Fulle, M., 1233, 1244, 1270
Fuller, C.E., 643
Fuller, G.M., 396, 428, 437
Fuller-Rowell, T.J., 1569, 1570, 1581, 1582
Fullerton, A.W., 299, 909, 911
Fullton, L.K., 935
Fung, A.K., 1616
Funsten, H.O., 1139
Furevik, B., 1615
Furia, M., 352, 1651
Furniss, I., 314, 329, 330, 636–638, 641, 869
Furth, H.P., 1011, 1031
Furusho, T., 497
Fusco-Femiano, R., 477, 496
Fuselier, S.A., 1231
Futamase, T., 467
Fynbo, J., 523

Gabriel, A.H., 217–219, 222, 1043, 1044, 1046, 1060–1063, 1077, 1086–1088
Gabriel, C., 871
Gabriel, M., 1060
Gaddis, L.R., 1321, 1322
Gadun, A.S., 1069, 1086, 1087
Gaensler, B.M., 736, 754
Gaetz, T.J., 941, 959
Gagné, M., 891, 893
Gaidos, J.A., 555, 559, 696
Gailitis, A., 1481, 1491
Gaines, G.A., 299
Gaisser, T.K., 689, 692, 695, 1231
Gaisser, T.R., 1004, 1005
Gal, R.R., 523–527
Galama, T.J., 510–518, 521, 524–528
Galeev, A.A., 691, 695, 1273
Galer, S.J., 1360, 1373
Gales, J.M., 328, 420, 636
Gallagher, D., 1188
Gallagher, J.S., 574, 576, 793, 809, 869
Gallais, P., 577, 869
Galloway, R.P., 438
Galvin, A.B., 893, 1001, 1003, 1122, 1137–1139, 1214
Gammie, C.F., 605
Gamow, G., 423, 436, 437
Ganapathy, R., 1379, 1401

INDEX OF CITED AUTHORS

Gancarz, A.J., 1374
Ganga, K., 330, 394, 405, 420, 466
Gangadharam, E.V., 1402
Gangpadhyay, P., 1214
Gänsicke, B.T., 809
Gaposchkin, C., 793
Gaposchkin, E.M., 1650
Gaposchkin, M., 1649
Gaposchkin, S., 810
Garanin, A.V., 1320
Garay, G., 860, 870
García Lopez, R.J., 439, 892
Garcia-Munoz, M., 137, 143, 150, 250, 252, 695, 990, 999, 1199, 1214
Garcia, M.R., 513, 525, 754, 808, 822, 855
García, R.A., 1061–1063
Gardiner, B.G., 1668
Gardner, D.J., 1188
Gardner, F.F., 1561, 1571, 1581
Gardner, I.S.K., 298
Gardner, L.D., 222, 1138, 1139
Gardner, M., 1291, 1293
Garignan, G.R., 1583
Garmire, A.B., 754
Garmire, G.P., 285, 547, 585, 586, 603, 604, 724, 736, 741, 754, 756, 757, 788, 821, 836
Garnavich, P.M., 395, 420, 468
Garner, H., 300
Garnett, J.D., 467
Garrard, T.L., 695
Garren, L., 1231
Garrison, D.H., 1360, 1362, 1363, 1371, 1373, 1374
Garton, W.R.S., 222
Garvin, J.B., 1322
Gary, S.P., 1133, 1137, 1140
Garzon, F., 871
Gaskell, C.M., 547
Gaskell, R.W., 1477
Gaspar, P., 1589, 1615
Gasperini, M., 366, 371
Gast, P.W., 1331, 1372, 1380, 1381, 1383, 1384, 1401, 1402
Gastaud, R., 869
Gates, W., 222
Gauld, B., 1004
Gault, D.E., 1188
Gauss, C.F., 228, 253
Gausted, J.E., 328, 329, 609, 639
Gautier, D., 428, 437, 1060, 1292, 1293, 1436, 1437, 1443, 1446, 1448, 1449
Gautier, T.N., 312, 329, 330, 577, 1430
Gauzit, J., 202
Gavrilova, E., 960
Gayley, K.G., 673, 903, 909, 910, 1215
Gazis, P.R., 1129, 1137
Ge, J.P., 481, 496
Gear, W.K., 560
Geballe, T.R., 524, 618, 636, 639, 640, 642, 643, 789, 837, 1440, 1449, 1476, 1477
Gebbie, K.B., 1087
Gebhardt, K., 456, 467, 541, 546, 547
Gee, P.S., 1231
Geer, A., 1657, 1668
Geer, S., 1231
Geese-Bähnisch, I., 1331, 1372
Geha, M., 465
Gehrels, N., 597
Gehrels, T., 166–169, 177, 524–527, 597, 604, 1233, 1272, 1376, 1426, 1430
Gehrz, R.D., 329, 870
Geiger, H., 120, 150

Geiss, J., 425, 427, 428, 434, 437, 438, 643, 673, 967, 986–988, 994, 996, 1001, 1004, 1118, 1122–1124, 1126, 1127, 1137–1140, 1148, 1156, 1159, 1209, 1214, 1215, 1231, 1278, 1285, 1286, 1288, 1291, 1292, 1333, 1338, 1351, 1371–1373, 1375, 1401, 1439, 1449, 1545, 1546, 1556,
Geissler, P.E., 1321, 1476, 1477
Geitel, H., 677, 695
Gekelman, W., 1023, 1024, 1031, 1033
Geldzahler, B.J., 833, 835
Gelly, B., 1062
Gemuend, H.P., 871
Genco, M.L., 1651
Gendron, E., 1273
Genetay, I., 1401
Genova, R., 662, 673
Gensheimer, F., 1291
Gensheimer, P., 1291
Gentemann, C., 1617
Genzel, R., 306, 329, 548, 570, 572, 573, 576, 577, 609, 610, 639–641, 643, 644, 869, 873
George, I.M., 855
George, J.S., 696, 697
George, S., 641
Georgelin, Y.M., 668, 673
Georgelin, Y.P., 668, 673
Geppert, U., 789
Gerakines, P.A., 624, 626, 637–639, 642, 644, 869, 872
Gérard, E., 637, 1265, 1269
Gerard, J.-C., 1271
Gerard, J.T., 1401, 1402
Gerbal, D., 397
Gerbaldi, M., 869
Gerber, H.-J., 1231
Gerbeth, G., 1491
Gerhard, O., 385, 394, 395
Gérin, M., 637
Gerlach, J.C., 187
Gerling, E.K., 1333, 1372
Gerling, S.K., 258, 259, 261, 270
Germain, B., 637, 1291
Germany, L.M., 516, 525
Gershberg, R.E., 892
Getman, V.S., 525
Getmantsev, G.G., 1526, 1555
Ghez, A.M., 859, 870
Ghielmetti, A.G., 1231
Ghigo, F.D., 855
Ghisellini, G., 524, 531, 535, 547, 549, 552–556, 558–560, 562
Ghosh, A., 695, 1371
Ghosh, P., 758, 827, 835
Ghosh, S.N., 201
Giacagla, G.E.O., 1649, 1651
Giacalone, J., 1002, 1154, 1156, 1159, 1160
Giacconi, R., 283, 285, 286, 496, 497, 532, 543, 546, 547, 549, 559, 562, 576–578, 582, 604, 722–724, 753, 754, 757, 759, 786, 788, 789, 809, 811, 820–824, 835, 836, 840, 854, 872, 893, 910
Giacometti, M., 394, 420, 466
Giaconni, R., 836
Giallongo, E., 466, 540, 547
Giampapa, M.S., 253, 254, 861, 870, 891, 893, 1085, 1139, 1140, 1160, 1161
Giannini, T., 628, 639, 641–643
Giard, M., 611, 636, 639
Giarrusso, S., 524, 559
Giavalisco, M., 498
Gibb, E., 623, 639
Gibbons, P.C., 636
Gibbons, R., 524
Giblin, T.W., 508, 525, 528, 758
Gibner, P.S., 695

Gibson, B.K., 377, 394–397, 496
Gibson, E.K., 1373
Gierasch, P., 1441, 1442, 1450
Gierliński, M., 848, 852, 854
Gies, D.R., 841, 854, 909
Giese, R.H., 1164, 1188
Gifanov, M., 820
Gilbert, A.D., 1481, 1492
Gilbert, G.K., 261, 270, 272, 275
Gilbert, W., 225, 226, 253
Giles, P.M., 1053, 1061, 1062
Gilfanov, M.R., 523, 819, 820, 960
Gill, E., 1650
Gille, J.C., 1660, 1662, 1668
Gillett, F.C., 330, 577, 612, 617, 623, 639, 644, 863, 868, 870, 920, 935
Gilli, R., 546
Gilliland, R.L., 395, 420, 468, 798, 799, 809
Gilman, D., 527, 528
Gilman, F.J., 1231
Gilman, P.A., 1041, 1058, 1061
Gillmor, C.S., 1496
Gilmore, G., 425, 427–429, 434, 437, 702–704, 711, 712, 714, 718, 719
Gilmour, J.D., 1375
Gilmozzi, R., 546
Gilra, D.P., 609, 639
Gimbel, J., 57, 58
Ginzburg, L., 695
Ginzburg, V.L., 552, 559, 678, 683, 685, 691, 694, 695, 811, 821
Gioia, I.M., 467, 482, 496, 497, 546, 560
Giommi, P., 524, 527, 546, 554, 559
Giordano, S., 222, 1134, 1137–1139
Giorgini, J.D., 1232, 1271
Giovanelli, R.G., 378, 393, 394, 1008, 1031, 1074, 1086, 1113, 1506, 1526
Giovannelli, F., 576
Girard, A., 1668
Gisler, G., 523
Gispert, R., 314, 328, 329, 1270
Giuli, R.T., 299
Gizon, L., 1053, 1061, 1062, 1069, 1086
Glaccum, W.J., 642, 1273
Gladman, B.J., 1239, 1270, 1453, 1456, 1476
Gladstone G.R., 1214
Glanz, J., 258, 270
Glassgold, A.E., 565, 576, 632, 633, 639, 642
Glassmeier, K.-H., 1226, 1232, 1233
Glatzmaier, G.A., 1322, 1481, 1491
Glazov, V.N., 1403
Gleason, J.D., 1375
Gleeson, L.J., 695
Glen, G., 729, 754
Glencross, W.M., 636–638, 641, 869
Glendenin, L.E., 1331, 1372
Glendenning, N.K., 725, 754
Glentzlin, M., 352, 1651
Gliem, F., 1001, 1138, 1214
Gloeckler, G., 150, 177, 425, 427, 428, 437, 438, 966, 967, 974, 986–989, 991, 992, 994–997, 1000–1004, 1032, 1122–1124, 1137–1140, 1159, 1201, 1209, 1214, 1215, 1439, 1449, 1555, 1556
Gloersen, P., 1591–1593, 1615, 1616
Gloria, K.A., 467
Gnedin, Y.N., 730, 754
Goad, M.R., 547
Gockel, A., 119, 150
Goddard, E.C., 113
Goddard, R.H., 60, 61, 113
Godfrey, P.D., 1292
Goebel, M., 1274
Goettel, K.A., 1400, 1402
Goetz, K., 1526

Goetz, B., 1272
Göğüş, E., 758
Gohin, F., 1593, 1615
Golay, M.J.E., 305, 329
Gold, R.E., 254, 525, 1002
Gold, T., 232, 253, 722, 726, 754, 759, 786, 1014, 1023, 1031, 1113, 1131, 1138, 1496, 1510, 1527, 1530, 1556
Goldberg, L., 222, 907, 909, 1068, 1086
Goldberg, R., 1580
Goldberger, M.L., 252
Golden, A., 757
Goldenberg, H.M., 1036, 1042, 1061
Goldhaber, G., 395, 420, 467
Goldman, P., 1322
Goldman, S.J., 97, 113
Goldoni, P., 523
Goldreich, P., 630, 637, 1059, 1061, 1237, 1270, 1425, 1429, 1459, 1466, 1476, 1496, 1527
Goldsmith, D., 1320, 1321
Goldsmith, D.W., 604, 638, 672
Goldsmith, P.F., 622, 636, 639, 641, 642, 1292, 1293
Goldstein, B.E., 254, 255, 1138, 1139, 1159, 1160, 1231
Goldstein, M.L., 1000, 1159
Goldstein, R., 270, 1231, 1291
Goldstein, R.B., 352
Goldstein, R.M., 265, 270, 560, 1246, 1270
Goldstein, W.C., 254
Goldsten, J., 525
Goldwurm, A., 523, 820
Golenetskii, S.V., 507, 508, 525, 527, 754, 755
Golisch, W.F., 1274
Golombek, M.P., 1320, 1371, 1395, 1402
Golovko, V.A., 1668
Golub, L., 234, 251, 253, 872, 892, 893, 908, 910, 1027, 1031, 1032, 1078, 1086
Gombogi, T., 1292
Gombosi, T.I., 1215, 1222, 1232, 1233, 1270, 1271
Gomez, A., 1272
Gomez, J.-L., 558
Gomez, M., 1061
Gonano, M., 1274
Gondhalekar, P.M., 298, 454, 467
González Hernández, I., 1053, 1061
González Riestra, R., 871, 872
Gonzalez, C., 908
Gonzalez, J.F., 524–526
González, R.A., 524
Gonzalez, W.D., 1016, 1031
González-Alfonso, E., 636–638, 644, 645
Gonzalez-Esparza, J.A., 1151, 1159
Gonzalez-Perez, N., 560
Gonzalez-Riestra, R., 935
Goobar, A., 395, 420, 467
Good, J., 603
Goode, P.R., 1050, 1060–1063
Goodman, J., 504, 521, 525
Goodman, M., 1231
Goodman, S.J., 524
Goodrich, C.C., 1215
Goodrich, R.W., 523, 526, 527, 861, 870
Goodwin, A.M., 1317, 1321
Goodwin, S.P., 382, 394
Gopakumar, A., 367, 371
Göpel, C., 1336, 1370, 1372
Gorchakov, E.V., 1004
Gordley, L.L., 1668
Gordo, J., 696
Gordon, H.R., 1587, 1601, 1615
Gordon, G.S. Jr., 1159
Gordon, S., 468

Gordon, W.E., 1572, 1581
Gorenstein, M.V., 421, 456, 466, 467, 480, 567
Gorenstein, P., 480, 496, 528, 567, 576, 605, 754, 788, 820, 821, 836, 872, 893, 894, 909, 910, 1401
Gorenstein, R, 788
Goret, P., 694
Gorgas, J., 467
Gorisse, M., 869
Gorman, G., 757
Gorn, L., 1233
Gorny, S.K., 935
Gorosabel, J., 523, 525, 527
Górski, K.M., 405, 410, 411, 413, 419–421
Gorsline, D.S., 1321
Gosling, B.E., 254
Gosling, J.T., 254, 969, 971, 986, 988, 1001, 1003, 1005, 1021, 1031, 1032, 1033, 1113, 1120–1122, 1128, 1136–1140, 1151, 1152, 1158–1161, 1231, 1527, 1541, 1556, 1557
Goss, W.M., 663, 672, 674, 755
Gott, J.R., 458, 467, 468, 476, 496
Gotthelf, E.V., 694, 696, 733, 734, 736, 741, 754, 755, 757, 948, 959, 960
Gottlieb, B., 1213
Gottlieb, D.M., 1037, 1041, 1063
Gottlieb, E.W., 759, 786
Goudfrooij, P., 524
Gough, D.O., 1038, 1043, 1045, 1048, 1055, 1058, 1060–1063
Gough, M.P., 1557
Gough, R., 855
Gouguenheim, L., 393, 395, 397
Gouiffes, C., 733, 754, 1062
Gould, R.J., 496, 552, 558, 583, 604
Goupil, M.J., 1060
Grab, C., 1231
Grabelsky, D.A., 809
Grad, H., 1506, 1527
Gradie, J., 1247, 1270
Gradsztajn, E., 437, 439
Grady, C.A., 642, 862, 865, 870, 901, 909
Grady, J.F., 300
Graf, H., 1401
Graf, S., 1292
Graham, G., 226, 253
Graham, J.A., 379, 393–397, 862, 870
Graham, J.R., 959
Graham-Smith, F., 755, 787
Granados, A., 466
Granahan, J.C., 1477
Grandel, J., 1593
Grandell, J., 1615
Grandi, P., 559
Grange, R., 299
Granier, C., 197, 202
Grant, C.E., 597, 604
Grard, R., 1282, 1292
Grard, R.J.L., 1232
Grasso, N., 888
Gratton, R., 376, 394
Grav, T., 527
Gravel, P., 393
Graves, J.E., 872
Gray, C.M., 1371
Gray, P.C., 1517, 1527
Greaves, J.S., 870
Grebowsky, J.M., 1534, 1555, 1556
Grec, G., 1061, 1062
Gredel, R., 523
Greeley, R., 1304, 1306, 1321–1323, 1375, 1402, 1476, 1477
Green, B.A., 1189
Green, D.A., 526, 732, 754
Green, I.M., 1001

Green, J.C., 299, 674
Green, R.F., 299, 300, 541, 546, 547
Green, S.F., 524, 1269
Greenberg, J.M., 612, 613, 623, 638, 639, 641–643, 673, 864, 871, 1232, 1237, 1241, 1242, 1248, 1261, 1270–1272, 1279, 1284, 1285, 1290, 1292
Greenberg, M., 1188
Greenberg, R., 1457, 1474, 1476
Greenfield, P., 547
Greenhill, L.J., 394
Greenhouse, M.A., 636, 637, 641, 869
Greenhow, J.S., 193, 202
Greenstadt, E.W., 981, 1001, 1531, 1556
Greenstein, G., 728, 730, 743, 754
Greenstein, J., 433, 437, 547
Greenstein, J.L., 132, 150, 530, 909
Greenwald, R.A., 1564, 1570, 1572, 1581
Greenwood, J.P., 1371
Gregorio-Hetem, J., 860, 870, 891
Gregory, P.C., 736, 754, 833, 835
Greiner, J., 505, 523, 525, 807, 809
Greisen, K., 692, 695
Greiveldinger, C., 733, 743, 754
Grenfell, T.C., 1616
Grenier, I.A., 604, 694, 753
Grenzel, R., 643
Grevesse, N., 428, 429, 434, 436, 438, 479, 490, 495, 940, 959, 1068, 1086, 1123, 1138, 1378, 1401
Grewing, M., 298, 299, 672, 1087
Gribbin, J., 394
Gridlay, J., 892
Griep, D.M., 1274
Griest, K., 466
Grieve, R., 270
Griffin W.G., 298
Griffin, M.J., 437, 636–638, 641, 642, 869, 1438, 1449
Griffin, W.G., 222
Griffith, R.B., 910
Griffiths, R.E., 831, 835, 837, 869
Grigorov, N.L., 999
Grigoryev, V., 1062
Grillmair, C.J., 393, 467, 546, 547
Grimm, R.E., 1321, 1322
Grindlay, J.E., 576, 753, 754, 784, 786, 806, 808, 809, 815, 821, 836, 909
Grineva, Yu.I., 217, 222
Gringauz, K.I., 247, 253, 1115, 1138, 1224, 1229, 1230, 1232, 1497, 1527, 1534, 1556, 1562, 1581, 1582
Gringel, W., 299
Grinin, V.P., 862, 870
Grinspoon, D.H., 1314, 1319, 1320, 1322, 1372, 1423
Grischenko, V.D., 1615
Grishchuk, L.P., 358, 371
Groegler, N., 1371
Groenewegen, M.A.T., 376, 394
Groezinger, U., 871
Grögler, N., 1372, 1373, 1375, 1401
Groh, K.H., 110, 113
Gronenschild, E.H.B.M., 809
Gronvall, H., 1616
Groom, D.E., 395, 420, 467, 1231
Groot, P.J., 511, 514, 517, 524, 525, 527, 528
Groote, D., 395
Grorod, P.A., 868
Gross, M., 351
Grossan, B., 523
Grosse A.V., 1373
Grosskreutz, C.L., 177
Grossmann-Doerth, U., 1069, 1086
Grosso, N., 888, 891
Groten, E., 1649

Groth, E.J., 393
Grothues, H.-G., 1265, 1271
Grotrian, W., 234, 253
Grove, J.E., 849, 854
Groves, G.V., 190, 202
Gruber, D.E., 694
Gruen, E., 871
Gruenwald, R., 428, 440
Gruenwaldt, H., 1214
Grumm, R.L., 1004
Grün, E., 673, 1166, 1169, 1170, 1176, 1179, 1180, 1184, 1185, 1187–1189, 1272, 1283, 1292, 1293
Grundahl, F., 391, 392, 394, 518
Gruntman, M.A., 1214
Grünwaldt, H., 1138, 1557
Gruzinov, A., 518, 525
Gry, C.J., 299, 595, 604, 615, 636–639, 641, 657, 675, 869, 1215
Gryaznov, G.M., 113
Grzedzielski, S., 1141, 1159, 1195, 1206, 1214, 1215
Guainazzi, M., 527
Guarnieri, A., 523
Güdel, M., 881, 885, 891
Gudmundsson, E.H., 729, 754
Guelin, M., 637
Guenther, D.B., 1060
Guérin, P., 265, 269
Guerrero, M.A., 935
Guest, J.E., 1301, 1321, 1322
Guest, S., 869
Guggisberg, S., 1338, 1353, 1372
Guhathakurta, M., 1138
Guhathakurta, P., 577
Guier, W.H., 1651
Guilloteau, S., 638
Guillout, P., 885, 886, 891
Guinan, E.F., 394, 891, 908
Guinot, B., 1633, 1650
Gulick, V.C., 1320
Gulkis, S., 419–421, 645, 1063, 1425, 1426, 1429
Gull, G.E., 315, 329, 330, 641, 642, 870, 1429
Gull, T.R., 298, 300, 527
Gummin, M.A., 299
Gundrum, T., 1491
Gunji, S., 960
Gunn, J.E., 444, 457, 468, 469, 475, 476, 496, 497, 541, 547, 548, 550, 559, 737
Gunnarsson, M., 1291
Gunnlaugsson, H.P., 1371
Guo, H., 330
Gupta, C., 1430
Gupta, M., 697
Gupta, S., 1477
Gurevich, A.V., 753, 1564, 1581
Gurian, I.A., 527
Gurman, J.B., 1086
Gurnett, D.A., 171, 177, 1123, 1132, 1138, 1158–1160, 1173, 1188, 1205, 1210, 1214, 1215, 1489, 1491, 1520, 1524, 1527, 1536, 1551, 1556
Gurney, J.J., 1403
Gürsel, Y., 370
Gursky, H., 286, 475, 477, 478, 496, 497, 532, 547, 559, 562, 576, 577, 583, 604, 753, 754, 757, 782, 786, 788, 789, 810, 815, 820, 821, 835, 836, 854
Gürtler, J., 863, 870
Gurtu, A., 1231
Guryan, Y.A., 525, 527, 755
Guseinov, O.H., 789, 822
Guseinov, O.K., 759, 787, 811
Guss, D.E., 972, 1000
Gussenhoven, M.S., 1556

Gustafson, B.A.S., 1164, 1183, 1185, 1187, 1188, 1273
Gustafsson, B., 397, 437, 642
Gustincic, J.J., 644
Guth, A., 414, 420
Gutierrez, C., 523
Guyeme, T.D., 1271
Guymer, T., 1600, 1604, 1611, 1615
Guzik, J.A., 1060
Guziy, S., 523
Guzzo, L., 487, 496, 497

Haardt, F., 524, 537, 547, 558, 559
Haas, M.R., 571, 577, 610, 620, 632, 639, 640, 645, 871
Habbal, S.R., 251, 253, 1137, 1138, 1140, 1160, 1215
Haber, D.A., 1063
Haber, H.E., 1231
Haberl, F., 736, 747, 748, 754–757, 830–832, 835
Haberle, R.M., 1402, 1423
Haberman, J.A., 1430
Habets, G.M.H.J., 778, 787, 837, 911
Habing, H.J., 330, 577, 604, 610, 638, 639, 644, 672, 864, 870, 872, 935, 1270
Hack, M., 299
Hackman, R., 261, 270
Hackwell, J.A., 329, 871
Hada, T., 1232
Haensel, P., 755
Haerendel, G., 182, 184–186, 187, 869, 1019, 1021, 1031–1033, 1233, 1515, 1519, 1523, 1526–1528, 1557, 1564, 1581, 1583
Haffner, L.M., 601, 602, 604
Hagen, J.P., 623, 625, 651, 673
Hagen, W., 643
Hagenaar, H.J., 255, 1113
Hagfors, T., 1573, 1581, 1582
Hagiwara, K., 1231
Hagmann, C., 1231
Hahn, G., 1274
Hahn, J.M., 1274, 1293
Hahn, O., 1320, 1328, 1330, 1372
Haigh, J.D., 1086
Haikala, L.K., 641
Hailey, C.J., 736, 754
Haiman, Z., 395
Hainaut, O.R., 524, 526, 1243, 1271, 1273, 1274
Haisch, B.M., 872, 878, 882, 891–893, 904, 905, 908–910, 1002, 1113
Hajduk, A., 1268, 1271
Hajduk, C., 871
Hajian, A.R., 935
Hakala, P., 527
Hakkila, J., 286, 523
Haldemann, A.F.C., 1402
Haldoupis, C., 1563, 1583
Hale, A., 1265, 1271
Hale, G.E., 229, 253, 1096, 1113
Halekas, J.S., 1479, 1491
Hall, A., 1452
Hall, C.M., 1375
Hall, D.T., 1207, 1214, 1216, 1469, 1477
Hall, G., 871
Hall, J.E., 298
Hall, J.S., 132, 150, 611, 639
Hall, P., 525, 560
Hall, T., 696
Hallenbeck, S., 868, 870
Haller, E.E., 330
Halliday, A.N., 1372, 1375
Halliday, I., 1164, 1188, 1366
Hallikainen, M., 1615
Halpern, D., 1586, 1615

Halpern, J.P., 521, 525–527, 547, 725, 743, 744, 746, 754, 755, 757, 758
Halpern, L., 123, 150
Hamaguchi, K., 892
Hamann, W.R., 935
Hamilton, D.C., 177, 1001, 1002, 1138, 1189, 1534, 1556
Hamilton, D.P., 1183, 1188, 1475, 1477
Hamilton, T.T., 758
Hamilton, W.O., 365, 371
Hammel, H.B., 871, 1273
Hammer, F., 467, 497, 498
Hammer, R., 1073, 1086
Hammergren, M., 393
Hammerschlag, A., 299
Hammerschlag-Hensberge, G., 835
Hammond, C.M., 254, 1160
Hamre, T., 1615
Hamuy, M., 375, 394
Han, M., 393–397
Han, S., 1062
Han, X., 853, 855
Han, Z., 788
Hanany, S., 394, 416, 420
Hanawa, T., 847, 849, 854, 855
Handy, B.N., 1032, 1067, 1086
Hänel, H., 1491
Hanel, R.A., 1427, 1429, 1430, 1436, 1437, 1439, 1443, 1449, 1477
Hanes, D.A., 394
Hanle, P.A., 57, 58
Hanlon, L., 528
Hanner, M.S., 638, 642, 867, 869, 870, 871, 1164, 1176, 1187–1189, 1232, 1244, 1246, 1270, 1271, 1274, 1284, 1292
Hansen, B., 392, 394, 395
Hansen, B.M.S., 814, 821
Hansen, C.J., 1466, 1477, 1478
Hansen, D.O., 1175, 1188
Hansen, G.B., 1477
Hansen, J.E., 1430
Hansen, M.M.S., 821
Hansen, S.H., 385, 394
Hansen, V.L., 1310, 1313, 1314, 1321
Hansen, W.L., 330
Hanski, M., 393, 397
Hanson, M.M., 778, 787
Hanson, W.B., 252, 1215, 1569, 1577, 1578, 1580, 1581, 1583, 1584
Hansteen, V.H., 1088
Hapgood, M.A., 1584
Hara, S., 960
Hara, T., 960
Haramundanis, K.L., 299
Harang, L., 1563, 1581
Harden, M., 1232
Harder, H., 1309, 1321
Harding, A.K., 695, 728, 729, 753, 754, 756, 757
Harding, P., 393, 395–397
Hardy, D.A., 1548, 1556, 1557
Hardy, E., 378, 381, 393, 396
Harford, J., 57, 58, 64, 86, 113, 259, 270
Hargraves, R.B., 1402
Hargraves, R.M., 1371
Hargreaves, J.K., 1580, 1584
Harland, D.M., 1323
Harmann, D., 397
Harmon, B.A., 855, 856
Harmon, J.K., 1220, 1232, 1246, 1267, 1269, 1271, 1300, 1321, 1400, 1402
Harnden, E.R., 724, 733
Harnden, F.R., 496, 497, 576, 754, 757, 872, 892, 893, 899, 904–910
Harper, C.L., 1326, 1366, 1367, 1372, 1374
Harper, C.T., 1372
Harper, G.M., 908

Harper, R.H., 1568, 1584
Harpold, D.N., 1430
Harries, J.E., 1658–1660, 1668
Harries, T.J., 909
Harris, A.I., 329, 640, 643
Harris, A.W., 658, 673, 1269, 1466, 1477, 1492
Harris, B., 1401
Harris, D.L., 1477
Harris, J.A., 1402
Harris, M.J., 528
Harris, N., 1615
Harris, S., 329, 330, 577, 924, 935
Harris, W.E., 394
Harrison, E.R., 413, 420
Harrison, F., 523
Harrison, F.A., 523, 526
Harrison, R.A., 1079, 1086
Harrison, S.R.K.F.A., 523
Harrison, T.E., 527
Harrison, V.A.W, 298
Hartig, G., 300
Hartke, G.J., 728, 730, 743, 754
Hartle, R.E., 1134, 1140, 1430
Hartley, K.F., 755
Hartman, D.H., 440
Hartman, J.W., 607, 761, 787, 814, 819–821
Hartman, R.C., 560, 696, 753–755, 757, 835
Hartmann, D.H., 505, 523–526, 587, 592, 604, 605, 607, 637, 639, 754
Hartmann, H.W., 809
Hartmann, J., 607, 639, 647, 673
Hartmann, L.W., 810, 835, 862, 870, 872, 891–893, 904, 909
Hartmann, W.K., 261, 270, 1253, 1266, 1268, 1270, 1271, 1274, 1297, 1321, 1323, 1348, 1349, 1359, 1364–1366, 1372, 1386, 1402, 1403
Hartquist, T.W., 610, 639
Hartree, D.R., 1561, 1581
Hartung, J.B., 1188
Hartwick, F.D.A., 452, 467
Harvey, B., 57, 58
Harvey, J.W., 1031, 1043, 1045, 1047, 1060, 1061, 1063, 1159
Harvey, K.L., 1027, 1032, 1063, 1087, 1139
Harvey, P.M., 642, 1273
Harwit, M., 306–309, 311, 312, 315, 318, 319, 329, 330, 576, 608, 609, 621, 622, 636, 639, 641–643
Harwood, A.S., 869
Harwood, R.S., 1668
Hasan, H., 299
Hasebe, N., 1004
Hasegawa, H., 1164, 1188
Hasegawa, T.I., 637, 960
Haser, L., 187, 641, 643, 869, 1526
Hashimoto, K., 960
Hashimoto, Y., 960
Hashimotodani, K., 960
Hasinger, G., 543, 546, 547, 756, 757, 817, 818, 820–822, 856
Haskin, L.A., 1380, 1402
Haslam, C.G.T., 672
Hass, M., 571, 871
Hassall, B.J.M., 821
Hasselmann, K., 1615, 1616
Hasselmann, S., 1615, 1616
Hassler, D.M., 1078, 1086–1088
Hatchett, S., 832, 835
Hathaway, D.H., 1038, 1054, 1062
Hatsukade, I., 495, 497, 947, 959
Hattori, M., 460, 467, 495
Hau, L.-N., 1506, 1527
Hauchecorne, A., 202
Hauck, S.A., 1322
Haugan, P.M., 1615

Haugan, S.V.H., 1086
Haugen, V.E.J., 1616
Hauser, M.G., 321, 329, 330, 419–421, 577, 642, 645, 705, 718, 1188, 1214
Häusler, B., 1523, 1527
Hawarden, T., 869
Hawking, S.W., 370
Hawley, J.F., 794, 808
Hawley, S.A., 550, 559
Hawley, S.H., 1140, 1161
Haxton, W.C., 440
Hayakawa, S., 314, 329, 498, 584, 604, 836, 959, 960
Hayami, Y., 697, 960
Hayashi, C., 1241, 1271
Hayashi, I., 959
Hayashi, T., 1004
Hayashida, K., 524, 527, 855, 959
Hayashida, N., 697
Hayden, R.J., 1322–1334, 1376
Hayes, K.G., 1231
Haynes, M.P., 393, 394
Hayward, T.L., 871
Hazard, C., 530, 547
Hazell, A.S., 869
Hazen, M.L., 396
Heacox, W.D., 429, 437
Head, J.W. III, 1298, 1301, 1303, 1310, 1317, 1320–1322, 1375, 1477
Head, W., 1373
Heap, S.R., 298–300, 394, 672, 1215
Hearn, A.G., 897, 909
Hearn, D.R., 821, 836
Hearth, L.W., 1429
Heasley, J.N., 391, 392, 394
Heath, E.P., 177
Heaviside, O., 1560, 1581
Heber, B., 1002
Hebrard, G., 299
Hechler, M., 113
Hecht, J., 1582
Heck, A., 298, 395–397
Heckman, T., 299
Hedgecock, P.C., 1032
Hedin, A.E., 1566, 1581, 1583
Heelis, R.A., 1577, 1578, 1581
Heemskerk, M.H.M., 910
Hefti, S., 1004, 1130, 1131, 1137, 1138, 1140, 1214
Hege, E.K., 1270
Heger, M.L., 607, 639
Heiken, G.H., 1322, 1323
Heiles, C., 595, 599, 604, 605, 615, 639, 653, 663–665, 672–674
Heiles, D., 577
Heimbach, P., 1604, 1615
Heindl, W.A., 694
Heinke, C.O., 786
Heinrich, M., 1477
Heinrich, W., 697
Heinrichsen, I., 871
Heirtzler, D., 1003
Heise, J., 523–528, 760, 789, 803, 809, 811, 822
Heiskanen, W.A., 1650
Heisler, J., 476, 497
Helbig, P., 389, 394, 457, 467
Helfand, D.J., 525, 564, 577, 724, 741, 753–755, 757, 758, 809, 877, 883, 891
Hellier, C., 804, 806, 809
Hellings, R.W., 371
Helliwell, R.A., 1581
Helmich, F.P., 626, 627, 637–640, 644, 863, 869, 870, 872, 873
Helmke, P.A., 1402
Helou, G., 570, 572, 577, 636

Helse, J., 803, 811
Helz, A.W., 1401
Hemenway, C.L., 1164, 1188
Hemmendinger, A., 1330, 1332, 1372, 1375
Hempe, K., 909
Henderson, J.T., 1566, 1581
Hendricks Jr., 1188
Hendry, M.A., 394
Heng, I.S., 370
Henke, T., 1122, 1138
Henkel, C., 394
Henney, W.J., 861, 870
Henning, R., 1060
Henning, T., 640, 870, 871
Hénoux, J.-C., 1126, 1138
Henrichs, H.F., 900, 909, 910
Henry, J.C., 644
Henry, J.P., 387, 393, 394, 473, 482, 485, 496, 497, 548, 576, 754
Henry, R.C., 582–584, 604, 673, 754
Hensberge, H., 894
Henschel, M., 1617
Henstock, D.R., 469
Henze, W., 286
Heppenheimer, P.A., 57
Heppenheimer, T.A., 58
Heppner, J.P., 187, 247, 253, 1018, 1032, 1582
Herant, M., 1086
Heras, A.M., 437, 610, 638, 639, 642–644, 869, 870, 872, 873
Herbig, G.H., 650, 673, 861, 870
Herbst, E., 637
Herbst, T., 636
Herbst, W., 872
Herbstmeier, U., 577, 605, 871
Hergenrother, C., 525
Herkenhoff, K.E., 1322, 1402
Herman, G., 1261, 1271
Herman, M., 1616
Herman, R., 308, 328
Hermsen, W., 286, 753, 755–758, 959
Hernandez, J.J., 1231
Hernanz, M., 527
Hernquist, L., 395, 496, 753
Herpe, G., 468
Herrera, R.G., 1477
Herreros, J., 1061
Herrick, D.L., 1319, 1321
Herrick, R.R., 1321
Herring, T.A., 467
Herrmann, U., 1293
Herrnstein, J.R., 383, 394
Herroros, J., M., 1061
Herrwerth, I., 1232, 1293
Herschel, W., 303, 1068, 1431, 1453
Herter, T.L., 330, 394, 644, 866, 870, 1429
Hertz, P., 527, 820, 822
Herzberg, G., 607, 638
Herzog, E., 475, 498
Herzog, L.F., 1315, 1331, 1372
Hesketh, W.D., 1668
Hess, D.C., 1371, 1372
Hess, P.C., 1299, 1314, 1321, 1333, 1340, 1342, 1372
Hess, V.F., 119, 150, 231, 253, 678, 695
Hess, W.N., 159, 177
Hesse, M., 251, 1025, 1032, 1159
Hesselman, G.C., 1610, 1615
Hesser, J.E., 394, 395
Hessman, F.V., 802, 809
Hester, J.J., 758, 869
Hetlinger, J., 300
Heubner, W.F., 1271

Heuerman, K.F., 1002
Heuring, F.T., 1651
Heuring, T.T., 1557
Hevesey, G., 1326, 1333, 1372
Hewagama, T., 328, 636, 642
Hewett, R., 870
Hewish, A., 759, 787, 840, 855, 1559, 1581
Hewitt, A., 546
Heyer, H.-H., 1269
Heyl, J.S., 753
Heymans, C., 466
Heyvaerts, J., 1013, 1032, 1083, 1086
Hibbitts, C.A., 1477
Hickam, H., 259, 270
Hidayat, B., 788, 909, 911
Hiesinger, H., 1321
Higbie, P.R., 1000
Higdon, J.C., 499, 525, 689, 695
Higdon, J.L., 526
Hikasa, K., 1231
Hilchenbach, M., 1138, 1212, 1214
Hildebrand, R.H., 315, 329, 611, 639
Hilditch, R., 809
Hildner, E., 221, 222, 1001, 1002, 1113, 1138, 1139
Hill, C.T., 693–695
Hill, F., 1036, 1047, 1052, 1060, 1062
Hill, H.A., 1038, 1062
Hill, J.R., 1232
Hill, R., 394, 1269
Hill, R.J., 393, 396, 733, 754
Hill, R.L., 395
Hill, T.W., 1541, 1556
Hill, V., 393, 439
Hillas, A.M., 696
Hillebrandt, W., 604
Hillenbrand, L.A., 859, 870
Hillier, D.J., 881, 892, 903, 909
Hillier, R.R., 722, 754
Hills, J., 523
Hiltner, W.A., 132, 150, 611, 639, 811, 821
Hinchliffe, I., 1231
Hind, J.R., 791, 793, 809
Hine, A.A., 1271
Hines, C.O., 196, 202, 1532, 1542, 1543, 1555, 1567, 1581
Hines, D.C., 330, 872
Hink, P.L., 696, 697
Hinshaw, G., 328, 410, 419–421
Hinshaw, J.C., 636
Hinson, D.P., 1477
Hintenberger, H., 434, 440, 1379, 1390, 1402
Hinteregger, H.E., 208, 209, 222, 1566, 1581
Hippelein, H., 523, 577
Hiraga, J., 960
Hirahara, M., 1557
Hirano, T., 855
Hirao, K., 1232
Hirao, T., 330
Hirata, K., 943, 959
Hirata, R., 909
Hirayama, T., 1111
Hiromoto, N., 330, 642
Hirshberg, J., 1122, 1138
Hirt, P., 1137
Hirth, G., 871
Hirth, W., 605
Hivon, E., 394, 420, 466
Hjellming, R.M., 753, 760, 787, 835, 840, 853, 855
Hjorth, J., 466, 523, 527
Ho, C., 524, 526, 576, 753
Ho, L.C., 546, 547, 854

Hoag, A., 458, 467
Hoare, M.G., 908
Hoban, S., 1231, 1291
Hobbs, L.M., 428, 650, 666, 667, 673–675
Hobson, M.P., 393, 395, 419
Hochedez, J.F., 1086
Hocke, K., 1582
Hodderson, L., 151
Hodges, R.R., 1232, 1292, 1293
Hodgkin, S.T., 892, 893
Hodgson, R., 229, 253
Hoegy, W.R., 1580
Hoeksema, J.T., 1060, 1062, 1063, 1086, 1139, 1147, 1159–1161
Hoeksema, T., 1061
Hoekstra, H., 525
Hoekstra, R., 299
Hoessel, J.G., 393–397, 869
Hoffman, J.A., 815, 821
Hoffman, J.H., 1232, 1292, 1293, 1578, 1581
Hoffman, R.D., 440
Hoffman, T.E., 352
Hoffmann, H.J., 1187
Hoffmann, W.F., 306, 307, 312, 329, 1272, 1430
Höflich, P., 377, 389, 394, 516, 525
Hofmann, E., 1615
Hofman-Wellenhof, B., 1650
Hofmeister, H., 1403
Hofmeyr, P.K., 1403
Höfner, H., 1138, 1528, 1583
Hogan, C.J., 395, 397, 420, 428, 438, 468, 1231
Hogerheijde, M.R., 607, 610, 625, 643, 644, 858, 870
Hogg, D.W., 521, 525
Hohenberg, C.M., 1140
Höhler, G., 1231
Holberg, J.B., 287, 299, 1430
Holder, G.P., 395
Holder, J., 960
Holeman, E., 1556
Holland, W.S., 870
Hollenbach, D.J., 610, 616, 620, 637–640, 643, 673, 675, 1269
Hollingworth, J., 1561, 1581
Holloway, J.R., 1321
Hollweg, J.V., 1001, 1085, 1087, 1131, 1138, 1140, 1160, 1215
Holm, A.V., 298, 797, 809
Holman, G.D., 1003
Holme, R., 1488, 1491
Holmes, A., 258, 270, 1327, 1330, 1372
Holmes, J.C., 1582
Holt, S.S., 497, 498, 547, 577, 696, 754, 757, 789, 820, 822, 834, 836, 837, 893, 946, 959
Holtzman, J.A., 869
Holweg, J.V., 253
Holweger, H., 1066, 1086
Holzer, T.E., 245, 253, 682, 695, 1001, 1088, 1134, 1138, 1139, 1191, 1193, 1199, 1201, 1214, 1215
Holzworth, R.H., 1527, 1582
Homan, D.C., 560
Homan, J., 817, 820, 821
Honda, K., 697
Honda, M., 429, 436, 439
Hones, E.W., 1016, 1022, 1031, 1032, 1541, 1544, 1545, 1556, 1557
Hong, S.S., 612, 639
Hongre, L., 1491
Honscheid, K., 1231
Hony, S., 618–620, 632, 633, 640, 643, 644
Hood, D., 300
Hood, L.L., 1479, 1491, 1492
Hoogeveen, G.W., 254, 1137, 1160
Hook, I.M., 395, 420, 467, 523
Hook, R., 524

Hoover W., 70
Hopmann, H., 76, 86, 96, 113
Hoppe, M.M., 981–983, 1001
Hoppe, P., 673, 1188, 1278, 1292
Hora, J.L., 1272, 1430
Horack, J.M., 523, 524, 526, 527
Horanyi, M., 1183, 1188, 1218, 1220, 1232, 1262, 1263, 1271
Horbury, T.S., 1002, 1155, 1159, 1160
Horita, R.E., 1555
Horn, J., 643
Horne, K., 547, 793, 799, 800, 808–810, 893
Horner, D.J., 497
Horner, J.P., 489
Hörnes, M., 1280, 1294
Hornstein, J.S., 1449
Hornung, K., 1188, 1293
Horowitz, N., 260, 270
Horwitz, J.L., 1556
Hörz, F., 1169, 1187, 1188
Hosemann, R., 1333, 1372
Hoshi, R., 854, 959
Hoshino, M., 1032
Houbolt, J.C., 85, 113
Houck, J.R., 312, 329, 330, 577, 868, 871, 935
Houck, T.E., 299
Houde, M., 637
Houdek, G., 1059, 1060, 1062
Houghton, J.T., 1615, 1617, 1660, 1668
Houpis, H.L.F., 1232, 1262, 1271
House, F.B., 1660, 1668
Houston, W.N., 1273
Houtermans, F.G., 258, 270, 1329, 1330, 1372
Houziaux, L., 298–300
Hovestadt, D., 143, 150, 990, 1000–1004, 1032, 1124, 1126, 1138, 1214, 1215
Hovis, W.A., 1615
Hovland, H.J., 1273
Howard, A., 1322
Howard, I.D., 671
Howard, R.A., 1000, 1004, 1052, 1062, 1086, 1113, 1136, 1138, 1140
Howarth, I.D., 797, 808, 834, 900, 909
Howe, J.E., 636, 639, 641, 642
Howe, R., 1050–1052, 1058, 1060–1063
Howell, D., 712, 718
Howell, S., 809
Howk, C., 525
Howk, J.C., 670–673
Hoyle, F., 433, 533, 437–440, 547, 553, 559, 613, 640, 829, 834, 990, 1002, 1008, 1009, 1014, 1023, 1031, 1032, 1113, 1376, 1506, 1527
Hoyle, R.A., 722, 754
Hoyt, W.G., 259, 270
Hristov, V.V., 394, 420, 466, 637
Hrivnak, B.J., 632, 640
Hsieh, K.C., 966, 969, 1001, 1002, 1138, 1214
Hsu, S., 1032, 1033
Hu, E.M., 496, 672
Hu, H.-Y., 1477
Hu, J.Y., 862, 870
Hu, X., 872
Hua, X.M., 910
Huang, C.Y., 1526
Huang, J., 661, 673
Huang, M., 810
Huang, X., 393
Hubbard, N.J., 1345, 1346, 1372, 1373, 1380, 1383, 1384, 1402
Hubbard, W.B., 1160, 1269, 1426, 1430, 1436, 1443, 1449
Hubble, E.P., 374, 394
Hubeny, I., 394, 799, 808–810, 908
Huber, M.C.E., 221, 222, 674, 1077, 1086–1088, 1138, 1139
Huchra, J.P., 393–397, 496, 547
Huchtmeier, W.K., 1291

Huckle, H.E., 837
Hudec, R., 505, 518, 523, 525
Hudgins, D.M., 619, 620, 640, 642, 644
Hudson, H.S., 526, 892, 971, 1002, 1048, 1063, 1070, 1088
Hudson, M.K., 1527, 1556, 1557
Huebner, W.F., 1233, 1236, 1257, 1261, 1271, 1272, 1274, 1288, 1292
Huensch, M., 892
Hufbauer, K., 203, 205, 216, 218, 222
Huggins, P.J., 330, 640, 642
Huggins, W., 303, 329
Hughes, D.W., 1249, 1250, 1268, 1271, 1272
Hughes, E.R., 835
Hughes, J.P., 484, 496–498
Hughes, M.P., 663, 673
Hughes, P.A., 518, 525
Hughes, S.M.G., 393–397
Hughes, V.A., 835
Hui, J., 1429
Huille, S., 439
Hulbert, E.O., 1566, 1582
Huldtgren, M., 636, 869, 871, 872
Hull, H., 1063
Hulleman, F., 736, 754
Hulse, A.R., 761, 787
Hulse, R.A., 286, 371
Hultqvist, B., 156, 172, 177, 1032, 1508, 1535, 1537, 1540, 1542, 1546–1550, 1556, 1563, 1569, 1581
Humason, M.L., 374, 375, 394
Humes, D.H., 1172, 1187, 1188
Hummer, D., 907, 909
Humphreys, R.A., 909
Humphries, C.M., 298
Hundhausen, A.J., 221, 222, 247, 251, 253, 254, 1001, 1113, 1126, 1128, 1136, 1138, 1139, 1152, 1160, 1161
Huneke, J.C., 1322, 1370, 1372, 1374, 1375
Huneke, J.L., 1341, 1348, 1353
Hunstead, R W., 855
Hunt, J., 1271
Hunt, L.K., 523
Hunten, D.M., 165, 177, 190, 198, 202, 1188, 1274, 1322, 1323, 1423, 1426, 1428, 1430, 1463, 1476
Hunter, R.H., 1375
Hunter, S.D., 684, 695, 696, 753, 755
Hunter, T., 641
Huntress, W.T., 1231, 1291
Huot, J.-P., 1273
Hurlburt, N.E., 1032, 1054, 1060, 1062, 1087
Hurley, K., 467, 504, 505, 508, 523–528, 736, 754, 755, 758, 787, 821
Hurley, P.M., 1374
Hurt, R., 1478
Hurwitz, M., 298, 299, 605, 669, 673
Hussain, L., 1348, 1371, 1374
Huster, E., 1331, 1372
Hut, P., 755, 783, 787, 1240, 1271
Hutchings, J.B., 298–300, 547, 672, 808, 820, 841, 854, 855
Hutsemékers, D., 468
Huuskonen, A., 1569, 1581
Huygen, E., 641, 644, 871, 872
Huygen, R., 638, 642, 869, 873
Hviid, S.F., 1371, 1402
Hwang, U., 696, 754, 946, 948, 959
Hya,W., 636
Hyett, B.J., 246, 255
Hynds, R.J., 255, 1000
Hynek, J.A., 1272
Hysell, D.L., 1564, 1582

Iacoangeli, A., 394, 420, 466
Ianna, P., 525
Ibanez, J.M., 558

Ibata, R.A., 392, 394
Ibbetson, P., 524
Iben, I., 642, 775, 787, 935
Ichimaru, S., 844, 855
Igenbergs, E.B., 1174, 1188, 1293
Iglesias, C.A., 731, 754, 1045, 1047, 1060, 1062
Iijima, T., 1538, 1539, 1556
Ikebe, Y., 492, 495–497, 576, 577
Ikeda, K., 959
Ikeda, M., 1615
Ilinskii, V.N., 525, 527, 755
Illarionov, A.F., 828, 835
Illiano, J.M., 1232, 1292, 1293
Illingworth, G.D., 393–397, 467, 498, 523, 526
Imbault, D., 869
Imhoff, C.L., 861, 870
Imori, M., 696
Impey, C.D., 467, 524, 559
Inan, U.S., 742, 755, 1570, 1581
Ingalls, R.P., 352
Ingersoll, A., 1426, 1429, 1430
Ingham, M.F., 233, 251
Inghram, M.G., 1331, 1372, 1374
Inhester, B., 1032, 1086
Innes, D.E., 1028, 1032, 1079, 1086
Innes, M., 269
Inomoto, M., 1032
Inoue, H., 498, 560, 757, 830, 835, 836, 855, 893, 956, 959
Inoue, M., 394, 836
Inouye, G.T., 1001
Intriligator, D.S., 996, 1002, 1426, 1430
Iorenas, V.A., 275
Iovlev, M., 1233
Ip, W.-H., 177, 1202, 1215, 1231, 1240, 1270, 1272, 1274, 1291, 1292, 1469, 1477
Ipatov, S.I., 1242, 1271
Ipavich, F.M., 691, 695, 1001–1003, 1032, 1124, 1556, 1137, 1138, 1214, 1556
Irvine, J.M., 433, 438
Irvine, W.M., 626, 640, 1284, 1286, 1290, 1292, 1445, 1449
Irwin, J.A., 489, 497
Irwin, M., 395, 420, 467
Isaac, G.R., 1043
Isaacman, R.B., 328, 420, 636, 645
Isaak, G.R., 1060–1062
Isenberg, P.A., 253, 1085, 1087, 1131, 1138, 1140, 1157, 1160, 1202, 1209, 1214, 1215, 1233
Isern, J., 527
Ishida, M., 803, 809, 959
Ishimaru, Y., 576
Ishino, M., 696
Ishitsuka, M., 1061
Isobe, N., 498
Isobe, S., 702, 718
Isobe, T., 577
Isotov, Y.I., 440
Israel, F.P., 870
Israel, G.L., 524, 736, 755, 756
Israel, M.H., 695
Israel, W., 370, 371
Israelevich, P.L., 1229, 1232
Istomin, V.G., 195, 202
Istomin, Ya.N., 753
Itoh, M., 524, 527, 576, 959
Itow, Y., 1061
Ivanov, A.B., 1322
Ivanov, B.A., 1320
Ivanov, E.N., 370
Ivanov, V.D., 223
Ivans, I.I, 396
Ives, J.C., 497, 824, 835

Ivison, R.J., 468
Iwamoto, K., 516, 524, 525
Iwaniszewska, C., 1476
Iwasaki, K., 1321
Iwasawa, K., 533, 536, 542, 546, 547, 753, 855
Iyer, B.R., 358, 367, 370, 371
Iyomoto, N., 498
Iyudin, A.F., 955, 959
Izenberg, N., 1321
Izmodenov, V.V., 1202, 1206, 1208–1211, 1214, 1215
Izotov, Y.I., 428, 440
Izsak, I.G., 1651
Izumiura, H., 640, 873

Jacchia, L.G., 1566, 1581, 1651
Jackson, A.A., 1179, 1188
Jackson, B.V., 245, 253
Jackson, J.C., 831, 835
Jackson, J.D., 1231, 1520, 1527
Jackson, P.D., 420, 421, 663, 675
Jackson, W.R., 754
Jacobi, R.B., 1370
Jacobs, H., 869
Jacobs, J.A., 1480, 1491
Jacobsen, J., 1457, 1466
Jacobsen, P., 542
Jacobson, S.B., 392, 396
Jacobson, T.S., 1478
Jacoby, G.H., 935
Jaeger, K., 525
Jaffe, A.H., 383, 385, 387, 388, 390, 394, 420, 466, 1328, 1372
Jaffe, D.T., 315, 329, 616, 640, 643
Jaffe, L.D., 274, 275
Jäger, C., 631, 640, 866, 870, 1331, 1333, 1338, 1348
Jäger, E., 1375
Jager, R., 510, 524, 526
Jagod, M.-F., 1449
Jagoutz, E., 1360, 1363, 1372, 1392, 1401–1403
Jahoda, K., 590, 603, 604, 694, 822
Jain, B., 395, 468
Jain, R.K., 855
Jakob, G., 869
Jakobsen, P., 523, 547, 589, 604
Jakosky, B.M., 177, 1321, 1323, 1376, 1423
Jakowksi, N., 1575, 1581
Jamar, C., 298, 1086, 1271, 1272
James, F., 1231
James, H.F., 1269
James, O.B., 1342, 1372
James, P.B., 1321
James, T.L., 1002
Jameson, R., 871
Jamieson, J.A., 326, 327, 329
Janardhan, P., 1291
Janes, D.M., 1476
Janes, K.A., 394
Janka, H.T., 521, 528
Jansen, D.J., 643
Jansen, F., 697, 822
Janssen, M., 419–421, 645, 1063
Janssen, P.A.E.M., 1616
Jär, C., 640
Jarnot, R.F., 1668
Jarvis, R.M., 396
Jaskulek, S., 1004
Jasniewicz, G., 935
Jau, A.W., 1584
Jaumann, R., 1402
Jauncey, D.L., 507, 559, 855
Jaunsen, A.O., 466, 523, 527
Jayaraman, S., 1187

Jayawardhana, R., 864, 870
Jean, C., 456, 467, 468
Jeans, J.H., 857, 870
Jedrzejewski, R., 547
Jefferies, S.M., 1061
Jefferts, K.B., 644
Jeffreys, H., 272, 275
Jeffries, R.D., 883, 892
Jegou, J.P., 197, 202
Jelinsky, P.S., 299
Jelinsky, S., 675
Jenkins, E.B., 246, 251, 287, 291–294, 298–300, 496, 584, 604, 608, 643, 652, 654, 657, 658, 660, 661, 672–675, 897, 910, 958, 959
Jennings, D.E., 1449
Jennings, R.E., 329, 330, 577, 935
Jennison, R.C., 531, 547
Jensen, B.L., 523
Jensen, M.D., 1577, 1582
Jerius, D., 577
Jerjen, H., 378, 394
Jernigan, J.G., 821
Jessberger, E.K., 1279, 1283, 1284, 1292, 1293, 1338, 1341, 1343, 1348, 1349, 1371, 1372
Jesse, W.P., 122, 151, 697
Jetzer, P., 395
Jewitt, D., 1236, 1239, 1243, 1244, 1247, 1258, 1264–1266, 1268, 1269, 1271–1273, 1281, 1292, 1293
Jewitt, D.C., 1271, 1272, 1293
Jger, C., 871
Jha, S., 395, 420, 468
Jhabvala, M., 1138
Ji, H., 1024, 1032, 1033
Jiang, X.J., 547
Jimenez, A., 1061
Jobes, F., 1033
Joblin, C., 618, 638, 640
Jocelyn, J.A., 645, 1002, 1088, 1137–1139
Jochum, K.P., 1402
Johannessen, J.A., 1615, 1616
Johannessen, O.M., 1593–1595, 1606, 1610, 1611, 1615, 1616
Johanssen, S., 439
Johansson, L.E.B., 1291
John, D., 637
Johnson, C.L, 1321, 1322
Johnson, C.Y., 1566, 1582
Johnson, D.A., 1354, 1371
Johnson, F.S., 298, 1215
Johnson, H.L., 304, 305, 329
Johnson, H.M., 872, 893, 910
Johnson, H.R., 909
Johnson, J., 396
Johnson, J.M., 1402
Johnson, J.R., 1371, 1402
Johnson, K.F., 1231
Johnson, M.C., 897, 909
Johnson, M.P., 755
Johnson, P., 1371, 1374, 1392, 1401
Johnson, R.A., 527
Johnson, R.E., 1258, 1271, 1462, 1477
Johnson, R.G., 1557, 1581
Johnson, S., 736
Johnson, T.H., 122, 150, 678, 695
Johnson, T.V., 1174, 1188, 1323, 1423, 1448, 1449, 1477, 1478
Johnson, W.N., 560, 854
Johnston, H., 814, 821
Johnston, H.M., 757
Johnston, K.J., 835
Johnston, M.D., 829, 835
Johnston, S., 753, 755, 786–789
Johnstone, A., 1226, 1232
Johnstone, R., 871
Johnstone, R.M., 498
Jokipii, J.R., 140, 150, 250, 251, 253, 254, 679, 687, 688, 690, 695, 696, 979, 987, 988, 993, 997, 1000, 1002, 1003, 1005, 1085, 1139, 1140, 1141, 1150, 1154–1156, 1158–1161, 1188, 1215
Jolliff, B., 1321
Joncas, G., 637
Jones, A., 638, 1061, 1188
Jones, A.P., 620, 638, 640, 673
Jones, B., 393, 525, 644
Jones, B.B., 222
Jones, B.W., 330
Jones, C., 285, 393, 475, 478–480, 482, 487, 489, 496, 497, 576–578, 754, 809, 835, 910
Jones, C.A., 823, 835
Jones, D., 1557
Jones, D.E., 1140, 1492
Jones, D.H.P., 755
Jones, D.L., 855
Jones, E.W. Jr., 1555
Jones, F.C., 686–688, 690, 695, 1002
Jones, G.O.L., 1547, 1556, 1557, 1581
Jones, H.P., 1074, 1080, 1085, 1086
Jones, J.A., 1139
Jones, J.H., 1363, 1372
Jones, K., 270
Jones, K.B., 1477
Jones, L.R., 497
Jones, R.L., 1662, 1668
Jones, T.E., 299
Jones, T.J.L., 222
Jones, T.W., 552, 559, 639, 696
Jones, W.P., 255
Jones, W.V., 695
Jonker, P.G., 525, 820
Joos, R., 1001, 1137, 1138
Jorda, L., 1236, 1265, 1270–1274
Jordan, C., 216, 217, 222, 878, 892, 893, 908, 1080, 1087
Jordan, P., 345, 351
Jordan, S.D., 1087
Jorgensen, H.E., 1243, 1244, 1274
Josberger, E.G., 1615
Joselyn, J.A., 1113, 1201, 1215
Joseph, C.L., 300
Joseph, R., 871
Joss, P.C., 781, 782, 787, 788, 814, 819, 821, 827, 836, 1239, 1271
Josties, F.J., 835
Joswiak, D.J., 1272
Joubert, M., 636–638, 641, 869
Joung, M.K., 638
Jourdain, E., 523
Journet, A., 352, 1651
Joy, A.H., 804, 806, 809
Joy, M., 546
Joy, S., 177, 1491, 1492
Joyce, R.R., 644
Juda, J.Z., 576
Juda, M., 564, 593, 595, 603, 604
Judge, D.L., 149, 1138, 1199, 1214, 1216
Judge, P.G., 904, 907, 909, 1078, 1087, 1088
Jugaku, J., 788, 821, 871
Juhasz, A., 1221, 1232
Julian, W.H., 1269, 1496, 1527
Julien, P., 285, 496, 754, 809
Juliusson, E., 126, 150
Jura, M., 298, 577, 672, 673, 865, 867, 868, 870
Jurgens, R.F., 352, 1232, 1270, 1271
Justtanont, K., 632, 633, 637, 640, 643, 873
Juszkiewicz, R., 387, 394

INDEX OF CITED AUTHORS

Kaas, A.A., 636, 869, 871, 872
Kaastra, J.S., 492, 497, 545, 547, 892, 910, 959
Kadnichansky, S.A., 1320
Kadomtsev, B.B., 1023, 1032
Kahabka, P., 780, 787, 808, 809
Kahane, C., 870
Kahler, S.W., 969, 972, 1000, 1002, 1003, 1152, 1159, 1160
Kahn, F.D., 589, 602, 604
Kahn, S.M., 604, 836, 893
Kaiser, M.E., 300
Kaiser, N., 387, 394, 397, 460, 467, 477, 497
Kaiser, T.R., 1560, 1582
Kaita, E., 420, 421
Kajino, T., 428, 430, 438
Kajita, T., 959, 1056, 1061, 1062
Kakar, R.K., 644
Kalas, P., 870
Kalata, K., 1086
Kalberla, P.M.W., 602, 604, 605, 672
Kalble, A., 299
Kaldeich, B.H., 639, 1087, 1272
Kaldeich-Schürmann, B., 1087
Kale, V.S., 1320
Kalen, J.D., 437
Kaler, J.B., 935
Kalkofen, W., 1073, 1087
Kallemeyn, P.H., 1402
Kallenbach, R., 1001, 1005, 1138, 1158, 1159, 1214
Kalligas, D., 366, 371
Kallman, T.R., 831, 834–837
Kalogera, V., 760, 761, 787, 789
Kaluzienski, J.L., 856
Kaluzienski, L.J., 836
Kamata, Y., 892
Kambe, E., 909
Kameda, J., 1061
Kamei, S., 697, 960
Kamél, L., 1273
Kamide, Y., 187, 1032
Kamijo, F., 613, 640
Kamionkowski, M., 384, 394
Kamisky, C.D., 1274, 1293
Kamoun, P.D., 1236, 1246, 1271
Kamp, I., 630, 640
Kamp, L.W., 1423
Kamperman, T., 869
Kamperman, T.M., 299, 869
Kanbach, G., 286, 694–696, 723, 737, 753, 755, 757, 758
Kandel, R., 1659, 1668
Kane, S.R., 465, 1000
Kaneda, H., 498
Kaneyuki, K., 1061
Kang, H.S., 440, 693, 696
Kaniovsky, A., 960
Kankelborg, C.C., 1032, 1085
Kantowski, R., 444, 466
Kapahi, V.K., 547
Kaper, J.A., 909
Kaper, L., 523, 528, 558, 789, 900, 909, 911
Kaplan, G., 1401
Kaplan, L.D., 260, 269, 270, 1381
Kappelmann, N., 299
Karamanolis, S., 113
Karatasos, K., 820
Karev, V.I., 222
Kargatis, V.E., 507, 526
Kargel, J.S., 1320, 1476, 1477
Karlen, D., 1231
Karnashov, A., 525
Karovska, M., 222, 577, 758, 1138, 1139

Karsh, J., 1004
Kasai, M., 467
Kashiwagi, T., 1004
Kashyap, V., 892
Kaspi, S., 545, 547
Kaspi, V.M., 726, 732, 736, 755, 757
Kasprzak, W.T., 1430
Kassiola, A., 458, 467
Kästle, H., 1002
Kastler, A., 189, 198, 202
Kastner, J.H., 645, 860, 870, 935
Katayama, K., 757
Kato, C., 1004
Kato, T., 604, 1087
Kattenberg, A., 1615
Katterloher, R.O., 869
Katz, J.I., 517, 526, 782, 787
Katz, N., 395, 496
Kaufman, H.R., 113
Kaufman, M.J., 616, 620, 621, 639, 640–642
Kaufman, T.M., 352
Kaufman, Y.J., 1616
Kaufmann, J.M., 1047, 1062
Kaufmann, R., 387, 394
Kaula, W.M., 268, 270, 1320, 1322, 1456, 1461, 1477, 1650, 1651
Kavelaars, J.J., 381, 394, 1270, 1476
Kavenson, N., 820
Kaw, P.K., 1011, 1031
Kawachi, A., 960
Kawada, M., 330, 637
Kawai, N., 524, 528, 745, 748, 753, 755–757, 833, 835, 836, 959
Kawamura, S., 370
Kawanomoto, S., 438
Kaye, M.J., 1401
Kayser, B., 1231
Kayser, D., 1582
Kayser, R., 446, 467, 468
Kayser, S.E., 1582
Kazanas, D., 560
Kazimirchak-Polonskaya, E.I., 1274, 1293
Keane, J.V., 624, 626, 639, 640, 644, 872
Keane, M., 525
Kearney, P.D., 1137
Keays, R.R., 1401
Keckhut, P., 202
Kecman, B., 1004
Kecskeméty, K., 1232, 1263, 1271
Kedziora-Chudczer, L.L., 559
Kedziora-Chudczer, R., 557
Keegan, R., 697
Keegstra, P., 420, 421
Keel, W.C., 541, 547
Keene, J., 616, 621, 636, 640, 645, 1292, 1293
Keeton, C.R., 394
Kehm, K., 1140
Kehoe, R., 523
Keiffer, H.B., 1423
Keil, K., 1373
Keil, S.L., 253, 1087
Keiser, G.M., 351
Keith, J.E., 1374
Keldysh, M.V., 265, 270
Keller, H.U., 638, 869, 1223, 1224, 1232, 1233, 1236, 1241, 1243, 1244, 1247, 1251, 1253–1259, 1262, 1263, 1266, 1267, 1270–1274, 1292, 1402
Keller, L.D., 959, 1293
Kellett, B.J., 724, 741, 755, 893, 1233
Kelley, D., 1232, 1271
Kelley, M.C., 1527, 1557, 1563, 1564, 1570, 1581, 1582

Kelley, R.L., 820, 825, 835
Kellogg, E., 478, 496, 497, 498, 576, 754, 757, 788, 789, 820, 821, 835, 836
Kellogg, P.J., 246, 253, 1526, 1530, 1556
Kelsall, T., 312, 329, 402, 419–421, 642, 645, 718, 1177, 1188, 1214
Kelskemety, K., 1263
Kelson, D.D., 378, 393–397, 467, 498, 523, 526
Kelvin, W.T., 277, 286
Kemper, F., 643
Kemurdzhian, A.L., 1322
Kendziorra, E., 723, 755, 757, 789, 836
Kenfer, A., 577
Kenidge, J., 1292
Kennedy, G.P., 57, 58
Kennedy, M.J., 1002
Kennedy, T.C., 1649
Kennefick, D., 370
Kennel, C.F., 979, 1002, 1419, 1491, 1492, 1496, 1499, 1500, 1510, 1514, 1527, 1549, 1556
Kennelly, A.E., 1560, 1582
Kennicutt, R.C. Jr., 393–397
Kent, S.M., 322, 329, 475, 497
Kenyon, S.J., 808, 809, 870
Keohane, J.W., 689, 694, 696
Keppler, E., 986, 1002, 1232
Kerkhof, O., 638, 639
Kernan, P.J., 393
Kerp, J., 602, 604, 605
Kerr, A.R., 644
Kerr, F.J., 639, 695
Kerr, R.A., 1264, 1272
Kerr, R.E., 1647, 1651
Kerr, Y., 1612, 1616
Kertzman, M.P., 695
Kessler, D.J., 1168, 1188
Kessler, J.E., 638
Kessler, M.F., 321, 329, 437, 559, 570, 576–578, 610, 636–638, 640–645, 856, 862, 869, 870, 872, 873, 1435, 1449
Kesson, S.E., 1351, 1374, 1385, 1402
Kester, D.J.M., 638, 644, 869, 872, 873
Kesteven, M.J., 855
Key, J., 1616
Khabibrakhmanov, I.Kh., 1233
Khachikian, E.Ye., 537, 547
Khanna, R., 1232
Khare, B.N., 1293, 1476, 1477
Kharyukova, V.P., 1322, 1402, 1403
Khatri, G., 1062
Khavenson, N., 820
Khokhlov, A., 377, 394
Khristiansen, G.B., 691, 696
Khudukon, B.Z., 1582
Khurana, K.K., 177, 1490–1492
Kiang, T., 1259
Kidder, L.E., 359, 371
Kidger, M.R., 560, 1272
Kieboom, K., 809
Kieda, D.B., 695
Kieffer, H.H., 165, 177, 1296, 1321, 1323, 1375, 1376, 1423
Kieffer, S.W., 1477, 1478
Kienzle, F., 1060
Kiepenheuer, K.O., 222, 233, 253, 1095, 1113
Kifune, T., 959, 960
Kii, T., 560, 835, 855, 959
Kikuchi, J., 1004
Kikuchi, K., 489, 496, 497, 576
Kikuchi, S., 820
Killeen, J., 1031
Killen, T.L., 1583
Killian, J.R., 43, 57, 58

Kim, A.G., 395, 420, 467
Kim, D.-W., 57, 567–569, 576, 577
Kim, M.Y., 395, 420, 467
Kim, P.D., 893
Kim, S.B., 959
Kim, S.J., 1286, 1287, 1293
Kim, V., 113
Kimble, R.A., 299, 300
King, A.R., 764, 773, 782, 787, 809, 810, 813, 820, 821, 837
King, D.L., 789, 837
King, G.A.M., 1567, 1571, 1582
King, I.R., 498, 719
King, J.W., 1160, 1582
King, K.J., 330, 638, 641, 869
King-Hele, D.G., 1575, 1582, 1650, 1651
Kingma, S., 935
Kinkel, U., 871
Kinney, A., 299, 536, 546, 547
Kintner, E.C., 498
Kinugasa, K., 696, 947, 948, 959, 960
Kin-Wing, 642
Kinzer, R.L., 695
Kiplinger, A.L., 809
Kippen, R.M., 286, 495, 516, 523, 525, 526, 758
Kippenhahn, R., 761, 772, 775, 787
Kirchengast, G., 1568, 1582
Kirches, S., 871
Kirk, R.L., 1322, 1402, 1461, 1477, 1478
Kirkham, B., 298, 672
Kirkman, D., 440, 469
Kirkpatrick, D.J., 526, 1478
Kirnozov, F.F., 1403
Kirsch, E., 1001
Kirshner, R.P., 395, 420, 468
Kirsten, T., 1333, 1372
Kissel, J., 1176, 1181, 1187, 1188, 1279, 1283, 1292, 1293
Kissel, S.E., 960
Kistler, L.M., 1003, 1526
Kita, R., 960
Kitamoto, K., 959
Kitamoto, S., 576, 833, 835, 836, 893
Kitamura, Y., 1257, 1272
Kitayama, M., 1232
Kivelson, M.G., 171, 177, 1489–1492
Kiziloglu, Ü., 753, 756, 820
Kjeldsen, H., 1059, 1060, 1062
Kjeldseth-Moe, O., 1069, 1078, 1087
Klaas, U., 871
Klaasen, K.P., 1188, 1430
Klafter, J., 1527
Klapdor-Kleingrothaus, H.V., 396
Klare, G., 809
Klarmann, J., 696, 697
Klarmann, M.H., 695
Klebesadel, R.W., 500, 501, 523, 524, 526, 528
Klecker, B., 970, 993, 997, 999, 1000, 1002, 1003, 1032, 1138, 1214, 1215
Kleczewski, M., 869
Klein, R.I., 834
Klein, L.W., 1113, 1159
Klein, R.I., 897, 908
Kleine, M., 1288, 1293
Kleiner, S.C., 636, 639, 642
Kleinknecht, K., 1231
Kleinmann, D.E., 305, 311, 329, 609, 640
Klemaszewski, J.E., 1477
Klepach, E.G., 696
Klepczynski, W.J., 835
Kletetschka, G., 1491

Kleuner, S.C., 641
Klimchuk, J.A., 1088
Klimenko, I.N., 1232
Klimov, S., 1233
Klimov, Y.G., 443, 467
Klinger, J., 1219, 1232, 1259–1262, 1270–1272
Klinglesmith, D.A., 298, 1231, 1291
Klioner, S.A., 351
Kliore, A.J., 1426, 1430, 1469, 1477
Klitzing, A., 1583
Klobuchar, J.A., 1583
Klock, W., 1293
Klose, S., 524
Klosko, S.M., 1651
Kloster, K., 1615, 1616
Klostermeyer, J., 1568, 1582
Klumpar, D.M., 1526, 1582
Klypin, A., 496
Knacke, R.F., 629, 643, 864, 865, 870–872, 1244, 1269, 1271, 1445, 1449
Knaflich, H., 202
Knapp, G.R., 565, 577, 589, 604
Knapp, J., 694
Knauth, D.C., 429, 436, 438
Kneib, J.-P., 466, 467, 498, 577
Knie, K., 601, 604
Kniffen, D.A., 696, 723, 725, 753–757
Knigge, C., 796, 809
Knight, C.A., 560
Knight, S., 1516, 1527, 1536, 1556
Knittle, E., 1321
Knollenberg, J., 1256, 1257, 1272
Knollenberg, R.G., 1423
Knop, R.A., 395, 420, 467, 523
Knott, S.F., 1375
Knowles, I.G., 1231
Knox, L., 420
Knude, J., 603
Knudsen, D.J., 1584
Knudsen, J.M., 1371, 1402
Knulst, J.C., 1615
Ko, S., 695
Ko, Y.-K., 1127, 1138
Kobayashi, K., 1061
Kobayashi, M., 1004
Kobayashi, N., 526, 892
Kobayashi, S., 518, 526
Kobayashi, T., 696
Kobayashi, Y., 642
Koblinsky, C.J., 1526, 1591, 1616
Koch, D.G., 329, 576, 636, 639, 641, 642, 754
Kochan, H., 1233, 1272
Kochanek, C.S., 383, 394, 456, 457, 466–468
Kochanski, G.P., 498
Kochansky, A., 193, 202
Kochavi, E., 1269
Koch-Miramond, L., 892
Koeberl, C., 1374
Koehler, J.A., 1584
Koempe, C., 870, 871
Koerner, D.W., 526, 630, 640, 643, 872
Koester, D., 904, 911, 1216
Kogut, A., 413, 417, 419–421
Koh, D.T., 786, 820, 834
Kohl, H., 1567, 1582, 1583
Kohl, J.L., 218, 222, 242, 253, 1086, 1134, 1135, 1137–1139
Kohler, S., 395
Köhler, Th., 468
Kohmura, T., 960
Kohmura, Y., 577
Kohno, T., 1004

Koike, C., 631, 640, 866, 867, 871
Kojima, H., 1527, 1555
Kokubun, S., 1032, 1555–1557
Kolá, A., 525
Kolaczek, B., 1649
Kolatt, T., 397, 505, 526
Kolb, E.W., 371
Kolb, U., 764, 780, 787, 788, 791, 810, 813, 821
Kolchinsky, P., 559
Kolda, C., 1231
Kolhörster, W., 119, 121, 150
Kollatschny, W., 547
Kollgaard, R.I., 496, 560
Kolodziejczak, J.J., 758
Kölzer, G., 1233
Komarek, T.A., 352
Komatsu, G., 1320
Komberg, B.V., 748, 753, 761, 786
Komen, G.J., 1604, 1616
Kömle, N.I., 1274
Komm, R.W., 1060, 1062
Kommers, J.M., 516, 521, 526, 755, 787, 821
Komori, Y., 696
Komossa, S., 547
Konar, S., 762, 785, 787
Kondo, I., 524, 1087
Kondo, Y., 295, 299, 560, 662, 663, 672, 673, 780, 787, 808, 861, 870, 871, 892, 907–909
Kondoh, K., 1004
Kondor, A., 1232, 1273
Kondratyev, K.YA., 1187
Kondratyeva, M.A., 999
Kondratyuk, Y.V., 85, 113
Konenkov, D., 789
Königl, A., 554, 558, 559
Konishi, T., 960
Konno, I., 1292
Konod, Y., 298
Kononov, A.Ya., 222
Konopliv, A.S., 1322
Kontor, N.N., 1004
Koo, B.C., 665, 673, 674
Kooi, J., 640
Koomen, M.J., 1004, 1113, 1138, 1140
Koons, H.C., 1527
Koopmans, L.V.E., 467
Koornneef, J., 298
Kopal, Z., 270, 271, 275
Kopeikin, S.M., 358, 371
Kopp, R.A., 1028, 1032, 1111, 1113
Koppes, C.R., 57, 58
Kopriva, D.A., 140, 150, 250, 253
Koptsevich, A.B., 744, 755
Kopylov, A.I., 528
Koratkar, A.P., 547, 576
Korendyke, C.M., 1138, 1140
Koresko, C.D., 523, 526, 527
Korff, S.A., 124, 125, 149, 150
Korista, K.T., 535, 547
Kormendy, J., 467, 535, 546, 547, 564, 577
Korneev, V.V., 222
Korolev, S.P., 64
Körösmezey, A., 1232, 1271
Koroteev, A.S., 113
Korotev, R., 1375
Korschinek, G., 604
Korth, A., 1002, 1232
Korzennik, S.G., 1063
Kosch, M.J., 1564, 1582
Koshiba, M., 959

Koshio, Y., 1061
Koshut, T.M., 286, 524–527
Kosirev, N.A., 895, 909
Koskinen, H., 1581
Koslovsky, B., 143, 150, 252, 439
Kosmodemyansky, A., 57, 58
Kosovichev, A.G., 1040, 1048, 1050–1052, 1054, 1055, 1058, 1060–1063, 1086
Kosowsky, A., 384, 394
Kosugi, T., 1087
Kóta, J., 1002, 1150, 1155, 1160
Kotani, T., 808, 836
Kotelnikov, V.A., 1320
Kotova, G.A., 1233
Kouchi, A., 1259–1261, 1272
Kourganoff, V., 1066, 1086
Kouveliotou, C., 286, 507, 508, 521, 523–528, 736, 754, 755, 758, 785, 787, 789, 819, 821
Kovalevsky, J., 371, 1634, 1649–1651
Kovner, I., 467
Kowal, C.T., 375, 394, 1453
Kowalski, M.P., 479, 497
Koyama, K., 498, 576, 689, 696, 753, 757, 835, 836, 855, 885, 892, 893, 951, 952, 959, 960
Kozai, Y., 1651
Kozasa, T., 1239, 1272, 1275
Kozlovsky, B., 439, 695, 696, 1001, 1003, 1159, 1214, 1556
Kozlowski, R.W.H., 1430
Kozma, C., 788
Kraemer, S., 300
Kraetschmer, W., 870, 871
Kraft, R.P., 396, 568, 577, 590, 603, 604, 791–793, 804, 806, 808, 809, 811, 820, 821
Krähenbühl, U., 1373, 1401
Krainev, M.B., 1002
Kramer, G., 299
Kramer, M., 362, 371, 372, 755
Kramm, J-R., 1232, 1272
Kramm, R., 1272, 1274
Krankowsky, D., 1232, 1286, 1292, 1293, 1372, 1583
Krasnobaev, K., 1213
Krasnopolsky, V.A., 275, 1231
Krassa, R.F., 1198, 1216
Krassner, J., 644
Krätschmer, W., 673, 1188
Kratz, K.-L., 393
Kraushaar, W.L., 285, 584, 603–605
Krauss, L.M., 393, 430, 438
Krautter, J., 299, 797, 798, 808, 809, 871, 885, 892, 893
Kreitz, P., 1231
Kremser, G., 1555, 1556
Krennrich, F., 684, 696
Kreplin, R.W., 285, 1086
Kreysa, E., 871
Krieger, A.S., 221–223, 254, 1031, 1033, 1119, 1134, 1139
Krige, J., 57, 58
Krijger, J.M., 755
Krimigis, S.M., 172, 177, 966, 983, 996, 999, 1000, 1002–1004, 1492, 1556
Krimm, H.A., 505, 518, 526
Kriss, G.A., 299, 547, 724, 755, 756, 829, 835, 884, 891
Krist, J.E., 869
Kristen, H., 527
Kristian, J., 375, 394, 469, 753, 760, 787
Krivov, A.V., 1188
Kroeger, R.A., 854
Krolik, J.H., 465, 524, 534, 547
Kroll, P., 525
Kronawitter, A., 386, 395
Kronberg, P.P., 488, 495, 835
Kronrod, M.A., 1320

Kroupa, P., 712, 718
Krueger, F.R., 1188, 1284, 1293
Krueger, H., 871, 1184, 1188
Krueger, V.L., 298
Kruk, J.W., 299, 674
Krumholz, M.R., 521, 526
Krupa, P., 1270
Krupp, N., 1002
Kruse, H., 1403
Krutov, V.V., 222
Krymsky, G.F., 690, 696
Krysko, A.A., 275
Kryuchkov, V.P., 1320
Krzeminski, W., 793, 809, 823, 835
Ksanfomality, L.V., 1233
Ksenofontov, L.T., 694
Ku, W.H., 960
Kubo, S., 756
Kubota, A., 577, 855
Kuchar, T.A., 328, 330
Kuchner, M., 527
Kuczera, H., 1174, 1188, 1293
Kudrin, L.V., 1320
Kudritzki, R.P., 892, 897, 909, 910
Kufl, H.-U., 871
Kuhi, L.V., 861, 869, 871, 904, 909, 910
Kuhn, J.R., 1050, 1055, 1062
Kühne, M., 1087
Kühr, H., 468, 560
Kührt, E., 1259, 1262–1264, 1266–1268, 1272, 1293
Kuijken, K., 525, 711, 718
Kuijpers, J., 893
Kuiper, G.P., 151, 223, 253, 255, 258, 260–262, 265, 269, 270, 1031, 1113, 1239, 1272, 1281, 1293, 1452, 1453, 1457, 1477
Kuiper, L., 733, 748, 755–757
Kuiper, T.B.H., 330, 642, 644
Kulikov, G.V., 691, 696
Kulikovsky, A., 1213
Kulkarni, S.R., 508, 513, 514, 516, 518, 521, 523–527, 664, 668, 673, 674, 736, 748, 753–757, 767, 787–789, 1269
Kulkarni, V.P., 675
Kulsrud, R.M., 1025, 1032, 1033
Kumagai, S., 960
Kumar, P., 521, 526, 1050, 1055, 1061, 1062
Kumar, S., 457, 1201, 1215
Kunde, V.G., 437, 1429, 1430, 1449, 1477
Kundić, T., 467, 526
Kundt, W., 728, 755
Kundu, A., 1272
Kung, D.T., 674
Kunieda, H., 498, 835, 855, 959
Kunigasa, K., 757
Kunitake, M., 1569, 1582
Kunitsyn, V.E., 1575, 1576, 1582
Kunkel, M., 577, 871
Kunneth, E., 1138
Kunow, H., 255, 987, 1002, 1005, 1158, 1159
Kunth, D., 428, 438
Kuntz, K.D., 594, 604, 605
Kunz, S., 1124, 1139
Kunze, D., 329, 576, 577, 638, 869, 873
Kupo, I., 1426, 1430
Kupperian, J.E. Jr., 285, 286, 298, 1196, 1215
Kurfess, J., 560
Kurki-Suonio, H., 430, 438
Kuroda, T., 1272
Kürster, M., 822, 882, 893
Kurt, V.G., 755, 1198, 1213–1215, 1527, 1556, 1581
Kurth, W.S, 1123, 1138, 1140, 1158–1160, 1188, 1205, 1206, 1210, 1214, 1215, 1491

Kurtz, J.P., 1373
Kurtz, N.T., 643
Kushky, E.R., 1320
Kussendrager, D., 869
Kuulkers, E., 808
Kuzmin, R.O., 1320, 1322
Kuzmin, V.A., 692, 697
Kuznetsov, A.V., 524, 528, 820, 960
Kwan, J., 534, 547, 642
Kwok, P.W., 753
Kwok, R., 1605, 1616
Kwok, S., 632, 640, 642, 934
Kylafis, N., 644
Kyrola, E., 1214, 1215
Kyte, F., 1187

La Dous, C., 794, 809
La Franca, F., 546
LaBelle, J., 1033, 1528
Labhardt, L., 395, 396
Labitzke, K., 199, 202
Labonte, B.J., 1052, 1060, 1062
Labrum, N.R., 1520, 1527
Lacey, C., 524
Lachieze-Rey, M., 495
Lachlan-Cope, T., 1584
Lacis, A.A., 437, 1429
Laclare, F., 1061
Lacombe, F., 869
Lacy, J.H., 623, 626, 636, 638, 640, 642, 644, 872
Lacy, M., 527
Lada, C.J., 644, 858, 859, 871
Laeverenz, P., 1138, 1214
Laframboise, J.G., 1584
Lagache, G., 574, 577, 636, 637, 869, 871
Lagage, P.O., 637, 690, 696
Lagardere, H., 1061
Lagerloef, G.S.E., 1612, 1616
Lagerros, J., 1232, 1233
Lago, M.T.V.T., 861, 871
Lagrange, A.M., 864, 868, 871, 1215
Lagrange-Henri, A.M., 869, 870
Lahav, O., 383, 390, 393, 395
Lahuis, F., 626, 637–640, 644, 869, 870, 872, 873
Lal, D., 434, 436, 438
Lal, N., 437, 1160
Lala, P., 1651
Lallement, R., 605, 667, 672, 673, 869, 1156, 1158–1161, 1193–1195, 1201, 1203, 1204, 1208, 1210, 1214–1216
Lamarre, J.M., 639, 1270
Lamb, D.Q., 500, 504, 507, 508, 520, 521, 526, 810
Lamb, F.K., 526, 768, 787, 817, 820, 821, 823, 826, 827, 833–835
Lamb, R.C., 696, 835
Lambeck, K., 1650
Lambert, D.L., 438, 439
Lame, N.J., 547
Lamers, H.J.G.L.M., 251, 253, 298, 299, 829, 835, 869, 873, 896–898, 901, 902, 908–911, 935
Laming, J.M., 891, 893
Lammer, H., 1483, 1492
Lammers, E., 1160
Lämmerzähl, P., 1224, 1232, 1233, 1292, 1293, 1583
Lamport, J.E., 1492
Lampton, M., 287, 299
Lamy, P.L., 1138, 1140, 1164, 1188, 1236, 1244, 1245, 1247, 1264, 1270, 1272, 1274
Lancaster, N., 1321
Landau, L., 721, 755
Landecker, T.L., 637
Landgraf, M., 673, 1188

Landriu, D., 869
Landsman, W.B., 662, 673, 1215
Landua, R., 1231
Lane, A.L., 298
Langacker, P., 1231
Lange, A.E., 329, 330, 394, 420, 466, 637
Langer, W.D., 610, 640, 645
Langhoff, S.R., 619, 640
Langille, R.C., 1575, 1583
Lanning, H., 787
Lanoix, P., 395
Lanz, T., 809, 908
Lanzerotti, L.J., 177, 1001, 1002, 1004, 1271, 1491, 1492
Laor, A., 547
Lapègue, J., 869
Laplace, P.S., 1459, 1477, 1568, 1582
Laques, P., 860, 871, 1270
Lara, L.M., 1271
Larcialis, R.L., 1269
Large, M.I., 722, 755
Lark, N.L., 1273
Larmour, J., 1561, 1582
Laros, J.G., 508, 523, 524–526, 528, 754
LaRosa, T.H., 1003
Larsen, C., 695
Larsen, J.C., 1668
Larsen, M.F., 1559, 1561, 1579, 1582
Larsen, R.M., 1060–1063
Larsen, S., 1402
Larson, D.E., 1526
Larson, H.P., 437, 1426, 1430
Larson, L.E., 1448
Larson, S., 1265, 1272, 1273
Larson, S.B., 505, 526
Larson, S.M., 1270, 1274, 1477
Larsson, B., 641
Lasby, C.G., 57, 58
Lasch, J.E., 1402
Lasenby, A.N., 393, 395, 419
Lasher, L.E., 1433, 1449
Laskarides, P., 820
Lasnier, P., 1604, 1616
Lasota, J.-P., 355, 371, 794, 796, 799, 809, 844, 855
Lassen, H., 1559, 1561, 1582
Lassen, K., 1031
Latter, W.B., 636
Lattes, C.M.G., 125, 150
Lattimer, J.M., 729, 755, 757
Lau, E.L., 1401, 1449
Lauer, T.R., 467, 547
Laufer, D., 1269
Laul, J.C., 1375, 1382, 1401, 1402
Launius, R.D., 57, 58
Laureijs, R.J., 641, 870, 871
Laurent, P., 820, 850, 855
Laurent-Muehleisen, S.A., 496
Lauroesch, J.T., 666, 673, 675
Lautman, D.A., 1188
Lavender, S.J., 1616
Lavrentiev, M.A., 1232
Lawler, J.E., 396
Lawrence, A., 534, 544, 547
Lawrence, D.J., 1401
Lawson, A.J., 560
Layer, T.R., 546
Layne, G.D., 1374
Lazarian, A., 1265, 1270
Lazaro, C., 809
Lazarus, A.J., 251, 254, 1000, 1131, 1137, 1139, 1140, 1153, 1159, 1160, 1215, 1231, 1492, 1557

Lazcano, A., 1260, 1280, 1293
Lazrek, M., 1043, 1044, 1061, 1062
Lazzati, D., 524
Le Bertre, T., 636
Le Bourlot, J., 616, 640
Le Brun, V., 397
Le Contel, J.M., 909
Le Fèvre, O., 397, 467, 468, 483, 497
Le Grand, P., 1616
Le Provost, C., 1615–1617, 1647, 1651
Le Roux, I.A., 691, 696, 1002
Le Traon, P.-Y., 1599, 1600, 1616
Le Vine, D.M., 1616
Lea, S.M., 478, 497, 826, 835
Leacock, R.J., 547
Leahy, D.A., 830, 834–836
Lean, J., 1062, 1095, 1113
Lebach, D.E., 453, 467
Lebedev, M.G., 1213
Lebedinsky, A.I., 274, 275
Lebofsky, L.A., 1188, 1243, 1244, 1272, 1274, 1476
Lebre, A., 636
Lebrun, F., 820
Lebrun, J.-C., 1062
Lecacheux, A., 1265, 1272, 1453
Lecacheux, J., 1270, 1271, 1429
Lecar, M., 496
Leckrone, D.S., 298, 672
Lecomte, B., 641
Ledlow, M.J., 496
Ledoux, P., 1043, 1062
Lee, A.T., 394, 420
Lee, B., 518, 521, 523, 526, 527
Lee, C.H., 786
Lee, D.-C., 1366, 1372, 1375
Lee, H.-K., 526
Lee, H.M., 866, 870
Lee, J.C., 395, 420, 467
Lee, L.C., 1032, 1135, 1139, 1159, 1521, 1528
Lee, M.A., 696, 697, 976, 979–982, 988, 989, 997, 998, 1000–1002, 1005, 1138, 1156, 1158–1161, 1202, 1209, 1214, 1215, 1233
Lee, M.G., 394
Lee, S., 1321
Lee, U., 822
Leech, K., 437, 466, 577, 638, 639, 643, 644, 869, 870, 872, 873, 1270, 1292
Leeks, S.J., 637, 642
Leer, E., 694, 999, 1134, 1139
Lees, J.F., 673
Lefevere, T., 1002
Léger, A., 617, 618, 638, 641
Legrand, J.-P., 214
Lehár, J., 467
Lehmacher, 1430
Lehman, D.H., 114
Lehman, M., 57, 58
Lehner, M.J., 466
Lehnert, M., 525
Lehtinen, K., 641
Lehtinen, N.G., 755
Leibacher, J.W., 1038, 1043, 1060–1062
Leibowitz, E., 524, 820
Leibrand, K., 1062
Leibundgut, B., 395, 420, 468, 524, 526, 788
Leighton, R.B., 264, 270, 305, 326, 330, 1036–1038, 1062, 1063, 1068, 1087
Leinert, C., 641, 859, 871, 1176, 1179, 1188
Leipold, M., 113
Leisawitz, D., 642, 718, 1214
Leising, M.D., 528

Leister, N.V., 1061
Leitherer, B., 576
Leitherer, C., 299
Leitinger, R., 1575, 1582
Lellouch, E., 437, 638, 869, 1270, 1291, 1292, 1435, 1449, 1467, 1477
Lemaire, J.F., 1497, 1535, 1555, 1562, 1582
Lemaire, P., 1060, 1071, 1086–1088
Lemaitre, G., 122, 127, 150
Lemen, J.R., 1086
LeMignant, D., 1273
Lemke, D., 577, 641, 862, 870, 871
Lemme, C., 1291
Lemmon, M.T., 1402, 1478
Lemoine, F.G., 1321, 1322
Lemoine, M., 429, 438, 440, 1215
Lémonon, L., 466
Lemons, D.S., 1137
Léna, P., 639, 869
Lenard, P., 1565, 1582
Lengyel-Frey, D., 1140
Lense, J., 340, 351
Lentz, G.A., 151
Lenz, D., 908
Lenzen, R., 871
Leonard, D.C., 523, 854
Leong, C., 496, 497
Lepine, J.R.D., 870
Lepping, R.P., 1000, 1122, 1139, 1148, 1159, 1160, 1484, 1489, 1492, 1526, 1556, 1557
Lequeux, J., 637, 639, 640, 668, 672
Lerch, F.J., 1651
Lerche, I., 249, 253, 691, 696
Leroy, C., 1649
Leske, R.A., 689, 696, 697, 975, 987, 994, 1000, 1002–1004
Lessard, R.W., 696
Lester, M., 1584
Lestrade, J.P., 286
Leuenhagen, U., 935
Leutwyler, H., 1137
Levallois, J.-J., 1650, 1649
Levasseur-Regourd, A.-C., 1164, 1188, 1232, 1270, 1272
Levenfish, K.P., 758
Levenson, N.A., 956, 959
Leverington, D., 1463, 1477
Levin, B.J., 1236, 1272
Levine, A.J., 57, 58
Levine, A.M., 826, 835
Levine, M.W., 352
Levine, P.D., 351
Levinson, A., 558
Levinson, R., 757, 788, 789, 821, 836
Levison, A., 552
Levison, H.F., 1239, 1242, 1269, 1270, 1272, 1293, 1433, 1449
Levitus, S., 1612, 1616
Levshakov, S.A., 428, 438
Lev-Tov, S.J., 755
Levy, D.H., 1272, 1274
Levy, E.H., 869, 870, 1160, 1188
Lévy, P., 1519
Levy, R.H., 1011, 1014, 1015, 1017, 1019, 1032
Lewin, B.W., 787
Lewin, W.H.G., 526, 576, 764, 782, 786–789, 808, 812, 814–817, 820–822, 836, 849, 855
Lewis, B.M., 904, 909
Lewis, C., 298
Lewis, G.F., 392, 394
Lewis, J.S., 1303, 1321, 1458, 1477
Lewis, R., 1292
Lewis, R.S., 636

Lewis, R.V., 1584
Ley, W.W., 262, 270
Li, A., 613, 641, 643, 864, 871, 1289
Li, F.K., 819, 821, 836
Li, H., 526
Li, H.R., 1273
Li, L.-X., 521, 526
Li, W., 643, 854, 1584
Li, X., 1138, 1533, 1556
Liang, E.P., 499, 503, 504, 526
Libbrecht, K.G., 1047, 1050, 1061–1063
Libby, W.F., 127, 150, 1331, 1372
Liberman, R., 1004
Licandro, J., 1265, 1272
Lichtenberg, G., 1274
Lichti, G.G., 286, 755, 959
Lichtman, S.W., 222, 286
Lidman, C., 395, 420, 466, 467, 498, 524–526
Liebermann, R., 57, 58
Liebert, J., 548
Liebes, S., 443, 467
Liedahl, D.A., 547, 833, 835, 836
Lielausis, O., 1491
Lieske, J.M., 1649
Lieu, R., 497, 603
Liewer, P.C., 671, 673
Liggett, T.J., 1269
Lighthill, M.J., 1059, 1062
Lightman, A.P., 828, 836, 852, 855
Ligon, D.T. Jr., 1401
Lijowski, M., 696, 697
Likkel, L., 1273
Liller, W., 576, 578, 786, 835, 1430
Lilley, A.E., 673
Lillie, C.F., 299
Lim, T., 636–638, 641–643, 869
Limongi, M., 395
Lin, R.P., 969, 1002, 1003, 1232, 1320, 1321, 1479, 1489, 1491, 1492, 1526
Lin, Y.C., 560, 696, 753, 755, 757
Lind, F.D., 1575, 1583
Lindal, G.F., 1321, 1437, 1449
Lindblad, B.-A., 1188
Linde, J.J., 1211
Linde, T.J., 673, 1188, 1215
Lindemann, F.A., 230, 253
Linden-Voernle, M., 528
Lindholm, 909
Lindler, D.J., 298, 300, 870
Lindsey, C., 1048, 1054, 1060–1062
Lindstrøm, E., 1611, 1616
Lindstrom, M., 1339–1341, 1371, 1372
Lindzen, R.S., 1568, 1580
Lineweaver, C.A., 383, 390, 395, 419–421
Lingenfelter, R.E., 439, 499, 507, 525, 526, 528, 689, 694–696, 755, 756
Link, R., 465, 467
Linke, R.A., 605
Linker, J.A., 1136, 1139, 1491
Linkert, D., 1188
Linkert, G., 1188, 1189
Linnell, A.P., 793, 809
Linsky, J.L., 298–300, 385, 395, 427, 438, 617, 641, 662, 672, 673, 872, 878, 891–894, 904, 907–910, 1204, 1208, 1209, 1211, 1215, 1216
Linsley, J., 150, 691, 695
Lions, L., 1582
Liou, J.C., 1187–1189
Lipa, J.A., 351
Lippincott, C.R., 1581
Lipschutz, M.E., 1372
Lipson, J., 1374

Lis, D.C., 637, 640, 1286, 1291–1293, 1448
Liseau, R., 610, 627, 636–639, 641, 642, 645, 869
Lissauer, J.J., 1264, 1265, 1273, 1477
Lisse, C.M., 1229, 1232, 1245, 1246, 1272, 1273
Liszt, H., 707, 718
Lites, B.W., 1068, 1087
Littenberg, L., 1231
Little, B., 1430
Litvinenko, Y.E., 1027, 1032
Liu, A.K., 1605, 1616
Liu, Q.Z., 813, 821
Liu, S., 1128, 1139
Liu, X.-W., 633, 636–638, 641, 643
Livesay, L.L., 114
Livesey, W.A., 1002
Livi, S., 1001, 1137–1139, 1214, 1233
Livingston, W.C., 650, 673, 1041, 1061, 1062, 1087, 1088
Livingstone, R.C., 1584
Livio, M., 380, 395, 396, 521, 523, 524, 526, 528, 808, 828, 835, 836, 854
Livshitz, M.A., 223
Lizano, S., 872
Ljung, B., 641
Llebaria, A., 1138, 1140
Lloyd, N.D., 1584
Lloyd-Davies, E.J., 488, 497
Loaring, N., 466
Löb, H.W., 110, 113
Lobo, J.A., 366, 371
Lockman, F.J., 58, 497, 595, 601, 602, 604, 663, 664, 672
Lockwood, M., 1547, 1556
Lodders, K., 636
Lodge, O.J., 442, 467
Loeb, A., 516, 518, 520, 523, 526–528, 1047, 1060
Loeser, R., 1086, 1088
Loewenstein, K., 420, 421
Loewenstein, M., 488, 490, 496, 497, 569, 577
Loewenstein, R.F., 642, 1273
Löfdahl, M.G., 1085
Lofgren, E.J., 150, 695
Logatchev, Y.I., 1004
Logsdon, J.M., 57, 58, 93, 114
Loh, E.C., 695
Loiacono, J.J., 300
Loidl, J., 187, 1528
Loinard, L., 637
Loken, C., 496, 498
Lollouch, E., 1423
Lommen, A.N., 748, 753, 755
Long, D.C., 467
Long, J.A., 869
Long, K.S., 299, 564, 577, 584, 604, 724, 754, 755, 793, 798, 799, 807–809, 899, 910
Longair, M.S., 373, 395, 462, 466, 495, 719, 869
Longanecker, G.R., 298
Longanecker, G.W., 298
Longdon, N., 1272
Longhi, J., 1304, 1306, 1307, 1321, 1344, 1358, 1362, 1372, 1375
Looper, M.D., 1002
Lopate, C., 992, 1002
Lopez, R.E., 1544, 1556
Lopper, M.D., 1003
Lord, 1583
Loredo, T.J., 1270
Lorentz, H.A., 227, 253
Lorenz, E., 1233
Lorenz, R.D., 1478
Lorenzetti, D., 616, 628, 636–639, 641–643, 869
Lorimer, D.R., 753, 786, 814, 821
Lory, M.L., 190, 198, 201
Lotko, W., 1527

Lottermoser, R.-F., 1026, 1027, 1032
Lou, W., 543
Lou, Y.Q., 1002
Loup, C., 640, 873
Lovas, I., 1086
Love, S.G., 1269, 1272
Lovell, A.J., 1292
Lovell, J.E.J., 559, 855
Lovering, J.F., 1371
Low, B.C., 1113
Low, F., 641, 1188
Low, F.E., 252
Low, F.J., 311, 305, 307, 312, 328–330, 577, 609, 640, 642, 868, 935, 1274, 1426, 1430
Lowder, D.M., 694
Lowe, S.T., 453, 468
Lowell, P., 260, 270, 1406, 1454
Lowman, P., 1401
Lowrance, P.J., 872, 873
Lozinskaya, T.A., 688, 696
Loznikov, V., 960
Lu, J.Y., 1216
Lu, L., 299, 674
Lubin, D., 440
Lubin, L.M., 526
Lubin, P.M., 329, 420, 421, 642, 645
Lubovich, D.A., 435, 438
Lucas, R.A., 527
Lucchitta, B.K., 1304, 1321
Lucey, P.G., 1301, 1321
Lucke, P.B., 653, 673
Lucy, L.B., 829, 835, 879, 892, 897, 899, 903, 908, 910
Ludtke, C., 300
Ludwig, G.H., 177
Ludwig, K.R., 1371
Luetzow-Wentzky, P., 871
Luginbuhl, C., 754
Lugmair, G.W., 1333–1336, 1343, 1355, 1360, 1371, 1373, 1375, 1385, 1401
Lugovsky, S.B., 1231
Lugten, J.B., 641, 643
Luhman, M.L., 640
Luhn, A., 970, 974, 1002
Lühr, H., 1032, 1033, 1527, 1582
Lui, A.T.Y., 1537, 1544, 1550, 1555, 1556
Luinge, W., 869
Luis, R., 1272
Lukasiak, A., 689, 695, 696, 1003
Lum, K.S.K., 960
Lum, R.K.L., 1401
Lume, K., 1253, 1269
Luminet, J-P., 463, 468, 547
Lund, N., 528
Lundgren, R.A., 1001, 1138
Lundin, K., 1581
Lundin, R., 1541, 1542, 1556, 1581, 1584
Lundquist, C., 1651
Lundqvist, P., 788
Lunine, J.I., 869, 870, 1293, 1476–1478, 1492
Luo, M., 695
Luppino, G.A., 397, 459, 467, 497
Lüst, R., 150, 182, 185, 187, 192, 1142, 1160, 1506, 1526, 1527
Lütjens, P., 187
Luton, J.M., 200, 201
Luttermoser, D.G., 909
Lutz, D., 329, 573, 576, 577, 638, 641, 869, 873
Luu, J.X., 1236, 1239, 1243, 1247, 1258, 1264, 1271, 1272, 1281, 1293
Lyard, F., 1651
Lygre, K., 1616
Lynas Gray, A.E., 908

Lynch, D.K., 642, 866, 871
Lynden-Bell, D., 386, 392, 393, 395, 477, 497, 531, 547, 549, 559, 701, 702, 718, 1244, 1425, 1429, 1466, 1476
Lynds, C.R., 650, 673
Lynds, R., 458, 467
Lyne, A.G, 371, 726, 748, 753, 755–757, 766, 784, 786–788, 814, 821
Lyon, E.F., 251
Lyon, J., 1215
Lyons, L.R., 1500, 1527, 1534, 1536, 1537, 1543, 1550, 1556, 1557
Lyot, B., 233, 234, 253, 1113
Lyttleton, R.A., 1236, 1238, 1272
Lyubimov, G.P., 1004
Lyzenga, D., 1615, 1616

Maas, D., 1283, 1293
Mac Donald, G.J.F., 1650
Mac Low, M., 665, 673
MacAlpine, G.M., 547
Macau, D., 298
Maccacaro, T., 496, 497, 537, 546–548, 560
Maccagni, D., 478, 496
Maccarone, M.C., 524, 527
Macchetto, D., 524
Macchetto, F.D., 298, 373, 395, 396, 524, 528
Macchietto, R., 371
MacCormack, E., 895, 908
MacDonald, G.J.F., 1456
MacDonald, H.M., 1560, 1582
MacDowall, R.J., 1000
MacFadyen, A.I., 521, 526
MacFarlane, J.J., 891, 899, 908–910
Macfarlane, M., 753
MacGregor, K.B., 902, 909, 1088
Mach, F., 335, 351
Machida, S., 1032, 1557
Mack, J.E., 603
Mack, P., 1231
Mackay, M.B., 530, 547
MacKenty, J.W., 547
MacKenzie, E., 1580
Mackey, C., 524
Mackie, G., 569, 576, 577
Mackin, R.J., 251–253
Mackwell, S.L., 1321
MacNeil, P.E., 352
Macomb, D.J., 524, 525
MacPherson, G.J., 1336, 1373
Macquart, J.-P., 557, 559
MacQueen, R.M., 222, 245, 253, 1001, 1113, 1121, 1138, 1139
MacRae, D.A., 704, 718
Macri, L.M., 379, 394–397
Madau, P., 552, 559
Madden, M.P., 1004
Madden, S.C., 577
Madders, K., 57, 58
Maddox, J., 390, 395
Madejski, G.M., 560
Mädler, J.H., 271, 275
Madore, B.F., 299, 376, 393–397
Madsen, M.B., 1371, 1402
Maechling, C.R., 1373
Maeda, K., 1563, 1567, 1568, 1582
Maeda, Y., 892
Maeder, A., 774, 787
Maeno, T., 696
Maezawa, K., 1032
Magain, P., 466, 468, 548
Magalhaes, J.A., 1402
Magee, D.K., 526
Magee-Sauer, K., 1292, 1293

Maggio, A., 878–889, 892, 894
Maggiore, M., 366, 371
Magnan, A., 1060
Magnani, L., 591, 604
Magorrian, J., 536, 546, 547
Magris, C., 637
Magueijo, J., 420
Maguire, W., 1430
Mahadevan, R., 855
Mahaffy, P.R., 428, 438, 1430, 1448
Mahoney, M.J., 871
Mahoney, W.A., 753
Maier, E.J., 1063, 1578, 1582
Maihara, T., 314, 329, 642
Maillard, J.-P., 641
Maillard, M., 202
Maki, J.N., 1402
Makida, Y., 696
Makino, F., 560, 577, 835, 836, 855, 959
Makino, J., 755
Makino, N., 467
Makishima, K., 496–498, 564, 565, 576, 577, 833, 835, 836, 854, 855, 893, 959, 960
Makiuti, S., 330
Malaise, D., 298, 1271, 1272
Malama, Y.G., 1202, 1206, 1208, 1210, 1213–1215
Malcom, H., 1002
Malfait, K., 629–632, 641, 643, 644, 862, 866, 869, 871, 872
Malhotra, R., 1490, 1492
Malhotra, S., 467
Malin, M.C., 1321, 1323, 1402
Malina, R.F., 299, 604, 605, 891
Malkan, M.A., 497, 526, 536, 547, 559, 870
Malkov, Yu.F., 547
Mall, G., 1001
Mall, U., 1001, 1138, 1214
Mallina, R.F., 298
Mallozzi, R.S., 286, 495, 508, 524, 526
Malmquist, K.G., 382, 395
Maloney, F.P., 394
Maltby, P., 1086, 1087
Malumuth, E., 298
Mamon, G., 576
Managadze, G.G., 1138, 1214
Manchado, A., 917, 935
Manchester, R.N., 737, 748, 753, 755, 757, 766, 784, 786–788
Mancinelli, R.L., 1321
Mandel, E., 497
Mandel, H., 299
Mandel'shtam, S.L., 222
Mandolesi, R., 869
Mandrou, P., 820
Mandzhavidze, N., 910
Mangano, M.L., 1231
Mangeney, A., 1060
Mangin, J.F., 352, 1651
Mangoldt, T., 1273
Manhès, G., 1370, 1372
Mankov, S., 1231
Mankovsky, A., 1556
Manley, D.M., 1231
Mann, A.K., 959
Mann, I., 1188, 1272
Manney, G.L., 1668
Mannheim, K., 552, 559
Manning, R.M., 1402
Mannings, V., 190, 640, 643, 865, 871, 1270, 1292
Manno, V., 57, 58, 1003
Manohar, A.V., 1231
Manring, E.R., 190, 193, 201, 202

Manson, A.H., 1572, 1582
Manvelyan, O.S., 1402
Manzo, G., 856
Mao, S., 523
Maoli, R., 389, 395, 468
Maoz, D., 456, 467, 547
Maoz, E., 383, 395
Maquire, W., 1477
Maran, S.P., 298, 300, 672, 1063
Marandino, G.E., 560
Marani, G.F., 458, 467
Maraschi, L., 537, 547, 551, 552, 554, 558–560, 773, 787, 809, 824, 836
Marchant, H., 1615
Marchis, F., 1245, 1273
March-Russell, J., 1231
Marcialis, R.L., 1476
Marck, J.-A., 355, 371
Marco, O., 1245, 1273
Marconi, M.L., 1272
Marcout, J., 935
Marcy, G.W., 864, 870, 871, 1060
Margon, B., 299, 524, 762, 833, 836
Mariani, F., 1137, 1232
Mariewicz, W.J., 1274
Marigo, P., 935
Marina Cairos, L., 523
Marioge, J.P., 1086
Maris, H.B., 1566, 1582
Mark, D., 299
Markert, T.H., 577, 754, 835, 946, 960
Markevitch, M., 498
Markiewicz, W.J., 1270, 1272
Marklund, G.T., 1536, 1537, 1556
Markov, M.S., 1320
Marković, D.M., 370
Markwardt, C.B., 754
Marley, M.S., 1320, 1430, 1436, 1443, 1449
Marlow, D.R., 467
Marouf, E.A., 1476
Marov, M.Ya., 1423
Marra J., 1616
Marsch, E., 554, 1087, 1088, 1133, 1138–1141, 1145, 1159, 1160, 1188, 1214
Marscher, A.P., 543, 547, 559, 560
Marsden, B.G., 1238, 1239, 1269, 1271, 1273, 1274, 1293, 1476
Marsden, D., 694, 736, 755
Marsden, P., 935
Marsden, P.L., 329, 330, 577
Marsden, R.G., 250, 253, 255, 1001, 1002, 1004, 1140, 1141
Marsden, R.J., 1159
Marsh, J.G., 1651
Marsh, K.A., 865, 871
Marsh, T.R., 792, 793, 795, 796, 799, 800, 809
Marshall, E., 113
Marshall, F.E., 497, 498, 525, 734, 755, 818, 822, 836, 846, 848, 856, 1232
Marshall, F.J., 586, 589, 604
Marshall, H.L., 758, 821
Marshall, J., 1304
Marshall, S., 523
Marshall, S.L., 466
Marshall, T.R., 1582
Marstad, N.C., 891
Marten, A., 1449
Marten, M., 1444, 1449
Martens, P., 1085
Martens, P.C.H., 752, 757
Martì, J.M., 558
Marti, K., 1292, 1333, 1373
Martic, M., 1059, 1062

Martin, C., 744, 755
Martin, C.D., 298, 299
Martín, E.L., 439, 893
Martin, J.H., 1557
Martin, J.O., 351
Martin, L.J., 1321
Martin, L., 1304
Martin, M., 870
Martin, P.D., 1477
Martin, P.G., 547, 612, 632, 641
Martin, R., 222, 465, 1138, 1139, 1273, 1291
Martin, S., 242, 253
Martin, S.F., 1027, 1032, 1136, 1137
Martin, X., 527
Martinez, R.R., 1372
Martinis, L., 394, 420, 466
Martín-Pintado, J., 637, 638
Martyn, D.F., 1564, 1567, 1580, 1582
Martz, D.E., 330
Marvin, U.B., 1403
Marzari, F., 1473, 1477
Masai, K., 498, 836, 942, 960
Masaike, A., 960
Masci, F.J., 548
Masetti, N., 523–525, 527, 528
Mas-Hesse, J.M., 908
Masi, S., 394, 420, 466
Maskell, K., 1615
Masnou, J.L., 286, 755, 758
Mason, G.M., 150, 252, 966, 969, 975, 988–1003, 1214
Mason, H.E., 1086
Mason, I.M., 755
Mason, J., 1188, 1233
Mason, K.O., 753, 755, 795, 796, 799, 800, 808, 809, 827, 831, 836, 837, 856
Mason, P.V., 420, 466
Massa, D.L., 394, 669, 674
Massaglia, S., 558
Massaro, E., 756
Massey, H.S.W., 58, 209, 214, 222, 1566, 1580
Massi, M., 869
Massie, S.T., 1426, 1429
Masson, P., 1322
Masterson, C., 696
Mather, J.C., 314, 328, 329, 402, 404, 419–421, 636, 642, 645
Mathers, C., 672
Matheson, P., 1477
Matheson, T., 395, 420, 467, 547, 854
Mathews, G.J., 440
Mathews, W.G., 497, 569, 576, 577
Mathez, G., 467, 468, 498
Mathieu, R., 859, 871
Mathis, J.S., 612, 641
Mathis, K.D., 697
Mathur, D., 1062
Mathur, S., 535, 547
Matijevic, J., 1402
Matilsky, T.A., 810, 835, 836
Matsakis, D.N., 351, 835
Matson, D.L., 1423, 1466, 1467, 1474, 1477
Matsubara, Y., 960
Matsuda, T., 828, 836
Matsuhara, H., 330, 637, 642
Matsui, N., 696
Matsui, W., 835
Matsumoto, H., 577, 696, 1232, 1527, 1555
Matsumoto, T., 314, 329, 330, 637, 642, 643
Matsunaga, H., 696
Matsuo, H., 329
Matsuoka, M., 497, 498, 830, 835, 836, 855, 959

Matsuoka, T., 960
Matsura, M., 696
Matsushita, K., 497, 563, 568, 569, 576, 577
Matsushita, S., 252, 1568, 1583
Matsuura, M., 959
Matsuura, S., 330, 637
Matsuzaki, K., 696, 757, 809
Matt, G., 546
Mattauch, J., 1330, 1373
Mattei, J.A., 809, 810
Matteucci, F., 435, 437, 438
Matthaeus, W.H., 1216
Matthews, H.E., 1292, 1293, 1449
Matthews, J.M., 1059, 1062
Matthews, K., 548, 757, 870
Matthews, L.D., 725, 736, 757
Matthews, M.S., 169, 177, 1061, 1271, 1272, 1274, 1477, 1323, 1375, 1376, 1401, 1402, 1423, 1449, 1476–1478, 1491, 1492
Matthews, T.A., 530, 531, 547
Matthews, U.E., 1293
Matthews, W.G., 483
Matthis, J.S., 637
Mattila, K., 577, 612, 641, 871
Mattison, E.M., 352
Matto, S., 1615
Mattoo, S., 1616
Mattox, J.R., 753, 755
Matveinen, H., 1580
Matz, S.M., 523, 757
Mätzler, C., 1616
Matzner, R.A., 438
Mauche, C.W., 804, 809, 810
Maucherat, A., 1086
Mauk, B.H., 177, 1477, 1492
Maunder, E.W., 229, 253
Maurer, J.C., 1583
Maurer, P., 1348, 1349, 1372, 1373, 1375
Maurice, S., 1401
Mauskopf, P.D., 394, 420, 466, 637
Maxson, C.W., 329, 872, 893, 910
Maxwell, J.C., 227, 253
May, A., 702, 718
May, J., 604
Mayall, N.U., 394
Mayeda, T.K., 1385, 1386, 1401
Mayer, C.H., 262, 270
Mayer, H.G., 1583
Mayer, T., 271
Mayer, W., 496, 753, 821, 836
Mayer-Hasselwander, H.A., 286, 696, 753, 755, 757, 758
Mayer-Hasselwander, H.H., 755
Mayhew, M., 1491
Maynard, C., 1580
Maynard, N.C., 1031, 1576, 1582
Mayne, K.I., 1374
Mayor, M., 859, 864, 869–871
Mayr, H.G., 1555
Maza, J., 394
Mazaudier, C., 1572, 1582
Maze, R., 123, 149
Mazelle, C., 1233, 1320, 1321, 1491, 1492
Mazets, E.P., 502, 504, 508, 525, 527, 736, 754, 755, 1224, 1232
Mazur, J.E., 972, 975, 1000, 1002, 1003
Mazure, A., 397
Mazzali, P.A., 524, 525
Mc Carthy, D., 1649
Mc Crea, I.W., 1584
Mc Donnell, J.A.M., 1188
Mc Dougall, W.A., 58
Mc Fadden, J.P., 1555

McAllister, A.H., 1161
McBreen, B., 466, 577
McCammon, D., 582, 590, 592, 603–605, 756, 584–586
McCarthy, D.D., 351, 835
McCarthy, D.K., 299
McCarthy, J.F., 329
McCarthy, T.S., 1403
McCaughrean, M.J., 629, 641, 861, 871, 872
McClain, C.R., 1615
McClain, E.F., 673
McClain, E.P., 1616
McClanahan, T., 525
McClean, I.S., 526
McClintock, J.E., 525, 762, 764, 765, 786, 787, 789, 812, 813, 820–822, 824, 826, 834, 836, 841–845, 854, 855
McClintock, W., 662, 673
McClure, J.P., 1584
McCluskey, G.E., 672
McCollough, M.L., 286
McCollum, B., 547, 560, 870
McComas, D.J., 254, 255, 1120, 1129, 1137–1139, 1159, 1160, 1231, 1557, 1582
McConnell, J.C., 1427, 1430, 1448
McConnell, M.L., 523
McCord, T.B., 260, 268, 270, 1400, 1402, 1423, 1462, 1477
Mccorkle, S., 1271
McCormick, P.M., 1661, 1668
McCoy, T.J., 1373
McCracken, C.W., 1188
McCracken, H.J., 468
McCracken, K.G., 1003
McCray, R.A., 656, 665, 672, 673, 829, 830, 832, 834–836
McCrea, I.W., 1584
McCrea, W.H., 1238, 1273
McCrosky, R.E., 1269
McCulloch, M.T., 1370
McCulloch, P.M., 855
McCullough, P.R., 674
McCurdy, H.E., 57, 58
McDavid, D., 909
McDonald, F.B., 139, 143, 149, 150, 437, 696, 983, 986, 990, 992, 1000, 1002–1004, 1120, 1123, 1139, 1142, 1156, 1159, 1199, 1160, 1215
McDonald, G.D., 1285, 1293, 1462, 1477
McDonald, L.M., 696
McDonald, S.W., 1269
McDonnell, J.A.M., 1166, 1169, 1187, 1188, 1224, 1232, 1293
McDougall, W.A., 57, 1496
McDowell, J.C., 546
McEentire, R.W., 1556
McElhinney, M.W., 1481, 1492
McEnery, J.E., 696
McEntire, R.W., 1004, 1477, 1492, 1555, 1556
McEwen, A.S., 1321, 1476, 1477
McEwen, D.J., 1584
McFadden, J., 1320, 1491, 1492
McFadden, J.P., 1526, 1527, 1578, 1580
McFadden, L., 1292, 1323, 1448, 1449
McFadden, L.-A., 1269, 1274
McFadden, P.L., 1492
McGee, J.J., 1347
McGetchin, T.R., 1348, 1373
McGhee, C.A., 1477
McGill, G.E., 1309, 1321, 1322
McGovern, P.J., 1322
McGrath, M.A., 1467, 1477
McGuire, R.E., 999, 1000, 1003
McHardy, I.M., 478, 497, 560, 893
McIlwain, C.E., 167, 177, 1527, 1535, 1536, 1543, 1555–1557
McIntosh, B.A., 1164, 1188
McIntosh, P.S., 254, 1000, 1113

McIntyre, M.E., 1058, 1061
McKay, C.P., 1293, 1306, 1321
McKay, D., 1293
McKay, D.J., 855
McKay, D.S., 1359, 1373
McKay, G.A., 1373, 1374, 1402
McKay, N.P.F., 755
McKay, T. A., 523
McKee, C.F., 565, 577, 589, 590, 593, 599, 601, 604, 608, 620, 638, 639, 641, 665, 672, 673, 675
McKee, J.D., 479, 497
McKee, S., 694
McKeith, C.D., 439, 672
McKellar, A., 607, 641
McKenzie, J.F., 695, 1001, 1085, 1135, 1137–1139
McKibben, R.B., 142, 150, 151, 992, 993, 1002, 1003, 1154, 1160
McKinney, C.M., 1649
McKinnon, W.B., 1249, 1273, 1303, 1321, 1433, 1449, 1470, 1471, 1476, 1478
McLandress, C., 1569, 1582
McLaughlin, M., 755
McLean, D.J., 1520, 1527
McLean, I.S., 526
McLennan, J.C., 677, 696
McLeod, B.A., 394, 467
McLeod, C.P., 1060–1062
McLeod, M.G., 1481, 1492
McMahon, R.G., 395, 420, 467
McMullen, R.A., 1032
McMullin, J., 1294
McNab, M.C., 1002
McNair, A., 1331, 1373
McNall, J.F., 299
McNamara, B.J., 505, 518, 527
McNamara, B.R., 475, 498
McNaron-Brown, K., 560, 854
McNaught, R.H., 1243, 1273
McNutt, D.P., 639
McNutt, R.L., 1211, 1215
McNutt, R.L. Jr., 1159, 1205, 1215
McPhate, J.B., 1293
McPherron, R.L., 1545, 1557
McSween, H.Y. Jr., 1304, 1321, 1362, 1371, 1373, 1374, 1402
McTiernan, J., 525
Meade, M.R., 299
Meadows, E., 195, 202
Meadows, E.B., 1582
Mebold, U., 605, 672
Médale, J.L., 1491
Medvedev, M.V., 518, 527
Meech, K.J., 1243, 1244, 1264, 1271, 1273, 1274
Meegan, C.A., 286, 499, 500, 505, 507, 523–528, 755, 787, 821
Meek, C.E., 1572, 1582
Meekins, J.F., 604, 754
Meeks, M.C., 644
Megie, G., 197, 202
Mehltretter, J.P., 1068, 1087
Mehringer, D.M., 637, 640, 1291, 1293
Meier, A., 1231, 1291, 1292
Meier, D.L., 855
Meier, M., 524
Meier, M.F., 1598, 1616, 1617
Meier, R., 872, 1285, 1286, 1288, 1293
Meier, R.R., 1199, 1200, 1215, 1216
Meijer, G., 644
Mein, N., 1073, 1087
Meira, L.G., 1615
Meissner, R., 1317, 1321
Meister, J., 437, 1137
Mekler, Y., 1430

Melchior, P., 1647, 1649–1651
Melchiorri, A., 394, 420, 466
Melentyev, V.V., 1615
Melioranskiy, A., 960
Mellier, Y., 395, 460, 461, 467, 468, 498, 465,
Melnichenko, Y.A., 1582
Melnick, G.J., 323, 329, 330, 525, 616, 620, 621, 623, 625, 636, 639, 641–643
Melnick, J., 528, 538, 548
Melnik, J., 935
Melnikov, V.N., 439
Melosh, H.J., 1247–1249, 1273, 1274, 1303, 1319, 1321, 1367, 1375, 1386, 1402, 1476
Melotte, P.J., 1452
Meloy, D.A., 293, 299, 667, 672, 958, 959
Melrose, D.B., 696, 1521, 1527
Melzner, F., 187, 869, 1526–1528, 1583
Menard, F., 638
Mende, S.B., 1515, 1527
Mendenhall, J.A., 591, 603
Méndez, M., 820
Mendis, D.A., 1173, 1187, 1218, 1221, 1225, 1232, 1233, 1262, 1269, 1270, 1273
Mendoza-Gomez, C., 1292
Meneguzzi, M., 434, 438
Meng, C.-I., 1541, 1557
Menon, T.K., 673
Menou, K., 369, 371, 845, 855
Menten, K.M., 641
Menzies, J.W., 465, 468, 835
Meredith, L.H., 1569, 1582
Mereghetti, S., 736, 741, 742, 755, 756, 820
Merényi, E., 1222, 1251, 1253, 1255, 1273, 1274
Merkulova, N., 547
Merrihue, C., 1333, 1373
Merrill, K.M., 631, 639, 641, 644
Merrill, P.W., 607, 631, 641
Merrill, R.B., 1373
Merrill, R.T., 1480, 1492
Merritt, D., 541, 546
Meserole, R., 1269
Meshik, A.P., 1140
Messina, A., 150, 547
Messina, D.C., 528
Mészáros, P., 509, 512, 517, 518, 521, 527, 528, 732, 756, 834, 836
Metcalfe, L., 466, 574, 577, 869, 870
Metcalfe, N., 525
Metik, L., 547
Metlov, V., 525
Metzger, A.E., 502, 504, 513, 527, 528, 1382, 1402
Metzger, M.R., 511, 513, 524, 527
Metzler, K., 1272
Meunier, N., 1069, 1087
Meusemann, H., 113
Mewaldt, R.A., 254, 673, 689, 696, 697, 975, 993, 995, 999, 1000, 1002–1005, 1555
Mewe, R., 497, 547, 809, 872, 875, 876, 879, 887, 891–894, 905, 910
Meyer, B., 187, 1139, 1526
Meyer, C., 1341, 1345, 1346, 1372, 1373
Meyer, D.M., 660, 672, 673
Meyer, F., 777, 779, 786, 787
Meyer, J.-P, 434, 678, 680, 681, 689, 1126
Meyer, J.P., 439, 696, 1003, 1139
Meyer, M.R., 641, 870
Meyer, P., 126, 133, 150, 232, 245, 253, 696, 967, 969, 1003, 1196, 1215
Meyer, R.D., 756
Meyer, S.S., 419, 420, 421, 645
Meyer, W., 300
Meyer-Hofmeister, E., 777, 779, 787, 855
Meylan, G., 466

Meylan, T., 809
Meynet, G., 774, 787
Mezger, P.G., 641
Mezzena, R., 371
Mi, G., 834
Micela, G., 877, 883, 891–894
Michalitsianos, A.G., 299
Michaud, G., 428, 438
Michel, E., 1060, 1062
Michel, F.C., 725, 727, 728, 756, 757
Michel, K.W., 1237, 1269
Michels, D.J., 1004, 1086, 1113, 1138, 1140
Michels, J., 222, 1138, 1139
Michelson, P.F., 351, 696, 755, 757, 822
Michika, D., 300
Michin, W.P., 92, 114
Mickelson, M.E., 1448
Middleditch, J., 753, 755, 787, 808
Middlehurst, B.M., 269, 270, 673, 674, 1477
Midgley, J.E., 246, 253
Migenes, V., 855
Miggenrieder, H., 1139
Miglio, L., 420, 466
Mignani, R., 744, 753, 756
Mignard, F., 467, 1478
Mignoli, M., 547
Mihalov, J.D., 1003, 1137
Mihara, T., 577
Mikhailov, A.V., 1567, 1582
Mikhailov, Z., 270
Mikic, Z., 1136, 1139
Mikkelsen, I.S., 1582
Mikusch, E., 1274
Milani, A., 1650
Miles, M.W., 1615, 1616
Miley, G.K., 330, 577, 760, 786, 789, 837, 840, 854
Milford, P.N., 1062, 1086
Millar, T.J., 438, 637, 644, 872
Miller, B.A., 1060
Miller, D., 645
Miller, G.S., 820
Miller, H.R., 547
Miller, J.A., 971, 976, 1003, 1426
Miller, J.S., 537, 546, 550, 559
Miller, L., 466, 518, 525
Miller, M.C., 756, 818, 821
Miller, R.E., 330, 523, 642
Miller, W.D., 1002, 1430
Milliard, B., 299
Millier, F., 1086
Milligan, J.E., 290, 291, 300
Millikan, R.A., 119, 150
Milliken, B.D., 1004
Millis, R.L., 1244, 1269, 1273, 1291
Millman, P.M., 1164, 1188
Mills, B.Y., 755
Milne, E.A., 897, 910
Milsztajn, A., 385, 395
Milward, G.H., 1568, 1582
Min, K., 1371
Minakov, A.A., 465, 466
Minato, J.R., 834
Mineo, T., 733, 748, 749, 756, 959
Miner, E.D., 177, 1434, 1449, 1476, 1477
Miner, E.O., 1492
Mineshige, S., 855
Minh, Y.C., 610, 637–639, 641, 643, 644
Mink, D., 329
Minnaert, M., 230, 253
Minniti, D., 466

Mino, Y., 359, 371
Minster, J.-F., 1331, 1348, 1373
Minton, R.B., 1460, 1477
Mioduszewski, A.J., 855
Mirabel, I.F., 570, 572, 573, 577, 852, 853, 855
Miralda-Escudé, J., 395, 461, 467
Mironova, E.N., 1213, 1214
Misawa, K., 1353, 1373
Misconi, N.Y., 1185, 1188
Misner, C.W., 357, 371
Misra, K.C., 1403
Mitchell, D., 1320, 1491, 1526
Mitchell, D.L., 1232, 1271, 1321, 1491, 1492
Mitchell, G.F., 626, 641
Mitchell, J., 696
Mitchell, J.K., 1262, 1273
Mitchell, R.J., 475, 478, 488, 490, 497
Mitman, K.E., 518, 524
Mitra, S.K., 156, 158, 177
Mitsuda, K., 577, 578, 817, 820–822, 835, 847, 849, 854, 855, 959, 960
Mitsui, T., 696
Mittaz, J.P.D., 497
Mittlefehldt, D.W., 1372
Mitton, S., 788
Miura, M., 1061
Miyake, W., 1232
Miyamoto, S., 835, 959, 960
Miyano, K., 959
Miyata, E., 956–958, 960
Miyoshi, M., 394
Miyoshi, S., 836
Mizumoto, Y., 960
Mizuno, D.R., 330
Mizuno, H., 1432, 1449
Mizuno, T., 577, 855
Mizutani, H., 1232, 1271
Mizutani, K., 642
Mo, J.-E., 935
Moalic, J.-P., 1061
Möbius, E., 696, 967, 974, 995, 1000, 1002, 1003, 1005, 1138, 1202, 1209, 1215, 1233
Mochizuki, K., 641
Mochkovitch, R., 521, 527
Mock, M., 827, 834
Modigliani, A., 222, 1138, 1139
Modisette, J.L., 249, 253, 299, 673, 909
Moe, O., 1086
Moebius, E., 1214, 1215, 1526
Moffat, A.F.J., 909
Moffet, R.J., 1581, 1582
Mögel, H., 1561, 1562, 1582
Mogro-Campero, A., 972, 1003
Mohan, V., 523
Mohanty, G., 696
Möhlmann, D., 1218, 1258, 1232, 1273
Mohr, J.J., 387, 395
Moiseev, A., 696
Mojzsis, S.J., 1374
Molaro, P., 393, 439, 673
Molendi, S., 485, 496, 524, 546, 559
Moles, M., 468
Molina, M.J., 1660, 1668
Molinari, S., 630, 636–639, 641, 642, 869
Moller, P., 523
Molnar, L.A., 789, 833, 836, 837
Molster, F.J., 631, 632, 640, 641, 644, 867, 870–873
Molvik, G.A., 1215
Monaghan, J., 759, 787
Monaldi, L., 527

Moneti, A., 860, 872
Monfils, A., 298
Mönig, K., 1231
Montenbruck, O., 1650
Montesinos, B.M., 908
Montgomery, M.D., 1122, 1127, 1128, 1136, 1137, 1139, 1556, 1557
Montmerle, T., 394, 396, 397, 433, 438, 636, 869, 871, 872, 884, 885, 891, 892, 894
Montoya, L.M., 523
Montroy, T., 394, 420, 466
Mook, D.E., 811, 821
Mooney, T.J., 860, 871
Moore, C.B., 202
Moore, E.M., 547
Moore, G.F., 1602, 1616
Moore, H.J., 1322, 1371, 1402
Moore, J.H., 934
Moore, J.M., 1474, 1477
Moore, J.V., 298
Moore, M.H., 1218, 1232
Moore, P., 1271
Moore, R., 1080, 1087
Moore, R.K., 1616
Moore, R.L., 242, 254, 892, 910, 1086
Moore, T.E., 1543, 1549, 1555–1557, 1578, 1582
Moore, W.J., 329
Moorhouse, A., 639
Moorwood, A.F.M., 301, 315, 328–330, 576, 577
Moos, H.W., 298, 299, 300, 673, 674, 1435, 1449
Moraal, H., 254, 992, 1002, 1003
Morales-Rueda, L., 805, 809
Morall, H., 250
Moran, E.C., 854
Moran, J.M., 394, 870
Moraru, D., 525
Morbey, C.L., 299
Moreels, G., 1231
Morel, A., 1587, 1601, 1615, 1616
Morel, P., 1060
Moreno, R., 637, 1291
Moresi, L.-N., 1314, 1322
Morfill, G.E., 643, 673, 690, 695, 978, 1004, 1174, 1183, 1188, 1272
Morfill, G.M., 1293
Morgan, E.H., 820, 822
Morgan, F.J., 222
Morgan, J., 1381, 1466, 1477
Morgan, J.F., 222
Morgan, J.W., 1401
Morgan, P., 1321
Morgan, T.H., 299, 909
Morgan, W.W., 475, 497
Mori, M., 960
Moriarty, P., 696
Moriarty-Schieven, G.H., 526
Morin, J., 836
Moritz, H., 1650
Moriya, M., 960
Moriyama, S., 1061
Morosov, A.I., 113
Moroz, V.I., 177, 1296, 1270, 1273, 1304, 1309, 1321–1323, 1423, 1527, 1556, 1581
Morris, D.J., 755, 959, 1003
Morris, P.W., 437, 638, 639, 644, 869, 870, 872, 873
Morris, R.V., 1371
Morris, S.L., 393, 496, 497, 547, 560
Morrison, D.A., 1169, 1188, 1296, 1321, 1373, 1477, 1478
Morrison, G.H., 1381, 1402
Morrison, M., 1062
Morrison, N.D., 870
Morrison, P., 133, 134, 150, 231, 254, 492, 496, 722, 756

Morrison, R., 582, 605
Morrow, C.A., 1047, 1060, 1062
Morse, J.A., 298, 299
Morton, D.C., 290, 291, 293, 299, 300, 657, 673, 674, 675, 760, 773, 787, 835, 895–897, 909, 910, 1196, 1197, 1215
Morton, W.A., 675
Moseley, S.H. Jr., 328, 329, 419–421, 636, 642, 645, 1188, 1273
Mosen, A.W., 1402
Moses, J.D., 1086, 1138
Moses, J.I., 1447, 1449
Moskalenko, E.I., 525
Moskalenko, I.V., 684–686, 697
Moskalyeva, L.P., 1322, 1402, 1403
Mosquet, N., 869
Mosser, B., 1060
Motch, C., 736, 748, 754, 756, 812, 821, 891
Motoki, M., 696
Motschmann, U., 1233
Mott-Smith, L.M., 121, 150
Motz, L., 523
Mouginis-Mark, P.J., 1321, 1362, 1373
Mouillet, D., 868
Mould, J.R., 379, 393–397, 869
Mouri, H., 640, 871
Mouschovias, T.Ch., 858, 871
Moustakas, L.A., 466
Moutou, C., 619, 641
Mozer, F.S., 1016, 1031, 1526, 1536, 1555, 1557
Muchmore, D., 1088
Muchmore, V.M., 1074
Muecke, A., 696
Muehlner, D., 314, 330
Mueller, B.E.A., 1269, 1273
Mueller, D., 314, 694
Mueller, I.I., 1650
Mueller, J., 1668
Mueller, J.L., 1615
Mueller, S., 1433, 1449
Mueller, T., 871
Muhleman, D.O., 1322
Mühlhäuser, K.-H., 1004, 1139, 1140
Muhli, P., 527
Mukai, K., 808, 809
Mukai, T., 1022, 1026, 1032, 1227, 1232, 1272, 1540, 1555, 1557
Mukherjee, R., 525, 684, 696
Mulchaey, J.S., 488, 496, 497, 526
Mulkay, M.J., 323, 329
Mullan, D.J., 901, 904, 910
Müller, D., 150
Muller, E., 558
Müller, E.A., 437, 1086
Muller, H., 1584
Müller, J., 346, 351, 352
Muller, J.M., 524, 527
Muller, R., 1068, 1087
Muller, R.A., 421
Müller, W., 120, 150
Mullis, C.R., 496
Mumma, M.J., 640, 1232, 1281, 1286, 1287, 1290, 1292, 1293
Munakata, K., 1004
Münch, G., 230, 250, 252, 270, 647, 649–651, 653, 673, 1425, 1430
Munch, T.A., 165, 177
Mundt, R., 862, 870–872
Mundy, L.G., 604, 643
Munk, W.H., 1647, 1650, 1651
Munoz, G., 137
Muñoz, J.A., 456, 463, 467
Muñoz-Caro, G., 613, 639
Munro, G.H., 220, 1567, 1582
Munro, R.H., 222, 1001, 1113, 1139

Muraishi, H., 960
Murakami, H., 323, 329, 330, 959, 1004
Murakami, M., 330
Murakami, T., 504, 524, 526–528, 736, 754–758, 787, 821, 835, 959
Muraki, Y., 960
Murayama, H., 1231
Murchie, S.L., 1371, 1402
Murdin, P., 755, 760, 789, 811, 822, 823, 833, 834, 837, 840, 841, 856
Murdock, T.L., 326, 330, 419–421, 645
Murowinski, R., 299
Murphy, D.C., 604
Murphy, D.W., 855
Murphy, E.M., 299, 674
Murphy, J.R., 1402
Murphy, N., 1492
Murphy, R.J., 523, 972, 1003
Murphy, T., 527
Murray, A., 754
Murray, A.G., 869
Murray, B.B., 268, 270, 1322, 1323, 1497
Murray, N., 1061
Murray, S., 285, 496, 497, 577, 754, 809
Murray, S.S., 497, 577, 786
Musatov, S., 253
Muschietti, L., 1517, 1526, 1527
Mushotzky, R.F., 474, 479, 481, 488, 490, 495–498, 543, 569, 577, 854, 855, 959
Musielak, Z.E., 892, 910, 1086
Muslimov, A.G., 729, 754, 822
Musmann, G., 1145, 1160, 1232
Musser, J.A., 694
Mutchler, M., 524
Mutschke, H., 640, 870

Naber, R.M., 465, 525, 528
Nadeau, D., 456, 462, 465, 467
Naeslund, M., 527
Nagase, F., 560, 578, 604, 757, 808, 814, 821, 826, 830, 831, 834–836, 891, 893, 959
Nagase, N., 822
Nagel, W., 834, 836
Nagy, A.F., 1232, 1271, 1448
Nair, A.D., 560
Naismith, R., 1561, 1562, 1580
Naito, T., 960
Nakada, T., 1231
Nakagawa, M., 959
Nakagawa, T., 330, 614, 641, 642
Nakagawa, Y., 1271
Nakahata, M., 959, 1061
Nakai, N., 394
Nakamoto, A., 1004
Nakamura, H., 833, 836, 851, 1343
Nakamura, K., 1231
Nakamura, N., 855, 1373
Nakamura, T., 525
Nakariakov, V.M., 1083, 1087
Nakayama, S., 1061
Nakazawa, K., 496, 1271
Naletto, G., 222, 1138, 1139
Namiki, M., 528
Nandra, K., 852, 855
Nandy, K., 298
Napier, W.M., 1238, 1269
Napiwotzki, R., 924, 935
Narain, U., 1082, 1087
Naranan, S., 496
Narayan, R., 371, 456, 462, 465–468, 558, 559, 753, 844, 845, 852, 855
Narayeva, M.K., 1321
Narcisi, R.S., 1562, 1566, 1582

Narita, M., 330
Narlikar, J., 754
Nash, D., 1467, 1477
Nasuti, 737
Natarajan, P., 513, 521, 527, 528
Nather, R.E., 753, 793, 797, 810, 1430
Natta, A., 629, 641, 643
Nauenberg, M., 760, 787
Navalgund, R., 1616
Navarro, J.F., 477, 493, 496–498
Naylor, D.A., 437, 641, 869, 1449
Naylor, T., 821, 854
Neal, C.R., 1375
Neff, D.H., 909
Neff, S.G., 299, 547
Negoro, H., 959
Neizvestny, S.I., 757
Nel, H.I., 755
Nelemans, G., 765, 774, 787
Nelson, B., 1478
Nelson, C.A., 466
Nelson, M.J., 753
Nelson, R.F., 893
Nelson, R.N., 1435, 1449
Nelson, R.W., 786, 820, 834
Nemiroff, R.J., 452, 467, 502, 527, 457, 458
Nenow, E.V., 1376
Nerney, S., 1156, 1157, 1161
Ness, N.F., 151, 248, 250, 251, 253–255, 1000, 1118, 1138, 1140, 1144, 1152, 1159–1161, 1214, 1232, 1301, 1320, 1321, 1479–1485, 1487, 1488, 1490–1492, 1501, 1527, 1528, 1530, 1557
Netterfield, C.B., 394, 420, 466
Netzer, H., 547
Neubauer, F.M., 1139, 1160, 1229, 1230, 1232, 1489, 1490, 1492
Neufeld, D.A., 620, 621, 633, 636, 639, 640–642
Neufeld, M.J., 57, 58
Neugebauer, G., 304, 306, 311, 312, 314, 318, 319, 326, 328–330, 523, 527, 548, 570, 577, 636, 757, 868, 870, 1430
Neugebauer, M., 248, 251–255, 896, 910, 1116, 1117, 1123, 1129–1131, 1134, 1135, 1138–1140, 1144, 1159, 1160, 1214, 1216, 1231–1233, 1291–1293
Neuhäuser, R., 736, 756, 757, 860, 871, 882, 884, 885, 891, 892
Neukum, G., 1188, 1303, 1322
Neumann, D.M., 489, 495–498
Neumann, G.A., 1300, 1321, 1322
Neupert, W.M., 216, 217, 222, 1086
Neuss, H., 187, 1526, 1528
Neustroev, V.V., 757
Nevalainen, J., 489, 498
Nevatia, J., 999
Nevkirk, L.L., 177
New, R., 1060–1062
Newberg, H.J.M., 395, 420, 468
Newburn, R.L., 870, 1187, 1216, 1226, 1232, 1233, 1270, 1292, 1293
Newcomer, F.M., 959
Newell, H.E., 42, 57, 58, 60, 67, 80, 90, 114, 198, 202
Newhall, X.X., 351, 352, 1650
Newkirk, G., 1045, 1062
Newkirk, G. Jr., 1062
Newkirk, G.A. Jr., 1139
Newkirk, R.W., 218, 222
Newman, J., 526
Newman, W.S., 1160
Newmann, J.A., 395
Newport, B.J., 695
Newton, H.W., 1101, 1113
Newton, I., 335, 352, 467
Newton, R.R., 1651
Ney, E.P., 150, 695

Nezvara, M., 525
Ng, C.K., 973, 976, 1003
Nguyen-Q-Rieu, A.J., 636–638, 641, 643, 869
Nicasto, I., 756
Nicastro, L., 524–528, 756
Nice, D., 755
Nice, D.J., 371
Nicholls, R.W., 222
Nichols, J.S., 758, 909
Nichols, R.H., 1375
Nicholson, F., 1231
Nicholson, P.D., 1270, 1435, 1449, 1476, 1477
Nicholson, S.B., 253, 304, 330, 1452, 1453, 1460
Nicolas, K.R., 1078, 1087
Nicolis, A., 366, 371
Nicolosi, P., 222, 1138, 1139
Nicolson, G.D., 559
Niebur, S.M., 688, 696
Niel, M., 523, 524, 526, 528, 820
Nielsen, E., 1572, 1581, 1582
Niemann, H.B., 438, 1430, 1448, 1583
Nier, A.O., 1328–1330, 1332, 1370, 1373, 1375
Nigam, R., 1050, 1059, 1062
Nightingale, R.W., 1032, 1085
Nikolayeva, O.V., 1320, 1321
Nikolsky, Yu.V., 1270
Nilakshi, N., 523, 524
Nilsson, C., 1167, 1168, 1188
Ninkov, Z., 467
Nisbet, E.G., 1321
Nisbet, E.J., 1317
Nishida, A., 1018, 1019, 1032, 1517, 1534, 1557
Nishijima, K., 960
Nishimura, J., 524, 527, 688, 696
Nishimura, T., 641, 642
Nishiuchi, M., 696
Nisini, B., 627, 628, 636–639, 641–643, 869
Nissen, P.E., 428, 429, 437–439
Nizovsky, H., 1616
Noack, E., 113
Nobili, A., 1650
Noble, L.M., 247, 254
Noci, G., 222, 1080, 1083, 1085, 1087, 1137–1139
Noda, M., 330
Noels, A., 429, 434, 438
Noerdgaard-Nielsen, H.U., 528
Nolan, P.L., 503, 504, 526, 696, 755, 757
Nolan, T., 1004
Noll, K.S., 396, 1274, 1462, 1477
Nollett, K.M., 393
Nolte, J.T., 242, 250, 254
Nomoto, K., 394, 524, 525, 778, 780, 787, 810, 960
Nonino, M., 546
Norberg, O., 1581
Nordberg, W., 202
Nordh, L., 636, 637, 641, 862, 869, 871, 872
Nordholdt, J.E., 1582
Nordlund, Å, 1058, 1063, 1068, 1087
Nordsieck, K.H., 298, 299, 641
Nordström, B., 393, 439
Nordtvedt, K., 335, 343–346, 349–352, 365, 371
Norman, C.A., 546, 548, 617, 636, 693, 696
Norman, K., 869
Norman, M., 1341, 1342, 1370, 1371, 1373
Norman, M.L., 496
Normile, D., 431, 438
Norris, J.E., 439
Norris, J.P., 467, 508, 526, 527, 822
Northcott, M.J., 872
Northrop, T., 174, 175, 177, 1419, 1498, 1499

Norton, T.D., 466
Nouël, F., 1650
Nousek, J.A., 547, 603–605
Novak, G., 639
Novak, R., 1292
November, L.J., 1068, 1087
Novick, R., 577, 754
Novikov, B., 820
Novikov, I.D., 759, 787, 789
Noyes, R.W., 222, 234, 242, 244, 255, 877, 892, 1037, 1038, 1046, 1062, 1086, 1113
Nozaki, M., 696
Nozette, S., 1321
Nugent, J., 836
Nugent, P., 393, 395, 420, 467
Null, G.W., 1436, 1449
Nulsen, P.E.J., 481, 496
Nunes, N.J., 395, 420, 467
Nunes, P., 1373
Nussbaumer, H., 674, 789
Nusser, A., 393
Nussinov, S., 525
Nuth, J., 868, 870
Nutter, S.L., 694
Nylund, S.R., 1555
Nyman, L.A., 604
Nyquist, L.E., 1340–1343, 1347, 1351, 1353, 1355–1358, 1360, 1362, 1363, 1365, 1371–1375, 1391, 1402
Nystrom, G.U., 352, 1086, 1138

Obayashi, Y., 1061
Oberc, P., 1225, 1232
Oberg, J.E., 57, 58
Oberhummer, H., 440
Oberli, F., 1343, 1374
Oberly, J.J., 298
Oberth, H., 61, 77, 114, 288, 299
O'Brien, P.T., 547
Ocampo, A.C., 1423, 1477
Occhialini, G.P.S., 123, 125, 149, 150
O'C Drury, L., 696
O'Ceallaigh, C., 1003
Ochsenbein, F., 935
Ockert-Bell, M.E., 1183, 1188
O'Connell, R.W., 475, 498
Oda, M., 760, 788, 821, 823, 836, 841, 855, 959, 960
Oda, N., 329
Oda, S., 960
Odegard, N.P., 329, 642, 898, 910, 1188
O'Dell, C.R., 652, 674, 861, 870–872
O'Dell, R.C., 629, 641
O'Dell, S.L., 553, 559, 758
Odewahn, S.C., 523, 524, 526
Odishaw, H., 57, 58, 149
Oegelman, H., 754, 809, 834
Oegerle, W.R., 299, 476, 498, 674
Oemler, A., 475, 492, 496, 498
Oerlemans, J., 1617
Oeschger, H., 437
Oetliker, M., 1002
Oey, M.S., 546
Ofek, E.O., 524
Offermann, D., 1578, 1583
Offutt, W.B., 1476
Ofman, L., 1087
O'Gallagher, J.J., 151, 1138
Ogawa, H.S., 1138, 1214
Ogawara, Y., 835, 855, 959, 960, 1079, 1087
Ögelman, H., 724, 737, 743, 753–757, 812, 821, 828, 1001
Oger, F., 1616

Ogilvie, G.I., 526
Ogilvie, K.W., 1001, 1124, 1127, 1129, 1137–1139, 1214, 1279, 1293
Ogino, T., 1557
Ogio, S., 960
Ogor, F., 1599
Oh, S., 394, 420
Ohashi, T., 496, 497, 576, 577, 820, 830, 835, 836, 893, 959
Ohl, R.G., 299
Ohnishi, T., 577, 855
Ohta, K., 960
Øieroset, M., 1032, 1556
Ojakangas, G.W., 1476
Ojha, R., 560
Oka, T., 1449
Okada, A., 1061
Okada, K., 577, 640, 871
O'Kane, P., 757
Okazaki, T., 1032
Oke, B., 530, 547
Oke, J.B., 469, 550, 559
Okuda, H., 314, 329, 330, 576, 577, 610, 641–643
Okulesski, B.A., 1491
Okumura, K., 330, 466, 577, 869
Okumura, Y., 484, 497, 498
Olano, C.A., 674
Olbert, S., 231, 255, 756
Olehy, D.A., 1402
Olinger, C.T., 1375
Olive, K.A., 435, 438, 440, 1231
Oliveira, C.L.M., 299, 547
Oliver, M.B., 298
Olkin, C.B., 1269
Olnon, F.M., 330, 577, 935
Olofsson, G., 622, 636, 637, 641, 862, 869, 871, 872
Olsen, L.F., 523
Olsen, S., 1581
Olson, G.L., 836, 878, 879, 891, 898, 908
Olson, R.A., 526
Olszewski, E., 834
O'Meara, J.M., 440
Omelchenko, A., 1233
Omholt, A., 1581
Omidi, N., 1528
Omont, A., 617, 632, 636–638, 641–643, 869, 1261, 1273
Omura, Y., 1517, 1527
Onaka, T., 330, 640
O'Neal, R.H., 1138
Ono, Y., 1024, 1032, 1033
Onsager, T.G., 1031
Oort, A.H., 1616
Oort, J.H., 652, 674, 719, 1235, 1238, 1273, 1281, 1293
Oosterbroek, T., 527, 736, 755, 756
Oparin, A.I., 1280, 1281, 1293
Opgenoorth, H.J., 1547, 1557
Öpik, E.J., 258, 264, 270, 1235, 1273, 1281, 1293, 1425, 1430
Oppenheimer, F., 131, 150, 695
Oppenheimer, J.R., 756, 759, 788, 839, 855
Oppenheimer, T.R., 721
Ordway, F.I., 57, 58, 71, 114
O'Reilly, J.E., 1601, 1616
Orfei, R., 636–638, 641, 869
Ori, A., 370
Orio, M., 809
Orito, S., 681, 696
Orlandini, M., 524, 525, 527, 754
Orlando, S., 892
Ormes, J.F., 695, 696
Oró, J., 1280, 1293
Orosei, R., 1261, 1273
Orosz, J.A., 841, 855

Orr, A., 527, 789
Orrall, F.Q., 222, 905, 910
Orren, M.J., 1403
Ortolan, A., 371
Orton, G.S., 1430
Osaki, Y., 1061
Osawa, K., 788, 821
Osborne, J.P., 803, 809, 810
Oschlies, K., 993, 1003
Oscoz, A., 523
O'Shea, B., 496
Osip, D.J., 1269, 1272, 1291
Osmer, P.S., 540, 547, 820, 821
Osmer, R., 788
Ostensen, R., 466, 527
Osterbrock, D.E., 475, 497
Osterman, S.N., 299, 546
Ostreicher, R., 299
Ostriker, J.P., 392, 393, 395, 440, 468, 474–496, 589, 590, 593, 599, 601, 604, 608, 641, 665, 673, 690, 695, 768, 772, 773, 776, 786, 788, 818, 821, 823, 827, 833–835, 978, 1000
Ostro, S.J., 1232, 1271
O'Sullivan, D., 697, 1003
Otani, C., 528, 855
O'Toole, S.J., 1060
Ott, J., 523
Ott, L.L., 527
Ott, S., 869
Ottemayer, D.R., 91, 114
Ottens, F., 1001, 1138
Otterman, J., 202
Oudmaijer, R.D., 376, 394
Oue, H., 959
Ovenden, M.W., 1459, 1477
Owen, F.N., 475, 481, 491, 496, 498
Owen, T.C., 270, 428, 437, 438, 1288, 1292, 1293, 1306, 1321, 1430, 1437, 1443, 1448, 1449, 1476, 1477
Owens, A., 515, 524, 527
Owocki, S.P., 891, 899–902, 909, 910
Oyama, Y., 959
Ozaki, M., 696, 753, 892, 959
Ozel, M.E., 754
Ozerov, V.D., 253, 1138

Pacher, T., 871
Pacholczyk, A.G., 678, 683, 696
Paciesas, W.S., 286, 500, 523–527, 856, 960
Pacini, F., 306, 329, 395, 722, 726, 756, 1496, 1527
Packham, C., 524, 454, 459
Paczyński, B., 454, 459, 467, 527, 500, 504, 508, 515, 517, 521, 761, 772, 775, 776, 778, 781, 788
Padgett, D., 643
Padovani, P., 540, 547, 548, 553, 554, 559, 560
Paerels, F., 807–809, 833, 835, 836
Paerli, R., 1140
Paestermann, T., 604
Pagani, L., 641
Pagano, I., 908
Page, C.G., 498
Page, D., 730, 737, 756
Page, D.E., 187, 253, 1003, 1004, 1140, 1141, 1159, 1213
Page, E., 1215
Page, L., 420
Page, T.L., 498
Pagel, B.E.J., 424, 428, 438
Paget, T.M., 217, 222
Pahl, M., 1372
Pahre, M., 524
Paige, D.A., 1466, 1477
Pain, R., 395, 420, 467

Pais, A., 339, 352
Paizis, C., 1002
Pajot, F., 639
Pakhomov, L.A., 1668
Pakull, M.W., 812, 821
Palanque-Delabrouille, NF., 454, 466
Palazzi, E., 523–528, 559
Palla, F., 639, 641, 643, 893
Pallavicini, R., 877, 880, 887–889, 891–894, 969, 1003
Pallé, P.L., 1061, 1062
Pallottino, G.V., 365, 371
Palme, C., 1403
Palme, H., 1378, 1382, 1383, 1385, 1389, 1390, 1401–1403
Palmer, D., 525
Palmer, I.D., 986, 988, 1003
Palmer, P., 1286, 1293
Palmieri, T.M., 603
Palumbo, G., 559
Palumbo, M.E., 623, 626, 642
Pan, H.-C., 835
Pan, V., 1362, 1372
Panagia, N., 376, 395, 396, 420, 467, 523, 672
Panasyuk, A., 222, 1138, 1139
Panasyuk, M.I., 999
Pandey, A.K., 523, 524
Panek, R.J., 797, 809
Paneth, F.A., 1329, 1370, 1374
Panfilov, A.S., 1321
Pankiewitz, G.S., 1224, 1232
Paolini, F., 559, 820, 854
Paolini, F.P., 604
Paolini, F.R., 286, 754, 786
Pap, J., 1061
Papadakis, M., 820
Papaloizou, J.C.B., 868
Papamastorakis, I., 1003, 1032, 1033, 1527
Papanastassiou, D.A., 1322, 1332, 1337, 1338, 1343, 1348, 1353, 1355, 1370, 1374–1376
Papike, J.J., 1298, 1321, 1323, 1340, 1342, 1344, 1346, 1347, 1373, 1374
Pappalardo, R.T., 1477
Pappe, N., 960
Paquet, P., 1650
Parady, B., 1557
Pardo, J.R., 641
Paresce, F., 299, 547, 589, 590, 605, 658, 663, 674
Parigger, P., 1583
Parish, H.F., 1581
Parizot, E., 434, 438
Park, C.G., 1562, 1580
Park, H.S., 518, 527
Park, J.H., 1668
Park, S., 594, 605
Parke, M., 1650
Parker, E.L., 527
Parker, E.N., 133, 134, 138–140, 150, 230, 232, 234, 235, 237, 239, 241, 242, 244–254, 682, 691, 696, 896, 910, 1003, 1008, 1009, 1011–1015, 1023, 1032, 1054, 1063, 1079, 1083, 1084, 1087, 1113, 1115, 1117, 1135, 1139, 1142, 1144, 1147, 1148, 1150, 1154, 1160, 1197, 1198, 1215, 1491, 1492, 1506, 1507, 1511, 1514, 1527, 1530, 1539, 1557
Parker, R., 1491
Parker, R.H., 527
Parker, T.J., 1273, 1304, 1321, 1402
Parkes, G., 837
Parkin, C.W., 1491
Parkinson C.L., 1615
Parkinson, C.E., 1615
Parkinson, W.H., 222
Parkinson, W.J., 222
Parmar, A.N., 524, 527, 755, 756, 822, 827, 834, 836, 837, 856, 893, 959
Parmar, P.S., 617, 642

Parmentier, E.M., 1299, 1314, 1315, 1319, 1321, 1340, 1342, 1372
Parnell, C.E., 1032
Parodi, B.R., 375, 377–379, 395
Parrot, M., 1570, 1583
Parry, I., 872
Parsignault, D.R., 809
Pasachoff, J.M., 234, 251, 253, 438
Pascale, E., 394, 420, 466
Paschmann, G., 982, 1000, 1003, 1019, 1020, 1022, 1031–1033, 1505, 1507, 1527, 1541, 1549, 1556, 1577, 1583
Pascu, 1453
Pasquini, L., 428, 438, 524
Pastoriza, M.G., 547
Paswaters, S.E., 1138, 1140
Patat, F., 524, 525
Patchett, B.E., 959
Patchett, P.J., 1371
Paterakis, G., 820
Paterson, W.R., 1491, 1555, 1556
Pati, M.E., 358, 371
Patino, A., 637
Patnaik, A., 1293
Patrick, T.J., 869
Patrón, J., 1061
Patsourakos, S., 1076, 1087
Patten, B.M., 636, 639, 641, 642, 883, 892
Patterson, C.C., 1330, 1331, 1374
Patterson, J., 793, 795, 804, 810
Patterson, J.H., 275, 1403
Patterson, J.R., 960
Patterson, T.N.L., 1197, 1215
Paturel, G., 382, 393, 395, 397
Paubert, G., 637, 1291, 1449
Paul, E.R., 689, 702, 719
Paul, J., 394, 396, 397, 523, 820
Paul, J.A., 286, 695, 755, 758
Paularena, K.I., 1160
Pauldrach, A.W.A., 892, 897, 901, 902, 909, 910
Paulikas, G.A., 1490, 1492
Pauls, H.L., 673, 1158, 1161, 1208, 1215, 1216
Pavlenko, Y.V., 439
Pavlis, E., 351
Pavlov, G.G., 725, 728–732, 736, 737, 741, 744, 746–748, 753–758, 855
Pavlova, T.G., 1372
Pawsey, J.L., 1561, 1571, 1581
Paxton, H.J.B., 222, 298
Payne, D.G., 553, 558
Payne-Gaposchkin, C., 793, 810
Peach, J.V., 820
Peacock, A., 757, 856
Peale, S.J., 1261, 1264, 1265, 1273, 1460, 1477, 1478
Pearce, J.A., 648, 674
Pearce, J.B., 1450
Pearl, J., 1427, 1430, 1449, 1467, 1477
Pearson, T.J., 469, 559
Peccei, R., 371
Pechernikova, G.V., 1322
Peckett, A., 1372, 1373
Pedersen, H., 513, 523, 524, 527, 528
Pedersen, J.P., 1610, 1616
Pedraz, S., 523
Peebles, P.J.E., 387, 413, 414, 395, 420
Peeters, E., 640, 642–644
Pei, Y.C., 329
Peimbert, A., 428, 438
Peimbert, M., 428, 438, 871
Peixoto, J.P., 1596
Pel, J.-W., 465
Pelaez, F., 523
Pelat, D., 547
Pellegrini, S., 498, 568, 576, 577
Pelling, M.R., 694
Pelling, R.M., 951, 960
Pelló, R., 466
Pelz, G., 871
Pen, U.-L., 489, 490, 491, 498
Pendergast, K.H., 823, 836
Pendleton, G.H., 524, 526
Pendleton, G.N., 286, 523, 524, 526, 527
Pendleton, Y.J., 608, 612, 626, 642, 1293, 1476, 1477
Pendleton, Y.L., 638
Pendray, G.E., 113
Penford, J.E., 547, 854
Peng, C.Y., 467
Penninx, W., 822
Penny, A.J., 1059, 1063
Pennypacker, C.R., 395, 420, 467
Penston, M.V., 298, 546–548, 871
Penton, S., 560
Penzias, A.A., 400, 420, 644
Pepin, R.O., 1392, 1401, 1402
Pequignot, D., 636, 637, 638, 641, 869
Pérault, M., 611, 636, 637, 869 871, 872
Percival, J.W., 753, 754
Peredo, M., 1160
Peres, G., 564, 577, 881, 888, 892, 1082, 1087
Pereslegina N.V., 1004
Pérez Hernández, F., 1062
Perez, A.M., 527
Perez, E., 547
Perez, M., 547
Pérez, M.R., 859, 869–872
Perez-Martinez, S., 637
Perez-Mercader, J., 351
Peria, W., 1526, 1555
Perkins, F., 1033
Perley, R.A., 496, 532, 547, 559
Perlman, E.S., 497, 550, 559
Perlmutter, S.J., 388–390, 395, 415, 420, 457, 467, 495, 523
Perna, R., 518, 526
Pernechele, C., 222, 1138, 1139
Perola, G.C., 546, 547, 559
Peron, I., 642
Perrier, F., 869
Perrine, C.D., 1452, 1453
Perry, M., 639, 643
Perry, P.M., 298
Persaud, J.L., 547
Persi, P., 636, 837, 869, 871, 872, 911
Persic, M., 386, 395
Person, M.J., 1478
Persson, S.E., 636
Pesce, J.E., 560
Pesses, M.E., 250, 254, 682, 696, 977, 978, 997, 1000, 1003, 1156, 1160, 1200, 1215
Peter, H., 1078, 1079, 1087
Peter, P., 394, 396, 397
Peters, B., 131, 150, 434, 438, 695
Peters, G., 285, 496, 754, 809
Peters, J., 547
Petersen, C.C., 1322
Peterson, B., 1293
Peterson, B.A., 466, 525, 541, 546–548, 755
Peterson, B.M., 444, 454, 458, 465, 468, 534, 541, 547
Peterson, D.D., 1003
Peterson, L.E., 527, 528, 837, 960, 1402
Peterson, R.C., 439
Peterson, W., 1526
Peterson, W.K., 1541, 1557

Pethick, C.J., 754, 755, 787, 835
Petit, M., 1580
Pettini, M., 669
Pettit, E., 304
Petitto, J.M., 1062
Petre, R., 694, 696, 724, 736, 741, 754, 756, 953, 959, 960, 1232
Petrik, J., 299
Petro, L.D., 524, 528, 820, 835
Petrosian, V., 458, 467, 499, 526
Petrou, N., 1061
Petrov, G.M., 1320
Petry, C., 524
Petschek, H.E., 235, 254, 1008–1011, 1017, 1025, 1031, 1032, 1113, 1499, 1500, 1509, 1514, 1527
Pettengill, G.H., 268, 270, 352, 1271, 1322
Petterson, J.A., 831, 835, 836
Pettersson, L.H., 1616
Pettini, M., 498, 524, 525, 528, 672, 674, 669
Pettit, E., 304, 330
Pettit, H.B., 202
Pezzuto, S., 639, 641–643
Pfaff, R., 1526, 1555, 1580, 1581, 1583
Pfahl, E.D., 756, 788, 835
Pfau, W., 870
Pfeffermann, E., 809
Pfeifer, R.H., 299
Pfeiffer, B., 393
Pfenniger, D., 639
Pfister, L., 1430
Pfotzer, G., 122, 150
Pham Van, J., 352, 1651
Phan, T.-D., 1021, 1032
Phelps, R., 393–397
Phillips, C.B., 1477
Phillips, F.G., 1293
Phillips, J.E., 1138
Phillips, J.L., 254, 255, 1137, 1138, 1140, 1148, 1157, 1159–1161, 1188, 1189
Phillips, J.P., 639
Phillips, M.M., 394, 395, 420, 468
Phillips, R.J., 1310, 1321, 1322, 1403, 1423
Phillips, T.G., 315, 330, 616, 621, 636, 637, 640–642, 1291–1293
Philpotts, J.A., 1380, 1402
Phinney, E.S., 523, 526, 559, 788
Phinney, G., 548
Phinney, S., 540, 548, 553, 559, 814, 821
Piacentini, F., 394, 420, 466
Pian, E., 499, 505, 509, 513, 516–518, 521, 523–528, 556, 559, 560
Picat, J-P., 468
Piccinotti, G., 479, 492, 498
Piccioni, A., 523
Picha, J., 1650
Pichel, W.G., 1608, 1610, 1615, 1616
Pichon, B., 369, 371
Pichon, C., 718
Pick, M., 1650
Pickering, W.H., 1452, 1453
Piddington, J.H., 1466, 1477
Piepke, A., 1231
Pierce, J.R., 235, 254
Pierce, M.J., 465, 467
Pieri, D.C., 1321
Piersimoni, A., 523
Pieters, C.M., 1298, 1321, 1354, 1374, 1375
Pietsch, W., 601, 605, 754, 755, 757, 789, 808, 836
Pietz, J., 594, 605
Piexoto, J.P., 1616
Piggot, C.H., 1374
Piggot, C.S., 1327, 1372
Pijpers, F.P., 1063

Pikington, J.D.H., 1649
Pilbratt, G.L., 576
Pilcher, C.B., 1466, 1477
Pilcher, C.D., 1429
Pilipp, W.G., 1128, 1139
Pilkington, J.D.H., 787, 855, 1581
Pilon, A.M., 1584
Pina, R., 870
Pineau des Forêts, G., 637, 638, 640
Pines, D., 787, 812, 822, 835
Pinkau, K., 755
Pinkney, J., 496, 546
Pinson, W.H., 1331, 1372, 1374
Pinsonneault, M.H., 436, 437, 439
Piotto, G., 713, 719
Piper, J.O.A., 1481, 1492
Pipher, J.L., 309, 311, 329, 330, 467, 644
Piran, T., 371, 499, 505, 509, 517, 518, 521, 524, 526–528, 684, 696
Pirenne, B., 468
Piro, L., 516, 523, 524, 525, 527, 528
Pirraglia, J.A., 1429, 1449
Pirronello, V., 1292
Pisano, D.J., 525
Pisarski, R., 960
Piscitelli, J.R., 1273
Piskunov, N., 1208, 1215
Pitteway, M.L., 1571, 1584
Pitts, R., 547
Pitz, E., 1188
Pivovaroff, M.J., 753
Pizzella, G., 370, 371
Pizzichini, G., 523, 524
Pizzo, V.J., 1137, 1138, 1151, 1159, 1160
Plait, P., 870
Planck, B., 1188
Plaskett, J.S., 648, 674
Platacis, E., 1491
Platt, J.R., 642
Platt, S.R., 639
Platt, T., 1602, 1616
Plaut, L., 704, 719
Plavec, M.J., 761, 770, 788
Plets, H., 644, 864, 872
Plez, B., 393
Ploeger, G., 869
Plokhotnichenko, V.L., 757
Ploner, S.R.O., 1069, 1086, 1087
Plucinsky, P.P., 595, 597, 599, 605, 745, 756, 959
Plume, R., 636, 639, 641, 642
Plummer, W., 328
Plunkett, S.P., 1138, 1140
Pneuman, G.W., 1032, 1111, 1113
Pocobelli, G., 1584
Podolak, M., 1219, 1232, 1274
Podosek, F.A., 1335, 1348, 1370, 1372, 1374, 1375
Podsiadlowsky, P., 764, 782, 788, 814, 820
Podwysocki, M., 1401
Poe, C.H., 902, 903, 910
Poeppel, W.G., 654, 674
Poetzel, R., 862, 872
Pogge, R.W., 535, 547, 548
Poggianti, B.M., 492, 498
Poglitsch, A., 643
Pogosyan, D., 394, 420, 466
Pohl, M., 560, 696, 757, 1232
Poincare, H., 1560, 1583
Poisson, E., 358, 367, 370, 371
Pokhunkov, A.A., 195, 202
Polamskey, C., 177
Poland, A.I., 1001, 1061, 1087, 1113, 1138, 1139

Polanskey, C., 1188, 1491, 1492
Polatidis, A.G., 469
Poletto, G., 222, 1138, 1139
Pollack, J.B., 437, 1187, 1309, 1321, 1423, 1456, 1459, 1477, 1478
Pollard, G.S.G., 835
Pollard, K., 465
Pollock, C.J., 1582
Polnau, E., 1372
Polomski, E., 546
Poloskov, S.M., 199, 202
Pols, O.R., 523, 778, 786
Polyakov, A.M., 351
Pomerantz, M.A., 1061
Ponin, A., 270
Ponman, T.J., 488, 490, 496–498
Ponnamperuma, C., 1430, 1477
Ponsot, B., 370
Pont, F., 376, 395
Pontecorvo, B., 1036, 1063
Pooley, G.G., 524, 760, 788, 855
Pope Pius XII, 258
Popecki, M.A., 975, 1003
Popescu, C.C., 393
Popov, G.A., 113
Popow, E., 871
Popowski, P., 466
Popp, B.D., 1061
Poppel, W.G.L., 672
Porco, C.C., 1426, 1430, 1476
Poretti, E., 891
Portegies Zwart, S.F., 786, 789
Porter, J.G., 242, 254
Pospieszalska-Surdej, A., 444, 468
Pospisilova, L., 1651
Potekhin, Y.A., 732, 756
Potemra, T.A., 1538, 1539, 1556
Potgieter, M.S., 1002, 1154, 1160
Potocki, K.A., 299
Pottasch, S.R., 330, 577, 615, 642, 644, 919, 923, 934, 935
Pottelette, R., 1508, 1518, 1522, 1527, 1528
Potter, M., 559
Potter, N.M., 1402
Pouliquen, D., 869
Pound, R.V., 341, 352
Pounds, C.A., 959
Pounds, K.A., 534, 547, 548, 820, 831, 836, 837, 852, 855
Poutanen, J., 546, 854
Powell, B.N., 1403
Powell, C.F., 125, 150
Powell, K.G., 1215, 1232
Poynter, R.L., 1160, 1214, 1215
Poynting, J.H., 227, 254
Praderie, F., 862, 869, 872
Pradhan, A.K., 960, 1214
Prakash, M., 755, 757
Prantzos, N., 395, 435, 437–440
Prato, L., 638
Pratt, M.R., 466
Pravdo, S.H., 497, 820, 822, 830, 834, 836
Pravec, P., 525
Preece, R.D., 286, 523, 524, 526, 527, 758
Preger, B., 525
Preibisch, T., 882, 884, 892, 893
Preite-Martinez, A., 935, 959
Premo, W.R., 1341, 1374
Pres, P., 893
Press, W.H., 393, 444, 457, 468
Preston, R.A., 855
Prestwich, A.H., 577
Pretzl, K., 395

Prialnik, D., 810, 1219, 1232, 1260, 1262, 1273, 1274
Price, C., 793, 810
Price, M., 636, 637, 641, 1322
Price, M.C., 869
Price, P.B., 966, 1003
Price, S.D., 311, 313, 314, 323, 326, 328, 330, 869, 873
Priedhorsky, W., 523
Priest, E.R., 894, 1011, 1025, 1027, 1028, 1032, 1083, 1086, 1159
Prieto, A., 468
Prieur, J.L., 467
Prieur, L., 1587, 1601, 1616
Prilutskii, O.F., 1233
Primas, F., 393, 429, 434, 439
Primdahl, F., 1577, 1583
Primini, F.A., 577, 821, 826, 836
Prince, T.A., 523, 784, 786, 788, 820, 834
Pringle, J.E., 526, 786, 794, 810, 823, 827–829, 831–833, 835, 836, 855
Prinja, R.K., 900, 909, 910
Prinn, R.G., 1446, 1449
Prishchep, V.L., 690, 696
Pritchard, R.S., 755
Pritchet, C.J., 393, 394
Pritchett, P.L., 1527
Probstein, R.F., 1222, 1232, 1236, 1241, 1261, 1270
Proctor, M.R.E., 1481, 1492
Prodi, G.A., 365, 371
Proffitt, C.R., 1060
Proga, D., 797, 798, 809, 810
Prölls, G., 1568, 1570, 1583
Pronik, V.I., 547
Pronin, A.A., 1320, 1321
Prosser, C.F., 379, 395, 859, 872, 892, 893
Protheroe, R.J., 691, 696
Provost, J., 1060–1063
Prudkoglyad, A., 960
Prunet, S., 394, 420, 466
Prusti, T., 636–639, 642, 862, 869, 871, 872
Przybilski, O., 86, 114
Psaltis, D., 818, 821
Ptak, R.L., 547
Ptuskin, V.S., 685, 687, 690–697, 1002
Puetter, R.C., 644
Puget, J.L., 328, 329, 574, 576, 577, 617, 636–638, 641, 644, 868, 869, 871, 872, 1270
Puget, P., 869
Pulley, O.O., 1566, 1582
Puls, J., 892, 901, 902, 909, 910
Purcell, E.M., 607, 612, 638, 642, 1196, 1197
Purcell, J.D., 1085, 1215
Purkins, T., 869
Putney, A., 788
Pye, J.P., 605, 883, 892, 893, 908–910, 1232
Pyle, E.J., 298
Pyle, J.A., 1668
Pyle, K.R., 695, 977, 980, 1003

Qi, C., 638
Qin, L., 757
Quast, G.R., 870
Quast, R., 467
Quataert, E., 371, 855
Quebatte, J., 1269
Quegan, S., 1581, 1582
Queloz, D., 864, 871
Quémerais, E., 1206, 1207, 1209–1211, 1214, 1215
Quenby, J.J., 1002
Querci, F., 909
Quick, J., 855
Quickenden, T.I., 1462, 1477

Quijano-Rico, M., 1403
Quimby, R., 395, 420, 467
Quinn P.J., 546
Quinn, H., 1231
Quinn, J., 696
Quinn, J.M., 1527
Quinn, P.J., 466
Quinn, T., 1242, 1270, 1273, 1292
Quinn, T.J., 1056, 1063
Quirico, E., 1462, 1477

Raab, F.J., 370
Rabben, H.-H., 187, 1526
Rabello-Soares, M.C., 1062
Raberanto, P., 1668
Rabii, B., 394, 420
Rabin, D.M., 1037, 1063 1080, 1086, 1087
Rabinowitz, D., 1233
Rachen, J.P., 693, 696
Radhakrishnan, V., 663, 674, 761, 762, 784, 788
Rädler, K.-S., 1490, 1492
Radocinski, R.G., 1004
Raeder, J., 1232, 1556, 1557
Raesenberg, R.D., 433
Raffelt, G., 1231
Rahe, J., 809, 1188, 1216, 1232, 1233, 1270, 1292, 1293
Raimond, E., 330, 577
Raine, D.J., 809, 820
Rairden, R.L., 177
Raiteri, C.M., 559, 560
Raitt, W.J., 1580
Raizer, Yu.P., 1175, 1188
Rajagopal, M., 731, 732, 744, 756
Ramachandran, R., 814, 821
Ramanamurthy, P.V., 744, 756, 757
Ramaprakash, A.N., 521, 526, 527
Ramaty, R., 143, 150, 252, 434, 436, 439, 695, 696, 910, 1001, 1003, 1159, 1214, 1556
Ramirez-Ruiz, E., 518, 527
Rammensee, W., 1386, 1401, 1402
Ramsay, G., 809
Ramsey, B., 1617
Rand, R.J., 668
Randall, B.A., 177
Randall, C.E., 298, 1293
Randich, S., 883, 892
Rank, D.M., 637
Ranta, H., 1580
Rao, S., 394, 420, 466
Rao, U.R., 977, 1003
Rapisada, M., 754
Rapisarda, M., 524
Rapp, R.H., 1651
Rappaport, S.A., 496, 753, 756, 764, 765, 772, 778, 781–783, 787–789, 810, 812, 814, 821, 822, 824, 827–829, 835–837, 935
Raschke, E., 1668
Rasio, F.A., 748, 756, 764, 765, 772, 781–784, 788
Rasmussen, A.P., 809, 959
Rasmussen, I., 871
Ratckiewicz, R., 1211, 1214
Ratcliffe, J.A., 1559, 1566, 1567, 1580, 1583
Ratke, L., 1272
Ratkiewicz, R., 1211, 1215
Ratner, M.I., 467
Raubertas, R.F., 1321
Rauch, M., 385, 395
Rauer, H., 637, 1274, 1291–1293
Raulin, F., 1430, 1477
Rauschenbakh, B.V., 57, 58
Rawley, L., 547

Rawlings, S.G., 556, 559
Rawson, D.M., 395–397
Ray, R.D., 1647, 1650
Ray, T.R., 862, 870–872
Rayleigh, 1560, 1583
Rayman, M.D., 110, 114
Raymond, J.C., 222, 242, 255, 477, 498, 795, 802, 809, 810, 822, 835, 893, 909, 1062, 1085, 1135, 1138, 1139
Rayner, D.P., 559
Reach, W.T., 329, 611, 636, 642, 657, 673, 674, 1188
Read, S.M., 433, 439, 642, 645
Read, W.G., 1668
Reader, P.D., 113
Readhead, A.C.S., 469, 526
Reading, D.H., 959
Reale, F., 577, 892, 1082, 1087
Reale, M.A., 299
Reames, D.V., 969, 970–974, 979, 986, 987, 994, 1000, 1003, 1004
Reasbeck, P., 1374
Reasenberg, R.D., 342, 352, 439, 1321
Reasoner, D.L., 1582
Rebar, F., 298
Reber, M., 1287, 1292, 1293
Rebhan, H., 1616
Rebka, G.A. Jr., 341, 352
Rebolo, R., 385, 395, 428, 429, 434, 439, 892
Rebull, L., 429, 439
Recondo-Gonzalez, M.C., 547
Rector, T.A., 523
Redfern, R.M., 754, 757, 1001
Reece, J.M., 308
Reed, R., 329
Reedy, R.C., 1402
Rees, D., 1581, 1582
Rees, M.H., 1569, 1583
Rees, M.J., 477, 498, 504, 509, 512, 517, 518, 523, 527, 528, 546, 552–554, 557–559, 737, 786, 823, 827, 833, 836, 854
Reese, Y., 1371, 1373, 1375
Reeves, E.M., 222, 1086
Reeves, H., 426, 427, 430, 432–434, 437–439
Reeves, R., 57, 58
Refdsal, S., 493
Reffert, M., 521
Refregier, A., 466
Refsdal, S., 443, 444, 446, 452, 454, 465, 466, 468, 498
Régulo, C., 1061, 1062
Reichart, D.E., 511, 517, 520, 521, 526, 527
Reiche, K.-U., 1138, 1214
Reichert, G.A., 547
Reid, G.C., 968, 1003
Reid, I.N., 703, 704, 718
Reid, M., 816, 821
Reid, R., 1371
Reidel, D., 1160, 1161, 1649
Reifenstein, E.C.III., 722, 757, 788
Reilly, E.F., 869
Reimer, O., 696, 697, 757
Reimers, D., 385, 395, 468, 882, 892, 906, 907, 909, 910
Reimold, W.U., 1373
Reindl, B., 396
Reinecke, J.P.L., 992, 1003
Reinhard, R., 1004, 1249, 1270–1274, 1293
Reinisch, B.W., 1571, 1583
Reipurth, B., 639
Reisig, G.H.R., 62, 63, 114
Reiss, D., 395, 420, 468, 525
Reiter, J., 1060, 1062
Reitsema, H.J., 1271, 1272, 1274
Rembor, K., 1271, 1272
Rembor, K.-M., 1236, 1271–1273

Rème, H., 1227, 1228, 1230, 1232, 1320, 1321, 1491, 1492
Remillard, R.A., 765, 787, 822, 835, 841, 843, 854–856
Remizov, A.P., 1232
Remsberg, E.E., 1668
Remy, M., 466, 468
Renaud, C., 1061, 1062
Renaud, N., 395
Renda, M., 560
Renk, B., 1231
Renn, J., 442, 443, 468
Renne, P.R., 1371
Rense, W.A., 210, 222
Renzini, A., 439, 488, 498, 576, 632, 642, 910
Reppin, C., 755, 757, 789, 836
Ress, M., 549
Retallack, P., 835
Rettenmund, U., 1231
Rettig, T.W., 1274, 1292, 1293
Reuss, M.K., 1002
Revelle, D.O., 1262, 1274
Rey, M., 1617
Rey, P., 1402, 1403
Reyniers, M., 632, 644
Reynolds, C.S., 370, 371
Reynolds, J.E., 559, 855
Reynolds, J.H., 1334, 1336, 1374, 1597
Reynolds, R.J., 590, 604, 652, 657, 663, 664, 668, 672–674
Reynolds, R.T., 268, 270, 1477
Reynolds, S., 537, 548
Reynolds, S.P., 554, 559, 689, 694, 696
Reynoso, E.M., 597, 605
Rezek, T., 525
Rheinisch, B.W., 1571
Rho, J., 736, 953, 960
Rhoades, C.E., 760, 788
Rhoads, J.E., 467, 523–525
Rhodes, E.J., Jr., 1060, 1062, 1063
Rhodes, J.M., 1402
Rhodes, L., 1038, 1043, 1045–1047
Ribas, I., 394
Ribbe, P.H., 1374
Rich, M., 299
Richard, O., 438
Richards, P.L., 314, 329, 330, 394, 420, 637, 871
Richardson, F.F., 1175, 1187
Richardson, I.G., 1022, 1032
Richardson, J., 821, 1206, 1215
Richardson, J.D., 1000, 1157, 1159, 1160, 1161
Richardson, R.S., 288, 299
Richer, H.B., 392, 395
Richer, J., 438, 439
Richmond, A.D., 1568, 1570, 1583
Richmond, M.W., 547
Richstone, D., 467, 535, 546, 547, 564, 577
Richter, I., 1188
Richter, K., 1225, 1232, 1233, 1244, 1267, 1273
Richtmyer, R.D., 149
Ricker, G.R., 526, 577, 959
Ricketts, M.J., 498, 836, 1232
Ricklefs, R.L., 351, 1650, 1651
Rickman, H., 1219, 1233, 1236, 1253, 1255, 1261, 1262, 1273, 1274, 1291
Ridgway, S.T., 871
Rieck, W., 1001, 1138
Ried, M.J., 870
Rieder, R., 1360, 1371, 1374, 1395–1397, 1401–1403
Riedinger, J.R., 329, 577, 870
Rieger, E., 187, 1526, 1527, 1528
Riegger, J., 299
Rieke, G.H., 305, 330, 1269

Rieke, M., 330, 872
Rieke, M.J., 1269
Ries, J.G., 351, 1650
Riess, A.G., 388, 394, 395, 420, 457, 468, 523, 854
Rietveld, M.T., 1582, 1583
Rigaud, F., 466
Rigaut, F., 1273
Righter, K., 1322, 1374, 1375
Rigopoulou, D., 329, 574, 576, 577
Rihelä, P., 1581
Rikitaki, T., 156, 177
Riley, P., 1136, 1138, 1139, 1151, 1160
Ringwood, A.E., 1351, 1374, 1384, 1385, 1386, 1388, 1389, 1393, 1401, 1402
Rinnert, K., 1576, 1583
Rio, Y., 869
Rishbeth, H., 1567, 1573, 1583
Ritter, H., 764, 780, 782, 787, 788, 791, 810, 813, 821, 822
Ritz, C., 1270
Rivellini, T.P., 1402
Rix, H.-W., 456, 467
Roark, T.P., 643
Robbins, D.E., 1129, 1138, 1139
Roberge, W.G., 329
Robert, T., 869
Roberts, B., 1069, 1087
Roberts, D.H., 560
Roberts, H., 438
Roberts, K.-H., 1481
Roberts, M.D., 960
Roberts, P.H., 247, 255, 1481, 1491, 1492
Roberts, W.O., 220, 222
Robertson, D.S. 453, 468, 560
Robertson, J.G., 468
Robillot, J.-M., 1061, 1062
Robins, M.O., 58, 222
Robinson, C.R., 524, 525, 526, 784, 787, 788, 854
Robinson, D.R.T., 396
Robinson, F.D., 869
Robinson, I.S., 1596, 1616
Robinson, M.S., 1301, 1321
Robinson, R., 298
Roble, R.G., 1583
Robley, R., 202
Robson, A., 329, 577, 870
Robson, E.I., 560
Robson, R., 1582
Roca Cortés, T., 1061–1063
Rocard, F., 1270
Roche, P., 809
Roche, P.F., 1271
Rochester, G.D., 125, 151
Rochus, P., 1086
Rockwood, C.C., 298
Roddier, C., 872
Roddier, F., 867, 872, 1043, 1061
Rodgers, A.J., 696
Rodgers, B., 633, 642
Rodgers, C.D., 1657, 1660, 1668
Rodionov, A.V., 1217, 1220, 1222, 1223, 1232, 1233, 1258, 1269
Rodono, M., 908
Rodrí-Franco, A., 638
Rodriguez Espinosa, J.M., 527, 547, 871
Rodríguez, L.F., 853, 855
Rodríguez-Fernandez, N.J., 638
Rodriguez-Kuiper, E.N., 644
Rodriguez-Pascual, P., 395, 547
Roe, P.L., 1215
Roeder, S.C., 444, 466
Roederer, J.G., 1004, 1492

Roelfsema, P.R., 437, 618, 638, 640, 642, 644, 869, 872, 873
Roellig, T.L., 330, 642, 643
Roelof, E.C., 222, 254, 983, 1001–1004, 1139, 1492, 1555
Roemer, E., 1236, 1273
Roesler, F., 300
Roettiger, K., 496, 498
Roger, A.S., 1584
Rogers, A.E.E., 467
Rogers, C., 632, 641
Rogers, F.J., 731, 754, 1045, 1047, 1060, 1062
Rogerson, J.B., 292, 293, 299, 300, 657, 662, 674
Rogovaya, S.I., 696
Rokke, L., 421
Rokutanda, E., 577
Rol, E., 520, 524, 528
Rolfe, E.J., 909, 1270, 1271, 1273
Rolfs, C., 440
Roll, P.G., 420
Romanek, C.S., 1373
Romanelli, P., 430, 438
Romani, P.N., 1440, 1444, 1448, 1449
Romani, P.R., 1448
Romani, R., 819, 821
Romani, R.W., 728, 729, 731, 744, 756, 783, 788, 853, 855
Romaniello, M., 376, 395
Romanishin, W., 547
Romano, D., 435, 438
Romeiser, R., 1615
Romeo, G., 394, 420, 466
Romer, A., 1326, 1374
Romero, J., 1061
Romishevskii, E.A., 1539, 1557
Romney, J.D., 672
Romoli, M., 222, 1138, 1139
Romoli, R., 1137
Rompolt, B., 253
Ronan, M.T., 1231
Ronca, L.B., 1320
Rönnmark, K., 1551, 1557
Rood, H.J., 475, 480, 498
Rood, R.T., 428, 435, 439
Roos, M., 1231
Roos-Serote, M., 437
Roques, J.P., 523, 820
Ros, C.L., 1113
Rosati, P., 473, 487, 498, 546
Roscoe, H.K., 644
Rose, D., 300
Rose, H.A., 1581
Rose, H.J., 696, 1401
Rose, J.L., 1328, 1374
Rose, L.A., 830, 836
Rose, R., 1232, 1271
Rosema, K.D., 1232, 1271
Rosen, L.P., 642
Rosen, N., 363, 371
Rosen, S.R., 802, 810
Rosenbauer, G., 1541, 1557
Rosenbauer, H., 1004, 1032, 1113, 1119, 1128, 1139, 1140, 1216, 1231, 1233, 1291, 1292, 1556
Rosenberg, L.J., 1231
Rosenberg, R.L., 139, 151, 1140
Rosenberg, W., 1063
Rosenblatt, E.I., 547
Rosenbluth, M.N., 1031
Rosenfeld, L., 607, 643, 721, 756
Rosenqvist, J., 1444, 1449
Rosenthal, D., 620, 621, 636, 642
Rosino, L., 393, 396
Rosner, R., 836, 872, 877, 882, 892–894, 907–910, 1081, 1086, 1088

Ross, C.L., 1001, 1138, 1139
Ross, F.E., 259, 262, 270
Ross, H.E., 85, 114, 85
Ross, M.N., 1322, 1402, 1461, 1477
Ross, P., 1583
Ross, R.R., 852, 855
Rossi, B., 121, 122, 151, 231, 251, 255, 286, 559, 604, 696, 722, 754, 756, 786, 820, 854, 1137
Rosso, C., 908
Rossow, W.B., 437, 1429
Rostopchina, A.N., 870
Roth, H., 249, 1061
Roth, I., 971, 976, 1004, 1527, 1556
Roth, M., 249, 252
Rothenbach, M., 1330, 1372
Rothenberg, A.M., 1402
Rothenflug, R., 495, 959
Rothermel, H., 755
Rothery, D.A., 1476
Rothschild, R.E., 507, 528, 694, 736, 753, 755, 756, 834, 836
Rott, M., 1272
Rottenberg, J., 269
Röttger, J., 1581
Röttger, R., 1563
Rouan, D., 869
Roueff, E., 640
Roukema, B.F., 463, 468
Roumeliotis, G., 1080, 1087
Roush, T.L., 1462, 1476, 1477
Roussel, H., 387, 395
Roux, A., 1491
Rovetti, F., 558, 559
Rowan-Robinson, M., 329, 330, 577, 612, 642
Rowell, G.P., 960
Rowland, F.S., 1660, 1668
Rowlands, D.D., 1322
Rowlands, S., 1332, 1375
Rownlee, B.D.E., 1272
Roxburgh, I.W., 1060, 1061, 1063
Roy, A., 869
Roy, A.E., 1459, 1477
Roy, M., 1649
Roy, N.L., 1188
Rubbert, B., 1273
Rubin, B.C., 786, 820, 834
Rubin, H., 1506, 1527
Rubin, M., 1062
Rubino, A., 1615
Rubinstein, I., 1616
Rucinski, D., 1212, 1215
Rucklidge, A.M., 1054, 1062
Ruderman, M.A., 499, 502, 528, 728, 752, 753, 756, 757, 784–786, 788, 813, 819, 821
Ruderman, M.S., 1213
Rudnick, L., 491, 498
Rudy, R.J., 644
Rüedi, I., 1087, 1088
Ruffert, M., 521, 528
Ruffini, R., 371, 760, 788
Ruhl, J.E., 394, 420, 466
Ruiz-Lapuente, P., 395, 420, 467, 808, 810
Rummel, R., 1650, 1651
Rumpl, W., 641
Rumsey, H., 270
Runcorn, S.K., 1187
Rupen, M.P., 853, 855
Ruppe, H.O., 80, 91, 99, 114
Russ, G.P., 1370
Russel, S.S., 1270
Russell, A.S., 1328, 1374

Russell, C., 1491
Russell, C.T., 966, 983, 1001–1003, 1021, 1022, 1031–1033, 1138, 1159, 1233, 1491, 1492, 1527
Russell, D., 1581
Russell, H.N., 1068, 1087, 1243, 1273
Russell, J., 1322
Russell, J.M., 1662, 1664, 1668
Russell, K.S., 1269
Russell, R.W., 315, 330, 616, 617, 642–644, 871
Russell, S., 871, 1292
Russell, S.S., 640
Russo, A., 150
Russo, A., 57, 58, 120, 221
Rust, D.M., 221, 222, 1032
Rüster, J.A., 1569
Rüster, R., 1581, 1583
Rutherford, E., 677, 697, 1326, 1327, 1374
Rutledge, R.E., 852, 855
Rutten, R.G.M., 524, 525, 528, 808
Rutten, R.J., 1074, 1086, 1087
Ruzhin, Y.Y., 1527
Ruzmaikin, A.A., 687, 697
Ryan, E.V., 869
Ryan, F.D., 367, 371
Ryan, J., 959
Ryan, J.M., 523
Ryan, L., 1004
Ryan, S., 872
Ryan, S.G., 429, 434, 439
Rybchinsky, R.E., 253, 1138
Rybicki, G., 907, 909, 910
Rycroft, M.J., 1187
Ryde, F., 788
Ryde, N., 632, 642
Ryder, G., 1321, 1343, 1348, 1354, 1371–1374
Rydgren, A.E., 299
Ryter, C., 328, 433, 439, 637, 811, 821
Ryu, D., 696
Ryzkov, Y.A., 113
Rzhiga, O.N., 1320

Sa Basu, 1580
Saba, J.L.R., 836, 1062, 1086, 1113
Sabalisck, N., 1272
Sabau-Graziati, L., 576
Sabine, E., 228, 255, 1535, 1557
Sablik, M.J., 1292
Sabo, A.P., 691, 696
Sacco, B., 286, 753, 755, 756, 758
Sachs, R.K., 414, 421
Sackett, P.D., 465, 468
Sackman, I.-J., 436, 439
Sadat, R., 395
Saddlemyer, L., 547
Sadler, D.H., 223
Sadler, E.M., 525, 526
Sadun, A.C., 547
Saeki, T., 696
Safi-Harb, S., 754
Safronov, V.S., 258, 270, 1237, 1264, 1269, 1273, 1281, 1293, 1322
Sagan, C., 262, 270, 534, 548, 1187, 1293, 1430, 1475, 1477, 1478
Sagar, R., 523, 524
Sagawa, E., 1555
Sagdeev, R.Z., 1002, 1215, 1224, 1225, 1233, 1236, 1241, 1253–1255, 1261, 1265, 1273, 1274, 1292, 1461, 1475, 1510, 1511, 1523, 1527
Saglia, R.P., 395
Saha, A., 376–379, 383, 393–396, 397
Saha, M.N., 897, 910
Saha, P., 396, 577
Sahade, J., 910
Sahnow, D.J., 299

Sahr, J.D., 1575, 1583
Sahu, K.C., 465, 511, 523, 524, 528
Sahu, M.S., 427, 439
Saint-Pé, O., 869, 1245, 1273
Saio, H., 702, 719
Saisse, M., 299, 869
Saito, Y., 748, 755, 757, 1032, 1555, 1557
Sakai, S., 221, 376, 393–396, 397
Sako, M., 834, 836
Sako, T., 960
Sakurazawa, K., 960
Salah, J.E., 1568, 1583
Salama, A., 437, 638, 642, 644, 869, 870, 872, 873, 1449
Salamanca, I., 547
Salamon, M.H., 559, 695, 1003
Salaris, M., 395
Sale, R.G., 754
Salinari, P., 330
Salo, H., 1219, 1233
Salomone, M., 1273
Salotti, L., 820
Salpeter, E.E., 549, 559, 604, 672, 722, 753, 759, 788, 811, 818, 820, 821, 836, 1573, 1583
Salter, C.J., 672
Salton, J., 820
Salucci, P., 395
Salvail, J.R., 1262, 1270
Salvati, M., 559
Salzer, J.J., 393, 394
Sam Lone, J., 869
Samain, E., 346, 352, 1630, 1651
Samarasinha, N.H., 1265, 1273
Sambruna, R., 547
Sambruna, R.M., 559, 560
Samland, M., 392, 396
Samuelson, F.W., 696
Samuelson, R.E., 1449
Sanchez, E., 1542, 1557
Sanchez, M., 1061
Sanchez-Bejar, V., 523
Sandage, A., 373, 378–382, 375, 384, 385, 388, 391–397, 531, 532, 547, 548, 701, 702, 718, 787, 788, 811, 820, 821
Sandage, R.R., 759
Sandel, B.R., 299, 1215, 1430, 1448
Sandel, W.R., 1215
Sanders, D.B., 538, 548, 570, 577, 1449
Sanders, W.T., 570, 584, 589, 590, 594, 603–605, 908
Sanderson, T.R., 977, 981, 1002, 1004
Sandford, M.C.W., 298
Sandford, S., 1187
Sandford, S.A., 636, 640, 642, 866, 867, 872
Sandquist, E.L., 777, 778, 788
Sandven, P., 1615
Sandven, S., 1606, 1615, 1616
Sandwell, D.T., 1321
Sanford, M.C.W., 298
Sanford, P.W., 820, 835–837
Sanford, R.F., 190, 202
Sanghellini, L., 935
Sanitt, N., 444, 468
Sanko, N.F., 1270
Sano, S., 754
Santangelo, A., 959
Santiago-Muñoz, E., 1139
Santos, P., 1272
Santos-Lleo, M., 546, 547
Sanuki, T., 696
Sanwal, D., 756, 758
Sanz, H.G., 1370
Saracco, P., 524
Saraceno, P., 627, 628, 636–639, 641–643, 869

Sarajedini, A., 393
Sarazin, C., 474, 477, 489, 497, 498
Sargent, A.I., 629, 630, 636, 640, 643, 865, 869, 871
Sargent, W.L.W., 395, 428, 438, 513, 528, 536, 547, 559
Sari, R., 517, 518, 523, 526, 528
Sarkar, I.C., 1321
Sarkar, S., 999
Sarris, E.I., 1555
Sarris, E.T., 982, 1004
Sartoretti, P., 1477
Sartori, G., 1583
Sasaki, M., 371, 696
Sasseen, T., 675
Sastry, G.N., 480, 498
Sathyaprakash, B.S., 367, 370
Sathyendranath, S., 1602, 1616
Sato, K., 439
Sato, N., 830, 831, 836, 959
Sato, S., 329, 330
Sato, Y., 960
Satoh, T., 1491
Sauer, K., 1233
Sauer, T., 468
Saumon, D., 392, 396
Saunders, R.D.E., 559
Saunders, R.S., 556, 1262, 1273, 1274, 1310, 1320–1323
Saur, J., 1490, 1492
Saust, A.B., 466
Sauvage, M., 577, 637, 869
Sauvageot, J.L., 959
Sauval, A.J., 1068, 1086, 1123, 1138
Sauvaud, J.A., 1232, 1491, 1492
Savage, B.D., 287, 291, 292, 297–299, 605, 609, 636, 637, 654, 655, 658, 659, 664, 665, 667–674, 1215
Savonije, G.J., 764, 768, 786, 788, 789, 824, 835, 836, 910
Sawada, K., 836
Sawchuk, W., 1556
Sawyer, S.R., 547
Sazonov, S.Y., 523, 528
Sazonov, Y.A., 1526, 1555
Scalo, J., 652, 674, 862, 872
Scaramuzzi, F., 394, 420, 466
Scarf, F.L., 247, 254, 1002, 1160, 1188, 1215
Scargle, J.D., 527
Scarsi, L., 725, 286, 753
Scearce, C.S., 253, 254, 1557
Schaack, D., 329
Schaaf, R., 728, 755
Schaber, G.G., 1310, 1320, 1322
Schachter, J.F., 559
Schade, D., 452, 467
Schaefer, B.E., 395, 420, 467, 505, 523, 528
Schaefer, K.G., 935
Schaeffer, O.A., 1348, 1374, 1643
Schaeidt, S.G., 437, 638, 869, 873
Schaifers, K., 1402
Scharf, C.A., 474, 496, 497
Scharlemann, E.T., 728, 752
Scharmer, G.B., 1066, 1087
Schatten, K.H., 1492
Schatzman, A., 225, 234, 249, 255
Schatzman, E., 255, 437
Scheick, X., 1062
Schein, M., 122, 151, 678, 697
Scheinen, N.B., 1373
Schenk, P.M., 1247, 1249, 1273, 1472, 1477
Scherb, F., 248, 251, 255, 1137
Scherrer, P.H., 1043, 1046, 1047, 1060–1063, 1085, 1161
Scherrer, R.J., 433, 439
Schiaparelli, G., 1236, 1273, 1406

Schieder, R., 636, 639, 641, 642
Schiff, L.I., 344, 352
Schiffer III, F.H., 298
Schilke, P., 640, 645
Schiller, H., 1602, 1616
Schiminovich, D., 299, 755
Schinder, P.D., 1477
Schindler, K., 251, 1032, 1543, 1557
Schindler, R.H., 1231
Schindler, S., 496, 497
Schinnerer, E., 548
Schirmer, A.F., 560
Schklovsky, J.S., 1196
Schlegel, D.J., 375, 396, 587, 588, 591, 592, 605
Schlegel, E.M., 959
Schlegel, K., 1563, 1567–1569, 1581–1584
Schleicher, D.G., 1265, 1269, 1273, 1291
Schleuning, 639
Schlickeiser, R., 552, 558
Schlike, P., 1293
Schloerb, F.P., 640, 1292
Schlosser, W., 1265, 1273
Schluessel, P., 1615, 1616
Schlüter, A., 240, 255, 1506, 1527
Schmadebeck, R.L., 528, 1401
Schmahl, E.J., 1086
Schmeider, B., 1073
Schmerling, E.R., 1575, 1583
Schmid, J., 1130, 1139
Schmider, F.-X., 1062
Schmidt, B.P., 395, 420, 468, 525, 528
Schmidt, E., 810
Schmidt, G.C., 1326, 1374
Schmidt, G.D., 810
Schmidt, H.U., 187, 786, 1232, 1233, 1271, 1272, 1230
Schmidt, K., 1272
Schmidt, M., 443, 454, 458, 468, 504, 521, 528, 530, 540, 547, 548, 559, 702, 718, 757
Schmidt, R., 468
Schmidt, W., 641, 1214, 1215
Schmidt, W.K.H., 528, 1003, 1271, 1272
Schmidt-Kaler, T., 1265, 1271
Schmidtke, P.C., 808
Schmieder, B., 1073, 1087
Schmitt, B., 1270, 1477
Schmitt, J., 1062, 1232
Schmitt, J.H.M.M., 594, 605, 871, 879–883, 885, 888, 889, 891–894, 908, 909
Schmitt, M., 1231
Schmitt, R.A., 1380, 1402, 1403
Schmutzler, T., 603
Schnabel, L., 1401
Schneeberger, D.M., 1321
Schneid, E., 696, 755
Schneider, D.P., 530, 548, 559
Schneider, E., 694, 1188
Schneider, G., 330, 865, 872, 873
Schneider, J., 468
Schneider, J.P., 201
Schneider, M., 1650
Schneider, P., 395, 445, 456–458, 460, 465, 466, 468
Schnetzler, C.C., 1380, 1374, 1402
Schnopper, H., 871
Schoenfelder, V., 959
Schofield, J.T., 1402
Schohn, C., 935
Scholer, M., 643, 977, 978, 981, 1002–1004, 1022, 1025–1027, 1032, 1138, 1214, 1215, 1506, 1509, 1512, 1517, 1527, 1528
Scholten, J., 604
Schommer, R.A., 394, 395, 420, 468, 810
Schönberner, D., 935

Schonfeld, E., 1373
Schönfelder, V., 755, 959
Schou, J., 1050–1052, 1054, 1060–1063, 1086
Schramm, D.N., 430–432, 438–440, 524, 694, 695, 1231
Schramm, T., 467
Schreier, E., 497, 577, 754, 757, 760, 770, 788, 789, 811, 820, 821, 823, 831, 832, 835, 836
Schrijver, C.J., 234, 242, 255, 910, 1028, 1032, 1069, 1077, 1078, 1082, 1083, 1085, 1087, 1113, 1601
Schrijver, J., 809
Schröder, A., 379, 381, 396
Schroeder, W., 151
Schubert, G., 1303, 1304, 1306–1310, 1312, 1313, 1320, 1322, 1400, 1402, 1458, 1477, 1491
Schubert, J., 871
Schuecker, P., 496, 497
Schühle, U., 1086–1088
Schulte, W., 1232, 1292, 1293
Schultz, R.A., 1321
Schulz, B., 466, 577, 871, 1269, 1444, 1449
Schulz, M., 1481, 1490, 1492, 1516, 1526
Schulz, N.S., 758
Schulz, R., 1265, 1273
Schulze, H., 1284, 1293
Schumacher, E., 1331, 1374
Schunk, R.W., 1570, 1580, 1583
Schüssler, M., 1086
Schuster, A., 230, 255, 1568, 1583
Schuster, K., 638
Schutte, W.A., 613, 623, 624, 631, 637–640, 642, 644, 689, 872, 873
Schutz, B.E., 1650
Schwab, A.P., 1373
Schwadron, N.A., 988, 996, 1001, 1004, 1135, 1137, 1140, 1159, 1233
Schwaller, H., 1401
Schwandt, C., 1373
Schwartz, D.A., 477, 478, 497, 577, 822
Schwartz, P.R., 1661, 1668
Schwartz, S.J., 1031
Schwarz, G., 1231, 1256, 1274
Schwarz, H.E., 917, 935
Schwarz, J., 497, 576
Schwarz, K.P., 1649
Schwarz, R., 525, 757, 809
Schwarz, U., 437
Schwarzmüller, J., 1401
Schwarzschild, M., 234, 255, 328
Schwehm, G., 113, 1188, 1232, 1270, 1272, 1293
Schweizer, F., 576
Schwengeler, H., 395
Schwenn, R., 1004, 1113, 1119, 1120, 1137–1141, 1145, 1151, 1159, 1160, 1188, 1231, 1291
Schwope, A.D., 736, 756, 757
Sciama, D.W., 583, 604
Sciambi, R.K., 1001, 1004
Scime, E.E., 254, 1137, 1140, 1160, 1128, 1133
Sciortino, S., 891–894
Sckopke, N., 1003, 1032, 1033, 1527, 1541, 1557
Scott D., 1231
Scott, D.H., 1322, 1375
Scott, D.M., 786, 820, 834
Scott, D.W., 1651
Scott, K., 1062
Scott, P.F., 787, 855, 1581
Scott, R.F., 1273
Scotti, J.V., 1248, 1274
Scoville, N.Z., 642
Scowen, P.A., 869
Scudder, J.D., 1025, 1032, 1080, 1087, 1139
Seaquist, E.R., 835
Seares, F.H., 253

Seaton, M.J., 51, 58, 201, 674
Sebesta, J.F., 1584
Sebo, K.M., 394, 395
Seboldt, W., 113
Sedlmayer, E., 908
Sedov, L.I., 942, 960
Seeber, G., 1650
Seek, J.B., 254, 1557
Segreto, A., 755
Segura, M., 1429, 1477
Sehnal, L., 1632, 1651
Seifert, S., 1402
Seiff, A., 1402
Seige, P., 1271, 1272
Seina, A., 1616
Seitz, S., 466
Seitzer, P., 1477
Sekanina, Z., 1187, 1232, 1236, 1237, 1241, 1247, 1259, 1264, 1266, 1270, 1272, 1274, 1275, 1293
Sekido, Y., 151
Sekiguchi, K., 547
Sekii, T., 1062
Sekii, T., 1060, 1063
Selesnick, R.S., 993, 1002–1004, 1555
Selivanos, A.M., 275
Selivanov, A.S., 1321
Sellgren, K., 611, 641, 642
Selvelli, P.L., 298
Sembach, K.R., 299, 601, 602, 604, 605, 655, 658, 662, 669, 670, 672, 674
Sembay, S., 959
Sembroski, G.H., 696
Sempere, M.J., 637
Sen, H.K., 1571, 1583
Senay, M., 1272, 1292, 1293
Senior, C., 1582
Senske, D.A., 1322
Sentman, D., 1570, 1583
Seo, E.S., 687, 696, 697
Serber, R., 122, 123, 125, 151
Sergeev, S.G., 547
Serio, S., 893, 894, 1003
Serkowski, K., 793, 809
Serlemitsos, P.J., 475, 478, 497, 498, 577, 754, 820, 822, 833, 834, 836, 959
Serra, G., 328, 329, 636–639, 641, 869
Serrano, A., 576
Serra-Ricart, M., 935, 1272
Setti, G., 396, 543, 548, 811, 821
Settle, M., 1373
Setty, C.S.G.K., 1583
Seuss, S.T., 254
Severino, G., 1086
Sevre, F., 870
Seward, F.D., 577, 724, 733, 736, 754, 757, 801, 872, 893, 899, 910
Seyfert, C.K., 548
Sfeir, D.M., 592, 595, 597, 599, 603, 605, 1193, 1216
Sforna, D., 394, 420, 466
Shafer, R.A., 328, 329, 420, 498, 554, 560, 636, 642, 645, 893
Shafranov, V.D., 1506, 1527
Shaham, J., 752, 786, 817, 820, 821
Shahbaz, T., 812, 821
Shakura, N.I., 532, 548, 549, 558, 559, 794, 810, 823, 836, 843, 855
Shalina, E.V., 1616
Shanklin, J.D., 1668
Shanks, T., 396, 466, 546
Shapiro P.R., 605
Shapiro, E.P.R., 577
Shapiro, I.C., 1168, 1188
Shapiro, I.I., 342, 351, 352, 466, 467, 560, 1269, 1271
Shapiro, M.M., 123, 150

Shapiro, P.R., 601, 665, 672, 674
Shapiro, S., 351
Shapiro, S.L., 771, 788, 812, 821, 828, 836
Shapiro, V.D., 1002, 1215, 1229, 1233, 1273
Shapley, A., 575, 576
Shapley, H., 648, 674
Shapovalova, A.I., 547
Shara, M.M., 807, 809, 810
Share, G.H., 504, 523, 528
Sharova, V.A., 1491
Sharp, R.D., 1231, 1557
Sharp, R.P., 270
Sharpe, M.R., 57, 58, 114
Sharples, R.M., 498
Shashkina, V.P., 1320
Shaver, P.A., 466, 468, 497, 546, 548
Shaviv, O., 810
Shay, M.A., 1025, 1026, 1032
Shaya, E.J., 660, 673
Shea, M.A., 1000, 1161
Shearer, A., 744, 757
Shearer, C.K., 1321, 1374
Sheeley, N.R., 969, 1000, 1004, 1063, 1087, 1113, 1122, 1134, 1135, 1138, 1140, 1151, 1161
Sheldon, R.B., 1001
Shelley, E.G., 1231, 1291, 1292, 1526, 1536, 1545, 1546, 1557
Shelton, H., 1188
Shelton, R.L., 602, 605
Shelton, W., 57, 58
Shelus, P.J., 351, 1650, 1651
Shemansky, D.E., 1214, 1430, 1448, 1469, 1477
Sheminova, V.A., 1086
Shenton, C.B., 222
Shenton, D.B., 222, 298
Shepherd, D.S., 526
Shepherd, G.G., 190, 202, 1582
Shepherd, L.R., 114
Sheppard, D.C., 755
Sherrer, P.H., 1086
Shervais, J.W., 1347, 1374, 1375
Shestakova, L.I., 1264, 1274
Shevchenko, V.I., 1273
Shibai, H., 330, 615, 640–642, 866, 871
Shibanov, Y.A., 730–732, 755–758
Shibata, K., 1028–1030, 1032
Shibata, M., 371
Shibata, R., 528
Shibata, S., 754, 756, 757
Shibazaki, N., 756, 821, 840, 842, 845, 853, 855
Shields, G.A., 536, 548
Shields, J.C., 547
Shigeyama, T., 525
Shih, C.-Y., 1338, 1340–1342, 1344, 1347, 1350, 1354, 1355, 1357, 1358, 1363, 1367, 1402
Shikaze, Y., 696
Shilepsky, A., 603
Shima, M., 429, 439
Shimizu, M., 1224, 1233
Shimizu, T., 234, 255
Shimmins, A.J., 530, 547
Shine, R.A., 255, 1032, 1085, 1087, 1088, 1113
Shing, L., 1086
Shinn, D.H., 1580
Shinohara, I., 1557
Shipley, A., 546
Shipman, H., 299
Shipman, R.F., 328
Shirai, H., 1004
Shirley, J.H., 1323
Shivanandan, K., 311, 330, 639

Shklovsky, I.S., 650, 651, 674, 759, 788, 811, 821, 823, 836, 840, 855, 935, 1527, 1556, 1581
Shklovsky, J.S., 1216
Shlesinger, M.F., 1519, 1527
Shlyapnikov, A., 523
Shoemaker, C.S., 1247, 1248, 1274
Shoemaker, D., 370
Shoemaker, E.M., 261, 270, 1274, 1470, 1477, 1478
Shore, S.N., 298, 909, 1215
Shoshany, Y., 1262, 1274
Shoub, E.C., 1080, 1087
Showalter, D.L., 1402
Showalter, M.R., 1187, 1449, 1477
Showman, A.P., 1490, 1492
Shrader, C., 547, 560, 854
Shrock, R.E., 1231
Shu, F., 857–859, 871, 872
Shuchman, R.A., 1616
Shuchman, S., 1615
Shukurov, A.M., 697
Shul'man, L.M., 1262, 1274
Shull, J.M., 298, 299, 547, 657, 662, 674
Shulman, S., 604
Shuman, F.G.D., 642
Shuman, S., 1004
Shure, M.A., 315, 330
Siarkowski, M., 879, 893
Sibeck, D.G., 1542, 1549, 1556, 1557
Sibille, F., 636, 637, 869, 871, 872
Sicardy, B., 1477
Siddiqi, A.A., 57, 58
Siddique, N., 1188
Sidher, S.D., 636–638, 641, 645, 869
Sidorenko, A.I., 1320
Siebenmorgen, R., 637, 644, 869, 870, 873
Sieber, W., 526
Siedentopf, H., 235, 243, 251
Siegmund, O.H.W., 222, 299, 1087, 1138, 1139
Sienkiewicz, R., 781, 788
Siewers, L., 370
Sigad, Y., 387, 396
Sigl, G., 692, 695
Signer, P., 437, 1137, 1329, 1375, 1378, 1402
Sigurdsson, S., 523, 527
Sijm, N., 869
Sikora, M., 552, 559, 560
Silber, M., 329
Silberberg, R., 687, 697
Silbermann, N.A., 379, 393–397, 547
Silk, J., 385, 394, 491, 497, 498, 857, 872
Sill, G.T., 262, 270, 1462, 1478
Sillanpaeae, A., 560
Silva, R.W., 1116, 1140
Silvaggio, P., 1465, 1476
Silverberg, R.F., 329, 419–421, 642, 645, 697, 1188
Silverstone, M.D., 873
Sime, D.G., 1138
Simnett, G.M., 972, 1002, 1004, 1087, 1138, 1140, 1159, 1160
Simon, A., 1563, 1583
Simon, G.W., 1037, 1038, 1062, 1063, 1068, 1087
Simon, M., 638, 687, 697
Simon, T., 862, 869, 872, 882, 883, 892, 893, 908
Simonson, E.A., 428, 438
Simonson, S.C., 639
Simpson, J.A., 125–130, 133–137, 139, 142, 144, 145, 148–151, 231, 232, 250, 252, 253, 255, 679, 689, 695, 697, 967, 969, 972, 983, 985, 999–1005, 1118, 1140, 1142, 1160, 1214, 1215, 1224, 1225, 1233, 1483, 1492
Sinclair, M.W., 855
Singer, R.B., 1321, 1402

Singer, S., 1556
Singh, K.P., 603, 886, 893
Singh, S., 1004
Sinton, W.M., 260, 270
Sion, E.M., 798, 809, 810
Sipior, M.S., 935
Siren, C., 1003
Sirou, F., 869
Siscoe, G.L., 992, 1002, 1004, 1032, 1131, 1140, 1159, 1542, 1549, 1557
Sitdikov, A., 820
Sitko, M.L., 547, 629, 642, 870
Sitte, K., 1175, 1187
Sittler, E., 1137
Sizemore, K.O., 298
Sjoekvist, M., 642
Sjogren, W.L., 1311, 1321, 1322
Sjöstrand, T., 1231
Skadron, G., 694, 999
Skalsky, A., 1233
Skibo, J.G., 854
Skilling, J., 687, 697
Skillman, D.R., 547
Skilman, T.L., 253
Skinner, C., 572, 577
Skinner, C.J., 636–640, 641
Skinner, G.K., 829, 834, 836
Skinner, S.L., 891, 892
Skobeltzyn, D., 123, 151
Skorov, Yu.V., 1262, 1274
Skoug, R.M., 1139
Skrutskie, M.F., 870, 1478
Skrypnik, G.I., 1320
Skumanich, A., 905, 910
Slade, M.A., 1232, 1271, 1301, 1321, 1402
Slattery J.C., 1175, 1188
Slattery, W.L., 1320, 1401
Slavin, J.A., 1160, 1557
Slavin, J.D., 594, 605, 673
Sleep, N.H., 1307, 1322
Slettebak, A., 837
Sloan, R.K., 270
Slonaker, R., 270
Smail, I., 464, 468, 492, 498
Smak, J.I., 544, 548
Smale, A.P., 753, 757, 808, 821, 822
Smarr, L.L., 578, 761, 788
Smart, D.F., 1000, 1161
Smartt, S., 524
Smerd, S.F., 1004
Smette, A., 458, 468, 542, 548, 1271, 1274
Smialkowski, A., 694
Smirnov, G.G., 1402, 1403
Smirnov, V.G., 1233, 1615
Smith E.J., 1159
Smith, A.F., 869
Smith, A.G., 547
Smith, A.M., 298, 672
Smith, B.A., 265, 270, 864, 872, 873, 1273, 1455, 1464, 1478
Smith, B.J., 330
Smith, B.W., 475, 477, 479, 497, 498, 589, 595, 601, 604, 672, 820, 959
Smith, C.R., 1583
Smith, D., 397, 1617, 1650
Smith, D.A., 151, 855
Smith, D.E., 1304, 1306, 1321, 1322, 1651
Smith, D.M., 1591, 1594, 1616
Smith, D.R., 388, 396
Smith, E.J., 149, 151, 252, 253, 1001, 1002, 1004, 1032, 1138, 1139, 1140, 1159–1161, 1492, 1556, 1557
Smith, E.K., 1562, 1583
Smith, E.V.P., 1041, 1063

Smith, F.G., 755
Smith, G.H., 396
Smith, G.L., 1658, 1668
Smith, G.R., 1430
Smith, H.A., 547, 577, 636, 637, 639, 641–643, 869
Smith, I.A., 524, 526, 754, 755, 787, 821
Smith, J.D., 695
Smith, L.F., 903, 910
Smith, L.J., 661, 665, 674, 766, 908
Smith, M.G., 525, 548
Smith, N.R., 1526, 1616
Smith, P., 524
Smith, P.H., 1402, 1464, 1478
Smith, P.L., 222, 1138, 1139
Smith, P.S., 547, 560
Smith, R., 525
Smith, R.C., 395, 420, 468
Smith, R.H., 1402
Smith, R.J., 466
Smith, R.K., 959
Smith, S., 498
Smith, S.D., 1660, 1668
Smith, S.M., 547
Smith, V.V., 425, 429, 430, 434, 437, 439
Smith, W.B., 352, 821
Smith, W.C., 1660
Smith, W.L., 1657, 1668
Smits, D., 855
Smoluchowski, R., 1259, 1274
Smoorenburg, C., 869
Smoot, G.F., 329, 394, 401, 410, 417, 419–421, 642, 645, 1231
Smrekar, S.E., 1322
Smyers, S.D., 643
Smythe, W.D., 1423, 1477
Smythe, W.R., 1330, 1332, 1372, 1375
Sneden, C., 391–393, 396, 397
Snell, R.L., 623, 636, 639, 641, 642
Snider, J.L., 341, 352
Snijder, H., 759
Snijders, M.A.J., 298, 546, 547
Snow, M., 672
Snow, T.D., 837
Snow, T.P., 293, 299, 300, 607, 642, 654, 659, 660, 674, 897, 909, 910
Snowden, S.L., 582, 585–592, 594, 597, 600, 601, 603–605, 756
Snyder, C.W., 177, 248, 254, 255, 896, 910, 1116, 1117, 1123, 1134, 1139, 1140, 1144, 1160, 1296, 1304, 1321–1323, 1375, 1376, 1423
Snyder, G.A., 1338, 1340, 1342, 1344, 1346, 1347, 1351, 1356–1358, 1371, 1375
Snyder, H., 788, 855
Soberman R.K., 1173, 1187
Soberman, G.E., 771, 788
Sobolev, V.V., 895, 910
Sobornov, O.P., 1403
Socker, D.G., 1138, 1140
Soddy, F., 1326, 1375
Soderblom, D.R., 872
Soderblom, L.A., 1322, 1371, 1402, 1423, 1452, 1471, 1476–1478
Sodré, L., 395
Sodroski, T.J., 611, 642
Soffel, M., 352
Soffitta, P., 527, 528
Sofia, S., 1041, 1042, 1063
Sofia, U.J., 299, 674
Sogawa, H., 640, 871
Soifer, B.T., 310–312, 329, 330, 548, 570, 577, 642, 644
Soika, J.J., 1570, 1583
Sokar, N., 935
Soker, D.G., 1087
Soker, N., 828, 835, 836, 935

Sokoloff, D.D., 697
Sokolov, O., 513
Sokolov, V.V., 528, 755
Solanki, S.K., 1068–1070, 1074, 1086–1088
Soldan, J., 518, 525
Soldner, J., 468
Solem, J.C., 1248, 1274
Solheim, B.H., 1582
Solinger, A.B., 497, 498
Solinger, A.G., 479
Sollerman, J., 774, 788
Solomatov, V.V., 1314, 1322
Solomon, P., 829, 835, 860, 871, 897, 910
Solomon, S.C., 1298–1300, 1303, 1309, 1310, 1312, 1314, 1317, 1321, 1322, 1375, 1319, 1358
Soltan, A., 497, 548
Sommer, M., 755
Sommerfeld, A., 1560, 1583
Somogyi, A.J., 1232
Sonett, C.P., 252–254, 977, 1004, 1085, 1131, 1139, 1140, 1159, 1160, 1161, 1386, 1402
Song, I., 868
Song, P., 1033
Song, X.Y., 1086
Songaila, A., 428, 439, 672, 673
Sonneborn, G., 299, 657, 674, 675, 893, 909, 1010, 1011, 1016, 1017, 1019–1021, 1506
Sonnerup, B.U.Ö., 1032, 1033, 1527
Sonnett, C.P., 149, 1158
Sonobe, T., 736, 757, 836, 959
Sonoda, T., 696
Sopata, L.M., 1269
Sorowka, A., 1372
Sotirelis, T., 1555
Soucail, G., 458, 459, 467, 468, 493, 498
Soufflot, A., 869
Southwood, D.J., 177, 1159, 1491
Soutoul, A., 687, 689, 695, 696
Soward, A.M., 255, 1480, 1492
Spaans, M., 622, 642, 643
Spada, G., 496
Spadaro, D., 222, 1138, 1139
Spadaro, S., 1139
Spakman, J., 869
Spanier, S., 1231
Sparke, L.S., 547
Sparks, W.B., 298, 383, 396, 524, 527
Speer, R.J., 218, 222
Speiser, T.W., 1501, 1527, 1537, 1539, 1557
Spencer, D.A., 1402
Spencer, J.H., 835
Spencer, J.R., 1243, 1244, 1462, 1269, 1272, 1477, 1478
Spencer, M., 1571, 1578, 1580
Spencer, N.W., 1583
Spergel, D.N., 433, 439
Spero, R.E., 370
Spettel, B., 1401–1403
Spinoglio, L., 627, 628, 637, 639, 641–643, 869, 872
Spinrad, H., 270, 523, 560
Spite, F., 393, 428, 439
Spite, M., 393, 397, 428, 434, 437–440
Spitzer, L., 287–290, 293, 294, 299, 300, 565, 578, 594, 605, 608, 610, 643, 647–649, 651–654, 657, 660, 665, 669, 671, 674, 675, 895, 910
Spohn, T., 1322, 1402, 1477
Spoon, H.W.W., 329, 576, 577, 638, 869
Sprague, A.L., 1428, 1430
Spreiter, J.R., 246, 255, 1215, 1496, 1539, 1557
Springer, L., 1063
Spruit, H.C., 780, 788, 813, 822, 855
Spudis, P.D., 1297, 1298, 1301, 1322, 1323, 1346, 1353, 1375

Spurny, P., 525
Spyromilio, J., 395, 420, 468, 671
Squires, G., 460, 467, 523
Squyres, S.W., 1304, 1321, 1322, 1472, 1478
Sramek, R.A., 453, 467
Srednicki, M., 1231
Sreekantan, B.V., 788, 821
Sreekumar, P., 525, 695, 696, 755–757
Srinivasan, G., 761, 762, 765, 774, 784–786, 788, 819, 822
Srinivasan, R., 696
Srokosz, M., 1616
Sromovsky, L.A., 1478
St Maurice, J.-P., 1583
St. Cyr, O.C., 1138
St. Germain, K., 1616
St. John, C.E., 1088
St. John, C.R., 1069
St. Maurice, J.-P., 1563, 1569
Stabell, R., 466
Stacey, G.J., 577, 615, 616, 620, 643
Stachel, J., 468
Stacy, J.G., 663, 675
Stadlbauer, T., 959
Staelin, D.H., 722, 757, 759, 788
Stahara, J.R., 1539
Stahara, S.S., 1215, 1557
Stahler, S.W., 859, 872
Stairs, I.H., 362, 363, 371, 755, 756
Stalio, R., 299
Stamnes, K., 1583
Standish, E.M., 371
Stanek, K.Z., 525
Stanev, T., 695, 1231
Stanford, S.A., 498, 560
Stanley, H.R., 1651
Stapelfeldt, K., 630, 643, 860, 869, 872
Staples G., 1617
Starck, J.L., 576, 869
Stark, A.A., 592, 605
Starobinsky, A.A., 414, 421
Starr, R., 525
Starrfield, S., 809
Stathakis, R., 525
Staubach, P., 1188
Staubert, R., 755, 757, 789, 836
Staudacher, T., 1372
Stauffer, J.R., 636, 639, 641, 642, 861, 865, 868, 870–872, 883, 891–893
Stebbins, J., 1457, 1466, 1478
Stebbins, R.T., 1036, 1062
Stecher, T.P., 290, 291, 300, 609, 643
Stecker, F.W., 553, 554, 559
Stecklum, B., 524
Steel, D.I., 1188
Steen, M., 1616
Steenberg, C.D., 995, 1000, 1004
Stefani, F., 1491
Steffen, K., 428, 1593, 1594, 1616
Steffen, M., 437, 1062, 1074, 1088
Stegner, K., 642
Steidel, C.C., 466, 474, 492, 498, 513, 527, 528
Steiger, R.H., 1331, 1333, 1338, 1348, 1375
Steigman, G., 429, 431, 433, 437, 439, 440
Stein, R.F., 1038, 1043, 1058, 1062, 1063, 1070, 1073, 1074, 1086
Stein, W.A., 496, 559, 608, 631, 639, 641, 643, 935
Steinberg, J.T., 1139, 1492, 1557
Steinberg, M., 908
Steinberger, J., 755
Steiner, M., 1074
Steiner, O., 1086, 1088
Steinhardt, P.J., 419, 495

Steinle, H., 959
Steinmayer, M., 869
Steinweg, A.D., 1577, 1583
Steinz, J.A., 329, 577, 870, 1449
Stel, F., 395
Stella, L., 524, 736, 755, 756, 809, 818, 822, 824, 828, 829, 836, 854, 856, 892
Stencel, R.E., 892, 904, 909, 910
Stenflo, J.O., 1086, 1088
Stenholm, B., 935
Stenmark, L., 642
Stenzel, R.L., 1023, 1031, 1033
Stepanov, D., 960
Stephen, E., 524
Stephens, I.B., 1273, 1274
Stern, A.S., 1471, 1478
Stern, D., 523, 560
Stern, D.P., 1481, 1492, 1496, 1527, 1557
Stern, R.A., 299, 593, 604, 605, 877, 883, 888, 891–894, 1086
Stern, S.A., 299, 1240, 1242, 1269, 1270, 1274, 1293, 1433, 1450
Sternberg, A., 329, 576, 617, 643
Sterzik, M.F., 871, 891
Stetson, P.B., 393–397
Stettler, A., 1338, 1340, 1371–1373, 1375, 1401
Stevens G., 1000
Stevens, I.R., 824, 831, 834, 835
Stevens, J., 1271
Stevens, J.A., 560
Stevenson, D.J., 1295, 1296, 1322, 1402, 1461, 1478, 1480, 1490–1492
Stewart, B., 1560, 1583
Stewart, G.C., 484, 496, 835, 837, 855, 959
Stiavelli, M., 396
Stickel, M., 554, 560, 577
Stickland, D., 298
Stier, M.T., 303, 330
Still, M., 102, 114
Still, M.D., 787, 809
Stinebring, D.R., 157, 177, 786
Stirpe, G.M., 547
Stix, T.H., 1571, 1583
St-Louis, N., 909
St-Maurice, J.-P., 1583
Stocke, J.T., 299, 466, 496, 497, 548, 559, 560
Stöcker, J., 187, 1526, 1528
Stocker, R., 300
Stockman, H.S., 800, 810
Stockton, A., 533, 548, 1604
Stoecker, J., 869
Stofan, E.R., 1322
Stoffelen, A., 1604, 1616
Stöffler, D., 1348, 1371, 1402
Stoffregen, W., 187, 1526, 1563, 1581
Stogryn, A., 1588, 1616
Stoker, C.R., 1321, 1402
Stokes, G.M., 675
Stolarik, J.D., 187
Stollberg, M., 786, 820, 834
Stollman, G.M., 817, 822
Stolovy, S.R., 1429
Stompor, R., 394, 420
Stone, E.C., 149, 696, 697, 966, 972–974, 987, 990, 992, 994, 1000–1004, 1156, 1159, 1201, 1212, 1214, 1216, 1555
Stone, E.J., 303, 330
Stone, E.S., 696
Stone, J., 695
Stone, J.M., 496, 810
Stone, R.G., 1549, 1557
Stone, S.S., 1531, 1550
Stoner, R.E., 547
Stooke, P.J., 1251, 1252, 1254, 1274

Storey, J.W.V., 330, 616, 643, 644
Storey, L.R.O., 158, 177, 1570, 1583
Storey, P.J., 935
Störmer, C., 174, 177, 229, 231, 255, 1497, 1527, 1583
Storrs, A.D., 1262, 1274
Story, T.R., 1233
Stoy, R.H., 935
Strachan, L., 222, 1138, 1139
Strafella, F., 639, 641
Strain, C.V., 298
Stranathan, R.K., 1328, 1374
Strangeway, R.J., 1230, 1233, 1526, 1527
Straniero, O., 395
Strassmeier, K.G., 894
Straumann, N., 387, 388, 394, 396
Strauss, M.A., 393, 396, 397
Strauss, R., 1649
Strazzulla, G., 642, 1271
Streander, K., 1063
Strebel, R., 1292
Streitmatter, R., 696
Strickman, M.S., 737, 755, 757
Strimpel, O., 476, 495
Stringfellow, G.S., 756
Strittmatter, P.A., 548
Strobel, D.F., 1429, 1477, 1492
Strohmayer, T., 285, 524, 526, 754, 755, 758, 787, 816, 818, 821, 822
Strom, K.M., 869, 870
Strom, R.G., 270, 524, 525, 528, 1303, 1306, 1320, 1322, 1323, 1423, 1476
Strom, S.E., 869, 870
Stromgren, B., 647, 652, 675
Strong, A.W., 684, 685, 695, 697
Strong, I.B., 499, 502–504, 526, 528, 999, 1001
Strong, J., 328
Strong, K.T., 1002, 1107, 1113, 1161
Stroud, W.G., 198, 202
Stubbe, P., 1564, 1583
Stubbemann, U., 1232, 1292
Stubbs, C., 395, 420, 466, 468, 525
Stucki, K., 1078, 1088
Stüdemann, W., 1004, 1556, 1578, 1584
Stuhlinger, E., 57, 58, 61, 66, 67, 69, 70, 71, 108–110, 113, 114
Stuhlmann, R., 1668
Stumpff, P., 1243, 1269
Sturm, E., 329, 576, 577
Sturner, S.J., 689, 697, 728, 729, 757
Sturrock, P.A., 1013, 1014, 1033, 1080, 1088, 1111–1113, 1134, 1140
Stutzki, J., 609, 616, 639, 643
Su, 1580
Suda, T., 959
Sudan, R.N., 691, 695
Suess, H.E., 1378, 1402
Suess, S.T., 255, 892, 910, 1156, 1157, 1160, 1161, 1191, 1216
Sugawara, T., 754
Sugiura, M., 1139, 1580
Sugiyama, T., 329
Sugizaki, M., 696, 736, 754, 757
Sukhanov, A.L., 1320
Suleiman, R.M., 1138, 1139
Sullivan, J., 300
Sullivan, J.D., 254
Sullivan, R., 1402, 1476, 1477
Sullivan, W., 57, 58
Summerhayes, C., 1616
Summers, A.L., 268, 270
Summers, M.E., 1492
Sumner, T., 605
Sun, W.-H., 547
Sündermann, J., 1403

Sunshine, J.M., 1354, 1375
Suntzeff, N.B., 394, 395, 420, 468
Sunyaev, R.A., 474, 498, 523, 524, 528, 532, 548, 549, 558, 559, 794, 810, 820, 823, 828, 835, 836, 843, 849, 851, 855, 944, 960
Suomi, V.E., 1655, 1668
Surdej, J., 444, 452, 454, 455, 456, 457, 461, 462, 466, 467, 468, 1271, 465,
Suri, A.N., 1000
Surkov, Y.A., 1311, 1322, 1395, 1396, 1398, 1399, 1402, 1403
Sussman, G.J., 370
Susukita, R., 960
Suszkewicz, E., 787
Sutantyo, W., 761, 779, 780, 783, 788
Sutherland, P.G., 728, 729, 754, 756
Sutherland, W., 466
Suto, H., 640, 871
Sutton, S., 1273
Suvanto, K., 1556
Suzuki, A., 959, 960
Suzuki, J., 696
Suzuki, K., 836, 855
Suzuki, M., 1231
Suzuki, N., 440
Suzuki, R., 960
Suzuki, S., 960
Suzuki, T., 525
Suzuki, T.-K., 437, 438
Svendsen, E.A., 1592, 1594, 1616
Svensson, R., 523, 546, 553, 560
Svestka, J., 673, 1188
Švestka, Z., 222, 234, 253, 255, 969, 1004, 1581
Swallow, E., 1270
Swanenburg, B.N., 284, 286, 755
Swank, J.H., 498, 753, 758, 816, 820, 822, 830, 834, 836, 837, 856, 879, 893, 899, 908
Swann, G.A., 1375
Swanson, P.N., 644
Swartz, D.A., 758
Swartz, K., 836
Swartz, M.M., 222
Swarup, G., 547
Swaters, R., 524
Sweet, P.A., 234, 255, 1008, 1009, 1011–1013, 1033, 1113, 1506, 1507, 1528
Sweetnam, D.N., 1449
Swift, C.T., 1616
Swindle, T., 1348, 1375
Swings, J-P., 468
Swings, P., 607, 643, 1281, 1292
Swinyard, B.M., 636–638, 641, 869
Swordy, S.P., 694, 1001
Sykes, M.V., 1181, 1188, 1266, 1269, 1274, 1430, 1457, 1478
Sylvester, R.J., 631, 636, 637, 643
Sylvester, B., 222
Sylwester, J., 222
Synnott, S.P., 1478
Syrovatskii, S.I., 552, 559, 678, 683, 695
Szabo, A., 1160, 1214, 1492
Szalai, S., 1233, 1274
Szalay, A., 299
Szatjno, M., 822
Szczepanski, J., 619, 643
Szczerba, R., 642
Szegö, K., 1213, 1217, 1218, 1221, 1226, 1230, 1232, 1233, 1251, 1253, 1271, 1273, 1274
Szemerey, I., 1232
Szendro, S., 1232
Szkody, P., 802, 809, 810
Szymanski, J., 523
Szymkowiak, A.E., 481, 497, 836, 959

Taam, R.E., 762, 764, 768, 777, 778, 782, 786–788, 821, 828, 836
Taam, R.W., 757
Tacconi, L., 329, 548, 576
Tacconi-Garman, L.E., 329, 576
Tackley, P.J., 1322
Tadhunter, C.N., 546, 548
Taffarello, L., 371
Tagliaferri, G., 559, 891, 892
Tagoshi, H., 371
Tahshiro, M., 491
Takahashi, E., 1320, 1322
Takahashi, K., 959, 1556
Takahashi, M., 748, 749, 757
Takalo, L.O., 560
Takano, M., 564, 578
Takano, S., 484, 498, 959
Takano, T., 1087
Takashima, T., 994, 1004
Takeda, H., 1373
Takeda, M., 692, 697
Takehana, N., 1004
Takeshima, T., 756, 960
Takita, M., 959
Talavera, A., 862, 869, 870, 872
Talent, D.L., 673
Talmadge, C., 369, 371
Talon, R., 524
Tambovtseva, L.V., 870, 1264, 1274
Tammann, G.A., 376, 378, 379, 381, 388, 393–397, 516, 528
Tamura, K., 745, 755
Tamura, T., 496, 497, 576, 577, 960
Tanabashi, M., 1231
Tanabé, T., 330, 640, 871
Tanaka, K., 696, 745
Tanaka, K.L., 1304, 1306, 1310, 1320, 1322, 1359, 1362, 1364–1366, 1375
Tanaka, M., 330, 637
Tanaka, T., 371, 1616
Tanaka, Y., 485, 495, 498, 535, 539, 548, 585, 589, 604, 605, 725, 757, 758, 765, 788, 814, 822, 835, 836, 840, 842, 845, 847, 849, 851–855, 958–960
Tananbaum, H., 285, 496, 497, 548, 577, 723, 737, 754, 757, 760, 788, 789, 809, 818, 820–822, 835, 836
Tanberg-Hanssen, E., 1100, 1113
Taniguchi, Y., 536, 547
Tanimori, T., 689, 697, 952, 959, 960
Tanré D., 1601, 1616
Tanvir, N.R., 377, 396, 524, 525, 527, 528
Tanzi, E.G., 559
Tapia, S., 793, 800, 810, 1272
Tapin, S.J., 1138
Tarasov, A.F., 1526, 1555
Tarbell, T.D., 1032, 1062, 1063, 1085–1088
Tardiff, A., 329
Tarenghi, M., 395
Tarle, G., 694
Tarter, C.B., 831, 836
Tashiro, M., 498, 577, 855
Tatrallyay, M., 1232
Tatsumoto, M., 1341, 1373, 1374
Tatsy, L.P., 1403
Tauber, J.A., 576, 641
Tauris, T., 761, 764, 782, 787, 789
Tavani, M., 524, 527, 728, 752, 754, 821
Tawara, Y., 498, 835, 892, 959
Tawb, J.A., 1598, 1615
Tayler, R.J., 438
Taylor A., 1188
Taylor, A.R., 637, 837, 893, 911
Taylor, B.G., 329, 577, 722, 724, 726, 757, 856, 870

Taylor, D.A., 1185, 1188
Taylor, D.J., 753
Taylor, F.W., 1413, 1423, 1668
Taylor, G.B., 469, 524
Taylor, G.J., 1322, 1373, 1375, 1376, 1402, 1403
Taylor, H.A., 1567, 1583
Taylor, H.P., 1401
Taylor, J.H., 157, 177, 285, 286, 360–362, 371, 372, 668, 675, 740, 755, 757, 761, 766, 786, 787
Taylor, L.A., 1371, 1374, 1375, 1379, 1388–1390, 1403
Taylor, M.J., 753, 754
Taylor, S.R., 1296, 1299, 1316, 1318, 1322, 1323, 1350, 1352, 1353, 1356, 1357, 1375, 1403
Taylor, W., 672
Teegarden, B.J., 150, 523, 524, 528, 972, 1003, 1004, 1215
Teerikorpi, P., 380, 382, 393, 395–397
Tegier, S.C., 1294
Tegmark, M., 410, 421
Telesco, C.M., 629, 643, 864, 870–872, 1269
Teller, E., 131, 149
Telljohann, U., 1061
Telting, J., 524, 525, 528, 902, 910
Temerin, M., 971, 976, 1004, 1526, 1555, 1556
Tennant, A.F., 758
Tenorio, L., 419, 420, 421
Teplitz, H.I., 526
Tera, F., 1322, 1338, 1347, 1349, 1371, 1372, 1375
Tera, T., 1370
Terada, K., 854
Teraesranta, H., 560
Terasawa, N., 439
Terasawa, R.G., 1540
Terasawa, T., 1232, 1517, 1528, 1557
Terekhov, O., 504, 523, 524, 528
Teresawa, T., 1032
Tereshchenko, E.D., 1582
Terlevich, R.J., 393, 438, 516, 528, 538, 548
Terrell, N.J., 823, 836
Terrile, R.J., 864, 872, 873, 1445, 1478
Terzian, Y., 604, 651, 672, 924, 935
Teschke, F., 1403
Testermann, L., 1074, 1085
Testud, J., 1568, 1583
Teuber, D.L., 352
Teukolsky, S.A., 771, 788, 812, 821
Texier, D., 636, 637, 638, 641, 869
Thacker, R., 1403
Thaddeus, P., 604, 644
Thatte, N., 329, 576
Thayer, M.R., 695, 1004
Thé, P.S., 869–872, 911
Théado, S., 436, 440
Theile, B., 1527
Thejll, P., 1214
Theureau, G., 382, 393, 395, 397
Thi, W.F., 617, 630, 643
Thiel, K., 1218, 1233
Thiele, U., 547
Thieleman, F.K., 393, 394, 430, 440
Thielemann, F., 960
Thim, F., 396
Thirring, H., 340, 351
Thoburn, J.A., 439
Tholen, D.J., 1243, 1270, 1271, 1273, 1274
Thomas, B.T., 1148, 1154, 1160, 1161
Thomas, G.E., 200, 202, 1195, 1198, 1199, 1216
Thomas, H., 959
Thomas, H.C., 786, 808
Thomas, J.A., 329
Thomas, J.O., 1583
Thomas, K., 1284, 1293

Thomas, M.L., 1232, 1271
Thomas, N., 1232, 1244, 1247, 1253–1255, 1257, 1258, 1272, 1274, 1402
Thomas, P.C., 1188
Thomas, P.G., 1303, 1304, 1306, 1321, 1322
Thomas, P.J., 1293
Thomas, R.J., 1087
Thomas, R.N., 608, 637, 1062
Thomas, S., 1617
Thomas, V.A., 1528
Thomas-Keprta, K.L., 1373
Thompson, A., 689, 695, 697, 1003
Thompson, A.R., 673
Thompson, C., 508, 524, 526, 723, 725, 741, 742, 744, 745, 757, 758, 785, 789, 819, 822
Thompson, D., 835
Thompson, D.J., 560, 696, 754–757
Thompson, D.T., 1273
Thompson, F.C., 1332, 1375
Thompson, G.I., 298
Thompson, I.B., 855
Thompson, M.J., 1048–1063
Thompson, R., 560
Thompson, R.I., 323, 330, 872
Thompson, W.R., 1477
Thomsen, B., 393, 523
Thomsen, M.F., 177, 1031, 1033, 1231, 1582
Thomson, J.J., 1326, 1375
Thomson, N.R., 1581
Thorburn, J., 439
Thorne, K.S., 357, 359, 365, 370, 371
Thorne, R.M., 1010, 1011, 1032, 1527
Thornley, M., 577
Thornton, D.D., 637
Thornton, G.J., 960
Thorsett, S.E., 371, 524, 526, 527
Thorstensen, J.R., 525, 808, 834, 835
Thronson, H.A., 643, 673
Thuan, T.X., 428, 440, 576
Thuillier, G., 1060
Thurman, S.W., 1402
Thurston, M.R., 892
Tidman, D.A., 235, 254
Tielens, A.G.G.M., 610, 613–618, 623–626, 631, 636–644, 675, 869, 871–873, 1241, 1269, 1270, 1274, 1293
Tilanus, R.P.J., 524, 1449
Tilav, S., 695
Tilgner, C., 871
Tilton, G.R., 1370, 1374
Timmermann, R., 617, 636, 643, 644
Timofeev, G.A., 1004
Timothy, A.F., 222, 254, 1139
Timothy, J.G., 300, 1086, 1087, 1138
Timothy, P., 299
Tindo, I.P., 215, 222, 223
Tingay, S.J., 853, 855
Tinney, C., 525, 1066
Tiphene, D., 869
Titarchuk, L.G., 822, 849, 850, 855
Titheridge, J.E., 1571, 1583
Title, A.M., 255, 1032, 1062, 1063, 1068, 1085–1088, 1113
Tittemore, W.C., 1472, 1477, 1478
Titus, R., 202
Tjin A Djie, H.R.E., 862, 869, 872
Tkachenko, A.Y., 523
Tkachenko, N.P., 1231
Tobar, M.E., 370
Tobias, S.M., 1058, 1063
Tobiska, W.K., 1430
Tokarz, S., 547
Tokunaga, A.T., 642, 643, 870, 1244, 1271, 1274, 1292

Tollestrup, E., 525
Tolls, V., 636, 639, 641, 642
Tomandl, D., 1187
Tomaney, A.B., 466
Tomasch, A.D., 694
Tomasko, M.G., 1402
Tombaugh, C.W., 1454
Tomczyk, S., 1046, 1060, 1063
Tommasi, E., 639, 641, 642
Tondello, G., 222, 1138, 1139
Toneri, T., 960
Tonks, W.B., 1367, 1375
Tonry, J., 394, 395, 420, 468
Toomre, J., 1060, 1062, 1063, 1087
Toon, O.B., 1423
Topka, K., 496, 577, 754, 872, 893, 909, 910, 1088
Toptygin, I.N., 690, 695, 977, 1004
Torbert, R.B., 1527, 1557, 1583
Torbet, E., 694
Torii, K., 696, 736, 757, 954, 960
Tornikoski, M., 560
Törnqvist, N.A., 1231
Torre, C., 1651
Torre, J.M., 352
Torres, C.A.O., 870
Torroni, V., 527
Tosi, M., 395, 435, 437–440
Tossman, B., 1002, 1004
Tosti, G., 560
Totani, T., 520, 528
Toth, I, 1232, 1236, 1244, 1245, 1247, 1272, 1273
Totsuka, Y., 959
Tousey, R., 206–208, 223, 298, 1068, 1085, 1088
Toussaint, F., 892
Tout, C., 718
Toutain, T., 1060, 1061
Towlson, W.A., 298, 637, 638, 641, 869
Townes, C.H., 330, 637, 643, 644
Townsend, J., 195, 202
Tozzi, G.P., 1272, 1292
Tozzi, P., 546
Trafton, L., 1425, 1430
Trafton, L.M., 298, 672
Trager, G.B., 1401
Trainor, J.H., 150, 1003, 1004, 1160, 1215
Trams, N.R., 636–639, 641, 643, 644, 869–870, 873
Tran, D., 577, 869
Tran, K.-V., 467
Traniguchi, Y., 536
Trapero, J., 661, 675
Trattner, K.J., 1002
Traub, W.A., 303, 330
Trauger, J.T., 869
Treffers, R., 1430
Trefftz, E., 1236, 1269
Treiman, A.H., 1240, 1321, 1358, 1361, 1362, 1375
Treiman, S., 255
Treiman, S.B., 151
Tremaine, S., 467, 497, 523, 546, 547, 1240, 1270, 1271, 1273, 1292
Tremonti, C., 438
Tretyakova, Ch.A., 999
Treuhaft, R.N., 453, 468
Treumann, R.A., 1022, 1033, 1500, 1503, 1504, 1506, 1508, 1519, 1526–1528, 1556
Treves, A., 559, 560, 748, 757, 787, 809, 836
Trexel, H., 1004
Trilling, D.E., 864, 872
Trimble, V., 500, 528
Trinchieri, G., 483, 498, 567, 575–578
Tripp, T.M., 674, 1477
Trippe, T.G., 1231

Trishchenko, A.P., 1668
Tritton, K.P., 548
Trivedi, S.S., 1000
Troland, T.H., 673
Trombka, J.I., 504, 525, 527, 528, 1401, 1402
Trotter, D.E., 222
Truemper, J., 547, 576, 605, 785, 786, 802, 810, 833, 908, 1215
Trujillo, C., 1271, 1272
Trümper, J., 100, 114, 485, 498, 585, 605, 724, 725, 732, 733, 736, 740, 742, 746–748, 750, 753, 755–758, 770, 789, 822, 833, 836, 851, 855, 959, 1232
Trumpler, R.J., 607, 643
Truong-Bach, R.J., 633, 636, 641, 643
Truran, A., 391, 392
Truran, J.W., 393, 397
Tsai, C., 1000
Tsao, C.H., 688, 697
Tsatoulis, C., 1616
Tsou, P., 1187
Tsuchiyama, A., 640, 871
Tsuda, T., 1582
Tsunemi, H., 497, 498, 757, 835, 836, 893, 942, 947, 948, 955, 959, 960
Tsuneta, S., 234, 255, 893, 1003, 1033, 1087
Tsunoda, R.T., 1564, 1584
Tsuru, T., 577, 888, 893, 960
Tsuruda, T.K., 1555
Tsuruta, S., 729, 752, 754, 757, 1028
Tsurutani, B.T., 1002, 1133, 1140, 1159, 1232, 1233, 1557
Tsurutani, T., 1531, 1549, 1550
Tsusaka, Y., 577
Tsusaki, D.M., 1401
Tsvetanov, Z.I., 547, 548
Tsygan, A.I., 822
Tu, C.-Y., 1133, 1140
Tuchiyama, A., 640
Tuck, A.F., 1668
Tucker, W.H., 479, 496, 498, 576, 584, 604, 724, 757, 836
Tuffs, R., 871
Tufte, S.L., 604, 674
Tufts, B.R., 1476, 1477
Tukahara, M., 754
Tulinov, G.F., 199, 202
Tull, R.G., 1430
Tumer, T., 754, 755
Tums, E., 966, 1001, 1004, 1138
Tuohy, I.R., 724, 741, 757, 822, 836
Turatto, M., 526, 528
Turck-Chièze, S., 1057, 1060, 1061, 1063
Turco, R.P., 1423
Turcotte, D.L., 1314, 1315, 1322, 1650
Turcotte, P., 560
Turkevich, A.L., 274, 275, 1374, 1378, 1402, 1403
Turneaure, J.P., 351
Turner, A., 377, 393–397, 868
Turner, B.E., 575, 578, 626, 644
Turner, E.L., 393, 452, 457, 466, 467, 468
Turner, G., 1333, 1338, 1347–1349, 1353, 1360, 1371, 1373, 1375
Turner, J., 870
Turner, M.J., 836
Turner, M.J.L., 498, 893, 959
Turner, M.S., 393, 419, 474, 498
Turner, T.J., 534, 547, 548, 855
Turnrose, B.E., 298
Turnshek, C., 548
Turnshek, D.A., 547
Turok, N., 392, 393
Turolla, R., 757
Turpie, K., 328, 636
Turtle, E.P., 1477

Turunen, E., 1580
Turyshev, S., 351
Tussolino, A.J., 151
Tutukov, A.V., 760, 789
Tuve, M.A., 1559, 1561, 1571, 1580, 1584
Tuzzolino, A.J., 137, 1233
Twicken, J.D., 1477
Twigg, L., 1063
Tylenda, R., 935
Tyler, G.L., 1322
Tylka, A.J., 999, 1003
Tyrén, F., 217, 222
Tyson, J.A., 459, 460, 466, 468, 469, 493, 498
Tyson, T., 397
Tytler, D., 393, 428, 437, 440
Tyuflin, Yu.S., 1320
Tzioumis, A.K., 559, 855

Uberoi, M.S., 251, 255
Uchida, Y., 893, 1087
Udomprasert, P., 526
Udry, S., 872
Ueda, I., 696
Ueda, Y., 528, 535, 548, 577, 855, 1032
Ueno, S., 892
Ugras, N.G., 645
Uitenbroek, H., 1074, 1087
Ulaby, F.T., 1590, 1616
Ullberg, C., 642
Ulmer, A., 754
Ulmer, M., 603, 733, 757, 823, 837
Ulmer, M.P., 497, 528
Ulmschneider, P., 894, 1069, 1073, 1082, 1086–1088
Ulrich, M.-H., 546, 560
Ulrich, R.K., 1038, 1043, 1060–1063
Umeda, H., 394, 525
Unavane, M., 714, 718
Underhill, A.B., 298, 895, 910
Underwood, J.H., 893
Unger, S., 641
Unger, S.J., 636–638, 641, 869
Ungerechts, H., 604
Ungstrup, E., 1231, 1232
Uno, S., 528, 577, 855
Unruh, D.M., 1373
Unsöld, A., 243, 255, 650, 673
Unti, T.W.J., 1131, 1140
Unwin, J.J., 1580
Unwin, S.C., 560
Urch, I.H., 249, 255
Urey, H., 258, 261, 270, 272, 275, 1329, 1338, 1375, 1378, 1402, 1403
Urizicker, A.E., 285
Urnov, A.M., 222
Urpin, V., 785, 789
Urry, C.M., 540, 547, 548, 554, 559, 560
Usher, P.D., 820
Ushimaru, N., 835
Utlaut, W.F., 1564, 1584
Utterback, N.G., 1188, 1293
Uyama, K., 329
Uznov, A.M., 222

Vacanti, G., 559
Vachon, P.W., 1609, 1616, 1617
Vagnetti, F., 540, 547
Vaiana, G.S., 223, 254, 860, 872, 876, 892–894, 905, 908–910, 1003, 1027, 1031, 1033, 1081, 1088
Vainstein, L.A., 222
Vaisberg, O.L., 1224, 1225, 1233

Vakulov, P.V., 1004
Vala, M., 619, 643
Valageas, P., 491, 498
Valana, G.S., 908
Valdes, F., 468
Valencia, G., 1231
Valentijn, E.A., 437, 571, 578, 638–640, 644, 869, 870, 873
Valentini, G., 523
Valenzuela, A., 1524, 1528
Vales, N., 1650
Valeur, H., 1616
Vali, H., 1373
Vallarta, M.S., 122, 127, 150
Vallée, J.P., 687, 697
Vallerga, J.V., 594, 603, 605, 667, 672, 675, 908, 1210, 1214, 1216
Valnicek, B., 222, 223
Valtaoja, E., 560
van Agthoven, H., 869
van Albada, T.J., 702, 718
Van Allen, B.T., 1533
van Allen, J.A., 57, 58, 130, 150, 151, 159, 167, 170, 174, 177, 1496, 1497, 1528, 1533, 1557, 1569, 1584
Van Ballegooijen, A.A., 255, 1085, 1113, 1138
van Bibber, K., 1231
van Citters, G.W., 753, 754, 1430
van Cleve, J.E., 871
Van de Hulst, H.C., 210, 223, 233, 243, 247, 255, 613, 644, 651, 675
van del Klis, M., 820
van den Ancker, M.E., 626–630, 641, 643, 644, 863, 865–867, 869, 870–873
van den Bergh, S., 516, 528, 702, 719
van den Heuvel, E.P.J., 528, 558, 576, 748, 752, 753, 757, 760–762, 764, 765, 767, 769–778, 780, 782, 784–789, 793, 807–812, 814, 821, 822, 828, 829, 835–837, 854, 855, 869, 871, 872, 911
van den Klis, M., 789
van den Oord, G.H.J., 910, 1087
van der Bliek, N.R., 644, 873
Van der Hooft, F., 753
van der Hucht, K.A., 294, 299, 642, 644, 869, 903, 911
van der Hulst, T., 869
van der Klis, M., 757, 762, 789, 813, 814, 816–822, 833, 837
van der Kolff, G., 1615
van der Kruit, P.C., 702, 719, 833
van der Lei, S., 869
van der Linden, R., 869
van der Meulen, R.D., 959
van der Raay, H.B., 1060, 1061
van der Steene, G., 525, 528
van der Tak, F.F.T., 626, 627, 644
van der Werf, P.P., 571, 578
van Dessel, E.L., 835, 1086
van Dijkhuizen, C., 869
van Dishoeck, E.F., 607, 609–611, 616, 617, 622, 625–627, 636–645, 673, 863, 869, 870, 872, 873
van Dokkum, P.G., 467, 492, 498, 526
van Duinen, R.J., 330, 577, 642, 935
van Eyken, A.P., 1573, 1583, 1584
van Gorkom, J.H., 548, 559
van Heijnsbergen, D., 644
van Hollebeke, M.A.I., 985, 986, 1002, 1004, 1160
Van Howe, L., 396
van Kerckhoven, C., 619, 640, 643, 644
van Kerkwijk, M.H., 736, 748, 754, 755, 757, 761, 762, 767, 768, 778, 789, 828, 829, 832, 833, 837
van Langren, M.F., 271
Van Loon, H., 199, 202
van Loon, J.T., 644, 871, 873
Van Nes, P., 977, 1004
van Nguyen, D., 869

van Paradijs, J., 499, 507, 509, 511, 522–528, 576, 736, 742, 753–758, 761, 762, 764, 765, 778, 782, 786–789, 812–814, 816, 819–822, 837, 844, 845, 853–856, 787, 789
van Patten, R.A., 351
Van Speybroeck, L.P., 220, 223, 562, 577, 578, 754, 1033
van Waerbeke, L., 395, 468, 460
van Winckel, H., 632, 641, 644, 665, 867, 871–873
van Woerden, H., 665, 675
van Zadelhoff, G.J., 643
Van, J.T.T., 576
Van, R., 959
Vân, T.T., 1270
Vandehei, T., 466
vanden Bout, P.A., 835, 1430
Vandenberg, D.A., 391, 393
Vandenberg, N.R., 560
Vandenbussche, B., 437, 625, 638, 641, 644, 869, 871–873
Vanderriest, C., 444, 457, 466, 468
Vanderspek, R.K., 525, 526
Vangioni-Flam, E., 434, 435, 437–440
Vanhouten, C., 300
Vaniman, D., 1296, 1322, 1323
Vanyan, L.L., 1491
Vanzani, V., 1477
Vanzi, L., 524
Vapillon, L., 1429
Vardya, M.S., 631, 644
Varendorff, M., 959
Varga, A., 1232
Varghese, P., 114
Varhalmi, L., 1233, 1274
Vartanian, M.H., 951, 952, 960
Vasavada, A.R., 1430
Vasilevskis, J., 1402
Vasisht, G., 736, 754, 757
Vasiyev, B.N., 222
Vasseur, G., 1568
Vassiliev, V.V., 696
Vasyliunas, V.M., 992, 1004, 1009, 1010, 1033, 1491, 1492, 1528
Vauclair, S., 428, 428, 436, 440
Vaughan, A.H., 881, 891, 892
Vaughan, B.A., 786, 814, 820, 822, 834
Vaughan, P.A., 298
Vaughan, R.M., 1402
Vedder, J.F., 1175, 1188
Vedder, P.W., 675, 908, 956, 960
Vedovato, G., 371
Vedrenne, G., 523, 524, 528, 819, 820
Veeder, G.J., 1244, 1274, 1476
Vegard, L., 189, 202
Veiga, R.E., 1668
Veillet, C., 351, 352, 1630, 1650, 1651
Veilleux, S., 525, 547, 548
Veis, G., 1649, 1651
Vèlez, R.A., 1402
Vellante, M., 1131, 1140
Velusamy, T., 640
Veneziano, G., 370, 371
Venis, T.E., 298
Venkatavaradan, V.S., 999
Venkatesan, D., 1556
Venkateswaran, S., 1215
Vennes, S., 1214
Ventura, G., 869
Ventura, J., 756–758, 812, 822
Verbunt, F., 748, 754, 755–757, 761, 771, 778, 780, 783, 788, 789, 812, 813, 820–822, 854, 856, 884, 893
Verigin, M.I., 1138, 1214, 1232, 1233
Vermeulen, R.C., 469, 560
Vernazza, J.E., 1068, 1070, 1073, 1082, 1086, 1088

Vernon, M.J., 1371, 1401
Vernov, S.N., 1004
Verschuur, G.L., 651, 665, 675
Verstraete, L., 618, 641, 644
Verter, F., 605
Vervack, R.J. Jr, 1430
Vessot, R.F.C., 341, 352
Vestergaard, M., 523
Vette, J.I., 177
Vetto, J.I., 163–165
Veverka, J., 1188, 1247, 1270, 1426, 1430
Vial, J.-C., 1086, 1087
Viangoni-Flam, E., 439
Vick, C.P., 86, 114
Vidal, J.L., 860, 871
Vidal, N.V., 823, 837
Vidal-Madjar, A., 299, 427, 428, 438, 440, 604, 657, 673–675, 869–871, 1215
Viegas, S.M., 428, 440
Vietri, M., 524, 693, 697, 818, 822
Viggiano, A.A., 1577, 1580
Vignes, D., 1320, 1321, 1481, 1491, 1492
Vigouroux, G., 352, 1651
Vigroux, L., 577, 869
Vikhlinin, A.A., 523, 577, 693, 820
Vilas, F., 166, 177, 1301, 1303, 1322, 1323, 1400–1403, 1423, 1491
Vilcsek, E., 1403
Vilenkin, A., 693, 694
Vilhu, O., 753, 883, 893
Villa, G., 475
Villa, I.M., 1371
Villa, J.G., 498
Villain, J., 57, 58
Villante, F.L., 385, 394
Villata, M., 559, 560
Villumsen, J.V., 466
Vincent, P., 1651
Vincent, R.A., 1569, 1581
Viner, M.R., 835
Vink, J.S., 901, 910, 959
Vinokur, M., 1215
Viola, V.E., 433, 439, 440
Viollier, M., 1668
Virdi, T.S., 1569, 1581, 1584
Viscuso, P.J., 643
Visser, H., 869
Visser, M., 368, 369, 371
Visvanatan, N., 787
Vitagliano, H.D., 298
Vitale, S., 371
Vitry, R., 870
Vittorio, N., 420, 466
Vityazev, A.V., 1303, 1322
Vivares, F., 869
Vladilo, G., 673
Vladimirova, G.A., 1232
Vlahos, L., 1087
Vnuchkov, C.A., 1491
Voelk, H., 871
Vogel, J.D., 1372
Vogel, P., 1231
Voges, W., 495–497, 605, 757, 789, 836, 891, 908, 1139
Vogt, H.H., 1402
Vogt, N., 809
Vogt, N.P., 394
Vogt, R., 678, 696
Vogt, R.E., 370, 1000
Voigt, G.-H., 1482, 1488, 1492
Voigt, H., 1492
Voit, G.M., 496

Völk, H., 187, 1188
Völk, H.J., 184, 690, 695–697, 1293
Volk, K.M., 640
Volk, S.J., 757
Volkoff, G., 721, 756, 788
Volkov, A.M., 1615
Volland, H., 1570, 1583, 1584
Vollmer, O., 150, 1001
von Braun, W., 57, 58, 64, 67–69, 71, 80, 82, 85, 114
von der Luhe, O., 1138
von Duinen, R.J., 935
von Helden, G., 632, 634, 644
von Helmholtz, H., 1280, 1293
von Hippel, T., 525
Von Humboldt, A., 255
von Maanen, A., 253
von Montigny, C., 696, 755
von Rosenvinge, T.T., 150, 696, 697, 966, 992, 1000, 1002–1004, 1555
von Soldner, J., 442
von Steiger, R., 395, 437–439, 986, 987, 1001–1002, 1004, 1086–1088, 1125–1127, 1134, 1135, 1137–1141, 1156, 1158–1161, 1214
von Weizsäcker, C.F., 1332, 1375
von Winckel, H., 935
von Zahn, U., 1429, 1430, 1578, 1583, 1584
Vonder Haar, T.H., 1655, 1668
Voors, R.H.M., 631, 641, 644, 871
Vorontsov, S.V., 1060
Voshage, H., 434, 440, 1402
Voss, L., 109, 114
Voss, R., 1231
Voughan, A.E., 755
Vrba, F., 754, 872
Vreeswijk, P.M., 465, 524–526, 528
Vriend, W.-J., 612, 613, 644
Vrtilek, S., 812, 817, 822
Vsekhsvyatskii, S.K., 1238, 1274
Vuillemin, A., 1060
Vuks, M.F., 189, 202
Vynckier, C., 864, 872

Waddington, C.J., 695
Wade, C., 787
Wade, C.M., 760, 840, 855
Wade, G.A., 909
Wade, N., 1584
Wade, R.A., 799, 808, 809, 810, 869
Wadhwa, M., 1360, 1375
Wadsley, J., 385, 397
Waelkens, B., 636
Waelkens, C., 630, 641, 643, 644, 866, 869, 871–873
Wagner, C.A., 1651
Wagner, R.M., 547, 548
Wagner, S.J., 547, 557, 560
Wagoner, R.V., 358, 366, 367, 371, 430, 440
Wahlgren, G., 298
Wahlund, J.-E., 1547, 1557
Waite, J.H. Jr., 1448, 1555–1557, 1582
Wakita, H., 1380, 1402, 1403
Wakker, B.P., 665, 675
Waldmeier, M., 220, 223, 242, 255
Waldron, W.L., 899, 908, 910
Walker R.J., 1491
Walker, A.B.C., 216, 223
Walker, A.R., 376, 397
Walker, B.M., 1403
Walker, D., 1344, 1375
Walker, D.D., 671
Walker, H., 871
Walker, H.J., 636, 637, 641
Walker, H.-P., 1556

Walker, M.A., 559, 755, 786, 787, 789
Walker, M.F., 792, 793, 810
Walker, R.G., 313, 330, 577, 868
Walker, R.J., 177, 1371, 1491, 1492, 1544, 1549, 1557
Walker, R.M., 866, 872
Walker, T.P., 430–434, 440, 695
Wallace, P.T., 755
Wallace, R.A., 673
Wallerstein, G., 673
Walling, E., 1330, 1372
Wallington, S., 468, 456
Wallis, M.K., 1199, 1200, 1216, 1232, 1261, 1274
Wallner, C., 604
Wallyn, P., 753
Walpole, P.H., 1002, 1492
Walraven, T., 1043, 1062
Walsh, D., 444, 454, 469
Walsh, J.R., 202
Walt, M., 1498
Walter, C.M., 1428, 1430
Walter, F., 523
Walter, F.M., 298, 672, 725, 736, 747, 757, 884, 893
Walters, H.B., 57, 58
Waltman, E.B., 855
Walton, C.C., 1616
Walton, N., 395, 420, 467
Wambacher, H., 123, 125, 149
Wambsganss, J., 458, 465, 467–469
Wampler, E.J., 439
Wamsteker, W., 395, 547, 672, 872, 908, 935, 1461, 1478
Wand, R.H., 1568, 1583, 1584
Wanders, I., 547
Wang, C., 1157, 1158, 1161, 1216
Wang, D., 1138, 1140
Wang, F.Y.-H., 724, 733, 734, 743, 744, 746, 754, 757
Wang, L., 516, 525, 528
Wang, M., 1601, 1615
Wang, M.-S., 1372
Wang, P., 891
Wang, P.H., 1668
Wang, Q.D., 594, 601, 602, 605, 754, 755, 757
Wang, T., 547
Wang, Y., 467
Wang, Y.C., 1206
Wang, Y.-M., 828, 834, 837, 1083, 1087, 1088, 1134, 1137, 1139, 1140, 1151, 1157, 1161
Wang, Z., 177, 636, 639, 641, 642
Wang, Z.-R., 757
Wänke, H., 1321, 1363, 1372, 1374, 1376, 1378–1389, 1393, 1394, 1397, 1398, 1401–1403,
Wannier, P.G., 642
Ward, A.K., 298
Ward, A.W., 1402
Ward, D.B., 315, 330
Ward, M.J., 537, 546, 548, 835
Ward, W.R., 1237, 1270, 1380, 1386, 1401
Wardle, J.F.E., 550, 556, 560
Ward-Thompson, D., 862, 868, 873
Wargau, W., 809
Wark, R.M., 526
Warnecke, J., 1491
Warner, B, 753, 792, 793, 795–797, 801, 802, 804, 808, 810, 813, 822, 1430
Warner, J.L., 1321
Warner, J.W., 373, 395, 719
Warner, P., 1584
Warren, H.P., 1032, 1085
Warren, M.S., 393
Warren, P.H., 1340, 1341, 1344, 1346, 1375, 1376, 1385, 1389, 1403

Warrick, R.A., 1598, 1617
Warwick, J.W., 1441, 1450, 1533, 1557
Warwick, R., 837
Wasilewski, P., 1320, 1321, 1491, 1492
Wasserburg, G.J., 1297, 1322, 1331–1334, 1336–1338, 1341, 1343, 1348, 1353, 1355, 1370–1372, 1374–1376
Wasserman, I., 820
Wasserman, L.H., 1269, 1430
Wasson, J.T., 1341, 1344, 1346, 1362, 1376, 1385, 1389, 1391, 1403
Watabe, T., 330
Watanabe, S., 1555, 1584
Watanabe, T., 1087
Waters, J.R., 788, 820, 821
Waters, J.W., 610, 621, 631, 1661, 1663, 1668
Waters, L.B.F.M., 610, 631, 637, 640, 641, 643, 644, 768, 789, 828, 829, 837, 866, 867, 869–873, 901, 902, 905, 910, 911
Waterscheid, R., 1582
Watkins, G., 1584
Watson, A.M., 869
Watson, D.M., 315, 330, 620, 643, 644
Watson, M.G., 547, 805, 809, 810, 826, 831, 833, 835, 837
Watts, J.W., 352
Waxman, E., 515, 516, 518, 525, 527, 528, 693, 697
Way, S.H., 1430
Wayland, J.R., 1001
Wayte, R.C., 650, 675
Wdowczyk, J.J., 694
Weaver, H.A., 299, 1247, 1272, 1274, 1286, 1287, 1292, 1293
Weaver, H.F., 328
Webb, G.M., 1216
Webb, J.K., 439
Webb, J.R., 560
Webber, J.C., 467
Webber, W.R, 149, 150, 434, 437, 440, 683, 688, 695–697, 1000, 1003, 1160, 1214–1216
Webbink, R.F., 761, 778, 782, 787, 789, 809
Weber, E.J., 249, 255, 902, 905, 911
Weber, F., 725, 757
Weber, H.W., 1402
Weber, L., 1556
Weber, M.A., 752, 757, 856
Webster, B.L., 760, 789, 811, 822, 823, 837, 840, 841, 856
Webster, R.L., 533, 548
Weckwerth, G., 1402
Weedman, D.W., 537, 547, 548
Weedman, R.L., 533
Weekes, T.C., 554, 558, 696
Wegmann, R., 1233
Wegner, G., 393, 394, 497
Wehinger, P.A., 1293
Wehrle, A.E., 551, 560
Wehrli, C., 1061
Wei, J., 523
Weidenbeck, M., 696
Weidenschilling, S.J., 270, 1237, 1238, 1241, 1272, 1274, 1477
Weidlich, U., 1273
Weidner, J.R., 1401
Weigel, A., 1372
Weigert, A., 761, 772, 775, 787
Weiland, J., 718, 1214
Weiland, J.L., 329, 642, 1188
Weimer, D.R., 1031
Weinberg, D.H., 395, 496
Weinberg, J.L., 1164, 1176, 1188, 1189
Weinberger, A.J., 865, 873
Weinreb, S., 607, 644
Weinstein, S.A., 1319, 1322
Weintraub, D.A., 870
Weisberg, J.M., 361, 362, 372, 672, 753
Weiser, H., 222

Weisman, P.R., 1423
Weiss, A., 393
Weiss, A.A., 1004
Weiss, L., 1159
Weiss, L.A., 1138, 1152, 1160, 1161
Weiss, R., 314, 329, 330, 370, 419–421, 645
Weiss, W., 1001, 1581
Weisskopf, M.C., 733, 758, 836
Weissman, P.R., 1232, 1238, 1247, 1261, 1268, 1270, 1271, 1273, 1274, 1289, 1293, 1323, 1429, 1448, 1449
Weissman, P.W., 1433, 1449
Weistrop, D., 300
Weitz C., 1321
Weitzel, N., 871
Welch, D., 466
Welch, W.J., 637, 672
Weldon, R.J., 1401
Weller, C.S., 1200, 1216
Wellgate, G.B., 755
Wells, D.C., 1430
Wells, H.W., 1562, 1564, 1580
Wells, L.D., 793, 810
Wells, M., 871
Wells, W.K., 1430
Welsh, B.Y., 299, 605, 663, 672, 675, 908, 1216
Welsh, W.F., 547
Welter, G.L., 834, 837
Welty, A.D., 756
Welty, D.E., 650, 655–657, 666, 667, 670, 673, 675
Wen, Z., 871, 872
Wenk, R.A., 468
Wensink, H., 1615
Wensink, J., 869
Wentz, F.J., 1597, 1598, 1617
Wenzel, K.-P., 255, 695, 697, 967, 1002, 1004, 1120, 1140, 1158–1160
Wenzel, W., 525
Werner, M.W., 575, 637, 640, 872
Werner, W., 299
Wescott, E., 1570, 1583
Wescott, E.M., 185, 187
Wesselius, P., 642, 644
Wesselius, P.R., 330, 577, 637, 638, 642–644, 868, 869, 872, 873, 935
West R.M., 1271
West, D.K., 298
West, K.A., 669, 674
West, R., 1271, 1272
West, R.A., 1437, 1440, 1450
West, R.G., 593, 605, 959, 1232
West, R.M., 1243, 1244, 1260, 1269, 1271, 1274
Westerlund, B.E., 872
Westin, J., 391, 392, 397
Westphal, J.A., 394, 469, 869
Wetherill, G.W., 1262, 1274, 1331, 1333, 1349, 1362, 1366, 1370, 1376, 1386, 1391, 1403
Wex, N., 372, 755
Wexler, H., 201
Weyman, R., 255
Weymann, R., 246, 896, 911, 1197, 1216
Weymann, R.J., 298, 468, 469, 547, 548, 672
Whalen, B., 1581
Whalen, B.A., 1555, 1578, 1584
Whang, Y.C., 247, 255, 1206, 1212, 1216, 1492
Wharton, L.E., 1583
Wharton, R.A. Jr., 1321
Wheatley, J.M., 672
Wheatley, P.J., 796, 810
Wheaton, W.A., 502, 528, 837
Wheelan, J., 788
Wheeler, J.A., 371, 754
Wheeler, J.C., 394, 516, 525, 528, 855

Wheeler, S.J., 1060
Whipple, A.L., 351, 1650
Whipple, E., 1527
Whipple, E.C., 1232
Whipple, F., 258, 270, 1179, 1189
Whipple, F.L., 1218, 1233, 1236, 1262, 1264, 1269, 1271, 1272, 1274, 1275, 1279, 1281, 1293
Whitaker, E.A., 1247, 1273
Whitcomb, S.E., 370
White, D.A., 473, 492, 498
White, G., 636
White, G.J., 637, 639, 641, 642, 643, 869
White, N.E., 736, 755, 758, 813, 814, 818, 822, 824, 826, 829–831, 834–837, 846–848, 852–854, 856, 879, 880, 886, 891, 893, 894, 959
White, O.R., 1062, 1073, 1085
White, R., 870
White, R.A., 496
White, R.E., 490, 496, 498, 666, 675
White, R.J. III, 547
White, R.L., 892, 899, 910
White, S.D.M., 393, 420, 474, 477, 479, 481, 489, 497, 498
White, S.M., 892
White, T.R., 855
White, W.A., 222
Whited, L.J., 1293
Whitehead, J.D., 1562, 1584
Whitehouse, D.R., 755
Whitelock, P., 395
Whitelock, P.A., 835
Whitney, A.R., 551, 560
Whitney, C.A., 934
Whittet, D.C.B., 329, 610, 623, 637–640, 642–644, 863, 869, 872, 873, 959, 960
Wiant, J.R., 351, 1650, 1651
Wichmann, R., 809, 871, 892, 893
Wick, G.A., 1615
Wickramasinghe, D.T., 612, 613, 644, 800, 810
Widing, K.G., 891, 1134, 1140
Wiedemann, V., 904, 911
Wiedenbeck, M.E., 689, 696, 697, 1002, 1004
Wielebinski, R., 372, 526, 754, 755
Wieler, R., 1124, 1140
Wieprecht, E., 638, 869
Wieringa, M.H., 526, 559
Wiescher, M., 440
Wiesmann, H., 1371–1375, 1402
Wiezorrek, E., 869
Wijers, R.A.M.J., 509, 512, 513, 515, 516, 518, 520, 521, 523–528, 820, 821
Wijnands, R., 762, 789, 814, 818, 820, 822
Wijnbergen, J.J., 869, 1078, 1080
Wikstøl, Ø., 1088
Wilcke, J.C., 226, 255
Wilcots, E.M., 673
Wilcox, J.M., 151, 250, 255, 1031, 1063, 1118, 1139, 1140, 1144, 1147, 1149, 1158, 1160, 1161
Wild, J.P., 968, 969, 1004
Wild, N., 1031, 1033
Wildeman, K.J., 869
Wildeman, T.R., 1373
Wildt, R., 230, 255, 262, 270, 1066, 1088, 1425, 1430
Wilhelm, K., 1032, 1078, 1083, 1086–1088, 1271, 1272, 1274
Wilhelms, D.E., 261, 270, 1297, 1320, 1322, 1323, 1337, 1350, 1353, 1354, 1376
Wilhjelm, J., 1031
Wilken, B., 1001, 1002, 1004, 1137–1140, 1214, 1492, 1556, 1578, 1584
Wilkening, L.L., 866, 867, 873, 1274
Wilker, B., 1578
Wilkerson, T.D., 1124, 1139
Wilkes, B.J., 546, 547, 548
Wilkes, M.V., 1580
Wilking, B.A., 858, 871
Wilkinson, A., 393
Wilkinson, D.T., 329, 419–421, 645
Wilkinson, E., 299
Wilkinson, G., 386
Wilkinson, M.I., 393, 397
Wilkinson, P.N., 457, 458, 467, 469
Will, C.M., 344, 345, 352–354, 358, 363, 364, 367, 368, 370–372
Will, G., 1491
Willems, F.J., 631, 644
Williams, A., 465
Williams, C.L., 527
Williams, C.R., 1320
Williams, D.A., 610, 613, 617, 619, 638, 639
Williams, D.J., 983, 1004, 1477, 1489, 1492, 1501, 1526, 1528, 1534, 1543, 1550, 1555–1557
Williams, G., 150, 835, 1238, 1275
Williams, G.V., 1476
Williams, I.S., 1373
Williams, Ip., 1274
Williams, J.G., 346, 351, 352, 1650
Williams, L.L., 1211, 1215, 1216
Williams, M.A., 1427, 1429
Williams, O.R., 523
Williams, P.J.S., 1556, 1557, 1568, 1569, 1573, 1580, 1581, 1584
Williams, P.M., 903, 911
Williams, R.E., 468, 548
Williams, S.D., 1062
Williamson, F.O., 584, 605
Willick, J.A., 387, 397
Willingale, R., 605, 833, 837
Willis, A.J., 299, 674, 834, 900, 903, 908, 911
Willis, J.J., 907, 1321
Willis, J.P., 1381, 1403
Willner, S.P., 525, 546, 617, 623, 642, 644
Wills, B.J., 536, 547, 548
Wills, D., 547, 548
Wills, F.O., 352
Wills, R.D., 286, 723, 755, 758
Willson, L., 1298
Willson, L.A., 904, 907, 911
Willson, R.C., 1041, 1048, 1061, 1063, 1070, 1088
Willson, R.L., 1062
Wilson, A., 868, 870, 1087
Wilson, A.S., 540, 546, 548
Wilson, B.G., 1062
Wilson, C.A., 786, 820, 834
Wilson, C.T.R., 123, 151, 677, 697
Wilson, G., 388, 397
Wilson, G.R., 1402
Wilson, H.W., 1331, 1373
Wilson, L., 907, 1321
Wilson, O.C., 649, 675, 736, 911
Wilson, P.D., 1475, 1478
Wilson, R., 222, 223, 287, 295, 296, 298, 300, 454, 467, 674, 1434, 1448
Wilson, R.B., 523–525, 527, 786, 820, 834
Wilson, R.H., 400, 420
Wilson, R.W., 605, 607, 644
Wilson, T.J., 1291
Wilson, T.L., 439, 1291
Wilson, W.J., 1661, 1668
Wimmer-Schweingruber, R.F., 1001, 1002, 1004, 1125, 1137, 1139, 1140
Winant, C.D., 394, 420
Windridge, D., 525
Winge, C., 547
Wingham, D., 1612, 1616, 1617
Winglee, R.M., 1003
Wink, J., 1291
Wink, J.E., 637
Winkert, G., 1004

Winkler, C., 523, 959
Winkler, P.F., 756, 960
Winkler, R., 1232, 1271
Winnberg, A., 1291
Winnewisser, G., 636, 639, 641, 642
Winningham, J.D., 1580
Winser, K.J., 1547, 1556, 1557
Winske, D., 1025, 1032, 1503, 1528, 1556
Winter, F.H., 57, 58
Winterhalter, D., 1140, 1491
Winterhoff, H.P., 1138
Winters, J.M., 935
Wisdom, J., 1458, 1478
Wiseman, A.G., 358, 367, 370–372
Wisniewski, W.Z., 1244, 1264, 1270, 1275
Wisotzki, L., 395, 468, 541, 548
Witasse, O., 395
Withbroe, G.L., 220, 222, 234, 242, 244, 255, 1086
Witt, A., 1188
Witt, A.N., 659, 673, 674
Witt, N., 1213
Witte, M., 967, 1004, 1156, 1159, 1204, 1209, 1215, 1216
Witteborn, F.C., 329, 637, 643, 1430
Witten, E., 430, 440
Wittman, D., 388, 397
Wittman, D.M., 460, 469
Witzel, A., 557, 560
Wkerman, 1616
Wlérick, G., 468
Wlotzka, F., 1402, 1403
Woźniak, P.R., 674
Woch, J., 1138, 1139
Wohl, C.G., 1231
Wöhler, M.F., 1280, 1294
Woiceshyn, P.M., 1426, 1430
Wold, M., 527
Wolf, B., 809
Wolf, C., 527
Wolf, J., 871
Wolf, R.A., 249, 252, 1543, 1555
Wolfe, A.M., 414, 420, 421, 458, 469, 543, 548
Wolfe, J.H., 139, 151, 978, 983, 985, 1004, 1116, 1119, 1140, 1145, 1161
Wolfe, R.F., 1470, 1477
Wolfendale, A.W., 694
Wolfenstein, L., 1231
Wolff, B., 1210, 1216
Wolff, C., 251
Wolfire, M.G., 637, 640, 653, 675
Wolfschmidt, G., 203, 221, 223
Wolfson, C.J., 1062, 1063, 1086
Wolfson, J., 1032
Wolinski, K.G., 753
Wollan, E.O., 120, 122, 149, 151, 697
Wolszczan, A., 363, 371, 788
Wolter, A., 496, 497, 550, 554, 560
Woltjer, L., 543, 548, 752, 811, 821,
Wolverton, M., 148, 151
Wong, H.-K., 1556
Wong, M.H., 1448
Wong, S.K., 861, 872
Woo, J., 836
Wood, A., 1371
Wood, B., 1138
Wood, B.E., 427, 438, 662, 673, 1204, 1208, 1209, 1215, 1216
Wood, D.O.S., 869, 902, 911
Wood, J.A., 427, 1326, 1381, 1403
Wood, J.H., 808
Wood, K.S., 497, 527, 822, 834, 904, 911
Wood, P., 935
Woodard, M.F., 1047, 1048, 1050, 1062, 1063
Wooden, D.H., 637, 643, 1430

Wooden, J., 1373–1375, 1402
Woodgate, B.E., 298, 300, 870
Woodney, L.M., 1286, 1287, 1294
Woodruff, R., 300
Woods, D.T., 1080, 1088
Woods, P.M., 526, 736, 754, 758
Woodsworth, A., 835
Woodward, C., 808
Woodworth, P.L., 1616, 1617
Woody, D.P., 314, 329, 330
Woolf, N.J., 328, 329, 496
Woolfson, M.M., 1456, 1476
Woolum, D.S., 1432, 1448
Woosley, S.E., 434, 436, 440, 515–517, 521, 524, 526, 528, 822, 856
Wootten, A., 1292, 1293
Wopenka, B., 1294
Workman, R.L., 1231
Woronow, A., 270
Worrall, D.M., 491, 498, 526
Wotzlaw, S., 86, 114
Woudt, P.A., 558, 788
Woweries, J., 1232, 1292, 1293
Wren, J., 523
Wrickramasinghe, N.C., 640
Wright, C.M., 617, 618, 620–622, 636–638, 643–645, 869, 872
Wright, E.L., 328, 329, 410, 419–421, 450, 614, 636, 642, 645, 786, 1188
Wright, J.W., 1571, 1584, 1589, 1617
Wroe, H., 298
Wu, B.H., 1135, 1139
Wu, C.-C., 935, 298
Wu, C.S., 1521, 1528
Wu, F.S., 1199, 1216
Wu, H., 547
Wu, J.H.P., 394, 420
Wu, K., 809
Wu, S.T., 1161
Wu, S.Y., 1616
Wucknitz, O., 466
Wuerker, R.F., 1188
Wukelic, G.E., 212, 223
Wulf, Th., 151, 677, 697
Wunsch, C., 1647, 1651
Wurm, K., 1236, 1275
Wurz, P., 1138, 1140, 1214
Wyckoff, S., 1287, 1293, 1294
Wygant, J.R., 1556
Wyller, A.A., 1571, 1583
Wynn, G.A., 805, 806, 810
Wynne-Jones, I., 675
Wynn-Williams, G.G., 300
Wyskocil, V., 1650
Wyss, W., 351
Wytrzyszczak, I.M., 1649

Xie, D.X., 1293
Xie, T., 616, 645
Xilouris, K., 755
Ximenez de Ferran, S., 329, 577
Xu Y.L., 1187
Xu, H., 496, 497, 576
Xu, S., 1361, 1371
Xu, W., 469, 560

Yadav, R.K.S., 523, 524
Yadigaroglu, I. A., 526, 729, 756
Yahil, A., 386, 393, 396, 397
Yajima, N., 696
Yakovlev, D.G., 729, 730, 745, 757, 758
Yamada, M., 959, 1023, 1024, 1032, 1033
Yamada, Y., 1004
Yamagami, T., 329, 696

Yamagami, Y., 524
Yamamoto, A., 696
Yamamoto, T., 1232, 1239, 1271, 1272, 1275, 1294, 1555–1557
Yamamura, I., 330, 637, 640, 641, 871
Yamaoka, K., 577, 855
Yamasaki, N., 577
Yamasaki, N.Y., 497
Yamashita, A., 560
Yamashita, K., 495, 497, 498, 604, 835, 836, 959, 960
Yamauchi, M., 959
Yamauchi, S., 696, 757, 892
Yamazaki, T., 577
Yamburenko, N., 960
Yanagimachi, T., 1004
Yanagita, S., 960, 1004
Yanai, K., 1373
Yanasak, N.E., 696, 697
Yancopoulos, S., 746, 758, 820
Yao, W.-M., 1231
Yaqoob, T., 855
Yatteau, J., 1557
Yau, A.W., 1537, 1548, 1551, 1555, 1557
Yee, H.K.C., 393, 467, 468, 496
Yegingil, I., 673
Yeh Eueng-Nan, 1615
Yeh, T., 1010, 1033
Yelle, R.V., 1426, 1430, 1433, 1450, 1466, 1478
Yentis, D.J., 834
Yeomans, D.K., 1232, 1259, 1270, 1271, 1275, 1293
Yeroshenko, Y.G., 275, 1491
Yi, I., 558, 559, 855
Yin, L., 1401
Yingst, A., 1321
Yionoulis, S.A., 1651
Yoder, C.F., 351, 1459, 1478, 1650
Yoder, J.A., 1602, 1616
Yoder, K.A., 1478
Yonezawa, T., 1567, 1584
York, D.G., 293, 299, 300, 496, 589, 590, 604, 653, 654, 657, 658, 661, 665, 672–675
Yorke, H.W., 576, 756, 868, 869
Yoshida, A., 515, 524, 528, 754, 756, 757, 836, 959
Yoshida, K., 696
Yoshida, M., 960
Yoshida, S., 159, 176, 177
Yoshida, T., 696, 960
Yoshii, Y., 437, 702, 704, 719
Yoshikoshi, T., 960
Yoshimura, K., 696
Yost, S.A., 526
Young, A.J., 371, 855
Young, A.T., 262, 270
Young, C.A., 523
Young, D.T., 1231, 1556, 1580–1583
Young, E., 330, 577, 869, 873
Young, J.W., 1269
Young, K., 1292, 1293
Young, L.A., 1430
Young, P., 444, 469
Young, R., 222
Young, R.E., 1441, 1447, 1450, 1548
Young, T.R., 525
Younger, P.F., 548
Youssef, S., 1231
Yu, G., 249, 255
Yu, J.T., 413, 420
Yu, Q., 392
Yuen, L., 330
Yui, Y.Y., 641
Yukawa, 125

Yumi, S., 1649
Yun, J., 610, 645
Yung, Y., 1291
Yung. Y.L., 1426, 1429
Yungelson, L.R., 760, 778, 786, 789, 808
Yunin, S., 960

Zaal, P.A., 644
Zand, J., 524
Zabriskie, F., 289, 300
Zaglauer, H.W., 364, 372
Zaharia, S., 1032
Zahn, J.P., 1060
Zähringer, J., 1372, 1378, 1403
Zaitsev, L., 1615
Zaitseva, S.Ye., 1402, 1403
Zajac, B.J., 639
Zakharov, V.E., 1514, 1520, 1528
Zall, P.A., 644
Zamorani, G., 533, 547, 548
Zamorano, J., 467
Zanda, B., 434, 440
Zane, S., 757
Zanetti, L.J., 1556
Zank, G.D., 1206, 1209, 1212
Zank, G.P., 648, 654, 673, 675, 1004, 1005, 1158, 1161, 1188, 1206, 1208, 1209, 1213, 1215, 1216, 1227, 1233
Zanstra, H., 925, 935
Zapaterio Osorio, M.R., 523, 892
Zaragoza, A., 194, 202
Zare, R.N., 1373
Zarnecki, J.C., 1269
Zaroubi, S., 419
Zasetsky, V.V., 275
Zaslavsky, G.M., 1527
Zatsepin, G.T., 692, 697
Zavattini, G., 524, 525, 528
Zavlin, B.E., 731, 732, 736, 737, 741, 743, 744, 748
Zavlin, V.E., 754, 756, 757, 758, 855
Zavody, A.M., 1668
Zawodny, J.M., 1668
Zayer, I., 1062, 1063
Zbijewski, W., 523
Zdziarski, A.A., 852, 854, 856
Zeh, H.D., 1402
Zehnpfennig, T., 1033
Zekanina, Z., 1188
Zel'dovich, Y.B., 413, 421, 474, 498, 549, 759, 787, 789, 811, 822
Zelenyi, L.M., 1011, 1031
Zelikoff, M., 298
Zendri, J.P., 371
Zenneck, J., 1560, 1584
Zepf, S.E., 395
Zepka, A., 755
Zerbini, S., 1649, 1650
Zhang, B., 729, 758
Zhang, C.-Y., 935
Zhang, H., 523
Zhang, J., 604
Zhang, M., 151
Zhang, S.N., 849, 851, 856
Zhang, W., 755, 818, 822, 959
Zhang, Y.F., 636, 639, 641, 642
Zhao, J., 1054, 1062, 1063
Zhao, Y., 1616
Zharikov, S.V., 513, 528, 755
Zharkova, V.V., 1054, 1055, 1062
Zheng, W., 299, 546
Zhigulev, V.N., 1539, 1557

Zhitnik, I.A., 222
Zhou, A., 523
Zhu, J., 523
Zhu, T., 757
Zhulin, I.A., 275
Zhuravlev, D.A., 999
Zhuzgov, L.L., 275
Zhuzgov, L.N., 1491
Zickgraf, F.-J., 754, 756
Zijlstra, A.A., 644
Zimbelman, J.R., 1321, 1322, 1323
Zimmer C., 1492
Zimmermann, H.U., 576, 756
Zimmermann, J.-P., 870
Zink, S.M., 1139
Zinn, R., 394
Zinnecker, H., 871, 882, 884, 892, 893
Zinner, E., 613, 636, 1169, 1188, 1284, 1291, 1294, 1335, 1370, 1373
Zirakashvili, V.N., 691, 696, 697
Zirin, H., 971, 1004
Zirker, J.B., 139, 222
Zmuda, A.J., 1538, 1557, 1569, 1584
Zmuidzinas, J., 621, 640, 645
Znajek R.L., 521, 523, 544, 546, 553, 558
Zobl, R., 757
Zohar, S., 265, 270
Zolcinski, M.-C., 893

Zolensky, M., 1187
Zombeck, M., 894
Zook A., 560
Zook, H.A., 1179, 1188, 1189, 1293, 1182,
Zou, Z.L., 547
Zuber, M.T., 1306–1309, 1321, 1322
Zubkov, B.V., 1233
Zucchino, P.M., 299
Zucker, M.E., 370
Zuckerman, B., 630, 643, 645, 870, 873
Zund, J., 1650
Zurbuchen, T.H., 696, 1000–1002, 1004, 1005, 1137, 1140, 1159
Zurek, W.H., 393
Zwaan, C., 771, 789, 872, 893, 910
Zwaan, M.A., 465
Zwally, H.J., 1322, 1615
Zwarthoed, G.A.A., 909
Zweibel, E., 786
Zwickl, J., 1004
Zwickl, R.D., 1138, 1231, 1549, 1556, 1557
Zwicky, F., 443, 454, 464, 469, 475, 498, 678, 694, 721, 752, 759, 786, 840
Zwitsky, D., 396
Zybin, K.P., 1581
Zycki, P.T., 546, 852, 854
Zygielbaum, A.I., 352
Zylstra, G., 528, 834–836
Zytkow, A.N., 505

Name index
(index of names cited in the text)

Acton, L.W., 1089
Acuña, M.H., 1307, 1491
Acuna, M., 146
Adams, D., 1191
Adams, J., 1431
Adams, W., 259
Adams, W.S., 649
Africano, A., 65
Akasofu, S.I., 156
Aldrin, E.E., 50, 74, 88, 1337, 1352
Alexander, D., 1089
Alexander, W., 1113
Alfvén, H., 131, 231, 234, 1110, 1419, 1498, 1530, 1533, 1535
Allan Poe, E., 28
Allen, D., 1412
Allen, D.A., 301
Alpher, 399
Altamimi, Z., 1640
Altwegg, K., 1277
Alvarez, L., 122, 261
Aly, J.-J., 1112
Ambarzumian, 652, 665
Ampére, 227, 228, 237, 238
Anderegg, M., 324, 328
Anderson, S.B., 748
Anderson, C., 122, 123, 126
Anderson, C.D., 124
Anderson, J.D., 146
Antiochos, S.B., 1112
Appenzeller, I., 861
Appleton, E., 278, 279
Appleton, E.V., 51, 1560
Ariosto, L., 26
Aristotle, 1096, 1535
Armstrong, N., 50, 74, 88, 258, 977, 1337, 1352
Arnaud, K., 495
Arnold, J., 1297
Arnold, J.R., 271
Arnold, W., 65
Arrhenius, S., 1280
Ashworth, D.G., 1166
Asimov, I., 39
Aston, F.W., 1326
Atterley, J., 28
Auger, P., 52, 54, 124
Aumann, H.H., 318, 319
Axford, I., 1174, 1213
Ayres, T., 878

Baade, W., 678, 702, 721
Babcock, H., 1090, 1093, 1096, 1106
Babcock, H.D., 142
Bahcall, 473, 541, 542, 750, 783
Bahnsen, A., 1551
Bain, J.G., 205
Baldwin, R., 259, 1295
Balmino, G., 1648

Balogh, A., 1141
Baluteau, J.-P., 315, 324
Banday, A.J., 399
Baos, Jr., A., 124
Baranov, V., 1200, 1208
Bardin, I.P., 135
Barlier, F., 1623
Barnard, E.E., 1452
Barnes, A., 146
Barnett, M., 278
Barsukov, V., 1297
Bartels, J., 156, 158, 183, 1102
Bates, D.R. 190
Baumgartner, W., 495
Bause, H., 180
Baz, F.E. , 1320
Beals, C.S., 261
Bechtold, J., 542
Becker, K.E., 34
Becker, W., 721
Becklin, E., 325
Beckwith, S., 325
Becquerel, H., 119
Bedinger, A., 193
Begemann, F., 1370, 1380
Behannon, K.W., 1491
Behrenfeld, M.J., 1603
Beichman, C.A., 319
Beintema, D.A., 319, 324
Bell, J., 722
Beloussov, V.V., 135
Bèmont, G., 1288, 1325
Bennett, C., 1282
Bennett, C.L., 419
Bergé-Nguyen, M., 1641
Berg, O., 1184
Berger, C., 1649
Berkner, L., 69, 135
Berman, 1364, 1365
Bertaux, J.L., 1206, 1213
Bethe, H., 124, 399
Beutler, H., 124
Beynon, W.J.G., 135
Biermann, L., 179, 232, 233, 236, 248, 1095, 1109
Bigelow, F., 1089
Bigg, 1466
Bignami, G., 723
Bildsten, 785
Bingham, C., 21
Bingham, R., 534
Birkeland, K., 230, 1544, 1569, 1120
Bjørgo, E., 1614
Blaauw, 649
Blackett, P.M.S., 135
Blaes, O., 538
Blamont, J., 180, 189, 193, 197, 1196
Bleeker, J., 328

Bless, R., 298
Bogard, D.D., 1325
Bogdanov, A., 29
Boggess, A., 291
Boggess, N., 419
Boggess, N.W., 319
Bohlin, R., 295
Bohm, D., 1502
Bohn, 1167
Bondi, 356
Bonestell, C., 259
Borgman, J., 316, 319, 324
Börner, 760
Boslough, J., 404
Bostick, W., 124
Bothe, W., 124
Boucher, C., 1640
Bouillé, F., 1640
Bowen, I., 918
Bowen, I.S., 119
Bowyer, S., 543, 823
Boyle, B., 540
Bradbury, R., 260
Brahe, T., 1235
Brahic, 1455
Brans, C., 345
Breit, G., 278
Bricard, J., 190, 198
Bringer, H., 105
Brode, R., 124
Broglio, L., 52, 53
Brownlee, D., 1166
Brown, H. 1330, 1426
Bruce, C.W., 1452
Brueckner, G.E., 1664
Brunt, S., 27
Bruzek, A., 1101
Bull, H.W., 65
Bullard, 156
Burlaga, L., 1105
Burnight, T.R., 100, 210
Burroughs, E.R., 29, 260
Burrows, C., 944
Burrows, D.N., 603
Byram, E.T., 211

Camilo, 766
Campbell, W., 916
Campbell, D., 1309
Cane, H.V., 999
Canright, R., 83
Cardan, 226
Carlqvist, P., 1110
Carr, M., 1304
Carrington, R., 1094, 1096–1098, 1120
Carroll, N., 1320
Carruthers, G., 291
Carter, L., 65
Cassinelli, J., 878
Cassinelli, J.P., 895
Cassini, G.D., 271, 1451
Cauchy, A., 1519
Cazenave, A., 1641
Celotti, A., 558
Celsius, 1535
Chandrasekhar, S., 102, 245, 652, 721
Chapman, S., 44, 45, 51, 69, 135, 156, 158, 183, 204, 229, 230, 234–236, 242, 248, 278, 1497, 1529, 1544
Charles, P., 791
Cheney, R., 1635

Cheng, E., 419
Chertok, B., 37
Chree, C., 1102
Christiansen, W., 205
Christy, J., 1433
Christy, J.W., 1454
Chubb, T.A., 211
Ciufolini, I., 347
Claeskens, J.-F., 441
Clarke, A., 39
Clarke, A.C., 66, 1109, 1655
Clay (Holland), J., 124
Clay, J., 120
Cleator, P.E., 65
Cleaver, A.V., 65
Clegg, P.E., 319, 324
Coblentz, W.W., 304
Code, A., 291, 124
Cohez, E., 723
Colas, F., 1260
Collins, M., 74, 1337
Columbus, C., 1089
Comastri, A., 558
Compton, A.H., 121, 122, 124, 126
Connerney, J.E.P., 1491
Copernicus, 226, 257, 826, 1191
Cornas Solà, J., 1463
Cote, A., 1320
Coulomb, J., 135, 227
Crippen, R.L., 94
Crommelin, A., 339
Crow, A.D., 65
Crutzen, P., 1660
Curie, M., 1325, 1326
Curie, P., 1325
Curtis, 529
Curtis, H., 916, 921
Cuvier, 258

d' Azambuja, H., 204
Damour, T., 350, 356, 357
Danielson, R., 289
Danjon, A., 135
Darton, N.H., 259
Darwin, C., 260, 1280
Davies, K., 404
Davis, D., 495
de Boer, K., 296, 297
de Fontenelle, B., 27
De Gaulle, C., 201
de Graauw, T., 324, 325, 328
de Jager, C., 193, 197, 203
de Jong, T., 319
de Lapace, P.-S., 1432
De La Rue, W., 1089
De Mairan, 1535
de Muizon, J., 863
de Sitter, W., 340, 346, 347, 356
de Vaucouleurs, G., 260, 263
Defoe, D., 27
Dellinger, H., 204
Dellinger, J.H., 279
Delsemme, A., 1281, 1282
Dermoff, S., 1166
Dershem, E., 124
Deslandres, H., 1090, 1100
Dessler, A., 259, 1426
di Bondone, G., 1279
Dicke, R., 345

Diner, D., 1410
Dirac, 122
Dodson, H., 1098
Dohnanyi, J., 1166
Dooling, 57
Dornberger, W., 34, 36, 62, 63
Dow, W., 41
Drake, 608
Du Fay, 226
Dungey, J., 1109, 1110
Durand, D., 1585
Durant III, F.C., 70
Duvall, T., 1045
Dwek, E., 419
Dyal, P., 317
Dyck, H.M., 317

Ebert, H., 1101
Eckart, C., 124
Eddington, A., 120, 339, 340, 343, 344, 351, 356, 442, 648
Edgeworth, K., 1186, 1433
Edison, 29
Edison, T., 608
Ehricke, K., 72
Einstein, A., 20, 227, 250, 335, 337–339, 342–345, 350, 351, 353, 356, 432, 441, 442, 445, 453, 464, 734, 946
Eisenhower, D.D., 43, 45, 46, 48, 71, 136, 328
El Baz, F., 1297
Ellerman, F., 1100
Ellison, M., 1098, 1101
Elsässer, H., 156, 233
Elvey, C., 135
Elvis, M., 529, 558
Emerson, J.P., 319, 324
Encrenaz, T., 148, 1431
Engel, R., 62
Esnault-Pelterie, R., 31, 32
Esposito-Farése, G., 369
Evans, J., 281
Eviatar, A., 1466
Eyraus, A., 28

Fabbiano, G., 561
Fabian, A., 558
Fahr, 1198
Fairbanks, W., 340
Falkowski, P.G., 1602, 1603
Fan, C.Y., 136
Faraday, M., 227, 228
Farquar, B., 1181
Fechtig, H., 1166, 1169
Feissel, M., 1640
Felgett, P., 304
Fenton, K.B., 129
Fermi, E., 131, 132, 432, 1328
Fernandez, J., 1187, 1433
Ferraro, 1497, 1529
Fiammarion, 1405
Field, G., 653
Fielder, G., 1295
Finger, H., 107
Fisk, L.A., 999, 1141, 1199
Fitton, B., 324
Fitzgerald, 228
Flammarion, C., 29
Fleming, J., 279, 327
Föppl, H., 180, 181
Flynn, 1212
Forbush, S., 120, 124, 1118
Forman, E., 65

Fossati, G., 558
Fountain, J., 1454
Franklin, 226, 228
Fraunhofer, J., 914
Freundlich, E., 338
Friedman, H., 41, 67, 100, 135, 205, 211, 307, 721, 722
Friichtenicht, J., 1173
Frisch, P.C., 647, 1212
Froman, D., 124
Frost, E.B., 443
Furniss, I., 324
Fyodorov, N., 30

Gaensler, B., 785
Gagarin, Y., 49, 50, 83, 85, 91, 94, 258
Gaidukov, L., 38
Gaignebet, J., 1630
Galilei, G., 225, 271, 336, 1091, 1431, 1451, 1453, 1473
Gamov, G., 423
Gamow, 399
Ganesh, S., 706
Ganswindt, 72
Gaposchkin, M., 1627
Gaspar, P., 1649
Gast, P., 1297
Gatland, K.W., 65
Gauss, F., 277, 1560
Gautier, T.N., 319
Gehrels, T., 146
Gehrz, R.D., 317
Geiss, J., 1370, 1526
Ghisellini, G., 549
Giacconi, R., 100, 218, 499, 549, 562, 722
Gilbert, G.K., 259, 1295
Gilbert, W., 1480
Gill, P.S., 124
Gillett, F.C., 318, 319, 608
Gilmore, G.F., 699, 704, 712
Gilruth, R., 82, 83, 85, 93
Ginzburg, V.L., 650
Giovanelli, R., 1099, 1109
Gladman, B., 1456
Glenn, J., 50
Glenn, J.H., 83
Glennan, K., 83
Gloeckler, G., 963
Glushko, V.P., 33, 34, 39, 44, 45, 64
Goddard, R., 31, 32, 60, 65, 72, 108, 279
Godwin, F., 27
Goebbels, J., 64
Golay, M.J.E., 305
Gold, T., 232, 282, 1110, 1496
Goldberg, L., 67, 205, 216, 218, 295
Goldin, D., 110, 1415
Goldsmith, D., 653
Goldstein, R., 1309
Golub, L., 878
Gonsales, D., 27
Goody, R.M., 317, 1409
Göring, H., 35
Górski, K.M., 399
Gosling, J., 1104
Gottlieb, 1041
Goudsmit, S., 124
Gould, B.A., 653
Graham, 1535
Greaves, W.M.H., 1100
Grec, 1043
Greenberg, L., 315
Greenberg, M., 1166

Greenstein, J., 67, 530
Gregg, P., 29
Griest, 454
Griffin, M.J., 324
Grindlay, J., 501
Gringauz, K., 1501
Gringauz, K.I., 136
Grocco, G.A., 65
Groetzinger, G., 124
Groom, S., 1607, 1614
Grossmann, M., 338
Grotrian, W., 204
Grottrup, H., 37, 38
Grün, E., 1163, 1169
Gryaznov, G.M., 109
Guest, J., 1301
Guinot, B., 1640, 1649
Gulkis, S., 321, 419
Gum, C., 653
Gupta, D., 531
Gursadian, G., 91
Gursky, H., 100
Gustafson, B., 1166

Haas, A., 124
Haas, K., 60
Habing, H.J., 319
Hacking, P., 325
Hackwell, J.A., 317
Haerendel, G., 180, 184–186, 1007, 1535
Hagfors, T., 1559
Haisch, B., 1107
Hale, 1098, 1100, 1452
Hale, E.E., 29
Hale, G.E., 204, 1090, 1109, 1112
Hale, W., 65
Hall, A., 1452
Hall, C.F., 144
Halley, E., 1235, 1277, 1280
Hammer, R., 1065
Hanner, 866, 1164
Hansen, S., 1102
Harkins, W., 124
Harnden, R., 878
Harries, J.E., 1653
Harrington, R., 1454
Hartmann, W.K., 8, 257
Harwit, M., 301, 303, 609
Hathaway, D., 1094, 1106
Hauchecorne, A., 197
Hauser, H., 419
Hauser, M.G., 319, 321
Hawking, S., 404
Hayakawa, S., 205, 314, 324
Hayden, 1332–1334
Hays, P.B., 1664
Hazard, C., 530
Head, J.W., 1295, 1297
Heaviside, O., 227, 278
Heckmann, O., 204
Heger, M.L., 647
Heiles, C., 653
Heilmeier, L., 180
Heisenberg, W., 124
Helmholtz, 227, 228
Hemenway, C., 1164
Heraclitus, 1102, 1112
Herbig, G., 650
Herring, D., 316
Herring, R., 317

Herschel, 1405
Herschel, J., 303, 608
Herschel, W., 260, 608, 913, 1451, 1457
Herter, T., 325
Hertz, 227, 1560
Hertz, H., 1101
Herzog, E., 124
Hess, V., 3, 119, 124, 678
Hevelius, 1235
Hewish, A., 531, 722
Hey, J.S., 1101, 1102
Hickman, C., 65
Hildebrand, R., 315
Hillberry, N., 124
Hiltner, W.A.,132
Himmler, H., 63, 64
Hinners, N., 1320
Hinshaw, G., 419
Hinteregger, H.E., 209
Hiorter, 1535
Hipparchus, 226
Hirayama, K., 1164
Hirsch, A., 32
Hitler, A., 63
Hoag, B., 124
Hobbs, L., 650
Hodgson, R., 1097, 1098
Hoffman, 356
Hoffmann, W.F., 306, 323
Holmes, A., 1330
Holt, J., 1562
Hopfield, J.J., 67, 124
Hopmann, H., 75, 76, 86, 109
Horner, D., 495
Houbolt, J.C., 85
Houck, J.R., 309, 315, 317, 319, 323, 325, 328
Houtermans, F., 1329
Hoyle, F., 282, 423, 1098, 1110
Hsue-shen Tsien, 65, 69
Hubbard, 1455
Hubble, E., 20
Huggins, W., 303, 914
Hughes, D., 124, 1166
Hulburt, E.O., 205, 278, 279, 307
Hulse, J., 285
Hulse, R., 359, 360
Hultberg, J., 1370
Hultqvist, B., 54, 156, 1496, 1529
Hundhausen, A., 1102, 1105
Hunten, D., 1409, 1425
Huntress, W.T., 1277
Hurlburt, N., 1097
Hurley, K., 502
Huygens, C., 27, 1451
Hynek, J.A., 67
Hynes, R., 812

Ihaddadene, S., 999
Infeld, 356
Ingersol, A.P., 146
Irm, A., 1506
Itokaura, I., 56

Jackson, P., 419
Jaffe, D.T., 315
Jamieson, J.A., 326
Jansky, K., 499, 650
Janssen, J., 1089, 1100
Janssen, M., 419
Jansson, 233

Jeffreys, 1425
Jennings, R.E., 317, 319, 324
Jessberger, E., 1181
Jesse, W., 123, 124
Jewitt, D., 1433
Johannessen, J.A., 1585
Johannessen, O.M., 1614
Johnson, H.L., 304, 305
Johnson, L., 82
Johnson, T., 122, 124
Jokipii, R., 1183
Jonest, H., 124
Jorda, L., 1235
Joselyn, J.A., 1100
Judge, D.L., 146
Jurgens, R., 1309

Kahn, 608
Kaita, E., 419
Kamp, L., 1412
Kant, E., 1432
Kaplan, J., 45
Kastler, A., 190, 1196
Kaufman, H., 109
Kaula, W., 1627
Keegstra, P., 419
Keldysh, M.V., 1297
Keller, H.U., 1235
Kelsall, T., 419
Kelvin, L., 227, 229, 277, 541, 1109, 1280
Kennedy, J., 50
Kennedy, J.F., 82, 111, 113, 258, 273, 1337
Kennelly, A.E., 278
Kepler, J., 26, 50, 226, 257, 947, 1235, 1399
Kerr, R.A., 1647
Kessler, M.F., 324, 328, 561
Kester, D., 615
Khayyam, O., 226
Khrushchev, N., 47
Kiepenheuer, K.O., 40, 203, 204, 233, 1095, 1101, 1102
Killian, J., 47
Kilmuk, P., 91
Kinard,W., 146
King, I., 702
King, K.J., 324
Kirchhoff, G., 914
Kissel, J., 1181, 1284
Kleinmann, D.E., 305, 306
Klemperer, W.B., 67
Klimpton, L., 136
Kliore, A.J., 146, 1414
Kogut, A., 419
Kondracki, H.C., 308, 309
Koornneef, J., 296
Korolev, S., 33, 34, 38, 44, 45, 48, 86, 91
Koshiba, M., 943
Kotelnikov, V., 1310
Kowal, C., 1454
Kraichnan, 1513
Kramers, 613
Krause, E., 41, 67, 68, 205, 279
Krüger, H., 1183
Kuiper, G., 66, 257, 258, 261, 275, 609, 1186, 1433, 1452, 1454, 1462, 1463
Kumar, V., 419
Kummerer, R., 419
Kuntz, K.D., 603
Künzel, H., 1098
Kupo, I., 1466
Kupperien, J.E., 291

Labhardt, L., 376
Laclaváre, G., 135
Lacques, 1454
Lacroute, P., 717
Lacy, J., 617
Ladenburg, R., 67
Lagerkvist, K., 1164
Lagrange, 271
Lamarck, 258
Lamb, D., 505
Lambeck, K., 1627
Lamont, 228
Landau, L., 721, 1514
Lang, F., 31
Lange, 967
Lange, A., 314
Langmuir, 1196
Langseth, M., 1299
Laplace, 271, 442, 1235
Larson, S., 1454
Lassell, W., 1451
Lasswitz, K., 29
Latham, G., 1298
Laurain, O., 1649
Laursen, V., 135
Lawden, D., 81
Lebedew, W., 91
Lecacheux, J., 1260, 1454
Lee, M., 1212
Lefebvre, M., 1623, 1630
Leighton, R.B., 1095
Leighton, R.W., 304
Leisawitz, D.T., 419
Lejay, P., 135
LeMay, C., 46
Lense, J., 347
Lepping, R.P., 1491
Le Traon, P.-Y., 1614
Le Verrier, U., 1431
Levy, D.H., 1247, 1248
Levy, E., 1183
Ley, W., 65
Lichtman, 211
Lindemann, F.A., 230
Lindsay, J., 216
Lineweaver, C., 419
Linsky, J., 878, 1208
Little, A.G., 1102
Little, M., 328
Löb, H., 110
Lockeyer, 233
Lockyer, J., 1099
Loebell, E., 65
Loewenstein, K., 419
Loewenstein, M., 495
Lofgren, E.J., 131
Logsdon, J.M., 93
Loidl, H., 180
Long, L., 1370
Lopate, C., 130
Lorentz, H., 237, 343, 344, 350
Löst, R., 179
Lovell, B., 205
Lövgren, T., 1535
Low, B.C., 1108
Low, F., 305–307, 317, 319, 320, 323, 609
Lowell, P., 29, 259, 266, 1414, 1432, 1454
Lubin, P., 419
Lucchitta. B., 1297
Lucian of Samosota, 26

Ludwig, G., 154, 1497
Lundquist, C., 1628
Lusser, D., 65
Lüst, R., 180, 192
Luu, J., 1433
Lyne, A.G., 784
Lyot, B., 1089, 1100
Lyttleton, 1192, 1281

Macchetto, D., 376
Mach, E., 335, 336
Macmillan, H., 201
Malin, M., 1304
Malina, F., 42, 65, 66, 69
Mandel'shtamm, S.L., 205, 210, 212, 215
Mandl, R.W., 443
Mange, P., 283
Manning, L., 65
Manring, E., 193
Marconi, G., 29, 277, 1560, 1579
Marius, S, 1451
Marov, M., 1297
Marsden, P.L., 319
Marsden, R., 228
Marsh, J.G., 1649
Martz, D.E., 305
Mason, G.M., 970, 999
Masursky, H., 264
Mather, J.C., 314, 319, 321, 404, 419
Matsumoto, T., 314, 324
Maunder, W., 1095, 1097, 1098, 1112
Maxwell, J.C., 237, 250, 335–337, 348
Mayer, J.R., 303
Mayer, T., 271
Maze, R., 123
Mazets, E., 1181
McCauley, J., 264, 1297
McCleese, D., 1410, 1419
McDermot, M., 27
McDonald, F.B., 137, 146, 148, 677
McDonnell, T., 1164, 1166, 1181
McDowell, F., 1370
McElroy, N.H., 83
McIlwain, C., 154, 1497, 1533
McIntosh, P., 1100
McNaught, R.H., 943
McNutt, D., 308, 309
McQueen, R., 219
Meade, M., 298
Medlock, J., 85
Megarian, G.K., 67
Mekler, Y., 1466
Melotte, P.J., 1452
Melnick, G., 325
Melzner, F., 180
Mendez, R., 920, 924
Menon, T.K., 651
Menzel, D., 205, 1425
Merrill, 648, 649
Merton, R., 328
Messier, C., 913, 948
Metcalf, T., 1113
Mewe, R., 878
Meyer, P., 136
Meyer, S., 419
Michelson, A., 335
Michin, V.P., 86
Miles, M., 1614
Miley, G., 319
Millikan, R.A., 120, 228

Millikan, C.B., 69
Mitchell, J.K., 1262
Montgomery, C., 124
Montmerle, T., 878
Moore, 916
Moore, P., 1407
Moore, W.J., 317
Moorwood, A.F.M., 315, 324
Morabito, L., 1460
Morley, E., 335
Moroz, V., 1419
Morris, E., 1297
Morton, D., 289, 290
Moseley, H., 419
Moskalenko, I., 686
Mould, J., 379
Muehlberger, W., 1320
Muehlner, D., 314
Mueller, G., 85, 88, 93
Mulchaey, J., 495
Münch, 651–653
Munch, G., 146
Munro, R., 1106
Murdock, T.L., 419
Murray, B., 1301, 1407
Mushotzky, R., 473
Muslimov, 729
Mutch, T., 1304, 1320
Myer, B., 180

Nadig, 1167
Nagase, F., 834
Nebel, R., 65
Neddermeyer, S., 124
Negenborn, L., 21
Neher, V., 120
Neivert, P., 1320
Ness, N.F., 138, 1479
Neu, E., 68
Neubauer, F.M., 1491
Neugebauer, G., 304, 305, 317, 319, 323, 325
Neugebauer, M., 1115
Neuss, H., 180
Newell, H., 42, 67, 216
Newell, H.A., 205
Newell, H.E., 67
Newton, 226, 237, 250, 337, 342, 1098, 1100
Newton, H., 1097, 1100
Newton, I., 335, 442, 1280
Ney, E.P., 131
Nichols, M.H., 67
Nicholson, S., 1452, 1454
Nicholson, S.B., 304, 1091
Nicolas, J., 1649
Nicolet, M., 135, 195, 279
Nielsen, U., 124
Nilsson, C., 999
Nixon, 93
Nordheim, L., 124
Nordvedt, K., 353
Northrop, T., 1498
Nottale, L., 444
Nyquist, L.E., 1325

Oberth, H., 30–32, 60–62, 69, 72, 373
Occhialini, G., 205
O'Day, M., 41
O'Dell, C.R., 100
Odeshaw, H., 136
Oersted, 227, 228

Ohman, Y., 135
Okano, S., 56
Okuda, H., 324
Omont, A., 706
Oort, J.H., 613, 652
Oosterbroek, T., 815
Öpik, E., 258, 264, 1235, 1281, 1425
Oppenheimer, R., 721
Oppenheimer, J.R., 124
Oster, L., 1102
Ostriker, J.P., 477, 777
Owen, 260

Pacini, F., 306
Paczyński, B., 506, 777
Padovani, P., 554, 558
Pallavicini, R., 875, 878
Panagia, N,, 376
Paolini, F., 100
Papanastassiou, D., 1370
Parker, 134, 139, 244, 1501, 1506, 1507
Parker, E.N., 896, 1104, 1111, 1115
Parker, G., 647
Parsons, J., 65
Paschmann, G., 1526
Patterson, C., 1330
Pavlov, G., 721
Payne-Scott, R., 1102
Pendray, G.E., 33, 65
Penzias, 499
Perelman, J.I., 64
Perrine, C.D., 917, 1452
Perryman, M., 718
Pesses, L., 977, 978, 997
Peters, B., 433
Petschek, H., 1509
Pettengill, G., 1309, 1310
Pettersson, L., 1614
Pettit, E., 304, 1100, 1452
Pfaall, H., 28
Phillips, T.G., 315
Phinney, S., 540
Picard, A., 120
Picard, J., 120
Pickering, W.H., 70, 1452, 1497
Pietsch, W., 723
Piggot, C.S., 1327
Pipher, J.L., 317, 325
Plendl, H., 204
Pluskin, V.S., 677
Pollack, J., 1409
Polyakov, V., 92
Pomerantz, 1043
Pomerantz, M., 124
Popper, K., 1192
Portabella, M., 1609, 1614
Porter, R.W., 67
Pospieszalska, A., 465
Pottasch, S.R., 319, 324, 913
Pound, R., 341
Poynting, J.H., 228, 232, 237
Prandtl, L., 42
Press, F., 1298
Price, S.D., 313, 327, 328
Ptolemy, C., 959

Ramsey, W., 233
Rausch, 1331
Ray, E., 1497
Rayman, M.D., 110

Reagan, R., 327
Reber, G., 650
Reece, J.M., 308
Rees, M., 535, 540
Reeves, E., 219
Reeves, H., 423
Refsdal, S., 447
Regener, E., 40, 59, 63, 204
Resitema, H., 1453, 1455
Rense, W., 41
Ress, M., 551
Rhziga, O., 1310
Riccioli, G.B. , 271
Richards, P., 314, 323
Richardson, L.F., 1519
Rickman, H., 1164
Riedel, W., 39, 51
Rieder, R., 1395
Rieke, G.H., 305
Roach, F., 135
Roberts, W.O., 203
Roche, A.E., 1663
Roellig, T., 325
Roman, N., 309
Romani, R., 950
Röntgen, W., 118, 277
Rosen, M.W., 59, 67, 68, 70, 71
Rosner, B., 878
Rosny, J., 32
Ross, F.E., 259
Ross, H.E., 85
Rosse, W.P., 948
Rossi, B., 100, 122, 124, 205, 549, 678, 722
Rottman, G.J., 1664
Routly, 649
Rowan-Robinson, M., 319, 324
Rowland, 227
Ruderman, M., 499, 502, 785
Rummel, R., 1648
Russell, R.W., 315
Russen, D., 27
Russo, A., 25, 57
Rutherford, E., 111, 228, 1326, 1328
Ryan, J., 1107
Ryle, M., 205, 531
Rynin, N.A., 33, 64

Saba, J., 1107
Sabine, C., 277
Sabine, E., 1097, 1112
Sacco, B., 723
Safronov, V., 258
Sagan, C., 1408, 1409, 1471
Sagdeev, R., 47, 1297, 1514
Saha, A., 376, 377
Sandage, A., 373, 376
Sanders, W.T., 603
Sandven, S., 1585
Sänger, E., 40
Santana, J., 419
Satin, A., 70
Saunders, R.S., 1310
Savage, 670
Savage, B.D., 287
Sawyer. C., 1102
Schaack, D., 320
Scharf, C., 495
Schein, M., 124
Scherrer, P.H., 1035
Schiaparelli, G., 29, 266

Schiff, L., 340, 344
Schlegel, K., 1559
Schlüter, A., 240
Schmelz, J., 1107
Schmidt, M., 529
Schmidt, O., 258
Schmitt, H., 1343
Schmitt, J., 878, 1320
Schneider, N.M., 1468
Scholer, M., 1495
Schremp, J., 124
Schrödinger, 250
Schuster, A., 230
Schwabe, 228
Schwabe, H., 1092, 1112
Schwarzschild, M., 289
Sciama, D., 444
Scobee, F.R., 1108
Scott, D.R., 1320
Seamans, R.C., 85
Seaton, 652
Secchi, P., 1089
Sedov, L., 45, 135
Sembach, K., 669, 670
Serber, R., 122, 124
Seyfert, K., 529
Shafer, R., 419
Shalina, E., 1614
Shapiro, I., 342, 349
Shapiro, M.M., 124
Shapley, A.H., 135
Shapley, H., 648
Shelton, I., 943
Shepard, A.B., 50, 83
Shepherd, G.G., 1664
Shepherd, L.R., 65, 108
Shesta, J., 68
Shields, G., 536
Shih, C.-Y., 1325
Shine, D., 1113
Shivanandan, K., 308, 325
Shoemaker, E., 260, 274, 275, 1295, 1297
Shoemaker, E.M., 1247, 1470
Shoemaker, G., 1473
Shonka, F., 124
Silverberg, R., 419
Silverstein, A., 72
Simpson, J.A., 134, 135, 137, 142, 146, 148, 231, 233, 677, 1181, 1224
Singer, F., 45
Singer, S.F., 70
Skinner, L., 65
Sjogren, W., 1300
Slater, G. 1093
Slipher, 529
Smith, 1657
Smith, A., 291
Smith, A.M.O., 65
Smith, B., 1471
Smith, B.A., 1455
Smith, D., 1638
Smith, E.J., 146
Smith, G., 320
Smith, T.E., 1248
Smoot, F.G., 321
Smoot, G., 404, 419
Smyth, M., 1102
Snell, A.H., 124
Snowden, S.J., 603
Snowden, S.L., 581
Snyder, C.W., 1116

Soberman, R.K., 146
Soifer, B.T. , 318, 319, 325
Soifer, T., 320
Solanki, S.K., 1065
Southworth, G.C., 1101
Spettel, B., 1379
Spitzer, L., 67, 99, 100, 205, 288, 289, 291, 297, 565, 608, 649, 652, 653, 660, 670
Spörer, G., 1094, 1095
St John, C., 204
Stacey, G., 325
Stagg, J., 1102
Stearns, J., 124
Stecher, T., 291
Steeghs, D. 808
Stern, 1535
Stern, B., 878
Stewart, B., 277
Stone, E., 608, 985
Stone, E.C., 137
Stone, E.J., 303
Stoney, J., 228
Störmer, 1089
Street, J.C., 124
Strom, R., 1301
Stromgren, B., 648, 651, 652
Strong, J., 306
Strong, K., 1107
Struve, O., 189, 648
Stuhlinger, E., 57, 59, 71
Sturrock, P., 1111
Summerfied, M., 65, 66
Suomi, V., 1655
Surdej, J., 441
Svensson, R., 558
Swann, G., 1320
Swann, W., 1098, 1109
Swann, W.F.G., 124
Sweet, P., 1111
Swings, J.-P., 465

Tagliaferri. G., 558
Tammann, G.A., 373
Tanaka, Y., 839
Tandberg-Hanssen, E., 1100
Taylor, A.H., 278
Taylor, F.W., 1405, 1663
Taylor, J., 359
Teller, E., 124
Teller, V., 1498
Tennyson, A., 260
Tenorio, L., 419
Terrile, R., 1455
Thomas, 200
Thomas, N., 1451
Thomson, J.J., 228, 229, 1326
Tichomirov, N., 33
Tielens, A.G.G.M., 607
Tikhonravov, M., 34
Titov, G., 50
Toftoy, H., 36, 37, 66, 205
Toksoz, N., 1298
Tolstoy, A., 29
Tombaugh, C., 1433, 1454
Tousey, R., 41, 67, 100, 205–207, 218, 219, 290
Townes, C.H., 315, 316, 323, 609
Trauger, J.T., 1468
Trendelenburg, E.A., 324
Treumann, R.A., 21, 1495, 1550
Treves, A., 558

Trimble, V., 541
Truax, R.C., 65
Trümper, J., 100, 851
Tsander, 33, 34, 64
Tsiolkovsky, K.E., 30, 31, 33, 60, 61, 72, 108
Tsunemi, H., 937
Tsuruta, S., 750
Tucker, G., 28
Tucker, W., 724
Tukachevsky, M., 34
Tupolev, A., 34
Tuve, M.A., 41, 67, 205, 278

Unavane, 714
Underwood, J., 219
Urey, H., 258, 261, 275, 1295, 1297, 1329, 1338, 1378
Urry, M., 554, 558
Ustinov, D., 38

Vachon, W.P., 1614
Vaiana, G., 219
Vaiana, G.S., 878
Vaisberg, O., 1181
Vallarta, M.S., 124
van Allen, J.A., 41, 42, 44, 47, 48, 51, 57, 59, 67, 68, 70, 136, 137, 146, 148, 153, 205, 259, 272, 283, 1497
van de Hulst, H., 54, 233
van der Klis, M., 762, 811
van Dishoeck, E.F., 607
van Duinen, R.J., 316, 317, 319, 324, 328
van Mieghem, J., 135
van Oort, J., 710
van Paradijs, J., 500, 811
Vasyliunas, V., 1519
Vedenov, 1514
Velikhov, 1514
Verne, J., 28, 30, 31, 64, 271
Vernov, S., 47
Vessot, R., 341
Vinogradov. A.P., 1297
Vogt, R., 129
Völk, H., 184
Vololff, G., 721, 759
Voltaire, 27
von Brauchitsch, 63
von Braun, W., 59, 61–72, 81, 83, 85, 93, 100, 105, 112, 203–205, 259, 1497
von Eötvös, R., 336
von Humboldt, A., 228, 1568
von Kármán, T., 42, 65, 69, 111
von Neumann, J., 43
von Steiger, R., 1115,1526

Waelkens, C., 857
Waldmeier, M., 1101, 1106
Walker Fillius, R., 146
Walker, R., 1297
Walker, R.G., 313, 314, 317, 319, 321, 328
Walter, F., 878
Wänke, H., 1377
Ward, D.B., 315
Wasserburg, G., 1297
Waters, J.W. , 1663
Watson-Watt, R.A., 1561
Weaver, H., 663, 1248
Webb, J.E., 85, 86, 112
Webster, R., 533

Wegener, A., 1481
Weinberg, J., 1164, 1166
Weiss, R., 314, 319, 321, 402, 419
Wells, H.G., 29, 260
Welty, D., 655, 659, 671
Wenger, S., 21
Wenzel, K.-P., 963
Werner, M.W., 325
Wernsing, W., 309
Wescott, E.M., 185
Wesselius, P.R., 319, 324
Wetherill, G., 1297, 1370
Wheeler, J.A., 124, 759
Whipple, F., 42, 67, 70, 258, 291, 1166, 1237, 1279, 1281, 1282, 1287
White, N.E., 823
Wijers, R.A.M.J. , 499
Wijnbergen, J., 316, 324
Wilcox, J.M., 138
Wildhusen, F., 207
Wilhelms, D., 1297
Wilkins, J., 27
Wilkinson, D.T., 321, 419
Will, C.M., 353
Williams, I., 1164
Williams, L., 1208
Willson, R.D., 1665
Wilson, O., 649
Wilson, P., 1638
Wilson, R., 295
Wilson, V.C., 124
Winckler, J., 1107
Winkler, J., 33
Winningham, J.D., 1665
Withbroe, G., 1106
Witte, M., 1204
Witteborn, F.C., 317
Wolf, M., 1163
Wolf, Th., 118
Wolfe, J.H., 146
Wollan, E.O., 123, 124
Woody, D.P., 314
Wordie, J.M., 135
Wright, E., 419
Wright, E.L., 321
Wright, N., 404
Wulf, 678
Wyld, J.H., 65

York, D., 659, 662, 671
Young, C.A., 1089
Young, D., 70, 83
Young, E.T., 319
Young, J., 1320
Young, J.W., 94

Zähringer, J., 1169
Zahn, U.V., 1415
Zajac, B., 308
Zakharov, 1514
Zamrani, G., 533
Zank, G., 1208
Zanstra, 917
Zel'Dovitch, 759
Zirin, 651
Zlotnicki, V., 1649
Zwicky, F., 67, 400, 444, 460, 678, 721

Subject index

1C 4663, 916
1PN, 355, 356, 358, 364,
2dF, 461, 465
2PN, 358
2.5PN, 358
3C 48, 530
3C 58, 953
3C 273, 306, 530, 538, 543
3C 279, 551, 553
3C 295, 530
3PN, 358
4U0352+52, 824
4U 1700-37, 832
23 Ori, 655–657
30 Doradus, 714
53 Arietis, 651
53 B, 408
53 B sky map, 408
0405–385, 557
2060 Chiron, 1454
$\frac{1}{4}$ keV, 590
$\frac{1}{4}$ keV background models, 585
$\frac{1}{4}$ keV halo, 601
$\frac{1}{4}$ keV shadows, 592
$\frac{1}{4}$ KeV SXRB, 582, 585, 586
$\frac{1}{4}$ KeV SXRB, 584
$\frac{3}{4}$ keV band, 594

A0620-00, 765
A1909 + 04, 833
A 31, 925
A 33, 925
A 36, 925
A 74, 932
α Centauri, 453, 1208
α Cygnus, 650
α-disks, 794
α Oph, 655
α Orionis, 306
α SCO, 647
α Vir, 656
A-1, 62
A-2, 62
A-3, 62
A-4, 76
A-4 production line, 38
A-4 rocket, 35, 62, 63, 66, 105
A-4 test rocket, 59
A-5, 62
AAS, 1009, 1013
AAVSO, 796, 802
Abell, 574
Abell 194, 486
Abell 370, 444, 458, 459
Abell 2199, 482
Abell 2218, 493

ABMDA, 326
ABRIXAS, 102
absolute omnidirectional intensity, 164
absolute visual magnitudes, 813
absorption cross-section, 583
absorption model, 585
absorption of soft X-rays by circumstellar matter, 830
absorption of X-ray source, 831
absorption spectrum of SSS, 807
abundance, 13
abundance measurements, 11
abundance of chemical elements in cosmic-ray sources, 689
abundance of hydrogen and helium, 918
abundance of light nuclides, 429
abundance of stable secondary isotopes, 686
abundance of the anomalous component, 993
abundance properties of halo ISM, 667
abundance ratios, 1445
abundance ratios in giant planets, 1439
abundance variations, 649
abundances in halo gas, 670
abundances in planetary nebulae, 929
abundances of cometary volatile material, 1286
abundances of siderophile elements, 1388
AC118, 460
accelerating jet model, 554
acceleration at the Earth's bow shock, 981
acceleration by supernova, 688
acceleration efficiencies, 998
acceleration in wave-particle interaction, 1519
acceleration mechanism, 985
acceleration mechanisms of cosmic rays, 688
acceleration mechanisms of galactic cosmic rays, 689
acceleration of ACRs, 996
acceleration of CIR particles, 985
acceleration of corotating particles, 988
acceleration of cosmic-ray particles, 693
acceleration of cosmic rays, 693
acceleration of heliospheric particle, 963
acceleration of particles at the Sun, 1107
acceleration of particles in reconnection, 1525
acceleration of protons, 140
acceleration of SEPs, 975
acceleration regions for ACRs, 998
ACCESS, 694
access to space, 288
accretion disk around neutron star, 828
accreting companion, 776
accreting Neutron stars, 768
accretion binaries, 564
accretion disk, 532, 792, 794
accretion-disk material, 799
accretion-disk models, 536
accretion-disk winds, 797
ACC, 1144
accretion flow, 828, 833
accretion from stellar wind, 773

accretion from wind, 829
accretion luminosity, 794
accretion of Mars, 1366
accretion onto white dwarf, 804
accretion wakes, 831
accretion-powered pulsars, 769
ACE, 681, 686, 694, 964–966, 971, 973, 974, 976, 994, 1104, 1112, 1144
ACES, 1630
ACI, 734
Acidalia Planitia, 1362
ACIS, 463, 733, 739
ACIS-I, 494
ACM, 1164
ACO, 409
acoustic holography, 1053
acoustic oscillations, 1039
acoustic shock waves, 1073
acoustic wave, 234
acoustic wave speed beneath sunspot, 1053
ACR, 17, 143, 964, 987, 990, 991, 993, 995–998, 1156, 1199, 1200, 1202, 1213, 1533,
ACRIM, 1041, 1050
ACRIM II, 1663, 1665
Active Cavity Radiometer, 1048
Active Cavity Radiometer Irradiance Monitor (see ACRIM)
active comets, 1244
Active Galactic Nuclei (see AGN)
Active Magnetospheric Particle Tracer Explorer (see AMPTE)
Active Microwave (see AMW)
active Sun, 1089
activity and topography of comet 1P/Halley, 1256
ADAF, 369, 558, 844, 845, 851
ADEOS, 1666, 1667
adiabatic deceleration, 1154
adiabatic invariants, 175
adiabatically cooled plasma, 603
ADONIS, 1245
ADOS, 1630
ADS, 501
advance rocket systems, 107
Advanced Ballistic Missile Defense Agency (see ABMDA)
Advanced Charge Composition (see ACC)
Advanced Composition Explorer (see ACE)
Advanced Cosmic Ray Composition Experiment (see ACCESS)
Advanced Microwave Scanning Radiometer (see AMSR)
Advanced Research Projects Agency (see ARPA)
Advanced Synthetic Aperture Radar (see ASAR)
Advanced Satellite for Cosmology and Astrophysics (see ASCA)
Advanced Very High Resolution Radiometer (AVHRR)
Advanced Satellite for Cosmology and Astrophysics (see ASCA)
advanced X-ray Astrophysics Facility (see AXAF)
advection-dominated accretion flow (see ADAF)
advertising in space, 13
AE, 651
AE Aqr, 793, 806
aerobee, 42, 59, 68, 71, 150, 153, 159, 170, 205, 211, 313, 309
aerobee High, 68
aerobee rocket, 283, 290, 309
aerodynamic braking, 98
AESTUS, 105
AFCRL, 309, 311, 313, 325
AFCRL/AGL, 328
AFGL, 313, 314, 317, 325, 326
AFRCL, 41, 1204
afterglow, 509
afterglow light-curve of GRB, 512
afterglow of GRB, 514, 515
afterglow polarization, 518
AGARD, 283
AGASA, 692

AGB, 14, 437, 612, 613, 631–635, 706, 774, 808, 867, 930, 931, 933
AGB stars, 15
AGB winds, 633
age of lunar highland rocks, 1334
age of the Earth, 11, 1325, 1327, 1330
age of the Solar System, 11
age of the Sun, 1035
age of the Universe, 390, 392
Agenda 21, 1585
ages of iron meteorites, 1329
ages of lunar mare basalts, 1349
AGN, 297, 462, 463, 563–565, 570, 571, 573, 574, 669, 852, 957
AGN NGC 1068, 572
agriculture, 1586
AGU, 1164, 1646
AGW, 1567–1569
AGW–TID relationship, 1568
Air Force Cambridge Research Laboratories (see AFCRL)
Air Force Geophysical Laboratory (see ATGL)
Air-glow, 189, 198, 1564
Air–Sea–Ice interaction, 1612
AIS, 33, 65
AJISAI, 1629
Akebono satellite, 1548
Akeno Giant Air Shower Array (see AGASA)
AKG, 28
AKR, 1520, 1521, 1551
AKR fine structure, 1521
Alba Patera, 1307
albedo, 257, 262, 1587
Alfvén adiabatic invariant, 174
Alfvén layer, 1540
Alfvén Mach number, 1010
Alfvén Radius, 768
Alfvén speed, 1009
Alfvén wave, 234, 241, 1131
Algae bloom, 1606, 1607
Algol, 760, 888, 889
algorithms estimating sea ice concentration, 1594
ALH77005, 1363, 1368
ALH84001, 1360, 1361
alkali metal cloud experiments, 189
alkali-suite rocks, 1346, 1347
all-sky survey, 475, 585, 586, 885
all-sky survey data, 881
all-sky X-ray survey, 478
All-Union Society for the Study of Interplanetary Flight, 64
ALMA, 462, 463
Along Track Scanning Radiometer (see ATSR)
ALOSI, 1630
Alpha Magnetic Spectrometer (see AMS)
alpha particles in the solar wind, 1129
Alpha Proton X-ray Spectrometer (see APXS)
ALSEP, 1479
Altas, 74
altimetric missions, 1645
altimetry, 1631, 1635
aluminous mare basalt magma, 1351
aluminous mare basalt (see AMB)
aluminous soil, 1353
AM CVn, 777
AM Her, 793, 803, 804, 1357, 1368
AM Her systems, 800, 801
Amalthea, 166, 1452, 1453, 1457, 1458
Amazonis, 258, 267
Amazonis Planitia, 1362
AMB, 1350, 1354
ambient radioactivity, 231
AMD, 1172, 1173
American Geophysical Union (see AGU)

American Interplanetary Society (see AIS)
American Rocket Society (see ARS)
amorphous ice, 1260
amorphous silicates, 612
AMPTE, 967, 995, 1021, 1191, 1202, 1209, 1502, 1504, 1534, 1544
AMPTE/IRM, 965, 1502, 1524
AMS, 694
AMSR, 1598
AMW, 1588, 1590, 1591, 1611, 1617
analyses of lunar samples, 1378
analysis of micrometeoroids, 1170
analytical precision, 1333
analyzing helioseismic data, 1047
ananke, 1452, 1453
ancient Martian crust, 1359
Andrastea, 1453
angular momentum loss, 877
angular power spectrum of CMB anisotropy, 409, 411, 413
anisotropy in CMB, 404
anisotropy of cosmic rays, 688
announcement of opportunity (see AO)
anomalous collisions, 1503
anomalous component of cosmic rays, 1157
anomalous cosmic radiation, 143
anomalous cosmic ray (see ACR)
anomalous hydrogen, 992
anomalous transport coefficients, 1525
Anomalous X-ray pulsars (see AXP)
ANS, 3, 316, 563, 683, 795, 815, 919
ANS-B, 316
Antarctic ozone hole, 5
Antennae, 565
Antennae galaxies, 573
antiparellel field merging, 1017
antiproton, 681
antiquity of Moon, 1338
AO, 316
aperiodic variability, 796, 816
APL, 41, 153
Apollo, 164, 259, 340, 1381, 1489
Apollo 11, 345
Apollo 11, 1337–1339, 1342, 1345, 1351, 1354, 1378, 1383
Apollo 11 landing on the Moon, 87
Apollo 11 Mare Basalts, 1352
Apollo 11 soil samples and soil breccias, 1379
Apollo 12, 1338, 1339, 1345, 1346, 1351, 1353, 1354, 1383
Apollo 12 Mare Basalts, 1352
Apollo 12 soil and rock samples, 1380
Apollo 14, 1339, 1342, 1345, 1348,-1353, 1354, 1382
Apollo 14 Mare Basalts, 1349, 1352, 1356
Apollo 14 samples, 1381
Apollo 15, 10, 1299, 1339, 1344, 1346, 1348–1351, 1353, 1357
Apollo 15 Mare Basalts, 1352
Apollo 15 samples, 1381, 1382
Apollo 16, 502, 1298, 1339, 1341, 1344, 1348, 1350
Apollo 16 samples, 1381
Apollo 17, 1183, 1300, 1338, 1339, 1342–1344, 1348, 1351, 1353, 1357, 1383
Apollo 17 Mare Basalts, 1352
Apollo 17 samples, 1381
Apollo Command and Service Module, 90
Apollo landing, 257
Apollo lunar samples, 1331, 1370
Apollo Lunar Surface Experiments Packages (see ALSEP)
Apollo mission, 4, 1124, 1137
Apollo program, 8, 1337
Apollo seismic data, 1301
Apollo Telescope Mount, 13, 219
Apollo-Saturn V rocket, 74
Apollo–Soyuz, 287
Apollo–Sojuz orbital rendezvous, 91
application technology satellites (see ATS)
Applied Physics Laboratory (see APL)
APT, 1656
APX, 1377, 1417
APXS, 1394, 1396, 1397
Aquila Rift, 597
Araneus diadematus, 88, 89
ARC, 109
Arctic ice area, 1595
ARGOS, 1630, 1643
Argyre, 267
Ariane, 59, 75, 93
Ariane 4 rocket, 322
Ariane program, 64
Ariane project, 105
Ariane rockets, 103
Ariane V, 72
Ariane V33, 718
Ariane-4, 105
Ariane-5 rocket system, 105
Arianespace, 105
Ariel, 77, 1453, 1459, 1472
Ariel 1, 51
Ariel 1 satellite, 213
Ariel 5, 475, 478, 479, 549, 563, 794
Ariel satellite, 51
Ariel V survey, 534, 537
Ariel V, 543, 824, 826
Aries rocket, 314
ARISTOTELES, 1644
Arizona crater, 261
ARPA, 48, 83, 313
ARS, 33, 1627
arsenic-doped silicon, 326
Arsia Mons, 1305
artificial cometary tail, 180
artificial comets, 1524
artificial Earth Satellite, 29
artificial little moon, 69
artificial plasma cloud, 181
artificial satellites, 259
AS&E, 562, 94
ASAR, 1610, 1617
ASCA, 383, 474, 475, 485–493, 535, 537, 539, 543, 561–563, 567, 569, 689, 725, 732, 734, 741, 743, 745, 746, 748, 803, 807, 830, 831, 833, 845, 849, 851, 879, 880, 885–889, 945, 946, 948, 950, 951, 954–956
ASCA CCD, 569, 948
ASCA SIS, 744
ashen light on Venus, 1412
ASI, 1629
Asp rockets, 282
asteriod-meteoriod astronomy, 146
Astérix, 52
Astérix satellite, 52
asteroid, 171, 349
asteroid 4 Vesta, 1361
asteroid belt, 155, 1163, 1426
Asteroids, Comets, Meteors (see ACM)
asteroid meteoroid detector (see AMD)
asteroid orbits, 1164
asteroid spin state, 1264
asteroid Vesta, 1386
asteroidal band, 1177
asteroidal objects, 1457
asteroids, relationship with comets, 1268
ASTRO, 298
Astro 1, 798
Astro 2, 798
ASTRO-E, 886, 889

Astro-E project, 103
Astro-E, 475, 488
ASTRO-F, 576, 609
astrobiology, 21, 1475
astroblemes, 261
ASTROD, 1056
astrometric shift, 453
astrometry, 16
astronomical observatory, 288
astronomical silicates, 866
astronomical spectroscopy, 1280
Astronomische Nederlandse Satelliet (see ANS)
astroparticle laboratory, 137
astrophysical constraints, 703
astrophysical ices, 1260
astrophysics and magnetospheric physics, 1553
astrophysics of hot plasmas, 565
astrophysics of isolated neutron stars, 725
asymmetric lens models, 449
Asymptotic Giant Branch (see AGB)
Atacama Large Millimetre Array (see ALMA)
Atlantic Research Corp. (see ARC)
Atlantis, 94
Atlas, 43, 57, 76, 77, 93, 1453
Atlas Agena, 98
Atlas Centaur, 80, 98
Atlas rocket, 308, 309
Atlas-Agena, 80
Atlas-Centaur, 80
ATM-Skylab, 878
atmosphere, 195
atmosphere of a planet, 162
atmosphere of Io, 1467
atmosphere of Mercury, 1407
atmosphere of Venus, 306
atmosphere sandwich structure, 196, 197
atmospheres of the giant planets, 1438
atmospheres of the terrestrial planets, 1405
atmospheric absorption in the infrared, 303
atmospheric abundance of helium and hydrogen, 929
atmospheric effects, 1470
Atmospheric Gravity Wave (see AGW)
atmospheric nuclear tests, 1523
atmospheric parameter, 1654
atmospheric sodium layer, 198
Atmospheric Structure Instrument, 1428
atmospheric tides, 1568
atmospheric windows, 302
ATN, 1664
atoll sources, 817, 818
ATS, 77
ATS-6, 1543
ATSR, 1597
Auger ionization, 898
AURA, 450, 455, 459
Aurigae, 651
aurora, 5, 225, 226, 228, 245, 1495, 1535
aurora borealis, 7
aurorae on Jupiter, 7
aurorae on Saturn, 7
aurorae, 158
auroral bremsstrahlung, 159
auroral bulk upflow, 1547
auroral E layer, 1563
auroral electrons, 1569
auroral ionosphere, 184
auroral kilometric radiation (see AKR)
auroral magnetosphere, 1517
auroral radiation, 155
auroral source region, 1521
auroral substorm, 1544

auroral zone, 1519, 1537
auroral zones at Uranus, 1488
Automatic Pixture Transmission (see APT)
AVHRR, 1590, 1596, 1597, 1605, 1610, 1617, 1656
AXAF, 101, 102, 463, 879, 886, 887, 889, 946
AXAF CCD, 463
axially symmetric lens models, 448
AXP, 726, 727, 736, 742
AXS J161730-505505, 954
Azimuthal drift violation, 1419
AZUR, 53

B0655 + 64, 363
B1534 + 12, 363
B1855 + 09, 363
B2127 + 11C, 363
β Cen, 657
β CMa, 658
β-meteoroids, 1179, 1181
β Pictoris, 864
β Pictoris analogues, 865
B supergiants, 901
B-1, 76
B-L coordinate system, 157
B-type companion, 763
BABI, 1337, 1355
back-scatter techniques, 1661
BAL, 539
balck-hole X-ray binaries, 849
Baldwin effect, 536
Ball Brothers Research Corporation (see BBRC)
ballerina skirt, 326, 1120
ballistic missiles, 42
Balloon Borne UV Stellar Spectrograph (see BUSS)
Balloon Infrared Platform (see BIRAP)
balloon observation, 315
balloon-borne telescope, 306, 312
barium cloud experiments, 180, 1523
barium clouds, 179, 181, 192
barograph, 40
Bartels diagram, 1103
baryon catastrophe, 567
baryon density, 385
baryon number, 430
baryonic dark matter, 712
baryonic density, 430, 435
baryonic fraction, 489
baryonic mass of clusters of galaxies, 489
baryonic material in clusters, 476
baryonic matter in the Universe, 1495
baryonic number, 431
basalt compositions, 1355
basalt fragments, 1353
Basalt Sources of Earth, Moon and Mars, 1368
Basaltic Achondrite Best Initial (see BABI)
basaltic meteorites, 1361
basaltic rock, 8
basaltic shergottites, 1364
basaltic soils on Mars, 265
basaltic volcanism, 265
basaltic/iherzolitic rocks, 1362
base coronal model for hot stars, 879
bathymetry, 1590
BATSE, 284, 499, 500, 505–508, 516, 518, 521, 733, 769, 827, 844
BB, 731
BBN, 369, 423–426, 429–431, 433, 435, 437, 657
BBN test for cosmology, 432
BBRC, 316
BBSO, 1048, 1050
BBXRT, 484, 485
BCF, 211

BD + 30, 929
BD + 30 3639, 932
Be star optical counterpart, 824
Be star polarization, 902
Be transient, 825
Be X-ray binaries, 766, 778
Be X-ray transient system, 829
Beagle 2 lander, 1421
beaming, 551, 552
BEB, 1626, 1628
BEC, 1628
behind bow shock, 1227
Belinda, 1453
Bell Labs H I survey, 592
BEN, 1508
benzene, 14
BeppoSAX, 285, 474, 475, 485, 489, 492, 510, 535, 538, 543, 561–563, 725, 732, 745, 748, 844, 851, 853, 880, 886–889, 947, 952
BeppoSAX X-ray satellite, 17
beryllium, 19, 433
beryllium-9, 429, 437
beryllium-10 in Antarctic ice, 1213
beryllium optics, 313, 327
BESS, 686
Beta Regio, 1313
BH, 774
BHC, 825
bi-stability limit for B supergiants, 901
Bianca, 1453
BIB, 327
BICE, 614
Biermann's conjecture, 233
Biermann–Chapman dilemma, 235, 247
big bang, 20, 423, 431, 584, 939, 943
big bang expansion, 542
Big Bang Nucleosynthesis (see BBN)
big-bear solar observatory, 1047
big-blue bump, 534, 536
big chill, 431
big crunch, 431
BIH, 1634
binary evolution models, 761
binary lenses, 454
binary pulsar and general relativity, 359
binary pulsars, 362, 363
binary pulsars and scalar-tensor gravity, 363
binary pulsars and alternative gravitational theories, 362
binary radio pulsars, 763, 766–768, 778
binaries with compact objects, 762
binary X-ray sources, 772
biological pump, 1600
BIPM, 1633
BIRAMIS, 1632,
BIRAP, 316
binary pulsar PSR 1913 + 16, 360
Birkeland currents, 1569
Birmingham whole-disk data, 1043
BiSON, 1047, 1056
Bison, 43
BL, 1539
BL Lac, 550, 554, 556
BL Lacerate objectives, 550
BL Lacertae, 554
BL Lacs, 558
black-body spectra from type 1 buest, 816
black-hole binaries, 839, 842, 849
black-hole gravity, 549
black hole in Cygnus X-1, 100
black hole X-ray binaries, 765
black holes, 13, 15, 531, 532, 760
black holes identified from mass functions, 841

black-body spectrum of CMBR, 321
black-body radiation spectrum, 302
blazar central engine, 558
blazar emission, 550
blazar evolution, 558
blazar flux, 550
blazar jets, 551
blazar-radiation process, 552
blazar-radio spectrum, 550
blazar sequence, 555, 556
blazars, 326, 535, 543, 544, 549, 550
blazars are variable, 553
blazers emit γ ray, 553
blinkers, 1079
Blocked Impurity Band (see BIB)
blue-background galaxy, 459
blue jets, 1570
Blue Streak, 51
blue supergiant, 760
BMD, 326
BMFT, 96
BMRC, 1604
BNCSR, 51
BNKL, 311
Boeing Delta II rocket, 98, 99
Bok globule, 860
bolometer, 306, 310, 321
bolometry, 305
bombardment of interstellar gas, 433
Bondi–Hoyle accretion radius, 829
BOOMERang 98, 383, 385, 388–390, 415, 416, 418
Bootstrap algorithm, 1594
boron, 19
boron-10 and 11, 429, 437
Bose–Einstein photon energy, 429
boundaries of the heliosphere, 1145, 1156, 1208
boundary layer (see BL)
boundary of the magnetosphere, 1227, 1530, 1541
bow-shock geometry, 1510
bow-shock nebula, 750
braiding, 1083
Brans-Dicke effect, 1045
Brans-Dicke theory of gravity, 1036
breakup of Miranda, 1473
bremsstrahlung, 759, 1107
bright emission from loops, 1082
bright halo galaxies, 567
bright LMC stars, 296
bright spot, 799
bright X-ray sources, 811
brightness curves of comets, 1218
brightness of the Sun, 234
brightness temperature, 1588, 1657
brightness temperature of the Sun, 207
British Interplanetary Society, 65
British National Committee for Space Research (see BNCSR)
Broad Absorption Line quasar (see BAL)
broadband electric noise (see BEN)
broadband GRB spectra, 503
broadband UV photometry, 292
brown dwarfs, 431
BSSI, 1367
bulk composition of Mars, 1391–1393
bulk composition of the Earth, 1393
Bulk Solar System Initial (see BSSI)
bulk speed of solar wind, 1120
bumper project, 42, 68
Bundesministerium Für Forschung und Technologie (see BMFT)
Buneman-wave, 1503
Buote, 567
Buran shuttle, 86

Bureau International del' Heure (see BIH)
Bureau International des Poids et Mesures (see BIPM)
Burst and Transient Source Experiment (see BATSE)
bursts, 816
BUSS, 295
BVSP, 1354, 1357

C11358 + 62, 464
C12244-02, 449
C(M)IR, 1120
C-type asteroids, 1253, 1590
CACTUS, 1631, 1646
CAL83, 807, 808
CAL187, 807
Caliban, 1453, 1456
calibrating cratering chronologies, 1354
Callisto, 96, 1183, 1247, 1249, 1453, 1462, 1463, 1470, 1472, 1473, 1487, 1490
Calypso, 1453, 1459
CAM, 609, 706
CAM-CVF, 609, 620
Canada–France–Hawaii Telescope (see CFHT)
Capella, 876
carbon, 19, 321
carbon cycle, 1600
carbon dioxide of Mars, 260
carbon monoxide, 321
carbon-rich objects, 631
carbonaceous chondrites, 867, 1163, 1379
Carme, 1452, 1453
Carrington map, 1055
Cas A, 581
CASDN, 52
Cassini, 97, 155, 1176, 1429, 1475, 1480
Cassini mission, 1186, 1431
Cassini spacecraft, 169, 1465, 1475
Cassini-Huygens, 1429
Cassini/Huygens mission, 1447
Cassiopeia-A, 651, 945, 952
Cassiopeia-A X-ray image, 946
CASTLES, 456
CASTOR, 1632, 1646
cataclysmic variable, 791
catalogs of clusters, 475
catalogue of nearby stars (see CNS)
Cayley Plains, 1339
CBE, 104, 263, 274, 376, 377, 444, 455, 459, 475, 476, 479–481, 488, 563, 666, 709–711, 725, 731, 733, 830, 831, 833, 807, 817, 886, 889, 945, 946, 956, 995, 1243, 1244, 1248, 1455, 1456, 1502
CCD, 104, 263, 274, 376, 377, 394, 444, 455, 459, 463, 468, 475, 476, 488, 525, 563, 565, 569, 666, 725, 731, 733, 807, 830, 831, 833, 886, 889, 945, 946, 948, 951, 956, 1243, 1244, 1248, 1272, 1275, 1455, 1456, 1478
CCE, 995, 1502
CDA, 1656
CDM, 385, 386, 476, 477
CDOM, 1601
CDS, 777, 1069, 1070, 1078, 1079, 1081
CEFSR, 69
Celescope, 292
celestial mechanics, 146, 226
celestial mechanics of gravitational swingby, 81
CELIAS, 1124, 1130
cellular structure of the plasma universe, 74, 1454, 1531
Cen A/NGC 5128, 565
Cen X-3, 760, 824, 832
CEN-Saclay, 723
Centaures, 180, 568
Centaurus Association, 663
center of mass of the Earth, 1638

central bar, 706
Central Bureau for the Study of Rockets, 64
central regions of clusters, 491
central star temperature, 926
central stellar cluster, 706
centrally dominant galaxy, 475
Centre d' Etudes et des Recherches en Géodynamique et Astrométrie (see CERGA)
Centre National d' Etudes Spatiales (see CNES)
centrifugal barrier, 829
CEOS, 1586, 1610, 1611, 1617, 1623
Cepheid distances, 377
Cepheids, 376
Cepheus Flare, 597
CERGA, 345, 1649
CERN, 54
CERN Bernas, 433
Cesium telluride photocathode, 296
CfA–Arizona Space Telescope Survey (see CASTLES)
CFHT, 444
CGRO, 103, 284, 499, 500, 505–508, 535, 549, 551, 553, 558, 563, 686, 725, 827, 828, 844
CGRO OSSE, 849
Challenger, 94, 1413
Challenging Mini-satellite Payload (see CHAMP)
Chameleon I, 597
CHAMP, 1632, 1643, 1644, 1646
Chandra, 102, 383, 442, 491, 494, 543, 545, 561, 563, 565, 568, 569, 575, 581, 686, 725, 729, 730, 732–734, 739, 804, 806, 889, 890, 947, 953
Chandra High Resolution Camera, 566
Chandra Interactive Analysis of Observations (see CIAO)
Chandra observatory, 494
Chandra X-ray Center, 566
Chandra X-ray Observatory (see CXO)
Chandra X-ray Telescope, 101
chaotic component of interstellar cloud velocities, 649
chaotic variability, 796
chaotization of electron orbits, 1011
Chapman layer, 1565
Chapman–Enskog expansion, 234
charge analyzer, 1124
charge energy and mass (see CHEM)
charge states of the anomalous component, 993
charge-exchange reactions, 1197
charge-state measurements, 993
charge-state spectra of solar wind types, 1126
charged dust dynamics, 1220
charged dust grains, 1182, 1183
charged particle, 120
charged particle composition, 146
charged particle spectrogram, 1543
charging of dust, 1220
Charon, 171, 1433, 1452, 1457
Charon and Pluto, 1469
Chassigny, 1364
Chattonella spp. bloom, 1608
CHEM, 995
chemical abundance in comets, 1290
chemical composition of disk ISM, 658
chemical composition of interstellar clouds, 649, 652
chemical composition of martian soil samples, 1396
chemical composition of Mercury, 1399
chemical composition of nebulae, 928
chemical evolution, 710
chemical evolution of terrestrial planets, 1377
chemical evolution of the Moon, 1377
chemical layering, 1318
chemical makeup of a planet, 1317
chemical mixing in ISM, 658
chemical pollution of the troposphere, 1666

chemical Universe, 302
chemistry of old stars, 708
chemistry of the atmosphere, 1653, 1660
chondritic uniform reservoir (see CHUR)
chlorophyll, 1602
chlorophyll algorithms, 1601
chlorophyll concentration, 1607
chlorophyll pigment concentration, 1607, 1608
CHON, 1283, 1284, 1288, 1289, 1290
chopped photometric channel (see CPC)
chopping, 306, 307
Christmas tree theory, 551
chromosphere, 234, 1068, 1070, 1075, 1076, 1125
chromospheres in red giant stars, 907
chromospheric activity, 862
chromospheric eruptions, 1098
chromospheric gas, 1078
chromospheric heating, 1071
chromospheric helium, 1080
chromospheric models, 1070
chromospheric spectrum, 1070
chronological evolution of Moon and Mars, 1369
chronology of terrestrial planets, 1368
Chryse Planitia, 1306
CHUR, 1360
Chwolson ring, 443, 446
CIAO, 566, 568
CIE, 956
CIR, 139, 140, 457, 458, 901, 963, 975, 978, 982, 983, 985, 987–999, 1118, 1119, 1122, 1126, 1145, 1151–1156, 1141
CIRA, 1566
circulation of the Martian atmosphere, 1419
circumstellar debris disks, 864
circumstellar disk, 858, 860, 865
circumstellar disks around YSOs, 629
circumstellar dust and gas, 630
circumstellar envelopes, 1261
circumstellar gas and dust, 630
circumstellar silicates, 630
CL0024 + 1654, 459
CL1358 + 62, 460
CLAES, 1663
Clandestine nuclear explosions, 501
CLASS, 457, 465
classes of comets, 1242
classical rocketry, 106
classical T Tauri Stars (see CTT)
classification of YSO, 859
Clementine, 347, 1298, 1300
Clementine spacecraft, 1178
climate-change research, 1586
climate history of Mars, 1421
climate monitoring, 1599
climate research, 20, 1591
climate-altering encounter, 648
climatic change on Mars, 1405
clinopyroxenites, 1365
CLIVAR, 1596
clocks freely falling in gravity, 337
clock rates near the Sun, 341
Clorophyll, 1590
close binary evolution, 774
close binary systems, 760
close-up photograph of Mars, 264
closed expander cycle, 76
closed expander cycle rocket engine, 76
cloud chemistry, 1422
cloud cover, 262
cloud layers on Venus, 1413
cloud photopolarimeter, 1412

cloud structure of giant planets, 1440
clouds and dust in terrestrial planet atmospheres, 1407
clouds of Venus, 262, 1408, 1409
cloudy sightlines, 654
Clover-Leaf, 449
CLS, 1643
clumping absorbing medium, 583
cluster cooling time, 481
cluster elemental abundances, 489
cluster evolution, 485, 489
cluster galaxy populations, 492
cluster II, 155, 1104
cluster images, 480
cluster incompleteness bias, 380
cluster membership, 884
cluster metal abundances, 474
cluster metallicities, 883
cluster radio sources, 491
cluster size, 479
cluster temperature, 474
cluster theory, 477
clusters are very old, 474
clusters dominated by dark matter, 473, 475
clusters of galaxies are X-ray objects, 385, 410, 411, 481,
CMB, 369, 374, 383, 385, 387–390, 400, 401, 406, 408, 409, 411, 415–417, 423, 429, 431
CMB COBE-DR, 410
CMB phenomenology, 400
CMBR, 321, 404
CMD, 390, 391
CME, 220, 221, 969, 971, 974–977, 1013, 1054, 1105, 1106, 1108, 1110–1112, 1121, 1122, 1125, 1136, 1143, 1144, 1148, 1152, 1153, 1155, 1158, 1505
CME, 14, 220, 245, 1013, 1105, 1107, 1121, 1143, 1144, 1579
CMIR, 1120
CNES, 52, 104, 105, 201, 1611, 1626, 1629, 1630, 1632, 1636–1639, 1641, 1644
CNES/MICROSCOPE, 1632
CNM, 663–665
CNO, 798, 808, 816
CNR, 52
CNR-Milano, 723
CNRS, 193
CNS, 312–314, 316, 319, 321, 399, 400, 401, 403, 413, 415, 416, 463, 476, 592, 609, 611, 614, 700, 702, 703, 707, 708, 718, 1177, 1178, 1193
CO_2 on Mars, 257
CO_2 on Venus, 257
co-rotating interaction region (see CIR)
coagulation due to Brownian motion, 1237
Coalsack, 597, 598
coastal-zone color scanner (see CZCS)
coastal-zone management, 1586
COBE, 19, 312, 399, 463, 592, 700
COBE basic results, 405
COBE data processing, 403
COBE DIRBE, 408
COBE-DM 2, 405
COBE-DMR, 409, 411–413, 417
COBE instrumental design, 402
COBE mission chronology, 404
COBE project, 402
COBE satellite, 402, 403
COBRAS, 1416
COG, 648, 649, 656, 666
cohesive nucleus model, 1262
cold-cloud temperature, 655
cold dark matter (see CDM)
cold dark-matter model, 415, 416
cold envelopes around YSOs, 622
cold trapping of atmospheric gases, 1467

cold Universe, 301
Cold War, 288
collapse mechanism, 1522
colliding winds of hot stars, 903
collinder 121, 653
collision between Earth and a Mars-sized object, 8
collision of the Earth, 1386
collision, Miranda, 1473
collision-dominated coma, 1222
collisional ionization, 661
collisional plasma, 1515
collisionless Boltzmann equation, 236
collisionless plasma, 228, 1133
collisionless plasma of near-Earth space, 1513
collisionless shock formation, 1511
collisionless shocks, 1510
collisionless shockwaves in hot magnetized plasmas, 1531
collisionopause, 1229, 1242
collison model of asteroids, 1164
colored dissolved organic matter (see CDOM)
color temperatures of hot stars, 289
color–magnitude diagram (see CMD)
colours on the surface of Io, 1471
Columbia, 94
column densities for interstellar hydrogen, 658, 13
Coma cluster, 18, 381, 482, 484
comet 2P/Encke, 1243, 1280
comet 9P/Tempel 1, 1280
comet 10P/Tempel 2, 1244
comet 19P/Borrelly comet 107P/Wilson–Harrington, 1280, 1242–1244, 1257, 1265, 1279
comet 21P/Giacobini–Zinner, 1279
comet 26P/Grigg–Skjellerup, 1279
comet 28P/Neujmin 1, 1244
comet 29P/Schwassmann–Wachmann 1, 1244
comet 46P/Wirtanen, 1280
comet 49P/Arend-Rigaux, 1244
comet 55P/Tempel-Tuttle, 1243
comet 73P/Schwassmann–Wachmann 3, 1280
comet 81P/Wild 2, 1280
comet 95P/Chiron, 1243, 1243
comet 107P/Wilson-Harrington, 1244
comet antisolar tail, 233
comet axis ratios, 1247
comet C/1980 E1, 1243
comet C/1995 O1, 1243
comet death, 1268
comet Encke, 1181, 1236
comet final decay, 1268
comet flyby missions, 1224
comet Giacobini-Zinner, 1124, 1181, 1227, 1228
comet Grigg-Skjellerup, 1227
comet Hale–Bopp, 626, 1236, 1265, 1277, 1283, 1285, 1286, 1289, 1290
comet Halley, 11, 180, 1181, 1186, 1217, 1222, 1223, 1227, 1229, 1230, 1236, 1249, 1251, 1286, 1288, 1289
comet Halley dust, 1176
comet Halley flyby, 1249
comet Halley nucleus properties, 1255
comet Halley's carbon grains, 1285
comet Hyakutake, 1259, 1266, 1283, 1285, 1286, 1289, 1290
comet ion, 248
comet-mineralogical composition, 1284
comet missions, 1181
comet-model calculations, 1261
comet-nucleus split, 1248
comet P/Schwassman-Wachmann 1, 1236
comet P/Wirtanen, 1236
Comet Rendezvous and Asteroid Flyby (see CRAF)
comet Shoemaker-Levy 9, 1247, 1428, 1446, 1456, 1458
comet-simulation experiments, 1219

comet tail, 232
comet Tempel 2, 1186
comet Tempel-Tuttle, 1181
comet-thermal parameters, 1262
comet trails, 1177, 1181
comet Whipple-Fedke, 180
comet Wild 2 dust, 1176
comet Wirtanen, 1221
cometary activity, 1263
cometary constituents, 1277, 1280
cometary dust, size and composition, 1180
cometary dynamics, 1240
cometary geometric albedos, 1247
cometary ice composition, 626
cometary ice inventory, 625
cometary ions, 1231
cometary nuclei, 1235, 1247, 1277, 1282
cometary nucleus, 13, 1236
cometary nucleus material, 1261
cometary nucleus microstructure, 1262
cometary nucleus surface, 1267
cometary orbits, 1240
cometary refractory matrix, 1266
cometary space missions, 1279
cometary surface, 1220, 1286
cometary volatiles, 1285, 1287, 1288
cometary water vapor, 1428
cometary X-ray emission, 1229
cometesimals, 1235, 1237, 1240, 1249
cometopause, 1229
cometospace, 11, 171, 1164, 1217, 1226
comets and the interstellar medium, 1289
comets in the Oort cloud, 1218
comets to dust, 1268
Comité d' Action Scientifique de la Défense Nationale (see CASDN)
Command and Data Acquisition (see CDA)
comets, relationship with asteroids, 1268
Commission Préparatoire Européenne pour la Recherche Spatiale (see COPERS)
Committee for Evaluating the Feasibility of Space Rocketry, 69
Committee for Planetary and Lunar Exploration (see COMPLEX)
Committee on Earth Observation Satellites (see CEOS)
Committee on Space Research (see COSPAR)
common envelope evolution, 776
communicaion satellite, 39, 48
communication with Martians, 29
compact binaries, 823
compact bodies, 355
compact objects and SNRs, 952
compact/X-ray binary sources, 15
companion stars, 767
comparative magnetospheres, 1554
comparative planetology, 1317
COMPLEX, 1421, 1428
complex databases, 321
complex dynamical structure, 195
composition of central star atmospheres, 929
composition of cometary dust, 1283
composition of interstellar grains, 612
composition of Martian mantle, 1393
composition of Martian meteorites, 1397
composition of terrestrial planet atmospheres, 1406
composition of the bulk Moon, 1385
composition of the solar wind, 1147
compositional zoning of carbonate minerals, 1360
compositions of Mars soil, 1391
Comptel, 684, 733, 955
Compton drag, 553
Compton Gamma Ray Observatory (see CGRO)
Compton rocket effect, 553

Comptonization, 849
conductive heating interface, 669
confinement time of cosmic rays, 685
conservative cvolution, 777
Consiglio Nazionale delle Ricerche (see CNR)
constellation-X, 575, 752
constellation-X mission, 494
continental drift, 261
continious viewing zone (see CVZ)
continuum emission from CVs, 792
continuum gamma-ray radiation, 684
CONTOUR, 1280
contour, 1236
convection away from the Sun, 1154
convection of plasma sheet particles, 1543
convection pattern of magnetospheric plasma, 1543
convection zone, 1041
convective heat flux, 1035
Convention to Combat Desertification, 1585
cool circumstellar material, 863
cool gas, 13
cool giant stars in binary systems, 905
cool-star dividing line, 904
cooling flows, 492
cooling neutron star, 729
cooling photodissociation, 315
coolstar winds, 907
coorbital satellite orbits, 1459
Copernican Revolution, 258, 260
Copernicus, 584, 655, 657, 658, 660–662, 824, 829, 897, 1339
Copernicus program, 293
Copernicus satellite, 289, 293, 294, 650, 654, 901,
COPERS, 212
copper-doped germanium, 304
Coprates, 266
Corbet diagram, 828
Cordelia, 13, 244, 250, 1075, 1076, 1081, 1453, 1460
corona, 13
corona of the Sun, 5
corona plus cool wind model, 897
coronae of late-type stars, 879
coronagraph, 1089
coronagraph image, 1136
coronagraphy, 233
coronal abundances, 886, 887
coronal activity in stars, 888
coronal diagnostic spectrometer (see CDS)
coronal emission in late-type stars, 881, 882
coronal emission lines, 234
coronal expansion, 243, 245–247
coronal gas, 653
coronal green-emission line, 233
coronal heat input, 244
coronal heating, 876
coronal heating theory, 877
coronal holes, 220, 1072, 1077, 1078, 1081, 1106, 1107, 1119, 1120, 1134, 1147, 1150, 1153
coronal loops, 1081, 1082
coronal-magnetic field, 1147
coronal-magnetic neutral line, 1142
coronal-mass ejection, (see CME)
coronal plus cool wind model, 899
coronal rocket observations, 210
coronal spectral lines, 233
coronal streamer, 247, 1120
coronal temperature, 235
coronal X-ray emission, 877

COROT, 1059
corotating events, 985

corotating ion events, 983, 987
Corotating interaction regions (see CIR)
Corotating merged interaction regions (see MIR)
Corporal, 65
COS, 298, 117, 122, 124
COS B, 103, 284, 553, 684, 723
cosmic abundance, 940
Cosmic Background Explorer (see COBE)
cosmic background radiation, 308
cosmic dust, 1170
cosmic-dust collection, 1172
cosmic-dust collection in deep-sea sediments, 1172
cosmic-dust panel, 1164
Cosmic lens cell sky survey (see CLASS)
cosmic-microwave background radiation (see CMBR)
cosmic-microwave background (see CMB)
cosmic mirage, 463
cosmic origins spectrograph (see COS)
cosmic-ray, 13
cosmic-ray albedo neutron decay, 162
cosmic-ray astrophysics, 677
cosmic-ray birthplace, 951
cosmic-ray composition, 689
cosmic-ray charged particles, 120
cosmic-ray energy spectra, 679
cosmic-ray flare, 232
cosmic-ray generation, 678
cosmic-ray instrument, 154
cosmic-ray instrumentation, 136
cosmic-ray intensity, 128, 225, 245
cosmic-ray intensity and large flares, 1101
cosmic-ray in the intensity in the inner heliosphere, 1155
cosmic-ray modulation, 250, 682, 1154
cosmic-ray neutron monitor, 231, 232
cosmic-ray nuclear component, 137
cosmic-ray observations, 678
cosmic-ray pressure, 691
cosmic-ray research, 138
cosmic ray spallation, 666
cosmic ray spectrum, 679, 691
cosmic-ray stream instability, 16, 230, 231, 690, 1196
cosmic-ray isotopes, 681
cosmic-ray isotope spectrometer (see CRIS)
cosmic-ray modulation, 1154
cosmic-ray propagation in galaxy, 685, 688
cosmic-ray transport, 686, 687
cosmic rays in astrophysical research, 683
cosmic rays in the heliosphere, 1141
cosmic virial theorem, 387
cosmic winds, 1141
cosmochemistry, 261
cosmogenic nuclides, 1329, 1335
cosmoid hypothesis, 1173
cosmological anisotropy to CMB, 417
cosmological constant, 373, 388, 416, 457
cosmological density, 457, 458
cosmological parameters, 4, 17, 431, 456,
Cosmos, 1168, 1593
COSPAR, 192, 199, 662, 1164, 1229, 1626, 1627, 1634
counting stars, 702
CPA, 1116
CPC, 318, 319, 1199
Crab Nebula, 581, 651, 948, 949
Crab Nebula images, 950
Crab Nebula pulsar, 759
Crab Pulsar, 733, 734
Crab-like pulsars, 732
CRAF, 1280
Crand, 162, 169
crater density ages, 1354

Crater Mound, 259
crater-to-projectile-diameter ratio, 1169
cratered uplands of Mars, 1305, 1354
cratering history of the Moon, 1349
cratering rates, 1364, 1469
craters, 10
craters on Mars, 263
creation of NASA, 48
Cressida, 1453
CRIS, 314, 966,
critical-velocity ionization effect, 1524
crops, 13
crust of Mars is intensely magnetized, 1485
crustal evolution, 1318
crustal formation, 1318
crustal formation on Venus, 1314
crustal magnesium suite rocks, 1342
crusts, 1262
cryocoolers, 327
Cryogenic Limb Array Etalon Spectrometer (see CLAES)
cryogenically cooled detectors, 310
cryogenically cooled rocket payloads, 317
cryomagma, 1471
Cryosat, 1612
cryovolcanic flows, 1472
crystallinity of ices in astrophysical sites, 1261
crystallization ages of Martian meteorites, 1359
crystallization in the lunar crust, 1343
CSAGI, 45, 135
CSHKP, 1111
CTB109, 953
CTT, 859, 860, 884, 885
CTV5, 51, 562, 759, 762, 771, 777, 779–781, 791, 793–797, 800, 806, 807, 812, 813
current filamentation, 1011
current shear, 1610
current system in the ionosphere, 183
curved plate analyzers, 1116
curved spacetime, 442
cutoff in the cosmic-ray spectrum, 692
CV AM Her, 800
CV EX Hya, 804
CV OY Car, 322, 799
CVF, 322, 609, 620
CV-like binaries, 763
CVs in globular clusters, 806
CVZ, 799
CXC, 947, 950
CXO, 463, 557, 807
cyclical solar atmosphere, 1093
cyclical Sun, 1091
cyclotron radius, 248
Cyg, 797
Cyg X-1, 760, 824, 840, 841, 844, 846, 849, 851, 852
Cyg X-3, 761, 832, 833
Cygnus A, 532
Cygnus Loop, 595, 598, 938, 956, 957
Cygnus OB7, 597
Cygnus Rift, 597
Cygnus Superbubble, 595, 598, 601
CZCS, 1600, 1601–1603, 1617, 1661

D1C, 1628
D1D, 921, 1628
D-region ionization, 1563
D/H, 428
D/H ISM, 427
DAC, 900–902
DaG 476, 1365, 1368
DAG, 1649
DAIEIN, 1624

DAO, 921
Dar al Gani 476, 1364
dark compact objects, 457, 458
dark gamma-ray bursts, 514
dark halos of spiral galaxies, 399, 712, 713
dark matter in the universe, 400
dark matters, 17, 587
DASA, 105
day-side auroral oval, 1547
Dayglow, 189
dayside ionosphere, 1508
DE 2, 1019
de Sitter precession, 346
Deacon rocket, 281
decoupling temperature of weak interaction, 432
deep impact, 1236, 1280
deep meteorology, 1422
deep space 1, 110, 1280
deep space network, 155
deep survey telescope, 593
Defense Meteorological Satellite Program (see DMSP)
deflector, 456
degeneracy splitting, 1047
Deimos, 1186, 1253, 1452, 1453
δ, 290
Delta 2910-rocket, 59, 316
δ Ori, 647
DEM, 886, 887
DENIS, 713
dense cores in molecular clouds, 857
dense interstellar clouds, 623
density of interstellar neutral helium, 1202
density of luminous matter, 431,
density of Mercury, 1400
density of the Moon, 1386
density of the solar wind, 243
density of the Universe, 432
density parameter, 384
Department of Defense (see DOD)
Department of Terrestrial Magnetism (see DTM)
depleted elements, 658
depletion of chalcophile elements in Martian mantle, 1394
depletion patterns, 655
depletions in disk and halo gas, 660
depth of convection zone, 1041
derivation of the Moon from the Earth, 1319
Descent Imager/Spectral Radiometer experiment, 1465
Desdemona, 1453
Despina, 1453
destruction of heavy nuclei, 433
detecting youngest stellar populations, 860
detector systems on Pioneer, 145
detectors of energetic particles, 965
detrimental effects of spaceflight, 1169
Deuterium, 427, 434, 657
Deuterium measurement, 428
Deuterons, 429
Deutsche Versuchanstalt für Luft- und Raumfahrt (see DFULR)
Devana Chasma, 1313
development of rockets and spaceships, 39
DFVLR, 97
DIADEM, 1605
DIADEME, 1626
Diamant, 52, 93, 1629
Diamant rockets, 52
diameter of Venus, 1408
DIB, 607
DIDSY, 1224
differential emission measure (see DEM)
differentiation, 1318
differentiation of the Moon, 1385

diffuse aurora, 1553
diffuse galactic background, 585
diffuse gamma-ray emission, 685
Diffuse Infrared Background Explorer, 402, 700
diffuse ionized gas, 595
diffuse interstellar-medium, 610
diffusion, 193
diffusion in the ionosphere, 1567
diffusion of sodium, 192
diffusion reconnection site, 1509
diffusive shock acceleration, 978
DIM, 321, 402, 403, 611, 700, 706–708, 795
dimming of the X-ray corona, 1108
Dione, 12, 169, 1453, 1454, 1459, 1462
dipolar geomagnetic field, 1497
dipole magnetic field, 226
dipole moment of Mars, 1485
DIRBE-corrected IRAS, 587
DIRBE/COBE, 705
DIRE, 358
direct integration of relaxed Einstein equations (see DIRE)
dirty snowball, 1218, 1259
dirty snowball hypothesis, 13
disaster assessment, 1586
disaster management, 1610
DISCO, 1045
discontinuity surfaces in the heliosphere, 1202
discoverer, 77
discovering interstellar gas, 648
discovery, 94
discovery of the heliosphere, 132
discovery scarp, 1302
discrete absorption components, 900
discrete interstellar clouds, 652
disk-galaxy stability, 710
disk-instability model (see DIM)
disparition brusque, 1100
displacement model, 589, 590
dissipational dark matter, 710
dissipational disk galaxy formation, 700
dissolving phase, 942
distance scale of the Universe, 441
distance to nebulae, 925
distances of nearby clusters, 379
distances to nebulae, 934
distances to PN, 922
distance–redshift relation, 444
distant galaxies, 447
distant heliosphere, 1144
distribution of free electrons in our Galaxy, 668
distribution of nearby clouds, 658
distribution of planetary nebulae, 921
distribution of sunspots, 1094
DIVA, 376, 384, 391, 718
dividing line (see DL)
Division of Planetary Science (see DPS)
diversity of Comets, 1289
DL, 321, 402, 403, 406, 408, 410, 413, 415, 416, 651
DMR data reduction, 406
DMSP, 904, 905, 907, 1592, 1617
DMSP SSM/I, 1595
DNM, 1245
DNMI, 1604
Docking Module, 90
DOD, 313, 1631, 1643
Doppler shift of the Probe-Orbiter signal, 1428
DORIS, 1631, 1634–1636, 1638–1640, 1646, 1649
double Neutron Stars, 778
double white dwarfs, 763
doubly ionized oxygen, 315
DPS, 1164

DQ Her, 793
DQ Her systems, 804
Draco, 601
Draco enhancement, 601
Draco Nebula, 588, 591, 594, 597, 598
draped magnetic field, 1229
dredge-up event, 931
drift diffusion coefficient, 1504
DS 2, 98
DTM, 41, 128
DUCMA, 1224, 1225
dunites, 1342
duration–intensity and hardness–intensity correlation of GRB, 507
dust, 13
dust above the surface of the Moon, 1184
dust accelerator, 1175
dust around carbon-rich objects, 631
dust as a source for cometary Volatiles, 1288
dust Belt around the Earth, 1167
dust boundaries, 1224, 1225
dust charging, 1221
dust coagulation, 1238
dust-column density, 1222, 1223
dust community, 1164
dust-condensation zone, 632
dust-continuum emission, 619
dust detectors, 1168, 1174
dust distribution, 1176
dust distribution in coma, 1225
dust-emission spectrum, 611
dust environment of the Earth, 1185
dust evolution, 1180
dust fragmentation, 1225
dust from Io, 1174
dust grains on bound orbits, 1179
dust impact rates, 1182
dust in outer solar System, 1172
dust in planetary nebulae, 927
dust in the F corona of the Sun, 1185
dust in the Martian environment, 1186
dust instruments, 1176
dust leaving the Moon, 1183
dust mantle, 1262
dust mass, 571
dust measurement, 1224
dust measurements by Galileo, 1182
dust meetings, 1164, 1165
dust particles, 1166
dust rings, 320
dust rings around stars, 1178
dust rings of Jupiter, 1183
dust rings of Saturn, 1183
dust storms on Mars, 1414
dust streams from Io's volcanoes, 1182
dust telescope, 1186
dust to comets, 1237
dust-to-gas mass ratio, 1181
dust/gas ratio in comet, 1283, 1284
dusty circumstellar material, 863
dusty heliosphere, 1163
dusty phenomena in the Jovian system, 1174
dusty plasma, 1563
dusty rings of Saturn, 1186
dusty starburst galaxies, 572
dusty winds, 904
dust–gas coupling, 1223
dwarf nova outbursts, 795
dwarf novae, 792
DXS, 603, 1191
dyfess thermometer for interstellar clouds, 657
dynamic evolution of the Milky Way, 16

dynamic phenomena in the heliosphere, 1151
dynamic structure of the chromosphere, 1073
dynamical effects of cosmic rays, 691
dynamical interaction in SFR, 859
dynamics explorer, 155
dynamics in the Milky Way, 699
dynamics of collisionless system, 709
dynamics of dust grains, 1179
dynamics of the atmosphere, 192
dynamics of the ionosphere, 1567
dynamics of the solar corona, 237, 241
dynamo effect, 10
dynamo models, 882
dynamo theory, 877
dynamo-generated magnetic fields, 881

E0102-72, 941
E2003 + 225, 803
E and S0 galaxies, 563, 567, 1566
e-neutrino physics, 1057
E/S0, 461
EA, 1362
early Amazonian volcanism, 1362
early atmosphere of Mars, 1306
early Earth, 11
early ideas, 1560
early infrared space astronomers, 323
early rocket observations, 311
early Universe, 3, 19, 302, 423, 573
early-type stars, 6, 9, 10, 13, 157, 335, 340, 342, 348, 350, 878, 880, 1315, 1367, 1406, 1453
EARTH, 1624
Earth airglow emission, 288
Earth gravity field, 1634
Earth history, 1316
Earth impact craters, 260
Earth interior, 1316
Earth observations from satellite, 1585
Earth Orbit Rendezvous Mode (see EOR)
Earth orbits, 76
Earth radiation budget observations, 1658
Earth Radiation budget experiment (see ERBE)
Earth Radiation experiment (see ERB)
Earth Radiation budget satellite (see ERBS)
Earth Resources and Technology Satellite (see ERTS)
Earth rotation, 1633
Earth rotation parameters, 1640
Earth science, 3
Earth surface, 1315
Earth system dust, 1176
Earth system science, 4
Earth's absolute sidereal orbital rate, 349
Earth's age, 1329
Earth's atmosphere by attenuation, 287, 981
Earth's Bow Shock Wave, 1510
Earth's bowshock, 1531
Earth's ionosphere, 1559
Earth's iron–nickle core, 349
Earth's magnetic field, 5, 1480, 1481
Earth's magnetosphere, 160, 161, 1496, 1529
Earth's mantle, 1387, 1388
Earth's plasma environment, 1525
Earth's radiation belts, 259, 1498
Earth's surface brightness, 310
Earth's total inertia, 348
Earth-observation microwave sensors, 1619
Earth-shepherded ring, 1179
Earth-Sun line, 13
Earth-to-orbit transportation, 93
Earth–ionosphere wave guide, 1570

Earth–Mars transfer propulsion system, 113
Earth–Mercury radar ranging, 350
Earth–Moon system, 6, 340, 343, 347
earthquake-influences on the ionosphere, 1570
EAS, 123, 1605, 1617
east–west asymmetry of cosmic rays, 121
east–west effect, 678
eccentricity, 825
ECD, 144, 145
ECHO, 1624
eclipse mapping, 879
eclipse of the Sun, 338
eclipses, 879
eclipsing binary, 792
eclipsing binary systems, 879
eclipsing binary UV light curves, 289
eclipsing binary X-ray source, 760
eclipsing systems, 799
ecliptic termination shock, 997
ECMWF, 1604, 1605, 1617
EDC, 145
Eddington limit, 536, 772
Eddington luminosity, 564
eddy diffusion, 193
Edgeworth–Kuiper Belt, 1178, 1238, 1239, 1244, 1433
EDI, 1505
EDOC, 1630
EET79001, 1363, 1364, 1368
EETA 79001, 1392
effects of seeing on spatial resolution, 714
efficiency of black hole formation, 544
EGB 1, 932
EGM, 1638
egocentric world of fantasy, 226, 383, 483, 486, 549, 561, 563–567, 569, 807, 877, 878, 882, 884, 888, 890
EGRET, 508, 555, 558, 684, 689, 725, 748
EGS, 1164
Einstein catalog, 563
Einstein Cross, 449
Einstein Extended Medium Sensitivity Survey (see EMSS)
Einstein Observatory, 475, 480, 533, 534, 543, 562, 804, 806, 807, 826, 860, 876, 881, 945, 956
Einstein orbiting observatory, 585
Einstein ring, 443, 446–449
Einstein satellite, 899, 905
Einstein SLEW survey, 550
Einstein Solid State Spectrometer, 878
Einstein X-ray observatory, 724
Einstein's equations in relaxed form, 357
EISCAT, 1524, 1563, 1568, 1572, 1573
Eistla region, 11
EIT, 1077–1079, 1081, 1084, 1097, 1108, 1120
ejecta blanket, 261
ejection ages, 1363
EK, 1239, 1242, 1258, 1259, 1268
EK belt, 1258, 1268
El Niño, 1591, 1597, 1600
El Niño southern oscillation (see ENSO)
ELAIS, 574
Elara, 1453
elastic aether, 227
ELDO, 54, 55, 90, 104, 201
electric currents in ionosphere, 1579
electric fluid, 227
electric rockets, 108
electric space propulsion, 110
electrically scanning microwave radiometer (see ESMR)
electromagnetically dominated dust, 1182
electrometer, 119
electron attachment, 1566

SUBJECT INDEX

electron bombardment ion thruster, 109
electron current detector (see ECD)
electron densities in D layer, 1562
electron density, 656
electron-density contours, 1576
electron density versus height relation, 1571
electron distribution in shock transition, 1513
electron distribution with distance from the Sun, 1128
electron-drift experiment (see EDI)
electron-energy ratio, 147
electron foreshock, 1512
electron-heat flux, 1133
electron inertia, 1025
electron magneto-hydrodynamics (see EMHD)
electronic magnetograph, 229, 682
electron precipitation event, 1536
electron-cyclotron maser, 1520
electron phase-space distribution, 1516
electron–positron annihilation, 429
electron–positron pairs, 556
electroscope, 118, 119
electrostatic accelerators, 1175
electrostatic dust accelerator, 1175
electrostatic inverse square, 227
electrostatic plate analyzer, 1578
elektron, 213
elektron satellite, 1551
elemental abundances in comet dust, 1283
elemental abundance in cometary volatiles, 1285
elemental abundances in disk clouds, 658
elemental composition of cosmic rays, 679
elementary flare bursts, 218
elements released by Comet Halley, 1287
ellipticity of the lunar orbit, 272
ELOISE, 216
ELS, 701, 702, 1586, 1617
ELS model, 701
Elsinore, 1473
elves, 1570
Elysium, 1307, 1309
EM, 889, 1586–1589
EMHD, 1023
emission from inner Solar System, 1198
emission in the galactic halo, 592
emission mechanisms of middle-aged pulsars, 744
emission of gamma rays from ms-pulsars, 748
emission properties of galaxies, 575
emissivity, 1587
EMSS, 459, 482, 487, 550
Enceladus, 12, 169, 1458, 1469, 1472
Enceladus and Dioner, 1459
energetic gamma-ray experiment telescope (see EGRET)
energetic ISM, 661
energetic neutral atoms, 1212
energetic proton event, 969
energetic storm particle (see ESP)
energetics of lower transition region, 1079
Energia, 86
energy balance, 1075
energy-balance temperature, 926
energy distribution of Be stars, 901
energy distributions of B giant stars, 903
energy input to ISM, 610
energy spectra of cosmic rays, 146
energy spectrum of cosmic rays, 126, 232, 679
energy spectrum of PSR BO656 + 14, 743
energy-generating solar core, 1047
ENSO, 1597, 1623, 1635
entropy per particle, 490
envelopes of YSOs, 625, 485

ENVISAT, 1593, 1600, 1610, 1617, 1637, 1646, 1666, 1667
EO satellites and sensors, 1618
EOLE, 1629, 1630
EOQ, 1608
EOR, 82, 83, 85, 1611, 1661, 1664
EOS-CHEM, 1666, 1667
EPB, 1387
Epimetheus, 1453, 1454, 1459
epoch analysis of plasma and magnetic pressures, 1021
EPS, 1658
equality of free-fall rates, 346
equations of motion, 358
equator-S, 1502, 1504
equatorial anomaly, 1563
equatorial electrojet, 278
equilibrium spin, 768
equilibrium vapour pressure (see EVP)
equivalence principle, 336, 337, 339, 341, 345, 349, 350
ERB, 1041, 1658, 1659
ERBE, 1658, 1659
ERBS, 1658, 1659, 1664
Eridanus enhancement, 595, 597, 600
Eridanus, 585, 599
ERNO, 90
EROS, 454
EROSDOC, 1630
erosion of cometary nuclei, 1266
ERS-1, 1593, 1599, 1604, 1631, 1635, 1642, 1646, 1654
ERS-1/2, 1605
ERS-2, 1597, 1600, 1604, 1631
ERS-SAR, 1611, 1604,
ERTS, 1593, 1617
ESA, 56, 90, 95, 99, 102–105, 201, 212, 294, 321, 324, 325, 327, 373, 376, 405, 417, 453, 461, 494, 534, 544, 545, 557, 563, 576, 609, 700, 717, 718, 723, 807, 860, 861, 900, 917, 919, 1045, 1056, 1077, 1084, 1108, 1120, 1121, 1169, 1181, 1183, 1204, 1249, 1250, 1277, 1279, 1280, 1377, 1394, 1408, 1414–1416, 1421, 1422, 1429, 1432, 1435, 1436, 1447, 1479, 1483, 1486, 1539, 1613, 1614, 1617, 1623, 1630, 1631, 1637, 1644–1646, 1648, 1653, 1654, 1658, 1662, 1667, 1668
ESA EXOSAT, 830
ESA-99, 1613
ESA-ESRIN, 1597, 1600
ESA-NASA, 428, 966, 1066
ESA-SP137, 402, 718, 1200, 1632
ESA/NERSC, 1611
ESA/STScI, 944
escape velocities from galaxies, 566
ESLAB, 504
ESMR, 1593
ESO, 450, 455, 881, 886, 889
ESOC, 102, 718
ESP, 977
ESP events in outer heliosphere, 977
ESRO, 54, 90, 105, 180, 185, 201, 212, 294, 295, 654, 919
ESRO II, 214
ESSA, 1655
ESTEC, 324, 325
η Car, 903
ETALON, 1629
eucrite parent body (see EPB)
eucrites, 1361, 1380
EUMETSAT, 1654, 1657, 1658, 1667, 1668
EUMETSAT Polar System (see EPS)
EUR 1624
Eureca satellite, 1169
EuroGOOS, 1605, 1617
Europa, 11, 12, 96, 166, 1183, 1453, 1458, 1459, 1460, 1462, 1463, 1469, 1474–1476, 1487
Europa-induced magnetic field, 1490
Europa internal structure, 1463

Europa's salt-rich ocean, 11
European Center for Medium-range Weather Forecasting, 1604
European Geophysical Society (see EGS)
European large-orbiting instrumentation for solar experiments (see ELOISE)
European Launcher Development Organization (see ELDO)
European Meteorological Satellite Organisation (see EUMETSAT)
European ionospheric scatter radar facility (see EISCAT)
European rocket and satellite programs, 294
European Southern Observatory (see ESO)
European Space Agency (see ESA)
European Space Agency's Space Research and Technology Center (see ESTEC)
European Space Operation Centre (see ESOC)
European Space Research Organisation (see ESRO)
European X-ray Observatory Satellite (EXOSAT)
EUV, 217, 287, 535–537, 542, 555, 581, 582, 590, 594, 601–603, 663, 724, 794, 796, 802, 903, 1070, 1071, 1078–1080, 1098, 1106, 1108, 1109, 1126, 1197, 1198, 1199, 1201, 1212, 1564, 1565, 1569, 1579
EUV background, 593
EUV helium glow, 1212
EUV spectroscopy, 802
EUVE, 535, 561, 590, 593, 594, 725, 732, 748, 802, 803, 880, 886, 887, 903, 905, 1191, 1202, 1209, 1210, 1229
event horizon, 545
event horizons of block holes, 21
Evershed effect, 1069
evolution after nuclear burning, 934
evolution of a Neutron Star, 783
evolution of binaries with compact objects, 759
evolution of CV and compact LMXBs, 780, 781
evolution of intermediate-mass YSO, 628
evolution of isotopic composition, 1336
evolution of low-mass, 774
evolution of low-mass stars, 885
evolution of massive binary, 777
evolution of neutron stars, 751
evolution of SNR, 942
evolution of the Earth, 1329
evolution of Venus, 1313
evolution of wide LMXB, 781
evolution scenarios, 1445
EVP, 793, 1465–1467, 1470
excited spin state, 1264, 1265
EXE, 484, 599
exoplanets, 447
EXOSAT, 484, 724, 833
exosphere, 1408
exotic dark matter, 431
expansion method, 924
expansion rate of the Universe, 345
expansion velocity, 924, 700
Experiénce de Recherche d'Objets Sombres (see EROS)
experimental gravitational physics, 340
exploration of the Heliosphere, 1144
exploration of the Solar System, 3, 45, 47, 77, 272, 316, 964–966, 1168, 1479, 1529, 1530, 1533
Explorer, 316
Explorer 1, 47, 71, 259, 1497
Explorer 4, 47
Explorer 5, 47
Explorer 8, 1168
Explorer 10, 247–249, 1115, 1116
Explorer 12, 1116
Explorer 14, 1116
Explorer 34, 1124
Explorer 35, 164, 1489
Explorer I, 136, 154, 159, 160
Explorer II, 154
Explorer III, 154

Explorer IV, 154
explosive events, 550, 1027, 1029
extension of the solar wind, 249
external resurfacing, Iapetus, 1474
exterrestrial radiation, 118
extinct radionuclides, 1335, 1336
extinction by interstellar dust, 304
extra-energy problem, 490
extra-geocoronal Lyman-alpha glow, 1197
extraction of energy from black hole, 521
extragalactic background, 582
extragalactic emission, 583
extragalactic gravitational lensing, 454
extragalactic GRB models, 502
extragalactic radio sources, 343
extragalactic research, 476
extragalactic source-counts, 322
extragalactic space, 4
extragalactic stars, 447
extrasolar dust rings, 1178
extraterrestrial dust grains, 1170
extraterrestrial grains, 1171
extreme ultraviolet (see EUV)
extreme ultraviolet explorer (see EUVE)
extreme ultraviolet imaging telescope (see EIT)
extremely red objects, 305

F region, 1566
F-1, 84
F/G, 1357
faculae, 1068
failed supernova model, 515
faint-background sources, 460
faint-object camera, 917, 1436
faint-object spectrograph, 455
FAIR, 326
FAME, 718
far infrared, 14
far and extreme ultraviolet spectrometers, 298
far-infrared absolute spectrometer (see FIRAS)
far infrared and submillimeter telescope (see FIRST)
far-infrared molecular lines, 315
far-infrared polarization, 315
far-infrared sky survey experiment (see FIRSSE)
far infrared spectroscopy, 315
far ultraviolet spectroscopic explorer (see FUSE)
far-UV (see FUV)
far-UV-X-ray spectrum, 207
Farley-Buneman instability, 1563, 1564
farside Moon, 273
FAST, 1518, 1521, 1536, 1552, 1553
fast motions, 551
fast solar wind, 1083, 1150, 1153
FB, 145, 817
FC, 145, 1116
FCCC, 1585, 1617
Ferroan anorthosites, 1340
FFB, 1342
FGH, 653
field-aligned current regions, 1539
field-aligned electric field, 1537
field-aligned electric potentials, 1524
field reconnection, 1110
field-theory of gravity, 339
filament eruptions, 1111
filamentary material, 657
filamentation of sheet current, 1011
Fillius, 169
FILM, 609
final evolution of HMXB, 779

fine dust, 1224
fine-structure lines in PDR models, 616
fine-structure transitions, 315
finite universe, 463
FIP, 652, 660, 680, 681, 886, 972, 991, 994, 1124, 1125, 1126, 1135, 1147
FIP effect, 1126
FIR, 615
FIRAS, 321, 402–404, 609, 611, 614
fireball model, 508
fireball/blast-wave model, 509
firework rockets, 31
FIRSSE, 326
FIRST, 324, 325, 462, 574, 576, 609, 860, 863, 868, 1432, 1448
FIRST-Herschel, 442
first-ionization potential (see FIP)
first landing by humans on Moon, 3
fission cell (see FC)
flare as a pulled reconnection event, 1030
flare images in hard X-rays, 214
flare loop sizes, 889
flare model, 215, 889, 1111
flare phenomenon, 210, 229
flares and reconnection, 1028
flares v. CMEs controversy, 971
flat-spectrum radio quasar (see FSRQ)
flatness of the observed universe, 401
flickering, 10, 796
flow of plasma into magnetosphere, 1550
FLRW, 446
fluffy aggregates, 1250
fluid-dynamic model, 1505
flux distribution of BATSE GRB, 507
flux tubes, 1069
flux-gate magnetometer, 146, 1433, 1577
fly-along infrared (see FAIR)
fly's eye detector, 692, 694
fly's eye experiment, 692
FNMOC, 1604
focal plane arrays, 318
focal plane crystal spectrometer (see FPCS)
footpoint motions, 1082
forbidden transitions, 234
forbush decrease, 128, 137, 231
forcing of the ionosphere, 1570
foreshock particles, 1512
forestry services, 1586
forests, 13
formation Low-Mass X-ray Binary, 780
formation of a disk, 11
formation of black-hole binaries, 854
formation of catenae, 1249
formation of cometesimals, 1238
formation of disk galaxies, 700
formation of galaxies, 434
formation of helium spectrum, 1081
formation of protoplanetary disks, 857
formation of stars, 857
formation of the Earth, 1319
formation of the lunar crust, 1384
formation of the Solar System, 1303, 1335
Fornax A/NGC , 565, 1316
Fornax cluster, 379, 381
fossil, 432
fossil radiation, 435, 584
four-color photometry, 311
Fourier-transform spectrophotometer, 314
FOV, 954, 956, 1197
FP, 322
FPCS, 480, 946, 956
FR-I, 554

Fra Mauro Formation, 1339
fractal-like cometary nucleus, 1250
fractionation, 1123
fragment mass, 1249
Framework Convention on Climate Change, (see FCCC)
Fraunhofer lines, 233
Freedom-7, 50
Freja, 1536, 1553
Freja satellite, 1577
Freya Montes, 1311
FRI, 491
friable sponge model, 1219
friable sponge surface model, 1218
Fruchter pulsar, 782
FSRQ, 553, 558
FTE, 1022
fundamental equation of stellar statistics, 704
fundamental harmonic emission, 1521
FUSE, 298, 427, 428, 669, 920
fusion reaction, 111
FUV, 287, 612, 615–617, 634
FWHM, 294, 409, 412, 417, 463, 585, 592, 654, 670, 997
FY, 1617

gabbronorites, 1342
GAIA, 376, 384, 391, 442, 453, 461, 465, 704, 712, 718
Gaidos, 555
galactic bar, 706, 707
galactic census, 461
galactic center, 306, 307, 310, 591
galactic-center X-ray bulge, 594
galactic corona, 602, 155, 159, 225, 230, 250, 677
galactic cosmic-ray (see GCR)
galactic cosmic-ray sources (see GCRS)
galactic disk, 595, 602, 763
galactic electrons, 321, 682
galactic evolution, 1123
galactic evolution model, 435, 502, 601, 668
galactic model of GRB, 506
galactic plane, 306
galactic potential, 710
galactic radio disk, 687
galactic sources of cosmic rays, 688
galactic wind, 653, 691
galactic wind termination shock, 690
Galatea, 1453, 1460
galaxies in collision, 320
galaxy and black hole, 544
galaxy clusters, 447
galaxy ecology, 566
galaxy evolution explorer (see GALEX)
galaxy in visible light, 118
GALCIT, 65, 69
GALEX, 298
GALILEO, 1643, 1646
Galilean moons, 12
Galilean satellites, 9, 81, 96, 155, 966, 1174–1176, 1182, 1183, 1191, 1298, 1427, 1428, 1451, 1479, 1480, 1487, 1505
Galileo entry probe, 428
Galileo Jupiter orbiter, 1413
Galileo mission, 1447
Galileo near-infrared mapping spectrometer, 1413
Galileo project, 97
Galileo spacecraft, 1463, 1469
Galileo's dust detector, 1177
Galileo's voyage to Jupiter, 96
gallium-doped germanium, 304
γ Vel, 291, 294
gamma-ray, 13
gamma-ray astronomy, 284, 683, 684

gamma-ray bursts (see GRB)
gamma-ray burst source, 19, 499, 549, 557
gamma-ray emission of SNRs, 689
gamma-ray-emitting CV, 806
gamma-ray era, 553
gamma-ray large area space telescope (see GLAST)
gamma-ray map of the Galaxy, 723
gamma-ray sources, 723
gamma-ray spectrometer, 10, 1382
gamma-ray spectrum for Galaxy, 684
gamma-ray universe, 10, 14, 284
Ganymede, 96, 166, 1183, 1247, 1249, 1453, 1459, 1460, 1462, 1463, 1469, 1470, 1472, 1474, 1487
Ganymede magnetic field, 1463
gas–dust interaction, 1222
gas dynamic model, 1223
gas dynamics laboratory (see GDL)
gas ejected from cometary nucleus, 1257
gas giant planets, 1456
gas imaging spectrometer, (see GIS)
gas in the diffuse ISM, 614
gas jets of ions, 1223
gas streams, 831
gas tracers, 632
gaseous condensations, 1207
Gassiot Committee, 51
gas chromatograph–mass spectrometer (see GCMS)
GCM, 1419, 1420
GCMS, 1391
GCR, 423, 424, 427, 429, 433–435, 437
GCRS, 681
GDL, 33, 34
GEM, 1629, 1634
GEM-T, 1638
Geminga, 723
Gemini, 83, 335, 337, 338, 353,
general relativity (see GR)
general circulation model (see GCM)
general structure of ISM, 601
general theory of relativity, 20, 339
generation of aurora, 1537
generation of detected X-rays, 722
Genesis mission, 1124
genesis of igneous rocks on the Moon, 1357
genesis rock, 1340
genetic relationship between Earth and Moon, 1385, 1386
GEO-600, 357, 365
geochemical differentiation of Mars, 1366
geochronology in the twentieth century, 1334
geochronology of rock systems, 1331
geodesics of curved space, 442
geodetic precession, 347
geodynamic models, 1314
GEOLE, 1630, 1631
geologic history of terrestrial planets, 1316
geologic record of Venus, 1312
geologic Time, 1325
geological history of Venus, 1314
geology of Mercury, 257
geology of the Earth, 1315
geology of Venus, 1309
geomagnetic activity, 226, 228, 1120
geomagnetic activity and sunspots, 228, 229
geomagnetic dipole, 245
geomagnetic disturbance, 226, 248, 1097, 1481
geomagnetic field, 225, 249
geomagnetic fluctuations, 225, 228
geomagnetic latitude effect, 120
geomagnetic storm, 230, 245, 246, 1101
geomagnetic trapping, 175

geomagnetic variability, 245
geomagnetically induced currents (see GIC)
geomagnetically trapped corpuscular radiation, 154
geomagnetism, 1253, 1480
geometric effects on acceleration of solar wind, 1134
geometrical optical phase, 1624
geometrization of gravity, 339
geometry of intermediate polar, 805
geometry of events, 339
GEOS, 1548
GEOS 1, 1551, 1545
GEOS 3, 1630, 1631, 1646
GEOS A, 1628
GEOS B, 1628
GEOS C, 1628
GEOSAT, 1600, 1630, 1634–1636, 1646
geosciences, 1623
geostationary meteorological satellite (see GMS)
Geostationary Operational Environmental Satellite (see GOES)
geostationary satellites, 1654
geostationary weather satellites, 1655
geosynchronous orbit, 295
geotail, 155, 994, 1502, 1504, 1540
GERB, 1658
germanium bolometer, 305
germanium detectors, 313
Gesellschaft für Weltraumforschung (see GFW)
GfW, 39
GFZ, 1638
GFZ-1, 1629, 1631
GHRS, 654, 662, 798, 799
giant cells, 1054, 1068
giant-impact theory, 6
giant-luminous arcs (see GLA)
giant-molecular clouds, 860, 1431
giant planets beyond Jupiter, 1432
giant-radio outbursts, 833
giant rocket, 81
GIC, 1579
GINGA, 484, 879, 888, 944, 947, 954
Ginga, 484, 537, 847, 849, 850
Ginga satellite, 504
Gioia, 482
Giotto, 105, 1176, 1181, 1227, 1229, 1250, 1253, 1257, 1265, 1277, 1282, 1283, 1288
Giotto encounter, 1222–1224
Giotto experience, 1225
Giotto flyby, 1230, 1243
Giotto probe, 1217, 1228
Giotto spacecraft, 11, 1236, 1279
Gipul Catena, 1247, 1249
GIRD, 33
GIRD-09, 34
GIRD-X, 34
GIS, 486, 725
GK, 314, 864
GLA, 444, 458–460
GLAST, 508, 521, 748, 752
GLI, 1601, 1617
global astrometric interferometer for astrophysics (see GAIA)
global change on Earth, 1420
global distribution of chlorophyll, 1601
global magnetic field at Mercury, 1482
global magnetic field model, 987
global magnetic fields, 1481
global modeling of magnetosphere, 1553
global navigation satellite system (see GLONASS)
global ocean data assimilation experiment (see GODAE)
global oscillation network group (see GONG)
global positioning system (see GPS)

global solar monitoring, 204
global surveyor, 1414
global telecommunication system (see GTS)
global tide gauge, 1647
global tide models, 1647, 763
globular cluster X-ray binaries, 447, 713, 782
GLONASS, 1575, 1640, 1643, 1646, 1649
GMIR, 1123, 1156
GMS, 1596
GNSS, 1643
GNSS/GALILEO, 1644
GOCE, 1611–1613, 1617, 1631, 1632, 1643, 1645, 1646, 1648
GODAE, 1603, 1617
Goddard high resolution spectrometer, 1208
Goddard Space Flight Centre (see GSFC)
GOES, 1091, 1596, 1654
GOES-1, 1655
GOES-10, 1655
GOES-East, 1655
GOES-West, 1655
Goldstone ground station, 82
GOLF, 1043, 1044–1046, 1057
GOME, 21, 1649
GONG, 1039, 1045–1048, 1050, 1052, 1053, 1056, 1091, 1093
GONG +, 1046, 1092, 1585, 1617
GOOS, 1585, 1605, 1615, 1617
Gossamer ring, 1183, 1183
Gould's belt, 653, 654
GP-A, 341
GP-B, 344, 347
GPS, 336, 337, 339–351, 353, 354, 356, 357, 361, 362, 366, 368, 369, 441–443, 461, 1421, 1575, 1598, 1617, 1631, 1633, 1634, 1638–1640, 1643, 1646, 1649
GR, 335
GRACE, 1611–1613, 1617, 1632, 1643, 1644, 1646
gradient drift instability, 1563
GRADIO, 1635
gradiometry, 1634
grain-formation and growth, 1241
grain-formation hypotheses, 658
grain-surface chemistry, 624
grains condense in stellar outflows, 661
grains of stellar origin, 1279
GRANAT, 844, 849
granites/felsites, 1344
granitic magmatism, 1345
granulation, 1068, 1070
GRASP, 367
grating spectrograph, 291
grating spectrometer, 315, 319
gravitating Universe, 17
gravitational collapse, 11, 944
gravitational contraction, 1042
gravitational deflection of light trajectories, 342
gravitational field, 335, 336
gravitational field equations, 340
gravitational field of the Sun, 235
gravitational field potential, 338
gravitational force, 335
gravitational free-fall, 351
gravitational inertial mass of celestial bodies, 343
gravitational internal binding energy, 18, 343
gravitational lens, 17
gravitational lens experiment, 447, 448
gravitational lens mirage, 443
gravitational lens simulator, 441, 445, 465, 474, 493
gravitational lensing amplification, 446
gravitational lensing in galaxy clusters, 458
gravitational lensing optical depth, 452
gravitational maneuvers, 81, 97

gravitational mass parameter, 3, 339, 347
gravitational potentional well, 17
graviational radiation analysis and stimulation package (see GRASP)
gravitational radiation back-reaction, 366
gravitational radiation from cosmological sources, 1056
gravitational runaway, 1237
gravitational shift of clock rates, 341
gravitational waveform, 358
gravitational-to-inertial mass ratio, 343, 345
gravitational-wave detection, 359
gravitational-wave observatories, 357, 365
gravitationally bound corona, 241
gravitomagnetic field, 347
gravitomagnetic forces, 347
gravitomagnetic interaction, 340, 348
gravitomagnetic precession, 347
graviton mass, 10, 368, 1638
gravity-field models, 1639
gravity missions, 1644
gravity of central stars, 932
gravity probe-A (see GP-A)
gravity probe-B (see GP-B)
gravity recovery and climate experiment (see GRACE)
grazing incidence monochromator, 208
GRB, 284, 499–506, 508, 510–518, 520, 522
GRB counterparts, 505
GRB GeV emission, 508
GRB hosts, 521
GRB millisecond varriability, 508
GRB optical counterpart, 510
GRB polarimetry, 518
GRB sources, host galaxies, and cosmology, 519
GRBs and supernovae, 517
great Nebula in Andromeda, 914
Great Red Spot (see GRS)
greenhouse effect, 9, 262, 1408, 1409
greenhouse gas, 20
Greisen–Zatsepin–Kuzmin effect, 679
GRGS, 1637–1639
GRIM, 1629, 1638
GRM/SST, 1635
ground-based astronomy, 663
ground-based spectra, 918
Group for the Study of Reaction Motion (see GIRD)
Groupement Astronautique Français, 39
GRS, 1427, 1431, 1433, 1441
GSFC, 144, 293, 295, 319, 321, 325, 402, 495, 887, 1626, 1629, 1634
GSPC, 830
GTS, 39, 1604
Guggenheim Aeronautical Laboratory of the California Institute of Technology (see GALCIT)
guiding centre approximation, 237
Gula Mons, 11
Gum Nebula, 595, 597, 653, 663
Gunn–Peterson effect, 541, 542
GX 5-1, 817

H1413 + 117, 449
H2252-035, 793
HAB, 1607
HAC, 612–614
Haebe stars, 859, 862, 865, 1243, 1287
Halley fleet, 1279
Halley flybys, 1219
Halley missions, 1282
Halley multicolour camera (see HMC)
Halley's comet, 232, 257
Halley-type comets, 1242

halo cloud, 670
halo emission, 591
HALOE, 1663–1666
Halogen Occultation Experiment (see HALOE)
HAO, 1121
hard X-ray light curve of intermediate polar, 805
hard X-ray monitoring, 214
hard X-ray pulse, 215
hard X-ray telescope (see HXT)
hard-to-soft GRB evolution, 507
hare-and-hounds exercises, 1048
harmful algae bloom (see HAB)
HAS, 1346
Hauptstrom rocket engine, 75
hazard monitoring, 1586
hazards from meteor streams, 1186
hazards to future satellites, 1169
haze cover of Mars, 263
HB, 817
HBB, 931
HBL, 556, 558
HBO, 818
HCS, 1142, 1151, 1152, 1154, 1155
HD 20366, 670
HD 28497, 656
HD 38268, 296
HD 50896, 656
HD 93521, 670, 671
HD 100340, 670
HD 154368, 670
HD 215733, 670
HD 226868, 760, 841
HDE226868, 840
HDM, 386
HDW 4, 932
^{4}He fractional mass abudance, 424
HE1104-1805, 449, 458
He-3-1475, 916
He/H ratio, 583
HEAO, 78, 316, 409, 585, 724
HEAO 1, 479, 480, 485, 537, 543, 549, 563, 585, 586, 741, 795, 804, 824, 830, 846, 876
HEAO 2, 480, 876, 1175
HEAO II, 100
heat sources for the Earth, 1317
heating and cooling white dwarf, 798
heavier elements, 19
heavyside layer, 278
heavy elements in Galaxy, 938
heavy ions in solar wind, 1123
heavy nuclei in cosmic radiation, 678
heavy-element abundance, 1035
heavy-element atmospheres, 744
HED, 1360, 1361
height profiles of ions, 1578
helen Planitia, 1312
helene, 1453, 1454, 1459
heliarc welding, 310
heliolatitude, 1142
heliomagnetic coordinate system, 1120
heliopause, 5, 53, 77, 80, 965, 983, 986, 1128, 1130, 1141, 1158, 1176, 1205
helios, 1132
helios 1, 1128
helios 2, 1128
helios missions, 1119, 1145
helios spacecraft, 1151
helioseismic observations, 1046
helioseismic space mission, 1045
helioseismology, 5, 13, 133, 249, 250, 1038, 1045, 1056, 1069, 1125, 1141, 1183, 1191, 1192, 1196

helioseismology by irradiance measurements, 1048
heliosphere, 137
heliosphere boundary, 1185, 1205
heliosphere structure, 1212
heliospheric concepts, 133
heliospheric hydrogen wall, 1208
heliospheric magnetic field, 1148, 1152
heliospheric model, 134
heliospheric physics, 19, 423, 1143
helium abundance, 1035, 1429
helium abundance interferometer, 1428
helium burning, 774
helium emission, 1081
helium flash ignition, 774, 1204
helium ions, 250
helium isotopic ratio, 427
helium mass fraction in Jupiter, 1446
helium mass fraction in Saturn, 1446
helium settling, 1047
helium to hydrogen ratio, 932
helium-3, 428, 435
helium-4, 428, 435
Hellas, 267, 1307
HEOS, 185, 1183
HEOS 2, 1167, 1175, 1176, 1230
Her X-1, 760
Herbig-Ae/Be stars, 859, 865
Herbig-Haro object HH30, 860, 861
Herkules, 86
Herschel space observatory, 462
Hertzsprung-Russell diagram, 17, 297
Hesperian age Martian meteorites, 1365
HESSI, 999, 1104, 1108, 1112, 1113
HETE, 521
heterodyne instruments for FIRST (see HIFI)
heterodyne instruments in space, 621
hiatuses in planetary exploration, 268
hidden quasars, 537
HIFI, 325, 635
high-energy astronomical observatories, 77, 78, 99, 795, 876
high-energy astrophysical observatory (see HEAO)
high-energy solar spectroscopic imager (see HESSI)
high-energy astronomy, 723
high-energy emission of isolated neutron stars, 732
high-energy pulsar emission models, 726
high-energy radiation, 875
high-energy transient experiment (see HETE)
high-field neutron star atmospheres, 731
high-latitude magnetosphere, 1535, 1537
high-latitude star-count data, 703
high-mass X-ray binary (see HMXB)
high-polarization quasar, 550
high-precision parallax collecting satellite, 700
high-redshift quasarology, 541
high-resolution Doppler imager (see HRDI)
high-resolution camera (see HRC)
high-resolution imager (see HRI)
high-resolution picture transmission (see HRPT)
high-resolution spectrograph, 298, 1078
high-resolution telescope, 1078
high-resolution telescope and spectograph (see HRTS)
high-resolution optical data, 666
high-resolution wind field model (see HIRLAM)
high-spectral-resolution observations of optical absorption lines, 650
high-spectral-resolution, 650, 667
high-speed mass loss, 290
high-speed stellar mass, 291
high velocity (see HV)
high-velocity clouds (see HVC)
high-velocity disk clouds, 661

high-Z elements, 939
higher-Z elements, 940
highlands alkali suite (see HAS)
highly ionized gas, 661, 668
highly ionized halo gas, 669
highly magnetized neutron star, 833
HIM, 664, 665, 669
Himalia, 1452, 1453
HIMS, 603
Hinotori, 1104
Hipparcos catalogue, 17, 704, 717
Hipparcos satellite, 453
Hipparcos space telescope, 14
Hipparcos, 16, 105, 376, 390, 391, 700, 708, 710–712, 716, 717, 718
HIRLAM, 1609
HIRS-12, 1657
HIS, 1230
HISTAR, 13
historical supernova explosions, 945
history of Moon's orbit, 272
history of our galaxy, 706, 1310
Hiten, 814, 1176
HLQ, 454, 457, 462
IIM7, 105
HM20, 75
HMC, 13, 1250–1255, 1258
HMF, 1148, 1152, 1152
HMXB, 760–766, 772, 777–779, 783, 785, 811, 812, 814, 818, 819, 823–826, 828–834
HMXB pulse period, 826
HMXB supergiant systems, 832
hodograph, 194, 196
hodographs of horizontal winds, 193
homogeneity and isotropy of the universe, 401
homogeneous leaky box model, 685
Hopkins ultraviolet telescope, 298, 798
Horizon 2000, 1448
Horizon 2000 program, 1277
host galaxy of GRB, 511, 521
hot-Big Bang model, 400
hot-bottom burning (see HBB)
hot chromospheric gas, 861
hot dark matter (see HDM)
hot-galactic halo, 602
hot-halo gas, 669
hot halos, 567
hot interstellar medium spectrometer (see HIMS)
hot ISM, 581
hot ISM evolution, 569
hot plasmas, 566
hot plasmas in galaxies, 565
hot-star outflows, 900
hot-star winds, 899
HPQ, 550, 555
HR Del, 797
HR diagram, 904
HRC, 463, 566, 568
HRDI, 1664
HRI, 480, 481, 491, 565, 724, 733, 734, 737, 739, 880, 956
HRPT, 1656, 1657
HRTS, 1067, 1078, 1079
HST, 7, 13, 14, 16, 101, 102, 288, 297, 298, 323, 373, 376–379, 382–384, 387, 388, 391, 392, 427, 442, 445, 449, 450, 454–457, 459, 460, 465, 474, 487, 492, 493, 511, 513, 521, 531, 534, 536, 538, 541, 542, 561, 563, 572, 615, 629, 654, 656–658, 662, 669–671, 702, 703, 706, 712–715, 718, 725, 729, 732, 733, 746, 750, 796, 798, 799, 806, 832, 857, 858, 860, 861, 864, 865, 903, 913, 917, 930, 938, 944, 958, 1078, 1191, 1202, 1204, 1208, 1236, 1239, 1244, 1246, 1283, 1432–1435, 1442, 1447, 1448, 1457, 1462, 1464, 1467, 1469, 1490, 1579
HST UV observation, 799
HST+COS, 298
HST-NICMOS, 865, 1359, 1362
HST/GHRS, 655, 658, 799
HST/GHRS spectrum of U Gem, 799
Hubble 5, 915
Hubble constant, 305, 374, 441, 442, 447, 451, 452, 455, 456
Hubble deep field, 574, 703
Hubble diagram, 374
Hubble diagram of clusters of galaxies, 378
Hubble diagram of supernovae of type 1a, 375
Hubble expansion, 342
Hubble Space Telescope (see HST)
Hulse–Taylor, 767, 778
Hulse–Taylor binary, 784
HUT spectra of U Gem, 798
Huygens descent probe, 1475
Huygens entry probe, 169
Huygens probe, 1465
HV, 538, 656
HVC, 601, 602, 651, 660, 661, 665
HW 5, 932
HXT, 1108
hybrid arrays of detectors, 327
hybrid star, 878
hydrocarbon dust, 612
hydrodynamic blast wave, 1104
hydrodynamic coronal expansion, 248
hydrodynamical simulations of accretion, 828
hydrodynamics of the solar corona, 19, 239, 315
hydrogen and helium differentiation, 1202
hydrogen and helium in nebulae, 918
hydrogen atmosphere models, 732
hydrogen burning, 774
hydrogen chloride on Venus, 262
hydrogen column density, 888, 889
hydrogen corona radius, 1224
hydrogen fluoride on Venus, 262
hydrogen geocorona, 288
hydrogen glow, 1195
hydrogen in intergalactic space, 541
hydrogen maser oscillator, 341
hydrogen–oxygen engine, 72
hydrogen–oxygen rocket motors, 75
hydrogen peroxide, 75, 1207
hydrogenated amorphous carbon (see HAC)
hydrostatic stratification of the Sun, 1057
hydrous silicates, 867
Hyperion, 1452, 1453, 1458
hypernova, 19
hypernova 1998bw, 19
hyperon, 17

IAF, 39, 45, 70, 1640
IAG, 1640
IAGA, 1208
Iapetus, 1453, 1458, 1458
IAU, 291, 534, 663, 702, 1013, 1164, 1640
IAUC, 943
IBM, 85
IC 342, 565
IC 4637, 932
IC 4997, 917
ICBM, 43, 44, 63, 106, 326, 1153, 1181, 1227, 1228, 1279, 1504, 1517
ice and snow, 626, 1588
ice clouds and land elevation satellite (see ICESat)
ice deposits at the lunar pole, 1390
ice deposits on Mercury, 625, 1405
ice-mass fluctuations, 1611
ice-motion algorithm, 1605
ice segregation, 624

icebreaker "Sibiria", 622, 1606
ICESat, 1611, 1612
ICI, 491, 1124, 1130, 1577
ICM, 491
ICPP, 1025
ICSU, 44, 45
icy conglomerate model, 1241, 1262
icy conglomerate nucleus, 1236
icy dirt ball, 1259, 1262
icy grains, 1241
icy Saturnian satellites, 97, 257, 1462
identification of interstellar clouds, 652
IDP, 867, 1171, 1172, 1269, 1284
IERS, 1640, 1643
IFCAI-Palermo, 723
IGBP, 1585, 1617
IGM, 474, 479, 488, 490, 541, 542
IGOS, 1610
IGS, 1644
IGY, 5, 44–46, 56, 69, 129, 134, 135, 153, 154, 192, 193, 280, 1091, 1529, 1537, 1542, 1544, 1566
IKI, 1523
ILMT, 461, 465
ILR, 707
image magnification, 446
images of comet Halley, 1251
images of Mars, 263
imaging photopolarimeter, 1426
imaging photopolarmetry, 146
imaging planetary nebulae, 917
imaging proportional counter, 879
imaging radars, 1589
imaging SAR, 1608
imaging Spectrograph, 298
IMAPS, 657
imbrium basin, 261
IMF, 712–714, 860, 884, 1014–1016, 1018, 1019, 1541, 1542, 1553
IMP, 137, 144, 248, 964, 965, 985, 990, 1127, 1132
IMP 1, 248, 249, 1144
IMP 4, 1124
IMP 5, 1199
IMP 6, 501, 1128
IMP 7, 974
IMP 8, 1144, 1191
impact basins on the Moon, 261
impact cataclysm, 1175, 1361
impact craters on Mars, 264
impact features, 261
impact-ionization detector, 1175, 1181
impact resetting of age lunar cataclysm, 259, 1347
impact-reset ages, 1361
impacts on Earth, 262
improved stratospheric and mesospheric sounder (see ISAMS)
improved TIROS operational system (see ITOS)
impulsive events, 970
IMS-HERS, 1231
in-situ ionospheric measurements, 1576
in the subsolar region, 1015
inactive comets, 1243
incoherent scatter radar (see ISR)
incoherent scatter technique, 1568, 1573
incoherently scattered radio waves, 1574
incompatible lithologies in lunar evolution, 1345
indium antimonide, 304
inertial frame, 337
infrared, 13
infrared space observatory, 14
infrared and microwave background, 321
infrared-array camera (see IRAC)

infrared-astronomical satellite (see IRAS)
infrared astronomy, 302, 315, 705
infrared cirrus, 320
infrared-detection techniques, 307
infrared emission from circumstellar dust, 631
infrared forbidden lines, 920
infrared galaxies, 320
infrared-imaging devices, 1456
infrared-imaging spectrometer (see IRIS)
infrared-interference spectrometer, 1427
infrared luminosity, 305
infrared-microwave excess, 262
Infrared Processing and Analysis Center (see IPAC)
infrared radiation from interstellar grains, 608
infrared radiation from the galaxy, 321
infrared radiometer, 1426
infrared-rocket astronomy, 303
infrared satellite, 308, 317
infrared-space missions, 609
infrared-space observatory (see ISO)
infrared-space astronomy, 301, 324
infrared spectra, 920
infrared spectra of Neptune, 1443
infrared spectra of Uranus, 1443
infrared-spectral domain, 302
infrared spectrograph (see IRS)
infrared telescope facility (see IRTF)
infrared telescope in space (see IRTS)
inhomogeneous electric fields, 182
initial perturbations, 401
inner envelopes around YSOs, 626
inner galaxy, 705, 1120, 1143, 1145, 1150, 1152
inner magnetosphere, 1533, 1534, 1543
inner-old disk, 707
inner-Oort cloud, 1240
inner regions of quasars, 539
inner-solar system dust, 1176
inner-source pickup ions, 987
Institute of Space and Astronautical Science (see ISAS)
instrument pointing system, 91
INTEGRAL, 103
integral, 729, 752
integral surface mass density, 710
integrated Earth system model, 1614
integratel global observing statergy (see IGOS)
integrated light-curves for millisecond pulsars, 749
intensity distribution of dust, 1258
interaction of a hot star with a patchy ISM, 652
intercloud absorption, 656
intercloud material, 656
intercontinental ballistic missiles (see ICBM)
intergalactic clouds, 458
intergalactic gas clouds, 441
intergalactic medium (see IGM)
Intergovermental Oceanographic Committee (see IOC)
Intergovernmental Panel on Climate Change (see IPCC)
interior of Mars, 1308
interior of Mercury, 1303
interior of the Moon, 1301
interior of Venus, 1312
intermediate polar (see IP)
intermediate-mass X-ray binaries, 762
intermediate-range and ballistic missiles (see IRBM)
intermediate-velocity clouds (see IVC)
internal disruption in a tokamak, 1023
internal gravity wave, 234
internal nuclear evolution, 782
internal shock senario, 557
internal structure of SL9, 1249
internal structures of giant planets, 1436

internal waves, 1611
International Astronomical Union (see IAU)
International Astronomical Union Circular (see IAUC)
International Astronautical Federation (see IAF)
International Atomic Time, 1633
International Cometary Explorer, 1124, 1181, 1279
International Conference on Plasma Physics (see ICPP)
International Council of Scientific Union (see ICSLI)
International Geophysical Year (see IGY)
International Geosphere and Biosphere Program (see IGBP)
International Halley Watch, 1279
International Liquid Mirror Telescope (see ILMT)
International Monitoring Platform (see IMP)
International Solar Polar Mission (see ISPM)
International Solar Terrestrial Physics (see ISTP)
International Space Technology, Inc (see ISTI)
International Space Station (see ISS)
International Sun–Earth (see ISEE)
International Terrestrial Reference System (see ITRS)
International Ultraviolet Explorer (see IUE)
International Union of Geodesy and Geophysics (see IUGG)
interplanetary dust, 1176, 1426
interplanetary dust complex, 1181
interplanetary dust particles (see IDP)
interplanetary gas emission, 1196
interplanetary helioseismology by irradiance
  measurements (see IPHIR)
interplanetary laser ranging (see IPLR)
interplanetary magnetic field, 244, 246, 250, 1117, 1118,
  1178, 1507, 1532
interplanetary medium, 132
Interplanetary Monitoring Platform (see IMP)
interplanetary network (see IPN)
interplanetary plasma clouds, 1152
interplanetary plasma of electrons and protons, 233
interplanetary radar ranging, 341
interplanetary scintillations, 1147
interplanetary shock, 969, 1104, 1129
interplanetary shock particles, 976
interplanetary shocks accelerate particles, 977
interplanetary space missions, 1142
interplanetary transponder laser ranging, 348, 351
interplanetary interstellar connection, 1202
interpretation of spectra, 934
interrelationship between plasma experiments in laboratory and space (see
  IPELS)
interspersed model, 589, 625, 647
interstellar absorbing grains, 289
interstellar amorphous silicates, 631
interstellar C IV absorption, 297
interstellar Ca, 6, 649, 656
interstellar cometary ice composition, 301, 626, 1187
interstellar dust (see ISD)
interstellar dust in the heliosphere, 1183
interstellar extinction, 5, 289, 296, 705, 1192, 1197
interstellar gas and dust, 607, 616
interstellar gas convective motion, 691
interstellar gas-origin theory, 992
interstellar glow, 1183, 1192, 1207
interstellar graphite, 291
interstellar H I, 651
interstellar H I opacity, 287
interstellar helium, 1192
interstellar helium is ionized, 1209
interstellar hydrocarbon grains, 613
interstellar hydrogen flow, 1195, 1211
interstellar ice, 613
interstellar matter, 126, 246, 581, 647, 1191
interstellar molecular hydrogen, 292
interstellar neutral atoms, 1197
interstellar neutral derivatives, 1201
interstellar neutral gas, 991
interstellar origin of comets, 1291
interstellar PAH, 617
interstellar PAH family, 620
interstellar pickup ions, 987, 996
interstellar reddening, 306
interstellar space, 249
interstellar theory, 653
interstellar wind, 1192, 1198
intertial mass of celestial bodies, 343
intraday variability, 557
intranetwork magnetic flux emergence, 1029
Inverness, 1473
inverse Evershed effect, 1069
inverse square law, 227
inverted V events, 11, 12, 96, 166, 1183, 1452, 1453, 1455, 1456,
  1459–1461, 1466, 1470, 1472, 1487, 1490, 1535
Io plasma torus (see IPT)
Io torus, 11, 167
Io's internal structure, 1461
Io's ionosphere, 1467, 1469
Io's orbital position, 1466
Io's spectrum, 1461
IOC, 800, 804, 806, 1585, 1617
ion and neutral chemistry, 1569
ion anisotropy, 1133
ion chemistry, 1566
ion composition, 1577
ion composition instrument (see ICI)
ion drift, 1568
ion drift density and temperature, 1577
ion foreshock, 1512
ion propulsion, 108
ion-release module (see IRM)
ion trap, 1115
ion-cyclotron mechanism, 13
ion-cyclotron waves, 1135
ionization cavity, 1192
ionization chamber, 121
ionization cone, 534
ionization parameter, 942
ionization states, 974
ionized cometary tails, 179
ionized rims, 668
ionosheath, 1227
ionosheric current systems, 1507
ionosonde, 1568, 1571
ionosondes in satellites, 158, 182, 1545, 1575
ionosphere 1560
ionosphere of giant planets, 1440
ionospheric electron density, 1562
ionospheric irregularities, 1563, 1564, 1579
ionospheric layers, 1561
ionospheric observational techniques, 1570
ionospheric physics, 158
ionospheric processes, 1564, 1566
ionospheric radio-wave propagation, 1579
ionospheric research, 41, 278
ionospheric structure, 1560
ions in Jupiter's magnetosphere, 1467
IP, 666, 667, 793, 800, 804, 971
IPAC, 321, 325
IPC, 480–482, 485, 486, 488, 724, 879, 880, 1152
IPCC, 1585, 1591, 1617
IPELS, 1025
IPHIR, 1048
IPLR, 348
IPN, 502, 505
IPS, 91, 1147

IPT, 1426, 1427, 1466, 1468, 1469, 1617
IR emission of galaxies, 573
IR observation of galaxies, 570
IRAC, 306, 312–314, 316, 318–321, 324, 409, 534, 538, 561, 565, 570–572, 575, 576, 585, 588, 591, 700, 702, 705–708, 718, 860, 862–864, 901, 920, 1177–1179, 1181, 1266
IRAS, 609–612, 614–616, 619, 620, 631
IRAS-LRS, 631
IRAS mission, 320
IRBM, 43, 44, 51, 52, 70
IRC, 305
IRIS, 995, 1047, 1427, 1434, 1437, 1439, 1440, 1443, 1461, 1467, 1502
IRM, 110, 1021
IRON, 635, 928
iron-atmosphere models, 732
iron-oxide minerals, 260
irradiance observations, 1041
irradiation ages of iron meteorites, 1329
irradiation history of meteorites, 1333
irregular satellites, 1457
IRS, 262, 575
IRTF, 1486
IRTS, 314, 323, 325, 609
ISAGEX, 1629
ISAMS, 1663
ISAS, 57, 321, 323–325, 575
ISD, 1183, 1187
ISEE, 964, 965, 977, 981, 1130, 1132, 1502, 1504
ISEE 1, 1020, 1539
ISEE 2, 1539
ISEE 3, 974, 1022, 1124, 1144, 1153, 1181, 1191, 1517, 1539, 1549
Ishtar Terra, 1311
ISM in the solar neighborhood, 665
ISM metallicity, 659
ISM models, 665
ISM O VI, 601
ISMIRAN, 136, 321, 322, 324, 325, 327, 460, 544, 561, 570–573, 575, 609, 610, 613, 616, 617, 622, 631, 634, 635, 702, 706–708, 712–715, 718, 857, 860, 862–865, 868, 920, 1432, 1434, 1435, 1438, 1440, 1441, 1444, 1446–1448
ISO, 14, 561, 700, 706, 857, 860, 1283
ISO highlights, 570
ISO spectroscopy, 865
ISO-CAM, 460, 611, 612, 617, 712, 714, 860, 862
ISO-LWS, 616–618, 622, 627, 628, 633, 635
ISO-PHOT, 612, 629, 860, 862–864
ISO-SWS, 611–614, 617, 620, 621, 623, 624, 626–635, 863, 867
ISO-SWS-LWS, 867
isochron, 1330
isochron rubidium-strontium, 1331
ISOGAL, 706
isothermal corona, 241
isotope dilution method, 975, 994, 1331, 1333
isotopes of primary cosmic-ray nuclei, 681
isotopic abundances in cometary volatiles, 1288
isotopic composition of cosmic rays, 689
isotopic composition of lithium, 428
isotopic ratios in comets, 1290
isotrophy of GRB locational, 506
isotropic background infrared radiation, 321
ISPM, 967
ISR, 1573, 1575, 1577
ISRO, 1611
ISS, 86, 91, 92, 94, 95, 1579
ISSI, 1526
ISTI, 109
ISTP, 1511, 1517
ITOS, 1655
ITOS-A, 1655
ITRE, 1643

ITRS, 1640
IUE, 294–297, 376, 427, 471, 534, 536, 561, 615, 654, 656, 658, 662, 668–670, 795, 796, 798, 799, 829, 832, 861, 864, 868, 876–880, 890, 900, 901, 904, 905, 919, 1048, 1191, 1283, 1432, 1434, 1435, 1440, 1447, 1408
IUE spectrum of HR Del, 797
IUE spectrum of TT Ari, 798
IUE UV, 797
IUGG, 1535, 1624
IVC, 601, 602, 670

J1808.4-3658, 814
Janus, 1453, 1454, 1457, 1459
Janus and Epimetheus, 1459
Japanese Space Programme, 56
JASON-1, 1632, 1637, 1646
JATO, 65
JBIS, 65
jet acceleration, 553
jet formation in the coma, 1222
jet-assisted take-off (see JATO)
jet-Propulsion Laboratory (see JPL)
jet-like emission, 551, 833
jets of matter–antimatter, 556
JFET, 320
JGM, 1638
JHU, 153
JILA, 897
Jodrell Bank ULA Astrometric Survey (see JVAS)
JOP, 1427
Journal of the British Interplanetary Society (see JBIS)
Jovian charged particles, 146
Jovian helium abundance, 1429
Jovian infrared thermal structural, 146
Jovian magnetic field, 1486
Jovian magnetopause, 250
Jovian ring, 1172, 1174
Jovian trapped radiation, 146
Jovian water vapor, 1426
JPL, 47, 65, 68–71, 97, 98, 110, 248, 264, 266, 267, 325, 1116, 1173, 1246, 1395, 1428, 1455, 1460, 1461, 1464, 1497, 1631
Juliet, 1453
junction field effect transistor (see JFET)
Juno, 154
Juno 1, 47, 77
Juno II, 77, 79
Juno rockets, 6, 10, 12, 43, 70, 76, 77, 93, 155, 157, 166, 232, 250, 257, 307, 315, 348, 453, 1182, 1183, 1425–1427, 1431, 1453, 1480, 1485, 1497
Jupiter and Io, 1456
Jupiter C reentry project, 70
Jupiter C satellite project, 70
Jupiter D/H, 428
Jupiter flyby, 166, 1433
Jupiter objectives of Galileo, 1428
Jupiter Orbiter-Probe, 1427
Jupiter relativistic electron beacon, 250
Jupiter system dust, 1176
Jupiter's atmosphere, 96
Jupiter's Galilean Moons, 11
Jupiter's inner magnetosphere, 1468
Jupiter's inner radiation belt, 167
Jupiter's magnetic field, 166, 1183, 1467, 1469, 1505
Jupiter-encounter trajectory, 168
JVAS, 457

K1-22, 932
K1-27, 925, 932
K2-2, 932
K5V, 765

K-6, 56
K-8, 56
KAMIOKA, 943
KAMIOKANDE, 943
KAO, 314, 315, 317, 324, 325, 609, 613, 616, 617, 620, 621, 632, 633, 920, 1187, 1236, 1264, 1283, 1285, 1289, 1291, 1426, 1428, 1454
kaon, 17
Kaputnik, 45, 47
Katyusha solid propellant rocket, 64
KB-29, 34
KBO, 692, 1186, 1187
Keck telescope, 861
Kegeldüse, 31, 33
Kepler's laws, 16
Kepler's SNR, 948
Kepler's supernova, 947
kerosene–oxygen, 72
kerosene–oxygen engines, 74
kilohertz QPOs, 818
kinematics, 716
kinematics of old stars, 708
kinetic age, 933
King crater, 1298
Kitt Peak National Observatory (see KPNO)
KNMI, 1604
komatiitic magma, 1385
Kopernikus, 105
KOSI, 1219
Kosmos, 78, 93
Kosmos satellite, 78
Kourou, 104
KPNO, 458
KREEP, 1340, 1341, 1344–1348, 1350, 1351, 1353–1357, 1368, 1381–1385, 1388, 1389
Kristall, 92
KTG, 712, 713
KTG IMF, 712
Kuiper Airborne Observatory (see KAO)
Kuiper Belt Object (see KBO)
Kvant 1, 92
Kvant 2, 92

laboratory studies of reconnection, 1023
Lacertids, 550
LAGEOS, 347, 1629, 1634, 1638, 1640
Lageos, 1315
LAGEOS2, 295, 1631, 1634
Lagrangian point, 13
Laika, 47, 71
λ Sco, 656
landing on a cometary surface, 1268
Landsat, 1315, 1462, 1475, 1593, 1596, 1664
large angle and spectrometric coronagraph (see LASCO)
Large Astronomical Satellite (see LAS)
large magellanic cloud (see LMC)
large particles, 1225
large probe cloud particle spectrometer, 1412
large-scale, 399, 1146
large-scale kinetic simulations (see LSK)
large-scale structure, 485, 602
large-scale structure of galaxy distribution, 400
Larissa, 1453, 1455
LAS, 295
LASCO, 1081, 1108, 1135
LASER interferometric observations of gravitational waves, 367
LASER interferometry, 20, 366
LASER ranging to the Moon, 1630
LASER scattering surface, 413, 417
LASER upgrading, 1628

LASP, 1208
late Amazonian volcanism, 1362
late M supergiants, 904
late stages of accretion, 885
late-type giants, 877, 878, 880
latitude dependence of fast neutrons, 127
latitude gradients, 142
latitude-height cross section of water vapour, 1662
lava flows, 258
laval nozzle, 242
LB, 1192, 1539
LBL, 554, 558
LDEF, 680, 1167, 1169, 1172
lead sulfide, 304, 326, 685
LEAM, 6, 1167, 1184
LECP, 966, 996
LECS, 889
Leda, 1453, 1454
LEGOS-GRGS, 1641
Leiden–Dwingeloo, 587
Leiden–Dwingeloo H1 survey, 592
LenGIRD, 34
Leningrad GIRD (see LenGIRD)
lens equation, 445
lens statistics, 444
lensed quasar images, 444
Lense–Thirring precession, 344
lensing galaxies, 455
Leonid meteor storms, 1181
LET, 144, 145
LEW86010, 1360
LEW88516, 1363, 1368
LGM 1, 722
LHB, 589, 592–599, 601, 603
lherzolitic shergottites, 1363, 1364
LIC, 1203, 1210, 1212
life cycle of dust and gas, 608
life cycle of interstellar gas and dust, 607
life on Mars, 260
life on other planets, 21
light curves of afterglow of GRB, 512
light deflection, 339
light deflection in the Solar System, 453
light elements, 423
light-gas gun, 1174
light-curve of first GRB, 501
lighthouse effect, 553
lightning and radio detector, 1428
lightweight optics, 327, 1386
light rare Earth elements (see LREE)
LIGO, 357, 365–369, 1396
limb darkening, 1066
limb-infrared monitor of the stratosphere (see LIMS)
limb sounding, 1660
limonite, 260
LIMS, 1662
Lindblad's ring, 653, 654
line broadening in nebulae, 916
linear filtering of COBE-DMR sky maps, 412
LINER, 564
lines in GRBs, 504
Liouville's theorem, 175
liquid hydrogen, 72
liquid oxygen, 72
liquid-fuel rocket, 33, 34, 35
liquid-fuel rocket boosters, 32
liquid-fuel rocket motors, 31
liquid-helium-cooled devices, 305
liquid-helium-cooled telescope, 308–310, 318, 323
liquid-neon-cooled telescope, 313

liquid-nitrogen-cooled telescope, 308
liquid-propellant rockets, 19, 60
liquid-solid surface transitions, 1474
LISA, 369, 1056
LISM, 590, 662, 663, 1123, 1141, 1143, 1156, 1187
Lithium-6, 435
Lithium-6 and -7, 428
Lithium-7, 437
lithosphere, 1316, 1318
lithospheric heat transfer mechanisms, 1303
Little Green Man 1 (see LGM1)
LLBL, 1508, 1539, 1541, 1549
LLR, 340, 345–350
LMC, 296, 297, 376, 384, 454, 464, 508, 584, 651, 669, 670, 714, 766, 807, 808, 829, 943
LMC X-3, 765, 782, 807, 814, 846
LMO, 1340, 1342, 1344, 1345, 1351, 1354, 1358, 1369
LMXB, 16, 435, 760–762, 764, 771, 772, 777, 779–782, 784, 785, 797, 799, 806, 807, 811, 813–817, 819, 823, 825, 832, 853, 932
LMXB mass donors, 813, 1461, 1463
LMXRB, 569
local bubble, 653, 658, 662, 1192, 1212
local cavity, 595
local cloud, 1194, 1212
local-cloud absorption, 1203
local-cloud velocity, 1203
local flow beneath sunspot, 1054
local fluff, 595, 597, 662, 663
local galactic magnetic field, 668
local galaxies, 447
local group, 712
local helio seismology, 1052
local hot bubble (see LHB)
local interstellar medium (see LISM)
local interstellar wind, 662
local ISM, 596, 598
local standard of rest (see LSR)
local thermal equilibrium (see LTE)
local volume mass-density, 711
localized solar flows, 1053
Lockman Hole, 590
Lockman hole deep survey, 574
Lockman hole shallow survey, 574
Loki solid-propellant rockets, 70
Loki's volcanic plume, 1467
long-duration exposure facility (see LDEF)
long-baseline interferometry, 20
long-period comets (see LP)
long-period pulsar, 826
long-term light curve, 792
long-wave radiative flux, 1659
loop I, 585, 591, 595, 599
loop I/LHB Ring, 597
loop I superbubble, 596
loop models, 887
LOR, 84, 85
LORAL, 109
Lorentz-invariance violation, 346
Lorentz-invariant gravity, 344
loss of magnetospheric ions, 1548
loss-cone cyclotron maser, 1521
losses of orbital angular momentum, 780
losses of trapped particles, 162
low-energy charged particle (see LECP)
low-energy component of cosmic rays, 1199
low-energy telescope (see LET)
low-frequency peak BL lacs (see LBLs)
low-field neutron star atmospheres, 731
low-latitude boundary layer (see LLBL)
low-mass companion, 431, 763

low-mass X-ray binary (see LMXB)
low-polarization quasar, 550
low-polarizaion quasar (see LPQ)
low-resolution spectrometer (see LRS)
Lowell planetary patrol, 263
lower atmosphere of Venus, 1079, 1411
lower-hybrid drift wave, 1503
low-1, 1046, 1056
LOX, 86, 1238, 1242, 1259
LP, 1238
LPL, 1454, 1455
LPQ, 550, 555
LREE, 1342, 1347, 1353, 1355, 1367
LRI, 1617
LRS, 318, 319, 609
LSD, 280, 281
LSK, 1540, 1549, 1553
LSR, 648, 667, 1198
LST, 100, 316
LTE, 897, 898, 924, 929
luminosity, 933
luminosity calibration, 376
luminosity function, 485
luminosity of central stars, 932
luminosity of CVs, 791
luminosity of single star, 774
luminosity variation, 1041
luminous clusters, 479
luminous dark matter, 441
luminous magnetic rotator, 903
luminous radio sources, 491
luminous weak-field neutron star, 819
luminous X-ray binaries, 273, 274, 852, 1337, 1381
Luna 1, 86
Luna 2, 247, 1489
Luna 3, 86, 247, 273
Luna 9, 79, 80
Luna 10, 1489
Luna 16, 1339, 1351, 1353, 1354
Luna 16 Mare Basalts, 1352
Luna 20, 1339, 1342, 1348
Luna 24, 1339, 1351, 1353, 1354
Luna 24, 1354
Luna 24 Mare Basalts, 1352
Luna missions, 1377, 1479
Luna-Venus, 80
lunar and planetary laboratory (see LPL)
lunar basalts, 1351
lunar-basalts composition, 1354
lunar bombardment, 1369
lunar chronology, 1338
lunar craters, 259, 261
lunar ejecta and meteoroids (see LEAM)
lunar explorer, 35, 1175, 1183
lunar farside, 1298
lunar fines, 1379, 1380
lunar frontside maria age, 1354
lunar heat flow, 1299
lunar highlands, 1384
lunar-highland composition, 1390
lunar-highland crust, 1340, 1388
lunar landscape, 9
lunar laser ranging (see LLR)
lunar linear rilles, 1300
lunar magma ocean (see LMO)
lunar magmatism, 1354, 1479
lunar mantle, 1388
lunar mapping, 257, 1388
lunar meteorite ages, 1342
lunar-mission mode, 83

lunar-mode debate, 85
lunar motion, 339
lunar nearside, 1339
lunar-non-mare basalts, 1388
lunar-orbit rendezvous mode (see LOR)
lunar orbiter, 275, 1183, 1337
lunar orgin, 6
lunar prospector, 347, 1298, 1350, 1390, 1489
lunar rock, 1170
lunar-sample analyses program, 1169, 1298, 1379
lunar sampling sites, 1352
lunar-soil samples, 10, 1379
lunar-thermal evolution, 1338
lunar-thermal history models, 1357, 1358
lunar volcanism, 1357
Lundquist's chart, 1628
Lunik, 78
Lunik 1, 1116, 1144
Lunik 2, 1116, 1144
Lunik 3, 1116
LWP, 919
LWR, 322, 609, 610, 620, 629, 862, 863, 919, 920
LWS, 609, 616–618, 620, 622, 627–629, 633, 635, 637
Lyman α absorption lines, 542
Lyman–Spitzer Telescope (see LST)
Lysithea, 1452, 1453

M2 tide, 1646
M2-9, 915
M4-S, 57
M-4S-3, 57
M8, 312
M17, 307, 315
M23–24, 596
M31, 564, 571
M33, 564
M81, 564
M82, 566, 572, 573
M86, 562, 569
M87, 529, 567, 1362
MACEK, 1632
MACHO, 385, 454, 458
macroscopic plasma dynamics, 1553
macrospicules, 1077, 1078, 1311–1313, 1412, 1414
MAD, 886
Magellan data, 1311
Magellan mission, 11, 454, 766
Magellanic clouds, 297, 564, 668, 778
magenatic enthalpy, 10, 228
MAGER, 1479
magnet ograns of the Sun, 1090
magnetars, 727
magnetic braking model, 726
magnetic canopy, 1074
magnetic cleanliness, 248
magnetic clouds, 1152
magnetic CVs, 800
magnetic declination, 226, 160
magnetic elements, 146, 766, 1069
magnetic field above the solar surface, 1097
magnetic field around nucleus of comet, 1230
magnetic field at Jupiter, 1485
magnetic field at Neptune, 1487
magnetic-field decay, 819
magnetic-field decay by accretion, 785
MAGnetic field Electron Reflectometer (see MAGER)
magnetic field in high-speed solar wind, 1155
magnetic field in neutron star X-ray binaries, 764
magnetic-field measurements, 10
magnetic field of Earth, 226

magnetic field of Mercury, 1400
magnetic field of Moon, 1479
magnetic field of sunspots, 229
magnetic field of the Earth, 156
magnetic field of white dwarf, 804
magnetic-field profiles, 1531
magnetic-field structure, 983
magnetic-field structure of solar corona, 1015, 1069, 1085
magnetic fields in early-type stars, 902
magnetic fields of Moons, 1489
magnetic fields of Neutron Stars Decay, 785
magnetic fields of planets, 1479, 1480
magnetic fields of Uranus and Neptune, 1488
magnetic fields on the Sun, 226, 1090
magnetic-flux systems, 1096
magnetic-flux tubes, 1069, 1075, 1082
magnetic foot, 1511
magnetic loops, 1077
magnetic moment, 157, 1419
magnetic parts of chromosphere, 1073
magnetic pile-up boundary, 1229
magnetic propellor, 806
magnetic properties of Neptune, 171
magnetic reconfigurations, 1108
magnetic reconnection, 5, 1111
magnetic reconnection experiment (see MRX)
magnetic regions on Sun, 1054
magnetic-rotator theory, 905
magnetic-shock profile, 158, 1511, 1545
magnetic-storm aurora, 158
magnetic storms on Earth, 229, 1530
magnetic Sun, 1095
magnetic topology of reconnecting magnetosphere, 1015
magnetic-field aligned electric potential drops, 1515
magnetism of planetary bodies, 156, 1118
magnetized stellar winds, 877
magnetobremsstrahlung, 651
magnetogram, 1074
magnetohydrodynamics (see MHD)
magnetometers on satellites, 159, 246, 248, 249, 1017, 1502, 1505, 1530
magnetopause crossing, 1021, 1508
magnetopause structure, 1506
magnetopause transport, 1502
magnetosheath, (see MS)
magnetosheath boundary layer (see MSBL)
magnetosheath plasma, 5, 71, 137, 185, 768, 1014, 1481, 1501, 1539, 1541
magnetosphere as plasma laboratory, 1495
magnetosphere of giant planets, 1440
magnetosphere of Jupiter, 11
magnetospheres in the universe, 1554
magnetospheric boundary layers, 1548
magnetospheric convection, 1507
magnetospheric dynamics, 1018
magnetospheric emission models, 725, 728
magnetospheric field, 153, 172, 1532, 1545
magnetospheric reconnection, 1014
magnetospheric research, 1551
magnetospheric substorm, 1532, 1537, 1544
magnetospheric tail, 1545
magnetotail, 5, 1481, 1541, 1549
magnetotail cross section, 1540, 1542
magnetotail during plasmoid ejection, 1022
magnetotail during substorms, 1016
magnetotail reconnection, 1022
magnetotail of Neptune, 1490
magnetotail of Uranus, 1385, 1490
magnitude of a comet, 1236
MAGSAT, 1577
main sequence star, 6

main telescope (see MT)
Maja Valls, 1306
Malmquist bias, 382
manned Moon project, 81
mantle on Venus, 1313
MAP, 400, 405, 417, 418, 442, 463, 549, 1169
mapping circumstellar debris disks, 864
mapping of Mars, 265
MARCI, 1418
mare basalt ages, 1352
mare basalt volcanism, 1354, 1379, 1383
mare Crisium, 1339
mare Fecunditatis, 1339
mare Imbrium, 9, 1299, 1339
mare Serenitatis, 1300, 1339
mare surface ages, 1354
mare Tranquilitatis, 8, 1339
mare volcanism, 1353
MARECS, 56
marine geoid, 1611, 1613
Mariner, 148, 265, 964, 965, 983, 985, 1408, 1434, 1479
Mariner 1, 1116
Mariner 2, 155, 165, 248, 249, 262, 265, 1105, 1106, 1116, 1117, 1144, 1409
Mariner 4, 98, 140, 155, 165, 263, 264, 266, 1405, 1414, 1484
Mariner 5, 155, 1118
Mariner 6, 264, 266, 1414
Mariner 7, 264, 1414
Mariner 8, 265
Mariner 9, 98, 257, 264, 266–268, 1414
Mariner 10, 165, 268, 1128, 1201, 1301, 1302, 1304, 1399, 1405, 1407, 1480, 1482, 1483
Mariner 10 magnetic field data, 1483
Mariner missions, 80
Mariner Jupiter–Saturn (see MJS)
markings of Venus, 259
markings of Mars, 9, 10, 29, 155, 157, 165, 232, 257, 259, 263, 348, 349, 1304, 1358, 1367, 1377, 1390, 1405, 1406, 1414, 1453, 1479, 1480, 1484
Mars, 13
Mars 2, 266, 1414
Mars 3, 266
Mars a dead world, 268
Mars and Phobos, 1456
Mars chronology, 1325
Mars Climate Observer, 98
Mars Climate Orbiter (see MCO)
Mars Color Imager (see MARCI)
Mars completely dead, 264
Mars Enviormental Survey (see MESUR)
Mars expedition, 113
Mars Express, 1421
Mars geological activity, 264
Mars global surveyor, 98, 264, 265, 267, 1304, 1306, 1308, 1309, 1358, 1362, 1417, 1418, 1480, 1485, 1486
Mars history, 268
Mars Observer (see MO)
Mars orbiter camera (see MOC)
Mars orbiter laser altimeter (see MOLA)
Mars Pathfinder, 1394, 1416
Mars Pathfinder rocks and soil, 1397
Mars Polar Lander, 98, 1414
Mars program, 79
Mars Project, 67, 79, 85
Mars research, 258
Mars Surveyor Program, 98
Mars system dust, 1176
Martian canals, 260
Martian chronology, 1358
Martian crater retention ages, 1364
Martian cratering rate, 1360, 1366

Martian dust environment, 1186
Martian dust storm, 265
Martian dynamo, 1485
Martian fluvial channel system, 267
Martian geological features, 266
Martian invasions, 29
Martian magma ocean, 1367
Martian magnetic fields, 1485
Martian megaregolith, 1307, 1358, 1359, 1364, 1370, 1377, 1386, 1391, 1396
Martian rocks, 6
Martian seasonal polar caps, 1398, 1420
Martian source terrains, 1362
Martian space science, 1414
Martian surface, 1306
Maser radiation, 1521
mass accretion, 843
mass density of the early Universe, 432
mass dependence of luminosity, 862
mass distribution, 441, 456
mass distribution in galactic disk, 709
mass estimates, 489
mass-flux distribution, 1224
mass loss, 934
mass loss from cool stars, 895, 903
mass loss from early-type stars, 896
mass loss from stars, 895
mass loss in close binaries, 770
mass loss of cometary nuclei, 1266
mass movements, 1473
mass of an X-ray cluster, 343, 387, 476
mass of ions, 1578
mass of the galaxy, 386
mass-spectrograph analysis, 1327
mass spectrometer, 1175, 1181, 1328, 1428
mass spectrometric isotopic analysis, 1331
mass spectrum of a dust particle, 1181
mass transfer in close binaries, 770
mass transfer in CV, 782
mass transfer in LMXB, 782
mass-losing Star, 775
mass-loss of non-binary cool stars, 907
mass-loss rates, 897
mass-to-light ratio, 386, 476, 715
Massachusetts Institute of Technology (see MIT)
massive black holes, 540, 549
massive central black hole, 539
massive compact objects, 441
massive neutrinos, 431
massive X-ray binaries, 766
mass–radius dependencies, 752
material loss rates, 1266
Mathilda, 257
matter content, formation, and acceleration of jets, 557
matter density, 385
matter reaction with antimatter, 111
Matthew principle, 328
Maunder minimum, 1095, 1097, 1118
Max Planck Institute for Extraterrestrial Research (see MPE)
MAXIMA, 385, 415, 416, 418
MAXIMA 1, 383, 388–390
MBB, 75, 94, 96, 105
MBM 12, 23, 24, 594, 597
MCO, 1414, 1418, 1420, 1421
MDI, 945, 1042, 1046, 1050, 1052, 1053, 1055, 1068, 1076, 1096
mean-free-path curve, 582
mean-overall denisty of Universe, 431
mean planetary radius (see MRP)
measoscale wind field, 1609
measurements of pollution in the troposphere (see MOPITT)
measurement of seismic radius, 1050

measuring solar disturbances locally, 1047
mechanism of reconnection, 1506, 1525
mechanism that powers quasars, 544
MECS, 888, 889
Mediterranean Forecasting System Pilot Project (see MFSPP)
medium profile spectrometer, 298
MEDOC, 1630, 1636
mega-maser, 572
MEM, 707, 1199
Menelaus, 50, 83, 157, 165, 257, 259, 268, 337–339, 342, 347–349, 1300, 1377, 1399, 1405, 1406, 1408, 1421, 1452, 1482, 1483
MER, 1199
mercury cadmium telluride, 326
Mercury evolution with time, 1303
Mercury orbiter, 1421
Mercury perihelion precession, 350
Mercury programme, 50
Mercury rotation period, 268
Mercury's evolution, 1483
Mercury's global magnetic field, 1483
Mercury's magnetic field, 1399
Mercury's perihelion, 339
Mercury's perihelion advance, 340
merged interaction region (see MIR)
merging potential, 1019
MERIS, 1601, 1617
mesogranulation, 1068
mesopause, 200
mesoscale ocean current, 1605
mesosphere, 200
messenger, 886, 889, 1408
Messerschmitt–Boelkow–Blohm Corporation (see MBB)
MESUR, 1415–1417
metal-abundance deficiency (see MAD)
metal mass-to-light ratio, 488
metallic hydrogen, 1437
metallicity effect of Cepheids, 377
metallicity of interstellar gas, 647
metallicity—kinematics distribution, 702
metal–silicate partition coefficients, 1386
Meteor Crater, 258
meteor studies, 1163
meteoriods, 146
meteorite crater, 258
meteorite research, 6, 11, 1163
meteorites from Mars, 9
meteorites from the Moon, 1378
meteoritic and other extraterrestrial material, 1164
meteoroid flux, 1170, 1180
meteoroid residuals, 1163, 1168, 1172, 1183
meteoroids in space are electrically charged, 1182
meteorological satellites, 1654, 1655
Meteosat system, 1657
Meteosat, 1656
Meteosat second generation (see MSG)
metis, 1453, 1458
METOP, 1611, 1617, 1667
METOP/EPS, 1596
metric field of gravity, 339
metric gravitational field, 346
metric theory of gravity, 341, 350
metrical gravitational field, 340
metrology of the Earth, 1645
MFSPP, 1605
Mg-suite (see MGS)
MG, 598, 599
MG0414 + 0534, 449, 456
MG1131 + 0456, 449
MG1654 + 1346, 449
MGS, 227, 228, 234, 236, 237, 250, 1109, 1340, 1341, 1357, 1397, 1398, 1485

MHD, 227, 248, 1426
MHD turbulence, 1132
Miams, 12
Michelsonn–Doppler interferometer (see MDI)
Michelson interferometer, 650
microcraters, 1169, 1170
microflares, 1079, 1135
microlensing, 458, 465, 712, 715
microlensing in the Milky Way, 453, 1179
microphysics of reconnection, 1025
microwave anisotropy probe (see MAP)
microwave brightness, 1593
microwave emissivity, 1594
microwave limb sounder (see MLS)
microwave limb sounding, 1666
microwave observation, 262
microwave radiometers, 402
microwave temperature structure, 19
mid-UV, 288
midcourse space experiment (see MSX)
MIE, 967
military interest in infrared detection, 326
military–industrial complex, 4, 14, 297, 321, 328, 564, 582, 584, 647, 653, 706, 721,
Milky Way C IV absorption, 297
Milky Way galaxy, 704, 705
Milky Way globular clusters, 713
millimetre wavelength techniques, 748, 1661
millisecond radio pulsars, 766
Mimas, 169, 1453, 1458
Mimas and Tethys, 1459
MIMOSA, 1632
mineralogy, 10
mineralogy of lunar crust, 1298
minimum orbital unmanned satellite of the Earth (see MOUSE)
minitrack transponder, 45, 70
Minkowski's star, 949
minor ACR, 994, 996
Minuteman, 43
MIPS, 139, 575, 1123, 1579
MIR, 139
Mir space station, 86, 92, 95, 113
Mir-Kvant Space Station, 944
Mir–Shuttle rendezvous, 95, 1452, 1453, 1457, 1461
MISR, 1601, 1617
missions to Mars, 98, 1116, 1415
MIT, 247, 292, 1115
Mittelwerk, 36
mixing diagram, 1384, 1398
mixing of the solar wind and ionospheric plasmas, 1553
MJS, 1426
MLS, 1663
MMS, 665, 1308, 1485, 1665
MO, 1416, 1417, 1485
MOC, 1308
model atmosphere analysis, 926
model B rocket, 30
model E rocket, 31
model nebula, 1237
model of a flux transfer event, 1022
model of Be star X-ray binary, 765
model of the heliosphere, 1157
modeling cosmic-ray diffusion, 685
modeling neutral-plasma interaction, 1199
modeling neutron-star atmospheres, 730
modeling of the lunar orbit, 348
modeling UV data, 655
modelling atmosphere, 924
modelling cometary nucleus, 1259
modelling interplanetary dust, 1166
models of cometary grains, 1284

models of dissipational disk galaxy formation, 700
models of dynamical black hole, 765
models of Milky Way galaxy formation, 701
models of the heliosphere, 1198
models of the Sun, 230
models of X-ray emission, 477, 878
MODIS, 1601, 1617
modulation of cosmic rays, 17
MODUST, 866, 867
Mögel-Dellinger events, 1563
MOLA, 1305, 1307, 1308
molecular abundances in molecular clouds and comets, 1289
molecular astrophysics, 315, 607, 612, 1279
molecular composition of cometary volatiles, 1287
molecular diffusion, 193
molecular emission from AGB winds, 633, 657
molecular hydrogen in interstellar medium, 293
molecular ring, 707
molecular species in comets, 1280
Molnija, 80
momentum problem, 903
monitoring of the solar wind, 1144
monitoring ozone, 5, 1144
Monoceros OB1, 597
Monoceros R2, 597
Monoceros–Gemini Enhancement, 585, 599
Monogem Ring, 585, 595, 597, 599, 600, 746, 1585
MONS, 1059
Montreal protocol, 6, 9, 26, 76, 163, 157, 259, 306, 340, 335, 341, 348, 350, 1297, 1337, 1367, 1377, 1378, 1452, 1453, 1585, 1630
Moon, 13
Moon and terrestrial planets, 1295
Moon before Apollo, 271
Moon chronology, 1325
Moon flight, 26
Moon frozen in time, 1356
Moon landing, 258
Moon rocks, 1337
Moon's lithosphere, 271, 1299
Moon's perigee precession, 340
Moon's shape, 272
Moon's silica mantle, 349
MOPITT, 1667
morphologic model, 1254
morophology of comet surface, 1266
morphology of cometary nuclei, 1236
morphology of the Earth, 1315
MORTEX, 1418, 1601, 1617
MosGIRD, 34
moss, 1082
motion and gravitational radiation, 356
mountains of Venus, 1311
MOUSE, 45, 1508, 1539
MPAE, 1251
MPE, 100, 880, 946, 948, 953, 955, 956
MPE-Garching, 723
MPR, 1307
MRX, 1024, 1539
MRX experiment, 1024
MS, 1021, 1509, 1512
MS0440 + 02, 460
MSBL, 1541
MSD, 1245
MSFC, 1106
MSG, 1657, 1658
MSIS, 1566
MSX, 144, 322, 609, 706, 742–744
MT, 144
MTOF, 1124
MTS, 1167

μ Cephei, 306
μ Col, 292, 656, 657, 671
μ Columbae, 651
mu meson, 231
μ neutrinos oscillate, 1056
multi-band imaging photometer for SIRTF (See MIPS)
multi-color blackbody disk, 847
multi-mission modular spacecraft (see MMS)
multi-wavelength comparisons, 575
multi-wavelength studies, 792
multiple images of, 459
multiple imaging, 445
multiplexed sensing chips, 327
multiply-imaged QSO, 455, 458
multiply-imaged quasars, 444
multiply-imaged quasi-stellar objects, 454
multi-stage rockets, 60, 69
muon, 17
MX-774, 43
MY, 1617
MyCnl 8, 914

N-1/L-3, 86
NACA, 48, 57, 67, 82, 136, 1653
NAD, 1624
Nadir, 1660
Naiad, 1453
nakhlites, 1362, 1364, 1365
NAL, 539
nanocrystal clusters, 634
nanoflares, 234, 284, 305, 501, 1116, 1135
NASA, 26, 32, 42, 48, 50, 51, 57, 61, 68, 72–74, 77, 80, 83, 85–89, 91, 93–96, 98–101, 107, 109, 110, 118, 136, 144, 147, 148, 155, 185, 215, 216, 220, 260, 264–267, 272–274, 284, 291, 294, 295, 298, 303, 306, 309, 313, 315–317, 319–321, 323–325, 327, 328, 341, 344, 347, 399, 402, 405, 417, 450, 455, 459, 487, 493, 494, 508, 534, 539, 544, 545, 557, 558, 562, 563, 575, 576, 581, 609, 671, 680, 694, 723, 724, 795, 807, 823, 858, 861, 864, 865, 887, 897, 905, 917, 938, 944, 947, 950, 966, 967, 1009, 1013, 1041, 1045, 1046, 1056, 1067, 1077, 1084, 1104, 1106, 1109, 1118, 1120, 1121, 1123, 1124, 1127, 1132, 1145, 1157, 1168, 1169, 1172–1174, 1178, 1180, 1183, 1185, 1204, 1206, 1229, 1248, 1279, 1280, 1298–1300, 1302, 1305–1308, 1311–1313, 1339, 1370, 1377–1379, 1394, 1408, 1409, 1413, 1415–1418, 1420, 1421, 1428, 1429, 1431–1436, 1447, 1448, 1454, 1455, 1460, 1461, 1464, 1465, 1483, 1486, 1497, 1539, 1594, 1608, 1623, 1627–1629, 1631, 1634, 1638, 1644, 1647, 1653, 1654, 1658, 1660–1664, 1667, 1668
NASA/GSFC, 1602
NASA/JPL, 1597, 1598, 1606
NASA–ESA, 221
NASDA, 57, 1611, 1654
National Advisory Committee for Aeronautics (see NACA)
National Aeronautics and Space Administration (see NASA)
National Aeronautics and Space Development Agency (see NASDA)
National Bureau of Standards (see NBS)
National Centres for Environmental Prediction (see NCEP)
National Oceanographic and Atmospheric Adminstraton (see NOAA)
National Radio Astronomy Observatory (see NRAO)
National Science Foundation (see NSF)
NATO, 43
natural dust, 1169
natural films, 1610
natural radioactivity, 1325
Naval Research Laboratory (see NRL)
Navy Navigation Satellite System (see NNSS)
NB, 817
NBO, 817
NBS, 279
NCEP, 1604, 1605
near-surface wind field, 1608
near-Earth dust environment, 1168

SUBJECT INDEX

near-Earth Satellites, 348
near-infrared camera and multi-object spectrograph (see NICMOS)
near-infrared intensity, 705
near-infrared mapping spectrometer (see NIMS)
nearby radio pulsars, 745
nearby stars, 447
nearest ISM, 662
nebula space velocity, 917
nebulae numbers and galactic longitude, 922
nebular abundances, 930
nebular continuum emssion, 927
nebulium lines, 916, 918
Nederlands Instituut voor Vliegtuigontwikkeling en Ruimtevaart (see NIVR)
negative hydrogen ion, 42, 171, 157, 230, 257, 335, 453, 1431, 1434, 1435, 1441, 1453, 1480, 1488,
NEI, 942, 956
Neptune atmospheric structure and composition, 1443
Neptune internal structure, 1442
Neptune ionosphere and magnetosphere, 1444
Neptune magnetic field, 1488
Neptune ring system, 1173
Neptune's magnetosphere, 172
Neptune's stratosphere, 1446
Nereid, 1453, 1454, 1462
NERVA, 107, 108
NESDIS, 1596, 1617
net flux radiometer, 316, 1073, 1076, 1428
Neukum–Wise (see NW)
neutral atomic oxygen, 315
neutral carbon lines, 315
neutral gas outflow, 1224
neutral helium glow, 1200
neutral hydrogen flow, 1210
neutral hydrogen, 1206
neutral material in galactic disk, 595
neutralization of an ion beam, 1524
neutrino annihilation, 521
neutrino-bremsstrahlung, 729
neutrino burst, 943
neutrino-burst detection, 943
neutrino emission from neutron star, 729
neutrino-fossil radiation, 429
neutrino interactions, 429
neutrino production, 1047
neutrinos and solar models, 1040
neutron intensity, 128
neutron monitor locations, 130, 759, 760
neutron-star crust, 785
neutron-star formation, 678
neutron-star hypothesis, 505
neutron-star laboratory, 833
neutron-star magnetosphere, 827
neutron star mass, 826
neutron-star orbit, 826
neutron-star spin period, 826
neutron-star structure, 725
neutron-star thermal radiation, 730, 764
neutron-star/black hole X-ray binaries, 721, 807
neutron-star X-ray binaries, 844
neutron–proton capture, 429, 430
neutrons in the atmosphere, 123
new elements, 914
new generation space telescope, 1448
new Quebec impact crater, 261
new technology telescope (see NTT)
Newton, 889, 890
Newton telescope, 102
Newtonian gravity, 335
Newtonian multipolar gravitational field, 347
Newtonian Solar System, 338

Next Generation Space Telescope (see NGST)
NFI, 510
NGC 246, 919, 925, 932
NGC 253, 566, 575
NGC 1068, 307
NGC 1360, 925
NGC 1535, 925, 932
NGC 1850, 715
NGC 2516, 882
NGC 3132, 932
NGC 3242, 914, 925, 932
NGC 4038, 573
NGC 4039, 573
NGC 4472, 569
NGC 4636, 568, 569
NGC 5307, 915
NGC 6210, 925, 932
NGC 6302, 929
NGC 6357, 312
NGC 6445, 929
NGC 6543, 919
NGC 6572, 932
NGC 6720, 925, 932
NGC 6741, 917
NGC 6853, 925, 932
NGC 7008, 932
NGC 7027, 920, 929, 932
NGC 7293, 925, 932
NGC 7662, 932
NGST, 298, 384, 442, 456, 463, 576, 609, 636, 868, 1432, 1448
NICMOS, 323, 1428
nightglow, 189
NII-88, 38, 44
Nike, 48
Nike Booster, 281
Nimbus, 1593, 1661
Nimbus 7, 104, 1321, 1595, 1598, 1600, 1660, 1663, 1664
Nimbus science, 1662
NIMS, 1413, 1428, 1429, 1461, 1462
NIR, 1600, 1601
nitrogen to oxygen ratio, 932
NIVR, 316, 320
NIXT, 1067, 1078, 1081
NKVD, 34
NML Cyg, 305
NML Tau, 305
NNSS, 1575, 1643
NOAA, 1054, 1596, 1617, 1635, 1654, 1656, 1657, 1668
NOAA-1, 1655
NOAA-3, 1655
NOAA-9, 1658
NOAA-10, 1658
NOAA-15, 1655
NOAA-AVHRR, 1607, 1611
Noachian meteorites, 1365
Noachian-age Martian rock, 1359
noctilucent clouds, 1563
non-equilibrium ionization (see NEI)
non-linear effects in high-frequency plasma, 1515
non-LTE, 929
non-LTE analysis, 929
non-LTE effect, 898
non-LTE model, 897
non-magnetic CVs, 795
non-military space program, 259
non-Newtonian gravity theories, 710
non-nuclear super-Eddington sources, 564
non-symmetric lenses, 444
non-thermal plasma radiation, 1521
non-wave heating, 1083

non-gravitational physics, 341
non-thermal galactic radio risk, 687
Nordvedt effect, 356, 1344
norite clast, 1342, 1343
Normal-Incidence X-ray Telescope (see NIXT)
normal-mode frequencies of Sun, 1052
NORSEX, 1592, 1594, 1617
North American datum (see NAD)
north-polar region of Mars, 1308
North Polar, 585
North-Polar Spur (see NPS)
not strongly absorbing (see NSA)
Nozomi spacecraft, 1176, 1186
NPS, 663
NRAO, 532, 950
NRAO VLA, 534
NRL, 41, 45, 67, 71, 205, 207, 278, 280–283, 290, 291, 307–309, 314, 317, 325, 326, 763, 774
NSA, 664
NSA gas, 664
NSC, 736
NSCAT, 1590, 1605, 1617
NSF, 303, 307, 1575
NTT, 455, 522, 832
nuclear "star" event, 125
nuclear emulsion, 125
nuclear-explosion craters, 261
nuclear fusion inside a star, 939
nuclear physics, 424
nuclear-powered rockets, 107
nuclear reaction cross-sections, 1047
nuclear-electric power generators, 109
nuclear-electric propulsion system, 108
nuclear-thermal rocket motor, 107
nuclear-thermal rockets, 108
nucleation scenario, 1432
nucleon-capture reactions, 423
nucleonic, 432
nucleonic component of cosmic rays, 231
nucleonic density, 19, 430, 431, 938
nucleosynthesis of elements, 1334
nucleosynthesis of light nuclei, 429
nucleosynthetic origins for the elements, 658
nucleus albedo, 1242
nucleus imaging, 1258
nucleus of comet 1P/Halley, 1256
nucleus of comet Halley, 1246, 1251
nucleus size, 1242, 1326
numerical simulation of shock waves, 1073
numerical simulations, 1058, 1068
numerical simulations of cluster formation, 474
nutrino capture cross-sections, 432
NW, 1359
NWP, 1586

OAO, 77, 291, 295, 316
OAO-2, 292, 293, 654, 658, 793, 795, 1282
OAO-3, 293
OAO-A, 292
OAO-B, 292, 293
OAO-C, 292, 293, 653, 823, 825, 829, 831, 832
OB star companion, 824
OBC, 296
Oberon, 1453, 1462
objective grating spectrometer (see OGS)
oblateness, 1042
oblateness of gravitational equipotentials, 1036
oblateness of the Sun's surface, 1050
obscured quasars, 537
obscuring atmosphere, 287

observation of jets, 1225
observation of the solar wind, 247, 667
observations of active comets, 1244
observations of corotating particles, 987
observations of inactive comets, 1243
observations of solar particles, 968
observing the Earth, 13
OCC, 1245
occultation of β Sco by Jupiter, 1426
occultations of stars by nuclei, 1246
ocean circulation, 1598, 1641
ocean-color remote sensing, 1602
ocean-color sensing, 1601
ocean-dynamic topography, 1599
ocean-primary production, 1603
oceanic-brightness temperature, 1612
oceanographic altimetry, 1637
oceanographic applications, 1586
oceanography, 1585, 1603, 1631
Oceans Observation System Development Panel (see OOSDP)
Oceanus Procellarum, 1339
OCI, 1601, 1617
OCM, 1601, 1617
OCTS, 1601, 1617
OCZ, 1123
ODIN, 622
Office National d'Etudes et de Recherches Aéronautiques et Spatialle (see ONERA)
Office of Naval Research (see ONR)
OGLE, 77, 454, 1167, 1168, 1175
OGO 3, 1132
OGO 5, 77, 1127, 1132, 1168, 1195, 1198
OGS, 879
OHB, 102
oil spill, 1590, 1610
OIMS, 33
OIR, 1411
OIRO, 317
OKB-456, 38
Olympus Mons, 1307
Olympus, 105
on-board computer (see OBC)
one-fluid, MHD-type flow, 1227, 1235, 1238–1240, 1258, 1264, 1268, 1281, 1283, 1285, 1291
ONERA, 1631, 1632, 1644
ONR, 70, 153
OOSDP, 1613, 1614, 1617
open corona, 1083
open magnetosphere, 1532
operation backfire, 50
Oph DC, 596
Ophelia, 1453, 1460
Ophichus, 658
optical afterglow of GRB, 510
optical gravitational lensing experiment (see OGLE)
optical high-velocity clouds, 661
optical identification of X-ray sources, 478
optical light-curve, 517
optical luminosities of LMXB, 813
optical morphology, 480
optical properties of HMXB, 812
optical signal of GRB, 517
optical spectra of LMXB, 812
optically violent variables (see OVV)
optimum geological resolution, 264
OR-2, 34
orbital angular momentum loss, 781
orbital change, 771
orbital period, 769, 825
orbital periods of LMXB, 813

orbital precession of Triton, 1471
orbital variation of optical brightness, 812
orbiting array of wide-angle light collectors (see OWL)
orbiting astronomical observatory (see OAO)
orbiting geophysical observatory (see OGO)
orbiting infrared observatory (see OIRO)
orbiting solar observatory (see OSO)
orbiting the Moon, 28
orbits and companion stars, 767
orbits of SP comets, 1242
ordnance rocket branch, 37
ORFEUS-SPA, 669
orientale Basin, 261
origin and evolution of the solar system, 11
origin of an LMXB, 761, 1281
origin of cosmic rays, 130
origin of hot plasma, 601
origin of KREEP, 1382
origin of light elements, 433
origin of lunar craters, 272
origin of minor ACRs, 998
origin of molecules in comets, 1278
origin of shocks, 899
origin of the cosmic rays, 952
origin of the Earth, 9
origin of the Moon, 6, 272, 1386
origin of the solar system, 12, 1277
origin of ultra-high-energy cosmic rays, 291, 307, 651, 657, 658, 693
Orion A and B molecular clouds, 597
Orion high-velocity outflow, 620
Orion Nebula, 15, 16, 311, 315, 860
Orion OB associations, 585, 599
Orion Peak 1 shock, 621
Orion region, 648
Orion shock, 622
Orion's Cloak, 662
Orion–Eridanus, 647, 661
Ørsted satellite, 1577
Orthopyroxenite, 1359, 1364
OSMI, 1601, 1617
OSO, 13, 77, 215–217, 291, 316, 1091, 1104, 1106
OSO 4, 1067, 1070
OSO 6, 1067, 1070
OSO 7, 501, 1104
OSO 8, 475, 478, 479, 684, 733, 816, 830, 831, 849, 1067, 1073, 1078
OSTC, 465
OT, 511, 513, 521
OTS, 56
out-flow of ion conics and beams, 1548
outer-convective zone (see OCZ)
outer-gap models, 728
outer-heliosphere Lyman-alpha emission, 1206
outer planets, 81
outer-radius luminosity of single star, 774
outer-solar atmosphere, 11, 1075
outer-solar system dust, 1176
outer-stellar atmospheres, 875
outgassing, 9
outside massive black holes strong-field region, 369
OVI, 595, 602
OVI plasma, 602
OVV, 550, 694
OWL, 694
oxygen isotopes, 1385
oxygen-bearing species in wam, dense gas, 621, 1656
ozone hole, 13

P0201 + 113, 453
P Cygni, 290, 291
PACS, 636

PAGEOS, 1625
PAH, 322, 570, 574, 610, 612, 616, 617, 619, 629, 630, 632, 867
pallet, 90
Palomar observatory sky survey, 475
Pamela experiment, 694
Pandora, 1453, 1460
panspermia, 16, 454, 1280
parallax effect, 1198
parallel electric fields, 1515, 1535
parameter retrieval, 1590
parameters of giant planets, 1437
parametrization of metric gravitational field, 343
parent bodies of meteorites, 1336
parent-comet mass, 1249
parental magma of nakhlites, 1362
Parker spiral model, 1117
partially ionized diffuse gas, 662
particle acceleration, 1517
particle accelerators in astrophysics, 138
particle and high energy physics, 123
particle diffusive shock acceleration, 690
particle drift in magnetic field, 1154
particle-environment monitor (see PEM)
particle signatures, 964
particle-size distribution, 1241
particle velocities as a function of size, 1237
particles in corotating events, 986
particles upstream of planetary bow shocks, 981
Pasiphaë, 1452, 1453
passive microwave (see PMW)
patatoid, 1358, 1359, 1377, 1395, 1396, 1414, 1417, 1634
Pathfinder Mars Mission, 98, 1397, 1398
paths of ion clouds, 184
PCA, 1563
PDR, 315, 615–617, 627–629, 634
PDR gas phase, 616
PDS, 889
Peenemünde archive, 37
PEM, 1664, 1665
pencil rocket, 56
PEOLE, 1629
Pepsios spectrometer, 338, 650
Per, 658
Per OB2, 597
perihelion measurement, 349, 825
period gap, 780
periodicity geomagnetism, 1102
Perseus, 658
Perseus arm, 651
Perseus Cluster, 484
perturbation of the lunar orbit, 348
petrogenetic processes, 1363
PG 1115 + 080, 449
PG 1159, 921
PG 1159 stars, 921
phaeophytin, 1602
phaethontis, 264
PHARAO, 1630
PHEBUS, 504
Phobos 2, 1186, 1377, 1396
Phoebe, 1452, 1453, 1474, 1475
photo conducting array camera and spectrometer (see PACS)
photo-dissociation regions, 615
photo-polarimeter radiometer, 1428
photoconductor, 310, 321
photodiodes, 321
photodissociation region, 572
photographic observation, 1625
photography of Mars, 266
photoionization of trace elements, 666

photoionization of wind, 831
photometric parallax, 703
photometry of solar flares, 1098
photon-counting spectrometers, 666
photon-dominated regions (see PDR)
photo-polarimeter radiometer (see PPR)
photon-photon scattering optical depth, 504
photosphere, 234, 1065, 1068, 1069
photospheric abundances, 1068
photospheric emission from cooling neutron stars, 730
photospheric fine structure, 1070
photospheric magnetic fields, 1096
photoswitch detector system (see PDS)
photosynthetic pigments, 1602
PHT, 609
PHT-S, 609
physical and chemical conditions in galaxies, 571
physical data for terrestrial planets, 1406
physical models of clusters, 476
physical Sun, 1109
physics of collisionless shocks, 1525
physics of gravitational lensing, 445
physics of interplanetary shocks, 725, 1104
physics of planetary magnetosphere, 160
physics of quasar evolution, 540
physics of solar activity, 1112
physics of star formation, 710
physics of the atmosphere, 1653
physics of the solar wind, 251
π, 290
pickup ion sources, 967, 987, 990, 995, 1192, 1200, 1209
pickup ions in outer heliosphere, 1157
piezoelectric microphones, 17, 47, 79, 80, 140, 143, 144, 148, 964, 965, 983, 985, 986, 1151, 1168, 1172–1174, 1181, 1183, 1192, 1202, 1408, 1425, 1426, 1431, 1433, 1479, 1530, 1533
Pioneer 8, 1132, 1176
Pioneer 9, 1132, 1176
Pioneer 10, 155, 166, 1128, 1176, 1199, 1454, 1466, 1480
Pioneer 10-11 spacecraft, 147
Pioneer 11, 155, 166, 168, 169, 1128, 1144, 1176, 1454, 1480, 1485, 1487
Pioneer II, 154
Pioneer III, 154
Pioneer IV, 154
Pioneer missions, 1144
Pioneer probes, 1412
Pioneer Venus (see PV)
Pioneer Venus Orbiter, 165, 1411
Pioneer 2, 136
Pioneer 5, 140
Pioneer-10/11, 143
Pitch-angle diffusion, 1419, 1498, 1500, 1501
PKS, 557, 741, 742
PL, 742–744, 841, 1678–1703, 1705–1707
Planck mission, 462
Planck Surveyor Mission, 494
PLANCK, 385, 390, 405, 442
plane-parallel photosphere, 6, 76, 1065
PLANET, 454
planet formation around old star, 867
planet formation in the early Solar System, 1237
planet-geological history, 1297
planet-internal structure, 1297
planet X, 1433
planetary atmospheres, 289, 976, 981, 1575
planetary development, 261
planetary dynamos, 21, 1480
planetary exploration beyond Jupiter, 1447
planetary geophysical processes, 1296
planetary geoscience, 1295
planetary interiors, 1296, 1482

planetary magnetic fields, 1490
planetary magnetism, 1479
planetary mantles, 1367
planetary material, 16
planetary nebula central stars, 933
planetary nebula M2-9, 915
planetary nebula NGC 5307, 913, 915
planetary nebulae and stellar evolution, 916
planetary satellites, 1451
planetary satellites, sizes and shapes, 1452
planetary science, 9
planetary spacecraft imagery, 264
planetary swingby, 81, 97
planetesimals, 11, 1329, 1431
planets beyond Neptune, 1432
plasma analyser, 146
plasma and field data from a rotational discontinuity, 1020
plasma and magnetic fields in interplanetary space, 138
plasma astro-physics, 1010
plasma boundaries, 1503
plasma boundaries and transport, 1501
plasma-cloud drift, 182
plasma component in space, 1497
plasma confinement, 1496, 1498
plasma convection in the magnetosphere, 1542
plasma cup, 247
plasma data, 1228
plasma density, 1210
plasma density versus geocentric distance, 1534
plasma detector, 248
plasma-drag gum, 1174
plasma-dynamical equations, 236, 237, 1225
plasma flow in the subsolar region, 1015
plasma frequency, 968
plasma heating, 1510
plasma in near Earth space, 1495
plasma in the magnetosphere, 1496
plasma properties, 1127
plasma reconnection, 1504, 1540
plasma turbulence, 1513, 1564
plasma velocities, 1078
plasma waves, 159, 1513, 1549
plasma-dynamic systems, 109
plasma-wave dynamical spectrum, 1511
plasma-filled tunnel, 599
plasma-wave instruments, 1132
plasma-wave modes in solar wind, 240, 241, 1027, 1132, 1544
plasma-wave subsystem (see PWS)
plasmapause, 1534, 1562
plasmoid formation, 1507
plasmoid-driven reconnection, 1030
plate-tectonic motions, 1638
plate tectonics, 6, 157, 171, 1316, 1319, 1433, 1453
PLAZMAG, 1229, 1230
Pluto and Charon, 1457
Pluto-Charon system, 1186, 1435, 1436
plutonic cumulates, 1342
PMIRR, 1418–1421
PMS, 884, 885
PMSE, 1563
PMW, 357, 359, 917, 921, 924, 925, 928–931, 1588
PODAAC, 1597, 1598
point-source catalogue, 570
point-spread function (see PSF)
pointing control system, 290
polar-cap absorption event (see PCA)
polar-cap models, 728
polar-coronal holes, 1120
polar dipole/collar, 1422
polar-ice deposits, 1390

polar-magnetic substorm, 1544
polar-mesopheric summer echoes (see PMSE)
polar-orbiting weather satellites, 1656
polar temperatures on Venus, 1410
polar termination shock, 997
polar wind, 1546
polar-cap ion-cloud experiments, 185
Polaris, 43
polarization of gravitational waves, 365, 366
POLDER, 1601, 1617
POLDER-2, 1601
polluting the planetary environment, 13
polyacetylenic chains, 14
polycyclic aromatic hydrocarbons (see PAH)
polytrope model, 250
poor clusters, 475
POPSAT, 1630, 1631
population II sources, 583
population II subdwarf system, 699
population of black-hole binaries, 853
population of the Oort cloud, 1239
porcupine Xenon beam experiment, 1523
Portia, 16, 1453
position-sensitive gas proportional counter, 945
position-sensitive proportional counter (see PSPC)
positrons, 682
post-AGB, 612, 616, 618, 619, 631–634
post-main sequence mass loss, 904
post-Newtonian solar system dynamics, 348
potassium clouds, 198
potassium on Io, 1466
potassium–argon dating method, 1332
potassium–argon chronometer, 1333
power emitted by, 303
power of blazar, 556
power radiated by quasar accretion, 542
Poynting–Robertson effect, 1179, 1187
PPR, 1428
PRARE, 1599, 1630, 1631
pre-imbrium protolith, 1350
pre-main sequence (see PMS)
pre-stellar cores, 862, 1590
prebiotic matter, 14
precession of a gyroscope, 344
precession of Mercury, 1042
precise determinations of gravitational fields, 716
precise positioning, 1640
precise-range and range-rate equipment (see PRARE)
preferred frames in gravity, 346
President's Science Advisory Committee (see PSAC)
pressure modulation, 1660
pressure modulator infrared radiometer (see PMIRR)
pressure tensor, 1025
pressure-balance structures, 1131
primary radiation, 17
primitive meteorites, 1335
primordial helium abundance, 19
primordial nucleosynthesis, 399, 429
primordial Universe, 423
Priroda, 92
pristine lunar high-land rocks, 1389
private, 65
PRODEX, 465
production and loss of ionosphere, 1564
production models for highly ionized gas, 669
Proga, 798
Prognoz, 1191
progress, 80
project orbiter, 45, 71
project Vanguard, 45

Prometheus, 1453, 1460, 1461
prominence eruption, 1099
prominences, 1077
prompt-optical emission, 517, 519
propagating shock models, 899
propagation of a distorted wavefront, 449
propagation of light rays, 451
propagation of nucleons, 141
proper motion, 16
properties of dark matter, 710
properties of gamma-ray emission, 503
properties of normal galaxies, 561
properties of the solar wind, 1152
properties of X-ray pulsars, 834
proplyds, 125, 860, 861
propulsion technology, 20
prostellar cloud, 15
Proteus, 76, 79, 80, 93, 95, 1453
proto-planets, 1456
proton flux, 247
proton-primary radiation, 122
proton rocket, 91
protons in the solar wind, 1128
protoplanetary nebula, 15
protosolar D/H, 428
protostar, 858
prototype nucleus, 1249
PS, 1539
PSAC, 47, 48
PSF, 465, 485, 489
PSPC, 17, 483, 485, 591, 724, 725, 746, 747, 882, 884, 885, 945, 1490
PSR, 763
PSR 0655 + 64, 362
PSR 1534 + 12, 362
PSR 1855 + 09, 362
PSR 1913 + 16, 360, 761, 767
PSR 1913 + 16 class, 767
PSR 1953 + 29, 767
PSR 1953 + 29 class, 767
PSR 2127 + 11C, 362, 767
Puck, 1453
pulled fields, 549, 1013
pulsar astronomy, 722
pulsar in the crab nebula, 953
pulsar slowdown, 726
pulsar-timing measurements, 157, 721, 825
pulsar-wind nebulae (see PWN)
pulsars in globular clusters, 668
pulsating neutron star, 949
pulsations, 769, 816, 825
pulse period of X-ray pulsar, 827
pulse period–orbital period distribution, 829
pulse profile of Crab Pulsar, 733
pulse-timing noise, 828
PUMA, 1224
PUMA-1, 1284
PUMA-2, 1284
Puppis-A, 953
Puppis window, 663
pushed fields, 1012
push–pull chopping, 304
PV, 1309, 1409, 1410, 1412, 1413, 1415, 1418
PVS, 1205
PW 1, 925, 932
PWN, 726, 745, 750

Q0957 + 561, 444, 457
Q2237 + 0305, 449, 456
QBO, 199, 1665
QCC, 612

QMD, 1344, 1345, 1347, 1357
QPO, 796, 797, 817, 818,
QSO, 305, 444, 454, 456–458, 462, 464, 465, 669
QSO J03.13, 455
quadruply ionized neon, 315
quantum mechanics, 232
quark-hardon transition, 431
quartz-monzodiorite clast, 443, 549, 1345
quartz-monzodiorites (see QMD)
quasar continuum, 532, 540, 544
quasar history, 529
quasar in full, 534
quasar-linear symmetry, 531
quasar luminosities, 531
quasar-luminosity function, 540
quasar mysteries, 531
quasar spectra, 531
quasar-X-ray spectrum, 441, 529, 537
quasars accelerate material, 531
quasars are heavily obscured, 533
quasi-biennial oscillation (see QBO)
quasi-linear plasma theory, 1513
quasi-periodic oscillations (see QPO)
quasi-Roche lobe overflow, 829
quasi-static corona, 242
quasi-stationary reconnection, 1135
quasi-stationary types of solar wind, 1125
QUE94201, 1364, 1365, 1368
quenched carbonaceous carbon (see QCC)
quiet Sun, 1072
quiet Sun spectrum, 1071
quiet-time suprathermal tails, 998

R136a, 296, 297
R144, 10, 297
R CCD, 44, 455
R Monocerotis, 311
R-5, 44
R-7, 44
R-7 ICBM, 45
radar altimeter, 1588, 1635, 1637
radar altimetry, 11
radar echo from cometary nucleus, 1246
radar measurement, 342, 1588
RADARSAT geophysical processor system (see RGPS)
radial magnetic field, 232
radial velocities of nebulae, 5, 432, 923, 1520
radial velocities, 922
radiation-belt electron depletion, 1500
radiation-belt electrons, 155, 1496, 1500
radiation belts, 154
radiation pressure, 1179
radiation-driven flow from an accretion disk, 798
radiation-harden detectors, 327
radiation-hydrodynamic simulation, 1074
radiative cooling, 669
radiative envelope, 775
radiative phase, 10, 683, 942
radio afterglow, 513
radio astronomy, 14, 650
radio bursts, 968
radio-continuum loops, 665
radio emission by relativistic electrons, 683
radio emission of quasars, 722
radio emission of supernova, 688
radio-frequency tracking data, 1626
radio interferometry, 549
radio loops I and IV, 662, 723, 724, 766
radio loops, 663
radio pulsars in globular clusters, 784

radio sky, 663
radio tomography, 1576
radio-wave intermodulation, 1564, 1570
radiosotope thermoelectric generator (see RTG)
radio waves, 13
radio window, 230, 549
radio-frequency ion thruster, 109
radio-loud quasars, 304, 532, 533, 540, 550
radio-quiet neutron stars, 747
radio-quiet quasar, 533
radio-silent neutron star, 737, 741
radiogenic growth, 1367
radioheligraph, 282
radiometric ages for basalts, 1369
radiometric ages for lunar highlands rocks, 1341
radiometric ages of lunar samples, 1340
radiometric chronology, 1325
radiometric dating of meteorites, 1335
radiometric methods, application, 262, 1334
radius of comet, 1236
RAE, 51
Raketenbau und Entwicklung, 37
random walk, 250
Ranger, 273, 1337
Ranger 1, 1116
Ranger 2, 1116
Ranger spacecraft, 79
rapid burster, 815
rare-earth concentrations in lunar rocks, 1380
RASS, 550, 585, 592, 880
rate of reconnection, 1009
ratio of nucleons to photons, 423
ray paths, 1040
RBCS, 485, 595, 598–600
RBE, 595, 597, 600
RCA, 166, 167, 169, 170
RCW103, 954, 1019, 1020, 1131
RD, 1131
RD-214, 46, 1340, 1344, 1345, 1347, 1353–1355, 1360, 1380, 1381, 1383, 1392
reacceleration of cosmic rays, 687
recombination, 1566
reconnection configurations, 1007, 1014, 1111, 1506, 1525
reconnection at the magnetopause, 1021, 1507
reconnection configurations on Sun, 1012
reconnection events, 1079
reconnection in the solar corona, 1028
reconnection model, 1545
reconnection rate, 1509
reconnection rates in the MRX experiments, 1024, 1552
reconnection theory of solar flares flares, 1110
reconnection, indirect evidence, 1018
reconnection, solar observations, 1027
reconnections, in-situ observations, 1019
recurrent particle increases, 983
recycled millisecond pulsars, 748
recycled pulsars, 761
recycling, 783
recycling model, 784
red color of Mars, 259, 260, 435
red giants, 878
red-rectangle, 620, 868
red-rectangle nebula, 867
red, oxidized basalt, 260
reddening of starlight, 653
redshift controversy, 544
redshift in GRBs, 508
redshift of quasars, 530
redstone, 43, 76, 77, 93
redstone Jupiter C rocket, 72

redstone missile, 69, 70
redstone rocket, 70
REE, 1340, 1344–1347, 1353–1355, 1360, 1368, 1374, 1380, 1383, 1392
reflecting layer, 278
reflection at air–sea interface, 1587
REFLEX, 487
refractory elements, 659
refueling the Moon vehicle, 82
region of bizarre emptiness (see RBE)
regions of hot plasma, 597
regular satellites of the planets, 1470
relative bulk motion, 552
relative abundances in cosmic rays, 680
relative abundances of chemical elements, 9
relative cluster distances, 378
relative impact fluxes on planets, 264
relativistic bulk motion, 554
relativistic electrons in interstellar space, 683
relativistic jets, 852
RELIKT, 404
remote sensing from orbit, 10
remote sensing of Hale–Bopp, 1282
remote sensing of Hyakutakie, 1282
reservoirs of comets, 1238
reset ages for lunar highland samples, 1348
reset eucrite ages, 1361
resonance wave heating, 242
resonances and shepherds, 1459
resource mapping, 1586
restricted, 340
resurfacing processes, 1470
retarding potential analyzers, 1577
RETDOC, 1630
returning a spacecraft from orbit, 83
Reynolds number, 197
RGPS, 1605, 1606
Rhea, 169, 1453, 1462
ρ Ophiuchus, 597
RIASS, 905
rich clusters, 475
ring-current particle population, 1534
RIT, 94
riverbed, 264
RL 10, 72
RL 10 engine, 72
RLO, 771
RMS, 409
RNII, 34
Robotic optical transient search experiment (see ROTSE)
robotic probe, 265
robotic systems, 289
robotics, 21, 761, 772, 833
Roche lobe overflow (see RLO)
Roche lobes, 770
rocket and satellite research panel, 42
rocket developments, 64
rocket flight, 30
rocket infrared astronomy, 31, 307
rocket observation of ultraviolet spectrum, 896
rocket propulsion for space flight, 30
rocket societies, 32
rocket spectrograph, 206, 208
rocket spectrometers, 60, 289
rocket weapons, 29
rocket-borne scientific experiments, 205
rockets for research and exploration, 67
Rockoons, 68, 159, 211, 280, 281
Röntgen satellite, 56, 585, 724, 899
Röntgen Satellite (see ROSAT)
Rosalind, 1453

ROSAT, 100, 383, 475, 483, 485, 491, 495, 535, 543, 563–565, 569, 582, 586, 724, 732, 741, 743–747, 802, 806–808, 845, 851, 860, 862, 879–886, 888–890, 899, 905, 945–948, 953, 955, 956, 1191, 1229
Rosat $\frac{1}{4}$ keV, 587
Rosat $\frac{1}{4}$ keV All-Sky Survey, 591
Rosat $\frac{1}{4}$ keV band, 592
Rosat $\frac{1}{4}$ keV map, 588
Rosat $\frac{3}{4}$ keV, 587
Rosat $\frac{3}{4}$ keV map, 594
Rosat all-sky survey (see RASS)
Rosat HRI, 733–735, 750
Rosat PSPC, 738, 880
Rosat WFC S1 band, 593
Rosat WFC survey, 593
Rosat Wide Field Camera, 590
Rosetta, 1236, 1268, 1277, 1280
Rosetta mission, 1218
Rossi X-ray Timing Explorer (see RXTE)
rotating neutron superfluid, 785
rotating nucleus model, 1220
rotating winds, 901
rotation in cluster stars, 883
rotation of a magnetized neutron star, 722
rotation of the Sun, 249, 250
rotation of Venus, 265
rotation rate of Sun, 1050
rotation states of comets, 1265
rotation-powered pulsars, 726, 727, 732, 739, 742
rotational angular moments, 157
rotational axis of Uranus, 170
rotational discontinuities (see RD)
rotational excitation diagram, 617
rotational modulation, 879
rotational properties of cometary nuclei, 1264
ROTSE, 518, 519
Routly–Spitzer effect, 650, 652, 663
ROVER, 107
Royal Aircraft Establishment (see RAE)
RPA-COPERNIS, 650, 1227
RS Canum Venaticorum (see RS CVn)
RS CVn, 876, 878, 881, 886–888
RTG, 144, 145
rubidium–strontium dating method, 1330
rubidium-strontium isochron, 1332
Rubis, 181
Russian lunar module, 86
Russian manned Moon programme, 85
Russian space-stations, 91
RX J0852.0-4622, 595, 597
RXJ0911.4 + 0551, 456
RXTE, 285, 492, 689, 725, 732, 734, 748, 796, 814, 816–818, 844, 849

S3-3, 1536
S216, 925
S-Band occultation, 146
S/1999+A169J1+A186, 1453
Sabaeus Sinus, 263
Sagittarius area, 1194
Salyut, 1168
Salyut space station, 91, 113
samarium–neodymium chronometer, 1333
SAMBA, 1416
SAMPEX, 965, 975, 993, 1191, 1533, 1662
SAMSO, 313
San Marco satellite, 53
Sanduleak-69 202, 944
SAO, 11, 317, 947, 950, 1500, 1589, 1591, 1593, 1604–1606, 1608–1610, 1617, 1627, 1629, 1631, 1648

SAR, 11, 1589, 1604
SAS, 585, 795
SAS 1, 100, 723, 823
SAS 2, 684
SAS 3, 563, 585, 586, 795, 815, 824, 826, 843
SAS-2, 284
SAS-3 C, 589
SAS-C, 762, 782, 814
satellite altimetry, 1642
satellite and ring systems of outer planets, 1186
satellite atmospheres, 1463
satellite CMB mission, 418
satellite dynamics, 1456
satellite geodesy, 1623, 1624
satellite interiors, 1462
satellite-laser ranging, 1626
satellite launcher, 42
satellite oceanography, 1586
satellite projects, 70
satellite to satellite Tracking, 1634
satellites of the outer planets, 6, 11, 93, 97, 155, 157, 168, 257, 307, 453, 1431, 1434, 1435, 1451, 1453, 1480, 1487
Saturn and its moons, 12
Saturn encounter, 169
Saturn flyby, 1433
Saturn flyby trajectory, 169
Saturn I, 73, 74
Saturn IB, 73, 90
Saturn magnetic field, 1487
Saturn Orbiter-Titan Probe (see SOTP)
Saturn rocket, 59, 81
Saturn-system dust, 74, 82, 87, 1176
Saturn's interior, 1436
Saturn's ring system, 1172
Saturn-Apollo Moon, 79
Saturn-Apollo Moon Project, 74
Saturn-Apollo Project, 79, 87
Saturnian co-orbitals, 1459, 1462
SAX J1808.4-3658, 819
SBUV, 1661
scalar gravity in the present Universe, 350
scalar-tensor coupling constant, 363, 368
Scandinavian Twin Auroral Radar Experiment (see STARE)
scanning multichannel microwave radiometer (see SMMR)
ScanSAR, 1609
SCaRaB, 1659
scarcity, 13
scattering radio waves in lower ionosphere, 1571
scatterometer, 1590, 1593, 1604
SCC, 1245
Science and Engineering Research Council (see SERC)
science in space, 90
science in space era, 134
science working team (see SWT)
scientific ballooning, 40
scientific dynasty, 290, 324
SCM, 1245
Sco X-1, 759, 760, 818, 819
SCORE, 48
Scorpius, 290
Scorpius–Centaurus Association, 650
Scout rocket, 51–53, 77, 93, 316, 1590, 1592, 1593, 1605
Sco–Cen Association, 668
Sco–Cen OB Association, 585, 595, 597
SCR, 229–236, 243, 245–247, 249, 1115
SDI, 327
SDO, 1056
SDSS, 461, 465, 1627, 1629
SE, 1627
SE I, 1627

SE II, 1627
sea-floor spreading, 1319
sea-floor topography, 1641, 1642
sea-ice monitoring, 1586
sea-ice motion vectors, 1606
sea level, 1590, 1598
sea-level change, 1612
sea-level variability, 1599
sea state, 1604
sea-surface salinity, 1611, 1612
sea-surface temperature, 1587, 1596–1598
sea-surface topography, 1600
Sea WiFS, 1617
SeaSaT, 1630–1632, 1634, 1637
SeaSaT User Research Group in Europe (see SURGE)
seasat, 1593, 1598
SeaWiFS, 1601, 1607, 1608
SEC, 295, 296
Second High Eccentricity Orbit Satellite (see HEOS 2)
second-rank tensor gravitational field, 336
sectoral modes, 1045
secondary-electron conducting (see SEC)
SED, 75, 105, 109, 355, 356, 358, 362–365, 554–556, 558, 857, 858, 860, 862, 865, 963, 964, 968, 969, 971, 972, 974, 976, 982, 985, 998
Sedov model, 10, 942,
seismic-space observations, 1048
selective chopper technique, 1660
self-calibrating radiometers, 1041
self-excited dynamo, 156
Semiorka, 76, 93
Semiorka rocket, 91
Semyorka, 44
sensor intercomparison for marine bilogical and interdisciplinary ocean studies (see SIMBIOS)
SEP abundances, 972
SEP composition, 972
SEP events, 967, 971
SEPICA, 966
SERC, 294, 320
SEREB, 52
Sergeant, 45
Sergeant rockets, 70
Series, 95, 585
Service Hydrographique et Océanographique de la Marine (see SHOM)
SEVIRI, 1658
Seyfert galaxies, 444, 534, 537, 538, 543
SFR, 825, 859, 860, 865, 884, 885
SG, 825
SGR, 508, 726, 736, 742, 1093
Sgr A, 307
Sgr B2, 307
Sh2-174, 932
Sh2-216, 932
shadow catalog, 592
shadowing, 584
shallow water bathymetry, 1610
shape and orientation of nucleus, 1251
shape model for 1P/Halley, 1253
shepherd moons, 11, 1362, 1363, 1365, 1460
shift of periastron time, 362
Shinsei, 57
Shklovskii method, 5, 923, 924, 975
shock-acceleration models, 998
shock heating, 941
shock-spike events, 246, 979
shock-acceleration theory for galactic cosmic rays, 976
shocked interstellar gas, 620, 667
shocks in hot-star winds, 899
shocks in interplanetary space, 977
shocks in the interstellar medium, 689

shocks of sound waves, 1071
SHOM, 1605
short-period comets (see SP)
shrapnel A, 93, 955
Shuttle Endeavour, 95
Shuttle Program, 94
SID, 1562, 1563
side-looking radar (see SLR)
siderophile element redistribution, 1387, 1388, 1393
Sif Mons, 11
signal communication by orbiting relay equipment (see SCORE)
silicate optical depth, 613
silicate spectra, 867
silicon-carbide optics, 327
silicon detectors, 313
SIM, 384, 391, 718
SIMBIOS, 1602, 1617
simulated lensed images, 449
singly-ionized carbon, 315
Sinope, 1452, 1453
SIRTF, 325, 544, 545, 574, 575, 609, 635, 636, 725, 868, 887, 966, 1448
SIS, 725, 744, 887, 966, 994
SISPRE, 52
size of the heliosphere, 1143
SIZEX, 1617, 1631
SKB, 38
sky distribution of X-ray bursters, 816
sky distributions of HMXBs LMXBs, 814
sky maps of HMXB and LMXB, 764, 314
skyhook balloon, 281, 678
Skylab, 13, 80, 87–89, 99, 113, 218, 220, 221, 877, 1067, 1078, 1094, 1104–1107, 1119, 1121, 1136
Skylab Space Station, 74, 1091
Skylark, 51, 294
SL9, 1248, 1249
SLAR, 1589, 1591
SLEW, 550
sloan-digital sky survey (see SDSS)
slow-polar rotation, 1050
slow shocks, 1025, 1026
slow-solar wind, 1083, 1118, 1126, 1128, 1135
SLR, 1593, 1617, 1634, 1639, 1640
SLV, 656, 660
small astronomical satellite (see SAS)
small-dust particles, 1180, 1183
small explorer (see SMEX)
small-interstellar grains, 1183
small-Magellanic cloud (see SMC)
small-scale star formation, 860
SMC, 584, 669, 766, 807
SMC X-1, 826
SMEX, 325
Smithsonian Astrophysical Observatory (see SAO)
SMM, 221, 877, 1041, 1048, 1050, 1067, 1069, 1074, 1079, 1105, 1107, 1108, 1121, 1136
SMMR, 1593–1595, 1600, 1617, 1661
SMMR-SSM/I, 1594
SMOS, 1611–1613, 1617
SN 1006, 951, 952
SN 1987A, 670, 944, 958
SNAP, 144, 389
SNAP-19, 145, 1661
SNC, 1359, 1362, 1378
SNOWMASS, 1046
SNR, 562, 564, 581, 589, 590, 594, 599, 600, 653, 663, 665, 688, 690, 726, 732, 736, 737, 739, 741, 742, 747, 937–942, 944, 947, 949, 952, 954–958, 1056
SO, 1056
SO82 B, 1078

Società Italiana per lo Studio della Propulsione a Reazione (see SISPRE)
Société pour l'Etude et la Réalisation (see SEREB)
Society for the Study of Interplanetary Communication, 33
sodium cloud, 191, 192, 195, 198, 199, 201
sodium-cloud experiment, 190
sodium-cloud temperature data, 199
sodium on Io, 1466
sodium trail, 193
sodium-twilight airglow, 189
SOFIA, 325, 634, 635, 863, 1454
soft gamma-ray repeater (see SGR)
soft X-ray events, 969
soft X-ray light curve, 800
soft X-ray puzzle, 801
soft X-ray spectrum of SSS, 808
soft X-ray telescope (see SXT)
soft X-ray transients, 813, 844, 845
SOHO, 13, 221, 234, 242, 877, 1043–1047, 1050, 1052, 1054, 1056–1058, 1066–1070, 1076–1079, 1081, 1083, 1093, 1096, 1104, 1108, 1112, 1120, 1121, 1124, 1130, 1135, 1144, 1147, 1191, 1201, 1210–1212, 1415, 1498, 1511, 1579
SOHO GOLF, 1055
SOHO MDI, 1048
SOI/MDI, 1045–1047
Sojourner, 1377, 1394, 1395, 1416
Sojourner roving vehicle, 74, 80, 98
Sojuz capsule, 13, 90, 93
SOL-STICE, 1664
solar-acoustic waves, 1046, 1104
solar-activity cycle, 1036, 1143
solar and heliospheric observatory, 13, 242, 1055, 1065, 1068, 1121, 1144
solar anomalous and magnetospheric particle explorer (see SAMPEX)
solar atmospheres, 13
solar B, 1067, 1104, 1112, 1113
solar B mission, 1066
solar-backscatter ultra violet (see SBUV)
solar breeze, 247, 1115
solar convection, 1068
solar-convective flow, 233–235, 239, 248, 1059, 1115, 1143
solar-corona stores particles, 968
solar-corpuscular radiation (see SCR)
solar-corpuscular stream, 158
solar-cosmic ray, 231, 250
solar cycle, 1095
solar-disk sextant, 1042
solar dynamo, 1091
solar eclipse, 1089
solar-eclipse observations, 339
solar eclipse of 1919, 441
solar eclipse of 1958, 281
solar-elemental abundances, 1068
solar-energetic particle ionic charge analyzer (see SEPICA)
solar-energetic particle propagation, 1196
solar-energetic particles, 155, 967
solar environment, 1193, 1194
solar-eruptive events, 1108
solar-eruptive phenomena, 971, 1098
solar-filament eruption, 1097, 1100
solar-flare research, 1104
solar-flare spectrum, 217
solar-flare theory, 1110
solar-flare X-ray spectrum, 216, 280
solar flares and coronal mass ejections, 1108
solar flares and prominences, 1100
solar flares at the Sun, 1102
solar granulation, 1068
Solar Heliospheric Observatory, 1045
solar-intensity distribution, 207
solar interior, 13, 1035, 1056
solar-irradiance measurements, 1659

solar-irradiance variations, 1048
solar irradiance, 13, 1041
solar-isotopic abundances, 1123
solar-magnetic activity, 905
solar-magnetic field, 225, 1036, 1120
solar-magnetic flux, 1038
solar maximum, 1153
solar-maximum mission (see SMM)
solar minimum, 1153
solar missions, 80, 1045
solar modulation in interplanetary space, 127
solar modulation of galactic cosmic rays, 141, 204
solar near-UV radiation, 204
solar-neutrino flux, 13, 1036
solar optical universal polarimeter (see SOUP)
solar orbiter (see SO)
solar-oscillation spectrum, 1069
solar oscillations, 1037, 1038
solar oscillations investigation with the Michelson Doppler Imager (see SOI/MDI)
solar photosphere, 207, 1065, 1066, 1068
solar planets, 447
solar probe mission, 1042
solar radiation, 1178
solar-radio emission, 1101
solar-radius variations, 1041
solar-rotation period, 1148, 1153
solar sailing, 20
solar-space research, 203
solar-spectral energy distribution, 278
solar spectrum, 209
solar spin-down, 1036
solar-stellar irradiance comparison experiment (see SOL-STICE)
solar-surface velocity, 335, 339, 344, 1046
solar system, 13
solar-system chronology, 1337
solar-system contribution to interstellar dust, 1179
solar-system dust cloud, 1187
solar-system dynamical model, 335, 342, 345
solar-system evolution, 1123
solar-system gravity, 345
solar-system history, 1319
solar-system N-body system, 346
solar-system ranging experiments, 349
solar tachocline, 1058
solar thermal stratification, 1068
solar-transient activity, 1158
solar ultraviolet measurement of emitted radiation (see SUMER)
solar ultraviolet spectra, 13
solar ultraviolet spectral irradiance monitor (see SUSIM)
solar UV flux, 5, 6, 14, 139, 160, 161, 164, 179, 204, 225, 233, 242, 245, 246, 248–250, 427, 983, 986, 1083, 1107, 1115, 1142, 1146, 1196, 1481, 1501, 1530, 1545
solar-wind abundances of heavy ions, 1126
solar-wind acceleration mechanisms, 1133
solar wind and interstellar composition, 1209
solar-wind bubble, 1192
solar-wind composition, 1123
solar-wind detectors, 1116
solar-wind dynamic pressure, 1157
solar-wind electrons, 1127
solar-wind instruments, 1123
solar-wind ion composition spectrometer (see SWICS)
solar-wind ions flow through the magnetosphere, 1549
solar-wind mass flux, 1210
solar-wind morphology, 1118
solar-wind parameters, 1124
solar-wind plasma transport, 1548
solar-wind pressure variation, 1211
solar-wind sensor, 1124

solar-wind sources, 1133
solar-wind speed, 242, 1117, 1122, 1134
solar-wind termination shock, 1197, 1200
solar-wind theory, 896
solar-wind torque, 249
solar-wind velocity structure, 1120
solar-wind velocity vectors, 1227
solar-wind-implanted rare gases, 1379
solar-wind–magnetospheric processes, 1570
solar X-radiation, 210
solar X-rays, 155
solar-stellar connection, 905
solar-terrestrial disturbances, 1102
solar-terrestrial paradigm, 1107
solar-terrestrial phenomena, 1097
solar-wind acceleration, 13
solid body of a planet, 162
solid-fuel rocket exhausts, 1171
solid methane, 624, 899, 945
solid-state detectors (see SSD)
solid-state greenhouse effect, 1474
solid-state imaging spectrometer, 725
Solwind, 1121, 1136
SONG, 1631, 1632, 1637
SOPRANE, 1605
SOTP, 1429
sound speed in solar core, 1048
sound speed of the Sun, 1047
sounding-rocket lauchings, 181
SOUP, 1067, 1068, 1225, 1238, 1239, 1241, 1242, 1259
source functions of the solar wind, 1148
source in relativistic motion, 553
source of dusty rings, 17, 818, 1183
Soutoul, 25, 689
SP, 1238
Space and Missile Systems Organization (see SAMSO)
space clock, 341
space debris, 1169
space dust dangerous, 1166
space-dust research centres, 1167
space-express spaceplane, 106
space-geodetic scheme, 1625
space-infrared telescope facility (see SIRTF)
space observatories, 1434, 1447
space oceanography navigation and geodynamics (see SONG)
space-plasma phenomena, 1142
space-plasma physics, 1495, 1532
space-plasma radiate, 1520
space race, 3, 25, 40
space research in America, 39
space research in Europe, 50
space rockets, 76
space science and Earth, 1645
Space-Science Board (see SSB)
space-science projects, 66
Space Shuttle, 7, 56, 59, 72, 93, 100, 798, 1147
Space-Shuttle main engine (see SSME)
Space Station Freedom, 95
Space Task Group (see STG)
Space Telescope European Coordinating Facility (see ST-ECF)
space telescope imaging spectrograph (see STIS)
space transportation, 59
space travel, 60
Space Ultraviolet-visible Observatory (see SUVO)
space weather, 14
space-weather forecasting, 14
space-based facilities, 13
space-based systems, 13
space-borne spectroscopy, 14, 349, 1310
space-borne infrared observatory, 570

SUBJECT INDEX

spacecraft Galileo, 96
Spacelab, 56, 91, 484, 485
Spacelab-2 Infrared Telescope, 322
spaceship Earth, 4
spacetime metric $g_{\mu\nu}$, 356
spacetime near black holes, 369
spaghetti-type engine, 19, 75
spallation cross-section, 426, 434
spallation reaction, 426
Spartan, 484, 485, 1147
SPAS, 94
spatial diffusion of cosmic rays, 687
spatial distribution and locations of GRBs, 504, 714
spatial structure of milky way galaxy, 702
special-sensor microwave imager (see SSMI)
Special Committee for the International Geophysical Year (see CSAGI)
spectra measured from space, 919
spectra of bright sources, 618
spectra of central star, 920
spectra of corona, 211
spectra of Jupiter, 1439
spectra of nebulae, 918
spectra of newly formed stars, 629
spectra of planetary nebulae, 320
spectra of Saturn, 1439, 1442
spectra of X-ray binaries, 819
spectral and photometric imaging receiver (see SPIRE)
spectral analysis, 569
spectral diagnostics, 217
spectral-energy distribution (see SED)
spectral-energy distribution of GRB afterglow, 514
spectral-energy distribution of normal galaxy, 574
spectral images of clusters, 485
spectral imaging of coronal emission lines, 1147
spectral models for termal emission, 594
spectral shape and luminosity, 850
spectral type of companion, 924
spectrally integrated solar output, 1091
spectroheliograph, 204, 229
spectrohelioscope, 229, 650
spectroscopic satellite, 292
spectroscopy during total eclipse, 233
spectroscopy of Ly-alpha glow, 1212
spectroscopy of nearby stars, 1203
spectrum of central star, 924
spectrum of highly obscured quasar Markarian 3, 538
spectrum of Martian atmosphere, 1395
spectrum of Saturn, 1441
spectrum of Vela X-1, 831
speculations about Venus, 262
speed of gravitational waves, 368
speed of light, 227, 337, 338
speed of light near gravitating bodies, 342
Spektr, 92
Spektrostratoskop, 1067, 1068
spherical harmonic coefficients, 1483
spherical harmonic on surface of the Sun, 766, 1040, 1077
spin evolution of accreting neutron stars, 768
spin-magnetic field in neutron star X-ray binaries, 764
spin modulation, 804
spin-up, 768, 827
spin–orbit coupling, 768, 827, 1045
spiral-arm structure Milky Way Galaxy, 435, 653, 668, 713
SPIRE, 324
splitting of cometary nuclei, 1247
sporadic E, 1562
sporadic radio missions, 171
spread-F phenomenon, 1564
sprites, 585, 1570
SPT, 109

Sputnik, 52, 71, 77, 85, 136, 213, 258, 269, 272, 291, 340, 654, 1529, 1575
Sputnik 1, 3, 44–47, 49, 257, 259
Sputnik 2, 46, 259
Sputnik 3, 1479
Sputnik I, 153, 1496
Sputnik II, 71
Sputnik III, 154
sputtered exospheres, 1469
SQUID, 347
SRC, 295
SRE, 1245
SRON, 324, 879
SRON-Leiden, 723
SS4, 76
SS6, 76
SS433, 833
SS Cyg, 793, 796
SSB, 136, 1427, 1612
SSC, 158
SSD, 945, 1124
SSD-ESTEC, 723, 1590, 1592, 1594, 1605, 1606, 1617
SSME, 74, 76, 93, 259
SSMI, 1594
SSS, 480, 481, 485, 780, 793, 807, 808, 878, 899, 945, 951, 952, 1587, 1596–1598, 1607, 1610, 1612, 1617, 1634
ST-ECF, 459
stability of solar constant, 1037
standard candles, 374
standard Earth (see SE)
standard HMXBs, 777
standing hydromagnetic waves, 1009
star counts in inner Galaxy, 14, 306, 704, 857
star-count analyses, 702
star-forming region (SFR)
star with a circumstellar disk, 306, 538, 858
star-forming region in Eagle Nebula, 14, 858
star spectra, 914
stardust mission, 1185, 1186
stardust, 1176, 1236, 1280
stardust model, 613
STARE, 1572
STARLETTE, 1629, 1634, 1638
startburst galaxy, 566
static Universe, 20
stationary plasma thruster (see SPT)
stationary states far from thermal equilibrium, 1519
statistical distribution in line strengths, 897
statistical distribution of cosmological perturbations, 417
statistical properties of galaxies, 575
steady-state ocean circulation, 1611, 1613
steady-state reconnection, 289, 1011
steady-state Universe, 282
STELLA, 1629, 1631, 1634
stellar birth-line, 859
stellar-bolometric magnitudes, 289
stellar-chemical abundance, 708
stellar-coronal emission, 880
stellar-coronal physics, 889
stellar-coronal research, 880
stellar distance, 289
stellar dynamics, 14, 709, 858, 930
stellar flares, 879, 888
stellar formation, 858
stellar IMF, 713
stellar-nucleo synthesis, 433
stellar origin of chemical elements, 423
stellar populations, 699
stellar-system tests of gravitational theory, 15, 246, 359, 435, 650, 772
stellar-wind laboratory, 581, 829

stellar winds and mass loss, 895
stellar-X-ray astronomy, 876, 890
stellar-X-ray emission, 876
STEREO, 1104, 1108, 1112, 1113
STG, 83
stingray nebula, 915
STIS, 298
stochastic acceleration, 976, 998
stochastic topology, 250
storm sudden commencement (see SSC)
Störmer theory, 173
STP72, 1200
Strategic Defense Initiative (see SDI)
stratigraphic column, 261
stratigraphic mapping planets, 261
stratigraphic markers, 261
stratoscope, 1067, 1068
stratoscope II, 289, 306
stratosphere of Neptune, 1443
stratosphere of Uranus, 1443
stratospheric chemistry, 1083, 1135, 1586
Stratospheric Observatory for Infrared Astronomy (see SOFIA)
strong-low-velocity (see SLV)
stripped nuclei, 126
Strömgren sphere, 1198
strong-equivalence principle, 355
strong-galactic binary X-ray sources, 763
strong-greenhouse effect, 262
strong-solar wind, 1205
strong v. weak gravity, 354
strong-field systems, 355
strong-field systems in general relativity, 354
strongly ionized interstellar medium, 1199
structure and composition of comets, 1281
structure and rotation of the solar interior, 1045
structure of radiation belts, 1500
structure of RCW103, 954
structure of stellar coronae, 880
structure of the atom, 228, 1326
structure of the heliospheric magnetic field, 1146
structure of the magnetopause, 1505
structure of the radiation belts, 1499
structure of the Universe, 399
STScI, 21, 450, 455, 459, 487, 944, 1248, 1262
submillimeter domain, 302
submillimetre observation, 461
submillimeter wave astronomy satellite (see SWAS)
subnuclei, 1241
subsolar-coronal abundances, 887
subsonic-wind termination shock, 1205
sudden ionospheric disturbance (see SID)
sudden mass loss, 772
SUISEI, 1227
suite-rock formation, 1344
SULEICA, 995
sulphur on Io, 11, 13, 76, 157, 303, 335, 338–340, 442, 453, 648, 929, 1466
SUMER, 1069, 1070, 1078, 1079, 1081
Sun-accreted heavy elements, 1041
Sun as emitter of SCR, 229
Sun born in a star cluster, 1240
Sun's image at sunset, 453
Sun's interior, 1042
Sun's magnetic field, 13
Sun's motion through interstellar medium, 228, 1193
Sun-as-a-star, 1043
Sun-as-a-star oscillation spectrum, 1044
Sun-Earth connection, 1109
Sun-Earth-Explorer spacecraft, 1539
Sun-follower, 207

Sun-synchronous, 319
Sun-synchronous orbit, 317
Suns's rotation, 1057
Sunspot cycle, 1041
Sunspot magnetic fields, 230
Sunspot minimum, 1068, 1069, 1102, 1119, 1142
Sunyaev–Zeldovich effect, 383, 494
Sun–comet–Giotto phase angle, 1251
super-Eddington accretion, 564, 565, 833
super-rotation, 1409, 1422
superbubbles, 662, 665
Super Dual Auroral Network (see SUPERDARN)
SUPERDARN, 1570, 1572
supergiant HMXB, 831, 878
supergranulation, 1054, 1068
supergranulation cells, 1077
superinsulation, 309, 310
superionization, 897
superionization boundaries, 898, 899
superionization of Be stars, 901
superionization of hot star winds, 897
superKamiokande, 1056
superluminal knot, 554
superluminal motion, 551
supermassive black holes, 535, 538, 565
supernova acceleration probe (see SNAP)
supernova and GRB, 516
supernova blast wave, 690
supernova event, 306, 759, 938, 940
supernova in Large Magellanic Cloud Cloud, 943
supernova remnant (see SNR)
supernovae explosion, 375, 433
supernova–GRB connection, 515, 581, 678, 937
supersoft sources, 807
supersonic collisions between clouds, 652
supersonic coronal expansion, 241
supersonic expansion, 240–242
suprathermal and low energy ion composition analyzer, 995
surface albedo, 1593
surface brightness distribution, 480
surface charging, 10, 1221
surface composition, 1461
surface composition of star, 931
surface differential rotation, 1050
surface environment of Venus, 265
surface features of Venus, 1309
surface geology, 1319, 1469
surface geology of Mercury, 1303
surface-magnetic field, 1050
surface-magnetic fields at Uranus, 1489
surface-magnetic field of Neptune, 1489
surface of Ariel, 1472
surface of cometary nuclei, 1262
surface of comets, 1218
surface of Enceladus, 1472
surface of Europa, 1474
surface of Io, 1471
surface of Mars, 258
surface of Mercury, 1301, 1302
surface of Miranda, 1473
surface of Titan, 1464, 1465
surface of Triton, 1471
surface of Venus, 11, 1312
surface parameters from satellite based Earth Observation, 1590
surface photometric data, 703
surface pressure on Mercury, 1407
surface temperatures for pulsars, 274, 274, 275, 745, 1337, 1378
SURGE, 1631
SUSIM, 1664

SUVO, 298
Svezola, 95
SWAN, 1201, 1210, 1211
SWAS, 323, 325, 609, 621–623
SWICS, 428, 966, 967, 986, 995, 1124, 1125, 1127, 1129, 1130, 1201, 1209
SWP, 919
SWT, 581–585, 588, 590, 592, 593, 595, 599, 601, 603, 1638, 1647
SXRB, 581–586, 588–593, 595, 601, 603
SXT, 1029, 1091, 1108
Sycorax, 1453, 1456
symmetry and rare radio emission jets, 540
symphony communication satellite, 96
synchronous rotation, 1457
synchrotron emission, 510
synchrotron nature of cosmic radio emission, 683
synchrotron nebula, 733
synchrotron radiation, 651, 1520
Syncom, 77
synthesis of chemical elements, 19
synthetic aperture radar (see SAR)
Syrtis Major, 263, 267
systematic rocket technology, 61

T2L2, 1630
T Tauri, 433
T Tauri star, 311, 862, 1050
T/P, 1605
TAD, 1568
TAI, 1633
tail of the magnetosphere, 1539
tail plasma sheet, 1519
tails of comets, 225
tails of gamma-ray emission, 508
TAMA, 357
TAMA 300, 365, 919, 1131
tangential discontinuities (see TD)
tangential magnetopause, 1506
target of opportunity (see ToO)
Taurus Littrow Valley, 1339
TD, 1131
TD-1, 54, 294
TD-1A, 214, 217, 654, 1575
tearing-mode instability, 1011
TEC, 1575, 1581
technological challenges, 289
tectonic activity on the Moon, 1298
tectonic canyon, 264
tectonic deformation of Venus, 1093, 1311
TEGP, 354, 355, 361, 363, 366–369
TEGP 8, 356
telecommunications, 13
telescopes, 99
telescopic observations, 1242
Telesto, 1453, 1459
television, 13
television and infrared observation satellite (see TIROS 1)
Telstar, 77, 889
temperature distributions in clusters, 489
temperature fluctuations of small grains, 612
temperature fluctuations of the CMB, 384
temperature of central star, 925, 932
temperature of CVs, 791
temperature of the atmosphere, 198
temporal variations of geocenter, 1640
temporal variations of gravity field, 1638
TENMA, 484, 879, 945
tenma, 484, 846, 850
tenuous atmospheres, 1465
tepid-wind model, 897

terminal velocities, 897, 1157, 1211
terrella experiment, 229, 230, 1569
terrestrial atmosphere, 3
terrestrial γ-ray flashes, 507
terrestrial geology and geophysics, 1315
terrestrial magnetosphere, 6, 226
terrestrial meteorite craters, 257, 258, 1315
terrestrial volcanic rocks, 12, 169, 1397, 1453, 1459
TES, 382, 1397, 1398, 1667
TGS, 743, 879
TH, 743
Thalassa, 1453
Tharsis, 258, 265, 267, 1307, 1308
Tharsis Montes, 1307
Tharsis province, 1305
Tharsis region, 12, 1304
Thebe, 1453
theoretical cosmology, 401
theories of reconnection, 1016
theories of stellar evolution, 16
theory of reconnection, 1008, 1010
theory, technology, and observations, 311
thermal and mechanical layering, 1318
thermal balance at the surface of a satellite, 1466
thermal bremsstrahlung, 502, 938
thermal emission, 941
thermal emission from hot plasma, 584
thermal emission from middle-aged pulsars, 742
thermal emission from W49B, 949
thermal equilibrium emission models, 603
thermal evolution models, 1307
thermal evolution of Mercury, 1303
thermal evolution of neutron stars, 729
thermal evolution of the Moon, 1351
thermal flux of Venus, 1312
thermal infrared (see TIR)
thermal-infrared spectrometer, 1461
thermal-infrared spectrum, 262
thermal processing, 624
thermal profiles of giant planets, 1439
thermal stability in multiphase medium, 653
thermal structure of Saturn, 1437
thermal structure of the chromosphere, 1073
thermal structure of the Sun, 1048
thermal structure of Titan's atmosphere, 1464
thermocouple, 303
thermocouple radiometry, 262
thermograph, 40
thermohaline circulation, 1612
thermonuclear flash model, 816
thermopile, 303
thermosphere, 200, 1411
thermosphere-ionosphere general circulation model (see TIGCM)
thermosphere of Venus, 1409
thin-lens approximation, 445
thin-film detectors, 1181
tholins, 1461, 1462
Thor, 43, 76, 77, 93
Thor-Delta rocket, 51, 96
three-body problem, 340
three-color photometry, 311
three-dimensional adaptive MHD model, 1229
three-dimensional heliosphere, 1141
three-dimensional nature of CIRs, 1151
thunderclouds and the ionosphere, 1570
TiCCE, 1169
TID, 1567, 1568
tidal breakups, 1247–1249
tidal bulge on Io, 1460
tidal disruption, 1247

tidal evolution of Uranian satellite system, 1457
tidal flexing, 1460
tidal force, 11
tidally disrupted comets, 1249, 1640
tides influence on ionospheric parameters, 1568
TIGCM, 1568
time and frequencies, 1632
time-of-flight (see ToF-E)
time synchronisation, 1630, 1633
time variability, 879
time variability of hot-star outflows, 900
time variation of Newton's G, 346
timescales for formation of cometesimals, 1238
Timocharis, 1299
tiny cold clouds, 667
tiny-scale atomic structure, 665
TIR, 1586, 1587, 1590, 1591, 1597, 1611, 1617
TIRL, 1356
TIROS, 1655
Tiros, 11, 12, 43, 74, 76, 77, 93, 97, 1453, 1454, 1457, 1463, 1469, 1475, 1476, 1593
TIROS 1, 1653, 1655
TIROS-N, 1596
TIROS-NOAA, 1596
Titan and Hyperion, 1459
Titan and Iapetus, 1458
Titan Centaur rocket, 80, 98
Titan IIIE-Centaur, 53
Titan's spectrum, 1023, 1464
Titania, 1453, 1462
titanium basalts, 1350
titanium depletions, 667, 1465
TMI, 1598, 1617
TNO, 1236, 1238–1240, 1242, 1244, 1259, 1268, 1433, 1448
TOA, 966, 995, 1124, 1174, 1175, 1658
ToF, 1124
TOF-E, 966
TOGA, 1641
Tokyo University Spherical Torus, 1024
tomographic mapping of the ionosphere, 1575
TOMS, 1661–1663
ToO, 597, 795
TOPAZ, 1605
top-of-the-atmosphere (see TOA)
TOPEX/Poseidon, 1599, 1600, 1605, 1632, 1635, 1637, 1638, 1646, 1647, 1065
topography and morphology of comet 1P/Halley, 1256
topography of cometary nucleus, 1266
topography of lunar crust, 1298
topologically closed Universe, 336
topology of magnetic field lines, 1540
topside ionosphere, 1546
total destruction cross-section, 248, 426
total mass of clusters of galaxies, 489
total ozone mapping spectrometer (see TOMS)
TR, 330, 430, 1065, 1228, 1557
TRACE, 877, 1067, 1078, 1081–1083, 1104, 1108, 1109, 1112
trace element patterns of mare basalts, 1383
tracer of photoionized gas, 669
tracer populations, 709
TRANET, 1626, 1635
trans-Neptunian objects (see TNO)
transient sources, 844
transient Sun, 1096, 1102
transistor, 77
TRANSIT, 1626, 1630, 1631
transit-time damping, 997
transition region and coronal explorer (see TRACE)
transition region between chromosphere and corona, 1074
transition region camera (see TRC)

transition-region models, 1079
transition region of chromosphere, 1076
transition-region spectral lines, 1078
transmission grating spectrometer (see TGS)
transmission lines, 1579
transport across magnetopause, 1541
transport in plasma, 1504
transport processes, 1552
transportation through space, 60
trapped acoustic waves, 1038
trapped instantaneous residual liquid (see TIRL)
trapped particle intensities, 163
travel to Mars, 112
traveling ionospheric disturbance, 1567
travelling atmospheric disturbance (see TAD)
travelling ionospheric disturbance (see TID)
TRC, 1067, 1078
TRIAD, 1539
TRIGA, 1378
trigonometic parallaxes, 924
triply ionized oxygen, 315, 1452, 1461, 1462, 1465, 1469, 1470, 1475
Triton's atmosphere, 1465
Triton's plumes, 1474
TRMM, 1598, 1612, 1613, 1617, 1666
troctolites, 1342, 1343
Trojan asteroids, 1459
tropical rainfall measuring mission (see TRMM)
troposphere of Neptune, 1443
troposphere of Uranus, 1443
TRW, 102, 144, 1175, 1206
TS + TH + PL, 743
TS-3, 1024
TSAS, 665
TT Ari, 798
Tully–Fisher relation, 380
TUNDE, 1226
tuning-fork chopper, 312
turbopause, 191, 195, 197
turbulent convection in the Sun, 1135
turbulent mixing layer, 669
turbulent resistivities, 1013
TV Col, 804
twilight glow, 189
twin-exhaust model, 553
twisted fields, 1014
Two Degree Field (see 2dF)
two-dimensional gas-dynamical simulation, 831
two-dimensional models, 902
two-fluid models of the solar wind, 1134
two-pole accretion, 802
two-shock models, 1211
two-stage missiles, 42
TX Cam, 305
Tycho catalogue, 886
Tycho's SNR, 947
type I X-ray bursts, 816

U Gem, 793, 797–799
U Gem WD, 793
UARRP, 41, 42
UARS, 1654, 1569, 1660–1665, 1667
UARS MLS, 1666
UARS science, 1665
UBVRI, 304
UCLA, 321
UDMH, 72, 75, 76, 105
UDMH–nitrogen tetroxide, 72
Uhuru, 100, 283, 478, 543, 549, 561, 723, 759, 760, 782, 794, 840
Uhuru satellite, 532
Uhuru X-ray catalogue, 800

UIB, 570, 573
UIR, 616–619
UK, 534
UKMO, 1604
UKS, 1502
ULECA, 1124
ULEIS, 966
ULET, 966, 967
ULF, 981
ULIG, 570
ULIRG, 14, 306, 534, 538, 570, 572–574
ultraheavy nuclei, 680, 692
ultraluminous infrared galaxies (see ULIRG)
ultra-low energy, 1124
ultra-low energy charge analyzer (see ULECA)
ultra-low energy isotope spectrometer (see ULEIS)
ultra-low energy telescope (see ULET)
ultra-low frequency (see ULF)
ultra-violet photometer, 1426
ultra-violet spectrometer, 1426
ultraviolet, 13
ultraviolet coronagraph spectrograph, 1135
ultraviolet astronomical satellite (see UVAS)
ultraviolet-blue-visual-red-infrared (see UBVRI)
ultraviolet continuum excess, 861
ultraviolet coronagraph spectrometer (see UVCS)
ultraviolet line spectra of YSOs, 862
ultraviolet observations of ISM, 608
ultraviolet photometry, 146, 919
ultraviolet spectrometer (see UVS)
ultraviolet spectroscopy from space, 287
Ulysses, 143, 146, 681, 694, 964, 965, 987, 996, 1104, 1120, 1124, 1128, 1130, 1132, 1135, 1136, 1146, 1148, 1154, 1157, 1168, 1175–1177, 1182, 1191, 1201, 1202, 1209, 1428, 1479, 1480, 1505
Ulysses mission, 1141, 1143
Ulysses spacecraft, 1104, 1146, 1149
UM673, 458
Umbriel, 1453, 1459, 1472
understanding blazars, 554
UNESCO, 1638
unidentified infrared (see UIR)
unidentified infrared bands (see UIB)
unified propulsion system (see UPS)
universal expansion, 399
universal Time1 (see UT1)
universality of free fall, 349, 351
Universe at large redshifts, 402
Universe is flat, 463
unstable elementary particles, 125
unsymmetrical dimethyl hydrazine (see UDMH)
upper atmosphere, 179, 180, 189
Upper atmosphere research, 42
Upper atmosphere research satellite (see UARS)
upper Atmosphere Rocket Research Panel (see UARRP)
upper superadiabatic boundary layer, 1058
UPS, 95–97
upstream of comets, 1226
Uranian satellites, 1462
Uranium–helium 4 dating, 1329
Uranium–lead age of the Earth, 1330
Uranium–lead dating, 155, 157, 169, 170, 257, 335, 434, 1327, 1328, 1431, 1434, 1441, 1453, 1480
Uranus atmospheric structure and composition, 1443
Uranus internal structure, 1442
Uranus ionosphere and magnetosphere, 1445
Uranus magnetic dipole axis, 1487
Uranus's rotation axis, 1441
Uranus–Neptune zone, 1283
Urca processes, 729
URKREEP, 1344–1347, 1355–1358, 1368, 1369

Ursa Major, 590
Ursa Minor Publishing Corporation, 1191
US Air Force (see USAF)
USAF, 43, 50, 326
USAF/MSX, 708
USGS, 261, 263, 264
USNO, 1454
USO, 1626, 1636
UT1, 1633, 1634
UTC, 1611
UV astronomical spectroscopy, 287
UV delay, 795, 796
UV-glow measurements, 1196
UV interstellar extinction, 293
UV-line profile, 797
UV observations from balloons, 288
UV observatories in space, 652
UV sky survey telescope (see UVSST)
UV-space astronomy, 289
UV-spectra of hot stars, 291
UV spectra of supernovae, 289
UV spectrograph, 219
UV spectrometer/spectroheliometer, 219
UV spectrometers rockets, 291
UV spectroscopy of CV binaries, 797
UV spectroscopy of non-solar objects, 295
UV spectrum of the Sun, 290
UV stellar photometer, 292
UV stellar spectrophotometry, 290
UVAS, 295
UVCS, 1081, 1083, 1084, 1135
UVS, 1206, 1426, 1427, 1435
UVSP, 1069
UVSST, 294
UX UMa, 796, 799

V2, 34, 35, 36, 50
V2 assembly line, 36
V2 legacy, 39
V2 Panel, 42, 205
V2 Rocket, 34, 37, 39, 63, 66, 204, 279, 288, 290
V2 Rocket Panel, 41
V1974 Cyg, 808
vacuum vessels, 309
vacuum-tight telescope cover, 309
Valles Marineris, 266, 267, 1304, 1307
value of H_0, 381
value of H_0 from physical methods, 382
Van Allen belt, 296, 1498, 1569
Van Allen radiatior belts, 1533
Vanguard 3, 1479
Vanguard Project, 46, 71
Vanguard satellite, 47
Vanguard satellite launcher, 42
variability of solar irradiance gravity oscillations (see VIRGO)
variation of galactic cosmic ray intensity, 231
variations of F2 layer, 1562
Vedra Valles, 1306
VeGa, 1176
Vega, 265, 1225, 1265, 1282, 1283, 1311, 1398
Vega 1, 265, 1224, 1250, 1253, 1279
Vega 2, 265, 1181, 1224, 1250, 1251, 1253, 1279, 1412
Vega craft, 1217
Vega mission, 1377, 1455
Vega phenomenon, 863, 864
vegetation on Mars, 260
Veil Nebula, 595, 653, 938
Vela 3, 1129
Vela 4, 1127
Vela Pulsar, 737

Vela satellite, 501, 504
Vela Sheet, 597
Vela SNR, 597, 599, 600, 653, 739, 828, 955
Vela-like pulsars, 735
Venera, 1310, 1312, 1398, 1408, 1412, 1479
Venera 4, 265
Venera 7, 265, 1405
Venera 8, 1311, 1399
Venera 10, 265
Venera 13, 11
Venera 14, 11
Venera 16, 1409
Venera mission, 1377
Venera probes, 9, 10, 155, 157, 165, 248, 257, 259, 262, 315, 342, 1309, 1377, 1398, 1405, 1406, 1408, 1409, 1411, 1421
Venus 2, 983
Venus 3, 247
Venus greenhouse effect, 1412
Venus greenhouse heating, 262
Venus has no intrinsic magnetic field, 1310
Venus large-scale gravity anomalies, 1311
Venus missions, 79
Venus Orbiter Radiometric Temperature-sounding Experiment (see VORTEX)
Venus Probe, 1116
Venus Sample Return, 1422
Venus Sample Return Mission, 1421
Venus' environment, 265
Venus's atmosphere, 1409, 1410, 1413
Venus's cloud layers, 1405
Venusian atmosphere, 258
Venusian cloud, 257, 262, 1411
Venusian greenhouse effect, 1409
Venusian polar dipole, 1411
Venusian rocks, 1398, 1399
Venusian soil, 1399
Venusian surface material, 1398
Venusian volcanism, 1310
Verein für Raumschiffahrt (see VfR)
Vergeltungswaffe 2, 35
verification of general relativity, 353
Véronique, 52
Véronique rocket, 52
vertical temperature profile radiometer (see VTPR)
vertically generalized production model, 1603
very important particle (see VIP)
very large array (see VLA)
very long baseline interferometry, 343, 551, 1361
very low Ti (see VLT)
VfR, 33, 34, 62
VFW, 90
VHF, 1572
VHK, 1349, 1350, 1368, 1596
vidicon, 296
Viking, 42, 59, 68, 71, 80, 159, 205, 1304–1306, 1308, 1359, 1394, 1414
Viking 1, 1358, 1377, 1390
Viking 2, 1358, 1377, 1390, 1391
Viking I, 98
Viking II, 98
Viking missions, 79, 342
Viking orbiter, 264
Viking satellite, 1536, 1551
violent explosive events, 1079
VIP, 1199
VIRGO, 357, 365, 367–369, 453, 531, 557, 950, 1045, 1046, 1050, 1055–1057
Virgo cluster, 379, 483, 562–564, 567, 568, 575
Virgo cluster spirals, 380
Virgo ellipticals, 565

viscous processes at the magnetopause, 1532
visible light, 13
visible/near-infrared (see VNIR)
VLA, 450, 531
VLA/NRAO, 450
VLBA, 453
VLBI, 343, 453, 456, 457, 551, 552, 557, 665, 1634, 1640
VLF, 155
VLF radio waves, 155
VLT, 1351, 1353, 1354, 1357
VNIR, 1586, 1590, 1591, 1601, 1607, 1611, 1617
VNIR/TIR, 1611
void in nearby ISM, 653
volcanic activity on Io, 1468, 1471
volcanism on Io, 1460
volcanism on Mars, 1309
volcanism on Venus, 264, 1313
volcano, 1461
volcanoes of Mars, 265
volcanoes on Io, 710, 1174
VORTEX, 512, 1410, 1418
Voskhod spacecraft, 80
Vostok, 80
Vostok 1, 49, 91
Vostok programme, 50
voyage to the moon, 26, 28
Voyager, 81, 140, 143, 148, 434, 681, 686, 694, 965, 994, 996, 1132, 1151, 1155, 1157, 1172, 1174, 1181, 1192, 1196, 1202, 1205, 1206, 1209, 1212, 1426, 1427, 1431, 1433, 1456, 1460–1464, 1479
Voyager 1, 155, 1211, 1434, 1454, 1455, 1466, 1480, 1487
Voyager 2, 155, 169, 1128, 1130, 1143, 1207, 1434, 1453, 1455, 1457, 1472, 1473, 1480, 1487, 1488
Voyager Interstellar Mission, 1145
Voyager mission, 1173
Voyager observations, 1174
Voyager radio data, 1205
Voyager radio emissions, 1212
Voyager spacecraft, 1104, 1144, 1145, 1158
VRO 42.22.01, 550
VTPR, 1655
Vulcain motor, 75
Vulcain rocket engine, 75
Vulpecula Rift, 597

W49B, 948, 949
WAC, 42
Wac Corporal, 65, 68
WAC Corporal rocket, 42
WAGS, 1568
Walén correlation, 1021
Walén relation, 1020
WAM, 1604, 1617
warm cloud, 656
warm disk, 670
warm gas surrounding YSOs, 627
warm highly ionized medium (see WHIM)
warm intermediate-velocity gas, 671
warm ionized material, 668
warm ionized medium (see WIM)
warm neutral and ionized gas, 670
warm-wind model, 897
WAT, 475, 491
water coolant, 15
water cycle in Martian atmosphere, 1419
water frost on Europa and Ganymede, 1462
water ice at the poles of Mercury, 1400
water ice on Sallisto, 1462
water in circumstellar and interstellar environment, 14
water ions, 1226
water vapor in giant planets, 1442

SUBJECT INDEX

water vapor in interstellar medium, 322
water vapor in stratospheres of giant planets, 1446
wave-heating mechanisms, 1083
wave-height data, 1604
wave motions, 1078
wave spectra, 1590
wave-like change of plasma parameters, 1567
wave-particle interactions, 1549
wave-particle interactions in substorms, 1550
waves in the solar wind, 1131
WCD, 902
WCRP, 763, 793, 799, 1585, 1596, 1617
WD companion, 763
weak gravitational lensing, 387
weak-interaction decoupling, 429
weak lensing, 460
weak lensing in galaxy clusters, 460
weak lensing in the universe, 460
weak-lined T Tauri stars (see WTT)
weakly interacting massive particles (see WIMPs)
weather forecasts, 13, 1586
weather phenomenon on Mars, 1419
WEDOC, 1630
WFC, 510, 515, 590, 593, 603, 724, 844, 853
WFC S1, 593
WFPC, 860, 965, 994, 1244, 1245, 1608
WFPC2, 376, 377, 449, 455, 1244, 1245, 1248
WFPC3, 463
WGS 84, 1640
WHIM, 539
white-dwarf atmospheres, 798
white-dwarf binaries, 791, 793, 800
white-dwarf interacting binary, 792
white-dwarf model atmosphere, 807
white-dwarf stars, 1209
white-light coronograph, 219
white-light flare, 1098
white sands, New Mexico, 66
WHK, 666
wide-angled tailed source (see WAT)
wide-field camera (see WFC)
wide-field planetary camera (see WFPC)
wide radio-pulsar binary, 781
wide-field infrared explorer (see WIRE)
WIM, 601, 602, 665, 668
WIMPs, 20,1604
WIND, 1104, 1112
wind and the advanced composition explorer, 964
wind-compressed disk (see WCD)
wind-field model, 1609
wind-imaging interferometer (see WINDII)
wind spacecraft, 1505, 1511, 1590
WINDII, 879, 1664
winds from hot stars, 897
winds of early-type stars, 879
winds of O stars, 900
winds with embedded shocks, 900
wind–neutron star interaction, 828, 908
WIRE, 325
WISCONSIN, 1191
Wisconsin, 585
Wisconsin all-sky survey, 586, 594
Wisconsin detector, 582
Wisconsin Survey, 582, 592
Wisconsin Ultraviolet Photo Polarimeter Experiment (see WUPPE)
WLV, 660
WMH, 667
WMO, 1585, 1617
WNM, 663, 664
WOCE, 825, 895, 903, 1596, 1638, 1641

Wolf–Rayet binaries, 778
Wolf–Rayet system, 832
Wolszczan binary radio pulsar, 778
Wolszczan's pulsar, 767
World Climate Research Programme (see WCRP)
World Meteorological Organisation (see WMO)
world ocean circulation experiments (see WOCE)
World Weather Watch (see WWW)
worldwide atmospheric gravity wave studies (see WAGS)
wormholes, 111
WR stars, 903
WR winds, 903
wTT, 859, 860, 884, 885
WUPPE, 298, 902
WWW, 1585, 1617

X2030 + 375, 834
χ Oph, 650, 658
χ Per, 291, 292
X-ray, 13
X-ray afterglow, 17, 510
X-ray and optical parameters, 473, 484, 543, 545, 563, 581, 816, 889
X-ray background, 543, 843, 762, 763, 811, 846
X-ray binaries (see XRB)
X-ray binaries and binary radio pulsar, 783
X-ray bright points (see XBP)
X-ray burst source, 815, 819
X-ray cluster emission, 480
X-ray color–color diagrams, 250, 818, 846
X-ray coronal emission, 884
X-ray-dark regions, 1119
X-ray detected radio-quiet neutron star candidates, 736
X-ray detector, 281, 899
X-ray emission across the HR diagram, 880
X-ray emission among PMS stars, 884
X-ray emission during large flares, 889
X-ray emission from black holes, 564
X-ray emission from clusters, 476, 478
X-ray emission from comet Hyakutake, 1229
X-ray emission from groups, 487
X-ray emission from LHB, 603
X-ray emission from pulsars, 724
X-ray emission from star-forming regions, 884
X-ray emission from stellar clusters, 882
X-ray emission from stellar flares, 888
X-ray emission from the Galaxy, 283
X-ray emission of hot stars, 881
X-ray emitters, 876
X-ray-emitting coronae, 10, 590, 877
X-ray emitting gas, 491
X-ray emitting SNRs, 942
X-ray enhancement, 585
X-ray evolving universe spectroscopy mission (see XEUS)
X-ray fluorescence (see XRF)
X-ray fluorescence spectrometer (see XRFS)
X-ray flux variations, 817
X-ray halo, 565
X-ray image of Cassiopeia-A, 947
X-ray image of CTB109, 953
X-ray image of solar corona, 877
X-ray images, 480
X-ray images of Tycho's SNR, 949
X-ray imaging spectroscopy, 488
X-ray light curves, 842, 845
X-ray luminosity, 479, 830, 834, 880
X-ray luminosity functions, 883
X-ray luminosity of the Sun, 881
X-ray luminosity v. spin-down energy loss, 747
X-ray luminous E and S0 galaxies, 569

X-ray monitoring of solar activity, 212
X-ray morphologies, 482
X-ray Multi-Mirror Mission (see XMM)
X-ray nebula structures, 737
X-ray Nova Monoceros, 765, 975
X-ray novae, 845
X-ray observation, 463, 477, 565
X-ray observations of galaxies, 561
X-ray photograph of the Sun, 282
X-ray photon spectra, 847, 848
X-ray picture of a coronal hole, 221
X-ray properties, 843
X-ray properties of galaxies, 563, 844, 953
X-ray pulsations, 814, 824
X-ray satellites, 17, 846
X-ray sources in external galaxies, 564
X-ray source in the wind, 830
X-ray spectra, 818, 845
X-ray spectra of a flare, 214
X-ray spectra of Cassiopeia-A, 945
X-ray spectra of Cygnus Loop, 957
X-ray spectra of SN1006, 952
X-ray spectra of soft X-ray transients, 851
X-ray spectrographic telescope, 219
X-ray spectrometer, 1382
X-ray spectroscopy of flares, 217
X-ray spectroscopy of stellar sources, 885
X-ray spectrum at high luminosities, 846
X-ray spectrum at lower luminosities, 850
X-ray spectrum in quiescent state, 851
X-ray spectrum of a large flare, 890
X-ray spectrum of Geminga, 744
X-ray spectrum of solar corona, 210
X-ray spectrum of the Sun, 212
X-ray stars, 100, 219, 840
X-ray window, 549
X-rays from stellar winds, 878
X-rays from the solar corona, 721
X-rays in space, 100
X-rays of non-solar origin, 282
X-rays reflected by Thomson scattering, 852
XBAC, 485
XBP, 1027, 1028, 1094
Xenon beam injection experiment, 1523
XEUS, 18, 102, 383, 463, 494, 545, 575, 752, 879, 886, 887, 889, 890
XMM, 102, 106
XMM-LSS, 463, 464

XMM-Newton, 18, 106, 442, 463, 475, 487, 488, 492, 494, 543, 557, 561–563, 565, 569, 574, 575, 725, 729, 730, 732, 741, 742, 744, 746, 748, 752, 804, 806–808, 834, 948
XMM-Newton X-ray observatory, 14, 557
XRB, 564
XRF, 1398
XRFS, 1391
XSPEC, 569
XUV, 1106
XUV spectroheliograph, 219

Yakutsk array, 692
yellow giants, 878
Yohkoh, 234, 877, 1028, 1029, 1067, 1079, 1081, 1104, 1108, 1112
Yohkoh spacecraft, 1093, 1094
Yohkoh/XST, 881
young neutron stars in supernova remnants, 732
young stars with circumstellar disks, 620, 622, 630
young stellar objects (see YSO)
YSO, 610, 612, 616, 623, 625, 627, 628, 631
YSO activity, 861

ζ Aur, 906
ζ Aur binaries, 906
ζ Oph, 292–294, 654, 655
ζ Pup, 291, 294
Z sources, 817
Zanstra method, 927
Zanstra temperature, 925
Zarya, 95
zero-point energy of empty space, 111
zodiacal cloud, 1177, 1187
zodiacal-dust cloud, 321
zodiacal-dust grains, 312
zodiacal dust in the Solar System, 1176
zodiacal-dust reradiation, 308
zodiacal light, 235, 1164, 1166, 1176, 1178, 1426
zodiacal-light brightness, 1168, 1176
zodiacal-light observations, 313, 1163
zonal harmonics, 1627
zonal modes, 1043, 1045
zonal-wind velocities of giant planets, 1441
zonally averaged sound speed, 1050
Zond, 273, 274, 1337
Zond 3, 983
Zond program, 79
zone of silence, 278
zone of stellar evolution, 904